COMPUTERIZED STUDY GUIDE FOR

Biology, fourth edition by Sylvia S. Mader.
Written and compiled by Jay Templin.

☆ **ONLY $15.95!**

■ **CUT STUDY TIME!**   ■ **IMPROVE YOUR GRADES!**

ORDER TOLL FREE 1-800-748-7734

HERE'S HOW IT WORKS

Test yourself on each chapter in the text! The program presents the correct answer and a short rationale for each question.

After taking the computerized test on a chapter, the program will create a study plan with page references for the questions missed. Look up the answers to the questions you missed, then retake the test until you master each chapter.

✔ **FOCUS YOUR STUDY TIME ONLY ON THE AREAS WHERE YOU NEED HELP!**

Available for MS-DOS and Macintosh computers.

The program is very easy to use. If not completely satisfied, simply return the program in 10 days for a full refund. *This Computerized Study Guide is not a duplicate of the printed study guide to accompany the text.*

Medi-Sim, Inc. - 660 S. 4th Street - Edwardsville, KS 66113 - 1-800-748-7734

Mailing Instructions: 1) Remove order form from book. 2) Fold order form on Fold line #1. 3) Bring edge up to fold #2. 4) Fold this flap down over edge and seal completely across with tape. 5) Tape both sides where indicated.

------------------------------ Fold #2 ------------------------------

Lori
442-7788

Computerized Study Guide for:
Biology

ISBN # 0-697-20849-4

W9-AYC-719

ORDER INFORMATION

PLEASE CHECK APPROPRIATE BOX

TYPE	QTY	PRICE
❑ IBM PC/PC COMPATIBLE - 3 1/2"	_____	@ $15.95 = _____
❑ IBM PC/PC COMPATIBLE - 5 1/4"	_____	@ $15.95 = _____
❑ MACINTOSH - 3 1/2"	_____	@ $15.95 = _____
SHIPPING & HANDLING		2.95
TOTAL		_____

------------------------------ Fold line #1 ------------------------------

Phone Orders to 1-800-748-7734

Method of payment: ❑ Check Enclosed ❑ Mastercard ❑ Visa ❑ Money Order Enclosed

Credit card account number   Expiration date

Ship to :

_____
Cardholder's signature

Telephone Number: _____

Name
_____

Address
_____

Telephone #
_____

City   State   Zip

Medi-Sim, Inc. - 660 S. 4th Street - Edwardsville, KS 66113 Phone: 1-800-748-7734

NO POSTAGE
NECESSARY
IF MAILED
IN THE
UNITED STATES

BUSINESS REPLY MAIL
FIRST- CLASS MAIL   PERMIT NO. 47   EDWARDSVILLE, KS

POSTAGE WILL BE PAID BY ADDRESSEE

MEDI-SIM, INC.
P.O. BOX 13267
EDWARDSVILLE KS 66113-9989

BIOLOGY

Wild horses gallop across the Patagonian plains of South America's Argentine coast at the foot of the Fitzroy Mountains. In 1832, Charles Darwin came here aboard the HMS Beagle, captained by Robert Fitzroy. During his 5-year, around-the-world voyage on the Beagle, Darwin gathered data that would later support his theory of evolution.

Much has been learned since Darwin's day about the 65-million-year evolution of the horse, which began with a much smaller ancestor that had many toes and teeth with low crowns. As grasslands replaced a forestlike environment, the horse evolved into a much larger animal with fewer toes and teeth with high crowns. Today, the horse has only one toe and is highly adapted for running fast and far, allowing it to seek new sources of food and water when needed.

BIOLOGY

Fourth Edition

Sylvia S. Mader

WCB **Wm. C. Brown Publishers**
Dubuque, Iowa • Melbourne, Australia • Oxford, England

Book Team

Editor *Kevin Kane*
Developmental Editor *Carol Mills*
Production Editors *Anne E. Scroggin/Ann Fuerste*
Designer *Mark Elliot Christianson*
Art Editor *Miriam J. Hoffman*
Photo Editor *Lori Gockel*
Permissions Editor *Karen Storlie*
Art Processor *Andréa Lopez-Meyer*

Wm. C. Brown Publishers
A Division of Wm. C. Brown Communications, Inc.

Vice President and General Manager *Beverly Kolz*
National Sales Manager *Vincent R. Di Blasi*
Assistant Vice President, Editor-in-Chief *Edward G. Jaffe*
Marketing Manager *Paul Ducham*
Advertising Manager *Amy Schmitz*
Managing Editor, Production *Colleen A. Yonda*
Manager of Visuals and Design *Faye M. Schilling*
Design Manager *Jac Tilton*
Art Manager *Janice Roerig*
Publishing Services Manager *Karen J. Slaght*
Permissions/Records Manager *Connie Allendorf*

Wm. C. Brown Communications, Inc.

Chairman Emeritus *Wm. C. Brown*
Chairman and Chief Executive Officer *Mark C. Falb*
President and Chief Operating Officer *G. Franklin Lewis*
Corporate Vice President, President of WCB Manufacturing *Roger Meyer*

Cover Credits

Biology Front cover photo on complete text and background photo for all parts © Galen Rowell/Mountain Light Photography
Part 1: The Cell Front cover inset photo © Carolina Biological Supply Co.
Part 2: Genetic Basis of Life Front cover inset photo © Gunter Ziesler/Peter Arnold, Inc.
Part 3: Evolution and Diversity Front cover inset photo © Douglas Faulkner/Photo Researchers, Inc.
Part 4: Plant Structure and Function Front cover inset photo © Astrid and Hahns-Frieder Michler/ Science Photo Library/Photo Researchers, Inc.
Part 5: Animal Structure and Function Front cover inset photo © Runk/Schoenberger/Grant Heilman
Part 6: Behavior and Ecology Front cover inset photo © Carl R. Sams II/Peter Arnold, Inc.

Copy Edited by Laura Beaudoin
Photo Research by Michelle Oberhoffer

The Credits section for this book begins on page C-1 and is considered an extension of the copyright page.

Library of Congress Catalog Card Number: 91-76060

ISBN *Biology* Casebound 0-697-12383-9
ISBN *Biology* Paper binding 0-697-15096-8
ISBN *Part 1: The Cell* 0-697-15098-4
ISBN *Part 2: Genetic Basis of Life* 0-697-15099-2
ISBN *Part 3: Evolution and Diversity* 0-697-15100-X
ISBN *Part 4: Plant Structure and Function* 0-697-15101-8
ISBN *Part 5: Animal Structure and Function* 0-697-15102-6
ISBN *Part 6: Behavior and Ecology* 0-697-15103-4
ISBN *Biology* Boxed set: 0-697-15097-6

Printed in the United States of America by Wm. C. Brown Communications, Inc., 2460 Kerper Boulevard, Dubuque, IA 52001

10   9   8   7   6   5   4   3   2   1

PUBLISHER'S NOTE TO THE INSTRUCTOR

Binding Option	Description	ISBN
Biology, casebound	The full-length text (chapters 1–50), with hardcover binding.	0-697-12383-9
Biology, paperbound	The full-length text, paperback covered and available at a significantly reduced price, when compared with the casebound version.	0-697-15096-8
Part 1 *The Cell*, paperbound	Part 1 features the first unit or chapters 1–9 of the text, covering the scientific method, evolution and unity of life, basic chemistry, cell biology, photosynthesis, and respiration. This paperback option is available at a significantly reduced price when compared with both the full-length casebound and paperbound versions.	0-697-15098-4
Part 2 *Genetic Basis of Life*	Part 2 features chapters 10–18 and covers cell reproduction, Mendelian and molecular genetics, gene activity, and recombinant DNA and biotechnology. This paperback is also available at a significantly reduced price when compared with the full-length versions of the text.	0-697-15099-2
Part 3 *Evolution and Diversity*	Part 3 features chapters 19–28 on Darwin and evolution, genetics in evolution and diversity. Paperbound, it is available for a fraction of the full-length casebound or paperbound prices.	0-697-15100-X
Part 4 *Plant Structure and Function*	Part 4 features chapters 29–32 on plant structure and function. Paperbound, it is also available for a fraction of the full-length casebound or paperbound prices.	0-697-15101-8
Part 5 *Animal Structure and Function*	Part 5 features chapters 33–44 on animal systems and reproduction and development. Paperbound, it also sells for a fraction of the full-length book price.	0-697-15102-6
Part 6 *Behavior and Ecology*	Part 6 features chapters 45–50 on behavior and ecology. Paperbound, it sells for a fraction of the full-length book price.	0-697-15103-4
Biology, the Boxed Set	The entire text, offered in an attractive, boxed set of all six paperbound "splits." It is available at the same price as the full-length casebound text.	0-697-15097-6

Binding Options and Recycled Paper

Biology—in all its numerous binding options (listed here)—is printed on **recycled paper stock**. All of its ancillaries, as well as all advertising pieces will be printed on recycled paper, subject to market availability.

Our goal in offering the text and its ancillary package on **recycled paper** is to take an important first step toward minimizing the environmental impact of our products. If you have any questions about recycled paper use, *Biology*, its package, any of its binding options, or any of our other biology texts, feel free to call us at 1-800-331-2111. Thank you.

Kevin Kane
Executive Editor
Life Sciences

BRIEF CONTENTS

CONTENTS

6
Membrane Structure and Function 84

7
Cellular Energy 104

8
Photosynthesis 121

9
Glycolysis and Cellular Respiration 137

PART 2

Genetic Basis of Life 153

10
Cell Reproduction: Binary Fission and Mitosis 154

11
Cell Reproduction: Meiosis 169

Contents

PART 3

Evolution and Diversity 297

19

Darwin and Evolution 298

20

Genetics in Evolution 315

21

Speciation in Evolution 331

22

Origin of Life 353

23

Viruses and Monera 365

24

Protists and Fungi 379

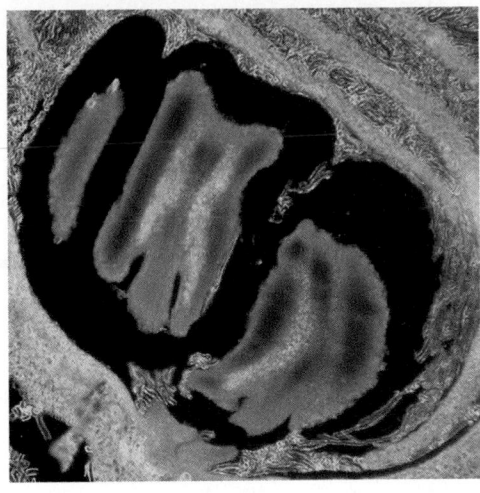

31

Reproduction in Plants 507

32

Growth and Development in Plants 522

PART 5

Animal Structure and Function 537

33

Animal Organization and Homeostasis 538

34

Circulatory System 552

35

Lymphatic System and Immunity 572

36

Digestive System and Nutrition 587

READINGS

THE **BIOLOGY** LEARNING SYSTEM

3

Basic Chemistry

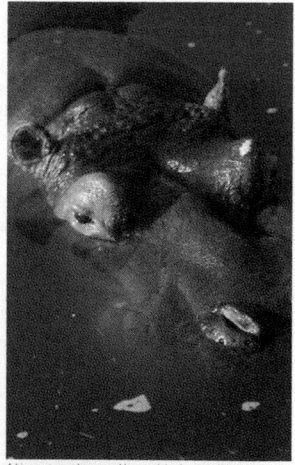

Your study of this chapter will be complete when you can

1. name and describe the subatomic particles of an atom, indicating which one accounts for the occurrence of isotopes;
2. describe and discuss the energy levels (electron shells) of an atom, including the orbitals of the first 2 levels;
3. draw a simplified atomic structure of any atom with an atomic number less than 20;
4. distinguish between ionic and covalent reactions, and draw representative atomic structures for ionic and covalent molecules;
5. tell which atom has been reduced and which has been oxidized in a particular reaction;
6. describe the chemical properties of water, and explain their importance for living things;
7. define an acid and a base; describe the pH scale, and state the significance of buffers.

A hippopotamus keeps cool by remaining in water. Water has many biological functions, both without and within organisms. Cells are largely composed of this inorganic molecule, which facilitates chemical reactions and helps maintain a normal body temperature because it is slow to heat.

24

Behavioral Objectives

Each chapter begins with a list of behavioral objectives designed to help students identify the major concepts of the chapter. Their study of the chapter is complete when they can satisfy these objectives.

Text Line Art

Graphic diagrams placed within textual passages help clarify difficult concepts and enhance learning.

HAVE YOU THANKED A GREEN PLANT TODAY? Plants dominate our environment, but most of us spend little time thinking about the various services they perform for us and for all living things. Chief among these is their ability to carry on photosynthesis, during which they use sunlight (*photo*) as a source of solar energy to produce food (*synthesis*):

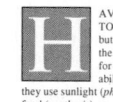

Other organisms, called algae, are also photosynthetic. Algae (chap. 24) are a diverse group, but many are water dwelling, microscopic organisms related to plants.

The organic food produced by photosynthesis is not only used by plants themselves, it is also the ultimate source of food for all other living things (fig. 8.1). Also, when organisms convert carbohydrate energy into ATP energy they make use of the oxygen (O_2) given off by photosynthesis. Nearly all living things are dependent on atmospheric oxygen derived from photosynthesis. At one time in the distant past, plant and animal matter accumulated without decomposing. This matter became the fossil fuels (e.g., coal, oil, and gas) that we burn today for energy. This source of energy, too, is the product of photosynthesis.

Photosynthesis is absolutely essential for the continuance of life because it is the source of food and oxygen for nearly all living things.

Sunlight

Photosynthesis is an energy transformation in which solar energy in the form of light is converted to chemical energy within carbohydrate molecules. Therefore, we will begin our discussion of photosynthesis with the energy source—sunlight.

Solar radiation can be described in terms of its energy content and its wavelength. The energy comes in discrete packets called **photons**. So, in other words, you can think of radiation as photons that travel in waves:

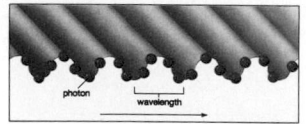

Figure 8.2*a* illustrates that solar radiation, or the **electromagnetic spectrum,** can be divided on the basis of wavelength—gamma rays have the shortest wavelength and radio waves have the longest

Figure 8.1
This squirrel is a herbivore. It feeds directly on plant material produced by a photosynthesizer. Carnivores, such as a hawk that may feed on this squirrel, are also dependent, although indirectly, on food produced by photosynthesizers.

wavelength. The energy content of photons is inversely proportional to the wavelength of the particular type of radiation; that is, short-wavelength radiation has photons of a higher energy content than long-wavelength radiation. High-energy photons, such as those of short-wavelength ultraviolet radiation, are dangerous to cells because they can break down organic molecules. Low-energy photons, such as those of infrared radiation, do not damage cells because they only increase the vibrational or rotational energy of molecules; they do not break bonds. But photosynthesis utilizes only the portion of the electromagnetic spectrum known as **visible light.** (It is called visible light because it is the part of the spectrum that allows us to see.) Photons of visible light have just the right amount of energy to promote electrons to a higher electron shell in atoms without harming cells.

We have mentioned previously that only about 42% of the solar radiation that hits the earth's atmosphere ever reaches the surface, and most of this radiation is within the visible-light range.

122

The Cell

Concept Summaries

At the ends of major sections within each chapter, concept summaries briefly highlight key concepts in the section, helping students focus their study efforts on the basics.

Bold-Faced Key Terms

Important terms are boldfaced and defined on first mention.

Readings

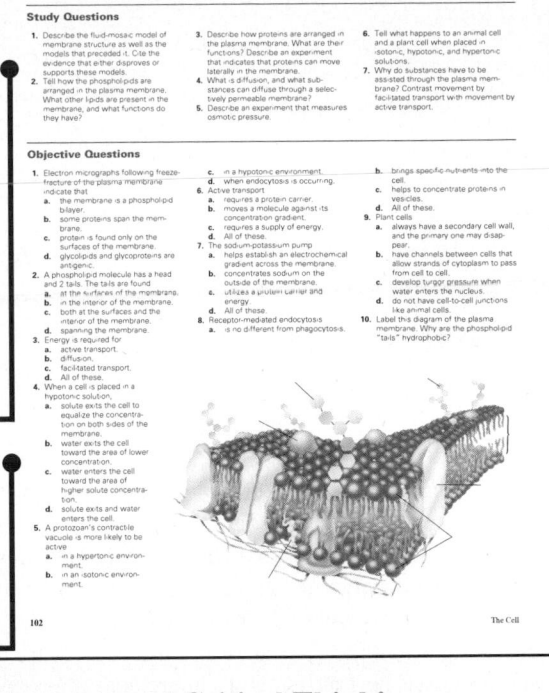

Throughout *Biology*, selected readings reinforce major concepts in the book. Most readings are written by the author, but a few are excerpted from popular magazines. A reading may provide insight into the process of science or show how a particular kind of scientific knowledge is applicable to the students' everyday lives.

Study Questions

These questions appear at the end of each chapter. They call for specific, short essay answers that challenge students' mastery of the chapter's basic concepts.

Objective Questions

Multiple-choice questions at the end of each chapter test basic recall of the chapter's key points. The answers to the objective questions are listed in appendix D.

Practice Problems

Practice problems at the end of passages in the genetics chapters help students master basic genetic quantifications. Answers to these problems are in appendix D.

Concepts and Critical Thinking

These end-of-chapter questions require students to apply the chapter's basic concepts to biological concerns. Writing the answers to these or the Study Questions fulfills any Writing-Across-the-Curriculum requirement.

Selected Key Terms

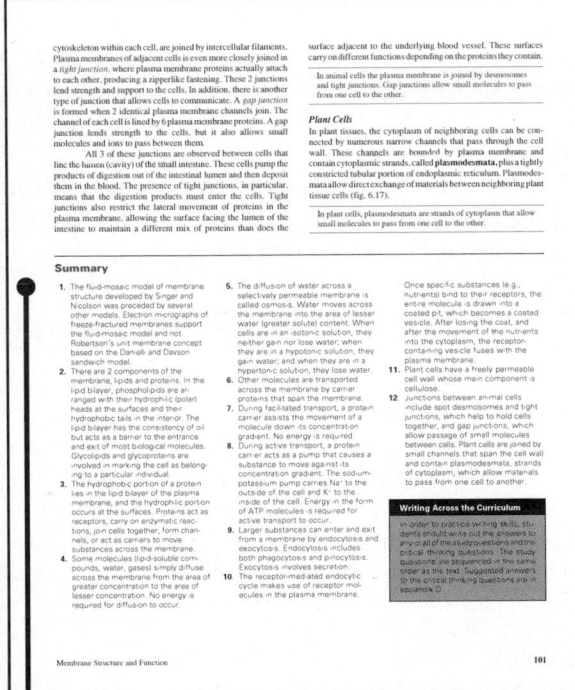

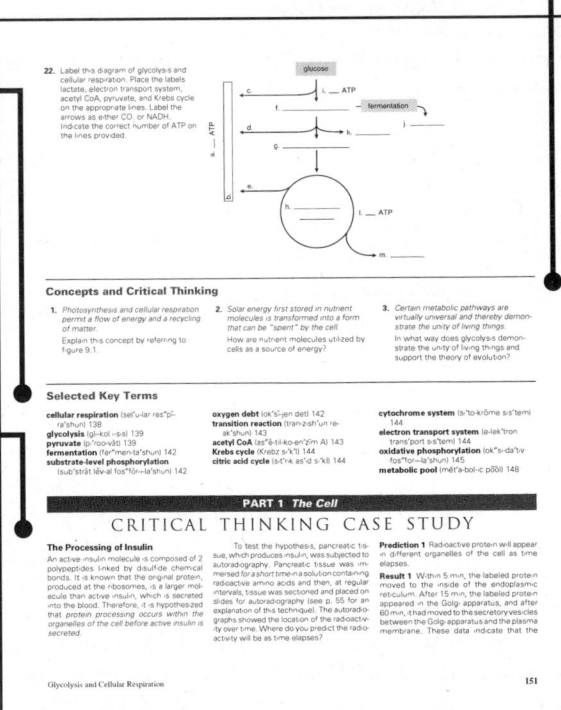

A selected list of bold-faced, key terms from the chapter appears at the end of each chapter. Each term is accompanied by its phonetic spelling and a page number indicating where the term is introduced and defined in the chapter.

Critical Thinking Case Studies

Each part ends with a case study designed to help students think critically by participating in the process of science. At many institutions, instructors are encouraged to develop the writing skills of their students. In such cases, instructors could require students to write out their answers to the questions in each case study. Suggested answers for each of these questions appear in the Instructor's Manual for the text.

Chapter Summaries

At the end of each chapter is a summary. This listing of important points should help students more readily identify important concepts and better facilitate their learning of chapter content.

Further Readings

A list of readings at the end of each part suggests references that can be used for further study of topics covered in the chapters of that part. The items listed in this section were carefully chosen for readability and accessibility.

PREFACE

Biology is an introductory college text that covers the basic concepts and principles of general biology. While the text is clear and straightforward enough to be read by liberal arts students, it is also comprehensive and authoritative, so it can just as easily be read by science majors. The text strives to use other forms of life, in addition to humans, as examples, making frequent reference to plants and invertebrates. The text takes a hierarchical approach, proceeding from the chemistry of the cell up to the organization of the biosphere.

Science Process and Critical Thinking

Biology stresses the process of science in many ways. Chapter 2 is entirely devoted to describing the scientific method with examples. And throughout the text, experiments that have led us to our present level of knowledge are described. *New to this edition are the "How Do You Do That?" boxes, which help to emphasize technological literacy.* The boxes are meant to take the mystery out of laboratory procedures and show students that everyone is capable of utilizing the methods by which data are collected.

Understanding the concepts of biology and engaging in the process of science encourages one to think critically. *New to this edition, each chapter ends with a list of the major concepts within the chapter and asks critical thinking questions based on those concepts.* Those questions allow students to see that thinking conceptually leads to predictions and deductions concerning particular aspects of biology. Suggested answers to the end-of-chapter critical thinking questions appear in appendix D. Each major part in the text ends with a critical thinking case study. As before, the case studies were written by Dr. Robert Allen and myself. They provide students with an opportunity to use scientific methodology as they think critically. Suggested answers to the case study questions appear in the Instructor's Manual for the text.

Writing Across the Curriculum

Students need a chance to practice their writing skills and at the same time test their ability to fulfill the behavioral objectives that begin each chapter. *This is provided by the "Writing Across the Curriculum Study Questions" at the close of the chapter,* which is based on an educational approach to improving the skills of students. Instructors who have their students write out the answers to the study questions and critical thinking questions will have answered the challenge of this new approach.

The objective questions prepare students to take examinations that contain recall-based questions. These, too, have been improved, and instructors will appreciate the addition of questions that require students to complete and/or label diagrams.

Organization of the Text

Each chapter in *Biology* introduces particular concepts and describes biological experiments that have contributed to our present level of knowledge. As in the third edition, the chapters are of a suitable length to be read in one sitting. *A new chapter on animal organization and homeostasis has been added to part 5.* This chapter includes a discussion of animal tissues. The text has the following parts, which have been revised as discussed.

Introduction

Chapter 1, which concerns the characteristics of life and introduces important biological concepts, was completely rewritten to increase student interest. The chapter shows that the unity of life is due to descent from a common ancestor and that diversity is due to adaptations to particular ways of life in an ecosystem.

Chapter 2 thoroughly explains scientific methods and gives examples of both experimental and observational biological research. The biological understanding of the word *theory* is also fully explained.

Part 1 The Cell

Part 1 concerns cell structure and function including energy metabolism. *Membrane protein diversity is more thoroughly discussed and illustrated in this edition.* The first of the 3 chapters devoted to energetics gives an overview. Instructors can use just this one chapter or proceed on to cellular respiration and/or photosynthesis.

Part 2 Genetic Basis of Life

There is a strong historical emphasis in this part, but practical aspects are not neglected. Students are given an opportunity to test their ability to do problems as they proceed. New advances in the field of genetics appear daily, and this part has been thoroughly updated to reflect our changing knowledge. *The very latest that is known about human genetic disorders is included. The biotechnology chapter includes a discussion of new techniques in gene therapy in humans and the human genome project.* The presentation is at an appropriate level for the beginning student.

Part 3 Evolution and Diversity

The evolution chapters were completely reorganized and rewritten with the valuable help of many instructors and scholars who specialize in this area. The geological time scale was completely updated. Chapter 22, concerning the origin of life, was also rewritten and now includes several hypotheses on this topic. The diversity chapters were revised and the most up-to-date classification system is used. Instructors will appreciate a new expanded table that compares the major features of the plant divisions.

Part 4 Plant Structure and Function

Four chapters are devoted to flowering plant anatomy and physiology. The first chapter provides a foundation for the others on nutrition and transport, reproduction, and growth and development.

Part 5 Animal Structure and Function

This part begins with a new chapter on animal organization and homeostasis. There is a discussion of animal tissues and organ systems before homeostasis is discussed in some detail. *The comparative section that begins each of the animal physiology chapters has been rewritten for clarity and improvement. A new extended section on AIDS is included in the reproduction chapter.*

Part 6 Behavior and Ecology

A new animal behavior chapter takes a more modern approach to this topic. The ecology chapters have a logical sequence from populations to communities to the biosphere. Environmental concerns are emphasized in the last 2 chapters. Some instructors may wish to begin the year's work with this part, which is certainly a workable alternative.

Aids to the Reader

Biology was written so that students can enjoy, appreciate, and come to understand the concepts of biology and the scientific process. The following text features are especially designed to assist student learning.

Text Introduction

The introduction section (chapters 1 and 2) lays a foundation upon which the rest of the text depends. The first chapter reviews the characteristics of life and presents the fundamental concepts of biology. The second chapter explains the scientific process in detail and gives examples of studies done by biologists. Various other experiments are described throughout the text.

Study Objectives

Each chapter begins with a list of objectives designed to guide students as they study the chapter. Their study of the chapter is complete when they can satisfy these objectives, therefore, they help students prepare for an examination.

Boxed Readings

Throughout *Biology,* selected boxed readings reinforce major concepts in the book. Most readings are of general interest and heighten student involvement by expanding on a topic discussed in the chapter.

"How Do You Do That?" Boxes

New to this edition are the "How Do You Do That?" boxes, which describe laboratory methods. They are designed to take the mystery out of the manner in which biological data are collected.

Drawings, Photographs, and Tables

The drawings, photographs, and tables in *Biology* have been designed to help students learn basic biological concepts as well as the specific content of the chapters, and are consistent with multicultural educational goals. Often it is easier to understand a given process by studying a drawing, especially when it is carefully coordinated with the text, as is the case here. The photographs were selected not only to please the eye but also to emphasize specific points in the text. The tables summarize and list important information, making it readily available for efficient study.

Chapter Summaries

The summary is a numbered series of statements that follow the organization of the chapter. The summary helps students identify the concepts of the chapter and facilitates their learning of chapter content.

Chapter Questions

Three kinds of questions—study questions, objective questions, and critical thinking questions—appear at the close of each chapter. They allow students to test their ability to fulfill the study objectives. *Writing across the curriculum* recognizes that students need an opportunity to practice writing in all courses. The study questions review the chapter, and their sequence follows that of the chapter. When students write out the answers to the study questions, they are writing while studying biology. The critical thinking questions are based on biological concepts that pertain to the chapter. They show that knowledge of a biological concept allows one to reason about some particular aspect of biology. Writing out the answers to the critical thinking questions also fulfills any writing across the curriculum requirements. The objective questions allow students to test their ability to answer recall-based, multiple choice questions. The objective questions have been expanded in this edition to include questions that require the completing or labeling of diagrams. Answers to the objective questions and critical thinking questions appear in appendix D.

Selected Key Terms

Each chapter ends with a selected key term list. Key terms are boldfaced in the chapter, defined in context, and also appear in the glossary. Especially significant key terms appear in the selected key term list. Each term is accompanied by its phonetic spelling and is referenced to the page on which it is introduced and defined.

Further Readings

The list of readings at the end of each part suggests references that can be used for further study of the topics covered in the chapters of that part. The items listed in this section were carefully chosen for readability and accessibility.

Critical Thinking Case Studies

Each part ends with a case study designed to help students think critically by participating in the process of science. At many institutions, instructors are encouraged to develop the writing skills of their students. In such cases, instructors could require students to write out their answers to the questions in each case study. Suggested answers for each of these questions appear in the Instructor's Manual.

Glossary

The glossary contains the terms that are boldfaced in the text. Many of the glossary entries are accompanied by a phonetic spelling and a definition, and each is referenced to the page on which it was first introduced and defined in the text.

ADDITIONAL AIDS

Instructor's Manual/Test Item File

Revised by Jean Helgeson, the Instructor's Manual/Test Item File is designed to assist instructors as they plan and prepare for classes using *Biology*. For each chapter in the text, the Instructor's Manual provides a chapter outline, key terms, an extended lecture outline, enrichment ideas, and a listing of selected films. The Test Item File contains approximately 40 multiple choice, true/false, and critical thinking essay questions per text chapter.

Suggested answers for the critical thinking case studies that appear at the end of each part in the text are placed at the end of the corresponding parts in the Instructor's Manual.

In addition, the Instructor's Manual includes an answer key for the Critical Thinking Case Study Workbook by Robert Allen, along with listings of the transparencies and micrograph slides available to instructors using *Biology*, fourth edition.

Student Study Guide

The Student Study Guide that accompanies the text was revised by Jay Templin. For each text chapter, there is a corresponding study guide chapter that includes a list of study objectives, a chapter review, vocabulary terms that are page-referenced to the text, learning activities with text page references that correlate to the study objectives and require students to apply textual information, critical thinking questions, and chapter test questions with an answer key providing the student immediate feedback.

Micrograph Slides

A boxed set of 99 slides includes all photomicrographs and electron micrographs printed in the text.

Laboratory Manual

The Laboratory Manual that accompanies *Biology* was thoroughly revised by Kenneth Kilborn and myself. Its 34 exercises provide enough variety to meet the needs of a broad spectrum of class designs. Student aids include a list of learning objectives at the beginning of each exercise and numerous full-color illustrations throughout. Each exercise includes an introduction, and ample space is provided for students to record their observations as the lab proceeds. A laboratory review, consisting of a series of questions, ends each exercise. Answers to the review questions are provided in an appendix.

Customized Laboratory Manual

The Laboratory Manual's 34 exercises are available as individual "lab separates," so instructors can custom-tailor the manual to their particular course needs. The separates, which are published in one color, can be individually selected at a greatly reduced price and will be collated and bound by **WCB** on request.

Laboratory Resource Guide

Helpful and thorough information regarding each lab preparation can be found in the Laboratory Resource Guide. Completely revised by Kenneth Kilborn, the guide is designed to help instructors make the laboratory experience a more meaningful one for the student. For handy reference, a list of suppliers is printed on the inside front cover. The Resource Guide is now divided into 3 parts. They are Laboratory Preparation and Instructions; Laboratory Exercises and Expected Results; and Answers to the Laboratory Review Questions.

Transparencies and Lecture Enrichment Kit

A full set of transparency acetates also accompanies the text. These feature key illustrations from the text in both 2 and full color. They are accompanied by a Lecture Enrichment Kit, which is a set of lecture notes featuring additional high-interest information about the pictured process or concept and not presented in the text.

Critical Thinking Case Study Workbook

Written by Robert Allen, this ancillary includes 30 additional critical thinking case studies of the type found in the text. Like the text case studies, they are designed to immerse students in the "process of science" and challenge them to solve problems in the same way biologists do. The case studies here are divided into 3 levels of difficulty (introductory, intermediate, and advanced) to afford instructors greater choice and flexibility. An answer key is printed in the Instructor's Manual.

Extended Lecture Outline Software

This instructor software features extensive outlines of each text chapter with a brief synopsis of each subtopic to assist in lecture preparation. Written in ASCII files for maximum utility, it is available in IBM, Apple, or Mac formats. It is free to all adopters, upon request.

Biology Art Masters

A set of 150 art masters consisting of one-color line art with labels can be used for additional transparencies or can be copied and used for student hand-outs.

The 7 packages include the following titles: Cell Biology, Genetics, Diversity, Plant Biology, Animal Biology, Ecology and Behavior, and Evolution.

Adopters of *Biology* can order one or all 7 packages for free.

WCB TestPak with Enhanced QuizPak and GradePak

WCB TestPak, a computerized testing service, provides instructions with either a mail-in/call-in testing program or the complete test item file on diskette for use with the IBM PC, Apple, or Macintosh computer. **WCB** TestPak requires no programming experience.

WCB QuizPak, a part of TestPak, provides students with true/false, multiple choice, and matching questions for each chapter in the text. Using this portion of the program will help students to prepare for examinations. Also included with the QuizPak is an on-line testing option to allow professors to prepare tests for students to take on the computer. The computer will automatically grade the test and update a gradebook file.

WCB GradePak, also a part of TestPak, is a computerized grade management system for instructors. This program tracks student performance on examinations and assignments. It will compute each student's percentage and corresponding letter grade, as well as the class average. Printouts can be made utilizing both text and graphics.

Other Titles of Related Interest from Wm. C. Brown Publishing

You Can Make a Difference
by Judith Getis

This short, inexpensive supplement offers students practical guidelines for recycling, conserving energy, disposing of hazardous wastes, and other pollution controls. It can be shrink wrapped with the text, at minimal additional cost (ISBN 0-697-13923-9)

How to Study Science
by Fred Drewes, Suffolk County Community College

This excellent new workbook offers students helpful suggestions for meeting the considerable challenges of a college science course. It offers tips on how to take notes; how to get the most out of laboratories; as well as how to overcome science anxiety. The book's unique design helps students develop critical thinking skills, while facilitating careful note-taking. (ISBN 0-697-14474-7)

The Life Science Lexicon
by William N. Marchuk, Red Deer College

This portable, inexpensive reference helps introductory-level students quickly master the vocabulary of the life sciences. Not a dictionary, it carefully explains the rules of word construction and derivation, in addition to giving complete definitions of all important terms. (ISBN 0-697-12133-X)

Biology Study Cards
by Kent Van De Graaff, R. Ward Rhees, and Christopher H. Creek, Brigham Young University

This boxed set of 300, 2-sided study cards provides a quick yet thorough visual synopsis of all key biological terms and concepts in the general biology curriculum. Each card features a masterful illustration, pronunciation guide, definition, and description in context. (ISBN 0-697-03069-5)

The Gundy-Weber Knowledge Map of the Human Body
by G. Craig Gundy, Weber State University

This 13-disk, Mac-Hypercard program is for use by instructors and students alike. It features carefully prepared computer graphics, animations, labeling exercises, self-tests, and practice questions to help students examine the systems of the human body. Contact your local Wm. C. Brown representative or call 1-800-351-7671.

The Knowledge Map Diagrams
1. Introduction, Tissues, Integument System (0-697-13255-2)
2. Viruses, Bacteria, Eukaryotic Cells (0-697-13257-9)
3. Skeletal System (0-697-13258-7)
4. Muscle System (0-697-13259-5)
5. Nervous System (0-697-13260-9)
6. Special Senses (0-697-13261-7)
7. Endocrine System (0-697-13262-5)
8. Blood and the Lymphatic System (0-697-13263-3)
9. Cardiovascular System (0-697-13264-1)
10. Respiratory System (0-697-13265-X)
11. Digestive System (0-697-13266-8)
12. Urinary System (0-697-13267-6)
13. Reproductive System (0-697-13268-4)

Demo—(0-697-13256-0)
Complete Package—(0-697-13269-2)

GenPak: A Computer Assisted Guide to Genetics
by Tully Turney, Hampden-Sydney College

This Mac-Hypercard program features numerous interactive/tutorial (problem-solving) exercises in Mendelian, molecular, and population genetics at the introductory level. (ISBN 0-697-13760-0)

Acknowledgments

Many persons have contributed to the success of previous editions and helped me make this edition our very best effort. My editor, Kevin Kane, directed the efforts of all. Carol Mills, my developmental editor, served as a liaison between the editor, me, and many other people. I especially want to mention Susan Murray and Anne Packard of Plymouth State College, who carefully read the galleys and made many helpful suggestions.

The production team at Wm. C. Brown was ever-faithful to their duties. Anne Scroggin and Ann Fuerste were the production editors; Miriam Hoffman the art editor; Lori Gockel the photo researcher; Michelle Oberhoffer the photo researcher; and Mark Christianson the designer. My thanks to each of them for a job well done!

The Contributors

With this edition, as with the last edition, there were several contributors to various sections of the book. They critically reviewed my revised chapters and told me how they could be improved. With their help, it was easier to make sure the content was sufficient, accurate, and up to date. The contributors were:

W. Dennis Clark (Ph.D. University of Texas), professor of botany at Arizona State University. Dr. Clark's expertise in the areas of chemistry, evolution, and physiology of plants was extremely helpful in the revision of the plant and energy chapters.

Thomas C. Emmel (Ph.D. Stanford University), professor of zoology at the University of Florida. He assisted invaluably by providing updated information in ecology, where he has published numerous articles and 2 books.

Robert M. Kitchin (Ph.D. University of California-Berkeley), professor of genetics at the University of Wyoming. He reviewed the genetics chapters and outlined new information and events taking place in this rapidly changing and expanding field.

The Ancillary Contributors

The ancillary package that accompanies *Biology* also involved the efforts of many instructors currently teaching biology.

Critical Thinking Case Study Book: Robert Allen (Ph.D. University of California-Los Angeles) is vice-president for instruction at Victor Valley College, whose special interests include research on the development of critical thinking skills in science. He is a frequent presenter on the subject for the National Science Foundation Chatauqua series.

Instructor's Manual-Test Item File: Jean Helgeson, (M. A. The University of Texas Southwestern Medical Center) is professor and coordinator of biology and health science at Collin County Community College.

Student Study Guide: Jay Templin (Ed.D. Temple University) is an instructor at Widener University who has authored numerous student and instructor's ancillaries for several college biology textbooks. He is also a co-author of the *College Board Achievement Test* in biology.

Laboratory Manual and Laboratory Resource Guide: Kenneth Kilborn (M. A. San Jose State University) has been teaching general biology and general botany in addition to 20 years of teaching in the biology laboratory at Shasta College.

QuizPak and The Lecture Enrichment Kit: Jane Aloi (Ph.D. University of California, Davis) is an ecologist teaching courses in introductory biology, zoology, and human anatomy at Saddleback College.

Reviewers

Many instructors of introductory biology courses around the country reviewed portions of the manuscript. Others assisted by taking the time to fill out survey forms that helped us make important decisions about content. With many thanks, we list their names here.

First Edition

A. Lester Allen *Brigham Young University*
William E. Barstow *University of Georgia*
Lester Bazinet *Community College of Philadelphia*
Eugene C. Bovee *University of Kansas*
Larry C. Brown *Virginia State University*
L. Herbert Bruneau *Oklahoma State University*
Carol B. Crafts *Providence College*
John D. Cunningham *Keene State College*
Dean G. Dillery *Albion College*
H. W. Elmore *Marshall University*
David J. Fox *University of Tennessee*
Larry N. Gleason *Western Kentucky University*
E. Bruce Holmes *Western Illinois University*
Genevieve D. Johnson *University of Iowa*
Malcolm Jollie *Northern Illinois University*
Karen A. Koos *Rio Hondo College*
William H. Leonard *University of Nebraska—Lincoln*
A. David Scarfe *Texas A & M University*
Carl A. Scheel *Central Michigan University*
Donald R. Scoby *North Dakota State University*
John L. Zimmerman *Kansas State University*

Second Edition

David Ashley *Missouri Western State College*
Jack Bennett *Northern Illinois University*
Oscar Carlson *University of Wisconsin-Stout*
Arthur Cohen *Massachusetts Bay Community College*
Rebecca McBride DeLiddo *Suffolk University*
Gary Donnermeyer *St. John's University (Minnesota)*
D. C. Freeman *Wayne State University*
Sally Frost *University of Kansas*
Maura Gage *Palomar College*
Betsy Gulotta *Nassau Community College*
W. M. Hess *Brigham Young University*
Richard J. Hoffmann *Iowa State University*
Trudy McKee
Brian Myres *Cypress College*
John M. Pleasants *Iowa State University*
Jay Templin *Widener University*

Third Edition

Wayne P. Armstrong *Palomar College*
Mark S. Bergland *University of Wisconsin-River Falls*
Richard Blazier *Parkland College*

William F. Burke *University of Hawaii*
Donald L. Collins *Orange Coast College*
Ellen C. Cover *Manatee Community College*
John W. Crane *Washington State University*
Calvin A. Davenport *California State University-Fullerton*
Robert Ebert *Palomar College*
Darrell R. Falk *Pt. Loma Nazarene College*
Jerran T. Flinders *Brigham Young University*
Sally Frost *University of Kansas*
Elizabeth Gulotta *Nassau Community College*
Madeline M. Hall *Cleveland State University*
James G. Harris *Utah Valley Community College*
Kenneth S. Kilborn *Shasta College*
Donald R. Kirk *Shasta College*
Jon R. Maki *Eastern Kentucky University*
Ric Matthews *Miramar College*
Joyce B. Maxwell *California State University-Northridge*
Leroy McClenaghan *San Diego State University*
Leroy E. Olson *Southwestern College*
Barbara Yohai Pleasants *Iowa State University*
David M. Prescott *University of Colorado*
Robert R. Rinehart *San Diego State University*
Mary Beth Saffo *University of California—Santa Cruz*
Walter H. Sakai *Santa Monica College*
Frederick W. Spiegel *University of Arkansas-Fayetteville*
Gerald Summers *University of Missouri-Columbia*
Marshall Sundberg *Louisiana St. University*
Kathy S. Thompson *Louisiana State University*
Anna J. Wilson *Oklahoma City Community College*
Timothy S. Wood *Wright State University*

Fourth Edition

Michael S. Gaines *University of Kansas*
Kerry S. Kilburn *West Virginia State College*
W. Sylvester Allred *Northern Arizona University*
Helen L. Grierson *Morehead State University*
Deborah K. Meinke *Oklahoma State University*
Robert Snetsinger *Queens University*
Gail Kingrey *Pueblo Community College*
Barbara Yohai Pleasants *Iowa State University*
Eugene Nester *University of Washington*

Survey Respondents

Sister Julia Van Demack *Silver Lake College*
Ken Beatty *College of the Siskiyous*
John W. Metcalfe *Potsdam College at the State University of New York*
Brenda K. Johnson *Western Michigan University*
Tom Dale *Kirtland Community College*
Marlene Kayne *Trenton State College*
Peter M. Grant *Southwestern Oklahoma State University*
Janet Carter *Delaware Technical & Community College*
Frank F. Escobar *Holyoke Community College*
John A. Chisler *Glenville State College*

Nancy Goodyear *Bainbridge College*
John F. Lyon *University of Lowell*
Rob Snetsinger *Queens University*
Fred Stevens *Schreiner College*
Paula H. Dedmon *Gaston College*
Bruce G. Stewart *Murray State College*
Leo R. Finkenbinder *Southern Nazarene University*
Mary P. Greer *Macomb Community College*
Kenneth M. Allen *Schoolcraft College*
Robert Cerwonica *SUNY-Potsdam*
Darcy Williams *Cecil Community College*
Dave McShaffrey *Marietta College*
Elizabeth L. Nebel *Macomb Community College*
Jal S. Parakh *Western Washington University*
Dwight Kamback *Northampton Community College*
David L. Haas *Fayetteville State University*
A. Quinton White *Jacksonville University*
Paul J. Hummer, Jr. *Hood College*
Dee Forrest *Western Wyoming College*
Linden C. Haynes *Hinds Community College*
John De Banzie *Northeastern State University*
Anne T. Packard *Plymouth St. College*
L. H. Buff, Jr. *Spartanburg Methodist College*
Dennis Vrba *North Iowa Area Community College*
Donald L. Collins *Orange Coast College*
Larrie E. Stone *Dana College*
Brian K. Mitchell *Longview College*
Dennis M. Forsythe *The Citadel*
Bonnie S. Wood *University of Maine at Presque Isle*
Eunice R. Knouso *Spartanburg Methodist College*
Larry C. Brown *Virginia State University*
Clyde M. Senger *Western Washington University*
James V. Makowski *Messiah College*
James R. Coggins *University Wisconsin-Milwaukee*
Roland Vieira *Green River Community College*
Donna Barleen *Bethany College*
S. M. Cabrini Angelli *Regis College*
W. Brooke Yeager, III *Penn State-Wilkes-Barre*
Diana M. Colon *Northwest Technical College*
Carolyn K. Jones *Vincennes University*
Rita M. O'Clair *University of Alaska Southeast*
Fred M. Busroe *Morehead State University*
Edward R. Garrison *Navajo Community College*
Joseph V. Faryniarz *Maitatuck Community College*
K. Dale Smoak *Piedmont Technical College*
Donald A. Wheeler *Edinboro University of Pennsylvania*
W. Lee Williams *Alamance Community College*
Eugene A. Oshima *Central Missouri State University*
Richard C. Renner *Laredo Junior College*
Peggy Rae Doris *Henderson State University*
Charles H. Owens *Virginia Highlands Community College*
Jeanette Oliver *Flathead Valley Community College*
E. L. Beard *Loyola University*
Monica Macklin *Northeastern State University*

1

A View of Life

Bengal tiger and cub. Life comes only from life, and reproduction passes life's characteristics from one generation to another.

Your study of this chapter will be complete when you can

1. list and explain 5 characteristics of life;
2. relate the concept of emergent properties to the increasingly complex levels of biological organization;
3. describe how populations of organisms interact within ecosystems;
4. explain how evolution accounts for both the unity and the diversity of life.

Figure 1.1
Examples of life's diversity, rockhopper penguins are birds in good standing, but they cannot fly. Their stubby forelimbs are modified into flippers for fast swimming. Only a little over a half meter tall, these penguins are named for their skill in leaping from rock to rock. *a.* Male and female rockhoppers court on the rocks of their breeding ground. *b.* Rockhoppers greet each other by braying; the sound has been likened to the screech of a rusty wheelbarrow. *c.* Yellow feathers above the eyes give a rockhopper a striking pair of eyebrows.

a.

b.

c.

ife on earth takes on a staggering variety of forms, often behaving in ways strange to us. For example, gastric-brooding frogs swallow their embryos and give birth to them later by throwing them up! Fetal sand sharks kill and eat their siblings while still inside their mother. Young black widow spiders spin a perfect web on their first try. Fringe-lipped bats swoop out of the black tropical night to attack mud-puddle frogs. Some *Ophrys* orchids look so much like female bees that male bees try to mate with them. Penguins are birds that cannot fly but can swim (fig. 1.1). Bacteria live out their entire lives in 15 minutes, while bristlecone pine trees outlive 10 generations of humans.

Categorizing the astounding array of life has always been difficult, and different types of classification systems have been and are in use today. Most modern biologists, however, use a system that divides organisms into 5 major groups, or **kingdoms:** protista, monera, fungi, plants, and animals. The reading in this chapter describes these kingdoms.

Despite life's diversity, all living things, be they pasteur roses, sidewinder snakes, or basketball players, are comprised of the same chemical elements that make up nonliving things and they obey the same laws of physics and chemistry as nonliving objects do. So, what is special about living things? Fortunately, all organisms share certain common characteristics. They give us insight into the nature of life and help us distinguish living things from nonliving things.

Characteristics of Life

Living Things Are Organized

The complex organization of living things (fig. 1.2) begins with the **cell,** the smallest, most basic unit of life. Most cells are microscopic specks, but some can be long and thin, as in the 3-foot-long fibers of human nerve cells. They can even be a blob the size of an apple, as with the ostrich's egg with its plentiful supply of yolk.

Some cells live by themselves as functioning organisms. Other cells sometimes cluster together in microscopic colonies. In multicellular organisms, similar cells combine to form a tissue, such as nerve tissue, for example. Tissues make up organs, as when various tissues combine to form the brain. Organs work together in systems; for example, the brain works with the spinal cord and a network of nerves to form the nervous system. Organ systems are joined within an individual (see fig. 3.1).

There are levels of biological organization that extend beyond the individual organism. All the members of one species in a particular area belong to a population. In your local forest there is a population of gray squirrels, for example. The populations of various animals and plants in the forest make up a community. The community of populations interacts with the physical environment and forms an ecosystem. How an ecosystem functions is discussed in more detail later in this chapter.

Figure 1.2
The high degree of organization of life shows up everywhere, from the patterns of veins on a dragonfly's wings to the arrangement of bundles of pine needles.

Emergent Properties

We have seen that biological organization includes these levels of organization: cells, tissues, organs, organ systems, individuals, populations, communities, and ecosystems. Each level of organization is more complex than the level preceding it and has properties beyond those of the former level. For example, when cells are broken down into bits of membrane and oozing liquids, these parts themselves cannot carry out the business of living. Slice a lump of coal and fit the pieces together, and you still have a lump of coal. Slice a frog and arrange the slices, and you have a former frog. It can no longer grow and reproduce.

In the living world, the whole is indeed more than the sum of its parts. Each new level of biological organization has emergent properties that are due to interactions between the parts making up the whole. All properties, even emergent properties, are governed by the laws of physics and chemistry.

Living things have levels of organization from cells to ecosystems. Each level of organization has emergent properties that cannot be accounted for by simply adding up the properties of previous levels.

Classification of Organisms

A golden-furred, rat-sized animal, discovered in Australia about 100 years ago, bores through desert sand like a mole but has a pouch like a kangaroo. Figuring out what to call such an oddity is the job of taxonomists. Taxonomy is the discipline of identifying and classifying organisms according to certain rules (*taxo,* to put into order; *nomy,* law or rule).

Taxonomists give each living thing a binomial (*bi,* two; *nomen,* name) in Latin, a custom started by the Swedish taxonomist Linnaeus. For example, 2 types of the mole mentioned are named *Notorcytes typhlops* and *Notorcytes caurinus.* The first word of the name is the genus and the second word is the specific epithet, which pinpoints a particular species within a genus.

Another aspect of taxonomy is the grouping of species into larger groups according to their shared characteristics. Taxonomists put similar species in the same genus, then group genera into families, families into orders, and so on as listed in table 1.A. All the species in a genus have fairly specific characteristics in common, whereas those organisms in the same kingdom have general characteristics in common. Most taxonomists now believe that the classification of an organism into progressively larger categories, defined by increasingly general characteristics, reflects the organism's evolutionary history. Species are defined by unique, recently evolved characteristics, but the traits that link all the organisms in a kingdom are ancient.

Taxonomists debate intensely over how many kingdoms to recognize and how to define them. Originally only 2 kingdoms were described—plants (Plantae) and animals (Animalia). Plants were literally organisms that were planted and immobile, while animals were animated and moved about. The discovery of photosynthesis gave an added characteristic to plants: they make their own food, whereas animals must eat. When microscopes were invented in the seventeenth century, scientists found many tiny creatures that weren't much like either plants or animals. What, for example, do you call a little cell that photosynthesizes like a plant but swims and eats like an animal?

These problems have inspired taxonomists to propose a number of classification systems. Some still retain just 2 kingdoms, and others use as many as 15 kingdoms. Such disagreements are good for progress in taxonomy but confusing to biology students. Therefore, we have elected to refer to just one classification system. Widely used today, this system divides living things into 5 kingdoms (table 1.B and fig. 1.A). Three of these kingdoms contain multicellular forms. These organisms differ in their mode of nutrition: plants make their own food through photosynthesis; animals ingest preformed food, and fungi live on preformed food but do not ingest it, secreting enzymes that break down food externally.

Table 1.A
Levels of Classification

	Example: Human	**Example: Corn**
Kingdom	Animalia	Plantae
Phylum (Division)	Chordata	Anthophyta
Class	Mammalia	Monocotyledonae
Order	Primates	Commelinales
Family	Hominidae	Poaceae
Genus	*Homo*	*Zea*
Species	*Homo sapiens*	*Zea mays*

Living Things Metabolize

A bedroom can easily turn into a giant heap of laundry, notebooks, magazines, and random doughnut boxes unless the occupant puts in energy to maintain order. Living things have the same problem. Maintaining their high degree of organization takes energy. Living things also need raw materials from their surroundings (their environment). **Metabolism** is the term we use for all the chemical and energy transformations that occur in cells as they use energy and nutrients to grow and make repairs.

The ultimate source of energy for all life on earth is the sun. Plants and plantlike organisms capture a small fraction of the energy that reaches the earth from the sun. They photosynthesize, a process that lets them store solar energy by converting carbon dioxide and water into energy-rich sugars. Animals get energy by eating plants; then other animals eat those animals (fig. 1.3).

Homeostasis

For metabolic processes to continue, living things need to keep themselves stable in temperature, moisture level, acidity, and other physiological factors. This is **homeostasis**—the maintenance of internal conditions within certain boundaries.

Living things can maintain homeostasis in a variety of ways. A chilly lizard may raise its internal temperature by basking in the sun on a hot rock. When the lizard starts to overheat, it scurries for cool shade. Biology students also need to maintain homeostasis. Consider a student who becomes so engrossed in her textbook that she forgets to eat lunch. In such an emergency, the liver responds by releasing stored sugar, giving her energy until she eats again.

An intake of materials and energy is needed if an organism's organization is to be maintained. The ultimate source of energy for life on earth is the sun.

Living Things Respond

Living things find energy and raw materials by interacting with their surroundings. Even one-celled organisms can respond to their environment. Some can move toward or away from light or chemicals by beating microscopic hairs or by snapping whiplike tails. Multicellular organisms can manage more complex responses. A maple seed can respond to cold weather, followed by

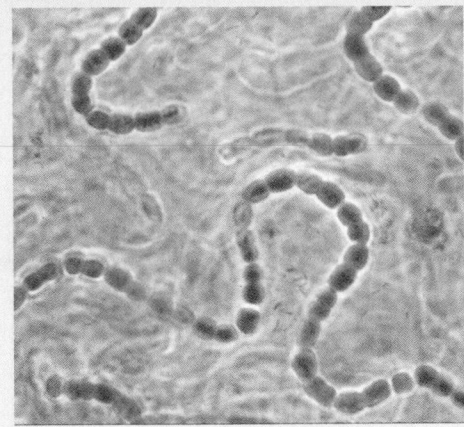

a.

Figure 1.A
Representatives from the 5 kingdoms recognized in this text: ***a.*** kingdom Monera: *Nostoc,* a cyanobacterium, magnification, X100; ***b.*** kingdom Protista: *Euglena,* a single-celled organism; ***c.*** kingdom Fungi: *Coprinus comatus,* a shaggy mane mushroom; ***d.*** kingdom Plantae: *Rosa* hybrid, a flowering plant; ***e.*** kingdom Animalia: *Felix caracal,* an African lynx (small cat).

Table 1.B
Kingdoms of Life

Kingdom	Organization	Type of Nutrition	Representative Organisms
Monera (monerans)	Small, simple single cell (sometimes chains or mats)	Absorb food (some photosynthetic)	Bacteria including cyanobacteria
Protista (protists)	Large, complex single cell (sometimes chains or colonies)	Absorb, photosynthesize, or ingest food	Protozoans, algae of various types
Fungi	Multicellular and filamentous with specialized complex cells	Absorb food	Molds and mushrooms
Plantae (plants)	Multicellular with specialized complex cells	Photosynthesize food	Mosses, ferns, woody and nonwoody flowering plants
Animalia (animals)	Multicellular with specialized complex cells	Ingest food	Sponges, worms, insects, fish, amphibians, reptiles, birds, and mammals

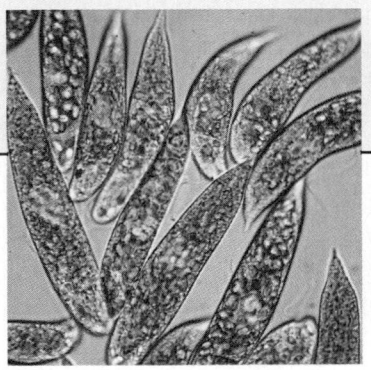

b.

c.

d.

e.

warm weather, by sprouting out of its seed coat—just plain warmth will not trigger this action. A vulture can smell meat a mile away and soar toward dinner. A monarch butterfly in Maryland can sense the approach of fall and begin its flight to Mexico for winter. And humans can detect the arrival of winter and head for the ski slopes, haul out sweaters, and buckle down to study for exams. All these responses make up what we call the **behavior** of an organism.

Living Things Reproduce

Life only comes from life. Every type of living thing can **reproduce,** or make a copy like itself (fig. 1.4). Bacteria, protozoa, and other one-celled organisms simply split in 2. Most multicellular organisms begin the reproductive process by pairing a sperm from one partner with an egg from the other partner.

The united egg and sperm have a long way to go to become a sperm whale or a yellow daffodil or a biology professor. The embryos develop into such complex adults thanks to blueprints inherited from their parents. These instructions for organization and metabolism are encoded in **genes.** Genes are made of long molecules of DNA shaped somewhat like spiral staircases with thousands of steps. All the cells of an organism have a copy of the very same genes.

Figure 1.3
All organisms require an outside source of nutrient materials and energy. Many animals actively seek, kill, and then, like all other animals, ingest their food. *a.* A lioness stalking her prey, a group of zebras, moves closer and closer with muscles taut. *b.* She explodes into her charge joined by other lionesses. The zebras, sensing danger, bolt away. *c.* One zebra of the herd is trapped and cannot escape the claws and jaws of the lionesses.

Figure 1.4

All organisms reproduce. **a.** A single-celled organism such as this *Amoeba proteus* simply divides in two. **b.** Multicellular plants produce seeds, which contain an embryo, an immature form that develops into the adult. This photograph shows the seed-containing fruit of a flowering dogwood. **c.** Multicellular animals go through developmental stages also. This fertilized queen wasp and workers are building a papery nest. Three stages in wasp development— eggs, larvae, and a pupa in its sealed cell—coexist in the nest. **d.** A hippopotamus and her young move into the water to cool off. Mammals give birth to live offspring that must be cared for.

a. b. c.

d.

Living Things Evolve

Occasionally cells make a mistake when they copy and pass on DNA. The change, or **mutation,** may cause an offspring to develop slightly differently. Most of the time, the change is for the worse, and the offspring dies or fails to reproduce well. The changed DNA disappears.

On rare occasions, however, the change in the DNA copy improves the offspring's chance of surviving and reproducing. For example, scientists in England noticed that a mutation was causing normally light-colored moths to be dark in color. In polluted areas, the mutant moths fared better because birds didn't eat them as often; dark bodies made them hard to spot against sooty trees.

Table 1.1
Characteristics of Living Things

Characteristic	Result
1. Are organized	Consist of cells
2. Metabolize	Maintain their organization
3. Respond to stimuli	Interact with the environment
4. Reproduce	Have offspring
5. Evolve	Adapt to the environment

Figure 1.5
The feeding patterns of organisms in an ecosystem can be described in terms of food chains that indicate "who eats whom." The food chain always begins with a photosynthesizing producer population followed by one or more consumer populations. The decomposers break down dead organisms and return the chemicals to the producer population. As one population feeds on another, energy flows through the system and gradually dissipates as heat, but chemicals (see inorganic nutrient supply) cycle back to the producers once again.

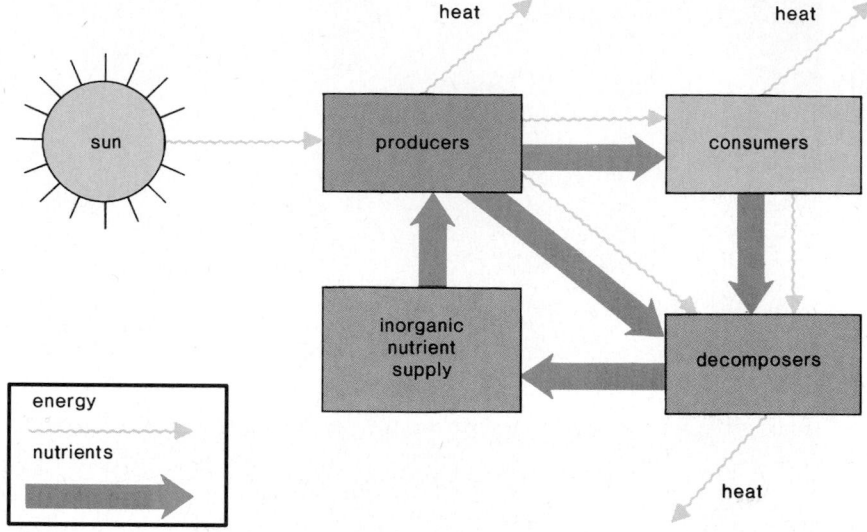

The process by which the characteristics of organisms change over time is called **evolution.** Evolutionary changes occur in groups of interbreeding individuals called **species.** Certain members of a species may inherit a genetic change that causes them to be better adapted or suited to a particular environment. These members can be expected to produce more surviving offspring, who also have the favorable characteristic. In this way the attributes of the species' members change over time. For example, the dark-colored moths just mentioned are better adapted to living in a polluted area than are the light-colored moths of the same species. The dark-colored moths will produce offspring in greater quantity than light-colored moths, and these offspring will also be dark in color.

An **adaptation** is any peculiarity of form or function or behavior that promotes the likelihood of a species' continued existence. If the environment should change, and members of a species no longer possess the genetic capability of adapting to the new environment, then extinction can follow.

> Living things reproduce and pass on genes that determine the form and function of their offspring. Genetic changes account for the ability of a species to evolve and adapt to a particular environment.

All organisms are adapted to particular places, climates, food sources, and ways of life. Hence the mind-boggling diversity of life on earth. For example, penguins are adapted to an aquatic existence in the antarctic (see fig. 1.1). Most birds have forelimbs proportioned for flying, but a penguin has stubby, flattened wings suitable for swimming. Their feet and tails serve as rudders in the water, but the flat feet also allow them to walk on land. Rockhopper penguins have a bill adapted to eating small shell fish. Their eggs—one, at most 2—are carried on their feet, where they are protected by a pouch of skin. This allows the birds to huddle together for warmth while standing erect and incubating eggs.

> Adaptations to different ways of life, including specific ways of reproducing and acquiring food, account for the great diversity of life forms.

Different as a penguin may be from a hummingbird, or from a house cat or a holly tree for that matter, they share basic characteristics (table 1.1).

Ecosystems

Adaptations include those features that allow populations to interact with themselves and with the physical **environment** within an **ecosystem** (fig. 1.5). All the ecosystems taken together make up the **biosphere,** the network of life on earth. For a description of 2 ecosystems within this network, turn to figures 1.6 and 1.7. Note the many ways that organisms can live within the 2 environments. Their specific adaptations allow them to play particular roles in their ecosystem. The diversity of organisms is best understood in terms of the many different ways in which organisms carry on their life functions within the ecosystem where they live, acquire energy, and reproduce.

The interactions between **populations** in an ecosystem tend to keep the system relatively stable. Although a forest or pond changes—trees fall, ducks come and go, seeds sprout—each ecosystem remains recognizable year after year. We say it is in dynamic balance. In many cases, even the extinction of species, and their replacement by new species through evolution, still permits the dynamic balance of the system to be maintained.

If the ecosystem is big enough, it needs no raw materials from the outside. (Individual organisms never escape this need, but ecosystems can.) A big ecosystem just keeps cycling its raw materials, like water and nitrogen. The only input it needs is energy.

Figure 1.6

In a coral reef, populations of organisms interact among themselves and with their marine environment. The reef consists of the skeletons of stony corals, colonial animals that form deposits of calcium carbonate. Only the outer layer of the reef is alive; the rest forms an inert structure full of nooks and crannies, where fish can hide from their predators. Some types of fish venture out of hiding only at night, when they feed on plankton, the microscopic organisms that drift through the ocean. The coral animals themselves often feed at night also, taking in whatever comes within reach of their extended tentacles. Other fish are active during the day. These hover near the surface of the coral, grazing on algae and worms or on shrimp and crabs if their jaws are able to crunch through shells. Parrot fish even grind the stony coral skeletons themselves. All the smaller fishes are prey to large carnivores like moray eels, jack fish, and barracudas.

Palau coral reef

Tubastraea coral

moray eel

red grouper

lionfish

Figure 1.7

A temperate deciduous forest is a terrestrial ecosystem found in the Northern Hemisphere where the climate is mild. Various types of insects and worms as well as bacteria and fungi live off the rich organic matter found in the soil. Some birds, like the robin and the tufted titmouse, feed on soil organisms. Other birds, like the red-eyed vireo and woodpeckers, prefer the insects that reside in the trees. Various small mammals, such as rabbits, squirrels, and white-footed mice, live mainly on vegetation, fruits, and insects. Wolves, bobcats, and gray foxes, which feed on these small animals, were more prevalent in days gone by; however, birds of prey like hawks and owls are still common and feed on these animals.

virgin forest at Porcupine Mountains State Park, Michigan

Eastern chipmunk

barn owl

millipede

bobcat

marsh marigolds

Organisms are members of populations within ecosystems, which are units of the biosphere. Their various adaptations help keep the system in dynamic balance and assure its continued existence.

The human population tends to modify ecosystems to suit human needs. We are now beginning to realize that this process can endanger us as well as other species.

The Human Population

People modify ecosystems for their own purposes: clearing forests for farms, clustering houses into towns, and, finally, covering square miles with the asphalt and skyscrapers of big cities. With each step, more of an area's original organisms are driven out. Big cities are populated largely by humans and the few types of life forms, such as rats and pigeons and office plants, that can tolerate a hostile environment. With such sparse, disconnected groups of living things, natural cycles do not function well.

Unlike the original ecosystem, humans rely on manufactured goods of various types and they supplement solar energy with fossil fuel (coal and oil) energy to maintain farms, towns, and cities. These inputs are eventually converted to outputs like trash and carbon dioxide, which become pollutants when they build up and cannot be dispersed easily.

We are beginning to learn that the human population is a part of the biosphere and is dependent upon the natural cycles that occur within the biosphere. As more and more of the biosphere is converted to towns and cities, fewer of the natural cycles are able to function adequately to sustain the human population. For example, the air is cleaner over forests, where many trees absorb carbon dioxide, than over cities with few trees. An increasing human population tends to destroy the very cycles that sustain it.

The Unity and Diversity of Life

Earlier we discussed the characteristics shared by living things. These common threads, the surprising unity of diverse living things, can be explained by descent from a common ancestor. Many kinds of evidence suggest that life began with single cells and that the present rainbow of organisms evolved from this common origin over hundreds of millions of years. In other words, the process of evolution explains the unity we observe in living things. The other striking thing about life on earth is its diversity. The same coral reef contains massive animals and mere pinpoints of life, rock-hard corals and jellyfish that are all frills and film. Yet, each body type suits a particular life-style. The process of evolution, which involves changes in the genetic material and then a sorting out of physical modifications suited to different environments, explains the diversity we observe in living things. This evolutionary process is the cornerstone of biology.

Evolution from a common ancestor explains the unity of life. Adaptations to different ways of living explain the diversity of life.

Summary

1. Although living things are diverse, they share certain characteristics in common.
2. All living things are organized; the cell is the smallest unit of life. Cells form tissues, tissues form organs, organs form systems in individuals, individuals are members of populations, and populations make up communities in ecosystems.
3. Each level of organization has emergent properties that cannot be accounted for by simply adding up the properties of the previous level.
4. Living things need an outside source of materials and energy. These are used during metabolism for repair and growth of the organism.
5. Organisms differ as to how they acquire materials and energy. Plants photosynthesize and make their own food; animals either eat plants or other animals. Therefore, there is a flow of energy from plants through all other living things.

6. Living things need a relatively constant internal environment and have various mechanisms and strategies for maintaining homeostasis.
7. Living things respond to external stimuli; taken together, these reactions constitute the behavior of the organism.
8. Living things reproduce and pass on genes, hereditary factors that control organization and metabolism. Changes in the hereditary material allow evolution to occur.
9. Organisms are adapted to various types of environments and to different ways of acquiring energy and reproducing.
10. Within an ecosystem, populations interact with each other and with the physical environment. Materials cycle through an ecosystem. Energy does not cycle.
11. Two examples of ecosystems, the coral reef and the temperate deciduous forest, show that the adaptations of organisms allow them to play particular roles within an ecosystem.

12. The human population is a part of the biosphere, and the diversity of the biosphere should be preserved in order to ensure the continuance of the human population.
13. The process of evolution explains both the unity and the diversity of life. Descent from a common ancestor explains why all organisms share the same characteristics, and adaptation to various ways of life explains the diversity of life forms.

Writing Across the Curriculum

In order to practice writing skills, students should write out the answers to any or all of the study questions and the critical thinking questions. The study questions are sequenced in the same order as the text. Suggested answers to the critical thinking questions are in appendix D.

Study Questions

1. What are the 5 common characteristics of life listed in the text?
2. What evidence can you cite to show that living things are organized?
3. Why do living things require an outside source of materials and energy?
4. What is passed from generation to generation when organisms reproduce? What has to happen to the hereditary material in order for evolution to occur?
5. Choose an organism from a temperate forest or coral reef and tell how it is adapted to its way of life.
6. What is an ecosystem, and why should human beings preserve ecosystems?
7. How does evolution explain both the unity and diversity of life?
8. What kingdoms are used in the classification system adopted by this text? What types of organisms are found in each kingdom?

Objective Questions

For questions 1–5, match the statements in the key with the sentences below.

Key:

   a. Living things are organized.
   b. Living things metabolize.
   c. Living things respond.
   d. Living things reproduce.
   e. Living things evolve.

1. Genes made up of DNA are passed from parent to child.
2. Zebras run away from approaching lions.
3. All organisms are composed of cells.
4. Cells use materials and energy for growth and repair.
5. There are many different kinds of living things.
6. Evolution from the first cell(s) best explains why
   a. ecosystems have populations of organisms.
   b. photosynthesizers produce food.
   c. diverse organisms share common characteristics.
   d. All of these.

7. Adaptation to a way of life best explains why living things
   a. display homeostasis.
   b. are diverse.
   c. began as single cells.
   d. are classified into 5 kingdoms.
8. Into which kingdom would you place a multicellular land organism that carries on photosynthesis?
   a. kingdom Monera
   b. kingdom Protista
   c. kingdom Plantae
   d. kingdom Animalia
9. Place the following labels on this diagram of chemical cycling and energy flow in an ecosystem: plants, animals, death and decay, nutrients (e.g., water, fertilizer) for plants.

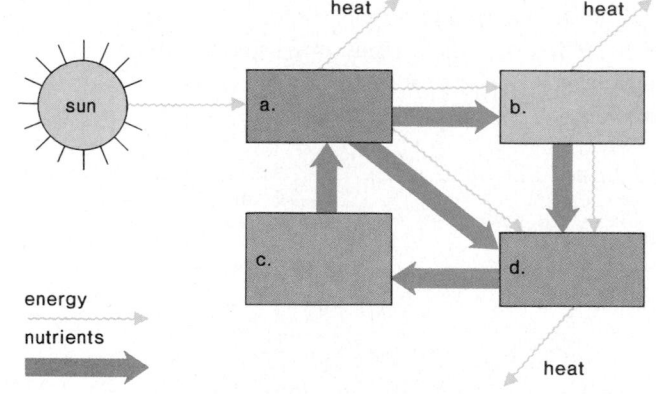

Concepts and Critical Thinking

1. *All living things evolved from a common ancestor.*

   How does this concept explain the unity of life forms?

2. *Organisms are adapted to particular ways of life.*

   How does this concept explain the diversity of life?

3. *Each level of biological organization has emergent properties.*

   How does this concept help explain the difference between the living and the nonliving?

Selected Key Terms

kingdom (king′dum) 3
cell (sel) 3
metabolism (me-tab′o-lizm) 4
homeostasis (ho″me-o-sta′sis) 4
behavior (be-hāv′yor) 5

reproduce (re″pro-dūs′) 5
gene (jēn) 5
mutation (mu-ta′shun) 7
evolution (ev″o-lu′shun) 8
species (spe′shēz) 8

adaptation (ad″ap-ta′shun) 8
environment (en-vi′ron-ment) 8
ecosystem (ek″o-sis′tem) 8
biosphere (bi′o-sfēr) 8
population (pop″u-la′shun) 8

2

The Scientific Method

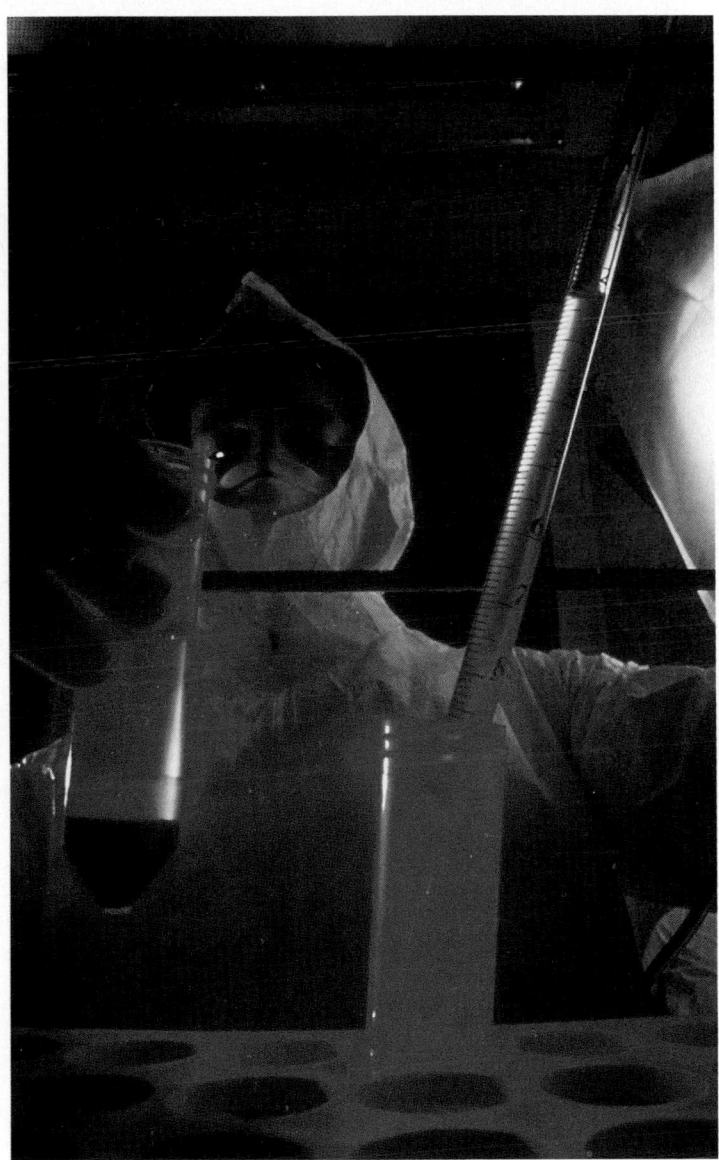

Many biological experiments are carried out in the laboratory where conditions can be rigidly controlled. This technician, who is testing blood for the presence of the HIV virus, is wearing protective clothing.

Your study of this chapter will be complete when you can

1. outline the basic steps of the scientific method;
2. explain how inductive and deductive reasoning are used during the steps of the scientific method;
3. explain why experimental results can falsify a hypothesis but cannot prove a hypothesis true;
4. give an example of an investigation that follows the steps of the scientific method, and identify the experimental variable and the dependent variable;
5. tell why investigators often report mathematical data and publish their results in scientific journals;
6. design a controlled experiment;
7. show that descriptive research also follows the scientific method;
8. explain what is meant by a "theory" in science, and give examples of scientific theories.

Science helps human beings understand the natural world and is concerned solely with information gained by observing and testing that world. Scientists, therefore, ask questions only about events in the natural world and expect that the natural world in turn will provide all the information needed to understand events. It is the aim of science, too, to be objective rather than subjective, even though it is very difficult to make objective observations and to come to objective conclusions because human beings are often influenced by their particular prejudices. Still, anything less than a completely objective observation or conclusion is not considered scientific. Finally, the conclusions of science are subject to change. Quite often in science, new studies, which might utilize new techniques and equipment, indicate that previous conclusions need to be modified or changed entirely.

> Scientists ask questions and carry on investigations that pertain to the natural world. The conclusions of these investigations are tentative and subject to change.

Scientific Method

Scientists, including biologists, employ an approach for gathering information that is known as the **scientific method.** Although the approach is as varied as scientists themselves, there are still certain processes that can be identified as typical of the scientific method. Figure 2.1 outlines the primary steps in the scientific method. On the basis of **data** (factual information), which may have been collected by someone previously, a scientist formulates a tentative statement, called a **hypothesis,** that can be used to guide further observations and experimentation. It is often said that this portion of the scientific method involves **inductive reasoning;** that is, a scientist uses isolated facts to arrive at a general idea that may explain a phenomenon. For example, scientists not only observed the prevalence of dark-colored peppered moths on trees in polluted areas, they also observed the prevalence of light-colored peppered moths on trees in nonpolluted areas. This caused them to formulate the hypothesis that predatory birds were responsible for the unequal distribution of these moths because the birds feed on the moths they can see (see fig. 20.10).

Once the hypothesis has been stated, **deductive reasoning** comes into play. Deductive reasoning begins with a general statement that infers a specific conclusion. It often takes the form of an "if ... then" statement: *If* predatory birds are responsible for

Figure 2.1

Steps involved in the scientific method. Inductive reasoning is used to formulate a hypothesis from accumulated scientific data, and deductive reasoning is used to decide what observations and experiments are appropriate in order to test the hypothesis.

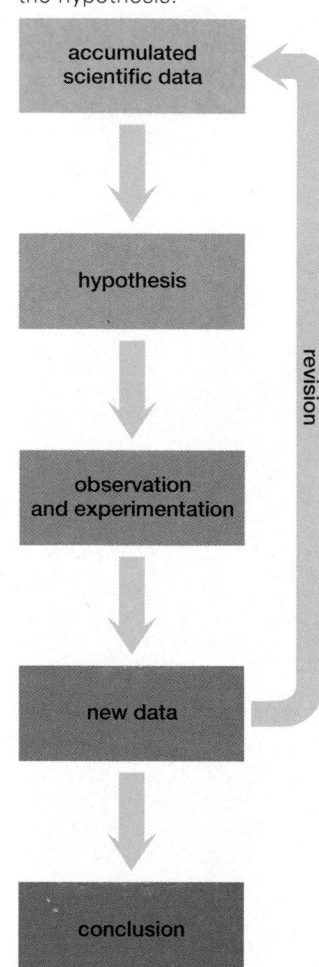

accumulated scientific data

hypothesis

observation and experimentation

new data

revision

conclusion

the unequal distribution of dark- and light-colored moths, *then* it will be possible to get evidence of birds feeding primarily on light-colored moths in polluted areas and on dark-colored moths in nonpolluted areas.

To see if this deduction was correct, the British scientist H.B.D. Kettlewell performed an experiment. He released equal numbers of the 2 types of moths in the 2 different areas. He observed that birds captured more dark-colored moths in nonpolluted areas and more light-colored moths in polluted areas. From the unpolluted area, Kettlewell recaptured 13.7% of the light-colored moths and only 4.7% of the dark form. From the polluted area, he recaptured 27.5% of the dark form and only 13% of the light-colored moths. Mathematical data like these are preferred because they are objective and cannot be influenced by the scientist's subjective feelings. These particular data supported the hypothesis (fig. 2.2).

Sometimes, as in this case, the data support the hypothesis; however, at other times, the data do not support the hypothesis, and it has to be rejected. Knowing that some future observation and/or experiment might falsify the hypothesis, a scientist never says that the data "prove" the hypothesis. In other words, a hypothesis cannot be proven true, but it can be shown to be false. Because of this feature, some think of science as what is left after alternative hypotheses have been rejected.

> The scientific method consists of making observations, formulating a hypothesis, testing, and coming to a conclusion. Hypotheses can be supported but not proven true. Hypotheses can be shown to be false, and then they are rejected.

Reporting the Experiment

It is customary for an investigator to report findings in a scientific journal so that the design and results of an experiment are available to all. This is especially desirable because all experimental results must be repeatable; that is, other scientists using the same procedures should get the same results. If they do not, the original data cannot be considered to support the hypothesis.

Often, too, the authors of a report suggest other experiments that could be done to clarify or broaden the understanding of the matter under study.

> The results of observations and experiments are published in scientific journals, where they can be examined by all. These results must be repeatable, that is, they should be obtainable by anyone doing the exact same procedure.

Figure 2.2

Analysis of experiment performed by Kettlewell.

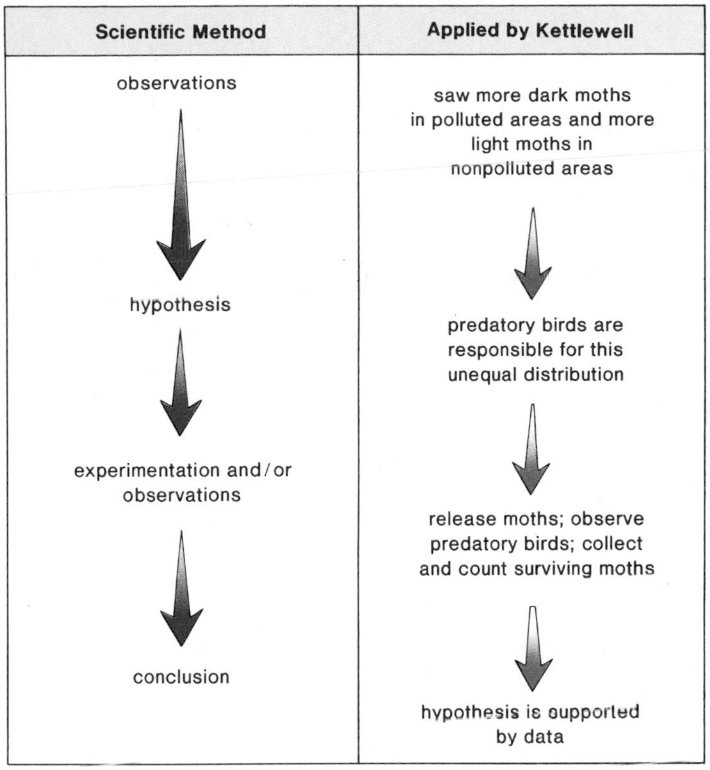

Scientific Method	Applied by Kettlewell
observations	saw more dark moths in polluted areas and more light moths in nonpolluted areas
↓	↓
hypothesis	predatory birds are responsible for this unequal distribution
↓	↓
experimentation and/or observations	release moths; observe predatory birds; collect and count surviving moths
↓	↓
conclusion	hypothesis is supported by data

Revision

Science is an ongoing endeavor. Because hypotheses are never proven true, they are reviewed constantly. Particularly when new procedures and instruments make it possible, the process of revision begins. Previous data are used as a starting point, and then new data is collected. The new data may require a new conclusion and the formulation of revised hypotheses, which will then be tested.

Controlled Experiment

Kettlewell performed his experiment in the field, but scientists often prefer to work in the laboratory, where they can oversee all elements of the experiment. These are controlled experiments.

Experimental Variable	Dependent Variable
Element of the experiment being tested	Result or change that occurs due to the experimental variable

Elements of an Experiment

When conducting a controlled experiment, all conditions should be held constant except the one being tested. This one element is called the **experimental variable;** in other words, the investigator deliberately varies this portion of the experiment. Then the investigator observes the effects of the experiment; in other words, he or she observes the **dependent variable:** To be absolutely sure that the results are due to the experimental variable and not due to some unknown factor, it is customary to have a separate sample called the control group. A **control group** undergoes all the steps of the experiment except the one being tested.

Design of an Experiment

As an example of a controlled experiment, suppose physiologists wish to determine if sweetener S is a safe additive for foods. On the basis of available information, they hypothesize that sweetener S has no effect on health at a low concentration but causes bladder cancer at a high concentration. They then decide to feed sweetener S to groups of mice at ever-greater percentages of the total intake of food:

Group 1: no sweetener S in food (control group)
Group 2: sweetener S in 5% of food
Group 3: sweetener S in 10% of food
↓
Group 11: sweetener S in 50% of food

The experimenters first place a certain number of randomly chosen, inbred (genetically identical) mice into the various groups--say 10 mice per group. It is hoped that if any of the mice are different from the others, random selection will distribute them evenly among the groups. The experimenters also make sure that all environmental conditions, such as availability of water, cage conditions, and temperature of surroundings are the same for all groups. The food for each group is exactly the same except for the amount of sweetener S. (The amount of sweetener S is the experimental variable.) After a designated period of time, the animals are killed and dissected to examine their bladders for evidence of cancer. (Evidence of bladder cancer is the dependent variable.) Figure 2.3 outlines the design of this experiment.

Results of an Experiment

Usually the data from experiments such as these are presented in the form of a table or graph (fig. 2.4). A statistical test may be run to determine if the difference in the number of cases of bladder cancer among the various groups is significant. After all, if a significant number of mice in the control group develop cancer, the results may be invalid. On the basis of the results, the experimenters try to develop a recommendation concerning the safety of adding sweetener S to the food of humans. They may determine that ever-larger amounts of the sweetener over 10% of food intake are expected to cause a progressively increased incidence of bladder cancer, for example.

Many scientists work in laboratories in which they carry out controlled experiments.

Descriptive Research

Scientists don't always gather data by experimenting. Much of the data they gather is purely observational, but even so, the steps described for the scientific method are still applicable: observations are made, a hypothesis is formulated, predictions are made on

Figure 2.3
Design of a controlled experiment. Genetically identical mice are randomly divided into a control group and the experimental groups. All groups are exposed to the same environmental conditions, such as housing, temperature, and water supply. Only the experimental groups are subjected to the test (experimental variable)--in this case the presence of sweetener S in the food. At the end of the experiments, all mice are examined for evidence of bladder cancer.

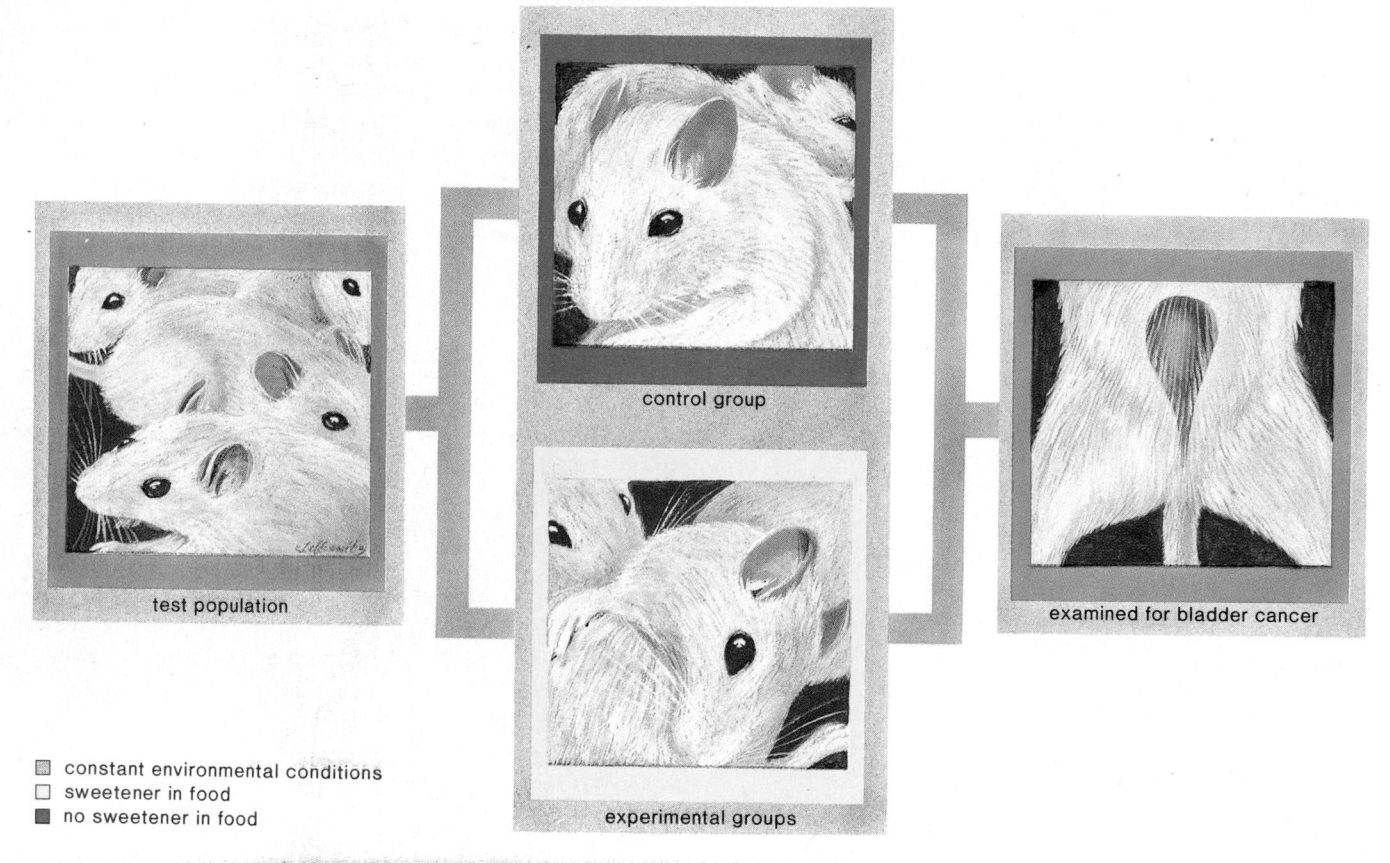

control group

test population

examined for bladder cancer

experimental groups

▨ constant environmental conditions
☐ sweetener in food
■ no sweetener in food

the basis of the hypothesis, and data are collected that support or disprove the hypothesis. For example, Kenneth E. Glander[1] was interested in determining the feeding pattern of mantled howling monkeys, which are herbivores that feed on plant products (fig. 2.5). He believed 2 of the hypotheses he found in the scientific literature could apply to the monkeys.

1. Most large herbivores try to obtain an optimal mix of nutrients.
2. Mammalian herbivores tend to avoid plant foods that contain toxins.

Glander felt that there was insufficient data to decide if one or both of these statements were pertinent to the feeding behavior of the howling monkeys. So he and his associates made new observations.

1. They marked off 2 test sites and identified every type of tree in each site. Altogether they identified 1,699 trees.

2. They observed the feeding behavior of a group of monkeys consisting of 2 adult males, 6 adult females, and 5 juveniles. A total of 1,853 hours of animal observation was accumulated for a period of one year. The average observation day lasted 12 hours and 11 minutes.

3. They performed chemical analyses of fresh plant samples from about 50 trees to determine their nutritional (amino acid) and toxin contents.

Glander found that the study group of monkeys obtained a majority of their food from 62 relatively rare tree species. During the 12-month study period the monkeys obtained all of their food from 331 different kinds of trees, or 19.5% of all trees available to them. Further, they tended to choose mature leaves and fruits in the wet season and new leaves and flowers in the dry season. The chemical analyses of the plant products showed that the howlers were maximizing their intake of water, digestible protein, total protein, and amino acids and minimizing their intake of fiber, ash, and toxic plant compounds. Glander concluded that the howlers chose those foods that gave the greatest net nutrient return. Glander stated, "A tropical

[1]Glander, K. E. 1981. Feeding patterns in mantled howling monkeys. In *Foraging behavior: ecological, ethological and psychological approaches,* ed. A. C. Kamil and T. D. Sargent, 231-57. New York: Garland STPM Press.

Figure 2.4

Scientists often acquire mathematical data, which they report in the form of a table or a graph. Mathematical data are more decisive and objective than visual observations. The data in this instance suggest that there is an ever greater chance of bladder cancer development if the food contains more than 10% sweetener S. The same experiment is repeated many times to test these results, and statistical analyses are conducted to see if the results are significant or simply due to chance alone.

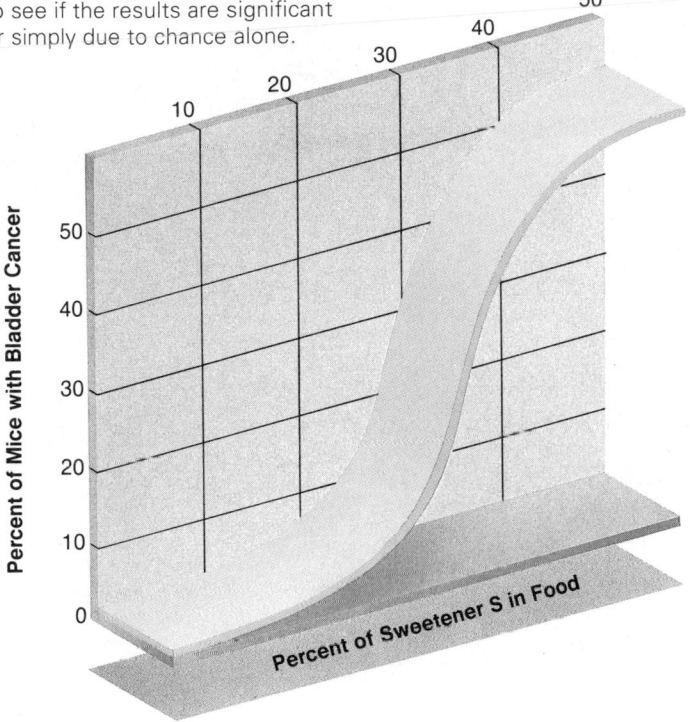

Figure 2.5

Mantled howling monkeys are herbivores that feed on plant products in South American forests. They tend to choose mature leaves and fruit in the wet season and new leaves and flowers in the dry season.

forest should not be viewed as a well-stocked larder waiting to be exploited, but rather as a spatially and temporally changing mosaic of items of varying value and availability."

This proposed model of the nature of the food sources for forest-dwelling herbivores is one of the most important outcomes of Glander's observations. A scientific **model** is a suggested explanation that can help direct future research. Again, the steps in the scientific method can be applied (fig. 2.6).

> A large proportion of scientific information is based on purely observational (descriptive) data, but the steps previously suggested for the scientific method still apply.

Theories and Principles

The ultimate goal of science is to understand the natural world in terms of **theories**, which are interpretations that take into account the results of many experiments and observations. In a movie, a detective may claim to have a theory about the crime. Or you may say that you have a theory about the won-lost record of your favorite baseball team. But in science, the word *theory* is reserved for a conceptual scheme that is supported by a large number of observations and has not yet been found lacking.

Figure 2.6

Analysis of Glander's study of howling monkeys.

Scientific Method	Applied by Glander
observations	studied scientific literature
↓	↓
hypothesis	feeding behavior may be influenced by nutrient value of food including avoidance of toxins
↓	↓
experimentation and/or observations	observe monkeys feeding and perform chemical analyses of food selected
↓	↓
conclusion	hypothesis is supported by data

Table 2.1
Unifying Theories of Biology

Name of Theory	Concept	Investigator
Cell	All organisms are composed of cells	Matthias Schleiden, 1838 Theodor Schwann, 1839
Biogenesis	Life comes only from life	Louis Pasteur, 1865
Homeostasis	The internal environment remains within a normal range	Claude Bernard, 1851
Evolution	All living things have a common ancestor and are adapted to a particular way of life	Charles Darwin, 1858
Gene	Organisms contain coded information that dictates their form, function, and behavior	Gregor Mendel, 1866 James Watson and Francis Crick, 1953

Table 2.1 lists some of the unifying theories of biology. You can see that in general the theories pertain to the characteristics of life we discussed in chapter 1. The theory of evolution enables scientists to understand the history of life, the variety of living things, and the anatomy, physiology, development, and even behavior of organisms. Because the theory of evolution has been supported by so many observations and experiments for over 100 years, some biologists refer to the "principle" of evolution, suggesting that this is the appropriate term for theories that are generally accepted as valid by an overwhelming number of scientists.

Sometimes scientists do experiments that they *expect* to be explainable by a certain theory:

> *The role of an experiment is illustrative. It allows a scientist to demonstrate the power of his theory, not as a collection of truths, but as a set of ideas. When an experiment succeeds, this shows that a certain way of describing the world has proved itself useful. When an experiment fails it shows that one's concepts were inadequate or confused.*[2]

For example, in 1975 David P. Barash[3] conducted a study of 2 mountain bluebird mating pairs, seemingly with the intention of using the theory of evolution to explain their behavior. He posted a model of a male bluebird 1 m from each nest while the resident male was foraging. Upon returning to its nest, each resident encountered the model in proximity to his female. This was done during 3 different stages of the nesting cycle. The behavior of the male toward the model and toward his female during the first 10 minutes of his discovery of the model was observed and noted. As a control, a model male robin was also posted 4 days after the initial

[2]Harre, R. 1981. *Great scientific experiments*. London: Phaidon Press Limited.

Figure 2.7
Results of an experiment with nesting mountain bluebirds. The graph shows that resident males are more likely to aggressively approach a male model and female mate before the nest is completed, less likely after the first egg is laid, and least likely after the eggs have hatched. The experimenter gives an evolutionary interpretation to the results: it is more adaptive for a male bluebird to be aggressive before mating in order to be sure the offspring will be his and less adaptive at the other stages mentioned.

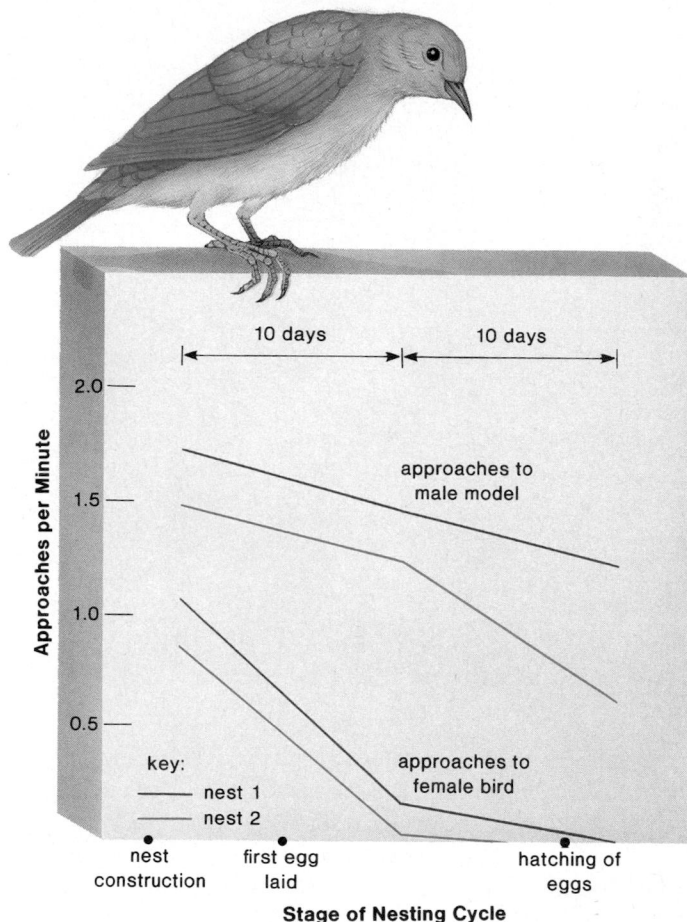

presentation of the bluebird model. The robin model elicited no response. The number of aggressive approaches toward the bluebird male model and female mate is given in graph form in figure 2.7. Aggression was most severe when the bluebird male model was presented before the first egg was laid, less severe once the first egg was laid, and least severe after the eggs had hatched.

In his report of this experiment, Barash explains how evolutionary theory can account for these results. It is adaptive for males to be especially aggressive before the eggs are laid because there is a limited number of nesting sites available to the birds, and he wants to be sure he is the one mating and passing on his genes. It is adaptive for males to be less aggressive after the first egg is laid; the reproductive process has begun and the male is more sure the

[3]Barash, D. P. 1976. The male response to apparent female adultery in the mountain bluebird, *Sialia currucoides:* An evolutionary interpretation. *American Naturalist* 110:1097-101.

offspring will be his. Finally, it would be maladaptive for the male to spend time and energy being too aggressive toward a rival and his mate after hatching because his offspring are already present.

Barash's discussion of his results seems to support the suggestion that scientists sometimes do experiments with the intention of explaining their results on the basis of a particular theory. He shows how his results can be given an evolutionary interpretation. It is adaptive for a male bluebird to be aggressive toward rivals, especially before mating, because it helps ensure that he will be the one to pass on genes. Barash shows that this is an observed characteristic among these birds. In doing so, he also shows that in addition to anatomy and physiology, behavior is subject to an evolutionary interpretation.

> It is the aim of science to formulate theories and principles that encompass a number of hypotheses concerning the natural world. The power of a theory is evident when it can be used to explain diverse observations and experiments.

Science and Social Responsibility

There are many ways in which science has improved life. The most obvious examples are in the field of medicine. The discovery of antibiotics, such as penicillin, and the vaccines for polio, measles, and mumps have increased the life span by decades. Cell biology research may someday help us understand the mechanisms that cause cancer. Genetic research has produced new strains of agricultural plants that have eased the burden of feeding our burgeoning population.

But science has also fostered technologies of grave concern to some of us. Even though we would not like a cancer patient to be denied the benefits of radiation therapy, we are concerned about the nuclear power industry and do not want nuclear wastes stored near our homes. Then, too, there may be undesirable side effects even from those technologies of which we approve. Not one of us wishes to go back to the days in which large numbers of people died of measles or polio. Yet as science has conquered one disease after another, the world's death rate has fallen and the human population has exploded. Few of us are willing to give up technology's gift of the private automobile, but we are concerned about the amount of air pollution that automobiles generate.

All too often we blame science for these side effects and say that our lives are infinitely more threatened than they were in "the good old days." We think that scientists should label research "good" or "bad" and be able to predict whether any resulting technology would primarily benefit or harm humanity. Yet science, by its very nature, is impartial and simply attempts to study natural phenomena. Science does not make ethical or moral decisions. When we wish to make value judgments, we must go to other fields of study to find the means to make those judgments. And these judgments must be made by all people. The responsibility for how we use the fruits of science, including a given technology, must rest with people from all walks of life, not upon scientists alone.

Scientists should provide the public with as much information as possible when such issues as nuclear power or recombinant DNA technology are being debated. Then, they, along with other citizens, can help make decisions about the future role of these technologies in our society. All men and women have a responsibility to decide how to use scientific knowledge so that it benefits the human species and all living things.

Summary

1. Scientific investigations always pertain to the natural world, and the conclusions of the investigations are always subject to change.

2. In general, scientists use the scientific method. Based on their own observations or on ones found in the scientific literature, inductive reasoning is used to formulate a hypothesis. Using deductive reasoning, a plan of action is devised and carried out. The results either support or fail to support the hypothesis.

3. Most scientists do controlled experiments. Particularly, in the laboratory (1) the environmental conditions are held constant, (2) there is a control group that goes through all the steps of the experiment except the one being tested, and (3) one or more experimental groups are involved.

4. In a controlled experiment, the experimental variable is that portion of the experiment that is being tested and the dependent variable is the results of the experiment.

5. Much of biological investigation is descriptive rather than experimental. Descriptive research requires that the investigator make careful observations. Descriptive research still follows the steps in the scientific method.

6. The goal of science is to understand the natural world in terms of theories, which can be used to explain diverse observations and experiments. Some theories are so well supported that many suggest they should be called principles.

7. Investigators sometimes make observations or do experiments with the purpose of explaining their results on the basis of a particular theory.

Writing Across the Curriculum

In order to practice writing skills, students should write out the answers to any or all of the study questions and the critical thinking questions. The study questions are sequenced in the same order as the text. Suggested answers to the critical thinking questions are in appendix D.

Study Questions

1. There is no standard scientific method but still it is possible to identify certain steps that are commonly accepted as comprising this method. What are these steps?
2. Apply these steps to Kettlewell's experiment regarding unequal distribution of dark- and light-colored moths.
3. Which of the steps identified in question 1 requires the use of inductive reasoning? Which requires the use of deductive reasoning?
4. Why do scientists prefer mathematical data and use charts and graphs to report such data?
5. For what purposes are scientific journals useful?
6. Design a controlled experiment using an example other than the one given in the text.
7. What is descriptive research, and how does it differ from experimental research?
8. What is the difference between a hypothesis and a theory? List 5 biological theories that pertain to all living things, and name some of the investigators that helped formulate these theories.
9. An investigator spills dye on a culture plate and then notices that the bacteria live despite exposure to sunlight. He decides to expose 2 culture plates to ultraviolet light--one plate contains bacteria and dye; the other plate contains only bacteria. The bacteria on both plates die. Fill in the right-hand portion of this diagram:

Scientific Method	Example
observations	a.
hypothesis	b.
experimentation and/or observations	c.
conclusion	d.

Objective Questions

1. Which of these is the second step in the scientific method?
   a. experimentation
   b. conclusion
   c. researching the scientific literature
   d. formulating the hypothesis
2. Which of these is the third step in the scientific method?
   a. experimentation
   b. conclusion
   c. researching the scientific literature
   d. formulating the hypothesis
3. With which of these steps in the scientific method do you associate inductive reasoning?
   a. experimentation
   b. conclusion
   c. researching the scientific literature
   d. formulating the hypothesis
4. With which of these steps in the scientific method would you associate deductive reasoning?
   a. experimentation
   b. conclusion
   c. researching the scientific literature
   d. formulating the hypothesis
5. Which is the experimental variable in this experiment?
   a. Conditions like temperature and housing are the same for all groups.
   b. The amount of sweetener S in food is varied.
   c. Two percent of the group who were fed food that was 10% sweetener S got bladder cancer, and 90% of the group who were fed food that was 50% sweetener S got bladder cancer.
   d. The data were presented as a graph.
6. Which is the control group in this experiment?
   a. All mice in this group died because a lab assistant forgot to give them water.
   b. All mice in this group received food that was 10% sweetener S.
   c. All mice in this group received no sweetener S in food.
   d. Some mice in all groups got bladder cancer; therefore, there was no control group.
7. Which is an example of descriptive research?
   a. Jones put broken egg shells in the nest of some gulls and watched their behavior.
   b. Smith measured the length of the twigs of all trees in the marked off area.
   c. Green put pesticide into one jar of amoebas but not into the other jar.
   d. Kettlewell counted how many moths the birds did not eat.
8. Which of these statements is correct?
   a. Because scientists speak of the "theory of evolution" they believe that evolution does not have much merit.
   b. A theory is simply a hypothesis that needs further experimentation and observation.
   c. Theories are hypotheses that have failed to be supported by experimentation and observation.
   d. The term theory in science is reserved for those hypotheses that have proven to have the greatest explanatory power.

Concepts and Critical Thinking

1. *Scientists use the scientific method to gather information about the natural world.*

   When utilizing the scientific method, hypotheses can be proven false but not true. Explain.

2. *There are several underlying theories in biology.*

   What is the difference between the use of the word theory in the everyday sense and in the scientific sense?

3. *All persons, not just scientists, should decide on the role of various technologies in society.*

   Science is not obligated to make policy decisions. Why not?

Selected Key Terms

scientific method (si"en-tif'ik meth'ud) 14
data (da'tah) 14
hypothesis (hi-poth'e-sis) 14
inductive reasoning (in-duk'tiv re'zun-ing) 14
deductive reasoning (de-duk'tiv re'zun-ing) 14

experimental variable (ek-sper" i-men'tal va're-ah-b'l) 15
dependent variable (de-pen'dent va're-ah-b'l) 15

control group (kon-trōl' grōōp) 15
model (mod'el) 17
theory (the'o-re) 17

Suggested Readings

Baker, J., and Allen, G. 1971. *Hypothesis, prediction, and implication in biology.* Reading, Mass.: Addison-Wesley.

Barash, D. P. 1976. The male response to apparent female adultery in the mountain bluebird, *Sialia currucoides:* An evolutionary interpretation. *American Naturalist.*

Glander, K. E. 1981. Feeding patterns in mantled howling monkeys. A. C. Kamil and T. D. Sargent, eds. *Foraging behavior: ecological, ethological and psychological approaches.* New York: Garland STPM Press.

Life: Origin and evolution: Readings from Scientific American. Intro. by C. E. Folsome. 1979. San Francisco: W. H. Freeman.

Scientific American Editors. 1978. *Evolution: A Scientific American book.* San Francisco: W. H. Freeman.

Volpe, E. P. 1985. *Understanding evolution.* 5th ed. Dubuque, Iowa: Wm. C. Brown Publishers.

The Cell

All organisms, including these wandering albatrosses, have the same levels of organization. In this part, we examine the structure and function of atoms, molecules, and cells—the first few levels of organization. Organization determines what an organism is like and how it acquires energy and carries on life's functions. These birds are performing a courtship ritual prior to reproducing.

3

Basic Chemistry

A hippopotamus keeps cool by remaining in water. Water has many biological functions, both without and within organisms. Cells are largely composed of this inorganic molecule, which facilitates chemical reactions and helps maintain a normal body temperature because it is slow to heat.

Your study of this chapter will be complete when you can

1. name and describe the subatomic particles of an atom, indicating which one accounts for the occurrence of isotopes;
2. describe and discuss the energy levels (electron shells) of an atom, including the orbitals of the first 2 levels;
3. draw a simplified atomic structure of any atom with an atomic number less than 20;
4. distinguish between ionic and covalent reactions, and draw representative atomic structures for ionic and covalent molecules;
5. tell which atom has been reduced and which has been oxidized in a particular reaction;
6. describe the chemical properties of water, and explain their importance for living things;
7. define an acid and a base; describe the pH scale, and state the significance of buffers.

Sometimes it is difficult to realize that life has a chemical basis. Maybe this is because we usually deal with whole objects, such as trees, dogs, and people, and do not often consider their minute composition. If we did consider it, though, we would soon learn that an organism is composed of organ systems with organs made up of tissues, that tissues contain cells made up of molecules, and that the molecules themselves are atoms bonded together (fig. 3.1). This shows that fundamentally, living things are made up of chemicals.

The structure and the function of living things are dependent upon chemicals. The success of the cheetah that darts across the plain to capture the gazelle is dependent upon the chemical reactions that occur throughout its body, such as those that allow its eyes to see and its muscles to contract. Also, a bittersweet vine can climb a fence only because of the chemical reactions that permit growth to occur.

As recently as the nineteenth century, scientists believed that only nonliving things, like rocks and metals, consisted of chemicals alone. They thought that living things, such as the cheetah and the bittersweet vine, were different and had a special force, called a vital force, necessary to being alive. Scientific investigation, however, has repeatedly shown that both nonliving and living things have only a chemical and physical basis. Therefore, it is appropriate for us to study chemical principles as an introduction to our study of life. In this chapter, we will consider the basic facts of chemistry, and in the next chapter we will take a look at some complex molecules that are especially associated with

Figure 3.1

Levels of organization from atom to organism. The chemical basis of life becomes apparent when we realize that all living things are fundamentally composed of small units of matter called atoms. Atoms join together to form molecules. Molecules are organized into cells. Cells of the same type make up tissues, and different types of tissues make up organs. Several organs are found within organ systems, and organ systems make up the organism.

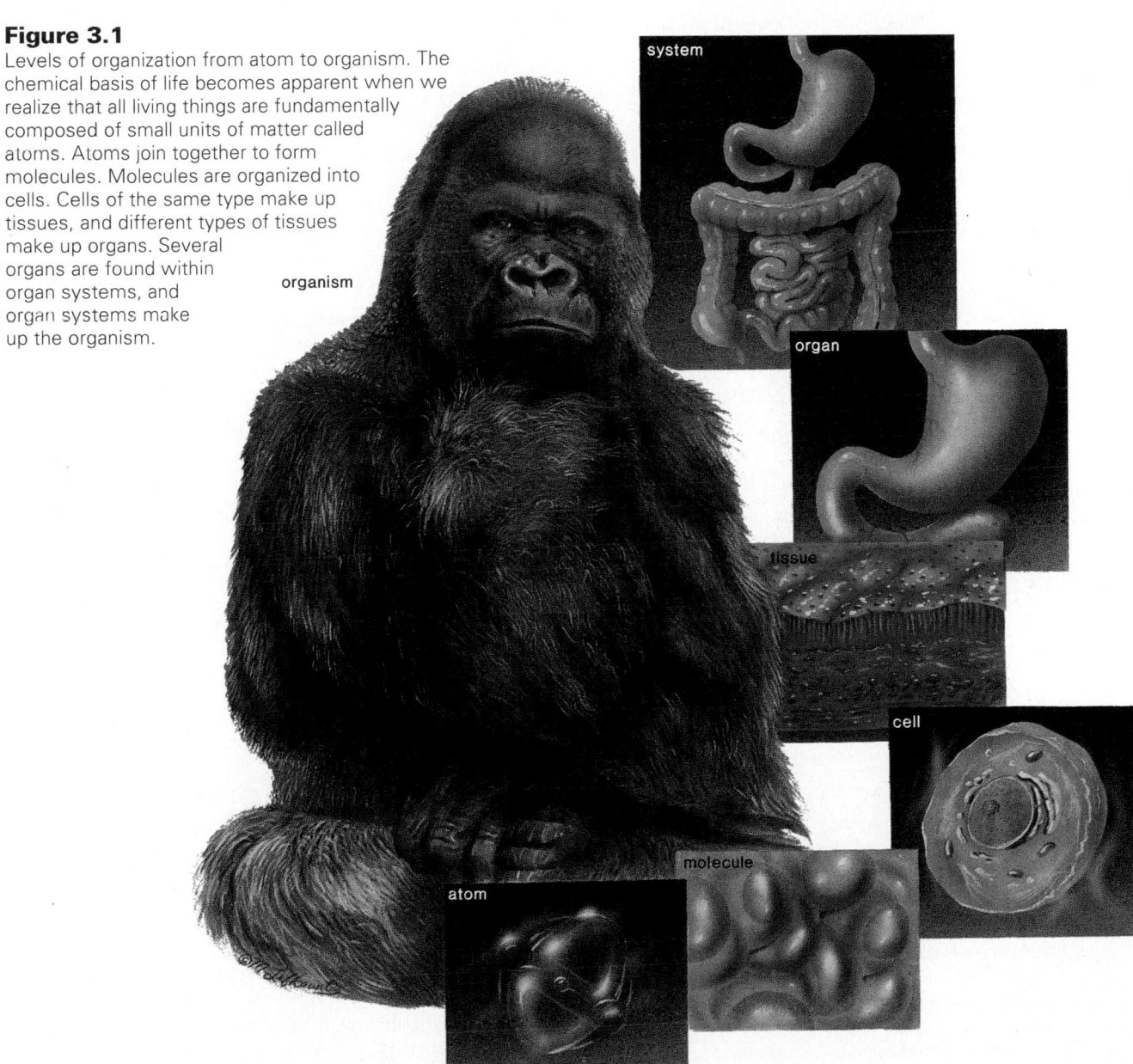

system

organism

organ

tissue

cell

molecule

atom

Table 3.1
Common Elements in Living Things

Element*	Atomic Symbol	Atomic Number	Atomic Weight[d]	Comment
Hydrogen	H	1	1	These elements make up most biological molecules.
Carbon	C	6	12	
Nitrogen	N	7	14	
Oxygen	O	8	16	
Phosphorus	P	15	31	
Sulfur	S	16	32	
Sodium	Na	11	23	These elements occur mainly as dissolved salts.
Magnesium	Mg	12	24	
Chlorine	Cl	17	35	
Potassium	K	19	39	
Calcium	Ca	20	40	
Iron	Fe	26	56	These elements also play vital roles in organisms.
Copper	Cu	29	64	
Zinc	Zn	30	65	

*A periodic table of the elements appears in appendix B.
[d]Average of most common isotopes.
Note: The atomic number gives the number of protons (and electrons in electrically neutral atoms). The number of neutrons is equal to the atomic weight minus the atomic number. There are no neutrons in a hydrogen atom.

living things. Then we will discuss how complex molecules are organized into the structures found within cells. In this way, not only the sameness but also the uniqueness of living things will begin to be apparent.

Living things are composed of chemicals organized to provide the structure and to perform the functions necessary to life.

Only 6 elements—carbon, hydrogen, nitrogen, oxygen, phosphorus, and sulfur—make up most (about 98%) of the body weight of organisms. The acronym CHNOPS helps us remember these 6 elements. The other elements noted in table 3.1 also play important roles in cells.

All living and nonliving things are matter composed of elements. Six elements in particular are commonly found in living things.

Matter

Matter refers to anything that takes up space and has weight.[1] It is helpful to remember that matter can exist as a solid, a liquid, or a gas. Then we can realize that not only are we matter but so too are the water we drink and the air we breathe.

[1]The term *atomic weight* will be used as if it was the same as atomic mass. Weight, however, is dependent on gravity, and therefore atomic weight (but not mass) would be different on the moon, for example, compared to the earth.

All matter, both nonliving and living, is composed of certain basic substances called **elements.** Most matter contains more than one type of element, but some matter contains only one type of element. For example, air contains more than one type of element, but the oxygen we respire contains only the element of that name. It is quite remarkable that there are only 92 naturally occurring elements (see appendix B). We know these are elements because any particular element cannot be broken down to substances with different properties (a property is a chemical or physical characteristic, such as denseness, smell, taste, and reactivity).

Atoms

In the early 1800s, the English scientist John Dalton proposed that elements actually contain tiny particles called **atoms.** He also deduced that there is only one type of atom in each type of element. You can see why, then, the name assigned to each element is the same as the name assigned to the type of atom it contains. Some of the names we use for the elements (atoms) are derived from English and some are derived from Latin. One or 2 letters create the *atomic symbol,* which stands for this name. For example, the symbol H stands for a hydrogen atom, and the symbol Na (for *natrium* in Latin) stands for a sodium atom. Table 3.1 gives the atomic symbols for the other elements (atoms) commonly found in living things.

Atoms Have Weight

From our discussion of elements, we would expect atoms to have a certain weight. The weight of an atom is in turn dependent upon the presence of certain subatomic particles. Although physicists have identified a number of subatomic particles, we will consider only the most stable of these: **protons, neutrons,** and **electrons.** Protons and neutrons are located within the nucleus of an atom, and electrons move about the nucleus. Figure 3.2 shows the arrangement of the subatomic particles in helium, an atom that has only 2 electrons.

Our concept of an atom has changed greatly since Dalton's day. If we could draw an atom the size of a football field, the nucleus would be like a gumball in the center of the field and the electrons would be tiny specks whirling about in the upper stands. Most of an atom is empty space. We should also realize that we can only indicate where the electrons are expected to be most of the time. In our analogy, the electrons may even stray outside the stadium at times.

Atoms contain protons, neutrons, and electrons that are arranged in a definite manner.

The subatomic particles are so light that their weight is indicated by special units called atomic mass units (table 3.2). Protons and neutrons each have only one atomic unit of weight. In comparison electrons have almost no weight; an electron weighs about 1/1,800 that of a proton or neutron. Therefore, it is customary to disregard the combined weight of the electrons when calculating the total weight, called the *atomic weight,* of an atom.

Figure 3.2

Model of helium (He), a simple atom. Atoms contain subatomic particles which are located where shown. Protons and neutrons are found within the nucleus, and electrons are outside the nucleus. **a.** The stippling shows the probable location of the electrons in the helium atom. **b.** The average location of electrons is sometimes represented by a circle, and each electron is represented by a small sphere.

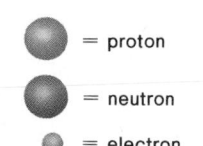

= proton

= neutron

= electron

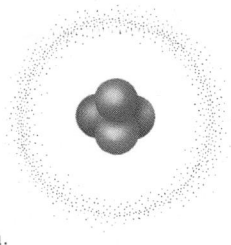

a.

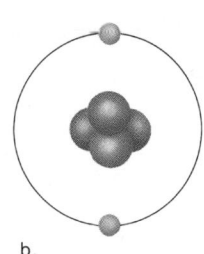

b.

Figure 3.3

All atoms of a particular element have the same atomic number, which tells the number of protons. This number is often written as a subscript in the lower left of the atomic symbol. The atomic weight (mass) is often written as a superscript at the upper left of the atomic symbol. For example, the carbon atom can be noted in this way:

atomic weight —— $^{12}_{6}C$ —— atomic symbol
atomic number ——

Isotopes are atoms of the same element that differ in weight. The number of protons stays the same, but the number of neutrons can vary. The above configuration is the most common isotope of carbon; other isotopes are

$$^{13}_{6}C \quad ^{14}_{6}C \quad ^{15}_{6}C$$

Table 3.2
Subatomic Particles

Name	Charge	Weight
Electron	One negative unit	Almost none
Proton	One positive unit	One atomic unit
Neutron	No charge	One atomic unit

All atoms of an element have the same number of protons. This is called the atom's *atomic number*. In table 3.1, atoms are listed according to increasing atomic number, as they are in the periodic table of the elements found in appendix B. This indicates that it is the number of protons (i.e., the atomic number) that makes an atom unique. The number of protons not only contributes to the physical properties (weight) of an atom but also *indirectly* determines the chemical properties, which we will discuss in a later section.

Isotopes

All atoms of an element have the same number of protons but may differ in the number of neutrons. Table 3.1 allows you to calculate only the usual number of neutrons for hydrogen, carbon, and oxygen. But any specific carbon atom, for example, may have a few more or a few less neutrons than the number indicated. Atoms that have the same atomic number and differ only by the number of neutrons are called **isotopes** (fig. 3.3). Most isotopes are stable, but a few are unstable and tend to break down to more stable forms. They emit radiation as they break down. These radioactive isotopes can be detected by physical or photographic means and are used by biologists as "labels" in biochemical experiments. For example, researchers used radioactive carbon dioxide ($^{14}CO_2$) to detect the sequential biochemical steps occurring during pho-

Figure 3.4

The thyroid gland is a butterfly-shaped organ located in front of the throat region of humans. It produces the hormone thyroxine, which contains iodine. Therefore, if a patient is administered radioactive iodine (atomic number 53) a scan of the thyroid viewed sometime later will show the distribution of the iodine throughout the tissue.

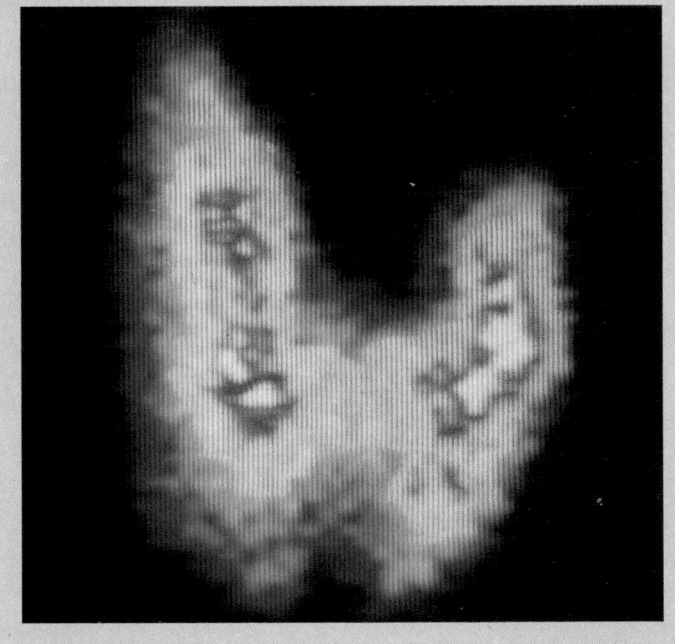

tosynthesis. Radioactive isotopes are sometimes used in medical diagnostic procedures also. For instance, because the thyroid gland uses iodine, it is possible to administer radioactive iodine and then scan the gland to determine the location of the iodine isotope (fig. 3.4).

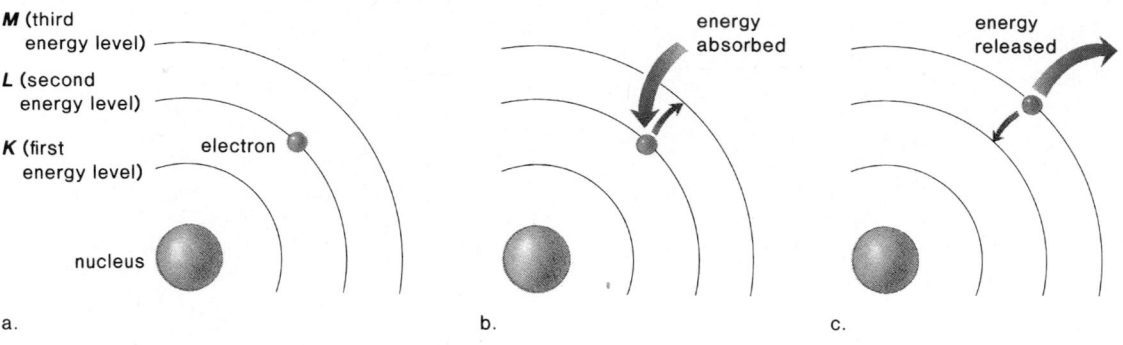

Carbon[14]

Most carbon atoms have 6 protons and 6 neutrons (shown as ^{12}C, carbon 12) and are not radioactive; however, ^{14}C (carbon 14) has 6 protons and 8 neutrons and is radioactive. With the release of radioactivity, ^{14}C spontaneously breaks down over a period of time into ^{14}N (nitrogen 14). A small amount of ^{14}C is present in all living things. Scientists can determine the age of a fossil because the rate at which ^{14}C breaks down is known. If the fossil still contains organic matter, the relative amounts of ^{14}C and ^{14}N are measured and the age is calculated. This method is unreliable for measuring fossil ages in excess of 20,000 years, however, because the radioactivity is so slight in such fossils that it is difficult to make an accurate determination.

Atoms Have Chemical Properties

Protons and electrons carry a charge; protons have a positive (+) electrical charge and electrons have a negative (-) electrical charge. When an atom is electrically neutral, the number of protons equals the number of electrons. Therefore, in such atoms, the atomic number tells you the number of electrons. For example, a carbon atom has an atomic number of 6 (table 3.1) and when electrically neutral, it has 6 electrons. Atoms have different chemical properties (the way they react with other atoms) because they each have a different number of electrons arranged about the nucleus.

> The chemical properties of atoms differ because the number and the arrangement of their electrons are different.

Atomic Energy Levels

All electrons have the same weight and charge, but they vary in energy content. **Energy** is defined as the ability to do work. We recognize this in everyday life by suggesting that those who are energetic do more work than those who are not. Electrons differ in the amount of their *potential energy,* that is, stored energy ready to do work. The relative amount of stored energy is designated by placing electrons at *energy levels,* also called *electron shells,* about the nucleus (fig. 3.5). Electrons with the least amount of potential energy are located in the shell closest to the nucleus, called the *K* shell. Electrons in the next higher shell, called the *L* shell, have more energy, and so forth as we proceed from shell to shell outside the nucleus. An analogy may help you appreciate that the farther electrons are from the nucleus, the more potential energy they possess. Falling water has potential energy, as witnessed by how it can turn a waterwheel connected to a shaft that transfers the energy to machinery for grinding grain, cutting wood, or weaving cloth, for example. The higher the waterfall, the greater the amount of energy released per unit amount of water.

Our analogy is not exact because the potential energy possessed by electrons is not due to gravity; it is due to the attraction between the positively charged protons and the negatively charged electrons. It takes energy to keep an electron farther away from the nucleus as opposed to closer to the nucleus. You have often heard that sunlight provides the energy for photosynthesis in green plants, but it may come as a surprise to learn that when a pigment such as chlorophyll absorbs the energy of the sun, electrons move to a higher energy level about the nucleus. When these electrons return to their previous level, the energy released is used to make food molecules (fig. 3.5*b* and *c*).

Electrons Occupy Orbitals

It is convenient to imagine electrons occupying concentric electron shells about a nucleus, as in figure 3.5; however, such depictions should be regarded as diagrams of convenience. In actuality, it is not possible to determine where rapidly moving electrons are from moment to moment. It is possible, however, to describe the pattern of their motion. These volumes of space where electrons are most often found are called **orbitals.**

At the first energy level, there is only a single spherical orbital, where at most 2 electrons are found about the nucleus (fig. 3.6). The space is spherical-shaped because the most likely location for each electron is a fixed distance in all directions from the

Figure 3.6

Electron orbitals. Each electron energy level (see fig. 3.5) contains one or more orbitals, a volume of space in which the rapidly moving electrons are most likely found. The nucleus is at the intersection of axes *x, y,* and *z.* **a.** The first energy level (electron shell) contains only one orbital, which has a spherical shape. Two electrons can occupy this orbital. **b.** The second energy level contains one spherical-shaped orbital and 3 dumbbell-shaped orbitals at right angles to each other. Since 2 electrons can occupy each orbital, there are a total of 8 electrons in the second electron shell. Each orbital is drawn separately here, but actually the second spherical orbital surrounds the first spherical orbital and the dumbbell-shaped orbitals pass through the spherical ones.

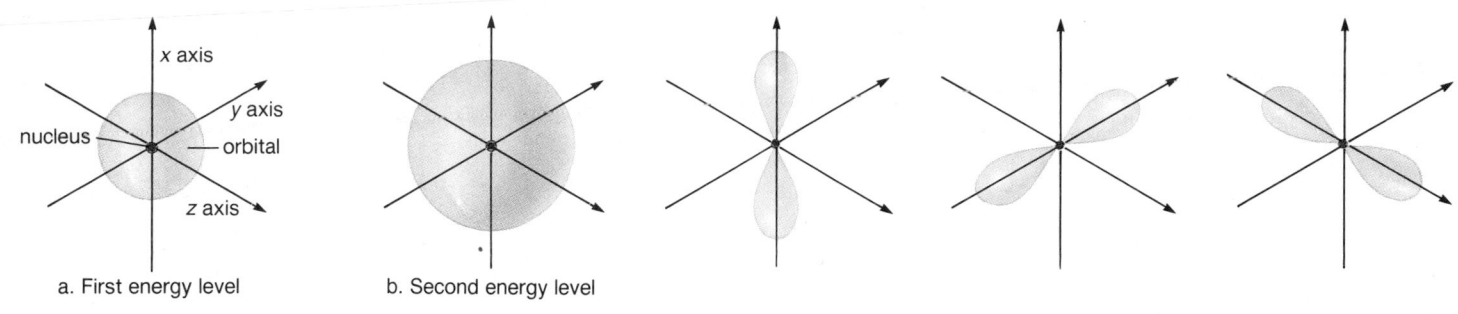

a. First energy level   b. Second energy level

nucleus. At the second energy level, there are 4 orbitals; one of these is spherical-shaped but the other 3 are dumbbell-shaped. This is the shape that allows the electrons to be most distant from one another. Since each orbital can hold 2 electrons, there are a maximum of 8 electrons in this shell. Higher shells can be more complex and contain more orbitals if they are inside another shell. If such a shell is the outer shell, it too has only 4 orbitals and a maximum number of 8 electrons.

Electron Configurations

Although it is a simplification to depict atoms by indicating only electron shells, it is a convenient way to keep track of the number of electrons in the outer shell (fig. 3.7). This is important because the number of electrons in the outer shell determines whether or not an atom reacts with another atom as well as the manner in which the atom reacts. If an atom has only one shell, this outer shell is complete when it has 2 electrons. Otherwise the *octet rule,* which states that the outer shell is complete when it has 8 electrons, holds. It is observed that atoms with 8 electrons in the outer shell do not react at all. They are said to be inert. Atoms that have fewer than eight electrons in the outer shell react with other atoms in such a way that after the reaction, each has a completed outer shell. Atoms can give up, accept, or share electrons in order to have a completed outer shell.

The manner in which atoms react with one another is determined by the number of electrons in the outer shell.

Compounds and Molecules

When 2 or more atoms of different elements react or bond together, a **compound** results. A **molecule** is the smallest part of a compound that still has the properties of that compound. Molecules can also form when 2 or more atoms of the same element react with one another.

Figure 3.7

Since the manner in which atoms react with one another is dependent upon the number of electrons in the outer shell, it is beneficial to have a convenient way to determine and to show this number. These examples of electrically neutral atoms will help you learn to do this.

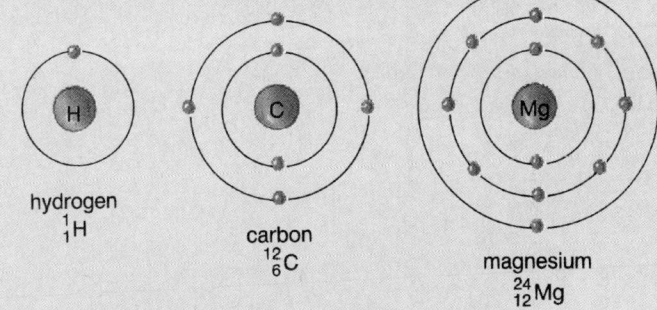

hydrogen
^1_1H

carbon
$^{12}_6C$

magnesium
$^{24}_{12}Mg$

Hydrogen, the smallest atom, has only one proton and one electron; therefore, the outer shell has only one electron. Carbon, on the other hand, has an atomic number of 6. It has 6 protons in the nucleus and 6 electrons in the shells (2 electrons in the first shell and 4 electrons in the second, or outer, shell). Each lower energy level is filled with electrons before the next higher energy level contains any electrons. Magnesium, with an atomic number of 12, has 3 shells (two electrons in the first shell, 8 electrons in the second shell, and 2 electrons in the third, or outer, shell).

Chemical Bonds

Electrons possess energy; therefore, the bonds that exist between atoms in molecules are energy relationships. Energy is required for a bond to form, and energy is released when a bond is broken. The provision of energy for bond formation and the utilization of energy released when a bond is broken are top priorities for organisms.

Figure 3.8

During an ionic reaction, an electron is transferred from one atom to another. **a.** Formation of sodium chloride, an ionic compound. Here a single electron passes from sodium (Na) to chlorine (Cl). At the completion of the reaction, each atom has 8 electrons in the outer shell, but each also carries a charge. Sodium now has a net charge of +1, symbolized as Na⁺, and chlorine has a net charge of -1, symbolized as Cl⁻. An ionic bond is the attraction between ions, atoms that carry a charge. **b.** Ionic bonding between Na⁺ and Cl⁻ causes the atoms to assume a 3-dimensional lattice in which each sodium ion is surrounded by 6 chlorine ions and each chlorine ion is surrounded by 6 sodium ions. **c.** The crystals of table salt can vary in size but their structure still is due to the lattice formation shown here. Magnification, ×30.

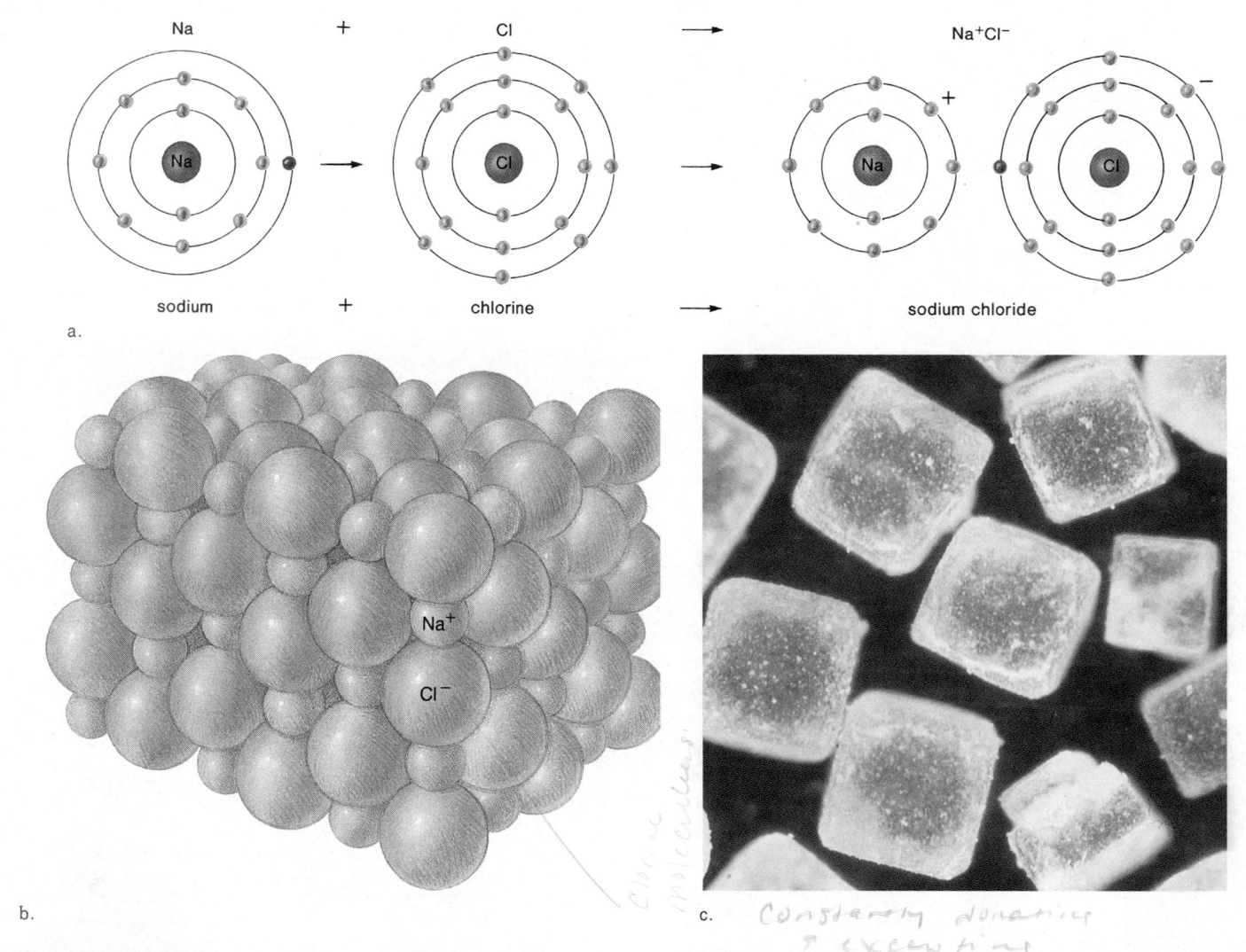

Na + Cl $\longrightarrow$ Na⁺Cl⁻

sodium + chlorine $\longrightarrow$ sodium chloride

a.

b.

c.

Ionic Bonding

Ionic bonds form when electrons are transferred from one atom to another. For example, sodium (Na), with only one electron in its third shell, tends to be an electron *donor* (fig. 3.8). Once it gives up this electron, the second shell, with 8 electrons, becomes its outer shell. Chlorine (Cl), on the other hand, tends to be an electron *acceptor*. Its outer shell has 7 electrons, so it needs only one more electron to have a completed outer shell. When a sodium atom and a chlorine atom come together, an electron is transferred from the sodium atom to the chlorine atom. Now both atoms have 8 electrons in their outer shells.

This electron transfer, however, causes a charge imbalance in each atom. The sodium atom has one more proton than it has electrons; therefore, it has a net charge of +1 (symbolized by Na⁺). The chlorine atom has one more electron than it has protons; therefore, it has a net charge of -1 (symbolized by Cl⁻). Such charged atoms are called **ions.** Sodium (Na⁺) and chlorine (Cl⁻) are not the only biologically important ions. Some, such as potassium (K⁺), are formed by the transfer of a single electron to another atom; others, such as calcium (Ca⁺⁺) and magnesium (Mg⁺⁺), are formed by the transfer of 2 electrons.

Figure 3.9

Killer whale emerges from the sea. Seawater and the blood of vertebrates are strikingly similar in the kinds of salts present and in the relative concentrations of these salts. Life is believed to have originated in the sea, and the first organisms were adapted to a seawater environment. Since then, life forms have also become adapted to the freshwater and land environments. Even so, the blood of these animals retains something of the same pattern of salts as the sea.

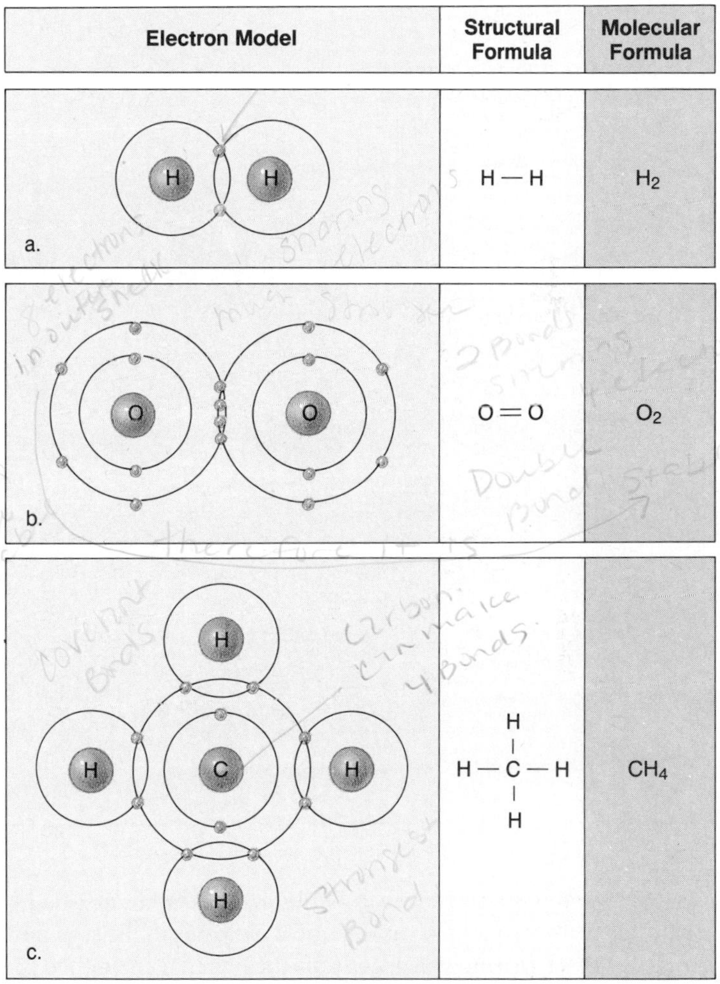

Ionic compounds are held together by an attraction between the charged ions called an **ionic bond.** For example, when sodium reacts with chlorine, an ionic compound called sodium chloride (Na^+Cl^-) results, and the reaction is called an ionic reaction. Sodium chloride is a salt commonly known as table salt because it is used at the table to season our food (fig. 3.8c). Salts can exist as dry solids, but when such a compound is placed in water, the ions separate as the salt dissolves. For example, Na^+Cl^- separates into Na^+ and Cl^-. Ionic compounds are most commonly found in this dissociated (ionized) form in biological systems because these systems are 70%–90% water (fig. 3.9).

> The transfer of electron(s) between atoms results in ions that are held together by an ionic bond, the attraction of negative and positive charges.

Covalent Bonding

A **covalent bond** results when 2 atoms *share* electrons in such a way that each atom has a completed outer shell. Consider the hydrogen atom, which has one electron in the first shell, a shell that is complete when it contains 2 electrons. If hydrogen is in the presence of a strong electron acceptor, it gives up its electron to become a hydrogen ion (H^+). But if this is not possible, hydrogen

Figure 3.10

Covalently bonded molecules. In a covalent bond, atoms share electrons. Sharing is represented as overlapping outer shells, and the shared electrons are counted as belonging to each atom. When this is done, each atom has a completed outer shell. *a.* A molecule of hydrogen (H_2) contains 2 hydrogen atoms sharing a pair of electrons. This single covalent bond can be represented in any of the 3 ways shown. *b.* A molecule of oxygen (O_2) contains 2 oxygen atoms sharing 2 pairs of electrons. This results in a double covalent bond. *c.* A molecule of methane (CH_4) contains one carbon atom bonded to 4 hydrogen atoms. Each hydrogen atom has a completed outer shell because the first shell only holds 2 electrons. By sharing with 4 hydrogens, the carbon atom has a completed outer shell containing 8 electrons.

Electron Model	Structural Formula	Molecular Formula
a.	H — H	H_2
b.	O = O	O_2
c.	H—C—H (with H above and below)	CH_4

can share with another atom and thereby have a completed outer shell. For example, a hydrogen atom can share with another hydrogen atom. In this case, the 2 orbitals overlap and the electrons are shared between them (fig. 3.10a). Because they share the electron pair, each atom has a completed outer shell. When a reaction results in a covalent molecule, it is called a covalent reaction.

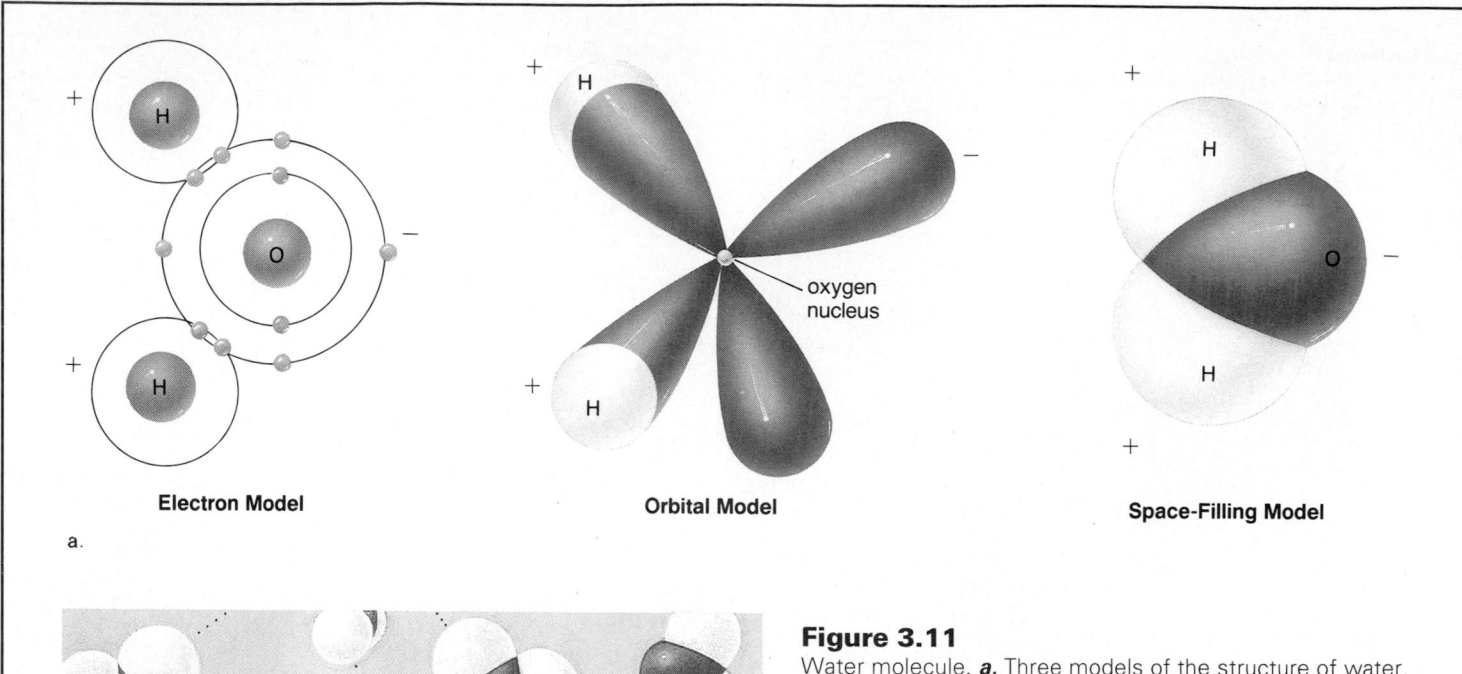

Electron Model **Orbital Model** **Space-Filling Model**

oxygen nucleus

a.

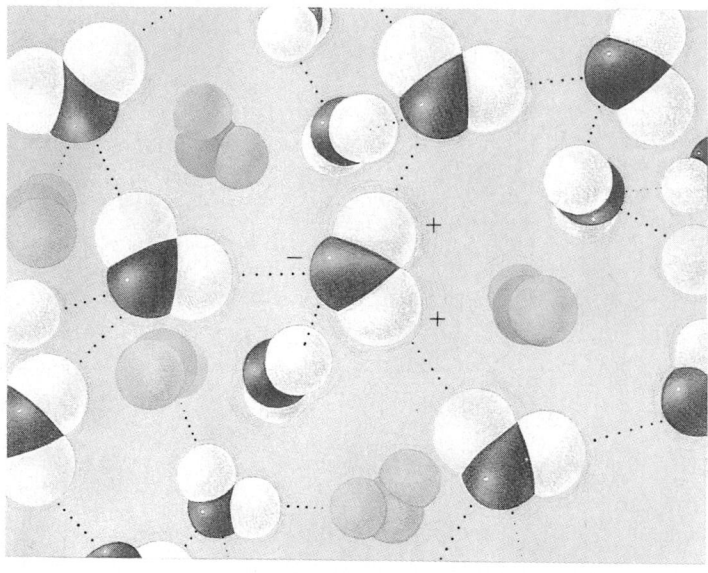

b.

Figure 3.11

Water molecule. *a.* Three models of the structure of water. The 2-dimensional electron model of the structure of water does not indicate how the molecule is orientated in space. The orbital model shows that when an atom of 2 energy levels (such as oxygen) combines with other atoms, the 4 orbitals of the outer shell (fig 3.6*b*) rearrange to give tear-shaped orbitals that point toward the corners of a tetrahedron; however, in water only 2 of the orbitals are utilized in covalent bonding. The space-filling model shows that the water molecule has a V shape. Water is a polar molecule, and because the oxygen attracts the electrons more strongly than do the hydrogens there is a partial positive charge on each hydrogen and a partial negative charge on the oxygen. *b.* Hydrogen bonding between water molecules. A hydrogen bond is the attraction of a partially positive hydrogen to a partially negative atom in the vicinity. Each water molecule can bond to 4 other molecules in this manner. When water is in its liquid state, some hydrogen bonds are forming and others are breaking at all times.

A more common way to symbolize that atoms are sharing electrons is to draw a line between the 2 atoms as in the structural formula H—H. In a molecular formula the line is even omitted and the molecule is simply written as H_2.

Double and Triple Covalent Bonds Like a single bond between 2 hydrogen atoms, a double bond can also allow 2 atoms to complete their octets. In a double covalent bond, 2 atoms share 2 pairs of electrons (fig. 3.10*b*). In order to show that oxygen gas (O_2) contains a double bond, the molecule can be written as O=O.

It is even possible for atoms to form triple covalent bonds as in nitrogen gas (N_2), which can be written as N≡N. Single covalent bonds between atoms are quite strong, but double and triple bonds are even stronger.

In a covalent molecule, atoms share electrons; not only single bonds but also double and even triple bonds are possible.

Polar Covalent Bonds

Normally, the sharing of electrons between 2 atoms is fairly equal, and the covalent bond is nonpolar. All the molecules in figure 3.10, including methane (CH_4), are nonpolar. In the case of water (H_2O), however, the sharing of electrons between oxygen and each hydrogen is not completely equal. The larger oxygen atom, with the greater number of protons, dominates the H_2O association and attracts the electron pair to a greater extent. This causes the oxygen atom to assume a slightly negative charge, showing that it is electronegative in relation to the hydrogen atoms. Each hydrogen atom assumes a slightly positive charge, showing that it is electropositive in relation to oxygen. The unequal sharing of electrons in a covalent bond creates a **polar covalent bond,** and the molecule itself is a polar molecule (fig. 3.11).

> The water molecule has an electronegative end and an electropositive end indicating that it is a polar molecule.

Hydrogen Bonding

Polarity within a water molecule causes the hydrogen atoms in one molecule to be attracted to the oxygen atoms in other molecules (fig. 3.11*b*). This attractive force creates a weak bond called a **hydrogen bond.** This bond is often represented by a dotted line because a hydrogen bond is easily broken. Hydrogen bonding is not unique to water. It occurs whenever an electropositive hydrogen atom is attracted to an electronegative atom in another molecule or parts of the same molecule.

Although a hydrogen bond is more easily broken than a covalent bond, many hydrogen bonds taken together are quite strong. Hydrogen bonds between parts of cellular molecules help maintain their proper structure and function. We will see that some of the important properties of water are also due to hydrogen bonding.

> A hydrogen bond occurs between a slightly positive hydrogen atom of one molecule and a slightly negative atom of another molecule or between parts of the same molecule.

Chemical Reactions

Chemical reactions are often indicated by a chemical equation. The atoms or molecules to the left of an arrow are the *reactants,* the substances that react with one another to give the product(s). The *product(s)* is placed to the right of the arrow. The reactants and the products are designated by formulas that give the number of atoms and possibly an indication of how they are arranged within the compound or molecule. For example, the equation for the formation of water from hydrogen gas and oxygen gas is written as follows:

$$2H_2 \;+\; O_2 \longrightarrow 2H_2O$$
$$\text{hydrogen} \quad \text{oxygen} \qquad \text{water}$$

Notice that all atoms present in the reactants are accounted for in the products; there are 4 hydrogen atoms and 2 oxygen atoms on both sides of the equation. The equation must be balanced in this way because mass can never be created and can never disappear.

Oxidation-Reduction Reactions

Oxidation-reduction reactions are an important type of reaction in cells, but the terminology was derived from studying reactions outside of cells. When oxygen combines with a metal, oxygen receives electrons and becomes negatively charged and the metal loses electrons and becomes positively charged.

Today, the terms **oxidation** and **reduction** are applied to many ionic reactions, whether or not oxygen is involved. Very simply, *oxidation refers to the loss of electrons, and reduction refers to the gain of electrons.* In our previous ionic reaction, $Na + Cl \rightarrow Na^+Cl^-$, the sodium has been oxidized (loss of electron) and the chlorine has been reduced (gain of electron).

The terms oxidation and reduction are also applied to certain covalent reactions. In this case, however, oxidation is the loss of hydrogen atoms, and reduction is the gain of hydrogen atoms. A hydrogen atom contains one proton and one electron; therefore, when a molecule loses a hydrogen atom, it has lost an electron, and when a molecule gains a hydrogen atom, it has gained an electron. We will have occasion to refer to this form of oxidation-reduction reaction again in chapter 5.

> When oxidation occurs, an atom is oxidized (loses electrons). When reduction occurs, an atom is reduced (gains electrons). These 2 processes occur concurrently in oxidation-reduction reactions.

Water

Properties of Water

Life evolved in water, and living things are 70%-90% water. The chemical properties of water are absolutely essential to the continuance of life (fig. 3.12). Water is a polar molecule—the oxygen end of the molecule is electronegative, and the hydrogen end is electropositive. Water molecules are hydrogen bonded one to the other (fig. 3.11). A hydrogen bond is much weaker than a covalent bond within a water molecule, but taken together, hydrogen bonds cause water molecules to cling together. Without hydrogen bonding between molecules, water would boil at $-80°$ C and freeze at $-100°$ C, making life as we know it impossible. But because of hydrogen bonding, water is a liquid at temperatures suitable for life. It boils at $100°$ C and freezes at $0°$ C. Water also has other important properties (table 3.3).

1. *Water is the universal solvent and facilitates chemical reactions both outside of and within living systems.* When a salt, such as sodium chloride (Na^+Cl^-), is put into water, the electronegative ends of the water molecules are attracted to the sodium ions, and the electropositive ends of the water molecules are attracted to the chlorine ions. This causes the sodium ions and the chlorine ions to separate and to dissolve in water:

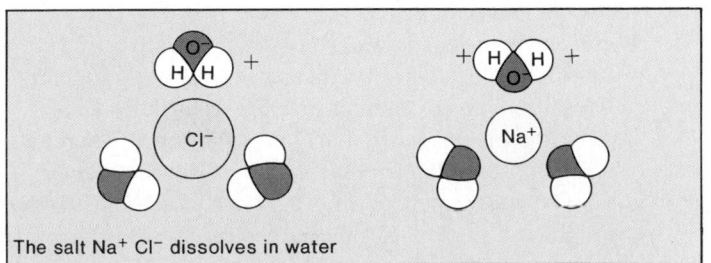

The salt $Na^+ Cl^-$ dissolves in water

Figure 3.12

a. The temperature of water changes slowly and a great deal of heat is needed for vaporization to occur. *b.* Ice is less dense than liquid water and therefore bodies of water freeze from the top down, making ice fishing possible after a hole is drilled through the ice. *c.* The heat needed to vaporize water can help animals in a hot climate to maintain internal temperatures.

a.

b.

c.

Water is also a solvent for larger molecules that contain ionized atoms or are polar molecules:

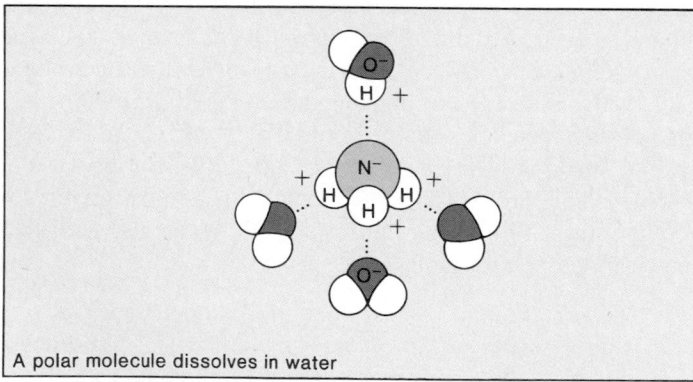

A polar molecule dissolves in water

When ions and molecules disperse in water, they move about and collide, allowing reactions to occur. Those molecules that can attract water are said to be hydrophilic, meaning "water loving." Nonionized and nonpolar molecules that cannot attract water are said to be hydrophobic, or "water hating."

2. *Water molecules are cohesive.* Water flows freely, yet water molecules do not break apart. They cling together because of hydrogen bonding. Water molecules also adhere to surfaces, particularly polar surfaces. Therefore, water can fill a tubular vessel and still flow so that dissolved and suspended molecules are evenly distributed throughout a system. For these reasons, water is an

excellent transport system both outside of and within living organisms. One-celled organisms rely on external water to transport nutrient and waste molecules, but multicellular organisms often contain internal vessels in which water serves to transport nutrients and wastes.

3. *The temperature of liquid water rises and falls more slowly than that of most other liquids.* A calorie of heat energy is needed to raise the temperature of 1 g of water 1° C.[2] This is about twice the amount of heat required for other covalently bonded liquids. The many hydrogen bonds that link water molecules help water absorb heat without a change in temperature. Because water holds heat so effectively, its temperature falls more slowly. This property of water is important not only for aquatic organisms but also for all living things. Water protects organisms from rapid temperature changes and helps them maintain their normal internal temperatures.

4. *Water tends to remain a liquid; it does not readily change to ice or steam.* Converting 1 g of liquid water to ice requires the loss of 80 calories of heat energy. Converting 1 g of water to steam requires an input of 540 calories of heat energy (fig. 3.13). Hydrogen bonds must be broken to change water to steam; this accounts for the very large amount of heat needed for evaporation. This

[2]A calorie (cal) is the amount of heat required to raise the temperature of 1 g of water 1° C. A kilocalorie (kcal) equals 1,000 calories and is the unit used to express the energy value of food.

Table 3.3
Water

Properties	Chemistry	Result
Universal solvent	Polarity	Facilitates chemical reactions
Adheres and is cohesive	Polarity; hydrogen bonding	Serves as transport medium
Resists changes in temperature	Hydrogen bonding	Helps keep body temperatures constant
Resists change of state (from liquid to ice and from liquid to steam)	Hydrogen bonding	Moderates earth's temperature Evaporation helps bodies remain cool
Less dense as ice than as liquid water	Hydrogen bonding changes	Ice floats on water

Figure 3.13
Water is the only common molecule that can be a solid, liquid, or gas according to the environmental temperature. At ordinary temperatures and pressure, water is a liquid and it takes a large input of heat to change it to steam. (When we perspire the heat of our bodies is causing water to vaporize, and therefore, sweating causes us to cool off.) In contrast, water gives off heat when it freezes and this heat will keep the environmental temperature higher than expected. Can you see why there are less severe changes in temperature along the coasts?

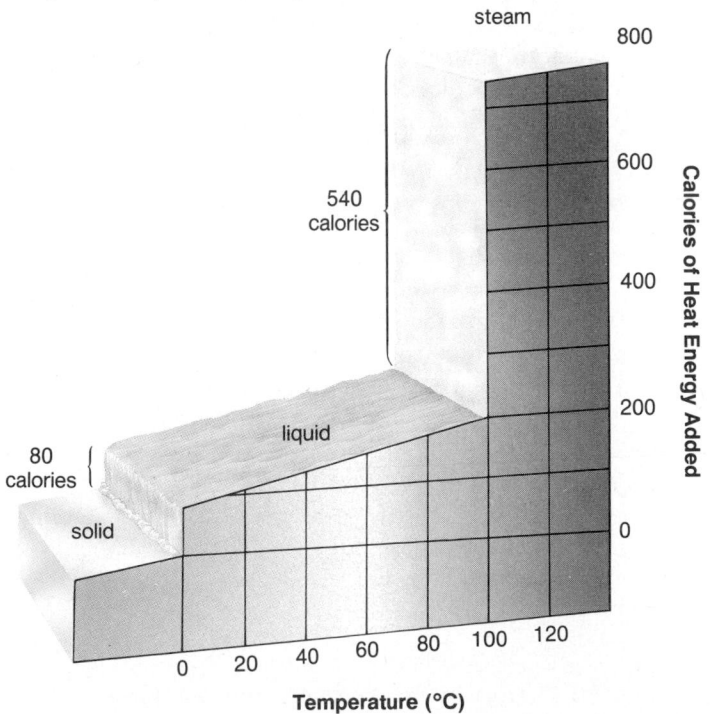

Figure 3.14
Most substances contract when they solidify, but water expands. Water is most dense at 4° C and becomes less dense as it either cools or warms. At 0° C water freezes, forming a lattice structure in which the hydrogen bonds are fixed. The water molecules in the lattice are further apart than in liquid water; therefore ice is less dense than liquid water. This means that ice can float on liquid water. When water vaporizes at 100° C, all the hydrogen bonds are broken, and the molecules move away from one another.

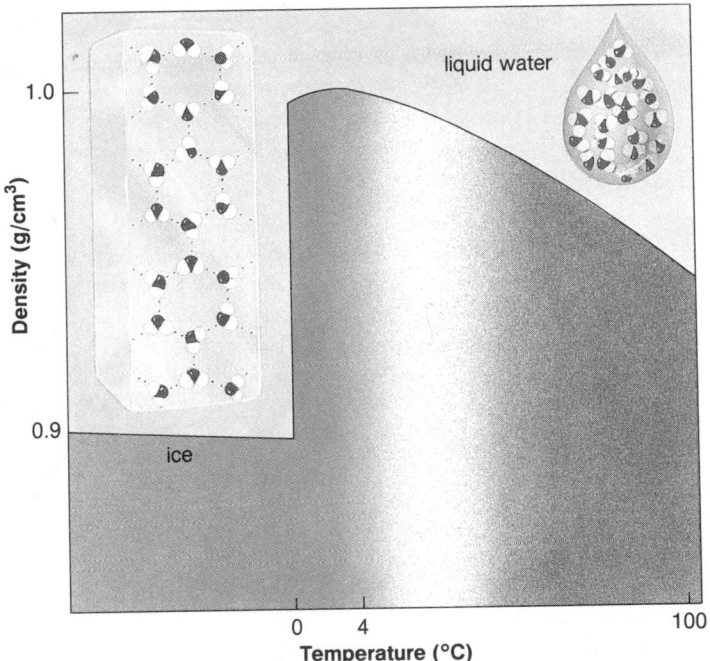

property of water helps moderate the earth's temperature so that it promotes the continuance of life. It also gives animals in a hot environment an efficient way to release excess body heat. When an animal sweats, body heat is used to vaporize the sweat, thus cooling the animal.

5. *Frozen water is less dense than liquid water.* As water cools, the molecules come closer together. They are densest at 4° C, but they are still moving about (fig. 3.14). At temperatures below 4° C, the water molecules cease moving, and hydrogen bonding becomes more rigid but also more open. This makes ice less dense than liquid water which is why ice floats on liquid water. Bodies of water always freeze from the top down. When a body of water freezes on the surface, the ice acts as an insulator to prevent the water below it from freezing. This protects many aquatic organisms so that they can survive the winter.

Water has unique properties that allow cellular activities to occur and that make life on earth possible.

Ionization of Water

Water has another property that has important consequences for living things. It tends to ionize or dissociate, releasing an equal number of hydrogen ions (H^+) and hydroxyl ions (OH^-):

$$H-O-H \rightleftharpoons H^+ + OH^-$$

water   hydrogen   hydroxyl
            ion   ion

This equation is written with a double arrow because the reaction is reversible. Just as water can release hydrogen and hydroxyl ions, these same ions can join together to form water. In fact, only a few water molecules at a time are actually dissociated.

There are many other substances that also dissociate and contribute ions to a given solution, which is a liquid containing a dissolved substance. **Acids** are molecules that release hydrogen ions when they dissociate. A strong acid, such as hydrochloric acid (HCl), dissociates almost completely when put into water:

$$HCl \longrightarrow H^+ + Cl^-$$

hydrochloric   hydrogen   chloride
acid   ion   ion

In contrast, **bases** are molecules that can take up hydrogen ions. When sodium hydroxide (NaOH) dissociates, the hydroxyl ion can combine with a hydrogen ion to form water. Sodium hydroxide is termed a strong base because it dissociates almost completely when put into water:

$$NaOH \longrightarrow Na^+ + OH^-$$

sodium   sodium   hydroxyl
hydroxide   ion   ion

pH Scale

Acids and bases affect the relative concentration of hydrogen ions and hydroxyl ions. Biologists use the **pH scale** because it indicates the relative concentrations of H^+ and OH^- in a solution (fig. 3.15). The pH scale range is from 0 to 14. The pH of pure water is 7; this is neutral pH. Acids lower the pH; any pH below 7 indicates that there are more hydrogen ions than hydroxyl ions. Bases increase the pH; any pH above 7 indicates that there are fewer hydrogen ions than hydroxyl ions. A pH change of one unit involves a 10-fold change in hydrogen ion concentration. Therefore, pH 6 has 10 times the $[H^+]$ of pH 7, and pH 5 has 100 times the $[H^+]$ of pH 7.

Buffers

Most organisms maintain a pH of about 7; for example, human blood has a pH of about 7.4. A much higher pH or much lower pH causes illness. Normally, pH stability is possible because organisms have built-in mechanisms to prevent pH changes. Buffers are the most important of these mechanisms. A **buffer** is a chemical or

Figure 3.15

The pH scale. The proportionate amount of hydrogen ions to hydroxide ions is indicated by the diagonal line. Any pH above 7 is basic, while any pH below 7 is acidic.

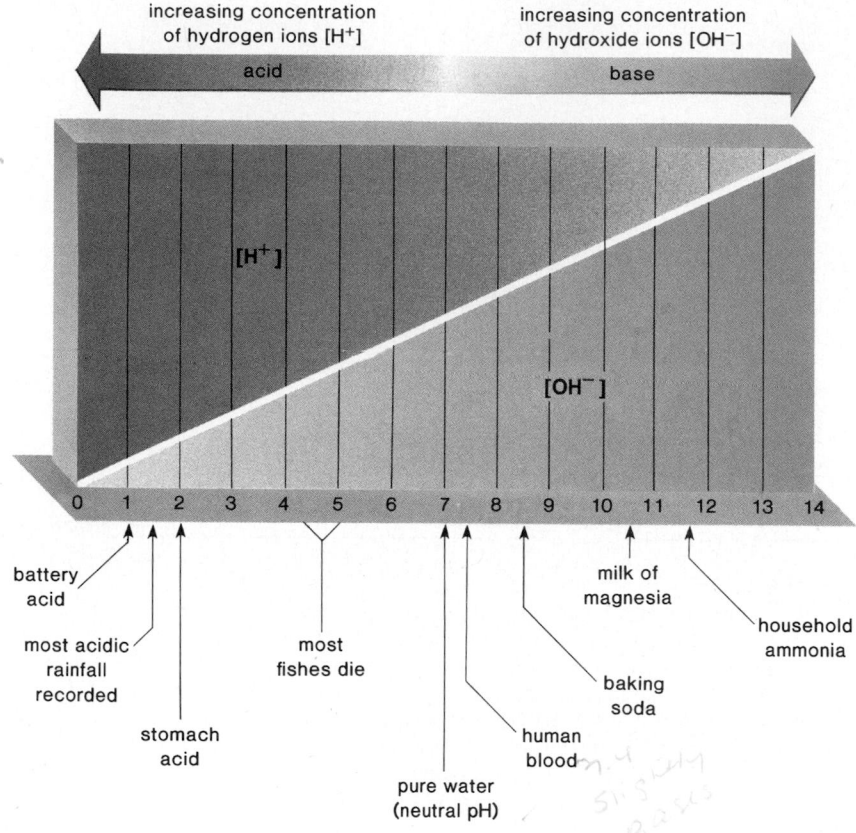

a combination of chemicals that can both take up and release hydrogen ions. Carbonic acid (H_2CO_3) helps buffer human blood because it is a weak acid that does not totally dissociate:

$$H_2CO_3 \rightleftharpoons H^+ + HCO_3^-$$

carbonic   hydrogen   bicarbonate
acid   ion   ion

When excess hydrogen ions are present in blood, the reaction goes to the left, and carbonic acid forms to maintain the pH. If hydroxyl ions are added to blood, water will form, effectively removing hydrogen ions. Other types of bases can directly combine with hydrogen ions. When a base removes hydrogen ions, the reaction goes to the right, and the pH is maintained.

It is possible to overcome an organism's buffering ability. Usually, however, buffers keep the pH within normal limits despite the many biochemical reactions that either release or take up hydrogen or hydroxyl ions.

Acids have a pH that is less than 7, and bases have a pH that is greater than 7. Organisms contain buffers that help maintain the pH within a normal range.

Acid Deposition

Normally rainwater has a pH of about 5.6 because the carbon dioxide in the air combines with water to form a weak solution of carbonic acid. Rain falling in the northeastern United States and southeastern Canada has a pH between 5.0 and 4.0. You have to remember that a pH of 4 is 10 times more acidic than a pH of 5 to appreciate the increase in acidity this represents.

There is very strong evidence now that this observed increase in rainwater acidity is a result of the burning of fossil fuels like coal and oil, as well as the gasoline derived from oil. When these substances are burned, sulfur oxides and nitrogen oxides are produced. These combine with water vapor in the atmosphere to form acids. These acids fall out of the atmosphere as rain or snow, in a process properly called wet deposition but more often called acid rain. Also, dry particles of sulfate salts and nitrate salts can fall out of the atmosphere, a process called dry deposition. The net result is that the burning of fossil fuels and related products leads to the presence of acids in the air, and these return to the earth sometime later.

These air pollutants can be carried in the wind to places far distant from their origin. Acid deposition in Canada and the northeastern United States is due to the burning of fossil fuels in factories and power plants located in the Midwest. Similarly, the Scandinavian countries are the recipients of air pollutants from England and other northern European countries. Unfortunately, regulations that require the use of tall smokestacks to reduce local air pollution only result in the pollutants being carried farther away.

Acid deposition has created a very serious situation in certain areas of the world. In the United States, vulnerable areas include not only the Northeast but also the Great Smokey Mountains, the lakes of Wisconsin and Minnesota, the Pacific Northwest, and the Colorado Rockies. The soil in these areas is thin and lacks limestone (calcium carbonate, $CaCO_3$), which can buffer acid deposition. The first cry of alarm came from Sweden, which reported as early as the 1960s that its lakes were dying. Today dying lakes are found across northeastern North America, Western Europe, and Eastern Europe. The acid rain leaches aluminum from the soil and carries this element into the lakes. The acid also allows mercury deposits in lake bottom sediments to be converted to soluble methyl mercury. Not only do the lakes become more acidic, they also show an accumulation of substances that are toxic to living things. Sweden now reports 15,000 of its lakes are too acidic to support any higher forms of aquatic life (fig. 3.A).

Factories, power plants, and automobiles put out sulfur dioxide (SO_2), nitrogen oxides (NO_x), and other pollutants.

a.

In the atmosphere, SO_2 is converted to sulfuric acid. NO_x is converted to nitric acid. Some of the pollutants quickly fall to the ground, helping to cause local environmental problems.

b.

Far from their source, the acids fall from the sky either in the form of dry particles or as wet deposition. They cause lakes to become sterile and statues and buildings to crumble and can seriously harm forests.

d.

Many of the acids, however, are carried hundreds of miles by the prevailing winds.

c.

Figure 3.A
The cause and effect of acid deposition.

Continued on next page

In the early 1980s, evidence began to accumulate that acid deposition was also damaging to forests. As of mid-1986, some 19 countries in Europe reported damage to their woodlands ranging from roughly 5%-15% of the forested area in Yugoslavia and Sweden to 50% or more in the Netherlands, Switzerland, and Germany. More than one-fifth of Europe's forests are now damaged.

These are not the only effects of acid deposition. Other effects include reduction of agricultural yields, damage to marble and limestone monuments and buildings, and even an increase in the illnesses of humans. Acid deposition is believed to be implicated in the increased incidence of lung cancer and possibly colon cancer along the East Coast of the United States. Tom McMillan, Canadian Minister of the Environment, says that acid deposition "is destroying our lakes, killing our fish, undermining our tourism, retarding our forests, harming our agriculture, devastating our heritage and threatening our health."

There are, of course, things that can be done.

1. Use alternative energy sources, such as solar, wind, hydropower, and geothermal energy whenever possible.
2. Use low-sulfur coal or remove the sulfur impurities from coal before it is burned.
3. Require the use of scrubbers to remove sulfur from factory and power plant emissions.
4. Require people to use mass transit rather than their own automobiles.
5. Reduce our energy needs through other means of energy conservation.

These measures and possibly others could be taken immediately. It is only necessary for us to determine that they are worthwhile.

This reading is based on these references: G. Tyler Miller, *Living in the Environment,* Wadsworth Publishing Company, Belmont, CA, 1988 and Lester R. Brown, et al., *State of the World,* W. W. Norton & Company, New York, NY, 1991.

Summary

1. Both living and nonliving things are composed of matter consisting of elements. Each element contains atoms of just one type. The acronym CHNOPS recalls the most common elements (atoms) found in living things.
2. Atoms contain subatomic particles. Protons and neutrons in the nucleus determine the weight of an atom. Electrons are outside the nucleus and have almost no weight.
3. The atomic number indicates the number of protons and the number of electrons in electrically neutral atoms. Protons have a positive charge and electrons have a negative charge. Before atoms react, the charges are equal.
4. Isotopes are atoms of the same type that differ by the number of neutrons. Radioactive isotopes are used as tracers in biological experiments and medical procedures.
5. Electrons occupy energy levels (electron shells) at discrete distances from the nucleus. When electrons absorb energy from an outside source, they move to a higher level, and when they drop back, they release energy. Certain biological processes are powered by this release of energy.
6. Electron shells contain orbitals. The first shell contains a single spherical-shaped orbital. The second shell contains 4 orbitals: the first is spherical-shaped and the others are dumb-bell-shaped.
7. The number of electrons in the outer shell determines the reactivity of an atom. The first shell is complete when it has 2 electrons; the second shell is complete when it has 8 electrons; other shells can hold more electrons, but if they are an outer shell, they are also complete with 8 electrons. This is the octet rule.
8. Most atoms, including those common to living things, do not have completed outer shells. This causes them to react with one another to form compounds and/or molecules. Following the reaction, the atoms have completed outer shells.
9. When an ionic reaction occurs, one or more electrons are transferred from one atom to another. The ionic bond is an attraction between the resulting ions.
10. When a covalent reaction occurs, atoms share electrons. A covalent bond is the sharing of these electrons. There are single, double, and triple covalent bonds.
11. In polar covalent bonds, the sharing of electrons is not equal; one end of the bond (molecule) is electronegative, and the other end is electropositive. A hydrogen bond is a weak bond that occurs between an electropositive hydrogen atom and an electronegative atom present in another molecule or another part of the same molecule. Hydrogen bonds help maintain the shape of cellular molecules.
12. Chemical equations are used to symbolize chemical reactions. An atom that has lost electrons (or hydrogen atoms) has been oxidized, and an atom that has gained electrons (or hydrogen atoms) has been reduced.
13. Water is a polar molecule, and hydrogen bonding occurs between water molecules. These 2 features account for the unique properties of water, which are summarized in table 3.3. These features allow cellular activities to occur and make life on earth possible.
14. Water dissociates to give an equal number of hydrogen ions and hydroxyl ions. This is termed neutral pH. In acidic solutions, there are more hydrogen ions than hydroxyl ions; these solutions have a pH less than 7. In basic solutions, there are more hydroxyl ions than hydrogen ions; these solutions have a pH greater than 7. Cells are sensitive to pH changes, and biological systems tend to have buffers that help keep the pH within a normal range.

Writing Across the Curriculum

In order to practice writing skills, students should write out the answers to any or all of the study questions and the critical thinking questions. The study questions are sequenced in the same order as the text. Suggested answers to the critical thinking questions are in appendix D.

Study Questions

1. Name the kinds of subatomic particles studied; tell their weight, charge, and location in an atom. Which of these varies in isotopes?

2. Define energy level and orbital. What is their relationship?

3. Draw a simplified atomic structure for a carbon atom that has 6 protons and 6 neutrons.

4. Draw an atomic representation for the molecule $Mg^{++}Cl_2^-$. Using the octet rule, explain the structure of the compound.

5. Tell whether CO_2 (O—C—O) is an ionic or covalent compound. Why does this arrangement satisfy all atoms involved?

6. Explain why water is a polar molecule. What is the relationship between polarity of the molecule and hydrogen bonding between water molecules?

7. Define oxidation and reduction and tell which has been oxidized and which reduced in this equation:
$$4\ Fe + 3\ O_2 \rightarrow 2\ Fe_2O_3$$
Satisfy yourself that this equation is balanced.

8. Name 5 properties of water and relate them to the structure of water including its polarity and hydrogen bonding between molecules.

9. Define an acid and a base. On the pH scale, which numbers indicate an acid, a base, and a neutral pH?

Objective Questions

1. An atom that has 2 electrons in the outer shell would most likely
   a. share to acquire a completed outer shell.
   b. lose these 2 electrons and become a negatively charged ion.
   c. lose these 2 electrons and become a positively charged ion.
   d. form hydrogen bonds only.

2. The atomic number tells you
   a. the number of neutrons in the nucleus.
   b. the number of protons in the atom.
   c. the weight of the atom.
   d. to what element the atom belongs.

3. An orbital is
   a. the same volume and charge as an electron shell.
   b. the same space and energy content as the energy level.
   c. the volume of space most likely occupied by an electron.
   d. what causes an atom to react with another atom.

4. A covalent bond is indicated by
   a. plus and minus charges attached to atoms.

   b. dotted lines between hydrogen atoms.
   c. concentric circles about a nucleus.
   d. overlapping electron shells or a straight line between atomic symbols.

5. An atom has been oxidized when it
   a. combines with oxygen.
   b. gains an electron.
   c. loses an electron.
   d. Both a and c.

6. In which of these are the electrons shared unequally?
   a. double covalent bond
   b. triple covalent bond
   c. hydrogen bond
   d. polar covalent bond

7. In the molecule

   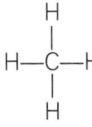

   a. all atoms have 8 electrons in the outer shell.
   b. all atoms are sharing electrons.
   c. carbon could accept more hydrogen atoms.
   d. All of these.

8. Which of these properties of water is not due to hydrogen bonding?
   a. stabilizes temperature inside and outside cell
   b. molecules are cohesive
   c. universal solvent
   d. ice floats on water

9. Acids
   a. release hydrogen ions.
   b. have a pH value above 7.
   c. take up hydroxyl ions.
   d. Both a and b.

10. Complete this diagram of a nitrogen atom by placing the correct number of protons and neutrons in the nucleus and electrons in the shells. Explain why the correct formula for ammonia is NH_3 and not NH_4.

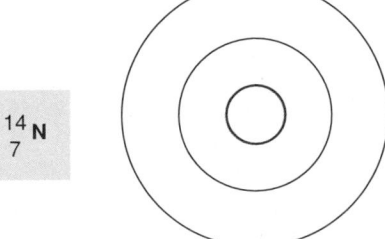

Concepts and Critical Thinking

1. *Life has a chemical and physical basis.*

Give an example from your knowledge of nutrition, medicine, or environmental effects to show that this concept has an everyday application.

2. *The organization of life begins with atoms and continues from molecules to cells to tissues to organism.*

Why does figure 3.1 suggest that the first 2 levels of organization in living things are atoms and molecules? Why aren't cells the first level of organization?

3. *Life is dependent upon the special properties of water.*

Show that living things are absolutely dependent upon a particular property of water by telling what would happen if water didn't have this property.

Selected Key Terms

atom (at'om) 26
electron (e-lek'tron) 26
isotope (i'so-tōp) 27
orbital (or'bɪ-tal) 28
compound (kom'pownd) 29

molecule (mol'e-kūl) 29
ionic bond (i-on'ik bond) 31
covalent bond (ko-va'lent bond) 31
polar covalent bond (po'lar ko-va'lent bond) 33
hydrogen bond (hi'dro-jen bond) 33

oxidation (ok"sɪ-da'shun) 33
reduction (re-duk'shun) 33
acid (as'id) 36
base (bās) 36
pH scale (pe āch skāl) 36
buffer (buf'er) 36

4

The Chemistry of Life

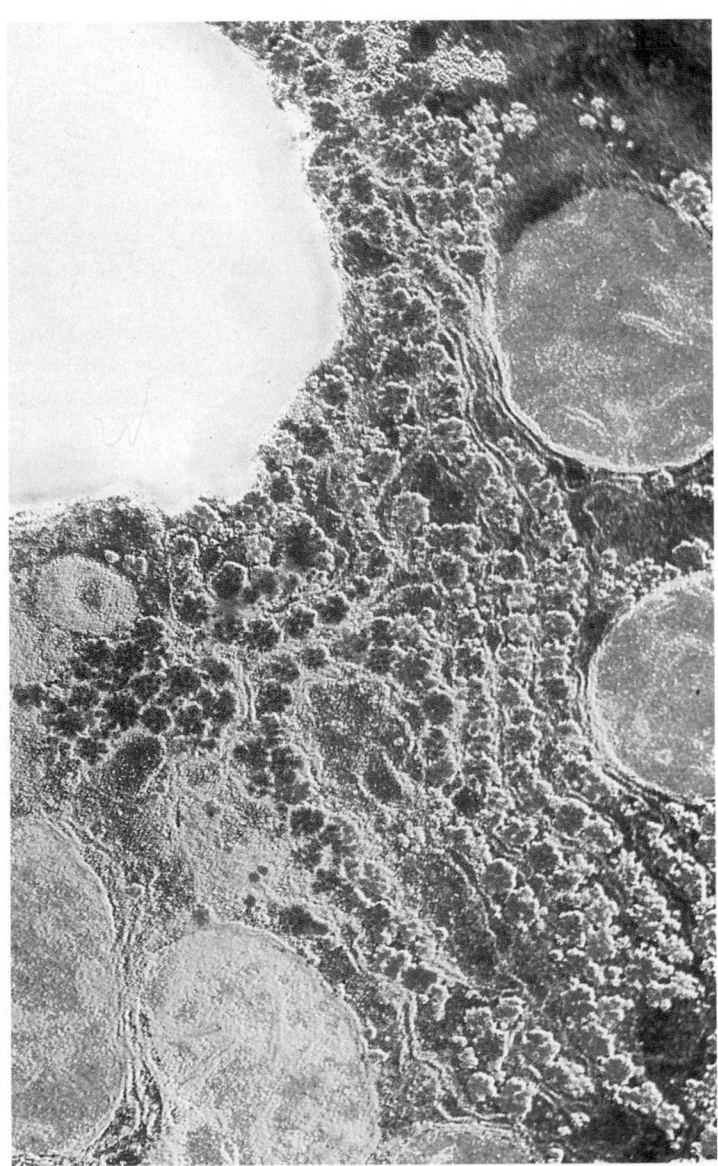

This electron micrograph of a liver cell has been colored to reveal its contents, which includes many types of organic molecules. Energy is stored in glycogen granules (red) and in a fat vacuole (yellow). Magnification, ×9,000.

Your study of this chapter will be complete when you can

1. explain and demonstrate the bonding patterns of carbon;
2. recognize the various functional groups found in cellular molecules;
3. distinguish between condensation of monomers and hydrolysis of polymers;
4. give examples of monosaccharides, disaccharides, and polysaccharides, and state their functions;
5. recognize the molecular and structural formulas for glucose;
6. give examples of various lipids, and state their functions;
7. recognize the structural formula for a saturated fatty acid and an unsaturated fatty acid and the structural formula for a fat;
8. give examples of proteins, and state their functions;
9. recognize an amino acid, and demonstrate how a peptide bond is formed;
10. relate the 4 levels of structure of a protein to the bonding patterns observed at each level;
11. give examples of nucleotides and nucleic acids, and state their functions;
12. state the components of a nucleotide, and tell how these monomers are joined to form a nucleic acid;
13. compare the structures of DNA and RNA.

Living things are highly organized; therefore, it is easy for us to differentiate between the plant, the animal, and the bacterium in figure 4.1. We might even be inclined to think that these organisms are so different that they must contain entirely different types of chemicals. But this is not the case. Instead, they and all living things contain the same classes of primary molecules: carbohydrates, proteins, lipids, and nucleic acids. But within these classes there is molecular diversity among organisms. For example, the plant, the animal, and the bacterium all utilize a carbohydrate molecule for structural purposes, but the exact carbohydrate is different in each of them.

The most common elements in living things are remembered by the acronym CHNOPS (see table 3.1), and of these, carbon, hydrogen, nitrogen, and oxygen constitute about 95% of your body weight. But we will see that both the sameness and the diversity of life are dependent upon the chemical characteristics of carbon, an atom whose chemistry is essential to living things.

The bonding of hydrogen, oxygen, nitrogen, and other atoms to carbon creates the molecules known as **organic** molecules. For a molecule to be organic, it must contain carbon and hydrogen. This is why carbon dioxide is termed an inorganic molecule. It is the organic molecules that characterize the structure and the function of living things, like the primrose, the blueshell crab, and the bacterium in figure 4.1. **Inorganic** molecules constitute nonliving matter, but even so, inorganic molecules, like salts (e.g., Na^+Cl^-), also play important roles in living things.

> Of the elements most common to living things, the chemistry of carbon allows the formation of varied organic molecules. This accounts for both the sameness and the diversity of living things.

Chemistry of the Cell and the Organism

Carbon has 4 electrons in its outer shell, and this allows it to bond with as many as 4 other atoms. Moreover, because these bonds are covalent, they are quite strong. Usually carbon bonds to hydrogen, oxygen, nitrogen, or another carbon atom. The ability of carbon to bond to itself makes carbon chains of various lengths and shapes possible. Long chains containing 50 or more carbon atoms are not unusual in living systems. Carbon can also share 2 pairs of electrons with another atom, giving a double covalent bond. Carbon-to-carbon bonding sometimes results in ring compounds of biological significance.

Small Organic Molecules

Carbon chains make up the skeleton or backbone of organic molecules. The small organic molecules in living things—sugars, fatty acids, amino acids, and nucleotides—all have a carbon backbone, but in addition they have characteristic functional groups (fig. 4.2). *Functional groups* are clusters of atoms with a certain pattern and that always behave in a certain way. Therefore, functional groups help determine the characteristics of organic molecules.

Figure 4.1
One thing these organisms (primrose, blueshell crab, and bacterium) have in common is the use of carbohydrate to provide structure. *a.* The *primrose* is held erect partly by the incorporation of the polysaccharide cellulose into the cell walls, which surround each cell. *b.* The shell of the *crab* contains chitin, a different type of polysaccharide. *c.* Most *bacterial* cells, like plant cells, are enclosed by a cell wall, but in this case the wall is strengthened by another type of polysaccharide known as peptidoglycan. Magnification, ×38,000.

a.

b.

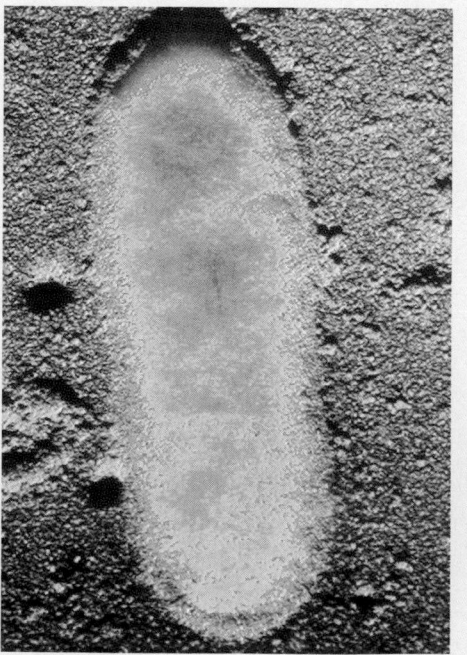

c.

One characteristic of an organic molecule of some importance is whether it is hydrophilic or hydrophobic. You will recall in chapter 3 that we showed how a polar molecule with plus and minus charges is **hydrophilic** and attracted to water because water is also a polar molecule. A hydrocarbon (molecule with only carbon and hydrogen) is always **hydrophobic,** but if there is a functional group that can ionize the molecule becomes hydrophilic:

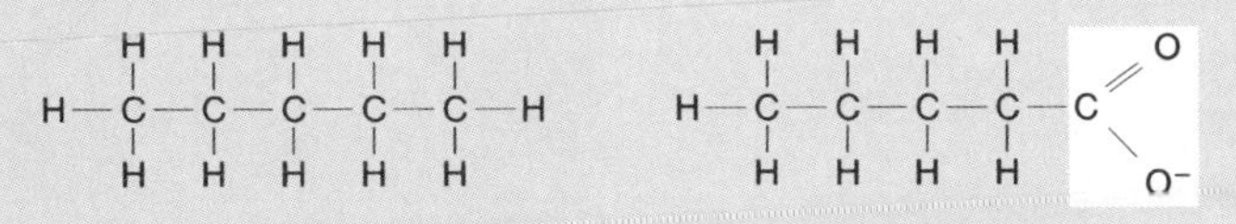

hydrocarbon
nonpolar
(hydrophobic)

acid in ionized form
polar
(hydrophilic)

The carboxyl (acid) functional group, —COOH, can give up a hydrogen ion (H^+) and become ionized as shown. Other functional groups (—OH, —CO, and —NH$_2$) are also polar even though they do not ionize. Polar molecules have nonbonding electrons that associate closely with water molecules. Since cells are 70%–90% water, the ability or inability to interact with water profoundly affects the function of organic molecules in cells.

Functional groups add diversity to organic molecules. Another factor contributing to the diversity of organic molecules is the existence of isomers. **Isomers** are molecules that have identical molecular formulas because they contain the same numbers and kinds of atoms, yet they are different molecules because the atoms in each isomer are arranged differently. For example, figure 4.3 shows 2 compounds with the molecular formula $C_3H_6O_3$. Each of these molecules is assigned its own chemical name because the molecules differ structurally.

Large Organic Molecules

Each of the small organic molecules already mentioned can be a subunit of a large organic molecule, often called a *macromolecule*. A subunit is called a *monomer*, and the macromolecule is called a **polymer.** Simple sugars (monosaccharides) are the monomers within polysaccharides; fatty acids and glycerol are found in fat, a lipid; amino acids join to form proteins; and nucleotides are the subunits of nucleic acids:

Macromolecule	Monomer
polysaccharide	monosaccharide
lipid (e.g., fat)	glycerol and fatty acid
protein	amino acid
nucleic acid	nucleotide

We will see that macromolecules in the same class can vary because of differences in monomer type or the way the monomers are joined. Also, different types of functional groups can be bonded to the same monomer backbone. This is why, for example, carbohydrates can have different characteristics and therefore play different roles in cells.

Figure 4.2
Molecules with the exact same type of backbone can still differ according to the type of functional group attached to the backbone. Many of these functional groups are polar, helping to make the molecule soluble in water. In this illustration, the remainder of the molecule (aside from the functional groups) is represented by an R.

Functional Groups		
Name	**Structure**	**Found in**
hydroxyl (alcohol)	R — OH	sugars
carboxyl (acid)	R — C (=O) (OH)	sugars fats amino acids
ketone	R — C (=O) — R	sugars
aldehyde	R — C (=O) (H)	sugars
amine (amino)	R — N (H) (H)	amino acids proteins
sulfhydryl	R — SH	amino acids proteins
phosphate	R — O — P (=O) (OH) — OH	phospholipids nucleotides nucleic acids

R = remainder of molecule

Figure 4.3
Isomers have the same molecular formula but different configurations. Both of these compounds have the formula $C_3H_6O_3$. *a.* In glyceraldehyde, oxygen is double-bonded to an end carbon. *b.* In dihydroxyacetone, oxygen is double-bonded to the middle carbon.

a. Glyceraldehyde

b. Dihydroxyacetone

There are only a few types of molecules in cells, but these molecules still have great variety and therefore play different roles in cells.

Figure 4.4

a. In cells, synthesis often occurs when monomers are joined together by condensation (removal of H_2O). *b.* Breakdown occurs when the monomers in a polymer are separated by hydrolysis (addition of H_2O).

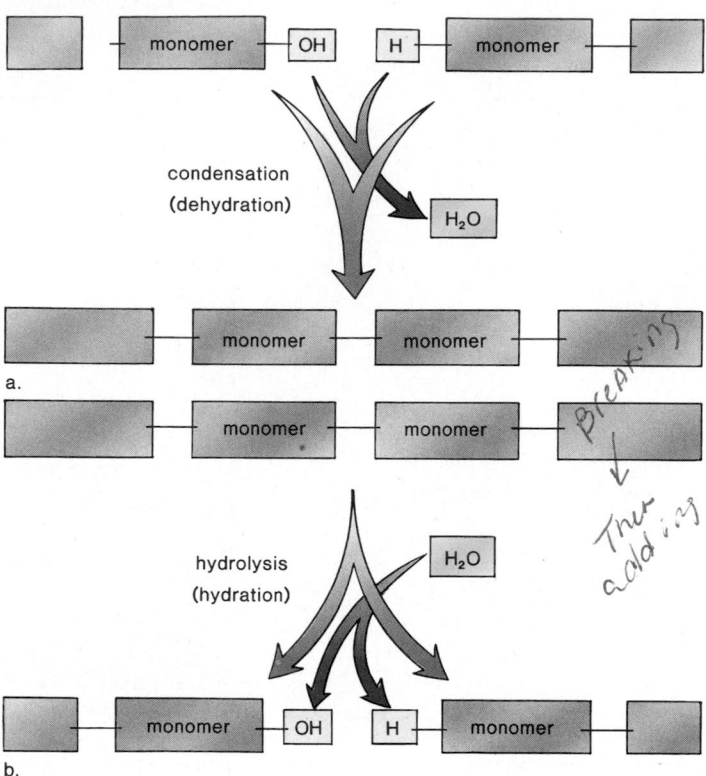

a.

b.

Figure 4.5

a. Glucose is a 6-carbon sugar that can exist as an open chain or as a ring compound; the ring is more common in cells. The shape of glucose is indicated in the space-filling model. Notice that the carbon atoms in glucose are numbered and that the colors allow you to decipher the manner in which the open chain becomes the ring. *b.* Fructose is an isomer of glucose. It too has the molecular formula $C_6H_{12}O_6$. Notice, though, that the atoms are bonded in a slightly different manner in fructose. This causes the open chain to form a slightly different ring and therefore also contributes to a different shape of the molecule.

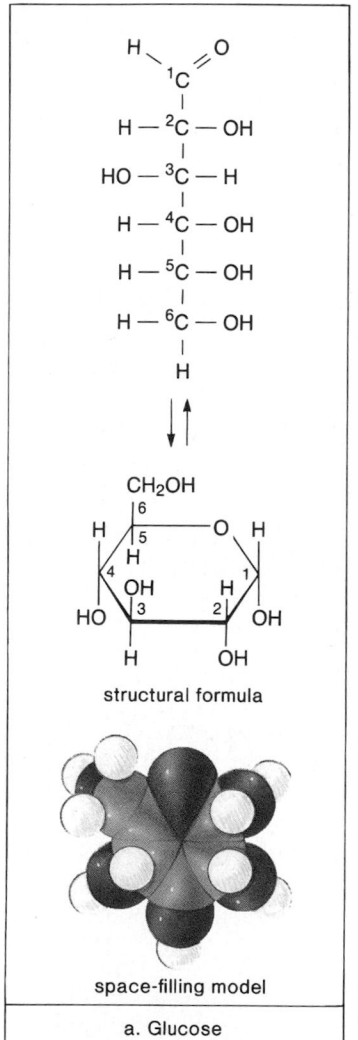

structural formula

space-filling model

a. Glucose

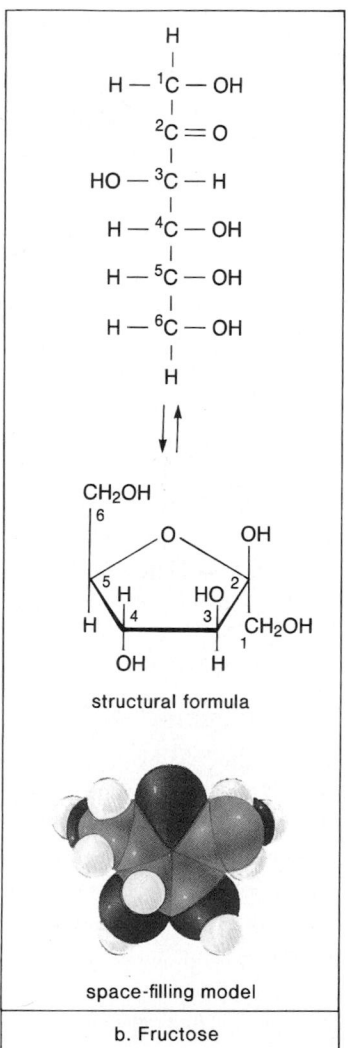

structural formula

space-filling model

b. Fructose

Condensation and Hydrolysis

All macromolecules are made within a cell in essentially the same way. Two monomers join together when a hydroxyl (—OH) group is removed from one monomer and a hydrogen (—H) is removed from the other (fig. 4.4*a*). This **condensation** of monomers is a *dehydration synthesis* because water is removed (dehydration) and a bond is made (synthesis). Condensation does not take place unless the proper enzyme (a molecule that speeds up a chemical reaction in cells) is present and energy is expended. When a bond is formed, energy is needed.

Polymers are broken down by **hydrolysis,** which is essentially the reverse of condensation; an —OH group from water attaches to one monomer and a —H attaches to the other (fig. 4.4*b*). This is a hydrolysis reaction because water (*hydro*) is used to break (*lyse*) a bond. When a bond is broken, energy is released, or made available.

Macromolecules are formed and are broken down in cells as each cell renews its many parts. Also, polysaccharide breakdown makes energy available to cells.

Condensation (the removal of water) joins monomers together to form macromolecules, and hydrolysis (the addition of water) breaks macromolecules apart into monomers.

Simple Sugars to Polysaccharides

Sugars and polysaccharides belong to a class of compounds called **carbohydrates.** Carbohydrates usually contain carbon and the components of water—hydrogen and oxygen—in a 1:2:1 ratio. The general formula for any carbohydrate is $(CH_2O)_n$; the *n* subscript stands for whatever number of these groups there are. For example, a carbohydrate with the molecular formula $C_6H_{12}O_6 = (CH_2O)_6$.

Figure 4.6

Sucrose is a disaccharide that contains glucose and fructose units. During condensation, a bond forms between the glucose and fructose molecules as the components of water are removed. During hydrolysis, the components of water are added as the bond is broken.

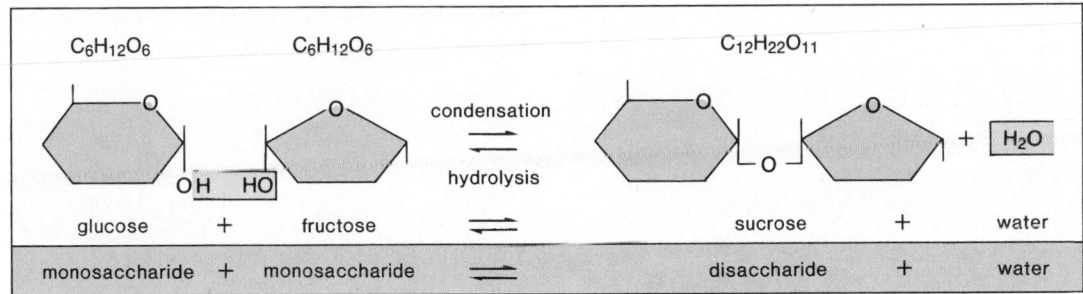

Simple sugars are *monosaccharides* (one sugar) with a carbon backbone of 3 to 7 carbon atoms. The best-known sugars are those that have 6 carbons (*hexoses*). **Glucose** is in the blood of animals, and *fructose* is frequently found in fruits. These sugars are isomers of one another. They all have the molecular formula $C_6H_{12}O_6$, but they differ in structure (fig. 4.5). Structural differences cause molecules to vary in shape, which is best seen by using space-filling models. Shape is very important in determining how molecules interact with one another.

There are 2 5-carbon sugars (*pentoses*) of significance. They are *ribose* and *deoxyribose,* which are found in the nucleic acids RNA and DNA, respectively (RNA and DNA are discussed later in the chapter).

A *disaccharide* contains 2 monosaccharides that have joined together by condensation. *Sucrose* is a disaccharide that contains glucose and fructose (fig. 4.6). Sugar is transported within the body of a plant in the form of sucrose, and this is the sugar we use at the table to sweeten our food. We acquire this sugar from plants, such as sugarcane and sugarbeets.

Lactose is a disaccharide that contains galactose and glucose and is found in milk. *Maltose* (composed of 2 glucose molecules) is a disaccharide of interest because it is found in our digestive tract as a result of starch digestion.

The most common *polysaccharides* in living things are starch, glycogen, and cellulose. Each of these is a polymer of glucose.

Starch and Glycogen

The structures of starch and glycogen differ only slightly (fig. 4.7). **Glycogen** is characterized by many side branches, which are chains of glucose that go off from the main chain. **Starch** has few of these chains.

Plant cells store extra carbohydrate as complex sugars or starches. When leaf cells are actively producing sugar by photosynthesis, they store some of this sugar within the cell in the form of starch granules (fig. 4.7b). Roots also store sugars as starch and so do seeds. Grains (e.g., the seeds of wheat, corn, and rice plants) are used by humans as a source of nourishment.

Animal cells store extra carbohydrate as glycogen, sometimes called "animal starch." After a human eats, the liver stores glucose as glycogen (fig. 4.7c). Between meals, the liver releases glucose to keep the blood concentration of glucose near the normal 0.1%.

Cells use sugars, especially glucose, as an immediate energy source. Glucose is stored as starch in plants and as glycogen in animals.

Cellulose and Chitin

Cellulose contains glucose molecules that are joined together differently than they are in starch and glycogen. The orientation of the bonds in starch and glycogen allows these polymers to form compact spirals, making these polymers suitable as storage compounds. The orientation of the bond in cellulose causes the polymer to be straight and fibrous, making it suitable as a structural compound.

As shown in figure 4.8, the long, unbranched polymers of cellulose are held together by hydrogen bonding to form microfibrils. Several microfibrils, in turn, make up a fibril. Layers of cellulose fibrils make up plant cell walls. The cellulose fibrils are parallel within each layer, but the layers themselves lie at angles to one another to give added strength.

Humans have found many uses for cellulose. Cotton fibers are almost pure cellulose, and we often wear cotton clothing. Furniture and buildings are made from wood, which contains a high percentage of cellulose. Most animals, however, including humans, cannot digest cellulose because the enzymes that digest starch are unable to break the linkage between the glucose molecules in cellulose. Even so, cellulose is a recommended part of our diet because it provides bulk (also called fiber or roughage) that helps the body maintain regularity of elimination. As mentioned, grains are a source of starch, but they are also a source of fiber.

Cattle and sheep receive nutrients from grass because they have a special stomach chamber, the rumen, where bacteria that can digest cellulose live. Unfortunately, cattle are often

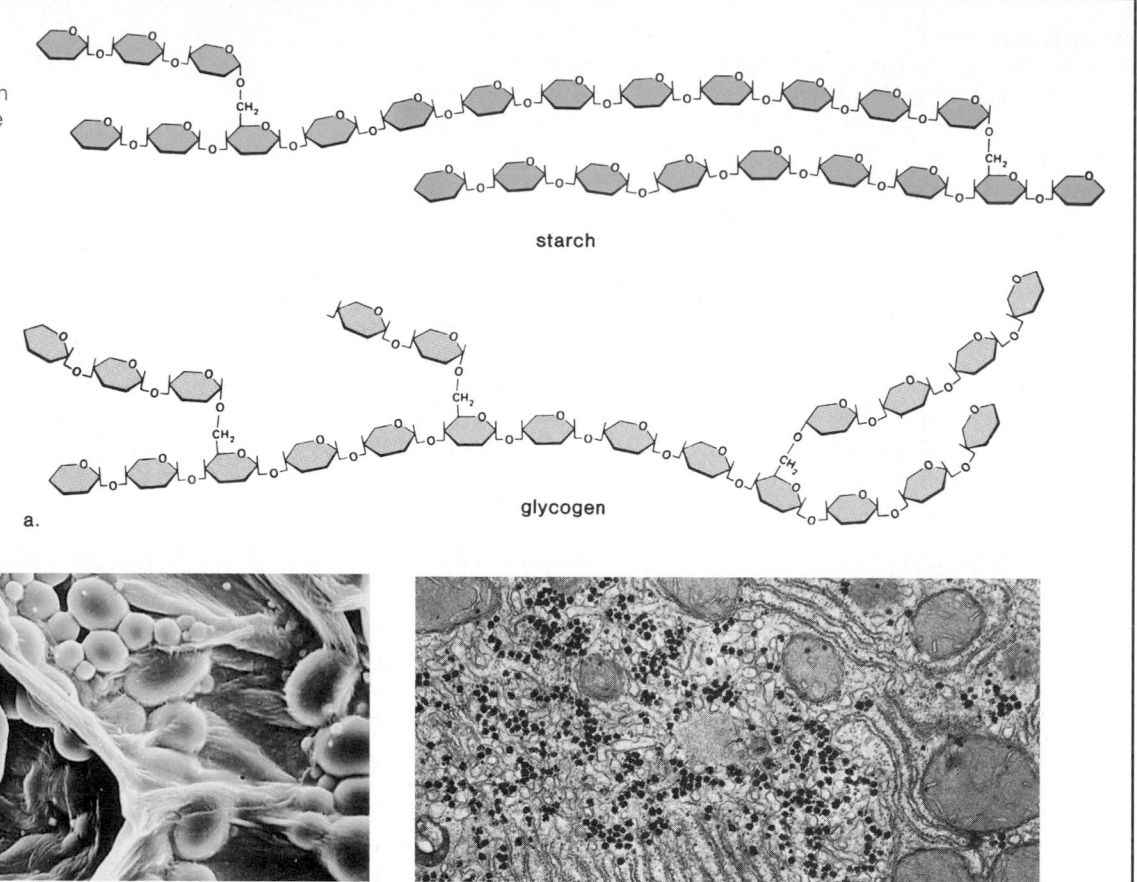

Figure 4.7
a. Starch and glycogen are polymers of glucose. *b.* Starch is the storage form of glucose in plant cells. This is a scanning electron micrograph of starch grains in a plant cell in which starch granules are highly visible. Magnification, X2,600. *c.* Glycogen is the storage form of glucose in animal cells. This is an electron micrograph of a liver cell in which glycogen granules are visible. Magnification, X24,000.

starch

glycogen

a.

b.

c.

Wheat

Humans commonly use only about 12 species of plants as food sources. Three of these are cereals: rice, corn, and wheat. A grain of wheat contains a seed and is composed of 3 parts: the embryonic plant, the endosperm (stored food), and the seed coat along with other protective layers (fig. 4.A). The embryonic plant is wheat germ, which is high in vitamins. The starchy endosperm is used to produce the flour from which our white breads are made. The seed coat is bran, composed mostly of cellulose, which cannot be digested by humans. For this reason, bran is an excellent roughage substance that adds bulk to the diet.

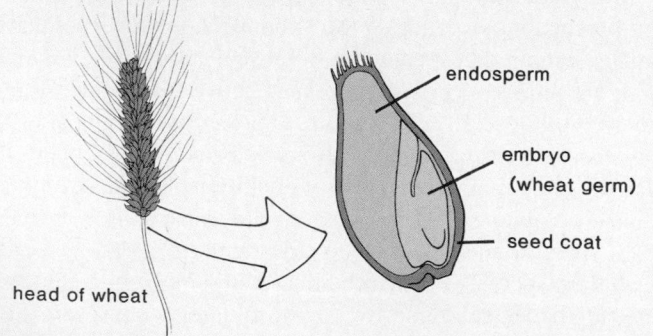

endosperm

embryo (wheat germ)

seed coat

head of wheat

kept in feedlots, where they are fed grains instead of grass. This practice wastes fossil fuel energy because it takes energy to grow the grain, to process it, to transport it, and finally to feed it to the cattle. Also, range-fed cattle move about and produce a leaner meat than do cattle raised in a feedlot. There is growing evidence that lean meat is healthier for humans than fatty meat.

Chitin, which is found in the exoskeleton of crabs and related animals like lobsters and insects, is also a polymer of glucose. Each glucose, however, has an amino group attached to it. The linkage between the glucose molecules is like that found in cellulose; therefore, chitin is not a digestible material. Recently, scientists have discovered how to turn chitin into thread that can be used as suture material. They hope to find other uses for treated chitin. If so, all those discarded crab shells that pile up beside crabmeat processing plants will not go to waste.

Fatty Acids to Lipids

A variety of organic compounds are classified as **lipids.** Many of these are insoluble in water because they lack any polar groups. The most familiar lipids are those found in fats and oils. Organisms use these molecules as long-term energy storage

Figure 4.8

Cellulose fibrils are present in plant cell walls. Magnification, X30,000. Each fibril contains several microfibrils, and each microfibril contains many chains of glucose hydrogen-bonded together. Finally, we have a close-up view of 3 cellulose polymers, each made up of glucose molecules.

fibril

microfibril

cell wall

plant cell

cellulose molecules

hydrogen bond

cellulose polymer

compounds. Phospholipids and steroids are also important lipids found in living things. For example, they are components of plasma membranes.

Lipids are quite varied in structure, but they tend to be insoluble in water.

Fats and Oils

Fats and **oils** contain 2 types of unit molecules: fatty acids and glycerol.

Each *fatty acid* consists of a long hydrocarbon chain with a carboxyl (acid) group at one end. Because the carboxyl group is a polar group, fatty acids are soluble in water. Most of the fatty acids in cells contain 16–18 carbon atoms per molecule, although smaller ones are also found. Fatty acids are either saturated or unsaturated (fig. 4.9). *Saturated* fatty acids have no double bonds between the carbon atoms. The carbon chain is saturated, so to

speak, with all the hydrogens that can be held. *Unsaturated* fatty acids have double bonds in the carbon chain wherever the number of hydrogens is less than 2 per carbon atom:

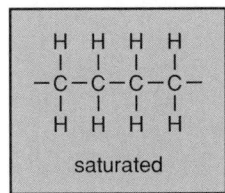

saturated

unsaturated

Glycerol is a compound with 3 hydroxyl groups (fig. 4.10). Hydroxyl groups are polar and therefore glycerol is soluble in water. When fat is formed, the acid portions of 3 fatty acids react with these hydroxyl groups so that fat and 3 molecules of water form. Again, the larger fat molecule is formed by condensation, and a fat can be hydrolyzed to its components. Since there are 3 fatty acids per one glycerol molecule, the fat molecule is sometimes called a triglyceride. Triglycerides are often referred to as neutral fats because they, unlike their component parts, lack polar groups that can hydrogen bond with water (p. 33). Therefore, for example, cooking oils do not mix with water even though they are both liquids. Even after shaking, the oil simply separates out.

Triglycerides containing unsaturated hydrocarbon chains melt at a lower temperature than those containing saturated chains. This is because a double bond makes a kink that prevents close packing among the chains (fig. 4.9). We can reason, then, that butter, which is a solid at room temperature, must contain saturated hydrocarbon chains and corn oil, which is a liquid even when placed in the refrigerator, must contain unsaturated chains. This difference is made use of by living things. For example, the feet of reindeer and penguins contain unsaturated triglycerides, and this helps protect these exposed parts from freezing.

Nearly all organisms use fats for long-term energy storage. Fats have mostly C—H bonds, making them a richer supply of chemical energy than carbohydrates, which have many C—OH bonds. Molecule per molecule, animal fat contains over twice as much energy as glycogen; gram per gram though, fat stores 6 times as much energy as glycogen. This is because fat droplets, being nonpolar, do not contain water. Small birds, like the broad-tailed hummingbird (fig. 4.11), store a great deal of fat before they start their long spring and fall migratory flights. About 0.15 g of fat per gram of body weight is accumulated each day. If the same amount of energy were stored as glycogen, a bird would be so heavy it would not be able to fly.

> Fats and oils are triglycerides (one glycerol plus 3 fatty acids). They are used as long-term energy storage compounds in plants and animals.

Waxes

In *waxes,* a long-chain fatty acid bonds with a long-chain alcohol. Waxes are also solid at normal temperatures because they have a high melting point. Being hydrophobic, they are also waterproof and resistant to degradation. In many plants, waxes form a protective covering that retards the loss of water for all exposed parts. In animals, waxes are involved in skin and fur maintenance. In humans, wax is produced by glands in the outer ear canal. Here its function is to trap dust and dirt, preventing them from reaching the eardrum.

Phospholipids

Phospholipids, as implied by their name, contain a phosphate group. A phosphate group is a polar group that can ionize:

$$
\underset{\text{nonionized phosphate}}{R-\overset{\overset{\displaystyle O}{\|}}{\underset{\underset{\displaystyle OH}{|}}{P}}-OH}
\qquad\qquad
\underset{\text{ionized phosphate}}{R-\overset{\overset{\displaystyle O}{\|}}{\underset{\underset{\displaystyle O^{-}}{|}}{P}}-O^{-}}
$$

Figure 4.9
Fatty acid structure. Fatty acids are long hydrocarbon chains ending in a carboxylic (acid) group. Stearate is a saturated fatty acid, and oleate is an unsaturated fatty acid.

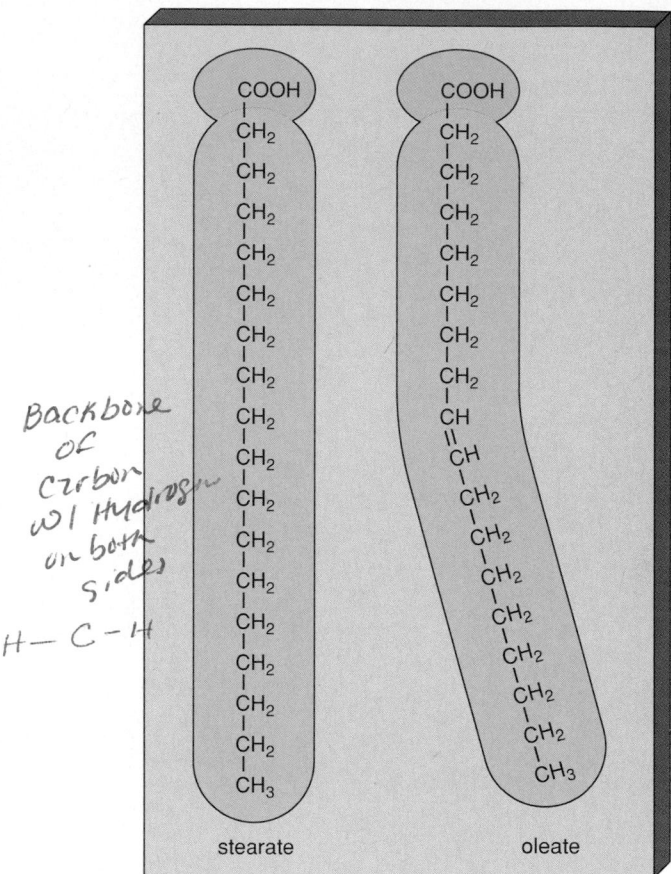

stearate   oleate

Essentially, phospholipids are constructed like neutral fats, except that in place of the third fatty acid, there is a phosphate group or a grouping that contains both phosphate and nitrogen. This group becomes the polar head of the molecule, while the hydrocarbon chains of the fatty acids become the nonpolar tails (fig. 4.12). When phospholipid molecules are placed in water, they form a sheet in which the polar heads face outward and the nonpolar tails face each other. This property of phospholipids means that they can form an interface or separation between 2 solutions such as the interior and exterior of a cell. The plasma membrane of cells is basically a phospholipid bilayer.

> Phospholipids have a polar head and nonpolar tails. They arrange themselves in a double layer in the presence of water, a property that makes them suitable as the plasma membrane of cells.

Steroids

Steroids are lipids that have an entirely different structure from neutral fats. Each steroid has a backbone of 4 fused carbon rings and varies from other steroids primarily by the type of functional

Figure 4.10

Formation of a neutral fat. Three fatty acids plus glycerol react to produce a fat molecule and 3 water molecules. A fat molecule plus 3 water molecules react to produce 3 fatty acids and glycerol.

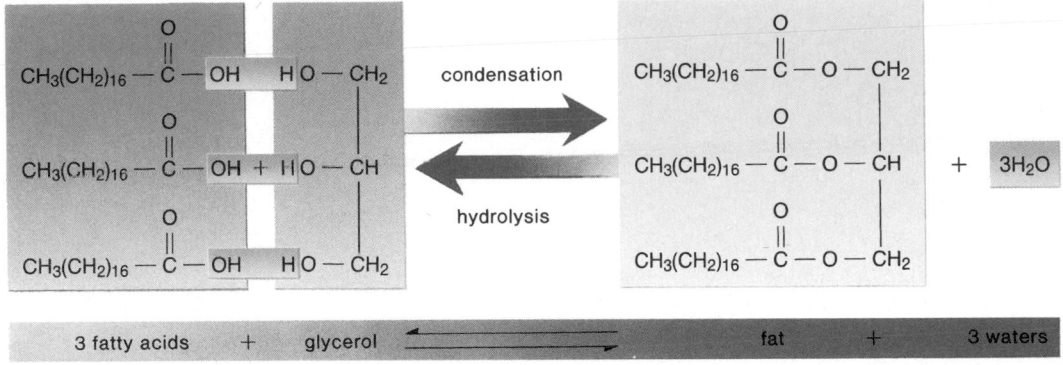

Figure 4.11

A broad-tailed humming-bird, like other small birds that migrate, stores a lot of energy as fat in order to have enough fuel for its long journey. Fat is more energy rich than glycogen, and it is found in concentrated droplets; therefore, it is an efficient way to store energy.

Figure 4.12

Structurally, phospholipids have a polar head and nonpolar tails (*left and center*). In water (*right*), the molecules arrange themselves as shown. The polar heads are attracted to water; the nonpolar tails are not.

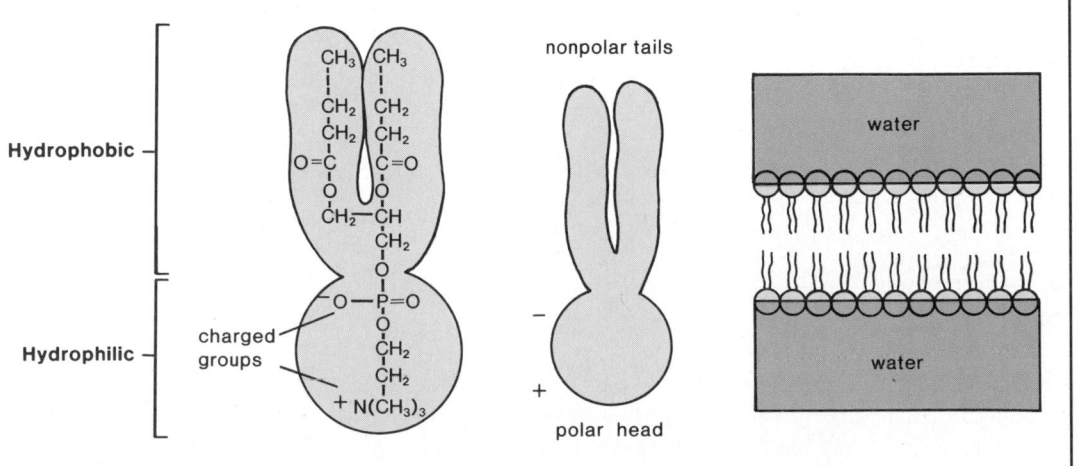

group attached to the ring (fig. 4.13). Cholesterol is the precursor of several other steroids, including the vertebrate hormones such as aldosterone, a hormone that helps regulate the sodium content of the blood, and the sex hormones, which help maintain male and female characteristics. Their functions vary due primarily to the different attached groups.

For many years, nutritionists have gathered evidence that a diet high in saturated fats and cholesterol can lead to reduced blood flow caused by the deposit of fatty materials on the linings of blood vessels. This problem is discussed at length in a reading in chapter 34.

> Steroids are ring compounds that have a similar backbone but vary according to the attached group. This causes them to have different functions in the bodies of humans and other animals.

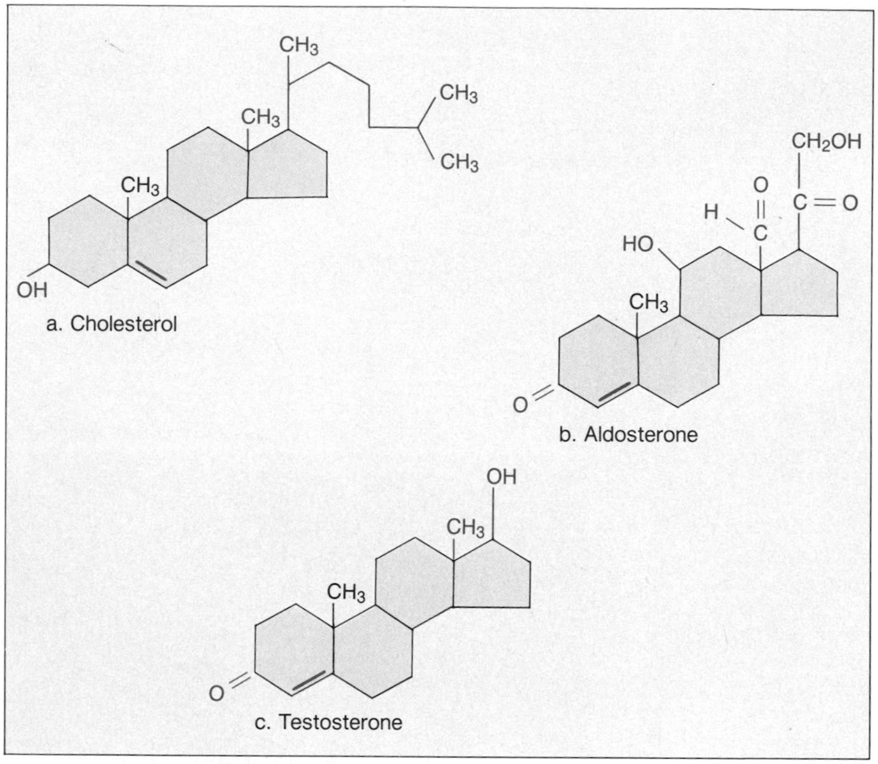

Amino Acids to Proteins

Amino acids are the monomers that condense to form **proteins,** which are very large molecules that have both structural and metabolic functions. For example, in animals the proteins myosin and actin are the contractile components of muscle; insulin is a hormone that regulates the sugar content of the blood; hemoglobin transports oxygen in the blood; and collagen fibers support many organs. Proteins are present in the membrane that surrounds each cell, and they also exist within the cell. The cellular proteins, called **enzymes,** are organic catalysts that speed up chemical reactions within cells.

In this section we will point out that the role a protein plays in cells is dependent on its biological properties, which in turn are dependent on its structure.

Amino Acids and Peptides

There are 20 different amino acids commonly found in cells, and figure 4.14 gives several examples. All amino acids contain 2 important functional groups: a carboxyl (acid), —COOH, group and an amino, —NH_2, group, both of which ionize at normal body pH in this manner:

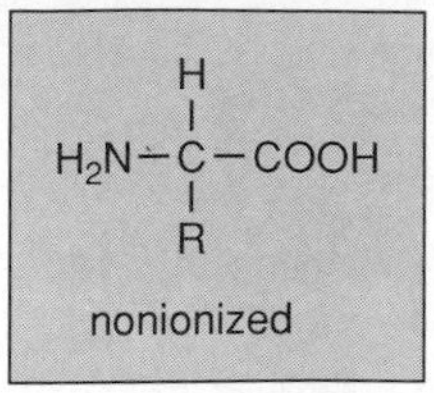

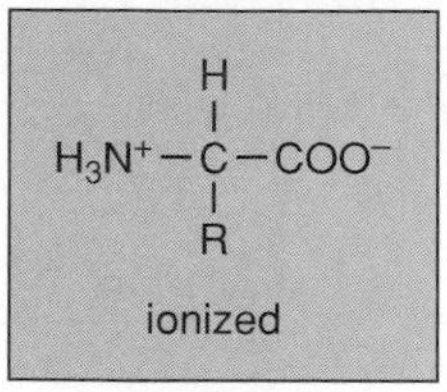

Amino acids differ in the nature of the R group (Remainder of the molecule), which ranges in complexity from a single hydrogen to complicated ring compounds. The unique chemical properties of

Figure 4.13
Steroid diversity. Like cholesterol in **a,** steroid molecules have 4 adjacent rings, but their effects on the body largely depend on the attached groups indicated in red. **b.** Aldosterone is involved in the regulation of sodium blood levels. **c.** Testosterone is the male sex hormone.

a. Cholesterol

b. Aldosterone

c. Testosterone

an amino acid depend on those of the R group. For example, some R groups are polar and some are not. Also, the amino acid cysteine has an R group that ends with a sulfhydryl, —SH, group that often serves to connect one chain of amino acids to another by a disulfide bond, —S—S.

A **peptide** is 2 or more amino acids joined together, and a *polypeptide* is a chain of many amino acids joined by peptide bonds. Since a protein can contain more than one polypeptide chain, you can see why a protein could have a very large number of amino acids.

Figure 4.15 shows how 2 amino acids are joined by a condensation reaction between the carboxyl group of one and the amino group of another. The resulting covalent bond between 2 amino acids is called a **peptide bond.** The atoms associated with the peptide bond share the electrons unevenly because both oxygen and nitrogen atoms are electronegative. The hydrogen attached to the nitrogen is electropositive. The polarity of the peptide bond means that hydrogen bonding is possible between parts of a polypeptide.

> Amino acids are joined by peptide bonds in polypeptides and proteins. Proteins have structural and metabolic functions in cells. Some proteins are enzymes that speed up chemical reactions.

Figure 4.14

Representative amino acids; the R groups are in the contrasting color. Some R groups are nonpolar and hydrophobic, some are polar and hydrophilic, and some are ionized and hydrophilic.

Levels of Structure

The final shape of a protein in large measure determines the function of a protein in the cell or body of an organism. An analysis of protein shape shows that proteins can have up to 4 levels of structure (table 4.1 and fig. 4.16).

The *primary structure* of a protein is the sequence of the amino acids joined together by peptide bonds. In 1953, Frederick Sanger determined the amino acid sequence of the hormone insulin, the first protein to be sequenced. He did it by first breaking insulin into fragments and then determining the amino acid sequence of the fragments before determining the sequence of the fragments themselves. It was a laborious 10-year task, but it established for the first time that a particular protein has a particular sequence of amino acids. Today, there are automated sequencers that tell scientists the sequence of amino acids in a protein within a few hours. Notice that since each amino acid differs from another by its R group, it is correct to say that proteins differ from one another by a particular sequence of the R groups. The fact that some of these are polar and some are not influences the final shape of the polypeptide.

The *secondary structure* of a protein comes about when the polypeptide takes a particular orientation in space. Linus Pauling and Robert Corey, who began studying the structure of amino acids in the late 1930s, concluded about 20 years later that polypeptides must have particular orientations in space. They said that the α (alpha) helix and the ß (beta) sheet were 2 possible patterns of amino acids within a polypeptide. They called it the "α" helix because it was the first pattern they discovered and the "ß" sheet because it was the second pattern they discovered.

Hydrogen bonding between the oxygen atoms and nitrogen atoms of amino acid monomers, in particular, stabilizes the α helix. In the ß sheet, or pleated sheet, the polypeptide turns back upon itself and hydrogen bonding occurs between these extended lengths of the polypeptide. Fibrous proteins are so-called because they are found in tough fibers. Keratin found in

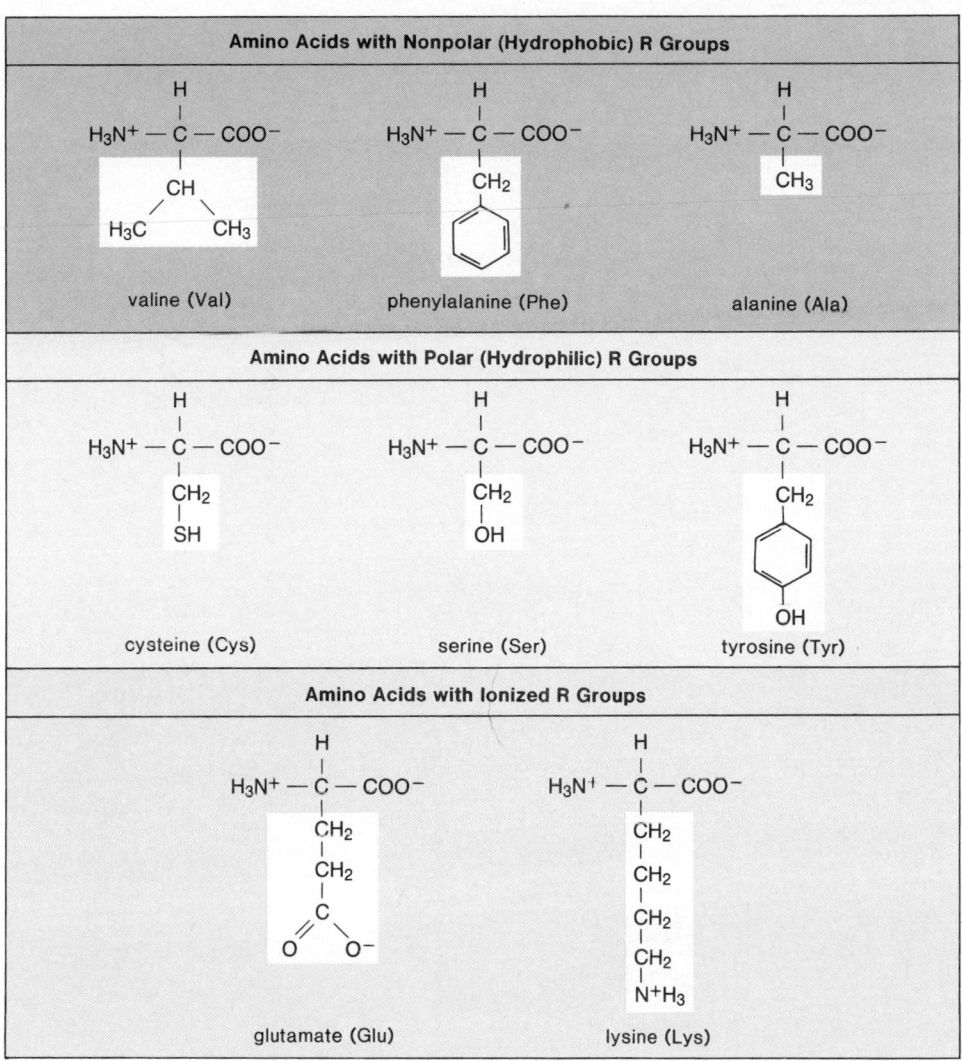

Amino Acids with Nonpolar (Hydrophobic) R Groups

valine (Val)

phenylalanine (Phe)

alanine (Ala)

Amino Acids with Polar (Hydrophilic) R Groups

cysteine (Cys)

serine (Ser)

tyrosine (Tyr)

Amino Acids with Ionized R Groups

glutamate (Glu)

lysine (Lys)

Figure 4.15

Synthesis of a peptide. The peptide bond forms as the components of water are removed from the carboxyl group of one amino acid and the amino group of the other amino acid. There is a partial negative charge on the oxygen and nitrogen and a partial positive charge on the hydrogen within the peptide bond. These charges are not shown.

Figure 4.16

Levels of protein structure for a globular protein. ***a.*** The primary structure is the sequence of amino acids. ***b.*** The secondary structure contains α helix segments and ß sheet segments. ***c.*** The tertiary structure is a twisting and turning of the polypeptide molecule. ***d.*** The quaternary structure contains more than one polypeptide each with its own levels of structure.

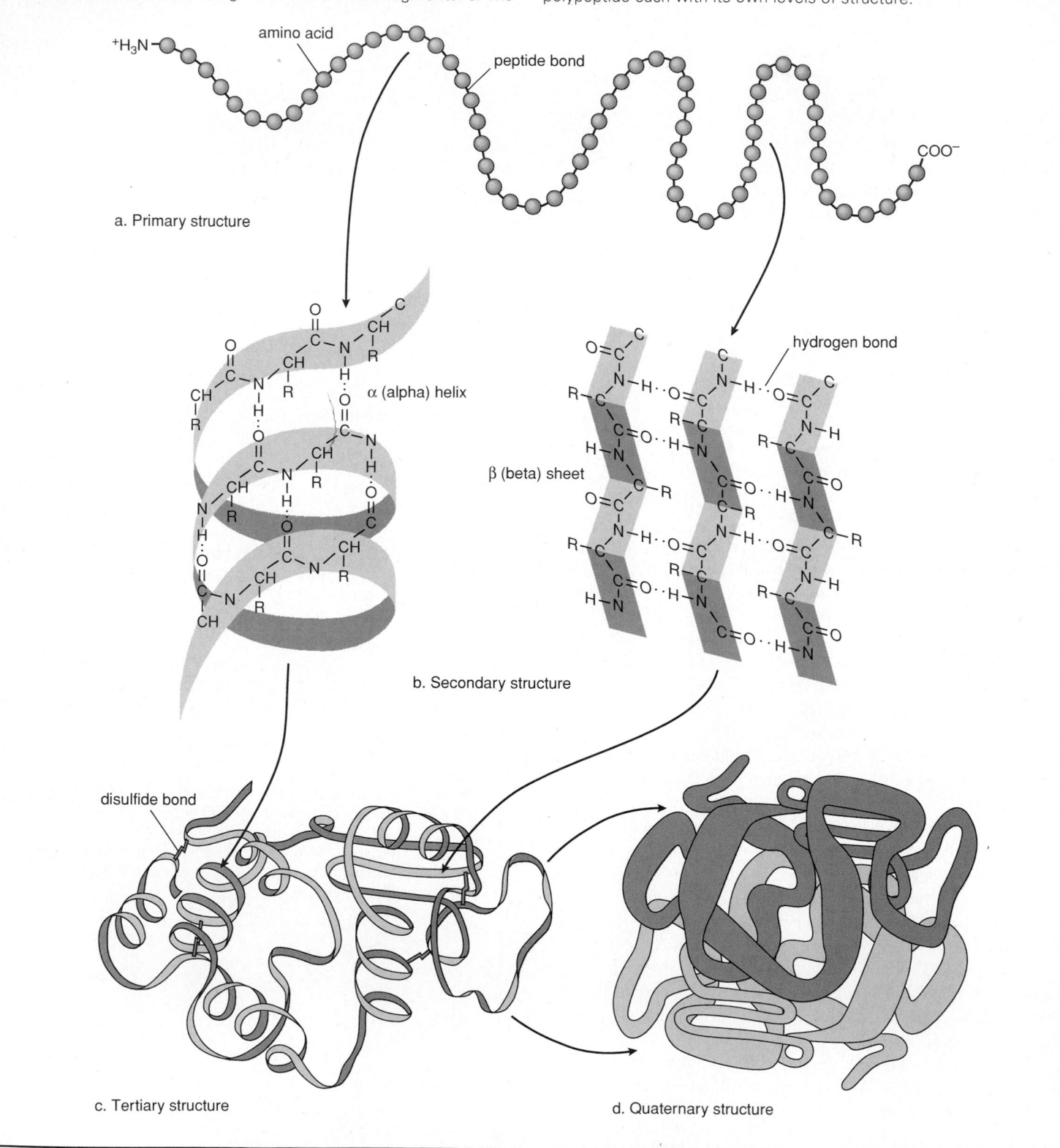

hair and wool is an example of a fibrous protein in which the polypeptides are an α helix; silk is an example of a fibrous protein in which the polypeptides are ß sheets.

Globular proteins, so-called because they have a globular shape, have regions of both alpha helix and beta sheet arrangements depending upon the particular amino acids in the primary structure. The polypeptide of a globular protein folds and twists into a *tertiary shape* that is maintained by various types of bonding between the R groups. Indeed, the folding and twisting is determined by those R groups that can bond with one another, giving stability to the shape of the molecule. Hydrogen bonds, ionic bonds, and covalent bonds are seen; when cysteines are brought close to one another, a disulfide, S—S, bond serves as a bridge between 2 amino acid monomers. The hydrophobic R groups, which do not react with other R groups, tend to be pushed together into a common inner region, where they are not exposed to water. (These are called hydrophobic interactions.)

The various enzymes of a cell are globular proteins that usually have only a tertiary level structure. Some proteins have more than one polypeptide chain, each with its own primary, secondary, and tertiary structures. Within the protein, these separate chains are arranged to give a fourth level of structure termed the *quaternary structure*. Various types of interactions between the polypeptide chains are observed. Hemoglobin is a much studied globular protein that has a quaternary structure of 4 polypeptides. Each polypeptide has a heme group associated with it that carries oxygen reversibly (fig. 37.10c).

A protein has up to 4 levels of structure that account for its final 3-dimensional shape. The shape of a protein determines its function in cells.

Denaturation of Proteins

Both temperature and pH can bring about a change in protein shape. For example, we are all aware that the addition of acid to milk causes curdling; heating causes egg white, a protein called albumin, to congeal, or coagulate. When a protein loses its normal configuration, it is said to be denatured. Denaturation occurs because the normal bonding patterns between parts of a molecule have been disturbed. Once a protein loses its normal shape, it is no longer able to perform its usual function.

If the conditions that caused denaturation were not too severe, and if these are removed, some proteins regain their normal shape and biological activity. This shows that the primary structure of a polypeptide forecasts its final shape (fig. 4.17).

The final shape of a protein is dependent upon the primary structure—the sequence of amino acids in the polypeptide(s).

Table 4.1
Levels of Protein Structure

Level of Structure	Description	Type of Bond
Primary	Sequence of amino acids	Covalent (peptide) bond between amino acids
Secondary	Alpha helix and beta sheet	Hydrogen bond between members of peptide bond
Tertiary	Folding and twisting of polypeptide	Hydrogen, ionic, covalent (S-S), hydrophobic interactions* between R groups
Quaternary	Several polypeptides	Hydrogen, ionic between polypeptide chains

*Strictly speaking, these are not bonds, but they are very important in creating and stabilizing tertiary structure.

Figure 4.17
Denaturation and reactivation of the enzyme ribonuclease. When a protein is denatured, it loses its normal shape and activity. If denaturation is gentle and if the conditions are removed, some proteins regain their normal shape. This shows that the normal conformation of the molecule is due to the various interactions between a set sequence of amino acids. Each type of protein has a particular sequence of amino acids.

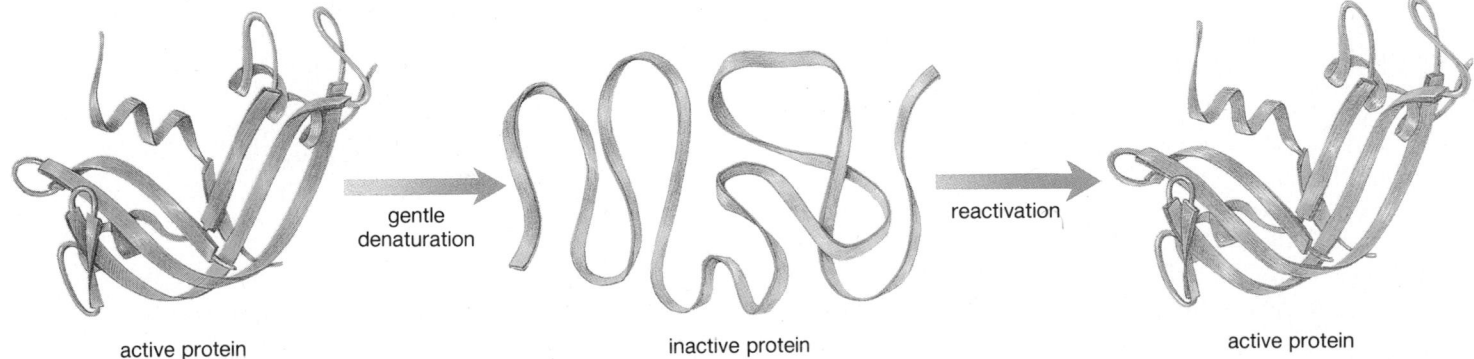

active protein   gentle denaturation   inactive protein   reactivation   active protein

Table 4.2
DNA Structure Compared to RNA Structure

	DNA	RNA
Sugar	Deoxyribose	Ribose
Bases	Adenine, guanine, thymine, cytosine	Adenine, guanine, uracil, cytosine
Strands	Double-stranded with base pairing	Single-stranded
Helix	Yes	No

Nucleotides to Nucleic Acids

Every **nucleotide** is a molecular complex of 3 types of unit molecules: phosphate (phosphoric acid), a pentose sugar, and a nitrogen-containing base:

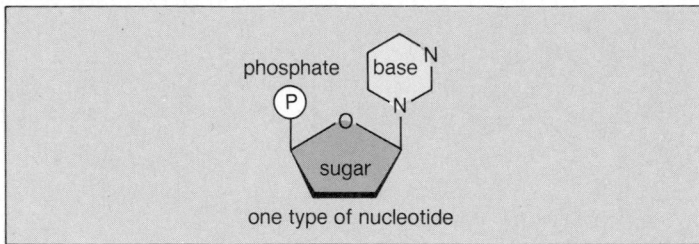

one type of nucleotide

Nucleotides have metabolic functions in cells. For example, some are components of coenzymes, which perform functions facilitating enzymatic reactions. ATP (adenosine triphosphate) is a nucleotide used in cells to supply energy for synthetic reactions and for various other energy-requiring processes. Nucleotides are also monomers in nucleic acids.

Nucleic acids are huge polymers of nucleotides with very specific functions in cells; for example, **DNA** (deoxyribonucleic acid) is the genetic material that stores information regarding its own replication and the order in which amino acids are to be joined to make a protein. Another important nucleic acid, **RNA** (ribonucleic acid), works in conjunction with DNA to bring about protein synthesis.

The nucleic acids DNA and RNA are polymers of nucleotides. DNA makes up genes and along with RNA controls protein synthesis within the cell.

In DNA the sugar is deoxyribose, and in RNA the sugar is ribose; the difference accounts for their respective names (table 4.2). There are 4 different types of nucleotides in DNA and RNA. Figure 4.18 shows the types of nucleotides that are present in DNA. The base can be one of the purines, adenine or guanine, which have a double ring, or one of the pyrimidines, thymine or cytosine, which have a single ring. These structures are called bases because their presence raises the pH of a solution. In RNA, the base uracil occurs instead of the base thymine.

Figure 4.18
All nucleotides found in DNA contain phosphate, the pentose sugar deoxyribose, and a nitrogen-containing organic base. **a.** The bases adenine and guanine are double-ringed purine bases. **b.** The bases thymine and cytosine are single-ringed pyrimidine bases.

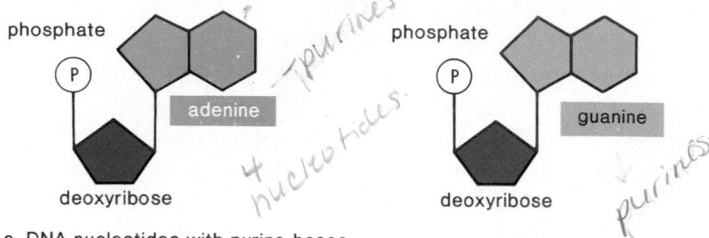

a. DNA nucleotides with purine bases

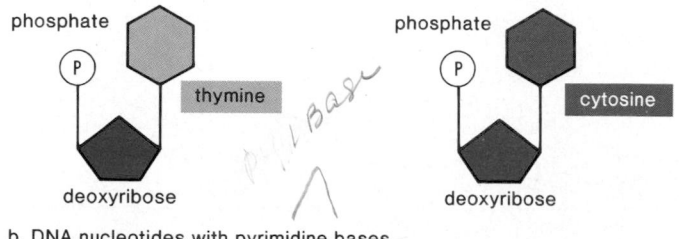

b. DNA nucleotides with pyrimidine bases

Figure 4.19
RNA is a single-stranded polymer of nucleotides. When the nucleotides join, the phosphate group (P) of one is bonded to the sugar of the next. The bases project to the side of the resulting sugar-phosphate backbone.

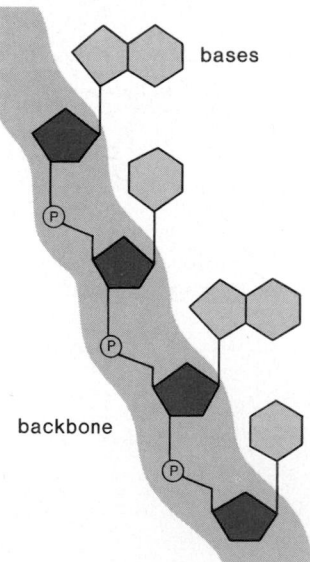

Nucleotides join together in a definite sequence when DNA and RNA form by condensation. The nucleotides form a linear molecule called a strand in which the backbone is made up of phosphate-sugar-phosphate-sugar, with the bases projecting to one side of the backbone. Since the nucleotides occur in a definite order, so do the bases.

Radioactive isotopes are often used today to trace the synthesis of macromolecules in cells and tissues. The first step is to expose the cells to a radioactively labeled molecule. For example, to detect the synthesis of DNA, ^{3}H- or ^{14}C-thymidine* can be used, and to detect the synthesis of RNA, ^{3}H- or ^{14}C-uridine* can be used. Other labeled molecules are available for protein and polysaccharide synthesis. After time has elapsed and synthesis has occurred, the labeled macromolecule is extracted from the cell by chemical means and placed in a vial that contains a special solution. The solution produces a light flash every time the labeled macromolecule gives off a radioactive particle. The number of flashes (called scintillations) are detected by placing the vial in a scintillation counter, an instrument that counts the flashes. The greater the amount of synthesis, the greater the radioactivity in the sample, and the greater the counts per minute (cpm). Figure 4.B shows the results of an experiment in which the amount of RNA synthesis was determined as a function of cell culture age. As cell cultures age, the amount of RNA synthesis decreases.

Autoradiography is a technique that is used to determine the placement of labeled molecules in cells. In this case, autoradiography would indicate where RNA synthesis is occurring. After exposure of cells (or tissues) to, say, radioactively labeled uridine, the cells are washed clean of any excess radioactivity before they are sectioned and placed on a slide. The slide is coated with an emulsion layer that contains silver bromide crystals. The radioactively labeled molecules in the cell (or tissues) interact with the silver bromide crystals in the emulsion, causing black grains to appear when the slide is photographically developed. It takes microscopic examination to see the location of the black grains in the emulsion and thereby tell which part of the cell took up the radioactively labeled molecules (fig. 4.C).

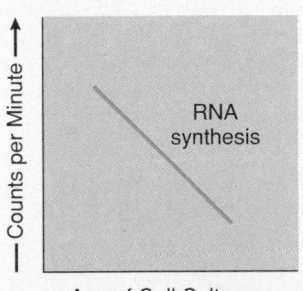

Figure 4.B
A scintillation chamber determines the relative amount of radioactivity in a number of samples, and a graph can be used to display the results.

scintillation chamber

*These molecules are nucleosides, that is, they lack the phosphate component but do contain the base and sugar of a nucleotide.

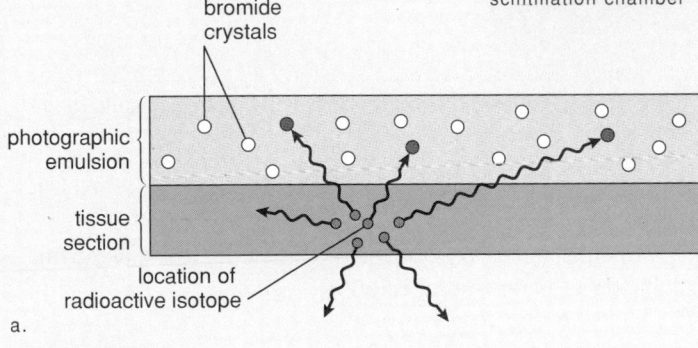
a.

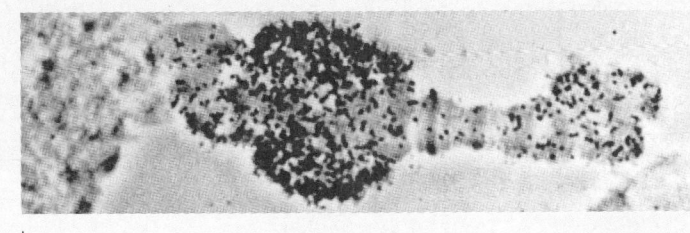
b.

Figure 4.C
a. Radioactive isotopes in uridine molecules within a tissue interact with silver bromide crystals in a photographic emulsion. *b.* When the emulsion is developed, black grains show up that indicate the location of RNA synthesis.

RNA is single-stranded (fig. 4.19), but DNA is double-stranded. The 2 strands of DNA twist about one another in the form of a double helix (fig. 4.20). The 2 strands are held together by hydrogen bonds between purine and pyrimidine bases. Thymine (T) in one strand is always paired with adenine (A) in the opposite strand, and guanine (G) is always paired with cytosine (C). This is called **complementary base pairing.**

If the DNA helix is unwound, it resembles a ladder (fig. 4.20c). The sides of the ladder are made entirely of phosphate and sugar molecules, and the rungs of the ladder are made only of the complementary paired bases. The bases can be in any order within a strand, but between strands, A is always paired with T and G is always paired with C, and vice versa. Therefore,

Figure 4.20

Overview of DNA structure. *a.* Double helix. *b.* Double helix showing complementary base pairing between bases. *c.* When the helix unwinds, the ladder structure of DNA is more evident. Sugar-phosphate molecules make up the sides of the ladder, and the bases, held together by hydrogen bonding, make up the rungs. One nucleotide has been boxed in the diagram.

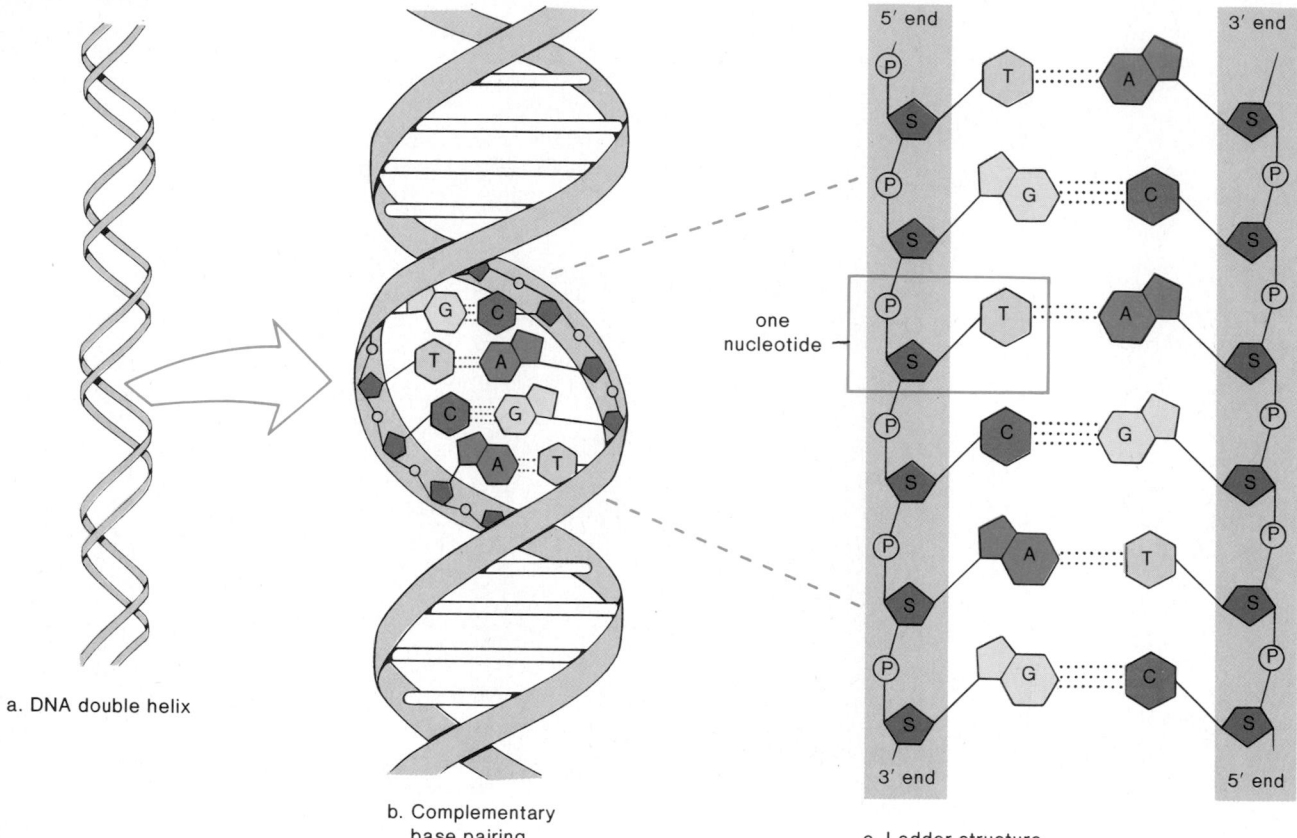

a. DNA double helix

b. Complementary base pairing

c. Ladder structure

regardless of the order or the quantity of any particular base pair, the number of purine bases always equals the number of pyrimidine bases.

DNA has a structure like a twisted ladder: sugar molecules and phosphate molecules make up the sides of the ladder, and hydrogen-bonded bases make up the rungs of the ladder. RNA differs from DNA in several respects.

The structure and the function of nucleic acids are discussed further in chapters 15 and 16 of this text.

Summary

1. The chemistry of carbon accounts for the diversity of organic molecules found in living things. Carbon can bond with as many as 4 other atoms and can also bond with itself to form both chains and rings. The former are called the backbone of the molecule.

2. Each type of small molecule in living things—simple sugars, fatty acids, amino acids, and nucleotides—has a carbon backbone and is characterized by the presence of functional groups.

Some functional groups are hydrophobic, and some are hydrophilic.

3. Differences in the backbone and attached functional groups cause molecules to have different shapes, and shape helps determine how a molecule interacts with other molecules in cells.

4. The large organic molecules in cells are macromolecules, polymers formed by the joining together of monomers. Enzymes carry out both condensation

(building up) and hydrolysis (breaking down) of macromolecules. For each bond formed, a molecule of water is removed, and for each bond broken, a molecule of water is added.

5. Monosaccharides, disaccharides, and polysaccharides are all carbohydrates. Therefore, the term *carbohydrate* includes both the monomers (e.g., glucose) and the polymers (e.g., starch, glycogen, and cellulose). Starch and glycogen are energy-storage

compounds, but cellulose has a structural function in plants.

6. Lipids include a wide variety of compounds that are insoluble in water. Fats and oils allow long-term energy storage and contain one glycerol and 3 fatty acids. Fats tend to contain saturated fatty acids, and oils tend to contain unsaturated fatty acids. In a phospholipid, one of the fatty acids is replaced by a phosphate group. In the presence of water phospholipids form a double layer because the head of the molecule is polarized and the tails are not. Waxes and steroids are also lipids.

7. Proteins are polymers of amino acids. Some proteins are enzymes, and proteins also have structural roles in cells and organisms.

8. A polypeptide is a long chain of amino acids joined by peptide bonds. There are 20 different amino acids in cells that differ only by their R groups. Polarity and nonpolarity are important aspects of the R groups.

9. A polypeptide has 3 levels of structure: the primary level is the sequence of the amino acids; the secondary level contains alpha (α) helices and beta (β) sheets held in place by hydrogen bonding between peptide bonds; and the tertiary level is the final folding and twisting of the polypeptide that is held in place by bonding and hydrophobic interactions between R groups. Proteins that contain more than one polypeptide have a quaternary level of structure.

10. The shape of a polypeptide influences its biological activity. A polypeptide can be denatured and lose its normal shape and activity but, if reactivation conditions are appropriate, it assumes both of these again. This shows that the final shape of a polypeptide is dependent upon the primary structure.

11. Nucleic acids are polymers of nucleotides. Each nucleotide has 3 components: a sugar, a base, and phosphate

(phosphoric acid). DNA, which contains the sugar deoxyribose, is the genetic material that stores information for its own replication and for the order in which amino acids are to be sequenced in proteins. DNA, with the help of RNA, controls protein synthesis.

12. Table 4.3 also summarizes our coverage of organic molecules in cells.

Table 4.3
Organic Molecules in Cells

	Categories	Examples	Functions
Carbohydrates	Monosaccharides 6-carbon sugar	Glucose	Immediate energy source
	Disaccharides 12-carbon sugar	Sucrose	Transport sugar in plants
	Polysaccharides polymer of glucose	Starch, glycogen Cellulose	Energy storage Plant cell wall structure
Lipids	Triglycerides 1 glycerol + 3 fatty acids	Fats, oils	Long-term energy storage
	Waxes fatty acid + alcohol	Cuticle	Protective covering for parts of plants
		Ear wax	Protective barrier
	Phospholipids like triglyceride except one fatty acid is replaced by a phosphate group	——	Plasma membrane component
	Steroids backbone of 4 fused rings	Cholesterol Testosterone	Plasma membrane component Male sex hormone
Proteins	Polypeptides polymer of amino acids; 3 levels of structure— some proteins have 4 levels	Enzymes Myosin and actin Insulin Hemoglobin Collagen	Speed up cellular reactions Muscle cell components Regulates sugar content of blood Oxygen carrier in blood Fibrous support of body parts
Nucleic Acids	Nucleic acids polymer of nucleotides	DNA RNA	Genetic material Protein synthesis
	Nucleotides	ATP Coenzymes	Energy carrier Assist enzymes

Study Questions

1. How are the chemical characteristics of carbon reflected in the characteristics of organic molecules?

2. Give examples of functional groups, and discuss the importance of their being hydrophobic or hydrophilic.

3. What molecules are monomers of the polymers studied in this chapter? How are monomers joined to give polymers, and how are polymers broken down to monomers?

4. Name several monosaccharides, disaccharides, and polysaccharides, and give a function of each. How are these molecules structurally distinguishable?

5. Name the different types of lipids, and give a function for each type. What is the difference between a saturated and unsaturated fatty acid? Explain the structure of a fat molecule by stating its components and how they are joined together.

6. How does the structure of a phospholipid differ from that of a fat? How do phospholipids form a double layer in the presence of water?

7. Draw the structure of an amino acid and of a dipeptide, pointing out the peptide bond.

8. Discuss the 4 levels of structure of a protein and relate each level to particular bonding patterns.

9. How is the tertiary structure of a polypeptide related to its primary structure? Mention denaturation as evidence of this relationship.

10. How are nucleotides joined to form nucleic acids? Name several differences between the structure of DNA and the structure of RNA.

Objective Questions

1. Which of these is not a characteristic of carbon?
  a. forms 4 covalent bonds
  b. bonds with itself
  c. is sometimes ionic
  d. forms long chains

2. The functional group COOH is
  a. acidic.
  b. basic.
  c. never ionized.
  d. All of these.

3. A hydrophilic group is
  a. attracted to water.
  b. a polar or ionized group.
  c. found in fatty acids.
  d. All of these.

4. Which of these is an example of hydrolysis?
  a. amino acid + amino acid → dipeptide + H_2O
  b. dipeptide + H_2O → amino acid + amino acid
  c. Both of these.
  d. Neither of these.

5. Which of these makes cellulose nondigestible?
  a. a polymer of glucose subunits
  b. a fibrous protein
  c. the linkage between the glucose molecules
  d. the peptide linkage between the amino acid molecules

6. A fatty acid is unsaturated if it
  a. contains hydrogen.
  b. contains double bonds.
  c. contains an acidic group.
  d. bonds to glycogen.

7. Which of these is not a lipid?
  a. steroid
  b. fat
  c. polysaccharide
  d. wax

8. The difference between one amino acid and another is found in the
  a. amino group.
  b. carboxyl group.
  c. R group.
  d. peptide bond.

9. The shape of a polypeptide is
  a. maintained by bonding between parts of the polypeptide.
  b. important to its function.
  c. ultimately dependent upon the primary structure.
  d. All of these.

10. Which of these is the peptide bond?

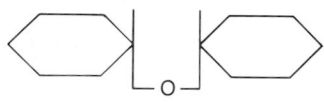

a.

$$\begin{matrix} O \\ \| \\ -C-O- \end{matrix} \qquad \begin{matrix} O \quad H \\ \| \quad | \\ -C-N- \end{matrix}$$
  b.   c.

11. Nucleotides
  a. contain a sugar, a nitrogen-containing base, and a phosphate molecule.
  b. are the monomers for fats and polysaccharides.
  c. join together by covalent bonding between the bases.
  d. All of these.

12. DNA
  a. has a sugar-phosphate backbone.
  b. is single-stranded.
  c. has a certain sequence of amino acids.
  d. All of these.

13. Label the following diagram using the terms monomer, hydrolysis, condensation, and polymer, and explain the diagram:

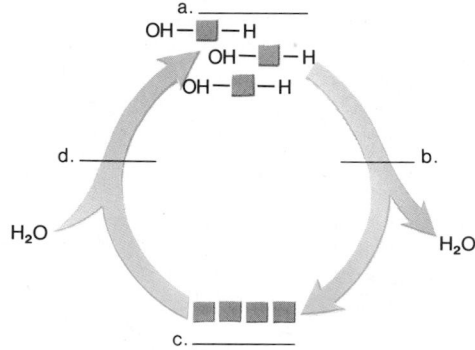

14. Label the levels of protein structure in this diagram of hemoglobin:

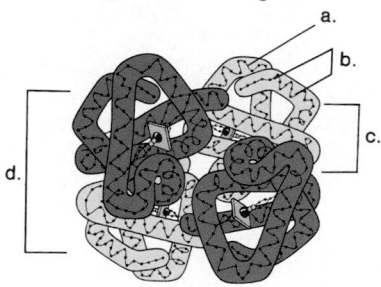

15. Complete the other strand of a DNA molecule.

| T | C | A | A | G | C | C | G | T | A | C | G |

Concepts and Critical Thinking

1. *The unity and the diversity of life begin at the molecular level of organization.*

   How do the organisms in figure 4.1 demonstrate this biological concept?

2. *The atoms and bonds within a molecule determine its chemical and physical properties.*

   Compare fats that largely contain either saturated or unsaturated fatty acid to demonstrate this concept.

3. *The properties of a molecule determine the role that the molecule plays in the cells or body of an organism.*

   Choose either a phospholipid or the protein keratin to demonstrate this biological concept.

Selected Key Terms

organic (or-gan'ik) 42
hydrophilic (hi"dro-fil'ik) 43
hydrophobic (hi"dro-fo'bik) 43
isomer (i'so-mer) 43
polymer (päl'e-mer) 43
condensation (kän-den-sa'shen) 44
hydrolysis (hi-drol'i-sis) 44

carbohydrate (kar"bo-hi'drāt) 44
lipid (lip'id) 46
phospholipid (fos"fo-lip'id) 48
steroid (ste'roid) 48
amino acid (ah-me'no as'id) 50
protein (pro'te-in) 50

enzyme (en'zīm) 50
peptide (pep'tīd) 50
nucleotide (nu'kle-o-tīd) 54
nucleic acid (nu-kle'ik as'id) 54
DNA 54
RNA 54

5

Cell Structure and Function

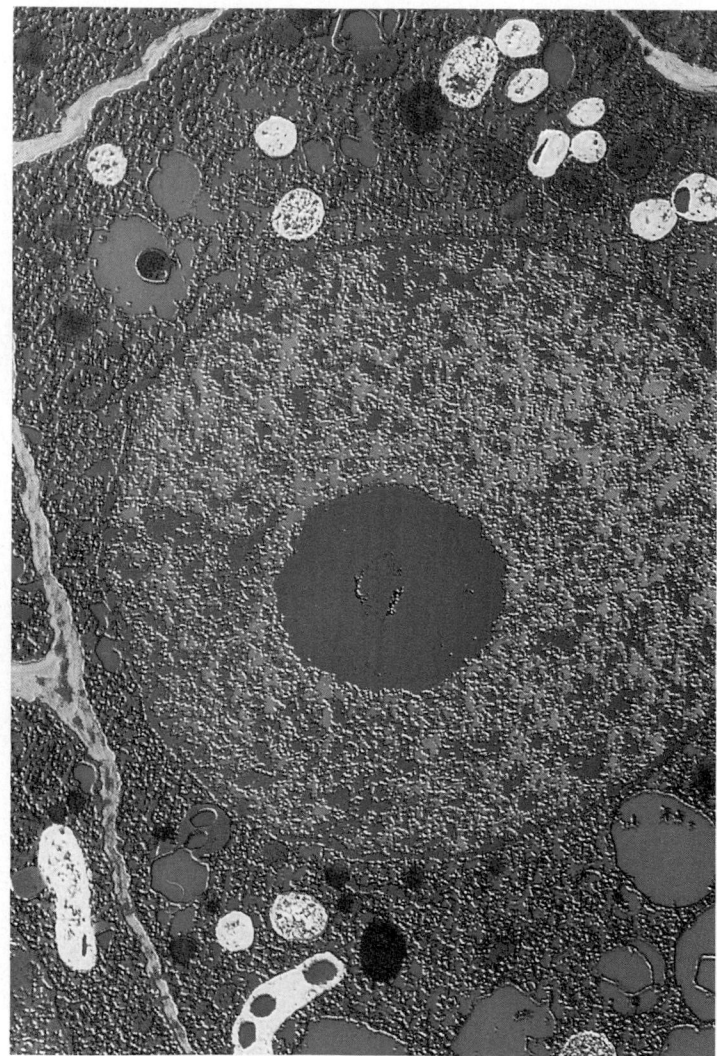

A cell in the root tip of a corn plant. The nucleus dominates the cell, signifying its importance in the life of the cell. The red granular material contains DNA, and the concentration of red contains RNA. These 2 nucleic acids are involved in protein synthesis within the cytoplasm, which lies outside the nucleus. The plant cell has many vacuoles; some appear yellow and others appear blue. The white outline of the cell is the cell wall, which maintains the shape of the cell. Magnification, X1,765.

Your study of this chapter will be complete when you can

1. state 2 tenets of the cell theory;
2. list several similarities and differences between prokaryotic cells and eukaryotic cells;
3. give several differences between bright-field light microscopy and transmission electron microscopy. Name several other types of microscopes that are available today;
4. explain, on the basis of cell volume to cell-surface relationships, why cells are so very small;
5. describe the structure of a prokaryotic cell, and give a function for each part mentioned;
6. describe the structure of the nucleus of a eukaryotic cell, and give a function for each part mentioned;
7. name the structures that form the endomembrane system, and tell how they are related to one another;
8. explain the relationship between chloroplasts and mitochondria; and describe the structure and function of each;
9. list 4 evidences for the endosymbiotic theory, which explains the origin of mitochondria and chloroplasts;
10. name the components of the cytoskeleton; describe the structure and the functions of each component;
11. contrast the structures of prokaryotic cells, eukaryotic animal cells, and eukaryotic plant cells.

**A**ll the organisms we see about us are made up of cells (fig. 5.1). The atoms and the molecules we studied previously were not alive, but the cell is. A **cell** is the smallest unit of living matter.

> All organisms are made up of cells, and a cell is the smallest unit of living matter.

A few cells, like a hen's egg or a frog's egg, are large enough to be seen by the naked eye, but most are not. This is the reason the study of cells did not begin until the invention of the first microscopes in the seventeenth century. It is not known exactly who was the first to see cells, but Anton van Leeuwenhoek of Holland is famous for observing tiny, one-celled living things that no one had seen before. Leeuwenhoek sent his findings to an organization of scientists called the Royal Society in London. Robert Hooke, an Englishman, confirmed Leeuwenhoek's observations and was the first to use the term *cell*. The tiny chambers he observed in the honeycomb structure of cork reminded him of the rooms, or cells, in a monastery. Naturally, then, he referred to the boundaries of these chambers as walls.

Although these early microscopists had seen cells, it was more than a hundred years—in the 1830s—before Matthias Schleiden published his theory that all plants are composed of cells and Theodor Schwann published a similar proposal concerning animals. These Germans based their ideas not only on their own work but on the work of all who had studied tissues under microscopes. Soon after, another German scientist, Rudolf Virchow, used a microscope to study the life of cells. He came to the conclusion that "every cell comes from a preexisting cell." By the middle of the nineteenth century, biologists clearly recognized that all organisms are composed of self-reproducing elements called cells.

> Cells are capable of self-reproduction. Cells come only from preexisting cells.

The previous 2 highlighted statements are often called the **cell theory,** but sometimes they are called the *cell doctrine*. Those who use the latter terminology want to make it perfectly clear that there are extensive data to support the cell theory and that it is universally accepted by biologists.

Cell Variety

Since the early days of cell study, it has become apparent that a cell carries out the activities we associate with living things, such as those needed for *growth* and *reproduction*. Further, a cell's organized parts have specific functions that permit it to carry on these activities. Improved microscopy has vastly extended our ability to view the internal organization of a cell, and today's standard biochemical techniques allow us to determine the function of a cell's parts.

Figure 5.1
All organisms, whether plants or animals, are composed of cells. This is not readily apparent because a microscope is usually needed to see the cells. *a.* Corn plant. *b.* Light micrograph of leaf showing many individual cells. *c.* Rabbit. *d.* Light micrograph of the intestinal lining showing that it too is composed of cells. The dark-staining bodies are nuclei.

a.

b.

c.

d.

Cell Biology: Technology and Techniques

The study of cells is dependent upon the use of microscopes to detect cellular components and biochemical techniques to decipher their functions.

Microscopes

In the *bright-field light microscope,* light rays passing through a specimen are brought to a focus by a set of glass lenses, and the resulting image is then viewed by the human eye. In the *transmission electron microscope,* electrons passing through a specimen are brought to a focus by a set of magnetic lenses, and the resulting image is projected onto a fluorescent screen or photographic film. We will use these statements to examine the differences between the 2 types of microscopes.

The Means of Illumination

Almost everyone knows that the magnifying capability of an electron microscope is greater than that of a light microscope. A light microscope can magnify objects a few thousand times, but an electron microscope can magnify them hundreds of thousands of times. The difference lies in the means of illumination. The paths of light rays and electrons moving through space are wavelike, but the wavelength of electrons is much shorter than the wavelength of light. This difference in wavelength not only accounts for the electron microscope's greater magnifying capability, it also accounts for its greater resolving power. The greater the resolving power, the greater the detail eventually seen. *Resolution* is the *minimum* distance between 2 objects before they are seen as one larger object. If oil is placed between the sample and the objective lens of the light microscope, the resolving power is increased, and if ultraviolet light is used instead of visible light, it is also increased. But typically, a light microscope can resolve down to 0.2 μm, while the electron microscope can resolve down to 0.0001 μm. If the resolving power of the average human eye is set at one, then

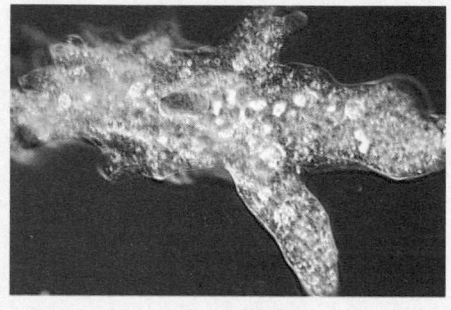

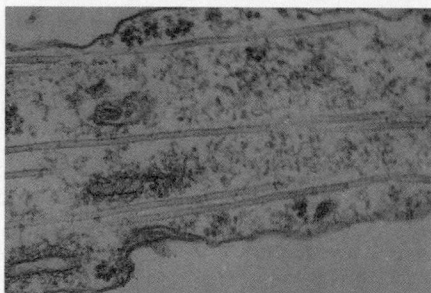

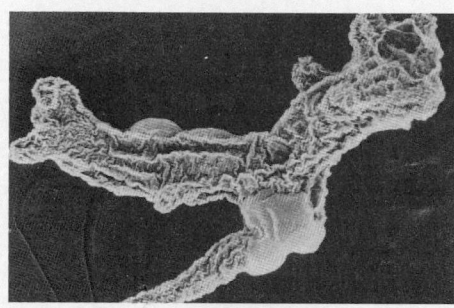

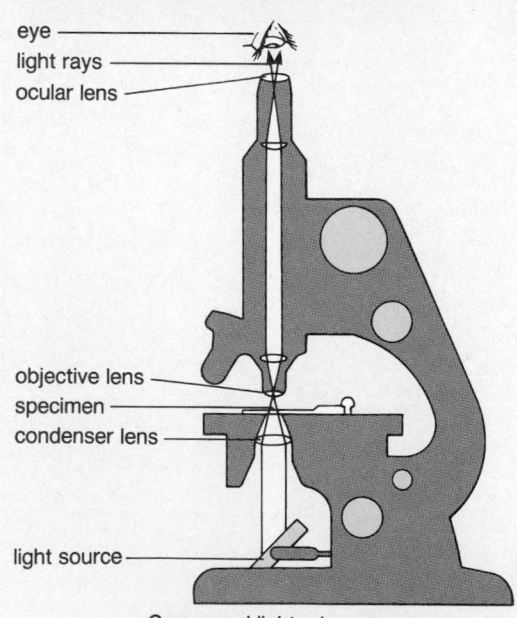

a. Compound light microscope

eye
light rays
ocular lens
objective lens
specimen
condenser lens
light source

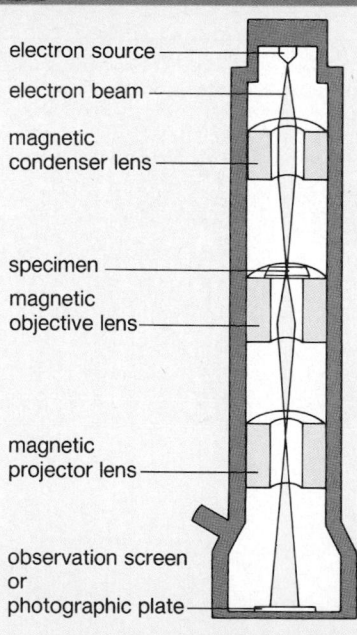

b. Transmission electron microscope

electron source
electron beam
magnetic condenser lens
specimen
magnetic objective lens
magnetic projector lens
observation screen or photographic plate

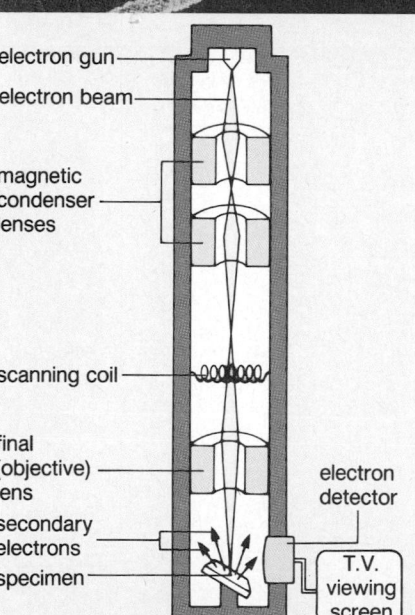

c. Scanning electron microscope

electron gun
electron beam
magnetic condenser lenses
scanning coil
final (objective) lens
secondary electrons
specimen
electron detector
T.V. viewing screen

Figure 5.A

a. Light micrograph of *Amoeba proteus,* a single-celled organism. A light microscope uses light to view a specimen. The specimen can be living if it is thin enough to allow light to pass through. A light microscope does not magnify or distinguish as much detail as the electron microscope. *b.* Transmission electron micrograph (TEM) of a portion of a pseudopodium, an extension by which *Amoeba proteus* moves. Cytoskeleton elements are visible. The transmission electron microscope uses electrons to "view" the specimen. The specimen is nonliving and must be thin enough to allow electrons to pass through. The magnification and the amount of detail seen are far greater than with the light microscope. *c.* Scanning electron micrograph (SEM) of *Amoeba proteus.* The scanning microscope scans the surface of the specimen with an electron beam. The secondary electrons given off are collected and the result is a 3-dimensional image of the specimen. Magnification, ×225.

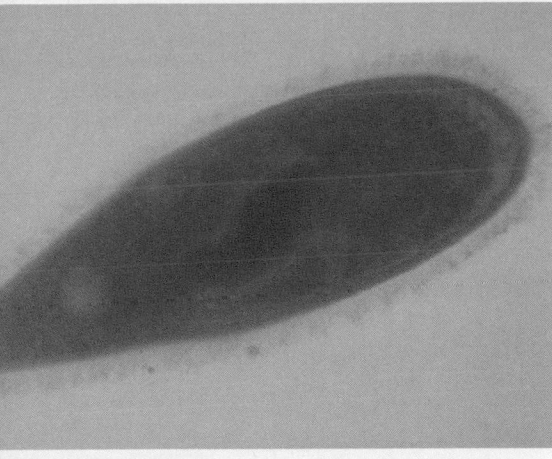

a.

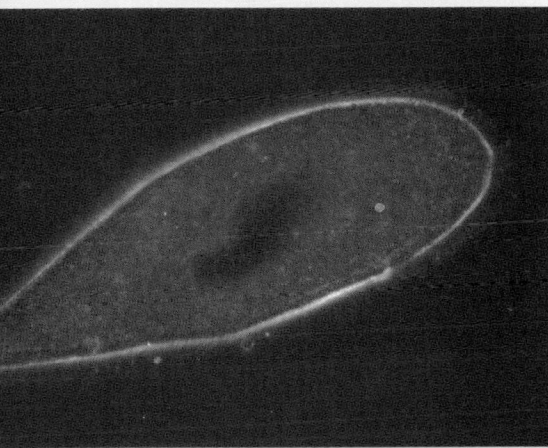

b.

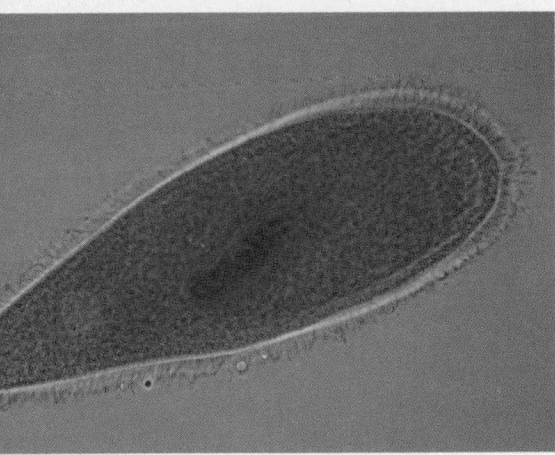

c.

Figure 5.B

Various light microscopes can be used to view a specimen under different conditions. Magnification, X240. *a.* Bright-field micrograph of a paramecium. *b.* Dark-field micrograph of a paramecium. *c.* Phase-contrast micrograph of a paramecium.

that of the typical light microscope is about 500 and that of the electron microscope is 100,000 (fig. 5.A).

The Lenses and the Means of Viewing the Object

Light rays can be bent (refracted) and brought to a focus as they pass through glass lenses, but electrons do not pass through glass. Electrons have a charge that allows them to be brought to a focus by magnetic lenses. The human eye utilizes light to see an object but cannot utilize electrons for the same purpose. Therefore, electrons leaving the specimen in the electron microscope are directed toward a screen or a photographic plate that is sensitive to their presence. Humans can then view the image on the screen or photograph.

The Specimen

Organisms that are quite small and thin can be viewed alive with a light microscope but not with an electron microscope. Electrons randomly scatter as they pass through air; therefore, it is necessary to create a vacuum inside an electron microscope. Even with the light microscope, specimens are usually prepared for observation. Cells are killed, fixed so that they do not decompose, and embedded into a matrix. The matrix strengthens the specimen so that it can be thinly sliced. These sections are stained with colored dyes (light microscopy) or with electron-dense metals (electron microscopy).

This lengthy treatment is necessary for a couple of reasons: (1) light rays and electrons pass through thin materials only, and (2) the dyes create differences in density (e.g., light and dark) that help distinguish detail. Some colored dyes bind preferentially to certain structures, and this aids in their identification. The specimen appears colored because white light is composed of many colors of light and the pigments of the dye absorb certain colors and not others. For example, a green object absorbs all the colors except green light, which then remains for us to see.

One of the primary disadvantages of viewing dead, treated cells is that you can never be sure that the living cell is similar or that an artifact (structure not present when the cell is alive) has not been introduced.

Other Microscopes

Notice that we made specific reference to the bright-field light microscope and to the transmission electron microscope. This is

obviously because there are other types of light microscopes and electron microscopes, each with their own particular advantage (fig. 5.B).

As mentioned, the light microscope has an advantage over the electron microscope because the untreated specimen can be viewed alive; however, special microscopes other than bright-field are needed to allow visualization of detail. *Phase-contrast* and *interference-contrast microscopy* capitalize on the fact that when light passes through a living cell, the phase of each light wave is altered according to the relative density of the cellular structure:

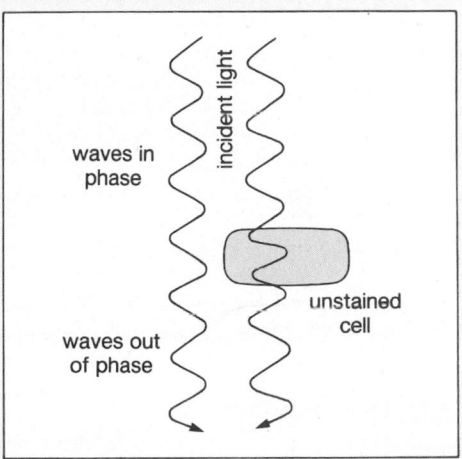

waves in phase

incident light

waves out of phase

unstained cell

Dark-field illumination, as the name suggests, illuminates the object against a dark background. The light has to be directed from the side; only scattered light enters the microscope lens, and as a result, the cell appears vividly lit on a black field. A similar arrangement uses ultraviolet light to detect fluorescence in a specimen.

The scanning electron microscope (SEM) permits the development of 3-dimensional images. A narrow beam of electrons is scanned over the surface of the specimen, which is coated with a thin metal layer. The metal gives off secondary electrons that are collected by a detector to produce an image on a television screen. The 1986 Nobel Prize in physics was won by Gerd Binnig and Heinrich Rohrer for the invention of the *scanning tunneling microscope* (STM). This microscope uses a tiny probe that is brought to within a nanometer of the specimen's surface. A voltage applied between the probe's tip and the specimen creates a flow of electrons that tunnels through the nanometer gap between them. As the probe scans across the

Continued on next page

sample, the tip is moved up and down by the experimenter to keep the tunneling current constant. A computer keeps track of the up-and-down movement of the probe and produces a 3-dimensional image.

The STM does not work particularly well for biological molecules and cells. In the past few years, however, researchers have developed other types of scanning probe microscopes that can be used to view cells successfully. For example, the *atomic force microscope* (AFM) places the probe's tip right up against the surface of the sample. The amount of force required to keep the probe's tip in this position as it moves up and down also allows a computer to produce a 3-dimensional image. The microscope works best with individual biomolecules; to obtain good pictures of cells, investigators plan to build a low-temperature atomic force microscope for samples that will be firmed up by freezing them in liquid nitrogen. Other microscopes derived from the STM are in development. They include one designed to monitor ion flow into and out of living cells and the near-field optical microscope, which should allow experimenters to view viruses and chromosomes in vivo.

Cell Fractionation

Cell fractionation is a means to separate cell components so that their individual biochemical composition and function can be determined (fig. 5.C). First, it is necessary to break open a large number of similar type cells. The cells are usually placed in a *homogenizer,* a tube that contains a close-fitting pestle; when the pestle is rotated, the cells are broken. Second, the "freed" contents of the cells are subjected to a spinning action known as *centrifugation.* At a low speed, large particles, like cell nuclei, settle out and are found in the sediment. Smaller particles are still in the fluid (supernatant), which can be poured into a fresh tube and subjected to centrifugation at a higher speed until the smallest particles have been separated out. The various cell fractions can then be biochemically analyzed.

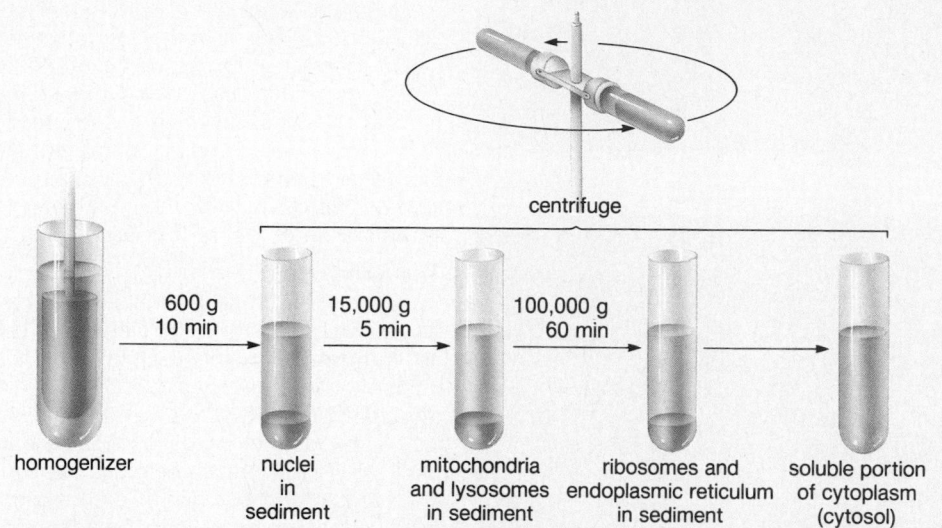

centrifuge

| 600 g 10 min | 15,000 g 5 min | 100,000 g 60 min |

homogenizer — nuclei in sediment — mitochondria and lysosomes in sediment — ribosomes and endoplasmic reticulum in sediment — soluble portion of cytoplasm (cytosol)

Figure 5.C

Cell fractionation. Cells are broken open mechanically by the action of the pestle against the sides of the tube. Contents are centrifuged at ever-increasing speed to separate out first larger and then smaller components of the cell. (g = gravity)

Although there are different types of cells, they all have a barrier called the **plasma membrane,** which separates the contents of the cell from the environment outside the cell. The structure and the function of the plasma membrane are discussed at greater length in chapter 6. For now, we will mention that the plasma membrane regulates the passage of molecules into and out of the cell. The cell takes in nutrient molecules, which are used as building blocks and as an energy source so necessary to growth and repair. The cell also gives off waste molecules, which are end products of metabolism.

The plasma membrane is thin and weak, and alone it cannot give much strength to a cell. Some cells, such as those of plants and microorganisms (organisms that cannot be seen without a microscope), are strengthened by the addition of a **cell wall,** which protects the plasma membrane.

All cells are surrounded by a plasma membrane that separates them from their environment. Some cells also have a cell wall.

Cells are divided into 2 major groups, chiefly on the basis of structure. **Eukaryotic cells** have a true nucleus (*eu*—true; *karyon*—nucleus), a membrane-bounded compartment that houses DNA within chromosomes, which are complex threadlike structures. The rest of the cell is also divided by membranes into compartments that perform various functions. Some of these are **organelles,** small membranous bodies whose structure suits their function.

The other major type of cell is called a **prokaryotic cell** because the cell lacks a true nucleus (*pro*—before; *karyon*—nucleus). The DNA is in a single chromosome located within a distinguishable region called the **nucleoid.** An internal membrane does not divide the cell into compartments, and there are few organelles.

The term **cytoplasm** refers to the cell contents found between the nucleus (or nucleoid) and the plasma membrane. The various structures within the cytoplasm are bathed by a semifluid medium known as **cytosol.** The cytosol in eukaryotic cells is now known to have an organized lattice of protein filaments called the cytoskeleton. The cytoskeleton is discussed on page 76.

There are 2 major types of cells: eukaryotic cells, which have a membrane-bounded nucleus, and prokaryotic cells, which do not have such a nucleus.

Figure 5.2

Biological measurements. **a.** It takes a microscope to see cells because they are so very small. You can see cells with a light microscope, but not with much detail. It takes an electron microscope to make out most of the organelles in the cells. Note that a prokaryote (bacterium) is smaller than the nucleus of a human liver cell. **b.** The units of measurement common to cell biology and their equivalents.

Cell Sizes

Cells are quite small. Most prokaryotic cells are from 1 μm-10 μm in diameter. Eukaryotic cells are much larger—10 μm-100 μm—but still too small to be seen by the naked eye (fig. 5.2). In fact, you could place as many as 10,000 prokaryotic cells and as many as 1,000 eukaryotic cells on the length of this 1 cm line —.

An explanation for why cells are so small and why the body of a plant or an animal is multicellular is found in a consideration of surface area-to-volume relationships. Suppose a cell is 1 mm in each dimension (height, width, depth); its surface area will be

$$6 (1 \times 1) = 6 \text{ mm}^2$$

The volume of the cell will be

$$1 \times 1 \times 1 = 1 \text{ mm}^3$$

If the linear dimensions of the cell are doubled, its surface area will be 24 mm² and its volume will be 8 mm³. Notice that as a result of the cell's doubling in size, the volume increased 8 times, but the surface area increased only 4 times.

Nutrients enter a cell and wastes exit a cell at the plasma membrane. A large cell requires more nutrients and produces more wastes than a small cell. Therefore, a large cell that is actively metabolizing cannot make do with proportionately less surface area than a small cell. Yet as previously mentioned, as a cell increases in size, surface area decreases dramatically in proportion to volume:

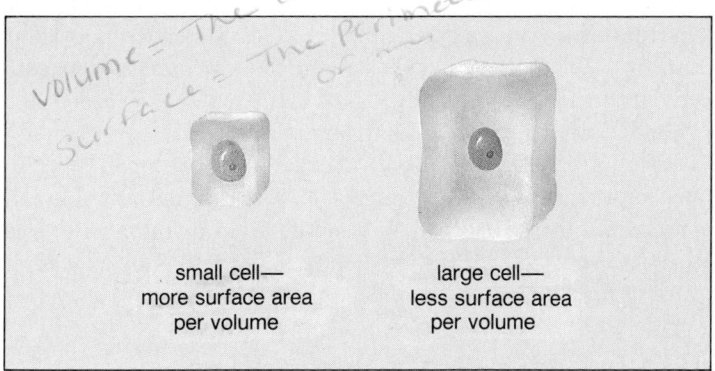

small cell—
more surface area
per volume

large cell—
less surface area
per volume

We would expect, then, that there would be a limit to how large an actively metabolizing cell can become. For example, consider that egg cells are among the largest known cells. A chick's egg is several centimeters in diameter, but the egg is not actively metabolizing and contains a large amount of storage material called yolk.

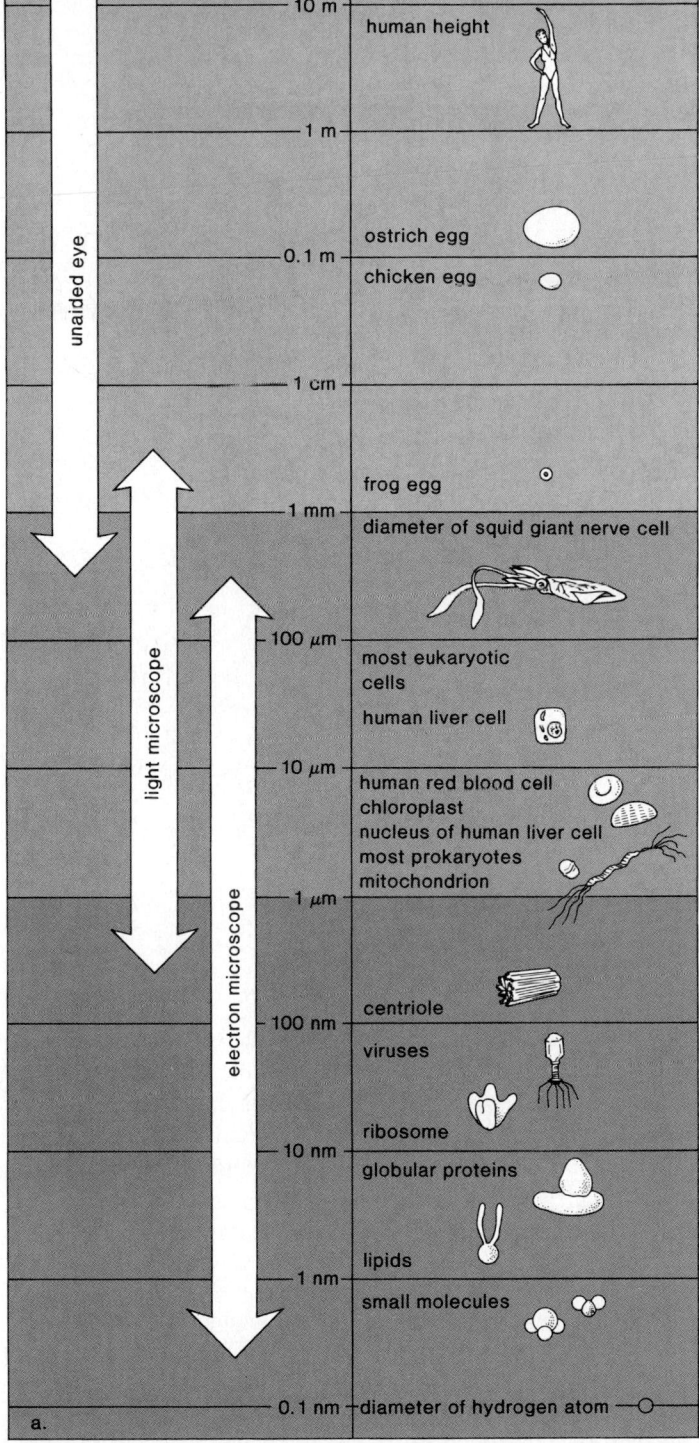

a.

Measurements Used in Cell Biology			
Unit	Symbol	Scale	Seen by
centimeter	cm	0.01 m	unaided eye
millimeter	mm	0.1 cm	unaided eye
micrometer	μm	0.001 mm	light microscope
nanometer	nm	0.001 μm	electron microscope

b.

Figure 5.3

Prokaryotic (i.e. bacterial) anatomy. A single chromosome is located within a region called the nucleoid. Ribosomes are the only cytoplasmic organelles. Prokaryotes usually have a cell wall and may also have a protective capsule. Magnification, X46,000.

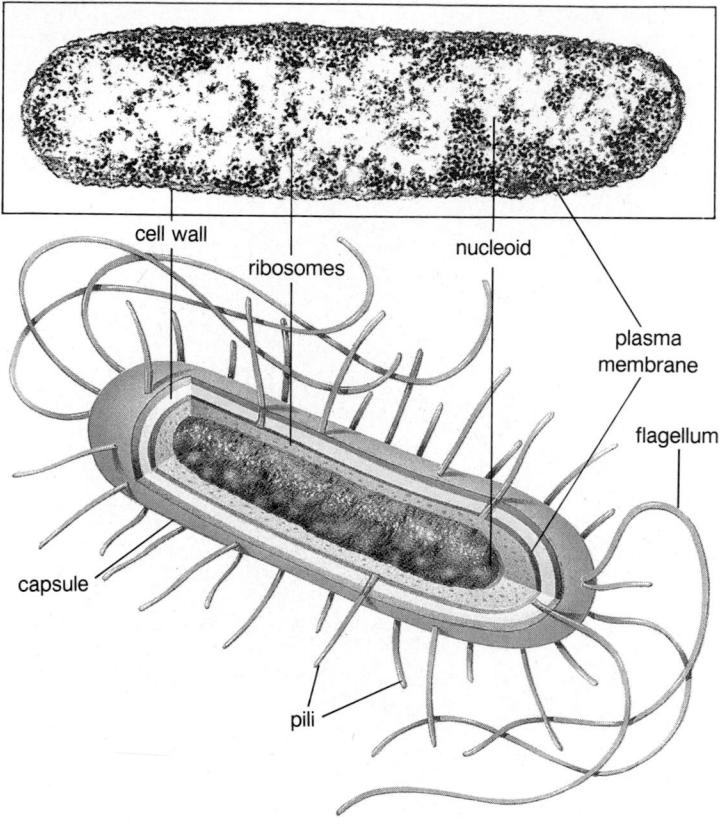

cell wall

ribosomes

nucleoid

plasma membrane

flagellum

capsule

pili

Once the egg is fertilized and metabolic activity begins, the egg divides repeatedly. Cell division provides the surface area needed for adequate exchange. Furthermore, cells that specialize in absorption have modifications to increase greatly the surface area per volume of the cell. The columnar cells along the surface of the intestinal wall have villi (villus, sing.) to increase their surface area. Modifications are necessary to increase surface area per volume.

A cell needs a surface area (plasma membrane) that can adequately exchange materials with the environment. Surface area to volume relationships require that cells stay small.

There are other factors limiting the size of cells. The nucleus is the control center of the cell, and it can control only a certain amount of cytoplasm. Also, not only must there be an exchange of materials between the cell and the external environment, but materials must be made available to various parts of the cytoplasm. A eukaryotic cell is 10 times larger than a prokaryotic cell, and we should expect to see modifications for the distribution of molecules inside the cell. The compartmentalization of eukaryotic cells enhances their ability to make molecules available where they are needed.

Prokaryotic and Eukaryotic Cell Differences

Bacteria are the only organisms that are prokaryotic cells. The bacteria are a very diverse group, and some members are photosynthetic. Among these are the **cyanobacteria,** which were formerly called blue-green algae.

Figure 5.3 illustrates the main features of prokaryotic anatomy. There is an exterior *cell wall* containing peptidoglycan, a large complex molecule consisting of polysaccharide polymers cross-linked by short chains of amino acids. In some bacteria, the cell wall is further surrounded by a gelatinous sheath, or *capsule.* Motile bacteria usually have long, very thin appendages called *flagella* that are composed of subunits of the protein called flagellin. The flagella, which rotate like propellers, rapidly move the bacterium in a fluid medium. Bacteria also have *pili,* which are short appendages that help bacteria attach themselves to an appropriate surface.

In prokaryotes, most genes are found within a single chromosome (loop of DNA) located within the nucleoid, but they may also have small accessory rings of DNA called *plasmids.* The cytoplasm lacks membranous organelles, but the plasma membrane may have folds and convolutions that extend into the cell. In addition, the photosynthetic cyanobacteria have light-sensitive pigments, usually within the membranes of flattened disks called *thylakoids.* Prokaryotic cytoplasm contains numerous granules known as *ribosomes,* which function in protein synthesis. These are smaller than the ribosomes of eukaryotic cells, and they are not attached to membranes.

Bacteria are prokaryotic cells with these constant features.	
Outer boundary:	cell wall
	plasma membrane
Cytoplasm:	ribosomes
	thylakoids (cyanobacteria)
Nucleoid:	chromosome (DNA only)

All cells aside from bacteria are eukaryotic. Eukaryotic organisms include algae, protozoa, and fungi, in addition to plants and animals. In this chapter, we will concentrate on the structure and the function of a typical animal cell (fig. 5.4) and a typical plant cell (fig. 5.5). Such drawings are composites based on data gathered from microscopic studies. In actuality, there is a great deal of variation among plant cells and animal cells.

The various organelles of eukaryotic cells have a structure that suits their function(s) (table 5.1). The cytoplasm is compartmentalized by these organelles and by other membranous structures. *Compartmentalization* keeps the cell organized and its various functions separate one from the other.

Eukaryotic Cell Organelles

The **nucleus** is a prominent structure in the eukaryotic cell. With staining, it is usually large enough to be observed through the light microscope, although electron micrographs show much more detail. The nucleus is of primary importance because it is the

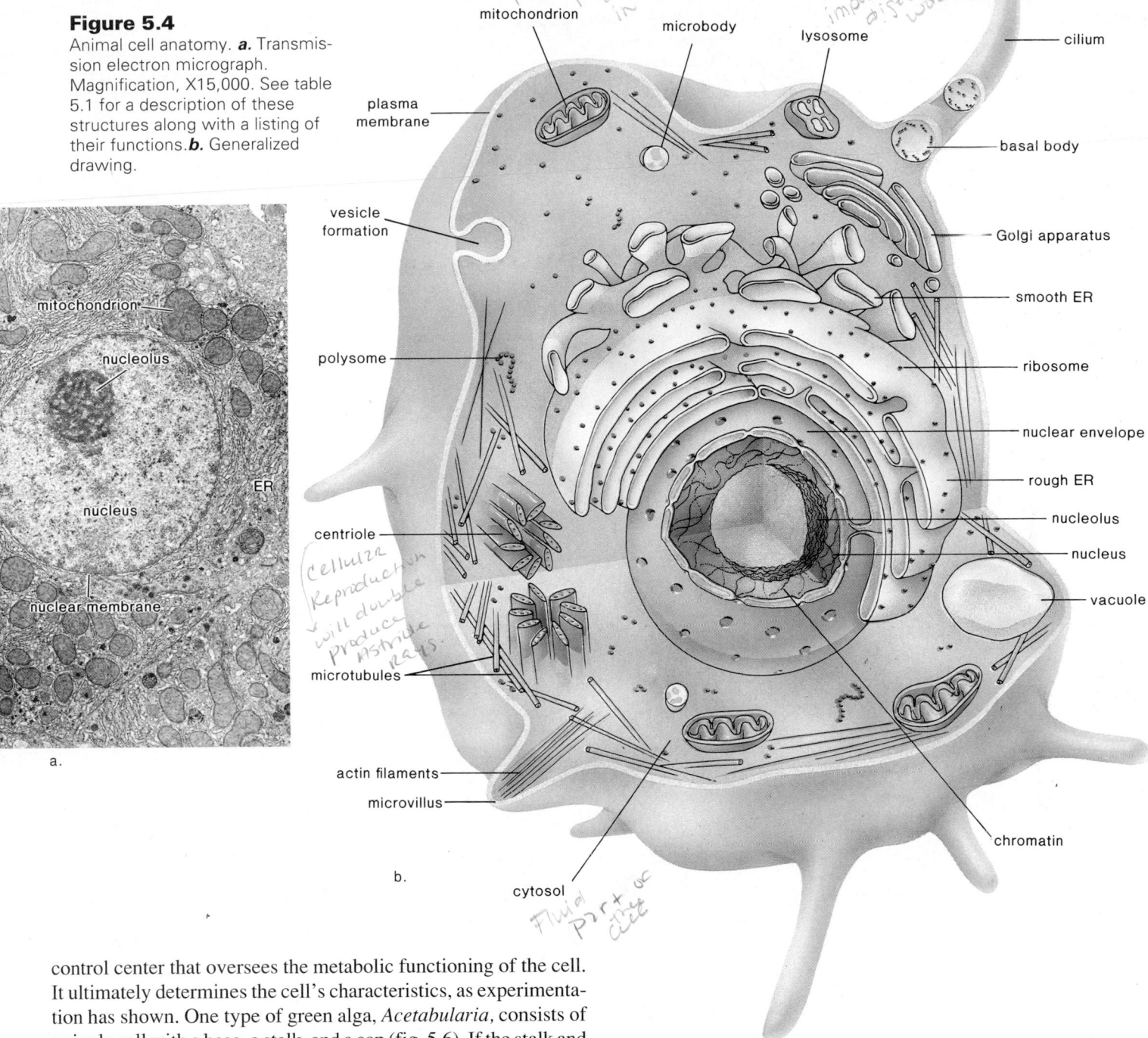

Figure 5.4
Animal cell anatomy. **a.** Transmission electron micrograph. Magnification, X15,000. See table 5.1 for a description of these structures along with a listing of their functions. **b.** Generalized drawing.

mitochondrion

microbody

lysosome

cilium

plasma membrane

basal body

vesicle formation

Golgi apparatus

smooth ER

polysome

ribosome

nuclear envelope

rough ER

nucleolus

centriole

nucleus

vacuole

microtubules

actin filaments

chromatin

microvillus

cytosol

a.

mitochondrion

nucleolus

ER

nucleus

nuclear membrane

b.

control center that oversees the metabolic functioning of the cell. It ultimately determines the cell's characteristics, as experimentation has shown. One type of green alga, *Acetabularia,* consists of a single cell with a base, a stalk, and a cap (fig. 5.6). If the stalk and the cap are removed from the base (fig. 5.6*b*), they die, but the nucleus-containing base develops into a new organism. The importance of the nucleus is further exemplified when the base is combined with the stalk of a different species. The regenerated cap resembles the species cap of the nucleus, not the species cap of the stalk cytoplasm. Similarly, the nucleus of a frog's egg can be removed and the nucleus of another species can be introduced. The tadpole and eventually the frog that develops is like that of the introduced nucleus.

The nucleus is the control center of the cell.

It is clear that the nucleus is the control center of the cell because it contains the hereditary material DNA (deoxyribonucleic acid). As mentioned in the previous chapter, DNA carries the instructions for the proper sequencing of amino acids in proteins. The proteins of a cell determine its structure and the functions it can perform. The nucleus directs what proteins will be present and in that way determines the structure and coordinates the activities of the cell.

When you look at the nucleus, even in an electron micrograph, you cannot see DNA molecules but you can see chromatin.

Figure 5.5

Plant cell anatomy. **a.** Generalized drawing. **b.** Transmission electron micrograph. Magnification, X20,000. See table 5.1 for a description of the structures along with a listing of their functions.

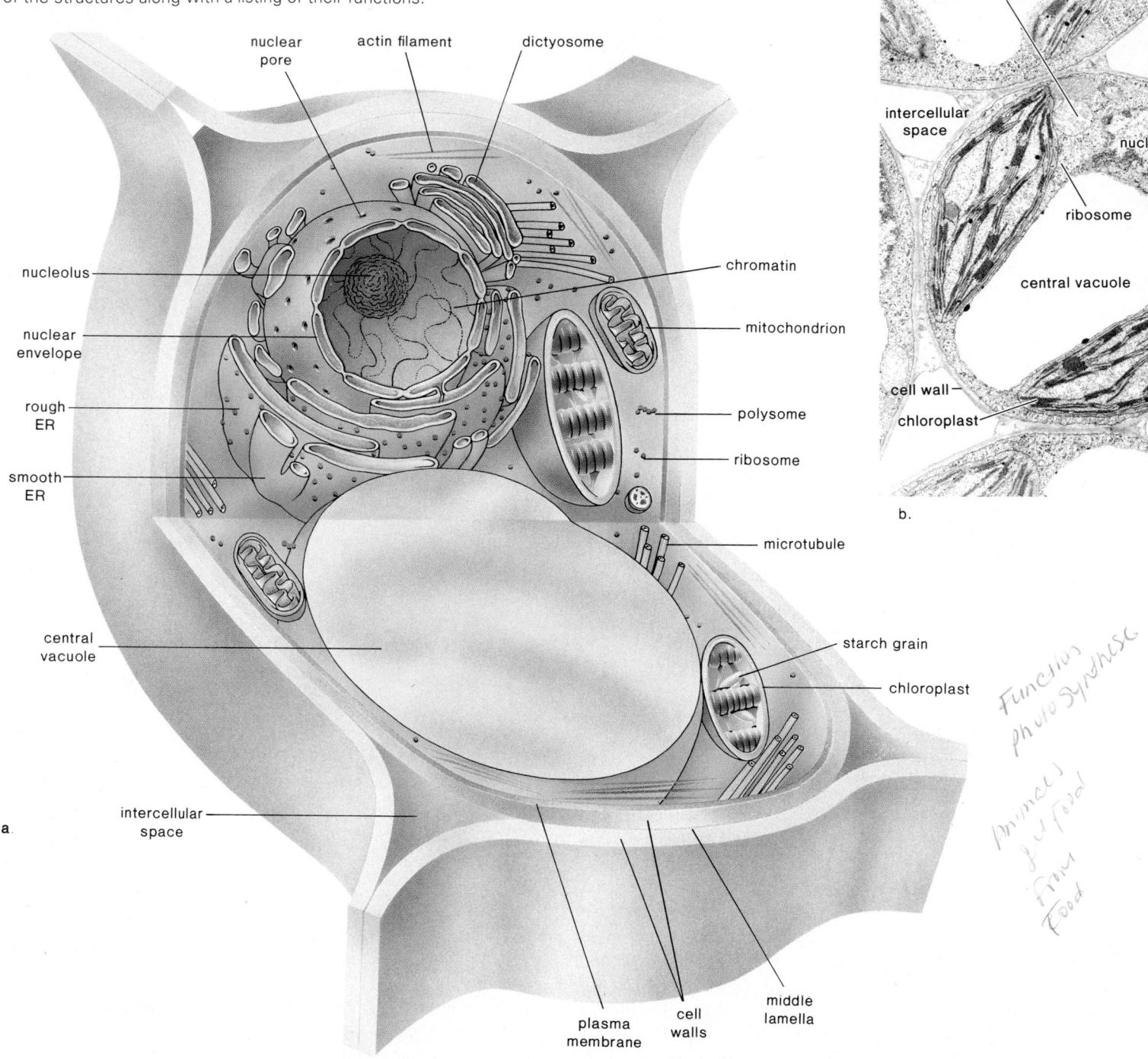

nuclear pore
actin filament
dictyosome
nucleolus
nuclear envelope
rough ER
smooth ER
central vacuole
intercellular space
a.
chromatin
mitochondrion
polysome
ribosome
microtubule
starch grain
chloroplast
plasma membrane
cell walls
middle lamella

mitochondrion
intercellular space
nucleus
ribosome
central vacuole
cell wall
chloroplast
b.

Functius Photosynthese
Mmanies J food from Food

Chromatin looks grainy, but actually it is a threadlike material that undergoes coiling into rodlike structures called **chromosomes** just before the cell divides. Chemical analysis shows that chromatin, and therefore chromosomes, contains DNA, much protein, and some RNA. Chromatin is immersed in a semifluid matrix called the *nucleoplasm*. Nucleoplasm has a different pH from the cytosol, and this suggests that it has a different composition.

Most likely, too, when you look at an electron micrograph of a nucleus, you will see one or 2 regions that look darker than the rest of the chromatin. These are **nucleoli,** where another type of RNA, called ribosomal RNA (rRNA), is produced. (Ribosomes are small bodies in the cytoplasm that contain ribosomal RNA and proteins.)

Table 5.1
Eukaryotic Structures in Animal Cells and Plant Cells

Name	Composition	Function
Cell wall*	Contains cellulose fibrils	Support and protection
Plasma membrane	Bilayer of phospholipid with embedded proteins	Selective passage of molecules into and out of cell
Nucleus	Nuclear envelope surrounding the nucleoplasm, chromatin, and nucleoli	Control of growth and reproduction
Nucleolus	Concentrated area of chromatin, RNA, and proteins	Ribosomal formation
Ribosomes (polysomes)	Protein and RNA in 2 subunits	Protein synthesis
Endoplasmic reticulum	Membranous flattened channels and tubular canals	Synthesis and/or modification of proteins and other substances, and transport by vesicle formation
Rough ER	Studded with ribosomes	Protein synthesis
Smooth ER	Having no ribosomes	Various; lipid synthesis in some cells
Golgi apparatus[d] (dictyosome)*	Stack of membranous saccules	Processing and packaging of molecules, (e.g., glycoproteins)
Vacuoles* and vesicles	Membranous sacs	Storage of substances
Lysosomes[d]	Membranous vesicles containing digestive enzymes	Intracellular digestion
Microbodies	Membranous vesicles containing specific enzymes	Various metabolic tasks
Mitochondria	Inner membrane (cristae) within outer membrane	Cellular respiration
Chloroplasts	Inner membrane (grana) within 2 outer membranes	Photosynthesis
Cytoskeleton	Microtubules, and actinfilaments	Shape of cell and movement of its parts
Cilia and flagella	9 + 2 pattern of microtubules	Movement of cell
Centrioles[d]	9 + 0 pattern of microtubules	Form basal bodies, which give rise to microtubules

*Plant cells
[d]Animal cells

Figure 5.6
The *Acetabularia* are single-celled algae up to 5 cm long which lend themselves to regeneration experiments. ***a.*** Photo of *Acetabularia mediterranea* commonly called mermaid's wineglass. ***b.*** Experiments to demonstrate that the nucleus is the control center of the cell. (1) In this experiment, the stalk and the cap which do not have a nucleus, die, but the base, with a nucleus, regenerates. (2) In this experiment, the regenerated cap resembles that of the species of the nucleus, not that of the species of the cytoplasm.

a.

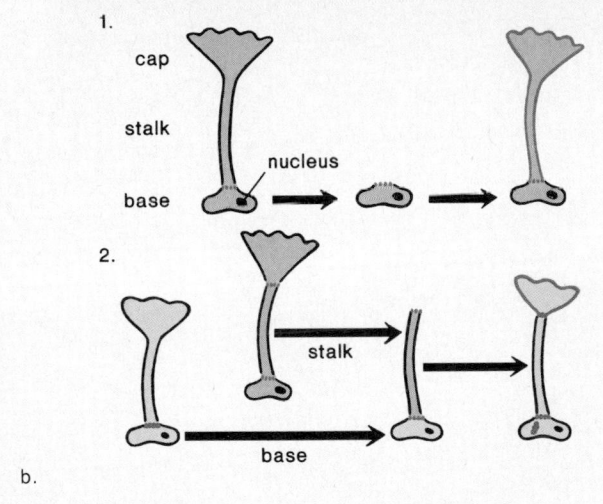

b.

The nucleus is separated from the cytoplasm by a double membrane known as the **nuclear envelope** (fig. 5.7). You might think that the presence of this envelope would prevent materials from passing to and from the nucleus, but electron micrographs show nuclear *pores* in the nuclear envelope. These pores are of sufficient size (100 nm) to permit the passage of ribosomal subunits. High-power electron micrographs show that the pores have a complex structure (fig. 5.7c). Therefore, they should not be regarded as simple openings. They have associated proteins that can regulate the passage of materials to and from the nucleus.

The structural features of the nucleus include

Chromatin (chromosomes):	DNA and proteins
Nucleolus:	chromatin and ribosomal subunits
Nuclear envelope:	double membrane with pores

Figure 5.7

Anatomy of the nucleus. **a.** The fluid portion of the nucleus is the nucleoplasm, where chromatin is found. The nucleolus is a special region of chromatin where rRNA is produced and joined with proteins from the cytoplasm to form the subunits of ribosomes. The nucleus is bounded by a double membrane which contains pores. **b.** Freeze-fracture (chap. 6) preparation of the nuclear envelope of a rat pancreas cell showing the existence of nuclear pores. Magnification, X145,000. **c.** Each nuclear pore contains a pore complex consisting of 8 granules and possible projections that block the opening. These complexes are believed to regulate the passage of materials into and out of the nucleus. Two proposed versions of the pore complex are shown.

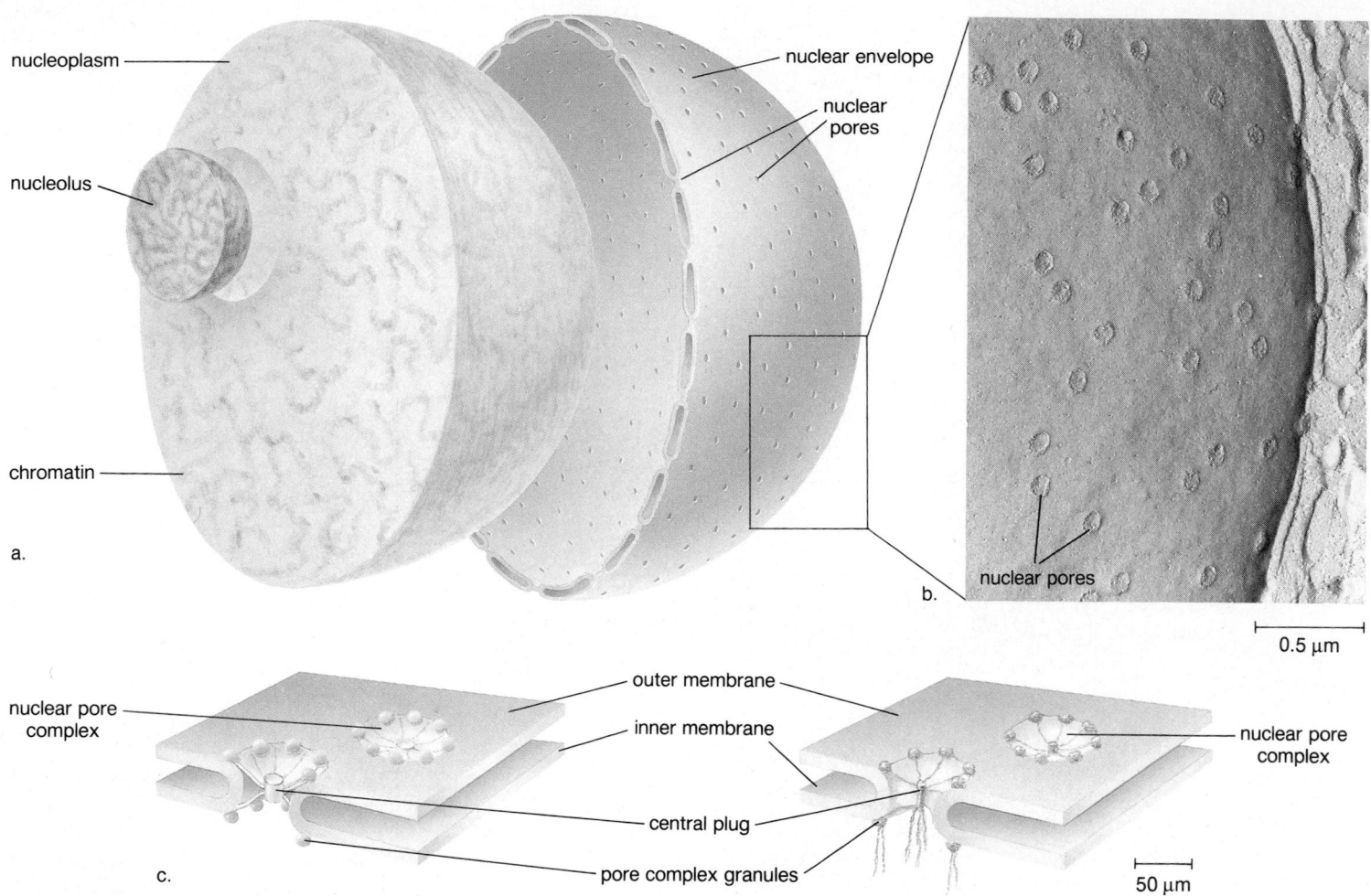

Ribosomes

Ribosomes are small particles composed of RNA and proteins. Unlike many of the organelles discussed in this chapter, ribosomes are found in both prokaryotes and eukaryotes. The ribosomes found in eukaryotic cells are about one-third larger than those found in prokaryotic cells. Both types of ribosomes are composed of 2 subunits, a large subunit and a small subunit. Each subunit has its own mix of proteins and rRNA. Ribosomes perform a very important function because they help carry out protein synthesis.

Several ribosomes synthesizing the same protein are called a *polysome*. Polysomes can lie free within the cytosol, and they can be attached to the endoplasmic reticulum, a membranous system of saccules and channels in the cytoplasm. Cells that produce proteins for secretion, such as those that produce digestive enzymes for the small intestine, contain a few million ribosomes and an extensive endomembrane system.

Ribosomes are small organelles that are involved in protein synthesis. A polysome contains a group of ribosomes, each one involved in producing a copy of the same protein.

Endomembrane System

The endomembrane system contains several membranous structures. Many types of enzymatic reactions occur all the time in cells, and the endomembrane system keeps each reaction restricted to a particular region. For example, the endomembrane system keeps the reactions that occur in the cytosol separate from those that occur in the endoplasmic reticulum and Golgi apparatus.

Figure 5.8

Rough ER. **a.** Electron micrograph of a mouse hepatocyte (liver) cell shows a cross section of many flattened vesicles with ribosomes attached to the side that abuts the cytosol. Magnification, ×75,000. **b.** The 3-dimensions of the organelle. **c.** Model of a single ribosome illustrates that each one is actually composed of 2 subunits. **d.** Method by which the endoplasmic reticulum acts as a transport system.

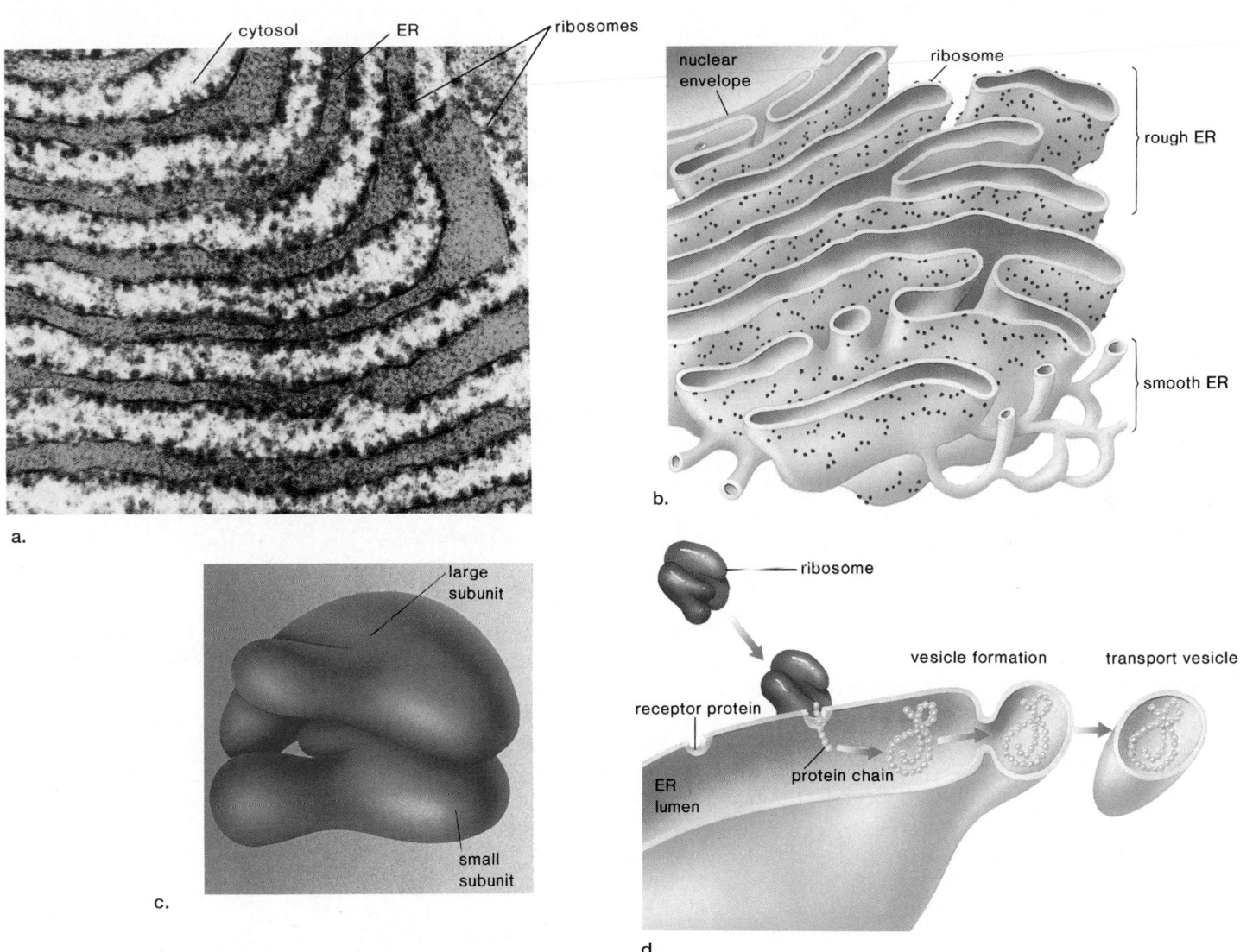

a.

b.

c.

d.

Endoplasmic Reticulum

The **endoplasmic reticulum** (*endoplasmic*—inside the cytoplasm; *reticulum*—network) is continuous with the nuclear envelope. Its complicated system of membranous channels and saccules may be studded with ribosomes, in which case it is called **rough ER,** or it may lack ribosomes, in which case it is called **smooth ER** (fig. 5.8).

Rough ER is involved in protein synthesis, modification of newly formed proteins, production of new membrane, and transport of these proteins and membrane to other locations within the cell. Proteins produced by polysomes attached to rough ER start entering the ER lumen even while protein synthesis is occurring (fig. 5.8d). Once inside the lumen they can be modified; for

example, they can be shortened or have a sugar chain attached to them. These proteins can be incorporated into ER membrane or they can move along until they reach smooth ER.

Smooth ER can have various functions depending on the particular cell. Sometimes it specializes in the production of steroid hormones. For example, it is abundant in cells of the testes and the adrenal cortex, both of which produce steroid hormones. In the liver it helps detoxify drugs; in muscle cells it acts as a storage site for calcium ions that are released when contraction occurs.

Regardless of any specialized function, smooth ER also forms *transport vesicles* (little vacuoles) in which large molecules are moved within the cell (fig. 5.8d). When transport vesicles fuse with another part of the endomembrane system, any contents of the vesicle are discharged and the receiving membrane is enlarged in this manner:

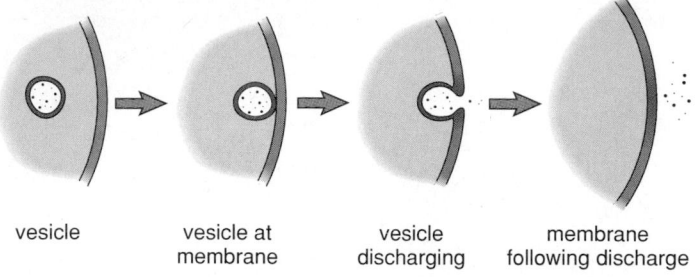

| vesicle | vesicle at membrane | vesicle discharging | membrane following discharge |

Often transport vesicles are on their way to the Golgi apparatus.

Rough ER specializes in protein synthesis. Smooth ER has a function suited to a particular type cell; sometimes it specializes in lipid synthesis. The ER is a transport system. Molecules moving through the system are eventually enclosed in vesicles that usually move to the Golgi apparatus.

Golgi Apparatus

In most cells the **Golgi apparatus** consists of a stack of 3 to 20 slightly curved saccules (flattened vacuoles) (fig. 5.9). In fact a stack of pancakes can be compared to the appearance of the Golgi apparatus except that there are vesicles at the ends of its saccules. The Golgi apparatus receives and processes molecules that are to be transported about the cell. It is involved in the packaging, storage, and distribution of these molecules about or from the cell.

In animal cells, the Golgi apparatus has a "forming face" nearest the endoplasmic reticulum and a "maturing face" pointing toward the plasma membrane. This terminology is consistent with the observation that vesicles are received from smooth ER at the forming face of the Golgi apparatus and vesicles leave the apparatus at the maturing face. After a vesicle is received at the forming face, its molecular contents are discharged into the first of the saccules. Thereafter, the molecules move through the Golgi in vesicles pinched off the saccules. In the meantime, the molecules undergo all sorts of modifications. A sugar chain or a phosphate or a sulfate group can be added, for example. Finally, the molecule is packaged in a vesicle that will carry it either to the plasma membrane for secretion or into a vesicle such as a lysosome (fig. 5.10).

The Golgi apparatus processes, packages, and distributes molecules about or from the cell. Therefore, it is involved in the secretion of molecules.

Researchers are trying to determine what causes molecules to be either secreted or kept in the cell. Just as letters are addressed, perhaps molecules carry some type of identification that determines their final destination in the cell. We do know that proteins that enter the rough ER lumen have a lead sequence of amino acids that causes the polysome to attach to the ER in the first place. Perhaps there are other types of amino acid sequences that act as signals to the Golgi apparatus. Or perhaps there are receptors inside the Golgi, and once a molecule attaches to a receptor its fate is determined.

Figure 5.9
Golgi apparatus in an animal cell. The Golgi receives protein-containing vesicles from the smooth ER at its forming face. These proteins are processed while they pass from saccule to saccule of the central region. Finally, they are repackaged in vesicles that leave the maturing face. Some of these vesicles fuse with the plasma membrane allowing secretion to occur, and others are lysosomes, which contain digestive enzymes. *a.* Electron micrograph from liver cells of a mammal. Magnification, X62,000. *b.* Artist representation of Golgi apparatus.

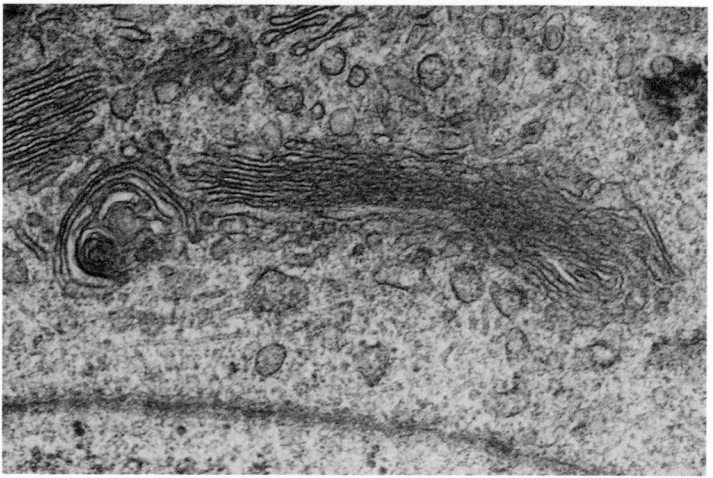

a.

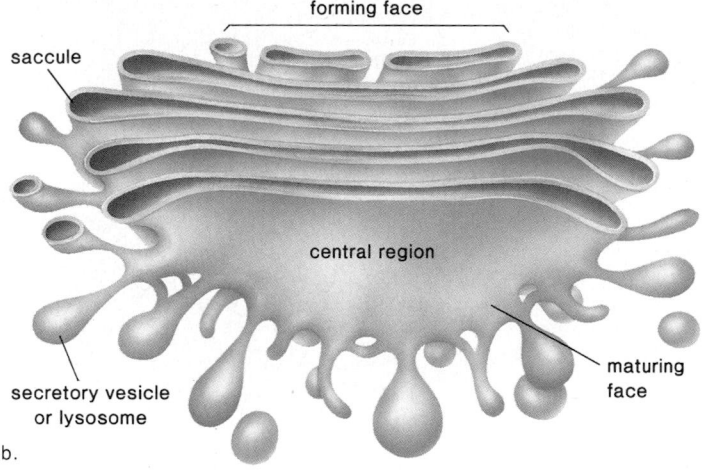

b.

In a plant cell, the Golgi apparatus has a different name and orientation, and it performs other functions. It is called a **dictyosome** and is found in special regions of the cell where its services seem to be needed. For example, the number of dictyosomes increases during cell division because they form vesicles containing polysaccharides needed for a cell plate. The cell plate develops into a cell wall, separating the 2 new cells. Most likely, dictyosomes also provide vesicles that fuse with developing plasma membrane.

In plant cells, dictyosomes provide the material to form new cell walls and new membrane following cell division.

Figure 5.10

Functions of the endomembrane system. Proteins made at the rough ER move through the lumen of the canals and tubules until they are transported in vesicles from the smooth ER to the Golgi apparatus. The Golgi apparatus produces vesicles that move to the plasma membrane and lysosomes, which contain digestive enzymes. Macromolecules, which enter a cell by vesicle formation, are broken down when the vesicle fuses with a lysosome to give a secondary lysosome.

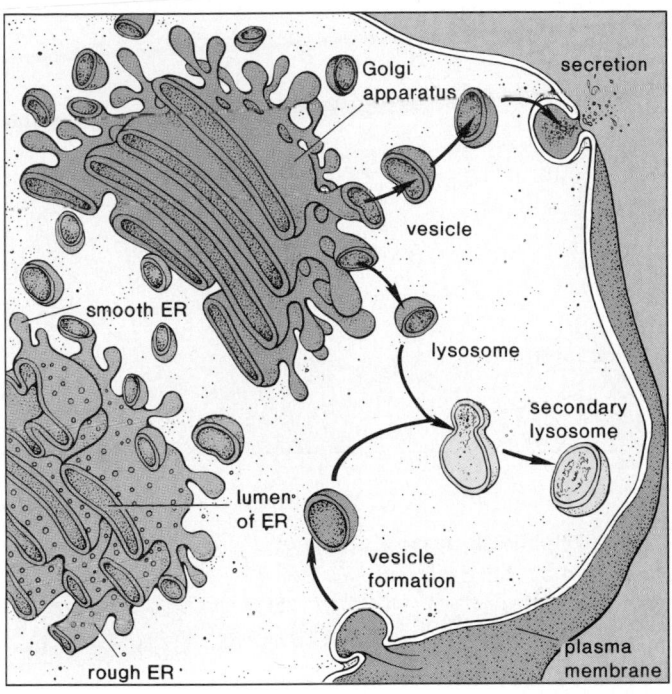

Figure 5.11

Lysosomes in a kidney cell of a mammal. The small dark body is a primary lysosome, which has not yet fused with an incoming vesicle. The larger bodies are secondary lysosomes, which have fused with such vesicles. The dark appearance of the lysosomes results from staining for a particular enzyme, acid phosphatase, whose presence is the test for this organelle. Magnification, X92,000.

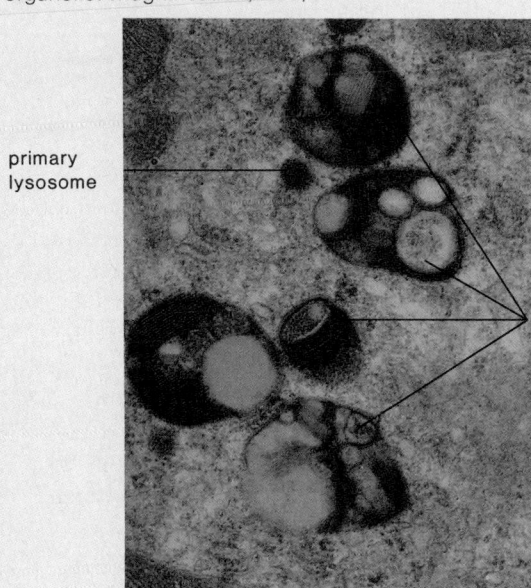

Lysosomes are membrane-bounded vesicles that contain specific enzymes. Lysosomes are produced by Golgi apparatuses, and their hydrolytic enzymes digest macromolecules from various sources.

Lysosomes

Lysosomes are membrane-bounded vesicles produced by the Golgi apparatus that contain hydrolytic digestive enzymes (fig. 5.11).

Sometimes macromolecules are brought into a cell by vesicle formation at the plasma membrane (fig. 5.10). A lysosome can fuse with such a vesicle and can digest its contents into simpler subunits that then enter the cytoplasm. Some white blood cells defend the body by engulfing bacteria that are then enclosed within vesicles. When lysosomes fuse with these vesicles, the bacteria are digested. It should come as no surprise then that even parts of a cell are digested by its own lysosomes (called autodigestion). Normal cell rejuvenation most likely takes place in this manner, but autodigestion is also important during development. For example, when a tadpole becomes a frog, lysosomes are utilized to digest away the cells of the tail. The fingers of a human embryo are at first webbed, but the fingers are freed from one another following lysosomal action.

Occasionally, a child is born with a metabolic disorder involving a missing or inactive lysosomal enzyme. In these cases, the lysosomes fill to capacity with a macromolecule that cannot be broken down. The cells become so full of these lysosomes that the child dies. Someday soon it may be possible to supply these children with the missing enzyme.

Microbodies

Microbodies are similar to lysosomes because they are vesicles bounded by a single membrane and because they contain specific enzymes (fig. 5.12). In this way, they also function to compartmentalize the cell. Although microbodies are believed to be quite varied, 2 types are noteworthy: peroxisomes and glyoxysomes.

Peroxisomes are microbodies that contain enzymes for transferring hydrogen atoms to oxygen, forming hydrogen peroxide (H_2O_2), a toxic molecule that is immediately broken down to water by the enzyme catalase. Peroxisomes are abundant in cells that are metabolizing lipids and in liver cells that are metabolizing alcohol. They are believed to help detoxify the alcohol.

Glyoxysomes have been observed in leaves that are carrying on photosynthesis. In this case, they contain enzymes that can metabolize some of the molecules involved in the photosynthetic process. They are also seen in germinating seeds, where they are believed to convert oils into sugars used as nutrients by the growing plant.

Cell Structure and Function

73

It is possible that the endoplasmic reticulum produces microbodies, but we know that their enzymes are produced by free ribosomes within the cytosol. Afterwards, the enzymes enter the microbodies directly from the cytosol.

Vacuoles

A **vacuole** is a large membranous sac; a vesicle is smaller. Animal cells have vacuoles, but they are much more prominent in plant cells. Typically, plant cells have one or 2 large vacuoles so filled with a watery fluid that they give added support to the cell. Most of the central area of the cell is occupied by vacuole, and the remaining cell contents are pushed to the sides (fig. 5.5).

Vacuoles are most often storage areas, but they can have specialized functions. Plant vacuoles contain not only water, sugars, and salts but also pigments and toxic substances. The pigments are responsible for many of the red, blue, or purple colors of flowers and some leaves. The toxic substances help protect a plant from herbivorous animals. (As long as the substance is contained within the vacuole, it will not harm the plant.) The vacuoles present in protozoans are quite specialized, and they include contractile vacuoles and digestive vacuoles (fig. 24.2).

The organelles of the endomembrane system are as follows:

Endoplasmic reticulum: synthesis and/or modification and transport of proteins and other substances
Rough ER: protein synthesis
Smooth ER: lipid metabolism, in particular
Golgi apparatus: processing and packaging of protein molecules
Lysosomes: intracellular digestion
Microbodies: various metabolic tasks
Vacuoles: storage areas

Figure 5.10 shows how certain of these organelles are related to each other.

Energy-Related Organelles

Life is possible only because of a constant input of energy to maintain the structure of the cell. Chloroplasts and mitochondria are the 2 eukaryotic membranous organelles that specialize in converting energy into a form that can be used by the cell.

Photosynthesis, which occurs in **chloroplasts,** is the process by which solar energy is converted to chemical-bond energy within carbohydrates. Photosynthesis can be represented by this equation:

light energy + carbon dioxide + water ⟶ carbohydrate + oxygen

Here the word *energy* stands for solar energy, the ultimate source of energy for cellular organization. Only plants, algae, and certain bacteria are capable of carrying on photosynthesis. (Bacteria, however, being prokaryotic, do not have chloroplasts.)

Cellular respiration, which occurs in **mitochondria,** is the process by which the chemical energy of carbohydrates is converted to that of ATP (adenosine triphosphate), the common

Figure 5.12
Peroxisome in a tobacco leaf. The crystalline, squarelike core of this microbody cross section is believed to consist of the enzyme catalase, which breaks down hydrogen peroxide to water. Microbodies carry out various metabolic reactions and are another means of compartmentalizing enzymatic reactions that occur in cells. Magnification, X155,000.

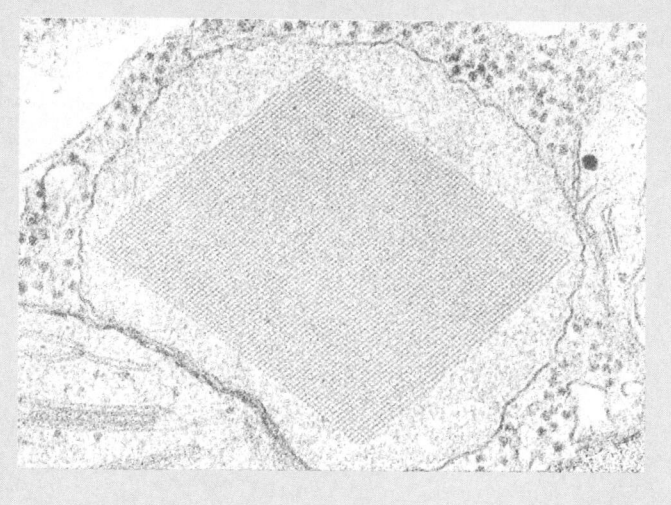

carrier of chemical energy in cells. Cellular respiration can be represented by this equation:

carbohydrate + oxygen ⟶ carbon dioxide + water + energy

Here the word *energy* stands for ATP molecules. When a cell needs energy, ATP supplies it. ATP energy is used for synthetic reactions, active transport, and all energy-requiring processes in cells. All organisms carry on cellular respiration, and all organisms except bacteria have mitochondria. Chloroplasts use the sun's energy to produce carbohydrates, and carbohydrate-derived products are broken down in mitochondria to build up ATP molecules.

Chloroplasts and mitochondria supply the eukaryotic cell with usable energy. Energy is needed for various cellular activities, all of them helpful for maintaining the structure of cells.

Chloroplasts

Chloroplasts are about 4 µm-6 µm in diameter and 1 µm-5 µm in length; they belong to a group of plant organelles known as plastids. Among the plastids are also the *amyloplasts,* which store starch, and the *chromoplasts,* which contain red and orange pigments.

A chloroplast is bounded by a double membrane, and inside there is even more membrane organized into flattened sacs called **thylakoids.** The thylakoids are piled up like stacks of coins, and each stack is called a **granum.** There are membranous connections between the grana called *lamellae* (fig. 5.13). The fluid-filled space about the grana is called the **stroma.**

Figure 5.13

Chloroplast structure. **a.** Electron micrograph. Magnification, X40,000. **b.** Generalized drawing in which the outer and inner membranes have been cut away to reveal the grana.

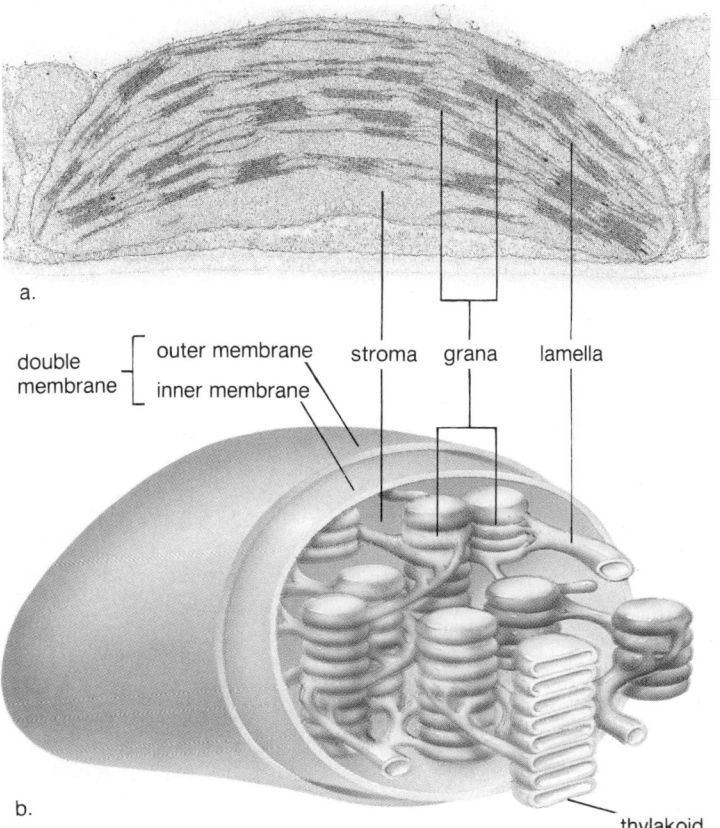

a.

double membrane — outer membrane / inner membrane   stroma   grana   lamella

b.

thylakoid

Figure 5.14

Mitochondrion structure. **a.** Electron micrograph. Magnification, X70,000. **b.** Generalized drawing in which the outer membrane and portions of the inner membrane have been cut away to reveal the cristae.

a.

double membrane — outer membrane / inner membrane   cristae   matrix

b.

Photosynthesis requires pigments that capture solar energy and enzymes that synthesize carbohydrate. The green pigments, called chlorophylls, are the major pigments found within the thylakoid membrane of grana. Enzymes are found within the stroma. This basically tells you how chloroplasts are organized to carry out photosynthesis. As mentioned previously, the cyanobacteria, which are prokaryotic, usually possess cytoplasmic thylakoids to which photosynthetic pigments are bound.

Chloroplasts have their own DNA and ribosomes and can produce certain proteins. They also reproduce themselves by division. In other words, a chloroplast is similar in many respects to a photosynthetic prokaryotic organism, an observation to which we will return.

Mitochondria

Most mitochondria are between 0.5 μm and 1.0 μm in diameter and 7 μm in length, although the size and the shape can vary.

Mitochondria, like chloroplasts, are bounded by a double membrane. The inner of these 2 membranes is folded to form little shelves, called **cristae,** which project into the **matrix,** an inner

space filled with a gel-like fluid (fig. 5.14). Analysis reveals that the matrix contains enzymes that break down carbohydrate-derived products, while ATP production occurs at the cristae.

Like chloroplasts, mitochondria also contain their own DNA and ribosomes and can produce a few of their own proteins. They also reproduce themselves by division.

Chloroplasts and mitochondria are membranous organelles whose structure is suited to compartmentalizing the processes that occur within them.

Origin of Energy-Related Organelles in Eukaryotic Cells

It is believed that the first eukaryotic cells did not have energy-related organelles; instead, they were acquired through the process of endosymbiosis. The **endosymbiotic theory** states that chloroplasts and mitochondria were originally prokaryotes that came to reside inside a eukaryotic cell, establishing a symbiotic relationship known as mutualism (p. 761). In other words, today's mitochondria may be derived from an aerobic (oxygen-using) bacterium that was taken up by a eukaryotic cell in exchange for a ready supply of nutrients. Chloroplasts may be derived from a photosynthetic bacterium that permitted its host to carry on photosynthesis (fig. 5.15).

The evidence for the endosymbiotic theory is as follows:

1. Mitochondria and chloroplasts are similar to bacteria in size and in structure.
2. Both organelles are bounded by a double membrane—the outer membrane may be derived from the engulfing vesicle, and the inner one may be derived from the plasma membrane of the original prokaryote.
3. Mitochondria and chloroplasts contain a limited amount of genetic material and are capable of self-reproduction. Their DNA is a circular loop like that of bacteria.
4. Although most of the proteins within mitochondria and chloroplasts are now produced by the eukaryotic host, they do have their own ribosomes and they do produce some proteins. Their ribosomes resemble those of bacteria.

The endosymbiotic theory states that mitochondria and chloroplasts are derived from bacteria that took up residence in a eukaryotic cell(s).

Cytoskeleton: Cell Shape and Movement

The **cytoskeleton** is a network of protein elements that extends throughout the cytosol in eukaryotic cells (fig. 5.16). Prior to the 1970s, it was believed that the cytosol was an unorganized mixture of biomolecules. High-voltage electron microscopes, which can penetrate thicker specimens, showed, however, that the cytosol was instead highly organized. It contains 3 types of protein elements: microtubules, actin filaments, and intermediate filaments.

The name *cytoskeleton* is convenient in that it allows us to compare the cytoskeleton to the bones and muscles of an animal. The bones and muscles give an animal structure and allow it to move. Similarly the cytoskeleton of a cell is responsible for its shape and movement. Various cells such as red blood cells, muscle cells, and nerve cells have a distinctive shape that allows easy recognition. The cytoskeleton maintains the shape of a cell, but at the same time it also allows movement, as when, for example, vesicles move from the Golgi apparatus to the plasma membrane or when chloroplasts circulate in large plant cells. The cytoskeletal elements seem to act as "highways" along which the organelles (and perhaps molecules) are propelled in a certain direction only. In addition, we know that cells have appendages, such as cilia, that can move or even at times make the entire cell move. Some cells can move by creeping in a forward direction.

Muscles are attached to bones, and it is possible that enzymes are attached to the cytoskeleton in a way that permits them to act in a sequential manner. The term *cytoskeleton* is inappropriate, however, if it makes us think of something rigid and nonchanging. On the contrary, the cytoskeleton appears to undergo

Figure 5.15
Endosymbiotic theory. **a.** Mitochondria may be derived from aerobic bacteria that were engulfed by a larger amoebalike eukaryote. **b.** Similarly, chloroplasts may be derived from photosynthetic bacteria that were engulfed by a larger amoebalike eukaryote.

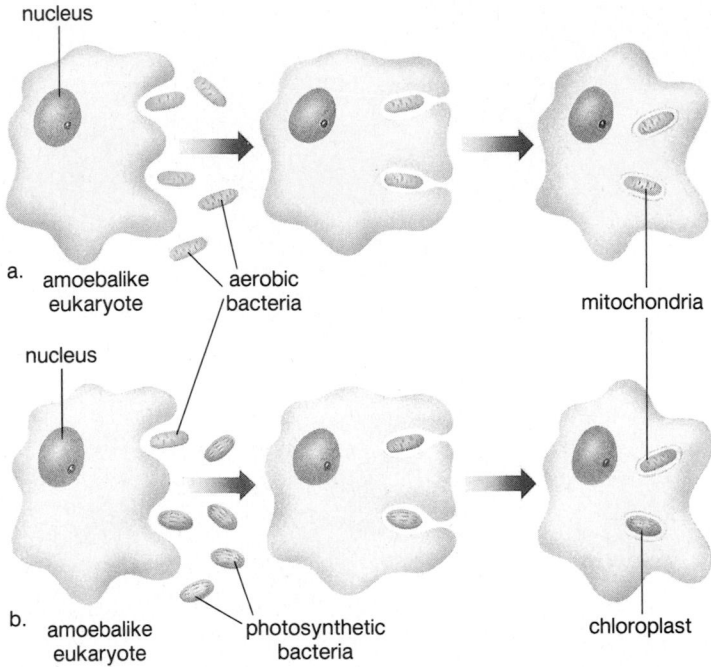

continuous and even rapid changes that allow the cell to change its shape as is necessary when cells creep forward or when embryonic cells become specialized during development.

In summary the cytoskeleton has these functions:

Maintenance of cell shape (e.g., red blood cells, nerve cells, and muscle cells have different shapes)
Change in cell shape (e.g., embryonic cells become specialized during development)
Movement of cell parts (e.g., organelles move about; cilia beat)
Movement of the cell (e.g., some cells creep; some are propelled by cilia or flagella)

The cytoskeleton contains 3 types of elements: microtubules, actin filaments, and intermediate filaments, which are responsible for cell shape and movement.

Microtubules

Microtubules are small cylinders about 25 nm in diameter and from 200 nm–25 μm in length. Their length can change because microtubules can assemble (polymerize) and disassemble (depolymerize). Microtubules are made of a globular protein called tubulin. When assembly occurs, 2 tubulin molecules come together as dimers and then the dimers form the hollow cylinder that is the microtubule (fig. 5.16). Microtubules have 13 rows of tubulin dimers. In many cells the regulation of microtubule assembly is under the control of a *microtubule organizing*

Figure 5.16

The cytoskeleton. **a.** Overview showing the general placement of the cytoskeletal elements. **b.** Intermediate filaments maintain the shape of the cell and give support to various structures. **c.** Microtubules radiate out from the microtubule organizing center and may serve as tracts for organelle movement. **d.** Actin filaments are present as bundles or a mesh network. They are often involved in cellular movements of all sorts.

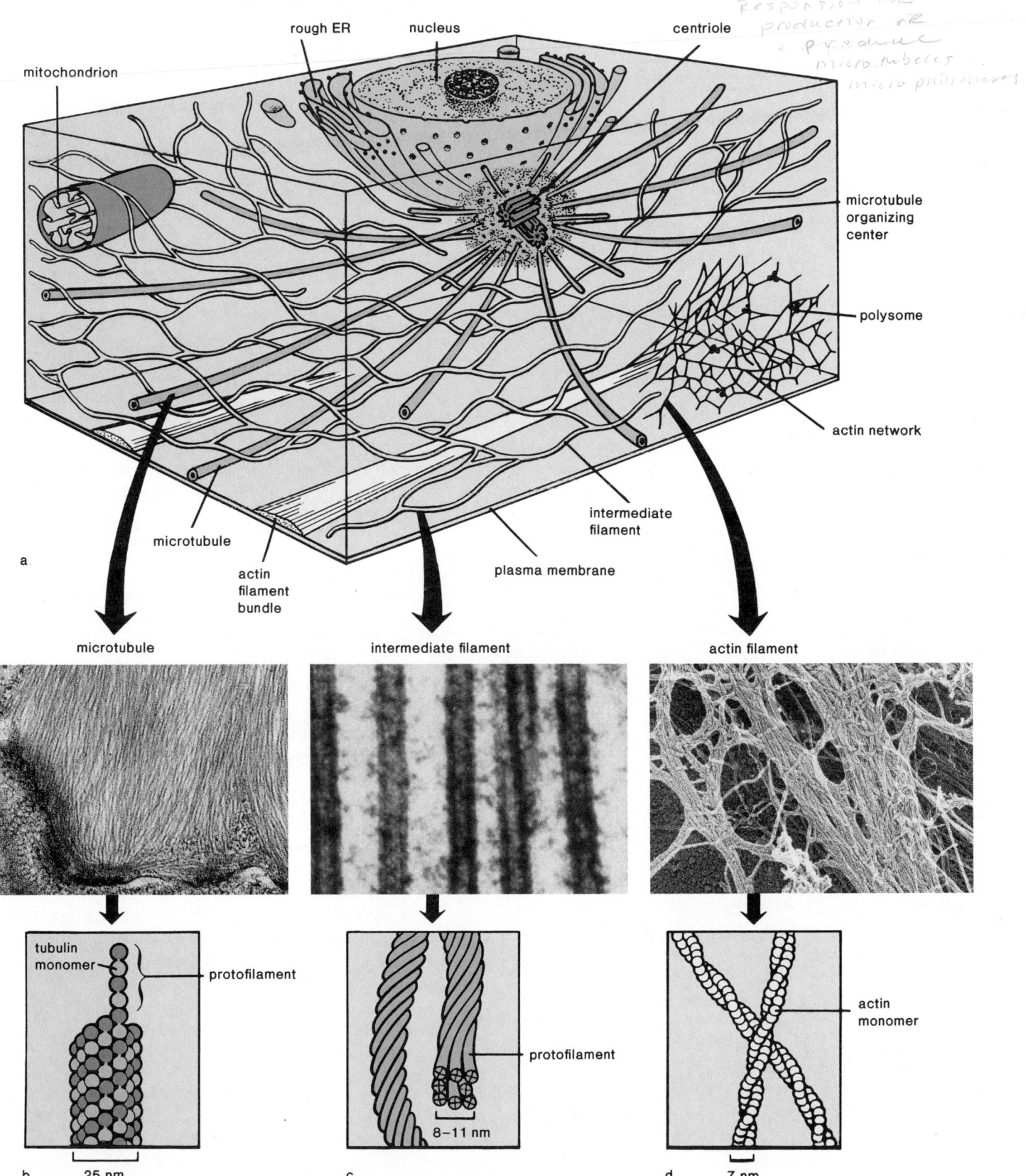

Responsible for production or + produce microtubules micro phillimeres

a.

- mitochondrion
- rough ER
- nucleus
- centriole
- microtubule organizing center
- polysome
- actin network
- intermediate filament
- plasma membrane
- microtubule
- actin filament bundle

microtubule

intermediate filament

actin filament

b. tubulin monomer — protofilament — 25 nm

c. protofilament — 8–11 nm

d. actin monomer — 7 nm

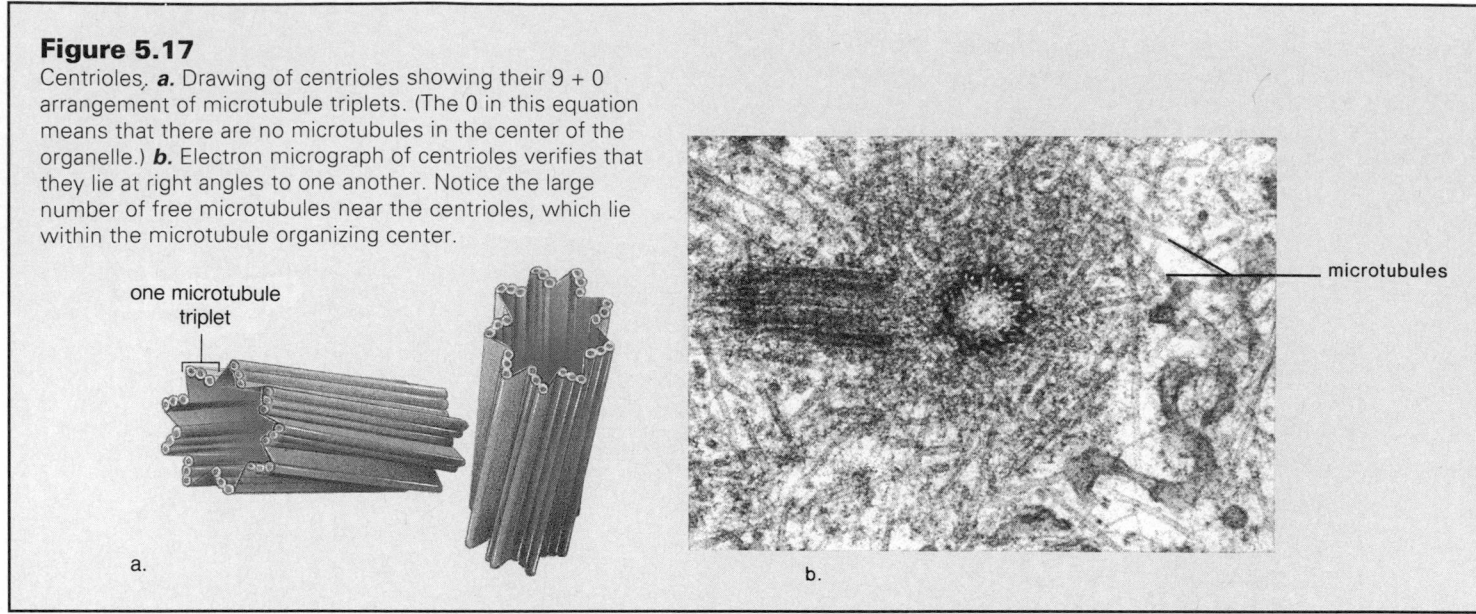

Figure 5.17
Centrioles. *a.* Drawing of centrioles showing their 9 + 0 arrangement of microtubule triplets. (The 0 in this equation means that there are no microtubules in the center of the organelle.) *b.* Electron micrograph of centrioles verifies that they lie at right angles to one another. Notice the large number of free microtubules near the centrioles, which lie within the microtubule organizing center.

one microtubule triplet

a.

microtubules

b.

center, which lies near the nucleus. The microtubules radiate out from the center, reinforcing the shape of the cell and acting as tracts along which organelles can move. The movement of vesicles from the Golgi apparatus to the plasma membrane is believed to be associated with microtubules. It is well known that during cell division, microtubules are involved in the movement of chromosomes (p. 160). During cell division, and indeed at all times, microtubules are forming and reforming as assembly and disassembly occur.

In animal cells, but not plant cells, the microtubule organizing center contains 2 centrioles lying at right angles to one another. **Centrioles** are short cylinders with a 9 + 0 pattern of microtubule triplets, that is, a ring having 9 sets of triplets with none in the middle (fig. 5.17). Before an animal cell divides, the centrioles replicate and the members of each pair are also at right angles to one another. During cell division the centriole pairs separate so that each new cell receives one pair.

Cilia and *flagella* are hairlike projections that can move either in an undulating fashion, like a whip, or stiffly, like an oar. Cells that have these organelles are capable of movement. For example, single-celled paramecia move by means of cilia; sperm cells move by means of flagella. The cells that line our upper respiratory tract are ciliated. The cilia sweep debris trapped within mucus back up into the throat, where it can be swallowed. This action helps keep the lungs clean.

Cilia are much shorter than flagella, but they have a similar construction and this construction is quite different from that of prokaryotic flagella. Eukaryotic cilia and flagella are membrane-bounded cylinders enclosing a matrix area. In the matrix are 9 microtubule doublets arranged in a circle around 2 central microtubules. This is called the 9 + 2 pattern of microtubules. Cilia and flagella move when the microtubule doublets slide past one another (fig. 5.18).

Each cilium and flagellum has a basal body lying in the cytoplasm at its base. Basal bodies have the same circular arrangement of microtubule triplets as centrioles and are believed to be derived from them. It is possible that basal bodies organize the microtubules within cilia and flagella, but we know that cilia and flagella grow by the addition of tubulin dimers to their tips.

Centrioles have a 9 + 0 pattern of microtubules and give rise to basal bodies that organize the 9 + 2 pattern of microtubules in cilia and flagella.

Actin Filaments

Actin filaments (formerly called microfilaments) are long, extremely thin fibers (about 7 nm in diameter) that occur in bundles or meshlike networks. The actin filament contains 2 chains of globular actin molecules twisted about one another in a helical manner (fig. 5.16). Actin filaments, like microtubules, can assemble and disassemble.

Actin filaments are well known for their role in muscle contraction, when they interact with other filaments composed of myosin. Actin filaments, however, assist movement in nearly all eukaryotic cells. For example, actin filaments are seen in the microvilli that project from intestinal cells. Their presence most likely accounts for the ability of microvilli alternately to shorten and extend into the intestine. In plant cells, they form the tracts that cause the cytoplasm and thus the chloroplasts to circulate or stream in a particular direction.

Working in conjunction with myosin molecules, actin filaments form a constricting ring that becomes ever smaller when an animal cell divides into 2 cells. Also, it has been discovered that the presence of a network of fibers lying beneath the plasma membrane probably accounts for the formation of pseudopodia, extensions that allow certain cells to move in an amoeboid fashion (fig. 5.19).

Figure 5.18

a. A sperm is propelled by a flagellum that contains microtubules (*b*). Magnification, ×500. *c.* A basal body with a 9 + 0 pattern of microtubule triplets is at the base of a flagellum. (Notice that there are no central microtubules within a ring of 9 triplets.) *d.* A flagellum has a 9 + 2 pattern of microtubule doublets. (Notice there are 2 central microtubule singlets with a ring of 9 doublets.) Compare the cross section of the basal body to the flagellum and note that in place of the third microtubule, the outer doublets have dynein side arms. Dynein is an enzyme that splits ATP. *e.* Movement of flagellum. In the presence of ATP, the dynein side arms reach out and connect to their neighboring doublet and push against it. If it were not for the 2 central microtubule singlets, each doublet would simply slide past its neighbor, but because of the radial spokes connecting them to the central microtubules, a bending occurs instead.

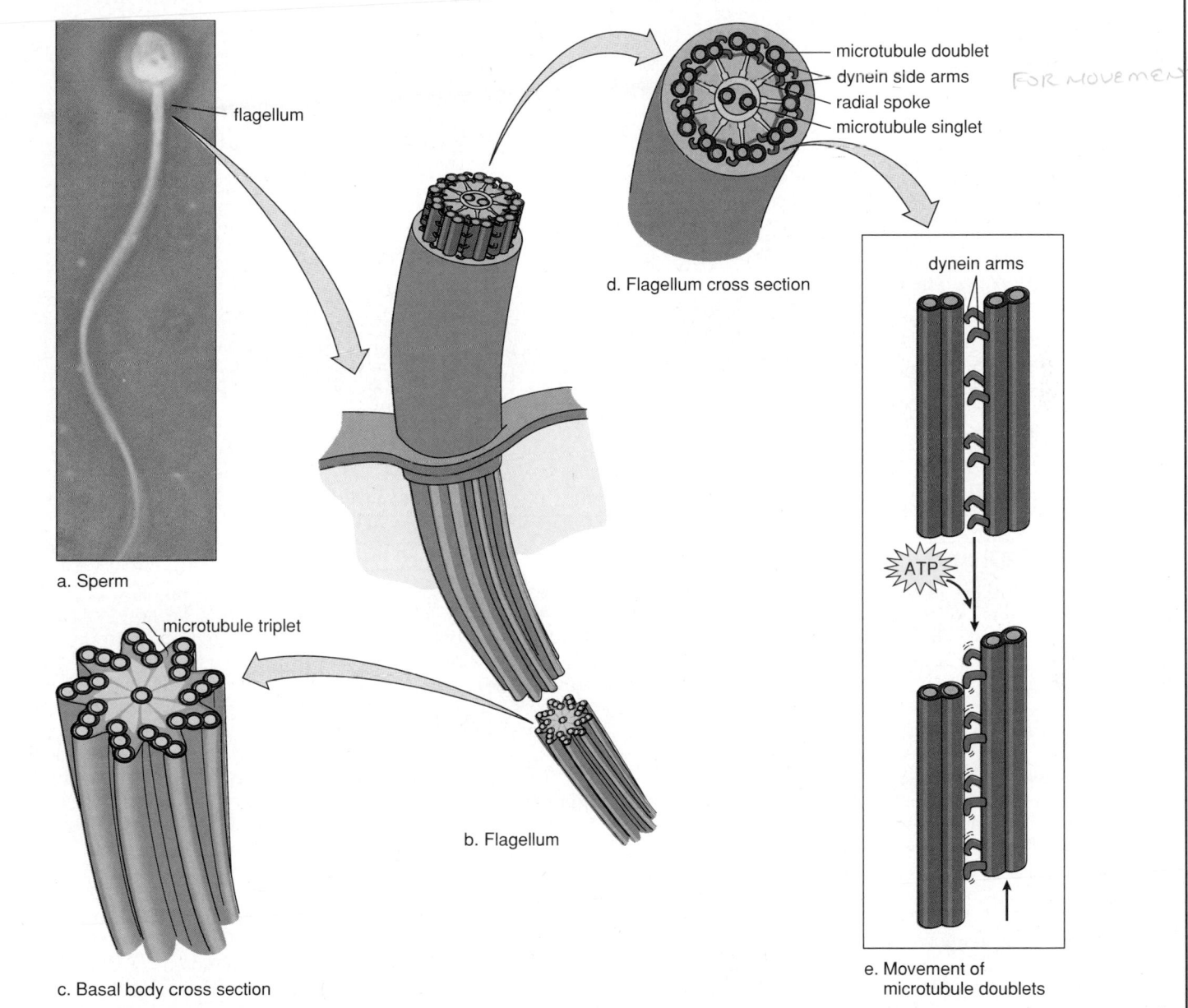

flagellum

microtubule doublet
dynein side arms
radial spoke
microtubule singlet

FOR MOVEMENT

d. Flagellum cross section

dynein arms

ATP

a. Sperm

microtubule triplet

b. Flagellum

c. Basal body cross section

e. Movement of microtubule doublets

Intermediate Filaments

Intermediate filaments (8 nm–11 nm in diameter) are intermediate in size between microtubules and microfilaments. They are ropelike polymers of fibrous proteins, but the specific type varies according to the tissue. In the skin, the filaments are made of the protein keratin, and they give great mechanical strength to cells. In all cells there are intermediate filaments that support the nuclear envelope.

Unlike the other 2 types of cytoskeletal elements, intermediate filaments do not assemble and disassemble. Therefore, they seem to be particularly important in maintaining the shape of cells and may have the function of supporting the other elements of the cytoskeleton.

Figure 5.19

a. Pseudopodium formation in a cell that moves in an amoeboid manner. There is a meshlike network of actin filaments beneath the plasma membrane. In one region the network changes from a gel state to a sol state as the actin filaments disassemble. When water from the rest of the cell moves into the region due to increased osmotic pressure, a pseudopodium forms. **b.** Reassembly of the actin filaments causes a return of the gel state and gives structure to the newly formed pseudopodium. The actin binding protein holds the rigid actin filaments together so that network results.

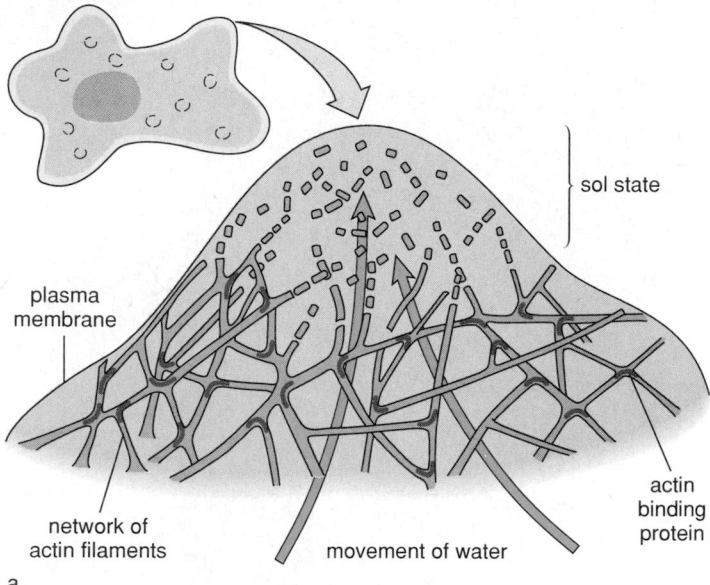

sol state

plasma membrane

network of actin filaments

movement of water

actin binding protein

a.

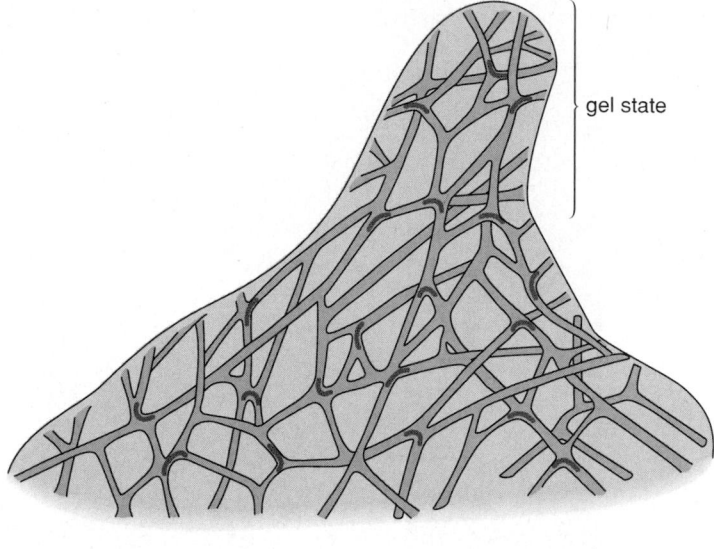

gel state

b.

Summary

1. All organisms are composed of cells, the smallest units of living matter. Cells are capable of self-reproduction, and existing cells come only from preexisting cells.

2. There are 2 major groups of cells: prokaryotic and eukaryotic. Both types have a plasma membrane and cytoplasm. Eukaryotic cells also have a nucleus and various organelles. Prokaryotic cells have a nucleoid that is not bounded by a nuclear envelope. They also lack most of the other organelles that compartmentalize eukaryotic cells. Table 5.2 summarizes the differences between prokaryotic and eukaryotic cells.

3. Cells are very small and are measured in small units of the metric system. The plasma membrane regulates exchange of materials between the cell and the external environment. Cells must remain small in order to have an adequate amount of plasma membrane per cell volume.

4. The nucleus of eukaryotic cells, represented by animal and plant cells, is bounded by a nuclear envelope containing pores. These pores serve as passageways between the cytoplasm and the nucleoplasm. Within the nucleus, the chromatin undergoes coiling into chromosomes at the time of cell division. The nucleolus is a special region of the chromatin where ribosomal RNA (rRNA) is produced and where proteins from the cytoplasm gather to form ribosomal subunits. These subunits are joined in the cytoplasm.

5. Ribosomes are organelles that function in protein synthesis. They can be bound to endoplasmic reticulum or exist free within the cytosol. Several ribosomes are involved in synthesizing the same type protein. The complex is called a polysome.

6. The endomembrane system includes the endoplasmic reticulum (both rough and smooth), the Golgi apparatus (dictyosome in plant cells), the lysosomes, and other types of vesicles and vacuoles. The endomembrane system serves to compartmentalize the cell and keep the various biochemical reactions separate from one another. Proteins that are made in the cytoplasm quite often enter the rough ER, where they may be modified before proceeding to the lumen of the smooth ER. The smooth ER has various metabolic functions depending on the cell, but it also forms vesicles that carry proteins and other substances to different locations, particularly to the Golgi apparatus. The Golgi apparatus processes proteins and repackages them into lysosomes, which carry out intracellular digestion, or into vesicles that fuse with the plasma membrane. Following fusion, secretion occurs. Also part of the endomembrane system are microbodies that have special enzymatic functions, and the large single plant cell vacuole, which not only stores substances but lends support to the plant cell when fluid-filled.

7. Cells require a constant input of energy to maintain their structure. Chloroplasts capture the energy of the

Table 5.2
Comparison of Prokaryotic Cells and Eukaryotic Cells

	Prokaryotic Cells	Eukaryotic Cells	
		Animal	Plant
Size	Smaller (1-10 μm in diameter)	Larger (10-100 μm in diameter)	
Plasma membrane	yes	yes	yes
Cell wall	usually (peptidoglycan)	no	yes (cellulose)
Nuclear envelope	no	yes	yes
Nucleolus	no	yes	yes
DNA	yes (single loop)	yes (chromosomes)	yes (chromosomes)
Mitochondria	no	yes	yes
Chloroplasts	no	no	yes
Endoplasmic reticulum	no	yes	yes
Ribosomes	yes (smaller)	yes	yes
Vacuoles	no	yes (small)	yes (usually large, single vacuole)
Golgi apparatus	no	yes	yes
Lysosomes	no	always	often
Microbodies	no	usually	usually
Cytoskeleton	no	yes	yes
Centrioles	no	yes	no (in higher plants)
9 + 2 cilia or flagella	no	often	no (in flowering plants) yes (in ferns, cycads, and bryophytes)

sun and carry on photosynthesis, which produces carbohydrate. Carbohydrate products are broken down in mitochondria as ATP is produced. This is an oxygen-requiring process called cellular respiration. It is believed that mitochondria and chloroplasts are derived from prokaryotes that took up residence in eukaryotic cells.

8. The cytoskeleton contains microtubules, intermediate filaments, and microfilaments. These are involved in maintaining cell shape and in the movement of cellular components. Microtubules are made up of 13 rows of tubulin dimers. They are found free within the cytosol but also are present in centrioles, cilia, and flagella. Centrioles give rise to basal bodies, which lie at the base of cilia and flagella. Actin filaments are 2 long chains of actin arranged in a helical manner. Intermediate filaments contain protein fibers and are intermediate in size between microtubules and microfilaments.

Writing Across the Curriculum

In order to practice writing skills, students should write out the answers to any or all of the study questions and the critical thinking questions. The study questions are sequenced in the same order as the text. Suggested answers to the critical thinking questions are in appendix D.

Study Questions

1. What are the 2 basic tenets of the cell theory?
2. What are the contrasting advantages of light microscopy and electron microscopy?
3. What similar features do prokaryotic cells and eukaryotic cells have? What is their major difference?
4. Contrast the size of prokaryotic and eukaryotic cells, and tell why cells are so small.

5. Roughly sketch a prokaryotic cell, label its parts, and state a function for each of these.
6. Describe an experiment that supports the concept of the nucleus as the control center of the cell.
7. Distinguish between the nucleolus, ribosomal RNA, and ribosomes.
8. Describe the structure and the function of the nuclear envelope and the nuclear pores.

9. Trace the path of a protein from rough ER to the plasma membrane.
10. Give the overall equations for photosynthesis and cellular respiration, contrast the 2, and tell how they are related.
11. Draw a diagram that explains the endosymbiotic theory.
12. What are the 3 components of the cytoskeleton? What are their structures and functions?

Objective Questions

1. The small size of cells is best correlated with
   a. the fact they are self-reproducing.
   b. their prokaryotic versus eukaryotic nature.
   c. an adequate surface area for exchange of materials.
   d. All of these.

2. Which of these is not a true comparison of the light microscope and the transmission electron microscope?

   Light————Electron

   a. Uses light to "view" object—uses electrons to "view" object.
   b. Uses glass lenses for focusing—uses magnetic lenses for focusing.
   c. Specimen must be killed and stained————specimen may be alive and nonstained.
   d. Magnification is not as great—magnification is greater.

3. Which of these best distinguishes a prokaryotic cell from a eukaryotic cell?
   a. Prokaryotic cells have a cell wall, but eukaryotic cells never do.
   b. Prokaryotic cells are much larger than eukaryotic cells.
   c. Prokaryotic cells have flagella, but eukaryotic cells do not.
   d. Prokaryotic cells do not have a membrane-bounded nucleus, but eukaryotic cells do have such a nucleus.

4. Which of these is not found in the nucleus?
   a. functioning ribosomes
   b. chromatin that condenses to chromosomes
   c. nucleolus that produces ribosomal RNA (rRNA)
   d. nucleoplasm instead of cytoplasm

5. Vesicles from the smooth ER most likely are on their way to the
   a. rough ER.
   b. lysosomes.
   c. Golgi apparatus.
   d. plant cell vacuole only.

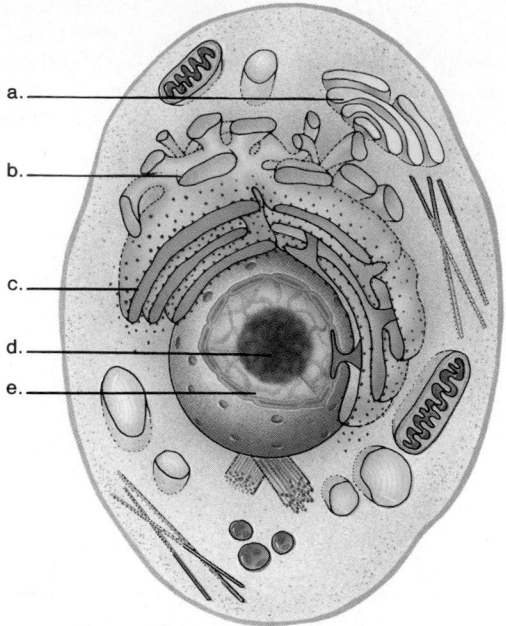

6. Lysosomes function in
   a. protein synthesis.
   b. processing and packaging.
   c. intracellular digestion.
   d. lipid synthesis.

7. Mitochondria
   a. carry on cellular respiration.
   b. break down ATP to release energy for cells.
   c. contain grana and cristae.
   d. All of these.

8. The endosymbiotic theory explains
   a. the origin of the nucleus.
   b. where the endomembrane system came from.
   c. why prokaryotic cells are different from eukaryotic cells.
   d. the origin of mitochondria and chloroplasts.

9. Which organelle releases oxygen?
   a. ribosomes
   b. Golgi apparatus
   c. mitochondria
   d. chloroplasts

10. Cilia and flagella contain
    a. microtubules.
    b. microfilaments.
    c. tubulin dimers.
    d. Both a and c.

11. Label these parts of the cell that are involved in protein synthesis and modification. Give a function for each structure.

12. Study the example given. Then for each other organelle listed, state another that is structurally and functionally related. Tell why you paired these 2 organelles.
    a. The nucleus can be paired with *nucleoli* because nucleoli are found in the nucleus. Nucleoli occur where chromatin is producing rRNA.
    b. mitochondria
    c. centrioles
    d. endoplasmic reticulum

Concepts and Critical Thinking

1. *All organisms are made up of cells.*

What data would you use to convince someone that all organisms are composed of cells? What data would you have to find to nullify the cell theory?

2. *Cells are compartmentalized and highly organized.*

Substantiate this concept by referring to table 5.1.

3. *Life begins at the cellular level of organization.*

Which organelles contribute to the ability of the cell to maintain its structure and grow? What are the functions of these organelles?

Selected Key Terms

cell (sel) 61
plasma membrane (plaz'ma mem'brān) 64
cell wall (sel wawl) 64
eukaryotic cell (u"kar-e-ot'ik sel) 64
organelle (or"gan-el') 64
prokaryotic cell (pro"kar-e-ot'ik sel) 64
nucleoid (nu'kle-oid) 64
cytoplasm (si'to-plazm") 64
cytosol (si'to-sol) 64
bacterium (bak-te're-um) 66

nucleus (nu'kle-us) 66
chromatin (kro'mah-tin) 68
chromosome (kro'mo-sōm) 68
nucleolus (nu-kle'o-lus) 68
nuclear envelope (nu'kle-ar en'vĕ-lōp) 69
ribosome (ri'bo-sōm) 70
endoplasmic reticulum (en"do-plas'mik rĕ-tik'u-lum) 71
Golgi apparatus (gol'je ap"ah-ra'tus) 72
lysosome (li'so-sōm) 73
microbody (mi"kro-bod'e) 73

chloroplast (klo'ro-plast) 74
mitochondrion (mi"to-kon'dre-an) 74
cytoskeleton (si"to-skel'ē-ton) 76
microtubule (mi'kro-tu'būl) 76
centriole (sen'tri-ōl) 78

6

Membrane Structure and Function

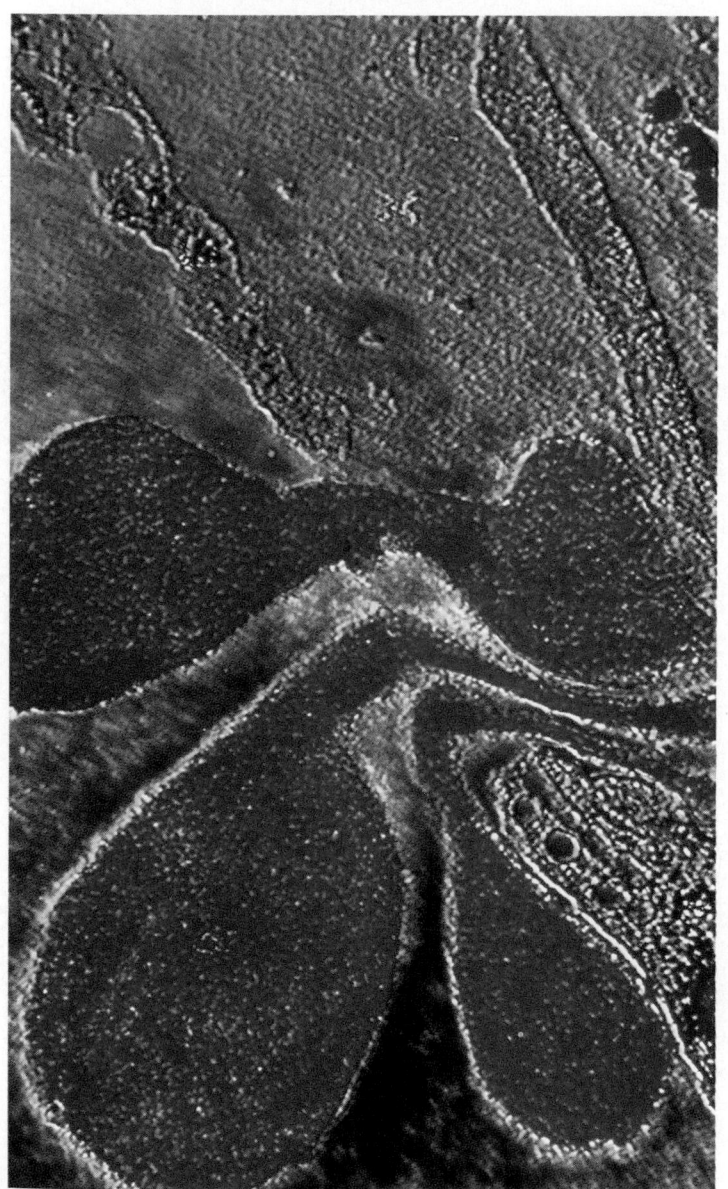

The plasma membrane is flexible. What greater proof than this photo of red blood cells squeezing through the walls of a small blood capillary?

Your study of this chapter will be complete when you can

1. contrast the "sandwich" model of membrane structure with the "fluid-mosaic" model, and cite the evidence supporting these models;
2. describe the arrangement of the lipid component, and give a function for each type of lipid in the membrane;
3. describe the arrangement of the protein component, and give several functions for these proteins;
4. define diffusion and osmosis, and explain their relevance to cell biology;
5. describe the appearance of a plant cell and an animal cell in isotonic, hypotonic, and hypertonic solutions;
6. name 2 types of transport by carrier proteins, and give an example for each;
7. contrast endocytosis and exocytosis. Name 3 types of endocytosis and differentiate among them;
8. describe the composition of the plant cell wall;
9. list and describe 3 types of junctions that occur between animal cells. Name and describe one type of junction between plant cells.

All cells have a plasma membrane that serves as an interface between the interior, which is alive, and the exterior, which is nonliving. An intact plasma membrane is absolutely essential to a cell, and if by chance the plasma membrane is disrupted, the cell loses its contents and dies. The plasma membrane was absolutely essential to the evolution of the first cell or cells. Only when a membrane is present can there be a cell at all.

The interior and exterior environments of a living cell are largely fluids, even if the cell is within a multicellular organism. Quite often the composition of intracellular fluid (i.e., cytosol) is not quite the same as that of the extracellular fluids. The plasma membrane functions to keep the intracellular fluid relatively constant. It regulates the entrance and exit of molecules into and out of the cell, and in this way the intracellular fluid remains compatible with the continued existence of the cell.

Studies of Membrane Structure

At the turn of the century, investigators noted that lipid-soluble molecules entered cells more rapidly than water-soluble molecules. This prompted them to suggest that lipids are a component of the plasma membrane. Later, chemical analysis disclosed that the plasma membrane contains phospholipids (fig. 6.1). In 1925, E. Gorter and G. Grendel measured the amount of phospholipid extracted from red blood cells and determined that there is just enough to form a bilayer around the cells.[1] They further suggested that the nonpolar (hydrophobic) tails are directed inward and the polar (hydrophilic) heads were directed outward:

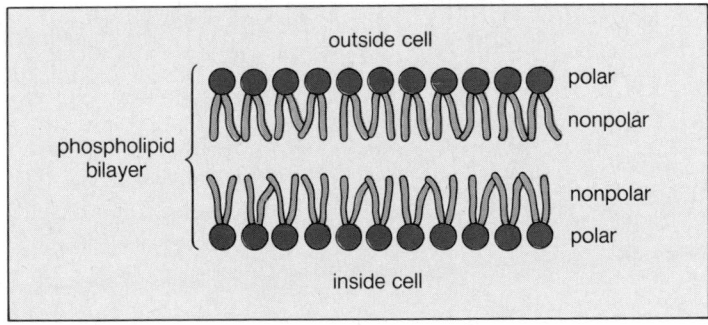

The presence of lipids cannot account for all the properties of the plasma membrane. For example, to account for the permeability of the membrane to certain nonlipid substances, J. Danielli and H. Davson suggested in the 1940s that globular proteins are also a part of the membrane. They proposed the "sandwich" model of membrane structure in which the phospholipid bilayer is a filling

[1]Later investigators found that Gorter and Grendel reported too low an amount of lipid and underestimated the surface area of red blood cells; however, since the 2 errors canceled each other, they still came to the correct conclusion.

Figure 6.1
The plasma membrane contains phospholipid molecules. *a.* Each phospholipid molecule is composed of glycerol bonded to 2 fatty acid chains and a chain that contains a phosphate group and a nitrogen group. *b.* The molecule arranges itself in such a way that the phosphate and nitrogen-containing chain forms a polar (hydrophilic) head, and the fatty acid chains form nonpolar (hydrophobic) tails. The presence of a double bond causes a "kink" in a fatty acid tail.

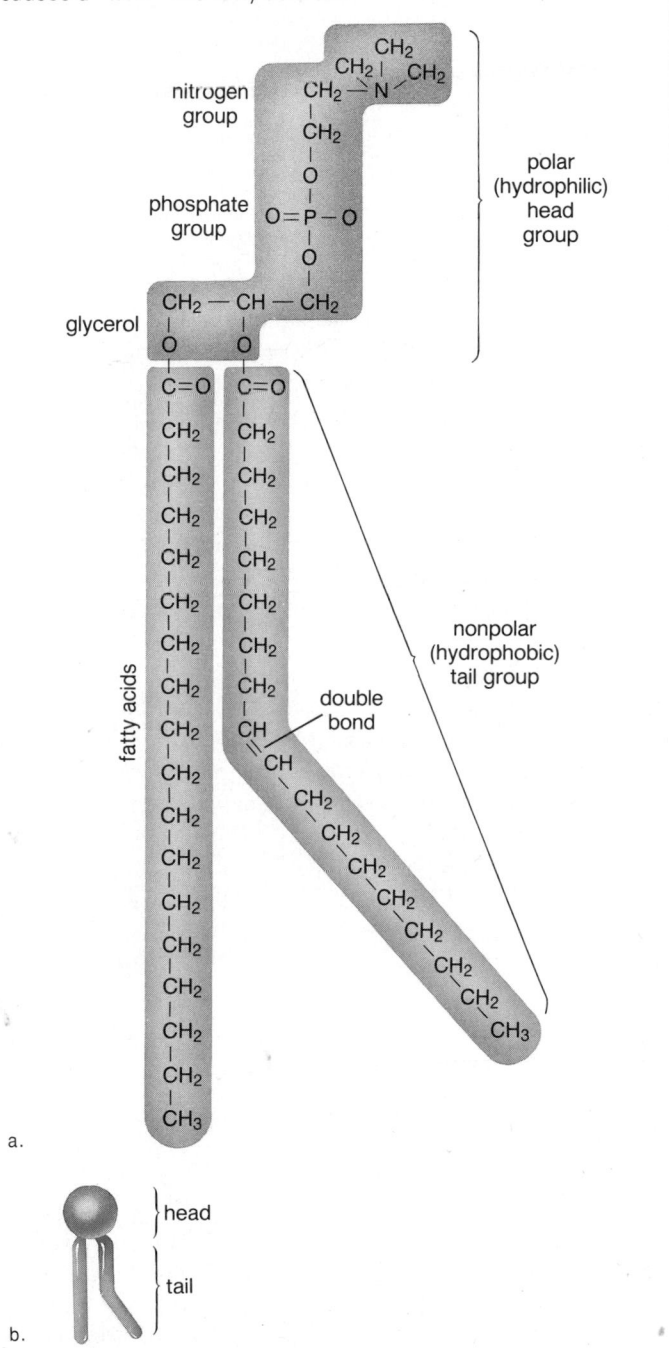

between 2 layers of proteins arranged to give channels through which polar substances may pass:

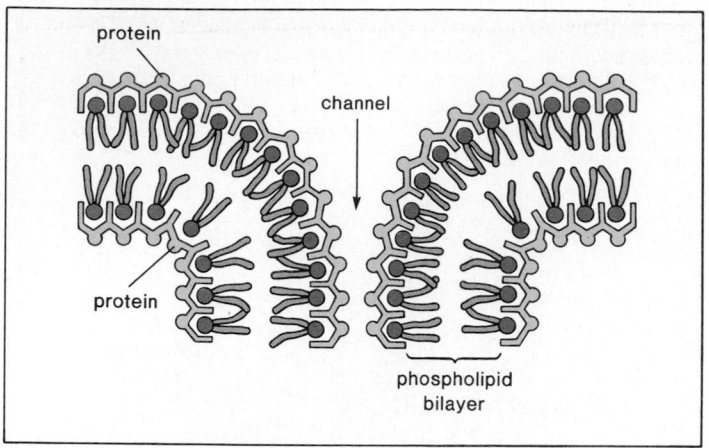

By the late 1950s, electron microscopy had advanced to allow viewing of the plasma membrane and other membranes in the cell. Since the membrane has a sandwichlike appearance (fig. 6.2), J. D. Robertson assumed that the outer dark layer (stained with heavy metals) contained protein plus the hydrophilic heads of the phospholipids. The interior was simply the hydrophobic tails of these molecules. Robertson went on to suggest that all membranes in various cells have basically the same composition. This proposal was called the "unit membrane" model.

This model of membrane structure was accepted for at least 10 years, even though investigators began to doubt its accuracy. For example, not all membranes have the same appearance in electron micrographs, and they certainly do not have the same function. The inner membrane of a mitochondrion is coated with rows of particles and functions in cellular respiration; it therefore has a far different appearance and function from the plasma membrane. Finally, in 1972, S. Singer and G. Nicolson introduced the **fluid-mosaic model** of membrane structure, which proposes that the membrane is a bilayer of phospholipids with the consistency of olive oil, in which protein molecules are either partially or wholly embedded. The proteins are scattered throughout the membrane, forming a mosaic pattern (fig. 6.3). The fluid-mosaic model of membrane structure is supported especially by electron micrographs of freeze-fractured membranes.

The fluid-mosaic model of membrane structure is widely accepted at this time.

Our perception of the plasma membrane has changed over the years. There have been a series of models, each one developed to suit the evidence available at that time. This illustrates that scientific knowledge is always subject to change, and modifications are made whenever new data are presented. A model is useful because it pulls together the available data and suggests new avenues of research.

Figure 6.2

Membrane structure. **a.** The red blood cell plasma membrane typically has a 3-layered appearance in electron micrographs; there are 2 outer dark layers and a central light layer. Magnification, X415,000. **b.** In 1960, Robertson's unit membrane model was widely accepted. This model proposed that the outer dark layers in electron micrographs were made up of protein and polar heads of phospholipid molecules, while the inner light layer was composed of the nonpolar tails. The Robertson model was challenged, particularly by the Singer and Nicolson fluid-mosaic model, which puts protein molecules within the lipid bilayer. **c.** A technique, called freeze-fracture, allows an investigator to view the interior of the membrane. Cells are rapidly frozen in liquid nitrogen and then fractured with a knife. The fracture often splits the membrane in 2, just in the middle of the lipid bilayer. Platinum and carbon are applied to the fractured surface to produce a faithful replica that is observed by electron microscopy. **d.** The electron micrograph is not as smooth as it would be if the unit membrane model was correct. Instead, the micrograph shows the presence of particles as is expected by the fluid-mosaic model.

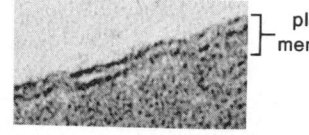

a. Electron micrograph of red blood cell plasma membrane

plasma membrane

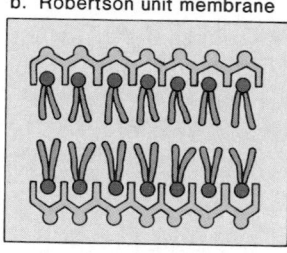

b. Robertson unit membrane

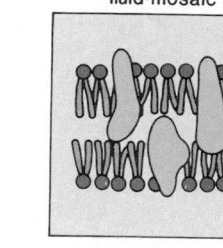

Singer and Nicolson fluid-mosaic model

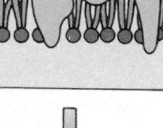

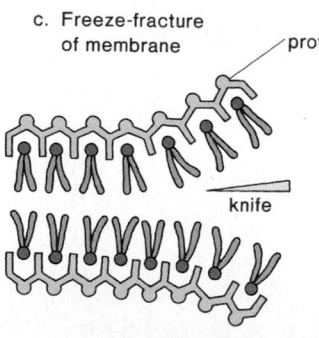

c. Freeze-fracture of membrane

protein

knife

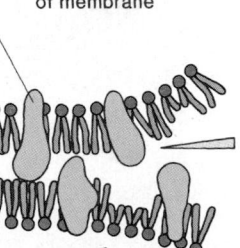

Freeze-fracture of membrane

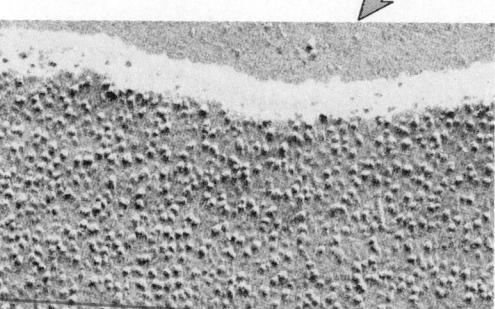

d. Electron micrograph of freeze-fractured membrane shows presence of particles

Figure 6.3

Fluid-mosaic model of the plasma membrane. The plasma membrane is composed of a phospholipid bilayer with embedded proteins. The hydrophilic heads of the phospholipids are at the surfaces of the membrane, and the hydrophobic tails make up the interior of the membrane. Note the asymmetry of the membrane; for example, carbohydrate chains project externally and cytoskeleton filaments attach to proteins on the cytoplasmic side of the plasma membrane.

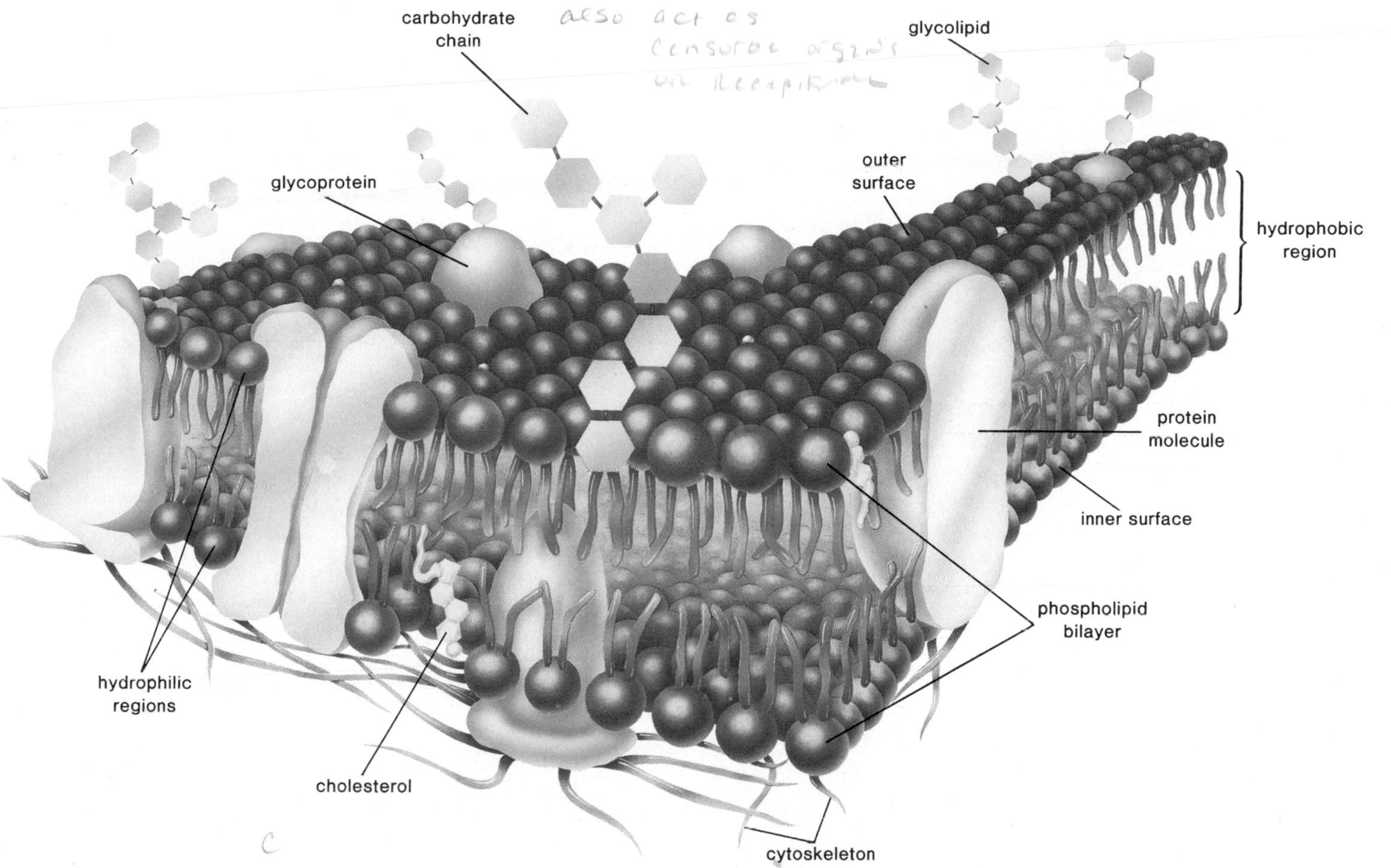

carbohydrate chain

also act as
censorce organ's
or Receptorne

glycolipid

glycoprotein

outer surface

hydrophobic region

protein molecule

inner surface

phospholipid bilayer

hydrophilic regions

cholesterol

cytoskeleton filaments

Fluid-Mosaic Model of the Plasma Membrane

The fluid-mosaic model of membrane structure has 2 components, lipids and proteins (fig. 6.3). The lipids form the matrix of the membrane, and the proteins carry out all of its functions.

Lipid Component

Most of the lipids in the plasma membrane are **phospholipids** (fig. 6.1). In a laboratory container, phospholipids spontaneously form a bilayer in which the hydrophilic (polar) heads face water on both sides and the hydrophobic tails avoid contact with water. The hydrophobic interior portions of the phospholipid bilayer are never exposed to water because it spontaneously forms a sphere—becoming a barrier just like the plasma membrane:

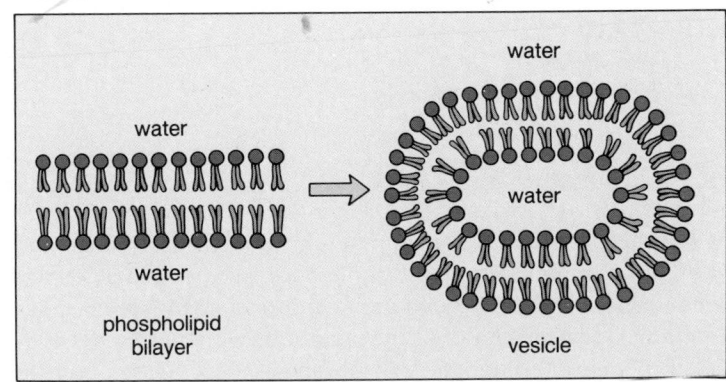

water

water

phospholipid bilayer

water

water

water

vesicle

At body temperature, the phospholipid bilayer of the plasma membrane has the consistency of olive oil. The greater the concentration of unsaturated fatty acid residues, the more liquid the bilayer. In each monolayer, the hydrocarbon tails wiggle, and the entire phospholipid molecule can move sideways, even exchanging places with its neighbor. (Phospholipid molecules rarely flip-flop from one layer to the other, because this would require the hydrophilic head to move through the hydrophobic center of the membrane.) The fluidity of a phospholipid bilayer means that cells are pliable. Imagine if they were not—the nerve cells in your neck would crack whenever you nodded your head!

The plasma membrane is asymmetrical. For example, the types of phospholipids in each half of the lipid bilayer can be different. Phospholipids can vary especially by the kind of nitrogen group they have, and certain kinds are more likely to be found in the outer half of the membrane than in the inner half. *Glycolipids* are another type of lipid found in the plasma membrane. They are like phospholipids except that the hydrophilic head is made up of a variety of sugars joined to form a straight or branching carbohydrate chain. As figure 6.3 shows, glycolipids contribute to the asymmetry of the plasma membrane because the carbohydrate chain is always directed externally, an observation that offers a clue to their function. It is possible that glycolipids function as part of a receptor or help mark the cell as belonging to a particular individual. They might also regulate the action of plasma membrane proteins involved in the growth of the cell, and if so they might have a role in the occurrence of cancer.

In animal cells, **cholesterol** is a major membrane lipid, equal in amount to phospholipids. Cholesterol has both a hydrophilic and a hydrophobic end and arranges itself in this manner:

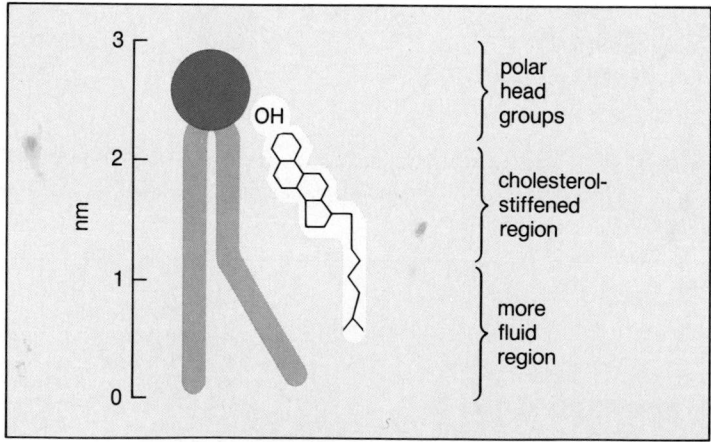

Cholesterol makes the membrane more impermeable to most biological molecules. Small molecules, like amino acids and sugars, do not pass through the membrane easily because they are polar, and the center of the membrane is nonpolar. Passage of molecules through the membrane is dependent upon the protein component.

The phospholipid bilayer portion of the plasma membrane forms a hydrophobic impermeable barrier that prevents the movement of polar (most biological) molecules through the plasma membrane.

Protein Component

The proteins associated with the plasma membrane are either attached to its inner surface or embedded in the phospholipid bilayer. Proteins often have hydrophobic and hydrophilic regions; hydrophobic regions are within the membrane, while the hydrophilic regions project from both surfaces of the bilayer.

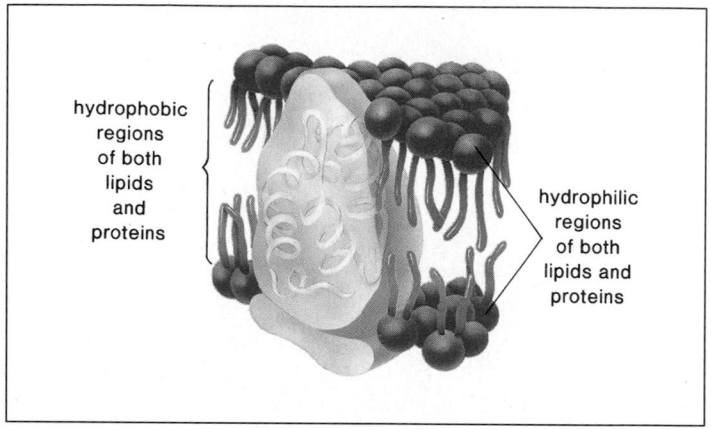

Although proteins are oriented so that a particular hydrophilic region always projects from either the outer or inner surface, they can move laterally in the fluid lipid bilayer. This has been demonstrated by the experiment explained in figure 6.4.

The protein composition of the plasma membrane varies among cells—red blood cell plasma membrane can have 20%-40% lipid and 60%-80% protein, for example. Also, the proteins on the outer and inner surfaces of the membrane can differ, and only those on the inner surface are attached to cytoskeletal filaments. This shows that the plasma membrane is asymmetrical, as does the placement of glycoproteins. *Glycoproteins,* like glycolipids, have an attached carbohydrate chain that only projects externally from the membrane.

Glycoproteins make cell-to-cell recognition possible and serve as the "fingerprints" of the cell. They enable the immune system to recognize its own organs and, unfortunately, are the cause of an immune system response to transplanted organs. As we will discuss in more detail, certain plasma membrane proteins are involved in the passage of molecules through the membrane. Some of these have a *channel* through which a substance simply can move across the membrane; others are *carriers* that combine with a substance and help it to move across the membrane. Still other proteins are *receptors;* each type of receptor has a specific shape that allows a specific molecule to bind to it. The binding of a molecule, such as a hormone, can influence the metabolism of the cell. Viruses often must attach to

Figure 6.4

Experiment to demonstrate that plasma membrane proteins move about. Mouse cells and human cells are exposed to their respective fluorescent antibodies. Proper laboratory treatment causes the cells to fuse. Immediately after fusion, microscopic observation shows that the mouse and human glycoproteins are still separated from one another. Forty minutes after fusion, the glycoproteins are completely intermixed. This shows that the proteins are able to move within the plasma membrane.

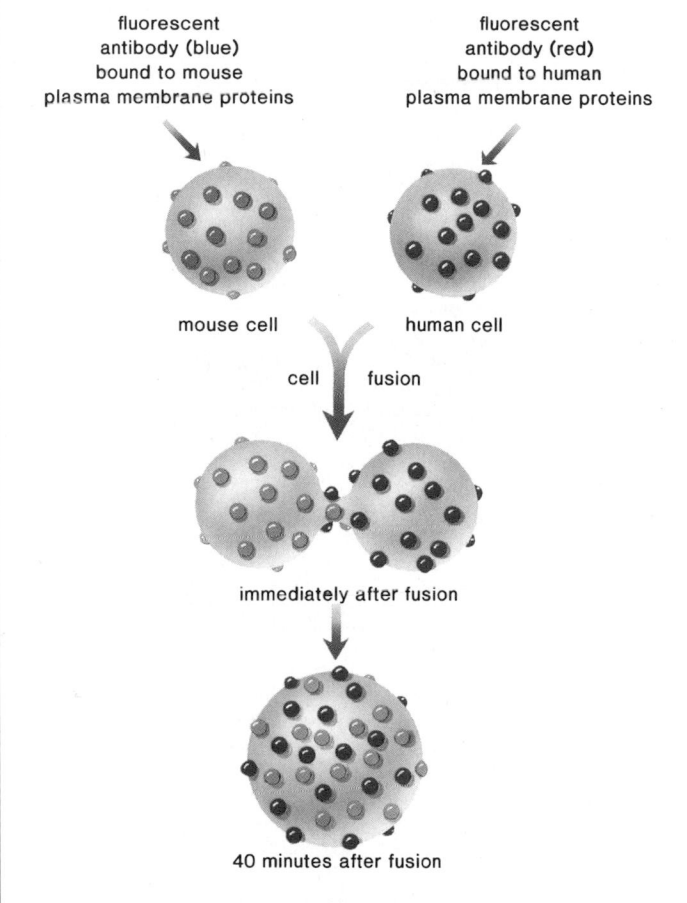

fluorescent antibody (blue) bound to mouse plasma membrane proteins

fluorescent antibody (red) bound to human plasma membrane proteins

mouse cell

human cell

cell fusion

immediately after fusion

40 minutes after fusion

receptors before they enter a cell. Some proteins have an *enzymatic function* and carry out metabolic reactions. Figure 6.5 depicts the various functions of membrane proteins.

Movement of Molecules into and out of Cells

If the plasma membrane were completely permeable, all substances would move across it with no difficulty. Since it can be observed that all substances cannot move across the plasma membrane, it is correct to say that the plasma membrane is semipermeable. It can further be observed, however, that proteins are often involved in allowing some selected substances to move across the

membrane. In addition, vesicle formation brings only certain macromolecules into cells. Because the plasma membrane is able to regulate which substances enter and exit the cell, it is said to be **selectively permeable.**

> The plasma membrane is selectively permeable—it has special mechanisms to regulate the passage of most molecules into and out of the cell.

Certain small molecules are simply able to diffuse across the plasma membrane.

Diffusion

Ions and molecules in air and in water constantly move about, seemingly at random. It can be demonstrated, however, that there is a net movement of molecules from a region of greater concentration to a region of lesser concentration. For example, when a few crystals of dye are placed in water, the dye molecules move in various directions, but their net movement is in a direction away from the concentrated source (fig. 6.6). The net movement of the water molecules is in the opposite direction. Eventually, the dye is evenly dissolved in the water, resulting in a colored solution. The dye and water molecules continue to move about, but there is no net movement in any one direction.

Diffusion is the movement of molecules from the area of greater concentration to the area of lesser concentration. Diffusion occurs spontaneously, and no energy is required to bring it about; however, diffusion does require a difference in concentrations.

A few substances freely diffuse across plasma membranes. For example, the respiratory gases, oxygen and carbon dioxide, diffuse into and out of cells. After you have taken in a breath of air, there is a greater concentration of oxygen in the alveoli of your lungs than there is in the blood capillaries surrounding the lungs (fig. 6.7). Oxygen diffuses down this *concentration gradient* from the lungs into the blood. Water is another molecule that simply diffuses across cell membranes. This has important biological consequences, which we will examine next.

> Only a few types of small molecules freely diffuse through the plasma membrane from the area of greater concentration to the area of lesser concentration.

Osmosis

The diffusion of water across a selectively permeable membrane has been given a special name: it is called **osmosis.** To illustrate osmosis, a thistle tube covered at one end by a selectively permeable membrane is placed in a beaker of distilled water. Inside the tube is a solution containing both a solute (e.g., sugar molecules) and a solvent (water) (fig. 6.8). Obviously, the lesser solute and greater water concentration is outside the tube, so there will be a net movement of water from outside the tube to inside through the membrane. The solute is unable to pass out of the tube because the membrane is impermeable to it; therefore, the level of the solution within the tube rises (fig. 6.8c). The eventual height of the solution

Figure 6.5

Membrane protein diversity. These are some of the functions performed by proteins found in the plasma membrane.

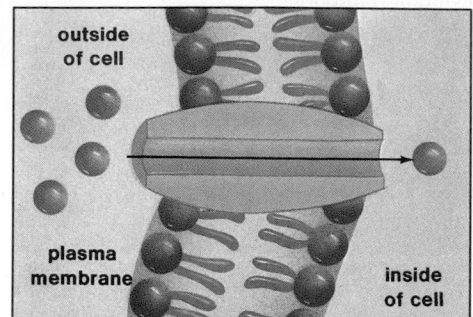

Channel Protein

A protein that allows a particular molecule or ion to cross the plasma membrane freely as it enters or exits the cell. Recently, it has been shown that cystic fibrosis, an inherited disorder, is caused by a faulty chloride (Cl^-) channel. When the channel is not functioning normally, a thick mucus collects in airways and in pancreatic and liver ducts.

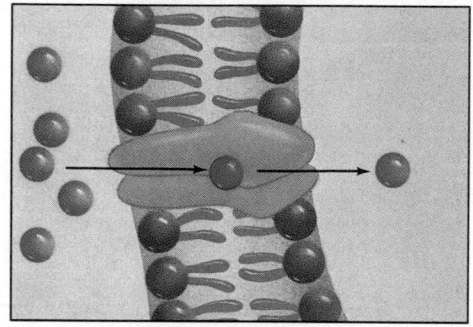

Carrier Protein

A protein that selectively interacts with a specific molecule or ion so that it can cross the plasma membrane to enter or exit the cell. The carrier protein that transports sodium (Na^+) ions and potassium (K^+) ions across the plasma membrane requires ATP energy. The inability of some persons to use up energy for sodium-potassium transport has been suggested as the cause of their obesity.

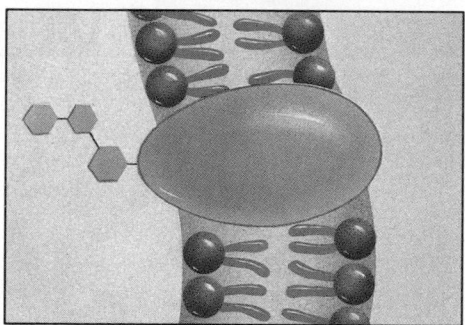

Cell Recognition Protein

A glycoprotein that identifies the cell. For example, the MHC (major histocompatibility complex) glycoproteins are different for each person, so organ transplants are difficult to achieve. Cells with foreign MHC glycoproteins are attacked by blood cells responsible for immunity.

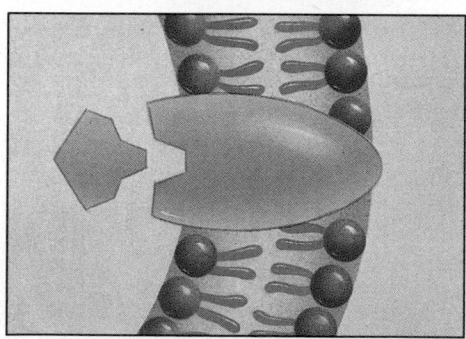

Receptor Protein

A protein that is shaped in such a way that a specific molecule can bind to it. Recently, it has been shown that Pygmies are short not because they do not produce enough growth hormone, but because their plasma membrane growth hormone receptors are faulty and cannot interact with growth hormone.

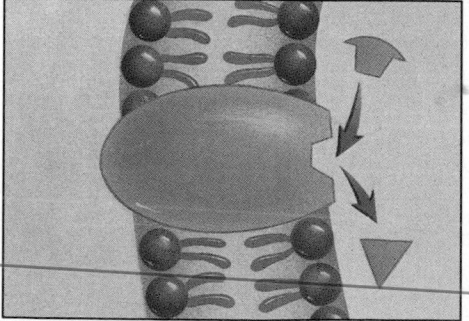

Enzymatic Protein

A protein that catalyzes a specific reaction. For example, there is a plasma membrane protein called adenylate cyclase that is involved in ATP metabolism. Polluted water can contain cholera bacteria, which release a toxin that interferes with the proper functioning of adenylate cyclase. So many sodium ions and so much water leave intestinal cells that the individual dies from severe diarrhea.

Figure 6.6

Process of diffusion is spontaneous and no energy is required to bring it about. **a.** When dye crystals are placed in water they are concentrated in one area. **b.** The dye dissolves in the water, and there is a net movement of dye molecules away from the area of concentration. There is a net movement of water molecules in the opposite direction. **c.** Eventually, the water and dye molecules are equally distributed throughout the container.

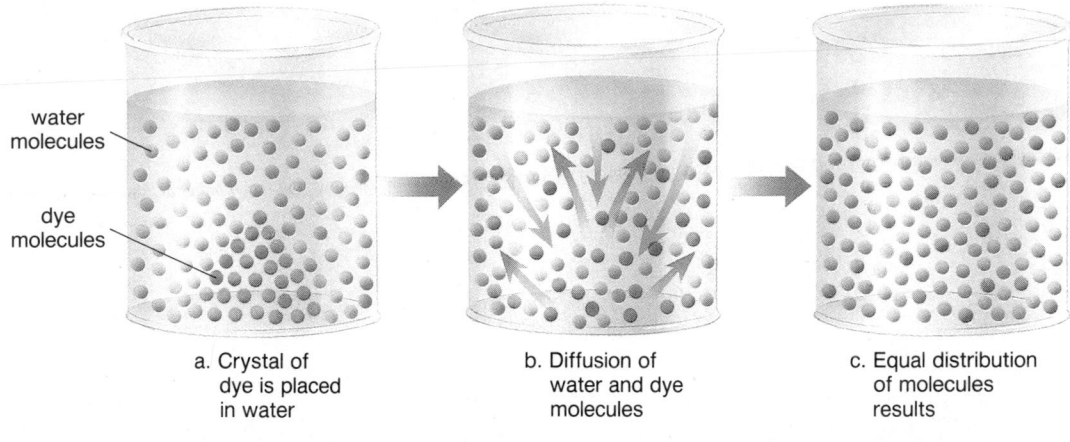

water molecules

dye molecules

a. Crystal of dye is placed in water

b. Diffusion of water and dye molecules

c. Equal distribution of molecules results

Figure 6.7

Oxygen (dots) diffuses into the capillaries because there is a greater concentration of oxygen in the alveoli (air sacs) of the lungs than in the capillaries.

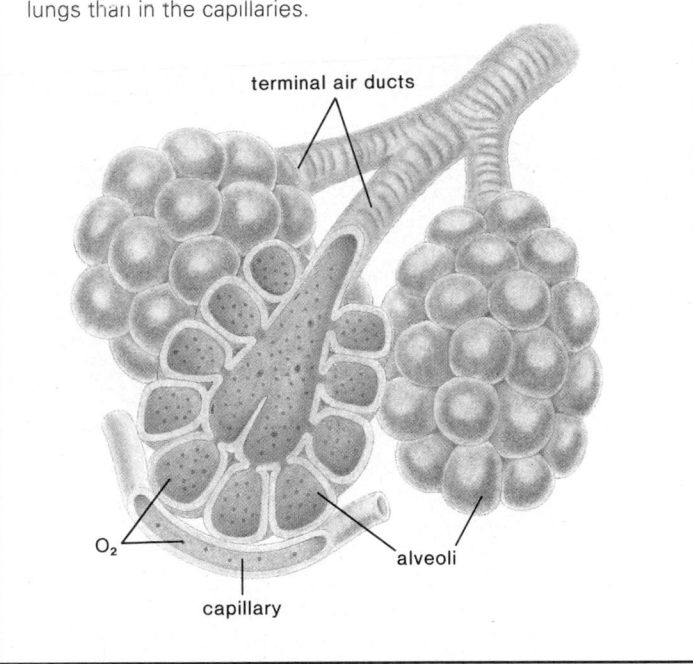

terminal air ducts

O₂

alveoli

capillary

in the tube indicates the degree of **osmotic pressure** caused by the flow of water from the area of greater water concentration to the area of lesser water concentration.

Notice the following elements in this illustration of osmosis:

1. A selectively permeable membrane separates a solution from pure water.

2. A difference in solute and water concentrations exists on the 2 sides of the membrane.

3. The membrane is impermeable (does not permit passage) to the solute particles.

4. The membrane is permeable (permits passage) to the water, which moves from the area of greater water (lesser solute) concentration to the area of lesser water (greater solute) concentration.

5. An osmotic pressure is present; the amount of liquid increases on the side of the membrane with the greater solute concentration.

These considerations will be important as we discuss osmosis in relation to cells placed in different solutions. The plasma membrane is not completely impermeable to solutes such as sugars and salts, but the difference in permeability between water and these solutes is so great that cells do have to cope with the osmotic movement of water.

Osmosis is the diffusion of water across a selectively permeable membrane. The presence of osmotic pressure is evident when there is an increased amount of water on the side of the membrane that has the greater solute concentration.

Tonicity

Cells can be placed in solutions that are isotonic, hypotonic, or hypertonic (table 6.1 and fig. 6.9). In an **isotonic solution,** a cell neither gains nor loses water because the concentration of solute and water is the same on both sides of the plasma membrane (*iso—* same as). Most animal cells normally live under isotonic conditions. For example, the tissue fluid that surrounds your cells is normally isotonic to them.

Figure 6.8

Osmosis demonstration. *a.* A thistle tube, covered at the broad end by a selectively permeable membrane, contains a sugar solution. The beaker contains only solvent (water). *b.* The solute (green circles) is unable to pass through the membrane, but the water passes through in both directions. There is a net movement of water toward the inside of the thistle tube, where the solute concentration is greater and the water concentration is lesser. *c.* In the end, the level of the solution rises in the thistle tube until a pressure equivalent to osmotic pressure builds.

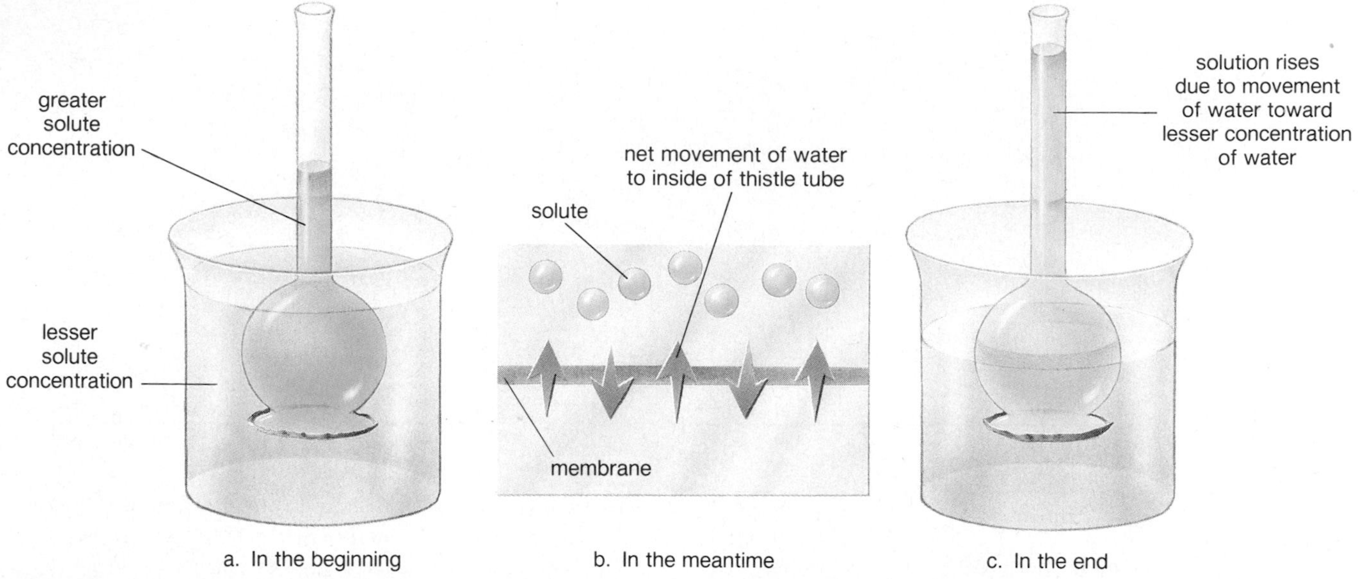

greater solute concentration

lesser solute concentration

net movement of water to inside of thistle tube

solute

membrane

solution rises due to movement of water toward lesser concentration of water

a. In the beginning

b. In the meantime

c. In the end

In a **hypotonic solution,** a cell tends to gain water because the concentration of solute is lesser (*hypo*—less than) and the concentration of water is greater than that of the cell. Plants and algae that live in freshwater ponds live under hypotonic conditions. The cytoplasm and central vacuoles gain water, and the plasma membrane pushes against the rigid cell wall. The resulting pressure, called **turgor pressure,** helps give internal support to the cell. Even plants that live on land are dependent upon turgor pressure to keep them from wilting. Freshwater protozoans are not protected from osmotic swelling by a rigid cell wall, but they have a contractile vacuole that rapidly pumps out excess water (fig. 6.10).

In a **hypertonic solution,** a cell tends to lose water because the concentration of solute is greater (*hyper*—more than) and the concentration of water is less than that of the cell. When plant cells are placed in a hypertonic solution, it is possible to see that the cytoplasm has lost water and has undergone plasmolysis because the plasma membrane pulls away from the cell wall. Some organisms, such as marine fishes, live in a hypertonic environment. The gills of these fishes extrude salt from the blood, and the blood remains isotonic to the cells.

Cells can die when placed in a solution of unfavorable tonicity. For example, red blood cells will lyse (burst) when placed in a very hypotonic solution and will undergo crenation (dry up) when placed in a very hypertonic solution (fig. 6.9*e* and *f*). Organisms that live in environments with unfavorable tonicities have evolved mechanisms to keep the tonicity of cellular cytoplasm within a normal range.

Table 6.1
Effect of Osmosis on a Cell

Tonicity of Solution	Concentrations		Net Movement of Water	Effect on Cell
	Solute	*Water*		
Isotonic	Same as cell	Same as cell	None	None
Hypotonic	Less than cell	More than cell	Cell gains water	Swells, turgor pressure
Hypertonic	More than cell	Less than cell	Cell loses water	Shrinks, plasmolysis

When a cell is placed in an isotonic solution, it neither gains nor loses water. When a cell is placed in a hypotonic solution (lesser solute concentration than isotonic) the cell gains water. When a cell is placed in a hypertonic solution (greater solute concentration than isotonic), the cell loses water and the cytoplasm shrinks.

Transport by Carriers

The presence of the plasma membrane impedes the passage of all but a few substances. Yet biologically useful molecules do enter and exit the cell at a rapid rate because there is a transport system

The Cell

Figure 6.9

Osmosis in plant and animal cells. The arrows indicate the net movement of water. In an isotonic solution, a cell neither gains nor loses water; in a hypotonic solution, a cell gains water; and in a hypertonic solution, a cell loses water.

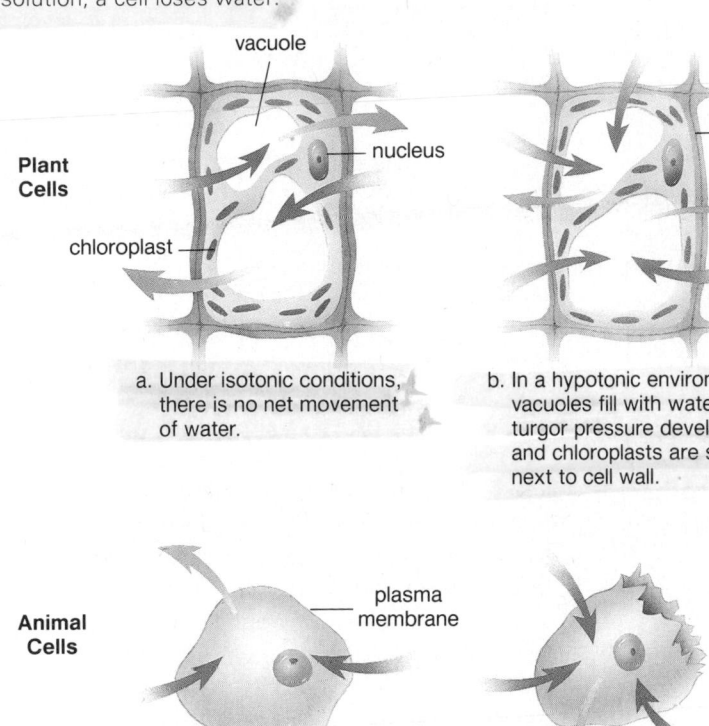
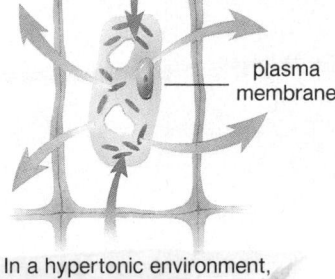

Plant Cells

vacuole

nucleus

chloroplast

cell wall

plasma membrane

a. Under isotonic conditions, there is no net movement of water.

b. In a hypotonic environment, vacuoles fill with water, turgor pressure develops, and chloroplasts are seen next to cell wall.

c. In a hypertonic environment, vacuoles lose water, cytoplasm shrinks (called plasmolysis), and chloroplasts are seen in center of cell.

Animal Cells

plasma membrane

d. Under isotonic conditions, there is no net movement of water.

e. In a hypotonic environment, water enters cell, which may burst (lyse) due to osmotic pressure.

f. In a hypertonic environment, water leaves the cell, which shrivels (crenation occurs).

Figure 6.10

Many organisms have evolved mechanisms to cope with the tonicity of the surrounding medium. A paramecium *a.* lives in fresh water, which is hypotonic to it. It has contractile vacuoles that expel water from the cell. Magnification, X107. *b.* Water from the cytoplasm flows into tubules, filling vacuole. *c.* The vacuole collapses, appearing to "contract," as water is released through a temporary opening in the plasma membrane.

contractile vacuole

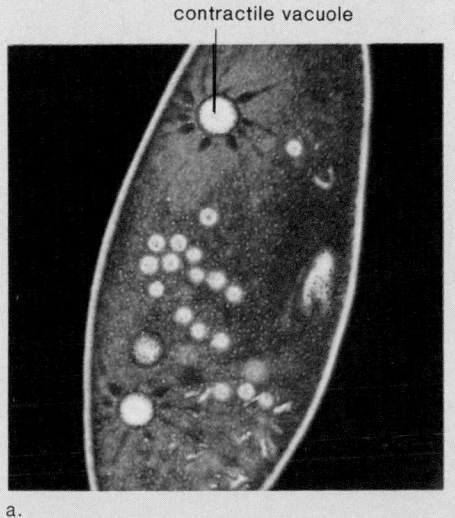
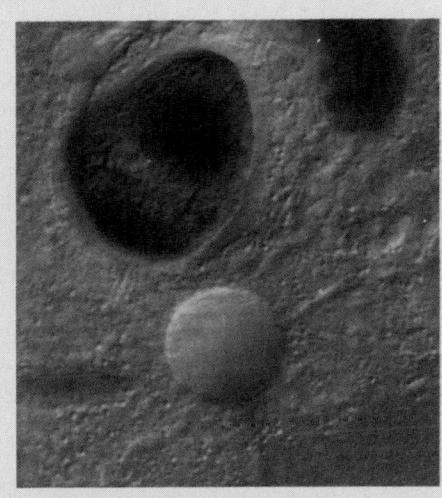
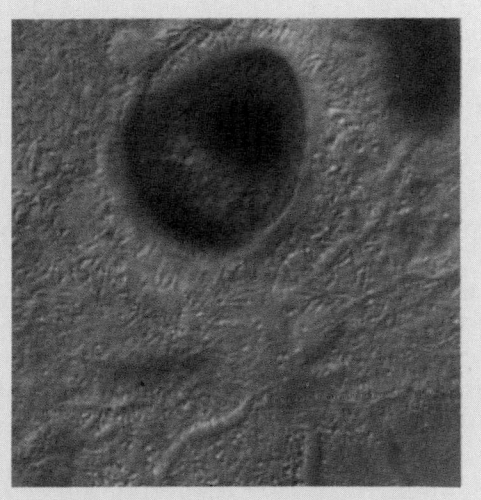

a.

b.

c.

Figure 6.11

Facilitated transport. A carrier protein speeds the rate at which the solute crosses the plasma membrane in the direction of decreasing concentration. Note that the protein carrier undergoes a change in shape (called a conformational change) as it moves a solute across the membrane.

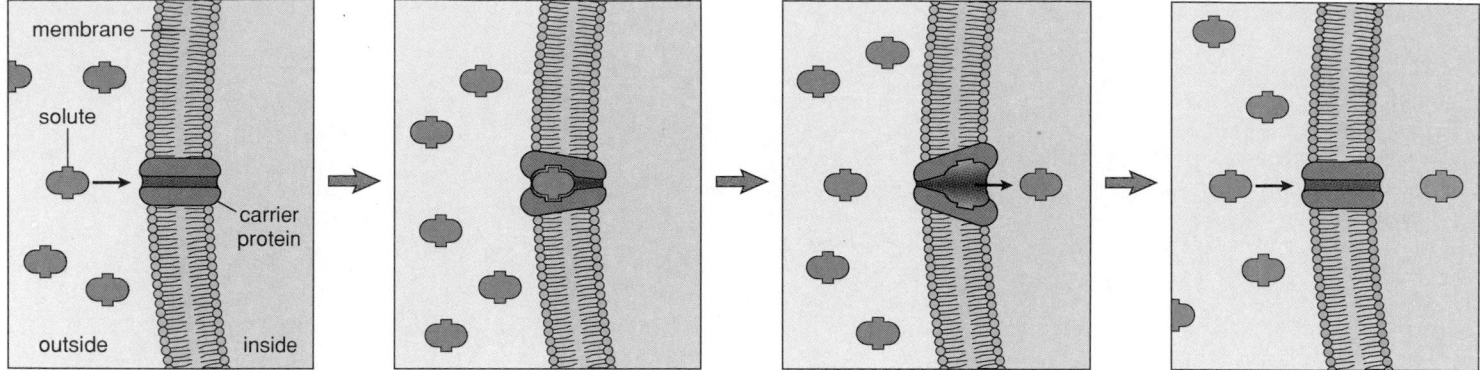

in the plasma membrane that delivers them in a timely manner. Some of the proteins that span the plasma membrane act as **carriers** to transport these molecules. Carrier proteins are specific, and each can combine with only a certain type of molecule, which is then transported across the plasma membrane. It is not completely understood how carrier proteins function, but after they combine with a substance, they are believed to undergo a *conformational change,* a change in shape that causes the substance to be moved across the membrane.

Some of the proteins in the plasma membrane are carriers that transport biologically useful molecules into and out of the cell.

Facilitated Transport

A model for **facilitated transport** suggests that after a carrier has assisted the movement of a molecule to one side of the membrane, it is free to assist the passage of another similar molecule (fig. 6.11). This form of transport is involved in carrying various sugars, amino acids, and nucleotides into or out of a cell according to concentration gradients. Because the molecule is still traveling down its concentration gradient, no significant energy input is required.

Active Transport

Due to **active transport,** molecules or ions accumulate either inside or outside the cell, even when this is the region of greater concentration. For example, the plasma membranes of plant root cells are able to extract inorganic ions from the soil due to active transport. We already mentioned that in marine fishes the cells in the gills are able to extrude salt so that the blood remains isotonic to body cells. In our body, the thyroid gland accumulates iodine and the kidneys actively reabsorb sodium ions (Na^+) so that water follows by osmosis. If this did not happen, we would be in constant danger of dehydration because of water loss by way of the kidneys. As described in this chapter's reading, the chambered nautilus uses a similar mechanism to maintain buoyancy.

Both a protein carrier and an expenditure of energy are needed for active transport. The necessary energy is produced by the breakdown of ATP (adenosine triphosphate). Therefore, it is not surprising that cells involved primarily in active transport, such as kidney cells, have a large number of mitochondria near the membrane where active transport is occurring. Proteins involved in active transport are often called *pumps* because just as a water pump uses energy to move water against the force of gravity, proteins use energy to move a substance against its concentration gradient.

One type of pump that is active in all cells, but especially associated with nerve and muscle cells, moves sodium ions (Na^+) to the outside of the cell and potassium ions (K^+) to the inside of the cell. These 2 events are presumed to be linked, and the carrier protein is called a **sodium-potassium pump.** ATP energy released by an ATPase enzyme (an enzyme that breaks down ATP) is believed to be necessary to bring about a change in shape of the carrier, allowing it to combine alternately with Na^+ and K^+ (fig. 6.12).

Sodium ions and potassium ions are charged particles; the action of the pump causes the plasma membrane to be positively charged on the outside and negatively charged on the inside. Therefore, the membrane has an electrical gradient; this, along with any concentration gradient, influences the movement of charged particles. Charged particles tend to move according to their *electrochemical gradient.*

The sodium gradient, established by the sodium-potassium pump, provides the energy for an animal cell to take up other types of molecules, such as sugars and amino acids. For example, there are carriers in the plasma membrane of an animal cell that cotransport both sodium ions and an amino acid. Because there are so many more sodium ions outside the cell than inside the cell, the carrier automatically transports sodium ions, and consequently the amino acid as well, even if there is

Chambered Nautilus

The chambered nautilus (fig. 6.A), a mollusk that takes its name from its multichambered shell, still occurs today in deep tropical oceans, but it was much more prevalent some 65 million years ago. Along with closely related forms, it dominated the oceans. The advantage these animals must have had was the ability to achieve buoyancy. Thus, they were no longer confined to the ocean floor. Recently, scientists discovered that as the nautilus grows, it empties liquid from each newly formed chamber of its spiral shell. This makes it light enough to float in water.

Research has been conducted into the puzzle of how the nautilus empties its chambers. A complex tubular organ, the siphuncle, contains blood vessels spirals in the same manner as the shell (b and c). Water from each newly formed chamber makes its way to these blood vessels, but by what means? The blood pressure within these vessels should cause water to *leave* the blood vessels, not enter them!

Electron micrographs have provided a possible explanation. They show that the wall of the siphuncle (d) has a folded tissue lining containing many small grooves (e). If the cells making up this tissue use active transport to carry solutes from the chamber liquid to the grooves, it would establish a concentration gradient causing water to follow passively by osmosis. This process is called *local osmosis* because it depends on the buildup of solutes by active transport in an isolated area, such as the grooves. The cells about these grooves do indeed contain many mitochondria that can supply the energy needed for active transport of the solute into the grooves (f) so that water will follow passively by means of osmosis. After being withdrawn, the water moves from the grooves of the tubular organ to collecting channels that funnel it to the blood vessels.

Although the nautilus can empty its chambers, it never attempts to fill them. This observation is consistent with the hypothesis that local osmosis accounts for how the chambered nautilus empties the chambers of its shell so that they contain only gas and not liquid.

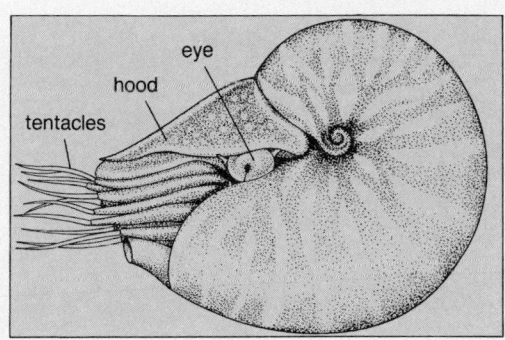

a. Chambered nautilus

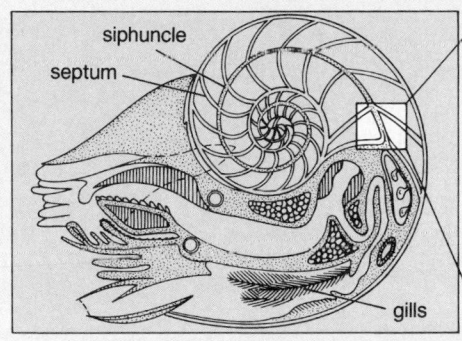

b. Longitudinal section of chambered nautilus

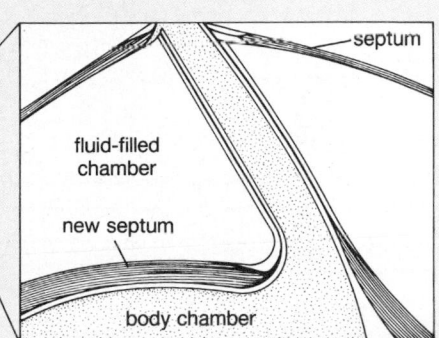

c. Longitudinal section of siphuncle

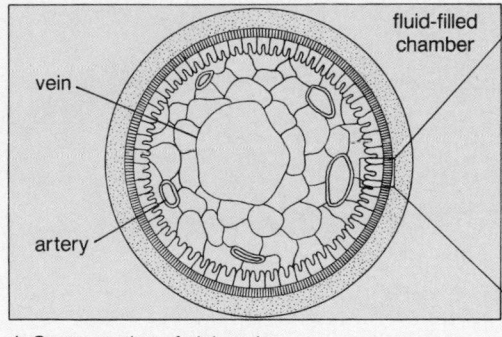

d. Cross section of siphuncle

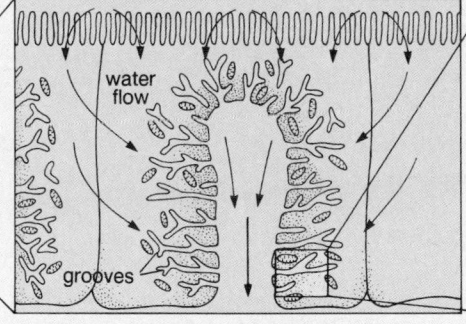

e. Siphuncular epithelium showing grooves

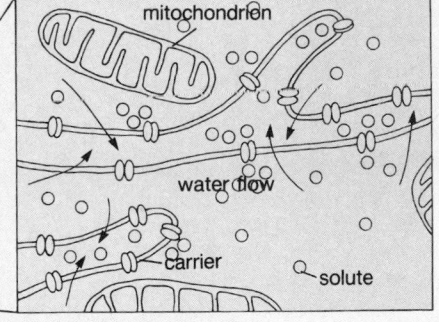

f. Enlargement of grooves

Figure 6.A
Aspects of chambered nautilus anatomy and physiology.

already a greater concentration of amino acid inside the cell than outside. The sodium-potassium pump establishes a gradient that is utilized for the uptake of nutrient molecules; therefore, we can understand why it apparently requires one-third of all our energy expenditure!

During facilitated transport (no energy required), small molecules follow their concentration gradients. During active transport (energy required), small molecules go against their concentration gradient.

Figure 6.12

The sodium-potassium pump. The same carrier protein transports sodium ions (Na+) to the outside of the cell and potassium ions (K+) to the inside of the cell because it undergoes an ATP dependent conformational change. The result is both a concentration gradient and an electrical gradient for these ions across the plasma membrane. Three sodium ions are carried outward for every 2 potassium ions carried inward; therefore, the inside of the cell is negatively charged compared to the outside.

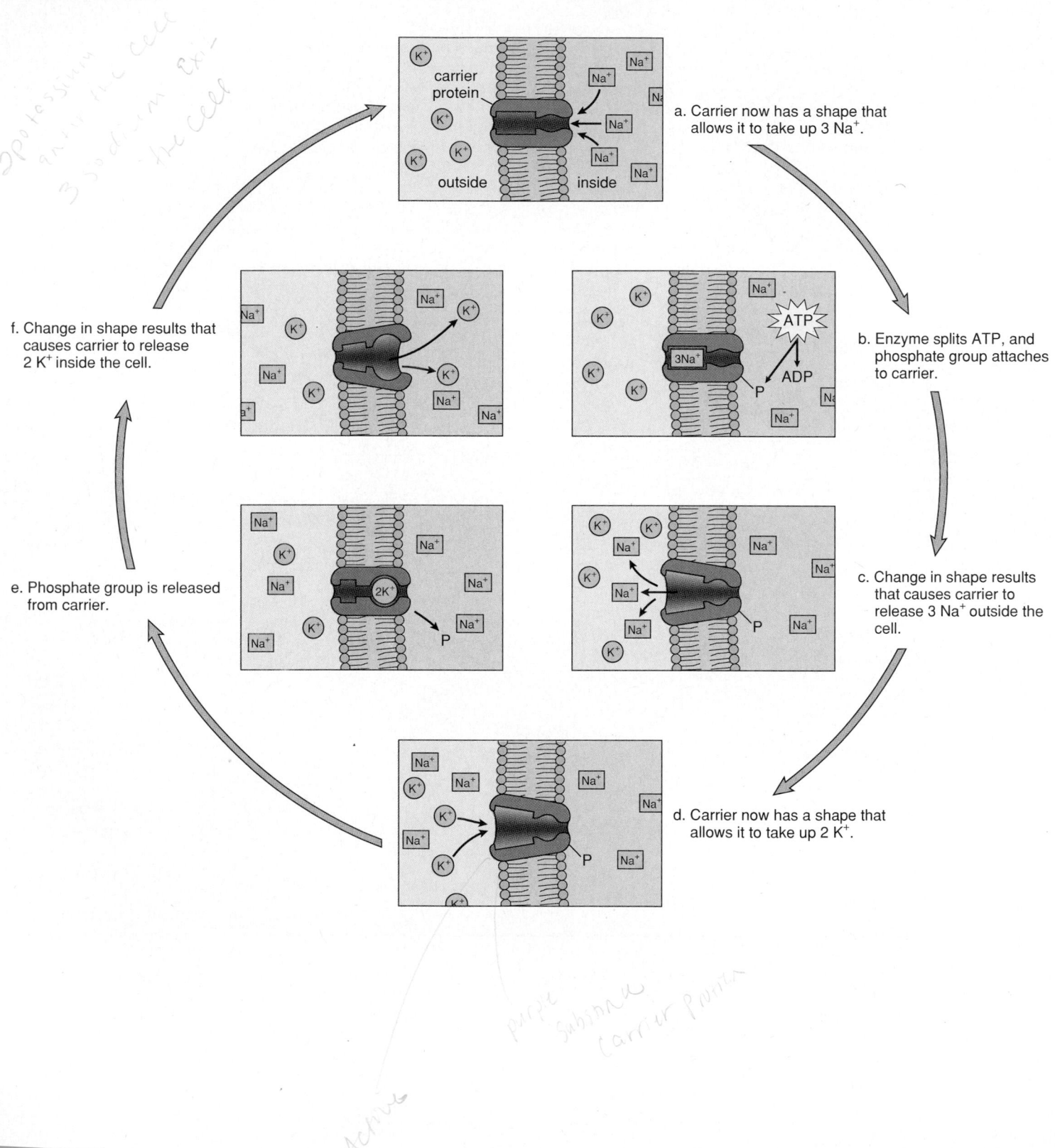

a. Carrier now has a shape that allows it to take up 3 Na+.

b. Enzyme splits ATP, and phosphate group attaches to carrier.

c. Change in shape results that causes carrier to release 3 Na+ outside the cell.

d. Carrier now has a shape that allows it to take up 2 K+.

e. Phosphate group is released from carrier.

f. Change in shape results that causes carrier to release 2 K+ inside the cell.

Endocytosis and Exocytosis

Some substances are too large to be transported by protein carriers; instead they are taken into the cell by **endocytosis:**

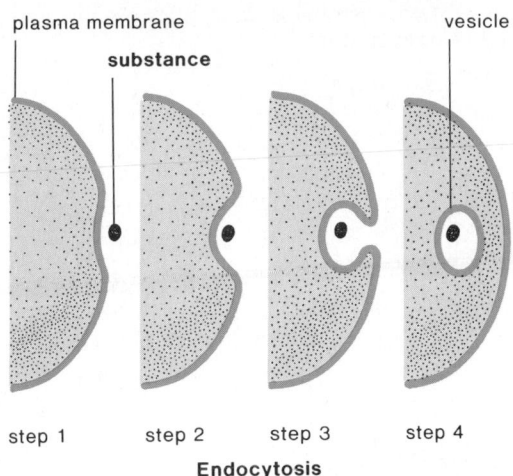

Endocytosis

When the material taken in is quite large, the process is called **phagocytosis** (cell eating), and the result is formation of a vacuole. Phagocytosis is common in amoeboid-type cells such as the cells known as macrophages, which take in bacteria and worn-out red blood cells (fig. 6.13).

When macromolecules are taken in by endocytosis, the process is called **pinocytosis** (cell drinking), and the result is formation of a vesicle. Both phagocytic vacuoles and pinocytic vesicles can fuse with lysosomes, whose enzymes digest their contents.

The opposite of endocytosis is **exocytosis.** During exocytosis, a vesicle fuses with the plasma membrane, discharging its contents:

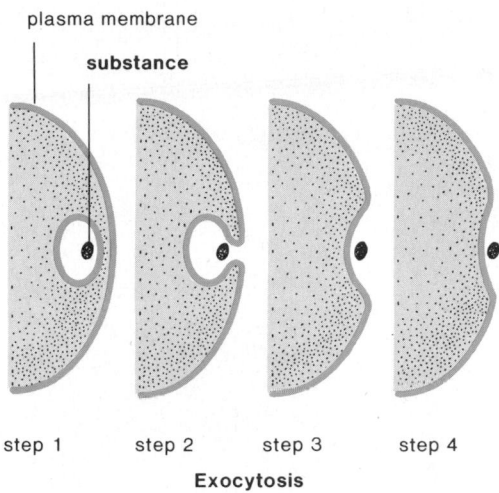

Exocytosis

Figure 6.13
Macrophages are large cells that act as scavengers. These "big eaters" take in (phagocytize) all sorts of debris, including worn out red blood cells, as shown here.

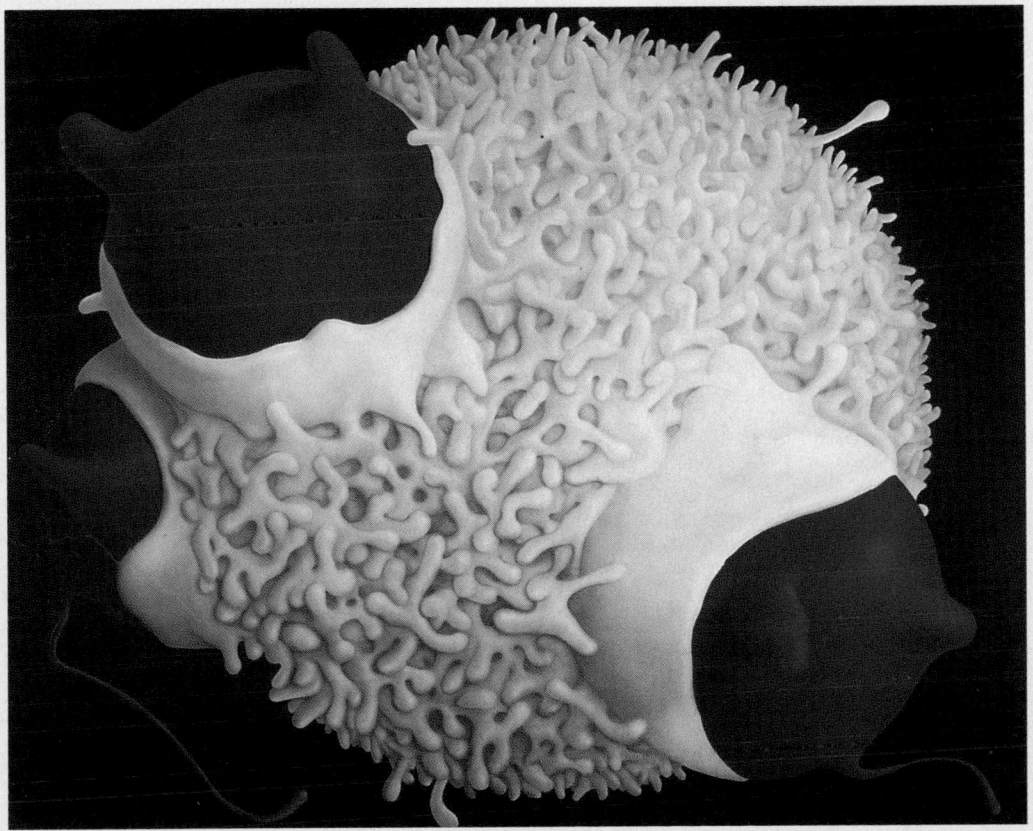

Figure 6.14

Receptor-mediated endocytic cycle takes up membrane when endocytosis occurs and returns it when exocytosis occurs. *a.* (1) The receptors in the coated pits combine only with a specific substance, called a ligand. (2) The vesicle that forms is at first coated with the structural protein clathrin but soon the vesicle loses its coat. (3) Ligands leave the vesicle. (4) When exocytosis occurs membrane is returned to the plasma membrane. *b.* Electron micrographs of a coated pit in the process of forming a vesicle.

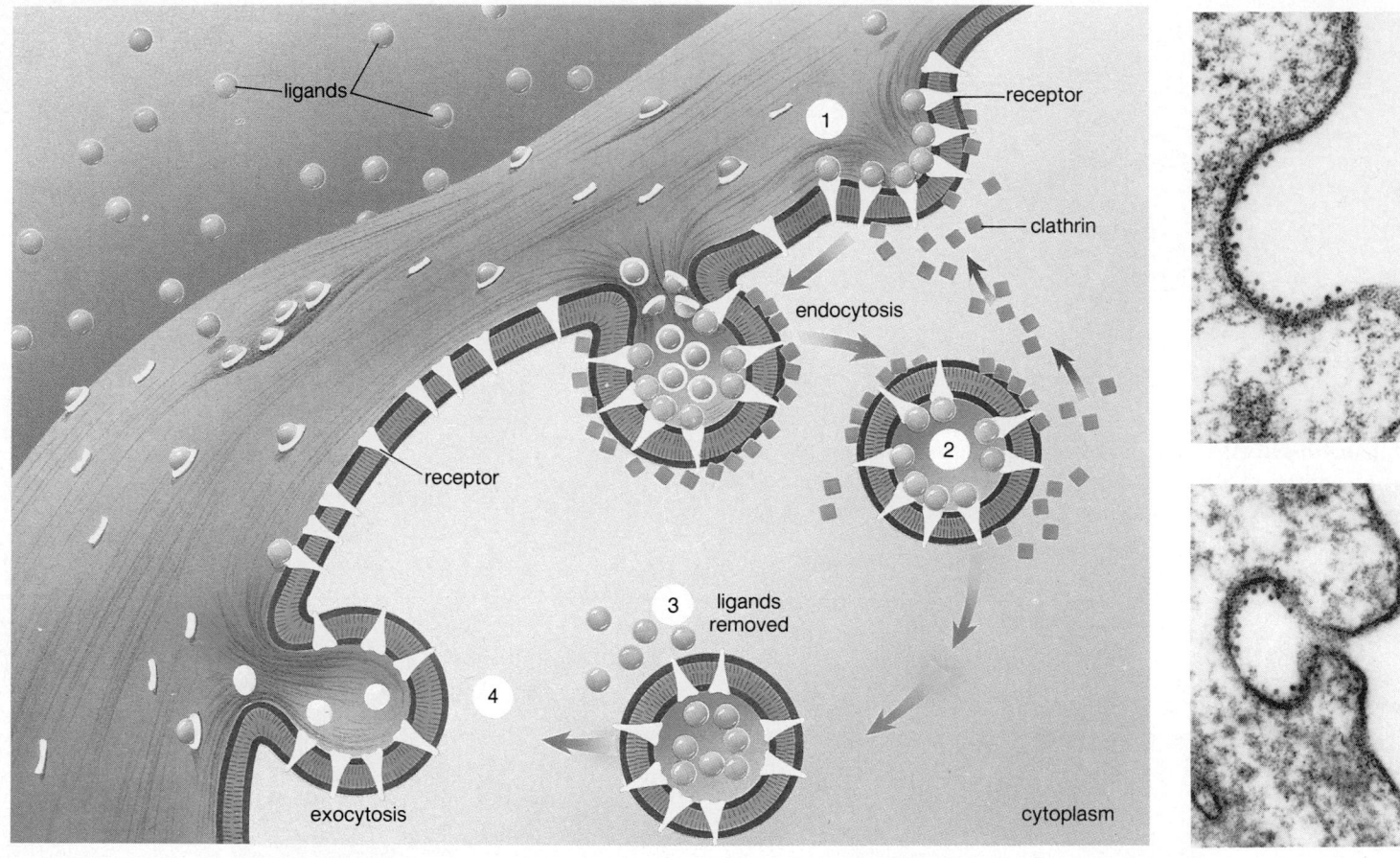

As noted in chapter 5, vesicles formed at the Golgi apparatus secrete cell products in this manner. Exocytosis adds plasma membrane to a cell, and endocytosis takes it away. It can be reasoned that the amount of plasma membrane lost by endocytosis must be replaced by exocytosis unless the cell is growing. It is possible in that case that exocytosis would occur simply for the purpose of adding on more plasma membrane.

Receptor-Mediated Endocytic Cycle

Receptor-mediated endocytosis is a form of pinocytosis that is very specific because it involves the use of plasma membrane receptors. A macromolecule that binds to a receptor is called a ligand (fig. 6.14). The binding of ligands to specific receptor sites causes the receptors to gather at one location before endocytosis occurs. This location is called a coated pit because there is a layer of fibrous protein, called clathrin, on the cytoplasmic side. Clathrin is a protein designed to form lattices around membranous vesicles. When a vesicle forms, it also is coated, but soon it loses its coat. At this point the ligands can directly enter the cell or else end up in lysosomes, which digest them to smaller molecules, which enter the cell. In any case, the receptors return to the plasma membrane and exocytosis occurs.

The importance of receptor-mediated endocytosis is exemplified by the occurrence of a genetic disease. Normally cells take up cholesterol, which is carried in the blood by a lipoprotein called low density lipoprotein (LDL). When cells need more cholesterol for membrane production they produce receptors for LDL. After LDL molecules bind to receptors, receptor-mediated endocytosis occurs. Later, the receptors are returned to the plasma membrane and lysosomes disengage cholesterol from LDL. This whole process goes awry in individuals who lack a gene or inherit a faulty gene for the LDL receptor. Because cholesterol is unable to enter their cells, it builds up and forms plaque on blood vessel walls leading to cardiovascular disease and heart attacks. Children with this genetic disorder have been known to have heart attacks even as early as 6 years old.

Table 6.2
Passage of Molecules into and out of Cells

Name	Direction	Requirements	Examples
Diffusion	Toward lesser concentration	—	Lipid-soluble molecules Water Gases
Transport			
Facilitated	Toward lesser concentration	Carrier	Sugars and amino acids
Active	Toward greater concentration	Carrier plus energy	Sugars, amino acids, and ions
Endocytosis			
Phago-cytosis	Toward inside	Vesicle formation	Cells and subcellular material
Pinocytosis	Toward inside	Vesicle formation	Macromolecules
Receptor-mediated	Toward inside	Vesicle formation	Macromolecules
Exocytosis	Toward outside	Vesicle fuses with plasma membrane	Macromolecules

Endocytosis and exocytosis are other ways that substances can enter and exit cells. Table 6.2 summarizes the various ways that molecules pass into and out of cells.

Modifications of the Cell Surface

The plasma membrane is the outer living boundary of the cell, but many cells have an additional component that is formed outside the membrane. For example, animal cells have an extracellular matrix and plant cells have a cell wall.

Extracellular Matrix of Animal Cells

An extracellular matrix is a meshwork of fibrous proteins and polysaccharides that is laid down by cells and is found in intercellular spaces. The carbohydrate chains of glycolipids and glycoproteins are a part of the extracellular matrix, but there are also various other types of macromolecules that are secreted by but are not part of the cell itself. Some of these molecules (e.g., unattached glycoproteins) form a gel in which other molecules (e.g., collagen) form fibers. Animal cells like to adhere, that is, stick to, a surface, and the extracellular matrix provides a suitable surface for adherence. The extracellular matrix not only keeps the cells stationary, it also helps them keep their usual shape so necessary to performing their usual functions.

Figure 6.15
Plant cell wall. All plant cells have a primary cell wall and some have a secondary cell wall. The cell wall, which lends support to the cell, is freely permeable.

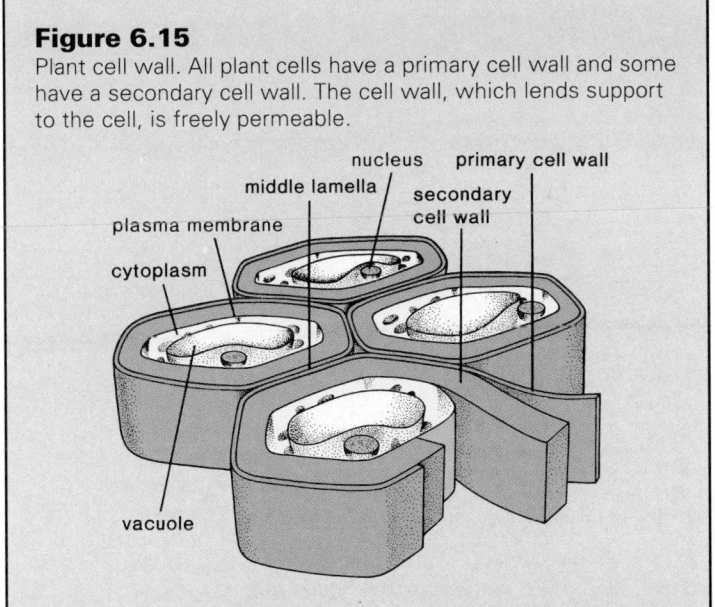

Cell Walls of Plant Cells

In addition to a plasma membrane, plant cells are surrounded by a porous cell wall that varies in thickness, depending on the function of the cell (fig. 6.15). All plant cells have a primary cell wall, whose best-known component is cellulose polymers united into thread-like microfibrils that form fibrils (fig. 4.8). The cellulose fibrils form a framework whose spaces are filled by noncellulose molecules. Among these are found pectic substances, which allow the wall to stretch when the cell is growing, and hemicelluloses, which harden the wall when the cell is mature. A layer of pectic substances, called the middle lamella, also occurs between plant cells. Some cells in woody plants have a secondary wall that forms inside the primary cell wall. The secondary wall has a greater quantity of cellulose fibrils than the primary wall, and these layers are laid down at right angles to one another. Lignin, a substance that adds strength, is another common ingredient of secondary cell walls in woody plants.

The plasma membrane in animals is supported by an extracellular matrix, and in plants it is supported by a cell wall. Both of these lend strength to the cell and do not interfere with the functions of the plasma membrane.

Junctions between Cells

The plasma membrane of cells sometimes interacts with other cells. The junctions that occur between cells are an example of such cellular interaction.

Animal Cells

Three types of junctions are seen between animal cells: spot desmosome, tight junction, and gap junction (fig. 6.16). In a *spot desmosome*, internal cytoplasmic plaques, firmly attached to the

Figure 6.16

Cell-to-cell junctions between animal cells. **a.** In a spot desmosome, filaments pass from one cell to the other at a region where plaques of dense material are reinforced by cytoskeleton filaments. **b.** In a tight junction, adjacent proteins are bonded directly together, preventing any passage of materials in the space between the cells. **c.** In gap junctions, donut-shaped proteins from each cell join to form tiny channels. These allow small molecules to pass from cell to cell. These junctions span the "gap" between cells.

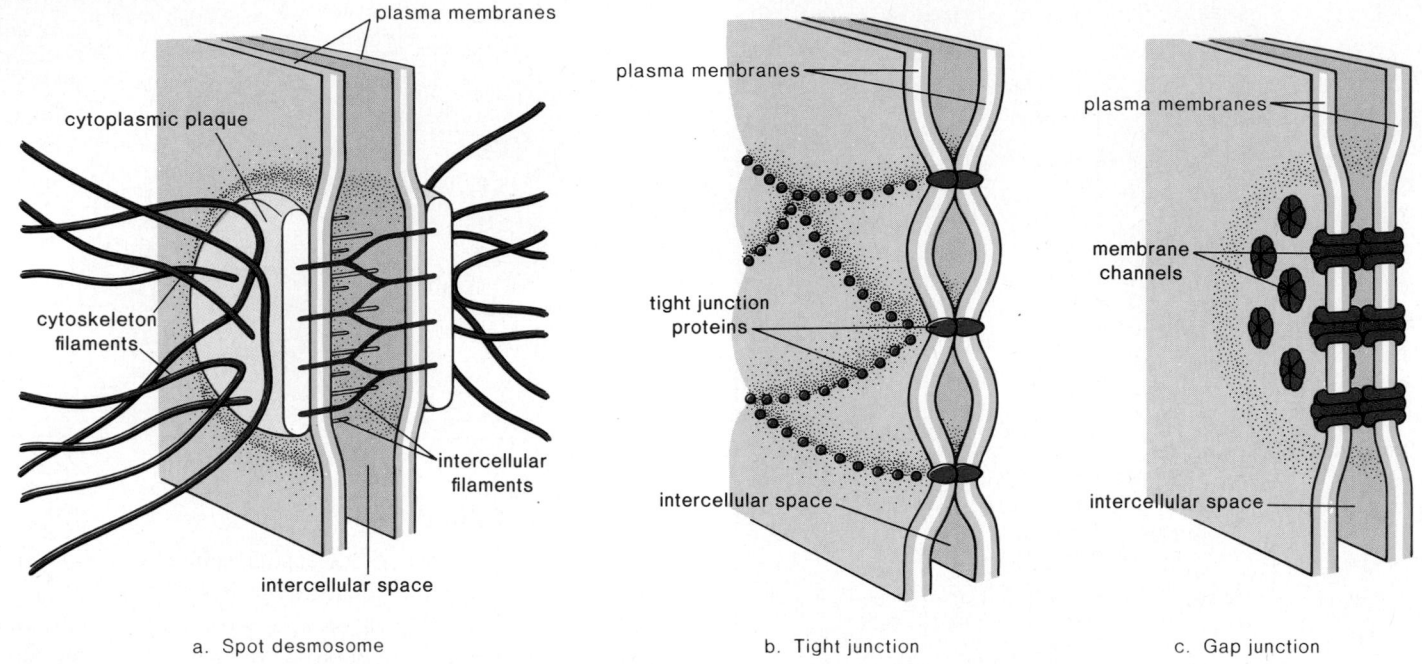

a. Spot desmosome

b. Tight junction

c. Gap junction

Figure 6.17

Cell-to-cell junction between plant cells. At the plasmodesmata, the plasma membrane of adjacent cells passes through the cell wall to form channels, which allow substances to pass from one cell to the other. **a.** Electron micrograph. **b.** Drawing.

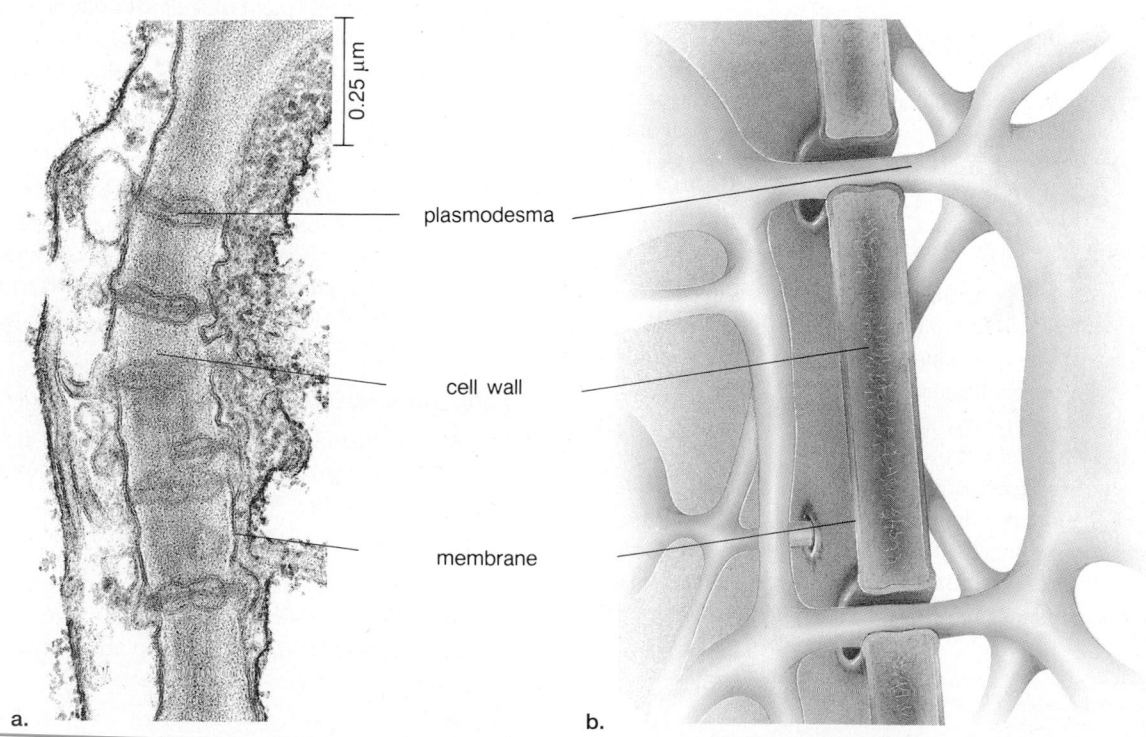

a.

b.

cytoskeleton within each cell, are joined by intercellular filaments. Plasma membranes of adjacent cells is even more closely joined in a *tight junction,* where plasma membrane proteins actually attach to each other, producing a zipperlike fastening. These 2 junctions lend strength and support to the cells. In addition, there is another type of junction that allows cells to communicate. A *gap junction* is formed when 2 identical plasma membrane channels join. The channel of each cell is lined by 6 plasma membrane proteins. A gap junction lends strength to the cells, but it also allows small molecules and ions to pass between them.

All 3 of these junctions are observed between cells that line the lumen (cavity) of the small intestine. These cells pump the products of digestion out of the intestinal lumen and then deposit them in the blood. The presence of tight junctions, in particular, means that the digestion products must enter the cells. Tight junctions also restrict the lateral movement of proteins in the plasma membrane, allowing the surface facing the lumen of the intestine to maintain a different mix of proteins than does the surface adjacent to the underlying blood vessel. These surfaces carry on different functions depending on the proteins they contain.

In animal cells the plasma membrane is joined by desmosomes and tight junctions. Gap junctions allow small molecules to pass from one cell to the other.

Plant Cells

In plant tissues, the cytoplasm of neighboring cells can be connected by numerous narrow channels that pass through the cell wall. These channels are bounded by plasma membrane and contain cytoplasmic strands, called **plasmodesmata,** plus a tightly constricted tubular portion of endoplasmic reticulum. Plasmodesmata allow direct exchange of materials between neighboring plant tissue cells (fig. 6.17).

In plant cells, plasmodesmata are strands of cytoplasm that allow small molecules to pass from one cell to the other.

Summary

1. The fluid-mosaic model of membrane structure developed by Singer and Nicolson was preceded by several other models. Electron micrographs of freeze-fractured membranes support the fluid-mosaic model and not Robertson's unit membrane concept based on the Danielli and Davson sandwich model.

2. There are 2 components of the membrane, lipids and proteins. In the lipid bilayer, phospholipids are arranged with their hydrophilic (polar) heads at the surfaces and their hydrophobic tails in the interior. The lipid bilayer has the consistency of oil but acts as a barrier to the entrance and exit of most biological molecules. Glycolipids and glycoproteins are involved in marking the cell as belonging to a particular individual.

3. The hydrophobic portion of a protein lies in the lipid bilayer of the plasma membrane, and the hydrophilic portion occurs at the surfaces. Proteins act as receptors, carry on enzymatic reactions, join cells together, form channels, or act as carriers to move substances across the membrane.

4. Some molecules (lipid-soluble compounds, water, gases) simply diffuse across the membrane from the area of greater concentration to the area of lesser concentration. No energy is required for diffusion to occur.

5. The diffusion of water across a selectively permeable membrane is called osmosis. Water moves across the membrane into the area of lesser water (greater solute) content. When cells are in an isotonic solution, they neither gain nor lose water; when they are in a hypotonic solution, they gain water; and when they are in a hypertonic solution, they lose water.

6. Other molecules are transported across the membrane by carrier proteins that span the membrane.

7. During facilitated transport, a protein carrier assists the movement of a molecule down its concentration gradient. No energy is required.

8. During active transport, a protein carrier acts as a pump that causes a substance to move against its concentration gradient. The sodium-potassium pump carries Na^+ to the outside of the cell and K^+ to the inside of the cell. Energy in the form of ATP molecules is required for active transport to occur.

9. Larger substances can enter and exit from a membrane by endocytosis and exocytosis. Endocytosis includes both phagocytosis and pinocytosis. Exocytosis involves secretion.

10. The receptor-mediated endocytic cycle makes use of receptor molecules in the plasma membrane.

Once specific substances (e.g., nutrients) bind to their receptors, the entire molecule is drawn into a coated pit, which becomes a coated vesicle. After losing the coat, and after the movement of the nutrients into the cytoplasm, the receptor-containing vesicle fuses with the plasma membrane.

11. Plant cells have a freely permeable cell wall whose main component is cellulose.

12. Junctions between animal cells include spot desmosomes and tight junctions, which help to hold cells together, and gap junctions, which allow passage of small molecules between cells. Plant cells are joined by small channels that span the cell wall and contain plasmodesmata, strands of cytoplasm, which allow materials to pass from one cell to another.

Writing Across the Curriculum

In order to practice writing skills, students should write out the answers to any or all of the study questions and the critical thinking questions. The study questions are sequenced in the same order as the text. Suggested answers to the critical thinking questions are in appendix D.

Study Questions

1. Describe the fluid-mosaic model of membrane structure as well as the models that preceded it. Cite the evidence that either disproves or supports these models.
2. Tell how the phospholipids are arranged in the plasma membrane. What other lipids are present in the membrane, and what functions do they have?
3. Describe how proteins are arranged in the plasma membrane. What are their functions? Describe an experiment that indicates that proteins can move laterally in the membrane.
4. What is diffusion, and what substances can diffuse through a selectively permeable membrane?
5. Describe an experiment that measures osmotic pressure.
6. Tell what happens to an animal cell and a plant cell when placed in isotonic, hypotonic, and hypertonic solutions.
7. Why do substances have to be assisted through the plasma membrane? Contrast movement by facilitated transport with movement by active transport.

Objective Questions

1. Electron micrographs following freeze-fracture of the plasma membrane indicate that
   a. the membrane is a phospholipid bilayer.
   b. some proteins span the membrane.
   c. protein is found only on the surfaces of the membrane.
   d. glycolipids and glycoproteins are antigenic.
2. A phospholipid molecule has a head and 2 tails. The tails are found
   a. at the surfaces of the membrane.
   b. in the interior of the membrane.
   c. both at the surfaces and the interior of the membrane.
   d. spanning the membrane.
3. Energy is required for
   a. active transport.
   b. diffusion.
   c. facilitated transport.
   d. All of these.
4. When a cell is placed in a hypotonic solution,
   a. solute exits the cell to equalize the concentration on both sides of the membrane.
   b. water exits the cell toward the area of lower concentration.
   c. water enters the cell toward the area of higher solute concentration.
   d. solute exits and water enters the cell.
5. A protozoan's contractile vacuole is more likely to be active
   a. in a hypertonic environment.
   b. in an isotonic environment.
   c. in a hypotonic environment.
   d. when endocytosis is occurring.
6. Active transport
   a. requires a protein carrier.
   b. moves a molecule against its concentration gradient.
   c. requires a supply of energy.
   d. All of these.
7. The sodium-potassium pump
   a. helps establish an electrochemical gradient across the membrane.
   b. concentrates sodium on the outside of the membrane.
   c. utilizes a protein carrier and energy.
   d. All of these.
8. Receptor-mediated endocytosis
   a. is no different from phagocytosis.
   b. brings specific nutrients into the cell.
   c. helps to concentrate proteins in vesicles.
   d. All of these.
9. Plant cells
   a. always have a secondary cell wall, and the primary one may disappear.
   b. have channels between cells that allow strands of cytoplasm to pass from cell to cell.
   c. develop turgor pressure when water enters the nucleus.
   d. do not have cell-to-cell junctions like animal cells.
10. Label this diagram of the plasma membrane. Why are the phospholipid "tails" hydrophobic?

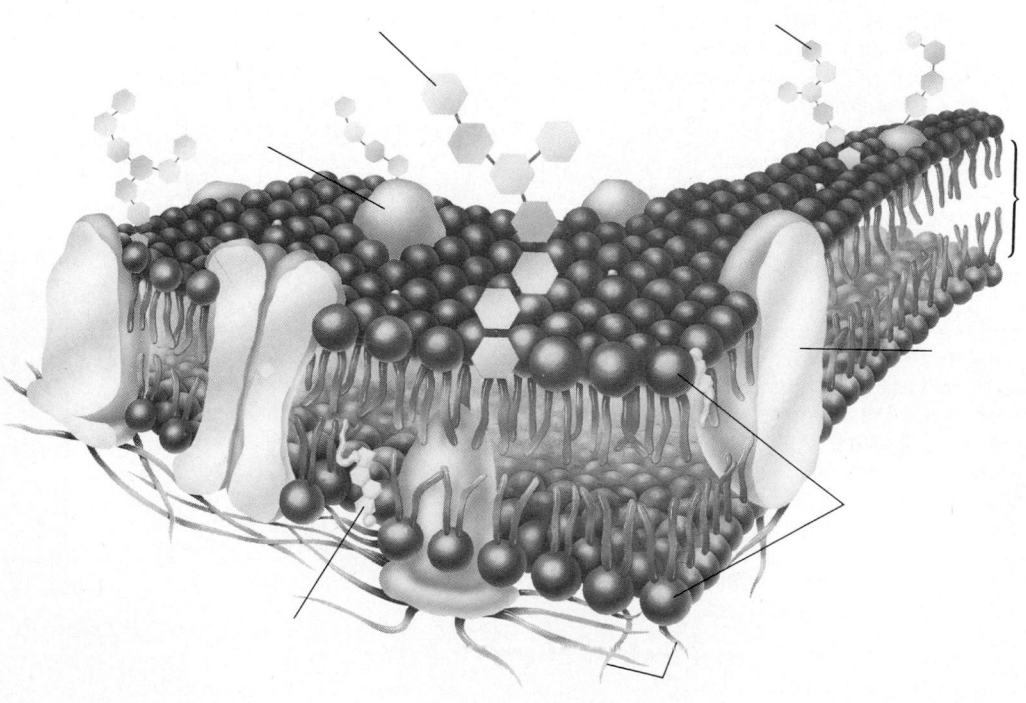

11. Write hypotonic solution or hypertonic solution beside each cell. Justify your conclusions.

a. _____

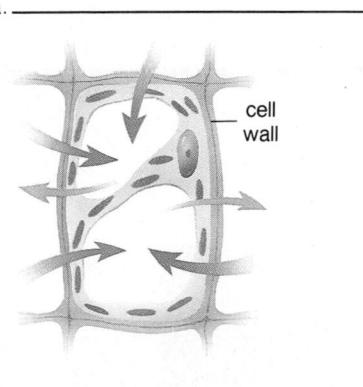

cell wall

b. _____

Concepts and Critical Thinking

1. *All cells have a membrane that separates their contents from the extracellular environment and serves to maintain their integrity.*

In a structural sense, how does the plasma membrane maintain the integrity of the cell? In a functional sense, how does the plasma membrane maintain the integrity of the cell?

2. *To remain alive, cells must have an extracellular environment that is compatible with their continued existence.*

How does the phenomenon of osmosis demonstrate this concept?

3. *Cells of multicellular organisms function as a unit.*

For cells to function as a unit, communication is necessary. How does the plasma membrane help distantly located cells communicate with each other? How does the plasma membrane help adjacently located cells communicate with each other?

Selected Key Terms

fluid-mosaic model (floo'id mo-za'ik mod'el) 86
phospholipid (fos"fo-lip'id) 87
cholesterol (ko-les'ter-ol) 88
selectively permeable (sĕ-lek'tiv-le per'me-ah-b'l) 89
diffusion (di-fu'zhun) 89
osmosis (oz-mo'sis) 89
osmotic pressure (os-mot'ik presh'ur) 91
isotonic solution (i"so-ton'ik sol-u'shun) 91

hypotonic solution (hi"po-ton'ik sol-u'shun) 92
turgor pressure (tur'gor presh'ur) 92
hypertonic solution (hi"per-ton'ik sol-u'shun) 92
carrier (kar'ē-er) 94
active transport (ak'tiv trans'port) 94
sodium-potassium pump (so'de-um po-tas'e-um pump) 94

endocytosis (en"do-si-to'sis) 97
phagocytosis (fag"o-si-to'sis) 97
pinocytosis (pi"no-si-to'sis) 97
exocytosis (ex"o-si-to'sis) 97

7

Cellular Energy

A lubber grasshopper feeds on leaves, acquiring the materials and energy it needs to maintain its organization. The organic molecules in cells are constantly being built up and broken down. Overall, there is a loss of useful energy, which must then be replaced from an outside source.

Your study of this chapter will be complete when you can

1. state 2 energy laws that have important consequences for living things;

2. explain, on the basis of these laws, why living things need an outside source of energy;

3. describe 2 types of metabolic reactions (pathways) that are found in all cells;

4. describe the structure and function of enzymes and the conditions that affect the yield of enzymatic reactions;

5. describe the regulation of enzymatic reactions (pathways) by the process of inhibition. Compare competitive inhibition, noncompetitive inhibition, and feedback inhibition;

6. describe the function of the coenzymes NAD^+ and FAD, and state their role in the production of ATP by the electron transport system in mitochondria. Contrast this electron transport system to the one in chloroplasts;

7. describe the function of the coenzyme $NADP^+$;

8. explain, with the aid of a diagram, how degradative pathways drive synthetic pathways, emphasizing the role of ATP and $NADP^+$. Tell why an outside source of matter and energy should be included in the diagram;

9. describe the structure and function of ATP and how it is produced by chemiosmotic phosphorylation.

W e can readily see that a butterfly's wing (fig. 7.1) is highly organized, but it takes a knowledge of cells to realize that they are also organized. All cells contain the same type of carbon-based molecules, and in eukaryotic cells, some of these molecules make up the organelles. It is the organelles (see fig. 5.4 and fig. 5.5) that show us that a cell is just as organized in its own way as the butterfly's wing. Such organization cannot be maintained without a supply of energy. As an analogy, consider that your room quickly becomes a total mess unless you keep doing work to keep it straight (organized). Similarly, a cell must continually use energy to first form macromolecules from unit molecules and then to arrange the macromolecules into its structural components.

Nature of Energy

Energy is defined as the capacity to do work, that is, to bring about a change. There are several different forms of energy. For example, you use your own mechanical energy to straighten up your room; light energy allows you to see what you are doing; electrical energy makes the motor turn in your vacuum cleaner; and heat energy maintains a comfortable room temperature.

Another form of energy, not as obvious as these, is chemical-bond energy, the energy that is stored in the bonds that hold atoms together. For example, the chemical-bond energy in food provides the energy for your muscles to do mechanical work. Just as chemical-bond energy in food can be transformed into the mechanical energy used by your muscles, other forms of energy

Figure 7.1
The complex organization of living things is dramatized by the pattern of color in a butterfly's wing. But the unseen cells that make up the wing are also highly organized, and the nectar the butterfly obtains from the flower is needed to maintain the organization of both the cells and the wing.

Celluar Energy

can be transformed into another. For example, electrical energy is transformed into light energy within a light bulb, and oil is transformed into electrical energy at a power plant.

A power plant does not create energy, it merely transforms it. Do transformations ever cause energy to be used up? We realize that matter is never created or destroyed. When we take materials and build a vacuum cleaner, we know that even if we throw the vacuum cleaner away, the materials are still not used up. On page 33, it was mentioned that chemical equations must always be balanced—the same number of atoms are present before and after a reaction—because mass cannot be used up. But what about energy? Can it ever be used up?

First Energy Law

As you have probably guessed, not only can energy never be created, it also can never be used up. The same amount of energy has always been present in the universe and will always be present. The first energy law states:

Energy can be transformed from one form into another, but it cannot be created or destroyed.

We have what appears to be a paradox, however, because it often seems as if energy has been "lost" when transformations occur. For example, cars need to be refueled and animals must continue eating. The answer to the paradox is that neither automobiles nor organisms are isolated. It is said that they are a *system* that has *surroundings*. The surroundings supply the gasoline (or the food), and when the gasoline is used to do work, it results in heat that is accepted by the surroundings:

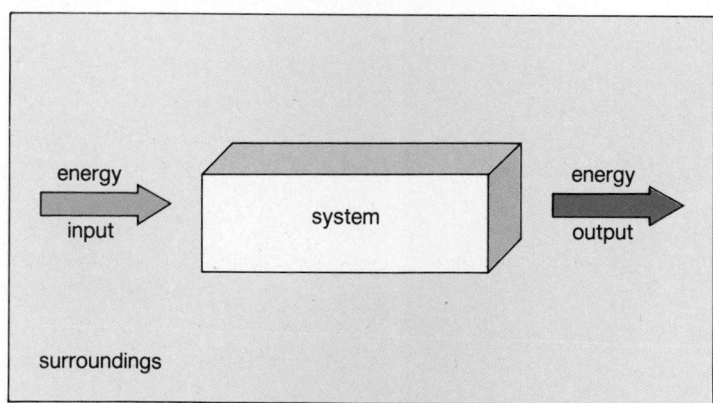

The first energy law can be interpreted to mean that if we put *X* amount of energy into a system (automobile, organism), then eventually *X* amount of energy will leave the system even if work has been done. Energy usually leaves in the form of heat.

Second Energy Law

Although energy was not lost in our examples, something seems to have been lost. To determine what this is, consider that gasoline can be burned to power a car, but the heat given off is no longer able to perform this task. The amount of *useful* energy, then, has changed. Similarly, once food energy is used to power

muscles, this energy is no longer available because it has been converted to heat. The second energy law states:

When one form of energy is transformed into another form, some useful energy is always lost as heat; therefore, energy cannot be recycled.

The second energy law implies that only processes that decrease the amount of useful energy occur naturally, or spontaneously. The word *spontaneous* does not mean that the process has to happen all at once, it simply means that it happens naturally over time. For example, your room tends to become messy, not neat; water tends to flow downhill, not up; and after death, organisms decay and eventually disintegrate. In other words, the amount of disorder (i.e., entropy) is always increasing in the universe. If this is the case, how do we account for an orderly room or the presence of highly organized structures, such as living cells and organisms? Obviously, these systems must have a continual input of energy, and this continual input of energy eventually increases the entropy of the universe.

While we have been drawing an analogy between nonliving systems (a car, your room) and a living system (cell or organism), there is a remarkable difference between the 2 types of systems. Although living things are continually taking in useful energy and putting out heat just like the nonliving system, the work performed in between these 2 events maintains their organization and even allows them to grow. A living organism represents stored energy in the form of chemical compounds:

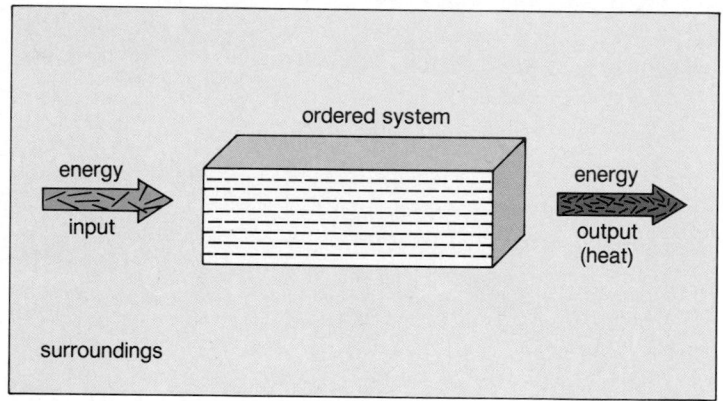

Implication for Living Things

Life does not violate the energy laws discussed earlier, since each living thing merely represents a temporary storage area for useful energy. The useful energy stored in one organism can be used by another to maintain its own organization. Because of this, energy flows through a community of organisms (fig. 7.2). As transformations of energy occur, useful energy is lost to the environment in the form of heat, until finally useful energy is completely used up. Since energy cannot recycle, there is a need for an ultimate source of energy. This source, which continually supplies all living things with energy, is the sun. The entire universe is tending toward disorder, but in the meantime, solar energy is sustaining all living things. Photosynthesizing organisms like plants and algae are able

Figure 7.2

The loss of useful energy in a community of organisms. About 2% of the solar energy reaching the earth is taken up by plants. This is the energy that allows them to make their own food. Herbivores obtain their food by eating plants, and carnivores obtain food by eating other animals. Whenever the energy content of food is used by organisms it is eventually converted to heat. With death and decay all the energy temporarily stored in organisms returns as heat to the atmosphere.

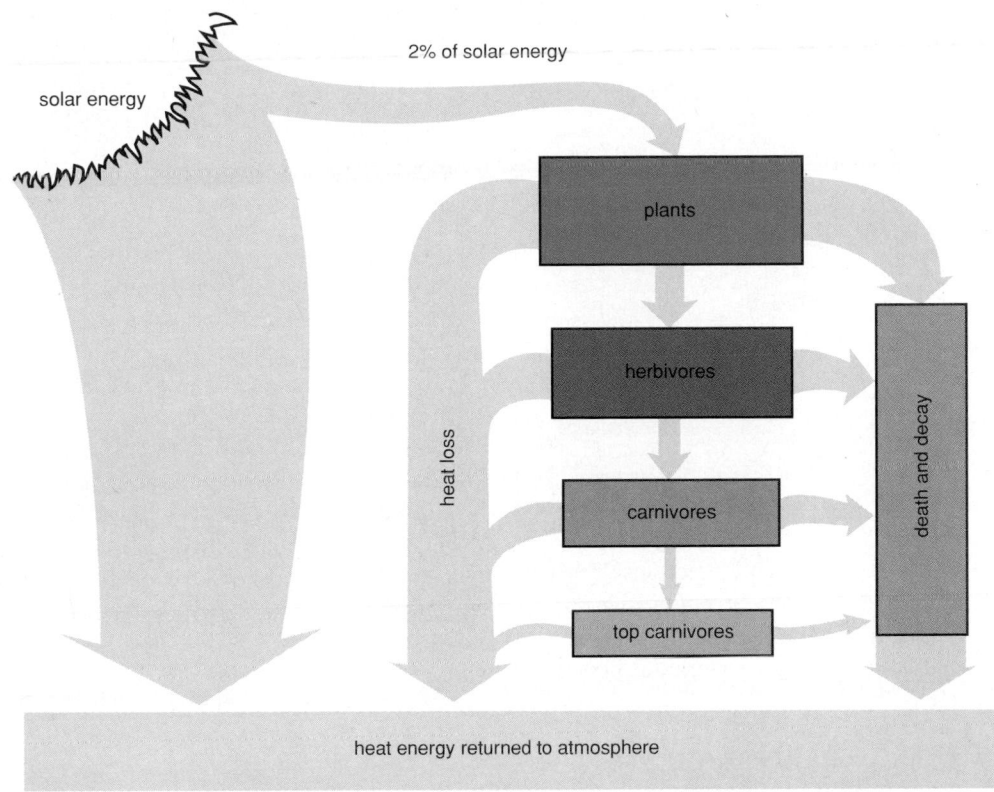

to capture less than 2% of the solar energy that reaches the earth. They use this energy to make their own organic food, which then becomes the food for all other types of living things also. Without photosynthesizing organisms, life as we know it could not exist.

Living things continually lose useful energy to the environment, but a new supply comes to them in the form of organic food produced by photosynthesizing organisms.

Energy-Balance Sheet

Only 42% of solar energy reaches the earth's surface; the rest is absorbed by or reflected into the atmosphere and becomes heat.

Of this usable portion, only about 2% is eventually utilized by plants; the rest becomes heat.

Of this, only 0.1%-1.6% is ever incorporated into plant material; the rest becomes heat.*

Of this, only 20% is eaten by herbivores; a large proportion of the remainder becomes heat.*

Of this, only 30% is ever eaten by carnivores; a large proportion becomes heat.*

Conclusion: Most of the available energy is never utilized by living things.

*Plant and animal remains became the fossil fuels that we burn today to provide energy. Eventually this energy becomes heat.

Cellular Metabolism

The food an organism eats becomes the nutrient molecules of its cells. Within cells, these molecules take part in a vast array of chemical reactions, collectively termed the **metabolism** of the cell. So far, we have stressed that nutrient molecules can be used as a source of energy. For this to occur, they have to be broken down, and in fact, they have to be oxidized. Recall that oxidation is the removal of an electron either alone or in the form of a hydrogen atom. An oxidation reaction is a *degradative* process that releases energy. In general, we can represent it like this:

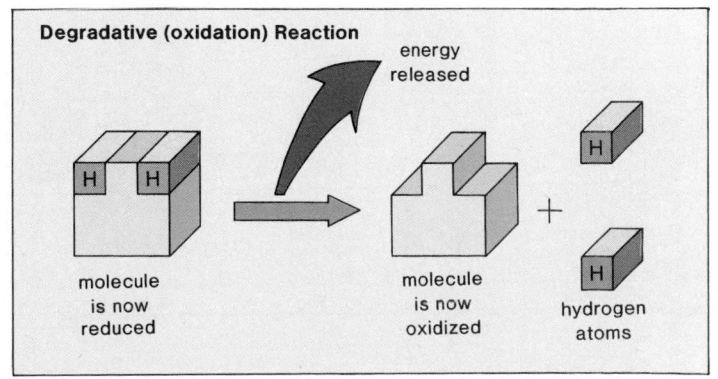

We have already indicated that a cell needs energy to maintain its structure and to grow. In other words, some of the nutrient molecules present in cells are used to form the structure of the cell. For example, amino acids are the unit molecules for proteins, and fatty acids are part of the lipids found in membranes. There are also *synthetic* reactions in which reduction, the gain of electrons, occurs. Reduction can occur when hydrogen atoms are added to a molecule:

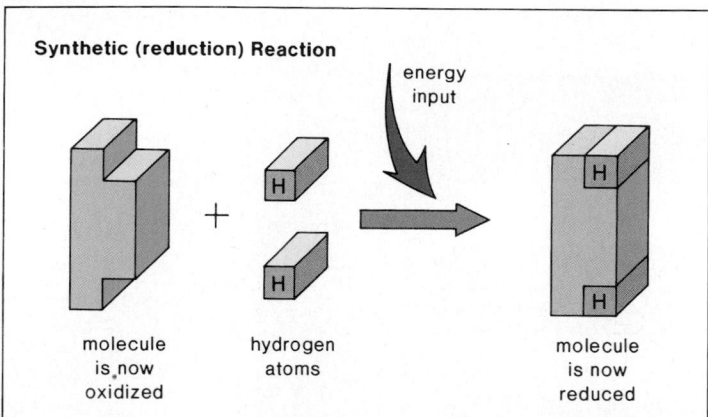

Synthetic (reduction) Reaction

energy input

molecule is now oxidized | hydrogen atoms | molecule is now reduced

Do you suppose it would be possible to take the energy released by degradative reactions and use it for synthetic reactions? Further, would it be possible to use the hydrogen atoms produced in certain degradative reactions for certain synthetic reactions? The answer to these questions is *yes,* and we are now going to explore how these processes occur inside cells.

Pathways

At any one time there are thousands of reactions occurring in cells. Compartmentalization, which we discussed in detail in chapter 5, helps to segregate different types of reactions but more segregation is needed. Within the different organelles, reactions are organized into *metabolic pathways.* Some of these pathways are largely degradative ones that release energy; for example, carbohydrates are broken down in mitochondria (p. 74). Other pathways are largely synthetic ones that require energy input; for example, lipids are synthesized in smooth ER.

In general, metabolic pathways can be represented by this diagram:

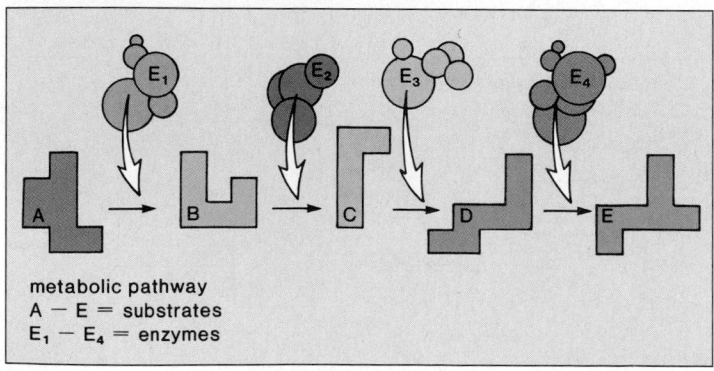

metabolic pathway
A — E = substrates
E_1 — E_4 = enzymes

In this pathway, A-E are products of the previous reaction and reactants of the next reaction, while E_1–E_4 identify different enzymes. A reactant in an enzymatic reaction is a **substrate** for that enzyme. A is the beginning substrate, and E is the end product of the pathway.

This simplified representation of a metabolic pathway does not indicate that the intermediate molecules (B-D) can also be starting points for other pathways. Degradative pathways can even interconnect with synthetic pathways, and in this way degradative pathways sometimes provide not only the energy but also the molecules needed for synthesis of cell components.

Some pathways (reactions) in cells are degradative ones that release energy and some are synthetic ones that require energy input.

Enzymes

As indicated previously, every reaction within a metabolic pathway requires an enzyme. **Enzymes** are organic catalysts, usually globular protein molecules, that speed up chemical reactions without being permanently changed by them. Some RNA molecules, particularly in the nucleus, can be enzymes, and these are called *ribozymes.* Every enzyme is very specific in its action and can speed up only one particular reaction or one type of reaction. Table 7.1 indicates that the name of an enzyme is often formed by adding *ase* to the name of its substrate. Some enzymes are named for the action they perform: for example, a dehydrogenase is an enzyme that removes hydrogen atoms from its substrate.

No reaction can occur in a cell unless its own enzyme is present and active. For example, if enzyme 2 (i.e., E_2) in the preceding diagram is missing or not functioning, the pathway shuts down at B. Since enzymes are so necessary in cells, their mechanism of action has been studied extensively.

Energy of Activation

Molecules frequently do not react with one another unless they are activated in some way. In the laboratory, activation is very often achieved by heating the reaction flask to increase the number of effective collisions between molecules. The energy that must be

Table 7.1
Enzymes Named for Their Substrate

Substrate	Enzyme
Lipid	Lipase
Urea	Urease
Maltose	Maltase
Ribonucleic acid	Ribonuclease
Lactose	Lactase

Figure 7.3

Enzymes speed up the rate of chemical reactions because they lower the required energy of activation. ***a.*** Energy of activation (E_a) that is required for a reaction to occur when an enzyme is not available. ***b.*** Required energy of activation (E_a) is much lower when an enzyme is available.

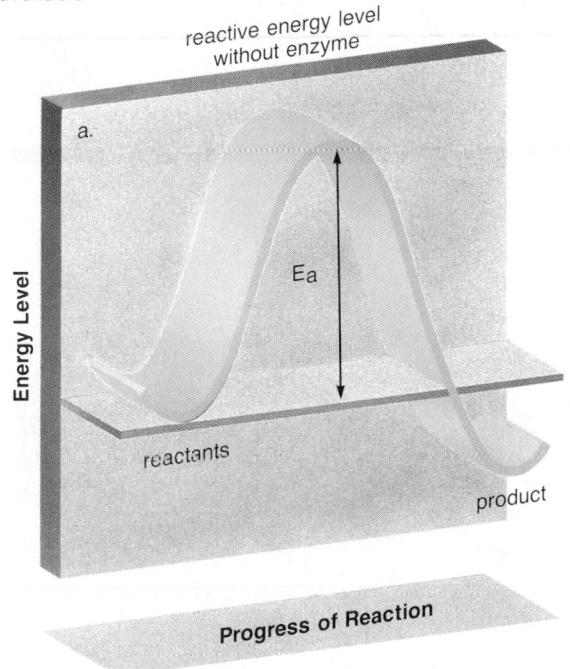

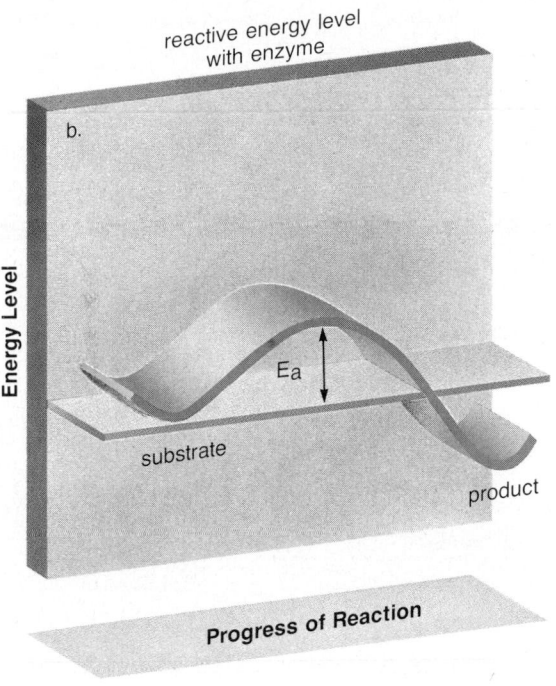

supplied to cause molecules to react with one another is called the *energy of activation* (E_a). Figure 7.3*a* and *b* compares E_a when an enzyme is not present to when an enzyme is present, illustrating that enzymes lower the energy of activation. For example, the hydrolysis of casein (the protein found in milk) requires an energy of activation of 20,600 Kcal/mole[1] in the absence of an enzyme, but only 12,600 Kcal/mole is required if the appropriate enzyme is present. Enzymes lower the energy of activation by forming an enzyme-substrate complex.

Enzyme-Substrate Complex

The following equation, which is pictorially shown in figure 7.4, is often used to indicate that an enzyme forms a complex with its substrate:

$$E + S \rightarrow ES \rightarrow E + P$$

| enzyme | substrate | enzyme-substrate complex | | product |

In most instances only one small part of the enzyme, called the **active site,** complexes with the substrate(s) (fig. 7.4). It is here that the substrate and enzyme fit together, seemingly like a key fits a lock; however, it is now known that the active site undergoes a slight change in shape in order to accommodate the substrate(s) more perfectly. This is called the *induced fit model* because as binding occurs, the enzyme is induced (undergoes a slight alteration) to achieve optimum fit. The change in shape of the active site facilitates the reaction that now occurs. After the reaction has been completed, the product(s) is released, and the active site returns to its original state. Only a small amount of enzyme is actually needed in a cell because enzymes are used repeatedly.

Some enzymes do more than simply complex with their substrate(s); they actually participate in the reaction. Trypsin digests protein by breaking peptide bonds (fig. 7.5). The active site of trypsin contains 3 amino acids with R groups that actually interact with members of the peptide bond first to break the bond and then to introduce the components of water. This illustrates that the formation of the enzyme-substrate complex is very important in speeding up the reaction.

Sometimes it is possible for a particular reactant(s) to produce more than one type of product(s). The presence or absence of an enzyme determines which reaction takes place. For example, if substance A can react to form either product B

---

[1] The Kcal (Kcalorie) is a common way to measure heat, and a mole is the molecular weight of a substance expressed in grams.

Figure 7.4

a. The enzyme and substrates before the reaction begins. *b.* Formation of the enzyme-substrate complex holds the substrates together in such a way that they can react. *c.* Following the reaction the enzyme returns to its prior configuration.

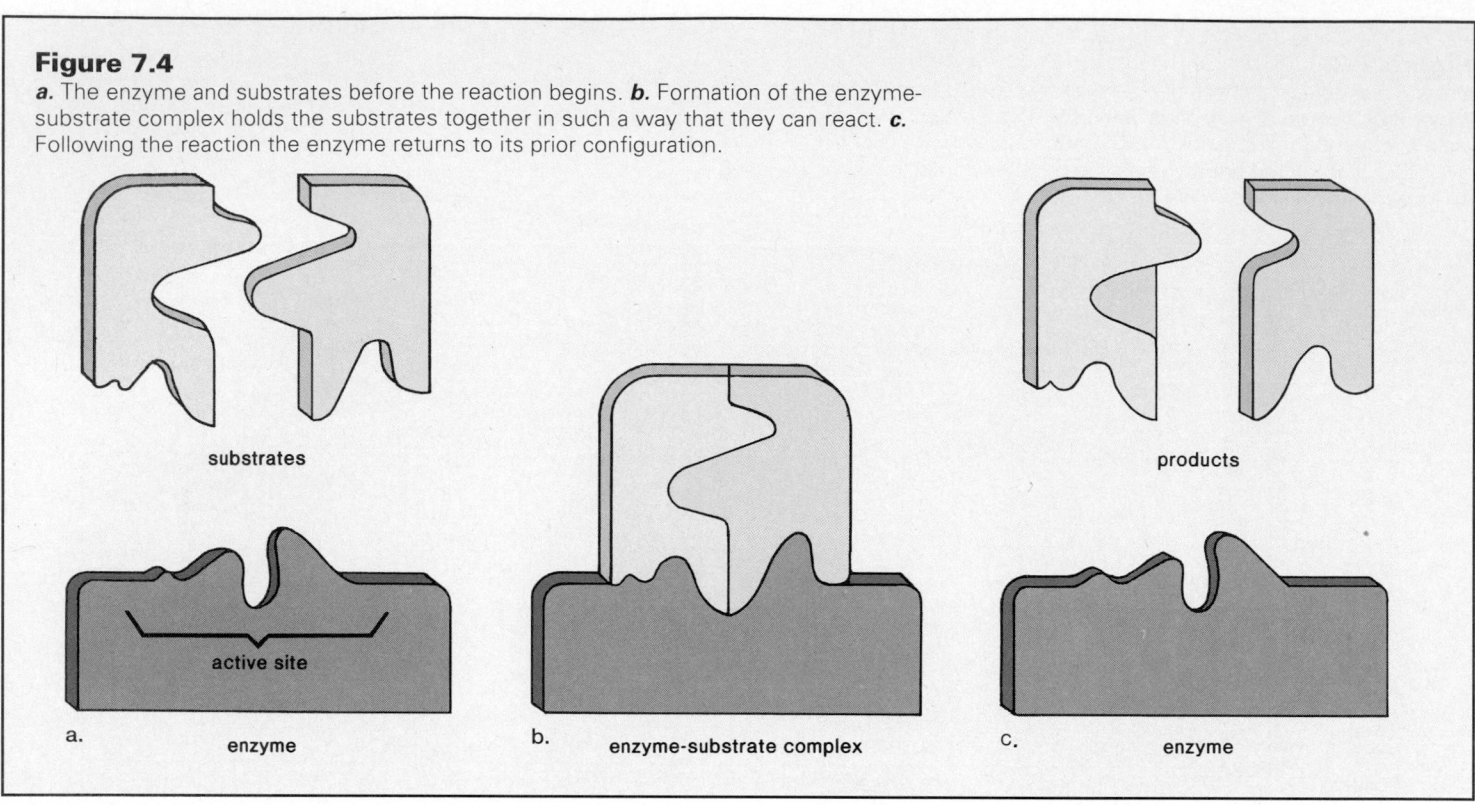

substrates

products

active site

a.   enzyme   b.   enzyme-substrate complex   c.   enzyme

Figure 7.5

Computer-generated model of the functional backbone of the enzyme trypsin. The 3 R groups shown in green are in the active site. They interact with members of the peptide bond in order to hydrolyze it.

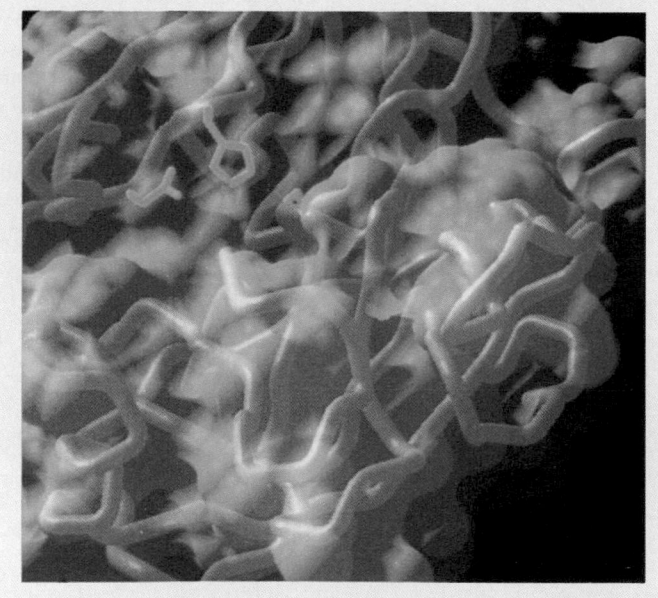

or product C, then the enzyme that is present and active—E_1 or E_2—determines which product is produced:

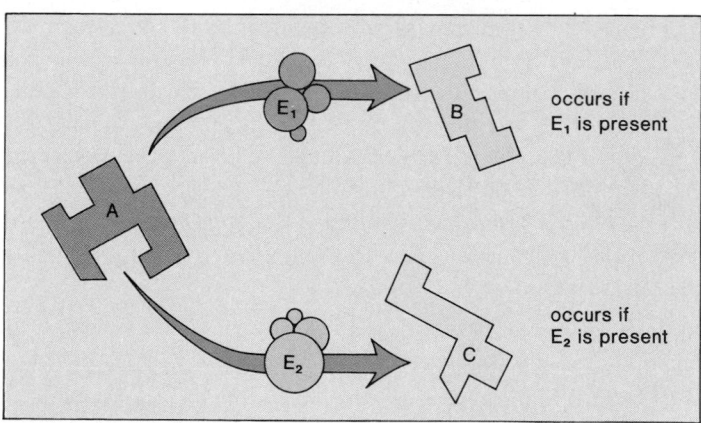

occurs if E_1 is present

occurs if E_2 is present

Enzymes are proteins that speed up chemical reactions by lowering the energy of activation. They do this by forming an enzyme-substrate complex.

Conditions Affecting Enzymatic Reactions

Enzymatic reactions proceed quite rapidly. For example, the breakdown of hydrogen peroxide into water and oxygen can occur 600,000 times a second when the enzyme catalase (an enzyme found in peroxisomes, p. 73) is present. How quickly an enzyme works, however, is affected by certain conditions.

Figure 7.6

Enzyme activity is affected by temperature and pH. **a.** At first, as with most chemical reactions, the rate of an enzymatic reaction doubles with every 10°C rise in temperature. In this graph, which is typical of a mammalian enzyme, the rate is maximum at about 40°C and then it decreases until it stops altogether, indicating that the enzyme is now denatured. **b.** Pepsin, an enzyme found in the stomach, acts best at a pH of about 2, while trypsin, an enzyme found in the small intestine, prefers a pH of about 8. Other pHs do not properly maintain the shape that enables enzymes to bind with their substrate.

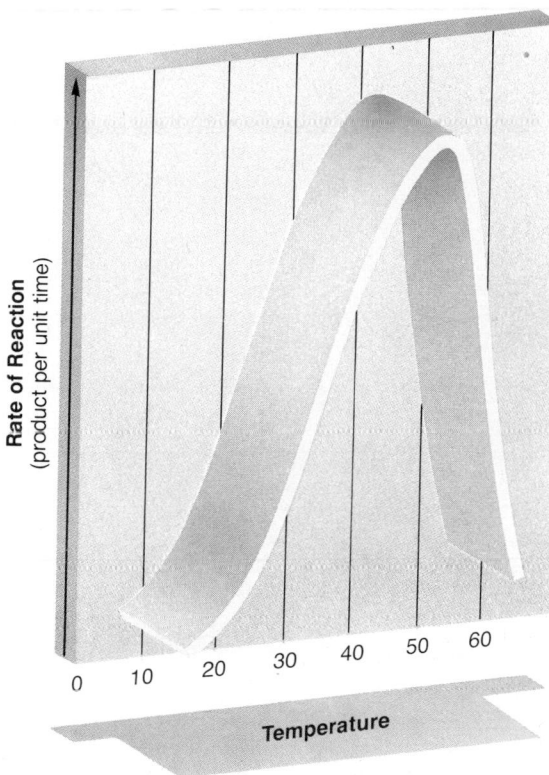

a.

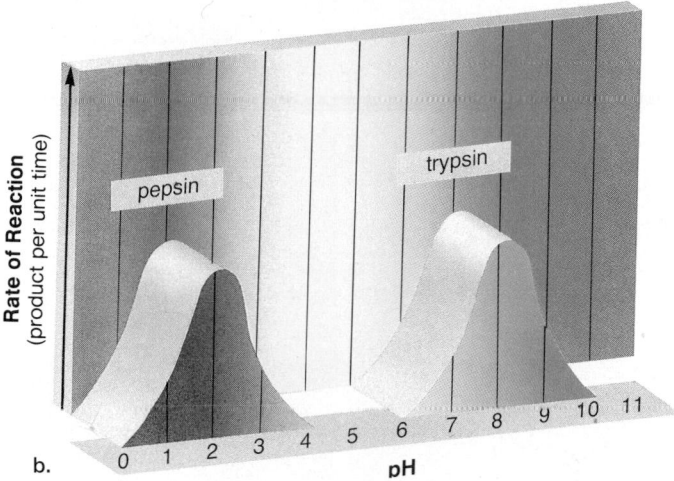

b.

dependent on interactions, such as hydrogen bonding, between R groups (p. 53). A change in pH can alter the ionization of these side chains and disrupt the normal interactions, and denaturation eventually occurs. Again, the enzyme is then unable to combine efficiently with its substrate.

Adequate substrate concentration and optimal temperature and pH all increase the speed of enzyme activity.

Substrate Concentration

Generally, an enzyme's activity increases with greater substrate concentration because there are more collisions between substrate molecules and the enzyme. More substrate molecules fill active sites for a larger proportion of time, and so more product results per unit time. But when the concentration of substrate is so great that the enzyme's active sites are almost continuously filled with substrate, the enzyme's rate of activity cannot increase. Maximum velocity has been reached.

Temperature and pH

A higher temperature generally results in an increase in enzyme activity. As the temperature rises, the movement of enzyme molecules and substrate molecules increases, and there are more effective collisions between them. If the temperature rises beyond a certain point, however, enzyme activity eventually levels out and then declines rapidly because the enzyme is **denatured** (fig. 7.6a). As shown in figure 4.17, an enzyme's shape changes during denaturation. Therefore, an enzyme can no longer bind substrate molecules efficiently.

A change in pH can also affect enzyme activity (fig. 7.6b). Each enzyme has an optimal pH that helps maintain its normal configuration. Recall that the tertiary structure of a protein is

Regulation of Enzyme Activity

While these factors affect enzyme activity, the cell also has built-in mechanisms to control directly both enzyme concentration and activity. First of all, cells are able to regulate whether an enzyme is present at all, but since this type of regulation involves control of protein synthesis, it is not discussed until chapter 17. Cells also have ways to control the level of activity of enzymes that have already been synthesized and are present in the cytoplasm.

Inhibition

Inhibition is a common means by which cells regulate enzyme activity. In *competitive inhibition,* another molecule is so close in shape to the enzyme's substrate that it can compete with the true substrate for the enzyme's active site. This molecule inhibits the reaction because only the binding of the true substrate results in a product. In *noncompetitive inhibition,* a molecule binds to an enzyme but not at the active site. This other binding site is called the *allosteric site* (*allo*—other; *steric*—space or structure). In this instance also, the molecule is an inhibitor because when it binds there is a shift in the 3-dimensional structure of the enzyme that prevents the substrate from binding to the active site. In cells, inhibition is usually reversible; that is, the inhibitor is not permanently bound to the enzyme.

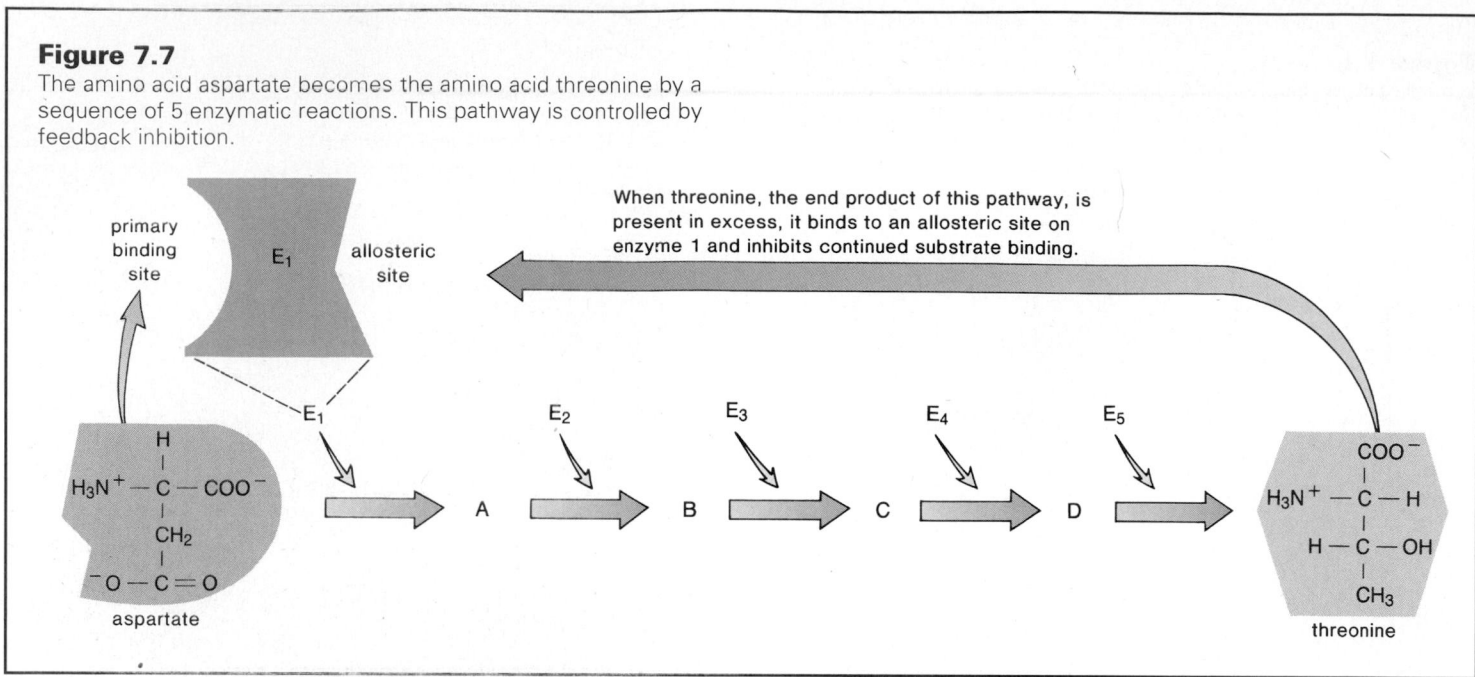

Figure 7.7

The amino acid aspartate becomes the amino acid threonine by a sequence of 5 enzymatic reactions. This pathway is controlled by feedback inhibition.

During competitive inhibition, an inhibitor binds to the active site; during noncompetitive inhibition, an inhibitor binds to an allosteric site.

Irreversible inhibition of enzymes also occurs, due to the presence of a poison. For example, penicillin causes the death of bacteria due to irreversible inhibition of an enzyme needed to form the bacterial cell wall. In humans, hydrogen cyanide irreversibly binds to a very important enzyme (cytochrome oxidase) present in all cells, and this accounts for its lethal effect on the human body.

Feedback Inhibition

The activity of almost every enzyme in a cell is regulated by feedback inhibition. **Feedback inhibition** is an example of a common biological control mechanism called negative feedback. Just as excessively high temperature can cause a furnace to shut off, so a product produced by an enzyme can inhibit its activity. When the product is in abundance, it binds competitively with its enzyme's active site (fig. 7.4c); as the product is used up, inhibition is reduced and more product can be produced. In this way, the concentration of the product is always kept within a certain range.

Most enzymatic *pathways* are also regulated by feedback inhibition, but in these cases the end product of the pathway binds at an allosteric site on the *first* enzyme of the pathway (fig. 7.7). This binding shuts down the pathway, and no more product is produced.

In feedback inhibition, the inhibitor is either the product of an enzymatic reaction that binds to the active site or the end product of a metabolic pathway that binds to an allosteric site on the first enzyme of the pathway.

Coenzymes

Many enzymes require a nonprotein *cofactor* to assist them in carrying out their function. Some cofactors are ions; for example, magnesium (Mg^{++}), potassium (K^+), and calcium (Ca^{++}) are often involved in enzymatic reactions. Some other cofactors, called **coenzymes,** are organic molecules that bind to enzymes and serve as carriers for chemical groups or electrons.

Usually coenzymes participate directly in the reaction. For example, the coenzyme tetrahydrofolate can carry a methyl group (CH_3). Sometimes it works in conjunction with the enzyme thymidylate synthase and donates a methyl group to form thymine, one of the bases found in DNA. In humans, this coenzyme is not present unless the diet includes the vitamin folic acid. **Vitamins** are relatively small organic molecules that are required in trace amounts in our diet and in the diet of other animals for synthesis of coenzymes that affect health and physical fitness.

The presence of other coenzymes in cells is similarly dependent on our intake of vitamins. For example,

Vitamin	Coenzyme
Niacin	NAD^+
B_2 (riboflavin)	FAD
B_1 (thiamine)	Thiamine pyrophosphate
Pantothenic acid	Coenzyme A (CoA)
B_{12}	Cobamide coenzymes

A deficiency of any one of these vitamins results in a lack of the coenzyme listed and therefore a lack of certain enzymatic actions. In humans, this eventually results in vitamin-deficiency symptoms. For example, niacin deficiency results in a skin disease called pellagra, and riboflavin deficiency results in cracks at the corners of the mouth.

Figure 7.8

Secondary structure of an enzyme that degrades alcohol, called alcohol dehydrogenase. NAD⁺ is a coenzyme for this enzyme, and therefore the enzyme has a binding site not only for its substrate (alcohol) but also for NAD⁺. The NAD⁺ binding site is in green and yellow; the substrate-binding site is in blue. A bound NAD⁺, in purple, lies close to the substrate-biding site.

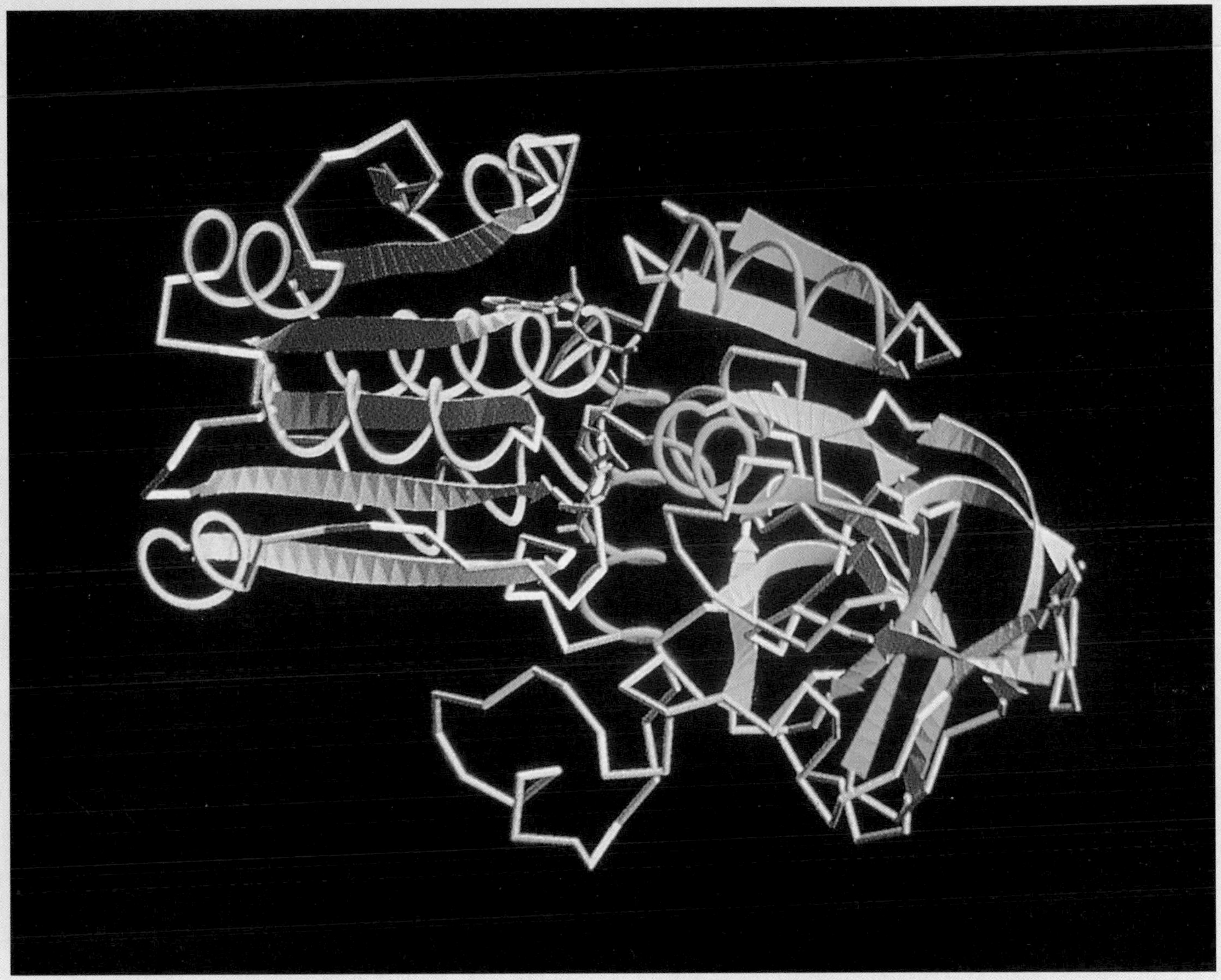

Coenzymes are nonprotein molecules that assist enzymes in performing their reactions. Coenzyme synthesis requires an intake of vitamins.

NAD⁺ and FAD

NAD⁺ (nicotinamide adenine dinucleotide) is a coenzyme that carries electrons and quite often works in conjunction with enzymes called dehydrogenases (fig. 7.8). An enzyme of this type removes 2 hydrogen atoms ($2e^- + 2H^+$) from its substrate; both electrons but only one hydrogen ion are passed to NAD⁺—NAD⁺ becomes NADH:

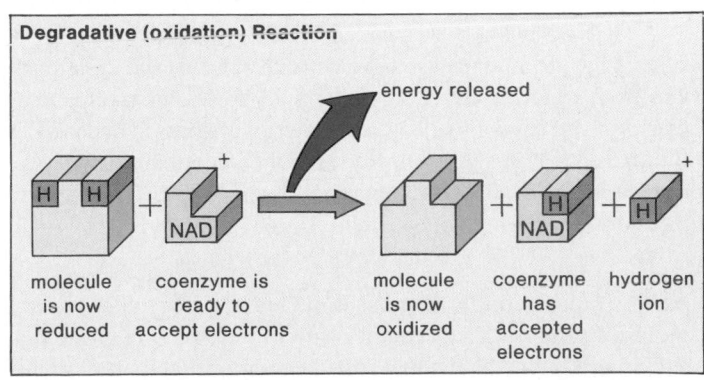

Degradative (oxidation) Reaction

energy released

molecule is now reduced + coenzyme is ready to accept electrons → molecule is now oxidized + coenzyme has accepted electrons + hydrogen ion

Figure 7.9

Electron transport system, in which electrons pass from carrier to carrier. With each oxidation reaction some energy is released and a portion of it is trapped for the purpose of producing ATP molecules.

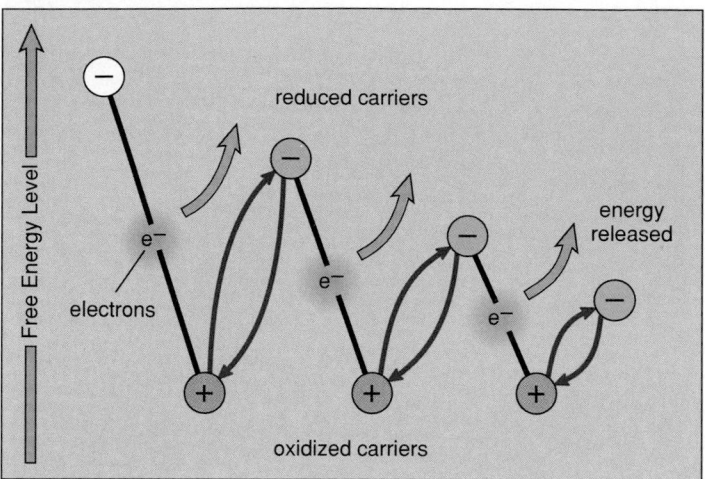

This is a degradative (oxidation) reaction and energy is released; but by comparing this equation to the similar one on page 107, you can see that much less energy has been lost because the red arrow is smaller. The energy has been transferred to NAD⁺ in the form of high-energy electrons. Another coenzyme called **FAD** (flavin adenine dinucleotide) also carries high-energy electrons, but it accepts both hydrogen ions (to become FADH$_2$).

Both NAD⁺ and FAD are involved in **cellular respiration,** a metabolic pathway by which carbohydrates are oxidized in a step-by-step manner to carbon dioxide and water. At various junctions along the way, these coenzymes accept electrons from certain substrates and carry them to an **electron transport system** consisting of membrane-bounded carriers that pass electrons from one carrier to another. High energy electrons are delivered to the system and low energy electrons leave it. Every time an electron is transferred, oxidation occurs and energy is released; this energy is ultimately used to produce ATP molecules (fig. 7.9).

NAD⁺ and FAD are electron carriers that take electrons to the electron transport system in mitochondria. This system uses the energy of oxidation to produce ATP molecules.

Chloroplasts also have an electron transport system for producing ATP. In this case the electrons that fuel the system are taken from water. Solar energy energizes these electrons, and as they pass from carrier to carrier, the energy is released and ATP is built up. In the end, the coenzyme NADP⁺ (nicotinamide adenine dinucleotide phosphate) accepts the electrons and becomes reduced to NADPH.

NADP⁺

NADP⁺ (nicotinamide adenine dinucleotide phosphate) has a structure similar to NAD⁺, but it contains a phosphate group that is lacking in NAD⁺. This may signal enzymes that its function is slightly different. It does carry electrons and a hydrogen ion just like NAD⁺, but its electrons are used to bring about reduction synthesis:

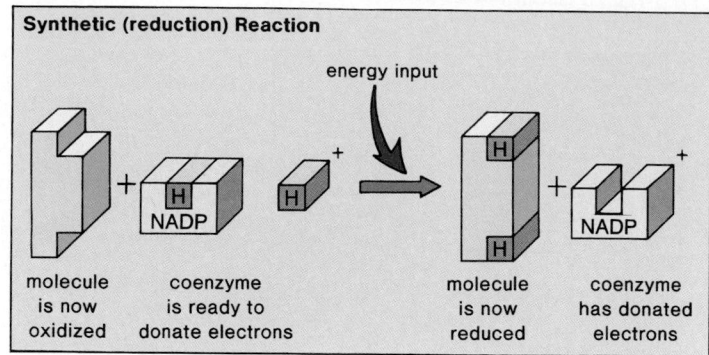

Synthetic (reduction) Reaction

During **photosynthesis** within chloroplasts, carbon dioxide is reduced to a carbohydrate. NADPH supplies the necessary electrons and ATP supplies the necessary energy to bring about this reduction.

NADPH also functions outside of chloroplasts. Within the cytosol, plant and animal cells use a special degradative pathway to produce NADPH. This NADPH is used during the synthesis of many necessary molecules in cells. Energy too is required, but for this purpose, ATP produced by mitochondria is utilized.

NADP⁺ is an electron carrier that participates along with ATP in synthetic reactions.

Metabolism Revisited

We stated it is possible to harness the energy and hydrogens provided by degradative reactions for synthetic reactions. This is possible *only* because ATP carries energy and NADPH carries electrons between degradative pathways and synthetic pathways (fig. 7.10).

Degradative pathways do run synthetic pathways as long as the cell receives an outside source of energy and matter. The energy laws discussed earlier in this chapter tell us that life is only possible because organisms (cells) are able to *store* temporarily some of the energy that flows through them.

Of all the organisms, only photosynthesizers such as plants are able to make organic food by utilizing *solar* energy. The green arrows in figure 7.10 illustrate the process of photosynthesis. In chloroplasts, solar energy is used to produce the ATP and the NADPH that are used to reduce carbon dioxide to carbohydrates. These and the other macromolecules produced by plants supply plants and eventually all other living things with a source of organic food.

Plants and animals use organic food as both an energy source and a source of unit molecules (blue arrows). When used as an energy source, carbohydrates are degraded (oxidized) to carbon dioxide and water, which are excreted. Heat is given off in the process, and some of the energy is used to produce ATP. Degradation also produces a supply of NADPH. Again ATP is a carrier for energy and NADPH is a carrier for electrons utilized

Figure 7.10

The degradative part of photosynthesis drives the synthetic part, and the degradative part of general metabolism drives the synthetic part. Degradation (oxidation) provides electrons and energy that are carried to synthetic reactions by NADPH and ATP, respectively.

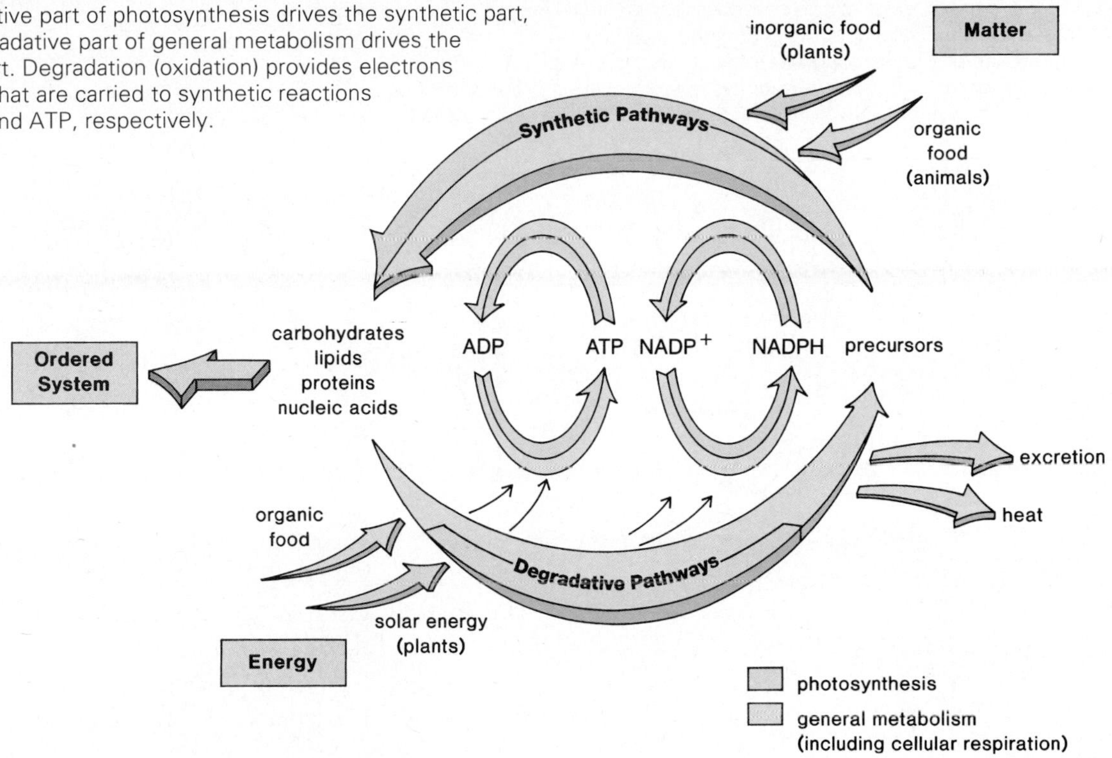

Matter

inorganic food
(plants)

organic
food
(animals)

Synthetic Pathways

carbohydrates
lipids
proteins
nucleic acids

Ordered
System

ADP   ATP   NADP⁺   NADPH   precursors

excretion

heat

organic
food

Degradative Pathways

solar energy
(plants)

Energy

☐ photosynthesis

☐ general metabolism
   (including cellular respiration)

Figure 7.11

ATP reaction. ATP, the common energy carrier in cells, is a high-energy molecule as represented by the wavy lines between the phosphate groups Ⓟ. When the last phosphate group is removed, the energy released is used by the cell for various purposes. ATP can be reformed if enough energy is supplied to rejoin ADP and Ⓟ.

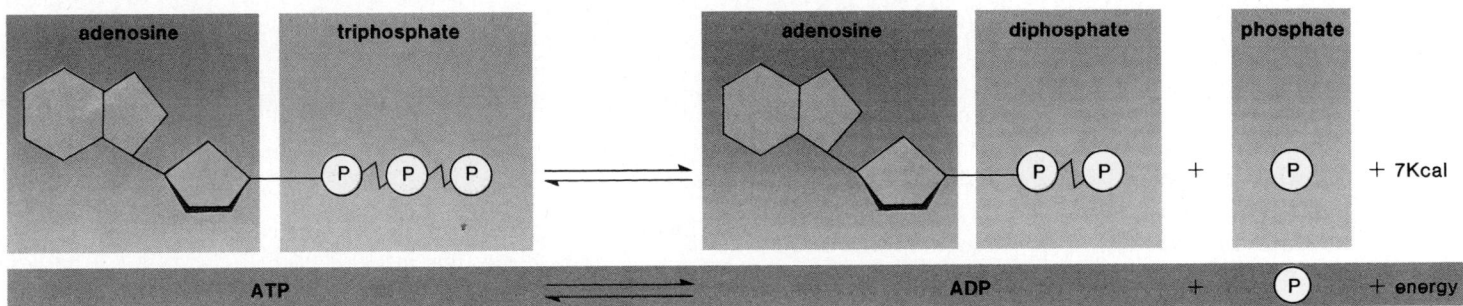

| adenosine | triphosphate | | adenosine | diphosphate | phosphate |

ATP ⇌ ADP + Ⓟ + energy

+ 7Kcal

in synthetic (reduction) reactions. In this way, the flow of energy through the organism (cells) is utilized for growth and maintenance of the organism.

The role of enzymes in metabolism must not be forgotten. They make "warm chemistry" possible. Degradative reactions occur spontaneously (naturally) but only if sufficient energy of activation is supplied. Enzymes lower the energy of activation, making it possible for these reactions to occur very quickly at body temperature. Since the enzymes involved in synthesis are not the same as those used for degradation, however, the cell needs the coenzyme NADP⁺ to carry electrons and ATP to carry energy from degradative to synthetic reactions.

ATP

ATP (adenosine triphosphate) is the common energy currency of cells; when cells require energy, they "spend" ATP (fig. 7.11). You may think that this causes our bodies to produce a lot of ATP, and it

Figure 7.12

The coupling of ATP buildup and breakdown to other reactions. Energy-releasing reactions are shown as gears with arrows going down; energy-requiring reactions are shown as gears with arrows going up. Degradation of food is coupled to the production of ATP (*far left*); breakdown of ATP releases energy to drive energy-requiring reactions (*right*). Very often this involves the transfer of a phosphate group from ATP to an intermediate molecule in the reaction (*lower right*).

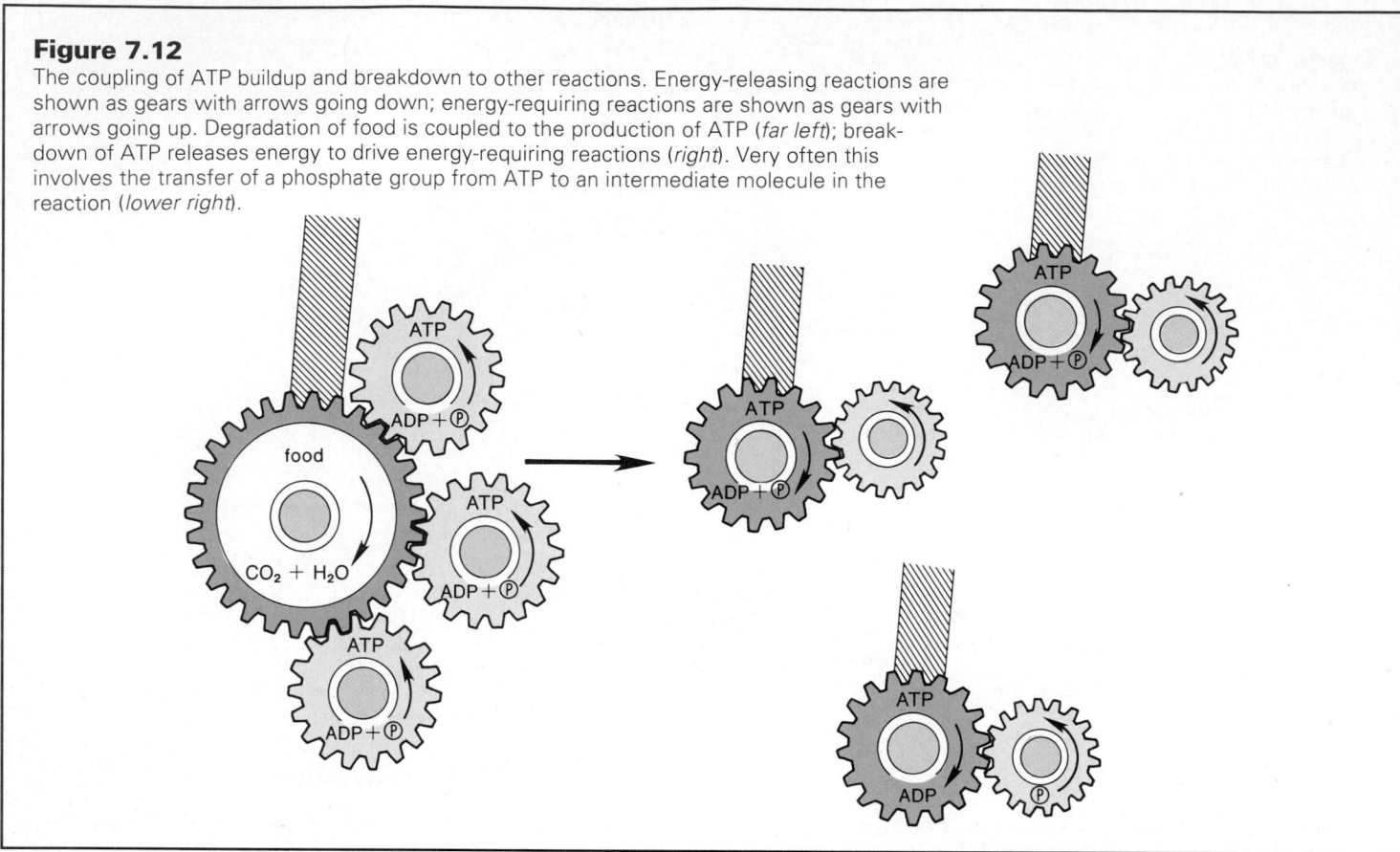

does; the average male needs to produce about 8 kg (over 17 lb) of ATP an hour! Obviously, it would be impossible to carry around even a few hours supply of ATP, and indeed the body has on hand only about 50 g (1.8 oz) at any time. The answer to the paradox is that ATP is constantly being recycled from **ADP** (adenosine diphosphate) and $\textcircled{P}$. The cell's entire supply of ATP is recycled about once each minute.

Structure of ATP

ATP is a nucleotide composed of the base adenine and the sugar ribose (together called adenosine) and 3 phosphate groups. ATP is called a "high-energy" compound because its phosphate groups are easily removed. Under cellular conditions, the amount of energy released is about 7.3 Kcal per mole. The wavy line between the phosphate bonds indicates the high potential energy of these bonds.[2]

　　The use of ATP as a carrier of energy derived from degradation has some advantages:

1.　The energy of degradation can be captured in a step-by-step manner that prevents waste (see fig. 7.10).
2.　It provides a common energy currency that can be used in many different types of reactions.
3.　When ATP becomes ADP + $\textcircled{P}$ the amount of energy released is just about enough for the biological purposes mentioned in the following section and so little energy is wasted.

[2]The high-energy content of ATP does not reside simply in the phosphate bond; rather it comes from the complex interaction of the atoms within the molecule. The use of the wavy line is for convenience only.

　　(1) and (3) are only possible because most reactions (e.g., ATP breakdown and synthetic buildup) are *coupled;* that is, they take place at the same time and at the same place and usually utilize the same enzyme (fig. 7.12). During the coupling process, a phosphate group is often transferred to an intermediate compound. For example, during active transport, a phosphate group is transferred to a membrane-bounded protein (see fig. 6.12).

Function of ATP

Recall that at various times we have mentioned at least 3 uses for ATP.

Chemical work. Supplies the energy needed to synthesize macromolecules that make up the cell.
Transport work. Supplies the energy needed to pump substances across the plasma membrane.
Mechanical work. Supplies the energy needed to cause muscles to contract, cilia and flagella to beat, chromosomes to move, and so forth.

　　All organisms, both prokaryotic and eukaryotic, use ATP. This illustrates the chemical unity of all living things.

ATP is a carrier of energy in cells. It is the common energy currency because it supplies energy for many different types of reactions.

Formation of ATP

Cells have 2 ways to make ATP. The first, called **substrate-level phosphorylation,** occurs in the cytosol. During this process, a high-energy substrate transfers a phosphate group to ADP, forming ATP:

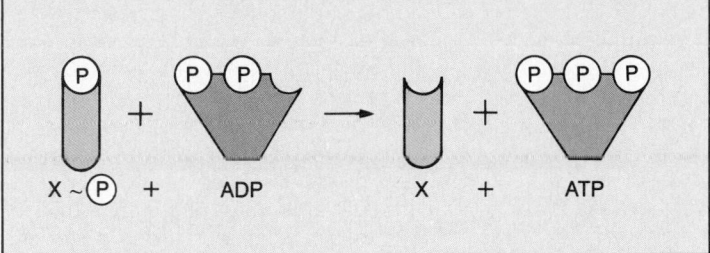

Substrate-level phosphorylation is not the primary way of forming ATP, however; most ATP is made within mitochondria and chloroplasts by way of the electron transport system mentioned earlier (fig. 7.9). In this case, the phosphorylation that occurs is called **chemiosmotic phosphorylation.**

Chemiosmotic Phosphorylation

For many years it was known that ATP synthesis was somehow coupled to the electron transport system, but the exact mechanism could not be determined. Peter Mitchell, a British biochemist, received a Nobel Prize in 1978 for his *chemiosmotic theory* of ATP production in both mitochondria and chloroplasts.

In mitochondria and chloroplasts, the carriers of the electron transport system are located within a membrane. According to the chemiosmotic theory, hydrogen (H⁺) ions tend to collect on one side of the membrane because they are pumped there by certain carriers. This establishes an electrochemical gradient across the membrane that can be used to provide energy for ATP production. Particles, called *ATP synthase complexes,* span the membrane. Each complex contains a channel protein that allows hydrogen ions to flow down their electrochemical gradient. The flow of hydrogen ions through the channel provides the energy for the ATP synthase enzyme to produce ATP from ADP + P (fig. 7.13). This process is called chemiosmotic phosphorylation because chemical and osmotic events have been joined to allow ATP synthesis.

The chemiosmotic synthesis of ATP was supported by a now-famous experiment performed by Andre Jagendorf of Cornell University utilizing chloroplasts (fig. 7.14). The experiment shows that ATP production is indeed tied only to a hydrogen ion gradient.

> Most of the energy for ATP synthesis is derived from a hydrogen ion gradient established across a membrane by an electron transport system.

Figure 7.13

The energy released by the passage of electrons down the electron transport system is used to pump hydrogen ions (H⁺) across a membrane in mitochondria and chloroplasts. This establishes an electrochemical gradient that is utilized for ATP synthesis. When the hydrogen ions flow back across the membrane through a channel protein in the membrane that is also an ATP synthase, ATP is synthesized.

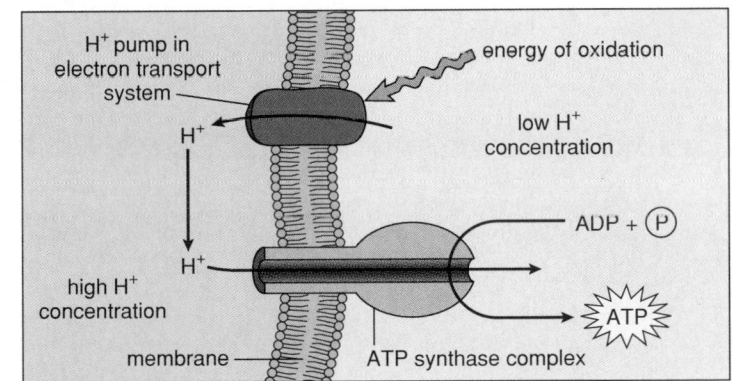

Figure 7.14

An experiment performed by Andre Jagendorf in 1966 supports the chemiosmotic theory. An electrochemical gradient was established by first soaking isolated chloroplasts in an acid medium and then abruptly changing the pH to basic. ATP production occurred after ADP and P were added. This all occurred in the dark, proving that ATP synthesis is not directly linked to the chloroplast electron transport system, which works only in the light.

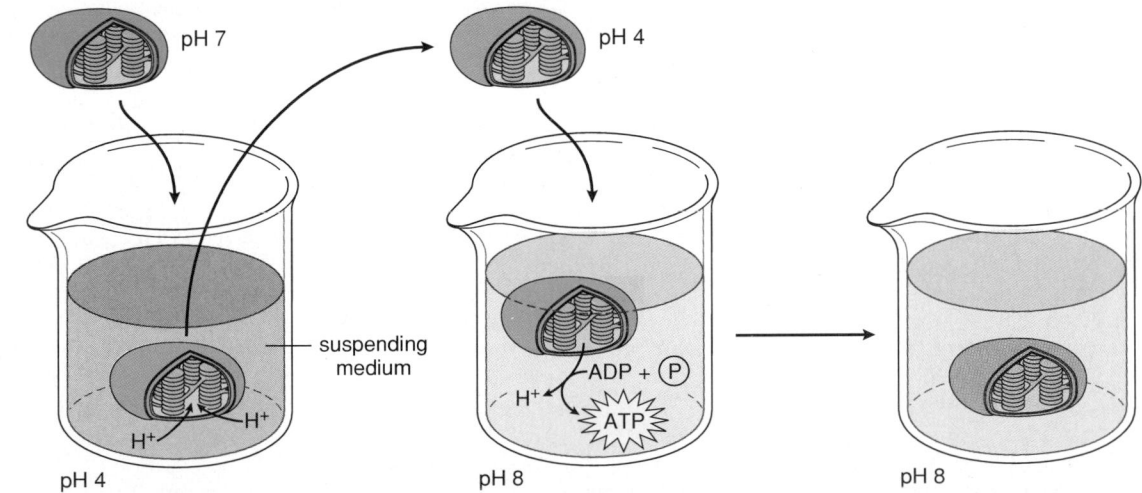

H⁺ diffuses into chloroplasts so that they become pH 4.

ADP and P are added to beaker. As H⁺ diffuses out of chloroplasts, ATP is produced.

Chloroplast and medium have same pH. H⁺ electrochemical gradient no longer exists and ATP production ceases.

Summary

1. There are 2 energy laws that are basic to understanding energy-use patterns in organisms (cells). The first states that energy cannot be created or destroyed, and the second states that some useful energy is always lost when one form of energy is transformed into another form.

2. In keeping with these laws, living things need an outside source of energy. Plants capture the energy of sunlight to make their own food, and animals eat either plants or other animals to obtain food.

3. Metabolism consists of 2 kinds of reactions, degradative and synthetic. During degradation, organic molecules derived from food are broken down to release energy; oxidation occurs and hydrogen atoms are removed from substances. During synthesis, an energy-requiring process, molecules derived from food accept hydrogen atoms and are reduced.

4. There are degradative metabolic pathways and synthetic metabolic pathways. Each reaction in a pathway requires its own enzyme. Enzymes speed up chemical reactions because they lower the energy of activation. They can do this because they form a complex with their substrate(s) at the active site.

5. Various factors affect the yield of enzymatic reactions, such as the concentration of the substrate(s), the temperature, and the pH. A high temperature or a pH outside the preferred range for that enzyme can lead to denaturation, a change in structure that prevents the enzyme from functioning.

6. The activity of most enzymes is regulated by inhibition. In competitive inhibition, a molecule competes with the substrate for the active site; during noncompetitive inhibition, a molecule binds to an allosteric site. During feedback inhibition, the product of an enzymatic reaction binds to the enzyme; for metabolic pathways, the end product usually binds to an allosteric site on the first enzyme in the pathway.

7. Many enzymes have cofactors or coenzymes that help them carry out a reaction. Coenzymes are nonprotein, organic molecules often derived at least in part from vitamins. NAD^+ and FAD are coenzymes that carry high-energy electrons to an electron transport system that is usually located in mitochondria. Here oxidation occurs each time one carrier passes electrons to the next one. At the same time, energy is released and is used to produce ATP molecules. Chloroplasts also contain an electron transport system that receives energized electrons taken from water.

8. NADPH is also a carrier of electrons, but it carries them to various synthetic reactions. These are reduction reactions that require hydrogen atoms and energy supplied by ATP.

9. Degradative pathways drive synthetic pathways. Degradative pathways supply the ATP and the NADPH that is needed to bring about synthesis. We must never forget, however, that in order for this to occur, the cell needs an outside source of energy and matter. Only a flow of energy through the organism allows it to sustain itself and to grow.

10. ATP is a high-energy molecule that serves as a common carrier of energy in cells. When ATP becomes ADP + $\circled{P}$, the right amount of energy is provided for many reactions. In order to produce ATP, the energy from carbohydrate breakdown is coupled to ATP production, and in order to pass this energy to synthetic reactions, coupling again occurs.

11. The chemiosmotic theory explains just how the electron transport system produces ATP. The carriers of this system deposit hydrogen ions (H^+) on one side of a membrane. When the ions flow down an electrochemical gradient in special channels, an enzyme utilizes the release of energy to make ATP from ADP and $\circled{P}$.

Writing Across the Curriculum

In order to practice writing skills, students should write out the answers to any or all of the study questions and the critical thinking questions. The study questions are sequenced in the same order as the text. Suggested answers to the critical thinking questions are in appendix D.

Study Questions

1. Explain the first energy law using the terms *system* and *surroundings* in your explanation. Explain the second energy law using the term *useful energy* in your explanation.

2. Discuss the importance of solar energy, not only to plants but to all living things.

3. Contrast a degradative reaction with a synthetic reaction.

4. Diagram a metabolic pathway and use this to explain the terms *substrate* and *product* and the need for various enzymes.

5. In what way do enzymes lower the energy of activation? What does the induced-fit model say about formation of the enzyme-substrate complex?

6. In what ways do substrate concentration, temperature, and pH affect the yield of enzymatic reactions?

7. Explain competitive inhibition, noncompetitive inhibition, and feedback inhibition of enzyme activity.

8. Contrast the function of NAD^+ and FAD to that of $NADP^+$.

9. In what way do degradative reactions drive synthetic reactions? Why is there still a need for an outside source of energy and matter?

10. Discuss the structure and the function of ATP, and outline the formation of ATP by chemiosmotic phosphorylation.

Objective Questions

1. Degradative reactions
   a. are often oxidation reactions.
   b. are the same as synthetic reactions.
   c. require a supply of NADPH molecules.
   d. Both a and c.
2. Enzymes
   a. make it possible for cells to escape the need for energy.
   b. are nonprotein molecules that help coenzymes.
   c. are not affected by a change in pH.
   d. lower the energy of activation.
3. An allosteric site on an enzyme is
   a. the same as the active site.
   b. where ATP attaches and gives up its energy.
   c. often involved in feedback inhibition.
   d. All of these.
4. A high temperature
   a. can affect the shape of an enzyme.
   b. lowers the energy of activation.
   c. makes cells less susceptible to disease.
   d. Both a and c.
5. Electron transport systems
   a. are found in both mitochondria and chloroplasts.
   b. contain oxidation reactions.
   c. are involved in the production of ATP.
   d. All of these.
6. The difference between NAD⁺ and NADP⁺ is
   a. only NAD⁺ production requires niacin in the diet.
   b. one contains high-energy phosphate bonds and the other does not.
   c. one carries electrons to the electron transport system and the other carries them to synthetic reactions.
   d. All of these.

7. ATP
   a. is used only in animal cells and not in plant cells.
   b. carries energy between degradative pathways and synthetic pathways.
   c. is needed for chemical work, mechanical work, and transport work.
   d. Both b and c.
8. Chemiosmotic phosphorylation is dependent upon
   a. the diffusion of water across a selectively permeable membrane.

   b. an outside supply of phosphate and other chemicals.
   c. the establishment of an electrochemical H⁺ gradient.
   d. the ability of ADP to join with ⓟ even in the absence of a supply of energy.
9. Label this diagram that gives an overview of metabolism. Label the arrows either degradative reactions or synthetic reactions. Label the long lines either macromolecules or small molecules. Label the short lines ADP + ⓟ, ATP, NADP, NADPH.

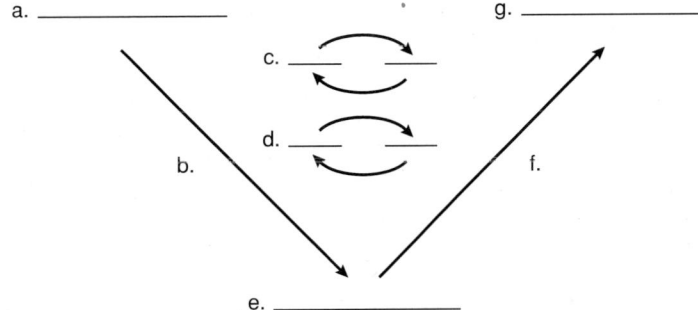

10. Label this diagram describing chemiosmotic phosphorylation.

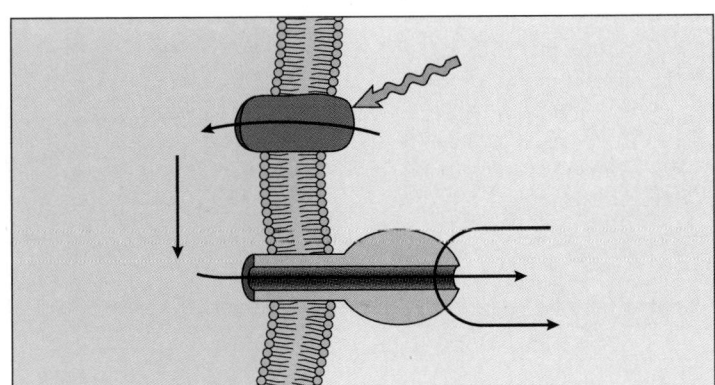

Concepts and Critical Thinking

1. *A constant input of energy is required to maintain the organization of the cell.*

   In reference to work performed by the rough ER and the Golgi apparatus, explain specifically why a cell needs energy to maintain its organization.

2. *In keeping with the first and second laws of thermodynamics, a cell takes in useful energy and gives off heat.*

   Explain why you would expect both parts of the ATP cycle to be responsible for the loss of useful energy through heat.

3. *The actions of enzymes allow the cell to maintain itself and grow.*

   Explain why no reaction occurs in a cell unless a specific enzyme is present.

Selected Key Terms

metabolism (mĕ-tab′o-lizm) 107
substrate (sub′strāt) 108
enzyme (en′zīm) 108
active site (ak′tiv sīt) 109
denatured (de-na′tūrd) 111
feedback inhibition (fēd′bak in″hĭ-bish′un) 112
coenzyme (ko-en′zīm) 112
vitamin (vi′tah-min) 112
NAD⁺ (nicotinamide adenine dinucleotide) (nik″o-tin′ah-mīd ad′ĕ-nīn di-nu′kle-o-tīd) 113

FAD (flavin adenine dinucleotide) (fla′vin ad′ĕ-nīn di-nu′kle-o-tīd) 114
cellular respiration (sel′u-lar res″pī-ra′shun) 114
electron transport system (e-lek′tron trans′port sis′tem) 114
NADP⁺ (nicotinamide adenine dinucleotide phosphate) (nik″o-tin′ah-mīd ad′ĕ-nīn di-nu′kle-o-tīd fos′fāt) 114
photosynthesis (fo″to-sin′thĕ-sis) 114

ATP (adenosine triphosphate) (ah-den′o-sēn tri-fos′fāt) 115
ADP (adenosine diphosphate) (ah-den′o-sēn di-fos′fāt) 116
substrate-level phosphorylation (sub′strāt lĕv-al fos″fōr-ĭ-la′shun) 117
chemiosmotic phosphorylation (kem″ĭ-os-mot′ik fos″fōr′ĭ-la′shun) 117

8

Photosynthesis

Green leaves are a site for photosynthesis. They are situated to absorb energy from the sun and carbon dioxide from the air. Solar energy is used to convert carbon dioxide to a carbohydrate, which serves as a food not only for plants but for the entire biosphere.

Your study of this chapter will be complete when you can

1. give at least 3 examples of the importance of photosynthesis to living things;
2. relate the visible light range to photosynthesis, and describe the role of chlorophyll;
3. describe the structure of chloroplasts, and give the function of the various parts;
4. explain the terms *light-dependent reactions* and *light-independent reactions*, and describe how the light-dependent reactions drive the light-independent reactions;
5. trace the noncyclic electron pathway, pointing out significant events as the electrons move;
6. trace the cyclic electron pathway, pointing out significant events;
7. describe the process of chemiosmotic ATP synthesis in chloroplasts;
8. describe the 3 stages of the Calvin cycle;
9. contrast C_3 and C_4 plants, and relate the phenomenon of photorespiration to the success of C_4 plants in hot, dry climates;
10. contrast a CAM plant with a C_4 plant, using the terms *partitioning in space* and *partitioning in time*.

 AVE YOU THANKED A GREEN PLANT TODAY? Plants dominate our environment, but most of us spend little time thinking about the various services they perform for us and for all living things. Chief among these is their ability to carry on photosynthesis, during which they use sunlight (*photo*) as a source of solar energy to produce food (*synthesis*):

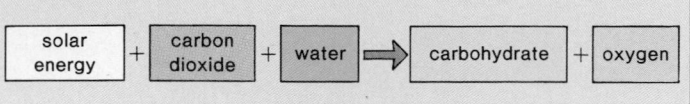

Other organisms, called algae, are also photosynthetic. Algae (chap. 24) are a diverse group, but many are water dwelling, microscopic organisms related to plants.

The organic food produced by photosynthesis is not only used by plants themselves, it is also the ultimate source of food for all other living things (fig. 8.1). Also, when organisms convert carbohydrate energy into ATP energy they make use of the oxygen (O_2) given off by photosynthesis. Nearly all living things are dependent on atmospheric oxygen derived from photosynthesis.

At one time in the distant past, plant and animal matter accumulated without decomposing. This matter became the fossil fuels (e.g., coal, oil, and gas) that we burn today for energy. This source of energy, too, is the product of photosynthesis.

Photosynthesis is absolutely essential for the continuance of life because it is the source of food and oxygen for nearly all living things.

Sunlight

Photosynthesis is an energy transformation in which solar energy in the form of light is converted to chemical energy within carbohydrate molecules. Therefore, we will begin our discussion of photosynthesis with the energy source—sunlight.

Solar radiation can be described in terms of its energy content and its wavelength. The energy comes in discrete packets called **photons.** So, in other words, you can think of radiation as photons that travel in waves:

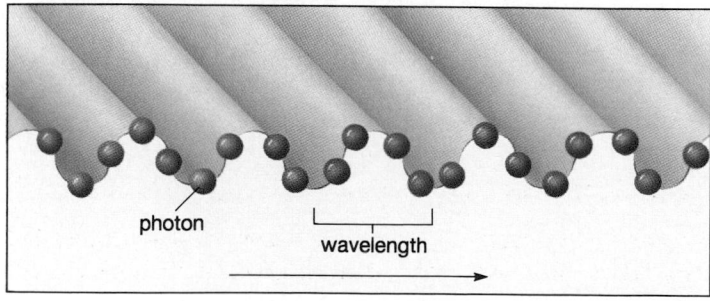

Figure 8.2*a* illustrates that solar radiation, or the **electromagnetic spectrum,** can be divided on the basis of wavelength—gamma rays have the shortest wavelength and radio waves have the longest

Figure 8.1
This squirrel is a herbivore. It feeds directly on plant material produced by a photosynthesizer. Carnivores, such as a hawk that may feed on this squirrel, are also dependent, although indirectly, on food produced by photosynthesizers.

wavelength. The energy content of photons is inversely proportional to the wavelength of the particular type of radiation; that is, short-wavelength radiation has photons of a higher energy content than long-wavelength radiation. High-energy photons, such as those of short-wavelength ultraviolet radiation, are dangerous to cells because they can break down organic molecules. Low-energy photons, such as those of infrared radiation, do not damage cells because they only increase the vibrational or rotational energy of molecules; they do not break bonds. But photosynthesis utilizes only the portion of the electromagnetic spectrum known as **visible light.** (It is called visible light because it is the part of the spectrum that allows us to see.) Photons of visible light have just the right amount of energy to promote electrons to a higher electron shell in atoms without harming cells.

We have mentioned previously that only about 42% of the solar radiation that hits the earth's atmosphere ever reaches the surface, and most of this radiation is within the visible-light range.

Figure 8.2

The electromagnetic spectrum, the components of visible light, and the absorption spectra of chlorophylls (**a**) and (**b**). Visible (white) light is actually made up of a number of different wavelengths of radiation; when it is passed through a prism (or through raindrops), we see these wavelengths as different colors of light. The chlorophylls, like other pigments, absorb only certain wavelengths; for example, the chlorophylls primarily absorb light in the blue-violet range and in the red-orange range. The chlorophylls do not absorb green light to any extent; therefore, light of this color is reflected to our eyes and we see chlorophyll as a green pigment. There are other pigments in plants, such as the carotenoids, that are yellow-orange and are able to absorb light in the violet-blue-green range. These pigments become noticeable in the fall when chlorophyll breaks down. To identify the absorption spectrum of a particular pigment, a purified sample is exposed to different wavelengths of light inside an instrument called a spectrophotometer. A spectrophotometer measures the amount of light that passes through the sample, but from this it can be calculated how much was absorbed. The amount of light absorbed at each wavelength is plotted on a graph, and the result is what we call the absorption spectrum in (**b**). When light is not absorbed, it is reflected or transmitted (**c**). How do we know that the peaks shown in chlorophyll's absorption spectrum indicate the wavelengths used for photosynthesis? Photosynthesis, of course, produces oxygen, and therefore we can use the rate of production of this gas as a means to measure the rate of photosynthesis at each wavelength of light. When such data are plotted, the resulting graph (called the action spectrum) is very similar to the absorption spectrum of the chlorophylls. Therefore, we are confident that the light absorbed by the chlorophylls does contribute extensively to photosynthesis.

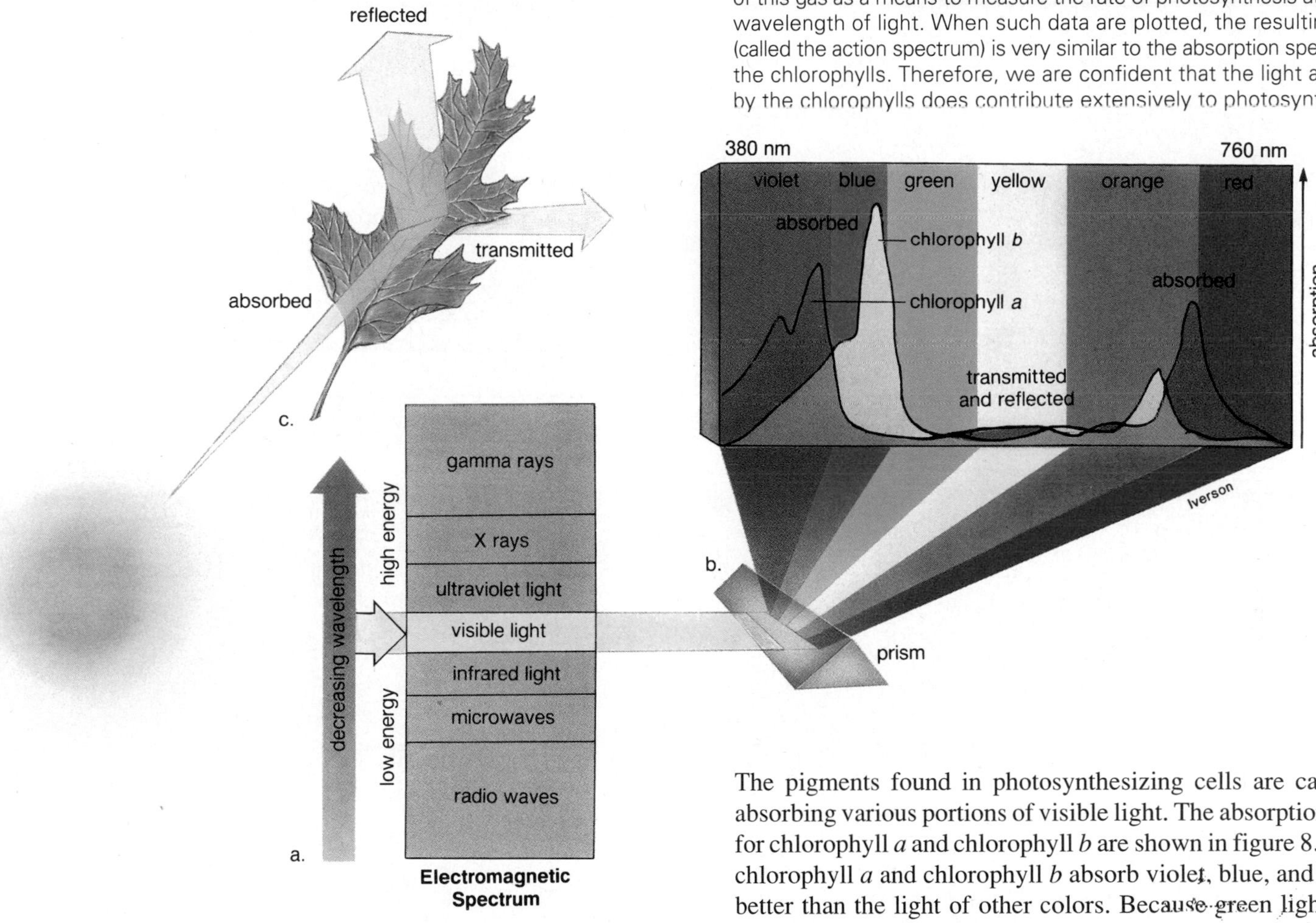

Higher energy wavelengths are screened out by the ozone layer in the atmosphere, and lower energy wavelengths are screened out by water vapor and carbon dioxide before they reach the earth's surface. The conclusion is, then, that both the organic molecules within organisms and certain life processes, such as vision and photosynthesis, are chemically adapted to the radiation that is most prevalent in the environment.

Photosynthesis utilizes the portion of the electromagnetic spectrum (solar radiation) known as visible light.

The pigments found in photosynthesizing cells are capable of absorbing various portions of visible light. The absorption spectra for chlorophyll *a* and chlorophyll *b* are shown in figure 8.2*b*. Both chlorophyll *a* and chlorophyll *b* absorb violet, blue, and red light better than the light of other colors. Because green light is only minimally absorbed and is primarily transmitted or reflected (fig. 8.2*c*), leaves appear green to us.

Chloroplast Structure and Function

The site of photosynthesis is the **chloroplast,** an organelle found in the cells of green plants and certain algae (fig. 8.3). In the chloroplast, a double membrane, or envelope, surrounds a large central space called the **stroma.** A membrane system within the stroma forms the **grana,** so called because they looked like piles of seeds (*grana*—seeds) to early microscopists.

Figure 8.3

Chloroplast structure. *a.* A chloroplast is bounded by a double membrane. Within, the grana contain pigments (e.g., chlorophyll) that absorb the energy of sunlight. The stroma contains enzymes that reduce carbon dioxide (CO_2). *b.* Electron micrograph of a chloroplast cross section showing the grana, their connecting stroma lamellae, and the stroma. Magnification, ×12,000. *c.* A granum is a stack of membranous sacs called thylakoids. Each thylakoid has a thylakoid space.

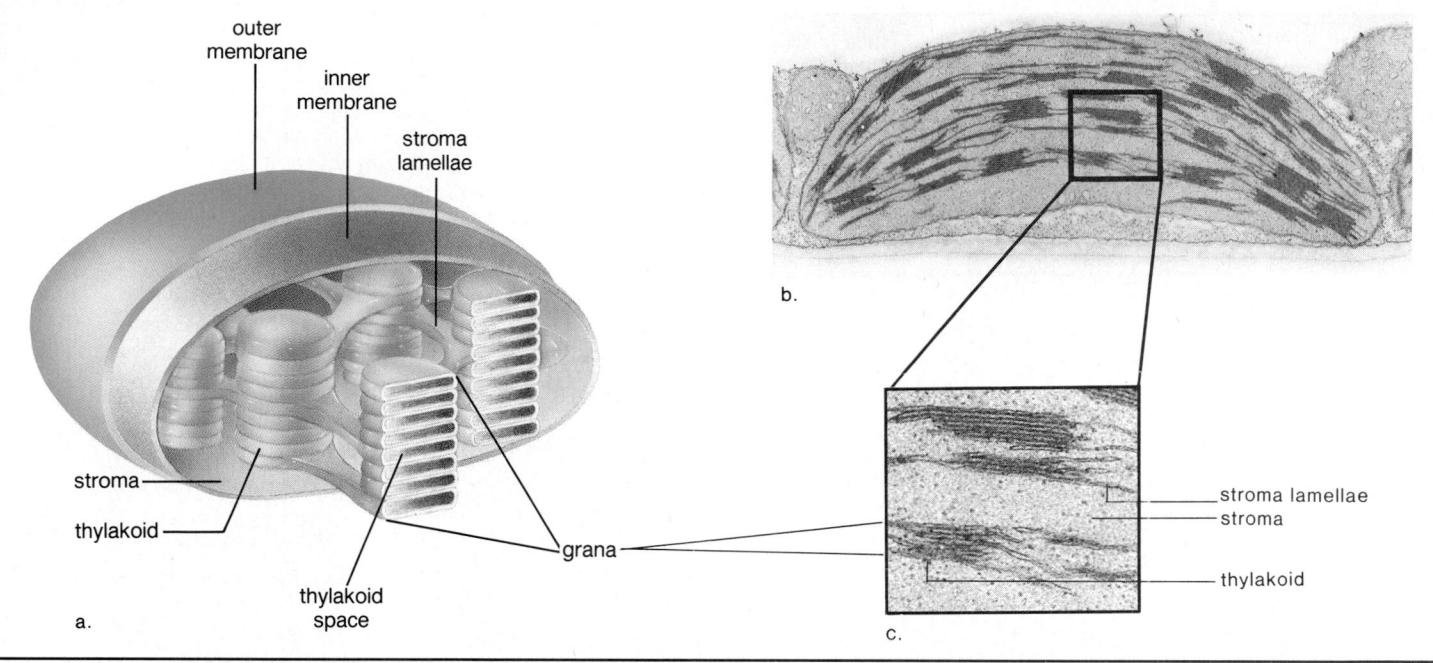

Figure 8.4

Photosynthesis overview. *a.* This overall equation for photosynthesis shows the reactants and the products. *b.* In the thylakoids, water is split and oxygen is released. ATP and NADPH are made. *c.* In the stroma, carbon dioxide (R—CO_2) is reduced to R—CH_2O using the ATP and the NADPH from the thylakoids.

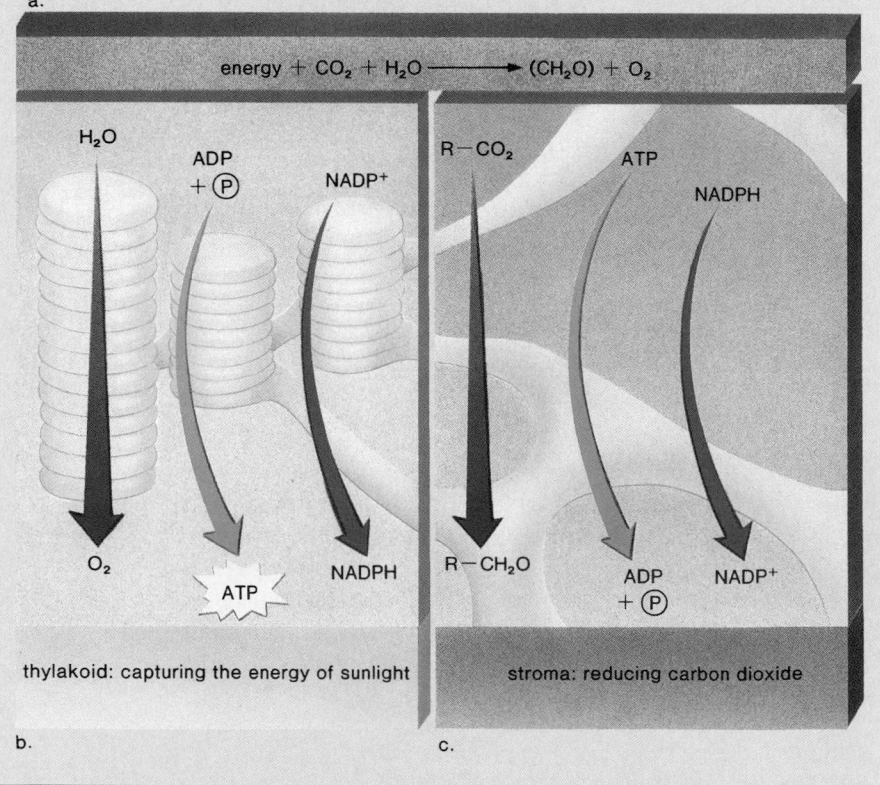

$$energy + CO_2 + H_2O \longrightarrow (CH_2O) + O_2$$

thylakoid: capturing the energy of sunlight

stroma: reducing carbon dioxide

The Cell

Grana

The electron microscope reveals that each granum is actually a stack of flattened sacs or disks, now called **thylakoids.** The grana are connected to one another by stroma lamellae, and therefore it is believed that all the thylakoids in a chloroplast are continuous with one another and surround a common interior space.

Chlorophyll and the yellow-orange pigments called carotenoids are found within the membranes of the thylakoids. These pigments are able to absorb the energy of the sun (fig. 8.2b); therefore, they make photosynthesis possible. Perhaps we should think of these pigments as the most important biomolecules there are because nearly all living things are dependent on the food produced by photosynthesis.

Thylakoids are the energy-generating system of chloroplasts not only because they contain chlorophyll but also because the cytochrome (electron transport) system associated with the production of ATP and NADPH is also present in the thylakoid membrane.

Stroma

The stroma contains an enzyme-rich solution. These enzymes incorporate carbon dioxide (CO_2) into organic molecules and reduce it to CH_2O. In other words, the carbohydrate produced by photosynthesis is formed within the stroma.

> Chloroplasts carry on photosynthesis, an energy-requiring reaction that results in carbohydrate synthesis. First the solar energy must be captured, and then carbon dioxide must be reduced to carbohydrate.

An overall equation for photosynthesis does not indicate that the process actually involves 2 sets of reactions (fig. 8.4). As described in this chapter's reading, the probability of 2 sets of reactions was suggested by F. F. Blackman in 1905 after he discovered that when light is being absorbed maximally, a rise in temperature still increases the rate of photosynthesis.

The first set of reactions, which takes place in the thylakoid, where chlorophyll and other pigments are located, is called the **light-dependent reactions** because they do not take place unless solar energy is available. As water is split and oxygen (O_2) is released, NADPH and ATP are made (fig. 8.4b).

The second set of reactions, which takes place in the stroma, where many enzymes are located, is called the **light-independent reactions** because solar energy is not required; that is, the reactions can occur in the dark as well as in the light. The light-independent reactions use the NADPH and ATP produced by the light-dependent reactions to reduce carbon dioxide (R—CO_2) to carbohydrate (R—CH_2O) (fig. 8.4c).

> The overall equation for photosynthesis does not indicate that the process involves the light-dependent reactions and the light-independent reactions. The light-dependent reactions, located in the thylakoid, capture the energy of sunlight needed by the light-independent reactions, located in the stroma, to reduce carbon dioxide to carbohydrate.

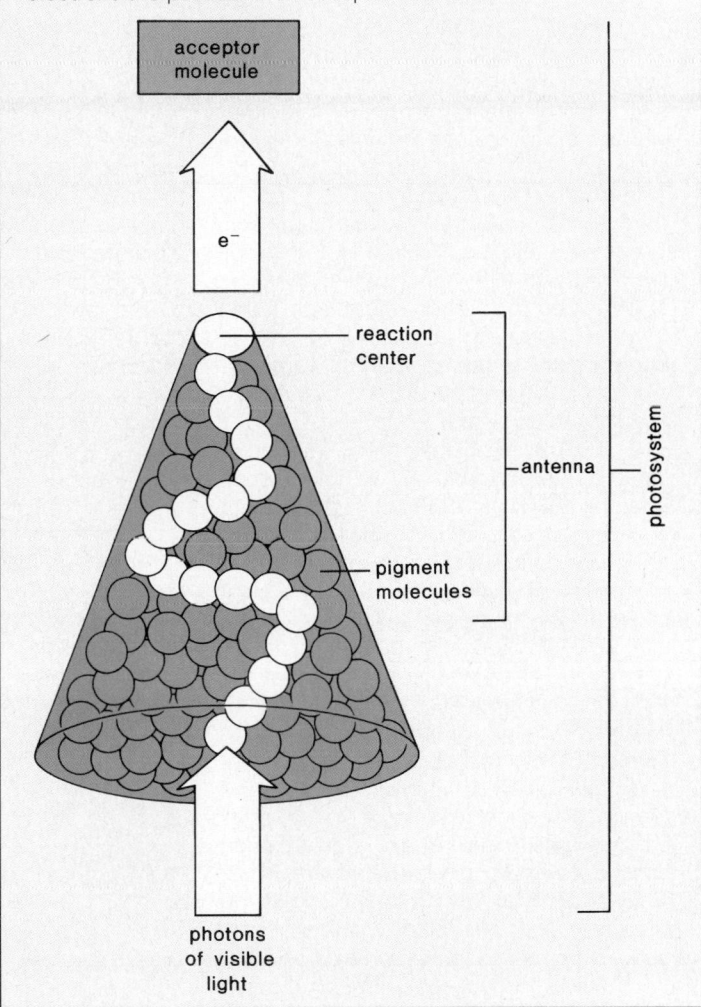

Figure 8.5
Components of a photosystem. An antenna consists of several hundred chlorophyll and carotenoid molecules plus a reaction-center chlorophyll *a* molecule. Light energy is absorbed in the form of photons and then passed from molecule to molecule (shown in yellow) until it reaches a reaction-center chlorophyll *a* molecule. Thereafter, energized electrons are passed to an acceptor molecule.

Light-Dependent Reactions

The light-dependent reactions require the participation of 2 **photosystems,** called Photosystem I and Photosystem II. Each photosystem contains pigment molecules that are part of a light-harvesting antenna (fig. 8.5). Like television antennae aimed to pick up signals, the leaves of a plant turn so the photosystem antennae can collect as much solar energy as possible. The pigment molecules in an antenna are chlorophyll molecules and carotenoid molecules. One chlorophyll *a* molecule is special; it is the reaction-center chlorophyll molecule.

Photosynthesis begins when the antenna of each photosystem absorbs photons of visible light and funnels the energy to their respective reaction center. The reaction-center chloro-

With the rise of the scientific method in the seventeenth century, investigators began to perform experiments that eventually led to our present-day knowledge of photosynthesis. The ancient Greeks believed that plants were "soil eaters" that somehow converted soil into plant material. Apparently to test this hypothesis, a seventeenth-century Dutchman named Jean-Baptiste Van Helmont planted a willow tree weighing 5 lb in a large pot containing 200 lb of soil. He watered the tree regularly for 5 years and then reweighed both the tree and the soil. The tree weighed 170 lb and the soil weighed only a few ounces less than the original 200 lb. Van Helmont concluded that the increase in weight of the tree was due primarily to the addition of *water*. He did not consider the possibility that air had anything to do with it.

Later, a number of other investigators did consider the role of the air. Joseph Priestley (English, 1733-1804) put a plant and a lighted candle under a bell jar that had trapped a certain amount of air. He found that after the candle had gone out, the plant could "renew" the air in such a way that the candle would burn if lighted again. The science of chemistry was just emerging at this time, and Jan Ingenhousz (Dutch, 1730-99) identified the gas given off by plants as *oxygen*. He also observed that the amount of oxygen given off is proportional to the amount of sunlight reaching the plant; in the shade, plants "injure" air and release carbon dioxide. It was Nicholas de Saussure (Swiss, 1767-1845) who found that an increase in the dry weight of a plant was dependent upon the presence of *carbon dioxide.*

Microscopes were increasingly used during those days, and Matthias Schleiden, a German, is credited with the conclusion in 1838 that plants are composed of cells. Julius Sachs (German, 1832-97) was a plant physiologist who studied plant cells under the microscope. He noted that the green substance named *chlorophyll* is confined to the chloroplasts. In one experiment with a whole plant, he coated a few leaves with wax. He observed that after exposure to sunlight, only the uncoated leaves, which could take in carbon dioxide, increase in *starch* content. (A similar experiment is found in the lab manual that accompanies this text.) Sachs must have been an excellent microscopist because he followed up this experiment by observing that the starch grains in chloroplasts increase in size when a leaf is photosynthesizing.

By the end of the nineteenth century, scientists knew that photosynthesis occurs in chloroplasts and that the basic reaction is:

---

carbon dioxide + water

sunlight $\downarrow$

carbohydrate + oxygen

In 1905, F. F. Blackman performed experiments in which he steadily increased the amount of light energy while the temperature was held constant; once the rate of photosynthesis was maximum, he increased the temperature. Now the rate of photosynthesis increased. This led him to conclude that only one part of photosynthesis has to do with light, and that there is a second part that requires temperature-sensitive enzymes. Today these 2 parts of photosynthesis are called the light-dependent reaction and the light-independent reaction.

By the 1940s, cell fractionation allowed investigators to study certain parts of cells. Robert Hill isolated chloroplasts from cells and discovered that oxygen is given off in the absence of CO_2. He suggested, therefore, that the oxygen released during photosynthesis comes not from carbon dioxide but from water. This was substantiated in 1941 by Samuel Ruben and Martin Kamen, who performed a study that utilized heavy oxygen ($^{18}O_2$) as a tracer. A plant was exposed first to carbon dioxide that contained heavy oxygen and then to water that contained heavy oxygen. Only in the latter instance did the oxygen given off contain this isotope. To balance the equation, it is necessary to show water on both sides of the equation for photosynthesis (fig. 8.A).

Cell fractionation, radioactive tracers, and electron microscopy continue to add to our knowledge of photosynthesis; some of the more recent experiments using these techniques are covered elsewhere in this chapter.

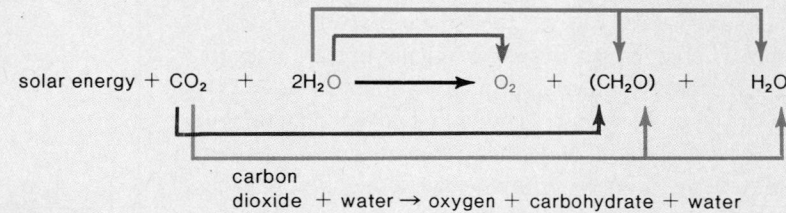

solar energy + CO_2 + $2H_2O$ ⟶ O_2 + (CH_2O) + H_2O

carbon dioxide + water → oxygen + carbohydrate + water

Figure 8.A
Overall equation for photosynthesis. The arrows show the relationship between the reactants and the products in the equation.

---

phyll *a* molecule of the Photosystem I antenna has an absorption spectrum that peak at a wavelength of around 700 nm and therefore is called *P700* (P—pigment). The reaction-center chlorophyll *a* molecule of the Photosystem II antenna has an absorption spectrum that peaks at a slightly shorter wavelength; it is called *P680*. The light energy received at a reaction center energizes electrons within the chlorophyll *a* molecule, and these electrons then pass to an acceptor molecule.

In a photosystem, the light-harvesting antenna absorbs solar energy and funnels it to a reaction-center chlorophyll *a* molecule, which then sends energized electrons to an acceptor molecule.

Two Electron Pathways

There are 2 possible pathways that electrons can take during the first phase of photosynthesis, but only one pathway, the *noncyclic electron pathway*, produces both ATP and NADPH. The other, the *cyclic electron pathway*, generates only ATP. Both pathways are diagrammed in figure 8.6.

Figure 8.6

Photophosphorylation (ATP formation) occurs in thylakoid membranes. In noncyclic photophosphorylation, electrons move from water to P680 to an acceptor molecule that passes them down a transport system to P700, which sends them to another acceptor molecule before they are finally sent to NADP⁺. Then NADP⁺ combines with a hydrogen (H^+) ion from stroma and becomes NADPH. In cyclic photophosphorylation (dotted lines), electrons pass from P700 to an acceptor molecule that sends them down the electron transport system before they return to P700.

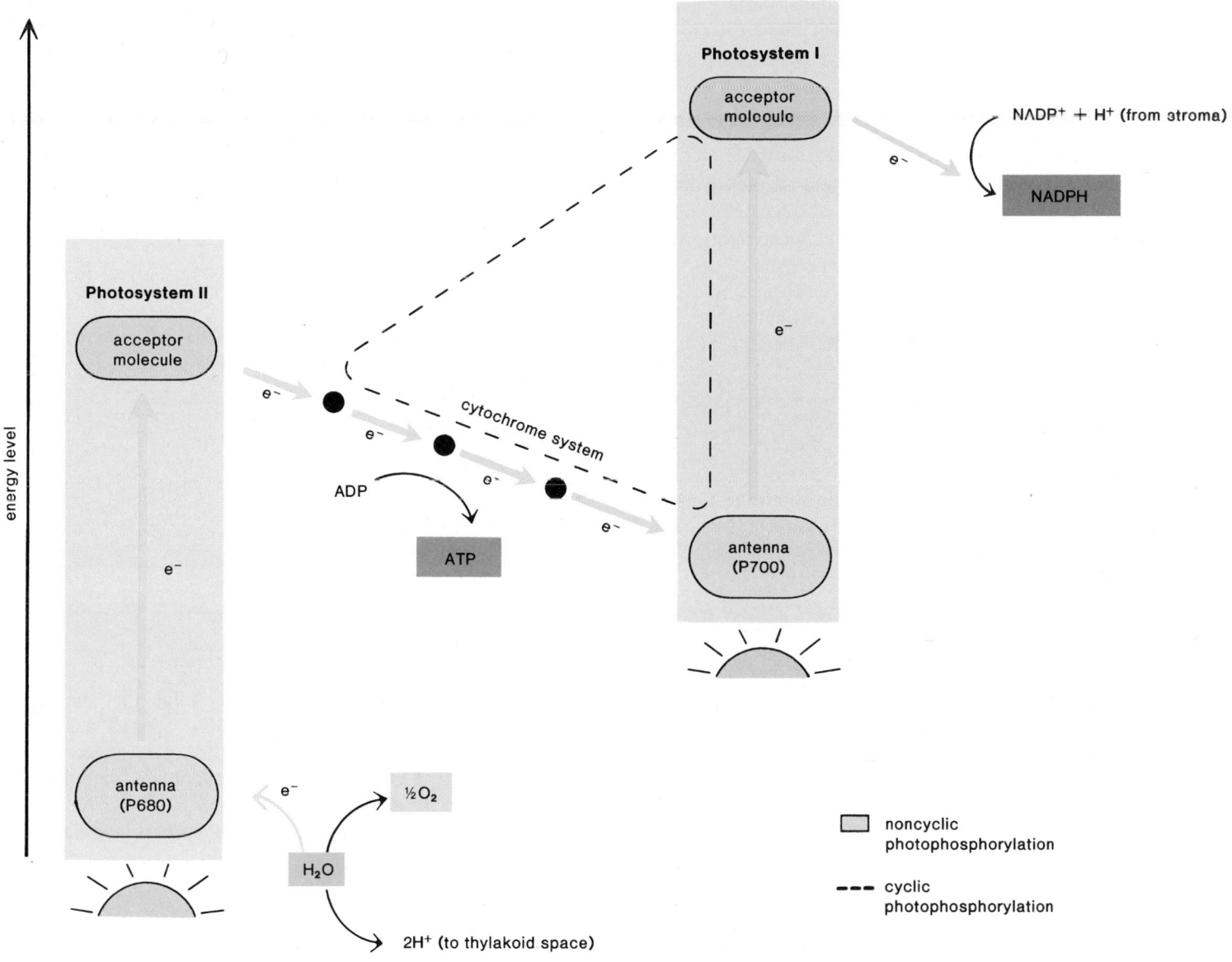

The light-dependent reactions of photosynthesis located in the thylakoid membrane include a noncyclic electron pathway (producing both ATP and NADPH) and a cyclic electron pathway (producing only ATP).

The Noncyclic Electron Pathway

From figure 8.6, you can see that each photosystem absorbs sunlight at the same time; however, it is easier to describe the events as if they occur in a sequential manner and as if they begin with Photosystem II.

1. As P680 absorbs solar energy (*lower left*), electrons (e⁻) become so highly energized that they can leave the thylakoid-bound chlorophyll *a* molecule. The "hole" left in the molecule is filled by electrons that come from the splitting of water:

$$H_2O \longrightarrow 2H^+ + 2e^- + \tfrac{1}{2}O_2$$

The freed oxygen (O_2) is the oxygen gas given off during photosynthesis.

2. The electrons that leave P680 are received by an acceptor molecule that sends them to an electron transport system consisting of a series of thylakoid-bound carriers, some of which are cytochrome molecules. For this reason, this electron transport system often is called a **cytochrome system.** High-energy electrons are delivered to the system and low-energy electrons leave it. As the electrons pass from one cytochrome molecule to another, energy is made available for ATP formation:

$$ADP + \text{P} + energy \longrightarrow ATP$$

3. Electrons leaving the cytochrome system are picked up by P700, the reaction-center chlorophyll a molecule of Photosystem I. As this antenna receives solar energy, the electrons are boosted to their highest energy level yet, and they move from the reaction center to an acceptor that passes them on to $NADP^+$, which also combines with hydrogen (H^+) ions from the stroma. In the end, NADPH results:

$$NADP^+ + 2e^- + H^+ \longrightarrow NADPH$$

These events make up the process of **noncyclic photophosphorylation** because it is possible to trace electrons in a one-way (noncyclic) direction from water (H_2O) to NADPH and because the energy from sunlight is used to produce ATP (photophosphorylation).

> The noncyclic electron pathway produces both ATP and NADPH, which are used in the stroma to reduce carbon dioxide. Oxygen is released as a by-product.

The Cyclic Electron Pathway

The other electron pathway in chloroplasts is the cyclic electron pathway. In this pathway, electrons leave P700 and eventually return to it (dotted line in figure 8.6) instead of passing to $NADP^+$. Before they return to P700, they pass down the cytochrome system, and ATP is produced as previously described. This pathway sometimes is called **cyclic photophosphorylation** because it is possible to trace electrons in a cycle from P700 to P700 and because solar energy is used to produce ATP (photophosphorylation). Only ATP is produced during cyclic phosphorylation, and there is no concomitant formation of NADPH.

It is believed that the cyclic electron pathway, and therefore Photosystem I, evolved very early (before Photosystem II) in the history of the earth. There are some photosynthetic bacteria even today that use the cyclic electron pathway only. These bacteria do not produce oxygen because they do not split water. Instead, they get the hydrogen atoms from other sources; for example, the photosynthetic sulfur bacteria take hydrogen atoms from hydrogen sulfide (H_2S). These bacteria release sulfur and not oxygen (O_2). There is no release of oxygen during cyclic photophosphorylation.

Table 8.1
Light-Dependent Reactions

Noncyclic Electron Pathway

Participant	Function	Result
Water	Splits to give O₂ H⁺ Electrons	Becomes [O₂] in atmosphere Collects in thylakoid space Passes to Photosystem II
Photosystem II	Absorbs solar energy	Supplies energized electrons to cytochrome system
Cytochrome system	Transports electrons	Releases energy for [ATP] production
Photosystem I	Absorbs solar energy	Supplies energized electrons to NADP⁺
NADP⁺	Accepts electrons (final acceptor)	Becomes [NADPH]

Cyclic Electron Pathway

Participant	Function	Result
Photosystem I	Absorbs solar energy	Supplies energized electrons to cytochrome system
Cytochrome system	Transports electrons	Releases energy for [ATP] production

☐ = products of light-dependent reactions

From this discussion, you may think that the cyclic electron pathway is vestigial and of little use to water-splitting photosynthesizers. The cyclic electron pathway, however, may be utilized whenever CO_2 is in such limited supply that carbohydrate is not being produced. At these times, all available $NADP^+$ is already NADPH and there is no more $NADP^+$ to receive electrons. Obviously, the energized electrons from Photosystem I have to go somewhere, and under these circumstances, the cyclic electron pathway acts as a sort of safety valve. Finally, the cyclic electron pathway provides another way to establish a chemiosmotic gradient for ATP production.

> The cyclic electron pathway utilizes only Photosystem I and produces only ATP. It probably evolved before Photosystem II.

The characteristics of the light-dependent reactions are summarized in table 8.1.

Chemiosmotic ATP Synthesis

The energy made available by the passage of electrons down the cytochrome system is coupled to the buildup of ATP in an indirect manner. Some of the carriers of the cytochrome system pump

The Cell

Figure 8.7

Chloroplast structure and function. **a.** Photosynthesis consists of light-dependent and light-independent reactions. The light-dependent reactions occur in the grana, where sunlight energy is captured, water (H_2O) is split and gives off oxygen (O_2), and ATP and NADPH are produced. The light-independent reactions occur in the stroma, where carbon dioxide (CO_2) is fixed and reduced after being incorporated into the Calvin cycle. Reduction uses the ATP and the NADPH from the light-dependent reactions. **b.** Chemiosmotic ATP synthesis. When water is split, the hydrogen (H^+) ions remain in the thylakoid space. Also, some of the cytochrome system carriers pump hydrogen (H^+) ions into the thylakoid space. When these flow out of the thylakoid space and down their electrochemical gradient through a special channel protein of the ATP synthase complex, ATP is produced.

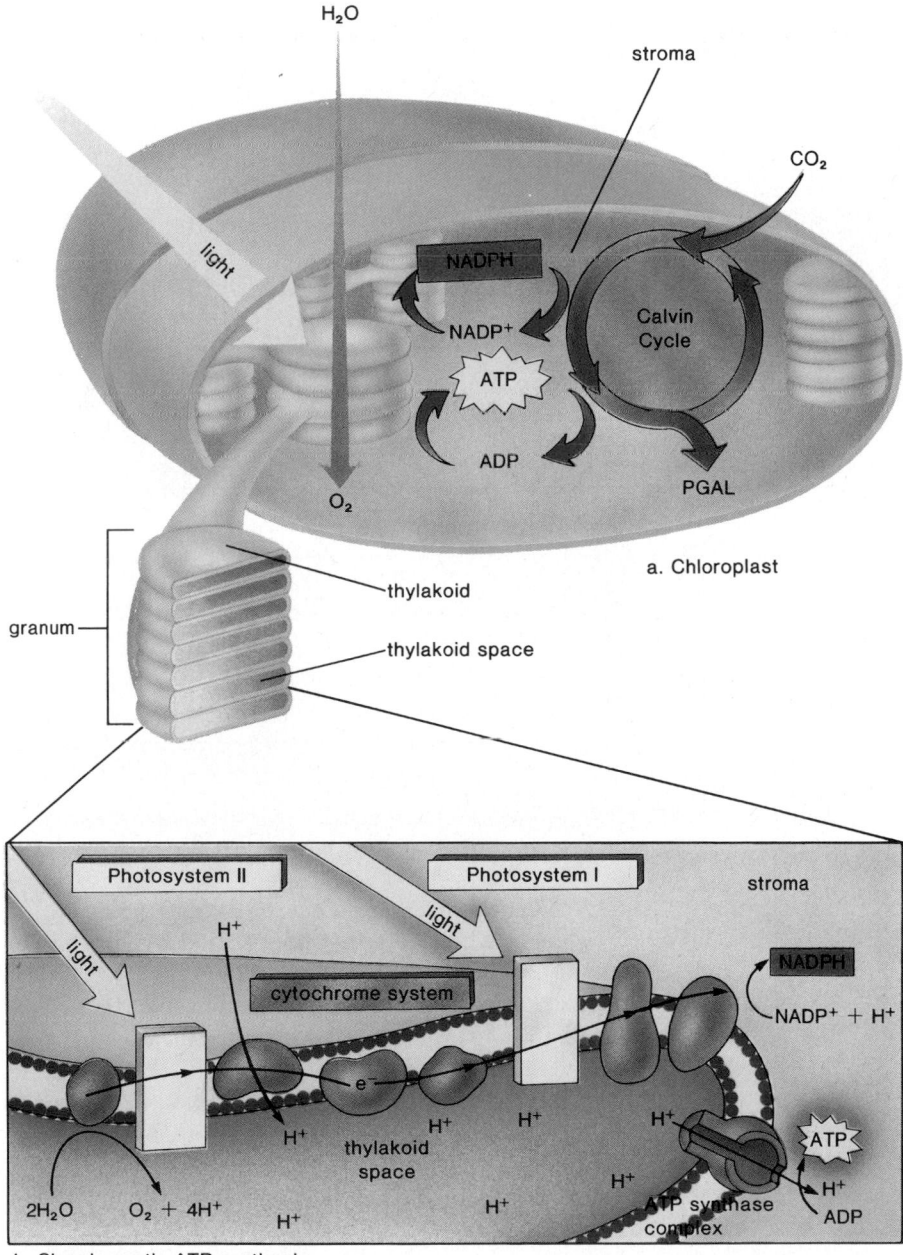

a. Chloroplast

b. Chemiosmotic ATP synthesis

hydrogen ions (H^+) from the stroma into the thylakoid space. The thylakoid space acts as a reservoir for hydrogen ions also because for every water molecule that is split at the beginning of noncyclic photophosphorylation, 2 hydrogen ions stay behind in the thylakoid space. The hydrogen ion taken up by $NADP^+$ comes from the stroma, not the thylakoid space.

Because of the large number of hydrogen ions in the thylakoid space compared to the stroma, an extreme electrochemical gradient is present. As previously discussed on page 117, when these hydrogen ions flow out of the thylakoid space by way of a channel protein present in a particle, called the *ATP synthase complex*, energy is provided for the ATP synthase enzyme to produce ATP from ADP + Ⓟ (fig. 8.7). This is called *chemiosmotic ATP synthesis* because chemical and osmotic events join to permit ATP synthesis.

Light-Independent Reactions

The light-independent reactions are the second stage of photosynthesis. They take their name from the fact that light is not directly required for these reactions to proceed. In this stage of photosynthesis, the NADPH and ATP produced by the noncyclic electron pathway during the first stage are used to reduce carbon dioxide: CO_2 becomes CH_2O within a carbohydrate molecule. This reduction process is a building-up or synthetic process because it requires the formation of new bonds. Hydrogen atoms (e^- + H^+) and energy are needed for reduction synthesis, and these are supplied by NADPH and ATP.

> The light-independent reactions use the ATP and NADPH from the light-dependent reactions to reduce carbon dioxide.

Calvin Cycle

The reduction of carbon dioxide occurs in the stroma of the chloroplast by means of a series of reactions known as the **Calvin cycle.** Melvin Calvin received a Nobel Prize in 1961 for his part in determining these reactions (fig. 8.8). The Calvin cycle includes carbon dioxide fixation, carbon dioxide reduction, and regeneration of RuBP.

Figure 8.8

a. To discover the reactions involved in carbon dioxide fixation and reduction, Calvin and his colleagues used the apparatus shown here. Algae (*Chlorella*) were placed in the flat flask, which was illuminated by 2 lamps. Radioactive carbon dioxide ($^{14}CO_2$) was added, and the algae were killed by transferring them into boiling alcohol (in the beaker below the flat flask). This treatment instantly stopped all chemical reactions in the cells. Carbon-containing compounds were then extracted from the cells and analyzed. **b.** By tracing the radioactive carbon, Calvin and his colleagues were able to trace the reaction series, or pathway, through which CO_2 was incorporated. They found that the radioactive carbon (shown in boxes) is first incorporated into RuBP and that the resulting molecule immediately splits to form two molecules of PGA. By gradually increasing the exposure of the algae to $^{14}CO_2$ before killing, they were able to identify the rest of the molecules in the cycle that is now called the Calvin cycle (see fig. 8.9).

a.

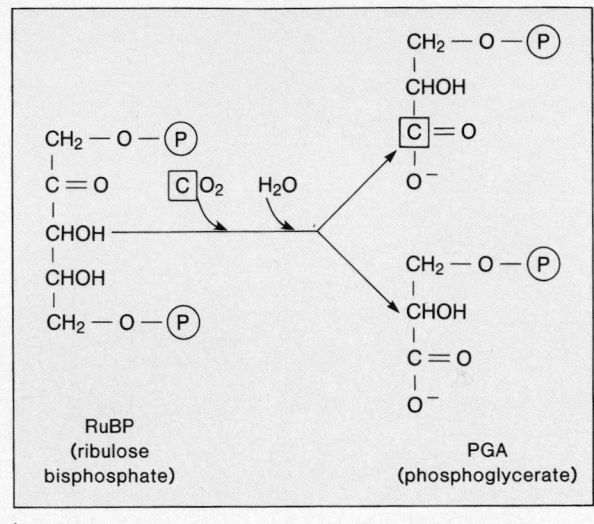

b.

Carbon Dioxide Fixation

Carbon dioxide (CO_2) fixation, the attachment of carbon dioxide to an organic compound, is the first event in the Calvin cycle (figs. 8.9 and 8.10). Carbon dioxide fixation occurs when a 5-carbon molecule called **RuBP** (ribulose bisphosphate) combines with carbon dioxide. The enzyme that speeds up this reaction is called RuBP carboxylase, or *rubisco*. Rubisco makes up about 20%–50% of the protein content in chloroplasts, and some say it is the most abundant protein in the world. The reason for its abundance may be that it is unusually slow (it processes only about 3 molecules of substrate per second compared to about 1,000 per second for a typical enzyme), and so there has to be a lot of it to keep the Calvin cycle going.

Carbon dioxide fixation occurs when carbon dioxide combines with RuBP. The enzyme for this reaction is rubisco.

After CO_2 combines with RuBP, the resulting 6-carbon molecule immediately splits to give 2 copies of the same 3-carbon (C_3) molecule.

Carbon Dioxide Reduction

During carbon dioxide reduction, R—CO_2 becomes R—CH_2O. (The R stands for the rest of the molecule.) The reduction process requires the NADPH and some of the ATP formed in the thylakoid membrane during the light-dependent reactions:

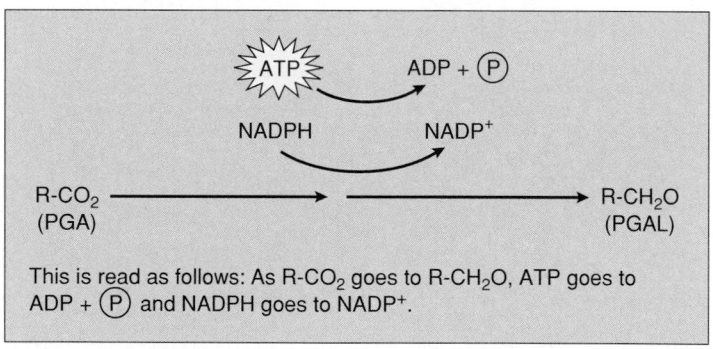

This is read as follows: As R-CO₂ goes to R-CH₂O, ATP goes to ADP + (P) and NADPH goes to NADP⁺.

The molecule *PGAL* (phosphoglyceraldehyde) is the end product of the Calvin cycle. For every 3 turns of the Calvin cycle (fig. 8.10), there is a net gain of one PGAL molecule. Two PGAL molecules combine to form glucose phosphate; therefore, glucose is often called the end product of photosynthesis.

Within the leaves of higher plants, glucose molecules are converted to the disaccharide sucrose for transport to other parts of the plant where glucose is used for the synthesis of macromolecules like cellulose and starch. In all organisms, glucose is also used as an energy source, as will be discussed in the next chapter.

Carbon dioxide reduction uses the NADPH and some of the ATP produced by the light-dependent reactions (fig. 8.7*b*).

The Cell

Figure 8.9

The Calvin cycle (simplified). RuBP combines with carbon dioxide (CO_2), forming a 6-carbon molecule (C_6), which immediately breaks down to 2 PGA (phosphoglycerate) molecules. PGA is reduced to PGAL (phosphoglyceraldehyde), the end product of this cycle. Outside the cycle, 2 PGA molecules can combine to give glucose-6-phosphate, a molecule that can be metabolized to other organic molecules. In the cycle, 5 out of every 6 PGAL are used to reform 3 RuBP.

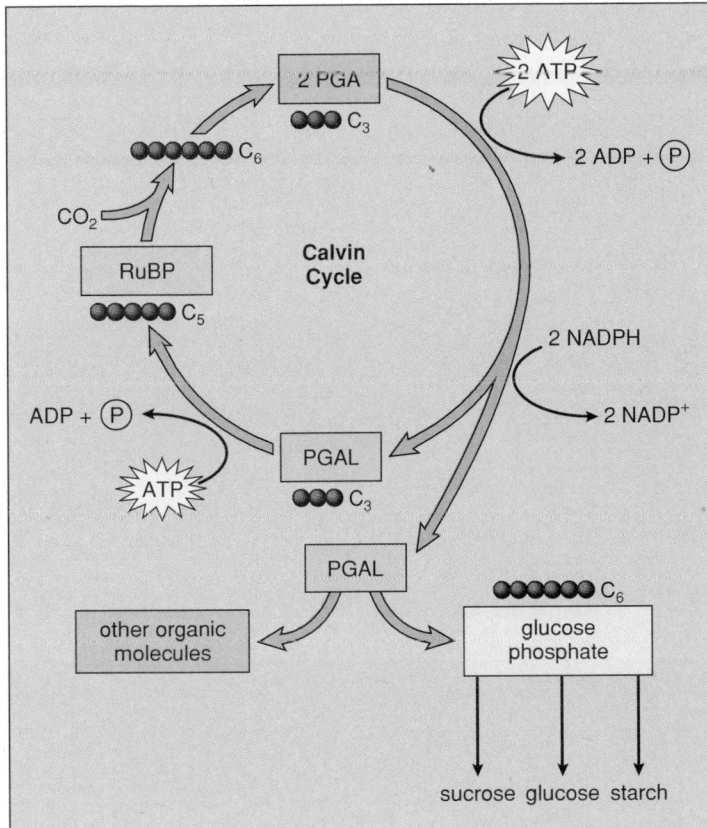

Regeneration of RuBP

For every 3 turns of the Calvin cycle, 5 molecules of PGAL are used to reform 3 molecules of ribulose bisphosphate (RuBP), the molecule the cycle started with:

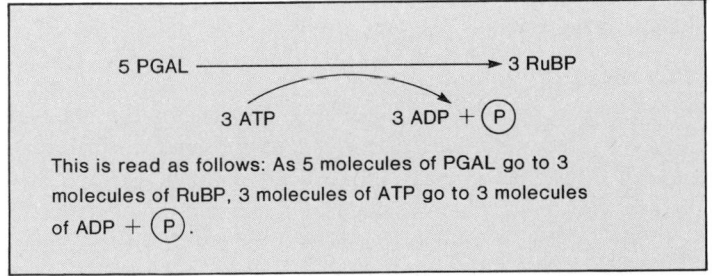

This reaction also utilizes some of the ATP produced by the light-dependent reactions. Altogether, the ATP and the NADPH consumed in producing glucose represents about a 30% energy-conversion rate. In other words, 30% of the energy available in ATP and NADPH is finally found in glucose.

Five out of 6 PGAL molecules of the Calvin cycle are used to regenerate 3 RuBP molecules so the cycle can start again. This regeneration process also uses up some of the ATP molecules produced by the light-dependent reactions.

Table 8.2 summarizes the light-independent reactions.

C_4 Plants

Some plants that live in hot, dry environments do not take up CO_2 from the atmosphere by attaching it to RuBP. Instead they use the enzyme PEP carboxylase (*pepco*) to fix CO_2 to PEP (phosphoenolpyruvate). These plants are called C_4 **plants** because the product of carbon dioxide fixation is oxaloacetate, a C_4 or 4-carbon molecule. The plants that do use the enzyme rubisco to fix CO_2 to RuBP are called C_3 **plants** because the first detected molecule following fixation is PGA, a C_3 molecule:

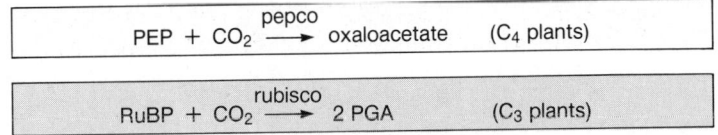

Since we have mentioned repeatedly that there is usually a close correlation between structure and function, you are probably not surprised that the structure of a leaf from a C_3 plant is different from that of a C_4 plant. In a C_3 plant, the *mesophyll cells* contain well-formed chloroplasts and are arranged in layers. In a C_4 leaf, not only the mesophyll cells but also the *bundle sheath cells* contain chloroplasts. Further, the mesophyll cells are arranged concentrically around the bundle sheath cells:

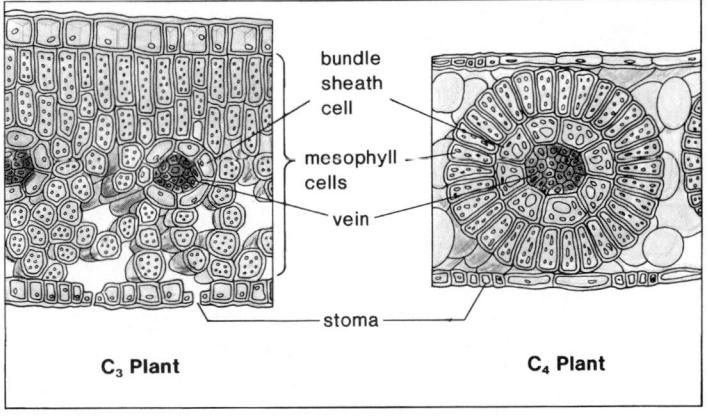

In a C_3 plant, all the mesophyll cells both fix CO_2 and produce glucose. In a C_4 plant, the mesophyll cells with chloroplasts only fix CO_2; the bundle sheath cells produce glucose by utilizing the Calvin cycle (fig. 8.11).

In a C_4 plant, PEP takes up CO_2 to give oxaloacetate in mesophyll cells and then a reduced molecule (malate) is pumped into the bundle sheath cells. After malate gives CO_2 to the Calvin cycle, the molecule pyruvate is returned to mesophyll cells. It takes 2 ATP molecules to change pyruvate to PEP once more. You would think that these "extra reactions" of the C_4 pathway

Figure 8.10

Calvin cycle (detailed). Per 3 turns of the cycle 5 molecules of PGAL (15 carbons) are used to reform 3 molecules of RuBP (15 carbons). Each RuBP molecule combines with one molecule of carbon dioxide (CO_2), forming a 6-carbon molecule (C_6), which immediately breaks down to 2 PGA molecules. Each molecule is reduced in 2 steps to PGAL, the end product of this cycle. Per 3 turns, one PGAL molecule leaves the cycle.

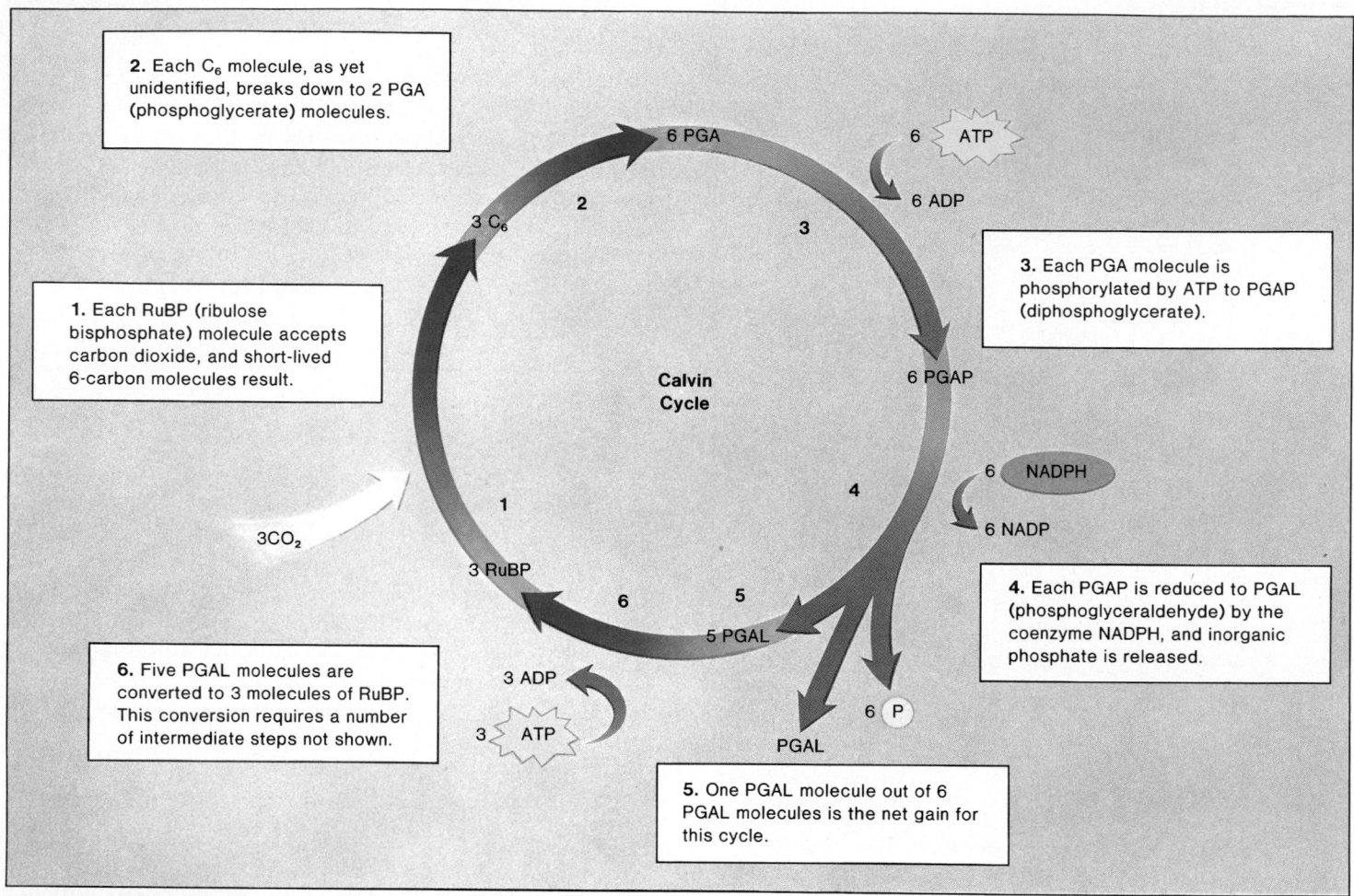

2. Each C_6 molecule, as yet unidentified, breaks down to 2 PGA (phosphoglycerate) molecules.

1. Each RuBP (ribulose bisphosphate) molecule accepts carbon dioxide, and short-lived 6-carbon molecules result.

3. Each PGA molecule is phosphorylated by ATP to PGAP (diphosphoglycerate).

4. Each PGAP is reduced to PGAL (phosphoglyceraldehyde) by the coenzyme NADPH, and inorganic phosphate is released.

6. Five PGAL molecules are converted to 3 molecules of RuBP. This conversion requires a number of intermediate steps not shown.

5. One PGAL molecule out of 6 PGAL molecules is the net gain for this cycle.

6 PGA · 6 ATP · 6 ADP · 6 PGAP · 6 NADPH · 6 NADP · 6 (P) · PGAL · 5 PGAL · 3 ADP · 3 ATP · 3 RuBP · 3 C_6 · $3CO_2$ · Calvin Cycle

Table 8.2

Light-Independent Reactions

Participant	Function	Result
RuBP	Takes up CO_2	CO_2 fixation
CO_2	Provides carbon atoms	Reduced to CH_2O
ATP	Provides energy for reduction of CO_2 and generation of RuBP	Broken down to ADP + (P)
NADPH	Provides hydrogen atoms for reduction	Oxidized $NADP^+$
PGAL	End product of photosynthesis	Glucose phosphate from 2 PGAL

would represent an energy drain for these plants, yet in hot, dry climates, the net photosynthetic rate of C_4 plants such as sugarcane, corn, and Bermuda grass is about 2 to 3 times that of C_3 plants such as wheat, rice, and oats (fig. 8.12).

Photorespiration

The explanation for the higher energy efficiency of C_4 plants in hot, dry climates compared to C_3 plants lies in the difference between *rubisco* and *pepco*, the respective enzymes for carbon dioxide fixation in C_3 and C_4 plants. Oxygen competes with CO_2 for the active site of rubisco but not for the active site of pepco. If there is an adequate supply of CO_2 inside the leaf, then most of RuBP is converted to 2 PGA molecules, and the C_3 pathway operates efficiently. But if the supply of CO_2 inside the leaf is inadequate, most of RuBP combines with O_2, giving one mol-

Figure 8.11

The C_4 pathway is superimposed on an electron micrograph of the cells of the mesophyll and the bundle sheath in corn. During the C_4 pathway, the enzyme called pepco fixes CO_2 to PEP forming oxaloacetate, which is reduced to a molecule (malate) that carries CO_2 to the Calvin cycle in the bundle sheath cells. Pyruvate then returns to the mesophyll, where ATP is used to reform PEP. Magnification, X8,300.

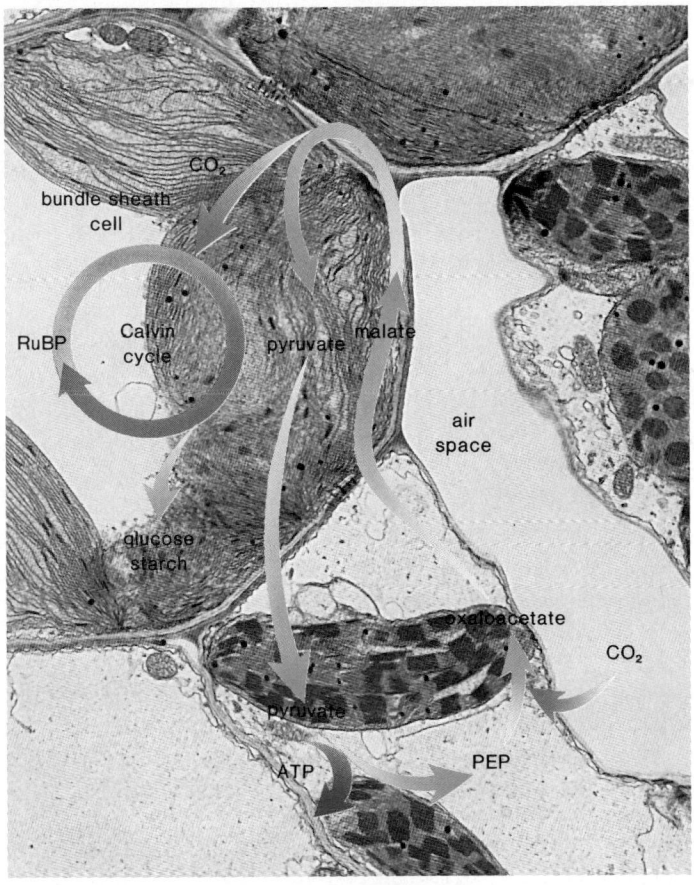

ecule of PGA and one molecule of phosphoglycolate; the latter rapidly breaks down to release CO_2:

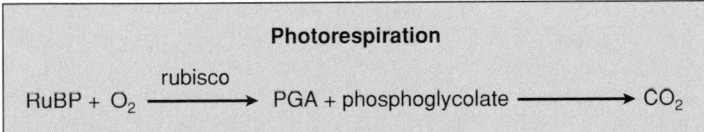

Photorespiration

$$RuBP + O_2 \xrightarrow{\text{rubisco}} PGA + \text{phosphoglycolate} \longrightarrow CO_2$$

This process is named photorespiration because in the presence of light (*photo*), oxygen is taken up and CO_2 is evolved (*respiration*). Obviously, this is just the reverse of the photosynthetic process: photorespiration wastes energy and does nothing to serve the needs of the plant.

Notice in the diagram on page 131 that there are little openings called stomata (stoma, sing.) in the surfaces of leaves through which water can leave and carbon dioxide can enter. If the weather is hot and dry, these openings close in order to

Figure 8.12

Percentage of grasses that have C_4 photosynthesis in areas of North America. The higher temperatures in southern locations apparently offer some advantage to these plants because they are more prevalent in these locations.

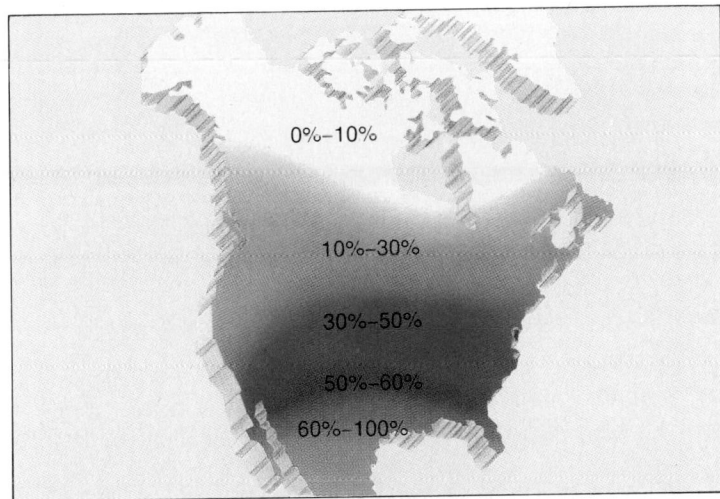

conserve water. Therefore, this is also the time when the concentration of CO_2 is the lowest in the leaf and when photorespiration will predominate in a C_3 plant. C_4 photosynthesis evolved as a way to circumvent this problem since pepco never binds to O_2—regardless of the CO_2 concentration in the leaf. The energy-requiring "extra reactions" of C_4 photosynthesis that occur in the mesophyll cells can deliver a constant supply of CO_2 to the Calvin cycle in the bundle sheath cells. This keeps the CO_2 concentration high in bundle sheath cells and prevents photorespiration from occurring to any extent, even if the weather is hot and dry.

When the weather is moderate, C_3 plants are not at a disadvantage, but when the weather becomes hot and dry, C_4 plants have a distinct advantage, and we can expect them to predominate (fig. 8.12). In the early summer, C_3 plants such as Kentucky bluegrass and creeping bentgrass predominate in lawns in the cooler parts of the United States, but by midsummer, crabgrass, a C_4 plant, begins to take over.

C_4 plants have an advantage over C_3 plants when the weather is hot and dry because their method of CO_2 fixation and their related structure prevent photorespiration from occurring to any extent.

CAM Plants

CAM plants use pepco (PEP carboxylase) to fix carbon dioxide at *night*, forming malate, which is stored in large vacuoles in their mesophyll cells until the next day. CAM stands for crassulacean-acid metabolism; the Crassulaceae are a family of flowering succulent plants, including the stonecrops—low-lying plants that live in warm, arid regions of the world (fig. 8.13). CAM was first

Figure 8.13
Most CAM plants, such as this stonecrop, are succulents that
have fleshy stems or leaves. They conserve water by keeping
their stomata closed during the day and can live under very
arid conditions.

Table 8.3
The 3 Pathways of Carbon Dioxide Fixation

Feature	C_3	C_4	CAM
Leaf structure	Bundle sheath cells lacking chloroplasts	Bundle sheath cells having chloroplasts	Large vacuoles in mesophyll cells
Enzyme utilized	Rubisco	Pepco	Pepco
Optimum temperature	15°C–25°C	30°C–40°C	35°C
Productivity rate (tons/ hectare/year)	22 ± 0.3	39 ± 17	Low—variable

Modified with permission from D. K. Northington and J. R. Goodin, *The Botanical World*. St. Louis, 1984, Times Mirror/Mosby College Publishing.

discovered in these plants, but now it is known to be prevalent among most succulent plants that grow in desert environments, including the cacti.

Whereas a C_4 plant represents partitioning in space—carbon dioxide fixation occurs in mesophyll cells and the Calvin cycle occurs in bundle sheath cells—CAM is an example of partitioning carbon dioxide fixation in time. The malate formed at night releases CO_2 during the day to the Calvin cycle within the same cell, which now has NADPH and ATP available to it from the light-dependent reactions. The primary reason for this partitioning again has to do with the conservation of water. CAM

plants open their stomata only at night, and therefore only at that time is atmospheric CO_2 available to pepco. When sunlight reaches the plant, the stomata close to conserve water and CO_2 cannot enter the plant.

Photosynthesis in a CAM plant is usually not as efficient as in a C_3 plant or a C_4 plant, but it does allow CAM plants to live under stressful conditions. Each of the 3 forms of carbon dioxide fixation does have an advantage. CAM plants can at least live under arid conditions; C_4 plants are more efficient photosynthesizers under hot, dry conditions; and C_3 plants are more efficient photosynthesizers below a temperature of 25°C.

> CAM plants use pepco to carry on carbon dioxide fixation at night when the stomata are open. During the day, the stored carbon dioxide can enter the Calvin cycle.

Table 8.3 contrasts the features of these 3 types of plants.

Summary

1. Photosynthesis (a) provides food either directly or indirectly for most living things, (b) produces oxygen, and (c) provides the energy present today in fossil fuels.

2. Photosynthesis uses solar energy in the visible-light range; the photons of this range contain the right amount of energy to energize the electrons of chlorophyll molecules. Specifically, chlorophylls *a* and *b* absorb violet-blue and red-orange wavelengths best. This causes chlorophyll to appear green to us.

3. A chloroplast is bounded by a double membrane and contains 2 main portions: the liquid stroma and the membranous grana made up of thylakoid sacs. The light-dependent reactions take place in the thylakoids, and the light-independent reactions take place in the stroma.

4. The noncyclic electron pathway of the light-dependent reactions begins when solar energy enters Photosystem II, where it activates the reaction center P680. Energized electrons leave P680 and go to an electron acceptor. The splitting of water replaces these electrons in P680, releases oxygen to the atmosphere, and gives hydrogen ions (H+) to the thylakoid space. The acceptor molecule passes electrons to Photosystem I by way of a cytochrome (electron transport) system. Light energy absorbed by Photosystem I is received by its reaction center P700. The energized electrons that leave this molecule are ultimately received by NADP+, which also combines with H+ from the stroma to become NADPH.

5. In the cyclic electron pathway, the energized electrons that leave Photosystem I are not sent to NADP+. Instead, they pass down the cytochrome system back to Photosystem I again. The cyclic pathway may occur only if there is no free NADP+ to receive the electrons; this is a way to prevent an energy overload of Photosystem I. It serves to generate ATP,

and in keeping with its name, it evolved first and is the only photosynthetic pathway present in certain bacteria today.

6. The energy made available by the passage of electrons down the cytochrome system allows carriers to pump hydrogen (H^+) ions into the thylakoid space. The buildup of hydrogen ions establishes an electrochemical gradient. When hydrogen ions flow down this gradient through the channel present in ATP synthase complexes, ATP is synthesized from ADP and $\textcircled{P}$.

7. The energy yield of the light-dependent reactions is stored in ATP and NADPH. These molecules are now used by the light-independent reactions to reduce carbon dioxide to carbohydrate.

8. During the light-independent reactions, the enzyme rubisco fixes carbon dioxide to RuBP giving a 6-carbon molecule that immediately breaks down to 2 C_3 molecules. This is the first stage of the Calvin cycle. During the second stage, $R—CO_2$ is reduced to $R—CH_2O$, and this step requires the NADPH and some of the ATP from the light-dependent reactions. For every 3 turns of the Calvin cycle, the net gain is one PGAL molecule; the other 5 PGAL molecules are used to reform 3 molecules of RuBP. This step also requires ATP for energy.

9. In C_4 plants, as opposed to the C_3 plants just described, the enzyme pepco fixes carbon dioxide to PEP to form a 4-carbon molecule, oxaloacetate, within mesophyll cells. A reduced form of this molecule is pumped into bundle sheath cells where carbon dioxide is released to the Calvin cycle. The resulting pyruvate returns to mesophyll cells, where ATP must be used to regenerate PEP. C_4 plants avoid photorespiration by a partitioning of pathways in space—carbon dioxide fixation in mesophyll cells and Calvin cycle in bundle sheath cells.

10. During CAM photosynthesis, pepco fixes carbon dioxide to PEP at night to produce malate. The next day, carbon dioxide is released and enters the Calvin cycle within the same cells. This represents a partitioning of pathways in time: carbon dioxide fixation at night and the Calvin cycle during the day. The plants that carry on CAM are desert plants, in which the stomata only open at night in order to conserve water.

Writing Across the Curriculum

In order to practice writing skills, students should write out the answers to any or all of the study questions and the critical thinking questions. The study questions are sequenced in the same order as the text. Suggested answers to the critical thinking questions are in appendix D.

Study Questions

1. Why is it proper to say that all living things are dependent on solar energy?
2. Contrast the electromagnetic spectrum with the absorption spectrum of chlorophyll. Why is chlorophyll a green pigment?
3. What are the inputs and outputs for the light-dependent reactions and for the light-independent reactions? Where do these reactions take place in the chloroplast?
4. Describe the structure of a photosystem. Explain the terms *P700* and *P680*.
5. Using figure 8.6, trace noncyclic and cyclic electron pathways, explaining the events as they occur.
6. Contrast the noncyclic and cyclic electron pathways and explain the differences noted.
7. Describe the chemiosmotic process of ATP production, mentioning the thylakoid membrane, the electron transport system, ATP synthase complex, and ATP synthase.
8. Describe the 3 stages of the Calvin cycle. What happens during the stage that utilizes ATP and NADPH from the light-dependent reactions? What specific molecules are involved?
9. Explain C_4 photosynthesis, contrasting the actions of rubisco and pepco.
10. Explain CAM photosynthesis, contrasting it to C_4 photosynthesis in terms of partitioning of a pathway.

Objective Questions

1. The absorption spectrum of chlorophyll
   a. approximates the action spectrum of photosynthesis.
   b. explains why chlorophyll is a green pigment.
   c. shows that some colors of light are absorbed more than others.
   d. All of these.
2. The final acceptor of electrons during the noncyclic electron pathway is
   a. Photosystem I.
   b. Photosystem II.
   c. ATP.
   d. $NADP^+$.
3. A photosystem contains
   a. pigments, a reaction center, and an electron acceptor.
   b. ADP, $\textcircled{P}$, and H^+.
   c. proto ns, photons, and pigments.
   d. Both b and c.
4. Which of these should not be associated with the electron transport system?
   a. cytochromes
   b. movement of H^+ into the thylakoid space
   c. photophosphorylation
   d. absorption of solar energy
5. Pepco has an advantage compared to rubisco. The advantage is that
   a. pepco is present in both mesophyll and bundle sheath cells, rubisco is not.
   b. rubisco fixes carbon dioxide only in C_4 plants, whereas pepco does it in both C_3 and C_4 plants.
   c. rubisco is subject to photorespiration but pepco is not.
   d. Both b and c.

6. The NADPH and ATP from the light-dependent reactions are used to
   a. cause rubisco to fix carbon dioxide.
   b. reform the photosystems.
   c. cause electrons to move along their pathways.
   d. convert PGA to PGAL.
7. CAM photosynthesis
   a. is the same as C_4 photosynthesis.
   b. is an adaptation to cold environments in the Southern Hemisphere only.
   c. is prevalent in desert plants that close their stomata during the day.
   d. occurs in plants that live in marshy areas.
8. Chemiosmotic phosphorylation depends on
   a. an electrochemical gradient.
   b. a difference in H^+ concentration between the thylakoid space and the stroma.
   c. ATP breaking down to ADP + Ⓟ .
   d. Both a and b.
9. Label this diagram of a chloroplast.

a. The light-dependent reactions occur in which part of a chloroplast?
b. The light-independent reactions occur in which part of a chloroplast?
10. Label this diagram using these labels: water, carbohydrate, carbon dioxide, oxygen, ATP, ADP +Ⓟ, NADPH, and NADP.

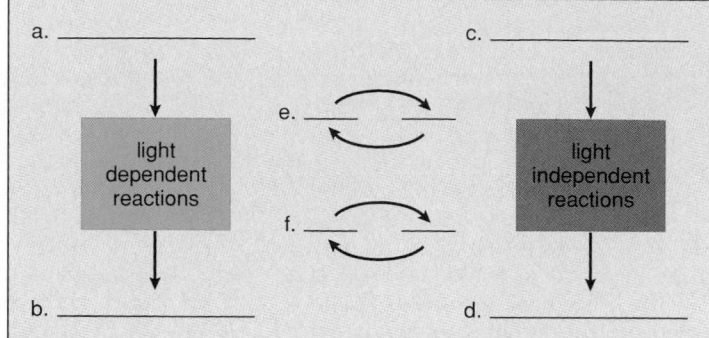

a. _____   c. _____

| light dependent reactions | e. _____ ⇄ _____ | light independent reactions |

f. _____ ⇄ _____

b. _____   d. _____

Concepts and Critical Thinking

1. *Virtually all living things are dependent on solar energy.*

   What data would you use to convince a friend that the lives of human beings are dependent on solar energy?

2. *Photosynthesis makes solar energy available to living things.*

   Tell how photosynthesis makes solar energy available by explaining the roles of chlorophyll, ATP, $NADP^+$, and CO_2 in photosynthesis.

3. *The cell and the organelles themselves are compartmentalized, and this assists their proper functioning.*

   Name 2 major compartments (separate areas) of a chloroplast, and tell how photosynthesis is dependent on keeping the areas separate.

Selected Key Terms

photon (fo'ton) 122
electromagnetic spectrum (e-lek"tro-mag-net'ik spek'trum) 122
visible light (viz'ĭ-b'l līt) 122
chloroplast (klo'ro-plast) 123
stroma (stro'mah) 123
granum (gra'num) 123
thylakoid (thi'lah-koid) 125
chlorophyll (klo'ro-fil) 125

light-dependent reaction (līt de-pen'dent re-ak'shun) 125
light-independent reaction (līt in"de-pen'dent re-ak'shun) 125
photosystem (fo"to-sis'tem) 125
noncyclic photophosphorylation (non-sik'lik fo"to-fos"for-ĭ-la'shun) 128
cyclic photophosphorylation (sik'lik fo"to-fos"for-ĭ-la'shun) 128
ATP synthase complex (sin'thase kom'pleks) 129

Calvin cycle (kal'vin ben'sun si'k'l) 129
carbon dioxide (CO_2) fixation (kar'bon di-ok'sīd fix-sa'shun) 130
RuBP 130
C_4 plant 131
C_3 plant 131
photo respiration (fo"to res"pi-ra'shun) 133
CAM 133

9

Glycolysis and Cellular Respiration

The powerful muscles of this cheetah allow it to dart forward to capture prey. Muscles are powered by ATP energy, and mitochondria are the organelles that convert the energy of other organic compounds to the energy of ATP.

Your study of this chapter will be complete when you can

1. describe the general function of cellular respiration;
2. describe both the anaerobic and aerobic fate of pyruvate;
3. discuss glycolysis, explaining the process of substrate-level phosphorylation and giving the inputs and outputs of the pathway;
4. describe lactate fermentation and alcoholic fermentation and the evolutionary significance of the fermentation process;
5. discuss the transition reaction, explaining the role of acetyl CoA and giving the inputs and outputs of the reaction;
6. discuss the Krebs cycle, explaining how it cycles and giving the inputs and outputs of the cycle;
7. discuss the electron transport system, and explain how it contributes to the formation of ATP;
8. describe the arrangement of electron carriers in the inner membrane of mitochondria, and explain the process of chemiosmotic ATP synthesis;
9. calculate the yield of ATP molecules per glucose molecule for glycolysis and cellular respiration;
10. discuss the concept of a metabolic pool and how the breakdown of carbohydrate, protein, and fat contributes to the pool.

It is common knowledge that human beings eat food, which they digest to nutrient molecules, but you may not be aware that most other organisms also use organic molecules as a source of building blocks and energy. Carbohydrate molecules (e.g., glucose) are a primary energy source because they are often broken down to acquire a supply of ATP molecules. We can write the overall equation for carbohydrate breakdown in this manner:

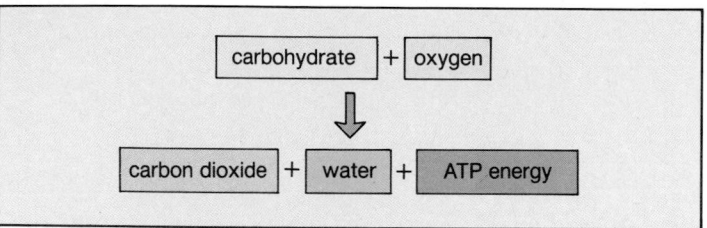

ATP, the energy currency of cells (p. 115), provides the energy that cells need for transport work, mechanical work, and synthetic work. Why do living things convert carbohydrate energy to ATP energy? Why not, for example, use glucose energy directly, thereby bypassing mitochondria? The answer is that the energy content of glucose is too great for individual cellular reactions; ATP contains the right amount of energy, and enzymes are adapted through evolution to couple ATP breakdown with energy-requiring cellular processes (fig. 7.12).

In cells, the energy of carbohydrates is converted to that of ATP molecules, the energy currency of cells.

You know from chapter 5 that chloroplasts produce the energy-rich carbohydrates that typically undergo cellular respiration in mitochondria. Cellular respiration, then, is the link between the energy captured by plants and the energy utilized by both plants and animals (fig. 9.1a). It is the means by which energy flows from the sun through all living things. This is true because, for example, plants produce the food eaten by animals.

As we discussed in chapter 7, there is always loss of usable energy when one form of energy is transformed into another form of energy. Eventually, the energy content of food is completely dissipated, but the chemicals themselves cycle. Note that when

Figure 9.1

a. Which of these 2 organisms is carrying on cellular respiration—the giraffe or the acacia tree? Both are, because both animals and plants require a source of ATP energy. **b.** The energy-rich carbohydrate molecules, broken down largely within the mitochondria, are originally derived from photosynthesis carried out within chloroplasts. The flow of *energy* among organisms (e.g., from plants to animals) goes one way only, but the *chemicals* can recycle because the carbon dioxide and water given off by the mitochondria are reused by the chloroplasts.

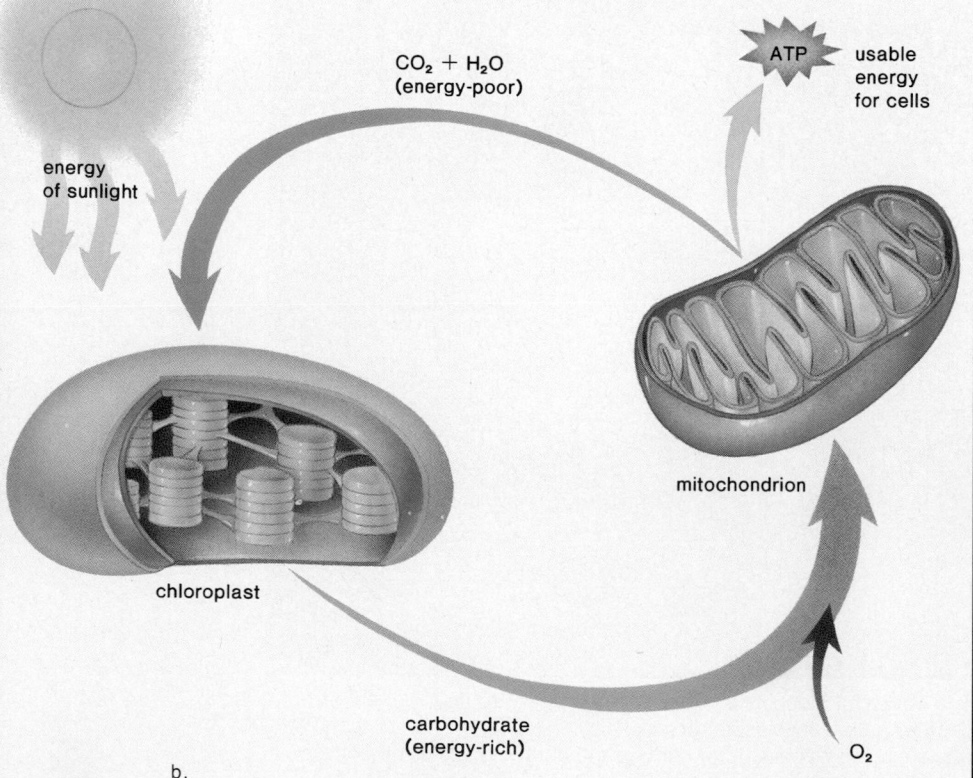

a.

b.

carbohydrate is broken down, the low-energy molecules carbon dioxide and water are given off. These are the raw materials for photosynthesis (fig. 9.1b).

> The energy content of carbohydrate was originally provided by solar energy.

Overview of Glucose Metabolism

We will especially be concerned with the breakdown of glucose because this molecule is the primary carbohydrate that cells break down for a constant supply of ATP molecules. The energy in the bonds of glucose is not released all at once; instead the energy is released slowly, step by step. *Metabolic pathways* (p. 108) are a series of enzymatic reactions that in this case allow the step-by-step breakdown of glucose. This gradual breakdown allows a concomitant gradual buildup of ATP molecules. If glucose is metabolized completely to carbon dioxide (CO_2) and water (H_2O), the cell realizes a maximum yield of 36 ATP molecules. The energy in 36 ATP molecules is equivalent to 39% of the energy that was available in glucose.

Glycolysis and Fermentation

Glycolysis is a metabolic pathway that breaks down glucose (C_6) to 2 molecules of **pyruvate** (C_3) (fig. 9.2). Glycolysis takes place in the cytosol outside mitochondria and is anaerobic—it does not require oxygen. Oxidation by removal of hydrogen atoms ($e^- + H^+$) does occur, though, and the energy of oxidation is used to generate 2 ATP molecules.

Pyruvate is a pivotal metabolite. What happens to pyruvate depends on whether oxygen is available to the cell. If oxygen is not available, anaerobic metabolism continues and fermentation occurs. **Fermentation** includes glycolysis plus the reduction of pyruvate to alcohol or to lactate. Yeast cells carry on alcoholic fermentation when they ferment sugars in fruits and grains to the alcohol in wine and beer. You may be aware that, in the absence of oxygen, our muscle cells carry on lactate fermentation. A cramping of muscles may occur when lactate builds up. With aerobic exercise, a common endeavor today, lactate fermentation is not expected and there should be no cramping of the muscles.

> **Glycolysis:** anaerobic process in cytosol; breakdown of glucose to pyruvate; net gain of 2 ATP per glucose molecule
> **Fermentation:** anaerobic process in cytosol; breakdown of glucose to alcohol or lactate; net gain of 2 ATP per glucose molecule

Glycolysis and Cellular Respiration

Pyruvate from glycolysis enters a mitochondrion when oxygen is available. **Cellular respiration** is the complete breakdown of pyruvate to carbon dioxide and water. Cellular respiration occurs in mitochondria and is aerobic—oxygen is required. Cellular respiration involves the conversion of pyruvate to an acetyl group, and 2 other

Figure 9.2

Overview of glucose metabolism. Glycolysis in the cytosol produces pyruvate, a key intermediary metabolite. If oxygen is not available, anaerobic metabolism continues and pyruvate is reduced to either alcohol or lactate depending on the type of cell. If oxygen is available, aerobic respiration begins in a mitochondrion and pyruvate is metabolized completely to carbon dioxide and water. The net gain of ATP for glycolysis (and fermentation) is 2 ATP; the net gain of ATP for cellular respiration is 34 ATP; the net gain of ATP for the complete breakdown of glucose is 36 ATP.

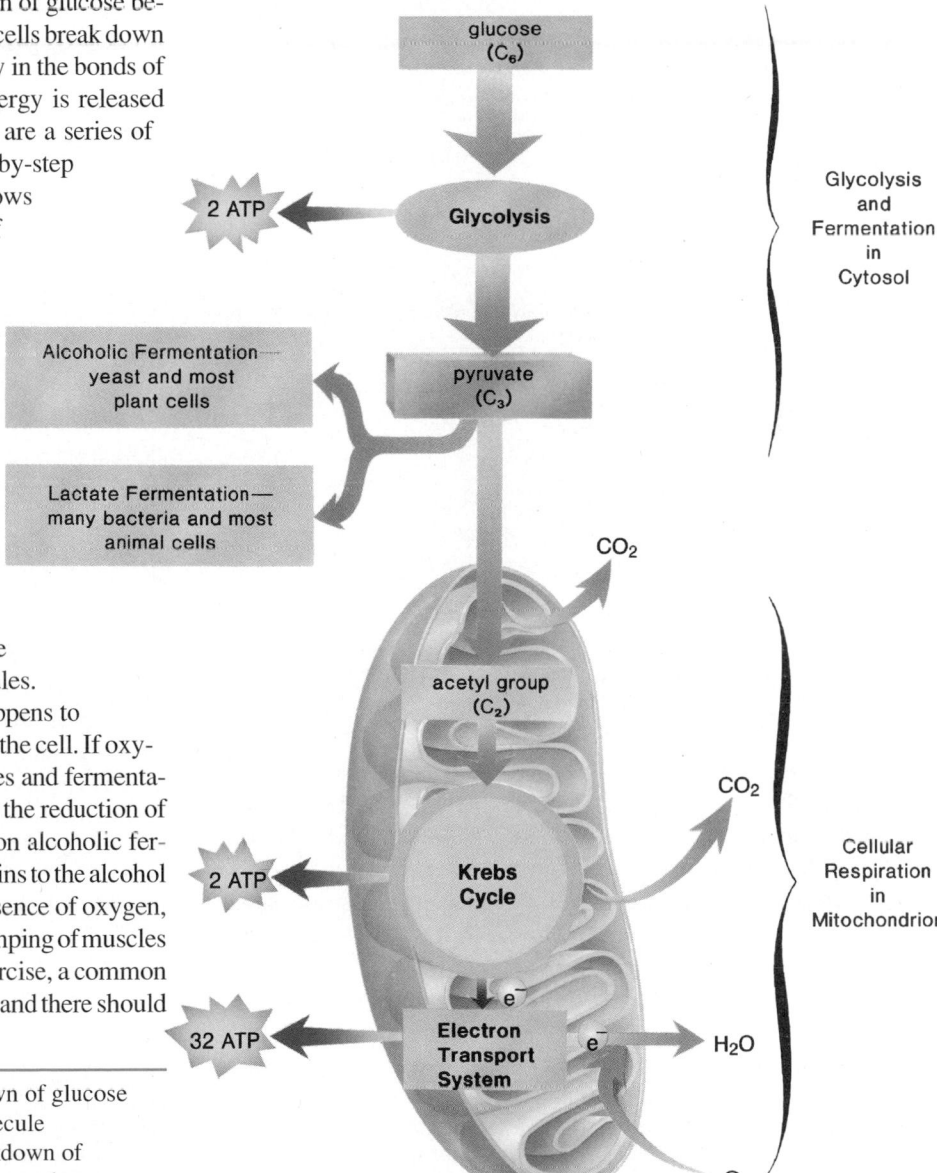

metabolic pathways shown in figure 9.2, namely the Krebs cycle and an electron transport system. At the completion of cellular respiration, glucose has been broken down completely to carbon dioxide and water. It is possible to write an overall equation for the complete breakdown of glucose as is done in figure 9.3, that is:

$$C_6H_{12}O_6 + 6O_2 \rightarrow 6CO_2 + 6H_2O$$

Figure 9.3

Complete glucose breakdown. Glucose enters a cell from the bloodstream and is metabolized to pyruvate which enters a mitochondrion. Oxygen from the bloodstream also enters a mitochondrion. Here pyruvate is metabolized completely to carbon dioxide and water, which exit the cell. The resulting ATP molecules stay behind in the cell to enable the cell to perform all sorts of energy-requiring processes.

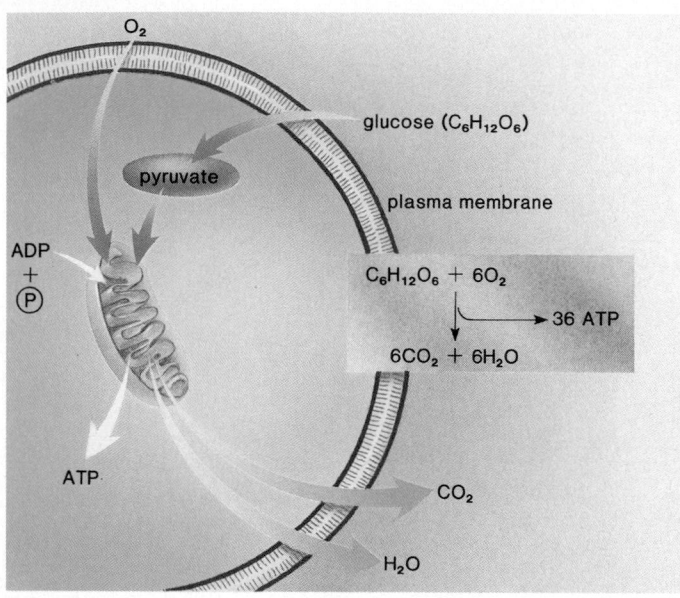

Unfortunately, the overall equation has limited usefulness because it does not remind us that complete glucose breakdown involves glycolysis, the conversion of pyruvate to an acetyl group, the Krebs cycle, and the electron transport system. The equation, however, allows us to discuss how the human body provides the reactants and deals with the end products of glucose breakdown. The air we breathe contains *oxygen* (O_2) and the food we eat contains *glucose* ($C_6H_{12}O_6$). The oxygen and the glucose enter the bloodstream, which carries them to the body's cells, where they diffuse into each and every cell. During cellular respiration in a mitochondrion, pyruvate is broken down to *carbon dioxide* (CO_2) and *water* (H_2O) as ATP is produced. All 3 of these then leave the mitochondrion. The ATP is utilized inside the cell for energy-requiring processes. Carbon dioxide diffuses out of the cell into the bloodstream. The bloodstream takes the carbon dioxide to the lungs, where it is exhaled. The water molecules produced, called metabolic water, play little if any role in humans. In other organisms, metabolic water helps prevent dehydration.

Even though the overall equation for glucose breakdown seems to suggest that oxygen combines directly with glucose, this is not the case at all (fig. 9.3). Instead, as expected, oxidation occurs by the removal of hydrogen atoms ($e^- + H^+$) from metabolites. We will see that these electrons (e^-) are carried by coenzyme NAD^+ (and FAD) to the electron transport system (p. 146), where they are eventually transferred to oxygen molecules. Reduced oxygen then takes up hydrogen (H^+) ions and becomes water.

Cellular respiration: aerobic respiration in mitochondria; breakdown of pyruvate to carbon dioxide and water; pyruvate → acetyl group, Krebs cycle, electron transport system; gain of 34 ATP per glucose molecule

Net maximum gain of ATP per glucose molecule breakdown to CO_2 and H_2O:

Glycolysis	2 ATP
Cellular respiration	34 ATP
	36 ATP

Glycolysis

Figure 9.4 illustrates the main reactions of glycolysis, the breakdown of glucose to 2 molecules of pyruvate. First, 2 ATP molecules are used to phosphorylate metabolites in order to activate them (steps 1 and 2). The term *glycolysis,* meaning the splitting of sugar, is appropriate because glucose is split into 2 smaller molecules (step 3). All reactions after this step are multiplied by 2.

Hydrogen atoms ($e^- + H^+$) are removed from a metabolite of the pathway and are picked up by NAD^+, a coenzyme of dehydrogenases (p. 114):

$$NAD^+ \xrightarrow{\text{2H}} NADH + H^+$$

Altogether, 2 NADH molecules result because step 4 occurs twice. The energy of oxidation results in high-energy phosphate bonds attached to certain substrates. These substrates are used to produce 4 ATP molecules (steps 5 and 7). Since 2 ATP molecules were used to get started, glycolysis yields a net gain of 2 ATP molecules per glucose molecule.

Glycolysis provides 2 NADH molecules and a net gain of 2 ATP per glucose molecule.

ATP production during glycolysis is called **substrate-level phosphorylation** because substrates have transferred high-energy phosphate bonds to ADP. This process is illustrated in figure 9.5, where the molecule PGAP gives up a high-energy phosphate to ADP and becomes PGA. ADP becomes ATP (Step 5 in fig. 9.4).

The end product of glycolysis is 2 molecules of pyruvate, a C_3 molecule (step 8 in fig. 9.4).

Altogether the inputs and outputs of glycolysis are as follows:

Glycolysis	
inputs	outputs
glucose	2 pyruvate
2 NAD$^+$	2 NADH
2 ADP + Ⓟ	2 ATP (net)

Figure 9.4

Glycolysis is a metabolic pathway that begins with glucose and ends with pyruvate. Net gain of 2 ATP molecules can be calculated by subtracting those expended from those produced. Text in boxes to the right explains the reactions.

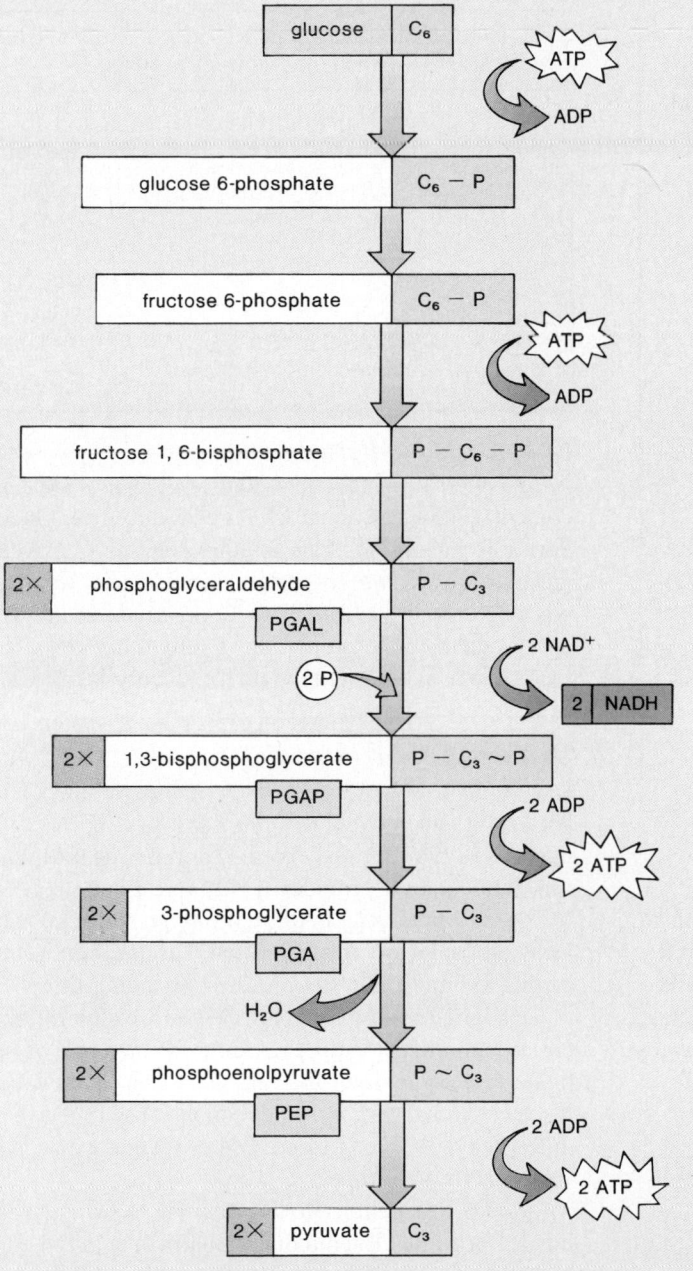

1. Phosphorylation of glucose by ATP produces an activated molecule.

2. Rearrangement, followed by a second ATP phosphorylation, gives fructose bisphosphate.

3. The 6-carbon molecule is split into 2 3-carbon PGAL. After this, each reaction has to occur twice per each glucose molecule.

4. Oxidation, followed by phosphorylation, produces 2 NADH molecules and gives 2 PGAP molecules—each with one high-energy phosphate bond, represented by a wavy line.

5. Removal of high-energy phosphate by 2 ADP molecules produces 2 ATP molecules and gives 2 PGA molecules. This pays back the original investment of 2 ATP used in steps 1 and 3.

6. Oxidation by removal of water gives 2 PEP molecules, each with a high-energy phosphate bond.

7. Removal of high-energy phosphate by 2 ADP molecules produces 2 ATP molecules. The net energy yield for the glycolysis pathway is 2 molecules of ATP.

8. Pyruvate is the end product of the glycolytic pathway. If cellular respiration occurs, pyruvate enters the mitochondria for further breakdown.

Figure 9.5

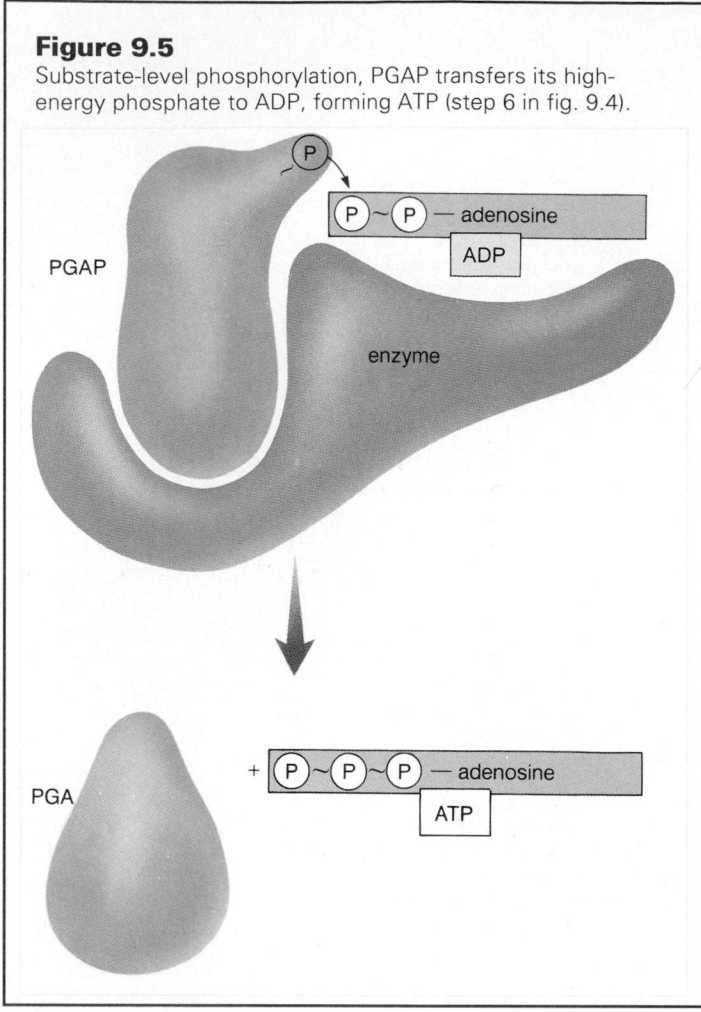

Substrate-level phosphorylation, PGAP transfers its high-energy phosphate to ADP, forming ATP (step 6 in fig. 9.4).

Fermentation

Fermentation includes glycolysis plus the reduction of pyruvate to either lactate or alcohol. During fermentation, pyruvate accepts the 2 hydrogen atoms that were removed from glycolytic metabolites earlier. In the 2 most familiar forms of fermentation, lactate fermentation and alcoholic fermentation, the end products are lactate (in many bacteria and in animal cells) or ethyl alcohol and carbon dioxide (in yeasts and in most plant cells):

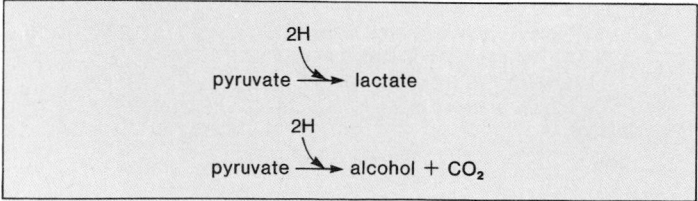

Recall that step 4 in glycolysis (fig. 9.4) requires the coenzyme NAD$^+$ and that following this step, ATP production occurs. Glycolysis is an **anaerobic** process, and if oxygen is not available, glycolysis is the only means of producing ATP. Therefore, it is extremely important to the life of a cell that glycolysis

continue when oxygen is not available. For glycolysis to continue, there must be a source of "free" NAD$^+$, that is, NAD$^+$ that is not carrying hydrogen atoms. During fermentation, NADH from step 4 of glycolysis (fig. 9.4) becomes free NAD$^+$ by passing on hydrogen atoms to pyruvate. Now glycolysis can continue and ATP production can occur (fig. 9.6).

The purpose of fermentation is to supply the glycolytic pathway with free NAD$^+$ when oxygen is not available.

The inputs and outputs of fermentation are as follows:

Fermentation	
inputs	outputs
glucose	2 lactate or 2 alcohol and 2 CO_2
2 ADP	2 ATP (net)

Energy Yield of Fermentation

A high-energy phosphate bond in an ATP molecule has an energy content of 7.3 Kcal, and 2 ATP are produced per glucose molecule during fermentation. This is equivalent to 14.6 Kcal. Complete glucose breakdown to carbon dioxide and water represents a possible energy yield of 686 Kcal per molecule. Therefore, the energy yield for fermentation is only $14.6/686 = 2.1\%$. This is much less than the energy yield for glycolysis followed by aerobic cellular respiration (p. 143).

Usefulness of Fermentation

Despite its low yield of ATP, fermentation does have its place, because it can provide a *rapid burst* of ATP energy. When our muscles work vigorously over a short period of time, as when we run, fermentation provides a way to produce ATP even though oxygen is temporarily in limited supply. At first, the blood carries away all the lactate formed in the muscles, but eventually lactate begins to build up in the muscles, changing the pH and causing muscle fatigue and discomfort. When we stop running, our body has a chance to catch up; the lactate is converted back to pyruvate by the removal of hydrogen atoms, which can now be transported to the electron transport system. In the meantime, we are said to be in **oxygen debt,** as signified by the fact that we keep on breathing very heavily for a time.

Fermentation allows yeast cells to grow and divide anaerobically for a time. Eventually, though, if the initial glucose level is high, they are killed by the very alcohol they produce. Presumably, human beings were delighted to discover this form of fermentation, as the ethyl alcohol produced by the process has been consumed in great quantity for thousands of years. In addition, we use the CO_2 produced to make bread rise.

Evolutionary Perspective

Glycolysis and fermentation have probably been present as long as there have been living things on earth. At first, there was no oxygen

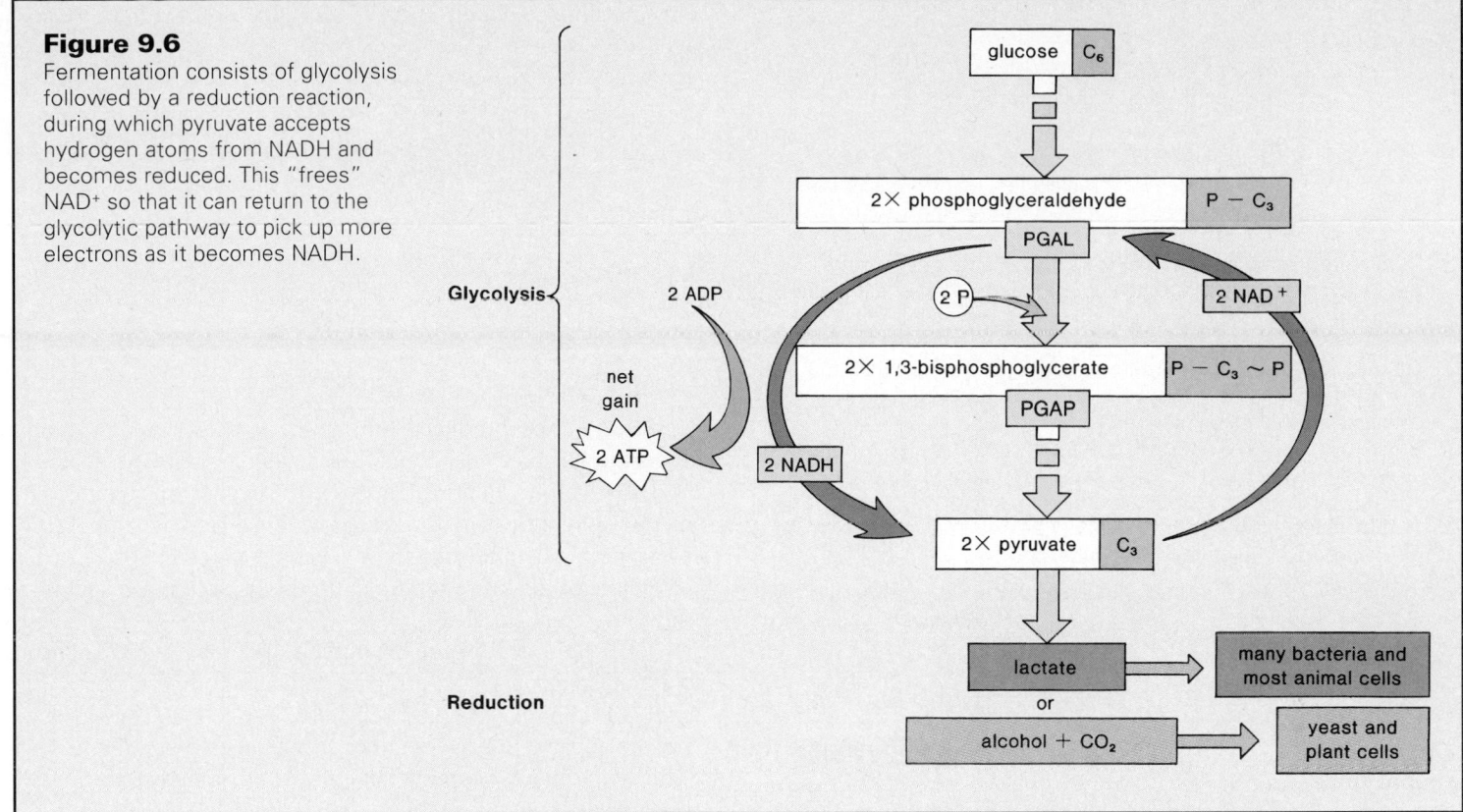

Figure 9.6
Fermentation consists of glycolysis followed by a reduction reaction, during which pyruvate accepts hydrogen atoms from NADH and becomes reduced. This "frees" NAD⁺ so that it can return to the glycolytic pathway to pick up more electrons as it becomes NADH.

in the atmosphere, but there were organic molecules in the oceans where the first cell(s) arose. Any cell that could utilize the energy of these molecules anaerobically had a competitive advantage, even if the method was not the most efficient. The presence of some form of fermentation in nearly all organisms shows how important this ancient competitive edge was to the continuance of life. Apparently, in the ancient past, only cells capable of fermentation survived, passing on the trait to future organisms.

Since fermentation is very inefficient compared to aerobic respiration, you might think it would have simply disappeared sometime during the history of life on earth. As with the 2 photosystems involved in photosynthesis, however, it appears that evolution simply builds on what has already been accomplished. In this instance, we see that pyruvate, the end product of glycolysis, can be metabolized either anaerobically or aerobically. Glycolysis, then, may have initially evolved as a part of fermentation, but today it is usually preliminary to aerobic respiration.

Cellular Respiration

Cellular respiration is aerobic respiration that occurs in mitochondria. During cellular respiration pyruvate from glycolysis is broken down completely to carbon dioxide and water. Altogether there is a gain of 36 ATP for the complete breakdown of glucose (fig. 9.7). This is far more than the 2 ATP that result from fermentation, an anaerobic process.

During cellular respiration, organic molecules are oxidized by the removal of hydrogen atoms ($e^- + H^+$). Most often, these hydrogen atoms are accepted by NAD⁺, which then becomes NADH. NADH delivers electrons (e^-) to an electron transport system, which passes them on to oxygen, which then becomes water (H_2O). Oxygen is the final acceptor for electrons during cellular respiration.

Cellular respiration requires 3 different events: the oxidation of pyruvate to an acetyl group (called the transition reaction), the Krebs cycle, and an electron transport system.

Transition Reaction

The **transition reaction** is so called because it connects glycolysis to the Krebs cycle (fig. 9.8). In this reaction, pyruvate (C_3) is oxidized and broken down to an acetyl group (C_2) and one molecule of CO_2. The acetyl group is transferred to coenzyme A (CoA)—a large molecule that contains a nucleotide and a portion of one of the B vitamins. This combination of acetyl group and CoA is called **acetyl CoA.** Acetyl CoA contains a high-energy bond, and therefore the molecule is called active acetate, AA (fig. 9.7). Oxidation of pyruvate results in an NADH molecule, which delivers electrons to the electron transport system.

During the transition reaction, pyruvate is broken down to an acetyl group and CO_2 is released. Oxidation of pyruvate results in an NADH molecule.

Figure 9.7

The overall equation for glucose breakdown shown does not indicate that the process requires these metabolic pathways plus the transition reaction.

Glycolysis: Glucose is broken down to 2 molecules of pyruvate (PYR) for a net gain of 2 ATP molecules. Electrons exit the pathways and are taken to the electron transport system by NADH.

Transition Reaction: Carbon dioxide (CO_2) and electrons are removed from pyruvate (PYR). Acetyl CoA (AA) results. The electrons are taken to the electron transport system by NADH.

Krebs Cycle: Acetyl CoA enters the cycle; carbon dioxide (CO_2) and electrons exit the cycle. There is a gain of 2 ATP per glucose molecule.

Electron Transport System: The system receives electrons (usually carried by NADH), and as they pass down the system from carrier to carrier (cytochromes are carriers), 32 ATP molecules are built up. Oxygen (O_2) is the final acceptor for electrons and water (H_2O) results.

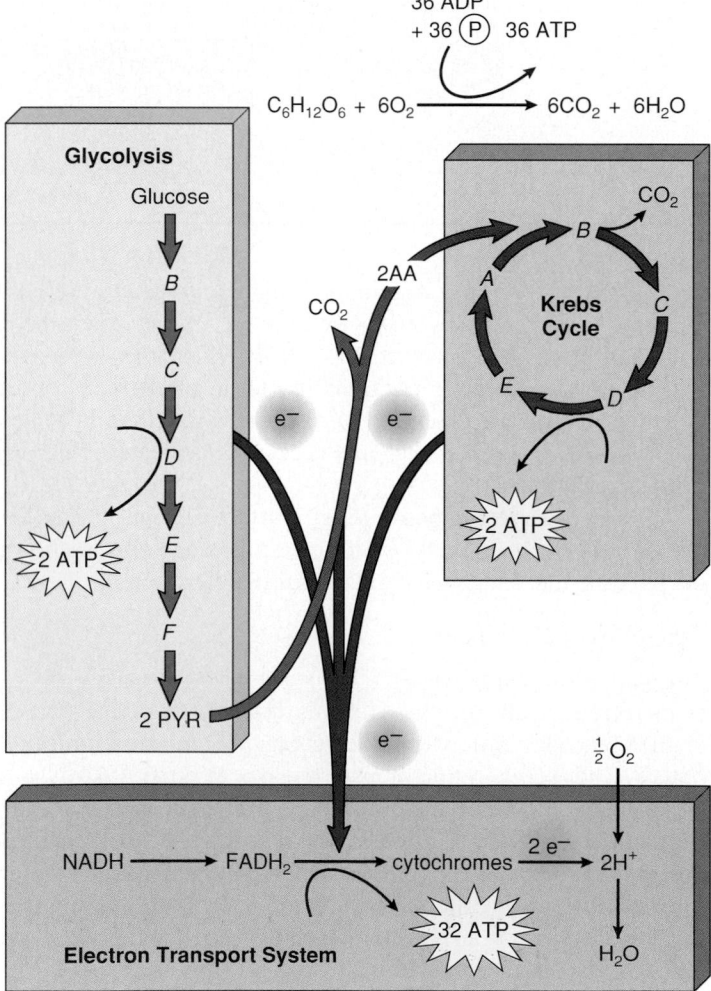

The transition reaction occurs twice for each original glucose molecule. Therefore, inputs and outputs of the transition reaction per glucose molecule are as follows:

Transition Reaction	
inputs	outputs
2 pyruvate 2 CoA 2 NAD$^+$	2 CO_2 2 acetyl CoA 2 NADH

Krebs Cycle

The acetyl CoA produced during the transition reaction then enters the **Krebs cycle,** a cyclical metabolic pathway located in the matrix of mitochondria (fig. 9.9). The Krebs cycle is named for Sir Hans Krebs, a British scientist who received the Nobel Prize for his part in determining these reactions in the 1930s. The first metabolite of the cycle is citrate; for this reason, it is also known as the **citric acid cycle.**

The C_2 acetyl group that enters the Krebs cycle is oxidized to 2 molecules of CO_2. During the oxidation process most of the hydrogen atoms ($e^- + H^+$) are donated to NAD$^+$, but in one instance they are donated to FAD, another coenzyme of oxidation in cells. Some of the energy of oxidation is used immediately to form ATP by substrate-level phosphorylation, as in glycolysis.

During the Krebs cycle, the C_2 acetyl group is oxidized to 2 molecules of CO_2. Oxidation results in NADH and FADH$_2$ molecules. Substrate phosphorylation yields an ATP molecule.

The Krebs cycle turns twice for each original glucose molecule. Therefore, the inputs and outputs of the Krebs cycle per glucose molecule are as follows:

Krebs Cycle	
inputs	outputs
2 acetyl groups 2 ADP + 2 P 6 NAD$^+$ 2 FAD	4 CO_2 2 ATP 6 NADH 2 FADH$_2$

Electron Transport System

The electrons that were removed from the substrates of glycolysis and the Krebs cycle are donated by NADH and FADH$_2$ to an **electron transport system** located on the cristae of mitochondria. Some of the electron carriers of the system are cytochrome molecules; therefore, the system is also called the **cytochrome system.**

Figure 9.10 of the electron transport system illustrates that high-energy electrons are delivered to the top of the system and low-energy electrons leave at the bottom. As a pair of electrons is passed from carrier to carrier, oxidation occurs, and the energy

Figure 9.8

Transition reaction. Inside mitochondria, the oxidation of pyruvate by NAD^+ is accompanied by the release of CO_2. The remaining 2-carbon (acetyl) group binds to coenzyme A (CoA) forming acetyl CoA. This reaction occurs twice per molecule of glucose broken down.

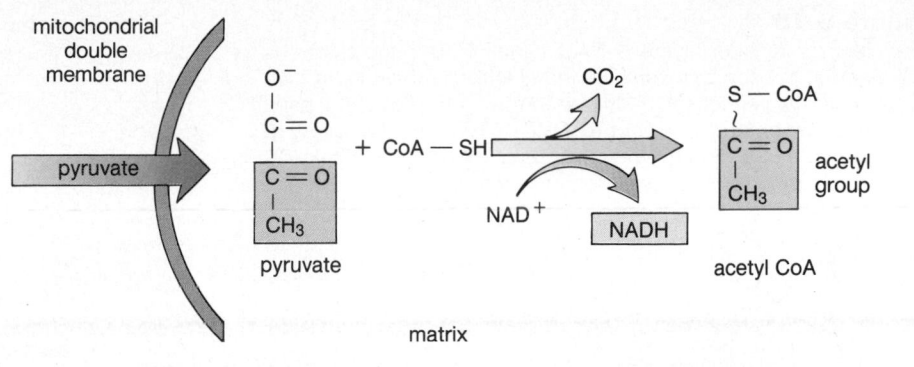

Figure 9.9

Krebs cycle. The net result of this cycle is the oxidation of an acetyl group to 2 molecules of CO_2. The energy of oxidation allows the immediate formation of one molecule of ATP. Most of the energy, however, resides in high-energy electrons, which are accepted by 3 molecules of NAD^+ and one molecule of FAD. NADH and $FADH_2$ then take the electrons to the electron transport system. (Keep in mind that the Krebs cycle turns twice per glucose molecule.)

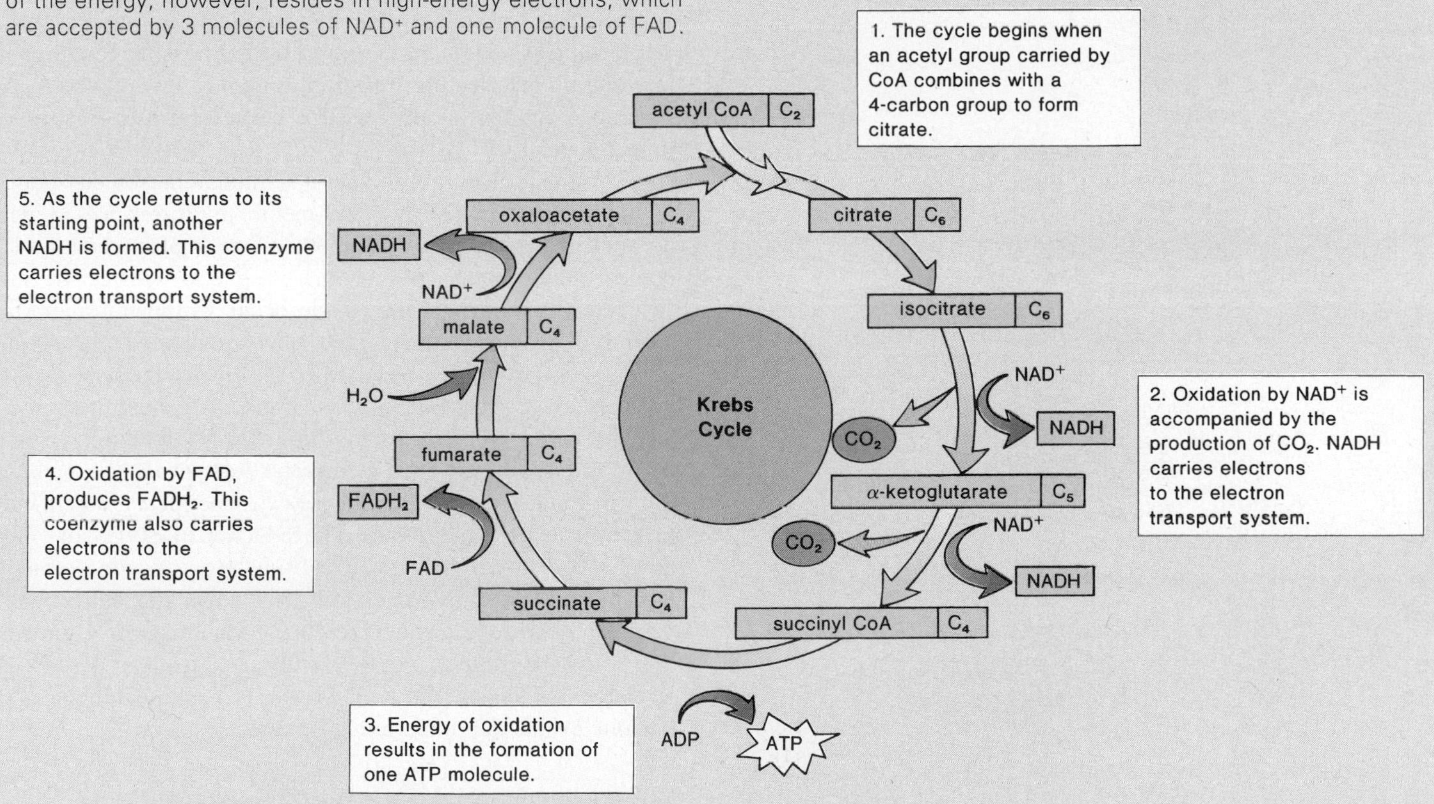

1. The cycle begins when an acetyl group carried by CoA combines with a 4-carbon group to form citrate.

5. As the cycle returns to its starting point, another NADH is formed. This coenzyme carries electrons to the electron transport system.

2. Oxidation by NAD^+ is accompanied by the production of CO_2. NADH carries electrons to the electron transport system.

4. Oxidation by FAD, produces $FADH_2$. This coenzyme also carries electrons to the electron transport system.

3. Energy of oxidation results in the formation of one ATP molecule.

Krebs Cycle

acetyl CoA C_2 — citrate C_6 — isocitrate C_6 — α-ketoglutarate C_5 — succinyl CoA C_4 — succinate C_4 — fumarate C_4 — malate C_4 — oxaloacetate C_4

released is used to form ATP molecules at 3 different sites. This process is sometimes called **oxidative phosphorylation** because oxygen receives the energy-spent electrons from the last of the carriers. An enzyme called cytochrome oxidase splits and reduces molecular oxygen (O_2) to water:

$$\tfrac{1}{2}O_2 + 2e^- + 2H^+ \longrightarrow H_2O$$

Oxygen is, then, the final acceptor for electrons during cellular respiration, an **aerobic** process.

Electrons removed by NAD^+ and FAD from the metabolites of glycolysis and the Krebs cycle are passed from one carrier to the next in the electron transport system. Each pair of electrons releases enough energy to allow the production of 3 ATP molecules.

Figure 9.10

The electron transport system. NADH and $FADH_2$ bring electrons to the electron transport system. As the electrons move down the system energy is released and used to form ATP. For every pair of electrons that enters by way of NADH, 3 ATP result. For every pair of electrons that enters by way of $FADH_2$, 2 ATP result. Oxygen, the final acceptor of the electrons, becomes water.

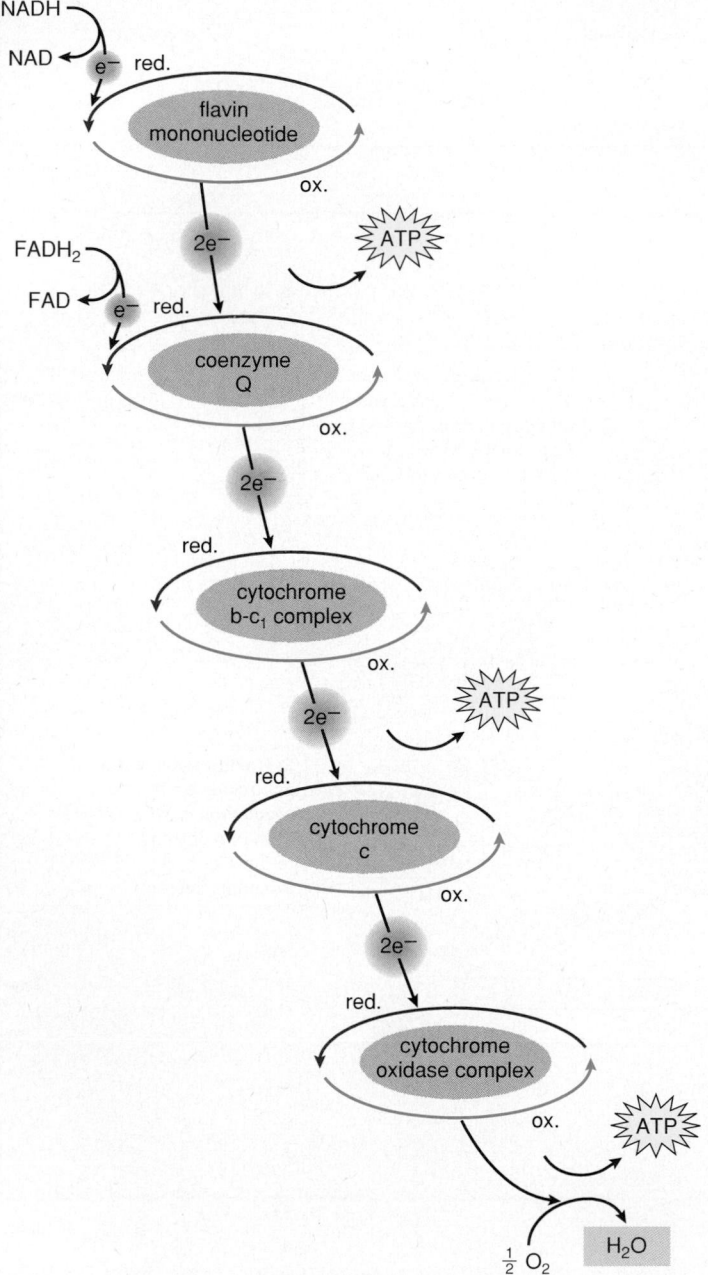

Recycling of NAD[+] and FAD

The coenzymes NAD[+] and FAD are constantly being reduced and oxidized inside the cell:

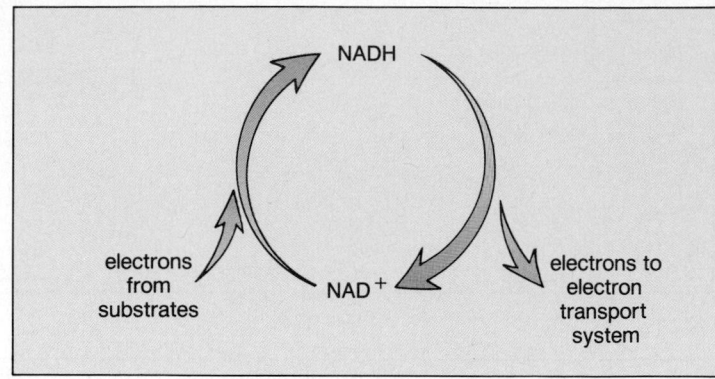

This means that the cell needs only a limited supply of these coenzymes because they are used repeatedly.

Chemiosmotic ATP Synthesis

ATP synthesis occurs as a result of the electron transport system. The *chemiosmotic theory* suggests how it is possible for the energy released by the electron transport system to be coupled to ATP buildup. Coupling is only possible because of the structure of a mitochondrion.

As discussed in chapter 5, a mitochondrion has a double membrane (fig. 9.11). The inner membrane forms cristae, which are shelflike projections that jut out into the matrix, the innermost compartment, which is filled with a gel-like substance. The transition reaction and the Krebs cycle occur within the matrix; the carriers of the electron transport system are located on the cristae.

Electrons are passed down the electron transport system and certain carriers of the electron transport system pump hydrogen (H^+) ions from the matrix into the intermembrane space between the outer and inner membrane. As hydrogen ions enter the intermembrane space, an extreme electrochemical gradient builds up across the inner membrane. The reduction of oxygen to water in the matrix removes hydrogen ions from the matrix and also adds to the gradient. When hydrogen ions flow down an electrochemical gradient by way of a channel protein present in a particle, called the ATP synthase complex, ATP is synthesized from ADP + ⓟ. This is called chemiosmotic ATP synthesis because chemical and osmotic events join to permit ATP synthesis.

Energy Yield From Glucose Metabolism

As we have suggested, it is possible to calculate the ATP yield for the complete breakdown of glucose to carbon dioxide and water. Table 9.1 shows the number of molecules of ATP produced by each part of the process.

Figure 9.11

Mitochondrion structure and function. **a.** Mitochondria have a double membrane with a space between the membranes. The inner membrane forms cristae, which project into the matrix, the innermost compartment. **b.** The transition reaction (PYR→ AA) and the Krebs cycle are located in the matrix. The carriers of the electron transport system are located on the cristae. The chemiosmotic theory says that some of the carriers pump H^+ into the intermembrane space. When these hydrogen ions flow down their electrochemical gradient through ATP synthase complexes, ATP is produced. Notice that the reduction of O_2 in the matrix contributes to the electrochemical gradient by lowering the number of H^+ in the matrix.

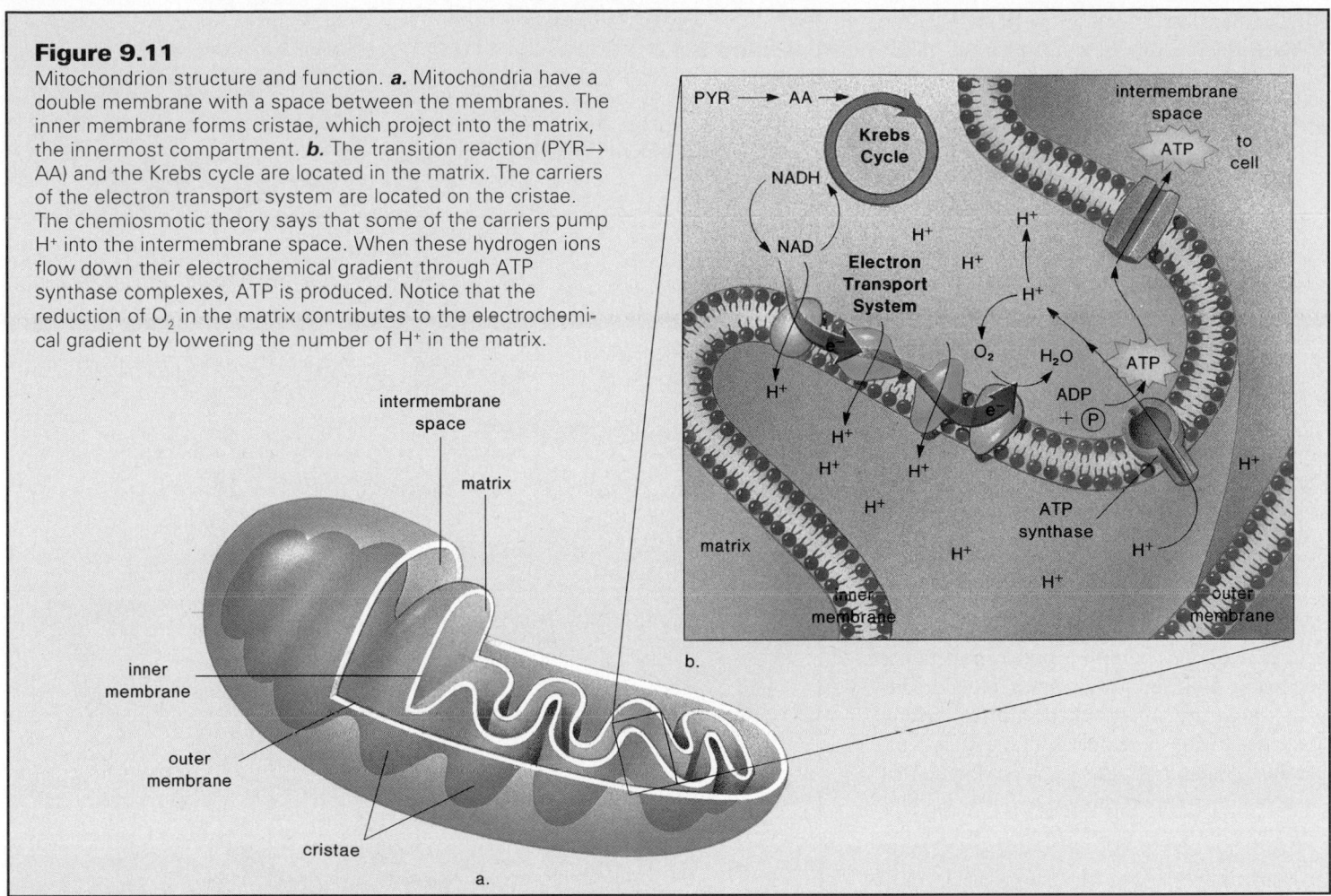

Table 9.1

Energy Yield of Glycolysis and Cellular Respiration per Glucose Molecule

Pathway	ATP Yield per Glucose Molecule		Total ATP
	Substrate-Level Phosphorylation	Oxidative Phosphorylation	
Glycolysis	2 ATP (net)*	2 NADH = 4 ATP+	6 ATP
Transition reaction		2 NADH = 6 ATP	6 ATP
Krebs cycle	2 ATP	6 NADH = 18 ATP	
		2 $FADH_2$ = 4 ATP	24 ATP
Total	4 ATP	32 ATP	36 ATP

*The only amount that is produced under anaerobic conditions.

+In most cells each NADH produced in the cytosol results in only 2 rather than 3 ATP because the electrons are shuttled to an FAD coenzyme. In the heart and the liver, a shuttle system delivers the electrons to an NAD^+ coenzyme, and in these cells each NADH does result in the production of 3 ATP.

Substrate-Level Phosphorylation

Per glucose molecule, 4 ATP molecules are formed directly by substrate-level phosphorylation: 2 during glycolysis and 2 during 2 turns of the Krebs cycle.

Oxidative Phosphorylation

Per glucose molecule 10 NADH molecules and 2 $FADH_2$ molecules take electrons to the electron transport system. For each molecule of NADH produced *inside* the mitochondria by the Krebs cycle, 3 ATP are produced by the electron transport system, but for each $FADH_2$, there are only 2 ATP. Figure 9.10 explains the reason for this difference: $FADH_2$ delivers its hydrogens to the transport system below the level of the first site for ATP production.

What about the ATP yield of NADH generated *outside* the mitochondria by the glycolytic pathway? NADH cannot cross mitochondrial membranes, but there is a "shuttle" mechanism, which allows its electrons to be delivered to the electron transport system inside the mitochondria. In most types of cells, the shuttle consists of an organic molecule, which can cross the outer membrane, accept the electrons, and deliver them to an FAD molecule in the inner membrane. This FAD, however, can produce only 2

ATP molecules; therefore, in most cells the NADH produced in the cytosol results in the production of only 2 ATP instead of 3 ATP.

Efficiency of Energy Transformation

It is interesting to consider how much of the energy in a glucose molecule eventually becomes available to the cell. The difference in energy content between the reactants (glucose and oxygen) and the products (carbon dioxide and water) is 686 Kcal. In other words, this is the total amount of energy available for the production of high-energy phosphate bonds. A high-energy phosphate bond has an energy content of 7.3 Kcal, and 36 of these are produced during glucose breakdown, so 36 phosphates are equivalent to a total of 263 Kcal. Therefore, 263/686, or 39%, of the available energy is transferred from glucose to ATP. The rest of the energy is lost in the form of heat. In birds and mammals, in particular, some of this heat is used to maintain a body temperature above that of the environment. In other organisms, including plants, there are few if any measures to conserve heat, and the body temperature generally fluctuates according to that of the external environment.

> The energy yield of 36 ATP molecules represents 39% of the total energy that is available in a glucose molecule.

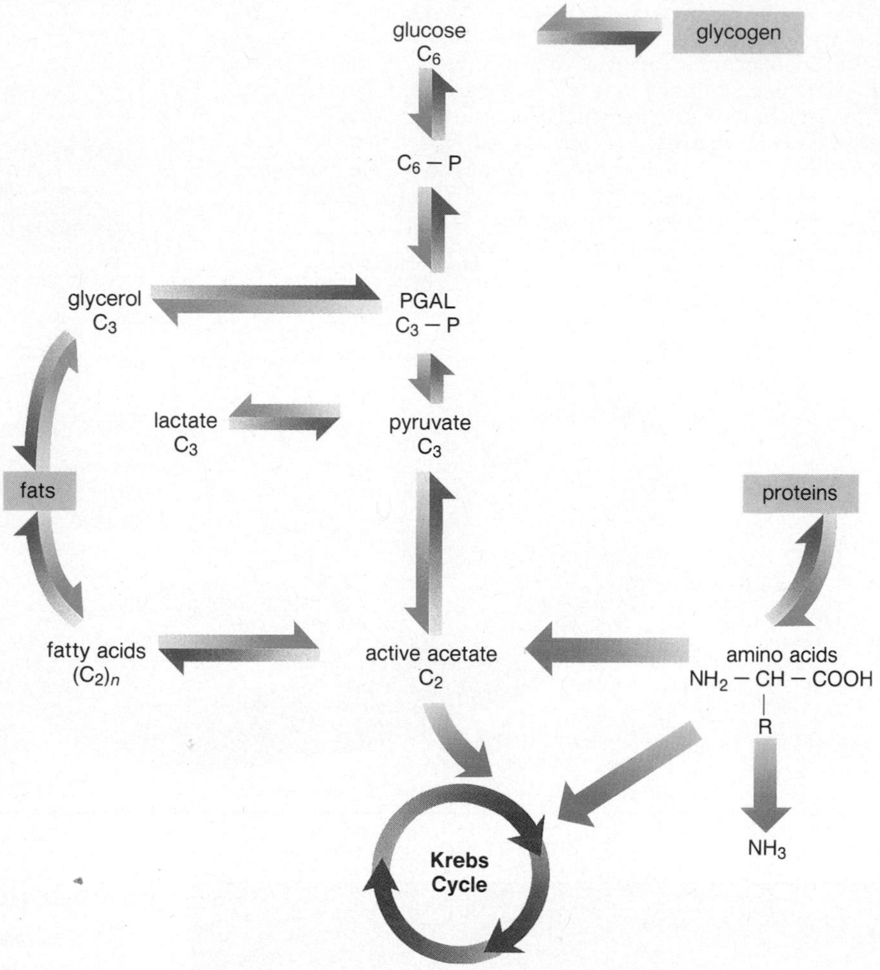

The Metabolic Pool and Biosynthesis

Figure 9.12 shows how the primary organic compounds of cells make up a metabolic pool. Cells can oxidize, or degrade, other molecules besides glucose to release energy. For example, a fat molecule is hydrolyzed to glycerol and 3 fatty acids when it is used as an energy source. Inside the cell, glycerol is easily broken down to PGAL and thereby can enter the glycolytic pathway. The fatty acids are degraded to acetyl groups, which can enter the Krebs cycle. If a fatty acid contains 18 carbons, it breaks down to give 9 acetyl groups. Calculation shows that these 9 groups would produce 108 ATP molecules, a very large number. Fats are a very efficient form of stored energy, especially since there are 3 long-chain fatty acids per fat molecule.

The energy yield of protein is equivalent to carbohydrates. Every protein is broken down to amino acids, which undergo deamination, the removal of the amino group. The number of carbons remaining now determines just where the molecule enters the Krebs cycle (fig. 9.12). The amino group is converted to ammonia (NH_3) and excreted.

Figure 9.12

The metabolic pool concept. Carbohydrates, fats, and proteins can be used as energy sources, and they enter degradative pathways at specific points. Degradation (green arrows) produces metabolites that can also be used for synthesis (blue arrows) of other compounds.

The substrates making up the pathways in figure 9.12 can also be used as starting materials for synthetic reactions. In other words, compounds that enter the pathways are oxidized to substrates that can be used for biosynthesis. This is the cell's **metabolic pool,** in which one type of molecule can be converted to another. In this way, carbohydrate intake can result in the formation of fat. PGAL molecules can be converted to glycerol molecules, and acetyl groups can be joined together to form fatty acids. Fat synthesis follows. This explains why you gain weight from eating too much candy, ice cream, and cake.

Some metabolites of the Krebs cycle can be converted to amino acids. Plants are able to synthesize all of the amino acids they need. Animals, however, lack some of the enzymes necessary for synthesis of all amino acids. Humans (and white rats), for example, can synthesize 9 of the common amino acids but cannot synthesize the other 11. The amino acids that cannot be synthesized must be supplied by the diet; they are called the essential amino acids. The nonessential amino acids are those that can be synthesized. It is quite possible for animals to eat large quantities of protein and still suffer from protein deficiency if their diet does not contain adequate quantities of the essential amino acids.

All the reactions involved in cellular respiration are part of a metabolic pool; the metabolites from the pool can be used either as energy sources or as substrates for various synthetic reactions.

There are other degradative pathways, aside from those discussed in this chapter or depicted in figure 9.12, that are needed to supply substrates for biosynthesis. For example, a pathway called the pentose phosphate shunt converts glucose to pentose (a 5-carbon sugar), rather than to pyruvate. NADPH is a by-product of the pentose phosphate shunt. The pentose sugars are needed for nucleotide formation, and the NADPH is used in synthetic reactions. Therefore, this pathway is also a very important one.

Summary

1. In cells, the energy of carbohydrate molecules, in particular, is converted to ATP energy.

2. Glucose metabolism often provides energy. Glycolysis anaerobically produces pyruvate, which either becomes reduced within the cytosol (fermentation) or becomes fully oxidized in mitochondria. Many bacteria and most animal cells carry on lactate fermentation; yeast and plant cells carry on alcoholic fermentation. Both kinds of fermentation only give a net yield of 2 ATP molecules.

3. Glycolysis is the breakdown of glucose to 2 molecules of pyruvate. This pathway is a series of enzymatic reactions that occurs in the cytosol. Oxidation by NAD^+ releases enough energy immediately to give a net gain of 2 ATP molecules by substrate-level phosphorylation. NADH either gives electrons to pyruvate or takes them to a mitochondrion.

4. Fermentation involves glycolysis, followed by the reduction of pyruvate by NADH to either lactate or alcohol and CO_2. The reduction process "frees" NAD^+ so that it can accept more hydrogen atoms from the glycolysis.

5. Although fermentation results in only 2 ATP molecules, it still serves a purpose: In vertebrates, it provides a quick burst of ATP energy for short-term, strenuous muscular activity. The accumulation of lactate puts the individual in oxygen debt because oxygen is needed when lactate is completely metabolized to CO_2 and H_2O. Fermentation evolved before aerobic respiration, and its continued presence exemplifies how evolution usually works—the newly evolved process is added onto those that have already evolved.

6. When glucose is completely oxidized, at least 36 ATP molecules are produced. Three subpathways are required: glycolysis, the Krebs cycle, and the electron transport system. Oxidation involves the removal of hydrogen atoms ($e^- + H^+$) from substrate molecules, usually by the coenzyme NAD^+.

7. Pyruvate from glycolysis can enter the mitochondrion, where the transition reaction takes place. During this reaction, oxidation occurs as CO_2 is removed. NAD^+ is reduced, and CoA receives the C_2 acetyl group that remains. Since the reaction must take place twice per glucose molecule, 2 NADH result.

8. The acetyl group enters the Krebs cycle, a cyclical series of reactions located in the mitochondrial matrix. Complete oxidation follows, as 2 CO_2 molecules, 3 NADH molecules, and one $FADH_2$ molecule are formed. The cycle also produces one ATP molecule. The entire cycle must turn twice per glucose molecule.

9. The final stage of pyruvate breakdown involves the electron transport system located in the inner membrane of mitochondria. The electrons received from NADH and $FADH_2$ are passed down a chain of carriers until they are finally received by oxygen, which combines with hydrogen ions to produce water. As the electrons pass down the chain, ATP is produced. The term *oxidative phosphorylation* is sometimes used for ATP production by the electron transport system.

10. Some of the electron carriers pump H^+ into the intermembrane space setting up an electrochemical gradient. H^+ flow down this gradient, and the energy released is used to form ATP molecules from ADP and Ⓟ.

11. In the mitochondria, for each NADH molecule donating electrons to the electron transport system, 3 ATP molecules are produced. In the cytosol, however, each NADH formed results in only 2 ATP molecules. This is because its hydrogen atoms must be shuttled across the mitochondrial outer membrane by a molecule that can cross it. In most cells, these hydrogen atoms are then taken up by FAD. Each molecule of $FADH_2$ results in the formation of only 2 ATP because the electrons enter the electron transport system at a lower level.

12. Once NAD^+ and FAD have delivered electrons to the electron transport system, they are "free" to turn around and accept more hydrogens. Therefore, the cell only needs a limited supply of these coenzymes because they are used repeatedly.

13. Of the 36 ATP formed by aerobic cellular respiration, 4 are the result of substrate-level phosphorylation and the rest are produced by oxidative phosphorylation. The energy for the latter comes from the electron transport system.

14. Carbohydrate, protein, and fat can be broken down by entering the degradative pathways at different locations. These pathways also provide metabolites needed for the synthesis of various important substances. Degradation and synthesis, therefore, both utilize the same metabolic pool of reactants.

Writing Across the Curriculum

In order to practice writing skills, students should write out the answers to any or all of the study questions and the critical thinking questions. The study questions are sequenced in the same order as the text. Suggested answers to the critical thinking questions are in appendix D.

Study Questions

1. What types of organisms carry on cellular respiration? What is the function of this process?
2. Glycolysis results in pyruvate molecules. Contrast the fate of pyruvate under anaerobic conditions and aerobic conditions.
3. What is the overall chemical equation for the complete breakdown of glucose to carbon dioxide and water? How does a human cell acquire the needed substrates, and what happens to the products?
4. What are the 3 pathways involved in the complete breakdown of glucose to carbon dioxide and water? What reaction is needed to join 2 of these pathways?
5. Outline the main reactions of glycolysis, emphasizing those that produce energy.
6. Contrast glycolysis with fermentation. What is the benefit of pyruvate reduction during fermentation? What types of organisms carry out lactate fermentation, and what types carry out alcoholic fermentation?
7. Give the substrates and products of the transition reaction. Where does it take place?
8. What happens to the acetyl group that enters the Krebs cycle? What are the other steps in this cycle?
9. What is the electron transport system, and what are its functions? Relate these functions to the location of this system.
10. Explain how, when, and where chemiosmotic phosphorylation takes place in mitochondria.
11. Calculate the energy yield of glycolysis and cellular respiration per glucose molecule. Distinguish between substrate-level phosphorylation and oxidative phosphorylation.
12. Give examples to support the concept of the metabolic pool.

Objective Questions

For questions 1-8, identify the pathway involved and match to the terms in the key.

Key:

a. glycolysis
b. Krebs cycle
c. electron transport system

1. carbon dioxide given off
2. water formed
3. PGAL
4. NADH becomes NAD$^+$
5. oxidative phosphorylation
6. cytochrome carriers
7. pyruvate
8. FAD becomes FADH$_2$
9. Which of these is *not* true of fermentation?
   a. net gain of only 2 ATP
   b. occurs in cytosol
   c. NADH donates electrons to electron transport system
   d. begins with glucose
10. The transition reaction
    a. connects glycolysis to the Krebs cycle.
    b. gives off CO$_2$.
    c. utilizes NAD$^+$.
    d. All of these.
11. The greatest contributor of electrons to the electron transport system is
    a. oxygen.
    b. glycolysis.
    c. the Krebs cycle.
    d. the transition reaction.
12. Substrate-level phosphorylation takes place in the
    a. glycolysis and the Krebs cycle.
    b. electron transport system and the transition reaction.

c. glycolysis and the electron transport system.
d. Krebs cycle and the transition reaction.

13. Fatty acids are broken down to
    a. pyruvate molecules, which take electrons to the electron transport system.
    b. acetyl groups, which enter the Krebs cycle.
    c. amino acids, which excrete ammonia.
    d. All of these.
14. Of the 36 ATP molecules that are produced during the complete breakdown of glucose, most are formed by
    a. chemiosmotic phosphorylation.
    b. the electron transport system.

c. substrate-level phosphorylation.
d. Both a and b.

For questions 15-20, match the items below to one of the locations named in the key.

Key:

a. matrix of the mitochondria
b. cristae of the mitochondria
c. between the inner and outer membranes of the mitochondria
d. in the cytosol

15. electron transport system
16. Krebs cycle
17. glycolysis
18. transition reaction
19. H$^+$ reservoir
20. ATP synthase complex
21. Label this diagram of a mitochondrian:

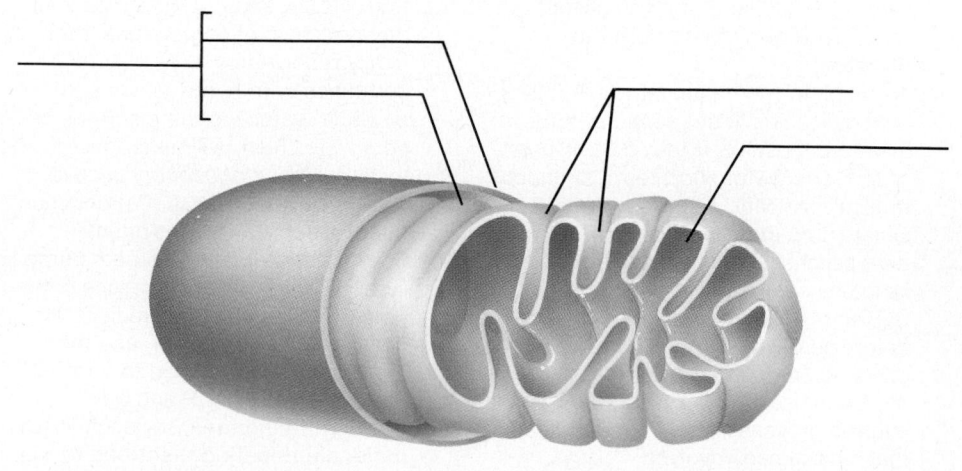

22. Label this diagram of glycolysis and cellular respiration. Place the labels lactate, electron transport system, acetyl CoA, pyruvate, and Krebs cycle on the appropriate lines. Label the arrows as either CO_2 or NADH. Indicate the correct number of ATP on the lines provided.

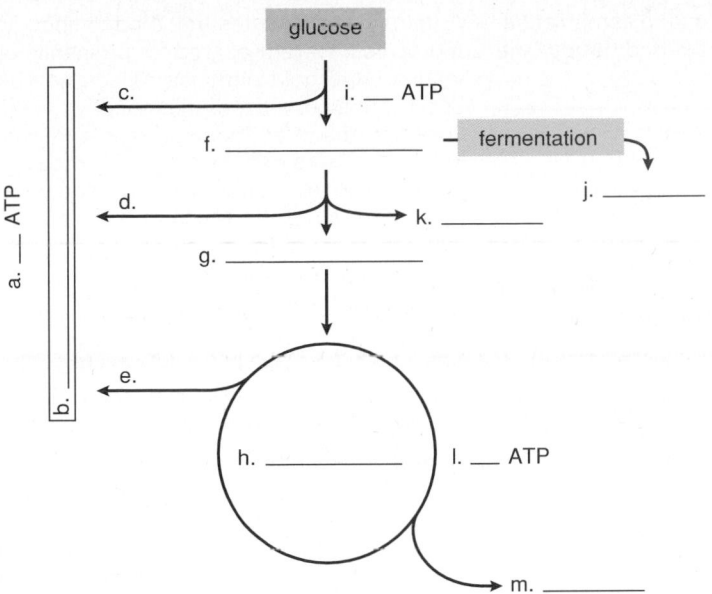

Concepts and Critical Thinking

1. *Photosynthesis and cellular respiration permit a flow of energy and a recycling of matter.*

Explain this concept by referring to figure 9.1.

2. *Solar energy first stored in nutrient molecules is transformed into a form that can be "spent" by the cell.*

How are nutrient molecules utilized by cells as a source of energy?

3. *Certain metabolic pathways are virtually universal and thereby demonstrate the unity of living things.*

In what way does glycolysis demonstrate the unity of living things and support the theory of evolution?

Selected Key Terms

cellular respiration (sel'u-lar res"pĭ-ra'shun) 138
glycolysis (gli-kol i-sis) 139
pyruvate (pi'roo-vāt) 139
fermentation (fer"men-ta'shun) 142
substrate-level phosphorylation (sub'strāt lĕv-al fos"fŏr-i-la'shun) 142

oxygen debt (ok'sĭ-jen det) 142
transition reaction (tran-zish'un re-ak'shun) 143
acetyl CoA (as"ĕ-til-ko-en'zīm A) 143
Krebs cycle (Krebz si'k'l) 144
citric acid cycle (sit'rik as'id si'kl) 144

cytochrome system (si'to-krōme sis'tem) 144
electron transport system (e-lek'tron trans'port sis'tem) 144
oxidative phosphorylation (ok"si-da'tiv fos"for-i-la'shun) 145
metabolic pool (mĕt'a-bol-ic pōol) 148

PART 1 *The Cell*

CRITICAL THINKING CASE STUDY

The Processing of Insulin

An active insulin molecule is composed of 2 polypeptides linked by disulfide chemical bonds. It is known that the original protein, produced at the ribosomes, is a larger molecule than active insulin, which is secreted into the blood. Therefore, it is hypothesized that *protein processing occurs within the organelles of the cell before active insulin is secreted.*

To test the hypothesis, pancreatic tissue, which produces insulin, was subjected to autoradiography. Pancreatic tissue was immersed *for a short time* in a solution containing radioactive amino acids and then, at regular intervals, tissue was sectioned and placed on slides for autoradiography (see p. 55 for an explanation of this technique). The autoradiographs showed the location of the radioactivity over time. Where do you predict the radioactivity will be as time elapses?

Prediction 1 Radioactive protein will appear in different organelles of the cell as time elapses.

Result 1 Within 5 min, the labeled protein moved to the inside of the endoplasmic reticulum. After 15 min, the labeled protein appeared in the Golgi apparatus, and after 60 min, it had moved to the secretory vesicles between the Golgi apparatus and the plasma membrane. These data indicate that the protein moved from the endoplasmic reticulum to the Golgi apparatus and then to the secretory vesicles.

I. In keeping with these results, would an autoradiograph produced after 60 min show any sign of remaining radioactivity? Why or why not?

If the tissue had been exposed to radioactive amino acids for a considerable length of time (instead of a short time), how might the results differ from those described?

These results tell us the pathway of protein movement, but they do not indicate where protein processing occurred. Where do you predict protein processing might occur?

Prediction 2 Protein processing might occur inside the endoplasmic reticulum or in the Golgi apparatus. Most likely active insulin is in the secretory vesicles.

Result 2 Chemical analysis of organelle contents showed that a small amino acid chain is removed from the protein as soon as it enters the endoplasmic reticulum. The remaining molecule is called proinsulin.

II. Molecules that cross membranes usually combine with receptor sites in a lock-and-key manner. What might have been the purpose of the small amino acid chain that was removed in the endoplasmic reticulum?

Researchers then turned their attention to the vesicles. Using electron microscopy they observed vesicles coated with protein bristles close to the Golgi saccules, and uncoated vesicles were some distance away. Then they continued the autoradiography experiment. What do you predict regarding the movement of protein via these vesicles?

Prediction 3 Protein will be found first in the coated vesicles near the Golgi and then later in the uncoated vesicles some distance away.

Result 3 The results showed radioactivity first in the coated vesicles and then about 30 min later in the uncoated vesicles.

III. Is it possible that the protein might move from one type vesicle to the other or that the coated vesicles might become uncoated vesicles? Why does the latter possibility seem more logical?

Another series of experiments was done to identify whether the protein in the vesicles was proinsulin or active* insulin.

Antibodies, molecules that bind to specific molecules, were prepared for proinsulin or insulin. Fine gold particles, which can be detected by electron microscopy, were linked to these antibodies. Pancreatic tissue slices were exposed *either* to proinsulin antibody or insulin antibody and then examined using the electron microscope. What do you predict in regard to binding of the antibodies to the vesicles?

Prediction 4 Since coated vesicles precede uncoated vesicles and proinsulin precedes active insulin, proinsulin antibody will bind to coated vesicles and insulin antibody will bind to uncoated vesicles.

Result 4 Results showed that proinsulin antibody binds only to the coated vesicles. Active insulin antibody binds only to the uncoated vesicles. These results strongly suggest that proinsulin is converted to insulin as coated vesicles develop into smooth vesicles.

IV. Why did the researchers prepare antibodies to proinsulin and insulin? Why not just continue with radiography?

Why did the researchers choose to apply the 2 different types of antibodies separately? Why not apply them together?

The results of these experiments provide strong support for the hypothesis that protein processing occurs within the organelles of the cell before active insulin is secreted. Understanding the complete pathway of the formation of the insulin molecule required a variety of techniques, each of which made an essential contribution to understanding the process.

Other Questions

1. Cell fractionation could be done following brief exposure of pancreatic tissue to radioactive amino acids. With reference to figure 5.C, what would be the results of this experiment? Assume three fractions: (a) endoplasmic reticulum; (b) Golgi apparatus; and (c) vesicles.
2. The researchers were careful to use the beta cells of the Isles of Langerhans in their experiment and not the entire pancreas? Why? What would it mean if they got similar results using the alpha cells?

Orci, L., J. D. Vassalli, and A. Perrelet. September 1988. The insulin factory. *Scientific American*, 85-94.

Suggested Readings for Part 1

Alberts, B., et al. 1989. *Molecular biology of the cell*. 2d ed. New York: Garland Publishing.

Allen, R. D. February 1987. The microtubule as an intracellular engine. *Scientific American*.

Baker, J. J. W., and Allen, G. E. 1981. *Matter, energy, and life*. 4th ed. Reading, Mass.: Addison-Wesley.

Berns, M. W. 1983. *Cells*. 3d ed. New York: Holt, Rinehart & Winston.

Cronquist, A. 1982. *Basic botany*. 2d ed. New York: Harper & Row.

Dautry-Varsat, A., and Lodish, H. F. May 1984. How receptors bring proteins and particles into cells. *Scientific American*.

de Duve, C. May 1983. Microbodies in the living cell. *Scientific American*.

Dickerson, E. March 1980. Cytochrome and the evolution of energy metabolism. *Scientific American*.

Dustin, P. August 1980. Microtubules. *Scientific American*.

Govindjee, and Coleman, William J. February 1990. How plants make oxygen. *Scientific American*.

Grivell, L. A. March 1983. Mitochondrial DNA. *Scientific American*.

Karplus, M., and McCammon, J. A. April 1986. The dynamics of proteins. *Scientific American*.

Lake, J. A. July 1981. The ribosome. *Scientific American*.

Miller, K. R. October 1979. The photosynthetic membrane. *Scientific American*.

Ostro, M. J. January 1987. Liposomes. *Scientific American*.

Porter, K. R., and Bonneville, M. A. 1973. *Fine structure of cells and tissues*. 4th ed. Philadelphia: Lea and Febiger.

Raven, H., et al. 1986. *Biology of plants*. 4th ed. New York: Worth.

Rothman, J. September 1985. The Compartmental organization of the Golgi apparatus. *Scientific American*.

Scientific American. October 1985. The molecules of life.

Sharon, N. November 1980. Carbohydrates. *Scientific American*.

Sheeler, P., and Bianchi, D. E. 1980. *Cell biology: Structure, biochemistry, and function*. New York: John Wiley and Sons.

Stryer, L. 1988. *Biochemistry*. 3d ed. San Francisco: W. H. Freeman.

Unwin, N. February 1984. The structure of proteins in biological membranes. *Scientific American*.

Wickramasinghe, H. Kumar. October 1989. Scanned-probe microscopes. *Scientific American*.

Yougan, D. C., and Mars, B. L. June 1987. Molecular mechanisms of photosynthesis. *Scientific American*.

Genetic Basis of Life

Like begets like because a genetic blueprint is passed from parent to offspring. We now know that the genes are segments of chromosomes and each is a DNA molecule. Soon the nucleotide sequence of human genes and other animals will be known. This will increase our ability to manipulate inheritance in the laboratory.

Cell Reproduction: Binary Fission and Mitosis

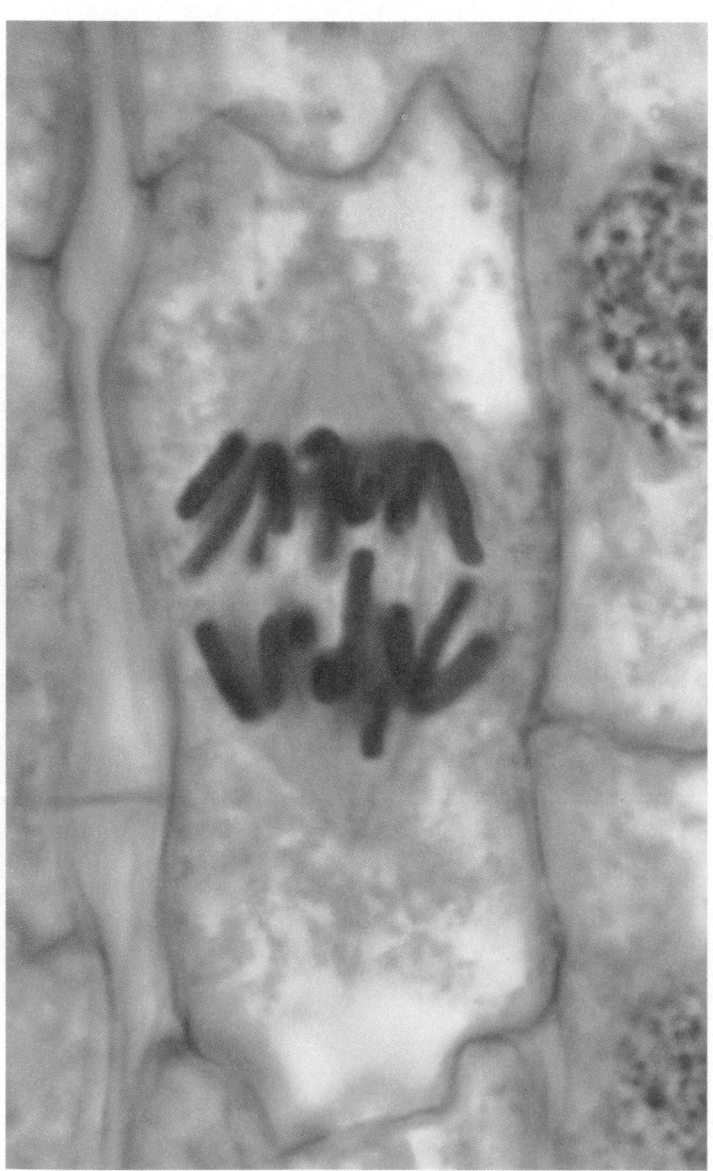

The mitotic spindle. During nuclear division, the chromosomes are attached to microtubules, which collectively assume the shape of a spindle. The spindle brings about the movement and distribution of a full set of daughter chromosomes to each of its poles.

Your study of this chapter will be complete when you can

1. relate binary fission and mitosis to the process of asexual reproduction among single-celled organisms and to growth and repair among multicellular forms;
2. list several benefits of binary fission and mitosis;
3. describe the prokaryotic chromosome and the process of binary fission;
4. describe a duplicated eukaryotic chromosome, and distinguish between the diploid and haploid number of chromosomes;
5. state the difference between chromatin and chromosome, and explain how they are related;
6. state the phases of the cell cycle, and describe what happens during each phase;
7. give evidence that the cell cycle is controlled, and contrast the behavior of normal cells with those of cancer cells;
8. draw a series of diagrams illustrating the stages of mitosis in animal cells, and tell what happens during each stage; describe cytokinesis in animal cells;
9. state at least one difference between animal and plant cell mitosis; describe cytokinesis in plant cells.

s we mentioned in the introductory chapter, the life cycle of all organisms includes reproduction. We have a tendency to think of reproduction in terms of multicellular organisms, even though reproduction always involves single cells. For example, all organisms begin life as a single cell—when that cell divides in single-celled organisms, it produces 2 new individuals that are like the parent. In other types of organisms, cell division is a part of the growth process that produces the multicellular form we recognize as the organism (fig. 10.1). Cell division is also important in multicellular forms for renewal and repair. For example, our own body is always producing new blood cells and skin cells to replace those that are worn out or damaged. In summary, then, simple cell division has these purposes:

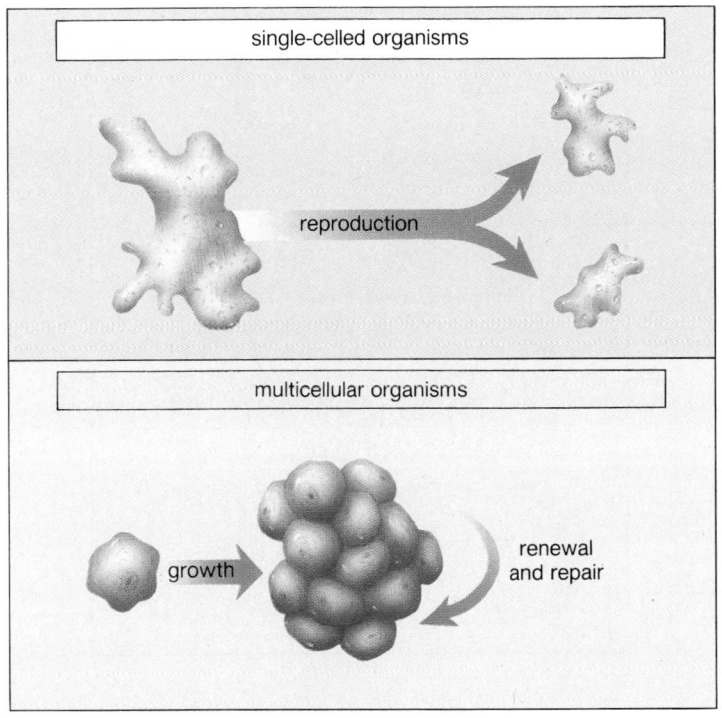

Cell division occurs in single-celled organisms when they reproduce. Simple cell division is also involved in the growth and in the renewal and repair of multicellular forms.

Cells reproduce, and in that way new cells arise. In fact, new cells can only come from preexisting cells. This is a tenet of the cell theory we discussed on page 61. The organization of a cell is such that it is very unlikely that the coming together of biomolecules would spontaneously produce a new cell. Instead, only cell division produces new cells. Why can't a cell simply grow larger? What is the benefit of cell division in the first place? Recall that we previously considered these questions on page 65. A small cell has a more adequate plasma membrane-to-volume ratio than a large cell. Everything that enters and exits a cell must cross the plasma membrane, and as the cell increases in size, there comes a point when the plasma membrane surface area is not adequate to meet the needs of the cell. Cell division keeps the surface area-to-volume ratio favorable for the efficient survival of the cell.

Figure 10.1

Cell reproduction occurs in both single-celled organisms and in multicellular organisms. ***a.*** When a single-celled organism divides, reproduction occurs. This is a photograph of a paramecium dividing. Magnification, X200. ***b.*** Life begins as a single cell in multicellular organisms. When this cell divides, a multicellular organism results. This is a photograph of a slipper-shell mollusk embryo after only a single division has occurred. Magnification, X140.

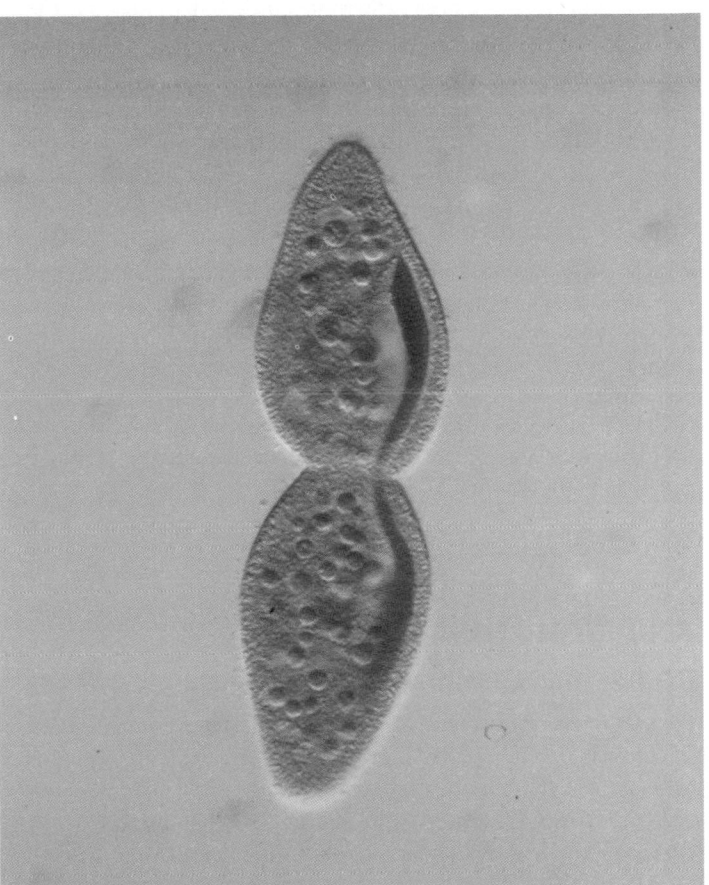

a.

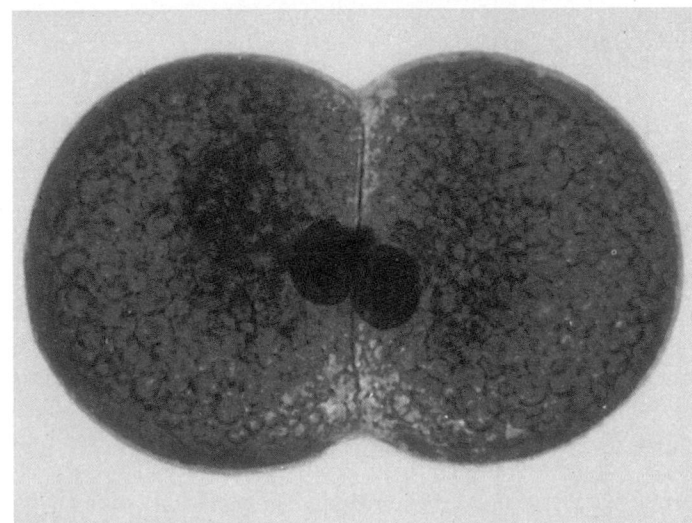

b.

New cells only come from preexisting cells. When growing cells divide, they maintain a favorable surface area-to-volume relationship.

Another advantage of cell division that we have not considered previously concerns only multicellular organisms. Cell division allows cells to specialize for various functions, such as digestion, respiration, and excretion. Obviously, it would wreak havoc if these specialized cells didn't give rise to their own kind. In other words, intestinal cells give rise to intestinal cells, and kidney cells give rise to kidney cells. Another way that biologists like to say this is, "like begets like." By what mechanism does this occur? We have already considered that the nucleus, which contains the chromosomes, controls the cell. Cell division must have a way by which copies of the chromosomes are passed to newly formed cells. Indeed, this chapter is principally about the way in which copies of chromosomes are passed from parent cell to daughter cells so that the daughter cells receive a copy of each and every chromosome. In this way the daughter cells receive a complete copy of the genetic material, DNA.

Cells always give rise to cells that are like themselves, that is, "like begets like." This occurs because the daughter cells receive a copy of the parent cell's genetic material.

Prokaryotic Chromosomes and Cell Division

Prokaryotic Chromosomes

Prokaryotes (i.e., bacteria) are single-celled organisms that lack a nucleus (p. 64). They have a single chromosome that is compacted into an electron-dense, irregularly shaped region called the **nucleoid,** which is not enclosed by a membrane. When stretched out, the chromosome is seen to be a circular loop attached to the inside of the plasma membrane. Its length may be about 500 times the length of the cell, which is why it needs to be compacted inside the cell.

The bacterial chromosome is not like the eukaryotic chromosome because it lacks associated proteins. It consists simply of the genetic material, DNA.

The prokaryotic chromosome is a single loop of DNA that is compacted into a region called the nucleoid.

Cell Division

Cell division in prokaryotes is termed **binary fission** because division (fission) produces 2 (binary) daughter cells that are identical to the original parent cell. Before division takes place, the DNA is replicated so that there are 2 chromosomes attached to the inside of the plasma membrane. Following replication, the 2 chromosomes separate by an elongation of the cell that pushes the chromosomes apart. When the cell is approximately twice its original length, the plasma membrane grows inward and a cell wall forms, dividing the cell into 2 approximately equal portions (fig. 10.2).

Figure 10.2

Binary fission in a prokaryote. **a.** The chromosome is a single loop of DNA that is attached to the plasma membrane. After the DNA replicates and the cell divides, each daughter cell contains its own chromosome. **b.** Following division, each daughter cell has its own nucleoid, a region where the chromosome is compacted in the cell. Magnification, X30,000.

a.

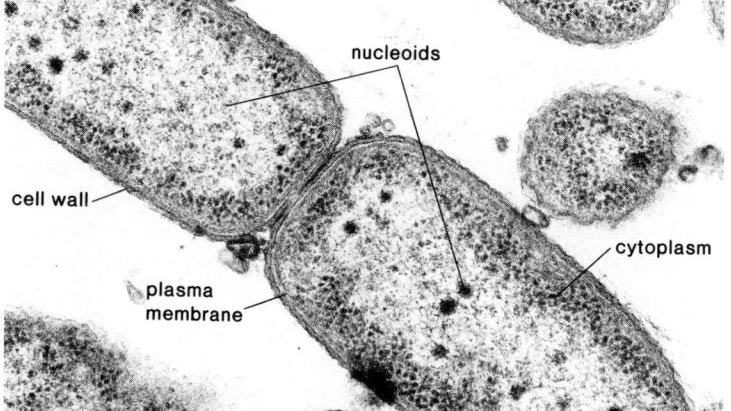

b.

Table 10.1		
Diploid Chromosome Number of Some Eukaryotes		
Type of Organism	**Name of Organism**	**Chromosome Number**
Fungi	*Aspergillus nidulans* (mold)	8
	Neurospora crassa (mold)	14
	Saccharomyces cerevisiae (yeast)	34
Plants	*Vicia faba* (broad bean)	12
	Zea mays (corn)	20
	Solanum tuberosum (potato)	48
	Nicotiana tabacum (tobacco)	48
Animals	*Felis domesticus* (cat)	38
	Equus caballus (horse)	64
	Gallus gallus (chicken)	78
	Pan troglodytes (chimp)	48
	Canis familiaris (dog)	78
	Rana pipiens (frog)	26
	Musca domestica (housefly)	12
	Homo sapiens (human)	46

Sometimes a prokaryotic cell has several nucleoids because several replications can be occurring at once. The third chromosome may begin at the site of the second, before the production of the second chromosome has even been completed.

Cell division in prokaryotes is by binary fission. DNA replicates before new plasma membrane and cell wall separate the cytoplasm.

Eukaryotic Chromosomes and Cell Division

Eukaryotic Chromosomes

Eukaryotes have a true nucleus (see fig. 5.4) that is bounded by a double membrane called the nuclear envelope. The nuclear envelope contains pores that allow substances to pass between the nucleoplasm and the cytoplasm (see fig. 5.7). The DNA (and associated proteins) within a nucleus that is not undergoing division is a tangled mass of thin threads called **chromatin.** At the time of division, chromatin becomes highly coiled and condensed, and it is easy to see the individual **chromosomes.** The reading for this chapter describes the manner in which the coiling and condensation take place. In addition to chromatin, the nucleus contains at least one nucleolus that is attached to, and formed by, special regions of particular chromosomes. Recall that the nucleolus is involved in the production of ribosomes, which are found in the cytoplasm.

When the chromosomes are highly coiled and condensed at the time of cell division, it is possible to photograph and count them. Each species has a characteristic chromosome number (table

10.1); the chromosome number for humans is 46. This is called the full or **diploid (2N) number** of chromosomes that occurs in all cells of the body. The diploid number includes 2 chromosomes of each kind. Half the diploid number, called the **haploid (N) number** of chromosomes, contains only one of each kind of chromosome. In the life cycle of many animals only sperm and eggs have the haploid number of chromosomes.

Eukaryotic chromosomes are located within the nucleus of the cell.

Cell Division

Cell division in eukaryotes involves nuclear division and **cytokinesis,** which is division of the cytoplasm. The typical life cycle of animals includes 2 types of nuclear division:

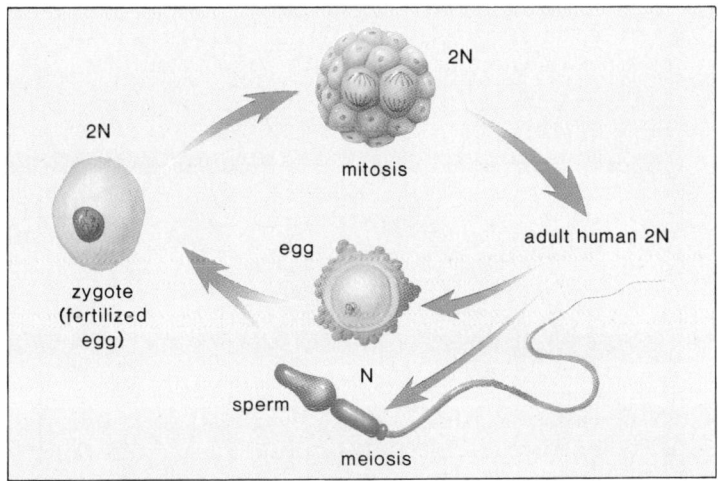

Mitosis is nuclear division in which the chromosome number stays constant. A 2N nucleus divides to give daughter nuclei that are also 2N. Mitosis is the type of nuclear division that is involved in growth and in the repair of the body. **Meiosis** is nuclear division in which the chromosome number is halved. A 2N nucleus divides to give daughter nuclei that are N. Meiosis is the type of nuclear division involved in the production of **gametes,** the egg and sperm. When the sperm fertilizes the egg, the resulting zygote again has the 2N number of chromosomes.

Before nuclear division takes place, DNA replicates; therefore, each chromosome is duplicated and has 2 identical parts called **sister chromatids** (fig. 10.3). Sister chromatids are genetically identical; that is, they contain exactly the same genes. Sister chromatids are constricted and are attached to each other at a region called the **centromere.**

Nuclear division has 4 stages (prophase, metaphase, anaphase, and telophase), which will be considered in depth later in the chapter. During one of these stages, the centromeres divide, and in that way a duplicated chromosome gives rise to 2 daughter chromosomes. These chromosomes, which have only one chromatid, are distributed equally to the daughter cells. In this way, each daughter cell gets a complete copy of the organism's genetic material.

What's in a Chromosome?

When early investigators decided that the genes were on the chromosomes, they had no idea of chromosome composition. By the mid-1900s, it was known that chromosomes are made up of both DNA and protein. Only in recent years, however, have investigators been able to produce models suggesting how chromosomes are organized.

A eukaryotic chromosome is more than 50% protein of 2 types: histones and nonhistones. The nonhistone proteins are acidic (have negatively charged R groups) and are only loosely associated with DNA.

The functions of the nonhistone proteins are being studied, and some are known to be regulators of gene activity. Much more is known about the histone proteins, which are basic (have positively charged R groups) and bind strongly to the negatively charged phosphate groups of DNA.

There are 5 different primary types of histone molecules designated H1, H2A, H2B, H3, and H4, and all 5 types are found in all tissues of an organism. Remarkably, the amino acid sequences of H3 and H4 vary little between organisms. For example, H4 of peas is only 2 amino acids different from H4 of cattle. This similarity suggests that mutations in the histone proteins have been selected against during the evolutionary process and that the histones therefore have very important functions.

A human cell contains 46 chromosomes, and the length of the DNA in each is about 5 cm on the average.

Therefore, a human cell contains at least 2 m of DNA. Yet all of this DNA is packed into a nucleus that is about 5 µm in diameter. The histones are responsible for packaging the DNA so that it can fit into such a small space. First the DNA double helix is wound at intervals around a core of 8 histone molecules (2 copies each of H2A, H2B, H3, and H4), giving the appearance of a string of beads (fig. 10.A*b*). Each bead is called a nucleosome, and the nucleosomes are said to be joined by "linker" DNA. This string is coiled tightly into a fiber that has 6 nucleosomes per turn (fig. 10.A*c*)—the H1 histone appears to mediate this coiling process. The fiber loops back and forth (fig. 10.A*d* and *e*) and can condense to give a highly compacted form (fig. 10.A*f*) characteristic of metaphase chromosomes.

An interphase nucleus shows darkly stained chromatin regions called heterochromatin and lightly stained chromatin regions called euchromatin (fig. 10.B). Heterochromatin is genetically inactive (fig. 10.A*e*). The mechanism of inactivation is unknown, but it may involve methylation (adding methyl groups—CH_3—to the bases of DNA). Euchromatin is genetically active (fig. 10.A*c*). It's believed that in this form, the nucleosomes allow access to the DNA, so that a gene can be turned on and can actively dictate protein synthesis in the cell.

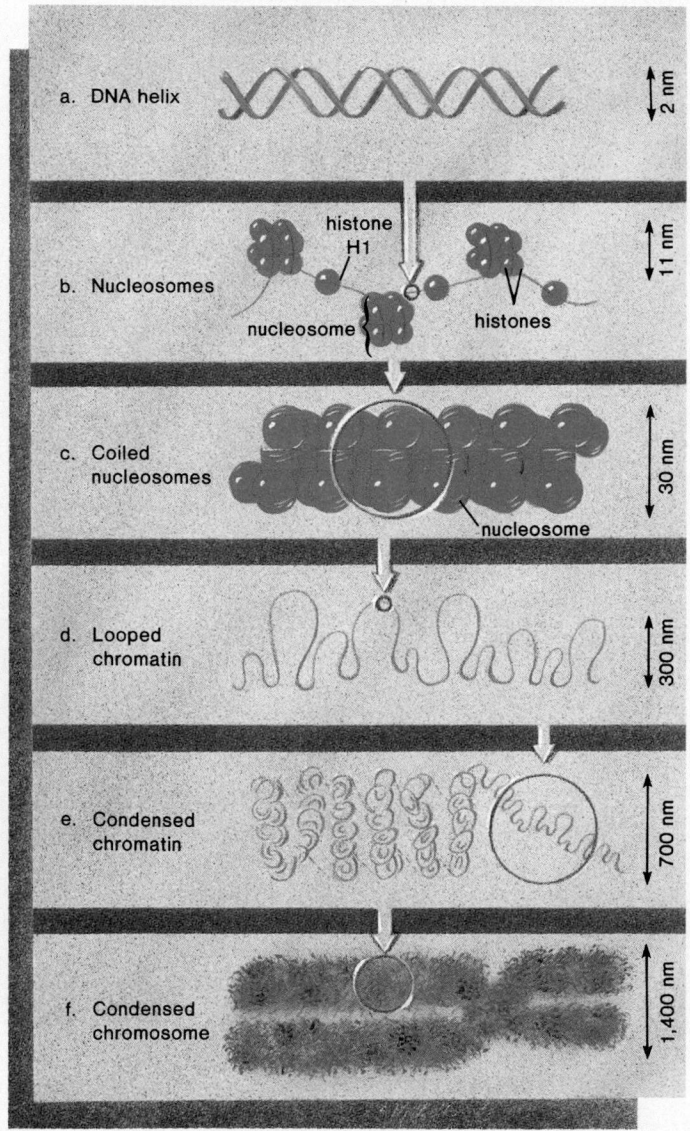

Figure 10.A

Levels of chromosome structure. Each drawing has a scale giving a measurement of length for that drawing. Notice each measurement represents an ever-increasing length; therefore, it would take a much higher magnification to see the structure in (*a*) compared to (*f*).

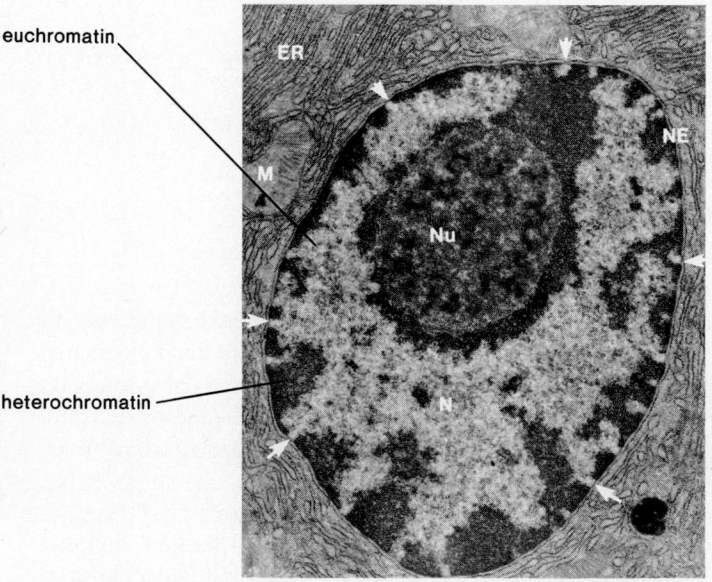

Figure 10.B

Eukaryotic nucleus. The nucleus contains chromatin, DNA at 2 different levels of coiling and condensation. Euchromatin is at the level of coiled nucleosomes (30 nm), and heterochromatin is at the level of condensed chromatin (700 nm). Arrows indicate nuclear pores. Magnification, X10,000.

Figure 10.3

Duplicated chromosomes. DNA replication has occurred, and each chromosome is held together at a region called the centromere. **a.** Electron micrograph of highly coiled and compacted chromosome, typical of a nucleus about to divide. Magnification, X8,400. **b.** A chromosome that is not as condensed. One chromatid is indicated by the box. A duplicated chromosome contains 2 sister chromatids, each containing copies of the same genes.

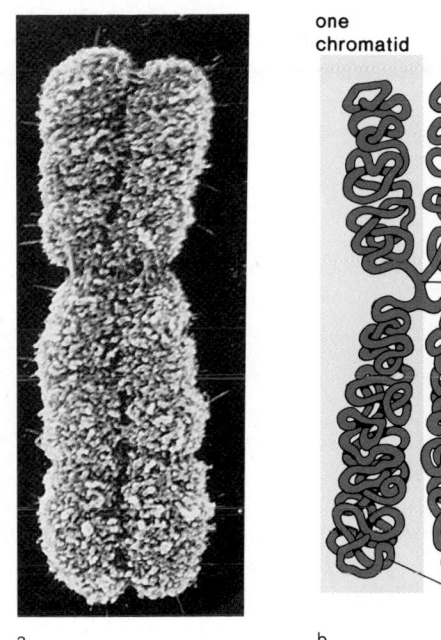

one chromatid

centromere

sister chromatids

a.   b.

DNA replicates prior to nuclear division, which consists of 4 stages.

Cell Cycle

By the 1870s, microscopy could provide detailed and accurate descriptions of chromosome movements during mitosis, but there was no knowledge of cellular events between divisions. Because there was little visible activity between divisions, the period of time was dismissed as a resting state termed **interphase.** When it was discovered in the 1950s that DNA replication occurs during interphase, the cell cycle concept was formulated.

Cells grow and divide during a cycle that has 4 phases (fig. 10.4). The entire cell division phase, including both mitosis and cytokinesis, is termed the *M phase* (M = mitosis). The period of DNA replication is termed the *S phase* (S = synthesis) of the cycle. The proteins associated with DNA in eukaryotic chromosomes (see reading) are also synthesized during this phase. There are 2 other phases of the cycle. The period of time following the M phase and before the S phase begins is termed the G_1 *phase*, and the period of time following the S phase and before the M phase begins is termed the G_2 *phase*. At first, not much was known about these phases, and they were thought of as G = gap phases. Now we know that during the G_1 phase, the cell grows in size and the cellular organelles increase in number. During the G_2 phase, synthesis of various enzymes and other types of proteins occurs in preparation for mitosis. Some biologists today prefer the designation G = growth for these 2 G phases. In any case, interphase consists of G_1, S, and G_2 phases.

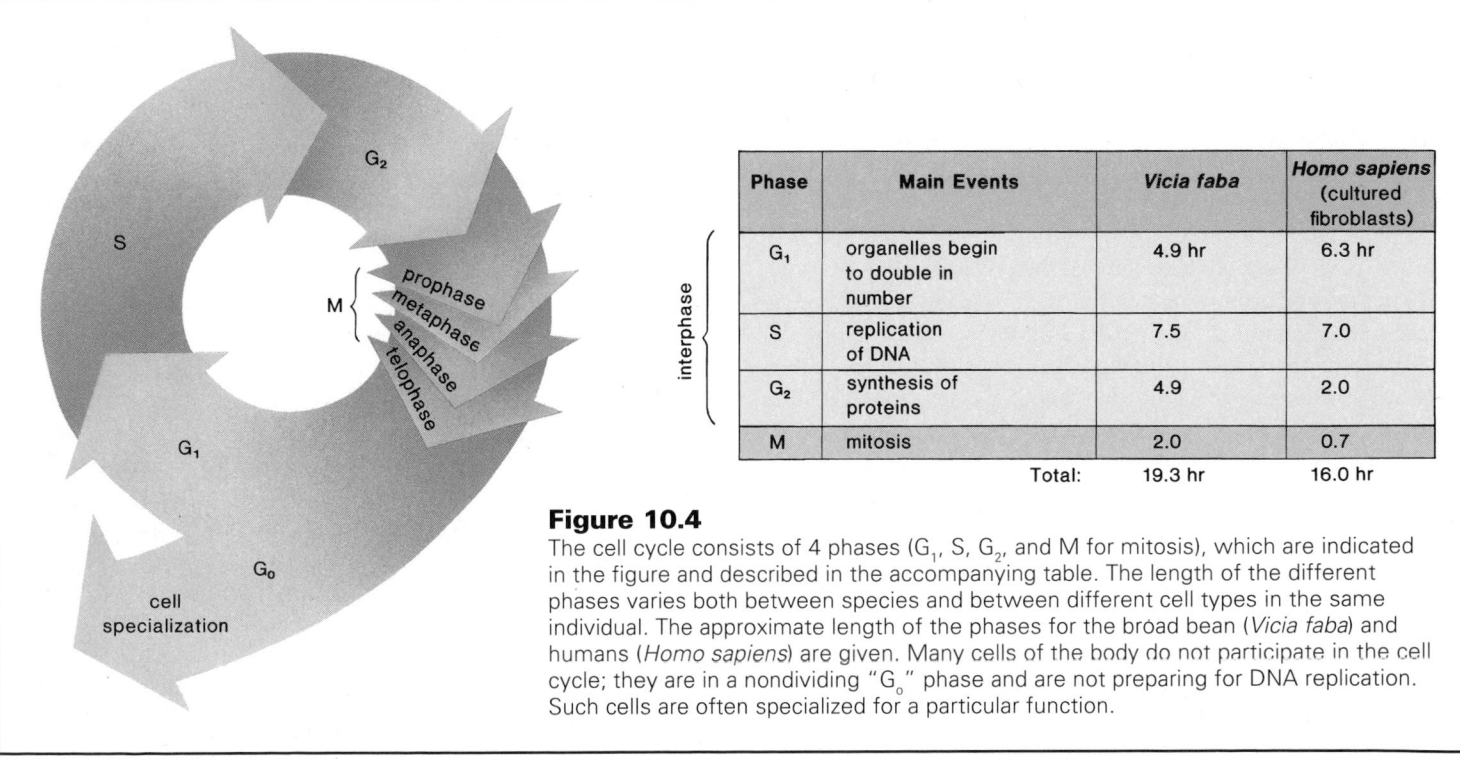

Phase	Main Events	*Vicia faba*	*Homo sapiens* (cultured fibroblasts)
G_1	organelles begin to double in number	4.9 hr	6.3 hr
S	replication of DNA	7.5	7.0
G_2	synthesis of proteins	4.9	2.0
M	mitosis	2.0	0.7
	Total:	19.3 hr	16.0 hr

Figure 10.4

The cell cycle consists of 4 phases (G_1, S, G_2, and M for mitosis), which are indicated in the figure and described in the accompanying table. The length of the different phases varies both between species and between different cell types in the same individual. The approximate length of the phases for the broad bean (*Vicia faba*) and humans (*Homo sapiens*) are given. Many cells of the body do not participate in the cell cycle; they are in a nondividing "G_0" phase and are not preparing for DNA replication. Such cells are often specialized for a particular function.

Some cells, such as skin cells, divide continuously throughout the life of the organism. Other cells that temporarily or permanently leave the cell cycle are said to be in a G_o phase. Cell specialization can cause cells to leave the cell cycle. Liver cells are specialized, but they will divide when repair of the organ is needed. Muscle and nerve cells in animals and sugar-transport cells in plants permanently lose the ability to divide.

Cells undergo a cycle that includes interphase (G_1, S, G_2) and mitosis, followed by cytokinesis.

Control of the Cell Cycle

Investigations are underway to determine what controls the cell cycle. For example, investigators want to know what causes a cell to continue to divide or what prevents it from doing so. It would appear that the ability to respond to outside influences is one factor. Cells have receptors for growth factors, and when growth factors are present cells divide. Also, cells of multicellular animals divide only if they are adhered to a solid surface. Without adhesion the cell's cytoskeleton cannot carry on mitosis and cytokinesis. Once normal cells make contact with their neighbors, they stop dividing, though. Cancer cells are just the opposite of normal cells; they will divide without adhesion to a surface, and they do not stop dividing even after making contact with neighboring cells.

Other investigators have centered their attention on factors within the cytoplasm that cause a cell to divide. First, we can note that the completion of early events is necessary to later events. For example, protein synthesis during the G_1 phase provides molecules necessary to DNA synthesis during the S phase. Otherwise, control seems to be exerted at 2 critical "checkpoints" during interphase. The first of these is when the G_1 phase becomes the S phase and DNA replication occurs. Investigators have found that when G_1 phase nuclei are placed in S phase cells, DNA replication begins, and when S phase nuclei are placed in G_2 cells, synthesis slows down. This suggests that S phase cytoplasm contains stimulatory substances that cause DNA replication to occur. Such stimulatory substances have been found for the second critical checkpoint during interphase—when G_2 becomes the M phase and mitosis begins. Those studying mitosis in newly fertilized frog's eggs have shown that the presence of a substance called maturation promoting factor (MPF) is necessary for mitosis to begin. MPF contains 2 molecules of interest, cdc2 protein and cyclin, another protein. Cyclin increases in quantity during interphase and influences the activity of cdc2, a kinase that transfers phosphate groups from ATP to various enzymes. The molecule cdc2 is involved in various events of mitosis such as condensation of chromatin, disintegration of the nuclear envelope, and formation of the mitotic spindle.

Geneticists have discovered that the cdc2 gene is remarkably similar in all eukaryotes. For example, the human gene for cdc2 will function in mutant yeast cells that do not have a functioning cdc2 gene! Much work remains to be done, but at least we are now aware that there are indeed stimulatory proteins that cause a cell to undergo mitosis.

Most likely the cytoplasm contains regulatory proteins that control the events of the cell cycle.

Mitosis and Cytokinesis

Mitosis is *division of the eukaryotic nucleus so that the daughter cell nuclei acquire the same number and kinds of chromosomes as the parent cell nucleus.* Before mitosis begins, each chromosome has duplicated and is composed of 2 chromatids; at the completion of mitosis each chromosome is composed of a single chromatid once more:

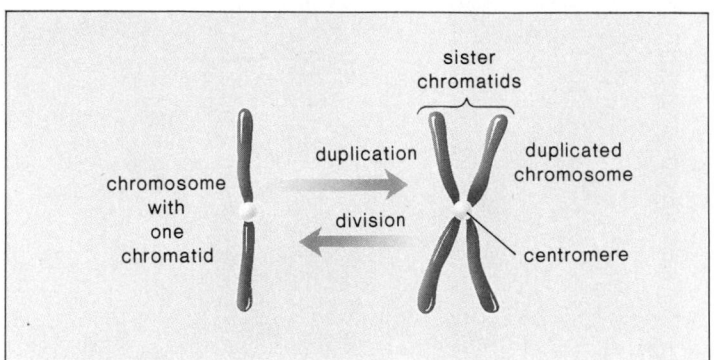

During mitosis, a microtubular structure, the mitotic **spindle,** forms and brings about an orderly distribution of chromosomes to the daughter cell nuclei. (There is no spindle during the process of binary fission in prokaryotic cells.) Mitosis, also called karyokinesis, is a continuous process that is arbitrarily divided into 4 stages for convenience of description. These stages are prophase, metaphase, anaphase, and telophase.

Mitosis in Animal Cells
Prophase

It is apparent during *prophase* that nuclear division is about to occur because the chromatin has condensed and the chromosomes are visible structures (see chapter reading; fig. 10.5). As the chromosomes continue to compact, the nucleolus disappears and the nuclear envelope fragments. *The chromosomes are duplicated and are composed of 2 sister chromatids held together at a centromere.* Counting the number of centromeres in diagrammatic drawings gives the number of chromosomes for the cell depicted.

The spindle begins to assemble as pairs of centrioles migrate toward the poles (opposite ends of spindle). Centrioles are short cylinders of microtubules (see fig. 5.17) found within a microtubule organizing center (MTOC). As its name implies, the MTOC is the organizing center for the spindle. (This accounts for the observation that plant cells do not have centrioles and yet they do have spindles during mitosis.) Short microtubules collectively called an **aster** appear and radiate from the centrioles. Between the centrioles we see the first indication of the polar microtubules, which will extend from one of the poles to the equator (midplane of spindle) or even beyond.

Figure 10.5

Prophase drawing (**a**) and micrograph (**b**) of whitefish (animal) embryonic cell. Magnification, X280. These events occur during prophase:

- chromatin condenses to (duplicated) chromosomes;
- nucleolus disappears;
- nuclear envelope fragments;
- centriole pairs migrate toward opposite poles;
- polar microtubules are assembling.

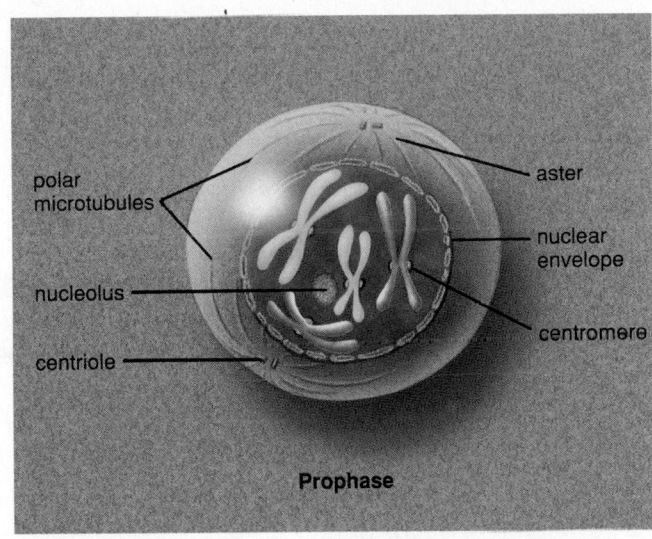
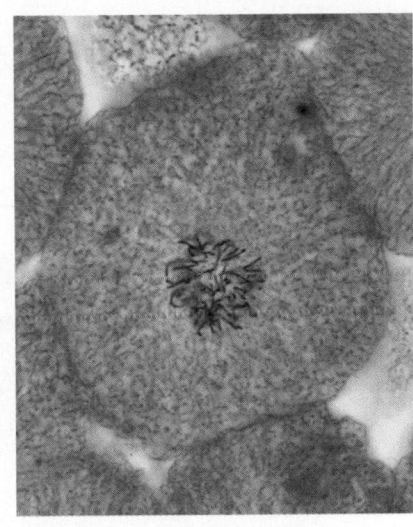

polar microtubules
aster
nuclear envelope
nucleolus
centromere
centriole

Prophase

a.

b.

Figure 10.6

Metaphase drawing (**a**) and micrograph (**b**) of whitefish (animal) embryonic cell. Magnification, X280. These events occur during metaphase:

- nuclear envelope fragmentation is complete;
- spindle formation is completed;
- duplicated chromosomes are aligned at equator;
- kinetochore microtubules of sister chromatids point to opposite poles.

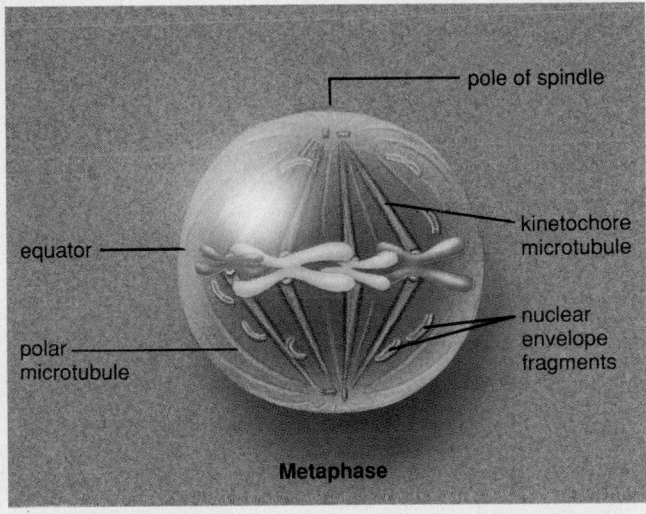

pole of spindle
equator
kinetochore microtubule
nuclear envelope fragments
polar microtubule

Metaphase

a.

b.

Figure 10.7

Anaphase drawing (**a**) and micrograph (**b**) of whitefish (animal) embryonic cell. Magnification, X500. These events occur during anaphase:

- centromeres divide;
- diploid set of daughter chromosomes move toward each pole;
- polar microtubules lengthen;
- cytokinesis begins.

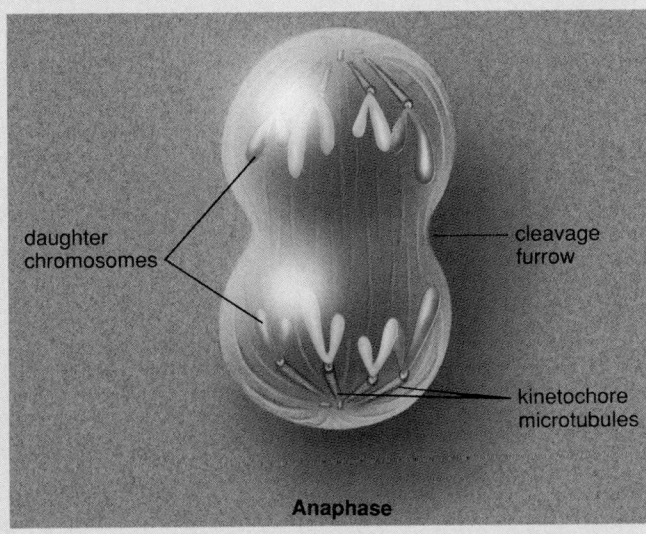
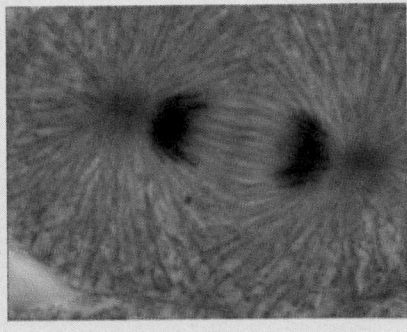

daughter chromosomes
cleavage furrow
kinetochore microtubules

Anaphase

a.

b.

Figure 10.8

Spindle apparatus anatomy and function. *a.* Artist's representation of the spindle apparatus from an animal cell. The microtubule organizing center (yellow) contains the centrioles with radiating fibers called asters. Polar fibers reach from the poles to the equator, where they overlap. The chromosomes (blue) attached to spindle by kinetochore fibers are moving toward the poles. *b.* Chromosome movement. Spindle fibers are composed of microtubules that can lengthen by the addition of tubulin dimers and shorten by the subtraction of tubulin dimers. It is believed that the polar spindle fibers lengthen at the equator due to the addition of tubulin subunits and that they also slide past one another in the region of overlap. This pushes the chromosomes apart. Also the kinetochore fibers shorten by disassembling at the kinetochore; this pulls the chromosomes toward the poles.

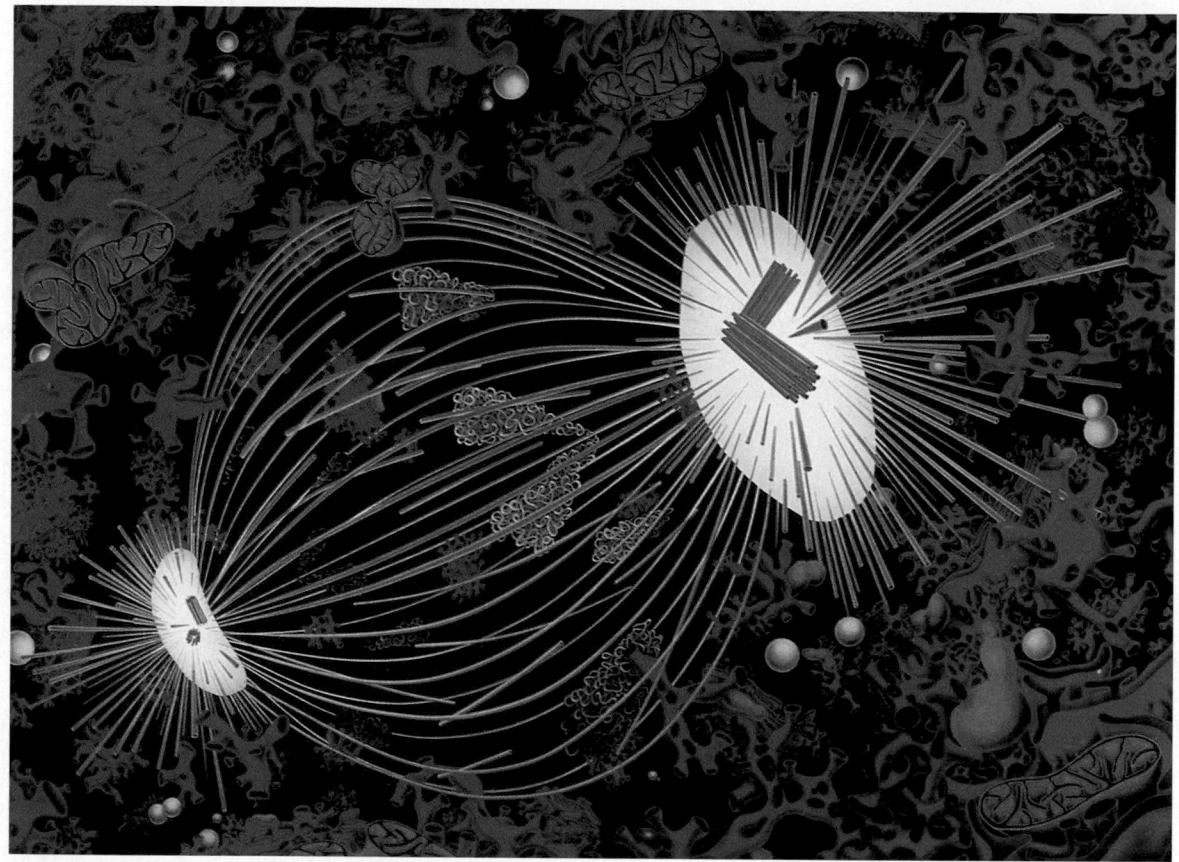

a.

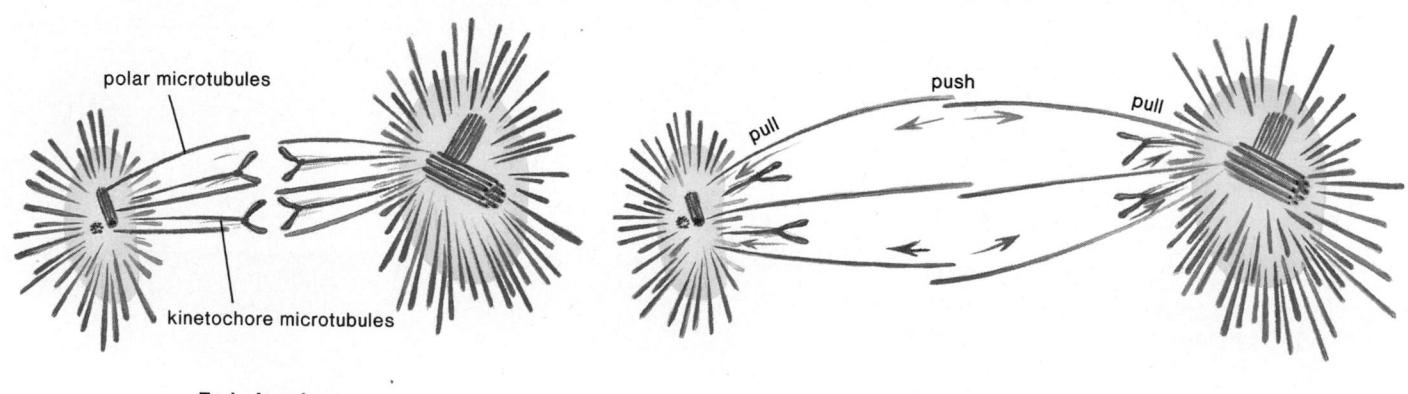

b.

Early Anaphase

Late Anaphase

Figure 10.9

Telophase drawing (**a**) and micrograph (**b**) of whitefish (animal) embryonic cell. Magnification, X280. These events occur during telophase:

- chromosomes uncoil to chromatin;
- nuclear envelopes reform;
- nucleoli reappear;
- daughter nuclei are diploid;
- cytokinesis is nearly complete.

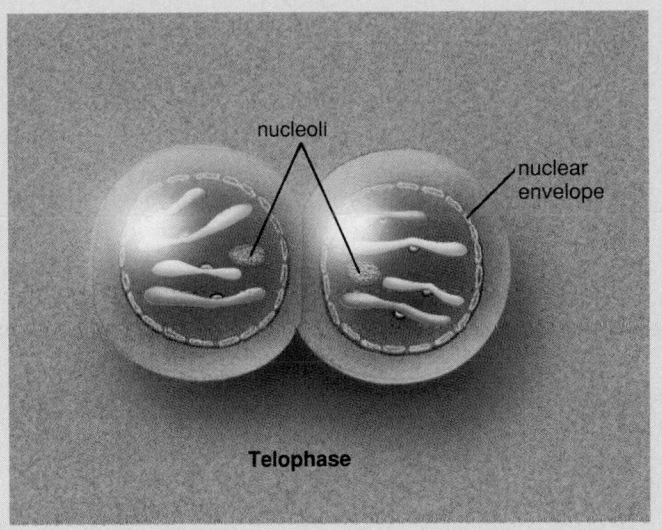

nucleoli

nuclear envelope

Telophase

a.

b.

During prophase, chromatin condenses to form the chromosomes, which have no particular orientation.

Metaphase

As *metaphase* begins, the fragmentation of the nuclear envelope that began in prophase continues, and the polar microtubules now extend to the equator of the spindle (fig. 10.6). Specialized structures called *kinetochores* develop on either side of each centromere. Microtubules called *kinetochore microtubules* extend from the kinetochores to the poles. The chromatids of a chromosome dangle from the centromere, but the kinetochores are orientated so that the sister chromatids face opposite poles.

Metaphase is recognized by the alignment of the chromosomes at the equator of a fully formed spindle.

Anaphase

At the start of *anaphase,* the centromeres of the duplicated chromosomes divide (fig. 10.7). Daughter chromosomes, each having a single chromatid, begin to move toward opposite poles. The mechanism by which the centromeres divide is unknown, but it has been suggested that the centromeres are merely a region where the DNA did not replicate. Replication in this area would then allow the chromatids to separate.

There seem to be 2 processes that account for the movement of the daughter chromosomes toward opposite poles. A lengthening of the polar microtubules acts to push the chromosomes away from the equator, and a shortening of the kinetochore microtubules acts to pull the chromosomes toward the poles (fig. 10.8).

As the chromosomes reach the poles, cytokinesis begins.

Anaphase is recognized by the movement of chromosomes toward each pole of the spindle.

Telophase

As the microtubules disassemble during *telophase,* new nuclear envelopes form around the daughter chromosomes (fig. 10.9) Each daughter nucleus contains the same number and kinds of chromosomes as the original parent cell. Remnants of the polar microtubules are still visible between the 2 nuclei.

The chromosomes become more diffuse chromatin once again, and a nucleolus appears in each daughter nucleus. Cytokinesis is nearly complete, and soon there will be 2 individual daughter cells.

Telophase is recognized by 2 clusters of daughter chromosomes within newly forming daughter nuclei.

Cytokinesis in Animal Cells

Cytokinesis, or cytoplasmic cleavage, usually accompanies mitosis. A cleavage furrow, which is an indentation of the membrane between the 2 daughter nuclei, begins as anaphase draws to a close. The cleavage furrow deepens as a band of actin filaments, called the contractile ring, slowly constricts the cell at the equator. A narrow bridge between the 2 cells can be seen during telophase, and then the contractile ring completely separates the cytoplasm and there are 2 daughter cells (fig. 10.10).

Cytokinesis in animal cells is accomplished by a furrowing process.

Figure 10.10
Cytokinesis in an animal cell. The single cell becomes 2 daughter cells by a furrowing process.

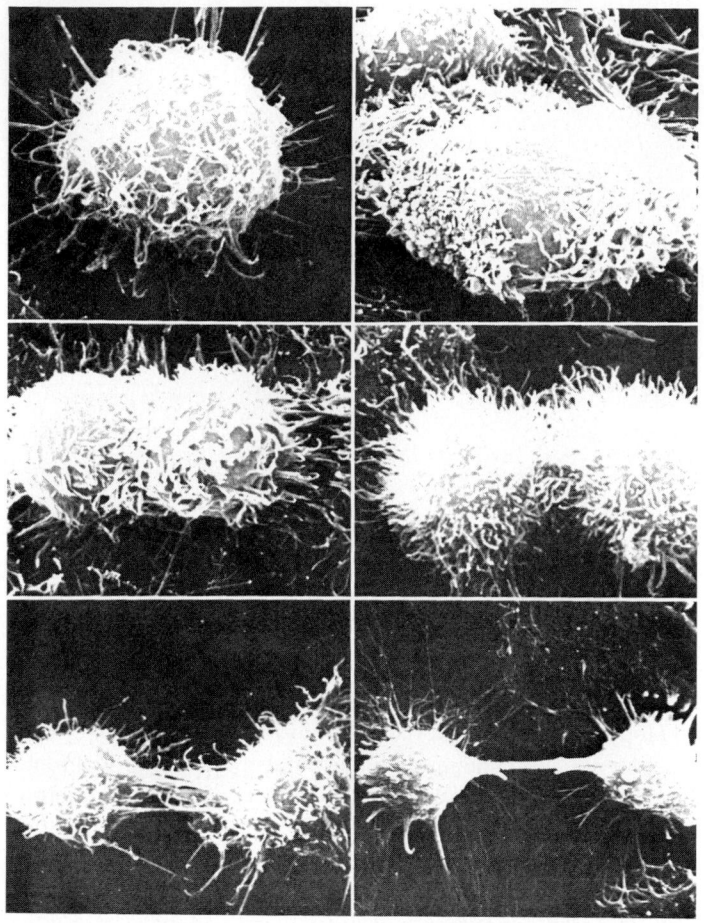

Mitosis in Plant Cells

Figure 10.11 illustrates mitosis in plants. Note that there are exactly the same stages in plant cells as in animals cells. As mentioned previously, plant cells have a microtubule organizing center, but there are no centrioles or asters.

Certain plant tissue, called meristem tissue, retains the ability to divide throughout the lifetime of a plant. Meristem tissue is found in root and shoot tips and in stems, and it accounts for the ability of trees to grow larger each growing season.

Cytokinesis in Plant Cells

Cytokinesis in plant cells occurs by a process different from that seen in animal cells. The rigid cell wall that surrounds plant cells does not permit cytokinesis by furrowing. Instead, vesicles, largely derived from the Golgi apparatus, travel down the polar microtubules to the region of the equator. These vesicles fuse, forming a **cell plate.** Their membrane completes the plasma membrane for both cells (fig. 10.12). They also release polysaccharides that signal the formation of plant cell walls. These walls are later strengthened by the addition of cellulose fibrils.

A spindle forms during mitosis in plant cells, but there are no centrioles or asters. Cytokinesis in plant cells involves the formation of a cell plate.

Importance of Binary Fission and Mitosis

Binary fission and mitosis are successful evolutionary adaptations that ensure daughter cells get equal amounts of genetic material and cytoplasm. Fission and mitosis differ chiefly in that binary fission does not utilize a spindle, whereas mitosis does utilize a spindle.

Figure 10.11
Mitosis in plant cells. Notice that there are no asters in plant cells because they lack centrioles. This is evidence that centrioles are not needed for a spindle apparatus to form. Magnification, X1,500.

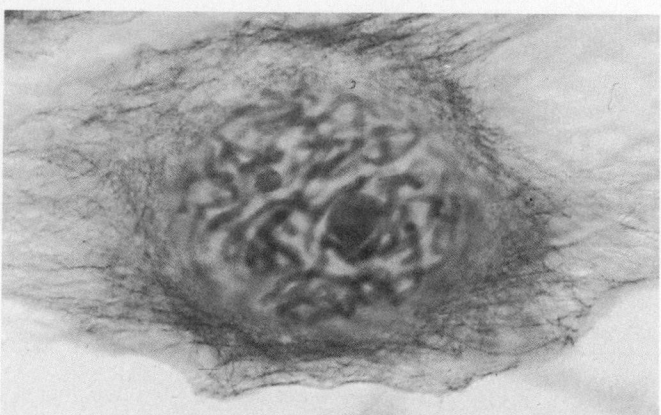

a. Prophase

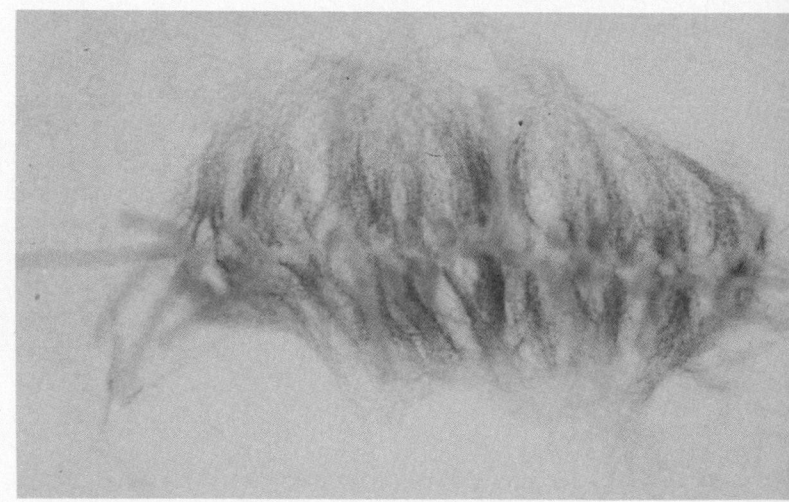

b. Metaphase

Genetic Basis of Life

Figure 10.12

Cytokinesis in a plant cell involves the formation of a cell plate. ***a.*** Electron micrograph showing the stage that corresponds to (***c***). CP = cell plate; CW = cell wall; Pl = plasma membrane; N = nucleus; SpF = spindle fiber. Magnification, X20,000. ***b.*** Vesicles containing polysaccharides are at the center of the cell. ***c.*** The cell plate forms at the cell equator and extends to the plasma membrane. ***d.*** Daughter cell plasma membranes are complete, and the cell wall has formed.

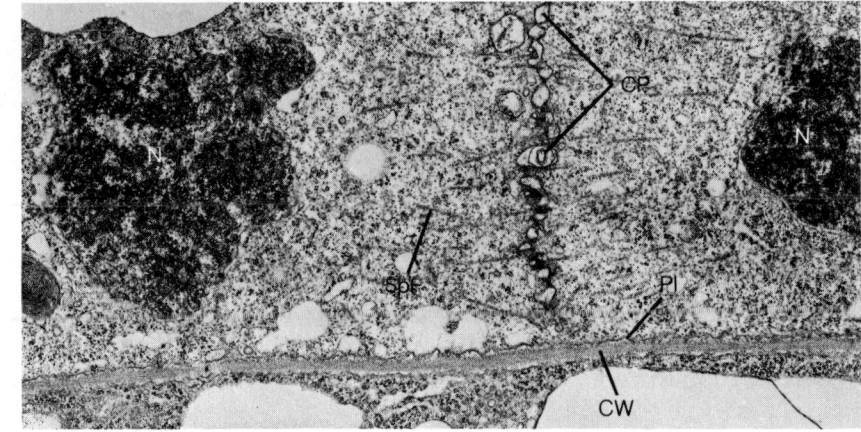

a.

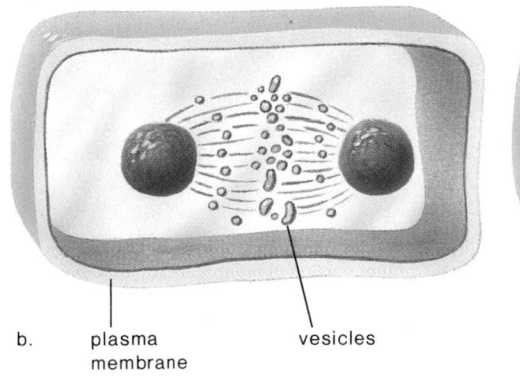

b. plasma   vesicles
    membrane

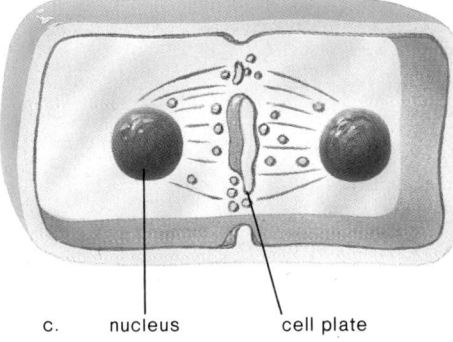

c. nucleus   cell plate

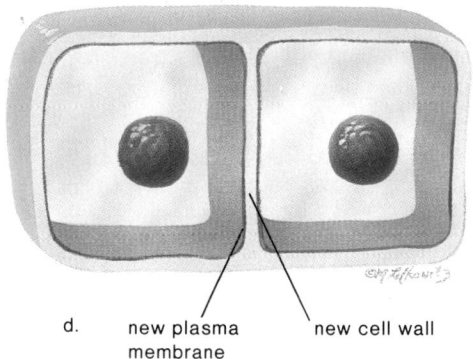

d. new plasma   new cell wall
    membrane

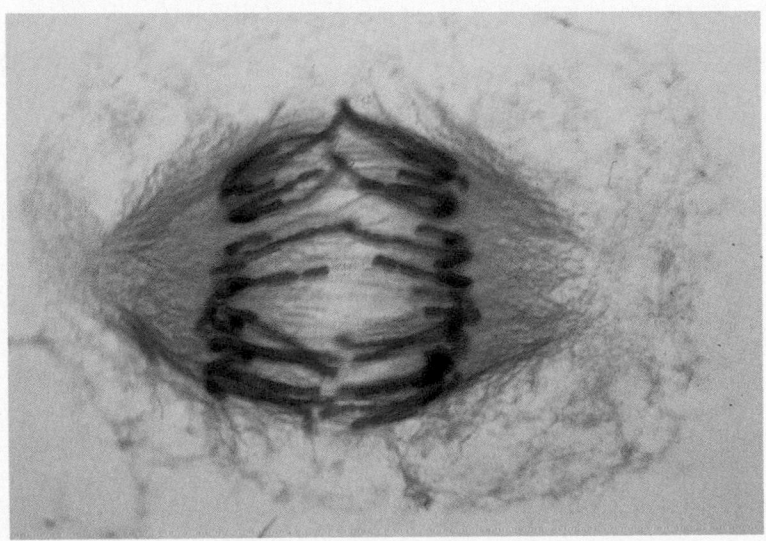

c. Anaphase

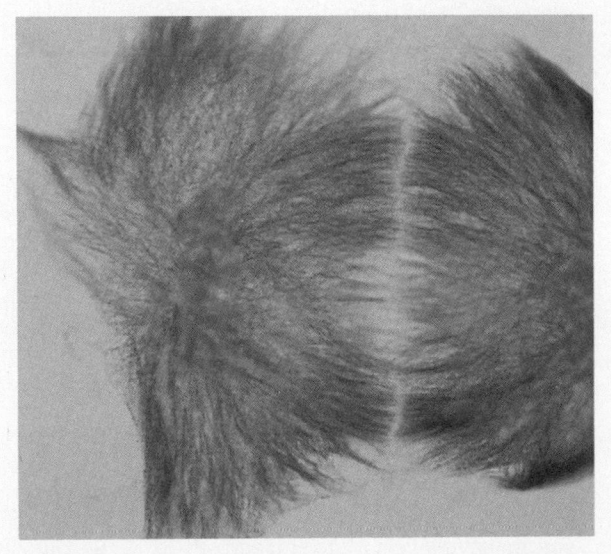

d. Telophase

Evolution of the Spindle

Observation of nuclear division in primitive eukaryotes, such as dinoflagellates and fungi, makes it possible to suggest the steps by which mitosis may have evolved. At first, division may have taken place in a manner similar to binary fission, except that the separating chromosomes were attached to the nuclear envelope instead of the plasma membrane. The purpose of the first involvement of microtubules may have been to support the nuclear envelope. Later, the microtubules became attached to the chromosomes themselves. As microtubules began to take a more active role in separating the chromosomes, the nuclear envelope became less important. Indeed, we have seen that the nuclear envelope plays no role in chromosome movement in the most advanced eukaryotes.

Asexual versus Sexual Reproduction

We have observed that single-celled prokaryotes reproduce by binary fission. Single-celled eukaryotic protists (e.g., protozoans and algae) reproduce by cell division that involves mitosis. Reproduction by binary fission and mitosis is termed **asexual reproduction** (table 10.2).

Although some multicellular forms do reproduce asexually, generally speaking, mitosis only occurs during the growth and repair of tissues. Multicellular forms usually carry on **sexual reproduction,** which requires sex cell (gamete) formation and fertilization. Meiosis, the form of cell division that reduces the chromosome number from diploid to haploid, is involved in sex cell formation and sexual reproduction. Meiosis is discussed in the next chapter.

Table 10.2
Functions of Cell Division

Type of Organism	Cell Division	Function
Prokaryotes (bacteria)	Binary fission	Asexual reproduction
Protists Some fungi	Mitosis and cytokinesis	Asexual reproduction
Some fungi Plants	Mitosis and cytokinesis	Primarily growth and repair; asexual reproduction in some
Animals	Meiosis and fertilization	Sexual reproduction

Prokaryotes reproduce asexually by binary fission, and single-celled eukaryotes reproduce asexually by mitosis. Mitosis in multicellular eukaryotes is usually involved in growth and repair of tissues, and meiosis occurs during sexual reproduction.

Summary

1. In single-celled organisms, simple cell division allows the organism to reproduce. In multicellular organisms, this type of cell division is primarily for growth and repair of tissues.
2. Cell division keeps the cell surface area-to-cell volume ratio at a level appropriate to the life of the cell. Its occurrence leads to specialization of parts.
3. The prokaryotic chromosome is a single long loop of DNA attached to the inside of the plasma membrane. Binary fission involves replication of DNA, followed by an elongation of the cell that pushes the chromosomes apart. Ingrowth of the plasma membrane and formation of new cell wall material divide the cell in 2.
4. Each eukaryote species has a characteristic number of chromosomes. The total number is called the diploid number, and half this number is the haploid number. Between nuclear divisions, the chromosomes are not distinct and are collectively called chromatin. Chromatin and chromosomes have levels of coiling and condensation that are explained in the chapter reading.
5. Replication of DNA precedes cell division. The duplicated chromosomes are each composed of 2 sister chromatids held together at a centromere.
6. Among eukaryotes, cell division involves nuclear division (karyokinesis) and division of the cytoplasm (cytokinesis). There are 2 types of nuclear division. Mitosis is simple nuclear division in which the chromosome number stays constant, and meiosis is nuclear division in which the chromosome number is reduced by one half. Only mitosis is considered in this chapter.
7. The cell cycle includes interphase and mitosis. Interphase has 3 phases. During the G_1 phase the organelles increase in number, during the S phase, DNA replication occurs, and during the G_2 phase various proteins are synthesized. Specialized cells leave the cycle.
8. Control of the cell cycle is under investigation. In cells grown in tissue culture, it appears that various proteins in the cytoplasm control the cycle.
9. Mitosis, nuclear division by which the daughter nuclei receive the same number and kinds of chromosomes as the mother cell, has 4 stages, which are described here for animal cells.

Prophase—Duplicated chromosomes are distinct, the nucleolus is disappearing, the nuclear envelope is fragmenting, and polar microtubules are forming as centriole pairs separate. Asters develop around the centrioles.

Metaphase—Spindle apparatus is fully formed. Duplicated chromosomes move to the equator and are attached to the poles by the kinetochore microtubules.

Anaphase—Centromeres divide and daughter chromosomes move toward the poles, perhaps by a push-pull mechanism. Push—polar microtubules elongate and slide past one another at the equator. Pull—kinetochore microtubules shorten at the kinetochores. Cytokinesis by furrowing begins.

Telophase—Nuclear envelopes reform, chromosomes begin changing back to chromatin, the nucleoli reappear, and spindle microtubules disappear. Cytokinesis completes the furrowing process that divides the cytoplasm.

10. Mitosis in plant cells is somewhat different from mitosis in animal cells because plant cells lack centrioles and therefore asters. Even so, the mitotic spindle forms and the same 4 stages are observed. Cytokinesis in plant cells involves the formation of a cell plate from which the plasma membrane and cell wall are completed.

11. Binary fission in single-celled prokaryotes and mitosis in single-celled eukaryotic protists and fungi allow these organisms to reproduce asexually. Mitosis in multicellular eukaryotes is primarily for the purpose of growth and repair of tissues. Multicellular forms generally carry on sexual reproduction, which requires meiosis, the type of cell division involved in gamete formation.

Study Questions

1. What benefits are associated with cell division?
2. Describe the prokaryotic chromosome and the process of binary fission.
3. A human cell contains at least 2 m of DNA, yet all of this DNA is packed into a nucleus that is about 5 μm in diameter. Explain.
4. Describe the cell cycle, including a reference to interphase.
5. How do cancer cells differ from normal cells in relation to the cell cycle?
6. Define the following words: chromosome, chromatin, chromatid, centriole, cytokinesis, and centromere.
7. Why is duplication of a chromosome required before mitosis occurs?
8. Describe the events that occur during the stages of mitosis.
9. Contrast cytokinesis in animal cells with that in plant cells.
10. Why is binary fission in prokaryotes and mitosis in single-celled eukaryotes a form of asexual reproduction?

Objective Questions

1. What feature in prokaryotes substitutes for the spindle in eukaryotes?
   a. centrioles with asters
   b. fission instead of cytokinesis
   c. elongation of plasma membrane
   d. looped DNA
2. How does a prokaryotic chromosome differ from a eukaryotic chromosome? A prokaryotic chromosome
   a. is shorter and fatter.
   b. is looped and in one piece.
   c. never replicates.
   d. All of these.
3. The diploid number of chromosomes
   a. is the 2N number.
   b. was in the parent cell and is in the 2 daughter cells following mitosis.
   c. varies according to the particular organism.
   d. All of these.

For question 4-6, match the descriptions that follow to the terms in the key.

   Key:
   a. centriole
   b. chromatid
   c. chromosome
   d. centromere

4. point of attachment for sister chromatids
5. found at poles; centers of asters
6. condensed chromatin
7. If a parent cell has 14 chromosomes prior to mitosis, how many chromosomes will the daughter cells have?
   a. 28
   b. 14
   c. 7
   d. Any number between 7 and 28.
8. In which stage of mitosis are the chromosomes moving toward the poles?
   a. prophase
   b. metaphase
   c. anaphase
   d. telophase
9. Interphase
   a. is the same as prophase, metaphase, anaphase, and telophase.
   b. includes phases G_1, S, and G_2.
   c. requires the use of polar microtubules and kinetochore microtubules.
   d. rarely occurs.

10. Cytokinesis
    a. is mitosis in plants.
    b. requires the formation of a cell plate in plant cells.
    c. is the longest part of the cell cycle.
    d. is half a chromosome.
11. Label this diagram of prophase.

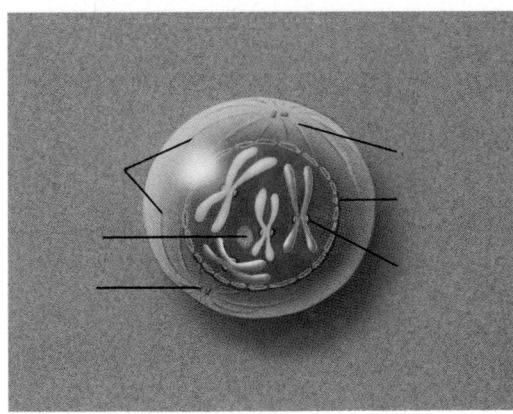

Concepts and Critical Thinking

1. *All cells come only from preexisting cells* (the principle of biogenesis).

   Cells can grow and reproduce. How does growth of a cell contribute to reproduction of a cell?

2. *Like begets like because the genetic material is copied and transmitted from parent to daughter cells* (the principle of heredity).

   In certain cells, certain genes are active and others are inactive. Considering this, why do only skin cells arise from skin cells, for example?

3. *Most eukaryotic cells reproduce by means of mitosis.*

   What are the benefits of dividing by mitosis for a cell?

Selected Key Terms

nucleoid (nu'kle-oid) 156
binary fission (bi'na-re fish'un) 156
chromatin (kro'mah-tin) 157
chromosome (kro'mo-sōm) 157
diploid (2N) number (dip'loid num'ber) 157
haploid (N) number (hap'loid num'ber) 157

cytokinesis (si"to-ki-ne'sis) 157
mitosis (mi-to'sis) 157
meiosis (mi-o'sis) 157
sister chromatid (sis'ter kro'mah-tid) 157
centromere (sen'tro-mēr) 157
interphase (in'ter-fāz) 159

spindle (spin'd'l) 160
aster (as'ter) 160
cell plate (sel plāt) 164
asexual reproduction (a-seks'u-al re"pro-duk'shun) 166
sexual reproduction (seks'u-al re"pro-duk'shun) 166

11

Cell Reproduction: Meiosis

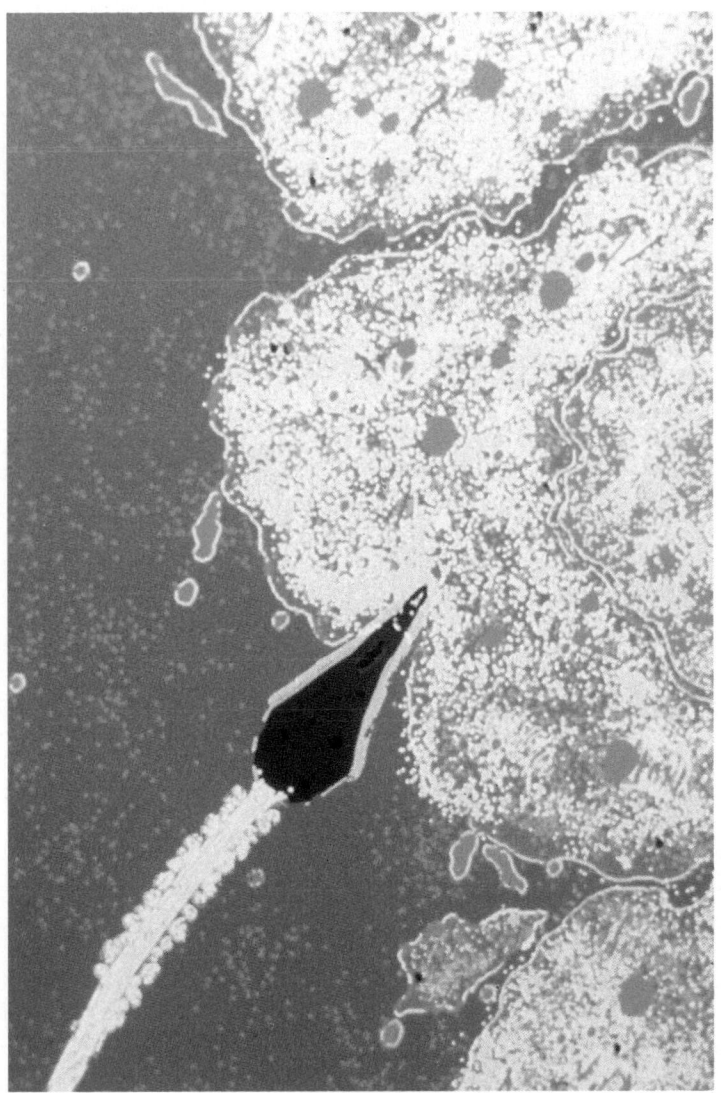

Sexual reproduction requires the production of gametes, the sperm and egg. A sperm is penetrating a layer of cells that surrounds the much larger egg in this scanning electron micrograph. Each sperm is tipped with an acrosome, a cap that opens to release an enzyme that slices through the coating of the egg. (The acrosome is colored mauve.) Magnification, X6,000.

Your study of this chapter will be complete when you can

1. state in general the role of meiosis in plant, animal, and fungal life cycles;
2. give an overview of the process of meiosis and its stages, emphasizing the main events;
3. describe the stages of meiosis I in detail;
4. describe synapsis (bivalent formation), and tell how crossing-over occurs;
5. describe the stages of meiosis II in detail;
6. describe the mammalian life cycle, and compare spermatogenesis to oogenesis;
7. describe the manner in which sexual reproduction brings about variation and contributes to the evolutionary process;
8. compare the process of meiosis to that of mitosis.

ome organisms reproduce asexually, but on occasion even these also reproduce sexually. Both asexual and sexual reproduction support the cell theory because the new life begins as a single cell.

Gamete formation and fusion to form a zygote are integral parts of **sexual reproduction** (fig. 11.1). Obviously, if the gametes contained the same number of chromosomes as the body cells, the number of chromosomes would double with each new generation. Within a few generations, the cells of the individual would be nothing but chromosomes! The early cytologists (biologists who study cells) realized this, and Pierre-Joseph van Beneden, a Belgian, was gratified to find in 1883 that the sperm and the egg of the worm *Ascaris* each contained only 2 chromosomes, while the cells of the embryo and the individual always have 4 chromosomes.

Figure 11.1

Life cycles of an animal, a plant, and a fungus. While meiosis always reduces the number of chromosomes, it occurs at a different point in each life cycle. Therefore, in animals, the haploid stage consists only of the gametes; in plants, there is both a diploid and haploid adult; in fungi, the adult is always haploid because the zygote undergoes meiosis.

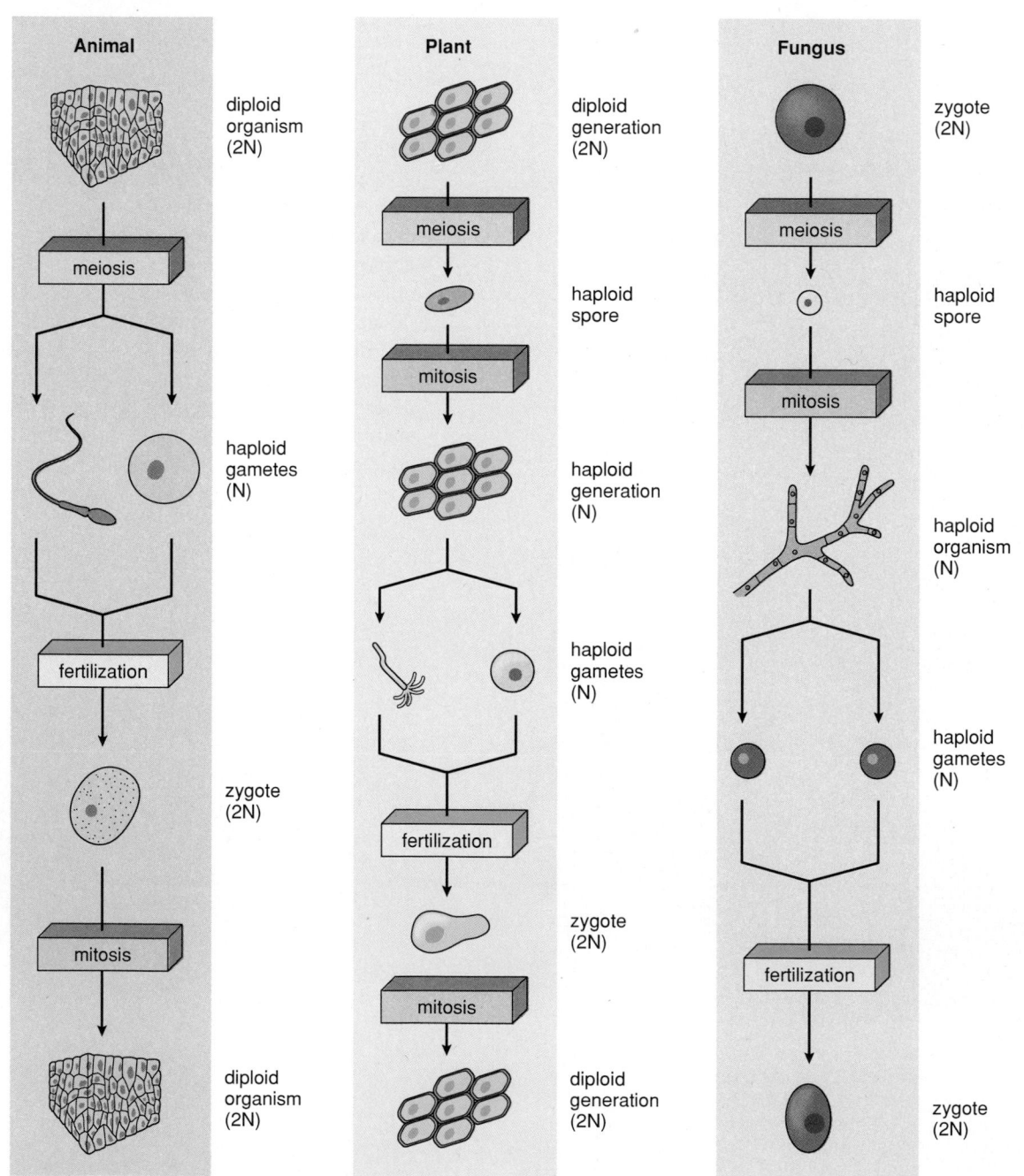

Meiosis is the type of nuclear division that reduces the chromosome number from the diploid (2N) number to the haploid (N) number. Meiosis occurs at different points during the life cycle of various kinds of organisms (fig. 11.1). In animals, it occurs during the production of the gametes. In plants, meiosis produces spores that divide mitotically to become a haploid generation. It is this generation that produces the gametes. In fungi (and some algae) meiosis occurs directly after zygote formation, and the adult is always haploid. This haploid adult produces the gametes.

Notice in figure 11.1 that all 3 types of life cycles have both a diploid stage and a haploid stage. In animals, the adult is always diploid and the haploid stage consists only of the gametes. In plants, there is both a diploid adult and a haploid adult that produces the haploid gametes. In fungi (and some algae), the zygote is the only diploid stage and the adult, which is always haploid, produces the haploid gametes.

Meiosis is a special type of nuclear division that reduces the diploid (2N) number of chromosomes to the haploid (N) number of chromosomes. Meiosis occurs at different points in the life cycle of organisms.

Overview of Meiosis

For the sake of simplicity, we presently will confine our discussion to meiosis in animals. Then, at the end of the chapter, we will discuss all 3 life cycles depicted in figure 11.1.

In a diploid cell, chromosomes come in pairs called **homologous chromosomes,** or **homologues.** Homologues are the same kind of chromosomes: they look alike, and they also carry genes for the same traits. One member of each pair was contributed by the male gamete, and the other was contributed by the female gamete:

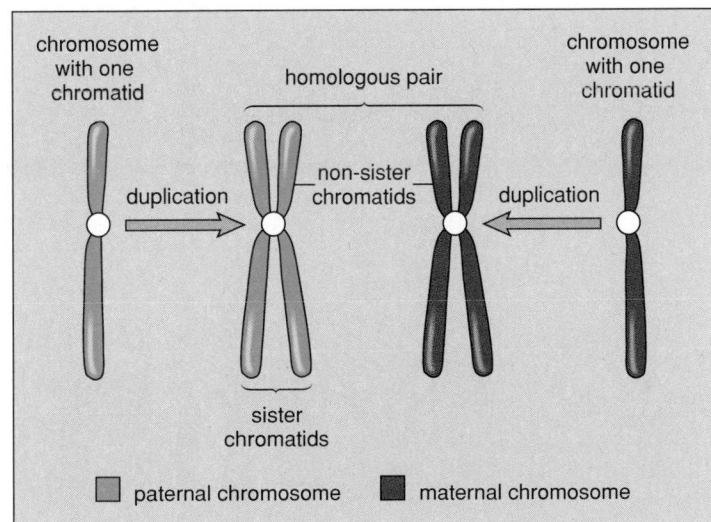

chromosome with one chromatid — homologous pair — chromosome with one chromatid

non-sister chromatids

duplication — duplication

sister chromatids

☐ paternal chromosome   ☐ maternal chromosome

Meiosis requires 2 nuclear divisions and produces 4 haploid daughter cells, each having one of each kind of chromosome and therefore half the total number of chromosomes present in the diploid parent nucleus. Figure 11.2 provides an overview of meiosis, indicating its 2 nuclear divisions, *meiosis I* and *meiosis II.*

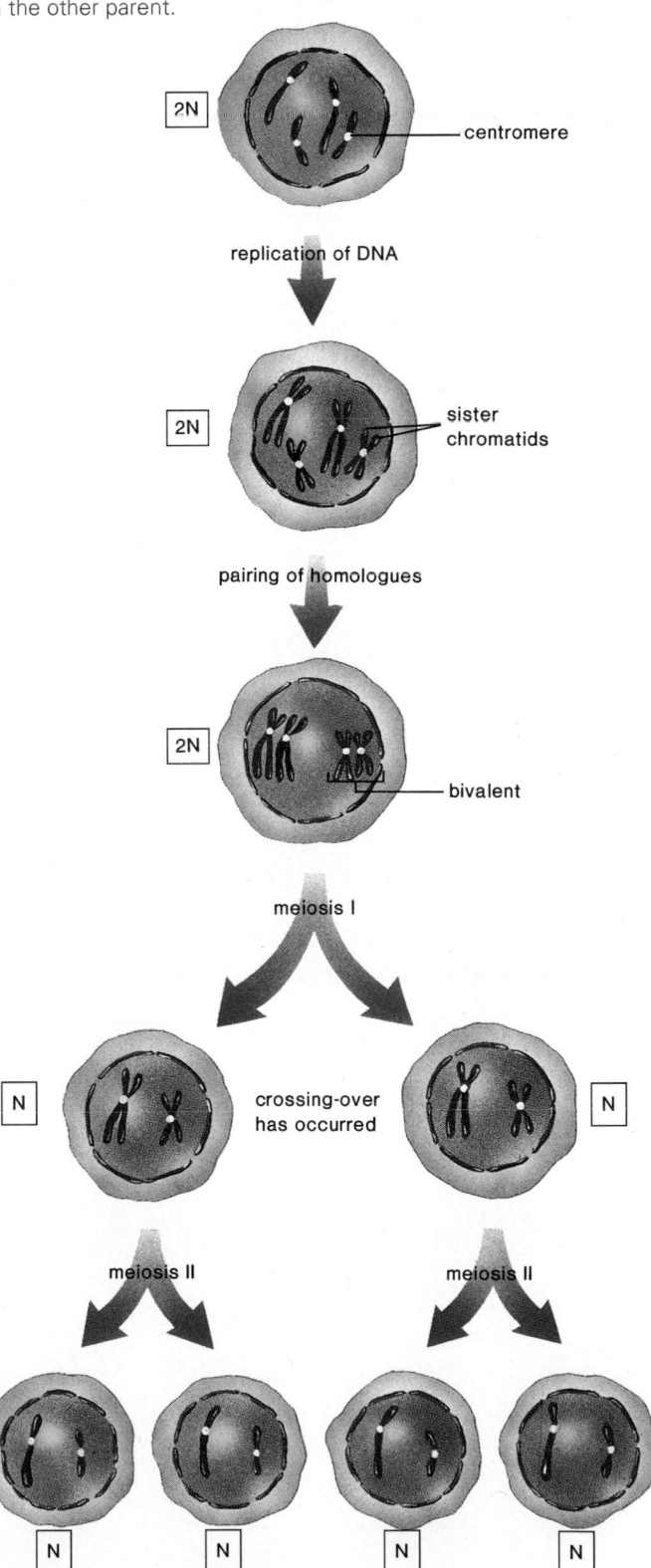

Figure 11.2

Overview of meiosis. The number of chromosomes in each nucleus can be determined by counting the number of centromeres. Following meiosis I there are 2 haploid daughter cells; following meiosis II there are 4 haploid daughter cells. Notice that no 2 daughter cells are alike because of crossing-over. The blue chromosomes were inherited from one parent and the red chromosomes were inherited from the other parent.

2N — centromere

replication of DNA

2N — sister chromatids

pairing of homologues

2N — bivalent

meiosis I

N — crossing-over has occurred — N

meiosis II — meiosis II

N   N   N   N

Prior to meiosis, DNA replication has occurred; therefore, each chromosome has 2 sister chromatids. Meiosis I reduces the number of chromosomes, and meiosis II produces daughter chromosomes.

During meiosis I, something new happens that does not occur in mitosis. The homologues actually come together and pair, forming a so-called **bivalent.** It is a bivalent because there are 2 chromosomes in close association. Sometimes the term **tetrad** is used because the bivalent contains 4 chromatids. **Crossing-over** of genetic material may occur between the nonsister chromatids of the tetrad. Due to crossing-over, the sister chromatids of a chromosome are no longer identical (see fig. 11.4).

Crossing-over temporarily holds the homologues together so that they interact with the spindle microtubules as if they were one chromosome. Eventually, the homologues of each bivalent separate, and this assures that one (duplicated) chromosome of each kind reaches the daughter nuclei of meiosis I. It is important for daughter nuclei to have one member from each pair of homologous chromosomes because only in that way can there be one copy of each kind of gene in the daughter nuclei.

The separation process has no restrictions; either homologue of a homologous pair may occur in a daughter nucleus with either homologue of any other pair.

Meiosis I is a form of nuclear division that results in cells that have only one copy of each kind of chromosome. Each homologous pair separates independently of all other pairs.

At the completion of meiosis I, the chromosomes in the daughter nuclei still consist of sister chromatids. Therefore, DNA replication does not occur between meiosis I and meiosis II. During meiosis II, the centromeres divide and a haploid set of daughter chromosomes moves toward each pole of the spindle. At the completion of meiosis II, there are 4 daughter cells, each with the haploid number of chromosomes.

Figure 11.3

Meiosis consists of 2 divisions, meiosis I and meiosis II, and each of these has 4 stages. Members of homologous pairs have different colors to indicate that one member of each pair is a maternal chromosome and the other is a paternal chromosome.

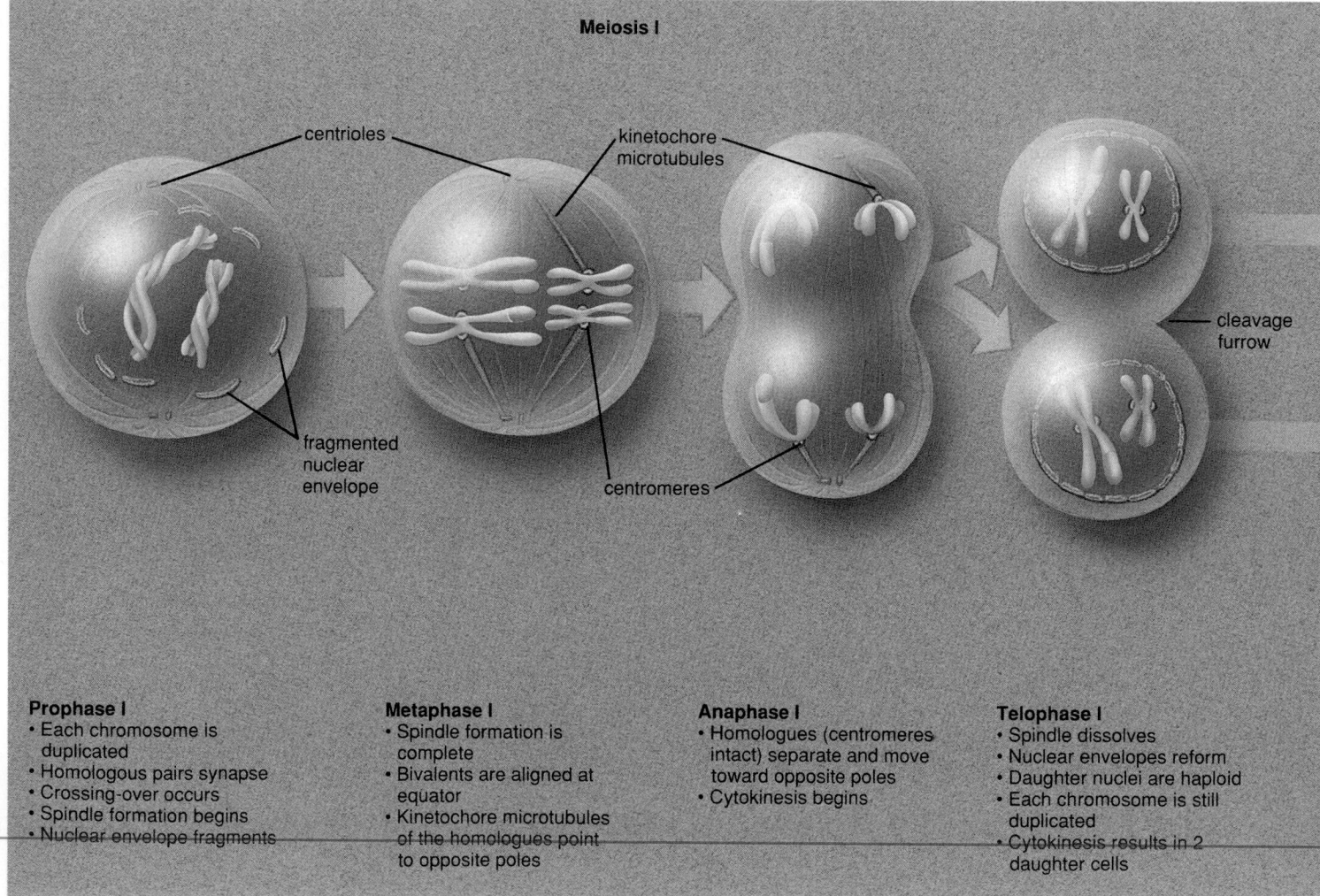

Meiosis I

Prophase I
- Each chromosome is duplicated
- Homologous pairs synapse
- Crossing-over occurs
- Spindle formation begins
- Nuclear envelope fragments

Metaphase I
- Spindle formation is complete
- Bivalents are aligned at equator
- Kinetochore microtubules of the homologues point to opposite poles

Anaphase I
- Homologues (centromeres intact) separate and move toward opposite poles
- Cytokinesis begins

Telophase I
- Spindle dissolves
- Nuclear envelopes reform
- Daughter nuclei are haploid
- Each chromosome is still duplicated
- Cytokinesis results in 2 daughter cells

During meiosis II, the centromeres holding sister chromatids together divide and daughter chromosomes move toward each pole. At the completion of meiosis there are 4 daughter cells each with the haploid number of chromosomes.

Meiosis I

Meiosis I is divided into the same 4 stages as mitosis, but the prophase stage is much more complicated (fig. 11.3).

Prophase I

During prophase I, replication has already occurred and the chromosomes consist of sister chromatids held together at a centromere. Only because of replication will there be tetrads permitting crossing-over between the nonsister chromosomes.

Before crossing-over occurs, each one of these duplicated but still extended chromosomes begins to pair with its homologue. During this process, called **synapsis,** the homologues line up side by side, and a nucleoprotein lattice (called the synaptomenal complex) appears between them (fig. 11.4). This lattice holds the members of a bivalent together in such a way that the DNA of the nonsister chromatids is aligned. Now crossing-over, an exchange of genetic material between nonsister chromatids, can occur. The lattice then begins to break down, and the homologues begin to move apart, but not too far apart because they are held together by **chiasmata** (chiasma sing.), regions where the nonsister chromatids are still attached due to crossing-over (fig. 11.5).

During prophase I, synapsis occurs and a nucleoprotein lattice develops between homologues. This close association assists crossing-over between nonsister chromatids of the bivalent.

Spindle microtubules begin to appear as the centriole pairs move apart. The nuclear envelope is fragmenting, and the nucleolus is disappearing. We should also mention that throughout all these events of prophase I, the chromosomes have been condensing so that by now they have the appearance of metaphase chromosomes. Prophase I is the longest stage of meiosis and takes most of the time required for the entire process.

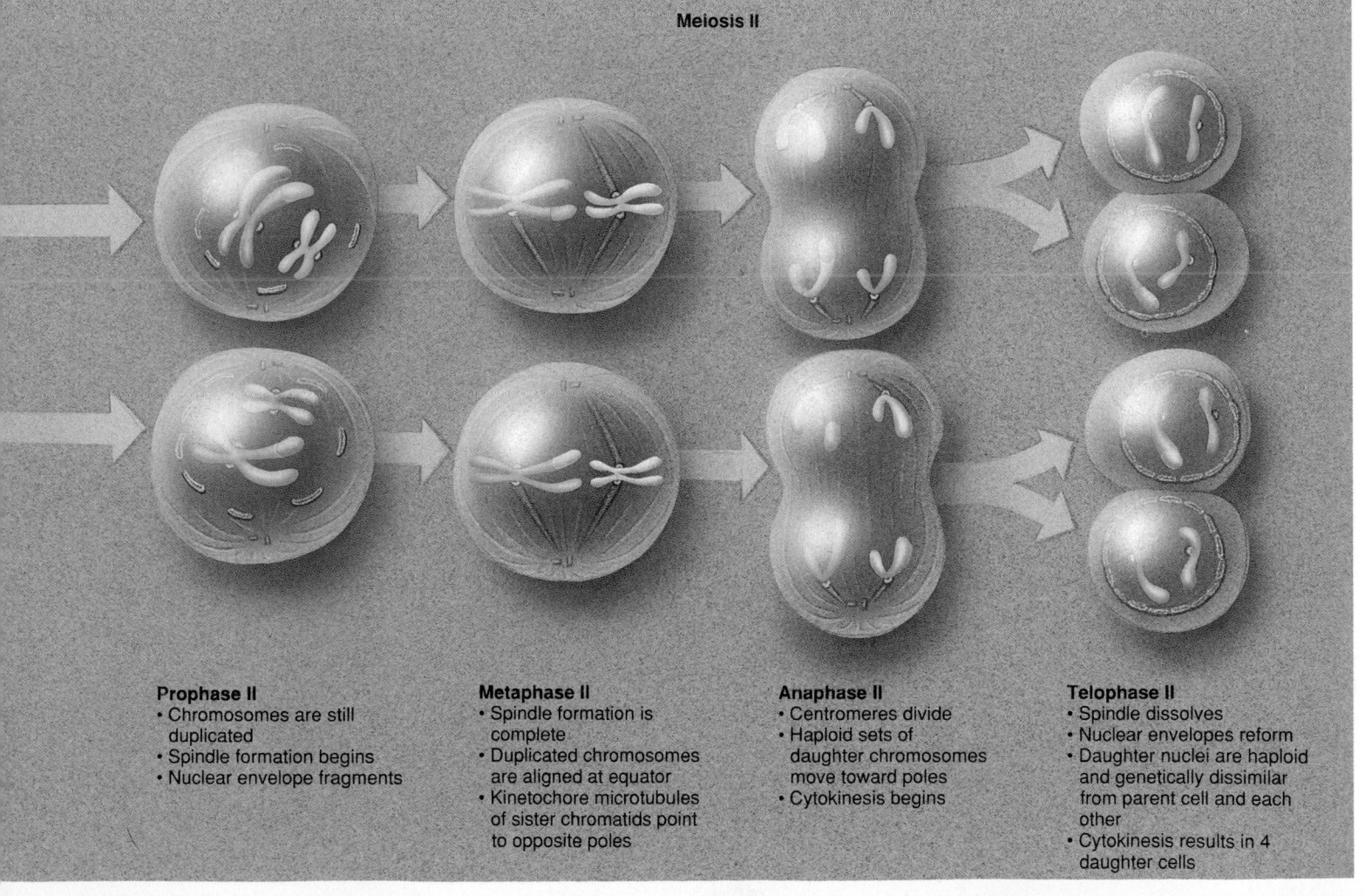

Meiosis II

Prophase II
- Chromosomes are still duplicated
- Spindle formation begins
- Nuclear envelope fragments

Metaphase II
- Spindle formation is complete
- Duplicated chromosomes are aligned at equator
- Kinetochore microtubules of sister chromatids point to opposite poles

Anaphase II
- Centromeres divide
- Haploid sets of daughter chromosomes move toward poles
- Cytokinesis begins

Telophase II
- Spindle dissolves
- Nuclear envelopes reform
- Daughter nuclei are haploid and genetically dissimilar from parent cell and each other
- Cytokinesis results in 4 daughter cells

Metaphase I

Spindle formation is now complete, but at the equator during metaphase I we find chromosome bivalents with this appearance:

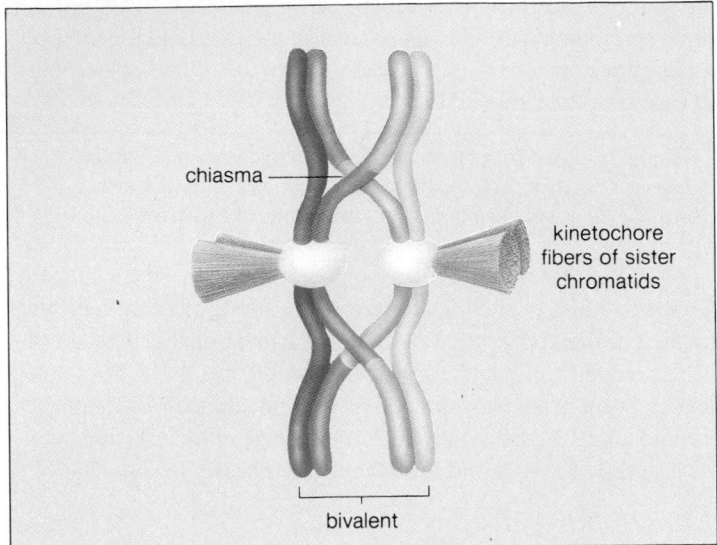

chiasma

kinetochore fibers of sister chromatids

bivalent

Notice that the *kinetochore microtubules* extend from only one side of the centromere, which holds the sister chromatids together, and that the bivalent is held together by chiasmata.

There is no set way for the chromosomes to align themselves at the equator of the spindle. The maternal homologue of one bivalent may be oriented toward one pole, and the maternal or paternal homologue of another bivalent may be aligned in like manner. Therefore, during meiosis I, the chromosomes assort independently of one another, and any possible combination of chromosomes can be found in the daughter cells.

Bivalents are present at the equator during metaphase I. The alignment is random, and therefore any possible combination of chromosomes can occur in the daughter cells.

Anaphase I

During anaphase I, the association between members of a bivalent is disrupted when the chiasmata, which mark points of exchange between the nonsister chromatids, cease to exist because crossing-over is completed (fig. 11.4d). Now homologues separate as the centromeres of each bivalent move to opposite poles. Due to this

Figure 11.4

Crossing-over occurs during prophase I. *a.* The homologous chromosomes pair up and a nucleoprotein lattice called the synaptonemal complex develops between them. This is an electron micrograph of the complex, which zippers the members of the bivalent together so that corresponding genes are in alignment. Magnification, X40,000. *b.* This diagrammatic representation shows only 2 places where nonsister chromatids 1 and 3 have come into contact. Actually, the other 2 nonsister chromatids most likely are also crossing-over. *c.* Chiasmata indicate where crossing-over has occurred. The exchange of color represents the exchange of genetic material. *d.* Following meiosis II, daughter chromosomes have a new arrangement of genetic material due to crossing-over, which occurred between nonsister chromatids in prophase I.

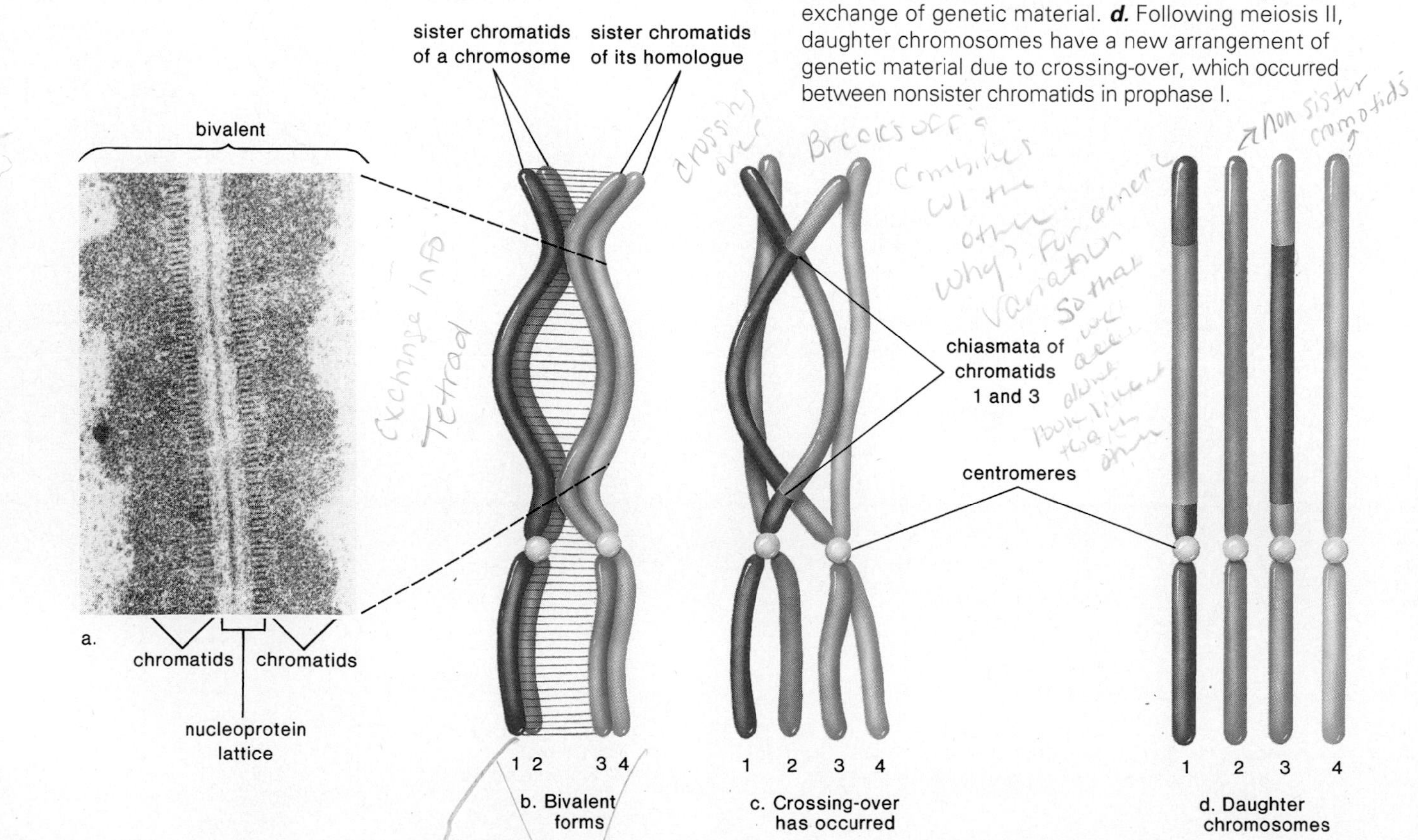

bivalent

sister chromatids of a chromosome   sister chromatids of its homologue

a.

chromatids   chromatids

nucleoprotein lattice

1 2   3 4

b. Bivalent forms

1   2   3   4

c. Crossing-over has occurred

chiasmata of chromatids 1 and 3

centromeres

1   2   3   4

d. Daughter chromosomes

Genetic Basis of Life

Figure 11.5

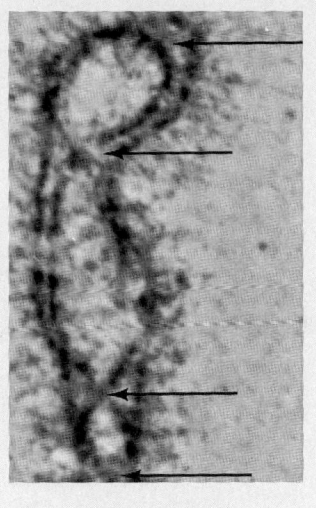

Chiasmata (arrows) of a chromosome bivalent, from a testis cell of a grasshopper. The chiasmata mark the places where crossing-over between nonsister chromosomes of the bivalent has occurred. The chiasmata hold the members of the bivalent together until separation of homologues occurs in anaphase I of meiosis. Magnification, X4,000.

Table 11.1
Stages of Meiosis I and Meiosis II

Stages	Meiosis I	Meiosis II
Prophase	Homologues, each composed of sister chromatids, synapse, forming bivalents.	Duplicated chromosomes are present; one from each pair of homologous chromosomes.
Metaphase	Bivalents are at the equator.	Duplicated chromosomes are at the equator.
Anaphase	Homologues separate and move to opposite poles.	Centromeres divide and daughter chromosomes move to opposite poles.
Telophase	At each pole there is one homologue from each pair of homologous chromosomes.	At each pole there is the haploid number and one of each kind of chromosome.

separation, the chromosome number is reduced from the diploid to the haploid number, and the daughter cells receive one of each kind of chromosome. Notice that at the end of anaphase I, each chromosome still consists of 2 chromatids.

Like mitosis, movement of chromosomes is due to a push provided by spindle elongation and a pull due to kinetochore microtubule disassembly (p. 162).

Homologues separate during anaphase I. This is the means by which the diploid number is reduced to the haploid number; it is also the reason the daughter cells receive one of each kind of chromosome.

Telophase I

In some species, there is a telophase I stage at the end of meiosis I. If there is a telophase, the nuclear envelope reforms and nucleoli appear. Also, this stage may or may not be accompanied by cytokinesis and daughter cell formation. In other species, the telophase I stage is omitted at the end of the first division.

Figure 11.3 shows only 2 of the 4 possible combinations of haploid chromosomes when the parent cell has 2 homologous pairs of chromosomes. Can you determine what the other 2 possible combinations of chromosomes are?

Interkinesis

Following telophase I (if this stage occurs), there is a short period called *interkinesis* before meiosis II begins. Interkinesis is similar to interphase between mitotic divisions except that DNA replication does not occur. Replication is unnecessary because the chromosomes are already duplicated.

Cytokinesis does not necessarily follow telophase I; however, there is a short period known as interkinesis before meiosis II begins. DNA does not replicate because the chromosome are already duplicated.

Meiosis II

The phases of **meiosis II** are referred to as prophase II, metaphase II, anaphase II, and telophase II (fig. 11.3). This second division of meiosis varies considerably from species to species, but it is essentially a mitosis-like division in which the centromeres divide during anaphase II and the daughter chromosomes move toward the poles. Additional variation is introduced, however, because there is independent assortment of the cross-over chromatids. At the end of telophase II, following cytokinesis, there are haploid cells. These cells are not genetically identical to the parent cell because (1) they are haploid and (2) crossing-over has resulted in different combinations of the genes on the daughter chromosomes. In other words, *genetic recombination* has occurred.

The haploid cells, which are the product of meiosis II, mature and become gametes in animals. In plants, they become spores that divide to give a haploid adult generation. Plants also have a diploid adult generation (fig. 11.1). In fungi and some algae, the zygote immediately undergoes meiosis, and therefore the adult is always haploid.

Meiosis II is similar to a mitotic division in that the centromeres divide and daughter chromosomes move toward the poles. Table 11.1 contrasts meiosis I and meiosis II.

Meiosis in Humans

Humans have a life cycle that is typical of all other mammals (fig. 11.1). Meiosis is confined to the germ line, those cells that produce the gametes and that remain separate and distinct from the somatic (body) cells. In females, the germ cells are located in the ovaries, where eggs are produced in a process called **oogenesis.** In males, the germ cells are located in the testes, where sperm are produced in a process called **spermatogenesis.** It is these gametes that carry genes from one generation to the next. As described in this chapter's reading, in 1868, Charles Darwin attempted unsuccessfully to explain how traits are passed from generation to generation by way of the gametes. Now, of course, it is known that the chromosomes within the gametes carry these traits to the new individual.

Oogenesis and Spermatogenesis

Figure 11.6 contrasts oogenesis with spermatogenesis. In females, the first meiotic cell division produces 2 cells, one of which is much smaller than the other. The smaller cell is called a **polar body** and remains attached to the larger cell. The larger cell, called the **secondary oocyte,** matures into the egg. The second meiotic division of the oocyte actually does not occur unless fertilization takes place. Regardless, complete oogenesis in females produces from each original cell only one egg and at least 2 nonfunctional polar bodies that disintegrate. In males, on the other hand, spermatogenesis results in 4 viable sperm from each original cell.

From the drawing on page 177, it is apparent that the sperm and egg are adapted to their functions. The sperm is a tiny flagellated cell that is adapted for swimming to the egg. The egg is a large cell that awaits the arrival of the sperm and contributes most of the cytoplasm and nutrients to the zygote. The sperm and egg represent the haploid stage in the human life cycle. When fertilization occurs, the resulting zygote has the diploid number of chromosomes. The zygote develops into the organism.

Meiosis in humans, like other mammals, occurs during spermatogenesis in males and oogenesis in females. The sperm and egg are the only haploid stages, and fertilization provides the zygote with the diploid number of chromosomes.

Figure 11.6

Spermatogenesis and oogenesis in mammals. Spermatogenesis produces 4 viable sperm, whereas oogenesis produces one egg and at least 2 polar bodies. In humans, both sperm and egg have 23 chromosomes each; therefore, following fertilization, the zygote has 46 chromosomes.

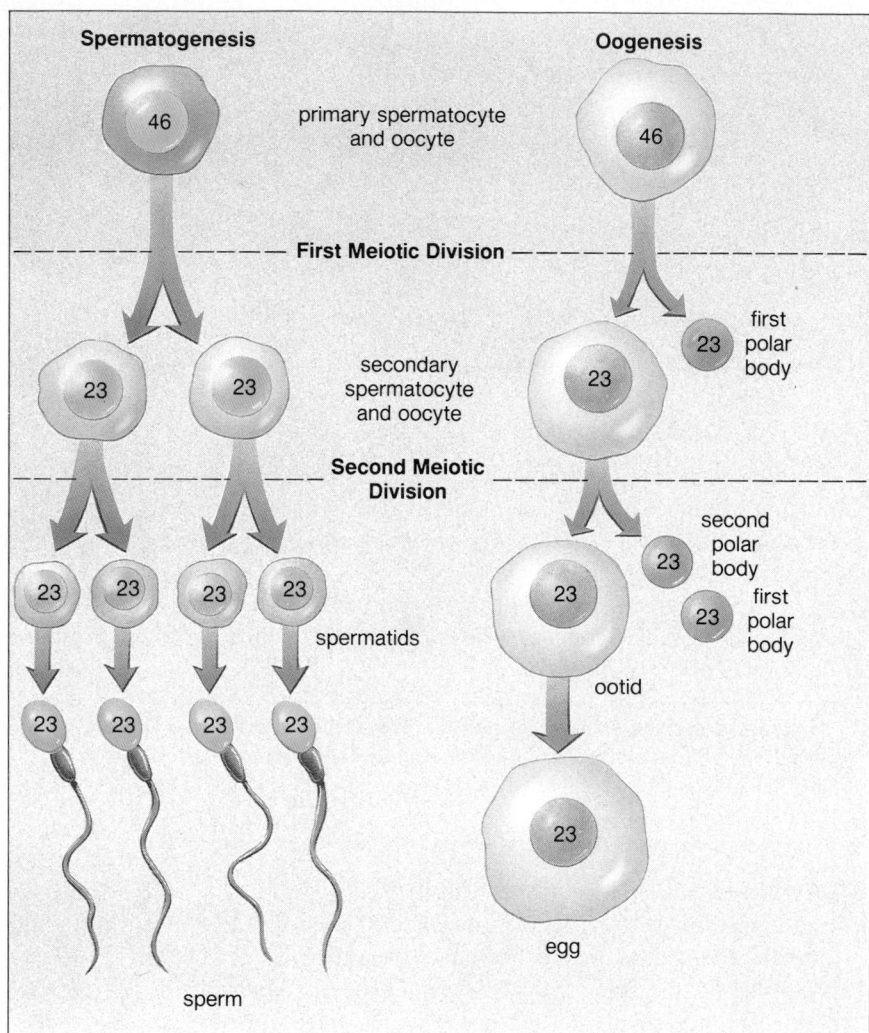

Importance of Meiosis

In all organisms, meiosis provides a way to keep the chromosome number constant generation after generation. Without meiosis, the chromosome number of the adult stages depicted in figure 11.1 would continually increase. Not only is the chromosome number halved precisely, each daughter cell receives a copy of each kind of chromosome. This ensures that each daughter cell receives one of each kind of gene.

Meiosis also helps ensure that genetic recombination occurs with each generation. As a result of *independent assortment of chromosomes,* the chromosomes are distributed to the daughter cells in various combinations. The total number of possible combinations is 2^N, where N is the haploid number of chromosomes. In humans, where N = 23, the number of possible chromosome combinations produced by meiosis is a staggering 2^{23}, or 8,388,608.

And this does not even consider the variations that are introduced due to *crossing-over*. For example, if we assume only one crossover occurs within each bivalent (actually many crossovers usually occur), then in humans crossing-over and independent assortment will generate 4^{23} different kinds of gametes, or 70,368,744,000,000.

This is just the variation introduced by meiosis, but if the entire sexual reproductive process is considered, there is another element that introduces variation. Due to *fertilization*, the chromosomes donated by the parents are combined, and in humans, this means that $(2^{23})^2$, or 70,368,744,000,000, chromosomally different zygotes are possible, even assuming no crossing-over. If crossing-over occurs once, then $(4^{23})^2$, or 4,951,760,200,000,000,000,000,000,000, genetically different zygotes are possible for every couple.

Genetic Basis of Life

Pangenesis

The egg and sperm cells are often referred to as *germ cells* because they are the beginnings, or germs, of new individuals. The germ cells represent the only connecting thread between successive generations. Accordingly, the mechanism of hereditary transmission must operate across this slender connecting bridge. In 1868, Charles Darwin proposed an ingenious theory to account for the transmission of traits by the germ cells (fig. 11.A). Darwin thought that all organs of the body sent contributions to the germ cells in the form of particles, or gemmules. These gemmules, or minute delegates of bodily parts, were supposedly discharged into the bloodstream and became ultimately concentrated in the gametes. In the next generation, each gemmule reproduced its particular bodily component. The theory was called *pangenesis,* or "origin from all," since all parental body cells were supposed to take part in the formation of the new individual.

Darwin's fanciful explanation of inheritance, however, has no foundation of factual evidence. Indeed, many observations argue against it. In 1909, W. E. Castle of Harvard University provided one of the earliest convincing observations refuting the theory of pangenesis. Castle successfully removed the ovaries of an albino guinea pig and grafted in their place the ovaries of a black guinea pig. The albino foster mother was then mated to an albino male. This pair produced several litters of young, and all the offspring were black. Ordinarily when albinos are crossed only albino offspring are produced. It is obvious, then, that the eggs produced by the transplanted ovaries (from the black guinea pig) were not

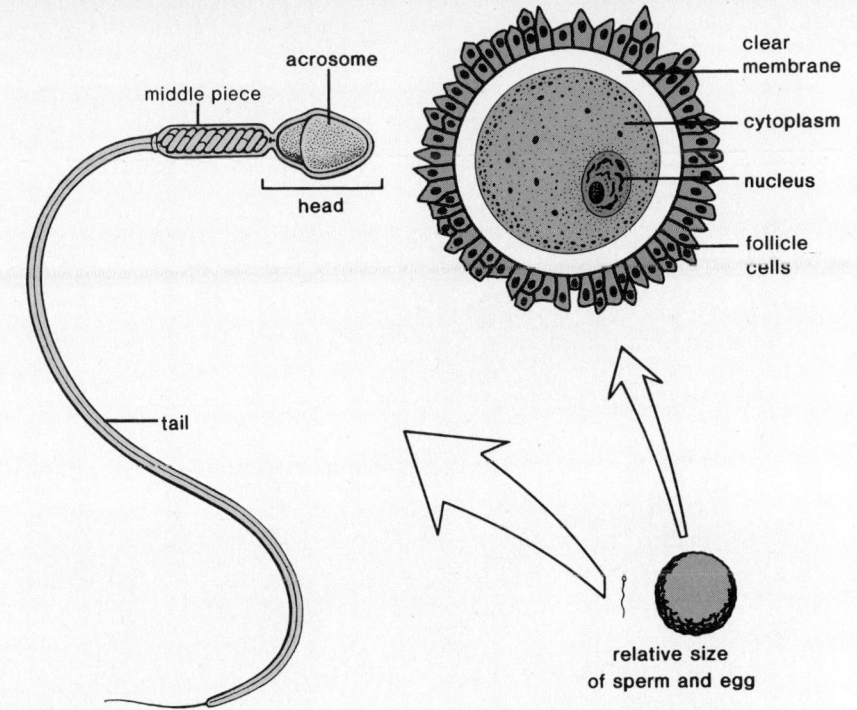

Figure 11.A
Darwin recognized that the egg and sperm must have a means of carrying information to the next generation, but he did not realize such information resides in the genes located on the chromosomes.

changed, or instructed otherwise, by their new residence in the albino foster mother guinea pig.

Notwithstanding Darwin's simplistic view, the mechanism of inheritance does reside in the gametes. Although the relatively large egg cell and the comparatively small sperm cell are as strikingly unlike in appearance as any 2 cells can be, they are equivalent in certain important components—their nuclei and, more

particularly, their *chromosomes.* The chromosomes are, in fact, the vehicles for transmitting the blueprints of traits from parents to offspring. Deoxyribonucleic acid (DNA) [is] ... the *chemical carrier* of the hereditary information.

From Peter J. Volpe, *Biology and Human Concerns,* 3d ed. Copyright © 1983 Wm. C. Brown Communications, Inc., Dubuque, Iowa. All Rights Reserved. Reprinted by permission.

If the diploid zygote becomes a diploid adult, some of this variability may remain hidden and not subject to direct testing by the environment. If there is a haploid stage as there is in plants and fungi, any variability present is tested immediately by the environment.

In any case, a sexually reproducing population has a tremendous storehouse of genetic recombinations, which may be advantageous for evolution, particularly when the environment is changing. Asexually reproducing organisms, like prokaryotes, depend primarily on mutations to generate variation. This is sufficient because they produce great numbers of offspring within a limited amount of time. Mutation is still the raw material for variation among

sexually reproducing organisms, but the shuffling of genetic material due to sexual reproduction may better allow adaptation to a changing environment. If the environment is not changing, the genetic makeup of the parents is most likely still adaptive.

Meiosis promotes genetic recombination; the more chromosomes there are, the greater the recombination possible.

Comparison of Meiosis and Mitosis

Figure 11.7 compares meiosis to mitosis. The differences between these nuclear divisions can be categorized as follows.

Figure 11.7

Comparison of meiosis and mitosis. For simplicity's sake crossing-over has been omitted from meiosis. (The blue chromosomes were inherited from one parent and the red chromosomes were inherited from the other parent.)

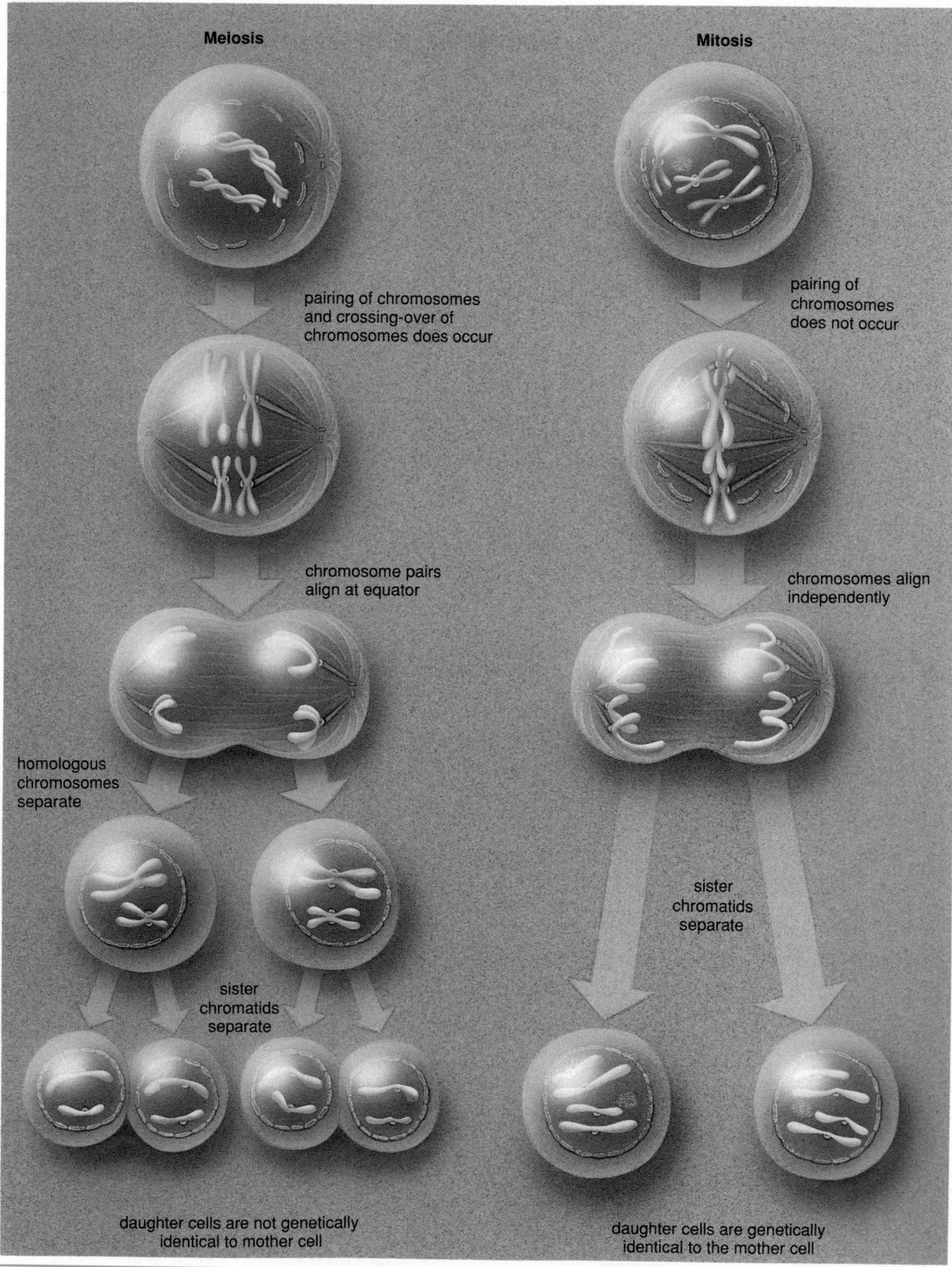

Meiosis

Mitosis

pairing of chromosomes and crossing-over of chromosomes does occur

pairing of chromosomes does not occur

chromosome pairs align at equator

chromosomes align independently

homologous chromosomes separate

sister chromatids separate

sister chromatids separate

daughter cells are not genetically identical to mother cell

daughter cells are genetically identical to the mother cell

Genetic Basis of Life

Occurrence

Meiosis occurs only at certain times in the life cycle of sexually reproducing organisms (fig. 11.1). Mitosis is more common because it allows growth and repair of body tissues in multicellular forms.

Process

The following are distinctive differences between the process of meiosis and mitosis.

1. DNA is replicated only once before both meiosis and mitosis; but there are 2 nuclear divisions during meiosis and only one nuclear division during mitosis.
2. Homologous chromosomes pair and undergo crossing-over during prophase I of meiosis but not during mitosis.
3. Paired homologous chromosomes (bivalents) align at the equator of metaphase I in meiosis; individual (duplicated) chromosomes align at the equator in metaphase of mitosis.

4. Homologous chromosomes (with centromeres intact) separate and move to opposite poles at anaphase I in meiosis; centromeres divide and daughter chromosomes separate and move to opposite poles in the anaphase of mitosis.

Daughter Nuclei and Cells

The genetic consequences of meiosis and mitosis are quite different.

1. Four daughter cells are produced from a single cell that undergoes meiosis; mitosis results in 2 daughter cells.
2. The 4 daughter cells formed by meiosis are haploid; the daughter cells produced by mitosis have the same chromosome number as the parent cell.
3. The daughter cells from meiosis are not genetically identical to each other or the parent cell because of crossing-over and independent separation of homologues. The daughter cells from mitosis are genetically identical to each other and to the parent cell.

Summary

1. Meiosis is involved in the sexual reproduction life cycle. It ensures that the chromosome number in offspring stays constant generation after generation.
2. In the animal life cycle, only the gametes are haploid; in plants and fungi, there is a multicellular haploid adult that produces the gametes.
3. The nucleus contains pairs of chromosomes, called homologous pairs (homologues). Due to the process of meiosis, the haploid daughter cells (gametes) receive one of each kind of chromosome.
4. Meiosis requires 2 cell divisions and results in 4 daughter cells. Replication of DNA takes place before meiosis begins. During meiosis I, the homologues pair (called a bivalent) and crossing-over between nonsister chromatids occurs before they independently separate. The daughter cells receive one member of each pair of homologous chromosomes. There is no replication of DNA during interkinesis. During meiosis II, the daughter chromosomes separate in a manner similar to mitosis. The 4 resulting daughter cells are not genetically identical to the mother cell; they are haploid, and due to crossing-over, their chromosomes carry a different arrangement of the genetic material.

5. Meiosis I is divided into 4 stages:

   Prophase I—Bivalents form, and crossing-over occurs as chromosomes condense; the nuclear envelope fragments.
   Metaphase I—Bivalents independently align at the equator.
   Anaphase I—Homologous chromosomes separate.
   Telophase I—Nuclei become haploid, having received one duplicated chromosome from each homologous pair.

6. Meiosis II is divided into 4 stages:

   Prophase II—Chromosomes condense and the nuclear envelope fragments.
   Metaphase II—Duplicated chromosomes align individually at the equator.
   Anaphase II—Daughter chromosomes separate.
   Telophase II—Four haploid daughter cells are genetically different from the mother cell.

7. During the life cycle of humans and many other animals, only the gametes are haploid and fertilization restores the diploid number. Meiosis is involved in spermatogenesis and oogenesis. Whereas spermatogenesis produces 4 sperm per meiosis, oogenesis produces one egg and 2 to 3 nonfunctional polar bodies.

8. Mutations are the primary source of genetic variation among asexually reproducing organisms. But among sexually reproducing organisms, meiosis produces variation by independent assortment of homologous chromosomes and crossing-over. Fertilization also contributes to variation. Variation is important to the process of evolution; it promotes the possibility of adaptation to a changing environment.

9. Meiosis I can be compared to mitosis in this manner:

Mitosis	Meiosis I
Prophase	
No pairing of chromosomes	Pairing of homologous chromosomes
Metaphase	
Duplicated chromosomes at equator	Bivalents at equator
Anaphase	
Chromatids separate	Homologous chromosomes separate
Telophase	
Daughter nuclei are diploid	Daughter nuclei are haploid

10. Meiosis II can be compared to mitosis in this manner:

Mitosis	Meiosis II
Metaphase	
Diploid number of duplicated chromosomes at equator	Haploid number of duplicated chromosomes at equator
Anaphase	
Chromatids separate	Chromatids separate
Telophase	
Two daughter cells	Four daughter cells

Study Questions

1. Why did early investigators predict that there must be a reduction division in the sexual reproduction process?
2. Compare the animal, plant, and fungal life cycles by indicating when meiosis occurs in each.
3. What are the events of meiosis that account for the production of 4 daughter cells with the haploid number of chromosomes having a different genetic makeup than in the mother cell?
4. Draw and explain a series of diagrams that illustrate the events of meiosis I.
5. Draw and explain a series of diagrams that illustrate synapsis and crossing-over. Indicate in your drawings what holds the sister chromatids together. Indicate what holds the homologues together.
6. Draw and explain a series of diagrams that illustrate the events of meiosis II.
7. Draw a diagram to illustrate the life cycle of humans, and compare spermatogenesis to oogenesis.
8. Construct a chart to describe the many differences between meiosis and mitosis.
9. What accounts for the genetic similarity between daughter cells and the parent cell following mitosis and the genetic dissimilarity between daughter cells and the parent cell following meiosis?
10. List the ways in which sexual reproduction contributes to variation among members of a population. What is the evolutionary significance of this variation?

Objective Questions

1. A bivalent (tetrad) is
   a. a homologous chromosome.
   b. the paired homologous chromosomes.
   c. a duplicated chromosome composed of sister chromatids.
   d. the 2 daughter cells after meiosis I.
2. If a mother cell has 12 chromosomes, then the daughter cells following meiosis will have
   a. 12 chromosomes.
   b. 24 chromosomes.
   c. 6 chromosomes.
   d. Any one of these.
3. At the equator during metaphase I of meiosis, there are
   a. singled chromosomes.
   b. unpaired duplicated chromosomes.
   c. bivalents (tetrads).
   d. 23 chromosomes.
4. At the equator during metaphase II of meiosis, there are
   a. singled chromosomes.
   b. unpaired duplicated chromosomes.
   c. bivalents (tetrads).
   d. always 23 chromosomes.
5. Gametes contain one of each kind of chromosome because
   a. the homologous chromosomes separate during meiosis.
   b. the chromatids never separate during meiosis.
   c. 2 replications of DNA occur during meiosis.
   d. crossing-over occurs during prophase I.
6. Crossing-over occurs between
   a. sister chromatids of the same chromosomes.
   b. 2 different bivalents.
   c. nonsister chromatids of a bivalent.
   d. 2 daughter nuclei.
7. During which stage of meiosis do homologous chromosomes separate?
   a. prophase II
   b. telophase II
   c. metaphase I
   d. anaphase I
8. Fertilization
   a. is also a source of variation during sexual reproduction.
   b. is fusion of the gametes.
   c. occurs in both animal and plant life cycles.
   d. All of these.

9. Which of these is not a difference between spermatogenesis and oogenesis in humans?

Spermatogenesis	*Oogenesis*
a. Occurs in males	Occurs in females
b. Produces 4 sperm per meiosis	Produces one egg per meiosis
c. Produces haploid cells	Produces diploid cells
d. Always goes to completion	Does not always go to completion

10. Which of these drawings represents meiosis I? How do you know?

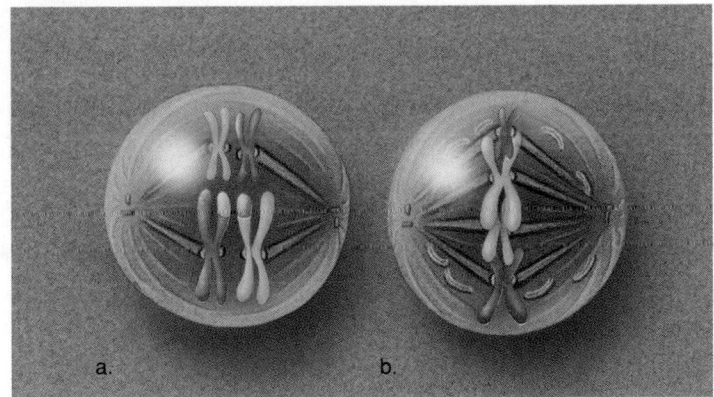

Concepts and Critical Thinking

1. *Methods of organismal reproduction support the cell theory.*

In what way do both asexual and sexual reproduction support the cell theory?

2. *Variation among offspring is greater for sexual reproduction than for asexual reproduction.*

How does asexual reproduction introduce variation among offspring?

3. *Meiosis is an essential part of the life cycle of eukaryotes.*

If the zygote undergoes meiosis, the adult is haploid. Haploidy has what advantage or disadvantage?

Selected Key Terms

homologous chromosome (ho-mol'o-gus kro'mo-sōm) 171
homologue (hom'o-log) 171
bivalent (or, tetrad) (bi-va'lent) (tet'rad) 172
crossing-over (kros'ing o'ver) 172

meiosis I (mi-o'sis wun) 173
synapsis (sĭ-nap'sis) 173
meiosis II (mi-o'sis tōō) 175
genetic recombination (jĕ-net'ik re"kom-bĭ-na'shun) 175

oogenesis (o"o-jen'ĕ-sis) 175
spermatogenesis (sper"mah-to-jen'ĕ-sis) 175
polar body (po'lar bod'e) 176
secondary oocyte (sek'on-dary o'o-sīt) 176

Mendelian Patterns of Inheritance

What causes a giraffe to have such a long neck? The genes inherited from its parents, of course. This may seem obvious to us, but it took many years of research to determine that the sperm and the egg carry heritory factors from parent to offspring. This is universally true, whether you are talking about giraffes, peas, or people.

Your study of this chapter will be complete when you can

1. discuss the ideas about heredity that were prevalent when Mendel began his experiments;
2. outline the general methodology that Mendel used for his experiments;
3. diagram the type of monohybrid cross and dihybrid cross Mendel did, using words to indicate the phenotype and letters to indicate the genotype;
4. state Mendel's law of segregation, and describe the reasoning he used to arrive at his law;
5. explain the use of a monohybrid testcross and a dihybrid testcross to determine the genotype of an individual;
6. state Mendel's law of independent assortment, and describe the reasoning he used to arrive at this law;
7. solve one-trait and 2-trait genetics problems using the Punnett square or the laws of probability.

Zebras always produce zebras, never bluebirds, and poppies always produce seeds for poppies, never dandelions. Almost everyone who observes such phenomena reasons that parents must pass hereditary information to their offspring. Many also observe, however, that offspring rarely resemble either parent exactly. After all, black-coated mice even on occasion produce white-coated mice, for example.

The laws of heredity must explain not only the stability but also the variation that is observed between generations of organisms. In the previous chapter, we concentrated on the passage of chromosomes via the egg and sperm from parent to offspring. The inheritance of chromosomes can explain genetic stability, and it can also explain variation if we can show that the chromosomes differ from one another in some way.

Gregor Mendel took the first step toward this end when he discovered certain laws of heredity in the early 1860s (fig. 12.1).

Gregor Mendel

Gregor Mendel was an Austrian monk who taught natural science at a local technical high school. Previously, he went to the University of Vienna to study science and mathematics. Various theories about heredity had been proposed before Mendel began his experiments. In particular, the blending theory of inheritance was believed by many at this time.

Blending Theory of Inheritance

When Mendel began his work, most plant and animal breeders believed in the *blending theory of inheritance* because it acknowledged that both sexes contribute equally to a new individual. They felt, however, that parents of contrasting appearance always produce offspring of intermediate appearance. Therefore, according to this theory, a cross between plants with red flowers and plants with white flowers would yield only plants with pink flowers. When red and white flowers reappeared in future generations, the breeders mistakenly attributed this to an instability in the genetic material.

The blending theory of inheritance presented a real problem to Charles Darwin (chap. 19), who wanted to give his theory of evolution a genetic basis. If populations contained only intermediate individuals and normally lacked variations, how could diverse forms evolve?

At the time Mendel began his study of heredity, the blending theory of inheritance was popular.

Mendel's Experiments

Most likely his background in mathematics caused Mendel to decide that his breeding experiments needed a statistical basis. He prepared for his experiments carefully and conducted preliminary studies with various animals and plants. He then chose to work with the garden pea *Pisum sativum* (fig. 12.2).

Figure 12.1

Mendel working in his garden. Mendel selected 22 varieties of the edible pea *Pisum* for his genetic experiments. He grew and tended the plants himself. For each experiment, he observed as many offspring as possible. For example, for a cross that required him to count the number of round seeds to wrinkled seeds, he observed and counted a total of 7,324 peas.

The garden pea was a good choice. The plants were easy to cultivate and had a short generation time. And although the peas normally self-pollinate, they could be cross-pollinated for the sake of the experiment. Many varieties of the peas were available, and Mendel chose 22 for his experiments. When these varieties self-pollinated, they were *true-breeding*—the offspring were like the parent plants and like each other. In contrast to his predecessors, Mendel studied the inheritance of relatively simple and distinguishable traits—seed shape, seed color, and flower color (table 12.1).

As Mendel followed the inheritance of individual traits, he kept careful records of the numbers of offspring that expressed each characteristic. And he used his understanding of the mathematical laws of probability to interpret the results. In other words, Mendel simply wanted to observe objective facts, and if he did have personal beliefs, he set them aside for the sake of the experiment. This is one of the qualities that make his experiments as applicable today as they were in 1860.

Mendel carefully designed his experiments and gathered mathematical data.

In the end, Mendel arrived at a particulate theory of inheritance that is described in detail following. As we shall see, a particulate theory of inheritance does account for the presence of variations among offspring, as is required for the occurrence of evolutionary change.

Figure 12.2

In the garden pea (*Pisum sativum*) each flower produces both male and female gametes. The pollen grains, which upon maturity contain sperm, are produced within the anthers of the stamens. The ovule, within the ovary of the pistil, eventually contains an egg. Following pollination—when pollen is deposited on the stigma (the top portion of the pistil)—the pollen grain develops a tube through which a sperm reaches the egg. Self-pollination is the rule in the garden pea because the sexual structures of the plant are entirely enclosed by petals. Sometimes Mendel allowed the plant to self-pollinate so that the male and female gametes of the same flower produced offspring. At other times, he performed cross-pollination, so that male and female gametes from different flowers produced offspring. He removed the pollen-producing anthers of one plant and brushed the stigmas with the anther from another plant.

Once the ovules had developed into seeds (peas), they could be observed or planted in order to observe the results of a cross. The open pod shows the results of a cross involving smooth or wrinkled seeds.

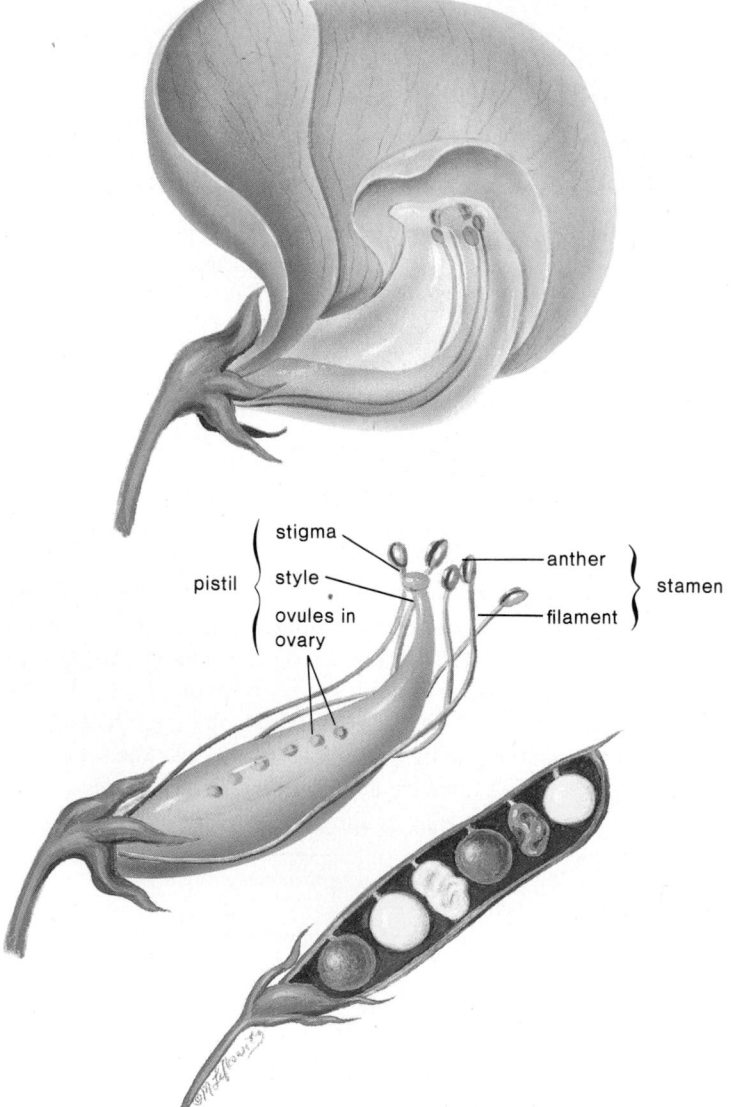

Table 12.1
Pisum Traits and Characteristics Studied by Mendel

Trait	Characteristics		F₂ Results*	
			Dominant	**Recessive**
Stem length	Tall	Short	787	277
Pod shape	Inflated	Constricted	882	299
Seed shape	Round	Wrinkled	5,474	1,850
Seed color	Yellow	Green	7,022	2,001
Flower position	Axial	Terminal	651	207
Flower color	Purple	White	705	224
Pod Color	Green	Yellow	428	152

*All of these give approximately a 3:1 ratio. For example,

$$\frac{787}{277} \cong \frac{3}{1}$$

Genetic Basis of Life

Figure 12.3

The monohybrid cross allowed Mendel to deduce that individuals have 2 discrete and separate genetic factors for each trait. In true-breeding plants that self-pollinate, the 2 factors are the same (see P generation). If these plants are crossed, all the offspring show the same characteristic (e.g. tall), indicating that it is dominant to the other, recessive characteristic (see F_1 generation). If these hybrid plants are allowed to self-pollinate, the recessive characteristic (e.g. short) reappears. On the average, 3 out of 4 F_2 plants show the dominant characteristic and one shows the recessive (see F_2 generation), giving a 3:1 ratio.

Mendel's Experimental Design and Results

P (parental) generation: Mendel crossed 2 true-brooding varieties that differed in one characteristic, such as length of stem.

 ×

TT   tt

F_1 (first filial) generation: All offspring were tall and therefore resembled only one of the parents. Mendel allowed these plants to self-pollinate.

Tt

F_2 (second filial) generation: Among the offspring 3/4 were tall but 1/4 were short. Therefore, Mendel said there are 2 factors (represented here by T and t) for each trait in the individual. These segregate into the gametes.

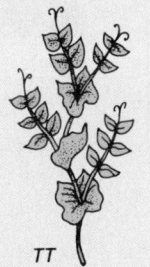   

TT   Tt   Tt   tt

key:
T = tall
t = short

Genotypes and Phenotypes
TT = tall
Tt = tall
tt = short

Monohybrid Inheritance

After ensuring that his pea plants were true-breeding, Mendel was then ready to perform a cross-pollination experiment between 2 strains. For these initial experiments, Mendel chose varieties that differed in only one trait. Crosses that involve only one trait are called *monohybrid crosses*. If the blending theory of inheritance were correct, then the cross should yield offspring with an intermediate appearance compared to the parents. For example, the offspring of a cross between a tall plant and short plant should be intermediate in height.

Mendel called the original parents the *P generation* and the first-generation offspring the *F_1 (for filial) generation* (fig. 12.3). He performed *reciprocal crosses:* first he dusted the pollen of tall plants on the stigmas of short plants, and then he dusted the pollen of short plants on the stigmas of tall plants. In both cases, all F_1 offspring resembled the tall parent.

Certainly, these results were contrary to those predicted by the blending theory of inheritance. Rather than being intermediate, the offspring were tall and resembled only one parent. Did these results mean that the other characteristic (i.e., shortness) had disappeared permanently? Apparently not, because when Mendel allowed the F_1 plants to self-pollinate, $^3/_4$ of the F_2 generation were tall and $^1/_4$ were short, a 3:1 ratio (fig. 12.3).

Mendel counted a lot of plants. For this particular cross, he counted a total of 1,064 plants, of which 787 were tall and 277 were short. In all crosses that he performed, he found a 3:1 ratio in the F_2 generation. The characteristic that had disappeared in the F_1 generation reappeared in $^1/_4$ of the F_2 offspring (table 12.1).

His mathematical approach led Mendel to interpret these results differently from previous breeders. He knew that the same ratio was obtained among the F_2 generation time and time again for each trait, and he sought an explanation for these results. A 3:1 ratio among the F_2 offspring was possible if:

1. the F_1 generation contained 2 separate copies of each hereditary factor, one of these being dominant and one being recessive;
2. the factors separated when the gametes were formed, and each gamete carried only one copy of each factor; and
3. random fusion of all possible gametes occurred upon fertilization.

In this way, Mendel arrived at the first of his laws of inheritance:

Mendel's law of segregation: Each organism contains 2 factors for each trait, and the factors segregate during the formation of gametes so that each gamete contains only one factor for each trait. When fertilization occurs, the new organism has 2 factors for each trait, one from each parent.

Today, we know that inheritance occurs by way of gametes, and that it is due to meiosis (see fig. 13.7) that each gamete carries only one factor for each trait. Today, too, we know that the genes within the gametes are unaffected by the somatic cells—Darwin's theory of pangenesis (p. 177) has been disproven.

Mendel's law of segregation is in keeping with a particulate theory of inheritance because individual and separate factors are passed on from generation to generation. It is the reshuffling of these factors that explains how variations come about and why offspring differ from their parents.

Modern Terminology

Figure 12.3 also shows how we now interpret the results of Mendel's experiments on inheritance of stem length in peas. Each pea plant is said to have 2 **alleles,** alternate forms of a gene that controls the length of the stem. In genetic notation, the alleles are identified by letters, the **dominant allele** (so named because of its ability to mask the expression of another allele) with an uppercase (capital) letter and the **recessive allele** with

Figure 12.4

Homologous chromosomes before and after replication. **a.** The letters represent alleles; that is, alternate forms of a gene. Each allelic pair, such as Gg or Zz, is located on homologous chromosomes at a particular gene locus. **b.** Following replication, each sister chromatid carries the same alleles in the same order.

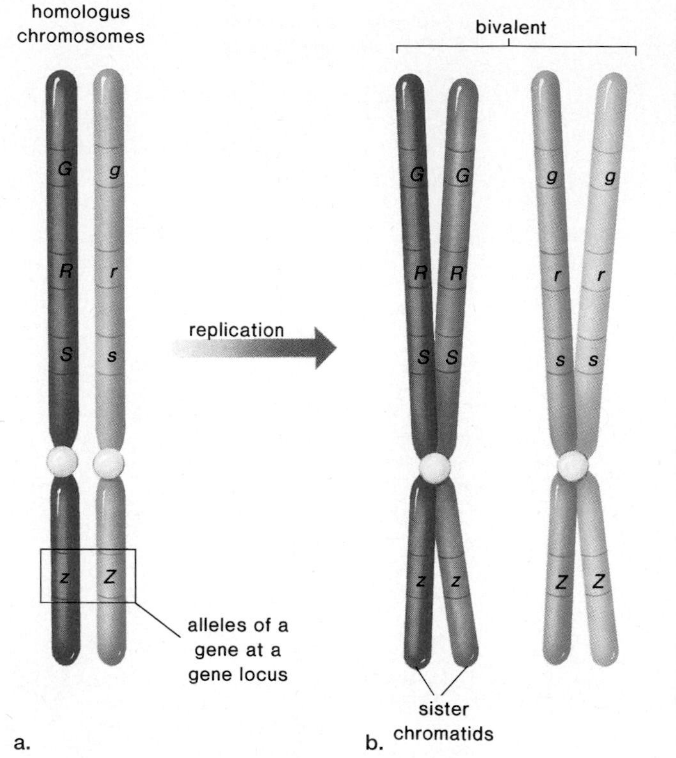

the same but lowercase (small) letter. For example, there is an allele for tallness (T) and an allele for shortness (t). One of these alleles occurs on each chromosome of a homologous pair at a particular location that is called the **gene locus** (fig. 12.4). During meiosis, first the homologous chromosomes segregate then the chromatids separate. Therefore, there is only one allele for each trait in the gametes (see fig. 13.7).

In Mendel's cross, the original parents (P generation) were true-breeding; therefore, the tall plants had 2 alleles for tallness (TT) and the short plants had 2 alleles for shortness (tt). When an organism has 2 identical alleles, as these had, we say it is **homozygous.** Because the first parents were homozygous, all gametes produced by the tall plant contained the allele for tallness (T), and all gametes produced by the short plant contained an allele for shortness (t).

After cross-pollination, all the individuals of the resulting F_1 generation had one allele for tallness and one for shortness (Tt). When an organism has 2 different alleles at a gene locus, we say that it is **heterozygous.** Although the plants of the F_1 generation had one of each type of allele, they were all tall. When only one allele in a heterozygous individual is expressed, we say that this allele is the dominant allele. The allele that is not expressed in a heterozygote is a recessive allele.

Genotype and Phenotype

It is obvious from our discussion that 2 organisms with different allelic combinations for a trait can have the same outward appearance (*TT* and *Tt* pea plants are both tall). For this reason, it is necessary to distinguish between the alleles present in an organism and the appearance of that organism.

The word **genotype** refers to the alleles an individual receives at fertilization. Genotype may be indicated by letters or by short descriptive phrases, as in table 12.2. Genotype *TT* is called homozygous dominant, and genotype *tt* is called homozygous recessive. Genotype *Tt* is called heterozygous.

The word **phenotype** refers to the physical appearance of the individual. The homozygous dominant individual and the heterozygous individual both show the dominant phenotype and are tall, while the homozygous recessive individual shows the recessive phenotype and is short.

When solving genetics problems, it is always necessary to know whether the genotype (genes of the individual) or the phenotype (appearance of the individual) is being considered.

We will now digress from Mendel's work with peas to show that these results are applicable to other organisms.

Monohybrid Genetics Problems

When solving a genetics problem, it is first necessary to know which characteristic is dominant. For example, the following key indicates that unattached earlobes are dominant over attached earlobes:

---

key: E = unattached earlobes
e = attached earlobes

Suppose a homozygous man with unattached earlobes reproduces with a woman who has attached earlobes. What types of earlobes will the children have? In row 1, P represents the parental generation, and the letters in this row are the genotypes of the parents. Row 2 shows that the gametes of each parent have only one type of allele for earlobes; therefore, all the children (F_1 offspring) will have a heterozygous genotype that will result in unattached earlobes:

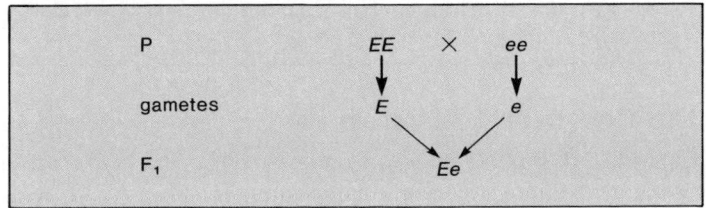

Table 12.2
Genotype versus Phenotype

Genotype	Genotype	Phenotype
TT	Homozygous dominant	Tall plant
Tt	Heterozygous	Tall plant
tt	Homozygous recessive	Short plant

If these children reproduce with someone having the same genotype, will their children have unattached earlobes? Previously

P	*Ee*	×	*Ee*
gametes	↓		↓
	½ *E*, ½ *e*		½ *E*, ½ *e*

there was only one possible type of gamete for each parent because they were homozygous. The F_1 offspring are heterozygous, however, so 2 types of gametes are possible: $\frac{1}{2}$ of the gametes will contain an *E*, and $\frac{1}{2}$ will contain an *e*. When designating the gametes, it is necessary to keep in mind that although the individual has 2 alleles for each trait, *each gamete has only one allele for each trait.* This is true of single-trait crosses as well as multiple-trait crosses. Practice Problems 1 will help you learn to designate the gametes.

Practice Problems 1*

---

1. For each of the following genotypes, give all genetically different gametes, noting the proportion of each for the individual.
   a. *WW*
   b. *Ww*
   c. *Tt*
   d. *TT*
2. For each of the following, state whether a genotype or a gamete is represented.
   a. *D*
   b. *GG*
   c. *P*

---

*Answers to Practice Problems appear in appendix D.

Punnett Square

Once the gametes have been designated, there are 2 possible ways to calculate the results of a cross. The first is to use a **Punnett square.** This simple method was introduced by a prominent poultry geneticist, R. C. Punnett, in the early 1900s. Figure 12.5 shows how the results of the cross under consideration (*Ee* X *Ee*) can be determined using a Punnett square. The results show that the expected proportions of offspring are $\frac{1}{4}$ *EE* to $\frac{1}{2}$ *Ee* to $\frac{1}{4}$ *ee*, giving a 1:2:1 ratio of genotypes. In genetics problems, however, you are usually asked for the ratio of phenotypes. In this example, $\frac{3}{4}$ have unattached earlobes and $\frac{1}{4}$ have attached earlobes, giving a 3:1 phenotypic ratio.

These are only expected proportions, not absolute numbers. For example, if a large number of heterozygous individuals produced 200 offspring, approximately 150 of these would have unattached earlobes and approximately 50 would have attached earlobes. In terms of genotypes, approximately 50 would be *EE*, about 100 would be *Ee*, and the remaining 50 would be *ee*.

Laws of Probability

Another method of calculating the expected ratios uses probability: *the chance, or probability, of 2 or more independent events occurring together is the product (multiplication) of their chance of occurring separately.*

In the cross just considered (*Ee* X *Ee*), what is the chance of obtaining either an *E* or an *e* from a parent?

The chance of $E = \frac{1}{2}$

The chance of $e = \frac{1}{2}$

Therefore, the probability of receiving these genotypes is as follows:

1. The chance of $EE = \frac{1}{2} \times \frac{1}{2} = \frac{1}{4}$
2. The chance of $Ee = \frac{1}{2} \times \frac{1}{2} = \frac{1}{4}$
3. The chance of $eE = \frac{1}{2} \times \frac{1}{2} = \frac{1}{4}$
4. The chance of $ee = \frac{1}{2} \times \frac{1}{2} = \frac{1}{4}$

Now we have to realize that *the chance of an event that can occur in 2 or more independent ways is the sum (addition) of the individual chances.*
Therefore,

The chance of offspring with unattached earlobes (EE, Ee, or eE) is $\frac{3}{4}$ (add 1, 2, and 3)

The chance of offspring with attached earlobes (ee) is $\frac{1}{4}$ (simply 4)

Because the gametes combine at random, it is necessary to observe a very large number of offspring before this 3:1 ratio can be obtained. Only in that way can all genetically different sperm have a chance to fertilize all possible types of eggs. This being the case, the most useful interpretation in humans is to consider the chances that a child will have a particular trait. Both methods tell us that each child has 3 chances in 4 (75%) of having unattached earlobes and one chance in 4 (25%) of having attached earlobes. You should also realize that *chance has no memory.* For example, if 2 heterozygous parents already have a child with attached earlobes, the next child still has a one-in-4 chance of having attached earlobes. Although this may not seem important as far as earlobes are concerned, it is important when calculating the chances of a child inheriting a genetic disease (chap. 14 concerns this topic).

Figure 12.5
Mendel's law of segregation illustrated in a cross between humans. Heterozygous individuals produce 2 types of gametes in equal amounts. Therefore, an offspring of 2 heterozygotes has one chance in 4 of showing the recessive phenotype.

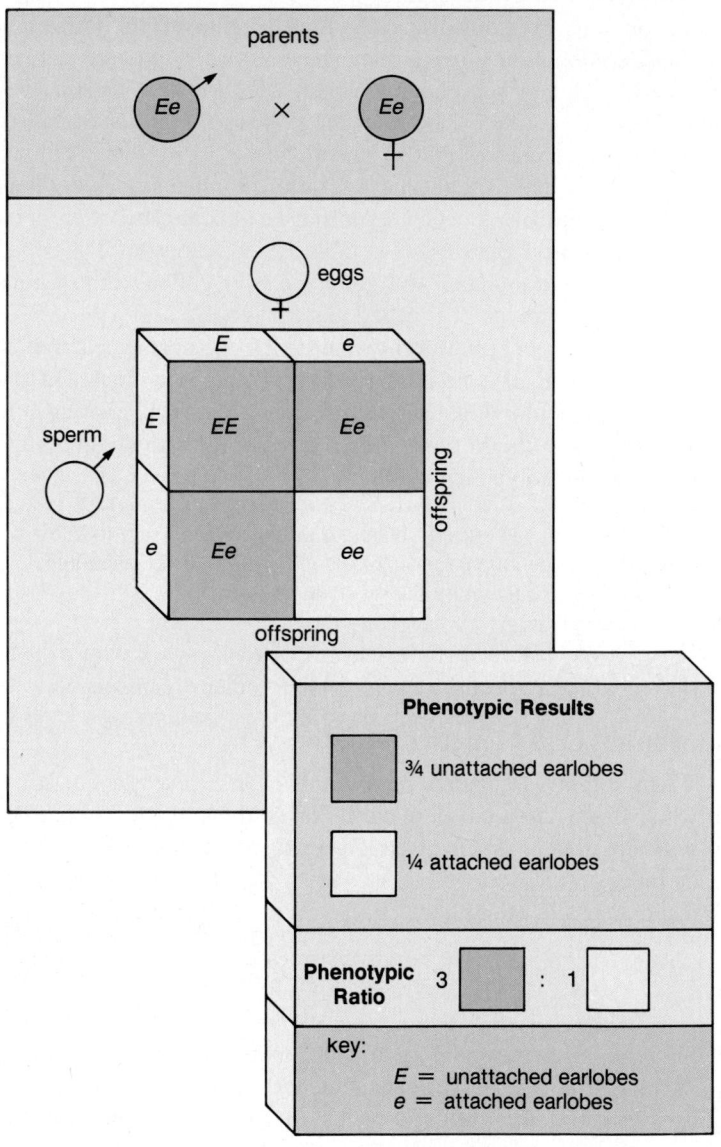

When solving a genetics problem, it is assumed that all genetically different sperm fertilize all genetically different eggs. The results may be expressed as a probable phenotypic ratio; it is also possible to state the chances of an offspring showing a particular phenotype.

The Monohybrid Testcross

To test his ideas about the segregation of alleles, Mendel crossed his F_1 generation tall (dominant phenotype) plants with true-

Genetic Basis of Life

Figure 12.6

Representation of a testcross to determine if the individual showing the dominant trait is homozygous or heterozygous. **a.** Because all offspring show the dominant characteristic, the individual is most likely homozygous, as shown. **b.** Because the offspring show a 1:1 ratio of phenotypes, the individual is heterozygous, as shown.

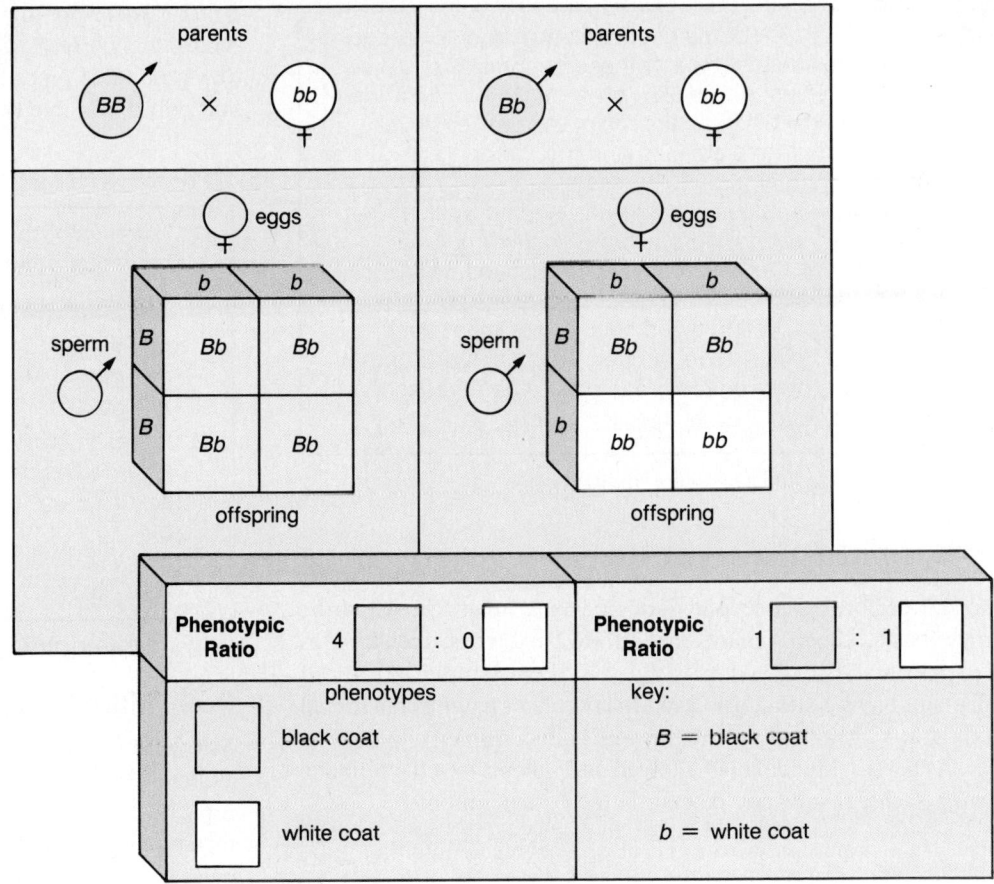

breeding short (recessive phenotype) plants. He reasoned that half the offspring should be tall, as you can see by using the following method to represent the cross:

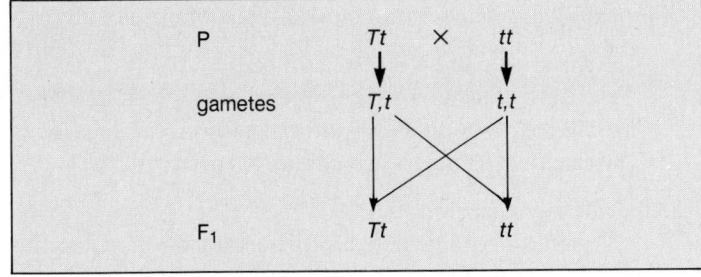

Mendel achieved the results he predicted when he did the cross. Therefore, his hypothesis that alleles segregate when gametes are formed was supported. It was Mendel's experimental use of simple dominant and recessive traits that allowed him to formulate and to test the law of segregation.

A monohybrid **testcross** is used to determine if an individual with the dominant phenotype is homozygous dominant or heterozygous for a particular trait. Since both of these genotypes

produce the dominant phenotype, it is not possible to determine the genotype by inspection. For example, in a certain strain of mice, black coat (*B*) is dominant over white coat (*b*). Therefore, the genotype of a black-coated mouse may be either *BB* or *Bb* and is represented as *B__*.

When doing a monohybrid testcross, an individual with the dominant phenotype is crossed with an individual with the recessive phenotype. For example, a black-coated mouse (*B__*) is crossed with a white-coated mouse (*bb*). Figure 12.6*a* shows that when the black-coated mouse is homozygous (*BB*), all offspring are expected to have black coats. Figure 12.6*b* shows that when the black-coated mouse is heterozygous (*Bb*), one-half of the offspring are expected to have white coats. The presence of offspring with white coats shows that the *B__* mouse is heterozygous with the genotype *Bb*.

The results of a testcross can indicate whether an individual who has the dominant phenotype is heterozygous or homozygous dominant. If any of the offspring of a testcross have the recessive phenotype, the parent with the dominant phenotype must be heterozygous.

Practice Problems 2*

1. In rabbits, if B = dominant black allele and b = recessive white allele, which of these genotypes could a white rabbit have? *Bb, BB, bb*
2. In horses, trotter (T) is dominant over pacer (t). A trotter is mated to a pacer, and the offspring is a pacer. Give the genotype of all horses.
3. In humans, freckles is dominant over no freckles. A man with freckles reproduces with a woman with freckles, but the children have no freckles. What chance did each child have for freckles?
4. In pea plants, yellow seed color is dominant over green seed color. Give the genotype of all plants that could possibly produce green peas when crossed with a heterozygote (Y = yellow, y = green).

*Answers to Practice Problems appear in appendix D.

Dihybrid Inheritance

Mendel performed a second series of experiments in which he crossed true-breeding plants that differed in 2 traits; such crosses are appropriately called *dihybrid crosses*. For example, he crossed tall plants having inflated pods with short plants having constricted pods (fig. 12.7). The F_1 plants showed both dominant characteristics. As before, Mendel then allowed the F_1 plants to self-pollinate. Two possible results could occur in the F_2 generation:

1. If the dominant factors (*TI*) always segregate into the gametes together, and the recessive factors (*ti*) always stay together, then there should be 2 phenotypes among the F_2 plants—tall plants with inflated pods and short plants with constricted pods.
2. If the 4 factors segregate into the gametes independently, then there should be 4 phenotypes among the F_2 plants— tall plants with inflated pods, tall plants with constricted pods, short plants with inflated pods, and short plants with constricted pods.

Figure 12.7 shows that Mendel observed 4 phenotypes among the F_2 plants, supporting the second hypothesis. Therefore, Mendel formulated his second law of heredity:

> **Mendel's law of independent assortment:** Members of one pair of factors segregate (assort) independently of members of another pair of factors. Therefore, all possible combinations of factors can occur in the gametes.

Today, we know that the law of segregation and the law of independent assortment hold because of the events of meiosis. There is only one allele for each trait in the gametes, and all possible combinations of alleles (located on different chromosomes) occur in the gametes because homologous chromosomes segregate and independently assort during meiosis (see fig. 13.7).

Let's use Mendel's law of independent assortment to designate the gametes for the F_1 genotype (*TtIi*). Only one letter of each kind can appear in a gamete, for example, there can be either a T or a t. The letter T or t may be present with either one of the other letters. Therefore, the genetically different gametes for this individual are *TI, Ti, tI, ti*. Depending on the genotype, there may be fewer possible types of gametes. For example, the genotype *TtII* has only 2 possible types of gametes: *TI* or *tI*.

Practice Problems 3*

1. For each of the following genotypes, give all possible gametes, noting the proportion of each for the individual.
   a. *ttGG*
   b. *TtGG*
   c. *TtGg*
   d. *TTGg*
2. For each of the following, state whether a genotype or a gamete is represented.
   a. *TT*
   b. *Tg*
   c. *IiCC*
   d. *TW*

*Answers to Practice Problems appear in appendix D.

Probability Revisited

Mendel always observed a 9:3:3:1 ratio among the F_2 generation of a dihybrid cross (fig. 12.7). He realized that this could only occur if each characteristic is inherited separately from any other. It is then possible to apply again the laws of probability mentioned on page 188. For example, we know the F_2 results for 2 separate monohybrid crosses are as listed here:

1. The chance of tallness = $^3/_4$
   The chance of shortness = $^1/_4$
2. The chance of inflated pods = $^3/_4$
   The chance of constricted pods = $^1/_4$

The probabilities for the 2-trait cross are, therefore, as follows:

The chance of tallness and inflated pods = $^3/_4 \times ^3/_4 = ^9/_{16}$
The chance of tallness and constricted pods = $^3/_4 \times ^1/_4 = ^3/_{16}$
The chance of shortness and inflated pods = $^1/_4 \times ^3/_4 = ^3/_{16}$
The chance of shortness and constricted pods = $^1/_4 \times ^1/_4 = ^1/_{16}$

And the phenotypic ratio is 9:3:3:1.

Again, since all genetically different male gametes must have an equal opportunity to fertilize all genetically different female gametes to even approximately achieve these results, a large number of offspring must be counted.

Dihybrid Genetics Problems

The fruit fly, *Drosophila melanogaster* (see fig. 13.9a), less than one-fifth the size of a housefly, is a favorite subject for genetic research because it has mutant characteristics that are easily determined. A "wild-type" fly has long wings and a gray body. There are mutant flies with short (vestigial) wings and black (ebony) bodies. The key for a cross involving these traits is L = long wing, l = short wing, G = gray body, and g = black body.

Genetic Basis of Life

Figure 12.7

The dihybrid cross shows that factors segregate into the gametes independently of other factors. In true-breeding plants that self-pollinate, the 2 factors for each trait are the same (see P generation). If these plants are crossed, the offspring show only dominant traits (see F₁ generation). If these F₁ plants are allowed to self-pollinate, all possible phenotypes appear among the offspring (see F₂ generation). On the average, 9 out of 16 offspring show both dominant characteristics, 3 show a dominant trait and a recessive trait, 3 show the other recessive trait and the other dominant trait, and one shows both recessive traits. This gives a 9:3:3:1 ratio of phenotypes. The blank represents that a dominant or a recessive allele can be present.

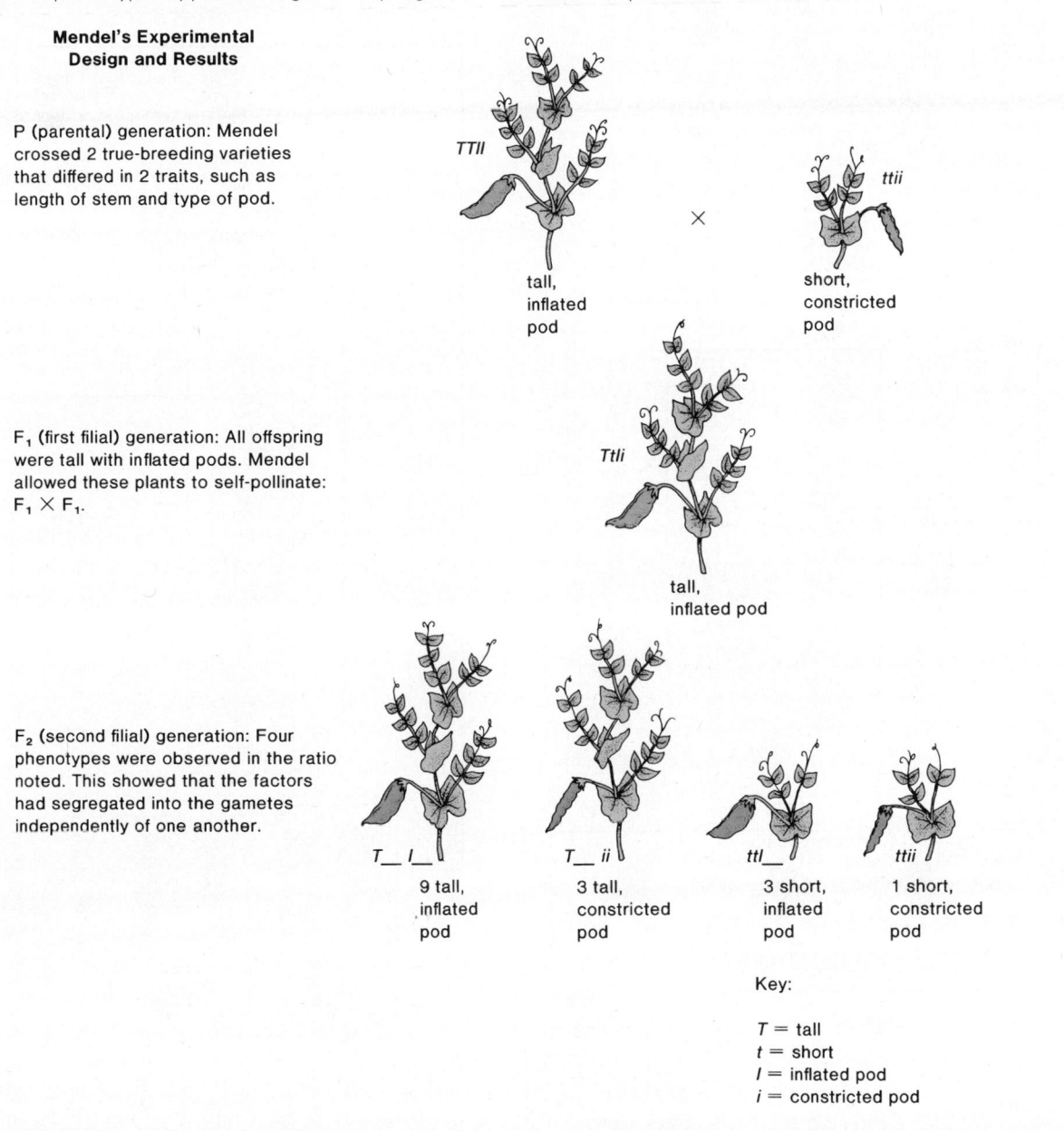

Mendel's Experimental Design and Results

P (parental) generation: Mendel crossed 2 true-breeding varieties that differed in 2 traits, such as length of stem and type of pod.

TTII

×

ttii

tall, inflated pod

short, constricted pod

F₁ (first filial) generation: All offspring were tall with inflated pods. Mendel allowed these plants to self-pollinate: F₁ × F₁.

TtIi

tall, inflated pod

F₂ (second filial) generation: Four phenotypes were observed in the ratio noted. This showed that the factors had segregated into the gametes independently of one another.

T__ I__

9 tall, inflated pod

T__ ii

3 tall, constricted pod

ttI__

3 short, inflated pod

ttii

1 short, constricted pod

Key:

T = tall
t = short
I = inflated pod
i = constricted pod

In figure 12.8, the flies of the P generation have only one possible type of gamete because both are homozygous. All the F₁ flies are heterozygous (*LlGg*) and have the same phenotype (long wings, gray body).

The Punnett square in figure 12.8 shows the expected results when the F₁ flies are crossed, assuming that all genetically different sperm have an equal opportunity to fertilize all genetically different eggs. Notice that $9/16$ of the offspring have long wings and a gray body; $3/16$ have long wings and a black body; $3/16$ have short wings and a gray body; and $1/16$ have short wings and a black body. This phenotypic ratio of 9:3:3:1 is expected whenever a heterozygote for 2 traits is crossed with another heterozygote for 2 traits and simple dominance is present in both genes.

Figure 12.8

Mendel's law of independent assortment illustrated in a fruit fly cross. Each F$_1$ fly (*LlGg*) produces 4 types of gametes because all possible combinations of alleles can occur in the gametes. Therefore, all possible phenotypes appear among the F$_2$ offspring.

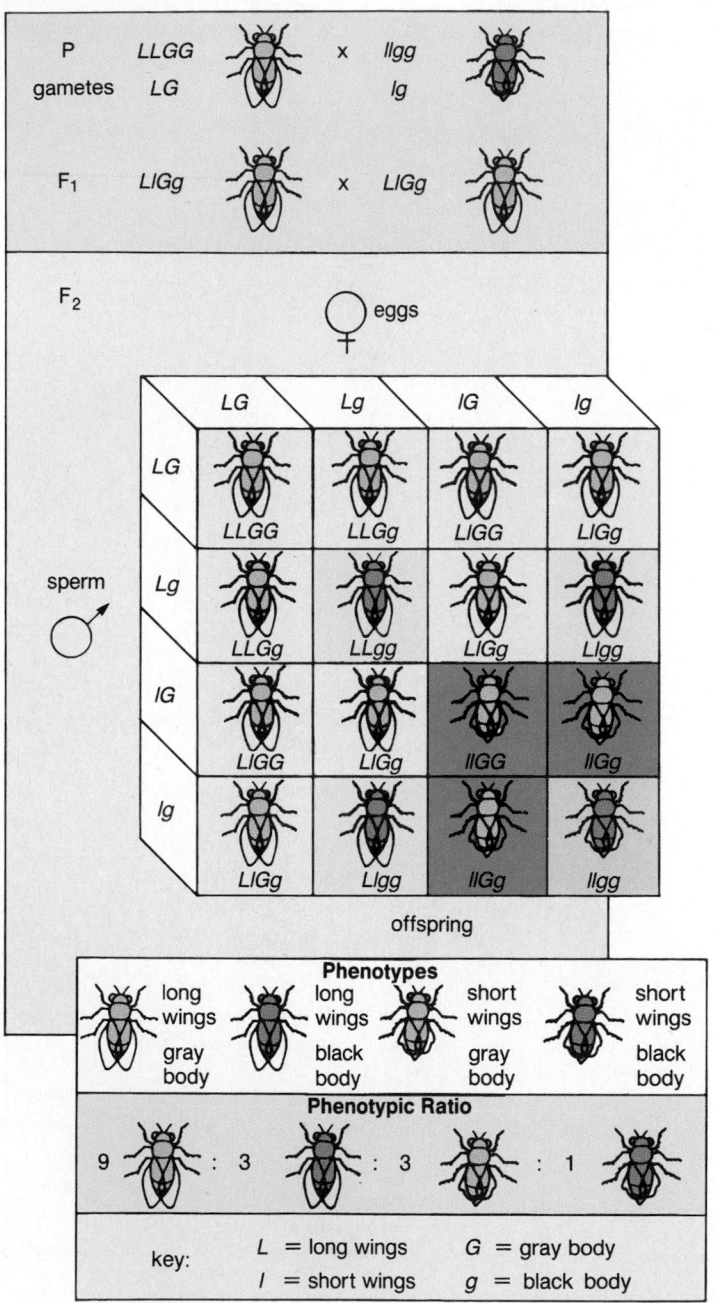

Figure 12.9

Representation of a dihybrid testcross to determine the genotype of a fly showing 2 dominant characteristics. If the fly is heterozygous dominant for both traits, and is crossed with another that is recessive for both traits as in the case illustrated here, the expected ratio of phenotypes is a 1:1:1:1. What would be the result if the test individual were homozygous dominant for both traits? Homozygous dominant for one trait but heterozygous for the other?

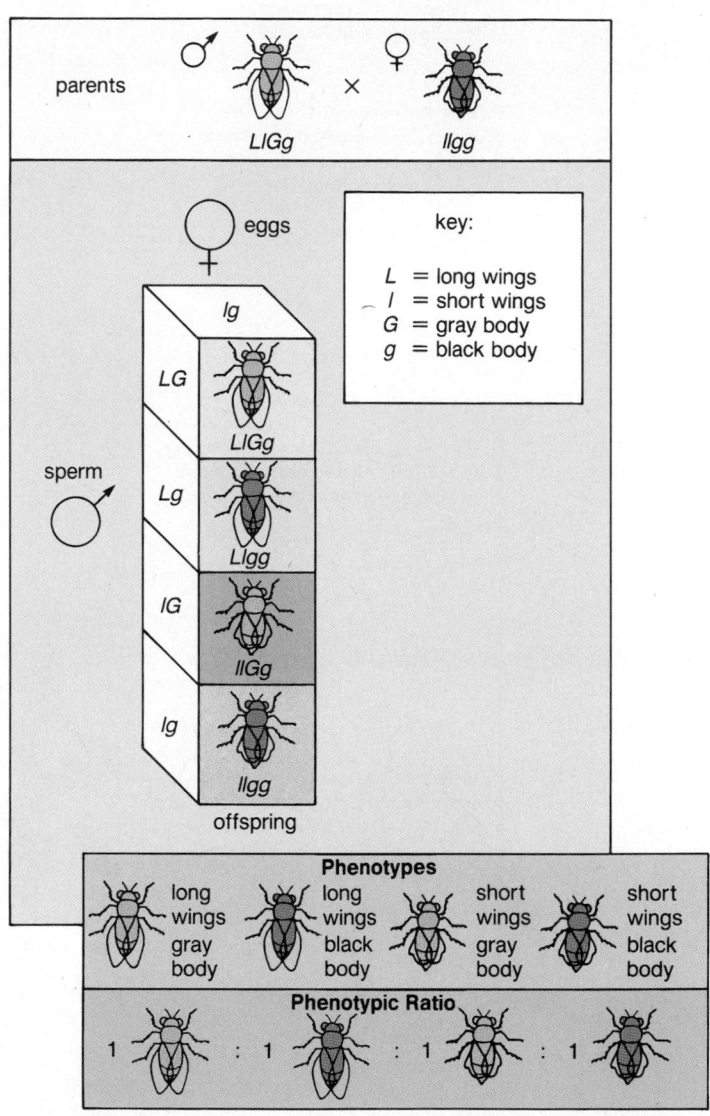

Dihybrid Testcross

A dihybrid testcross is used to determine if an individual is homozygous dominant or heterozygous for either of the 2 traits. Since it is not possible to determine the genotype of a long wing, gray body fly by inspection, for example, the genotype may be represented as *L__G__*.

When doing a dihybrid testcross, an individual with the dominant phenotypes is crossed with an individual with the recessive phenotypes. For example, a long wing, gray body fly is crossed

with a short wing, black body fly. A long wing, gray body fly heterozygous for both traits will form 4 different types of gametes. The homozygous fly with short wings and a black body can form only one kind of gamete:

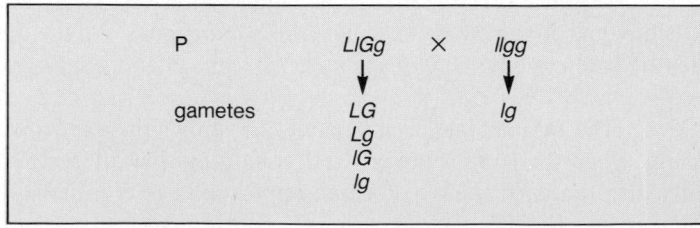

Genetic Basis of Life

Table 12.3
Common Crosses Involving Simple Dominance

Examples	Phenotypic Ratios
$Tt \times Tt$	3:1 (dominant to recessive)
$Tt \times tt$	1:1 (dominant to recessive)
$TtYy \times TtYy$	9:3:3:1 (9 dominant; 3 mixed; 3 mixed; 1 recessive)
$TtYy \times ttyy$	1:1:1:1 (all possible combinations in equal number)

Figure 12.9 shows that the expected offspring are $^{1}/_{4}$ long wings and a gray body; $^{1}/_{4}$ long wings and a black body; $^{1}/_{4}$ short wings and a gray body; and $^{1}/_{4}$ short wings and a black body. This is a 1:1:1:1 phenotypic ratio. The presence of offspring with short wings and a black body shows that the $L\_G\_$ fly is heterozygous for both traits and has the genotype $LlGg$.

If the $L\_G\_$ fly is homozygous for both traits, then no offspring will have short wings and/or a black body. If the $L\_G\_$ fly is heterozygous for one trait but not the other, what is the expected phenotypic ratio among the offspring?

Table 12.3 gives examples of crosses we have studied thus far. When these types of crosses are performed, these phenotypic ratios are observed.

Practice Problems 4*

1. In horses, B = black coat, b = brown coat, and T = trotter, t = pacer. A black pacer mated to a brown trotter produces a black trotter offspring. Give all possible genotypes for this offspring.
2. In fruit flies, long wings (L) is dominant over short wings (l), and gray body (G) is dominant over black body (g). In each instance, what are the most likely genotypes of the previous generation if a student gets the following phenotypic results?
    a. 1:1:1:1
    b. 9:3:3:1
3. In humans, short fingers and widow's peak are dominant over long fingers and continuous hairline. A heterozygote reproduces with a heterozygote. What is the chance of any one child's having the same phenotype as the parents?

*Answers to Practice Problems appear in appendix D.

Summary

1. At the time Mendel began his hybridization experiments, the blending theory of inheritance was popular. This theory stated that whenever the parents were distinctly different, the offspring would be intermediate between them.
2. Mendel, unlike preceding plant breeders, decided to do a statistical study, most likely because he had a mathematical background.
3. Mendel chose the garden pea for several reasons: many true-breeding varieties were available, they were easy to cultivate, had a short generation time, and could be cross-pollinated even though they normally self-pollinated. He studied only distinctly different traits, kept careful mathematical records, and interpreted his results mathematically.
4. When Mendel performed his monohybrid crosses, he found that the F_1 plants resembled only one of the parents; however, the characteristic of the other parent reappeared in about $^{1}/_{4}$ of the F_2 plants. Mendel saw that these 3:1 results were possible if the

F_1 hybrid contained 2 factors for each trait, one being dominant and the other recessive, and the factors segregated into the gametes.
5. Mendel's law of segregation states that there are 2 factors for each trait in the individual, and these segregate into gametes that then have only one factor for each trait.
6. Both the Punnett square and the laws of probability can be used to calculate the expected phenotypic ratio of a cross. In practice, a large number of offspring must be counted in order to observe the expected results, because only in that way can it be ensured that all possible types of sperm have fertilized all possible types of eggs.
7. Because humans do not produce a large number of offspring, it is best to use a predicted ratio as a means of determining the chances of an individual inheriting a particular characteristic.
8. Mendel also crossed his F_1 plants with homozygous recessive plants. The results indicated that the recessive factor was present in the F_1 plants (i.e.,

that it was heterozygous). Today, we call this a testcross, because it is used to test whether an individual showing the dominant characteristic is homozygous dominant or heterozygous.
9. When Mendel performed his dihybrid crosses, the F_1 individuals were dominant in both traits, but there were 4 phenotypes among the F_2 offspring. This allowed Mendel to deduce the law of independent assortment, which states that the members of one pair of factors segregate independently of those from another pair. Therefore, all possible combinations of factors can occur in the gametes.
10. The Punnett square and laws of probability for a dihybrid cross similar to Mendel's show that $^{9}/_{16}$ of the F_2 offspring have the 2 dominant traits, $^{3}/_{16}$ have one dominant trait with one recessive trait, $^{3}/_{16}$ have the other dominant trait with the other recessive trait, and $^{1}/_{16}$ have both recessive traits, giving a 9:3:3:1 ratio.
11. The dihybrid testcross allows you to test whether an individual showing 2 dominant characteristics is homozy-

gous dominant for both traits or for one trait only or is heterozygous for both traits.

12. Table 12.3 lists the most typical crosses and the expected results. Learning the table saves the trouble of working out the results repeatedly for each of these crosses.

Study Questions

1. How did Mendel overcome the difficulty of not being able to see the factors? In what way did he set aside any personal beliefs he may have had?
2. How did the monohybrid crosses performed by Mendel refute the blending theory of inheritance?
3. Using the monohybrid cross Mendel did as an example, trace his reasoning to arrive at the law of segregation.
4. Define these terms: allele, dominant allele, recessive allele, genotype, homozygous, heterozygous, and phenotype.

5. Use a Punnett square and the laws of probability to show the results of a cross between 2 individuals who are heterozygous for one trait. What are the chances of an offspring having the dominant phenotype? the recessive phenotype?
6. In what way did a monohybrid test-cross support Mendel's law of segregation?
7. How is a monohybrid testcross used today?
8. Using the dihybrid cross Mendel did as an example, trace his reasoning to arrive at the law of independent assortment.

9. Use a Punnett square and the laws of probability to show the results of a cross between 2 individuals who are heterozygous for 2 traits. What are the chances of an offspring being homozygous dominant for both traits? recessive for both traits?
10. If an individual is a heterozygote for 2 traits, what would the results of a testcross be? If the individual is heterozygous in only one trait? If the individual is homozygous dominant in both traits?

Objective Questions

For questions 1-4, match the cross with the results in the key:

Key:
- **a.** 3:1
- **b.** 9:3:3:1
- **c.** 1:1
- **d.** 1:1:1:1

1. *TtYy* X *Tt Yy*
2. *Tt* X *Tt*
3. *Tt* X *tt*
4. *TtYy* X *ttyy*
5. Which of these could be a gamete?
   - **a.** *GgLl*
   - **b.** *GLl*
   - **c.** *Gl*
   - **d.** None of these.
6. Which of these properly describes a cross between an individual who is homozygous dominant for hairline but heterozygous for finger length and an individual who is recessive for hairline and fingers? (*Key: W* = widow's peak, *w* = continuous hairline, *S* = short fingers, *s* = long fingers)
   - **a.** *WwSs* X *WwSs*
   - **b.** *WWSs* X *wwSs*
   - **c.** *Ws* X *ws*
   - **d.** *WWSs* X *wwss*
7. In peas, yellow seeds (*Y*) is dominant over green seeds (*y*). In the F$_2$ generation of a monohybrid cross in which a dominant homozygote is crossed with a recessive homozygote, you would expect
   - **a.** plants that produce 3 yellow seeds to every green seed.
   - **b.** plants with one yellow seed for every green seed.
   - **c.** only plants with either the genotype *YY* or *yy*.
   - **d.** Both a and c.

8. In humans, pointed eyebrows are dominant over smooth eyebrows. Mary's father has pointed eyebrows, but she and her mother have smooth. What is the genotype of the father?
   - **a.** *BB*
   - **b.** *Bb*
   - **c.** *bb*
   - **d.** Any one of these.
9. In guinea pigs, smooth coat (*S*) is dominant over rough coat (*s*) and black coat (*B*) is dominant over white coat (*b*). In the cross *SsBb* X *SsBb*, how many of the offspring will have a smooth black coat?
   - **a.** 9 only
   - **b.** about 9 of every 16
   - **c.** only one out of 16
   - **d.** 6 of every 16

Genetic Basis of Life

10. In horses, B = black coat, b = brown coat, and T = trotter, t = pacer. A black trotter that has a brown pacer offspring is
   a. BT.
   b. $BbTt$.
   c. $bbtt$.
   d. $BBtt$.

11. In tomatoes, red fruit (R) is dominant over yellow fruit (r) and tallness (T) is dominant over shortness (t). A plant that is $RrTT$ is crossed with a plant that is $rrTt$. What are the chances of an offspring being heterozygous for both traits?
   a. none
   b. $1/2$
   c. $1/4$
   d. $9/16$

12. If a plant with an $RrTt$ genotype is crossed with a plant that is $rrtt$,
   a. All the offspring will be tall with red fruit.
   b. 75% will be tall with red fruit.
   c. 50% will be tall with red fruit.
   d. 25% will be tall with red fruit.

Additional Genetics Problems

1. If a man homozygous for widow's peak (dominant) reproduces with a woman homozygous for continuous hairline (recessive), what are the chances of the children's having a widow's peak? a continuous hairline?

2. John has unattached earlobes like his father, but his mother has attached earlobes. What is John's genotype?

3. In humans, the allele for short fingers is dominant over that for long fingers. If a person with short fingers who had one parent with long fingers reproduces with a person with long fingers, what are the chances of each child's having short fingers?

4. In a fruit fly experiment (see key on page 192), 2 gray-bodied fruit flies produce mostly gray-bodied offspring, but some offspring have black bodies. If there are 280 offspring, how many do you predict will have gray bodies

and how many will have black bodies? How many of the 280 offspring do you predict will be heterozygous? If you wanted to test whether a particular gray-bodied fly was homozygous dominant or heterozygous, what cross would you do?

5. In rabbits, black color is due to a dominant allele B and brown color to its recessive allele b. Short hair is due to the dominant allele S and long hair is due to its recessive allele s. In a cross between a homozygous black, long-haired rabbit and a brown, homozygous short-haired one, what would the F_1 generation look like? the F_2 generation? If one of the F_1 rabbits reproduced with a brown, long-haired rabbit, what phenotypes and in what ratio would you expect?

6. In horses, black coat (B) is dominant over brown coat (b) and being a trotter (T) is dominant over being a pacer (t).

A black horse that is a pacer is crossed with a brown horse that is a trotter. The offspring is a brown pacer. Give the genotypes of all these horses.

7. The complete genotype of a gray-bodied, long-winged fruit fly is unknown. When this fly is crossed with a black-bodied, short-winged fruit fly, the offspring all have a gray body but about half of them have short wings. What is the genotype of the gray-bodied, long-winged fly? (See key on page 192.)

8. In humans, widow's peak (pointed) hairline is dominant over continuous (smooth) hairline, and short fingers are dominant over long fingers. If an individual who is heterozygous for both traits reproduces with an individual who is recessive for both traits, what are the chances of a child's also being recessive for both traits?

Concepts and Critical Thinking

1. *The laws of heredity explain genetic stability and variation between generations.*

   Explain why you are similar in appearance to your parents but do not resemble either one exactly.

2. *The laws of heredity are the same for all organisms.*

   Why would you expect the same laws of heredity to apply to both plants and animals?

3. *Genes, carried by way of gametes from generation to generation, are not affected by the phenotype.*

   Explain why an individual, deformed by accident, can have children who have no such deformities.

Selected Key Terms

13

Chromosomes and Genes

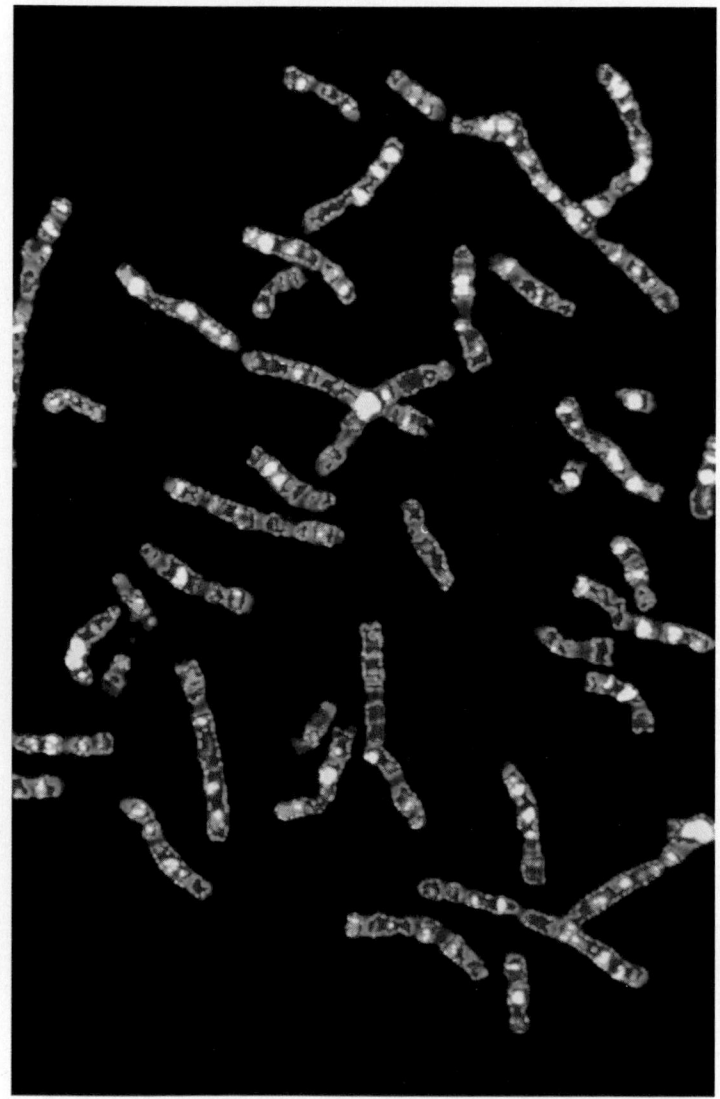

The genes composed of DNA, are in the chromosomes, which when stained have specific banding patterns. The banding patterns help distinguish one chromosome from the other, and the bands also serve as landmarks. Sometimes, it is possible to associate particular genes with particular bands. Magnification, X3,000.

Your study of this chapter will be complete when you can

1. recognize and solve genetics problems involving degrees of dominance, multiple alleles, and polygenes;
2. discuss, with examples, the relative influence of the genotype and the environment on the phenotype;
3. support the chromosome theory of inheritance by listing similarities in the behavior of chromosomes and genes;
4. explain the normal sex chromosome makeup of human males and females and the manner in which sex is determined;
5. describe the experimental results that allowed investigators to determine the existence of X-linked genes;
6. solve genetics problems involving sex-linked genes;
7. explain why linked genes do not obey Mendel's laws, and solve genetic linkage problems;
8. use the results of crosses involving linked genes to determine the order of genes on a chromosome;
9. give examples of chromosome mutations caused by a change in chromosome number, and tell how such mutations could occur;
10. give examples of chromosome mutations caused by a change in chromosome structure, and tell how such mutations could occur.

n this chapter, we go beyond Mendel's experiments first to examine varied relationships between alleles and then to show that the genes are on the chromosomes. Mendel never tried to determine where in the cell his hereditary factors might be located. That came later.

Beyond Mendel's Experiments

Inheritance that follows the patterns explained by Mendel's experiments is sometimes called "simple Mendelian inheritance" and for good reason. The dominance and recessive relationship between alleles is actually quite varied, and we have space here to present only a few examples. It is hoped that this discussion will allow you to see that inheritance depends on the genome (all the genes) and not individual alleles or even allelic pairs.

Degrees of Dominance

The terms *dominant* and *recessive* are best applied, as Mendel originally intended, to the phenotype. A dominant characteristic is one that is seen in the heterozygous F_1 generation, and a recessive characteristic is one that is not seen in the F_1 generation (see fig. 12.3). Mendel always observed complete dominance of one characteristic over another in his experiments. If Mendel had chosen other characteristics, he most likely would have observed **incomplete dominance** on occasion. In a cross between a true-breeding, red-flowered snapdragon strain and a true-breeding, white-flowered strain, the offspring have pink flowers (fig. 13.1). This is not an example that supports the blending theory of inheritance prevalent when Mendel did his work, however. If these F_1 plants self-pollinate, the F_2 generation has a phenotypic ratio of 1 red-flowered:2 pink-flowered:1 white-flowered plant. Because the parental phenotypes reappear in the F_2 generation, it is obvious that we are still dealing with an example of particulate inheritance of the type described by Mendel.

There are also examples of **codominance** on the phenotype level. An individual with the blood type AB is exhibiting codominant characteristics. We know this because there are some individuals who are blood type A and others who are blood type B. The recessive individual has blood type O and exhibits neither dominant characteristic (p. 225).

If we turn from the phenotype to the genotype, it is possible to examine dominance in terms of allele activity. Quite often, when heterozygous individuals are examined biochemically, the level of gene-directed protein product is in between that of the 2 homozygotes. In other words, the dominant allele is directing the production of a protein, but the recessive allele is not or at least not an effective protein. For example, we can explain the results depicted in figure 13.1 by assuming that the allele R is directing the production of red pigment, while the allele r is not. In peas it has been shown that the allele R (for round) causes effective starch enzyme, while the allele r (for wrinkled) results in defective starch enzyme. In the heterozygote, there is enough effective enzyme to give the normal amount of starch and therefore the normal phenotype. In the

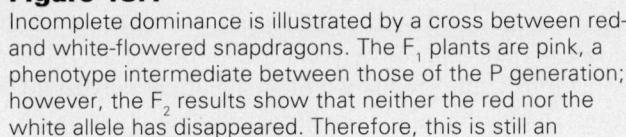

Figure 13.1
Incomplete dominance is illustrated by a cross between red- and white-flowered snapdragons. The F_1 plants are pink, a phenotype intermediate between those of the P generation; however, the F_2 results show that neither the red nor the white allele has disappeared. Therefore, this is still an example of particulate inheritance, though the heterozygote *Rr* is pink.

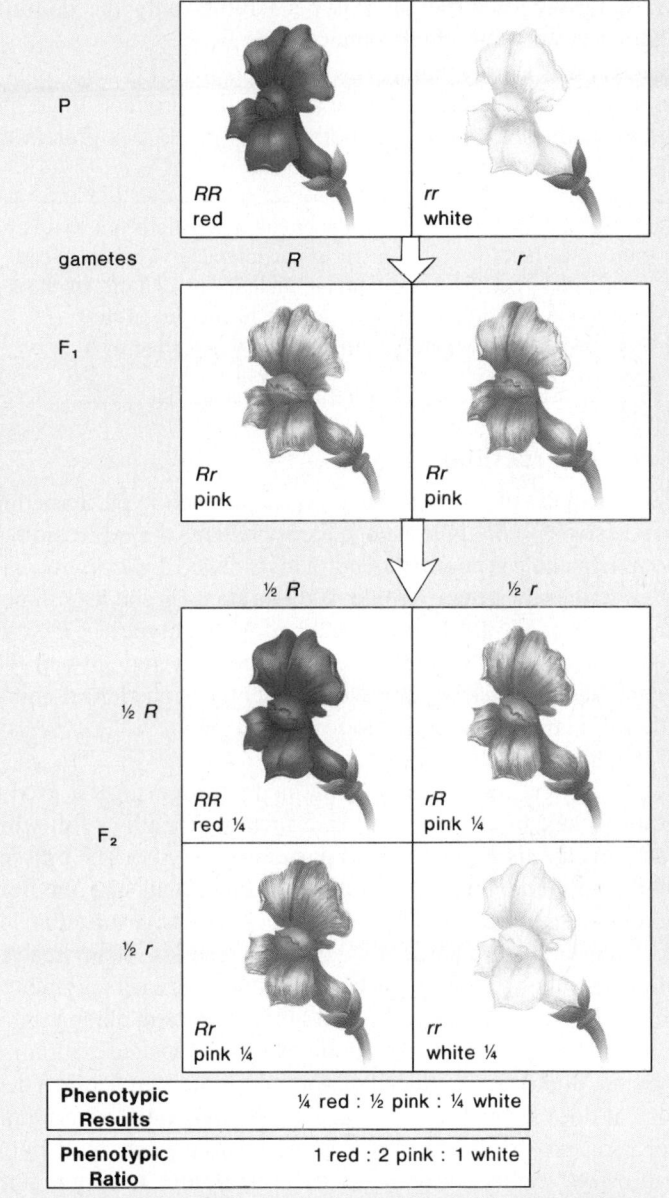

Phenotypic Results	¼ red : ½ pink : ¼ white
Phenotypic Ratio	1 red : 2 pink : 1 white

homozygous recessive, excess nonconverted sugar leads to an uptake of water and a stretching of the seedcoat. When the usual drying of a seed occurs, wrinkling results.

The phenotype can be drastically altered when an individual inherits an allele that codes for a defective product. (The term *pleiotropy* is used to describe an allele that affects more than one characteristic.) Individuals with Marfan syndrome tend to be tall and thin with long legs, arms, and fingers. They are nearsight-ed, and the wall of the aorta is weak, causing it to enlarge and, eventually, to split. All of these effects are caused by an inability to produce normal connective tissue. The difficulty stems from an allele that directs the production of an abnormal polypeptide for collagen. The normal allele produces a normal polypeptide, but when the abnormal and normal polypeptides combine, the end result is abnormal collagen. Although both alleles are active in the heterozygote, the phenotype exhibits only the dominant characteristic—abnormal connective tissue.

Most likely you will now have no difficulty deducing that sometimes both alleles produce an effective product. In an indi-vidual with blood type AB, both alleles produce a protein that appears on the red blood cells.

> On the phenotype level, a dominant characteristic is one that is seen in the heterozygote; the recessive characteristic is not seen. On the genotype level, a dominant allele is one that directs the production of a product that affects the phenotype of the heterozygote more than the product, if any, directed by a recessive allele.

Gene Interactions

Pairs of alleles often interact to produce the phenotype. Sometimes a recessive pair of alleles at one locus prevents the expression of a dominant allele at another locus. (This is called *epistasis,* meaning a "covering-up".) For example, William Bateson and R.C. Punnett deduced that purple-flowered peas had the genotype $C\_\_P\_\_$. The genotypes $C\_\_pp$ or $ccP\_\_$ were white-flowered instead. The homozygous recessive *pp* prevents the expression of the dominant allele *C,* and the homozygous recessive *cc* prevents the expression of the dominant allele *P.*

These results can be explained by observing that produc-tion of plant pigments involves a metabolic pathway in which each reaction is catalyzed by a particular enzyme. The presence of one dominant allele results in a functional enzyme for a particular reaction; the presence of 2 recessives results in a nonfunctioning enzyme for a particular reaction. If any reaction of the metabolic pathway is blocked, the plant lacks pigment. A somewhat similar situation occurs in humans and other animals. If an individual inherits the allelic pair (*aa*) that causes albinism, they are unable to produce the normal pigment melanin. There-fore, it does not matter what genes they may have inherited for color of eyes and hair, for example. These genes will not be expressed, and the individual will be an albino, lacking pigment in all parts of the body (fig. 13.2).

Sometimes, alleles at another locus seem to modify the expression of a dominant allele at another locus. (This is called *dominance modification.*) How modifying genes work is not known. The spotted pattern in Holstein cattle is determined by a single pair of alleles, but the relative amounts of black and white are determined by modifying genes. Also, in humans there may be only 2 alleles for eye color, *B* for brown eyes, and *b* for blue eyes. The presence of modifying genes, however, could explain why

Figure 13.2
An albino koala. The inheritance of 2 recessive alleles prevented the production of melanin and resulted in these animals that have no color in the skin, hair, or eyes.

there are also shades of gray and green in addition to brown and blue eyes. The modifying genes affect the amount of pigment actually deposited in the iris and the pattern of its deposition.

Recently, investigators have suggested that dominance modification can explain a newly discovered phenomenon called *parental imprinting of genes.* Recall that Mendel found that recipro-cal crosses gave the same results; it did not matter whether a factor came from the male or female parent, he still got the same results. In contrast to Mendel's finding, in rare instances, it does seem to matter which parent contributes a particular allele. In mice, for example, if both copies of chromosome 11 are from the male parent (investiga-tors are able to bring this about experimentally), the mouse is oversized. If both copies of chromosome 11 are from the female parent, the mouse is undersized. It appears that a certain allele is expressed to a greater extent if inherited from the father and to a lesser extent if inherited from the mother. Imprinting has also been ob-served in humans (fig. 13.3). When the allele that causes Huntington

Figure 13.3

Parental imprinting of genes. Contrary to Mendel's experiments, it appears that in certain rare instances it does matter whether an allele is inherited from a father or a mother. It is assumed that these alleles are modified differently during spermatogenesis and oogenesis, and this is what causes imprinting to occur.

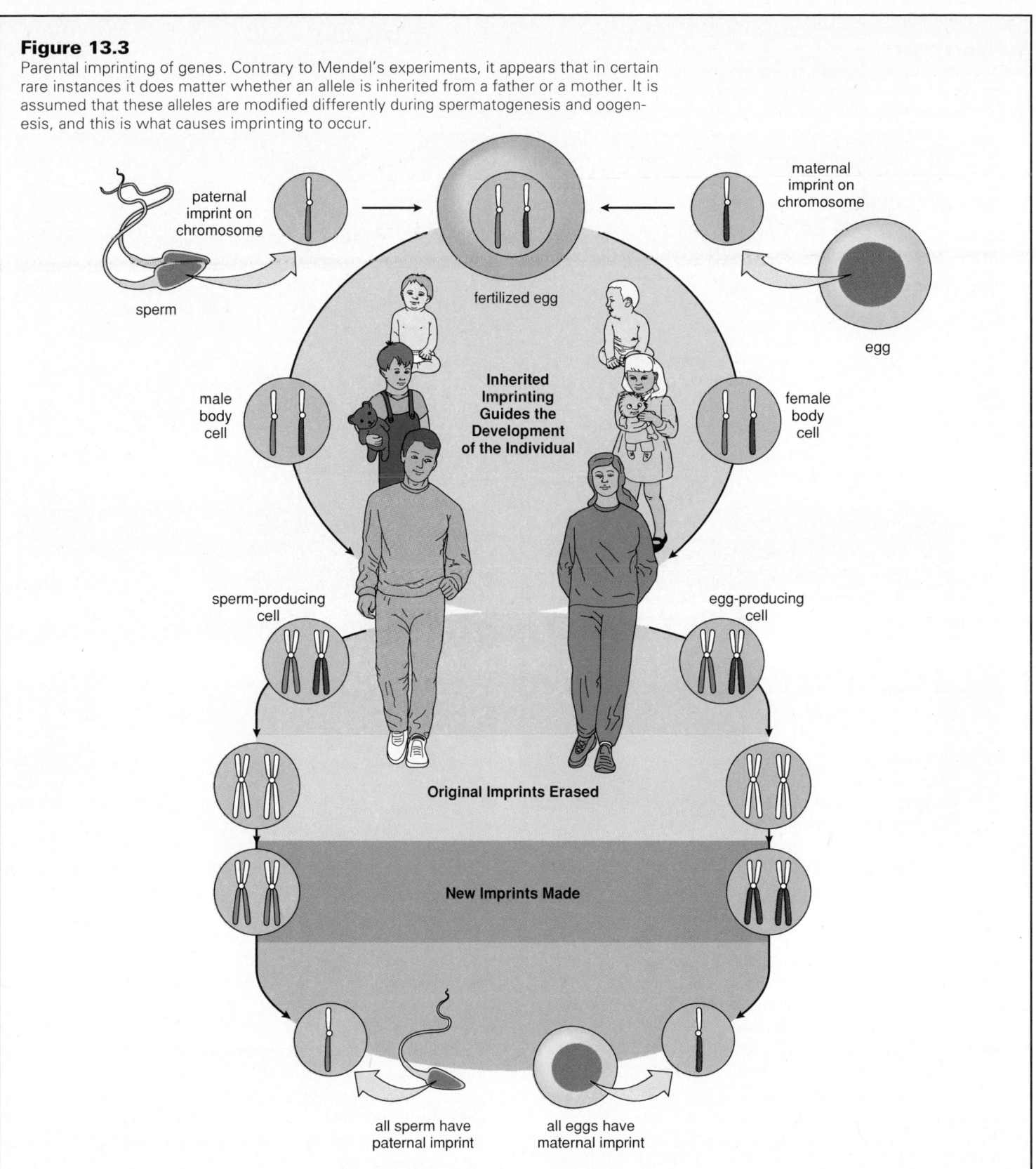

paternal imprint on chromosome

sperm

maternal imprint on chromosome

egg

fertilized egg

Inherited Imprinting Guides the Development of the Individual

male body cell

female body cell

sperm-producing cell

egg-producing cell

Original Imprints Erased

New Imprints Made

all sperm have paternal imprint

all eggs have maternal imprint

Figure 13.4

Inheritance by multiple alleles explains inheritance of coat color in these rabbits. Because there are 4 alleles that range in dominance (each individual has only 2), there are 4 different genotypes and phenotypes. The blank line means any allele less dominant.

genotype: C_____
phenotype: wild type of agouti coat

genotype: c^{ch}_____
phenotype: chinchilla or light gray coat

genotype: c^h_____
phenotype: himalayan coat (albino with black extremities)

genotype: cc
phenotype: albino

disease (p. 223) is inherited from a father, the genetic disorder occurs earlier in life than when the allele is inherited from a mother. This can be explained by assuming that the allele is imprinted (i.e., modified) differently in each sex before it is passed on.

Gene interactions often occur and result in the phenotype.

Multiple Alleles

Sometimes there are more than 2 alleles for a given chromosome locus in which case a trait is controlled by **multiple alleles.** Each individual has only 2 of all the available alleles.

In rabbits there are 4 alleles for coat color, and there is a dominance sequence that may be described like this:

$$C > c^{ch} > c^h > c$$

Therefore, there are 4 genotypes and phenotypes, as shown in figure 13.4.

Inheritance by multiple alleles causes a trait to exhibit more than 2 possible phenotypes. ABO blood type is also controlled by multiple alleles, and there are 4 possible phenotypes, as described preceding and on page 226.

Polygenic Inheritance

Sometimes a trait is controlled by several pairs of alleles, and each allele has the same degree of influence. This is called **polygenic**

inheritance. For example, H. Nilsson-Ehle studied the inheritance of seed color in wheat. When he crossed plants that produced white and dark red seeds, the F_1 seeds were intermediate in color between those of the parents. But if the F_1 plants were allowed to self-pollinate, the F_2 seeds had 7 phenotypes ranging from white to dark red:

⬦	aabbcc	(all recessive alleles)
⬦	Aabbcc	(any one dominant allele)
⬦	AAbbcc	(any 2 dominant alleles)
⬦	AABbcc	(any 3 dominant alleles)
⬦	AABBcc	(any 4 dominant alleles)
⬦	AABBCc	(any 5 dominant alleles)
⬦	AABBCC	(all dominant alleles)

Notice that each allele seems to have a small but equal quantitative effect, and that this accounts for the observed range in F_2 phenotypes. If we examine matters a little more closely, as in figure 13.5, we find that few individuals have the original F_1 phenotypes and most have one of the intermediate phenotypes. If we graph the F_2 results the proportions of the various phenotypes result in a bell-shaped curve.

Figure 13.5

Polygenic inheritance. **a.** The population shows continuous variation with a distribution illustrated by a bell-shaped curve. The range in phenotypes can be explained by assuming that the trait is controlled by several pairs of alleles and each allele adds quantitatively to produce the phenotypes. **b.** In this cross, the trait is controlled by 3 pairs of alleles. The Punnett square illustrates the resulting range in genotypes when a triple heterozygote reproduces with another triple heterozygote. The colorless circles represent recessive alleles and the colored circles represent dominant alleles.

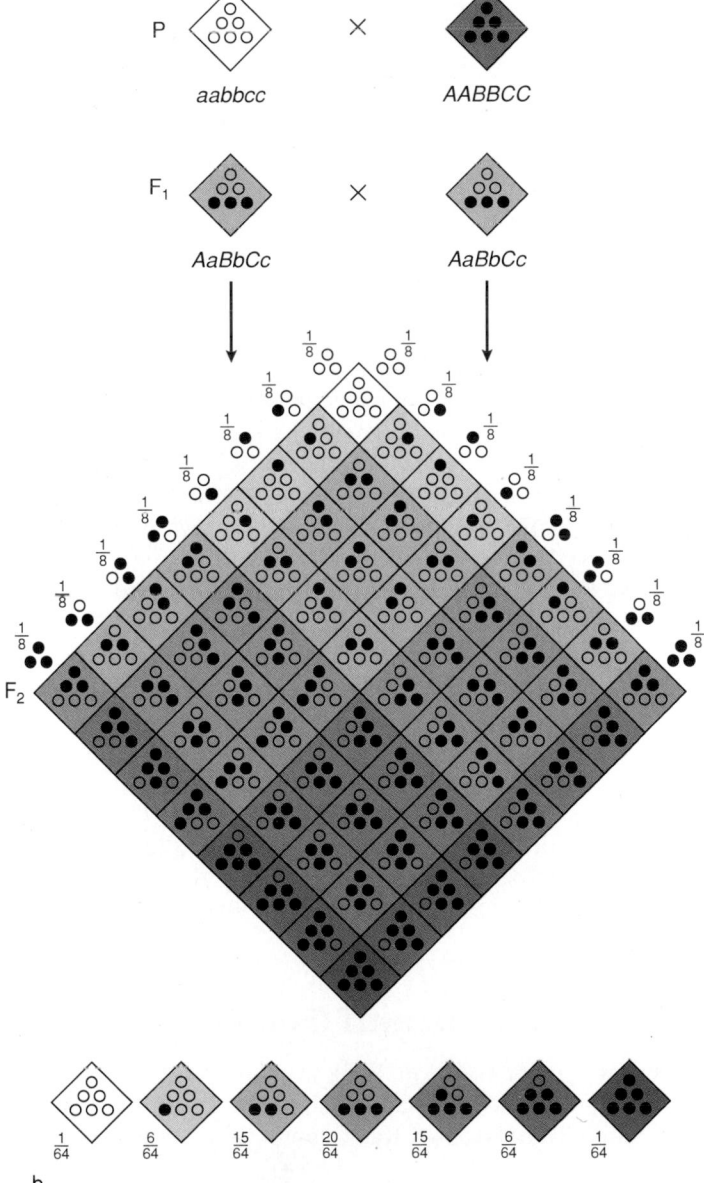

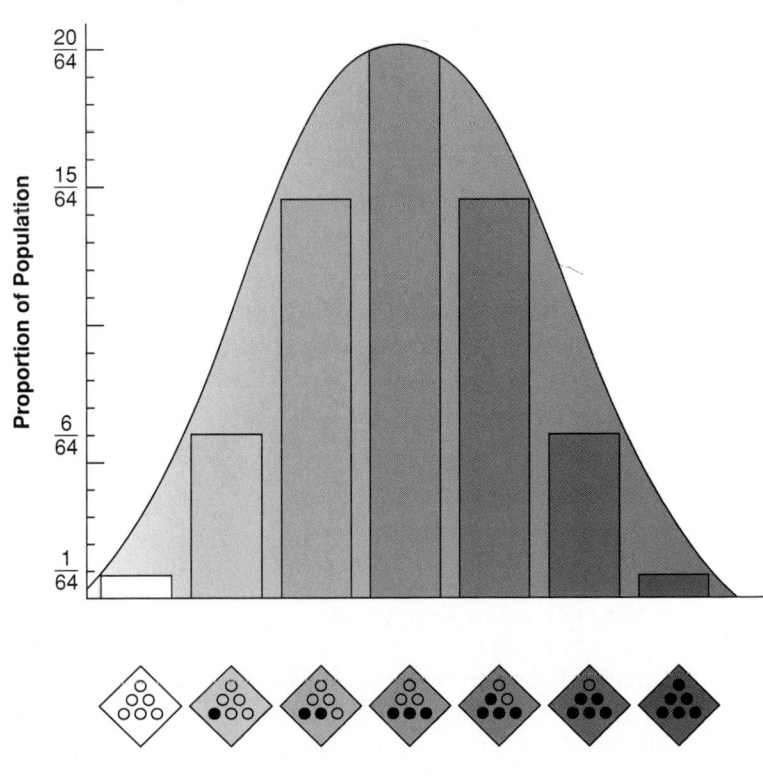

a.

b.

Polygenic inheritance accounts for the range in phenotypes we observe for such characteristics as ear size in corn plants or skin color in humans.

> In polygenic inheritance, a trait is controlled by several genes and each allele adds quantitatively to the phenotype. The variations are continuous—there is a range of intermediate phenotypes—and the phenotypic distribution follows a bell-shaped curve.

Polygenic traits are typically influenced by the environment; for example, the height and the weight of plants and animals are in part due to the type of nurture they receive.

Practice Problems 1*

1. What genotypes and phenotypes are possible among the offspring if a rabbit with the genotype Cc^h is mated to a rabbit with the genotype $C^{ch}c^h$?
2. What are the results of a testcross involving incomplete dominance when snapdragons with pink flowers are crossed with white-flowered ones?
3. Breeders of dogs note various colors among offspring that range from white to black. If the coat color is controlled by 3 pairs of alleles, how many different shades are possible?
4. Investigators note that albino tigers usually have crossed eyes. What 2 inheritance patterns can account for a phenotype such as this?

*Answers to Practice Problems appear in appendix D.

Genes and the Environment

There is no doubt that both the genotype and the environment affect the phenotype. The relative importance of genetic and environmental influences on the phenotype can vary, but in some instances the environment seems to have an extreme effect.

In the water buttercup, *Ranunculus peltatus* (fig. 13.6), the submerged part of the plant has a different appearance from that part above water. Apparently, the presence or absence of an aquatic environment dramatically influences the phenotype.

In humans, if the drug thalidomide is taken during the second month of pregnancy, a seriously malformed baby either without arms or legs or with stublike appendages can result. This phenotype can be due to a rare recessive mutation in humans, but in this instance it is caused by the environmental effect of thalidomide—the genotype is normal!

> The relative effect of genes and the environment on the phenotype varies; there are examples of extreme environmental influence.

Chromosomes and Genes

Mendel was rediscovered in 1900, at a time when the scientific community was ready to appreciate the work he had done so many years before. Three botanists, Karl Correns (German), Hugo de Vries (Dutch), and Erich Tschermak von Seysenegg (Austrian), working independently, discovered the laws of segregation and independent assortment. Upon surveying the scientific literature, they found that Gregor Mendel had already worked out these laws before them.

We could say that Mendel was ahead of his time to explain why his work lay unknown in dusty scientific libraries and archives for nearly 35 years. The scientific community caught up to Mendel because the cellular basis of inheritance had been worked out in the intervening years. The behavior of chromosomes in mitosis had been described by 1875, and meiosis had been described in the 1890s. By 1902, both Theodor Boveri (German) and Walter S. Sutton (American) had independently noted the parallel behavior of genes and chromosomes and had proposed the **chromosome theory of inheritance,** which states that the genes are located on the chromosomes. This theory is supported by the following observations (fig. 13.7):

1. Both chromosomes and factors (now called alleles) exist in pairs in diploid cells.
2. Both homologous chromosomes and alleles of each pair segregate during meiosis so that the gametes have one-half the total number.

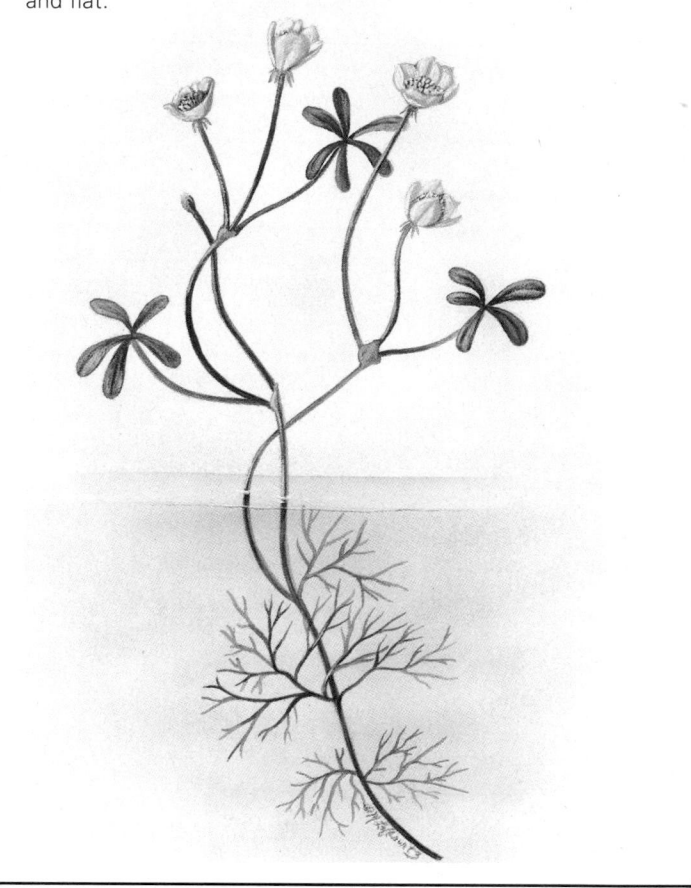

Figure 13.6
Environmental influence on the phenotype of the water buttercup is obvious. The submerged leaves are thin and finely divided, whereas those that lie above water are broad and flat.

Genetic Basis of Life

Figure 13.7

Mendel's laws of segregation and independent assortment have a chromosomal basis. The law of segregation holds because homologous chromosomes, and therefore alleles, segregate during meiosis I. Consequently, there is only one chromosome and one allele of each kind in the gametes. The law of independent assortment holds because homologous chromosomes and therefore alleles, segregate independently during meiosis I.

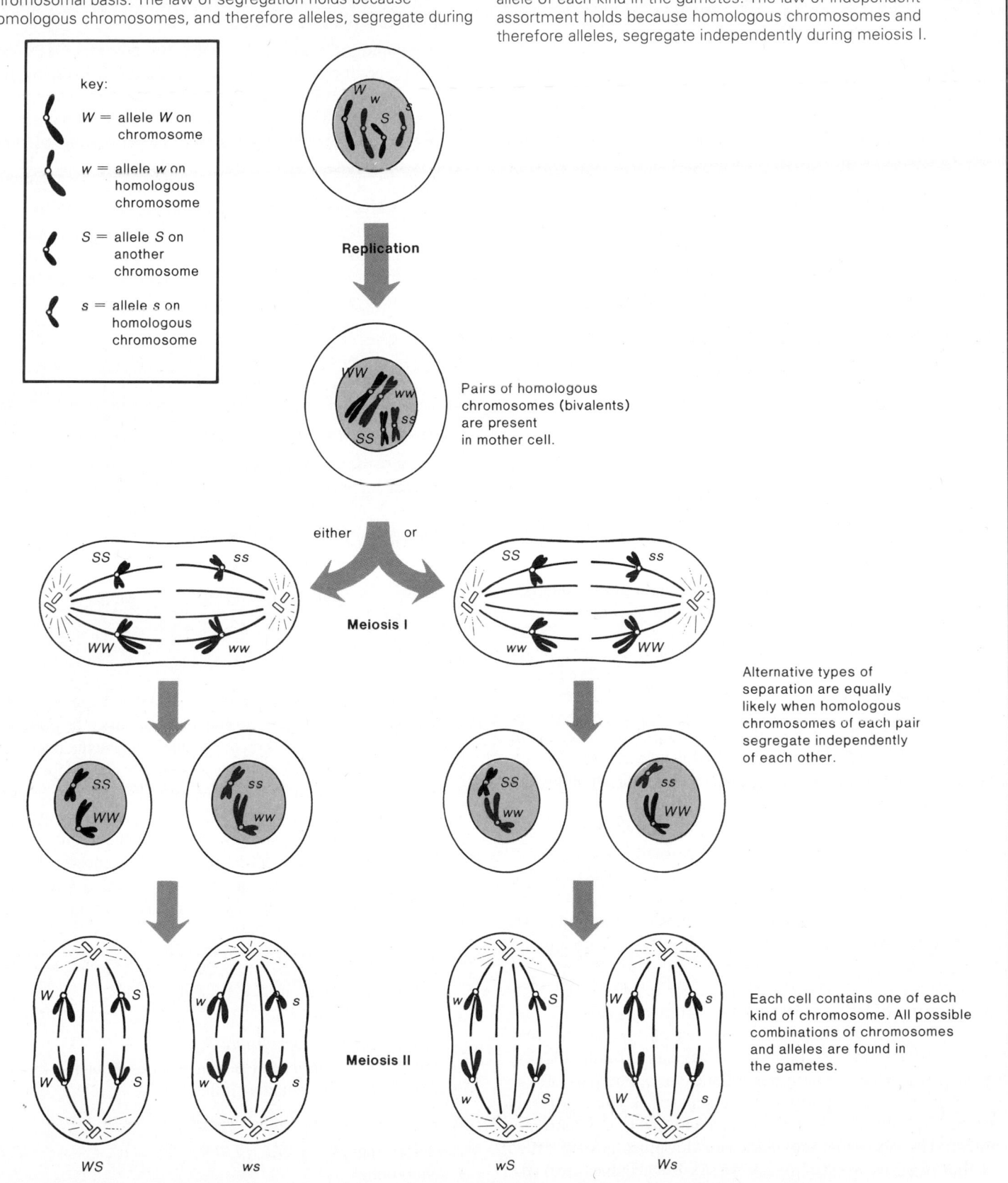

key:

W = allele W on chromosome

w = allele w on homologous chromosome

S = allele S on another chromosome

s = allele s on homologous chromosome

Replication

Pairs of homologous chromosomes (bivalents) are present in mother cell.

either   or

Meiosis I

Alternative types of separation are equally likely when homologous chromosomes of each pair segregate independently of each other.

Meiosis II

Each cell contains one of each kind of chromosome. All possible combinations of chromosomes and alleles are found in the gametes.

WS   ws   wS   Ws

Figure 13.8

In this Punnett square, the sperm and eggs are shown as carrying only a sex chromosome. (Actually, they also carry 22 autosomes.) The offspring are either male or female, depending on whether they receive an X or Y chromosome from the male parent.

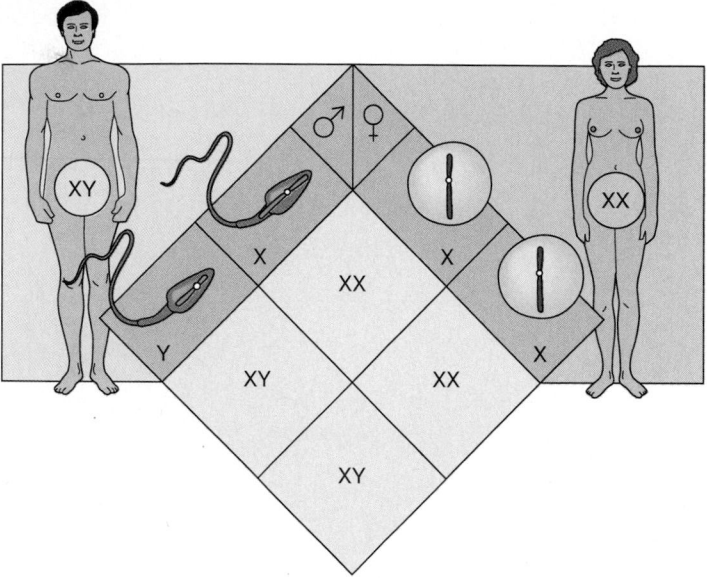

3. Both homologous chromosomes and alleles of each pair segregate independently so that the gametes contain all possible combinations.

4. Fertilization restores both the diploid chromosome number and the paired condition for alleles in the zygote.

The chromosome theory of inheritance states that the genes are on the chromosomes; therefore, they behave similarly during meiosis and at fertilization.

Chromosome Sex Determination

In many animal species, chromosomes differ between the sexes. All but one of the pairs of chromosomes in males and females are the same: these are called the **autosomes.** The pair of chromosomes that are different are called the **sex chromosomes.** The sex chromosomes in human females are XX, and those in males are XY. Because males can produce 2 different types of gametes—those that contain an X and those that contain a Y—normally they determine the sex of the new individual (fig. 13.8). Although the chances of any human couple having a boy or a girl are 50:50, actually more boys than girls are born. For reasons that are not clear, the ratio of boys to girls is 106:100 among newborns.

As you would suspect, the sex chromosomes carry genes that determine sex. It has taken 30 years to finally identify the specific gene that determines maleness as opposed to female-ness. This and other findings are discussed in the reading in chapter 14, "Active Y Chromosomes and Inactive X Chromosomes." The genes that are on the sex chromosomes are called **sex-linked genes.** Most of the known sex-linked genes are on the X chromosome and so are called **X-linked genes.** Some—perhaps most—of the genes on the X chromosome control traits that have nothing to do with sex.

Discovery of X-Linkage

The hypothesis that genes are on the chromosomes was substantiated by experiments performed by a Columbia University group of *Drosophila* geneticists, headed by Thomas Hunt Morgan. These flies are even better subjects for genetic studies than garden peas: they can be easily and inexpensively raised in simple laboratory glassware; females mate only once and then lay hundreds of eggs during their lifetimes; and the generation time is short, taking only about 10 days when conditions are favorable (fig. 13.9).

Drosophila flies have the same sex chromosome pattern as humans, and this facilitates our understanding of a cross performed by Morgan. Morgan took a newly discovered mutant male with white eyes and crossed it with a red-eyed female:

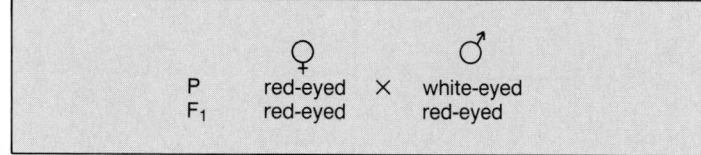

From these results, he knew that red eyes are dominant over white eyes. He then crossed the F_1 flies. In the F_2 generation, there was the expected 3 red-eyed:1 white-eyed ratio, but it struck him as odd that all of the white-eyed flies were males:

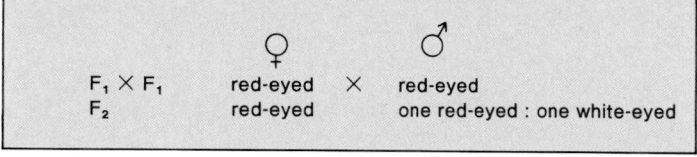

Obviously, a major difference between the male flies and the female flies was their sex chromosomes. Could it be possible that an allele for eye color was on the Y chromosome but not on the X? This idea could be quickly discarded because normal females have red eyes and they have no Y chromosome. Perhaps an allele for eye color was on the X, but not the Y, chromosome. Figure 13.9*b* indicates that this explanation would match the results obtained in the experiment. These results support the chromosomal theory of inheritance by showing that the behavior of a specific allele corresponds exactly with that of specific chromosome, the X chromosome in *Drosophila*.

Some alleles are linked to the X chromosome, and the Y chromosome is blank for these alleles.

X-Linked Genetics Problems

Recall that when solving autosomal genetics problems, the key and genotypes are represented as follows:

Key: *Genotypes:*

L = long wings LL, Ll, ll
l = short wings

Figure 13.9

Drosophila and *Drosophila* cross. **a.** Both males and females have 4 chromosomes, 3 pairs of autosomes and one pair of sex chromosomes. Males are XY and females are XX. **b.** When solving X-linked genetics problems, males do not have a superscript attached to the Y chromosome because they lack the gene for eye color on that chromosome. Among the offspring in this cross, 50% of the male offspring have white eyes. Because the male offspring receive a Y chromosome from the male parent, they express whichever X chromosome is inherited from the female parent. In contrast, 50% of the female offspring are red-eyed heterozygotes—when an X^r is received from the female parent, it is masked by an X^R received from the male parent.

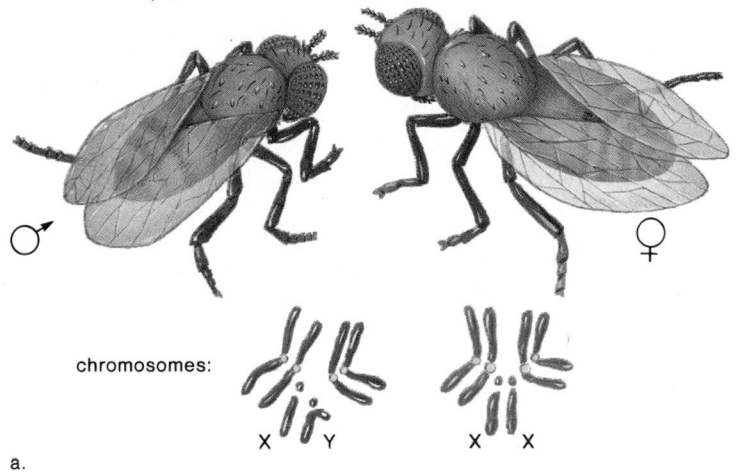

a.

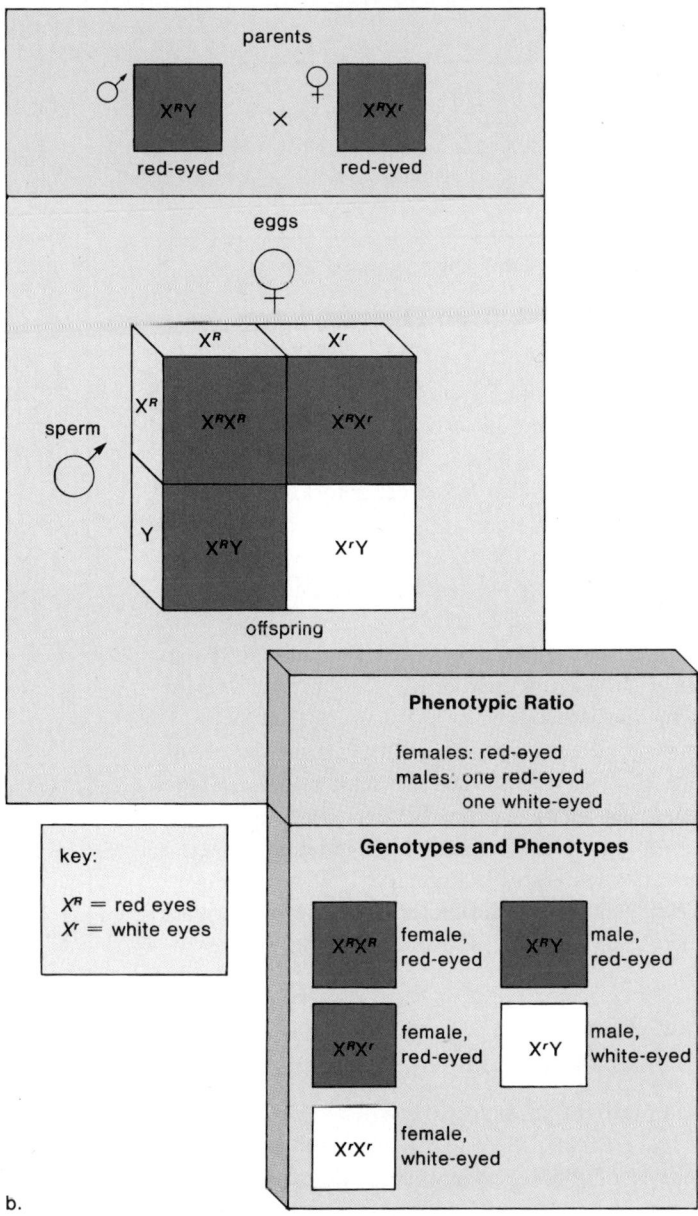

b.

As noted in figure 13.9, however, the key for an X-linked gene shows the allele attached to the X:

Key:

X^R = red eyes
X^r = white eyes

Notice, too, that there are 3 possible genotypes for females but only 2 for males. Females can be heterozygous with the genotype X^RX^r, in which case they are carriers. Carriers usually do not show an abnormality, but they are capable of passing on an allele for an abnormality. Males cannot be carriers; if the normal allele is on the single X chromosome, they appear normal, and if the mutated allele is on the single X chromosome, they show the abnormality.

Notice in figure 13.9 that white-eyed male offspring received the recessive allele from the female parent. The contribution of the male parent was a Y, which does not have an allele for eye color. Some other ways in which it is possible to recognize X-linked recessive traits are as follows:

1. More males than females are afflicted.
2. For a female to have the characteristic, the male parent must also have it. The female parent must have it or be a carrier.
3. The characteristic often skips a generation from the F_1 to the F_2 (see also fig. 14.14).
4. If a female has the characteristic, all of the male offspring will have it.

Practice Problems 2*

1. Using the key for *Drosophila* eye color shown in figure 13.9, give the 3 possible genotypes for females and the 2 possible genotypes for males. What are the possible gametes for these males?
2. Which *Drosophila* cross would produce white-eyed males? in what ratio?
   a. $X^RX^R \times X^rY$
   b. $X^RX^r \times X^RY$
3. A woman is color blind (recessive). What are the chances of her sons being color blind? If she reproduces with a man with normal vision, what are the chances of her daughters being color blind? being carriers?

Practice Problems 2*
Continued

4. Both parents are right-handed (autosomal: *R* = right-handed, *r* = left-handed) and have normal vision. Their son is left-handed and color blind. Give the genotype of each person.

*Answers to Practice Problems appear in appendix D.

Genetic Linkage and Chromosome Mapping

Drosophila probably has thousands of different genes controlling all aspects of its structure, biochemistry, and behavior. Yet it has only 4 chromosomes. This paradox led investigators like Sutton to believe that each chromosome must carry a large number of genes. For example, it is now known that genes controlling eye color, wing type, body color, leg length, and antennae type are all located on chromosome 2 (fig. 13.10). These genes are said to form a **linkage group** because they are found on the same chromosome.

It is easy to predict that crosses involving linked genes will not give the same results as those involving unlinked genes. For example, suppose you are doing a cross between a gray-bodied, red-eyed heterozygote and a black-bodied fly with purple eyes. These genes are on chromosome 2; therefore, they are linked (fig. 13.10). Using the following,

Key:

G = gray body   *R* = red eyes[1]
g = black (ebony) body   *r* = purple eyes

you predict that the results will be 1:1 instead of 1:1:1:1 as in the case for unlinked genes. The reason, of course, is that linked alleles tend to stay together and do not segregate independently as predicted by Mendel's laws because they are on the same chromosome. Under these circumstances, the heterozygote forms only 2 types of gametes and produces offspring with only 2 phenotypes (fig. 13.11*a*).

When you do the cross, however, you find that a very small number of offspring show recombinant phenotypes (i.e., those that are different from the original parents). Specifically, you find that 47% of the offspring have gray bodies and red eyes, 47% have black bodies and purple eyes, 3% have gray bodies and purple eyes, and 3% have black bodies and red eyes (fig. 13.11*b*). What happened?

[1]*Drosophila* has several genes for eye color. Red/white is on the X chromosome; red/purple is on chromosome 2.

Figure 13.10

Each homologous pair of chromosomes carries a number of genes. The alleles on a chromosome form a linkage group because they tend to go together into the gametes. This simplified chromosome map shows the relative positions of some of the genes on *Drosophila* chromosome 2. The distances between the genes (the numbers = map units) are proportional to the percentage of crossing-over events that occurs between the various alleles. For example, the crossing-over frequency between gray body and long wings would be 48.5 - 31.0 = 17.5. This means that 17.5% of all gametes would carry a recombinant chromosome. Notice that the chromosome theory of inheritance includes the concept that alleles are arranged linearly along a chromosome at specific loci.

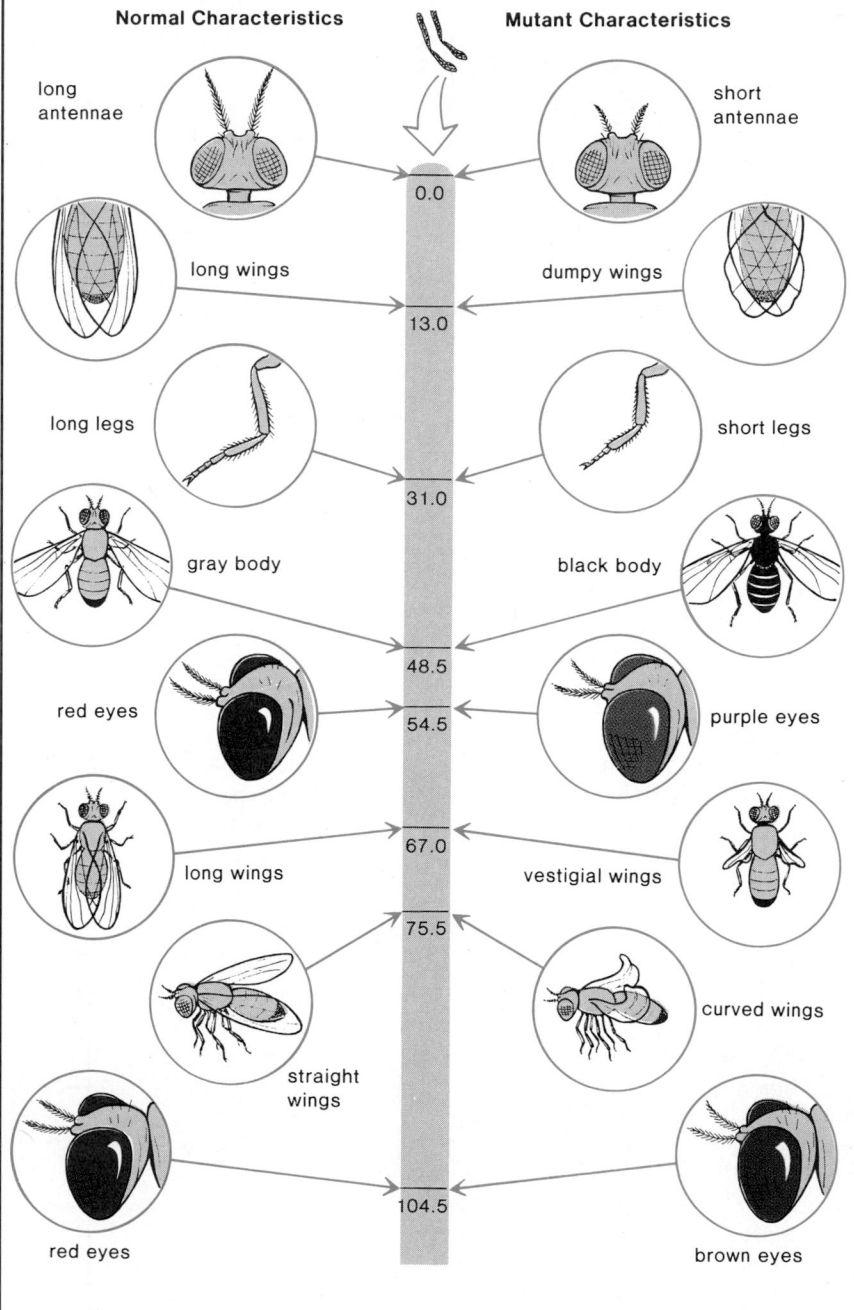

Genetic Basis of Life

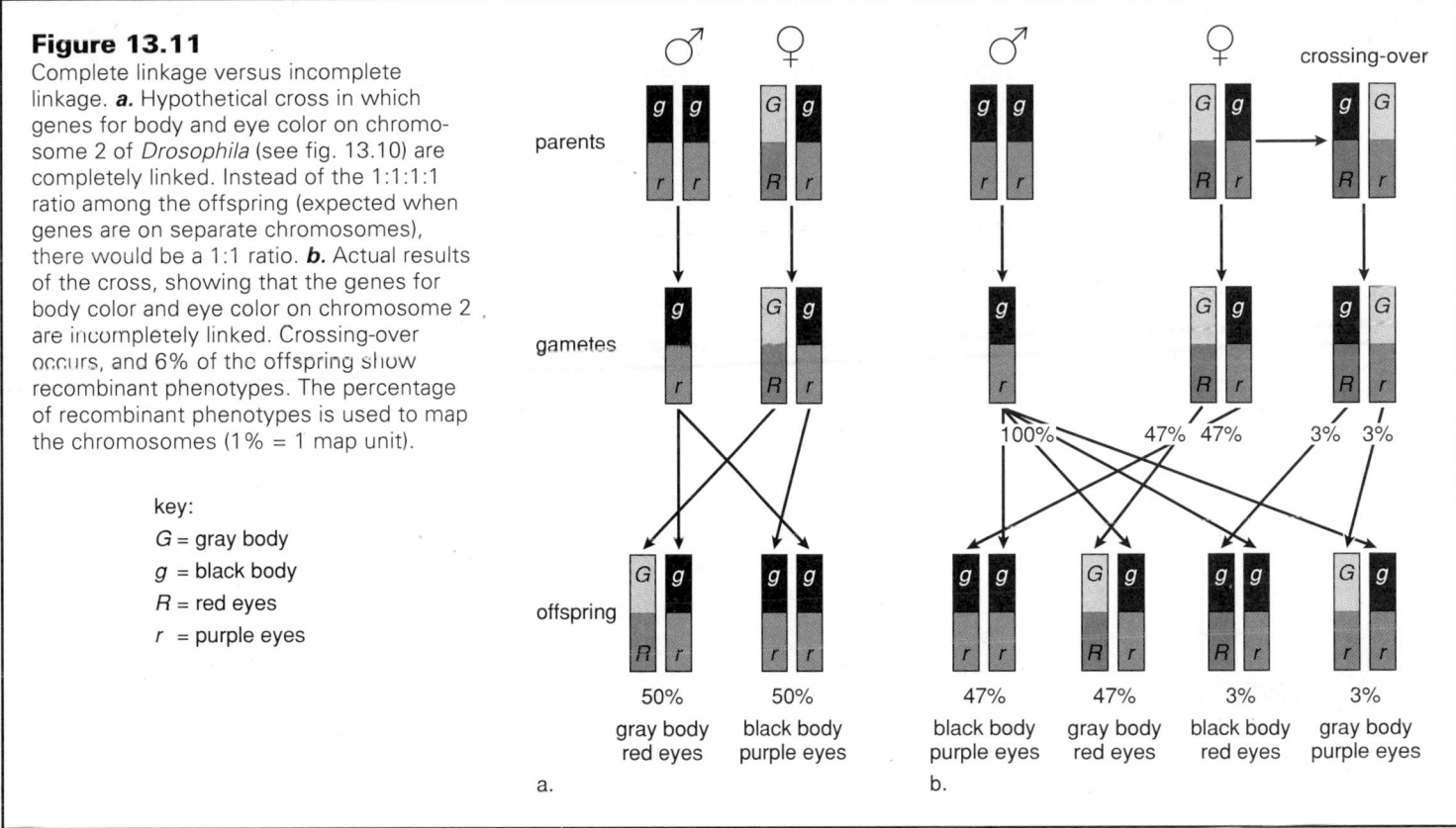

Figure 13.11
Complete linkage versus incomplete linkage. **a.** Hypothetical cross in which genes for body and eye color on chromosome 2 of *Drosophila* (see fig. 13.10) are completely linked. Instead of the 1:1:1:1 ratio among the offspring (expected when genes are on separate chromosomes), there would be a 1:1 ratio. **b.** Actual results of the cross, showing that the genes for body color and eye color on chromosome 2 are incompletely linked. Crossing-over occurs, and 6% of the offspring show recombinant phenotypes. The percentage of recombinant phenotypes is used to map the chromosomes (1% = 1 map unit).

key:
G = gray body
g = black body
R = red eyes
r = purple eyes

Recall that crossing-over can occur between homologous chromosomes when they are paired during meiosis, and that crossing-over produces recombinant gametes (fig. 13.12). In this instance, crossing-over produced a very small number of *recombinant gametes* (fig. 13.11*b*) and *recombinant phenotypes*. Take a look at figure 13.10 again and notice that these 2 sets of alleles are very close together. Doesn't it stand to reason that the closer together 2 genes are, the less likely they are to cross over? This is exactly what various crosses have repeatedly shown.

> All the genes on one chromosome form a linkage group that tends to stay together, except when crossing-over occurs.

Chromosome Mapping

You can use the percentage of recombinant phenotypes to map the chromosomes, because there is a direct relationship between the frequency of crossing-over and the percentage of recombinant phenotypes. In our example, a total of 6% of the offspring are recombinants, and for the sake of mapping the chromosomes, it is assumed that 1% of crossing-over = 1 map unit. Therefore, the allele for black body and the allele for purple eyes are 6 map units apart.

Suppose you want to determine the order of any 3 genes on the chromosomes. To do so, you can perform crosses that tell you the map distance between all 3 pairs of alleles. Considering just the mutant alleles that follow, if you know that

1. the distance between the black-body and purple-eye alleles = 6 map units,
2. the distance between the purple-eye and vestigial-wing alleles = 12.5 units, and
3. the distance between the black-body and vestigial-wing alleles = 18.5 units, then the order of the alleles must be as shown here:

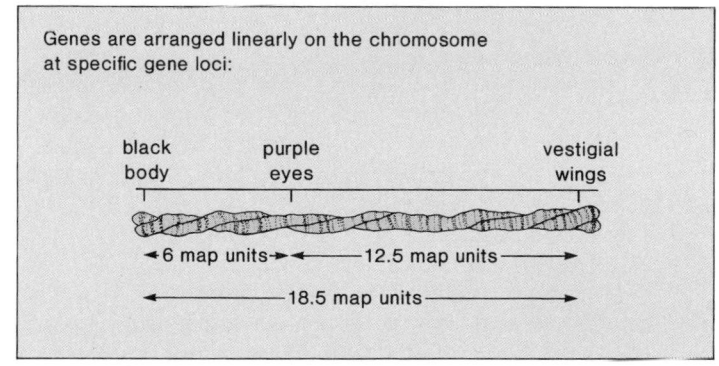

Genes are arranged linearly on the chromosome at specific gene loci:

black body — purple eyes — vestigial wings
◄6 map units► ◄——12.5 map units——►
◄———————18.5 map units———————►

Of the 3 possible orders (any one of the 3 genes can be in the middle), only this order gives the proper distance between all 3 allelic pairs. Because it is possible to map the chromosomes, we know that genes are indeed located on chromosomes, they have a definite location, and they occur in a definite order.

Various methods are used to map human chromosomes. These will be discussed in the next chapter.

Figure 13.12

Crossing-over causes an exchange of alleles between nonsister chromatids of a bivalent during meiosis I. Whenever crossing-over occurs, the gametes will contain recombinant chromosomes.

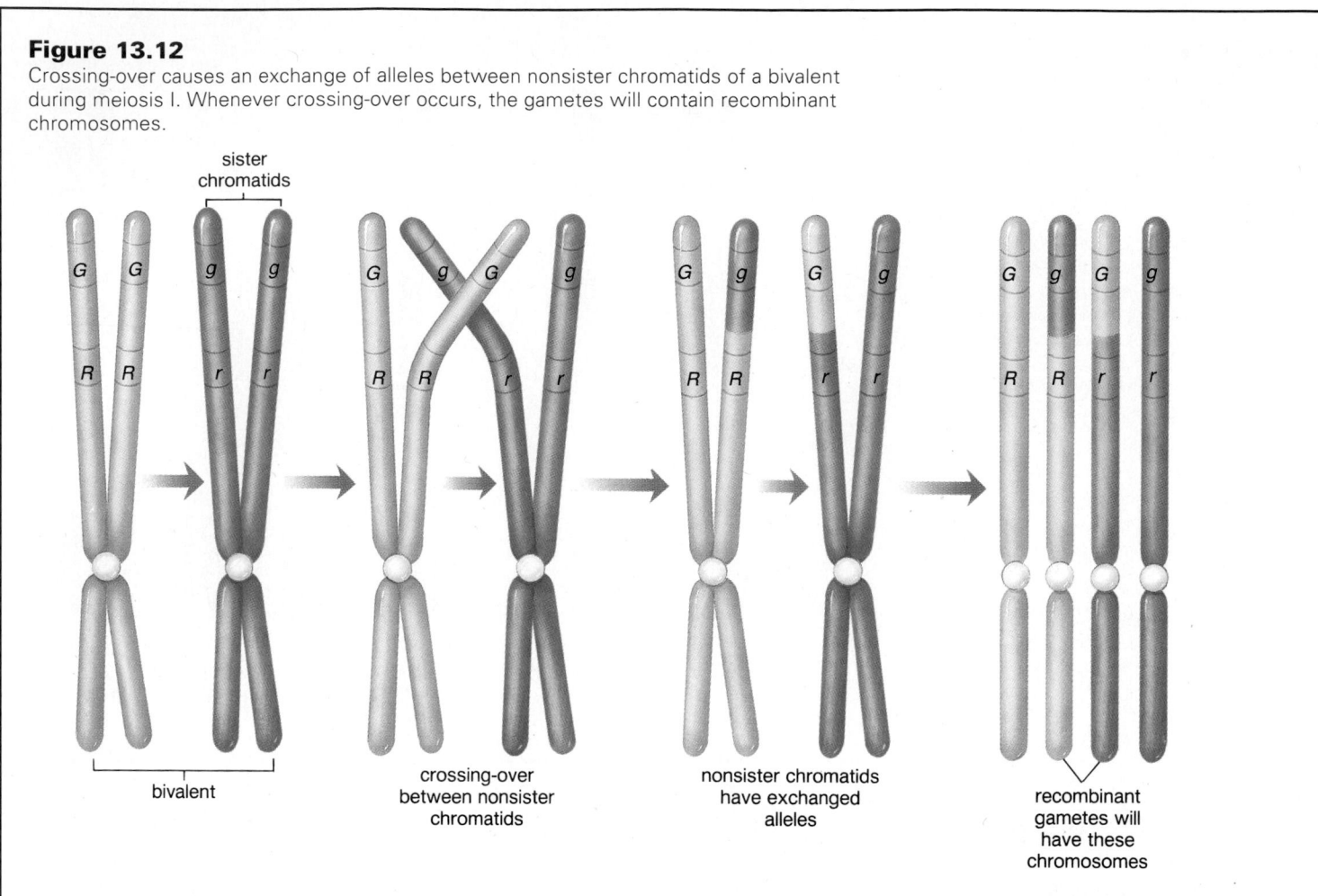

sister chromatids

bivalent

crossing-over between nonsister chromatids

nonsister chromatids have exchanged alleles

recombinant gametes will have these chromosomes

Practice Problems 3*

1. When *AaBb* individuals are allowed to self-breed, the phenotypic ratio is just about 3:1. What ratio was expected? What may have caused the observed ratio?

2. In a cross involving a heterozygote in which long wings and red eyes are on one member of a homologous pair of chromosomes and dumpy wings and purple eyes are on the other, what total percentage of recombinant phenotypes are expected according to figure 13.10?

3. In 2 sweet pea strains, *B* = blue flowers, *b* = red flowers, and *L* = long pollen grains, *l* = round pollen grains. In a cross between a heterozygous plant with blue flowers and long pollen grains and a plant with red flowers and round pollen grains, 44% of offspring are blue, long; 44% are red, round; 6% are blue, round; and 6% are red, long. How many map units separate these 2 sets of alleles?

4. Investigators performed crosses that indicated bar-eye and garnet-eye alleles are 13 map units apart, scalloped-wing and bar-eye alleles are 6 units apart, and garnet-eye and scalloped-wing alleles are 7 units apart. What is the order of these alleles on the chromosome?

*Answers to Practice Problems appear in appendix D.

The crossing-over frequency is proportional to the recombinant phenotypic frequency, which in crosses involving linked genes indicates the distance between genes on the chromosomes.

Chromosome Mutations

Mutations are permanent gene or chromosome changes that can be passed to offspring if they occur in cells that become gametes. Like crossing-over and recombination of chromosomes during meiosis and gamete fusion during fertilization, mutations increase the amount of variation among offspring. *Chromosome mutations* include changes in chromosome number and changes in chromosome structure.

Changes in Chromosome Number

Polyploidy

Some mutant eukaryotes have more than 2 sets of chromosomes. They are called **polyploids** (*poly*—many; *ploid*—sets). Polyploid organisms are named according to the number of sets of chromosomes they have. Triploids (3N) have 3 of each kind of chromosome, tetraploids (4N) have 4 sets, pentaploids (5N) have 5 sets, and so on.

Figure 13.13

Many food sources are specially developed polyploid plants. Among these are seedless (**a**) watermelons and (**b**) bananas. They are infertile (seeds poorly developed) triploids, but they can be propagated by asexual means. Polyploidy makes plants and their fruits larger, such as these (**c**) jumbo McIntosh apples.

a.

b.

c.

Polyploidy is a major evolutionary mechanism in plants, and it is estimated that 47% of all flowering plants are polyploids. Among these are many of our most important crops, such as wheat, corn, cotton, sugarcane, and also fruits such as watermelons, bananas, and apples (fig. 13.13). Also many attractive flowers, such as chrysanthemums and day lilies, are polyploids.

It appears from its common occurrence that polyploidy must be advantageous to plants. Hybrids, like polyploids, are often more vigorous than parental species; however, unlike polyploids, hybrids are infertile because the odd number of chromosomes do not pair successfully during meiosis. Doubling the chromosome number following hybridization allows a hybrid to be fertile:

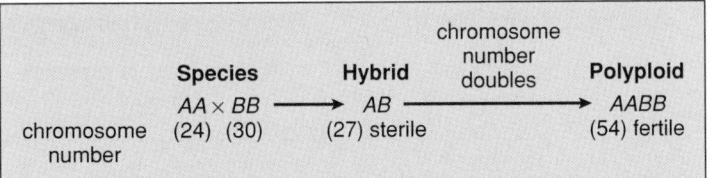

Hybridization followed by doubling of the chromosome number contributes to plant evolution by allowing new species to arise. This mode of speciation generally does not work for animals because they have sex chromosomes. Not only do abnormal sex chromosome numbers cause anatomical and physiological difficulties, but XXYY males would always have sons.

Polyploids (having more than one set of chromosomes) are common in plants. Hybridization followed by doubling the chromosome number is believed to have contributed to speciation in plants.

Monosomy and Trisomy

Monosomy occurs when an individual has only one of a particular type of chromosome (2N - 1), and *trisomy* occurs when an individual has more than 2 of a particular type of chromosome (2N + 1). The usual cause of monosomy and trisomy is nondisjunction during meiosis (fig. 13.14). Nondisjunction can occur when either the members of a homologous pair or the daughter chromosomes fail to separate and instead go into the same daughter cell. Nondisjunction is more common during meiosis I than during meiosis II. Nondisjunction can also occur during mitosis.

Monosomy and trisomy occur in both plants and animals. In animals, autosomal monosomies and trisomies are generally lethal, although a trisomic individual is more likely to survive than a monosomic one. The survivors are characterized by a distinctive set of physical and mental abnormalities called a syndrome. The most common autosomal trisomy among humans is trisomy 21 (Down syndrome), which is discussed in chapter 14.

Monosomy (2N - 1) and trisomy (2N + 1) sometimes occur. These changes in chromosome number generally cause abnormalities, particularly in animals.

Changes in Chromosome Structure

There are various agents in the environment, such as radiation, certain organic chemicals, or even viruses, that can cause chromosomes to break. Ordinarily when breaks occur in chromosomes, the 2 broken ends reunite just as they were before. Sometimes, however, the broken ends of one or more chromosomes do not rejoin in

Figure 13.14

Nondisjunction can occur during meiosis I if the members of a homologous chromosome pair fail to separate, or during meiosis II if the daughter chromosomes fail to separate completely. The resulting zygotes following fertilization with a normal gamete are either N/ms/1 (monosomic, a lethal condition) or N+1 (trisomic, sometimes lethal).

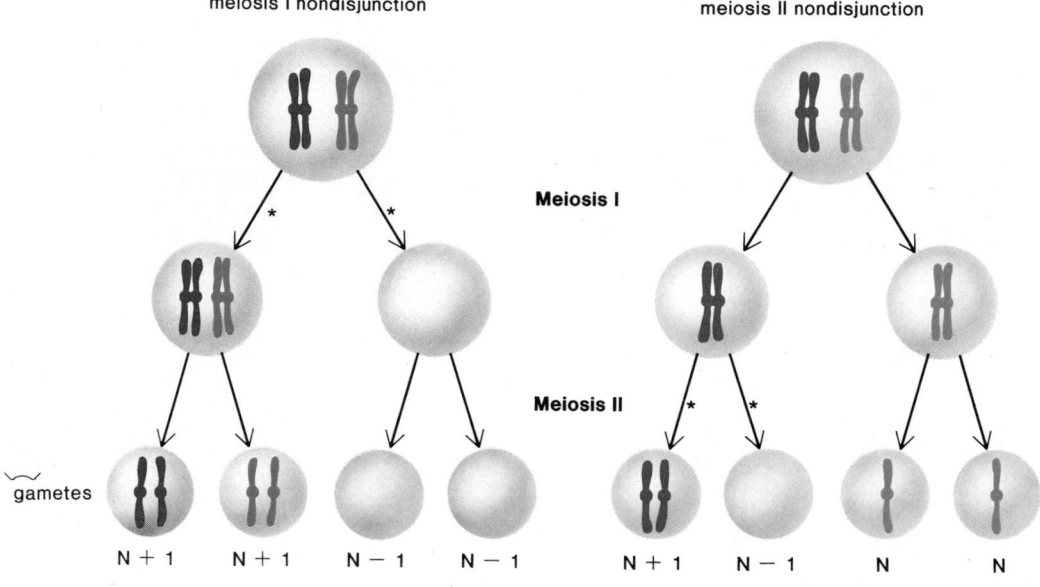

the same pattern as before, and this results in a change in chromosome structure. The existence of such mutations can be readily detected in chromosomes taken from larval salivary gland cells of *Drosophila* (fig. 13.15). In these cells, the chromosomes duplicate repeatedly, but instead of separating, they remain side by side to produce what are called giant polytene chromosomes. The large size of these chromosomes makes it possible to see some of the types of chromosome mutations we will be discussing.

An *inversion* occurs when a segment of a chromosome is turned around 180° (fig. 13.16). You might think this is not a problem because the same genes are present, but the new position might lead to altered gene activity. Also, crossing-over might be prevented because the chromosome pair cannot align properly during synapsis; if crossing-over does occur, duplications and deletions might result.

A *translocation* is the movement of a chromosomal segment from one chromosome to another, nonhomologous chromosome. Translocation heterozygotes usually have reduced fertility, again due to production of abnormal gametes. A translocation can also cause a Down syndrome child.

A *deletion* occurs when an end of a chromosome breaks off or when 2 simultaneous breaks lead to the loss of a segment. Deletions are usually lethal when homozygous (both members of a pair of chromosomes affected), although exceptions have been detected in corn and other organisms. Even when heterozygous (one member of a pair of chromosomes affected), a deletion often causes abnormalities. An example of a heterozygous deletion in humans is cri du chat (cat's cry) syndrome. The affected individual has a small head, is mentally retarded, and has facial abnormalities.

Abnormal development of the glottis and larynx results in the most characteristic symptom—the infant's cry resembles that of a cat.

A *duplication* is the presence of a chromosome segment more than once in the same chromosome. There are several ways a duplication can occur. A broken segment from one chromosome can simply attach to its homologue, but the presence of a duplication in the middle of the chromosome is more likely due to unequal crossing-over. This can occur when homologues are slightly mispaired and simultaneously produce a gene duplication and a deletion:

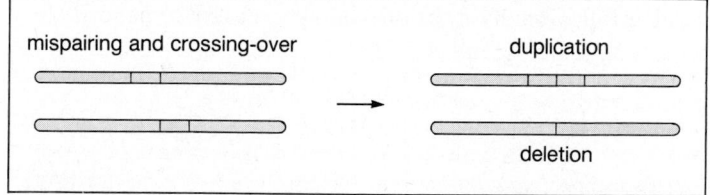

Multiple copies of genes that occur because of duplication can now mutate differently and thereby provide additional genetic variation for the species. For example, there are several closely linked genes for human globin (globin is a part of hemoglobin, which is present in red blood cells and carries oxygen), which may have arisen by a process of duplication followed by different mutations. The evidence for this is the great similarity in the amino acid sequences of the different globins.

Chromosome mutations can be the result of structural changes. Although most of these lead to abnormal gametes and offspring, some no doubt have played a role in evolutionary history.

Figure 13.15

Giant polytene chromosomes from salivary gland cells of *Drosophila* are composed of many hundreds of chromatids that lie side by side. Each of these chromosomes has a characteristic banding pattern; the darker bands are places where DNA is more condensed. Sometimes it is possible to make out chromosomal mutations in these chromosomes; for example, a duplication leads to the increased size of a particular band. Magnification, X240.

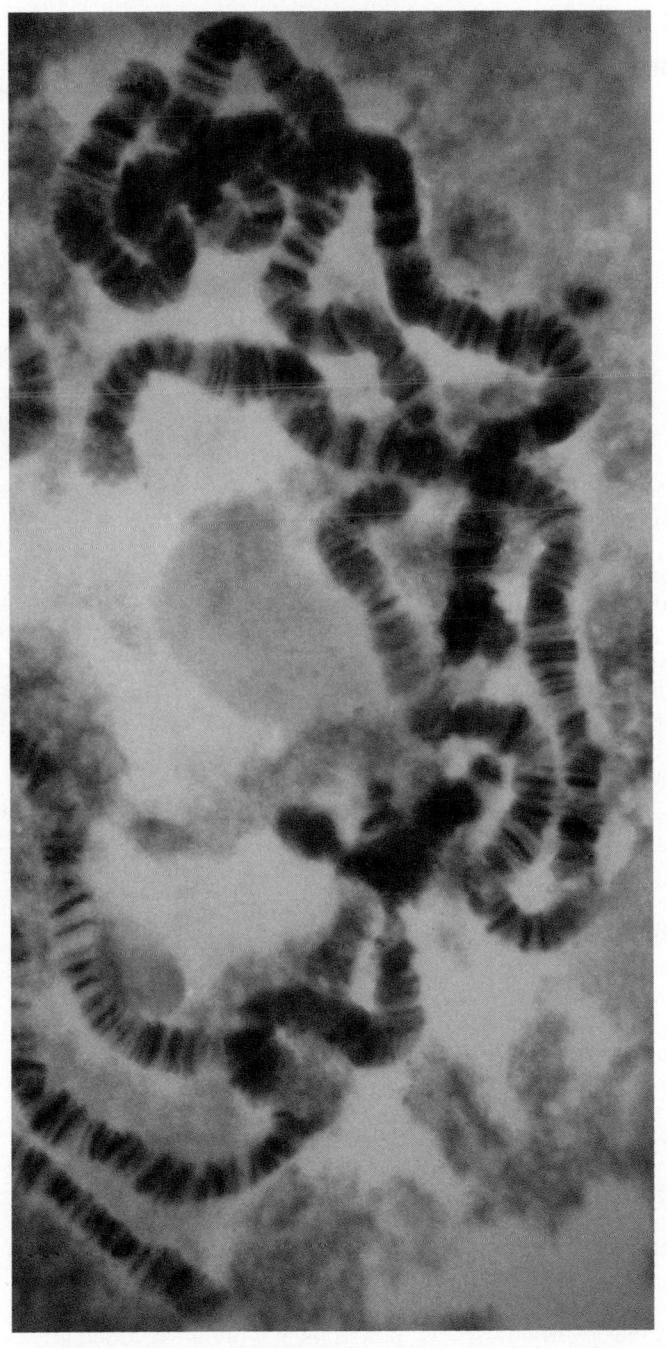

Figure 13.16

Types of chromosome mutations. *a.* Inversion is when a piece of chromosome breaks loose and then rejoins in the reversed direction. *b.* Translocation is the exchange of chromosome pieces between nonhomologous pairs. *c.* Deletion is the loss of a chromosome piece. *d.* Duplication occurs when the same piece is repeated within the chromosome.

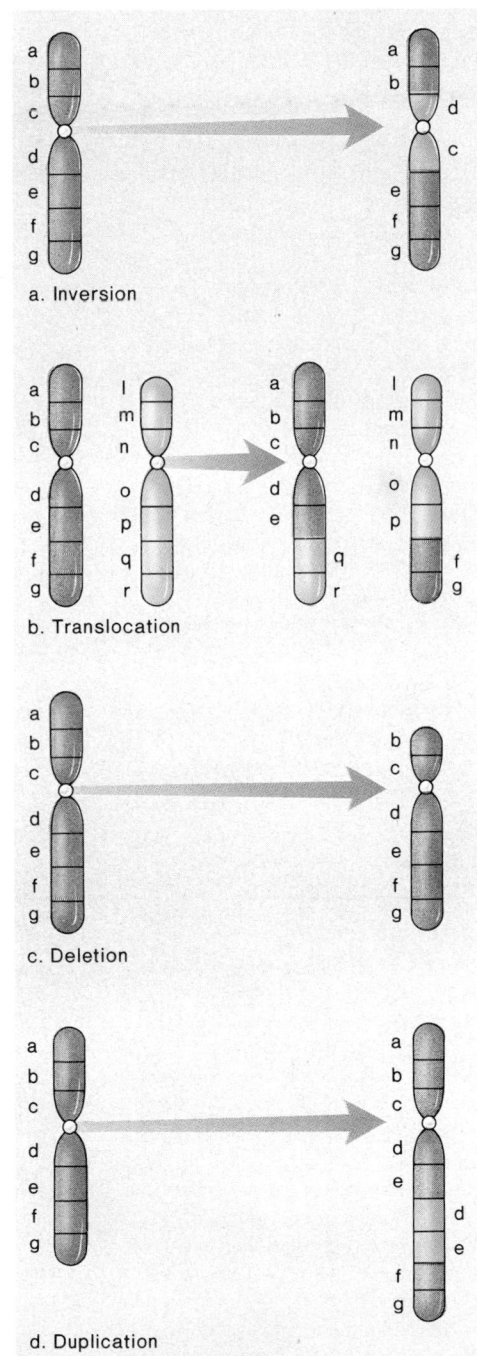

a. Inversion

b. Translocation

c. Deletion

d. Duplication

Summary

1. Different degrees of dominance have been observed. With incomplete dominance, the F_1 individuals are intermediate between the parental types; this does not support the blending theory because the parental phenotypes reappear in F_2. With codominance, the F_1 individuals express both alleles.

2. On the genotype level, the action of a dominant allele can make up for a noneffective recessive allele (simple dominance) or can cause an effect that drastically alters the phenotype (pleiotropy).

3. Genes interact. The homozygous recessive condition can mask the effect of a dominant allele at another locus (epistasis). Modifiers can alter the effect of a dominant allele (dominance modification). The latter may explain parental imprinting of genes.

4. Some genes have multiple alleles, although each individual organism has only 2 alleles. Coat color in rabbits is controlled by multiple alleles, and 4 different type coats are possible.

5. Some traits are controlled by many genes and are therefore called polygenic. Each allele adds quantitatively to the phenotype. Therefore, there are many intermediate F_2 phenotypes. The distribution of these continuous variations follows a bell-shaped curve.

6. The relative influence of the genotype and the environment on the phenotype can vary. There are examples of extreme environmental influence.

7. The chromosome theory of inheritance says that the genes are located on the chromosomes, which accounts for the similarity of their behavior during meiosis and fertilization.

8. Sex determination in animals is dependent upon the chromosomes. Usually females are XX and males are XY.

9. Solid experimental support for the chromosome theory of inheritance came when Morgan and his group were able to determine that the white-eye allele in *Drosophila* is on the X chromosome.

10. Genes on the X chromosome are called X-linked genes. Such genes are not on the Y chromosomes. Therefore, when doing X-linked genetics problems, it is the custom to indicate the sexes by using sex chromosomes and to indicate the alleles by superscripts attached to the X. The Y is blank because it does not carry these genes.

11. All the genes on a chromosome form a linkage group. Linked genes do not obey Mendel's laws because they tend to go into the gametes together. Crossing-over can cause recombinant gametes and recombinant phenotypes to occur. The percentage of recombi-nant phenotypes is used to measure the distance between gene pairs and to map the chromosomes.

12. Chromosome mutations fall into 2 categories: changes in chromosome number and changes in chromosome structure.

13. Polyploidy occurs when the eukaryotic individual has more than 2 complete sets of chromosomes. Monosomy is a deficiency in an individual chromo-some ($2N - 1$); trisomy is an excess in an individual chromosome ($2N + 1$).

14. Changes in chromosome structure include inversions, translocations, deletions, and duplications. Although these generally disrupt synapsis and segregation, there are instances in which they seem to have played a positive role in the evolutionary history of organisms.

Writing Across the Curriculum

In order to practice writing skills, students should write out the answers to any or all of the study questions and the critical thinking questions. The study questions are sequenced in the same order as the text. Suggested answers to the critical thinking questions are in appendix D.

Study Questions

1. Explain inheritance by multiple alleles. List the human blood types and give the possible genotypes for each.

2. Compare the F_2 phenotypic ratio for a cross involving a completely dominant allele, an incompletely dominant allele, and codominant alleles.

3. Show that if a dihybrid cross between 2 heterozygotes involves an epistatic gene, you do not get a ratio of 9:3:3:1.

4. Explain why traits controlled by polygenes show continuous variation that can be measured quantitatively, have many intermediate forms, and give a distribution in the F_2 generation that follows a bell-shaped curve.

5. If you were studying a particular population of organisms, how would you recognize that several traits are being affected by a pleiotropic gene?

6. Give examples to show that the environment can have an extreme effect on the phenotype despite the genotype.

7. What is the chromosome theory of inheritance? List the ways in which genes and chromosomes behave similarly during meiosis and fertilization.

8. How is sex determined in humans? Which sex determines the sex of the offspring?

9. How did a *Drosophila* cross involving an X-linked gene help investigators show that certain genes are carried on certain chromosomes?

10. Show that a dihybrid cross between a heterozygote and a recessive homozygote involving linked genes does not give the expected 1:1:1:1 ratio. What is the significance of the small percentage of recombinants that occurs among the offspring?

11. What are the 2 types of chromosome mutations? What 2 types of changes in chromosome number were discussed? What 4 types of changes in chromosome structure were discussed?

12. In what way may polyploidy have assisted plant evolution? Why is polyploidy not generally found in animals? Why would you expect a monosomy to be more lethal than a trisomy?

Objective Questions

For questions 1-4, match the statements that follow to the items in the key:

Key:

   a. multiple alleles
   b. incomplete dominance
   c. polygenes
   d. epistatic gene
   e. pleiotropic gene

1. A cross between oblong and round squash produced oval squash.
2. Although most people have an IQ of about 100, IQ generally ranges from about 50 to 150.
3. Investigators noted that whenever a particular species of plant had narrow instead of broad leaves, it was also short and yellow, instead of tall and green.
4. In rabbits, the allele *C* is dominant to the allele *c^{ch}*, which is dominant to the allele *c^h*, which is dominant to the allele *c*. Each individual has only 2 of these alleles.
5. When a white-eyed *Drosophila* female occurs

   a. both parental flies could have red eyes.
   b. the female parent could have red eyes but the male parent has to have white eyes.
   c. both the male and female parent must have white eyes.
   d. Both a and b.

6. Investigators found that a cross involving the mutant genes *a* and *b* produced 30% recombinants, a cross involving *a* and *c* produced 5% recombinants, a cross involving *c* and *b* produced 25% recombinants. Which is the correct order of the genes?

   a. *a, b, c*
   b. *a, c, b*
   c. *b, a, c*

7. A boy is color blind (sex-linked recessive) and has a continuous hairline (autosomal recessive). Which could be the genotype of his mother?

   a. *bbww*
   b. *X^bYWw*
   c. *bbX^wX^w*
   d. *X^BX^bWw*

8. Which 2 of these chromosome mutations are most likely to occur when homologous chromosomes are undergoing synapsis?

   a. inversion and translocation
   b. deletion and duplication
   c. deletion and inversion
   d. duplication and translocation

9. Investigators do a dihybrid cross between 2 heterozygotes and get about a 3:1 ratio among the offspring. The reason must be due to

   a. polygenes.
   b. pleiotropic genes.
   c. linked genes.
   d. epistatic genes.

10. In snakes, the recessive genotype *cc* causes the animal to be albino despite the inheritance of the dominant allele (*B* = black). What would be the results of a cross between 2 snakes with the genotype *CcBB?*

   a. all black snakes
   b. all white snakes
   c. 9:3:3:1
   d. 3 black:1 albino

Additional Genetics Problems

1. Chickens that are homozygous for the frizzled trait have feathers that are weak and stringy. When raised at low temperatures, the birds have circulatory, digestive, and hormonal problems. What inheritance pattern explains these results?
2. Chickens having black feathers are crossed with chickens having white feathers, and the result is chickens that appear to have blue feathers. What inheritance pattern explains these results, and what key do you suggest for this cross?
3. In radish plants, the shape of the radish may be long (*LL*), round (*ll*), or oval (*Ll*). If oval is crossed with oval, what proportion of offspring will also be oval?
4. If blood type in cats were controlled by 3 codominant and multiple alleles, how many genotypes and phenotypes would occur? Could the cross AC X BC produce a cat with AB blood type?
5. In *Drosophila, S* = normal, *s* = sable body; *W* = normal, *w* = miniature wing. In a cross between a heterozygous normal fly and a sable-bodied, miniature-winged fly the results were 99 normal flies, 99 with a sable body and miniature wings, 11 with a normal body and miniature wings, and 11 with a sable body and normal wings. What inheritance pattern explains these results? How many map units separate the genes for sable body and miniature wing?
6. In *Drosophila,* the gene that controls red eye color (dominant) versus white eye color is on the X chromosome. What are the expected phenotypic

results if a heterozygous female is crossed with a white-eyed male?
7. Bar-eye in *Drosophila* is dominant and sex-linked. What phenotypic ratio is expected for reciprocal crosses between pure-breeding flies?
8. In *Drosophila*, a male with bar-eyes and miniature wings is crossed with a female who has normal eyes and normal wings. Half the female offspring have bar-eyes and miniature wings and half have bar-eyes and normal wings. What is the genotype of the female parent?
9. In cats, *S* = short hair, *s* = long hair; *X^c* = yellow coat, *X^C* = black coat, *X^CX^c* = tortoiseshell (calico) coat. If a long-haired yellow male is crossed with a tortoiseshell female homozygous for short hair, what are the expected phenotypic results?

Concepts and Critical Thinking

1. *Genes often work together to produce the phenotype.*

   To demonstrate that the phenotype is controlled by the entire genome, give examples of human traits controlled by more than one pair of alleles.

2. *Genes are located on the chromosomes.*

   Give evidence that the genes are on the chromosomes and that they are linearly arranged on the chromosomes.

3. *Genes on sex chromosomes determine the sex of the individual.*

   What evidence suggests that alleles on the Y chromosome determine maleness in humans? Considering the degrees of dominance discussed in the chapter, how do you think these alleles might relate to sex-determining alleles on the X chromosome?

Selected Key Terms

incomplete dominance (in″kom-plēt′ dom′ĭ-nans) 197
codominance (ko-dom″ĭ-nans) 197
multiple allele (mul′ti-p′l ah-lēl) 200
polygenic inheritance (pol″ĕ-jēn′ik in-her′ĭ-tans) 200

chromosome theory of inheritance (kro′mo-sōm the′o-re uv in-her′ĭ-tans) 202
autosome (aw′to-sōm) 204
sex chromosome (seks kro′mo-sōm) 204

sex-linked gene (seks linkt jēn) 204
X-linked gene (eks linkt jēn) 204
linkage group (lingk′ij grōōp) 206
mutation (mu-ta′shun) 208
polyploid (polyploidy) (pol′e-ploid) 208

14

Human Genetics

Genes and the traits they control are passed via chromosomes from parents to children. Perhaps one day it will be possible to treat or even cure many genetic disorders prenatally so that all parents can have children free of known genetic disorders.

Your study of this chapter will bc complete when you can

1. describe how a karyotype is prepared, what it consists of, and how it is used;
2. describe Down syndrome, its symptoms, its causes, and its relation to the age of the mother;
3. list and describe different types of chromosome abnormalities seen in humans;
4. recognize 3 patterns of inheritance (autosomal dominant, autosomal recessive, X-linked recessive) when viewing a human pedigree chart;
5. give examples and describe the most common genetic disorders in humans;
6. solve genetics problems concerning modes of inheritance in humans;
7. tell how it is possible to test for genetic disorders prior to the birth of a child.

he principles of genetics we have studied are not applicable just to peas and fruit flies. They are applicable to all organisms, including ourselves. Sometimes we tend to think that academic research is of little direct value, but in this instance we have a perfect example that in the end such research can improve our lives and the lives of our children.

In this chapter, we will consider the relevance of classical genetics to humans. The relevance of biochemical genetics to humans is considered in chapter 18.

Human Chromosomes

To view clearly and count the chromosomes, cells can be treated and photographed just prior to dividing, as described in figure 14.1 for white blood cells. Metaphase chromosomes can then be cut out of the photograph and arranged in pairs. The members of a pair not only have the same size and shape, they also have the same banding pattern. In the resulting display of paired chromosomes, called a **karyotype,** autosomal chromosomes are usually ordered by size and numbered from largest to smallest; the sex chromosomes are identified separately.

Karyotyping of an adult or child indicates whether the chromosome number and structure are normal. Karyotyping can also be done before birth to determine if the chromosome number and structure of a developing child is normal. The reading on page 226 discusses the procedures for retrieving embryonic and fetal cells so that they can be tested.

Autosomal Chromosome Abnormalities

Autosomal chromosome abnormalities in humans are either changes in number of individual chromosomes or changes in chromosome structure.

Down Syndrome

The most common autosomal chromosome number abnormality in humans is trisomy 21, also called **Down syndrome,** because it was first described by John L. Down in 1866 (table 14.1). This syndrome is generally characterized by mental retardation, a distinc-

Figure 14.1

Human karyotype preparation (simplified version). In order to get the banding shown here the chromosomes are prepared for microscopy by Giemsa staining. The bands help researchers identify and analyze the chromosomes. This is the karyotype of a male because X and Y chromosomes are present.

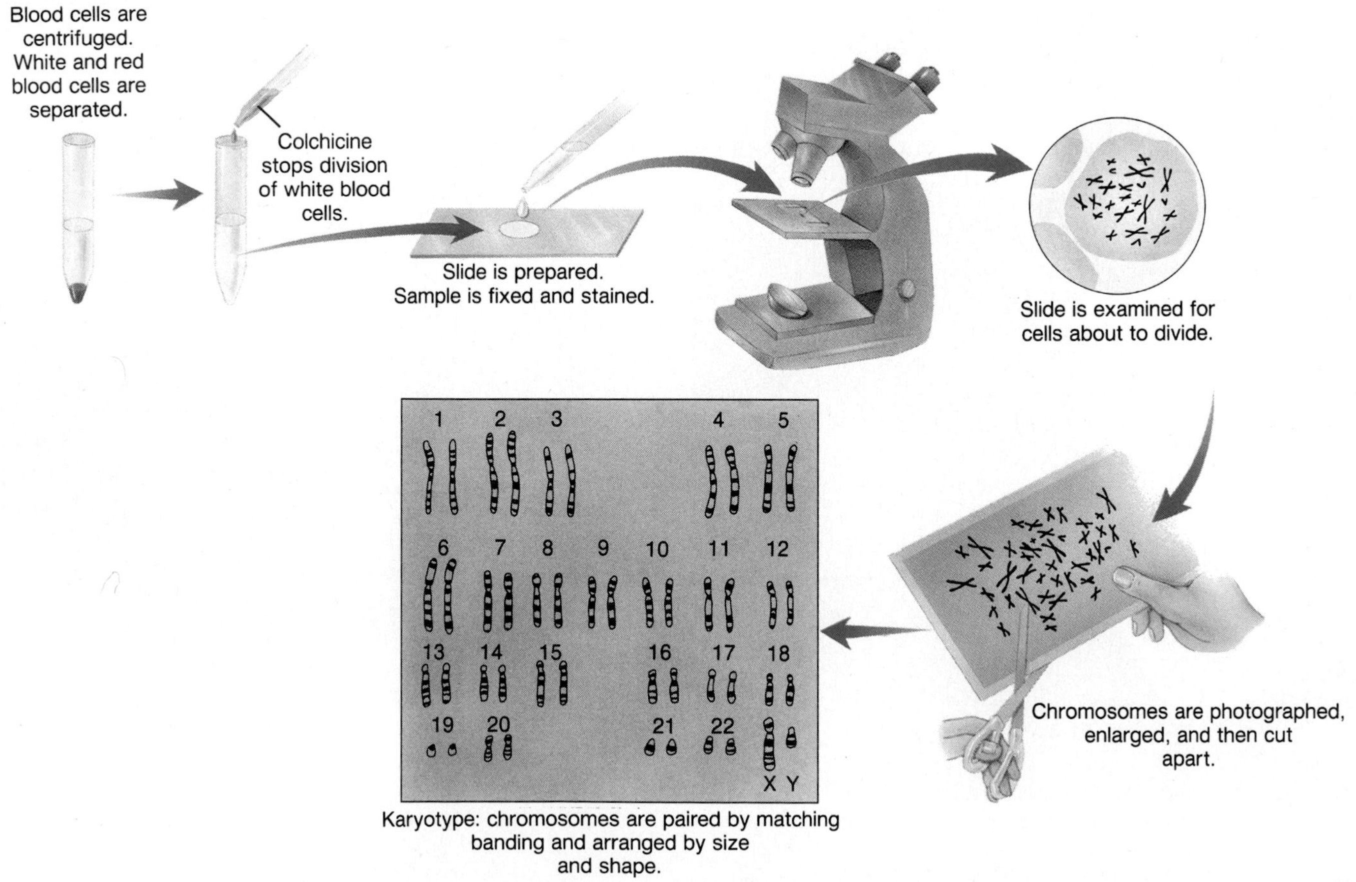

Blood cells are centrifuged. White and red blood cells are separated.

Colchicine stops division of white blood cells.

Slide is prepared. Sample is fixed and stained.

Slide is examined for cells about to divide.

Chromosomes are photographed, enlarged, and then cut apart.

Karyotype: chromosomes are paired by matching banding and arranged by size and shape.

Table 14.1
Frequency and Effects of the Most Common Chromosome Abnormalities in Humans

Syndrome	Sex	Chromosomes	Frequency		Expected Viability and Fertility
			Abortuses	**Births**	
Down	M or F	Trisomy 21	1/40	1/700	15 years, rarely reproductive
Patau	M or F	Trisomy 13	1/33	1/15,000	< 6 months
Edward	M or F	Trisomy 18	1/200	1/5,000	< 1 year
Turner	F	XO	1/18	1/5,000	Sterile
Metafemale	F	XXX (or XXXX)	0	1/700	Limited
Klinefelter	M	XXY (or XXXY)	0	1/2,000	Sterile
XYY male	M	XYY	?	1/2,000	—

From Robert F. Weaver and Philip W. Hedrick, *Genetics.* Copyright © 1989 Wm. C. Brown Publishers, Dubuque, Iowa. All Rights Reserved. Reprinted by permission.

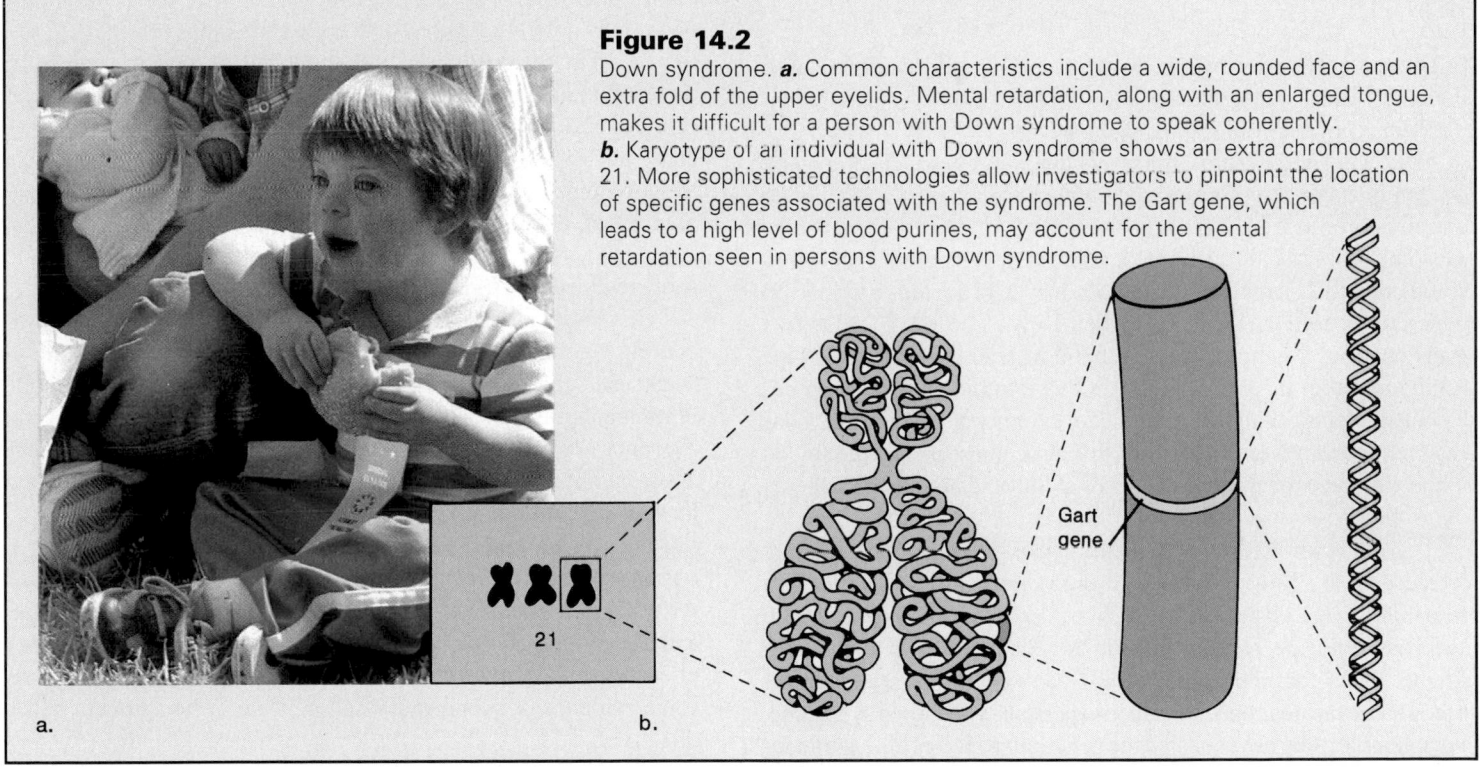

Figure 14.2
Down syndrome. *a.* Common characteristics include a wide, rounded face and an extra fold of the upper eyelids. Mental retardation, along with an enlarged tongue, makes it difficult for a person with Down syndrome to speak coherently.
b. Karyotype of an individual with Down syndrome shows an extra chromosome 21. More sophisticated technologies allow investigators to pinpoint the location of specific genes associated with the syndrome. The Gart gene, which leads to a high level of blood purines, may account for the mental retardation seen in persons with Down syndrome.

Gart gene

21

a.

b.

tive palm print, and a common facial appearance that includes an extra fold of the eyelids (fig. 14.2). Although most persons with Down syndrome live only until the middle teens, some live much longer and with a certain degree of independence. There is no cure possible for Down syndrome, but it is hoped that eventually there may be a treatment for some of its symptoms.

Persons with Down syndrome usually have 3 copies of chromosome 21 because nondisjunction (see fig. 13.14) resulted in a gamete with 2 copies instead of one. Chances of Down syndrome rapidly increase with parental age starting at about age 35 in women (fig. 14.3) and age 55 in men. A recent study has found that nondisjunction is more likely to occur during meiosis I rather than meiosis II, and the incidence of nondisjunction is the same for younger and older women. Previously, it was thought there might be a tendency toward nondisjunction as a woman ages because her eggs age as she does—a woman is born with all the eggs she will ever have and they remain in suspended animation until one matures each month. In keeping with their findings, the researchers suggest that older women

Human Genetics

Figure 14.3

Frequency of Down syndrome in children as it relates to the age of the mother.

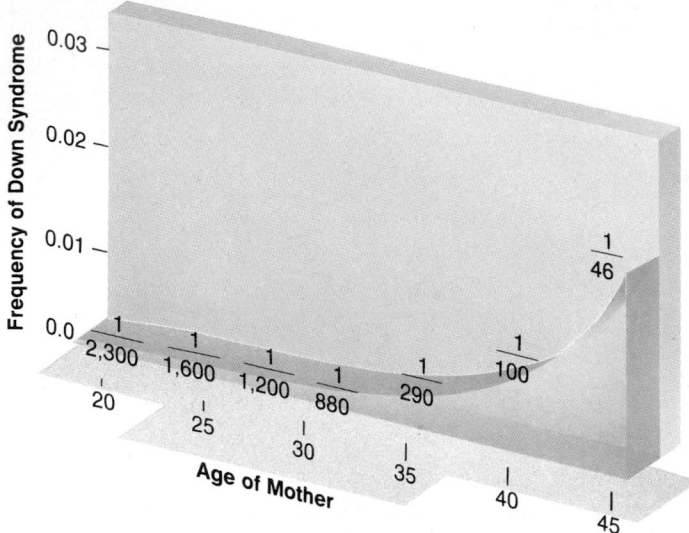

might be more likely to bring a Down syndrome child to term because their body fails to recognize and abort abnormal embryos. This explanation has not been substantiated, however.

Down syndrome is always due to an extra chromosome 21, but in about 5% of cases, the extra chromosome is attached to another chromosome, often chromosome 14. This abnormal chromosome arose because a translocation occurred between chromosomes 14 and 21 in one of the parents or even in a relative who lived generations earlier. Therefore, when Down syndrome is due to a translocation, it is not age related and instead tends to run in the family of either the father or the mother. Amniocentesis followed by chromosome analysis can indicate whether an unborn child has Down syndrome, and therefore this procedure is recommended whenever the expectant mother is older than 35 and/or the father is older than 55 or when the condition runs in the family.

It is known that the genes causing Down syndrome are located on the lower third of chromosome 21 (fig. 14.2b), and researchers have directed their efforts to the discovery of which specific genes are responsible for the characteristics of the syndrome. Thus far, they have discovered several genes that may account for various conditions seen in persons with Down syndrome. For example, they have located the genes most likely responsible for the increased tendency toward leukemia, cataracts, Alzheimer disease, and mental retardation. The gene thought to be responsible for mental retardation, dubbed the *Gart* gene, causes an increased level of purines in the blood. It is hoped that someday a way to control the expression of the Gart gene before birth will be found so that at least this symptom of Down syndrome will not appear.

Other Autosomal Abnormalities

Recall that a chromosome deletion is responsible for a syndrome known as cri du chat, in which the infant has a cry that sounds like a cat and shows other abnormalities, including mental retardation.

Chromosome analysis shows that a portion of one chromosome 5 is missing (deleted), although the other chromosome 5 is normal.

Two other trisomies, besides that of chromosome 21, are seen with some regularity in humans. Individuals with trisomy 13 (Patau syndrome) and trisomy 18 (Edward syndrome) have an average life span of less than one year. Heart and nervous system defects prevent normal development.

> There are several autosomal chromosome abnormalities in humans. Trisomy 21 (Down syndrome) is the most common of these.

Sex Chromosome Abnormalities

Sex chromosome abnormalities (table 14.1) are due to nondisjunction in the sex chromosomes. Nondisjunction of sex chromosomes during oogenesis can lead to an egg with either 2 X chromosomes or no X chromosome. Nondisjunction of the sex chromosomes during spermatogenesis can result in a sperm that has no sex chromosome, both an X and a Y chromosome, 2 X chromosomes, or 2 Y chromosomes. Assuming that the other gamete is normal, the zygote could develop into an individual with one of the conditions noted in figure 14.4.

Turner syndrome is a monosomy of the X chromosome; these individuals are designated as XO—the O signifies the absence of a second sex chromosome. Turner females are short, have a broad chest, and may have congenital heart defects. Because the ovaries never become functional, Turner females do not undergo puberty or menstruate, and there is a lack of breast development (fig. 14.5a). Although no overt mental retardation is reported, Turner females show reduced skills in interpreting spatial relationships.

When an egg with 2 X chromosomes is fertilized by an X-bearing sperm, a **metafemale** with 3 X chromosomes results. It might be supposed that the XXX female would be especially feminine, but this is not the case. Although there is in some cases a tendency toward learning disabilities, most metafemales have no apparent physical abnormalities except perhaps menstrual irregularities, including early onset of menopause.

When an egg with 2 X chromosomes is fertilized by a Y-bearing sperm, a male with **Klinefelter syndrome** (XXY) results. Affected individuals are males, but the testes are underdeveloped and there may be some breast development (fig. 14.5b). These phenotypic abnormalities are not apparent until puberty, although some evidence of subnormal intelligence may be apparent before this time.

XYY males may also result from nondisjunction during spermatogenesis. Affected males are usually taller than average, suffer from persistent acne, and tend to have barely normal intelligence. At one time, it was suggested that these men were likely to be criminally aggressive, but it has since been shown that the incidence of such behavior among them is no greater than among XY males.

> One sex chromosome monosomy (Turner syndrome) and 3 trisomies (Klinefelter syndrome, metafemale, and XYY male) are known in human beings.

Figure 14.4

Nondisjunction of sex chromosomes during oogenesis followed by fertilization with normal sperm results in the conditions noted (*left*). Nondisjunction of sex chromosomes during spermatogenesis followed by fertilization of normal eggs results in the conditions noted (*right*).

Oogenesis

XX

meiosis I
or
meiosis II

eggs   XX

X   Y   X

or

XO   XXY   XXX
Turner   Klinefelter   metafemale

Spermatogenesis

XY

meiosis I   meiosis II

nondisjunction

sperm

XY   XX   YY

X   X   X   X

XO   XXY   XXX   XYY
Turner   Klinefelter   metafemale   XYY male

Figure 14.5

a. A female with Turner syndrome (XO) is distinguished by a thick neck, short stature, and immature sexual features. *b.* A male with Klinefelter syndrome (XXY) has immature sex organs and shows breast development.

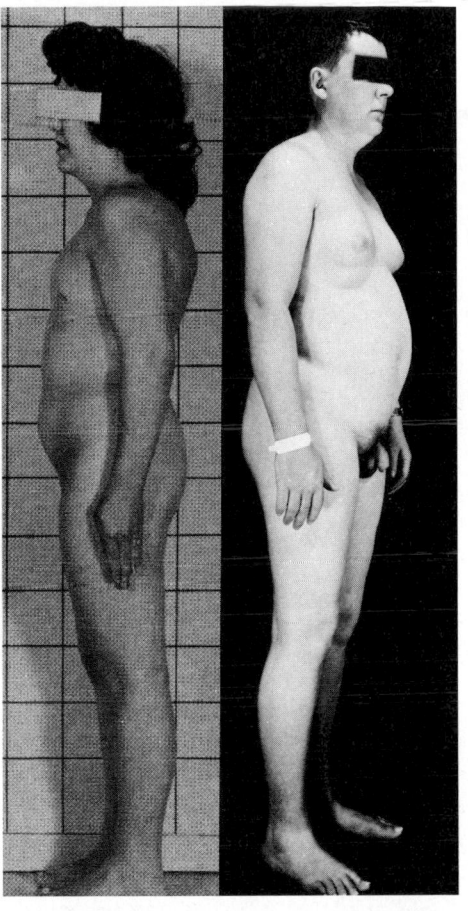

a.   b.

Examination of these abnormal sex chromosome patterns indicates how sex is determined in humans. Turner syndrome (XO) produces a person who is more female than male. Klinefelter syndrome (XXY) produces a person who is more male than female. Clearly, the Y chromosome is not neutral in humans. Rather, it bears genes that cause maleness, as discussed in the reading on page 220.

Human Genes

The patterns of inheritance in humans are the same as those we previously studied in chapters 12 and 13 and as we will describe in the rest of this chapter.

Pedigree Charts

Genetic counselors often construct pedigree charts to predict whether a couple is likely to pass on a genetic disorder. A pedigree chart can indicate whether the characteristic is dominant, recessive, or X-linked recessive. For example, examine these 2 possible patterns of inheritance:

pattern I   pattern II

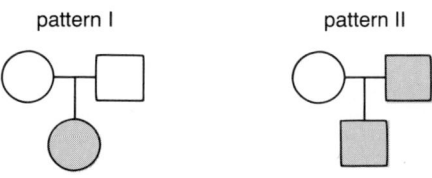

In both patterns, males are designated by squares and females by circles. Shaded circles and squares are affected individuals. A line between a square and a circle represents a union. A vertical line going downward leads, in these patterns, to a single child. (If there are more children they are placed off a horizontal line.) Which pattern of inheritance do you suppose represents an autosomal dominant characteristic and which represents an autosomal recessive characteristic?

In pattern I, the child is affected but neither parent is; this can happen if the characteristic is recessively inherited. What are the chances that any offspring from this union will be affected? Because the parents are heterozygous, the chances are 1 in 4, or 25% (see table 12.3). Notice that the parents are **carriers** because they have a normal phenotype, but they are capable of having a child with a genetic disorder. See figure 14.6 for other ways to recognize an autosomal recessive pattern of inheritance.

Active Y Chromosomes and Inactive X Chromosomes: The Functions of Sex Chromosomes in Body Cells

Although the Y chromosome in mammals has few gene loci, it has a powerful effect on sex determination. For years, it has been proposed that there must be a gene located on the Y chromosome that brings about maleness. Embryos begin life with no evidence of their sex, but about the third month of development, males can be distinguished from females. Researchers have recently found a gene called *Sry* gene (short for sex-determining region Y gene), which they believe is the best candidate for the gene that switches development to the male pathway. In other words, the embryo is basically female and will automatically develop into a female unless *Sry* gene promotes the development of testes instead of ovaries. When the researchers injected XX mice embryos with a small fragment of Y chromosome containing the *Sry* gene, they became males with testes and male behavior. In the past, investigators have found two other genes that they believed determined maleness—the H-Y+ gene and the testes-determining factor—but these genes are now known to fail this same test.

The H-Y+ allele directs the synthesis of a specific kind of protein molecule, called the H-Y antigen, that is present in the plasma membranes of virtually all cells of a male, but not in those of a female. It is called an antigen because females produce antibodies against it. To test for maleness, it is possible to suspend a sample of white blood cells in a solution that contains some of these antibodies. If the cells carry H-Y antigen, indicating the person is a male, the antibodies bind with them. The testes-determining factor probably plays a contributory role during development. When this gene is lacking from the Y chromosome, the individual is a female even though the chromosomal inheritance is XY.

Aside from identifying gene loci on the sex chromosomes that cause a person to be male or female, biologists have been interested in how males manage with only one X chromosome while females have 2 X chromosomes. That question turned out to have a rather unexpected answer.

Years ago, M. L. Barr observed a consistent difference between nondividing cells taken from female and male mammals, including humans. Females, but not males, have a small, darkly staining mass of condensed chromatin adhering to the inner edge of the nuclear envelope (fig. 14.A). This dark-staining spot is called a Barr body after its discoverer.

In 1961, Mary Lyon, a British geneticist, proposed that the Barr body is a condensed, inactive X chromosome. Her hypothesis has been tested and found to be valid. One of the X chromosomes is inactivated in the cells of female embryos, but which one of the 2 is determined by chance. About 50% of the cells have one X chromosome active, and 50% have the other X chromosome active. The female body, therefore, is a mosaic, with "patches" of genetically different cells. For example, human females who are heterozygous for an X-linked recessive form of ocular albinism have patches of pigmented and nonpigmented cells at the back of the eye. Women heterozygous for Duchenne muscular dystrophy have patches of normal muscle tissue and degenerative muscle tissue (the normal tissue increases in size and strength to make up for the defective tissue), and women who are heterozygous for hereditary absence of sweat glands have patches of skin lacking sweat glands. Inactivation of an X chromosome does not occur in female oocytes, most likely because both are needed for normal development of the immature egg.

The female calico cat (fig. 14.A) also provides phenotypic proof of the Lyon hypothesis. In these cats, an allele for black coat color is on one X chromosome and a corresponding allele for orange coat color is carried on the other X chromosome. The patches of black and orange in the coat can be related to whichever X chromosome is in the Barr bodies of the cells found in the patches.

The existence of Barr bodies also explains why metafemales and persons with Klinefelter syndrome don't show a greater degree of abnormality than they do. The extra X chromosomes are inactivated early in development and form Barr bodies.

Barr body

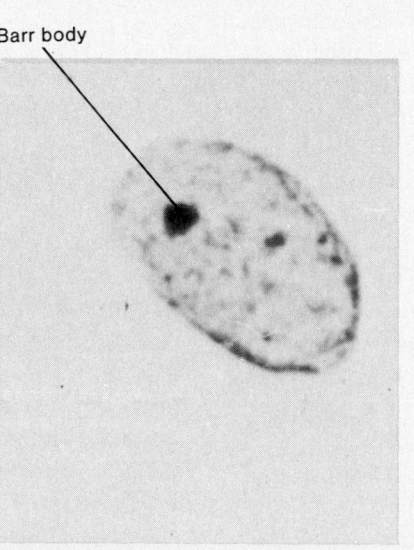

a.

b.

Figure 14.A

a. Electron micrograph illustrates that a Barr body is a dark-staining spot, which can be viewed microscopically. Magnification X2,500. *b.* The coat of a calico cat is orange and black. The alleles for coat color occur at a gene locus on the X chromosome. Presumably, the black patches occur where cells have Barr bodies carrying the allele for orange, and the orange patches occur where cells have Barr bodies carrying the allele for black coat. The white areas are due to a separate gene.

Genetic Basis of Life

Figure 14.6
Sample pedigree chart for an autosomal recessive genetic disorder. The listing tells how to recognize this type of inheritance.

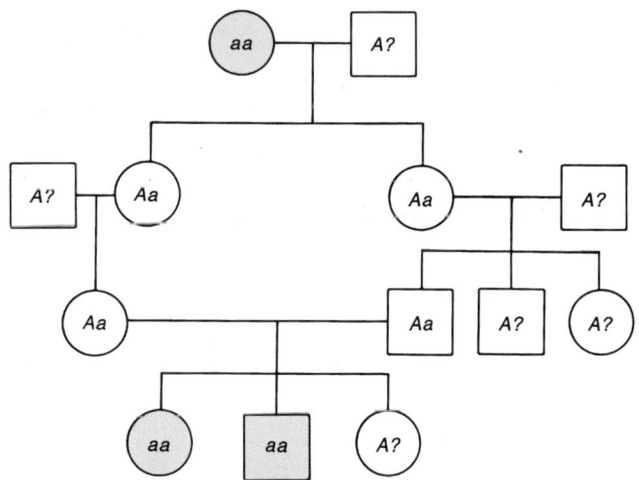

Autosomal Recessive Genetic Disorders
- Most affected children have normal parents.
- Heterozygotes have a normal phenotype.
- Two affected parents will always have affected children.
- Affected individuals who have noncarrier mates will have normal children.
- Close relatives who reproduce are more likely to have affected children.
- Both males and females are affected with equal frequency.

key:
aa = affected
Aa = carrier
      (appears normal)
AA = normal

Figure 14.7
Sample pedigree chart for an autosomal dominant genetic disorder. The listing tells how to recognize this type of inheritance.

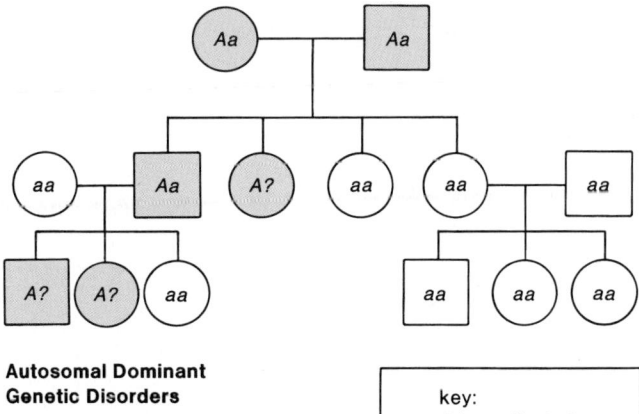

Autosomal Dominant Genetic Disorders
- Affected children usually have an affected parent.
- Heterozygotes are affected.
- Two affected parents can produce an unaffected child.
- Two unaffected parents will not have affected children.
- Both males and females are affected with equal frequency.

key:
AA = affected
Aa = affected
aa = normal

In pattern II, the child is affected, as is one of the parents. When a characteristic is dominant, an affected child usually has at least one affected parent. Of the 2 patterns, this one shows a dominant pattern of inheritance. What are the chances that any offspring from this union will be affected? Because this is a heterozygote and homozygous recessive cross, the chances are 50% (see table 12.3). Figure 14.7 gives other ways to recognize an autosomal dominant pattern of inheritance.

Autosomal Recessive Inheritance

There are many autosomal recessive disorders, and 3 of the better known will be discussed. Others that are well known are albinism, galactosemia (accumulation of galactose in the liver and mental retardation), thalassemia (production of abnormal amounts of hemoglobin), and xeroderma pigmentosum (lack of ability to repair ultraviolet-induced damage). The homozygous recessive phenotype is more likely to occur among the children of closely related individuals, which helps explain why these disorders are sometimes more prevalent among members of a particular ethnic group.

Cystic Fibrosis
Cystic fibrosis is the most common lethal genetic disease among Caucasians in the United States. About one in 25 Caucasians is a carrier and about one in 1,800 children born to this group has the disorder (table 14.2). In these children the mucus in the digestive tract and lungs is particularly thick and viscous. Thick mucus impedes the secretion of pancreatic juices and food cannot be properly digested; large, frequent, and foul-smelling stools occur. Breathing is difficult and lung infections are frequent. New treatments have raised the average life expectancy today to 17-20 years of age.

In the past few years, much progress has been made in our understanding of cystic fibrosis. The cystic fibrosis gene, which is located on chromosome 7, has been isolated and found to code for a protein involved in the passage of chloride (Cl⁻) ions through the membrane. It seems the protein is a chloride ion channel that regulates itself. The protein is called the cystic fibrosis transmembrane conductance regulator (CFTR). Ordinarily, after chloride ions have passed through the membrane, water follows. It is believed that lack of water in the lungs causes the mucus to be so thick.

Scientists quickly developed a commercially available test capable of detecting the mutation that causes cystic fibrosis. Then they found that only 70%-75% of those with cystic fibrosis have this particular mutation, and there are many other mutations that account for the other 25%-30% of defective CFTR genes. Because of this, the test is available only to those who request it. There are those who believe it might be simpler to test for the presence of CFTR in plasma membranes rather than to test chromosome 7 for the mutation.

The possibility of using gene therapy to cure cystic fibrosis seems more of a reality after scientists inserted a normal CFTR gene into diseased cells growing in culture. The treated cells did gain the ability to regulate passage of the chloride ion across the plasma membrane. There have been various proposals for delivering the gene to cells that line the airways. For example, an aerosol can be prepared that contains viruses stripped of their ability to

Table 14.2
Some Human Genetic Disorders

Name	Chromosome	Incidence among Newborns	Status
Autosomal Recessive Disorders			
Cystic fibrosis	7	One in 2,500 Caucasians	Allele located; chromosome test now available;* treatment and cure a possibility soon
Tay-Sachs disease	15	One in 3,600 eastern European Jews	Biochemical test now available*
Phenylketonuria (PKU)	12	One in 5,000 Caucasians	Biochemical test now available
Autosomal Dominant Disorders			
Neurofibromatosis (NF)	17	One in 3,000	Allele located; chromosome test now available*
Huntington disease	4	One in 20,000	Chromosome test now available*
Incomplete dominance			
Sickle-cell disease	11	One in 500 American blacks	Chromosome test now available*
X-linked recessive			
Hemophilia A	X	One in 1,500 live male births	Treatment available
Duchenne muscular dystrophy	X	One in 5,000 live male births	Allele located; biochemical test of muscle tissue available; treatment available soon

*Prenatal testing is done.

cause disease and engineered to contain the CFTR gene. When the aerosol is inhaled, it is hoped that the viruses will enter the airway cells, bringing with them the gene that can cure cystic fibrosis.

Tay-Sachs Disease

Tay-Sachs disease is the best-known genetic disease among U.S. Jewish people, most of whom are of central and eastern European descent. At first, it is not apparent that a baby has Tay-Sachs disease. Development begins to slow down between 4 and 8 months of age, however, as neurological impairment and psychomotor difficulties become apparent. Ophthalmologic examination reveals a characteristic red spot and a yellowish accumulation in the region of the retina called the fovea centralis. The child gradually becomes blind and helpless, develops uncontrollable seizures, and eventually becomes paralyzed. There is no treatment or cure for Tay-Sachs disease, and most affected individuals die by the age of 3 or 4.

So-called late-onset Tay-Sachs disease occurs in adults. The symptoms are progressive mental and motor deterioration, depression, schizophrenia, and premature death. The gene for late-onset Tay-Sachs disease has been sequenced, and this form of the disorder apparently is due to one changed pair of bases in the DNA.

Tay-Sachs disease results from a lack of the enzyme hexosaminidase A (Hex A) and the subsequent storage of its substrate, a glycosphingolipid, in lysosomes. More and more lysosomes build up in many body cells, but the primary sites of storage are the cells of the nervous system (fig. 14.8). This accounts for the onset and the progressive deterioration of psychomotor functions.

There is a biochemical test to detect carriers of Tay-Sachs disease. The test uses a sample of blood serum, white blood cells, or tears to determine whether Hex A activity is present. Affected individuals have no detectable Hex A activity. Carriers have about half the level of Hex A activity found in normal individuals. Prenatal diagnosis is also possible following either amniocentesis or chorionic villi sampling as described in the reading on page 226.

Phenylketonuria (PKU)

Phenylketonuria (PKU) is a familiar example of a genetic disorder caused by the lack of an enzyme. In this case, the homozygote is unable to metabolize the amino acid phenylalanine, and an abnormal breakdown product, phenylketone, accumulates in the urine. Today newborns are routinely tested for the absence of the necessary enzyme. Those who test positive for PKU are placed on a diet low in phenylalanine until the brain is fully developed. If treatment is not begun very soon after birth, severe mental retardation develops and the person may never walk or talk properly. Severe convulsions and seizures occur.

Genetic Basis of Life

Figure 14.8

Tay-Sachs disease. **a.** When Hex A is present, glycosphingolipids are broken down. **b.** When Hex A is absent, these lipids accumulate in lysosomes and lysosomes accumulate in the cell. **c.** Electron micrograph of cell crowded with lysosomes.

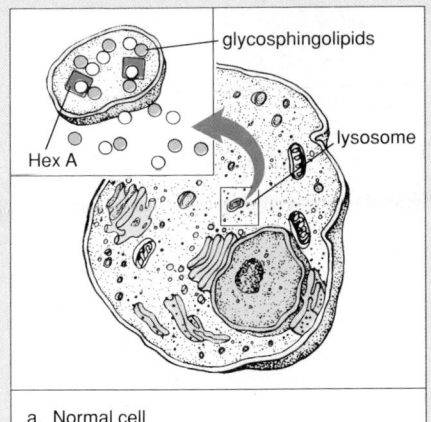

a. Normal cell

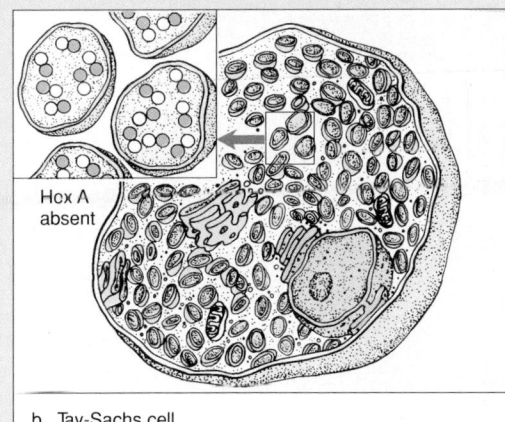

b. Tay-Sachs cell

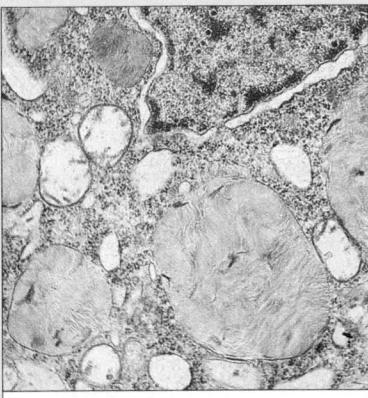

c. Electron micrograph of Tay-Sachs cell

In order to avoid the high risk of having a microcephalic child—one with an abnormally small head and severe mental retardation—a PKU woman should resume her limited diet several months before becoming pregnant. The diet must not include the artificial sweetener Nutrasweet® because it contains phenylalanine.

> There are several autosomal recessive disorders in humans. Among these are cystic fibrosis, Tay-Sachs disease, and phenylketonuria (PKU).

Autosomal Dominant Inheritance

There are many autosomal dominant disorders. Only 2 of the better known are discussed here. Others that are well known are Marfan syndrome (p. 197), achondroplasia (dwarfism), brachydactyly (abnormally short fingers), porphyria (inability to metabolize porphyrins from hemoglobin breakdown), and hypercholesterolemia (elevated levels of cholesterol in blood).

Neurofibromatosis (NF)

Neurofibromatosis (NF), sometimes called the elephant man disease,[1] is one of the most common genetic disorders, occurring roughly in one in 3,000 newborns. It is found equally in every racial and ethnic group throughout the world.

At birth, or later, the affected individual may have 6 or more large tan spots on the skin. Such spots may increase in size and number and get darker. Nerve cells proliferate and become small-to-large benign tumors called neurofibromas, which occur under the skin, in the muscles, and in the central nervous system. Many children with NF have learning disabilities and may be overactive.

[1]Although neurofibromatosis is commonly associated with Joseph Merrick, the severely deformed nineteenth-century Londoner depicted in *The Elephant Man,* researchers today believe Merrick actually suffered from a much rarer disorder called Proteus syndrome.

This genetic disorder shows variable expressivity; in most cases, symptoms are mild, and patients live a normal life. In other cases the effects are severe. The tumors can cause skeletal deformities, such as a large head, and eye and ear deformities leading to blindness and hearing loss. Eventually an early death may result.

The NF gene is known to be on chromosome 17, and researchers have developed a test for diagnosing the disorder. The gene has been isolated and found to code for a protein now called NF1, which normally suppresses abnormal cell division. In other words, it is a tumor-suppressor gene whose mutation leads to the development of tumors.

Huntington Disease

Huntington disease (HD), although its incidence is only one in 20,000, is one of the more common inherited brain disorders. Most individuals who inherit the allele appear normal until middle age. Then, a slow degeneration of brain cells sets in. The most recognizable symptoms are involuntary jerking movements of the body, including facial muscles, and slurred speech. Later, common symptoms include difficulty swallowing and loss of balance. Mood swings are accompanied by impaired reasoning and memory problems. Eventually, the victim is incapacitated, and death occurs often due to pneumonia or heart failure.

It is known that the gene that causes Huntington is on chromosome 4, but so far, despite great effort by many, the gene has not been located precisely. Although there is a fairly reliable test for Huntington, few may wish to have this information because no treatment is available.

Because the gene has not been isolated, it is not known what the gene codes for. It has been found, however, that the brain of a person with Huntington produces more than the usual amount of quinolinic acid, an excitotoxin that can overstimulate certain nerve cells. Perhaps quinolinic acid leads to the death of nerve cells

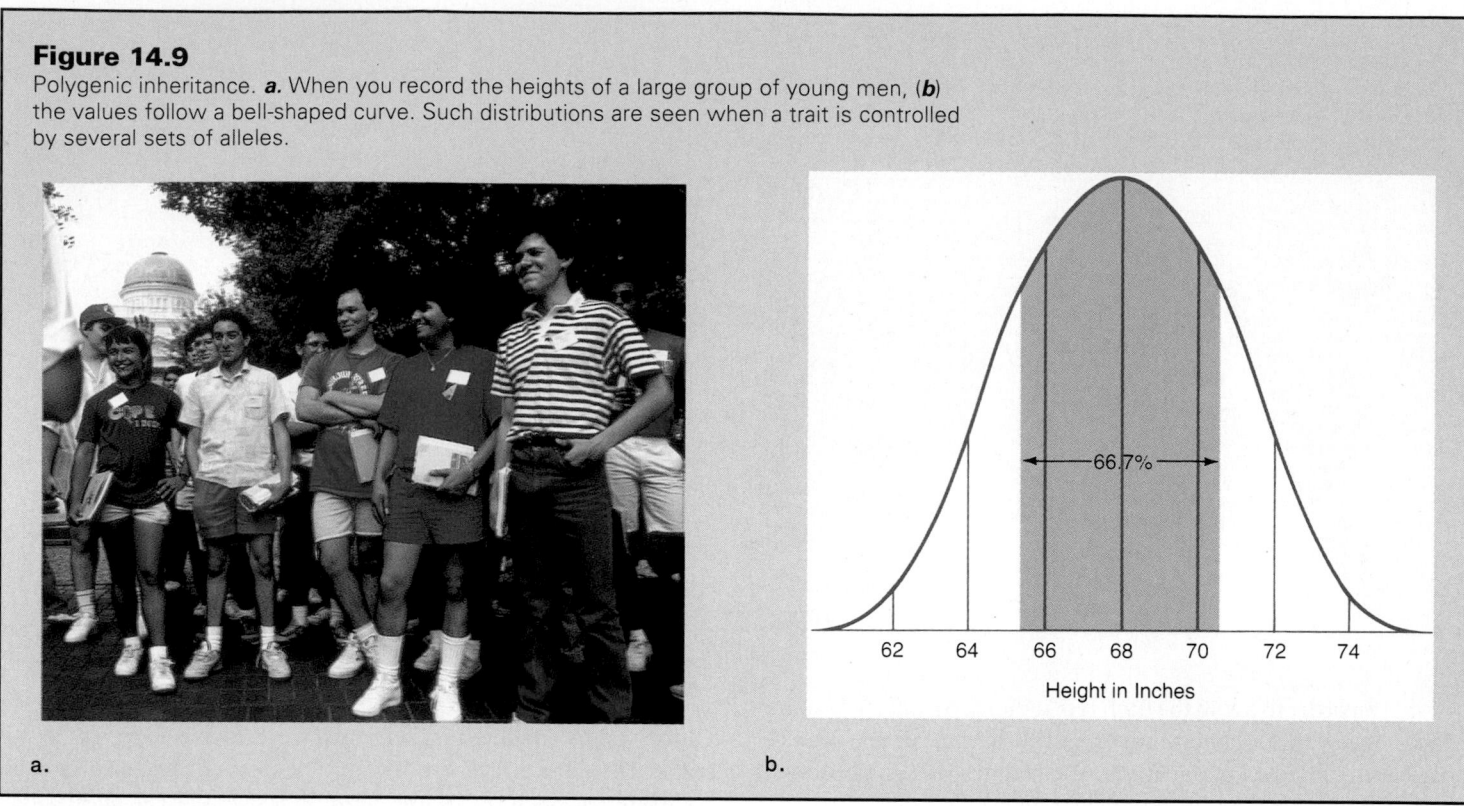

Figure 14.9
Polygenic inheritance. *a.* When you record the heights of a large group of young men, (*b*) the values follow a bell-shaped curve. Such distributions are seen when a trait is controlled by several sets of alleles.

66.7%

Height in Inches

a.

b.

and the subsequent symptoms of Huntington disease. Researchers are looking for chemicals that can block quinolinic acid's action or inhibit quinolinic acid synthesis.

There are several autosomal dominant disorders in humans. Among these are neurofibromatosis (NF) and Huntington disease.

Practice Problems 1*

1. Both a man and a woman are heterozygous for Tay-Sachs disease. What are the chances that any child born to them will have Tay-Sachs disease?
2. A man who is heterozygous for Huntington disease reproduces with a normal woman. What are the chances the child will develop Huntington disease later in life?
3. A child has PKU, but neither parent has PKU. What is the genotype of the parents?
4. A child has neurofibromatosis. The mother appears normal. What is the genotype of the father?

*Answers for Practice Problems appear in appendix D.

Polygenic Inheritance

As discussed in the previous chapter, 2 or more sets of alleles can affect the same trait, sometimes in an additive fashion. Polygenic inheritance causes the distribution of phenotypes according to a bell-shaped curve, with most individuals exhibiting the average phenotype (fig. 14.9). The more genes that control the trait, the more continuous the distribution.

Skin Color

Just how many pairs of alleles control skin color is not known, but a range in colors can be explained on the basis of 3 or even 2 pairs.

Black = *AABB*
Dark = *AABb* or *AaBB*
Medium brown = *AaBb* or *AAbb* or *aaBB*
Light = *Aabb* or *aaBb*
White = *aabb*

When a black person has children with a white person, the children have medium-brown skin but 2 medium-brown individuals can produce children who range in skin color from black to white. If a medium-brown person reproduces with a white person, the very darkest individual possible is medium-brown, but a white child is also possible (fig. 14.10).

Behavioral Inheritance

Is behavior primarily inherited, or is it shaped by environmental influences? This nature (inherited) versus nurture (environment) question has been asked for a long time, and twin studies are employed to attempt to find the answer. Twins can be identical (derived from the same fertilized egg) or fraternal (derived from 2 separate eggs). Identical twins have inherited exactly the same chromosomes and genes, while fraternal twins have no more genes in common than do any other brother and sister.

Genetic Basis of Life

Figure 14.10
Inheritance of skin color. This white man (aabb) and medium-brown woman (AaBb) had fraternal twins, one was white and one was medium-brown.

Figure 14.11
Oskar Stohr (*right*) and Jack Yufe (*left*) are identical twins who were reared separately but who exhibit remarkably similar behavior.

Table 14.3
ABO Blood Types

Phenotype	Genotype
A	*AA, AO*
B	*BB, BO*
AB	*AB*
O	*OO*

Twin studies have been conducted to see to what extent behavior is inherited. It has been found that fraternal twins raised in the same environment are not remarkably similar in behavior, whereas identical twins raised separately are sometimes remarkably similar. For example, Oskar Stohr was raised as a Catholic by his grandmother in Nazi Germany; Jack Yufe was raised by his Jewish father in the Caribbean (fig. 14.11). Yet these two men "like sweet liquers . . . store rubber bands on their wrists, read magazines from back to front, dip buttered toast in their coffee and have highly similar personalities."[2]

Responses to a questionnaire designed to provide information about behavioral traits showed that identical twins reared separately tend to have a more similar personality than fraternal twins reared together. Altogether, the data seem to show that about 50% of the *differences* in personality traits was due to polygenic inheritance and 50% was due to environmental influence.

Polygenic Disorders
A number of serious genetic disorders, such as cleft lip or palate, clubfoot, congenital dislocation of the hip, and certain neural tube defects, are traditionally believed to be controlled by a

[2]Holden, C. 1980. Identical twins reared apart. *Science* 207:1323-28.

combination of genes. Researchers who studied the inheritance of cleft palate in a large family in Iceland report finding a cleft palate gene on the X chromosome. This might be another cause of the condition.

Multiple Alleles
ABO Blood Type
Three alleles for the same gene control the inheritance of ABO blood types. These alleles determine the presence or absence of antigens on the red blood cells.

A = type A antigen on red blood cells
B = type B antigen on red blood cells
O = no A or B antigens on the red blood cells

Each person has only 2 of the 3 possible alleles, and both *A* and *B* are dominant over *O*. Therefore, as table 14.3 shows, there are 2 possible genotypes for type A blood and 2 possible genotypes for type B blood. On the other hand, alleles *A* and *B* are fully expressed in the presence of the other. Therefore, if a person inherits one of each of these alleles, that person will have type AB blood. Type O blood can only result from the inheritance of 2 *O* alleles.

An examination of possible matings between different blood types sometimes produces surprising results; for example,

Parents: *AO* X *BO*
Children: *AB, OO, AO, BO*

Therefore, from this particular mating, every possible phenotype (types AB, O, A, B blood) is possible.

Karyotyping of fetal or embryonic cells can indicate whether a developing child has one of the chromosome abnormalities listed in table 14.1. Also, the appropriate chromosome can be tested for the presence of a mutant allele for cystic fibrosis, neurofibromatosis, and sickle-cell disease, among others (see table 14.2). Chapter 18 discusses the type of test done. Biochemical tests are available for dozens of enzymes such as hexosaminidase A, the missing enzyme in persons with Tay-Sachs disease.

Amniocentesis allows testing on fetal cells, but this procedure cannot be done until the sixteenth week of pregnancy (fig. 14.B*a*). A long needle is passed through the abdominal wall to withdraw a small amount of amniotic fluid along with fetal cells. Since there are only a few cells in amniotic fluid, testing must be delayed for 4 weeks until the cells have grown and multiplied in cell culture and there are enough cells for testing purposes.

Chorionic villi sampling can be done as early as the fifth week of pregnancy (fig. 14.B*b*). The doctor inserts a long, thin tube through the vagina into the uterus. With the help of ultrasound, which gives a picture of the uterine contents, the tube is placed between the lining of the uterus and the chorion, a membrane that surrounds the fetus and has projections called chorionic villi. Suction is used to remove a sampling of the chorionic villi cells.

Screening eggs for genetic defects is a new technique (fig. 14.B*c*). Preovulation eggs are removed by aspiration after a telescope with a fiber-optic illuminator, called a laparoscope, is inserted into the abdominal cavity through a small incision in the region of the navel. The prior administration of FSH ensures that several eggs are available for screening. Only the chromosomes within the first polar body are tested because if the woman is heterozygous for a genetic defect and it is found in the polar body, then the egg must be normal. Normal eggs undergo in vitro fertilization and are placed in the prepared uterus. At present, only one in 10 attempts results in a birth, but it is known ahead of time that the child will be normal.

Blood typing sometimes can aid in paternity suits. A blood test of a supposed father, however, only can suggest that he *might* be the father, not that he definitely *is* the father. For example, it is possible, but not definite, that a man with type A blood (having genotype *AO*) is the father of a child with type O blood. On the other hand, a blood test sometimes can prove definitely that a man is not the father. For example, a man with type AB blood cannot possibly be the father of a child with type O blood. Therefore, blood tests can be used legally only to exclude a man from possible paternity.

The Rh Blood Factor

The Rh blood factor is inherited separately from types A, B, AB, or O type blood. In each instance, it is possible to be Rh positive (Rh$^+$) or Rh negative (Rh$^-$). When you are Rh positive, there is a particular antigen on the red blood cells, and when you are Rh negative, it is absent. It can be assumed that the inheritance of this antigen is controlled by a single allele pair in which simple dominance prevails: the Rh-positive allele is dominant over the Rh-negative allele. Complications arise when an Rh-negative woman reproduces with an Rh-positive man and the child in the womb is Rh positive. Under certain circumstances, the woman may begin to produce antibodies that will attack the red blood cells of this baby or of a future Rh-positive baby.

Skin color is an example of a human trait controlled by polygenes. ABO blood type is an example of a human trait controlled by multiple alleles.

Degrees of Dominance

The field of human genetics also has examples of incomplete dominance and codominance. For example, when a curly-haired Caucasian person reproduces with a straight-haired Caucasian person, their children will have wavy hair. We have already mentioned that the multiple alleles controlling blood type are codominant. An individual with the genotype AB has the blood type AB, and there are individuals who have either blood type A or blood type B.

Sickle-Cell Disease

Sickle-cell disease is an example of a human disorder that is controlled by incompletely dominant alleles. Individuals with the genotype Hb^AHb^A are normal, those with genotype Hb^SHb^S have sickle-cell disease, and those with the genotype Hb^AHb^S have sickle-cell trait, a condition in which the cells are sometimes sickle shaped. Two individuals with sickle-cell trait can produce children with all 3 phenotypes, as indicated in figure 14.12.

Among regions of malaria-infested Africa, infants with sickle-cell disease die but infants with sickle-cell trait have a better chance of survival than the normal homozygote. Their sickle cells give protection against the malaria-causing parasite that uses red blood cells during its life cycle. The parasite dies when potassium leaks out of the red blood cells as they become sickle-shaped. The protection afforded by sickle-cell trait keeps the allele prevalent in populations exposed to malaria. As many as 60% of blacks in malaria-infested regions of Africa have the allele. In the United States about 10% of the black population carries the allele.

The red blood cells in persons with sickle-cell disease cannot easily pass through small blood vessels. The sickle-shaped cells either break down or they clog blood vessels, and the individual suffers from poor circulation, anemia, and sometimes internal hemorrhaging. Jaundice, episodic pain in the abdomen and joints, poor resistance to infection, and damage to internal organs are all symptoms of sickle-cell disease.

Persons with sickle-cell trait do not usually have any difficulties unless they experience dehydration or mild oxygen deprivation. Although a recent study found that army recruits

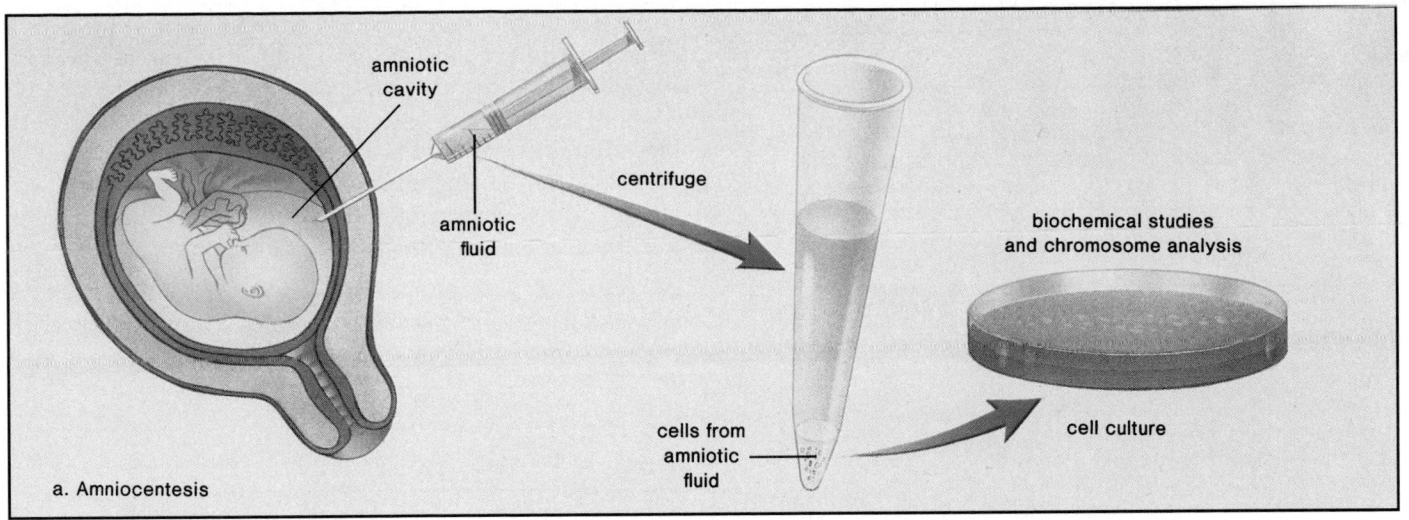

a. Amniocentesis

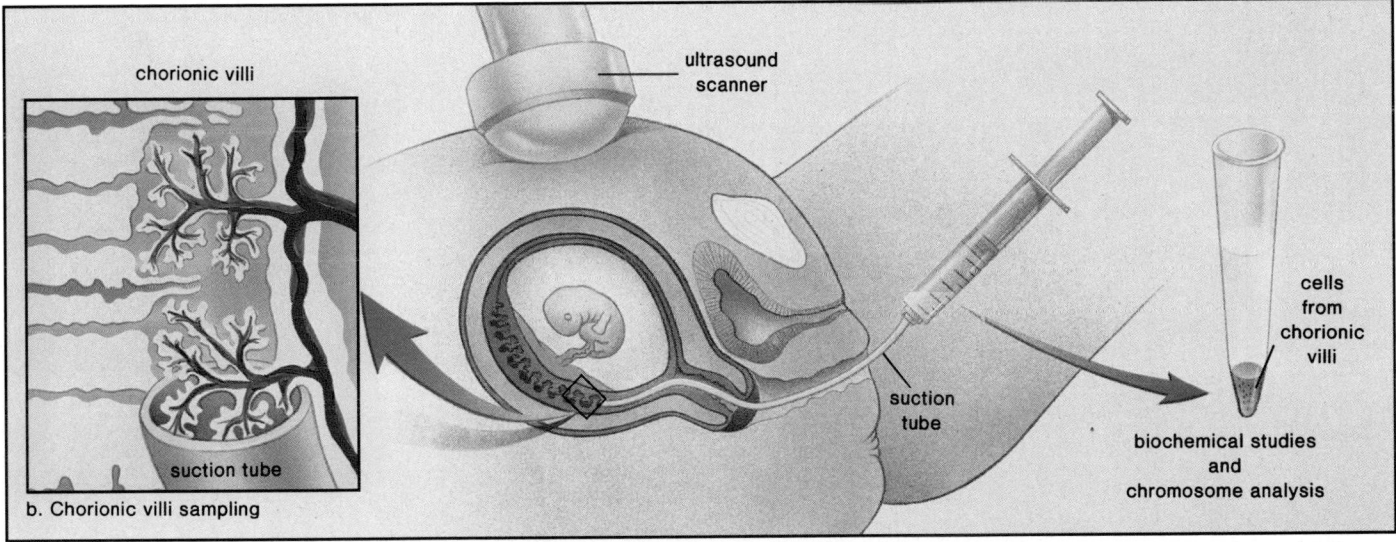

b. Chorionic villi sampling

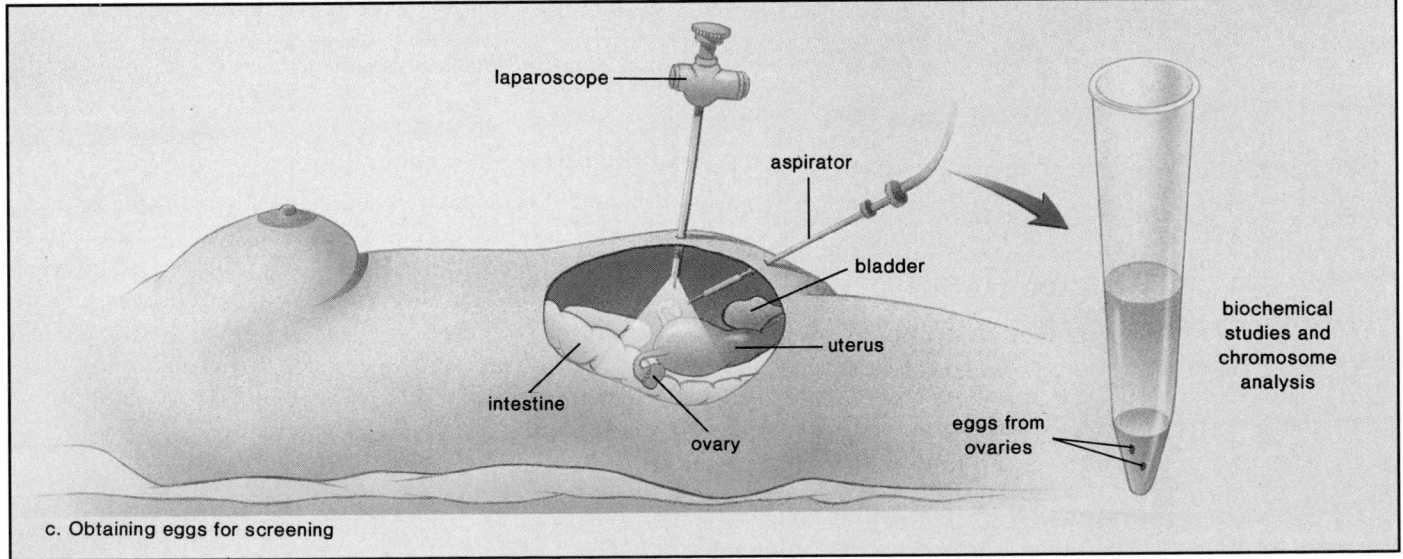

c. Obtaining eggs for screening

Figure 14.B

Technique for obtaining (**a**) fetal cells, (**b**) embryonic cells, and (**c**) eggs for genetic and biochemical testing.

Figure 14.12

a. Inheritance of sickle-cell disease, a condition more prevalent among blacks. In this example, both parents have the sickle-cell trait and are therefore carriers. Each child, therefore, has a 25% chance of having sickle-cell disease or of being perfectly normal and a 50% chance of having the sickle-cell trait. *b.* Sickled cells. Individuals with sickle-cell disease have sickled red blood cells, which tend to clump as illustrated here. Magnification, X2,500.

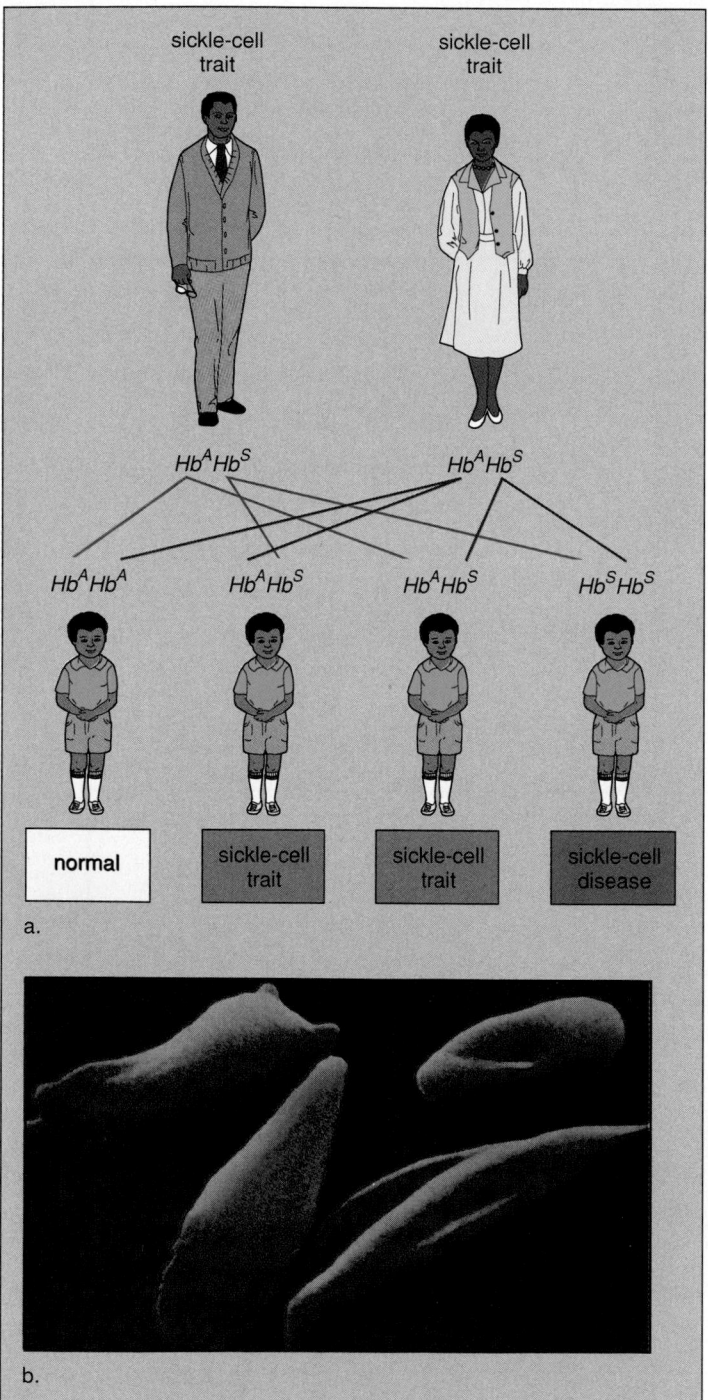

sickle-cell trait

sickle-cell trait

$Hb^A Hb^S$ $Hb^A Hb^S$

$Hb^A Hb^A$ $Hb^A Hb^S$ $Hb^A Hb^S$ $Hb^S Hb^S$

| normal | sickle-cell trait | sickle-cell trait | sickle-cell disease |

a.

b.

with sickle-cell trait were more likely to die when subjected to extreme exercise, previous studies of athletes do not substantiate these findings. At present, most investigators believe that no restrictions on physical activity is needed for persons with sickle-cell trait.

Innovative therapies are being attempted in persons with sickle-cell disease. For example, these individuals, like everyone else, produce fetal hemoglobin during development.

Sickle-cell disease is an example of a human genetic disorder that is controlled by incompletely dominant alleles.

Practice Problems 2*

1. What is the genotype for sickle-cell trait? Reproduction with what genotype would produce a child with sickle-cell disease?
2. What is the darkest child that could result from a mating between a light individual and a white individual?
3. What is the lightest child that could result from a mating between 2 medium-brown individuals?
4. From the following blood types, determine which baby belongs to which parents:

Mrs. Doe	Type A
Mr. Doe	Type A
Mrs. Jones	Type A
Mr. Jones	Type AB
Baby 1	Type O
Baby 2	Type B

5. Prove that a child does not have to have the blood type of either parent by indicating what blood types *might* be possible when a person with type A blood reproduces with a person with type B blood.

*Answers for Practice Problems appear in appendix D.

X-Linked Recessive Inheritance

Some human characteristics are X-linked recessive. Among these are a number of genetic disorders. Three of these will be discussed here. Others that are well known are Lesch-Nyhan syndrome (compulsive self-mutilation and mental retardation), one type of manic depressive psychosis (mood swings from extreme elation to severe depression), and ichthyosis (skin disorder that causes scaling).

Color Blindness

In humans there are 3 genes involved in distinguishing color because there are 3 different types of cones, the receptors for color vision. Two of these are X-linked genes: one affects the green-sensitive cones, and the other affects the red-sensitive cones.

Genetic Basis of Life

Figure 14.13

Cross involving X-linked genes. The male parent is normal, but the female parent is a carrier; an allele for color blindness is located on one of her X chromosomes. Therefore, each son stands a 50-50 chance of being color blind.

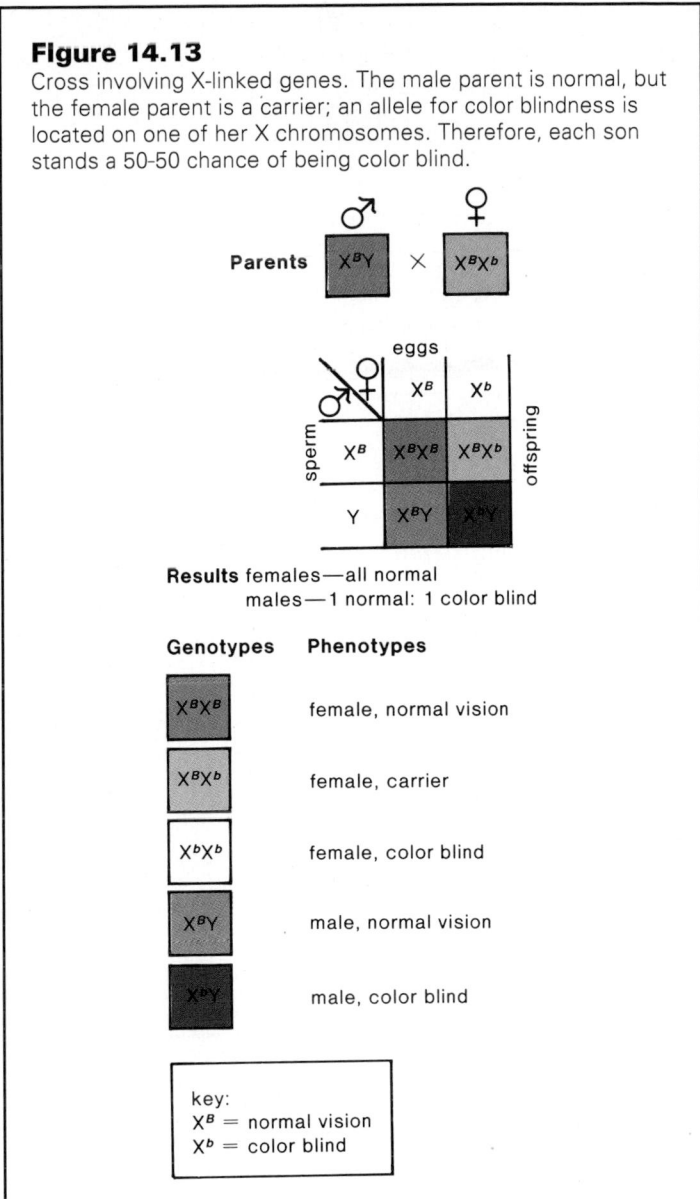

Figure 14.14

Sample pedigree chart for a recessive X-linked genetic disorder. This listing tells how to recognize this type of inheritance.

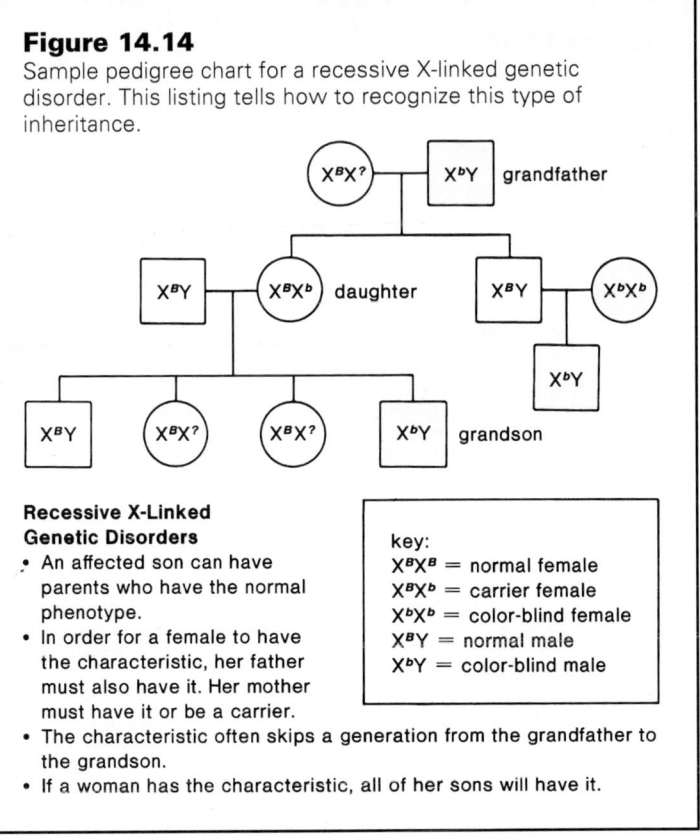

Recessive X-Linked Genetic Disorders

- An affected son can have parents who have the normal phenotype.
- In order for a female to have the characteristic, her father must also have it. Her mother must have it or be a carrier.
- The characteristic often skips a generation from the grandfather to the grandson.
- If a woman has the characteristic, all of her sons will have it.

key:
X^BX^B = normal female
X^BX^b = carrier female
X^bX^b = color-blind female
X^BY = normal male
X^bY = color-blind male

About 6% of men in the United States are color blind due to a mutation involving green perception, and about 2% are color blind due to a mutation involving red perception.

Let us consider a particular cross for this trait. If a heterozygous woman reproduces with a man with normal vision, what are the chances of their having a color-blind daughter? a color-blind son?

$$\text{Parents } X^BY \times X^BX^b$$

Figure 14.13 illustrates the use of the Punnett square when solving X-linked problems such as this one. Notice that when indicating the results of a cross involving an X-linked gene, you give the phenotypic ratios for males and females separately.

Figure 14.14 gives a pedigree chart for an X-linked recessive disorder and lists ways to recognize this pattern of inheritance.

Hemophilia

About one in 1,500 males is born with hemophilia in the United States (table 14.2). Most have hemophilia A, caused by the absence or minimal presence of a particular clotting factor called factor VIII. **Hemophilia** is called the bleeder's disease because the affected person's blood does not clot. Not only do hemophiliacs bleed externally after an injury, they also suffer from internal bleeding, particularly around joints. Hemorrhages can be stopped by transfusions of fresh whole blood (or plasma) or concentrates of the clotting protein. Unfortunately, some hemophiliacs have contracted AIDS after receiving concentrated blood from donors, but this is less likely to occur if hemophiliacs use a purified form of the concentrate from donors who have been tested.

At the turn of the century, hemophilia was common in the royal families of Europe, and all of the affected males could trace their ancestry to Queen Victoria of England (fig. 14.15). Because none of the queen's ancestors or relatives was affected, it seems that the recessive allele she carried arose by mutation either in Victoria or one of her parents. Her daughters Alice and Beatrice were carriers and introduced the allele into the ruling houses of Russia and Spain. Alexis, the last heir to the Russian throne before the Russian Revolution, was a hemophiliac. The current British royal family has no hemophiliacs because Victoria's eldest son, King Edward VII, did not receive the allele and therefore could not pass it on to any of his descendants.

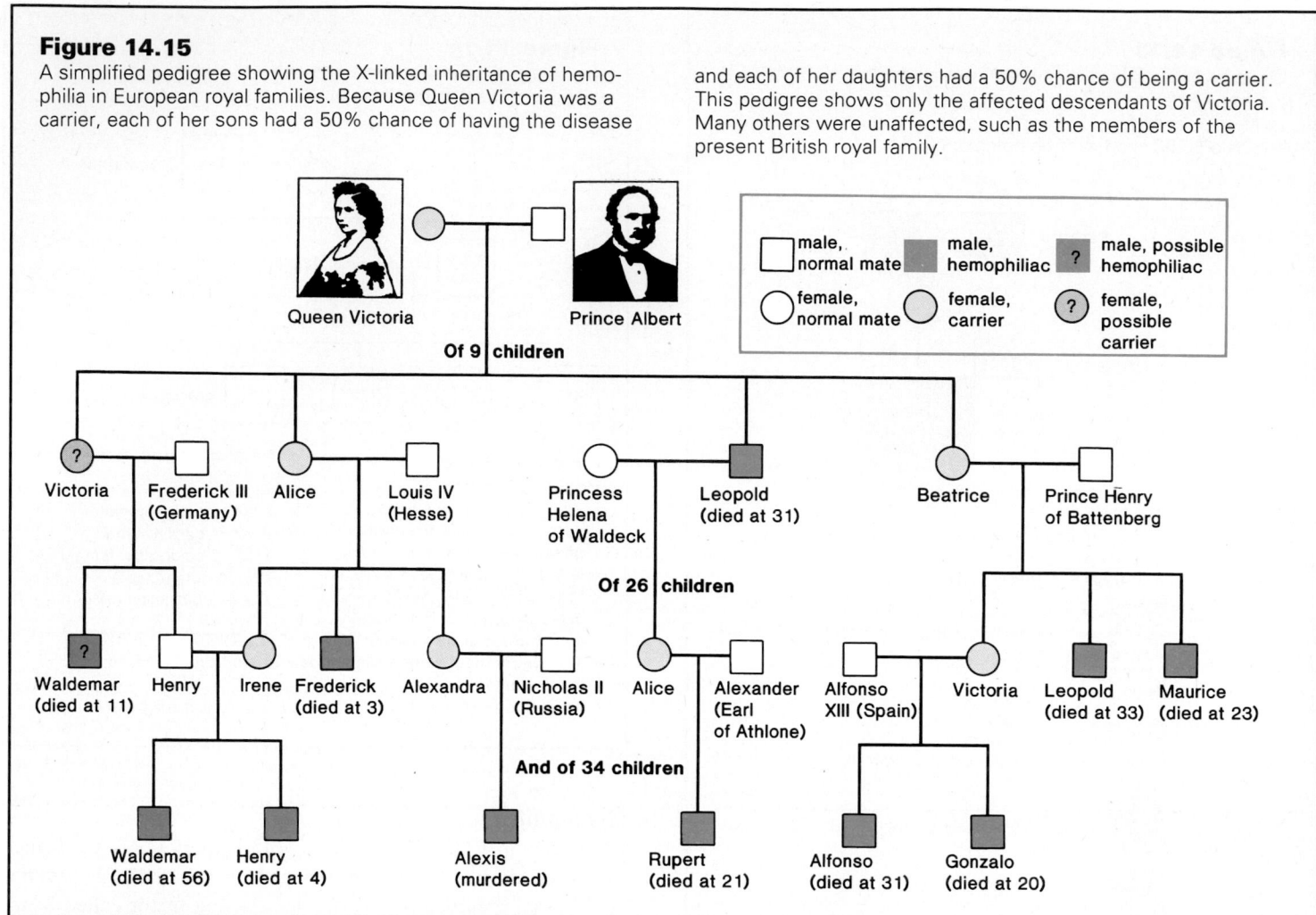

Figure 14.15

A simplified pedigree showing the X-linked inheritance of hemophilia in European royal families. Because Queen Victoria was a carrier, each of her sons had a 50% chance of having the disease and each of her daughters had a 50% chance of being a carrier. This pedigree shows only the affected descendants of Victoria. Many others were unaffected, such as the members of the present British royal family.

Muscular Dystrophy

Muscular dystrophy, as the name implies, is characterized by a wasting away of the muscles. The most common form, **Duchenne muscular dystrophy,** is an X-linked recessive disorder that occurs in about one in 5,000 live male births. Symptoms, such as waddling gait, toe walking, frequent falls, and difficulty in rising may appear as soon as the child starts to walk. Muscle weakness intensifies until the individual is confined to a wheelchair. The condition also affects the muscles that allow us to breathe and the heart; death usually occurs during the teenage years, therefore, affected males are rarely fathers. The recessive allele remains in the population by passage from carrier mother to carrier daughter.

The allele for muscular dystrophy has been isolated, and it is now known that the absence of a plasma membrane protein, called dystrophin, is the cause of the disorder. To detect carriers, they are tested for the normal amount of dystrophin. The complete lack of dystrophin causes calcium to leak into a muscle cell, and this promotes the action of an enzyme that dissolves muscle fibers.

When the body attempts to repair the tissue, the formation of fibrous tissue occurs, cutting off the blood supply so that more and more muscle cells die.

Cell therapy is being attempted as a treatment for muscular dystrophy. There are immature muscle cells called myoblasts that reside under the plasma membrane of most muscle cells. These are available to repair any muscle cell that becomes damaged. Myoblasts can be extracted from a donor, such as a father or brother who is normal, and grown in vitro until they have multiplied in number. When myoblasts are injected into the muscles of someone with muscular dystrophy they apparently will fuse with muscle cells as they normally do. Now the muscle cell is expected to be capable of making dystrophin. Clinical trials appear hopeful and are being expanded.

There are several X-linked recessive abnormalities in humans. Among these are color blindness, hemophilia, and Duchenne muscular dystrophy.

Practice Problems 3*

1. Both the mother and father of a hemophilic son appear to be normal. From whom did the son inherit the gene for hemophilia? What is the genotype of the mother, the father, and the son?

2. A man is color blind. If he reproduces with a woman who is homozygous normal, what is the chance that sons will be color blind? daughters will be color blind? will be carriers?

3. A father is right-handed (R = right-handed, r = left-handed) and has normal vision. A son is left-handed and color blind. Give all possible genotypes of the mother.

4. Both a man and woman have normal vision. A woman has a color-blind daughter. What can you deduce about the girl's father?

*Answers to Practice Problems appear in appendix D.

Other Inheritance Patterns

Fragile-X Syndrome

Fragile-X syndrome is one of the most common genetic causes of mental retardation, second only to Down syndrome. It is called fragile-X syndrome because under the right laboratory conditions, a lesion can be seen in the X chromosome. Nevertheless, the condition is caused by a particular allele on the X chromosome. The allele has been identified and is called FMR-1 for fragile X mental retardation-1. In persons with fragile-X syndrome, certain portions of the gene are duplicated many times, and it is believed that this may cause the gene to be disfunctional.

The manner in which fragile-X syndrome is inherited does not follow the usual X-linked recessive pattern of inheritance. For example, males with the defective allele can have normal intelligence, but their daughters can have fragile-X positive sons. Also, investigators have found a number of girls with the fragile-X lesion who are subnormal in intelligence. It appears that fragile X may be an example of parental imprinting of genes, which was discussed in the previous chapter. In other words, the activity of the allele is dependent on the sex that passes the gene on.

Sex-Influenced Traits

Some genes not located on the X chromosome or Y chromosome are expressed differently in the 2 sexes, and therefore they are referred to as **sex-influenced traits.** Pattern baldness is caused by an autosomal allele that is dominant in males due to the presence of testosterone, the male sex hormone. Heterozygous women with adrenal tumors develop pattern baldness, but hair returns when the tumor is removed. This hair loss can be explained by the fact that an abnormally large amount of testosterone is produced by the malfunctioning adrenal gland. The adrenal glands normally produce some testosterone, even in women.

Another sex-influenced trait is index finger length. An index finger equal to or longer than the fourth finger is dominant in females but recessive in males.

Summary

1. It is possible to treat and photograph the chromosomes of a cell so that they can be cut out and arranged in pairs. The resulting karyotype can be used to diagnose chromosome abnormalities.

2. Down syndrome (trisomy 21) is the most common autosomal chromosome abnormality. The occurrence of this syndrome, which is related to the mother's age, can be detected by amniocentesis. Most often, Down syndrome is due to nondisjunction during gamete formation, but in a small percentage of cases, there has been a translocation between chromosomes 14 and 21, in which chromosome 21 is attached to chromosome 14.

3. Turner syndrome (XO) is a monosomy for the X chromosome. There are several trisomies: metafemales, which are XXX, Klinefelter syndrome, which is XXY, and XYY males.

4. Amniocentesis and chorionic villi testing allow physicians to recover fetal cells, whose chromosome content can be analyzed for any possible abnormalities. Biochemical (enzyme) disorders can also be detected.

5. When studying human genes biologists often construct pedigree charts to show the pattern of inheritance of a characteristic within a family. The particular pattern indicates the manner in which a characteristic is inherited. Sample charts are given for autosomal recessive, autosomal dominant, and X-linked recessive patterns.

6. Cystic fibrosis, lysosomal storage diseases, and PKU are autosomal recessive disorders that have been studied in detail.

7. Neurofibromatosis (NF) and Huntington disease are autosomal dominant disorders that have been well studied.

8. Skin color is an example of a human trait controlled by polygenes. ABO blood type is an example of a human trait controlled by multiple alleles.

9. Sickle-cell disease is a human disorder that is controlled by incompletely dominant alleles.

10. Color blindness, hemophilia, and muscular dystrophy are X-linked recessive disorders. Some traits, like pattern baldness, are sex-influenced traits and are controlled by autosomal genes.

Writing Across the Curriculum

In order to practice writing skills, students should write out the answers to any or all of the study questions and the critical thinking questions. The study questions are sequenced in the same order as the text. Suggested answers to the critical thinking questions are in appendix D.

Study Questions

1. What does the normal human karyotype look like?
2. What are the characteristics of Down syndrome? What is the most frequent cause of this condition and how might it arise due to a translocation?
3. What is the only known sex chromosome monosomy in humans? Name and describe 3 sex chromosome trisomies.
4. How might you distinguish an autosomal dominant from an autosomal recessive trait when viewing a pedigree chart?
5. Describe the symptoms of cystic fibrosis, Tay-Sachs, and PKU. For any one of these, what are the chances of 2 carriers having an affected child?
6. Describe the symptoms of neurofibromatosis (NF) and Huntington disease. How do they illustrate variable expressivity? For either of them, what are the chances of a heterozygote and a normal individual having an affected child?
7. Why is the inheritance of skin color believed to be an example of polygenic inheritance?
8. Why is ABO blood type an example of inheritance by multiple alleles?
9. Explain how sickle-cell disease is inherited. What are the symptoms of sickle-cell trait and sickle-cell disease? If 2 persons with sickle-cell trait reproduce, what are the chances of having a child with sickle-cell trait? with sickle-cell disease? What race of people is more likely to have this condition? Why?
10. Explain how color blindness, hemophilia and Duchenne-type muscular dystrophy are inherited. What are the symptoms of these conditions? What are the chances of having an affected male offspring if the mother is a carrier and the father is normal?
11. What is a sex-influenced trait? Give 2 examples of such traits.

Objective Questions

For questions 1-3, match the conditions in the key with the descriptions below:

Key:
   a. Down syndrome
   b. Turner syndrome
   c. Klinefelter syndrome
   d. XYY

1. Male with underdeveloped testes and some development of the breasts.
2. Trisomy 21
3. XO female
4. Down syndrome
   a. is always caused by nondisjunction of chromosome 21.
   b. shows no overt abnormalities.
   c. is more often seen in children of mothers past the age of 35.
   d. Both a and c.
5. A person has a genetic disorder. Which one of these is inconsistent with autosomal recessive inheritance?
   a. Both parents have the disorder.
   b. Both parents do not have the disorder.
   c. All the children (males and females) have the disorder.
   d. Both a and c.
6. A male has a genetic disorder. Which one of these is inconsistent with X-linked recessive inheritance?
   a. Both parents do not have the disorder.
   b. Only males in a pedigree chart have the disorder.
   c. Only females in a pedigree chart have the disorder.
   d. Both a and c.

For questions 7-10, match the conditions in the key with the descriptions below:

Key:
   a. cystic fibrosis
   b. Huntington disease
   c. hemophilia
   d. Tay-Sachs

7. autosomal dominant
8. most often seen among Jewish people
9. X-linked recessive
10. thick mucus in lungs and digestive system
11. Determine if the characteristic possessed by the darkened males and females is an autosomal dominant, autosomal recessive, or X-linked recessive condition:

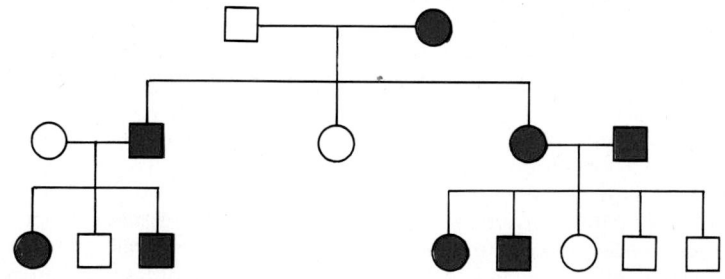

232

Genetic Basis of Life

Additional Genetics Problems

1. A hemophiliac man reproduces with a homozygous normal woman. What are the chances that their sons will be hemophiliacs? that their daughters will be hemophiliacs? that their daughters will be carriers?

2. A son with cystic fibrosis is born to a couple who appear to be normal. What are the chances that any child born to this couple will have cystic fibrosis?

3. A man who is heterozygous for Huntington disease reproduces with a normal woman. What are the chances that a child will inherit the allele for Huntington disease? If the child does inherit the allele, what are the chances that she or he will pass it on?

4. A man has type AB blood. What is his genotype? Could this man be the father of a child with type B blood? If so, what blood types could the child's mother have? (p. 225)

5. A woman with white skin has medium-brown skin parents. If this woman married a light man, what is the darkest skin color possible for their children? the lightest? (p. 224)

6. What is the genotype of a man who is color blind (X-linked recessive) and has a continuous hairline? If this man has children by a woman who is homozygous dominant for normal color vision and widow's peak, what will be the genotype and phenotype of the children? (pp. 228, 195)

7. Determine if the characteristic possessed by the darkened squares (males) and circles (females) is dominant, recessive, or sex-linked recessive. Write in the genotypes for the starred individuals:

*Answers to Additional Genetics Problems appear in appendix D.

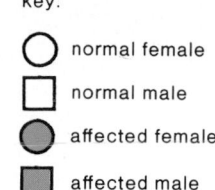

key:
- ○ normal female
- □ normal male
- ● affected female
- ■ affected male

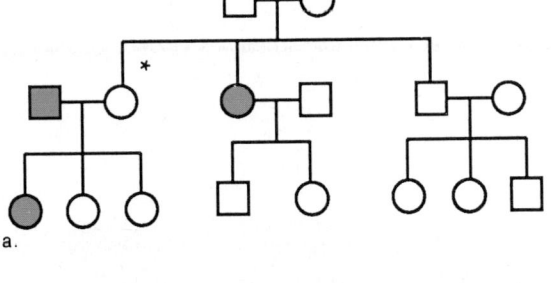

a.

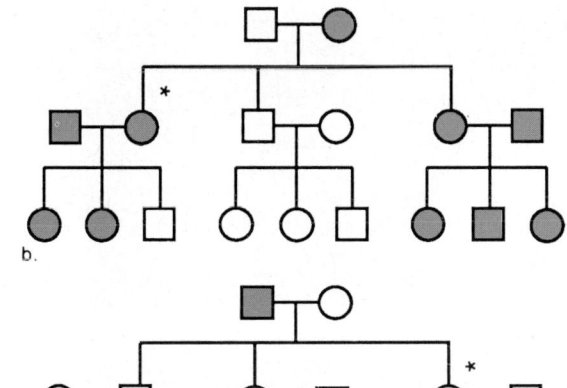

b.

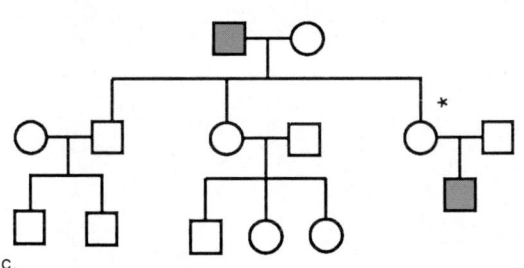

c.

Concepts and Critical Thinking

1. *The principles of genetics are the same for all organisms.*

   What are some similarities and differences between plant genetics and animal genetics? Considering that there are some differences, does this concept still hold?

2. *Inheritance plays a significant role in determining our mental and physical characteristics.*

   Without regard to ethical considerations, design an experiment to solve the nature-nurture question.

3. *Genetic diseases, like other conditions, are sometimes preventable and treatable.*

   What measures, if any, should be taken to prevent genetic diseases in the next generation?

Selected Key Terms

karyotype (kar'e-o-tīp) 216
Down syndrome (down sin'drōm) 216
Turner syndrome (tur'ner sin'drōm) 218
metafemale (met"ah-fe'māl) 218
Klinefelter syndrome (klīn'fel-ter sin'drōm) 218
XYY male (eks wi wi māl) 218

carrier (ka'ē-er) 219
cystic fibrosis (sis'tik fi-bro'sis) 221
Tay-Sachs disease (ta saks' dī-zēz) 222
phenylketonuria (PKU) (fen"il-ke"to-nu're-ah) 222
neurofibromatosis (NF) (nu"ro-fy brŏ mah tō'sis) 223

Huntington disease (HD) (hunt'ing-tun dī-zēz') 223
sickle-cell disease (sik'l sel dī-zēz) 227
hemophilia (he"mo-fil'e-ah) 229
Duchenne muscular dystrophy (du-shen' mus'ku-lar dis'tro-fe) 230
sex-influenced trait (seks in-flu'enst trāt) 231

15

DNA: The Genetic Material

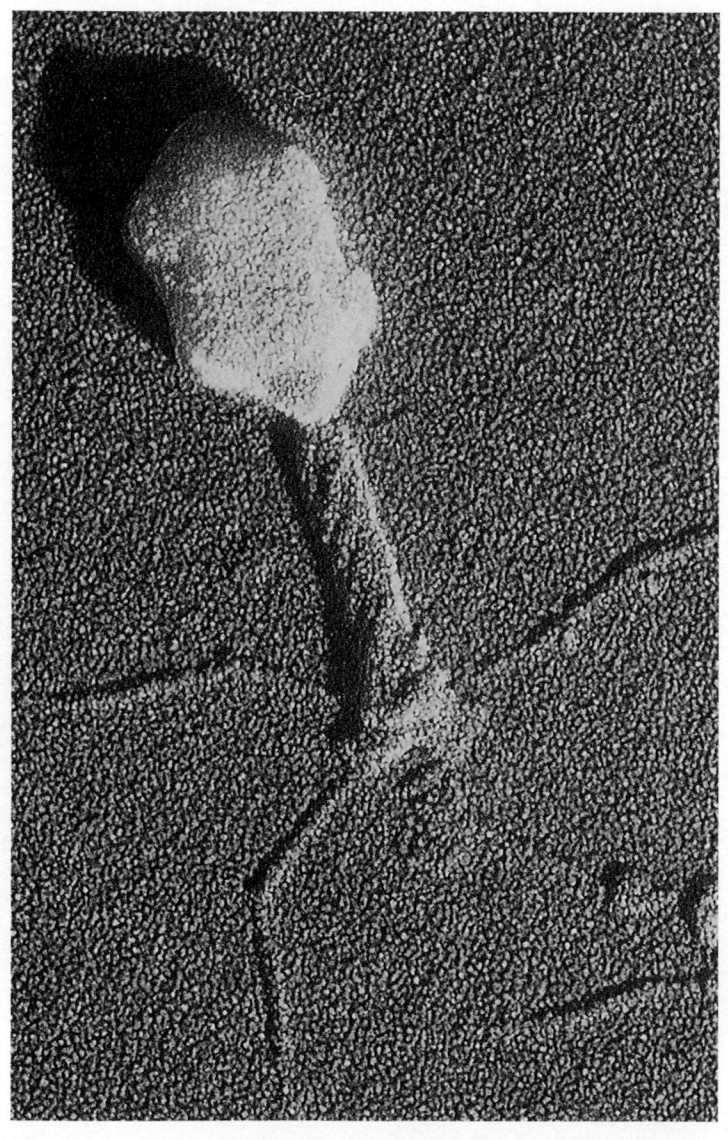

How do you prove that DNA is the genetic material? Investigators turned to T viruses, which are composed only of a nucleic acid core within a protein coat. They showed that the DNA, located within the hexagonal head and not the protein coat, coded for new viruses. Magnification, X60,000.

Your study of this chapter will be complete when you can

1. list the requirements for a substance to serve as the genetic material;
2. describe the transformation experiment of F. Griffith, including his surprising results and conclusion;
3. tell how O. Avery and colleagues showed that DNA is the transforming substance;
4. describe the experiment of Hershey and Chase with T2 viruses, and tell how it showed that DNA is the genetic material;
5. explain Chargaff's rules for DNA and tell why they are significant;
6. describe the Watson and Crick model of DNA, and tell how it fits the Chargaff and the Franklin data;
7. describe the semiconservative manner in which DNA replicates;
8. tell how Meselson and Stahl demonstrated that DNA replication is semiconservative;
9. contrast the process of DNA replication in prokaryotes and eukaryotes;
10. discuss new findings regarding DNA biochemistry.

ven though Morgan and his colleagues were able to confirm that the genes are on the chromosomes and were even able to map the *Drosophila* chromosomes, they still didn't know just what genes consisted of. You can well imagine, then, that the search for the genetic material was of utmost importance to biologists at the beginning of the twentieth century. They knew that this material must be

1. able to *store information* that is used to control both the development and the metabolic activities of the cell or organism;
2. stable so that it can *be replicated* with high fidelity during cell division and be transmitted from generation to generation;
3. able to *undergo rare changes* called **mutations** to generate genetic variability that is acted upon during evolution.

The genetic material must be able to store information, be replicated, and undergo mutations.

Search for the Genetic Material

Knowledge about the chemistry of DNA was absolutely essential to come to the conclusion that this is the genetic material. In 1869, the Swiss chemist Friedrich Miescher removed nuclei from pus cells (these cells have little cytoplasm) and found that they contained a chemical he called *nuclein*. Nuclein, he said, was rich in phosphorus and had no sulfur, properties that distinguished it from protein. Later, other chemists did further work with nuclein and said that it contained an acidic substance they called **nucleic acid.** Soon it was realized that there are 2 types of nucleic acids: **DNA** (**deoxyribonucleic acid**) and **RNA** (**ribonucleic acid**).

When it was discovered early in the twentieth century that nucleic acids contain 4 types of nucleotides, an unfortunate idea called the *tetranucleotide* (4-nucleotide) *hypothesis* arose. It said that DNA was composed of repeating units, and each unit always had just one of each of the 4 different nucleotides. In other words, DNA could not vary between species and therefore could not be the genetic material!

Transformation of Bacteria

In 1931, the bacteriologist Frederick Griffith performed an experiment with a bacterium (*Streptococcus pneumoniae,* or pneumococcus for short) that causes pneumonia in mammals. He noticed that when these bacteria are grown on culture plates, some, called S strain bacteria, produce shiny, smooth colonies and others, called R strain bacteria, produce colonies that have a rough appearance. Under the microscope, S strain bacteria have a mucous (polysaccharide) coat but R strain bacteria do not. When Griffith injected mice with the S strain of bacteria, they died, and when he injected mice with the R strain, they did not die (fig. 15.1). In an effort to

Figure 15.1

Griffith's transformation experiment *a.* S strain pneumococcus bacteria are virulent and kill the host, a mouse. *b.* R strain bacteria are not virulent and do not kill the mouse. *c.* Heat-killed S strain bacteria are not virulent and do not kill the mouse. *d.* If heat-killed S strain bacteria and R strain bacteria are both injected into a mouse, it dies because the R strain bacteria have been transformed into the virulent S strain. Source: R. Sager and J. Ryan, *Cell Heredity,* Copyright ©1961 John Wiley and Sons, Inc., New York, NY.

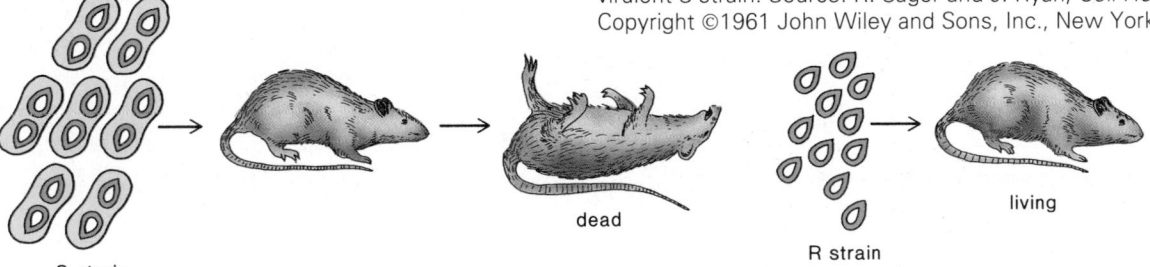

Figure 15.2

Bacteria and bacteriophages. Bacteria and bacteriophages were the experimental material of choice for determining the physical and chemical characteristics of the genetic material. Recall that bacteria are prokaryotes (see fig. 5.3), one-celled organisms whose genetic material is found within the nucleoid. The most intensely studied species of bacterium is *Escherichia coli*, which normally lives within the human gut. A few thousand of these bacteria can be placed in a liquid medium containing only a few salts and an energy source such as the sugar glucose and within a few hours, there will be as many as 3×10^9 bacteria per ml. Bacteriophages (or phages), viruses that attack bacteria, consist only of a protein coat surrounding a nucleic acid core. Enormous populations of a phage can be easily obtained—up to 10^{11} per ml or more in an infected bacterial culture is not unusual. The T virus, depicted here in close-up (**a**) and attacking an *E. coli* cell (**b**) is designated T2 (the T simply means "type"). There are other types of bacteriophages that attack *E. coli*, and they too have proven to be good experimental material. Magnification, X50,000.

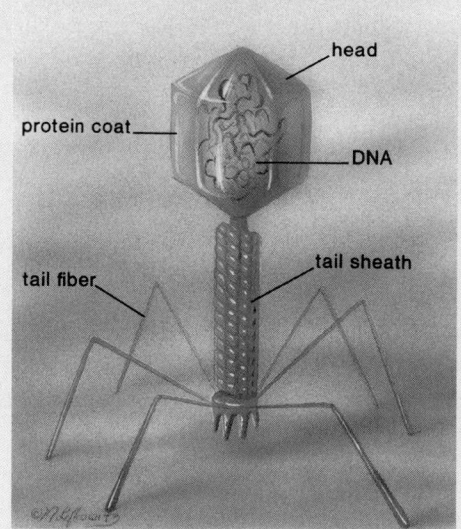

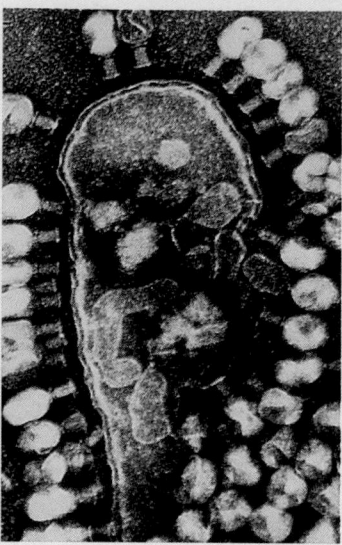

a.

b.

determine if the smooth coat alone was responsible for the virulence (ability to kill) of the S strain bacteria, he injected mice with heat-killed S strain bacteria. The mice did not die.

Finally, Griffith injected the mice with a mixture of heat-killed S strain and live R strain bacteria. Most unexpectedly, the mice died and living S strain bacteria were recovered from the bodies! Griffith concluded that some substance necessary to the synthesis of a mucous coat and therefore virulence must have passed from the dead S strain bacteria to the living R strain bacteria so that the R strain bacteria were *transformed:*

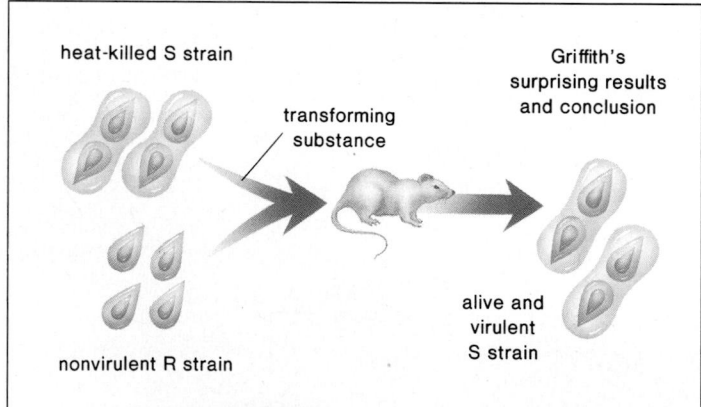

This change in the phenotype of the R strain bacteria must be due to a change in their genotype. Indeed, couldn't the transforming substance that passed from S strain to R strain be genetic material? Reasoning such as this prompted investigators at the time to begin looking for the transforming substance to determine the chemical nature of the hereditary material.

The Transforming Substance

Obviously, it is not convenient to look for the transforming substance in mice, so it is not surprising that the next group of investigators, led by Oswald Avery, worked in vitro (in laboratory glassware). After 16 years of research, this group published a paper that demonstrated that the transforming substance is DNA. Their evidence included the following observations:

1. DNA alone from S strain bacteria caused R strain bacteria to be transformed. And the DNA they used was pure—99.98% pure!
2. Enzymes that degrade proteins had no effect upon transformation, nor did RNase, an enzyme that digests RNA.
3. Enzymatic digestion of the transforming substance with DNase, an enzyme that digests DNA, did prevent transformation.
4. The molecular weight of the transforming substance was so great that it must contain about 1,600 nucleotides! Certainly this is enough for some genetic variability.

These experiments showed not only that DNA is the hereditary material, but also that DNA controls the biosynthetic properties of a cell. While these experiments seem quite convincing, remember that at the time, biologists knew proteins were very complex, but they still were not sure about DNA. Some thought that perhaps these results pertained only to this experiment.

Bacterial cells can be transformed when DNA passes from one to the other. Transformation changes the biosynthetic properties of the transformed cell.

Genetic Basis of Life

Figure 15.3

Hershey and Chase Experiment. Bacteriophage T2 virus is composed of a DNA core and protein coat only. It was reasoned that whichever of these enters a bacterium and controls viral replication is the genetic material. ***a.*** In this experiment, ^{32}P was used to label viral DNA. The coats were removed by agitation in a blender, and the radioactively labeled DNA entered the cell. Because replication proceeded normally, DNA is the genetic material. ***b.*** In this experiment, ^{35}S was used to label the protein coat. When the cells were agitated in a blender, the radioactively labeled protein coats were removed. Because replication proceeded normally, protein is not the genetic material.

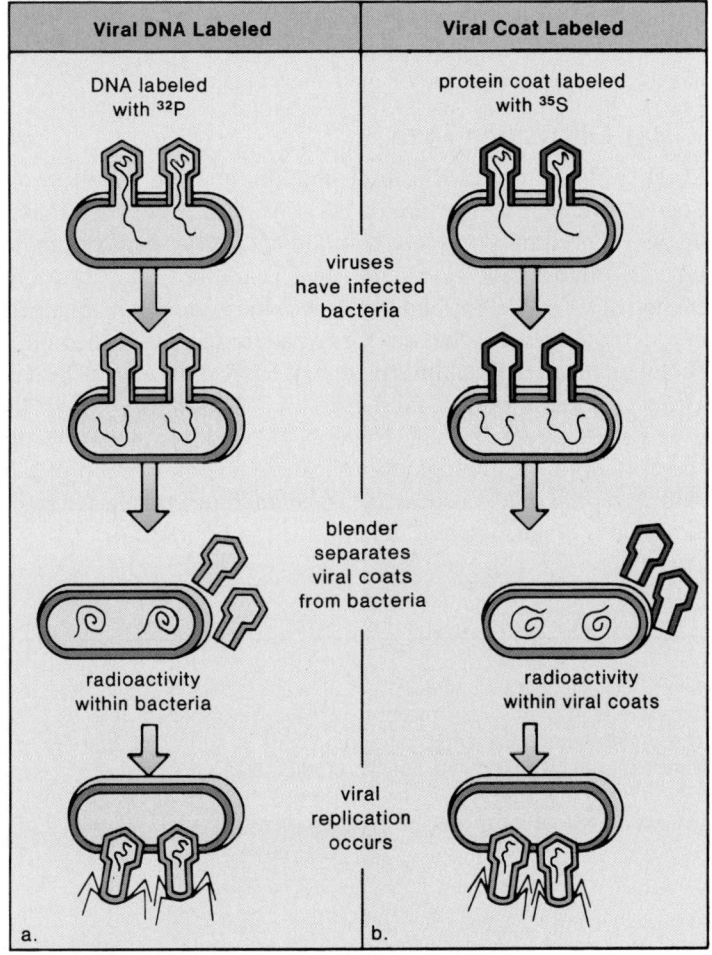

There is a chemical distinction between DNA and protein. Phosphorus (P) is generally absent, but sulfur (S) *is present in amino acids* of protein. On the other hand, sulfur (S) is absent but phosphorus (P) *is present in DNA* in high amounts. Therefore, it was possible for Hershey and Chase to prepare 2 batches of phages: one that had the DNA core radioactively labeled with ^{32}P, and another that had the protein coat radioactively labeled with ^{35}S. Radioactive phages were allowed to attach briefly to *Escherichia coli* bacterial cells; once the infection process had started, most of the adhering phage coats were sheared from the bacterial cells by agitation in a kitchen blender. Centrifugation then caused the bacterial cells to collect as a pellet at the bottom of the tube. In the one experiment (fig. 15.3a), they found that most of the ^{32}P-labeled DNA remained in the pellet:

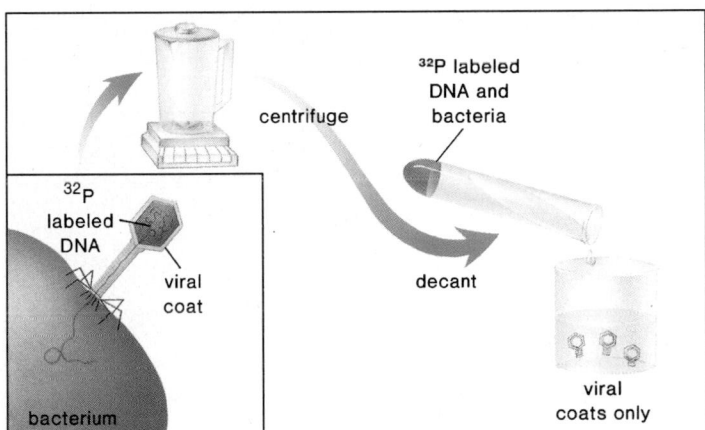

In the other experiment (fig. 15.3b), they found that most of the ^{35}S-labeled protein remained in the phage coats. These results indicated that the DNA of the virus (and not the protein) enters the host, where viral replication takes place. Therefore, DNA is the genetic material of bacteriophage T2. It directs protein coat synthesis and allows replication to occur.

Biologists, in general, were persuaded that DNA is the genetic material by this experiment, possibly because recent chemical studies of the composition of DNA had finally shown that the tetranucleotide hypothesis was not true.

The Hershey and Chase experiment showed that DNA and not protein is the genetic material of the T2 bacteriophage.

Genetic Material of Viruses

In the twentieth century, many geneticists who were interested in determining the chemical nature of the genetic material began to work with **bacteriophages,** viruses that attack bacteria (fig. 15.2). In 1952, 2 experimenters, Alfred D. Hershey and Martha Chase, chose a bacteriophage known as T2 as their experimental material. They decided to see which of the bacteriophage components— protein or DNA—entered bacterial cells and directed reproduction of the virus (see fig. 23.3). Whichever did this, they reasoned, was the genetic material.

Structure of DNA

During the same period of time that biologists were using viruses to show that DNA is the genetic material, biochemists were busy trying to determine its structure.

Chargaff's Rules

With the development of new chemical techniques in the 1940s, it was possible for Erwin Chargaff to analyze in detail the base content of DNA. For at least 20 years, the tetranucleotide hypothesis had been endorsed by many. This hypothesis stated that DNA was composed of repeating units, each unit having 4 nucleotides (one nucleotide for each of the 4 bases): **adenine (A), thymine (T), cytosine (C),** and **guanine (G).** According to this hypothesis, DNA would contain approximately 25% of each kind of base in every species.

A sample of Chargaff's data is shown in table 15.1. You can see that while some species—for example, *E. coli* and *Zea mays* (corn)—do have approximately 25% of each base in their DNA, most do not. Chargaff did find that the amount of A is always equal to the amount of T and the amount of G is always equal to the amount of C, regardless of the species.

Table 15.1
Chargaff's DNA Data: Base Composition in Various Species (%)

Species	A	T	G	C
Homo sapiens	31.0	31.5	19.1	18.4
Drosophila melanogaster	27.3	27.6	22.5	22.5
Zea mays	25.6	25.3	24.5	24.6
Neurospora crassa	23.0	23.3	27.1	26.6
Escherichia coli	24.6	24.3	25.5	25.6
Bacillus subtilis	28.4	29.0	21.0	21.6

Note: In any particular species listed, the amount of A = T and the amount of G = C. Between species, however, the A,T and G,C percentages differ. For example, in humans (*Homo sapiens*) the A,T percentage is about 31%, but in fruit flies (*Drosophila melanogaster*) the percentage is about 27%.

Chargaff's data showed the following:

1. The base composition of DNA differs from species to species. For example, *Bacillus subtilis,* a type of bacterium, does not have the same amount of A, T, C, or G as *E. coli* (table 15.1).
2. In each species, however, the percent of A equals the percent of T and the percent of G equals the percent of C. Further, 50% of the bases are **purines** (A + G) and 50% are **pyrimidines** (T + C). These relationships are called Chargaff's rules.

The first finding means that when comparisons between species are made, DNA does have the *variability* required of the genetic material. The second finding means that within the species, DNA has the *constancy* required of the genetic material.

X-Ray Diffraction Data

M. H. F. Wilkins and Rosalind Franklin at King's College in London studied the structure of DNA using X rays. They found that if a concentrated, viscous solution of DNA is made, it can be separated into fibers. Under the right conditions, the fibers are enough like a crystal (a solid substance whose atoms are arranged in a definite manner) that an X-ray pattern can be formed on a photographic film. Franklin's picture of DNA showed that DNA is a helix (fig. 15.4). The helical shape is indicated by the crossed (X) pattern in the center of the photograph. The dark portions at the top and bottom of the photograph indicate that some portion of the helix is repeated. It was possible for Franklin to estimate certain dimensions of the molecule, as indicated in figure 15.5.

Figure 15.4
a. When a crystal is X-rayed, the way in which the beam is diffracted reflects the pattern of the molecules in the crystal. The closer together two repeating structures are in the crystal, the farther from the center the beam is diffracted. *b.* The diffraction pattern of DNA produced by Rosaline Franklin. The crossed (X) pattern in the center told investigators that DNA is a helix and the dark portions at the top and the bottom told them that some feature is repeated over and over again. Watson and Crick figured that this feature was the hydrogen-bonded bases.

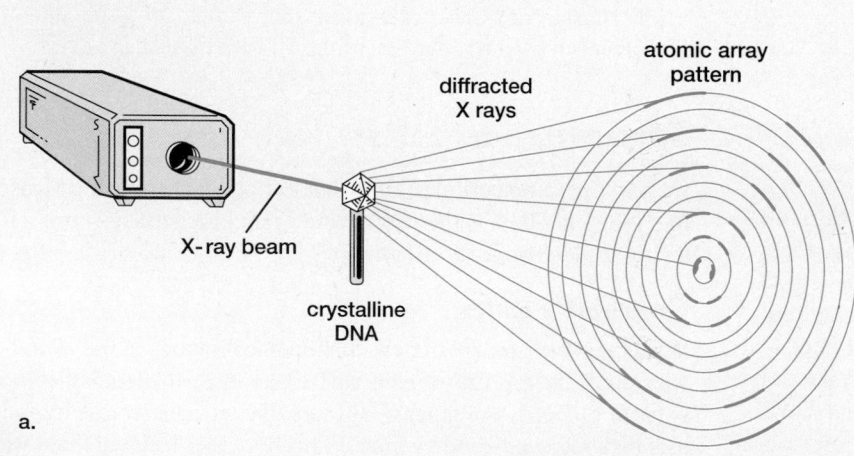
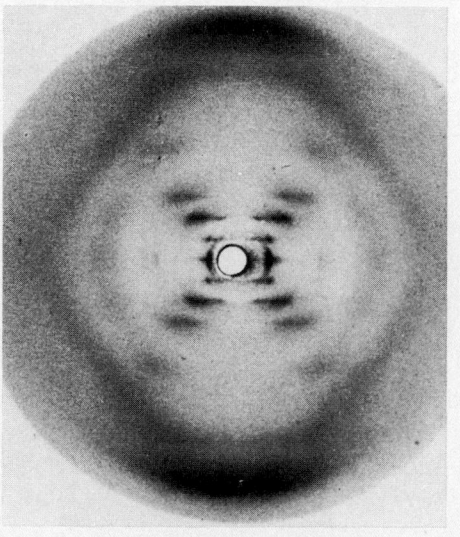

a.

b.

Figure 15.5

Watson and Crick model of DNA. **a.** A space-filling model of DNA. Notice the close stacking of the paired bases, as determined by the X-ray diffraction pattern of DNA. (Color code for atoms: yellow = phosphate; dark blue = carbon; red = oxygen; turquoise = nitrogen, and white = hydrogen.) **b.** Diagram of DNA double helix shows that the molecule resembles a twisted ladder. Sugar-phosphate backbones make up the sides of the ladder, and hydrogen-bonded bases make up the rungs of the ladder. Complementary base pairing dictates that A is bonded to T and G is bonded to C. Notice that the two strands of the molecule are antiparallel, that is the sugar-phosphate groups are oriented in different directions. This introduces complications, which are described in the reading on page 242.

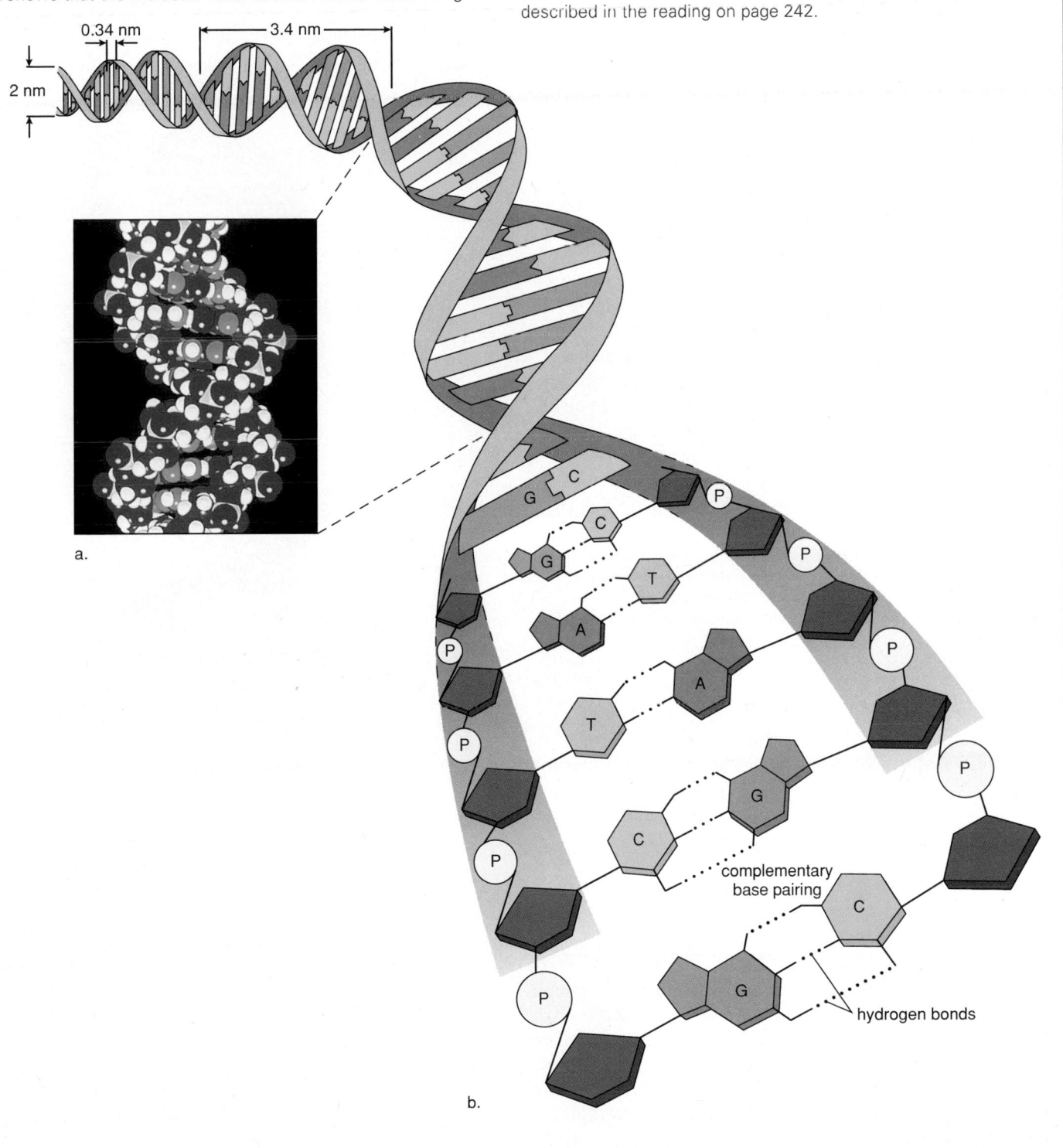

X-ray diffraction data showed that DNA is most likely a helix with certain definite molecular measurements.

Watson and Crick Model

In the early 1950s, James Watson, an American, was on a postdoctoral fellowship at Cavendish Laboratories in Cambridge, England. There he met the biophysicist Francis H. C. Crick. Using the data we have just presented, they constructed a model of DNA (fig. 15.5).

Watson and Crick knew, of course, that DNA is a polymer of nucleotides (see figs. 4.18 and 4.19), but they did not know how the nucleotides were arranged within the molecule. This is what they decided.

1. The Watson and Crick model shows that DNA is a double helix with sugar-phosphate backbones on the outside and paired bases on the inside. This arrangement fits the mathematical measurements provided by the X-ray diffraction data for the spacing between the base pairs (.34 nm) and for a complete turn of the double helix (3.4 nm).
2. Chargaff's rules said that A = T and G = C. The model shows that A is hydrogen bonded to T and G is hydrogen bonded to C. This so-called **complementary base pairing** means that a purine is always bonded to a pyrimidine. Only in this way will the molecule have the width (2 nm) dictated by its X-ray diffraction pattern, since 2 pyrimidines together are too narrow and 2 purines together are too wide.

The Watson and Crick double helix model of DNA is like a twisted ladder. The sugar-phosphate backbones of the 2 DNA strands make up the sides of the ladder, and the hydrogen-bonded bases make up the rungs, or steps, of the ladder.

Variability of Bases

The Watson and Crick model also allows for variability of the bases, as dictated by Chargaff's data. The paired bases occur in any order:

A···T	T···A	C···G	T···A
T A	C G	A T	A T
C···G	A···T	T···A	C···G

The variability that can be obtained is overwhelming. For example, it has been calculated that the human chromosome contains on the average about 140 million base pairs. Since any of the 4 possible nucleotides can be present at each nucleotide position, the total number of possible nucleotide sequences is 4 X 140 X 10^6, or over 500 million possible arrangements for one chromosome. No wonder each species has its own base percentages!

Figure 15.6

Semiconservative replication. After the DNA molecule unwinds, each old strand serves as a template for the formation of the new strand. Complementary nucleotides available in the cell pair with those of the old strand and then are joined together to form a new strand. After replication is complete, there are two daughter DNA molecules. Each is composed of an old strand and a new strand. Notice that each daughter molecule has the same sequence of base pairs as the parent molecule had before unwinding occurred.

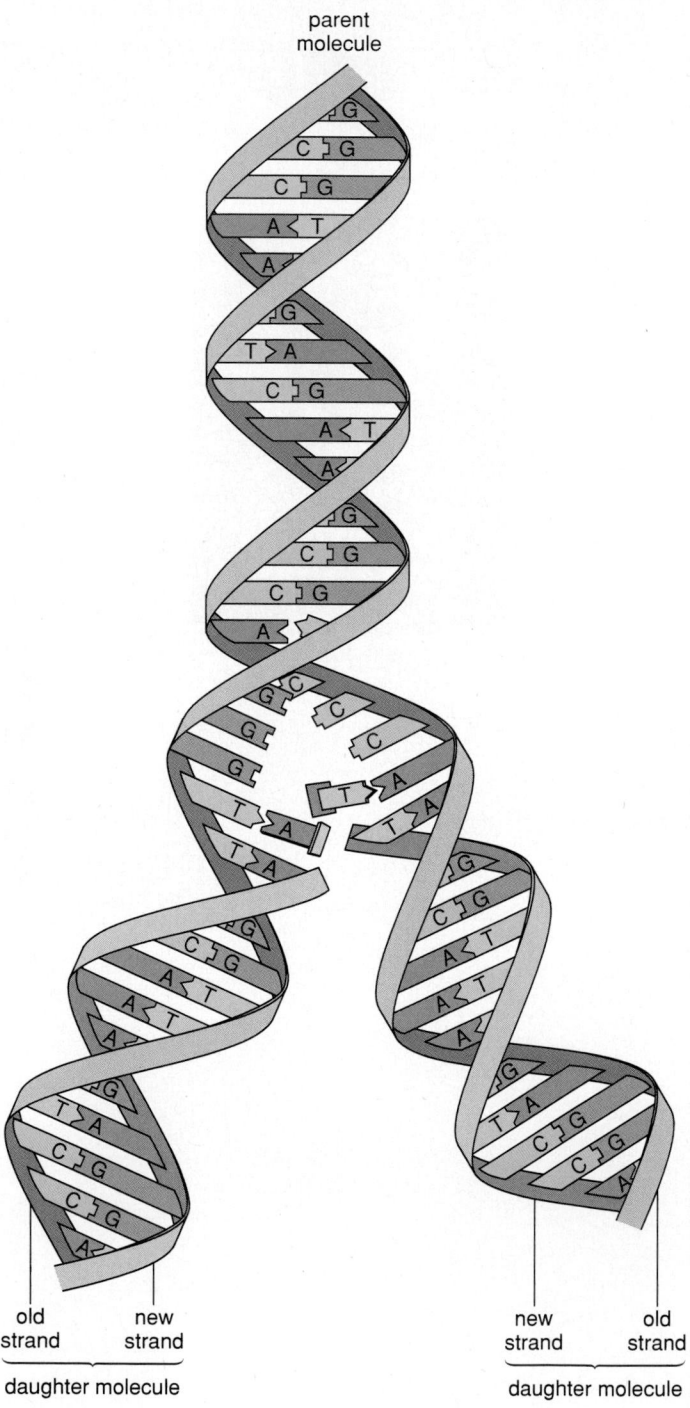

parent molecule

old strand — new strand
daughter molecule

new strand — old strand
daughter molecule

Genetic Basis of Life

Figure 15.7

Meselson and Stahl's experimental design and results. **a.** When a dense salt solution of CsCl is centrifuged at very high speed for 2-3 days, a density gradient forms which can separate DNA molecules according to their density. **b.** In their experiment Meselson and Stahl grew bacteria in ^{15}N until the DNA molecules were heavy. Then they removed the bacteria and placed them in a medium containing ^{14}N. After one generation the DNA molecules were hybrid and therefore intermediate in density. After two generations half of the DNA molecules were light. This showed that DNA replication is semiconservative.

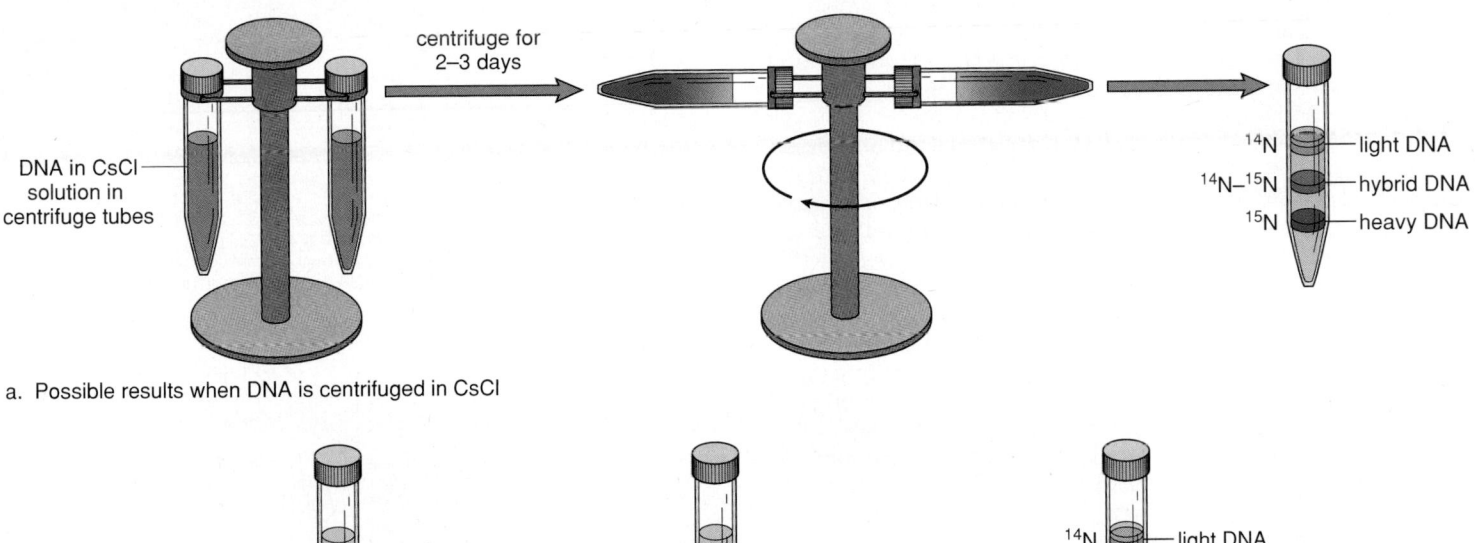

a. Possible results when DNA is centrifuged in CsCl

results when bacteria are grown in ^{15}N

results after bacteria are transferred to ^{14}N and one cell generation has occurred

results after bacteria are grown in ^{14}N for another cell generation

b. Results of Meselson and Stahl's experiment

DNA Replication

The Watson and Crick model suggests that DNA can be replicated by means of complementary base pairing. During replication, each old DNA strand of the parent molecule serves as a template for a new strand in a daughter molecule (fig. 15.6). A template is most often a mold used to produce a shape complementary to itself. Replication requires the following steps:

1. *Unwinding.* The old strands that make up the parent DNA molecule are unwound and "unzipped" (i.e., the weak hydrogen bonds between the paired bases are broken). There is a special enzyme called helicase that unwinds the molecule.
2. *Complementary base pairing.* New complementary nucleotides, always present in the nucleus, are positioned by the process of complementary base pairing.
3. *Joining.* The complementary nucleotides become joined together to form new strands. Each daughter DNA molecule contains an old strand and a new strand. Steps 2 and 3 are carried out by the enzyme **DNA polymerase.**

Notice in figure 15.6 that each daughter DNA molecule has the same sequence of bases as the parent DNA molecule had originally. Although DNA replication can be easily explained in this manner, it is actually a complicated process. Some of the more precise molecular events are explained in this chapter's reading.

Semiconservative Replication

DNA replication is termed **semiconservative replication** because one of the old strands is conserved, or present, in each daughter double helix. Semiconservative replication was experimentally confirmed by Matthew Meselson and Franklin Stahl in 1958.

These investigators knew that it would be possible to centrifuge DNA molecules in a gradient that can separate them on the basis of density (fig. 15.7a). A DNA molecule in which both strands contained heavy nitrogen (^{15}N with an atomic weight of 15) is most dense. A DNA molecule in which both strands contained light nitrogen (^{14}N with an atomic weight of 14) is least dense. A hybrid DNA molecule in which one strand is heavy and one is light has an intermediate density.

Meselson and Stahl first grew bacteria in a medium containing ^{15}N so that only heavy DNA molecules were present in the cells.

Aspects of DNA Replication

Watson and Crick realized that the strands in DNA had to be antiparallel to allow for complementary pairing of the bases. This opposite polarity of the strands introduces complications for DNA replication, as we will now see. First, it is important to take a look at a deoxyribose molecule in which the carbon atoms are numbered (fig. 15.A*a*). Use the structure to see that one of the strands of DNA (*b*) runs in the 5′ → 3′ direction and the other runs in the 3′ → 5′ direction. You can tell the 5′ (read as 5 "prime") end because that's the last phosphate.

During replication (*b*), one nucleotide is joined to another. Each new nucleotide already has a phosphate group at the 5′ carbon atom and it is joined to the 3′ carbon atom of a sugar already in place. Therefore it is said that *DNA polymerase synthesizes the replicated strand in the 5′ → 3′ direction*. This presents a problem at the replication fork (*c*), where DNA is unwound and unzipped and where replication is occurring.

At a replication fork, only one of the new (daughter) strands runs in the 5′ → 3′ direction (the template for this strand, of course, runs in the 3′ → 5′ direction). The new strand running in the 5′ → 3′ direction can be synthesized continuously and is called the *leading strand*. But what about the situation when the 5′ → 3′ parental strand is serving as the template? Synthesis of the new strand must also be in the 5′ → 3′ direction, and therefore synthesis has to begin at the fork. (Although synthesis is in the 5′ → 3′ direction, this new daughter strand in the end runs from 3′ → 5′, opposite to its template—do you see why?) Because replication of the 5′ → 3′ parental strand must begin repeatedly as the DNA molecule unwinds and unzips, replication is discontinuous; indeed, replication of this strand results in segments called Okazaki fragments, after the Japanese scientist who discovered them. Discontinuous replication takes more time than continuous replication, and therefore the new strand in this case is called the *lagging strand*.

The fact that DNA polymerase can only join a nucleotide to the free 3′ end of another nucleotide presents another problem: *DNA polymerase cannot start the synthesis of a new DNA chain* at the origin of replication. (In the lagging strand there are many origins of replication.) Here, RNA polymerases lay down a short amount of RNA, called an RNA primer, that is complementary to the DNA strand being replicated. Now DNA polymerase can add DNA nucleotides in the 5′ → 3′ direction. Later, while proofreading, DNA polymerase removes the RNA primer and replaces it with complementary DNA nucleotides. Another enzyme, called DNA ligase, joins the 3′ end of each fragment to the 5′ end of another.

Figure 15.A

DNA replication (in depth). *a.* Structure of deoxyribose, showing where nitrogen base and phosphate groups are attached. *b.* Template strand and replicated strand which always grows from the 5′ end toward the 3′ end. *c.* Both parental strands are templates for a daughter strand. Replication is a continuous process for one daughter strand and a discontinuous process for the other daughter strand.

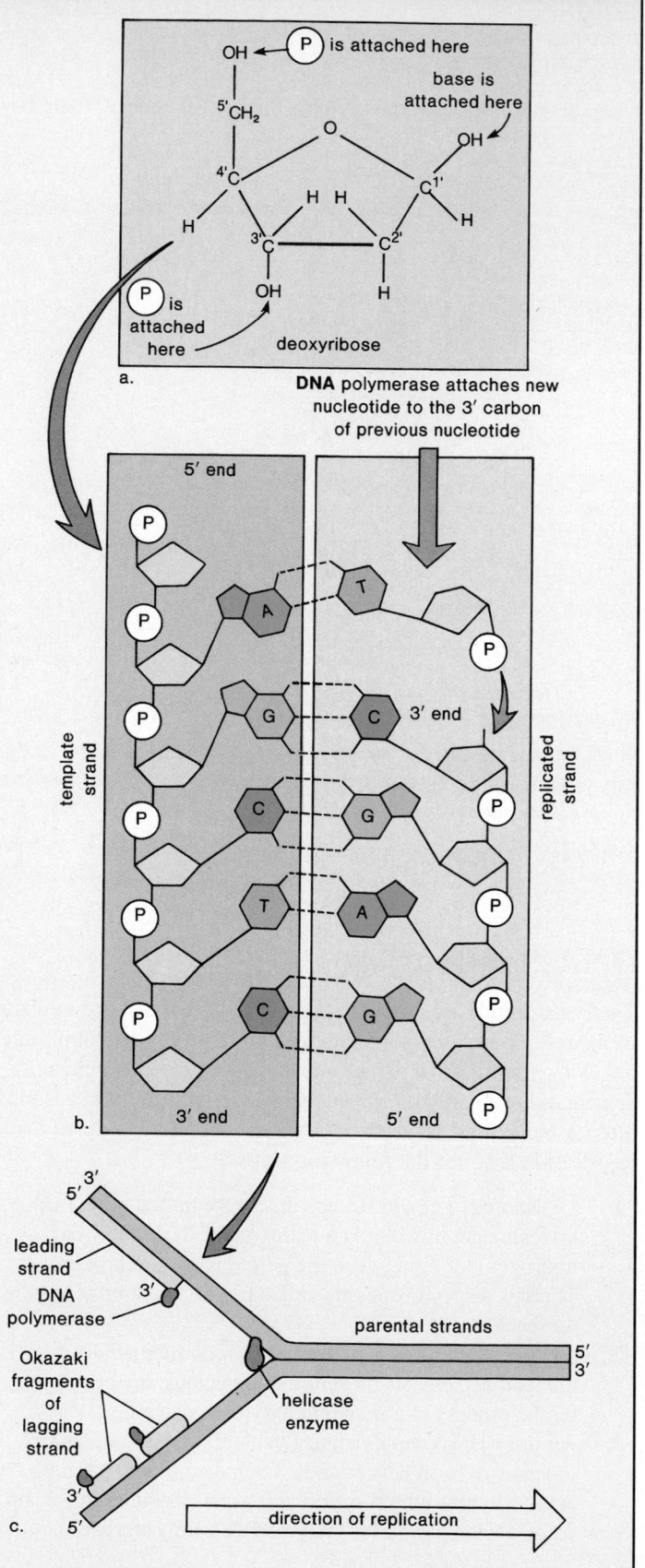

Genetic Basis of Life

Then, they switched the bacteria to a medium containing [14]N. After one cell generation, only hybrid DNA molecules were in the cells. After 2 cell generations, half of the DNA molecules were light and half of the DNA molecules were hybrid. These were exactly the results to be expected if DNA replication is semiconservative.

> DNA replication is semiconservative. The parent DNA molecule unwinds and unzips. Then, each old strand serves as a template for a new strand. Because of complementary base pairing, the daughter DNA molecules are exactly like each other and the parent DNA molecule.

Accuracy of Replication

Three enzymatic processes carried out by a DNA polymerase complex are responsible for the high fidelity of DNA replication. First, the complex chooses a nucleotide triphosphate from the cellular pool that is complementary to a template nucleotide. Then the nucleotide triphosphate is converted to a nucleotide monophosphate that is aligned with the template nucleotide.[1] Only if the fit is stable is the nucleotide incorporated into the daughter strand. If the fit is not stable, the nucleotide is restored to its triphosphate form and released. A mismatched nucleotide slips through this selection process only once per 100,000 base pairs.

Such errors can still be caught, however, because the DNA polymerase complex also carries out a proofreading function. The mismatched nucleotide causes a pause in replication, and during this time, the mismatched nucleotide is excised from the daughter strand. After proofreading has occurred, the error rate is only one mistake per 10 million base pairs.

The errors that slip through nucleotide selection and proofreading cause a gene mutation to occur. Actually it is of benefit for mutations to occur occasionally because variation is the raw material for the evolutionary process. Gene mutations are covered in more detail in chapter 16.

Prokaryotic versus Eukaryotic Replication

Recall that bacteria have a single circular loop of DNA (see fig. 10.2) that must be replicated before the cell divides. In these organisms, replication begins at an origin of replication and then spreads out in both directions. Therefore, replication is bidirectional:

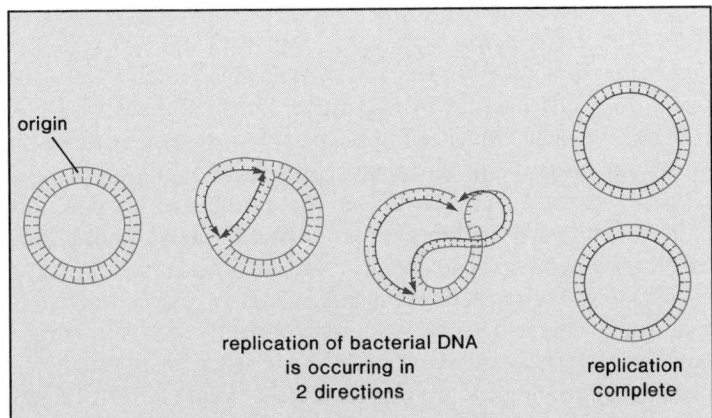

origin

replication of bacterial DNA is occurring in 2 directions

replication complete

[1]When nucleotide monophosphates are cleaved from nucleotide triphosphates, energy is provided for DNA synthesis.

Bacterial cells are able to replicate their DNA at a rate of about 10^6 base pairs per minute, and about 40 min are required to replicate the complete chromosome. Because bacterial cells are able to divide as often as once every 20 min, it is possible for a new round of DNA replication to begin even before the previous round is completed!

In eukaryotes, DNA replication begins at numerous origins of replication along the length of the chromosome and the so-called replication bubbles spread bidirectionally until they meet. Notice that there is a V shape wherever DNA is being replicated. This is called a *replication fork*.

Although eukaryotes replicate their DNA at a slower

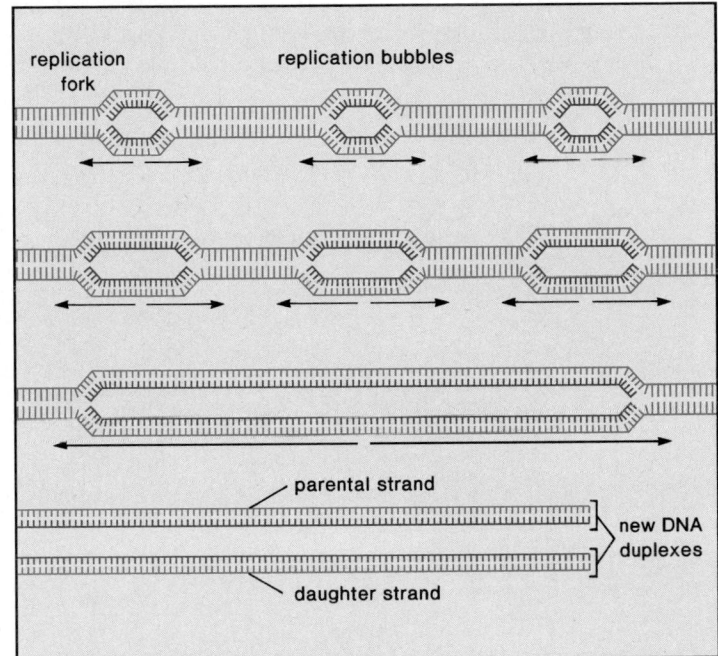

replication fork

replication bubbles

parental strand

new DNA duplexes

daughter strand

rate—500–5,000 base pairs per minute—there are many individual origins of replication. Therefore, eukaryotic cells complete the replication of the diploid amount of DNA (in humans over 6 billion base pairs) in a matter of hours!

> Generally speaking, DNA replication is rapid and accurate. If an error in base pairing does take place, then a mutation has occurred.

Other Aspects of DNA Biochemistry

There have been several unexpected discoveries regarding DNA biochemistry. We will discuss some of these.

B-DNA and Z-DNA

The Watson and Crick model showed that DNA is a right-handed, smooth double helix with 10 stacked bases for each complete turn of the molecule. This form of DNA is called B-DNA. But other forms of DNA have been found.

In 1979, Alexander Rich, Andrew Wang, and their colleagues discovered a new form of DNA—a left-handed double helix that is not smooth. The backbones of this form of DNA zigzag as they make their turns; therefore, it is called Z-DNA. Z-DNA is narrower

Figure 15.8

Two forms of the double helix. **a.** B-DNA is the form that corresponds to the structure proposed by Watson and Crick. B-DNA is said to be a right-handed spiral because the backbone (red line connecting red phosphate groups) goes up from left to right. B-DNA has both major and minor grooves. **b.** Z-DNA is a left-handed spiral with a backbone that zig-zags as it goes from right to left. Z-DNA has only one type of groove.

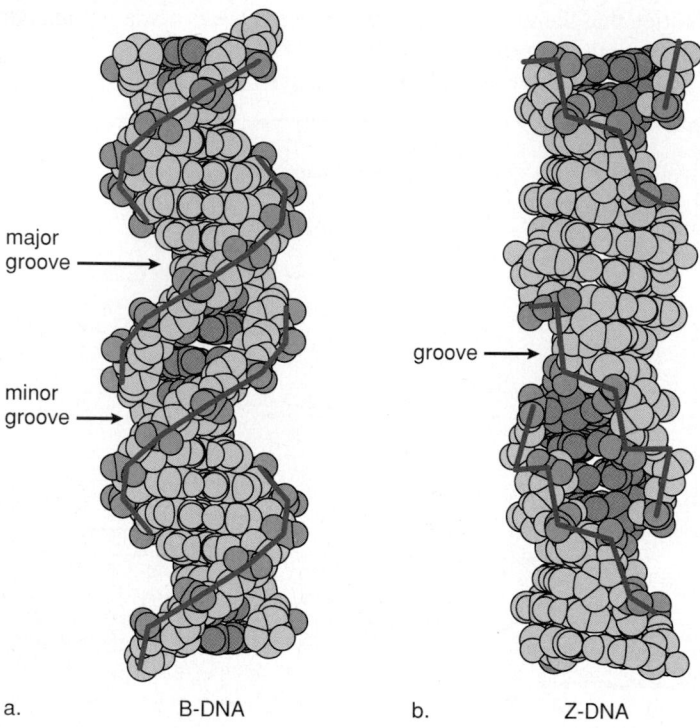

major
groove →

minor
groove →

groove →

a.   B-DNA   b.   Z-DNA

(1.8 nm) than B-DNA (2.0 nm), and it contains 12 nucleotide base pairs per turn. It is also grooved differently from B-DNA. Whereas B-DNA contains both major (wide) grooves and minor grooves, Z-DNA contains only one kind of groove (fig. 15.8).

Z-DNA has been found to occur naturally in cells. For example, the interband regions of *Drosophila* polytene chromosomes contain Z-DNA (see fig. 13.15). Stability of the molecule is greatly improved if it has undergone methylation (the addition of—CH$_3$ groups) and if there are proteins associated with it. All of these observations suggest that the Z-DNA configuration is involved in regulating the activity of genes. Perhaps the transformation of B-DNA to Z-DNA or vice versa causes genes to be turned on or off. Turned-on genes are directing protein synthesis, and turned-off genes are not doing so.

Organization of DNA in Cells

Classical geneticists who made chromosome maps would have predicted that the **genome** (all the genes of an organism) consists of DNA that directs the synthesis of cytoplasmic proteins. It appears, however, that at most 70% of eukaryotic DNA seems to function in this way. The rest either has no function or has a function we haven't been able to determine yet.

Figure 15.9

a. Barbara McClintock reported forty years ago that her corn experiment results could only be explained by the presence of movable genetic elements, now called jumping genes. She was so ahead of her time that the scientific community did not recognize her achievement for many years. She received the Nobel Prize in 1983. **b.** An example of Indian corn showing kernels that have unusual colors. This was the experimental material that prompted McClintock to suggest that there were movable genetic elements.

a.

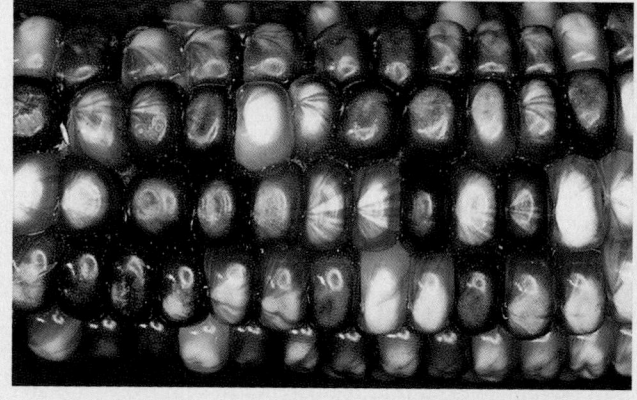

b.

Repetitive DNA Sequences

Genes that direct the synthesis of cytoplasmic proteins are separated from one another by repetitive DNA. **Repetitive DNA** contains the same sequence of base pairs repeated many times. In *highly repetitive DNA,* the same sequence of 5-15 base pairs is repeated 100,000-1 million times. This number of base pairs is much too short to be multiple copies of the same gene. Since highly repetitive DNA is found in the region of the centromere, it's speculated it has a structural role here and in all places where it is found. One interesting aspect of highly repetitive DNA is that the number of repeats between genes is inherited in a Mendelian fashion and is unique to the person. One's so-called DNA fingerprint is based on this finding (p. 291).

In *moderately repetitive DNA*, the same sequence of 1,000-1,500 base pairs is repeated only 10-3,000 times. Some of these sequences seem to be multiple copies of genes for ribosomal RNA, ribosomal proteins, and histones. It can be speculated that the many copies of these genes allow a faster production rate of these substances than would otherwise be possible.

Intervening DNA Sequences

Eukaryotic genes in particular are interrupted by noncoding DNA sequences. The portions of DNA that are expressed (code for protein) are called exons, and the portions of DNA that are not expressed (do not code for protein) are called introns. The next chapter discusses exons and introns in more detail.

Movable Genetic Elements

Classical geneticists always assumed that the genes were fixed in a definite order on the chromosomes. It came as a surprise, then, when it was proven that organisms have movable genetic elements called **transposons.** Such elements are often called "jumping genes" because they are able to move from one location to another in the genome. The movement of DNA from one place to another is called *transposition*.

All transposons contain a gene that allows them to move, but some contain other types of genes as well. For example, in bacteria, transposons contain one or more genes that make a bacterium resistant to antibiotics. This is of some concern because transposons have even been known to move between bacterial cells.

The first jumping genes were discovered in corn by Barbara McClintock in the 1940s. She called them controlling elements because when inserted into a new location they altered the activity of nearby genes, including those, for instance, that affected the color of kernels (fig. 15.9). For this discovery she received the Nobel Prize in 1983. Transposons have now been found in fruit flies and humans, in addition to bacteria and corn.

New findings regarding DNA biochemistry have shown that DNA is more variable in its form and function than originally thought.

Summary

1. Early work on the biochemistry of DNA wrongly suggested that it was composed of units each having only one of each of the 4 different nucleotides—A, T, C, or G. This so-called tetranucleotide hypothesis meant that DNA lacked the variability necessary for the genetic material.

2. Frederick Griffith injected strains of pneumococcus into mice and observed that smooth (S) strain bacteria are virulent but rough (R) strain bacteria are not. When heat-killed S strain bacteria were injected along with live R strain bacteria, however, virulent S strain bacteria were recovered from the dead mice. Griffith said that the R strain had been transformed by some transforming substance passing from the dead S strain to the live R strain.

3. Twenty years later, Oswald Avery and his colleagues reported that the transforming substance is DNA. They showed that purified DNA (and not protein, for example) is capable of bringing about the transformation.

4. Hershey and Chase turned to bacteriophage T2 as their experimental material. In 2 separate experiments, they labeled the protein coat with ^{35}S and the DNA with ^{32}P and then showed that the radioactive P alone is largely taken up by the bacterial host and that reproduction of viruses proceeds normally. This convinced most people that DNA is the genetic material.

5. In the meantime, work on the biochemistry of DNA was progressing. Chargaff did a chemical analysis of the molecule and found that A = T and G = C and that the amount of purine equals the amount of pyrimidine. These relationships are known as Chargaff's rules.

6. Franklin prepared an X-ray photograph of DNA that showed it is helical, has repeating structural features, and has certain dimensions.

7. Watson and Crick built a model of DNA in which the sugar-phosphate molecules made up the sides of a twisted ladder and the complementary paired bases were the rungs of the ladder. This model was in keeping with Chargaff's rules and the dimensions provided by the Franklin X-ray diffraction pattern.

8. The Watson and Crick model immediately suggested a method by which DNA could be replicated. The 2 strands unwind and unzip, and each parental strand acts as a template for a new (daughter) strand. In the end, each new duplex is like the other and like the parental duplex. The actual process is more complicated and involves many enzymes. The enzyme DNA polymerase joins the nucleotides together and proofreads them to make sure the bases have been paired correctly.

9. Replication is semiconservative since each new duplex contains an old (parental) strand and a new (daughter) strand. Meselson and Stahl demonstrated this by the following experiment: bacteria were grown in heavy nitrogen (^{15}N) and then switched to light nitrogen (^{14}N). The density of the DNA following replication was intermediate between these 2, as measured by centrifugation of the molecules through a salt gradient.

10. Replication in prokaryotes is bidirectional from one point of origin and proceeds until there are 2 copies of the circular chromosome. Replication in eukaryotes is also bidirectional, but there are many points of origin and many bubbles (places where the DNA strands are separate and replication is occurring). Replication occurs at the ends of the bubbles—at replication forks.

11. Since the time of Watson and Crick, it has been discovered that DNA is more variable than realized and that not all of

it functions as genes in the classical sense. As examples of the former, we now know of the existence of Z-DNA and transposons, and as examples of the latter, we now know there is highly repetitive DNA with a function yet to be determined.

Study Questions

1. List and discuss the requirements for genetic material.
2. What is the tetranucleotide hypothesis, and why did this hypothesis hinder the acceptance of DNA as the genetic material?
3. Describe Griffith's experiments with pneumococcus, his surprising results, and conclusion.
4. How did Avery and his colleagues demonstrate that the transforming substance is DNA?
5. Describe the experiment of Hershey and Chase, and explain how it shows that DNA is the genetic material.
6. What is the difference between a base, a nucleotide, and a nucleic acid? Which of these pertains to Chargaff's rules and what are the rules?
7. What is an X-ray diffraction pattern? What information was provided by DNA's diffraction pattern?
8. Describe the Watson and Crick model of DNA structure. How did it fit the data provided by Chargaff and the X-ray diffraction pattern?
9. Explain how DNA replicates semiconservatively. What role does DNA polymerase play? What role does helicase play?
10. How did Meselson and Stahl demonstrate semiconservative replication?
11. List and discuss differences between prokaryotic and eukaryotic replication of DNA.
12. What are Z-DNA, repetitive DNA, and transposons, and why are they significant?

Objective Questions

For questions 1-4, match the names in the key to the statements below:

Key:

    a. Griffith
    b. Chargaff
    c. Meselson and Stahl
    d. Hershey and Chase

1. A = T and G = C.
2. Only the DNA from T2 enters the bacteria.
3. R strain bacteria became an S strain through transformation.
4. DNA replication is semiconservative.
5. If 30% of an organism's DNA is thymine then
    a. 70% is purines.
    b. 20% is guanine.
    c. 30% is adenine.
    d. Both b and c.

6. If you grew bacteria in heavy nitrogen and then switched them to light nitrogen, how many generations after switching would you have some light/light DNA?
    a. never, because replication is semiconservative
    b. first generation
    c. second generation
    d. only the third generation
7. If one strand of DNA has the base sequence ATCGTA, what will the complementary strand have?
    a. TAGCAT
    b. ATCGTA
    c. CAGTCT
    d. All of these.
8. The double helix model of DNA resembles a twisted ladder in which the rungs of the ladder are

    a. a purine paired with a pyrimidine.
    b. A paired with G and C paired with T.
    c. sugar-phosphate paired with sugar-phosphate.
    d. Both a and b.
9. Cell division requires that the genetic material be able to
    a. store information.
    b. be replicated.
    c. undergo rare mutations.
    d. All of these.
10. In a DNA molecule, the
    a. bases are covalently bonded to the sugars.
    b. sugars are covalently bonded to the phosphates.
    c. bases are hydrogen bonded to one another.
    d. All of these.

Genetic Basis of Life

11. In the following diagram, blue stands for heavy DNA (contains ^{15}N) and red stands for light DNA (does not contain ^{15}N). Label all 3 strands as heavy or light DNA, and explain why the diagram is in keeping with the semiconservative replication of DNA.

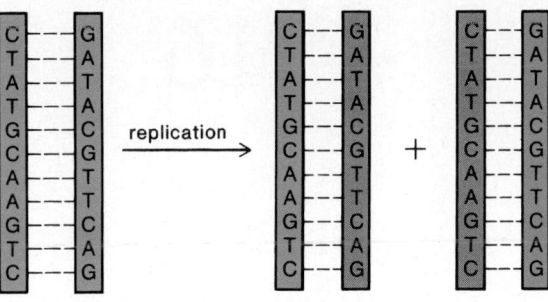

Concepts and Critical Thinking

1. *DNA is the genetic material that dictates the form, the function, and the behavior of organisms.*

   What would have been the results of the Hershey and Chase experiment if protein and not DNA were the genetic material?

2. *The genetic material has to be capable of replication.*

   To explain the importance of DNA replication, associate the necessity of replication with the cell cycle and the life cycle.

3. *Evolution is dependent on the ability of the genetic material to mutate.*

   Explain the importance of mutation to the evolutionary process.

Selected Key Terms

mutation (mu-ta'shun) 235
nucleic acid (nu-kle'ik as'id) 235
DNA (deoxyribonucleic acid) (de-ok"se-ri"bo-nu-kle'ik as'id) 235
RNA (ribonucleic acid) (ri"bo-nu-kle'ik as'id) 235
bacteriophage (bak-te're-o-fāj") 237

adenine (ad'ĕ-nīn) 238
thymine (thi'min) 238
cytosine (si'to-sin) 238
guanine (gwan'in) 238
purine (pu'rin) 238
pyrimidine (pi-rim'i-din) 238
complementary base pairing (kom"plĕ-men'tă-re bās pār'ing) 240

DNA polymerase (de'en-a pol-im'er-ās) 241
semiconservative replication (sem"e-kon-ser'vah-tiv rĕ"plĭ-ka'shun) 241
genome (je'nom) 244
transposon (trans-po'zun) 245

16

Gene Activity

In a eukaryotic cell. DNA located in the nucleus *(lower right)* controls protein synthesis, which occurs in the cytoplasm. An RNA copy (yellow line) of DNA moves from the nucleus to the ribosomes (blue particles) in the cytoplasm, where it will be translated into the amino acid sequence of a specific protein.

Your study of this chapter will be complete when you can

1. list and discuss the early studies that led to the recognition of gene activity;
2. draw and explain a diagram that outlines the central dogma of biology;
3. list the biochemical differences between RNA and DNA;
4. show that the DNA code is triplet, degenerate, and almost universal;
5. describe the process by which RNA is made complementary to DNA;
6. list and discuss 2 ways that mRNA is processed before it leaves the eukaryotic nucleus;
7. describe the roles of ribosomes, mRNA, tRNA, and amino acids during protein synthesis;
8. determine the mRNA codons, possible tRNA anticodons, and the sequence of amino acids in the resulting protein when given a DNA coding strand and a table of codons;
9. discuss the different types of gene mutations, the rate of mutation, and the manner in which DNA is protected from mutation.

The genes are composed of DNA, a molecule that lies within the nucleus of a eukaryotic cell (fig. 16.1). From its headquarters, DNA controls the metabolism of the cell and ultimately brings about the phenotype of the individual. How does DNA manage to store the necessary information and how does it exert its control? Just as it took some time for investigators to determine that DNA is the genetic material, so it also took some time to discover exactly what a gene is and does. Knowledge about the action of genes preceded that of their makeup.

The Activity of Genes

In the early 1900s, the English physician Sir Archibald Garrod suggested that there is a relationship between inheritance and metabolic diseases. He introduced the phrase *inborn error of metabolism* to dramatize this relationship. Garrod observed that family members often had the same disorder, and he said this inherited defect could be caused by the lack of a particular enzyme in a metabolic pathway. Since it was known at the time that enzymes are proteins, Garrod was among the first to hypothesize a link between genes and proteins.

One Gene-One Enzyme Hypothesis

Many years later, in 1940, George Beadle and Edward Tatum performed a series of experiments on *Neurospora crassa,* the red bread mold, which reproduces by means of spores. Normally, the spores become mold capable of growing on minimal medium (containing only a sugar, mineral salts, and the vitamin biotin) because it can produce all the enzymes it needs. In their experiments, Beadle and Tatum used X rays to induce mutations in asexually produced haploid spores. Some of the X-rayed spores could no longer become mold capable of growing on minimal medium; however, growth was possible on medium enriched by certain metabolites. In the example given in figure 16.2, the mold grows only when supplied with enriched medium that includes all metabolites or C and D alone. Since C and D are a part of this hypothetical pathway

$$A \xrightarrow{1} B \xrightarrow{2} C \xrightarrow{3} D$$

in which the numbers are enzymes and the letters are metabolites, it is concluded that the mold lacks enzyme 2. Beadle and Tatum further found that each of the mutant strains has only one defective gene leading to one defective enzyme and one additional growth requirement. Therefore, they proposed that each gene specifies the synthesis of one enzyme. This is called the *one gene-one enzyme hypothesis.*

One Gene-One Polypeptide Hypothesis

The one gene-one enzyme hypothesis suggests that a gene mutation causes a change in the structure of a protein. To test this idea, Linus Pauling and Harvey Itano decided to see if the hemoglobin in the red blood cells of persons with sickle-cell disease has a structure different from that of the red blood cells of normal

Figure 16.1

DNA location and structure. DNA is highly compacted in chromosomes, but it is extended as chromatin during interphase. It is during this time that DNA can be extracted from a cell and its structure studied.

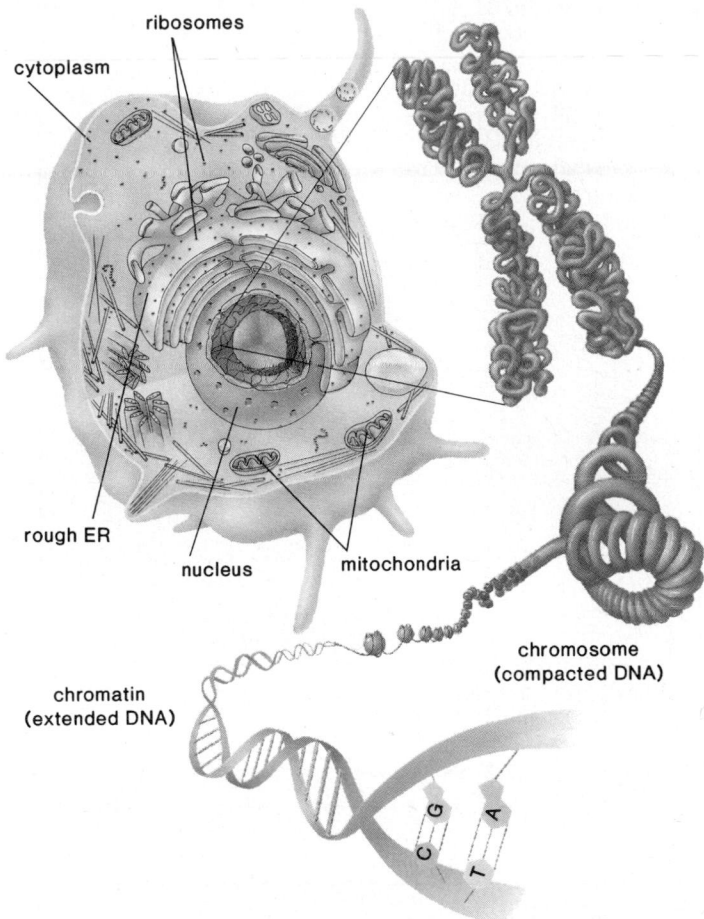

individuals (fig. 16.3). Recall that proteins are polymers of amino acids, some of which carry a charge. These investigators decided to see if there was a charge difference between normal hemoglobin (Hb^A) and sickle-cell hemoglobin, (Hb^S). To determine this, they subjected hemoglobin collected from normal individuals, sickle-cell trait individuals, and sickle-cell disease individuals to electrophoresis, a procedure that separates molecules according to their size and charge (p. 290). Here is what they found:

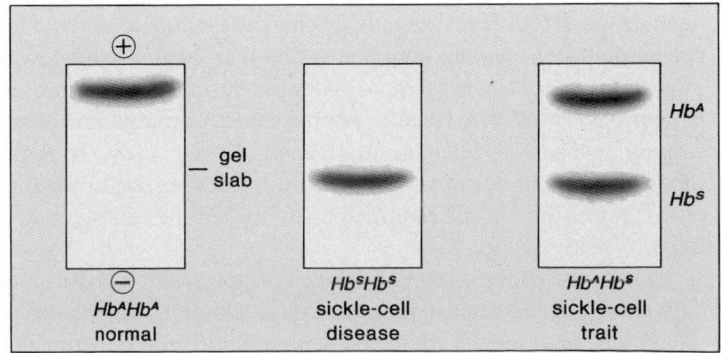

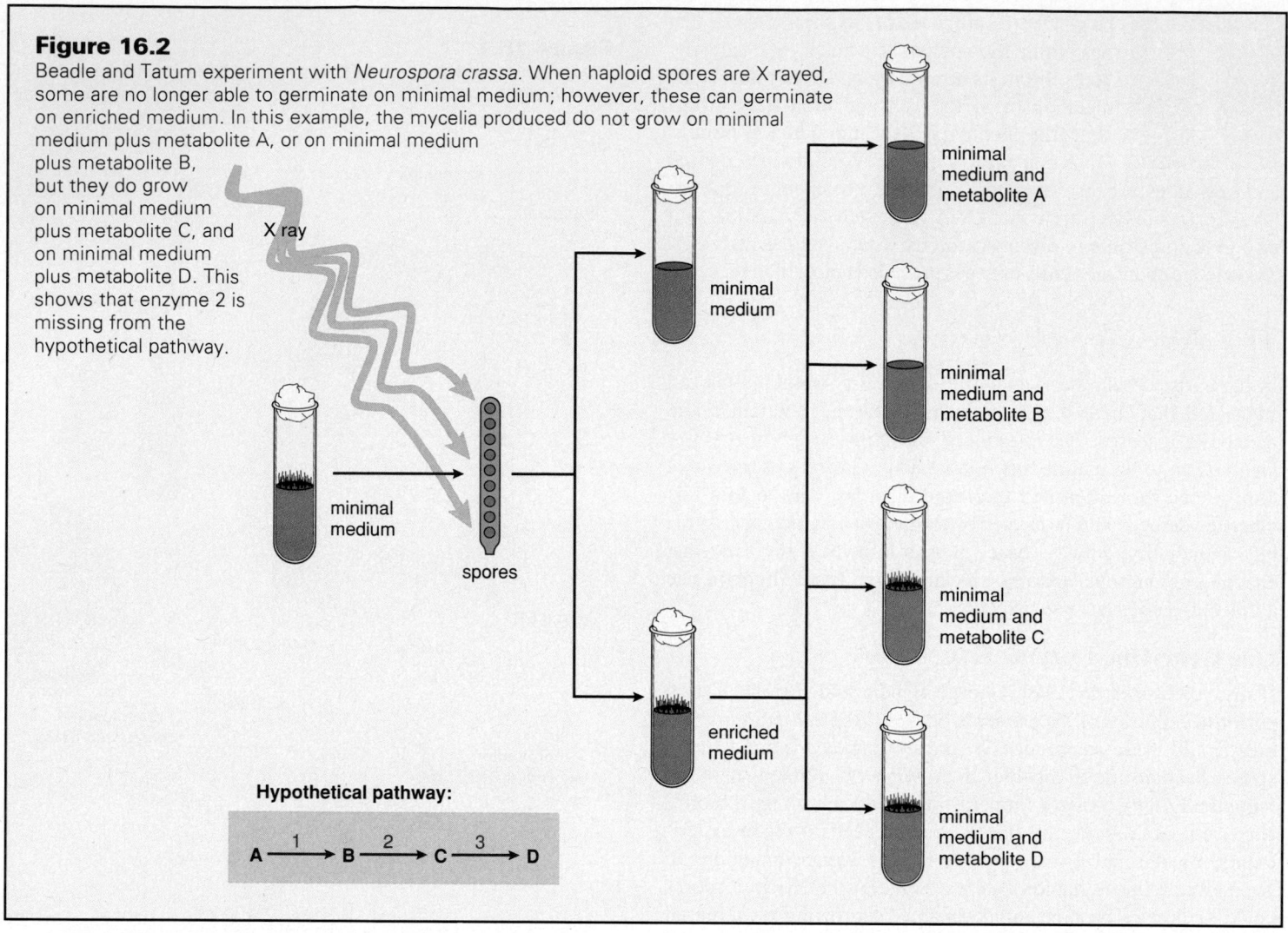

Figure 16.2

Beadle and Tatum experiment with *Neurospora crassa*. When haploid spores are X rayed, some are no longer able to germinate on minimal medium; however, these can germinate on enriched medium. In this example, the mycelia produced do not grow on minimal medium plus metabolite A, or on minimal medium plus metabolite B, but they do grow on minimal medium plus metabolite C, and on minimal medium plus metabolite D. This shows that enzyme 2 is missing from the hypothetical pathway.

X ray

minimal medium

spores

minimal medium

enriched medium

minimal medium and metabolite A

minimal medium and metabolite B

minimal medium and metabolite C

minimal medium and metabolite D

Hypothetical pathway:

A $\xrightarrow{1}$ B $\xrightarrow{2}$ C $\xrightarrow{3}$ D

As you can see, there is a difference in migration rate toward the positive pole between normal hemoglobin and sickle-cell hemoglobin. Further, hemoglobin from those with sickle-cell trait separates into 2 distinct bands, one corresponding to that for Hb^A hemoglobin and the other corresponding to that for Hb^S hemoglobin. Pauling and Itano therefore demonstrated that a mutation leads to a change in the structure of a protein.

Several years later, Vernon Ingram was able to determine the structural difference between Hb^A and Hb^S, where normal hemoglobin Hb^A contains negatively charged glutamate, and sickle-cell hemoglobin contains nonpolar valine. This explains the slower migration of Hb^S toward the positive pole during electrophoresis. It also causes Hb^S to be less soluble and to precipitate out of solution, especially when environmental oxygen is low. At these times, the Hb^S molecules stack up into long, semirigid rods that push against the plasma membrane and distort the red blood cell into the sickle shape.

Hemoglobin contains 2 types of polypeptide chains and only one type is affected in persons with sickle-cell trait; therefore, there must be a gene for each type of polypeptide. A refinement of the one gene—one enzyme hypothesis was needed, and it was replaced by the *one gene—one polypeptide hypothesis.*

Each gene contains the information for making one polypeptide of a protein, a molecule that may contain one or more different polypeptides. A mutated gene can lead to a defective polypeptide chain.

The Central Dogma of Molecular Biology

DNA is a linear polymer of nucleotides, and proteins are linear polymers of amino acids. This colinearity suggests that the nucleotide sequence of DNA somehow determines the order of amino acids in proteins. A DNA molecule, however, cannot directly control protein synthesis because DNA is found in the nucleus (or the nucleoid in prokaryotes) and protein synthesis occurs in the cytoplasm. Therefore, there must be a molecule that acts as a go-between, and the most likely candidate is RNA, a nucleic acid found in both the nucleus and the cytoplasm.

Genetic Basis of Life

Figure 16.3

Sickle-cell disease in humans. **a.** Scanning electron micrograph of normal (*left*) and sickled (*right*) red blood cells. Magnification, X2,500. **b.** Portion of the chain in normal hemoglobin Hb^A and in sickle-cell hemoglobin Hb^S. Although the chain is 146 amino acids long, the one change from glutamate to valine in the sixth position results in sickle-cell disease. **c.** Glutamate has a polar R group, while valine has a nonpolar R group, and this causes Hb^S to be less soluble and to precipitate out of solution. The Hb^S molecules stack up in long, semirigid rods that push against the plasma membrane and distort the red blood cell into sickle shape.

a.

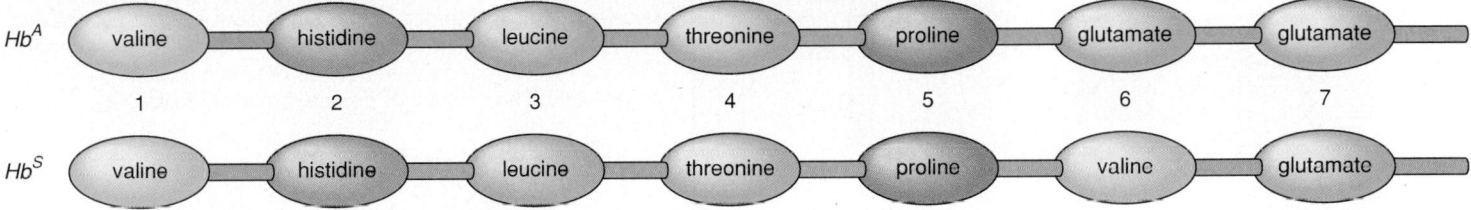

b.

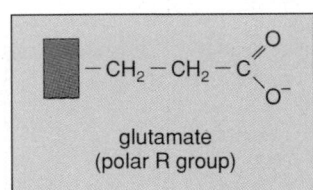
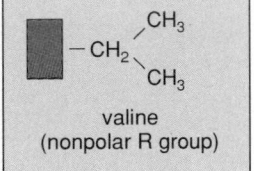

c.

RNA

Like DNA, RNA is a polymer of nucleotides (fig. 16.4). The nucleotides in RNA, however, contain the sugar ribose and the bases adenine (A), cytosine (C), guanine (G), and uracil (U). In other words, the base uracil replaces the thymine found in DNA (table 16.1). Finally, RNA is single-stranded and does not form a double helix in the same manner as DNA.

There are 3 different classes of RNA, each with specific functions in protein synthesis:

Messenger RNA (mRNA): takes a message from DNA in nucleus to ribosomes in cytoplasm.
ribosomal RNA (rRNA): along with proteins makes up the ribosomes, where proteins are synthesized.
transfer RNA (tRNA): transfers amino acids to the ribosomes.

Gene Expression

The process by which a gene produces a product, usually a protein, is called *gene expression*. The central dogma of molecular biology explains the manner in which genes are expressed:

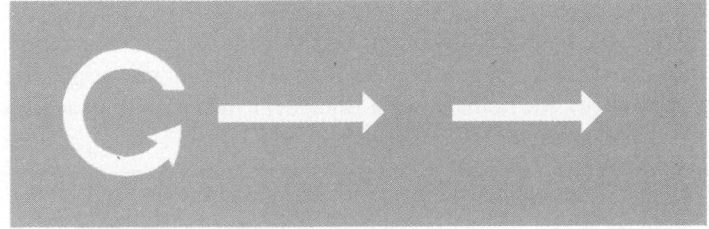

This diagram indicates that DNA not only serves as a template for its own replication, it is also a template for RNA formation. Most often it is mRNA that is produced. The process by which a mRNA

Figure 16.4

Like DNA, RNA is a polymer of nucleotides. RNA, however, is single-stranded, the pentose sugar is ribose, and uracil replaces thymine as one of the pyrimidine bases.

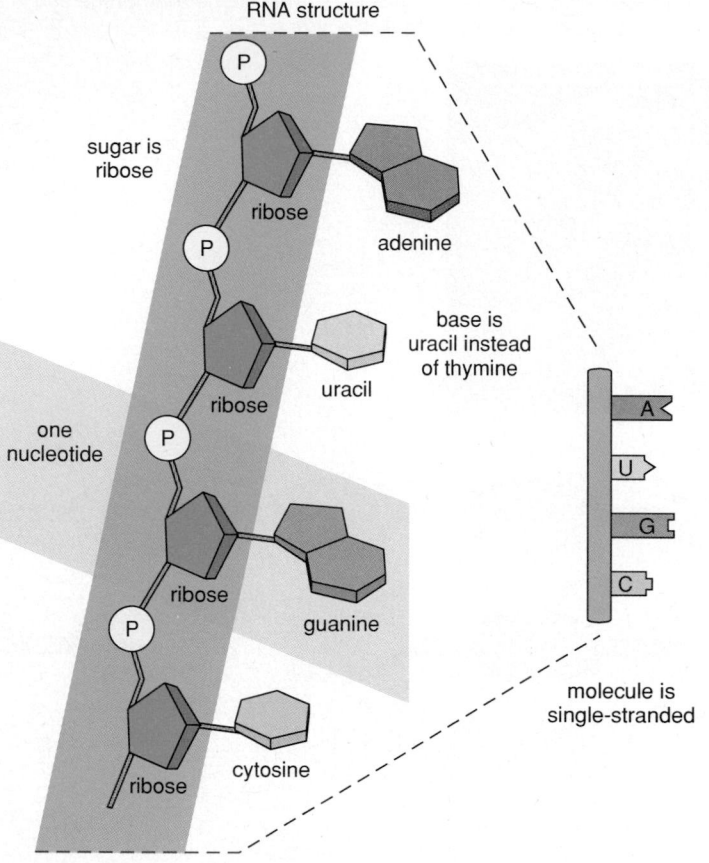

RNA structure

P

sugar is ribose

ribose

adenine

base is uracil instead of thymine

ribose

uracil

one nucleotide

P

ribose

guanine

P

ribose

cytosine

molecule is single-stranded

A

U

G

C

Table 16.1
DNA Compared to RNA

	DNA	RNA
Function	Genes; controls protein synthesis	Helper to DNA involved in protein synthesis
Sugar	Deoxyribose	Ribose
Bases	Adenine, guanine, thymine, cytosine	Adenine, guanine, uracil, cytosine
Strands	Double-stranded with base pairing	Single-stranded
Helix	Yes	No

Figure 16.5

Gene expression requires transcription and translation. Transcription occurs when DNA acts as a template for RNA synthesis. Translation occurs when the sequence of codons (groups of 3 bases) carried by a type of RNA called messenger RNA (mRNA) dictates the sequence of amino acids in a polypeptide. The product of gene expression is often a protein.

DNA double helix

DNA

transcription

G C C A T G A C C

C G G U A C U G G

mRNA

codon 1   codon 2   codon 3

translation

$$O \qquad\quad O \qquad\quad O$$
$$\| \qquad\quad \| \qquad\quad \|$$

polypeptide   – N – C – C – N – C – C – N – C – C –

R₁   R₂   R₃

copy is made of a portion of DNA is called **transcription.** Following transcription, mRNA will have a sequence of bases that is complementary to that of DNA (fig. 16.5). Then, mRNA moves into the cytoplasm. Photographic data shows radioactively labeled RNA moving from the nucleus to the cytoplasm, where protein synthesis occurs.

The central dogma of molecular biology also says that mRNA directs the synthesis of a polypeptide. During **translation,** the sequence of bases in mRNA dictates the sequence of amino acids in a protein.

Gene expression requires both transcription and translation. These terms are apt. Transcribing a document means making a copy of it, and translating a document means putting it in a different language.

Gene expression includes the processes of transcription and translation. During transcription, DNA serves as a template for the formation of complementary RNA. During translation, the sequence of bases in RNA determines the sequence of amino acids in a protein.

The Genetic Code

The central dogma of molecular biology suggests that the sequence of nucleotides in DNA and in RNA directs the order of amino acids in a protein. It would seem that there must be a code for each of the 20 amino acids found in proteins. But can 4 nucleotides provide enough combinations to code for 20 amino acids? If each code word, called a **codon,** were made up of 2 bases, such as AG, there could be only 16 codons—not enough to code for 20 amino acids. But if each codon were made up of 3 bases, such as AGC, there would be 64 codons—more than enough to code for 20 different amino acids:

Number of Bases in Genetic Code	Number of Different Amino Acids That Can Be Specified
1	4
2	16
3	64

It is no surprise, then, to learn that the genetic code is a **triplet code** and that each codon therefore consists of 3 nucleotide bases.

Cracking the Code

In 1961, Marshall Nirenberg and J. Heinrich Matthei performed an experiment that laid the groundwork for cracking the genetic code. First, they found a cellular enzyme could be used to construct a synthetic RNA (one that does not occur in cells), and then they found that the synthetic polymer could be translated in a cell-free system (a test tube that contains "freed" cytoplasmic contents of a cell). Their first synthetic RNA was composed only of uracil, and the protein that resulted was composed only of the amino acid phenylalanine. Therefore, the codon for phenylalanine was known to be UUU. Later, a cell-free system was developed by Nirenberg and Philip Leder in which only 3 nucleotides at a time were translated; in that way it was possible to assign an amino acid to each of the RNA codons (fig. 16.6).

A number of important properties of the genetic code can be seen by careful inspection of figure 16.6:

1. The genetic code is degenerate. This means that most amino acids have more than one codon; leucine, serine, and arginine have 6 different codons, for example. The degeneracy of the code is probably a protective device that reduces the potentially harmful effects of mutations.
2. The genetic code is unambiguous. Each triplet codon has only one meaning.
3. The code has "start" and "stop" signals. There is only one start signal but 3 stop signals.

The triplet genetic code is made of 64 3-base code words called codons. Except for the stop codons, all the codons code for amino acids.

Universality of the Genetic Code

The genetic code given in figure 16.6 is just about universally used by living things. This suggests that the code dates back to the very first organisms on earth and that all living things are related. Once the code was established, changes in it would have been very disruptive, and therefore it has remained unchanged over eons.

Figure 16.6

These messenger RNA (mRNA) codons are complementary to those in DNA. The DNA code is degenerate; more than one codon exists for most of the 20 amino acids. Notice that in this chart, each of the codons is comprised of a letter from the first, second, and third positions. For example, find the square in which C (from the first position) and A (from the second position) come together and then look across to the right; you will notice the letters for the third position of the codons for histidine and glutamine.

First Base	Second Base				Third Base
	U	C	A	G	
U	UUU phenylalanine	UCU serine	UAU tyrosine	UGU cysteine	U
	UUC phenylalanine	UCC serine	UAC tyrosine	UGC cysteine	C
	UUA leucine	UCA serine	UAA *stop*	UGA *stop*	A
	UUG leucine	UCG serine	UAG *stop*	UGG tryptophan	G
C	CUU leucine	CCU proline	CAU histidine	CGU arginine	U
	CUC leucine	CCC proline	CAC histidine	CGC arginine	C
	CUA leucine	CCA proline	CAA glutamine	CGA arginine	A
	CUG leucine	CCG proline	CAG glutamine	CGA arginine	G
A	AUU isoleucine	ACU threonine	AAU asparagine	AGU serine	U
	AUC isoleucine	ACC threonine	AAC asparagine	AGC serine	C
	AUA isoleucine	ACA threonine	AAA lysine	AGA arginine	A
	AUG (*start*) methionine	ACG threonine	AAG lysine	AGG arginine	G
G	GUU valine	GCU alanine	GAU aspartate	GGU glycine	U
	GUC valine	GCC alanine	GAC aspartate	GGC glycine	C
	GUA valine	GCA alanine	GAA glutamate	GGA glycine	A
	GUG valine	GCG alanine	GAG glutamate	GGG glycine	G

Exceptions to the universality of the genetic code are found in some prokaryotes and mitochondria. Recall that mitochondria have genes and carry on protein synthesis also. As it turns out, the code in mitochondria is slightly different from that utilized in the nucleus. This is consistent with the endosymbiotic theory, which says mitochondria were once independent organisms (see fig. 5.15). It also means that eukaryotes have a type of cytoplasmic inheritance—inheritance dependent on genes outside the nucleus. It seems that these genes are inherited only through the maternal line because the egg and not the sperm contributes organelles to the new individual. Investigators have been using this knowledge to explain certain genetic diseases that are inherited only from one's mother and to establish evolutionary relationships among human races.

Transcription

Transcription is the first step required for gene expression, the process by which a gene product is made. Most often this gene product is a protein, but we should note that the molecules tRNA and rRNA are also transcribed off DNA templates. These molecules are also gene products. Just now we are interested in the formation of mRNA, which carries genetic information to the ribosomes, where protein synthesis occurs.

Messenger RNA

During transcription, a mRNA molecule is formed that has a sequence of bases complementary to a portion of one DNA strand; wherever A, T, G, or C is present in the DNA template, U, A, C, or G is incorporated into the mRNA molecule (fig. 16.7). A segment of the DNA helix unwinds and unzips, and complementary RNA nucleotides pair with DNA nucleotides of the strand that is to be transcribed. When these RNA nucleotides are joined together by an **RNA polymerase,** an mRNA molecule results. This molecule now carries a sequence of codons that will be used to order the sequence of amino acids in a protein.

Transcription begins at a region of DNA called a promoter. A promoter is a special sequence of DNA bases where RNA polymerase attaches and the transcribing process begins. A promoter is at the start end of the gene to be transcribed. Some genes are on one of the DNA strands, and some are on the other strand.

Elongation of the mRNA molecule occurs as long as transcription proceeds. Only the newest portion of a RNA molecule is bound to the DNA, and the rest dangles off to the side. Finally, RNA polymerase comes to a terminator sequence at the other end of the gene being transcribed. The terminator causes RNA polymerase to stop transcribing the DNA and to release the mRNA molecule, now called a RNA transcript.

Many RNA polymerase molecules can be working to produce a RNA transcript at the same time (fig. 16.8). This allows the cell to produce many thousands of copies of the same mRNA molecule and eventually many copies of the same protein within a shorter period of time than otherwise.

RNA Processing

Since the advent of modern molecular techniques, investigators can compare the structure of various eukaryotic RNA transcripts and their corresponding genes. They do this by first isolating the mRNA and its corresponding gene. Then, they separate the DNA molecule into single strands and allow the mRNA to bind to its complementary strand. If the 2 molecules are indeed colinear, then the mRNA should bind along the entire length of its template DNA. Much to their surprise, researchers have found that much of human template DNA does not bind to its mRNA. The segments of a gene

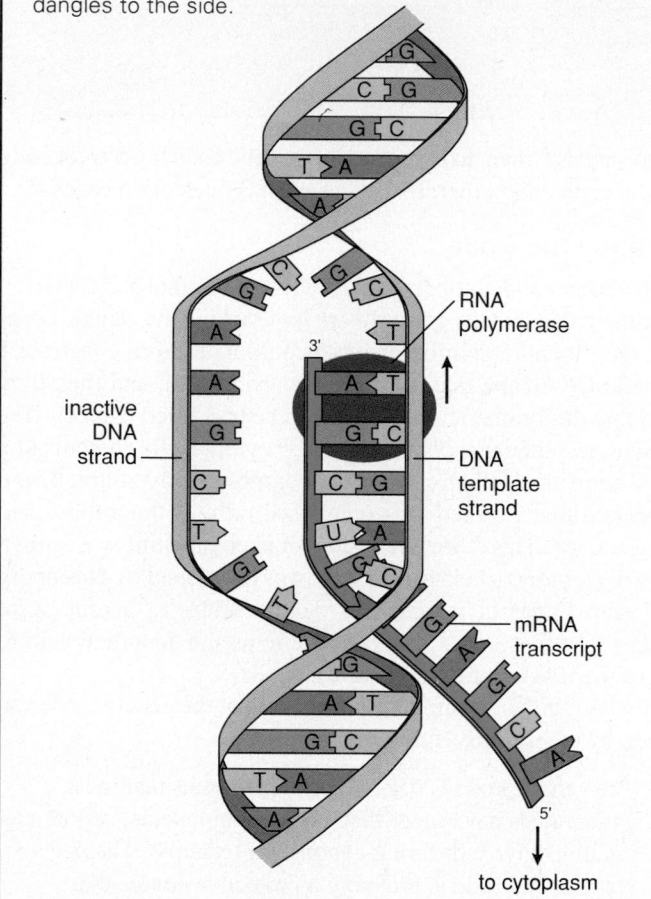

Figure 16.7
During transcription, complementary RNA is made from a DNA template. At the point of attachment of RNA polymerase, the DNA helix unwinds and unzips, and complementary RNA nucleotides are joined together. This is called transcription because the nucleotide sequence in DNA is transcribed into RNA nucleotides. This is the message that mRNA will take to the cytoplasm. After RNA polymerase has passed by, the DNA strands rejoin and the RNA transcript dangles to the side.

that do not bind to mRNA and therefore do not code for protein are called intervening sequences, or **introns.** The segments of a gene that do bind to mRNA and therefore do code for protein are called **exons** because they are expressed.

By comparing the mRNA molecule present in the nucleus with that in the cytoplasm, it can be shown that both exons and introns are present in the *primary mRNA* transcript, but only exons are present in the *mature mRNA* transcript that leaves the nucleus and enters the cytoplasm. The introns are removed from the primary mRNA transcript by a process called *RNA processing* or RNA splicing (fig. 16.9).

Since the discovery of split (interrupted) genes in eukaryotes, 2 essential questions have been asked: How is processing carried out? and What is the function of introns in the first place? It was discovered that splicing of RNA may be done by *spliceosomes,*

Figure 16.8

a. Numerous RNA transcripts extend from a horizontal gene in an amphibian egg cell. Magnification, X65,000. **b.** The strands get progressively longer because transcription begins to the left. The dark dots along the DNA are RNA polymerase molecules that are joining RNA nucleotides together. The dots at the end of the strands are spliceosomes involved in RNA processing (*see* fig. 16.9).

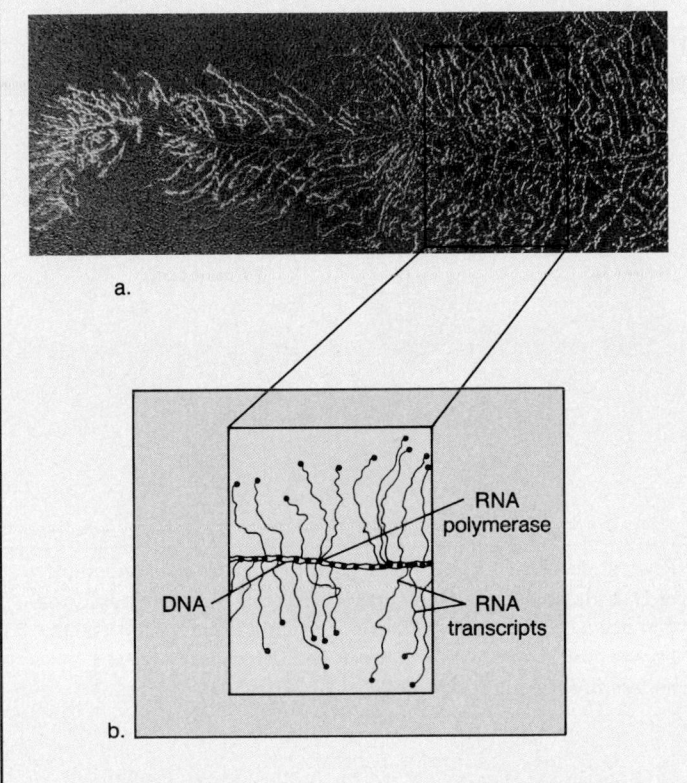

Figure 16.9

mRNA processing in eukaryotes. **a.** DNA contains both exons (coding sequences) and introns (noncoding sequences). **b.** Both of these are transcribed and are present in primary mRNA. **c.** During processing a cap and a poly A tail (a series of adenine nucleotides) are added to the molecule. Also, there is excision of the introns and a splicing together of the exons. This is accomplished by complexes called spliceosomes. **d.** Now the mature mRNA molecule is ready to leave the nucleus.

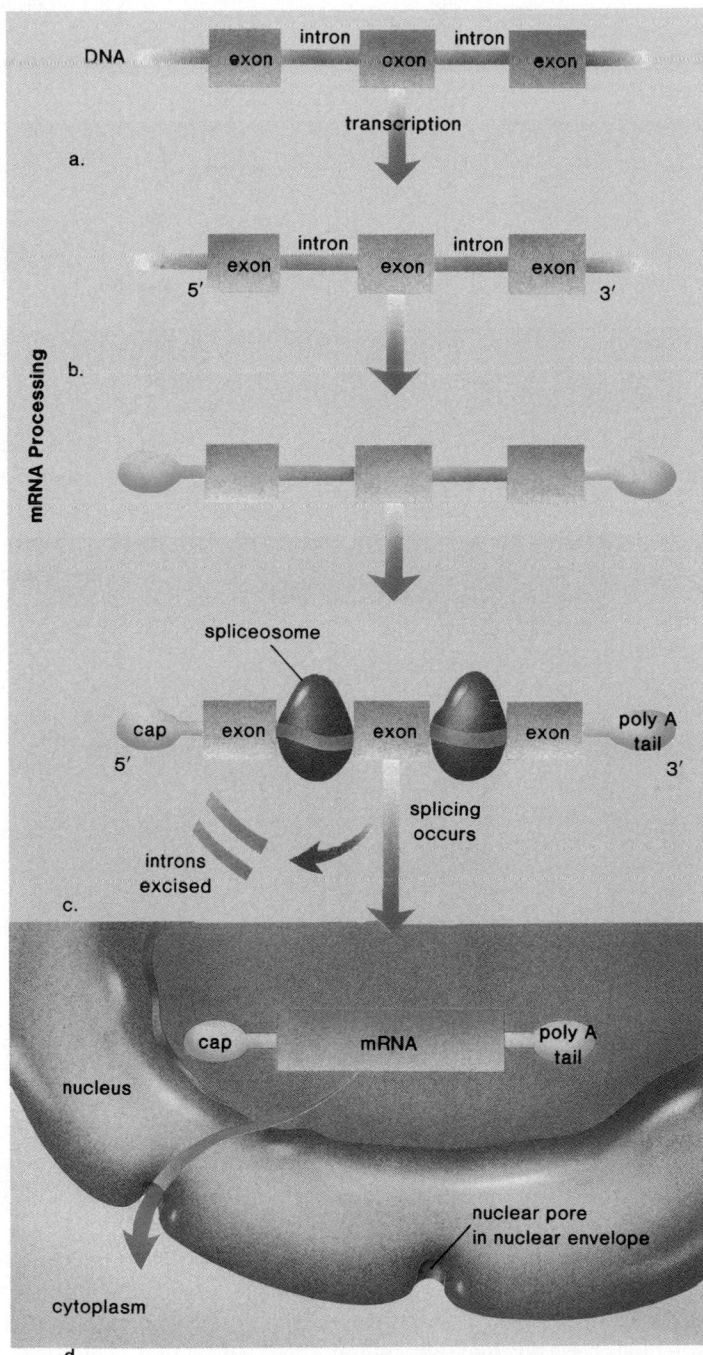

a complex that contains several kinds of ribonucleoproteins. A spliceosome cuts the primary mRNA and then rejoins the adjacent exons (fig. 16.9).

There has been much speculation about the possible role of introns in the eukaryotic genome. It's possible that introns allow crossing-over within a gene during meiosis. It's also possible that introns divide a gene up into regions that can be joined in different combinations to give novel genes and protein products, a process that perhaps facilitates evolution.

Some researchers are trying to determine whether introns exist in more primitive eukaryotes and in prokaryotes. They have found that the more primitive the eukaryote, the less likely a gene is to be interrupted by introns and the introns that do exist are shorter. At first it was thought that introns do not exist at all in prokaryotes, but an intron has been discovered in the gene for a tRNA molecule in *Anabaena*, a cyanobacterium. This particular intron belongs to a class of introns called "self-splicing." Self-splicing introns, which have the enzymatic capability of splicing themselves out of an RNA transcript, were discovered in the early

Figure 16.10

Polysome structure. **a.** Several ribosomes move along an mRNA at a time. They function independently of each other so that several polypeptides can be made at the same time. **b.** Electron micrograph of a polysome. Magnification, X400,000.

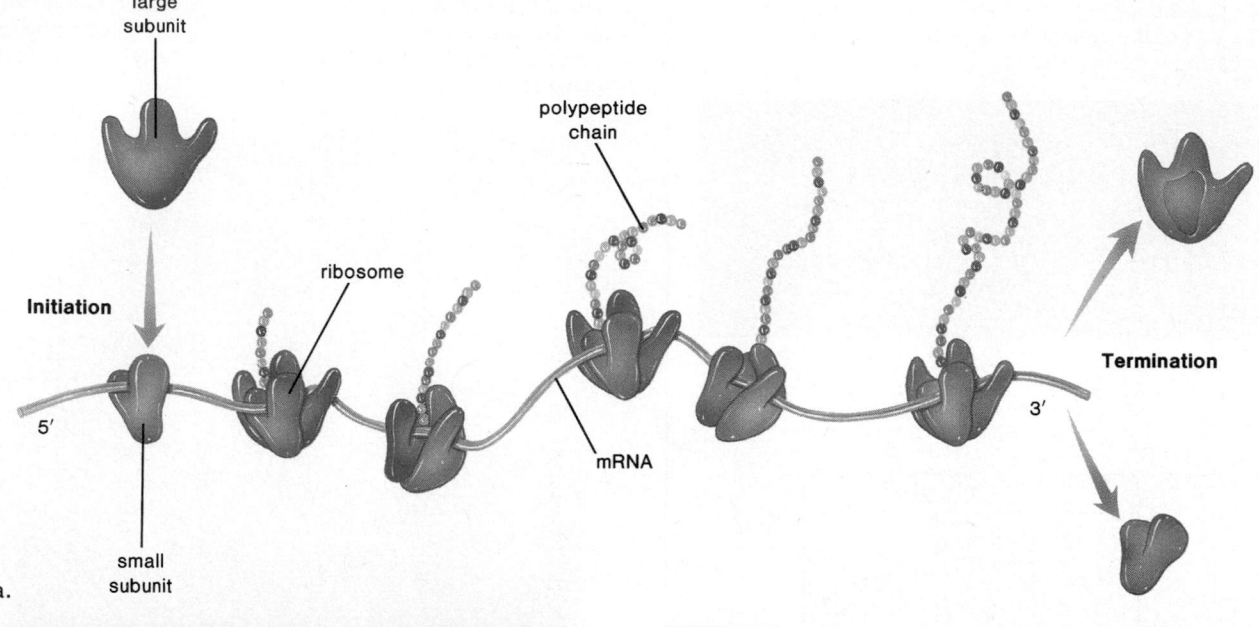

a.

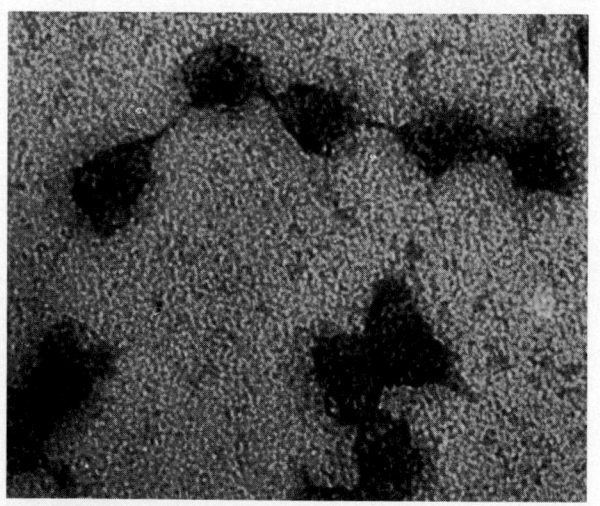

b.

RNA molecules (mRNA, rRNA, and tRNA) are transcribed off of DNA templates. mRNA carries a copy of the genetic information needed for protein synthesis. Particularly in eukaryotes, the primary mRNA transcript is processed before it becomes a mature mRNA transcript.

Translation

Translation is the second step by which gene expression leads to protein synthesis. During translation, the sequence of codons in mRNA directs the sequence of amino acids in a protein. Two other types of RNA are needed for protein synthesis. rRNA is contained in the ribosomes, where the codons of mRNA are read, and tRNA carries amino acids to the ribosomes so that protein synthesis can occur.

Ribosomal RNA

Ribosomal RNA (rRNA) molecules, along with a variety of proteins, make up the ribosomes, which are tiny organelles that are attached to endoplasmic reticulum and also occur free within the cytoplasm of eukaryotic cells (fig. 16.1).

Prokaryotic cells contain about 10,000 ribosomes, but eukaryotic cells contain many times this number. In both types of organisms, ribosomes are composed of 2 subunits that don't join together until protein synthesis begins. The small subunit contains one rRNA molecule and several different types of proteins, and the large subunit contains 2 rRNA molecules and several different types of proteins. Among these proteins is the enzyme that joins amino acids together by means of a peptide bond.

1980s. The finding of these so-called ribozymes did away with the belief that only proteins can function as enzymes. Ribozymes, however, are restricted in their function since each one cleaves only RNA at specific locations.

This discovery of ribozymes in prokaryotes is being used to substantiate the belief that RNA could have been the first genetic material and the first enzyme in the history of life. For many years, scientists have puzzled over which came first—DNA, which is the genetic material, or proteins, which are enzymes. Now it appears that this is an unnecessary dilemma. Possibly RNA could have fulfilled both functions in the first cell or cells.

Genetic Basis of Life

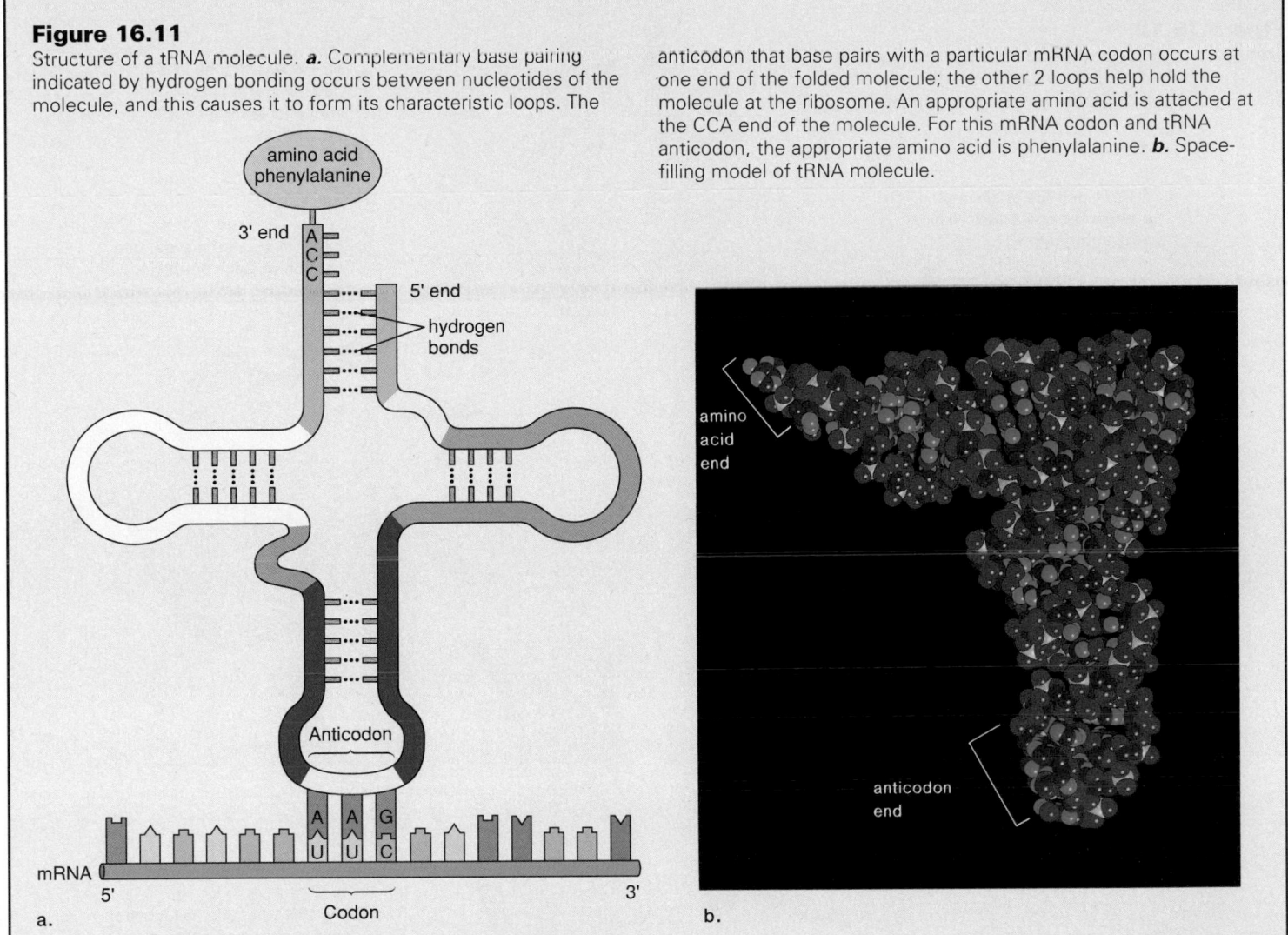

Figure 16.11

Structure of a tRNA molecule. *a.* Complementary base pairing indicated by hydrogen bonding occurs between nucleotides of the molecule, and this causes it to form its characteristic loops. The anticodon that base pairs with a particular mRNA codon occurs at one end of the folded molecule; the other 2 loops help hold the molecule at the ribosome. An appropriate amino acid is attached at the CCA end of the molecule. For this mRNA codon and tRNA anticodon, the appropriate amino acid is phenylalanine. *b.* Space-filling model of tRNA molecule.

The function of ribosomes is to ensure that protein synthesis occurs in an orderly fashion and that the amino acids carried by tRNA molecules are attached in the order directed by the mRNA codons.

Polysomes

As soon as the initial portion of mRNA has been translated by one ribosome, it is free to have another ribosome attach to it. Therefore, several ribosomes are often attached to and translating the same mRNA. The entire complex is called a **polysome** (fig. 16.10).

Transfer RNA

Transfer RNA (*tRNA*) molecules transfer amino acids to the ribosomes. A tRNA molecule is a single-stranded nucleic acid that doubles back on itself to give regions where complementary bases are hydrogen bonded to one another. The structure of a tRNA molecule is generally drawn as a flat cloverleaf, but the space-filling model shows the molecule's actual shape (fig. 16.11).

There is at least one tRNA molecule for each of the 20 amino acids found in a protein. The amino acid binds to one end of the molecule where there is a CCA sequence. The other end of the molecule contains an **anticodon,** a group of 3 bases that is complementary to a specific codon of mRNA. For example, a tRNA that has the anticodon AAG binds to the codon UUC and carries the amino acid phenylalanine. Because the anticodons of tRNA molecules bind to the codons of a mRNA molecule, the amino acids become sequenced in a polypeptide according to the information provided by a gene. In other words, tRNA molecules translate one language (nucleic acids) into another language (protein).

One area of active research is to determine how the correct amino acid becomes attached to the correct tRNA molecule. So far, it is known that the task is carried out by a number of amino acid activating enzymes, called *tRNA synthetases.* Somehow a tRNA synthetase recognizes which amino acid

Figure 16.12

Translation. tRNA-amino acid molecules arrive at the ribosome, and the sequence of mRNA codons dictates the order in which amino acids become incorporated into a polypeptide.

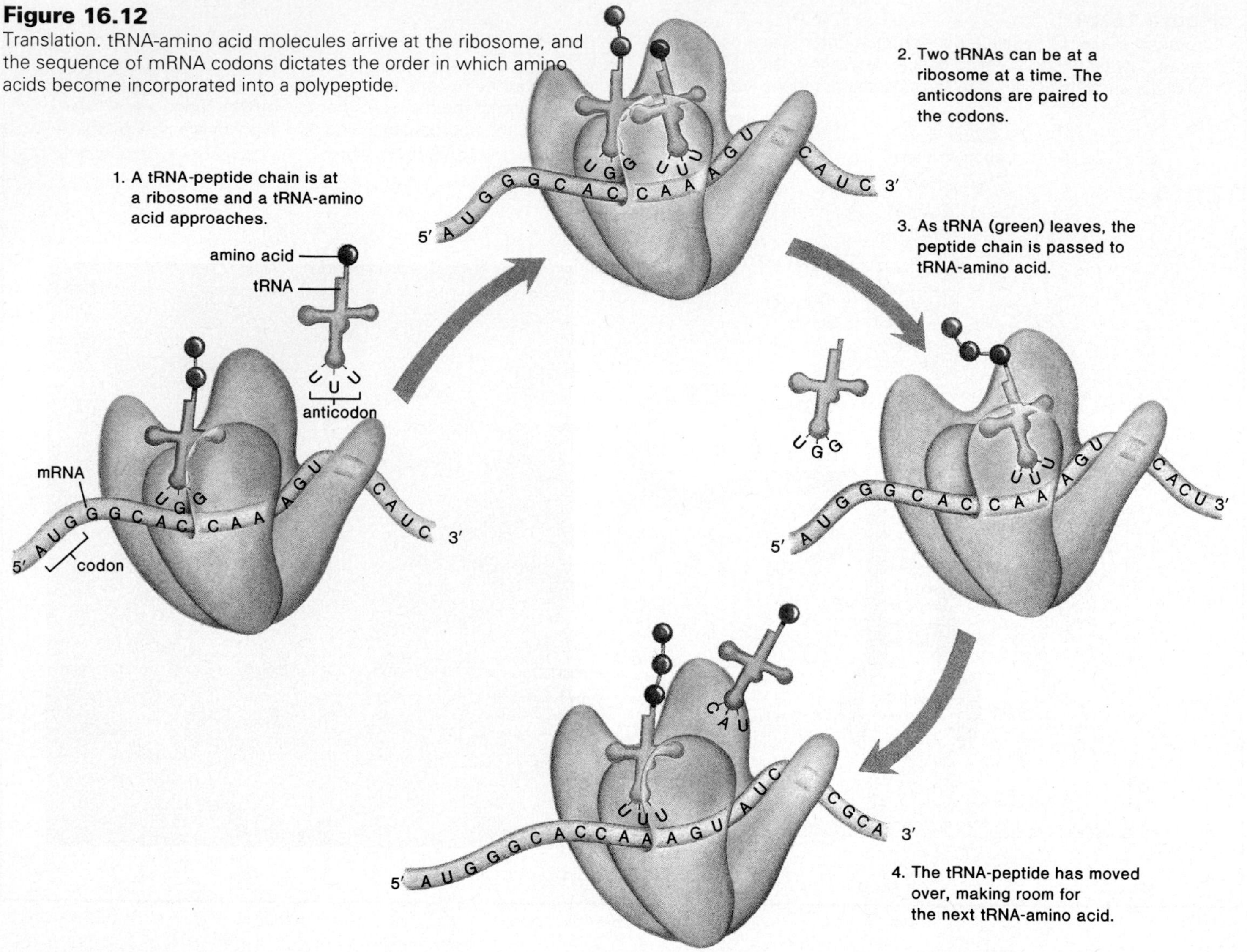

1. A tRNA-peptide chain is at a ribosome and a tRNA-amino acid approaches.

amino acid

tRNA

anticodon

mRNA

codon

5′ AUGGGCACCAAAGU CAUC 3′

2. Two tRNAs can be at a ribosome at a time. The anticodons are paired to the codons.

5′ AUGGGCACCAAAGU CAUC 3′

3. As tRNA (green) leaves, the peptide chain is passed to tRNA-amino acid.

5′ AUGGGCACCAAAGU CACU 3′

4. The tRNA-peptide has moved over, making room for the next tRNA-amino acid.

5′ AUGGGCACCAAAGU AUC CGCA 3′

should be joined to which tRNA molecule. This is an energy-requiring process that utilizes ATP. Once the amino acid-tRNA complex is formed, it travels through the cytoplasm to a ribosome, where protein synthesis is occurring.

Translation Process

The process of translation must be extremely orderly so that the amino acids of a polypeptide are sequenced correctly. Protein synthesis involves 3 steps: initiation, elongation, and termination.

1. Initiation of translation: A small ribosomal subunit attaches to the mRNA in the vicinity of the *start codon* (AUG). The first or initiator tRNA pairs with this codon. Then a large ribosomal subunit joins to the small subunit, and translation begins.

2. Chain elongation: Each ribosome contains 2 sites, the P (for polypeptide) site and the A (for amino acid) site:

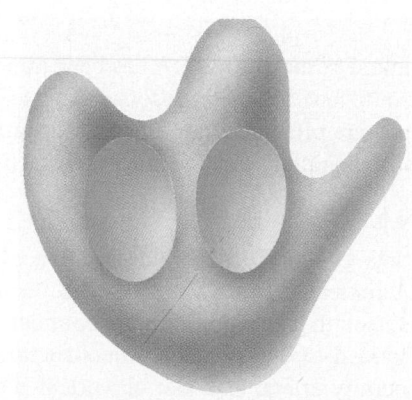

During elongation, a tRNA with attached polypeptide chain is at the P site and an tRNA-amino acid complex is just arriving at the A site (fig. 16.12). The polypeptide chain is transferred and attached by a peptide bond to the newly arrived amino acid. An enzyme (peptidyl transferase), which is a part of the larger ribosomal subunit, and energy are needed to bring about this transfer. Now the tRNA molecule at the P site leaves.

Then *translocation* occurs: the mRNA along with the peptide-bearing tRNA moves from the A site to the empty P site. This makes it seem as if the ribosome has moved forward 3 nucleotides, especially since there is a new codon now located at the empty A site.

The complete cycle—pairing of new tRNA-amino acid complex, transfer of peptide chain, translocation—is repeated at a rapid rate (about 15 times each second in *E. coli*).

3. Chain termination: Termination of polypeptide chain synthesis occurs at a *stop codon,* codons that do not code for an amino acid. The polypeptide chain is enzymatically cleaved from the last tRNA, and it leaves the ribosome, which dissociates into its 2 subunits (fig. 16.10).

During translation, the codons of mRNA base pair with the anticodons of tRNA molecules carrying specific amino acids. The order of the codons determines the order of the tRNA molecules and the sequence of amino acids in a polypeptide.

Gene Expression in Review

The following list, along with table 16.2 and figure 16.13, provides a brief summary of the events involved in gene expression that results in a protein product.

1. DNA contains genetic information. The sequence of its bases determines the sequence of amino acids in a protein.
2. During transcription, one strand of DNA serves as a template for the formation of messenger RNA (mRNA). The bases in mRNA are complementary to those in DNA; every 3 bases is a codon that codes for an amino acid.
3. Messenger RNA is processed before it leaves the nucleus, during which time the introns are removed.
4. Messenger RNA carries a sequence of codons to the ribosomes, which are composed of ribosomal RNA (rRNA) and proteins.
5. Transfer RNA (tRNA) molecules, each of which is bonded to a particular amino acid, have anticodons that pair complementarily to the codons in mRNA.
6. During translation, tRNA molecules and their attached amino acids arrive at the ribosomes and the linear sequence of codons of the mRNA determines the order in which the amino acids become incorporated into a protein.

Definition of a Gene and Gene Mutations

Classical (Mendelian) geneticists thought of a gene as a particle on a chromosome. To molecular geneticists, however, a gene is a sequence of DNA nucleotide bases that codes for a product. Most often the gene product is a polypeptide chain. A gene product can also be rRNA or tRNA, however. These RNA molecules are involved in protein synthesis, but they do not contain codons.

Gene Mutations

Early geneticists understood that genes undergo mutations, but they didn't know what causes mutations. It is apparent today that a *gene mutation* is a change in the nucleotide sequence of a gene. Gene mutations can be contrasted with chromosome mutations considered earlier (p. 208), which are changes in chromosome number or in the structure of a chromosome.

Frameshift Mutations

The term *reading frame* applies to the sequence of codons because they are read from some specific starting point as is this sentence: THE CAT ATE THE RAT. If the letter C is deleted from this sentence and the reading frame is shifted, we read THE ATA TET HER AT—something that doesn't make sense. *Frameshift mutations* occur most often because one or more nucleotides are either inserted or deleted from DNA. The result of a frameshift mutation can be a completely nonfunctional protein because the sequence of codons is completely altered. If by chance 3 adjacent nucleotides are inserted or deleted, the protein will be normal except for the addition or loss of one amino acid from the polypeptide chain. This observation lends support to the conclusion that the genetic code is a triplet code.

Point Mutations

Point mutations involve a change in a single nucleotide and therefore a change in a specific codon. When one base is substituted for another, the results can be variable. For example, in figure 16.14, if UAC is changed to UAU, there is no noticeable effect, because both of these codons code for tyrosine. Therefore, this is called a silent mutation. If UAC is changed to UAG, however, the result could very well be a drastic one because UAG is a stop codon. If this substitution occurs early in the gene, the resulting protein may be too short and may be unable to function. Such an effect is called a nonsense mutation. Finally, if UAC is changed to CAC, then histidine is incorporated into the protein instead of tyrosine. This is a missense mutation. A change in one amino acid may not have an effect if the change occurs in a noncritical area or if the 2 amino acids have the same chemical properties. In this instance, however, the polarities of tyrosine and histidine differ; therefore, this substitution most likely will have a deleterious effect on the functioning of the protein. Recall that the occurrence of valine instead of glutamate in the beta (ß) chain of hemoglobin results in a sickle-cell disease (fig. 16.3), for example.

Figure 16.13

Gene expression leads to the formation of a product, most often a protein. Transcription, which occurs in the nucleus, and translation, which occurs in the cytoplasm at the ribosomes, are 2 steps required for gene expression.

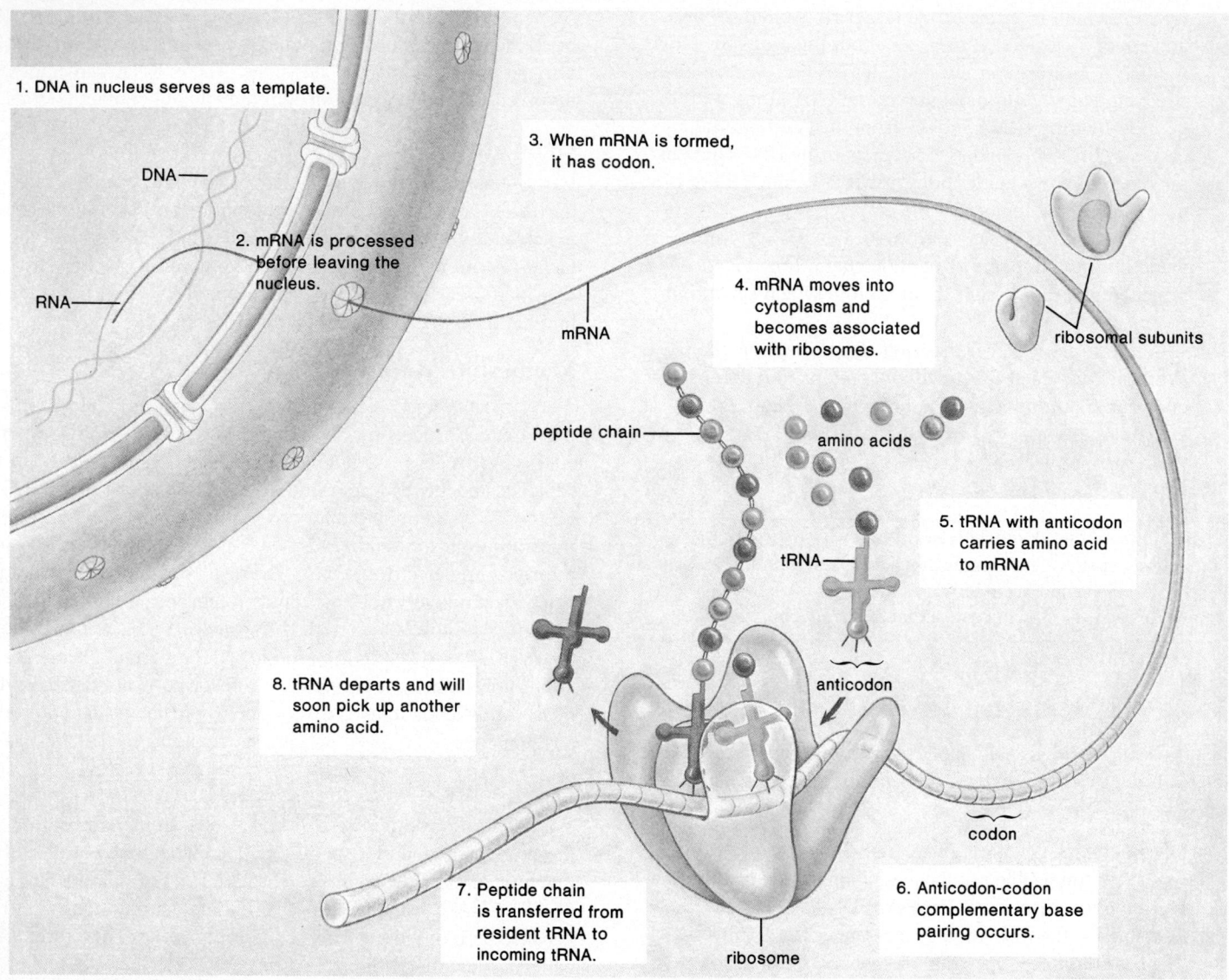

1. DNA in nucleus serves as a template.

DNA

RNA

2. mRNA is processed before leaving the nucleus.

3. When mRNA is formed, it has codon.

mRNA

4. mRNA moves into cytoplasm and becomes associated with ribosomes.

ribosomal subunits

peptide chain

amino acids

5. tRNA with anticodon carries amino acid to mRNA

tRNA

anticodon

8. tRNA departs and will soon pick up another amino acid.

7. Peptide chain is transferred from resident tRNA to incoming tRNA.

ribosome

codon

6. Anticodon-codon complementary base pairing occurs.

Rate of Gene Mutations

Per cell cycle, gene mutations don't occur very often; a frequency of 10^{-8} to 10^{-5} per generation is often quoted. There are several mechanisms that protect against the occurrence of mutations. The bases are on the interior of the DNA molecule, and the supercoiling of the molecule in eukaryotes also lends stability. During replication, DNA polymerases proofread the new strand against the old strand and detect any mismatched pairs, which are then replaced with the correct nucleotides. In the end, there is usually only one mistake for every one billion nucleotide pairs replicated.

Organisms, including humans, are continually exposed to **mutagens,** environmental substances that cause mutations, such as radiation (e.g., radioactive elements and X rays) and organic chemicals (e.g., certain pesticides and cigarette smoke). If these mutagens bring about a mutation in the gametes, then the offspring of the individual may be affected. On the other hand, if the mutation occurs in the body cells, then cancer may be the result. Ultraviolet (UV) radiation is a mutagen that everyone is exposed to; it easily penetrates the skin and breaks the DNA of underlying tissues. Wherever there are 2 thymine molecules next to one another, ultraviolet radiation may cause them to bond together, forming thymine dimers.

Table 16.2
Participants in Gene Expression

Name of Molecule	Special Significance	Definition
DNA	Genetic information	Sequence of DNA bases
mRNA	Codons	Complementary sequence of RNA bases in threes
tRNA	Anticodon	Sequence of 3 bases complementary to codon
rRNA	Ribosomes	Site of protein synthesis
Amino acids	Building blocks for protein	Transported to ribosomes by tRNAs
Protein	Enzyme, structural protein, or secretory product	Amino acids joined in a predetermined order

Figure 16.14

Point mutation. The effect of a base alteration can vary. Starting at far left, if the base change codes for the same amino acid, there is no noticeable effect; if the base change codes for a stop codon, the resulting protein will be incomplete; and if the base change codes for a different amino acid, a faulty protein is possible.

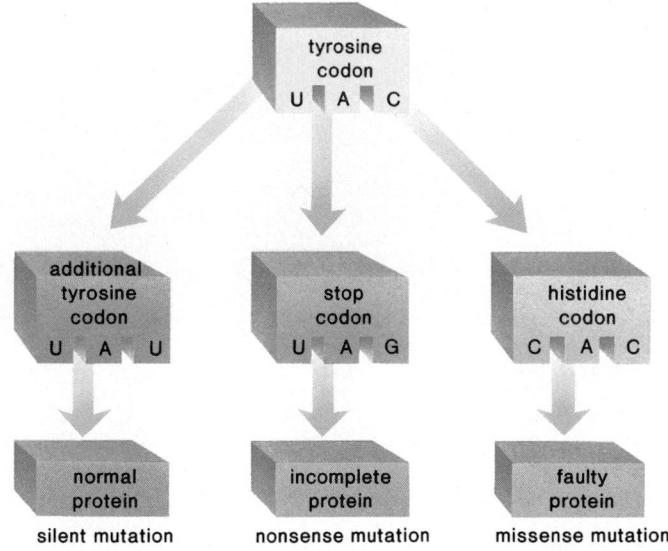

Usually, these dimers are removed from damaged DNA by special enzymes called repair enzymes. Repair enzymes constantly monitor DNA and repair any irregularities. One enzyme excises a portion of DNA that contains the dimer;

Figure 16.15

Repair enzymes. Several types of enzymes are involved in repair of a DNA thymine dimer. *a.* One enzyme carries out excision of the faulty section of DNA. *b.* Another carries out repair replication using the other strand as a template. *c.* Another is responsible for rejoining the new and old sections of DNA into a complete whole.

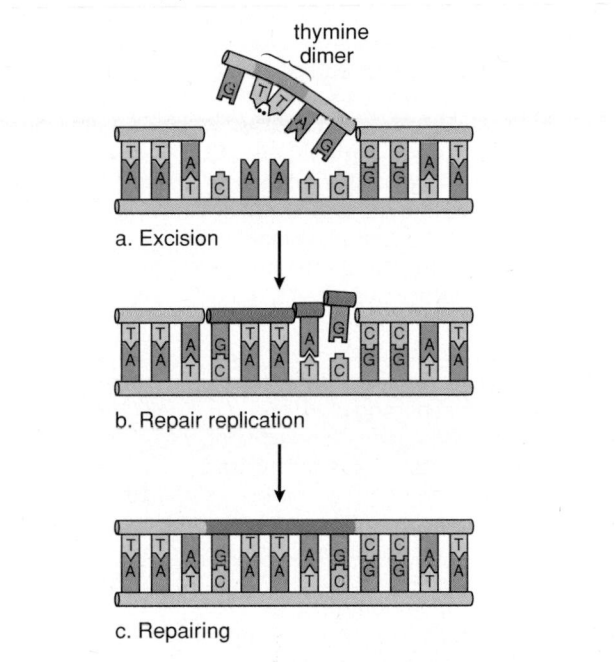

a. Excision

b. Repair replication

c. Repairing

another makes a new section by using the other strand as a template; and still another seals the new section in place (fig. 16.15). The importance of repair enzymes is exemplified by individuals with the condition known as xeroderma pigmentosum. They lack some of these repair enzymes, and as a consequence these individuals have a high incidence of skin cancer.

Significance of Mutations

Although most mutations we observe are harmful, they still provide the raw material for evolutionary change. Gene mutations, along with recombinations of genes, produce a new genotype whose phenotype may be better adapted to the environment. Better-adapted individuals are expected to have more surviving offspring than those less adapted and in this way the new genotype becomes prevalent. Therefore, evolution occurs.

A gene mutation is an alteration in the nucleotide sequence of a gene. The spontaneous rate of mutation is low because DNA polymerases proofread the new strand against the old strand during replication and because DNA repair enzymes constantly monitor and repair any irregularities.

Summary

1. Several investigators contributed to recognition of gene activity. Garrod is associated with the phrase "inborn error of metabolism" because he suggested that some of his patients had inherited their inability to carry out certain enzymatic reactions.

2. Beadle and Tatum X-rayed spores of *Neurospora crassa* and found that certain of the subsequent cultures lacked a particular enzyme needed for growth on minimal medium. Since they found that the mutation of one gene results in the lack of a single enzyme, they suggested the one gene—one enzyme hypothesis.

3. Pauling and Itano found that the chemical properties of the beta chain of sickle-cell hemoglobin differ from those of normal hemoglobin, and therefore the one gene—one polypeptide chain hypothesis was formulated. Later, Ingram showed that the biochemical change is due to the substitution of the amino acid valine (nonpolar) for glutamate (polar).

4. The central dogma of molecular biology says that (1) DNA is a template for its own replication and is a template for RNA formation during transcription, and (2) the complementary sequence of nucleotides in mRNA orders the correct sequence of amino acids of a polypeptide during translation (i.e., DNA→RNA→protein).

5. RNA differs from DNA in these ways: (1) the pentose sugar is ribose, not deoxyribose; (2) the base uracil replaces thymine; (3) RNA is single-stranded.

6. DNA has a code to indicate the proper sequence of amino acids in a protein. It is a triplet code, and each codon (code word) consists of 3 bases. The code is degenerate; that is, more than one codon exists for most amino acids. There are also one start and 3 stop codons.

7. The genetic code is just about universal; one exception is mitochondrial DNA, which has been found to have different codons for certain amino acids.

8. Transcription requires template binding (RNA polymerase begins transcription at a promoter region), RNA chain initiation (first 2 nucleotides join), chain elongation (chain lengthens), and chain termination (RNA is released because of stop signal).

9. mRNA is processed following transcription, and introns are removed. In eukaryotes, introns are excised by spliceosomes.

10. Translation requires mRNA, rRNA, and tRNA. Ribosomal RNA (rRNA) is found in the ribosomes, and transfer RNA (tRNA) is found in the cytosol. Each tRNA has an anticodon at one end and an amino acid at the other; there are amino acid activating enzymes that ensure that the correct amino acid is attached to the correct tRNA. When a tRNA binds with its codon at the ribosome, the correct amino acid is positioned for inclusion in a polypeptide.

11. Translation requires these steps. During initiation, mRNA, the first (initiator) tRNA, and the ribosome all come together in the proper orientation at a start codon. During elongation, as the tRNAs recognize their codons, the growing peptide chain is transferred by peptide bonding to the next amino acid in the chain. During termination at a stop codon, the polypeptide is cleaved from the last tRNA. The ribosome dissociates, and the mRNA leaves.

12. Many ribosomes move along an mRNA at a time. Collectively these are called a polysome.

13. In molecular terms, a gene is a sequence of DNA nucleotide bases, and a gene mutation is a change in this sequence. Frameshift mutations result when a base is added or deleted and the result is a nonfunctioning protein. Point mutations can range in effect depending on the particular codon change. Gene mutation rates are rather low because DNA polymerase proofreads the new strand during replication and because there are repair enzymes that constantly monitor the DNA.

14. Gene mutations are the raw material for the evolutionary process, and along with recombination, may lead to greater adaptation to an environment and a change in the most prevalent phenotype of the species.

Writing Across the Curriculum

In order to practice writing skills, students should write out the answers to any or all of the study questions and the critical thinking questions. The study questions are sequenced in the same order as the text. Suggested answers to the critical thinking questions are in appendix D.

Study Questions

1. What made Garrod think there were inborn errors of metabolism?
2. Explain Beadle and Tatum's experimental procedure.
3. How did Pauling and Itano know that the chemical properties of *Hb*^{S} differed from those of *Hb*^{A}? What change does a substitution of valine for glutamate cause in the chemical properties of these molecules?
4. Draw a diagram for the central dogma of biology and tell where the various events occur in a eukaryotic cell.
5. What are the chemical differences between RNA and DNA?
6. How did investigators reason that the code must be a triplet code and in what manner was the code cracked? Why is it said that the code is degenerate, unambiguous, and universal?
7. What are the specific steps that occur during transcription of RNA off a DNA template?
8. How is mRNA processed before leaving the eukaryotic nucleus?
9. Compare the functions of mRNA, rRNA, and tRNA during protein synthesis. What are the specific events of translation?
10. Why is a frameshift mutation always expected to result in a faulty protein, while a substitution of one base for another may not always result in a faulty protein?

Genetic Basis of Life

Objective Questions

For questions 1-4, match the investigator to the phrase in the key:

Key:

   a. one gene—one enzyme hypothesis

   b. inborn error of metabolism

   c. one gene—one polypeptide hypothesis

   d. triplet code

1. Sir Archibald Garrod
2. Francis Crick
3. George Beadle and Edward Tatum
4. Linus Pauling and Harvey Itano
5. Considering the following pathway, if Beadle and Tatum found that *Neurospora* can not grow if metabolite A is provided but can grow if B, C, or D is provided, what enzyme would be missing?

       1   2   3

   A——→B——→C——→D

   a. enzyme 1
   b. enzyme 2
   c. enzyme 3
   d. All of these.

6. The central dogma of biology
   a. states that DNA is a template for all RNA production.
   b. states that DNA is a template only for DNA replication.
   c. states that translation precedes transcription.
   d. pertains only to prokaryotes because humans are unique.
7. If the sequence of bases in DNA is TAGC, then the sequence of bases in RNA will be
   a. ATCG.
   b. TAGC.
   c. AUCG.
   d. Both a and b.
8. RNA processing is
   a. the same as transcription.
   b. an event that occurs after RNA is transcribed.

   c. the rejection of old, worn-out RNA.
   d. Both b and c.
9. During protein synthesis, an anticodon on tRNA pairs with
   a. DNA nucleotide bases.
   b. rRNA nucleotide bases.
   c. mRNA nucleotide bases.
   d. other tRNA nucleotide bases.
10. If the DNA codons are CAT CAT CAT, and a guanine base is added at the beginning, then which would result?
   a. G CAT CAT CAT
   b. GCA TCA TCA T
   c. frameshift mutation
   d. Both b and c.
11. This is a segment of a DNA molecule. What are (a) the mRNA codons, (b) the possible tRNA anticodons, and (c) the sequence of amino acids in the polypeptide?

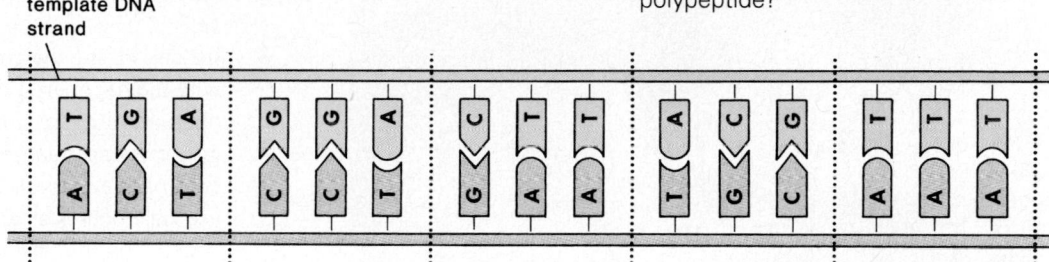

template DNA strand

Concepts and Critical Thinking

1. *DNA stores genetic information.*

   What is the genetic information stored by DNA, and where is it stored in the molecule?

2. *The genetic code is almost universal.*

   What is the evolutionary significance of a universal genetic code?

3. *There is a flow of information from DNA to RNA to protein.*

   DNA and a protein are colinear molecules. Explain.

Selected Key Terms

messenger RNA (mRNA) (mes'en-jer ar'en-a) 251
ribosomal RNA (rRNA) (ri'bo-sōm"al ar'en-a) 251
transfer RNA (tRNA) (trans'fer ar'en-a) 251

transcription (trans-krip'shun) 252
translation (trans"-lā'shun) 252
triplet code (trip'let kōd) 253
RNA polymerase (ar'en-a pol-im'er-ās) 254

intron (in'tron) 254
exon (eks'on) 254
polysome (pol'e-sōm) 257
anticodon (an"tǐ-ko'don) 257

17

Regulation of Gene Activity

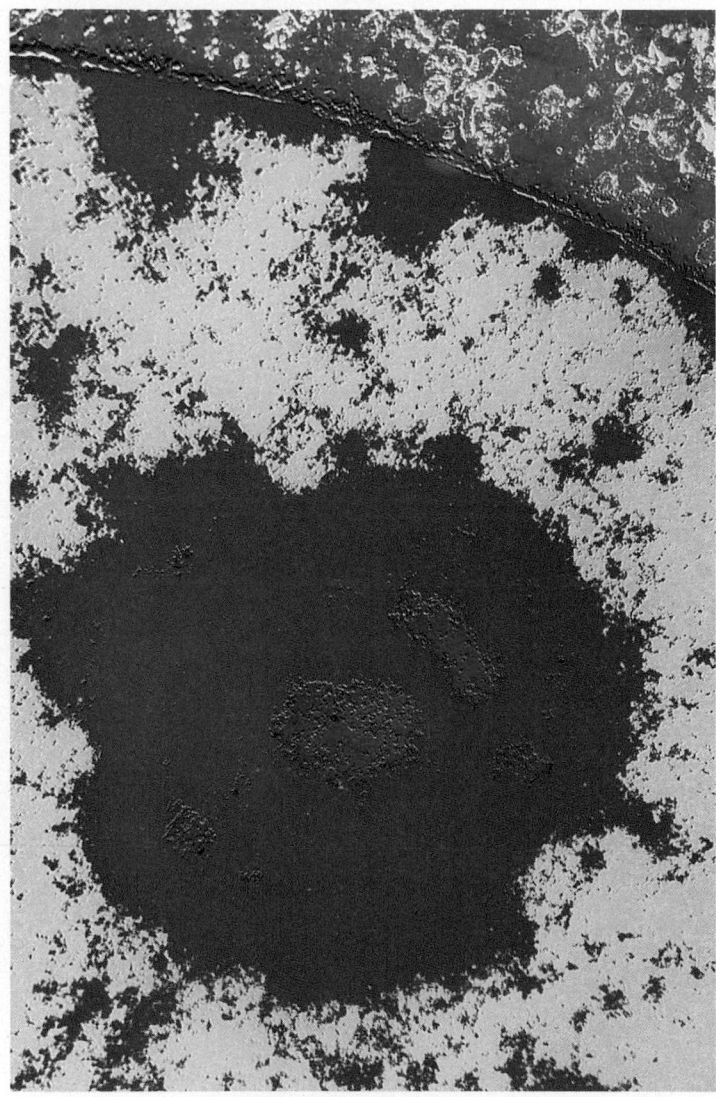

Colored electron micrograph of a section through a nucleus. The central red body is the nucleolus where rRNA is made and ribosomal subunits are assembled. Ribosomal subunits (red areas) are awaiting export to the cytoplasm (blue). Regulation of the expression of genes is not confined to transcriptional control, it also involves how many ribosomes are functioning in the cytoplasm. Magnification, X12,825.

Your study of this chapter will be complete when you can

1. explain the necessity for control of gene expression in prokaryotes and in eukaryotes;
2. list the levels of control of gene expression in eukaryotic cells;
3. list and define the components of an operon, and explain why the *lac* operon is an inducible operon and the *trp* operon is a repressible operon;
4. give evidence that genes are regulated at all levels of control in eukaryotes;
5. use Barr bodies as an example that heterochromatin is genetically inactive;
6. use giant (polytene) chromosome puffs as an example that euchromatin is genetically active;
7. describe cancer as a failure in genetic control, and relate the development of cancer as requiring initiation and promotion;
8. discuss the discovery of oncogenes and tumor suppressor genes and their possible functions;
9. describe the development of cancer as a multistep process;
10. list some recommendations for avoiding the development of cancer.

ells don't need the same enzymes (and possibly other proteins) all the time. Suppose, for example, the environment is supplying the amino acid tryptophan—wouldn't it be disadvantageous to produce the enzymes needed for tryptophan synthesis? Indeed, a medium containing tryptophan can be inoculated with 2 strains of *E. coli.* The first strain is capable of turning off its genes that code for tryptophan-producing enzymes, and the second, a mutated strain, is unable to do so. After a few days, the culture contains only cells of the first type—they have overgrown the cells of the second type. It takes energy to carry out transcription and translation, and using energy unnecessarily puts the mutated cells at a disadvantage. It's to be expected, then, that prokaryotic cells normally have a way to regulate gene expression.

Or, let's consider eukaryotic cells in a multicellular organism. Each mature cell is specialized for a particular task: for example, in humans, cells lining the gut produce a lot of digestive enzymes, while muscle cells need the contractile proteins actin and myosin. Numerous experiments of the type discussed in this chapter's reading indicate that each type of specialized cell contains a full complement of genes. In other words, all human cells have genes for digestive enzymes and muscle proteins, but they are expressed only in certain cells of the body. This tells us that eukaryotic cells also must have a way to regulate the action of genes.

Cells are able to regulate the expression of genes, so that at any particular time, only certain ones are fully turned on and producing a protein product.

Regulation of Gene Action

Regulation of gene activity pertains to all the steps involved in protein synthesis and protein function. In eukaryotic cells, there are 4 primary levels of control of gene activity (fig. 17.1).

1. **Transcriptional control:** A number of mechanisms serve to control which genes are transcribed and/or the rate at which transcription occurs.
2. **Posttranscriptional control:** Differential processing of mRNA (see fig. 16.9) and also the speed with which mRNA leaves the nucleus can affect the amount of gene expression finally realized.
3. **Translational control:** The life expectancy of mRNA molecules (how long they exist in the cytoplasm) can vary as can their ability to bind ribosomes. It is also possible that some mRNAs may need additional changes before they are translated at all.
4. **Posttranslational control:** Even after translation has occurred, the polypeptide product may have to undergo additional changes before it is biologically functional. Also, a functional protein is subject to feedback control in the manner described in figure 7.7.

Figure 17.1

Levels at which control of gene expression occurs in eukaryotic cells. Transcriptional and posttranscriptional control occur in the nucleus; translational and posttranslational control occur in the cytoplasm.

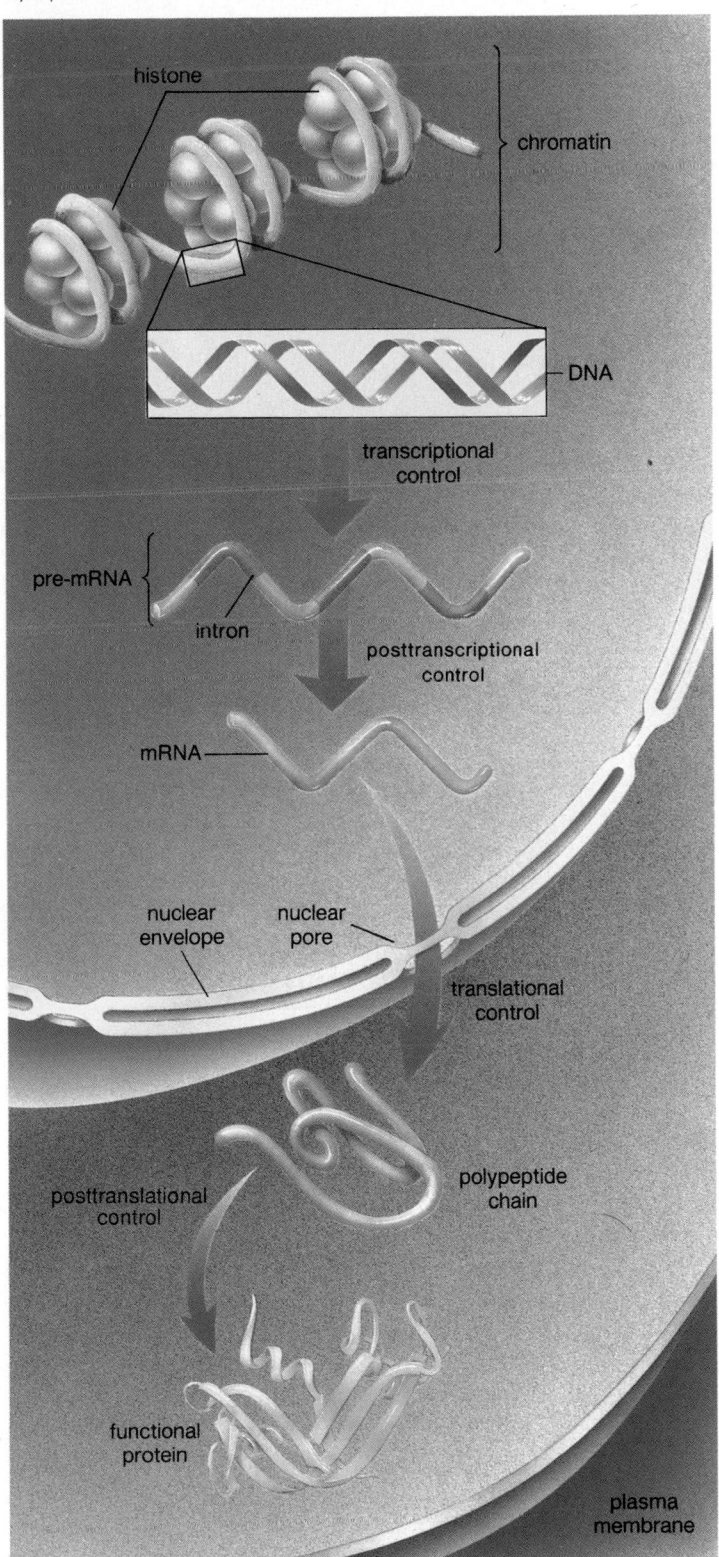

A multicellular organism begins life as a zygote that undergoes cell division to produce many cells. These cells then become differentiated and assume a particular structure and function. You might think at first that a parceling out of genes brings about differentiation, that is, different cells receive different genes. But, after you have studied this chapter, you will most likely hypothesize that the same genes are *present* in all cells and differentiation comes about because different genes are *active* in different cells.

Investigators have tested these 2 hypotheses for many years. First, they showed that the amount of DNA remains constant in most cells (exceptions are the sex cells [gametes], which have half the amount of DNA). This fact suggests that the hypothesis that genes are parceled out in different cells is incorrect. Then, investigators progressed to more challenging experiments to show that the nuclei in multicellular organisms are totipotent. *Totipotent* means "all powerful" and this expression is used to indicate that any cell in a multicellular organism is capable of acting like a zygote—it contains all the genetic information required to bring about complete development of the organism.

Adult plant tissue can be dissociated to give single cells, and if these cells are cultured in nutrient medium, they will multiply and form an undifferentiated mass called a *callus*. If the right mix of hormones is added, the callus will go on to develop into a mature plant. This experiment, which has been carried out with carrot, tobacco, and several other plants, supports the hypothesis that adult plant cells are totipotent. Therefore, differentiation must be the result of regulation of gene expression rather than the parceling out of genes.

Totipotency has been demonstrated in animals, too. For example, when a nucleus from an intestinal cell of a tadpole is transplanted into a frog's egg, the nucleus of which has been destroyed, development of a tadpole occurs, sometimes followed by further development into an adult frog (fig. 17.A). Of course, you may argue that support for totipotency would be stronger if the nucleus had been taken from the intestinal cell of a fully grown frog. This is true, and as yet such experiments have never resulted in adult frogs. Apparently, all the genes are present in fully differentiated cells, but we just don't know how to turn on the genes that are normally inactive.

A more direct way to determine that different types of cells are expressing different types of genes is to look at the mRNAs. For example, the mRNAs found in liver cells are expected to be different from those found in kidney cells if these cells vary in the genes that have been turned on. It's possible to use nucleic acid hybridization techniques to see if this is the case. During nucleic acid hybridization, a hybrid molecule is formed between mRNA and DNA. The technique is as follows: DNA from an organism is heated so that the hydrogen bonds between the 2 strands break and the strands separate. This is called denaturation. Now, a large excess of RNA from a particular cell type is added and the amount of hybridization between DNA and RNA is noted (fig. 17.B). In one such experiment, it was found that 4.5% of the total DNA forms hybrids with nuclear RNA from mouse liver cells and 4.0% of the total DNA forms hybrids with nuclear RNA from mouse kidney cells. If these RNA molecules were different in the 2 cells, then when both types of RNA are added to denatured DNA, a total of 8.5% should hybridize. The actual results were less than this: 7.5% hybridize, which means that some of the RNA molecules were the same but most were different. Similar results were obtained when liver cells were compared to muscle and brain cells. These experiments support the hypothesis that different genes are active in different cells of a multicellular organism and that certain genes are turned on and certain others are turned off in specialized cells.

Genetic Control in Prokaryotes

We will consider 2 levels of control in prokaryotes: transcriptional control and translational control.

Transcriptional Control

Usually, regulation of gene expression in prokaryotes occurs at the level of transcription. When a gene is to be transcribed, RNA polymerase attaches to a special DNA sequence called a **promoter**; thereafter, this enzyme moves along the DNA, joining mRNA nucleotides together. If RNA polymerase can be prevented from attaching to the promoter, then transcription will not occur.

Obviously, a barrier molecule would prevent RNA polymerase from attaching to the promoter. Such barrier molecules exist; they are rather bulky protein molecules called repressors. When a **repressor** binds to a short sequence of DNA called an **operator,** it prevents RNA polymerase from attaching to the promoter, which is located adjacent to the operator. A repressor is coded for by a **regulator gene,** which is separate from the genes being regulated:

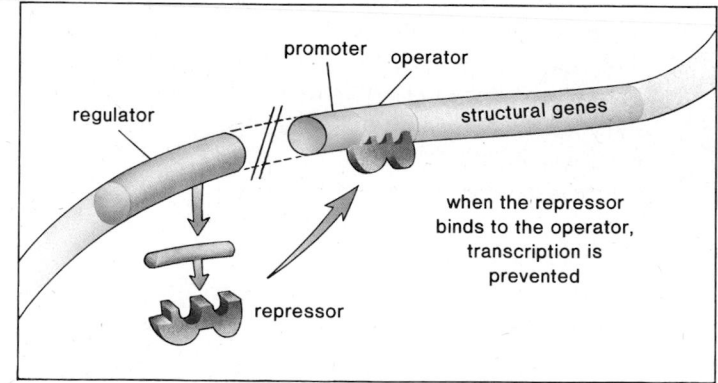

when the repressor binds to the operator, transcription is prevented

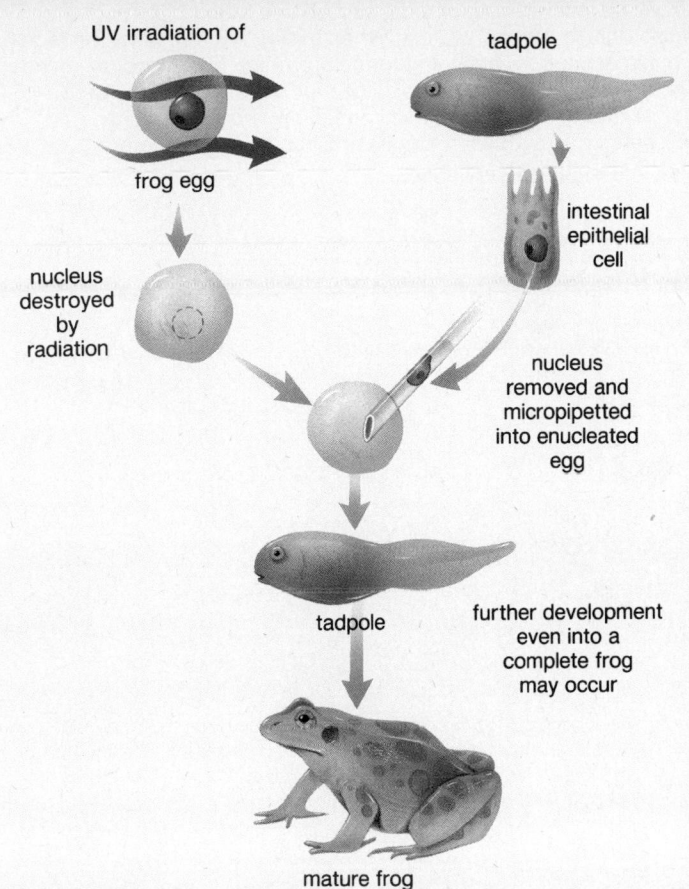

Figure 17.A

Totipotency experiment. The haploid nucleus of a frog's egg is destroyed by ultraviolet radiation (*left*). Now it can receive a diploid nucleus taken from an intestinal cell of a tadpole (*right*). In some cases, the reconstituted cell develops into an adult frog.

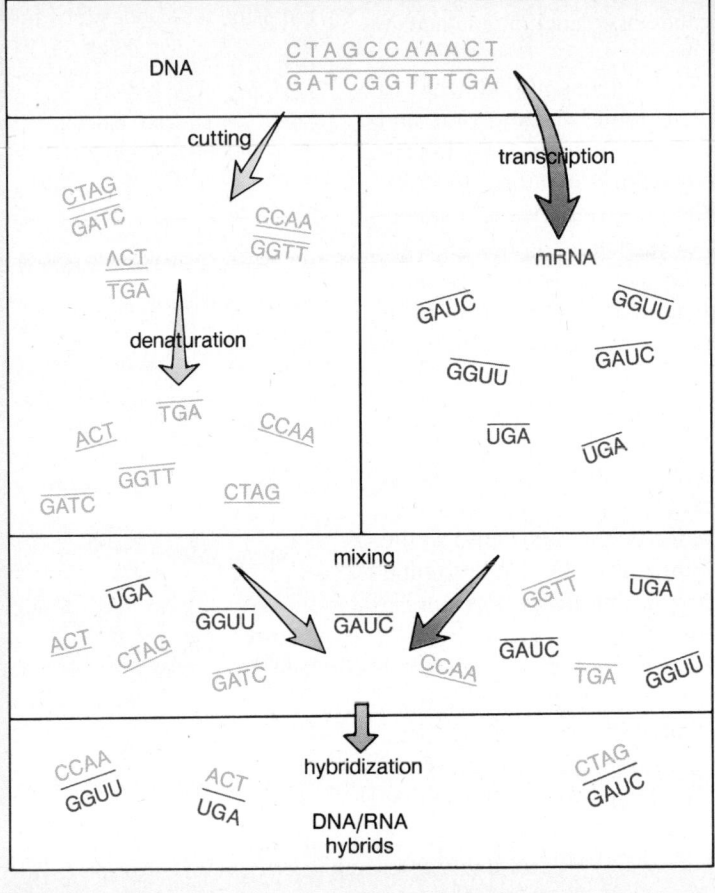

Figure 17.B

DNA/RNA hybridization. An organism's DNA is cut into fragments and denatured (*above*). Transcribed mRNA is obtained from a particular cell type. The 2 are mixed (*middle*). RNA and DNA hybridize whenever they find a complementary strand (*below*). Many of the DNA strands do not hybridize because they were not transcribed in this cell.

The elements that have just been described make up a unit of gene operation called an operon. The **operon** model explains gene regulation in prokaryotes. It suggests that several genes, located in sequence on a chromosome, are controlled by the same regulatory factors. Altogether, an operon includes these elements:

Regulator gene: a gene that codes for a repressor protein or activator protein. A repressor binds to the operator and prevents RNA polymerase from binding to the promoter.

Promoter: a short sequence of DNA where RNA polymerase first attaches when a gene is to be transcribed.

Operator: a short sequence of DNA where the repressor binds, preventing RNA polymerase from attaching to the promoter. Often called the on/off switch of transcription.

Structural genes: one to several genes of a metabolic pathway that are transcribed as a unit. (These genes are called structural genes because they determine the structure of proteins in a metabolic pathway.)

> In prokaryotes, the operon model helps explain regulation of structural gene transcription.

The *Lac* Operon

In 1961, the French microbiologists François Jacob and Jacques Monod published the results of a long series of experiments showing that *E. coli* is capable of regulating the expression of genes necessary for lactose metabolism. They coined the word *operon* and later received a Nobel Prize for their investigations.

When *E. coli* is denied glucose and is given the milk sugar lactose instead, it immediately begins to make the 3 enzymes needed for the metabolism of lactose. These 3 enzymes are encoded by genes designated by *lac z, lac y,* and *lac a. Lac z* codes for an enzyme called ß-galactosidase, which breaks down the disaccharide lactose to glucose and galactose; *lac y* codes for a permease that facilitates the entry of lactose into the cell; and *lac a* codes for an enzyme called transacetylase, which has an accessory function in lactose metabolism.

The 3 genes are adjacent to one another on the chromosome and are under the control of a single promoter and a single operator (fig. 17.2). The regulator, located some distance ahead of the promoter, codes for a *lac* operon repressor that ordinarily binds to the operator and prevents transcription of the *z, y,* and *a* genes. But when *E. coli* is switched to a medium containing lactose, the lactose binds to the repressor, and the repressor undergoes a change in shape that *prevents* it from binding to the operator. Because the repressor is unable to bind to the operator, the promoter is able to bind to RNA polymerase. RNA polymerase carries out transcription, and the 3 enzymes are produced.

Since the 1960s, it's been discovered that matters are a little more complex than originally thought. When glucose is missing, there is a helper complex—cAMP—connected to a protein called catabolite *activator* protein, CAP—which attaches to the promoter and assists the binding of RNA polymerase to the promoter.

Because the presence of lactose brings about production of enzymes, it is called an **inducer** of the *lac* operon: the enzymes are said to be inducible enzymes, and the entire unit is called an **inducible operon.**

The *Trp* Operon

Jacob and Monod found that other operons in *E. coli* usually exist in the on rather than off condition. For example, the prokaryotic cell ordinarily produces 5 enzymes that are needed for the synthesis of the amino acid tryptophan. If tryptophan is present in the medium, however, these enzymes are no longer produced. In this operon model, the regulator codes for a repressor that ordinarily is unable to attach to the operator. The repressor has a binding site for tryptophan, and if tryptophan is present it binds to the repressor. Now, a change in shape allows the repressor to bind to the operator. The enzymes are said to be repressible, and the entire unit is called a **repressible operon.** Tryptophan is called the **corepressor.**

Figure 17.2

The *lac* operon, a model for an inducible operon, presented by Jacob and Monod. **a.** The regulator gene codes for a repressor that is normally active. When active, the repressor binds to the operator and prevents RNA polymerase from attaching to the promoter. Therefore, transcription of the three structural genes does not occur. **b.** When lactose (or more correctly, a derivative of lactose) is present, it binds to the repressor and this changes the repressor's shape, preventing it from binding to the operator. Now RNA polymerase binds to the promoter; transcription and translation of the 3 structural genes follow.

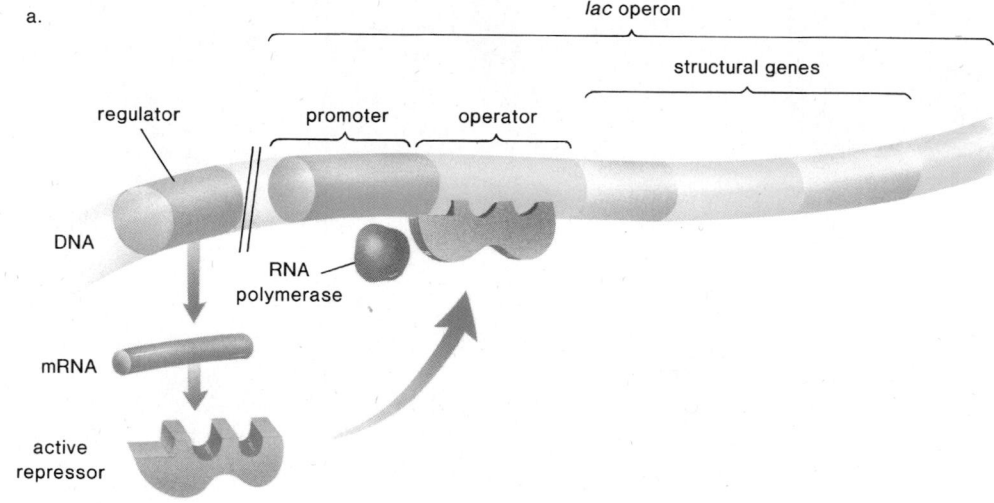

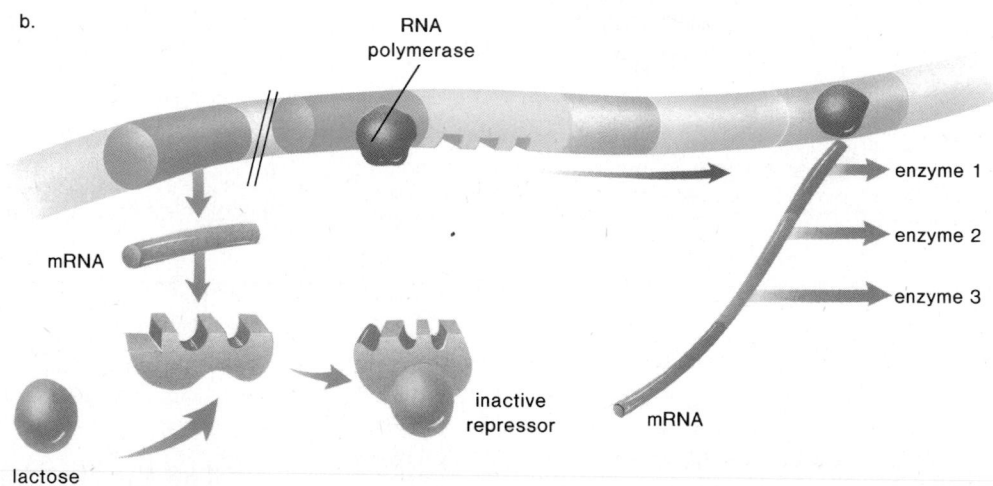

In the *lac* operon, the repressor ordinarily binds to the operator but is unable to do so when lactose is present. In the *trp* operon, the repressor ordinarily does not bind to the operator but is able to do so when tryptophan is present. Therefore, some operons are inducible and some are repressible.

Translational Control

In 1983, Nancy E. Kleckner of Harvard University conducted a series of experiments showing that antisense RNA can control the expression of a gene in bacteria. Antisense RNA is transcribed

Genetic Basis of Life

from the DNA strand that is complementary to a mRNA template DNA and is ordinarily inactive. When bacteria transcribe both sense mRNA and antisense RNA, the 2 bind together by complementary base pairing, forming an RNA duplex. This prevents ribosomes from translating the mRNA into protein:

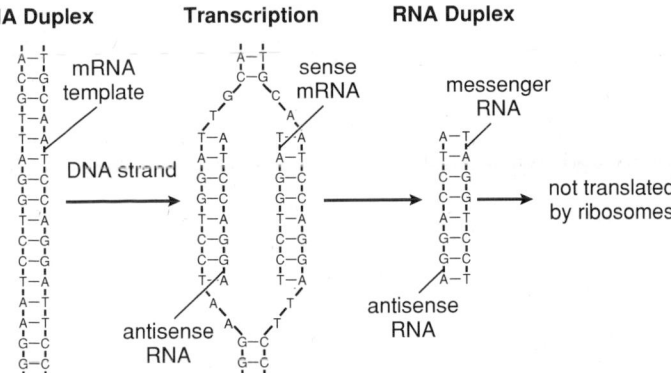

Later investigators have suggested that inhibition of gene expression with antisense RNA is universal among bacteria. As yet, they have not shown that this is a normal means by which eukaryotic cells regulate gene expression. Laboratory-produced antisense RNA, however, can be injected into eukaryotic cells to control the activity of various sense mRNA molecules. Drug companies are very interested in the possibility of producing this type of product to control the activity of viruses and even cancer genes in humans.

Genetic Control in Eukaryotes

As mentioned previously and as described in figure 17.1, there are 4 levels of genetic control in eukaryotic cells: transcriptional control, posttranscriptional control, translational control, and posttranslational control.

Transcriptional Control

We might expect transcriptional control in eukaryotes (1) to involve the organization of chromatin and (2) to include regulatory proteins such as those we have just observed in prokaryotes.

Chromatin and the Regulation of Transcription

The reading "What's in a Chromosome," on page 158, describes how DNA in eukaryotes first winds around histone molecules and how this beaded string then coils, giving a fiber that is then compacted to varying degrees. Chromatin that is highly compacted and visible with the light microscope during interphase is called **heterochromatin.** More diffuse chromatin is called **euchromatin** (fig. 17.3).

To demonstrate that heterochromatin is genetically inactive, we can refer to the **Barr body,** a highly condensed structure. In mammalian females, one of the X chromosomes is found in the Barr body (p. 220); chance alone determines which of the X chromosomes is condensed. For a given gene, if a female is heterozygous, 50% of her cells have Barr bodies containing one allele and the other 50% have the other allele. This causes the body

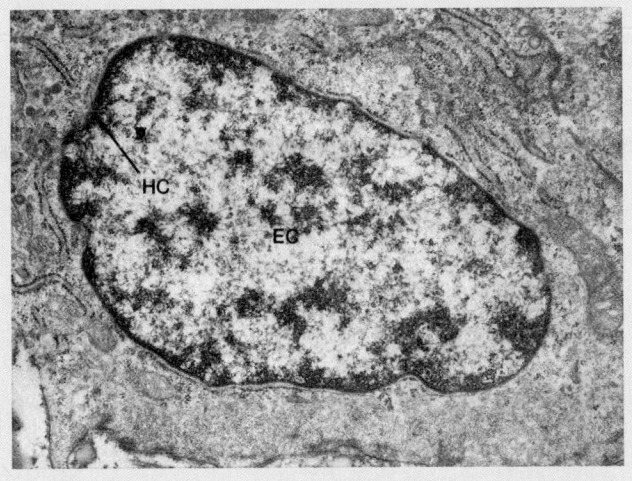

Figure 17.3
Electron micrograph of a nucleus that contains some heterochromatin (HC), highly compacted and genetically inactive DNA. Euchromatin (EC) is more diffuse and more likely to be genetically active. Magnification, X25,400.

of heterozygous females to be mosaic, with patches of genetically different cells. The mosaic is exhibited in various ways: some females show columns of both normal and abnormal dental enamel; some have patches of pigmented and nonpigmented cells at the back of the eye; some have patches of normal muscle tissue and degenerate muscle tissue. Barr bodies are not found in the cells of the female gonads, where the genes of both X chromosomes appear to be needed for development of the immature egg. Likewise, it is clear that the inactivation of one X chromosome in the female zygote lowers the dosage of gene products to that seen in males. Most likely, development has become adjusted to this lower dosage, and a higher dosage would bring about abnormalities. We certainly know that in regard to other chromosomes, any imbalance causes a greatly altered phenotype.

In contrast to heterochromatin, euchromatin is genetically active. In the salivary glands and other tissues of *Drosophila* larvae, the chromosomes duplicate and reduplicate many times without dividing mitotically. The homologues, each consisting of about 1,000 sister chromatids, synapse together to form giant chromosomes (fig. 13.15) called *polytene chromosomes* (*poly*—many; *tene*—thread). Polytene chromosomes are visible, genetically active, interphase chromosomes, whereas the chromosomes that appear during mitosis are highly compacted and genetically inactive.

A useful feature of polytene chromosomes is their banding pattern because it is possible to associate a particular gene with a particular band. It is observed that as a *Drosophila* larva develops, first one and then another of the chromosomes' bands bulge out, forming chromosome puffs (fig. 17.4). The use of radioactive uridine, a label specific for RNA, indicates that DNA is being actively transcribed at *chromosome puffs*. It appears that the

Figure 17.4

Structure of giant polytene chromosome. **a.** Electron micrograph of an insect, the midge (*Chironomus*), chromosome which has a puff in one region. **b.** Artist's interpretation of a puffed region. **c.** Puffs contain loops of DNA, where enzymes can transcribe mRNA off the DNA template. Supporting this interpretation is the observation that varying regions of the chromosomes exhibit puffs as development of *Drosophila* and other insects proceeds.

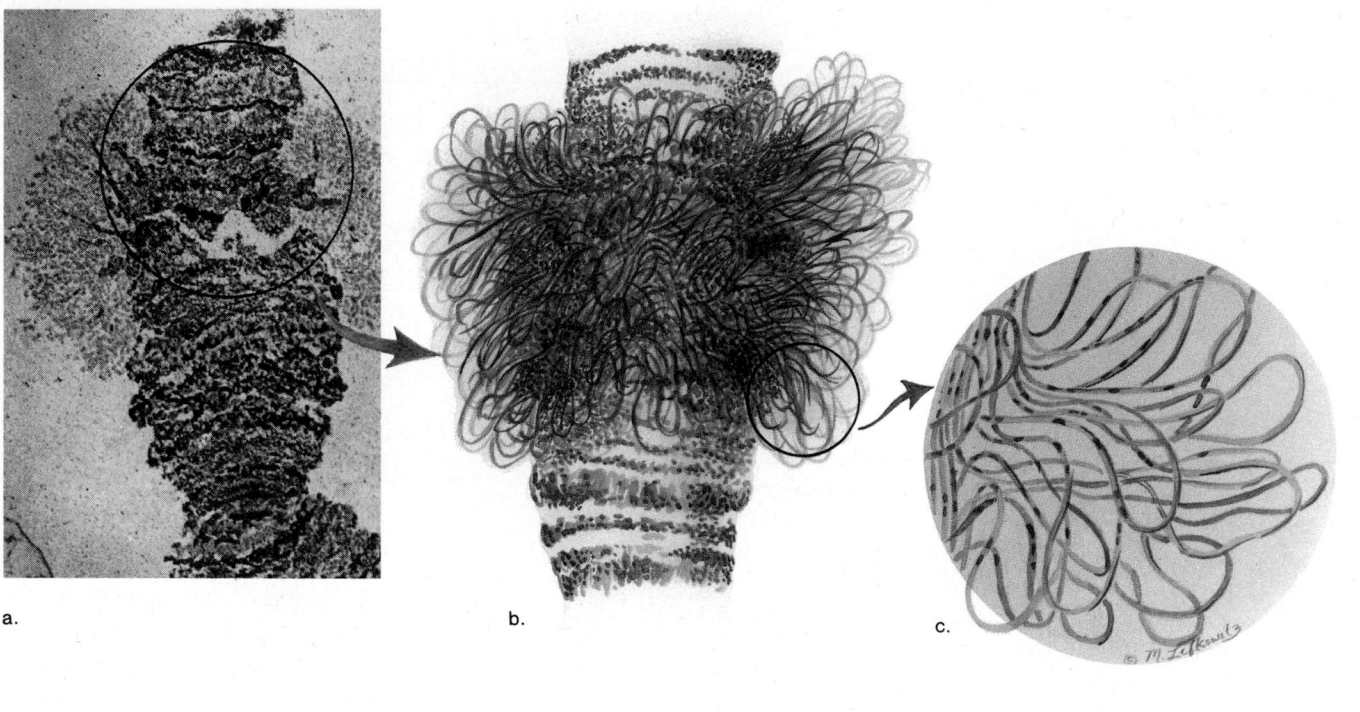

a. b. c.

chromosome is decondensing at the puffs so that transcription enzymes have access to a section of DNA. Histone molecules may regulate this change in chromosome structure.

It is of interest, too, that the hormone ecdysone, which plays an important role in the life cycle of insects, has been shown to induce formation of chromosome puffs, which always appear at specific regions of specific chromosomes. This hormone may bind to DNA and in that way bring about transcription (see fig. 42.3*b*).

Varying the amount of chromatin is another form of transcriptional control. When the gene product is either tRNA or rRNA, only an increased amount of transcription permits an increase in the amount of gene product. In *Xenopus* (a frog) germ cells that are producing eggs, the number of nucleoli actually increases to nearly 1,000. This means that there is a total of over one million copies of rRNA genes—much more than normal. This ensures an immense number of ribosomes in the egg cytoplasm to support protein synthesis during the early stages of development.

Transcriptional control in eukaryotes may involve changes in chromatin structure and amount.

Regulatory Proteins and Transcription

Although no operons like those of prokaryotic cells have been found in eukaryotic cells, investigations suggest that transcription is controlled by DNA-binding proteins called transcription factors.

Every cell contains many different types of transcription factors, and a different combination is believed to regulate the activity of any particular gene. After the right combination of transcription factors bind to DNA, an RNA polymerase attaches to DNA and begins the process of transcription.

As cells mature they become specialized. Specialization is determined by which genes are active in a cell and therefore, perhaps, by which transcription factors are present. Signals received from inside and outside the cell could turn on genes that code for certain transcription factors.

There is evidence of regulatory proteins in eukaryotic cells, but there are no definite operons as for prokaryotic cells.

Posttranscriptional Control

In eukaryotes, mRNA molecules (and other RNAs) are processed before they leave the nucleus and pass into the cytoplasm (see fig. 16.9). Differential excision of introns and splicing of mRNA can vary the type of mRNA that leaves the nucleus. For example, both the hypothalamus and thyroid gland produce a hormone called calcitonin. The mRNA that leaves the nucleus is not the same in both types of cells, however; radioactive labeling studies show that they vary because of a difference in mRNA splicing (fig. 17.5). Evidence of different

Figure 17.5

Variable RNA splicing can bring about different gene products. Notice that the introns that have been excised from the mRNA on the left are different from those that have been excised from the mRNA on the right.

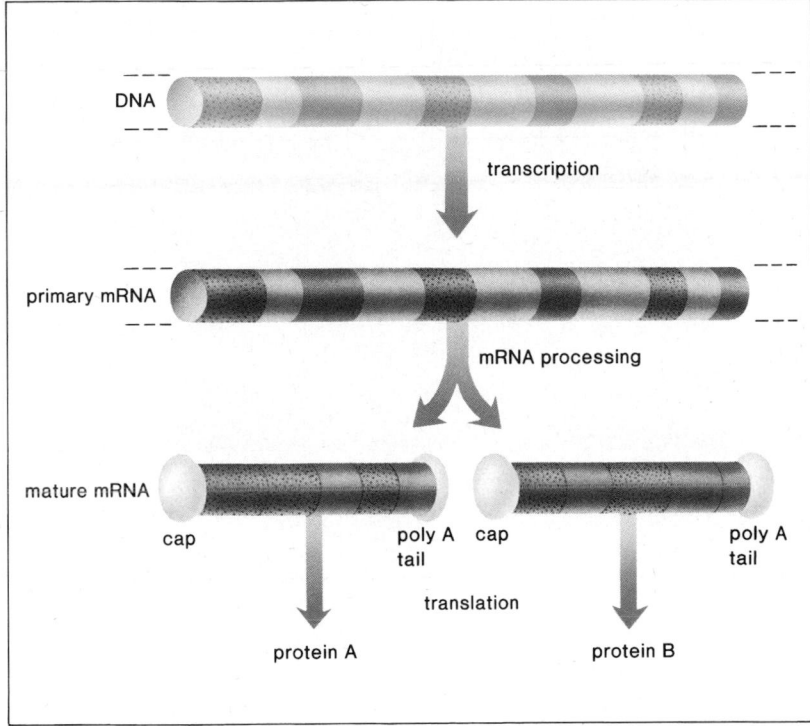

all this time. The agents of mRNA destruction are ribonucleases, enzymes that have been found to be associated with ribosomes. Every mRNA has a 5' noncoding segment and a 3' noncoding segment; evidently differences in these segments influence how long the mRNA avoids being degraded.

Hormones seem to cause the stabilization of certain mRNA transcripts. For example, the hormone prolactin promotes milk production in mammary glands, but it does so primarily by affecting the length of time the mRNA for casein, a major protein in milk, is translated. Similarly, in estrogen-treated amphibian cells, a mRNA for a protein called vitellogenin lasts 3 weeks compared to 16 hours in untreated cells. There is evidence to suggest that estrogen binds to and thereby interferes with the action of a ribonuclease.

> Translational controls directly affect whether an mRNA is translated and the amount of gene product eventually produced.

Posttranslational Control

Some proteins are not active immediately after translation. For example, bovine proinsulin is at first biologically inactive. After the single long polypeptide folds into a 3-dimensional structure, a sequence of about 30 amino acids is enzymatically removed from the middle of the molecule. This leaves 2 polypeptide chains that are bonded together by disulfide (S—S) bonds and results in an active protein.

Finally, the metabolic activity of genes is often under feedback control (see fig. 7.7).

> Posttranslational control affects the activity of a protein product, whether or not it is functional, and the length of time it is functional.

patterns of mRNA splicing are found in other cells, such as those that produce neurotransmitters, muscle regulatory proteins, and antibodies.

The speed of transport of mRNA from the nucleus into the cytoplasm can ultimately affect the amount of gene product realized per unit time following transcription. There is evidence that there is a difference in the length of time it takes various mRNA molecules to pass through a nuclear pore.

> Posttranscriptional control in eukaryotes involves differential mRNA processing and factors that affect the length of time it takes mRNA to travel to the cytoplasm.

Translational Control

The eggs of frogs and certain other animals contain mRNA that is not translated at all until fertilization occurs. These mRNAs are called *masked messengers* because apparently the cell is unable to recognize them as available for translation. As soon as fertilization occurs, unmasking occurs, and there is a rapid burst of specific gene product synthesis.

Obviously, the longer a particular mRNA molecule remains in the cytoplasm, the more product there will eventually be. During maturation, mammalian red blood cells eject their nucleus, and yet they continue to synthesize hemoglobin for several months thereafter. This means that the necessary mRNAs are able to persist

Cancer—A Failure in Genetic Control

Cancer cells exhibit characteristics that indicate they have experienced a severe failure in the control of gene expression.

Cancer cells exhibit uncontrolled and disorganized growth. Normal cells divide only about 50 times, but cancer cells enter the cell cycle (see fig. 10.4) repeatedly and never differentiate.

In tissue culture, normal cells grow in only one layer because they adhere to the glass, and they stop dividing once they make contact with their neighbors, a phenomenon called *contact inhibition.* Cancer cells have lost contact inhibition and grow in multiple layers, most likely because of cell-surface changes (fig. 17.6). In the body, a cancer cell divides to form a growth, or tumor, that invades and destroys neighboring tissue (fig. 17.7). The cells are disorganized because they don't differentiate into the tissues of the organ; therefore, they can never help fulfill the function of the organ. To support their growth, cancer cells release a growth factor

Figure 17.6

Transformation from normal cell to cancerous cell. *a.* Normal fibroblasts are flat and extended in cell culture. *b.* After being infected with a cancer-causing virus, the cells become round and cluster together in piles.

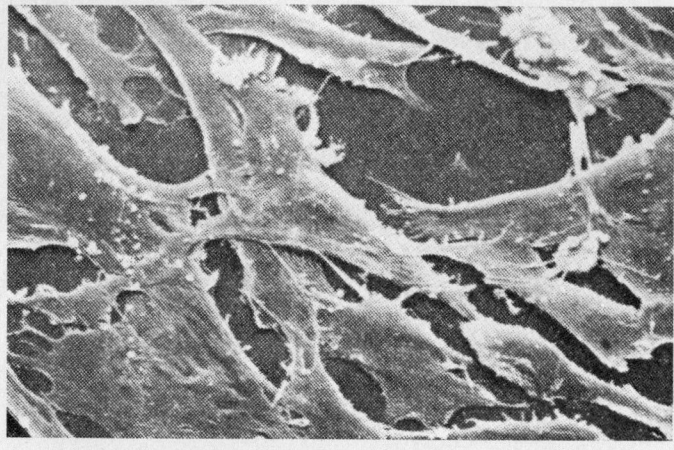

a.

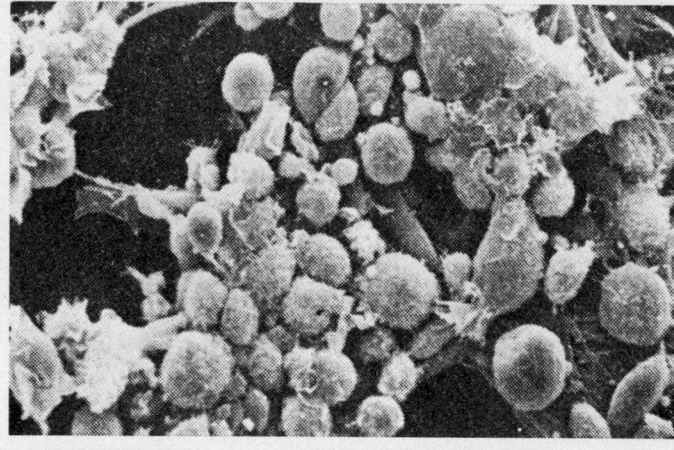

b.

that causes neighboring blood vessels to branch into the cancerous tissue. This phenomenon has been termed *vascularization*, and some modes of treatment are aimed at preventing it.

Cancer cells detach from the tumor and spread around the body. Cancer cells produce proteolytic enzymes that allow them to invade underlying tissues. They are also motile, most likely because they lack bundles of intact microfilaments and because they have a disorganized internal cytoskeleton. After traveling through the blood or lymph, cancer cells start new tumors wherever they relocate. This process is called *metastasis*. If the original tumor is found before metastasis occurs, the chances of a cure are greatly increased. This is the rationale for the early detection of cancer. A tumor is considered benign (non-cancerous) until metastasis occurs. *Benign tumors* are those that remain in one place; *malignant tumors* are those that metastasize.

Cancer cells grow and divide uncontrollably, and they metastasize, forming new tumors wherever they relocate.

Figure 17.7

In the body, cancer cells form a tumor, a disorganized mass of cells undergoing uncontrolled growth. As a tumor grows, it invades underlying tissues. Some of the cells leave this primary tumor and move through layers of tissue into blood or lymphatic vessels. These vessels transport the cells to new locations where they start new tumors elsewhere in the body. Carcinoma is cancer that begins in epithelial tissue; sarcoma begins in connective tissue.

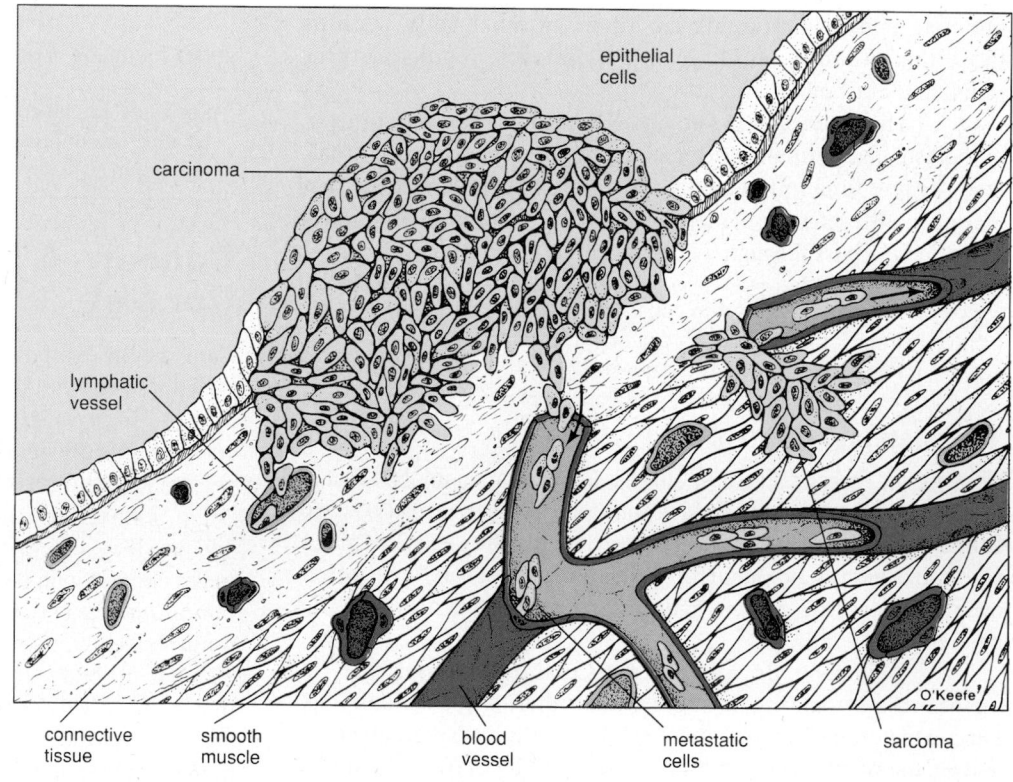

Genetic Basis of Life

Figure 17.8

Control of cell reproduction involves both oncogenes and tumor suppressor genes. Oncogenes promote cell proliferation, and tumor suppressor genes suppress cell proliferation. Some of the proteins coded for by these genes are plasma membrane receptors that receive incoming signals. Reception of a signal activates certain chemical reactions in cells. Oncogenes may be tyrosine kinases that add phosphate to the amino acid tyrosine in cellular proteins and tumor suppressor genes may possibly be tyrosine phosphatases that remove phosphate from tyrosine in cellular proteins.

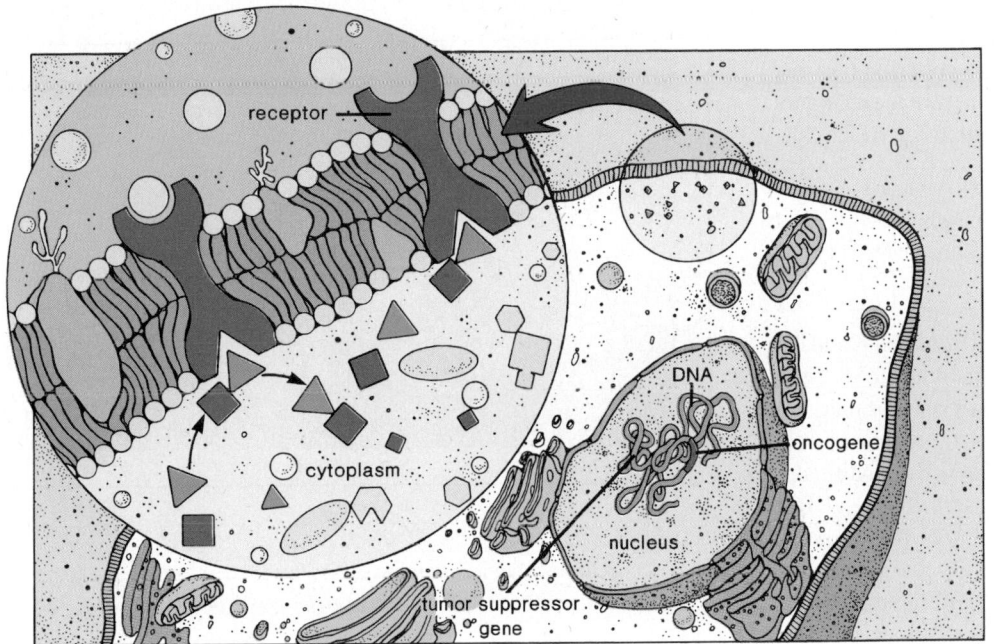

O'Keefe

Causes of Cancer

The development of cancer can be viewed as a 2-step process involving initiation and promotion.

During initiation, a mutation occurs that makes cancerous growth a possibility at a later date. **Carcinogens,** agents that contribute to the development of cancer, are *initiators*. For example, the papilloma virus that causes genital warts is now associated with the development of cervical cancer. Radiation, including ultraviolet (UV) radiation, radon (a radioactive gas made by the natural decay of radium), and X rays, damages the normal bonding patterns between DNA nucleotides. Chemical carcinogens, such as pesticides, are known to bring about base sequence changes in DNA. Cigarette smoke, which contains numerous carcinogens, plays a significant role in the development of lung cancer.

A *promoter* of cancer is any influence that triggers a cell to start uncontrolled growth. It's possible that promotion simply involves a second change in the DNA brought about by another carcinogen. In other words, cumulative DNA changes can finally result in uncontrolled growth. Or, a promoter may provide the environment that causes transformed cells to form a tumor. For example, there is some evidence to suggest that a diet rich in saturated fats and cholesterol is a possible promoter for cancer. Considerable time may elapse between initiation and promotion, and this is why cancer is more apt to develop in older rather than younger individuals.

One model suggests that carcinogenesis involves a gene mutation or a chromosome mutation (initiation) and a second influence that triggers cancerous growth (promotion).

Effect of Mutated Genes

Carcinogens are now believed to cause oncogenes to become active and tumor suppressor genes to become inactive (fig. 17.8). **Oncogenes** code for proteins that promote uncontrolled cell division and disorganized growth of cells. **Tumor suppressor genes** (also called antioncogenes) code for proteins that ordinarily suppress cell division and in this way promote organized growth of cells.

Oncogenes In 1911, Peyton Rous reported that a virus, later named Rous sarcoma virus (RSV), is capable of causing sarcoma, a type of cancer, in chickens. It wasn't until 1966 that the scientific community realized the value of Rous's contribution and awarded him a Nobel Prize. In the meantime, it was clear that RSV is an RNA virus. After the virus infects a cell, it inserts a DNA transcript of its RNA genome, which includes a gene known as *src* (for sarcoma gene), into the host chromosome. Investigators prepared radioactive copies of the *src* gene, knowing that these copies would bind to the host chromosome wherever the *src* gene was located. To their surprise, they found that binding occurred in 2 locations—one where viral DNA had inserted and another in a normal part of the chromosome! This showed that the *src* gene was a normal gene in chickens, but it had become a cancer-causing gene only because it was being carried by the virus. It's possible that the *src* gene is under the control of a viral promoter that has turned it on, for example.

The *src* gene is now known to be an oncogene. In its normal, nonmutated state, it is a proto-oncogene, a gene that can be transformed into an oncogene. Oncogenes are not alien to the cell; they are normal, essential genes that have undergone a mutation. Many oncogenes, including several that cause human cancers, have been isolated and studied and found to be mutated forms of genes involved in the regulation of cellular growth. There are numerous elements that affect the growth and the development of cells. These include growth factors that act on the surface of the cell, receptors for these substances, proteins that carry signals from the receptors, and nuclear functions that regulate the response to these influences. Oncogenes are genes that cause one of these elements to malfunction in such a way that the cell divides repeatedly to produce a tumor.

It is of extreme interest to investigators that the src gene codes for an enzyme called a tyrosine kinase. The enzyme adds a phosphate group to tyrosine, an amino acid of proteins, and in some unknown way this promotes cell division:

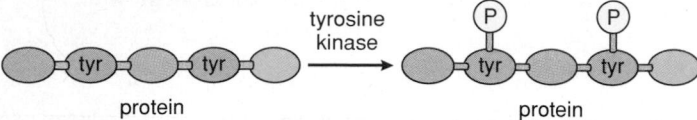

tyr = tyrosine

At least 40 different genes have been found to code for tyrosine kinases. Some of these enzymes are also plasma membrane receptors for growth factors, which are hormones that stimulate a cell to divide.

> Oncogenes are mutated copies of proto-oncogenes, normal genes that code for proteins that promote cell division. Some of these proteins are tyrosine kinases, enzymes that add phosphate to the amino acid tyrosine in proteins.

Tumor Suppressor Genes Throughout the 1970s, Alfred Knudson at the M. D. Anderson Hospital and Tumor Institute in Houston studied the inheritance of retinoblastoma, a cancer of the retina. We now know that retinoblastoma cells have a specific region of both copies of chromosome 13 visibly missing. The absence of a particular gene, called RB, is the specific cause of retinoblastoma. When RB is absent a cell becomes cancerous, and when it is present a cell does not become cancerous. RB is a tumor suppressor gene. Experiments support this hypothesis. Laboratories maintain cancerous retinoblastoma cells in cell culture, and when these cells are supplied with an RB gene, they are no longer cancerous. Similarly, the RB gene has reversed malignancy in cultured cells that cause kidney cancer in children and others that cause prostate cancer in adult males.

Investigators have found another tumor suppressor gene they call p53. This gene is responsible for the Li-Fraumeni syndrome, characterized by a high rate of lethal cancers among family members. When a person inherits only one normal allele, there is a 50% chance of developing cancer by age 30 because of the high chance it will also mutate.

The function of the RB gene has been studied. It codes for a phosphorylated nuclear protein, which may be an inhibitor of cell division. The possibility exists that some tumor suppressor genes code for tyrosine phosphatases, enzymes that remove phosphate groups from tyrosine. Tyrosine phosphatases have been isolated from cells, and some of these are receptors for signaling molecules, as are some oncogene products. Whether tyrosine phosphatases are growth inhibitors or not is not known, but when a cell line known to contain a tyrosine phosphatase gene was also imbued with a tyrosine kinase oncogene, the cells did not become cancerous. In this one case a tyrosine phosphatase has been shown to have antioncogenic effects.

> Tumor suppressor genes are antioncogenes that code for proteins which inhibit cell division. The possibility exists that some of these genes code for tyrosine phosphatases, but this is not yet known.

Figure 17.9
Data suggest that diet can influence the development of cancer. Fresh fruits, especially those high in vitamins A and C, and vegetables, especially those in the cabbage family, are believed to reduce the risk of cancer.

Rather than a 2-step process, it now appears that cancer is a multistep process. An examination of cancerous colon cells suggests that at least 5 mutations are required before cancer develops. One of these causes the oncogene called RAS to be activated. RAS codes for a protein that regulates the activity of a tyrosine kinase. The cells have also lost tumor suppressor genes from chromosomes 5, 18, and 17.

> The development of cancer usually requires several mutations. Some of these mutations cause proto-oncogenes to become oncogenes, and some of these cause tumor suppressor genes to become ineffective.

Prevention of Cancer

There is clear evidence that the risk of certain types of cancer can be reduced by adopting protective behaviors. In general, the avoidance of carcinogenic chemicals and excessive radiation is helpful. Specifically, avoiding excessive sunlight reduces the risk of skin cancer, for example, and not smoking cigarettes and cigars reduces the risk of lung cancer and other types of cancer as well. Cigarette smoke contains many carcinogenic chemicals that can cause mutations. Smokeless tobacco products such as snuff or chewing plugs are associated with cancer of the mouth.

There is also an emphasis on dietary considerations to help prevent cancer (fig. 17.9). Guidelines suggest that the following are helpful:

Lower total fat intake. A high-fat intake has been linked to development of breast, colon, and prostate cancer.

Cut down on consumption of salt-cured, smoked, and nitrite-cured foods. Processed meats, such as lunch meats and breakfast meats, for example, bacon and sausage, fall into this category. Some of the chemicals added to these foods may be promoters of cancer.

Eat more high-fiber foods such as those in whole-grain cereals, fruits, and vegetables. Studies have indicated that a high-fiber diet protects against colon cancer.

Increase consumption of foods rich in vitamins A and C. These vitamins are antioxidants that prevent the formation of free radicals (organic ions that have a nonpaired electron) that can possibly damage DNA. Dark-green, leafy vegetables, carrots, and various fruits contain beta-carotene, a precursor of vitamin A. Citrus fruits contain vitamin C.

Increase consumption of vegetables of the cabbage family. This incudes cabbage, broccoli, and cauliflower. These foods also seem to protect against the development of cancer.

Be moderate in the consumption of alcohol. People who drink and smoke are at an unusually high risk of cancer of the mouth, larynx, and esophagus.

Summary

1. In prokaryotes, genes are turned on and off as appropriate for environmental conditions. In eukaryotes, specialized cells have different genes turned on. In any case, both types of cells are able to regulate the expression of genes.

2. The following levels of control of gene expression are possible in eukaryotes: transcriptional control, posttranscriptional control, translational control, and posttranslational control.

3. Regulation in prokaryotes usually occurs at the level of transcription. The operon model developed by Jacob and Monod says that a regulator gene codes for a repressor, which sometimes binds to the operator. When it does, RNA polymerase is unable to bind to the promoter and transcription of structural genes cannot take place.

4. The *lac* operon is an example of an inducible operon because when lactose, the inducer, is present it binds to the repressor. The repressor is unable to bind to the operator, and transcription of structural genes takes place.

5. The *trp* operon is an example of a repressible operon because when tryptophan, the corepressor, is present, it binds to the repressor. The repressor is then able to bind to the operator; and transcription of structural genes does not take place.

6. All 4 levels of control have been observed in eukaryotes. First we note that DNA is coiled and compacted to various degrees. An aspect of transcriptional control is therefore whether chromatin is diffuse enough to allow transcription to take place.

7. Highly compacted heterochromatin is genetically inactive, as exemplified by Barr bodies. Less-compacted euchromatin is genetically active, as exemplified by polytene chromosome puffs.

8. Gene amplification is the replication of just one or a few genes such that there are more copies of these genes (i.e., rRNA genes in the oocyte) than were originally present in the zygotic nucleus.

9. Regulatory proteins as well as DNA sequences called enhancers play a role in controlling transcription in eukaryotes.

10. Posttranscriptional control refers to variations in mRNA processing and the speed with which a particular mRNA molecule leaves the nucleus.

11. Translational control affects mRNA translation and the length of time it is translated. Posttranslational control affects whether or not a protein is active and how long it is active.

12. Development of cancer is a failure in genetic control. Cancer cells have 2 main characteristics: they exhibit uncontrolled and disorganized growth, and they metastasize.

13. The development of cancer is often described as a 2-step process involving initiation and promotion. Carcinogens, such as radiation and certain chemicals, are initiators and may also be promoters. Promoters can also be agents that affect the environment of the cell.

14. Oncogenes are mutated copies of proto-oncogenes, normal genes that code for proteins that promote cell division. Some of these are tyrosine kinases, enzymes that add phosphate to the amino acid tyrosine in proteins.

15. Tumor suppressor genes are antioncogenes that code for proteins that inhibit cell division. Some of these may be tyrosine phosphatases, enzymes that remove phosphate from the amino acid tyrosine in proteins.

16. There has been an emphasis of late on the proper diet to prevent cancer. The intake of leafy green vegetables, fruits, high-fiber grains, and cabbage family vegetables is encouraged. The intake of fat, processed meats, and alcohol is discouraged.

Writing Across the Curriculum

In order to practice writing skills, students should write out the answers to any or all of the study questions and the critical thinking questions. The study questions are sequenced in the same order as the text. Suggested answers to the critical thinking questions are in appendix D.

Study Questions

1. How do prokaryotes and eukaryotes differ regarding their need for control of gene expression?
2. What are the 4 levels of control of gene expression in eukaryotes? In what part of the cell do you find each level?
3. Name and state the function of the 4 components of operons.
4. Explain the operation of the *lac* operon in a way that stresses it is an inducible operon.
5. Explain the operation of the *trp* operon in a way that stresses it is a repressible operon.
6. Explain how Barr bodies show that heterochromatin is genetically inactive.
7. Explain how giant (polytene) chromosome puffs show that euchromatin is genetically active.
8. What do regulatory proteins do in eukaryotic cells? What are enhancers?
9. Give examples of posttranscriptional, translational, and posttranslational control in eukaryotes.
10. What 2 characteristics are shared by all cancer cells? Describe the development of cancer as a 2-step process.
11. What is an oncogene and what is the function of oncogenes in cells?
12. What is a tumor suppressor gene, and what is the function of tumor suppressor genes in cells?

Objective Questions

1. Which type of prokaryotic cell would be more successful as judged by its growth potential?
   a. One that is able to express all its genes all the time.
   b. One that is unable to express any of its genes any of the time.
   c. One that expresses some of its genes some of the time.
   d. One that only divides when all types of amino acids and sugars are present in the medium.
2. Which of these is mismatched?
   a. posttranslational control—nucleus
   b. transcriptional control—nucleus
   c. translational control—cytoplasm
   d. posttranscriptional control—nucleus
3. When lactose is present
   a. the repressor is able to bind to the operator.
   b. the repressor is unable to bind to the operator.
   c. transcription of *lac y*, *lac z*, and *lac a* genes occurs.
   d. Both b and c.
4. When tryptophan is present
   a. the repressor is able to bind to the operator.
   b. the repressor is unable to bind to the operator.
   c. transcription of structural genes occurs.
   d. Both b and c.
5. RNA processing varies in different cells. This is an example of
   a. transcriptional control of gene expression.
   b. posttranscriptional control of gene expression.
   c. translational control of gene expression.
   d. posttranslational control of gene expression.
6. A scientist adds radioactive uridine (label for RNA) to a culture of cells and examines an autoradiograph. Which type of chromatin will be labeled?
   a. heterochromatin
   b. euchromatin
   c. Both a and b.
   d. Neither a nor b.
7. If Barr bodies were genetically active, females heterozygous for an X-linked gene would
   a. be mosaics.
   b. not be mosaics.
   c. die.
   d. Both a and c.
8. Which of these might cause a proto-oncogene to become an oncogene?
   a. exposure of cell to radiation
   b. exposure of cell to certain chemicals
   c. viral infection of the cell
   d. All of these.
9. A cell is cancerous. Where might you find an abnormality?
   a. only in the nucleus
   b. only in the plasma membrane receptors
   c. only in cytoplasmic reactions
   d. in any part of the cell concerned with growth and cell division
10. Viruses are apt to contribute to the development of cancer when they
   a. cause an infectious disease.
   b. place an oncogene in the genome of the host.
   c. are retroviruses.
   d. Both b and c.
11. a. Label the diagram below of a *lac* operon. Note that transcription is not occurring.
    b. What one art change is needed to have the diagram represent a *trp* operon when no transcription is occurring?

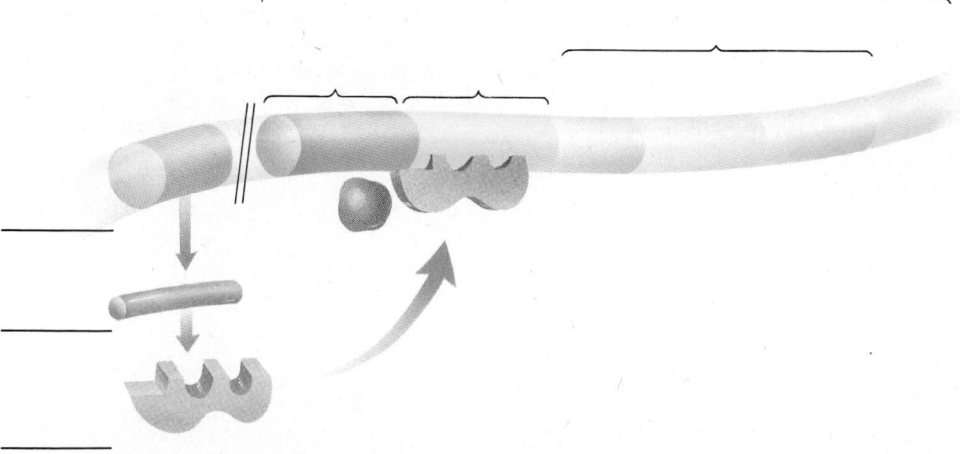

Concepts and Critical Thinking

1. *Cells are able to regulate the expression of genes so that only certain ones are expressed at any particular time.*

   What firsthand evidence is there that only certain genes are active in certain cells?

2. *Genes are controlled at various levels that involve both transcription and translation.*

   Define gene expression, and then explain why it is appropriate to speak of posttranslational control of genes.

3. *Faulty regulation of genes that promote cell division and/or genes that inhibit cell division cause the development of cancer.*

   Why would you expect normal cells to have opposing genes and opposing enzymes involved in the regulation of cell division?

Selected Key Terms

promoter (pro-mo′ter) 266
repressor (re-pres′or) 266
operator (op′er-a-tor) 266
regulator gene (reg′u-la″tor jēn) 266
operon (op′er-on) 267
structural gene (struk′tūr-al jēn) 267
inducer (in-dūs′er) 268

inducible operon (in-dūs′i-b′l op′er-on) 268
repressible operon (re-pres′ĭ-b′l op′er-on) 268
corepressor (ko″re-pres′or) 268
heterochromatin (het″er-o-kro′mah-tin) 269

euchromatin (u-kro′mah-tin) 269
Barr body (bahr bod′e) 269
carcinogen (kar-sin′o-jen) 273
oncogene (ong′ko-jēn) 273
tumor suppressor gene (tu′mor se′pres′or jēn) 273

18

Recombinant DNA and Biotechnology

DNA can be extracted from a cell and manipulated using standard laboratory techniques. This discovery has led to the commercial availability of gene products and the hope that one day there will be a cure for genetic disorders.

Your study of this chapter will be complete when you can

1. describe how the use of a vector permits cloning of a gene;
2. describe the process by which recombinant DNA is prepared;
3. describe various other ways DNA can be manipulated in the laboratory;
4. categorize and list some of the biotechnology products currently on the market;
5. describe ways in which bacteria have been engineered to perform various services;
6. explain how plants lend themselves to genetic manipulation and discuss any problems involved;
7. list some of the ways plants can be modified as a result of recombinant DNA technology;
8. describe a method for genetically engineering animals and state any problems involved;
9. outline procedures by which gene therapy may be conducted in humans;
10. describe the genetic and physical map of a human chromosome and tell how such maps are determined.

endel's original research was rediscovered in 1900, and in the intervening years, geneticists have made startling advances in their understanding of the molecular mechanisms of heredity. Recombinant DNA technology has even led to a new era of biotechnology. Biotechnology, the use of a natural biological system to produce a product or to achieve an end desired by humans, is not new. Plants and animals have been bred to express particular phenotypes since the dawn of civilization. The biochemical capabilities of microorganisms have also been exploited for a very long time. For example, the rising of bread and the fermentation of fruit for wine are dependent on the presence of yeast cells.

Today, however, biotechnology is first and foremost an industrial process that provides products due to our ability to **genetically engineer** bacteria (fig. 18.1). The product can be an organic chemical of interest, a protein that is useful as a vaccine, or a drug to promote human health. Engineered bacteria need not be confined to a chemical plant or laboratory; they may soon be released into the environment to clean up pollutants, increase the fertility of the soil, or kill insect pests. Biotechnology even extends beyond unicellular organisms; it is now possible to alter the genotype and subsequently the phenotype of plants and animals. An agricultural revolution of unequaled magnitude is not inconceivable even in the near future. And there is no need to stop here; routine gene therapy in humans may be just over the horizon. Let's begin our discussion with the most basic biotechnology procedure, the cloning of a gene.

Cloning of a Gene

Genetic engineering can produce cells that contain a foreign gene and are capable of producing a new and different protein. Often, **vectors** are used to carry the foreign gene into a bacterium, a yeast cell, a plant cell, or an animal cell as a part of recombinant DNA.

Vectors

Recombinant DNA (rDNA) contains DNA from 2 or more different sources. To make rDNA, a technician often begins by selecting a vector, the means by which rDNA is introduced into a host cell. One common type of vector is a plasmid. **Plasmids** are small accessory rings of DNA found in some bacteria that carry genes not present in the bacterial chromosome. The reading "The Discovery of Plasmid and Viral Vectors," tells how plasmids were discovered by investigators studying the sex life of the intestinal bacterium *Escherichia coli*.

Plasmids that are used as vectors have been removed from bacteria and have had a foreign gene inserted into them (fig. 18.2*a*). Treated cells take up plasmids, and after they enter, bacteria and plasmids reproduce. Eventually, there are many copies of the plasmid and therefore many copies of the foreign gene. The gene is now said to have been **cloned.**

The bacteriophage known as lambda is commonly used as a vector to carry rDNA into bacterial cells (fig. 18.2*b*). Viruses are often the vector of choice for animal cells, also. After a virus has

Figure 18.1
Biotechnology is an industrial endeavor. *a.* Laboratory procedures must be adapted to mass-produce the product. *b.* Microbes are grown in huge tanks, called fermenters because they were first used for yeast fermentation to produce wine. *c.* The product is purified. *d.* The product is packaged.

a.

b.

c.

d.

attacked a cell, the DNA is released from the virus and enters the cell proper. Here it may direct the reproduction of many more viruses. Each virus derived from a viral vector contains a copy of the foreign gene. Therefore, viral vectors also allow cloning of a particular gene.

Recombinant DNA

The introduction of foreign DNA into vector DNA to produce rDNA requires 2 enzymes (fig. 18.3). The first enzyme cuts plasmid (or viral) DNA open and the second seals foreign DNA into this opening.

The first enzyme needed to make rDNA cuts a DNA molecule into discrete pieces. Such enzymes occur naturally in some bacteria, where they stop viral reproduction by cutting up viral DNA. They are called **restriction enzymes** because they *restrict* the growth of viruses. In 1970, Hamilton Smith, at Johns Hopkins University, isolated the first restriction enzyme; now hundreds of different restriction enzymes have been iso-

Figure 18.2

Gene cloning using bacteria and viruses. **a.** A plasmid is removed from a bacterium and is used to make recombinant DNA. After the recombined plasmid is taken up by a host cell, replication produces many copies. **b.** Viral DNA is removed from a virus and is used to make recombinant DNA. The virus containing the recombinant DNA infects a host bacterium. Cloning is achieved when the virus reproduces and then leaves the host cell.

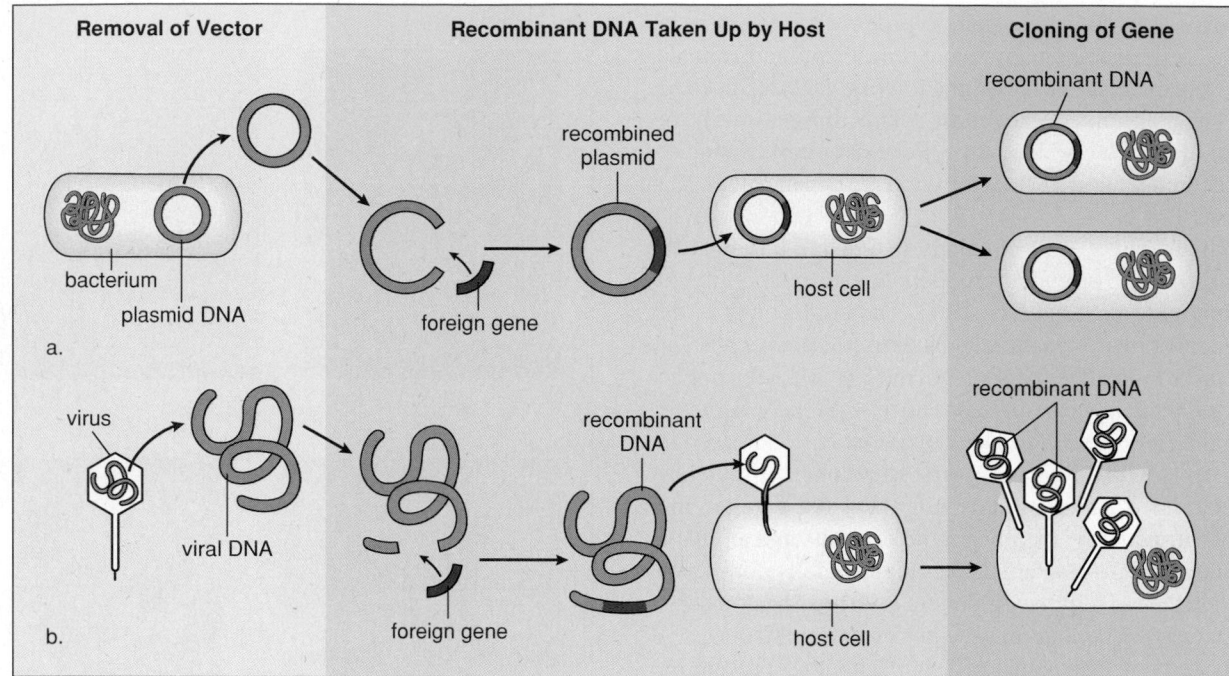

lated and purified. Each one cuts DNA at a specific cleavage site. For example, the restriction enzyme called *Eco*RI always cuts double-stranded DNA when it has this sequence of bases and in this manner:

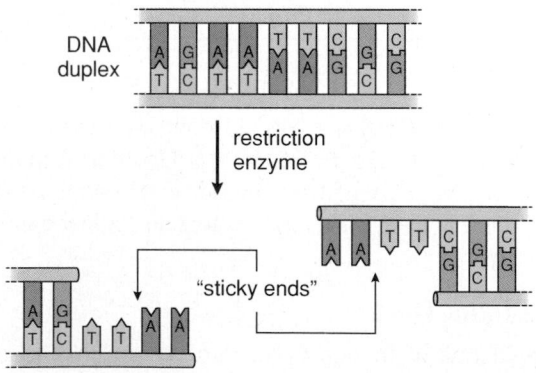

Notice there is now a gap into which a piece of foreign DNA can be placed if it ends in bases complementary to those exposed by the restriction enzyme. To assure this, it is only necessary to cleave the foreign DNA with the same type of restriction enzyme. The single-stranded but complementary ends of the 2 DNA molecules are called "sticky ends" because they can bind by complementary base pairing. They therefore facilitate the insertion of foreign DNA into vector DNA.

The second enzyme needed for preparation of rDNA is **DNA ligase,** a cellular enzyme that seals any breaks in a DNA molecule. Genetic engineers use this enzyme to seal the foreign piece of DNA into the vector. DNA splicing is now complete; an rDNA molecule has been prepared.

The Product

Bacterial cells take up a recombined plasmid, especially if they are treated with calcium chloride to make them more permeable. Thereafter, if the inserted foreign gene is replicated and actively expressed, the investigator can recover either the cloned gene or a protein product (fig. 18.3).

In order for mammalian gene expression to occur in a bacterium, the gene has to be accompanied by the proper regulatory regions. Also, the gene should not contain introns because bacterial cells do not have the necessary enzymes to process primary mRNA. It's possible to make a mammalian gene that lacks introns, however. The enzyme called reverse transcriptase can be used to make a DNA copy of mature mRNA. This DNA molecule, called *complementary DNA (cDNA),* does not contain introns. Alternatively, it is possible to manufacture small genes in the laboratory. A machine called a DNA synthesizer joins together the correct sequence of nucleotides, and this resulting gene also lacks introns. A gene can have nucleotides added to it so that it can be inserted into a vector that has been cleaved by a restriction enzyme.

Genetic Basis of Life

Figure 18.3

Recombinant DNA technology, in detail, utilizing a plasmid to clone a human gene. (The human gene is not shown to scale.) Human DNA and plasmid DNA are cleaved by the same type of restriction enzyme and spliced together by using the enzyme DNA ligase.

After the host bacterium takes up the plasmid, DNA cloning occurs. Also, the protein product is produced. The investigator can retrieve either the DNA or its product for further analysis or use.

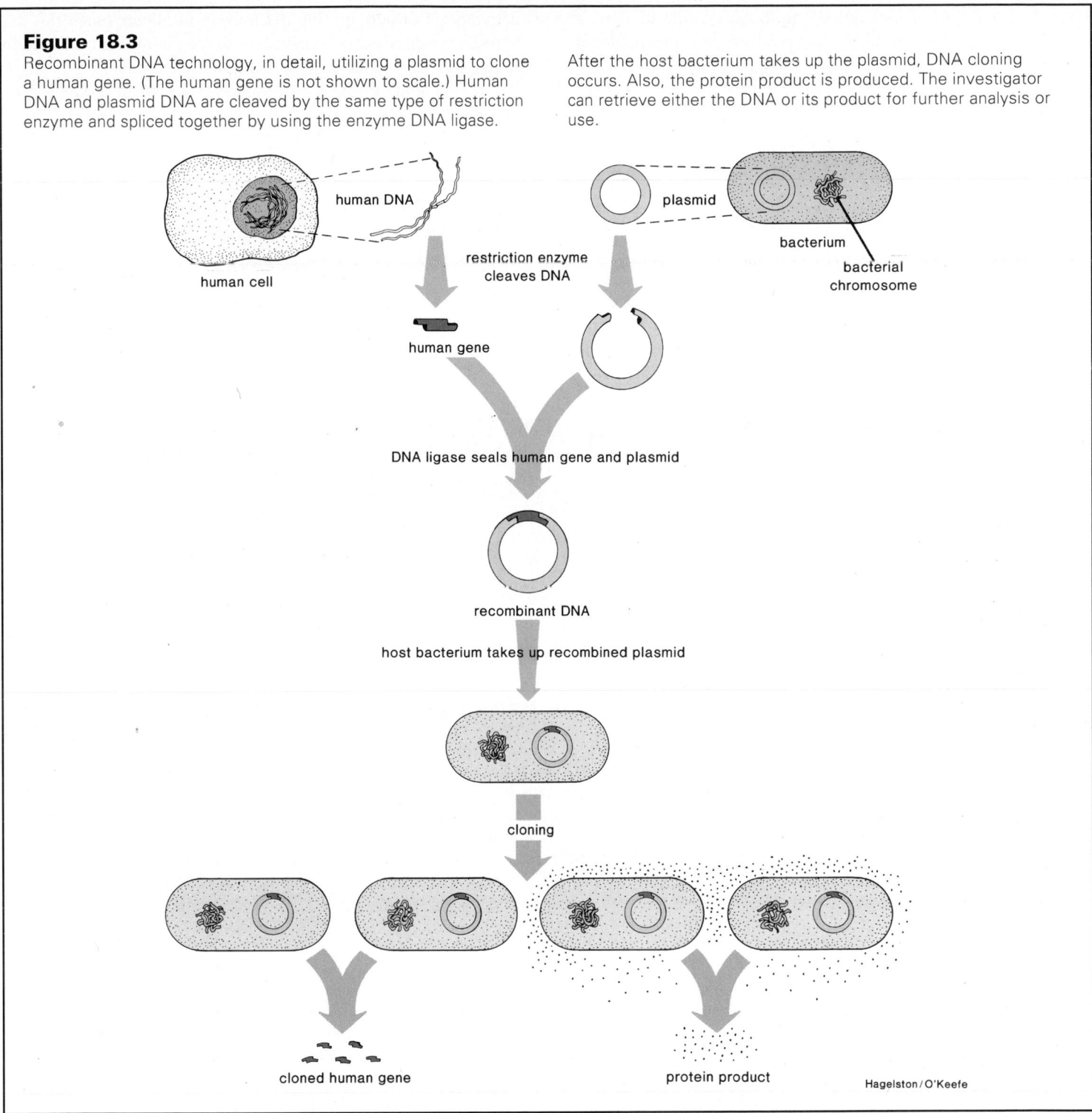

human DNA

human cell

plasmid

bacterium

bacterial chromosome

restriction enzyme cleaves DNA

human gene

DNA ligase seals human gene and plasmid

recombinant DNA

host bacterium takes up recombined plasmid

cloning

cloned human gene

protein product

Hagelston/O'Keefe

Yeast cells are eukaryotes, and they can carry out mRNA processing. Therefore, another methodology is to use bacteria to clone mammalian genomic DNA (gDNA) and then transfer the resulting genes into yeast cells, where gene expression will occur and a protein product will result. Yeast cells even have plasmids by which this is possible.

Biotechnology Products

Table 18.1 shows at a glance the types of biotechnology products now available. There are 3 categories in the table: hormones and similar types of proteins, DNA probes, and vaccines. The production of monoclonal antibodies by bacteria is in its experimental stage. In the meantime, monoclonal antibodies, while still considered a biotechnology product, are produced by the procedure described in chapter 34.

Hormones and Similar Types of Proteins

One impressive advantage of biotechnology is it allows mass production of proteins that are very difficult to obtain otherwise. For example, human growth hormone was previously extracted from the pituitary glands of cadavers, and it took 50 glands to obtain enough for one dose. Now human growth hormone produced by biotechnology is used to treat dwarfs. Insulin was previously extracted from the pancreas glands of slaughtered cattle and pigs; it was expensive and sometimes caused allergic reactions in recipients. Now insulin produced by biotechnology is used to treat diabetics. And few of us knew of tPA (tissue plasminogen

activator), a protein present in the body in minute quantities that activates an enzyme to dissolve blood clots. Now tPA produced by biotechnology is used to treat heart attack victims.

A study of the list of prospective protein products indicates that some of the most troublesome and serious afflictions in humans may soon be treatable: clotting factor VIII will be available for hemophilia; human lung surfactant will be available for premature infants with respiratory distress syndrome; atrial natriuretic factor may be helpful to many with hypertension. And the list will grow, because bacteria (or other cells) can be engineered to produce virtually any protein (fig. 18.4).

Hormones are also produced for use in animals. It is no longer necessary to feed steroids to farm animals; they can be given growth hormone. These animals then produce a leaner meat that is more healthful for humans. When cows are given bovine growth

The Discovery of Plasmid and Viral Vectors

It wasn't until 1946 that investigators began to study bacterial genetics in earnest. This is surprising because bacteria make wonderful laboratory subjects. They are simply organized and easily broken apart, making it easy to get at and analyze their biochemical machinery. They also have relatively few genes, and because one cell produces so many descendants, it's easy to detect very rare mutants. But investigators did not believe bacteria could be prime candidates for genetic analysis because no routine sexual method had been observed by which recombinants were produced. It was the study of recombinant offspring that had allowed Mendel to deduce his laws of genetics and Morgan to map the *Drosophila* chromosome.

In 1946, Joshua Lederberg and Edward Tatum went looking for recombinants among bacteria. They mixed 2 strains of *E. coli* together; one strain required the growth factors biotin and methionine and the other required threonine and leucine. The mixture was plated on a minimal medium that lacked all 4 growth factors because they reasoned that only those bacteria that had experienced genetic recombination would grow. A small but significant number of colonies were obtained when cell-to-cell contact had occurred. When each strain was exposed to only extracts from the other strain, no colonies resulted; therefore, this was not simply another example of *transformation,* a process first observed by Griffith in 1928 (p. 235).

Subsequently, it was discovered that these bacteria were undergoing a form of *conjugation* that allows genes to be transferred from one to the other. In the conjugation process, one bacterium, the "male," develops one or more sex pili (pilus, sing.), surface structures that make contact with a "female." (It is now believed that as the sex pili retract, a conjugation bridge of some sort forms through or on which DNA passes from one cell to another.) In 1953, William Hayes found that after conjugation, the "females" had become "males" and could also form sex pili. Later, Lederberg decided to designate bacteria capable of forming pili as F+ (fertility positive) and those that were incapable as F-, and it was said that the F+ bacteria had passed a fertility factor (F-factor) into the F- bacteria. This is what caused them to become F+ and capable of forming sex pili (fig. 18.A).

It turned out that the F-factor was a plasmid (a small extrachromosomal ring of DNA)—the very first plasmid to be discovered in bacteria. The F-plasmid includes genes responsible for allowing bacteria to carry out conjugation. Occasionally, however, it was noted that an F-plasmid can transfer genes from the main bacterial chromosome, also. Such F-plasmids are no longer independent of the host chromosome; instead, they are integrated into the host chromosome. The bacterial genes that are transferred bring about recombinant genotypes, and this explains the results observed by Lederberg and Tatum. Strains that have the inte-

grated F-plasmid are called Hfr strains because they bring about a *high frequency* of recombination.

François Jacob and Elie Wollman thought of a way to use Hfr strains to map the *E. coli* chromosome. During conjugation, the integrated F-plasmid breaks open and a single strand begins to move through the conjugation bridge, dragging with it the bacterial genes that are attached. It takes about 90 minutes for the entire bacterial chromosome to be transferred. Jacob and Wollman could interrupt this mating process (they used a kitchen blender to force the mating pairs apart) after different time intervals, then they could determine which genes per unit time had been transferred and therefore the order of the genes in the *E. coli* chromosome! This is called the interrupted mating technique for mapping the *E. coli* chromosome.

Many other types of plasmids are now known. Among the most prevalent and well studied are the R-plasmids which confer resistance to antibiotics. In most cases, antibiotic resistance is possible because the R-plasmid codes for enzymes that inactivate the drug. These plasmids can be transferred from one bacterium to another, and this allows more and more bacteria to become resistant. The indiscriminate use of antibiotics selects for the survival of those bacteria that have the R-plasmid and ensures that most bacteria are resistant. Therefore, antibiotics should only be prescribed and taken when absolutely necessary.

hormone (bGH) they produce 25% more milk than usual, which should make it possible for dairy farmers to maintain fewer cows and to cut down on their overhead expenses.

DNA Probes

Biotechnology contributes greatly to diagnostics by making DNA probes available. A **DNA probe** is a specific sequence of single-stranded DNA, often radioactive, that binds by complementary base pairing to a gene of interest. (Probes can be denatured DNA from cells, cDNA, made by a DNA synthesizer, or mRNA.) As listed in table 18.1 and discussed in the following reading, there are many applications for DNA probes. These include detection of disease microorganisms, inherited disorders, and various cancers. Probes are especially useful following DNA amplification by the polymerase chain reaction (PCR), which is also described in the reading.

Vaccines

Vaccines are used to make people immune to an infectious organism so they do not become ill when exposed to it. In the past, vaccines were made from treated bacteria or viruses. Both bacteria and viruses have surface proteins, and a gene for just one of these can be placed in a plasmid. When bioengineered bacteria produce many copies of the surface protein, these copies can be used as a vaccine. A vaccine for hepatitis B is now available, and ones for malaria and AIDS are in experimental stages. Malaria has been a scourge on humankind for many thousands of years, and AIDS is a deadly disease known to modern humans.

Vaccines are available through biotechnology for the inoculation of farm animals. There are vaccines for such illnesses as hoof-and-mouth disease and scours. These animal

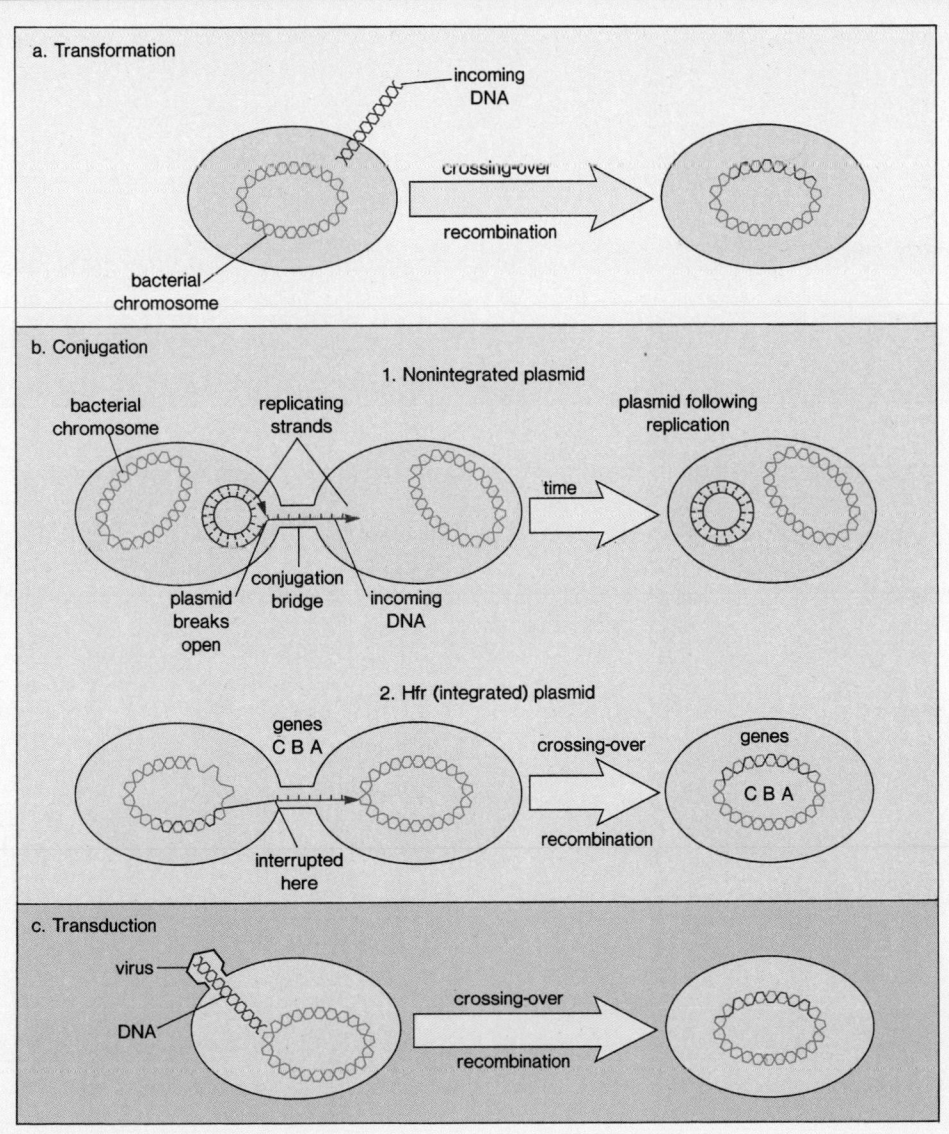

Later, still another way for genetic recombination of bacteria was discovered. In 1952, Joshua Lederberg and Norton Zinder found when working with *Salmonella typhimurium* that recombination was occurring without cell-to-cell contact and even in the presence of an enzyme that could destroy free DNA. This ruled out the possibility of transformation as the mechanism of transfer and it was discovered that transfer was being accomplished by a virus. The transfer of genetic information between bacteria via viruses is called *transduction*. It occurs because the viral genome contains host genes that were accidently picked up during the reproductive cycle of the virus within the host cell.

Figure 18.A

Recombinant bacterial genotypes can be achieved in 3 ways. *a.* Transformation. The cell simply takes up DNA that is added to the medium, and is incorporated into the chromosome. *b.* Conjugation. (1) An independent nonintegrated plasmid such as the F-plasmid breaks open, and one strand enters a F⁻ bacterium by way of a conjugation bridge. (2) If the plasmid has been integrated into the host chromosome as in Hfr plasmids, it can bring host genes with it into the Hfr bacterium. *c.* Transduction. A virus can introduce bacterial genes picked up from a former host. When these genes are incorporated into the bacterial chromosome, a recombinant genotype occurs.

illnesses were once a severe drain on the time, energy, and resources of farmers because they caused an untold number of animal illnesses and deaths each year.

Transgenic Organisms

Free-living organisms in the environment that have had a foreign gene inserted into them are called **transgenic organisms.**

Transgenic Bacteria

Naturally occurring bacteria have been bioengineered to perform a service in the environment. For example, in agriculture, engineered bacteria can be used to promote the health of plants. Certain bacteria that normally live on plants have been changed from frost-plus to frost-minus bacteria. Whereas before they promoted frost damage of plants, now they prevent it. Also, a

Table 18.1
Representative Biotechnology Products Available Now or Being Developed

Hormones and Similar Types of Proteins		DNA Probes	Vaccines
Treatment of Humans	*For*	*Diagnostics in Humans*	*Use in Humans*
Insulin	diabetes	Various diseases (e.g., sexually transmitted diseases)	AIDS
Growth hormone	pituitary dwarfism		Herpes (oral and genital)
tPA, tissue plasminogen activator	heart attack	Inherited disorders (e.g., cystic fibrosis, sickle-cell disease, hemophilia, Huntington disease, Duchenne muscular dystrophy)	Hepatitis A, B, and C
Interferons	cancer		Malaria
Erythropoietin	anemia		Lyme disease
Interleukin-2	cancer		Whooping cough
Clotting factor VIII	hemophilia		
Human lung surfactant	respiratory distress of newborn		
Atrial natriuretic factor	high blood pressure		
Tumor necrosis factor	cancer		
Ceredase	Gaucher disease		

Figure 18.4
Possible biotechnology scenario. *a.* A protein is removed from a cell and the amino acid sequence is determined. (This protein is growth hormone releasing factor.) From this the sequence of nucleotides in DNA can be deduced. *b.* The DNA synthesizer can be used to string nucleotides together in the correct order. *c.* A small section of the gene can be used as a DNA probe to test fetal cells for an inborn error of metabolism. (The manufactured gene could be placed in bacterium to produce more of the protein, which then could possibly be used in treatment of the defect.)

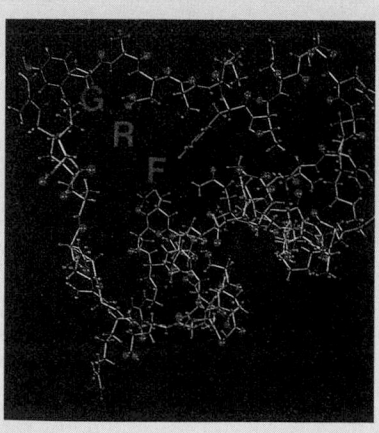
a.

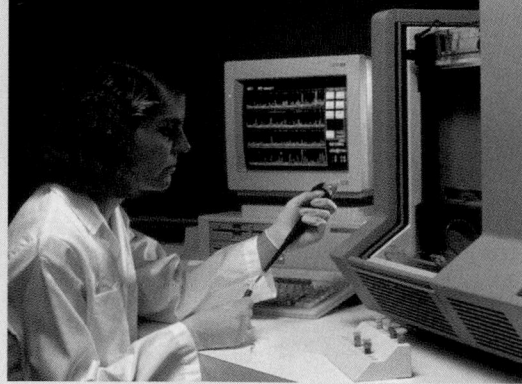
b.

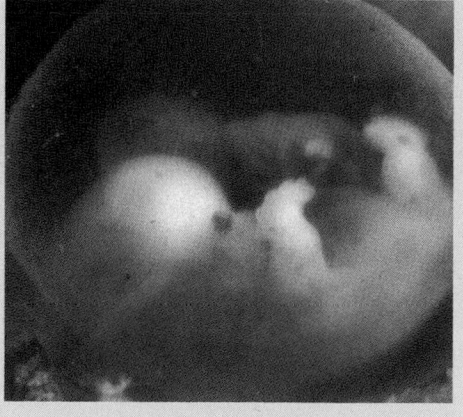
c.

Polymerase Chain Reaction

The polymerase chain reaction (PCR) is a way to make multiple copies of a single gene, or any specific piece of DNA, in a test tube. Further, the process is very specific—the targeted DNA sequence can be less than one part in a million of the total DNA sample! This means that a single gene among all the human genes can be amplified (copied) using PCR.

The process takes its name from DNA polymerase, the enzyme that carries out DNA replication in a cell. The reaction is a chain reaction because DNA polymerase is allowed to carry out replication over and over again until there are a million or more copies of the targeted DNA. The polymerase chain reaction does not replace DNA cloning. Cloning provides many more copies of a gene than this, and it still is used whenever a large quantity of a gene or a protein product is needed.

Before carrying out PCR, it is necessary to have available *primers*—sequences of about 20 bases that are complementary to the bases on either side of the "target DNA." The primers are needed because DNA polymerase does not start the replication process—it continues or extends the process. After the primers bind complementarily to the DNA strand, DNA polymerase copies the target DNA and only the target DNA between the primers. Therefore, PCR is very specific.

PCR has been in use for several years, but a recent advance has been the introduction of automated PCR machines allowing almost any laboratory to carry out the procedure.

Automation became possible after a thermostable DNA polymerase was extracted from a bacterium. Using this enzyme means that there is no need to add more DNA polymerase each time a high temperature is used to separate double-stranded DNA so that replication can occur once again.

Figure 18.B shows that after PCR amplification, it becomes a lot easier to use a probe to detect that target DNA is present. (The binding of the probe is detectable because the probe is labeled either radioactively or with a fluorescent dye.) Therefore, because of the polymerase chain reaction, DNA probes are increasingly used for all sorts of purposes, including those discussed in the following paragraphs.

Genetic Disorders

If a probe for a mutant gene is available, PCR analysis (PCR amplification followed by the use of a probe) can be used to determine if a fetal or an embryonic cell contains the mutant gene. Therefore, following in vitro fertilization, PCR analysis can be used to ensure the implantation of only normal embryos—ones that lack the faulty gene.

Infectious Diseases

Blood tests currently used to diagnose human illness often detect the presence of antibodies to the pathogen rather than the pathogen itself. For example, the blood test for AIDS does not detect HIV but confirms the presence of antibodies to HIV in the blood. PCR analysis can be used to detect the presence of the pathogen before the immune system has even begun an antibody response. It then becomes possible to begin treatment of a disease like AIDS earlier than before.

Cancer Diagnosis

It has long been the custom for physicians to save the tissues of persons who have died from various illnesses. Some of these are cancer patients. PCR makes it possible to analyze the DNA of deceased patients and of living patients who currently have a particular type of cancer. Using this method, for example, investigators found that 90% of pancreatic tumors contain a particular mutated gene.

Human Genome Project

PCR analysis makes it possible for investigators to gather linkage data from the sperm of a single individual. No longer is it necessary to depend on finding a couple who has had a large number of children. PCR also allows scientists to determine the order of restriction fragments by determining whether a particular sequence is found in any 2 fragments.

Evolutionary Relations of Organisms

Scientists used PCR to amplify DNA segments from a quagga, an extinct zebralike animal whose only remains consisted of bits of dried skin. This allowed them to determine that the quagga was a zebra rather than a horse. In other studies, DNA sequences from a 7,000-year-old mummified human brain and from a 17-20-million-year-old plant fossil were analyzed. Recently, PCR was used to amplify mitochondrial DNA sequences from different human populations. These data were used to construct a human phylogenetic tree that suggests an African origin for all humans.

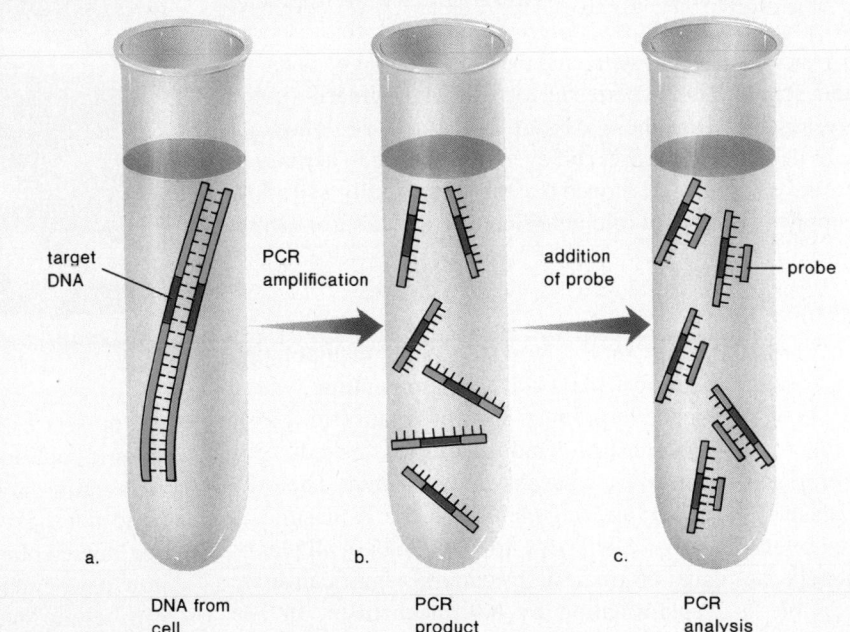

Figure 18.B

PCR analysis. ***a.*** DNA is removed from a cell and placed in a test tube along with appropriate primers, DNA polymerase, and a supply of nucleotides. Primers (not shown) are sequences of about 20 nucleotides complementary to the nucleotides on either side of target DNA. DNA polymerase cannot start a replication process, it can only continue or extend it. Therefore, after the primers are in place, DNA polymerase copies only the targeted DNA sequence. ***b.*** Following PCR amplification, many copies of target DNA (red) are present. ***c.*** Binding of a labeled DNA probe (blue) allows the scientist to determine that a particular DNA segment was indeed present in the original sample.

bacterium that normally colonizes the roots of corn plants now has been given genes (from another bacterium) that code for an insect-killing toxin.

Bacteria can be used for *bioremediation*, which is a term that means pollution cleanup. There are naturally occurring bacteria that can degrade most any type of chemical or material. For example, bacteria were used to help clean up the beaches of Alaska after a massive oil spill. These were naturally occurring bacteria that eat oil, but there are genetically engineered bacteria that could have done a better job (fig. 18.5).

Organic chemicals often are synthesized by having catalysts act on precursor molecules or by using bacteria to carry out the synthesis. Today, it is possible to go one step further and to manipulate the genes that code for these enzymes. For example, biochemists discovered a strain of bacteria that is especially good at producing phenylalanine, an organic chemical needed to make aspartame, the dipeptide sweetener better known as NutraSweet. ™ They isolated, altered, and formed a vector for the appropriate genes so that various bacteria could be geared to produce phenylalanine.

Many major mining companies already use bacteria to obtain various metals. Genetic engineering may enhance the ability of bacteria to extract copper, uranium, and gold from low-grade sources. At least 2 mining companies plan to test genetically engineered organisms with enhanced bioleaching capabilities.

Ecological Considerations

Many individuals are very concerned about the deliberate release of genetically engineered microbes (GEMs) into the environment. Ecologists point out that these bacteria might displace those that normally reside in an ecosystem, and the effects could be deleterious. Others rely on past experience with GEMs, primarily in the laboratory, to suggest that these fears are unfounded. Tools now are available to detect, to measure, and even to disable cell activity in the natural environment. It is hoped that these eventually will allow GEMs to play a significant role in agriculture and environmental protection.

Transgenic Plants

Plants, in particular, lend themselves to genetic manipulation because it is possible to grow plant cells in tissue culture, where each cell can be stimulated to produce an entire plant (fig. 18.6).

Plant cell walls must be removed to give "naked" cells called protoplasts before they will take up a plasmid. The only possible plasmid for bioengineering plant cells, the Ti plasmid, is transferred by its host, *Agrobacterium,* to many but not all plants. Unfortunately the plants of greatest agricultural significance are not susceptible to transformation by this mechanism. In one experiment to show that transgenic plants are possible, a gene for the enzyme luciferase was transplanted from a firefly into a tobacco plant. Whenever the plant was sprayed with luciferin it glowed, proving that the inserted gene was indeed present and active (fig. 18.7).

Since plant cells of interest do not take up the Ti plasmid, new techniques have been developed to introduce foreign DNA into plant cells. For example, it is possible to treat

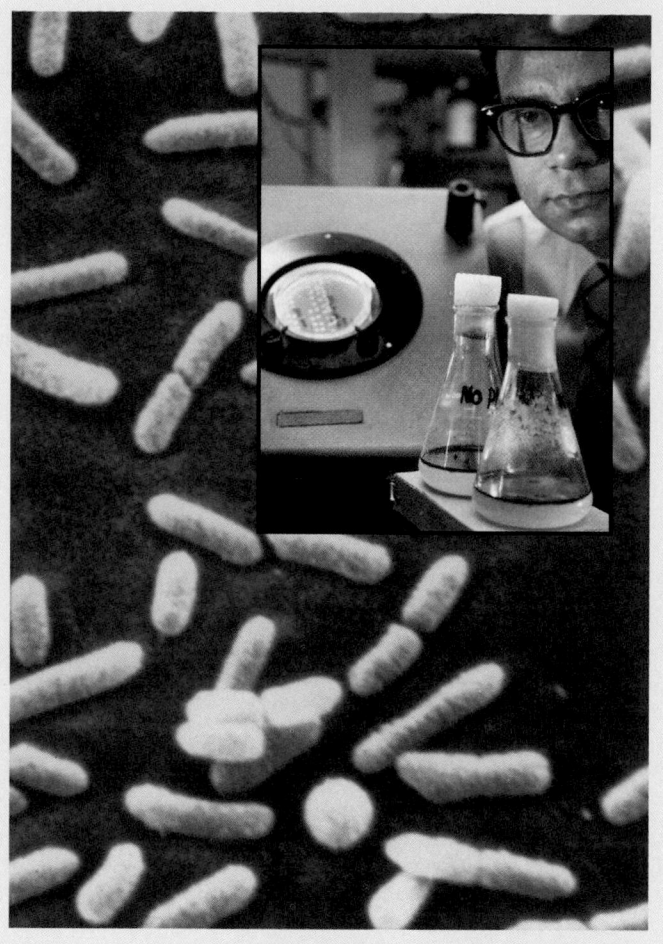

Figure 18.5
Bacteria capable of decomposing oil have been engineered and patented by the investigator, Dr. Chakrabarty. In the inset, the flask toward the rear contains oil and no bacteria; the flask toward the front contains the bacteria and is almost clear of oil. Now that engineered organisms (e.g., bacteria and plants) can be patented, there is an even greater impetus to create them.

protoplasts with an electric current while they are suspended in a liquid containing foreign DNA. The electric current makes tiny self-sealing holes in the plasma membrane through which genetic material can enter. Presently, about 50 types of genetically engineered plants that resist either insects, viruses, or herbicides now have entered small-scale field trials. The major crops that can be improved in this way are soybean, cotton, alfalfa, and rice; however, even genetically engineered corn may reach the marketplace by the year 2000.

There are ecological considerations about bioengineered plants also. Some are concerned that if herbicide resistant plants are available, farmers will tend to use more herbicide and the environment will be degraded. If plants are resistant to insects because they produce a toxin, perhaps resistant insects will evolve.

Figure 18.6
Cloning of entire plants from tissue cells. ***a.*** Sections of carrot root are removed, and thin slices are placed in a nutrient medium. ***b.*** After a few days, the cells form a callus, a lump of undifferentiated cells. ***c.*** After several weeks, the callus begins sprouting carrot plants. ***d.*** Eventually the cloned carrot plants can be moved from culture medium to potting soil.

a.

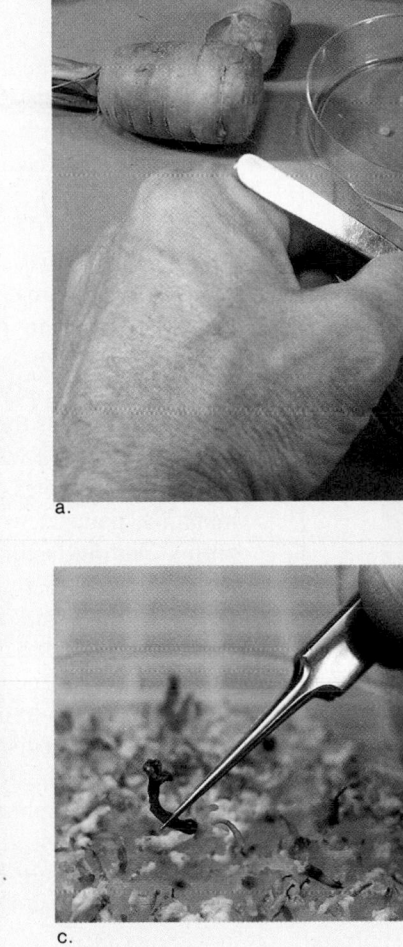

b.

c.

d.

On the other hand, it is hoped that one day even more agriculturally significant plants will be produced by bioengineering. The following types of plants are needed:

Plants that are more nutritious and in which the seed contains all the amino acids required by humans. Protein-enhanced beans, corn, soybeans, and wheat are now being developed.

Plants that require less fertilizer and are able to make use of nitrogen from the atmosphere. (Plants ordinarily take nitrogen from the soil only.) Certain bacteria, however, can make use of atmospheric nitrogen. If the required genes of bacteria were cloned, perhaps they could be transferred to plants.

Plants that have an increased ability to grow under unfavorable environmental conditions such as lack of water or in salty soil. Less irrigation of crops would then be needed. First, it will be necessary to identify which genes in plants might allow them to do this.

Transgenic Animals

Genetic engineering of animals has begun. Animal cells will not usually take up plasmids, but it is possible to microinject foreign genes into eggs before they are fertilized. The most common procedure attempted today is to microinject bovine growth hormone (bGH) into the eggs of various animals. It is hoped that the gene will establish itself and be transmitted to all the cells of the developing organism and even be passed along to the next generation of offspring. This procedure has been used in fishes, chickens, cows, pigs, rabbits, and sheep in the hope of producing bigger varieties. Bioengineered trout are expected to be released in ponds soon.

Figure 18.7

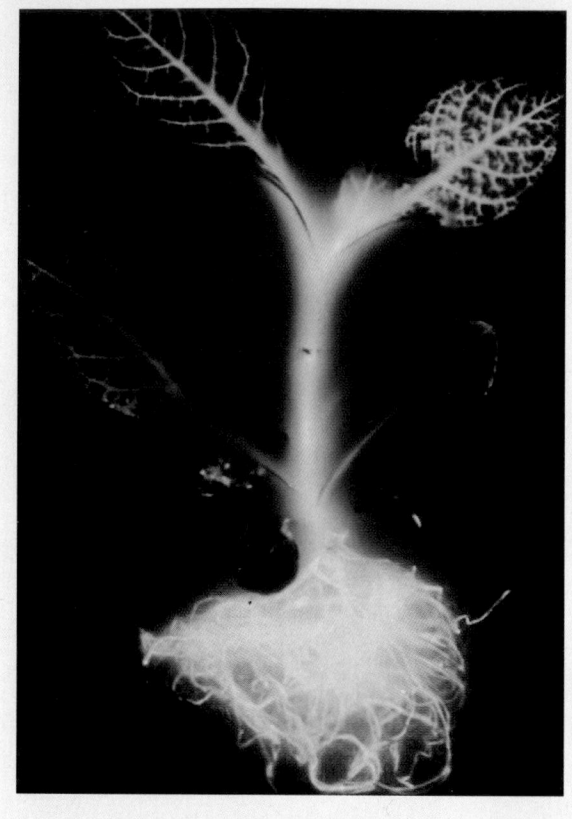

Bioengineered plant. This transgenic plant glows when sprayed with luciferin because its cells contain the protein luciferase, a firefly enzyme that acts on the chemical luciferin that then emits light.

Transgenic farm animals are being developed to produce biotechnology products. A transgenic calf has been produced that carries a gene for the production of human lactoferin in cow milk. Human lactoferin is a protein that is involved in iron transport and has antibacterial activity. Similarly a sheep has been bioengineered to produce tPA (see table 18.1) in her milk. And there is now a pig that can produce human hemoglobin with the hope that one day artificial blood for humans will be a reality.

Gene Therapy in Humans

Gene therapy provides the body with healthy genes to replace defective genes. Gene therapy also includes the use of genes to treat other human ills such as diabetes and AIDS.

Ex Vivo Therapy

During ex vivo therapy, cells are removed from a patient, treated, and returned to the patient. A retrovirus (fig. 23.4), which has RNA genes instead of DNA genes, is used as a vector to carry healthy genes into the cells of the patient, where they become incorporated into the genome. When a retrovirus is used for gene therapy, it has been equipped with recombinant RNA. After recombinant RNA enters a human cell, such as a bone marrow stem cell, reverse transcription occurs and then recombinant DNA enters a human chromosome (fig. 18.8).

During a clinical trial, children with severe combined immune deficiency syndrome (SCID),[1] a genetic disorder, are having the gene they need introduced into their lymphocytes. They lack an enzyme that is involved in the maturation of T and B cells; as a precaution, all participants in the trial will also receive the enzyme directly. In another clinical trial, investigators will treat the patients' liver cells so that these cells will remove cholesterol from the blood. The participants in this trial are inclined toward heart attacks early in life because they have too much cholesterol in their blood.

In Vivo Therapy

If in vivo therapy is used, no cells are removed from the patient. In vivo procedures are using viruses, laboratory-grown cells, or even synthetic carriers to introduce genes directly into the patient. For example, an adenovirus that contains a gene to treat cystic fibrosis patients has been placed in an aerosol spray. When inhaled by laboratory animals the virus enters the cells lining the lungs.

Perhaps it will be possible to use in vivo therapy to cure hemophilia, diabetes, Parkinson's disease, or AIDS. To cure hemophilia, patients would get regular doses of cells containing normal clotting-factor genes. Or cells could be placed in organoids, artificial organs that can be implanted in the abdominal cavity (fig. 18.9). To cure Parkinson's disease, dopamine-producing cells could be grafted directly into the brain. These procedures will use laboratory-grown cells that have been stripped of antigens that might cause them to be attacked by the immune system. To cure AIDS, decoys that display the HIV CD4 receptor could be injected into the body. The HIV virus would combine with these cells instead of T4 lymphocytes.

Some investigators are pursuing gene therapy by direct injection of a protein-DNA complex that is expected to enter liver cells for the purpose of treating diseases associated with the liver, including cancer.

The Human Genome Project

The *human genome project* is an international effort to map the various types of human chromosomes and to sequence their 3 billion base pairs. The project is enormous; the amount of paper needed to record the results will require on the order of 200 books the size of an encyclopedia volume. At first, investigators will prepare a separate genetic (linkage) map and a physical (base-pair) map for each chromosome. Then the 2 maps will be synchronized.

The Genetic Map

The human genome is believed to have 100,000 genes, and as yet, each chromosome has been assigned at most about 200 genes. Before DNA technology, some genes were assigned to

[1]SCID is often called the "bubble-baby" disease after David, a young person who lived under a plastic dome to protect him from infection.

Figure 18.8

Retroviruses as vectors. ***a.*** A retrovirus has an RNA chromosome. When this chromosome enters a cell, reverse transcription occurs. In this way, it is possible to retrieve viral DNA. ***b.*** A foreign human gene can now be inserted into viral DNA. ***c.*** Transcription inside a host cell produces an RNA copy of the recombinant DNA. ***d.*** A virus now is used to carry the recombinant RNA into a human host cell. Following reverse transcription, recombinant DNA is inserted into the human genome.

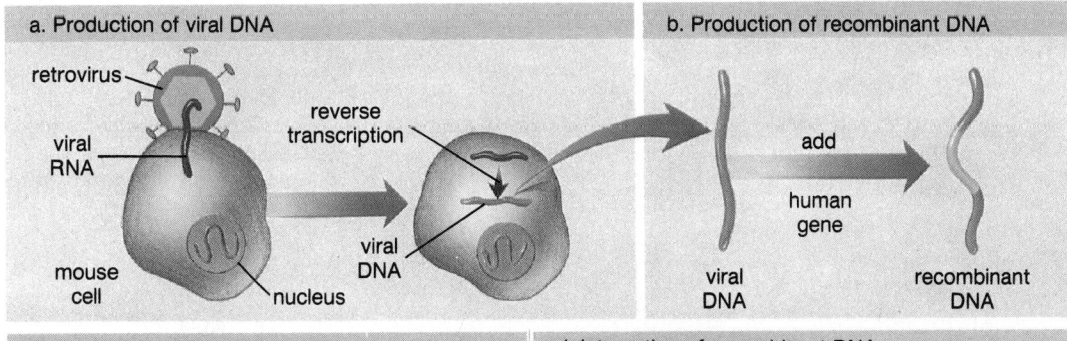

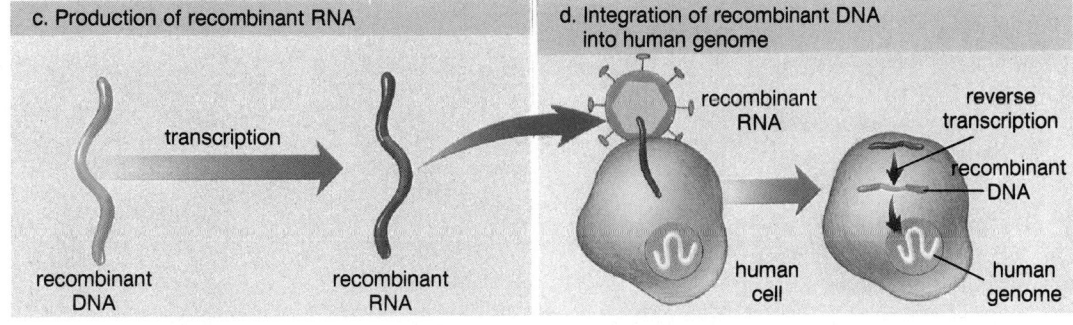

Figure 18.9

Cross section of an organoid. This organoid was made by treating Gore-Tex fibers with a growth factor that stimulates blood vessel formation. Once implanted in the abdominal cavity of a rat, blood vessels extend from the fibers to the animal's natural liver. In this cross section, note the presence of abundant vessels (red color) lined with endothelial cells and surrounded by layers of smooth muscle. These vessels could serve as a means to transfer proteins, made by genetically engineered endothelial cells, into the general circulation.

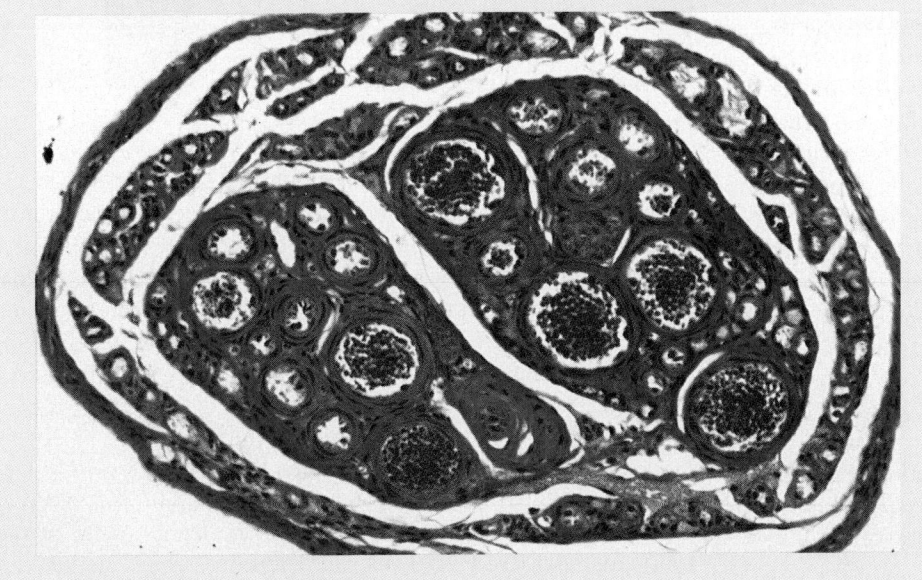

chromosomes on the basis of human and mouse cell fusion results. When human and mouse cells are mixed together in a laboratory dish in the presence of a specially treated virus, they fuse. As the hybrid cells grow and divide, some of the human chromosomes are lost, and eventually the daughter cells contain only a few human chromosomes. The particular human chromosomes can be recognized because each one has a distinctive banding pattern (fig. 14.1). Analysis of the proteins isolated from the cells has allowed scientists to determine which human genes should be associated with these human chromosomes.

Faster methods are now available to assign genes to chromosomes. For example, if a cDNA copy of a gene is available, it can be attached to a fluorescent dye and used as a probe to determine to which chromosome, and indeed to which band of this chromosome, it belongs. Base pairing between probe and chromosome will occur right on a microscope slide, and location of fluorescence allows us to see where the gene is located:

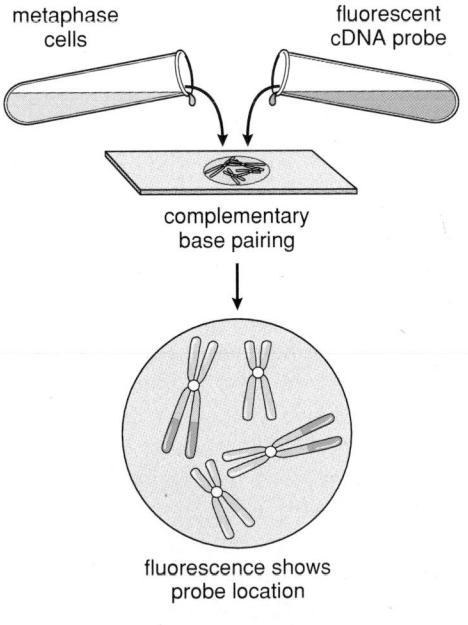

The genetic map of a chromosome will tell the sequence of the genes on that chromosome. Figure 18.10 shows a preliminary genetic map for chromosome 17, one of the smaller chromosomes. This map, like the one in figure 13.11 for *Drosophila*, is based on linkage data. Recall that the closer 2 gene loci are, the less likely their alleles will cross-over (recombine) during meiosis. A 1% cross-over rate is assigned a distance of one map unit.

In the past, linkage data for human genes were obtained by studying the results of human matings. This method of gathering such data has severe drawbacks. Human couples tend to have few children, and the generation time is lengthy. Obviously, a faster method is desirable, and fortunately it has been found. This method involves the use of DNA markers.

DNA Markers

Recall that a significant portion (about 30%) of the human genome consists of repetitive DNA that does not code for protein (p. 244). A **DNA marker** is a variation usually within highly repetitive DNA that is inherited in a Mendelian fashion. The number of DNA repeat sequences between any 2 genes varies, but that distinctive number is inherited. This means, for example, that Susan might have 10 repeats of say ATCCAGGAATG between gene 1 and gene 2 while Bob has 30 repeats between gene 1 and gene 2. The pattern of repeats between all the genes is called a person's DNA fingerprint because it is so distinctive (fig. 18.11). The accompanying reading describes how you take a DNA fingerprint. The process of getting a DNA fingerprint involves the necessity of performing gel electrophoresis.

Gel Electrophoresis Gel electrophoresis allows fragments of DNA to be separated according to their charge and size. The entire genome is fragmented by the use of one or more restriction enzymes, and the fragments are placed in a well at one end of a *gel* (a gelatin-like material) slab. When an electric current is passed through the gel, *electrophoresis* (*phoresis*—being carried) occurs. DNA fragments are negatively charged, and therefore they all move toward the positive pole, but the shorter fragments move the fastest and the longer fragments move the slowest (fig. 18.11). Gel electrophoresis is so sensitive that it can separate DNA fragments that differ in length from each other only by a single nucleotide.

The DNA fragments are now denatured into single-stranded molecules and blotted onto a membrane (called Southern blotting after E. M. Southern, who invented the technique), where they are exposed to a probe, say a probe for the DNA sequence ATCCAGGAATG mentioned previously. The probe locates the position and therefore the size of a fragment that contains this sequence. It's been discovered that the length of a fragment is related to the number of repeats. Referring back to our example, Susan's repetitive DNA fragment with 10 repeats will be shorter and will move farther on the gel than Bob's repetitive DNA fragment with 30 repeats. This would be just one difference between the DNA fingerprints of these 2 individuals. Other fragment differences related to repeats will be observed also. Since we observe these differences between people after treating DNA with restriction enzymes, they are called *restriction fragment length polymorphisms* (*RFLPs*, pronounced *riflips*). Polymorphism means many structural variations.

Figure 18.10

Genetic map for chromosome 19 (as known in 1990) and density bars. **a.** Gene density bar shows the amount of progress toward identifying the estimated genes per band of the chromosome. The light color means 1% progress and the dark color means 10% progress. **b.** Chromosome 17. The light bands are Giemsa (p. 216) negative, and the dark bands are Giemsa positive. The centromere is shown as a constriction and the diagonal shading is centric heterochromatin. Representative genes are TP53, also called p53, the tumor suppressor gene (p. 274); NF1 the neurofibromatosis gene (p. 223); HOX2@, a homeobox cluster; COL1A1 is a gene for collagen (p. 57). **c.** Polymorphism density bar indicates the number of known polymorphic loci per band. Red = 50 known markers, and light blue = 0 known markers. Lines connect representative markers (i.e. D17S34) to a location on the chromosome.

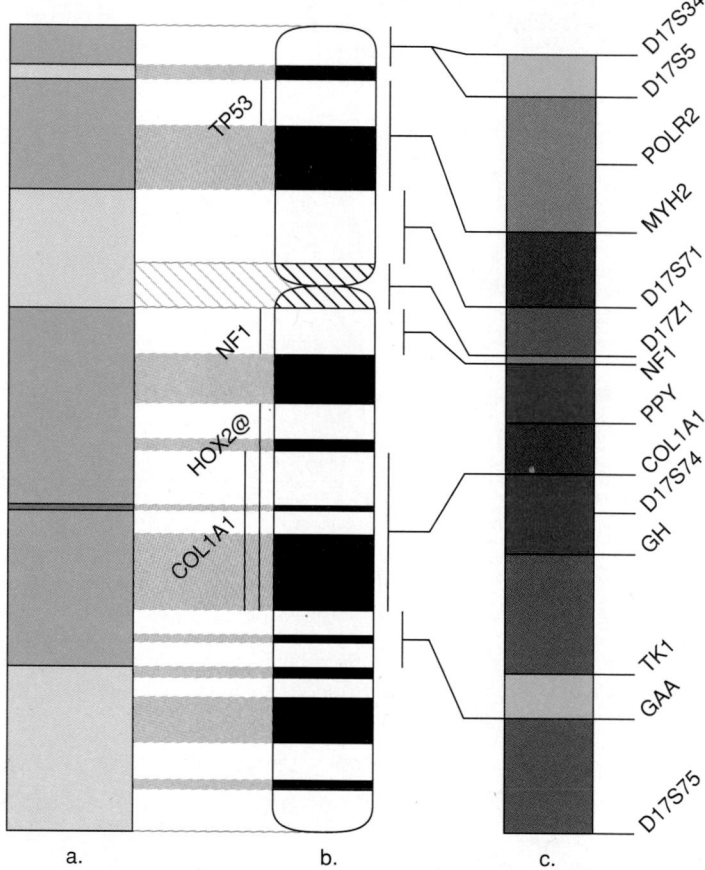

a.　　　　b.　　　　c.

RFLPs RFLPs are markers (variations) that are sometimes linked to a particular gene on a chromosome. Genetic disorders run in families, and at times it is discovered by studying the RFLPs of that family if a marker is always inherited with a disorder. Sometimes such a marker is seen in an entire population, and therefore a test for the marker, that is, RFLP analysis, is appropriate to all the members of the population to determine who has inherited a mutated allele.

Today's linkage map of a chromosome (fig. 18.10) includes not only genes, it also includes RFLP markers. PCR (see reading, p. 285) contributes to the gathering of linkage data because it can amplify the DNA from the many sperm of one individual and RFLP analysis can determine which markers are

Genetic Basis of Life

Figure 18.11

RFLP analysis. The placement of restriction enzyme cuts differs between individuals. In a given segment, individuals I and II have the same restriction sites but individual III has one more site. This could be caused by a difference in the number of DNA repeat sequences between genes. Electrophoresis of the restriction fragments separates them according to their length because short fragments migrate farther than long fragments. The placement of the fragments on the gel slab is detected by the use of a probe for the DNA repeat. If investigators know that the additional restriction enzyme cut is always inherited with (linked to) a genetic disorder, then individual III has the disorder. Also, the pattern, particularly if obtained by using several different restriction enzymes, is called a person's DNA fingerprint.

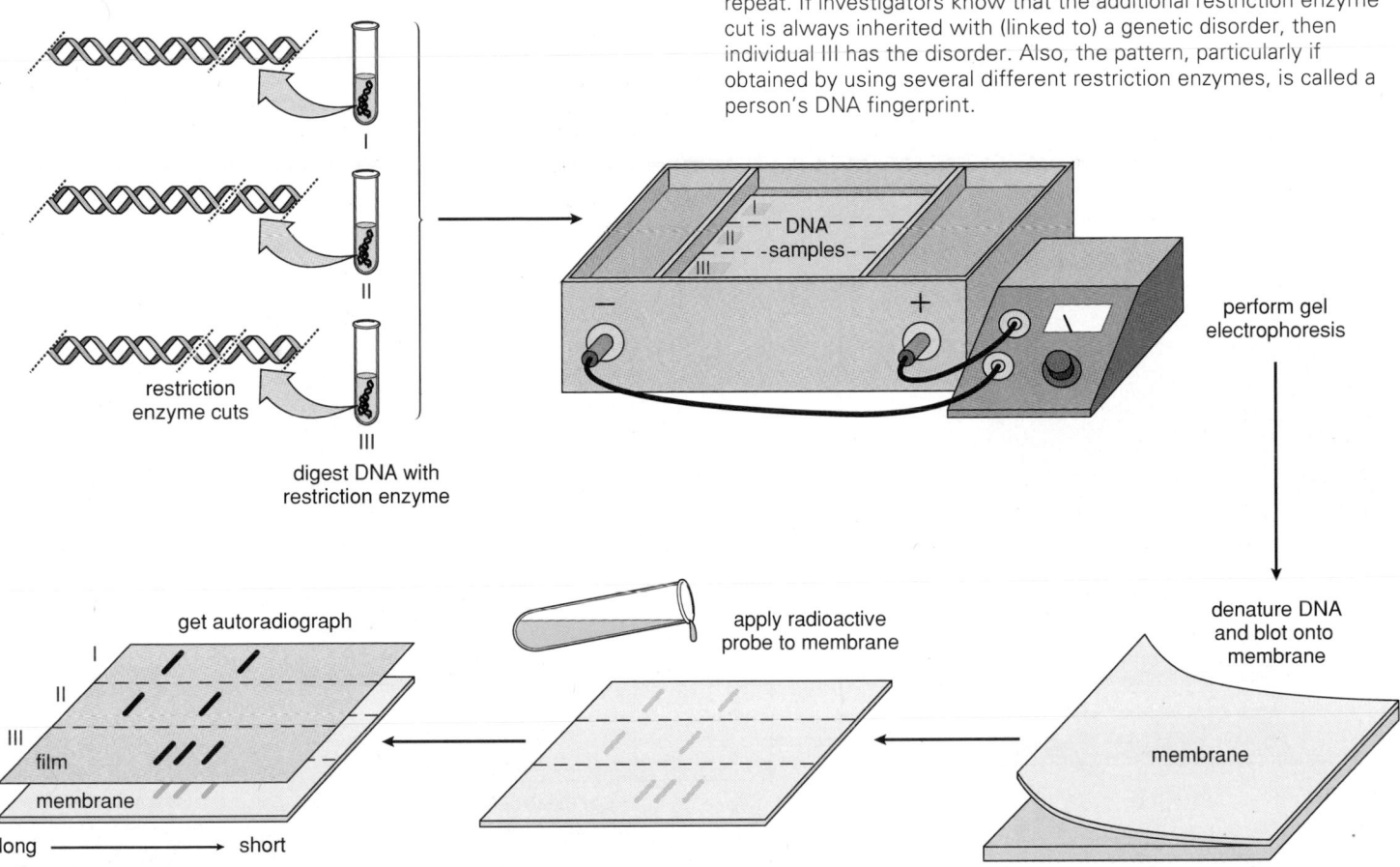

How Do You Take a DNA Fingerprint?

Since it was first used in a criminal case in 1988, DNA fingerprinting has established the guilt or innocence of suspects in over 100 criminal trials. If the DNA fingerprint of biological tissue (blood, hair, or semen) taken from the scene of a crime is the same as that of a suspect, then he or she is almost certainly guilty of the crime.

To take the DNA fingerprint, repetitive DNA is isolated from the genome and then treated with a restriction enzyme like *Hae* III or *Alu* I, which cuts the repetitive DNA into fragments of varying lengths. The length of the restriction fragments are peculiar to the person—no other person has the same series of different-sized fragments. Gel electrophoresis is used to separate the restriction fragments. The shorter the fragment, the farther it moves along in the gel. Therefore, in the end the fragments are separated according to their length from long to short. These double-stranded DNA fragments are denatured into single strands and blotted onto a membrane, where they remain in place. The next step is to incubate the membrane in a solution containing a radioactively labeled DNA probe. The radioactive probe searches for complementary sequences among the DNA fragments on the membrane. When a match occurs, the probe hybridizes, or irreversibly binds, to its target by base pairing. The probe remains in place even after the membrane is washed to remove excess, unbound probe. The membrane is then placed in direct contact with a sheet of X-ray film. The radioactivity in the labeled fragments exposes the sheet of X-ray film, and the result is a pattern of 15-40 dark bands unique to the individual. This is the DNA fingerprint (fig 18.11).

With the exception of identical twins, the probability that 2 people selected at random would have the same DNA fingerprint is less than one in a billion—even as low as one in 10^{19}! This is why DNA fingerprints can be used to tell the guilty from the innocent. They are not always accepted in a court of law because there is no uniform procedure and standard for taking the fingerprint. Once this is worked out, DNA fingerprinting will be a positive step forward for the criminal justice system.

DNA fingerprinting can also be used to identify individuals under other circumstances. For example, through DNA fingerprinting, parents have been matched with children many years after a separation. The U.S. military is planning to keep biosamples from its military personnel with the intention of using DNA fingerprinting to positively identify remains if necessary.

Figure 18.12

DNA sequencing. **a.** An example of DNA gel electrophoresis in which the investigator sees orange bands because the gel has been stained with a DNA-binding dye that fluoresces orange in ultraviolet light. The length of the DNA at each band can be determined by comparing the bands observed to those made previously by DNA molecules whose lengths were already known. **b.** Each test tube contains all 4 nucleotides, DNA polymerase, radioactive primer, and one of 4 analogues that stop replication at one of the bases in DNA. The length of the resulting fragments will vary in each test tube. It is possible to determine the sequence of the bases in DNA by simply reading up from the bottom. Move horizontally whenever necessary to determine the next nucleotide.

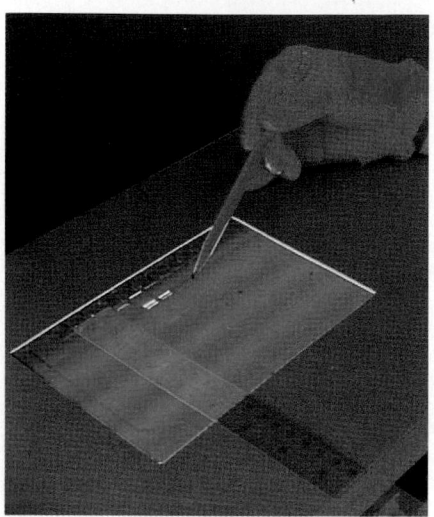

a.

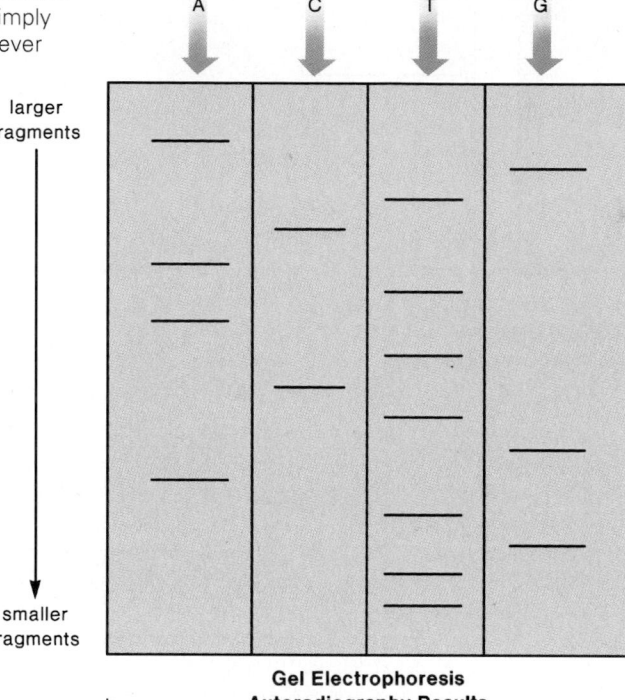

**Gel Electrophoresis
Autoradiography Results**

b.

linked to which alleles. Now we don't have to wait for children to be born or to study recombination in so few examples to determine how genes are linked.

The Physical Map

The physical map of a chromosome shows the sequence of DNA bases for that chromosome. One way to sequence DNA was developed by Frederick Sanger, who shared a Nobel Prize with Walter Gilbert for his work in this area. First, the DNA is divided up into fragments by using restriction enzymes. Then, each of these fragments is individually sequenced after being denatured into single strands. A copy of the strand to be sequenced is placed in 4 test tubes (fig. 18.12). The test tube contains everything necessary for DNA synthesis (DNA polymerase and all 4 types of nucleotides) and a radioactive primer that labels the 5′ end of the newly replicated strand of DNA. Four test tubes are needed because a different nucleotide analogue (a dideoxyribonucleoside triphosphate) is added to each. These stop DNA synthesis after they are in position because the

next phosphate group cannot attach to the analogue. For example, in the first test tube, replication produces synthesized DNA strands of different lengths, each length representing a place where an A (adenine) analogue is attached. In this instance, gel electrophoresis followed by autoradiography tells the investigator what these newly synthesized different length fragments are. Comparing the results of electrophoresis for all 4 test tubes allows the investigator to determine the sequence of the bases (fig. 18.12) in the fragment. Eventually the entire genome will be sequenced.

Relating the Maps

A way has been found to link the genetic map to the physical map. Investigators have discovered that there are segments of DNA that reoccur every few hundred base pairs and that contain a unique sequence of bases. These segments, known as sequence-tagged sites, or STSs, occur within markers. By identifying STSs on the physical map, it is possible to relate the physical map to the genetic map.

Summary

1. Biotechnology, the use of natural biological systems to produce products or achieve ends desired by humans, is being expanded greatly by DNA technology.
2. To achieve a genetically engineered cell, a vector is first prepared. Both plasmids and viruses are used as vectors to carry a foreign gene into a cell. When the plasmid replicates or the virus reproduces, the foreign gene is cloned.
3. The preparation of recombinant DNA requires restriction enzymes, which are used to cleave plasmid DNA and to cleave foreign DNA. The "sticky ends" produced facilitate the insertion of foreign DNA into vector DNA. The foreign gene is sealed into the vector DNA by DNA ligase.
4. Many other laboratory procedures are available to facilitate recombinant DNA technology. For example, it is possible to produce cDNA using reverse transcriptase; to manufacture a gene using a DNA-synthesizing machine; and to attach regulatory genes to structural genes before they are inserted into plasmid or viral DNA.
5. Transformed cells produce such products of interest to humans as hormones, DNA probes, and vaccines.
6. Transgenic organisms have also been made. Bacteria have been produced to promote the health of plants, perform bioremediation, extract minerals, and produce chemicals.
7. Plants lend themselves to genetic manipulation because it is possible to grow an adult plant from a cell that originally was maintained in tissue culture.
8. It is hoped that plants having the following characteristics may be developed using recombinant DNA technology: natural resistance to various diseases and pests, ability to fix atmospheric nitrogen, increased ability to grow in arid and salty soils, and production of seeds possessing all the essential amino acids.
9. Genetic engineering of animals is being intensely investigated. Genetic engineering of animal zygotes has not as yet led to uniform results, and the genetic engineering of the human zygote has not yet begun.
10. Human gene therapy has just begun. A retrovirus may serve as a vector that carries a normal gene into bone marrow stem cells removed from a patient with a genetic disorder. Once the normal gene has been incorporated into the DNA of the stem cells, they can be injected back into the patient. Alternatively, organoids can be made that can possibly carry bioengineered cells.
11. The human genome project will produce a genetic map and a physical map for each chromosome.
12. The genetic map of a chromosome will give the order of the genes and the markers on that chromosome. A marker is a variation that exists usually within highly repetitive DNA, and it usually concerns the number of repeat sequences between genes. Sometimes a variation is linked to a genetic disorder and can be used to tell if a disorder has been inherited. The variations can be detected by RFLP analysis.
13. The physical map of a chromosome will tell the sequence of base pairs in the chromosome. Laboratory techniques are available to determine DNA sequences.

Writing Across the Curriculum

In order to practice writing skills, students should write out the answers to any or all of the study questions and the critical thinking questions. The study questions are sequenced in the same order as the text. Suggested answers to the critical thinking questions are in appendix D.

Study Questions

1. What are the similarities and differences between using a plasmid and using a virus as a vector for gene cloning?
2. What is the methodology for producing recombinant DNA to be used in gene cloning?
3. List and explain other types of laboratory procedures that are used in recombinant DNA experiments.
4. Categorize and give examples of types of biotechnology products available today.
5. Why are plants good candidates for genetic engineering, and what are the problems involved in using this method to achieve agriculturally significant plants?
6. In what ways is it hoped that plants will soon be modified?
7. What types of bacteria have been produced by genetic engineering to provide plants with desirable characteristics? Give a pro and a con for releasing these into the environment.
8. Describe an experiment to produce transgenic animals. Why is this procedure not expected any time soon in humans?
9. Outline a method of gene therapy in humans.
10. What is the difference between the genetic map and the physical map of the human genome? How does a DNA marker differ from a gene?

Objective Questions

1. Which of these is a true statement?
   a. Both plasmids and viruses can serve as vectors.
   b. Plasmids can carry recombinant DNA but viruses cannot.
   c. Vectors carry only the foreign gene into the host cell.
   d. All of these statements are true.

2. Using this key, put the steps in the correct order to achieve a plasmid carrying recombinant DNA.

   Key:

   Step 1 use restriction enzymes
   Step 2 use DNA ligase
   Step 3 remove plasmid from parent bacterium
   Step 4 introduce plasmid into new host bacterium

   a. 1,2,3,4
   b. 4,3,2,1
   c. 3,1,2,4
   d. 2,3,1,4

3. Which of these is a benefit to having insulin produced by biotechnology?
   a. It can be mass-produced.
   b. It is nonallergenic.
   c. It is less expensive.
   d. All of these.

4. Why is it easier to engineer plants genetically as compared to animals?
   a. Plants can be grown from a single cell in tissue culture.
   b. *E. coli* plasmids can serve as vectors in plants.
   c. Agriculturally significant plants have been engineered.
   d. All of these.

5. Restriction fragment length polymorphisms (RFLPs)
   a. identify individuals genetically.
   b. are the basis for DNA fingerprints.
   c. can be subjected to gel electrophoresis.
   d. All of these.

6. Which of these would you not expect to be a biotechnology product?
   a. modified enzyme
   b. DNA probe
   c. protein hormone
   d. steroid hormone

7. What is the benefit of using a retrovirus as a vector in gene therapy?
   a. It is not able to enter cells.
   b. It incorporates the foreign gene into the host chromosome.
   c. It eliminates a lot of unnecessary steps.
   d. All of these.

8. Gel electrophoresis
   a. measures the size of plasmids.
   b. tells whether viruses are infectious.
   c. measures the charge and size of proteins and DNA fragments.
   d. All of these.

9. Which of these is incorrectly matched?
   a. protoplast—plant cell engineering
   b. base analogs—sequencing DNA
   c. retro-virus—gene therapy
   d. DNA ligase—mapping human chromosomes

10. After a plasmid has been cloned
   a. you can remove the cloned foreign gene and sequence it.
   b. it does not express itself and no protein product results.
   c. you can use it to carry out human gene therapy.
   d. All of these.

11. The restriction enzyme called Eco RI has cut double-stranded DNA in this manner. The piece of foreign DNA to be inserted ends in what bases, (a) to the left? (b) to the right?

 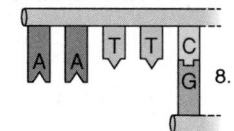

12. Use these labels where needed to complete this diagram of a possible gene therapy procedure in humans:

Most of the viral genes are removed and a human gene is inserted
Recombinant RNA is repackaged in viral coats
Transcription of DNA occurs
Reverse transcription occurs

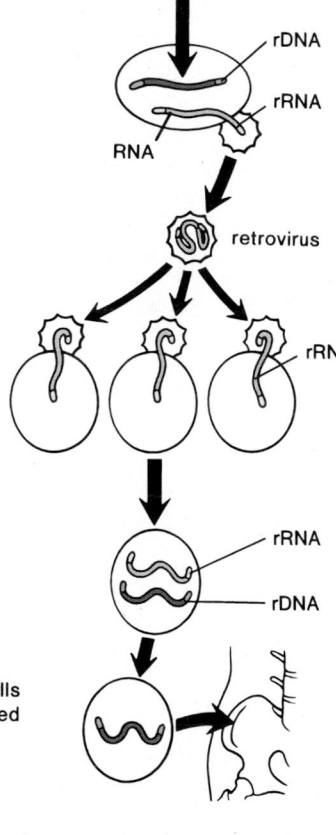

1. Viral gene infects mouse cells.

2.

3. Viral DNA is extracted.

4.

5.

6.

7. Virus infects patient's blood cells taken from bone marrow.

8.

9. Bone marrow cells carrying corrected human gene are reinjected into the patient.

Concepts and Critical Thinking

1. *Organisms are chemical and physical machines.*

   Provide examples to show that DNA can be manipulated, and tell why you think some people would find DNA manipulation disturbing.

2. *DNA is the common genetic material of all organisms.*

   How does recombinant DNA technology support our belief in this concept?

PART 2 *Genetic Basis of Life*

CRITICAL THINKING CASE STUDY

Chromosome Puffs

Chromosome puffs (fig. 17.4) have been observed in cells of the salivary glands in the midge fly *Chironomus*. It was hypothesized sometime ago that *chromosome puffs are regions of the chromosome where specific genes are active.*

Since gene activity leads to specific protein products, the hypothesis suggests that it should be possible to find specific proteins associated with particular puffs. Researchers have been fortunate in finding 2 species of midge fly whose salivary gland secretions differ. The salivary gland cells of *Chironomus pallidivittatus* produce a secretion that contains protein granules. The salivary gland cells of *Chironomus tentans* produce a secretion that does *not* contain protein granules. In which species do you expect to find a chromosome puff in salivary gland cells producing a secretion?

Prediction 1 When secretion is produced, a puff will be apparent in *C. pallidivittatus* because its secretions have protein granules. A corresponding puff will not be present in *C. tentans*.

Result 1 Observation reveals that is the case and supports the hypothesis that puffs are sites where specific genes are active.

I Why wouldn't it be sufficient just to witness puffing in *C. pallidivittatus*? What role does *C. tentans* play in the experiment?

To test the hypothesis further, hybrid offspring were produced by crossing the 2 midge fly species. In hybrid flies, one member of each homologous pair of chromosomes is inherited from *C. pallidivittatus* and the other is inherited from *C. tentans*. Which homologous pair member will show a chromosome puff in salivary gland cells producing a secretion?

Prediction 2 A chromosome puff will be seen on the chromosome inherited from *C. pallidivittatus* but not on the chromosome inherited from *C. tentans*.

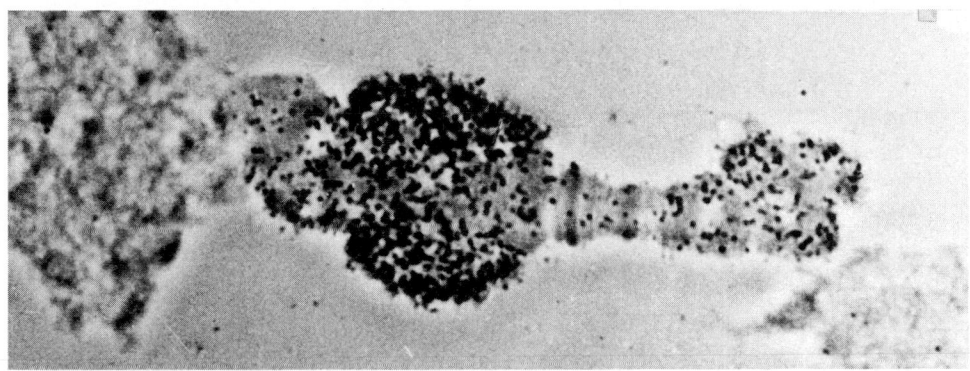

a.

b.

Figure A

A comparison of chromosome puffs. *a.* Chromosome puffs treated only with radioactive uridine. *b.* Chromosome puffs treated with actinomycin D and radioactive uridine. Magnification X4,200.

Result 2 Careful observation supports the prediction. A puff was apparent on the homologue inherited from *C. pallidivittatus* when secretion was produced. No puff was observed on the other homologue inherited from *C. tentans*.

II Why did researchers follow up Prediction 1 and Result 1 with Prediction 2 and Result 2?
Also, what would you predict about the concentration of granules in secretions of the hybrid midge fly?

So far we have supported the hypothesis by showing there is a relationship between chromosome puffs and a protein product. A much stronger argument for gene activation can be made if it is determined that mRNA is synthesized at chromosome puffs. Audioradiography, a technique described on page 55, was used to determine if this is the case. Radioactive uridine (a precursor of the base uracil) was injected into a midge fly. Salivary glands were removed and sectioned. The sections were used to prepare radiographs in which dark spots indicate the presence of RNA molecules. Where do you predict that dark spots will appear in the radiographs?

Prediction 3 Many dark spots will appear over chromosome puffs. None will appear on unpuffed regions of the chromosomes.

Result 3 As shown in part *a* of figure A, the concentration of dark spots is much greater on the puffed region of the chromosome. These results suggest that mRNA synthesis is occurring predominantly at the puff.

III Figure A shows some dark spots over the rest of the chromosome also. What could this mean?

Actinomycin D is an inhibitor of RNA synthesis. What do you predict will be the results of autoradiography when a midge fly is exposed to radioactive uridine and actinomycin D at the same time?

Prediction 4 When treated with both actinomycin D and radioactive uridine, no RNA synthesis is apparent at the puffs.

Result 4 As shown in part *b* of figure A, the number of dark spots is much fewer than before. The implication is that actinomycin D has inhibited RNA synthesis at the puff.

IV Why was this last experiment performed? Does comparing parts *a* and *b* of figure A seem more convincing than part *a* alone? Why?

The results of these experiments provide strong support for the hypothesis that chromosome puffs are regions of the chromosome where specific genes are active.

Other Questions

1. If 2 hybrid midge flies were crossed, how might their offspring differ as to amount of protein product in salivary secretions?

2. Could these experiments also be used to support the hypothesis that specific genes are active only at certain times? What other experiments could be done to support this hypothesis? (See also p.270 of the text.)

Reference: Beerman, W., and Clever, U. April 1964. Chromosome puffs. *Scientific American.*

Suggested Readings for Part 2

Antebi, E., and Fishlock, D. 1986. *Biotechnology: Strategies for life.* Cambridge, Mass.: The MIT Press.

Bishop, J. M. March 1982. Oncogenes. *Scientific American.*

Chilton, M. June 1983. A vector for introducing new genes into plants. *Scientific American.*

Cohen, S. N., and Shapiro, J. A. February 1980. Transposable genetic elements. *Scientific American.*

Darnell, J. E. October 1983. The processing of RNA. *Scientific American.*

Dickerson, R. E. December 1983. The DNA helix and how it is read. *Scientific American.*

Doolittle, R. F. October 1985. Proteins. *Scientific American.*

Drlica, K. 1984. *Understanding DNA and gene cloning.* New York: John Wiley and Sons.

Elkington, J. 1985. *The gene factory: Inside the science and business of biotechnology.* New York: Carroll and Graf Publishers.

Feldman, M., and Eisenback, L. November 1988. What makes a tumor cell metastatic? *Scientific American.*

Glover, D. M. 1984. *Gene cloning: The mechanics of DNA manipulation.* New York: Chapman and Hall.

Grivell, L. A. March 1983. Mitochondrial DNA. *Scientific American.*

Holliday, R. June 1989. A different kind of inheritance. *Scientific American.*

Kieffer, G. H. 1987. *Biotechnology, genetic engineering, and society.* Reston, Va.: National Association of Biology Teachers.

Klug, W. S., and Cummings, M. R. 1986. *Concepts of genetics.* 2d ed. Westerville, Ohio: Charles E. Merrill Publishing Co.

Kornbert, R. D., and Klug, A. February 1981. The nucleosome. *Scientific American.*

McIntosh, J. R., and McDonald, K. L. October 1989. The mitotic spindle. *Scientific American.*

Mendel, G. 1965. Experiments in plant hybridization. In *The origins of genetics,* by C. Stern and E. Sherwood. San Francisco: W. H. Freeman Co.

Mullis, K. B. April 1990. The unusual origin of the polymerase chain reaction. *Scientific American.*

Murray, A. W., and Szostak, J. W. November 1987. Artificial chromosomes. *Scientific American.*

Neufeld, P. J., and Colman, N. May 1990. When science takes the witness stand. *Scientific American.*

Nomura, M. January 1984. The control of ribosome synthesis. *Scientific American.*

Patterson, D. August 1987. The causes of Down syndrome. *Scientific American.*

Prescott, D. 1988. *Cells.* Boston: Jones and Bartlett.

Ptashne, M. January 1989. How gene activators work. *Scientific American.*

Radman, M., and Wagner, R. August 1989. The high fidelity of DNA duplication. *Scientific American.*

Ross, J. April 1989. The turnover of messenger RNA. *Scientific American.*

Sapienza, C. October 1990. Parental imprinting of genes. *Scientific American.*

Shepard, J. F. May 1982. The regeneration of potato plants from leaf-cell protoplasts. *Scientific American.*

Smith-Klein, C., and Kish, V. 1988. *Principles of cell biology.* New York: Harper & Row.

Stahl, F. W. February 1987. Genetic recombination. *Scientific American.*

Steitz, J. A. June 1988. "Snurps." *Scientific American.*

Tompkins, J. S., and Rieser, C. June 1986. Special report: Biotechnology. *Science Digest.*

Torrey, J. G. July-August 1985. The development of plant biotechnology. *American Scientist.*

Verma, I. M. November 1990. Gene therapy. *Scientific American.*

Weinberg, R. A. November 1983. A molecular basis of cancer. *Scientific American.*

———. September 1988. Finding the antioncogene. *Scientific American.*

Weintraub, H. M. January 1990. Antisense RNA and DNA. *Scientific American.*

White, R., and Lalouel, J. February 1988. Chromosome mapping with DNA markers. *Scientific American.*

Evolution and Diversity

Mammals evolved on land, but various types have returned to water. The manatee, a close relative of the elephant, lives in warm water and eats sea grass. Its proboscis is actually a reduced trunk. Many kinds of evidence reviewed in this part support the belief that all life forms are related and indeed can even trace their ancestry to the first cell or cells. Common descent explains why all living things share the same characteristics of life. Adaptation to particular environments explains the diversity of life.

19

Darwin and Evolution

Fossils provide the best evidence that previous forms differ from present forms and that evolution therefore occurs. Evolution of the vertebrates is now reasonably understood because they provide good fossils, like these fishes of the Eocene epoch.

Your study of this chapter will be complete when you can

1. contrast the pre-Darwinian world view with the post-Darwinian world view;
2. describe the guiding principles behind eighteenth-century taxonomy;
3. describe the catastrophe hypothesis of Cuvier;
4. describe Lamarck's theory of evolution, and tell which aspects of his theory have been falsified;
5. describe Charles Darwin's background and the path of the voyage of the HMS *Beagle;*
6. tell how the study of geology and fossils influenced Darwin's thinking;
7. tell how the study of biogeography influenced Darwin's thinking;
8. describe natural selection as a process that results in adaptation to the environment;
9. tell how the fossil record, biogeography, comparative anatomy, comparative embryology, and comparative biochemistry support a hypothesis of common descent.

harles Darwin was only 22 in 1831 when he accepted the position of naturalist aboard the HMS *Beagle*, a British naval ship about to sail around the world (fig. 19.1). The captain was hopeful that Darwin would find evidence of the biblical account of creation. The results of Darwin's observations were just the opposite, however, as we shall examine later in the chapter.

Table 19.1 tells us that prior to Darwin, people had an entirely different way of looking at the world. Their mindset was determined by deep-seated beliefs held to be intractable truths. To turn from these beliefs and accept the Darwinian view of the world required an intellectual revolution of great magnitude. This revolution was fostered by changes in both the scientific and the social realms. Here, we will only touch on a few of the scientific contributions that helped bring about the new world view.

Figure 19.1

Map (*a*) shows the journey of the HMS *Beagle* around the world. Notice the encircled colors are keyed to the photographs, which show us what Darwin observed. As Darwin traveled along the east coast of South America he saw (*b*) a bird called a rhea, which he later noted looked like the African ostrich, and (*c*) the sparse vegetation of the Patagonian Desert. Along the west coast he saw

(*d*) the Andes Mountains, with strata containing fossil animals, and (*e*) the lush vegetation of a rain forest. In the Galápagos Islands he saw (*f*) marine iguanas that have large claws to help them cling to rocks and blunt snouts for eating seaweed and (*g*) a finch that has the ecological niche of a woodpecker. Lacking a long tongue to ferret out insects from trees, it uses a long stick instead.

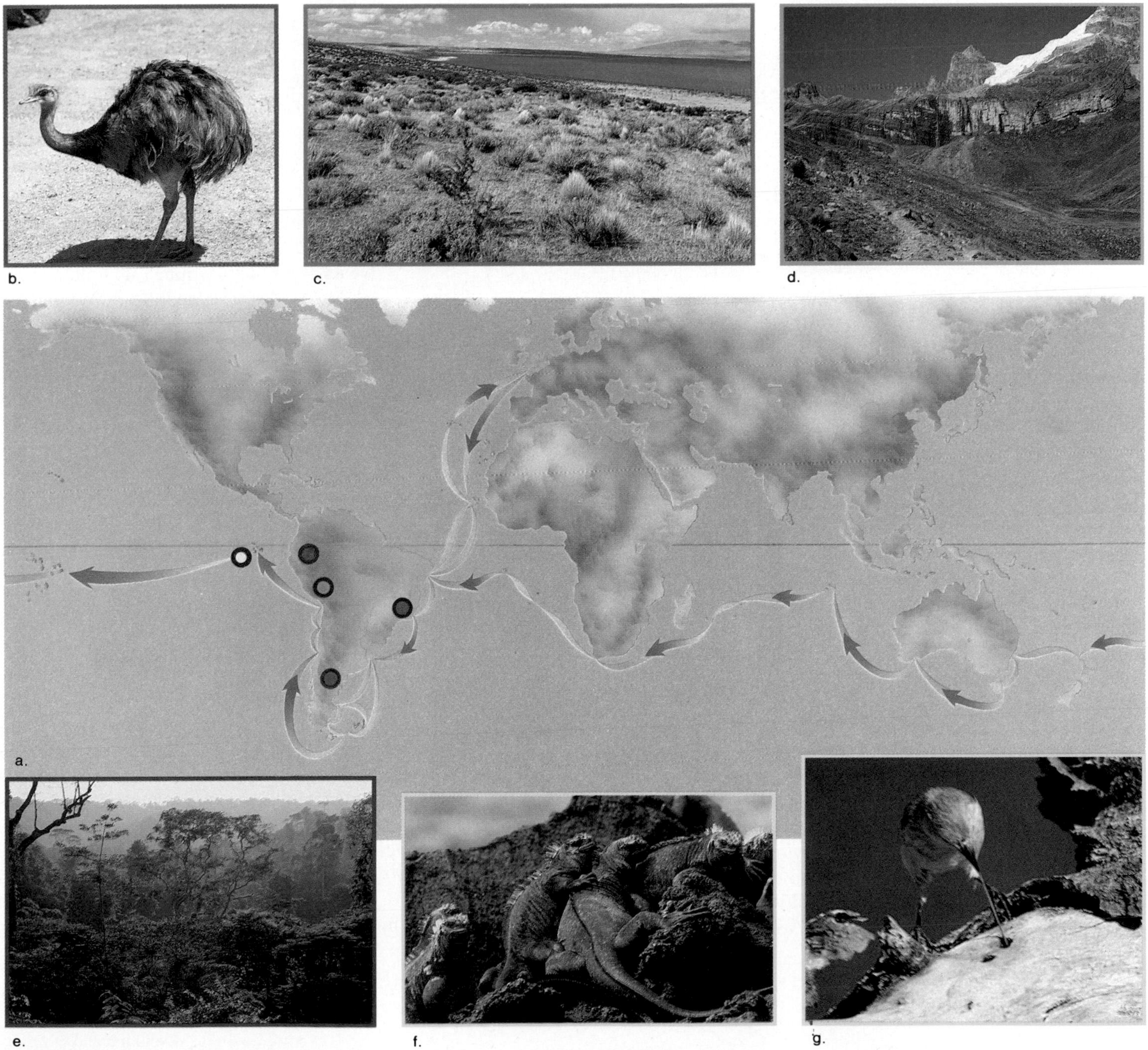

b.

c.

d.

a.

e.

f.

g.

Although it is often believed that Darwin forged this change in world view by himself, biologists during the preceding century had slowly begun to accept the idea of **evolution,** that is, that species change with time. As discussed in the first chapter, evolution explains the unity and diversity of life (p. 8). Living things share common characteristics because they are descended from common ancestors. Living things are diverse because each species is adapted to its habitat and way of life. We will see that the history of evolutionary thought is a history of ideas about descent and adaptation. Darwin himself used the expression "descent with modification," by which he meant that as descent occurs so does diversification. He saw the process of adaptation as the means by which diversification comes about.

Evolution before Darwin

Mid-Eighteenth Century

Taxonomy, the science of classifying organisms, was an important endeavor during the mid-eighteenth century. Chief among the taxonomists was Carolus Linnaeus (1707-1778), who gave us the binomial system of nomenclature (p. 4) and who developed a system of classification for all known plants. Linnaeus, like other taxonomists of his time, believed in the ideas of *special creation* and *fixity of species* (fig. 19.2). He thought each species had an "ideal" structure and function and also a place in the *scala naturae,* a sequential ladder of life. The simplest and most material of beings was on the first rung of the ladder, and the most complex and most spiritual of beings was on the last rung. Human beings occupied the last rung of the ladder.

These ideas were Christian dogma, which can be traced to the works of the famous Greek philosophers Plato and Aristotle. Plato said that every object on earth was an imperfect copy of an ideal form, which can be deduced upon reflection and study. To Plato, individual variations were imperfections that only distract the observer. Aristotle saw that organ-

Table 19.1
Contrast of World Views

Pre-Darwinian View	Post-Darwinian View
1. Earth is relatively young—age is measured in thousands of years.	1. Earth is relatively old—age is measured in billions of years.
2. Each species is specially created; species don't change, and the number of species remains the same.	2. Species are related by descent—it is possible to piece together a history of life on earth.
3. Adaptation to the environment is the work of a creator, who decided the structure and function of each type of organism. Any variations are imperfections.	3. Adaptation to the environment is the interplay of random genetic variations and environmental conditions.
4. Observations are supposed to substantiate the prevailing world view.	4. Observation and experimentation are used to test hypotheses, including hypotheses about evolution.

Figure 19.2
Linnaeus believed that each species was created separately and that classification should reveal God's divine plan. Variations were imperfections of no real consequence.

isms were diverse, and some were more complex than others. His belief that all organisms could be arranged in order of increasing complexity became the *scala naturae* just described.

Linnaeus and other taxonomists wanted to describe the ideal characteristics of each species and also wanted to discover the proper place for each species in the *scala naturae.* Therefore, Linnaeus did not even consider the possibility of evolutionary change for most of his working life. Even so, there is evidence he did eventually perform hybridization experiments, which made him think that at least species, if not higher categories of classification, might change with time.

Georges Louis Leclerc, better known by his title, Comte de Buffon (1707-1778), was a French naturalist who devoted many years of his life to writing a 44-volume natural history that described all known plants and animals. He provided evidence of descent with modification, and he even speculated on various mechanisms. Throughout his writings, he mentioned that the following factors could influence evolutionary change: direct influences of the environment, migration, geographical isolation, overcrowding, and the struggle for existence. Buffon seemed to vacillate, however, as to whether or not he believed in evolutionary descent, and often he professed to believe in special creation and the fixity of species.

Erasmus Darwin (1731-1802), Charles Darwin's grandfather, was a physician and a naturalist. His writings on both botany and zoology contained many comments, although they were mostly in footnotes and asides, that suggested the possibility of common descent. He based his conclusions on changes undergone by animals during development, artificial selection by humans, and the presence of vestigial organs (organs that are believed to have been functional in an ancestor but are reduced and nonfunctional in a descendant). Like Buffon, Erasmus Darwin offered no mechanism by which evolutionary descent might occur.

Figure 19.3

Artist's recreation of a *Mastodon* known only from the fossil record. One of George Cuvier's first contributions was to compare the structure of a *Mastodon* to a living elephant. Given only a single fossil bone, he was able to suggest the anatomy of an entire *Mastodon*. He said these animals were shorter than modern elephants but had massive, pillarlike legs. The skull was lower and flatter and of generally simpler construction than in modern elephants.

In the mid-eighteenth century there was intense interest in classifying known plants and animals. Some of the naturalists (e.g., Buffon and Erasmus Darwin) thought descent was possible, but they offered no definite mechanism by which evolution might occur.

Figure 19.4

Lamarck is most famous for suggesting that the long neck of the giraffe is due to continual stretching to reach food. Each generation of giraffes stretched a little farther and this was passed to the next generation. With the advent of modern genetics, it became possible to explain why this so-called inheritance of acquired characteristics would not be possible.

Late Eighteenth Century

George Cuvier and Catastrophism

In addition to taxonomy, biologists prior to Darwin were quite interested in comparative anatomy. Explorers and collectors traveled the world and brought back not only currently existing species to be classified but also fossils, remains of once-living organisms, to be studied. George Cuvier (1769-1832), a distinguished vertebrate (animals with backbones) zoologist, was the first to use comparative anatomy to develop a system of classifying animals. He also founded the science of **paleontology,** the study of fossils, and suggested that a single fossil bone was all he needed to deduce the entire anatomy of an animal (fig. 19.3).

Because Cuvier was a staunch advocate of special creation and fixity of species, he faced a real problem when geological evidence of a particular region showed a succession of life forms in the earth's strata (layers). To explain these observations, he hypothesized that a series of local catastrophes or mass extinctions had occurred whenever a new stratum of that region showed a new mix of fossils. After each catastrophe, the region was repopulated by species from surrounding areas, and this accounted for the appearance of new fossils in the new stratum. The result of all these catastrophes was the appearance of change with time. Some of his followers even suggested that there had been worldwide catastrophes and that after each of

these events, God created a new set of species. This explanation of the history of life came to be known as **catastrophism.**

Lamarck's Theory of Evolution

Jean Baptiste de Lamarck (1744-1829) was the first biologist to support common descent clearly and to link diversity with adaptation to the environment. Lamarck's ideas about descent were entirely different from those of Cuvier, perhaps because he was an invertebrate (animals without backbones) zoologist. Lamarck concluded after studying the succession of life forms in strata that more complex organisms are descended from less complex organisms. He mistakenly said, however, that increasing complexity was the result of a natural force—a tendency toward perfection—that was inherent in all living things. (As an analogy to the inanimate world, objects fall as a result of the force of gravity.)

To explain the process of adaptation to the environment, Lamarck proposed as a mechanism a hypothesis that was prevalent in his day. He supported the idea of **inheritance of acquired characteristics.** One example that he gave—and for which he is most famous—is that the long neck of a giraffe developed over time because animals stretched their necks to reach food high up in trees and then passed on a long neck to their offspring (fig. 19.4). The inheritance of acquired characteristics has never been substan-

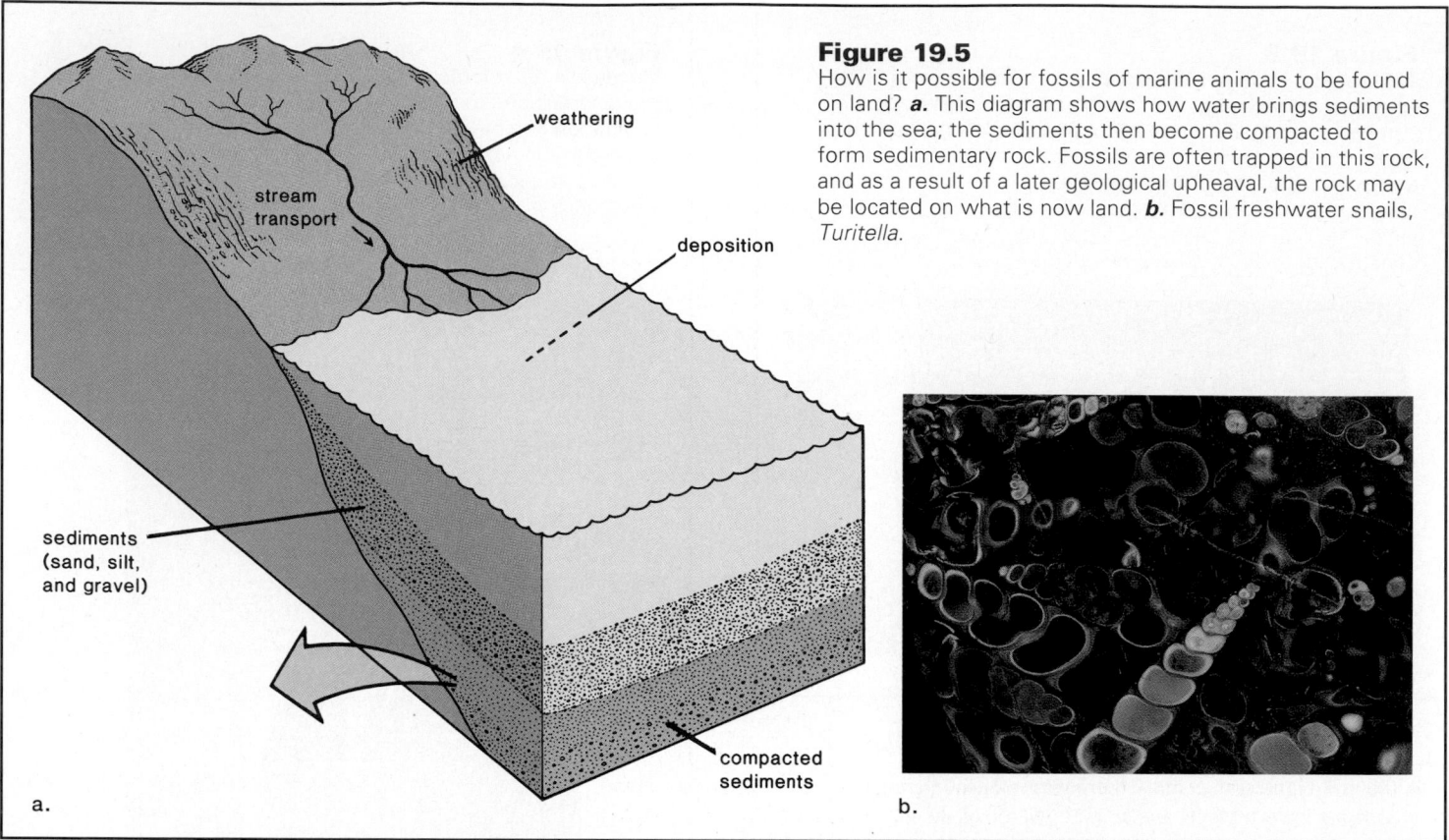

Figure 19.5

How is it possible for fossils of marine animals to be found on land? *a.* This diagram shows how water brings sediments into the sea; the sediments then become compacted to form sedimentary rock. Fossils are often trapped in this rock, and as a result of a later geological upheaval, the rock may be located on what is now land. *b.* Fossil freshwater snails, *Turitella.*

weathering

stream transport

deposition

sediments (sand, silt, and gravel)

compacted sediments

a.

b.

tiated by experimentation. The molecular mechanism of inheritance explains why. Phenotypic changes acquired during an organism's lifetime do not result in genetic changes that can be passed to subsequent generations.

Lamarck stated that more complex organisms evolved from less complex organisms, but he believed they were striving for perfection. He also said that organisms are adapted to their environments, but he believed in the inheritance of acquired characteristics.

Darwin's Theory of Evolution

When Darwin signed on as naturalist aboard the HMS *Beagle*, he had a suitable background for the position. He was an ardent student of nature and had been a collector of insects since his early years. His sensitive nature prevented him from studying medicine, and he went to divinity school at Cambridge instead. Even so, he attended many lectures in both biology and geology, and he was also tutored in these subjects by a friend and teacher, the Reverend John Henslow. As a result of arrangements made by Henslow, Darwin spent the summer of 1831 doing field work with Adam Sedgwick, a geologist at Cambridge, and it was Henslow who recommended Darwin for the post aboard the HMS *Beagle*. The trip was to take 5 years, and the ship was to traverse the Southern

Hemisphere (fig. 19.1), where life is most abundant and varied. Along the way, Darwin encountered forms of life very different from those in his native England.

Evolutionary Change

Although it was not his original intent, Darwin began to gather evidence that common descent does occur and that adaptation to various environments results in diversity.

Geology and Fossils

Darwin took Charles Lyell's *Principles of Geology* on the voyage, which presented arguments to support a theory of geological change proposed by James Hutton. In contrast to the catastrophists, Hutton believed the earth was subject to slow but continuous cycles of erosion and uplift. Weather causes erosion; thereafter dirt and rock debris are washed into the rivers and transported to oceans. These loose sediments are deposited in thick layers, which are converted eventually into sedimentary rocks (fig. 19.5). Then, sedimentary rocks, which often contain fossils, are uplifted from below sea level to form land. Hutton concluded that extreme geological changes can be accounted for by slow, natural processes given enough time. Lyell went on to suggest that these slow changes occurred at a uniform rate. Hutton's general ideas about slow and continual geological change are still accepted today, although modern geologists realize that rates of change have not always been uniform. Darwin was not taken by the idea of uniform

Figure 19.6
a. A giant armadillo-like glyptodont, known only by the study of its fossil remains. Darwin found such fossils and came to the conclusion that this extinct animal must be related to living armadillos. The glyptodont weighed 2,000 kg. *b.* A modern armadillo weighs about 4.5 kg.

Figure 19.7
The Patagonian hare has the face of a guinea pig and is native to South America, which has no native rabbits. The hare has long legs and has adaptations similar to those of rabbits.

change, but he was convinced, as was Lyell, that the earth's massive geological changes are the result of slow processes and that the earth, therefore, must be very old.

On his trip, Darwin observed massive geological changes firsthand. When he explored what is now Argentina, he saw raised beaches along the coast for great distances. When he got to the Andes, he was impressed by their great height. In Chile, he found marine shells inland, well above sea level, and witnessed the effects of an earthquake that caused the land to rise several feet. At the same time Darwin made geological observations, he collected fossil specimens. For example, on the east coast of South America, he found the fossil remains of a giant ground sloth and an armadillo-like animal (fig. 19.6). Darwin relied on the supposition that the earth must be very old, and he began to think that there would have been enough time for descent with modification to occur. Therefore, living forms must be descended from extinct forms known only from the fossil record. It would seem that species were not fixed and instead they change over time.

Darwin's geological observations were consistent with those of Hutton and Lyell. He began to think that the earth was very old and there would have been enough time for descent with modification to occur. Present-day species must be related to extinct species known only by their fossil remains.

Biogeography

Biogeography is the study of the geographic distribution of life forms on earth. Darwin could not help but compare the animals of South America to those with which he was familiar. For example, instead of rabbits he found the Patagonian hare (*Dolichotis patagonium*) in the grasslands of South America. The Patagonian hare has long legs and ears but the face of a guinea pig, a rodent native to South America (fig. 19.7). Did the Patagonian hare resemble a rabbit because the 2 types of animals were adapted to the same type of environment? Both animals ate grass, hid in bushes, and moved rapidly using long hind legs. Did the Patagonian hare have the face of a guinea pig because of common descent?

As he sailed southward along the continent of South America, Darwin saw how similar species replaced each other. For example, the greater rhea (a type of ostrich) found in the north was replaced by the lesser rhea in the south. Therefore, Darwin reasoned that related species could be modified according to the environment. When he got to the Galápagos Islands he found further evidence of this. The Galápagos Islands are a small group of volcanic islands off the western coast of South America. The few types of plants and animals found there were slightly different from species Darwin had observed on the mainland, and even more important, they also varied from island to island.

Tortoises Each of the Galápagos Islands seemed to have its own type of tortoise, and Darwin began to wonder if this could be correlated with a difference in vegetation between islands (fig. 19.8). Long-necked tortoises seemed to inhabit only dry areas, where food was scarce, most likely because the longer neck was helpful in reaching cacti. In moist regions with relatively abundant ground foliage, short-necked tortoises were found. Had an ancestral tortoise given rise to these different types, each adapted to a different environment?

Finches Although the finches on the Galápagos Islands seemed to Darwin like mainland finches, there are many more types (fig. 19.9). There are ground-dwelling finches with different-sized beaks, depending on the size of the seeds they feed on and a cactus-

Figure 19.8

Each of the Galápagos Islands had its own form of tortoise. Darwin wondered if all of the tortoises were descended from a common ancestor. **a.** The tortoises with dome shells and short necks fed at ground level and were from well-watered islands where grass was available. **b.** Those with shells that flared up in front had long necks and were able to feed on tall treelike cacti. They were from arid islands where prickly pear cactus was the main food source. Only on these islands were the cacti treelike.

a.

b.

Figure 19.9

It occurred to Darwin after returning home that all of the Galápagos finches were descended from a common ancestor. Each of the present-day 13 species of finches has a bill adapted to a particular way of life. **a.** For example, the large, tree-dwelling finch (*Camarhynchus psittucula*) grinds fruit and insects with a parrotlike bill. **b.** The small, ground-dwelling finch (*Geospiza fuliginosa*) has a pointed bill and eats tiny seeds and ticks picked from iguanas. **c.** The woodpecker-type finch (*Camarhynchus pallidus*) has a stout, straight bill, which chisels through tree bark to uncover insects, but because it lacks the woodpecker's long tongue it uses a tool—usually a cactus spine or a small twig—to ferret out insects.

a.

b.

c.

eating finch with a more pointed beak. The beak size of the tree-dwelling finches also varies but according to the size of their insect prey. The most unusual of the finches is a woodpecker-type finch. This bird has a sharp beak to chisel through tree bark but lacks the woodpecker's long tongue to probe for insects. To make up for this, the bird carries a twig or cactus thorn in its beak and uses it to poke into crevices. Once an insect emerges, the finch drops this tool and seizes the insect with its beak.

Some time later, Darwin speculated whether all these different species of finches could have descended from a type of mainland finch. In other words, a mainland finch was the common ancestor to all the types on the Galápagos Islands. Had speciation occurred because the islands allowed isolated populations of birds to evolve independently? Could the present-day species have resulted from accumulated changes occurring within each of these isolated populations?

Evolution and Diversity

Biogeography had a powerful influence on Darwin and made him think that adaptation to the environment is a process that accounts for diversification; one species can give rise to many species, each adapted differently.

Once Darwin decided that adaptation develops over time (instead of being the instant work of a creator), he began to think about a mechanism by which adaptation might occur.

Natural Selection

Both Charles Darwin and Alfred Russel Wallace, as explained in this chapter's reading, proposed natural selection as a mechanism for evolutionary change. **Natural selection** brings about adaptation to the environment, but it has no particular goal because environmental conditions are constantly changing. Therefore, perfect adaptation is not a probable outcome of natural selection.

Natural selection is a process in which preconditions (1-3) may result in certain consequences (4-5):

1. The members of a population have heritable variations.
2. In a population, many more individuals are produced each generation than can survive and reproduce.
3. Some individuals have adaptive characteristics that enable them to survive and reproduce better than do other individuals.
4. An increasing proportion of individuals in succeeding generations have the adaptive characteristics.
5. The result of natural selection is a population adapted to its local environment.

Variations

Darwin emphasized that the members of a population vary in their functional, physical, and behavioral characteristics (fig. 19.10). In the old world view (table 19.1), variations were imperfections that should be ignored since they were not important to the description of a species. For Darwin, variations were essential to the natural selection process. Darwin suspected but did not have the evidence we have today that the occurrence of variations is completely random; they arise by accident and for no particular purpose. New variations are just as likely to be harmful as helpful to the organism.

The variations that make adaptation to the environment possible are those that are passed on from generation to generation. The science of genetics was not yet helpful, and Darwin was never able to determine the cause of variations or how they are passed on. Today, we realize that genes determine the phenotype of an organism, and that mutations and recombination of alleles during sexual reproduction can cause new variations to arise.

Struggle for Existence

In Darwin's time, a socioeconomist, Thomas Malthus, stressed the reproductive potential of human beings. He proposed that death and famine were inevitable because the human population tends to increase faster than the supply of food. Darwin applied this concept to all organisms and saw that the available resources were not sufficient for all members of a population to survive. He calculated the reproductive potential of elephants. Assuming a life span of

Figure 19.10
Variations occur among individuals of a human population. For Darwin, variations were highly significant and were required for natural selection to result in greater adaptation to the environment.

about 100 years and a breeding span of from 30-90 years, a single female probably bears no fewer than 6 young. If all these young survive and continue to reproduce at the same rate, after only 750 years, the descendants of a single pair of elephants will number about 19 million!

Each generation has the same reproductive potential as the previous generation. Therefore, there is a constant struggle for existence, and only certain members of a population survive and reproduce each generation.

Fitness

Fitness is the ability of an organism to survive and reproduce in its local (immediate) environment. The most fit individuals are the ones that survive, that capture a disproportionate amount of resources, and that convert these resources into a larger number of viable offspring. Since organisms vary anatomically and physiologically and the challenges of local environments vary, what determines fitness varies for different populations. For example, among western diamondback rattlesnakes (*Crotalus atrox*) living on lava flows, the most fit are those that are black in color. But among those living on desert soil, the most fit are those with the typical light and dark brown coloring. Background matching helps an animal both capture prey and avoid being captured; therefore, it is expected to lead to survival and increased reproduction.

Darwin noted that when humans carry out artificial selection, they select the animals that will reproduce. For example, prehistoric humans probably noted desirable variations among

Figure 19.11
Artificial selection of dogs. All dogs are descended from the wolf (*Canis lupus*), which began to be domesticated about 14,000 years ago. In evolutionary terms, the process of diversification has been exceptionally rapid. Several factors may have contributed: (1) the wolves under domestication were separated because human settlements were separate; (2) humans in each tribe selected for whatever traits appealed to them. Artificial selection of dogs continues even today.

Shetland sheepdog

Boston terrier

bloodhound

red chow

French bulldog

dalmation

beagle

Chihuahua

Figure 19.12

Humans have developed several types of vegetables from a single species of *Brassica oleracea:* (**a**) Chinese cabbage, (**b**) brussels sprouts, and (**c**) kohlrabi. Darwin believed that artificial selection provided a model by which to understand natural selection. With natural selection, however, the environment provides the selective force.

a. b. c.

wolves and selected particular animals for breeding. Therefore, the desirable traits increased in frequency in the next generation. This same process was repeated many times over. The result today is that there are many varieties of dogs all descended from the wolf (fig. 19.11). In a similar way, several varieties of vegetables can be traced to a single ancestor (fig. 19.12).

In nature, interactions with the environment determine which members of a population reproduce to a greater degree than other members. In contrast to artificial selection, the end result of natural selection is not predesired. Natural selection occurs because certain members of a population happen to have a variation that allows them to survive and/or capture resources. For example, any variation that increases the speed of a hoofed animal helps it escape predators and live longer; a variation that reduces water loss helps a desert plant survive; and one that increases the sense of smell helps a wild dog find its prey. Therefore, we expect organisms with these traits to have increased fitness. *Differential reproduction* occurs when the most fit organisms reproduce and leave more offspring than the less fit.

Adaptation

An **adaptation** is a trait that helps an organism be more suited to its environment. We can especially recognize an adaptation when unrelated organisms living in a particular environment display similar characteristics. For example, manatees, penguins, and sea turtles all have flippers, which help them move through the water. Similarly, fishes, including sea skates, have flattened bodies for the same function (fig. 19.13).

Natural selection results in the adaptation of populations to their specific environments. Because of differential reproduction generation after generation, adaptive traits are disproportionately represented in each succeeding generation. As we shall see in

Figure 19.13

Animals living in the same type of environment have the same adaptations. Manatees, penguins, and sea turtles all have flippers. Fishes, including sea skates, have flattened bodies.

the next chapter, there are other processes of evolution aside from natural selection; however, natural selection is the only process that results in adaptation to the environment.

Origin of Species

After the HMS *Beagle* returned to England in 1836, Darwin waited over 20 years to publish his ideas. During the intervening years, he used the scientific process (fig. 2.1) to test his hypothesis that today's diverse life forms arose by descent from a common

Alfred Russel Wallace

Alfred Russel Wallace (1823-1913) is best known as the English naturalist who independently proposed natural selection as a process to explain the origin of species (fig. 19.A). Like Darwin, Wallace was a collector at home and abroad. Even at 14, while learning the trade of surveying, he became interested in botany and started collecting plants. While he was a school teacher at Leicester in 1844-45, he met Henry Walter Bates, an entomologist, who got him interested in insects. Together they went on a collecting trip to the Amazon, which lasted for several years. Wallace's knowledge of the world's extensive flora and fauna was much expanded by a tour he made of the Malay Archipelago from 1854-62. After studying the animals of every important island, he divided the islands into a western group, with animals like those of the Orient, and a eastern group, with animals like those of Australia. The dividing line between the islands of the archipelago is a narrow but deep strait that is now known as Wallace's Line.

Like Darwin, Wallace was a writer of articles and books. As a result of his trip to the Amazon, he wrote 2 books, entitled *Travels on the Amazon and Rio Negro* and *Palm Trees of the Amazon.* In 1855, during his trip to Malay, he wrote an essay called "On the Law Which Has Regulated the Introduction of New Species." In the essay, he said that "every species has come into existence coincident both in time and space with a preexisting closely allied species." It's clear, then, that by this date he believed in the origin of new species rather than the fixity of species. Later, he said that

Figure 19.A

Wallace was far from England in the Malay Archipelago when he formulated a hypothesis of natural selection to explain the origin of new species. His proposal matched that of Charles Darwin almost exactly, and a joint paper explaining natural selection was read to the Linnean society in 1858.

he had pondered for many years about a mechanism to explain the origin of species. He, too, had read Malthus's treatise on human population increases, and in 1858, while suffering an attack of malaria, the idea of "survival of the fittest" came upon him. He quickly completed an essay discussing a natural selection process, which he chose to send to Darwin for comment. Darwin was stunned upon its receipt. Here

before him was the hypothesis he had formulated as early as 1844 but had never dared to publish. In 1856 he had begun to work on a book that would supply copious data to support natural selection as a mechanism for evolutionary change. He told his friend and colleague Sir Charles Lyell that even Wallace's "terms now stand as heads of my chapters."

Darwin suggested that Wallace's paper be published immediately, even though he as yet had nothing in print. Lyell and others who knew of Darwin's detailed work substantiating the process of natural selection suggested that a joint paper be read to the Linnean society. The title of Wallace's section was "On the Tendency of Varieties to Depart Indefinitely from the Original Type." Darwin's half was an abstract of a paper he had written in 1844 and an abstract of his book *On the Origin of Species,* which was published in 1859.

When Wallace arrived back in England, he supported himself by selling specimens, lecturing, and writing books and popular articles. In 1870 he wrote a book entitled *Contributions to the Theory of Natural Selection.* In a book entitled *Geographical Distribution of Animals* (1876), he divided the world into 6 biogeographical regions. The land areas in each region have characteristic plants and animals, and the fossils of each region often resemble the living organisms only of that region. Wallace said the regions occur because they are separated by impassable barriers. It would seem that Wallace, too, was very much influenced by a study of biogeography, as was Darwin.

ancestor and that natural selection is the mechanism by which species change and even new species can arise. He did not make his findings known until after he received a paper from another naturalist, Alfred Russel Wallace, who independently proposed natural selection as a mechanism for the origin of species. This chapter's reading tells more about this.

Evidence of Common Descent

Many different lines of evidence support the hypothesis of common descent. This is significant, because the more varied the evidence supporting a hypothesis, the more certain it becomes. Darwin cited many of the evidences we will discuss, except he had no knowledge, of course, of the biochemical evidence that became available after his time.

In each case, we will predict what is to be expected, assuming either the special creation hypothesis of Darwin's day or his own common descent hypothesis. The special creation hypothesis (table 19.1) proposes that organisms are perfectly adapted to their environments because they were created by a fully rational, beneficent, omnipotent creator. The common descent hypothesis proposes that the numerous features of organisms are a consequence of their individual evolutionary histories, and that therefore organisms will not necessarily be designed ideally for their present way of life and environment.

Fossil Record

The fossil record is the history of life as recorded by remains from the past. Fossils are at least 10,000 years old and include such items as fossilized pieces of bone, impressions of plants pressed into

Figure 19.14

The fossil record does contain transitional fossils. **a.** This is *Archaeopteryx,* a transitional link between reptiles and birds. It had feathers and wing claws. Most likely, it was a poor flier. Perhaps it ran over the ground on strong legs and climbed up into trees with the assistance of these claws. **b.** It also had a feather-covered reptilian-type tail, which shows up well in this artist's representation of the animal.

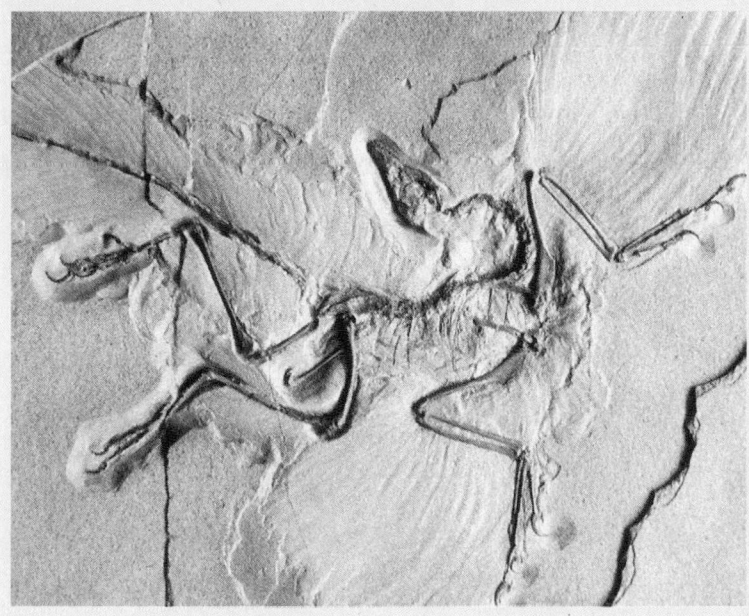

a.

b.

shale, and even insects trapped in tree resin (which we know as amber). Over the last 2 centuries, paleontologists have studied fossils all over the world and have pieced together the story of past life (see also chap. 21).

The fossil record is rich in information. One of its most striking patterns is the succession of life forms over time. Catastrophists offered an explanation for the extinction and subsequent replacement of one group of organisms by another group, but they never could explain successive changes that link groups of organisms historically. Particularly interesting are the fossils that serve as transitional links between groups. For example, the famous fossil *Archaeopteryx* is intermediate between reptiles and birds (fig. 19.14). The dinosaur-like skeleton of this fossil has reptilian features, including jaws with teeth and a long, jointed tail, but *Archaeopteryx* also had feathers and wings. Other intermediate fossils among the fossil vertebrates include the amphibious fish *Eustheopteron,* the reptile-like amphibian *Seymouria,* and the mammal-like reptiles, or therapsids. These fossils allow us to deduce the following line of descent for the vertebrates:

Sometimes, the fossil record allows us to trace the history of one particular organism, such as the modern-day horse, *Equus* (fig. 21.19). *Equus* evolved from *Hyracotherium,* which was about the size of a dog. This fossil animal had cusped, low-crowned molars and 4 toes on each front foot and 3 toes on each hind foot. In contrast, *Equus* has flattened, high-crowned molars and a single toe on each foot. In all cases, living organisms closely resemble the most recent fossils in their line of descent. Underlying similarities, however, allow us to trace a line of descent over vast amounts of time.

A special creation hypothesis would predict that the species living today are exactly the same ones that were originally created, that is, the perfectly adapted ones. Therefore, there would be no historical pattern of change in the fossil record. Extinctions could occur, but there would be no other discernible pattern. The evidence is opposite to this and supports the common descent hypothesis. Fossils can be linked over time because they do show a similarity in form, despite observed changes.

Biogeography

As previously mentioned, Darwin noted that South America lacked rabbits, even though the environment was quite suitable to them. This contradicted the prevailing special creation view that perfectly adapted organisms would have been created wherever there was a suitable environment. And it made Darwin consider an alternative hypothesis; namely, that organisms arise and disperse from the place of their origin. In other words, there were no rabbits in South America because rabbits had originated somewhere else and they had no means to reach South America.

Figure 19.15

There are many types of marsupials in Australia, each adapted to a different way of life. All of the marsupials in Australia presumably evolved from a common ancestor who entered Australia some 60 million years ago.

wombat (*Phascdomys*) (woodchucklike)

flying phalanger (*Petaurus*) (squirrellike)

kangaroo (*Macrapus*) (ungulatelike)

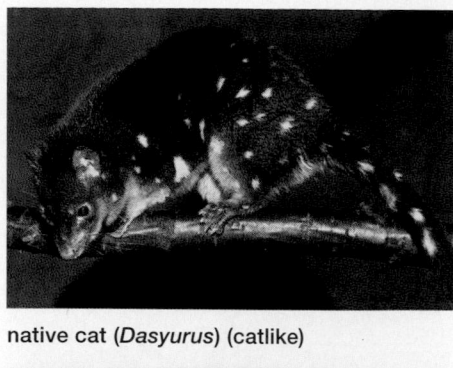

native cat (*Dasyurus*) (catlike)

Tasmanian wolf (*Thylacinus*) (wolflike)

Physical factors, such as the location of continents, often determine where a population can spread. For example, at one time in the history of the earth, South America, Antarctica, and Australia were all connected (fig. 21.15). Marsupials (pouched mammals) arose at this time, and their fossils are now found in all three areas. When Australia separated and drifted away, the marsupials diversified into many different types suited to specific environments (fig. 19.15). These marsupials have features and ways of life that are similar to placental mammals (development occurs in the uterus) in other parts of the world. Placental mammals, which evolved later, could not reach South America until the isthmus of Panama was formed, and they did not reach Australia until they were introduced by humans.

These observations are consistent with the hypothesis of common descent. It is quite plausible that a particular line of descent in a particular geographical region could give rise to many forms, each adapted to a different environment. We recognize that these many forms are related because they are anatomically similar. In our example here, they are all marsupials, and none are placental mammals.

Comparative Anatomy

Darwin was able to show that a common descent hypothesis offers a plausible explanation for anatomical similarities between organisms. The special creation hypothesis predicts that perfectly adapted vertebrates would have little similarity in the *structures* of their forelimbs, which are used for different *functions*. Vertebrate forelimbs are used for flight (birds and bats), orientation during swimming (whales and seals), running (horses), climbing (arboreal lizards), or swinging from tree branches (monkeys). Yet, all vertebrate forelimbs contain the same sets of bones organized in similar ways, despite their dissimilar functions (fig. 19.16). The best explanation for this unity is that the basic forelimb plan originated with a common ancestor, and then the plan was modified in the succeeding groups as each continued along its own evolutionary pathway. Such structures are called **homologous structures.**

Vestigial structures are anatomical features that are fully developed and functional in one group of organisms but reduced and functionless in similar groups. Most birds, for example, have well-developed wings used for flight. Some bird species, however,

Figure 19.16

Bones in the vertebrate forelimbs. Although the same bones are present, the specific design details of the limbs are different. This unity of plan is evidence of a common ancestor. (The bones are color coded; the green color signifies union of the yellow- and blue-coded bones.)

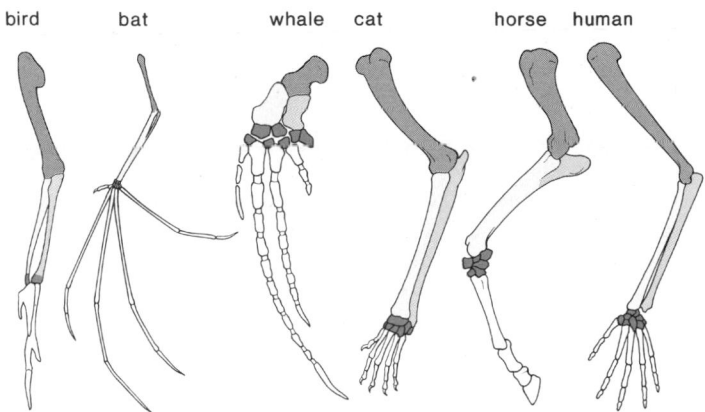

bird   bat   whale   cat   horse   human

have wings that are greatly reduced, and they do not fly. Similarly, snakes have no use for hindlimbs, and yet some have remnants of a pelvic girdle and legs. Humans have a tail bone but no tail.

The special creation hypothesis predicts that perfectly adapted organisms would have no nonfunctional structures, such as wings that do not allow flight. The presence of vestigial structures is explained by the common descent hypothesis, however. Vestigial structures occur because organisms inherit their anatomy from their ancestors; they are traces of an organism's evolutionary history.

Comparative Embryology

The unity of plan shared by vertebrates extends to their embryological development (fig. 19.17). At some time during development, all vertebrates have a supporting dorsal rod, called a notochord, and exhibit paired pharyngeal pouches bordering gill clefts. In fishes and amphibian larvae, these pouches and clefts become functioning gills. In humans, the first pair of pouches becomes the cavity of the middle ear and the eustachian tube. The second pair becomes the tonsils, while the third and fourth pairs become the thymus and parathyroid glands.

The special creation hypothesis does not predict any particular pattern of development among animals. The presence of a pattern, however, is consistent with the common descent hypothesis. Organisms that share a unity of plan also share similar stages of development. Why should terrestrial vertebrates develop and then modify structures like pharyngeal pouches that have lost their original function? Because animals tend to repeat the developmental stages of their ancestors.

Comparative Biochemistry

Almost all living organisms use the same basic biochemical molecules, including DNA, ATP, and many identical or nearly identical enzymes. Further, organisms utilize the same DNA triplet code and

Figure 19.17

(*a*) A chick and (*b*) a pig embryo at comparable early development stages have many features in common, although eventually they are completely different animals. This is evidence that they evolved from a common ancestor.

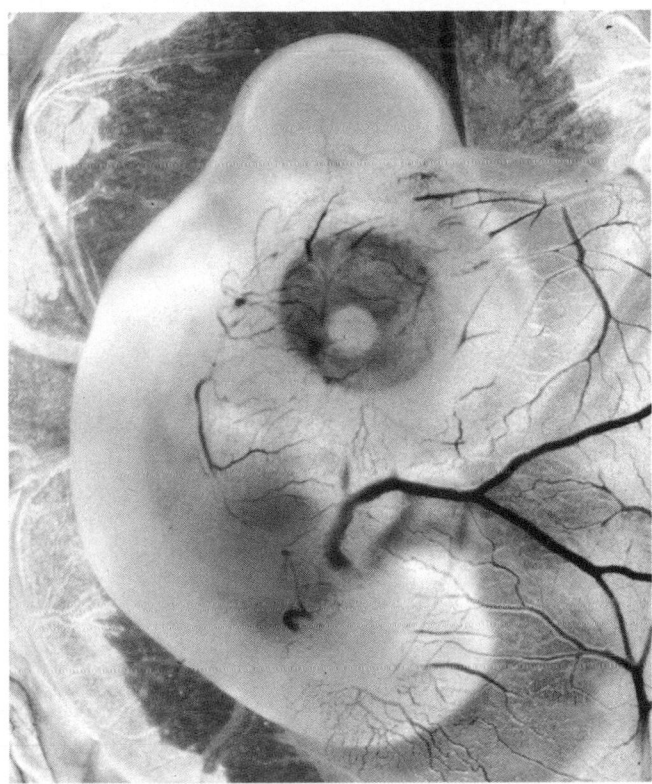

a.

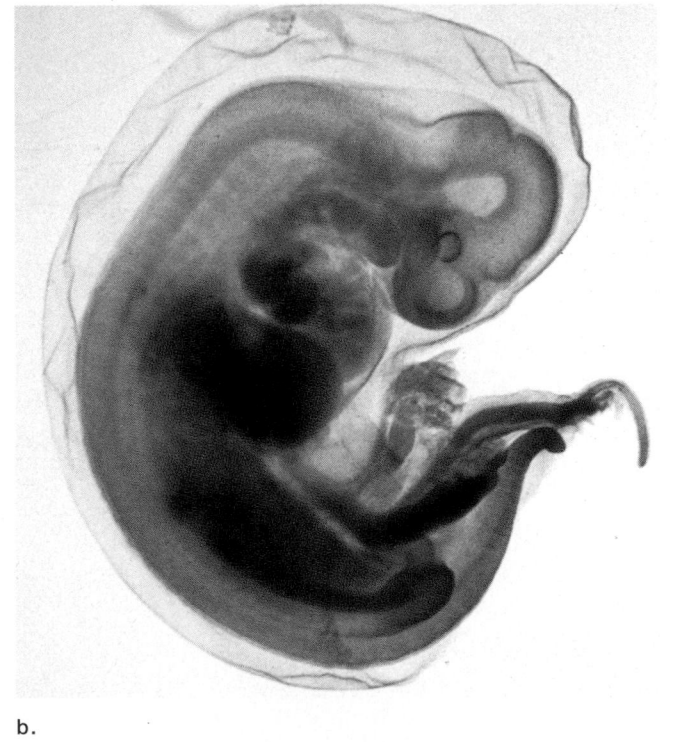

b.

Table 19.2
Percent Difference in Nucleotide Sequences of DNA between Selected Pairs of Animal Species

Human/chimpanzee	2.5
Human/gibbon	5.1
Human/green (Old World) monkey	9.0
Human/capuchin (New World) monkey	15.8
Human/lemur	42.0
Mouse/rat	30.0

From *DARWIN TO DNA, MOLECULES TO HUMANITY* by G. Ledyard Stebbins. Copyright © 1982 by W. H. Freeman and Company. Reprinted with permission.

the same 20 amino acids in their proteins. Organisms even share the same introns and hypervariable regions. There is obviously no functional reason why these elements need be so similar. But their similarity can be explained by descent from a common ancestor.

When the degree of similarity in DNA base sequences or the degree of similarity in amino acid sequences of proteins is examined, the data are as expected, assuming common descent. For example, investigators have determined the DNA base sequence differences between a number of organisms. Table 19.2 shows that humans and chimpanzees have only a 2.5% difference in DNA base sequences and that humans and lemurs have a 42.0% difference in DNA base sequences. This can be understood by assuming that humans and chimpanzees share a more recent common ancestor than do humans and lemurs. These data are consistent with other data regarding the anatomical similarity of these animals and, therefore, their relatedness.

The special creation hypothesis predicts no particular pattern of DNA base sequence similarity between organisms except perhaps according to their adaptations. Instead, organisms have remarkable similarities in nonfunctional DNA elements and have DNA sequence differences according to the degree of their anatomical relatedness. The pattern of DNA base sequence differences (and the amino acid sequence differences) among organisms is supportive of the common descent hypothesis.

Darwin discovered that many lines of evidence support the hypothesis of common descent. Since his time, it has been found that biochemical evidence also supports the hypothesis. Strength is lent to a hypothesis when it is supported by many different lines of evidence.

Evolution is no longer considered a hypothesis. It is one of the great unifying theories of biology. In science, the word *theory* is reserved for those conceptual schemes that are supported by a large number of observations and have not yet been found lacking. The theory of evolution has the same status in biology that the atomic theory has in chemistry.

Summary

1. Charles Darwin formulated hypotheses concerning evolution after taking a trip around the world as naturalist aboard the HMS *Beagle,* from 1831-36. His hypotheses were that common descent does occur and that natural selection results in adaptation to the environment.

2. The pre-Darwinian world view was entirely different from the post-Darwinian world view, as is shown in table 19.1.

3. A century before Darwin's trip, classification of organisms had been a main concern of biology. In keeping with the pre-Darwinian world view, Linnaeus thought that classification should describe the ideal features of species and reveal God's divine plan. Gradually, some naturalists, such as Comte de Buffon and Erasmus Darwin, began to put forth tentative suggestions that descent does occur.

4. George Cuvier and Jean Baptiste de Lamarck, contemporaries in the late eighteenth century, differed sharply on evolution. To explain the fossil record of a region, Cuvier proposed that a whole series of catastrophes (extinctions) and repopulations from other regions had occurred. Lamarck said that descent with modification does occur and that organisms do become adapted to their environments; however, he relied on the then commonly held but erroneous beliefs (*scalae naturae* and inheritance of acquired characteristics) to substantiate and provide a mechanism for evolutionary change.

5. Darwin's trip involved 2 primary types of observations. His study of geology and fossils caused him to concur with Lyell that the observed massive geological changes were caused by slow, continuous changes. Therefore, he concluded that the earth is old enough for descent with modification to occur.

6. Darwin's study of biogeography, including the animals of the Galápagos Islands, caused him to conclude that adaptation to the environment can cause diversification, including the origin of new species.

7. Natural selection is the mechanism Darwin proposed for how adaptation comes about. Members of a population exhibit random but inherited variations. (In contrast to the old world view, variations are highly significant.) Relying on Malthus's ideas regarding overpopulation, Darwin stressed that there was a struggle for existence. The most fit organisms, those possessing characteristics that allow them to acquire more resources, survive, and reproduce more than the less fit. In this way, natural selection results in adaptation to a local environment.

8. The hypothesis that organisms share a common descent is supported by many lines of evidence. The fossil record, biogeography, comparative

anatomy, comparative embryology, and comparative biochemistry all support the hypothesis. A hypothesis is greatly strengthened when many different lines of evidence support it.

9. Today, the theory of evolution is one of the great unifying theories of biology because it has been supported by so many different lines of evidence and has never been found lacking.

Writing Across the Curriculum

In order to practice writing skills, students should write out the answers to any or all of the study questions and the critical thinking questions. The study questions are sequenced in the same order as the text. Suggested answers to the critical thinking questions are in appendix D.

Study Questions

1. In general, contrast the pre-Darwinian world view with the post-Darwinian world view.
2. Cite naturalists who made a contribution to biology in the mid-eighteenth century, and state their beliefs about evolutionary descent.
3. How did Cuvier explain the succession of life forms in the earth's strata?
4. What is meant by the inheritance of acquired characteristics, a hypothesis which Lamarck used to explain adaptation to the environment?

5. What reading did Darwin do, and what observations did he make regarding geology?
6. What observations did Darwin make regarding biogeography? How did these influence his conclusions about the origin of new species?
7. What are the essential features of the process of natural selection as proposed by Darwin?
8. Distinguish between the concepts of fitness and adaptation to the environment.

9. How do data from the fossil record, biogeography, comparative anatomy, comparative embryology, and biochemical studies support the common descent hypothesis?
10. How do data from each field mentioned in question 9 dispute a special creation hypothesis?

Objective Questions

1. According to the theory of acquired inheritance,
   a. if a man loses his hand, then his children will also be missing a hand.
   b. changes in phenotype are passed on by way of the genotype to the next generation.
   c. organisms are able to bring about a change in their phenotype.
   d. All of these.
2. Why was it helpful to Darwin to learn that Lyell thought the earth was very old?
   a. An old earth has more fossils than a new earth.
   b. It meant there was enough time for evolution to have occurred slowly.
   c. Evolution doesn't occur without upheavals.
   d. Darwin said that natural selection occurs slowly.

3. All the finches on the Galápagos Islands are
   a. completely unrelated.
   b. descended from a common ancestor.
   c. now in competition with one another because they all feed on seeds.
   d. Both a and c.
4. Organisms
   a. compete with other members of their species.
   b. differ in fitness.
   c. are adapted to their environment.
   d. All of these.
5. DNA nucleotide differences between organisms
   a. indicate how closely related organisms are.
   b. indicate that evolution occurs.
   c. explain why there are phenotypic differences.
   d. All of these.

6. If evolution occurs, we would expect different biogeographical regions with similar environments to
   a. all contain the same mix of plants and animals.
   b. each have its own specific mix of plants and animals.
   c. have plants and animals that have similar adaptations.
   d. Both b and c.
7. The fossil record offers direct evidence for common descent because you can
   a. see that the types of fossils change over time.
   b. sometimes find common ancestors.
   c. trace the ancestry of a particular group.
   d. All of these.
8. Organisms adapted to the same way of life will
   a. have structures that show they share a unity of plan.

   b. have structures that need not indicate a unity of plan.

   c. live in the same biogeographical region.

   d. Both a and c.

9. Offer explanations for the following observations.

   a. Similar fossils of Triassic period reptiles are found in Antarctica, South Africa, and India.

   b. Amphibians, reptiles, birds, and mammals all have pharyngeal pouches at some time during development.

   c. Cacti and euphorbs exist on different continents, but both have spiny, water-storing, leafless stems.

   d. The tomato, potato, and jimson-weed are members of the nightshade family with similar flowers and fruits. The plants have other characteristics that are quite diverse.

Concepts and Critical Thinking

1. *All organisms share certain common characteristics because they are descended from a common ancestor.*

What evidence do you have that organisms as different as bacteria and humans share a common ancestor?

2. *Adaptation to various environments accounts for the diversity of life.*

The presence of vestigial structures suggests that the process of adaptation is not purposeful. Why?

3. *The scientific method has been used to test the unifying theories of biology.*

Use figure 2.1 to show that the scientific method can be used to test the concept of evolutionary descent.

Selected Key Terms

evolution (ev″o-lu′shun) 300
paleontology (pā′lē-ŏn-tol-ogē) 301
catastrophism (kah-tas′tro-fizm) 301

inheritance of acquired characteristics (in-her′ĭ-tans uv ah-kwīrd′ kar′ak-ter-is′tiks) 301

biogeography (bi″o-je-og′rah-fe) 303
adaptation (ad″ap-ta′shun) 307

20

Genetics in Evolution

Did the beautiful tail of the peacock evolve because of sexual selection—female preference for males with decorative feathers? If so, the tail affects a male's biological fitness, which is measured by the number of surviving offspring.

Your study of this chapter will be complete when you can

1. state the sources of variation in a population of sexually reproducing diploid organisms;
2. explain the Hardy-Weinberg law, along with its conditions;
3. list and discuss the agents of evolutionary change;
4. contrast the effects of gene flow and genetic drift on the gene pool of a population;
5. give examples of genetic drift, the founder effect, and a bottleneck;
6. contrast the effects of genetic drift and natural selection on the gene pool of a population;
7. state the steps required for adaptation by natural selection in modern terms;
8. distinguish between disruptive, stabilizing, and directional selection by giving examples;
9. tell how variation is maintained in a population.

arwin, you will recall, stressed that the members of a population are subject to natural selection. Today, we define a population as all the members of a single species occupying a particular area at the same time (fig. 20.1). A population could be all the green frogs in a frog pond, all the field mice in a barn, or all the English daisies on a hill. The members of a population will reproduce with one another to produce the next generation. With the rediscovery of Mendel in the early 1900s and the rise of modern genetics, it became possible to study evolution by examining the genetics of a population. Only then did it become clear that Darwin's theory of natural selection was explainable in terms of genetics.

Population Genetics

Although Darwin emphasized that the members of a population vary from one another, he could never identify the sources of variation, that is, the raw material for evolutionary change.

Gene Mutations

Gene mutations provide new alleles, and therefore they are the ultimate source of variation. A gene mutation is an alteration in the DNA nucleotide sequence of an allele. We know, for example, that a change in a single DNA base spells the difference between a normal hemoglobin molecule and sickle-cell hemoglobin. Also, a change in a regulatory gene can increase or decrease the expression of a structural gene.

Mutations occur at random. A mutation is like a shot in the dark; any particular gene can mutate at any particular time. This means that the occurrence of a mutation is not tied to its adaptive value, and the chance of a particular mutation is not expected to increase because the mutation is beneficial to the organism. You might think, for example, that it is exposure to a drug that causes bacteria to become resistant. The experiment described in figure 20.2, however, shows that drug-resistant mutations occur in bacteria by chance, before they are exposed to the drug.

Rather than being beneficial, a mutation can be neutral in its effect or even harmful to an organism in its present environment. Many times observed mutations seem harmful, such as those that cause human genetic disease. This may be because members of a population are so adapted to their present environment that only nonbeneficial changes are apparent. As discussed later, however, some mutations may remain hidden in a diploid organism to become an important source of variation should the environment change. Or the same mutations may arise again when conditions have changed and they are now favorable mutations.

Chromosome Mutations

Some chromosome mutations are simply the alteration in the number of chromosomes inherited. The nondisjunction of chromosome 21 during meiosis causes a human to inherit 3 copies of this chromosome, for example.

Figure 20.1
a. A population of penguins in the Antarctica. *b.* A population of sunflowers in Romania. In each population all individuals contribute alleles to a common gene pool. While they may appear to be very similar in appearance, genetic and phenotypic variations are present, as they are in all populations.

a.

b.

Figure 20.2

In 1950, Joshua Lederberg devised and performed this replica-plating experiment, which provides evidence that a drug-resistant mutation occurs in a bacterium before it is exposed to the drug, not as a response to exposure to the drug. **_a._** Replica-plating technique. Bacterial colonies each derived from a single bacterium are grown on an agar medium in a petri plate. A piece of sterile velvet mounted on a solid support is pressed onto the surface of the plate. When the velvet is pressed onto the surface of another agar plate, colonies arise having positions identical to those of the first plate. This replica plating can be done again and again. **_b._** Replica-plating experiment. Colonies growing on a master plate are replica-plated onto a streptomycin-free agar plate and an agar plate that contains streptomycin. A few colonies of resistant bacteria occur in the same position on each of the plates. Since the colonies on the master plate had never been exposed to the streptomycin, the drug-resistant mutation had arisen by chance before exposure to the streptomycin. Colonies that contain mutant cells are shown in color.

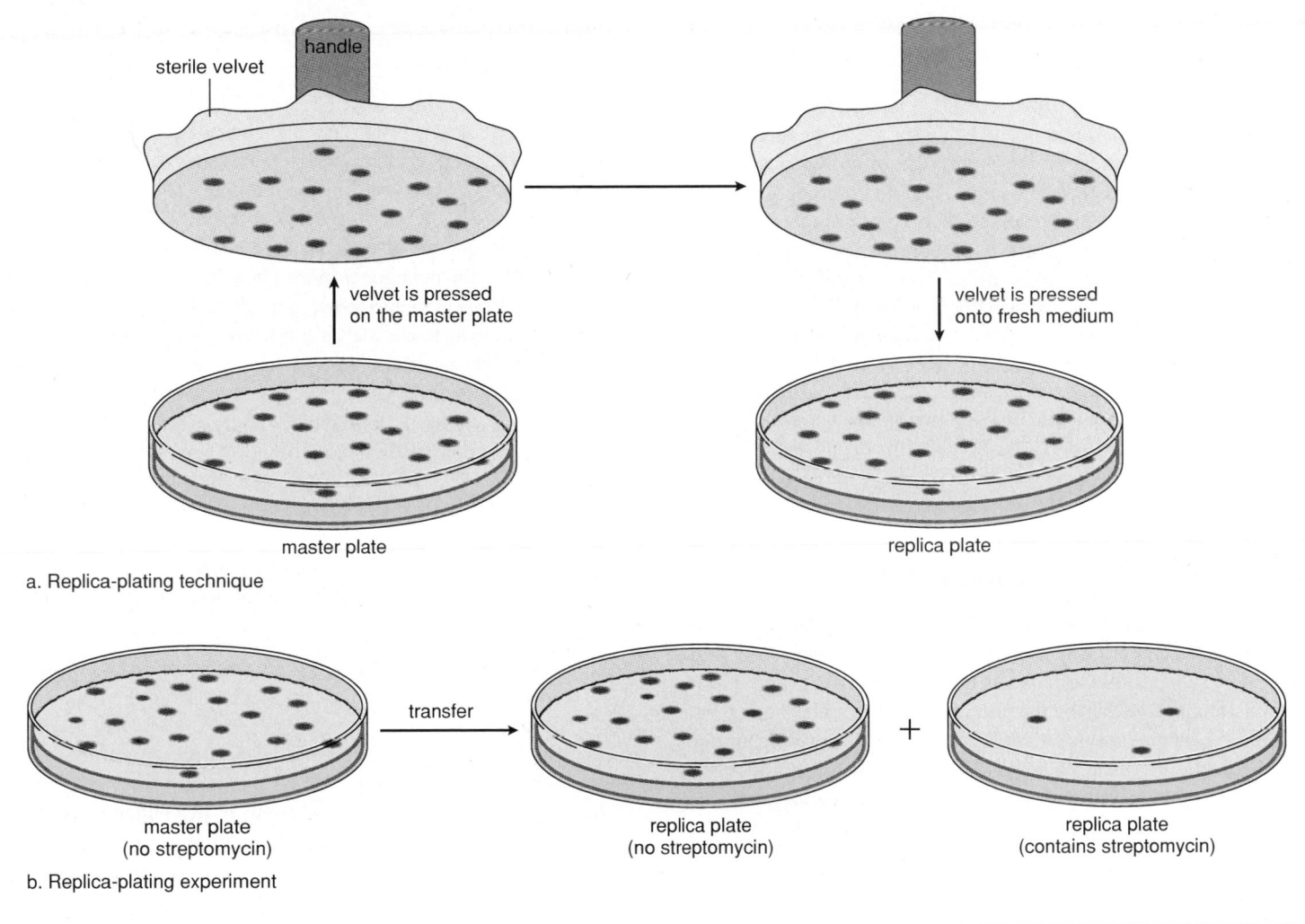

a. Replica-plating technique

b. Replica-plating experiment

Other chromosome mutations are an alteration in the arrangement of the alleles on the chromosomes. Various rearrangements of a chromosome are depicted in figure 13.16. Inversions and translocations can result in altered allele activity if the allele thereby comes under the control of a different regulatory gene. Duplication of an allele followed by a change in base sequence of one copy introduces a new allele while still retaining the old allele in the human genome. The best studied example of such an event concerns the globin alleles that occur at different loci but have similar DNA sequences. Adult hemoglobin contains 2 chains of α globin and 2 chains of ß globin. Besides the ones found in hemoglobin, humans have several other versions of both globin alleles. Most of these versions are expressed only during development, but some are expressed in adults.

Recombination

In sexually reproducing organisms, recombination of alleles and chromosomes is an important source of variation. During meiosis, crossing-over between nonsister chromatids and an independent assortment of chromosomes produce unlike gametes. During fertilization, random gamete union occurs, and this also contributes to a genotype that is unlike those of the parents. Altogether, each population has a storehouse of allelic and chromosomal combinations that potentially exceeds the number of individuals in the population (p. 176).

The entire genotype, and not individual alleles, is subjected to the natural selection process. A new and different combination of alleles may therefore have great selective value. For example, in a population of snails, stripes and brown color combined might make animals less visible in a woodland habitat. If stripes are controlled by one allele and brown color is controlled by another allele, it is a combination of the 2 alleles that will be selected for. Recombination may at some time bring the 2 alleles together so that the combined phenotype can be subjected to natural selection.

Along the same line, consider that most evolutionary changes in structure, function, and behavior are polygenic (see fig. 20.7a). The traits involved are controlled by polygenes, that is, genes that are at more than one locus. Certain combinations of the several alleles might be more adaptive in a particular environment than other combinations. The most favorable combination might not occur until just the right alleles are brought together by recombination. Once the improbable but most favorable phenotype occurs, it can become the prevalent phenotype in the population because of the natural selection process.

Variation is the raw material for evolutionary change. Only gene mutations result in new alleles. Chromosome mutations and recombination, however, contribute greatly to the production of variant genotypes.

The Hardy-Weinberg Law

Not only are variations created, they are also preserved and passed on from one generation to the next. The various alleles at all the gene loci in all individuals make up the **gene pool** of the population. It is customary to describe the gene pool of a population in terms of gene frequencies. For example, suppose it is known that one-fourth of all flies in a *Drosophila* population are homozygous dominant for long wings, one-half are heterozygous, and one-fourth are homozygous recessive for short wings. Therefore, we can describe the population in this manner:

$$\tfrac{1}{4} LL \ + \ \tfrac{1}{2} Ll \ + \ \tfrac{1}{4} ll$$

Given a population of 100 individuals, we have

$$25\,LL, \quad 50\,Ll, \quad \text{and} \quad 25\,ll$$

What is the number of the allele L and the allele l in the population?

Number of L allele:

$LL \ (2\ L \ \times \ 25) = 50$
$Ll \ (1\ L \ \times \ 50) = 50$
$ll \ \ (0\ L) \qquad\qquad 0$
$\qquad\qquad\qquad\quad 100\ L$

Number of l allele:

$LL \ \ (0\ l) \qquad\qquad\quad 0$
$Ll \ \ (1\ l \ \times \ 50) = 50$
$ll \ \ (2\ l \ \times \ 25) = 50$
$\qquad\qquad\qquad\quad 100\ l$

To determine the frequency of each allele, calculate its percentage of the total number of alleles in the population; in each case, $100/200 = 50\% = .50$. The sperm and eggs produced by this population will also contain these alleles in these frequencies. Assuming random mating (all possible gametes have an equal chance to combine with any other), we can calculate the ratio of genotypes in the next generation by using a Punnett square. In this Punnett square, the frequency of each type of sperm is shown on the horizontal and the frequency of each type of egg is shown on the vertical:

eggs	sperm		Results:
	0.5 L	0.5 l	$0.25LL + 0.5Ll + 0.25ll$
0.5 L	0.25 LL	0.25 Ll	$(\tfrac{1}{4} LL + \tfrac{1}{2} Ll + \tfrac{1}{4} ll)$
0.5 l	0.25 Ll	0.25 ll	

There is an important difference between this Punnett square and one used for a cross between individuals; here the sperm and eggs are those produced by the members of a population—not those produced by a single male and female. The results of the Punnett square show that there has been no change in frequencies of these alleles in the next generation. Therefore, sexual reproduction alone cannot bring about a change in allele frequencies. Also, the dominant allele need not increase from one generation to the next. Dominance does not cause an allele to become a common allele.

The potential constancy, or equilibrium state, of gene pool frequencies was independently recognized in 1908 by G. H. Hardy, an English mathematician and W. Weinberg, a German physician. This chapter's reading shows you how to use the Hardy-Weinberg equation to calculate allele frequencies. In nonmathematical terms, a *Hardy-Weinberg equilibrium* of allele frequencies in a gene pool will remain in effect in each succeeding generation of a sexually reproducing population as long as the 5 conditions described following are met:

1. No mutations: allele changes do not occur or else changes in one direction are balanced by changes in the opposite direction.
2. No gene flow: migration of alleles into or out of the population does not occur.
3. Random mating: individuals pair up by chance and not according to their genotypes or phenotypes.
4. No genetic drift: the population is very large, and changes in allele frequencies due to chance alone are insignificant.
5. No selection: no selective force favors one genotype over another.

In real life, these conditions are rarely, if ever, met, and allele frequencies in the gene pool of a population do change from one generation to the next. Therefore, evolution has occurred. The significance of the Hardy-Weinberg law is that it tells us what factors cause evolution—those that violate the conditions listed. Evolution can be detected by noting any deviation from a Hardy-Weinberg equilibrium of allele frequencies in the gene pool of a population.

The accumulation of small changes in the gene pool over a relatively short period of 2 or more generations is called microevolution (fig. 20.3). This chapter concerns microevolution, and the next concerns macroevolution, which includes speciation.

The Hardy-Weinberg Equation

 Because individuals live and reproduce within populations, it is at the population level that evolutionary change occurs. When biologists began to recognize this fundamental premise, they became interested in population genetics. G. H. Hardy and W. Weinberg recognized that the binomial expression

$$p^2 + 2pq + q^2$$

could be used to calculate the genotype and allele frequencies of a population directly. Assign the letter p to the frequency of the dominant allele and the letter q to the frequency of the recessive allele in a gene pool. The sperm and eggs produced by this population will contain these alleles in the same frequencies. Random union of gametes will result in these genotypes:

sperm

		p	q
eggs	p	p^2	pq
	q	pq	q^2

Results:

p^2 = homozygous dominant individuals
q^2 = homozygous recessive individuals
$2pq$ = heterozygous individuals

To use the binomial expression to calculate the genotype frequencies in a population, it is necessary to know the frequency of each allele in the population. For example, suppose the dominant allele A has a frequency of $p = 0.7$ and the recessive allele a has a frequency of $q = 0.3$. Then the genotype frequencies of the population would be

p^2 (homozygous dominant)	= 0.49
q^2 (homozygous recessive)	= 0.09
$2pq$ (heterozygous)	= 0.42
	1.00

Notice that the sum $p + q$ (frequencies of the 2 alleles) must equal 1.00 and the sum $p^2 + q^2 + 2pq$ (frequencies of the genotypes) must also equal 1.00.

When the frequency of a recessive phenotype is known, it is possible to calculate the genotype frequencies and the allelic frequencies. For example*, suppose by inspection we determine that 16% of a human population has a continuous hairline caused by a recessive allele. This automatically means that 84% of the population has the dominant phenotype, a widow's peak. Of these, how many have the homozygous dominant genotype? How many have the heterozygous genotype?

To answer these questions, first convert 16% to a decimal. Then we know that $q^2 = 0.16$ and, therefore, $q = .4$. Since $p + q = 1.0$, we know that $p = 0.6$ and, therefore, p^2 (frequency of the population that is homozygous dominant) = 0.36. To determine the frequency of the heterozygote, we simply realize that thus far we have accounted for only 0.52 of the population ($p^2 + q^2 = 0.52$). The remainder, $2pq$ (1-0.52), or 0.48 will have the heterozygous genotype. Or, if you prefer, calculate that $2pq = 0.48$. In summary, we have found that

homozygous recessive	= 0.16	= 16% have a continuous hairline
homozygous dominant	= 0.36	= 84% have a widow's peak
heterozygous	= 0.48	

We have considered only examples involving 2 alleles, but it is possible to use more complicated equations to calculate expected frequencies when there are more than 2 possible alleles at a gene locus.

*Additional problems are given on page 300.

The Hardy-Weinberg equilibrium provides a baseline by which to judge whether or not evolution has occurred. Any change of allele frequencies in the gene pool of a population signifies that evolution has occurred.

Agents of Evolutionary Change

The list of conditions for genetic equilibrium implies that the opposite conditions can cause evolutionary change. The conditions that can cause a deviation from the Hardy-Weinberg equilibrium are mutation, gene flow, nonrandom mating, genetic drift, and natural selection. Only natural selection results in adaptation to the environment.

Mutation

As mentioned, mutations provide new alleles, and therefore they underlie all the other mechanisms that provide variation, the raw material for evolutionary change. Due to the elaborate DNA replication and DNA repair mechanisms discussed in chapter 15,

the mutation rates of individual genes are low (about 10^{-5} per gene per cell cycle), but each organism has many genes, and a population has many individuals. Therefore, per individual and per population, mutations are common, not rare, events, as we can see from the following calculations:

For an individual human: Since there are about 100,000 genes (200,000 alleles) in humans, the average mutation rate per human gamete is (2×10^5 alleles) $\times$ (10^{-5} mutation per gene) = 2 mutations per human gamete.

For a human population: Since there are about 252 million people in the United States, the mutation rate per generation is (2.5×10^8 individuals) $\times$ (2 mutations per gamete) = 5×10^8 new mutations in each generation.

Therefore, even a modest mutation rate produces a tremendous amount of genetic variation in a population.

In 1966, R.C. Lewontin and J. L. Hubby set out to determine the molecular polymorphism (variation) of 18 different enzymes in natural populations of *Drosophila pseudoobscura.*

Figure 20.3

Microevolution diagram. Each bear represents a large number of bears in a population. **a.** Initially, 25% of the population had light coats and 75% had heavy coats. **b.** Several generations later, 37.5% of the population have light coats and 62.5% have heavy coats. There has been a change in the frequency of genotypes (shown) and alleles (not shown), and therefore microevolution has occurred.

Initial Generation

Genotype Frequencies		Phenotype Frequencies	
aa	.25	light coat	25%
AA	.25	heavy coat	75%
Aa	.50		

a.

Several Generations Later

aa	.375	light coat	37.5%
AA	.395	heavy coat	62.5%
Aa	.23		

b.

They extracted various enzymes and subjected them to electrophoresis (p. 291). They concluded that a fly population is polymorphic at no less than 30% of all its gene loci and that an individual fly is likely to be heterozygous at about 12% of its loci. Similar results have been found in studies on many species, demonstrating that high levels of molecular variation are the rule in natural populations.

Many of these mutations will not affect the phenotype, and therefore they will not ordinarily be detected. As discussed previously, harmful genotypic effects may be easiest to detect when a population is already well adapted. In a changing environment, however, even a seemingly harmful mutation can be a source of variation that can help a population become adapted to a new environment. For example, the water flea *Daphnia* ordinarily thrives at temperatures around 20°C and cannot survive at a temperature of 27°C or greater. There is a mutation, however, that requires *Daphnia* to live at temperatures between 25°C and 30°C. The adaptive value of this mutation is entirely dependent on environmental conditions.

> Mutations cause a gene pool to have multiple alleles of each gene.

Gene Flow

Gene flow, also called gene migration, is the movement of alleles between populations by, for example, the migration of breeding individuals. Figure 20.4 shows populations of the rat snake *Elaphe obsoleta*. Since the populations are adjacent to one another, there is a constant gene flow between them. Adult plants are not able to migrate, but their gametes are often either windblown or carried by insects. The wind, in particular, can carry pollen for long distances and be a factor in causing gene flow between plant populations.

Gene flow can increase the variation within a population by introducing novel alleles that were produced by mutation in some other population. Continued gene flow between populations makes their gene pools similar and reduces the possibility of allele frequency differences between populations that might be due to natural selection and genetic drift. Gene flow between the *Elaphe* populations in figure 20.4 causes them all to belong to the same species. Indeed, gene flow between populations can prevent speciation from occurring, as we shall see in the next chapter.

> Gene flow tends to decrease the diversity between populations and causes their gene pools to become similar.

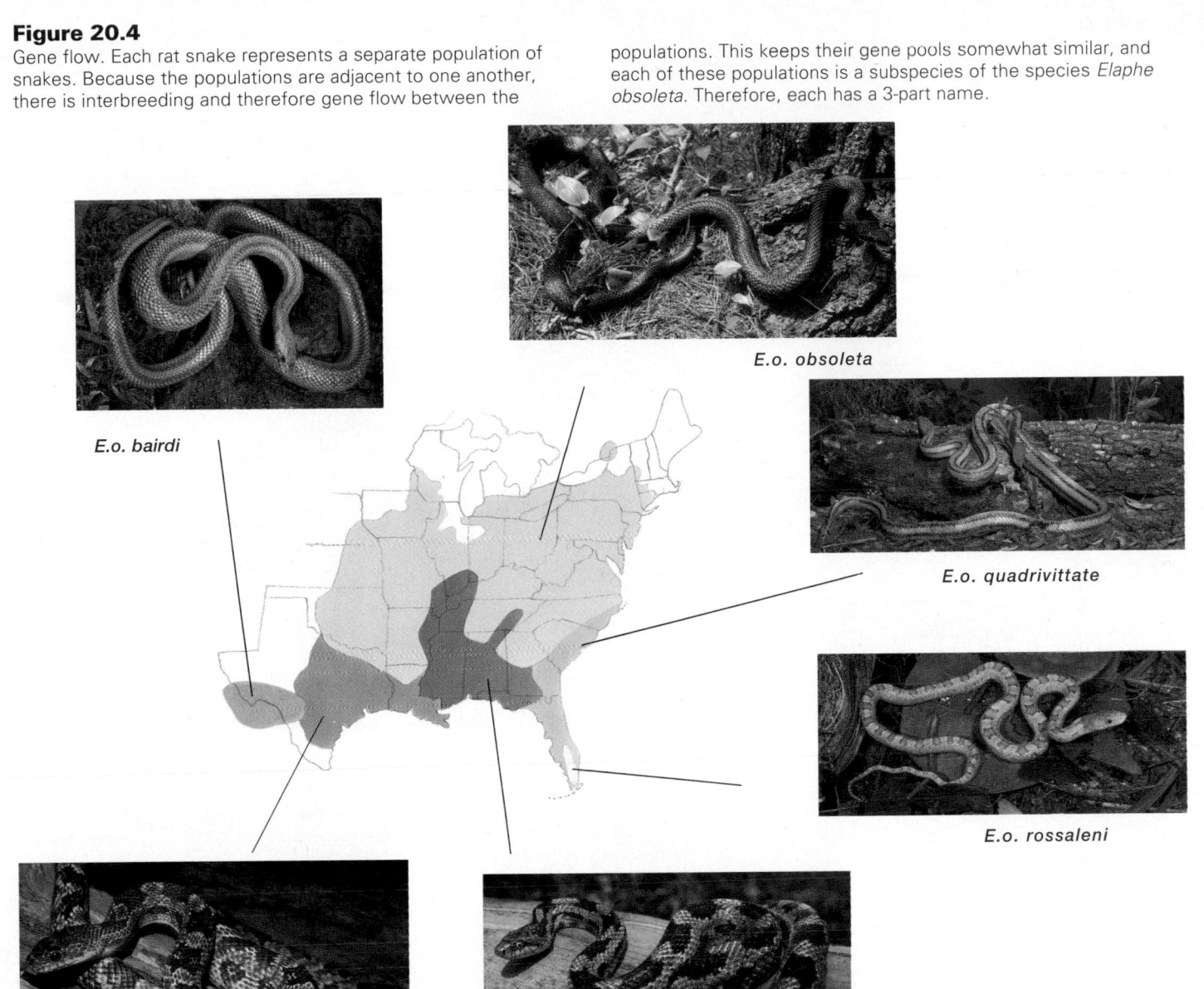

Figure 20.4
Gene flow. Each rat snake represents a separate population of snakes. Because the populations are adjacent to one another, there is interbreeding and therefore gene flow between the populations. This keeps their gene pools somewhat similar, and each of these populations is a subspecies of the species *Elaphe obsoleta*. Therefore, each has a 3-part name.

E.o. bairdi

E.o. obsoleta

E.o. quadrivittate

E.o. rossaleni

E.o. lindheimeri

E.o. spiloides

Nonrandom Mating

Random mating occurs when individuals pair up by chance and not according to their genotypes or phenotypes. Inbreeding, or mating between relatives to a greater extent than by chance, is therefore an example of *nonrandom mating*. This can happen if dispersal is so low that mates are likely to be siblings. Inbreeding does not change allele frequencies, but it does decrease the proportion of heterozygotes and increase the proportions of both homozygotes at all gene loci. In a human population, inbreeding increases the frequency of recessive abnormalities.

Assortative mating occurs when individuals tend to mate with those that have the same phenotype with respect to some characteristic. For example, in humans tall women seem to prefer to mate with tall men. Assortative mating causes the population to subdivide into 2 phenotypic classes, between which there is reduced gene exchange. Homozygotes for the gene loci that control the trait in question increase in frequency, and heterozygotes for these loci decrease in frequency. Other loci remain in Hardy-Weinberg equilibrium, except for those very closely linked to the loci governing the trait.

Figure 20.5

Genetic drift experiment. At the beginning of this experiment each of 107 *Drosophila* populations had 8 heterozygous flies (bw 75/bw, bw = brown eyes) of each sex. From the many offspring in each generation, only 8 females and 8 males were chosen at random to be the parents of the next generation. Nineteen generations later, 50% of the populations had fixed alleles.

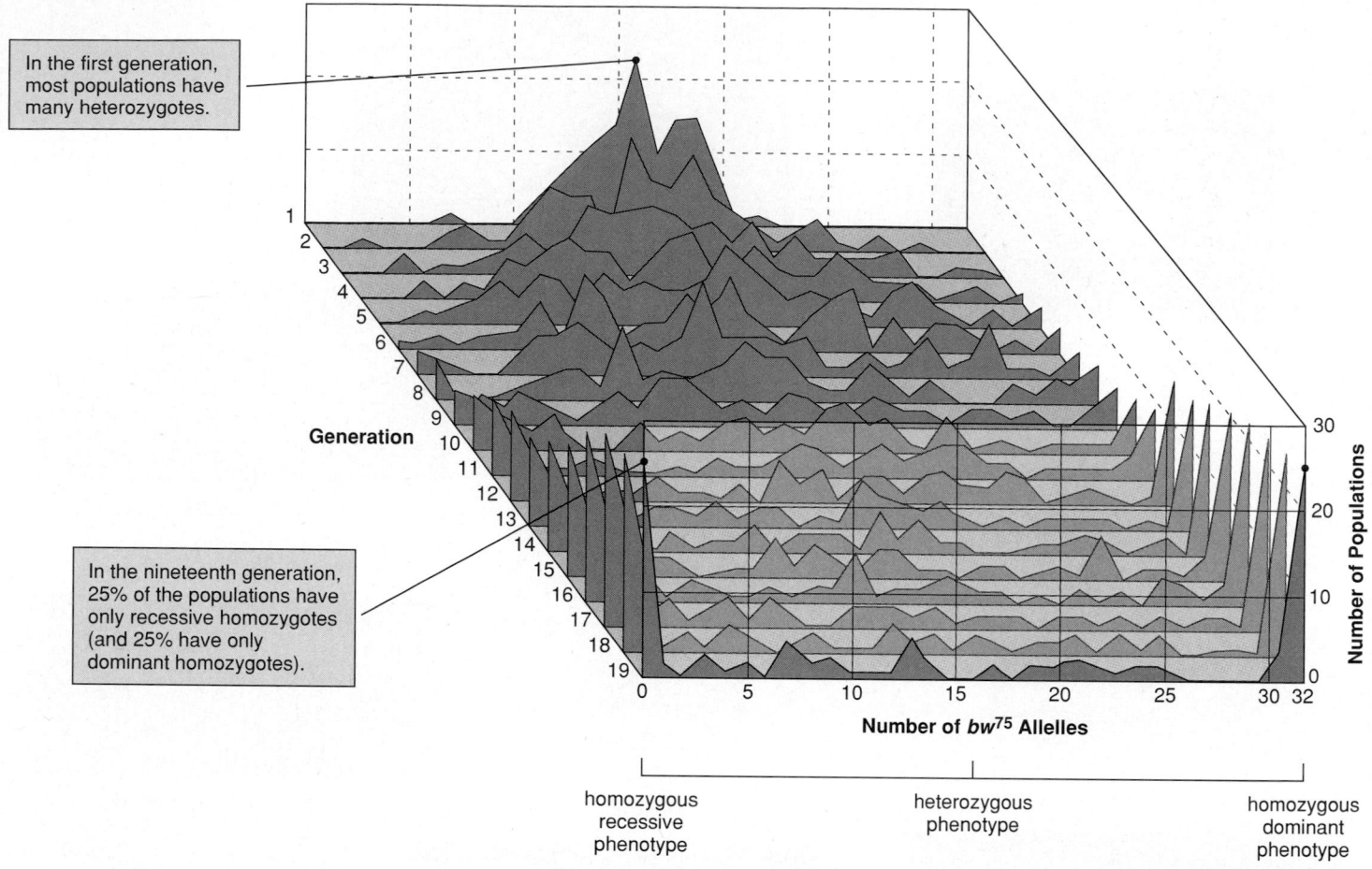

In the first generation, most populations have many heterozygotes.

In the nineteenth generation, 25% of the populations have only recessive homozygotes (and 25% have only dominant homozygotes).

Generation

Number of bw⁷⁵ Alleles

Number of Populations

homozygous recessive phenotype

heterozygous phenotype

homozygous dominant phenotype

Nonrandom mating involves inbreeding and assortative mating. The former results in increased frequency of homozygotes for all loci, and the latter results in increased frequency of homozygotes for only certain loci.

Genetic Drift

Genetic drift refers to changes in allele frequencies of a gene pool due to chance. Although genetic drift occurs in both large and small populations, a larger population is expected to suffer less of a *sampling error* than a smaller population. Suppose you had a large bag of 1,000 green balls and 1,000 blue balls, and you randomly drew 10%, or 200, of the balls. Because there is a large number of balls of each color in the bag, you can reasonably expect to draw 100 green balls and 100 blue balls or at least a ratio close to this. But suppose you had a bag of only 10 green balls and 10 blue balls and you drew 10%, or only 2 balls. The chances of drawing one green ball and one blue ball are now considerably less than before.

When a population is small, there is a greater chance that some rare genotype might not participate at all in the production of the next generation. Suppose there is a small population of flowers in which only 5% of individuals carry a rare allele. If by chance these flowers die out, the next generation would certainly have a change in allele frequencies. When genetic drift leads to the loss of one of a pair of alleles, the other allele then becomes *fixed* in the population. The experimental results given in figure 20.5 illustrate the occurrence of fixity of one allele or the other in small populations due to genetic drift.

1. The experiment began with 107 *Drosophila* populations, each in its own culture bottle. Each bottle contained 8 heterozygous flies of each sex. There were no homozygous recessive or homozygous dominant flies.

2. From the many offspring, the experimenter chose at random 8 males and 8 females. These were the parents for the next generation and so forth for 19 generations.

3. For the first few generations, most populations still contained many heterozygotes. (The exact composition of

Figure 20.6

Genetic drift. Various populations of the California cypress (**a** and **b**) differ from one another in shape of tree, type of bark, color of foliage, and size of cones. It is hypothesized that genetic drift caused these isolated populations to become dissimilar from each other.

a.

b.

the population can be calculated by using the Hardy-Weinberg equation, page 319. The recessive characteristic was brown eyes.)

4. By the nineteenth generation, 25% of the populations contained only homozygous recessive flies and 25% of the populations contained only homozygous dominant flies.

In keeping with this experiment, after random genetic drift has continued for a sufficient number of generations, most populations will have become fixed for some alleles.

Simultaneous drift occurring in a number of small, isolated populations causes their gene pools to diverge. The experiment in figure 20.5 considered only one allelic pair. If many pairs had been considered, it is expected that each of the populations would have a different gene pool composition after 19 generations. Genetic drift is a random process, and therefore it is not likely to produce the same results in several populations. In California, there are a number of cypress groves, each a separate population (fig. 20.6). All the trees belong to the same species, but still the phenotypes within each grove are more similar to each other than they are to the phenotypes in the other groves. Some groves have longitudinally shaped trees, and others have pyramidally shaped trees. The bark is rough in some

colonies and smooth in others. The leaves are gray to bright green or bluish, and the cones are small or large. Because the environmental conditions are similar for all groves, and no correlation has been found between phenotype and environment across groves, it is hypothesized that these variations between populations are due to genetic drift.

In nature, 2 situations, called the founder effect and the bottleneck effect, lead to small populations in which genetic drift affects gene pool frequencies.

Founder Effect

The **founder effect** is an example of genetic drift in which rare alleles, or combinations of alleles, occur at a higher frequency in a population isolated from the general population. After all, founding individuals contain only a fraction of the total genetic diversity of the original gene pool. Which particular alleles are carried by the founders is dictated by chance alone. The Amish of Lancaster, Pennsylvania, are an isolated group that was begun by German founders. Today, as many as one in 14 individuals carries a recessive allele that causes an unusual form of dwarfism (affecting only lower arms and legs) and polydactylism (extra fingers). In the population at large, only one in 1,000 individuals has this allele.

Figure 20.7

Variation in a population and modes of natural selection. *a.* Many traits, such as height in humans, are polygenic, and the distribution curve is bell-shaped. Most of the population is near the average value rather than the extremes. *b.* Modes of natural selection affect the distribution curve of a polygenic trait. In the diagrams, natural selection is acting against the phenotypes in the gray areas and is selecting for the phenotypes indicated by the arrows. In stabilizing selection, the average phenotype is favored, and the curve constricts because the extreme phenotypes are being eliminated. In disruptive selection, the average phenotype is being eliminated because the extreme phenotypes are favored; therefore, 2 distribution curves are eventually seen. In directional selection, one extreme phenotype is eliminated and the other is favored; therefore, the curve shifts in that direction.

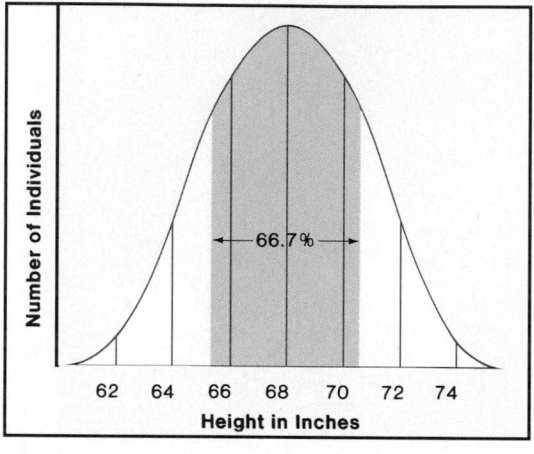

a.

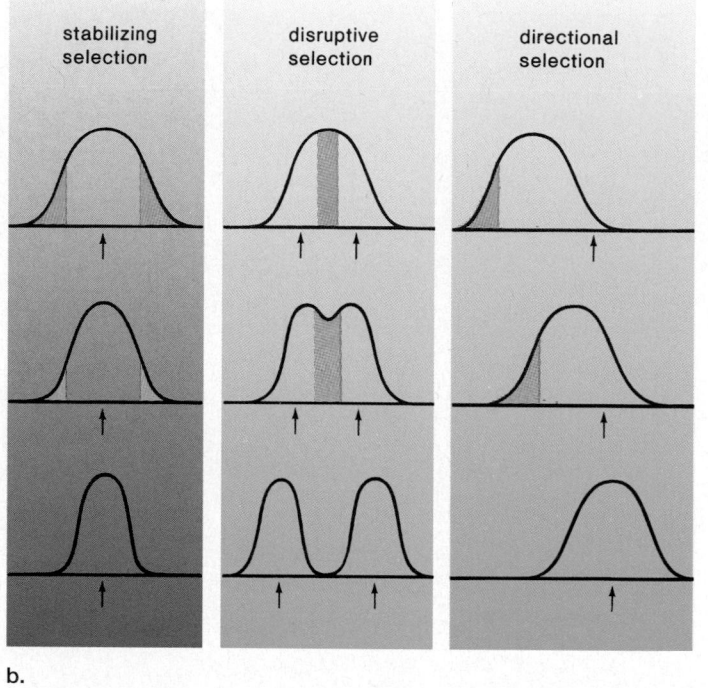

b.

All the Amish with this genetic syndrome can trace their ancestry back to a single couple (the Kings) who immigrated to Pennsylvania in 1744. They and their offspring had by chance large families, and therefore the rare allele became concentrated in the population.

Bottleneck Effect

Sometimes a population is subjected to near extinction because of a natural disaster (e.g., earthquake or fire) or because of slaughter by humans. Again, chance alone may determine which individuals survive these unfavorable times, which act like a bottleneck, preventing the majority of types of genotypes from participating in the production of the next generation.

The large genetic similarity found in cheetahs is believed to be due to a bottleneck. In a study of 47 different enzymes, each of which can come in several different forms, all the cheetahs had exactly the same form. This demonstrates that genetic drift can cause certain alleles to be lost from a population. Exactly what caused the cheetah bottleneck is not known. It is speculated that perhaps cheetahs were slaughtered by nineteenth-century cattle farmers protecting their herds, or were captured by Egyptians as pets 4,000 years ago, or were decimated by a mass extinction tens of thousands of years ago. Today, cheetahs suffer from relative infertility because of the intense inbreeding that occurred after the bottleneck.

Genetic drift is exaggerated when founder effects and bottleneck effects occur. In these instances, only a few types of genotypes contribute to the gene pool of a population, which is much smaller than the original population.

Natural Selection

Natural selection causes the allele frequencies of a gene pool to change, and when it does the members of a population become more adapted to their local environment. In chapter 19, we discussed how Darwin came to view natural selection as a mechanism involving several steps. Here we restate the steps in the context of population genetics.

Evolution by natural selection requires the following:

1. *Variation.* The members of a population differ from one another.
2. *Inheritance.* Many of these differences are inheritable, genetic differences.
3. *Differential reproduction.* Because of these differences, some phenotypes are more fit and reproduce to a greater extent than the other members.
4. *Accumulation of adaptive traits.* With each generation, alleles contributing to reproductive success increase in frequency.

Figure 20.8

Stabilizing selection favors an average human birth weight of about 7 lb (3 kg). The curve shows the percent mortality per birth weight in pounds. The highest mortality is seen when birth weight is very low; mortality rises again at weights above about 8 lb (3.5 kg). A high percentage of the population has a birth weight close to the optimal birth weight; therefore, the average birth weight is very close to optimal birth weight.

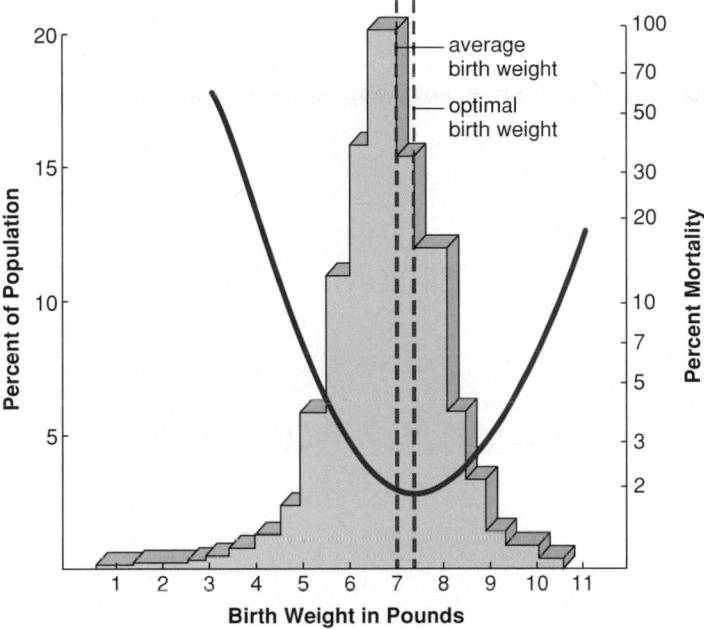

Figure 20.9

Disrupting selection results in polymorphism in a *Cepaea* snail population. Thrushes (birds) are more apt to feed on dark, unbanded snails in low vegetation areas (light circles) and light-banded snails in forest areas (dark circles). As a result of these conflicting selection pressures, due to a wide geographical range, both phenotypes are maintained in the population.

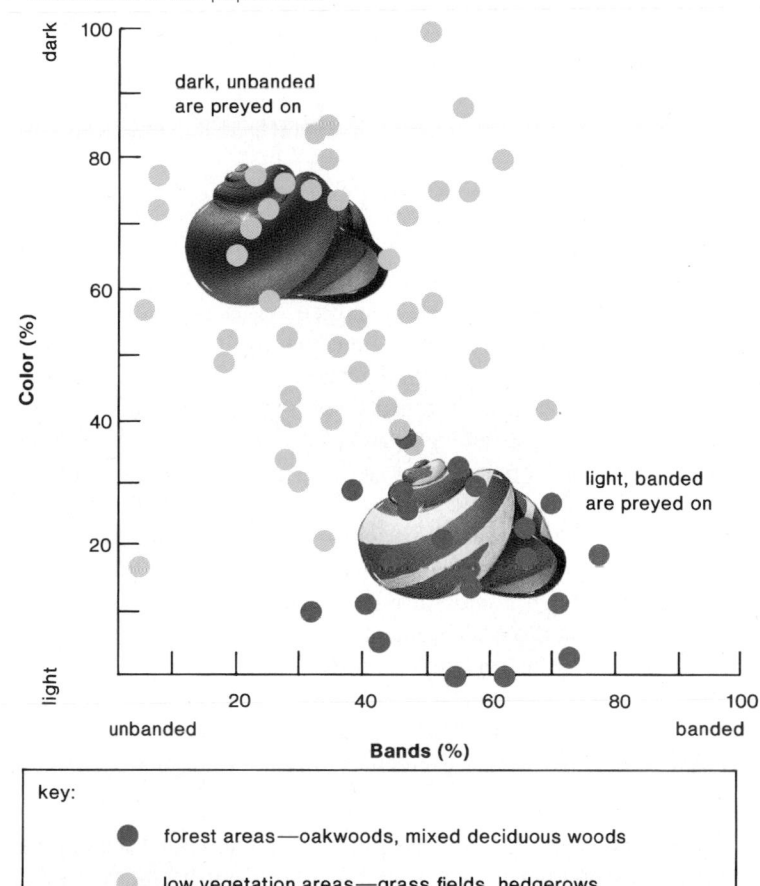

key:

● forest areas—oakwoods, mixed deciduous woods

● low vegetation areas—grass fields, hedgerows

Types of Selection

Most of the traits on which natural selection acts are polygenic and controlled by many allelic pairs. Such traits have a range of phenotypes whose frequency distribution usually resembles a bell-shaped curve (fig. 20.7). Three types of natural selection have been described for any particular trait: stabilizing selection, disruptive selection, and directional selection.

Stabilizing Selection

Stabilizing selection occurs when an intermediate phenotype is favored. It can improve adaptation of the population to those aspects of the environment that remain constant.

With stabilizing selection, extreme phenotypes are selected against, and individuals near the average are favored. As an example, consider that the birth weight of human infants ranges from 2-11 lb. The death rate is higher for infants who are at the extremes of this range and lowest for babies who have a birth weight of 7-8 lb. Most babies have a birth weight near this amount, which gives the best chance of survival (fig. 20.8). Similar results have been found in other animals, also.

Disruptive Selection

In **disruptive selection,** the 2 extreme phenotypes for a given trait are favored over any intermediate phenotype (fig. 20.9). For example, the shells of British land snails (*Cepaea nemoralis*) can

be light-colored or dark-colored and have light or dark bands. The snails have a wide geographic range that includes low vegetation areas (grass fields and hedgerows) and forested areas. Song thrushes feed on snails that are most easily seen. In low vegetation areas, they tend to prey on dark-colored, unbanded shells, and in forested areas, they tend to prey on light-colored, banded shells (fig. 20.9). Since selection pressure varies across the geographical range, polymorphism (2 or more forms) in this snail population is favored.

Directional Selection

Directional selection occurs when an extreme phenotype is favored. It can improve adaptation within a population when an environment is changing.

Industrial Melanism Before the Industrial Revolution in England, collectors of the peppered moth (*Biston betularia*) noted that most moths were light-colored, although occasionally a dark-colored (melanistic) moth was captured. Several decades after the Industrial Revolution, however, black moths made up 99% of the moth population in air-polluted areas.

Moths rest on the trunks of trees during the day (fig. 20.10); if they are seen by predatory birds, they are eaten. As long as the trees in the environment are light in color, the light-colored moths live to reproduce. But when the trees turn black from industrial pollution, the dark-colored moths survive and reproduce to a greater extent than the light-colored moths. The dark-colored phenotype, then, becomes the more frequent one in the population. Thereafter, if pollution is reduced and the trunks of the trees regain their normal color, the light-colored moths should increase in number. An experiment by H.B.D. Kettlewell of Oxford University showed that when their coloring matches the trees moths are more likely to avoid being eaten by predatory birds, which act as a selective agent.

Bacteria and Insects Indiscriminate use of antibiotics and pesticides causes populations of bacteria and insects to become resistant to these chemicals. The mutation that permits resistance is already present before exposure (fig. 20.2); the chemicals are merely acting as a selective agent. Those that survive exposure have offspring that are resistant, and in this way the population becomes resistant.

Another example of directional selection is the human struggle against malaria, a disease caused by an infection of the liver and the red blood cells (fig. 24.7). The *Anopheles* mosquito transmits the disease-causing protozoan *Plasmodium vivax* from person to person. In the early 1960s, international health authorities thought that malaria would soon be eradicated. A new drug, chloroquine, was more effective than the quinine that had been used previously against *Plasmodium,* and DDT (an insecticide) spraying had reduced the mosquito population. But in the mid-1960s, *Plasmodium* showed signs of chloroquine resistance, and worse yet, mosquitoes were becoming resistant to DDT. A few drug-resistant parasites and a few DDT-resistant mosquitoes had survived and multiplied, making the fight against malaria more difficult than ever. Another avenue has now been taken—a vaccine against malaria is being developed.

Figure 20.10
Directional selection results in a change from predominantly light-peppered moths to predominantly dark-peppered moths in a *Biston betularia* population. **a.** The light form is visible on soot-blackened oak trees and will be eaten by birds under these conditions. Therefore, the dark phenotype becomes the primary phenotype in the population. **b.** The dark form is visible on light, lichen-coated oak trees and will be eaten by birds under these conditions. Therefore, the light phenotype becomes the primary phenotype in the population.

a.

b.

There are 3 major types of natural selection:

1. Stabilizing selection, common in relatively unchanging environments, favors the intermediate phenotype so that it increases in frequency at the expense of the extreme phenotypes.
2. Disruptive selection favors extreme phenotypes at the expense of the intermediate. This type of selection may be common when geographically diverse populations are subjected to different environmental circumstances.
3. Directional selection favors one of the extreme phenotypes so that this phenotype becomes increasingly more common in the population. Directional selection is likely in a changing environment.

Figure 20.11

A male widowbird flying over his territory in the African savanna. After cutting off the tail feathers of some birds and gluing them onto the tail feathers of other birds, Malte Andersson found that females tend to choose a mate on the basis of tail length. The reproductive success of males, therefore, is largely dependent on the length of their tails, a feature that does not seem to help them establish territories.

Sexual Selection

We have noted that among the conditions that can cause evolutionary change, only natural selection results in adaptation to the environment (p. 324). If fitness is defined only in terms of number of surviving offspring, however, natural selection doesn't necessarily have to result in adaptation to the environment. A consideration of **sexual selection,** the preference of one mate for another, will substantiate this statement.

Many animal species, especially among birds and mammals, exhibit sexual dimorphism (fig. 43.2a). It's possible that the greater strength and ferocity of males may help them win contests or establish territories so that they have a better chance of mating with more females. In some instances, it seems that female preference might be the selection pressure controlling male attributes, such as the multicolored face of male mandril baboons or the splendid tail of male peacocks. Malte Andersson performed an experiment to see if this was the case in regard to the long black tail of male widowbirds (fig. 20.11).

Among widowbirds, males compete for territories in which mottled brown and short-tailed females then nest. Andersson allowed females to choose between 3 experimental groups of males:

1. Males with tails of normal lengths of 50 cm (half of the males had their feathers clipped, and then they were glued back on).
2. Males with tails clipped to 14 cm.
3. Males with tails lengthened to 75 cm (the feathers clipped from the second group were glued onto the tails of the third group).

At the beginning of the experiment, the same number of females were nesting in all territories, and at the end of the experiment, significantly more females were nesting in the territories of males with longer tails who did not seem disadvantaged by having a longer tail.

These results suggest that female sexual preference can help bring about directional selection of certain male characteristics. These characteristics, then, contribute to the fitness of the males independent of their possible adaptive value.

Maintenance of Variation

A population always shows some genotype variation. The maintenance of variation is beneficial because populations that lack variation may not be able to adapt to new conditions and may become extinct. How can variation be maintained in spite of selection constantly working to reduce it?

First, we must remember that the forces that work to promote variation are still at work: mutation still creates new alleles, and recombination still recombines these alleles during gametogenesis and fertilization, for example. Second, gene flow might still occur. If the receiving population is small and exhibits a high degree of homozygosity, gene flow can be a significant source of new alleles. Finally, we have seen that a range of phenotypes may be reduced but is not entirely eliminated by selection (fig. 20.9). And disruptive selection even promotes polymorphism in a population. There are other ways variation is maintained also.

Diploidy and the Heterozygote

We mentioned that only alleles that are exposed (cause a phenotypic difference) are subject to natural selection. In diploid organisms, this makes the heterozygote a potential protector of recessive alleles that otherwise would be weeded out of the gene pool. For example, consider these gene pool frequency distributions:

Frequency of Allele *a* in Gene Pool	Genotype Frequencies		
	AA	*Aa*	*aa*
0.9	0.01	0.18	0.81
0.1	0.81	0.18	0.01

Notice that even when selection reduces the recessive allele from a frequency of 0.9 to 0.1, the frequency of the heterozygote remains the same. The heterozygote remains a source of the recessive allele for future generations. In a changing environment the recessive allele may be favored by natural selection.

Sickle-Cell Disease

In certain regions of Africa, the contribution of the heterozygote to the maintenance of variation is exemplified even more by sickle-cell disease because the heterozygote phenotype is more fit than either of the homozygote phenotypes; therefore, all 3 genotypes are maintained in the population. The relative fitness of the heterozygote and the 2 homozygotes will determine their percentages in the population. The optimum ratio will be the one that correlates with the production of the most offspring. When the ratio of 2 or more phenotypes remains the same in each generation, it is called balanced polymorphism.

As discussed on page 249, individuals with sickle-cell disease have the genotype Hb^SHb^S and tend to die at an early age. Individuals who are heterozygous and have sickle-cell trait (Hb^AHb^S) are better off, but their red blood cells may still at times become sickle-shaped in the low oxygen environment of the tissues. Geneticists studying the distribution of sickle-cell disease in Africa have found that the recessive allele (Hb^S) has a higher frequency (0.2 to as high as 0.4 in a few areas) in regions with malaria (fig. 20.12). Further research has shown that the frequency of the genotype Hb^SHb^S is maintained because the heterozygote has an advantage over the normal phenotype in these regions. The malarial parasite flourishes in normal homozygotes but dies in heterozygotes because potassium ions leak out of the red blood cells when they become sickle-shaped. For this reason more heterozygotes survive than do the normal homozygotes in regions with malaria.

Figure 20.12
The red color shows the areas where malaria was prevalent in Africa, the Middle East, and Southern Europe, in 1920 before eradiation programs began; the blue color shows the areas where sickle-cell disease most often occurred. The overlap of these two distributions (purple) suggested that there might be a causal connection.

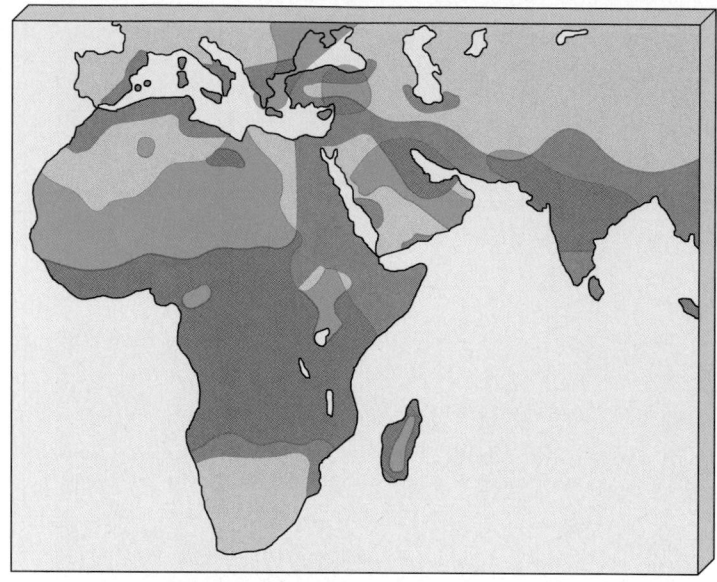

Even after populations become adapted to their environments, variation is promoted and maintained by various mechanisms. In diploid organisms, the heterozygote genotype can help maintain variation.

Summary

1. The principles of population genetics allow scientists to explain evolution in terms of genetics.
2. The members of a population vary from one another. Primary sources of variation are gene mutations, which produce new alleles; chromosome mutations, which can rearrange alleles on a chromosome; and recombination, which can bring certain alleles together into the same genotype. Most traits of evolutionary significance are polygenic.
3. The Hardy-Weinberg equilibrium is a constancy of gene pool frequencies that remains from generation to generation if certain conditions are met. The conditions are no mutation, no gene flow, random mating, no genetic drift, and no selection.
4. Since these conditions are rarely met, a change in gene pool frequencies is likely. When gene pool frequencies change, evolution has occurred. Deviations from a Hardy-Weinberg equilibrium allow us to determine when evolution has taken place.
5. The agents of evolutionary change are mutation, gene flow, nonrandom mating, genetic drift, and selection.
6. The rate of mutation per gene per generation is low, but since individuals have many genes and a population has many individuals, the mutation rate in a population is an adequate source of new alleles. Genotype variations may only be of evolutionary significance should the environment change.
7. Gene flow occurs when a breeding individual (animals) migrates to another population or when gametes and seeds (plants) are carried into another population. Constant gene flow between 2 populations causes their gene pools to become similar.
8. Nonrandom mating occurs when relatives mate (inbreeding) and assortative mating occurs. Both of these cause an increase in homozygotes.
9. Genetic drift occurs when allele frequencies are altered by chance, that is, by sampling error. Genetic drift can cause the gene pools of 2 isolated populations to become dissimilar as some alleles are lost and others are fixed. Genetic drift is particularly evident when founders start a new population or after a bottleneck, when severe inbreeding occurs.
10. The process of natural selection can now be restated in terms of population genetics (p. 324). As adaptation occurs, there is a change in gene pool frequencies. Only natural selection results in adaptation to the environment.

11. Three types of selection occur: stabilizing (the average human birth weight is near the optimum birth weight for survival); disruptive (*Cepaea* snails vary because a wide geographic range causes selection to vary); and directional (dark-colored peppered moths are prevalent in polluted areas and light-colored peppered moths are prevalent in nonpolluted areas). Sexual selection can bring about directional selection without regard to adaptation.

12. Despite constant natural selection, variation is maintained. Mutation and recombination still occur; gene flow between small populations can introduce new alleles; and natural selection itself sometimes results in variation. In sexually reproducing diploid organisms, the heterozygote acts as a repository for recessive alleles whose frequency is low. In regard to sickle-cell disease, the heterozygote is more fit in areas with malaria and therefore maintains the homozygotes in a population.

Study Questions

1. What are the sources of variation in a population of sexually reproducing diploid organisms?
2. What is the Hardy-Weinberg law?
3. What are the 5 agents of evolutionary change?
4. Which of these agents would cause the gene pools of 2 populations to become dissimilar? similar?
5. Give an example of genetic drift, the founder effect, and a bottleneck.
6. Why would you expect natural selection rather than genetic drift to increase the adaptation of a population?
7. State the steps required for adaptation by natural selection in modern terms.
8. Distinguish among disruptive, stabilizing, and directional selection by giving examples.
9. Does sexual selection influence fitness? Why or why not?
10. List ways in which variation is maintained in a population.

Objective Questions

1. Assuming a Hardy-Weinberg equilibrium, 21% of a population is homozygous dominant, 50% is heterozygous, and 29% is homozygous recessive. What percentage of the next generation is predicted to be homozygous recessive?
   a. 21%
   b. 50%
   c. 29%
   d. 25%
2. When evolution occurs
   a. adaptation to the environment always occurs.
   b. gene frequencies always change.
   c. phenotype changes are obvious.
   d. All of these.
3. A human population has a higher-than-usual percentage of individuals with a genetic disease. The most likely explanation is

   a. gene flow.
   b. natural selection.
   c. genetic drift.
   d. All of these.
4. The offspring of better-adapted individuals are expected to make up a larger proportion of the next generation. The most likely explanation is
   a. mutation.
   b. gene flow.
   c. natural selection.
   d. genetic drift.
5. The continued occurrence of sickle-cell disease in parts of Africa with malaria is due to
   a. continual mutation.
   b. gene flow between populations.
   c. fitness of the heterozygote.
   d. disruptive selection.
6. During the evolution of humans, brain size gradually increased. This is an example of

   a. disruptive selection.
   b. stabilizing selection.
   c. directional selection.
   d. All of these.
7. Which of these are necessary to natural selection?
   a. variations
   b. differential reproduction
   c. inheritance of differences
   d. All of these.
8. The average phenotype increases its frequency in a population due to
   a. directional selection.
   b. disruptive selection.
   c. stabilizing selection.
   d. sexual selection.
9. When a population is small, there is a greater chance of
   a. gene flow.
   b. genetic drift.
   c. natural selection.
   d. mutations occurring.

10. The following diagrams represent a distribution of genotypes (phenotypes) in a population. Under (*a*), draw another diagram to show that disruptive selection has occurred. Under (*b*), draw another diagram to show that stabilizing selection has occurred. Under (*c*), draw another diagram to show that directional selection has occurred.

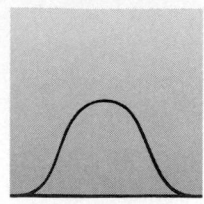 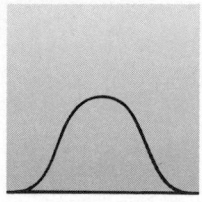 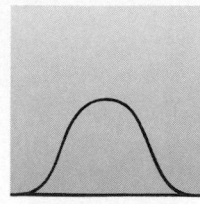

a. Disruptive selection   **b.** Stabilizing selection   **c.** Directional selection

Population Genetic Problems

1. If 1% of a human population has the recessive phenotype, what percentage has the dominant phenotype, assuming a Hardy-Weinberg equilibrium?

2. Four percent of the members of a population of pea plants are short (recessive characteristic). What are the frequencies of both the recessive allele and the dominant allele? What are the genotype frequencies in this population, assuming a Hardy-Weinberg equilibrium?

Concepts and Critical Thinking

1. *The process of evolution has a genetic basis.*

On page 177, we discussed pangenesis and Darwin's support of the hypothesis. Show that this hypothesis cannot be supported, by referring to the mechanism by which traits are passed from one generation to the next.

2. *Selection is the process by which populations accumulate adaptive traits.*

Reconcile this statement with your knowledge that selection acts at the level of the individual.

3. *Evolution involves several mechanisms, not just natural selection.*

Divide these mechanisms into those that increase variation and those that reduce variation between populations. Justify the items on each of your lists.

Selected Key Terms

gene pool (jēn pool) 318
gene flow (jēn flō) 320
genetic drift (jĕ-net′ik drift) 322

disruptive selection (dis-rup′tiv sĕ-lek′shun) 325
directional selection (di-rek′shun-al sĕ-lek′shun) 326

sexual selection (sek′shu-al sĕ-lek′shun) 327

21

Speciation in Evolution

The giant panda (*Ailuropoda melanoleuca*) is specialized for sitting upright on its hindquarters while it eats bamboo with large, flattened teeth set in powerful jaws. Molecular studies rather than anatomical and behavioral considerations have allowed scientists to deduce that the giant panda belongs in the bear family. Most likely its evolutionary line began about 20 million years ago, perhaps due to a fusion of chromosomes.

Your study of this chapter will be complete when you can

1. explain the biological definition of a species;
2. discuss the importance of reproductive isolating mechanisms, and give examples of both premating and postmating isolating mechanisms;
3. explain and contrast allopatric and sympatric speciation;
4. tell what taxonomy is and cite the categories of classification;
5. discuss systematics and the construction of phylogenetic trees;
6. tell what a cladogram is, and explain how this diagram differs from a phylogenetic tree based on systematics;
7. in general, describe the history of life on earth as revealed by the fossil record;
8. discuss continental drift, and tell how it affected the history of life on earth;
9. discuss the patterns of evolution that recur in the fossil record;
10. describe the phylogenetic history of *Equus,* and relate the history to anagenesis and adaptive radiation.

In this chapter, we will show how the microevolutionary mechanisms discussed in chapter 20 can lead to speciation, that is, the origin of species. Speciation is also a main event in macroevolution, because only following speciation events do lineages diverge from one another. Macroevolution considers the broad patterns of evolution that have occurred during the history of life.

Speciation

Sometimes it is very difficult to tell one species from another (fig. 21.1). Before we consider the origin of species, then, it is first necessary to define a species. For Linnaeus, whom we discussed in chapter 19, one species was separated from another by morphology, that is, their physical traits differed. Darwin saw that similar species, such as the 3 flycatcher species in figure 21.1, are related by common descent. The field of population genetics (see chap. 20) has produced the *biological definition of a species*. The members of one species interbreed and have a shared gene pool. Each species is reproductively isolated from every other species. The flycatchers in figure 21.1 are separate species because they do not interbreed.

Gene flow occurs between the populations of a species but not between populations of different species. For example, the human species has many populations that certainly differ in physical appearance. We know, however, that all humans belong to one species because the members of these populations can produce fertile offspring. On the other hand, the red maple and the sugar maple are separate species. Each species occurs over a wide geographical range in the eastern half of the United States and is made up of many populations (fig. 21.2). The members of each species' populations, however, do not hybridize in nature. Therefore, these 2 plants are related but separate species.

Reproductive Isolating Mechanisms

For 2 species to be separate, they must be reproductively isolated; that is, gene flow must not occur between them. A *reproductive isolating mechanism* is any structural, functional, or behavioral characteristic that prevents successful reproduction from occurring. Table 21.1 lists the mechanisms by which reproductive isolation is maintained.

Premating Isolating Mechanisms

Premating isolating mechanisms are those that prevent reproduction attempts. These include the following:

Habitat Isolation. When 2 species occupy different habitats, even within the same geographic range, they are less likely to meet and to attempt to reproduce. This is one of the reasons the flycatchers in figure 21.1 do not mate and the maples in figure 21.2 do not exchange pollen. In tropical rain forests, many animal species are restricted to a particular level of the forest canopy, and in this way they are isolated from similar species.

Figure 21.1

Three closely related flycatcher species. Although the 3 species are nearly identical, we would still expect to find some structural feature that distinguishes them. We know they are separate species because they are reproductively isolated—the members of each species reproduce only with each other. Each species has a characteristic song, and also helpful is the fact that each has its own particular habitat during the mating season.

least flycatcher (*Empidonax minimus*)

Acadian flycatcher (*Empidonax virescens*)

Trail's flycatcher (*Empidonax trailii*)

Evolution and Diversity

Figure 21.2

Habitat isolation. The red maple and the sugar maple do not interbreed—pollination between them is never successful. Therefore, the 2 species remain separate in the same locales. The upper illustration shows distribution of the red maple, and the lower shows distribution of the sugar maple. The anatomical differences between the 2 plants are also shown. Reprinted by permission from *North American Trees*, Third Edition, by Robert Preston, Jr., © 1976 by The Iowa State University Press, 2121 South State Avenue, Ames, Iowa 50010.

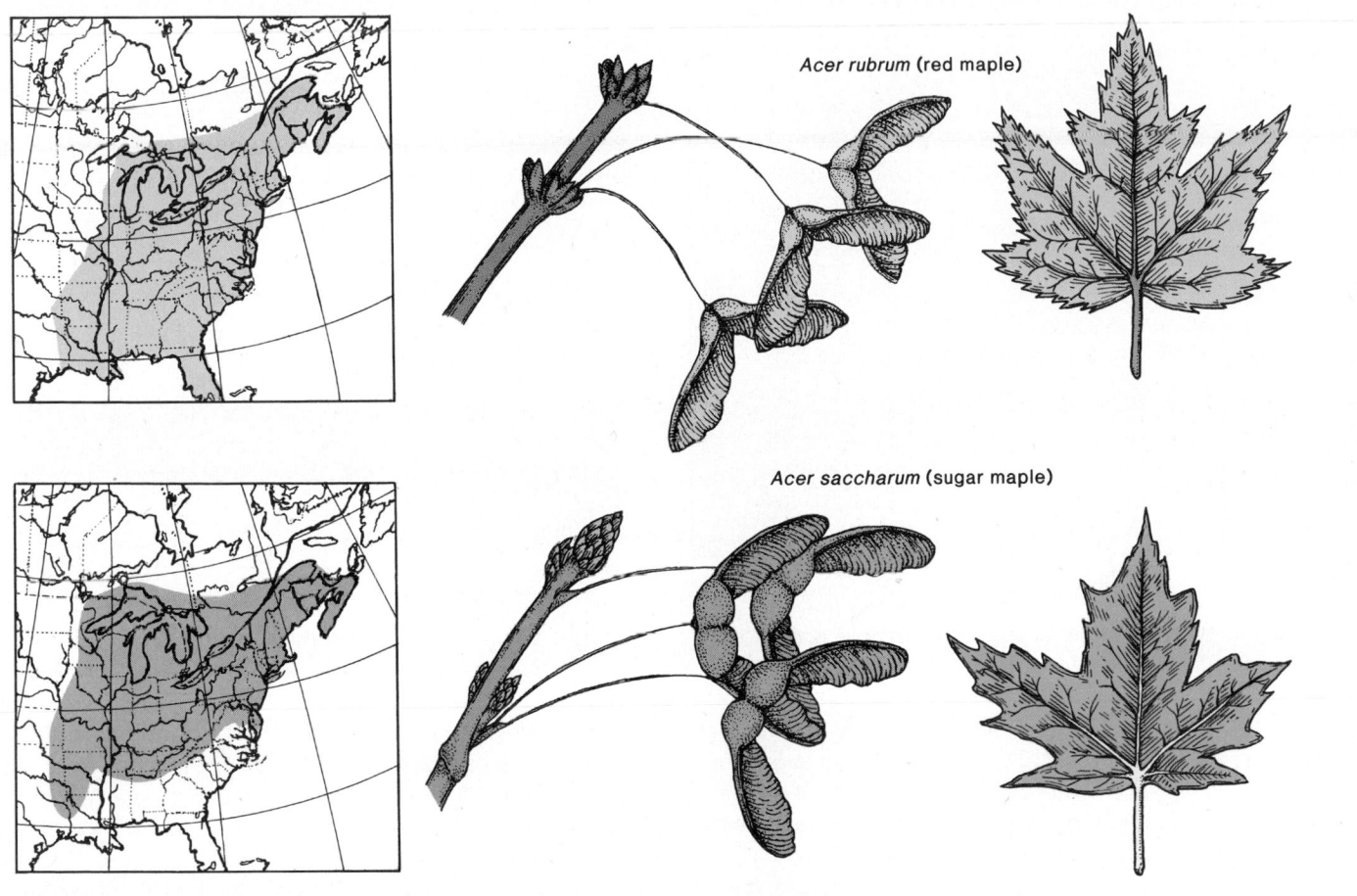

Acer rubrum (red maple)

Acer saccharum (sugar maple)

Temporal Isolation. Two species can occur in the same locale, but if each reproduces at a different time of year, they do not attempt to mate. For example, *Reticulitermes hageni* and *R. virginicus* are 2 species of termites. The former has mating flights in March through May, whereas the latter mates in the fall and winter months. Similarly, the frogs featured in figure 21.3 have different time periods of most active mating.

Behavioral Isolation. Many animal species have courtship patterns that allow males and females to recognize one another. Male fireflies are recognized by females of their species by the pattern of their flashings; similarly, male crickets are recognized by females of their species by their chirping. Many males recognize females of their species by sensing chemical signals called pheromones. For example, female gypsy moths secrete chemicals from special abdominal glands. These chemicals are detected downwind by receptors on male antennae.

Mechanical Isolation. When animal genitalia or plant floral structures are incompatible, reproduction cannot occur. Inaccessibility of pollen to certain pollinators can prevent cross fertilization in plants, and the sexes of many insect species have genitalia that do not match. For example, male dragonflies have claspers that are suitable for holding only the females of their own species.

Postmating Isolating Mechanisms

Postmating isolating mechanisms prevent hybrid offspring from developing or breeding, even if reproduction has been successful. These include the following:

Gamete Isolation. Even if the gametes of 2 different species meet, they may not fuse to give a zygote. In animals, the sperm of one species may not be able to survive in the reproductive tract of another species, or the egg may have

Table 21.1
Reproductive Isolating Mechanisms

Isolating Mechanism	Example
Premating	
Habitat	Species at same locale occupy different habitats
Temporal	Species reproduce at different seasons or different times of day
Behavioral	In animals, courtship behavior differs or they respond to different songs, calls, pheromones, or other signals
Mechanical	Genitalia unsuitable for one another
Postmating	
Gamete isolation	Sperm cannot reach or fertilize egg
Zygote mortality	Hybrid dies before maturity
Hybrid sterility	Hybrid survives but is sterile and cannot reproduce
F_2 fitness	Hybrid is fertile but F_2 hybrid has lower fitness

Figure 21.3
Temporal isolation. Five species of frogs of the genus *Rana* are all found at Ithaca, New York. The species remain separate because the period of most active mating is different for each and because whenever there is an overlap, different breeding sites are used. For example, pickerel frogs are found in streams and ponds on high ground, leopard frogs in lowland swamps, and wood frogs in woodland ponds or shallow water.

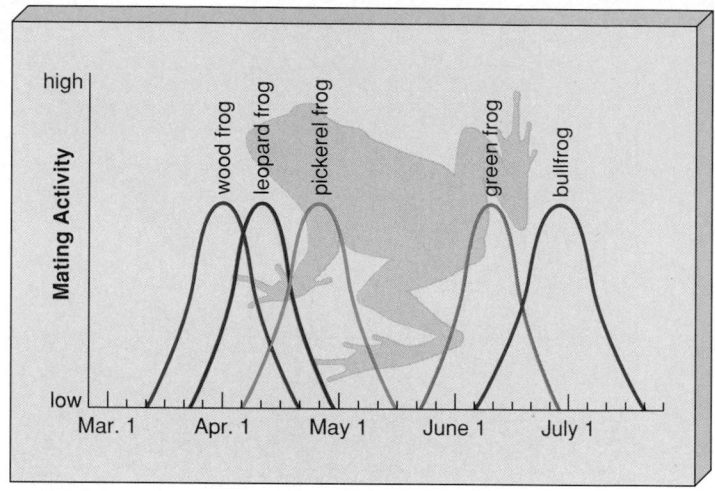

receptors only for sperm of its species. In plants, the stigma controls which pollen grains successfully complete pollination and which will not.

Zygote Mortality, Hybrid Sterility, and F_2 Fitness. If by chance the 2 frog species in figure 21.3 do form interspecies zygotes, the zygotes fail to complete development or else the offspring are frail. As is well known, a cross between a horse and a donkey gives a mule, which is usually sterile—it cannot reproduce. In some cases, mules are fertile, but the F_2 generation is not fertile. This has also been observed in both evening primrose and cotton plants.

The members of a biological species are able to breed and produce fertile offspring only among themselves. There are several mechanisms that keep species reproductively isolated from one another.

Origin of a Species

Gene flow is experienced between populations of the same species and is not experienced between populations of different species. Whenever reproductive isolation develops between groups of populations, **speciation** has occurred. Ernst Mayr, at the American Museum of Natural History, proposed one model of speciation after examining situations such as that depicted for the rat snake *Elaphe obsoleta* in figure 20.4. If *E.o. bairdi* became geographically isolated from the other populations, gene flow would stop, and different variations due to mutations, genetic drift, or natural selection could build up, causing reproductive isolation to occur.

Mayr called this model **allopatric speciation** (*allo*—different; *patric*—country) and said that postmating isolation would occur before premating isolation.

For example, the Baltimore oriole, which inhabits eastern forests, was geographically separated for many years from the Bullock's oriole, which inhabits western woodlands (fig. 21.4). This separation occurred because the great plains of the United States had few trees in which these birds could nest. Once humans settled in the plains, they planted trees, and both types of orioles then came into contact. The birds hybridized, but the hybrids seem less viable; nonhybrids are becoming more prevalent in regions of overlap between the birds. When the birds develop a premating isolating mechanism, speciation will be complete.

Allopatric speciation occurs when populations of a single species become separated by a geographical barrier. Without gene flow, genetic differences accumulate and the 2 sets of populations may become reproductively isolated.

Sympatric (*sym*—same; *patric*—country) **speciation** is a second model of speciation. With sympatric speciation, a population develops into 2 or more reproductively isolated groups without prior geographic isolation. The best evidence for this type of speciation is found among plants, where speciation can occur by means of polyploidy (p. 208). There can be a multiplication of the chromosome number in certain plants of a single species. Or, there can be hybridization between 2 species, followed by a doubling of the chromosome number (fig. 21.5). Polyploid plants are reproduc-

Evolution and Diversity

Figure 21.4

The male Baltimore oriole (**a**) and the male Bullock's oriole (**b**). These 2 types of orioles were once geographically separated but now are in contact and interbreed. The hybrids are less fertile than the nonhybrids, and it is therefore hypothesized that speciation is underway.

a.

b.

Figure 21.5

Spartina anglica is a vigorous cordgrass with 122 chromosomes. It arose by hybridization and doubling of the chromosome number. *S. townsendii* is a sterile hybrid of a cross between *S. alterniflora* and *S. maritima*. The odd number of chromosomes (61) cannot pair during meiosis. Doubling of the chromosome number allows synapsis to occur and meiosis to proceed normally.

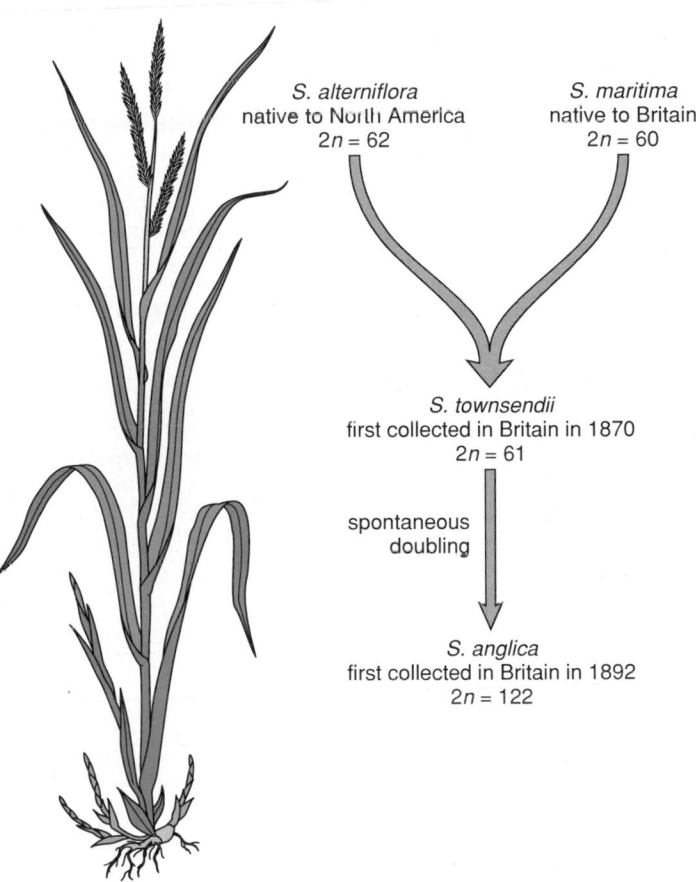

tively isolated by a postmating mechanism: polyploid plants can only reproduce successfully with other like polyploids, and backcrosses with diploid parents are sterile.

Chromosome rearrangements possibly occur among animals and cause speciation. The giant panda is bearlike, and it has been found that every bear chromosome can be matched with an arm from a giant panda chromosome. Giant pandas have 42 chromosomes, while bears have 74 chromosomes. At least some of the giant panda chromosomes are like 2 bear chromosomes fused together. Apparently chromosome fusion occurred as the giant panda line arose from a common ancestor with bears some 20 million years ago.

Sympatric speciation occurs when members of a single population develop a genetic difference, such as a chromosomal mutation, that prevents them from reproducing with the parent type.

Figure 21.6

Allopatric versus sympatric speciation. ***a.*** Allopatric speciation occurs after a geographic barrier prevents gene flow between populations that originally belonged to a single species. ***b.*** Sympatric speciation occurs when members of a population achieve immediate reproductive isolation without any prior geographic barrier.

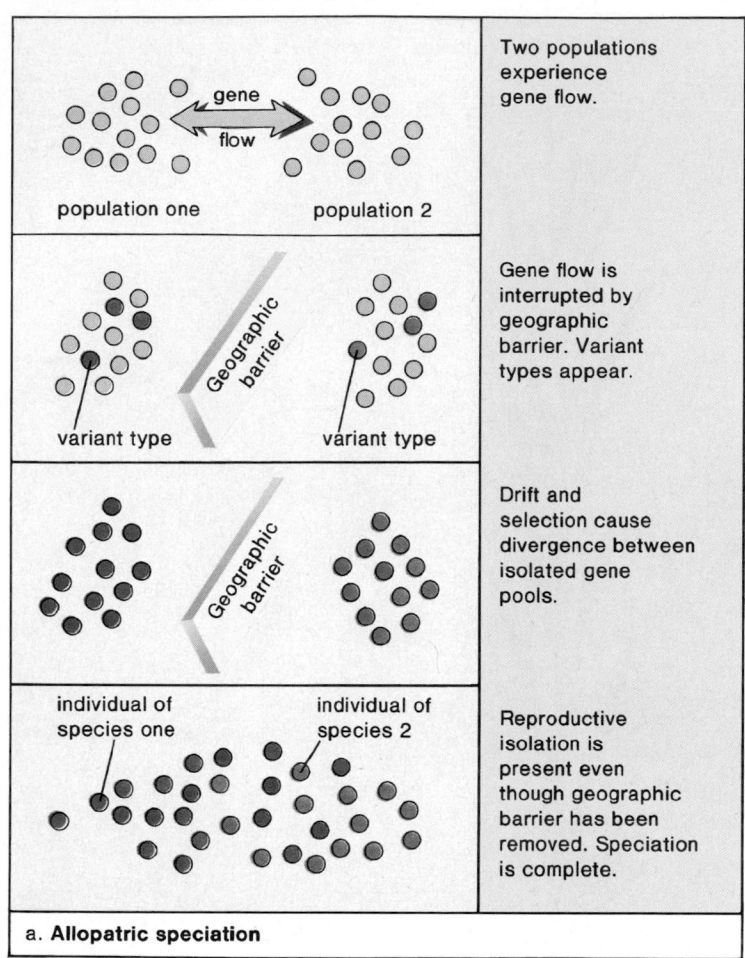

a. Allopatric speciation

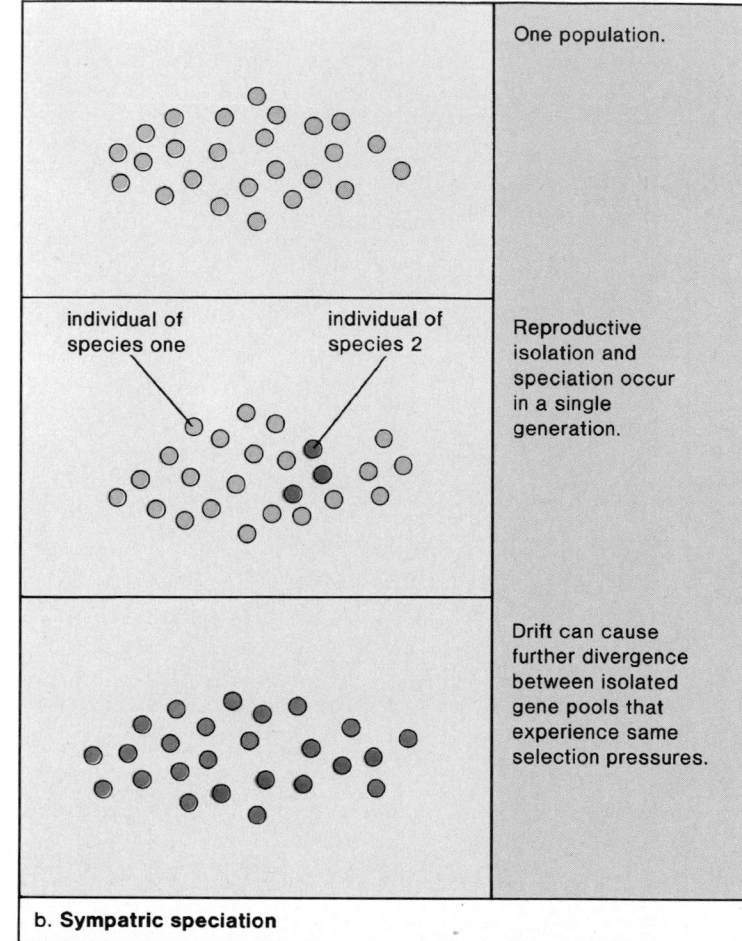

b. Sympatric speciation

Figure 21.6 contrasts allopatric speciation with sympatric speciation. Allopatric speciation is by far the more common means of speciation.

Allopatric Speciation on Island Chains

Allopatric speciation is exemplified very well by 13 species of finches that live on the Galápagos Islands (see fig. 19.9). All these birds are believed to be descended from a type of mainland finch that invaded the islands. As the parent population increased in size, daughter populations were established on the various islands (fig. 21.7). These daughter populations were subjected to the founder effect and the process of natural selection. Because of natural selection, each population became adapted to a particular habitat on its island. In time, the various populations became so genotypically different that now, when they by chance reside on the same island, they do not interbreed and are therefore separate species. There is evidence that the finches use beak shape to recognize members of the same species during courtship. Rejection of suitors with the wrong type of beak is a behavioral type of premating isolating mechanism.

The Galápagos finches are an example of **adaptive radiation.** On the species level, adaptive radiation is the rapid development from a single ancestral species of many new species, which then spread out and become adapted to various ways of life. (Adaptive radiation can also occur during macroevolution, p. 394.) Adaptive radiation occurred because there were few if any birds on the islands when the Galápagos finches arrived. Adaptive radiation among the finches is evident because each type of finch has a bill specialized for eating a particular type of food. For example, there are ground-dwelling finches whose beaks differ according to the size of seed they feed on and tree-dwelling finches whose beaks differ according to the size of their insect prey.

Figure 21.7
Adaptive radiation of Galápagos finches through a series of isolations on several islands.

A small number of mainland finches arrive at one of the Galápagos Islands (*a*). Genetic drift and natural selection begin.

Finches from population *a* are now on two different islands (*b* and *c*). Gene flow does not occur between the populations and each is subjected to genetic drift and different selection pressures.

Finches from population *c* migrate to the same island as population *b*. They are now separate species and interbreeding does not occur.

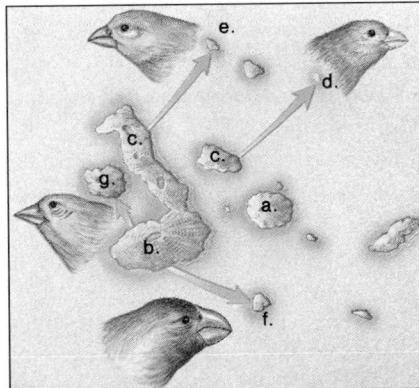

Finches from species *a, b, c* migrate to the other islands and each new population (*d, e, f, g*) begins to adapt to the local environment.

Eventually there are many species of finches on one island and each is adapted to a particular way of life.

Figure 21.8

Adaptive radiation in Hawaiian honeycreepers. More than 20 species evolved from a single species of a finch-like bird that colonized the Hawaiian Islands. This illustration shows only 6 of the more extreme adaptive forms.

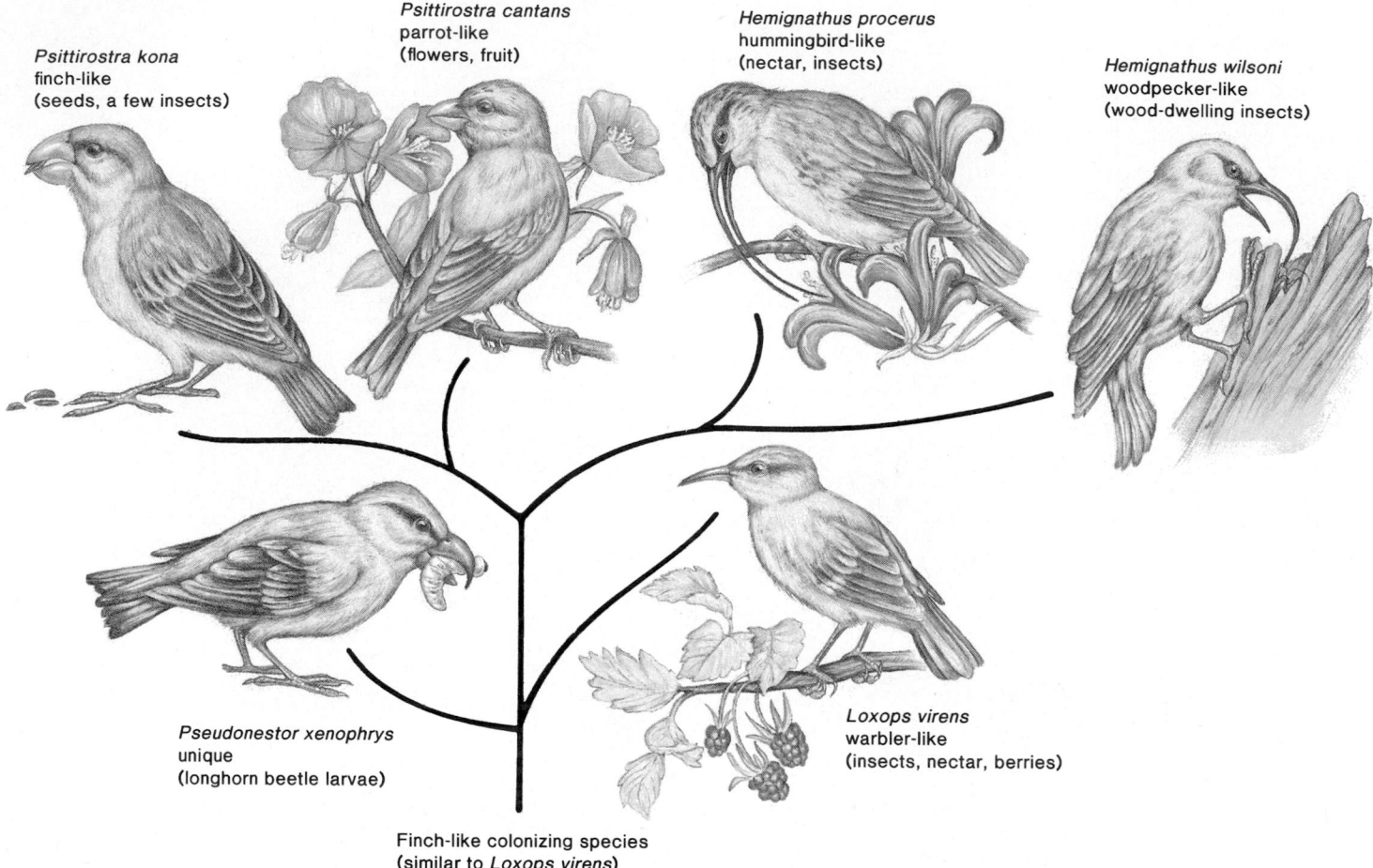

Psittirostra kona
finch-like
(seeds, a few insects)

Psittirostra cantans
parrot-like
(flowers, fruit)

Hemignathus procerus
hummingbird-like
(nectar, insects)

Hemignathus wilsoni
woodpecker-like
(wood-dwelling insects)

Pseudonestor xenophrys
unique
(longhorn beetle larvae)

Loxops virens
warbler-like
(insects, nectar, berries)

Finch-like colonizing species
(similar to *Loxops virens*)

Other island chains also provide examples of adaptive radiation. On the Hawaiian Islands, there is a wide variety of honeycreepers that are descended from a common goldfinch-like ancestor that arrived from Asia or North America about 5 million years ago. Today, honeycreepers have a range of beak sizes and shapes for feeding on various food sources, including seeds, fruits, flowers, and insects (fig. 21.8).

Adaptive radiation has also been observed in plants. On the island of New Caledonia, there are more than 6 species of *Dacrydium,* each adapted to a particular microclimate. For example, *D. araucariodes,* with narrow, incurvate leaves, is apparently adapted to a dry environment, whereas *D. taxoides,* with broad leaves, seems adapted to a moist environment. It is hypothesized that the softwood *Dacrydium* plants radiated on New Caledonia because of a lack of competition from hardwood trees.

Tempo of Speciation

All of the honeycreepers pictured in figure 21.8 evolved in about 5 million years. Is this a short time or a long time? The answer depends on your field of specialty. To a population geneticist who calculates gene pool changes from generation to generation, this is a long time. To a paleontologist who studies the fossil record, this is a very short time.

The *gradualist model* of evolutionary change suggests that the mechanisms of microevolution lead to speciation, and therefore the lineage (sequence of species from ancestor to descendant) of any group of organisms should be portrayed as in figure 21.9a. Species are expected to diverge after they come into existence, but phenotypic differences may not immediately show up in fossils. New species are recognized by the reproductive isolation of their gene pools, and the origin of new species may not coincide with phenotypic differences.

Evolution and Diversity

Figure 21.9

Tempo of speciation. **a.** In the gradualist model, change occurs constantly as descendants from a common ancestor gradually become new species. **b.** In the punctuated equilibrium model, rapid speciation is separated by periods in which no change occurs.

a. Gradualist model

b. Punctuated equilibrium model

Opposing this model of evolutionary change is the *punctuated equilibrium model*, supported in particular by Niles Eldredge, of the American Museum of Natural History, and Stephen Jay Gould, of Harvard University, both of whom are paleontologists. This model says that a species tends to remain the same for a long time, and then suddenly, new species appear (fig. 21.9*b*). In other words, an equilibrium phase (long periods without change) is punctuated by a rapid burst of change during which new species arise. According to this model, in which species are recognized by phenotypic differences, there are few intermediate forms in the fossil record because such forms exist only temporarily during a period of rapid change before a quiet period returns.

Table 21.2
Hierarchy of Classification

Category	Description		
Species	Reproductively isolated populations with phenotypic similarities		
Genus	Contains related species	More characteristics in common	Characteristics in common are more distinctive
Family	Contains related genera		
Order	Contains related families		
Class	Contains related orders		
Phylum (animals) Division (plants)	Contains related classes		
Kingdom	Contains related phyla		

The present controversy has stimulated much research. For example, Peter Grant tagged 1,500 medium ground finches on Daphne Major, one of the Galápagos Islands. In 1977, there was a severe drought and the majority of these birds died. The birds that survived were of a larger size than those that died. This suggests that an ecological crises is quite capable of causing a drastic change in species anatomy. Perhaps on a larger scale, speciation itself is apt to occur only after periods of crises. On the other hand, Peter Sheldon, of Trinity College in Dublin, has analyzed about 15,000 fossils of trilobites (extinct arthropod) from shale deposits in Wales. He finds a gradual change in the average number of tail ridges and feels that the necessity to assign species names to fossils artificially makes it seem as if there is no gradual change.

Macroevolution

Major phenotypic changes are the province of **macroevolution.** As we learned in chapter 1, the full classification of an organism includes the categories listed in table 21.2.

Macroevolutionists study the evolutionary relationships between groups of organisms above the species level. They are concerned with the broad patterns of evolution, since life began, among major groups of organisms. The study of macroevolution, then, is related to the science of taxonomy.

Taxonomy

Taxonomy, the science of classifying organisms, began with Linnaeus, who developed the binomial system of naming organisms (see reading p. 4) that recognizes species by their physical characteristics. He believed that the members of species reflected a divine idea or type. Modern taxonomists also rely heavily on anatomical features; however, they maintain that members of a species have phenotypic traits in common because they belong to a group of organisms that is reproductively isolated from other groups in nature. In other words, they define a species in evolutionary terms.

Taxonomists group species into ever-larger categories (table 21.2). A named category is called a *taxon* (taxa, pl.); for example, animalia, chordata, mammalia, etc. are taxa. Taxonomists don't always agree as to the methodology of classifying organisms, and therefore there are different schools of taxonomy. Systematics and cladistics are two of these schools of taxonomy.

Systematics

The goal of **systematics** is to classify organisms in a way that reflects phylogeny, or the evolutionary history of a group of organisms. The results are frequently portrayed in diagrams called **phylogenetic trees.** The tree usually has this appearance:

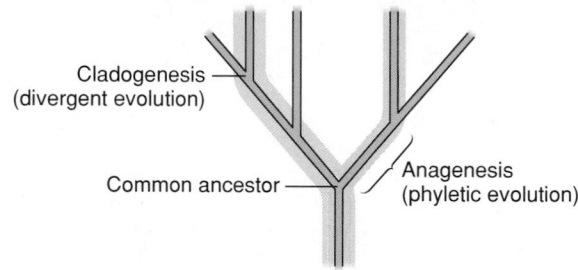

This tree has several lineages, and 2 are in color. *Cladogenesis* (also called divergent evolution), is a branching point, and *anagenesis* (also called phyletic evolution), is gradual change, occurring in between cladogeneses. Both events are presumed to involve speciation. The lineages in this tree have common ancestors; those that give rise to a divergence in particular may be known from the fossil record. In other words, the tree represents the history of a group of organisms. An actual phylogenetic tree based on the way organisms are classified is shown in figure 21.10. Notice how the most closely related organisms have a more recent common ancestor.

The fossil record and comparative anatomy, including embryological evidence, provide much information for the classification of organisms. Systematics relies heavily on homology to decipher relationships between species because homologous structures were inherited from a common ancestor. Figure 19.16 shows that the forelimbs of vertebrates are homologous because they contain the same sets of bones organized in the same general way. *Homologous structures* have the same structure but may differ in their function. In contrast, *analogous structures* have the same function but differ as to structure. The wing of an insect and bat are analogous structures; both are for flight, but they are constructed differently.

Molecular Systematics Amino acid sequences in certain proteins like hemoglobin and cytochrome c have been determined for various animals, as have DNA nucleotide differences between animals. When 2 lineages first diverge from a common ancestor, the genes and proteins of the lineages are nearly identical. But as time goes by, each lineage accumulates its own changes in gene and protein structure. Therefore, the degree of differences in these molecules provides some indication of how long ago the 2 lineages diverged from a common ancestor. In other words, DNA and

Evolution and Diversity

Figure 21.10

Phylogenetic trees are based on the way animals are classified. The classification and the tree tell the phylogenic history of the organism. A species is most closely related to other species in the same genus and then to genera in the same family and so forth from order to class to phylum to kingdom.

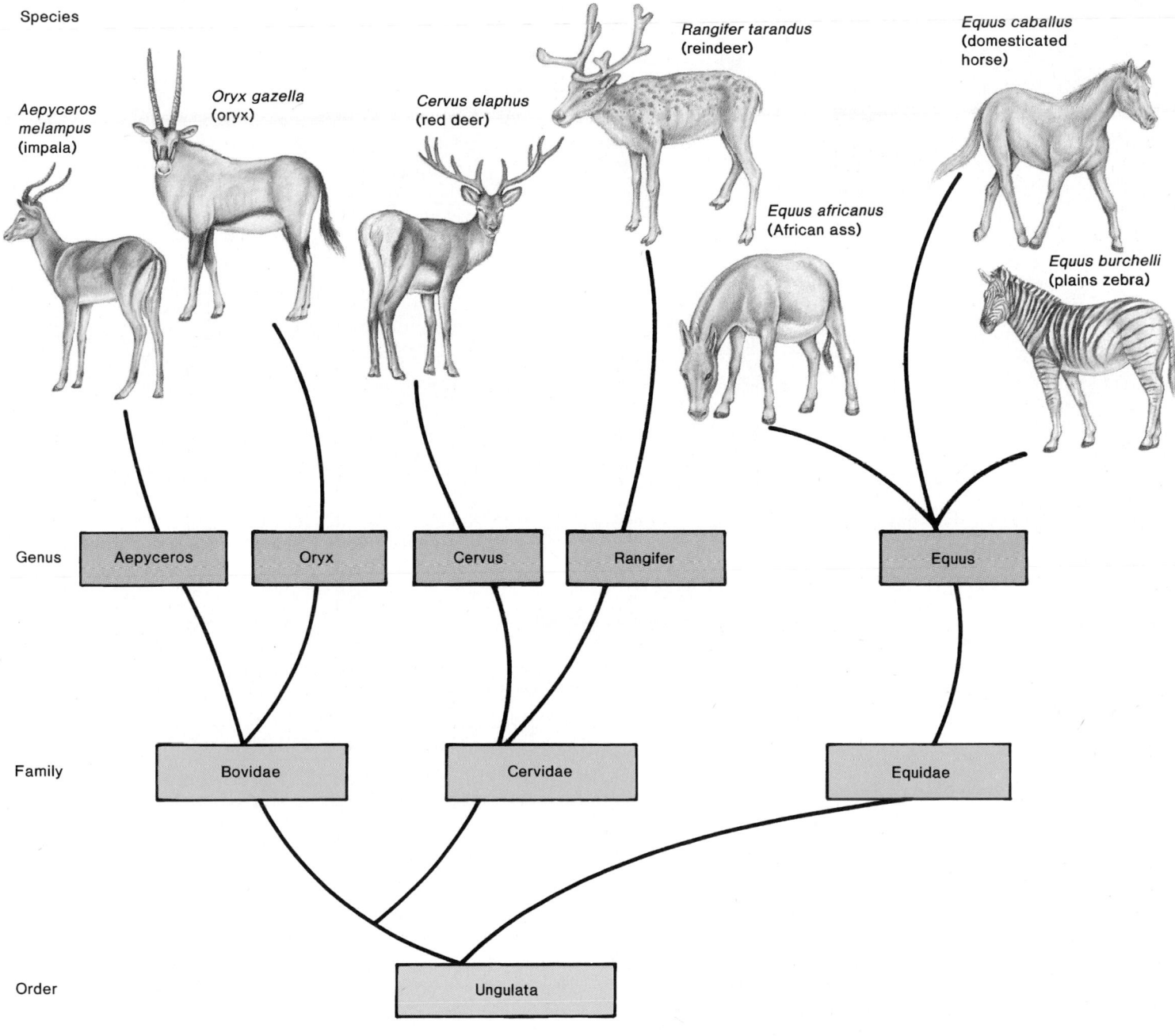

Species

Aepyceros melampus (impala)

Oryx gazella (oryx)

Cervus elaphus (red deer)

Rangifer tarandus (reindeer)

Equus africanus (African ass)

Equus caballus (domesticated horse)

Equus burchelli (plains zebra)

Genus: Aepyceros | Oryx | Cervus | Rangifer | Equus

Family: Bovidae | Cervidae | Equidae

Order: Ungulata

amino acid differences can be used as a "molecular clock" to indicate evolutionary time. Figure 21.11 shows the results of one such study of DNA differences. Biochemical data and the fossil record can be used as independent ways to measure the length of time 2 lineages have been diverging.

Phylogenetic trees are a way to show the evolutionary history of a group of organisms and how they are related by way of common ancestors.

Figure 21.11

Phylogenetic tree of certain primate species based on a biochemical study of their genomes. The length of the branches indicates the relative number of nucleotide pair differences that were found between groups. With the help of the fossil record it is possible to suggest a date at which each group diverged from the other.

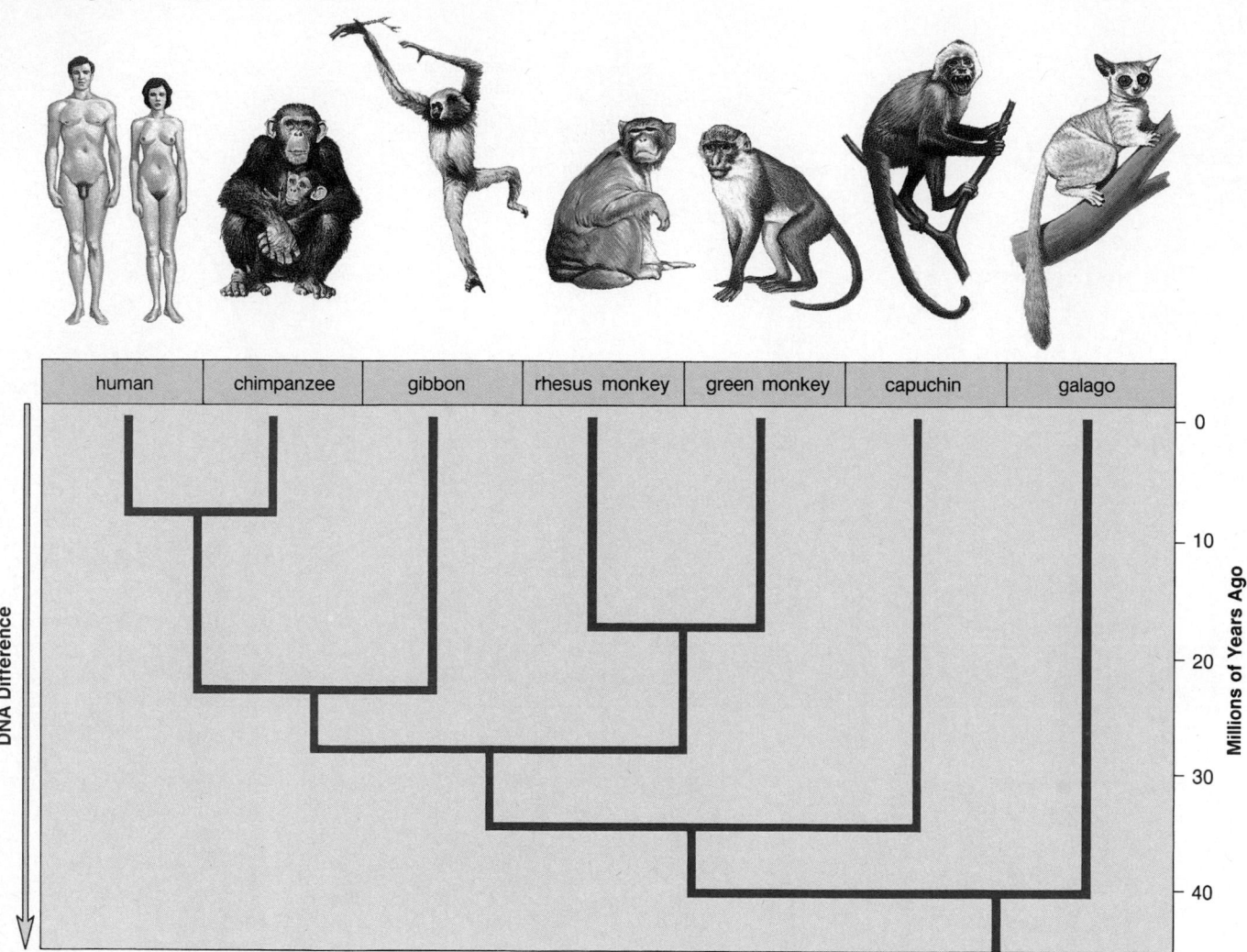

Cladistics

Cladistics differs greatly from systematics. Cladists place the common ancestor and all the organisms evolved from that common ancestor in one taxon. They use homologous structures and other similarities to decide which organisms are most closely related. They do not take into account how much the members of a taxon may have diverged from one another.

The results of a cladist's investigation are portrayed in a *cladogram.* In a cladogram, branches separate the taxons, and each branch represents the origin of some unique homology shared by its members. The sequence of the branching tells you the sequence in which these innovations occurred in the history of life, but the spacing does not represent actual lengths of time. Figure 21.12*a* gives a cladogram for vertebrate classes except fishes. It says that crocodiles, birds, and dinosaurs belong to one taxon, say a class,

and that birds and dinosaurs belong together in a lower taxonomy category, such as an order in that class. The similarities of crocodiles, birds, and dinosaurs include a 4-chambered heart and the behavioral habits of territoriality, nesting, and parental care of the young. Notice that the cladogram does not take into account that modern-day birds are quite different from modern-day reptiles. No one would have any difficulty telling a bird from a crocodile!

The traditional systematic view regarding the relationship of these vertebrates is shown in figure 21.12*b*. According to this view, turtles, snakes, crocodiles, and dinosaurs are all in the same class. Birds are descended from a common ancestor with dinosaurs, but they are placed in a separate class because they are a major group of organisms today. Modern-day birds, in contrast to reptiles, have feathers and can fly; they maintain a constant body temperature and lack teeth.

Figure 21.12

Cladistic and traditional views of reptile, bird, and mammal phylogeny. **a.** In a cladogram, organisms are grouped according to their similarities; the branching tells the order in which unique characteristics appeared. In this cladogram, birds are in the same group as crocodiles. **b.** In a phylogenetic tree, organisms are grouped according to their similarities and also according to their differences. Since birds are very different from reptiles, they are placed in a separate group.

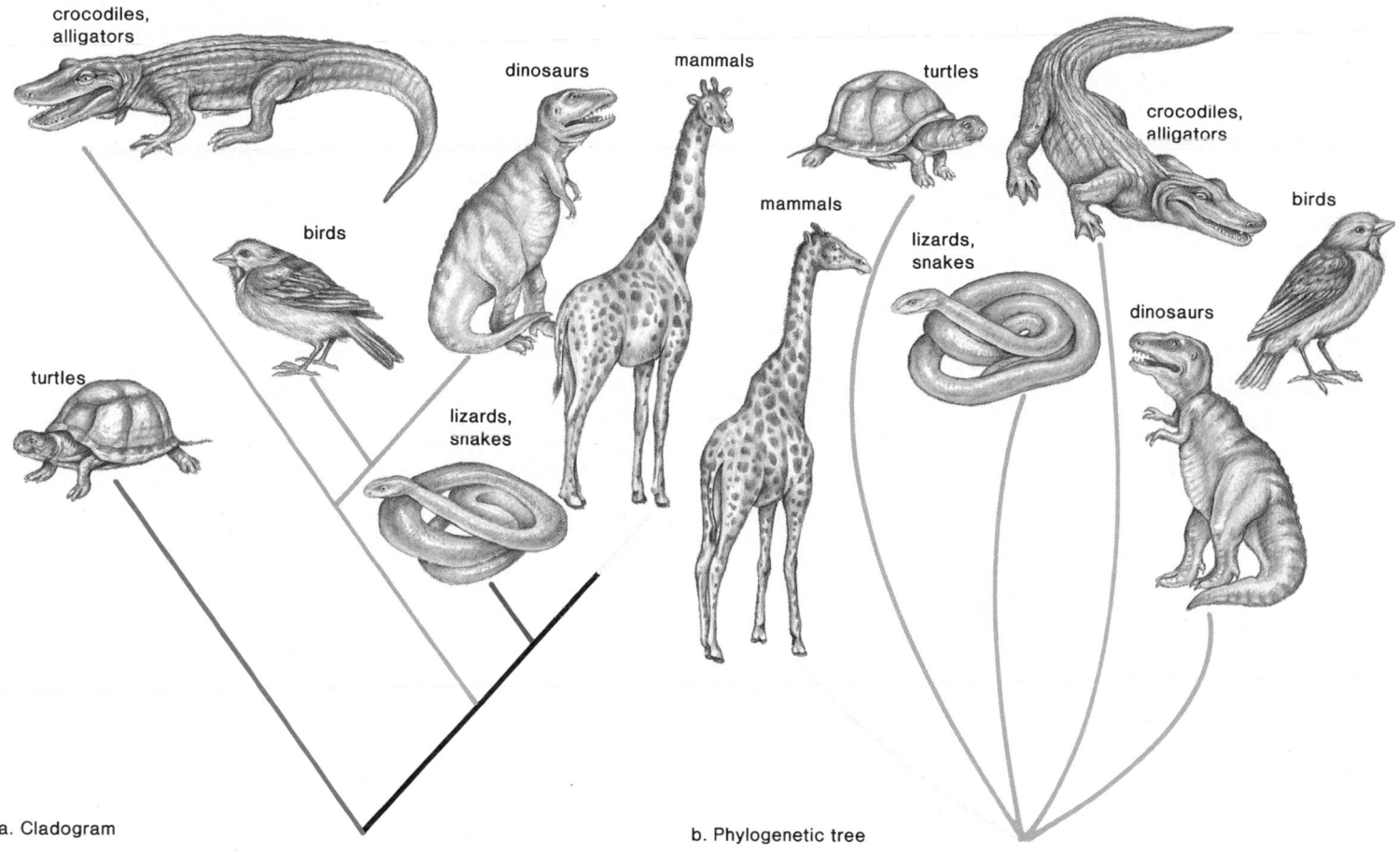

a. Cladogram

b. Phylogenetic tree

Cladists believe that their method of classification is more objective and testable. They simply identify observable similarities and do not make value judgments about the degree of divergence of organisms. Systematists maintain that cladists do make value judgments, such as choosing to ignore the unique presence of feathers in birds.

Fossil Record

Macroevolutionists, in particular, study the fossil record to trace the lineage of organisms throughout the history of life (fig. 21.13). Recall that *fossils* are the remains of past life or any other direct evidence of past life over 10,000 years ago. Usually when an organism dies, it is either consumed by scavengers or it undergoes bacterial decomposition. Occasionally, it is buried quickly and in such a way that decomposition is never completed or is completed so slowly that the organism leaves an imprint, or a cast, that clearly indicates its external and perhaps its internal structure.

Preserved fossils are most often, but not always, found in sedimentary rock (see fig. 19.5). Weathering produces sediments that are carried by streams and rivers into the oceans and other large bodies of water. There, they slowly settle and are converted into sedimentary rock. Later, sedimentary rocks are uplifted from below sea level to form new land, and researchers are able to search for fossil remains trapped in the rocks.

History of Life

During the last century, geologists and paleontologists (those who study fossils) noticed that fossil deposits are in recognizable layers, or strata (see fig. 19.1*d*), and each layer has its own mix of fossils. Boundaries between the strata, where one mix of fossils gives way to another, provide the basis for dividing geological time, as shown in table 21.3. We now know that many of these abrupt transitions were due to mass extinctions. **Mass extinctions** are large-scale losses of up to 55% of taxa at the family level. When all the species classified in all the genera of a family become extinct, the family becomes extinct. This happens to 55% of all the families during a mass extinction. The mass extinctions between the eras are of special significance. The extinction at the end of the Permian period, which was due to continental drift, is described later in this chapter. The dinosaurs became extinct at the end of the Cretaceous period. Walter and Luis Alvarez and their colleagues found that Cretaceous clay contains an abnormally high level of iridium, an element that is rare on earth but more common in asteroids. They hypothesize that a comet or asteroid collided with the earth, causing an explosion much like that of an atomic bomb. A cloud of dust would have mushroomed into the atmosphere, shading out the sun and causing plants to freeze and die. Mass extinction of animals and plants would have followed.

Notice that the names of the largest geological time divisions, the eras, pertain to life forms:

Eras:	Cenozoic	Modern life
	Mesozoic	Middle life
	Paleozoic	Ancient life

The dates for the geological time scale are determined by techniques described in "How Do You Date Fossils." We now know that the Solar System is 4.6 billion years old and that the history of life began about 3.5 billion years ago. A large portion of this time was devoted to the evolution of prokaryotes because eukaryotic cells didn't evolve until about 1.5 billion years ago. It was during the intervening 2 billion years that most of the metabolic pathways evolved. Single-celled organisms remained in the oceans and increased in complexity until multicellular organisms evolved about 650 million years ago. Fossils of organisms with skeletons are much easier to detect than those without skeletons; therefore, the history of life is better documented from this time on.

There was a great radiation of life forms in the Cambrian period, and most of the animal phyla arose at this time, including the first aquatic vertebrates. Plants invaded the land environment about 400 million years ago, and they were followed by fungi, insects, and finally vertebrates. The dominance

of reptiles lasted from 300 million to 65 million years ago, when the dinosaurs died out. Flowering plants and mammals had their origins some 200 million years ago. It was not until the dinosaurs vanished, however, that mammals became abundant. Hominids do not appear until about 3 million years ago, which is very recent compared to how long life has been evolving. If the history of the earth is measured using a 24-hour time scale that starts at midnight, humans do not appear until one-half minute before the next midnight (fig. 21.14).

Continental Drift

The evolution of life on earth has been dramatically affected by *continental drift,* a phenomenon proposed in 1915 by Alfred Wegener, a German meteorologist. Since then it has been concluded that the continents are on plates that move, or drift, relative to one another. There are ridges in the oceans where volcanic activity forms new crust, and this crust spreads slowly away from the ridges until it is consumed, often at oceanic trenches. Plates are mammoth pieces of crust that act like conveyor belts, carrying along the drifting continents. The study of the movement of plates is called plate tectonics.

Geological upheavals are common wherever the plates meet one another; this is where mountain building, volcanoes, and earthquakes occur. For example, the San Andreas Fault in California is at the boundary of 2 plates, where earthquakes are a common occurrence. Mount St. Helens, which erupted in 1979, is also near the edge of a plate.

Continental drift has affected the evolution of life on earth. About 230 million years ago, near the end of the Permian period, all the land masses came together and formed a supercontinent called Pangaea (*pangaea*—all land) (fig. 21.15). There was less shoreline, and the ocean basins increased in depth so that altogether the amount of shallow water habitat close to the

Figure 21.13

Fossils can be used to trace the history of life. Those featured here are placed in the order in which they appear in the fossil record. *a.* Prokaryotic cell fossils are about 3.5 billion years old. This is a filamentous form. Magnification X1,000. *b.* Eukaryotic cell fossils are about 1.5 billion years old. Magnification, X250. *c.* Invertebrate brachiopod (lamp shell) fossils are about 395 million years old. *d.* Fern fossils date from the Carboniferous period, which lasted from 350 to 286 million years ago. *e.* Vertebrate dinosaur fossils are about 190 million years old. *f.* Placental mammals date as early as 65 million years ago, but this fossil rabbit is from the Pleistocene epoch, which began about 2 million years ago.

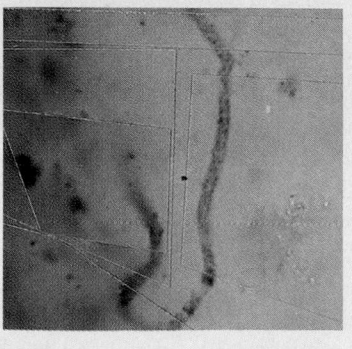
a.

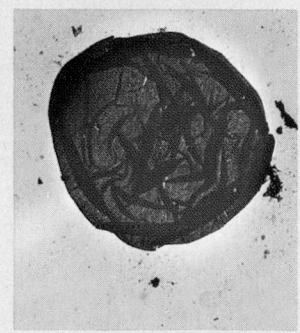
b.

c.

How Do You Date Fossils?

There are 2 ways to date fossils, the remains of past life or any other direct evidence of past life. The relative dating method determines the relative order of fossils but does not determine the actual date they were formed. The relative dating method is possible because fossil-containing sedimentary rocks occur in layers—the top layers are younger than the lower layers. Therefore, the fossils in each successive layer are older than the fossils in the layer above.

The absolute dating method assigns an actual rather than a relative date to a fossil. All radioactive isotopes have a particular half-life, the length of time it takes for half of the isotope in a specimen to decay and become nonradioactive (table 21.A). If the fossil has organic matter, half of the ^{14}C in it will be gone in 5,730 years. This is not very helpful unless we know how much ^{14}C was in the fossil to begin with. To make this determination, it is reasoned that organic matter always begins with the same amount of ^{14}C. (In reality, it is known that the ^{14}C levels in the air—and therefore the amount ingested by animals— can vary from time to time.) Now it is only necessary to use a radiation counter to compare the ^{14}C radioactivity of the fossil to that of a modern sample of organic matter. The older the fossil, the lower the counts per minute of radiation. In fact, after 50,000 years there is not enough ^{14}C radioactivity left to measure the age of a fossil accurately.

^{14}C is the only isotope listed in table 21.A that is contained within organic matter, but it is possible to use the others listed in the table to measure the age of a rock containing a fossil. This will indicate the age of the fossil also because as figure 19.5 shows, the 2 have most likely formed together. It is possible to measure ^{40}K, a common constituent of rocks, and its decay product, the gas argon (Ar), which is usually trapped inside the rock. The ratio of ^{40}K to ^{40}Ar in the rock can then be used to determine its age. For example, if the ratio is 1:1, then half of the ^{40}K has decayed and the rock is 1.3 billion years old. The isotopes ^{87}Rb, ^{232}Th, and ^{238}U are of no use for dating minerals less than about 100 million years. They have such a long half-life, no perceptible decay will have occurred in a length of time shorter than this.

Table 21.A
Principal Decay Series Used for Dating Rock

Parent Isotope	Half-life	Ultimate Stable Product	Effective Age Range
Rubidium, ^{87}Rb	47 billion yr (b.y.)	Strontium, ^{87}Sr	>100 million yr (m.y.)
Thorium, ^{232}Th	13.9 b.y.	Lead, ^{208}Pb	>200 m.y.
Uranium, ^{238}U	4.5 b.y.	Lead, ^{206}Pb	>100 m.y.
Potassium, ^{40}K	1.3 b.y.	Argon, ^{40}Ar	>100,000 yr
Uranium, ^{235}U	0.71 b.y	Lead, ^{207}Pb	>100 m.y.
Carbon, ^{14}C	5,730 yr	Nitrogen, ^{14}N	0-50,000 yr

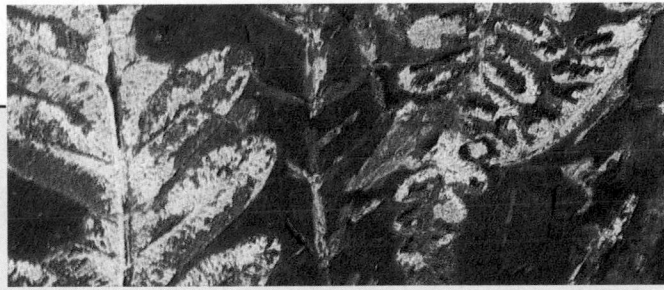

d.

e.

f.

Table 21.3

The Geological Time Scale: Major Divisions of Geological Time with Some of the Major Evolutionary Events of Each Geological Period

ERA	PERIOD	EPOCH	MILLIONS OF YEARS AGO	PLANT LIFE	ANIMAL LIFE
Cenozoic (from the present to 66.4 million years ago)	Quaternary	Holocene	0–0.01	Destruction of tropical rain forests by humans accelerates extinctions	AGE OF HUMAN CIVILIZATION
		Pleistocene		SIGNIFICANT MAMMALIAN EXTINCTION	
			0.01–2	Herbaceous plants spread and diversify	Modern humans appear
	Tertiary	Pliocene	2–5	Herbaceous angiosperms flourish	First hominids appear
		Miocene	5–24	Grasslands spread as forests contract	Apelike mammals and grazing mammals flourish; insects flourish
		Oligocene	24–37	Many modern families of flowering plants evolve	Browsing mammals and monkeylike primates appear
				SIGNIFICANT MAMMALIAN EXTINCTION	
		Eocene		SIGNIFICANT MAMMALIAN EXTINCTION	
			37–58	Subtropical forests with heavy rainfall thrive	First horses appear; all modern orders of mammals are represented
		Paleocene	58–66	Angiosperms diversify	Primitive primates, carnivores and insectivores appear
Mesozoic (from 66.4 to 245 million years ago)	Cretaceous		MASS EXTINCTION: DINOSAURS AND MOST REPTILES		
			66–144	Flowering plants spread; coniferous trees decline	Placental mammals appear; modern insect groups appear
	Jurassic		MASS EXTINCTION		
			144–208	Gymnosperms flourish	AGE OF DINOSAURS Birds appear
	Triassic		MASS EXTINCTION		
			208–245	Cycads and gingkos appear; forests of gymnosperms and ferns dominate	First mammals appear; first dinosaurs appear
Paleozoic (from 245 to 570 million years ago)	Permian		MASS EXTINCTION		
			245–286	Conifers appear	Reptiles diversify; amphibians decline
	Carboniferous		286–360	Age of great coal-forming forests: clubmosses, horsetails, and ferns flourish	AGE OF AMPHIBIANS
	Devonian		MASS EXTINCTION		
			360–408	First seed ferns appear	AGE OF FISHES First insects, first amphibians appear
	Silurian		408–438	Low-lying primitive vascular plants appear on land	First jawed fishes appear
	Ordovician		MASS EXTINCTION		
			438–505	Marine algae flourish	Invertebrates spread and diversify; Jawless fishes, first vertebrates, appear
	Cambrian		MASS EXTINCTION		
			505–570	Marine algae flourish	AGE OF INVERTEBRATES
Precambrian time (from 570 to 4600 million years ago)			700	Multicellular organisms appear	
			1500	First complex (eukaryotic) cells appear	
			3100–3500	First prokaryotic cells in stromatolites appear	
			4600	Earth forms	

Figure 21.14

The outer ring of this diagram shows the history of the earth as it would be measured on a 24-hour time scale starting at midnight. (The inner ring shows the actual years starting at 4 1/2 billion years ago.) When the history of the earth is measured as if it had all happened in 24 hours, the Cambrian period does not start until 8 P.M.! This means that a very large portion of life's history was devoted to the evolution of single-celled organisms. The first multicellular organisms probably did not appear until just before 8 P.M., and humans were not on the scene until less than a minute before midnight.

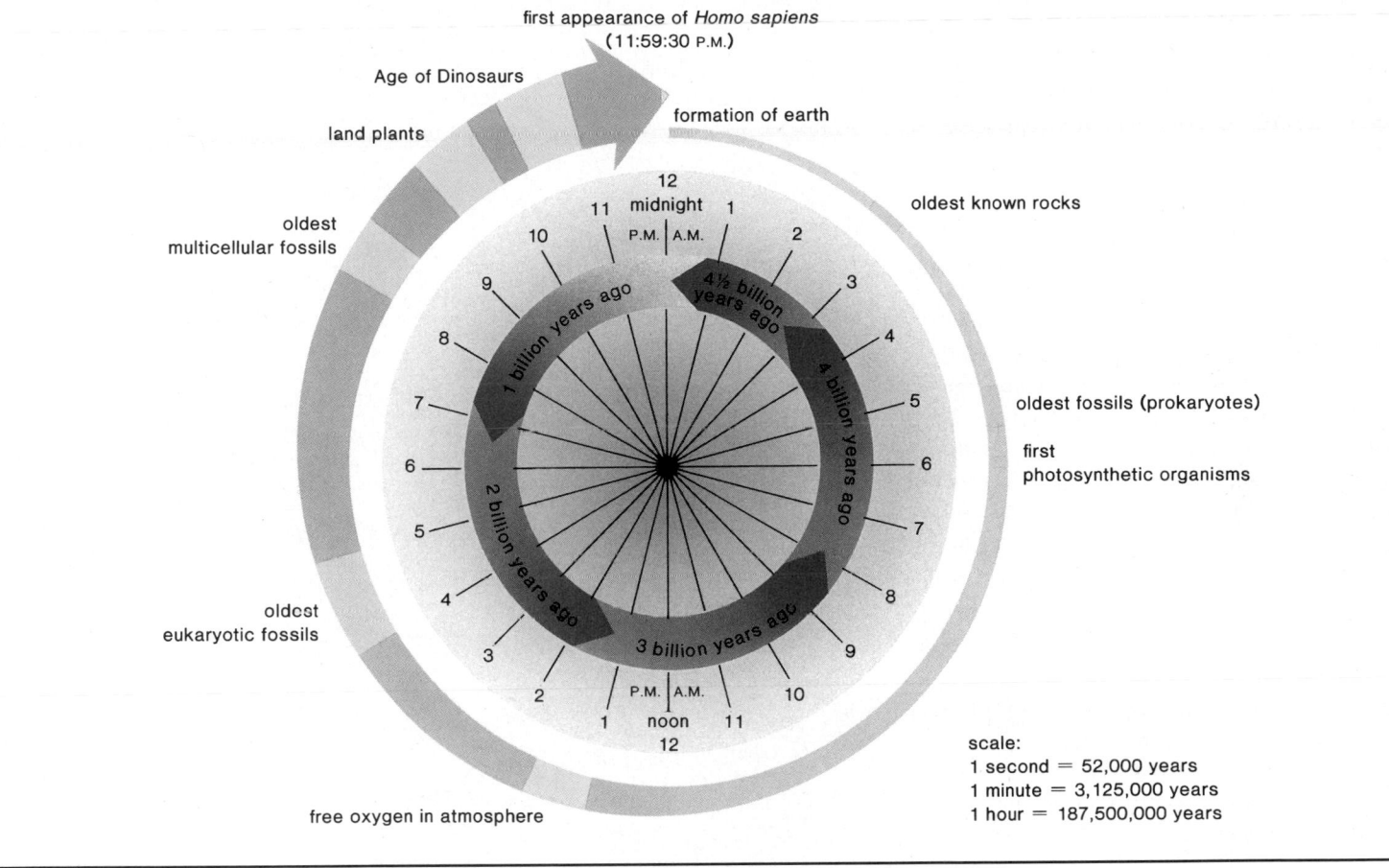

continents was greatly reduced. This is where most marine life is found; therefore, most marine species became extinct most likely because of habitat reduction. Due to the joining of the continents, land species that had been isolated from one another were competing for food and habitat. Matters were worsened by a climatic change due to altered oceanic currents.

During the early Mesozoic era, Pangaea began to break up and geographical isolation between continents occurred once more. As the continents drifted apart (fig. 21.15b–d), each became a separate evolutionary area, and the plants and the animals of the different continents diverged. Continental drift also helps explain the distribution of fossils on earth. Fossils of reptilian ancestors have a wider distribution than fossils of mammalian ancestors. This is because reptiles were evolving when Pangaea was forming, whereas mammals didn't start to diversify until after the continents started to move apart. Why were marsupials endemic only to Australia and South America

until recent times? It was possible for animals to migrate between South America and Australia by way of Antarctica when marsupials arose.

The fossil record tells the history of life on earth, which has been very much influenced by continental drift.

Patterns of Evolution

The fossil record shows various patterns of evolution (fig. 21.16), and we will discuss a few of these.

Cladogenesis

Cladogenesis (also called divergent evolution) occurs when a lineage splits and the 2 lines of descent give rise to new and different species. It is the process of cladogenesis that has resulted in the great diversity of past and living organisms. With adaptive radiation, diversification occurs as descendants of a single species become adapted to varied environments. The fossil record shows that amphibians, reptiles, and mammals underwent adaptive radiation in turn. The reptiles included some forms that were herbivores, some that were carnivores, some that could fly, and even some that returned to the water. Mammals arose while the reptiles were still dominant on earth. When the reptiles became extinct, mammals underwent adaptive radiation, moving into the environments once filled by the reptiles (fig. 21.17).

Convergent Evolution

Convergence occurs when 2 markedly different lineages give rise to organisms that have similar physical and functional traits because they are adapted to the same type of environment. Both euphorbia and cacti are adapted similarly to a hot, dry environment, and they both are succulent, spiny, flowering plants (fig. 21.18). Botanists, however, know they are not closely related because the details of flower structure are very different. Similarity due to convergence results in analogous structures (not homologous structures).

Anagenesis and Stasis

The fossil record often shows anagenesis (also called phyletic evolution)—evolutionary change that occurs gradually. On occasion, however, some lineages don't seem to change at all; this is called stasis. The horseshoe crab (*Limulus polyphemus*) and the bony fish called the coelacanth (*Latimeria,* see fig. 28.6*a*), are known as living fossils because they have changed so little for so great a period of time.

Evolutionary Trends

Sometimes it seems as if there is an overall evolutionary trend in a lineage. For example, during the phylogenetic history of *Equus,* we can note (1) increased size, (2) increased grinding surface of the molar teeth, and (3) a reduced number of toes (fig. 21.19). The history begins with *Hyracotherium* (known colloquially as *Eohippus*), a dog-sized mammal with a small head and small, low-crowned molars with cusps. The feet, with 4 toes on each front foot and 3 toes on each hind foot, were padded. *Hyracotherium* was adapted to the forestlike environment of the Eocene, an epoch of the Tertiary period. This small animal could have hidden among the trees for protection, and the low-crowned teeth were appropriate for browsing on leaves. In the Miocene and Pliocene epochs, however, grasslands began to replace the forests. Now the ancestors of the modern horse, *Equus,* were subjected to selective pressure for the development of strength, intelligence, speed, and

Figure 21.15

History of the placement of today's continents. *a.* About 200-250 million years ago, all the continents were joined into a supercontinent call Pangaea. *b.* When the joined continents of Pangaea first began moving apart, there were 2 large continents called Laurasia and Gondwanaland. *c.* By 65 million years ago, all the continents had begun to separate. This process is continuing today. *d.* North America and Europe are presently drifting apart at a rate of about 2 cm per year.

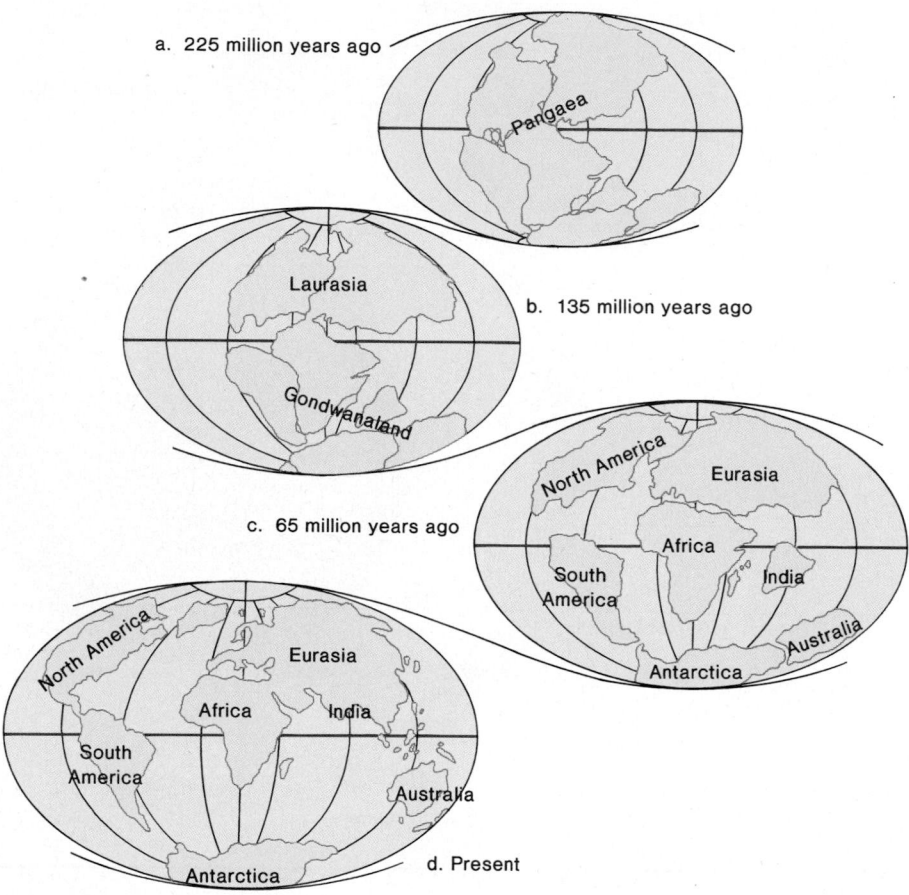

durable grinding teeth. A large size provided the strength needed for combat, a large skull made room for a larger brain, elongated legs ending in hooves gave speed to escape enemies, and the durable grinding teeth enabled the animals to feed efficiently on grasses.

By picking and choosing only certain fossils, it is possible to trace the lineage of the modern horse as if only anagenesis were involved. Closer examination, however, reveals that as *Hyracotherium* evolved into *Equus,* there were also periods of adaptive radiation. Only by choosing certain fossils does the evolution of the horse seem to be one of just anagenesis.

A study of the fossil record reveals certain patterns of evolutionary change. Divergent evolution and convergent evolution are 2 of these. Evolutionary trends are also seen.

Figure 21.16

Two patterns of evolution. *a.* During divergent evolution, one lineage splits and the resultant species have different adaptations. Adaptive radiation occurs when many splittings occur within a relatively short period of time. Divergent evolution accounts for the diversity of life. The species in these lineages will have homologous structures. *b.* During convergent evolution, different lineages produce species that have similar adaptations. The species in these lineages will most likely have analogous structures.

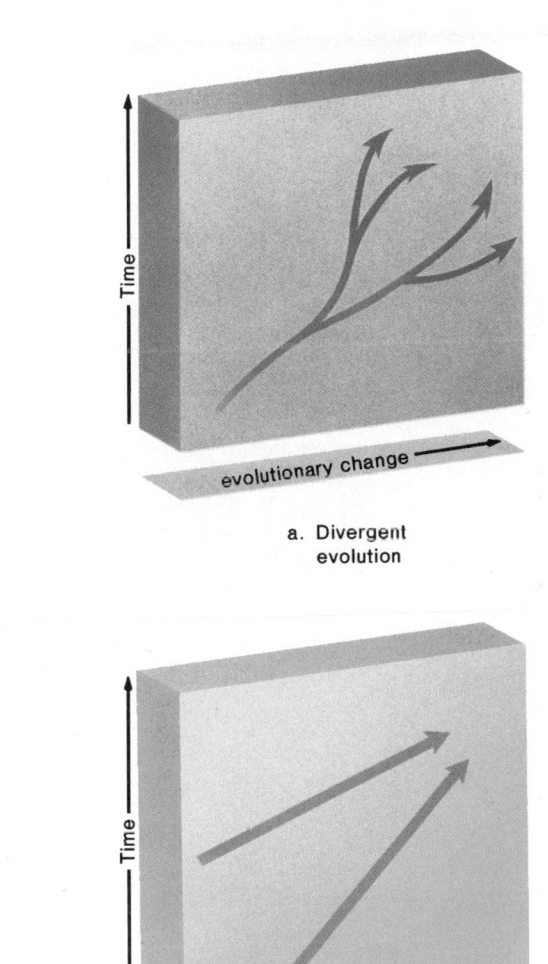

a. Divergent
evolution

b. Convergent
evolution

Figure 21.17

Adaptive radiation among reptiles and mammals. Both reptiles and mammals evolved into the types of forms shown; however, the mammals did not begin their adaptive radiation until the reptiles had declined. Therefore, each group shows similar adaptations because they were subjected to the same selective pressures.

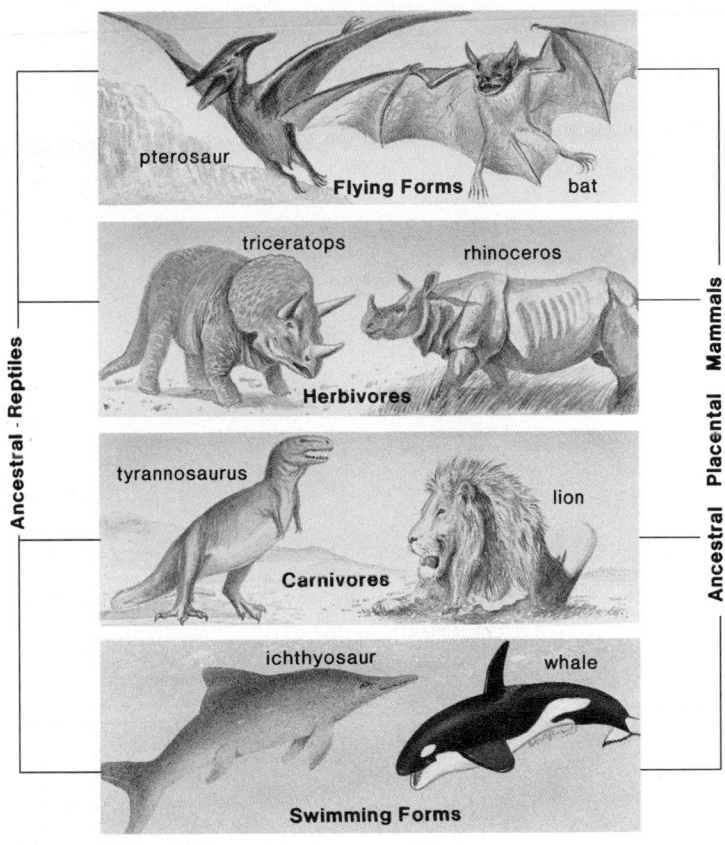

Figure 21.18

Cacti and euphorbia evolved on different continents, and yet they are both succulent flowering plants with spines. Cacti are adapted to living in North American deserts, and euphorbia are adapted to living in African deserts.

Figure 21.19

Evolution of the horse during the Tertiary and Quaternary periods (table 21.3). The modern horse evolved from the dog-sized *Hyracotherium,* but there were a number of adaptive radiations along the way. The other lineages became extinct, and only *Equus* is alive today. Whereas *Hyracotherium* was adapted to a forest environment, *Equus* is adapted to a grassland environment.

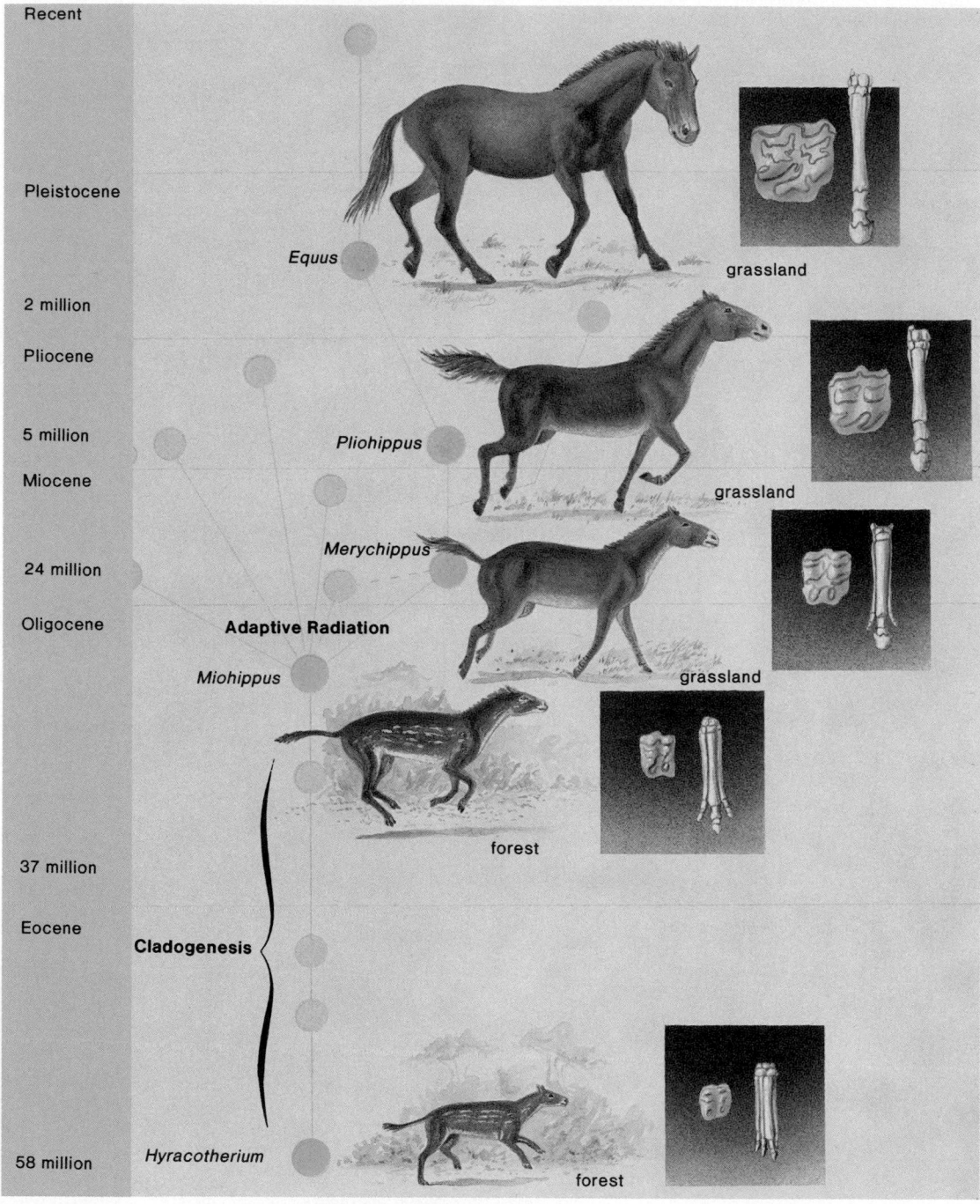

Evolution and Diversity

Summary

1. The biological definition of a species recognizes that populations of the same species breed only among themselves and are reproductively isolated from other species.

2. Reproductive isolating mechanisms prevent gene flow between species. Premating isolating mechanisms (habitat, temporal, behavioral, and mechanical) prevent mating from being attempted. Postmating isolating mechanisms (gamete isolation and zygote mortality, hybrid sterility, and F_2 fitness) prevent hybrid offspring from surviving and/or reproducing.

3. Allopatric speciation requires geographic isolation before reproductive isolation occurs. Adaptive radiation, as exemplified by the Galápagos finches, is a form of allopatric speciation. It occurs because the opportunity exists for new species to adapt to new habitats.

4. Sympatric speciation does not require geographic isolation for reproductive isolation to develop. The occurrence of polyploidy in plants is an example of this type of speciation.

5. Macroevolution is the study of evolutionary change above the species level. Members of a species share phenotypic similarities because species are reproductively isolated from one another.

6. The science of taxonomy recognizes certain categories of classification. A taxon is a named category.

7. Systematics is classification of organisms in keeping with their phylogenetic history. Systematists use both similarities (e.g., homologous structures) and differences (e.g., divergent evolution) both to classify organisms and to construct phylogenetic trees.

8. An analysis of protein structure or DNA structure can show how closely related 2 lineages are. The more differences in amino acid sequence or DNA nucleotide sequences there are, the longer ago they shared a common ancestor. This information can be used to construct phylogenetic trees, which serve as a check on those constructed by conventional means.

9. Cladistics is classification of organisms according to their shared similarities. A taxon includes the common ancestor who introduced a unique structure and all the descendants who share this and other similarities. Divergent evolution is not taken into account in this method of classification.

10. The fossil record provides evidence of history of life in general and the lineages of organisms in particular. The earth is about 4.6 billion years old, and the history of life began about 3.5 billion years ago. Because multicellular forms didn't appear until about 700 million years ago, most of life's history was devoted to the evolution of single-celled organisms. Humans do not appear in the record until quite recently (2-3 million years ago).

11. The continents are on massive plates that move, carrying the continents with them. This is called continental drift. Mammals didn't start to diversify until the supercontinent Pangaea began to break up. Therefore, each continent has its own particular types of mammals.

12. The fossil record gives evidence of mass extinctions, 2 of which have been studied in detail. The extinction at the end of the Permian period, the time when Pangaea was forming, was catastrophic. Loss of habitat caused the extinction of many forms of marine life. The extinction at the end of the Cretaceous period, when the dinosaurs died out, may have been due to asteroid or comet bombardment.

13. The fossil record reveals many patterns of evolution. Two significant ones are cladogenesis (divergent evolution) and convergent evolution. Divergent evolution accounts for the diversity of life, such as occurs during adaptive radiation. These diverse life forms share homologous structures. Convergent evolution occurs when 2 lineages produce species that are similarly adapted. These species will most likely have analogous structures.

14. When the phylogenetic history of one species is studied, trends may be discerned. In the phylogenetic history of the horse, there was an increase in size, the development of teeth suitable to grazing, and a reduction in number of toes to only one toe. The presence of trends might make it seem as if only anagenesis occurred; however, adaptive radiation also occurred during the phylogenetic history of the horse.

Writing Across the Curriculum

In order to practice writing skills, students should write out the answers to any or all of the study questions and the critical thinking questions. The study questions are sequenced in the same order as the text. Suggested answers to the critical thinking questions are in appendix D.

Study Questions

1. What is the biological definition of a species?
2. What is a reproductive isolating mechanism? Give examples of both premating and postmating isolating mechanisms.
3. How does allopatric speciation occur? sympatric speciation?
4. What is taxonomy? What are the categories of classification?
5. What is the relationship between systematics and phylogenetic trees?
6. What is a cladogram, and how does a cladogram differ from a phylogenetic tree?
7. In general terms, describe the history of life on earth as revealed by the fossil record.
8. What is continental drift, and how has it affected the history of life on earth?
9. What are some of the patterns of evolution that are seen in the fossil record?
10. In general terms, describe the history of *Equus*, and relate the history to anagenesis and adaptive radiation.

Objective Questions

1. The biological definition of a species depends on
   a. anatomical and developmental differences between 2 groups of organisms.
   b. the geographic distribution of 2 groups of organisms.
   c. differences in the adaptations of 2 groups of organisms.
   d. reproductive isolation of 2 groups of organisms.

2. Which of these is a premating isolating mechanism?
   a. habitat isolation
   b. temporal isolation
   c. gamete isolation
   d. Both a and b.

3. Male moths recognize females of their species by sensing chemical signals called pheromones. This is an example of
   a. gamete isolation.
   b. habitat isolation.
   c. behavioral isolation.
   d. mechanical isolation.

4. Allopatric but not sympatric speciation requires
   a. reproductive isolation.
   b. geographic isolation.
   c. hybridization first.
   d. spontaneous differences in males and females.

5. The many species of Galápagos finches were each adapted to eating different foods. This is an example of
   a. anagenesis.
   b. adaptive radiation.
   c. sympatric speciation.
   d. All of these.

6. Phylogenetic trees are based on
   a. fossil evidence.
   b. anatomical and developmental evidence.
   c. biochemical evidence.
   d. All of these.

7. Which best describes strata?
   a. sedimentary rock layers that contain fossils
   b. sedimentary rock layers that all date from the same historical time
   c. molten rock that contains radioactive material and is dangerous to the health of humans
   d. All of these.

8. Continental drift helps explain the occurrence of
   a. mass extinctions.
   b. distribution of fossils on earth.
   c. geological upheavals like earthquakes.
   d. All of these.

9. Which of these is mismatched?
   a. cladogenesis—divergent evolution
   b. convergence—recent common ancestor
   c. anagenesis—gradual change
   d. adaptive radiation—divergent evolution

10. The following diagram represents one species. (a) Use the labels population and gene flow where appropriate. (b) What changes would you make to this diagram in order to symbolize 2 species?

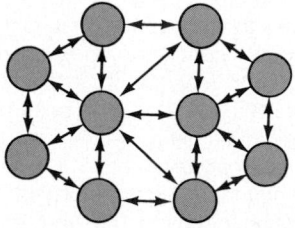

Concepts and Critical Thinking

1. *The diversity of life is dependent upon the process of speciation.*

   If only new species evolve, then of what use are the higher taxa (classification categories)?

2. *Every group of organisms has a phylogenetic history.*

   Why would you expect every group of organisms to have a phylogenetic history?

3. *Adaptation to the local environment is the driving force behind evolutionary change.*

   How do you recognize adaptive radiation in the phylogenetic history of a group of organisms?

Selected Key Terms

speciation (spe"se-a'shun) 334
allopatric speciation (al"o-pat'rik spe"se-a'shun) 334
sympatric speciation (sim-pat'rik spe"se-a'shun) 334
adaptive radiation (ah-dap'tiv ra-de-a'shun) 336
macroevolution (mak"ro-ev"o-lu'shun) 340
systematics (sis-tem-mat'iks) 340
phylogenetic tree (fi'lo-jĕ-net-ik trē) 340

22

Origin of Life

Field researchers examine rocks near North Pole, Australia, for evidence of the earth's earliest forms of life.

Your study of this chapter will be complete when you can

1. trace the steps by which chemical evolution produced the protocell;
2. describe any supporting experiments or data that support the occurrence of chemical evolution;
3. describe the differences between the protocell and a true cell;
4. explain why the protocell likely was a heterotrophic fermenter;
5. describe how genes may have first formed;
6. give dates for the oldest prokaryotic fossils and the evolution of the eukaryotic cell;
7. state the significance of a long prokaryotic evolution;
8. explain how the presence of atmospheric oxygen changed the world so that now life comes only from life;
9. discuss the importance of multicellularity and the evolution of sexual reproduction;
10. explain the 5-kingdom system of classification, and discuss the criteria by which organisms are placed in these kingdoms.

Figure 22.1

A model for the origin of life.

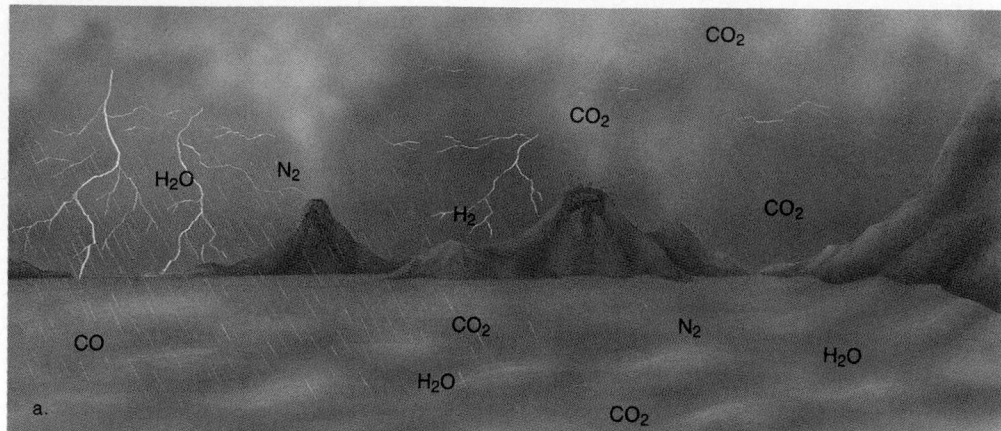

a. The primitive atmosphere contained gases, including water vapor that escaped from volcanoes; as the latter cooled, some gases were washed into the ocean by rain.

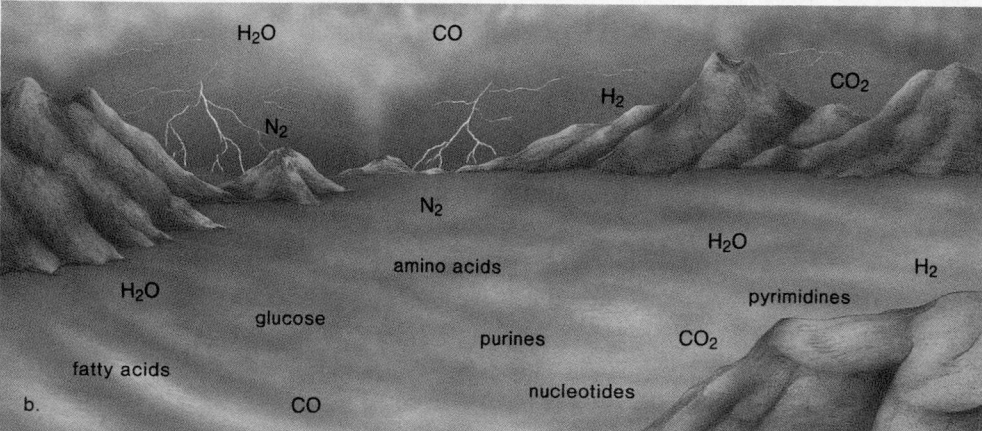

b. The availability of energy from volcanic eruption and lightning allowed gases to form simple organic molecules.

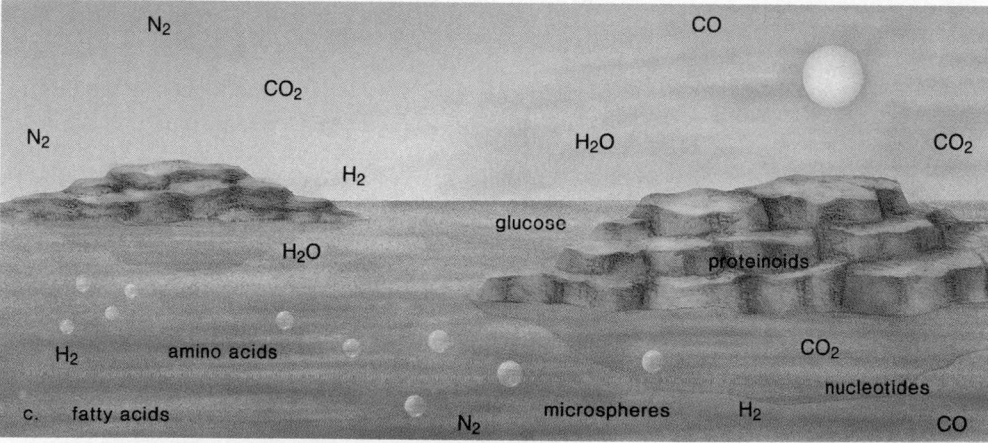

c. Amino acids that splashed up onto rocky coasts could have polymerized into polypeptides (proteinoids) that became microspheres when they reentered the water.

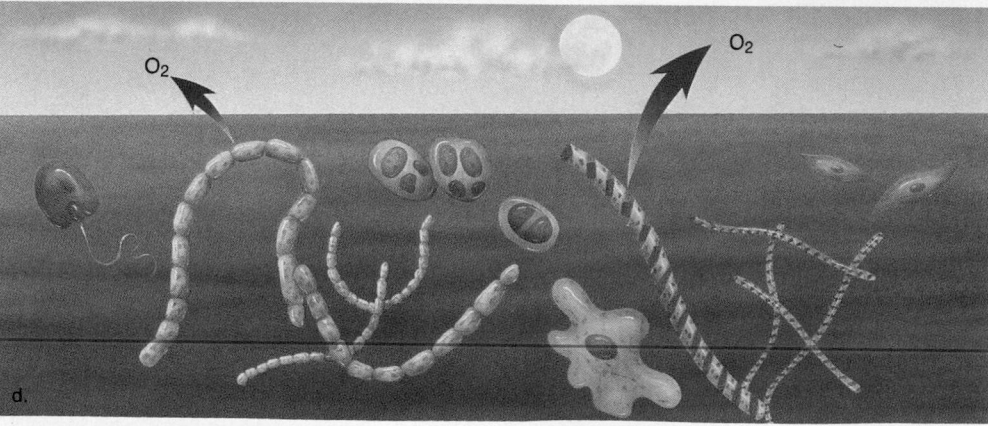

d. Eventually, various types of prokaryotes and then eukaryotes evolved. Some of the prokaryotes were oxygen-producing photosynthesizers.

Today we do not believe that life arises spontaneously from nonlife, and we say that "life comes only from life." But if this is so, how did the first form of life come about? We can assume that the first form of life was a single cell (or cells) that could grow, reproduce, and mutate. Since it was the very first living thing, it had to come from nonliving chemicals. Could there have been an increase in the complexity of the chemicals—could a **chemical evolution** have produced the first cell(s)? This chapter will review the evidence for such a chemical evolution; it is based on our knowledge of the primitive earth (fig. 22.1) and on experiments that have been performed in the laboratory.

Chemical Evolution

The sun and the planets probably formed from aggregates of dust particles and debris about 4.6 billion years ago. Intense heat produced by gravitational energy and radioactivity caused the earth to become stratified into several layers. Heavier atoms of iron and nickel became the molten liquid core, and dense silicate minerals became the semiliquid mantle. The unstable mantle caused the thin crust to move continually, so there were no stable landmasses during the time life was evolving.

Primitive Atmosphere

The *primitive atmosphere* was not the same as today's atmosphere. Alexander Oparin, a Soviet biochemist, proposed in the 1920s that the earliest atmosphere contained a lot of hydrogen (H_2) because it is the most abundant element in our Solar System. Later, it was suggested that lightweight atoms, including hydrogen, would have been lost as the earth formed because its gravitational field was not strong enough to hold them. It is now thought that the primitive atmosphere was produced after the earth formed by outgassing from the interior, particularly by volcanic action. In that case, the primitive atmosphere would have consisted mostly of water vapor (H_2O), nitrogen (N_2), and carbon dioxide (CO_2), with only small amounts of hydrogen (H_2) and carbon monoxide (CO). The primitive atmosphere, with little if any free oxygen, was a *reducing atmosphere* as opposed to the *oxidizing atmosphere* of today. This was fortuitous because oxygen (O_2) attaches to organic molecules, preventing them from joining to form larger molecules.

The primitive atmosphere was a reducing atmosphere that contained little if any oxygen gas (O_2).

At first the earth was so hot that water was present only as a vapor that formed dense, thick clouds. Then as the earth cooled, water vapor condensed to liquid water, and rain began to fall (fig. 22.1*a*). It rained in such quantity that the oceans of the world were produced.

Small Organic Molecules Evolve

The atmospheric gases, dissolved in rain, were carried down into newly forming oceans. The remaining steps shown in table 22.1 took place in and about the ocean, where life may have originated.

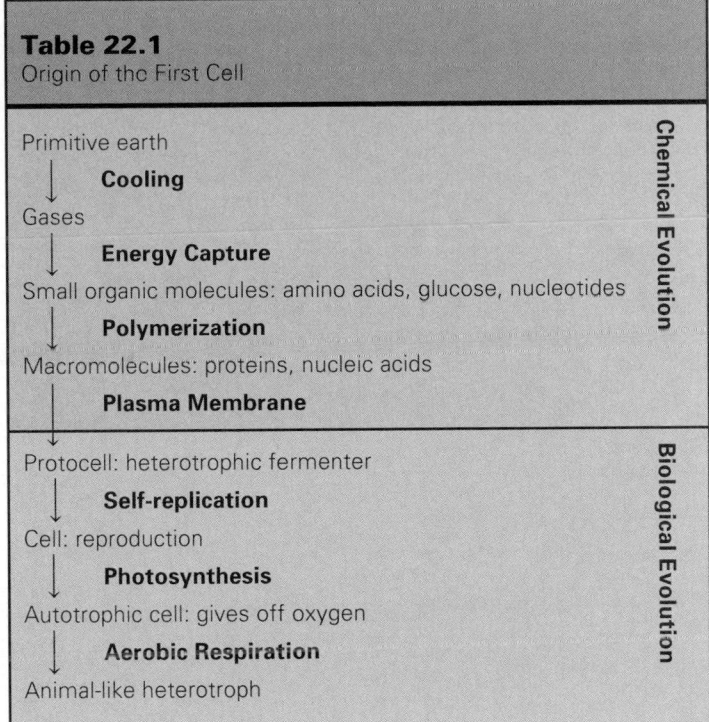

Table 22.1
Origin of the First Cell

Chemical Evolution
Primitive earth
↓ **Cooling**
Gases
↓ **Energy Capture**
Small organic molecules: amino acids, glucose, nucleotides
↓ **Polymerization**
Macromolecules: proteins, nucleic acids
Plasma Membrane

Biological Evolution
Protocell: heterotrophic fermenter
↓ **Self-replication**
Cell: reproduction
↓ **Photosynthesis**
Autotrophic cell: gives off oxygen
↓ **Aerobic Respiration**
Animal-like heterotroph

Oparin also suggested that organic molecules could have been produced from the gases of the primitive atmosphere in the presence of strong outside *energy sources* present on the primitive earth. These energy sources included heat from volcanoes and meteorites, radioactivity from the earth's crust, powerful electric discharges in lightning, and solar radiation, especially ultraviolet radiation (fig. 22.1*b*).

In 1953, Stanley Miller provided support for Oparin's ideas through an ingenious experiment (fig. 22.2). Miller placed a mixture resembling a strongly reducing primitive atmosphere [methane (CH_4), ammonia (NH_3), hydrogen (H_2), and water (H_2O)] in a closed system, heated the mixture, and circulated it past an electric spark (simulating lightning). After a week's run, Miller discovered that a variety of amino acids and organic acids had been produced. Since that time, other investigators have achieved similar results by utilizing other, less-reducing combinations of gases dissolved in water.

These experiments support the hypothesis that the primitive gases did react with one another to produce small organic compounds that accumulated in the ancient ocean. Neither oxidation (there was no free oxygen) nor decay (there were no bacteria) would have destroyed these molecules, and they would have accumulated in the oceans for hundreds of millions of years. With the accumulation of these small organic compounds, the oceans became a thick, hot **organic soup** containing a variety of organic molecules.

Cooling caused water vapor to turn to rain, which formed the oceans. It's possible that atmospheric gases reacted with one another under the influence of an outside energy source to produce simple organic molecules.

Figure 22.2

In Miller's experiment, gases (CH_4, methane; NH_3, ammonia; H_2, hydrogen; H_2O, water) were admitted to the apparatus, circulated past an energy source (electric spark), and cooled to produce a liquid that could be withdrawn. Upon chemical analysis, the liquid was found to contain various small organic molecules.

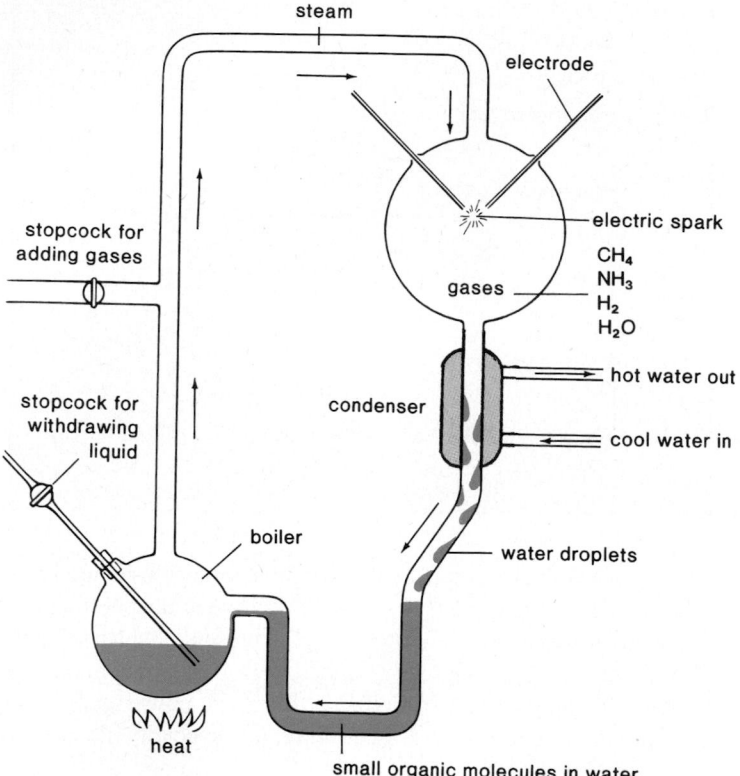

Macromolecules

The newly formed organic molecules likely polymerized to form still larger molecules and then macromolecules (fig. 22.1c). There are 3 primary hypotheses concerning this stage in the origin of life. One is the *RNA-first hypothesis,* which suggests that only the macromolecule RNA was needed at this time to progress toward formation of the first cell or cells. Thomas Cech (University of Colorado) and Sidney Altman (Yale University), shared a Nobel Prize in 1989 because they discovered that during transcript processing (p. 255), RNA is both a substrate and an enzyme. Other types of ribozymes (RNA enzymes) have been discovered to be involved in translation. It would seem, then, that RNA could have carried out the processes of life commonly associated today with DNA (the genetic material) and proteins (enzymes). Some viruses today have RNA genes; therefore, the first genes could have been RNA. And the first enzymes also could have been RNA molecules, since we now know that ribozymes exist. Those who support this hypothesis say that it was an "RNA world" some 4 billion years ago.

Another hypothesis is termed the *protein-first hypothesis.* Sidney Fox, now at Southern Illinois University, has shown that amino acids polymerize abiotically when exposed to dry heat. He suggests that amino acids collected in shallow puddles along the rocky shore and the heat of the sun caused them to form **proteinoids,** small polypeptides that have some catalytic properties. When proteinoids are returned to water, they form **microspheres,** structures composed only of protein that have many properties of a cell. It's possible that the first polypeptides had enzymatic properties and some proved to be more capable than others. Those that led to the first cell or cells had a selective advantage. This hypothesis assumes that DNA genes came after protein enzymes arose. After all, it is protein enzymes that are needed for DNA replication.

The third hypothesis is put forth by Graham Cairns-Smith of Glasgow University. He believes that clay was especially helpful in causing polymerization of both proteins and nucleic acids at the same time. Clay attracts small organic molecules and contains iron and zinc, which may have served as inorganic catalysts for polypeptide formation. In addition, clay has a tendency to collect energy from radioactive decay and to discharge it when the temperature and/or humidity changes. This could have been a source of energy for polymerization to take place. Cairns-Smith suggests that RNA nucleotides and amino acids became associated in such a way that polypeptides were ordered by and helped synthesize RNA. It is clear that this hypothesis suggests that both polypeptides and RNA arose at the same time.

Small organic molecules polymerized to produce macromolecules. This could have occurred on heated rocks or in clay.

Plasma Membrane

Before the first true cell arose, there would have been a **protocell,** a structure that has a lipid-protein membrane and carries on energy metabolism (fig. 22.3). Sidney Fox has shown that if lipids are made available to microspheres, they tend to become associated with them so that the outer boundary is a lipid-protein membrane. Phospholipid molecules automatically form droplets called **liposomes** in a liquid environment. Perhaps the first membrane formed in this manner. In that case, the protocell could have contained only RNA, which functioned as both genetic material and enzymes.

Some researchers support the work of Oparin, who was one of the original researchers in this area. As early as 1938, Oparin showed that under appropriate conditions of temperature, ionic composition, and pH, concentrated mixtures of macromolecules tend to give rise to complex units called *coacervate droplets.* Coacervate droplets have a tendency to absorb and incorporate various substances from the surrounding solution. Eventually, a semipermeable-type boundary may form about the droplet.

Heterotroph Hypothesis

The protocell would have had to carry on nutrition so that it could grow. Nutrition was no problem because the protocell existed in the ocean, which at that time was an organic soup containing simple organic molecules that could have served as building blocks and energy. Therefore, the protocell likely was a **heterotroph,** an organism that takes in preformed food. Notice that this suggests that heterotrophs are believed to have preceded **autotrophs,** organisms that make their own food.

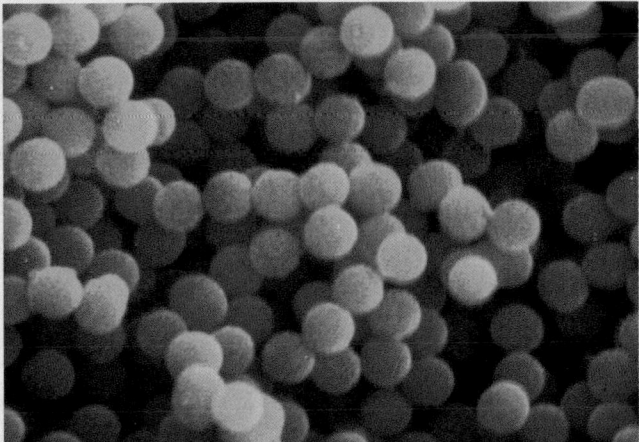

a.

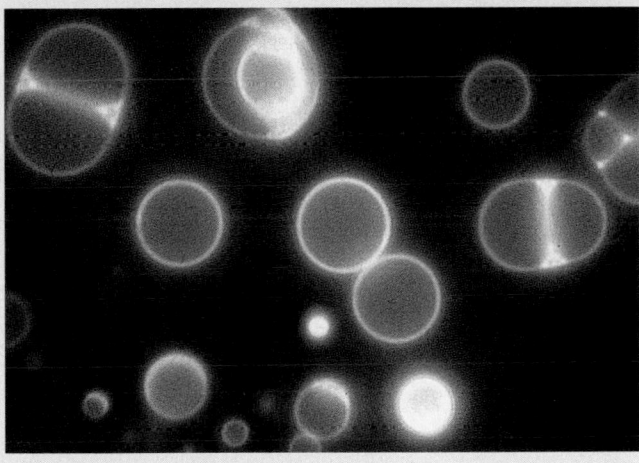

b.

At first, the protocell may have used preformed ATP, but as this supply dwindled, natural selection favored any cells that could extract energy from carbohydrates in order to transform ADP to ATP. Glycolysis is a common metabolic pathway in living things, and this testifies to its early evolution in the history of life. Since there was no free oxygen, we can assume that the protocell carried on a form of fermentation.

It seems logical that the protocell at first had limited ability to break down organic molecules and that it took millions of years for glycolysis to evolve completely. It is of interest that Fox has shown that a microsphere from which the protocell may have evolved has some catalytic ability and that Oparin found that coacervates do incorporate enzymes if they are available in the medium.

The protocell is hypothesized to have had a membrane boundary and to have been a heterotrophic fermenter with some degree of enzymatic ability.

Biological Evolution

Once a protocell arose, biological evolution would have begun. Biological evolution depends on the presence of genes that can replicate, mutate, and control the characteristics of the cell.

First Replication System

A true cell is a membrane-bounded structure that can self-replicate. It contains genetic material and has enzymes composed of proteins. All cells today have genes composed of DNA, and indeed the central dogma of genetics states that the flow of information is always from DNA → RNA → protein. It is possible that this sequence developed in stages.

According to the RNA-first hypothesis, RNA would have been the first to evolve and the first true cell would have had RNA genes. These genes would have directed and enzymatically carried out protein synthesis. As mentioned, RNA enzymes called ribozymes have been discovered. Also, today we know there are viruses that have RNA genes. These viruses have a protein enzyme called reverse transcriptase that uses RNA as a template to form DNA. Perhaps with time, reverse transcription occurred within the protocell, and this is how DNA genes arose. Once there were DNA genes, then protein synthesis would have occurred in the usual manner.

According to the protein-first hypothesis, proteins, or at least polypeptides, were the first of the 3 (i.e., DNA, RNA, and protein) to arise. Only after the protocell developed sophisticated enzymes did it have the ability to synthesize DNA and RNA from small molecules provided by the ocean, an organic soup. Researchers point out that a nucleic acid is a very complicated molecule, and the likelihood that RNA arose *de novo* (on its own) is minimal. It seems more likely that enzymes were needed to guide the synthesis of nucleotides and then nucleic acids.

Cairns-Smith proposes that polypeptides and RNA evolved simultaneously, so the first true cell would have contained RNA genes that could have replicated because of the presence of proteins. This eliminates the baffling chicken-and-egg paradox: which came first, proteins or RNA? Are 2 uncertain events, however, more certain when they occur simultaneously? Certainly not; therefore, Cairns-Smith's hypothesis is not attractive to many at this time.

Once the protocells acquired genes, they became true cells, capable of evolving into other types of cells. Since all organisms today use the same replication system and enzymatic pathways, it is likely that they have a common origin from the first cell or cells.

Prokaryotic Cells

The first fossils, dated about 3.5 billion years ago, are presumed to be prokaryotic cells. Some of these are found in fossilized *stromatolites* (*stroma*—bed; *lithos*—stone), which are pillarlike structures

Figure 22.4

The oldest prokaryotic fossils date from 3.5 billion years ago. **_a._** This fossil is a string of cells resembling cyanobacteria of today. Magnification X500. **_b._** Prokaryotic fossils are also found in the layers of fossilized stromatolites. **_c._** Living stromatolites are located in shallow waters off the shores of western Australia.

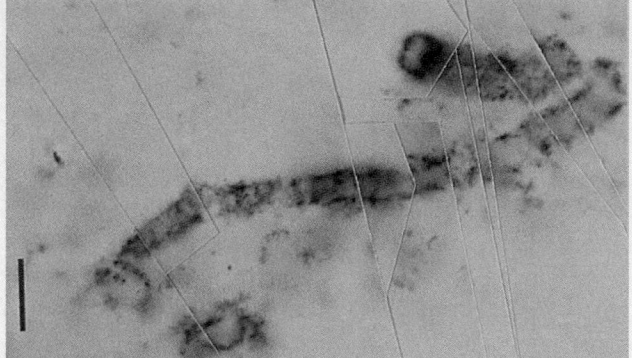

a.

b.

c.

composed of sedimentary layers containing communities of photosynthetic microorganisms (fig. 22.4). Stromatolites containing living cells exist even today in shallow waters off the west coast of Australia.

Autotrophism most likely arose because once the supply of organic molecules in the organic soup began to decline, natural selection favored cells that could use solar energy to produce carbohydrates. Some of the first autotrophs may have given off oxygen as they photosynthesized.

> Some early prokaryotes are believed to have been photosynthesizers.

Oxidizing Atmosphere

Once oxygen-producing photosynthesizers (cyanobacteria) evolved, the atmosphere became an oxidizing one instead of a reducing one (fig. 22.1d). This is an example of the profound effect that life has on the physical condition of the earth. As discussed in this chapter's reading, there are those who have been so struck with this realization they have formulated a new hypothesis. The Gaia hypothesis states that living things function in such a way as to preserve the earth for life.

In contrast to a reducing atmosphere, which promotes the buildup of organic molecules, an oxidizing one tends to break down organic molecules. Also, oxygen in the upper atmosphere forms ozone (O_3), which filters out the ultraviolet rays of the sun. Before the formation of this **ozone shield,** the amount of ultraviolet radiation reaching earth could possibly have been used to create new molecules. Because of oxygen's effects on the atmosphere, it is believed that today there could not be a chemical evolution leading to the origin of life. Instead, as the cell theory states, there is substantial evidence that life comes only from life.

Although the ozone shield may help prevent the origin of life today, it does serve an important function. Without the ozone shield, life would never have been able to exist on land because the ultraviolet radiation reaching earth was sufficient to destroy land-dwelling organisms. This is why there is such concern today about pollutants, which act to break down the ozone shield (p. 801).

Evolution and Diversity

The Gaia hypothesis (after the classical Greek word for our Mother Earth) maintains that life keeps the climate and the chemical composition of the earth suitable for life. For example, photosynthesizing organisms put oxygen (O_2) into the atmosphere, and its level has been maintained about 20% for the last half-billion years. If, by chance, this level rises to 30%, there would be disastrous fires any time lightning occurred. What keeps the level so constant? Those who support the Gaia hypothesis point to the presence of bacteria that produce methane (CH_4), which combines with oxygen to form carbon dioxide (CO_2). It is these bacteria, they say, that keep atmospheric levels of oxygen suitable for all other living things.

To take another example, the sun has been heating up since its formation, yet the temperature of the earth always has been about the same. About 4 billion years ago, when the sun was perhaps 25% less luminous, the primitive atmosphere contained more ammonia (NH_3) and methane, which acted like the glass of a greenhouse to trap the heat of the sun near the earth. Today, carbon dioxide put into the atmosphere by all living things plays a more prominent role in creating a greenhouse effect. As the sun heats up, less and less of a greenhouse effect is needed to maintain a suitable temperature for life. How has this been regulated up until now? Those who support the Gaia hypothesis point out that there are microorganisms in the oceans with calcium carbonate ($CaCO_3$) shells (see fig. 22.A). When carbon dioxide enters an ocean, it becomes bicarbonate and is utilized by these organisms to form their shell. When these organisms die, their shell sinks to the bottom of the ocean; in this way, carbon dioxide in effect is removed from the atmosphere. This explanation is suitable for today's world, but what about the 2 billion years before phytoplankton evolved? Perhaps photosynthesizing bacteria played a role in removing carbon dioxide, but this has not been shown as yet.

While many biologists agree that living things play a role in maintaining suitable

Figure 22.A

Coccolithophore. This microorganism has calcium carbonate ($CaCO_3$) plates embedded in its cell wall and exemplifies marine organisms that take calcium carbonate out of the sea. Calcium carbonate forms after carbon dioxide (CO_2) enters from the air; therefore, this action helps to decrease the amount of carbon dioxide in the atmosphere. The sea is full of these tiny organisms that are only 60 µm in diameter (a human hair is 100–200 µm thick).

conditions for life on earth, they doubt there is anything purposeful about it. Quite often those who support the Gaia hypothesis see the earth as a giant geophysiological organism that is purposefully maintaining itself. In this way, the Gaia hypothesis seems more like a religion than a science, and therefore it would be difficult to devise ways to test the hypothesis.

Those who refute the Gaia hypothesis point out that inorganic processes also play a role in maintaining the chemical composition and the temperature of the earth. For example, oxygen no doubt failed to build up quickly in the atmosphere because much of it was taken up as iron was oxidized—there is much iron in the earth's crust. Also, carbon dioxide is washed out of the atmosphere by rain and in the process becomes carbonic acid. Carbonic acid combines with calcium silicate in rocks to give sediments containing carbon compounds. In other words, inorganic and organic processes have contributed to keeping the earth suitable for life.

Another argument against the Gaia hypothesis is that while some life forms have benefited from changed conditions, other forms have not. For example, when oxygen was put into the atmosphere, it may have helped aerobic organisms to evolve, but the populations of anaerobic organisms could only decline. The history of life contains many species of organisms that became extinct due to changing conditions. New ones that are more suited to the present conditions then evolve.

The Gaia hypothesis challenges us to see the earth at a geophysiological level of organization. Life is not passively acted upon by physical processes; it has contributed to changing and to shaping the world we now live in.

The evolution of photosynthesizing organisms caused oxygen to enter the atmosphere. Oxygen breaks down organic molecules and forms an ozone shield that protects the earth from ultraviolet radiation.

Eukaryotic Cells

The presence of oxygen in the atmosphere meant that most environments were no longer suitable for anaerobic prokaryotes, and they began to decline in importance. The photosynthetic cyanobacteria proliferated, and eventually the atmosphere contained a stable amount of oxygen. As figure 22.5 shows, a large portion of the history of life was devoted to prokaryotic evolution, during which time various metabolic pathways, including aerobic respiration, came into existence.

The eukaryotic cell, which evolved about 1.5 billion years ago, is nearly always aerobic and contains a nucleus and all the other organelles studied in chapter 5. Most likely the eukaryotic cell acquired its organelles gradually. It may be that the nucleus and at least a primitive form of mitosis arose once there were several chromosomes. As discussed on page 76, the mitochondria of the eukaryotic cell probably were once free-living aerobic prokaryotes, and the chloroplasts probably were free-living photosynthetic prokaryotes. The theory of endosymbiosis says that the nucleated cell engulfed these prokaryotes, which then became organelles. It's been suggested that flagella (and cilia) also arose by endosymbiosis. First, slender undulating prokaryotes could have attached themselves to a host cell in order to take advantage of food leaking from the host's outer membrane. Eventually, these prokaryotes were drawn inside the host cell and became the flagella and cilia we know today.

Prokaryotes were alone on the earth for 2 billion years, during which time many biochemical pathways evolved. The first eukaryotes evolved about 1.5 billion years ago.

Multicellularity

It is not known when multicellularity began but the very first multicellular forms were most likely microscopic. It's possible that the first multicellular organisms practiced sexual reproduction. Among protists today, we find colonial forms in which some cells are specialized to produce gametes needed for sexual reproduction. Separation of germ cells, which produce gametes from somatic cells, may have been an important first step toward the complex macroscopic animals that suddenly appeared 700 million years ago. It's also possible that the increase in complexity correlates with an increase in atmospheric oxygen at this time. Macroscopic animals with muscular tissue require a significantly higher concentration of oxygen than simpler forms. In any case the Cambrian era begins with a burst of diversity, particularly in animals.

Figure 22.5
Some of the major events that occurred during Precambrian history. The entire Precambrian period takes up seven-eighths of the history of life on earth! (BYA = billions of years ago)

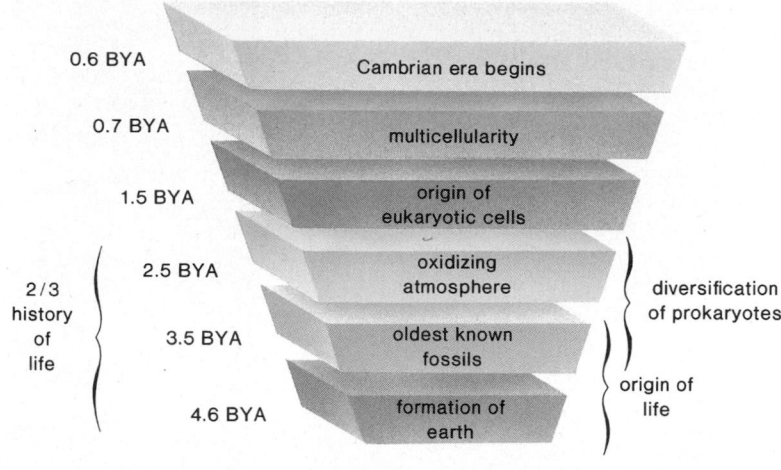

Diverse, complex organisms appear in the fossil record about 700 million years ago, most likely as a result of the evolution of multicellularity and sexual reproduction.

Five Kingdoms

The many diverse forms of life can be classified in various ways. This text recognizes 5 kingdoms, which differ according to the criteria in table 22.2. The prokaryotic bacteria are placed in the kingdom Monera (chap. 23). Their mode of nutrition is diverse, and all forms of nutrition are seen except ingestion. Remember that prokaryotes were evolving about 2 billion years before eukaryotes came on the scene. During this time they developed all sorts of metabolic pathways. Therefore, it is not surprising that some are autotrophic and are either chemosynthetic or photosynthetic. Some are heterotrophic and are either saprophytic or parasitic. A saprophyte breaks down dead organic matter and then absorbs the nutrient molecules.

The other kingdoms contain the eukaryotes. Single-celled eukaryotes and all the algae are placed in the kingdom Protista (chap. 24). Among the single-celled eukaryotes are the protozoans, which are heterotrophic by ingestion or else parasitic. The water molds, which resemble fungi, are saprophytic. The algae, of course, are photosynthetic.

The kingdoms Plantae, Animalia, and Fungi contain only multicellular forms. Fungi (chap. 24) such as molds and mushrooms are mainly saprophytic, although some are parasites. Only some bacteria and most fungi are saprophytes. They are the decomposers that release nutrients when they break down dead organic matter (fig. 1.5).

Plants (chap. 25), of course, are photosynthetic. They are distinguishable from algae because they are multicellular and adapted to a land environment. Notably, they protect the zygote from drying out by retaining it within an internal organ. Animals (chap. 26-28) are the only organisms with a nervous system. Typically, they move about in search of food and are heterotrophic by ingestion. Figure 22.6 diagrams the 5-kingdom system of classification, indicating in a general way how the organisms might be related.

Table 22.2
Classification Criteria for 5 Kingdoms

	Monerans	Protists	Fungi	Plants	Animals
Type of Cell	Prokaryotic	Eukaryotic	Eukaryotic	Eukaryotic	Eukaryotic
Complexity	Single cell	Single cell	Multicellular	Multicellular	Multicellular
Type of Nutrition	Autotrophic by various Heterotrophic by various	Photosynthetic Heterotrophic by various	Heterotrophic saprophytes	Photosynthetic	Heterotrophic by ingestion
Motility	Sometimes by flagella	Sometimes by flagella (or cilia)	Nonmotile	Nonmotile	Motile by contractile fibers
Life Cycle	Asexual usual	Various	Haplontic	Alternation of generations	Diplontic
Internal Protection of Zygote	No	No	No	Yes	Yes
Nervous System	None	Conduction of stimuli in some forms	None	None	Present

Summary

1. The unique conditions of the primitive earth allowed a chemical evolution to occur, which produced the protocell.
2. The gases of the primitive atmosphere probably escaped from volcanoes and included water vapor, nitrogen, and carbon dioxide, with small amounts of hydrogen and carbon monoxide.
3. The primitive atmosphere was a reducing atmosphere with no free oxygen gas.
4. At first the earth was very hot, but when it cooled, the water vapor condensed and the rains began to fall. The rains carried the gases into the newly formed oceans.
5. In the presence of an outside energy source, such as ultraviolet radiation, lightning, or radioactivity, the gases reacted with one another to produce small organic molecules such as amino acids and glucose. This scenario is supported by experiments performed by Stanley Miller and others.
6. All of these organic molecules made the oceans an organic soup.
7. The next step was the formation of macromolecules by polymerization. Three hypotheses are offered. The RNA-first hypothesis is supported by the discovery of ribozymes, RNA molecules that are present in today's cells and are both substrate and enzyme. The protein-first hypothesis is supported by the observation that amino acids will polymerize in a preferred fashion when heated to drying. Cairns-Smith has suggested that amino acid polymerization and nucleotide polymerization could have occurred in clay, where radioactivity would have supplied the necessary energy.
8. Polymerized amino acids, called proteinoids, become microspheres when placed in water. Microspheres have many properties similar to cells and may have evolved into protocells, structures that have a lipid-protein boundary and carry on energy metabolism.
9. Phospholipids readily form spherical liposomes, and perhaps this was the origin of the plasma membrane.
10. Many years ago, Oparin showed that coacervate droplets automatically form when concentrated mixtures of macromolecules are held at the right temperature, ionic composition, and pH.
11. The protocell must have been a heterotrophic fermenter living on the preformed organic molecules in the organic soup.
12. The protocell was not a true cell until it contained genes. The RNA-first hypothesis suggests that the first genes were RNA molecules. The protein-first hypothesis suggests that protein enzymes were needed before nucleic acids could form. Cairns-Smith suggests that polypeptides and RNA became associated in clay in such a way that the polypeptides catalyzed RNA formation. So, the first cell would have contained both genes and protein enzymes at the same time.
13. The earliest fossil evidence for cells, dated about 3.5 billion years ago, is found in fossilized stromatolites. These fossils are believed to be photosynthesizing bacteria. Eventually,

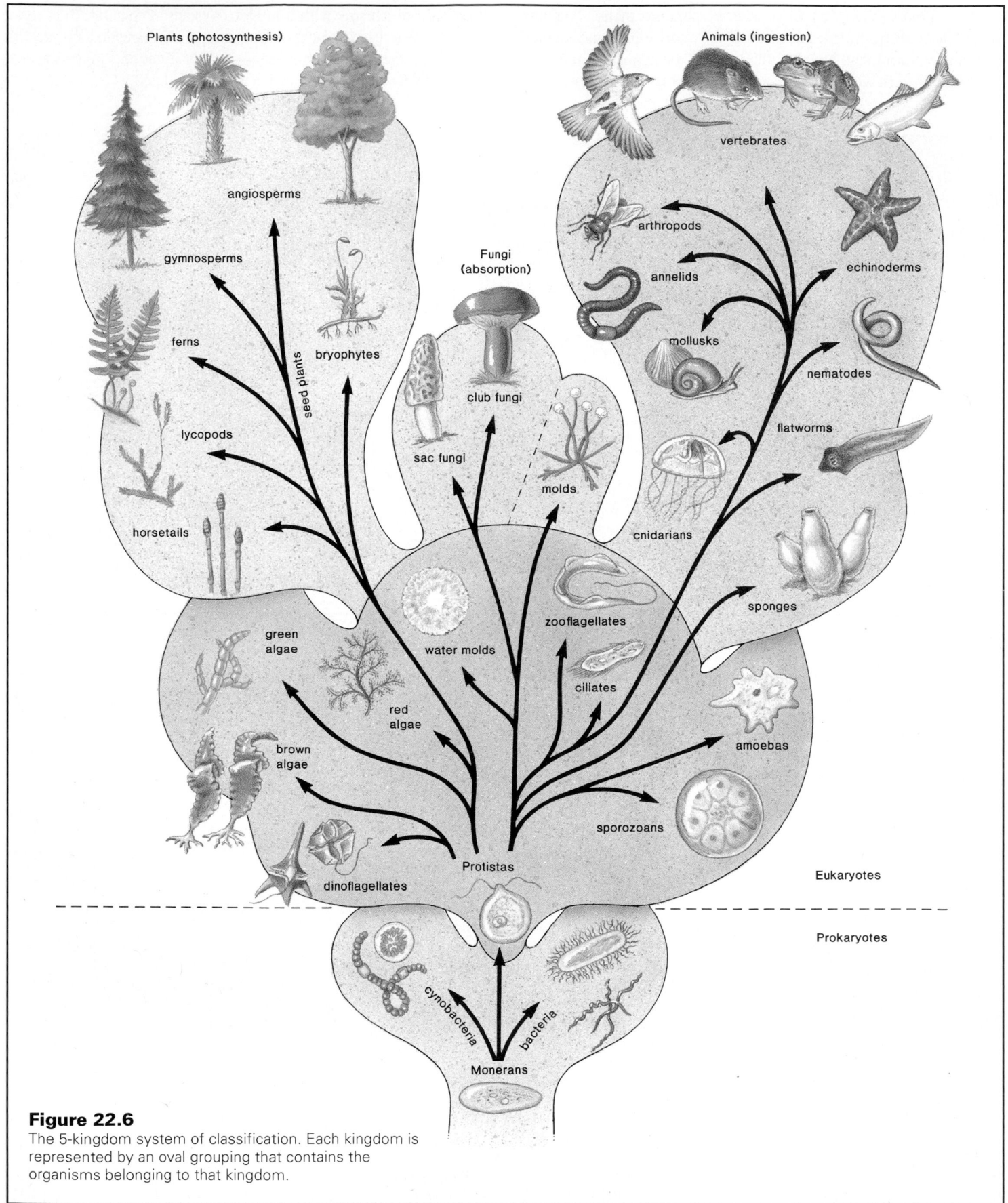

Plants (photosynthesis)

angiosperms

gymnosperms

ferns

bryophytes

seed plants

lycopods

horsetails

Fungi (absorption)

club fungi

sac fungi

molds

Animals (ingestion)

vertebrates

arthropods

annelids

echinoderms

mollusks

nematodes

flatworms

cnidarians

sponges

green algae

water molds

zooflagellates

ciliates

red algae

amoebas

brown algae

sporozoans

Protistas

dinoflagellates

Eukaryotes

Prokaryotes

cynobacteria

bacteria

Monerans

Figure 22.6

The 5-kingdom system of classification. Each kingdom is represented by an oval grouping that contains the organisms belonging to that kingdom.

the presence of oxygen in the atmosphere changed it from a reducing to an oxidizing one.

14. By 1.5 billion years ago, aerobic respiration had evolved and so had eukaryotes. Prokaryotes existed alone on the surface of the earth for at least two billion years. During this time all the metabolic pathways found in modern cells developed.

15. The presence of oxygen in the atmosphere also allowed the formation of the ozone shield, which filters out ultraviolet rays. This, along with the fact that oxygen attaches to small organic molecules, accounts for the inability of protocells to originate by chemical evolution today.

16. Multicellularity and sexual reproduction arose about 700 million years ago, and the Cambrian record has a rich supply of complex life forms.

17. The 5-kingdom system of classification utilizes various criteria to place organisms in the kingdoms Monera, Protista, Fungi, Plantae, and Animalia (see table 22.2 and figure 22.6).

Study Questions

1. Trace the steps by which a chemical evolution may have produced a protocell.
2. Describe Stanley Miller's experiment, and discuss its significance.
3. Contrast the RNA-first hypothesis and the protein-first hypothesis. If polymerization occurred in clay, what macromolecules would have resulted at the same time?
4. How might a plasma membrane have evolved? What is a protocell?
5. Why is it likely the protocell was a heterotrophic fermenter?
6. How might the first replication system have come about? What is a true cell?
7. The earliest fossils are found in stromatolites. What kind of fossils are they? How old are they? What are stromatolites?
8. Relate the diversity of prokaryotes to the 2 billion years they existed before eukaryotes.
9. Contrast the primitive atmosphere with today's atmosphere. Why is life unlikely to evolve today?
10. What is the importance of multicellularity and the evolution of sexual reproduction?
11. Describe the 5-kingdom system of classification, and list the criteria by which organisms are placed in a particular kingdom.

Objective Questions

1. Which of these gives a possible sequence of organic chemicals leading up to the protocell?
   a. inorganic gases, amino acids, polypeptide, microsphere
   b. inorganic gases, nucleotides, nucleic acids, genes
   c. water, salts, protein, oxygen
   d. Both a and b.
2. Which of these did Stanley Miller place in his experimental system to show that organic molecules could have arisen from inorganic molecules on the primitive earth?
   a. microspheres
   b. purines and pyrimidines
   c. the primitive gases
   d. All of these.
3. Which of these is the chief reason the protocell was probably a fermenter?
   a. It didn't have any enzymes.
   b. The atmosphere didn't have any oxygen.
   c. Fermentation provides the most amount of energy.
   d. All of these.
4. Which of these is a true statement?
   a. Eukaryotes evolved before prokaryotes.
   b. Prokaryotes evolved before eukaryotes.
   c. The true cell evolved before the protocell.
   d. Prokaryotes didn't evolve until 1.5 billion years ago.
5. Most of the history of life concerns the evolution of
   a. prokaryotes.
   b. eukaryotes.
   c. photosynthesizers.
   d. plants and animals.
6. What was happening during prokaryotic evolution?
   a. development of organelles
   b. development of specialization
   c. development of metabolic pathways
   d. All of these.
7. Which of these is a true statement?
   a. The primitive atmosphere was an oxidizing one and today's is a reducing one, making photosynthesis possible.
   b. The primitive atmosphere was 20% oxygen, just like it is today.
   c. The reducing primitive atmosphere contributed to the origin of life, and the oxidizing one of today would hinder it.
   d. It took so long for prokaryote evolution because the primitive atmosphere screened out the ultraviolet radiation from the sun.
8. Evolution of the DNA→RNA→ protein system was a milestone because the protocell
   a. was a heterotrophic fermenter.
   b. could now reproduce.
   c. needed energy to grow.
   d. All of these.
9. The significance of liposomes, phospholipid droplets, is it shows
   a. the first plasma membrane contained protein.
   b. a plasma membrane could have easily evolved.
   c. a biological evolution produced the first cell.
   d. that there was water on the primitive earth.

10. Multicellularity is properly associated with the origin of
    a. sexual reproduction.
    b. complex organisms.
    c. the end of Precambrian time.
    d. All of these.
11. Explain the significance of the following diagram, which shows a eukaryotic cell engulfing aerobic bacteria and photosynthetic bacteria.

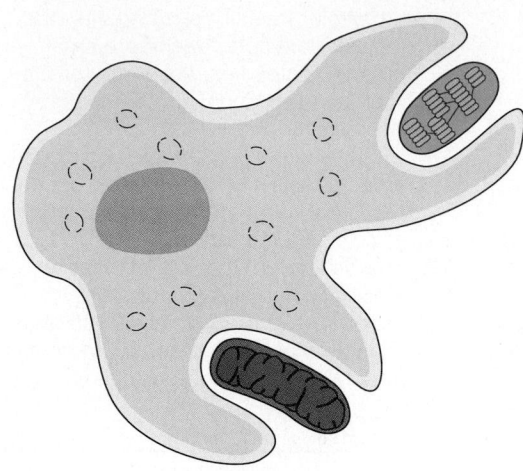

Concepts and Critical Thinking

1. *Life is a physical and chemical phenomenon.*

   How do all 3 hypotheses about the origin of life support this concept?

2. *Sexual reproduction promotes the evolutionary process.*

   What relationship might the evolution of sexual reproduction have to the diverse life forms found in the Cambrian era?

3. *Cells come only from preexisting cells.*

   Reconcile this concept with the belief that a chemical evolution produced the first cells.

Selected Key Terms

chemical evolution (kem′ĭ-kal ev″o-lu′shun) 355
organic soup (or-gan′ik sōōp) 355
proteinoid (pro′te-in-oid″) 356
microsphere (mi-kro-sfēr) 356
protocell (pro′to-sel) 356
liposome (lip′o-som) 356
heterotroph (het′er-o-trōf″) 356
autotroph (aw′to-trōf″) 356
ozone shield (o′zōn shēld) 358

23

Viruses and Monera

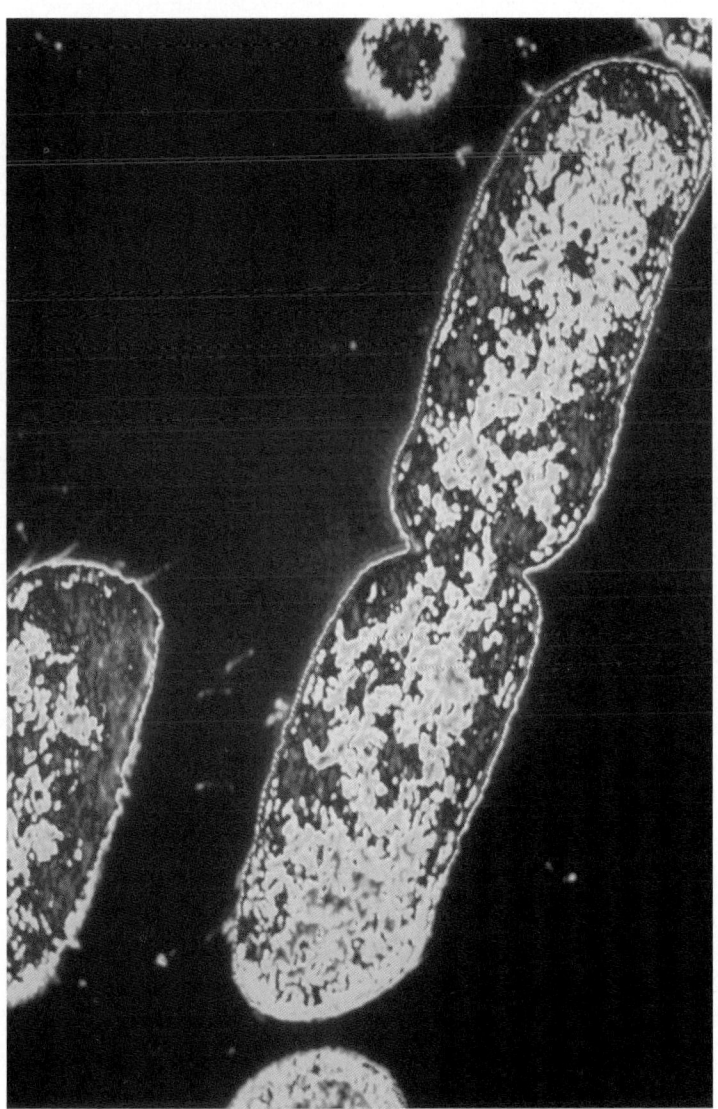

Most bacteria don't cause illness, but *Salmonella typhimurium* is a bacterium that causes food poisoning. Infections of poultry and eggs are common, and inadequately cooked pork and beef are also potential sources, as are oysters, which are farmed near sewage outlets. Clinical symptoms include headache, fever, abdominal pain, and diarrhea. Magnification, ×12,500.

Your study of this chapter will be complete when you can

1. describe in general the structure of viruses, including bacteriophages and animal viruses;
2. describe the bacteriophage lytic and lysogenic life cycles;
3. compare the life cycle of animal viruses to bacteriophage life cycles;
4. state how RNA retroviruses differ from other viruses, and discuss their significance to human beings;
5. compare viroids to viruses;
6. describe the anatomical characteristics of prokaryotic cells;
7. tell how bacteria reproduce, and relate this to their means of achieving genetic variation;
8. explain how bacteria differ in their tolerance and need for oxygen;
9. list the different types of autotrophic bacteria, and specifically contrast the primitive and advanced types of photosynthesis;
10. describe the nutrition of most heterotrophic bacteria, and give examples of different types of symbiotic bacteria;
11. classify the different types of monerans;
12. describe the anatomy and physiology of the cyanobacteria, and discuss their historical significance.

We will discuss living organisms from the simple to the complex and from the primitive (earliest evolved) to the most advanced (most recently evolved). How living organisms may be related can be discussed only in the broadest of terms, since detailed information is often lacking. It is also important to remember that no living group of organisms is the direct ancestor of another living group of organisms, although it is possible for 2 living groups to have shared a common ancestor.

It is curious that we begin our discussion with viruses when they are not even included in the classification table found in appendix A. We begin with viruses only because they are on the borderline between living and nonliving things.

Viruses

In 1892, Dimitri Ivanowsky, a Russian biologist, performed experiments with tobacco plants infected with tobacco mosaic disease, a condition that takes its name from the wrinkled and mottled appearance of the infected leaves. After transmitting the disease to healthy plants by rubbing them with juice extracted from diseased plants, Ivanowsky passed the infective extract through a fine-meshed porcelain filter. To his surprise, the filtrate was still infective. This meant that the disease-causing agent was smaller than any known bacteria. Disease-causing agents that could pass through filters came to be known as filterable viruses and later simply as viruses.

In 1935, W. M. Stanley discovered that viruses are remarkably different from bacteria and that they have a noncellular organization. Today, we know that they are tiny particles (10 nm-300 nm in diameter) composed of at least 2 parts: an *outer protein coat* and an *inner nucleic acid core*. The coat, called a capsid, is sometimes surrounded by an outer membranous envelope. The nucleic acid of the core may be either DNA or RNA. Although viruses cannot be seen with the light microscope, their structure can be studied with the electron microscope (fig. 23.1). Notice how the capsid is made up of repeating protein subunits.

Viruses are capable of reproduction but only within living cells; therefore, they are called obligate parasites. In the laboratory, some animal viruses are maintained by injecting them into live

Figure 23.1
Viruses (micrographs, *below*, and drawings, *above*) *a.* The tobacco mosaic virus is an RNA virus that attacks tobacco. Magnification, X100,000. The drawing shows how the nucleic acid core circles within a capsid composed of individual protein subunits. *b.* The adenovirus is a DNA virus that causes respiratory and intestinal infections in humans. Magnification, X250,000. *c.* T4 bacteriophage, a virus that attacks bacteria, has a complex structure. Magnification, X120,000. *d.* The influenza virus is an RNA virus whose protein capsid is enclosed by a membranous envelope. Magnification, X160,000.

RNA

protein subunits of capsid

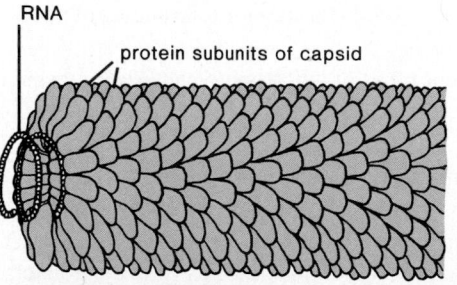

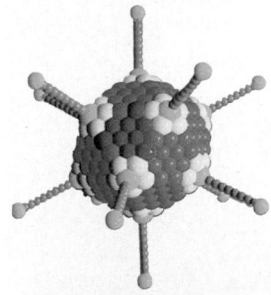

a.

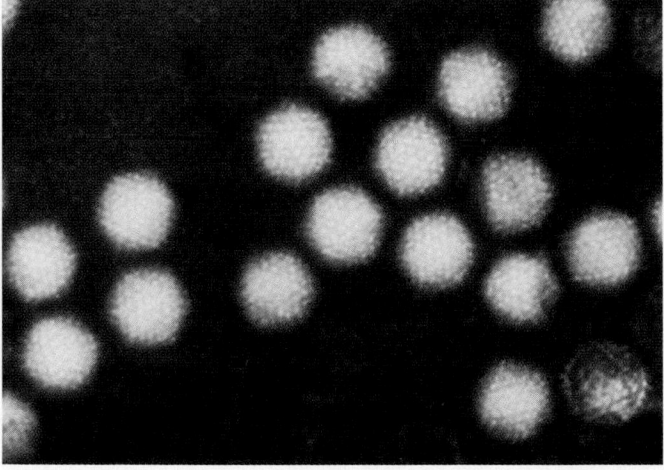

b.

chick embryos (fig. 23.2). Outside living cells, viruses are nonliving and can be stored in much the same way as chemicals are stored. Therefore, it is valid to ask if viruses should be considered alive.

Viruses typically have a specific host range. Certain ones attack only plants; others attack only animals, and the viruses called bacteriophages attack only bacteria. The human disease-causing viruses attack only certain types of cells. It is now known that there must be a match between a protein in the outer coat or envelope of the virus and the cell's outer surface for a virus to gain entry to the cell.

Viruses are noncellular obligate parasites that always have a protein coat and a nucleic acid core.

Life Cycles

Bacteriophages

Two types of bacteriophage life cycles, termed the **lytic cycle** and the **lysogenic cycle** have been carefully studied. The bacteriophage called lambda can enter either cycle (fig. 23.3).

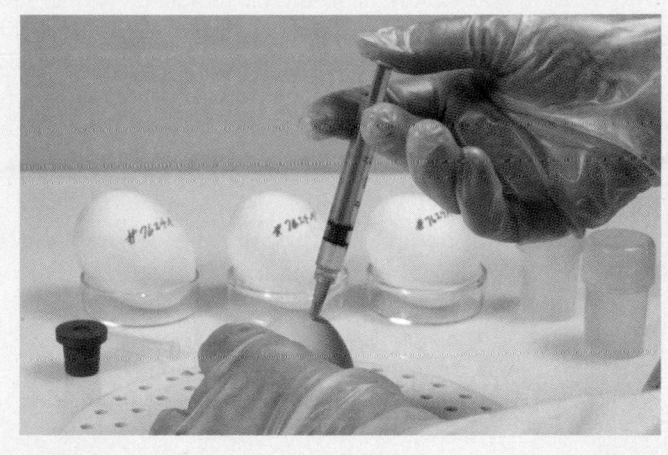

Figure 23.2
Inoculation of live chick eggs with viral particles. A virus reproduces only inside a living cell, not because it uses the cell as nutrients, but rather because it takes over the machinery of the cell.

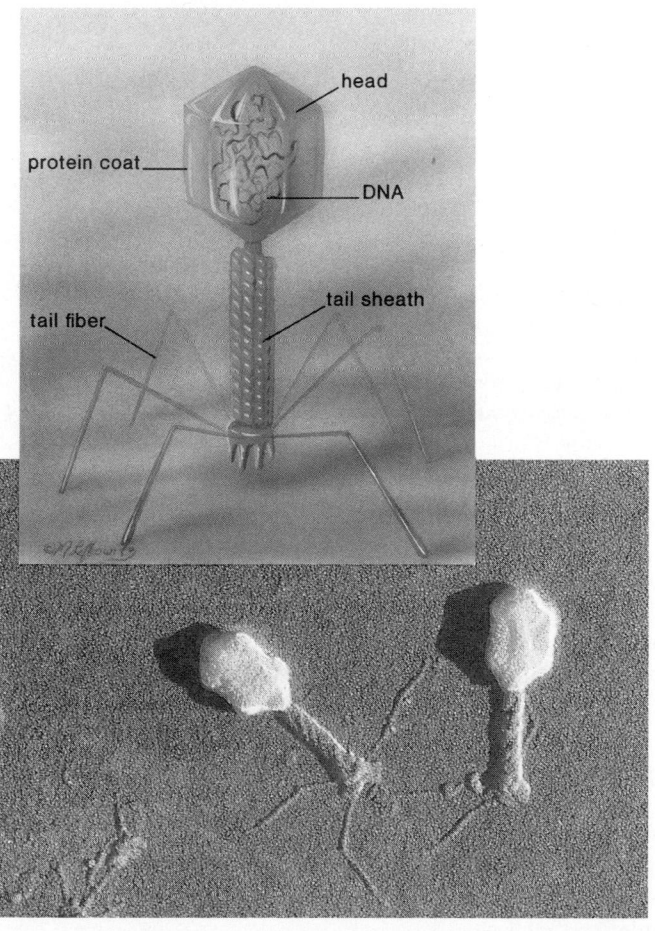

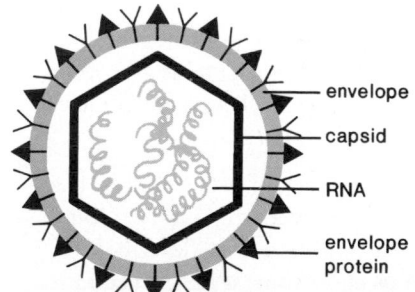

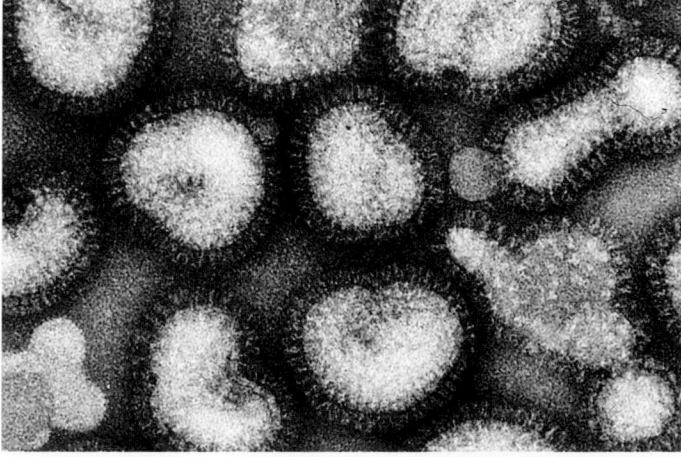

c.

d.

Lytic Cycle When the bacteriophage lambda collides with an *E. coli* cell, it binds to the surface and injects its DNA. Once inside, the viral DNA brings about disintegration of host DNA and takes over the operation of the cell. Viral DNA replication, utilizing the nucleotides within the host cell, produces many copies of viral DNA. Transcription occurs, and mRNA molecules utilize host ribosomes to bring about the production of multiple copies of coat proteins. Viral DNA and capsids are *assembled* to produce about 100 viral particles. In the meantime, viral DNA has directed the synthesis of lysozyme, an enzyme that digests the cell wall. The host cell bursts and the viral particles are released.

Lysogenic Cycle During the lysogenic cycle, viral DNA is integrated into the bacterial DNA. In this stage the viral DNA is called a *prophage*. The prophage is replicated along with the host DNA, and all subsequent cells, called *lysogenic cells,* carry a copy of the prophage. Certain environmental factors, such as ultraviolet radiation, can induce the lytic cycle. The prophage leaves the bacterial chromosome; replication of viral DNA, production of capsids, assembly, and cell lysis follow.

> During the lytic cycle of a bacteriophage, the bacterial cell dies when the viral particles burst from the cell. During the lysogenic cycle, viral DNA is integrated into bacterial DNA for an indefinite period of time.

Animal Viruses

Animal viruses sometimes have a membranous outer envelope. Such viruses enter a cell by endocytosis, and *uncoating* releases viral nucleic acid from the capsid. These viruses leave a cell by exocytosis, or *budding,* and in this way, they acquire the membranous envelope (fig. 23.4*a*). The envelope often contains glycoproteins that interact with the next host cell plasma membrane, permitting entry of the virus into this cell.

DNA Animal Viruses After a DNA virus with an envelope has gained entry to the cell and uncoating has occurred, transcription, translation, and assembly occur as we have just described for bacteriophages. Then, as each virus buds from the cell, it acquires its envelope. The DNA codes for coat protein and also envelope glycoprotein (fig. 23.4*a*).

RNA Animal Viruses Some animal viruses have RNA genomes; that is, only RNA is enclosed within the capsid. In most of these viruses, the single strand of RNA serves as a template for the production of double-stranded RNA. This unique molecule then serves as a template for multiple replicas of the genetic material and for the transcription of mRNA molecules. Special enzymes, termed *RNA replicase* and *transcriptase,* perform these tasks.

Other RNA viruses called **retroviruses** (fig. 23.4*b*) have the enzyme called *reverse transcriptase,* which carries out RNA → DNA transcription. Following replication, the resulting double-stranded DNA (called *cDNA* because it is complementary to viral RNA) is integrated into the host chromosome for an indefinite period of time before reproduction of the virus and budding from the host cell occurs. Because integration occurred, RNA viruses sometimes carry host genes into a new host cell.

Figure 23.3

Lytic versus lysogenic life cycle. In the lytic cycle, a bacteriophage reproduces within the host bacterial cell and then the cell is lysed (broken open), allowing the viral particles to escape. In the lysogenic cycle, the DNA of a bacteriophage is integrated into the host DNA and becomes a prophage. Thereafter, the prophage is replicated along with the bacterial chromosome and is passed to all the daughter cells. When and if the viral DNA leaves the chromosomes, the lysogenic cycle can be followed by the lytic cycle.

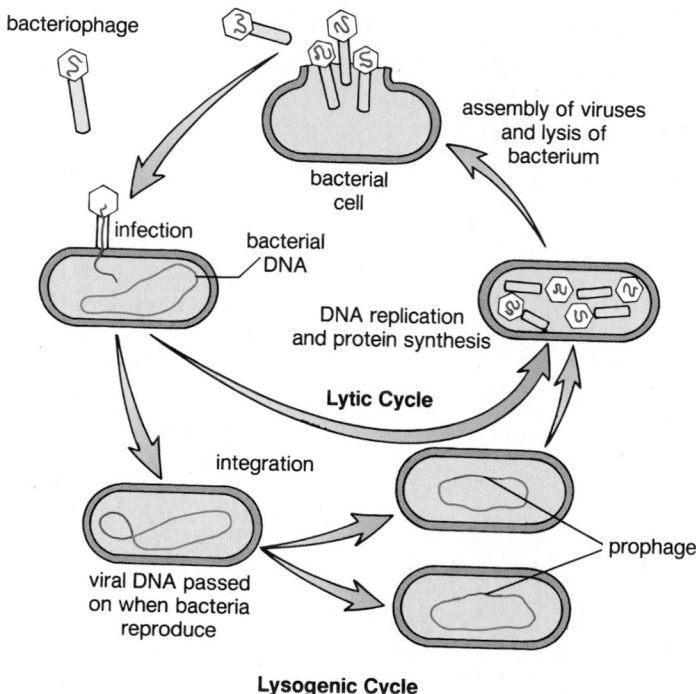

Retroviruses are of extreme interest because these are the viruses that bring cancer-causing oncogenes into a host cell. Also, the AIDS viruses are retroviruses.

> Animal viruses sometimes have a membranous envelope that they acquire by budding from the host cell. Retroviruses are RNA viruses that carry out RNA → DNA transcription and integrate this cDNA into the host cell.

Viroids

Viroids are very unusual infectious particles that can be compared to viruses. **Viroids** differ from viruses in that they consist only of a short chain of naked RNA. The number of nucleotides present is not sufficient to code for a capsid and there is no evidence that the viroid RNA is ever translated into a protein. We know of their presence only because they cause disease in plants; the first viroid to be recognized is the cause of potato spindle tuber disease. Viroids also have attacked coconut trees in the Philippines and chrysanthemums in the United States. The manner in which they cause disease is not known, but it is hypothesized that they are reproduced by the host cell and that they interfere with gene regulation in these cells.

Figure 23.4

Life cycles of animal viruses. *a.* DNA virus. After entering the host cell by endocytosis, the virus is uncoated. The DNA then codes for proteins, some of which are capsid (coat) proteins and some of which are envelope proteins. Assembly follows replication of the DNA. When the virus exits by budding, it is enclosed by an envelope made up of host cell plasma membrane lipids and viral envelope proteins.

b. RNA retrovirus. The life cycle includes steps not seen in (*a*). The RNA genes are transcribed to cDNA (complementary DNA), which is integrated into the host DNA. Transcription produces many copies of the RNA genes, which also serve to direct the synthesis of 3 types of proteins: the enzyme reverse transcriptase, capsid (coat) protein, and envelope protein. Again, the virus buds from the host cell.

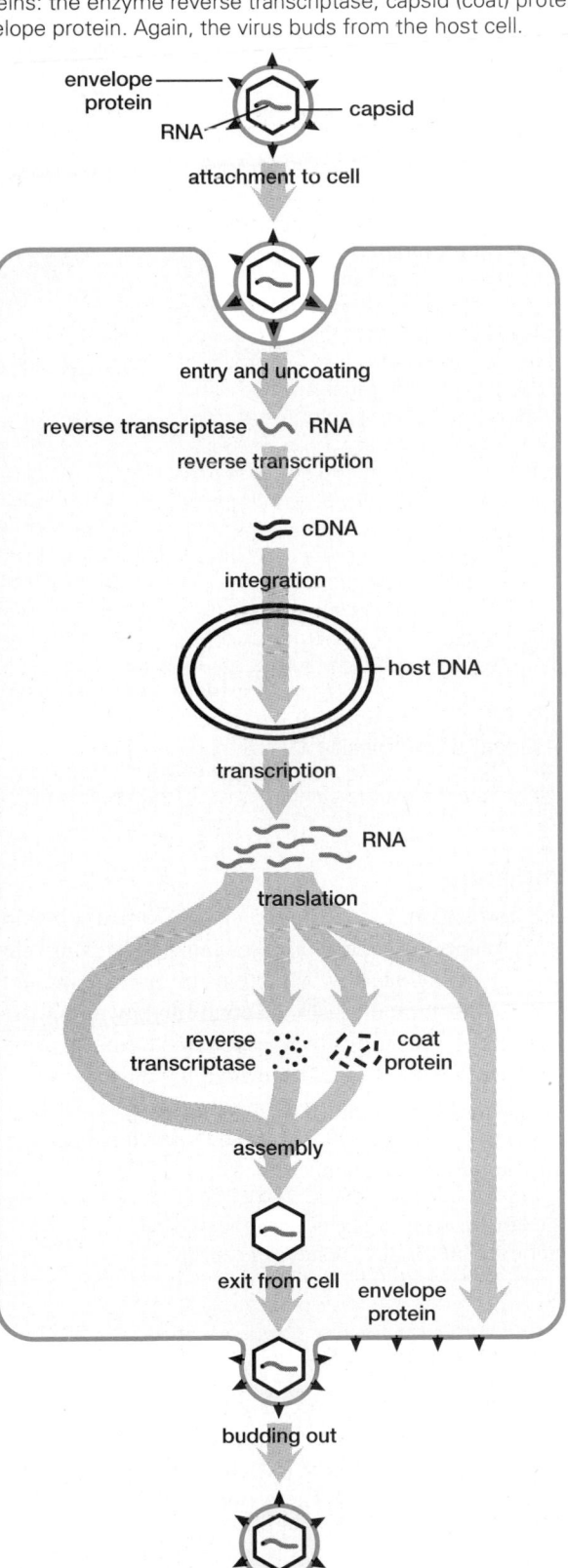

a. DNA virus

b. RNA retrovirus

An antibiotic is a chemical that selectively kills bacteria when it is taken into the body as a medicine. Since the introduction of the first antibiotics in the 1940s, there has been a dramatic decline in deaths due to pneumonia, tuberculosis, and other infectious diseases.

Most antibiotics are produced naturally by soil microorganisms. Penicillin is made by the fungus *Penicillium;* streptomycin, tetracycline, and erythromycin are all produced by the bacterium *Streptomyces.* Sulfa, a chemotherapeutic agent rather than an antibiotic, is an analogue of a bacterial growth factor and can be produced in the laboratory.

A few antibiotics are metabolic inhibitors specific for bacterial enzymes. This means that they poison bacterial enzymes without harming host enzymes. Penicillin blocks the synthesis of the bacterial cell wall; streptomycin, tetracycline, and erythromycin block protein synthesis; and sulfa prevents the production of a coenzyme.

There are problems associated with antibiotic therapy. Some patients are allergic to antibiotics, and their reaction to them may even be fatal. Antibiotics not only kill off disease-causing bacteria, they also reduce the number of beneficial bacteria in the intestinal tract. The latter may have

Figure 23.A
Penicillium chrysogenum, from which the antibiotic penicillin is prepared.

held in check a pathogen that now is free to multiply and invade the body. The use of antibiotics sometimes prevents natural immunity from occurring, leading to recurring antibiotic therapy. Most important, perhaps, is the growing resistance of certain strains of bacteria to antibiotics. While penicillin used to be 100% effective against hospital strains of *Staphylococcus aureus,* today it is far less effective. Tetracycline

and penicillin, long used to cure gonorrhea, now have a failure rate of more than 20% against certain strains of gonococcus. Most physicians believe that antibiotics should only be administered when absolutely necessary. Some believe that if antibiotic use is not strictly limited, then resistant strains of bacteria will completely replace present strains and antibiotic therapy will no longer be effective at all. They are very much opposed to the current practice of adding antibiotics to livestock feed in order to make animals grow fatter because resistant bacteria are easily transferred from animals to humans.

The development of antiviral drugs has lagged far behind the development of antibiotics. Viruses lack most enzymes and instead utilize the metabolic machinery of the host cell. Rarely has it been possible to find a drug that successfully interferes with viral reproduction without also interfering with host metabolism. One such drug, however, called vidarabine, was approved in 1978 for treatment of viral encephalitis, an infection of the nervous system. Acyclovir (ACV) seems to be helpful in treating genital herpes, and the drugs zidovudine (AZT) and dideoxyinosine (DDI) are being used with AIDS patients.

Viral Infections

Viruses are best known for causing infectious diseases in animals and plants. In plants, infectious diseases can only be controlled by burning those plants that show symptoms of disease. In animals, especially humans, viral diseases are controlled by administering vaccines and, only recently, by the administration of antiviral drugs. Some well-studied viral diseases in humans are influenza, mumps, measles, polio, rabies, and infectious hepatitis. As mentioned previously, retroviruses cause AIDS and have been implicated in the development of cancer.

Viruses cause diseases in plants and animals. They have also been implicated in the development of cancer.

Kingdom Monera

Kingdom Monera is one of the 5 kingdoms recognized in this text (see table 22.2 and appendix A). A kingdom is the largest category of classification. Kingdom Monera contains the **prokaryotes,** the first types of cells to evolve. These organisms provide the first

fossils—dated about 3.5 billion years ago—and the fossil record indicates that the prokaryotes were the only type of living organisms for at least 2 billion years (see fig. 22.5).

In our classification system, the prokaryotes are the **bacteria** and are divided into 2 groups, the archaebacteria and the eubacteria. It is important to realize that the eubacteria include the cyanobacteria, which were formerly called blue-green algae (fig. 23.5). Monerans are small, single cells (*moneres*—single) that generally range in size from 1 μm–10 μm in length (see fig. 5.2) and from 0.2 μm–0.3 μm in width. Since they are microscopic, it is not always obvious that they are abundant in air, water, and soil and on most objects. It has even been suggested that the combined number of all bacteria exceeds that of any other type of organism on earth.

Kingdom Monera includes the various types of bacteria, which are microscopic, single-celled prokaryotes.

Prokaryotic Structure

Prokaryotic cells lack a nucleus and do not have the cytoplasmic organelles found in eukaryotic cells, other than ribosomes (see fig. 5.3). The eubacteria have a cell wall composed of the unique

Figure 23.5

Electron micrograph following freeze-fracture of the cyanobacterium cell *Anabaena cylindrica*. Typical of prokaryotic cells, there is no membrane-bounded nucleus or the other organelles that are found in eukaryotic cells. Cyanobacteria carry out photosynthesis in the same manner as plants, but there are no chloroplasts. The chlorophyll is found in an extensive array of thylakoids (photosynthetic membranes).

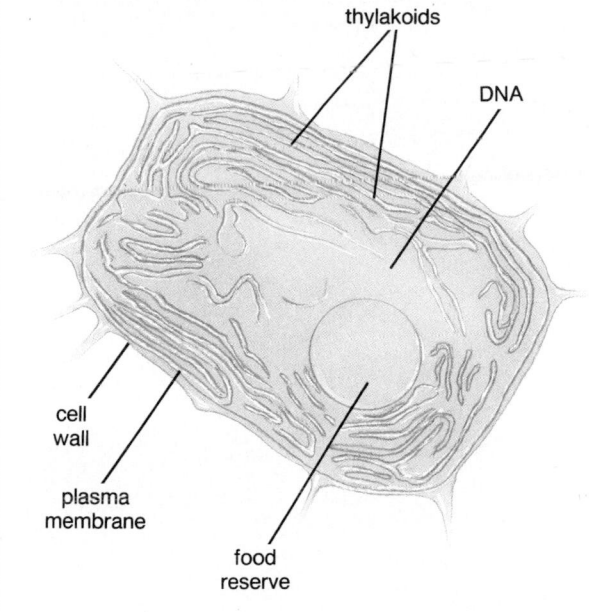
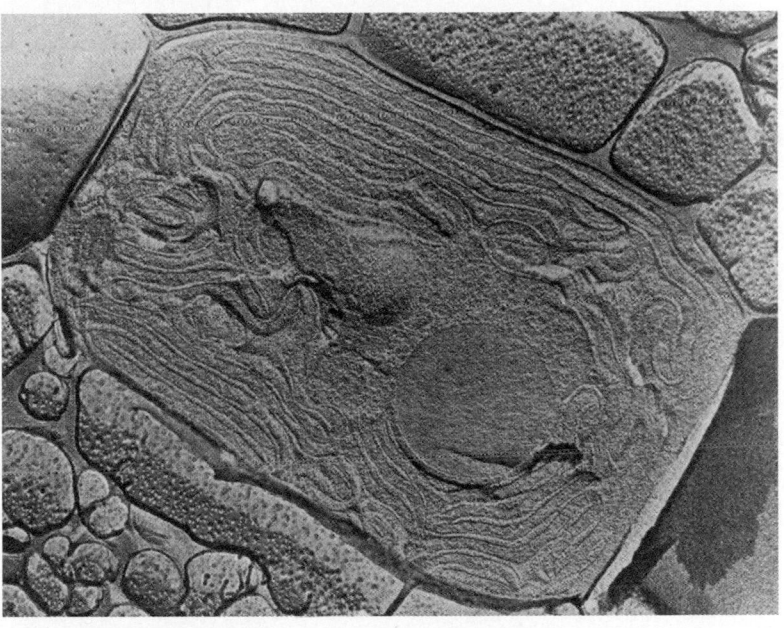

molecule peptidoglycan. The cell wall may be surrounded by a gelatinous sheath or capsule. Some bacteria move by means of flagella, and some adhere to surfaces by means of pili.

Bacteria occur in 3 basic shapes (fig. 23.6): *rod* (bacillus), *round* or spherical (coccus), and *spiral* (i.e., helical shape called a spirillum). The rods and bacilli may form chains of a length typical of the particular bacterium.

Prokaryotic cells lack a nucleus and most of the other organelles found in eukaryotes. They occur in 3 basic shapes.

Prokaryotic Reproduction

Prokaryotes reproduce asexually by means of **binary fission,** a process depicted in figure 10.2. The single circular chromosome replicates, and then the 2 copies separate as the cell enlarges. Newly formed plasma membrane and cell wall separate the cell into 2 cells. Mitosis, which requires the formation of a spindle apparatus, does not occur in prokaryotes.

As discussed in the reading on page 282, genetic recombination can occur among bacteria in 3 ways. *Conjugation* has been observed between bacteria when the so-called male cell passes DNA to the female by way of a sex pilus. *Transformation* occurs when a bacterium picks up from the medium DNA released by dead bacteria. During *transduction,* bacteriophages carry portions of bacterial DNA from one cell to another.

Since prokaryotes have no simple method for achieving genetic recombination, as do plants and animals, *mutation* becomes a most important avenue for evolutionary change. With a generation time as short as 20 minutes under favorable conditions, mutations are generated and distributed throughout a population more quickly than in eukaryotes. Also, prokaryotes are haploid, and so mutations are immediately subjected to natural selection for any possible beneficial effect.

Prokaryotes reproduce asexually by binary fission. Mutations are the chief means of achieving genetic variation.

When faced with unfavorable environmental conditions, some bacteria form **endospores.** A portion of the cytoplasm, along with a copy of the new chromosome, rounds up within the plasma membrane and a new, thicker cell wall forms around the developing endospore (fig. 23.7). When the endospore is mature, it is released from the bacterial cell in which it formed. Endospores are amazingly resistant to extremes in temperature, drying, and harsh chemicals, including acids and bases. When conditions are again suitable for growth, the endospore absorbs water, breaks out of the inner shell, and becomes a typical bacterial cell, reproducing once again by binary fission.

Figure 23.6
Scanning electron micrographs of bacteria. *a.* Spherical-shaped bacteria. Magnification, X25,500.
b. Rod-shaped bacteria. Magnification, X11,000. *c.* Spiral-shaped bacteria with flagella used for
locomotion. See figure 5.3 for a generalized drawing of bacteria. Magnification, X9,000.

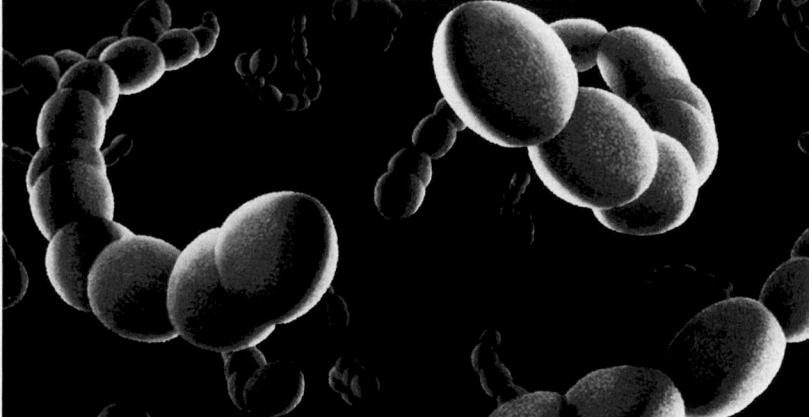

a.

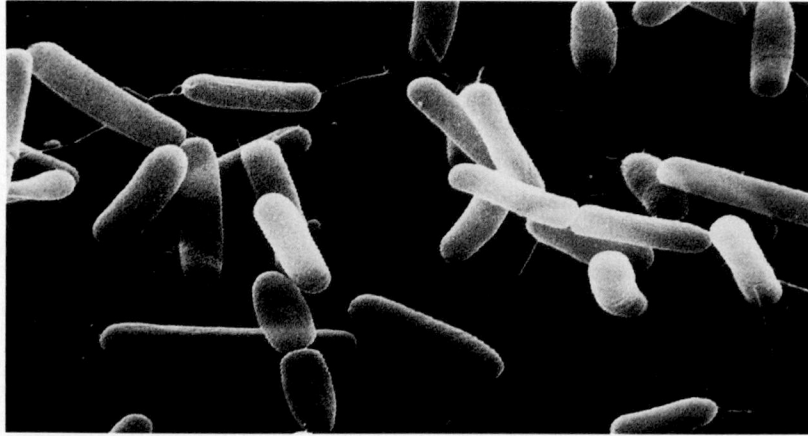

b.

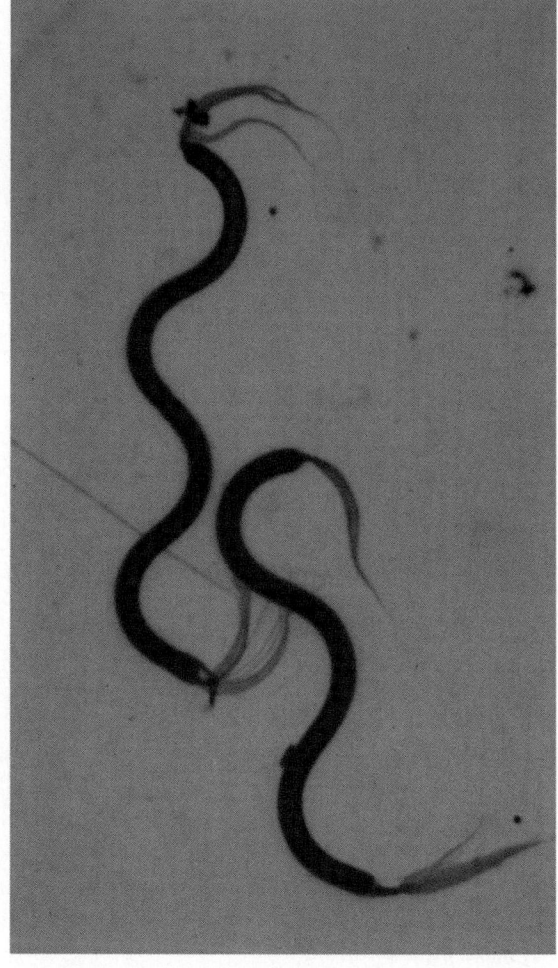

c.

Prokaryotic Metabolism

Some bacteria are *obligate anaerobes* and are unable to grow in the
presence of oxygen. A few serious illnesses, such as botulism, gas
gangrene, and tetanus, are caused by anaerobic bacteria. Some
other bacteria, called *facultative anaerobes,* are able to grow in
either the presence or the absence of oxygen. Most bacteria,
however, are *aerobic* and, like animals, require a constant supply
of oxygen to carry out cellular respiration.

Every type of nutrition is found among bacteria except for
ingestion of whole food. There are both autotrophic and heterotrophic bacteria.

Autotrophic Bacteria

Some autotrophic bacteria are photosynthetic; they use light as a
source of energy to produce their own food. These photosynthetic
bacteria can be divided into 2 types—those that are primitive
(evolved first) and those that are more advanced (evolved later):

Primitive	Advanced
Photosystem I only	Photosystems I and II
Do not give off O_2	Do give off O_2
Unique type of chlorophyll	Type of chlorophyll found in plants

The green sulfur bacteria and the purple sulfur bacteria carry
on the primitive type of photosynthesis. They do not give off oxygen
because they do not use water as an electron donor; instead, they use
hydrogen gas (H_2) and hydrogen sulfide (H_2S). These bacteria usually
live in anaerobic conditions such as the muddy bottom of marshes and
cannot photosynthesize in the presence of oxygen. The cyanobacteria
(see figs. 23.5 and 23.11) carry on the advanced type of photosynthesis in the same manner as plants (chap. 8).

Figure 23.7

Clostridium botulinum containing an endospore. This organism causes the type of food poisoning known as botulism. Sterilization, a process that kills all living things, even endospores, is used whenever all bacteria must be killed. Sterilization can be achieved using an autoclave, a container that maintains steam under pressure. Magnification, ×55,000.

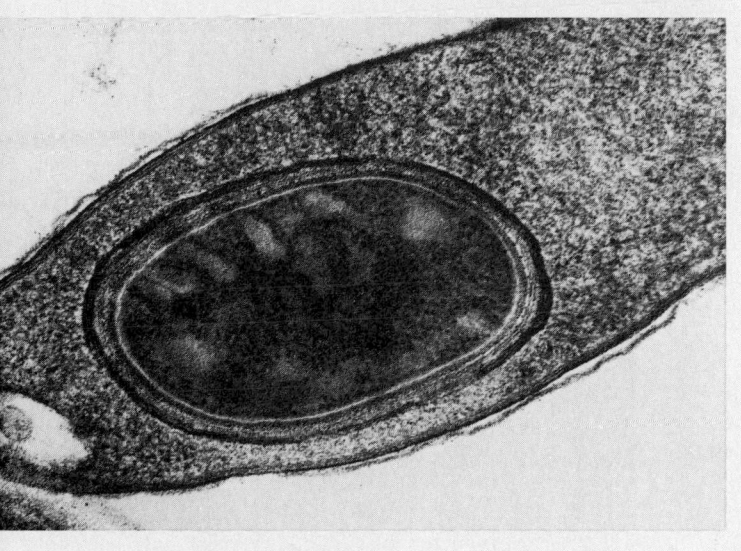

Figure 23.8

Giant tubeworms are a part of the community of organisms found near deep ocean ridges and vents. These communities have been given such names as Dandelions, the Rose Garden, and the Garden of Eden.

Some autotrophic bacteria are *chemoautotrophs,* which oxidize inorganic compounds to obtain the necessary energy to produce their own food. Among the inorganic compounds oxidized by specific bacteria are ammonia, nitrites, and sulfides, and they trap the small amount of energy released from these oxidations to use in the reactions that synthesize carbohydrates. Chemoautotrophs that support entire communities 2.5 km below sea level, where light never penetrates, have been discovered. At the mid-ocean ridge system, hot minerals spew out of the inner earth, providing hydrogen sulfide to chemosynthetic bacteria. The organic molecules produced by the bacteria support the growth of giant tubeworms (fig. 23.8), clams, and crabs. There is evidence these bacteria also reside within the body of the worms.

Nitrifying bacteria are also chemoautotrophs that oxidize ammonia (NH_3) to nitrites (NO_2^-) and nitrites to nitrates (NO_3^-). We will mention these important bacteria again when we discuss the nitrogen cycle.

Autotrophic bacteria include those that are photosynthetic and those that are chemosynthetic. Among the photosynthetic bacteria, there are those that evolved first and do not give off oxygen and those (cyanobacteria) that evolved later and do give off oxygen.

Heterotrophic Bacteria

The majority of bacteria are aerobic **saprophytes.** They break down dead organic matter by secreting digestive enzymes and then they absorb the nutrient molecules. These so-called **decomposers** can break down such a large variety of molecules that there probably is no natural organic molecule that cannot be digested by at least one bacterial species. The decomposing bacteria play a critical role in recycling matter in ecosystems and in making inorganic molecules available to photosynthesizers (see fig. 1.5). The metabolic capabilities of various heterotrophic bacteria are exploited by human beings, who use them to perform services ranging from the digestion of sewage and oil to the production of such products as vitamins and even antibiotics. Also, by means of gene splicing, bacteria can now be used to produce such substances as human insulin and growth hormone.

Heterotrophs may be free-living or **symbiotic** (table 23.1), forming mutualistic, commensalistic, or parasitic relationships. Some bacteria carry on **nitrogen fixation.** They reduce atmospheric nitrogen (N_2) to ammonia (NH_3) and are *mutualistic,* living in the root nodules of soybean, clover, and alfalfa plants and providing them with organic nitrogen (fig. 23.9). Plants are unable to fix atmospheric nitrogen, so these bacteria perform a useful function for plant nutrition. The intestinal bacteria in humans release vitamins K and B_{12}, which we can use to help produce blood components, and in the stomachs of cows and goats other bacteria digest cellulose, enabling these animals to feed on grass. *Commensalistic* bacteria live on organisms and cause them no harm, but *parasitic* bacteria are responsible for a wide variety of plant and animal diseases. Common human infections caused by parasitic bacteria include strep throat, diphtheria, typhoid fever, and gonorrhea.

The majority of bacteria are free-living heterotrophs (saprophytic decomposers) that contribute significantly to recycling matter through ecosystems. Many are also symbiotic heterotrophs, including those that cause disease.

Table 23.1
Symbiotic Relationships

Name	Definition	Examples
Mutualism	Both organisms benefit	Bacteria in plant nodules Dinoflagellates in coral animals
Commensalism	One organism gains and the other is unharmed	Bacteria on human skin
Parasitism	One organism gains and the other is harmed	Bacterial, fungal, and protozoan infections

Classification of Bacteria

Bacteria are a diverse group of organisms that vary widely in structure and metabolism; this diversity has made it more difficult to classify them according to evolutionary lines. Recently, however, comparisons of amino acid sequence in proteins and of base sequences in DNA and RNA have begun to indicate possible relationships. These metabolic studies indicate that prokaryotes split very early into 2 types: the archaebacteria and the eubacteria.

Archaebacteria

Most likely archaebacteria were the earliest prokaryotes. Their cell walls and plasma membranes do not have the same composition as the same structures in the eubacteria. Some believe that they should even be placed in a different kingdom.

Many **archaebacteria** are able to live in the most extreme environments, perhaps representing the kinds of habitats that were available when the earth first formed. The methanogens are anaerobic and live in swamps and marshes, producing methane, known as marsh gas. They also live in the guts of organisms, including humans. The halophiles live where it is salty, such as the Great Salt Lake in Utah. Curiously, a type of rhodopsin pigment (related to the one found in our own eyes) allows halophiles to carry on a primitive type of photophosphorylation for ATP production. The thermoacidophiles live where it is both hot and acidic. Those that live in the hot sulfur springs of Yellowstone National Park obtain energy by oxidizing sulfur.

Eubacteria

Most of our previous discussions pertain to the most common types of **eubacteria,** which can be classified in many different ways. One way to classify eubacteria is to consider their reaction to the *Gram stain*. After applying this stain, the gram-positive bacteria appear purple and the gram-negative bacteria appear red (fig. 23.10). This difference is dependent on the construction of the cell wall. The gram-positive cell wall has one thick layer, while the gram-negative wall has several thinner layers. Some

Figure 23.9
While there are some free-living bacteria that carry on nitrogen fixation, those of the genus *Rhizobium* invade the roots of legumes with the resultant formation of nodules. Here the bacteria convert atmospheric nitrogen to an organic nitrogen that the plant can use. These are nodules on the roots of a soybean plant.

Figure 23.10
Photomicrograph of bacteria that have been subjected to Gram staining. The gram-positive bacteria are purple, and the gram-negative bacteria are red. These bacteria are *Staphylococcus aureus* and *Escherichia coli,* respectively.

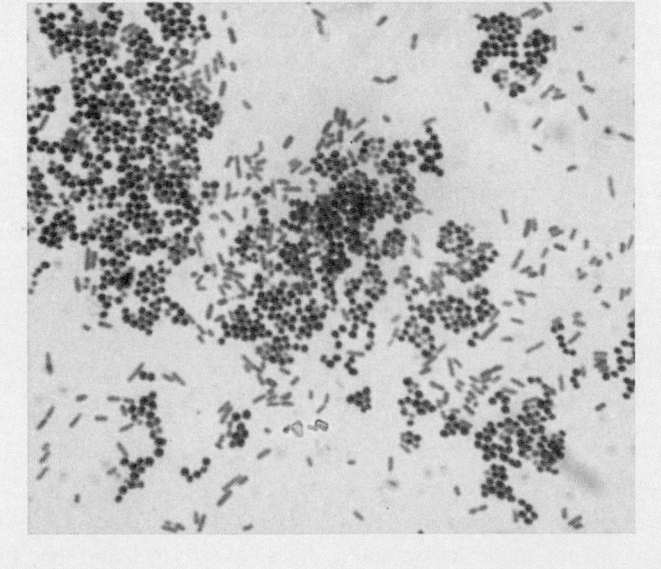

Figure 23.11

Diversity among the cyanobacteria. *a.* In *Gloeocapsa,* single cells are grouped together in a common gelatinous sheath. Magnification, X2,300. *b.* Filaments of cells occur in *Ocillatoria.* Magnification, X200.

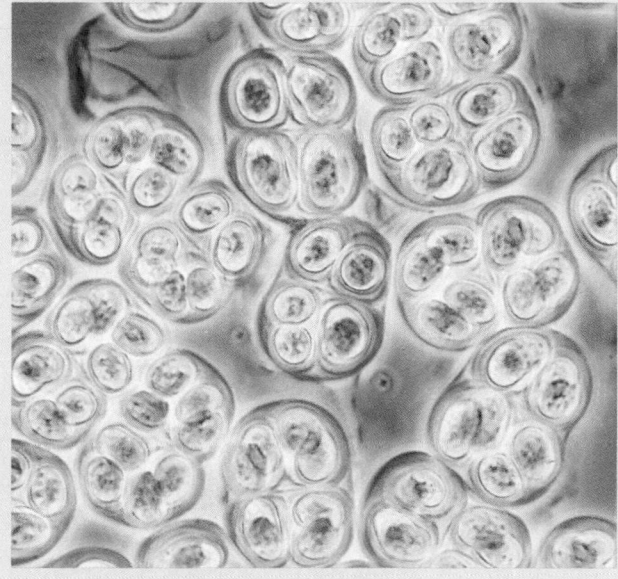

a.

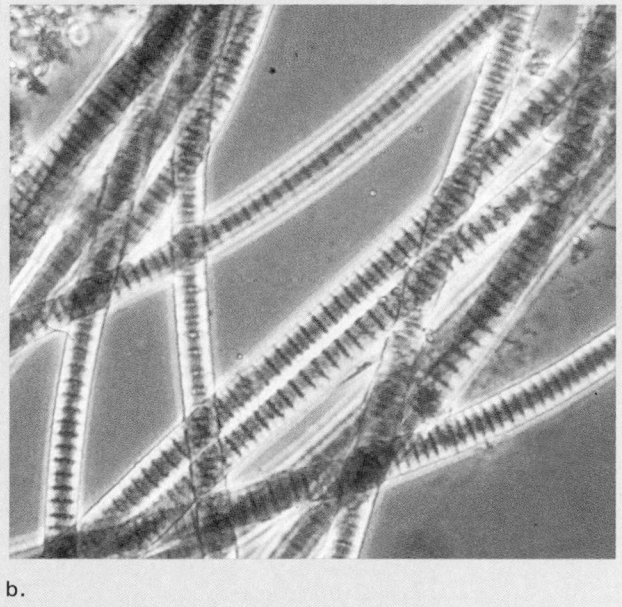

b.

Figure 23.12

Lichens contain fungal cells and photosynthetic cells (either cyanobacterial cells or green algal cells) in close association. The photosynthesizer provides food, but whether or not the fungus contributes to the relationship is debatable. This lichen, *Peltigera apathosa,* contains cyanobacterial cells and illustrates that lichens can superficially resemble low-lying plants. The cyanobacteria lie just beneath the surface, surrounded by fungal hyphae.

Among the photosynthetic bacteria, the cyanobacteria are of special importance; therefore, we will discuss them further here.

Cyanobacteria

Cyanobacteria, formerly called blue-green algae, are the most prevalent photosynthetic bacteria (fig. 23.11). The cyanobacteria carry on photosynthesis in a manner similar to plants; they possess chlorophyll *a* and evolve oxygen. They also have other pigments that can mask the color of chlorophyll, giving them, for example, not only blue-green but also red, yellow, brown, or black colors. We mentioned in chapter 22 that the cyanobacteria are believed to be responsible for first introducing oxygen into the primitive atmosphere.

Cyanobacteria may be unicellular, filamentous, or colonial. The filaments and colonies are not considered multicellular because each cell is independent of the others. Cyanobacteria lack any visible means of locomotion, although some glide when in contact with a solid surface and others oscillate (sway back and forth). Some cyanobacteria have a special advantage because they possess heterocysts, which are thick-walled cells without nuclei where nitrogen fixation occurs. The ability to photosynthesize and also to fix atmospheric nitrogen (N_2) means that their nutritional requirements are minimal.

Cyanobacteria are common in fresh water, soil, and on moist surfaces, but they are also found in habitats devoid of higher forms of life, such as hot springs. They are symbiotic with a number of organisms, such as liverworts, ferns, and even at times invertebrates like corals. In association with fungi, they form **lichens** that can grow on rocks. A lichen is a symbiotic relationship in which the cyanobacterium provides organic nutrients to the fungus, while the fungus possibly protects and furnishes inorganic nutrients to its partner (fig. 23.12). It is also

gram-negative bacteria, such as the rickettsias, chlamydias, and mycoplasmas, are much smaller than average bacteria. The rickettsias, which cause Rocky Mountain spotted fever and typhus, and the chlamydias, which cause a sexually transmitted urethritis, actually grow inside other cells. The mycoplasmas live in the body fluids of plants and animals, but they are unusual in that they don't have cell walls.

possible that the fungus is parasitic on the algal component. Lichens help transform rocks into soil; other forms of life then may follow. It is presumed that cyanobacteria were the first colonizers of land during the course of evolution.

Cyanobacteria are ecologically important in still another way. If care is not taken in the disposal of industrial, agricultural, and human wastes, phosphates and nitrates drain into lakes and ponds, resulting in a "bloom" of these organisms. The surface of the water becomes turbid and light cannot pen-etrate to lower levels. When a portion of the cyanobacteria die off, the decomposing bacteria use up the available oxygen, causing fishes to die from lack of oxygen.

Cyanobacteria are photosynthesizers that sometimes can also fix atmospheric nitrogen. They first introduced oxygen into the atmosphere and probably first colonized land. Even today, in association with fungi, they form lichens, which are important soil formers.

Summary

1. All viruses have at least 2 parts: an outer protein coat (capsid) and inner nucleic acid core.

2. Viruses reproduce inside living cells; therefore, they are obligate parasites. Since they are parasites, they cause diseases in plants and animals.

3. Bacteriophages exhibit 2 primary life cycles—lytic and lysogenic. During the lytic cycle, the bacteriophage takes over the operation of the cell. The cycle consists of (a) attachment and release of viral DNA into the cell, (b) disintegration of host DNA, (c) viral DNA replication, (d) protein coat formation, (e) assembly of virus, (f) production of lysozyme, and (g) release of new viruses from lysed host.

4. In the lysogenic cycle, viral DNA is integrated into bacterial DNA for an indefinite period of time, but it can undergo the lytic cycle when stimulated.

5. Animal viruses have an outer membranous envelope and enter cells by endocytosis. Uncoating occurs, which releases the viral nucleic acid; after replication, coat protein formation, and assembly, the viruses bud from the cell.

6. There are RNA animal viruses that never have a DNA stage; however, the RNA retroviruses possess an enzyme, reverse transcriptase, that causes the production of cDNA, which becomes integrated into host DNA. These viruses sometimes carry cancer-causing oncogenes. The AIDS virus is also a retrovirus.

7. Viroids, which have been known to cause disease in certain plants, consist of only a short chain of naked RNA.

8. Kingdom Monera contains single-celled prokaryotes, that is, the bacteria. Prokaryotic cells lack a nucleus and most other cytoplasmic organelles found in eukaryotic cells. They do have a cell wall composed of peptidoglycan.

9. Bacteria have 3 basic shapes: rod (bacillus), round (coccus), and spiral (spirillum).

10. Bacteria reproduce asexually by binary fission, and since they do not have a regular means of genetic recombination, their chief method for achieving genetic variation is mutation.

11. Some bacteria form endospores, which are extremely resistant to destruction; their genetic material can thereby survive unfavorable conditions.

12. Bacteria differ in their tolerance and need for oxygen. There are obligate anaerobes, facultative anaerobes, and aerobic bacteria.

13. Some bacteria are autotrophic and are either photosynthetic or chemosynthetic. There are the primitive photosynthesizers that do not give off oxygen, and the advanced photosynthesizers (cyanobacteria) that do give off oxygen. Chemosynthetic bacteria oxidize inorganic compounds, like sulfides, to acquire energy to make their own food. Surprisingly, these bacteria support communities at mid-oceanic ridges.

14. Most bacteria are aerobic heterotrophs and are saprophytic decomposers that are absolutely essential to the cycling of nutrients in ecosystems. Their metabolic capabilities are so vast that they are used by humans both to dispose of substances and to produce substances.

15. Many heterotrophic bacteria are symbiotic. The mutualistic nitrogen-fixing bacteria live in nodules on the roots of legumes. Some symbiotes, however, are parasitic and cause plant and animal diseases.

16. Bacteria are divided into the archaebacteria and the eubacteria. The archaebacteria have such a different type of cell wall and plasma membrane from the other bacteria that they should perhaps be in a different kingdom. They live under harsh conditions—anaerobic marshes, salty lakes, and hot sulfur springs.

17. Most bacteria are eubacteria and can be classified in various ways. One way is to consider which eubacteria are gram positive and which are gram negative.

18. Of special interest are the cyanobacteria, which were the first organisms to photosynthesize in the same manner as plants. When symbiotic with fungi, they form lichens.

Writing Across the Curriculum

In order to practice writing skills, students should write out the answers to any or all of the study questions and the critical thinking questions. The study questions are sequenced in the same order as the text. Suggested answers to the critical thinking questions are in appendix D.

Evolution and Diversity

Study Questions

1. Describe the general anatomy of viruses and tell why they are obligate parasites.
2. Describe both the lytic and lysogenic cycles of bacteriophages.
3. How do animal viruses differ in structure and life cycle from bacteriophages?
4. What are the unique features of retroviruses?
5. How do viroids differ from viruses?
6. What is the general anatomy of bacteria and how do they reproduce?
7. How do bacteria achieve genetic recombination, and why is the occurrence of mutations the more common source of variations?
8. How do some bacteria manage to survive the severest of unfavorable environments?
9. How do bacteria differ in their tolerance and need for oxygen?
10. What are the differences between the primitive type of photosynthesis and the advanced type of photosynthesis found in bacteria?
11. What are chemosynthetic bacteria, and where have they been found to support whole communities?
12. Discuss the importance of heterotrophic bacteria, and give examples of their symbiotic relationships with other types of organisms.
13. How are bacteria classified? Why might archaebacteria be placed in a separate kingdom? Discuss the importance of cyanobacteria in the history of the earth.

Objective Questions

1. Which of these are found in all viruses?
   a. membranous envelope and capsule
   b. capsid and nucleic acid
   c. tail fibers and head
   d. All of these.
2. Which step in the lytic cycle follows attachment of virus and release of DNA into the cell?
   a. production of lysozyme
   b. disintegration of host DNA
   c. assemblage
   d. DNA replication
3. Which of these is a true statement?
   a. Viruses carry with them their own ribosomes for protein formation.
   b. New viral ribosomes form after viral DNA enters the cell.
   c. Viruses use the host ribosomes for their own ends.
   d. Viruses do not need ribosomes for protein formation.
4. During the lysogenic cycle, viral DNA
   a. is lysed and disappears.
   b. becomes incorporated into host DNA.
   c. replicates to give many copies.
   d. bursts from the cell.
5. RNA retroviruses have a special enzyme that
   a. disintegrates host DNA.
   b. polymerizes host DNA.
   c. transcribes viral RNA to cDNA.
   d. translates host DNA.
6. Viroids
   a. have a more complicated structure than viruses.
   b. unlike viruses do not cause disease.
   c. are simpler than viruses and do cause disease.
   d. Both a and b.

7. The organisms in Kingdom Monera are
   a. prokaryotes and bacteria.
   b. either archaebacteria or eubacteria.
   c. cyanobacteria and viruses.
   d. Both a and b.
8. Facultative anaerobes
   a. require a constant supply of oxygen.
   b. are killed in an oxygenated environment.
   c. do not always need oxygen.
   d. are photosynthetic.
9. Cyanobacteria, unlike other types of bacteria that photosynthesize, do
   a. not give off oxygen.
   b. give off oxygen.
   c. not have chlorophyll.
   d. not have a cell wall.

10. Chemosynthetic bacteria
    a. are autotrophic.
    b. use the rays of the sun to acquire energy.
    c. oxidize inorganic compounds to acquire energy.
    d. Both a and c.
11. Heterotrophic bacteria
    a. always cause disease.
    b. serve many useful functions.
    c. are usually saprophytic decomposers.
    d. Both b and c.
12. Label this diagram of the lytic and lysogenic cycles.

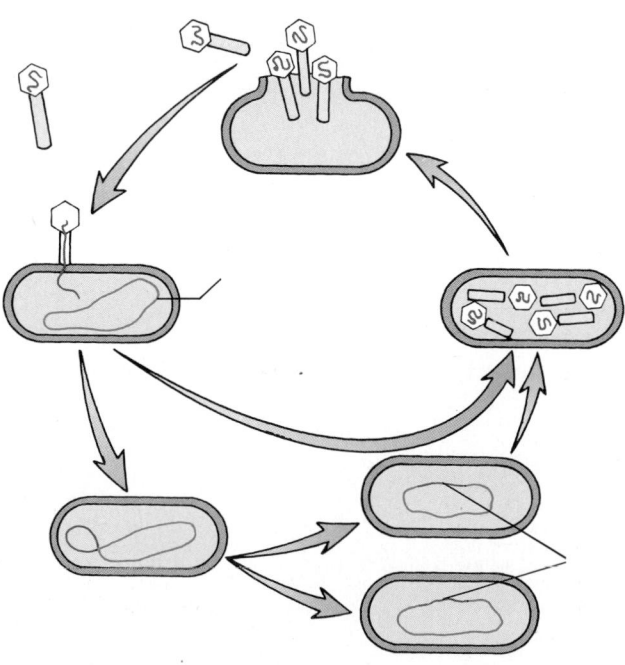

Concepts and Critical Thinking

1. *Viruses are obligate parasites.*

Explain the meaning of this statement.

2. *Monerans are a diverse group of organisms.*

What evidence do you have that Monerans are diverse?

3. *Life evolved from the simple to the complex.*

How can you reconcile this concept with the former concept?

Selected Key Terms

lytic cycle (lit′ik si′kl) 367
lysogenic cycle (li-so-jen′ik si′kl) 367
retrovirus (ret″ro-vi′rus) 368
viroid (vi′roid) 368
prokaryotes (pro′kar-e-otz) 370
bacterium (bak-te′re-um) 370

endospore (en′do-spōr) 371
chemoautotroph (ke″mo-aw′to-trof) 373
saprophyte (sap″ro-fit′) 373
symbiotic (sim″bi-ot′ik) 373
nitrogen fixation (ni′tro-jen fik-sa′shun) 373

archaebacterium (ar″ke-bak-te′re-um) 374
eubacterium (u″bak-te′re-um) 374
cyanobacterium (si″ah-no-bak-te′re-um) 375
lichen (li′ken) 375

24

Protists and Fungi

Amanita muscaria is a type of mushroom with a yellow, orange, or red cap speckled with cream or white flakes. The cap contains metabolic poisons that have been used to kill insects and intoxicate humans. In the Near East, it was used during religious ceremonies and may even have been considered a god at one time.

Your study of this chapter will be complete when you can

1. list and describe the types of organisms that are placed in the kingdom Protista;
2. tell how the protozoans are grouped, and describe a member of each group;
3. describe the life cycle of *Plasmodium vivax,* the cause of malaria;
4. list the different types of algae in the kingdom Protista;
5. describe the 3 types of life cycles found among living organisms; for each, name an alga that has this life cycle, and describe it specifically;
6. tell how the green algae are grouped, and give an example for each group; contrast the manner in which these examples reproduce;
7. describe the 3 types of seaweeds, and contrast their characteristics; tell how *Fucus* reproduces;
8. list, describe, and state the importance of euglenoids, dinoflagellates, and diatoms;
9. describe the funguslike protists, and tell why they are called funguslike;
10. describe the anatomical features of organisms in kingdom Fungi, and show the basis on which they are classified;
11. describe the life cycle of a black bread mold in detail and the life cycles of a sac fungus and a club fungus in general.

Classification

Kingdom PROTISTA
    Phylum Sarcodina: amoeboid protozoans
    Phylum Ciliophora: ciliated protozoans
    Phylum Zoomastigina: flagellated protozoans
    Phylum Sporozoa: parasitic protozoans
    Phylum Chlorophyta: green algae
    Phylum Pyrrophyta: dinoflagellates
    Phylum Euglenophyta: *Euglena* and relatives
    Phylum Chrysophyta: diatoms
    Phylum Rhodophyta: red algae
    Phylum Phaeophyta: brown algae
    Phylum Myxomycota: slime molds
    Phylum Oomycota: water molds
Kingdom FUNGI
    Division Zygomycota: black bread molds
    Division Ascomycota: sac fungi
    Division Basidiomycota: club fungi
    Division Deuteromycota: imperfect fungi (i.e., means of
      sexual reproduction not known).

The protists and fungi are eukaryotes, as are plants and animals (fig. 22.6). Their cells contain a nucleus and all the other organelles we studied in chapter 5. Protists are usually unicellular forms, however, and even the multicellular ones lack the tissue differentiation seen in more complex organisms. This makes it reasonable to assume that protists evolved before those found in the kingdoms Fungi, Plantae, and Animalia. We will rest content with this basic observation; it is not our intent to explore the possible evolutionary relationships among the protists themselves or to say for certain which of them may have given rise to fungi, plants, and animals.

Kingdom Protista

The **protists** are grouped according to their mode of nutrition and other characteristics into the following categories: (1) protozoans, which are the animal-like protists—they are motile and ingest their food like animals; (2) algae, which are the plantlike protists because they are photosynthetic like plants; and (3) slime molds and water molds, which are the funguslike protists (fig. 24.1). Slime molds are heterotrophic and produce windblown spores like fungi, but unlike fungi, they ingest their food. Water molds are heterotrophic and saprophytic like the fungi. A **saprophyte** carries on external digestion and absorbs the resulting nutrients across the plasma membrane. Unlike the fungi, water molds produce swimming zoospores.

    Protists, even those that are only one celled, should not be considered simple organisms. Each cell is organized to cope with the environment; alone it can carry out all the functions assumed by various tissues and organs in more complex organisms.

Figure 24.1

Dark-field micrographs illustrating protist diversity. ***a.*** Several paramecia. They are representative of the protists that are animal-like in that they ingest their food. Magnification, X63. ***b.*** *Volvox* colonies with young escaping. These organisms are representative of the protists that are plantlike in that they carry on photosynthesis. This particular protist is colonial, being a loose association of cells that show some specialization. Magnification, X16. ***c.*** A slime mold plasmodium moves slowly along the forest floor using dead organic matter as its food. In this way, slime molds are like fungi, which also feed on dead organic matter.

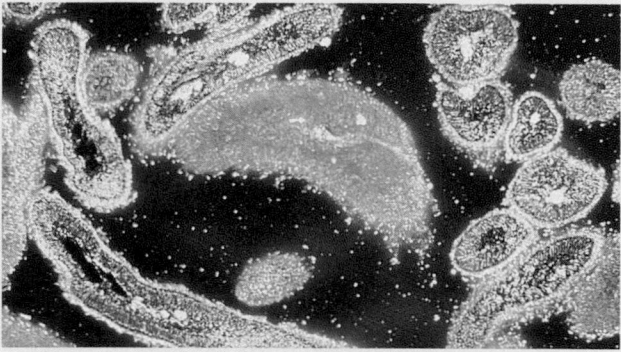

a.

b.

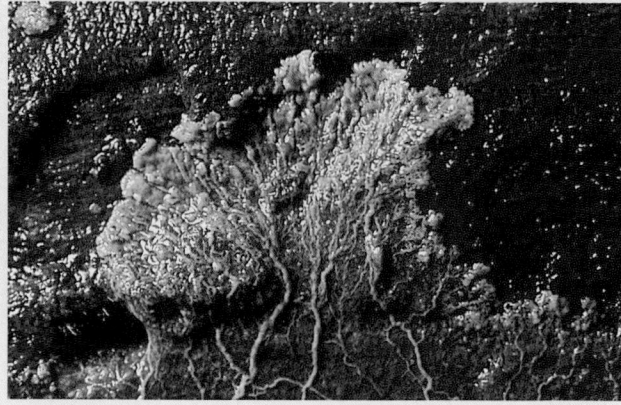

c.

The protists are usually single-celled organisms that are grouped according to their mode of nutrition. There are the heterotrophic protozoans, the photosynthetic algae, and slime molds and water molds, which somewhat resemble fungi.

Protozoans

Protozoans are small (2µm–1,000µm), usually colorless unicellular organisms that lack a cell wall. Like animals, they are heterotrophic and characteristically motile. Some are colonial; a colony is a loose association of cells, each of which must still take care of its own physiological needs. On occasion, some cells of a colony specialize for reproduction.

In addition to the usual organelles, protozoans have specialized vacuoles. For example, they take in food by simple diffusion, transport, or phagocytosis. They digest food in food vacuoles. Also, freshwater protozoans continually gain water by osmosis and eliminate it by means of "contractile" vacuoles. These vacuoles expand with water drawn from the cytoplasm and then appear to "contract," releasing the water through a temporary opening in the plasma membrane (fig. 6.10).

Although asexual reproduction involving binary fission and mitosis is the rule, many protozoans also reproduce sexually during some part of their life cycle. Sexual exchange is preceded by meiosis.

The protozoans we will study can be placed in 4 groups according to their type of locomotor organelle (table 24.1).

Amoeboids

Typically, the *amoeboids* (phylum Sarcodina) move about and feed by means of cytoplasmic extensions called **pseudopodia.** In *Amoeba proteus,* a commonly studied member of this group, the pseudopodia form when the cytoplasm streams forward in a particular direction (fig. 24.2). When amoebas feed, they **phagocytize;** the pseudopodia surround and engulf the prey, which may be algae, bacteria, or other protozoans.

Amoeba proteus is found in fresh water, but most amoebas are marine and are either radiolarians or foraminiferans. The radiolarians float near the ocean surface despite an internal skeleton composed of silica or strontium sulfate. Their skeletons are intricate, exquisite, and of almost infinite variety. The foraminiferans have an external calcareous skeleton (made up of calcium carbonate) (fig. 24.3). When they feed, pseudopodia extend through holes in the skeleton called foramina. These protozoa are more often found in the ooze of the ocean floor. They have been so populous through the ages that following a land upheaval, their accumulated shells formed the White Cliffs of Dover along the southern coast of England.

Ciliates

The *ciliates* (phylum Ciliophora), such as those in the genus *Paramecium* (fig. 24.4), are the most complex of the protozoans. Hundreds of cilia project through tiny holes in the outer covering, or pellicle. An array of organelles performs special functions. For example, when a paramecium feeds, food is

Table 24.1		
Types of Protozoans		
Protozoans	**Locomotor Organelles**	**Example**
Amoeboids	Pseudopodia	*Amoeba*
Ciliates	Cilia	*Paramecium*
Zooflagellates	Flagella	*Trypanosoma*
Sporozoans	No locomotor organelles	*Plasmodium*

Figure 24.2

Amoeba proteus is a protozoan that moves by formation of pseudopodia. Note also the unique organelles, including the food vacuoles and the contractile vacuole. This artist's representation is based on electron micrographs. Arrows indicate the flow of cytoplasm as a newly formed pseudopodium takes shape.

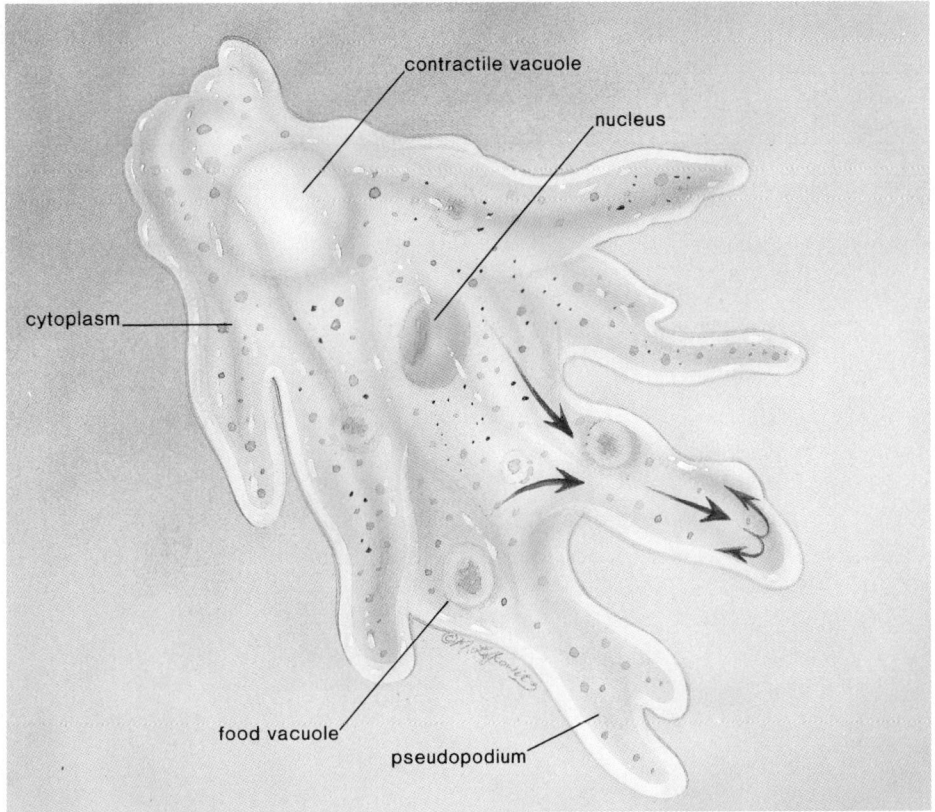

contractile vacuole
nucleus
cytoplasm
food vacuole
pseudopodium

swept down a gullet, below which food vacuoles form. Following digestion, the soluble nutrients are absorbed by the cytoplasm, and the indigestible residue is eliminated at the anal pore.

Ciliates have 2 types of nuclei: a large macronucleus and one or more small micronuclei. The macronucleus controls the normal metabolism of the cell, while the micronuclei are con-

cerned with reproduction. Following meiotic division of the micronuclei, 2 ciliates may exchange these in a sexual process called conjugation (fig. 24.5).

The diversity of ciliates is quite remarkable. The barrel-shaped didinia expand to consume paramecia much larger than themselves. Suctorians have an even more dramatic way of getting food. They rest quietly on a stalk until a hapless victim comes along. Then they promptly paralyze it and use their tentacles like straws to suck it dry. *Stentor* may be the prettiest ciliate. It resembles a giant blue vase decorated with stripes when viewed under the microscope.

Zooflagellates

Protozoans that move by means of flagella (phylum Zoomastigina) are called *zooflagellates* to distinguish them from unicellular algae that also have flagella. The flagella of these eukaryotic cells have the characteristic 9 + 2 microtubular structure discussed on page 79.

Figure 24.3
Shell of foraminiferan, an amoeboid that has existed for more than 500 million years, time enough for hundreds of lineages to have evolved, diversified, and become extinct. In large shells like this one, each story of the shell is multichambered. Not only cytoplasm but also symbiotic algae are within each chamber. The algae use the foraminiferan's waste products to make food, which they share with their host. Without the algae, a foraminiferan could not grow as large, live as long, or produce as many offspring.

Figure 24.5
Two ciliates (*Paramecium*) undergoing conjugation, during which they exchange nuclear material. Magnification, X40.

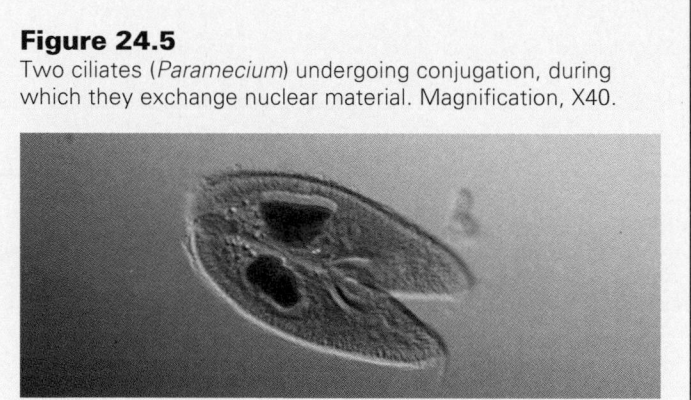

Figure 24.4
The principal structures found in *Paramecium caudatum.* Despite its complexity, a paramecium is a single-celled organism. Trichocysts are poisonous, threadlike darts used for defense and capturing prey. Arrows show path of food vacuoles from formation to discharge of nondigestible particles.

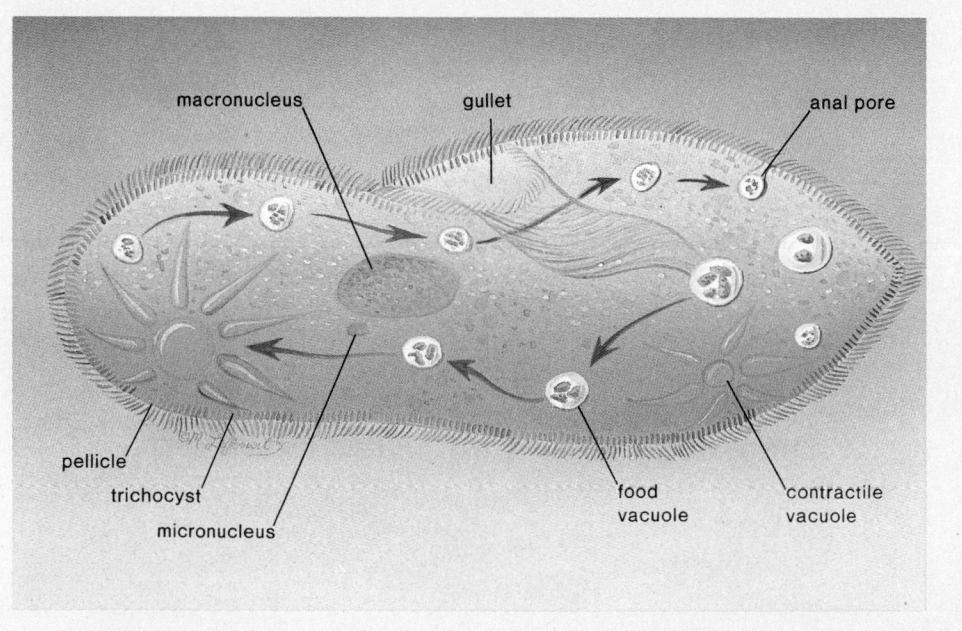

macronucleus   gullet   anal pore

pellicle

trichocyst

micronucleus

food vacuole

contractile vacuole

Evolution and Diversity

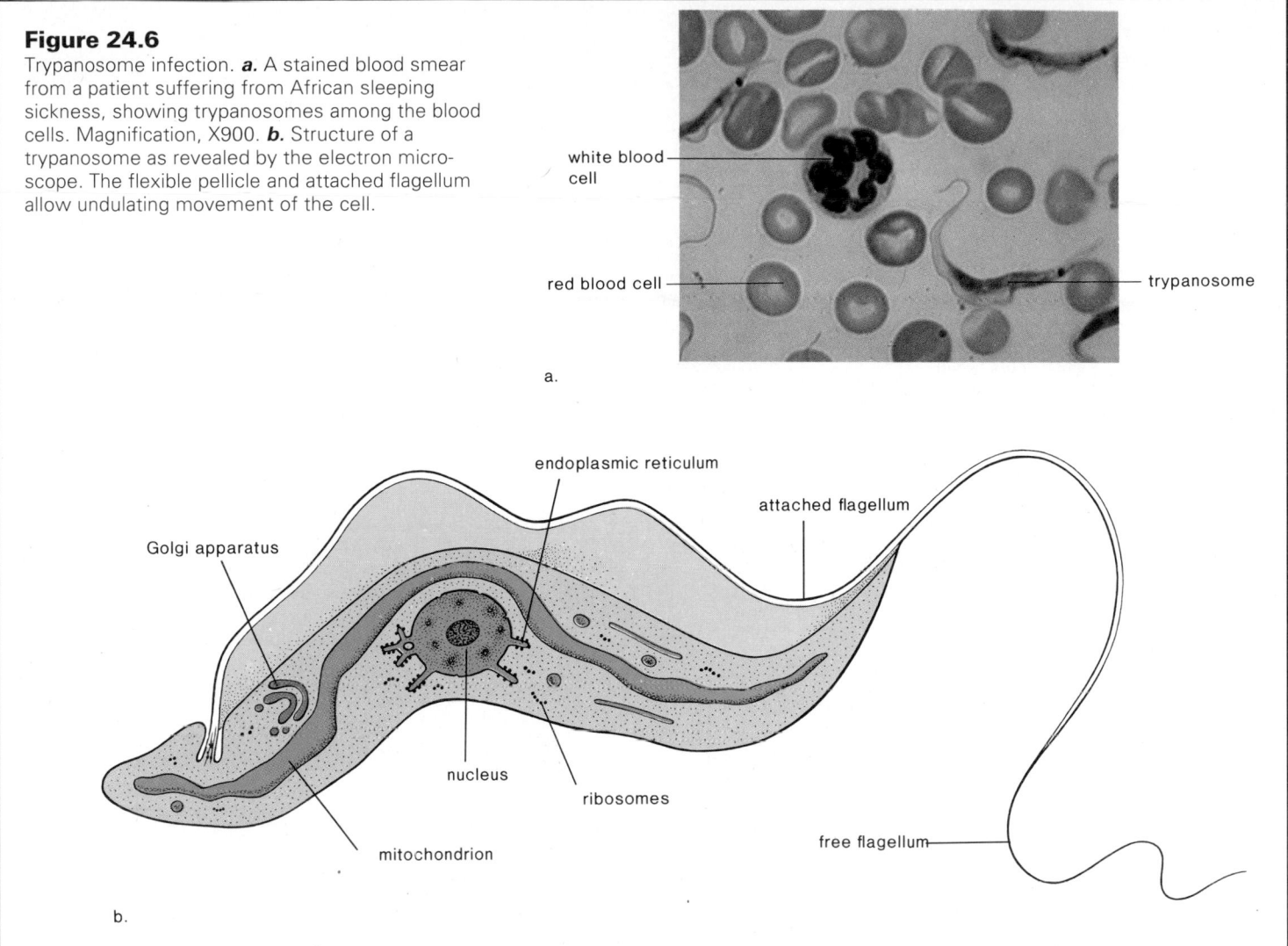

Figure 24.6
Trypanosome infection. **a.** A stained blood smear from a patient suffering from African sleeping sickness, showing trypanosomes among the blood cells. Magnification, X900. **b.** Structure of a trypanosome as revealed by the electron microscope. The flexible pellicle and attached flagellum allow undulating movement of the cell.

white blood cell

red blood cell

trypanosome

a.

endoplasmic reticulum

attached flagellum

Golgi apparatus

nucleus

ribosomes

mitochondrion

free flagellum

b.

Many zooflagellates enter into symbiotic relationships (see table 23.1). *Trichonympha collaris* lives in the gut of termites; it contains a bacterium that enzymatically converts the cellulose of wood to soluble carbohydrates easily digested by the insect. The trypanosomes, which cause African sleeping sickness, are transmitted to the blood of vertebrates by the tsetse fly (fig. 24.6). The tsetse fly, which becomes infected when it takes a blood meal from a diseased animal, passes on the disease when it feeds on another victim. The white blood cells in an infected animal accumulate around the blood vessels leading to the brain and cut off circulation. The lethargy characteristic of the disease is caused by an inadequate supply of oxygen for normal brain alertness.

Sporozoans

The *sporozoans* (phylum Sporozoa) are also parasitic protozoans whose complicated life cycle almost always involves the formation of infective spores. The most widespread human parasite is *Plasmodium vivax*, the causative agent of one type of malaria. When a human is bitten by an infected *Anopheles* mosquito, the parasite eventually invades the red blood cells. The chills and fever of malaria occur when the infected cells burst and release toxic substances into the blood (fig. 24.7).

Malaria is still a major killer of humans, despite extensive efforts to control it. As discussed on page 325, resurgence of this disease is caused primarily by the development of insecticide-resistant strains of mosquitoes and by parasite resistance to antimalarial drugs. Biotechnicians are presently trying to develop a vaccine utilizing a specific protein from the protozoan.

Protozoans are generally single cells that are heterotrophs and ingest food as animals do. They are grouped according to their mode of locomotion.

Algae

The term **algae** is used for aquatic organisms that photosynthesize as do terrestrial plants. Algae produce the food that maintains communities of organisms both in oceans and in bodies of fresh

Figure 24.7

Life cycle of *Plasmodium vivax*, a protozoan that causes one type of malaria. *Asexual reproduction in humans:* In the liver, spindle-shaped sporozoites produce asexual spores that invade and reproduce inside the red blood cells. After a time, some of these spores become gametocytes and then gametes if taken up by another mosquito. *Sexual reproduction in mosquitos:* The gametes mate, and a wormlike zygote encysts in the stomach wall where it divides to produce sporozoites. Sporozoites find their way to the salivary glands of the mosquito, and the cycle repeats.

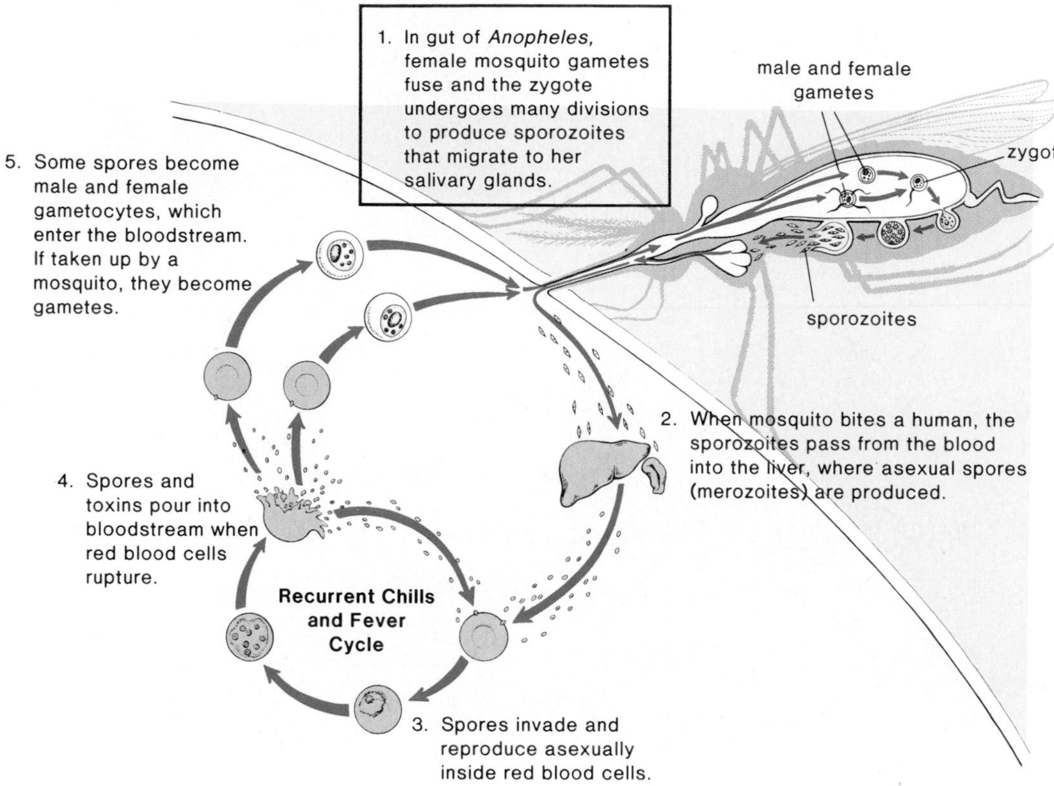

1. In gut of *Anopheles*, female mosquito gametes fuse and the zygote undergoes many divisions to produce sporozoites that migrate to her salivary glands.

male and female gametes

zygote

sporozoites

5. Some spores become male and female gametocytes, which enter the bloodstream. If taken up by a mosquito, they become gametes.

4. Spores and toxins pour into bloodstream when red blood cells rupture.

Recurrent Chills and Fever Cycle

2. When mosquito bites a human, the sporozoites pass from the blood into the liver, where asexual spores (merozoites) are produced.

3. Spores invade and reproduce asexually inside red blood cells.

water. They are commonly named for the type of pigment they contain; therefore, there are green, golden brown, brown, and red algae. All algae contain green chlorophyll, but they may also contain other pigments that mask the color of the chlorophyll. Algae are grouped according to pigmentation and biochemical differences, such as the chemistry of the cell wall and the chemical compound used to store excess food.

Life Cycles

A **life cycle** is the entire sequence of developmental phases from zygote formation to gamete production. There are 3 types of life cycles exhibited by living organisms, and we find all these types among the algae: in the **haplontic cycle,** the adult is haploid; in the **alternation of generations cycle,** a haploid generation alternates with a diploid generation; in the **diplontic cycle,** the adult is always diploid. These cycles are diagrammed in figure 24.8 and are further contrasted in table 24.2.

In the haplontic and alternation of generations cycles, meiosis produces **spores,** structures that mature to a haploid generation. Plants, which are studied in the next chapter, use the alternation of generations life cycle. In the diplontic cycle,

meiosis produces gametes, the only haploid structures found within the cycle. Animals, including human beings, follow the diplontic cycle.

In general, an organism exhibits one of 3 life cycles (haplontic, alternation of generations, or diplontic). All 3 of these are seen among the algae.

Green Algae

The green algae (phylum Chlorophyta) are single-celled, colonial, filamentous, and multicellular. It is sometimes suggested that the green algae are ancestral to the first plants because both of these groups possess chlorophylls *a* and *b*, both store reserve food as starch, and both have cell walls that contain cellulose.

Flagellated Green Algae *Chlamydomonas* is a single-celled green alga that has been studied in detail using the electron microscope (fig. 24.9*a*). It has a definite cell wall and a single, large, cup-shaped chloroplast that contains a pyrenoid, a dense body where starch is stored. A red-pigmented eyespot is sensitive to light and helps bring the organism into the light, where photo-

Figure 24.8

Organisms have one of 3 types of life cycles. **a.** The haplontic cycle is typical of algae and fungi. Notice that the adult is haploid, the gametes are sometimes isogametes (look alike), and the only diploid part of the cycle is the zygote, which undergoes meiosis to produce spores. **b.** The alternation of generations cycle is typical of plants. Notice that there are 2 generations. The sporophyte, the diploid generation (2N), produces the spores by meiosis. The gametophyte, the haploid generation (N), produces gametes. **c.** The diplontic cycle is typical of animals. Notice that the adult is always diploid, and meiosis produces heterogametes (egg and sperm). The earliest cycle may have been the haplontic, which could have led to both the alternation of generations cycle and the diplontic cycle.

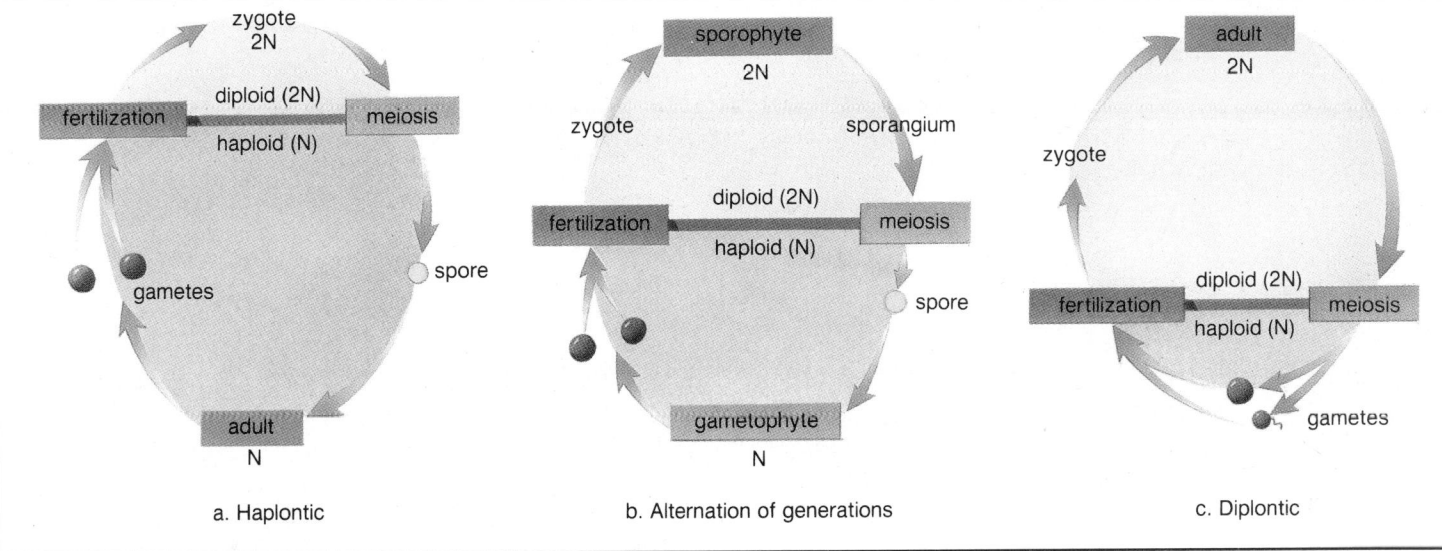

a. Haplontic   b. Alternation of generations   c. Diplontic

Table 24.2 Life Cycles		
Cycle	**Chromosome Number in Adult(s)**	**Spores**
Haplontic	Haploid only	Yes
Alternation of generations	Haploid ⟷ Diploid	Yes
Diplontic	Diploid only	No

synthesis can occur. Two flagella project from the anterior end of this alga and move the cell freely toward the light. *Chlamydomonas* has both animal-like and plantlike characteristics in that it is motile and yet makes its own food; it is believed that such an organism is a good example of what the most primitive protist was like.

Chlamydomonas has the haplontic life cycle (fig. 24.9b). Usually this protist practices asexual reproduction and the adult divides to give *zoospores* (flagellated spores) that resemble the parent cell. During sexual reproduction, gametes of 2 different strains come into contact and join to form a zygote. A heavy wall forms around the zygote, and it becomes a resistant zygospore able to survive until conditions are favorable for germination. Then, it produces 4 zoospores by meiosis. The gametes shown in figure 24.9b are *isogametes:* that is, they look exactly alike.

The life cycle of *Chlamydomonas* illustrates the primary differences between asexual and sexual reproduction (table 24.3). Sexual reproduction is simply reproduction that involves the use of gametes. Distinct and separate sexes and heterogametes (dissimilar gametes, such as egg and sperm) are not required. Sexual reproduction aids the process of evolution because it offers a means to produce variations in addition to mutations.

Colonial Green Algae   A number of *colonial* (loose association of cells) forms occur among the flagellated green algae. A *Volvox* colony is a hollow sphere with thousands of cells arranged in a single layer surrounding a watery interior. The cells of a *Volvox* colony, each of which resembles a *Chlamydomonas* cell, cooperate in that the flagella beat in a coordinated fashion. Some cells are specialized for reproduction, and each of these can divide asexually to form a new daughter colony (fig. 24.10a). This daughter colony resides for a time within the parental colony, but then it leaves the parental colony by releasing an enzyme that dissolves away a portion of the parental colony, allowing it to escape. Sexual reproduction involves **heterogametes**—large, nonmotile eggs and small, flagellated sperm (fig. 24.10b).

Filamentous Green Algae   Filaments are end-to-end chains of cells that form after cell division occurs in only one plane.

Spirogyra, found in green masses on the surfaces of ponds and streams, has chloroplasts that are ribbonlike and are arranged in a spiral within the cell (fig. 24.11). These species practice a form of sexual reproduction called **conjugation,** during which the

Figure 24.9

a. The structure of *Chlamydomonas,* a motile green alga.
b. *Chlamydomonas* has the haplontic life cycle. During sexual reproduction, mature adults produce gametes, which fuse to give a zygote. The zygote becomes a thick-walled zygospore, which enters a period of dormancy. Upon germination, meiosis produces zoospores, which grow to become haploid adults. These individuals usually reproduce asexually, with each of the original zoospores giving rise to many individuals.

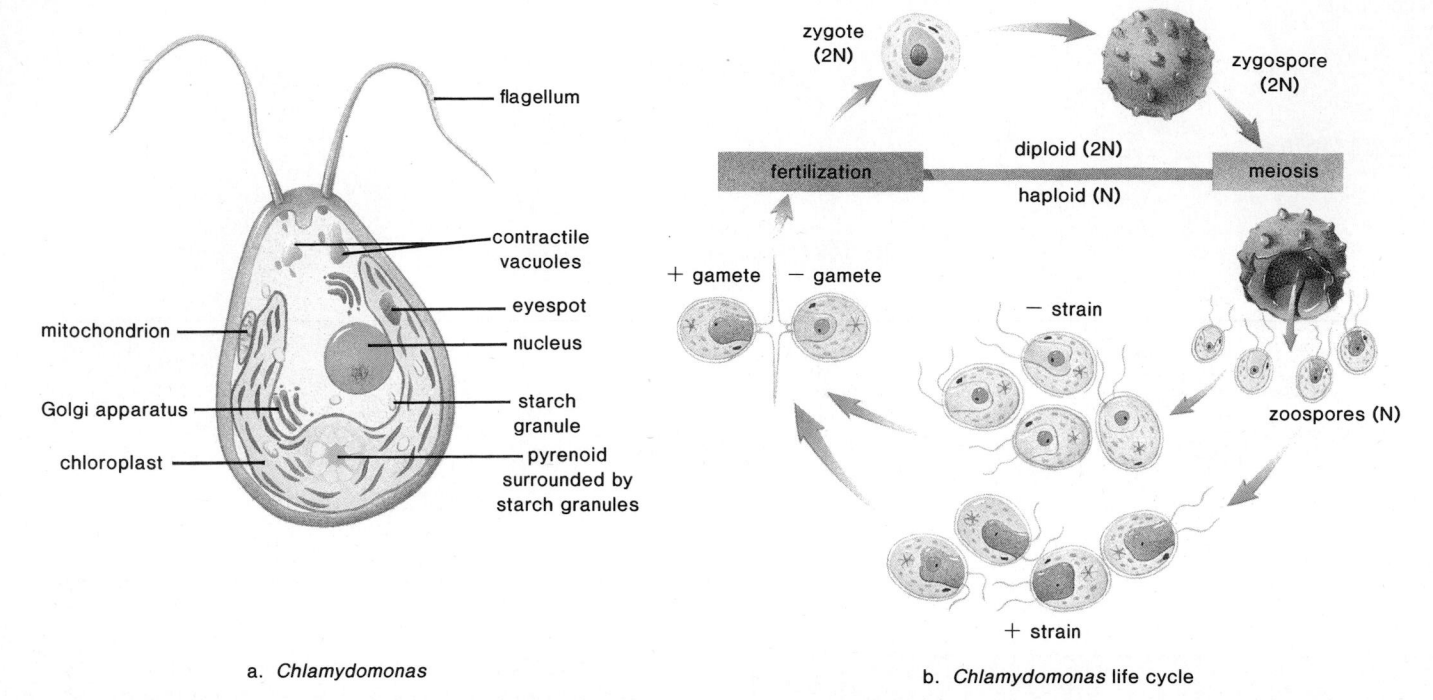

a. *Chlamydomonas*

b. *Chlamydomonas* life cycle

Table 24.3
Asexual Reproduction versus Sexual Reproduction

	Asexual	Sexual
Number of parents	One	Two
Gametes	No	Yes
Recombination of genes	No	Yes

---

contents of the cells act as isogametes. The 2 filaments line up next to each other, and the contents of the cells of one filament move into the cells of the other filament, forming diploid zygotes. These zygotes survive the winter, and in the spring, they undergo meiosis to produce new haploid filaments.

Among the filamentous green algae, there are forms that use heterogametes during sexual reproduction. For example, the genus *Oedogonium* contains filamentous algae in which the cells are cylindrical with netlike chloroplasts. Members of this genus produce sperm and egg for the purpose of sexual reproduction.

Multicellular Green Algae Multicellular *Ulva* is commonly called sea lettuce because of a leafy appearance (fig. 24.12). *Ulva* shows alternation of generations in the same way as terrestrial plants except that (1) the spores are flagellated, (2) there are isogametes, and (3) both generations look exactly alike. In plants, the spores are not flagellated; there are heterogametes, and one generation is typically dominant over (longer lasting than) the other.

> Green algae are a diverse group that have some of the same characteristics as plants. The haplontic life cycle is typical but *Ulva* has a life cycle that has 2 distinct generations, like that of plants.

Seaweeds

Multicellular green algae (such as *Ulva*), red algae, and brown algae are all seaweeds. All seaweeds have chlorophyll, but the green pigment can be masked by red or brown pigments. At one time it was believed that accessory pigments helped determine the depth at which algae were found, but recent studies have shown that there is no difference in the overall depth distribution of red, brown, and green seaweeds.

Figure 24.10
The genus *Volvox* contains colonial green algae. ***a.*** The adult often contains daughter colonies, which are asexually produced by special cells. Magnification, X30. ***b.*** During sexual reproduction, colonies produce definite sperm and eggs. Some produce either sperm or eggs, and others produce both sperm and eggs.

daughter colonies
sperm
vegetative cell
sperm
zygote
egg

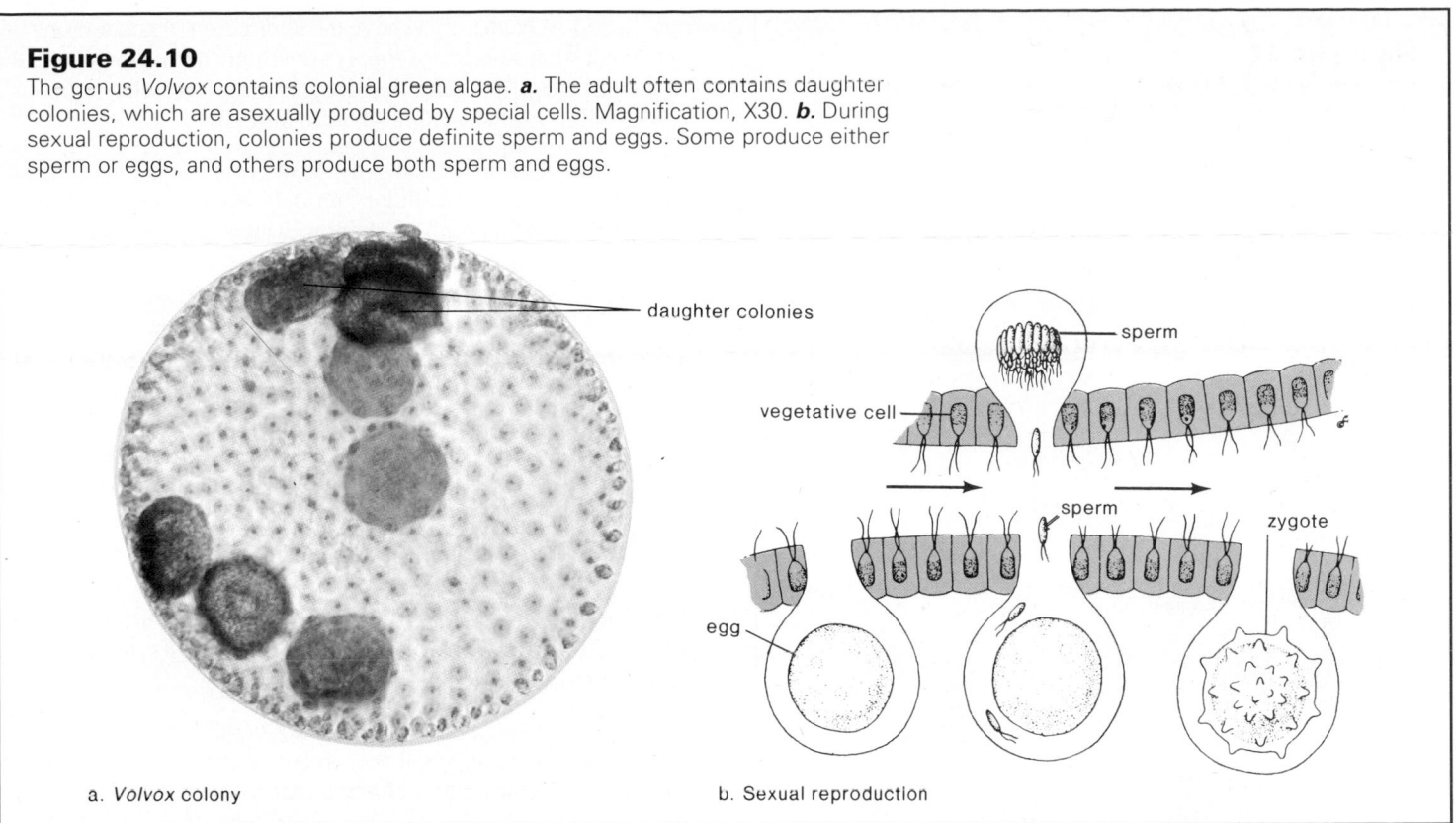

a. *Volvox* colony

b. Sexual reproduction

Figure 24.11
Members of the genus *Spirogyra* are filaments of cells.***a.*** Anatomy of one cell. ***b.*** Micrograph depicting conjugation between cells of 2 filaments. Magnification, X350.

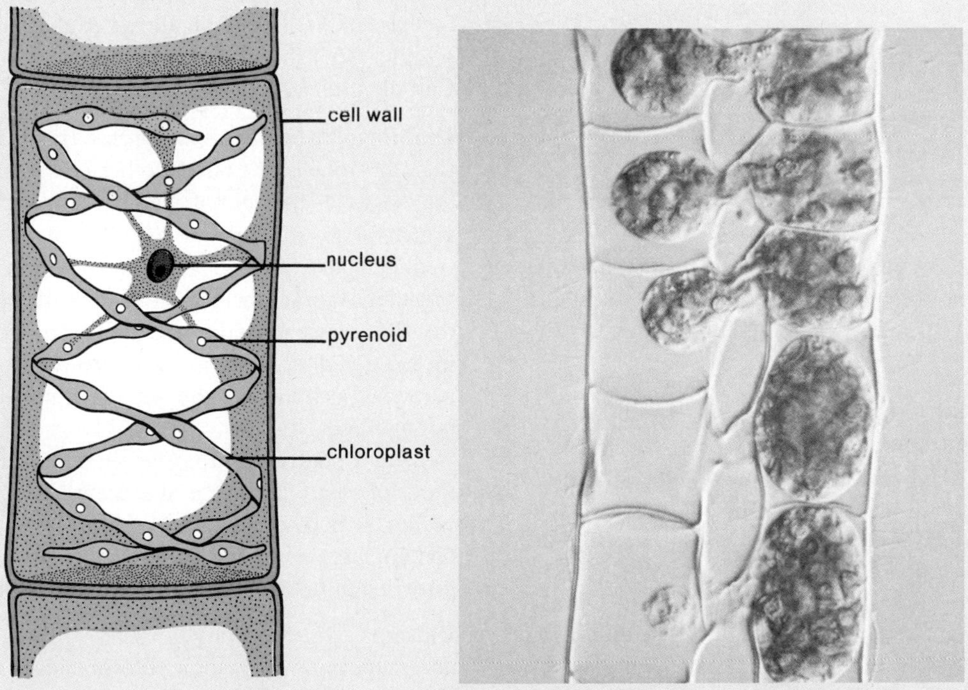

cell wall
nucleus
pyrenoid
chloroplast

a.

b.

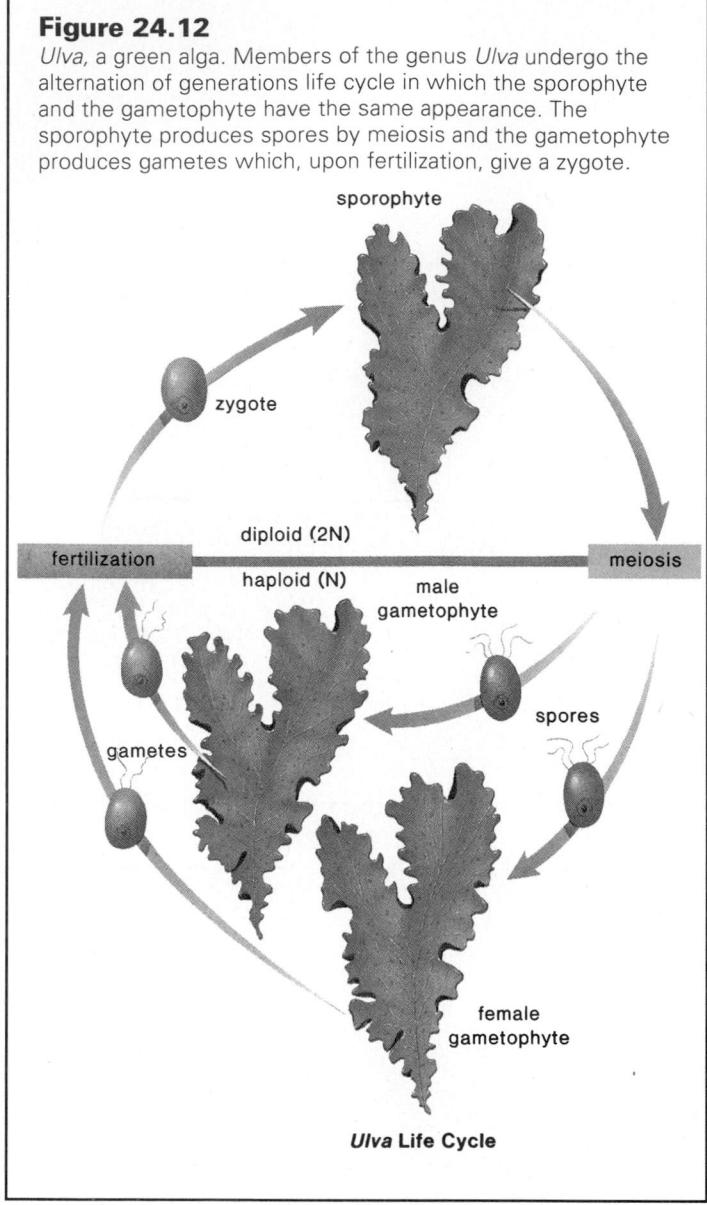

Figure 24.12

Ulva, a green alga. Members of the genus *Ulva* undergo the alternation of generations life cycle in which the sporophyte and the gametophyte have the same appearance. The sporophyte produces spores by meiosis and the gametophyte produces gametes which, upon fertilization, give a zygote.

sporophyte

zygote

diploid (2N)

fertilization   meiosis

haploid (N)

male gametophyte

spores

gametes

female gametophyte

***Ulva* Life Cycle**

Although the seaweeds are multicellular, as are plants, this does not mean that they are closely related to plants. Multicellularity most likely arose independently in several different lines.

Brown Algae Brown algae (phylum Phaeophyta) range from small forms with simple filaments to large forms between 50 m and 100 m long (fig. 24.13). Large brown algae are often observed along the rocky shoreline in the north temperate zone, where they are pounded by waves as the tide comes in and are exposed to dry air at low tide. These plants are firmly anchored by holdfasts, and when the tide is in, their broad flattened blades are buoyed by air bladders. When the tide is out, they do not dry out because their cell walls contain a mucilaginous, water-retaining material.

Most brown algae have the alternation of generations life cycle, but some species of *Fucus* are unique in that they have the diplontic cycle (fig. 24.8). In this cycle meiosis produces gametes and the adult is always diploid, as in animals.

Red Algae Like the brown algae, the red algae (phylum Rhodophyta) are multicellular, but they occur chiefly in warmer seawaters, growing both in shallow waters and as deep as light penetrates. Some forms of red algae are filamentous, but more often they are complexly branched, with the branches having a feathery, flat, and expanded or ribbonlike appearance. Coralline algae are red algae that have cell walls impregnated with calcium carbonate. In some instances, they contribute as much to the growth of coral reefs as do coral animals.

The seaweeds are multicellular protists that are grouped according to the type of accessory pigment they contain.

Other Types of Algae

There are several groups of exclusively unicellular algae in the kingdom Protista. They also contain chloroplasts and carry out photosynthesis in a manner similar to plants.

Euglenoids *Euglenoids* (phylum Euglenophyta) are freshwater organisms with both animal-like and plantlike characteristics (fig. 24.14). Their animal-like characteristics include motility and a flexible body wall. Euglenoids move by means of 2 flagella, typically one much longer than the other, that project to the anterior. Because euglenoids are bounded by a flexible pellicle instead of a rigid cell wall, they can assume different shapes as the underlying cytoplasm undulates or contracts.

Their plantlike characteristics sometimes include the possession of chloroplasts. A photoreceptor shaded by an eyespot at the base of one flagellum allows euglenoids to judge the direction of light. After they move toward light, photosynthesis takes place. Carbohydrates are stored as starch.

Dinoflagellates *Dinoflagellates* (phylum Pyrrophyta) have 2 flagella—one is free but the other is located in a transverse groove that encircles the animal. The beating of these flagella causes the organism to spin like a top. The cell wall, when present, is frequently divided into polygonal plates of cellulose closely joined together. At times there are so many of these organisms in the ocean that they cause a condition called "red tide." The toxins given off in these red tides cause widespread fish kills and can cause paralysis in humans who eat shellfish that have fed on the dinoflagellates.

Usually the dinoflagellates are an important source of food for small animals in the ocean. They also live as symbiotes within the body of some invertebrates. For example, corals usually contain large numbers of these organisms and this allows them to grow much faster than would otherwise be possible.

Diatoms *Diatoms* (phylum Chrysophyta) have a golden brown accessory pigment in their chloroplasts that can mask the color of chlorophyll (fig. 24.15). The structure of a diatom is often com-

Evolution and Diversity

Figure 24.13

Diversification among the brown algae, marine organisms. Both *Laminaria,* a type of kelp, and *Fucus,* known as rockweed, are examples of brown algae that grow along the shoreline. In deeper waters, giant kelps, such as *Nereocystis* and *Macrocystis,* often form spectacular underwater forests. Individuals of *Sargassum natans* sometimes break off from their holdfasts and form floating masses, where life forms congregate in the ocean. Brown algae not only provide food and habitat for marine organisms, they are harvested for human food and for fertilizer in several parts of the world. They are also a source of algin, a pectinlike material that is added to ice cream, sherbert, cream cheese, and other products to give them a stable, smooth consistency.

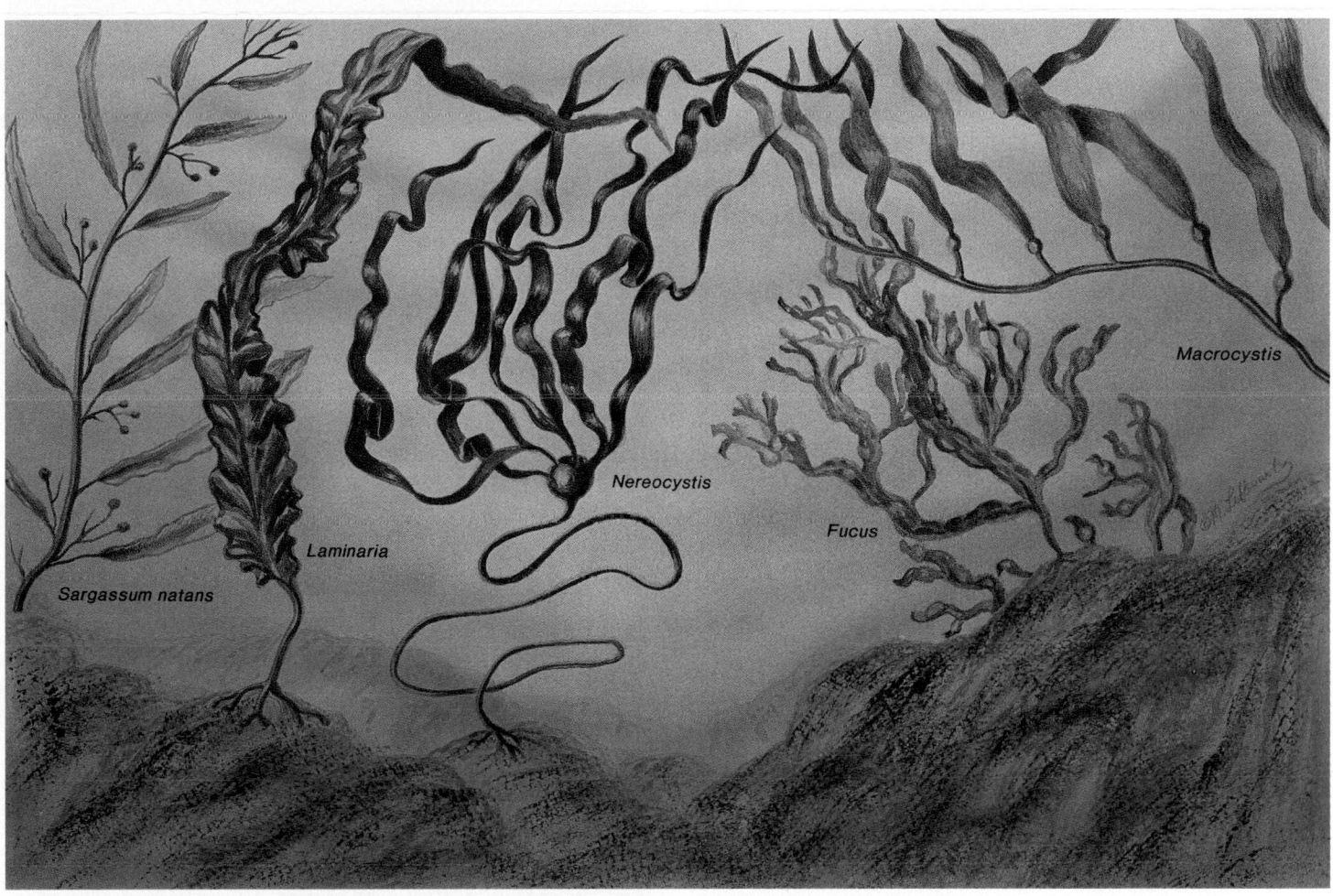

pared to a box because the cell wall has 2 halves, or valves, with the larger valve acting as a "lid" for the smaller valve. When diatoms reproduce, each receives one old valve. The new valve fits inside the old one.

The cell wall has an outer layer of silica, a common ingredient of glass. The valves are covered with a great variety of striations and markings that form beautiful patterns when observed under the microscope. Diatoms are the most numerous unicellular algae in the oceans. As such, they serve as an important source of food for other organisms. Their remains, called diatomaceous earth, accumulate on the ocean floor and are mined for use as filtering agents, soundproofing materials, and scouring powders.

Slime Molds and Water Molds

Slime molds and water molds have funguslike characteristics in that they live on dead organic matter and produce spores.

Slime Molds

There are 2 types of *slime molds* (phylum Myxomycota): cellular and acellular (fig. 24.1c). Both types have an amoeboid stage that lives on rotting logs or dead agricultural crops. In cellular slime molds, the total mass is composed of individual amoeboid cells, but in acellular slime molds, the mass is multinucleated and is called a *plasmodium*. In another part of their life cycle, slime molds form fruiting bodies, stalks that produce windblown spores. The spores germinate to give cells that join to form a zygote, the beginning of a new cycle.

Water Molds

Some *water molds* (phylum Oomycota) live in the water, where they parasitize fishes, forming furry growths on their gills. In spite of their common name, others live on land and parasitize insects and plants. A water mold was responsible for the 1840s

Figure 24.14

Euglena anatomy. *Euglena* is typical of protozoans that have both animal-like and plantlike characteristics. A very long flagellum propels the body, which is enveloped by a flexible pellicle made of protein. A photoreceptor allows *Euglena* to find light, after which photosynthesis can occur in its numerous chloroplasts. In addition to the pyrenoids, which store starch, there are starch granules in the cytoplasm.

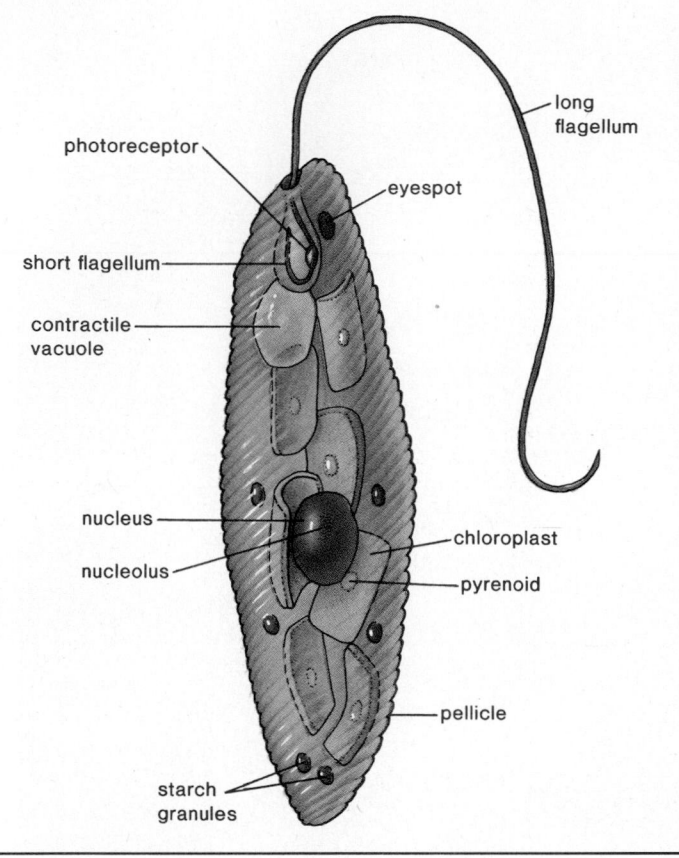

Figure 24.15

Diatom diversity. These unicellular algae may be various colors but they all have a unique golden brown pigment in addition to chlorophyll within the chloroplasts. These beautiful patterns are markings on their silica-embedded walls. Magnification, X63.

Evolution and Diversity

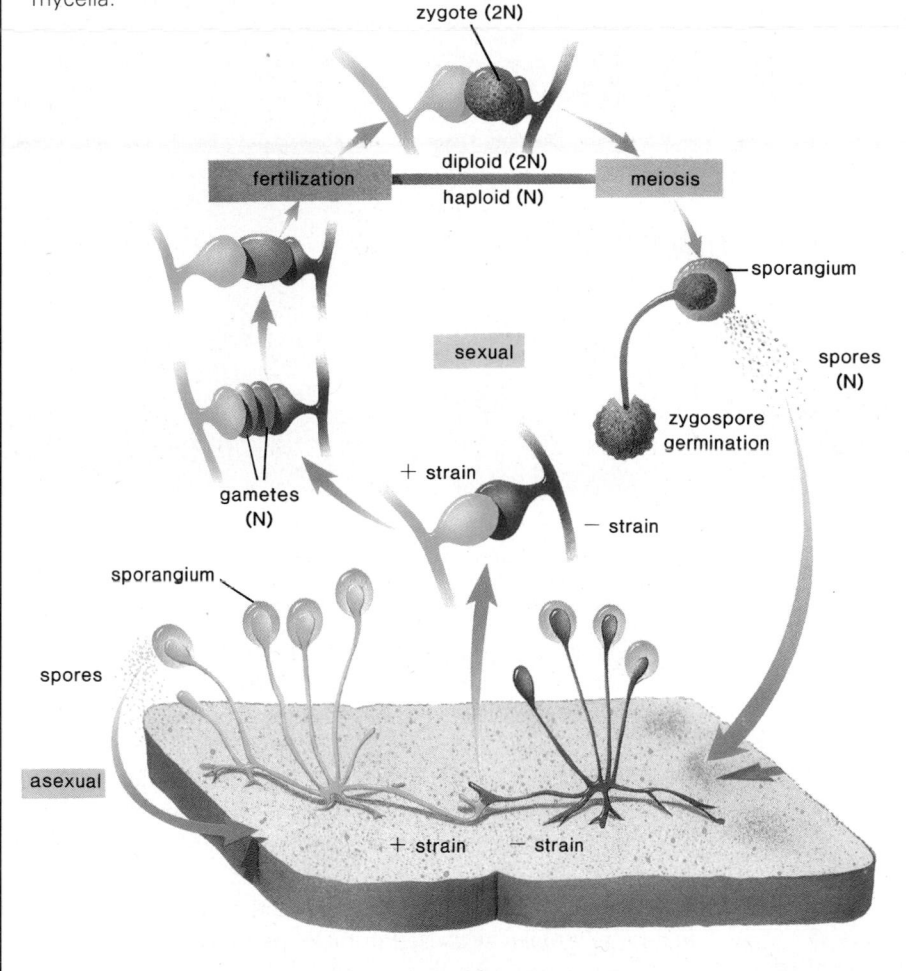

Figure 24.16
Life cycle of the black bread mold *Rhizopus*. Asexually, a mycelium gives rise to spore-producing sporangia (*lower left*). Sexually, the tips of hyphae from opposite mating strains can fuse, giving a zygote, the only diploid (2N) portion of the cycle (*above*). After a period of dormancy, meiosis is followed by germination and production of a sporangium. Windblown spores, an adaptation to land, produce mycelia.

zygote (2N)

fertilization

diploid (2N)

haploid (N)

meiosis

sporangium

sexual

spores (N)

zygospore germination

gametes (N)

+ strain

− strain

sporangium

spores

asexual

+ strain   − strain

hypha is an elongated cylinder containing a mass of cytoplasm and many haploid nuclei, which may or may not be separated by cross walls. A collection of hyphae is called a **mycelium.**

Fungi reproduce in accordance with the haplontic life cycle (fig. 24.8), and most are adapted to life on land in that they produce windblown spores. Also, most fungi reproduce both asexually and sexually; classification is largely based on the mode of sexual reproduction.

Fungi are saprophytic heterotrophic eukaryotes composed of hyphae (a mycelium). Fungi produce spores during both sexual and asexual reproduction, and the major groups of fungi are distinguishable on the basis of sexual reproduction.

Black Bread Molds

Black bread molds (division Zygomycota) belonging to the genus *Rhizopus* are often used as an example of this group. These molds exist as a whitish or grayish haploid mycelium on bread or fruit (fig. 24.16). During asexual reproduction, some hyphae grow upright, bearing a spherical *sporangium* within which thousands of spores are formed.

During sexual reproduction, hyphae of 2 different strains (usually called plus and minus) reach out to one another and form a diploid zygote that darkens as it enlarges into a *zygospore*. After remaining dormant for several months, the zygospore germinates. Now the nucleus of the zygospore undergoes meiosis to produce a short haploid hypha, which immediately forms a sporangium; windblown spores are subsequently released.

potato famine, which caused many Irish to come to the United States. Most water molds are saprophytic and live off dead organic matter, however.

Water molds have a threadlike body as do fungi, but their cells are diploid, not haploid, and their life cycle is diplontic (fig. 24.8). Also, the cell walls are largely composed of cellulose, quite unlike the fungi.

Kingdom Fungi

Fungi (fungus, sing.), like bacteria, are most commonly saprophytic decomposers that assist in the recycling of nutrients in ecosystems. Some fungi are parasitic, causing serious diseases in plants and animals. The body of all fungi, except unicellular yeast, is made up of filaments called **hyphae.** A

Sac Fungi

There are many different types of sac fungi (division Ascomycota) (fig. 24.17), most of which produce asexual spores called *conidia.* During sexual reproduction, sac fungi form spores called ascospores within saclike cells called asci. In most species, the asci are supported within a **fruiting body,** a collection of specialized hyphae. In cup fungi, the fruiting body takes the shape of a cup.

Yeasts are sac fungi that do not form fruiting bodies. In fact, yeasts are different from all other fungi in that they are unicellular and most often reproduce asexually by budding. Yeasts, as you know, carry out fermentation as follows:

glucose → carbon dioxide + alcohol

Figure 24.17

Sac and club fungi. **a.** Colorful cup fungi (ascomycetes). **b.** Bracket fungi (basidiomycetes) growing on a tree trunk. **c.** Mushrooms (basidiomycetes) of the genus *Mycena* have bell-shaped caps.

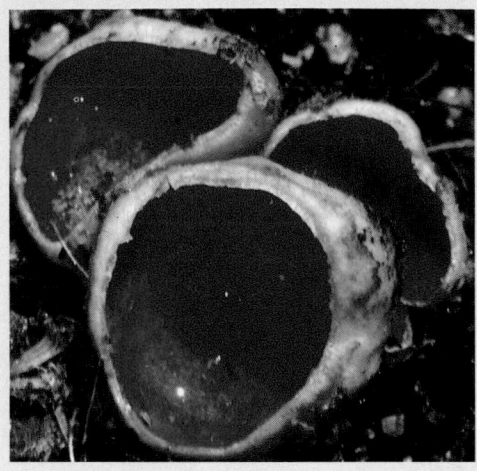

a.

b.

c.

Figure 24.18

Lichen structure. Lichens come in many shapes and sizes, but each is a symbiotic relationship between a fungus and an alga. Whether the fungus contributes to the relationship is debatable, and it may even be parasitic on the alga. **a.** Diagrammatic longitudinal section of a crustose (flat) lichen. **b.** Photo of a crustose lichen. Lichens are important soil formers.

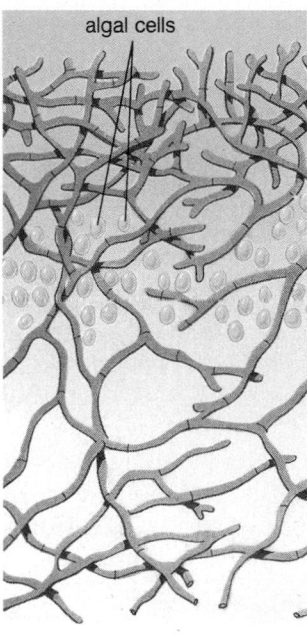

algal cells

a.

b.

In baking, the carbon dioxide from this reaction makes bread rise. The alcohol is of primary importance in the production of wines and beers. The carbon dioxide is retained for beers and sparkling wines; it is released for still wines.

Blue-green molds are sac fungi. The mold *Penicillium* grows on many different organic substances, such as bread, fabrics, leather, and wood. It is used by humans to provide the characteristic flavor of Camembert and Roquefort cheeses; more important, it produces the antibiotic penicillin. Another mold, the red bread mold *Neurospora,* was used in the experiments that helped decipher the function of genes.

Unfortunately, sac fungi are also the cause of chestnut tree blight and Dutch elm disease. These conditions have resulted in the death of most chestnut trees and elm trees in the United States. Also, powdery mildew, apple scab, and ergot, a disease of cultivated cereals, are caused by sac fungi.

A **lichen** is a symbiotic relationship between an alga and a fungus (fig. 24.18). The fungal portion of lichens is usually a sac fungus, while the algal part of this symbiotic relationship is usually a green alga. Lichens can live on bare rock or in poor soil, and they are able to survive great temperature extremes and dryness in all regions of the world. Reindeer moss is a lichen that is an important food source for arctic animals.

Club Fungi

Among the club fungi (division Basidiomycota) (fig. 24.17*b* and *c*), asexual reproduction is accomplished by formation of conidia. As a result of sexual reproduction, members of this group form club-shaped structures called *basidia,* often within fruiting bodies.

Figure 24.19

Life cycle of a mushroom. Beginning at lower right, hyphae from 2 different strains fuse and produce a mycelium in which there are haploid nuclei from both strains. The mycelium gives rise to a mushroom consisting of a stalk and cap. The gills on the underside of the cap are lined with basidia (club-shaped structures). After nuclei fuse to give a diploid nucleus, meiosis in each basidium produces basidiospores, which are windblown. If conditions are favorable, each basidiospore germinates into hyphae. Scanning electron micrograph of a basidium and basidiospores (*upper right*).

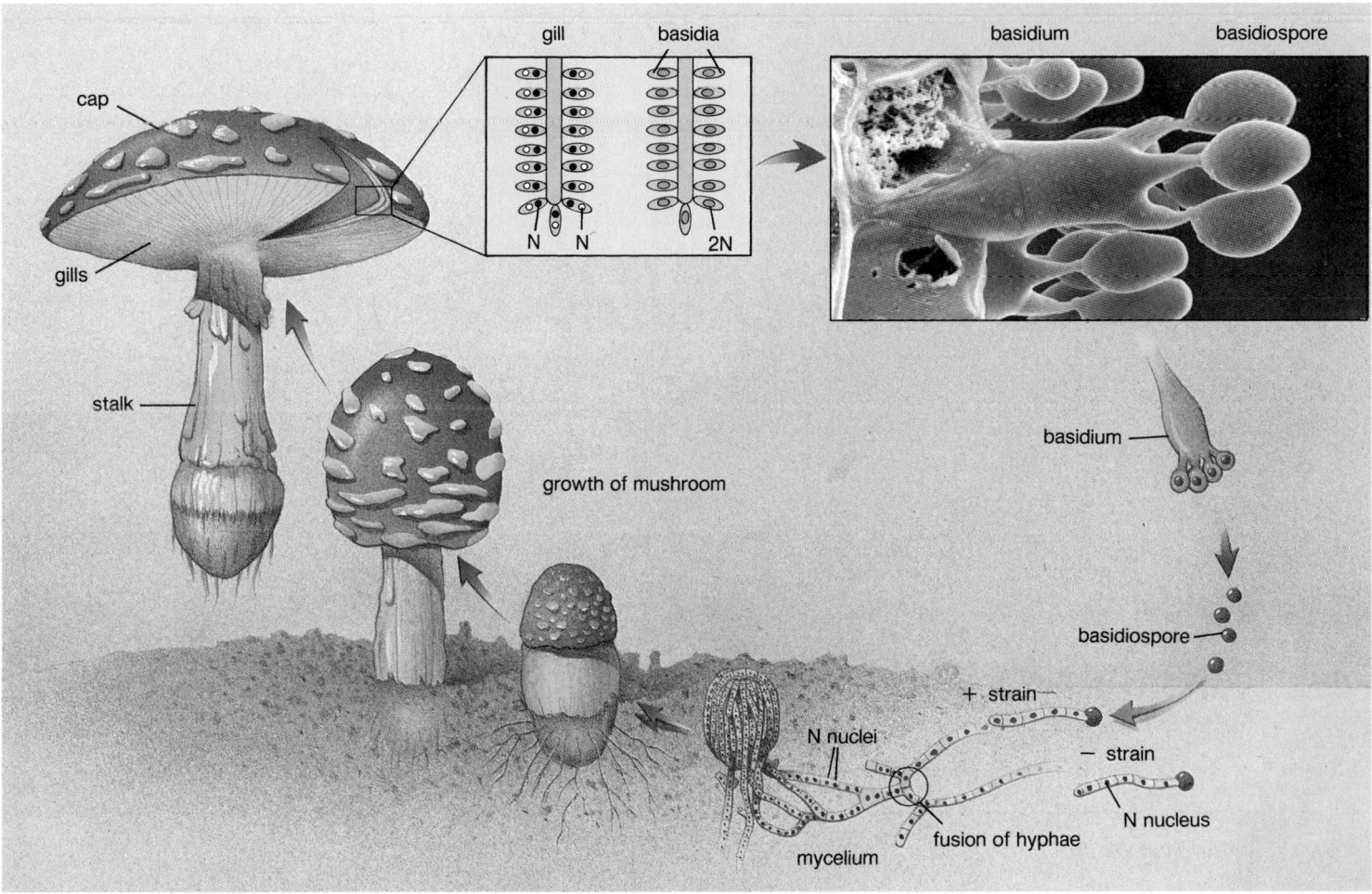

The mushroom, puffball, and bracket (shelf) fungi are club fungi; the visible portions of these are actually fruiting bodies, and the mycelia lie beneath the surface. On the underside of a mushroom cap, the basidia project from the gills. Within each basidium, a diploid nucleus undergoes meiosis to produce basidiospores, which are windblown (fig. 24.19).

Club fungi are economically important. Mushrooms are commercially raised and sold as a delicacy. Rusts and smuts are parasitic club fungi that attack grains, resulting in great economic loss and necessitating expensive control measures. They do not have conspicuous fruiting bodies, and the vegetative hyphae produce spores at various stages of the life cycle. On the other hand, the mycelia of club fungi that lie beneath the soil often form beneficial symbiotic relationships with plants, notably pine trees. These *mycorrhizae* (fungal roots) help the trees garner nutrients from the soil and grow at a faster rate. Therefore, foresters make sure the tree roots are exposed to the fungi before they are planted.

Other Types of Fungi

Some fungi (division Deuteromycota) cannot be assigned to a definite group because the sexual portion of their life cycle has not been observed. For this reason, these fungi are sometimes called "imperfect fungi." The fungi that cause ringworm and athlete's foot belong to this group, as does the yeast *Candida albicans*, which causes thrush, a mouth infection, and moniliasis, a fairly common vaginal infection in females who take birth-control pills.

During sexual reproduction, the black bread molds produce spores in sporangia, the sac fungi produce spores in saclike cells, and the club fungi produce spores in club-shaped structures. The sac and club fungi typically have fruiting bodies. The sexual life cycles for certain fungi that parasitize humans are unknown.

Summary

1. Protists and fungi are eukaryotes. The eukaryotic cell has a nucleus and other organelles, and it divides mitotically.

2. Kingdom Protista contains (1) protozoans, which, like multicellular animals, are heterotrophic and ingest food, (2) algae, which carry on photosynthesis in the same manner as plants, and (3) slime molds and water molds, which have some of the characteristics of fungi.

3. The protozoan cell contains vacuoles that are specialized to carry on functions like digestion of food and removal of excess water (contractile vacuole).

4. The protozoans are grouped according to the type of locomotor organelle: the amoeboids move and feed by forming pseudopodia; the ciliates move by means of cilia; the zooflagellates have flagella; the sporozoans have no locomotor organ and are parasitic (e.g., malaria).

5. Algae are aquatic organisms that photosynthesize, as do higher plants. They are often named for the type of pigment they contain. The 3 different types of life cycles (haplontic, alternation of generations, and diplontic) (fig. 24.8) are found among the algae.

6. Green algae possess chlorophylls *a* and *b*, store reserve food as starch, and have cell walls of cellulose, as do higher plants. Green algae are divided into those that are flagellated (e.g., *Chlamydomonas*, which is haplontic with isogametes, and *Volvox*, which is colonial); filamentous (e.g., *Spirogyra*, which undergoes conjugation, and *Oedogonium*, which is haplontic with heterogametes); and multicellular sheets (e.g., *Ulva*, which exhibits alternation of generations).

7. Some algae are seaweeds. These include the green alga *Ulva*, the brown algae, such as rockweeds and kelps, and the red algae, which are more delicate than the brown algae. It is of interest that the brown alga *Fucus* undergoes the diplontic life cycle.

8. There are 3 types of algae that are exclusively unicellular: euglenoids, which have both plantlike and animal-like characteristics; dinoflagellates; and diatoms, which are golden brown algae that are important to food chains in the ocean.

9. The slime molds and water molds are protists with funguslike characteristics. Slime molds, both cellular and acellular, have an amoeboid stage and then form fruiting bodies, which produce spores. Although most water molds, with threadlike bodies, are saprophytic, they are better known for parasitizing fishes in aquatic habitats and parasitizing higher plants, most notably causing the Irish potato famine of the 1840s.

10. Kingdom Fungi contains organisms that are usually saprophytic decomposers but also cause diseases in plants and animals. The body is made up of hyphae, collectively called a mycelium. Fungi are classified according to specific features of the sexual life cycle.

11. The black bread molds exemplify the haplontic life cycle. The spores are windblown, an adaptation to land. As a part of their sexual life cycle, sac fungi and club fungi form fruiting bodies, conspicuous structures in which windblown spores are produced by meiosis. Other fungi, sometimes termed imperfect fungi because their sexual life cycle is not known, cause some well-known diseases in humans.

Writing Across the Curriculum

In order to practice writing skills, students should write out the answers to any or all of the study questions and the critical thinking questions. The study questions are sequenced in the same order as the text. Suggested answers to the critical thinking questions are in appendix D.

Study Questions

1. Describe the eukaryotic cell, and tell how it may have evolved.
2. What are the 3 types of protists, and what are their distinguishing characteristics?
3. Describe and diagram the protozoan cell.
4. In what manner are the protozoans grouped? Give an example of each group.
5. Describe the life cycle of *Plasmodium vivax*, the causative agent of malaria.
6. What are the 3 types of life cycles found among living things? How do they differ? Show that *Chlamydomonas* and black bread mold have the haplontic cycle; that *Ulva* undergoes alternation of generations. What type of cycle does *Fucus* have?
7. Contrast how *Volvox, Spirogyra,* and *Oedogonium* reproduce, mentioning anything that is particularly distinctive.
8. List the distinguishing features of the euglenoids, dinoflagellates, and diatoms, mentioning any anatomical and ecological aspects that are of importance.
9. Why are slime molds and water molds sometimes called the funguslike protists?
10. Describe the anatomical features of the fungi, and tell how fungi are classified.
11. Describe the life cycle of the black bread mold.
12. Tell how sac and club fungi reproduce sexually.

Evolution and Diversity

Objective Questions

1. Which of these organisms is not classified as a protist?
   a. protozoans
   b. algae
   c. slime molds
   d. fungi
2. Sexual reproduction of the malarial parasite occurs in
   a. the mosquito.
   b. humans.
   c. red blood cells.
   d. Both b and c.
3. Which of these is not a green alga?
   a. *Volvox*
   b. *Fucus*
   c. *Spirogyra*
   d. *Oedogonium*
4. Which of these is the most sturdy of the seaweeds?
   a. green
   b. red
   c. brown
   d. golden brown
5. Euglenoids
   a. have both plantlike and animal-like characteristics.
   b. carry on photosynthesis.
   c. have a flexible body wall.
   d. All of these.
6. Fungi have a body that is
   a. composed of hyphae.
   b. colonial.
   c. composed of tissues.
   d. flagellated.

7. Which of these is mismatched?
   a. black bread mold—zoospores
   b. sac fungi—fruiting body
   c. club fungi—conidia
   d. slime mold—sporangia
8. Yeasts are
   a. tiny protozoans.
   b. colorless algae.
   c. a type of water mold.
   d. a type of fungus.
9. Give the kingdom and phylum for each of these organisms:
   a. *Chlamydomonas* (see p. 384 and appendix A)

   b. *Neurospora* (see p. 392 and appendix A)
   c. *Amoeba proteus* (see p. 381 and appendix A)
10. Classification is usually based on some feature that distinguishes the members of a group. What feature is used to classify the following:
    a. protozoans
    b. algae
11. Label the following diagram of the *Chlamydomonas* life cycle, and explain why it represents sexual reproduction:

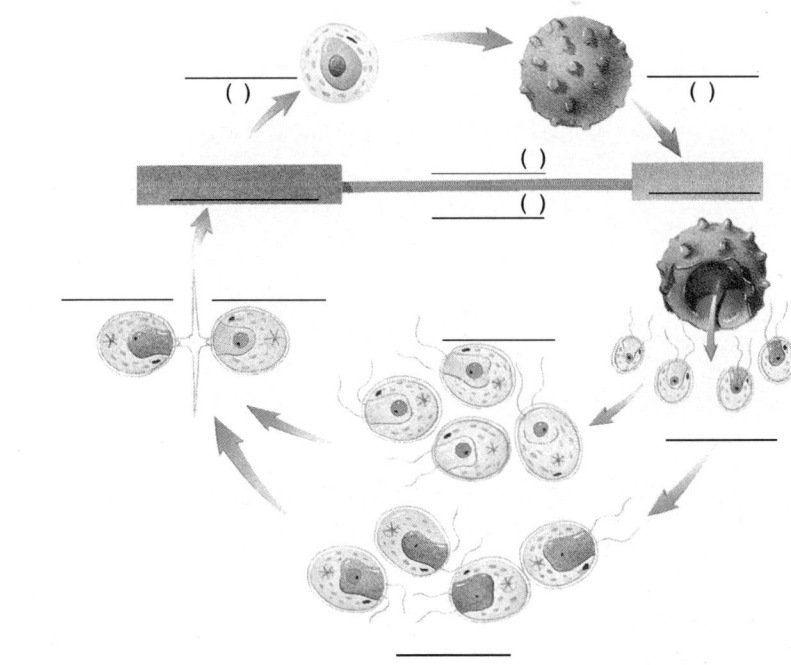

Concepts and Critical Thinking

1. *Protists are a diverse group of organisms.*

   Use nutrition to divide the protists into three groups.

2. *Each organism has one of three types of life cycles.*

   Use the occurrence of meiosis and haploidy to distinguish the various life cycles.

3. *Fungi are the most complex saprophytic organisms.*

   Considering their way of life why is this mode of nutrition suitable to fungi but not to plants and animals?

Selected Key Terms

saprophyte (sap′ro-fit) 380
protozoan (pro-ti-zo′en) 381
algae (al′-jā) 383
haplontic cycle (hap-lon′tik si′kl) 384

alternation of generations cycle (awl″ter-na′shun uv jen″e-ra′shunz si′kl) 384
diplontic cycle (dip-lon′tik si′kl) 384
spore (spōr) 384
heterogamete (het″er-o-gam′ēt) 385

conjugation (kon″ju-ga′shun) 385
fungus (fun′gus) 391
hypha (hi′fa) 391
mycelium (mi-se′le-um) 391
fruiting body (frōot′ing bod′e) 391

25

Plant Diversity

Do all plants produce flowers? Of course not. This is *Equisteum sylvaticum,* a type of horsetail. In the spring, these plants bear strobili, conelike structures where spores are produced. It is spores, not seeds, that disperse the species.

Your study of this chapter will be complete when you can

1. list the characteristics shared by all plants;
2. describe the bryophytes and give the most significant stages of the moss life cycle;
3. describe the structure and the most likely life cycle for the rhyniophytes, the first vascular plants;
4. list the nonseed vascular plants alive today and describe the life cycle of the fern;
5. describe the life cycle of seed plants, and tell how their life cycle is adapted to a terrestrial environment;
6. list and describe the 4 divisions of gymnosperms, and give the life cycle of the pine tree;
7. contrast dicots with monocots, and give the life cycle of flowering plants;
8. discuss the human and ecological relevance of bryophytes, nonseed vascular plants, conifers, and flowering plants;
9. compare the various plant life cycles, and give reasons why angiosperms are more diverse and widespread than the other plant groups.

Classification

Plants can be distinguished from algae even by a most cursory examination (fig. 25.1); however, it is possible to more specifically list the characteristics of plants (table 25.1).

Characteristics of Plants

Plants are photosynthetic organisms adapted to living on land. (There are aquatic plants today, but they are secondarily adapted to living in the water. The same can be said for a number of reptiles, such as sea turtles, and mammals, such as whales.) Green algae contain many different types of organisms, some of which resemble plants. In the classification system adopted by this text, however, plants have suitable adaptations for a land existence; for

Figure 25.1

Representatives of the dominant types of land plants on earth today. **a.** Most plants today are flowering plants. This is a tiger lily. **b.** Mosses are low lying and are usually found in moist locations. This is hairy-cap moss. **c.** Ferns were much more prominent in ancient times. This is bracken fern. **d.** The gymnosperms are seed plants, as are flowering plants. This is a conifer in which the seeds are borne in cones.

a.

c.

b.

d.

Table 25.1
Characteristics of Plants

1. Contain chlorophylls *a* and *b* and carotenoids. Store reserved food as starch, and have cellulose cell walls.
2. Lack organized muscle fibers.
3. Are multicellular and have cells specialized to form tissues and organs.
4. Have sex organs with an outer layer of nonreproductive cells to prevent desiccation of developing gametes.
5. Protect the developing embryo from drying out by providing it with water and nutrients within the female reproductive structure.
6. Have a life cycle that is described as alternation of generations.

example, they all protect the embryo from desiccation, or drying out. The green alga *Ulva,* which is macroscopic with specialized parts, lacks these adaptations and therefore is not a plant.

A land existence offers some advantages to plants. One advantage is the great availability of light for photosynthesis—even clear water filters light. Another advantage is that carbon dioxide is present in higher concentrations and diffuses more readily in air than in water.

The land environment, however, requires some adaptations to deal with the constant threat of desiccation. For example, in flowering plants, the body, gametes, zygote, and embryo are all protected from drying out. The leaves are covered by a waxy cuticle interrupted on the underside by stomata, which are small openings that open and close to regulate gas exchange and water loss. Flowering plants have an internal skeleton to oppose the force of gravity and to lift the leaves toward the sun and a vascular system that transports water up to and nutrients down from the leaves.

> Plants are multicellular photosynthesizers that are adapted to living on land. All plants protect their embryos from desiccation.

Alternation of Generations Life Cycle

All plants have a 2-generation life cycle known as alternation of generations:

1. The **sporophyte**, the diploid generation, produces haploid spores by meiosis. Spores develop into a haploid generation.
2. The **gametophyte**, the haploid generation, produces gametes that unite to form a diploid zygote.

In plants, one of the generations is dominant; in other words, it is larger and exists for a longer period of time.

> All plants have a life cycle that shows an alternation of generations; some have a dominant gametophyte and some have a dominant sporophyte.

Figure 25.2
A liverwort, *Marchantia.* Male and female gametophytes are flat, lobed structures that fork as they grow. The umbrella-like structures rising from the body of the plants are gametophores that in the male contain antheridia, where sperm are produced, and in the female contain archegonia, where eggs are produced.

Nonvascular Plants

The nonvascular plants are restricted to the bryophytes. All other plants have vascular tissue.

Bryophytes

The **bryophytes** (division Bryophyta, 24,000 species[1]) include liverworts and mosses. Although liverworts, with flattened, lobed bodies, are widespread and well-known, they represent but a small fraction of the total number of bryophyte species (fig. 25.2). Most species of liverworts are "leafy" and look somewhat like mosses, but examination shows that the body of a liverwort has distinct top and bottom surfaces, with numerous **rhizoids** (rootlike hairs) projecting into the soil. In contrast, a moss has a stemlike structure with radially arranged leaflike structures (fig. 25.1*b*). Rhizoids anchor the plant and absorb minerals and water from the soil. Because bryophytes do not have vascular tissue, *they lack true roots, stems, and leaves.* Instead, they have rhizoids, stemlike structures, and leaflike structures.

Moss Life Cycle

In mosses, the gametophyte is dominant—it is the longer-lasting generation (fig. 25.3*a*). In some mosses, there are separate male and female gametophytes (fig. 25.4). At the tip of a male gametophyte are **antheridia,** in which flagellated sperm are produced. After a rain or heavy dew, the sperm swim to the tip of a female gametophyte, where eggs have been produced within the **archegonia.** Antheridia and archegonia are both multicellular structures,

[1]The approximate number of known species is given beside each plant division name.

Figure 25.3

Life cycles of plants. All plants have a life cycle that shows an alternation of generations, but they differ as to which generation is dominant. **a.** In bryophytes, the gametophyte is dominant; therefore, more space is allotted to this generation in the diagram. **b.** In vascular plants, the sporophyte is dominant; therefore, more space is allotted to this generation in the diagram. **c.** Heterospory is seen in vascular plants that produce seeds. Note there are separate male and female gametophytes.

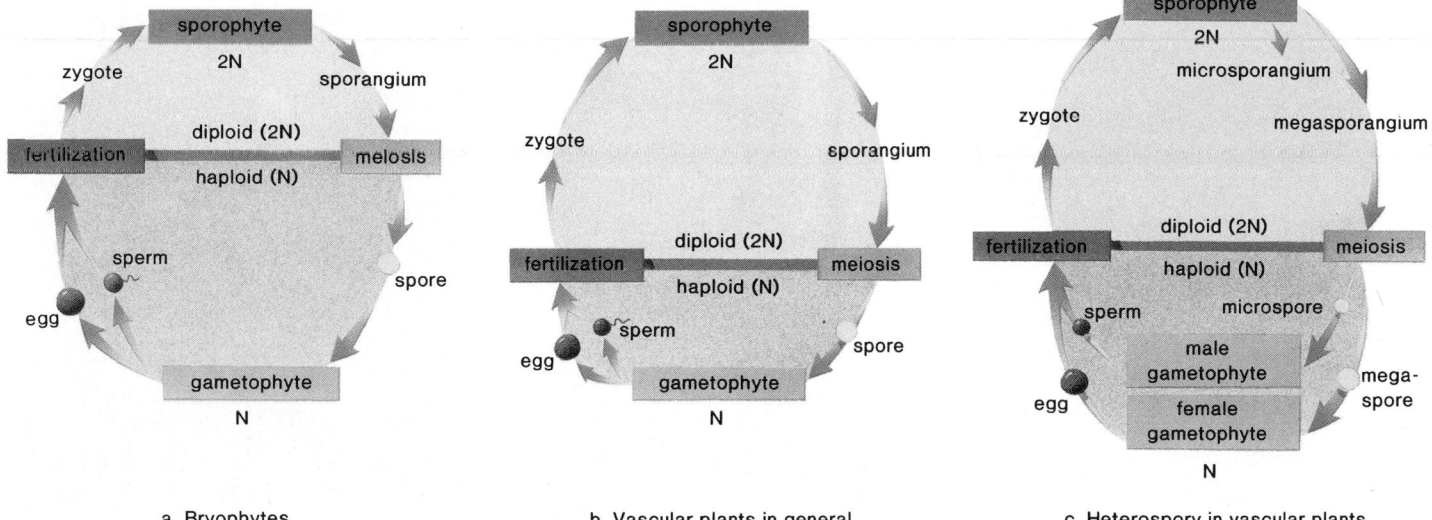

a. Bryophytes   b. Vascular plants in general   c. Heterospory in vascular plants

and each has an outer layer of jacket cells that protects the enclosed gametes from desiccation. After an egg is fertilized, the developing sporophyte is retained within the archegonium. The sporophyte, which is dependent—indeed parasitic—on the gametophyte, consists of a *foot,* which grows down into the gametophyte tissue, a *stalk,* and an upper *capsule,* or *sporangium,* where meiosis occurs and haploid spores are produced. In some species of mosses, a hoodlike covering is carried upward by the growing sporophyte. When this covering and the capsule lid fall off, the spores are mature and are ready to escape. The release of spores is controlled by one or 2 rings of "teeth" that project inward from the margin of the capsule. The teeth close the opening when the weather is wet but curl up and free the opening when the weather is dry. This appears to be the mechanism that allows spores to be released at times when they are most likely to be dispersed by the wind.

When a spore lands on an appropriate site, it germinates. The first row of cells branches, giving an algalike structure, is called a *protonema.* After about three days of favorable growing conditions, new moss plants can be seen at intervals along the protonema. Each of these consists of rootlike rhizoids and an upright shoot.

Adaptation of Bryophytes

Bryophytes are capable of living where low mineral availability prohibits invasion by other types of plants. There is no vascular tissue for water transport, and the plants are generally low-lying. There must be external water or mist at least periodically for sperm to swim from the antheridia to the eggs in the archegonia. The embryo is protected from drying out, however, by remaining within the archegonium. Also, the organism is dispersed to new locations as windblown spores.

Importance of Bryophytes

Bryophytes colonize bare rock, and they slowly convert the rocks to soil that can be used for the growth of other organisms.

Sphagnum, also called bog or peat moss, has commercial importance. This moss has special nonliving cells that can absorb moisture, which is why peat moss is often used in gardening to improve the water-holding capacity of the soil. In some areas, like bogs, where the ground is wet and acidic, dead mosses, especially sphagnum, accumulate and do not decay. This accumulated moss, called peat, can be used as fuel.

The bryophytes include the inconspicuous liverworts and mosses, plants that have a dominant gametophyte. Bryophytes lack vascular tissue, and fertilization requires an outside source of moisture. Windblown spores disperse the species.

Vascular Plants

All the other plants we will study are called the **vascular plants** because they have vascular tissue: **xylem** conducts water and minerals up from the soil, and **phloem** transports organic nutrients from one part of the body to another. The sporophyte, or diploid generation, is dominant (fig. 25.3*b*). This is the generation that has vascular tissue. Xylem, with its strong-walled cells, supports the body of the plant against the pull of gravity. The tallest organisms in the world are vascular plants—the redwood trees of California. Another advantage of having a dominant sporophyte relates to its being diploid. If a faulty gene is present, it can be masked by a functional gene. Vascular plants are a diverse and widely distributed group.

Figure 25.4

a. Moss life cycle. The male gametophyte (a leafy shoot) produces swimming sperm in antheridia. They need external water to reach egg-producing archegonia within the female shoot. Following fertilization, the diploid (2N) zygote develops into the sporophyte, which produces haploid (N) spores by meiosis. These spores develop into separate male and female gametophytes. *b.* *Polytrichum* sporophyte and gametophyte.

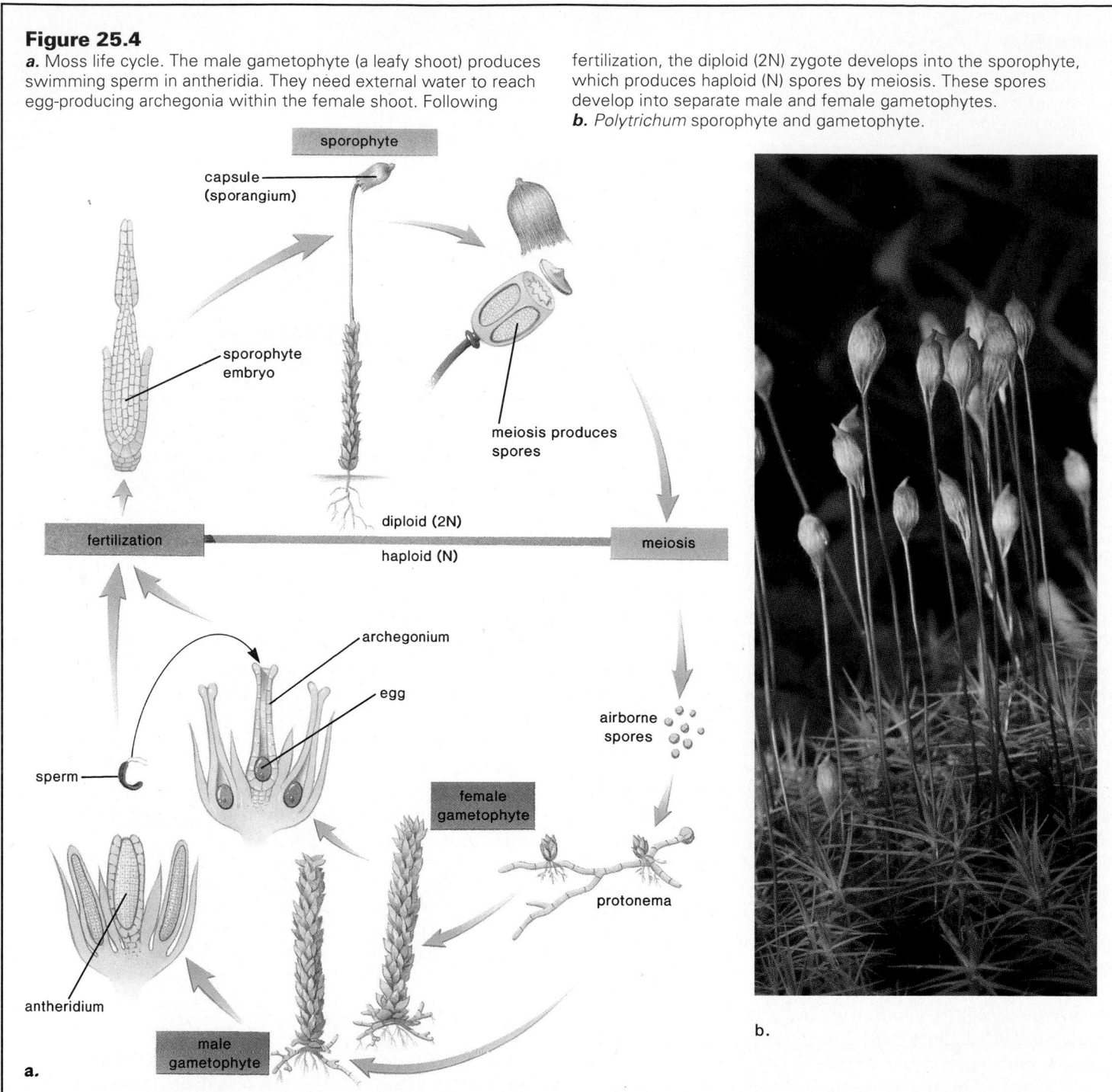

An extinct group of plants, the *rhyniophytes*, are known from the fossil record of the Silurian period (table 25.2). These plants in the genus *Rhynia* fulfill the qualifications of an ancestral prototype for all vascular plants. The sporophyte has a horizontal, underground rhizome with erect branches above and rhizoids below (fig. 25.5). The stem does have vascular tissue, but the plant has no true roots or leaves. Sporangia are at the ends of some of the branches.

The vascular plants evolved during the Silurian period. They are the most diverse and widely distributed of the plants.

Seedless Vascular Plants

Some vascular plants do not produce seeds. They disperse the species by producing windblown spores. The seedless vascular plants include the whisk ferns (division Psilophyta, 10–13 species),

Table 25.2

The Geological Time Scale: Some Major Evolutionary Events of Plants

Era	Period	M.Y.A*	Major Biological Events—Plants
Cenozoic	Quaternary	2	Increase in number of herbaceous plants
	Tertiary		Dominance of land by angiosperms — Age of Angiosperms
		66	
Mesozoic	Cretaceous	144	Angiosperms spread
	Jurassic		First angiosperms appear
			Age of Gymnosperms
		208	
	Triassic		Dominance of land by gymnosperms and ferns
		245	
Paleozoic	Permian	286	Land covered by forests of seedless vascular plants
	Carboniferous		Age of great coal-forming forests, including club mosses, horsetails, and ferns
		360	
	Devonian		Expansion of seedless vascular plants over land — Swamp forest
		408	
	Silurian		Seedless vascular plants appear on land
		438	
	Ordovician		First plant fossils
		505	
	Cambrian		Unicellular marine algae abundant — Age of Algae
		570	

*M.Y.A. = millions of years ago.

Figure 25.5

Rhynia major, the simplest and earliest-known vascular plant, thrived about 350–400 million years ago. The leafless rhizome and branches were green and carried on photosynthesis. There were nonvascular rhizoids instead of roots. The terminal sporangia apparently released their spores by splitting longitudinally.

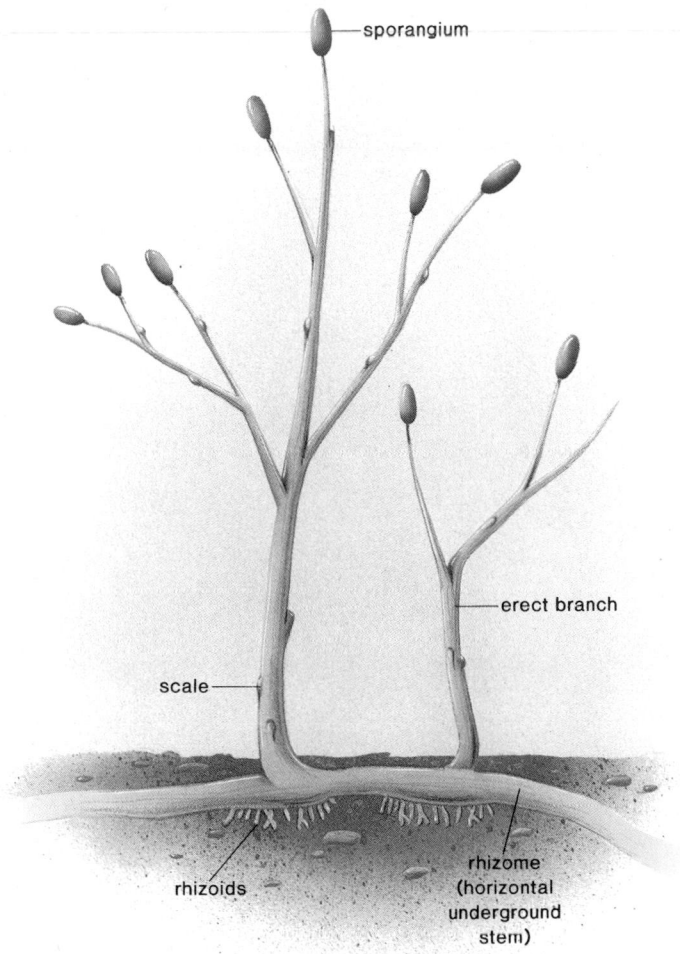

club mosses (division Lycophyta, 900 species), and horsetails (division Sphenophyta, 15 species) (fig. 25.6). Among these, the whisk fern most resembles the rhyniophytes because it has a branched stem and lacks true roots and leaves. The stem is green and carries on photosynthesis. Sporangia are located on the branches. The gametophyte is separate from the sporophyte and water dependent. In fact, the life cycle is very close to that of the fern, which is also a seedless, vascular plant.

Ferns

Ferns (division Pterophyta, 12,000 species) vary in appearance. Many are low-lying, but there are also tall tree ferns in the tropics. Figure 25.7 shows the life cycle of a common fern of the temperate zone, with a horizontal stem (rhizome) from which hairlike roots project downward and large leaves (fronds) project upward. Young fronds grow in a curled-up form called fiddleheads, which unroll as they grow. The fronds are often subdivided into a large number of leaflets.

Figure 25.6

The Paleozoic swamp forests (see fig. 25.9) contained the seedless vascular plants including whisk ferns, lycopod trees, and treelike giant horsetails. Today the whisk ferns (**a**) are represented by *Psilotum;* the club mosses (**b**) are represented by *Lycopodium;* and the horsetails (**c**) are represented by *Equisetum.* Their giant relatives died out during the Permian period.

a.

c.

b.

The sporophyte is the dominant generation in a fern plant. Sporangia develop in clusters called *sori* (sorus, sing.) (fig. 25.8a), which in some genera are protected by a covering, the *indusium.* Within the sporangia, meiosis occurs and spores are produced. As a band of thickened cells on the rim of a sporangium, known as *annulus,* dries out, it moves backward, pulling the sporangium open, and the spores are released (fig. 25.8b). The gametophyte is a tiny (1 cm–2 cm), heart-shaped structure called a *prothallus.* The antheridia and archegonia develop on the underside of a prothallus. Typically, the archegonia are at the notch and antheridia are toward the tip between the rhizoids. Fertilization takes place when moisture is present because the flagellated sperm must swim in external water from the antheridia to the egg within the archegonia. The resulting zygote begins its development inside an archegonium, but the embryo soon outgrows the available space. As a distinctive first leaf appears above the prothallus and as the roots develop below it, the sporophyte becomes visible. Often the sporophyte tissues and the gametophyte tissues are distinctly different shades of green. The young sporophyte grows and develops into the familiar fern plant.

Adaptation of Ferns Ferns have roots and leaves in addition to a stem. The well-developed leaves fan out, capture solar energy, and photosynthesize. There is a water-dependent gametophyte, which lacks vascular tissue and is separate from the sporophyte. Flagellated sperm require an outside source of water in which to swim to the eggs

Figure 25.7

Fern life cycle. The heart-shaped prothallus (gametophyte, much enlarged here) produces swimming sperm in antheridia. They need external water in order to reach egg-producing archegonia. Following fertilization, the diploid (2N) zygote develops into a sporophyte—the familiar fern plant with large fronds. Spores produced by meiosis are released from sori located on the underside of the fronds. Each can germinate to give a gametophyte.

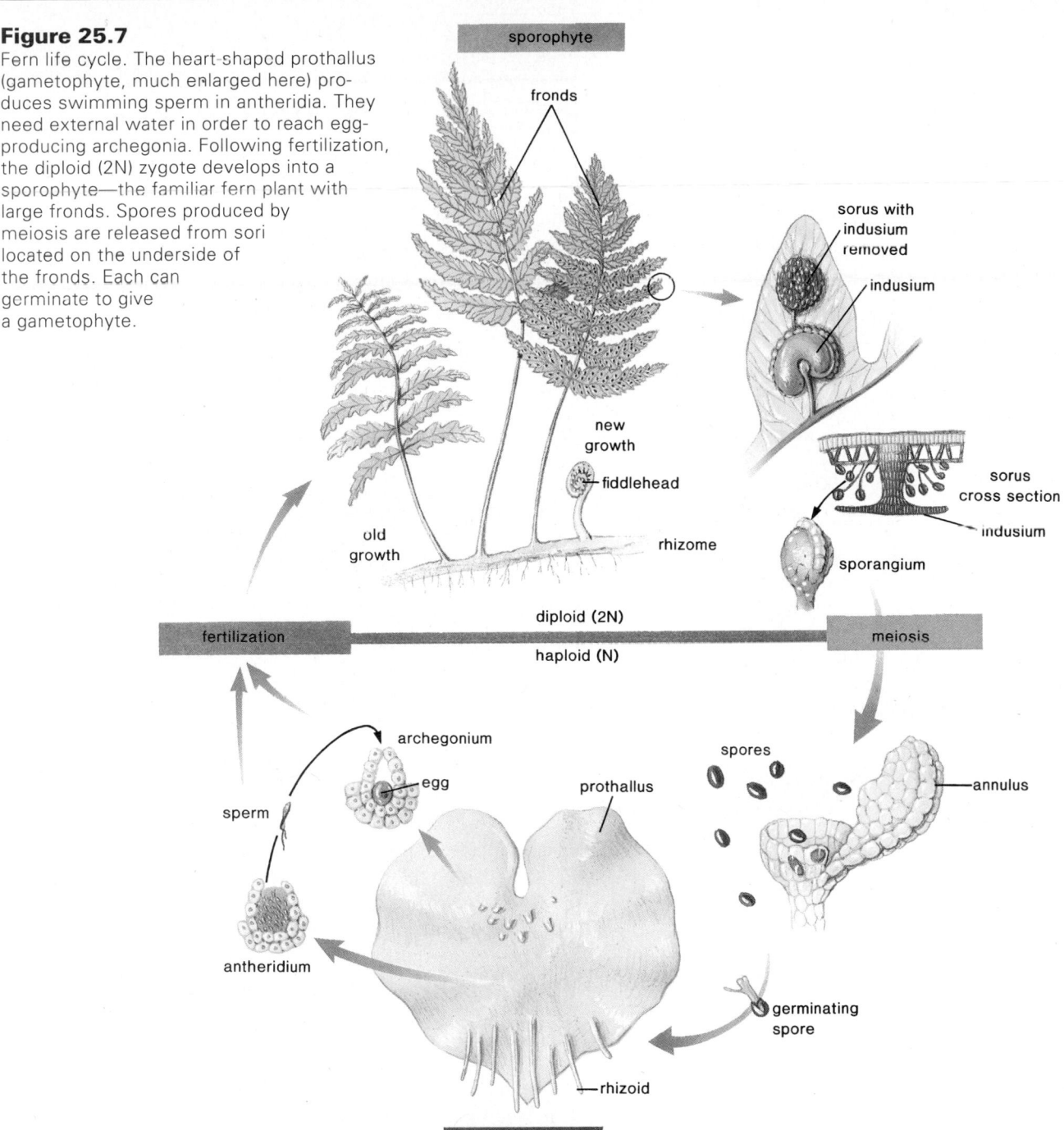

in the archegonia (fig. 25.7). Once established, some ferns, like the brachen fern *Pteridium aquilinum,* can spread into drier areas by means of vegetative reproduction. As the rhizomes grow horizontally in the soil, the fiddleheads grow up as new fronds.

Importance of Seedless Vascular Plants

During the Carboniferous period (table 25.2), horsetails, club mosses, and ferns were abundant, very large, and treelike (fig. 25.9). For some unknown reason, a large number of these plants died and did not decompose completely. Instead, they were compressed and compacted to form the coal we still mine and burn today. (Oil was formed similarly, but most likely it formed in marine sedimentary rocks that included animal remains.)

In the nonseed vascular plants, such as ferns, there is a dominant vascular sporophyte, which produces windblown spores. These plants have an independent nonvascular gametophyte, and flagellated sperm swim in external water to reach the egg.

Figure 25.8

Fern sporophyte anatomy. *a.* A photomicrograph of the underside of a leaflet showing sori. *b.* A scanning electron micrograph of sporangia within a sorus. When the rim (annulus) contracts, a sporangium breaks open, and the spores are released (see fig. 25.7). Magnification, X180.

a.

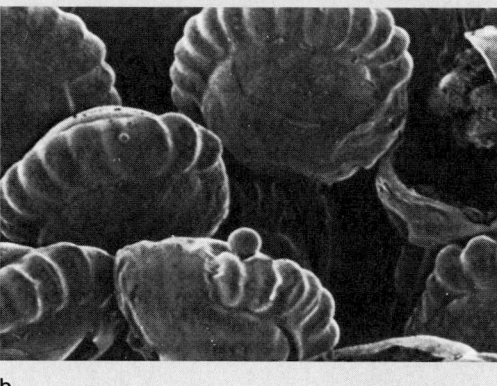

b.

Table 25.3
Seed Plant Terminology

Term	Definition
Heterospores	Two different kinds of spores, one of which develops into male gametophyte and the other into female gametophyte.
Megasporangium	A sporangium that produces megaspores.
Megaspore	A spore that develops into a female gametophyte.
Ovule	Reproductive structure that holds the megasporangium and then the female gametophyte (which produces an egg) and then becomes seed.
Microsporangium	A sporangium that produces microspores.
Microspore	A spore that becomes a pollen grain.
Pollen grain	A male gametophyte generation, which has sperm nuclei and develops a pollen tube.
Pollination	Transfer of pollen from microsporangium to female gametophyte.
Fertilization	Fusion of sperm nucleus with egg nucleus.
Seed	A mature ovule, which contains embryonic sporophyte and stored food.

Seed Plants

As Pangaea (see fig. 21.15) formed during the late Paleozoic era, mountain ranges arose and deserts appeared on their leeward side. Land that had been swampy became much drier (table 25.2). A mass extinction occurred, and among those that all but vanished were the nonseed vascular plants that had made up the great swamp forests of the Carboniferous period. Then the seed plants, with their many adaptations to life on land, came into their own. During the Mesozoic era, the more primitive gymnosperms dominated, but by the Cenozoic era, the **angiosperms** had become the dominant plants on earth. Some of these plants were trees with extensive root systems and enlarged stems—a result of the ability of vascular tissue to produce *secondary growth,* page 478.

The seed plants, to a greater extent than any of their predecessors, produce *heterospores* (fig. 25.3c), called microspores and megaspores, instead of homospores (table 25.3). Each microspore develops into a **pollen grain,** which is the immature male gametophyte, while it is still retained within a

microsporangium. After they are released, pollen grains develop into mature, sperm-bearing male gametophytes. Therefore, the *pollen grain* in seed plants *replaces external swimming sperm* in nonseed plants. **Pollination** is the transfer of the male gametophyte, pollen, to the vicinity of the female gametophyte. Then, the sperm (male gamete) is delivered to the female gametophyte through a pollen tube; therefore, no external water is required for fertilization to occur.

A megasporangium is located within an ovule and produces only one functional megaspore. Megaspores develop into egg-bearing female gametophytes that are still retained within ovules. After fertilization, the zygote becomes an embryonic plant enclosed within a seed. While nonseed plants are dispersed by means of spores, the seeds serve as the dispersal package in seed plants. A **seed** contains an embryonic sporophyte and stored food enclosed by a protective seed coat. Seeds are resistant to adverse conditions, such as dryness or temperature extremes.

There are 2 groups of seed plants, the gymnosperms ("naked seeds") and the angiosperms ("enclosed seeds"). Angiosperm seeds are enclosed in fruits; gymnosperm seeds are not.

The seed plants didn't begin to flourish until the Mesozoic era, when geological upheavals brought about a change in the climate that fostered their diversification.

Figure 25.9

Carboniferous swamps are believed to have contained plants with fernlike foliage (*left*), treelike club mosses (*left*), and treelike horsetails (*right*).

Gymnosperms

The **gymnosperms** are not believed to be closely related and therefore they are placed in 4 separate divisions.

> Gymnosperms:
>   cycads (division Cycadophyta)
>   ginkgo (division Ginkgophyta)
>   gnetae (division Gnetophyta)
>   conifers (division Coniferophyta)

Cycads (100 species) are cone-bearing, palmlike plants found today mainly in tropical and subtropical regions (fig. 25.10*a*). During the Mesozoic era, cycads were so numerous that this period is sometimes referred to as the Age of Cycads and Dinosaurs, and it is believed that dinosaurs fed on cycad seeds and foliage. Today, there are a limited number of cycad species, and only one occurs in the United States.

Only one species of *ginkgo,* the maidenhair tree (fig. 25.10*b*), survives today. The maidenhair tree was largely restricted to ornamental gardens in China until it was discovered that it does quite well in polluted areas. Because female trees produce rather smelly seeds, it is the custom to use only male trees, propagated vegetatively, in city parks. The maidenhair tree is recognized by its fan-shaped leaves with a forked vein pattern.

The *gnetae* (70 species) contain only 3 genera of plants. They look quite different from one another, though they share certain anatomical features (related to their gametophyte and vascular tissue) that make them seem more related to angiosperms than other gymnosperms. Only one genus is found in the United States: *Ephedra* is a many-branched shrub with small, scalelike leaves that is found largely in desert regions. *Welwitschia* belongs to a genus of very unusual plants found in Africa. Most of the plant exists underground. Only a few centimeters of trunk occurs above ground, but the trunk can be over a meter wide! There are only 2 enormous, straplike leaves, which grow for hundreds of years. These may cover most of the exposed portion of the trunk (fig. 25.10*c*).

Among the first seed-producing plants were the gymnosperms, which produce naked seeds. The 4 divisions of these plants are probably not closely related.

Conifers

The largest group of gymnosperms is the cone-bearing **conifers** (550 species), which include pine, cedar, spruce, fir, and redwood trees (fig. 25.11). These trees have needlelike leaves that are well adapted to withstand not only hot summers but also cold winters and high winds. Most conifers are evergreen (leaves are green the entire year).

Pine Life Cycle The life cycle illustrated in figure 25.12 is a good example of a conifer's life history. The sporophyte is dominant, and its sporangia are located on the scales of the cones. There are 2 types of cones—male and female.

Typically, the male cones are quite small and develop near the tips of lower branches. Each scale of the male cone has 2 or more microsporangia on the underside. Within these sporangia, each meiotic cell division produces 4 *microspores.* Each microspore develops into a pollen grain, which is the *male gametophyte.* The pollen grain has 2 wings and is carried by the wind. Pine trees release so many pollen grains during pollen season that everything in the area may be covered with a dusting of yellow, powdery pine pollen.

The mature female cones are larger than the male pine cones and are located near the top of the tree. Each scale of the female cone has 2 ovules that lie on the upper surface. Each ovule is surrounded by a thick, layered coat, having an opening at one end. The megasporangium is within the ovule, where meiosis produces 4 **megaspores.** Only one of these spores develops into a *female gametophyte,* with 2-6 archegonia, each containing a single large egg lying near the ovule opening.

During pollination, pollen grains are transferred from the male to the female cones. Once enclosed within the female cone, the pollen grain develops a pollen tube that slowly grows toward the ovule. The pollen tube discharges 2 nonflagellated sperm. One of these fertilizes an egg in the ovule and the other degenerates. Fertilization takes place 15 months after pollination, and is an entirely separate event from pollination, which is simply the transfer of pollen.

After fertilization, the ovule matures and becomes the seed, composed of the embryo, reserve food, and a seed coat. Finally, in the third season, the female cone, by now woody and hard, opens to release winged seeds. When a seed germinates, the sporophyte embryo develops into a new pine tree, and the cycle is complete.

Figure 25.10

Representative of the lesser-known gymnosperm divisions. *a.* Cycads resemble palm trees but are gymnosperms that produce cones. Male plants have pollen cones and female plants have seed cones. *b.* Ginkgos exist only as a single species, the maidenhair tree. This photograph of a female plant features the seeds. Male plants have pollen cones. *c.* There are 3 genera of gnetophytes—this is *Welwitschia mirabilis,* a very unusual plant of Africa. In this photograph, we see seed cones at the edges of the 2 enormous straplike leaves.

b.

a.

c.

a.

Figure 25.11

Conifers are usually evergreen and produce pollen cones and seed cones. *a.* Hemlock tree. *b.* Seed cones.

b.

Evolution and Diversity

Adaptation of Gymnosperms Gymnosperms have well-developed roots and stems. Many are tall trees that can withstand heat, dryness, and cold. The reproductive pattern of conifers has several important innovations not found in the plants we have considered so far. Transfer of pollen grains by wind and growth of the pollen tube eliminate the requirement of surface water for swimming sperm. Enclosure of the dependent female gametophyte in a cone protects it during its development and shelters the developing zygote as well. Finally, the embryo is protected by the seed and is provided with a store of nutrients that supports development for the first period of its growth following germination. All of these factors increase the chance for reproductive success on land.

Importance of Gymnosperms Conifers grow on large areas of the earth's surface and are economically important. They supply much of the wood used for construction of buildings and production of paper. They also produce many valuable chemicals, such as those extracted from resin, a waxy substance that protects the conifers from attack by fungi and insects.

Perhaps the oldest and largest trees in the world are conifers. Bristlecone pines in the Nevada mountains are known to be more than 4,500 years old, and a number of redwood trees in California are 2,000 years old and more than 90 m tall.

A conifer is the most typical example of a gymnosperm. In the conifer life cycle, windblown pollen grains replace swimming sperm. Following fertilization, the seed develops from the ovule, a structure that has been protected within the body of the sporophyte plant. The seeds are uncovered and dispersed by the wind.

Angiosperms

Angiosperms (division Anthophyta, 235,000 species), or flowering plants, are an exceptionally large and successful group of plants. They range in size from tiny pond-surface plants only 0.5 mm in diameter to very large trees. The oldest angiosperm fossils come from the Cretaceous period (table 25.2) of the Mesozoic era, but flowering plants didn't diversify until the Cenozoic era. The continents drifted to their present locations during the Cenozoic era, and the climate grew colder than that of the Mesozoic era. Perhaps the angiosperms were better able to diversify during these changing times, and this led to their present dominance. They have 2 innovations, both related to the evolution of the **flower,** where the reproductive structures are located. The flower attracts insects and birds that aid in pollination (pollination by wind also occurs), and it produces seeds enclosed by fruit. There are many different types of **fruits,** some of which are fleshy (e.g., apple) and some of which are dry (e.g., peas in a pod). Fruits are sometimes specialized to aid in dispersal of seeds. Fleshy fruits may be eaten by animals, which transport the seeds to a new location and then deposit them when they defecate.

Angiosperms have well-developed vascular and supporting tissues that make them well adapted to terrestrial life. Their xylem tissue contains xylem vessel elements as well as tracheids. Other vascular plants, including virtually all gymnosperms, have only tracheids in their xylem. Whereas the gymnosperms are softwood trees, **woody** angiosperms are hardwood trees. Nonwoody angiosperms are called **herbaceous** plants.

The angiosperms became the dominant land plants in the Cenozoic era. They have flowers, which attract pollinators and produce seeds enclosed by fruits. Also, their vascular tissue is more complex than that of the gymnosperms.

Angiosperms are divided into 2 classes: **dicotyledons** (dicots) and **monocotyledons** (monocots) (see fig. 29.6). The dicots are either woody or herbaceous, and they have flower parts usually in 4s and 5s, net leaf veins, vascular bundles arranged in a circle within the stem, and 2 cotyledons, or seed leaves. Dicot families include many familiar plant groups, such as the buttercup, mustard, maple, cactus, pink, pea, and rose families. The rose family includes roses, apples, plums, pears, cherries, peaches, strawberries, raspberries, and a number of other shrubs. The monocots are almost always herbaceous and have flower parts in 3s, parallel leaf veins, scattered vascular bundles in the stem, and one cotyledon, or seed leaf. Monocot families include the lily, palm, orchid, iris, and grass families. The grass family includes wheat, rice, corn (maize), and other agriculturally important plants.

Flowering Plant Life Cycle

A flower consists of several kinds of highly modified leaves that are arranged in concentric rings and attached to a modified stem tip, the receptacle (fig. 25.13). The sepals, which form the outermost ring, are frequently green and quite similar to ordinary foliage leaves. They enclose the flower before it opens. Next are the petals, which are often large and colorful. Their color often helps attract pollinators. Within the petals, the stamens form a whorl around the pistil. **Stamens** are the pollen-producing portion of the flower. Each stamen has a slender filament with an **anther** at the tip. In the anther, meiosis within microsporangia produces microspores. Each microspore divides internally and becomes the male gametophyte, or pollen grain. The pollen grains are either blown by the wind or carried by pollinators to the pistil. The **pistil** of most flowers has 3 parts—the **stigma, style,** and **ovary.** The ovary contains from one to many **ovules,** depending on the species of plant. Each ovule contains a megasporangium, where meiosis results in 4 megaspores, only one of which survives. The surviving megaspore develops into the embryo sac, which has 8 haploid nuclei. At this point, the female gametophyte is ready to be fertilized (see fig. 31.5).

When pollen grains are transferred to the stigma, each develops a pollen tube, which penetrates the style and ovary. Two sperm nuclei travel to an opening in the ovule. One sperm nucleus fertilizes the egg nucleus, and the other unites with 2 other nuclei (the polar nuclei) of the female gametophyte to form endosperm, which is food for the embryo. This so-called double fertilization is unique to angiosperms. The mature ovule, or seed, contains an embryo and food enclosed by a protective seed coat. The wall of the ovary and sometimes adjacent parts develop into a fruit that surrounds the seeds. Therefore, angiosperms are said to have *covered seeds.*

In angiosperms, the reproductive structures are located in the flower, which consists of highly modified leaves.

Figure 25.12

Life cycle of a pine tree. The mature sporophyte (pine tree) has female pine cones, which produce megaspores that develop into female gametophytes, and male pine cones, which produce microspores that develop into male gametophytes (mature pollen grains). Following fertilization, the immature sporophyte is present in seeds located on the female cones.

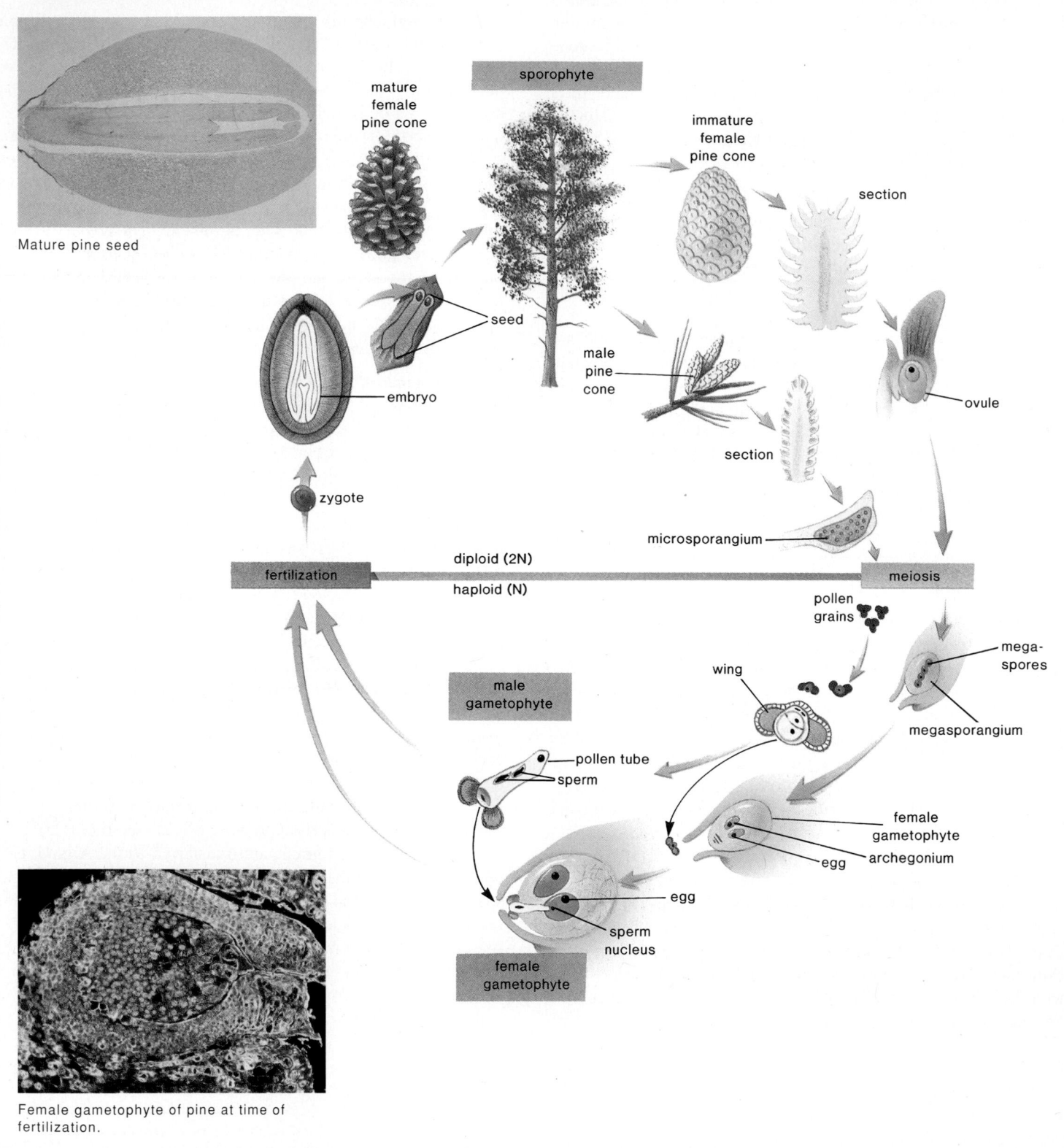

Mature pine seed

Female gametophyte of pine at time of fertilization.

Figure 25.13

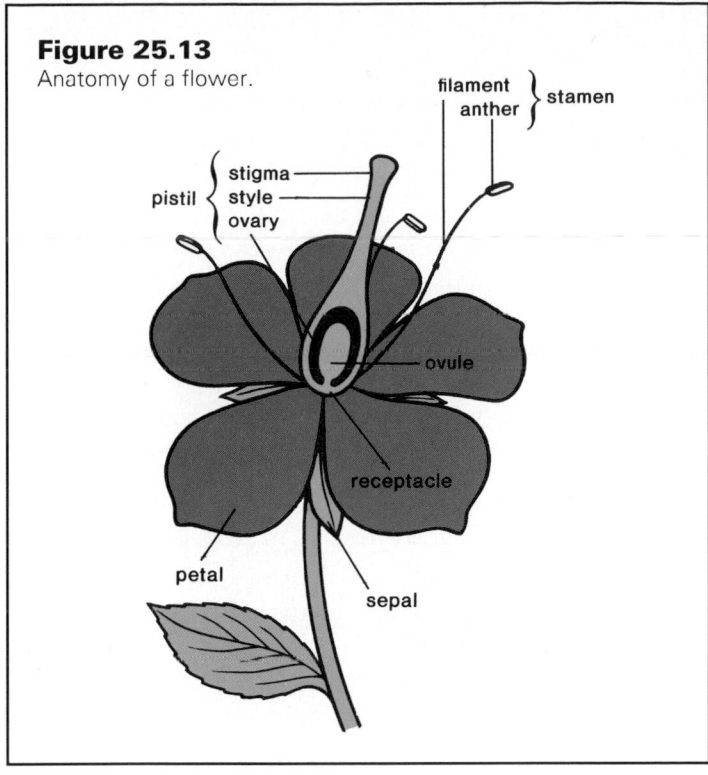

Anatomy of a flower.

Figure 25.14

The relative importance of the haploid (N) generation and the diploid (2N) generation among plants. In mosses, the gametophyte is dominant and larger than the sporophyte; in ferns, the sporophyte is dominant and separate; in conifers and flowering plants, the sporophyte is dominant and the gametophyte is dependent on the sporophyte.

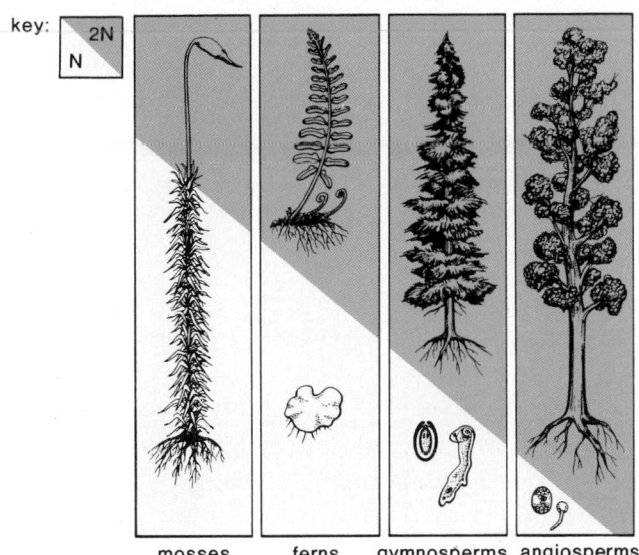

Adaptation of Angiosperms

Angiosperms have true roots, stems, and leaves. The vascular tissue is well developed, and the leaves are generally broad. Angiosperms are found in all sorts of habitats; some have even returned to the water.

The reproductive organs are in the flowers, which often attract animal pollinators; their ovules are located in ovaries that develop into fruits. Therefore, angiosperms produce covered seeds. Fruits often help with dispersal of seeds.

Importance of Angiosperms

The angiosperms are the major producers in most terrestrial ecosystems. They provide food that sustains most of the animals on land, including humans. Still, it has been observed that humans use relatively few types of plants for food. Grasses, alfalfa, and clover are also used as food, or forage, for livestock.

Humans use angiosperms for many other functions. Hardwood becomes the timber used for construction and for furniture. It is also used for fuel (firewood), particularly in poor countries. Flax and cotton are sources of natural fibers for making cloth, and cellulose can be treated to yield rayon. Plant oils are not only used in cooking but also in making perfumes and medicines. Spices are from various parts of plants: peppercorns are small berrylike structures from a vine; cinnamon comes from the bark of a tree; and cloves are dried flower buds. Various drugs come from angiosperm plants, including menthol from eucalyptus, quinine from quinine bark, morphine from the juice of the poppy, and marijuana from the leaves and flowers of the plant *Cannabis sativa*.

Extinctions

Many flowering plants are now on the verge of extinction. This is not caused by a change in climate or bombardment by comets but by the activities of humans. There are many arguments for attempting to preserve all plants, some of which are discussed in this chapter's reading.

Comparisons between Plants

Table 25.4 compares the plants we have studied in this chapter. The bryophytes lack vascular tissue and are unable to transport water any distance; they are low-lying. The first primitive vascular plants in the genus *Rhynia* had a stem but no roots or leaves. These structures are generally not well developed in the primitive vascular plants. A notable exception is the well-developed leaves (fronds) of the ferns. Gymnosperms and angiosperms have well-developed roots and stems. Gymnosperms have needlelike leaves and angiosperms have broad leaves. These groups have many types of tall trees, which transport water some distance from the roots to the leaves.

Figure 25.14 shows the relative importance of the haploid (N) generation and the diploid (2N) generation in plants. The gametophyte is dominant and larger than the sporophyte in bryophytes. In primitive vascular plants, including ferns, the small gametophyte is separate from the dominant sporophyte. The gametophyte lacks vascular tissue and fares best in moist habitats. In gymnosperms and angiosperms, the gametophyte is

Table 25.4
Comparison of Living Plant Divisions

Division	Examples	Habitat
Nonvascular Plants (dominant gametophyte; nonseed)		
Bryophyta (bryophytes)	Mosses, liverworts	Moist and shaded common; sometimes dry areas and even bare rocks
Seedless Vascular Plants (dominant sporophyte; nonseed)		
Psilophyta	Whisk ferns	Tropical and subtropical
Lycophyta	Club mosses	Moist and shaded common in temperate zone; also tropical to arctic and also dry areas
Sphenophyta	Horsetails	Moist and damp sites, by streams along edge of woods
Pterophyta	Ferns (including bracken fern, lady fern, staghorn fern)	Moist, warm shaded common; sometimes dry areas
Seed Plants (dominant sporophyte; seeds)		
Gymnosperms (naked seeds)		
Cycadophyta	Cycads	Tropical and subtropical
Ginkgophyta	Ginkgo	Gardens and parks
Gnetophyta	*Gnetum, Ephedra, Welwitschia*	Moist tropics; desert sites; desert sites
Coniferophyta	Conifers (including pines, firs, yews, redwoods, spruces)	Widespread distribution, particularly Northern Hemisphere
Angiosperms (covered seeds)		
Anthophyta		
Monocotyledonae (monocots)	Grasses, lilies, pineapple	Widespread distribution; some aquatic
Dicotyledonae	Most flowering herbs, shrubs, and trees	Widespread distribution; some aquatic

Sporophyte Description	Gametophyte Description	Reproductive Strategy
In the moss, a stalk with spore-bearing capsule	In moss, small, stemlike structure with attached leaflike structures, rhizoids; antheridia and archegonia at tips of shoots	In gametophyte, flagellated sperm swim in external water from antheridia to archegonia; sporophyte produces windblown homospores that disperse the species
Rhizome with erect branches, no root or leaves; sporangia borne on stems	Independent, with antheridia and archegonia	In gametophyte, flagellated sperm swim in external water from antheridia to archegonia; sporophyte produces windblown homospores that disperse the species
In *Lycopodium,* rhizome with upright branches and roots; small leaves cover stems and branches; sporangia borne in terminal cones	Independent, with antheridia and archegonia	In gametophyte, flagellated sperm swim in external water from antheridia to archegonia; sporophyte produces windblown homospores that disperse the species
In *Equisetum,* rhizome with hollow, upright, jointed stem; small scalelike leaves at joints and roots; sporangia borne in terminal cones	Independent, with antheridia and archegonia	In gametophyte, flagellated sperm swim in external water from antheridia to archegonia; sporophyte produces windblown homospores that disperse the species
Rhizome with roots and well-developed leaves called fronds; sporangia in sori on underside of fronds	Independent, heart-shaped prothallus with antheridia and archegonia	In gametophyte, flagellated sperm swim in external water from antheridia to archegonia; sporophyte produces windblown homospores that disperse the species
Woody stem short and bulbous, large leaves resembling palm fronds; massive male and female cones	Dependent female gametophyte within ovule; male gametophyte is mature pollen grain	Sporophyte produces heterospores in male and female cones; windblown pollen; seeds disperse the species
Woody tree with long side branches, fan-shaped, deciduous leaves; microsporangia in cones, naked ovules on stalks	Dependent female gametophyte within ovule; male gametophyte is mature pollen grain	Sporophyte produces heterospores; windblown pollen; seeds disperse the species
In *Gnetum,* trees and climbing vines with large leathery leaves. In *Ephedra,* a profusely branched shrub with small, scalelike leaves. In *Welwitschia,* massive, woody, disk-shaped stem with straplike leaves	Dependent female gametophyte within ovule; male gametophyte is mature pollen grain	Sporophyte produces heterospores; windblown pollen; seeds disperse the species
Woody tree with thick trunk and needlelike or scalelike evergreen leaves (some are shrubby); reproductive organs in male and female cones	Dependent female gametophyte within ovule; male gametophyte is mature pollen grain	Sporophyte produces heterospores in male and female cones; windblown pollen; seeds disperse the species
Trees, shrubs, herbs most with broad leaves; herbs are perennials or annuals; reproductive organs in flower	Dependent female gametophyte within ovule; male gametophyte is mature pollen grain	
Flower parts in 3s or multiples of 3s; parallel-veined leaves, one cotyledon		
Flower parts in 4s, 5s, or multiples of these; net-veined leaves, 2 cotyledons		Sporophyte produces heterospores in flowers; windblown pollen or animals assist pollination; seeds disperse the species; seeds enclosed by fruit

microscopic and dependent on the dominant sporophyte. Along with these changes, plants became less dependent on external water for reproduction. The nonseed plants (bryophytes and primitive vascular plants) have flagellated sperm that swim in external water from antheridia to eggs in archegonia. In seed plants, the pollen is generally windblown or carried by animals to the female gametophyte, and then the sperm are delivered to the egg by a pollen tube. There is no need of external water for fertilization to occur.

The gymnosperms and angiosperms are widely distributed on land. The delicate heterospores, gametophytes, gametes, zygotes, and embryos (seeds) are enclosed within coverings that protect them from drying out.

The Living Library of Plants

The California condor, the Maryland darter, the Florida panther, and other animals struggling to survive are not the only endangered species. Largely because of man's encroachment, many, perhaps dozens of American plant species are disappearing each year. Indeed, botanists estimate that some 3,000 of the 22,000 species of higher plants native to the United States may be facing extinction. Around the world, as many as 40,000 plant species are in trouble [fig. 25.A].

Now help is on the way, at least for America's vegetation, in the form of the Center for Plant Conservation, which has its headquarters at Harvard University's Arnold Arboretum. Started with a grant of $500,000, the center has an unprecedented program, by far the most comprehensive to date, that aims to preserve every kind of threatened plant in the United States.

Through a network of 18 affiliated botanical gardens and horticultural research facilities in 14 states, the center coordinates the collection of threatened species. By growing these rare plants, the center expects eventually to reintroduce some into their natural habitats and to satisfy the needs of both researchers and collectors.

Figure 25.A
Texas bladderpod, *Lesquerella pallida*, is a potential source of seed oil, but it may soon become extinct.

Collectors have contributed to the near extinction of several species. One victim is the Knowlton cactus, the first endangered species cataloged by the center. Says Donald Falk, the center's administrative director, "Collectors will go out and decimate populations, uprooting the cactus to send it back to live on windowsills."

Why spend money and energy to save, say, the frostweed or the small whorled pogonia? Medical benefits alone, says Frank Thibodeau, the center's scientific director, could justify the center's efforts: "Well over a quarter of all prescription medicines in the United States are based on plant products." He points, for example, to antitumor alkaloids found in the Madagascar periwinkle that are now used in the treatment of childhood leukemia and Hodgkin's disease. "The question," says Thibodeau, "is whether you're willing to bet that there isn't another important drug out there among those 3,000 plants or whether you're willing to hold the plants long enough to study them."

Then, too, some of the plants may have as yet undiscovered characteristics important to agriculture: for example, resistance to disease or drought. Using new recombinant DNA techniques, scientists look forward to identifying the genes that confer these traits and transferring them from wild plants to crop plants. By preserving the endangered species, says Falk, "we're building a genetic library." Thibodeau considers the library essential "even if it turned out that these plants have no other identifiable value."

Summary

1. Plants are photosynthetic organisms adapted to a land existence. Among the various adaptations, all plants protect the developing embryo from desiccation.
2. All plants have the alternation of generations life cycle. Some types of plants have a dominant gametophyte; others have a dominant sporophyte.
3. The bryophytes, which include the liverworts and mosses, are nonvascular plants and therefore lack true roots, stems, and leaves.
4. In the moss life cycle, the gametophyte is dominant. Antheridia on the male shoot produce swimming sperm that need external water to reach the eggs in the archegonia of the female shoots. Following fertilization, the dependent sporophyte consists of a foot, stalk, and capsule within which windblown spores are produced by meiosis. Each spore germinates to give either a male or female gametophyte shoot.
5. Mosses colonize rocks, helping to create soil. Humans use sphagnum in gardens to hold moisture, and it becomes a fuel called peat in areas with wet and acidic bogs.
6. Vascular plants arose during the Silurian period of the Paleozoic era. The extinct rhyniophytes may be ancestral vascular plants. These plants had photosynthetic stems (no leaves or roots) with sporangia at their tips. Most likely, the life cycle was similar to today's ferns. The sporophyte, which is diploid and has vascular tissue, is the dominant generation in vascular plants.
7. The seedless vascular plants include the whisk ferns, club mosses, and horsetails. While rather small and

limited in diversity today, they were more diverse and quite large during the Carboniferous period. Their life cycle resembles that of the fern, which is also a seedless vascular plant.

8. In ferns, the separate and water-dependent gametophyte (the heart-shaped prothallus) produces swimming sperm in antheridia and eggs in archegonia. Following fertilization, the zygote develops into the sporophyte, which has large fronds. On the underside of the fronds are sori, each containing several sporangia. Here meiosis produces windblown spores, each of which develops into a prothallus.

9. The Mesozoic era saw many geological changes as Pangaea formed and then broke apart. A mass extinction occurred that paved the way for the diversification of the seed plants. Seed plants that are trees have especially well-developed roots and stems due to secondary growth of vascular tissue. Seed plants produce heterospores and separate male and female gametophytes. The pollen grain replaces external swimming sperm, and the female gametophyte is retained within the ovule that develops into the seed.

10. Three divisions of gymnosperms (meaning naked seeds) contain little-known plants: cycads, ginkgos, and gnetae. Conifers, a fourth division, are the best known of the gymnosperms.

11. In conifers, male and female cones are produced by the sporophyte plant. On the underside of a male cone scale, there are 2 sporangia that produce microspores; each becomes a pollen grain, the male gametophyte. On the upper surface of a female cone scale, there are two ovules, where meiosis produces one megaspore that develops into the female gametophyte. After windblown pollination, the pollen grain develops a tube through which sperm reach the egg-containing female gametophyte. After fertilization, the ovule matures to be the seed.

12. Conifers are of great economic importance. They are a source of softwood and valuable chemicals.

13. Angiosperms (meaning covered seeds) are more diverse than the other types of plants. Perhaps their success is due to climatic changes in the Cenozoic era.

14. In a flower, the microsporangia develop within the anther portion of a stamen, and the megasporangia develop within ovules located in the ovary of the pistil. Pollination brings the mature male gametophyte (pollen grain) to the pistil and the pollen tube brings the sperm to the ovule within the ovary. Angiosperms exhibit double fertilization: one sperm nucleus fertilizes the egg, and the other joins with the polar nuclei to form the endosperm, which is food for the embryo. The ovule develops into the seed, and the ovary becomes the fruit.

15. Angiosperms have complex vascular tissue and are found in various habitats. Their reproductive organs are found in flowers. Animal pollination increases the chance of appropriate fertilization, and fruit production helps with the dispersal of seeds.

16. Angiosperms provide most of the food that sustains terrestrial life, and they are the source of many products used by humans.

Writing Across the Curriculum

In order to practice writing skills, students should write out the answers to any or all of the study questions and the critical thinking questions. The study questions are sequenced in the same order as the text. Suggested answers to the critical thinking questions are in appendix D.

Study Questions

1. What are the characteristics that define plants?
2. Draw a diagram to describe the life cycle of the moss, and point out significant features of this cycle.
3. What plants comprise the seedless vascular plants, and at what time in the history of the earth were they larger and more abundant than today?
4. Draw a diagram to describe the life cycle of the fern and point out significant features of this cycle.
5. List 3 innovations that are observed in the life cycle of seed plants but are not seen in the life cycle of ferns.
6. List and describe the 4 divisions of gymnosperms.
7. Describe the life cycle of the pine, being careful to point out the 3 innovations you mentioned in question 5.
8. Describe innovations found in flowering plants. Which are directly related to the life cycle of these plants?
9. For each group of plants studied, list one feature with ecological relevance and one with human relevance.
10. Draw a diagram that shows the increasing dominance of the sporophyte generation among the 4 groups of plants studied. Relate your diagram to dependence on external water for fertilization.

Objective Questions

1. Which of these are characteristics of plants?
   a. multicellular with specialized tissues and organs
   b. photosynthetic and contain chlorophylls *a* and *b*
   c. protect the developing embryo from drying out
   d. All of these.
2. In the moss life cycle, the sporophyte is
   a. leafy green shoots.
   b. the heart-shaped prothallus.
   c. a foot, stalk, and capsule.
   d. the dominant generation.
3. The rhyniophytes
   a. are a flourishing group of plants today.
   b. had large leaves like today's ferns.
   c. had sporangia at the tips of their branches.
   d. All of these.
4. You are apt to find ferns in a moist location because they have
   a. a water-dependent sporophyte generation.
   b. flagellated spores.
   c. swimming sperm.
   d. All of these.

5. Which of these is mismatched?
  a. pollen grain—male gametophyte
  b. ovule—female gametophyte
  c. seed—immature sporophyte
  d. pollen tube—spores
6. In the life cycle of the pine tree, the ovules are found on the
  a. needlelike leaves.
  b. female pine cone.
  c. male pine cone.
  d. All of these.
7. Which of these is mismatched?
  a. anther—produces microsporangia
  b. pistil—produces pollen
  c. ovule—becomes seed
  d. ovary—becomes fruit
8. Which of these plants contributed the most to our present-day supply of coal?
  a. bryophytes
  b. seedless vascular plants
  c. conifers
  d. angiosperms
9. Which of these is found in seed plants?
  a. complex vascular tissue
  b. pollen grain to replace swimming sperm
  c. retention of female gametophyte within the ovule
  d. All of these.

10. Label the following diagram of alternation of generations.

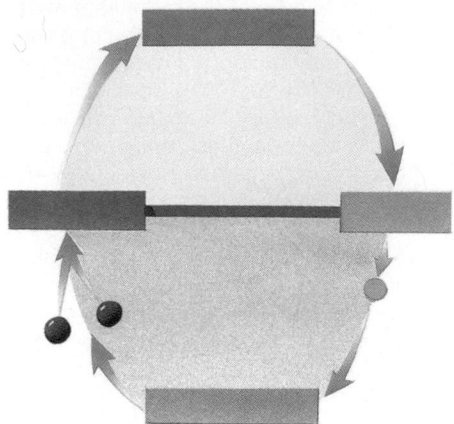

11. Complete this table to compare a moss to a fern:

	Moss	Fern
Description of Gametophyte	a.	b.
Description of Sporophyte	c.	d.

Which plant has the sporophyte dominant? How do you know?

12. Complete this table to compare the pine and flower life cycle:

	Pine	Flower
Location of Pollen Sacs	a.	b.
Location of Ovules	c.	d.

Which plant produces fruit? From what structure?

Concepts and Critical Thinking

1. *Plants are one of the major groups of organisms alive today.*

Should all organisms that photosynthesize be placed in the same kingdom? Why or why not?

2. *Plants have a phylogenetic history.*

Construct a simplified evolutionary tree for plants. Explain your tree.

3. *Plants are adapted to living on land.*

Vascular plants have roots, stems, and leaves. How does each organ contribute to a plant's adaptation to life on land?

Selected Key Terms

bryophyte (bri'o-fīt) 398
rhizoid (ri'zoid) 398
antheridium (an"ther-id'ĭ-um) 398
archegonium (ar"ke-go'nĕ-um) 398
xylem (zi'lem) 399
phloem (flo'em) 399

angiosperm (an'je-o-sperm) 404
pollen grain (pol'en grān) 404
pollination (pol"ĭ-na'shun) 404
seed (sēd) 404
gymnosperm (jim'no-sperm) 405
conifer (ko'nĭ-fer) 405

fruit (frōōt) 407
woody (wood'e) 407
herbaceous (her-ba'shus) 407
dicotyledon (di-kot"ĭ-le'don) 407
monocotyledon (mon"o-kot"ĭ-le'don) 407

26

Animals: Sponges to Roundworms

It seems remarkable that a gorgonian coral such as this sea fan (*Gorgonia ventilina*) is a colony of polyp animals. The polyps are embedded in calcareous cylindrical filaments that branch and anastomose, giving the entire colony a plantlike appearance.

Your study of this chapter will be complete when you can

1. list the characteristics of animals;
2. draw a phylogenetic tree that shows the 9 major animal phyla;
3. compare these phyla in terms of body plan, type of symmetry, number of germ layers, level of organization, and presence of coelom;
4. describe the way of life and the anatomical features of sponges; indicate those features that are not shared by other animals;
5. describe the cnidarian way of life, their 2 body forms, and other anatomical features using a hydra as an example;
6. describe the way of life and the anatomical features of a free-living flatworm, such as a planarian;
7. describe the life cycle of tapeworms, emphasizing anatomical changes that accompany the parasitic way of life;
8. describe the way of life and anatomical features of roundworms.

Classification

Whereas plants are multicellular photosynthetic organisms, animals are *multicellular heterotrophic organisms,* which must take in preformed food (fig. 26.1). Unlike the fungi, which rely on external digestion, animals ingest their food and have a central cavity where it is digested. Animals follow the diplontic life cycle (see fig. 24.8), in which the adult is always diploid. In this cycle, meiosis is necessary to produce haploid gametes, which join to give a zygote that develops into the adult. Table 26.1 lists the characteristics of animals.

Animals are traditionally divided into **vertebrates**—animals with backbones—and **invertebrates**—animals without backbones. The great majority of animal species—97%—are invertebrates, and many of these live in the sea, where early animal evolution occurred. The evolutionary history of animals (see table 28.1) indicates that there was an explosion of invertebrate diversification at the start of the Cambrian period. All of the major invertebrate phyla existing today are represented in the fossil record of this time. There is evidence of earlier animals, but their relationship to the Cambrian fossils is uncertain. Apparently, the Cambrian diversification was so rapid it cannot be traced in the fossil record.

Diversification of invertebrates is apparent in the fossil record of the Cambrian period.

Table 26.1
Characteristics of Animals

1. Are heterotrophic and usually acquire food by ingestion followed by digestion.

2. Typically have organized muscle fibers.

3. Are multicellular, and most have cells specialized to form tissues and organs.

4. Have a life cycle that is described as diplontic, in which the adult is always diploid.

5. Practice sexual reproduction and produce an embryo that undergoes stages of development.

Figure 26.1

The animal kingdom is diverse. A hydra (Magnification, X20) (*a*) and a bobcat (*b*) are both multicellular heterotrophic organisms, which must take in preformed food. Note the radial symmetry of the aquatic hydra, which has tentacles for capturing small prey. Cats stalk their prey, and it is important not to be seen. The black-spotted brown coat of a bobcat blends in well with dense vegetation. Locating prey by sound may be required, and the bobcat has ear tufts, which are thought to help hearing.

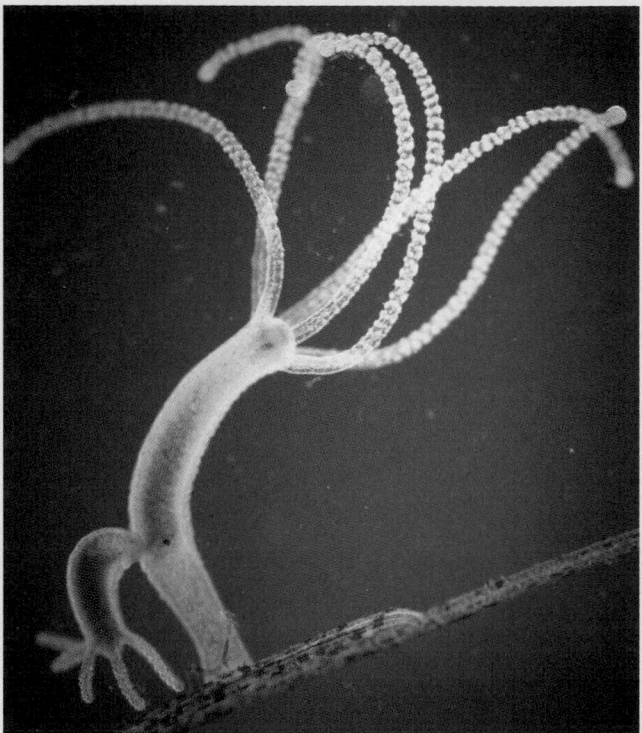

a.

b.

Figure 26.2

Phylogenetic tree of the animal kingdom. All animals are believed to be descended from protists; however, the sponges may have evolved separately from the rest of the animals.

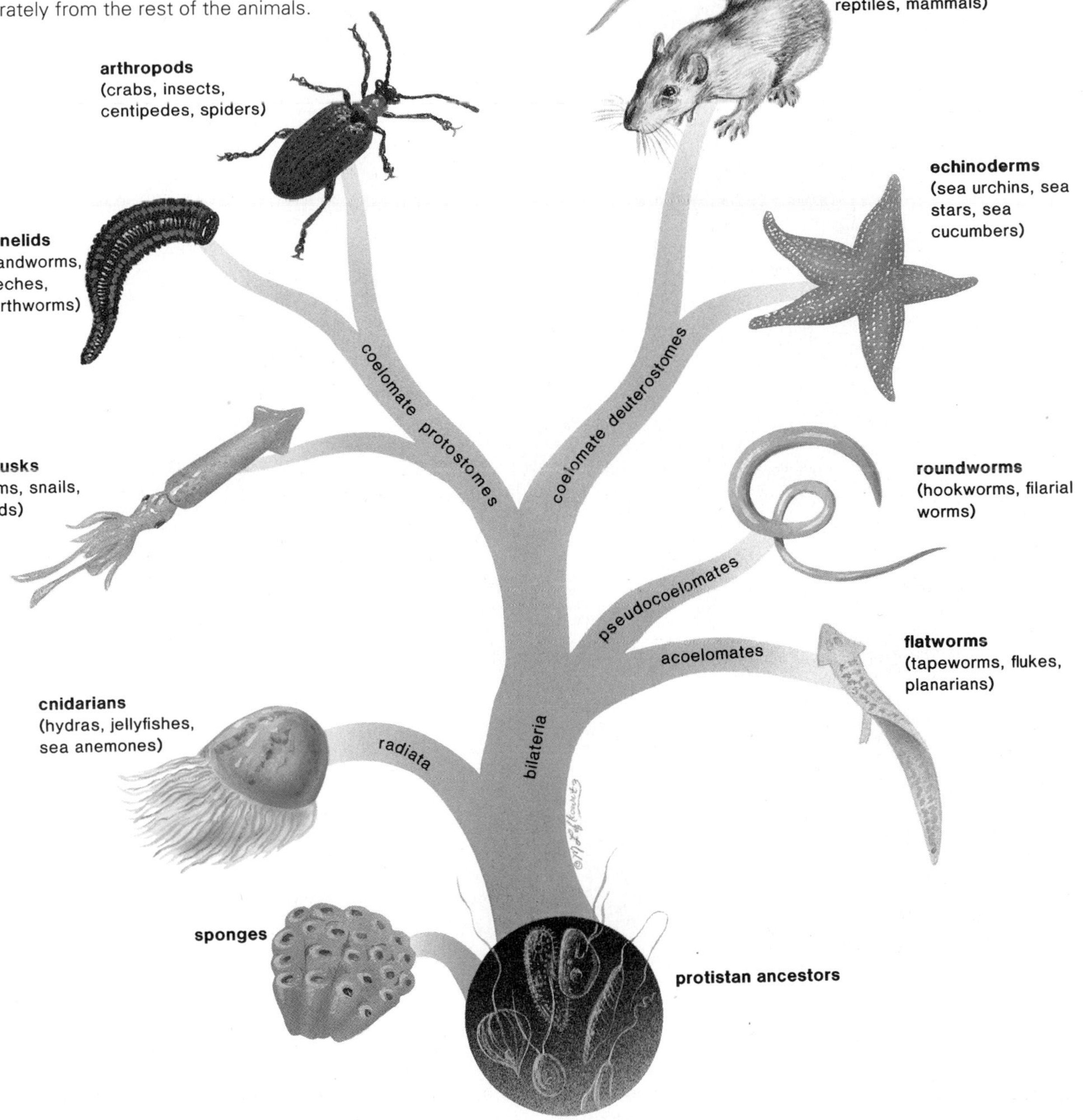

chordates
(fishes, amphibians, reptiles, mammals)

arthropods
(crabs, insects, centipedes, spiders)

echinoderms
(sea urchins, sea stars, sea cucumbers)

annelids
(sandworms, leeches, earthworms)

coelomate protostomes

coelomate deuterostomes

mollusks
(clams, snails, squids)

roundworms
(hookworms, filarial worms)

pseudocoelomates

acoelomates

flatworms
(tapeworms, flukes, planarians)

cnidarians
(hydras, jellyfishes, sea anemones)

radiata

bilateria

sponges

protistan ancestors

Without an adequate fossil record, biologists have developed possible phylogenetic trees, such as the one in figure 26.2, based on a study of present-day animal groups. Although classification systems differ, most authorities recognize approximately 30 animal phyla. We will consider only 9—those that most authorities believe are the major ones. The animals we will study are the product of the evolutionary process, which has continued since all the major phyla first evolved.

Sponges

Most often, sponges (phylum Porifera, 4,300 species[1]) are found attached to a substratum in shallow coastal waters. Sponges are **sessile filter feeders;** they stay in one spot and filter their food from the water that enters the central cavity through *pores* in the body wall. The body wall has 2 layers of cells; *epidermal cells,* which make up the outer layer, and *collar cells,* which make up the inner

[1]The approximate number of known species is given beside each animal phylum name.

Figure 26.3

Generalized sponge anatomy. Epidermal cells form an outer layer of cells, and collar cells line the central cavity and the canals. The collar cells (enlarged) have flagella, which beat, moving the water through pore cells as indicated by the arrows. Food particles in the water are trapped by the collar cells and are digested within their food vacuoles. Amoeboid cells transport nutrients from cell to cell; spicules make up an internal skeleton of some sponges.

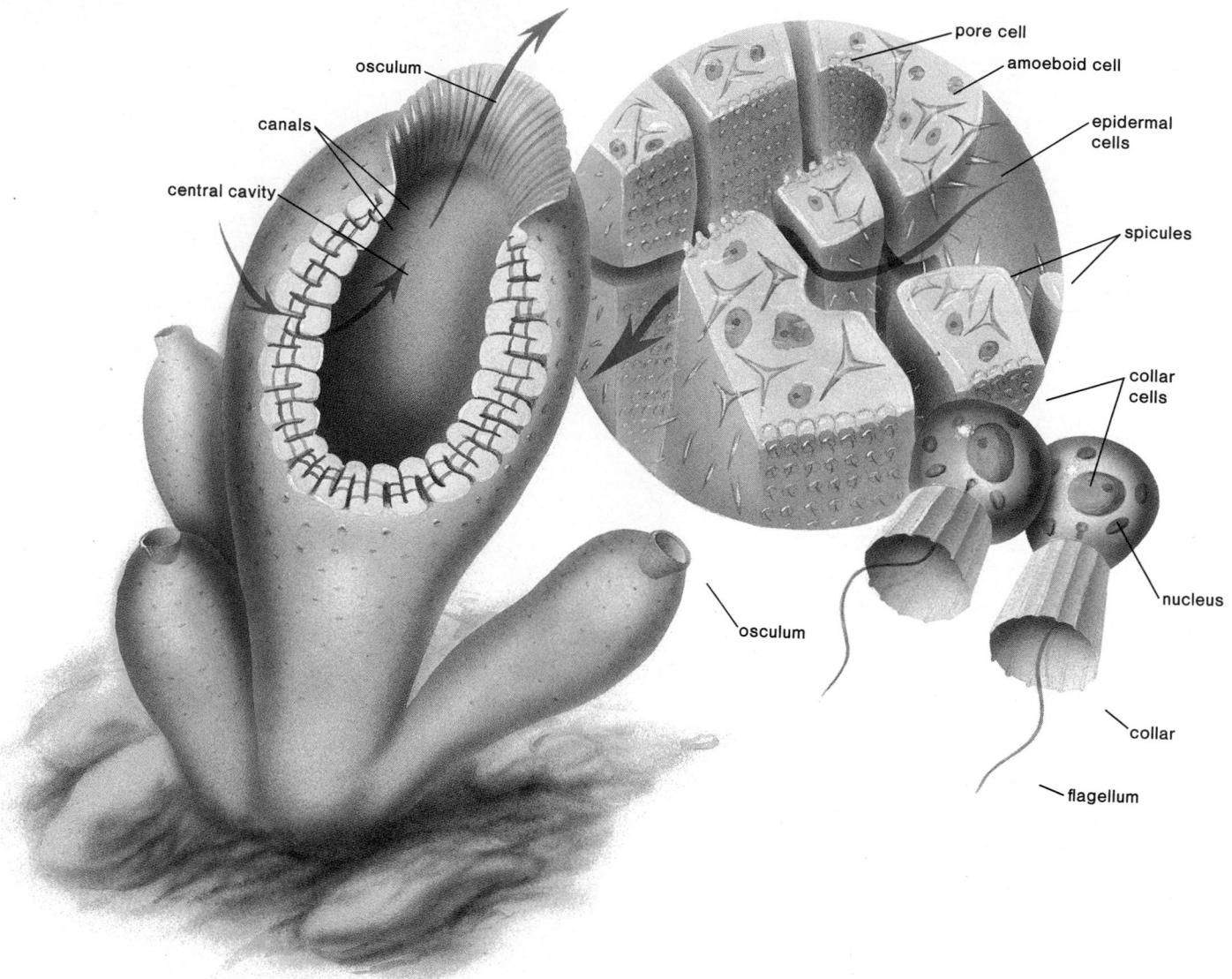

layer (fig. 26.3). Water enters the sponge through the pores and moves to a central cavity. The flagella of the collar cells beat to keep the water moving, and it exits from the sponge by way of the *osculum,* which is a single opening larger than the pores. Although it may seem that sponges can't do much, even a simple one only 10 cm tall is estimated to filter as much as 100 l of water each day. The food particles carried in the water are trapped by the collar cells, which pass them on to the *amoeboid cells* for digestion. The amoeboid cells also act as a circulatory device to transport nutrients from cell to cell, and they produce the skeletal fibers and gametes. Fertilization results in a zygote that develops into a ciliated larva (independent embryonic stage) capable of swimming to a new location. Sponges also reproduce asexually by budding, a process

that can produce large colonies of sponges. It is not surprising, then, that sponges can regenerate, or regrow, an entire organism from a small portion.

Many sponges are simple and small, with pores leading directly from the outside water into the central cavity. Other sponges are quite large and more complex, with canals leading to internal pores. Sponges are classified according to the type of skeletal material they contain. **Spicules,** which act as an *internal skeleton* in some sponges, are made of either calcium carbonate or silicate. Other sponges have skeletal fibers made of spongin, a fibrous protein. Natural sponges are prepared by beating spongin-containing sponges until all of the living cells are removed and only the skeleton remains. Commercial sponges, however, are largely synthetic.

Comments on Sponges

In many ways, sponges resemble a colony of protozoan cells more than a multicellular animal. Like a colony, sponges have a flexible body organization. In 1907, H. V. Wilson cut sponges into small pieces and squeezed the pieces inside bags so that individual cells were separated from one another. These cells, if left undisturbed, moved about, found other cells of their type, and reaggregated to give small but normally functioning sponges. A sponge cell can even change its structure and function to be like another cell in the organism. Such plasticity has not been demonstrated in any other type of animal.

Sponges have a cellular level of organization and don't have true tissues. Also, they don't have fully developed muscle and nerve fibers, and movement is limited to beating of the flagella, constriction of the osculum, and larval-stage swimming. For these reasons, sponges are believed to be an early branch in animal evolution, and they are not believed to have contributed to the evolution of more complex animals. It is possible that they and the other truly multicellular animals evolved independently.

> Sponges have a cellular level of organization. Movement is limited to beating of the flagella, constriction of the osculum, and larval-stage swimming. Sponges are classified according to the type of skeletal material they contain.

Cnidarians

Cnidarians (phylum Cnidaria, 11,000 species) are tubular or bell-shaped animals that, like sponges, reside mainly in shallow coastal waters, except for the oceanic jellyfishes.

Cnidarians have 2 tissue layers. The outer tissue layer is a protective *epidermis*. The inner layer, the *gastrodermis,* secretes digestive juices into the internal cavity, called the **gastrovascular cavity.** This cavity serves as a location for digestion (gastro) and also as a means to transport (vascular) nutrients to cells that line the cavity. Some of the cells that make up the tissue layers have muscle fibers, and between them is a jellylike packing material, the **mesoglea.** Within the mesoglea, nerve cells form a *nerve net*, a connecting network throughout the body. Having both muscle fibers and nerve fibers, these animals are capable of movement, for example, the movement of their tentacles.

Unlike sponges, cnidarians capture their prey. Most cnidarians have special stinging cells that are fluid-filled capsules containing a long, spiral, threadlike fiber called a **nematocyst.** When the trigger of a stinging cell is touched, the nematocyst is discharged. Some of these nematocysts trap the prey, and others have spines that penetrate and inject a paralyzing substance. Once the prey has been subdued, the tentacles about the cnidarian's mouth maneuver it into an internal cavity, where it is digested by enzymes. Cnidarians also use their nematocysts for defense.

Cnidarians are **radially symmetrical;** that is, splitting them lengthwise through the center always results in 2 equal halves because their bodies are arranged around a central axis. Most other animals are **bilaterally symmetrical;** only one lengthwise cut through the center gives 2 equal but opposite halves. This gives them definite right and left halves, which the cnidarians do not have:

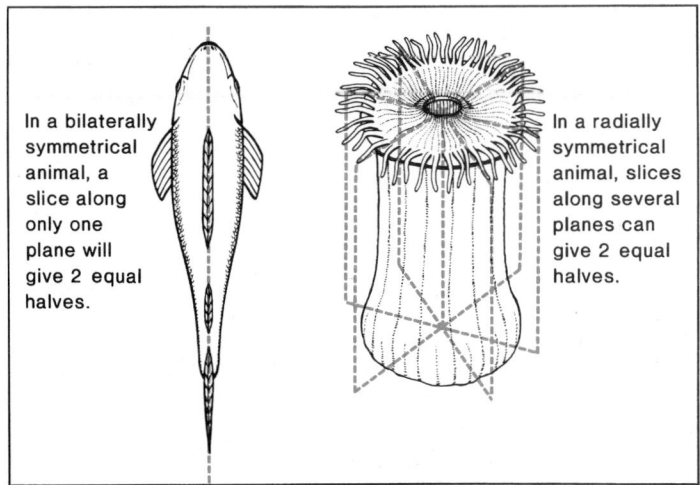

In a bilaterally symmetrical animal, a slice along only one plane will give 2 equal halves.

In a radially symmetrical animal, slices along several planes can give 2 equal halves.

Two basic body forms are seen among the cnidarians (fig. 26.4). The medusa is a bell-shaped, jellyfish form that swims about by contractions of the bell and with the mouth directed downward. In the more slender and usually sessile polyp, the mouth is usually directed upward. Some cnidarians pass through both the medusan and the polypoid forms during their life. (In such cases, the polyp is an asexual stage while the medusa is a sexual stage in the life cycle.) In other cnidarians, one stage is dominant and the other is reduced or, in yet other species, absent altogether.

Figure 26.4

The 2 body forms of cnidarians. The drawings indicate the manner in which the 2 tissue layers (epidermis and gastrodermis), separated by the mesoglea, surround the central gastrovascular cavity. **a.** The polyp, with the mouth side upward, is usually attached to a surface. **b.** The medusa, with the mouth side downward, is the free-swimming form.

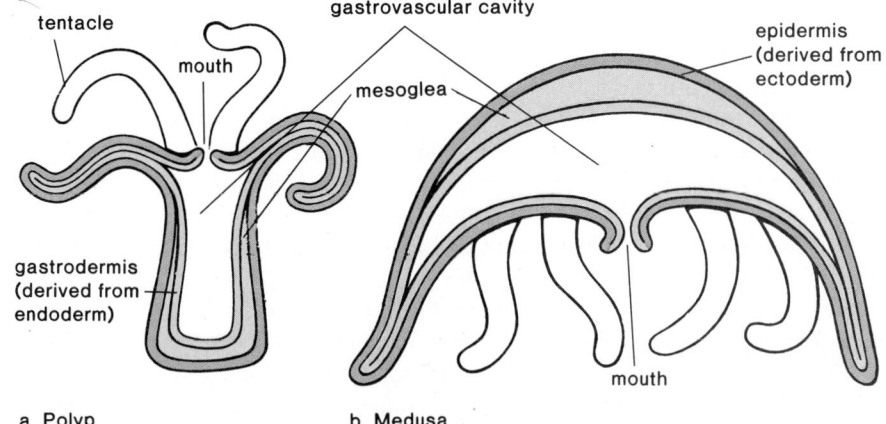

Figure 26.5

Generalized hydra anatomy. Two tissue layers are separated by mesoglea. The cells contain muscle fibers, and the mesoglea contains a nerve net; therefore, the animal can move. In fact, it is carnivorous and uses its nematocysts to stun and its tentacles to capture prey. Inset shows that before discharge, a nematocyst is tightly coiled within a stinging cell, and after discharge, it is extended.

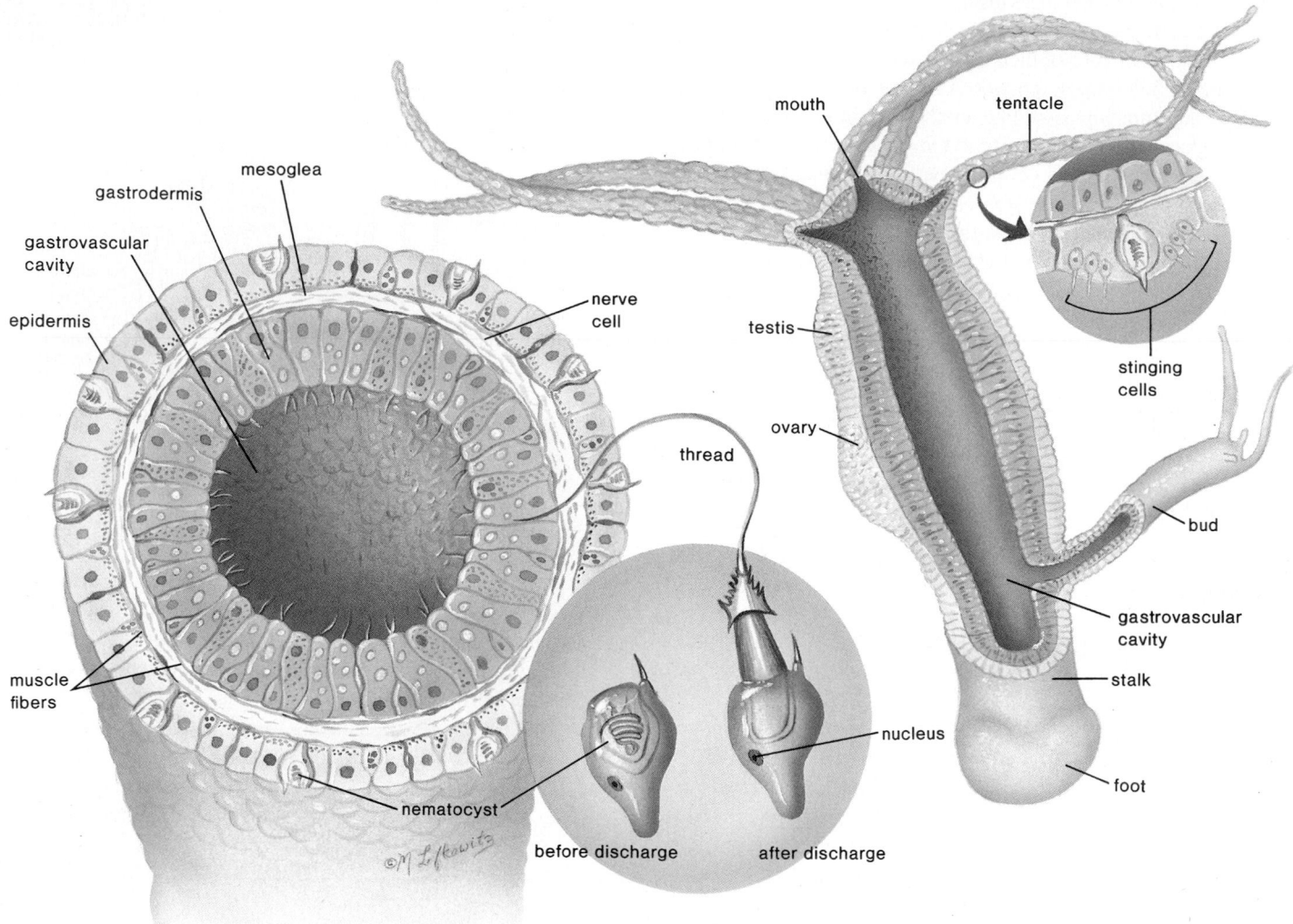

Hydrozoans

In most hydrozoans (class Hydrozoa), the polypoid stage is dominant. This is certainly the case in **hydras,** which are freshwater cnidarians. Even though most hydrozoans are marine, freshwater hydras are usually studied to show the typical cnidarian anatomy (fig. 26.5).

Freshwater hydras can reproduce both asexually and sexually. They reproduce asexually by forming buds, which are small outgrowths from the side of the animal. When hydras reproduce sexually, sperm from a testis swim to an egg within an ovary. Fertilization and early development occur within the ovary, after which the embryo is encased within a hard, protective shell that allows it to survive until conditions are optimum for it to emerge and to develop into a new polyp.

One of the most unusual hydrozoans is the Portuguese man-of-war (*Physalia*), which is a colony of polyps although it looks as if it might be an odd-shaped medusa. Some specialized polyps are gas-filled floats that provide buoyancy—they keep the colony afloat. Others are specialized for feeding or for reproduction. The Portuguese man-of-war also has stinging polyps armed with numerous nematocysts. Swimmers who accidentally come upon a Portuguese man-of-war can receive painful, even serious, injuries from these stinging polyps.

Other Cnidarians

There are 2 other classes of cnidarians. One of these includes the true *jellyfishes*, in which the medusan stage is the dominant phase, and the other contains the *sea anemones* and *corals*, which are polyps having no medusan stage (fig. 26.6). Sea anemones and corals are more often found in warm waters, where the anemones are the "flowers of the sea" and the corals are famous for building reefs. Coral polyps are rather small and inconspicuous, but they secrete hard and limy skeletons that remain after the animals die and degenerate. New generations of live corals grow on the skeletal remains of past generations. Live reef-building corals are found in

Figure 26.6

Cnidarian diversity. ***a.*** Medusae of *Aurelia.* A thick layer of jelly (mesoglea) gives medusae buoyancy and accounts for their common name, jellyfishes. ***b.*** Living coral polyps. Each polyp is an individual but has a calcareous skeleton that is joined to its neighbor's skeleton. The polyps feed on zooplankton but also contain symbiotic algae that contribute to their nutrition. ***c.*** Sea anemone. These animals are called the "flowers of the sea," but despite their appearance, they are carnivorous animals.

a.

b.

c.

sunlit waters because their walls contain symbiotic dinoflagellates that carry on photosynthesis. Coral formations provide sheltered habitats for a great variety of organisms.

Comments on Cnidarians

Cnidarians have the tissue level of organization. They have 2 tissue layers, the epidermis and the gastrodermis. These are derived from 2 embryonic *germ layers,* the ectoderm and endoderm, respectively. (One of the first events during animal development is the establishment of layers of cells called germ layers.) Some animals, like the cnidarians, have only 2 germ layers, whereas more complex animals have 3 layers: ectoderm, endoderm, and mesoderm, which lies between the first 2 layers.

Some cells of the epidermis and gastrodermis contain muscle fibers, and the mesoglea has nerve cells arranged to give a nerve net. Cnidarians are capable of motion.

The *radially symmetrical* cnidarians have a **sac body plan;** the mouth must serve as both an entrance for food and an exit for wastes. More complex animals are usually bilaterally symmetrical and have a **tube-within-a-tube body plan,** with both a mouth and an anus:

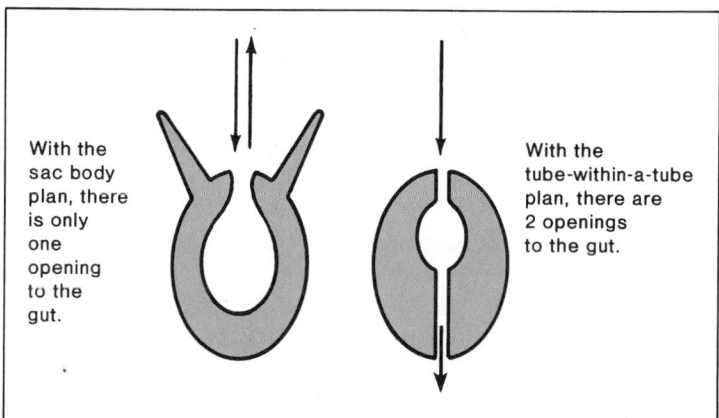

With the sac body plan, there is only one opening to the gut.

With the tube-within-a-tube plan, there are 2 openings to the gut.

The cnidarians are radially symmetrical and have a sac body plan. There are 2 tissue layers, the epidermis and gastrodermis, derived from the ectoderm and endoderm, embryonic germ layers. Nematocysts are a unique feature of cnidarians.

Flatworms

The flatworms (phylum Platyhelminthes, 13,000 species) take their name from the ribbonlike appearance of their body. There are free-living flatworms as well as parasitic groups. We will look at both groups.

Whereas the adult cnidarians are radially symmetrical, the flatworms are *bilaterally symmetrical.* Also, flatworms have *3 tissue layers.* There is a solidly packed mesoderm layer between the ectoderm and endoderm, and this has led to an increase in complexity. They have the organ level of organization to carry on the functions of the animal; however, they have a gastrovascular cavity with only one opening, the same as the cnidarians, and they have no specialized respiratory or circulatory structures. Carbon dioxide and oxygen diffuse through the body surface or through the lining of the gastrovascular cavity.

Planarians

Most *planarians* (class Turbellaria) are marine animals, but freshwater ones are commonly studied as representative of the group.

Planarians show **cephalization,** that is, a definite head region with a brain and sense organs (fig. 26.7). The head often has lateral extensions that function as sense organs to detect potential food sources and enemies. Planarians also have 2 pigmented eyespots that are sensitive to light changes. Inside, a concentration of nerve cells functions as a primitive brain, and a lengthwise pair of nerve cords are joined by connecting nerves. Planarians are said to have a *ladder-type nervous system* because the nerve cords plus the connecting nerves look like a ladder. When planarians move, the action of ventrally located ciliated cells seems to be aided by rhythmical muscle contractions.

A muscular pharynx is extended through the ventrally placed mouth when a worm is feeding on dead or living small animals. A strong sucking action tears off chunks of food. These chunks are then drawn into the branching gastrovascular cavity, which is extensive enough to make a circulatory system unnecessary for food transport. A water-regulating organ, consisting of a series of interconnecting canals, runs the length of the body on each side. Water drawn into the canals by ciliary action within bulbous cells exits at pores located in the body wall. The beating of the cilia in each bulbous cell looks like the flickering of a flame, so this organ is called a *flame-cell system.*

When planarians are cut crosswise, regeneration occurs—each half can develop into another worm. Planarians are **hermaphroditic** animals; that is, each individual has both male and female reproductive organs. Zygotes develop directly into small worms—there are no larval stages.

Figure 26.7

Planarian anatomy. Flatworms are bilaterally symmetrical and have nerve cords plus a primitive brain. The presence of eyespots in a head region further demonstrates cephalization. When the pharynx is extended as shown, food is sucked up into the gastrovascular cavity, which extends throughout the body.

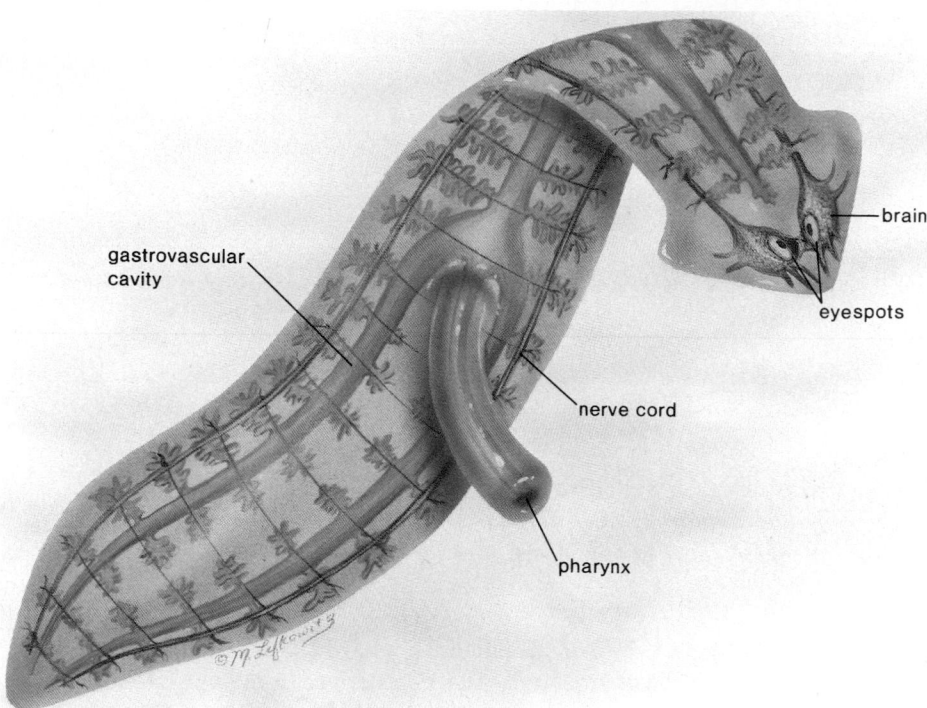

Flukes

Flukes (class Trematoda) are parasites inasmuch as they derive nourishment from a living host organism. Some flukes are external parasites, but the majority are endoparasites; that is, they live within the body of their host. Flukes often have 2 suckers by which they attach to and feed from host tissues.

Flukes that infect humans are commonly named for the organ they invade. For example, there are liver flukes and blood flukes. It is estimated that nearly half of all the people living in the tropics are infected by blood flukes of the genus *Schistosoma*. The accumulation of vast quantities of fluke eggs within the human body causes the symptoms of schistosomiasis—general weakness, anemia, diarrhea, and severely reduced resistance to other infections. Schistosome transmission depends on human feces reaching water, subsequent infection of the snail that is the intermediate host, and contact between resulting larvae and human skin. Rice-paddy agriculture in tropical countries, where human feces are used for fertilizer and people work in the water planting rice plants, creates a nearly perfect environment for transmission of schistosome flukes.

Tapeworms

Tapeworms (class Cestoda) are long (6 m–20 m), flattened worms that as adults live in the intestines of vertebrate animals, including humans. They cause problems because they excrete toxic wastes, absorb nutrients through the body wall, and sometimes interfere with passage of food through the intestinal tract.

Instead of a head, a tapeworm has a structure called a *scolex*. The scolex has hooks and suckers that allow the worm to attach to the host's intestinal wall (fig. 26.8). The rest of the body is made of subunits called *proglottids*. When a proglottid is mature, it primarily contains male and female reproductive organs. Thousands (more than 100,000 in some cases) of eggs are stored in the uterus, which expands until it fills the entire proglottid. Such "ripe" proglottids detach and pass out with the host's feces, scattering fertilized eggs on the ground. If pigs or cattle happen to ingest these, larvae develop and eventually become encysted in muscle that humans may eat in poorly cooked or raw meat. The cyst wall is digested, and a bladder worm (cysticercus) develops into a new tapeworm that attaches to the intestinal wall. To prevent tapeworm infections, meat, especially pork, should be cooked thoroughly.

Comments on Flatworms

Although flatworms have a *sac body plan,* they are more complex than cnidarians. They have *3 germ layers;* the presence of a mesoderm enables them to have the organ level of organization. The free-living forms exhibit *cephalization* and *bilateral symmetry,* both of which are typical of actively moving predaceous animals.

Evolution and Diversity

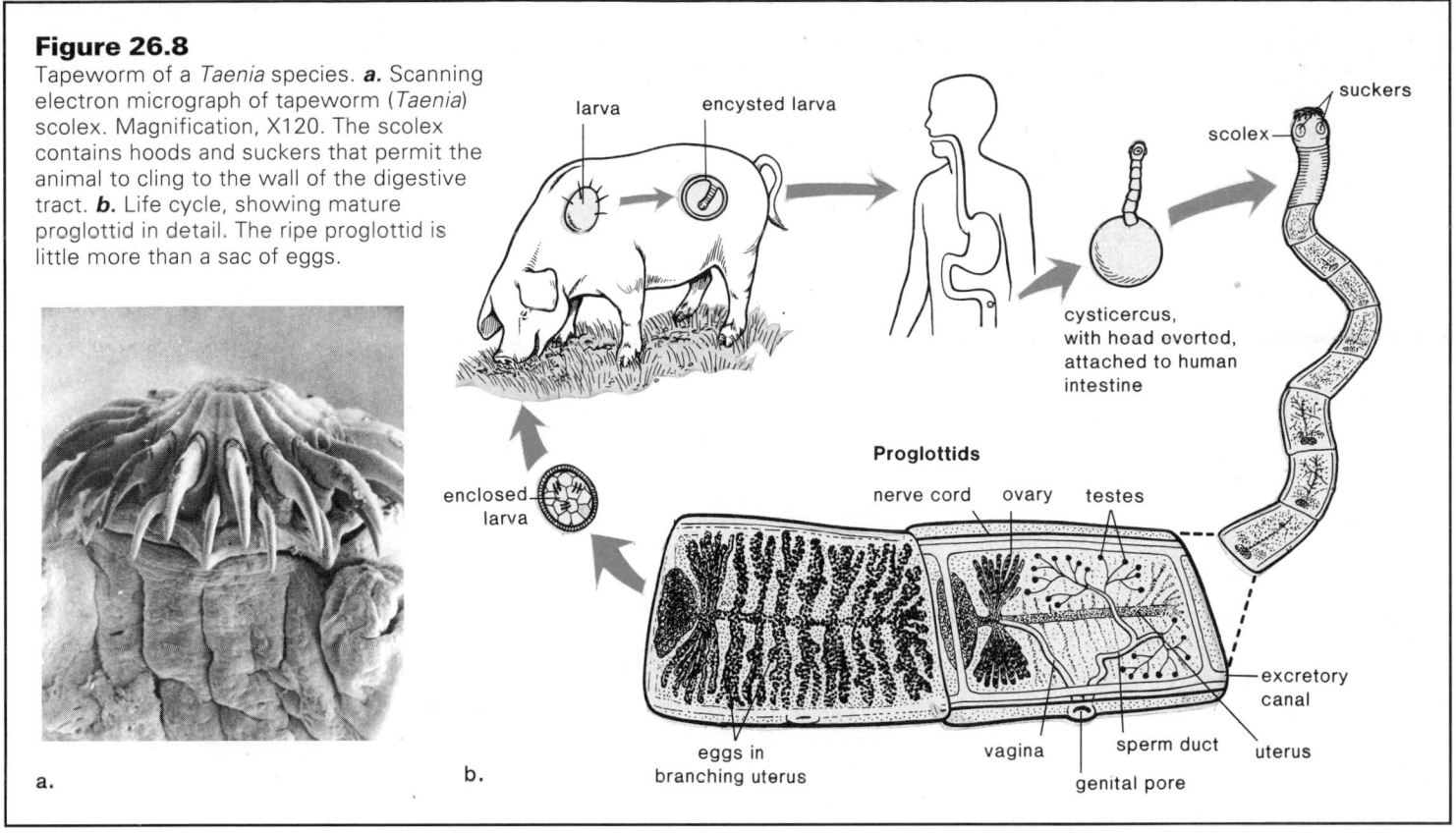

Figure 26.8
Tapeworm of a *Taenia* species. *a.* Scanning electron micrograph of tapeworm (*Taenia*) scolex. Magnification, X120. The scolex contains hoods and suckers that permit the animal to cling to the wall of the digestive tract. *b.* Life cycle, showing mature proglottid in detail. The ripe proglottid is little more than a sac of eggs.

larva

encysted larva

suckers

scolex

cysticercus, with head everted, attached to human intestine

enclosed larva

Proglottids

nerve cord   ovary   testes

eggs in branching uterus

vagina

genital pore

sperm duct

excretory canal

uterus

a.

b.

Although flatworms have a sac body plan, they are more complex than cnidarians because they have 3 germ layers and possess true organs. The free-living forms exhibit cephalization and bilateral symmetry.

The parasitic worms illustrate the modifications that occur in parasitic animals. Concomitant with the loss of predation, there is an absence of cephalization; the anterior end notably carries hooks and/or suckers for attachment to the host. There is an extensive development of the reproductive system at the expense of the other organs. Well-developed nerves and a gastrovascular cavity are not needed because the animal no longer seeks out and digests prey; instead, it acquires nutrients from its host. Both flukes and tapeworms utilize a secondary or intermediate host to transport the species from primary host to primary host. The primary host contains the sexually mature adult; the secondary host(s) contains the larval stage or stages.

Roundworms

Roundworms (phylum Nematoda, 12,000 species), as their name implies, have a smooth outside wall, indicating that they are not *segmented*. These worms, which are generally colorless and less than 5 cm long, occur almost anywhere—in the sea, in fresh water, and in the soil—in such numbers that thousands of them can be found in a very small area. They have various life-styles: some are carnivores with mouthparts to attack their prey; some are herbivores with stylets to pierce plant cell walls; and some feed on dead organic matter that they predigest by secreting digestive juices. Some are parasites of plants and can cause extensive agricultural damage. Still others are parasites of animals; in humans, they cause serious medical conditions.

The roundworms, like the animals we will study in chapter 27, have the *tube-within-a-tube body plan*. The digestive tract is the inner tube within the rest of the animal, which is the outer tube. Roundworms also have a body cavity—a **pseudocoelom**—that is incompletely lined with mesoderm (fig. 26.9).

Ascaris

Ascaris, an intestinal parasite of vertebrates, including humans, is a commonly studied representative of roundworms. Females (20 cm-35 cm) tend to be larger than males, but both sexes move by means of a *whiplike motion* because all their muscles run lengthwise. The reproductive system is the most well developed of all the organ systems (fig. 26.10). Mating produces embryos that mature in the soil, and therefore the parasite is limited to warmer environments. When the larvae in their protective covering are swallowed, they escape and burrow through the intestinal wall. Then they move to the liver, heart, and lungs. Within the lungs, further development takes place, and after about 10 days, the larvae migrate up the windpipe to the throat, where they are swallowed, allowing them once again to reach the intestine. After the mature

Figure 26.9

Presence of a **coelom**. A coelom is a body cavity surrounding the digestive organ or system. **a.** Flatworms are **acoelomate**: they have no body cavity, and the mesoderm is packed solidly between the ectoderm and endoderm. **b.** The roundworms are **pseudocoelomate**: they have a body cavity in which the organs lie loose. Apparently, this is disadvantageous to an increase in size, and most of these worms are small. **c.** In coelomates, the coelom develops as a cavity within the mesoderm; therefore, it is completely lined with mesoderm. Such a coelom is often called a "true coelom." The organs in a **true coelom** are held in place by mesenteries, assuring a more stable arrangement with less crowding. Further, the gut is muscular (muscles are derived from mesoderm) and shows specialization of parts not seen in the pseudocoelomates. In the true coelomates (all the other animals we will study), the internal organs are more complex.

The coelom serves other functions as well. It allows the organs and the body wall to move independently. This means an animal can stretch and bend without putting a strain on the internal organs. The coelom is fluid filled, and this fluid protects and cushions the internal organs. In some animals, the fluid aids in the movement of materials, such as metabolic wastes; in others, this function is taken over by blood vessels. Not only metabolic wastes but also sex cells may be deposited into the coelomic cavity before they are transported away by ducts. The gastrovascular cavity of acoelomates (cnidarians and flatworms) and the fluid-filled coelom of soft-bodied coelomates can act as a hydrostatic skeleton. It offers some resistance to the contraction of muscles and yet permits flexibility, so that the animal can change shape and perform a variety of movements.

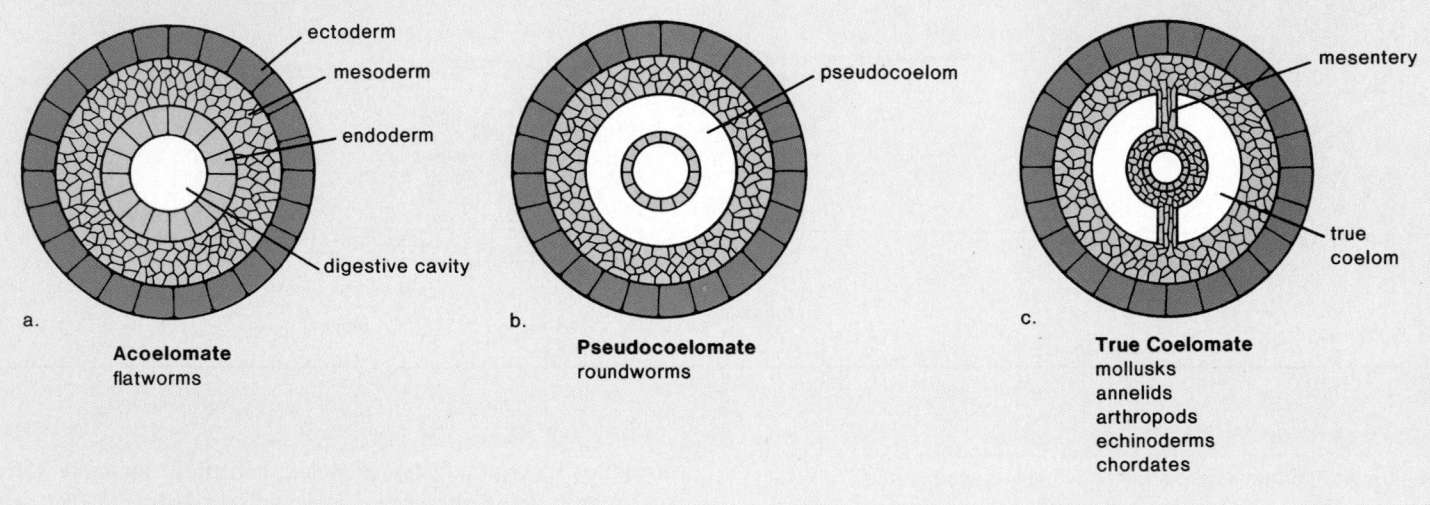

a. **Acoelomate** flatworms

b. **Pseudocoelomate** roundworms

c. **True Coelomate** mollusks annelids arthropods echinoderms chordates

worms mate, the female produces embryo-containing eggs that pass out with the feces. As with other roundworms, to complete this life cycle, eggs must reach the mouth of the next host; therefore, proper sanitation is the best means to prevent infection with *Ascaris* and other parasitic roundworms.

Other Roundworm Infections

Trichinella is a roundworm that encysts in the muscles of animals including humans (fig. 26.11). When humans eat poorly cooked pork, for example, they may contract an infection called *trichinosis*, in which permanent cysts cause muscle damage.

Elephantiasis is caused by a type of roundworm called a filarial worm, which utilizes the mosquito as a secondary host. Because the adult worms reside in lymphatic vessels, draining of fluid is impeded, and the limbs of an infected human may swell to a monstrous size (fig. 26.12). When a mosquito bites an infected person, it passes larvae to new hosts.

Other roundworm infections are more common in the United States. Children frequently acquire a pinworm infection. Although hookworm is primarily seen in southern states, it is judged by some to be the most troublesome parasitic intestinal worm of humans. An infection can be very debilitating because the worms attach themselves to the intestinal wall and feed on blood.

Comments on Roundworms

Roundworms have a pseudocoelom. They lack a circulatory system, but the fluid-filled coelom distributes gases and nutrients about the animal. The coelom also has the other advantages as noted in figure 26.9.

Roundworms have a tube-within-a-tube plan, as do all the rest of the animal phyla. The tube-within-a-tube plan leads to specialization of parts, as we shall see. Roundworms are nonsegmented, a feature that will appear in both branches of animal evolution.

Roundworms possess 2 features not seen previously in animals: a body cavity and a tube-within-a-tube body plan. The body cavity is a pseudocoelom rather than a true coelom. The tube-within-a-tube plan, in contrast to the sac plan, has both a mouth and an anus.

Evolution and Diversity

Figure 26.10

Roundworm anatomy. **a.** Anatomy of female *Ascaris*. The cross section shows the pseudocoelomate arrangement. There is mesodermally derived muscle tissue inside the epidermis, but there is no mesodermal tissue adjacent to the intestinal wall. **b.** A photograph of a roundworm.

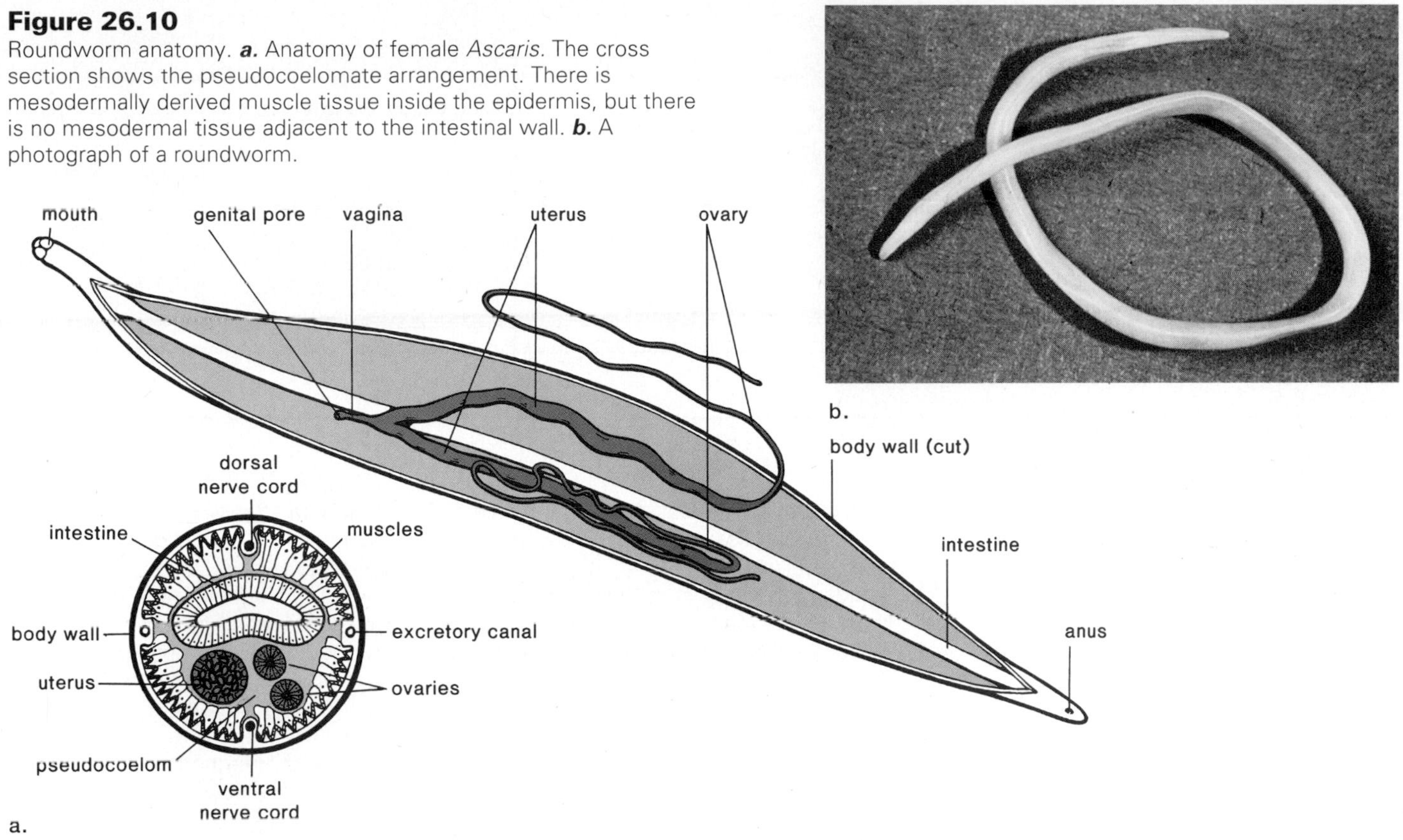

mouth · genital pore · vagina · uterus · ovary

b.

body wall (cut)

intestine

anus

dorsal nerve cord · muscles

intestine

body wall · excretory canal

uterus · ovaries

pseudocoelom

ventral nerve cord

a.

Figure 26.11

Trichinella larva embedded in a muscle. When meat like this is eaten raw or poorly cooked, *Trichinella* larvae may infect the consumer. Magnification, ×400.

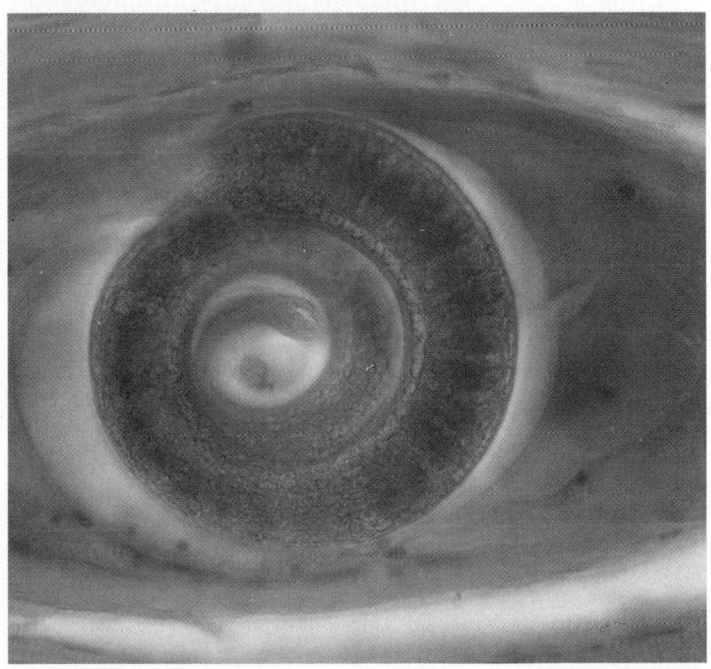

Figure 26.12

An infection from a filarial worm causes elephantiasis, a condition in which the individual experiences extreme swelling in regions where the worms have blocked the lymphatic vessels.

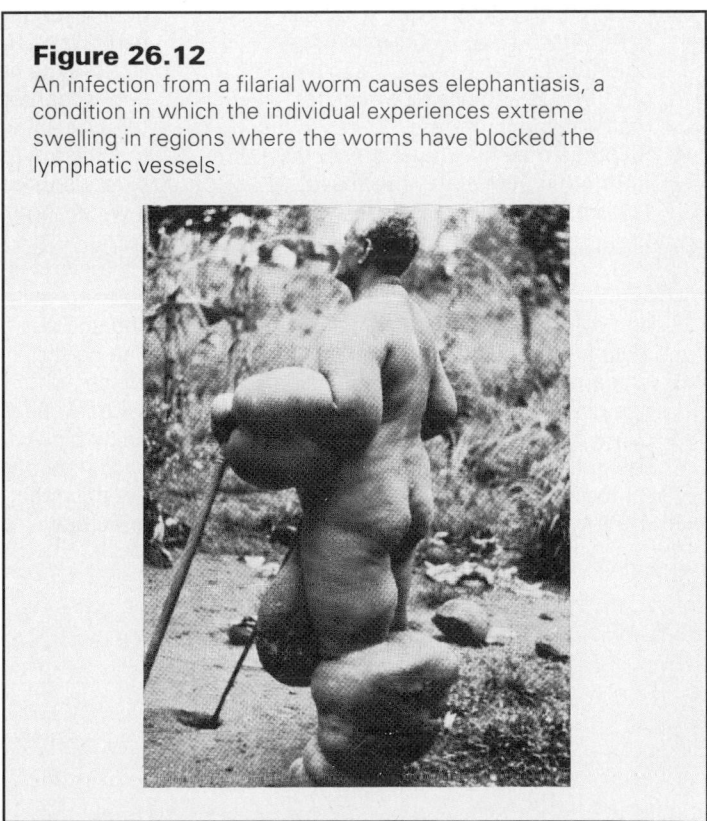

Animals: Sponges to Roundworms

Table 26.2 presents a comparison of the phyla we have discussed in this chapter.

Table 26.2
Comparison of Phyla

	Sponges	Cnidarians	Flatworms	Roundworms
Body plan	_____	Sac	Sac	Tube-within-a-tube
Symmetry	Radial or none	Radial	Bilateral	Bilateral
Germ layers	_____	2	3	3
Level of organization	_____	Tissues	Organs	Organs
Body cavity	_____	_____	_____	Pseudocoelom

Summary

1. Animals are multicellular organisms that are heterotrophic and ingest their food. They follow the diplontic life cycle. Typically, they have the power of motion by means of contracting fibers.

2. It's possible to construct an evolutionary tree for animals, but this is largely based on a study of today's forms. All the major phyla studied today evolved during an explosive and short period of diversification at the start of the Cambrian period.

3. Sponges are sessile filter feeders. They contain these types of cells: an outer layer of epidermal cells; pore cells, which admit water, and collar cells, which keep water moving; and amoeboid cells, which aid circulation of nutrients and produce skeletal material and gametes.

4. Sponges may have evolved separately from other animals as they have features that set them apart. They lack tissues and don't have fully developed muscle and nerve fibers. Movement is limited to beating of flagella, constriction of the osculum, and larval-stage swimming.

5. Cnidarians are carnivores. An epidermis covers the animal. Some cells contain muscle fibers, and a nerve net lies in the mesoglea. They possess tentacles to capture prey and nematocysts to stun it. The prey is digested within the gastrovascular cavity, which is lined by a tissue called the gastrodermis.

6. Cnidarians can exist in the polypoid form or the medusan form, or they can alternate between the 2. Hydras and sea anemones have a dominant polypoid stage; jellyfishes have a dominant medusan stage.

7. Cnidarians possess the tissue level of organization. They are radially symmetrical and have a sac body plan.

8. Flatworms may be free-living or parasitic. Freshwater planarians exemplify the features of flatworms in general and free-living forms in particular. They have muscles and a ladder-type nervous system, and they show cephalization. They take in food through an extended pharynx leading to a gastrovascular cavity, which extends throughout the body. There is a water-regulating organ that contains flame cells.

9. Flukes and tapeworms are parasitic. Tapeworms have a scolex with hooks and suckers for attaching to the host. The body is made up of proglottids, which, when mature, contain thousands of eggs. If these eggs are taken up by pigs or cattle, the larvae become encysted in their muscles. If humans eat this meat, they too may become infected.

10. In general, flatworms are bilaterally symmetrical, have 3 germ layers, and have the sac body plan. Mesoderm accounts for their organ level of development.

11. Roundworms are mostly small and very diverse; they are present most everywhere in great numbers. The parasite *Ascaris* is representative of the group. Infections can be caused by *Trichinella,* whose larval stage encysts in the muscles of humans. Elephantiasis is caused by a filarial worm that blocks lymphatic vessels.

12. Roundworms have 2 features not seen previously: a body cavity and a tube-within-a-tube body plan. The body cavity is a pseudocoelom rather than a true coelom.

Writing Across the Curriculum

In order to practice writing skills, students should write out the answers to any or all of the study questions and the critical thinking questions. The study questions are sequenced in the same order as the text. Suggested answers to the critical thinking questions are in appendix D.

Study Questions

1. What are the characteristics that separate animals from plants? from fungi?
2. What does the phylogenetic tree in figure 26.2 tell you about the evolution of the animals studied in this chapter?
3. List the types of cells found in a sponge, and describe their functions.
4. What features make sponges different from the other organisms placed in the animal kingdom?
5. Explain how cnidarians carry on their carnivorous life-style.
6. What are the 2 body forms found in cnidarians? Explain how they function in the life cycle of various types of cnidarians.
7. What features suggest that free-living planarians are predaceous?
8. Describe the parasitic flatworms, and give the life cycle of a tapeworm.
9. Describe the anatomy of a roundworm.
10. Compare the animals studied in this chapter in terms of body plan, symmetry, germ layers, level of organization, and presence of a coelom.

Objective Questions

1. Which of these is not a characteristic of animals?
   a. heterotrophic
   b. diplontic life cycle
   c. have contracting fibers
   d. single cells, colonial, or multicellular
2. The evolutionary tree of animals shows that
   a. cnidarians evolved from sponges.
   b. flatworms evolved from roundworms.
   c. both sponges and cnidarians evolved from protists.
   d. All of these.
3. Which of these sponge characteristics is not typical of animals?
   a. They practice sexual reproduction.
   b. They have the cellular level of organization.
   c. They do not have fully developed muscle fibers.
   d. Both b and c.
4. Which of these is mismatched?
   a. sponges—spicules
   b. tapeworms—proglottids
   c. cnidarians—nematocysts
   d. roundworms—cilia
5. Flukes and tapeworms
   a. show cephalization.
   b. have well-developed reproductive systems.
   c. have well-developed nervous systems.
   d. are plant parasites.

6. The presence of mesoderm
   a. restricts the development of a coelom.
   b. allows the development of a coelom.
   c. is seen only in acoelomates.
   d. Both a and c.
7. *Ascaris* is a parasitic
   a. roundworm.
   b. flatworm.
   c. hydra.
   d. sponge.
8. Label the following diagram, and explain how it pertains to the life cycle of cnidarians:

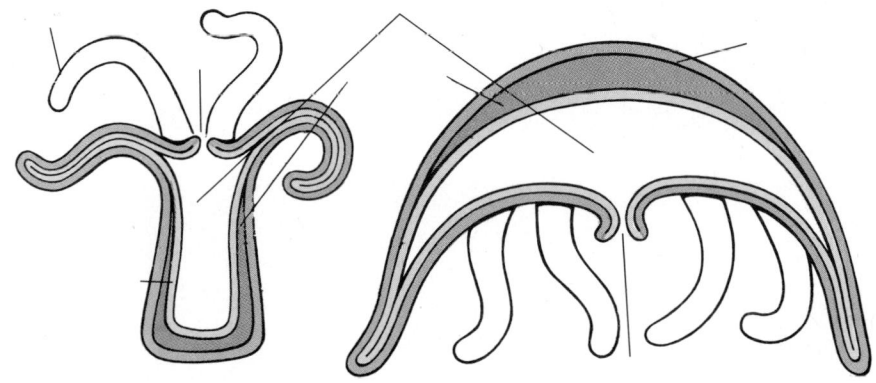

9. Under the phyla headings given, write which of these terms apply:

Terms: radial symmetry pseudocoelom
        tissue level of organization 3 germ layers
        tube-within-a-tube

Sponges   **Cnidarians**   **Flatworms**   **Roundworms**

10. Under the names of the animals given, write which of these features apply:

Terms: proglottids uterus
        eyespots collar cells
        stinging cells gastrovascular cavity

Sponges   **Planarians**   **Tapeworms**   **Roundworms**

Concepts and Critical Thinking

1. *Animals are one of the major groups of organisms alive today.*

Contrast the plant cell with the animal cell to show that the characteristics of animals are reflected in cell morphology.

2. *Animals have a phylogenetic history.*

A common ancestor for flatworms and roundworms might have what anatomical features?

3. *Parasites are adapted to their way of life.*

What types of adaptations are helpful to an internal parasite?

Selected Key Terms

vertebrate (ver'tĕ-brāt) 416
invertebrate (in-ver'tĕ-brāt) 416
sessile filter feeder (ses'il fil'ter fēd'er) 417
spicule (spik'ūl) 418
cnidarian (ni-dah're-an) 419
gastrovascular cavity (gas"tro-vas'ku-lar kav'ĭ-te) 419

mesoglea (mes"o-gle'ah) 419
nematocyst (nem'ah-to-sist) 419
radially symmetrical (ra'de-al-e sĭ-met're-kal) 419
bilaterally symmetrical (bi-lat'er-al-e sĭ-met're-kal) 419
hydra (hi'drah) 420

sac body plan (sak bod'e plan) 421
tube-within-a-tube body plan (tūb within' ah tūb bod'e plan) 421
cephalization (sef"al-i-za'shun) 421
hermaphroditic (her-maf"ro-dit'ik) 422
pseudocoelom (su"do-se'lom) 423
coelom (se'lom) 424

27

Animals: Mollusks to Chordates

Moths are flying insects. Like butterflies, the wings have overlapping scales that rub off when touched. Usually there is a long tongue for sucking nectar from flowers. This moth can feed during the day because its bright colors warn would-be predators it doesn't taste very good.

Your study of this chapter will be complete when you can

1. list the embryological differences between protostomes and deuterostomes, and list the phyla in each group;
2. describe the general characteristics of mollusks; contrast the anatomy of the snail, clam, and squid, indicating how each is adapted to its way of life;
3. describe the characteristics of annelids; contrast the anatomy of earthworms and predaceous polychaetes;
4. describe the characteristics of arthropods, and list the 5 major classes, giving examples of each;
5. describe the anatomy of the grasshopper, indicating how it is adapted to a terrestrial existence;
6. describe the characteristics of echinoderms, in particular, the sea stars;
7. list the 3 chordate characteristics, and tell which animals are the primitive chordates.

Classification

Phylum Mollusca: soft-bodied, unsegmented animals
   Class Polyplacophora: chitons
   Class Monoplacophora: *Neopilina*
   Class Gastropoda: snails and slugs
   Class Bivalvia: clams and mussels
   Class Cephalopoda: squids and octopuses
Phylum Annelida: segmented worms
   Class Polychaeta: sandworms
   Class Oligochaeta: earthworms
   Class Hirudinea: leeches
Phylum Arthropoda: chiton exoskeleton, joint-legged
   animals
   Class Crustacea: lobsters, crabs, barnacles
   Class Arachnida: spiders, scorpions, ticks
   Class Chilopoda: centipedes
   Class Diplopoda: millipedes
   Class Insecta: grasshoppers, termites, beetles
Phylum Onychophora: small, sluglike animals with legs;
   internal features of both annelids and arthropods
Phylum Echinodermata: marine; spiny, radially sym-
   metrical animals
   Class Crinoidea: sea lilies and feather stars
   Class Asteroidea: sea stars
   Class Ophiuroidea: brittle stars
   Class Echinoidea: sea urchins and sand dollars
   Class Holothuroidea: sea cucumbers
Phylum Chordata: dorsal supporting rod (notochord) at
   some stage; dorsal hollow nerve cord; pharyngeal
   pouches or gill slits
   Subphylum Urochordata: tunicates
   Subphylum Cephalochordata: lancelets

Figure 27.1

Artist's interpretation of the shallow seas of the Cambrian era. Apparently, all the major phyla of animals arose during a short few tens of millions of years. The animals depicted here are found as fossils in the Burgess Shale, a formation of the Rocky Mountains of British Columbia that originate in the sea. Not all of the Cambrian animals survive today. Some have become extinct, which was the fate of most species. The rest have been evolving since they first arose.

ll the major phyla of animals arose in the Cambrian seas (fig. 27.1). Many groups remained in the water, but some ventured onto land and became very successful there. In this chapter, we will have an opportunity to contrast animal adaptations suitable to living in water with adaptations suitable to living on land.

The phyla we will study are divided into the protostomes and deuterostomes. There are 3 major differences in the embryological development of these 2 groups of animals (fig. 27.2). First, cleavage of the very first cells occurs differently. In **protostomes,** spiral cleavage occurs, and daughter cells sit in grooves formed by the previous cleavages. It can also be noted that the fate of these cells is fixed and determinate in protostomes; each can only contribute in one particular way to development. In **deuterostomes,** radial cleavage occurs, and the daughter cells sit right on top of the previous cells. The fate of these cells is indeterminate; that is, if they are separated from one another, each cell can go on to become a complete organism.

As development proceeds, a hollow sphere forms, and the indentation that follows produces an opening called the blastopore. In protostomes, the blastopore becomes the mouth (*proto*—first;

Figure 27.2

Coelomate animals are divided into 2 groups—the protostomes and the deuterostomes—based on the embryological evidence given here.

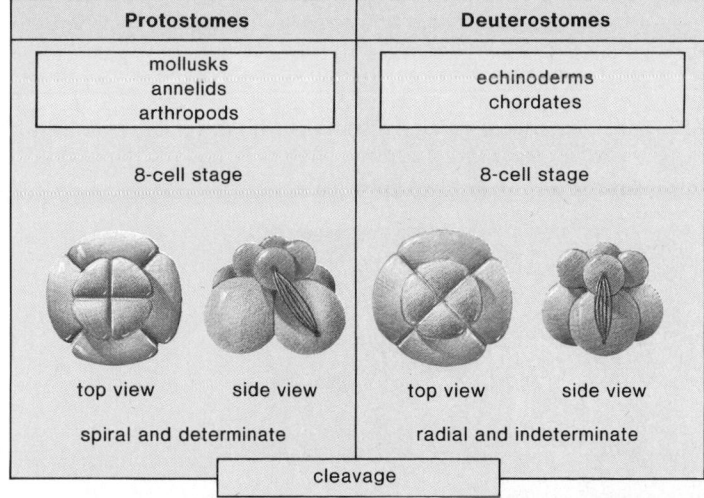

Protostomes	Deuterostomes
mollusks annelids arthropods	echinoderms chordates
8-cell stage	8-cell stage
top view · side view	top view · side view
spiral and determinate	radial and indeterminate
cleavage	

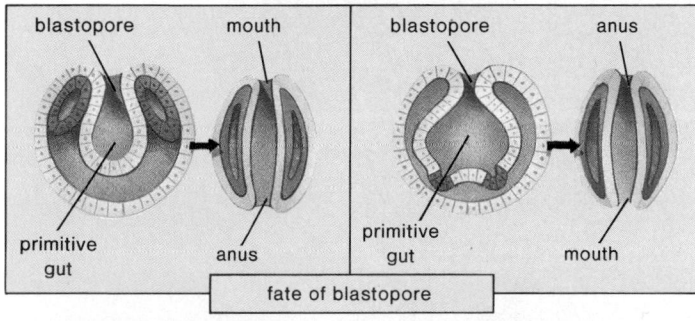

blastopore · mouth · blastopore · anus · primitive gut · anus · primitive gut · mouth

fate of blastopore

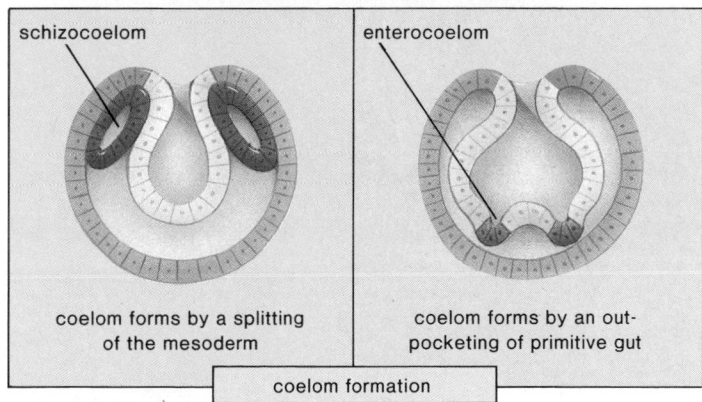

schizocoelom · enterocoelom

coelom forms by a splitting of the mesoderm · coelom forms by an out-pocketing of primitive gut

coelom formation

stome—mouth); in deuterostomes, this opening becomes the anus, and only later does a new opening form the mouth (*deutero*—second; *stome*—mouth).

Finally, the coelom develops differently in the 2 groups. In protostomes, the mesoderm arises from cells located near the embryonic blastopore, and a splitting occurs that produces the coelom, called a schizocoelom. In deuterostomes, the coelom arises as a pair of mesodermal pouches from the wall of the primitive gut. The pouches enlarge until they meet and fuse, forming an enterocoelom.

The phylogenetic tree (see fig. 26.2) indicates which phyla are the protostomes and which are the deuterostomes.

Protostomes	Deuterostomes
mollusks	echinoderms
annelids	chordates
arthropods	

The coelomate animals are divided into 2 groups. In the protostomes, there is spiral cleavage, the blastopore becomes the mouth, and there is a schizocoelom. In the deuterostomes, there is radial cleavage, the blastopore becomes the anus, and there is an enterocoelom.

Mollusks

There are over 50,000 living species of **mollusks** (phylum Mollusca), more than twice the number of vertebrate species! You can well imagine, then, that there must be quite a number of life-styles among the mollusks, which include such animals as squids, octopuses, clams, scallops, oysters, snails, and slugs. All these animals share 3 common features: a visceral mass, a mantle, and a foot.

The *visceral mass* contains the internal organs, including a highly specialized digestive tract, paired kidneys, and reproductive organs. The phylum name comes from the Latin word *mollis,* meaning soft, which refers to the visceral mass. The *mantle* is a covering that lies to either side of but does not completely enclose the visceral mass. It may secrete a shell and/or contribute to the development of gills or lungs. The space between the folds of the mantle is called the mantle cavity. The *foot* is a muscular organ that may be adapted for locomotion, attachment, food capture, or a combination of functions. Another feature often present is a *radula,* an organ that bears many rows of teeth and is used to obtain food.

The earliest mollusks are presumed to have had a nervous system consisting of a *nerve ring and 2 longitudinal nerve cords,* one serving the foot and the other the visceral mass (fig. 27.3). The *coelom is much reduced* and is largely limited to the region around the heart. In most mollusks, the heart pumps blood through vessels that empty into open spaces, called *sinuses,* before being collected in vessels and returned to the heart. This arrangement is called an *open circulatory system* because the blood is not always enclosed within blood vessels. Hemocyanin, rather than hemoglobin, is the common blood pigment in the animals of this phylum. Hemoglobin is an iron-containing red pigment, and hemocyanin is a copper-containing blue pigment. Both molecules carry oxygen.

Snails and Relatives

Snails, whelks, conchs, periwinkles, and sea slugs are representative *gastropods* (class Gastropoda) that are found usually in marine habitats, although they sometimes inhabit freshwater environments. In addition, garden snails (fig. 27.4) and slugs are gastropods adapted to terrestrial habitats. Many gastropods are herbivores and use their radula to scrape food from surfaces. Others are carnivores, using their radula to bore through surfaces such as bivalve shells to obtain food.

Figure 27.3

a. Relationship of parts in a hypothetical mollusk. **b.** Photograph of a chiton (class Polyplacophora), a mollusk whose anatomy may have changed very little from its primitive state. On the upper surface is a row of 8 overlapping plates. Below is a flat foot used for creeping along or clinging to rocks. The chiton scrapes algae and other plant food from rocks with its well-developed radula.

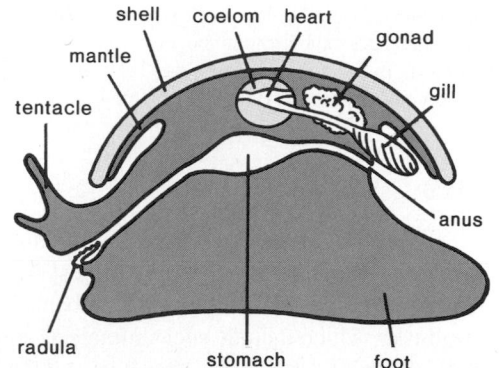

a.

b.

Figure 27.4

Garden snail anatomy. **a.** Garden snail from Papua, New Guinea. **b.** Snails have 3 obvious divisions of the body: a head with 2 pairs of tentacles, one having eyes at the tips; a flat, long muscular foot; and a visceral mass surrounded by a shell. The mantle serves as a lung. The embryo of a snail is at first bilaterally symmetrical, with the anus posterior. Then the body twists, and the gut, nerve cords, and many other structures are swung forward, so that the anus comes to lie just behind the mouth. **c.** Garden snails are adapted to a terrestrial life; compare their appearance with this nudibranch (sea slug), a marine gastropod that lacks shell, gills, and mantle cavity. It has outgrowths that function as gills.

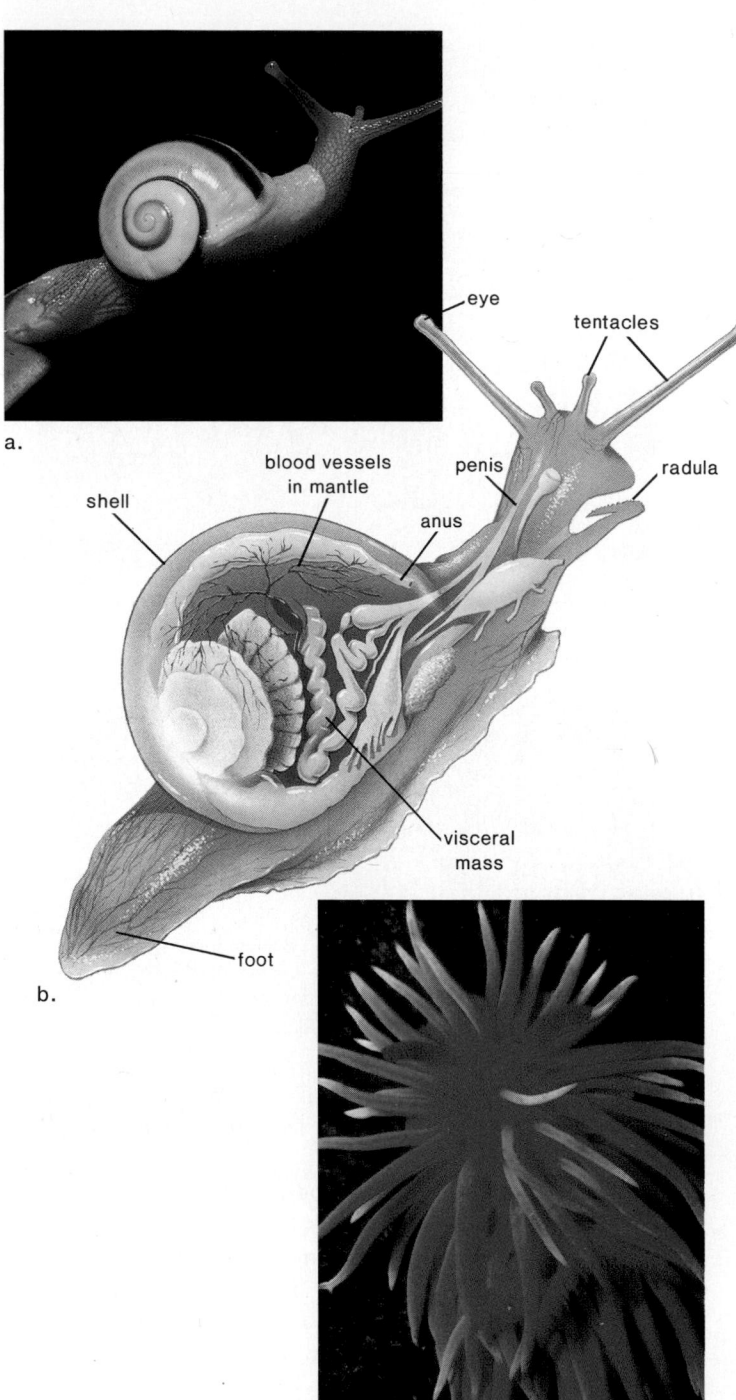

Gastropods have an elongated, flattened foot for crawling. In most, there is a head region and a coiled shell to protect the visceral mass—nudibranches (sea slugs) and terrestrial slugs, however, lack a shell. Although these animals begin life bilaterally symmetrical, the larva undergoes a *torsion*, or twisting, that brings the anus and mantle cavity downward, then forward and around to a position above the head. Torsion positions the visceral mass squarely above the foot.

In aquatic gastropods, gills are found in the mantle cavity, but in those adapted to land, the mantle is richly supplied with blood vessels and functions as a lung when air is moved in and out

Figure 27.5

a. Bivalve anatomy is exemplified by the anatomy of a clam. The shell and the mantle on one side have been removed. Notice the 3-ganglion nervous system and the lack of cephalization, a characteristic that is consistent with filter-feeding animals. ***b.*** This diagram of gill structure shows the flow of water through the mantle cavity. Water (blue arrows) enters the mantle cavity via the incurrent siphon and passes through pores entering the gills before exiting via the excurrent siphon. In the meantime, food particles trapped in mucus on the gills pass to the mouth (red arrows).

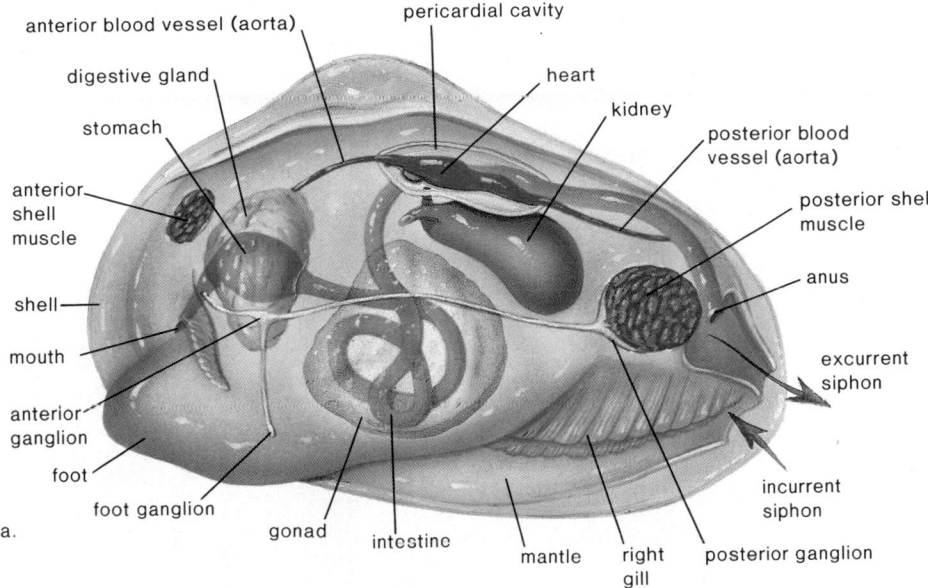

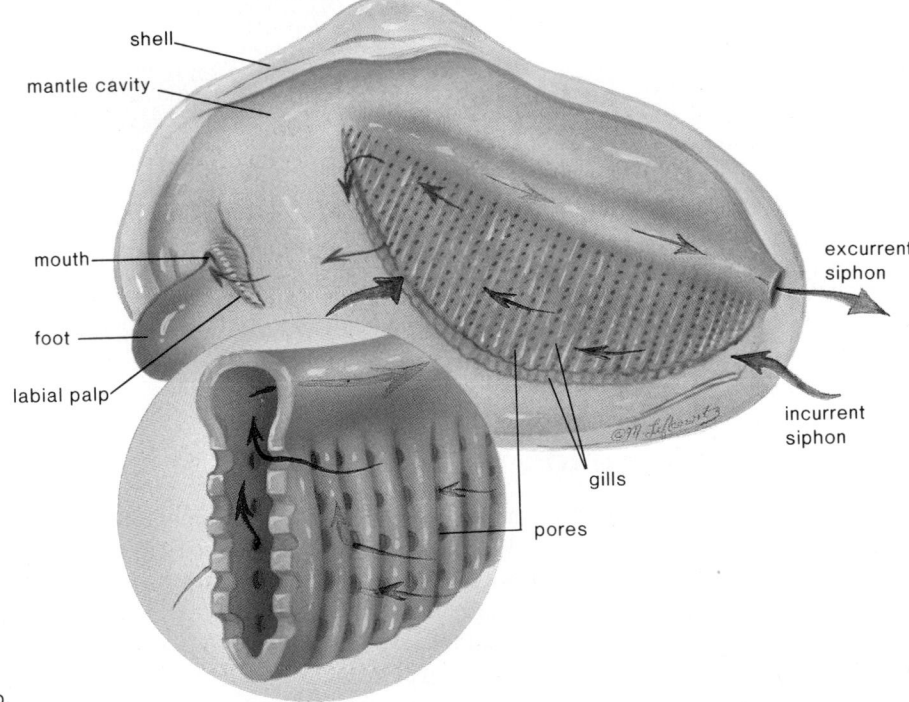

through respiratory pores. As another adaptation to life out of water, terrestrial gastropod development does not include the swimming larval stage found in aquatic species.

Bivalves

Clams, oysters, scallops, and mussels are all *bivalves* (class Bivalvia). The term *bivalve* refers to their 2 shells, which are hinged and closed by powerful muscles. Clams use their hatchet-shaped foot for burrowing in sandy or muddy soil, and mussels use their foot for production of threads, which attach them to nearby objects. Scallops both burrow and swim; rapid clapping of the valves releases water in spurts and causes the animal to move forward.

The bivalves have large gills that hang down in the mantle cavity. These are used for both respiration and for gathering food. They are *filter feeders* and glean food from the water that enters and exits the mantle by way of *siphons* located at the posterior end. Debris trapped on the *gills* is swept toward the mouth by ciliary action.

Bivalves have no head. Their nervous system consists of *3 pairs of ganglia* (ganglion, sing.), or nerve centers located anteriorly, posteriorly, and in the foot. These ganglia are connected by nerves (fig. 27.5).

Squids and Relatives

Cephalopods (class Cephalopoda) include squids (fig. 27.6), octopuses, and nautiluses, all of which are fast-moving, marine, swimming organisms. Both squids and octopuses can squeeze their mantle cavity so that water is forced out, propelling them rapidly backward by a sort of *jet propulsion.* Cephalopods are adapted to a predatory way of life. The foot has evolved into tentacles around the head (*cephalopod*—head-footed) that are used to seize prey. The powerful, parrotlike beak is used to tear apart the prey. They have well-developed sense organs, including focusing *camera-type eyes* that are very similar to those of vertebrates. Cephalopods, particularly octopuses,

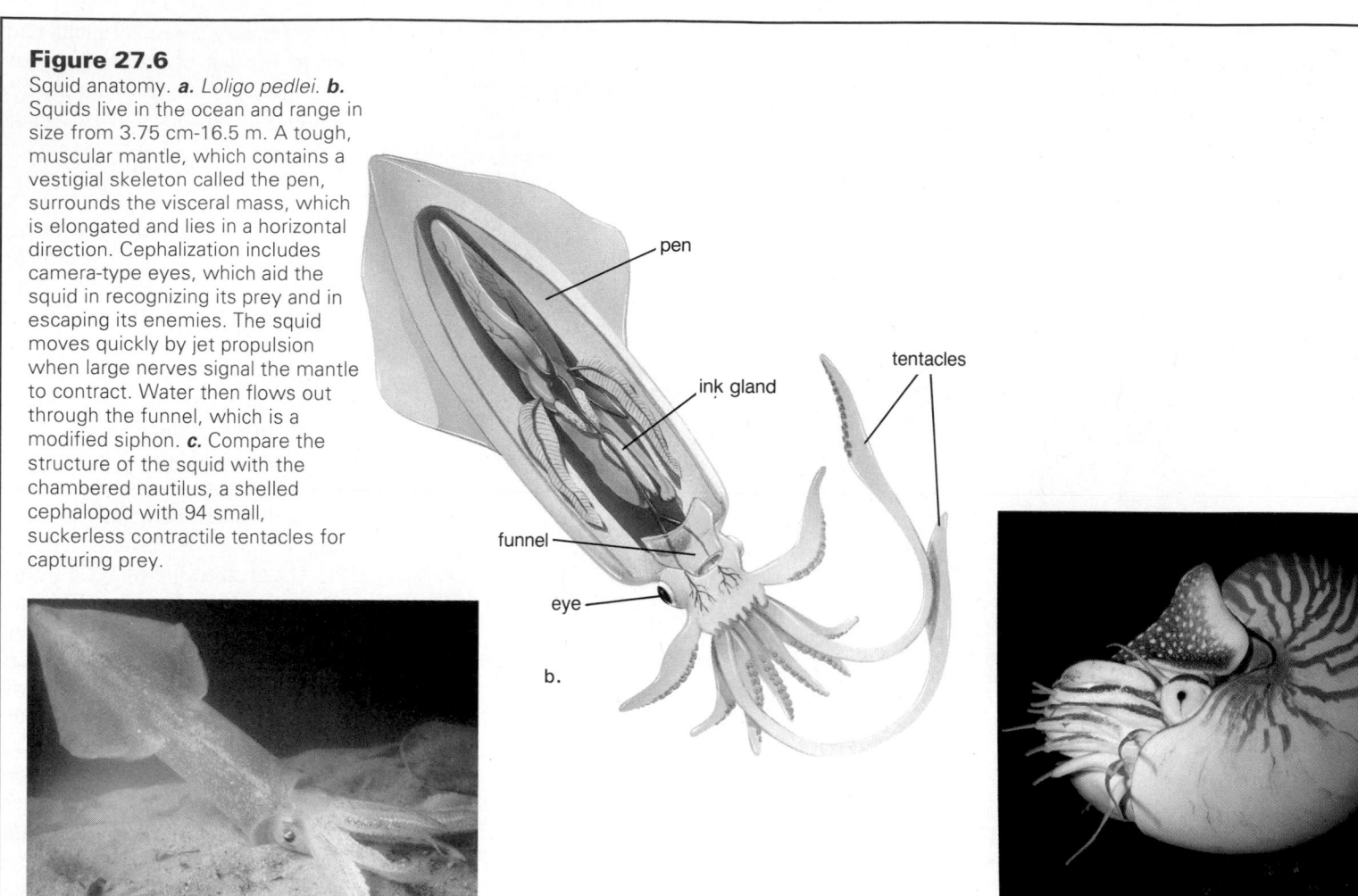

Figure 27.6

Squid anatomy. *a. Loligo pedlei. b.* Squids live in the ocean and range in size from 3.75 cm–16.5 m. A tough, muscular mantle, which contains a vestigial skeleton called the pen, surrounds the visceral mass, which is elongated and lies in a horizontal direction. Cephalization includes camera-type eyes, which aid the squid in recognizing its prey and in escaping its enemies. The squid moves quickly by jet propulsion when large nerves signal the mantle to contract. Water then flows out through the funnel, which is a modified siphon. *c.* Compare the structure of the squid with the chambered nautilus, a shelled cephalopod with 94 small, suckerless contractile tentacles for capturing prey.

have well-developed brains and show a remarkable capacity for learning. For protection, they possess *ink sacs,* from which they can squirt a cloud of brown or black ink. This action often leaves a would-be predator completely confused. Nautiluses are enclosed in shells, but squids have a shell that is reduced and internal. Octopuses lack shells entirely.

> Mollusks are a very diverse group, with a body plan composed of 3 parts: the foot, the mantle, and the visceral mass. The mostly herbivorous gastropods have a flattened foot, the filter-feeding bivalves have a 2-part shell, and the predaceous cephalopods have tentacles about the head.

Comments on Mollusks

The 3-part plan of the molluscan body—visceral mass, mantle, and foot—is unique to this animal phylum. A protective shell is particularly useful in slow-moving animals, and therefore we see

that the predaceous and fast-moving cephalopods have a reduced shell. The mollusks are quite diverse, with modifications in methods of feeding and locomotion. The foot is modified variously in the different groups.

Annelids

Annelids (phylum Annelida, 8,700 species), represented by earthworms and many aquatic species, are **segmented** worms, as is externally evident by the rings that encircle the body. Internally, partitions divide the *well-developed, fluid-filled coelom,* that acts as a **hydrostatic skeleton** (see fig. 26.9). Segmentation of the hydrostatic skeleton makes it more efficient because it allows regional differences in pressure and muscle action. Annelids have both circular and lengthwise muscles. When lengthwise muscles contract, segments of the body shorten; when circular muscles contract, segments of the body elongate.

Figure 27.7

a. A sandworm, such as *Neanthes,* has fleshy lobes, the parapodia, on each body segment. They are used for swimming and as respiratory organs. Numerous chitinous bristles grow out from the parapodia, and hence they are polychaetes, or "many bristled." **b.** This worm is a predator; its small prey is captured by a pair of strong chitinous jaws, which evert with a part of the pharynx when the worm is feeding. **c.** Compare the diagram in (**b**) with the head region of this sedentary tubeworm, which extends long ciliated tentacles into the water. The cilia create currents, filter food particles, and move these particles toward the mouth.

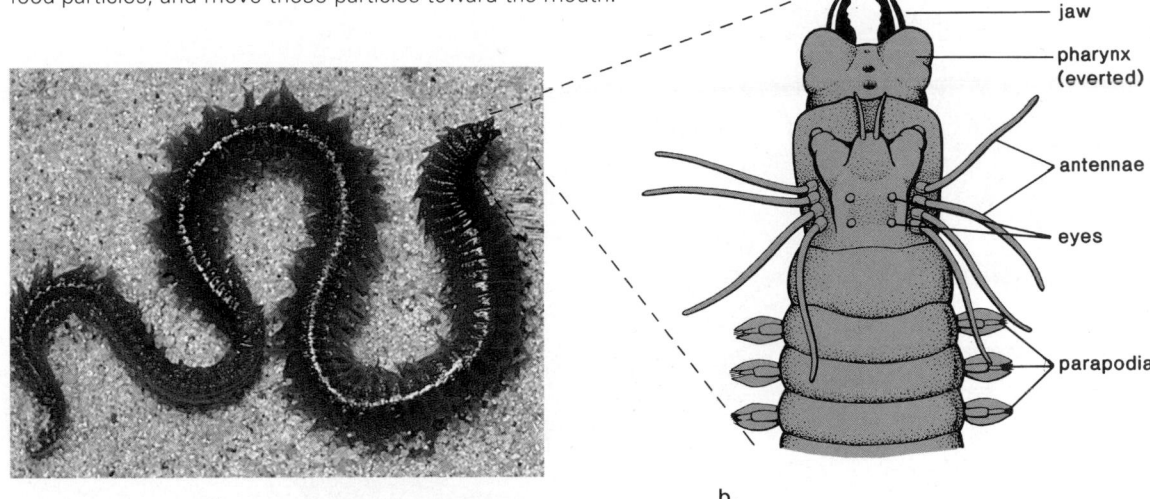

jaw

pharynx (everted)

antennae

eyes

parapodia

a.

b.

c.

Annelids are *bilaterally symmetrical and have the tube-within-a-tube body plan.* The digestive tract shows specialization of parts along its length; for example, there may be a pharynx, stomach, and accessory glands. There is an extensive closed circulatory system, in which blood is always enclosed within blood vessels that run the length of the body and branch to every segment. The nervous system consists of a *brain, connected to a ventral solid nerve cord, with a ganglion in each segment.* The excretory system consists of *paired* **nephridia,** which are coiled tubules in each segment that collect waste material from the coelom and excrete it through openings in the body wall.

Marine Annelids

Most annelids are marine and belong to the class Polychaeta, named for the presence of many setae (fig. 27.7). The *setae* are bristles that anchor the worm or help it move. In *polychaetes,* the setae occur in bundles on *parapodia,* which are paddlelike appendages found on most segments. Some polychaetes tend to be active, and aided by their parapodia, they wander about in search of prey. These worms have a definite head with sense organs. Other polychaetes are sedentary tubeworms, with tentacles around their head. Water currents, created by the action of cilia, bring debris within reach of the tentacles, to which it adheres. These worms are continuous filter feeders. Because the parapodia have a rich supply of blood vessels, they assist the process of gas exchange in polychaetes.

The polychaetes have breeding seasons, and only during these times do the worms have sex organs. Internal sex organs develop each year and produce either sperm or eggs. In some worms, the gametes can escape by way of the nephridia, whereas in others, the body literally breaks apart to release them. The zygote develops into a larval stage.

Earthworms

Earthworms (class Oligochaeta, meaning few setae) do not have a well-developed head or parapodia (fig. 27.8). Their setae protrude in clusters directly from the surface of their body. The placement of setae and other anatomical features indicate that the worms are segmented (table 27.1).

Earthworms reside in soil, where there is adequate moisture because the *body wall must remain moist for gas exchange purposes.* They feed on vast quantities of soil that contains living and decaying organic matter. Much of what earthworms eat passes through their body and is deposited as small piles of dirt (worm casts) on the soil surface. This activity is an important one for soil turnover.

Earthworms are *hermaphroditic,* but they still practice cross fertilization (fig. 27.8). Two worms lie parallel to each other, facing in opposite directions. The fused mid-body segment, called a *clitellum,* secretes mucus, protecting the sperm from drying out as they pass between the worms. After the worms separate, the clitellum of each produces a slime tube, which is moved along over

Figure 27.8

Earthworm anatomy. The drawing shows the internal anatomy of the anterior part of an earthworm's body. The closed circulatory system has 5 pairs of hearts. Earthworms are hermaphroditic; each worm has both male and female reproductive structures. The small sketch shows the location of the clitellum. During mating 2 worms are held together by mucus secreted by the clitellum.

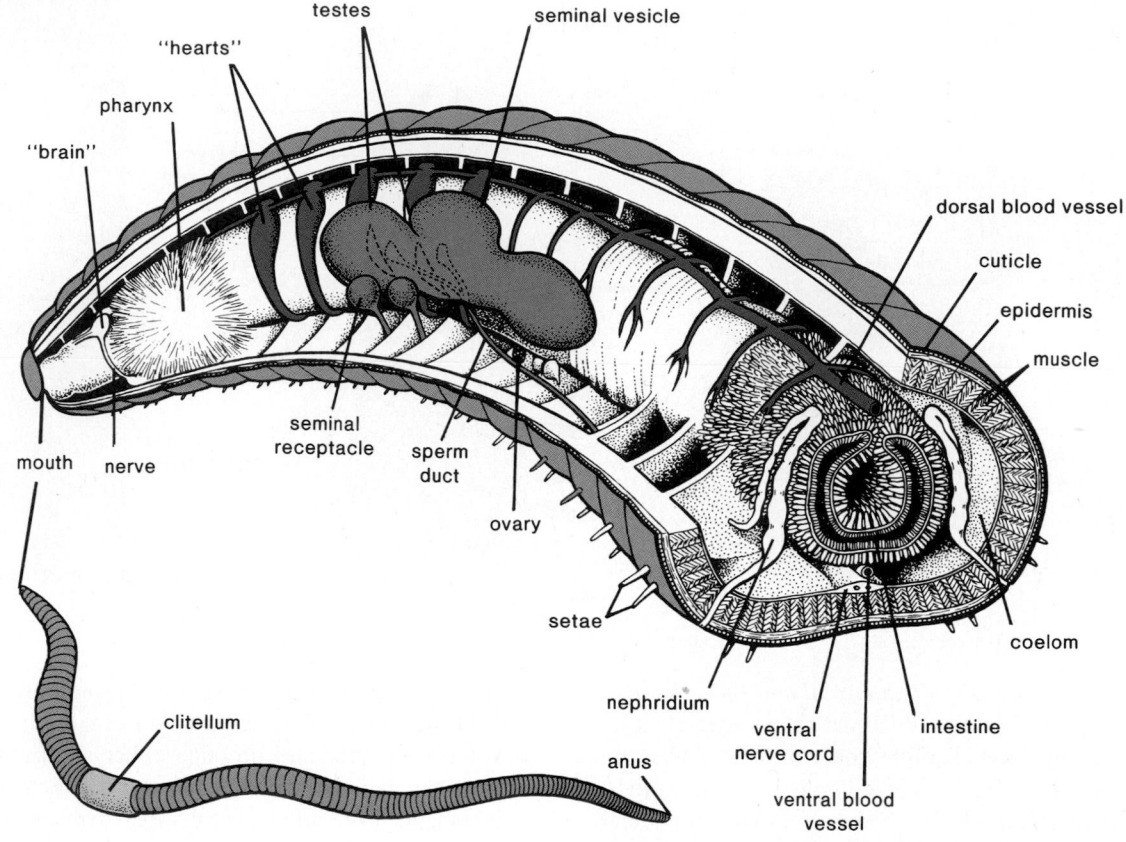

the head by muscular contractions. As it passes, eggs and the sperm received earlier are deposited and fertilization occurs. The slime tube then forms a cocoon to protect the worms as they develop. There is no larval stage.

Leeches

Leeches (class Hirudinea) have the same body plan as other annelids, but among their modifications are 2 suckers—a small, oral one around the mouth and a large, posterior one. While some leeches are free-living, the ones most remembered are the blood-sucking species. Leeches are able to keep blood flowing and prevent clotting by means of a substance in their saliva known as hirudin, a powerful anticoagulant.

Comments on Annelids

Annelids have a well-developed coelom and are segmented. Segmentation is an important element to have evolved in animals. Annelids have a hydrostatic skeleton, and the division of the coelom into a series of individual segments allows body parts to move independently of one another. This is especially significant, but evidence of segmentation is also seen in other systems of the body (table 27.1).

Table 27.1
Segmentation in the Earthworm

1. Body rings
2. Coelom divided by septa
3. Setae on each segment
4. Ganglia and lateral nerves in each segment
5. Nephridia in each segment
6. Branch blood vessels in each segment

Annelids have a ventral solid nerve, and we shall see that this contrasts with vertebrates, which have a dorsal hollow nerve cord.

Annelids are segmented worms. Most of their organ systems show evidence of segmentation. These worms have a well-developed coelom divided by septa, a closed circulatory system, and a ventral solid nerve cord. Marine annelids have adaptations for aquatic life, and earthworms are adapted to terrestrial living.

The protostomes (annelids, arthropods, and mollusks) share embryological features that indicate they are related, but how they are related is not known. Mollusks are not segmented except for *Neopilina gelatheae* (see fig. 27.A). This animal was presumed to be extinct for the past 500 million years, but 10 living specimens were dredged up from a depth of more than 3,500 m in the Pacific Ocean near Costa Rica in 1952. There is a segmental arrangement of gills, nephridia, and muscles, which could indicate the animal is similar to a possible common ancestor of both mollusks and annelids. On the other hand, it is possible that the segmentation seen in *Neopilina* arose independently in the molluskan line and that both mollusks and annelids are descended from nonsegmented flatworms.

A possible direct relationship between arthropods and annelids can also be found. *Peripatus* (phylum Onychophora) (see fig. 27.A) is called the walking worm because it has 14–400 pairs of short, stumpy legs and possesses organs with annelid or arthropod characteristics. Its excretory organs and musculature are more like those of annelids, but its respiratory and circulatory systems are more like those of arthropods. It has a cuticle of chitin and modified appendages that serve as jaws. Although the appendages of *Peripatus* are not jointed, the fossil record does contain wormlike animals with jointed appendages. It's possible that arthropods and annelids share a common ancestor, or perhaps arthropods are descended directly from annelids.

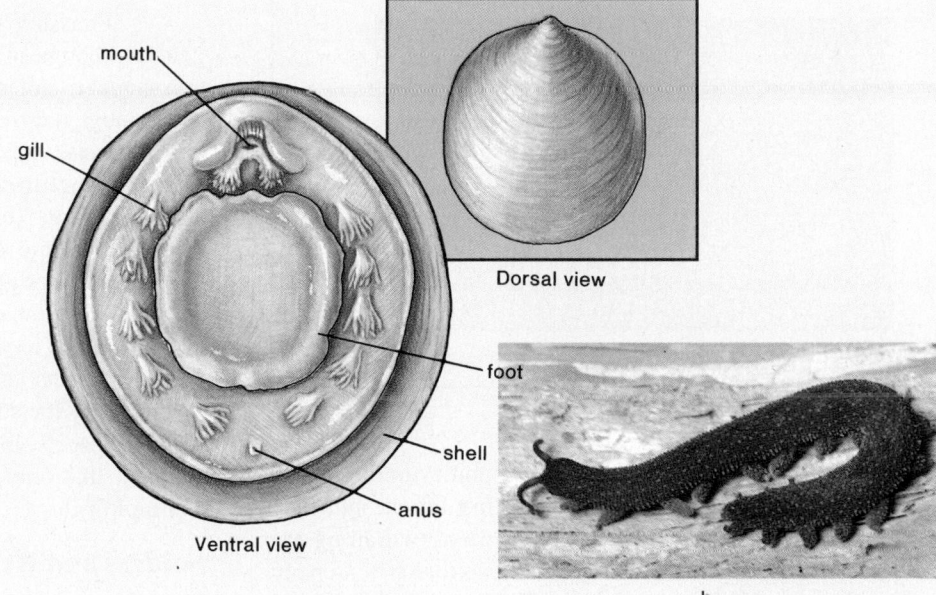

Figure 27.A
a. *Neopilina gelatheae,* a mollusk. *b. Peripatus,* an onychophoran.

Arthropods

Arthropods (phylum Arthropoda, 1,000,000 species) show such diversity and are adapted to so many different habitats that they are often said to be the most successful of all animals. Arthropod means literally jointed foot. Actually, they have **jointed appendages** that are freely movable. Their body is segmented, and at one time, there was a pair of similar appendages on each body segment (fig. 27.9). In modern arthropods, some segments have fused into regions, such as a head, thorax, and abdomen (fig. 27.10*f*). The appendages may be specialized for such functions as walking, swimming, reproducing, and eating.

Arthropods have a well-developed nervous system. There is a brain and a *ventral solid nerve cord.* The head bears various types of sense organs, including eyes of 2 types—compound and simple. The **compound eye** is composed of many complete visual units grouped in one structure. Each visual unit contains a separate lens and a light-sensitive cell.

The **exoskeleton** (external skeleton) of arthropods is composed primarily of **chitin,** a strong, flexible, nitrogenous polysaccharide. Because it is hard and nonexpandable, arthropods must **molt,** or shed, the exoskeleton as they grow larger. Before molting, the body secretes a new, larger exoskeleton, which is soft and wrinkled, underneath the old one. After enzymes partially dissolve and weaken the old exoskeleton, the animal breaks it open and wriggles out. The new exoskeleton then quickly expands and hardens.

Lobsters and Relatives

Members of the class Crustacea (40,000 species) are named for their hard shells. Although their anatomies are extremely diverse, the head usually bears a pair of compound eyes and 5 pairs of appendages. There are 2 pairs of antennae, which have various functions (sensory, locomotion, or feeding) according to the group. Behind the antennae are 3 pairs of mouthparts for chewing, grinding, or filter feeding.

In decapods, which include shrimps, crayfish, lobsters, and crabs, the thorax bears 5 pairs of walking legs (fig. 27.11). The first pair may be modified as claws. Typically, there are gills situated above the walking legs.

Copepods and krill are small crustaceans (less than 2 mm long) that live in the water, where they feed on algae. In the marine environment, they serve as food for fishes, sharks, and whales. They are so numerous that despite their small size, some believe that they are harvestable as food. Barnacles have a heavy calcareous shell that is actually a modified exoskeleton. Stalked (goose) barnacles are attached by a stalk, and stalkless (acorn) barnacles are attached directly by their shells. You see barnacles

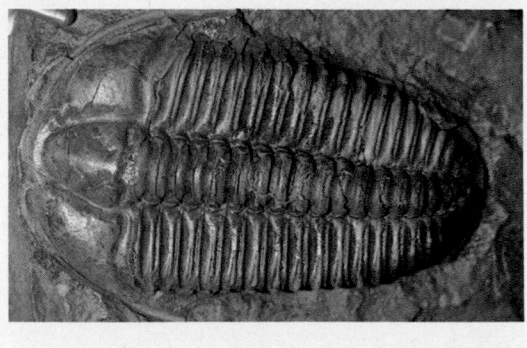

Figure 27.9
Fossil trilobite, an arthropod that has been extinct for about 225 million years. Note the regular segmentation of this animal, named for the 3 lobes seen here. There are so many trilobites in the fossil record that 3,900 species have been recognized, and even the developmental stages of some species have been studied.

on wharf pilings, ship hulls, seaside rocks, and even the bodies of whales. They begin life as free-swimming larvae, but they undergo a metamorphosis that transforms their swimming appendages to cirri, which are feathery structures that are extended and allow them to filter feed when they are submerged.

Insects

Insects (class Insecta, 900,000 species) include more known species than all other animal species combined. Many live on the land, where *wings* enhance the insects' ability to survive by providing a new way of escaping enemies, finding food, facilitating mating, and dispersing the species.

The grasshopper is often studied as a representative insect (fig. 27.12). It has 3 pairs of legs, one pair of which is suited to jumping. There are 2 pairs of wings. The forewings are tough and leathery, and when folded back at rest they protect the broad, thin hindwings. A large tympanum for the reception of sound waves is on the lateral surface of the abdomen.

The grasshopper's digestive system is suitable for a grass diet. The food is first subjected to the grinding action of mouthparts and the chemical action of saliva. It is then temporarily stored in the crop before being ground finely in the gizzard. Chemical digestion is completed by enzymes secreted by the gastric caeca. A *cecum* is a cavity open at one end only. Undigested remains pass into the intestine and leave the body by way of the anus.

Respiration occurs when air enters small tubules called **tracheae** through openings in the exoskeleton called spiracles. The tracheae branch and rebranch, ending in moist areas adjacent to cells, where the actual exchange of gases takes place. Excretion is carried out by **Malpighian tubules,** which extend into the hemocoel, a network of open spaces where blood flows around the organs. (The hemocoel is the coelomic cavity in arthropods.)

Malpighian tubules form a solid, nitrogenous waste that passes out of the animal by way of the digestive tract. Solid waste allows the animal to conserve water.

The grasshopper's reproductive system is adapted to life on land. The male passes sperm to the female by way of a penis, and fertilization is internal. In this way, both gametes and zygotes are protected from drying out. The female deposits the fertilized eggs in the ground with her ovipositor.

Grasshoppers undergo incomplete **metamorphosis,** a gradual change in form as the animal matures. The immature grasshopper, called a nymph, is recognizable as a grasshopper, even though it differs somewhat in shape and form from the adult. Other insects, such as butterflies, undergo complete metamorphosis, involving drastic changes in form. At first, the animal is a wormlike larva (caterpillar) with chewing mouthparts. It then forms a case, or cocoon, about itself and becomes a *pupa*. During this stage, the body parts are completely reorganized; the *adult* then emerges from the cocoon. This life cycle allows the larvae and adults to make use of different food sources.

Insects also show remarkable behavior adaptations, exemplified by the social systems of bees, ants, termites, and other colonial insects. Insects are so numerous and so diverse that the study of this one group is a major specialty in biology called entomology.

Spiders and Relatives

Spiders, scorpions, ticks, and mites are *arachnids* (class Arachnida). Spiders are predators, usually spinning a web in which to trap insects. They have 6 pairs of appendages. The first pair are modified as fangs, with ducts from poison glands. The second pair have become basal parts for chewing. The other 4 pairs are walking legs that end in claws.

The body of a spider doesn't show much obvious segmentation. The head and thorax are fused into a cephalothorax, and there is also an abdomen. Spiders don't have compound eyes; instead, the eyes are simple. Spiders are adapted to living on land, as witnessed by the presence of *book lungs,* which are air pockets within a cavity surrounded by blood.

Scorpions, which are more common in warm climates, are also predators. Ticks and mites are often parasites. Ticks suck the blood of vertebrates and are sometimes transmitters of diseases, such as Rocky Mountain spotted fever or Lyme disease. Chiggers are the larvae of certain mites and feed on the skin of vertebrates.

Centipedes and Millipedes

Centipedes (class Chilopoda) have a body composed of a head and trunk. The trunk has many segments, and each segment has a pair of walking legs. They are carnivorous animals, and the head bears antennae and mouthparts with jaws.

Millipedes (class Diplopoda) have the same segmented organization as centipedes, but some segments are fused and therefore they appear to have 2 pairs of walking legs on each segment. Millipedes dwell in the soil, feeding on dead organic matter.

Evolution and Diversity

Figure 27.10

There are 5 major classes of arthropods. ***a.*** Class Crustacea is represented by this crab with a large carapace and 5 pairs of legs. The first pair are pinching claws. ***b.*** Class Arachnida is represented by this land-dwelling scorpion, with 4 pairs of legs, poisonous pincers, and stinging tail. ***c.*** Class Chilopoda is represented by this centipede, a carnivorous animal, with a pair of appendages on every segment. ***d.*** Class Diplopoda is represented by this millipede, a scavenger that seems to have 2 pairs of appendages on each segment because every 2 segments are fused. Class Insecta is represented by (***e***) a swallowtail butterfly with wings and (***f***) an ant with clearly defined head, thorax, and abdomen. Insects have 3 pairs of legs.

d.

a.

b.

c.

e.

f.

Figure 27.11

Decapods are large crustaceans including lobster, crayfishes, shrimps, and crabs, which have 5 pairs of walking legs. **a.** Photograph of a shrimp. **b.** As in other crustaceans, the head bears compound eyes and 2 pairs of antennae (first antennae and second antennae). The rostrum conceals the 3 pairs of mouthparts. Swimmerets are attached to the abdomen, whose last 2 segments bear the uropods and the telson, which make up a fan-shaped tail for swimming backwards. **c.** Notice how the first pair of walking legs are claws in the lobster.

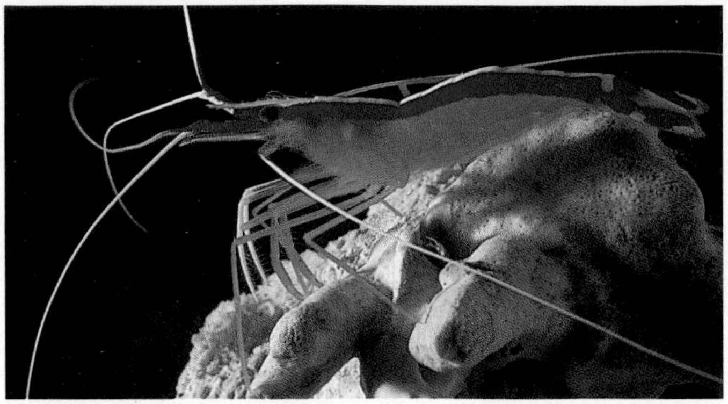

a.

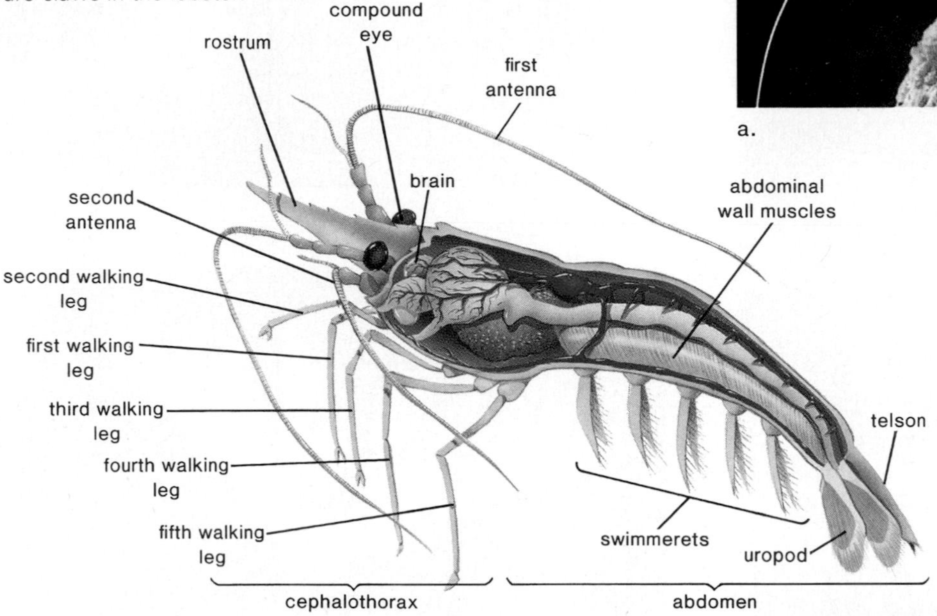

b.

Figure labels: rostrum, compound eye, first antenna, second antenna, brain, second walking leg, first walking leg, third walking leg, fourth walking leg, fifth walking leg, cephalothorax, abdominal wall muscles, telson, swimmerets, uropod, abdomen

c.

Comments on Arthropods

Based on number of species and dispersal of these species into various habitats, arthropods are the most successful phylum in the animal kingdom. Arthropods are segmented, and they also have a rigid, but jointed, skeleton. The presence of these 2 features allows complex body movements and adaptability to the land environment. We will see that this winning combination is seen not only in arthropods, which are protostomes, but also in chordates, which are deuterostomes. The 2 features, therefore, evolved in both lineages.

The segmentation first observed in annelids has led to specialization of body parts in arthropods. In comparison to annelids, arthropods, which also have a ventral solid nerve cord, show marked cephalization. The head carries sense organs; arthropods are especially known for the presence of compound eyes.

Arthropods are the most numerous and varied of all the animal phyla. They are segmented, with an external skeleton and jointed appendages. Several groups (insects, arachnids, centipedes, and millipedes) contain species that are adapted to terrestrial life.

Echinoderms

Echinoderms (phylum Echinodermata, 6,000 species) are a diverse marine group having *radial symmetry,* usually with a 5-part body organization. The endoskeleton (internal skeleton) of echinoderms consists of spine-bearing plates. The spines stick out through the delicate skin, which accounts for their name. Echinoderm comes from the Greek words meaning spiny skin.

Figure 27.12

The anatomy of a female grasshopper illustrates many adaptations to life on land. **a.** Externally, the hard skeleton prevents loss of water. There are spiracles, openings in the skeleton, which admit air into tracheae (air tubes for respiration). The tympanum uses air waves for sound reception, and the hopping legs and wings are for locomotion. The ovipositor deposits eggs in soil. **b.** Internally, the digestive system is adapted to digesting grass. The Malpighian tubules excrete a solid nitrogenous waste (uric acid). A seminal receptacle receives sperm from the male, which has a penis. Internal fertilization prevents the gametes from drying out.

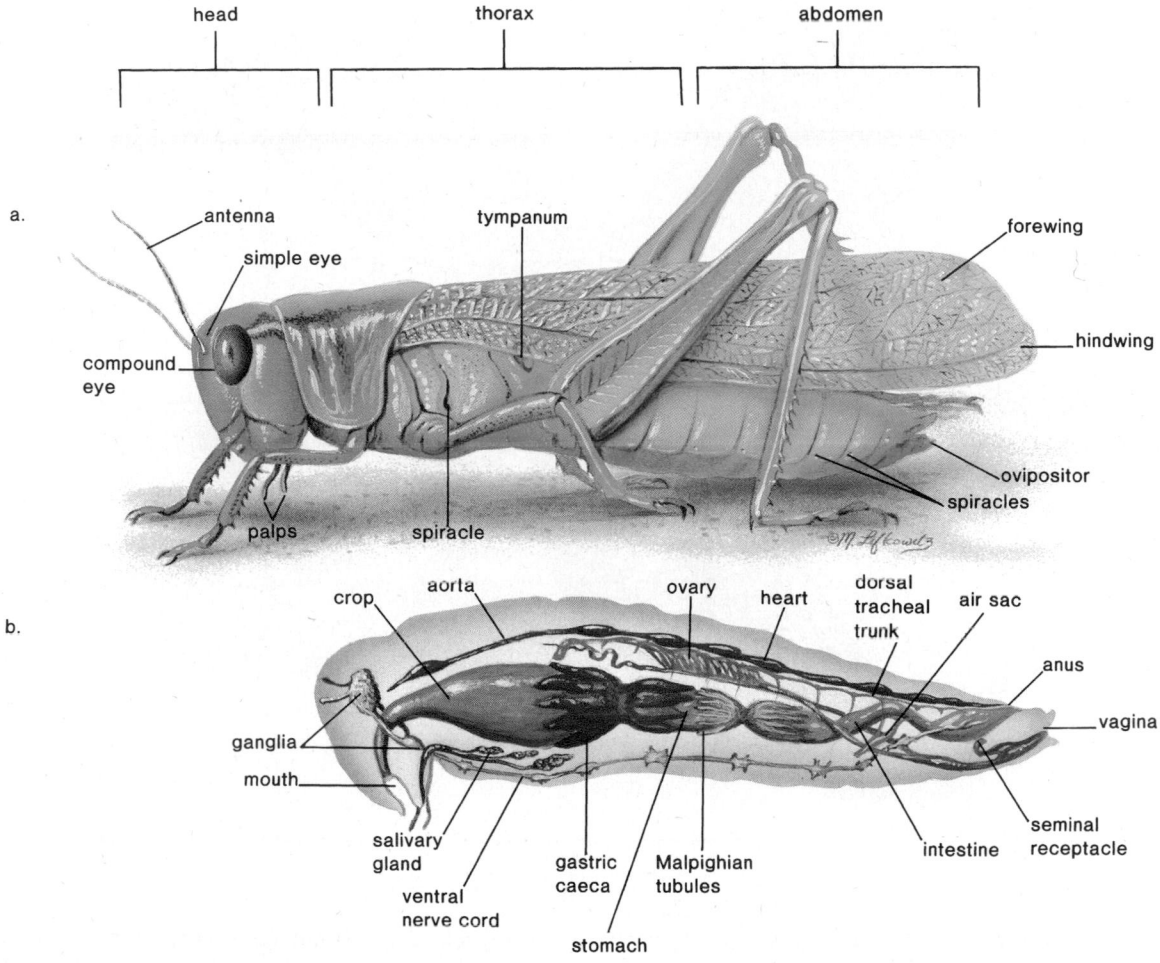

Echinoderms are deuterostomes, with a coelom that develops similarly to that of the chordates (fig. 27.2). Coelomic fluid circulates substances, and wandering amoeboid cells clean up particulate wastes. Gas exchange takes place through *skin gills,* which are tiny fingerlike extensions of the skin. Echinoderms have a central *nerve ring and nerve branches* extending into the multiple body divisions. They are capable of coordinated but slow responses and body movements.

Movement is facilitated by *tube feet,* which are part of a *water vascular system* (fig. 27.13). Water entering a *sieve plate* goes into major canals extending into each portion of the body. Small canals to the tube feet branch off the major canals. Each of the small canals ends in an expanded region called an *ampulla.* Contraction and expansion of the ampullae control the tube feet. By alternating the attachment and release of the tube feet, the animal moves.

Sexes are separate in echinoderms. They shed gametes into the water, where fertilization occurs. The zygotes become free-swimming, bilaterally symmetrical larvae that undergo metamorphosis to become radially symmetrical adults.

Sea Stars and Other Echinoderms

Sea stars (starfishes) (class Asteroidea) are the most familiar echinoderms. The body of most sea star species is flattened and has a central disk with 5, or a multiple of 5, sturdy arms (rays) extending outward from it. Each arm has rows of tube feet in a groove that extends along its ventral surface.

Sea stars commonly feed on clams, oysters, and other bivalve mollusks. To feed, a sea star positions itself over a bivalve and attaches some of its tube feet to each side of the shell. By working its tube feet in alternation, it pulls the shell open. A very small crack is enough for the sea star to push the larger, lower part of its stomach

Figure 27.13

Sea star anatomy. Aside from the water vascular system shown here in color, gonads and digestive glands occur in each arm. The 2-part stomach is in the central disk with the anus on the upper dorsal surface and the mouth hidden beneath, on the ventral surface.

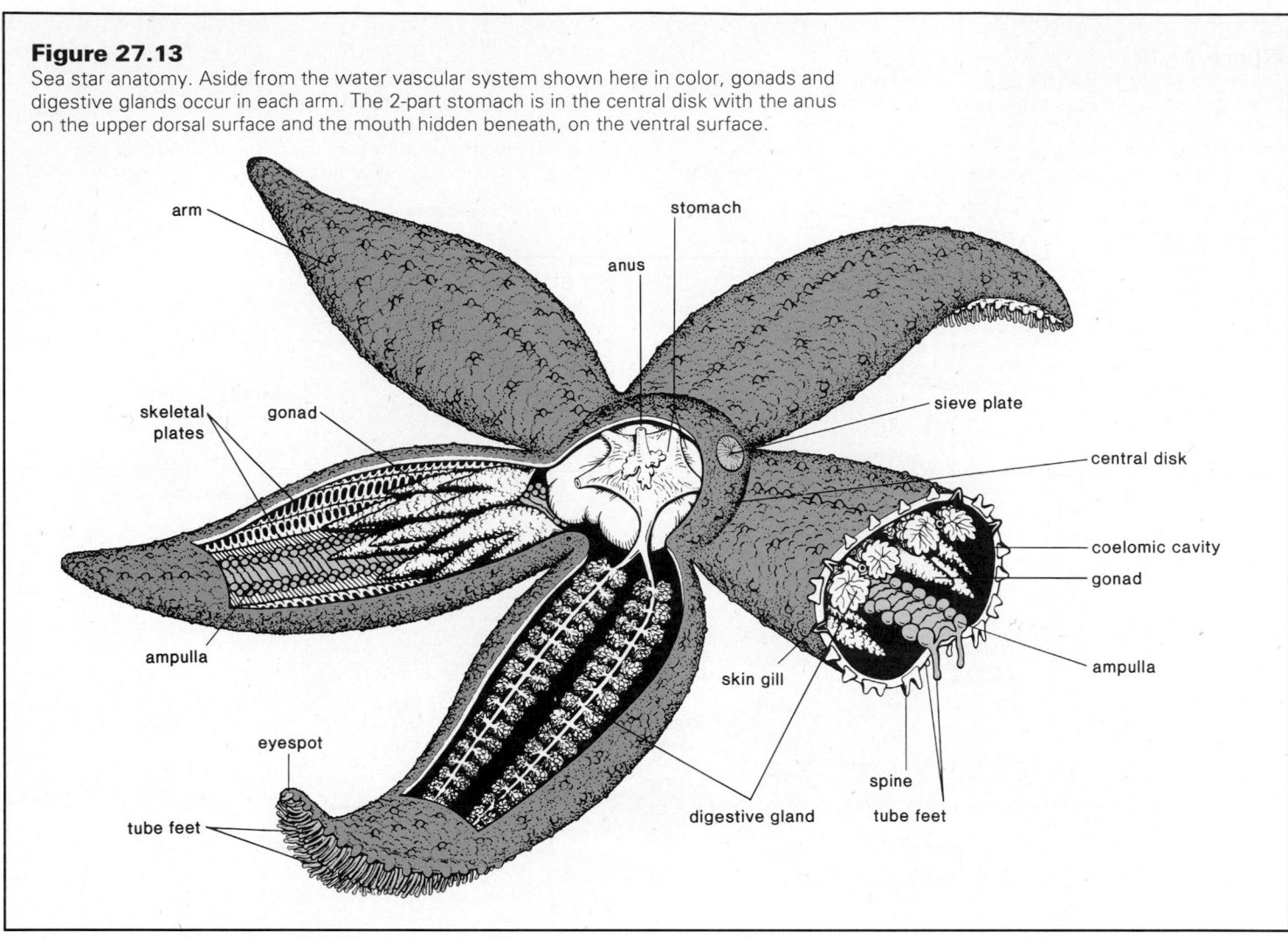

out of its mouth and through the crack, so that it contacts the soft parts of the bivalve. The stomach secretes enzymes, and digestion begins even while the bivalve is attempting to close its shell. Later, partly digested food is taken into the sea star's body, where digestion continues in the upper part of the stomach using enzymes from the digestive glands found in each arm.

There are several other classes of echinoderms. Brittle stars (class Ophiuroidea) have a central disk from which long arms radiate. The tube feet function mainly in chemoreception, and they move quickly by using their arms to push themselves along. Sand dollars and sea urchins (class Echinoidea) both have spines, which they use for locomotion, defense, and burrowing. Sand dollars have a fused, single, flattened plate with pores that form a 5-part flowerlike pattern. Special tube feet modified for respiration extend through these pores.

Sea cucumbers (class Holothuroidea) have a long, leathery body that resembles a cucumber. They lack arms, but they have tube feet with suckers that form feeding tentacles around the mouth.

The tentacles are inserted one at a time into the mouth, where food is scraped off. Sea lilies (class Crinoidea) are an ancient group with many species in the fossil record of the Carboniferous period; today they might be becoming extinct. They have a flowerlike appearance because their branched arms, used for filter feeding, are attached to a stalk.

Comments on Echinoderms

Based on embryological evidence, echinoderms are believed to be closely related to the chordates. Most likely these 2 groups share a common bilateral ancestor, but the echinoderms went on to become radially symmetrical as an adaptation to a sessile way of life.

Echinoderms are radially symmetrical and have a spiny skin. They move by tube feet, which are part of their water vascular system. Their other body systems are rather primitive.

Table 27.2
Comparison of Animal Phyla: Mollusks to Chordates

	Symmetry	Germ Layers	Body Plan	Coelom	Organs	Segmentation
Mollusks	Bilateral	3	Tube-within-a-tube	Coelom (reduced)	Yes	No
Annelids	Bilateral	3	Tube-within-a-tube	Coelom	Yes	Yes
Arthropods	Bilateral	3	Tube-within-a-tube	Coelom (reduced)	Yes	Yes
Echinoderms	Radial (as adults)	3	Tube-within-a-tube	Coelom	Yes	No
Chordates	Bilateral	3	Tube-within-a-tube	Coelom	Yes	Yes

Chordates

Among the chordates (phylum Chordata, 47,200 species), there are several species of filter-feeding marine invertebrates and the rest are vertebrates, animals in which the notochord has been replaced by a vertebral column. We will study only the invertebrate chordates (tunicates and lancelets) in this chapter. The next chapter considers the vertebrates, namely the fishes, amphibians, reptiles, birds, and mammals.

Chordates have all those advanced features we have discussed, such as bilateral symmetry, a well-developed coelom, and segmentation (table 27.2). To be considered a chordate, an animal must also have the following 3 basic characteristics at some time in its life history:

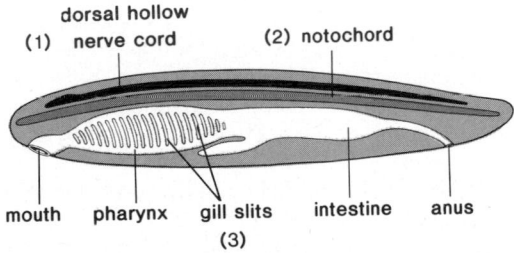

1. A dorsal hollow nerve cord. By hollow, it is meant that the cord contains a canal filled with fluid.
2. A dorsal supporting rod called a notochord. This is replaced by the vertebral column in vertebrates.
3. Gill slits or pharyngeal pouches. Gill slits are seen only in chordates that breathe by means of gills. All chordates, however, at some time during embryological development have pharyngeal pouches. Instead of becoming gills, these pouches are modified for other purposes in terrestrial chordates.

Invertebrate Chordates

A tunicate (subphylum Urochordata), or sea squirt, is a thick-walled squat sac with 2 openings, an incurrent siphon and an excurrent siphon (fig. 27.14). Inside the central cavity of the animal are numerous gill slits, the only chordate feature retained by the

Figure 27.14
Tunicate anatomy. **a.** The siphons of a living tunicate. A common name for these animals is sea squirts because water squirts from the excurrent siphon when the animals are irritated. **b.** The tunicate body. The numerous gill slits in the central cavity are the only chordate feature retained by the adult.

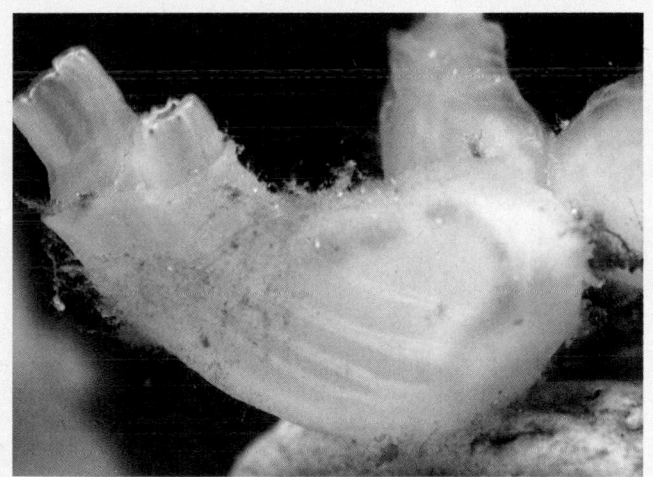

a.

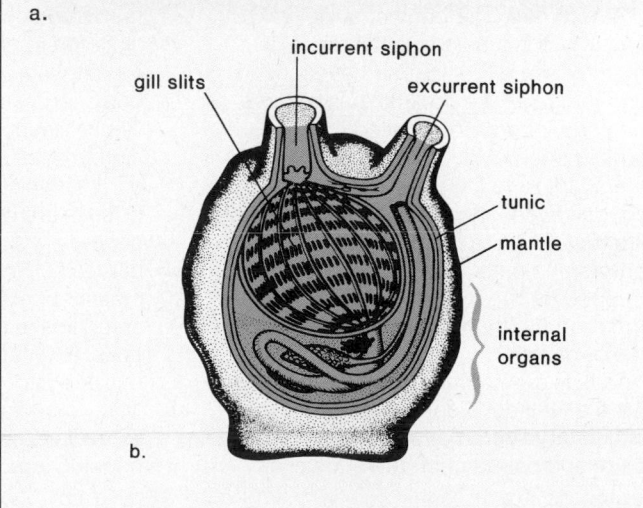

b.

adult. The larva of the tunicate, however, has a tadpole shape and possesses the 3 chordate characteristics. It has been suggested that such a larva may have become sexually mature without developing the other adult tunicate characteristics. If so, it may have evolved into a fishlike vertebrate.

A lancelet (subphylum Cephalochordata) is a chordate that shows the 3 chordate characteristics as an adult (fig. 27.15). In addition, segmentation is present, as witnessed by the fact that the muscles are segmentally arranged and the nerve cord gives off periodic branches.

> The invertebrate chordates include the tunicates and the lancelets. A lancelet is the best example of a chordate that possesses the 3 chordate characteristics as an adult.

Comments on Chordates

Lancelets are segmented, but they do not have a well-developed internal skeleton as do the vertebrates. Segmentation along with a rigid, but jointed, skeleton evolved only in arthropods and vertebrates. As mentioned previously, this combination seems to be adaptive to living on land. Among the vertebrates, the reptiles, birds, and mammals evolved on land and do not need external water for reproductive purposes.

Figure 27.15
The lancelet *Branchiostoma* (amphioxus). **a.** Diagrammatic illustration showing that lancelets retain the 3 chordate characteristics as adults. Lancelets are filter feeders that take food from the water passing out their numerous gill slits. **b.** Photograph of several individuals, in which segmentation of the musculature is obvious. Sometimes lancelets swim freely. Other times, they filter feed while partially buried in the bottom soil. Photograph by Carolina Biological Supply Company.

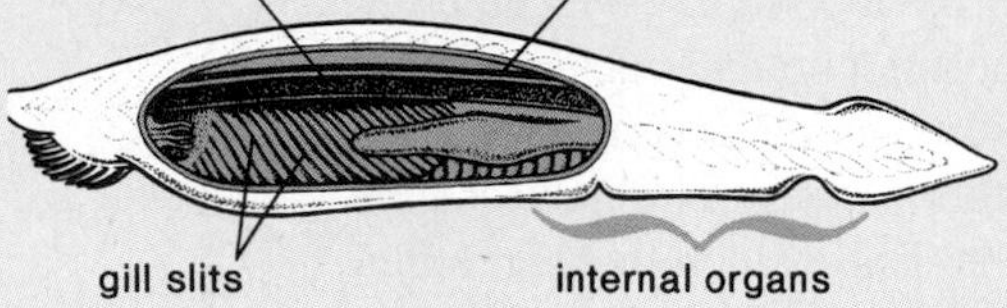

notochord   nerve cord

gill slits   internal organs

a.    b.

Summary

1. There are 2 groups of coelomate animals. In the protostomes, the blastopore becomes the mouth, and in the deuterostomes, it becomes the anus. Other differences between the 2 are spiral cleavage and a schizocoelom in protostomes versus radial cleavage and an enterocoelom in deuterostomes.

2. Mollusks, annelids, and arthropods are protostomes. The body of a mollusk typically contains a visceral mass, a mantle, and a foot. Some have a head and a radula. The mollusk has a nerve ring and 2 cords, a reduced coelom, and an open circulatory system. Snails are adapted to life on land, squids to an active life, and clams to a sedentary life. Bivalves are filter feeders. Water enters and exits by siphons; food trapped on the gills is swept toward the mouth. The nervous system has 3 major ganglia.

3. Annelids are segmented worms. They have a well-developed coelom, closed circulatory system, ventral solid nerve cord, and paired nephridia. Polychaetes (many bristles) are marine worms that have parapodia. They may be predators, with a defined head region, or they may be filter feeders, with ciliated tentacles to filter food from the water. Earthworms are oligochaetes (few bristles) that use the body wall for gas exchange. Leeches are also part of this phylum.

4. Arthropods (the most varied and numerous of animals) have an external skeleton of chitin (necessitating molting) and jointed appendages. Their body segments are often fused. Cephalization, with compound and simple eyes and a ventral solid nerve cord, is typical. The 5 major arthropod classes are crustaceans (e.g., lobsters, shrimps, krill, sow and pill bugs, and barnacles), insects (e.g., butterflies, grasshoppers, bees, and beetles), arachnids (e.g., spiders, scorpions, ticks, and mites), centipedes, and millipedes.

5. Many insects are adapted to life on land. Grasshoppers have wings and hopping legs, a tympanum for sound reception, a specialized digestive system for a grass diet, excretion of solid, nitrogenous waste by Malpighian tubules, tracheae for respiration, internal fertilization, and incomplete metamorphosis.

6. Echinoderms and chordates are deuterostomes. Echinoderms (e.g., sea stars, sea urchins, sea cucumbers, and sea lilies) have radial symmetry and an internal skeleton with spines. Notable are their tiny skin gills, nerve ring with branches, and water vascular system for locomotion. Each arm of a sea star contains branches from the nervous, digestive, and reproductive systems.

7. Chordates (tunicates, lancelets, and vertebrates) have a notochord, dorsal hollow nerve cord, and pharyngeal pouches at one time in their life history. Lancelets are the only chordate to have the 3 characteristics in the adult stage. In vertebrates, the notochord is replaced by the vertebral column.

8. Table 27.2 compares anatomical features of the animal phyla studied in this chapter.

Study Questions

1. Contrast the development of protostomes and deuterostomes in 3 ways. Tell which phyla belong to each group.
2. Contrast the phyla with respect to symmetry and segmentation.
3. Name at least 3 members of each phylum studied in this chapter.
4. What are the 3 characteristics of all mollusks? Contrast the way of life of the garden snail, clam, and squid, pointing out their different adaptations.
5. What are the general characteristics of annelids? How is the earthworm adapted to its way of life?
6. What are the general characteristics of arthropods? Name the 5 major classes, and give an example of each.
7. List different ways in which the grasshopper is adapted to a terrestrial habitat.
8. What are the general characteristics of echinoderms? Explain how the water vascular system works in sea stars.
9. What 3 characteristics do all chordates have sometime in their life history? What happens to the notochord in vertebrates?

Objective Questions

1. Which of these does not pertain to a protostome?
   a. blastopore becomes the anus
   b. spiral cleavage
   c. schizocoelom
   d. annelids, arthropods, and mollusks
2. Which of these is an adaptation to a terrestrial way of life in the snail?
   a. using the mantle instead of gills for respiration
   b. cephalization with antennae and eyes
   c. flat foot for crawling
   d. torsion so that anus is above the head
3. Which of these is mismatched?
   a. clam—gills
   b. lobster—gills
   c. grasshopper—book lungs
   d. polychaete—parapodia
4. Which of these is an incorrect statement?
   a. Spiders are carnivores in the phylum Arthropoda.
   b. Clams are filter feeders in the phylum Mollusca.
   c. Earthworms are scavengers in the phylum Arthropoda.
   d. Lancelets are filter feeders in the phylum Chordata.

5. A radula is a unique organ for feeding found in
   a. mollusks.
   b. annelids.
   c. arthropods.
   d. echinoderms.
6. Which of these is mismatched?
   a. crayfish—walking legs
   b. clam—hatchet foot
   c. grasshopper—wings
   d. earthworm—many tube feet
7. Which of these is mismatched?
   a. mollusk—schizocoelom
   b. insects—hemocoel
   c. echinoderms—enterocoelom
   d. chordates—schizocoelom

8. The tube feet of echinoderms
   a. are their head.
   b. are a part of the water vascular system.
   c. help pass sperm to females during reproduction.
   d. All of these.
9. Which of these is not a chordate characteristic?
   a. dorsal supporting rod, the notochord
   b. dorsal hollow nerve cord
   c. pharyngeal pouches
   d. vertebral column

10. Under the phyla headings given, write which terms apply.

    Terms: tissue level of organization
    organ system level of organization
    segmentation
    coelom
    cephalization in some representatives
    radial symmetry
    complete gut

Mollusks   **Annelids**   **Arthropods**   **Chordates**

11. Contrast a clam with an earthworm by writing beneath each the terms that apply.

Terms: annelid
ventral solid nerve cord
gills
open circulatory system
hatchet foot

mollusk
3 ganglia
closed circulatory system
hydrostatic skeleton
setae

Clam   **Earthworm**

12. Label on these diagrammatic drawings only those features that indicate grasshoppers are adapted to living on the land. Tell why you selected these features.

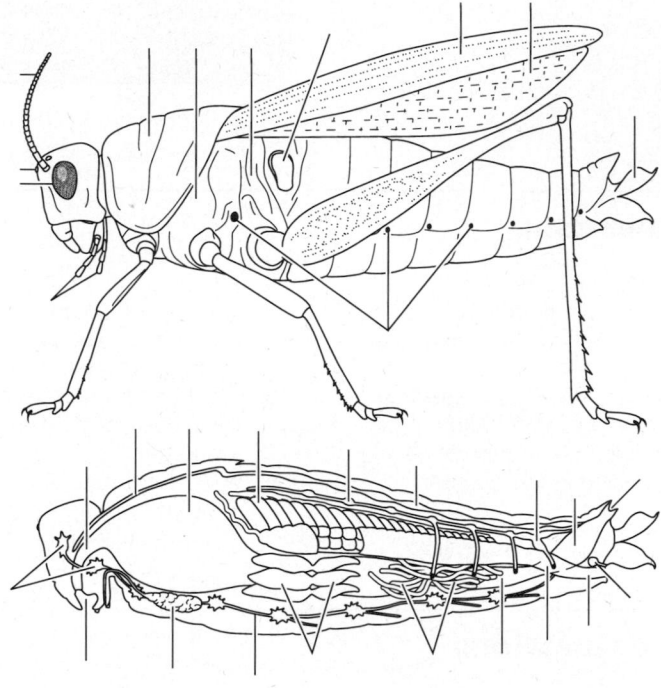

Concepts and Critical Thinking

1. *Segmentation leads to specialization of parts.*

Compare external features of the earthworm to that of the grasshopper to demonstrate this concept.

2. *Protostomes and deuterostomes both evolved similar features.*

Instead of pointing out differences between these 2 lineages, as in figure 26.2, list and discuss some similarities found in both arthropods and chordates.

3. *Insects are very well adapted to a land environment.*

List and discuss grasshopper features that adapt it to a life on land.

Selected Key Terms

protostome (pro′to-stōm) 430
deuterostome (du′ter-o-stōm″) 430
mollusk (mol′usk) 431
segmented (seg-ment′ed) 434
hydrostatic skeleton (hi″dro-stat′ik skel′ĕ-ton) 434

nephridium (nĕ-frid′e-um) 435
arthropod (ar′thro-pod) 437
jointed appendage (joint′ed ah-pen′dij) 437
exoskeleton (ek″so-skel′e-ton) 437

chitin (ki′tin) 437
molt (mōlt) 437
Malpighian tubule (mal-pig′ĭ-an tu′būl) 438
trachea (tra′ke-a) 438
metamorphosis (met″ah-mor′fo-sis) 438
echinoderm (e-kin′o-derm) 440

28

Vertebrate Diversity and Human Evolution

The female orangutan peers out at us, and we recognize right away a primate animal in our own mammalian order. The face is flat, and so the eyes are able to look straight ahead. The skull is of a size to hold a large brain. Orangutans are highly intelligent with a good memory for where and when to find various food sources in tropical rain forests.

Your study of this chapter will be complete when you can

1. tell why vertebrates are placed in the phylum Chordata, and tell how they differ from invertebrate chordates;
2. draw an evolutionary tree for the vertebrates;
3. list the 3 living classes of fishes, and trace the evolution of amphibians from fishes;
4. give examples of modern-day amphibians, and tell how amphibians are adapted to life in the water and on land;
5. give examples of modern-day reptiles, and tell how they are well adapted to life on land;
6. give examples of modern-day birds, and tell how they are adapted for flight;
7. list and give examples of the 3 types of modern-day mammals;
8. give the complete classification of humans; distinguish (with examples) among prosimians, anthropoids, hominoids, hominids, and humans;
9. list anatomical features of *Dryopithecus,* and tell how most are retained by modern-day apes but not humans;
10. draw and discuss a hypothetical phylogenetic tree for hominids;
11. contrast the australopithecines, *Homo habilis, Homo erectus, Homo sapiens neanderthalis,* and *Homo sapiens sapiens.*

Classification

Kingdom ANIMALIA
    Subphylum Vertebrata: vertebrates
        Class Agnatha: jawless fishes (lampreys, hagfishes)
        Class Chondrichthyes: cartilaginous fishes (sharks, rays)
        Class Osteichthyes: bony fishes
            Subclass Dipnoi: lungfishes
            Subclass Crossopterygii: lobe-finned fishes
            Subclass Actinopterygii: ray-finned fishes
        Class Amphibia: frogs, toads, salamanders
        Class Reptilia: snakes, lizards, turtles
        Class Aves: birds
        Class Mammalia: mammals

In this chapter, we will study the animals most closely related to us and trace our immediate ancestry. Even though we have a strong interest in other vertebrate animals, it is well to keep in mind that most animal species are invertebrates and that it is our bias that makes us see a world populated mostly by vertebrates.

Vertebrate Evolution

Vertebrates are in the phylum Chordata and show at some time in their life history the 3 chordate characteristics (notochord, dorsal hollow nerve cord, and pharyngeal pouches). The embryonic notochord, however, is generally replaced by a *vertebral column* composed of individual vertebrae. The main axis of the internal and jointed skeleton consists not only of a vertebral column but also of a skull that encloses and protects the brain. There is an extreme degree of cephalization, with complex sense organs. The eyes develop as outgrowths of the brain. The ears are primarily equilibrium devices in aquatic vertebrates, but they also function as sound-wave receivers in land vertebrates.

Vertebrates belong to the phylum Chordata and have the 3 chordate characteristics at some time during their life. The notochord is replaced by the vertebral column.

Vertebrates have *bilateral symmetry* and the coelomic, *tube-within-a-tube body plan,* and they are *segmented.* There is a *closed circulatory system,* in which blood is contained entirely within blood vessels. The kidneys are important excretory and water-regulating organs that conserve or rid the body of water as necessary. The sexes are generally separate, and reproduction is usually sexual.

Occasionally, similar adaptations are seen in both invertebrates and vertebrates; both insects and birds have wings, and both fishes and squids have fins. Such structures are analogous rather than homologous, even though they serve the same function, because their basic plans are completely different and because they did not evolve from a common ancestor.

We have mentioned previously that the Cambrian rocks contain fossils of all the major animal phyla (see fig. 27.1). But only invertebrate chordates are found; there is no record of vertebrate beginnings at this time. The phylogenetic tree of vertebrates begins with the fishes, which evolved later (fig. 28.1).

Fishes

There are 3 classes of living *fishes* (22,000 species): the jawless fishes (class Agnatha), the cartilaginous fishes (class Chondrichthyes), and the bony fishes (class Osteichthyes). The earliest vertebrate fossils are jawless fishes of the Ordovician period. These fishes had heavy, bony armor and were probably filter feeders. Living representatives of the *jawless fishes* are hagfishes and lampreys (fig. 28.2). They are cylindrical, up to a meter long, and have smooth, scaleless skin. The hagfishes are scavengers, feeding mainly on dead fishes. Parasitic lampreys have a round, muscular mouth that serves as a sucker. They use this sucker to attach themselves to another fish and to suck nutrients from the host's circulatory system. Marine parasitic lampreys gained entrance to the Great Lakes when a canal from the St. Lawrence River was deepened. The lamprey population grew quickly and caused extensive reduction in the trout population of the Great Lakes in the early 1950s.

The first jawed vertebrates in the fossil record are *placoderms,* a class of fishes of the Devonian period that has become extinct (fig. 28.3; table 28.1). Jaws are a distinct advantage because they allow the development of an active, herbivorous or carnivorous life-style. It is not surprising, then, that these fishes also had paired pectoral fins and pelvic fins. Paired fins allow a fish to balance and to maneuver well in the water. Unknown early jawed fishes were ancestral to the placoderms and to today's jawed fishes.

Cartilaginous Fishes

Sharks, rays, and skates are *cartilaginous fishes* (fig. 28.4); they have a *skeleton of cartilage* instead of bone. Their body is covered with small, toothlike scales called denticles, which is why a shark's skin feels like sandpaper. The menacing teeth of sharks and their relatives are simply larger, specialized versions of these scales.

Rays and skates have horizontally flattened bodies. They swim slowly along the sea bottom and feed on animals that they dredge up. Sharks have beautifully streamlined bodies and are fast-swimming predators in the open sea.

Some people believe sharks attack and eat most anything in the water. While this may be true of some sharks, there are other sharks that are not active predators of large prey. Instead, they are filter feeders, like the basking sharks and whale sharks that ingest tons of small crustaceans, collectively called *krill.*

Bony Fishes

The earliest *bony fishes* are thought to have evolved in freshwater habitats because their earliest fossils are found in rocks from lakes and rivers. In addition to gills, they possessed simple, *saclike lungs* connected to the anterior end of their digestive tract. These lungs probably functioned as supplementary gas exchange surfaces that allowed them to breathe air when the stagnant water was oxygen deficient. These fishes gave rise to the groups of bony fishes alive today.

Figure 28.1

Phylogenetic tree of vertebrates. Extinct jawless fishes are the oldest vertebrates in the fossil record. Although they had a well-developed, sometimes armored head, they were most likely bottom feeders. Fishes with jaws, which appeared during the late Silurian period, evolved into the cartilaginous fishes and the bony fishes of today. The ray-finned fishes, which are the most common fishes today, are descended from bony fishes with lungs; amphibians came from fleshy-finned fishes and were dominant during the Carboniferous period. Stem reptiles also appeared around this time, and from them arose the dinosaurs, which ruled during the Jurassic period, and the first mammals, which did not come into their own until the Tertiary period. Birds may very well be of dinosaurian origin.

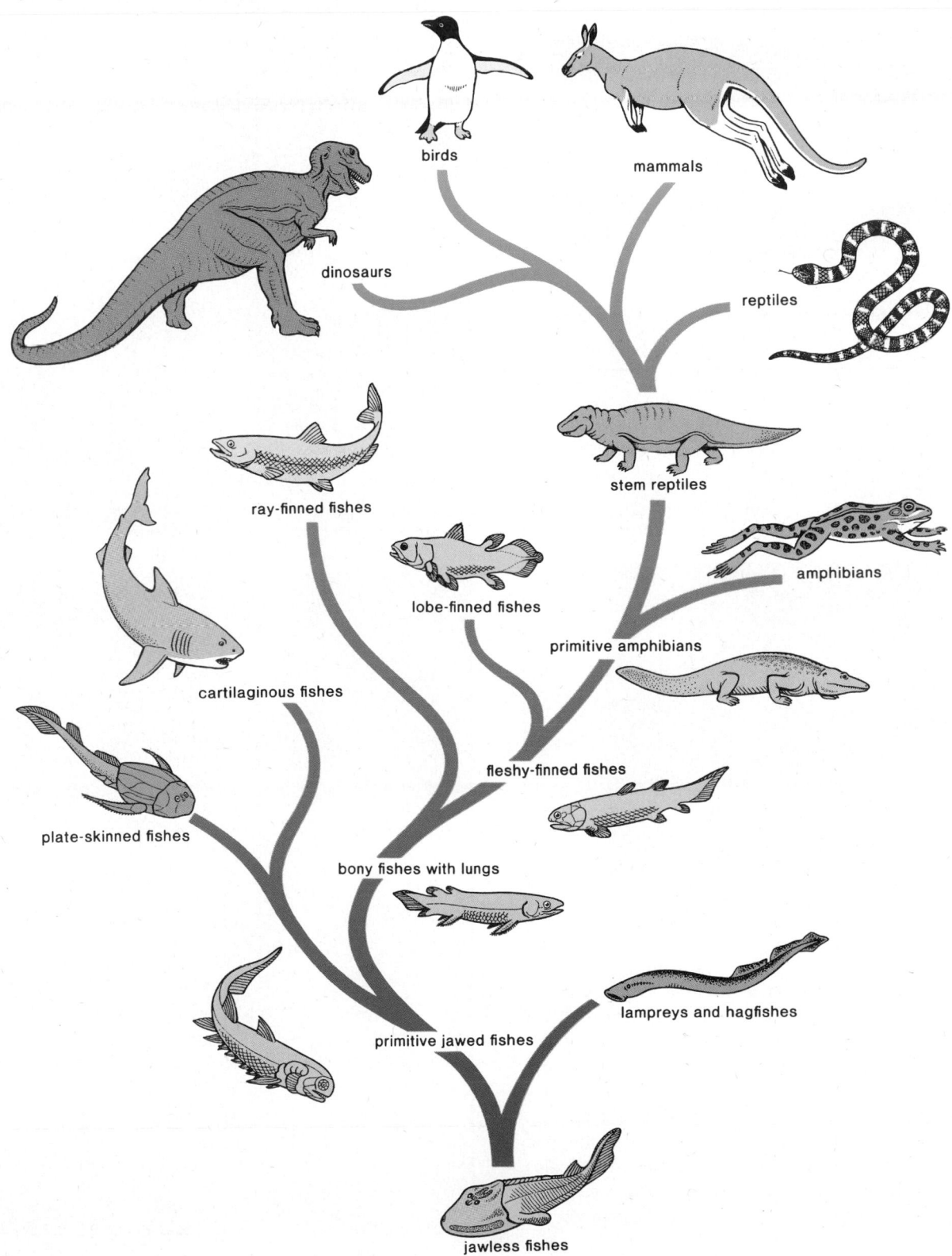

Figure 28.2

The jawless fishes of today are not at all like their fossil ancestor (see fig. 28.1). ***a.*** The hagfishes are scavengers with sensory tentacles around the mouth and a row of slime glands along each side. ***b.*** Parasitic lampreys attach themselves to fishes with their suckerlike mouth, rasp away the flesh with their horny teeth, and suck blood.

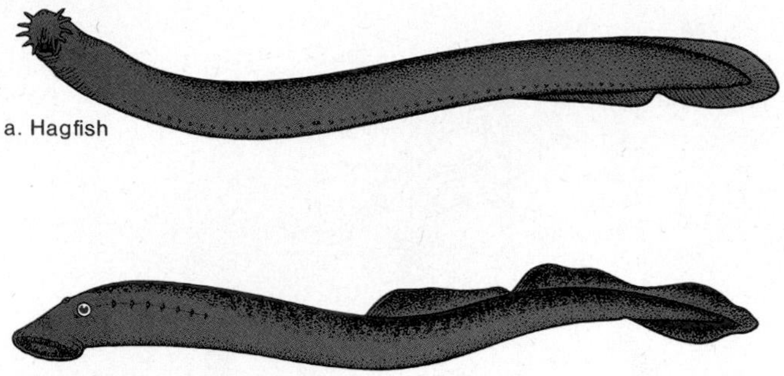

a. Hagfish

b. Lamprey

Figure 28.3

Painting of an extinct placoderm, the first type of jawed vertebrate in the fossil record. The appearance of jaws, which most likely evolved from the first anterior pair of gill arches, was a revolutionary event in the evolutionary history of vertebrates.

There are 2 groups of bony fishes alive today, the ray-finned fishes and the fleshy-finned fishes. The *ray-finned fishes* (subclass Actinopterygii), which have fins supported by spine-like rays, include most of today's bony fishes (fig. 28.5). In these fishes, the ancient lung eventually became a swim bladder. By secreting gases into or absorbing gases from this bladder, a fish can change its relative density and thereby move up or down in the water. Modern bony fishes have broad, flattened scales covering some or all of their body surfaces. Their gills are located in an enclosed gill chamber that is covered and protected

Table 28.1

The Geological Time Scale: Some Major Evolutionary Events of Animals

Era	Period	M.Y.A.*	Major Biological Events—Animals
Cenozoic	Quaternary	2	Age of Human Civilization
Cenozoic	Tertiary		First hominids appear
Cenozoic	Tertiary		Dominance of land by mammals and insects — Age of Mammals
Mesozoic	Cretaceous	66	Dinosaurs become extinct
Mesozoic	Jurassic	144	First mammals and birds appear
Mesozoic	Triassic	208	First dinosaurs appear — Age of Dinosaurs
Paleozoic	Permian	245	Reptiles appear and expand
Paleozoic	Carboniferous	286	Decline of amphibians
Paleozoic	Devonian	360	Insects appear — Age of Amphibians
Paleozoic	Devonian		First amphibians move onto land — Age of Fishes
Paleozoic	Silurian	408	
Paleozoic	Ordovician	438	First fishes (jawless) appear
Paleozoic	Cambrian	500	Invertebrates dominate the seas
Paleozoic	Cambrian	570	Trilobites abundant — Age of Invertebrates

*M.Y.A. = millions of years ago.

Figure 28.4
Cartilaginous fishes. *a.* The southern sting ray has a flattened shape and moves slowly across the bottom dredging up food. *b.* The bull shark is a streamlined predator that moves swiftly through the water to capture its prey.

a.

b.

Figure 28.5
Ray-finned fishes. Most fishes today are bony, ray-finned fishes adapted to living in a great variety of habitats. For example, trout and stickleback are in fresh water; tuna and swordfish are in the open seas; parrot and angelfish live among the coral reefs. *a.* Queen angelfish. *b.* Rainbow trout.

a.

b.

by a hard, bony flap, the operculum. Thus, they have only one gill opening on each side instead of the separate openings for each gill slit that are found in sharks.

Today, the *fleshy-finned fishes* are represented by 2 subclasses, the lungfishes (subclass Dipnoi) and only one species of lobe-finned fishes (subclass Crossopterygii). There are 3 genera of lungfishes—one each in Australia, Africa, and South America—that regularly breathe air to supplement gas exchange in their gills. The single species of lobe-finned fishes is the coelocanth, long thought to be extinct but now known to exist off the southeast coast of Africa (fig. 28.6). The coelocanth does not have lungs.

Figure 28.6

The fleshy-finned fishes of the Paleozoic era underwent an adaptive radiation that included the lobe-finned fishes and the ancestors of the amphibians. ***a.*** The coelocanth is a living fossil, a lobe-finned fish that is alive today. Most likely, it resembles the fleshy-finned fishes that gave rise to the amphibians. Note the tuft at the end of its tail; no other fish has such a tail pattern. ***b.*** Comparison of the limbs of a flesh-finned fish and the limbs of a primitive amphibian. These amphibians must have been awkward walkers, but legs were an improvement for locomotion, as is made clear by the side drawings.

a.

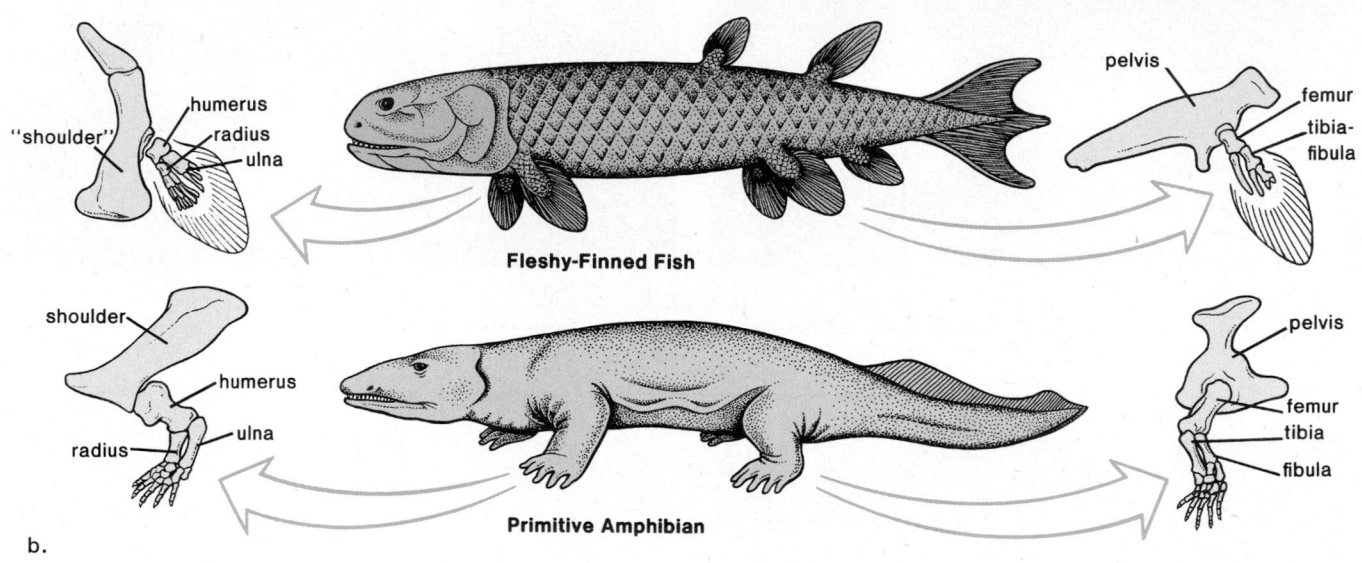

b.

The fleshy-finned fishes are of interest because they are thought to have been ancestral to the amphibians. Most likely, they could breathe air while they crawled clumsily on stumpy fins from pond to pond during the Devonian period, a time of alternating floods and droughts. Land plants and insects had evolved before land vertebrates, so there was an excellent food supply available on land for any animal that could remain ashore long enough to utilize it. These animals could avoid more vigorous competition in the water. Figure 28.6 compares the features of an ancestral fleshy-finned fish with those of a primitive amphibian.

The most primitive fishes today, the lampreys and hagfishes, are jawless. Sharks, skates, and rays are cartilaginous fishes, all other types of fishes are bony fishes. Amphibians are descended from an ancient group of bony fishes that, unlike most today, had fleshy fins and lungs.

Amphibians

Improved locomotion on land and other adaptations for terrestrial life led to a diversification of **amphibians** (class Amphibia, 3,900 species) during the Carboniferous period, now known as the Age of Amphibians (table 28.1). Many of these became extinct; amphibians are mainly represented today by newts and salamanders, frogs (fig. 28.7), and toads. These animals are adapted to a dual life. They have small, relatively *inefficient lungs,* and most of them also depend on gas exchange through the skin. For skin to function efficiently as a gas exchange surface, it must be quite thin and kept moist. The most important factor binding amphibians to water, however, is their method of reproduction. Amphibians, like most fishes, shed their eggs and sperm into the water, where *external fertilization* takes place. The eggs are not protected against desiccation when exposed to air because they are only enclosed by a jelly coat. When young amphibians hatch from their egg, they are tadpoles (aquatic larvae with gills) that feed and grow in the water. After they undergo *metamorphosis,* amphibians emerge from the water as adults that breathe air (fig. 28.7). The life history of these animals is divided between water and land, as is accurately described by their class name, Amphibia, which means double life.

A few amphibians have special reproductive adaptations that enable them to reproduce on land. For example, some frogs carry their developing eggs in fluid-filled pouches. The

Figure 28.7

Frog metamorphosis, during which the animal changes from an aquatic to a terrestrial organism. *a.* Hatching of eggs. *b.* Tadpoles breathe by means of gills. *c.* Development of legs. *d.* Frogs breathe by means of lungs and their skin.

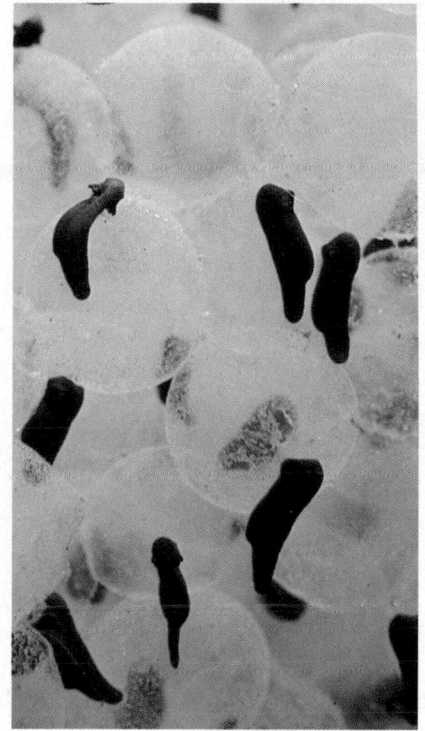

a.

b.

c.

d.

Figure 28.8

Drawing of *Seymouria,* an animal known only as a fossil, which is closely related to a possible evolutionary link between amphibians and reptiles. Inset is a fossil head of *Seymouria.*

young metamorphose very quickly, becoming tiny adults that are ready to live and grow on land.

Some time before the Permian period, amphibians that were probably related to *Seymouria* gave rise to the reptiles (fig. 28.8). Reptiles were to become the dominant organisms on earth during the next geological era, the Mesozoic era.

Metamorphosis allows amphibians to switch from an aquatic existence to a land existence. There they are restricted to a moist environment because they use their skin for respiration and because they must return to the water for reproduction. Reptiles are descended from amphibians.

Reptiles

Reptiles (class Reptilia, 7,000 species) are represented today largely by turtles, crocodiles and alligators, lizards, and snakes (fig. 28.9). All of them have thick, scaly skin that is keratinized and impermeable to water. Because the skin is also impermeable to gases, their lungs must be more efficient.

The reptiles' most outstanding adaptation to living on land is their means of reproduction. They have internal fertilization and lay eggs that are protected by leathery shells. This

Figure 28.9

Reptilian diversity. *a.* A snake has tough, scaly skin, which resists drying out. *b.* The slow-moving turtle has an external shell in addition to scaly skin.

a.

b.

shelled egg eliminated the need for a swimming larval stage during development. The egg contains *extraembryonic membranes,* which protect the embryo, remove nitrogenous wastes, and provide the embryo with oxygen, food, and water. These membranes are not part of the embryo itself and are disposed of after development is complete. One of the membranes, the amnion, is a sac that fills with fluid and provides a "private pond," within which the embryo develops (fig. 28.10).

The earliest reptiles, called stem reptiles, gave rise to several lines of descent during the Mesozoic era, called the Age of Reptiles (table 28.1). One line produced the *dinosaurs,* which underwent adaptive radiation and dominated the earth for more than 100 million years before their relatively sudden extinction, possibly due to asteroid or comet bombardment of the earth (p. 342). Birds may be closely related to dinosaurs, although their poor fossil record makes it difficult to trace their history. *Archeopteryx,* the oldest known fossil bird, had several reptilian characteristics, including jaws with teeth and a long, jointed reptilian tail. The bird was about the size of a crow and had rather feeble wings for its size, but it was well feathered.

There is another branch from the stem reptiles of interest. Mammal-like reptiles called *therapsids* gave rise to the true mammals early in the Jurassic period (fig. 28.11). The therapsids had mammal-like limbs, skulls, and teeth, and some of them resembled large dogs and were aggressive predators.

Reptiles are well adapted to a land environment. They have scaly skin which prevents loss of water, and they can reproduce on land because they lay a shelled egg. Both mammals and birds are descended from reptilian ancestors.

Birds

Birds (class Aves, 8,900 species) are the only modern animals that have feathers, which are actually modified reptilian scales. Scales on the legs and feet of birds are other reminders of their reptilian ancestry, as are the claws at the ends of their toes.

Bird feathers are of 2 types. *Contour feathers* are attached to the wings in such a way that they overlap to produce a broad, flat surface beneficial for flight. *Down feathers* provide excellent insulation against body heat loss. This is important because birds maintain a constant, relatively high body temperature, which permits them to be continuously active even in cold weather.

There are many orders of birds, including birds that are *flightless* (ostriches), *web footed* (penguins), *divers* (loons), *fish eaters* (pelicans), *waders* (flamingos), *broad billed* (ducks), *birds of prey* (hawks), *vegetarians* (fowl), *shore dwellers* (sandpipers), *nocturnal* (owls), *small nectar feeders* (hummingbirds), and *songbirds,* the most familiar birds.

Nearly every anatomical feature of birds can be related to their ability to fly. Their forelimbs are wings. They have hollow, very light bones with internal air spaces that are a part of the respiratory system. A bird's head is light because a light, horny beak has replaced jaws equipped with teeth. A slender neck connects the head to a rounded, compact torso. The breastbone is enlarged and has a keel, to which the strong flight muscles are attached.

Flight requires well-developed sense organs. Birds have particularly acute vision and excellent muscle reflexes. They can land precisely on small tree branches and swoop from great heights to capture small animals that are spotted while cruising high above the unsuspecting prey.

Evolution and Diversity

Figure 28.10

The reptilian egg was a major advancement because it allowed these animals to reproduce on land. **a.** Baby crocodile hatching out of its shell. Note that the shell is leathery and flexible, not brittle like birds' eggs. **b.** Inside the egg, the embryo is surrounded by extra embryonic membranes. The chorion aids gas exchange; the yolk sac provides nutrients; the allantois stores waste; and the amnion encloses fluid, which prevents drying out and provides protection.

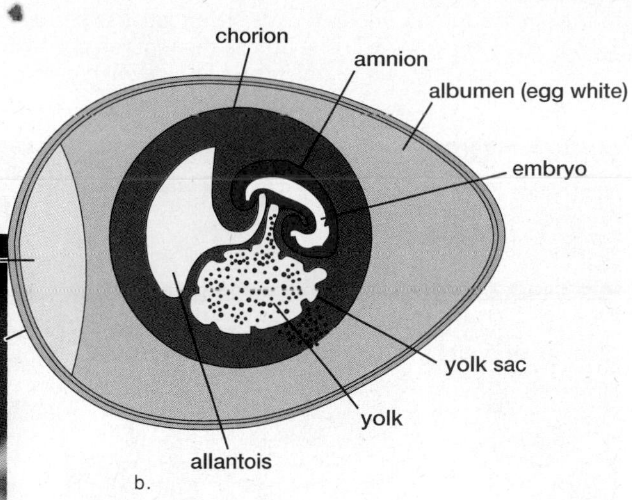

chorion

amnion

albumen (egg white)

embryo

air space

egg shell

yolk sac

yolk

allantois

b.

a.

Figure 28.11

Therapsid, a Jurassic reptile that had many mammalian features, including their characteristic means of locomotion and differentiated teeth. Most likely, therapsids chewed their food before swallowing, which would have provided quick energy to maintain a warm temperature.

Birds have internal fertilization and lay *hard-shelled eggs*. Many newly hatched birds require parental care before they are able to fly away and seek food for themselves (fig. 28.12). Complex hormonal regulation and behavior responses are involved in bird behavior. Some species migrate long distances twice a year, apparently using the sun and stars as a compass and even the earth's magnetic field to find their way on cloudy days.

Birds can maintain a constant internal temperature. All organ systems are adapted to allow birds to fly.

Mammals

The chief characteristics of **mammals** (class Mammalia, 4,500 species) are hair and milk-producing mammary glands (fig. 28.13). Hair provides insulation against heat loss and helps maintain a constant body temperature so mammals can be active in cold weather. Other mammalian characteristics include limbs for running, well-developed sense organs, and an enlarged brain. All of these are important for an active life.

Mammary glands enable females to feed (nurse) their young without deserting them to find food. Nursing also creates a bond between mother and offspring that helps ensure parental care while the young are helpless. In most mammals, the young are born alive after a period of development in the uterus, a part

Figure 28.12
Unlike *Archaeopteryx*, modern birds have toothless beaks and an entirely feathered tail instead of a long, bony tail. They retain the reptilian egg, except that a bird egg is hard-shelled, not leathery. Also, birds are usually devoted parents. Here, a song sparrow feeds its young.

Evolution and Diversity

Figure 28.13
Mammals are animals that have hair and mammary glands. The mammary glands allow females to nurse their young. Nursing not only nourishes the young, it also builds a strong mother-offspring relationship.

of the female reproductive tract. Internal development shelters the young and allows the female to move actively about while the young are maturing.

Mammals evolved during the Mesozoic era from therapsid, a mammal-like reptile (fig. 28.11). Some of the earliest mammalian groups, represented today by the *monotremes* and *marsupials,* are not abundant today. Mammals remained relatively small in size and number until the reptiles declined. Then the marsupials, and later the *placental mammals,* radiated into many of the habitats previously occupied by the large reptiles. Mammals are classified according to means of reproduction.

Monotremes

Monotremes (subclass Prototheria; order Monotremata) are represented by the duckbilled platypus and the spiny anteater, both of which are found in Australia. The female duckbilled platypus lays her eggs, which resemble reptilian eggs, in a burrow in the ground. She incubates the eggs and after hatching, the young lick up milk that seeps from modified sweat glands on the abdomen. The spiny anteater has a pouch on the belly side formed by swollen mammary glands and longitudinal muscle. The egg moves from the cloaca to this pouch where hatching takes place and the young remain for about 53 days. Then they stay in a burrow, which the mother periodically visits and where she nurses them.

Marsupials

The young of **marsupials** (subclass Metatheria; order Marsupialia) begin their development inside the female's body, but they are born in a very immature condition. Newborns crawl up into a pouch on their mother's abdomen. Inside the pouch, they attach to nipples of mammary glands and continue to develop.

Today, marsupial mammals are found mainly in Australia; only a few marsupials, such as the American opossum, are found outside that continent. In Australia, marsupials underwent adaptive radiation for several million years without competition from placental mammals, which arrived there only recently. Among the herbivorous marsupials, koalas are tree-climbing browsers and kangaroos are grazers. The Tasmanian "wolf" or "tiger," thought to be extinct, was a carnivorous marsupial about the size of a collie dog.

Figure 28.14
Most mammals are placental mammals, which develop in the uterus of the female parent. Here they receive oxygen and nutrients by way of a special structure called the placenta, an area of exchange between maternal blood and fetal blood.

Placental Mammals

Developing **placental mammals** (subclass Eutheria) are dependent on the placenta, an organ of exchange between maternal blood and fetal blood. Nutrients are supplied to the growing offspring, and wastes are passed to the mother for excretion (fig. 28.14). While the fetus is clearly parasitic on the female, in exchange, she is free to move about as she chooses while the fetus develops.

Placental mammals evolved during the Tertiary period and entered the habitats left vacant when the dinosaurs became extinct. Table 28.2 lists the major orders of placental mammals, including those that are adapted to life in the water, on land, and in the air.

Placental mammals are far more numerous than monotremes (egg laying) and marsupials (pouched) mammals. They are adapted to life in the air, in water, and on land. Humans are classified as primates.

Primate Evolution

Primates evolved during the Cenozoic era at a time when angiosperm trees were becoming prevalent (table 28.3). Among living mammals, the first primates resembled the living tree shrews of Southeast Asia, which are rat-sized animals with a snout, claws, and sharp front teeth (fig. 28.15). With time, primates evolved many characteristics suitable to living in trees.

Limbs Typically, in primates, both forelimbs and hindlimbs are highly mobile and both hands and feet have 5 long dexterous digits (fingers and toes). The thumb and big toe are opposable; that is, they can close to meet the other digits. Therefore, these animals can easily reach out and bring food, such as fruits, to the mouth. Claws are replaced by nails, which offer protection but still allow a tree limb to be freely grasped and released.

Vision In primates, the snout is considerably shortened and the eyes are in the front of the head. These animals have excellent binocular (3-dimensional) vision, allowing them to make accurate judgments about the distance and the position of adjoining tree limbs.

Brain All primates have well-developed brains; apes and humans have the most complex brains of any mammals. The visual portion of the brain is proportionately large, as are those centers responsible for hearing and touch.

Table 28.2
Some Major Orders of Placental Mammals

Insectivora (moles, shrews)	Primitive; small, sharp-pointed teeth
Chiroptera (bats)	Digits support membranous wings
Carnivora (dogs, bears, cats, sea lions)	Canine teeth long; teeth pointed
Rodentia (mice, rats, squirrels, beavers, porcupines)	Incisor teeth grow continuously
Lagomorpha (rabbits, hares, pikas)	Chisel-like incisors; hind legs longer than front legs; herbivorous
Perissodactyla (horses, zebras, tapirs, rhinoceroses)	Large, long-legged, one or 3 toes, each with hoof; grinding teeth
Artiodactyla (pigs, cattle, camels, buffalos, giraffes)	Medium to large; two or 4 toes, each with hoof; many with antlers or horns
Cetacea (whales, porpoises)	Medium to very large; forelimbs paddlelike; hind limbs absent
Primates (lemurs, monkeys, gibbons, chimpanzees, gorillas, humans)	Mostly tree-dwelling; head freely movable on neck; 5 digits, usually with nails; thumbs and/or large toes usually opposable

Adapted from *Essentials of Biology,* Second Edition, by Willis H. Johnson, Louis E. Delanney, Thomas A. Cole, Austin E. Brooks, copyright © 1974 by Saunders College Publishing, reprinted by permission of the publisher.

Table 28.3
Classification of Primates

Category	Animals
Phylum Chordata Subphylum Vertebrata Class Mammalia Subclass Eutheria	
Order Primates **Suborder Prosimii** (prosimians)	Lemurs Tarsiers
Suborder Anthropoidea (anthropoids)	Monkeys Apes Humans
Superfamily Hominoidea (hominoids)	*Dryopithecus* Modern apes Humans
Family Hominidae (hominids)	*Australopithecus* *Homo habilis* *Homo erectus* *Homo sapiens*
Genus Homo (humans)	*Homo habilis* *Homo erectus* *Homo sapiens*

Figure 28.15
A tree shrew is a small animal with feet adapted for climbing in trees, where it feeds on fruits and insects. Notice that the ears are similar to the ears of the primates—rounded with folded edges. A tree shrew's brain is relatively large, but the olfactory portion is surprisingly small. Tree shrews may resemble the ancient ancestors of modern primates.

Care of Offspring Primates have one offspring at a time, and the period of postnatal maturation is prolonged, giving the immature young adequate time to learn complex behavior patterns.

Prosimians

Prosimians, a term that means premonkeys, evolved about 55 million years ago. The living members of the group include lemurs and tarsiers (fig. 28.16). Lemurs, which have a squirrel-like appearance, are confined largely to the island of Madagascar. Their face is long and snoutlike; some digits have claws and some have nails. The big toe and thumb are mobile but not opposable. The tarsiers, which are found in the Philippines and East Indies, are curious, mouse-sized creatures, with enormous eyes suitable for their nocturnal way of life. They have a flattened face, and their digits terminate in nails.

This once-successful group of animals became largely extinct by the end of the Eocene epoch (table 28.4)

Anthropoids

The *anthropoids* include the monkeys and apes, as well as human beings.

Monkeys

Monkeys evolved from ancestral prosimians about 30 million years ago, when the weather was warm and vegetation was like that of a tropical rain forest. Monkeys became divided into the Old World and the New World monkeys when South America separated from

Figure 28.16
Evolution of primates took place in the Cenozoic era, which includes the Tertiary and Quaternary periods. Prosimians include the lemurs (shown here) and the tarsiers; the monkeys include the New World and the Old World monkeys; apes include the gibbons and great apes (orangutans, gorillas, and chimpanzees).

Anthropoids

Prosimians

Hominoids

Hominids

gibbons

Old World
monkeys

great apes

modern
prosimians

New World
monkeys

humans

hominids

hominoids

anthropoids

prosimians
primates

insectivores

Millions of
Years Ago

0

10

20

30

40

50

60

70

Table 28.4
Evolution of Primates during Cenozoic Era

Millions of Years ago	Epoch	Evolutionary Event
3	Pleistocene	Late hominids and humans
6	Pliocene	Hominids
27	Miocene	Apes
38	Oligocene	Monkeys
56	Eocene	Prosimians
66	Paleocene	First primates

Africa (p. 348). The New World monkeys today have long prehensile (grasping) tails and flat noses, and Old World monkeys, which lack such tails, have protruding noses. Two of the well-known New World monkeys are the spider monkey and the capuchin, the "organ grinder's" monkey. Some of the better-known Old World monkeys are now ground dwellers, such as the baboon and the rhesus monkey, which has been used in medical research.

Primates evolved from shrewlike insectivores and became adapted to living in trees, as exemplified by skeletal features, good vision, and aspects of their reproduction. Prosimians were the first primates to evolve, and they were followed by monkeys, the first of the anthropoids.

Hominoids
The hominoids include the apes and humans (fig. 28.16).

Apes
About 25 million years ago, the weather was changing. The continents were drifting into their present locations, and as a result, the weather was becoming somewhat cooler and drier. In places the tropical forests were giving way to grasslands.

At this time, apes became abundant and widely distributed in Africa, Europe, and Asia. Among these Miocene apes, members of the genus *Dryopithecus* are of particular interest because they are thought to be a possible hominoid ancestor—an ancestor that led to all other apes and eventually to humans. **Dryopithecines** were forest dwellers that probably spent most of the time in the trees. The bones of their feet, however, indicate that when they did walk on the ground, they walked on all 4s virtually all the time, using the knuckles of their hands to support part of their weight. Such "knuckle-walking" is retained by modern apes but not by humans, who stand erect. The skull of *Dryopithecus* has a sloping brow, prominent eyebrow ridges (called supraorbital ridges), and a rectangular-shaped jaw with long canine teeth. In general, these are features that were retained by apes but not humans (fig. 28.17)

There are 4 types of living apes: gibbon, orangutan, gorilla, and chimpanzee (fig. 28.18). The *gibbon* is the smallest of the apes, with a body weight ranging from 5 1/2 kg-11 kg. Gibbons have extremely long arms that are specialized for swinging between tree limbs. The *orangutan* is large (75 kg) but nevertheless spends a great deal of time in trees. The *gorilla,* the largest of the apes (180 kg), spends most of its time on the ground. *Chimpanzees,* which are at home both in trees and on the ground, are the most humanlike of the apes in appearance and are frequently used in psychological experiments.

Biochemists have compared the similarity of chromosome banding, base sequence in DNA, and amino acid sequence in proteins between humans and living apes. This data indicates that humans share a more recent common ancestor with chimpanzees than any of the other living apes. Therefore, it is now believed the gibbon separated from the human lineage about 17 million years ago; orangutan separated at least 12 million years ago, the gorilla separated about 9 million years ago; and the chimpanzee did not separate until about 6 million years ago. Notice that this means that the apes separated from the human lineage during the Miocene epoch.

The first hominoid (ancestor to all apes and humans) may have been *Dryopithecus,* whose characteristics are somewhat like those of today's apes.

Hominids

Just where in Africa the first hominid (ancestor leading to humans) lived and exactly what the ancestor was like are not known. This ancestor, however, probably had a bipedal posture (walking on 2 limbs) rather than a quadrupedal posture like modern apes (fig. 28.17). It's probable that bipedalism arose as an energetic strategy for gathering food in trees that were becoming distant from one

Figure 28.17

a. As outlined in the table, the Dryopithecines had primitive characteristics that could have led to both modern-day apes and humans. Compare the modern human skeleton (**b**) to the modern ape (gorilla) skeleton (**c**). Humans are bipedal while modern-day apes are knuckle-walkers. In the ape, the pelvis is very long and tilts forward, whereas in humans it is short and upright. The ape shoulder girdle is more massive than that of humans, and the head hangs forward because of the angle of attachment of the vertebral column to the skull. In apes, the foramen magnum (hole in skull through which the spinal cord passes) is well to the rear of the skull; in humans it is almost directly in the bottom center of the skull. Also, in humans the spine has an S-shaped curve, allowing a better weight distribution and improved balance when the body is upright.

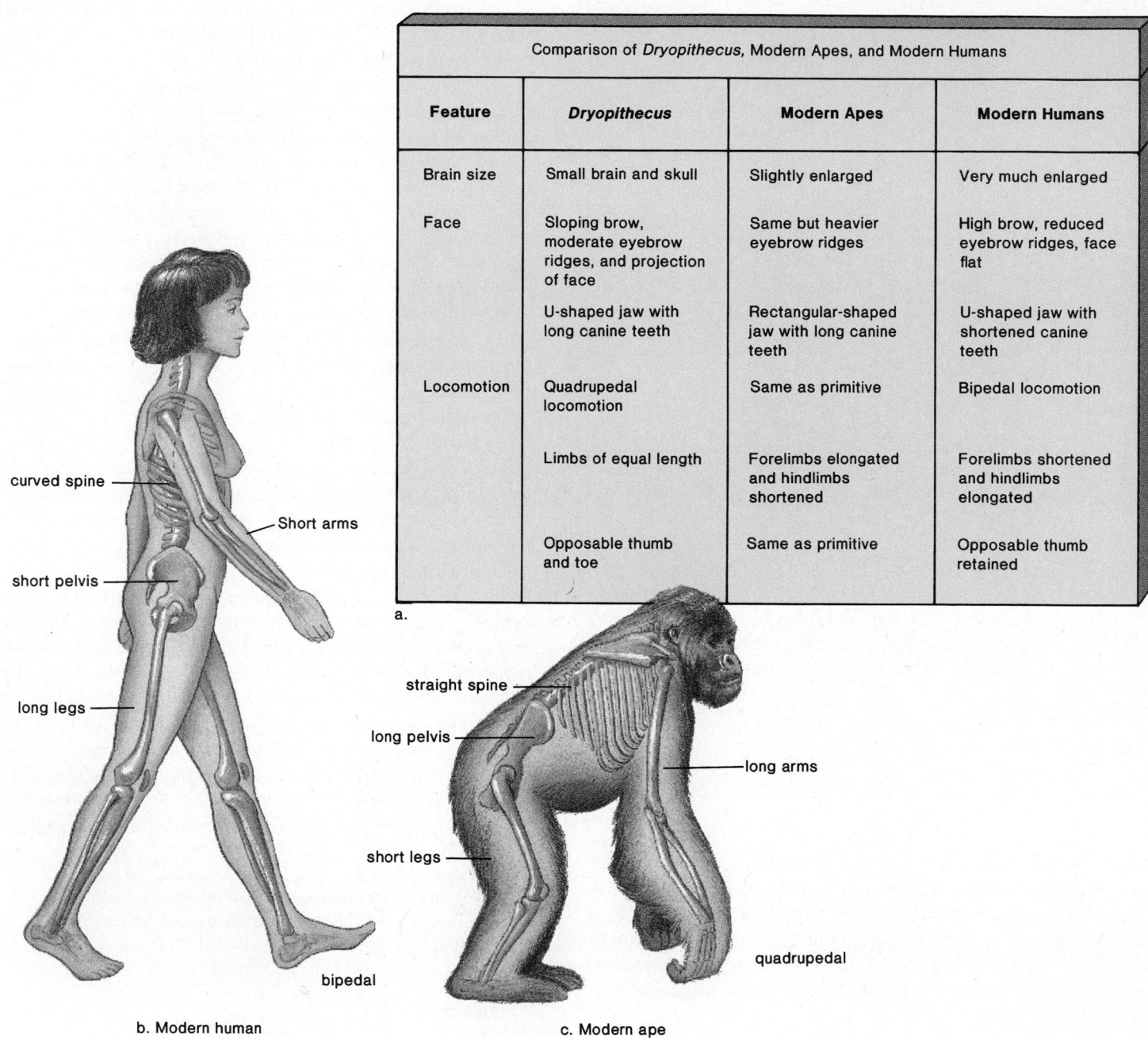

Comparison of *Dryopithecus*, Modern Apes, and Modern Humans			
Feature	***Dryopithecus***	**Modern Apes**	**Modern Humans**
Brain size	Small brain and skull	Slightly enlarged	Very much enlarged
Face	Sloping brow, moderate eyebrow ridges, and projection of face	Same but heavier eyebrow ridges	High brow, reduced eyebrow ridges, face flat
	U-shaped jaw with long canine teeth	Rectangular-shaped jaw with long canine teeth	U-shaped jaw with shortened canine teeth
Locomotion	Quadrupedal locomotion	Same as primitive	Bipedal locomotion
	Limbs of equal length	Forelimbs elongated and hindlimbs shortened	Forelimbs shortened and hindlimbs elongated
	Opposable thumb and toe	Same as primitive	Opposable thumb retained

a.

curved spine

Short arms

short pelvis

long legs

bipedal

b. Modern human

straight spine

long pelvis

long arms

short legs

quadrupedal

c. Modern ape

Figure 28.18

Ape diversity. *a.* Of the apes, gibbons are the most distantly related to humans. They dislike coming down from trees, even at watering holes. They will extend a long arm into the water and then drink collected moisture from the back of the hand. *b.* Orangutans are solitary except when they come together to reproduce. Their name means "forest man"; early Malayans believed that they were intelligent and could speak but did not because they were afraid of being put to work. *c.* Gorillas are terrestrial and live in groups in which a silver backed male, such as this one, is always dominant. *d.* Of the apes, chimpanzees sometimes seem the most humanlike.

a.

b.

c.

d.

Figure 28.19

Evolution of hominids. Some authorities believe that *Australopithecus afarensis* is a common ancestor for the australopithecines and for humans, as shown here. Notice that the australopithecines and humans coexisted for a time. The numbers are the cranial capacities.

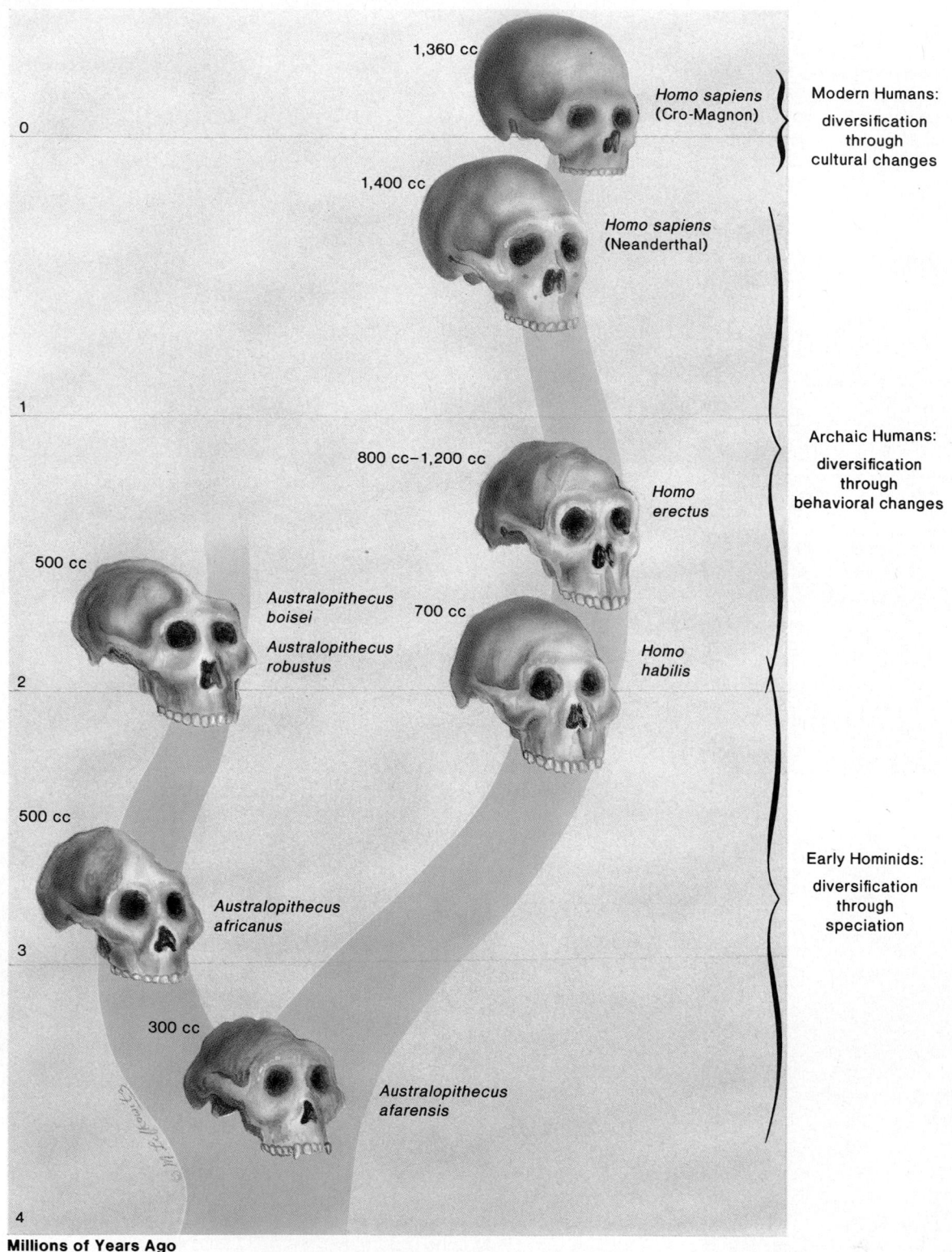

1,360 cc

Homo sapiens (Cro-Magnon)

Modern Humans: diversification through cultural changes

0

1,400 cc

Homo sapiens (Neanderthal)

1

800 cc–1,200 cc

Homo erectus

Archaic Humans: diversification through behavioral changes

500 cc

Australopithecus boisei

Australopithecus robustus

700 cc

Homo habilis

2

500 cc

Australopithecus africanus

Early Hominids: diversification through speciation

3

300 cc

Australopithecus afarensis

4

Millions of Years Ago

another. At the time that hominids evolved, the weather was becoming colder and drier; forests were disappearing and grasslands were becoming more prevalent in Africa.

The first hominid (ancestor leading to humans) arose at a time when a change in weather reduced the size of the African forests, making it advantageous to move on the ground.

At one time scientists attempted to arrange the hominid fossils in a straight line from the most primitive to the most advanced. Now it is reasoned that speciation gave rise to at least 2 hominid lineages (fig. 28.19). One of these lineages contains only australopithecines, and the other contains archaic humans and modern humans.

Australopithecines

Four species of *Australopithecus* (Southern Apeman) have been identified—*A. afarensis, A. africanus, A. robustus,* and *A. boisei.* Most authorities believe that A. afarensis is ancestral to the other *australopithecines* and to the humans (fig. 28.19).

The fossils of A. afarensis are dated from 5-3 million years ago. The nearly complete skeleton of one fossil, known as Lucy, is 1.0 m-1.4 m in length. The overall anatomy forms a connecting link between apes and humans. The skull has a sloping brow, and the face projects forward although the canines are reduced in size. The brain size of 400 cc is about that of today's chimpanzee. Cheek teeth are large, with thick enamel indicating a diet that may have included nuts and seeds in addition to fruit.

The arms are somewhat longer than modern humans, while the legs are shorter. Features of the forelimbs and shoulders suggest an ability to climb well in trees, and the fingers are long and curved, that is, suitable to grasping tree limbs. Nevertheless, the hand was most likely dexterous.

Lucy could walk erect! The foramen magnum is well to the rear of the skull, the pelvic girdle is short, and the knees angle toward the midline as they do in humans. Also, the big toe is not opposable. The foot is long and the toes are curved, however. Fossilized footprints that could have been made by Lucy suggest that her stride was short, and, therefore, walking and running would have been slow.

Researchers are intrigued by the observation that the australopithecine skeletons are dimorphic—occurring in 2 sizes. Presumably, the males were significantly larger than the females. It's speculated that perhaps the males did not range far, and one or 2 formed the nucleus of a social group.

A. africanus, which is dated between 2.5 and 2.0 million years ago, differs little from A. afarensis (fig. 28.20). The brain was slightly larger, and the cheek teeth were slightly bigger.

The other 2 species of australopithecines, A. robustus (found in southern Africa) and A. boisei (found in East Africa) are considerably larger. Lucy is believed to have weighed 50 kg, and A. boisei probably weighed over 70 kg. Brain volume in A. boisei is 500 cc or over. The jaws are massive and robust, and the cheek teeth are very large with very thick enamel. There is a midline skull crest for the attachment of large chewing muscles.

Figure 28.20
Australopithecus africanus. While at one time these hominids were thought to have been hunters, a study of cut marks left on animal bones suggests that they were most likely scavengers; they sought and confiscated the kills of coexisting predators and scavengers.

Weather was becoming even more cooler and drier around 2.5 million years ago. Do the anatomical features of A. robustus and A. boisei mean that their food was lower in quality compared to the earlier australopithecines? Larger animals can survive on lower-quality food even though they must spend more time eating.

A. robustus and A. boisei were contemporaries of Homo habilis.

Bipedalism arose before any notable increase in brain size. *Australopithecus afarensis* is most likely a common ancestor to an australopithecine lineage and a human lineage.

Archaic Humans

Humans are distinguishable by certain traits. Among these are a bipedal gait, a large brain, the making of tools, the eating of meat and green plants (omnivorous), pair bonding between male and female, the ability to use language, and a highly developed culture.

Homo habilis

Fossils identified as **Homo habilis** (handy man) are dated at 2 million years ago (fig. 20.19). These people were somewhat taller than the australopithecines at 1.5 m, but their brain size (700 cc) was half again as large. In other words, there was an increase in brain size relative to body mass, which might indicate an increase in intelligence.

The face was large and deep, with big incisors and canines and large cheek teeth. The pelvic and leg bones resemble modern humans more closely when compared to the australopithecines. *Homo habilis* people stood erect and had limbs fully adapted for walking. Their hands appear dexterous, but the fingers were still curved, indicating these people spent at least part of their time in trees.

Homo habilis people are especially known for the stone tools they made and used, as the species name handy man implies. The tools were made by knocking off flakes from a small core of stone. They used both the flakes and the core on a wide range of objects like wood, plants, and the carcasses of animals. They were meat eaters who seem to frequent particular sites where water and food were available. (These were not home bases, however, that is, a place to which a human returns and shares food after foraging or hunting.) Some of the meat they ate probably was the result of hunting, and some was probably scavenged from other animals such as lions and hyenas.

Homo habilis people lived at least 2 million years ago, and they had a significantly larger brain per body size than did the australopithecines. These people also made tools used to butcher animals.

Homo Erectus

Homo erectus people evolved in Africa about 1.5 million years ago and then later spread throughout Eurasia (fig. 28.21). This geographical expansion indicates that these people were quite successful at adapting to a wide range of environments. When they first arose, the brain size was just about 800 cc, but by 500,000 years ago the brain size was 1,200 cc.

An almost complete fossil of a 12-year-old boy has been found and dated at 1.6 million years ago. This fossil tells us that early *Homo erectus* was taller than *Homo habilis*. The fossil has a length of 1.6 m, but the boy would have grown to 1.8 m had he lived. The limbs are more robust but even so look quite like ours of today. The skull is longer than that of *Homo habilis* but does not have as high a brow as we do. The face is smaller than *Homo habilis,* but the eyebrow ridges are still more distinctive than ours, especially in males.

The teeth have decreased in size and massiveness. Both plants and animals were used for food, and judging from the smaller face and teeth, it is more likely that *Homo erectus* would have eaten prepared food.

Homo erectus people made tools of a better quality than the tools of predecessors—these tools had symmetrical bifaces and were made from flakes that were fashioned before they were chipped from the stone. How much meat they ate is questionable, and some researchers do not believe they were big-game hunters. Circumstantial evidence suggests that they could have hunted big game such as elephants, antelope, and bears. For example:

Social cooperation was present. Skeletons of old and diseased individuals are found, indicating that frail and sick people were allowed to live beyond the time of usefulness. Because humans are relatively small, they need to cooperate in order to kill large animals.
They had a wide geographic range. Hunters need to spread out in order to find sufficient game.

Figure 28.21
Homo erectus people. This drawing shows these people had the use of fire, made stone tools, and may have been big-game hunters.

The newborn's brain was small. The adult pelvis of *Homo erectus* is too narrow to allow a large-brained baby to pass through. The brain of the newborn most likely increased in size after birth, which could only happen if the lactating mother received good nutrition (such as hunting might provide).

Homo erectus people did have knowledge of fire, and this may be the reason they were able to live in colder climates. Apparently they were able to compete with other animals for caves and may even have built houses. During the one million years they were the prevalent human species, however, technological advancement was extremely slow. This suggests that although they may have had language, their communicative skills were primitive. Most likely their behavior was largely dominated by biological constraints rather than by culture.

Homo erectus people had a large brain, they walked erect, and they used fire. They may even have been big-game hunters.

Neanderthals

The **Neanderthal** people (*Homo sapiens neanderthalensis*) take their name from Germany's Neander Valley, where one of the first Neanderthal skeletons was discovered. The earliest fossils are 130,000 years old, and their latest fossils are only 32,000 years old. Neanderthals evolved and lived in Eurasia during the last Ice Age.

The Neanderthals had massive eyebrow ridges, and the nose, the jaws, and the teeth protruded far forward (fig. 28.22). The forehead was low and sloping, and the lower jaw sloped back without a chin. Surprisingly, the Neanderthal brain was, on the average, slightly larger than that of modern humans. The brain was 1,400 cc, whereas that of most modern humans is 1,360 cc.

Figure 28.22

The Neanderthals. This drawing shows that the nose and the mouth of these people protruded from the face, and the muscles were massive. They made stone tools and perhaps were hunters.

Figure 28.23

Cro-Magnon people are the first to be designated *Homo sapiens sapiens*. Their toolmaking ability and other cultural attributes, such as their artistic talents, are legendary.

The Neanderthals were heavily muscled, especially in the shoulders and the neck. The bones of the limbs were shorter and thicker than those of modern humans. It is hypothesized that a larger brain than that of modern humans was required to control the extra musculature.

Neanderthals seem to have been more culturally advanced than *Homo erectus* people. Most lived in caves, but those in the open may have built houses. They buried their dead with flowers and tools.

> The Neanderthals were more massive than modern humans, and they had a larger brain. Perhaps they were adapted to a cold climate because they lived in Eurasia during the last Ice Age.

Modern Humans

A fossil named for **Cro-Magnon,** France, where these remains were first found, is dated at about 30,000 years ago. Everyone agrees that the Cro-Magnons had a thoroughly modern appearance. They also had an advanced form of stone technology that included the making of compound tools; stone flakes were fitted to a wooden handle. They were the first to throw spears, enabling them to kill animals from a distance. They also were the first to make knifelike blades. They were such accomplished hunters that some researchers believe they were responsible for the extinction of many larger mammals, such as the giant sloth, the mammoth, the saber-toothed tiger, and the giant ox during the Upper Pleistocene epoch.

Cro-Magnons hunted cooperatively, and perhaps they had a language. They are believed to have lived in small groups, the men going out to hunt by day, while the women remained at home with the children. The ability to speak may have led to the development of a more advanced culture. The Cro-Magnons sculpted small figurines out of reindeer bones and antlers. They also painted beautiful drawings of animals on cave walls in Spain and France (fig. 28.23).

Origin of Modern Humans

A number of fossils dated about 100,000 years ago, which have a more modern appearance than the Neanderthals, have been found in Europe, Israel, and Africa. It is not known how these fossils are related among themselves and with the Neanderthals. Perhaps early on there were many migrations from Africa, and modern humans arose in several locations in Eurasia. Or, perhaps modern humans had a single place of origin and then as they spread out, various races came into being. (All modern races of humans are classified as *Homo sapiens sapiens*). It's possible that humans simply replaced the Neanderthals without interbreeding with them, but this is of some dispute.

The hypothesis that modern humans had a single place of origin is consistent with a recent study on mitochondrial DNA. Mitochondrial DNA is passed to offspring by the mother only, and therefore there is no possibility of DNA changes being due to genetic recombinations. The number of mitochondrial DNA mutations that have accumulated in a race over time can serve as an evolutionary clock: the greater the number of mutations among the members of a race, the older it is. Rebecca Cann and her colleagues from the University of California, Berkeley, used restriction fragment length polymorphism (p. 290) to study the mitochondrial DNA of populations from Africa, Asia, Europe, Australia, and New Guinea. The results showed that mitochondrial DNA has been evolving the longest time in Africa and can be traced to a single

African woman who lived about 150,000 years ago. The popular phrase "mitochondrial Eve" is misleading. There were many other women who lived at the same time and had children, but eventually their lines died out. The important conclusion to this study (and another study based on blood Rh factor) is that humans most likely originated in Africa. The data seem to suggest that humans began migrating out of Africa about 100,000 years ago, and then they founded the populations that became the various races of humans. There has been much controversy about these conclusions as discussed in the reading for this chapter.

It is hypothesized that modern humans arose in Africa about 150,000 years ago. From there they migrated into Eurasia and replaced the Neanderthals. The Cro-Magnons, dated about 30,000 years ago, are modern humans in all respects.

More about Eve and Our Ancestors

The Eve theory suggests that the origin of modern-looking people can be traced to a common African ancestor who lived as recently as 140,000 to 280,000 years ago. Molecular biologists at the University of California, Berkeley, led by the late Allen C. Wilson, claimed to have traced our ancestry through genetics—specifically, by using blood samples of people living today from around the world and studying their mitochondrial DNA, passed along only by women.

Mitochondrial DNA was chosen for study because it represents a kind of molecular clock inside all of us, says Mark Stoneking, Pennsylvania State University, who aided in the research. The passing of time could be measured by counting the number of mutations found in people's DNA. Mutations are known to occur at a rate of about one every 6,000 years.

So they theorized that the identity and birthplace of the oldest common ancestor would be revealed in whoever's blood samples held the largest number of mutations. People with the fewest mutations would be on the youngest branches of the human family tree; those with the greatest number would represent the oldest branches.

Using a computer program to construct the family tree, they concluded the common ancestor was between 140,000 and 280,000 years old and that the oldest lineage was African.

No anthropologist would dispute that pre-humans originated in Africa. The traditional belief has us descending from earlier ape-like humans called Homo erectus, who left Africa a million years ago and spread to China, Europe and Indonesia. The descendants of these first world travelers are known from fossils as Neanderthal, Beijing Man and Java Man. Paleoanthropologist Milford Wolpoff, one of the strongest opponents of the Eve theory, says we evolved from these species on our respective continents with our respective racial features. He calls it the "multi-regional hypothesis."

"Our goal is to describe a theory that synthesizes everything known about modern human fossils, archeology and genes," Wolpoff, University of Michigan, Ann Arbor, writes in *Scientific American*. "The Eve theory cannot do so."

The original Eve theory doesn't account for any mixing of genes with Neanderthal, Beijing Man or any other group, and therein lies the rub. "They claim (there was) a mass movement of modern Homo sapiens from Africa and that they didn't interbreed—that they just somehow took over," says anthropologist Bob Sussman, Washington University, St. Louis.

Like Wolpoff, Sussman argues that modern humans would have had to wipe out Neanderthal, Beijing Man and other "archaic" humans. "That just didn't make sense from the fossil record or what we know about population movement. . . ."

But the real trouble for the Eve theory began when Sussman commissioned a biologist and statistical expert to check the math used by the Eve group. The result was the discovery of error that showed the calculations were wrong.

"I showed that they estimated it incorrectly," says Eve-buster Allen Templeton, Washington University, St. Louis. "When you look at the set of evolutionary trees and whether they're compatible with the data, they have no way to determine where the common ancestor is from. . . ."

Stoneking sighs and concedes the mistake, but suggests rumors of the Eve theory's death may be premature. A re-analysis by others at Penn State with a new statistical test supports the age of the common ancestor, even though it doesn't prove conclusively that Eve originated in Africa.

Stoneking is still convinced Eve was African but has modified his theory to suggest that modern humans came out of Africa and perhaps interbred with their more archaic cousins.

And to get in his two cents, he asks what would happen if statistical tests were applied to the fossil record?

"I think the fossil evidence of modern humans is still up in the air. So much rests in the interpretation of so few specimens," Stoneking says.

So what will settle the debate? Says Stoneking: "The Eve theory really seems to get people worked up and going. I think it boils down to who can yell the longest and loudest."

Summary

1. Vertebrates are in the phylum Chordata and have the 3 chordate characteristics as embryos. As adults, the notochord is replaced by the vertebral column.

2. Vertebrates include fishes, amphibians, reptiles, birds, and mammals.

3. The earliest fishes, the jawless fishes, are represented today by hagfishes and lampreys. Sharks, skates, and rays are cartilaginous, and the bony fishes include all other well-known fishes.

4. The original bony fishes gave rise to 2 groups: today's ray-finned fishes and fleshy-finned fishes. The fleshy-finned

fishes are represented today by 3 species of lungfishes and the coelacanth, a lobe-finned fish.

5. Fleshy-finned fishes gave rise to the amphibians during the Devonian period (Mesozoic era). Amphibians, represented today by frogs and salamanders, must return to water to reproduce.

6. Reptiles (e.g., today's turtles, crocodiles, snakes, and lizards) evolved from amphibians. They lay a shelled egg, which allows them to reproduce on land.

7. One main group of ancient reptiles, the stem reptiles, produced a line of descent that evolved into both dinosaurs and birds. A different line of descent from stem reptiles evolved into mammals.

8. Birds are feathered, which helps them maintain a constant body temperature. They are adapted for flight: their bones are hollow, their shape is compact, their breastbone is keeled, and they have well-developed sense organs.

9. Mammals are vertebrates with hair and mammary glands. The former helps them maintain a constant body temperature, and the latter allows them to nurse their young. Monotremes lay eggs; marsupials have a pouch in which the newborn matures; and the placental mammals, which are far more varied and numerous, retain offspring inside the uterus until birth.

10. Mammals remained small and insignificant while the dinosaurs existed, but when the latter became extinct during the Cretaceous period (end of Mesozoic era), mammals became the dominant land organisms.

11. Primates are adapted to an arboreal existence. Prosimians (e.g., today's lemurs and tarsiers) are more primitive than the anthropoids, which include monkeys, apes, and humans.

12. During the Miocene epoch (Tertiary period, Cenozoic era), when the climate was becoming cooler and drier, apes arose. Dryopithecus could be the hominoid ancestor that gave rise to all the apes and humans. The dryopithecines had a sloping brow, heavy eyebrow ridges, and distinct muzzle, because the molars were large and the canine teeth were large and pointed. Most likely, they were knuckle-walkers and had an opposable thumb. Most of these features are seen today in apes but not in humans.

13. In the early Pliocene epoch, when savanna was replacing forests, a hominid ancestor (leading only to humans) began to walk erect.

14. The australopithecines (A. afarensis, A. africanus, A. robustus, and A. boisei) were hominids. They had a small brain but walked erect. The teeth of australopithecines indicate they were vegetarians.

15. It is possible that A. afarensis was a common ancestor to both the australopithecine lineage and the Homo lineage.

16. The brain size in the Homo lineage did increase to a remarkable degree. Homo habilis had an increased brain size compared to body mass that most likely indicates an increase in intelligence. Homo habilis people walked erect; made tools and butchered animals; therefore, they were meat eaters.

17. Homo erectus were the first hominids to migrate out of Africa into Europe and Asia. At first the brain size was 800 cc, but by the time they became extinct the brain size was 1,200 cc. These people made better tools, had fire, and may have been big-game hunters.

18. Two fossils are designated as Homo sapiens: Neanderthals and Cro-Magnon. Neanderthals are now thought to have been more modern than previously supposed. What happened to them is not definitely known. Cro-Magnon evolved into modern humans and definitely had culture.

Study Questions

1. What are the 3 chordate features, and which is replaced by the vertebral column in vertebrates?

2. Draw a simplified evolutionary tree of vertebrates using figure 28.1 as a guide.

3. What are the 3 classes of living fishes, and how are they related? The amphibians evolved from what type of fish?

4. In what ways are amphibians well adapted to a land existence, and in what ways are they ill adapted?

5. In what ways are reptiles better adapted to a land existence than amphibians?

6. In what ways are birds adapted to flying?

7. What are the 3 subclasses of mammals, and what are their primary characteristics?

8. List representative examples of all classes of vertebrates.

9. Give the complete classification of humans using table 28.3 as a guide.

Distinguish between prosimians, anthropoids, hominoids, hominids, and humans.

10. Describe Dryopithecus, and tell in what ways the anatomy of modern apes and modern humans differ from that of this fossil.

11. Draw a phylogenetic tree for hominids, and describe each group in the tree.

12. Discuss the phylogenetic relationship of erect walking, brain size, and tool making, and tell how these aspects relate to the hominids in your tree.

Objective Questions

1. Which of these features is replaced by the vertebral column in vertebrates?
   a. notochord
   b. dorsal hollow nerve cord
   c. gill pouches
   d. All of these.
2. Amphibians arose from
   a. cartilaginous fishes.
   b. jawless fishes.
   c. ray-finned fishes.
   d. bony fishes with lungs.
3. Which of these is not a feature of amphibians?
   a. dry skin that resists desiccation
   b. metamorphosis from a swimming form to a land form
   c. small lungs necessitating a supplemental way to achieve gas exchange
   d. reproduction in the water
4. Which of these are not reptiles?
   a. snakes
   b. penguins
   c. crocodiles
   d. lizards
5. Which of these is a true statement?
   a. In all mammals, offspring develop within the female.
   b. All mammals have hair and mammary glands.
   c. All mammals are land-dwelling forms.

6. d. All of these are true.
6. Which of these gives the correct evolutionary order?
   a. anthropoids—prosimians— hominids—hominoids
   b. prosimians—hominoids—homi- nids—anthropoids
   c. hominids—anthropoids— prosimians—humans
   d. prosimians—anthropoids— hominoids—hominids
7. Which of these features of the dryopithecines is seen in apes today?
   a. face that projects forward
   b. eyebrow ridges
   c. knuckle-walking
   d. All of these.
8. *Homo habilis* most likely existed at the same time as the
   a. dinosaurs.
   b. australopithecines.
   c. Cro-Magnon.
   d. All of these.
9. Which of these is mismatched?
   a. australopithecines—walked erect
   b. *Homo habilis*—large brain
   c. *Homo erectus*—had fire
   d. *Homo sapiens*—flat face
10. The hominid ancestor most likely
    a. made tools.
    b. had a large brain.
    c. walked erect.
    d. All of these.

11. Place the common names of the vertebrate classes on the lines provided in this phylogenetic tree.

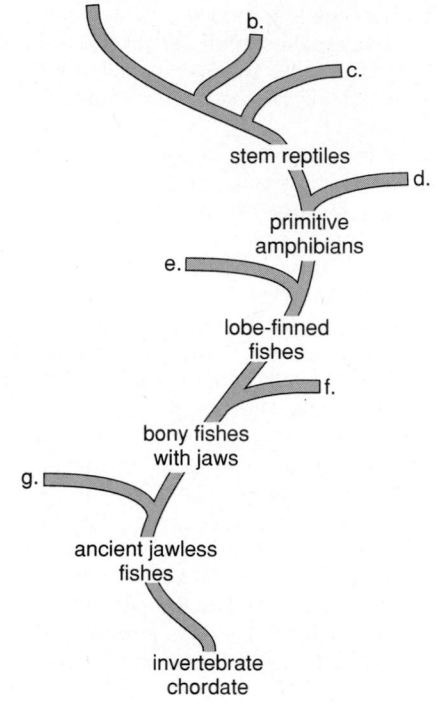

12. Place the checkmarks for the categories that apply to the particular animals listed.

	Primate	Prosimian	Anthropoid	Hominoid	Hominid
Lemurs					
Monkeys					
Living apes					
Dryopithecines					
Australopithecines					
Humans					

13. Fill in the blanks to give the classification of humans. Consult the classification system in appendix A if needed.

Kingdom _____

Phylum _____

Subphylum _____

Class _____

Order _____

Superfamily _____

Family _____

Genus and Species _____

Concepts and Critical Thinking

1. *Arthropods and vertebrates are highly successful groups of animals.*

   What criteria might you use to support this concept? What features do arthropods and vertebrates have in common?

2. *Reptiles are adapted to reproduce on land.*

   Compare the adaptations of reptiles and angiosperm for reproduction on land.

3. *Humans have a phylogenetic history.*

   Why is it incorrect to say that humans evolved from modern apes?

Selected Key Terms

amphibian (am-fib′e-an) 452
reptile (rep′tīl) 453
mammal (mam′al) 455
monotreme (mon′o-trēm) 457
marsupial (mar-su′pe-al) 457

placental mammal (plah-sen′tal mam′al) 458
primate (pri′māt) 458
dryopithecine (dri″o-pith′ĕ-sīn) 461
australopithecine (aw″strah-lo-pith′ĕ-sĭn) 465

Homo habilis (ho′mo hah′bi-lĭs) 465
Homo erectus (ho′mo ĕ-rek′tus) 466
Neanderthal (ne-an′der-thawl) 466
Cro-Magnon (kro-mag′non) 467

PART 3 *Evolution and Diversity*

CRITICAL THINKING CASE STUDY

Polyploidy

In 1917, O. Winge observed that species of *Chrysanthemum* differ by their diploid chromosome number. Their diploid chromosome numbers were 18, 36, 54, 72, and 90, all of which are multiples of 18. This caused Winge to hypothesize that *new plant species can be formed by polyploidy, the addition of an entire set of chromosomes.* In this case, 18 chromosomes (the diploid number in the first species listed) are added each time as the chromosome number increases from 18 to 90. New plant species that arise by polyploidy are called polyploids.

A study of hybridization, a cross between 2 species, has contributed to our knowledge about polyploidy. The genus *Triticum* (common wheat) contains 14 species that can be divided into 3 groups based on chromosome number:

Group	Diploid Number	Haploid Number	Number of Species
Einkorn	2N = 14	N=7	6 species
Emmer	2N = 28	N=14	5 species
Vulgare	2N = 42	N=21	3 species

Hybrids are fertile when the chromosomes from the 2 parents can form homologous pairs during meiosis, and they are infertile when homologous pairs cannot form during meiosis. What do you predict will be the result of crosses between *Triticum* species of the same group? between species of different groups?

Prediction 1 Crosses between *Triticum* species of the same group will be fertile because these species have the same haploid number of chromosomes. Crosses between species of the different groups will not be fertile because homologous pairs of chromosomes cannot form during meiosis.

Result 1 Subsequent crossing experiments have shown that when species are crossed within a group, fertile hybrids are produced. When species from the Einkorn group are crossed with species from the Vulgare group, partially fertile hybrids result. Crosses between the other groups are not fertile.

I Explain why a hybrid resulting from Einkorn × Vulgare could possibly produce gametes. Why might fertility be reduced? Explain why a hybrid resulting from Einkorn × Emmer is sterile (not fertile).

Crosses between species of 2 different genera usually produce hybrids that are sterile. For example, when a radish (N = 9) is crossed with a cabbage (N = 9), the hybrid is sterile even though both parents have the same haploid number of chromosomes (see fig. 28.A). Do you predict that a doubling of the chromosome number will result in a fertile hybrid?

Prediction 2 If there is a doubling of the chromosome number, the hybrid will be fertile because now there will be homologous pairs of chromosomes and meiosis can proceed normally.

Result 2 A hybrid from the cross radish × cabbage does become fertile when the chromosome number doubles. Many new species have been produced by artificial

radish 18	radish 9	radish 18	cabbage 18
	cabbage 9	cabbage 18	

a.   b.   c.   d.

Figure A
Karyotypes of (*a*) radish, (*b*) diploid radish-cabbage hybrid, (*c*) tetraploid radish-cabbage hybrid, and (*d*) cabbage. Source: After Karpechenko, as appeared in Th. Dobzhansky, *Genetics and the Origin of Species,* 1937.

hybridization and subsequent doubling of the chromosome number. (Such plants are called allopolyploids.)

II Which of the plants in figure 28.A would be sterile? Why? Suppose by nondisjunction a radish plant (N = 9) produced gametes in which N = 18. Following self-fertilization, 2N = 36. In such a plant, called an autopolyploid, how many chromosomes might try to pair with one another during meiosis? Might this also cause fertility problems?

There are 2 primary mechanisms by which autopolyploids are formed: nondisjunction of chromosomes during meiosis and a somatic doubling of chromosomes during mitosis. Although it seems unlikely that nondisjunction would occur in sperm and egg at the same time, the enormous number of plant gametes that are formed increases the chances of such an occurrence. Polyploids may arise when somatic cells, with a doubled chromosome number, produce new plants by vegetative (asexual) reproduction (p. 515). Do you predict that growth habit and/or climate will correlate with number of polyploid species per genera?

Prediction 3 Growth habit is more likely to correlate with number of polyploid species per genera than climate.

Result 3 The data shown following was collected on number of polyploids per genera. It shows that the perennial growth habit is conducive to polyploidy in both temperate and tropical climates.

Percent of Genera Exhibiting Polyploidy

Growth and Habitat	26%-50% Polyploid	75% Polyploid
Woody, temperate	27	2
Woody, tropical	8	0
Perennial herbs, temperate	37	15
Perennial herbs, tropical	13	13

Source: Data from G. L. Stebbins, *Chromosomal Evolution in Higher Plants*, Addison-Wesley Publishing Company, Reading, MA, 1971.

III Explain why these results support the prediction that habit of growth and not climate correlate with number of polyploid species per genera.
Speculate why the perennial growth habit (nonwoody plants that grow each season) and not the woody growth habit would promote polyploidy.
The chemical colchicine can prevent cell division after chromosomes have duplicated in preparation for mitosis. Would use of the chemical promote polyploidy?

Only plants have been discussed thus far since polyploidy does not often occur in animals. A doubling of the sex chromosomes increases the chances of gametes with an abnormal sex chromosome number. On the basis of the findings in plants, do you predict that there will be polyploid species in groups of animals that reproduce asexually?

Prediction 4 There will be polyploid species in groups of animals that reproduce asexually.

Result 4 Polyploid species are found in groups of animals that reproduce by parthenogenesis (i.e., offspring develop from the unfertilized egg). This type of asexual reproduction is found in invertebrates and in fishes, amphibians, and reptiles.

IV What abnormal gametes and abnormal offspring might result if a cell undergoing meiosis has 2 X chromosomes and 2 Y chromosomes?
Do the observations on polyploidy in animals strengthen a possible relationship between polyploidy and growth habit in plants?

Since Winge's initial hypothesis in 1917, varied and extensive observational data have supported the hypothesis concerning polyploids. Hybridization experiments and formation of artificial polyploids have established that polyploidy can be a mechanism for speciation.

Other Questions

1. Why might you expect fewer polyploids among annuals (live only one season) than among perennials?
2. Normally during meiosis in males the X and Y chromosomes pair even though they have different shapes. How do you know?

References

Grant, V. 1981. *Plant speciation,* 2d ed. New York: Columbia Univ. Press

Mayr, E. 1963. *Populations, species, and evolution.* Cambridge, Mass.: Harvard Univ. Press.

Stebbins, G. L. 1971. *Chromosomal evolution in higher plants.* Reading, Mass.: Addison-Wesley Publishing Co.

Suggested Readings for Part 3

Beehler, B. M. December 1989. The birds of paradise. *Scientific American.*

Bonatti, E. March 1987. The rifting of continents. *Scientific American.*

Brock, T. D., and Madigan, M. T. 1988. *Biology of microorganisms.* 5th ed. Englewood Cliffs, N.J.: Prentice Hall.

Cairns-Smith, A. G. June 1985. The first organisms. *Scientific American.*

del Pino, E. M. May 1989. Marsupial frogs. *Scientific American.*

Dickerson, R. E. September 1981. Chemical evolution and the origin of life. *Scientific American.*

Dodson, E. O., and Dodson, P. 1985. *Evolution: Process and product.* 3d ed. Boston: Prindle, Weber, and Schmidt.

Eckert, R., and Randall, D. 1983. *Animal physiology.* 2d ed. San Francisco: W. H. Freeman.

Feder, M. E., and Burggren, W. W. November 1985. Skin breathing in vertebrates. *Scientific American.*

Futuyma, E. J. 1986. *Evolutionary biology.* 2d ed. Sunderland, Ma. Sinauer.

Gallo, R. C. December 1986. The first human retrovirus. *Scientific American.*

Gosline, J. M., and De Mont, M. E. January 1985. Jet-propelled swimming in squids. *Scientific American.*

Grant, P. R. 1981. Speciation and the adaptive radiation of Darwin's finches. *American Scientist* 69:653.

Hadley, N. F. July 1986. The arthropod cuticle. *Scientific American.*

Hay, R. L., and Leaky, M. D. February 1982. The fossil footprints of Laetoli. *Scientific American.*

Hickman, Z. P., and Roberts, L. S. 1988. *Integrated principles of zoology.* 8th ed. St. Louis: C. V. Mosby Co.

Hirsch, M. S., and Kaplan, J. C. April 1987. Antiviral therapy. *Scientific American.*

Hogle, J. M. March 1987. The structure of poliovirus. *Scientific American.*

Koehl, M. A. R. December 1982. The interaction of moving water and sessile organisms. *Scientific American.*

Margulis, L. 1982. Early life. Boston: Science Books International.

Mossman, D. J., and Sarjeant, W. A. S. January 1983. The footprints of extinct animals. *Scientific American.*

Raven, P. H., et al. 1986. *Biology of plants.* 4th ed. New York: Worth.

Rensberger, B. April 1982. Evolution since Darwin. *Science 82.*

Scientific American. 1978. *Evolution.* San Francisco: W. H. Freeman.

Simons, K., et al. February 1982. How an animal virus gets into and out of its host cell. *Scientific American.*

Stebbins, G. L., and Ayala, F. J. July 1985. The evolution of Darwinism. *Scientific American.*

Vidal, G. February 1983. The oldest eukaryotic cells. *Scientific American.*

Volpe, E. P. 1981. *Understanding evolution.* 4th ed. Dubuque, Iowa: Wm. C. Brown Publishers.

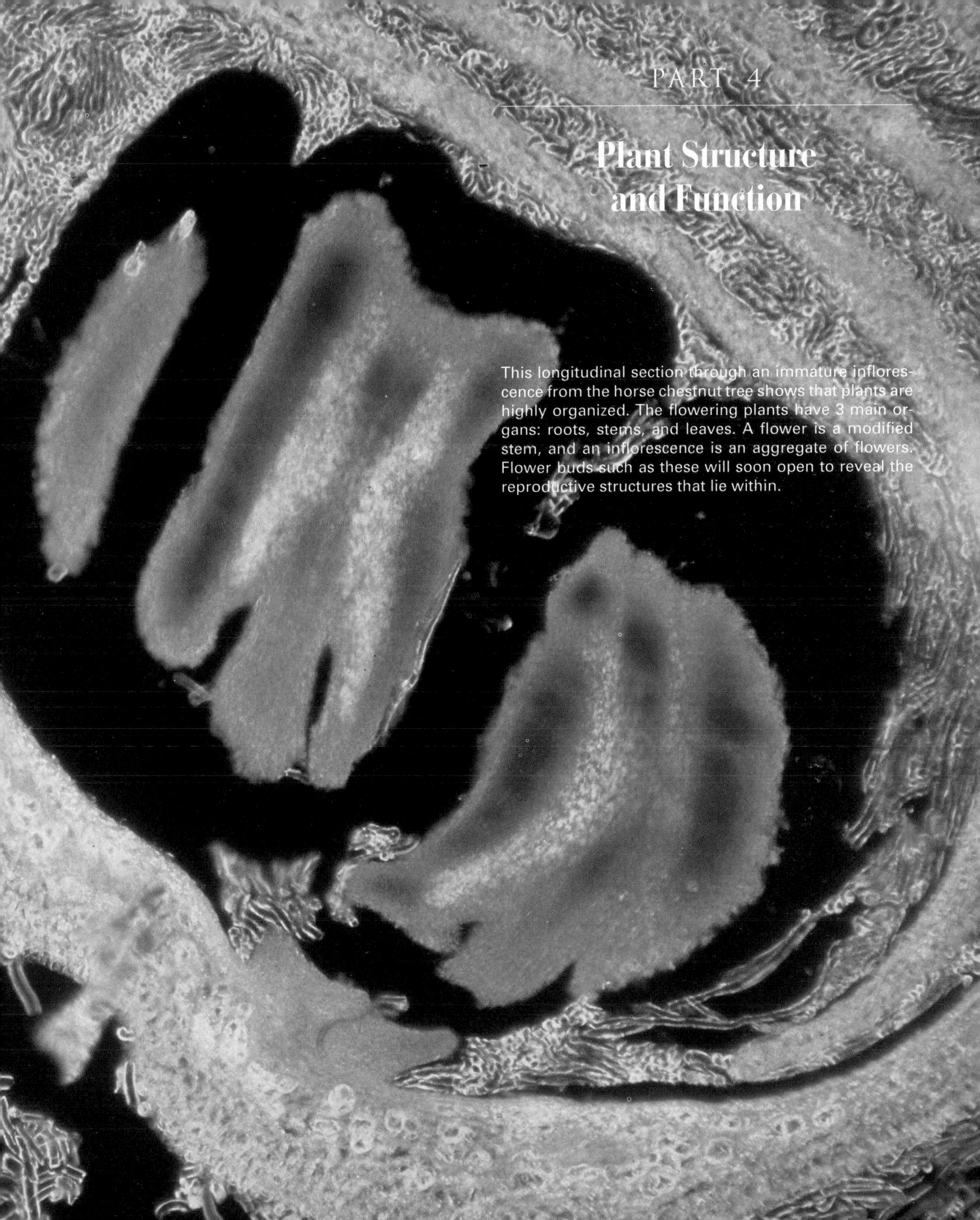

Plant Structure and Function

This longitudinal section through an immature inflorescence from the horse chestnut tree shows that plants are highly organized. The flowering plants have 3 main organs: roots, stems, and leaves. A flower is a modified stem, and an inflorescence is an aggregate of flowers. Flower buds such as these will soon open to reveal the reproductive structures that lie within.

Plant Structure

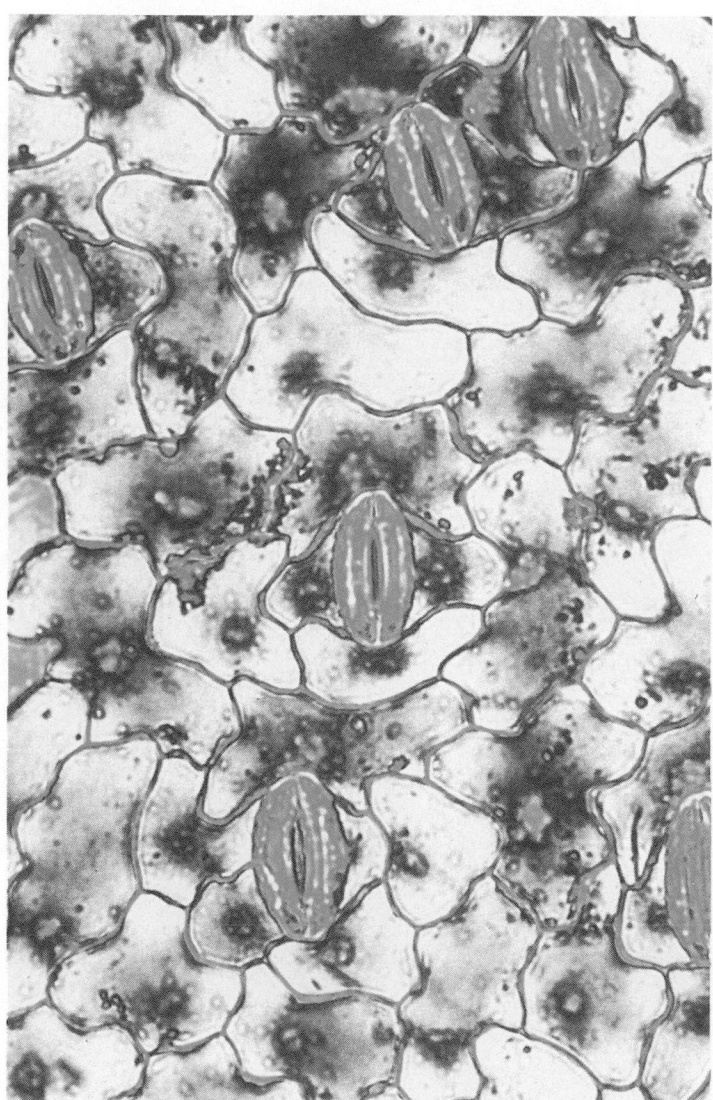

A layer of epidermal cells resembling puzzle pieces covers a leaf. Interspersed are the stomata, little openings surrounded by the sausage-shaped guard cells. When a plant is water-stressed, the stomata close, and this helps prevent further loss of water. When a plant is photosynthesizing, the stomata open, and this allows carbon dioxide to enter the leaf. Magnification, X250.

Your study of this chapter will be complete when you can

1. describe the overall organization of a flowering plant, distinguishing between the shoot system and the root system;

2. describe the structure of epidermis (dermal tissue) and the modification of epidermis in various parts of the plant body;

3. describe the structure and function of the parenchyma, collenchyma, and sclerenchyma cells in ground tissue;

4. describe the structure and function of xylem and phloem in vascular tissue;

5. contrast monocots and dicots in 3 ways;

6. name the various types of meristem tissue, and state their specific functions;

7. give 3 functions of roots; describe the structure and function of the 3 zones within a root tip; describe the cross section of herbaceous dicot and monocot roots;

8. describe 3 patterns of root growth;

9. give 3 functions of stems; describe the cross section of a herbaceous dicot and a monocot stem;

10. explain how secondary growth of stems and roots occurs, and describe the structure and function of a woody stem;

11. describe several ways in which stems can be modified;

12. describe the structure and function of tissues found in a C_3 dicot leaf; contrast C_3 leaves with C_4 leaves;

13. describe several leaf modifications.

n this chapter, we will study the organization and structure of flowering plants, which are the most complex, recently evolved, and widely distributed plants (see chap. 25).

Organization of a Plant Body

The body of a flowering plant is divided into 2 portions: the *root system* and the *shoot system* (fig. 29.1). The root system anchors the plant in the soil, and the stem system holds the leaves aloft to catch the rays of the sun. Both the roots and the stem transport water, inorganic nutrients, and organic nutrients. The roots absorb water and minerals, and then these are transported in the stem to the leaves, which receive carbon dioxide from the air and carry on photosynthesis. The sugar produced by photosynthesis is first transported in the stem to the rest of the plant. Finally it reaches the root, where it is stored as starch.

Tissue Types in Plants

Plants grow throughout their entire life because there is **meristem** (embryonic) **tissue** located in the shoot and root apexes. As apical meristem cells divide, they give rise to 3 types of primary meristematic tissue. Each of these supplies cells for a primary meristem tissue (p.478), one of the specialized tissue types in the body of a plant:

1. *Dermal tissue*—forms the outer protective covering of a plant
2. *Ground tissue*—fills the interior of a plant
3. *Vascular (transport) tissue*—conducts water and nutrients in a plant

Figure 29.1

Organization of plant body. The body of the plant is divided into 2 main portions: the root system below ground and the shoot system containing the stems and leaves above ground. A plant grows lengthwise by the production of new cells at the terminal bud (shoot tip) and root tip. The body of a plant contains 3 types of tissues. Dermal tissue covers the entire body of a nonwoody plant. Ground tissue makes up the bulk of the plant, and vascular tissue conducts water and nutrients to all body parts.

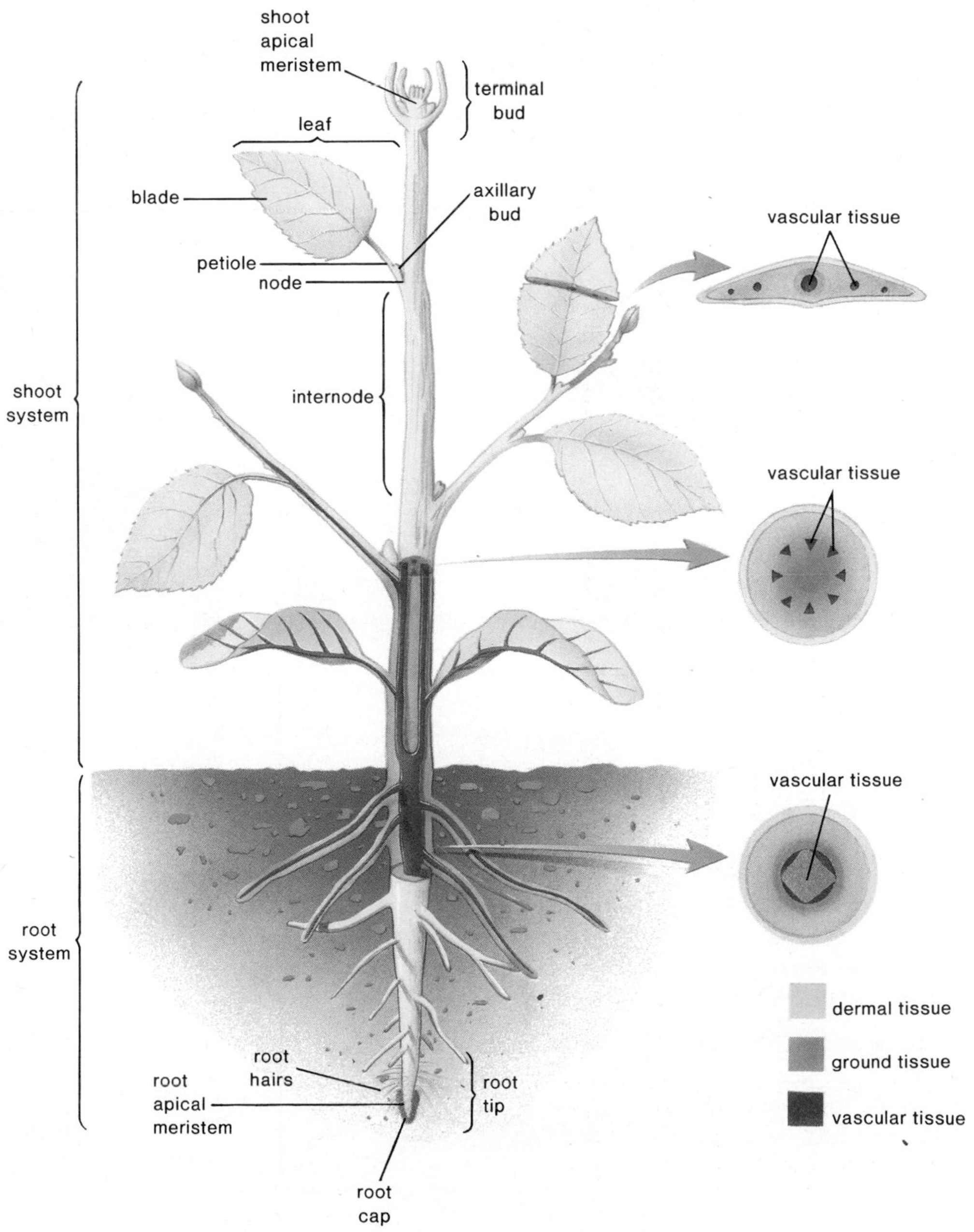

Figure 29.2

Epidermis is modified in different parts of a plant. **a.** Leaf epidermis contains stomata for gas exchange. Magnification, X230. **b.** Root epidermis has root hairs to absorb water. Magnification, X30. **c.** Cork replaces epidermis in older, woody stems. Magnification, X500.

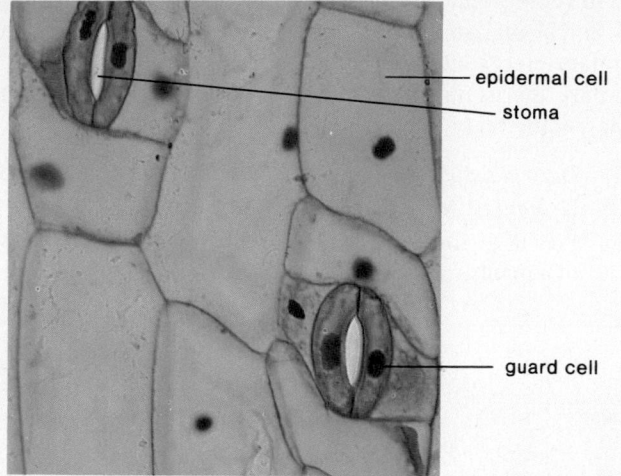

epidermal cell
stoma

guard cell

a.

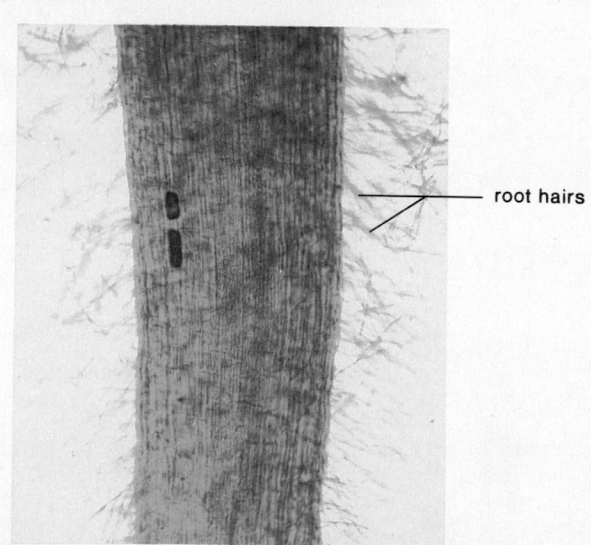

root hairs

b.

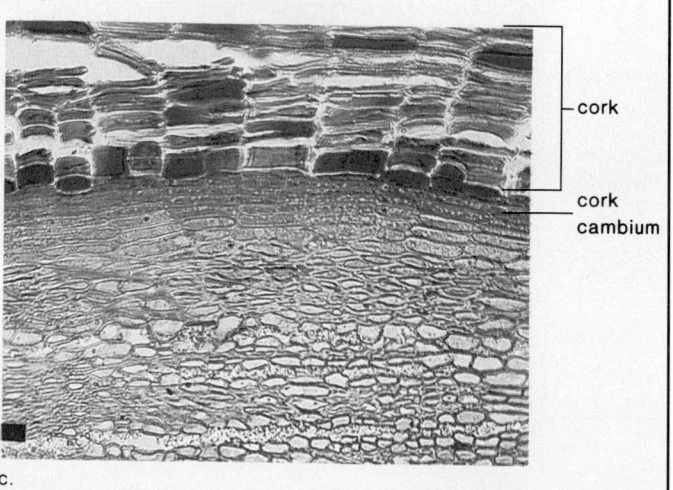

cork

cork cambium

c.

Figure 29.3

Ground tissue cells. **a.** Parenchyma cells are the least specialized of the plant cells. This type of parenchyma cell is found in stems and roots. Those in the leaf contain chloroplasts. Magnification, X150. **b.** Collenchyma cells. Notice how much thicker the walls are compared to those of parenchyma cells. Collenchyma cells give flexible support. Magnification, X255. **c.** Sclerenchyma cells have very thick walls and are nonliving—their only function is to give strong support. Magnification, X340.

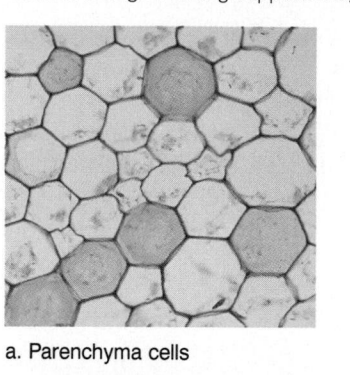
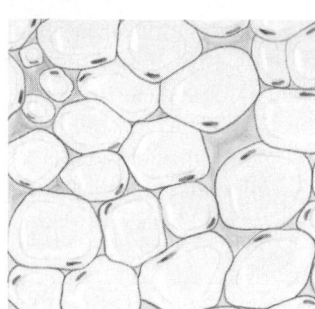

a. Parenchyma cells

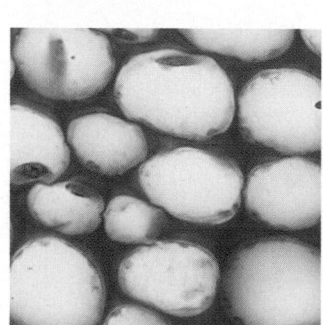
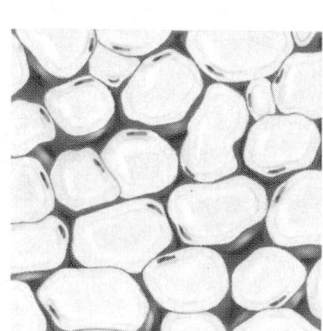

b. Collenchyma cells

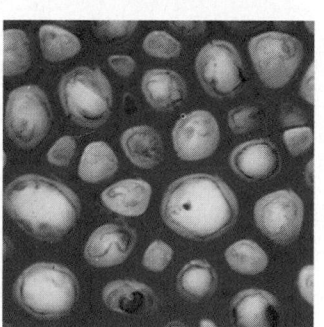
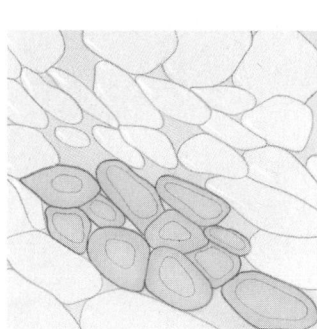

c. Sclerenchyma cells (fibers)

Dermal Tissue

The entire bodies of nonwoody (herbaceous) and young woody plants are covered by a layer of **epidermis,** which contains epidermal cells. Epidermal cells are closely packed, and the epidermis protects the inner tissues of a plant. Also, the walls of epidermal cells that are exposed to air are covered with a waxy cuticle to minimize water loss. In addition to the epidermal cells, the epidermis may also contain specialized cells, such as guard cells (fig. 29.2a). The guard cells surround microscopic pores called **stomata** (stoma, sing.). When the stomata are open, gas exchange can occur.

Plant Structure and Function

Figure 29.4

Xylem structure. **a.** General organization of xylem (*far left*) followed by an external view of vessel elements stacked one on top the other and a longitudinal view of several tracheids. Tracheids and vessel elements usually conduct water in the direction shown by the arrows. **b.** Photomicrograph of xylem vessels. Magnification, X225.

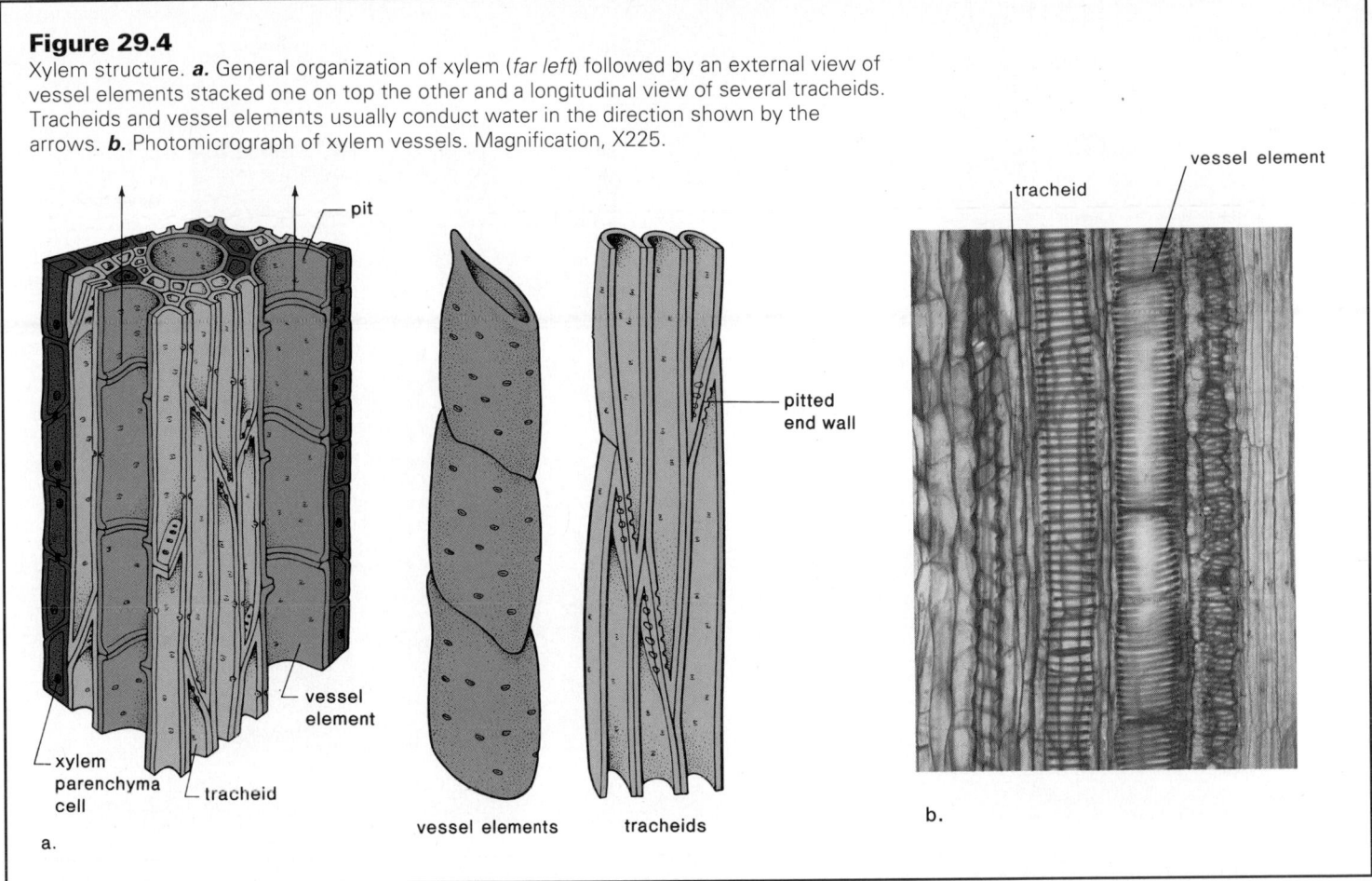

pit

vessel element

tracheid

pitted end wall

vessel element

xylem parenchyma cell

tracheid

vessel elements

tracheids

a.

b.

The epidermal cells of roots have long, slender projections called root hairs (fig. 29.2b). The hairs increase the surface area of the root for absorption of water and minerals; they also help to anchor the plant firmly in place.

In older woody plants, the epidermis of the stem is replaced by cork tissue. **Cork,** the outer covering of the bark of trees, is made up of dead cork cells that may be sloughed off (fig. 29.2c). New cork cells are made by meristematic tissue called cork cambium.

Ground Tissue

Ground tissue forms the bulk of a plant and contains parenchyma, collenchyma, and sclerenchyma cells (fig. 29.3). **Parenchyma** cells correspond best to the typical plant cell depicted in figure 5.5. These are the least specialized of the cell types and are found in all the organs of a plant. They may contain chloroplasts and carry on photosynthesis, or they may contain colorless plastids that store the products of photosynthesis. Parenchyma cells can divide and give rise to more specialized cells, as when roots develop from stem cuttings placed in water.

Collenchyma cells are like parenchyma cells except they have thicker primary walls (see fig. 6.15). The thickness is uneven and usually involves the corners of the cell. Collenchyma cells

often form bundles just beneath the epidermis and give flexible support to immature regions of a plant body. The familiar strands of celery stalks are composed mostly of collenchyma cells.

Sclerenchyma cells have thick secondary cell walls (see fig. 6.15), usually impregnated with lignin, which is an organic substance that makes the walls tough and hard. Most sclerenchyma cells are nonliving; their primary function is to support mature regions of a plant. Two types of sclerenchyma cells are fibers and sclereids. Although fibers occasionally occur in ground tissue, most are found in vascular tissue, which is discussed next. Fibers are long and slender and may occur in bundles that are sometimes commercially important. Hemp fibers can be used to make rope, and flax fibers can be woven into linen. Flax fibers, however, are not lignified, which is why linen is soft. Sclereids, which are shorter than fibers and more varied in shape, occur in seed coats and nut shells. They also give pears their characteristic gritty texture.

Vascular (Transport) Tissue

There are 2 types of vascular (transport) tissue. **Xylem** transports water and minerals from the roots to the leaves, and **phloem** transports organic nutrients, usually from the leaves to the roots. Xylem contains 2 types of conducting cells: *tracheids* and *vessel elements* (fig. 29.4). Both types of conducting cells are hollow and

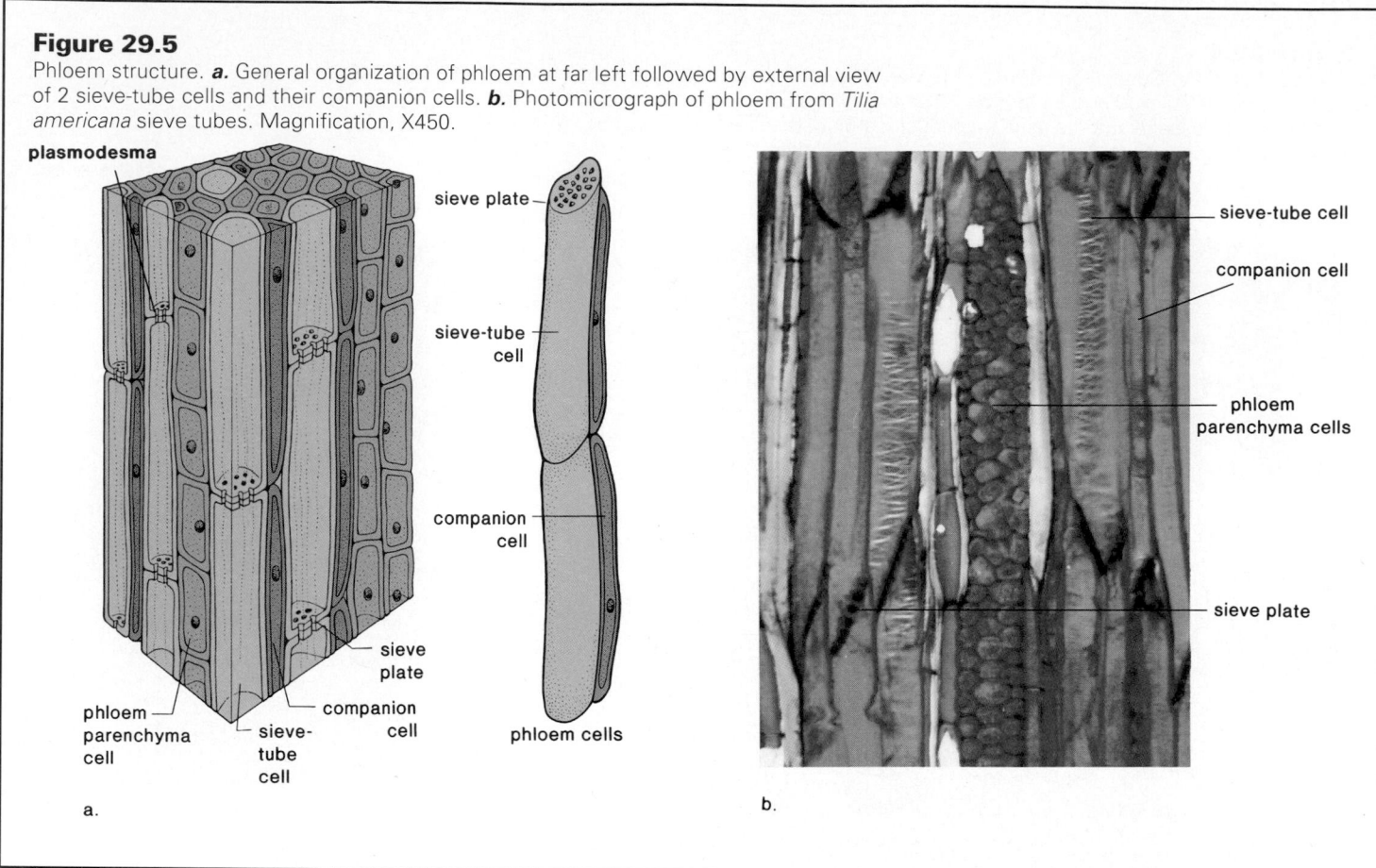

Figure 29.5

Phloem structure. *a.* General organization of phloem at far left followed by external view of 2 sieve-tube cells and their companion cells. *b.* Photomicrograph of phloem from *Tilia americana* sieve tubes. Magnification, X450.

plasmodesma

sieve plate

sieve-tube cell

companion cell

sieve plate

companion cell

phloem parenchyma cell

sieve-tube cell

phloem cells

a.

sieve-tube cell

companion cell

phloem parenchyma cells

sieve plate

b.

nonliving, but the vessel elements are larger, lack transverse end walls, and are arranged to form a continuous pipeline for water and mineral transport. The elongated tracheids, with tapered ends, form a less obvious means of transport, but water can move across the end walls and sidewalls because there are pits, or depressions where the secondary wall does not form.

The conducting cells of phloem are sieve-tube cells, each of which has a companion cell (fig. 29.5). *Sieve-tube cells* contain cytoplasm but no nuclei. These cells have channels in their end walls that in cross section make them resemble a sieve. Plasmodesmata extend from one cell to another through this so-called *sieve plate*. The smaller *companion cells* are more generalized cells and are closely connected to sieve-tube cells by numerous plasmodesmata. The nucleus of the companion cell controls and maintains the life of both cells.

It is important to realize that vascular tissue (xylem and phloem) extends from the root to the leaves and vice versa. In the roots, the vascular tissue is located in the vascular cylinder; in the stem, it forms vascular bundles; and in the leaves, it is found in leaf veins.

Monocots and Dicots

Flowering plants are divided into the *monocots* and *dicots,* depending on the type of cotyledon(s) or seed leaf (leaves) that the embryonic plant has. **Cotyledons** provide nutrient molecules for growing embryos before the true leaves begin photosynthesizing. Some embryos have one cotyledon, and plants with such embryos are known as monocotyledons, or monocots. Other embryos have 2 cotyledons, and plants with these embryos are known as dicotyledons, or dicots. Adult monocots and dicots have several other structural differences, as illustrated in figure 29.6.

Primary and Secondary Growth

All plants show primary growth, which increases the length of a plant. Primary growth occurs because shoot and root apical meristems are actively producing new cells. The following 3 primary meristems develop from apical meristem cells and give rise to the tissues noted (fig. 29.7).

1. Protoderm: produces dermal (epidermal) tissue
2. Ground meristem: produces ground tissue
3. Procambium: produces primary vascular tissue

Some dicots, notably woody trees, also have secondary growth, which increases the diameter of stems and roots; it makes them thicker. Secondary growth is associated with these lateral meristems;

1. Vascular cambium: produces secondary vascular tissue
2. Cork cambium: produces new cork cells

As discussed on page 477, cork replaces epidermis in woody plants.

Plant Structure and Function

Figure 29.6

Flowering plants are either monocots or dicots. Three features used to distinguish monocots from dicots are the number of cotyledons, arrangement of vascular bundles, and pattern of leaf veins. Not discussed in this chapter are the number of flower parts: monocots have flower parts in 3s, dicots in 4s or 5s.

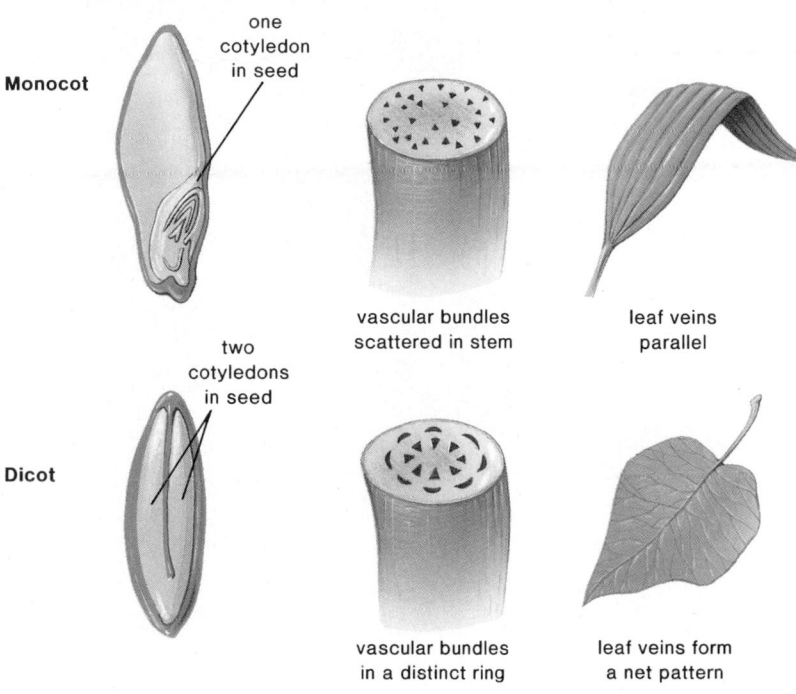

Monocot — one cotyledon in seed — vascular bundles scattered in stem — leaf veins parallel

Dicot — two cotyledons in seed — vascular bundles in a distinct ring — leaf veins form a net pattern

Figure 29.7

Diagram of shoot tip and root tip illustrating the arrangement of plant meristem tissues. Apical meristem gives rise to 3 primary meristems: protoderm, ground meristem, and procambium. Vascular cambium and cork cambium are present in those parts of the plant body where secondary growth occurs. Secondary growth increases the girth of the plant.

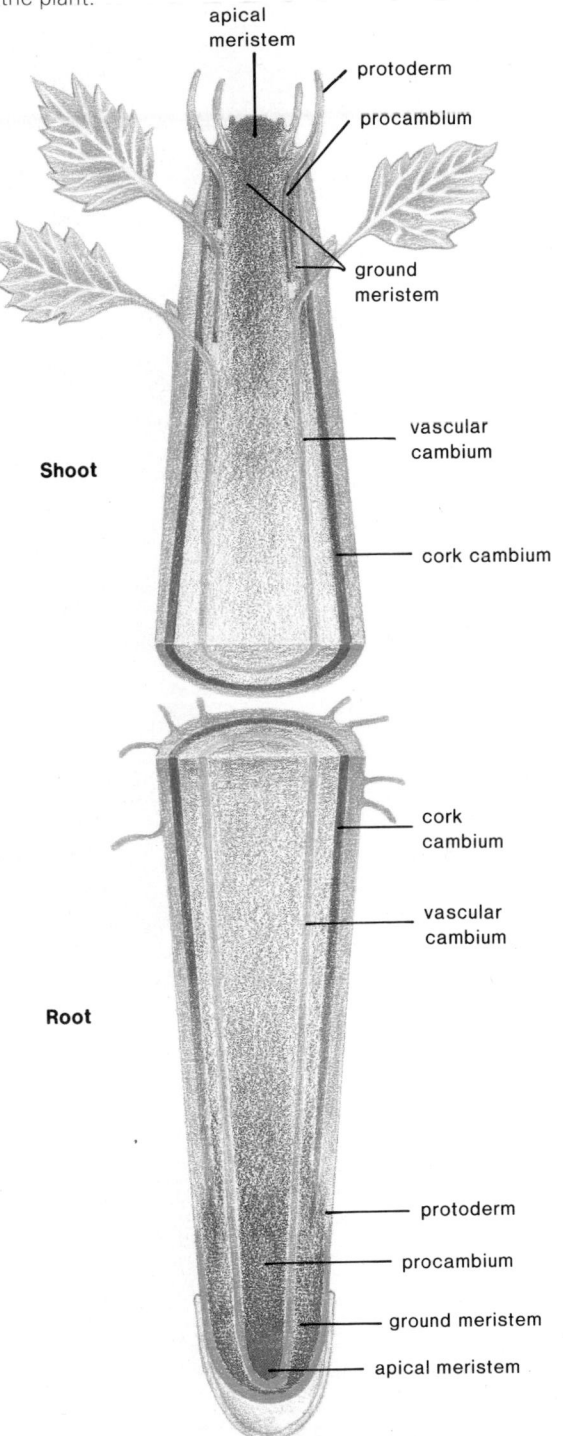

Shoot — apical meristem, protoderm, procambium, ground meristem, vascular cambium, cork cambium

Root — cork cambium, vascular cambium, protoderm, procambium, ground meristem, apical meristem

Root System

The **root** system anchors a plant in the soil, absorbs water and minerals from the soil, and stores the products of photosynthesis received from the leaves. The entire root system of a plant is quite extensive. A single 4-month-old rye plant has a root system that is over 600 km in length. There are literally millions of branched roots, and the plant has about 14×10^9 root hairs. The total absorptive surface area of all the roots and root hairs is almost 640 sq m.

Growth of Dicot Roots

Figure 29.8a, a longitudinal section of a dicot root, reveals zones where cells are in various stages of differentiation as *primary growth* occurs. The root apical meristem is in the *zone of cell division*. Here cells are continuously added to the root cap below and the zone of elongation above. The root cap is a protective cover for the root tip. The cells in the root cap have to be replaced constantly, because they are ground off as the root pushes through rough soil particles. In the *zone of elongation*, the cells get longer as they become specialized. In the *zone of maturation*, the cells are mature and fully differentiated. This zone is recognizable even in a whole root because root hairs are borne by many of the epidermal cells.

Figure 29.8

Dicot root tip. **a.** Root tip is divided into 4 zones, best seen in a longitudinal section such as this. **b.** Vascular cylinder of a dicot root contains the vascular tissue. Xylem is typically star shaped, and phloem lies between the points of the star. Magnification, X300. **c.**

Casparian strip. Because of the Casparian strip (waxy substance), water and minerals must pass through the cytoplasm of endodermal cells. In this way, endodermal cells regulate the passage of minerals into the vascular cylinder.

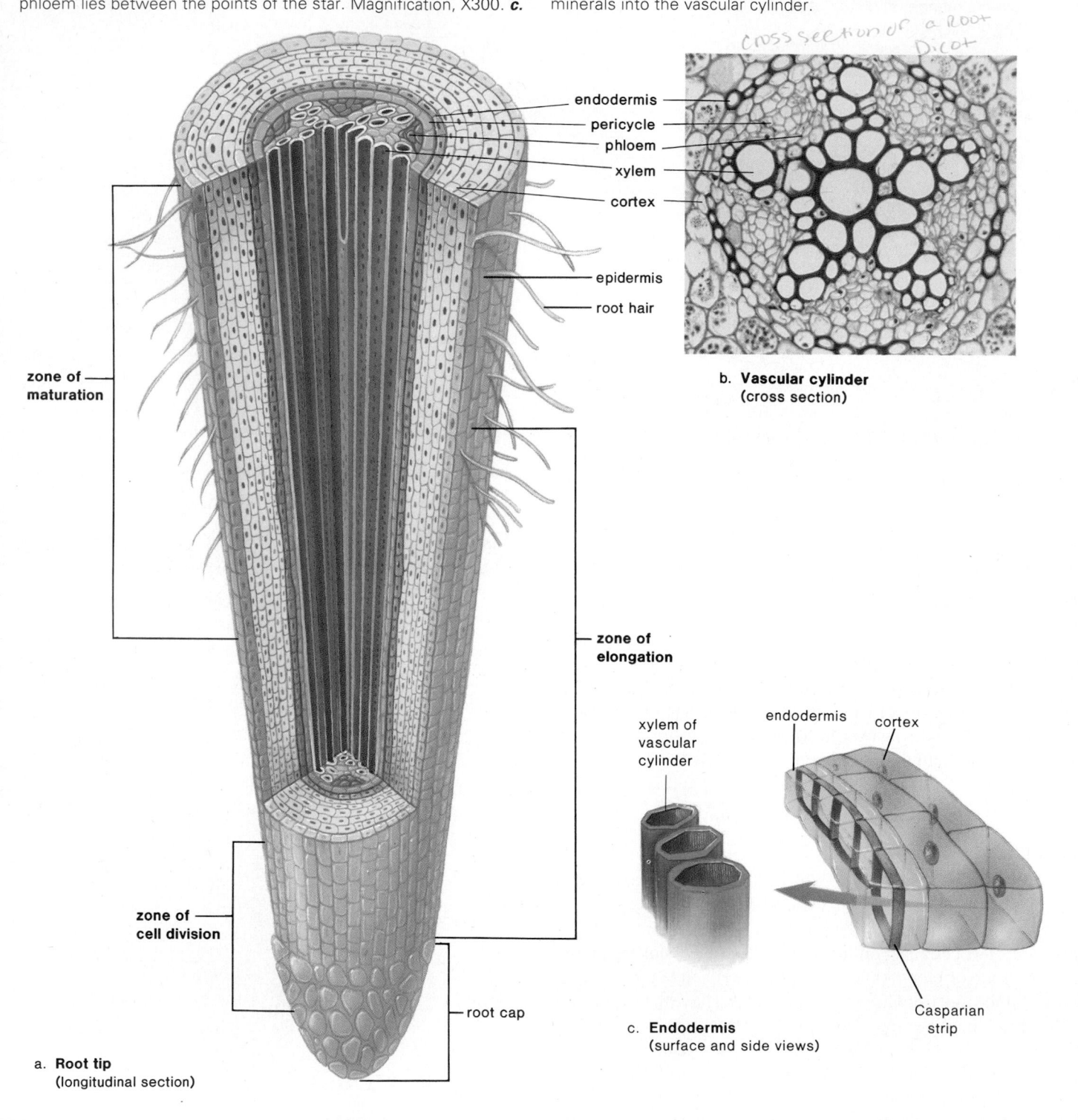

cross section of a Root Dicot

endodermis
pericycle
phloem
xylem
cortex

epidermis
root hair

b. Vascular cylinder
(cross section)

zone of maturation

zone of elongation

zone of cell division

root cap

a. Root tip
(longitudinal section)

xylem of vascular cylinder

endodermis   cortex

Casparian strip

c. Endodermis
(surface and side views)

Plant Structure and Function

Dicot Root Tissues

Figure 29.8*a* also shows a cross section of a root at the region of maturation. These specialized tissues are identifiable:

Epidermis The epidermis, which forms the outer layer of the root, consists of only a single layer of cells. The majority of epidermal cells are thin-walled and rectangular, but in the zone of maturation, many epidermal cells have root hairs. These project as far as 5 mm–8 mm into the soil particles.

Cortex Moving inward, next to the epidermis, large, thin-walled parenchyma cells that make up the *cortex* of the root are found. These irregularly shaped cells are loosely packed, and it is possible for water and minerals to move through the cortex without entering the cells. The cells contain starch granules, and the cortex functions in food storage.

Endodermis The **endodermis** is a single layer of rectangular endodermal cells that forms a boundary between the cortex and the inner vascular cylinder. The endodermal cells fit snugly together and are bordered on 4 sides (but not the 2 sides that contact the cortex and the vascular cylinder) by a strip of waxy material known as the **Casparian strip** (fig. 29.8*c*). This strip does not permit water and mineral ions to pass between adjacent cell walls. Therefore, the only access to the vascular cylinder is through the endodermal cells themselves, as shown by the arrow in figure 29.8*c*. It is said that the endodermis regulates the entrance of minerals into the vascular cylinder.

Vascular Tissue The **pericycle**, the first layer of cells within the *vascular cylinder*, has retained its capacity to divide and can start the development of branch or secondary roots (fig. 29.9). The main portion of the vascular cylinder, though, contains vascular tissue. The xylem appears star shaped in dicots because several arms of tissue radiate from a common center (fig. 29.8*b*). The phloem is found in separate regions between the arms of the xylem.

Growth of Monocot Roots

Monocot roots often have pith, which is centrally located ground tissue. In a monocot root, pith is surrounded by a ring of alternating xylem and phloem bundles (fig. 29.10). They also have pericycle, endodermis, cortex, and epidermis. Typically, monocot roots do not undergo secondary growth.

The root system of a plant absorbs water and minerals that cross the epidermis and cortex before entering the endodermis, the tissue that regulates the entrance of molecules into the vascular cylinder.

Root Diversity

In some plants, notably dicots, the first or primary root grows straight down and remains the dominant root of the plant. This so-called *taproot* is often fleshy and stores food (fig. 29.11*a*). Carrots, beets, turnips, and radishes have taproots that we consume as vegetables.

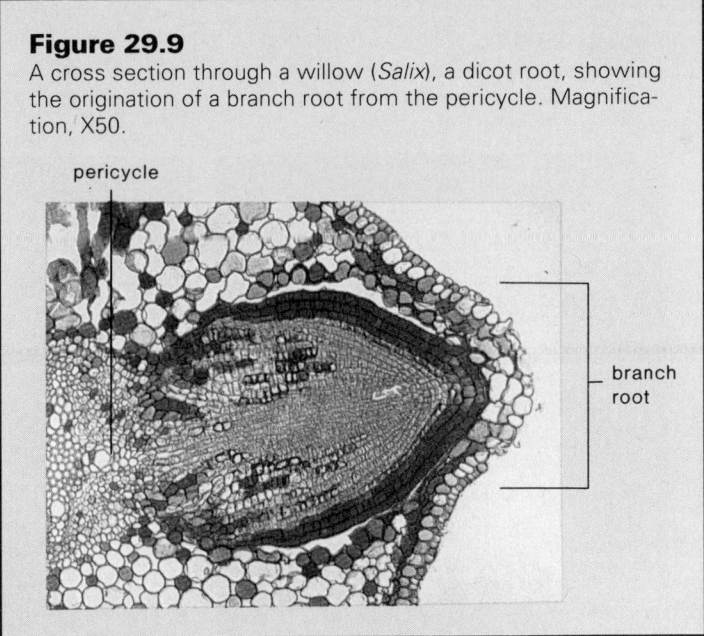

Figure 29.9
A cross section through a willow (*Salix*), a dicot root, showing the origination of a branch root from the pericycle. Magnification, X50.

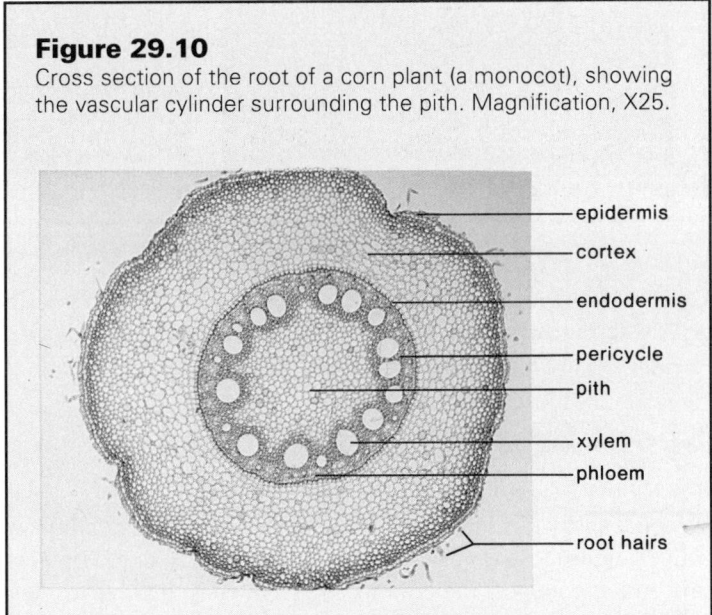

Figure 29.10
Cross section of the root of a corn plant (a monocot), showing the vascular cylinder surrounding the pith. Magnification, X25.

In other plants, notably monocots, a number of slender roots develop; there is no single, main root. These slender roots and their lateral branches make up a *fibrous root* system (fig. 29.11*b*). Most everyone has observed the fibrous root systems of grasses and has noted how these roots can hold the soil.

Sometimes a root system develops from an underground stem or from the base of an aboveground stem. These are *adventitious roots*, whose main function may be to help anchor the plant. If so, they are called *prop roots*. Mangrove plants have large prop

Figure 29.11

Three types of roots. *a.* A taproot may have secondary roots in addition to a main root. *b.* A fibrous root has many secondary roots with no main root. *c.* Adventitious roots, such as prop roots, extend from the stem.

a.   b.   c.

roots that spread away from the plant to help anchor it in marshy soil, where mangroves are typically found. Corn plants also develop prop roots for support (fig. 29.11*c*).

Shoot System

The shoot system of a plant includes the stem and the leaves. First, we will consider the growth of the stem, and in the next section, we will examine the leaf. A **stem** supports **leaves** (leaf, sing.), flowers, and fruits; conducts material to and from roots and leaves; and helps store water and the products of photosynthesis.

Even highly modified stems can be recognized by the presence of nodes, internodes, buds, and leaves (fig. 29.1). A *node* is the point on a stem at which leaves or buds are attached, and an *internode* is the segment of a stem between nodes.

Primary Growth of Stems

The terminal bud at the tip of each growing stem contains shoot apical meristem (fig. 29.12). Whereas root-tip meristem is protected by a root cap, shoot-tip meristem is protected by newly formed leaves within a bud. Shoot-tip meristem produces cells that become the stem and the leaves. At first, the nodes are very close together, and then internode growth occurs so that the nodes are

Figure 29.12
Terminal bud anatomy. The shoot apical meristem is surrounded by 2 sets of leaves, a younger pair and an older pair. Axillary buds are seen at the axes of the older pair of leaves. Axillary buds can give rise to side branches when they are active. Magnification, X100.

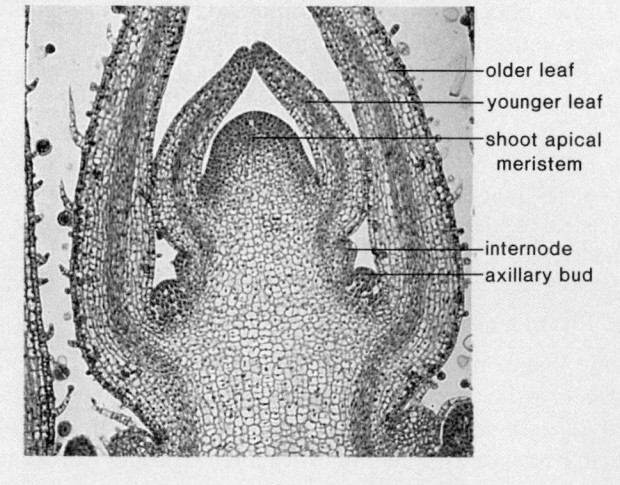

older leaf
younger leaf
shoot apical meristem
internode
axillary bud

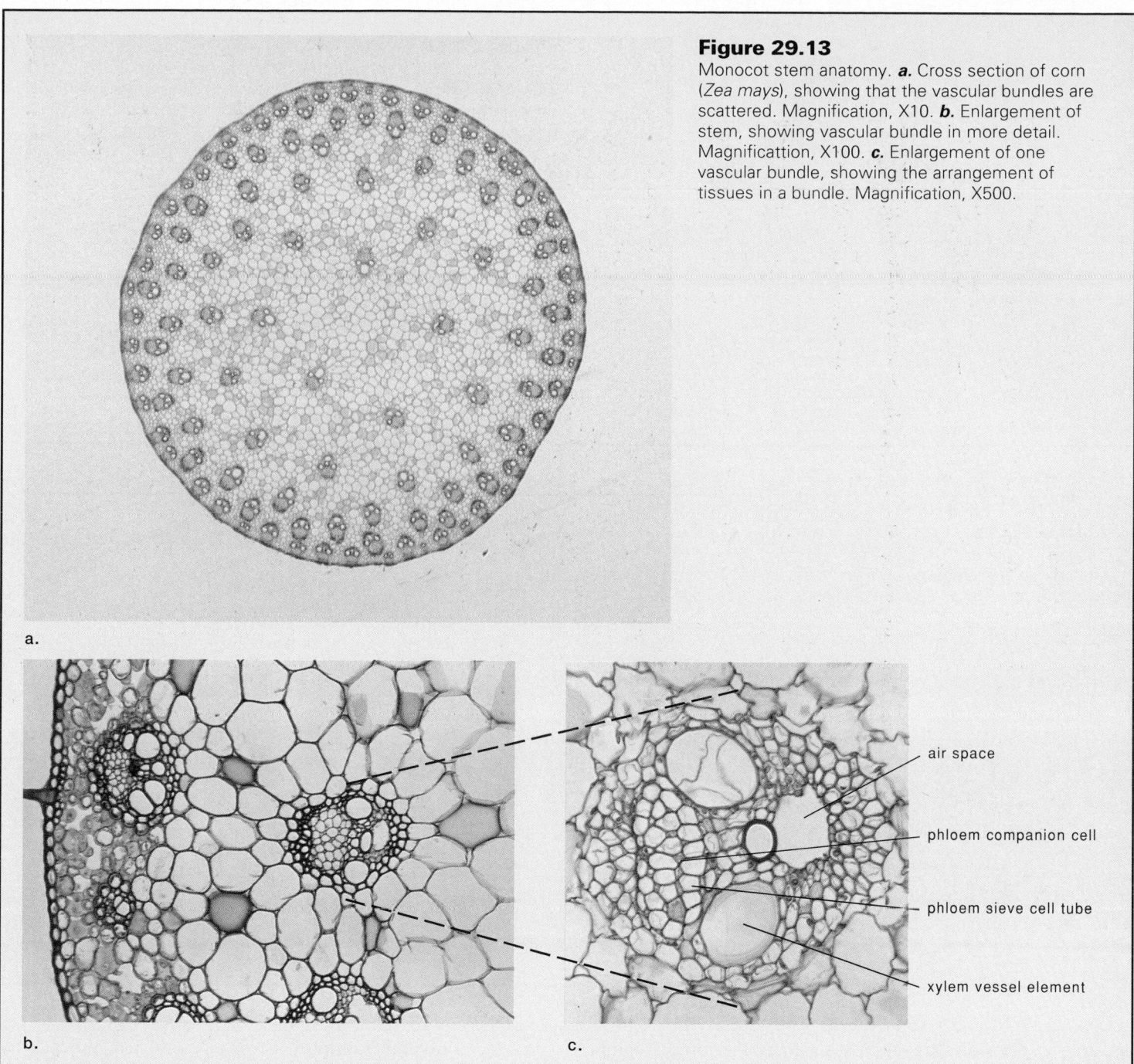

Figure 29.13
Monocot stem anatomy. *a.* Cross section of corn (*Zea mays*), showing that the vascular bundles are scattered. Magnification, X10. *b.* Enlargement of stem, showing vascular bundle in more detail. Magnificattion, X100. *c.* Enlargement of one vascular bundle, showing the arrangement of tissues in a bundle. Magnification, X500.

a.

b.

c.

air space

phloem companion cell

phloem sieve cell tube

xylem vessel element

distant from one another. This pattern of growth complicates the growth of stems, and it is not possible to divide the entire stem into zones of cell division, elongation, and maturation as with the root.

 Axillary buds, which are usually dormant but may develop into branch shoots, are seen at the axes of mature leaves. Inactive buds are covered by protective bud scales. In the temperate zone, the terminal bud stops growing in the winter and is then protected by bud scales also. In the spring, when growth resumes, these scales fall off and leave a scar. You can tell the age of a stem by counting these bud scale scars.

Herbaceous Stems

Mature nonwoody stems, called *herbaceous stems,* exhibit only primary growth. The outermost tissue of herbaceous stems is the epidermis, which is covered by a waxy cuticle to prevent water loss. These stems have distinctive *vascular bundles,* where xylem and phloem are found. In each bundle, xylem is typically found toward the inside of the stem and phloem is found toward the outside.

 In the monocot stem, the vascular bundles are scattered throughout the stem, and there is no well-defined cortex or well-defined pith (fig. 29.13). In the dicot herbaceous stem, the bundles

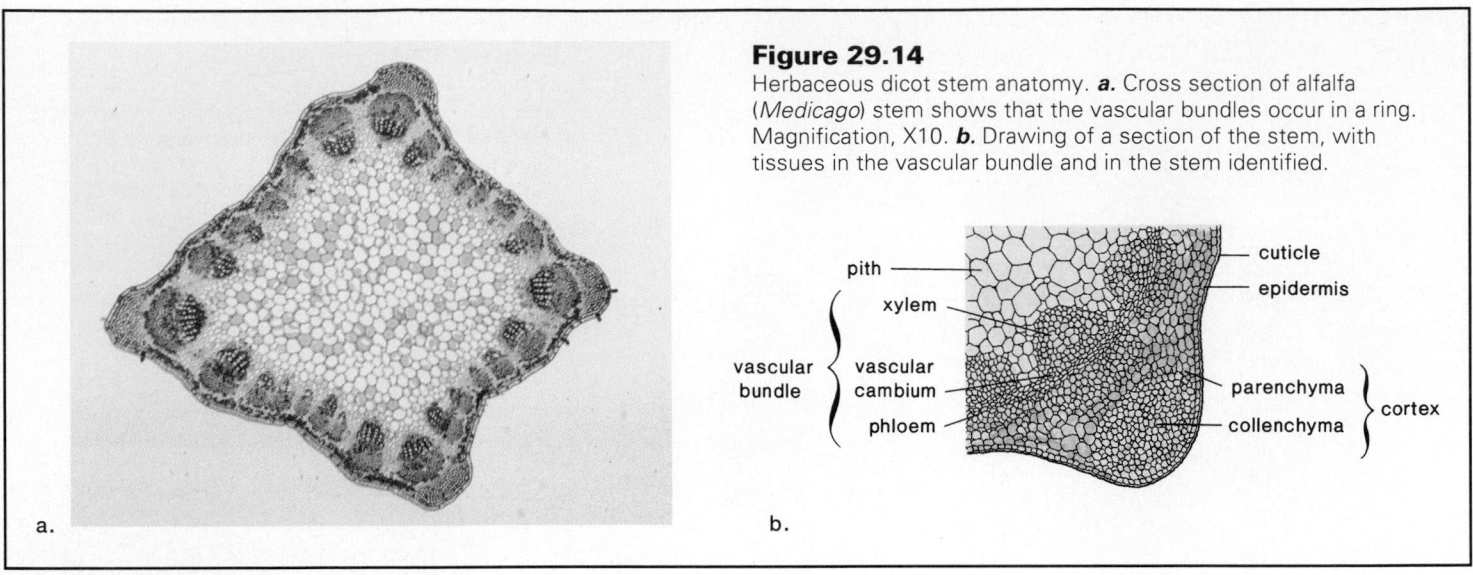

Figure 29.14
Herbaceous dicot stem anatomy. **a.** Cross section of alfalfa (*Medicago*) stem shows that the vascular bundles occur in a ring. Magnification, X10. **b.** Drawing of a section of the stem, with tissues in the vascular bundle and in the stem identified.

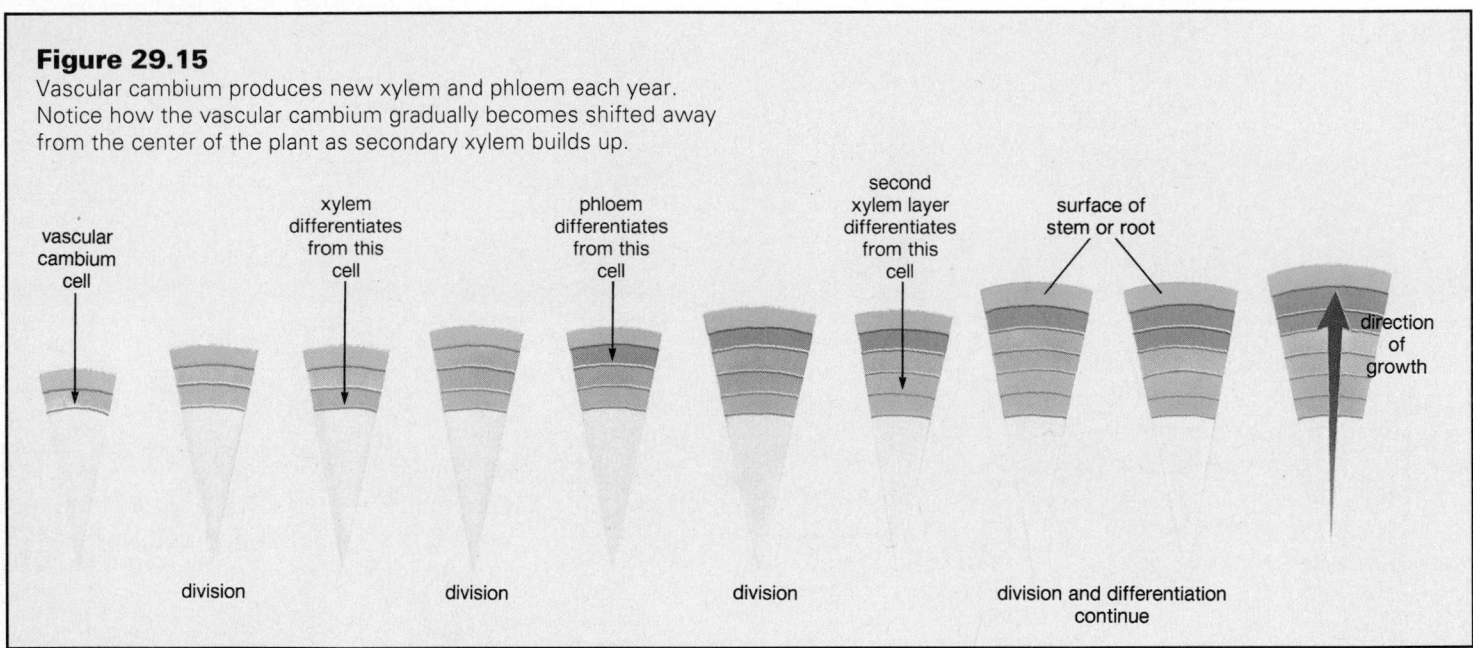

Figure 29.15
Vascular cambium produces new xylem and phloem each year. Notice how the vascular cambium gradually becomes shifted away from the center of the plant as secondary xylem builds up.

are arranged in a distinct ring that separates the cortex from the central pith (fig. 29.14). The cortex is sometimes green and carries on photosynthesis, and the pith may function as a storage site.

Secondary Growth of Stems

Secondary growth of stems is seen primarily in woody plants, such as trees, that live for many years. Almost all trees, other than conifers, are dicots. Primary growth in woody plants occurs for a short distance behind the apical meristem, at which point secondary growth begins. Recall that secondary growth occurs because of meristematic tissues called vascular cambium and cork cambium.

Vascular cambium begins as meristematic cells between the xylem and phloem of each vascular bundle. Then these cells join to form a ring of meristematic tissue. The division of this meristematic tissue takes place mostly in a plane parallel to the surface of the plant. The secondary tissues produced by the vascular cambium, called secondary xylem and secondary phloem, therefore add to the girth of the stem instead of to its length (fig. 29.15).

Cork cambium is located beneath the epidermis. When it begins to divide, it produces tissue that disrupts the epidermis and replaces it with cork cells. Cork cells are impregnated with suberin, a fatty substance that makes them waterproof. Dead cork only allows gas exchange in pockets of loosely arranged cells called *lenticels,* which are not impregnated with suberin.

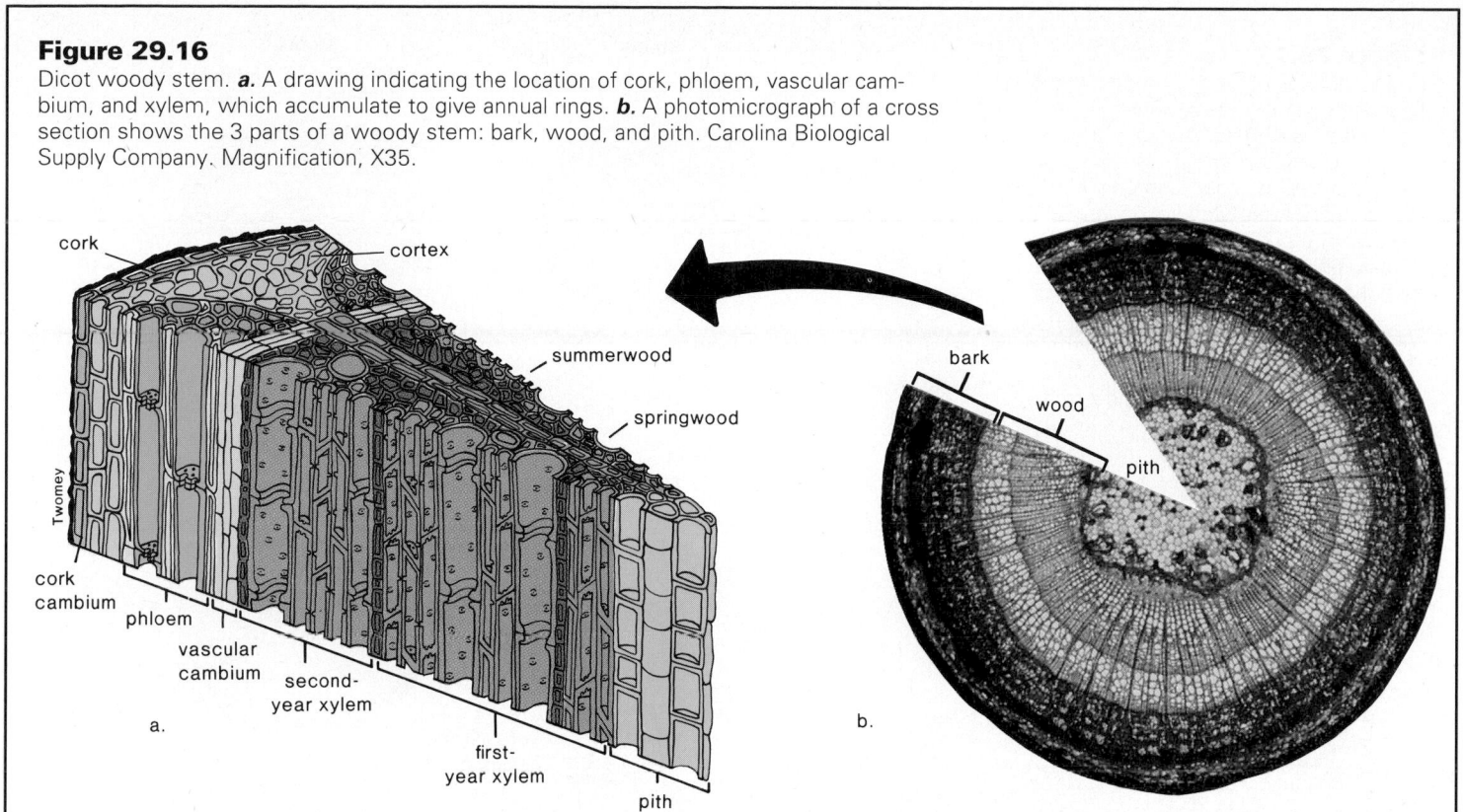

Figure 29.16

Dicot woody stem. **a.** A drawing indicating the location of cork, phloem, vascular cambium, and xylem, which accumulate to give annual rings. **b.** A photomicrograph of a cross section shows the 3 parts of a woody stem: bark, wood, and pith. Carolina Biological Supply Company. Magnification, X35.

Labels in figure a: cork, cortex, summerwood, springwood, Twomey, cork cambium, phloem, vascular cambium, second-year xylem, first-year xylem, pith

Labels in figure b: bark, wood, pith

Woody Stems

As a result of secondary growth, a woody stem has an entirely different type of organization than that of a dicot herbaceous stem (fig. 29.16). After secondary growth has continued for a time, it is no longer possible to make out individual vascular bundles. Instead, a woody stem has 3 distinct areas: the bark, the wood, and the pith. The bark contains cork, cork cambium, and phloem. Although secondary phloem is produced each year by vascular cambium, phloem does not build up for many seasons. Before it disappears it produces new cork cambium. Because phloem is in the bark, even partial removal can seriously damage a tree.

The wood of trees that grow in seasonal climates contains rings of secondary xylem called growth rings because there is one for each season of growth. In temperate regions with one growing season per year, these are called annual rings. It is easy to tell where one ring begins and another ends. In the spring, when moisture is plentiful, the xylem cells are much larger than later in the summer, when moisture is scarcer. In large trees, only the more recently formed layer of xylem, the sapwood, functions in water transport. The older, inner part, called the heartwood, becomes plugged with deposits, such as resins, gums, and other substances. Heartwood may help support a tree, although some trees stand erect and live for many years after the heartwood has rotted away.

Like stems, the roots of woody plants also show secondary growth. Figure 29.17 illustrates that the secondary growth of roots arises and occurs in about the same manner as that of the stem.

Types and Uses of Stems

Stem modifications are illustrated in figure 29.18. Aboveground horizontal stems, called **stolons** or runners, produce new plants where nodes touch the ground. The strawberry plant is a common example of this type of propagation.

Underground horizontal stems, **rhizomes,** may be long and thin, as in sod-forming grasses, or thick and fleshy, as in iris. Rhizomes survive the winter and contribute to asexual reproduction because each node bears a bud. Some rhizomes have enlarged portions called tubers, which function in food storage. For example, potatoes are tubers. The eyes of potatoes are buds that mark the nodes.

Corms are bulbous underground stems that lie dormant during the winter, just as rhizomes do. They also produce new plants the next growing season. Gladiolus corms are referred to as bulbs by laypersons, but the botanist reserves the term *bulb* for a structure composed of modified leaves.

Aboveground, vertical stems can also be modified. For example, cacti have succulent stems modified for water storage. The tendrils of grape plants that twine around a support structure are modified stems.

Humans use stems for many purposes. The stem of the sugarcane plant is a primary source of table sugar. The spice cinnamon and the drug quinine are derived from the bark of 2 different plants. The softwood of gymnosperms is used for numerous purposes, including lumber for construction and the

Figure 29.17

Cross sections of woody roots and stems. Both stems and roots experience secondary growth. *a.* Stems. A young stem in which the vascular cambium is just forming a ring (*above*); an older stem in which there is secondary xylem and secondary phloem (*below*). The primary xylem remains but the primary phloem has disappeared. *b.* Roots. A young dicot root in which the vascular cambium is located between the primary xylem and phloem (*above*); an older root in which there is secondary xylem and secondary phloem (*below*). The primary xylem remains but there are only remnants of the primary phloem.

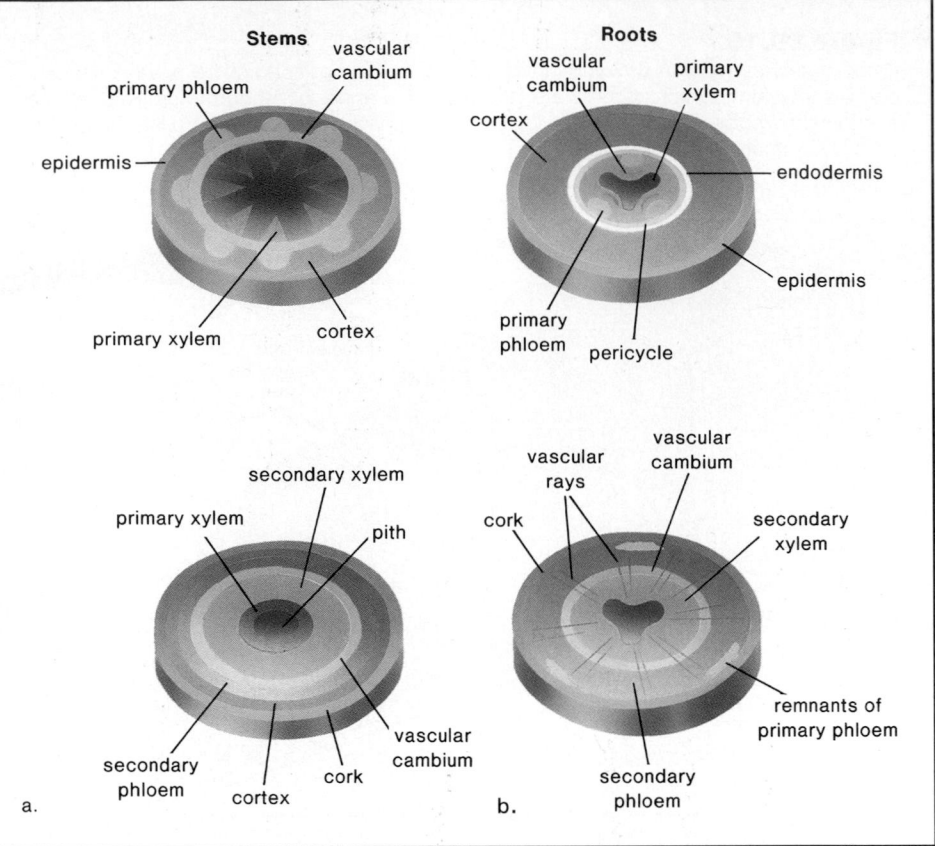

Figure 29.18

Stem modifications. *a.* A strawberry plant has aboveground, horizontal stems called stolons. Every other node produces a new shoot system. *b.* The underground horizontal stem of an iris is a fleshy rhizome. *c.* The underground stem of a potato plant has enlargements called tubers. We call the tubers potatoes. *d.* Corms are stems covered by papery leaves.

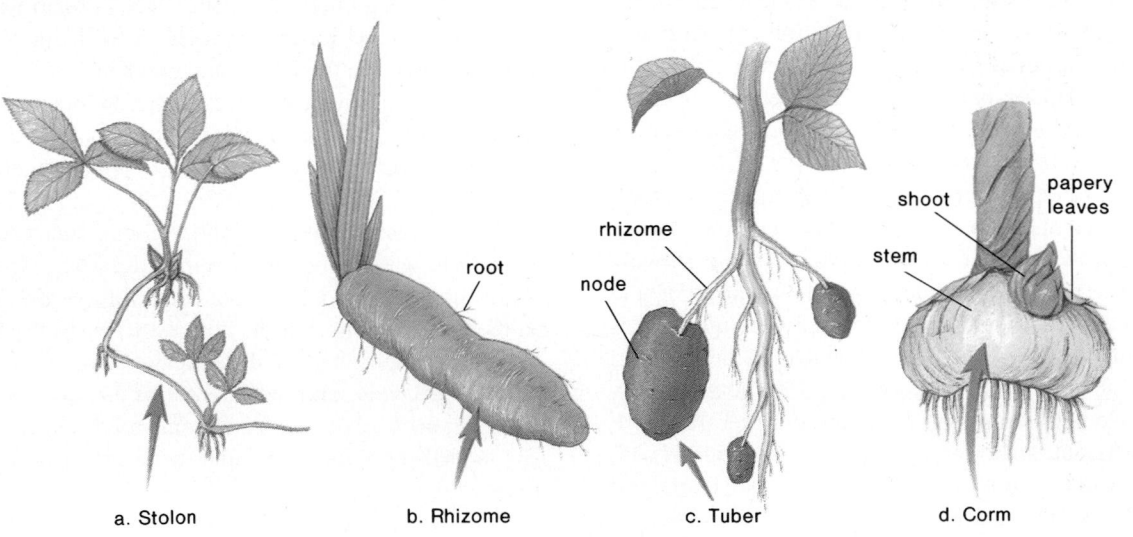

a. Stolon   b. Rhizome   c. Tuber   d. Corm

How Do You Tell What a Tree Has Been Through?

ach year a tree adds a new growth ring to its trunk. In the spring, growth is fast and the wood is light in color; in the summer, growth is slower and the wood is dark in color. Counting the dark rings tells you the age of the tree, but studying tree rings to determine their shape, thickness, color, and evenness lets you know what a tree has been through (fig. 29.A). Dates of past forest fires can be pinpointed, a year in which caterpillars stripped trees of foliage can be determined, and years in which smog damage occurred may be revealed by the rings.

A great deal can also be learned from tree rings about past climatic conditions. For many years, A. E. Douglass, Harold C. Fritts, and others associated with the Laboratory of Tree-Ring Research at the University of Arizona studied ring widths in trees from arid sites. They discovered that very significant statistical relationships exist between growth of trees and climatic data. With the aid of computers and statistical analyses, they developed techniques that take into account subtle climatic and other environmental variables. They were able to reconstruct relatively precise histories of climatic fluctuations and changes dating back hundreds of years. Today, extrapolations of tree ring data are made to try to determine climates dating back to prehistoric times.

Trees do not have to be cut down to examine the rings. A simple instrument called an *increment borer,* which consists primarily of a rigid metal cylinder, is driven into the trunk of a tree, and a core of wood is removed. The hole is then plugged to prevent disease organisms from entering the tree, and the rings revealed by the core are then examined and analyzed.

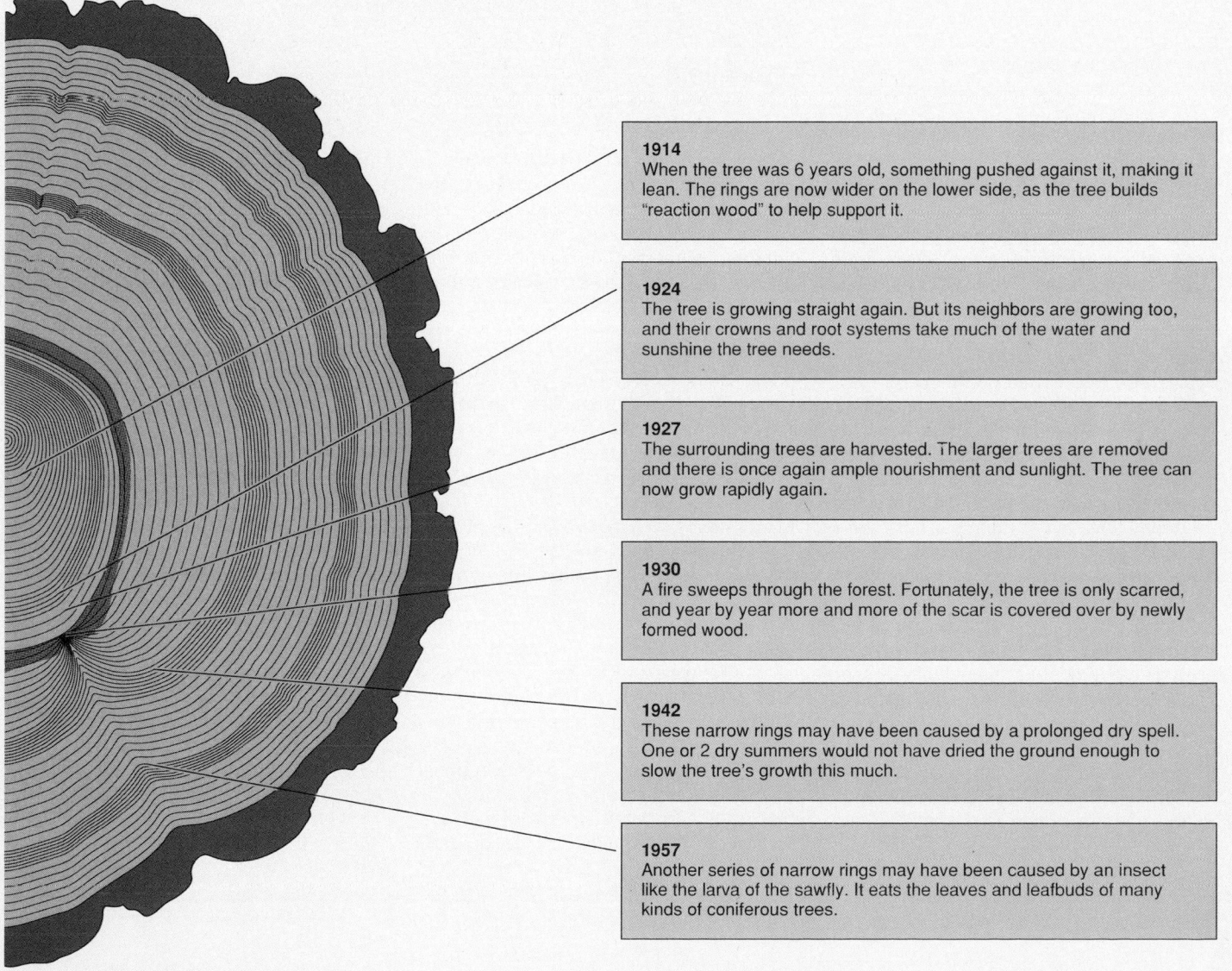

1914
When the tree was 6 years old, something pushed against it, making it lean. The rings are now wider on the lower side, as the tree builds "reaction wood" to help support it.

1924
The tree is growing straight again. But its neighbors are growing too, and their crowns and root systems take much of the water and sunshine the tree needs.

1927
The surrounding trees are harvested. The larger trees are removed and there is once again ample nourishment and sunlight. The tree can now grow rapidly again.

1930
A fire sweeps through the forest. Fortunately, the tree is only scarred, and year by year more and more of the scar is covered over by newly formed wood.

1942
These narrow rings may have been caused by a prolonged dry spell. One or 2 dry summers would not have dried the ground enough to slow the tree's growth this much.

1957
Another series of narrow rings may have been caused by an insect like the larva of the sawfly. It eats the leaves and leafbuds of many kinds of coniferous trees.

Figure 29.A
Life history of a tree planted in 1908 and cut down in 1970. The anatomy of the tree rings correlate to the events listed by date. Source: Data from St. Regis Paper Company, New York, NY., 1966.

Paper is Secondary Xylem

Over 230 kg per year, or over ½ kg per day, is a close approximation of the paper used annually by each person in America. Figured on a per capita basis, we use paper products at nearly twice the rate of the next highest consumer, Canada.

Immediately, of course, we think of paper as a means of communication; historically, that was nearly its exclusive use for over a thousand years. Paper appears to have been invented in China around A.D. 105. Paper remained an exclusively Chinese product for about 500 years until it appeared in Japan. It wasn't until another 500 years had passed that it found its way westward into Egypt, and finally Europe in the twelfth century. Its use for other than communication purposes was virtually nil until the mid-1800s when paper bags and boxes were invented. Around the turn of this century, we find milk cartons, cups, plates, food wrappers, and all kinds of things being made of paper.

What is this wonderful stuff made of? Plants, of course! Actually, the fibers used in manufacturing paper come from rags, straw, grasses, old newspapers, and a number of other sources. But the ultimate source is always plant material, and most of it, better than 90%, is wood. Cellulose, the basic material for paper, is a material in the cell walls of secondary xylem, which is wood. You may have the feeling that paper manufacturing is very complicated. Actually the basic process is quite simple and hasn't changed much in its nearly 2,000-year history. There are 4 steps: (1) make the pulp, (2) make the stock (optional), (3) make wet paper, and (4) dry it out.

Pulp is simply a suspension of pulverized wood cells. In the trade these cells are called fibers, but as plant biologists we know that most of them are really tracheids and vessels. This suspension of fibers is made in either of 2 ways. The logs with the bark removed may simply be ground against a grinding stone in the presence of water. Paper made from this kind of pulp is relatively inexpensive but weak and generally

Figure 29.B
Paper machine take-up spool at Potlatch Corporation, Lewiston, Idaho.

of poor quality. Newsprint, for example, is primarily of this type. . . . Alternatively, the logs may be fragmented into chips and then pulverized, using hot sulfites, soda, or sulfates, depending on the various processes. These chemicals remove impurities (especially lignin) from the chips, leaving only the cellulose; consequently the yield is only about 40%–50% of the original wood, but the product is of much finer quality. The pulp is then bleached so that the paper will be white instead of wood colored. It can be used directly to make paper, but usually it passes through an intermediate step during which it is made into stock.

Pulp is refined into stock simply by further grinding. The purpose is to roughen and fray the extracted fiber fragments and to make them more uniform in size. As a result of this treatment, the fibers adhere to one another more readily, and the strength and quality of the paper are improved. Also dyes, strengtheners, and other additives may be introduced at this point.

The third step is to make wet paper out of the pulp or stock. This is accomplished by applying the pulp or stock to a wire-screen filter, draining off the liquid and

thereby leaving only the wood fibers, which form a thin, matted network on the surface of the screen. In times past this was done by simply immersing a screen in a vat that contains the stock, and then drawing the screen upward, causing the wood fibers to collect on it. This was done by hand, and one man could produce perhaps 750 sheets per day. Today you'll find manual operation replaced by a revolving wire-screen belt up to 30 feet wide which moves at rates up to 88 feet per second or about 60 miles per hour. This belt picks up the stock and drains off the water rapidly by using a vacuum on the underside. . . . At such rates one machine can turn out in less than 2 seconds the equivalent of one man working all day using the techniques of the early 1800s.

The final step is to press the paper and dry it. This is accomplished by squeezing the paper between heated rotating drums. Finally, the paper is collected in giant rolls [fig. 29.B].

"Paper Is Secondary Xylem" from *Botany: A Human Concern*, Second Edition, by David L. Rayle and Hale L. Wedberg, copyright © 1980 by Saunders College Publishing, reprinted by permission of the publisher.

Plant Structure and Function

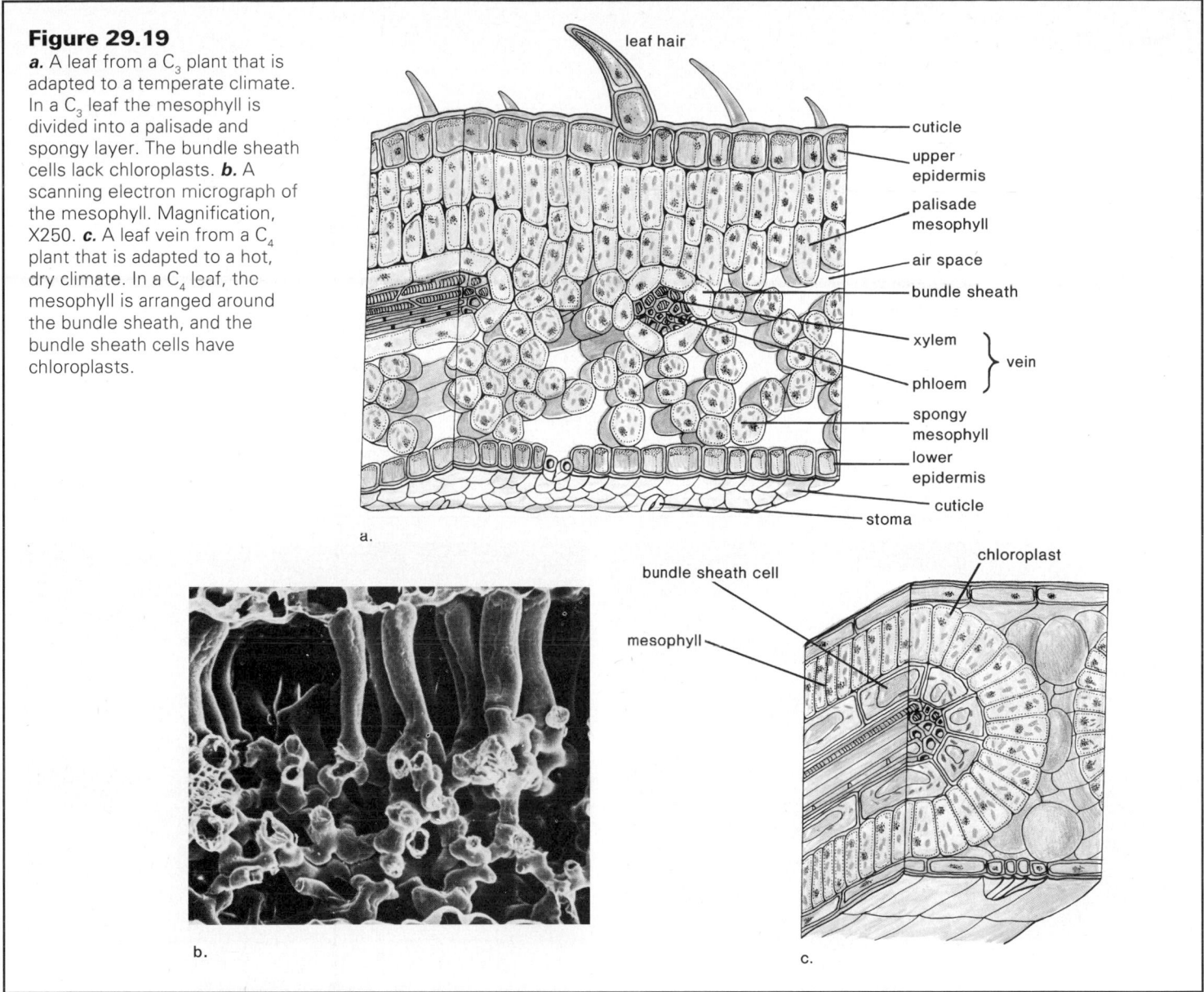

Figure 29.19
a. A leaf from a C_3 plant that is adapted to a temperate climate. In a C_3 leaf the mesophyll is divided into a palisade and spongy layer. The bundle sheath cells lack chloroplasts. **b.** A scanning electron micrograph of the mesophyll. Magnification, X250. **c.** A leaf vein from a C_4 plant that is adapted to a hot, dry climate. In a C_4 leaf, the mesophyll is arranged around the bundle sheath, and the bundle sheath cells have chloroplasts.

production of paper (see this chapter's reading) and rayon. Hardwood from angiosperms is used for flooring, furniture, and interior finishes.

> Within the shoot system, the stem transports water and nutrients between the leaves and the roots. Stem anatomy differs according to whether the plant is a monocot or dicot and whether it is herbaceous or woody. There are also various types of stem modifications.

Leaves

Leaves are the organs of photosynthesis in vascular plants. A leaf usually consists of a flattened *blade* and a *petiole*, which connects the blade to the stem. The blade may be single or composed of several leaflets. Externally, it is possible to see the pattern of the *leaf veins*, which contain vascular tissue. Leaf veins have a net pattern in dicot leaves and a parallel pattern in monocot leaves (fig. 29.6).

Figure 29.19*a* shows a cross section of a typical dicot leaf of a temperate zone plant. At the top and bottom is a layer of epidermal tissue that often bears protective hairs and/or glands that produce irritating substances. These features may prevent the leaf from being eaten by insects. Regardless of whether these additions are present, the epidermis always has an outer, waxy cuticle that keeps the leaf from drying out. Unfortunately, the cuticle also prevents gas exchange because it is not gas permeable. The epidermis, however, particularly the lower epidermis, contains openings called stomata, which allow gases to move into and out of the leaf. Epidermal cells do not ordinarily contain chloroplasts;

Figure 29.20

Classification of leaves. **a.** The cottonwood tree has a simple leaf. **b.** The shagbark history has a pinnately compound leaf. **c.** The honey locust has a twice pinnately compound leaf. **d.** The buckeye has a palmately compound leaf.

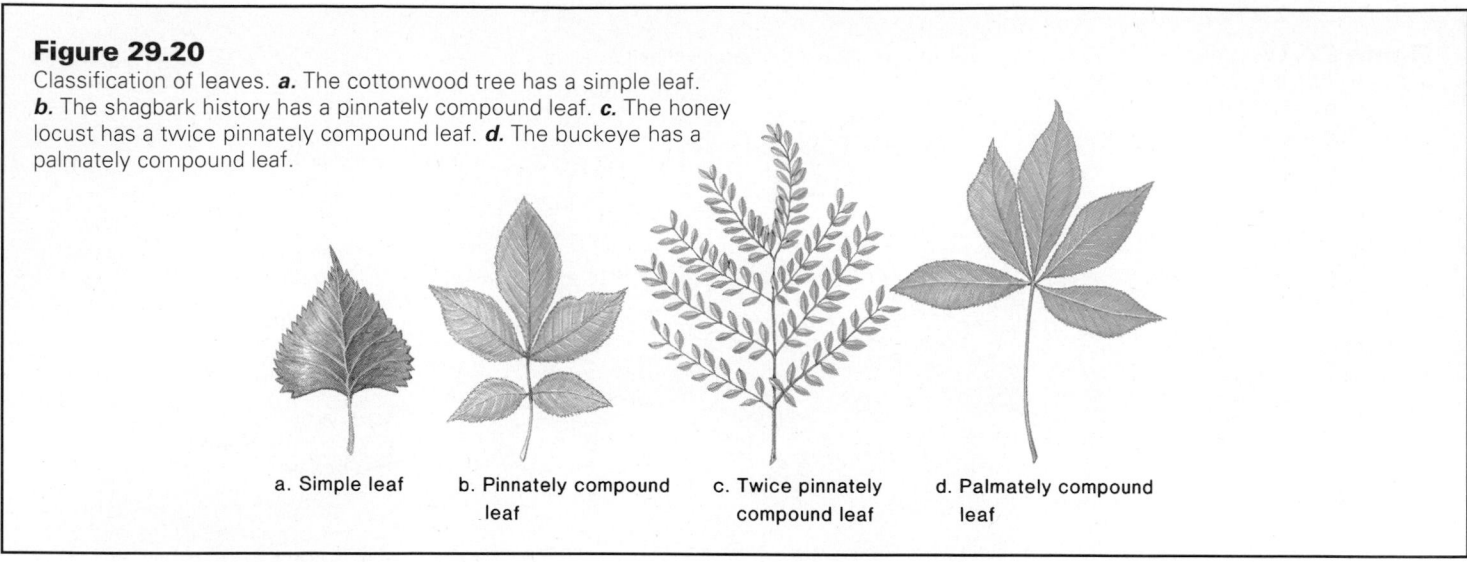

a. Simple leaf

b. Pinnately compound leaf

c. Twice pinnately compound leaf

d. Palmately compound leaf

a.

b.

Figure 29.21

Leaf modifications. **a.** The spines of a cactus plant are leaves modified to protect the fleshy stem from animal consumption **b.** The tendrils of a cucumber are leaves modified to attach the plant to a physical support. **c.** The leaves of the Venus's-flytrap are modified to serve as a trap for insect prey. When triggered by an insect, the leaf snaps shut. Once shut, the leaf secretes digestive juices, which break down the soft parts of the insect's body.

c.

Plant Structure and Function

however, the guard cells surrounding the stomata do contain chloroplasts. These guard cells regulate the opening and closing of the stomata, as discussed on page 498.

The body of a leaf is composed of **mesophyll** tissue, which has 2 layers of cells: the **palisade layer** containing elongated cells, and the **spongy layer** containing irregular cells bounded by air spaces. The parenchyma cells of these layers have many chloroplasts and carry on most of the photosynthesis for the plant. The loosely packed arrangement of the cells in the spongy layer increases the amount of surface area for gas exchange.

The cells of the mesophyll are protected from drying out by a thin film of water that is constantly evaporating. Loss of water by evaporation at the leaves is called transpiration. At least 90% of water taken up by the roots is eventually lost by transpiration. This means that the total amount of water lost by a plant over a long period of time is surprisingly large. For example, a single *Zea mays* (corn) plant loses somewhere between 135 l and 200 l of water through transpiration during a growing season. If this water loss is multiplied by the number of corn plants in a heavily planted cornfield, one can understand why farming requires so much water. The amount of water required by different agricultural crops to produce a given quantity of food can also be compared. Millet requires about 225 kg of water for every kilogram of food produced, wheat requires about 500 kg of water per kilogram of food, and potatoes require about 800 kg of water per kilogram of food.

Leaf Veins

A cross section of a leaf shows that leaf veins consist of a strand of xylem and a strand of phloem surrounded by a **bundle sheath.** The bundle sheath differs in C_3 and C_4 plants. As discussed on page 131, these terms refer to the number of carbon atoms in the first detected molecule after the start of photosynthesis. The bundle sheath cells in the leaves of C_4 plants characteristically have chloroplasts, and some of the mesophyll cells are often closely packed around the bundle sheath in a radial fashion (fig. 29.19c). The bundle sheath cells in C_3 plants do not contain chloroplasts, and mesophyll tissue is divided into the palisade and spongy layers (fig. 29.19a).

Within the shoot system, the leaves carry on photosynthesis. Leaf anatomy differs according to whether the plant is a C_3 or C_4 plant. There are also various types of leaf modifications.

Types and Uses of Leaves

In a leaf, the blade can be simple and undivided or it can be compound, with 2 or more separate leaflets making up the blade (fig. 29.20). The leaflets of a compound leaf can be arranged along the length of a central stalk (pinnately compound) or attached to the end of the petiole (palmately compound). Aside from these blade patterns, there are also various patterns of vascular arrangement in leaves and innumerable combinations of overall leaf shape, margin, and base modifications. Leaves are adapted to environmental conditions. Plants that usually live in the shade tend to have broad, wide leaves, and those that live where it is dry tend to have reduced leaves with sunken stomata. The leaves of a cactus are the spines attached to the succulent stem (fig. 29.21a). Other succulents, however, have leaves adapted to hold moisture.

Leaves can also be specialized for food storage. An onion *bulb* is made up of leaves surrounding a short stem. In a head of cabbage large leaves overlap one another. The petiole of a leaf can be thick and fleshy, as in celery and rhubarb.

Climbing leaves, such as those of peas and cucumbers, are modified into tendrils that can attach to nearby objects (fig. 29.21b). The leaves of a few plants are specialized for catching insects. The leaves of a sundew have sticky epidermal hairs that trap insects and then secrete digestive enzymes. The Venus's-flytrap has hinged leaves that snap shut and interlock when an insect triggers sensitive hairs (fig. 29.21c). The leaves of a pitcher plant resemble a pitcher and have downward-pointing hairs that lead insects into a pool of digestive enzymes. Insectivorous plants such as these commonly grow in marshy regions, where the supply of soil nitrogen is severely limited. The digested insects provide the plants with a source of organic nitrogen.

Summary

1. The body of a flowering plant is divided into the root system and the shoot system, which contains stems and leaves. Both systems contain 3 types of tissues: dermal, ground, and vascular (table 29.1).
2. Dermal tissue contains the epidermis, which is modified in different organs of the plant. In the roots, epidermal cells bear root hairs; in the leaves the epidermis contains guard cells. Cork replaces epidermis in woody plants.
3. Ground tissue contains parenchyma cells, which are thin-walled and capable of photosynthesis when they contain chloroplasts. If they contain only colorless plastids, they serve as storage cells. Collenchyma cells have thicker walls for flexible support. Sclerenchyma cells are hollow, nonliving, support cells with secondary walls.
4. Vascular tissue consists of xylem and phloem. Xylem contains xylem vessel elements and tracheids, which are elongated and tapered with pitted end walls. Xylem transports water and minerals. Phloem contains sieve-tube cells, each of which has a companion cell. Phloem transports organic nutrients.
5. The primary growth of roots and shoots results in a lengthening of the plant and occurs when apical meristem produces 3 types of primary meristem, one for each of the types of tissues. Secondary growth, which causes a stem or root to increase in diameter, is due to the activity of vascular cambium and cork cambium.
6. A root anchors a plant, absorbs water and minerals, and stores the products of photosynthesis. A root tip shows 3 zones in the region of primary growth: the zone of cell division (apical meristem), the zone of elongation, and the zone of maturation.

Table 29.1
Vegetative Organs and Major Tissues

	Roots	Stems	Leaves
Function	Absorb water and minerals Anchor plant Store materials	Transport water and nutrients Support leaves Help store materials	Carry on photosynthesis
Tissue			
Epidermis*	Root hairs absorb water and minerals	Protect inner tissues	Stomata carry on gas exchange
Cortex†	Store products of photosynthesis and water	Carry on photosynthesis if green	———
Endodermis†	Regulate passage of minerals into vascular cylinder	Regulate passage of minerals into vascular tissue if present	Regulate passage of minerals into vascular tissue if present
Vascular‡	Transport water and nutrients	Transport water and nutrients	Transport water and nutrients
Pith†	Store products of photosynthesis and water	Store products of photosynthesis	———

Mesophyll†
  Spongy layer
  Palisade layer
			Carry on gas exchange and photosynthesis

Plant tissues belong to one of 3 tissue systems.

*Dermal tissue   †Ground tissue   ‡Vascular tissue

7. A cross section of a herbaceous dicot root shows the epidermis (protects), cortex (stores food), endodermis (regulates movement of minerals), and vascular cylinder (vascular tissue). In the vascular cylinder of a dicot, the xylem appears star-shaped and the phloem is found in separate regions, between the arms of the xylem. In contrast, a monocot root has a ring of vascular tissue with alternating bundles of xylem and phloem surrounding a pith.

8. Three types of roots are taproots, fibrous roots, and adventitious roots.

9. Stems support leaves, conduct materials to and from roots and leaves, and help store plant products. Primary growth of a stem is due to the activity of the apical meristem. In cross section, a nonwoody dicot has epidermis, cortex tissue, vascular bundles in a ring, and an inner pith. Monocot stems have scattered vascular bundles, and the cortex and pith are not well defined.

10. Secondary growth of a temperate woody stem is due to vascular cambium, which produces new xylem and phloem every year, and cork cambium, which produces new cork cells when needed. Cork replaces epidermis in woody plants. The cross section of a woody stem shows bark, wood, and pith. The bark contains cork and phloem. Wood contains annual rings of xylem. Roots undergo secondary growth in the same manner as stems.

11. Stems are modified in various ways, such as horizontal aboveground and underground stems. Corms and some tendrils are also modified stems.

12. A leaf, the organ of photosynthesis, has a blade and a petiole. The leaf veins form a net in dicot leaves and are parallel in monocot leaves.

13. A cross section of a leaf shows the epidermis, with stomata mostly below. Mesophyll tissue within the leaf has a palisade and spongy layer in C_3 leaves. In C_4 leaves, some mesophyll cells surround the bundle sheath cells, which contain well-formed chloroplasts.

14. One way to classify leaves is to note whether they are simple or compound. If compound, they may be either pinnately or palmately compound. Leaves are variously modified. The spines of a cactus are leaves. Other succulents have fleshy leaves. An onion is a bulb with fleshy leaves, and the tendrils of peas are leaves. The Venus's-flytrap has leaves that trap and digest insects.

Writing Across the Curriculum

In order to practice writing skills, students should write out the answers to any or all of the study questions and the critical thinking questions. The study questions are sequenced in the same order as the text. Suggested answers to the critical thinking questions are in appendix D.

Study Questions

1. Contrast the root and shoot systems of a plant in as many ways as possible.
2. Contrast an epidermal cell with a cork cell. These cells occur in what type of plant tissue? Tell how epidermis is modified in various organs of a plant.
3. Contrast the structure and function of parenchyma, collenchyma, and sclerenchyma cells. These cells occur in what type of plant tissue?
4. Contrast the structure and function of xylem and phloem. Xylem and phloem occur in what type of plant tissue?
5. List 3 differences between monocots and dicots.
6. Contrast primary growth with secondary growth of stems and roots.
7. Name and state the function of the main tissues within each plant organ.
8. Name and discuss the zones of a root tip. Trace the path of water and minerals from the root hairs to xylem. Be sure to mention the Casparian strip.
9. Describe 3 basic types of roots, and give examples of other root modifications.
10. Describe the cross section of a monocot, a herbaceous dicot, and a woody stem.
11. Discuss the adaptation of stems by giving several examples.
12. Describe the structure and organization of a C_3 dicot leaf, a C_4 plant.
13. Give several ways in which leaves are specialized.

Objective Questions

1. Which of these is an incorrect contrast between monocots and dicots?

   monocots   dicots

   a. one cotyledon—2 cotyledons
   b. leaf veins parallel—net veined
   c. vascular bundles in a ring—vascular bundles scattered
   d. All are incorrect.

2. Which of these types of cells is most likely to divide?
   a. parenchyma
   b. meristem
   c. epidermis
   d. xylem

3. Which of these cells in a plant is apt to be nonliving?
   a. parenchyma
   b. collenchyma
   c. sclerenchyma
   d. epidermal

4. Root hairs are found in the zone of
   a. cell division.
   b. elongation.
   c. maturation.
   d. All of these.

5. Cortex is found in
   a. roots, stems, and leaves.
   b. roots and stems.
   c. roots only.
   d. stems only.

6. Between the bark and the wood in a woody stem, there is a layer of meristem called
   a. cork cambium.
   b. vascular cambium.
   c. apical meristem.
   d. the zone of cell division.

7. Which part of a leaf carries on most of the photosynthesis of a plant?
   a. epidermis
   b. mesophyll
   c. epidermal layer
   d. guard cells

8. Annual rings are the number of
   a. internodes in a stem.
   b. rings of vascular bundles in a monocot stem.
   c. layers of secondary xylem in a stem.
   d. Both b and c.

9. The Casparian strip is found
   a. between all epidermal cells.
   b. between xylem and phloem cells.
   c. on 4 sides of endodermal cells.
   d. within the secondary wall of parenchyma cells.

10. Which of these is a stem?
   a. taproot of carrots
   b. stolon of strawberry plant
   c. spines of cacti
   d. Both b and c.

11. Complete this simplified drawing of a root so that you can label endodermis, phloem, xylem, cortex, and epidermis:

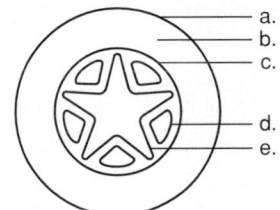

12. Complete this simplified drawing of a woody stem so that you can label cork, phloem, vascular cambium, xylem (wood), pith, and bark:

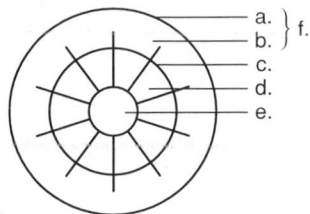

13. Complete this simplified drawing of a leaf so that you can label upper epidermis, palisade mesophyll, leaf vein, spongy mesophyll, and lower epidermis:

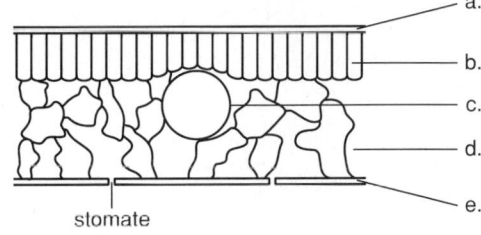

stomate

Concepts and Critical Thinking

1. *Structure suits the function.*

How does the structure of a dicot leaf suit its function of carrying on photosynthesis?

2. *Living things are organized.*

Show that a plant is organized on all levels of its structure.

3. *Organisms are adapted to the environment.*

In what ways do xylem rings (wood) help a plant live on land?

Selected Key Terms

meristem tissue (mer″ĭ-stem tish′u) 475
epidermis (ep″ĭ-der′mis) 476
stoma (**stomata**, pl.) (stō-ma) 476
cork (kork) 477
parenchyma (pah-reng′kĭ-mah) 477

xylem (zi′lem) 477
phloem (flo′em) 477
endodermis (en″do-der′mis) 481
Casparian strip (kas-par′e-on strip) 481
cambium (kam′be-um) 484

rhizome (rī′zōm) 485
mesophyll (mes′o-fil) 491
palisade layer (pal-a-sād la′er) 491
spongy layer (spun′je la′-er) 491
bundle sheath (bun′d′l shēth) 491

30

Nutrition and Transport in Plants

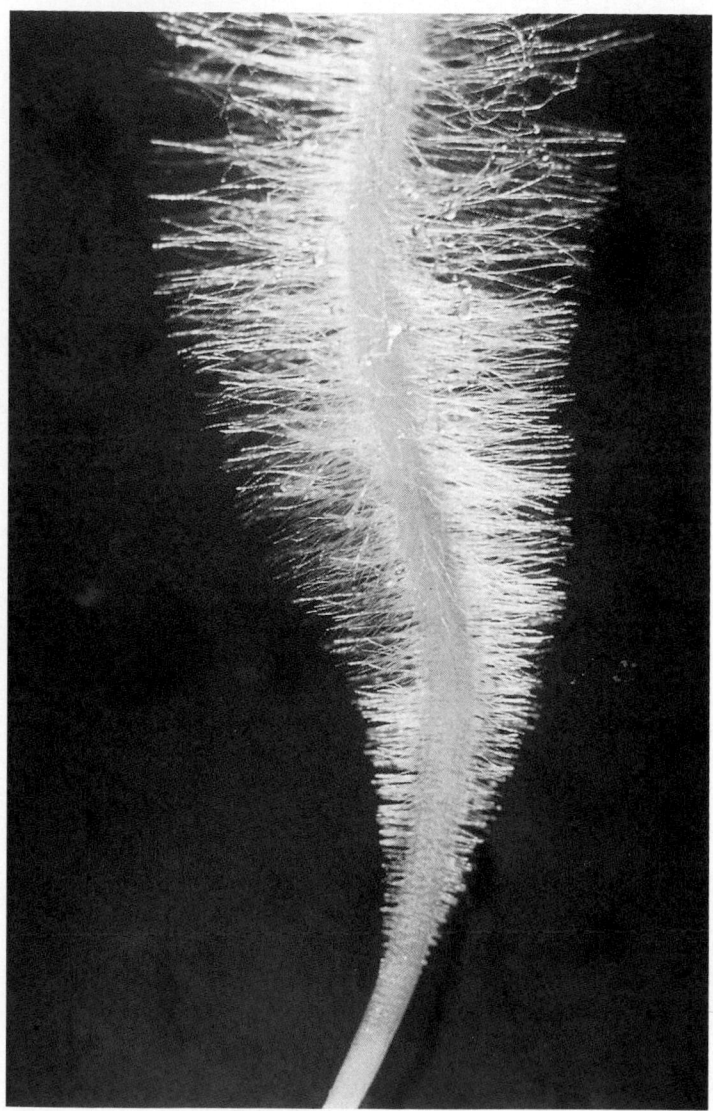

The root of a radish seedling has many projections called root hairs. These increase the surface area of a root, facilitating the uptake of water and minerals. Thereafter, water and minerals are transported from the roots to the rest of a plant. Nitrogen, potassium, phosphorus, and magnesium are some of the minerals that enter a plant via root hairs.

Your study of this chapter will be complete when you can

1. list the major plant nutrients;
2. trace 2 pathways of water from the soil to the vascular cylinder of a dicot root;
3. relate water availability to soil type, and contrast field capacity with permanent wilting point;
4. explain the mechanism by which water enters root cells, and relate root pressure to guttation;
5. describe the cohesion-tension model of xylem transport;
6. explain the mechanism by which stomata open and close;
7. distinguish between macroelements and microelements needed by a plant, and describe how mineral requirements are determined;
8. explain how acid deposition affects availability of minerals to a plant;
9. describe the mechanism of mineral uptake, and trace the path of minerals from the soil to the xylem within the vascular cylinder;
10. describe some adaptations of roots for nutrient uptake;
11. discuss the pressure-flow model of phloem transport.

To grow, plants require only water, carbon dioxide, oxygen, and certain minerals. The minerals required by plants are absorbed by roots from the soil. After being dissolved in water, the minerals are transported from the roots to the leaves in vascular tissue, which is continuous throughout the body of the plant (fig. 30.1). Water carried to the leaves is needed for photosynthesis, although much is lost by evaporation through the stomata. The carbon dioxide required for photosynthesis enters a leaf through the stomata.

We don't usually think plants need oxygen because it is available to them as a by-product of photosynthesis. Plants, however, do occasionally need to take oxygen from the atmosphere via the stomata and from small pockets of oxygen in the soil via root hairs. Because roots don't photosynthesize, these pockets are their only supply of oxygen. Oxygen is needed for plant cells to carry on cellular respiration, a process that makes ATP available to them (see chap. 9).

In this chapter, we will concentrate on how plants acquire water and minerals from the soil and how they transport these substances. We will also see how they transport the products of photosynthesis.

Figure 30.1
Leaves and roots are regions of a flowering plant body that are specialized to interact with the environment. The roots take up water and minerals. The leaves carry on gas exchange, but at the same time, they lose water to the environment.

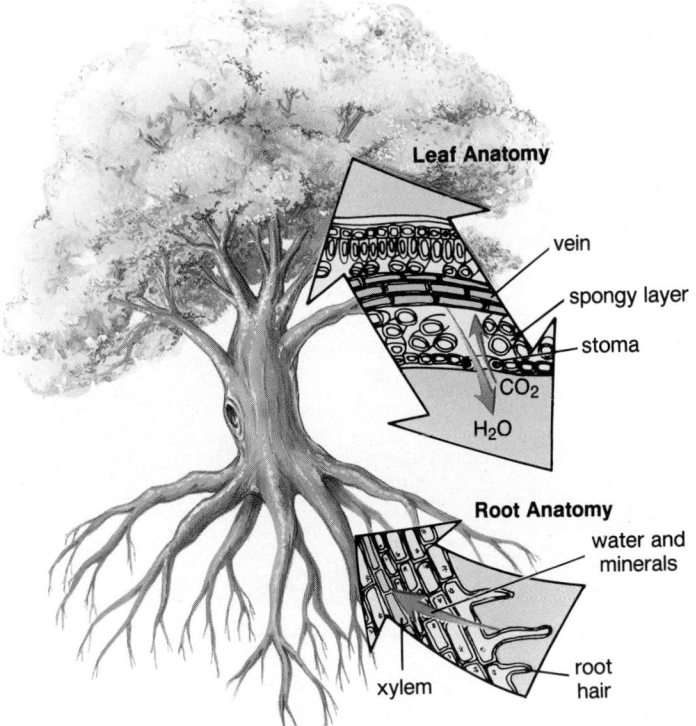

Water Uptake and Transport

Water Uptake

As figure 30.2 shows, water can enter the root of a flowering plant from the soil simply by diffusing *between* the cells of the epidermis and cortex via the porous cell walls. Eventually, however, the *Casparian strip* (see fig. 29.8c) prevents further progress between the cells, and the water (and any substances dissolved in the water) must enter the endoderm cells if it is to reach the xylem. Alternatively, water can move directly into the root-hair cells by osmosis and then progress from cell to cell across the cortex and endodermis until it reaches the xylem. This pathway is facilitated by the joining of plant cells to one another by plasmodesmata.

Regardless of the route by which water crosses the epidermis and cortex, it must at some time enter root cells. The cytoplasm of root cells usually has a higher concentration of solutes than does soil water. Therefore, water enters root cells passively by osmosis.

Figure 30.2
Water and minerals can travel one of 2 pathways across the cortex to the xylem of the vascular cylinder. Pathway A shows that water and minerals can travel between the cells of the cortex via porous walls but they must eventually enter endodermal cells because of the Casparian strip (see also fig. 29.8). In this way, the endodermal cells control the passage of materials into the vascular cylinder and xylem. Pathway B shows that water and minerals can immediately enter a root hair and thereafter move from cell to cell to finally enter the xylem.

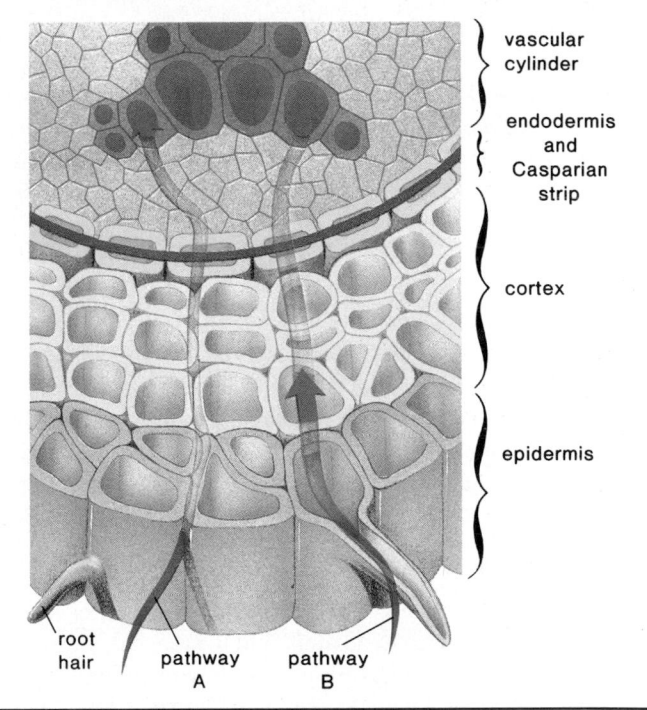

Figure 30.3

Availability of soil water. ***a.*** Root hairs absorb water from the pore spaces in soil. If the pores are large, as in sand, the water drains quickly. If the pores are small, as in clay, water does not drain because it adheres to the many small soil particles. ***b.*** Graph illustrating field capacity and permanent wilting point. Sand has a low field capacity because water drains from the soil. Clay has a high field capacity, but much of the water is unavailable to the plant. Loam has a perfect combination of qualities for plants. It is composed of 20% clay, 40% silt (particles of intermediate size), and 40% sand. Sandy, silty, and clay loam have proportionately more sand, silt, or clay, respectively.

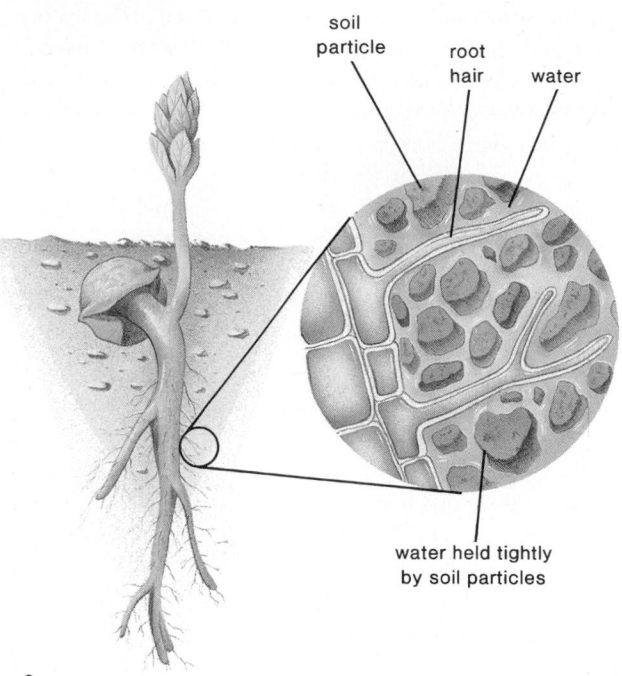

a.

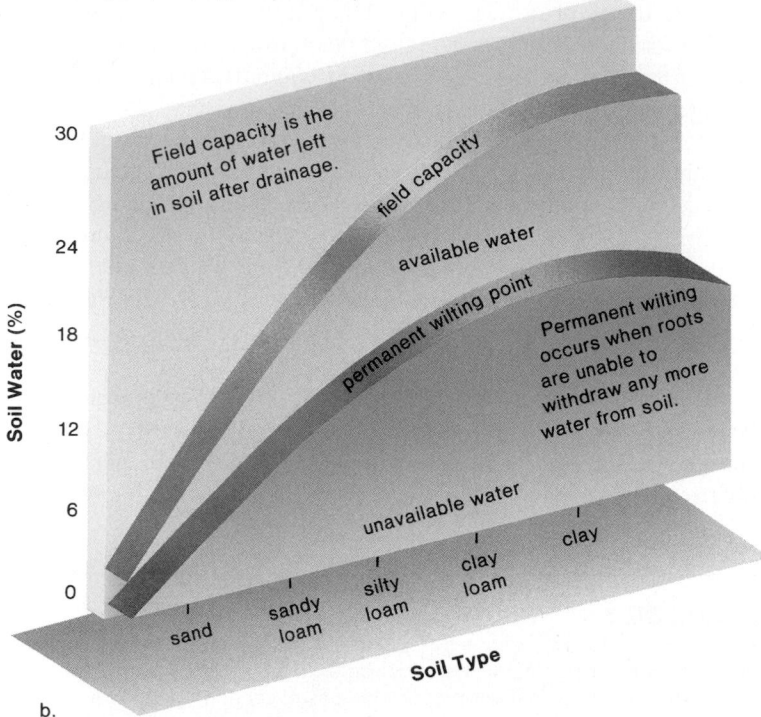

b.

Water Availability

After a rain or irrigation, gravity causes water in the soil to drain away. The speed with which this happens can be correlated with the size of the *soil pores*, the spaces between the particles of the soil. Sandy soil, which drains rapidly, has the largest pores because its particles are rather large. Clay soil, which drains slowly, has the smallest pores because its particles are very tiny. The water remaining in the soil after drainage is referred to as the *field capacity* of the soil. As you can see in figure 30.3, sand has a much lower field capacity than clay.

Another factor to consider is the availability of the water in the soil. Available water is water the roots are able to take up. The point at which water is no longer available to the roots is called the permanent wilting point of the soil. At this point, the soil, especially clay soil, still contains water, but it is held so tightly in small pores and on tiny particles, that the roots are unable to remove it (fig. 30.3*b*).

Water Transport

Once water enters xylem, it is transported to all parts of a plant. Transport does not need to be rapid, but the mechanism employed must be capable of transporting materials over long distances. Recall from the previous chapter (p. 477) that xylem contains 2 types of conducting cells: tracheids and vessel elements. The tracheids have pitted end walls but the vessel elements do not, and

Figure 30.4

Drops of guttation water on the edges of a strawberry leaf. Guttation, which occurs at night, is thought to be due to root pressure.

instead they form vessels that are continuous, completely hollow pipelines from the roots to the leaves. The vessels, therefore, offer the better route for water transport from the roots to the leaves.

Plant Structure and Function

Figure 30.5

Cohesion-tension model of xylem transport. **a.** This scanning electron micrograph shows how vessel elements join to form a hollow xylem vessel, which reaches from the roots to the stems. Magnification, X400. **b.** Water (H_2O) enters a plant at the root hairs and evaporates (transpires) at the leaves. Each xylem vessel is completely full of water. Transpiration exerts a pull on this water column that causes it to move upward.

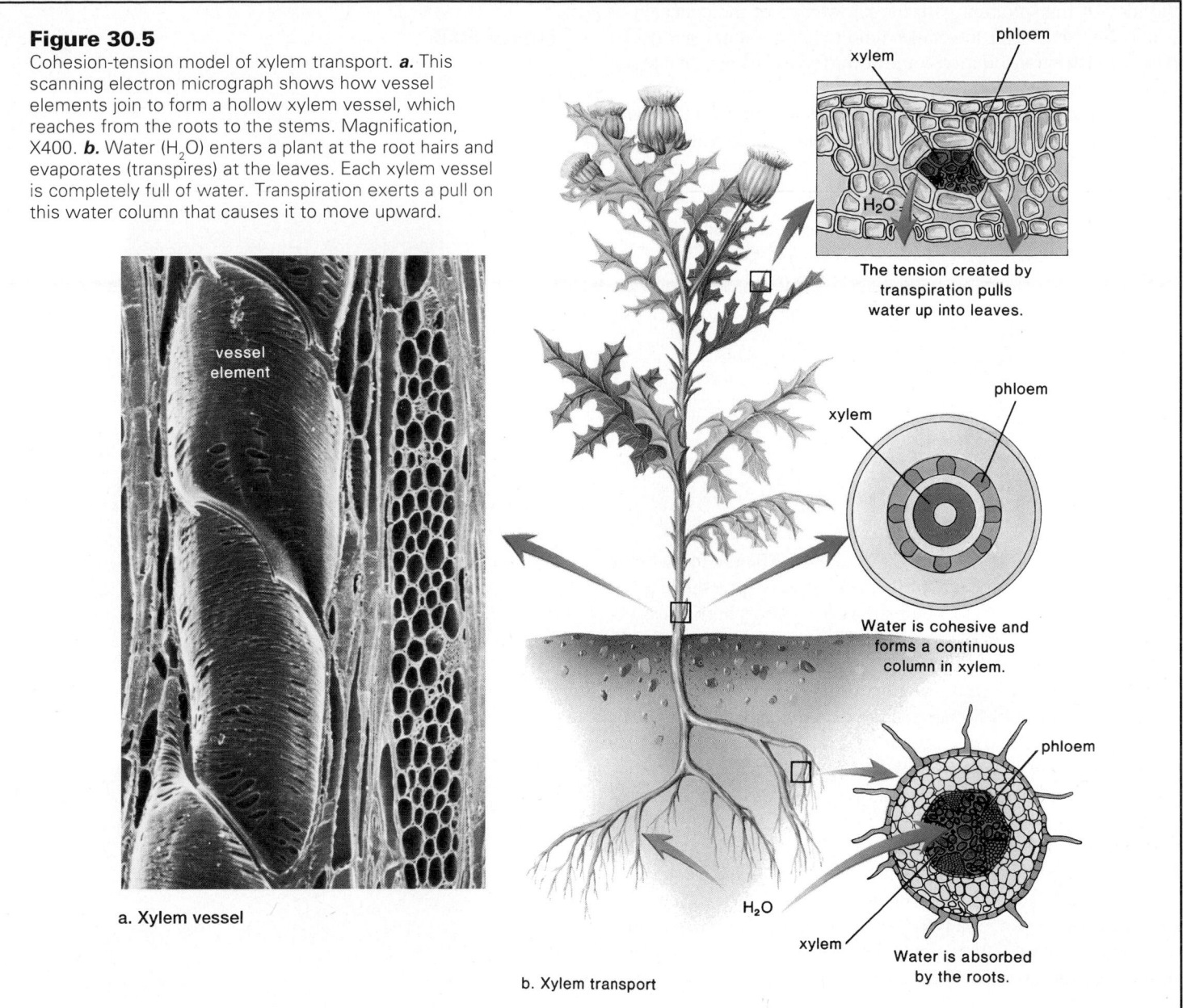

a. Xylem vessel

The tension created by transpiration pulls water up into leaves.

Water is cohesive and forms a continuous column in xylem.

Water is absorbed by the roots.

b. Xylem transport

When water enters root cells by osmosis, it creates root pressure. **Root pressure** tends to push xylem sap (water and minerals) upward. This can be shown by attaching a glass tube to the cut stem of a potted plant and watching the water rise. Water will rise as much as a meter above a cut tomato stem and several times that height in some kinds of vines. Root pressure is also responsible for **guttation,** when drops of water are forced out of vein endings along the edges of leaves (fig. 30.4). Although root pressure may contribute to the upward movement of water in xylem vessels, it is not nearly as important as transpirational pull, which will be discussed next.

Cohesion-Tension Model of Xylem Transport

The **cohesion-tension model** of water transport explains how water is transported to great heights, against gravitational force. First, we must realize that water molecules have a great tendency to cling together because they are polar. The *cohesive property* of water means that a column of water can be pulled without breaking. In fact, it is easier to pull apart the molecules in fine wires made of some common metals than it is to pull apart a thin column of water in a small-diameter, airtight tube. Second, we must also remember that water is continuously lost by **transpiration** (evaporation) through the stomata in the leaves. When water evaporates from the mesophyll cells in a leaf, these cells become less turgid. This loss of turgidity creates an osmotic gradient that causes water to move out of the xylem and into these cells. Transpiration therefore creates a *tension* that can pull water upward in the xylem (fig. 30.5).

Stephen Hales, an English botanist who worked at the beginning of the eighteenth century, first suggested that transpiration pulls water up through plants. In 1914, Henry Dixon coupled the idea of transpiration with his knowledge of the cohesion of water. Because water molecules tend to cling to one another, he postulated that transpiration could be the force causing water to rise in the hollow, continuous xylem vessels.

The following experiment shows that transpiration aids the movement of water in the xylem. A tube is filled with mercury and then quickly inverted and placed in a pan of mercury:

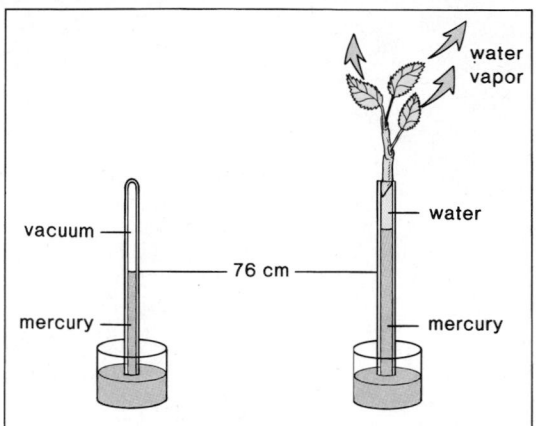

Atmospheric pressure (the weight of air) is sufficient to maintain the height of the mercury at 76 cm. (This height expressed in terms of water is equal to 10.4 m.) This shows that atmospheric pressure, even combined with root pressure, could not raise water in xylem to the height of the tallest trees (120 m). But if a twig with leaves is placed in the top of a tube that contains water above mercury, the mercury column rises higher than 76 cm. The explanation for this phenomenon is that transpiration of water at the leaves exerts a pulling force on the contents of the tube and raises the height of the mercury above that expected. Consistent with the belief that water is pulled upward by transpiration rather than pushed upward by atmospheric pressure is the observation that water moves in the branches of a tree earlier in the day than water in the trunk.

Although biologists in general accept the cohension-tension model of water transport, there are many specific questions yet to be answered. For example, how do aquatic plants or plants living in the desert or in cold climates manage to overcome varied difficulties? They must have particular adaptations for water transport, and these are areas of active research.

There is an important consequence of the way water is transported in plants. As mentioned in chapter 29, transpiration usually results in a loss of at least 90% of the water taken in at the roots. When the ground is dry and the plant is under water stress, the stomata close (fig. 30.6a). Now the plant loses little water because the leaves are protected against water loss by the waxy cuticle of the upper and lower epidermis (fig. 29.19a). When the stomata are closed, however, carbon dioxide (CO_2) cannot enter the leaves and plants are unable to photosynthesize. Photosynthesis therefore requires an abundant supply of water so that the stomata can remain open and allow carbon dioxide to enter.

Movement of water in a plant is dependent upon the physical and chemical properties of water. Transpiration of water at the leaves causes water to flow upward in the xylem.

Figure 30.6

Opening of stomata. **a.** Colored scanning electron micrograph of an open stoma. **b.** When a stoma opens, first potassium (K^+) ions and then water enter guard cells. This increases the turgor pressure of guard cells so that the stoma opens. **c.** When a stoma closes, first potassium ions and then water exit guard cells. This decreases the turgor pressure of guard cells so that the stoma closes.

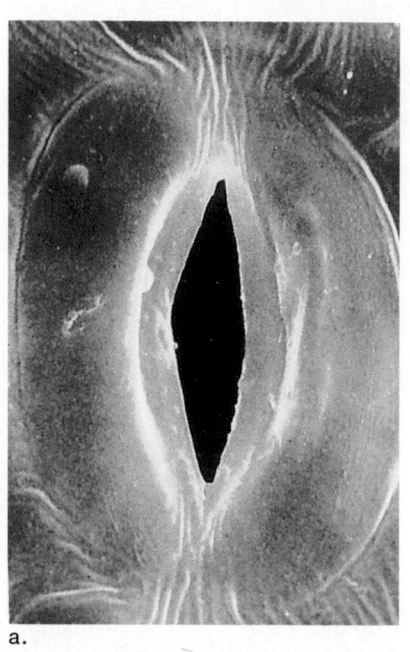

a.

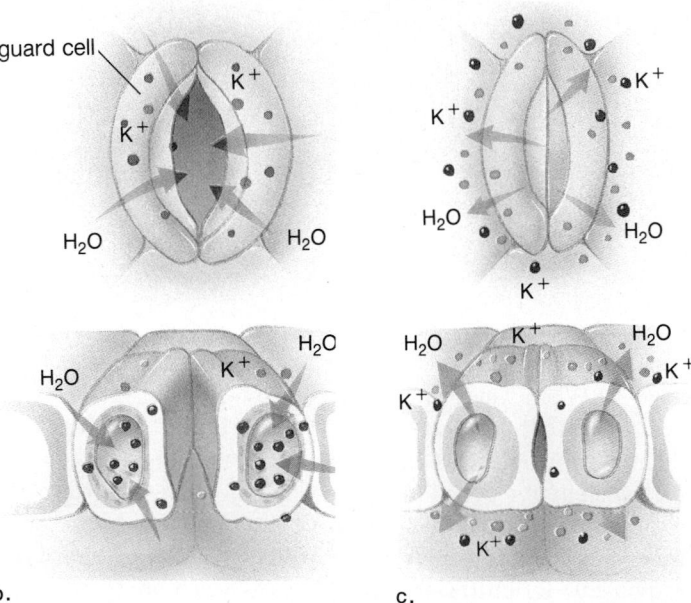

b.   c.

Opening and Closing of Stomata

Each stoma has 2 **guard cells** with a pore between them. When water enters the guard cells and *turgor pressure increases*, a stoma opens; when water exits the guard cells and they lose turgor pressure, the stoma closes. Notice in figure 30.6a that the guard

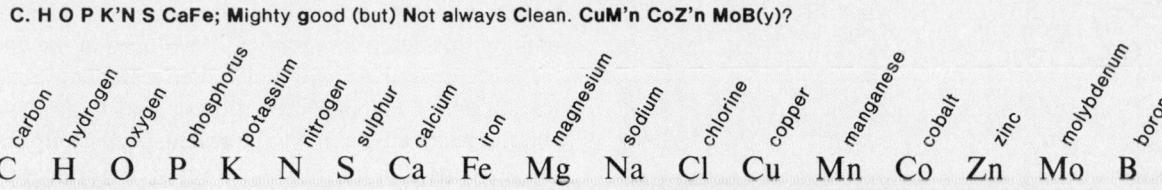

Figure 30.7

Essential elements for plant nutrition. (Note: Many plants do not require sodium. A few plants and most bacteria require cobalt.)

C. H O P K'N S CaFe; Mighty good (but) Not always Clean. CuM'n CoZ'n MoB(y)?

carbon	hydrogen	oxygen	phosphorus	potassium	nitrogen	sulphur	calcium	iron	magnesium	sodium	chlorine	copper	manganese	cobalt	zinc	molybdenum	boron
C	H	O	P	K	N	S	Ca	Fe	Mg	Na	Cl	Cu	Mn	Co	Zn	Mo	B

cells are attached to each other at their ends and that the inner walls are thicker than the other walls. When water enters, a guard cell's radial expansion is restricted because of cellulose microfibrils in the walls, but lengthwise expansion of the outer walls is possible. When the guard cells expand lengthwise, they buckle out from the region of their attachment and the stoma opens.

Botanists have been trying to determine what causes guard cells to take up or lose water. Because guard cells have chloroplasts, it was at first assumed that whenever photosynthesis was occurring, the accumulation of sugar in guard cells would cause water to enter. In 1968, R. A. Fischer offered a different explanation. He floated small strips of lower epidermis tissue from a broad-bean plant (*Vicia fava*) on various concentrations of potassium chloride. He found that guard cells take up potassium (K^+) ions *in the light if the air lacks carbon dioxide (CO_2)*. Once the guard cells take up potassium ions, water enters, and stomata open. The connection between low carbon dioxide concentration and uptake of potassium ions is not known. It's possible, however, that photosynthesis in guard cells leads to ATP production. ATP could then provide the energy needed for active transport of potassium ions into the guard cells. When guard cells are not photosynthesizing, potassium ions exit, causing the guard cells to lose water and the stomata to close. Still unexplained is the observation that some variegated plants do not have chloroplasts in their guard cells; however, they still open and close their stomata.

Stomata close even when a plant is in full sunlight if the leaves are water stressed. This is caused by the hormone abscisic acid (ABA), which is produced by cells in wilting leaves. ABA is believed to cause membrane leakage, leading to a loss of potassium ions and the closing of stomata.

It has also been observed that stomata open and close on a daily basis even if plants are kept in the dark. Behavior that occurs regularly every 24 hours is called a circadian rhythm. Biological clocks, which are discussed on page 530, are known to control circadian rhythms.

If water is available, stomata tend to open, and the plant photosynthesizes. They close when a plant is water stressed, even though light may be present for photosynthesis.

Table 30.1
Essential Elements for Plants

Element	Chemical Symbol	Form Available to Plant	Percent of Plant's Dry Weight
Macroelements			
Hydrogen	H	H_2O	6
Carbon	C	CO_2	45
Oxygen	O	O_2, CO_2, H_2O	45
Nitrogen	N	NO_3^-	1.5
Potassium	K	K^+	1.0
Calcium	Ca	Ca^{++}	0.5
Magnesium	Mg	Mg^{++}	0.2
Phosphorus	P	$H_2PO_4^-$, $HPO_4^=$	0.2
Sulfur	S	$SO_4^=$	0.1
Microelements			
Chlorine	Cl	Cl^-	0.01
Iron	Fe	Fe^{++}, Fe^{+++}	0.01
Boron	B	$BO_3^=$, $B_4O_7^=$	0.002
Manganese	Mn	Mn^{++}	0.005
Zinc	Zn	Zn^{++}	0.002
Copper	Cu	Cu^+, Cu^{++}	0.006
Molybdenum	Mo	$MoO_4^=$	0.00001

Mineral Requirements and Uptake

Mineral Requirements

Certain elements are considered essential to the health of plants; these can be remembered by the mnemonic given in figure 30.7. These elements are divided into macroelements and microelements as indicated in table 30.1. Studies show that the elements classified as macroelements are required in greater concentration than those classified as microelements.

Table 30.2
Functions of Some Essential Elements for Plant Nutrition

Element	Function
Nitrogen	Constituent of amino acids, proteins, coenzymes, nucleic acids, and chlorophyll
Phosphorus	Constituent of sugar phosphates, nucleotides, nucleic acids, phospholipids, coenzymes
Potassium	Not part of any known organic molecule; coenzyme and osmoregulator
Sulfur	Constituent of sulfur-containing amino acids, coenzyme A, and vitamins
Magnesium	Constituent of chlorophyll; coenzyme, especially in reactions involving ATP
Calcium	Constituent of cell wall, middle lamella, and membranes; cofactor for some enzymes; may detoxify large concentrations of heavy metals
Iron	Cofactor in enzymes of chlorophyll synthesis and other enzymes; constituent of cytochromes and ferredoxin, where it functions in electron transport

From D. K. Northington and J. R. Goodin, *The Botanical World*. Copyright © 1984 Times Mirror/Mosby College Publishing, St. Louis, MO. Reprinted by permission.

Carbon, hydrogen, and oxygen make up 96% of a plant's dry weight. Carbon dioxide is the source of carbon for a plant, and water is the source of hydrogen. Oxygen can come from atmospheric oxygen, CO_2, or water.

The other essential elements listed in table 30.1 are minerals that come from the soil. Nitrogen (N) is a common atmospheric gas, but plants are unable to make use of it. Instead, most plants are dependent upon a supply of nitrate (NO_3) in the soil; chapter 49 discusses how the presence of nitrate in the soil is dependent upon the work of bacteria. As discussed on page 502, however, some plants have root nodules containing mutualistic bacteria that fix atmospheric nitrogen and make nitrogen compounds available to their host.

Table 30.2 outlines the functions of some of these elements in plant nutrition.

Determination of Mineral Requirements

If you burn a plant, its nitrogen component is given off as ammonia and other gases, but most other mineral elements remain in the ash. The presence of a particular element in the ash, however, does not necessarily mean that the plant normally requires it. For example, plants take up [239]plutonium and the naturally occurring [90]strontium, elements they do not require. These radioactive elements are subsequently found in their ashes. This uptake of radioactive elements by plants is unfortunate because it intensifies the concerns and worries of a nuclear power plant mishap. Complicating the situation still further is the knowledge that if cows feed on plants containing [90]strontium, their milk will contain this radioactive element in even greater concentrations. Therefore, humans are likely to receive a concentrated dose of radioactivity when they drink the milk.

Because analysis of the ash does not allow an investigator to determine the mineral requirements of a plant, a method for making this determination was developed at the end of the nineteenth century. This method is called water culture, or **hydroponics** (fig. 30.8). Hydroponics allows plants to grow well if they are supplied with all the mineral nutrients they need. The investigator omits a particular mineral and observes the effect on plant growth. If growth suffers, it can be concluded that the omitted mineral is a required nutrient. This method has been more successful for macronutrients than for micronutrients. For studies involving the latter, the water and the mineral salts used must be absolutely pure, and purity is difficult to attain, because even instruments and glassware can introduce micronutrients. Then, too, the element in question may already be present in the seedling used in the experiment. These factors complicate the determination of essential plant micronutrients by means of hydroponics.

Mineral Availability

As mentioned previously, the essential minerals that a plant needs are normally taken from the soil. Location of minerals in the soil is critical because if a root does not come within a few millimeters of an ion, no uptake occurs. The downward movement of minerals through soil is influenced by the quantity and the pH of water. Sometimes water makes mineral ions available to roots, but often it leaches, or removes, mineral ions from the soil zone in which roots grow. The lower the pH, the greater the leaching power of water.

Soil particles, particularly clay and any organic matter in soil, contain negative charges that attract positively charged ions, such as calcium (Ca^{++}), potassium (K^+), and magnesium (Mg^{++}). This attraction keeps these ions at a soil level where they are available to plants. In acidic soils, however, hydrogen ions (H^+) replace these other positive ions so that the useful ions float free and are easily leached by water:

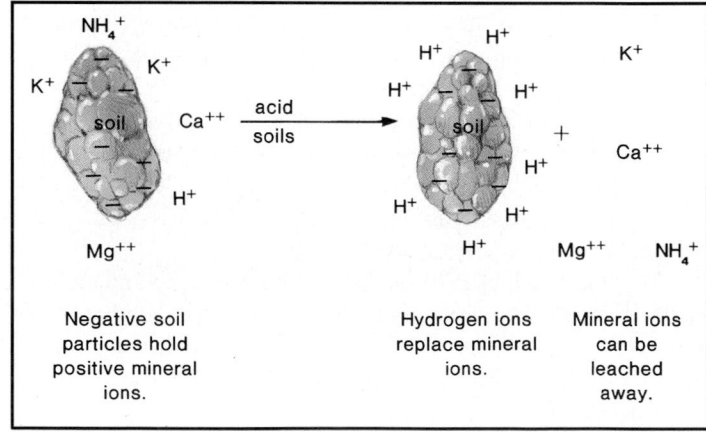

Negative soil particles hold positive mineral ions.

Hydrogen ions replace mineral ions.

Mineral ions can be leached away.

On the other hand, in very acidic soils, aluminum (Al^{+++}) and iron (Fe^{+++}) ions become available to plants. These mineral ions are insoluble above a pH of 4, but they become soluble at a more

Figure 30.8

Some plant nutrient deficiencies are easily diagnosed when plants are grown in a complete nutrient solution with elimination of one nutrient at a time. Sunflower plants respond rapidly to a deficiency of nitrogen (*a*), phosphorus (*b*), and calcium (*c*). Nutrient-deficient plants are shown on the left in each photograph; healthy plants grown in a complete nutrient solution are shown on the right.

a.

b.

c.

Figure 30.9

Bromine (Br) ion absorption by a potato tuber. As respiration increases, bromine ion absorption also increases. This shows that active transport is involved in nutrient uptake by plants.

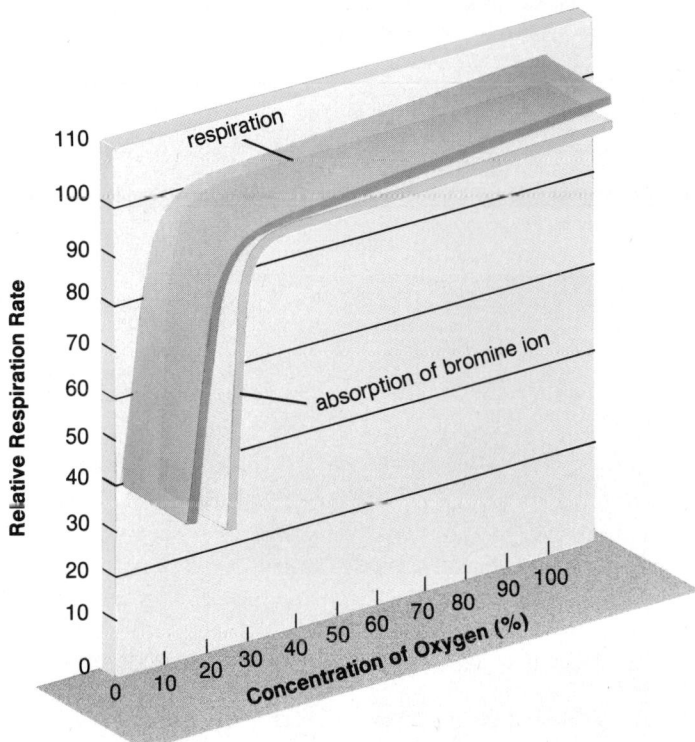

It is clear, then, that acid deposition has 2 effects that are deleterious to plants: it causes the removal of the mineral ions that plants need for proper nutrition, and it makes available other mineral ions that are toxic at high levels. The death of trees in certain portions of the United States is, in part, the result of acid deposition (see fig. 50.8).

Mineral Uptake

Like water, mineral ions can move past the epidermis and through the cortex by way of porous cell walls (fig. 30.2). Because of the Casparian strip, however, they eventually must enter the cytoplasm of the endodermal cells if they are to proceed farther and enter the vascular cylinder. Plasma membranes are freely permeable to water but not to all mineral ions. Therefore, mineral ions that cannot pass through the plasma membranes of endodermal cells can go no farther. In this way, the endodermis regulates the entrance of minerals into the vascular cylinder.

Mineral ions are often actively absorbed into the cytoplasm at the root-hair region of a young root. Only when particular mineral ions are highly concentrated in soil water, such as after a fertilizer application, do mineral ions enter root cells by simple diffusion. Figure 30.9 presents data from one experiment showing that active transport is most likely involved in the uptake of minerals by a plant. The rate of respiration and the rate of bromine (Br) ion absorption by thin disks of potato increased as more

acidic pH. They displace calcium, potassium, and magnesium on soil particles in the same way as hydrogen does. High levels of aluminum and iron in soils are toxic to plants.

Figure 30.10

Symbiosis and mineral nutrition. ***a.***
Nodules on soybean roots that contain
nitrogen-fixing bacteria. Their red color is
caused by a pigment produced while
nitrogen fixation is occurring. ***b.*** Loblolly
pine root systems. Roots on the left
were not treated and those on the right
were inoculated with a fungus. The
fungus-covered mycorrhizal roots absorb
minerals and water more efficiently.
Photograph by Carolina Biological Supply
Company.

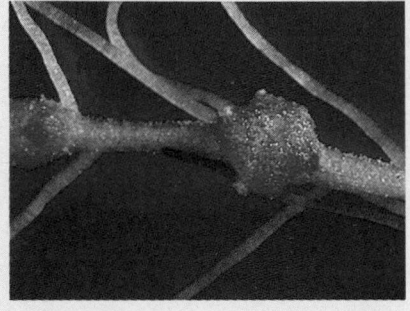

a.

b.

Figure 30.11

Aphids are small insects that remove nutrients from phloem by means of a hypodermic-
like mouthpart called a stylet. ***a.*** Aphid with stylet in place. ***b.*** When the aphid's body is
removed, phloem sap is available to the experimenter.

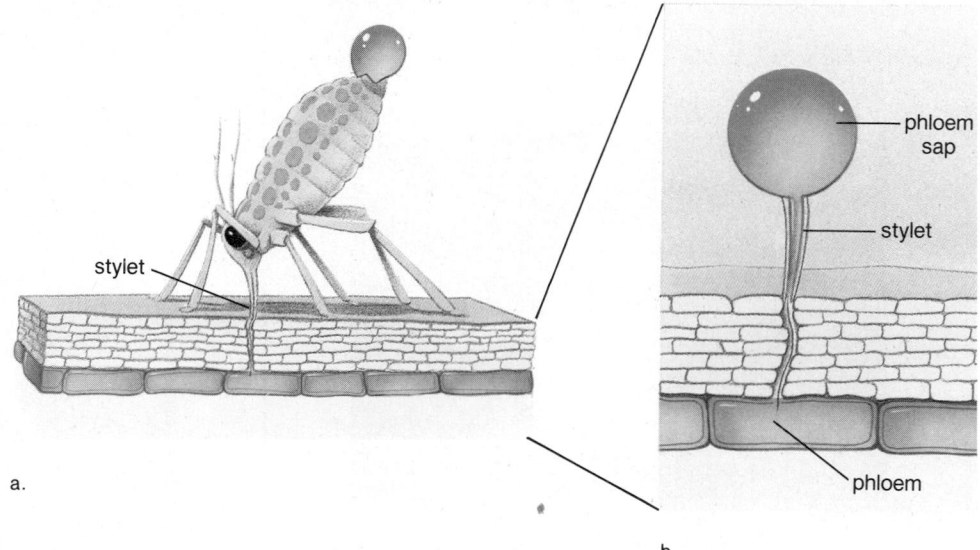

a.

b.

oxygen was made available by bubbling the gas through the culture
solution. This result suggests that energy is used to take up the
bromine ions.

Plants possess an astonishing ability to concentrate ions,
that is, to absorb them until the ions are many times more concentrated in the plant than in the surrounding medium. Experiments
have shown that the concentration of certain minerals in roots is as
much as 10,000 times greater than in the surrounding soil. Nutrition in animals is also dependent on this ability of plants to
concentrate minerals.

Once mineral ions have entered root cells, they diffuse
down a concentration gradient and enter the xylem within the
vascular cylinder. From there, they are distributed to the various
organs of the plant.

Minerals follow the same path as water. They are taken up
primarily by root hairs, move across the cortex, and then enter
the xylem. From there, they are distributed to other plant organs.

Root Adaptations for Mineral Uptake

Two symbiotic relationships are known to assist roots in supplying
mineral nutrients to the rest of a plant. Plants are unable to make use
of atmospheric nitrogen (N_2) because they do not have the cellular
enzymes to break the $N \equiv N$ bond. Therefore, plants are usually
dependent on the nitrate (NO_3^-) that roots remove from the soil. But
some roots, such as those of the legumes, soybean and alfalfa, are
infected by bacteria in the genus *Rhizobium*, which can break the
$N \equiv N$ bond and reduce nitrogen to NH_4^+ for incorporation into
organic compounds. (The reduction of N_2 to NH_4^+ is the process
called nitrogen fixation.) The bacteria live in **nodules** and are
supplied with carbohydrates by the host plant (fig. 30.10*a*). The
bacteria, in turn, furnish their host with nitrogen compounds. This
is such a useful service that researchers are now trying to find ways
to make more plants receptive to infection by nitrogen-fixing
bacteria. Gene-splicing techniques are being developed to introduce the bacterial genes that control nitrogen fixation (called *nif*
genes) into plant chromosomes. Less fertilizer would then be
needed to grow crops.

The second symbiotic relationship involves fungi. The
fungus grows around the root, often penetrating even to the level
of the cortex. The fungus increases the surface area available for
mineral and water uptake and breaks down organic matter, releasing nutrients that the plant can use. In return, the root furnishes the
fungus with sugars and amino acids. These "fungus roots," or
mycorrhizae, assist the growth and development of plants, but
they seem to be particularly essential to some plants. Orchid seeds,
which are quite small and contain limited nutrients, do not germinate until a mycorrhizal fungus has invaded their cells.
Nonphotosynthetic plants, such as Indian pipe, use their fungus
roots to extract nutrients from nearby trees by way of mycorrhizal

hyphae that connect the roots of the tree and the plant. To encourage growth, some tree seedlings are deliberately infected with mycorrhizal fungi before they are transplanted (fig. 30.10b).

As a special adaptation, some plants have poorly developed roots or no roots at all because minerals and water are supplied by other mechanisms. The carnivorous plants discussed on page 491 have poorly developed roots and rely on their leaves to absorb compounds containing minerals. **Epiphytes** are "air plants"; they do not grow in soil but on larger plants, which give them support. They do not get nutrients from their host, however. Some epiphytes have roots that absorb moisture from the atmosphere, and many catch rain and minerals in special pockets at the base of their leaves. Other epiphytes are parasitic plants, which attach to a host by way of rootlike projections called haustoria. The haustoria of mistletoe penetrate only the xylem of the host to remove water and minerals. Other plants, such as dodders, broomrapes, and pinedrops, send their haustoria into both the xylem and phloem of the host.

Transport of Organic Substances

As long ago as 1679, Marcello Malpighi suggested that bark is involved in translocating sugars from leaves to roots. He observed the results of removing a strip of bark from around a tree, a procedure called **girdling.** If a tree is girdled below the level of the majority of leaves, the bark swells just above the cut and sugar accumulates in the swollen tissue. We know today that when a tree is girdled, the phloem is removed but the xylem is left intact. Therefore, the results of girdling suggest that phloem is the tissue that transports sugars. Radioactive tracer studies have confirmed this. When ^{14}C-labeled CO_2 is supplied to mature leaves, radioactively labeled sugar is soon found moving down the stem into the roots. This labeled sugar is found mainly in the phloem, not in the xylem. Radioactive tracer studies have also confirmed the role of phloem in transporting other substances, such as amino acids, hormones, and even mineral ions. Hormones are transported from their production sites to target areas, where they exert their regulatory influences. In the autumn, before leaves fall, mineral ions are removed from the leaves and are taken to other locations in the plant.

Chemical analysis of phloem sap has shown that its main component is sucrose, and the concentration of nutrients is usually 10%–13% by volume. It is difficult to take samples of sap from the phloem without injuring the phloem, but this problem is solved by using aphids (fig. 30.11), small insects that are phloem feeders. The aphid drives its stylet, which is a sharp mouthpart that functions like a hypodermic needle, between the epidermal cells and withdraws sap from a sieve-tube cell. If the aphid is anesthetized using ether, its body can be carefully cut away, leaving the stylet. Phloem can then be collected and analyzed by the researcher.

Pressure-Flow Model of Phloem Transport

The conducting cells of phloem are sieve-tube cells lined up end-to-end with their sieve plates abutting (p. 478). Cytoplasm extends through the sieve plates of adjoining cells to form a continuous sieve-tube system that extends from the roots to the leaves and vice

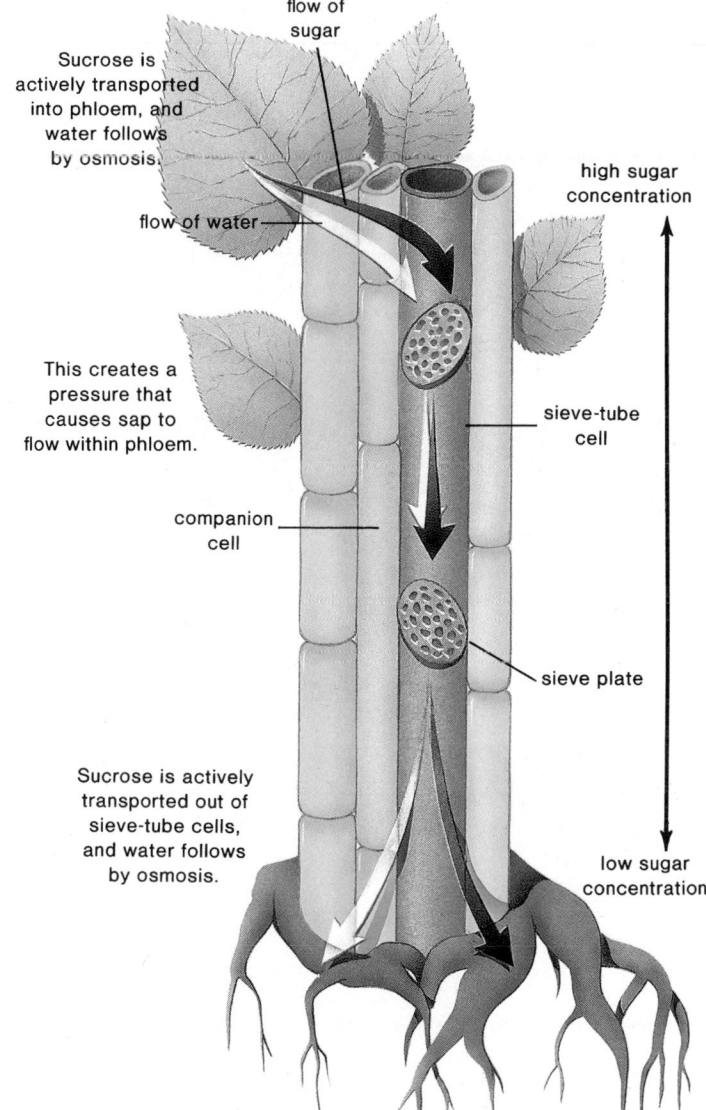

Figure 30.12
Pressure-flow model of phloem transport. Sucrose and water enter sieve-tube cells at a source. Sieve-tube cells form a continuous pipeline from a source to a sink, where sucrose and water exit sieve-tube cells.

flow of sugar

Sucrose is actively transported into phloem, and water follows by osmosis.

flow of water

high sugar concentration

This creates a pressure that causes sap to flow within phloem.

sieve-tube cell

companion cell

Sucrose is actively transported out of sieve-tube cells, and water follows by osmosis.

sieve plate

low sugar concentration

versa. An explanation of phloem transport must account for the movement of fairly large amounts of organic material for long distances in a relatively short period of time. The movement rate of radioactively labeled ^{14}C sugar has been determined by analysis of sap withdrawn from 2 areas of the stem. Materials appear to move through the phloem at a rate of 60 cm–100 cm per hour and possibly up to 300 cm per hour.

The **pressure-flow model** is a current explanation for the movement of organic materials in phloem (fig. 30.12). During the growing season, the leaves are photosynthesizing and producing sugar. Sucrose is actively transported into phloem, and water follows passively by osmosis. Transport and osmosis are possible because sieve-tube cells have a living plasma membrane. Also, the energy needed for sucrose transport is provided by the companion cells. The buildup of water within the sieve-tube cells at the leaves creates a *pressure*.

At the roots (or other places in the plant) sucrose is transported out of the phloem and water follows passively by osmosis. This exit of sucrose and water at the roots means that the pressure created at the leaves causes a *flow* of water from the leaves (higher pressure) to the roots (lower pressure). As the water flows, it brings the sucrose with it.

This model can be supported by the following experiment: Two bulbs are connected by a glass tube. The first bulb contains solute at a higher concentration than the second bulb. Each bulb is bounded by a selectively permeable membrane, and the entire apparatus is submerged in distilled water:

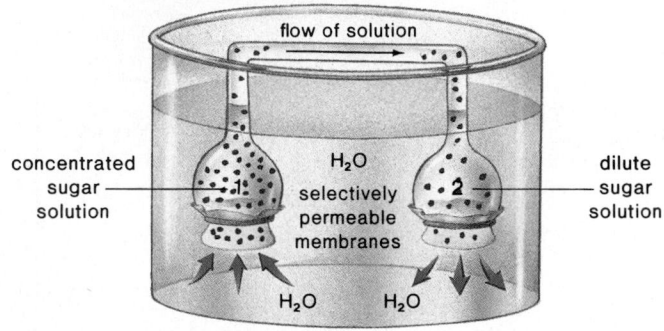

Distilled water flows into the first bulb because it has the higher concentration of solute. The entrance of water creates *pressure*, which causes a *flow* toward the second bulb. The pressure flow

not only drives water toward the second bulb, it also provides enough drive to force water out through the membrane of the second bulb, even though the second bulb contains a higher concentration of solute than the distilled water.

The pressure-flow model of phloem transport can account for any direction of flow in the sieve-tube cell system if we consider that the direction of flow is from *source to sink*. In young seedlings, the cotyledons, containing reserve food, are a major source of sucrose and roots are a sink. Therefore, the flow is from the cotyledons to the roots. In older plants, the most recently formed leaves that are actively photosynthesizing are a source of sucrose for the shoot tip (a sink), and the lower leaves are a source of sucrose for the root (a sink). When a plant is in fruit, it has been observed that phloem flow is monopolized by the fruits, little goes to the rest of the plant, and vegetative growth is slow.

The pressure-flow model of phloem transport suggests that phloem sap can move either up or down as appropriate for the plant at a particular time in its life cycle.

As with xylem transport, there are many questions to be answered regarding phloem transport. For example, what factors aside from source-to-sink relationships affect the direction and the rate of phloem transport? Do plant hormones have an effect? These and many other questions are now being addressed by plant physiologists.

Summary

1. To grow, plants require water, carbon dioxide, oxygen, and certain minerals.
2. Water can enter the root by moving between the cells until it reaches the Casparian strip, after which it passes through an endodermal cell before entering xylem. Water can also enter root hairs, then pass through the cells of the cortex and endodermis to reach xylem.
3. Much of the water in soil is not available to plants. In sandy soils, it drains quickly, and in clay, it adheres to the tiny particles. Field capacity is the amount of water in a soil after drainage. At the permanent wilting point, no more water is available to the plant.
4. Water enters root cells by osmosis, creating root pressure. Root pressure does not contribute greatly to the upward movement of water in xylem, but it does cause guttation.
5. The cohesion-tension model of xylem transport states that transpiration creates a tension that pulls water upward in xylem. This means of transport works only because water molecules are cohesive. Most of the water taken in by a plant is lost through stomata by transpiration. Only when there is plenty of water do stomata remain open, allowing carbon dioxide to enter the leaf and photosynthesis to occur.
6. Stomata open when guard cells take up water. They stretch lengthwise, and this causes them to buckle out. Water enters the guard cells after potassium (K^+) ions have entered. Uptake of K^+ has been observed when there is a low level of CO_2 inside leaves. Perhaps photosynthesis in guard cells leads to ATP production and active transport of K^+.
7. Plant nutrition requires 16 different elements. Carbon, hydrogen, and oxygen make up 96% of a plant's dry weight. The other necessary elements are taken up by the roots as mineral ions. Even nitrogen, which is present in the atmosphere, is taken as NO_3^- from the soil.
8. You can determine mineral requirements by hydroponics, which is growing plants in a solution. The solution is varied by the omission of one mineral. If the plant grows poorly, then the missing mineral is required for growth.
9. Roots can only take up the minerals that are available in soil. Acid deposition reduces the availability of some minerals and increases the availability of other minerals. Ions such as calcium, potassium, and magnesium are displaced from soil particles by hydrogen ions and are leached from the soil. Acid deposition increases the solubility of aluminum and iron and makes them available in toxic amounts.

10. Root hairs use active transport to take in mineral ions. After entering root hairs, minerals move across the cortex and endodermis and finally enter the xylem.

11. Plants have various adaptations that assist them in acquiring nutrients. Legumes have nodules infected with the bacterium *Rhizobium,* which makes nitrogen compounds available to these plants. Many other plants have mycorrhizae, or fungus roots. The fungus gathers nutrients from the soil,

and the root provides the fungus with sugars and amino acids. Some plants have poorly developed roots. Most epiphytes live on, but do not parasitize trees, whereas mistletoe and some other plants parasitize their host.

12. The pressure-flow model of phloem transport states that sucrose is actively transported into phloem at a source and water follows by osmosis. The resulting increase in pressure creates a flow that moves water and sucrose to a sink.

Study Questions

1. Give 2 pathways by which water and minerals can cross the epidermis and cortex of a root. What feature allows endodermal cells to regulate the entrance of molecules into the vascular cylinder?

2. Tell how field capacity and permanent wilting point relate to water availability in soils. Why do soils differ as to water availability?

3. Describe and give evidence for the cohesion-tension theory of water transport.

4. What events precede the opening and closing of stomata by guard cells?

5. Name the elements that make up most of a plant's body. Why are some elements called microelements and some macroelements?

6. Give reasons why a plant needs nitrogen, phosphorus, and potassium.

7. Briefly describe 2 methods used to determine the mineral nutrients of a plant.

8. Explain how acid deposition affects availability of minerals to a plant and why this can lead to plant death.

9. Contrast the manner in which water and minerals are absorbed by root cells.

10. Name 2 symbiotic relationships that assist plants in taking up minerals and 2 types of plants that tend not to take up minerals by roots from soil.

11. What data are available to show that phloem transports organic compounds? Explain the pressure-flow model of phloem transport.

Objective Questions

1. Which of these is not a nutrient for plants?
   a. water
   b. carbon dioxide gas
   c. mineral ions
   d. nitrogen gas

2. The Casparian strip affects
   a. how water and minerals move into the vascular cylinder.
   b. how water but not minerals move.
   c. how minerals but not water move.
   d. neither the flow of water nor of minerals into a plant.

3. Field capacity is
   a. the same as permanent wilting point.
   b. the amount of water in soil available to plants.
   c. the amount of water in soil after drainage takes place.
   d. higher for sandy soils than clay soils.

4. Stomata are usually open
   a. at night, when the plant requires a supply of oxygen.

   b. during the day, when the plant requires a supply of carbon dioxide.
   c. whenever there is excess water in the soil.
   d. All of these.

5. Which of these is not a mineral ion.
   a. NO^-
   b. Mg^+
   c. CO_2
   d. Al^{+++}

6. Acid deposition affects the availability of
   a. water to plants.
   b. oxygen.
   c. minerals.
   d. All of these.

7. The pressure-flow model of phloem transport states that
   a. phloem sap always flows from the leaves to the root.
   b. phloem sap always flows from the root to the leaves.
   c. water flow brings sucrose from a source to a sink.
   d. Both a and c.

8. Root hairs do not play a role in
   a. oxygen uptake.
   b. mineral uptake.
   c. water uptake.
   d. carbon dioxide uptake.

9. Explain why this experiment supports the hypothesis that transpiration can cause water to rise to the top of tall trees.

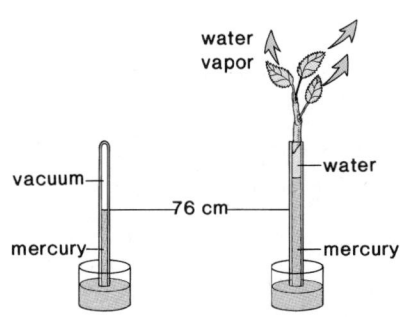

10. Label H$_2$O and K$^+$ in these diagrams. What is the role of K$^+$ in the opening of stomata?

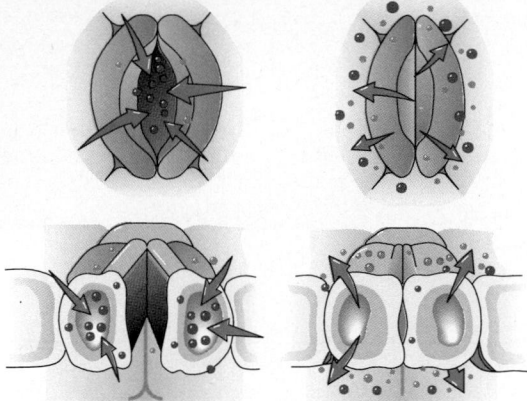

11. Explain why there is a flow of solution from the first bulb to the second bulb when this experiment is done.

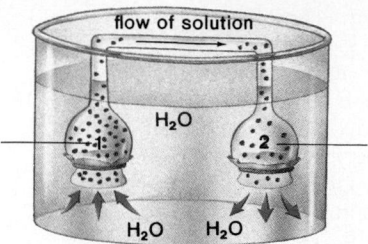

Concepts and Critical Thinking

1. *Living things are physical and chemical machines.*

(a) How does transport of water in xylem support this concept? (b) How does the opening and closing of a stoma support this concept? (c) How does transport of organic substances in phloem support this concept?

2. *Plants, like other living things, are highly organized.*

How does the structure of xylem contribute to the coordination of plant processes and parts?

Selected Key Terms

root pressure (root presh'ur) 497
guttation (gut-ta'shun) 497
cohesion-tension model (ko-he'zhun ten'shun mod'el) 497

transpiration (tran"spi-ra'shun) 497
guard cell (gahrd sel) 498
hydroponics (hi"dro-pon'iks) 500
nodule (nod'ūl) 502
mycorrhiza (mi"ko-ri'zah) 502

epiphyte (ep'i-fīt) 503
girdling (ger'd'l-ing) 503
pressure-flow model (presh'ur flo mod'el) 503

31

Reproduction in Plants

The reproductive organs of a lily are in the flower. They consist of the pistil, a central vaselike structure, and the surrounding 6 stamens (filaments with enlarged heads). Sexual reproduction leads to the production of seeds that contain a new embryonic plant.

Your study of this chapter will be complete when you can

1. draw and explain a diagram that shows the life cycle of flowering plants;
2. state and describe the parts of a flower;
3. describe the development of a female gametophyte in flowering plants from the megaspore mother cell to the production of an egg;
4. describe the development of the male gametophyte in flowering plants from the microspore mother cell to the production of sperm;
5. distinguish between pollination and fertilization, and explain the events of double fertilization in flowering plants;
6. give a classification of fruits, with examples of different types of fruits;
7. describe different means of dispersing seeds and fruits;
8. describe the development of a dicot embryo during seed formation;
9. contrast the structure and the germination of a bean seed with those of a corn kernel;
10. describe different ways of propagating plants asexually.

he flowering plants, or angiosperms, are the most diverse and widespread of all the plants. Their life cycle is like that of the gymnosperms, except that the seeds are covered within fruits. Whereas the seeds of conifers develop on the scales of cones, those of angiosperms develop within a flower (see chap. 25). The structure of the flower allows it to produce seeds within fruits. The evolution of the flower has no doubt contributed heavily to the enormous success of angiosperms in another way. It permits pollination to take place not only by wind, but also by animals. Flowering plants that rely on animals have a symbiotic relationship with their specific pollinator. The flower provides nutrients for a pollinator, such as a bee, fly, beetle, bird, or even bat. When this animal visits flowers to gather food, it carries pollen from one flower to another, allowing pollination to occur.

Life Cycle of Flowering Plants

Flowering plants, like other seed plants, exhibit an alternation of generations life cycle in which the diploid generation, the sporophyte, alternates with a haploid generation, the gametophyte. In flowering plants the sporophyte produces haploid heterospores, called microspores and megaspores (fig. 31.1). A microspore develops into a *male gametophyte* called a pollen grain, which is either windblown or carried by an animal to the vicinity of the female gametophyte. When a pollen grain matures, it contains sperm, which travel by way of a pollen tube to the female gametophyte, which is called an embryo sac. Therefore, flowering plants, like gymnosperms, are not dependent on an outside source of water for fertilization to take place.

A megaspore develops into a *female gametophyte*, the embryo sac, which is microscopic and retained within the body of the sporophyte. Each female gametophyte produces an egg, which is

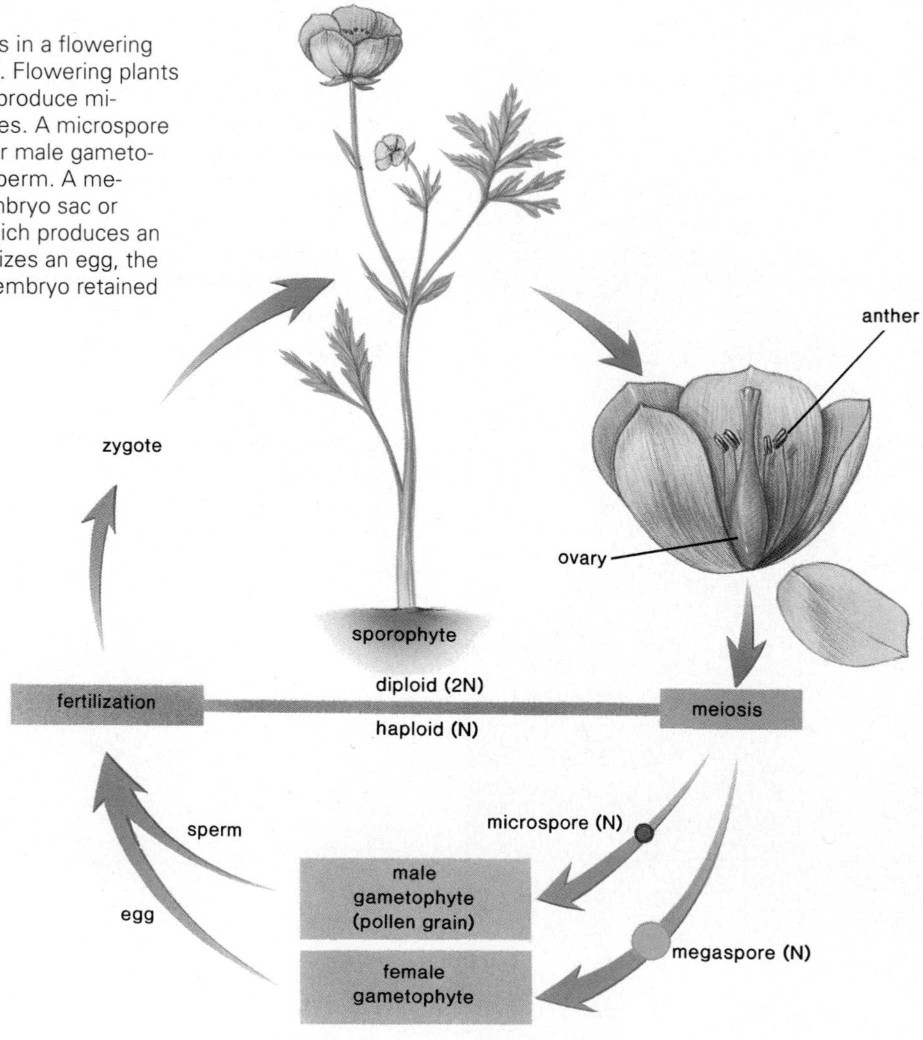

Figure 31.1

Alternation of generations in a flowering plant (see also fig. 25.3c). Flowering plants are heterosporous; they produce microspores and megaspores. A microspore becomes a pollen grain or male gametophyte, which produces sperm. A megaspore becomes the embryo sac or female gametophyte, which produces an egg. When a sperm fertilizes an egg, the zygote develops into an embryo retained within a seed.

zygote

anther

ovary

sporophyte

diploid (2N)

haploid (N)

fertilization

meiosis

sperm

microspore (N)

male gametophyte (pollen grain)

egg

megaspore (N)

female gametophyte

fertilized by a sperm released by a mature pollen grain. The zygote becomes an embryonic plant enclosed within a seed. A seed contains the embryonic plant of the next sporophyte, plus stored food.

Figure 31.1 diagrams the life cycle of a flowering plant. In this life cycle, the following occurs:

1. The dominant generation is represented by the diploid sporophyte, which bears flowers.
2. The flower produces microspores and megaspores, which develop into the male and female gametophytes, respectively.
3. The mature male gametophyte is the pollen grain, which releases sperm, one of which fertilizes the egg. The female gametophyte is retained within the flower and produces an egg.
4. Fertilization results in an embryo, which eventually is enclosed within a seed covered by fruit.

In flowering plants, the flower is a reproductive structure that produces microspores and megaspores. These develop into separate gametophytes, which produce sperm and egg. Fertilization results in a seed covered by fruit.

The Flower

A flower develops within a bud. In many plants, the same shoot apical meristem that previously formed leaves suddenly stops producing leaves and starts producing a flower. In other plants, axillary buds develop directly into flowers. Flower structures are modified leaves attached to a short stem tip called a *receptacle*

(fig. 31.2). In monocots, flower parts occur in 3s and multiples of 3. In dicots, flower parts are in 4s or 5s and multiples of 4 or 5. The **sepals,** which are the most leaflike of all the flower parts, are green, and they protect the bud as the flower develops within. When the flower opens, there is an outer whorl of sepals and an inner whorl of **petals,** whose color accounts for the attractiveness of many flowers. The size, shape, and color of a flower are attractive to a specific pollinator. Wind-pollinated flowers often have no petals at all.

The reproductive structures of a flower lie within the petals. At the very center of a flower is the **pistil,** which is often vaselike in appearance. A pistil may be simple or compound. A simple pistil contains a single reproductive unit called a **carpel.** A carpel usually has 3 parts: the **stigma,** an enlarged sticky knob; the **style,** a slender stalk; and the **ovary,** an enlarged base. A compound pistil has multiple carpels, which are often fused:

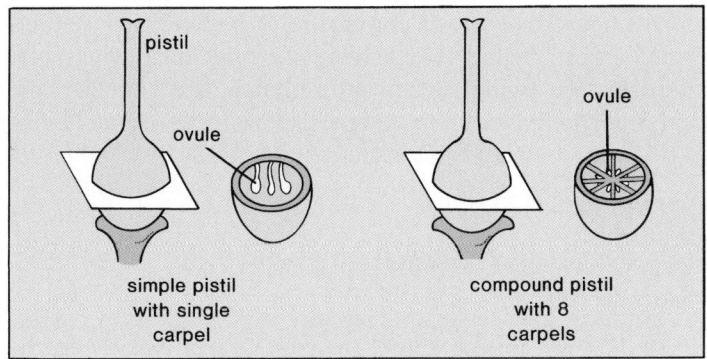

Figure 31.2

Parts of a typical flower, a sporophyte structure. Microspores are produced in the anthers, and megaspores are produced in ovules.

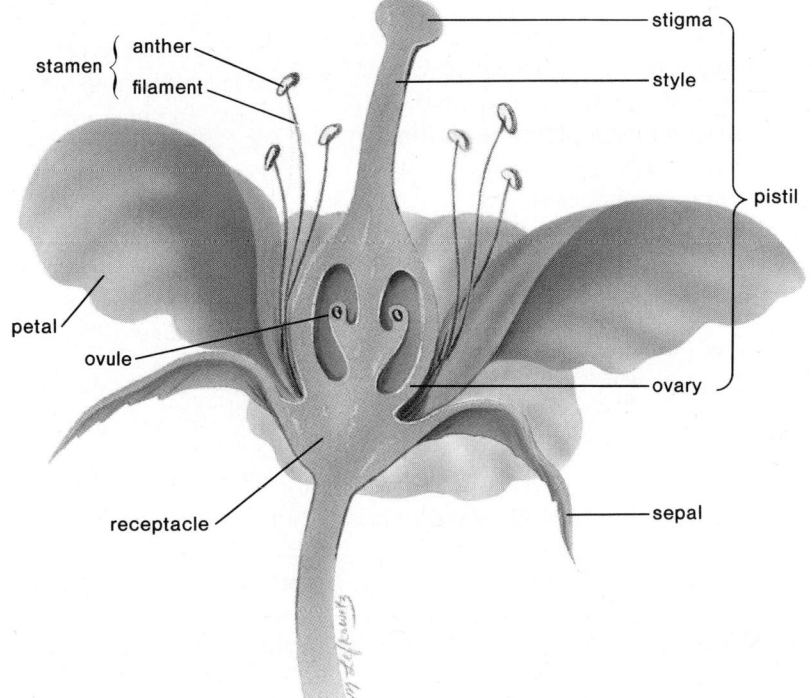

The ovary contains a number of **ovules,** which play a significant role in the production of megaspores and therefore female gametophytes. Grouped about the pistil are a number of **stamens,** each of which has 2 parts: the **anther,** a saclike container, and the **filament,** a slender stalk. Pollen grains develop in the anther.

Sometimes the pistil is called the female part of the flower and the stamens are called the male part of the flower, but this is not strictly correct. The pistil and stamens do not produce gametes; they produce megaspores and microspores, respectively. A microspore matures into a male gametophyte, which produces sperm, and a megaspore matures into a female gametophyte, which produces an egg.

Development of the Gametophytes

Within a flower, there is a megasporangium in each ovule of the ovary. The megasporangium contains a megaspore mother cell, which undergoes meiosis, producing 4 haploid megaspores. Three of these disintegrate, leaving one functional megaspore, whose nucleus divides mitotically until there are 8 nuclei of the *embryo sac or female gametophyte.* Cell walls form later to give 7 cells, one of which is binucleate:

1 egg cell associated with
2 synergid cells
1 endosperm mother cell with 2 polar nuclei
3 antipodal cells

Figure 31.3 is a micrograph of the embryo sac. Notice that the antipodal cells are at one end, and the egg cell is in between the synergid cells at the other end of the sac. The entire embryo sac is enclosed by 2 protective layers of cells called the integuments, which have a small opening, the micropyle, near the egg cell.

Male gametophytes are produced in the stamens. An anther has 4 pollen sacs or microsporangia (fig. 31.4). A pollen sac contains many microspore mother cells, each of which undergoes meiosis to produce 4 haploid microspores. A microspore divides mitotically, forming 2 cells enclosed by a finely sculptured wall. This structure, called the *pollen grain,* is the *immature male gametophyte* and contains a *tube cell* and a *generative cell.* Either now or later, the generative cell divides mitotically to give 2 sperm. The walls separating the pollen sacs in the anther break down when the pollen grains are ready to be released.

In flowering plants, an egg is found within the embryo sac, the female gametophyte that is located within an ovule. Two sperm cells are found within the pollen grain, which is the male gametophyte.

Pollination

Pollination is simply the transfer of pollen from the anther to the stigma; it should not be confused with fertilization. As mentioned previously, and as considered in this chapter's reading, pollination is brought about by the wind or with the assistance of a particular pollinator, that is, an animal that carries the pollen from the anther to the stigma.

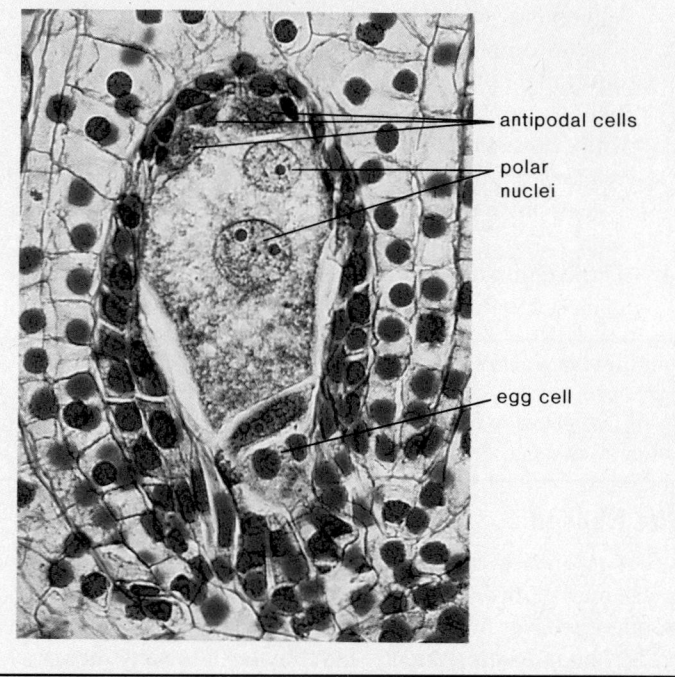

Figure 31.3
Embryo sac in a lily ovule. An embryo sac is the female gametophyte of flowering plants. It contains 8 nuclei arranged in 7 cells, one of which is the egg. The 2 polar nuclei located in the large central cell have not yet migrated to the center of the sac. Magnification, X300.

antipodal cells

polar nuclei

egg cell

Fertilization

When a pollen grain lands on the stigma of the same species, it germinates, forming a pollen tube. The germinated pollen grain, containing a tube nucleus and 2 sperm nuclei, is the mature male gametophyte. As it grows, the pollen tube passes between the cells of the stigma and the style to reach the micropyle of the ovule. Now *double fertilization* occurs. One sperm nucleus unites with the egg nucleus, forming a 2N zygote, and the other sperm nucleus migrates and unites with the polar nuclei, forming a 3N endosperm cell. The zygote divides mitotically to become the **embryo,** a young sporophyte plant, and the endosperm cell divides mitotically to become the endosperm.

Flowering plants practice double fertilization. One sperm nucleus unites with the egg nucleus, producing a zygote, and the other unites with the polar nuclei, forming a 3N endosperm cell.

Table 31.1 and figure 31.5 summarize the angiosperm life cycle.

Seeds and Fruits

With double fertilization, a number of processes begin. (1) The endosperm cell divides, forming the endosperm. **Endosperm** is a nutrient material for the developing embryo and sometimes for the young seedling as well. (2) The **zygote** develops into an embryo, as we will discuss in detail later. (3) The integuments of the ovule

Figure 31.4

Development of pollen grains. **a.** A mature anther showing that the walls between the pollen sacs have opened, allowing the pollen grains to be released. Magnification, X75. **b.** Photomicrograph of various pollen grains showing their diversity. The appearance of pollen grains varies with the species. Magnification, X200. **c.** Electron micrograph of goldenrod pollen grains. Magnification, X630.

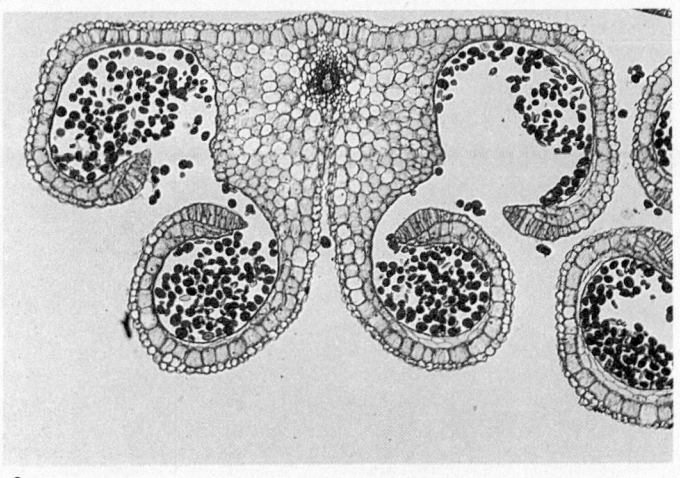

a.

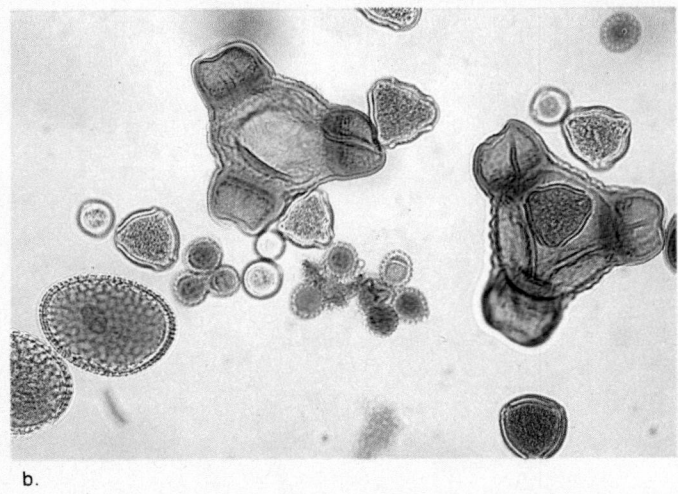

b.

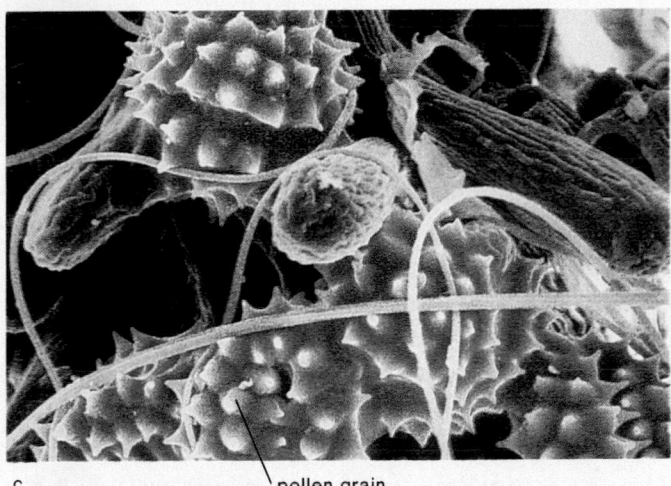

c. `pollen grain`

harden and become the seed coat. A **seed** is a structure formed by the maturation of the ovule; it contains a sporophyte embryo plus stored food. The ovary, and sometimes other floral parts, develop into a fruit. Although peas, beans, tomatoes, and cucumbers are commonly called vegetables, botanists categorize them as fruits. A **fruit** is a mature ovary that usually contains seeds.

Kinds of Fruits

As fruit develops from an ovary, the ovary wall thickens to become the *pericarp*. Most fruits are *simple fruits;* they are derived from an individual ovary, either simple or compound. In fleshy simple fruits, the pericarp is at least somewhat fleshy (table 31.2). Peaches and plums are good examples of fleshy simple fruits. An apple develops from a compound ovary, but much of the flesh comes from the receptacle, which grows around the ovary. It's more obvious that a tomato comes from a compound ovary, because in cross section, you can see several seed-filled cavities (fig. 31.6a).

Dry fruits have a dry pericarp. Legumes, such as peas (fig. 31.6b) and beans, produce a fruit that splits along 2 sides, or seams. In the case of peas, we eat the seeds, but in the case of beans, we eat the fruit and the seeds. Not all dry fruits split at maturity. The pericarp of a grain is tightly fused to the seed and cannot be separated from it. A corn kernel is actually a grain, as are the fruits of wheat, rice, and barley plants.

Figure 31.5

Life cycle of a flowering plant. Each ovule in the ovary of a pistil contains a megaspore mother cell that produces one functional haploid megaspore by meiosis. A megaspore divides mitotically 3 times, and the resulting structure with 8 nuclei in 7 cells is the female gametophyte (embryo sac). One of these cells is an egg. An anther contains many microspore mother cells, each of which produces 4 haploid microspores by meiosis. Each microspore divides mitotically, becoming a 2-celled pollen grain. When a pollen grain germinates, it contains a tube nucleus and 2 sperm nuclei. This is the mature male gametophyte. Following pollination, double fertilization occurs; one sperm nucleus unites with the egg nucleus, giving a zygote, and the other unites with the polar nuclei, giving a 3N endosperm nucleus. The zygote becomes an embryo, and the endosperm nucleus becomes stored food contained within a seed. Germination of the seed gives a sporophyte plant.

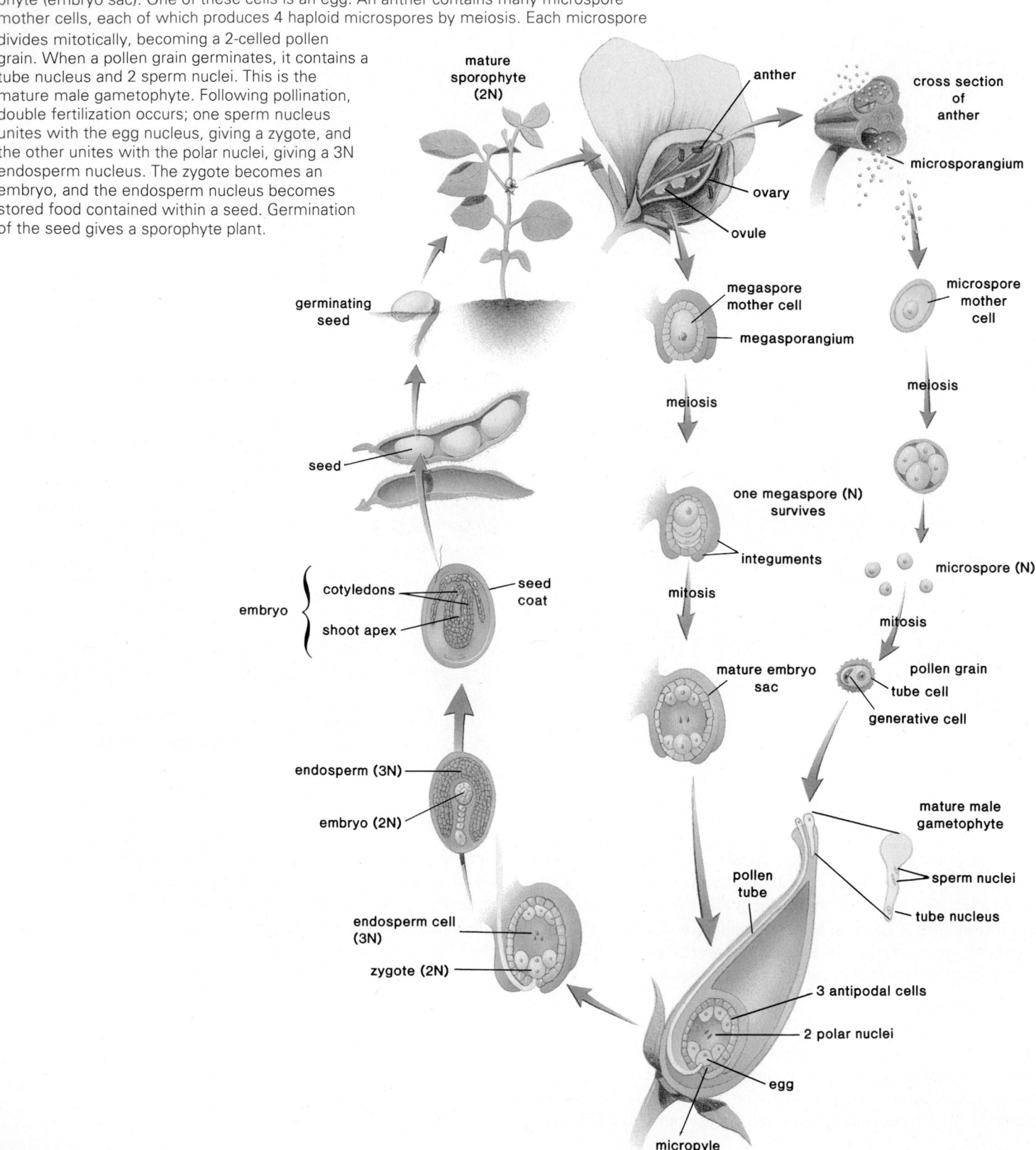

mature sporophyte (2N)

anther

cross section of anther

microsporangium

ovary

ovule

megaspore mother cell

microspore mother cell

megasporangium

germinating seed

meiosis

meiosis

seed

one megaspore (N) survives

integuments

microspore (N)

mitosis

mitosis

cotyledons

seed coat

embryo

shoot apex

mature embryo sac

pollen grain

tube cell

generative cell

endosperm (3N)

embryo (2N)

mature male gametophyte

sperm nuclei

tube nucleus

pollen tube

endosperm cell (3N)

zygote (2N)

3 antipodal cells

2 polar nuclei

egg

micropyle

Plant Structure and Function

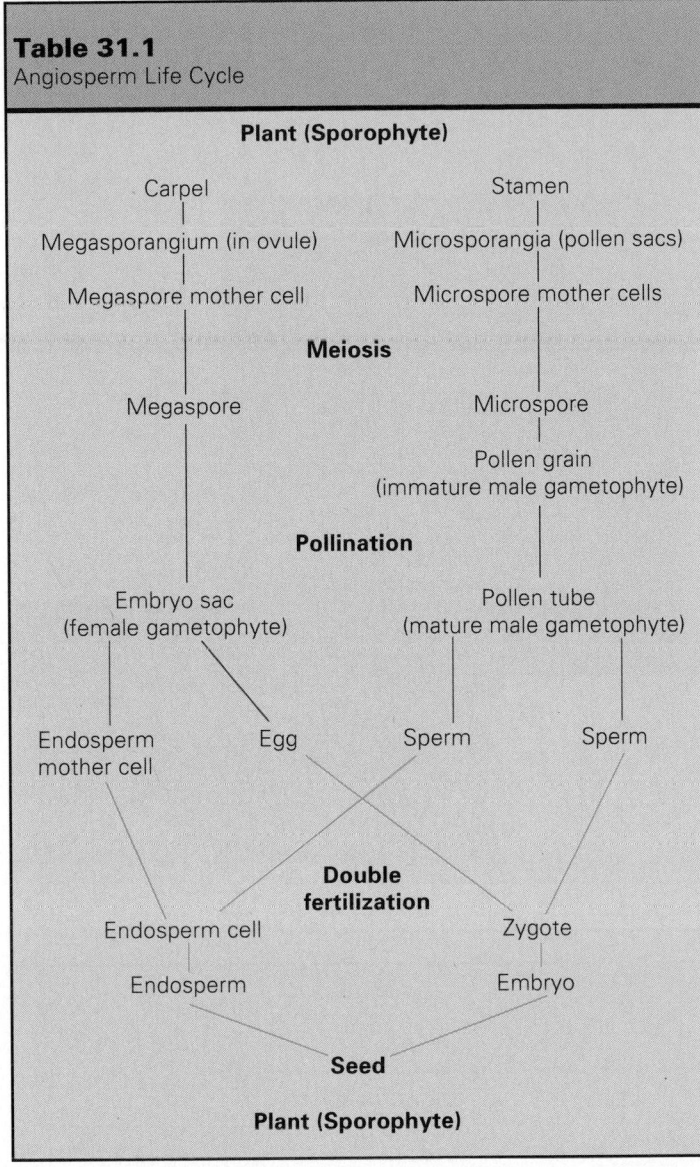

Table 31.1
Angiosperm Life Cycle

Plant (Sporophyte)

Carpel   Stamen

Megasporangium (in ovule)   Microsporangia (pollen sacs)

Megaspore mother cell   Microspore mother cells

Meiosis

Megaspore   Microspore

Pollen grain
(immature male gametophyte)

Pollination

Embryo sac   Pollen tube
(female gametophyte)   (mature male gametophyte)

Endosperm   Egg   Sperm   Sperm
mother cell

**Double
fertilization**

Endosperm cell   Zygote

Endosperm   Embryo

Seed

Plant (Sporophyte)

From T. Elliot Weier, et al., *Botany*, 6th ed. Copyright © 1982 John Wiley & Sons, Inc., New York, NY. Reprinted by permission of the author.

Some fruits are compound fruits; they develop from several individual ovaries. A blackberry is an aggregate fruit in which each berry is derived from the separate ovaries of a single flower (fig. 31.6c). The strawberry is also an aggregate fruit, but each ovary becomes a one-seeded fruit called an achene. The flesh of a strawberry is from the receptacle. In contrast, a pineapple comes from the fruit of many individual flowers attached to the same fleshy stalk. As the ovaries mature, they fuse to form a large, multiple fruit.

Seed Dispersal

For plants to be widely distributed, they have to disperse their seeds, preferably long distances from the parent plant. Wind, water, or various animals are all agents of dispersal for angiosperms.

The hooks and spines of clover, bur, cocklebur, and stickseed attach to the fur of animals and the clothing of humans. Birds and mammals sometimes eat fruits, including the seeds, which are then defecated (passed out of the digestive tract with the feces) some distance from the parent plant. Squirrels and other animals gather seeds and fruits, which they bury some distance away. Even seeds that are forgotten or abandoned aid the dispersal of the plant that produced them.

The fruit of the coconut palm can be dispersed by ocean currents. It may land many hundreds of kilometers away from the parent plant.

Some seeds, such as the approximately one million produced by an orchid, are so small and light that they need no special adaptation to carry them far away. The somewhat heavier dandelion fruit uses a tiny "parachute" for dispersal (fig. 31.7). The winged fruit of a maple tree, which contains 2 seeds, has been known to travel up to 10 km from its parent. Some seeds are forcibly ejected from a fruit that remains attached to the parent. Touch-me-not has seed pods that swell as they mature. When the pods finally burst, the ripe seeds are hurled out.

Development of the Embryo

As the integuments of the ovule become the seed coat, the zygote develops into an embryo. Early divisions of the zygote produce 2 parts: the upper part is the embryo, and the lower part is the suspensor, which anchors the embryo and transfers nutrients to it from the sporophyte plant. Soon the cotyledons, or seed leaves, can be seen. Figure 31.8 shows the development of a dicot embryo with its 2 cotyledons. At this point, the dicot embryo is heart-shaped. Later, when it becomes torpedo-shaped, it is possible to distinguish the shoot apex and the root apex. These contain apical meristems, the tissues that bring about primary growth in a plant; the shoot apical meristem is responsible for aboveground growth, and the root apical meristem is responsible for underground growth (see fig. 29.1).

Monocots, unlike dicots, have only one cotyledon. Another important difference between monocots and dicots is the manner in which nutrient molecules are stored in the seed. In a monocot, the cotyledon rarely stores food; rather, it absorbs food molecules from the endosperm and passes them to the embryo. During the development of a dicot embryo, the cotyledons usually store the nutrient molecules that the embryo uses. Therefore, in figure 31.8 we can see that the endosperm seemingly disappears. Actually, it has been taken up by the 2 cotyledons.

Regardless of whether or not the endosperm is taken up by the cotyledon(s), a mature seed contains an embryonic sporophyte plant and stored food. It is surrounded by a **seed coat,** which is derived from the integuments of the ovule.

Some seeds do not germinate (begin to grow) until they have been dormant for a period of time. For seeds, dormancy is the time when growth does not occur, even though conditions may be favorable for growth. Seed dormancy may not occur in the moist tropics, but dormancy in the temperate zone may not

Figure 31.6

Fruit diversity. In general, fruits develop from the ovary and seeds develop from the ovules. Often the other parts disappear. ***a.*** Developing fruit of tomato plant and a tomato cut crosswise revealing ripened compound ovary and seeds. ***b.*** Developing fruit of pea plant and dry fruit of pea plant developed from a simple ovary. Each pea developed from an ovule. ***c.*** Flower of blackberry with several individual ovaries and aggregate fruit of blackberry. Each cluster is a berry from a single flower.

a.

b.

Table 31.2
Kinds of Fruits

Name	Description	Example
Simple Fruits	*Develop from an individual ovary*	
Fleshy	Pericarp is usually fleshy	
Drupe	From simple ovary with one seed (pit) and soft "skin"	Peach, plum, olive
Berry	From compound ovary with many seeds	Grape, tomato
Pome	From compound ovary; flesh is from accessory flower parts	Apple, pear
Dry	Pericarp is dry	
Follicle	From simple ovary that splits open down one side	Milkweed, peony
Legume	From simple ovary that splits open down both sides	Pea, bean, lentil
Capsule	From compound ovary with capsules that split in various ways	Poppy
Achene	From simple ovary with one-seeded small fruit; pericarp easily removed	Sunflower
Nut	From simple ovary with one-seeded fruit; hard pericarp	Acorn, hickory nut, chestnut
Grain	From simple ovary with one-seeded small fruit; pericarp completely united with seed coat	Rice, oat, barley
Compound Fruits	*Develop from a group of individual ovaries*	
Aggregate fruits	Ovaries are from a single flower	Blackberry, raspberry, strawberry
Multiple fruits	Ovaries are from separate flowers clustered together	Pineapple

c.

Figure 31.7
What you might think of as a dandelion seed is actually a one-seeded fruit. The dandelion inflorescence resembles a single flower, but actually it is a composite of many individual flowers called florets. Dandelion fruits are adapted to dispersal by the wind. Each fruit has a tiny "parachute," which keeps it afloat in the breeze.

be broken until seeds have been exposed to a period of cold weather. In deserts and dry tropical areas, germination does not occur until there is adequate moisture. This requirement helps ensure that seeds do not germinate until the most favorable growing season has arrived. The first sign of the onset of germination is the bursting of the seed coat, due to the uptake of water and swelling of the cells. Water, bacterial action, and even fire act on the seed coat, allowing it to become permeable to water. Germination takes place if there is sufficient water, warmth, and oxygen to sustain growth.

Structure and Germination of Seeds

To illustrate the main features of seeds, we will contrast a bean seed (dicot) with a corn kernel (monocot). A bean seed has a small white scar called a *hilum,* where the ovule was attached to the ovary wall (fig. 31.9*a*). Right next to the hilum is the *micropyle,* which is visible as a small pore.

If the cotyledons of the seed are parted, you can see a rudimentary plant. The embryonic shoot bears young leaves and is called a **plumule.** The **epicotyl** is the portion of the stem above the attachment of the cotyledons, and the **hypocotyl** is the portion of stem below the attachment of the cotyledons. The **radicle** is the part of the embryo that contains the root apical meristem and

becomes the first (primary) root of the seedling. All of these parts become more obvious when the seed germinates. **Germination** is the resumption of growth by a seed or other reproductive plant structure. As the dicot seedling emerges from the soil, the shoot is hook-shaped to protect the delicate plumule (fig. 31.9*b*).

Corn kernels are actually fruits, and the outer covering is the pericarp (fig. 31.10*a*). The bulk of the food-storage tissue is endosperm, and the cotyledon does not play a storage role. The plumule and radicle are enclosed in protective sheaths called the coleoptile and coleorhiza, respectively. The plumule and the radicle burst through these coverings when germination occurs (fig. 31.10*b*).

Asexual Reproduction in Plants

Asexual reproduction, also known as *vegetative propagation,* is common in plants. In vegetative propagation, a portion of one plant gives rise to a completely new plant. Both plants then have identical genes. Some plants have aboveground, horizontal stems, called stolons or runners, and others have underground stems, called *rhizomes,* that produce new plants. For example, strawberry plants grow from the nodes of stolons and violets grow from the nodes of rhizomes. White potatoes can be propagated in a similar

Figure 31.8

Stages in development of a dicot embryo and seed. **a.** The single-celled zygote lies beneath the endosperm nucleus. **b.** and **c.** The endosperm is a mass of tissue surrounding the embryo. The embryo itself is actually made up of the cells above the suspensor. **d.** The embryo becomes heart-shaped as the cotyledons begin to appear. **e.** There is progressively less endosperm as the embryo differentiates and enlarges. As the cotyledons bend around, the embryo takes on a torpedo shape. **f.** The mature seed has a seed coat, which protects the enclosed embryo. The embryo consists of the epicotyl, represented here by the shoot apex; the hypocotyl, and the radicle, which forms the first seedling root.

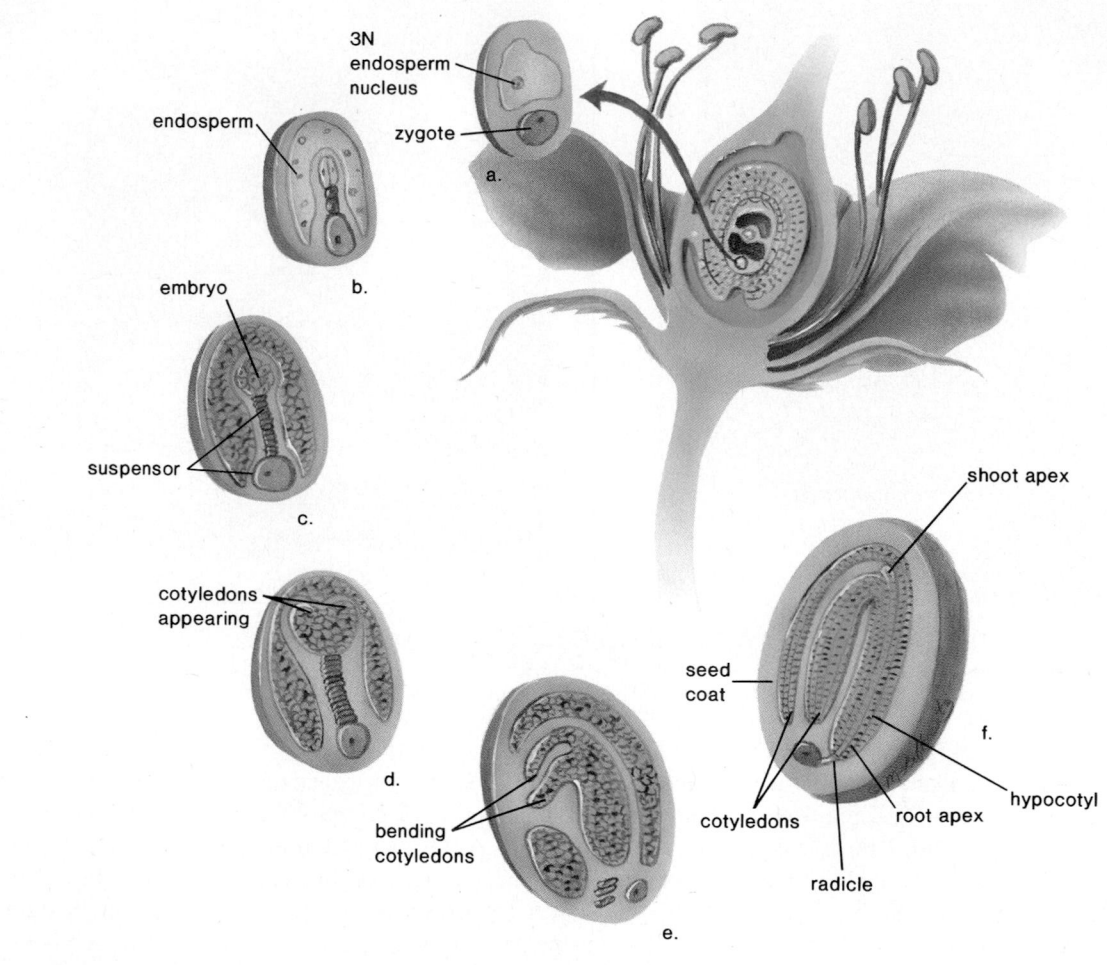

manner. White potatoes are actually portions of underground stems, and each eye is a bud that will produce a new potato plant if it is planted with a portion of a swollen tuber. Sweet potatoes are modified roots; they can be propagated by planting sections of the root. You may have noticed that the roots of some fruit trees, such as cherry and apple trees, produce "suckers," small plants that can be used to grow new trees.

Asexual reproduction has a great deal of commercial importance. Once a plant variety with desired characteristics has been developed by a sexual means such as hybridization, new plants can be supplied to gardeners and farmers by vegetative propagation. Cuttings can be taken from the plant, and the cut end can be treated to encourage it to grow roots, or a cutting can be grafted to the stem of a plant that has a root. Budding is the form of grafting most often used commercially. In this procedure, the axillary buds are grafted onto the stem of another plant.

Today, entire plants can be produced by tissue culture, a technique that may eventually replace the older methods thus far discussed. Usually an embryonic *tissue* is removed from a plant and placed in a special *culture* medium. After the tissue has grown for a while, it is subdivided to produce many identical plants from a very small number of starting cells.

Plants reproduce both asexually and sexually. Asexual reproduction occurs when a portion of one plant gives rise to an entirely new plant. Grafting, particularly budding, and tissue-culture propagation have commercial importance today.

Figure 31.9

A common garden bean. *a.* Seed structure. *b.* Germination and development of the seedling.

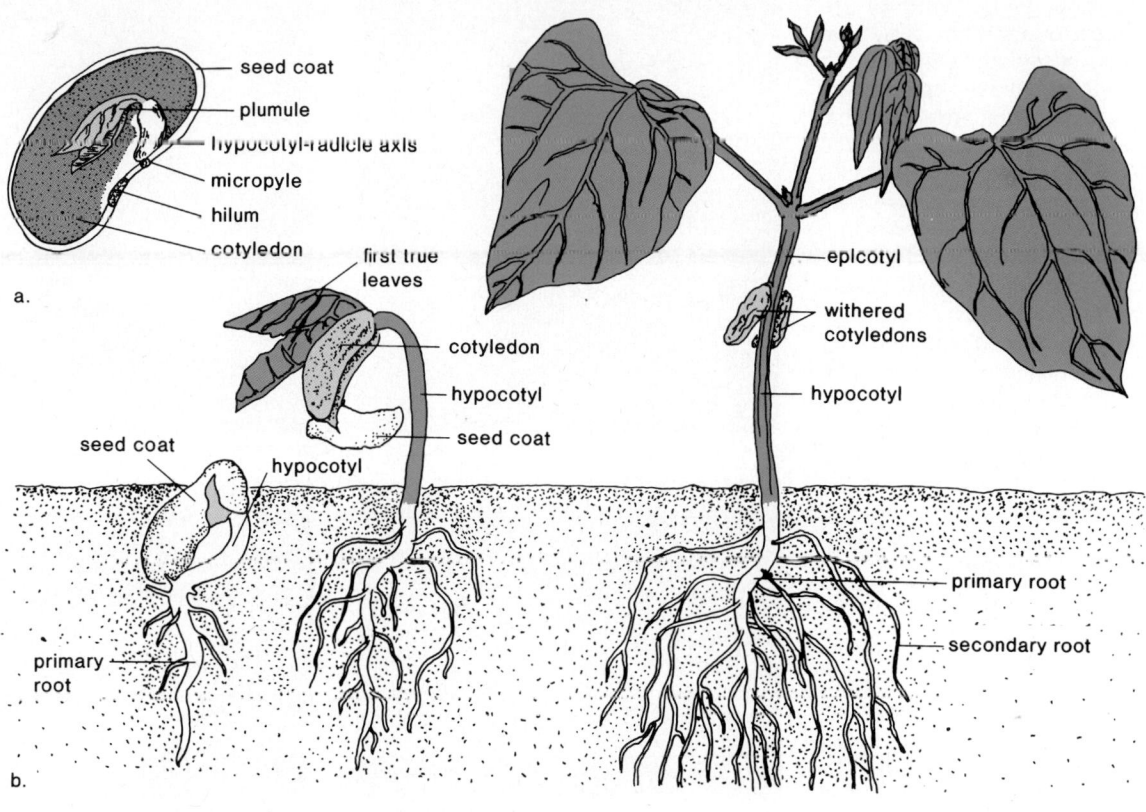

Figure 31.10

Corn. *a.* Grain structure. *b.* Germination and development of the seedling.

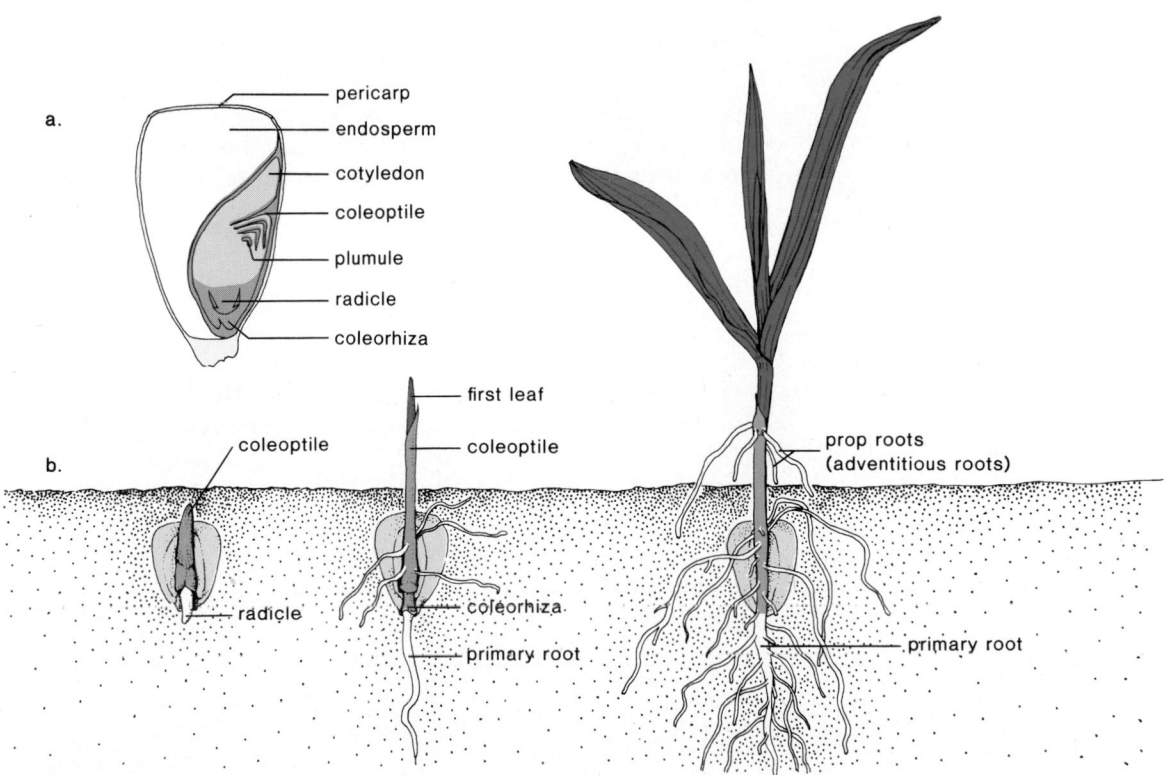

Coevolution of Plants and Their Pollinators

A plant and its pollinator(s) are adapted to one another. They have a mutualistic relationship in which each benefits—the plant uses its pollinator to ensure that cross-pollination takes place, and the pollinator uses the plant as a source of food. This mutualistic relationship came about through the process of coevolution; that is, the codependency of the plant and the pollinator is the result of suitable changes in structure and function in each. The evidence for coevolution is observational. For example, floral coloring and odor are suited to the sense perceptions of the pollinator, the mouthparts of the pollinator are suited to the structure of the flower, the type of food provided is suited to the nutritional needs of the pollinator, and the pollinator forages at the time of day that specific plants are open. The following are examples of such coevolution.

Bee-Pollinated Flowers

There are now 20,000 different species of bees that pollinate flowers. The best-known pollinators are the honeybees (fig. 31.A*a*). Bee eyes see a spectrum of light that is different from the spectrum seen by humans. The bee's visible spectrum is shifted so that they do not see red wavelengths but do see ultraviolet wavelengths. Bee-pollinated flowers are usually brightly colored and are predominantly blue or yellow; they are not entirely red. They may also have ultraviolet shadings called honey guides, which highlight the portion of the flower that contains the reproductive structures (see fig. 40.4). The mouthparts of bees are fused into a long tube that contains a tongue. This tube is an adaptation for sucking up nectar provided by the plant, usually at the base of the flower.

Bee flowers are delicately sweet and fragrant to advertise that nectar is present. The honey guides often point to a narrow floral tube large enough for the bee's feeding apparatus but too small for other insects to reach the nectar. Bees also collect pollen as food for their larvae. Pollen clings to the hairy body of a bee, and the bees also gather it by means of bristles on their legs. They then store the pollen in pollen baskets on the third pair of legs. Bee-pollinated flowers are sturdy and irregular in shape because they often have a landing platform

a.

b.

c.

pollen

d.

Figure 31.A

Flowers have adaptations that make them attractive to their pollinators. *a.* A bee-pollinated flower is a color other than red (bees cannot detect this color) and has a landing platform where the reproductive structures of the flower brush up against the bee's body. *b.* A butterfly pollinated flower is often a composite, containing many individual flowers. The broad expanse provides room for the butterfly to land, after which it lowers its proboscis into each flower in turn. *c.* Hummingbird-pollinated flowers are curved back, allowing the bird to insert its beak to reach the rich supply of nectar. While doing this, the bird's forehead and other body parts touch the reproductive structures. *d.* Bat-pollinated flowers are large, sturdy flowers that can take rough treatment. Here the head of the bat is positioned so that its bristly tongue can lap up nectar.

where the bee can alight. The landing platform requires the bee to brush up against the anther and stigma as it moves toward the floral tube to feed. One type of orchid (genus *Ophrys*) has evolved a unique adaptation. The flower resembles a female bee and when the male of that species attempts to copulate with the flower, the bee receives pollen.

Moth- and Butterfly-Pollinated Flowers

Contrasting moth- and butterfly-pollinated flowers emphasizes the close adaptation between pollinator and flower. Both moths and butterflies have a long, thin, hollow proboscis, but they differ in other characteristics. Moths usually feed at night and

Plant Structure and Function

have a well-developed sense of smell. The flowers they visit are visible at night because they are lightly shaded (white, pale yellow, or pink) and have a strong, sweet perfume, which helps attract moths. Moths hover when they feed, and their flowers have deep tubes with open margins that allow the hovering moths to reach the nectar with their long proboscis. Butterflies are active in the daytime and have good vision but a weak sense of smell. Their flowers have bright colors, even red because butterflies can see the color red, but the flowers tend to be odorless. Unable to hover, butterflies need a place to land. Flowers that are visited by butterflies often have flat landing platforms on which butterflies can alight (fig. 31.A*b*). The flowers also tend to be composites, with many individual flowers clustered in a head. Each flower has a long, slender floral tube, accessible to the long, thin, butterfly proboscis.

Bird- and Bat-Pollinated Flowers

In North America, the most well-known bird pollinators are the hummingbirds. These tiny animals have good eyesight but do not have a well-developed sense of smell. Like moths, they hover when they feed. Typical flowers pollinated by hummingbirds are red, with a slender floral tube and margins that are curved back and out of the way. And although they produce copious amounts of nectar, the flowers have little odor. As a hummingbird feeds on nectar with its long, thin beak, its head comes into contact with the stamens and pistil (fig. 31.A*c*).

Bats are adapted to gathering food in various ways, including feeding on the nectar and pollen of plants. Bats are nocturnal and have an acute sense of smell. Those that are pollinators also have keen vision and a long, extensible, bristly tongue. Typically, bat-pollinated flowers open only at night and are light-colored or white. They have a strong, musty smell similar to the odor that bats produce to attract one another. The flowers are generally large and sturdy and are able to hold up when a bat inserts part of its head to reach the nectar. While the bat is at the flower, its head is dusted with pollen (fig. 31.A*d*).

Coevolution

These examples are evidence of coevolution, but how did coevolution come about? About 200 million years ago, when seed plants were just beginning to evolve and insects were not as diverse as today, wind alone was used to carry pollen. Wind pollination, however, is a hit-or-miss affair. Perhaps beetles feeding on vegetative leaves were the first insects to carry pollen directly from plant to plant by chance. This use of animal motility to achieve cross-fertilization no doubt resulted in the evolution of flowers, which have features, such as the production of nectar, to attract pollinators. Then, if beetles developed the habit of feeding on flowers, other features, such as protection of ovules within ovaries, may have evolved.

As cross-fertilization continued, more and more flower variations likely developed, and pollinators became increasingly adapted to specific angiosperm species. Today, there are some 235,000 species of flowering plants and over 700,000 species of insects. This diversity suggests that the success of angiosperms has contributed to success of insects and vice versa.

Summary

1. Angiosperms produce flowers and often have a symbiotic relationship with an animal pollinator. They also produce seeds that are covered by fruits.
2. Flowering plants exhibit an alternation of generations life cycle that includes heterospores and separate male and female gametophytes. The pollen grain is the male gametophyte. The female gametophyte is called the embryo sac and remains within the body of the sporophyte plant.
3. Typical parts of a flower are sepals, which are green in color and form an outer whorl; petals, often colored, which form an inner whorl; the pistil, which is in the center and contains the carpels, each consisting of stigma, style, and ovary; and the stamens, each having a filament and anther, which are around the base of the carpels. The ovary contains ovules.
4. Each ovule contains a megaspore mother cell, which divides meiotically to produce 4 haploid megaspores, only one of which survives. This megaspore divides mitotically to produce the embryo sac (female gametophyte), which usually has 8 nuclei. The 2 center nuclei are the polar nuclei, and one of the 3 cells next to the micropyle is an egg cell.
5. The anthers contain microspore mother cells, each of which divides meiotically to produce 4 haploid microspores. Each of these divides mitotically to give a 2-celled pollen grain. One cell is the tube cell, and the other is the generative cell. The generative cell later divides to give 2 sperm. The pollen grain is the male gametophyte. After pollination, the pollen grain germinates, and as the pollen tube grows, the sperm nuclei travel to the embryo sac. Pollination is simply the transfer of pollen from anther to stigma.
6. Angiosperms practice double fertilization. One sperm nucleus unites with the egg nucleus, forming a 2N zygote, and the other unites with the polar nuclei, forming a 3N endosperm cell.
7. After fertilization, the endosperm cell divides to form the endosperm. The zygote becomes the sporophyte embryo. The ovule matures into the seed (its integuments become the seed coat). The ovary becomes the fruit.
8. There are different types of fruits. Simple fruits are derived from a single ovary (which can be simple or compound). Some simple fruits are fleshy, such as a peach or apple. Others are dry, such as peas, nuts, and grains. Aggregate fruits develop from a number of ovaries of a single flower, and compound fruits develop from multiple ovaries of separate flowers.
9. Flowering plants have several ways to disperse seeds. Seeds may be blown by the wind, attached to animals that carry them away, eaten by animals that defecate them some distance away, or adapted to water transport.
10. As the ovule is becoming a seed, the zygote is becoming an embryo. After the first several divisions, it is possible to discern the embryo and the suspensor. The suspensor attaches the embryo to the ovule and supplies it with nutrients. The dicot embryo becomes first heart-shaped and then torpedo-shaped. Once you can see the 2 cotyledons, it is possible to distinguish the shoot apex and the root apex, which contain the apical meristems. In dicot seeds, the cotyledons frequently take up the endosperm.

11. Prior to germination, you can distinguish a bean (dicot) seed's 2 cotyledons and plumule, which is the shoot that bears leaves. Also present are the epicotyl, hypocotyl, and the radicle. In a corn kernel (monocot), the endosperm, cotyledon, plumule, and radicle are visible.

12. Many flowering plants reproduce asexually. Sometimes the nodes of stems (either above ground or underground) give rise to entire plants. Cuttings of plants will produce entire plants, assuming that the cutting can be encouraged to develop roots. Grafting and budding also allow the production of new plants. Tissue culture is the latest method for asexually producing plants.

Study Questions

1. Name 2 unique features of flowering plant reproduction.
2. Draw a diagram that illustrates the life cycle of seed plants, and tell how it applies to the flowering plants. Why don't flowering plants require a source of outside water for pollination?
3. Draw a diagram of a flower and name the parts.
4. Describe the development of a female gametophyte from the megaspore mother cell to the production of an egg.
5. Describe the development of the male gametophyte from the microspore mother cell to the production of sperm.
6. What is the difference between pollination and fertilization?
7. Distinguish between simple, fleshy and simple, dry fruits. Give an example of each type. What is an aggregate fruit? a multiple fruit?
8. Name several mechanisms of seed and fruit dispersal.
9. Describe the sequence of events as a dicot zygote becomes a seed. Describe the dormant and germinated seed of a dicot; of a monocot.
10. In what ways do plants reproduce asexually without human assistance? with human assistance?

Objective Questions

1. In flowering plants, the
   a. gametes become the gametophyte generation.
   b. spores become the sporophyte generation.
   c. sporophyte organism produces spores.
   d. Both a and b.
2. The flower part that contains ovules is the
   a. carpel.
   b. stamen.
   c. sepal.
   d. petal.
3. The megaspore mother cell and the microspore mother cell
   a. both produce pollen grains.
   b. both divide meiotically.
   c. both divide mitotically.
   d. produce pollen grains and embryo sacs, respectively.
4. A pollen grain is a
   a. haploid structure.
   b. diploid structure.
   c. first diploid and then haploid structure.
   d. first haploid and then diploid structure.
5. Which of these is mismatched?
   a. polar nuclei—plumule
   b. egg and sperm—zygote
   c. ovule—seed
   d. ovary—fruit
6. Which of these is not a fruit?
   a. walnut
   b. pea
   c. green bean
   d. peach
7. Animals assist with
   a. both pollination and seed dispersal.
   b. only pollination.
   c. only seed dispersal.
   d. only asexual propagation of plants.
8. A seed contains
   a. a zygote.
   b. an embryo.
   c. stored food.
   d. Both b and c.
9. Which of these is mismatched?
   a. plumule—leaves
   b. cotyledon—seed leaf
   c. epicotyl—root
   d. pericarp—corn kernel

Plant Structure and Function

10. Gametes are involved in
- **a.** both the asexual and the sexual reproduction of plants.
- **b.** the asexual reproduction of plants.
- **c.** the sexual reproduction of plants.
- **d.** neither the asexual nor the sexual reproduction of plants.

11. Label the following diagram of alternation of generations in an-giosperms:

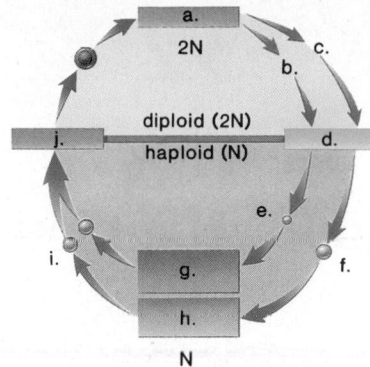

Concepts and Critical Thinking

1. *Living things have a reproductive strategy.*

What is the reproductive strategy of flowering plants?

2. *Living things have life cycles that promote survival of the species.*

How does the life cycle of flowering plants promote the possibility of gametophyte and sporophyte survival?

3. *Coevolution increases the fitness of both parties.*

How does coevolution between plants and their pollinators increase the fitness of the pollinators?

Selected Key Terms

pistil (pis′til) 509
ovary (o′vah-re) 509
ovule (o′vūl) 510
stamen (sta′men) 510
anther (an′ther) 510

pollination (pol″ĭ-na′shun) 510
endosperm (en′do-sperm) 510
zygote (zi′gōt) 510
seed (sēd) 511
fruit (frōot) 511

plumule (plōō′mul) 515
epicotyl (ep″ĭ-kot′il) 515
hypocotyl (hi″po-kot′il) 515
radicle (rad′ĭ-k′l) 515

Growth and Development in Plants

These bean seedlings are bending toward the light because cells on the shady side of the stem have elongated. Plants often respond to outside stimuli by unequal growth patterns, which are controlled by plant hormones. The phototropism of plants has survival value because it enhances their ability to catch the rays of the sun.

Your study of this chapter will be complete when you can

1. name the 5 major types of hormones, state which are stimulatory and which are inhibitory, and give a primary function for each type;
2. describe 2 experiments that contributed to understanding the role that auxin plays in phototropism;
3. indicate the manner in which plant hormones function according to gibberellin research;
4. describe a tissue culture experiment demonstrating that plant hormones interact;
5. describe 3 types of tropisms that occur in response to external stimuli; indicate any involvement of plant hormones in these responses;
6. list some movements of plants that occur as a response to internal stimuli, and provide evidence that biological clocks function in plants;
7. discuss the relationship of photoperiodism to flowering in certain plants;
8. explain the phytochrome conversion cycle, and suggest possible functions of phytochrome in plants.

Plants often respond to external and internal stimuli with changes in their growth patterns. Some of the external factors that regulate growth are light, day length, gravity, and temperature. Among the principal internal factors that regulate growth are plant hormones.

Hormones

A plant **hormone** is an organic molecule synthesized by the plant that has physiological and/or developmental effects at very low concentrations. Some plant hormones are synthesized in meristematic tissue and transported in vascular tissue (cytokinins and abscisic acid); others are synthesized and utilized in the same tissues (ethylene) or move directly from tissue to tissue without entering vascular tissue (auxins).

Each naturally occurring hormone has a specific chemical structure. Other chemicals, some of which differ only slightly from the natural hormones, also affect the growth of plants. These and the naturally occurring hormones are sometimes grouped together and called plant growth regulators. This chapter's reading discusses the various uses of plant growth regulators.

Hormones that Promote Growth

Notice in table 32.1 that 3 groups of hormones (auxins, gibberellins, and cytokinins) promote activities associated with growth. We will consider these hormones first.

Auxins

The most common naturally occurring **auxin** is indolacetic acid (IAA), which is produced in shoot meristem, in young leaves, and in flowers and fruits.

Early researchers observed that plants bend toward light, a phenomenon called *phototropism*. Around 1881, Charles Darwin and his son Francis reported on experiments they had performed with grass and oat seedlings. At first, each seedling is covered by a sheath called the *coleoptile*. The leaves soon break through this covering of the tiny shoot (fig. 32.1). The Darwins found that if the coleoptile is kept intact, the seedlings bend toward a unidirectional light source. But if the tip of the seedling is cut off or is covered by a black cap, the seedling does not respond to light. They concluded that some influence that causes bending is transmitted from the coleoptile tip to the rest of the shoot.

Building on the results of the Darwins and others, Frits W. Went performed still more experiments in 1926 (fig. 32.2). He cut off the tips of coleoptiles and placed them on agar (a gelatin-like material). After about an hour, he removed the tips

Table 32.1
Plant Hormones

Type	Primary Example	Notable Function
Growth Promoters		
Auxins	Indolacetic acid (IAA)	Cell elongation
Gibberellins	Gibberellic acid (GA)	Stem elongation
Cytokinins	Zeatin	Cell division
Growth Inhibitors		
Abscisic acid	Abscisic acid (ABA)	Dormancy
Ethylene	Ethylene	Abscission

Figure 32.1

a. An oat seedling is at first protected by a hollow sheath called a coleoptile. Later the leaves break through this sheath. *b.* Experiments with oat seedlings show that the coleoptile is necessary and must be exposed to light in order for oat seedlings to bend toward the light. If the tip of a seedling is cut off or is covered by a black cap, the seedling will not bend toward unidirectional light.

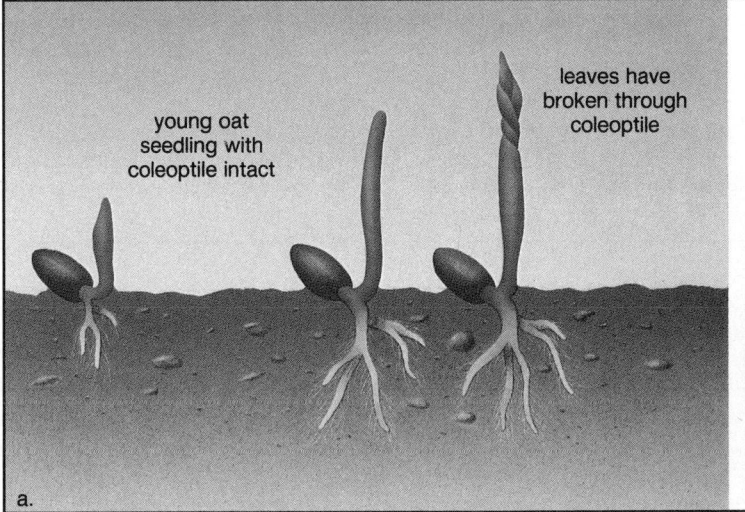
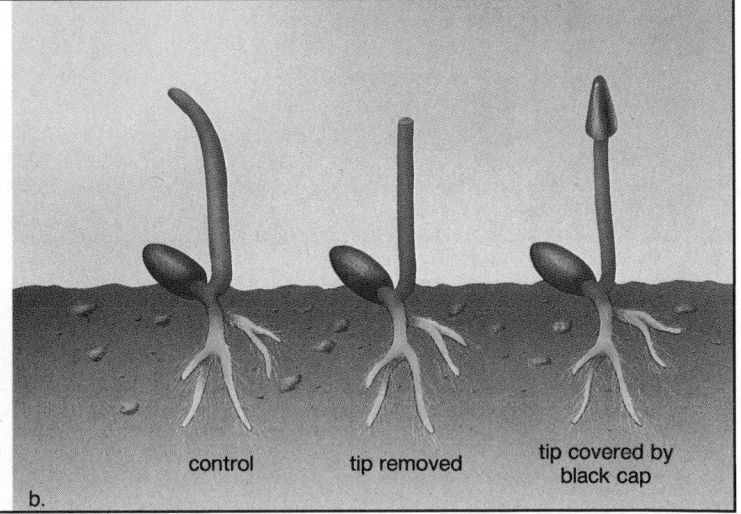

Plant Growth Regulators

N ow that the formulas for many plant hormones are known, it is possible to synthesize them as well as related chemicals in the laboratory. Collectively, these substances are known as plant growth regulators. Many scientists hope plant growth regulators will bring about an increase in crop yield, just as fertilizers, irrigation, and pesticides have done in the past.

Since auxin was first discovered, researchers have found many agricultural and commercial uses for it. Auxins cause the base of stems to form new roots quickly, so that new plants are easily started from cuttings. When sprayed on apples, pears, and citrus fruits shortly before harvest, auxins prevent fruit from dropping too soon. Because auxins inhibit the growth of lateral buds, potatoes sprayed with an auxin do not sprout and thus have a longer storage life.

In high concentrations, some auxins are used widely in agriculture as herbicides to prevent the growth of broad-leaved plants. In addition to their use in weed control, the synthetic auxins known as 2,4D and 2,4,5T were used as defoliants during the Vietnam War. Even though 2,4D has been known for over 35 years, we do not yet know how it works. Apparently, it is structurally different enough from natural auxin that a plant's own enzymes cannot break it down. Its concentration rises until metabolism is disrupted, cellular order is lost, and the cells die.

The other plant hormones studied in this chapter also have agricultural and commercial uses. Gibberellins are used to stimulate seed germination and seedling growth of some grains, beans, and fruits. They also increase the size of some mature plants. Treatment of sugarcane with as little as 2 oz per acre increases the cane yield by more than 5 metric tons. The application of either auxins or gibberellins can cause an ovary and accessory flower parts to develop into fruit, even though pollination and

Figure 32.A
Effect of gibberellic acid (GA_3) on Thompson seedless grapes (*Vitis vinifera*). Control grapes (*left*). GA_3 was sprayed at bloom and at fruit set (*right*). Almost all grapes sold in stores are now treated with gibberellic acid.

fertilization have not taken place. In this way, it is sometimes possible to produce seedless fruits and vegetables or bigger, more uniform bunches with larger fruit, as shown in figure 32.A.

Because cytokinins retard the aging of leaves and other organs, they are sprayed on vegetables to keep them fresh during shipping and storage. Such treatment of holly, for example, allows it to be harvested many weeks prior to a holiday.

Plant growth regulators are making it possible to grow plants from a few cells in laboratory glassware. It may even be possible to develop new varieties of food plants with particular characteristics, such as tolerance to heat, cold, toxins, or drought. Using gene-splicing techniques, it may eventually be possible to create plants capable of utilizing atmospheric nitrogen.

Several synthetic inhibitors are used to oppose the action of auxins, gibberellins, and cytokinins normally present in plants. Some of these can cause leaf and fruit drop at a time convenient to the farmer. Remov-

ing leaves from cotton plants aids harvesting of cotton, and thinning the fruit of young fruit trees results in larger fruit as the trees mature. Retarding the growth of other plants sometimes increases their hardiness. For example, an inhibitor has been used to reduce stem length in wheat plants, so that the plants do not fall over in heavy winds and rain.

The commercial uses of ethylene were greatly increased with the development of ethylene-releasing compounds. Ethylene gas is injected into airtight storage rooms to ripen bananas, honeydew melons, and tomatoes. It will also degreen oranges, lemons, and grapefruit when the rind would otherwise remain green because of a high chlorophyll level. When sprayed on certain fruit and nut crops, ethylene increases the chances that the fruit will detach when the trees are shaken at harvest time.

Today, fields and orchards are often sprayed with synthetic growth regulators just as they are sprayed with pesticides.

and cut the agar into small blocks. When an agar block was placed to one side of a tipless coleoptile, the shoot would bend away from that side. The bending occurred even though the seedlings were not exposed to light.

Went concluded that the agar blocks contained a chemical that had been produced by the coleoptile tips. It was this chemical, he decided, that had caused the shoots to bend. He

named the chemical substance auxin after the Greek word *auxein*, which means to increase. It can be shown that when a plant is exposed to unidirectional light, auxin is transported to the shady side. Elongation of the cells on this side brings about the characteristic bending toward the light (fig. 32.3). Auxin causes the affected cell to degrade some of the polysaccharides in the cell wall. The cell then elongates because it is less able to

Figure 32.2

This experiment shows that a chemical produced by the coleoptile tip causes a seedling to bend. *a.* Coleoptile tips are cut off and are placed on agar (a gelatin-like material). After a time the tips are removed and the agar is cut into small blocks. *b.* A block is placed to one side of a tipless coleoptile. *c.* Bending occurs even in the absence of a light stimulus.

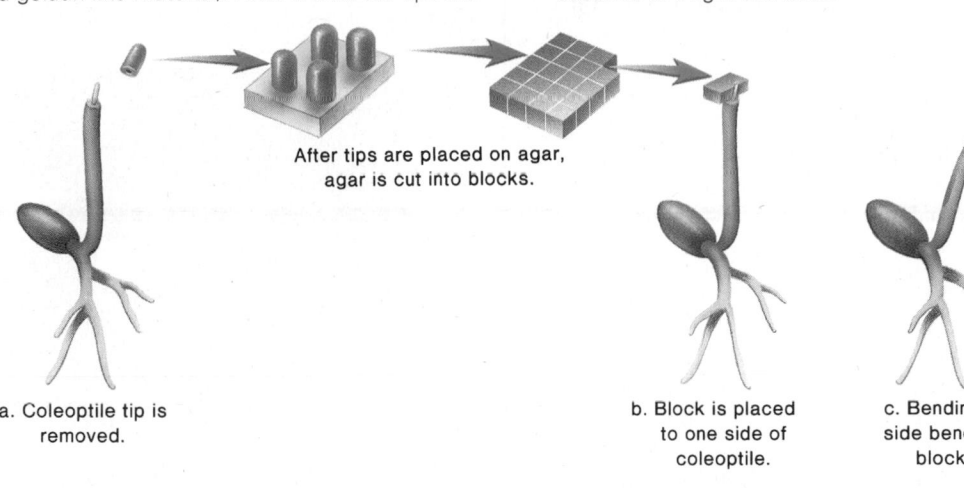

After tips are placed on agar, agar is cut into blocks.

a. Coleoptile tip is removed.

b. Block is placed to one side of coleoptile.

c. Bending side beneath block.

Figure 32.3

Auxin is transported from the illuminated side to the shaded side of a coleoptile as indicated by the arrows. This unequal concentration of auxin causes the cells on the shaded side to elongate. Elongation of the cells results in the bending of a seedling toward the light source.

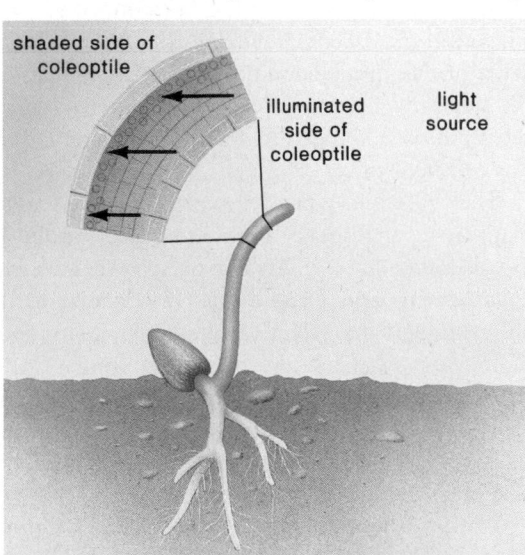

shaded side of coleoptile

illuminated side of coleoptile

light source

resist the expansion caused by osmotic movement of water into the cell. The direction of cell growth is dependent on where the wall is weakest.

Other Effects of Auxins Auxin has been found to affect many other aspects of plant growth. Auxin produced in a terminal bud at the apex of a plant prevents other bud development and growth for some distance from the apex. When a terminal bud is removed deliberately or accidently, the nearest axillary buds begin to grow, and the plant branches (fig. 32.4).

Figure 32.4

The trunk of this Jeffrey pine tree, which would normally be single, is forked because of the earlier removal of the terminal bud.

The application of a weak solution of auxin to a woody cutting causes roots to develop. Auxin production by seeds also promotes the growth of fruit. As long as auxin is concentrated in leaves or fruits rather than in the stem, leaves and fruits do not fall off. Therefore, trees can be sprayed with auxin to keep mature fruit from falling to the ground.

Auxin is also involved in the response of stems to gravity. This will be discussed later in more detail.

Researchers studying the bending of coleoptiles toward light discovered the presence of a hormone called auxin. Auxin is also involved in apical dominance and in the response of stems to gravity.

Gibberellins

We know of about 70 gibberellins that chemically differ only slightly. The most common of these is GA_3 (the subscript designation distinguishes it from other gibberellins). **Gibberellins** are likely present in newly developing plant organs because they are growth promoters that bring about cell division and enlargement of the resulting cells. When gibberellins are applied externally to plants, the most obvious effect is *stem elongation* (fig. 32.5).

Gibberellins were discovered in 1926, the same year that Went performed his classic experiments with auxin. Kurosawa, a Japanese scientist, was investigating a fungal disease of rice plants called "foolish seedling disease." The plants elongated too quickly, causing the stem to weaken and the plant to collapse. Kurosawa found that the fungus infecting the plants produced an excess of a chemical he called gibberellin, named after the fungus (*Gibberella fujukuroi*).

It wasn't until 1956 that gibberellic acid was isolated from a flowering plant rather than from a fungus. Since minute amounts of externally applied gibberellic acid caused genetically dwarfed pea plants to grow to a normal size, it was thought that dwarfed plants contained little gibberellin. Surprisingly, they contained more gibberellin than normal plants. For some reason, they were unable to utilize the gibberellin present in their cells. It is often the balance of hormones in a plant cell that produces an effect rather than the presence or absence of a particular hormone.

If applied at the appropriate concentration, gibberellins cause most dicots and monocots to grow taller and faster. A cabbage plant grows to 2 m, and bush beans become pole beans instead. Apparently, gibberellins, as well as the auxins, help regulate general plant development. The dormancy of seeds and buds can be broken by an application of gibberellin, and flowering can be induced in a mature plant.

Gibberellin research provides an example of how a plant hormone can act as a chemical messenger. Barley seeds have a large, starchy endosperm, which must be broken down into sugars to provide energy for growth. After the embryo produces gibberellin, cells inside the seed coat synthesize an enzyme that breaks down the starch. It is hypothesized that gibberellin turns on the gene that codes for the necessary enzyme.

Figure 32.5
The plant on the right was treated with gibberellin; the plant on the left was not treated. Gibberellins are often used to promote stem elongation in economically important plants, but the exact mode of action still remains unclear.

Gibberellins cause stems to elongate and dwarf plants to grow to normal size. This hormone also causes barley seeds to produce an enzyme that breaks down endosperm.

Cytokinins

The **cytokinins** are a class of plant hormones that promote cell division; *cytokinesis means cell division*. These substances are derivatives of the purine adenine, one of the bases in DNA and RNA. A naturally occurring cytokinin called *zeatin* has been extracted from corn kernels. Kinetin, a synthetic cytokinin, also promotes cell division.

The cytokinins were discovered as a result of attempts to grow plant tissue and organs in culture vessels in the 1940s (fig. 32.6). It was found that cell division occurs when coconut milk (a liquid endosperm) and yeast extract are added to the culture medium. Although the effective agent or agents could not be isolated, they were collectively called cytokinins. Not until 1967 was the naturally occurring cytokinin zeatin isolated from coconut milk. Cytokinins have been isolated from various seed plants, where they occur in the actively dividing tissues of roots and also in seeds and fruits.

Plant tissue culturing is now common practice, and researchers are well aware that the ratio of auxin to cytokinin and the acidity of the culture medium determine whether or not the plant tissue forms an undifferentiated mass, called a callus, or differentiates to form roots, vegetative shoots, leaves, or floral shoots (fig. 32.6). Researchers have reported that chemicals called oligosaccharins (chemical fragments released from the cell wall) are also effective in directing differentiation. They hypothesize that the function of auxin and cytokinin is actually to activate enzymes that release these more specific chemical messengers from the cell wall. They similarly believe that all

Plant Structure and Function

Figure 32.6

Interaction of hormones. It is now clear that plant hormones rarely act alone; it is the relative concentrations of these hormones that produce an effect. The modern emphasis is to look for an interplay of hormones when a growth response is studied. For example, tissue culture experiments have revealed that auxin and cytokinin interact to affect differentiation during development. ***a.*** In a tissue culture that has no added amounts of these 2 hormones, tobacco strips develop into a callus of undifferentiated tissue. ***b.*** If the usual ratio of auxin to cytokinin is changed slightly, the callus produces roots. ***c.*** Change the ratio again, and vegetative shoots and leaves may be produced. ***d.*** Yet another ratio results in floral shoots. These results illustrate that it is the *relative* proportions of hormones that probably bring about specific effects. To attribute a particular growth response to only one hormone is no doubt an oversimplification. Most likely, it will be many years before the interactions of hormones in determining the growth responses of plants are well understood.

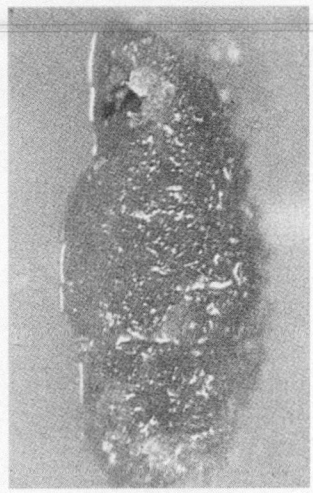

a.　　　　b.　　　　c.　　　　d.

plant hormones probably have such a function, which may explain the many and various effects of any specific plant hormone.

Other Effects of Cytokinins When a plant organ, such as a leaf, loses its natural color, it is most likely undergoing an aging process called *senescence*. During senescence, large molecules within the leaf are broken down and transported to other parts of the plant. Senescence does not always affect the entire plant at once; for example, as some plants grow taller, they naturally lose their lower leaves. It has been found that senescence can be prevented by the application of cytokinins. Not only can cytokinins prevent death of plant organs, they can also initiate growth. Lateral buds begin to grow despite apical dominance when cytokinin is applied to them.

Cytokinins promote cell division, prevent senescence, and initiate growth. The interaction of hormones is well exemplified by the effect of varying ratios of auxin and cytokinins on differentiation of plant tissues.

Plant Growth Inhibitors

As is evident in table 32.1, 2 types of plant hormones, ethylene and abscisic acid, inhibit growth. Hormones are expected to be present and active whenever the plant faces conditions that are unfavorable for continued growth, such as lack of water or cold temperatures.

Ethylene

In the early 1900s, it was common practice to ready citrus fruits for market by placing them in a room with a kerosene stove. Because heat alone did not have the same effect, researchers finally realized that an incomplete combustion product of kerosene, namely **ethylene,** is responsible for the ripening of fruits. Ethylene ripens fruits by increasing the activity of enzymes that soften fruits. For example, it stimulates the production of *cellulase*, an enzyme that hydrolyzes the cellulose of plant cell walls.

Because it is a gas, ethylene moves freely through the air—a barrel of ripening apples can induce ripening of a bunch of bananas even some distance away. Its presence in the air can also retard the growth of plants in general. Home owners who use natural gas to heat their homes sometimes report difficulties in growing houseplants, for example. Ethylene is also present in automobile exhaust, and it's possible that plant growth in general is affected by the exhaust that enters the atmosphere. It only takes 1 part of ethylene per 10 million parts of air to bring about inhibition of plant growth.

Other Functions of Ethylene High concentrations of auxin stimulate the production of ethylene in shoot meristem. This means that certain plant responses attributed to auxin may really be due to ethylene. For example, ethylene is probably needed for apical dominance.

Figure 32.7

Abscission layer. Before a leaf falls, a special band of cells called the abscission layer develops at the base of the petiole, the leaf stem. Here the hormone ethylene promotes the breakdown of plant cell walls so that the leaf finally falls. Here also, a layer forms a leaf scar, which protects the plant from possible invasion by microorganisms.

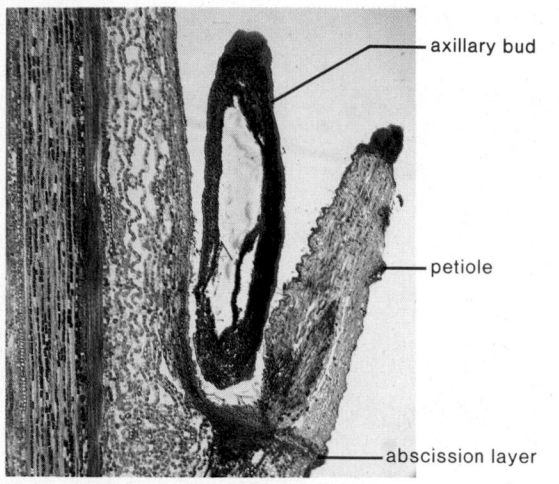

Figure 32.8

Stomata (stoma, sing.) are opened and closed by guard cells. Because of transpiration, a plant ordinarily loses a great deal of water by way of the stomata; therefore, it is beneficial for the stomata to be closed when a plant is under water stress. The hormone ABA is believed to bring about the closing of the stomata when water stress occurs. As discussed in chapter 30, stomata close when K^+ and then H_2O leave the guard cell.

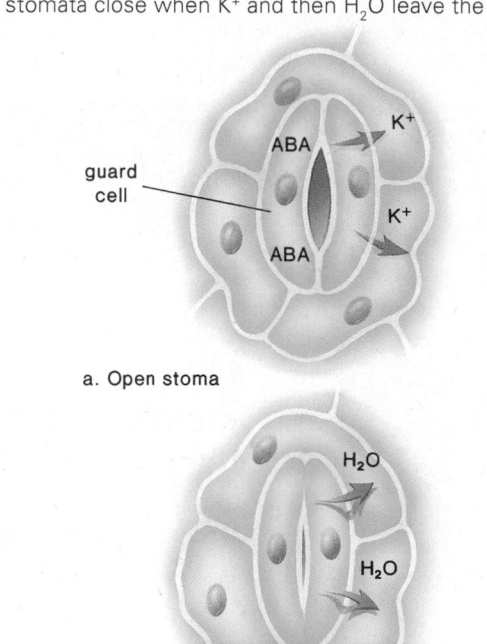

a. Open stoma

b. Closed stoma

Ethylene also inhibits stem elongation and prevents the maturation of young leaves. This effect can have survival value, as when young seedlings are germinating (fig. 31.9). If the soil is compact, the dicot shoot produces more ethylene, leading to a thickening of the hook, which helps the young plant push up through the soil.

Ethylene is also involved in **abscission,** the dropping of leaves, fruits, and flowers from a plant. As mentioned, a relative decrease of auxin and perhaps gibberellin in these areas of the plant compared to the stem probably initiates abscission (fig. 32.7). But once the process of abscission has begun, ethylene is produced. This is believed to stimulate such enzymes as cellulase, which causes leaf, fruit, or flower drop.

Abscisic Acid

Abscisic acid (ABA) is sometimes called the stress hormone because it initiates and maintains seed and bud dormancy and brings about the closure of stomata. Although the external application of abscisic acid promotes abscission, this hormone is no longer believed to function naturally in this process.

Dormancy occurs when a plant organ readies itself for adverse conditions by stopping growth (even though conditions at the time are favorable for growth). For example, it is believed that abscisic acid moves from leaves to vegetative buds in the fall, and thereafter these buds are converted to winter buds. A winter bud is covered by thick and hardened scales. In the spring, dormancy is broken when seeds germinate, and buds begin to send forth new

growth. A reduction in the level of abscisic acid and an increase in the level of gibberellins are believed to break seed and bud dormancy.

Abscisic acid is called the stress hormone not only because it promotes dormancy, but also because it brings about the closing of stomata when a plant is under water stress (fig. 32.8). In some unknown way, ABA causes K^+ to leave guard cells. Thereafter, the guard cells lose water, and the stomata close. Notice that this is the opposite sequence of events to those that cause stomata to open (see fig. 30.6).

Ethylene and abscisic acid are plant growth inhibitors. Ethylene ripens fruits and is involved in abscission. Abscisic acid maintains dormancy and causes stomata to close.

Plant Movements

As noted at the beginning of this chapter, plants respond to both external and internal stimuli by changing their growth patterns. These growth patterns often result in movement of some part of the plant.

Figure 32.9

When an upright plant is turned on its side, in time the roots bend down and the stem bends up. Roots therefore demonstrate positive gravitropism, and the stem demonstrates negative gravitropism.

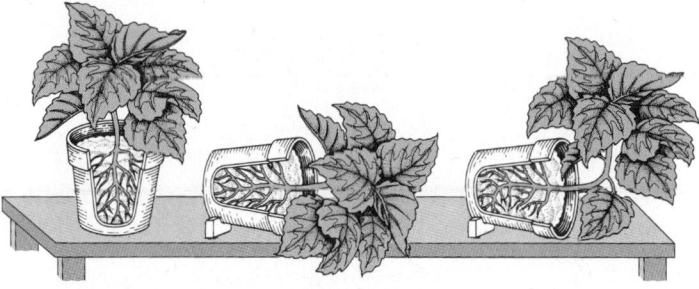

External Stimuli

Plant movement toward or away from an external stimulus is called a **tropism.** Three well-known tropisms are as follows:

1. Phototropism: a movement in response to a light stimulus
2. Gravitropism: a movement in response to gravity
3. Thigmotropism: a movement in response to touch

Phototropism

As discussed, plant stems bend toward the light (fig. 32.1). This is called positive **phototropism.** Bending away from light is called negative phototropism. Roots, depending on the species examined, are either insensitive to light or exhibit negative phototropism.

Response to a stimulus requires a receptor, and in this case, the photoreceptor is believed to be a yellow pigment related to the vitamin riboflavin. Following detection of light, the plant hormone auxin migrates from the bright side to the shady side of a stem. The cells on that side elongate faster than those on the bright side, causing the stem to bend toward the light (fig. 32.3).

Gravitropism

When an upright plant is placed on its side, the stem displays negative **gravitropism** because it grows opposite to the direction of gravity (fig. 32.9). Roots display positive gravitropism because they grow in the same direction as gravity. This difference in growth response is explained in this manner:

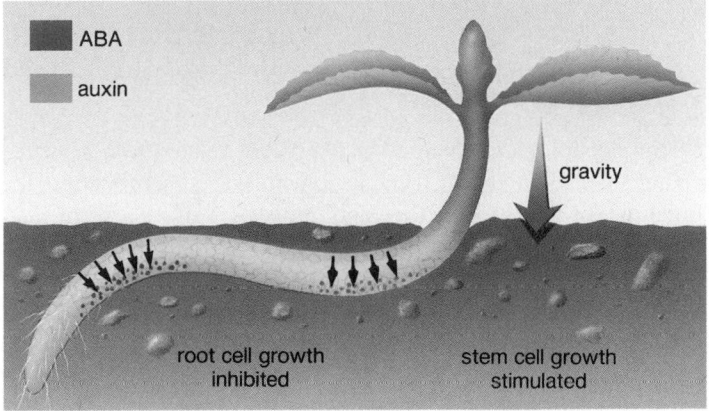

Figure 32.10

Coiling response of a morning glory (*Ipomea*) plant.

Gravity causes stem cells to grow more quickly (so they become more elongated) and inhibits the growth of root cells, leading to the bending we observe.

A hormone could accumulate on the lower side of roots and stems to cause these gravitropic bendings. In the case of stems, it can be demonstrated that auxin accumulates on the lower side, and in the case of roots, it is believed that the hormone ABA, which inhibits growth, accumulates on the lower side.

How do plants detect a change in gravity? There is evidence that roots perceive gravity by the movement of starch grains in special root-cap cells called statocytes. Due to gravity, the starch grains accumulate on the lower side of these cells. How and why this causes a redistribution of hormones is not known.

Thigmotropism

The response of a plant or a plant part to contact is called **thigmotropism** (*thigma* in Greek means touch). The presence of this response in plants is seen in the coiling of tendrils and the twining of a climbing plant stem. The response occurs because the cells in contact with an object, such as a pole, grow less and those on the opposite side elongate (fig. 32.10).

The Venus's-flytrap is a plant with specialized lobed leaves that snap shut when they are touched by an insect (see fig. 29.21c). While the leaves are closed, glands on the leaf surface secrete enzymes that digest the insect's body. The leaf closes when the outer epidermal cells expand rapidly due to an increase in turgor pressure. When the inner epidermal cells expand, the leaf opens again. There are other examples of touch responses due to turgor pressure changes. For example, leaflets of the sensitive plant (*Mimosa pudica*) fold when they are touched because of turgor pressure changes in specialized cells at the base of the leaflets and in the leaf itself.

Internal Stimuli

Some plant movements occur primarily in response to internal stimuli, including the following examples:

1. Helical movements—Although plants appear to be growing straight up, many actually spiral as they grow.
2. Nodding movements—Members of the legume family, such as garden beans, nod from side to side as the seedling pushes up through the soil.
3. Twining movements—In some plants, the visible twining of tendrils as they grow (even when they have not made contact with an object) is due to auxin. In other plants, similar movements are due to ethylene.
4. Contraction movements—Contractile roots of bulbs pull them deeper into the ground to a point where there is no difference in daytime and nighttime temperatures.
5. Nastic "sleep" movements—Plant organs, such as leaves and flowers, sometimes bend up and then down. For example, prayer plants have leaves that fold up and reopen daily due to turgor pressure changes (fig. 32.11). These movements are controlled by a "biological clock."

Circadian Rhythms in Plants

Organisms exhibit periodic fluctuations that correspond to environmental changes. For example, your temperature and blood pressure tend to change with the time of day, and you become sleepy at a certain time of night. The prayer plant we mentioned earlier displays rhythmic "sleep" behavior (fig. 32.11). A biological rhythm with a 24-hr cycle is called a **circadian rhythm** (Latin *circa* = about; *dies* = day; i.e., about a day).

Circadian rhythms tend to persist, even if the appropriate environmental cues are no longer present. For example, on a transcontinental flight, you will likely suffer jet lag, and it will take several days to adjust to the time change because your body will still be attuned to the day-night pattern of your previous environment. The internal mechanism by which a biological rhythm is maintained in the absence of appropriate environmental stimuli is termed a **biological clock.** Typically, if organisms are sheltered from environmental stimuli, their circadian rhythms continue, but the cycle is extended a bit. In prayer plants, the sleep cycle changes from 24 to 26 hr. Therefore, it is believed that biological clocks are synchronized by external stimuli to 24-hr rhythms. The length of daylight compared to the length of darkness, called the photoperiod, sets the clock. Temperature has little or no effect. This is adaptive because the photoperiod indicates seasonal changes better than temperature changes. Spring and fall, in particular, can have both warm and cold days.

Figure 32.11
A prayer plant. (*Maranta leuconeura*) *a.* The plant at noon. *b.* The same plant at 10 P.M., after "sleep" movements of its leaves have occurred.

a.

b.

There are other examples of circadian rhythms in plants. For example, stomata usually open in the morning and close at night, and some plants secrete nectar at the same time of the day or night. But the primary usefulness of the biological clock appears to be to measure day-length changes, so that a plant is able to respond to the changing seasons.

Photoperiodism

Many physiological changes in plants are related to a seasonal change in day length. Such changes include seed germination, the breaking of bud dormancy, and the onset of senescence. A physiological response prompted by changes in the length of day or night is called **photoperiodism.** In some plants, photoperiodism influences flowering: violets and tulips flower in the spring, and asters and goldenrods flower in the fall.

Plant Structure and Function

Figure 32.12

Day length (night length) effect on 2 types of plants. **a.** Short-day (long-night) plant. (1) When the day is shorter (the night is longer) than a critical length, this type of plant flowers. (2) The plant does not flower when the day is longer (night is shorter) than the critical length. (3) It also does not flower if the longer-than-critical-length night is interrupted by a flash of light. **b.** Long-day (short-night) plant. (1) When the day is shorter (the night is longer) than a critical length, this type plant does not flower. (2) The plant flowers when the day is longer (the night is shorter) than a critical length. (3) It also flowers if the slightly longer-than-critical-length night is interrupted by a flash of light.

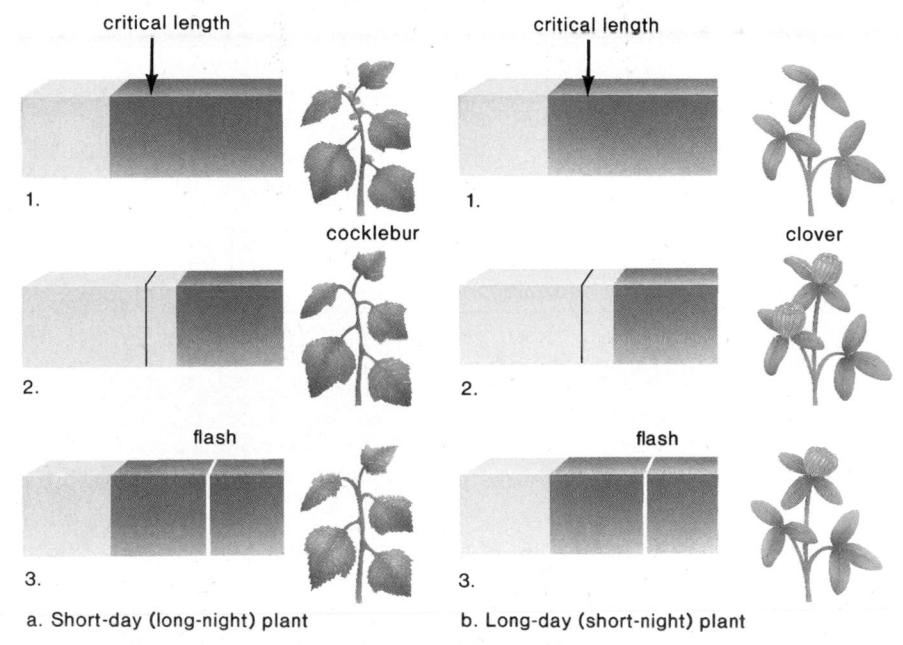

a. Short-day (long-night) plant

b. Long-day (short-night) plant

length shortens to 14 hours or less. We now know that some plants may require a specific sequence of day lengths in order to flower.

In 1938, K. C. Hammer and J. Bonner began to experiment with artificial lengths of light and dark that did not necessarily correspond to a normal 24-hour day. They discovered that the cocklebur, a short-day plant, flowers as long as the dark period is continuous for $8\frac{1}{2}$ hours, regardless of the length of the light period. Further, if this dark period is interrupted by a brief flash of light, the cocklebur does not flower. (Interrupting the light period with darkness has no effect.) Similar results have also been found for long-day plants. They require a dark period that is shorter than a critical length, regardless of the length of the light period. If a slightly longer-than-critical-length night is interrupted by a brief flash of light, however, long-day plants flower. We must conclude, then, that the length of the dark period controls flowering, not the length of the light period. Of course, in nature, short days always go with long nights and vice versa.

> Researchers have discovered that short-day plants require a period of darkness that is longer than a critical length to flower, and long-day plants require a period of darkness that is shorter than a critical length to flower.

In the 1920s, when U.S. Department of Agriculture scientists working in Beltsville, Maryland, began to study photoperiodism in more detail, they decided to grow plants in a greenhouse, where they could artificially alter the photoperiod. This work led them to conclude that plants can be divided into 3 groups:

1. **Short-day plants**—flower when the day length is shorter than a critical length. (Good examples are cocklebur, poinsettia, and chrysanthemum.)
2. **Long-day plants**—flower when the day length is longer than a critical length. (Good examples are wheat, barley, clover, and spinach.)
3. **Day-neutral plants**—flowering is not dependent on day length. (Good examples are tomato and cucumber.)

Further, we should note that both a long-day plant and a short-day plant can have the same critical length (fig. 32.12). Spinach is a long-day plant that has a critical length of 14 hours; ragweed is a short-day plant with the same critical length. Spinach, however, flowers in the summer when the day length increases to 14 hours or more, and ragweed flowers in the fall, when the day

Phytochrome

If flowering is dependent on day and night length, plants must have some way to detect these periods. Many years of research by U.S. Department of Agriculture scientists led to the discovery of a plant pigment called phytochrome. **Phytochrome** is a blue-green leaf pigment that alternately exists in 2 forms. As figure 32.13 indicates,

P_r (phytochrome red) absorbs red light (of 660 nm wavelength) and is converted to P_{fr}.

P_{fr} (phytochrome far-red) absorbs far-red light (of 730 nm wavelength) and is converted to Pr.

Direct sunlight contains more red light than far-red light; therefore, Pfr is apt to be present in plant leaves during the day. In the shade and at sunset, there is more far-red light than red light; therefore, Pfr is converted to Pr as night approaches. There is also a slow metabolic replacement of Pfr by Pr during the night.

It was thought for some time that this slow reversion might provide a means for the plant to measure the length of the night. It is now assumed, however, that the active form of phytochrome, Pfr, signals a biological clock that brings about flowering.

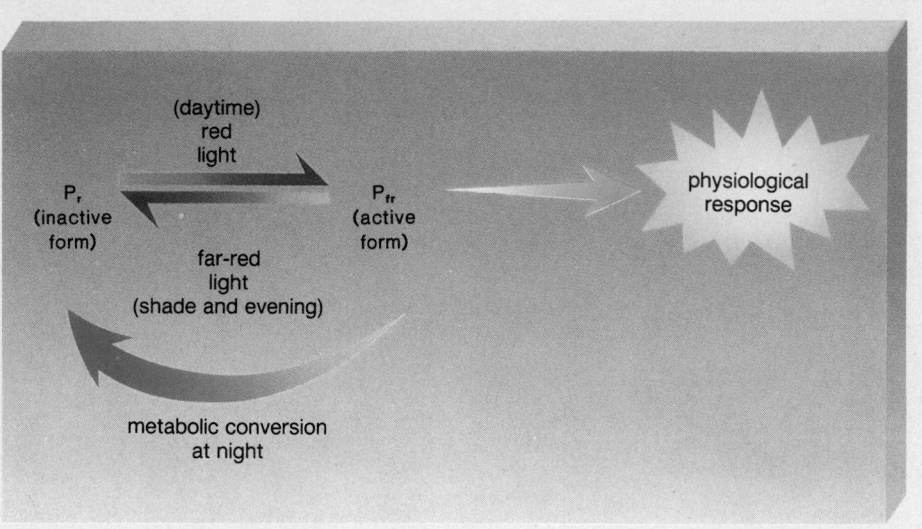

Figure 32.13
The $P_r \rightleftarrows P_{fr}$ conversion cycle. The inactive form P_r is prevalent during the night. At sunset or in the shade when there is more far-red light, P_{fr} is converted to P_r. Also during the night, metabolic processes cause P_{fr} to be converted into P_r. P_{fr}, the active form of phytochrome, is prevalent during the day because at that time there is more red light than far-red light.

(daytime) red light

P_r (inactive form)

P_{fr} (active form)

physiological response

far-red light (shade and evening)

metabolic conversion at night

Just how this happens is unknown. Hormones are probably involved, but this has not been proven. At one time, researchers thought they might find a special flowering hormone called florigen, but such a hormone has never been discovered.

Phytochrome is a leaf pigment that alternates between 2 forms (Pfr during day and Pr during night), apparently this conversion allows a plant to detect photoperiod changes that may result in flowering.

Other Functions of Phytochrome

The $Pr \rightleftarrows Pfr$ conversion cycle is now known to control other growth functions in plants. It promotes seed germination and inhibits stem elongation, for example. The presence of Pfr indicates to some seeds that sunlight is present and conditions are favorable for germination. This is why some seeds must be partly covered with soil when planted. Germination of other seeds is inhibited by light, so they must be planted deeper. Following germination, the presence of Pr indicates that stem elongation may be needed to reach sunlight. Seedlings that are grown in the dark etiolate, that is, the stem increases in length and the leaves remain small (fig. 32.14). Once the seedling is exposed to sunlight and Pr is converted to Pfr, the seedling begins to grow normally—the leaves expand and the stem branches.

Figure 32.14
Phytochrome has other functions besides regulating flowering. For example, if far-red light is prevalent, as it is in the shade, the stem of a seedling elongates but the leaves remain small (*left-hand side of photo*); however, if red light is prevalent, as it is in bright sunlight, the stem does not elongate but the leaves expand. These effects are due to phytochrome.

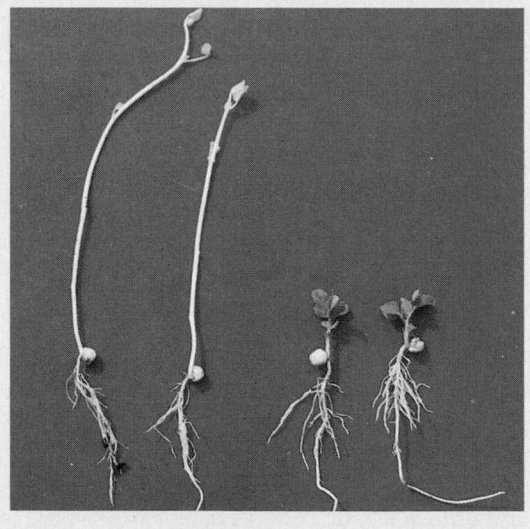

Summary

1. Both stimulatory and inhibitory hormones help control certain growth patterns of plants. There are hormones that stimulate growth (auxins, gibberellins, cytokinins) and hormones that inhibit growth (ethylene and abscisic acid).

2. The auxins, which promote cell elongation, control phototropism. When a plant is exposed to light, auxin moves laterally from the bright to the shady side of a stem. Auxins also influence apical dominance, fruit maturation, leaf and fruit drop, and gravitropism.

3. Gibberellins promote stem elongation. External application of gibberellins can cause dwarf plants to reach normal size. They also help break seed and bud dormancy.

4. Cytokinins cause cell division, the effects of which are especially obvious when plant tissues are grown in culture. They can also prevent leaf senescence.

5. Ethylene causes fruits to ripen and promotes abscission. Abscisic acid is no longer believed to be involved in abscission, but it does promote bud dormancy and causes stomata to close when a plant is under water stress. ABA may also be involved in root negative gravitropism.

6. It is typical for plant hormones to interact, and therefore it is probably an oversimplification to suggest that each has specific functions that do not involve other hormones.

7. Plants respond to external and internal stimuli by growth responses that bring about movement of plant parts.

8. Tropisms are responses to external stimuli. Positive phototropism occurs when auxin moves to the shady side of a stem. Negative gravitropism occurs when auxin moves to the lower side of a stem. Thigmotropism occurs when a plant part makes contact with an object. Tendrils and stems coil, and Venus's-flytrap leaves snap shut.

9. Several plant movements, such as helical, nodding, twining, contraction, and nastic movements, are believed to be responses to internal stimuli. Also, plants exhibit circadian rhythms, which are believed to be controlled by a biological clock. For example, bean leaves have rhythmic sleep patterns.

10. Photoperiodism is seen in some plants. For example, short-day plants flower only when the days are shorter than a critical length, and others, called long-day plants, flower only when the days are longer than a critical length. Actually, research has shown that it is the length of darkness that is critical. Interrupting the dark period with a flash of white light prevents flowering in a short-day plant and induces flowering in a long-day plant.

11. Phytochrome is a pigment that responds to both red and far-red light and is involved in flowering. Daylight causes phytochrome to exist as Pfr, but during the night, it is reconverted to Pr by metabolic processes. Phytochrome probably communicates with a biological clock that in some unknown way brings about flowering.

12. In addition to being involved in the flowering process, Pfr promotes seed germination, leaf expansion, and stem branching. When Pr dominates, the stem elongates and grows toward sunlight.

Study Questions

1. Name 3 types of hormones that promote growth processes, and give several specific functions for each.
2. Name 2 types of hormones that inhibit growth processes, and give several specific functions for each.
3. Name 3 types of tropisms, and indicate the role of hormones in these plant movements.
4. Describe 2 experiments that led to the discovery of auxin.
5. Give experimental evidence to suggest that hormones interact when they bring about an effect.
6. List and describe some movements of plants that occur as a response to internal stimuli.
7. What is a biological clock, how does it function, and what is its primary usefulness in plants?
8. Define photoperiodism, and discuss its relationship to flowering in certain plants.
9. What is the phytochrome conversion cycle, and what are some possible functions of phytochrome in plants?

Objective Questions

For questions 1–5, match the items below with the hormones in the key.

Key:

    a. auxin
    b. gibberellin
    c. cytokini
    d. ethylene
    e. abscisic acid

1. One rotten apple can spoil the barrel.
2. Cabbage plants bolt (grow tall).
3. Stomata close when a plant is water stressed.
4. Sunflower plants all point toward the sun.
5. Coconut milk causes plant tissues to undergo cell division.

6. Which of these is a correct statement?
    a. Both stems and roots show positive gravitropism.
    b. Both stems and roots show negative gravitropism.
    c. Only stems show positive gravitropism.
    d. Only roots show positive gravitropism.

7. Short-day plants
    a. are the same as long-day plants.
    b. are apt to flower in the fall.
    c. do not have a critical photoperiod.
    d. All of these.

8. A plant requiring a dark period of at least 14 hours
    a. will flower if a 14-hour night is interrupted by a flash of light.
    b. will not flower if a 14-hour night is interrupted by a flash of light.
    c. will not flower if the days are 14 hours long.
    d. Both b and c.

9. Phytochrome
    a. is a plant pigment.
    b. is present as Pfr during the day.
    c. may communicate with a biological clock.
    d. All of these.

10. Circadian rhythms
    a. require a biological clock.
    b. do not exist in plants.
    c. are involved in the tropisms.
    d. All of these.

11. Label the arrows in the following diagram, and explain what causes oat seedlings to bend toward the light:

12. Label the arrows in the following diagram, and explain what causes stomata to close when a plant is water stressed:

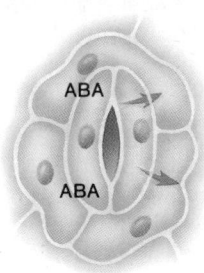

Concepts and Critical Thinking

1. *Physiological processes are coordinated in organisms.*

Give an example to show that plant hormones are involved in coordinating physiological processes.

2. *Biological clocks help organisms maintain adaptive behaviors.*

(a) Give examples to show that circadian rhythmic behavior in plants is adaptive. (b) Use figure 32.13 and figure 45.2 to suggest that flowering is controlled by a biological clock system.

Selected Key Terms

hormone (hor′mōn) 523
auxin (awk′sin) 523
gibberellin (gib-ber-el′in) 526
cytokinin (si″to-ki′nin) 526
ethylene (eth′ĭ-lēn) 527

abscisic acid (ABA) (ab-sis′ik as′id) 528
tropism (tro′pizm) 529
phototropism (fo-tot′ro-pizm) 529
gravitropism (grav″ĭ-tro′pizm) 529
thigmotropism (thig-mot′ro-pizm) 529

circadian rhythm (ser″kah-de′an rith′m) 530
biological clock (bi-o-loj′ē-kal klok) 530
photoperiodism (fo″to-pe′re-od-izm) 530
phytochrome (fī′to-krom) 531

PART 4 *Plant Structure and Function*

CRITICAL THINKING CASE STUDY

Leaf Senescence

The senescence (aging) and death of older leaves is a part of a plant's life cycle. It is hypothesized that *senescence of leaves is controlled by hormones and is accompanied by the breakdown of organic macromolecules.*

To test the hypothesis, researchers have determined if the amount of macromolecules decreases in *attached* leaves as a function of age. Do you predict that macromolecules decrease in attached leaves as they age?

Prediction 1 In keeping with the hypothesis, it is predicted that macromolecules decrease in attached leaves as they age.

Result 1 As shown in figure A, RNA and protein decrease in attached leaves as a function of age.

> I Why does it seem logical that the amount of RNA (fig. A*a*) would decrease in leaves as they age before the amount of protein (fig. A*b*) decreases?
> Is protein synthesis outstripping protein breakdown before age 47 days or after age 47 days? How do you know? Is this consistent with the hypothesis?

In other experiments, researchers investigated whether the amino acid content within *detached* leaves increases. Following detachment the petiole of the leaf is placed in water. Even so, the characteristics of senescence, such as a yellowing of the leaves, become evident quickly. In keeping with the hypothesis, do you predict that amino acid content will increase in detached leaves?

Prediction 2 Yes, it is predicted that the amino acid content of detached leaves will increase.

Result 2 The amino acid content of detached leaves does increase in detached leaves.

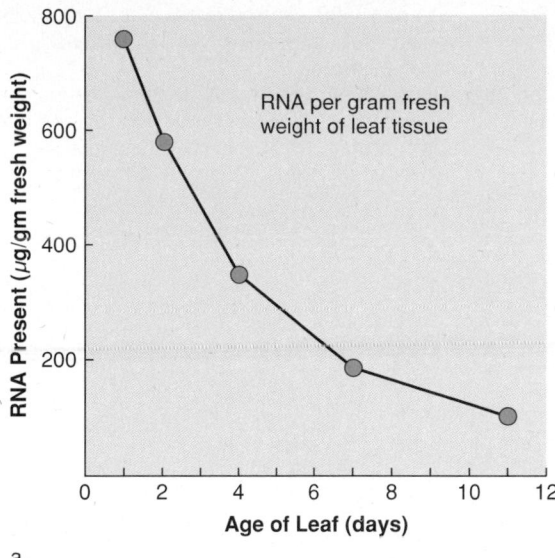

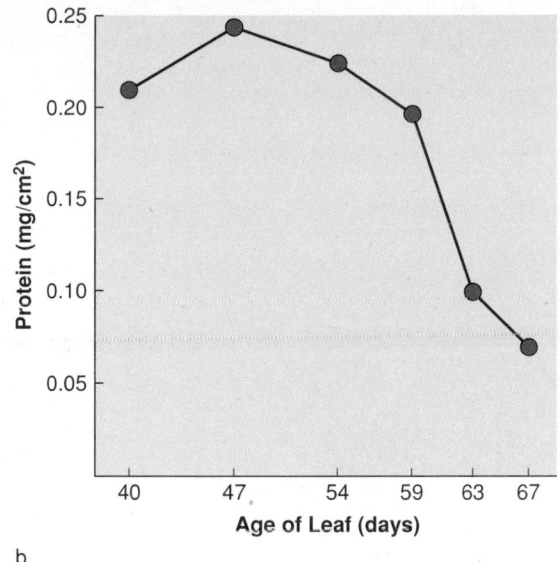

a.

b.

Figure A

Decrease in RNA content and leaf protein with senescence. **a.** Changes in RNA content of attached leaves of pea (*Pisum sativum*). **b.** Changes in protein content of attached leaves of *Perilla frutescens*.

II Add a curve to figure A*b* to indicate amino acid content of the leaf. Is your curve consistent with the observation that protein synthesis at first increases and only later decreases?
If amino acids are being translocated out of the intact leaf, how would the shape of your curve change?

Roots sometimes develop at the base of the petiole in detached leaves. In these instances, senescence is immediately reversed in detached leaves, that is, protein synthesis begins again. Cytokinins are known to be hormones that are synthesized and released by roots. Do you predict that the application of cytokinins to detached leaves will prevent senescence from occurring?

Prediction 3 If cytokinins are applied to detached leaves, senescence will be prevented from occurring.

Result 3 Application of kinetin (a synthetic cytokinin) causes detached leaves to remain green.

III In intact plants, older, larger leaves undergo senescence before younger, smaller leaves. Which of these leaves do you predict are receiving more cytokinins from the root by way of xylem?
Application of cytokinin to one part of a leaf causes metabolites to accumulate in that part. In intact plants, cytokinins may be causing young leaves or older leaves to act as sinks for phloem transport?
In intact plants, could the older leaves be a source of metabolites to the younger leaves? Why do you say so?

Other investigations have indicated that senescence of detached leaves is markedly decreased in the light when photosynthesis is occurring. Researchers decided to test if glucose, a product of photosynthesis, delays senescence. If glucose delays senescence, will the application of DCMU, an inhibitor of glucose production, delay or promote senescence of detached leaves in the light?

Prediction 4 If glucose production in the light delays senescence, the application of an inhibitor of sugar production will promote senescence.

Result 4 When the inhibitor DCMU is applied to detached leaves in the light, senescence is unaffected.

IV What do these results indicate about the effect of light on senescence? about the effect of glucose production in the light on senescence? about the effect of some other product of photosynthesis, such as ATP, on senescence? How could you test whether ATP production in the light affects senescence?

These results give support to the hypothesis that senescence of leaves is controlled by hormones and is accompanied by the breakdown of organic macromolecules. Further investigation is needed to determine if cytokinins cause the translocation of nutrient molecules from aging leaves into young leaves.

Other Questions

1. Would you expect auxins and gibberellins to promote senescence of leaves? Why? Would you expect abscisic acid and ethylene to promote senescence of leaves? Why?

2. What part of a woody plant never undergoes senescence? Does this lend support to the supposition that cytokinins prevent senescence? Why?

Reference

Street, H. E., and H. Opik. 1976. *The physiology of flowering plants.* 2d ed. London: Edward Arnold Publishers Ltd.

Suggested Readings for Part 4

Albersheim, P., and Darvill, A. G. September 1985, Oligosaccharins. *Scientific American.*

Barrett, S. C. H. September 1987, Mimicry in plants. *Scientific American.*

Bold, H. C. 1987. *Morphology of plants and fungi.* 5th ed. New York: Harper & Row, Publishers.

Brill, W. J. March 1977. Biological nitrogen fixation. *Scientific American.*

Epel, D. November 1977. The program of fertilization. *Scientific American.*

Heslop-Harrison, Y. February 1978, Carnivorous plants. *Scientific American.*

Jansen, W., and Salisbury, F. B. 1971. *Botany: An ecological approach.* Belmont, Calif.: Wadsworth.

Niklas, K. J. July 1987, Aerodynamics of wind pollination. *Scientific American.*

Raven, P. H., et al. 1986. *Biology of plants.* 4th ed. New York: Worth.

Rayle, D., and Wedberg, H. L. 1980. *Botany: A human concern*. Boston: Houghton Mifflin.

Rost, R., et al. 1984. Botany: *A brief introduction to plant biology*. 2d ed. New York: Wiley.

Salisbury, F. B., and Ross, C. W. 1985. *Plant physiology*. 3d ed. Belmont, Calif.: Wadsworth.

Shepard, J. F. May 1982. The regeneration of potato plants from leaf-cell protoplasts. *Scientific American*.

Stern, K. 1991. *Introductory Plant Biology*. 5th ed. Dubuque, Iowa: Wm. C. Brown Publishers.

Tippo, O., and Sterm, W. L. 1977. *Humanistic botany*. New York: W. W. Norton and Co.

Zimmerman, M. H. March 1963. How sap moves in trees. *Scientific American*.

Animal Structure and Function

Land snails have organ systems that carry on functions necessary to life. This part studies how the organ systems maintain homeostasis, the relative constancy of the internal environment. The circulatory system carries nutrients from the digestive system and oxygen from the respiratory system to the cells. Then it carries metabolic wastes from the cells to the excretory system. The antennae of the snail send messages to the nervous system which coordinates the functions of all the other systems.

33

Animal Organization and Homeostasis

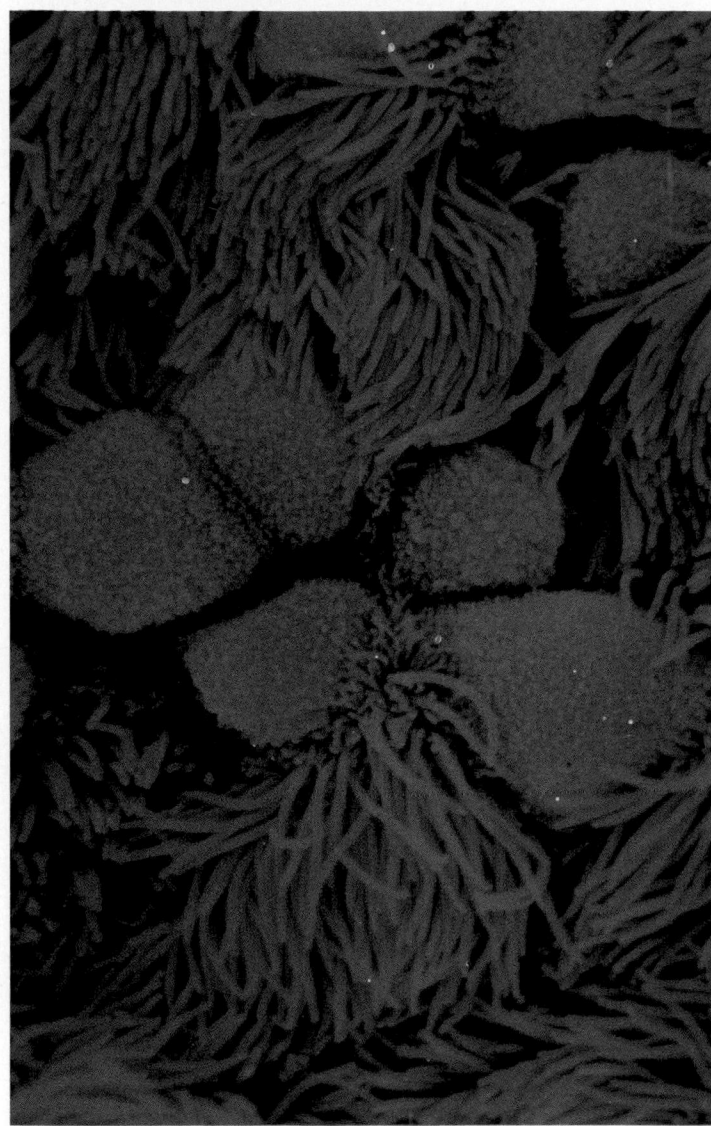

An animal's body is highly organized. This is a scanning electron micrograph of the tissue that lines the trachea, an organ that is part of the respiratory system. Some of the cells have cilia, which sweep impurities away from the lungs and toward the throat, where they can be swallowed. Magnification, X3,000.

Your study of this chapter will be complete when you can

1. name the 4 types of tissues in animals, and give a general function for each;
2. describe the structure and function of 3 types of epithelium along with their possible modifications;
3. describe the structure and function of the various types of connective tissue;
4. describe the structure and function of 3 types of muscular tissue;
5. describe the structure and function of 2 types of cells found in nervous tissue;
6. describe the structure of human skin, and tell how this illustrates the composition of an organ;
7. name the various organ systems in humans, associate each with a particular life process, and tell the specific functions of each system;
8. name the body cavities in a human, and tell which organ systems are in each cavity;
9. define homeostasis and tell how control by negative feedback works;
10. describe how body temperature is controlled in humans, and explain why body temperature fluctuates above and below a mean.

nimals, like plants, have levels of organization (fig. 33.1). In this chapter, we begin at the level of cells. The same type of cells form a tissue, and different types of tissues make up organs. Several organs are found within an organ system, and organ systems make up the organism.

Note that the structure and function of an organ system are dependent upon the structure and function of the organ, tissue, and cell type contained therein. For example, the structure and the function of the skeletal muscle system are the same as that of the skeletal muscles, the muscular tissue, and the muscle cells.

Types of Tissues

There are 4 major types of **tissue** in animals: *epithelial tissue* covers body surfaces and lines body cavities; *connective tissue* binds and supports body parts; *muscular tissue* causes body parts to move; and *nervous tissue* responds to stimuli and transmits impulses from one body part to another.

The embryological development of an animal begins with a single cell that divides to produce 3 fundamental layers, called the germ layers. The different tissues we have just mentioned arise from these germ layers:

Embryonic Germ Layer	Vertebrate Adult Structures
Ectoderm (outer layer)	Epidermis of skin; epithelial lining of mouth and rectum; nervous system
Mesoderm (middle layer)	Notochord; skeleton; muscular system; dermis of skin; circulatory system; excretory system; reproductive system—including most epithelial linings; outer layers of respiratory and digestive systems
Endoderm (inner layer)	Epithelial lining of digestive tract and respiratory tract; associated glands of these systems; epithelial lining of the urinary bladder

Epithelial Tissue

Epithelial tissue, also called epithelium, forms a continuous layer, or sheet, over body surfaces and most of the inner cavities. There are 3 types of epithelial tissue. *Squamous epithelium* is composed of flat cells; *cuboidal epithelium* contains cube-shaped cells; and in *columnar epithelium,* the cells resemble pillars or columns because they have an oblong shape. Figure 33.2 describes the structure and function of epithelium in vertebrates. Any epithelium can be simple or stratified. Simple means that the tissue has a single layer of cells, and *stratified* means that the tissue has layers piled one on top of the other. One type of epithelium is pseudostratified—it appears to be layered, but actually true layers do not exist because each cell touches a baseline.

Epithelial tissues have various functions. Epithelial cells can have hairlike extensions called cilia, which bend and move materials in a particular direction. In the human body, the ciliated

Figure 33.1
Levels of organization in the human body. A cell is composed of molecules; a tissue is made up of cells; an organ is composed of tissues, and the organism contains organ systems.

organism

system

organ

tissue

cell

molecule

epithelium lining the respiratory tract sweeps impurities toward the throat so they do not enter the lungs. An epithelium sometimes secretes a product and is described as glandular. A gland can be a single epithelial cell, such as the mucus-secreting globlet cells in the lining of the human intestine, or a gland can contain numerous cells. *Exocrine glands* secrete their product into ducts, and *endocrine glands* secrete their product directly into the bloodstream.

Epithelium forms the skin of many animals. The skin of earthworms and snails is glandular and produces mucous that lubricates the body, helping to ease movement through a dry environment. In roundworms, annelids, and arthropods, an outer

Figure 33.2

Type of epithelial tissues in vertebrates. Epithelial tissues are classified according to shape of cell, whether they are simple or stratified, and whether they have cilia. Epithelial tissues have functions associated with protection, absorption, and secretion.

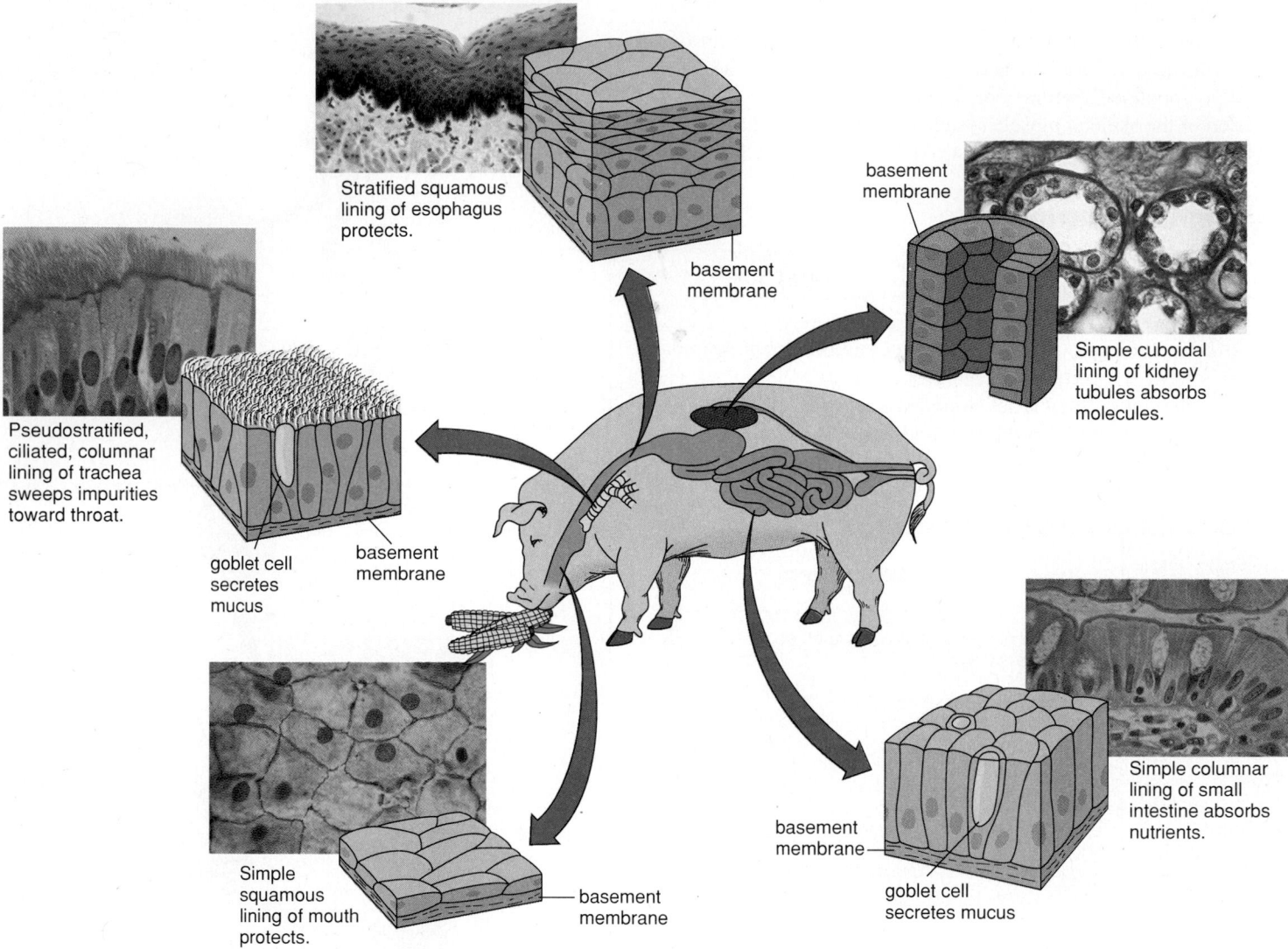

Stratified squamous lining of esophagus protects.

basement membrane

basement membrane

Simple cuboidal lining of kidney tubules absorbs molecules.

Pseudostratified, ciliated, columnar lining of trachea sweeps impurities toward throat.

goblet cell secretes mucus

basement membrane

Simple squamous lining of mouth protects.

basement membrane

basement membrane

goblet cell secretes mucus

Simple columnar lining of small intestine absorbs nutrients.

nonliving and protective cuticle is produced by epithelium. In terrestrial vertebrates, skin cells contain keratin, a substance that makes the skin protective against the possible loss of water.

Epithelial tissue cells are packed tightly and joined to one another in one of 3 ways: spot desmosomes, tight junctions, and gap junctions, (see fig. 6.16). Epithelium is often attached to a basement membrane, which is glycoprotein reinforced by fibers supplied by an underlying connective tissue.

> Epithelial tissue is classified according to the shape of the cell. There can be one or many layers of cells, and the epithelium can be ciliated and/or secretory.

Connective Tissue

Connective tissue binds structures together, provides support and protection, fills spaces, stores fat, and forms blood cells. It provides the source cells for muscular and skeletal cells in animals that can regenerate lost parts.

Connective tissue cells are separated widely by a *matrix*, a noncellular material found between cells.

Loose Connective Tissue and Fibrous Connective Tissue

The cells of loose and fibrous connective tissues, called **fibroblasts,** are located some distance from one another and are separated by a jellylike matrix that contains white collagen fibers and yellow elastic fibers. Collagen fibers provide flexibility and strength; elastic fibers provide elasticity.

Figure 33.3
Loose connective tissue. *a.* Loose connective tissue has plenty of space between components. This type of tissue supports epithelial tissue and is found within organs. It also surrounds organs and binds them to one another. Magnification, X250. *b.* Adipose tissue looks like white ghosts because the fat has been washed out during preparation of the tissue. Magnification, X250.

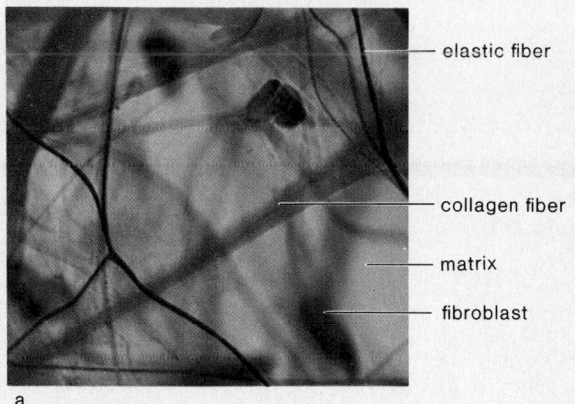

elastic fiber

collagen fiber

matrix

fibroblast

a.

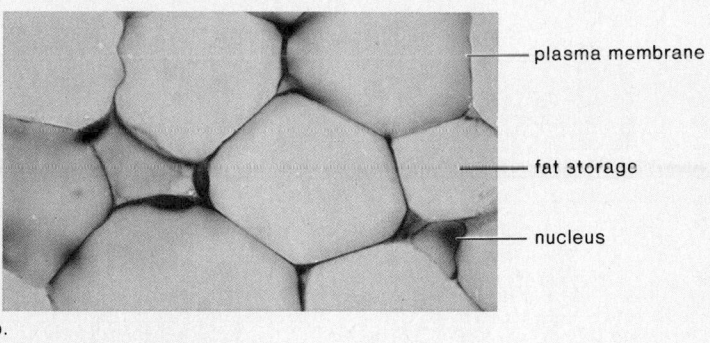

plasma membrane

fat storage

nucleus

b.

Loose connective tissue lies beneath epithelium in the skin and also beneath most internal vertebrate organs (fig. 33.3*a*). Its presence in lungs, arteries, and the urinary bladder allows these organs to expand. It forms a protective covering encasing many internal organs, such as muscles, blood vessels, and nerves.

Adipose tissue is a type of loose connective tissue in which the fibroblasts enlarge and store fat and in which the intercellular matrix is reduced (fig. 33.3*b*). In vertebrates, it is found particularly beneath the skin, around the kidneys, and on the surface of the heart. It insulates the body and provides padding.

Fibrous connective tissue contains many collagenous fibers that are packed closely together. This type of tissue has more specific functions in vertebrates than does loose connective tissue. For example, fibrous connective tissue is found in **tendons,** which connect muscles to bones, and in **ligaments,** which connect bones to other bones at joints.

Loose connective tissue and fibrous connective tissue contain fibroblasts separated by a matrix, which contains collagen and elastic fibers.

Cartilage and Bone

Cartilage and bone are rigid connective tissues in which structural proteins (cartilage) or calcium salts (bone) are deposited in the intercellular matrix (fig. 33.4).

In **cartilage,** the cells lie in small chambers called lacunae (lacuna, sing.), separated by a matrix that is strong yet flexible. There are various types of cartilage, which are classified according to type of collagen and elastic fiber found in the matrix. In some vertebrates, notably sharks and rays, the entire skeleton is made of cartilage. In humans, the fetal skeleton is cartilage, but it is later replaced by bone. Cartilage is retained at the ends of long bones, at the end of the nose, in the framework of the ear, in the walls of respiratory ducts, and within intervertebral disks.

In **bone,** the matrix of calcium salts is deposited around protein fibers. The minerals give bone rigidity, and the protein fibers provide elasticity and strength, much as steel rods do in reinforced concrete.

In compact bone, bone cells (osteocytes) are located in lacunae that are arranged in concentric circles around tiny tubes called Haversian canals. Nerve fibers and blood vessels are in these canals. The latter bring the nutrients that allow bone to renew itself. The nutrients can reach all of the cells because there are minute canals (canaliculi) containing thin processes of the osteocytes that connect them with one another and with the Haversian canals.

The ends of a long bone contain spongy bone, which has an entirely different structure. Spongy bone contains numerous bony bars and plates separated by irregular spaces. Although lighter than compact bone, spongy bone still is designed for strength. Just as braces are used for support in buildings, the solid portions of spongy bone follow lines of stress.

Cartilage and bone are support tissues. Cartilage is more flexible than bone because the matrix is rich in protein and not calcium salts, like that of bone.

Blood

Blood is a connective tissue in which the cells are separated by a liquid called plasma. In vertebrates, blood cells are of 2 types: red blood cells (erythrocytes), which carry oxygen, and white blood cells (leukocytes), which aid in fighting infection (fig. 33.5). Also present in plasma are platelets, which are important to the initiation of blood clotting. Platelets are not complete cells; rather, they are fragments of giant cells found in the bone marrow.

Figure 33.4

Anatomy of a long bone. **a.** A long bone is encased by fibrous connective tissue except where it is covered by cartilage at the ends. The central shaft is composed of compact bone, but the ends are spongy bone, which can contain red bone marrow. A central medullary cavity contains yellow bone marrow. **b.** Photomicrograph of cartilage. Magnification, X250. **c.** Micrograph of compact bone. Magnification, X320.

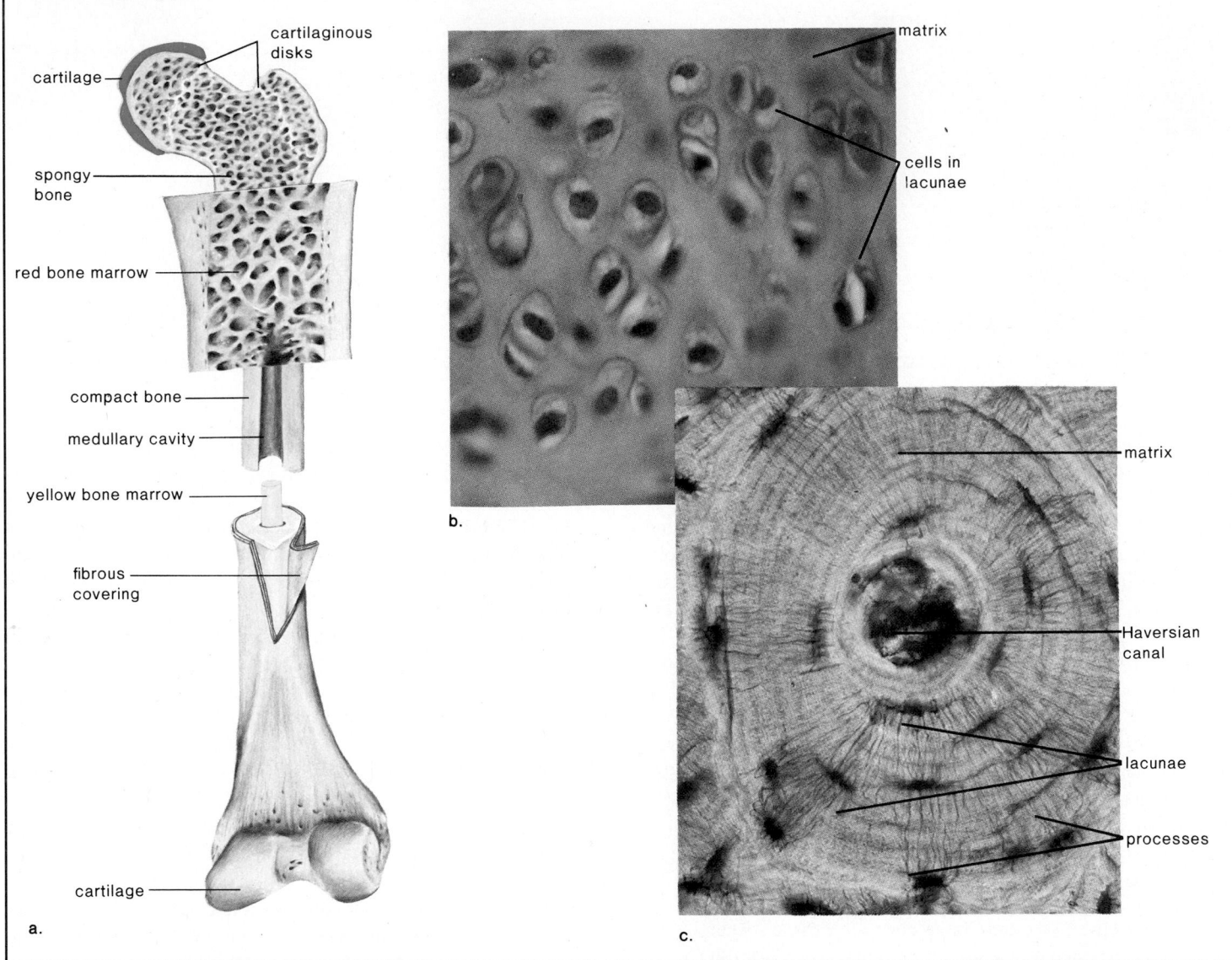

Blood is unlike other types of connective tissue in that the intercellular matrix (i.e., plasma) is not made by the cells. Plasma is a mixture of different types of molecules that enter the blood at various locations.

Blood is a connective tissue in which the matrix is plasma.

Muscular Tissue

Muscular (contractile) tissue is composed of cells called *muscle fibers*. Muscle fibers contain actin filaments and myosin filaments, whose interaction accounts for the movements we associate with animals.

Some muscle fibers are striated and some are smooth (fig. 33.6). *Striated muscle fibers* have light and dark bands perpendicular to the length of the cell. These bands reflect the placement of actin filaments and myosin filaments in the cell. *Smooth muscle fibers* are spindle-shaped cells that lack striations. They form layers in which the thick middle portion of one cell is opposite the thin ends of adjacent cells. Consequently, the nuclei form an irregular pattern in smooth muscular tissue.

Both striated and smooth muscular tissue assist locomotion in many invertebrates. In vertebrates, *smooth muscle* moves materials through internal organs such as the digestive

Animal Structure and Function

Figure 33.5

Blood, a liquid tissue. Blood is classified as connective tissue because the cells are separated by a matrix—plasma. Plasma, the liquid portion of blood, usually contains several types of cells. This shows the cells in human blood (red blood cells, white blood cells, and platelets, which are actually fragments of a larger cell).

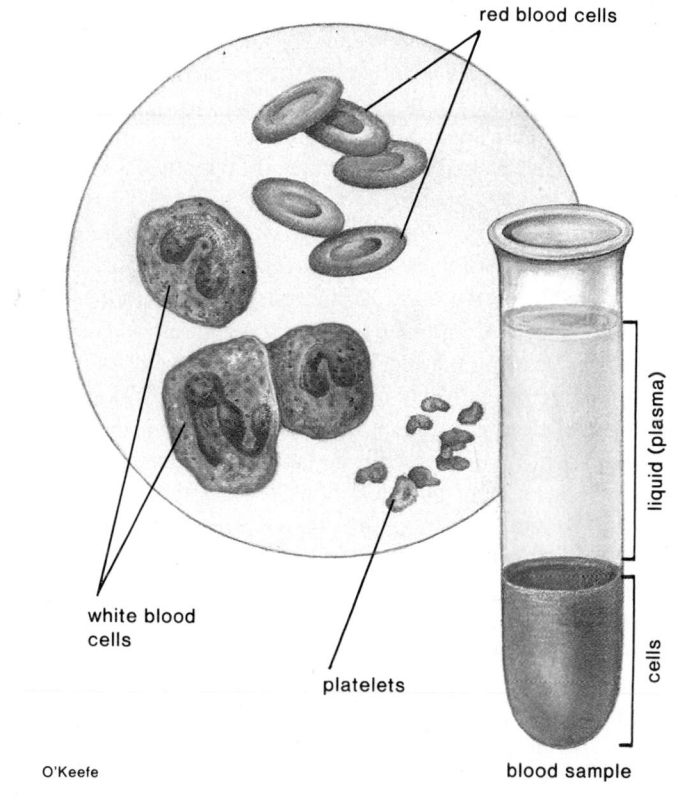

Figure 33.6

Muscular tissue. How do you distinguish an animal from a plant? One way is to detect rapid motion—only animals have contractile fibers that permit movement. *a.* Smooth muscle. Note the single nucleus and the lack of striations. Magnification, X250. *b.* Skeletal muscle is found attached to the skeleton in vertebrates. It is a type of striated muscle. Magnification, X100. *c.* Cardiac muscle pumps the heart. Note the branching of the fibers, the central position of the nuclei, and the presence of intercalated disks, which are folded plasma membranes between adjacent individual cells. Magnification, X160.

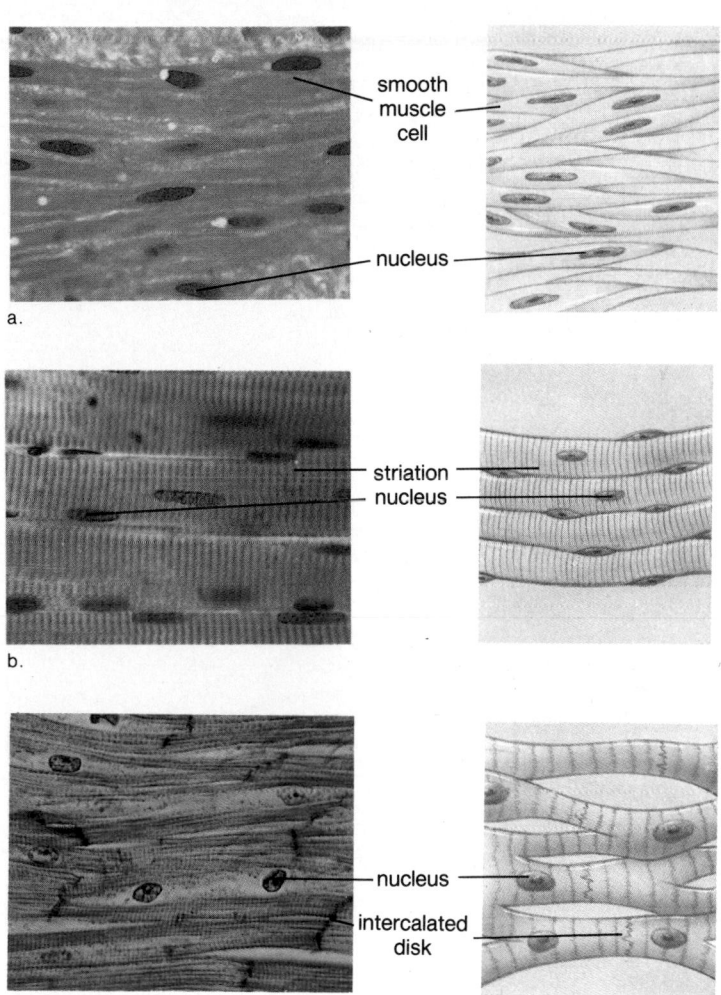

and reproductive tracts and the blood vessels. Only striated muscular tissue makes up the *skeletal muscles*, which are under voluntary control and are attached to the bones of the skeleton. Skeletal muscle cells are cylindrical and quite long—they arise during development when several cells fuse, giving one multinucleated cell. The nuclei are placed at the periphery of the cell, just inside the plasma membrane. A type of striated muscular tissue, called *cardiac muscle*, is found in the heart. Cardiac muscle cells differ from skeletal muscle cells in that they have a single, centrally placed nucleus. The cells are branched and seemingly fused one with the other, and the heart appears to be composed of one large interconnecting mass of muscle cells. Actually, cardiac muscle cells are separate and individual, but they are bound end to end at intercalated disks, areas of folded plasma membrane between the cells.

All muscle tissue contains actin and myosin filaments; these form a striated pattern in skeletal and cardiac muscle but not in smooth muscle.

Nervous Tissue

Nervous tissue contains nerve cells called **neurons** (fig. 33.7). A neuron is a specialized cell that has 3 parts: (1) dendrites, which conduct impulses (send a message) to the cell body; (2) the cell body, which contains most of the cytoplasm and the nucleus of the neuron; and (3) the axon, which conducts impulses away from the cell body. Long axons and dendrites are called nerve fibers, and they are bound together by connective tissue to form nerves. Neurons are specialized to detect environmental stimuli and then to conduct signals to other neurons and to muscle or glands, thereby bringing about a coordinated response to the stimulus.

Figure 33.7

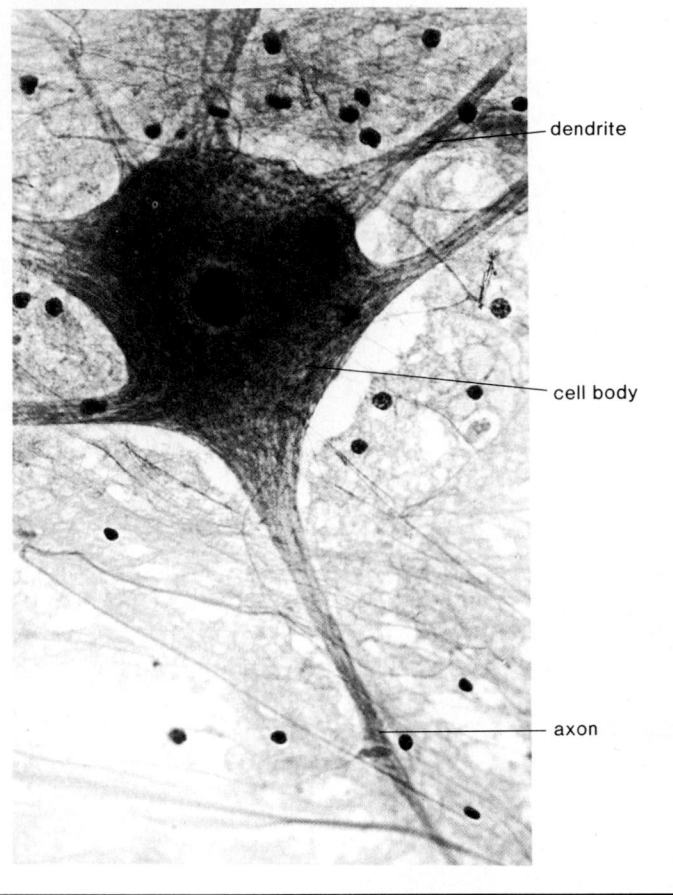

Micrograph of neuron. Conduction of the nerve impulse is dependent on neurons, each of which has the 3 parts indicated. A dendrite takes nerve impulses to the cell body, and an axon takes them away from the cell body. Magnification, X200.

dendrite

cell body

axon

In addition to neurons, nervous tissue contains neuroglial cells. These cells maintain the tissue by supporting and protecting neurons. They also provide nutrients to neurons and help to keep the tissue free of debris.

Nerve cells, called neurons, have fibers (processes) called axons and dendrites. Long fibers are found in nerves.

Organs and Organ Systems

In most animals each **organ** consists of several types of tissues (fig. 33.1). For example, human skin contains epithelium, connective tissue, nervous tissue, and blood (fig. 33.8). Each organ is also specialized to perform specific functions. In humans, the skin covers the body, protecting underlying parts from physical trauma, microbial invasion, and water loss. The skin also helps to regulate body temperature, and because it contains receptors (sense organs), the skin helps us to be aware of our surroundings and to communicate with others.

Some authorities speak of the skin as an organ within the integumentary system. They maintain that the hair follicles, oil and sweat glands, receptors, and skin are separate organs, and these organs work together to perform various functions.

Human Skin

Skin has an outer epidermal layer (the epidermis) and an inner layer (the dermis). Beneath the dermis, there is a subcutaneous layer that binds the skin to underlying organs.

The **epidermis** is the outer, thinner layer of the skin. It is a stratified squamous epithelium whose cells are derived from the basal cells, which undergo continuous cell division. As newly formed cells are pushed to the surface, they gradually flatten and harden. Eventually, they die and are sloughed off. Hardening is caused by cellular production of a waterproof protein called keratin. Over much of the body, keratinization is minimal, but the palm of the hand and the sole of the foot have a particularly thick outer layer of dead keratinized cells arranged in spiral and concentric patterns. We call these patterns fingerprints and footprints.

Particular cells in the epidermis, called melanocytes, produce melanin, the pigment responsible for skin color in dark-skinned persons. When you sunbathe, the melanocytes become more active, producing melanin in an attempt to protect the skin from the damaging effects of the ultraviolet (UV) radiation in sunlight.

The **dermis** is a layer of fibrous connective tissue that is deeper and thicker than the epidermis. It contains elastic fibers and collagen fibers. The collagen fibers form bundles that interlace and run, for the most part, parallel to the skin surface. There are several types of structures in the dermis. A hair, except for the root, is formed of dead, hardened epidermal cells; the root is alive and resides in a hair follicle found in the dermis. Each follicle has one or more oil (sebaceous) glands that secrete sebum, an oily substance that lubricates the hair and the skin. A smooth muscle called the arrector pili muscle is attached to the hair follicle in such a way that when contracted, the muscle causes the hair to stand on end. When you are frightened or are cold, goose bumps develop due to the contraction of these muscles.

Sweat (sudoriferous) glands are quite numerous and are present in all regions of the skin. A sweat gland begins as a coiled tubule within the dermis, but then it straightens out near its opening. Some sweat glands open into hair follicles, and others open onto the surface of the skin.

Small receptors are present in the dermis. There are different receptors for touch, pressure, pain, and temperature. The fingertips contain the most touch receptors, and these add to our ability to use our fingers for delicate tasks. The dermis also contains nerve fibers and blood vessels. When blood rushes into these vessels, a person blushes, and when blood is reduced in them, a person turns blue.

The subcutaneous layer, which lies below the dermis, is composed of loose connective tissue, including adipose tissue. Adipose tissue helps to insulate the body by minimizing both heat gain and heat loss. A well-developed subcutaneous layer gives a rounded appearance to the body. Excessive development of this layer accompanies obesity.

Figure 33.8

Human skin anatomy. Skin contains 3 layers: epidermis, dermis, and subcutaneous.

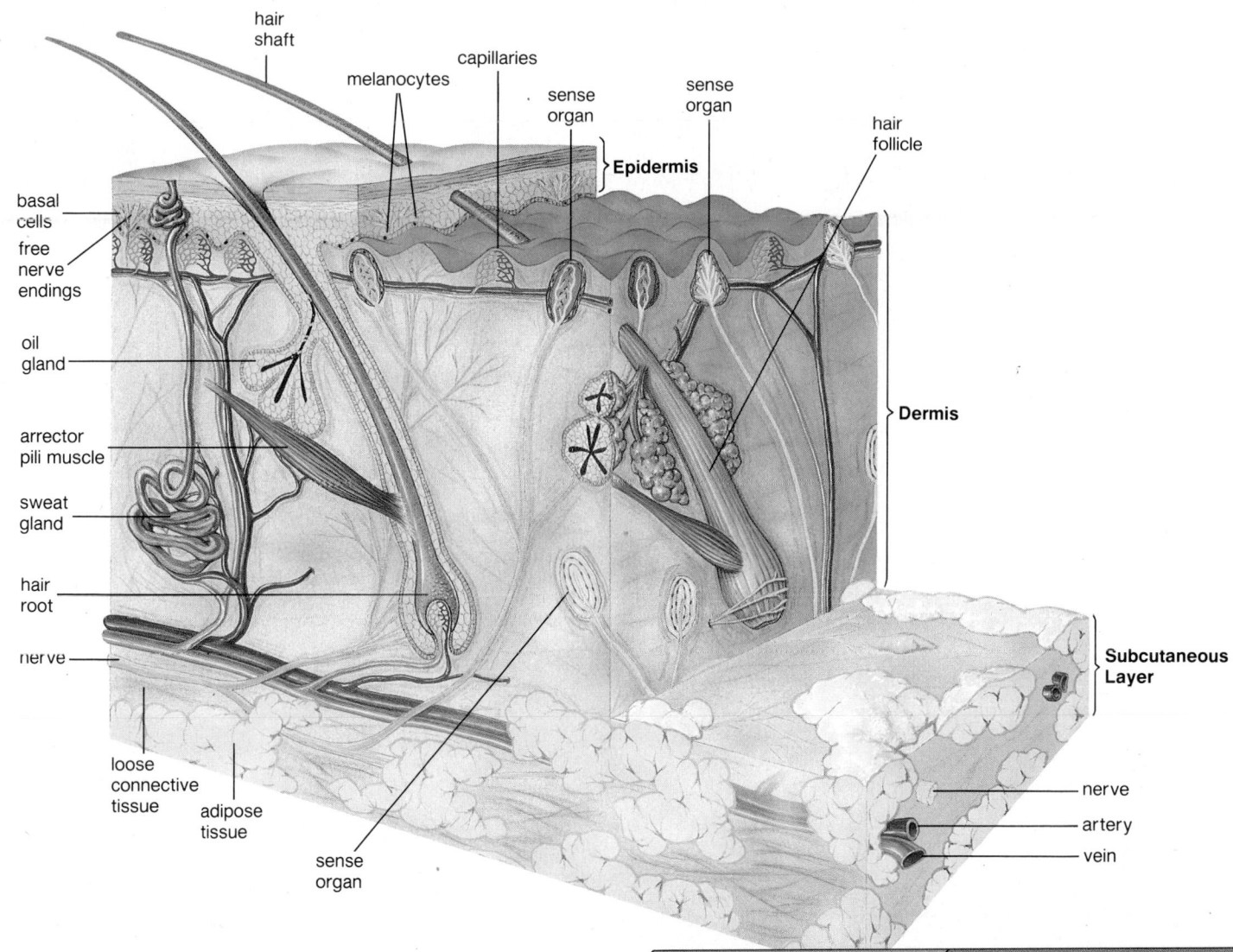

Each animal organ has a structure that suits its function. In human skin, the hardened cells of epidermis provide a protective covering; the dermis contains receptors, blood vessels, and nerves; and the adipose tissue of the subcutaneous layer insulates the body.

Organ Systems

In most animals, individual organs function as part of an *organ system*, the next higher level of animal organization. Figure 33.9 pictorially shows the organ systems within the human body. These same systems are found in all vertebrate animals. The organ systems carry out the life processes that are common to all animals and indeed to all organisms:

Life Processes	Human Systems
Acquire materials and energy (food)	Skeletal system Muscular system Digestive system
Maintain body shape	Muscular system Skeletal system
Exchange gases	Respiratory system
Transport materials	Circulatory system
Excrete wastes	Excretory system
Protect the body from disease	Lymphatic system
Coordinate body activities	Nervous system Endocrine system
Produce offspring	Reproductive system

Figure 33.9
Human organ systems and
their functions.

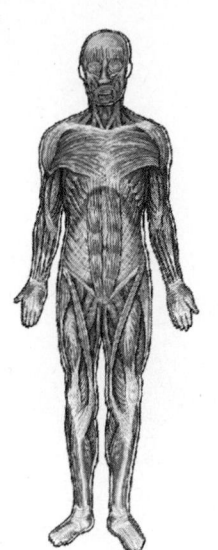 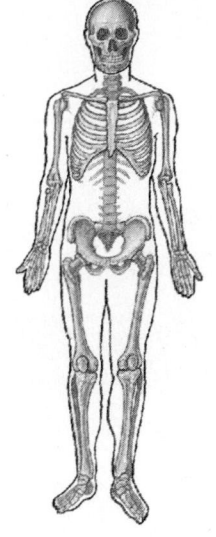 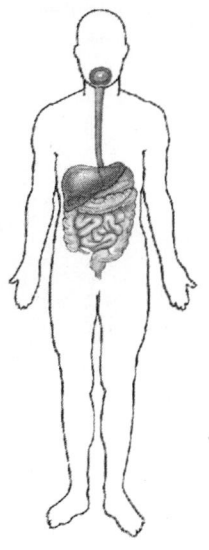 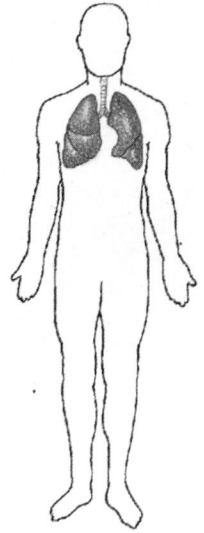 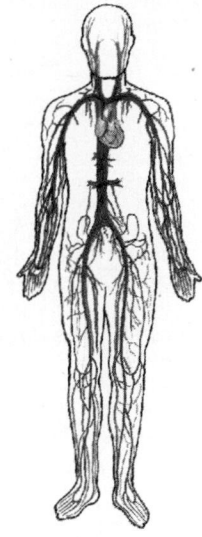

Name of System	Muscular system	Skeletal system	Digestive system	Respiratory system	Circulatory system
Life Process	Acquires materials and energy	Acquires materials and energy	Acquires materials and energy	Exchanges gases	Transports material
Specific Functions	Produces body movements; produces body heat; maintains posture and supports the body	Provides rigid framework for body movement; supports and protects internal parts; produces blood cells; stores minerals	Digests food into small molecules and absorbs these molecules	Exchanges gas between external environment and blood; maintains blood pH by excreting carbon dioxide	Transports nutrients and oxygen to and metabolic wastes from cells; distributes hormones; protects against injury and microbes

Body Cavities

Each organ system has a particular location within the human body. There are 2 main body cavities: the smaller dorsal body cavity and the larger ventral body cavity (fig. 33.10). The brain and the spinal cord are in the dorsal body cavity.

During development, the ventral body cavity develops from the *coelom*. In humans and other mammals, the coelom is divided by a muscular diaphragm that assists breathing. The heart, a pump for the closed circulatory system, and the lungs are located in the upper (thoracic or chest) cavity. The major portion of the digestive system, the entire excretory system, and much of the reproductive system are located in the lower (abdominal) cavity. The major organs of the excretory system are the paired kidneys. The accessory organs of the digestive system are the liver and pancreas. Each sex has characteristic sex organs.

Homeostasis

Claude Bernard pointed out in 1959 that while an animal lives in an external environment, the cells of the body lie within an internal environment. The *internal environment* is a fluid, called tissue fluid, that bathes the cells of the body. Bernard said that the relative stability of the internal environment allows animals to live in an external environment that can vary considerably. Later, Walter Cannon, an American physiologist, introduced the term **homeostasis.** He said there is a dynamic interplay between events that tend to change the internal environment and those that counter this possibility. To achieve homeostasis, the composition of blood and tissue fluid, the body temperature, and the blood pressure, for example, must stay within a normal range.

Excretory system	**Lymphatic system**	**Nervous system**	**Endocrine system**	**Reproductive system**
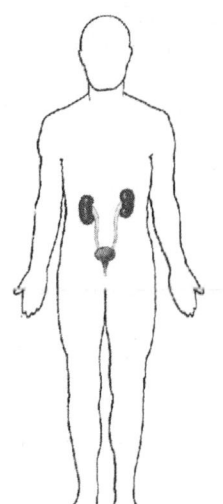	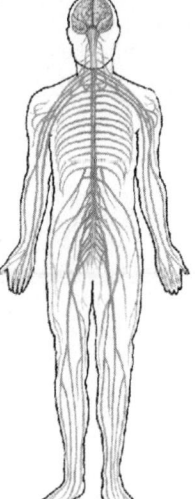	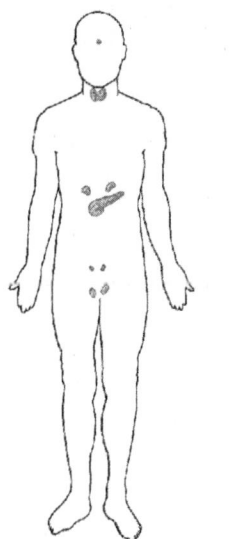		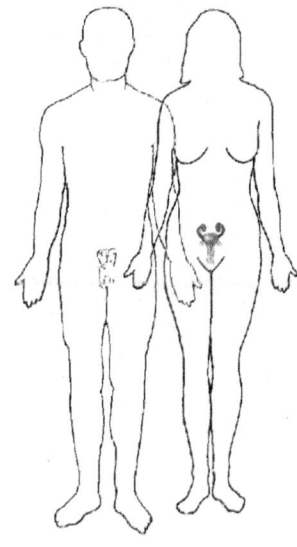
Excretes wastes	*Protects the body from disease*	*Coordinates body activities*	*Coordinates body activities*	*Produces offspring*
Maintains volume and chemical composition of blood and tissue fluid	Transports excess tissue fluid to bloodstream; transports fat to blood; helps provide immunity against disease	Along with endocrine system regulates body systems; learning and memory	Secretes hormones that regulate body metabolism, growth, and reproductive system	Male: produces hormones and sperm; transfers sperm to female. Female: produces hormones and egg; provides site for fertilization of egg, implantation, and development of embryo and fetus

---

The internal environment of an animal's body consists of tissue fluid, which bathes the cells.

In the chapters that follow, we will see that most organ systems of the human body contribute to homeostasis (fig. 33.9). The digestive system takes in and digests food, providing nutrient molecules that enter the blood and replace the nutrients that are constantly being used by the body cells. The respiratory system adds oxygen to the blood and removes carbon dioxide. The amount of oxygen taken in and carbon dioxide given off can be increased to meet body needs. The liver and the kidneys contribute greatly to homeostasis. For example, immediately after glucose enters the blood, it can be removed by the liver and stored as glycogen. Later, glycogen is broken down to replace the glucose used by the body cells; in this way, the glucose composition of the blood remains constant. The hormone insulin, secreted by the pancreas, regulates glycogen storage. The kidneys are also under hormonal control as they excrete wastes and salts, substances that can affect the pH level of the blood.

Although homeostasis is, to a degree, controlled by hormones, it is ultimately controlled by the nervous system. Homeostatic systems have 3 elements: *receptors* (sensors) that react to a stimulus; an *integrating center* that evaluates information from the sensors; and *effectors* (muscles and glands) that carry out a response to the stimulus (fig. 33.11). Maintaining proper temperature and blood pressure levels requires receptors that detect a change and signal an integrating center in the brain. The center then directs the muscles by way of nerves to bring

Figure 33.10

Mammalian body cavities. There is a dorsal cavity, which contains the cranial cavity and the vertebral canal. The brain is in the cranial cavity, and the spinal cord is in the vertebral canal. There is a well-developed ventral cavity, which is divided by the diaphragm into the thoracic cavity and the abdominal cavity. The heart and lungs are in the thoracic cavity, and most other internal organs are in the abdominal cavity.

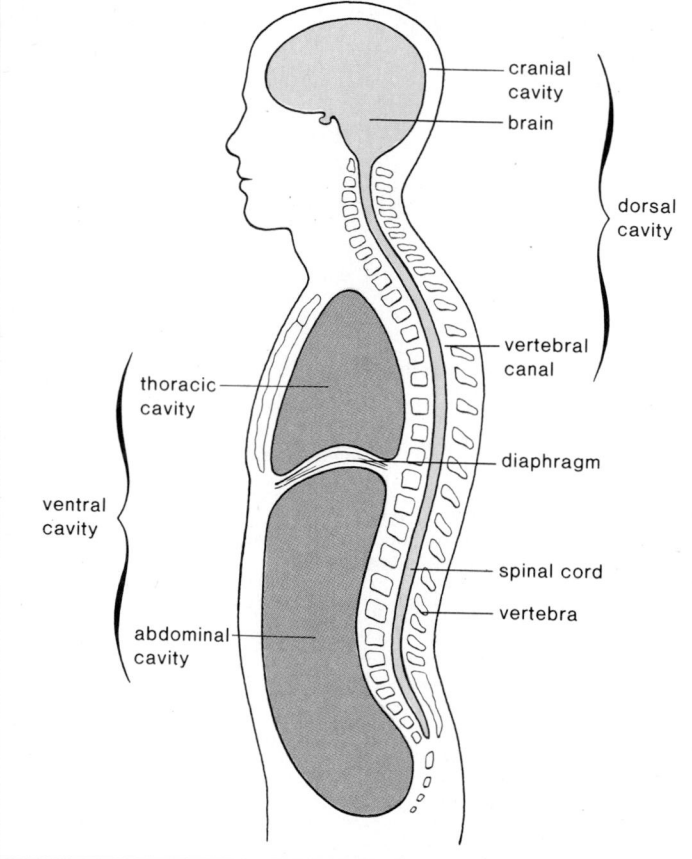

Figure 33.11

Negative feedback control. A receptor (sense organ) is stimulated by a change, such as low temperature, and signals an integrating center (in the brain) that directs effectors (muscles) to react. The response, constriction of blood vessels in the skin, and perhaps shivering, raises the temperature, negating the original stimulus.

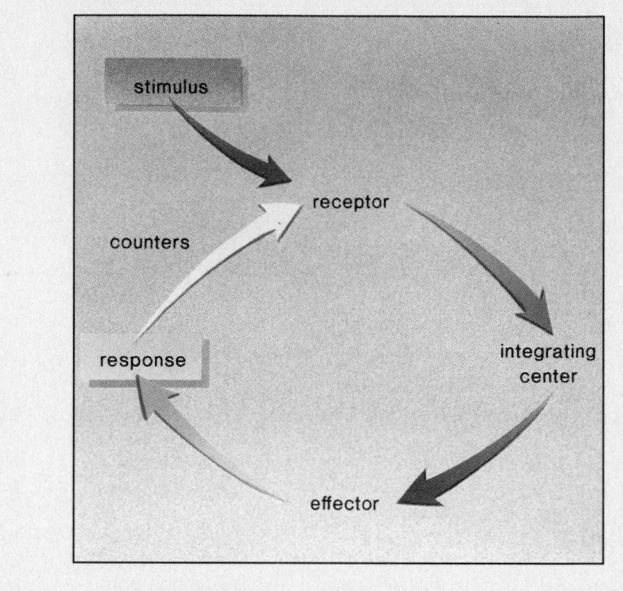

The receptor and the integrating center for body temperature are located in the hypothalamus, a part of the brain. The receptor is sensitive to the temperature of the blood, and when the temperature falls below normal, the receptor signals the integrating center, which directs (via nerve impulses) the muscles (the effectors) of blood vessels in the skin to constrict. This conserves heat. Also, the arrector pili muscles pull hairs erect, and a layer of insulating air is trapped next to the skin. If body temperature falls even lower, the integrating center sends nerve impulses to the skeletal muscles, and shivering occurs. Shivering generates heat, and gradually body temperature rises to 37°C and perhaps higher.

During the period of time the body temperature is normal, the receptor and the integrating center are not active, but once body temperature is higher than normal, they are activated. Now the integrating center directs the blood vessels of the skin to dilate. This allows more blood to flow near the surface of the body, where heat can be lost to the environment. The integrating center also activates the sweat glands, and the evaporation of sweat also helps lower body temperature. Gradually, body temperature decreases to 37°C and perhaps lower. Once body temperature is below normal the cycle begins again.

Homeostasis of internal conditions is a self-regulating mechanism that results in slight fluctuations above and below a mean. For example, body temperature rises above and drops below a normal temperature of 37°C.

---

about an adaptive response. Very often homeostatic systems utilize a **negative feedback loop.** For example, if the stimulus (input) was a decrease in body temperature, the response (output) is a rise in body temperature. It is a feedback loop because the output is fed back into the system; it is a negative feedback loop because the output is counter to and cancels the input. **Positive feedback loops** also sometimes occur. In these instances, a choice of events increases the likelihood of a particular response. Blood clotting, for example, requires a number of steps, each leading to the next step until a blood clot forms.

Body Temperature Control

Body temperature control in humans is dependent on a homeostatic system that is self-regulated by a negative feedback loop. This type of homeostatic regulation results in fluctuation between 2 levels (fig. 33.12).

Animal Structure and Function

Figure 33.12

Temperature control. When the body temperature rises, the regulator center directs the blood vessels to dilate and the sweat glands to be active. Now, the body temperature lowers. Then, the regulator center directs the blood vessels to constrict, hairs to stand on end, and even shivering to occur if needed. Now, the body temperature rises again. Because the receptors are sensitive only to a change in the internal environment, the body temperature fluctuates above and below normal.

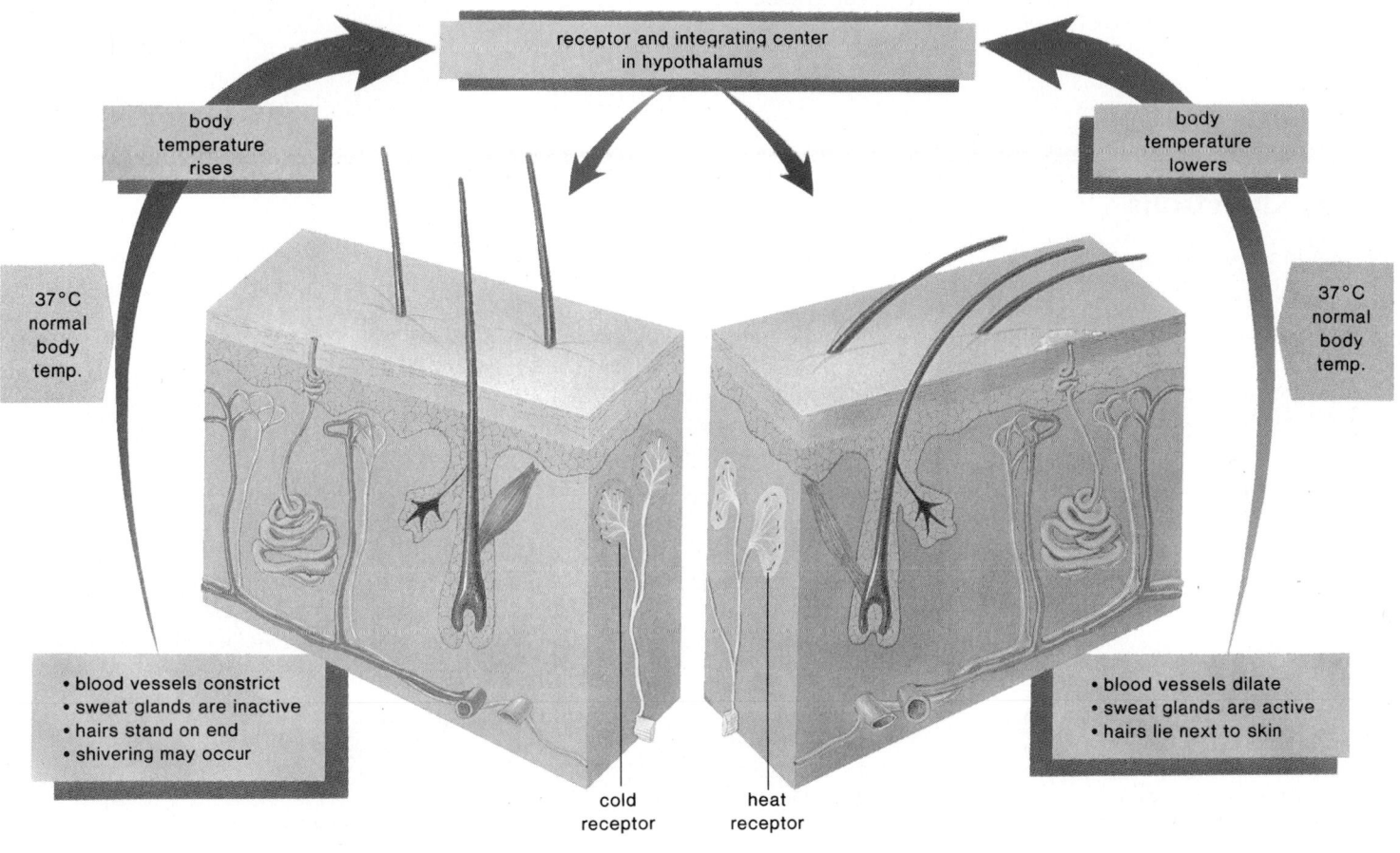

receptor and integrating center
in hypothalamus

body temperature rises

body temperature lowers

37°C normal body temp.

37°C normal body temp.

• blood vessels constrict
• sweat glands are inactive
• hairs stand on end
• shivering may occur

• blood vessels dilate
• sweat glands are active
• hairs lie next to skin

cold receptor

heat receptor

Summary

1. Tissues arise from the embryonic germ layers: ectoderm, mesoderm, and endoderm. Ectoderm produces nervous tissue; mesoderm produces muscle and certain systems (circulatory, excretory, reproductive); endoderm produces the linings of the digestive and respiratory tracts and associated organs. Epithelial tissue comes from all 3 layers.

2. Epithelial tissue covers the body and lines cavities. There is squamous, cuboidal, and columnar epithelium. Each type can be simple or stratified, glandular, or have modifications like cilia. Epithelial tissue protects, absorbs, secretes, and excretes.

3. Connective tissue has a matrix between cells. Loose connective tissue and fibrous connective tissue

contain fibroblasts and fibers (collagen and elastic). Adipose tissue is a type of loose connective tissue; tendons and ligaments are fibrous connective tissue.

4. Cartilage (protein matrix) and bone (calcium matrix) are rigid connective tissues. Blood is connective tissue in which plasma is the matrix.

5. Muscular (contractile) tissue can be smooth or striated (both skeletal and cardiac). In humans, smooth muscle is in the wall of internal organs; skeletal muscle is attached to bone, and cardiac muscle is in the heart.

6. Nervous tissue contains neurons having 3 parts: dendrites, cell body, and axon. Neuroglial cells support and protect neurons.

7. Organs contain various tissues. Skin is an organ that has 3 tissue layers: epidermis (stratified squamous epithelium); dermis (fibrous connective epithelium); and subcutaneous (loose connective epithelium). Receptors, hair follicles, blood vessels, and nerves are in the dermis.

8. Organ systems contain several organs. Organ systems of humans carry out the life processes that are common to all organisms and also have specific functions (see fig. 33.9).

9. The human body contains 2 main cavities. The dorsal cavity contains the brain and spinal cord. The ventral cavity contains the thoracic (heart and lungs) and abdominal cavity (most other internal organs).

10. Homeostasis refers to the relative constancy of the internal environment, which is very often maintained by negative feedback loops in which the response negates the stimulus. As an example, consider body temperature control in humans. When the body is cold, receptors signal an integrating center that directs the muscles of the blood vessels (effectors) to constrict. When the body is warm, receptors signal an integrating center that directs the blood vessels to dilate. Also, sweat glands are activated. It can be seen that this type of regulation causes fluctuations above and below a mean.

Study Questions

1. Name the 4 major types of tissues.
2. Describe the structure and functions of 3 types of epithelial tissue.
3. Describe the structure and functions of 6 types of connective tissue.
4. Describe the structure and functions of 3 types of muscular tissue.
5. Nervous tissue contains what types of cells?
6. Describe the structure of skin, and state at least 2 functions of this organ.
7. In general terms, describe the location of the human organ systems.
8. Tell how the various systems of the body contribute to homeostasis.
9. What is the function of receptors, the integrating center, and effectors in a negative feedback loop? Why is it called negative feedback?

Objective Questions

1. Which of these is mismatched?
   a. epithelial tissue—protection and absorption
   b. muscular tissue—contraction and conduction
   c. connective tissue—binding and support
   d. nervous tissue—conduction and message sending
2. Which of these is not epithelial tissue?
   a. simple cuboidal and stratified columnar
   b. bone and cartilage
   c. stratified squamous and simple squamous
   d. All are epithelial tissue.
3. Which tissue is more apt to line a lumen?
   a. epithelial tissue
   b. connective tissue
   c. nervous tissue
   d. muscular tissue
4. Tendons and ligaments are
   a. connective tissue.
   b. associated with the bones.
   c. found in vertebrates.
   d. All of these.
5. Which tissue has cells in lacunae?
   a. epithelial tissue
   b. cartilage
   c. bone
   d. Both b and c.

6. Cardiac muscle is
   a. striated.
   b. involuntary.
   c. smooth.
   d. Both a and b.
7. Which of these components of blood fights infection?
   a. red blood cells
   b. white blood cells
   c. platelets
   d. All of these.
8. Which of these body systems contribute to homeostasis?
   a. digestive system and excretory system
   b. respiratory system and nervous system
   c. nervous system and endocrine system
   d. All of these.

9. With negative feedback,
   a. the output cancels the input.
   b. there is a fluctuation above and below the average.
   c. there is self-regulation.
   d. All of these.
10. When a person is cold, the blood vessels
    a. dilate and the sweat glands are inactive.
    b. dilate and the sweat glands are active.
    c. constrict and the sweat glands are inactive.
    d. constrict and the sweat glands are active.
11. Identify each of these tissues, and tell whether the tissue is a type of epithelial tissue, muscular tissue, nervous tissue, or connective tissue.

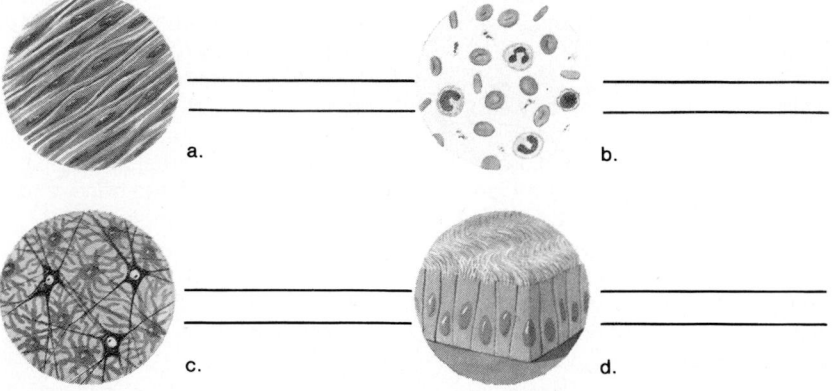

a.   b.

c.   d.

Concepts and Critical Thinking

1. *An animal's body has levels of organization.*

   How is each level of organization more than the sum of its parts?

2. *All organisms carry out certain life processes.*

   Review figure 7.10 and associate parts of the figure with appropriate life processes listed on page 545.

3. *The internal environment of organisms stays relatively constant.*

   How does each of the life processes (except producing offspring) help keep the internal environment relatively constant?

Selected Key Terms

epithelial tissue (ep"ĭ-the'le-al tish'u) 539
connective tissue (kŏ-nek'tiv tish'u) 540
fibroblast (fi'bro-blast) 540
adipose tissue (ad'ē-poz tish'u) 541
tendon (ten'don) 541

ligament (lig'ah-ment) 541
cartilage (kar'tĭ-lij) 541
bone (bōn) 541
blood (blud) 541
muscular (contractile) tissue (mus'ky-lar tish'u) 542

neuron (nu'ron) 543
epidermis (ep"ĭ-der'mis) 544
dermis (der'mis) 544
homeostasis (ho"me-o-sta'sis) 546
negative feedback (neg'ah-tiv fēd'bak) 548

34

Circulatory System

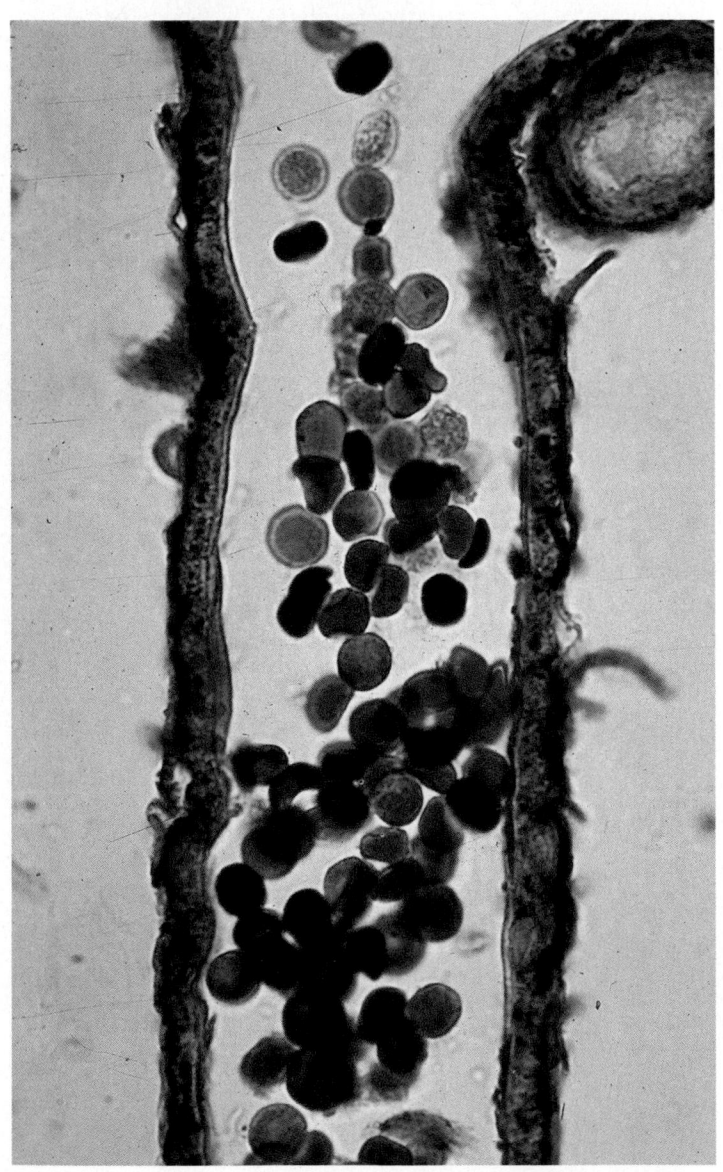

Vertebrates have a closed circulatory system—the blood is always contained within vessels and never runs free. Exchange of gases and nutrients for wastes has to take place across one-cell-thick capillary walls.

Your study of this chapter will be complete when you can

1. compare the circulatory systems of invertebrate and vertebrate animals;
2. compare the structure and function of the 3 types of blood vessels;
3. trace the path of blood through the mammalian heart and about the body;
4. describe how the heart beats, and tell how the heartbeat is controlled;
5. list the factors that control the pressure and velocity of blood in the vessels;
6. list the components of human blood plasma, and describe the function and the source of each;
7. describe the structure and function of human blood cells;
8. list the 3 steps necessary for the clotting of blood;
9. draw and explain a diagram depicting capillary exchange within tissues;
10. list the ABO blood types and describe a laboratory procedure for typing blood.

Figure 34.3

Vertebrate circulatory system. The heart is a pump that keeps the blood moving. **a.** Blood leaving the heart moves from an artery to arterioles to capillaries to venules and then returns to the heart by way of a vein. **b.** Arteries have well-developed walls with a thick middle layer of elastic tissue and smooth muscle. **c.** Capillary walls are one cell thick. Exchanges take place across their thin walls. **d.** Veins have flabby walls, particularly because the middle layer is not thick. Veins have valves that point toward the heart.

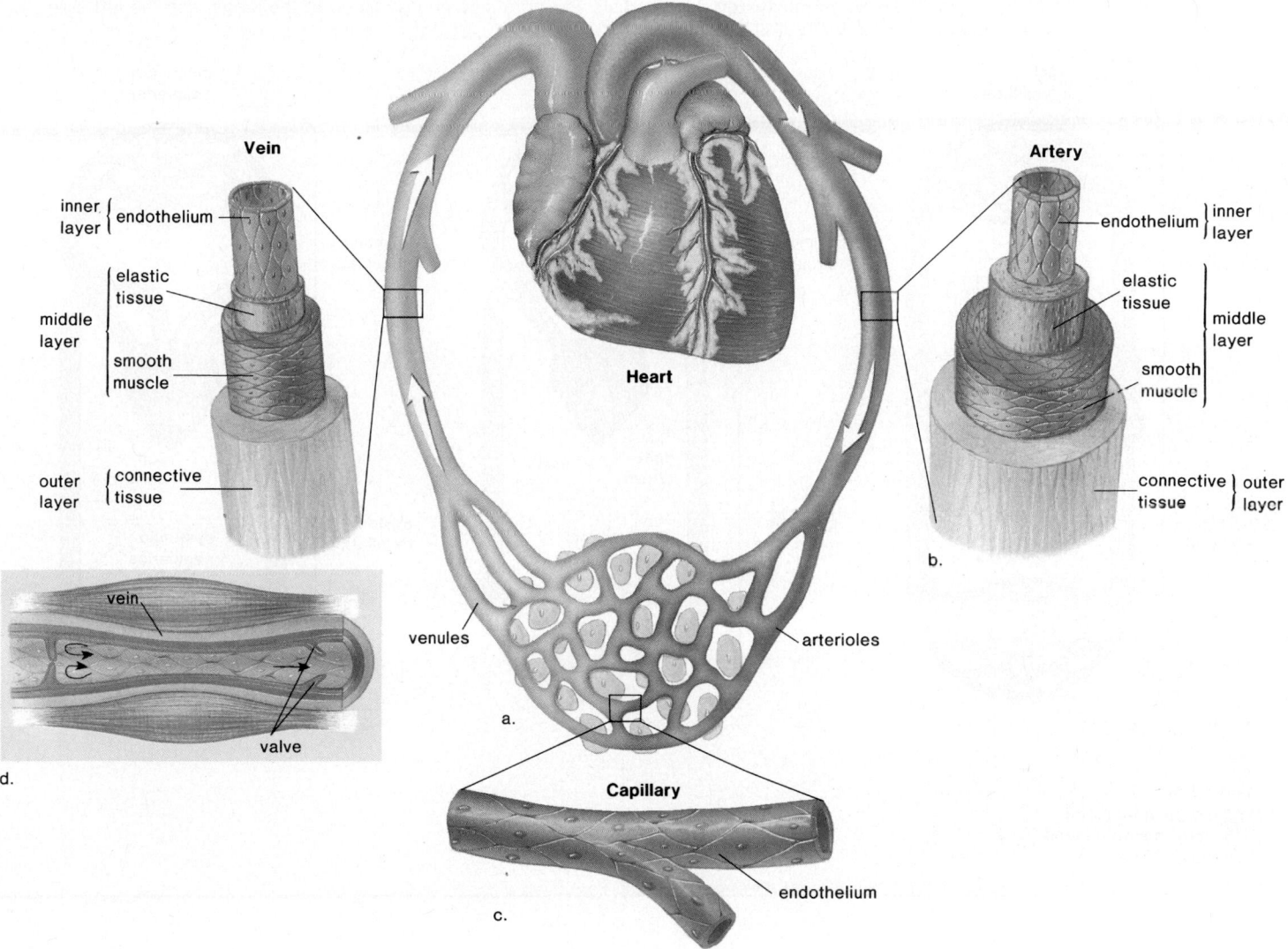

Vertebrates

All vertebrate animals have a closed circulatory system called a **cardiovascular system** (fig. 34.3). It consists of a strong, muscular heart, in which the *atria* (atrium, sing.) primarily receive blood and the muscular *ventricles* pump blood out through the blood vessels. There are 3 kinds of blood vessels: **arteries,** which carry blood away from the heart; **capillaries,** which exchange materials with tissue fluid; and **veins,** which return blood to the heart.

Arteries, the strongest of the blood vessels, are resilient; they are able to expand and constrict because their walls have substantial layers of both elastic tissue and smooth muscle. Arterioles are small arteries whose constriction can be regulated by the nervous system. Arteriole constriction and dilation affect blood pressure. The greater the number of vessels dilated, the lower the blood pressure.

Arterioles branch into capillaries, which are extremely narrow, microscopic tubes with a wall composed of only one layer of cells. Capillary beds (many capillaries interconnected) are present in all regions of the body; consequently, a cut to any body tissue draws blood. Capillaries are the most important part of a closed circulatory system because exchange of nutrient and waste molecules takes place across their thin walls.

The distribution of blood in the various capillary beds is regulated by **sphincters,** circulatory muscles that open and close tubular structures. In this instance, when the sphincter

Figure 34.4

Comparison of circulatory systems in vertebrates. **a.** In a fish, the blood moves in a single loop. The heart has a single atrium and ventricle and pumps the blood into the gill region, where gas exchange takes place. Blood pressure created by the pumping of the heart is dissipated after the blood passes through the gill capillaries. This is a disadvantage of this single-loop system. **b.** Amphibians have a double-loop system in which the heart pumps blood to both the lungs and the body itself. The system is not very efficient because oxygenated blood and deoxygenated blood mix in the single ventricle. **c.** The pulmonary and systemic systems are completely separate in birds and mammals since the heart is divided by a septum into a right and left half. The right side pumps blood to the lungs, and the left side pumps blood to the body proper.

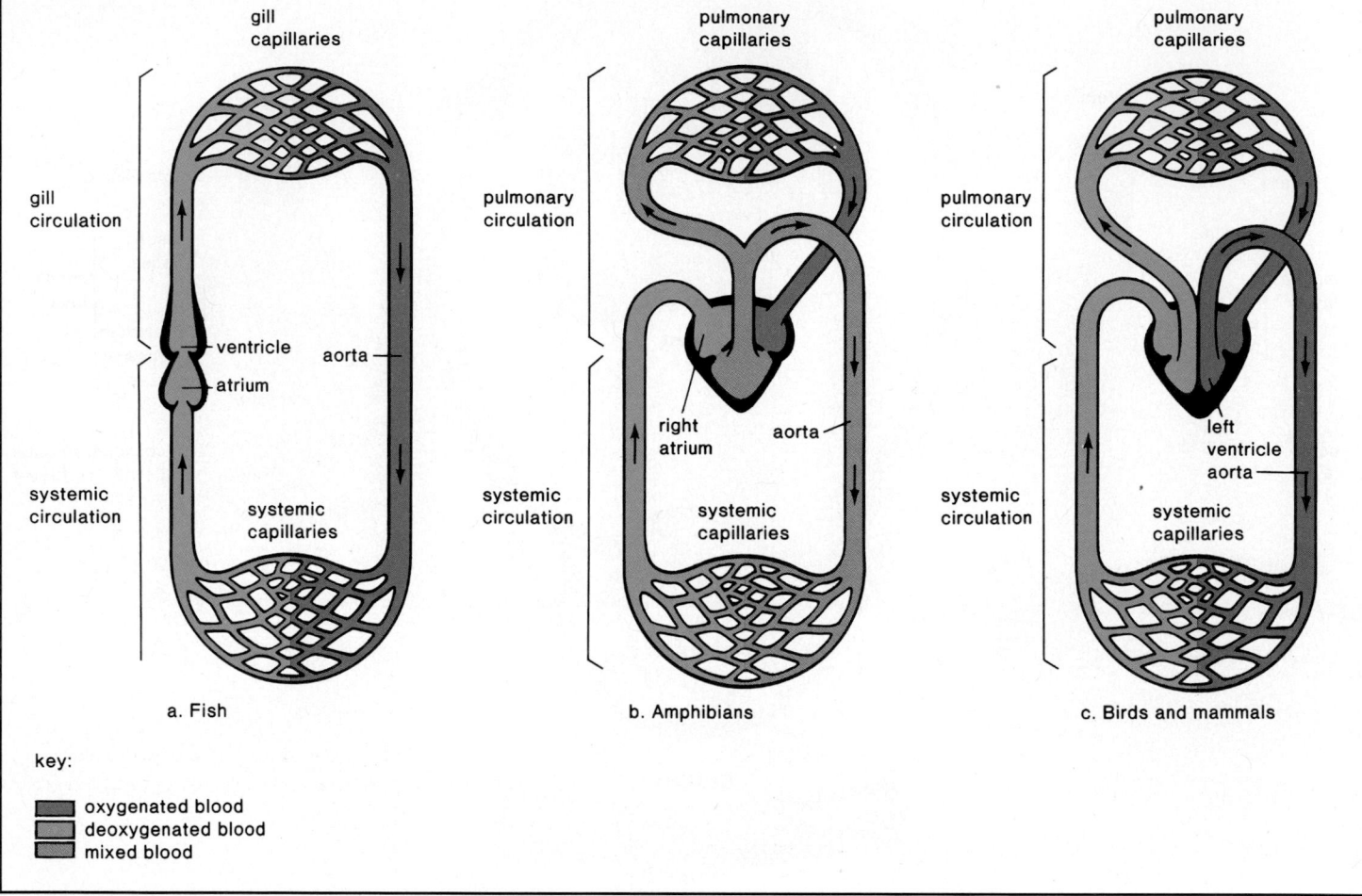

key:

- oxygenated blood
- deoxygenated blood
- mixed blood

relaxes, blood can enter a capillary bed. When the sphincter contracts, a capillary is closed and no blood can enter. Contraction and relaxation of sphincters is controlled by the nervous system. Not all capillary beds are open at the same time. After an animal has eaten, the capillary beds around the digestive tract are usually open, and during muscular exercise, the capillary beds of the skeletal muscles are open.

Venules and veins collect blood from the capillary beds and take it to the heart. First the venules drain the blood from the capillaries, and then they join together to form a vein. The wall of a vein is much thinner than that of an artery because the muscle and elastic fiber layers are poorly developed. Valves within the veins point, or open, toward the heart, preventing a backflow of blood.

Comparison of Circulatory Pathways

Among vertebrate animals, there are 3 different types of circulatory pathways. In fishes, blood follows a one-circuit (single-loop circulatory) pathway through the body. The heart has a single atrium and a single ventricle (fig. 34.4*a*). The pumping action of the ventricle sends blood under pressure only to the gills, where it is oxygenated. After passing through gill capillaries, there is little blood pressure left to distribute the oxygenated blood to the tissues.

In contrast, the other vertebrates have a 2-circuit (double-loop circulatory) pathway. The heart pumps blood to the lungs, called *pulmonary circulation,* and pumps blood to the tissues, called *systemic circulation.* This double pumping action assures adequate blood pressure and flow to both circulatory loops.

Animal Structure and Function

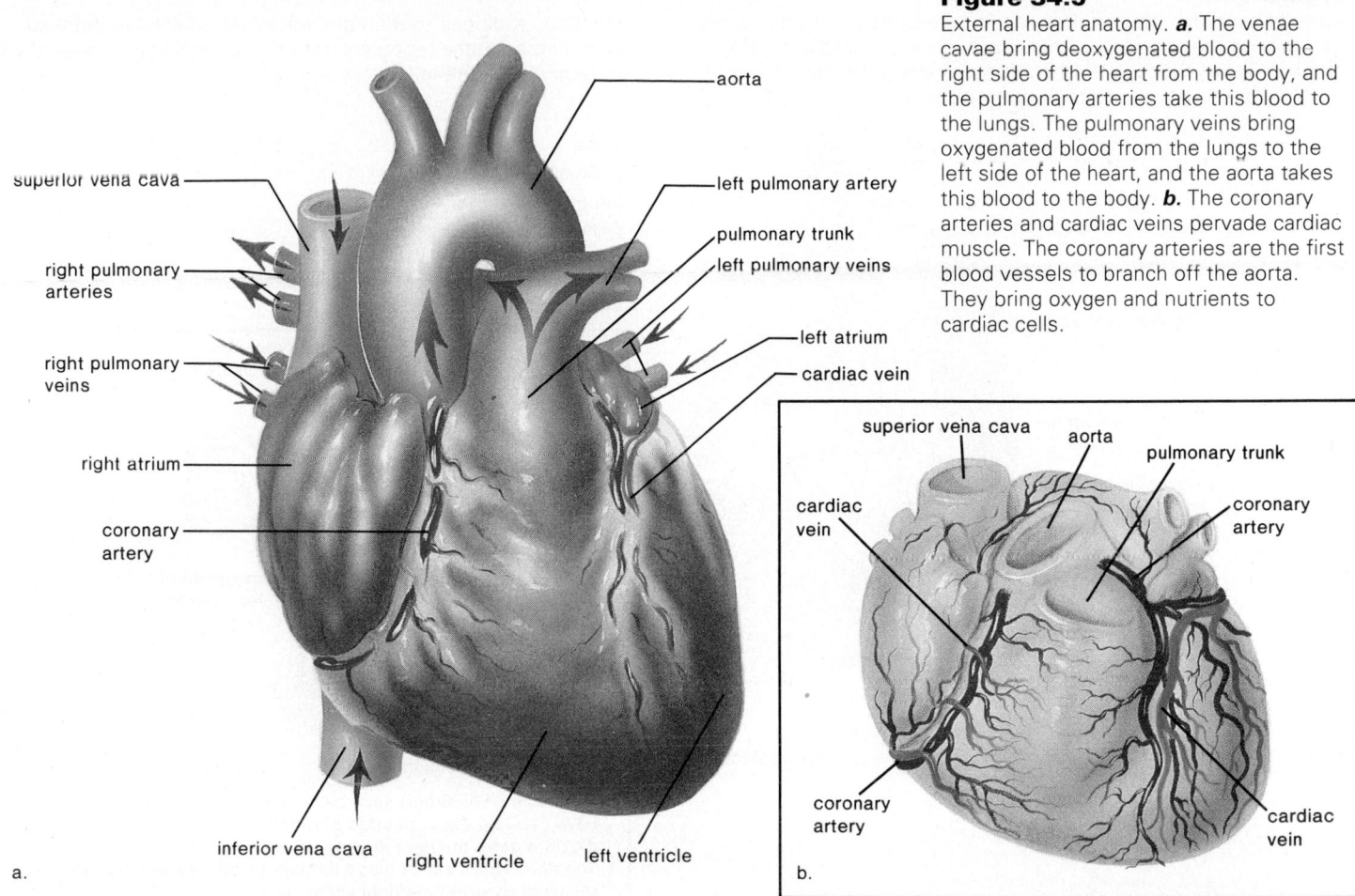

Figure 34.5
External heart anatomy. **a.** The venae cavae bring deoxygenated blood to the right side of the heart from the body, and the pulmonary arteries take this blood to the lungs. The pulmonary veins bring oxygenated blood from the lungs to the left side of the heart, and the aorta takes this blood to the body. **b.** The coronary arteries and cardiac veins pervade cardiac muscle. The coronary arteries are the first blood vessels to branch off the aorta. They bring oxygen and nutrients to cardiac cells.

In amphibians, the heart has 2 atria, but there is only a single ventricle (fig. 34.4b). The hearts of other vertebrates are partially (most reptiles) or completely (some reptiles, all birds, and mammals) divided into right and left halves (fig. 34.4c). The right ventricle pumps blood to the lungs, and the left ventricle pumps blood to the rest of the body. This arrangement increases the likelihood of adequate blood pressure for both the pulmonary and systemic circulations.

> In mammals, birds, and reptiles, the heart is divided into a right side (pumping deoxygenated blood) and a left side (pumping oxygenated blood); this ensures adequate blood pressure for both the pulmonary and systemic circulations.

Human Circulation

William Harvey discovered in the seventeenth century that the blood circulates in humans and other animals (see following reading). He realized that human circulation resembles that of other mammals (fig. 34.4c). The functions of the circulatory system include (1) transporting gases, nutrients, and wastes about the body, (2) clotting to prevent loss of blood from injured blood vessels, and (3) fighting infection or invasion of the body by microorganisms.

The Human Heart

The heart is a cone-shaped muscular organ about the size of a fist (fig. 34.5). It is located between the lungs directly behind the sternum and is tilted so that the apex is directed to the left. The major portion of the heart is called the *myocardium* and consists largely of cardiac muscle tissue (fig. 33.6c). Epithelial tissue lines the heart chambers, and the organ is surrounded by a membranous *pericardial sac*. Normally, this sac contains a small quantity of liquid to lubricate the heart.

Internally, the right and left sides of the heart are separated by a *septum* (fig. 34.6). The heart has valves that direct the flow of blood and prevent its backward flow. The valves that lie between the atria and the ventricles are called the *atrioventricular valves*. These valves are supported by strong fibrous strings called *chordae tendineae*. There are also *semilunar valves,* which resemble half moons, between the ventricles and their attached vessels.

Figure 34.6

Internal heart anatomy. **a.** The venae cavae empty into the right atrium, and the pulmonary trunk leave from the right ventricle. The pulmonary veins enter the left atrium, and the aorta leaves from the left ventricle. Note that the muscles attaching the chordae tendineae to the right ventricular wall have been cut. **b.** A diagrammatic representation that allows you to trace the path of blood through the heart.

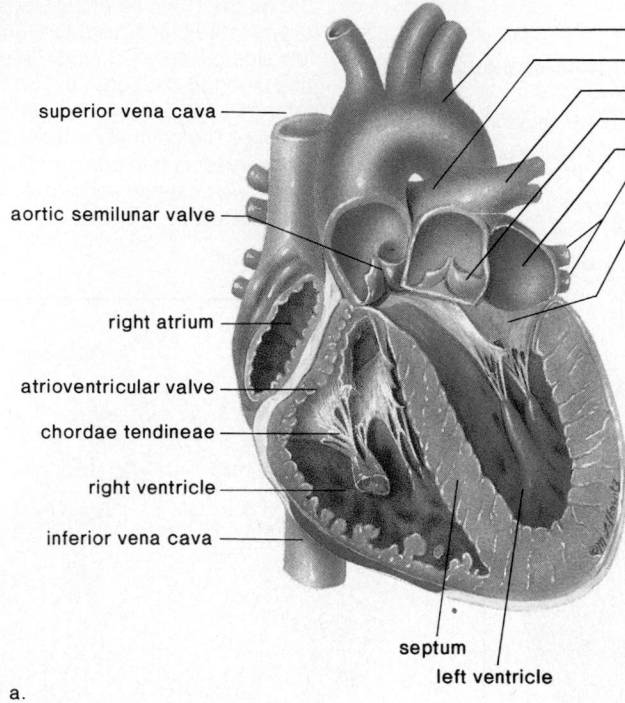

a.

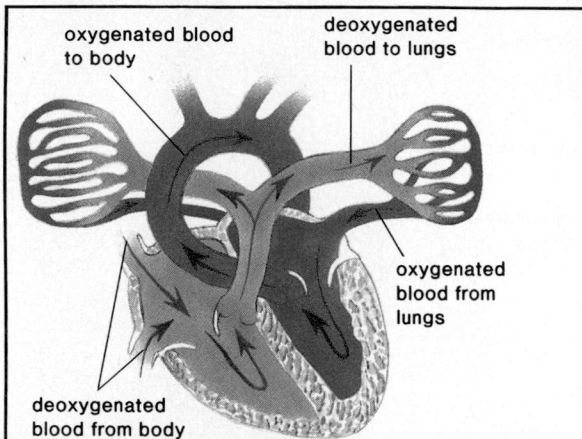

b. Path of blood through the heart

- The superior (anterior) vena cava and the inferior (posterior) vena cava carrying deoxygenated blood (low in oxygen and high in carbon dioxide) enter the right atrium.
- The right atrium sends blood through an atrioventricular valve (the tricuspid valve) to the right ventricle.
- The right ventricle sends blood through the pulmonary semilunar valve into the pulmonary trunk and the pulmonary arteries to the lungs.
- The pulmonary veins carrying oxygenated blood (high in oxygen and low in carbon dioxide) from the lungs enter the left atrium.
- The left atrium sends blood through an atrioventricular valve (the bicuspid, or mitral, valve) to the left ventricle.
- The left ventricle sends blood through the aortic semilunar valve into the aorta to the body proper.

The Heartbeat

The heart contracts, or beats, about 70 times a minute, and each heartbeat lasts about 0.85 sec. Each heartbeat, or *cardiac cycle,* consists of the following elements (the word **systole** refers to contraction of heart muscle, and the word **diastole** refers to relaxation of heart muscle):

Cardiac Cycle

Time	Atria	Ventricles
0.15 sec	Systole	Diastole
0.30 sec	Diastole	Systole
0.40 sec	Diastole	Diastole

First the atria contract (while the ventricles relax), and then the ventricles contract (while the atria relax) and then all chambers rest. The short systole of the atria is appropriate since the atria send blood only into the ventricles. It is the muscular ventricles that actually pump blood out into the circulatory system proper. When the word *systole* is used alone, it usually refers to the left ventricular systole.

When the heart beats, the familiar lub-DUPP sound is heard as the valves of the heart close. The *lub* is caused by vibrations of the heart when the atrioventricular valves close, and the *DUPP* is heard when vibrations occur due to the closing of the semilunar valves. These valves aid circulation of the blood through the heart because they permit only one-way flow and do not allow a backward flow of blood. The *pulse* is a wave of fibrillation, which passes down the walls of the arterial blood vessels when the aorta expands following ventricle systole. Because there is one arterial pulse per ventricular systole, the arterial pulse rate can be used to determine the heart rate.

The contraction of the heart is intrinsic to the heart; it does not require outside nervous stimulation. The reason for this is a unique type of tissue called nodal tissue, with both muscular

Figure 34.7

Control of the cardiac cycle. **a.** The SA node sends out a stimulus that causes the atria to contract. When this stimulus reaches the AV node, it signals the ventricles to contract by way of the Purkinje fibers. **b.** A normal ECG indicates that the heart is functioning properly. The P wave occurs as the atria contract; the QRS wave occurs as the ventricles contract; and the T wave occurs when the ventricles are recovering from contraction. **c.** Abnormal ECGs: sinus tachycardia is an abnormally fast heartbeat due to a fast pacemaker; ventricular fibrillation is irregular heartbeat due to irregular simulation of the ventricles, and mitral stenosis occurs because the bicuspid (mitral) valve is obstructed.

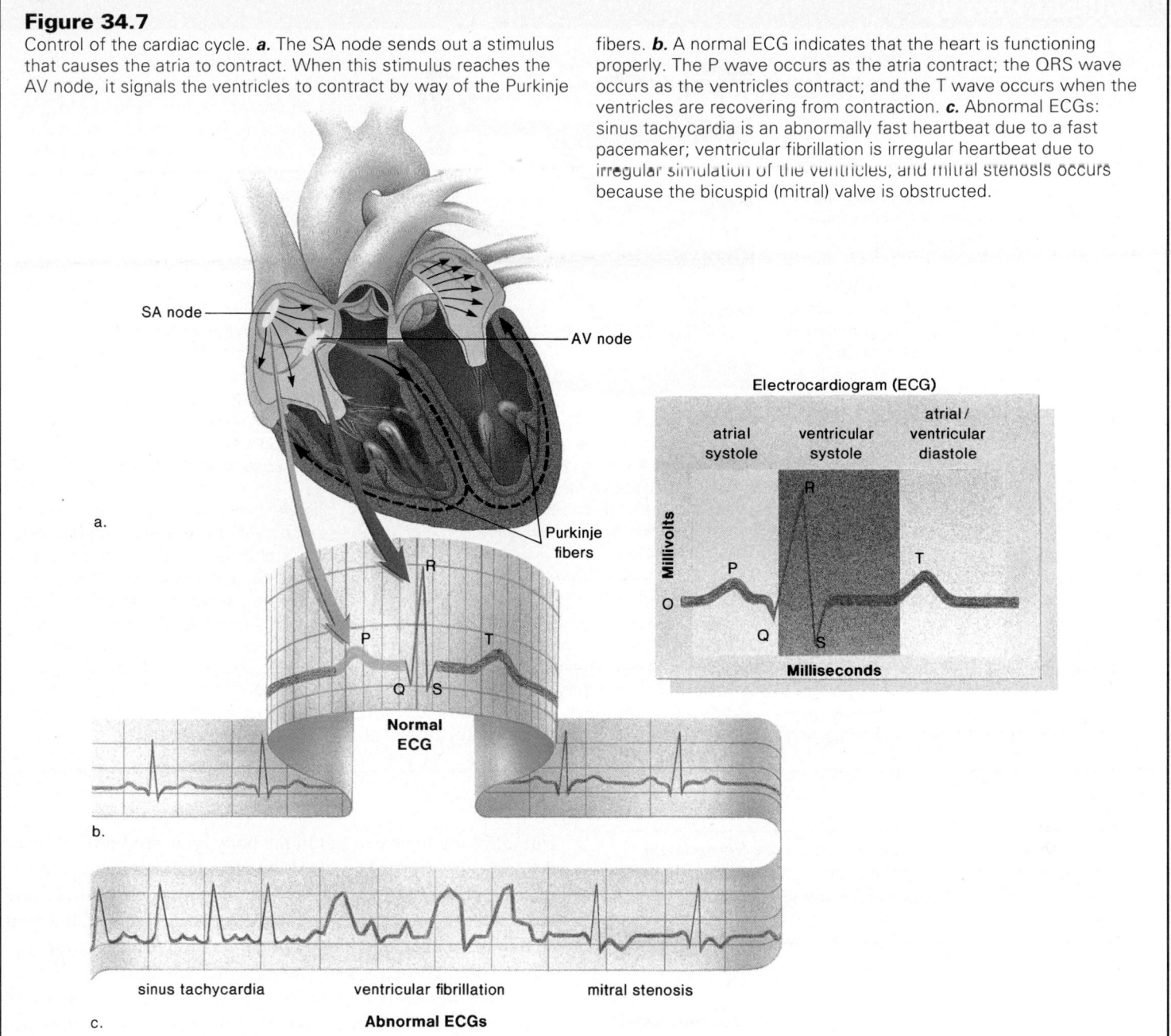

and nervous characteristics, located in 2 regions of the heart. The first of these, the *SA (sinoatrial) node,* is found in the upper dorsal wall of the right atrium; the other, the *AV (atrioventricular) node,* is found in the base of the right atrium very near the septum (fig. 34.7). The SA node initiates the heartbeat and automatically sends out an excitation impulse every 0.85 sec to cause the atria to contract. When the impulse reaches the AV node, it signals the ventricles to contract by way of specialized fibers called *Purkinje fibers.* The **SA node** is called the *pacemaker* because it usually keeps the heartbeat regular.

The rate of the heartbeat is also under nervous control. A cardiac center in the medulla oblongata of the brain can alter the beat of the heart by way of the autonomic nervous system. This system is made up of 2 divisions: the parasympathetic system, which promotes the functions we associate with normal activities, and the sympathetic system, which brings about the responses we associate with times of stress. The parasympathetic system slows the heartbeat, and the sympathetic system increases the heartbeat. Various factors, such as the relative need for oxygen or the level of blood pressure, determine which of these systems is activated.

William Harvey (1578-1657) was the first to offer proof that the blood circulates in the body of humans and other animals. Harvey was an English scientist of the seventeenth century, a time of renewed interest in the collection of facts, the use of the hypothesis, experimentation, and respect for mathematics. The seventeenth century was the time of the scientific revolution.

After many years of research and study, Harvey hypothesized that the heart is a pump for the entire circulatory system and that blood flows in a circuit. In contrast to former anatomists, Harvey dissected not only dead but also live organisms and observed that when the heart beats, it contracts, forcing blood into the aorta. Had this blood come from the right side of the heart? To do away with the complication of the lungs (pulmonary circulation), Harvey turned to fishes and noticed that the heart first receives and then pumps the blood forward. He observed that blood in the fetus passes directly from the right side of the heart through the septum to the left side. He felt confident that in mature higher organisms, all blood moves from the right to the left side of the heart by way of the lungs.

Harvey then wanted to show an intimate connection between the arteries and

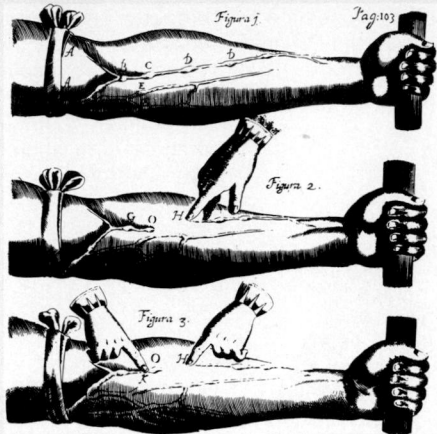

Figure 34.A

To show that the veins return blood to the heart, Harvey tied a ligature (*Figure 1*) above the elbow to observe the accumulation of blood in the veins. Blood can be forced past a valve from H to O (*Figure 2*), but not in the opposite direction (*Figure 3*). Therefore, it can be deduced that blood ordinarily moves toward the heart in the veins.

veins in the tissues of the body. Again using live lower organisms, he demonstrated that if an artery is slit, the whole blood system empties, including arteries and veins. He measured the capacity of the

left ventricle in humans and found it to be 2 oz. Since the heart beats 72 times a minute, in one hr the left ventricle forces into the aorta no less than $72 \times 60 \times 2 = 8{,}640$ oz $= 540$ lb, or 3 times the weight of a heavy man! Could so much blood be created and consumed every hour? The same blood must return again and again to the heart.

Harvey also studied the valves in the veins and suggested their true purpose. By the use of ligatures, he demonstrated that a tight ligature on the arm causes the artery to swell on the side of the heart, and a slack ligature causes the vein to swell on the opposite side (fig. 34.A). He said, "This is an obvious indication that the blood passes from the arteries into the veins . . . . and there is an anastomosis of the 2 orders of vessels."

Harvey's methods showed how fruitful research might be done. He established that physical and mechanical evidence could provide data for a theory of circulation. He erred, however, when he speculated on the function of the heart and lung. He thought the heart heated the blood, and the lung served to cool it or control the degree of heat. His basic method, however, contributed to the scientific revolution and set an example for others to follow.

The conduction system of the heart includes the SA node, the pacemaker, which results in atrial contraction, and also the AV node and the Purkinje fibers, which result in ventricular contraction.

Vascular Pathways

Human circulation includes 2 circular pathways, the **pulmonary circuit** and the **systemic circuit** (fig. 34.8).

Pulmonary Circuit

Deoxygenated blood from all regions of the body collects in the right atrium and then passes into the right ventricle, which pumps it into the pulmonary trunk. The pulmonary trunk divides into the pulmonary arteries, which carry blood to arterioles and capillaries in the lungs. After passing through lung capillaries located around the alveoli (see fig. 37.8), oxygenated blood returns to the left atrium of the heart through pulmonary venules and veins.

Systemic Circuit

The **aorta** and the **venae cavae** (vena cava, sing.) serve as the major pathways for blood in the systemic circuit. To trace the

path of blood to any organ in the body, you need only mention the aorta, the proper branch of the aorta, the organ, and the vein returning blood to the vena cava. In the systemic circuit, arteries contain oxygenated blood and have a bright red color, but veins contain deoxygenated blood and appear in purplish color.

The *coronary arteries* are extremely important because they serve the heart muscle itself (fig. 34.5). (The heart is not nourished by the blood in its chambers.) The coronary arteries arise from the aorta just above the aortic semilunar valve. They lie on the exterior surface of the heart, where they branch into arterioles and then capillaries. The capillary beds enter venules, which join to form the cardiac veins, and these empty into the right atrium. The coronary arteries have a very small diameter and may become blocked, causing a heart attack, as discussed in the reading, "Cardiovascular Disease."

A **portal system** is one that begins and ends in capillaries. One place in the human body where a portal system is found is between the small intestine and the liver. Blood passes from the capillaries of intestinal villi into venules that join to form the hepatic portal vein, a vessel that connects the intestine with the liver. The hepatic vein leaves the liver and enters the inferior vena cava.

Figure 34.8

Blood vessels in the pulmonary and systemic circuits. The blue-colored vessels carry deoxygenated blood, and the red-colored vessels carry oxygenated blood; the arrows indicate the flow of blood. In order to trace blood from the right to the left side of the heart, you must begin at the lung capillaries. In order to trace blood from the gut capillaries to the right atrium, you must consider the hepatic portal vein and hepatic vein.

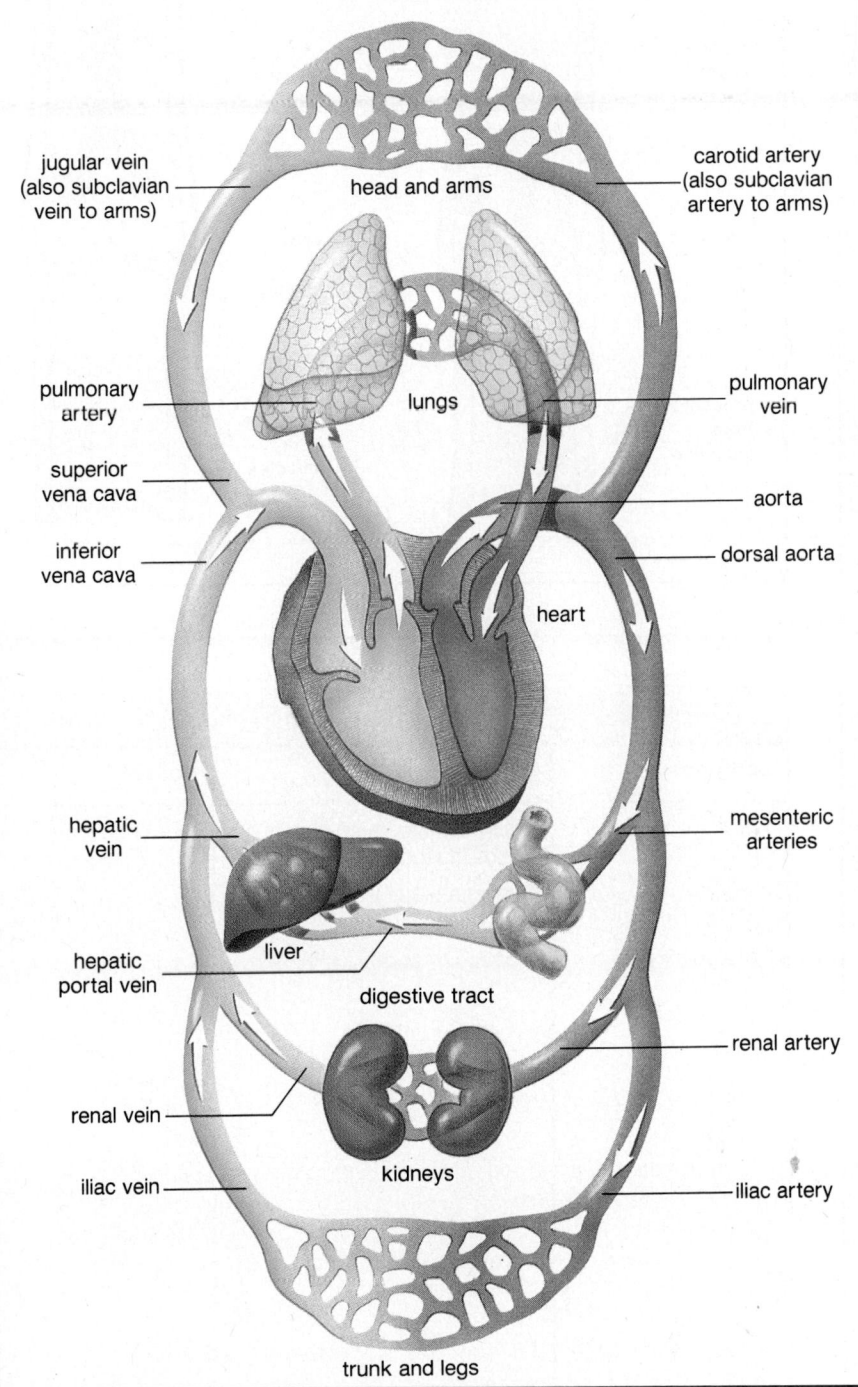

The pulmonary circuit takes deoxygenated blood to the lungs and oxygenated blood to the heart. The systemic circuit takes blood through the aorta to the vena cavae by numerous routes.

Systemic Blood Flow

When the left ventricle contracts, blood is forced into the arteries under pressure. *Systolic pressure* results from blood being forced into the arteries during ventricular systole, and *diastolic pressure* is the pressure in the arteries during ventricular diastole. Human **blood pressure** is measured with a sphygmomanometer, which has a pressure cuff that permits measurement of the amount of pressure required to stop the flow of blood through an artery. Blood pressure is normally measured on the brachial artery, which is in the upper arm, and is stated in millimeters of mercury (mm Hg). A blood pressure reading consists of 2 numbers, for example, 120/80— which represents systolic and diastolic pressures, respectively. For a man under age 45, a reading above 130/90 indicates *hypertension* (high blood pressure) and beyond age 45, a reading above 140/95 is considered hypertensive. High blood pressure is associated with cardiovascular disease, as discussed on page 566.

As blood flows from the aorta into the various arteries and arterioles, blood pressure falls (fig. 34.9). Also, the difference in pressure between systolic and diastolic gradually decreases until it disappears in the capillaries, where there is an even, steady flow of blood. Blood pressure in the venules is very low (5–10 mm Hg) and even lower in the vena cava (2 mm Hg). This low a pressure cannot move blood back to the heart, especially from the limbs of the body. When skeletal muscles near veins contract, they put pressure on the collapsible walls of the veins and on blood contained in these vessels. Veins, however, have *valves* that prevent the backward flow of blood, and therefore pressure from muscle contraction is sufficient to move blood through veins toward the heart (fig. 34.10). *Varicose veins,* abnormal dilations in superficial veins, develop when the valves of the veins become weak and ineffective due to a backward pressure of the blood.

In the human circulatory system, the beat of the heart supplies the pressure that keeps blood moving in the arteries; pressure drops off in the capillaries; and skeletal muscle contraction pushes blood in the veins back to the heart.

Human Blood

Blood has 2 main portions: the liquid portion, called plasma, and cells. **Plasma** contains many types of molecules, including nutrients, wastes, salts, and proteins (table 34.1). The salts and proteins buffer the blood, effectively keeping the pH near 7.4. They also maintain

Figure 34.9

Blood pressure in different parts of the human circulatory system. The pulse pressure (includes systolic and diastolic pressures) decreases as blood flows into smaller arteries and disappears as blood enters capillaries. The slow and even movement of the blood through the capillaries facilitates exchange of molecules there. The blood pressure in the veins is so low that it cannot account for the movement of blood in these vessels. Skeletal muscle contraction pushes the blood along from valve to valve in the veins.

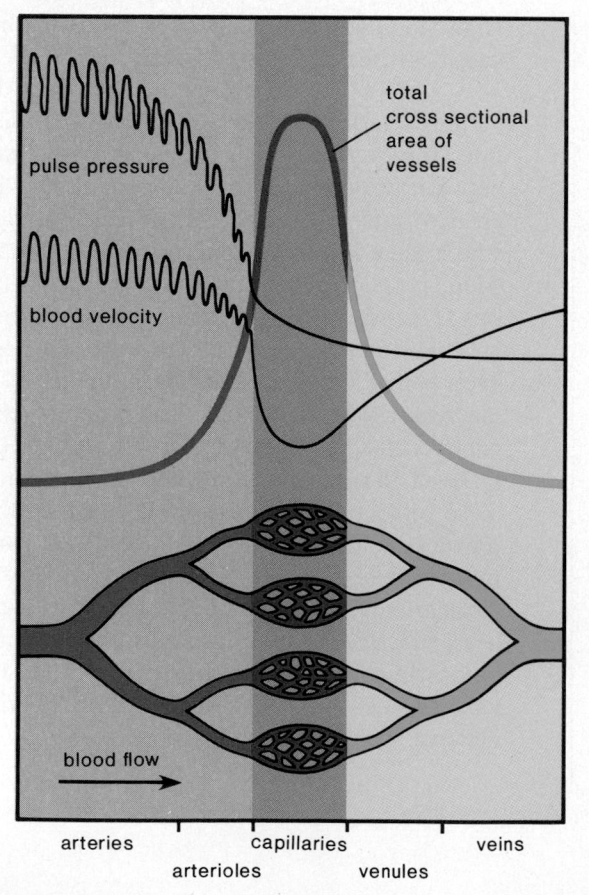

Figure 34.10

Cross section of a valve. *a.* When a valve is open, a result of pressure on the veins exerted by skeletal muscle contraction, the blood flows toward the heart in veins. *b.* Valves close when external pressure is no longer applied to them. Closure of the valves prevents the blood from flowing in the opposite direction.

blood moves toward heart

valve is open

pressure from muscles

blood falls back

valve is closed

no pressure from muscles

a.

b.

Table 34.1
Components of Plasma

Component	Function	Source
Water	Maintains blood volume and transports molecules	Absorbed from intestine
Proteins	Maintain blood osmotic pressure and pH	
Albumin	Transports organic molecules	Liver
Fibrinogen	Clotting	Liver
Globulins	Fight infection	Lymphocytes
Gases		
Oxygen	Cellular respiration	Lungs
Carbon dioxide	End product of metabolism	Tissues
Nutrients		
Fats, glucose, amino acids, etc.	Food for cells	Absorbed from intestinal villi
Salts	Maintain blood osmotic pressure and pH; aid metabolism	Absorbed from intestinal villi
Wastes		
Urea and uric acid	End products of metabolism	Liver
Hormones, vitamins, etc.	Aid metabolism	Varied

Plasma is 90%–92% water, 7%–8% plasma proteins, and not quite 1% salts. All other components are present in even smaller amounts.

Animal Structure and Function

blood's osmotic pressure so that water has an automatic tendency to enter blood capillaries. The plasma proteins assist in transporting large organic molecules in blood. For example, the lipoproteins that transport cholesterol are globulins, and bilirubin, a breakdown product of hemoglobin, is transported by albumin.

Types of Blood Cells

Blood cells are of 3 types: red blood cells, or **erythrocytes**, white blood cells, or **leukocytes**; and platelets, or **thrombocytes.**

Red Blood Cells

Red blood cells are small biconcave disks that lack a nucleus and contain the respiratory pigment hemoglobin. There are 4–6 million red blood cells per mm³ of whole blood, and each one of these cells contains about 200 million hemoglobin molecules. **Hemoglobin** contains 4 globin protein chains, each associated with heme, an iron-containing group. Iron combines loosely with oxygen, and in this way oxygen is carried in the blood. If there is an insufficient number of red blood cells or the cells do not have enough hemoglobin, the individual suffers from anemia and has a tired, run-down feeling.

Red blood cells are manufactured continuously in the red bone marrow of the skull, ribs, vertebrae, and ends of the long bones. The growth factor erythropoietin, which stimulates the production of red blood cells, is now a biotechnology product (p. 284). It is helpful to persons with anemia and is also sometimes abused by athletes who want to increase performance.

Before they are released from the bone marrow into blood, red blood cells lose their nucleus and acquire hemoglobin (fig. 34.11a). Possibly because they lack a nucleus, red blood cells live about 120 days. They are destroyed chiefly in the liver and the spleen, where they are engulfed by large phagocytic cells. When red blood cells are destroyed, hemoglobin is released. The iron is recovered and is returned to the red bone marrow for reuse. The heme portions of the molecules undergo chemical degradation and are excreted by the liver as bile pigments in the bile. The bile pigments are primarily responsible for the color of feces.

Oxygen is transported to the tissues in combination with hemoglobin, a pigment found in red blood cells.

White Blood Cells

White blood cells differ from red blood cells in that they are usually larger and have a nucleus, they lack hemoglobin, and without staining, they appear white in color. With staining, white blood cells appear light blue unless they have granules that bind with certain stains. The *granular leukocytes* have a lobed nucleus and may be neutrophils, which have granules that do not take up a dye; eosinophils, which have granules that take up the red dye eosin; and basophils, which have granules that take up a basic dye staining them a deep blue. *Agranular leukocytes* (with no granules) have a circular or indented nucleus. There are 2 types—the larger monocytes and the smaller lymphocytes (fig. 34.11a and b). The newly discovered stem cell growth factor (SGF) can be used to increase the production of all white blood cells, and there are also various

specific stimulating factors such as GM-CSF (granulocyte-macrophage colony-stimulating growth factor) that can be used to stimulate the production of specific stem cells. These growth factors should be helpful to patients with low immunity, such as AIDS patients.

When microorganisms enter the body due to an injury, the response is called an *inflammatory reaction* because there is swelling and reddening at the injured site. The damaged tissue has released kinins, which cause vasodilation, and histamines, which cause increased capillary permeability. **Neutrophils,** which are amoeboid, squeeze through the capillary wall and enter the tissue fluid, where they phagocytize foreign material. Monocytes appear and are transformed into **macrophages,** which are large phagocytizing cells that release white blood cell growth factors (see fig. 35.4). Soon there is an explosive increase in the number of leukocytes. The thick, yellowish fluid called pus contains a larger proportion of dead white blood cells that have fought the infection.

Lymphocytes also play an important role in fighting infection. Certain lymphocytes called T cells attack cells that contain viruses. Other lymphocytes called B cells produce antibodies. Each B cell produces just one type of antibody, which is specific for one type of antigen. An **antigen,** which is most often a protein but sometimes a polysaccharide, is found in the outer covering of a parasite or is present in its toxin. When **antibodies** combine with antigens, the complex is often phagocytized by a macrophage. An individual is actively immune when a large number of B cells are all producing the specific antibody needed for a particular infection. Chapter 35 deals with immunity and explores this topic in detail.

The inflammatory response is a "call to arms"—it marshals white blood cells to the site of invasion by microorganisms. Neutrophils and monocytes (macrophages) are phagocytic; B lymphocytes produce antibodies to combine with disease-causing antigens.

Platelets

Strictly speaking, platelets (thrombocytes) are not cells. Platelets result from fragmentation of certain large cells, called megakaryocytes, in the bone marrow (fig. 34.11). **Platelets** are involved in the process of blood clotting, or coagulation. Also necessary to the process are fibrinogen and prothrombin, proteins manufactured and deposited in blood by the liver. If clotting occurs in a test tube, a fluid called *serum* collects above the clot. Serum has the same composition as plasma except that it lacks prothrombin and fibrinogen.

Blood Clotting When a blood vessel in the body is damaged, platelets clump at the site of the puncture and partially seal the leak. They and the injured tissues release a clotting factor called *prothrombin activator* that converts prothrombin to thrombin. This reaction requires calcium (Ca^{++}) ions. Thrombin, in turn, acts as an enzyme that severs 2 short amino acid chains from each fibrinogen molecule. These activated fragments then join end to end, forming

Figure 34.11

a. Blood cell formation in red bone marrow. Multipotent stem cells give rise to specialized stem cells that produce the various types of blood cells. Each stem cell is under the control of specific growth factors. ***b.*** Human blood cells, their abundance, size, and function. (See next page.)

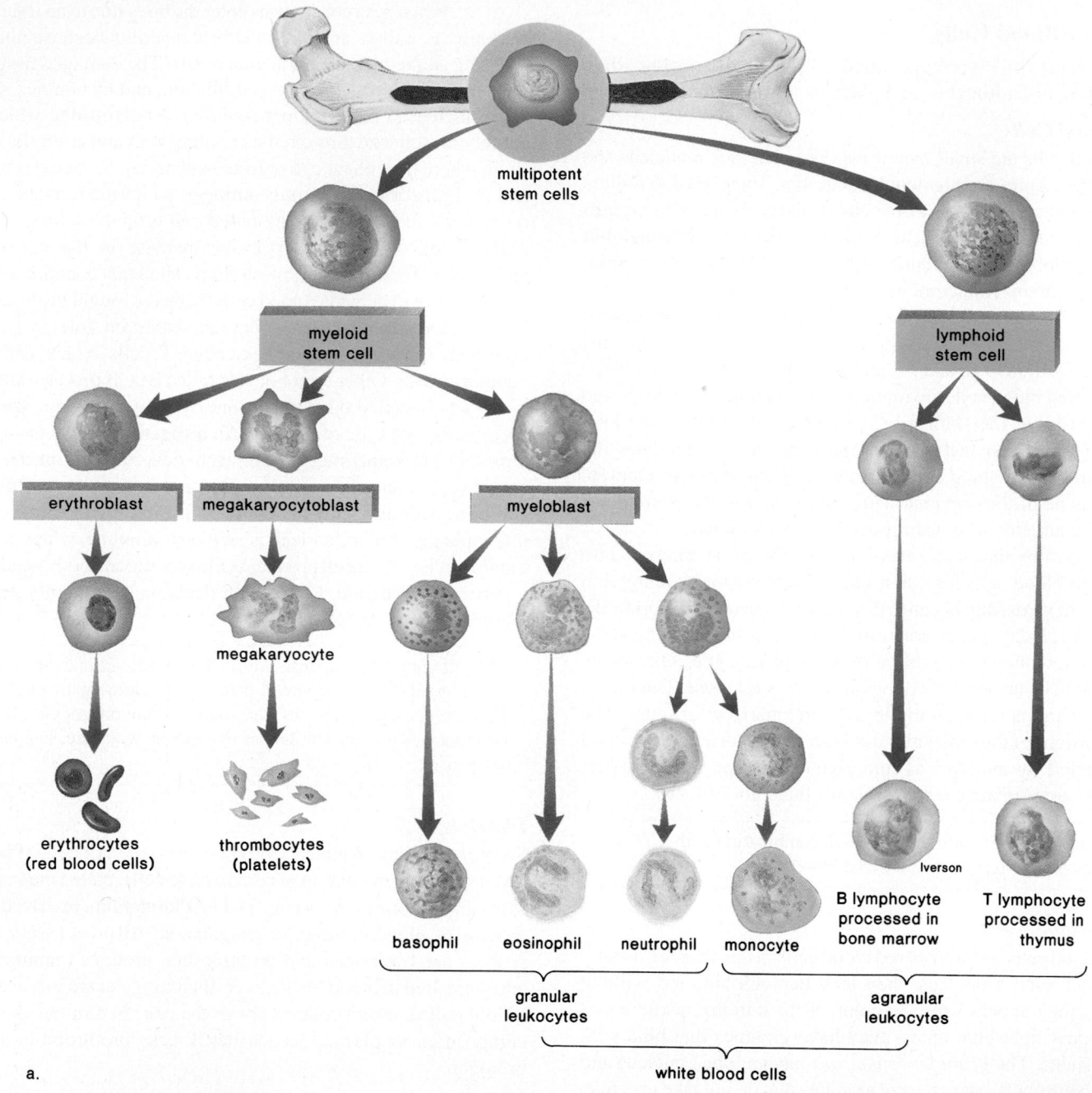

multipotent
stem cells

myeloid
stem cell

lymphoid
stem cell

erythroblast

megakaryocytoblast

myeloblast

megakaryocyte

erythrocytes
(red blood cells)

thrombocytes
(platelets)

basophil

eosinophil

neutrophil

monocyte

Iverson

B lymphocyte
processed in
bone marrow

T lymphocyte
processed in
thymus

granular
leukocytes

agranular
leukocytes

white blood cells

a.

long threads of *fibrin*. Fibrin threads wind around the platelet plug in the damaged area of the blood vessel and provide the framework for the clot. Red blood cells also are trapped within the fibrin threads; these cells make a clot appear red.

A fibrin clot is only temporarily present. As soon as blood vessel repair is initiated, an enzyme called plasmin destroys the fibrin network and restores the fluidity of plasma. This is a protective measure because a blood clot can act as a thrombus or an embolus. In either case, it interferes with circulation and even can cause death of tissues in the area.

The steps necessary for blood clotting upon injury are quite complex, but they can be summarized as in figure 34.12.

A blood clot consists of platelets and red blood cells entangled within fibrin threads.

Human Blood Cells			
Component	Number (per mm³ of blood)	Size	Function
Erythrocyte (red blood cell)	5,000,000	7.5 µm	Transport oxygen, assists CO_2 transport
Leukocyte (white blood cell)	7,000	8–20 µm	Fight infection
Thrombocytes (platelets)	200,000–400,000	2–4 µm	Aid blood clotting
White Blood Cell Types	% of Total White Blood Cells	Size	Function
Granular leukocytes			
Neutrophils	55–65	9–12 µm	Phagocytize primarily bacteria
Eosinophils	2–3	9–12 µm	Phagocytize and destroy antigen-antibody complexes
Basophils	0.5	9–12 µm	Congregate in tissues; release histamine when stimulated
Agranular leukocytes			
Lymphocytes	25–33	8–10 µm	B cell produces antibodies in blood and lymph; T cell destroys virus infected cells
Monocytes	4–7	12–20 µm	Become macrophages— phagocytize bacteria and viruses

b.

Figure 34.12

Blood clotting. **a.** Platelets and damaged tissue cells release prothrombin activator, which acts on prothrombin in the presence of calcium to produce thrombin, which acts on fibrinogen to form fibrin. **b.** A scanning electron micrograph showing a red blood cell caught in the fibrin threads of a clot. Magnification, X3,000.

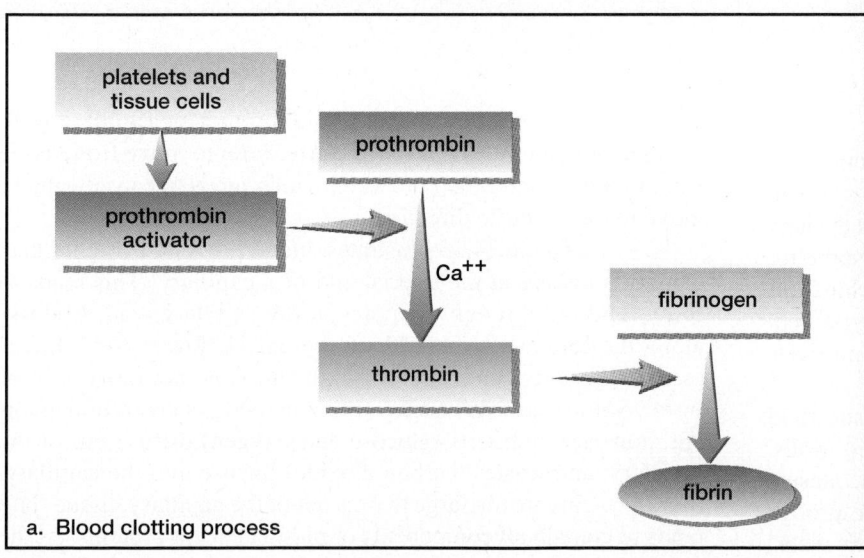

a. Blood clotting process

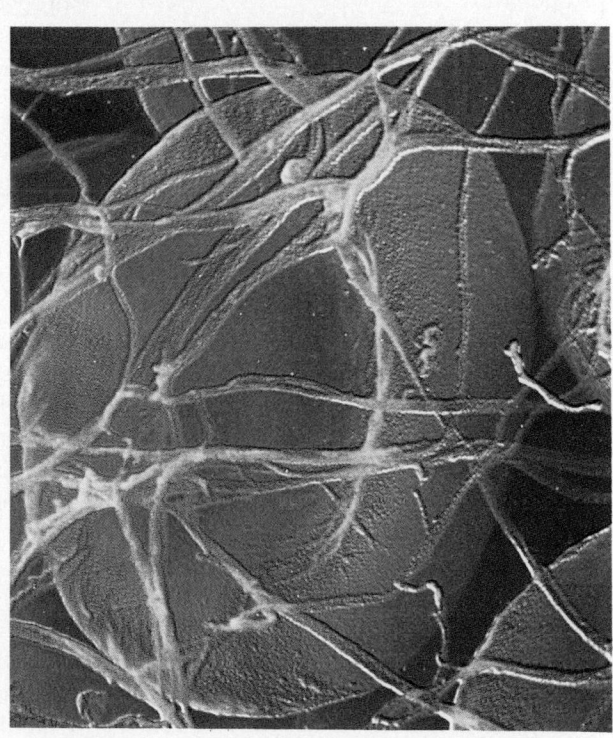

b. Blood clot

ecause sudden cardiac death happens once every 72 sec in the United States, it is well to identify the factors that predispose an individual to cardiovascular disease. The risk factors for cardiovascular disease include the following:

Male sex
Family history of heart attack under age 55
Smoking more than 10 cigarettes a day
Severe obesity (30% or more overweight)
Hypertension
Unfavorable HDL and LDL cholesterol blood levels
Impaired circulation to the brain or the legs
Diabetes mellitus

Hypertension (high blood pressure) is well recognized as a major factor in cardiovascular disease, and 2 controllable behaviors contribute to hypertension. Smoking cigarettes, including filtered cigarettes, causes hypertension, as does obesity. It is best to never take up the habit of smoking cigarettes, but most of the detrimental effects can be reversed if you stop smoking. Since it is very difficult for obese individuals to lose weight, it is recommended that weight control be a lifelong endeavor.

Investigators have identified several behaviors that may help to reduce the possibility of heart attack and stroke. Exercise seems to be critical. Sedentary individuals have a risk of cardiovascular disease that is about double that of those who are very active. One physician, for example, recommends that his patients walk for one hour, 3 times a week. Stress reduction also is desirable. The same investigator recommends everyday mediation and yogalike stretching and breathing exercises to reduce stress.

Another behavior that is now well known is the adoption of a diet that is low in saturated fats and cholesterol because such a diet is believed by many to protect against the development of cardiovascular disease. Cholesterol is ferried in the blood by 2 types of plasma proteins called LDL (low-density lipoprotein) and HDL (high-density lipoprotein). LDL (called "bad" lipoprotein) takes cholesterol to the tissues from the liver, and HDL (called "good" lipoprotein) transports cholesterol out of the tissues to the liver. When the LDL level in blood is abnormally high or the HDL level is abnormally low, cholesterol accumulates in the cells. When cholesterol-laden cells line the arteries, plaque develops, which interferes with circulation (fig. 34.B).

Cholesterol guidelines have been established by the National Heart, Lung, and Blood Institute. According to the institute, everyone should know his or her cholesterol blood level. Individuals with a borderline high cholesterol blood level (200–239 mg/100 ml) should be tested further if they already have heart disease or if they have 2 known risk factors of cardiovascular disease (see list). Individuals with a high cholesterol blood level (240 mg/100 ml) always should be tested further. Persons with an LDL cholesterol level of over 130 mg/100 ml should be treated if they have other risk factors, and those with an LDL cholesterol level of 160 mg/100 ml should be treated even if this is the only risk factor.

Persons with a total-to-HDL cholesterol ratio higher than 4.5 also are considered to be at risk. Heart attack has occurred in individuals who have a normal total cholesterol level but who also have an unfavorable total-to-HDL cholesterol ratio. For example, if a person's total cholesterol blood level is 200, but the HDL level is only 25 mg/100 ml, then the total-to-HDL cholesterol ratio is 8.0, and circulatory difficulties most likely will develop.

First and foremost, treatment for unfavorable cholesterol levels consists of adopting a diet that is low in saturated fat and cholesterol (see p. 000). Although the prescribed diet does not lower cholesterol blood level in all persons, it is expected to do so for most individuals. If diet alone does not bring down the cholesterol blood level, drugs can be prescribed. Some of the drugs act in the intestine to remove cholesterol, and others act in the body to prevent its production. These drugs reduce the blood level of cholesterol, but their long-term side effects are not completely known and may be serious. Considering this, some investigators do not recommend these drugs to lower cholesterol blood levels.

A diet low in saturated fat and cholesterol can lower the total cholesterol blood level and the LDL level of some individuals, but this diet most likely will not raise the

Capillary Exchange

No cell is far away from one or more capillaries because capillaries are found throughout every tissue in the body. The cross-sectional area of capillaries is so large that blood flows very slowly through the capillaries (fig. 34.9). This slow rate allows adequate time for exchange of materials between blood and tissue fluid. **Tissue fluid** is the internal environment of cells, and the circulatory system performs an important homeostatic function by providing nutrients to and removing wastes from tissue fluid.

Much of the exchange between blood and tissue fluid occurs by diffusion through the one-cell-thick capillary walls. While lipid-soluble substances can pass freely through the plasma membrane of cells, water-soluble substances usually diffuse through pores present in junctions between the cells of the capillary walls.

Two forces control movement of fluid through the capillary wall: osmotic pressure, which tends to cause water to move from tissue fluid to blood, and blood pressure, which tends to cause water to move in the opposite direction.

As figure 34.13 illustrates, blood pressure is higher than osmotic pressure at the arterial end of a capillary. This tends to force fluid out through the pores in the capillary wall. Midway along the capillary, where blood pressure is lower, the 2 forces essentially cancel one another, and there is no net movement of water. Solutes now diffuse according to their concentration gradient, however, nutrients (glucose and oxygen) diffuse out of the capillary, and wastes (carbon dioxide) diffuse into the capillary. Since proteins are too large to pass out of the capillary, tissue fluid tends to contain all components of plasma except proteins. At the

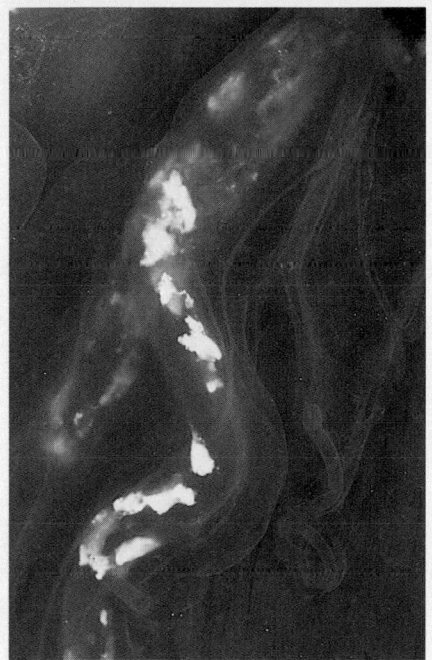

a.

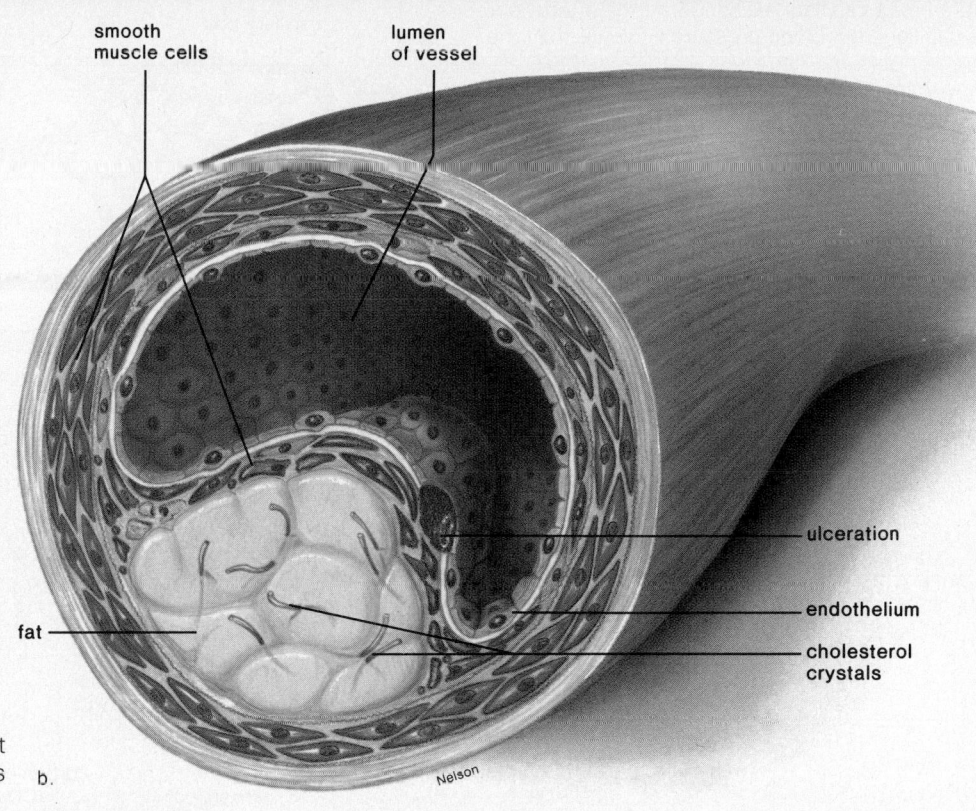

smooth
muscle cells

lumen
of vessel

ulceration

endothelium

cholesterol
crystals

fat

b.

Nelson

Figure 34.B

a. Plaques (*yellow*) in the coronary artery of a heart patient. *b.* Cross section of an artery that is partially blocked by plaque. The plaque bulges out into the lumen of the artery, obstructing blood flow. *b.* From Kent M. Van De Graaff and Stuart I. Fox, *Concepts of Human Anatomy and Physiology*, 2d ed. Copyright © 1989 Wm. C. Brown Publishers, Dubuque, Iowa. All Rights Reserved. Reprinted by permission.

HDL level. Aside from certain drugs that apparently can raise HDL level, exercise is sometimes effective.

There is nothing that can be done about some of the cardiovascular risk factors, such as male gender and family history. Other risk factors, however, likely can be controlled if the individual believes it is worth the effort. It is clear that the 4 great admonitions for a healthy life—heart-healthful diet, regular exercise, proper weight maintenance, and refraining from smoking—all contribute to acceptable blood pressure and cholesterol blood levels.

venule end of a capillary, where blood pressure has fallen, osmotic pressure is greater than blood pressure, and water tends to move into the capillary. Almost the same amount of fluid that left the capillary returns to it, although there is always some excess tissue fluid collected by the lymphatic capillaries (fig. 34.14).

Oxygen and nutrient molecules exit a capillary near the arterial end; carbon dioxide and waste molecules enter a capillary near the venous end.

Blood Typing

Although there are at least 12 well-known blood type identification systems, the ABO system and the RH system are most often used to determine blood type.

The ABO System

Before the twentieth century, blood transfusions were attempted, although sometimes the results were dire and even caused the death of the recipient. Concerned about these occurrences, a newly established physician in Vienna, Karl Lansteiner, began to examine the effect of mixing different samples of blood. In the end, he and his associates determined that there are 4 major blood groups among humans. They designated the types of blood as A, B, AB, and O. Table 34.2 shows that blood type is dependent upon the presence of antigens, specific glycoproteins, on red blood cells. Type O blood has neither the A antigen nor the B antigen on red blood cells; the other types of blood do have one or both of the antigen(s) present. These are antigens only to a recipient and not to the person with that type blood.

Figure 34.13

Diagram of a capillary exchange. At the arterial end of a capillary, the blood pressure is higher than the osmotic pressure. Therefore, water exits here. At the venous end of a capillary, the osmotic pressure is higher than the blood pressure. Therefore, water enters here. In between the 2 ends of the capillary, nutrients, gases, and wastes diffuse in or out according to their concentration gradients.

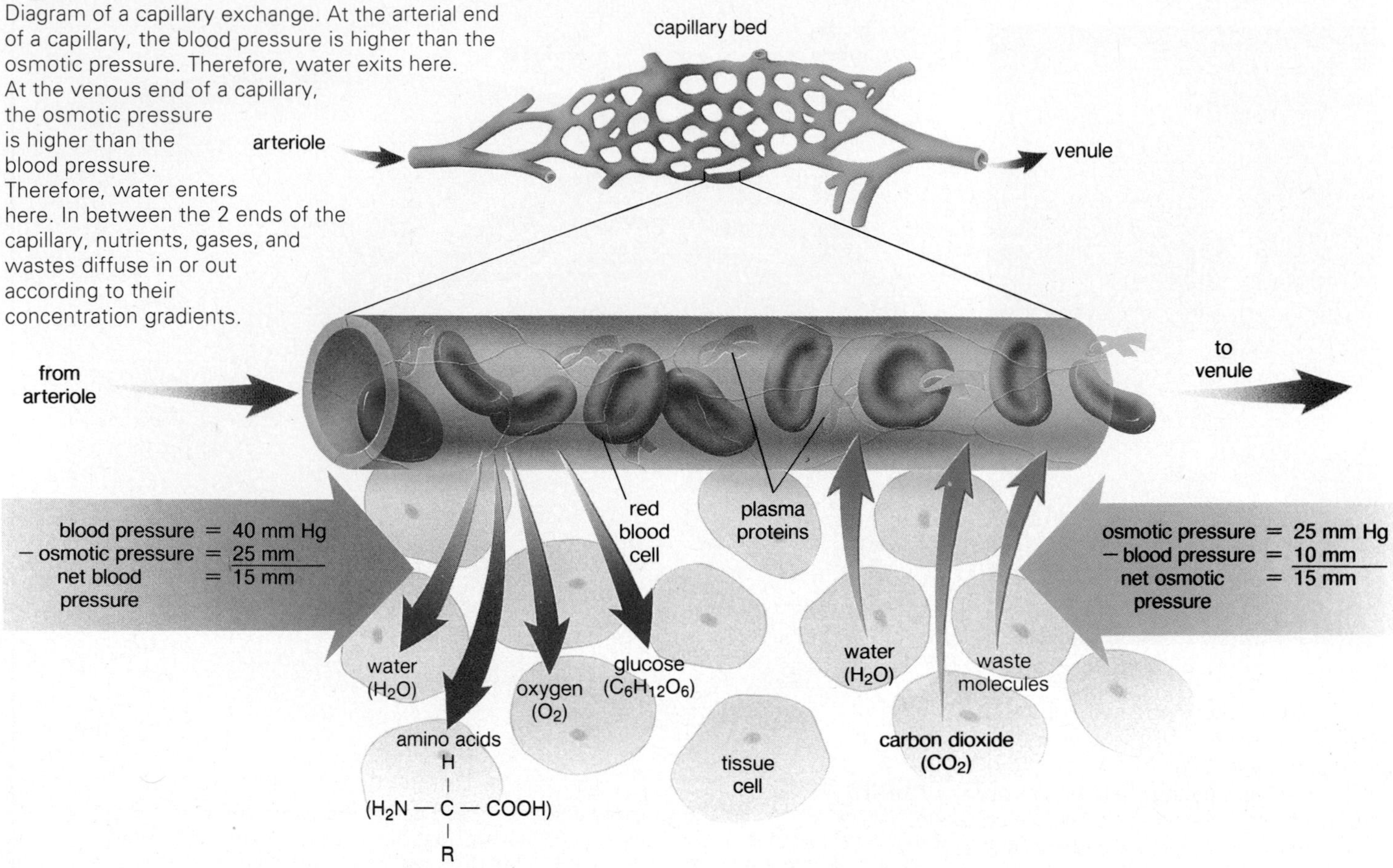

capillary bed

arteriole

venule

from arteriole

to venule

blood pressure = 40 mm Hg
− osmotic pressure = 25 mm
net blood pressure = 15 mm

osmotic pressure = 25 mm Hg
− blood pressure = 10 mm
net osmotic pressure = 15 mm

red blood cell

plasma proteins

water (H_2O)

oxygen (O_2)

glucose ($C_6H_{12}O_6$)

water (H_2O)

waste molecules

amino acids

H
|
(H_2N — C — COOH)
|
R

tissue cell

carbon dioxide (CO_2)

Within the plasma, there are antibodies to the antigens that are not present on the red blood cells. Therefore, for example, type A has an antibody called anti-B in the plasma. This is reasonable because if the A antigen and anti-A antibody are present, for example, **agglutination,** or clumping of red blood cells, occurs. For a recipient to receive blood from a donor, the recipient's plasma must not have an antibody that causes the donor's cells to agglutinate. For this reason it is important to determine each person's blood type. Figure 34.15 demonstrates a way to use the antibodies derived from plasma to determine the blood type. If clumping occurs after a sample of blood is exposed to a particular antibody, the person has that type of blood.

The Rh System

Another important antigen in matching blood types is the Rh factor. Persons with this particular antigen on the red blood cells are Rh positive (Rh^+); those without it are Rh negative (Rh−). Rh-negative individuals do not normally make antibodies to the Rh factor, but they do make them when exposed to the Rh factor. It is possible to extract these antibodies and use them for blood type testing (fig. 34.15).

The Rh factor is particularly important during pregnancy. If the mother is Rh negative and the father is Rh positive, the child may be Rh positive. The child's Rh-positive red blood cells begin to

Figure 34.14

Lymphatic vessels. Arrows indicate that lymph is formed when lymphatic capillaries take up excess tissue fluid. Lymphatic capillaries lie near blood capillaries.

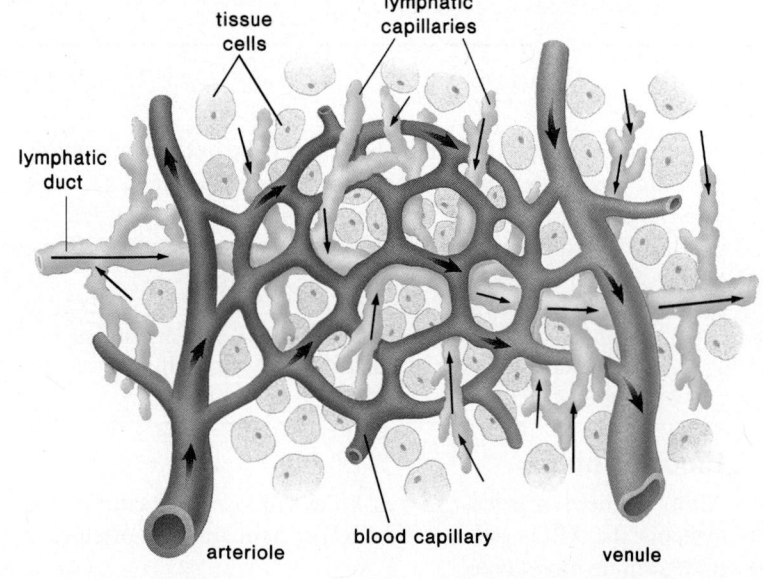

tissue cells

lymphatic capillaries

lymphatic duct

arteriole

blood capillary

venule

Animal Structure and Function

Table 34.2
Blood Groups

Blood Type	Antigen or Red Blood Cells	Antibody in Plasma	% U.S.* Black	% U.S.* Caucasian
A	A	Anti-B	25	41
B	B	Anti-A	20	7
AB	A,B	None	4	2
O	None	Anti-A and anti-B	51	50

*Blood type frequency for other races is not available.

leak across the mother's circulatory system because placental tissues normally break down before and at birth. This causes the mother to produce Rh antibodies. If the mother becomes pregnant with another Rh-positive baby, Rh antibodies may cross the placenta and destroy this child's red blood cells. This is called *hemolytic disease of the newborn*, and the problem is solved by giving Rh-negative women an Rh-immune globulin injection just after the birth of an Rh-positive child. This injection contains Rh-antibodies, which attack the baby's red blood cells before these cells stimulate the mother's immune system to produce its own antibodies.

Human blood consists of liquid plasma, which transports nutrient and waste molecules, and cells. Red blood cells carry oxygen, and white blood cells fight infection. Blood clotting involves the platelets, and blood typing is based on types of antigens on the red blood cells.

Figure 34.15

The standard test to determine ABO and Rh blood type consists of putting a drop of anti-A antibodies, anti-B antibodies, and anti-Rh antibodies on a slide. To each of these a drop of the person's blood is added. *a.* A possible reaction to any one of these is shown. Top, no reaction; therefore, no antigen to that antibody is present. Bottom, agglutination occurs; therefore, antigen is present. *b.* Several possible results. For example, in the top row there is no agglutination with either anti-A or anti-B, but there is agglutination with anti-Rh. Therefore, the person's blood type is O⁺.

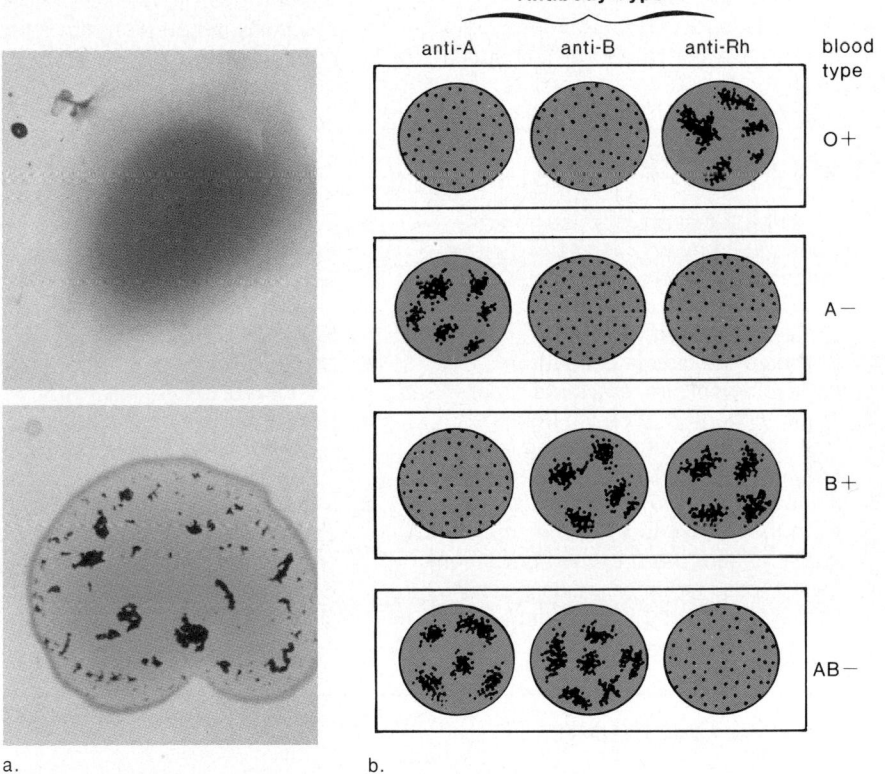

Summary

1. Some invertebrates do not have a transport system. The presence of a gastrovascular cavity allows diffusion alone to supply the needs of cells in cnidarians and flatworms.

2. Other invertebrates do have a transport system. Insects have an open system and earthworms have a closed one.

3. Vertebrates have a closed system in which arteries carry blood away from the heart to capillaries, where exchange takes place, and veins carry blood to the heart.

4. Fishes have a single circulatory loop because the heart with the single atrium and ventricle pumps blood only to the gills. The other vertebrates have both a pulmonary and systemic circulation. Amphibians have 2 atria but a single ventricle. Birds and mammals, including humans, have a heart with 2 atria and 2 ventricles, in which oxygenated blood is always separate from deoxygenated blood.

5. The heartbeat in humans begins when the SA node (pacemaker) causes the 2 atria to contract, sending blood through the atrioventricular valves to the 2 ventricles. The AV node causes the 2 ventricles to contract, sending blood through the semilunar valves to the pulmonary trunk and the aorta. Now all chambers rest. The heart sounds, lub-DUPP, are caused by the closing of the valves.

6. Blood pressure created by the beat of the heart accounts for the flow of blood in the arteries, but skeletal muscular contraction is largely responsible for the flow of blood in the veins, which have valves preventing a backward flow.

7. Blood has 2 main parts: plasma and cells. Plasma contains mostly water (92%) and proteins (8%) but also nutrients and wastes.

8. The red blood cells contain hemoglobin and function in oxygen transport.

9. Defense against disease depends on the various types of leukocytes. Neutrophils and monocytes are phagocytic and are especially responsible for the inflammatory reaction. Lymphocytes are involved in the development of immunity to disease.

10. The platelets and 2 plasma proteins, prothrombin and fibrinogen, function in blood clotting, an enzymatic process that results in fibrin threads.

11. Blood clotting is a complex process that includes 3 major events: platelets and injured tissue release prothrombin activator, which enzymatically changes prothrombin to thrombin; thrombin is an enzyme that causes fibrinogen to be converted to fibrin threads.

12. When blood reaches a capillary, water moves out at the arterial end, due to blood pressure. At the venule end, water moves in, due to osmotic pressure. In between, nutrients diffuse out and wastes diffuse in.

13. In the ABO blood system there are 4 types of blood: A, B, AB, and O, depending on the type of antigen present on the red blood cells. If the red blood cells have a particular antigen, they clump when exposed to the corresponding antibody. Table 34.2 tells which types of antibody are present in the plasma for the various ABO blood groups.

Study Questions

1. Describe transport in those invertebrates that have no circulatory system; those that have an open circulatory system; and those that have a closed circulatory system.

2. Compare the circulatory systems of a fish, an amphibian, and a mammal.

3. Trace the path of blood in humans from the right ventricle to the left atrium; from the left ventricle to the kidneys and return to the right atrium; from the left ventricle to the small intestine and return to the right atrium.

4. Define these terms: pulmonary circulation, systemic circulation, portal system.

5. Describe the beat of the heart, mentioning all the factors that account for this repetitive process. Describe how the heartbeat affects blood flow; what other factors are involved in blood flow?

6. Discuss the life cycle and function of red blood cells.

7. How are white cells classified? What is the function of neutrophils, monocytes, and lymphocytes?

8. Name the steps that take place when blood clots. Which substances are present in blood at all times, and which appear during the clotting process?

9. What forces operate to facilitate exchange of molecules across the capillary wall?

10. What are the 4 ABO blood types in humans? For each, state the antigen(s) on the red blood cells and the antibody(ies) in the plasma.

11. Problems can arise during childbearing if the mother is which Rh type and the father is which Rh type? Explain why this is so.

Objective Questions

1. Which one of these would you expect to be part of a closed, but not an open, circulatory system?
   a. ostia
   b. capillary beds
   c. hemocoel
   d. heart

2. In a one-circuit circulatory system, blood pressure
   a. is constant throughout the system.
   b. drops significantly after gas exchange has taken place.
   c. is higher at the intestinal capillaries than at the gill capillaries.
   d. cannot be determined.

3. In which of the following animals is the blood entering the aorta incompletely oxygenated?
   a. frog
   b. chicken
   c. monkey
   d. fish

4. Which of these factors has little effect on blood flow in arteries?
   a. the heartbeat
   b. blood pressure
   c. total cross-sectional area of vessels
   d. skeletal muscle contraction

5. In humans, blood returning to the heart from the lungs returns to the
   a. right ventricle.
   b. right atrium.
   c. left ventricle.
   d. left atrium.

6. Systole refers to the contraction of the
   a. major arteries.
   b. SA node.
   c. atria and ventricles.
   d. All of these.

7. A baby born to which of these couples is most likely to suffer from fetal erythroblastosis?
    a. Rh-positive mother and Rh-negative father
    b. Rh-negative mother and Rh-negative father
    c. Rh-positive mother and Rh-positive father
    d. Rh-negative mother and Rh-positive father

8. During blood typing, agglutination indicates that the
    a. plasma contains certain antibodies.
    b. red blood cells carry certain antigens.
    c. plasma contains certain antigens.
    d. red blood cells carry certain antibodies.

9. Water enters capillaries on the venule side as a result of
    a. active transport from tissue fluid.
    b. an osmotic pressure gradient.
    c. increased blood pressure on this side.
    d. higher red blood cell concentration on this side.

10. Which of these associations is incorrect?
    a. white blood cells—infection fighting
    b. red blood cells—blood clotting
    c. plasma—water, nutrients, and wastes
    d. platelets—blood clotting

11. The last step in blood clotting
    a. requires calcium ions.
    b. occurs outside the bloodstream.
    c. converts prothrombin to thrombin.
    d. converts fibrinogen to fibrin.

12. Label this diagram of the heart.

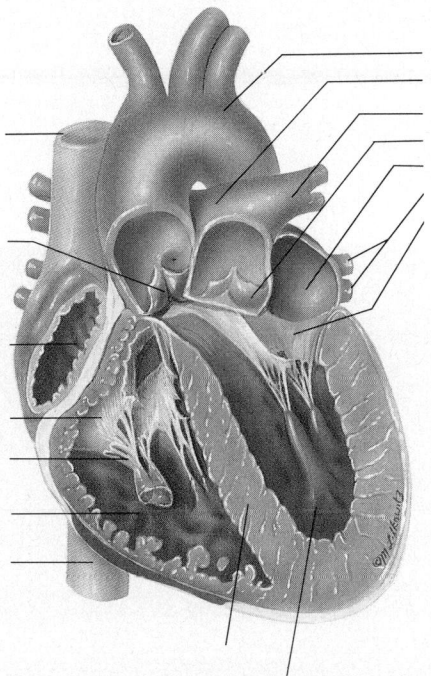

Concepts and Critical Thinking

1. *The internal environment of most animals remains relatively constant.*

How does the composition of tissue fluid stay relatively constant in vertebrates?

2. *A circulatory system is necessary to the life of a complex animal.*

Why is a circulatory system imperative for complex animals?

3. *Animals are physical and chemical machines, for example, energy must be exerted in order to keep the blood circulating.*

What type of energy keeps the blood circulating? How do animals acquire the necessary energy?

Selected Key Terms

hemoglobin (he″mo-glo′bin) 554
artery (ar′ter-e) 555
capillary (kap′ĭ-lar″e) 555
vein (vān) 555
systole (sis′to-le) 558
diastole (di-as′to-le) 558

pulmonary circuit (pul′mo-ner″e ser′kut) 560
systemic circuit (sis-tem′ik ser′kut) 560
aorta (a-or′tah) 560
vena cava (ve′nah ka′vah) 560
plasma (plaz′mah) 562
erythrocyte (ĕ-rith′ro-sīt) 563

leukocyte (lu′ko-sīt) 563
thrombocyte (throm′bō-sīt) 563
neutrophil (nu′tro-fil) 563
macrophage (mak′ro-fāj) 563
lymphocyte (lim′fo-sīt) 563
tissue fluid (tish′u floo′id) 566

35

Lymphatic System and Immunity

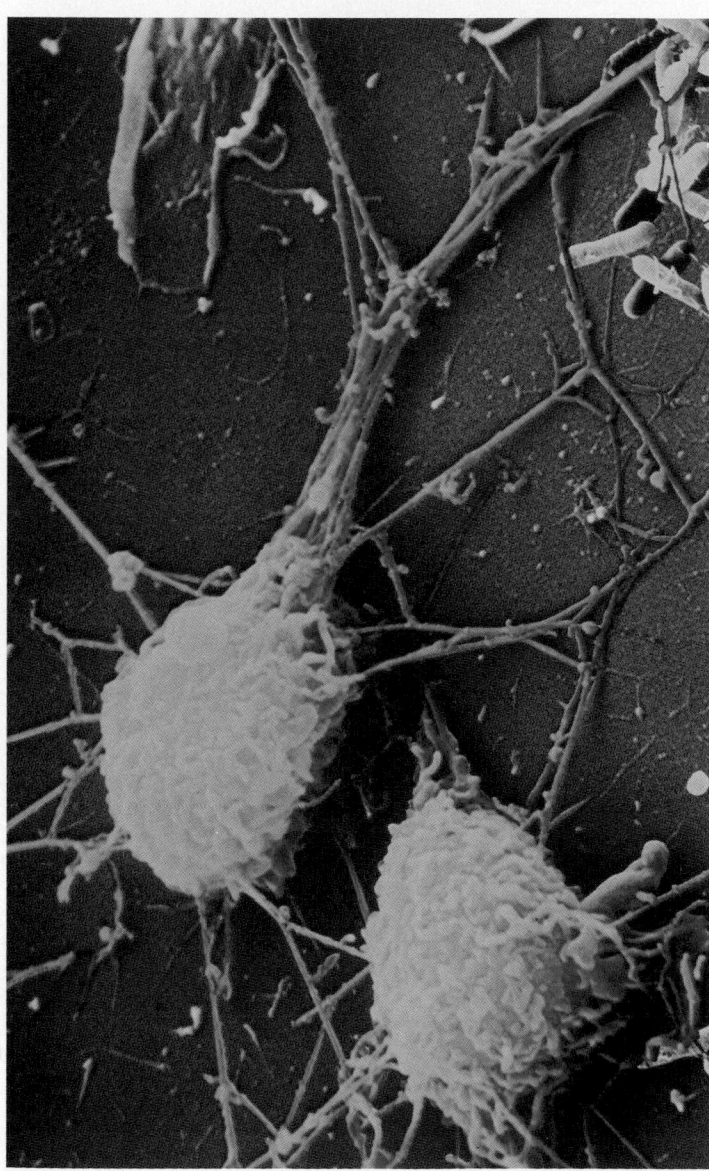

Immunity is dependent on the work of the white blood cells. Here the largest of the white blood cells, called a macrophage, is about to engulf some *E. coli* bacteria. Macrophages can consume a bacterium in less than 1/100 of a second.

Your study of this chapter will be complete when you can

1. name 3 functions of the lymphatic system;
2. describe the structure and the function of the lymphatic vessels and the lymphoid organs;
3. name 3 general ways the body defends itself against infections, and give examples of each;
4. contrast the maturation, the structure, and the function of B and T lymphocytes;
5. explain the clonal selection theory as it pertains to B cells and T cells;
6. describe the structure and the function of an antibody;
7. describe the different types of T cells, and describe the function of each type;
8. contrast active immunity with passive immunity, and tell why the former lasts longer;
9. tell how monoclonal antibodies are produced, and list ways in which they are used today;
10. name and discuss 3 types of immunological side effects.

The Lymphatic System

The mammalian **lymphatic system** consists of lymphatic vessels and the lymphoid organs. This system, which is closely associated with the cardiovascular system, has 3 main functions: (1) lymphatic vessels take up excess tissue fluid and return it to the bloodstream; (2) lymphatic capillaries absorb fats at the intestinal villi and transport them to the bloodstream (p. 596); and (3) the lymphatic system helps to defend the body against disease.

Lymphatic Vessels

Lymphatic vessels are quite extensive; every region of the body is supplied richly with lymphatic capillaries (fig. 35.1). The construction of the larger lymphatic vessels is similar to that of cardiovascular veins, including the presence of valves. Also, the movement of lymph within these vessels is dependent upon skeletal muscle contraction. When the muscles contract, the lymph is squeezed past a valve that closes, preventing the lymph from flowing backwards.

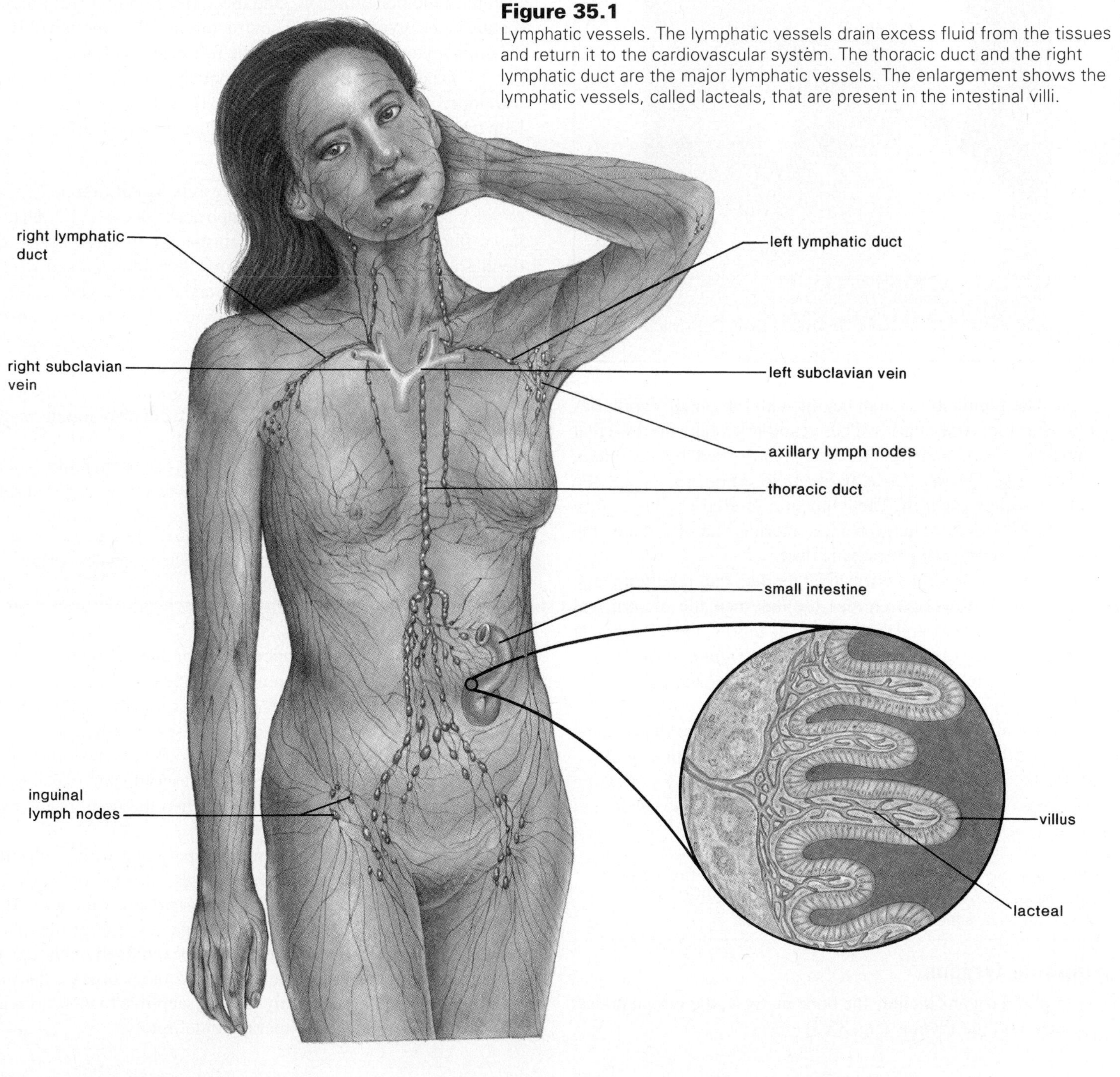

Figure 35.1

Lymphatic vessels. The lymphatic vessels drain excess fluid from the tissues and return it to the cardiovascular system. The thoracic duct and the right lymphatic duct are the major lymphatic vessels. The enlargement shows the lymphatic vessels, called lacteals, that are present in the intestinal villi.

right lymphatic duct

right subclavian vein

inguinal lymph nodes

left lymphatic duct

left subclavian vein

axillary lymph nodes

thoracic duct

small intestine

villus

lacteal

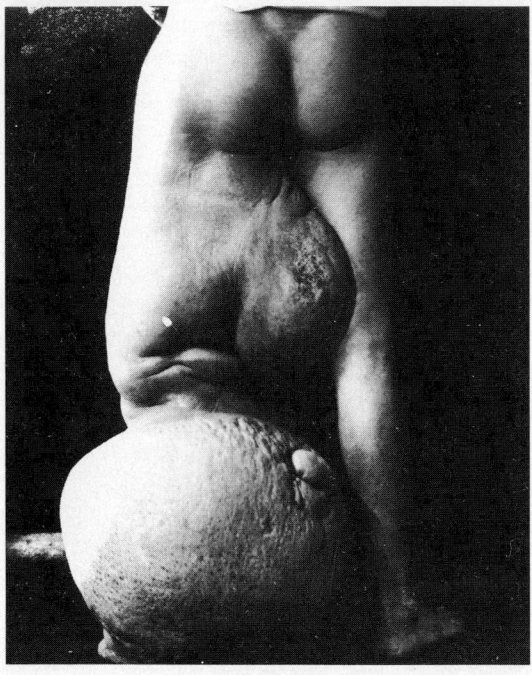

Figure 35.2
Elephantiasis. When the lymphatic vessels are blocked due to an infection by filarial worms, extreme edema results. Because tissue fluid is not drained away by the lymphatic vessels, the leg is swollen and the skin is thickened.

The lymphatic system begins with lymphatic capillaries that lie near blood capillaries. These capillaries take up fluid that has diffused from and has not been reabsorbed by the blood capillaries (fig. 34.14). Once tissue fluid enters the lymphatic vessels, it is called **lymph.** The lymphatic capillaries join to form lymphatic vessels that merge before entering one of 2 ducts: the thoracic duct or the right lymphatic duct.

The *thoracic duct* is much larger than the right lymphatic duct. It serves the lower extremities, the abdomen, the left arm, and the left side of the head and the neck. In the thorax, the left thoracic duct enters the left subclavian vein. The *right lymphatic duct* serves only the right arm and the right side of the head and the neck. It enters the right subclavian vein.

A malfunction in the lymphatic system can result in edema, which is localized swelling caused by the accumulation of tissue fluid. In the tropics, infection of lymphatic vessels by a parasitic worm causes elephantiasis, a condition in which a limb swells and supposedly resembles the limb of an elephant (fig. 35.2).

The lymphatic system is a one-way system. Lymph flows from a capillary to ever-larger lymphatic vessels and finally to a lymphatic duct, which enters a subclavian vein.

Lymphoid Organs

The lymphoid organs include the bone marrow, the lymph nodes, the spleen, and the thymus (fig. 35.3).

Bone Marrow

In the adult, bone marrow is present only in the bones of the skull, the sternum, the ribs, the clavicle, the spinal column, and the ends of the femur and the humerus (p. 666). Bone marrow contains connective tissue cells called reticular cells, which produce a network of reticular fibers. Reticular cells and developing blood cells are packed about thin-walled venous sinuses. Mature blood cells enter the bloodstream at these sinuses.

Radioactive tracer studies have shown that the bone marrow is the site of origination for all the types of blood cells (see fig. 34.11), including both the granular and agranular leukocytes. In other words, the bone marrow contains lymphoid tissue that produces lymphocytes. The B (for bone marrow) lymphocytes mature in the bone marrow, and the T (for thymus) lymphocytes mature in the thymus gland. The structure and the function of B and T lymphocytes are discussed in the following section.

The bone marrow also contains monocytes that have developed into resident macrophages. These large phagocytic cells help to cleanse the marrow and the adjacent blood sinuses.

Lymph Nodes

At certain points along lymphatic vessels, small (about 2.5 cm), ovoid or round structures called lymph nodes occur. A lymph node has a fibrous connective tissue capsule. Connective tissue also divides a node into nodules. Each nodule contains a sinus (open space) filled with many lymphocytes and macrophages. As lymph passes through the sinuses, it is purified of infectious organisms and any other debris.

While nodules usually occur within lymph nodes, they also can occur singly or in groups. The *tonsils* are composed of partly encapsulated lymph nodules. There are also nodules called *Peyer's patches* within the intestinal wall.

The lymph nodes occur in groups in certain regions of the body. For example, the inguinal nodes are in the groin and the axillary nodes are in the arm pits.

Lymph nodes are divided into sinus-containing lobules, where the lymph is cleansed by phagocytes.

The Spleen

The spleen is located in the upper left abdominal cavity just beneath the diaphragm. The spleen is constructed the same as a lymph node. Outer connective tissue divides the organ into lobules that contain sinuses. In the spleen, however, the sinuses are filled with blood instead of lymph. Especially since the blood vessels of the spleen can expand, this organ serves as a blood reservoir and makes blood available in times of low pressure or when the body needs extra oxygen in the blood.

A spleen nodule contains red pulp and white pulp. Red pulp contains red blood cells, lymphocytes, and macrophages. The white pulp contains only lymphocytes and macrophages. Both types of pulp help to purify the blood that passes through the spleen. If the spleen ruptures due to injury, it can be removed. Although the functions of the spleen are duplicated by other organs, the individual is expected to be slightly more susceptible to infections and may have to take antibiotic therapy indefinitely.

Animal Structure and Function

Figure 35.3

Lymphoid organs. The bone marrow is the site of lymphocyte and monocyte (macrophage) production and B cell maturation. The thymus is the site of T cell maturation. (A child is shown because the thymus is larger in children than in adults.) The lymph node is the site of lymphocyte and macrophage accumulation (the tonsils are modified lymph nodes). The spleen is the site of lymphocyte, macrophage, and red blood cell accumulation.

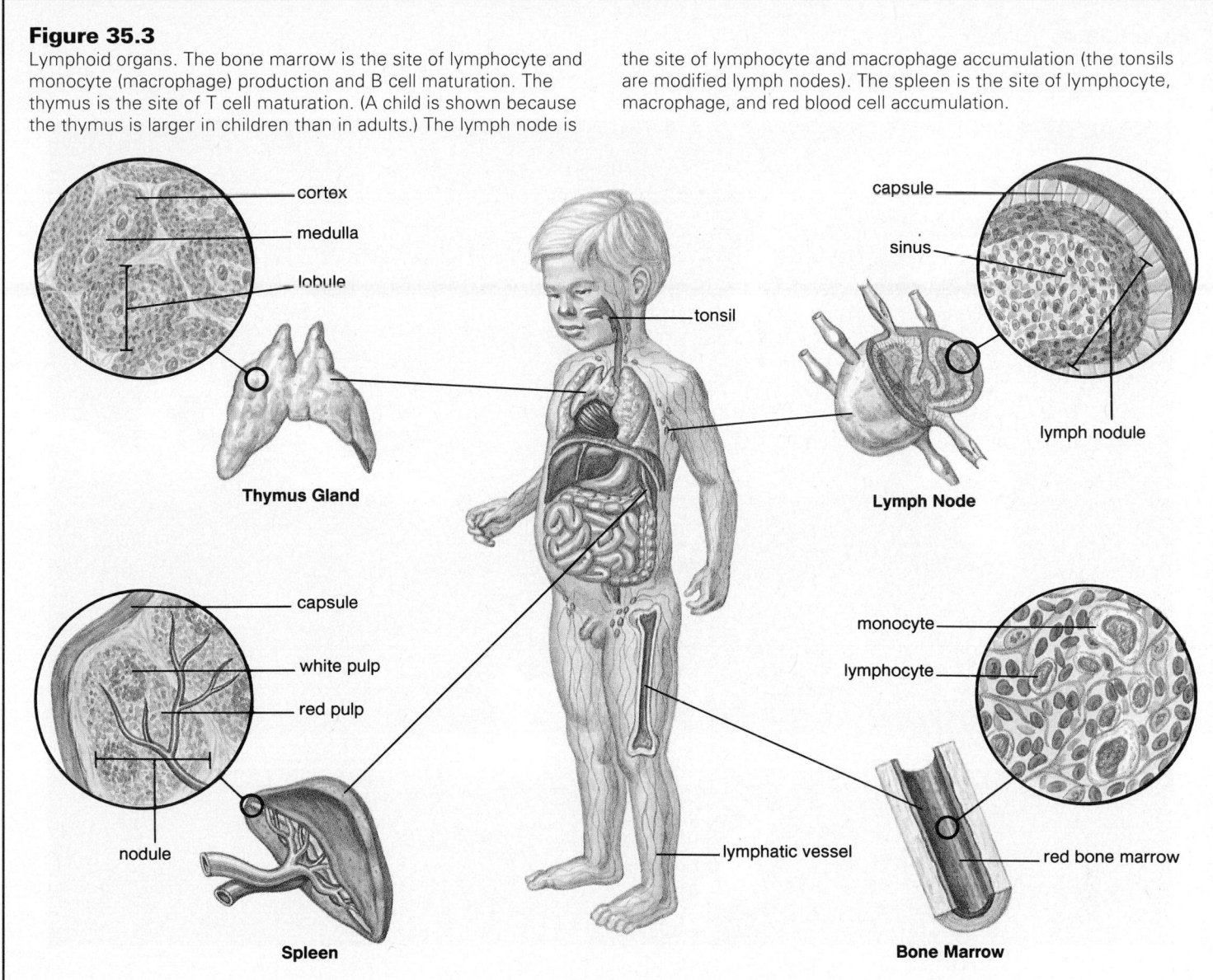

The spleen is divided into sinus-containing lobules, where the blood is cleansed by phagocytes.

The Thymus

The thymus is located along the trachea behind the sternum in the upper thoracic cavity. This gland varies in size, but it is larger in children than in adults. The thymus also is divided into lobules by connective tissue. The T lymphocytes mature in these lobules.

The thymus secretes thymosin, a molecule that is believed to be an inducing factor; that is, it causes pre-T cells to become T cells. Thymosin also may have other functions in immunity.

The thymus is divided into lobules, where T lymphocytes mature.

Immunity

Immunity is the ability of the body to protect itself from foreign substances and cells, including infectious microbes. The *first line of defense* is available immediately because it involves mechanisms that are nonspecific. The *second line of defense* takes a little longer to act because it is highly specific and contains mechanisms that are tailored to a particular threat.

General Defense

The environment contains many types of organisms that are able to invade and to infect the body. There are 3 general defense mechanisms that are useful against all types of organisms: barriers to entry, phagocytic white blood cells, and protective proteins.

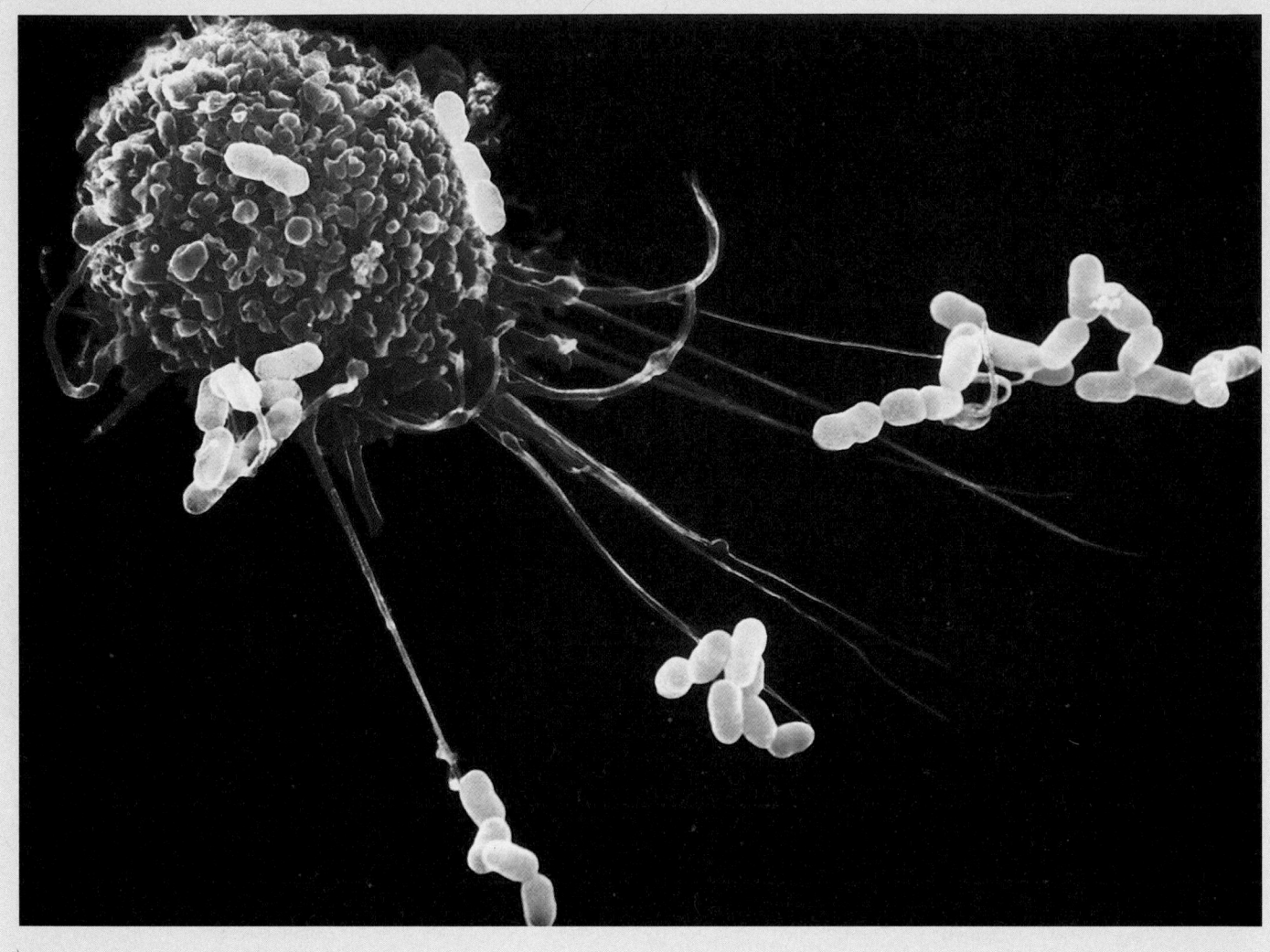

Barriers to Entry

The skin, along with the mucous membrane lining the respiratory and digestive tracts, is a mechanical barrier to entry by bacteria and viruses. The secretions of the oil glands in the skin contain chemicals that weaken or kill bacteria. The respiratory tract is lined by cells that sweep mucous and trapped particles up into the throat, where they can be swallowed. The stomach has an acidic pH that inhibits the growth of many types of bacteria. A mix of bacteria that normally reside in the intestine and other organs, such as the vagina, prevent pathogens from taking up residence.

Phagocytic White Blood Cells

If microbes do gain entry to the body, the inflammatory reaction (p. 563) and other nonspecific forces come into play. For example, neutrophils and monocyte-derived macrophages are phagocytic white blood cells that engulf some bacteria upon contact (fig. 35.4). Infections may be accompanied by fever, which is a protective response because phagocytes function better at a higher-than-normal body temperature.

Protective Proteins

The **complement system** is a series of proteins produced by the liver that are present in the plasma. When the first protein is activated, a cascade of reactions occurs. Every protein molecule in the series activates another in a predetermined sequence. Certain complement proteins form pores in bacterial cell walls and membranes. Then fluids and salts enter the bacterium until it bursts (fig. 35.5). Some simply coat bacteria while others attract phagocytes to the scene. Although complement is a general defense mechanism, it also plays a role in specific defense, as we will see later.

Figure 35.5

Action of the complement system. The complement system is a number of proteins always present in the plasma. When activated, some of these form pores in bacterial cell wall membranes, allowing fluids and salts to enter, until the cell eventually bursts.

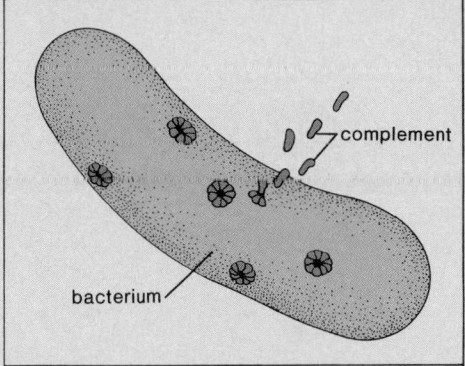

1. Complement proteins form pores in the bacterial cell wall and plasma membrane.

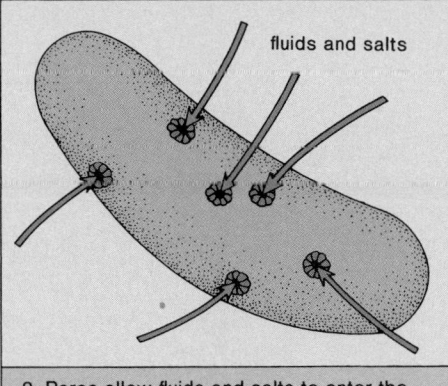

2. Pores allow fluids and salts to enter the bacterium.

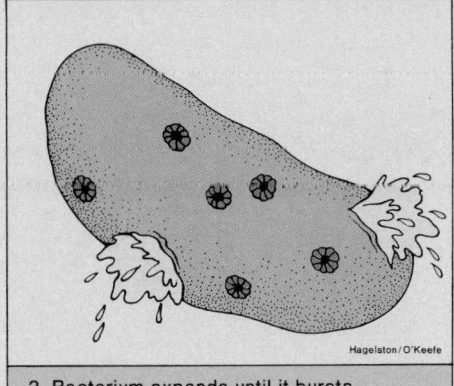

3. Bacterium expands until it bursts.

When most viruses infect a tissue cell, the infected cell produces and secretes interferon. **Interferon** binds to receptors on noninfected cells, and this action causes these cells to prepare for possible attack by producing substances that interfere with viral replication.

A cell that has bound interferon is protected against any type of virus; therefore, interferon should be very useful in preventing viral infection. Interferon is specific to the species, however; only human interferon can be used in humans. It used to be quite a problem to collect enough interferon for clinical and research purposes, but now interferon is a biotechnology product.

> The first line of defense against disease is nonspecific. It consists of barriers to entry, phagocytic white blood cells, and protective proteins.

Specific Defense

Sometimes, we are threatened with an invasion by microorganisms that cannot be counteracted successfully by the general defense mechanisms. In such cases, the immune system is activated to provide a specific defense. The **immune system** consists of lymphocytes and monocytes and also the lymphoid organs and lymphatic vessels where these white blood cells are found in high concentration.

The immune system allows us to develop an immunity against a specific antigen. **Antigens** are usually protein (or polysaccharide) molecules that specific lymphocytes recognize as foreign to the body. Antigens occur on bacteria and viruses, but they also can be part of a foreign cell or a cancerous cell. Ordinarily, we do not become immune to our body's own normal cells; therefore, it is said that the immune system is able to distinguish self from nonself. Immunity is primarily the result of the action of B

lymphocytes and T lymphocytes, which have different functions. **B lymphocytes,** also called B cells, become plasma cells that produce **antibodies,** proteins that are capable of combining with and inactivating antigens. These antibodies are secreted into the blood and the lymph. In contrast, **T lymphocytes,** also called T cells, do not produce antibodies. Instead, certain T cells directly attack cells bearing antigens they recognize. Other T cells regulate the immune response.

Lymphocytes are capable of recognizing an antigen because they have receptor molecules on their surface. The shape of the receptors on any particular lymphocyte are complementary to the shape of one specific antigen. It is often said that the receptor and the antigen fit together like *a lock and a key*. It is estimated that during our lifetime, we encounter a million different antigens, so we need the same number of different lymphocytes for protection against those antigens.

> There are 2 types of lymphocytes. B cells produce and secrete antibodies that combine with antigens. Certain T cells directly attack antigen-bearing cells, and others regulate the immune response.

The Action of B Cells

The receptor on a B cell is called a membrane-bound antibody because it is structured like an antibody. When a B cell encounters a bacterial cell or a toxin bearing an appropriate antigen, it is activated; that is, it has the potential to produce many **plasma cells** that will secrete antibodies against this antigen (fig. 35.6). (For the B cell to realize this potential, it must be stimulated by a helper T cell [p. 583].) All of the plasma cells derived from one parent lymphocyte are called a clone, and a clone produces the same type of antibody. Notice that a B cell does not clone until its antigen is present. The *clonal selection theory* states that the antigen selects which B cell will produce a clone of plasma cells.

Figure 35.6

Clonal selection theory as it applies to B cells. Antigens have different shapes. In this diagram the shape of the antigen causes it to attach to the receptor of the middle B cell and not those of the B cells on the right or the left. The antigen and receptor fit together in a lock-and-key manner because their shapes are complementary. The B cell will divide if stimulated by a helper T cell (fig. 35.10) and produce many plasma cells that secrete specific antibodies against this antigen by the fifth day. Memory cells that retain the ability to secrete these specific antibodies at a future time are also produced.

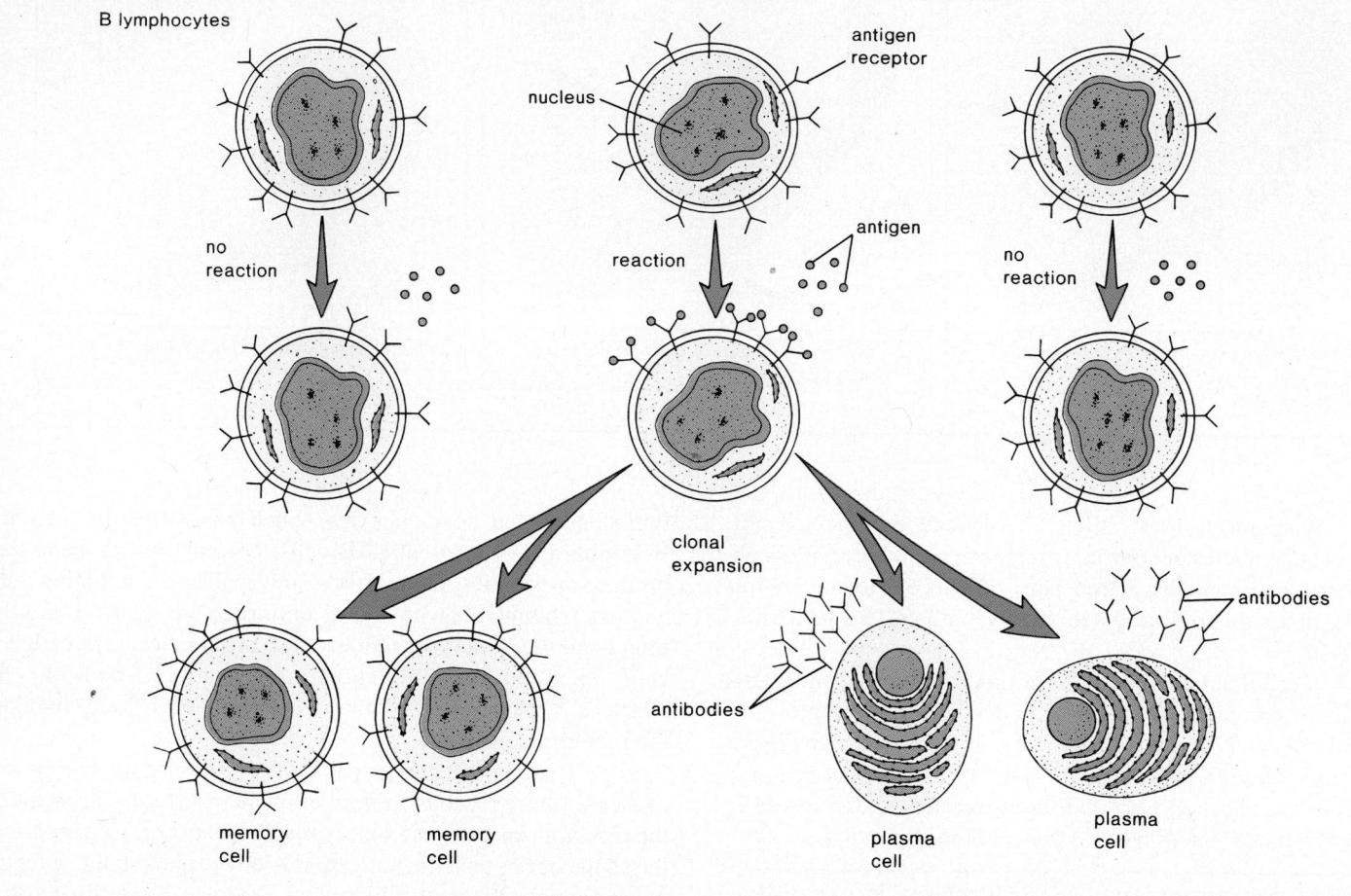

Once antibody production is sufficient, the antigen disappears from the system, and the development of plasma cells ceases. Some members of a clone do not participate in antibody production, however; instead, they remain in the bloodstream as **memory B cells.** Memory B cells are capable of producing the antibody specific to a particular antigen for some time. As long as these cells are present, the individual is said to be actively immune: future antibody production is possible because the memory cells can produce more plasma cells if the same antigen invades the system again.

Defense by B cells is called *antibody-mediated immunity* because B cells produce antibodies. It also is called humoral immunity because these antibodies are present in the bloodstream.

B cells are responsible for antibody-mediated immunity. After they recognize an antigen, they (if stimulated by a helper T cell) divide to produce both antibody-secreting plasma cells and memory B cells.

Antibodies The most common type of antibody (IgG) is a Y-shaped protein molecule having 2 arms. Each arm has a long "heavy" chain and a short "light" chain of amino acids. These chains have *constant regions,* where the sequence of amino acids is set, and *variable regions,* where the sequence of amino acids varies to produce a particular shape (fig. 35.7). The antigen binds to the antibody at the variable regions of one arm in a lock-and-key manner. In other words, the variable regions form an antibody-binding site that is specific for a particular antigen.

The constant regions are not identical among all the antibodies. Instead, they are the same for different classes of antibodies. Most antibodies found in the blood belong to the class IgG (immunoglobulin G) (table 35.1).

The antigen-antibody reaction can take several forms, but quite often the antigen-antibody reaction produces complexes of antigens combined with antibodies. Such an antigen-antibody complex, sometimes called the immune complex, marks the anti-

Animal Structure and Function

Figure 35.7

Structure of the most common antibody (IgG). ***a.*** Computer model of an antibody. ***b.*** Schematic drawing. Each arm of an antibody contains a light (short) amino acid chain and a heavy (long) amino acid chain. In the variable regions, the amino acid sequence varies between antibodies so that the binding sites have a shape that is complementary to the shape of a particular antigen. An antigen binds to a variable region in a lock-and-key manner.

antigen-binding site

antigen-binding site

light chains

heavy chains

constant regions

variable regions

a.

b.

gen for destruction by other forces. For example, the complex may be engulfed by neutrophils or macrophages, or it may activate a portion of blood serum called the complement system, page 576.

> An antibody combines with its antigen in a lock-and-key manner. When several antigens bind with antibodies, an antigen-antibody complex forms.

The Actions of T Cells

There are 4 different types of T cells: cytotoxic T cells, helper T cells, memory T cells, and suppressor T cells. All 4 types look alike but can be distinguished by their functions.

Cytotoxic T (T_C) cells sometimes are called **killer T cells.** In immune individuals, they attack and destroy cells bearing a foreign antigen, such as virus-infected cells or cancerous cells. These T cells have storage vacuoles that contain a chemical called perforin because it perforates cell membranes. The perforin molecules form a pore in the membrane that allows water and salts to enter. The cell under attack then swells and eventually bursts (fig. 35.8).

It often is said that T cells are responsible for *cell-mediated immunity,* characterized by destruction of antigen-bearing cells. Of all the T cells, only T_C cells are involved in this type of immunity.

Helper T (T_H) cells regulate immunity by enhancing the response of other immune cells. In response to an antigen, they enlarge and secrete lymphokines, including interferon and the interleukins.[1] **Lymphokines** are stimulatory molecules that cause

[1]For example, the lymphokine that stimulates B cell to secrete antibodies is called interleukin-4.

Table 35.1
Types of Antibodies

Classes	Description
IgG	Main antibody type in circulation; attacks microorganisms and their toxins
IgA	Main antibody type in secretions, such as saliva and milk; attacks microorganisms and their toxins
IgF	Antibody responsible for allergic reactions
IgM	Antibody type found in circulation; largest antibody, with 5 subunits
IgD	Antibody type found primarily as a membrane-bound immunoglobulin

T_H cells to clone and other immune cells to perform their functions. For example, T_H cells stimulate macrophages to phagocytize antibodies and B cells to manufacture antibodies. Because the AIDS virus attacks T_H cells, it inactivates the immune response. This is discussed in chapter 43.

When an activated T_H cell divides, the clone contains **suppressor T (T_S) cells** and **memory T (T_M) cells.** Once there is a sufficient number of T_S cells, the immune response ceases. Following suppression, a population of T_M cells persists, perhaps for life. These cells are able to secrete lymphokines and to stimulate macrophages and B cells whenever the same antigen enters the body once again.

Figure 35.8

Cell-mediated immunity. Scanning electron micrograph showing cytotoxic T cells destroying a cancer cell. During the killing process, vacuoles in a T cell fuse with the plasma membrane and release pore-forming proteins. Fluids and salts enter through the pores, and the target cell eventually bursts.

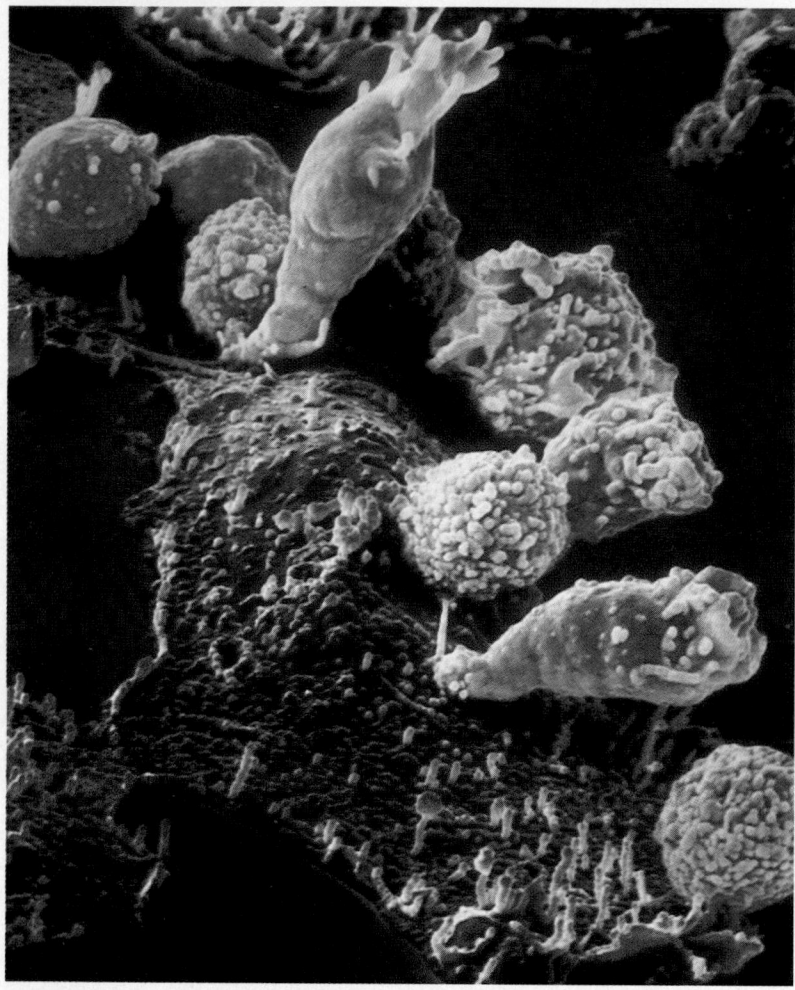

T_C cells are responsible for cell-mediated immunity; T_H cells promote the immune response; T_S cells suppress the immune response; and T_M cells maintain immunity.

The Activation of Cytotoxic and Helper T Cells T cells have receptors just as B cells do. Unlike B cells, however, T_C cell and T_H cells are unable to recognize an antigen that simply is present in lymph or blood. Instead, the antigen must be presented to them by an antigen-presenting cell (APC). When an APC, such as a macrophage, engulfs a bacterium or a virus, the APC enzymatically breaks it down to peptide fragments that are antigenic (have the properties of an antigen). The antigenic peptide fragment is linked to an **MHC** (major histocompatibility complex) **protein,** and together they are displayed to a T cell at the plasma membrane. The importance of MHC proteins first was recognized when it was discovered that they contribute to the specificity of tissues and make it difficult to transplant tissue from one person to another. In other words, the donor and the recipient must be histo-(tissue) compatible (the same or nearly so) for a transplant to be successful without the administration of immunosuppressive drugs.

There are 2 types of MHC proteins, known as MHC I and MHC II. Most cells of the body display MHC I–type proteins and only cells of the immune system, that is, macrophages, B cells, and some T cells display MHC II-type proteins. This allows cells of the immune system to recognize one another.

Figure 35.9*a* shows a macrophage and a B cell presenting an antigen to a T_H cell. Once a T_H cell recognizes an antigen, it undergoes clonal expansion, producing T_S cells and T_M cells that also can recognize this same antigen. While well-known APCs are macrophages and B cells, actually any cell in the body can be an APC. For example, figure 35.9*b* shows a virus-infected cell presenting an antigen to a T_C cell. Once a T_C cell recognizes an antigen, it attacks and destroys any cell that is infected with the same virus.

In order for T_C and T_H cells to recognize an antigen, the antigen, along with an MHC protein, must be presented to them by an APC.

Table 35.2 and figure 35.10 summarize our discussion of B cells and T cells.

Immunotherapy

The immune system can be manipulated to help people avoid or recover from diseases. Some of the techniques to do this have been utilized for a long time, and some are relatively new.

Induced Immunity

Active immunity, which provides long-lasting protection against a disease-causing organism, develops after an individual is infected with a virus or a bacterium. In many instances today, however, it is not necessary to suffer an illness to become immune because it is possible to be artificially immunized against a disease using vaccines. **Vaccines,** traditionally are bacteria and viruses (antigens) that have been treated so that they are no longer virulent (able to cause disease), but new methods of producing vaccines are being developed. For example, it is possible to use the recombinant DNA technique to mass-produce a protein that can be used as a vaccine. This method is being used to prepare a vaccine against hepatitis B.

After a vaccine is given, it is possible to determine the amount of antibody present in a sample of serum—this is called the *antibody titer*. After the first exposure to an antigen, a primary response occurs. For a period of several days, no antibodies are present; then, there is a slow rise in the titer, followed by a gradual

Animal Structure and Function

How Do You Test for an HIV Infection?

It is now well recognized that acquired immune deficiency syndrome (AIDS) is caused primarily by a human immunodeficiency virus (HIV-1), which infects T$_H$ lymphocytes. Even though the virus attacks the immune system, B cells do begin to produce antibodies within weeks or months after the infection begins. All of the commonly used tests detect the presence of these antibodies by having them react to their antigens. They use plastic beads that are coated with either inactivated virus or just specific viral envelope proteins.

At the start of the test, an antigen coated bead is placed in a small tube along with the specimen (human serum or plasma) to be tested (fig. 35.A). If the specimen contains HIV-1 antibodies, they will bind to the bead and remain bound when the bead is washed and rinsed. These antibodies are designated as Ab$_1$ because they are the first antibodies attached to the bead.

The next step is to get a color reaction that can be read photometrically by a laboratory instrument. First, a conjugate designated as Ab$_2$ – Enz is added to the tube. The conjugate is made up of antihuman globulin attached to an enzyme. The Ab$_2$ – Enz will bind to Ab$_1$ on the bead. After washing and rinsing, a substrate is added that will react with the enzyme and cause a color change. For example, one of the commonly used enzymes is peroxidase derived from horseradish. When hydrogen peroxide and o-phenylenediamine are added, the hydrogen peroxide releases oxygen that causes the o-phenylenediamine to go from a clear to a yellow-orange color. The intensity of the color reaction is compared to a known negative test result (no HIV-1 antibody is present in the specimen) and a known positive test result (HIV-1 antibody is present in the specimen).

The test we have described is called an HIV-1 antibody *enzyme-linked immunoassay* (ELISA). Note that since this is a test for HIV-1 antibody, it cannot detect persons who are infectious but have not yet produced HIV-1 antibodies. Nor can it reliably detect persons infected with HIV-2, another HIV that can cause AIDS. On the other hand, false positive results are possible in persons suffering from diseases other than AIDS. To guard against false positive results, it is common practice to do a more specialized and expensive test on specimens that give a positive ELISA test. (This test resembles the procedure described for DNA fingerprinting on page 592.) Patients who give a positive result with both the HIV-1 antibody ELISA and with the confirmatory test are told they have an HIV infection and that they should seek immediate treatment.

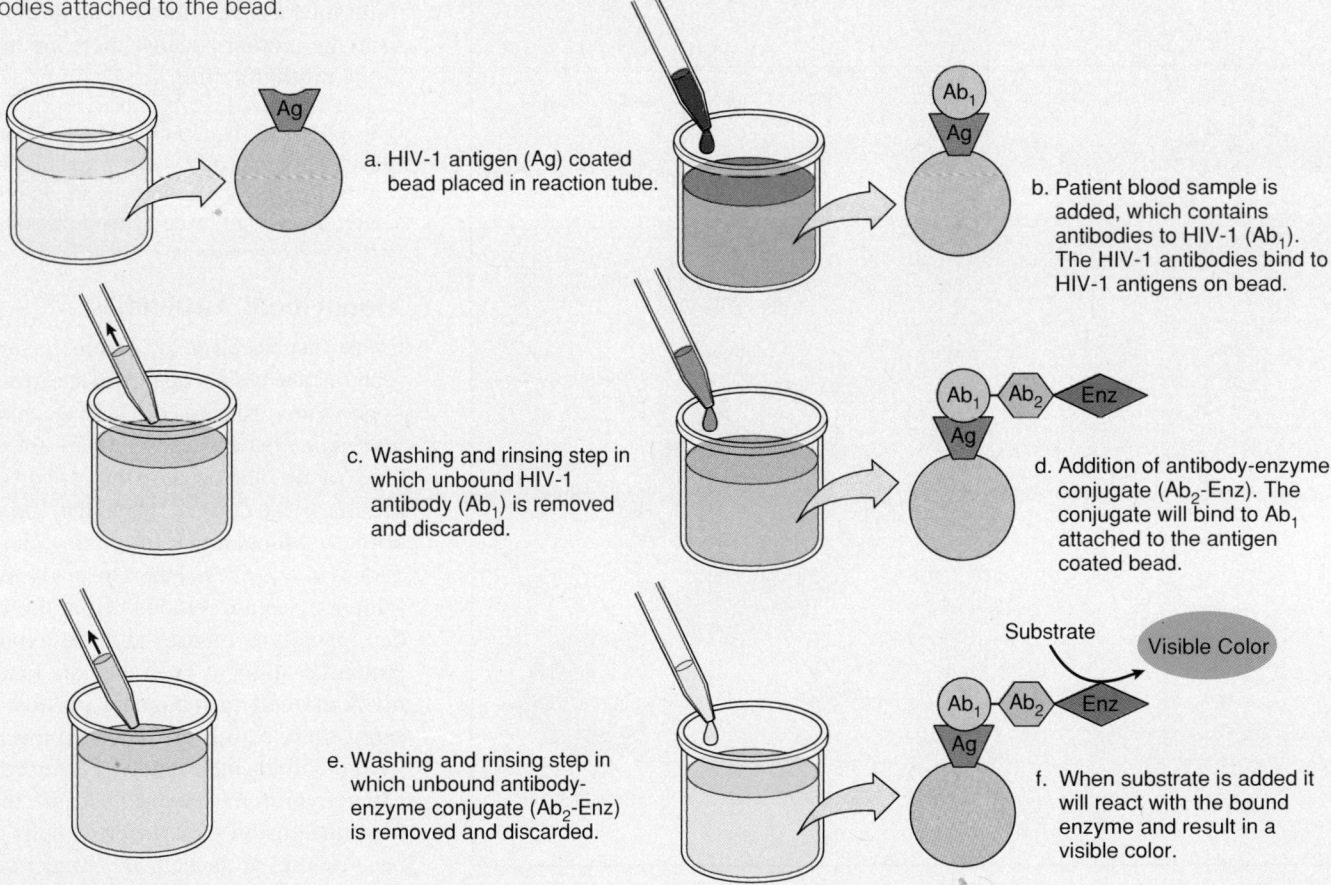

a. HIV-1 antigen (Ag) coated bead placed in reaction tube.

b. Patient blood sample is added, which contains antibodies to HIV-1 (Ab$_1$). The HIV-1 antibodies bind to HIV-1 antigens on bead.

c. Washing and rinsing step in which unbound HIV-1 antibody (Ab$_1$) is removed and discarded.

d. Addition of antibody-enzyme conjugate (Ab$_2$-Enz). The conjugate will bind to Ab$_1$ attached to the antigen coated bead.

e. Washing and rinsing step in which unbound antibody-enzyme conjugate (Ab$_2$-Enz) is removed and discarded.

f. When substrate is added it will react with the bound enzyme and result in a visible color.

Figure 35.A

A positive test for an HIV infection has these steps. In the first step, HIV-1 antibodies (Ab$_1$) in the specimen bind to HIV-1 antigens on a bead. In the second step, antibody-enzyme conjugate (Ab$_2$) binds to Ab$_1$. The enzyme brings about a color change that signifies the test is positive.

Figure 35.9

T cell activation. *a.* Either a macrophage or a B cell presents an antigen to a helper T cell. To accomplish this, the antigen has to be digested to peptides that are combined with an MHC II protein. The complex is presented to the T cell. In return, the helper T cell produces and secretes lymphokines that stimulate T cells and other immune cells. *b.* Cells infected with a virus present one of the viral proteins along with an MHC I protein to a cytotoxic T cell. This causes the cytotoxic T cell to attack and to destroy any cell infected with the same virus (see fig. 35.8).

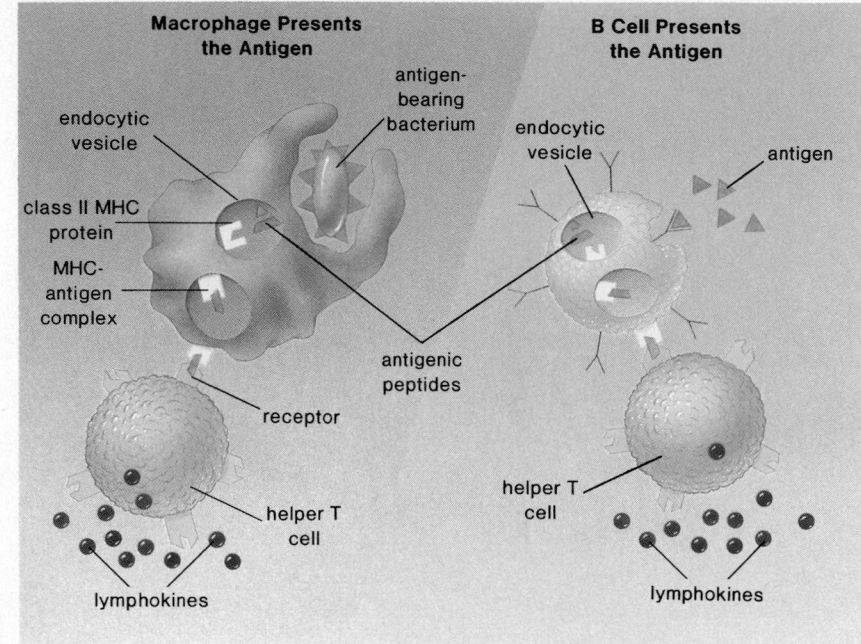

Macrophage Presents the Antigen

endocytic vesicle

antigen-bearing bacterium

class II MHC protein

MHC-antigen complex

antigenic peptides

receptor

helper T cell

lymphokines

B Cell Presents the Antigen

endocytic vesicle

antigen

helper T cell

lymphokines

a.

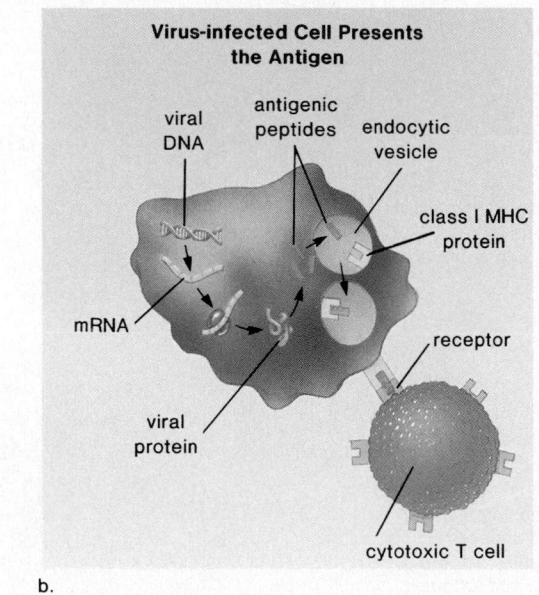

Virus-infected Cell Presents the Antigen

viral DNA

antigenic peptides

endocytic vesicle

class I MHC protein

mRNA

receptor

viral protein

cytotoxic T cell

b.

decline (fig. 35.11). After a second exposure, a secondary response may occur. If so, the titer rises rapidly to a level much greater than before. The second exposure in that case often is called the *"booster"* because it boosts the antibody titer to a high level. A good secondary response can be related to the number of plasma cells and memory cells in the serum. Upon the second exposure, these cells already are present, and antibodies can be produced rapidly.

Passive immunity occurs when an individual is given antibodies (immunoglobulins) to combat a disease. Since these antibodies are not produced by the individual's B cells, passive immunity is short-lived. For example, newborn infants possess passive immunity because antibodies have crossed the placenta from the mother's blood. These antibodies soon disappear, however, so that within a few months, infants become more susceptible to infections. Breast-feeding prolongs the passive immunity an infant receives from the mother because there are antibodies in the mother's milk.

Vaccines can be used to make people actively immune. Passive immunity is short-lived because the antibodies are administered to and not made by the individual.

Monoclonal Antibodies

Every plasma cell derived from the same B cell secretes antibodies against a specific antigen, as previously discussed. These are *monoclonal antibodies* because all of them are the same type (mono) and because they are produced by plasma cells derived from the same B cell (clone). Monoclonal antibodies can be produced in vitro (in laboratory glassware). B lymphocytes are removed from the body (today, usually a mouse) and are exposed to a particular antigen. Then they are fused with a myeloma cell (a malignant plasma cell) because these cells, unlike normal plasma cells, live and divide indefinitely. The fused cells are called hybridomas; *hybrid* because they result from the fusion of 2 different cells and *oma* because one of the cells is a cancer cell.

Animal Structure and Function

Figure 35.10
Summary of B cell and T cell functions. Both cells are produced in the bone marrow, but only B cells mature there. T cells mature in the thymus. An antigen activates the appropriate B cell, which divides and differentiates to give a clone of antibody-producing plasma cells and memory cells. An antigen activates a T cell (either cytotoxic T or helper T) if it is presented with an MHC II protein by an APC such as a macrophage. Cytotoxic T cells then attack and destroy antigen-bearing cells; helper T cells divide to give a clone of helper T cells that produce lymphokines and also suppressor T and memory T cells.

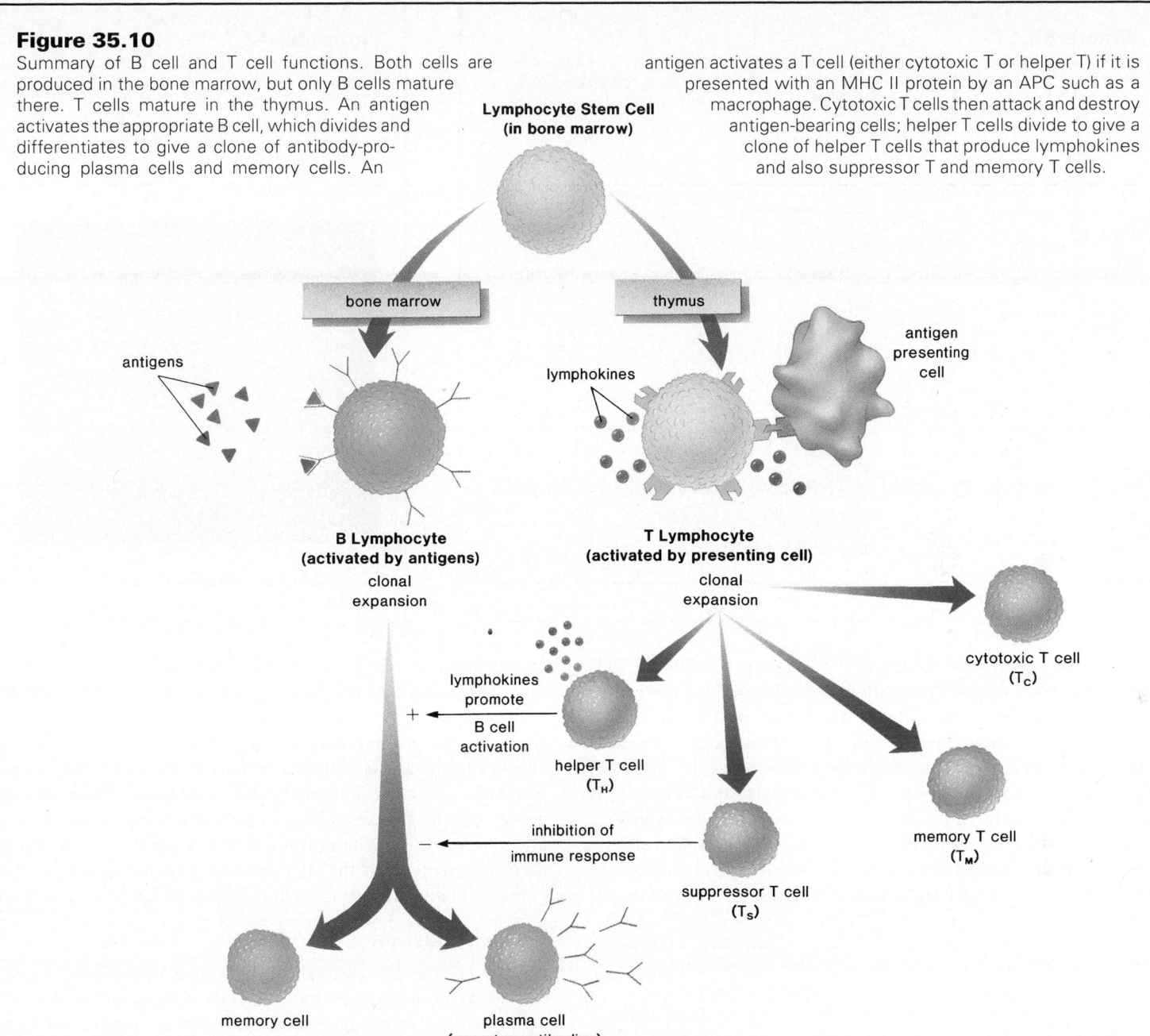

Table 35.2
Some Properties of B Cells and T Cells

Property	B Cells	T Cells	
		Cytotoxic	**Helper**
Type of immunity	Antibody mediated	Cell mediated	_____
Antigen recognition	Direct recognition	Must be presented by APC	Must be presented by APC
Response	Become antibody-producing plasma cells	Search and destroy antigen-bearing cells	Secrete lymphokines; stimulate other immune cells
Final response	Memory cells	_____	Suppressor and memory cells

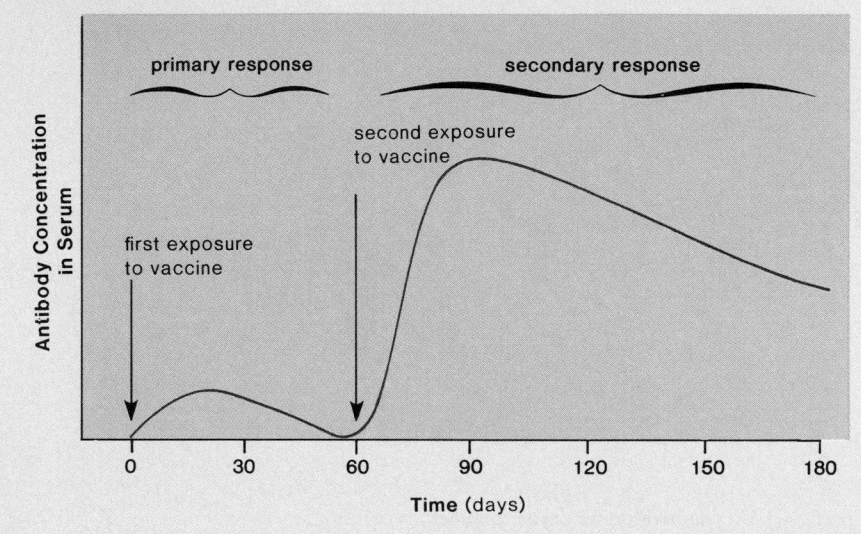

Figure 35.11

Development of active immunity due to immunization. The primary response after the first exposure of a vaccine is minimal, but the secondary response that may occur after the second exposure shows a dramatic rise in the amount of antibody present in serum.

Figure 35.12

Monoclonal antibody use. These cells are infected with herpes simplex type 1 virus (HSV-1), which can be detected by a monoclonal antibody specific for the virus and tagged with a fluorescent dye. When the cells fluoresce, it shows that the virus is inside the cells.

At present, monoclonal antibodies are used for quick and certain diagnosis of various conditions. For example, a particular hormone is present in the urine of a pregnant woman. A monoclonal antibody can be used to detect the hormone and so indicate that the woman is pregnant. Monoclonal antibodies also are used for identifying infections (fig. 35.12). They are so accurate they even can sort out the different types of T cells in a blood sample. Because monoclonal antibodies can distinguish between cancerous and normal tissue cells, they are used to carry radioactive isotopes or toxic drugs only to tumors so these can be destroyed selectively.

Monoclonal antibodies are produced in pure batches—they react specifically with just one type of molecule (antigen); therefore, they can distinguish one cell, or even one molecule, from another.

Immunological Side Effects and Illnesses

The immune system protects us from disease because it can tell self from nonself. Sometimes, however, the immune system is underprotective, as when an individual develops cancer, or is overprotective, as when an individual has allergies.

Allergies

Of the 5 varieties of antibodies (table 35.1)—IgA, IgD, IgE, IgG, and IgM—IgE antibodies cause allergies. IgE antibodies are found in the bloodstream, but they, unlike the other types of antibodies, also reside in the membrane of *mast cells* found in the tissues. Some investigators contend that mast cells are basophils that have left the bloodstream and have taken up residence in the tissues. In any case, when the *allergen,* an antigen that provokes an allergic reaction, attaches to the IgE antibodies on mast cells, these cells release histamine and other substances that cause mucus secretion and airway constriction, resulting in the characteristic symptoms of allergy. On occasion, basophils and other white blood cells release these chemicals into the bloodstream. The increased capillary permeability that results from this can lead to fluid loss and shock.

Tissue Rejection

Both T_C cells and/or antibodies can bring about disintegration of foreign tissue introduced into the body. Organ rejection can be controlled in 2 ways: careful selection of the organ to be transplanted and the administration of immunosuppressive drugs. It is best if the transplanted organ has the same type of MHC proteins as those of the recipient, because T_C cells learn to recognize foreign MHC proteins. The immunosuppressive drug cyclosporine has been in use for some years. An experimental drug, FK-506, eventually may replace cyclosporine as the drug of choice for transplant patients. In more than 100 patients taking FK-506, the rate of organ rejection was one-sixth that of cyclosporine.

Autoimmune Diseases

Certain human illnesses are due to the production of antibodies that act against an individual's own tissues. In myasthenia gravis, autoantibodies attack the neuromuscular junctions so that the muscles do not obey nervous stimuli. Muscular weakness results. In multiple sclerosis (MS), antibodies attack the myelin of nerve fibers, causing various neuromuscular disorders. A person with systemic lupus erythematosus (SLE) forms various antibodies to different constituents of the body, including the DNA of the cell nucleus. The disease sometimes results in death, usually due to kidney damage. In rheumatoid arthritis, the joints are affected. When an autoimmune disease occurs, a viral infection of tissues often has set off an immune reaction to the body's own tissues. There is evidence to suggest that type I diabetes is the result of this sequence of events, as well as heart damage following rheumatic fever.

Summary

1. The lymphatic system has various functions. Lymphatic capillaries collect excess tissue fluid, which moves in lymphatic veins to blood circulatory veins. Lacteals absorb the products of fat digestion, and lymphoid organs (bone marrow, thymus, lymph nodes, and spleen) help defend the body from disease.

2. The general defense of the body consists of barriers to entry, phagocytic white blood cells, and protective proteins.

3. The immune response is specific to a particular antigen and requires 2 types of lymphocytes, both of which are produced in the bone marrow. B cells mature in the bone marrow, and T cells mature in the thymus.

4. B cells are responsible for antibody-mediated immunity. When the shape of an antigen fits the shape of a B-cell receptor, that B cell undergoes clonal expansion, producing antibody-secreting plasma cells and memory cells. In keeping with the clonal selection theory, the antigen selects the B cell that proliferates.

5. An antibody is a Y-shaped molecule that has 2 binding sites. Each antibody is specific for a particular antigen.

6. For a T cell to recognize an antigen, the antigen must be presented by an APC along with an MHC protein. (MHC I in macrophages and MHC II for virus-infected cells.)

7. There are 4 types of T cells: T_C cells kill cells on contact; T_H cells stimulate other immune cells; T_S cells suppress the immune response; T_M cells are memory cells.

8. AIDS is an HIV infection of T_H cells, and therefore it destroys the cells that are needed to mount an immune response.

9. Immunity can be fostered by immunotherapy. Vaccines are available to promote active immunity, and antibodies sometimes are available to provide an individual with short-term passive immunity. Monoclonal antibodies are used for various purposes.

10. Immunity has certain undesirable side effects. Allergies are due to an overactive immune system that forms antibodies to substances not normally recognized as foreign. T_C cells attack transplanted organs. Autoimmune illnesses occur when antibodies against the body's own cells form.

Writing Across the Curriculum

In order to practice writing skills, students should write out the answers to any or all of the study questions and the critical thinking questions. The study questions are sequenced in the same order as the text. Suggested answers to the critical thinking questions are in appendix D.

Study Questions

1. What is the lymphatic system, and what are its 3 functions?
2. Describe the structure and the function of the bone marrow, lymph nodes, the spleen, and the thymus.
3. Distinguish between the general defense and the specific defense of the body against disease.
4. Contrast B and T cells in as many ways as possible.
5. What is the clonal selection theory?
6. Explain the process that allows a T cell to recognize an antigen.
7. List the 4 types of T cells, and state their functions.
8. Discuss allergies, tissue rejection, and autoimmune disease as they relate to the immune system.
9. Relate active immunity to the presence of plasma cells and memory cells.
10. How is active immunity achieved? How is passive immunity achieved?

Objective Questions

1. Complement
   a. is a general defense mechanism.
   b. is a series of proteins present in the plasma.
   c. plays a role in destroying bacteria.
   d. All of these.

2. Which one of these does not pertain to T cells?
   a. have specific receptors
   b. have cell-mediated immunity
   c. stimulate antibody production by B cells
   d. have no effect on macrophages

3. Which one of these does not pertain to B cells?
   a. have passed through the thymus
   b. specific receptors
   c. antibody-mediated immunity
   d. synthesize and liberate antibodies

4. The clonal selection theory says that
   a. an antigen selects out certain B cells and suppresses them.
   b. an antigen stimulates the multiplication of B cells that produce antibodies against it.
   c. T cells select out those B cells that should produce antibodies regardless of antigens.
   d. T cells suppress all those B cells except the ones that should multiply and divide.
5. Plasma cells are
   a. the same as memory cells.
   b. formed from blood plasma.
   c. B cells that are actively secreting antibody.
   d. inactive T cells carried in the plasma.
6. For a T cell to recognize an antigen, it must interact with
   a. complement.
   b. a macrophage.
   c. a B cell.
   d. All of these.
7. Antibodies combine with antigens
   a. at variable regions.
   b. at constant regions.
   c. only if macrophages are present.
   d. Both a and c.
8. Which one of these is mismatched?
   a. Helper T cells—help complement react
   b. Killer T cells—active in tissue rejection
   c. Suppressor T cells—shut down the immune response
   d. Memory T cells—long-living line of T cells
9. Vaccines are
   a. the same as monoclonal antibodies.
   b. treated bacteria or viruses or one of their proteins.
   c. MHC proteins.
   d. All of these.
10. The theory behind the use of lymphokines in cancer therapy is
   a. if cancer develops, the immune system has been ineffective.
   b. lymphokines stimulate the immune system.
   c. cancer cells bear antigens that should be recognizable by T_K cells.
   d. All of these.

Concepts and Critical Thinking

1. *The body defends its integrity.*

   How does your study of immunity support this concept?

2. *Multitiered mechanisms maintain homeostasis.*

   How does your study of immunity support this concept?

3. *Organs belong to organ systems.*

   Argue that red bone marrow is a part of the skeletal, circulatory, and lymphatic systems.

Selected Key Terms

lymphatic system (lim-fat′ik sis′tem) 573
immunity (i-mūn-i-tee) 575
interferon (in″ter-fer′on) 577
antigen (an′tĭ-jen) 577
B lymphocyte (be lim′fo-sīt) 577
antibody (an′tĭ-bod″e) 577
T lymphocyte (te lim′fo-sīt) 577
plasma cell (plaz′mah sel) 577
memory B cell (mem′o-re be sel) 578
cytotoxic T (T_c) cell (si″to-tok′sik te sel) 579
helper T (T_H) cell (hel′per te sel) 579
suppressor T (T_s) cell (sŭ-pres′or te sel) 579
memory T (T_M) cell (mem′o-re te sel) 579
MHC protein (em āch se pro′tein) 580
vaccine (vak′sēn) 580

Animal Structure and Function

36

Digestive System and Nutrition

A long-nosed tree snake is feeding on an anole. Snakes are discontinuous eaters—few eat more than once a week; many eat about once a month. Snakes can eat large prey because their jaws and their entire digestive tract are expandable. A snake's curved teeth hold onto the prey animal and pull it back into the esophagus.

Your study of this chapter will be complete when you can

1. contrast the dentition and digestive tracts of mammalian herbivores and mammalian carnivores;
2. contrast the characteristics of a continuous and a discontinuous feeder;
3. contrast the characteristics of an incomplete and complete digestive tract;
4. name the parts of the human digestive system and describe, in general, their structure and their function;
5. outline the digestion of carbohydrates (e.g., starch), proteins, and fats in humans by listing the names and functions of the appropriate enzymes and by telling where in the human digestive tract these enzymes function;
6. list the contributions of the liver and pancreas to the digestive process in humans;
7. name 6 functions of the liver;
8. give the names and the functions of hormones that affect the flow of digestive juices;
9. in general, tell how an animal's body makes use of the products of digestion;
10. describe a balanced diet; describe how to achieve a balanced diet and how to avoid too much fat, sugar, and salt in the diet.

Animals are heterotrophic organisms that must take in preformed food. The many adaptations animals exhibit to acquire and to digest food are extremely varied and can be related to a wide variety of available food. Necessarily, we will be able to discuss only a few examples. When animals take in food, they acquire the energy needed to carry on daily activities and the building blocks needed to grow and/or repair tissues.

Animal Feeding Modes

Some animals are omnivores; they eat both plants and animals. Others are herbivores; they feed only on plants. Still others are carnivores; they eat only other animals (fig. 36.1).

Among mammals, the dentition differs according to mode of nutrition (fig. 36.2). Humans, as well as raccoons, rats, and brown bears, are omnivores, in which dentition is nonspecialized and adequate to deal with both a vegetable diet and a meat diet. An adult human has 32 teeth. One-half of each jaw has teeth of 4 different types: 2 chisel-shaped *incisors* for biting; one pointed *canine* for tearing; 2 fairly flat *premolars* for grinding; and 3 *molars,* well-flattened for crushing.

Among herbivores, the koala of Australia is famous for its diet of only eucalyptus leaves, and likewise many other mammals are browsers, feeding off bushes and trees. Grazers, like the horse, feed off grasses. The horse has sharp, even incisors for neatly clipping off blades of grass and large, flat premolars and molars for grinding and crushing up the grass. Extensive grinding and crushing disrupts plant cell walls, allowing bacteria located in the cecum

---

Figure 36.1

Feeding modes of invertebrates. *a.* Some invertebrates are omnivores, as are earthworms. As an earthworm burrows, it takes up soil and organic material by the joint action of its liplike prostomium and muscular pharynx. *b.* Some invertebrates are herbivores, as are grasshoppers. There are mouthparts for chewing the food before it enters the pharynx. *c.* Some invertebrates are carnivores, as are spiders. Wolf spiders, jumping spiders, and tarantulas hunt for prey, which they seize and immobilize with their fangs and pedipalps, anterior appendages. Other spiders trap prey in webs made of silk. Digestive juices are secreted into the body of an insect and then it is sucked dry.

a.

b.

c.

Figure 36.2

Composition of the teeth of several mammals. ***a.*** Horses are herbivores that graze on grasses. Note the sharp incisors, reduced canines, and large, flat premolars and molars. ***b.*** Lions are carnivores that prey on other animals. Note the pointed incisors, enlarged canines, and jagged premolars and molars. ***c.*** Humans are omnivores that have nonspecialized teeth.

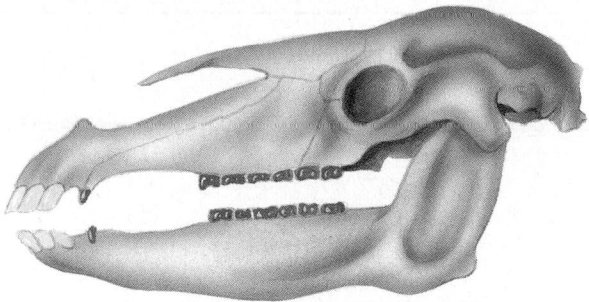

a. Horse

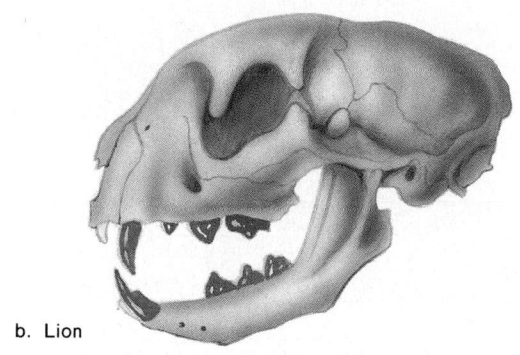

b. Lion

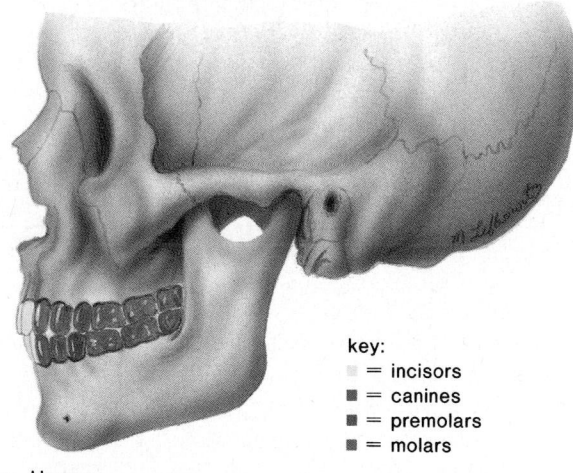

key:
- ☐ = incisors
- ■ = canines
- ■ = premolars
- ■ = molars

c. Human

to get at and digest cellulose. Other mammalian grazers, like the cow and deer, are ruminants. In contrast to horses, they graze quickly and swallow partially chewed grasses into a special part of the stomach called a rumen. Here, microorganisms start the digestive process and the result, called cud, is regurgitated at a later and more convenient time. The cud is chewed before being swallowed for complete digestion.

Many mammals, including dogs, toothed whales, and polar bears, are carnivores. A carnivore, like a lion, uses pointed incisors and enlarged canine teeth to tear off pieces small enough to be quickly swallowed. Meat is rich in protein and fat and easier to digest than plant material. Therefore, the digestive system of carnivores is shorter and doesn't have the specialization seen in herbivores.

The dentition of a mammal reflects its type of diet. Omnivores have nonspecialized teeth. Herbivores have large, flat molars that grind food; carnivores have incisors and canines to tear off chunks of food.

Continuous Feeders versus Discontinuous Feeders

The clam is a continuous feeder, which is often called a *filter feeder* (fig. 36.3*a*). Water is always moving into the mantle cavity by way of the incurrent siphon (slitlike opening) and depositing particles on the gills. The size of the incurrent siphon permits the entrance of only small particles, which adhere to the gills. Ciliary action moves suitably sized particles to the labial palps, which force them through the mouth into the stomach. Digestive enzymes are secreted by a large digestive gland, but amoeboid cells present throughout the tract are believed to complete the digestive process by intracellular digestion.

Marine fanworms (p. 435) are sessile filter feeders that live in a tube and extend feathery tentacles to capture food. The baleen whale is an active filter feeder. The baleen, a curtainlike fringe, hangs from the roof of the mouth and filters small shrimp called krill from the water. The baleen whale filters up to a ton of krill in a single feeding.

The squid is an example of a discontinuous feeder (fig. 36.3*b*). The body of a squid is streamlined, and the animal moves rapidly through the water using jet propulsion (forceful expulsion of water from a tubular funnel). The head of a squid is surrounded by 10 arms, 2 of which have developed into long, slender tentacles whose suckers have toothed, horny rings. These tentacles seize prey (fishes, shrimps, and worms) and bring it to the squid's bearlike jaws, which bite off pieces pulled into the mouth by the action of a radula, a toothy tongue. An esophagus leads to a stomach and cecum, where digestion occurs. The stomach, supplemented by the cecum, holds food until digestion is complete. Discontinuous feeders, whether they are carnivores or herbivores, require such a storage area.

Continuous feeders, such as sessile filter feeders, do not need a storage area for food; discontinuous feeders, such as herbivores and carnivores, do need a storage area for food.

Digestive Tracts

Generally, animals have some sort of digestive tract or gut. The gut can be simple or complex. A simple gut has few, if any, specialized parts, and a complex gut does have specialized parts.

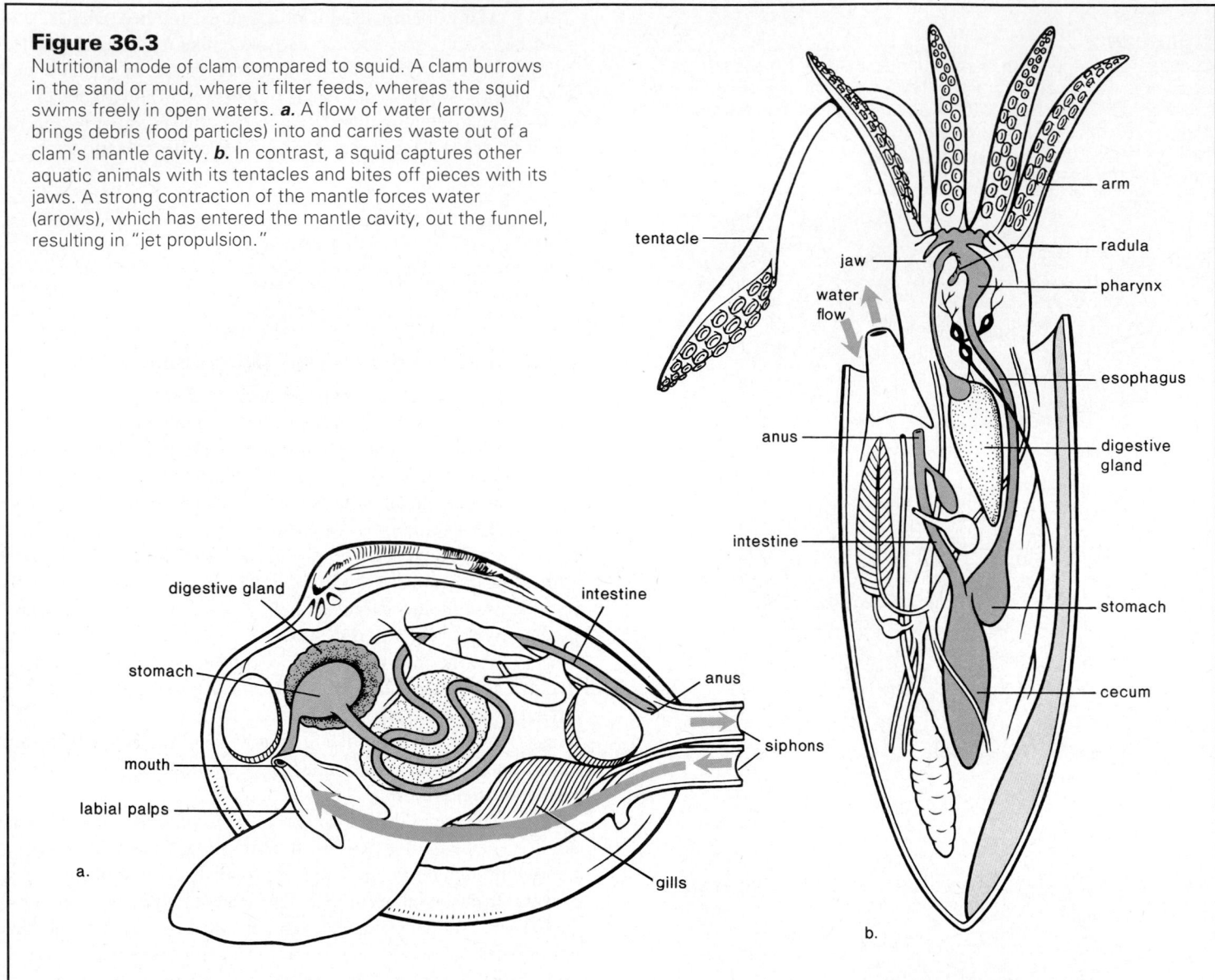

Figure 36.3
Nutritional mode of clam compared to squid. A clam burrows in the sand or mud, where it filter feeds, whereas the squid swims freely in open waters. **a.** A flow of water (arrows) brings debris (food particles) into and carries waste out of a clam's mantle cavity. **b.** In contrast, a squid captures other aquatic animals with its tentacles and bites off pieces with its jaws. A strong contraction of the mantle forces water (arrows), which has entered the mantle cavity, out the funnel, resulting in "jet propulsion."

Incomplete Gut versus Complete Gut

An *incomplete gut* has a single opening, usually called a mouth. The planarian is an example of an animal with an incomplete gut (fig. 36.4). It is carnivorous and feeds largely on smaller aquatic animals. Its digestive system contains only a mouth, a pharynx, and an intestine. When the worm is feeding, the pharynx actually extends beyond the mouth. It wraps its body about the prey and uses its muscular pharynx to suck up minute quantities at a time. Digestive enzymes present in the *gastrovascular cavity* allow some extracellular (outside the cells) digestion to occur. Digestion is finished intracellularly by the cells that line the cavity, which branches throughout the body. No cell in the body is far from the intestine, and therefore diffusion alone is sufficient to distribute nutrient molecules.

The digestive system of a planarian is notable for its lack of specialized parts. It is saclike because the pharynx serves not only as an entrance for food but also as an exit for nondigestible material. Specialization of parts does not occur under these circumstances.

Recall that the planarian has some parasitic relatives. The tapeworm (see fig. 26.8), for example, has no digestive system at all—it simply absorbs nutrient molecules from the intestinal juices that surround its body. The body wall is highly modified for this purpose: it has millions of microscopic fingerlike projections that increase the surface area for absorption.

In contrast to the planarian, an earthworm has a complete gut. A *complete gut* has a mouth and anus. Earthworms feed mainly on decayed organic matter in dirt (fig. 36.5). The muscular *pharynx* draws in food with a sucking action. The *crop* is a storage area that has expansive thin walls. The *gizzard* has thick muscular walls for churning and grinding the food. Digestion is extracellular—in the intestine. The surface area of digestive tracts is often increased for absorption of nutrient molecules, and in the earthworm there is an intestinal fold called the typhlosole. Undigested remains pass out of the body at the anus.

Specialization of parts is obvious in the earthworm because the pharynx, crop, gizzard, and intestine each has a particular function in the digestive process. A complete gut leads to a digestive system with specialization of parts.

In contrast to the incomplete, saclike gut, the complete gut with both a mouth and an anus has many specialized parts.

Human Digestive Tract

The human gut is complete, and there is a complex digestive system (fig. 36.6). Each part of the system has a specific function (table 36.1).

Mouth

Food is chewed in the mouth, and it is also mixed with saliva. We have already mentioned that human dentition is nonspecialized because humans are omnivores (fig. 36.2*c*). There are 3 pairs of *salivary glands* that send their juices by way of ducts to the mouth. Saliva contains the enzyme *salivary amylase,* which begins the process of starch digestion. The disaccharide maltose is a typical end product of salivary amylase digestion:

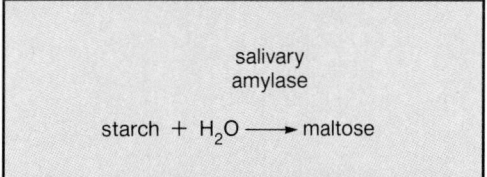

salivary
amylase

starch + H₂O ⟶ maltose

While in the mouth, food is manipulated by a muscular tongue, which has touch and pressure receptors similar to those in the skin. *Taste buds,* chemical receptors that are stimulated by the chemical composition of food, are also found primarily on the tongue and also on the surface of the mouth and the

Figure 36.4

Planarians are rather simple aquatic organisms. *a.* They have a very extensive nonspecialized digestive tract, the gastrovascular cavity. Incomplete digestive tracts have little specialization of parts because the single opening serves as both entrance and exit. *b.* The planarian relies on intracellular digestion to complete the digestive process. Phagocytosis produces a vacuole, which joins with an enzyme-containing lysosome. Notice that the food particles are always surrounded by membrane; therefore, they are never part of the cytoplasm. The digested products pass from the vacuole before any nondigestible material is eliminated at the plasma membrane.

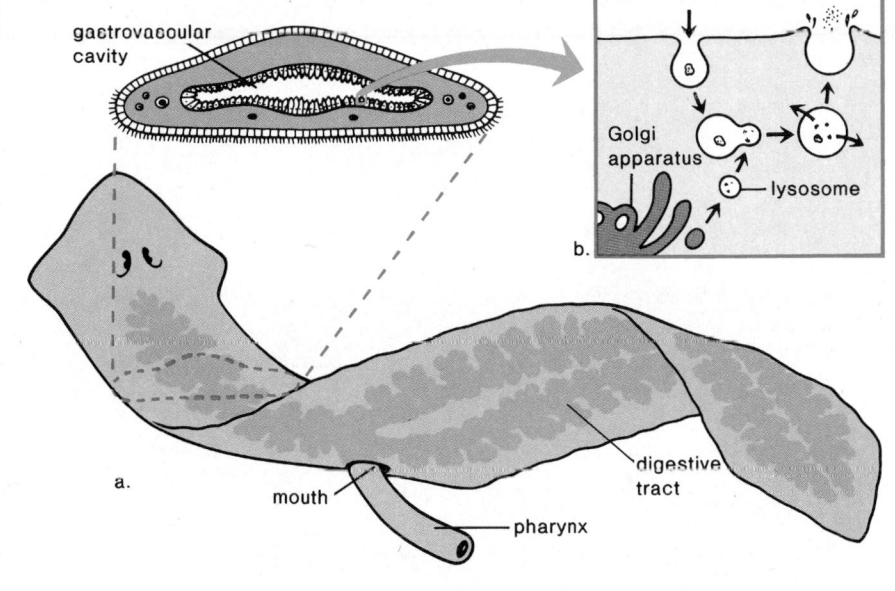

Figure 36.5

The earthworm has a complete digestive tract with both a mouth and an anus and many other specialized parts. The absorptive surface of the intestine is increased by an internal fold called the typhlosole.

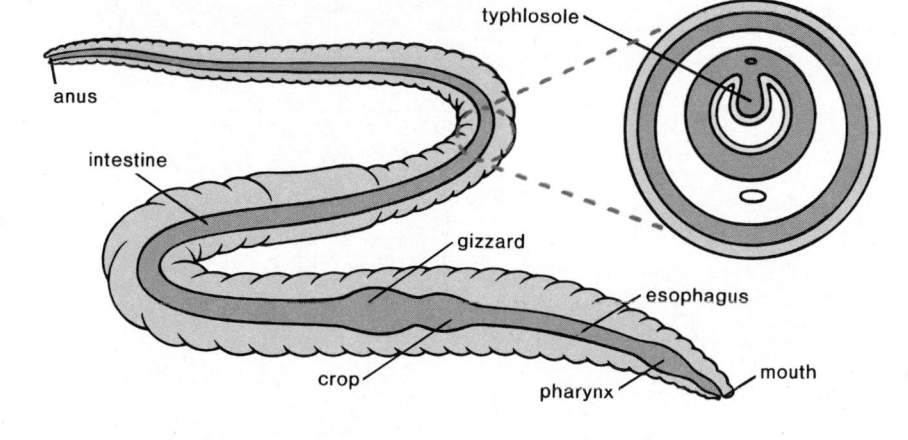

Figure 36.6

Human digestive system. The enlargement shows the layers that make up the wall of the small intestine. The *mucous membrane layer* is made up of columnar epithelium overlying loose connective tissue. The epithelium has deep folds called villi (fig. 36.10), which increase the absorptive surface of the small intestine. The *submucosal layer* is a dense connective tissue layer containing nerve fibers, blood vessels, and lymphatic vessels. The products of digestion enter these vessels. The *smooth muscle layer* has both circular and longitudinal muscular tissue. Rhythmic contraction of these muscles brings about peristalsis. In the *serous membrane layer* a thin sheet of connective tissue underlies a thin outermost sheet of squamous epithelium. This membrane is part of the peritoneum, which lines the entire abdominal cavity.

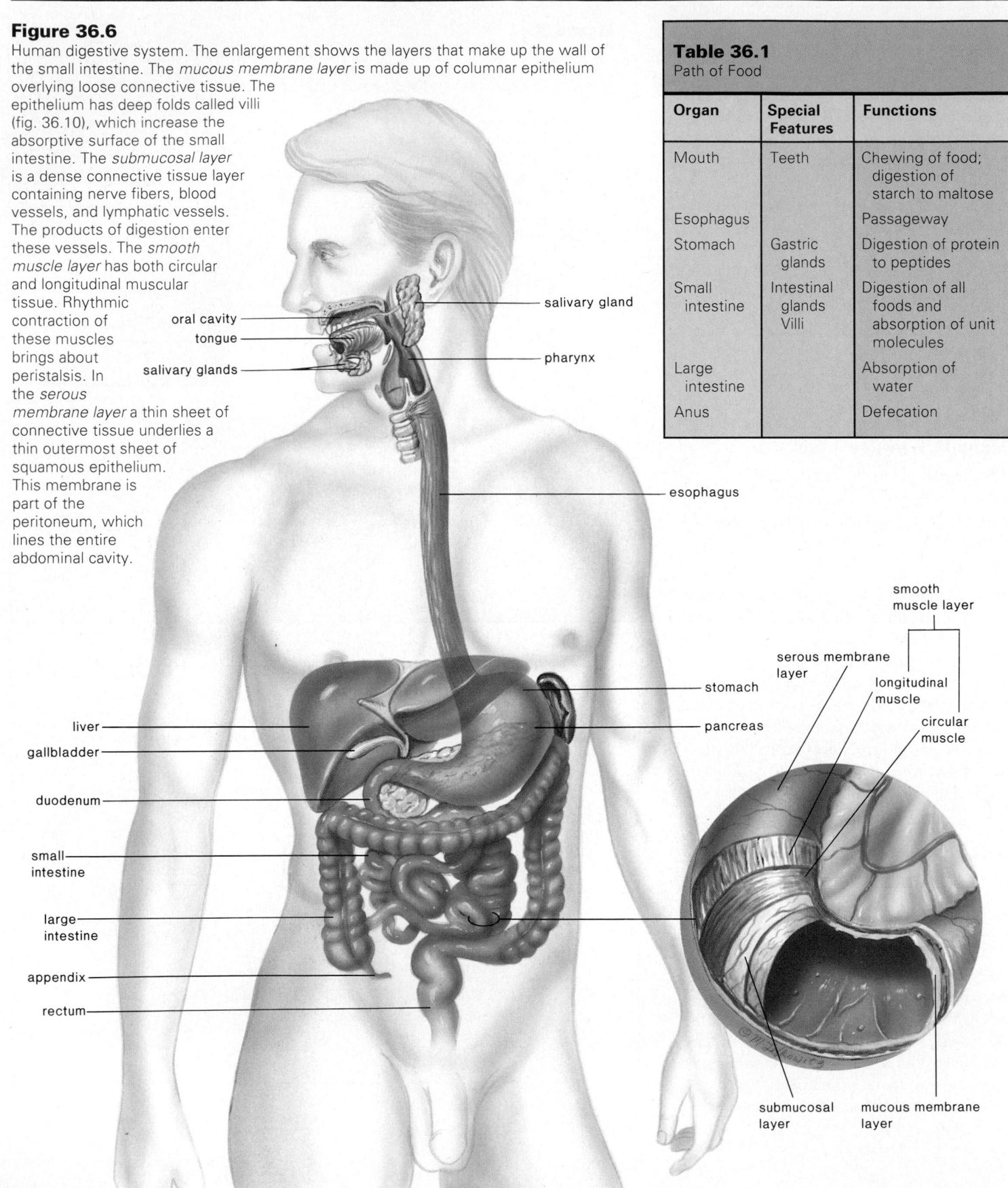

Table 36.1
Path of Food

Organ	Special Features	Functions
Mouth	Teeth	Chewing of food; digestion of starch to maltose
Esophagus		Passageway
Stomach	Gastric glands	Digestion of protein to peptides
Small intestine	Intestinal glands Villi	Digestion of all foods and absorption of unit molecules
Large intestine		Absorption of water
Anus		Defecation

salivary gland

oral cavity

tongue

pharynx

salivary glands

esophagus

smooth muscle layer

serous membrane layer

longitudinal muscle

circular muscle

stomach

pancreas

liver

gallbladder

duodenum

small intestine

large intestine

appendix

rectum

submucosal layer

mucous membrane layer

Animal Structure and Function

Figure 36.7

Swallowing. Respiratory and digestive passages diverge in the pharynx. During swallowing, the epiglottis covers the opening to the trachea and prevents food from entering this airway.

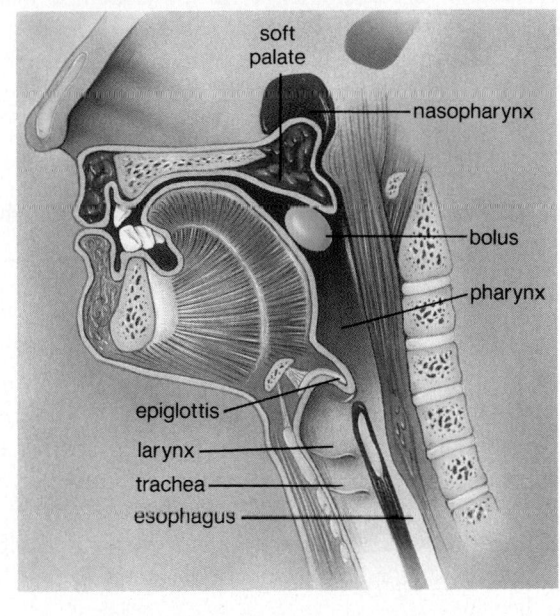

Figure 36.8

Peristalsis in the digestive tract. Rhythmic waves of muscle contraction move material along the digestive tract. The 3 drawings show how a peristaltic wave moves through a single section of gut over time.

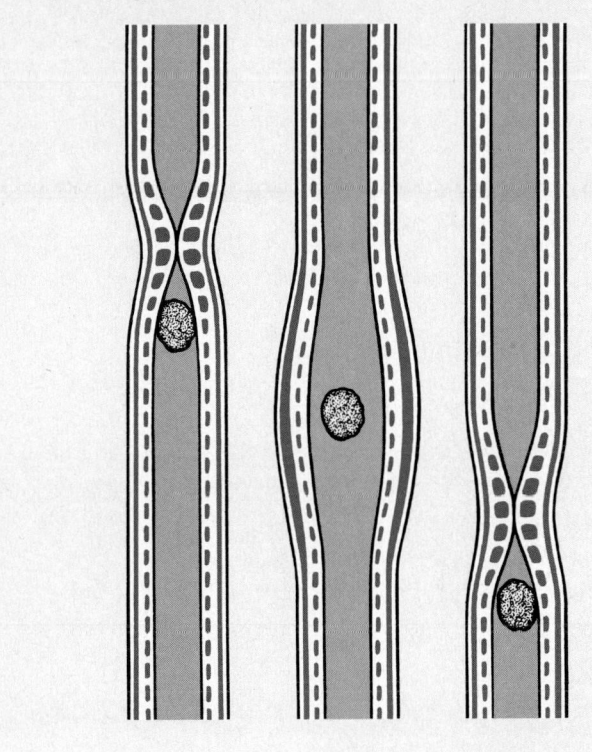

wall of the pharynx. After the food has been thoroughly chewed and mixed with saliva, the tongue starts the process of swallowing by pushing the food (bolus) back to the pharynx.

Pharynx and Esophagus

The digestive and respiratory passages cross in the pharynx (fig. 36.7). Therefore, during swallowing, the path of air to the lungs could be blocked if food enters the trachea (windpipe). Normally, however, a flap of tissue called the **epiglottis** covers the opening into the **trachea** as muscles move the bolus through the pharynx into the esophagus, a tubular structure that takes food to the stomach. When food enters the **esophagus,** the rhythmic contraction of the gut wall, **peristalsis,** begins (fig. 36.8). Peristalsis pushes the bolus down the esophagus to the stomach.

Stomach

Usually, the stomach stores up to 2 liters of partially digested food. Therefore, humans can periodically eat relatively large meals and spend the rest of their time at other activities. But the stomach is much more than a mere storage organ, as was discovered by William Beaumont in the mid-nineteenth century. Beaumont, an American doctor, had a French Canadian patient, Alexis St. Martin. St. Martin had been shot in the stomach, and when the wound healed, he was left with a fistula, or opening, that allowed Beaumont to collect gastric (stomach) juices produced by *gastric glands* and to look inside the stomach to see what was going on there. Beaumont was able to determine that the muscular walls of the stomach contract vigorously and mix food with juices that are secreted whenever food enters the stomach (fig. 36.9). He found that *gastric juice* contains *hydrochloric acid* (HCl) and a substance

active in digestion. (This substance was later identified as pepsin.) He also found that the gastric juices are produced independently of the protective mucous secretions of the stomach. Beaumont's work, which was very carefully and painstakingly done, pioneered the study of the physiology of digestion.

So much hydrochloric acid is secreted by the stomach that it routinely has a pH of about 2.0. Such a high acidity usually is sufficient to kill bacteria and other microorganisms that might be in food. This low pH also stops the activity of salivary amylase, which functions optimally at the near-neutral pH of saliva, but it promotes the activity of pepsin. *Pepsin* is a hydrolytic enzyme that acts on protein to produce peptides:

$$\text{protein} + H_2O \xrightarrow{\text{pepsin}} \text{peptides}$$

By now the stomach contents have a thick, soupy consistency and are called *chyme.* At the base of the stomach is a narrow opening controlled by a *sphincter.* Whenever the sphincter relaxes, a small quantity of chyme squirts through the opening into the **duodenum,** the first part of the *small intestine* (fig. 36.6). When chyme enters the duodenum, it sets off a neural reflex that causes

Figure 36.9

This scanning electron micrograph provides a greatly enlarged view of what Beaumont saw when he looked through the opening in St. Martin's stomach. The wall of the stomach has folds called rugae (Ru), which disappear when the stomach is full. The arrows indicate the openings to the gastric glands, which produce gastric secretions, including hydrochloric acid and pepsinogen, a precursor that becomes pepsin in the stomach. The wall of the stomach has the same 4 layers as the small intestine, but with modifications. Notice how large the mucous membrane layer (Mu = mucous membrane layer) is because of the presence of mucosal muscle (MM). Su = submucosal layer; ML = smooth muscle layer; Se = serous membrane layer. Magnification, X55.

Kessel, R. G., and Kardon, R. H.: *Tissues and Organs: A Text-Atlas of Scanning Electron Microscopy.* © 1979 by W. H. Freeman and Co.

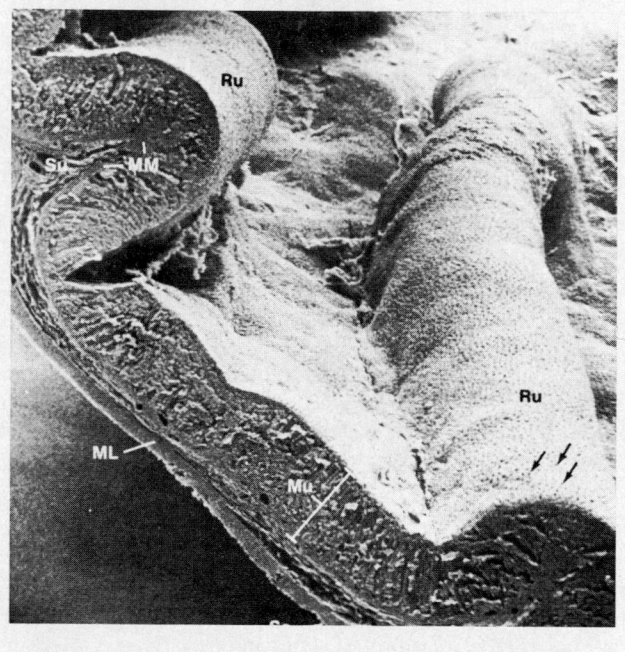

the muscles of the sphincter to contract vigorously and to close the opening temporarily. Then the sphincter relaxes again and allows more chyme to squirt through. The slow manner in which chyme enters the small intestine allows digestion to be more thorough than it otherwise would be.

Normally, a thick layer of mucus protects the wall of the stomach and first part of the duodenum. But if by chance gastric juice does penetrate the mucus, pepsin starts to digest the stomach or duodenal lining and an *ulcer* results. An ulcer is an open sore in the wall caused by the gradual disintegration of tissues. It is believed that the most frequent cause of an ulcer is oversecretion of gastric juice due to too much nervous stimulation. Evidence is growing, however, that suggests a bacterial infection may be involved.

Before food enters the small intestine, it passes through the mouth, esophagus, and stomach. In the mouth, salivary amylase digests starch to maltose; in the stomach, pepsin digests protein to peptides.

Small Intestine

The human *small intestine* is a coiled tube about 3 m long in a living person. The mucous membrane layer has ridges and furrows that give it an almost corrugated appearance. On the surface of these ridges and furrows are small, fingerlike projections called **villi.** Cells on the surfaces of the villi have minute projections called microvilli. The villi and microvilli greatly increase the effective surface area of the small intestine. If the small intestine were simply a smooth tube, it would have to be 500 m–600 m long to have a comparable surface area.

When the chyme enters the duodenum, proteins and carbohydrates are only partly digested, and fat digestion still needs to be carried out. Considerably more digestive activity is required before these nutrients can be absorbed through the intestinal wall. Two important accessory glands, the liver and the pancreas, send secretions to the duodenum (fig. 36.6). The liver produces **bile,** which is stored in the **gallbladder** and sent to the duodenum by way of a duct. Bile looks green because it contains pigments that are products of hemoglobin breakdown. This green color is familiar to anyone who has observed the color changes of a bruise. Hemoglobin within the bruise is breaking down into the same types of pigments found in bile. Bile also contains bile salts, which are *emulsifying* agents that break up fat into fat droplets so that they mix with the water:

$$\text{fat} \xrightarrow{\text{bile salts}} \text{fat droplets}$$

Emulsified fat is more easily acted on by enzymes.

The pancreas sends pancreatic juice into the duodenum, also by way of a duct. While the pancreas is an *endocrine* gland when it produces and secretes insulin into the bloodstream, it is an *exocrine* gland when it produces and secretes pancreatic juice into the duodenum. Pancreatic juice contains sodium bicarbonate ($NaHCO_3$), which neutralizes the chyme and makes the pH of the small intestine slightly basic. Pancreatic juice also contains digestive enzymes that act on every major component of food. *Pancreatic amylase* digests starch to maltose; *trypsin* and other enzymes digest protein to peptides; and *lipase* digests fat droplets to glycerol and fatty acids:

$$\text{starch} + H_2O \xrightarrow{\substack{\text{pancreatic} \\ \text{amylase}}} \text{maltose}$$

$$\text{protein} + H_2O \xrightarrow{\text{trypsin}} \text{peptides}$$

$$\text{fat droplets} + H_2O \xrightarrow{\text{lipase}} \text{glycerol} + \text{fatty acids}$$

Table 36.2
Digestive Enzymes

Reaction	Enzyme	Produced by	Site of Occurrence
Starch + H_2O → maltose	a. Salivary amylase b. Pancreatic amylase	a. Salivary glands b. Pancreas	a. Mouth b. Small intestine
Maltose + H_2O → glucose*	Maltase	Intestinal cells	Small intestine
Protein + H_2O → peptides	a. Pepsin b. Trypsin	a. Gastric glands b. Pancreas	a. Stomach b. Small intestine
Peptides + H_2O → amino acids*	Peptidases	Intestinal cells	Small intestine
Fat + H_2O → glycerol + fatty acids*	Lipase	Pancreas	Small intestine

*Absorbed by villi.

Figure 36.10
Anatomy of intestinal lining. *a.* The products of digestion are absorbed by villi, fingerlike projections of the intestinal wall. *b.* Each villus contains blood vessels and a lacteal.

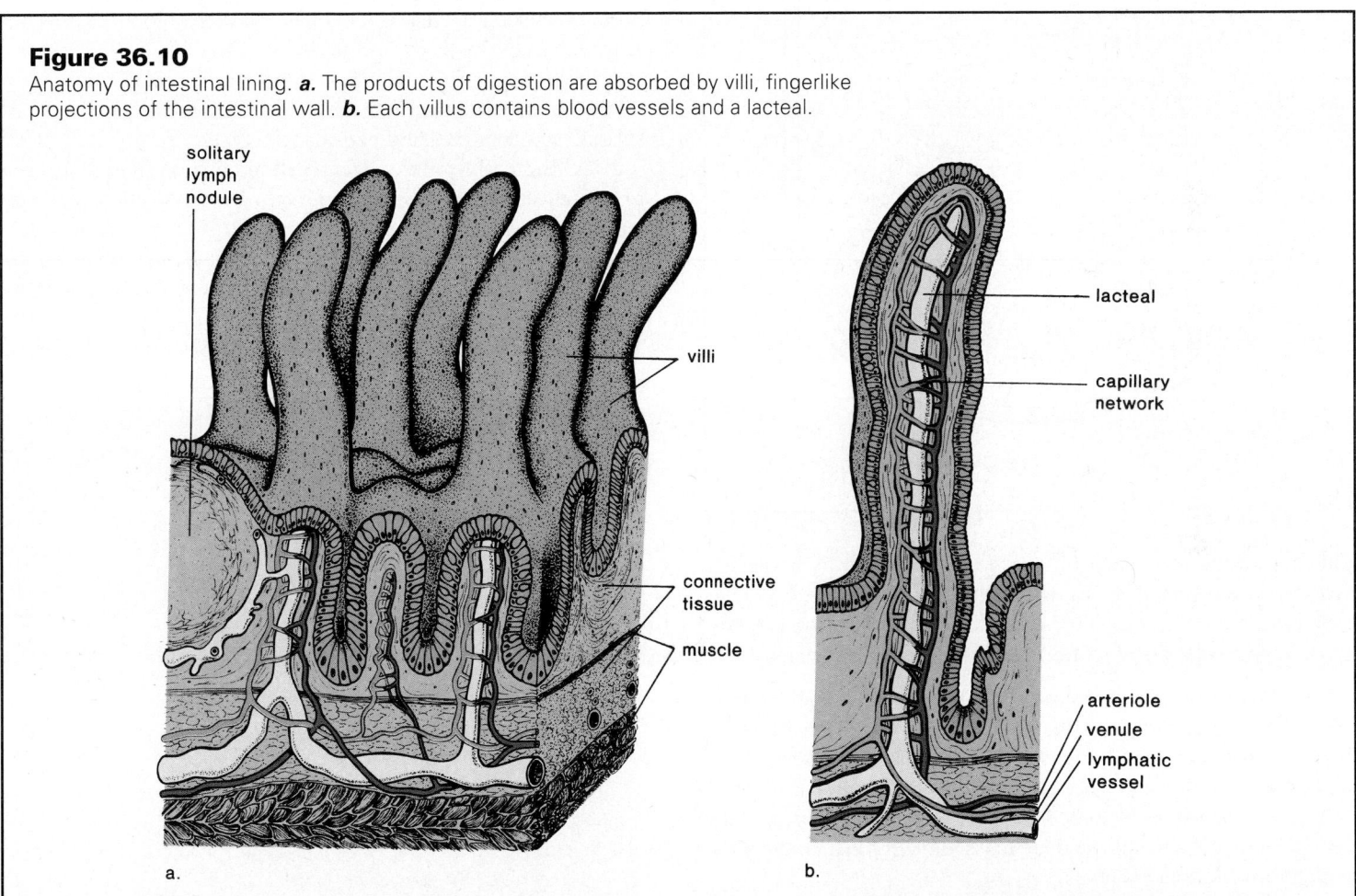

The epithelial cells of the villi produce *intestinal enzymes*, which remain attached to the plasma membrane of microvilli. These enzymes complete the digestion of peptides and sugars. Peptides, which result from the first step in protein digestion, are digested by peptidases to amino acids. Maltose, which results from the first step in starch digestion, is digested by maltase to glucose:

peptidases
peptides + H_2O ⟶ amino acids

maltase
maltose + H_2O ⟶ glucose

Figure 36.11

Hormonal control of digestive gland secretions. Especially after eating a protein-rich meal, gastrin produced by the lower part of the stomach enters the bloodstream and thereafter stimulates the upper part of the stomach to produce more digestive juices. Acid chyme from the stomach causes the duodenum to secrete secretin and CCK-PZ. Secretin and CCK-PZ stimulate the pancreas to release pancreatic juice, and CCK-PZ alone stimulates the gallbladder to release bile.

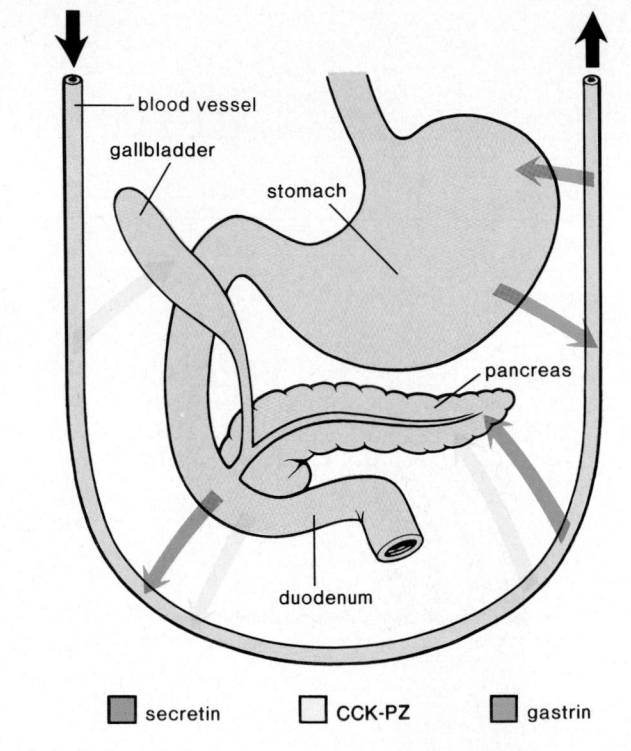

The wall of the small intestine is lined by villi. The epithelial cells of the villi produce intestinal enzymes that finish the digestive process. Sugars and amino acids enter blood vessels, and reformed fats enter the lacteals of the villi.

Control of Secretions Under normal circumstances, the digestive secretions are released *only* when food is present in the digestive tract. Saliva in the mouth and gastric juices in the stomach flow primarily in response to the taste, smell, or sometimes even the thought of food. Also, food in the stomach stimulates sense receptors that signal the brain to stimulate, by way of nerves, the gastric glands in the stomach.

Perhaps the major factor in gastric secretory regulation, however, is the hormone *gastrin,* which is produced by the stomach itself. When proteins contact the stomach mucosa, gastrin-producing cells are stimulated to release gastrin into the bloodstream (fig. 36.11). As soon as gastrin circulates through blood vessels and reaches the acid- and enzyme-secreting cells of the stomach lining, those cells respond by secreting large quantities of HCl and pepsin. As the stomach empties, both the neural reflexes and gastrin release subside, and less HCl and pepsin are secreted.

Similarly, some duodenal cells produce the hormone *secretin,* which stimulates the pancreas to release pancreatic juice, especially the sodium bicarbonate component. Other duodenal cells release a hormone called *CCK-PZ* (cholecystokinin-pancreozymin), which stimulates the release of bile. CCK-PZ stimulates the gallbladder to empty its contents through a duct into the duodenum. This same hormone also stimulates the pancreas, especially pancreatic enzyme secretion, as its full name suggests (fig. 36.11)

Liver

Blood vessels from the large intestine and the small intestine merge to form the hepatic portal vein, which leads to the liver:

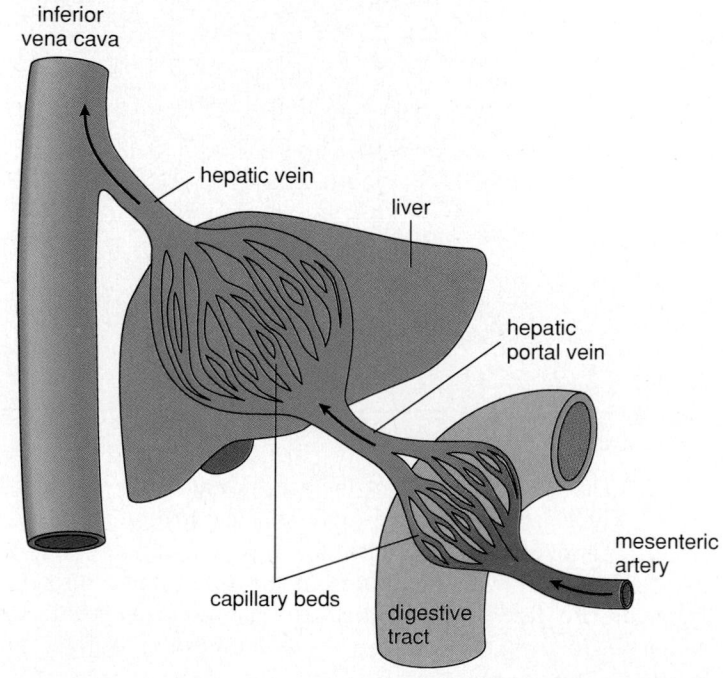

Other disaccharides, each of which is acted upon by a specific enzyme, are digested in the small intestine. Table 36.2 reviews reactions required for the digestion of food and the various digestive enzymes discussed in this chapter.

Absorption The small intestine is specialized for absorption. Molecules are absorbed by the huge number of villi that line the intestinal wall (fig. 36.10). Each villus contains blood vessels and a lymphatic vessel called a **lacteal.** Sugars and amino acids enter villi cells and then are absorbed into the bloodstream. Glycerol and fatty acids enter villi cells and are reassembled into fat molecules, which move into the lacteals.

Absorption continues until almost all products of digestion have been absorbed. Absorption involves active transport and requires an expenditure of cellular energy.

Food is largely made up of carbohydrate (starch), protein, and fat. These very large macromolecules are broken down by digestive enzymes to small molecules that can be absorbed by intestinal villi. This table indicates the steps needed for carbohydrate digestion (starch and maltose), protein digestion (protein and peptides), and fat digestion (fat) and shows that they are all hydrolytic reactions.

Animal Structure and Function

The liver has numerous functions, including the following:

1. Detoxifies the blood by removing and metabolizing poisonous substances.
2. Makes the blood proteins.
3. Destroys old red blood cells and converts hemoglobin to the breakdown products in bile (bilirubin and biliverdin).
4. Produces bile, which is stored in the gallbladder before entering the small intestine, where it emulsifies fats.
5. Stores glucose as glycogen and breaks down glycogen to glucose between meals to maintain a constant glucose concentration in the blood.
6. Produces urea from amino groups and ammonia.

Here we are interested in the last 2 functions listed. The liver helps maintain the glucose concentration in blood at about 0.1% by removing excess glucose from the hepatic portal vein and storing it as glycogen. Between meals, glycogen is broken down and glucose enters the hepatic vein. Glycogen is sometimes called *animal starch* because both starch and glycogen are made up of glucose molecules joined together (see fig. 4.7). If by chance the supply of glycogen and glucose runs short, the liver converts amino acids to glucose molecules. Amino acids contain nitrogen in the form of amino groups, whereas glucose contains only carbon, oxygen, and hydrogen. Therefore, before amino acids can be converted to glucose molecules, *deamination,* or the removal of amino groups from the amino acids, must take place. By an involved metabolic pathway, the liver converts these amino groups to urea, the most common nitrogenous waste product of humans, and after its formation in the liver, it is transported by the bloodstream to the kidneys, where it is excreted.

Blood from the small intestine enters the hepatic portal vein, which goes to the liver, a vital organ that has numerous important functions as listed.

Liver Disorders Jaundice, hepatitis, and cirrhosis are 3 serious diseases that affect the entire liver and hinder its ability to repair itself. When a person is jaundiced, there is a yellowish tint to the skin due to an abnormally large amount of bilirubin in the blood. In *hemolytic jaundice,* red blood cells are broken down in abnormally large amounts; in *obstructive jaundice,* there is an obstruction of the bile duct or damage to the liver cells. Obstructive jaundice often occurs when crystals of cholesterol precipitate out of bile and form gallstones.

Jaundice can also result from *viral hepatitis,* a term that includes hepatitis A, hepatitis B, and hepatitis C. Hepatitis A is most often caused by eating shellfish from polluted waters. Hepatitis B and C are commonly spread by blood transfusions, kidney dialysis, and injection with inadequately sterilized needles. All 3 types of hepatitis can be spread by sexual contact.

Cirrhosis is a chronic disease of the liver in which the organ first becomes fatty. Liver tissue is then replaced by inactive fibrous scar tissue. In alcoholics, who often get cirrhosis of the liver, the condition most likely is caused by the excessive amounts of alcohol the liver is forced to break down.

Large Intestine

About 1.5 liters of water enter the digestive tract daily as a result of eating and drinking. An additional 8.5 liters enter the digestive tract each day carrying the various substances secreted by the digestive glands. About 95% of this water is reabsorbed into cells of the *large intestine,* or *colon.* This water reabsorption is essential. Failure to reabsorb water can result in *diarrhea,* which can lead to serious dehydration and ion loss, especially in children.

The large intestine functions both in water conservation and ion regulation. The colon absorbs some sodium and other ions from the material passing through it. At the same time, colon cells excrete metallic ions into the wastes leaving the body. The role that the large intestine plays in excretion is discussed on page 620. *Vitamin K,* produced by intestinal bacteria, is also absorbed in the colon.

The last 20 cm of the large intestine is the rectum, which terminates in the anus, an external opening. Digestive wastes (feces) eventually leave the body through the rectum and anus. Feces are about 75% water and 25% solid matter. Almost one-third of this solid matter is made up of intestinal bacteria. The remainder is undigested plant material, fats, waste products (such as bile pigments), inorganic material, mucus, and dead cells from the intestinal lining.

Appendicitis and Polyps Two serious medical conditions are associated with the large intestine. The small intestine joins the large intestine in such a way that there is a blind end on one side (fig. 36.6). This blind sac, or cecum, has a small projection about the size of the little finger, called the appendix. In the case of *appendicitis,* the appendix becomes infected and so filled with fluid that it may burst. The appendix should be removed before it bursts, because a burst appendix can lead to a serious, generalized infection of the abdominal lining, called peritonitis.

The colon is subject to the development of polyps, which are small growths arising from the epithelial lining. Polyps, whether they are benign or cancerous, can be removed individually. The cause of colon cancer is not known, but a low-fat, high-fiber diet, which promotes regularity, is recommended as a protective measure.

Nutrition

To be sure that the essential nutrients are included in the diet, it is necessary to eat a balanced diet. A balanced diet is obtained by eating a variety of foods from the 4 food groups pictured in figure 36.12. Some authorities maintain there are only 3 food groups because milk products and meats provide the same type of nutrients.

Food largely consists of proteins, carbohydrates, and fats. Therefore, we will begin by considering these substances. Vitamins and minerals are also important nutrients and will be discussed later in this chapter.

Proteins

Foods rich in protein include red meat, fish, poultry, dairy products, legumes, nuts, and cereals. Following digestion of protein, amino acids enter the bloodstream and are transported to the tissues. Most of these amino acids are incorporated into structural proteins found in muscles, skin, hair, and nails. Others are used to synthesize such proteins as hemoglobin, plasma proteins, enzymes, and hormones.

Protein formation requires 20 different types of amino acids. Of these, 9 are required from the diet because the body is unable to produce them. These are termed the **essential amino acids.** The body produces the other 11 amino acids by simply transforming one type into another type. Some protein sources, such as meat, are *complete;* they provide all types of amino acids. Vegetables and grains supply us with amino acids, but each vegetable or grain alone is an *incomplete* protein source because at least one of the essential amino acids is absent. It is possible to combine foods, however, to acquire all of the essential amino acids. For example, the combinations of cereal with milk or beans with rice provide all the essential amino acids.

A complete source of protein is absolutely necessary to ensure a sufficient supply of the essential amino acids.

Even though in this country we emphasize protein intake, it does not take very much protein to meet the daily requirement. In the United States, the *required dietary (daily) allowances (RDAs)* are determined by the National Research Council, a part of the National Academy of Sciences. The RDA for the reference woman (120 lb) is 44 g of protein a day. For the reference man (154 lb), it is 56 g of protein a day. A single serving of roast beef (3 oz) provides 25 g of protein, and a cup of milk provides 8 g.

Unfortunately, in this country, the manner in which we meet our daily requirement for protein is not always the most healthful way. The foods that are richest in protein are apt to be richest in fat.

Foods richest in protein also tend to be richest in fat.

While it is very important to meet the RDA for protein, consuming more actually can be detrimental. Calcium loss in the urine has been noted when dietary protein intake is over twice the RDA. Everything considered, it is probably a good idea to depend on protein from plant origins (e.g., whole-grain cereals, dark breads, legumes) to a greater extent than is the custom in this country.

Carbohydrates

The quickest, most readily available source of energy for the body is carbohydrates, which can be complex, as in breads and cereals, or simple, as in candy, ice cream, and soft drinks. As

Figure 36.12
The 4 food groups. A variety of foods from each group should be eaten daily.

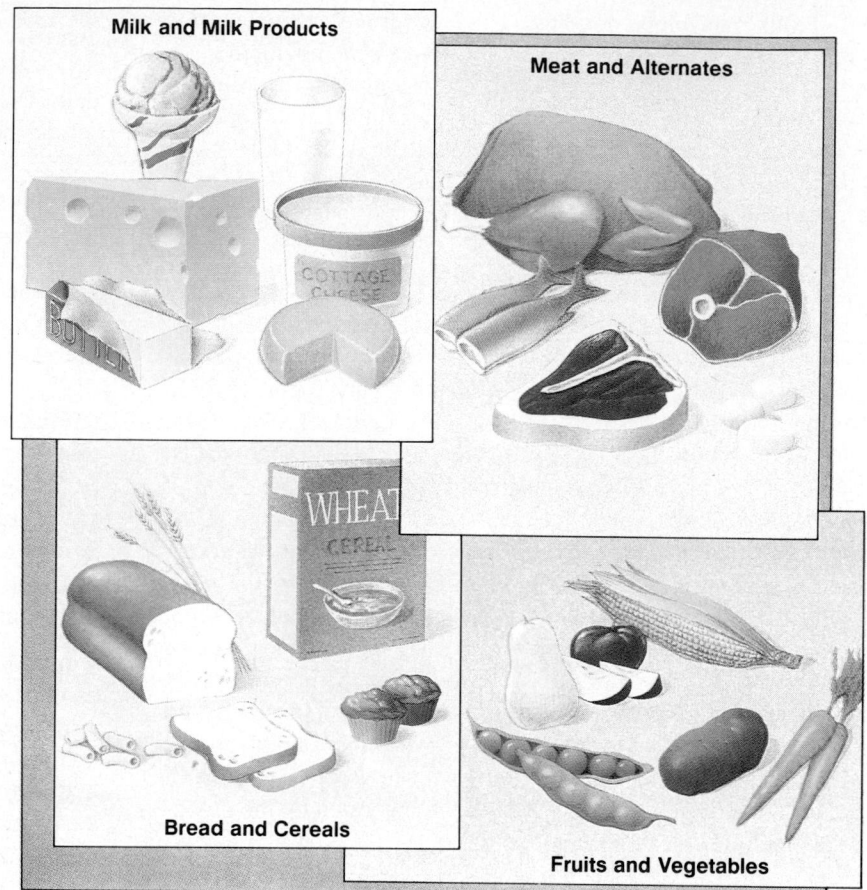

mentioned previously, starches are digested to glucose, which is stored by the liver in the form of glycogen. Between eating, blood glucose is maintained at about 0.1% by the breakdown of glycogen or by the conversion of amino acids to glucose. If necessary, these amino acids are taken from the muscles, even from heart muscle. To avoid this situation, it is suggested that the daily diet contain at least 100 g of carbohydrate. As a point of reference, a slice of bread contains approximately 14 g of carbohydrate.

Carbohydrates are needed in the diet to maintain the blood glucose level.

Actually, the dietary guidelines produced jointly by the U.S. Department of Agriculture and the Department of Health and Human Services recommend that we increase the proportion of carbohydrates per total energy content of the diet from 46% to 58%. Further, it is assumed that these carbohydrates are complex and not simple. Simple carbohydrates (e.g., sugars) are labeled "empty calories" by some dieticians because they contribute to energy needs and weight gain and are not a part of food that supply other nutritional requirements. Table 36.3 gives suggestions on how to cut down on your consumption of dietary sugars (simple carbohydrates).

In contrast to simple sugars, complex carbohydrates are likely to be accompanied by a wide range of other nutrients and by fiber, which is nondigestible plant material. Insoluble fiber, such as that found in wheat bran, has a laxative effect and therefore may reduce the risk of colon cancer. Soluble fiber, such as that found in oat bran, may possibly reduce cholesterol in the blood because it combines with cholesterol in the gut and prevents it from being absorbed.

While the diet should have an adequate amount of fiber, a high-fiber diet can be detrimental. Some evidence suggests that the absorption of iron, zinc, and calcium is impaired by a high-fiber diet.

Complex carbohydrates along with fiber are considered beneficial to health.

Carbohydrates provide most of the dietary Kcalories.[1] There are only 4 Kcalories per gram of carbohydrate as compared to 9 Kcalories per gram of fat, but since carbohydrates are the bulk of the diet, they provide the most calories.

Lipids

Our discussion of lipids is divided into 2 parts: fats and cholesterol.

Fats

Fats are present not only in butter, margarine, and oils but also in foods high in protein. After being absorbed, the products of fat digestion are transported by the lymph and the blood to the tissues. The liver can alter ingested fats to suit the body's needs, except it is unable to produce the fatty acid linoleic acid. Since this is required for phospholipid production, linoleic acid is considered an *essential fatty acid.*

Fats have the highest Kcaloric content, but they should not be avoided entirely because they contain the essential fatty acid linoleic acid.

While we need to ingest some fat to satisfy our need for linoleic acid, recent dietary guidelines suggest that we should reduce the amount of fat per total energy content of the diet from 40% to 30%. Dietary fat has been implicated in cancer of the colon, the pancreas, the prostate, and the breast (fig. 36.13). Many animal studies have shown that a high-fat diet stimulates the development of mammary tumors, while a low-fat diet does not. It also has been found that women who have a high-fat diet are more likely to develop breast cancer. Surprisingly, it has been discovered that a reduction in the amount of linoleic acid in the diet helps to prevent breast cancer. Linoleic acid is found in corn, safflower, sunflower, and other common plant oils but is not abundant in olive oil or in fatty fishes and marine animals.

There is very strong evidence that women who have a diet high in fat are more apt to develop breast cancer.

---

[1]The amount of heat required to raise 1 g of water 1°C

Table 36.3
Recommendations for Improved Diet

The Less-Fat Recommendations:
1. Choose lean meat, poultry, fish, dry beans, and peas. Trim fat off.
2. Eat eggs and such organ meats as liver in moderation. (Actually, these are high in cholesterol rather than fat.)
3. Broil, boil, or bake, rather than fry.
4. Limit your intake of butter, cream, hydrogenated oils, shortenings, coconut oil.

The Less-Salt Recommendations:
1. Learn to enjoy unsalted food flavors.
2. Add little or no salt to foods at the table and add only small amounts of salt when you cook.
3. Limit your intake of salty prepared foods such as pickles, pretzels, and potato chips.

The Less-Sugar Recommendations:
1. Eat less sweets such as candy, soft drinks, ice cream, and pastry.
2. Eat fresh fruit or canned fruit without heavy syrup.
3. Use less sugar—white, brown, raw—and less honey and syrups.

Source: American Dietetic Association, based on *Dietary Guidelines for Americans 1980,* U.S. Departments of Agriculture and Health, Education, and Welfare.

Fat is the component of food that has the highest energy content (9.3 Kcal/g compared to 4.1 Kcal/g for carbohydrate). Raw potatoes, which contain roughage, have about 0.9 Kcalories per gram, but when they are cooked in fat, the number of Kcalories jumps to 6 Kcalories per gram. Another problem for those trying to limit their Kcaloric intake is that fat is not always highly visible; for example, butter melts on toast or potatoes. Table 36.3 gives suggestions for cutting down on the amount of fat in the diet.

As a nation, we have increased our consumption of fat from vegetable sources and have decreased our consumption from animal sources, such as red meat and butter. Most likely, this is due to recent studies linking diets high in saturated fats and cholesterol to hypertension and heart attack.

Cholesterol

The risk of cardiovascular disease includes many factors, which are discussed on page 566. One of these factors, according to the National Heart, Lung, and Blood Institute, is a cholesterol blood level of 240 mg/100 ml or higher. If the cholesterol level is this high, additional testing can determine how much of each of 2 important subtypes of cholesterol is in the blood. Cholesterol is carried from the liver to the cells (including the endothelium of the arteries) by plasma proteins called *low-density lipoprotein (LDL)* and is carried away from the cells to the liver by *high-density lipoprotein (HDL).* LDL cholesterol apparently contributes to the formation of plaque, which can clog the arteries, while HDL cholesterol protects against the development of clogged arteries. It has been found that a diet low in saturated fats prevents the

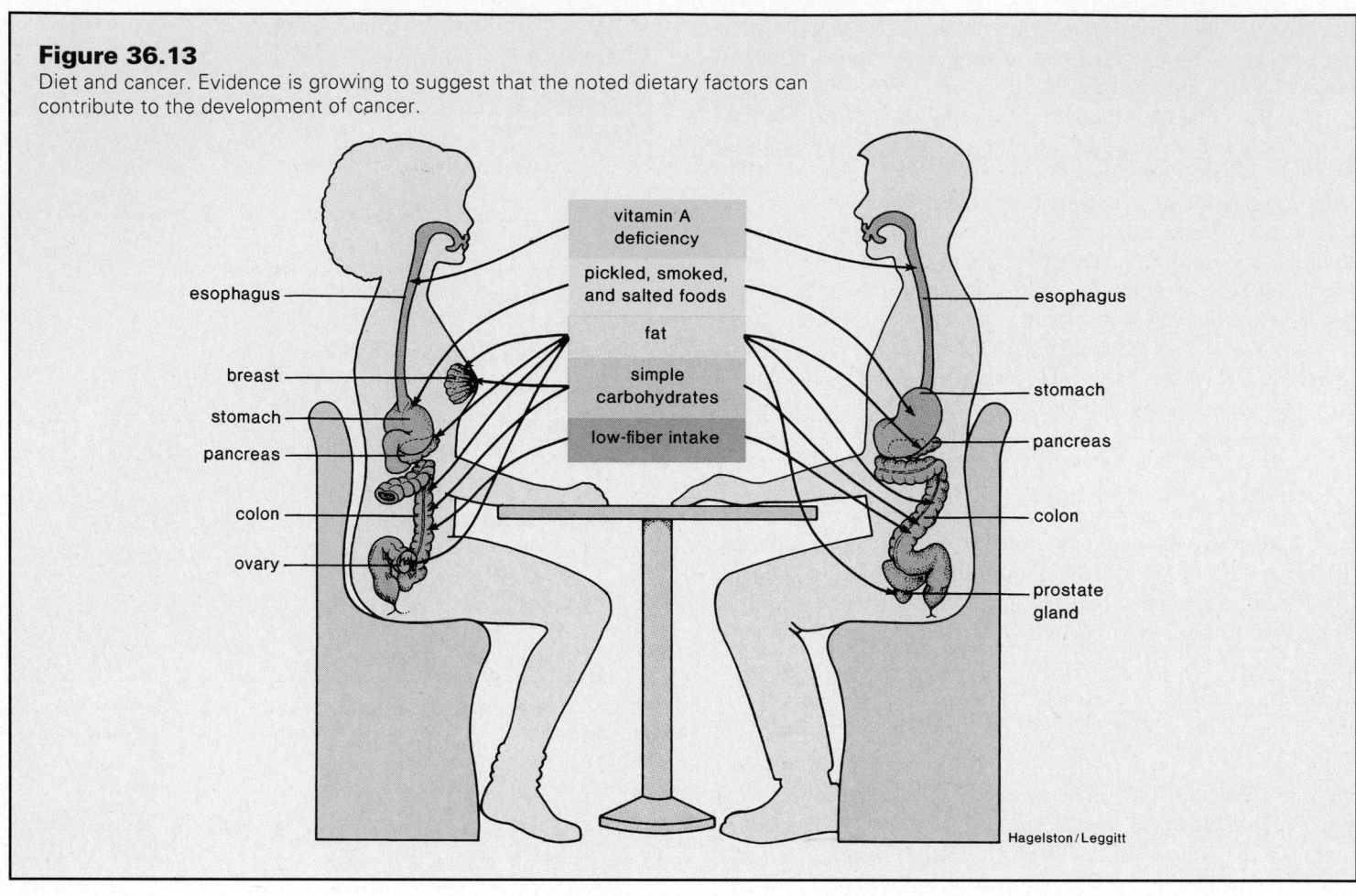

Figure 36.13

Diet and cancer. Evidence is growing to suggest that the noted dietary factors can contribute to the development of cancer.

accumulation of LDL cholesterol in the body. Therefore, the suggestions in regard to fat in table 36.3 are also recommended for cholesterol management. It may also be helpful to restrict the intake of cholesterol itself. To do this, white fish, poultry, and shellfish are recommended; cheese, egg yolks, and liver are not recommended. Egg whites can be substituted for egg yolks for both cooking and eating.

Vitamins and Minerals

Vitamins

Vitamins are organic compounds (other than carbohydrates, fats, and proteins) that the body is unable to produce but needs for metabolic purposes. Therefore, vitamins must be present in the diet; if they are lacking, various symptoms develop (fig. 36.14). Many substances are advertised as vitamins, but in reality there are only 13 vitamins (table 36.4). The table also gives the RDA vitamin values for a "typical" female and a "typical" male. In general, carrots, squash, turnip greens, and collards are good sources for vitamin A. Citrus fruits, other fresh fruits, and vegetables are natural sources of vitamin C. Fortified milk is the primary source of vitamin D, and whole grains are a good source of B vitamins.

It is not difficult to acquire the RDAs for vitamins if your diet is balanced (fig. 36.12), because each vitamin is needed in small amounts only. Many vitamins are portions of coenzymes. For example, niacin is part of the coenzyme NAD^+, and riboflavin is a part of FAD. Coenzymes are needed in only small amounts because each one can be used repeatedly.

The National Academy of Sciences suggests that we eat more fruits and vegetables to acquire a good supply of vitamins C and A because these 2 vitamins may help guard against the development of cancer. Nevertheless the intake of excess vitamins by way of pills is discouraged because this practice can possibly lead to illness. For example, excess vitamin C can cause kidney stones and can also be converted to oxalic acid, a molecule that is toxic to the body. Vitamin A taken in excess can cause peeling of the skin, bone and joint pains, and liver damage. Excess vitamin D can cause vomiting, diarrhea, and kidney damage. Megavitamin therapy, therefore, should always be supervised by a physician.

A properly balanced diet includes all the vitamins and minerals needed by most individuals to maintain health.

Figure 36.14
Illnesses due to vitamin deficiency. **a.** Bowing of bones (rickets) due to vitamin D deficiency. **b.** Dermatitis of areas exposed to light (pellagra) due to niacin deficiency. **c.** Bleeding of gums (scurvy) due to vitamin C deficiency. **d.** Fissures of lips (cheilosis) due to riboflavin deficiency.

a.

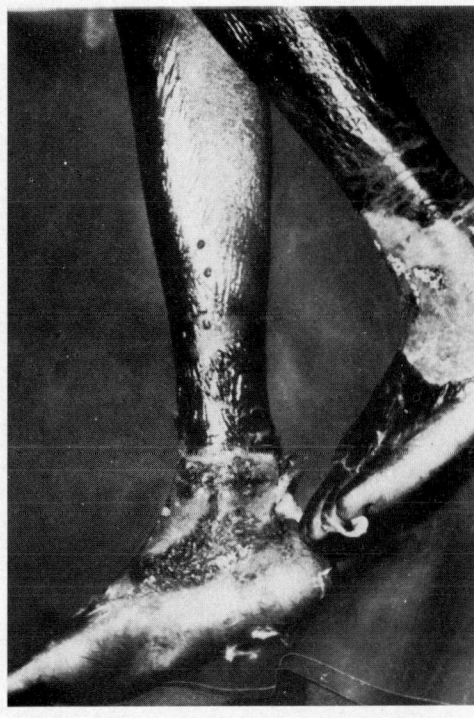
b.

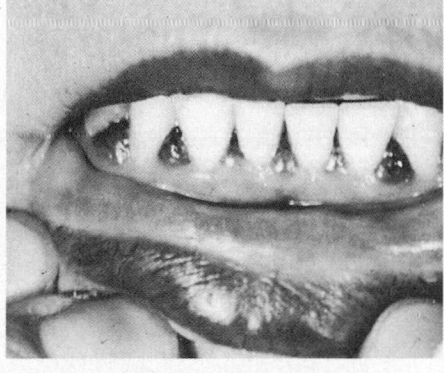
c.

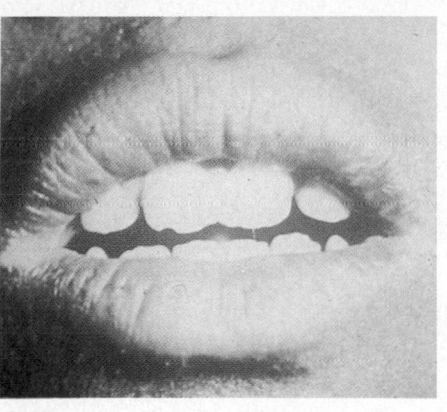
d.

Minerals

In addition to vitamins, various minerals are also required by the body. Some minerals (calcium, phosphorus, potassium, sulfur, sodium, chlorine, and magnesium) are recommended in amounts more than 100 mg per day. These macrominerals serve as constituents of cells and body fluids and as structural components of tissues. For example, calcium is needed for the construction of bones and teeth and also for nerve conduction and muscle contraction. Other minerals (iron, manganese, copper, iodine, cobalt, and zinc) are recommended in amounts of less than 20 mg per day.

These microminerals are more likely to have very specific functions. For example, iron is needed for the production of hemoglobin, and iodine is used in the production of thyroxin, a hormone produced by the thyroid gland. As research continues, more and more elements have been added to the list of those considered essential. During the past 3 decades, very small amounts of molybdenum, selenium, chromium, nickel, vanadium, silicon, and even arsenic have been found to be essential to good health.

Some individuals do not receive enough iron (especially women), calcium, magnesium, or zinc in their diets. Adult females need more iron in their diet than males (RDA of 18 mg compared to 10 mg) because they lose hemoglobin each month during menstruation. Stress can bring on a magnesium deficiency, and a vegetarian diet may lack zinc, which is usually obtained from meat. A varied and complete diet, however, usually supplies the RDAs for minerals.

Calcium

There is much interest in calcium supplements to counteract the possibility of developing *osteoporosis,* a degenerative bone disease that afflicts an estimated one-fourth of older men and one-half of older women in the United States. These individuals have porous bones that tend to break easily because they lack sufficient calcium. In 1984, a National Institute of Health conference on osteoporosis advised postmenopausal women to increase their intake of calcium to 1,500 mg and all others to 1,000 mg (compared with the RDA of 800 mg).

Table 36.4
Vitamins: Their Role in the Body and Food Sources

Vitamins	Role in Body*	Good Food Sources*	Deficiency[†]	Excess[†]
Fat-Soluble Vitamins				
Vitamin A	Assists in the formation and maintenance of healthy skin, hair, and mucous membranes; aids in the ability to see in dim light (night vision); essential for proper bone growth, tooth development, and reproduction	Deep yellow/orange and dark green vegetables and fruits (carrots, broccoli, spinach, cantaloupe, sweet potatoes); cheese, milk, and fortified margarines	Blindness, especially night blindness	Headache, vomiting, peeling of skin, liver damage, anorexia, swelling of long bones
Vitamin D	Aids in the formation and maintenance of bones and teeth; assists in the absorption and use of calcium and phosphorus	Milks fortified with vitamin D; tuna, salmon, or cod liver oil; also made in the skin when exposed to sunlight	Bone deformities, rickets in children	Vomiting, diarrhea, weight loss, kidney damage
Vitamin E	Protects vitamin A and essential fatty acids from oxidation; prevents cell membrane damage	Vegetable oils and margarine, nuts, wheat germ and whole grain breads and cereals, green leafy vegetables	None known in humans	Relatively nontoxic
Vitamin K	Aids in synthesis of substances needed for clotting of blood; helps maintain normal bone metabolism	Green leafy vegetables, cabbage, and cauliflower; also made by bacteria in intestines of humans, except for newborns	Possibly severe bleeding	Synthetic forms at high doses may cause jaundice
Water-Soluble Vitamins				
Vitamin C	Important in forming collagen, a protein that gives structure to bones, cartilage, muscle, and vascular tissue; helps maintain capillaries, bones, and teeth; aids in absorption of iron; helps protect other vitamins from oxidation	Citrus fruits, berries, melons, dark green vegetables, tomatoes, green peppers, cabbage, and potatoes	Scurvy (degeneration of skin, teeth, blood vessels)	Possibility of kidney stones
Thiamin	Helps in release of energy from carbohydrates; promotes normal functioning of nervous system	Whole-grain products, dried beans and peas, sunflower seeds, nuts	Beriberi (nerve changes, edema, heart failure)	None reported
Riboflavin	Helps body transform carbohydrate, protein, and fat into energy	Nuts, yogurt, milk, whole-grain products, cheese, poultry, leafy green vegetables	Cheilosis (cracks at corners of mouth), lesions of eye	None reported
Niacin	Helps body transform carbohydrate, protein, and fat into energy	Nuts, poultry, fish, whole-grain products, dried fruit, leafy green vegetables, beans; can be formed in the body from tryptophan, an essential amino acid found in protein	Pellagra (skin and gut lesions; nervous and mental disorders)	Flushing, burning, and tingling around neck, face, and hands
Vitamin B_6	Aids in the use of fats and amino acids; aids in the formation of protein	Sunflower seeds, beans, poultry, nuts, leafy green vegetables, bananas, dried fruit	Irritability, convulsions, kidney stones	None reported
Folic acid	Aids in the formation of hemoglobin in red blood cells; aids in the formation of genetic material	Dark green leafy vegetables, nuts, beans, whole-grain products, fruit juices	Anemia, gastrointestinal disturbances, red tongue	None reported
Pantothenic acid	Aids in the formation of hormones and certain nerve-regulating substances; helps in the metabolism of carbohydrate, protein, and fat	Nuts, beans, seeds, dark green leafy vegetables, poultry, dried fruit, milk	Sleep disturbances, fatigue, impaired coordination, nausea (rare in humans)	None reported
Biotin	Aids in the formation of fatty acids; helps in the release of energy from carbohydrate	Occurs widely in foods, especially eggs. Made by bacteria in the human intestine	Fatigue, depression, nausea, dermatitis, muscular pains	None reported
Vitamin B_{12}	Aids in the formation of red blood cells and genetic material; helps the functioning of the nervous system	Milk, yogurt, cheese, fish, poultry, and eggs; not found in plant foods unless fortified (such as in some breakfast cereals)	Pernicious anemia, neurological disorders	None reported

Studies have shown, however, that calcium supplements cannot prevent osteoporosis after menopause even when the dosage is 3,000 mg a day. In postmenopausal women, bone-eating cells called osteoclasts are known to be more active than bone-forming cells called osteoblasts (p. 665). Until now, the most effective defense against osteoporosis in older women has been exercise, which encourages the work of osteoblasts, and estrogen replacement. Recently, however, studies have shown that the drug etidronate disodium, which inhibits osteoclast activity, is effective in osteoporotic women when administered at the proper dosage.

Young women can guard against the possibility of developing osteoporosis when they grow older by forming strong, dense bones before menopause. Eighteen-year-old women are apt to get only 679 mg of calcium a day when the RDA is 800 mg. They should consume more calcium-rich foods, such as milk and dairy products. Calcium supplements may not be as effective. A cup of milk supplies 270 mg of calcium, while a 500 mg tablet of calcium carbonate provides only 200 mg. The rest is just not taken up by the body; it is not in a form that is *bioavailable*. An excess of bioavailable calcium can lead to kidney stones.

Dietary calcium and exercise, plus estrogen therapy if needed, are the best safeguards against osteoporosis.

Sodium

The recommended amount of sodium intake per day is 400–3,300 mg, and the average American takes in 4,000–4,700 mg, mostly as salt (sodium chloride). In recent years, this imbalance has caused concern because high-sodium intake has been linked to hypertension in some people. About one-third of the sodium we consume occurs naturally in foods; another third is added during commercial processing; and we add the last third either during home cooking or at the table in the form of table salt.

Clearly it is possible for us to cut down on the amount of sodium in the diet by reducing the amount of salt we eat. Table 36.3 gives recommendations for doing so.

Summary

1. All mammals have teeth. Herbivores need teeth that can clip off plant material and grind it up. Also, the herbivore's stomach contains bacteria that can digest cellulose.

2. Carnivores need teeth that can tear and rip meat into pieces. Meat is easier to assimilate, so the digestive system of carnivores has less specialization of parts and a shorter intestine than that of herbivores.

3. Some animals are continuous feeders (e.g., clams, which are sessile filter feeders); others are discontinuous feeders (e.g., squid). Discontinuous feeders need a storage area for food.

4. Some animals (e.g., planarians) have an incomplete digestive tract, which shows little specialization of parts. Other animals (e.g., earthworms) have a complete digestive tract, which does show specialization of parts.

5. In the human digestive tract, food is chewed and manipulated in the mouth, where salivary glands secrete saliva. Saliva contains salivary amylase, which begins carbohydrate digestion.

6. Food then passes down the esophagus to the stomach. The stomach stores and mixes food with mucus and gastric juice to produce chyme. Pepsin begins protein digestion here.

7. Chyme gradually enters the duodenum where bile, pancreatic juice, and the intestinal secretions are found. Enzymes in the small intestine hydrolyze all of the organic nutrients. Table 36.2 summarizes the enzymes involved in digesting food.

8. Most nutrient absorption takes place in the small intestine, but some water and minerals are absorbed in the colon. Digestive wastes leave the colon by way of the anus.

9. The liver produces bile, which is stored in the gallbladder. The liver is also involved in the processing of absorbed nutrient molecules and in maintaining the blood concentration of nutrient molecules, such as glucose. The liver converts ammonia to urea and breaks down toxins.

10. The pancreas produces digestive enzymes and hormones (insulin and glucagon) involved in the control of carbohydrate metabolism in the body.

11. There are 3 hormones that regulate digestive-tract secretions: gastrin, which stimulates acid- and enzyme-secreting cells in the stomach; secretin, which stimulates the pancreas to release sodium bicarbonate; and CCK-PZ, which stimulates the gallbladder to release bile and the pancreas to release digestive enzymes.

12. A balanced diet is required for good health. Food should provide us with all the necessary vitamins, minerals, amino acids, fatty acids, and an adequate amount of energy.

Writing Across the Curriculum

In order to practice writing skills, students should write out the answers to any or all of the study questions and the critical thinking questions. The study questions are sequenced in the same order as the text. Suggested answers to the critical thinking questions are in appendix D.

Study Questions

1. Contrast the incomplete with the complete gut using the planarian and earthworm as examples.

2. Contrast a continuous with a discontinuous feeder using the clam and squid as examples.

3. Contrast the dentition of the mammalian herbivore with the mammalian carnivore using the horse and lion as examples.

4. List the parts of the human digestive tract, anatomically describe them, and state the contribution of each to the digestive process.

5. State the location and describe the functions of both the liver and pancreas.

6. Assume that you have just eaten a ham sandwich. Discuss the digestion of the contents of the sandwich.

7. Discuss the absorption of the products of digestion into the circulatory system.

8. What are gastrin, secretin, and CCK-PZ? Where are they produced and what are their functions?

9. Give reasons why carbohydrates, fats, proteins, vitamins, and minerals are all necessary to good nutrition.

Objective Questions

1. Animals that feed discontinuously
  a. must have digestive tracts that permit storage.
  b. are able to avoid predators by limiting their feeding time.
  c. exhibit extremely rapid digestion.
  d. Both a and b.

2. In which of the following types of animals would you expect the digestive system to be more complex?
  a. those with a single opening for the entrance of food and exit of wastes
  b. those with 2 openings, one serving as an entrance and the other as an exit
  c. only those complex animals that also have a respiratory system
  d. Both b and c.

3. The typhlosole within the gut of an earthworm compares best to which of these organs in humans?
  a. teeth in the mouth
  b. esophagus in the thoracic cavity
  c. folds in the stomach
  d. villi in the small intestine

4. Which of these animals is a continuous feeder with a complete gut?
  a. planarian
  b. clam
  c. squid
  d. lion

5. The most common food digested in the human stomach is
  a. carbohydrate.
  b. fat.
  c. protein.
  d. nucleic acid.

6. Assuming normal body temperature, which of these combinations is most likely to result in complete digestion?
  a. fat, bile, sodium bicarbonate, lipase
  b. fat, bile, sodium bicarbonate, pepsin
  c. fat, HCl, maltase
  d. fat, bile, HCl, sodium bicarbonate, trypsin

7. Which association is incorrect?
  a. protein—trypsin
  b. fat—lipase
  c. maltose—pepsin
  d. starch—amylase

8. Most of the absorption of the products of digestion takes place in humans across the
  a. squamous epithelium of the esophagus.
  b. convoluted walls of the stomach.
  c. fingerlike villi of the small intestine.
  d. smooth wall of the large intestine.

9. The hepatic portal vein is located between the
  a. hepatic vein and the vena cava.
  b. mouth and the stomach.
  c. pancreas and the small intestine.
  d. small intestine and the liver.

10. Bile in humans
  a. is an important enzyme for the digestion of fats.
  b. is made by the gallbladder.
  c. emulsifies fat.
  d. All of these.

11. Which of these is not a function of the human liver?
  a. produce bile
  b. store glucose
  c. produce urea
  d. make red blood cells

12. The large intestine in humans
  a. digests all types of food.
  b. is the longest part of the intestinal tract.
  c. absorbs water.
  d. is connected to the stomach.

13. Label this diagram of the digestive tract.

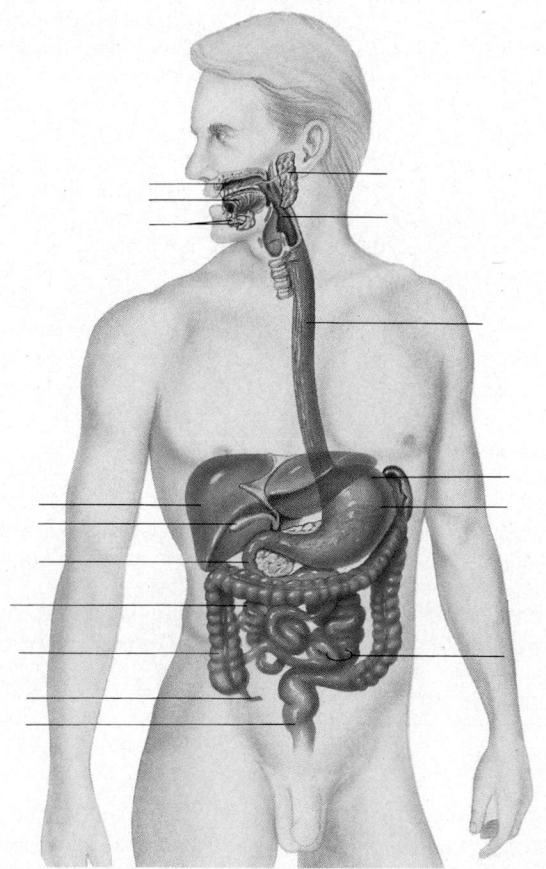

14. Predict and explain the digestive results per test tube for this experiment.

Test tube 1:

Test tube 2:

Test tube 3:

Test tube 4:

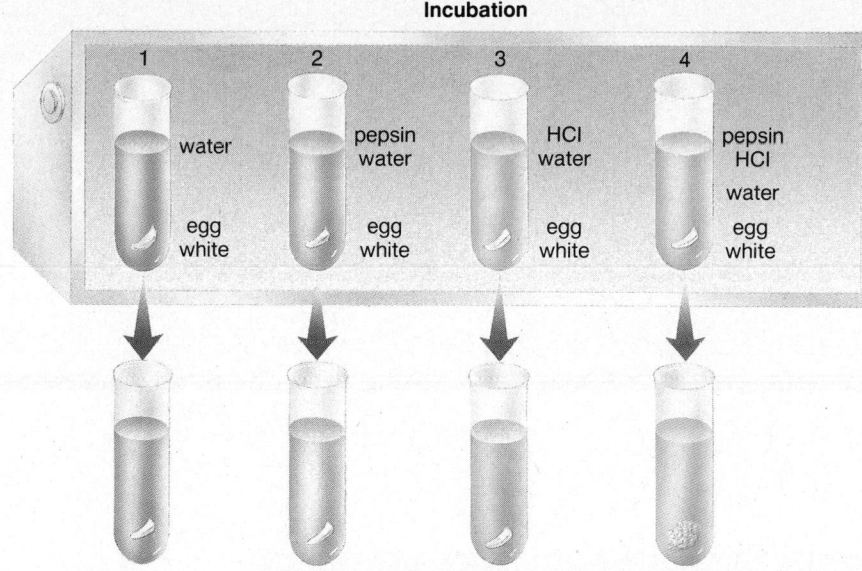

Incubation

Concepts and Critical Thinking

1. *All systems of the animal's body contribute to homeostasis.*

Tell several ways the human digestive system contributes to homeostasis.

2. *Organisms are adapted to obtaining a share of available resources.*

Animals vary according to how they obtain and process food. Choose either the grasshopper, earthworm, clam, or squid to support this concept.

3. *Forms of life depend on each other for materials and energy.*

Review figure 1.5, and explain how heterotrophs (e.g., animals) are dependent on autotrophs (e.g., plants) and how autotrophs in turn are dependent on heterotrophs.

Selected Key Terms

omnivore (om'nĭ-vōr) 588
herbivore (her'bĭ-vōr) 588
carnivore (kar'nĭ-vōr) 588
epiglottis (ep"ĭ-glot'is) 593

esophagus (ĕ-sof'ah-gus) 593
peristalsis (per"ĭ-stal'sis) 593
duodenum (du"o-de'num) 593
villus (vil'lus) 594
bile (bīl) 594

gallbladder (gawl'blad-der) 594
lacteal (lak'te-al) 596
essential amino acid (ĕ-sen'shal ah-me'no as'id) 598
vitamin (vi'tah-min) 600

37

Respiratory System

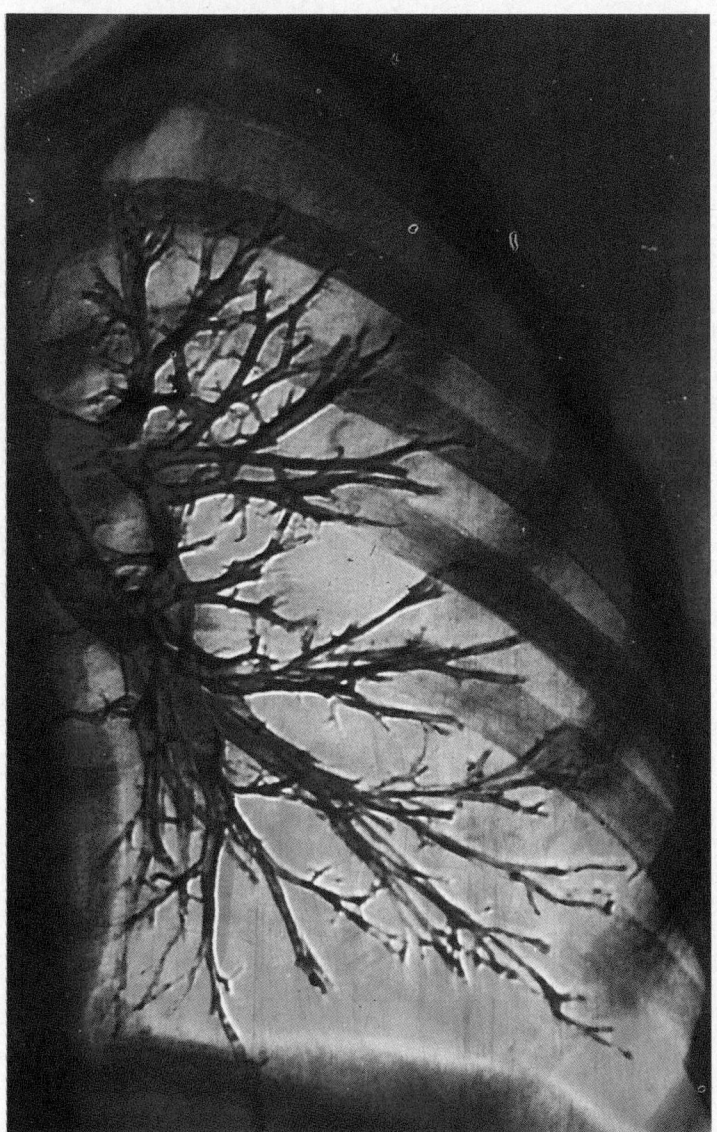

The human lung lies within the rib cage and is serviced by a branching pulmonary artery. The rib cage not only protects the lungs, it also assists breathing. When the rib cage moves up and out, the lungs are pulled open, and when it moves down and in, the lungs recoil.

Your study of this chapter will be complete when you can

1. compare the oxygen-containing capacities of water and air;
2. describe the mechanism of gas exchange in a hydra and a planarian;
3. compare the respiratory organs of aquatic and terrestrial animals;
4. compare the various methods for ventilating the lungs among terrestrial vertebrates;
5. compare the incomplete method of ventilation used by amphibians, reptiles, and mammals with the complete method used by birds;
6. describe the path of air in humans, and describe in general the structure and function of all organs mentioned;
7. describe how the breathing rate is controlled in humans;
8. give the equations applicable to the exchange of gases in the lungs and tissues, and tell how the gases are transported in the blood;
9. name 3 types of hemoglobin found in humans;
10. tell how the respiratory tract protects itself from infection, and relate emphysema and lung cancer to cigarette smoking.

Animals do not have a storage area for gases and must continually acquire oxygen and rid the body of carbon dioxide. Without a constant supply of oxygen an animal dies. This was first observed in 1772 by the English scientist Joseph Priestley, who collected oxygen by heating red mercuric oxide. His contemporary, the Frenchman Antoine Lavoisier, correctly deduced that both combustion by a lit candle and respiration by an animal remove oxygen from the air. This was one of the first times that a physiological process was explained with reference to a nonliving mechanism.

We know today that oxygen is the final acceptor for electrons during cellular respiration, a metabolic process that supplies ATP energy for repair, growth, and movement.

Respiratory Gas-Exchange Surfaces

Gas exchange takes place by the physical process of diffusion. For diffusion to be effective, the gas-exchange region must be (1) moist, (2) thin, and (3) large in relation to the size of the body. Some animals are small and shaped in a way that allows the surface of the animal to be the gas-exchange surface. Other animals are complex and have a specialized gas-exchange surface. The effectiveness of diffusion is enhanced by vascularization, and delivery to cells is promoted when the blood contains a respiratory pigment such as hemoglobin.

Aquatic Environments

It is more difficult for animals to obtain oxygen from water than from air. Water fully saturated with air contains about 5% the amount of oxygen than does the same volume of air. Also, water is more dense than air. Therefore, aquatic animals expend more energy for breathing than do terrestrial animals. Fishes use up to 25% of their energy output, while terrestrial mammals use only 1%–2% of their energy output for breathing.

Hydras and planarians have a large surface area in comparison to their size (fig. 37.1). This makes it possible for most of their cells to exchange gases directly with the environment. In hydras, the outer layer of cells is in contact with the external environment and the inner layer can exchange gases with the water within the gastrovascular cavity. In planarians, the flattened body permits cells to exchange gases with the external environment.

A tubular shape also provides a surface area adequate enough for gas-exchange purposes. In addition to a tubular shape, polychaete worms (p. 435) have extensions of the body wall called parapodia, which are vascularized. Quite often aquatic animals have **gills**, which are finely divided and vascularized outgrowths of either the outer or inner body surface. Among mollusks, such as clams and many snails, water is drawn into the mantle cavity, where it passes through the gills. In decapod crustaceans the gills are located in brachial chambers covered by the exoskeleton. Water is kept moving by the action of specialized appendages located near the mouth.

Figure 37.1

Animal shapes and gas exchange. Some small aquatic animals use the body surface for gas exchange. This works because the body surface is large compared to the size of the animal. *a.* In hydras every cell is near a source of oxygen. The inner layer of cells exchanges gases with the water in the gastrovascular cavity, and the outer layer of cells exchanges gases with the external environment. *b.* In planarians the body is flattened and most cells can carry out gas exchange with the external environment. Planarians do not have a circulatory system.

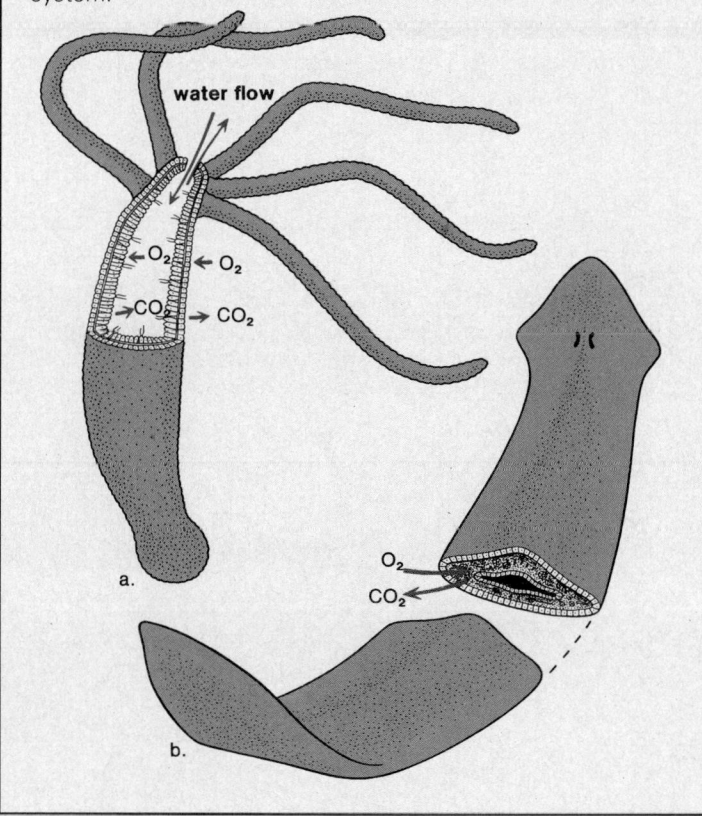

Among vertebrates, the gills of fishes are outward extensions of the pharynx (fig. 37.2). Ventilation is brought about by the combined action of the mouth and gill covers, or opercula (operculum, sing.). When the mouth is open, the opercula are closed and water is drawn in. Then the mouth closes and the opercula open, drawing the water from the pharynx through the gill slits located between the gill arches. On the outside of the gill arches, the gills are composed of *filaments* that are folded into platelike *lamellae*. In the capillaries of each lamella, the blood flows in a direction opposite to the movement of water across the gills. This *countercurrent flow* increases the amount of oxygen that can be taken up; as the blood in each lamella gains oxygen, it encounters water having a still higher oxygen content. The countercurrent mechanism allows about 80%–90% of the initial dissolved oxygen in water to be extracted.

Small aquatic animals sometimes use the body surface for gas exchange, but many larger ones have localized gas-exchange surfaces known as gills.

Terrestrial Environments

Air is a rich source of oxygen compared to water; however, it does have a drying effect on respiratory surfaces. A human loses about 350 ml of water per 24 hours when the air has a relative humidity of 50%.

The earthworm (see fig. 27.8) is an example of an invertebrate terrestrial animal that uses its body surface for respiration. An earthworm expends much energy to keep its body surface moist by secreting mucus and by releasing fluids from excretory pores. Further, the worm is behaviorally adapted to remain in damp soil during the day, when the air is driest.

Insects and certain other terrestrial arthropods have a respiratory system known as a tracheal system (fig. 37.3). Oxygen enters the trachea at spiracles, which are valvelike openings on each side of the body. The trachea branch and then rebranch, ending in tiny channels, the tracheoles, that are in direct contact with the body cells. Larger insects have a ventilation system to keep the air moving in and out of the trachea. Many have air sacs located near major muscles; contraction of these muscles causes the air sacs to empty, and relaxation causes the air sacs to expand and draw air in. The tracheal system of insects is effective in delivering oxygen to the cells so that the insect circulatory system plays no role in gas transport.

Terrestrial vertebrates, in particular, have evolved vascularized outgrowths from the lower pharyngeal region known as **lungs**. The lungs of amphibians are simple, saclike structures

Figure 37.2

Anatomy of gills in detail. **a.** The operculum (folded back) covers and protects several layers of delicate gills. **b.** Each layer of gills has 2 rows of gill filaments. **c.** Each filament has many thin, platelike lamellae. Gases are exchanged between capillaries inside the lamellae and the water that flows between the lamellae. **d.** Blood in the capillaries flows in the direction opposite to that of the water. Blood takes up 90% of the oxygen in the water as a result of this countercurrent flow.

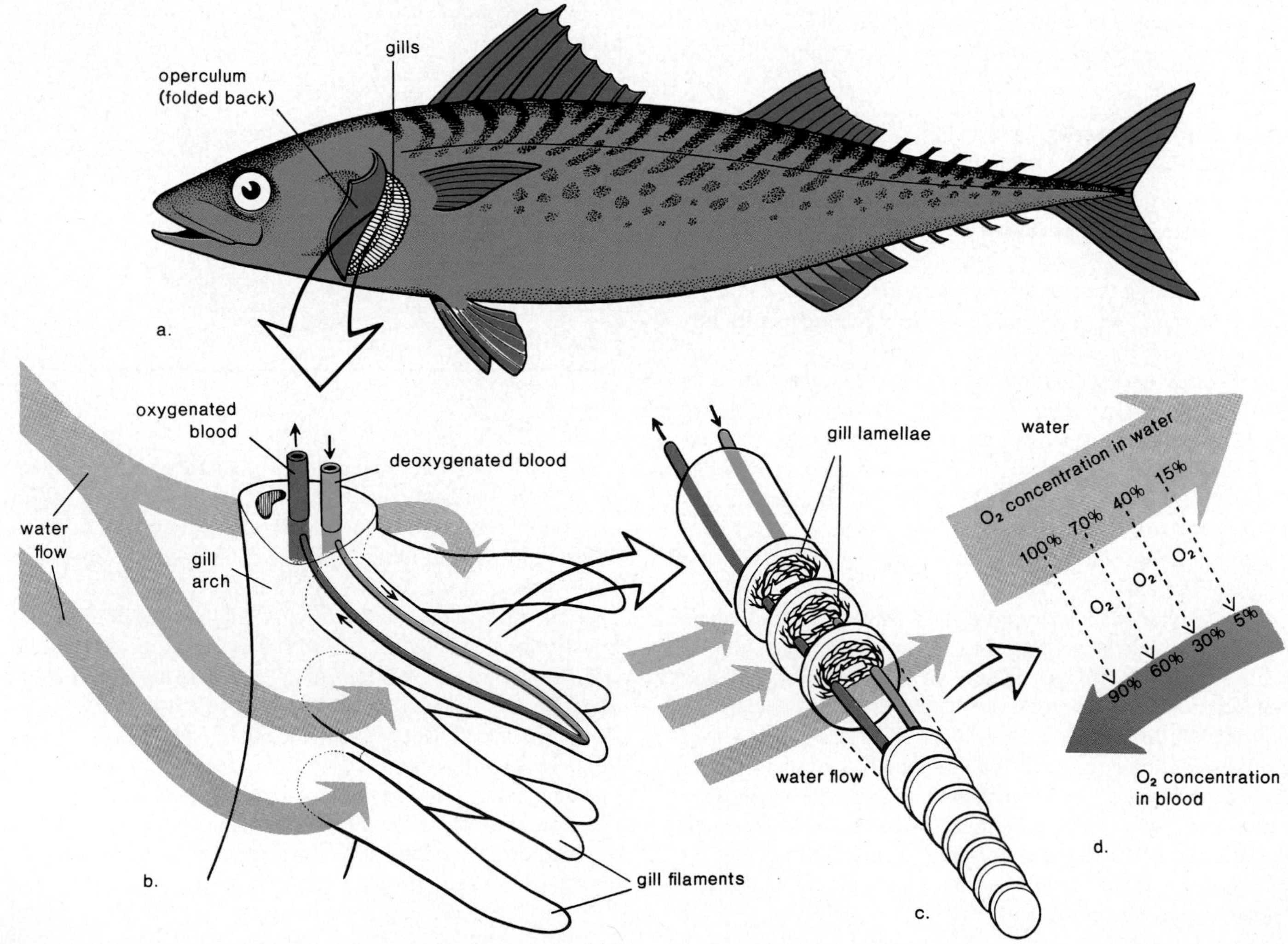

Animal Structure and Function

Figure 37.3

Tracheal system of insects. **a.** A system of air tubes extends throughout the body of an insect and they, rather than blood, carry oxygen to the cells. **b.** Air enters the tracheae at openings called spiracles. From here, it moves to the smaller tracheoles, which take it to the cells, where gas exchange takes place.

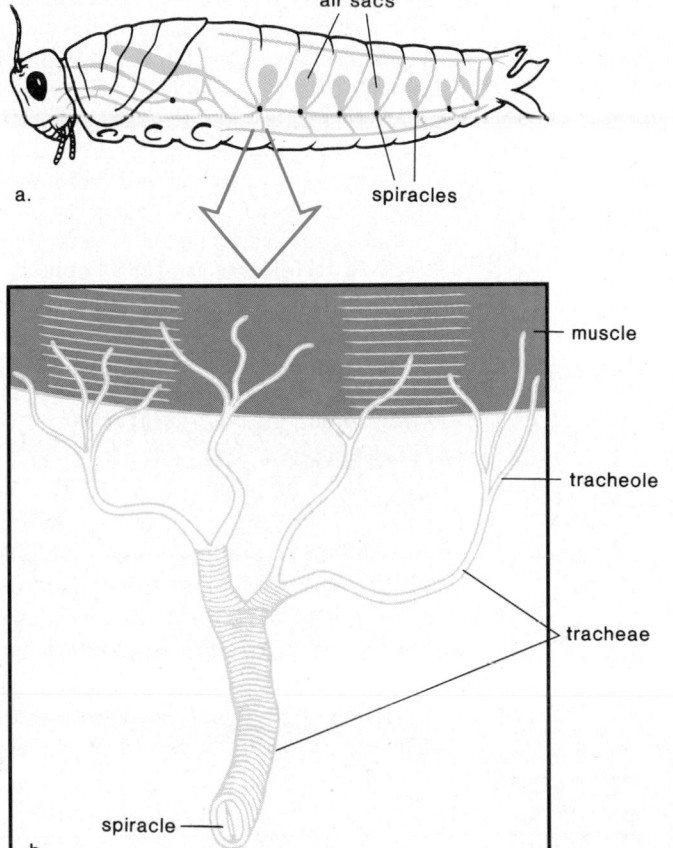

(fig. 37.4*a*). Many amphibians possess a short trachea, which divides into 2 bronchi that open into the lungs. Most amphibians respire to some extent through the skin, which is kept moist by the presence of mucus produced by numerous glands on the surface of the body. During the winter in temperate climates, amphibians burrow in the mud, and all gas exchange occurs by way of the skin.

The inner lining of the lungs is more finely divided in reptiles than in amphibians (fig. 37.4*b*). The lungs of birds and mammals are even more elaborately subdivided into small passageways and spaces. It has even been estimated that human lungs have a total surface area that is at least 50 times the skin's surface area. To keep the lungs from drying out, air is moistened as it moves along the tubes leading to the lungs.

Terrestrial vertebrates ventilate the lungs by moving air in and out of the respiratory tract. Frogs use positive pressure to force air into the respiratory tract. With the nostrils firmly shut, the floor of the mouth rises and pushes the air into the lungs. Reptiles, birds, and mammals use negative pressure (fig. 37.5). The lungs first expand, and then air comes rushing in. Reptiles have jointed ribs that can be raised to expand the lungs. Mammals have a rib cage that is lifted up and out and a muscular **diaphragm** that is flattened. Both of these actions increase the volume of the lungs, so that air is thereby drawn in, during the process called **inhalation,** or inspiration. After inhalation, exhalation occurs. During **exhalation** (also called expiration), air is pushed out of the lungs. In reptiles, lowering the ribs exerts a pressure that forces air out. In mammals, the rib cage is lowered, and the diaphragm rises, forcing the air out of the lungs.

Figure 37.4

Respiration in amphibians compared to reptiles. **a.** The amphibian lung is small and saclike—the moist, thin skin is often used as an auxiliary gas-exchange surface. Amphibians use positive pressure to push air into the lungs. **b.** The reptilian lung has more convolutions—the tough, scaly skin protects the animal from drying out. Reptiles use negative pressure to draw air into the lungs.

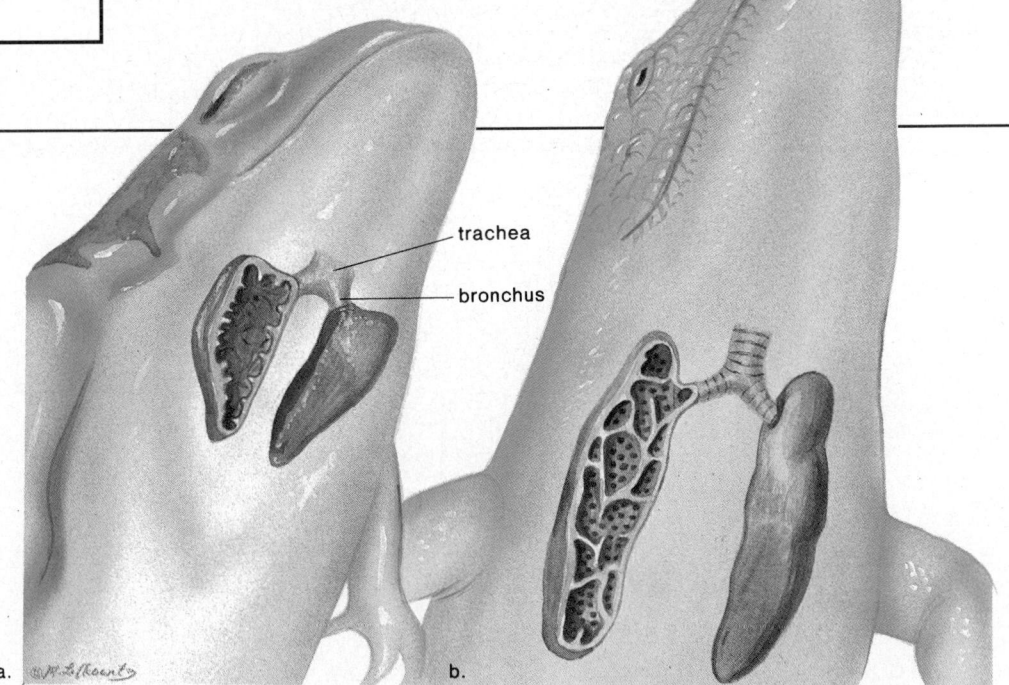

The lungs of amphibians, reptiles, and mammals are not completely emptied and refilled during each breathing cycle. Because of this *incomplete ventilation* method, the air coming in mixes with used air still in the lungs. While this does help conserve water, it also decreases gas-exchange efficiency. The high oxygen requirement of flying birds, however, necessitates a method of *complete ventilation* (fig. 37.6). Incoming air is carried past the lungs by a bronchus that takes it to a set of posterior air sacs. The air then passes forward through the lungs into a set of anterior air sacs. From here, it is finally expelled. Fresh, oxygen-rich air passes through the lungs in a one-way direction only and does not mix with used air.

A few terrestrial animals use the body surface for gas exchange, but most utilize a specialized region. Insects have a tracheal system, while terrestrial vertebrates depend on lungs, which are ventilated variously.

Human Respiratory System

Path of Air

The human respiratory system includes all structures that conduct air to and from the lungs (fig. 37.7; table 37.1). The lungs lie deep within the **thoracic cavity**, where they are protected from drying out. As air moves through the nose, pharynx, trachea, and bronchi to the lungs, it is fil-

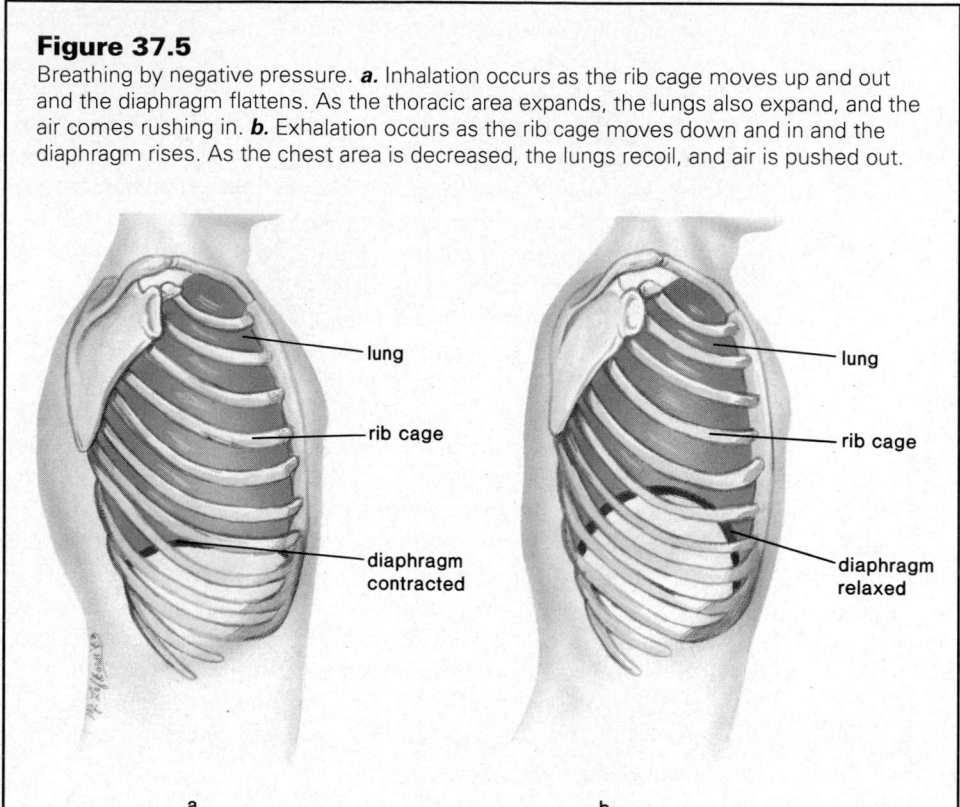

Figure 37.5
Breathing by negative pressure. *a.* Inhalation occurs as the rib cage moves up and out and the diaphragm flattens. As the thoracic area expands, the lungs also expand, and the air comes rushing in. *b.* Exhalation occurs as the rib cage moves down and in and the diaphragm rises. As the chest area is decreased, the lungs recoil, and air is pushed out.

lung

rib cage

diaphragm contracted

lung

rib cage

diaphragm relaxed

a.   b.

Figure 37.6
The lungs of amphibians, reptiles, and mammals are incompletely ventilated, whereas those of birds are completely ventilated. *a.* Inhalation. There is residual "used" air (blue) in the lungs of amphibians, reptiles, and mammals. In birds, the posterior air sacs are cleared of residual air. *b.* Exhalation. Most, but not all, used air leaves the lungs of amphibians, reptiles, and mammals. All used air leaves the lungs of birds because fresh air (pink) enters by its own pathway, as shown in (*a*).

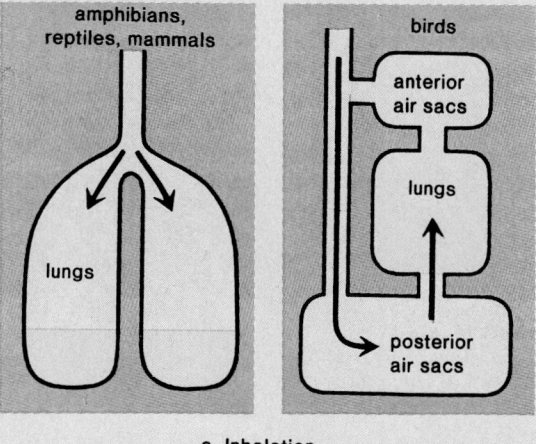
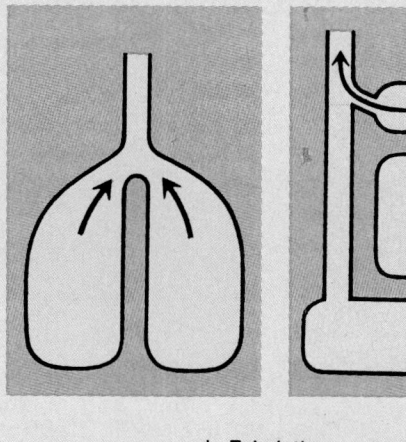

a. Inhalation   b. Exhalation

fresh air   used air

Figure 37.7

Human respiratory system. The path of air can be traced from the nose and mouth to the alveoli. The enlargement shows that alveoli are air sacs surrounded by capillaries. When the concentration of oxygen is higher in the sacs than in the blood, oxygen diffuses into the blood. When the concentration of carbon dioxide is higher in the blood than in the sacs, carbon dioxide diffuses out of the blood.

Table 37.1
Path of Air

Structure	Function
Nasal cavities	Filter, warm, and moisten
Pharynx (throat)	Connection to larynx
Glottis	Permits passage of air
Larynx (voice box)	Sound production
Trachea (windpipe)	Passage of air to thoracic cavity
Bronchi	Passage of air to each lung
Bronchioles	Passage of air to each alveolus
Alveoli	Air sacs for gas exchange

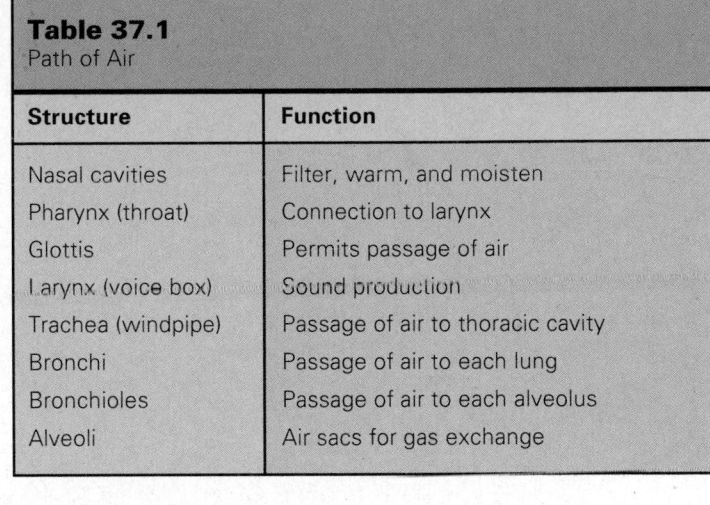

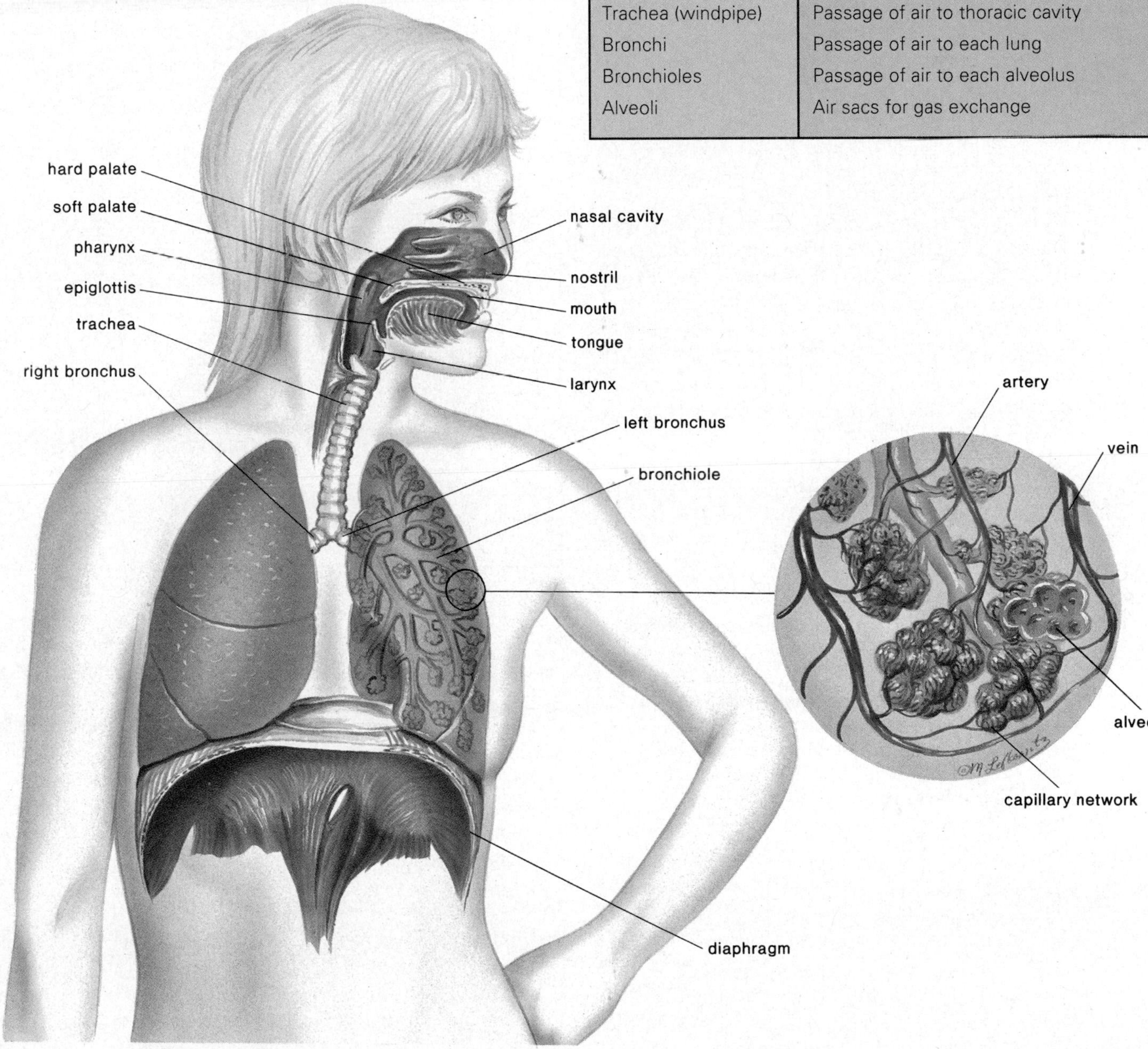

hard palate

soft palate

pharynx

epiglottis

trachea

right bronchus

nasal cavity

nostril

mouth

tongue

larynx

left bronchus

bronchiole

artery

vein

alveolus

capillary network

diaphragm

tered so that it is free of debris, warmed, and humidified. By the time the air reaches the lungs, it is at body temperature and is saturated with water. In the nose, hairs and cilia act as a screening device. In the trachea and the bronchi, cilia beat upward, carrying mucus, dust, and occasional bits of food that "went down the wrong way" into the throat, where the accumulation may be swallowed or expectorated.

The hard and soft palates separate the nasal cavities from the mouth, but the air and food passages cross in the **pharynx.** This may seem inefficient, and there is danger of choking if food accidentally enters the trachea, but this arrangement does have the advantage of letting you breathe through the mouth in case your nose is plugged up. In addition, it permits greater intake of air during heavy exercise, when greater gas exchange is required.

Air passes from the pharynx through the **glottis,** an opening into the **larynx,** or voice box. At the edges of the glottis, embedded in mucous membrane, are the *vocal cords.* These elastic ligaments vibrate and produce sound when air is expelled past them through the glottis from the larynx.

The larynx and the **trachea** are permanently held open to receive air. The larynx is held open by the complex of cartilages that form the Adam's apple. The trachea is held open by a series of C-shaped, cartilaginous rings that do not completely meet in the rear. When food is being swallowed, the larynx rises, and the glottis is closed by a flap of tissue called the **epiglottis.** A backward movement of the soft palate covers the entrance of the nasal passages into the pharynx. The food then enters the esophagus, which lies behind the larynx (see fig. 36.7).

The trachea divides into 2 **bronchi** (bronchus, sing.), which enter the right and left lungs; each then branches into a great number of smaller passages called **bronchioles.** The 2 bronchi resemble the trachea in structure, but as the bronchial tubes divide and subdivide, their walls become thinner and rings of cartilage are no longer present. Each bronchiole terminates in an elongated space enclosed by a multitude of air pockets, or sacs, called **alveoli** (alveolus, sing.), which make up the lungs.

In humans, air moves through the nose, pharynx, and larynx to the trachea, which is held open by cartilaginous rings. Two bronchi divide into bronchioles, which finally take air into the lungs. Each bronchiole terminates in a large number of alveoli, which are covered by an extensive network of capillaries.

Breathing

Humans breathe using the same mechanism employed by all other mammals. The volume of the thoracic cavity and lungs is increased by muscle contractions that lower the diaphragm and raise the ribs (fig. 37.5). These movements create a negative pressure in the thoracic cavity and lungs, and air then flows into the lungs. When rib and diaphragm muscles relax, air is exhaled as a result of increased pressure in the thoracic cavity and lungs.

Increased carbon dioxide (CO_2) and hydrogen ion (H^+) concentrations in the blood are the primary stimuli that increase breathing rate. The chemical content of the blood is monitored by

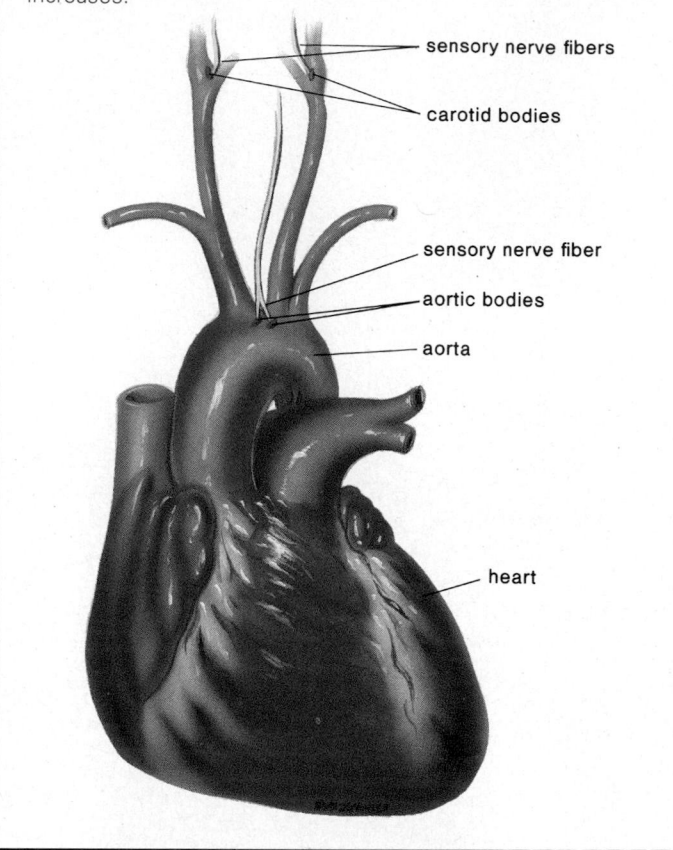

Figure 37.8
When chemoreceptors within the aortic and carotid bodies are stimulated by a rise in the H^+ concentration of the blood, caused by an increased amount of CO_2, nerve impulses travel to the respiratory center of the brain and the breathing rate increases.

sensory nerve fibers

carotid bodies

sensory nerve fiber

aortic bodies

aorta

heart

chemoreceptors called the *aortic* and *carotid bodies,* which are specialized structures located in the walls of the aorta and the carotid arteries (fig. 37.8). These receptors are very sensitive to changes in CO_2 and H^+ concentrations, but they are only minimally sensitive to a lower O_2 concentration. Information from the chemoreceptors goes to the respiratory center in the medulla oblongata of the brain, which then increases the breathing rate when CO_2 or H^+ concentration increases. This respiratory center is itself sensitive to the chemical content of the blood reaching the brain.

Gas Exchange and Transport

Diffusion primarily accounts for the exchange of gases between the air in the alveoli and the blood in the pulmonary capillaries (fig. 37.9). Atmospheric air contains little CO_2, but blood flowing into the pulmonary capillaries is almost saturated with the gas. Therefore, CO_2 diffuses out of the blood and into the alveoli. The pattern is the reverse for oxygen: blood coming into the pulmonary capillaries is oxygen poor and the alveolar air is oxygen rich; therefore, O_2 diffuses into the capillaries.

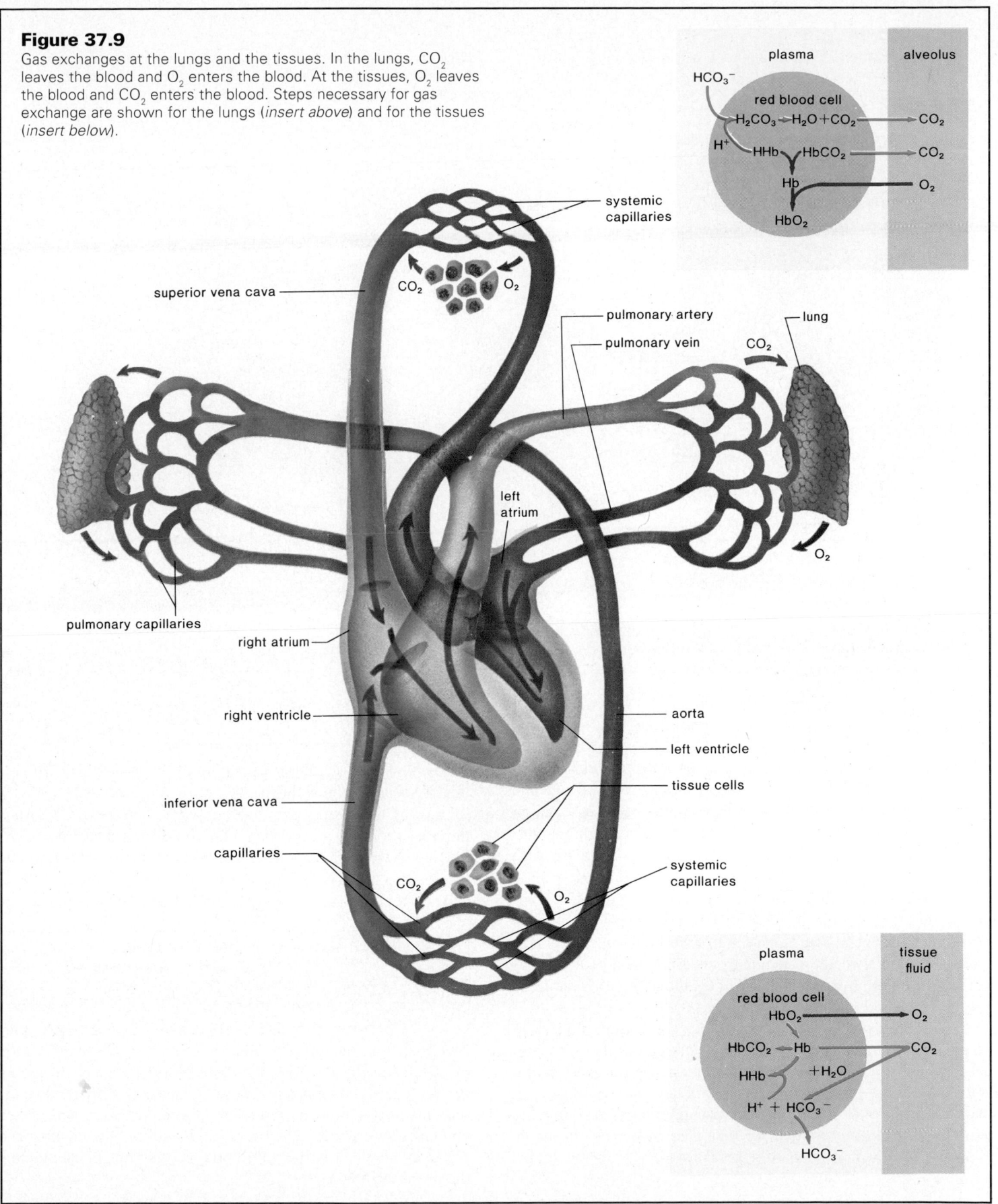

Figure 37.9
Gas exchanges at the lungs and the tissues. In the lungs, CO_2 leaves the blood and O_2 enters the blood. At the tissues, O_2 leaves the blood and CO_2 enters the blood. Steps necessary for gas exchange are shown for the lungs (*insert above*) and for the tissues (*insert below*).

plasma

alveolus

HCO_3^-

red blood cell

$H_2CO_3 \rightarrow H_2O + CO_2$ CO_2

H^+ HHb $HbCO_2$ CO_2

Hb O_2

HbO_2

systemic capillaries

superior vena cava

CO_2 O_2

pulmonary artery

pulmonary vein

lung

CO_2

left atrium

O_2

pulmonary capillaries

right atrium

right ventricle

aorta

left ventricle

tissue cells

inferior vena cava

capillaries

CO_2 O_2

systemic capillaries

plasma

tissue fluid

red blood cell

HbO_2 O_2

$HbCO_2 \leftarrow Hb$ CO_2

HHb $+H_2O$

$H^+ + HCO_3^-$

HCO_3^-

Figure 37.10

Hemoglobin, the respiratory pigment. **a.** Red blood cells move single file through capillaries. **b.** Each one is a biconcave disk containing many molecules of hemoglobin. **c.** Hemoglobin contains 4 polypeptide chains, 2 of which are alpha (α) chains and 2 of which are beta (ß) chains. The plane in the center of each chain represents an iron-containing heme group. Oxygen combines loosely with iron when hemoglobin is oxygenated.

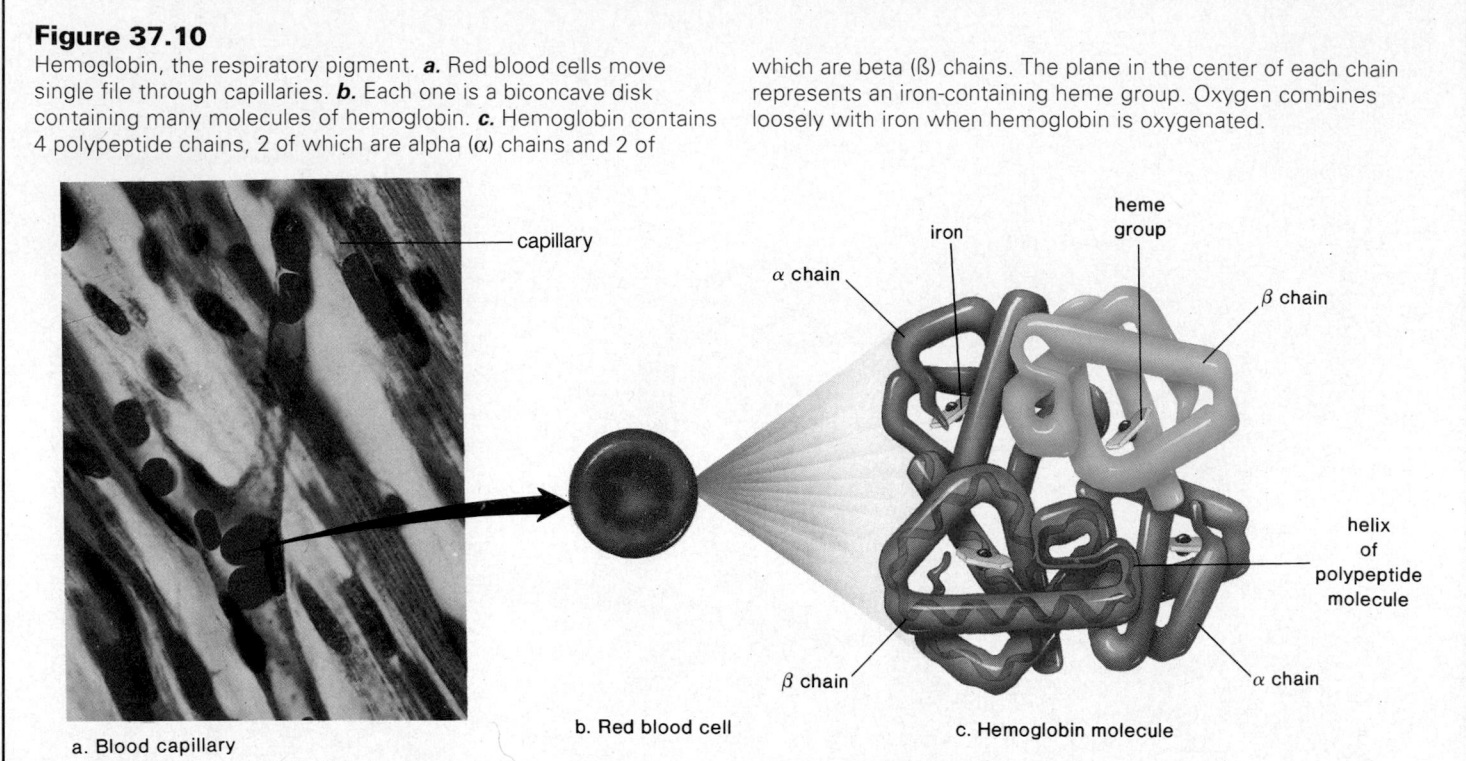

a. Blood capillary

b. Red blood cell

c. Hemoglobin molecule

Transport of O_2 and CO_2

Most oxygen entering the blood combines with **hemoglobin** (Hb) in red blood cells to form oxyhemoglobin (HbO_2):

$$Hb + O_2 \rightarrow HbO_2$$
$$\text{oxyhemoglobin}$$

Each hemoglobin molecule contains 4 polypeptide chains, and each chain is folded around an iron-containing group called heme (fig. 37.10). It is actually the iron that forms a loose association with oxygen. Since there are about 280 million hemoglobin molecules in each red blood cell, each cell is capable of carrying more than one billion molecules of oxygen.

Carbon monoxide, present in automobile exhaust, combines with hemoglobin more readily than oxygen, and it stays combined for several hours regardless of the environmental conditions. Accidental death or suicide from carbon monoxide poisoning occurs because the hemoglobin of the blood is not available for oxygen transport.

Oxygen-binding characteristics of hemoglobin can be studied by examining oxyhemoglobin dissociation curves (fig. 37.11). These curves show the percentage of oxygen-binding sites of hemoglobin that are carrying oxygen at various oxygen partial pressures (PO_2). A partial pressure of a gas is simply the amount of pressure exerted by that gas among all the gases present. At the normal partial pressures of O_2 in the lungs, hemoglobin becomes practically saturated with O_2, but at the partial pressures in the tissues, oxyhemoglobin quickly gives up much of its oxygen:

$$HbO_2 \longrightarrow Hb + O_2$$

The acid pH and warmer temperature of the tissues also promote this dissociation (breakdown).

In the tissues, some hemoglobin combines with carbon dioxide to form *carbaminohemoglobin*. Most of the carbon dioxide, however, is transported in the form of the **bicarbonate ion.** First carbon dioxide combines with water, forming *carbonic acid*, and then this dissociates to a hydrogen ion and the bicarbonate ion:

$$CO_2 + H_2O \longrightarrow H_2CO_3 \longrightarrow H^+ + HCO_3^-$$
$$\text{carbonic} \qquad \text{bicarbonate}$$
$$\text{acid} \qquad \text{ion}$$

Carbonic anhydrase, an enzyme in red blood cells, speeds up this reaction. The released hydrogen ions, which could drastically change the pH of the blood, are absorbed by the globin portions of hemoglobin, and the bicarbonate ions diffuse out of the red blood cells to be carried in the plasma. Hemoglobin, which combines with a hydrogen ion, is called reduced hemoglobin and can be symbolized as HHb. HHb plays a vital role in maintaining the pH of the blood.

The Risks of Smoking and Benefits of Quitting

Based on available statistics, the American Cancer Society informs us that smoking carries a high risk. Among the risks of smoking are the following:

Shortened life expectancy A 25-year-old who smoke 2 packs of cigarettes a day has a life expectancy 8.3 yrs shorter than a nonsmoker. The greater the number of packs smoked, the shorter the life expectancy.

Lung cancer The first event appears to be a thickening of the cells that line the bronchi. Then there is a loss of cilia so that it is impossible to prevent dust and dirt from settling in the lungs. Following this, cells with atypical nuclei appear in the thickened lining. A disordered collection of cells with atypical nuclei may be considered to be cancer in situ (at one location). The final step occurs when some cells break loose and penetrate the other tissues, a process called metastasis. This is true cancer (fig. 37.A*b*).

Cancer of the larynx, mouth, esophagus, bladder, and pancreas The chances of developing these cancers are from 2 to 17 times higher in cigarette smokers than in nonsmokers.

Emphysema Cigarette smokers have 4 to 25 times greater risk of developing emphysema. Damage is seen in the lungs of even young smokers. Smoking causes the lining of the bronchioles to thicken. If a large part of the lungs is involved, the lungs are permanently inflated and the chest balloons out due to this trapped air. The victim is breathless and has a cough. Since the surface area for gas exchange is reduced, not enough oxygen reaches the heart and brain. The heart works furiously to force more blood through the lungs, which may lead to a heart condition. Lack of oxygen for the brain may make the person feel depressed, sluggish, and irritable.

Coronary heart disease Cigarette smoking is the major factor in 120,000 additional U.S. deaths from coronary heart disease each year.

Reproductive effects Smoking mothers have more stillbirths and low-birthweight babies who are more vulnerable to disease and death. Children of smoking mothers are smaller and underdeveloped physically and socially even seven years after birth.

In the same manner, the American Cancer Society informs smokers of the benefits of quitting. These benefits include the following:

Risk of premature death is reduced Do not smoke for 10 to 15 years, and the risk of death due to any one of the cancers mentioned approaches that of the nonsmoker.

Health of respiratory system improves The cough and excess sputum disappear during the first few weeks after quitting. As long as cancer has not yet developed, all the ill effects mentioned can reverse themselves and the lungs can become healthy again. In patients with emphysema, the rate of alveoli destruction is reduced and lung function may improve.

Coronary heart disease risk sharply decreases After only one year the risk factor is greatly reduced, and after 10 years an exsmoker's risk is the same as that of those who never smoked.

The increased risk of having stillborn children and underdeveloped children disappears Even for women who do not stop smoking until the fourth month of pregnancy, such risks to infants are decreased.

People who smoke must ask themselves if the benefits of quitting outweigh the risks of smoking.

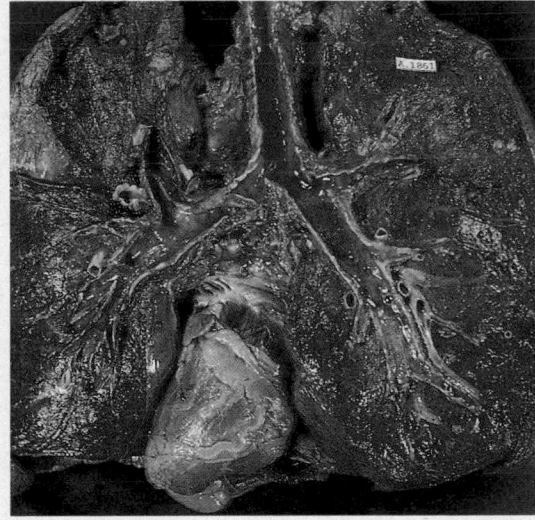
a.

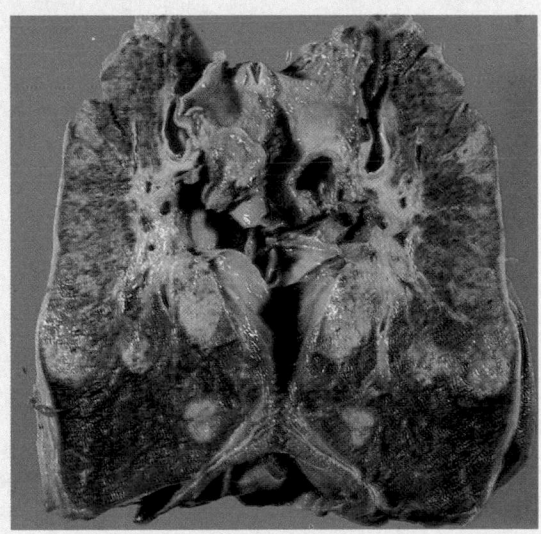
b.

Figure 37.A
a. Normal lungs with heart in place. Notice the healthy red color. *b.* Lungs of a heavy smoker. Notice how black the lungs are except where cancerous tumors are located.

As blood enters the pulmonary capillaries, most of the carbon dioxide is present in plasma as the bicarbonate ion. The little free carbon dioxide remaining begins to diffuse out, and the following reaction is driven to the right:

$$H^+ + HCO_3^- \longrightarrow H_2CO_3 \longrightarrow H_2O + CO_2$$

Carbonic anhydrase also speeds up this reaction, during which hemoglobin gives up the hydrogen ions it has been carrying, HHb becoming Hb.

Gas exchange in the lungs and the tissues is dependent upon the process of diffusion. Oxygen is transported in the blood by hemoglobin, and carbon dioxide is carried as the bicarbonate ion.

Types of Hemoglobin

Muscle cells contain an oxygen-binding pigment called *myoglobin*, which has a higher affinity for oxygen than does hemoglobin. Myoglobin tends to hold oxygen until the PO_2 falls very low. Myoglobin provides an excellent reserve source of oxygen for muscle cells when they are contracting and metabolizing rapidly.

Fetal hemoglobin has a higher affinity for oxygen than adult hemoglobin at the PO_2 levels found in the placenta. This facilitates transfer of oxygen from mother's blood to fetal blood. After birth, fetal hemoglobin is gradually replaced with the adult type of hemoglobin.

Respiration and Health

The full length of the respiratory tract is lined with a warm, moist mucous membrane that is constantly exposed to environmental air. The quality of this air, determined by the microorganisms and pollutants it contains, can affect our health.

Common Respiratory Infections

Droplets from one single sneeze can be loaded with billions of bacteria or viruses. The pseudostratified, ciliated, columnar epithelium of the trachea (p. 540) is protected by the production of mucus and by the constant beating of the cilia. If the number of infective agents is large and/or our resistance is reduced, however, a respiratory infection such as a cold or the flu can result.

Lung Disorders

Pneumonia and tuberculosis are 2 serious infections of the lungs that ordinarily can be controlled by antibiotics. Two other illnesses, emphysema and lung cancer, are not due to infections; in most instances, they are due to cigarette smoking, as discussed in this chapter's reading.

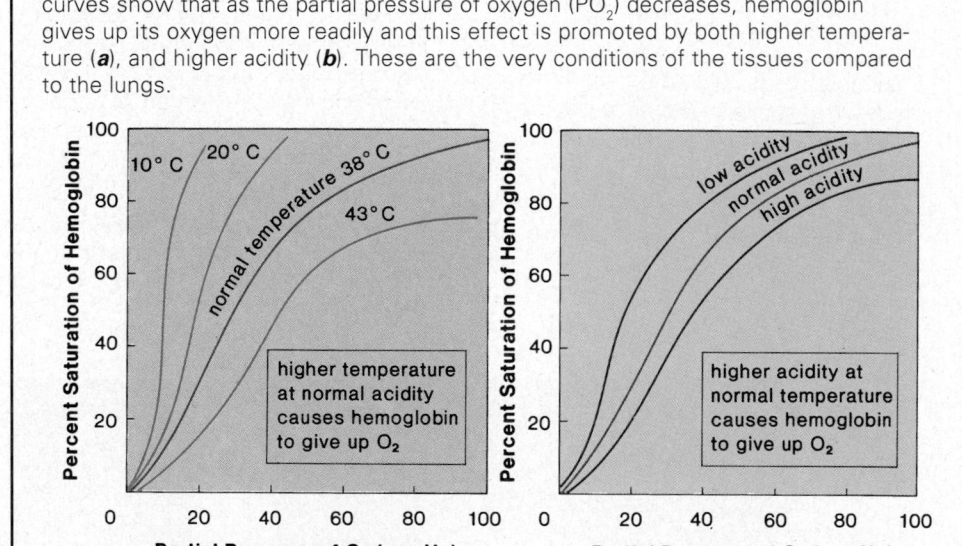

Figure 37.11

Properties of hemoglobin as revealed by hemoglobin dissociation curves. These curves show that as the partial pressure of oxygen (PO_2) decreases, hemoglobin gives up its oxygen more readily and this effect is promoted by both higher temperature (**a**), and higher acidity (**b**). These are the very conditions of the tissues compared to the lungs.

Summary

1. Some aquatic animals like hydras and planarians use their entire body surface for gas exchange.

2. Most animals have a localized, special gas-exchange area. Most aquatic animals pass water over gills. On land, insects utilize tracheal systems and vertebrates have lungs.

3. Lungs are found inside the body, where water loss is reduced. To ventilate the lungs, some vertebrates use positive pressure, but most inhale, using muscular contraction to produce a negative pressure that causes air to rush into the lungs. When the breathing muscles relax, air is exhaled.

4. Birds have a series of air sacs that allow a one-way flow of air over the gas-exchange area. This is called a complete ventilation system.

5. Table 37.1 lists the structures found in the human respiratory system.

6. Humans breathe by negative pressure, as do other mammals. During inhalation, the rib cage goes up and out, and the diaphragm lowers. The lungs expand and air comes rushing in.

7. During exhalation, the rib cage goes down and in, and the diaphragm rises. Therefore, air goes out.

8. The rate of breathing is dependent upon the amount of carbon dioxide in the blood, as detected by chemoreceptors such as the aortic and carotid bodies.
9. Gas exchange in the lungs and tissues is brought about by diffusion. Hemoglobin transports oxygen in the blood; carbon dioxide is mainly transported in plasma as the bicarbonate ion. The enzyme carbonic anhydrase found in red blood cells speeds up the formation of the bicarbonate ion.
10. The respiratory tract is subject to infections. Two disorders of the lungs, emphysema and cancer, are usually due to cigarette smoking.

Study Questions

1. Compare the respiratory organs of aquatic animals to those of terrestrial animals.
2. How does the countercurrent flow of blood within gill capillaries and water passing across the gills assist respiration in fishes?
3. Why don't insects require a blood respiratory pigment, and why is it beneficial for the body wall of earthworms to be moist?
4. Explain the phrase, "breathing by using negative pressure."
5. Contrast incomplete ventilation in humans with complete ventilation in birds.
6. Name the parts of the human respiratory system and list a function for each part.
7. The concentration of what gas in the blood controls the breathing rate in humans? Explain.

Objective Questions

1. One problem faced by terrestrial animals with lungs, but not by freshwater aquatic animals with gills, is
   a. gas exchange involves water loss.
   b. breathing requires considerable energy.
   c. oxygen diffuses very slowly in air.
   d. All of these.
2. Which of these exemplifies that not all active animals require that the circulatory system transport gases?
   a. mouse
   b. dragonfly
   c. trout
   d. sparrow
3. Birds have a more efficient lung than humans because the flow of air is
   a. the same during both inhalation and exhalation.
   b. in only one direction through the lungs.
   c. never backed up like in human lungs.
   d. not hindered by a larynx.

4. Which animal breathes by positive pressure?
   a. fish
   b. human
   c. bird
   d. frog
5. Which of these is a true statement?
   a. In lung capillaries, carbon dioxide combines with water to give carbonic acid.
   b. In tissue capillaries, carbonic acid breaks down to carbon dioxide and water.
   c. In lung capillaries, carbonic acid breaks down to carbon dioxide and water.
   d. In tissue capillaries, carbonic acid combines with hydrogen ions to form the carbonate ion.
6. Air comes into the human lungs because
   a. atmospheric pressure is less than the pressure inside the lungs.
   b. atmospheric pressure is greater than the pressure inside the lungs.
   c. although the pressures are the same inside and outside, the partial pressure of oxygen is lower within the lungs.

   d. the residual air in the lungs causes the partial pressure of oxygen to be less than outside.
7. If the digestive and respiratory tracts were completely separate in humans, there would be no need for
   a. swallowing.
   b. external nares.
   c. an epiglottis.
   d. a diaphragm.
8. To trace the path of air in humans you would place the trachea
   a. directly after the nose.
   b. directly before the bronchi.
   c. before the pharynx.
   d. Both a and c.
9. In humans, the respiratory center
   a. is stimulated by carbon dioxide.
   b. is located in the medulla oblongata.
   c. controls the rate of breathing.
   d. All of these.
10. Carbon dioxide is carried in the plasma
   a. in combination with hemoglobin.
   b. as the bicarbonate ion.
   c. combined with carbonic anhydrase.
   d. All of these.

11. Label this diagram of the human respiratory system:

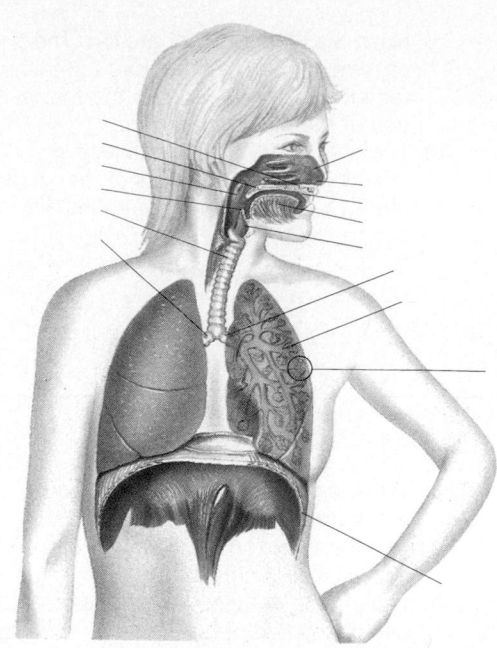

Concepts and Critical Thinking

1. *All systems of the animal's body contribute to homeostasis.*

Tell 2 ways the human respiratory system contributes to homeostasis.

2. *Organisms have localized boundaries for exchanging materials with the environment.*

Compare the similarities and differences between gills and lungs and the obvious localized boundaries with the environment.

3. *The structure of an organ suits its function.*

How does the structure of the lungs suit their function?

Selected Key Terms

gills (gilz) 607
tracheal system (tra′ke-al sis′tem) 608
lung (lung) 608
diaphragm (di′ah-fram) 609
inhalation (in″hah-la′shun) 609
exhalation (eks″hah-la′shun) 609

pharynx (far′inks) 612
glottis (glot′is) 612
larynx (lar′inks) 612
trachea (tra′ke-ah) 612
epiglottis (ep″ĭ-glot′is) 612
bronchus (brong′kus) 612

bronchiole (brong′ke-ōl) 612
alveolus (al-ve′o-lus) 612
hemoglobin (he″mo-glo-bin) 614
bicarbonate ion (bi-kar′bo-nāt i′on) 614
carbonic anhydrase (kar-bon′ik an-hi′drās) 614

38

Excretory System

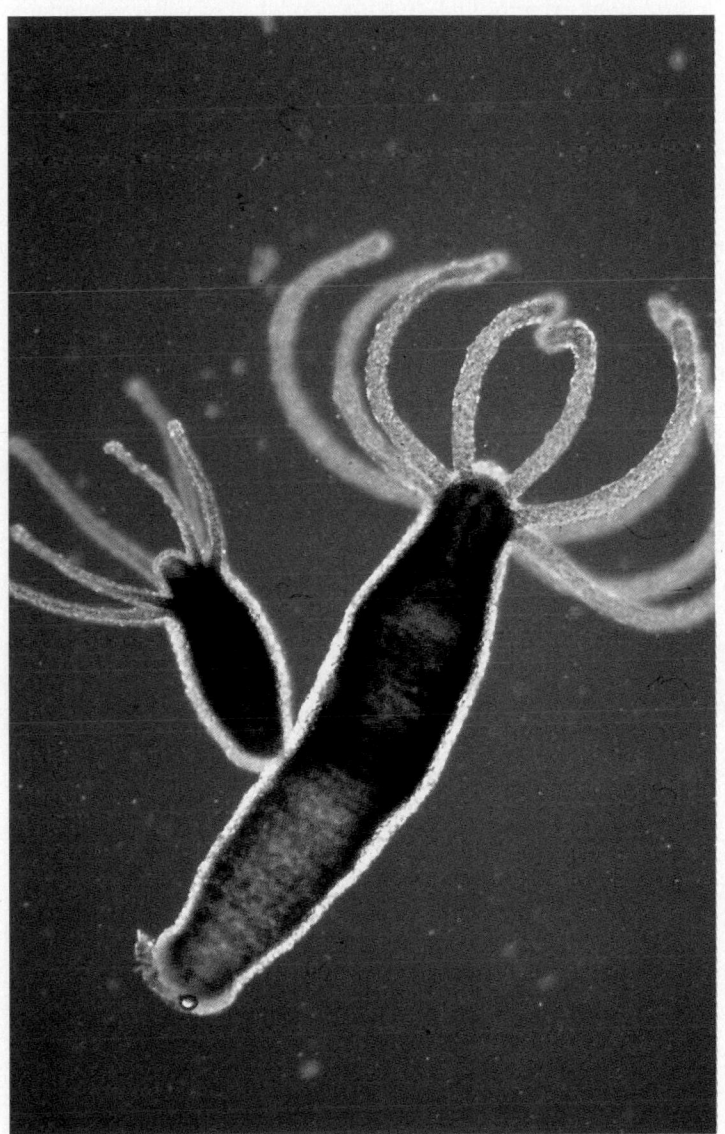

Some aquatic animals, like the hydra pictured here, don't have an excretory system. The body wall has only 2 layers of cells, and water enters and exits a large, central, fluid-filled cavity. Each cell excretes metabolic wastes directly into water, which washes them away. Magnification, X8.

Your study of this chapter will be complete when you can

1. relate the excretion of ammonia, urea, and uric acid to the animal's environment;
2. give examples of how various animals regulate the water and salt balance of the body;
3. contrast the manner in which marine bony fishes and freshwater bony fishes regulate the water and salt content of the blood;
4. compare the operation of planarian flame cells, earthworm nephridia, insect Malpighian tubules, and human kidneys;
5. trace the path of urine in humans, and describe in general the structure and function of each organ mentioned;
6. list the parts of the kidney nephron, and relate these to the macroscopic anatomy of the kidney;
7. describe the 3 steps in urine formation, and relate these to the parts of the nephron;
8. describe how water excretion is regulated and how the pH of the blood is adjusted by the kidneys;
9. tell, in general, how an artificial kidney machine works.

ike the digestive and respiratory systems, the excretory system plays a major role in maintaining homeostasis. Excretion is the elimination of molecules that have taken part in metabolic reactions and should not be confused with defecation, which is the elimination of nondigested material from the gut.

Physiology of Excretion

The excretory system is largely responsible for ridding the body of nitrogenous and other waste products (fig. 38.1). It also helps regulate the amount of water and ions in body fluids.

Nitrogenous Wastes

The breakdown of various molecules, including nucleic acids and amino acids, results in nitrogenous wastes. For simplicity's sake, however, we will limit our discussion to amino acid metabolism. Amino acids, derived from protein in food, can be used by cells for synthesis of new body protein or other nitrogen-containing molecules. The amino acids not used for synthesis are oxidized to generate energy or are converted to fats or carbohydrates that can be stored. In either case, the *amino groups* ($-NH_2$) must be removed because they are not needed for any of these purposes (fig. 38.2). Once the amino groups have been removed from amino acids, they may be excreted from the body in the form of ammonia, urea, or uric acid, depending on the species (table 38.1). Removal of amino groups from amino acids requires a fairly constant amount of energy. The energy requirement for the conversion of amino groups to either ammonia, urea, or uric acid, however, differs, as indicated in figure 38.2.

Ammonia

Amino groups removed from amino acids immediately form *ammonia* (NH_3) by addition of a third hydrogen ion. Therefore, little or no energy is required to convert an amino group to ammonia. Ammonia is quite toxic and can only be used as a nitrogenous excretory product if a good deal of water is available to wash it from the body. The high solubility of ammonia permits this means of excretion in bony fishes, aquatic invertebrates, and amphibians whose gills and skin surfaces are in direct contact with the water of the environment.

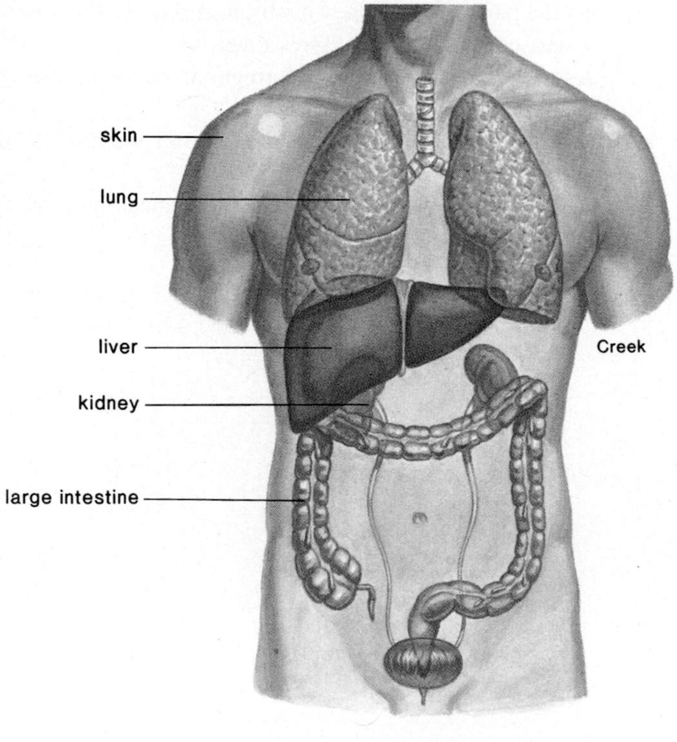

Figure 38.1

The internal environment of cells (blood and tissue fluid) in humans and other animals stays relatively constant because the blood is continually refreshed by certain organs of the body.

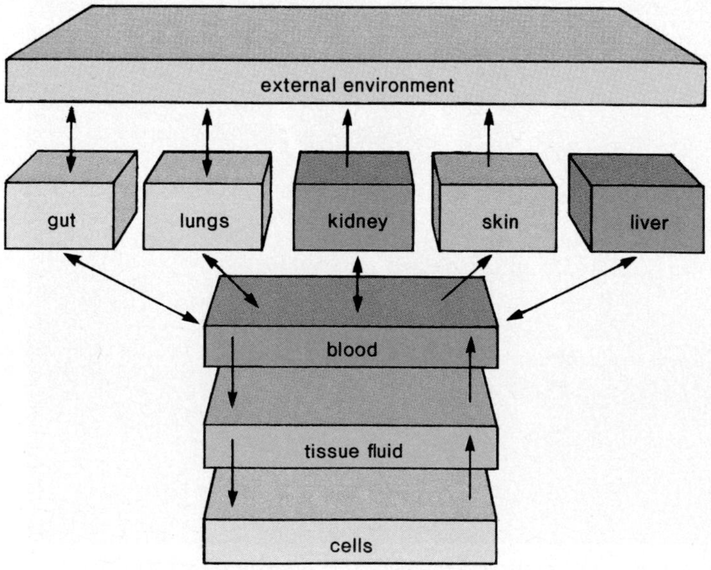

a. These organs are involved in excretion—ridding the body of metabolic wastes. In particular, the gut (large intestine) excretes heavy metals, the lungs excrete carbon dioxide, and the kidneys and skin excrete nitrogenous wastes. The liver removes toxic substances from the blood and converts them to molecules that are excreted by the kidneys. *b.* Diagrammatic representation of exchanges that occur in the body and with the external environment. The arrows that point toward the external environment represent pathways of excretion.

a.

b.

Animal Structure and Function

Figure 38.2

Nitrogenous wastes from protein breakdown. Proteins are hydrolyzed to amino acids, whose breakdown results in a carbon skeleton and amino groups. The carbon skeleton can be used as an energy source, but the amino groups must be excreted as either ammonia, urea, or uric acid. The energy and water requirements for these vary as indicated.

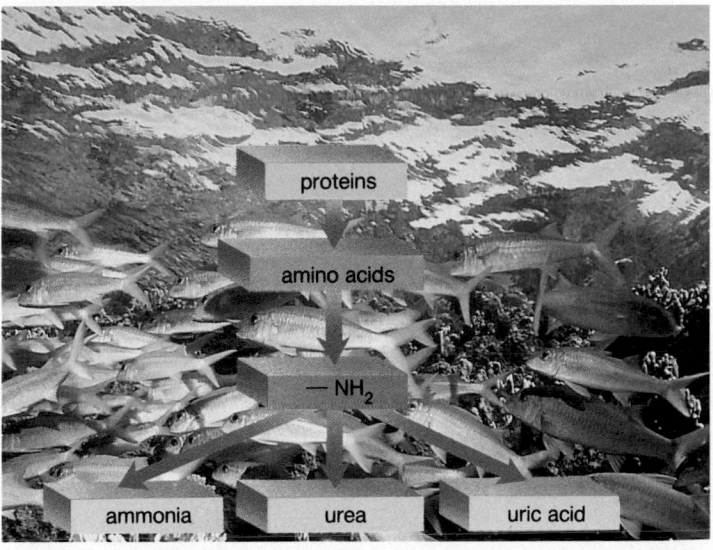

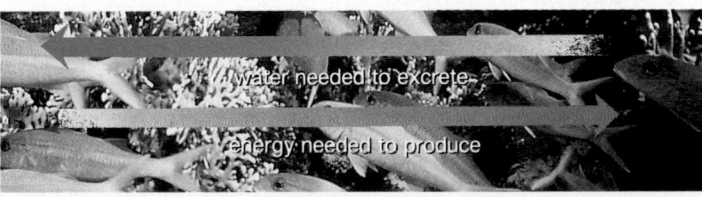

Table 38.1
Nitrogenous Waste Excretion

Product	Habitat	Animals
Ammonia	Water	Aquatic invertebrates Bony fishes Amphibian larvae
Urea	Land	Adult amphibians Mammals
Uric acid	Land	Insects Birds Reptiles

Urea

Terrestrial amphibians and mammals usually excrete **urea** as their main nitrogenous waste. Urea is much less toxic than ammonia and can be excreted in a moderately concentrated solution. This allows body water to be conserved, an important advantage for terrestrial animals with limited access to water.

Production of urea, however, requires the expenditure of energy. Urea is produced in the liver by a set of energy-requiring enzymatic reactions known as the *urea cycle*. In the cycle, carrier molecules take up carbon dioxide and 2 molecules of ammonia finally releasing urea.

Uric Acid

Uric acid is excreted by insects, reptiles, birds, and some dogs (e.g., dalmatians). Uric acid is not very toxic and is poorly soluble in water. Poor solubility is an advantage for water conservation because uric acid can be concentrated even more readily than can urea. In reptiles and birds, a dilute solution of uric acid passes from the kidneys to the *cloaca,* a common reservoir for the products of the digestive, urinary, and reproductive systems. After water is absorbed by the cloaca, the uric acid passes out with the feces.

Embryos of reptiles and birds develop inside shelled eggs that are completely enclosed. The production of insoluble, relatively nontoxic uric acid is advantageous for shelled embryos because all nitrogenous wastes are stored inside the shell until hatching takes place.

Uric acid is synthesized by a long, complex series of enzymatic reactions that require expenditure of even more ATP than does urea synthesis. Here again, there seems to be a trade-off between the advantage of water conservation and the disadvantage of energy expenditure for synthesis of an excretory molecule.

> Animals excrete nitrogenous wastes derived from protein breakdown as either ammonia, urea, or uric acid. Ammonia requires the most amount of water to excrete; uric acid requires the most amount of energy to produce. Habitat plays a primary role in determining which type of molecule is excreted.

Osmotic Regulation

In addition to their role in excretion of nitrogenous wastes, excretory organs also have the important function of regulating the water and salt balance of the body. Figure 38.3 shows that among animals, only marine invertebrates and cartilaginous fishes, such as sharks and rays, have body fluids that are nearly isotonic to seawater. These organisms have little difficulty maintaining their normal salt and water balance. Surprising, though, is the observation that while they are isotonic, the body fluids of cartilaginous fishes do not contain the same amount of salt as seawater. The answer to this paradox is that their blood contains a concentration of urea high enough to match the tonicity of the sea! For some unknown reason, this amount of urea is not toxic to them.

The body fluids of all bony fishes have only a moderate amount of salt. Apparently, their common ancestor evolved in fresh water, and only later did some groups invade the sea. Marine bony fishes (fig. 38.4a) are therefore prone to water loss and could become dehydrated. To counteract this, they drink seawater almost constantly. On the average, marine bony fishes swallow an amount of water estimated to be equal to 1% of their body weight every hour. This is equivalent to a human drinking about 700 ml of water every hour around the clock. While they get water by drinking, this habit also causes these fishes to acquire salt. Instead of forming a hypertonic urine, however, they actively transport sodium (Na⁺) and chloride (Cl⁻) ions into the surrounding seawater at the gills.

It is easy to see that the osmotic problems of freshwater bony fishes are exactly opposite to those of marine bony fishes (fig. 38.4*b*). The body fluids of freshwater bony fishes are hypertonic to fresh water, and they are prone to gain water. These fishes never drink water but instead eliminate excess water through production of large quantities of dilute (hypotonic) urine. They discharge a quantity of urine equal to one-third their body weight each day. Because they tend to lose salts, they actively transport salts into the blood across the membranes of their gills.

Excretory organs help regulate the water and salt balance of the body. For example, marine bony fishes drink water constantly and excrete salt from the gills. Freshwater bony fishes never drink water and excrete a dilute urine.

The difference in adaptation between marine and freshwater bony fishes makes it remarkable that some fishes actually can move between the 2 environments during their life cycle. Salmon, for example, begin their lives in freshwater streams and rivers, move to the ocean for a period of time, and finally return to fresh water to breed. These fishes alter their behavior and their gill and kidney functions in response to the osmotic changes they encounter when moving from one environment to the other.

Like marine bony fishes, some animals that evolved on land are also able to drink seawater despite its high toxicity. Birds

Figure 38.3

Comparison of relative ion concentration of animal body fluids and seawater (blue horizontal plane above) and fresh water (blue horizontal plane below). For example, marine invertebrates are the only animals with fluids with the ionic concentration of seawater, and freshwater invertebrates are the only animals with fluids that approach the ionic concentration of fresh water.

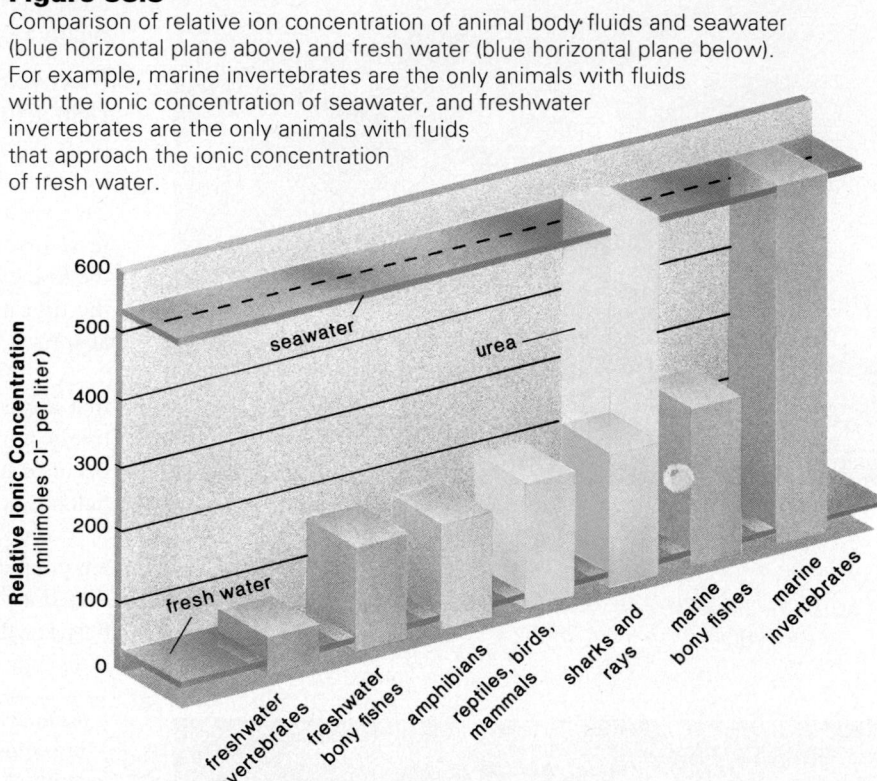

Figure 38.4

Water and salt balances in bony fishes. The black arrows represent passive transport from the environment, and the colored arrows represent active transport by the fishes to counteract environmental pressures. *a.* Marine bony fish. *b.* Freshwater bony fish.

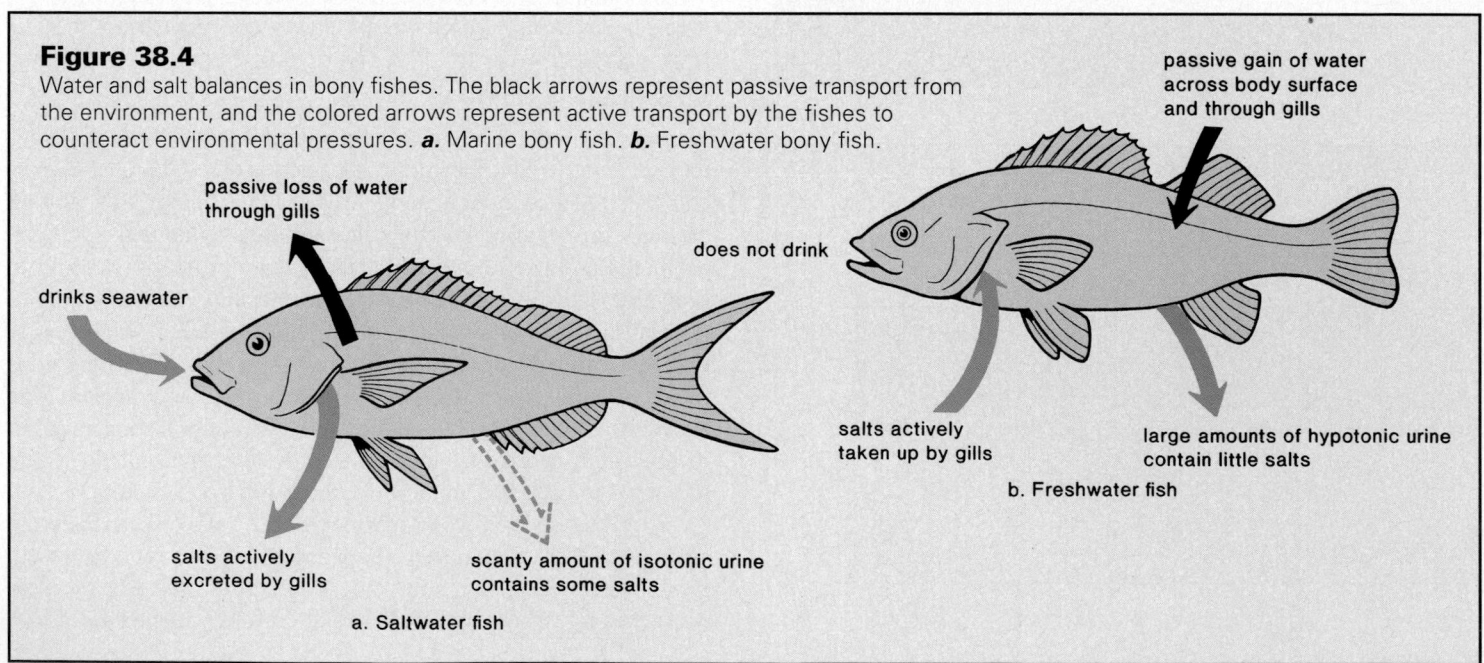

passive loss of water through gills

drinks seawater

salts actively excreted by gills

scanty amount of isotonic urine contains some salts

a. Saltwater fish

passive gain of water across body surface and through gills

does not drink

salts actively taken up by gills

large amounts of hypotonic urine contain little salts

b. Freshwater fish

Animal Structure and Function

Figure 38.5

a. The flame cell excretory system in planarians. Two or more tracts of branching tubules run the length of the body and open to the outside by pores. At the ends of side branches there are small flame cells, bulblike cells whose beating cilia causes fluid to enter the tubules that remove excess fluid from the body. **b.** The earthworm nephridium. The nephridium has a ciliated opening, the nephridiostome, that leads to a coiled tubule surrounded by a capillary network. Urine can be temporarily stored in the bladder before being released to the outside via a pore termed a nephridiopore. Most segments contain a pair of nephridia, one on each side.

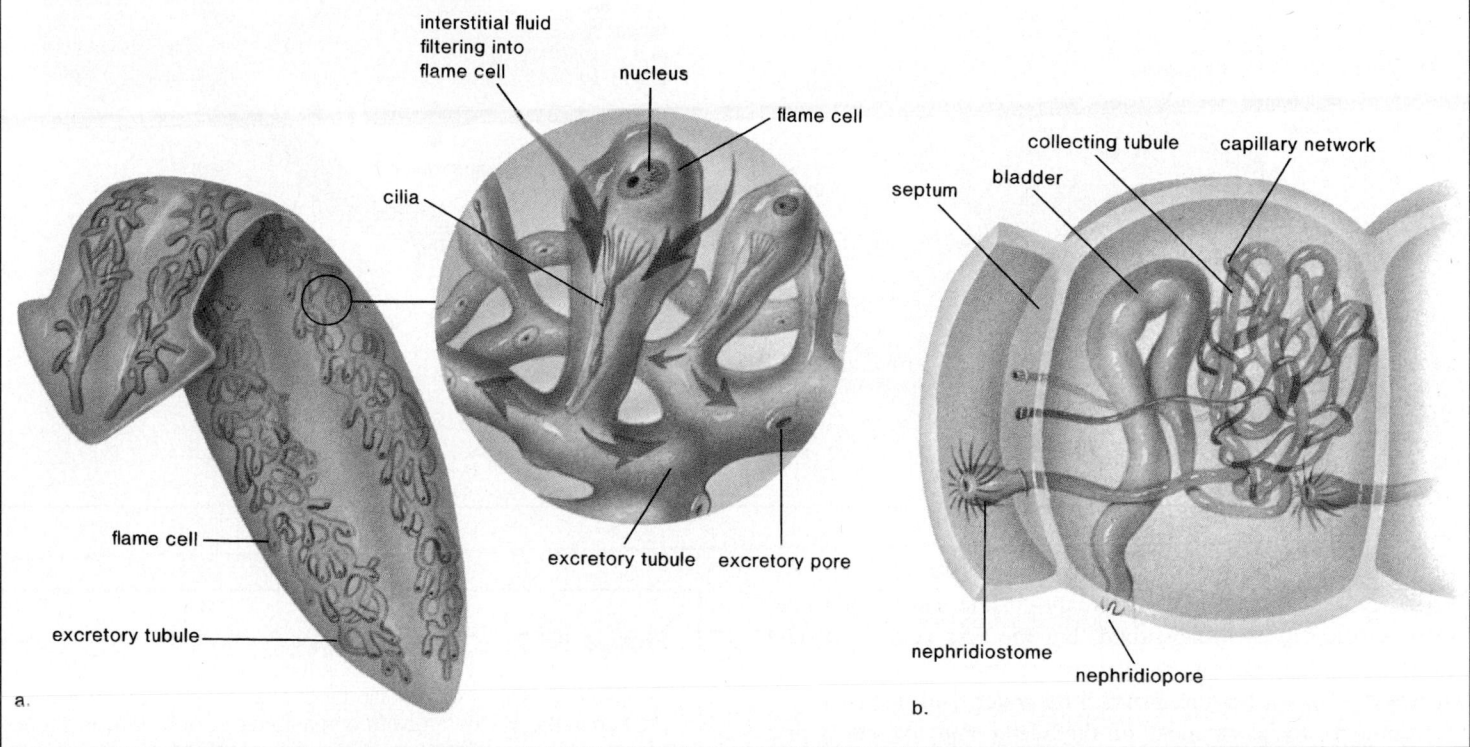

and reptiles that live near the sea have a nasal salt gland that can excrete large volumes of concentrated salt solution. Mammals that live at sea, like whales, porpoises, and seals, most likely can concentrate their urine enough to drink salt water. Humans cannot do this, and they die if they drink only seawater.

Most terrestrial animals need to drink water occasionally, but the kangaroo rat manages to get along without drinking water at all. It forms a very concentrated urine, and its fecal material is almost completely dry. These abilities allow it to survive using metabolic water derived from the breakdown of nutrient molecules alone.

Organs of Excretion

Most animals have tubular organs that function in the excretion of nitrogenous wastes and osmotic regulation. Sometimes the primary function of these organs is osmotic regulation.

Flame Cells in Planarians

Planarians have 2 strands of branching excretory tubules that open to the outside of the body through excretory pores (fig. 38.5a). Located along the tubules are bulblike *flame cells*, each of which contains a cluster of beating cilia that looks like a flickering flame under the microscope. The beating of flame-cell cilia propels fluid through the excretory canals and out of the body. The system is believed to function in osmotic regulation and also in excreting wastes.

Nephridia in Earthworms

The earthworm's body is divided into segments, and nearly every body segment has a pair of excretory structures called **nephridia** (nephridium, sing.). Each nephridium is a tubule with a ciliated opening (the nephridiostome) and an excretory pore (the nephridiopore) (fig. 38.5b). As fluid from the body cavity is propelled through the tubule by beating cilia, certain substances are reabsorbed and carried away by a network of capillaries surrounding the tubule. This process results in the formation of urine that contains only metabolic wastes, salts, and water.

Each day, an earthworm excretes a lot of water and may produce a volume of urine equal to 60% of its body weight. Its excretion of ammonia is consistent with this finding.

Malpighian Tubules in Insects

Insects have a unique excretory system consisting of long, thin tubules,(i.e., **Malpighian tubules)** attached to the gut (fig. 38.6). Water and uric acid simply flow from the surrounding hemolymph

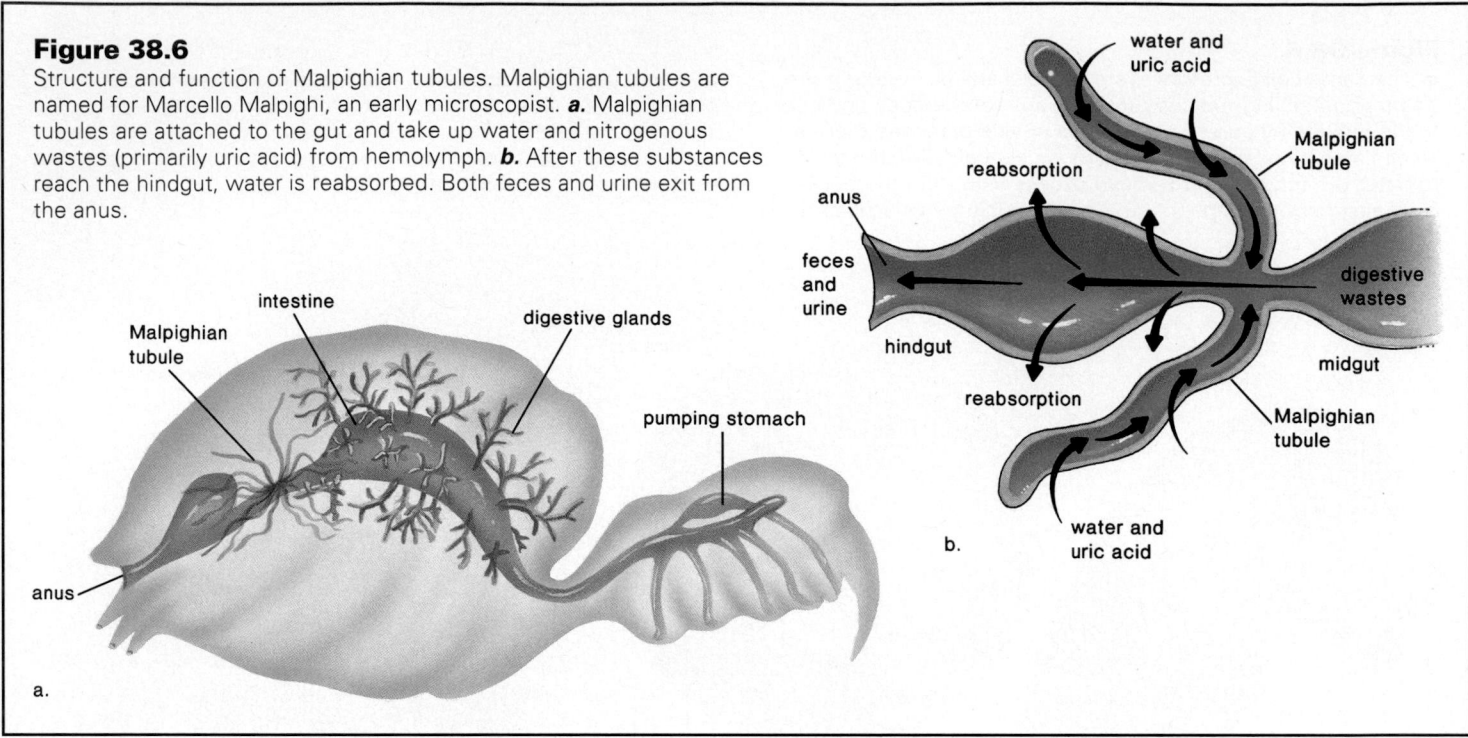

Figure 38.6

Structure and function of Malpighian tubules. Malpighian tubules are named for Marcello Malpighi, an early microscopist. *a.* Malpighian tubules are attached to the gut and take up water and nitrogenous wastes (primarily uric acid) from hemolymph. *b.* After these substances reach the hindgut, water is reabsorbed. Both feces and urine exit from the anus.

into these tubules before moving into the gut. Here, water and other useful substances are reabsorbed, but the uric acid eventually passes out of the gut. Insects that live in water or that eat large quantities of moist food reabsorb little water. But insects in dry environments reabsorb most of the water and excrete a dry, semisolid mass of uric acid.

Most animals have a primary excretory organ for excreting nitrogenous wastes and regulating water and salt balance. Examples are flame cells in planaria, nephridia in earthworms, and Malpighian tubules in insects and spiders.

Human Kidney

The human **kidneys** are bean-shaped, reddish brown organs, each about the size of a fist. They are located one on either side of the vertebral column just below the diaphragm, where they are partially protected by the lower rib cage. They are a part of the human urinary system (fig. 38.7). This system is composed not only of the kidneys, which make **urine,** but also of organs that conduct urine from the body. Each kidney is connected to a **ureter,** a duct that carries urine from the kidney to the **urinary bladder,** where it is stored until it is voided from the body through the single **urethra.** In males, the urethra passes through the penis, and in females, it opens ventral to the opening of the vagina.

Structure

If a kidney is sectioned longitudinally, 3 major parts can be distinguished (fig. 38.8). The outer region is the *cortex*, which has a somewhat granular appearance. The *medulla* lies on the inner side

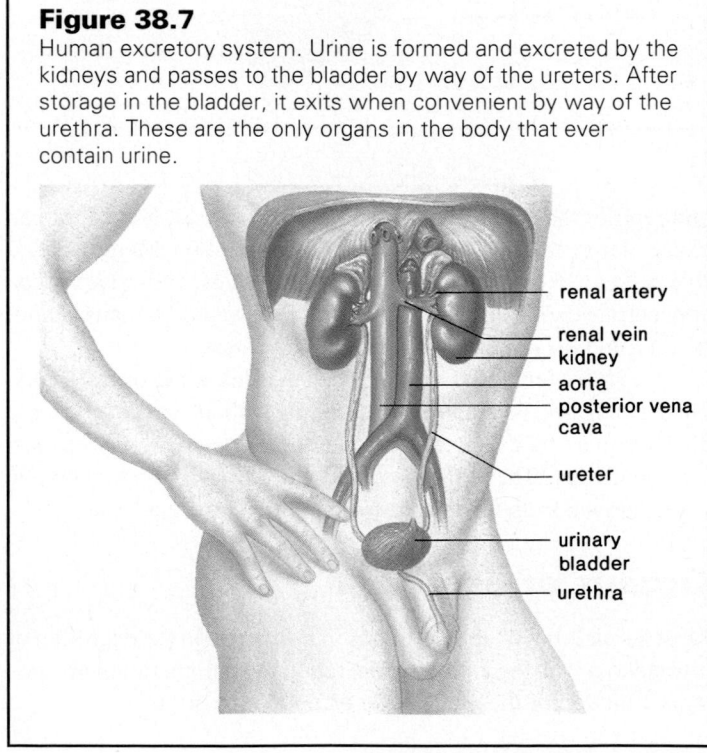

Figure 38.7

Human excretory system. Urine is formed and excreted by the kidneys and passes to the bladder by way of the ureters. After storage in the bladder, it exits when convenient by way of the urethra. These are the only organs in the body that ever contain urine.

of the cortex and is arranged in a group of pyramid-shaped regions, each of which has a striped appearance. The innermost part of the kidney is a hollow chamber called the *renal pelvis*. Urine formed in the kidney collects in the renal pelvis before entering the ureter.

Animal Structure and Function

Figure 38.8

Macroscopic and microscopic anatomy of the kidney. *a.* Longitudinal section of kidney showing location of cortex, medulla, and renal pelvis. *b.* An enlargement of one renal pyramid showing the placement of nephrons.

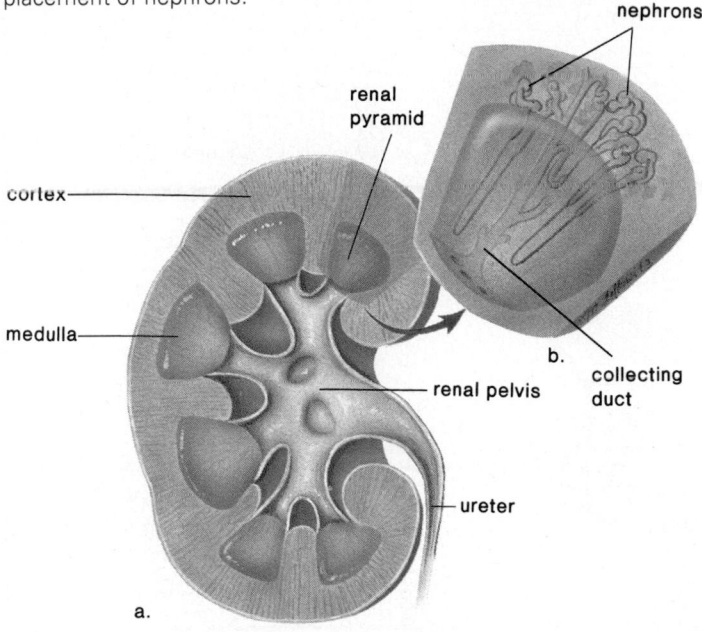

Nephrons

Microscopically, each kidney is composed of about one million tiny tubules called **nephrons.** Some nephrons are located primarily in the cortex, but others dip down into the medulla. Each nephron is made of several parts (fig. 38.9).

The blind end of a nephron is pushed in on itself to form a cuplike structure called the **glomerular capsule** (formerly called Bowman's capsule). Next, there is a region known as the **proximal** (near the glomerular capsule) **convoluted tubule,** which leads to a narrow U-turn known as the **loop of the nephron** (formerly called loop of Henle). This is followed by the **distal** (far from the glomerular capsule) **convoluted tubule.** Several distal convoluted tubules enter one **collecting duct.** The collecting duct transports urine down through the medulla and delivers it to the pelvis. The loop of the nephron and the collecting duct give the pyramids of the medulla their striped appearance (fig. 38.8).

Each nephron has its own blood supply (fig. 38.9). After the renal artery leaves the aorta to enter the kidney, it branches into numerous small arteries. These small arteries pass to all parts of the kidney and give off tiny arterioles, one for each nephron. Each arteriole, called an *afferent arteriole,* divides to form a capillary tuft, the **glomerulus,** which is surrounded by the glomerular capsule. The glomerular capillaries drain into an *efferent arteriole,* which subsequently branches into a second capillary network around the tubular parts of the nephron. These capillaries, called *peritubular capillaries,* lead to venules that join the renal vein, a vessel that enters the inferior vena cava.

Urine Formation

Human nephrons function to produce urine somewhat like earthworm nephridia in that they exchange molecules with the blood (table 38.2). Urine production requires 3 distinct processes (fig. 38.10):

1. Pressure filtration at the glomerular capsule
2. Reabsorption, including selective reabsorption, at the proximal convoluted tubule in particular
3. Tubular secretion at the distal convoluted tubule in particular

Pressure Filtration

When blood enters the glomerulus, blood pressure is sufficient to cause small molecules, such as nutrients, water, salts, and wastes, to move from the glomerulus to the inside of the glomerular capsule, especially since the glomerular walls are 100 times more permeable than the walls of most capillaries elsewhere in the body. The molecules that leave the blood and enter the glomerular capsule are called the *glomerular filtrate.* Blood proteins and blood cells are too large to be part of this filtrate, so they remain in the blood as it flows into the efferent arteriole. Glomerular filtrate has the same composition as tissue fluid, and if this composition were not altered in other parts of the nephron, death from loss of water (dehydration), loss of nutrients (starvation), and lowered blood pressure would quickly follow. *Selective reabsorption,* however, prevents this from happening.

Reabsorption

Reabsorption from the nephron to the blood takes place through the walls of the proximal convoluted tubule. Nutrients, water, and even some waste molecules diffuse passively back into the peritubular capillary network. Osmotic pressure further influences the movement of water. The nonfilterable proteins remain in the blood and exert an osmotic pressure. Also, after sodium (Na^+) is actively reabsorbed, chlorine ($Cl–$) follows along, and these 2 ions together increase the osmotic pressure of the blood.

Selective reabsorption of nutrient molecules is brought about by active transport. The cells of the proximal convoluted tubule have numerous microvilli, which increase the surface area, and numerous mitochondria, which supply the energy needed for reabsorption. Reabsorption is selective since only molecules recognized by carrier molecules are actively transported through the tubule into the interstitial spaces. From there these molecules diffuse into the peritubular capillary network.

Glucose is an example of a molecule that ordinarily is reabsorbed completely because there is a plentiful supply of carrier molecules for it. However, every substance has a maximum rate of transport, and after all its carriers are in use, any excess in the filtrate will appear in the urine. For example, as reabsorbed levels of glucose approach 400mg/100ml plasma, the rest will appear in the urine. In diabetes mellitus, there is an excess of glucose in the filtrate because the liver fails to store glucose as glycogen.

Tubular Secretion

Tubular secretion is the means by which other nonfilterable wastes are added to the fluid as it passes through the distal convoluted

Figure 38.9

The human nephron. Each kidney contains over one million nephrons. The term *proximal* means nearer; *distal* means farther. In this case, the proximal convoluted tubule is proximal (closer) to the glomerular capsule and the distal convoluted tubule is farther from the glomerular capsule. Arrows indicate flow of blood to and away from the glomerulus.

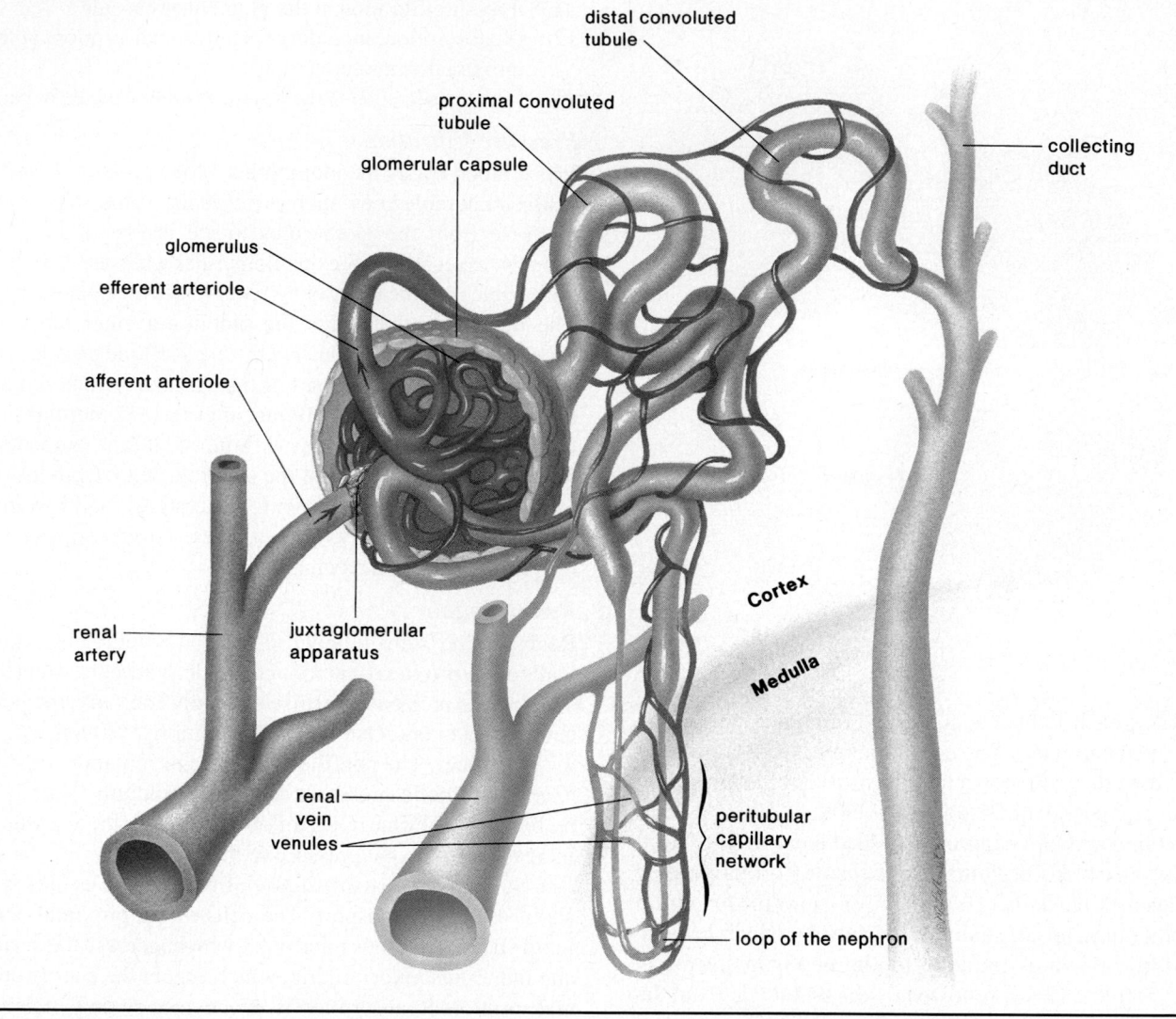

tubule. Toxic substances, such as foreign acids and bases that have been absorbed in the gut, are eliminated by tubular secretion. Penicillin and a number of other substances also are excreted in this way. But the process of tubular secretion is not as important to urine formation as are the first 2 steps studied.

Each step in urine formation is associated with a particular part of the nephron. Pressure filtration occurs at the glomerular capsule; reabsorption occurs at the proximal convoluted tubule; tubular secretion occurs at the distal convoluted tubule. The loop of the nephron plays the primary role in the reabsorption of water so that the urine of humans is hypertonic.

Control of Water Excretion

Reptiles and birds rely primarily on the gut to reabsorb water, but mammals rely on the kidneys. A *countercurrent mechanism* enables mammals to excrete a hypertonic urine. The arrangement of the loop of the nephron in relation to the collecting duct is essential to the process. A concentration gradient is established in the medulla by the extrusion of salt from the ascending limb of the loop of the nephron and by diffusion of urea from the collecting duct. The contents of the descending limb of the loop of the nephron are exposed to an increasing osmotic concentration that draws water from the descending limb into the medulla. From there the water enters the blood capillaries.

Figure 38.10

Steps in urine formation (simplified). *a.* The steps are noted at the location of the nephron where they occur. *b.* The molecules involved in the steps and the processes are listed.

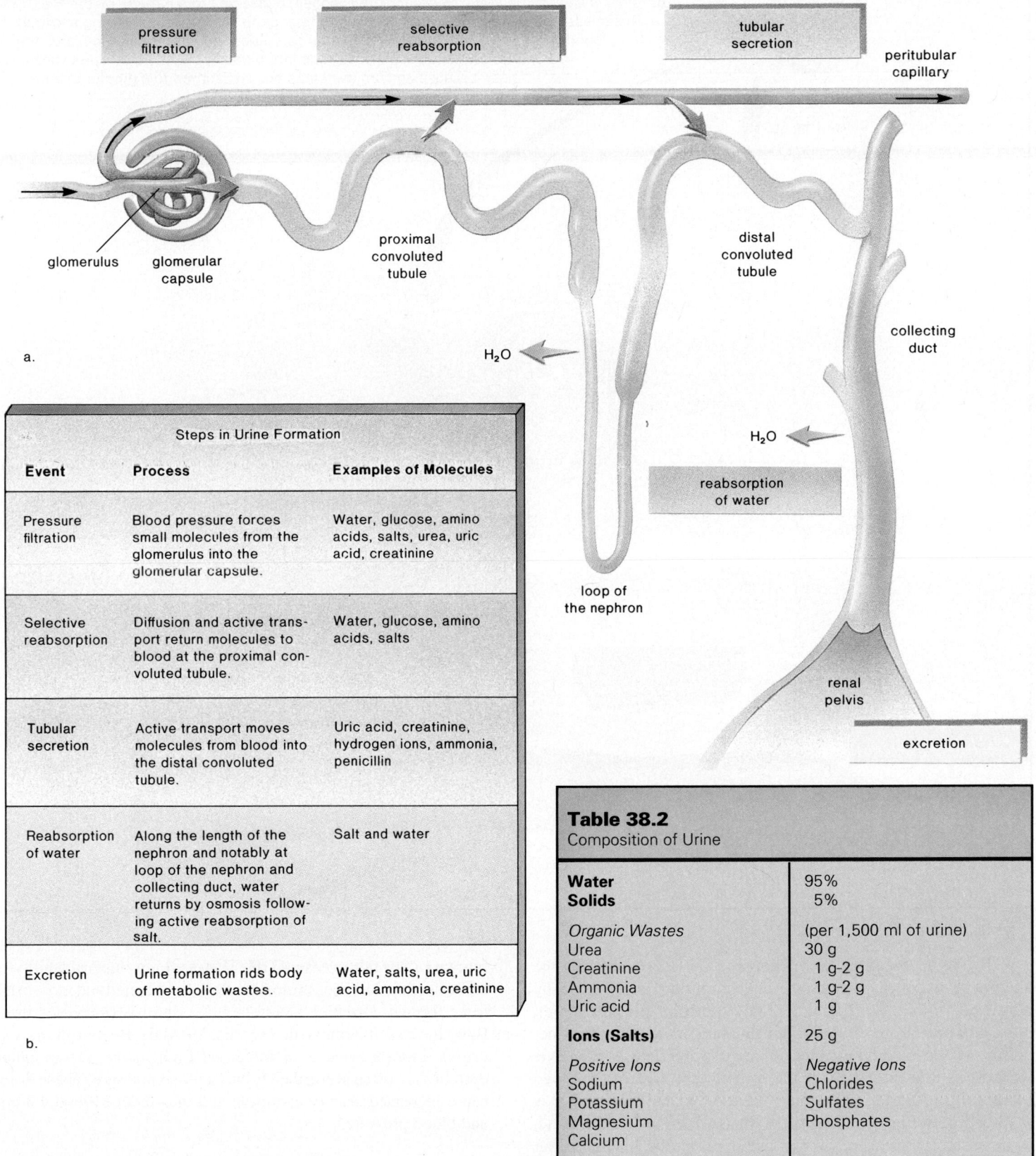

Steps in Urine Formation

Event	Process	Examples of Molecules
Pressure filtration	Blood pressure forces small molecules from the glomerulus into the glomerular capsule.	Water, glucose, amino acids, salts, urea, uric acid, creatinine
Selective reabsorption	Diffusion and active transport return molecules to blood at the proximal convoluted tubule.	Water, glucose, amino acids, salts
Tubular secretion	Active transport moves molecules from blood into the distal convoluted tubule.	Uric acid, creatinine, hydrogen ions, ammonia, penicillin
Reabsorption of water	Along the length of the nephron and notably at loop of the nephron and collecting duct, water returns by osmosis following active reabsorption of salt.	Salt and water
Excretion	Urine formation rids body of metabolic wastes.	Water, salts, urea, uric acid, ammonia, creatinine

b.

Table 38.2
Composition of Urine

Water	95%
Solids	5%
Organic Wastes	(per 1,500 ml of urine)
Urea	30 g
Creatinine	1 g-2 g
Ammonia	1 g-2 g
Uric acid	1 g
Ions (Salts)	25 g

Positive Ions	*Negative Ions*
Sodium	Chlorides
Potassium	Sulfates
Magnesium	Phosphates
Calcium	

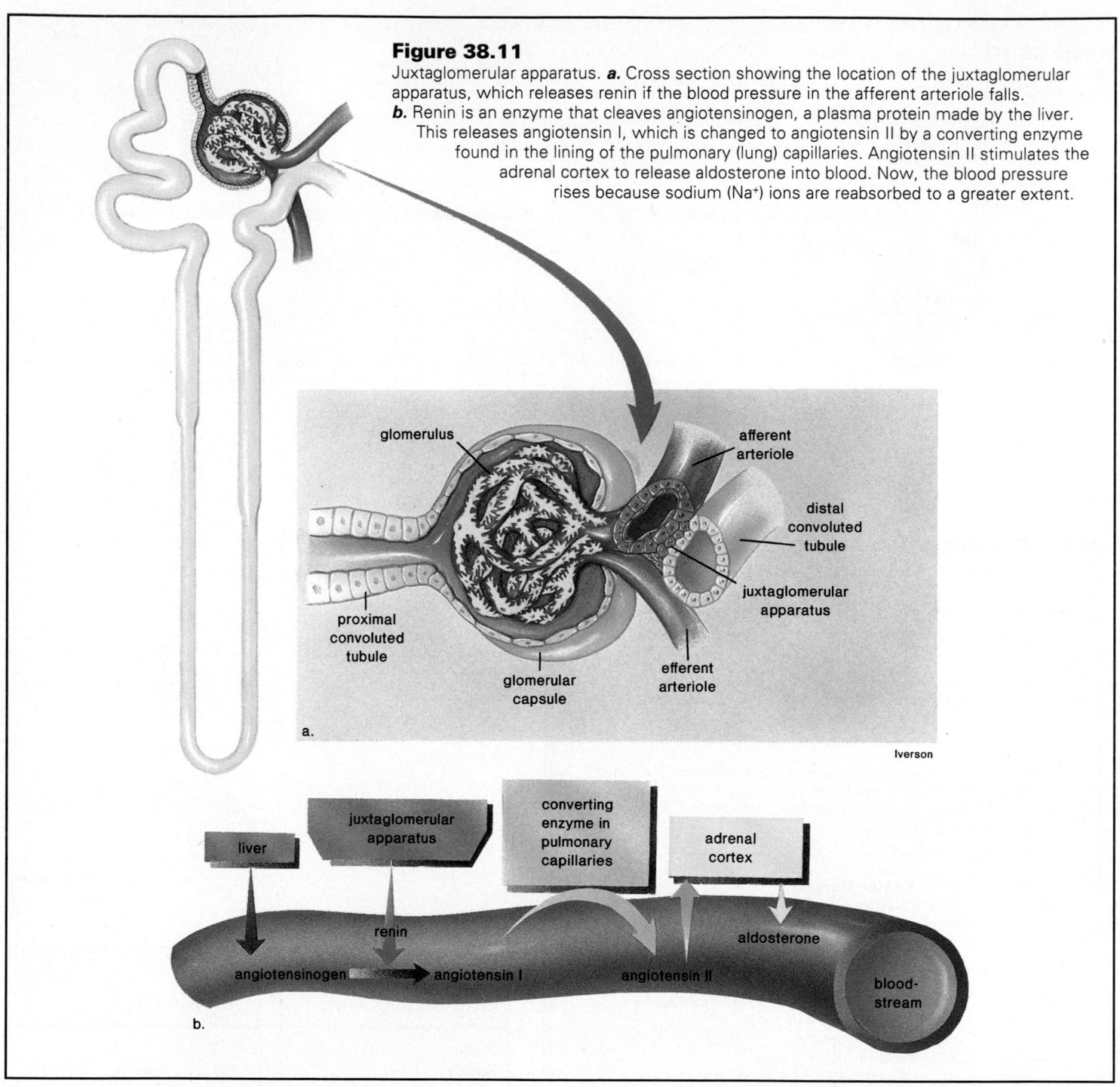

Figure 38.11

Juxtaglomerular apparatus. ***a.*** Cross section showing the location of the juxtaglomerular apparatus, which releases renin if the blood pressure in the afferent arteriole falls.
b. Renin is an enzyme that cleaves angiotensinogen, a plasma protein made by the liver. This releases angiotensin I, which is changed to angiotensin II by a converting enzyme found in the lining of the pulmonary (lung) capillaries. Angiotensin II stimulates the adrenal cortex to release aldosterone into blood. Now, the blood pressure rises because sodium (Na⁺) ions are reabsorbed to a greater extent.

The descending limb automatically loses water, but the amount of water that leaves the collecting duct is regulated by hormonal action. The hormone **ADH (antidiuretic hormone),** released by the posterior lobe of the pituitary, increases the permeability of the collecting duct so more water leaves it and is reabsorbed into the blood. If the osmotic pressure of the blood increases, the posterior pituitary releases ADH, more water is reabsorbed, and consequently there is less urine. On the other hand, if the osmotic pressure of the blood decreases, the posterior pituitary does not release ADH. The resulting impermeability of the collecting duct causes more water to be excreted and more urine to be formed. Drinking alcohol causes diuresis (increased urine flow) because it inhibits the secretion of ADH. Beer drinking also causes diuresis because of increased fluid intake. Drugs called diuretics are often prescribed for high blood pressure. These drugs cause increased urinary excretion and thus reduce blood volume and blood pressure.

How Do You Do Hemodialysis?

Hemodialysis is a way to remove nitrogenous wastes and regulate the pH of the blood when the kidneys are unable to perform these functions due to disease or injury. Dialysis occurs when small molecules pass through a membrane that will not allow the passage of large molecules. This is just the process that occurs when a patient is hooked up to a kidney machine (fig. 38.A). After the administration of an anticoagulant, blood from an artery moves through a tube to a cellulose acetate membrane that is con-

tinually washed by a solution called the dialysate. Later the blood is returned to the patient by way of a vein.

During dialysis, large substances like proteins and red blood cells cannot pass through the membrane, but small molecules like urea, glucose, and ions can pass through the dialyzing membrane. The composition of the dialysate is maintained at normal levels, so that if the blood lacks glucose and ions they will pass from the dialysate to the blood. On the other hand, nitrogenous wastes, toxic chemicals and

drugs pass from the blood to dialysate. In the course of a 6-hour dialysis, from 50 g–250 g of urea can be removed from a patient, an amount that greatly exceeds the urea clearance rate of normal kidneys. Most patients undergo treatment only about twice a week.

There are drawbacks to hemodialysis, and they include the need to use an anticoagulant, possible damage to red blood cells, risk of infection, limitation on fluid and protein intake, and the time it takes.

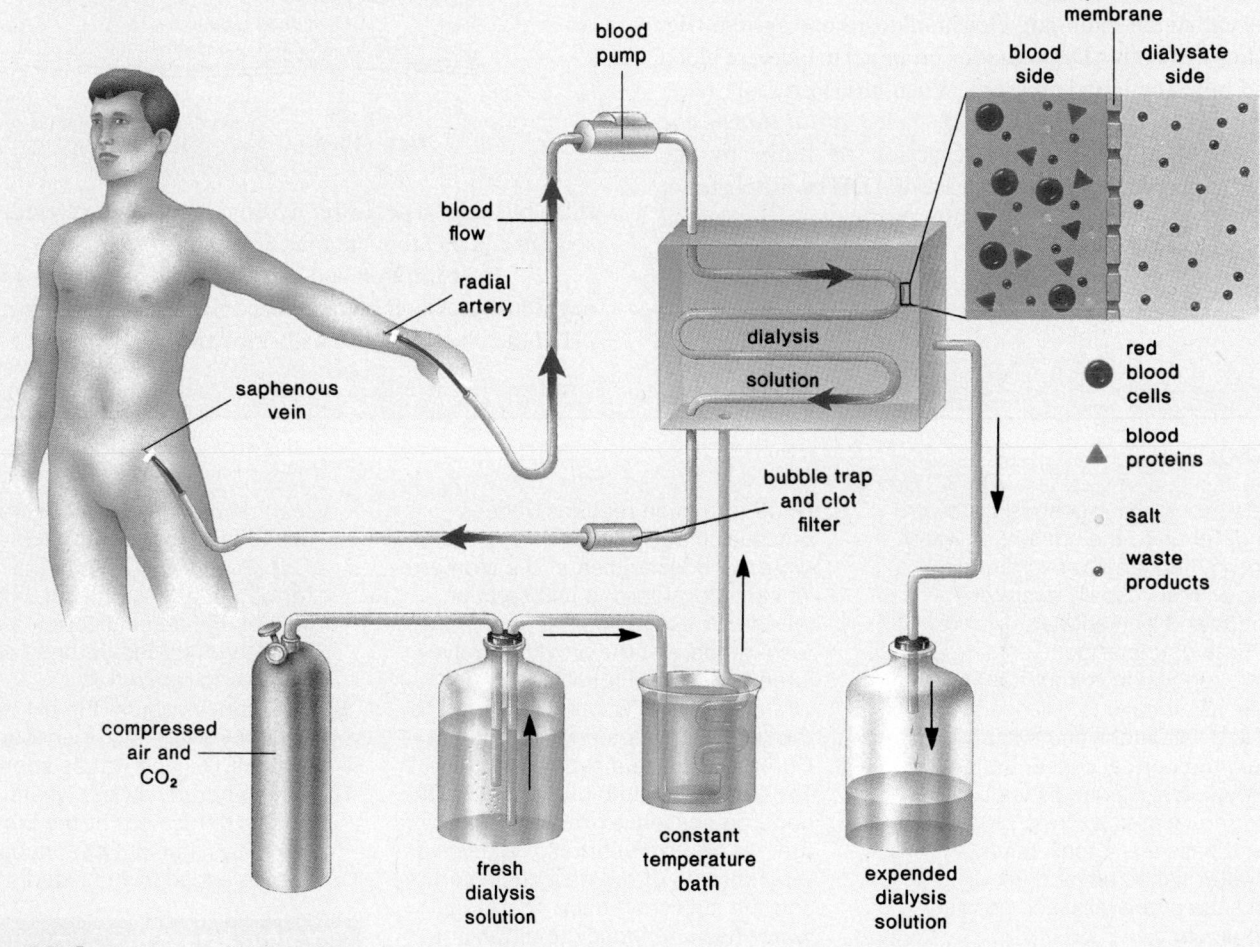

Figure 38.A

Diagram of an artificial kidney. As the patient's blood circulates through dialysis tubing, it is exposed to a dialysis solution. Wastes exit from blood into the solution because of a preestablished concentration gradient. In this way, blood is not only cleansed, the pH also can be adjusted.

The hormone **aldosterone,** which is secreted by the adrenal cortex, primarily maintains the sodium (Na^+) and potassium (K^+) balance of the blood. It causes the distal convoluted tubule to reabsorb Na^+ and to excrete K^+. The increase of Na^+ in the blood causes water to be reabsorbed, leading to an increase in blood volume and blood pressure.

Blood pressure is constantly monitored within the *juxtaglomerular apparatus,* a region of contact between the afferent arteriole and the distal convoluted tubule (fig. 38.11). The afferent arteriole cells in the apparatus secrete the enzyme *renin* when the blood pressure is insufficient to promote efficient filtration in the glomerulus. Renin converts a large plasma protein called angiotensinogen to *angiotensin.* Angiotensin then stimulates the adrenal cortex to release *aldosterone,* which allows Na^+ to be reabsorbed and blood volume and pressure to rise. This sequence of events is called the renin-angiotensin-aldosterone system (fig. 42.10). Notice that both ADH and aldosterone act to increase blood volume and raise the blood pressure. When blood pressure rises, the heart produces a peptide hormone called *atrial natriuretic factor (ANF)* that inhibits the secretion of renin by the juxtaglomerular apparatus and the release of ADH by the posterior pituitary. This is an example of how hormones serve as checks and balances to one another.

Adjustment of Blood pH

The kidneys help maintain the pH level of the blood within a narrow range, and the whole nephron takes part in this process. The excretion of hydrogen ions and ammonia, together with the reabsorption of sodium and bicarbonate ions, is adjusted to keep the pH within normal bounds. If the blood is acidic, hydrogen ions are excreted in combination with ammonia, while sodium and bicarbonate ions are reabsorbed. This will restore the pH because sodium ions promote the formation of hydroxyl ions, while bicarbonate takes up hydrogen ions when carbonic acid is formed:

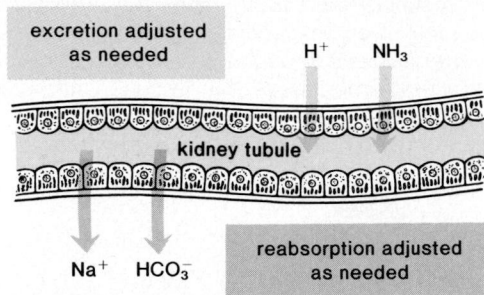

If the blood is basic, fewer hydrogen ions are excreted, and fewer sodium and bicarbonate ions are reabsorbed.

Reabsorption and/or excretion of ions (salts) by the kidneys illustrates their homeostatic ability: they maintain not only the pH of the blood but also its osmolarity.

Summary

1. Animals excrete nitrogenous wastes, which differ as to the amount of water and energy required to excrete them. Aquatic animals usually excrete ammonia, and land animals excrete either urea or uric acid.

2. Osmotic regulation is important to animals. Most have to balance their water and salt intake and excretion to maintain the normal concentration in the body fluids. Marine fishes constantly drink water, excrete salts at the gills, and pass an isotonic urine. Freshwater fishes never drink water, take in salts at the gills, and pass a hypotonic urine.

3. Animals often have an excretory organ. Earthworm nephridia exchange molecules with the blood in a manner similar to vertebrate kidneys. Malpighian tubules in insects take wastes and water from the hemocoel to the gut, where only water is reabsorbed.

4. Kidneys are a part of the human urinary system. Microscopically, each kidney is made up of nephrons, each of which has several parts and its own blood supply.

5. Urine formation requires 3 steps: pressure filtration, when nutrients, water, and wastes enter the glomerular capsule; selective reabsorption, when nutrients and some water are reabsorbed into the proximal convoluted tubule; and tubular secretion, when additional wastes are added to the distal convoluted tubule.

6. Humans excrete a hypertonic urine. The ascending limb of the loop of the nephron actively extrudes salt so that the medulla is hypertonic relative to the contents of the descending limb and the collecting duct. Therefore, water has a tendency to diffuse out of these.

7. Two hormones are involved in maintaining the water content of the blood. The hormone ADH, which makes the collecting duct more permeable, is secreted by the posterior pituitary in response to an increase in the osmotic pressure of the blood.

8. The hormone aldosterone is secreted by the adrenal cortex after the low Na^+ content of the blood and the

resultant low blood pressure have caused the kidneys to release renin. The presence of renin leads to the formation of angiotensin, which causes the adrenal cortex to release aldosterone. Aldosterone causes the kidneys to retain Na^+.

9. The kidneys adjust the pH of the blood by excreting or conserving H^+, NH_3, HCO_3, and Na^+ as appropriate.

10. During hemodialysis, waste molecules diffuse out of the blood into the dialysis fluid. Other substances can be added to the blood.

Writing Across the Curriculum

In order to practice writing skills, students should write out the answers to any or all of the study questions and the critical thinking questions. The study questions are sequenced in the same order as the text. Suggested answers to the critical thinking questions are in appendix D.

Study Questions

1. Relate the 3 primary nitrogenous wastes to the habitat of animals.
2. Contrast the osmotic regulation of a marine bony fish with that of a freshwater bony fish.
3. Give examples of how other types of animals regulate their water and salt balance.
4. Describe how the excretory organs of the earthworm and the insect function.
5. Give the path of urine, and give a function for each structure mentioned.
6. Describe the macroscopic anatomy of a human kidney, and relate it to the placement of nephrons.
7. List the parts of a nephron, and give a function for each structure mentioned.
8. Describe how urine is made by telling what happens at each part of the nephron.
9. What role do ADH and aldosterone play in regulating the tonicity of urine? How does this effect blood pressure?
10. How does the nephron regulate the pH of the blood?

Objective Questions

1. Which of these is mismatched?
   a. insects—excrete uric acid
   b. humans—excrete urea
   c. fishes—excrete ammonia
   d. birds—excrete ammonia
2. One advantage of urea excretion over uric acid excretion is that urea
   a. requires less energy to form.
   b. can be concentrated to a greater extent.
   c. is not a toxic substance.
   d. requires less water to excrete.
3. Freshwater bony fishes maintain water balance by
   a. excreting salt across their gills.
   b. periodically drinking small amounts of water.
   c. excreting a hypotonic urine.
   d. excreting wastes in the form of uric acid.
4. Animals with which of these are most likely to excrete a semisolid, nitrogenous waste?
   a. nephridia
   b. Malpighian tubules
   c. human kidneys
   d. All of these.
5. In which of these human structures are you least apt to find urine?
   a. large intestine
   b. urethra
   c. ureter
   d. bladder

6. Excretion of a hypertonic urine in humans is associated with
   a. the glomerular capsule.
   b. the proximal convoluted tubule.
   c. the loop of the nephron.
   d. the distal convoluted tubule.
7. The presence of ADH causes an individual to excrete
   a. sugars.
   b. less water.
   c. more water.
   d. Both a and c.
8. In humans, water is
   a. found in the glomerular filtrate.
   b. reabsorbed from the nephron.
   c. in the urine.
   d. All of these.

9. Pressure filtration is associated with
   a. the glomerular capsule.
   b. the distal convoluted tubule.
   c. the collecting duct.
   d. All of these.
10. In humans, glucose
    a. is in the filtrate and urine.
    b. is in the filtrate and not in urine.
    c. undergoes tubular secretion and is in urine.
    d. undergoes tubular secretion and is not in urine.
11. Label this diagram of a nephron, and give the steps for urine formation in the boxes:

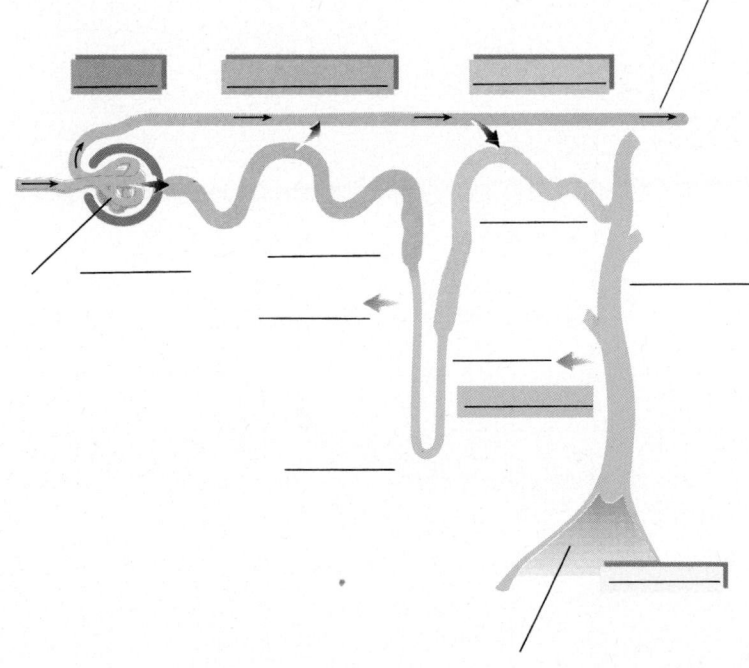

Concepts and Critical Thinking

1. *The excretory system plays a primary role in homeostasis.*

   Relate the need of excretion to figure 7.10 of the text.

2. *Animals utilize countercurrent mechanisms to increase blood concentrations of substances.*

   Tell how the countercurrent mechanism in the gills of fishes (p. 607) and in the kidneys of mammals (p. 626) helps adapt them to their environments.

3. *Structure suits the function.*

   Why would you expect the proximal convoluted tubule to be lined with cells, which have many mitochondria and microvilli?

Selected Key Terms

urea (u-re′ah) 621
uric acid (u′rik as′id) 621
nephridium (nĕ-frid′e-um) 623
ureter (u-re′ter) 624
urinary bladder (u′rĭ-ner″e blad′der) 624
urethra (u-re′thrah) 624
nephron (nef′ron) 625

glomerular capsule (glo-mer′u-lar kap′sūl) 625
proximal convoluted tubule (prok′sĭ-mal kon′vo-lūt-ed tu′būl) 625
loop of the nephron (lo͞op uv the nef′ron) 625

distal convoluted tubule (dis′tal kon′vo-lūt-ed tu′būl) 625
collecting duct (kŏ-lekt′ing dukt) 625
glomerulus (glo-mer′u-lus) 625
antidiuretic hormone (ADH) (an″tĭ-di″u-ret′ik hor′mōn) 628
aldosterone (al″do-ster′ōn) 630

Animal Structure and Function

39

Nervous System

Your study of this chapter will be complete when you can

1. state, in general, the overall function of the nervous system;
2. compare the organization and complexity of various invertebrate nervous systems with that of the human nervous system;
3. describe, in general, the structure of a neuron; name 3 types of neurons, and state their specific functions;
4. describe the nerve impulse as an electrochemical change that is recorded as the action potential by the oscilloscope;
5. describe the structure and function of a synapse, including transmission across a synapse;
6. describe the structure and function of the peripheral nervous system;
7. draw a diagram depicting a spinal reflex, and explain the function of all parts included;
8. state the parts of the autonomic nervous system; cite similarities and differences in structure and function of its 2 divisions;
9. describe, in general, the structure and function of the central nervous system;
10. list the major parts of the brain, and give a function for each part;
11. discuss, in general, current research into learning and memory;
12. list several well-known excitatory and inhibitory neurotransmitters in the peripheral and central nervous systems;
13. discuss and give examples of the effects of drugs at synapses, and list 5 criteria for drug abuse.

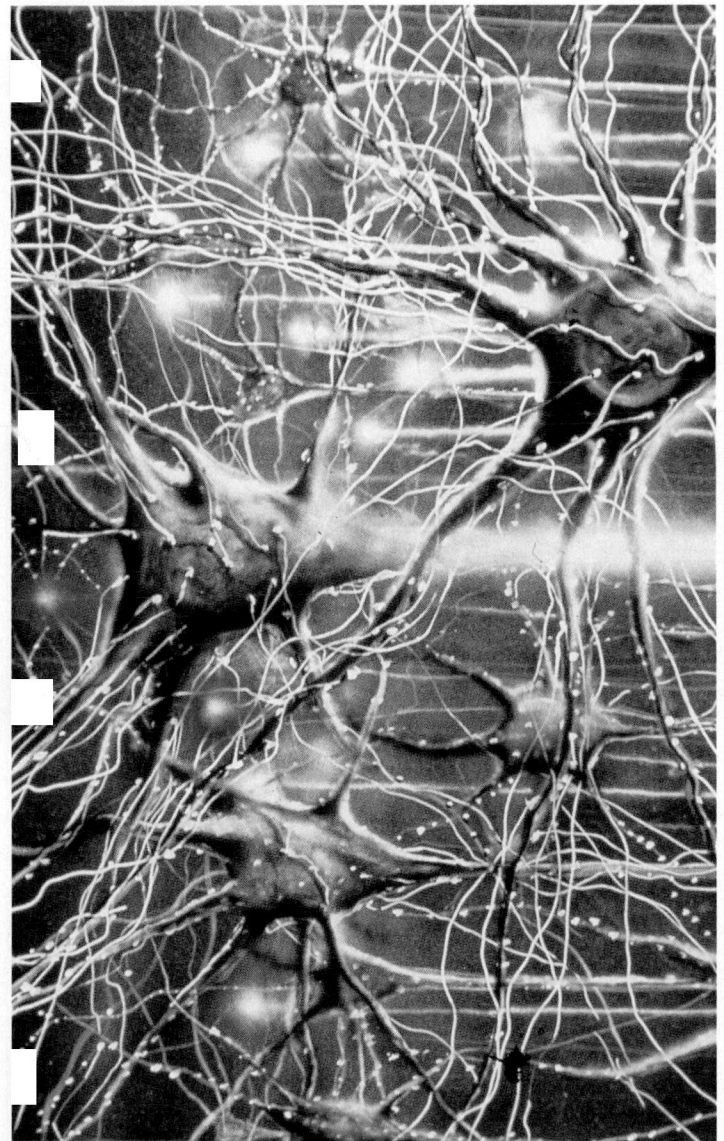

A dramatic representation of neurons with many processes that reach out in all directions. When neurons communicate with one another, they release transmitter substances that are represented here as bursts of color.

he ability to respond to stimuli is a characteristic of all living things. In complex animals, such as vertebrates, this ability depends on the nervous, endocrine (hormonal), sensory, and musculoskeletal systems. Working together, the nervous and endocrine systems coordinate the actions of all the other systems of the body to produce effective behavior and to keep the internal environment within safe limits for life. Because hormones are transported in the blood, it may require seconds, minutes, hours, or even longer for these chemical messengers to produce their effects. The nervous system, on the other hand, communicates rapidly, requiring only thousandths of a second. The nervous system receives and processes information before sending out signals to the muscles and glands for an appropriate response. In this way, the nervous system integrates and controls the other systems of the body.

Evolution of Nervous Systems

A comparative study of animal nervous systems indicates the steps that may have led to the centralized nervous system found in vertebrates (fig. 39.1). Hydras have a simple nervous system that looks like a net of threads extending throughout the radially symmetrical body. The net is actually composed of neurons located in the mesoglea that contact one another and muscle fibers within epidermal cells. Experiments with sea anemones and jellyfishes suggest that cnidarians may have 2 nerve nets. A fast-acting one allows major responses, particularly in times of danger, and the other coordinates slower and more delicate movements.

Planarians are bilaterally symmetrical, and they have a more complex nervous system that resembles a ladder. Cephalization is present—there is a brain, a concentration of neurons at the anterior end of the animal that receives sensory information from photoreceptors in the eyespots and sensory cells in the auricles. Two longitudinal nerve cords allow a rapid transfer of information from anterior to posterior. Transverse nerves between the nerve cords keep the movement of the 2 sides coordinated. Therefore, planarians have the rudiments of a central nervous system (brain and nerve cords) and a peripheral nervous system (transverse nerves).

The annelids and arthropods have what is usually considered the typical invertebrate nervous system. The **central nervous system** consists of a brain and a single, ventral solid nerve cord; the **peripheral nervous system** consists of nerves that contain both sensory and motor fibers. The ventral solid nerve cord has a ganglion in each segment, which apparently controls the muscles of that segment. Even so, the brain, which normally receives sensory information, controls the activity of the ganglia so that the entire animal is coordinated.

Vertebrates also have a central nervous system and a peripheral nervous system (fig. 39.1d). There is a vast increase, however, in the number of neurons. For example, an insect's entire

---

Figure 39.1

Evolution of the nervous system. **a.** The nerve net of a hydra, a cnidarian. **b.** The paired nerve cords with cross connections of a planarian, a flatworm, have a ladder appearance. **c.** The earthworm, a segmented worm, has a central nervous system consisting of the brain and a ventral solid nerve cord. It also has a peripheral nervous system consisting of nerves. **d.** A rabbit, like other vertebrates, has a dorsal, hollow nerve cord in its central nervous system.

a. Hydra

nerve net

b. Planarian

"brain"
eyespot
auricle
nerve cord
transverse nerve

c. Earthworm

lateral nerve
ventral solid nerve cord
brain

d. Rabbit

dorsal hollow nerve cord
brain

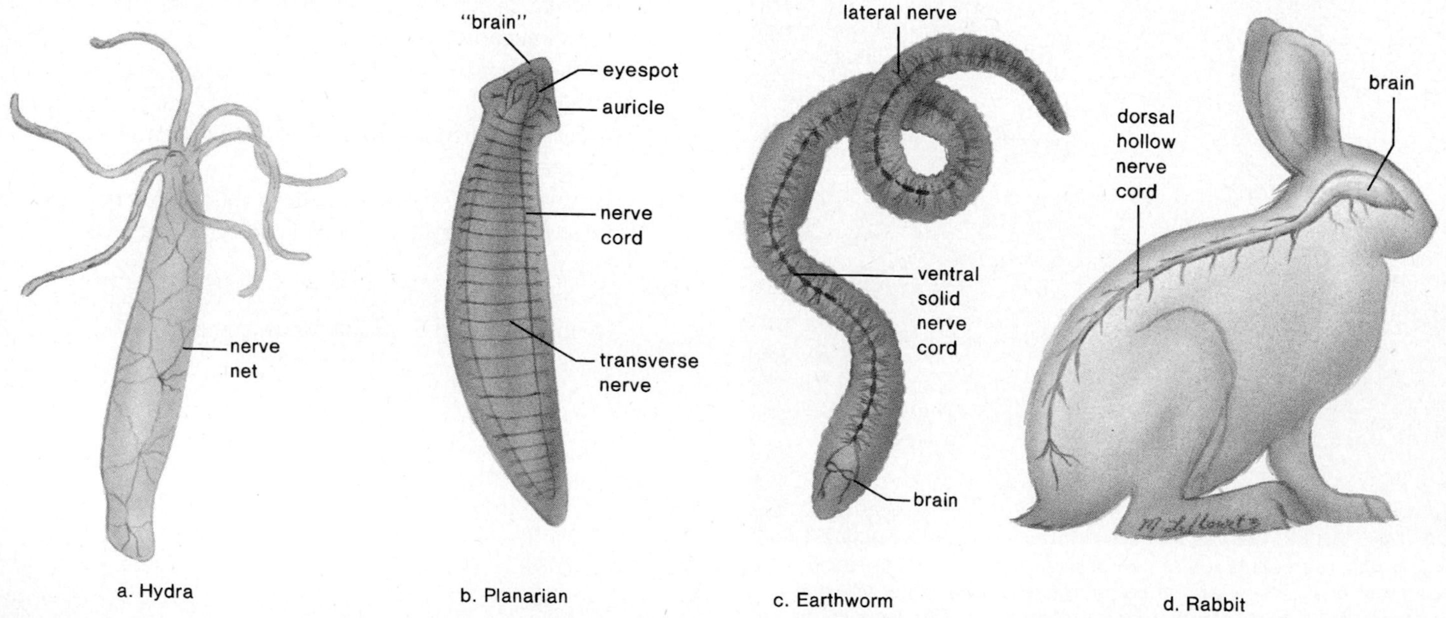

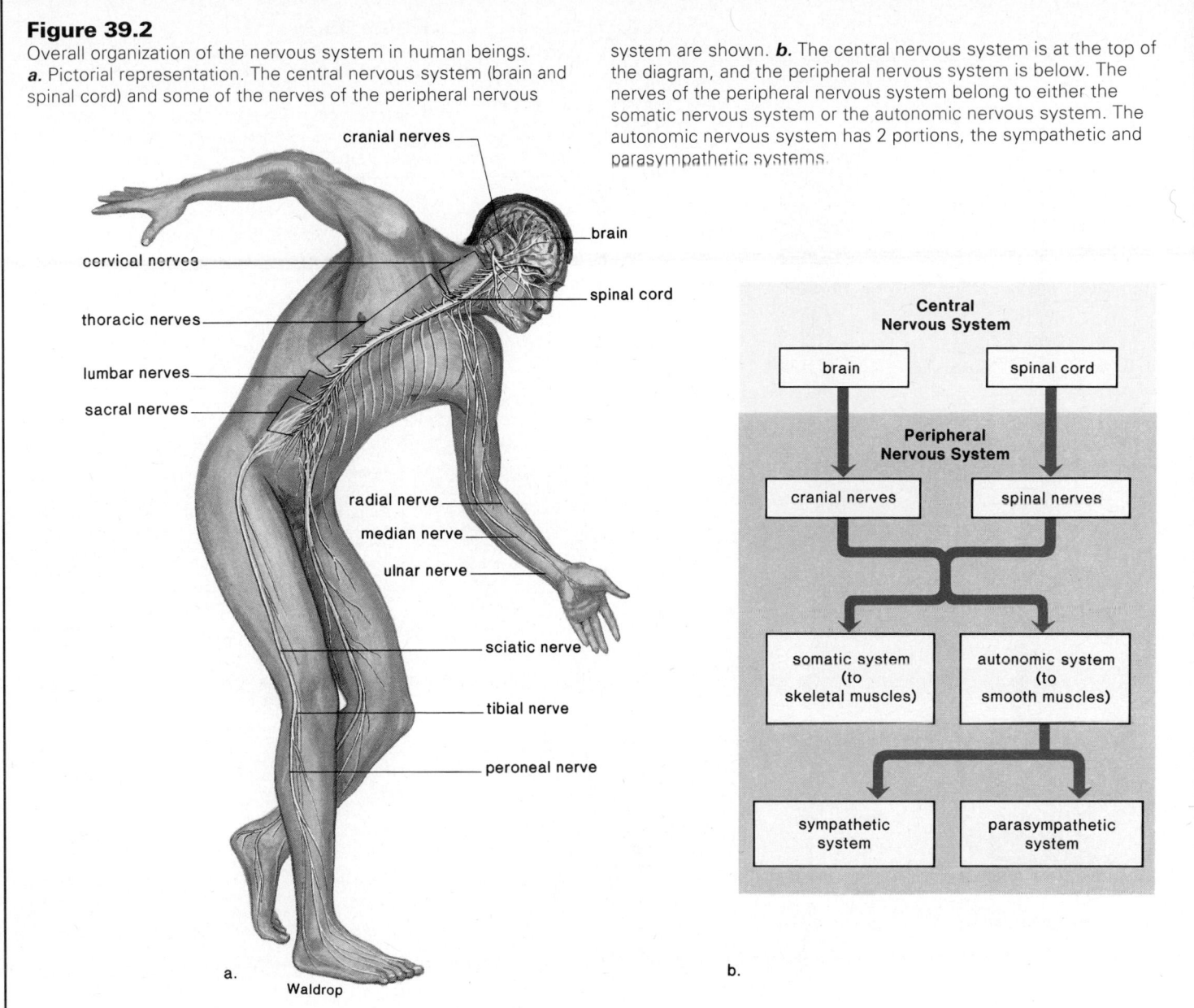

Figure 39.2
Overall organization of the nervous system in human beings.
a. Pictorial representation. The central nervous system (brain and spinal cord) and some of the nerves of the peripheral nervous system are shown. *b.* The central nervous system is at the top of the diagram, and the peripheral nervous system is below. The nerves of the peripheral nervous system belong to either the somatic nervous system or the autonomic nervous system. The autonomic nervous system has 2 portions, the sympathetic and parasympathetic systems.

nervous system may contain a total of about one million neurons, while a vertebrate nervous system may contain many thousand to several billion times that number.

> In complex animals the nervous system has 2 parts: the central nervous system and the peripheral nervous system.

Human Nervous System

In the human nervous system, the *central nervous system* includes the brain and spinal cord (dorsal nerve cord), which lie in the midline of the body, where the skull protects the brain and the vertebrae protect the spinal cord (fig. 39.2). The *peripheral nervous system* contains both cranial nerves, which originate in the brain, and spinal nerves, which project from either side of the spinal cord. The peripheral nervous system is further divided into the somatic system and the autonomic (or visceral) system. The somatic system contains nerves that control skeletal muscles, skin, and joints. The autonomic system contains nerves that control the glands and smooth muscles of the internal organs. Before we continue our discussion of the nervous system, we must first examine the anatomy and functions of nerve cells, or neurons.

Neurons

Neurons are cells that vary in size and shape, but they all have 3 parts: the dendrite(s), the cell body, and the axon (fig. 39.3). The **dendrites** receive information from other neurons and generally conduct nerve impulses *toward* the cell body. The **axon,** on the

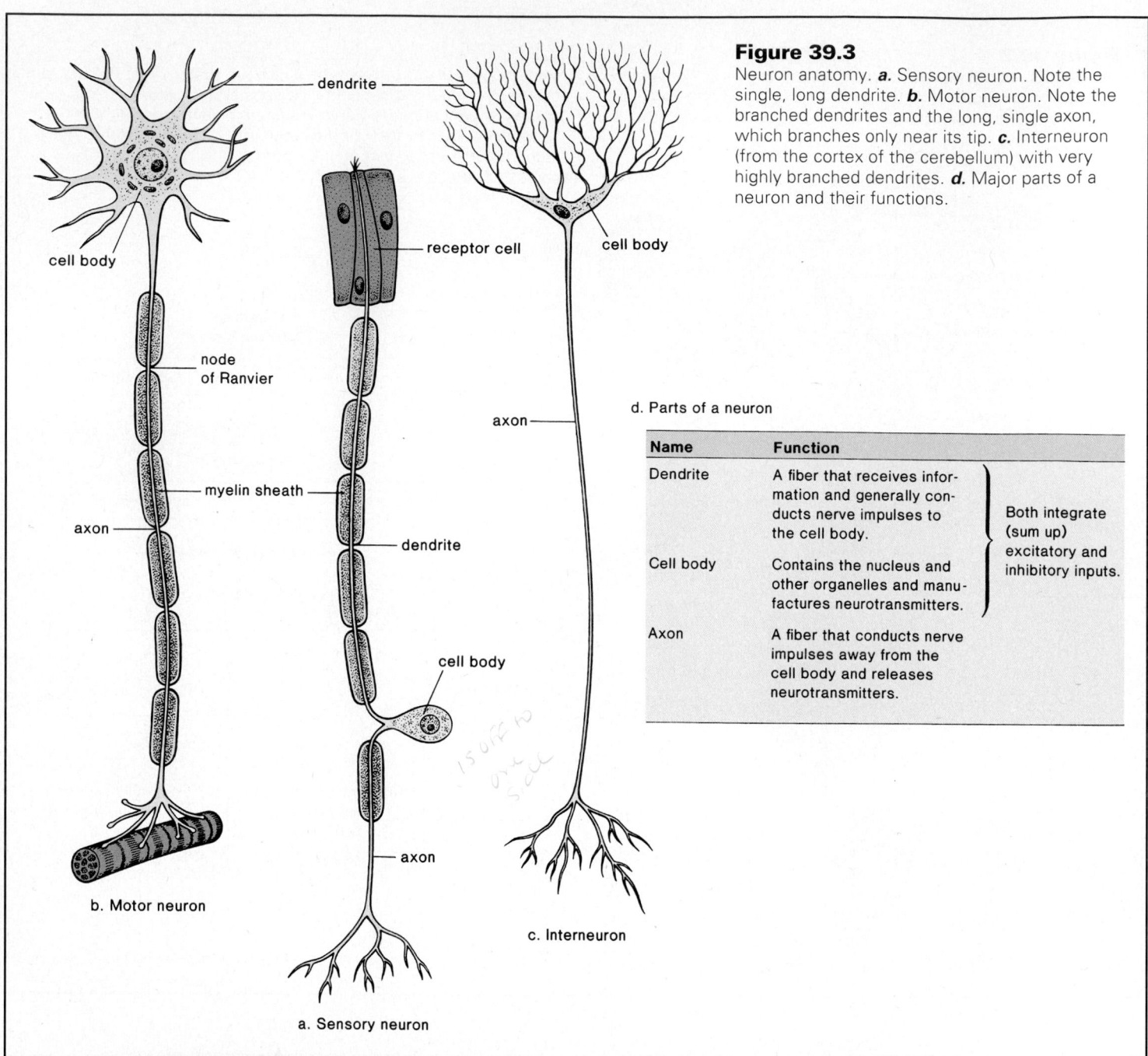

Figure 39.3

Neuron anatomy. **a.** Sensory neuron. Note the single, long dendrite. **b.** Motor neuron. Note the branched dendrites and the long, single axon, which branches only near its tip. **c.** Interneuron (from the cortex of the cerebellum) with very highly branched dendrites. **d.** Major parts of a neuron and their functions.

d. Parts of a neuron

Name	Function	
Dendrite	A fiber that receives information and generally conducts nerve impulses to the cell body.	Both integrate (sum up) excitatory and inhibitory inputs.
Cell body	Contains the nucleus and other organelles and manufactures neurotransmitters.	
Axon	A fiber that conducts nerve impulses away from the cell body and releases neurotransmitters.	

b. Motor neuron

c. Interneuron

a. Sensory neuron

other hand, conducts nerve impulses *away* from the cell body. The cell body contains the nucleus and other organelles typically found in cells. One of the main functions of the cell body is to manufacture neurotransmitters, which are chemicals stored in secretory vesicles at the ends of axons. When neurotransmitters are released, they influence the excitability of nearby neurons.

Three types of neurons are shown in figure 39.3. *Sensory neurons,* each with a long dendrite and a short axon, take messages from sense organs to the central nervous system. *Motor neurons,* each with a long axon and short dendrites, take messages from the central nervous system to muscle fibers or glands. Because motor

neurons cause muscle fibers or glands to react, they are said to innervate these structures. The third type of neuron, called an **interneuron**, is only found within the central nervous system. It conveys messages between various parts of the central nervous system, such as from one side of the brain or spinal cord to the other or from the brain to the cord, and vice versa. An interneuron has short dendrites and either a long axon or a short axon.

The dendrites and axons of these neurons are sometimes called fibers, or processes. Most long fibers, whether dendrite or axon, are covered by a white **myelin sheath** formed from the membranes of the tightly spiraled Schwann cells surrounding these fibers.

Animal Structure and Function

Schwann cells are one of the several types of glial cells in the nervous system. Neuroglial cells service the neurons and have supportive and nutritive functions.

Nerve Impulse

Italian investigator Luigi Galvani discovered in 1786 that a nerve can be stimulated by an electric current. But it was realized later that the speed of the nerve impulse is too slow to be due simply to the movement of electrons or current within a nerve fiber. In the early 1900s, Julius Bernstein, at the University of Halle, Germany, suggested that the nerve impulse is an electrochemical phenomenon involving the movement of unequally distributed ions on either side of a neuron plasma membrane. It was not until 1939, however, that the investigators developed a technique that enabled them to substantiate this hypothesis. A. L. Hodgkin and A. F. Huxley, English neurophysiologists, received the Nobel Prize in 1963 for their work in this field. They and a group of researchers, headed by K. S. Cole and J. J. Curtis, at Woods Hole, Massachusetts, managed to insert a very tiny electrode into the giant axon of the squid *Loligo* (fig. 39.4). This internal electrode was then connected to a voltmeter and an *oscilloscope,* an instrument with a screen that shows a trace or pattern indicating a change in voltage with time (fig. 39.4*d*). *Voltage* is a measure of the electrical potential difference between 2 points, which in this case is the difference between 2 electrodes, one placed inside and another placed outside the axon. (An electrical potential difference across a membrane is called the membrane potential.) When a potential difference exists, we can say that a plus and minus pole exist; therefore, an oscilloscope indicates the existence of polarity and records polarity changes.

Resting Potential

When the axon is not conducting an impulse, the oscilloscope records a membrane potential equal to about −65 mV (millivolts), indicating that the inside of the neuron is more negative than the outside (fig. 39.4*b*). This is called the **resting potential** because the axon is not conducting an impulse.

The existence of this polarity can be correlated with a difference in ion distribution on either side of the axomembrane (plasma membrane of axon). As figure 39.5*a* shows, there is a higher concentration of sodium ions (Na^+) outside than inside the axon and a higher concentration of potassium ions (K^+) inside than outside the axon. The unequal distribution of these ions is due to the action of the sodium-potassium pump. This pump is an active transport system in the plasma membrane that pumps sodium ions out of and potassium ions into the axon. The work of the pump maintains the unequal distribution of sodium and potassium ions across the axomembrane.

The pump is always working because the axomembrane is somewhat permeable to these ions and they tend to diffuse toward their lesser concentration. Since the axomembrane is more permeable to potassium ions than to sodium ions there are always more positive ions outside the axomembrane than inside and this accounts for the polarity recorded by the oscilloscope. There are also large, negatively charged proteins in the axoplasm

Figure 39.4

The original nerve impulse studies utilized giant squid axons and a voltage recording device known as an oscilloscope. *a.* The squid axons shown produce rapid muscle contraction so that the squid can move quickly. *b.* These axons are so large (about 1 mm in diameter) that a microelectrode can be inserted inside them. When the axon is not conducting a nerve impulse, the electrode registers and the oscilloscope records a resting potential of −65 mV. *c.* When the axon is conducting a nerve impulse, the *threshold* for an action potential has been achieved, and there is a rapid change in potential from −65 mV to +40 mV (called depolarization), followed by a return to −65 mV (called repolarization). *d.* Enlargement of action potential of nerve impulse.

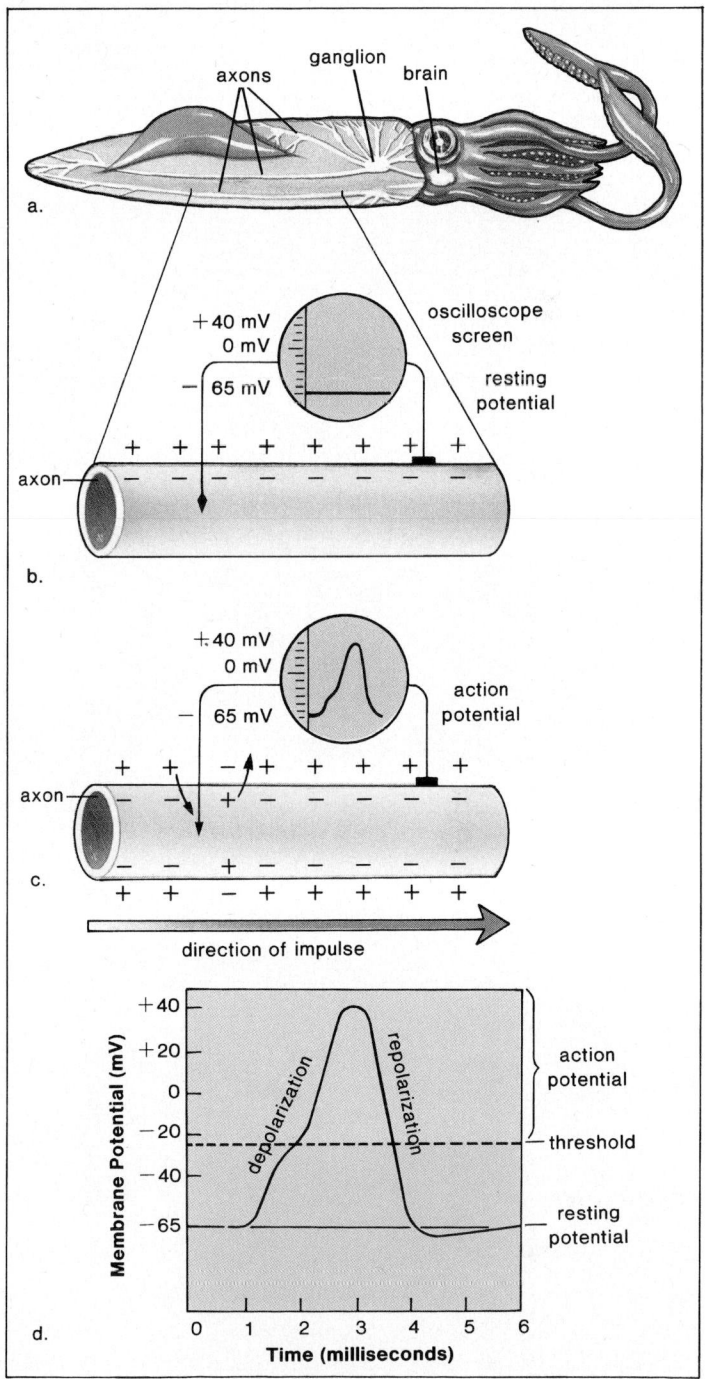

Figure 39.5

Action potential and resting potential. The action potential is the result of an exchange of sodium (Na⁺) ions and potassium (K⁺) ions, and it is recorded as a change in polarity by an oscilloscope (as shown on the right). So few ions are exchanged for each action potential that it is possible for a nerve fiber to conduct nerve impulses repeatedly. Whenever the fiber rests, the sodium-potassium pump restores the original distribution of ions.

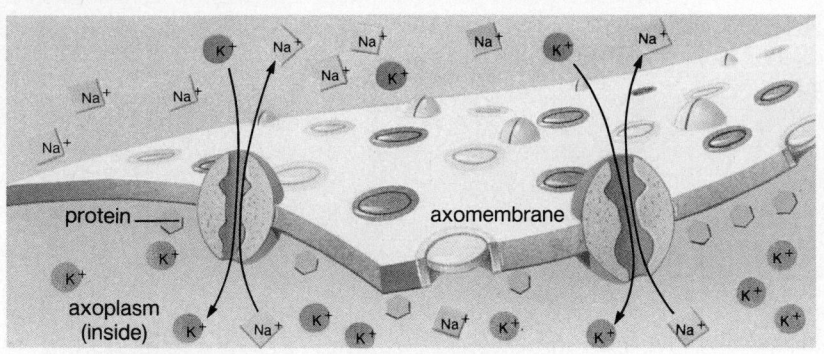

a. Resting potential

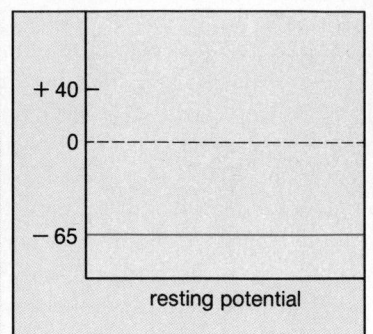

resting potential

When a neuron is not conducting a nerve impulse, the sodium (Na⁺) and potassium (K⁺) gates are closed. The sodium-potassium pump maintains the uneven distribution of these ions across the axomembrane. The oscilloscope registers a resting potential of −65 mV inside compared to outside.

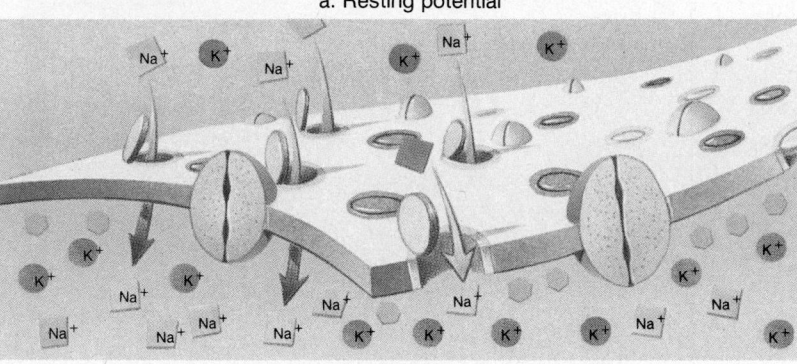

b. Sodium gates open

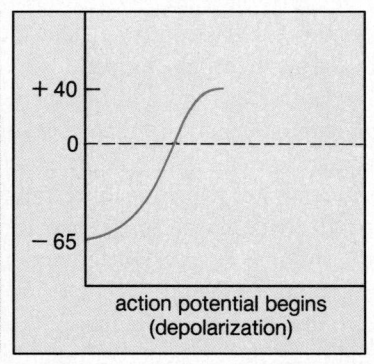

action potential begins (depolarization)

An action potential begins when the sodium gates open and sodium ions move to the inside. The oscilloscope registers a depolarization as the axoplasm reaches +40 mV compared to tissue fluid.

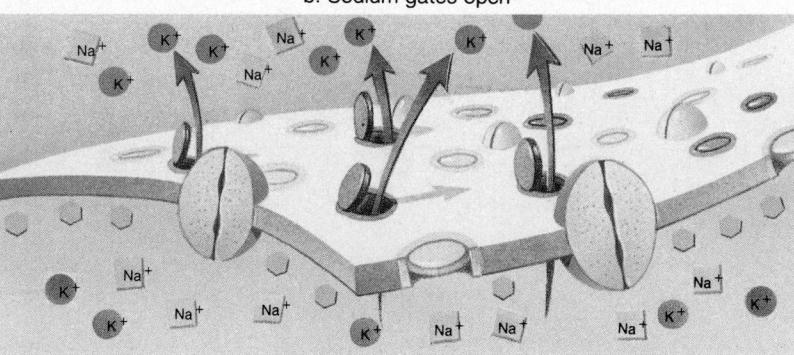

c. Potassium gates open

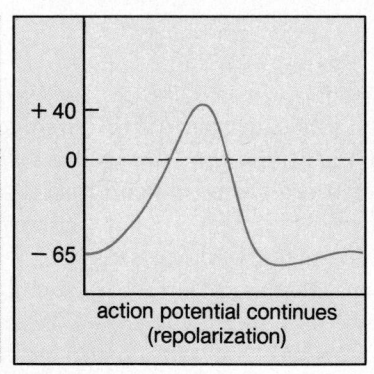

action potential continues (repolarization)

Recovery occurs as the sodium gates close and the potassium gates open, allowing potassium ions to move to the outside. The oscilloscope registers repolarization as the axoplasm again becomes −65 mV compared to tissue fluid.

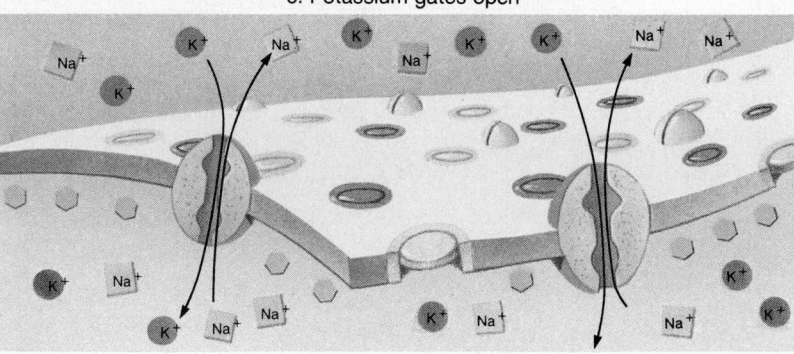

d. Sodium-potassium pump

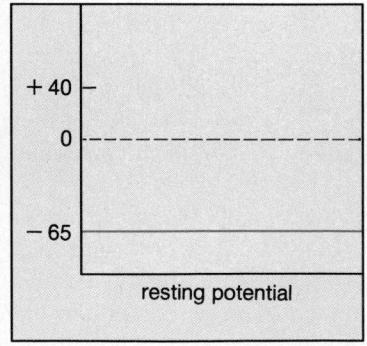

resting potential

The oscilloscope registers −65 mV again, but the sodium-potassium pump is working to restore the original sodium and potassium ion distribution illustrated in (a). The sodium and potassium gates are now closed but will open again in response to another stimulus.

(cytoplasm of axon); altogether, then, the oscilloscope records that the axoplasm is −65 mV compared to tissue fluid. This is the resting potential.

Action Potential

If the axon is stimulated to conduct a nerve impulse by either an electric shock, a sudden change in pH, or a pinch, there is a rapid change in the polarity recorded as a trace on the oscilloscope screen. This change in polarity is called the **action potential** (fig. 39.5). First, the trace goes from −65 mV to +40 mV (called *depolarization*), indicating that the axoplasm is now more positive than tissue fluid. Then the trace returns to −65 mV again (called *repolarization*), indicating that the inside of the axon is negative again.

The action potential is due to special protein-lined channels in the axomembrane, which can open to allow either sodium or potassium ions to pass through (fig. 39.5*b* and *c*). These channels have gates called "sodium gates" and "potassium gates." During the depolarization phase of the action potential, the sodium gates open and sodium rushes into the axon. Once this phase is complete, repolarization occurs. During repolarization, the potassium gates open and potassium rushes out of the axon.

Notice that at the completion of an action potential, the original ion distribution has been altered somewhat (fig. 39.5*d*). There are now more sodium ions inside the axon than before, and there are more potassium ions outside the axon than before. The sodium-potassium pump is able to restore the former distribution, however.

The oscilloscope records changes at only one location in a fiber, but actually the action potential travels along the length of a fiber. It is self-propagating because the ion channels are prompted to open whenever the membrane potential decreases in an adjacent area. In invertebrates, some fibers are larger than others and a few axons are called giant axons because they are so large. The nerve impulse travels a thousand times faster in giant axons compared to thin fibers (25 m/s compared to 0.025 m/s). In vertebrates the speed of travel is improved not by increase in size but because most long fibers have a myelin sheath (fig. 39.3). There are gaps in the myelin sheath called nodes of Ranvier, where one Schwann cell ends and the next begins. The action potential simply jumps from one node of Ranvier to the next and may reach speeds of 200 m per sec. This is called **saltatory** (saltatory—jumping) **conduction.**

All neurons transmit the same type of nerve impulse—an electrochemical change that is self-propagating along the fiber(s).

Transmission across a Synapse

In 1897, the English scientist Sir Charles Sherrington and others noted 2 important aspects of nerve impulse transmission between neurons. First, an impulse passing from one vertebrate nerve cell to another always moves in only one direction. Second, there is a very short delay in transmission of the nerve impulse from one neuron to another. This latter observation led to the hypothesis that there is a minute space between neurons. Sherrington called the region where the impulse moves from one neuron to another a **synapse,** meaning "to clasp." Synapses occur between the end of an axon and a dendrite or between the end of an axon and a cell body (fig. 39.6).

A synapse has 3 components: a presynaptic membrane, a gap now called the synaptic cleft, and a postsynaptic membrane. As mentioned previously, there are vesicles at the ends of axons where transmitter substances, called **neurotransmitters,** are stored. When nerve impulses reach a presynaptic membrane, the vesicles fuse with the membrane and discharge their contents into the synaptic cleft. The discharged neurotransmitter molecules diffuse across the synaptic cleft and bind in a lock-and-key manner to the postsynaptic membrane at *receptor sites*. This binding process alters the membrane potential of the postsynaptic membrane in the direction of either excitation or inhibition. If excitation occurs, the axoplasm becomes less negative, and if inhibition occurs, the axoplasm becomes more negative compared to tissue fluid. For example, if a neurotransmitter causes only potassium gates to open, potassium will exit and the axoplasm will become even more negative.

Summation and Integration

A dendrite or cell body is on the receiving end of many synapses. Whether the neuron fires (initiates a nerve impulse) or not depends on the summary, or net, effect of all the excitatory and inhibitory neurotransmitters it receives. If the amount of excitatory neurotransmitter received is sufficient to raise the membrane potential above the threshold level (fig. 39.4*d*), the neuron fires. If the amount of excitatory neurotransmitter received is insufficient, only local excitation occurs. For this reason, the dendrites and the cell body are the parts of a neuron where *integration* (a summing up) occurs (fig. 39.3). This is the way that the nervous system fine tunes its response to the environment.

Neurotransmitters

Acetylcholine (ACh) and **norepinephrine** (NE) are the primary excitatory neurotransmitters and are active in both the peripheral and the central nervous systems. Examples of inhibitory substances, so far discovered only in the central nervous system, are given on page 647.

Once neurotransmitters have been released into a synaptic cleft, they remain active for only a short time. In some neurons, the cleft contains enzymes that rapidly inactivate the neurotransmitter. For example, the enzyme acetylcholinesterase (AChE), or simply cholinesterase, breaks down acetylcholine. A single molecule of AChE catalyzes the breakdown of 25,000 molecules of acetylcholine per second. The breakdown products are then taken up into the presynaptic neuron and used in the synthesis of new acetylcholine molecules. The enzyme monoamine oxidase breaks down norepinephrine after it is absorbed.

Transmission of a nerve impulse across a synapse is dependent on neurotransmitter substances that alter the potential difference across the postsynaptic membrane.

Figure 39.6

Diagrammatic representation of a synapse at 3 different magnifications. **a.** Typically, a synapse is located wherever an axon is close to a dendrite or cell body. Drawing based on a photomicrograph shows that there are several synaptic endings per axon because of terminal branching of the axon. **b.** Drawing based on low-power electron micrographs shows that an axon ending has numerous synaptic vesicles, each filled with a neurotransmitter. This drawing makes it clear that a synapse contains a cleft (space) between the axon and the dendrite. **c.** Drawing based on high-power electron micrographs shows that a synapse consists of the presynaptic membrane, the synaptic cleft, and the postsynaptic membrane. When a nerve impulse reaches the synaptic vesicles, they move toward and fuse with the presynaptic membrane, discharging their contents. The neurotransmitter diffuses across the cleft and combines with a receptor. A nerve impulse may follow.

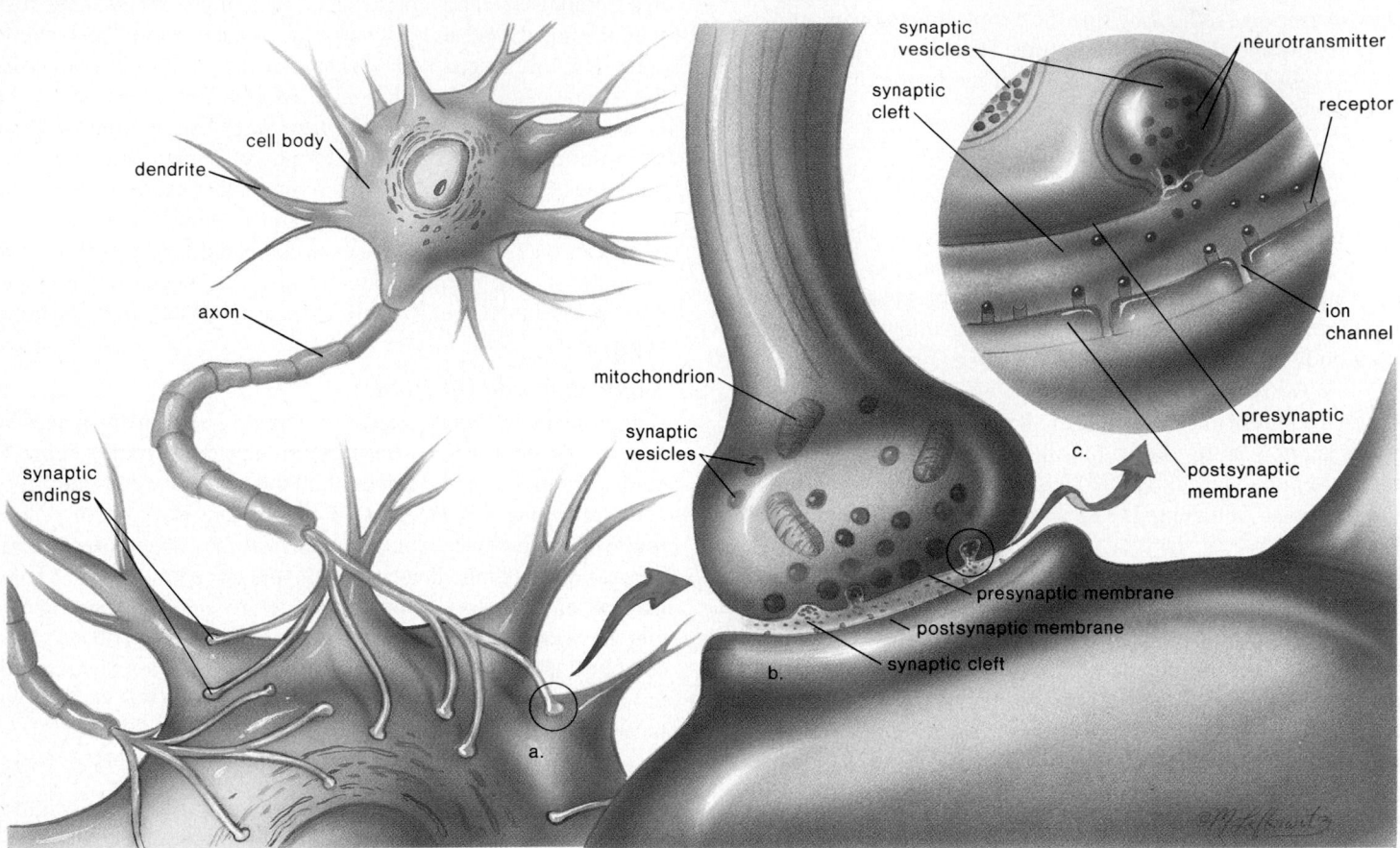

Peripheral Nervous System

The peripheral nervous system (PNS) contains **nerves** (fig. 39.2), structures that contain only long dendrites and/or long axons. Each of these fibers is surrounded by a myelin sheath (fig. 39.3), and therefore these nerves have a white glistening appearance. There are no cell bodies in nerves because cell bodies are found only in the central nervous system (CNS) or in the ganglia. **Ganglia** (ganglion, sing.) are collections of cell bodies within the PNS.

Humans have 12 pairs of cranial nerves and 31 pairs of spinal nerves. *Cranial nerves* are either sensory nerves (having long dendrites of sensory neurons only), motor nerves (having long axons of motor neurons only), or mixed nerves (having both long dendrites and long axons). All cranial nerves, except the vagus nerve, control the head, neck, and face. The *vagus nerve* controls the internal organs.

Spinal nerves are all mixed nerves that take impulses to and from the **spinal cord** (fig. 39.7). Their arrangement shows that humans are segmented animals: there is a pair of spinal nerves for each segment. Spinal nerves project from the spinal cord, which is a part of the central nervous system. The spinal cord is a thick, whitish nerve cord that extends longitudinally down the back, where it is protected by the vertebrae (vertebra, sing.). The cord contains a tiny, central canal filled with cerebrospinal fluid, gray matter consisting of cell bodies and short fibers, and white matter consisting of myelinated fibers.

In the PNS, cranial nerves take impulses to and/or from the brain and spinal nerves take impulses to and from the spinal cord.

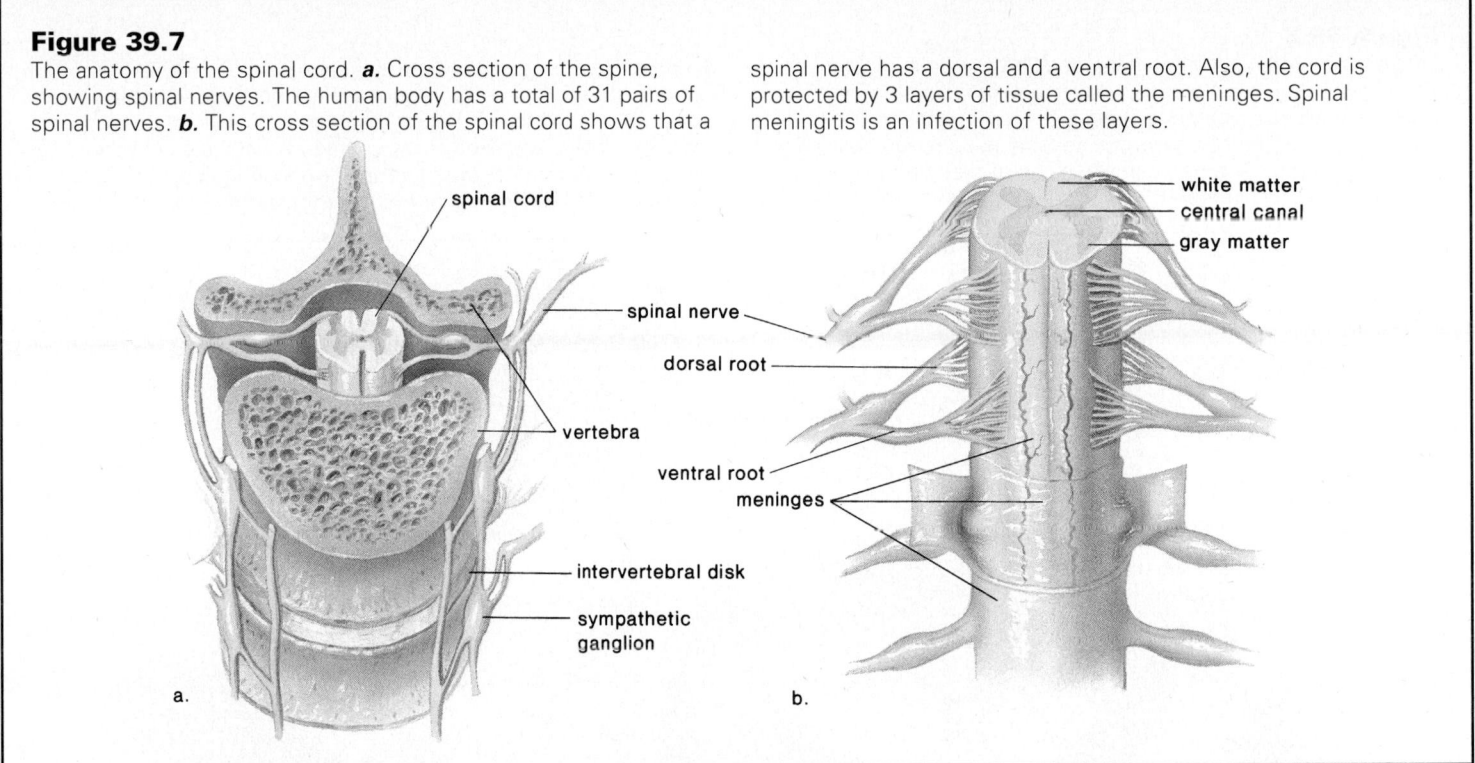

Figure 39.7
The anatomy of the spinal cord. *a.* Cross section of the spine, showing spinal nerves. The human body has a total of 31 pairs of spinal nerves. *b.* This cross section of the spinal cord shows that a spinal nerve has a dorsal and a ventral root. Also, the cord is protected by 3 layers of tissue called the meninges. Spinal meningitis is an infection of these layers.

Somatic Nervous System

The *somatic nervous system* includes all nerves that serve the musculoskeletal system and the exterior sense organs, including those in the skin. Exterior sense organs are **receptors,** which receive environmental stimuli and then initiate nerve impulses. Muscle fibers are **effectors,** which bring about a reaction to the stimulus. Receptors are discussed in chapter 40, and muscle effectors are discussed in chapter 41.

Reflex Arc

Reflexes are automatic, involuntary responses to changes occurring inside or outside the body. In the somatic nervous system, outside stimuli often initiate a reflex action. Some reflexes, such as blinking the eye, involve the brain, but others, such as withdrawing the hand from a hot object, do not necessarily involve the brain. Figure 39.8 illustrates the path of the second type of reflex action involving the spinal cord and a spinal nerve, called a *spinal reflex,* or *reflex arc.* Various other reactions occur; the person may look in the direction of the object, jump back, and utter appropriate exclamations. This whole series of responses is explained by the fact that the sensory neuron stimulates several interneurons, which take impulses to all parts of the central nervous system, including the cerebrum, which in turn makes the person conscious of the stimulus and his or her reaction to it.

> The reflex arc is a major functional unit of the nervous system. It allows us to react to internal and external stimuli.

Autonomic Nervous System

The **autonomic nervous system** is a part of the PNS (fig. 39.9). It is made up of motor neurons that control the internal organs automatically and usually without conscious intervention. There are 2 divisions to the autonomic nervous system (table 39.1): the **sympathetic nervous system** and the **parasympathetic nervous system.** Both of these (1) function automatically and usually subconsciously in an involuntary manner; (2) innervate all internal organs; and (3) use 2 motor neurons for each impulse. The cell body of the first motor neuron is located in the CNS, and the cell body of the second motor neuron is located in a ganglion. The axon that occurs before the ganglion is called the preganglionic fiber, and the axon that occurs after the ganglion is called the postganglionic fiber.

Sympathetic Nervous System

Preganglionic fibers of the sympathetic nervous system arise from the middle, or thoracic-lumbar, portion of the cord and almost immediately terminate in ganglia that lie near the cord (fig. 39.9). The sympathetic nervous system is especially important during emergency situations and may be associated with the "fight or flight response." For example, it inhibits the digestive tract but dilates the pupil of the eye, accelerates the heartbeat, and increases the breathing rate. It is not surprising, then, that the neurotransmitter released by the postganglionic axon is *norepinephrine* (NE), a chemical close in structure to epinephrine (adrenalin), a well-known heart stimulant.

Figure 39.8

Diagram of a reflex arc shows the detailed composition of a spinal nerve. When the receptors in the skin are stimulated, nerve impulses (see arrows) move along a sensory neuron to the spinal cord. (Note that the cell body of a sensory neuron is in a ganglion outside the cord.) The nerve impulses are picked up by an interneuron, which lies completely within the cord, and pass to the dendrites and cell body of a motor neuron, which lies ventrally within the cord. The nerve impulses then move along the axon of the motor neuron to an effector, such as a muscle fiber that contracts. The brain receives information concerning sensory stimuli by way of other interneurons having long fibers in tracts that run up and down the cord within the white matter.

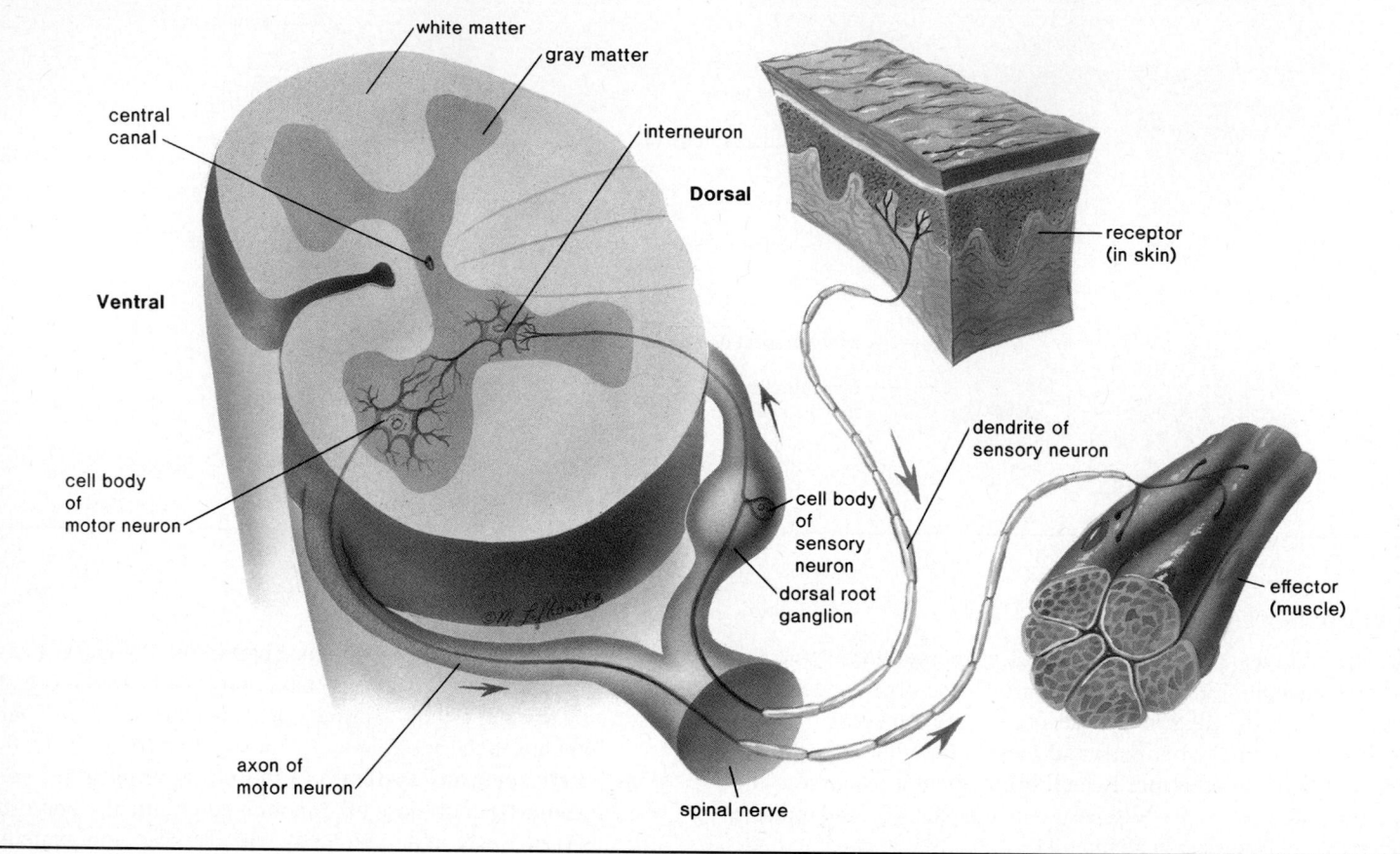

Parasympathetic Nervous System

In the parasympathetic nervous system, the preganglionic fibers arise from the brain and the bottom (sacral) portion of the cord system. This system is therefore often referred to as the craniosacral portion of the autonomic nervous system. The preganglionic fibers terminate in ganglia that lie near or within the organ (fig. 39.9). The parasympathetic system, sometimes called the "housekeeper system," promotes all the internal responses we associate with a relaxed state; for example, it causes the pupil of the eye to contract, promotes digestion of food, and retards the heartbeat. The neurotransmitter utilized by the parasympathetic system is *acetylcholine* (ACh).

> The autonomic nervous system controls the functioning of internal organs without conscious control. The sympathetic nervous system brings about those responses we associate with "fight or flight." The parasympathetic nervous system brings about those responses we associate with normally restful activities.

Central Nervous System

The central nervous system (CNS) consists of the spinal cord and brain (fig. 39.10). The CNS is protected by bone: the brain is enclosed within the skull, and the spinal cord is surrounded by vertebrae. Also, both the brain and the spinal cord are wrapped in 3 layers of protective membranes known as the *meninges;* spinal meningitis is a well-known infection of these coverings. The interior of the brain contains spaces known as *ventricles.* Like the central canal of the spinal cord, these are filled with cerebrospinal fluid.

The spinal cord and brain contain gray matter and white matter. As mentioned, the *gray matter* is made up of cell bodies and short fibers. The **white matter** consists of the long, myelinated fibers of interneurons. Some of these are grouped together in bundles called *tracts,* which run between the brain and the spinal cord. The tracts cross so that the left side of the brain controls skeletal muscles on the right side of the body and vice versa.

Animal Structure and Function

Figure 39.9

Structure and function of the autonomic nervous system. The sympathetic fibers arise from the thoracic and lumbar portions of the cord; the parasympathetic fibers arise from the brain and sacral portion of the cord. Each system innervates the same organs but has a contrary effect. For example, the sympathetic system speeds up the beat of the heart, whereas the parasympathetic system slows it down.

Table 39.1
Sympathetic versus Parasympathetic System

Sympathetic	Parasympathetic
Fight or flight	Normal activity
Norepinephrine is neurotransmitter	Acetylcholine is neurotransmitter
Postganglionic fiber is longer than preganglionic	Preganglionic fiber is longer than postganglionic
Preganglionic fiber arises from middle portion of cord	Preganglionic fiber arises from brain and lower portion of cord

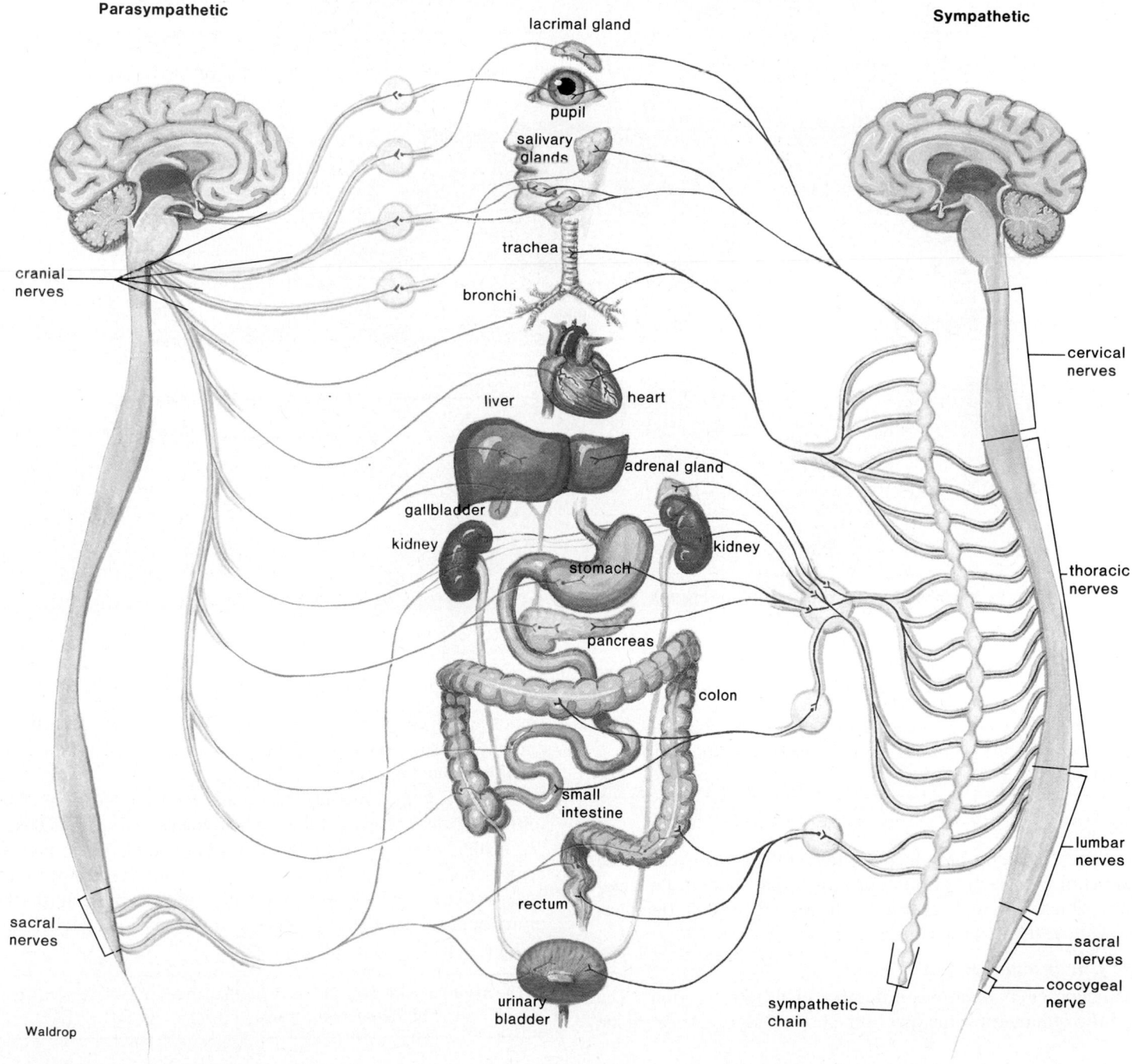

Parasympathetic

Sympathetic

lacrimal gland
pupil
salivary glands
trachea
bronchi
liver
heart
adrenal gland
gallbladder
kidney
kidney
stomach
pancreas
colon
small intestine
rectum
urinary bladder

cranial nerves
sacral nerves
Waldrop

cervical nerves
thoracic nerves
lumbar nerves
sacral nerves
coccygeal nerve
sympathetic chain

Nervous System

643

Figure 39.10

The human brain. Note how large the cerebrum is, compared to the rest of the brain.

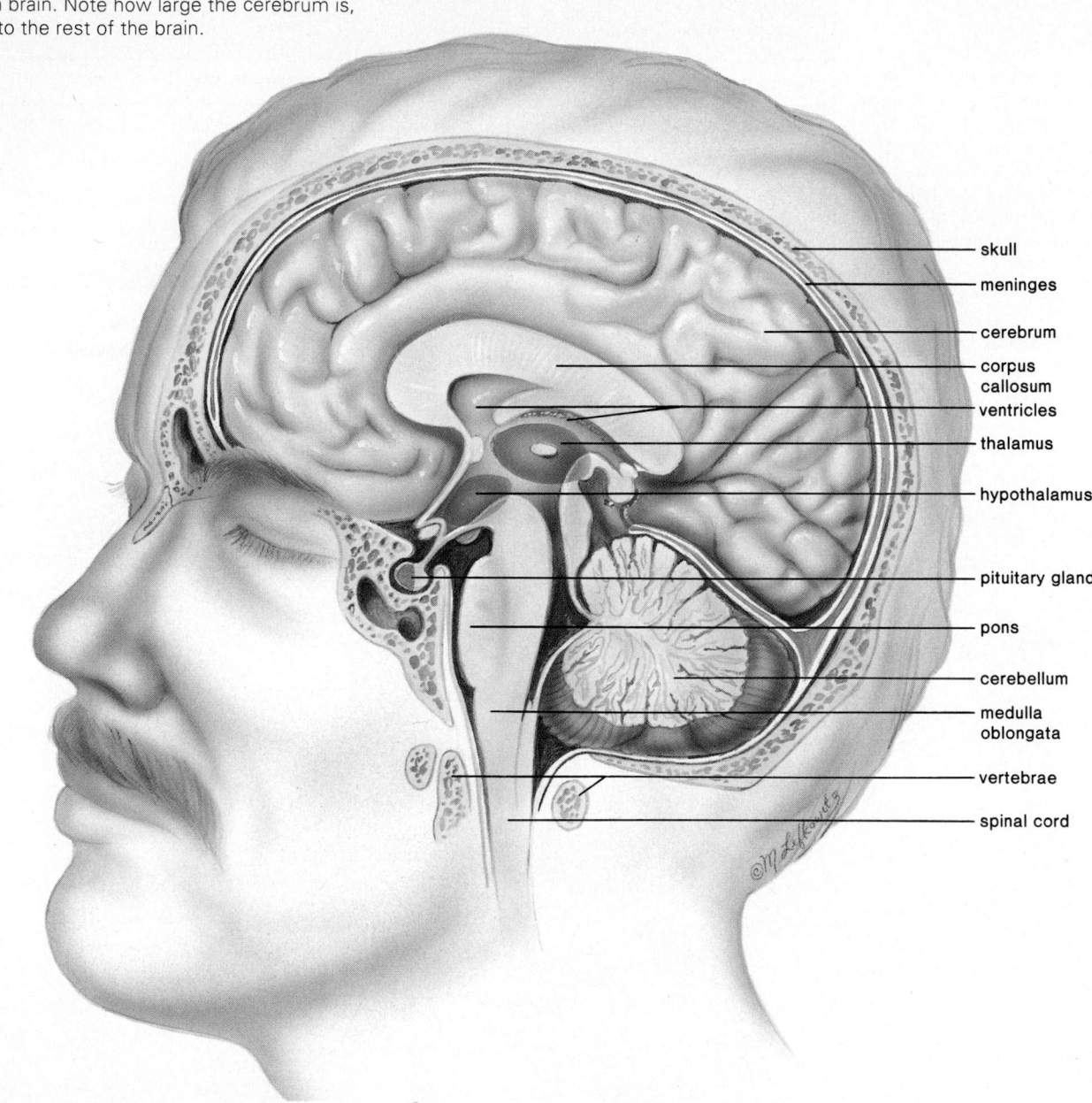

- skull
- meninges
- cerebrum
- corpus callosum
- ventricles
- thalamus
- hypothalamus
- pituitary gland
- pons
- cerebellum
- medulla oblongata
- vertebrae
- spinal cord

The CNS consists of the brain and the spinal cord and is the place where sensory information is received and motor control is initiated.

The Brain

The largest and most prominent portion of the human brain is the **cerebrum** (fig. 39.10). Consciousness resides only in the cerebrum, which we will discuss later; the rest of the brain functions below the level of consciousness.

The Unconscious Brain

The unconscious brain has a number of different regions. The **medulla oblongata** is the part of the brain that lies closest to the spinal cord. It contains centers for regulating heartbeat, breathing, and vasoconstriction (blood pressure) and also reflex centers for vomiting, coughing, sneezing, hiccupping, and swallowing.

The **hypothalamus** forms the lower walls and floor of the third ventricle. This part of the brain is concerned with homeostasis, or the constancy of the internal environment, and contains centers for regulating hunger, sleep, thirst, body temperature, water balance, and blood pressure. The hypothalamus controls the pituitary gland and thereby serves as a link between the nervous and the endocrine systems.

The medulla oblongata and the hypothalamus are both concerned with control of the internal organs.

Figure 39.11

Comparison of brain sizes among vertebrates. Note the enlargement in the size of the cerebrum counterclockwise from bass to human.

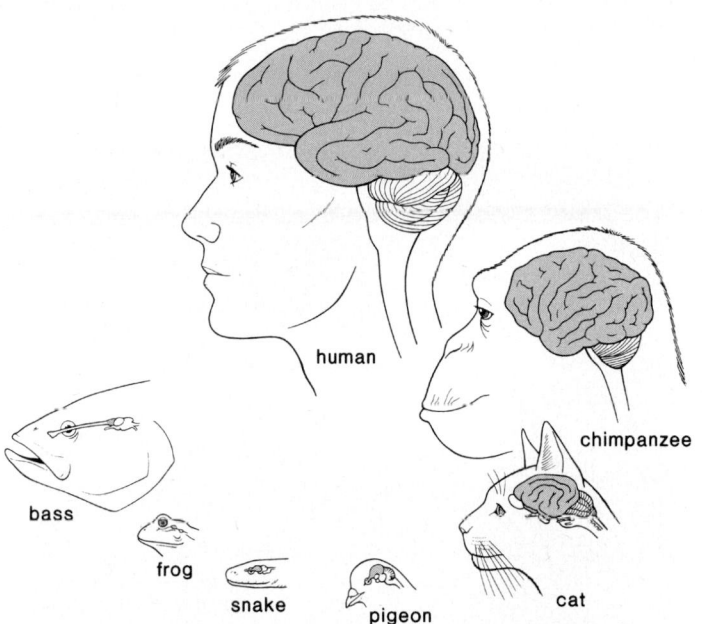

human

chimpanzee

bass

frog

snake

pigeon

cat

The *midbrain* and *pons* contain tracts that connect the cerebrum with other parts of the brain. The midbrain controls head and eyeball movements in response to visual and auditory stimuli. The pons helps the medulla regulate the breathing rate.

The **thalamus** is an egg-shaped structure in the third ventricle. It is the last portion of the brain for sensory input before the cerebrum. It serves as a central relay station for sensory impulses traveling upward from other parts of the spinal cord and the brain to the cerebrum. It receives all sensory impulses (except those associated with the sense of smell) and channels them to appropriate regions of the cortex for interpretation. The thalamus is sometimes called the gatekeeper to the cerebrum because it monitors the sensory data to be sent to the cerebrum. The thalamus allows us to ignore extraneous sensory information and pay attention instead to more important matters.

The **cerebellum,** a bilobed structure that resembles a butterfly, is the second largest portion of the brain (fig. 39.10). It is located dorsal to the pons and medulla oblongata. The cerebellum functions in muscle coordination by integrating impulses received from higher centers to ensure that all the skeletal muscles work together to produce smooth and graceful motions. The cerebellum is also responsible for maintaining normal muscle tone and transmitting impulses that maintain

Figure 39.12

The limbic system (*rear*), which includes portions of the cerebrum, thalamus, and hypothalamus, is concerned mainly with emotion and memory. The convoluted cortex of the cerebrum (*front*) is divided into 4 lobes: frontal, temporal, parietal, and occipital. It is possible to map the cerebrum because each particular area has a particular function.

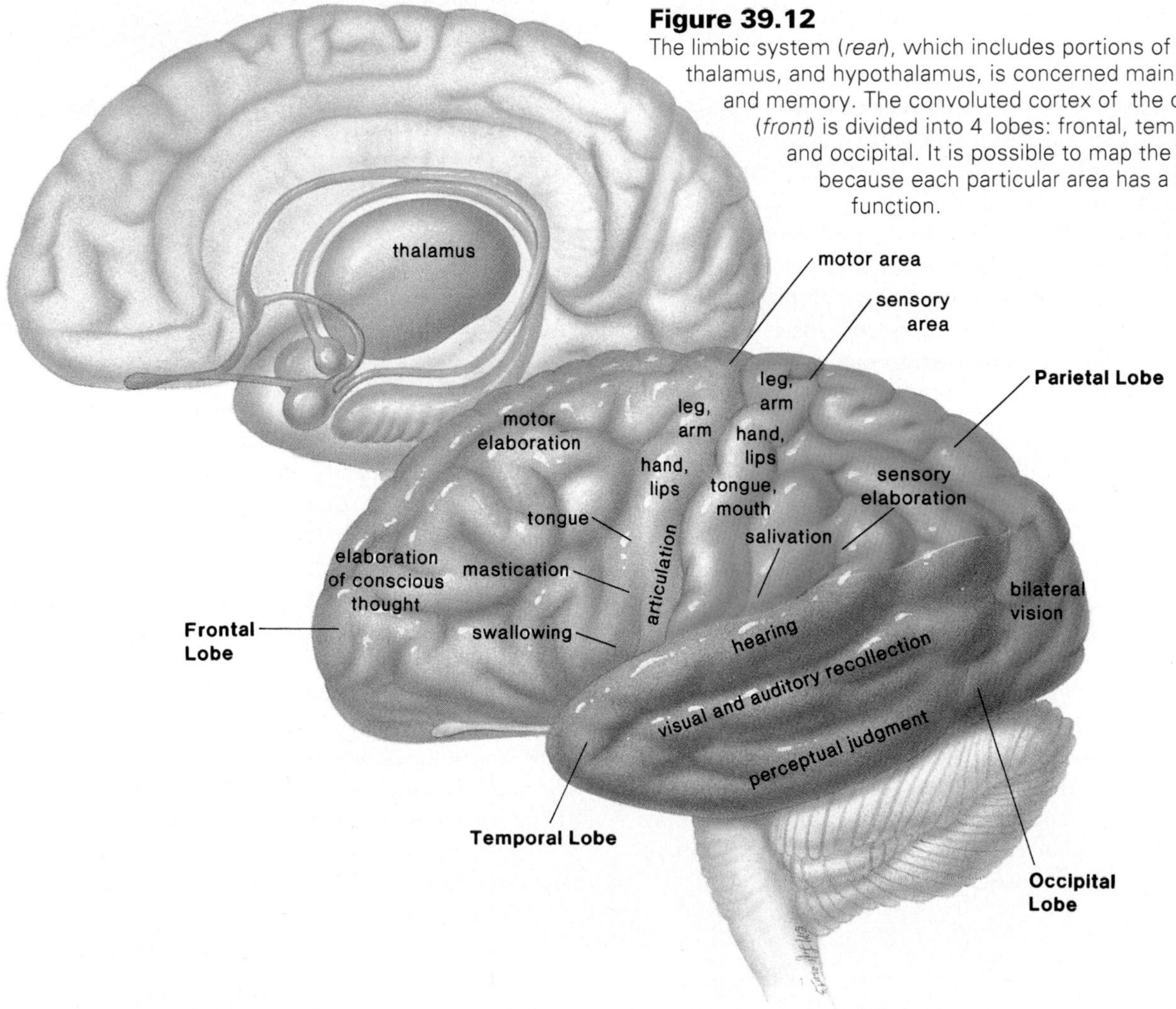

thalamus

motor area

sensory area

Parietal Lobe

leg, arm

leg, arm

hand, lips

motor elaboration

hand, lips

tongue, mouth

sensory elaboration

salivation

tongue

articulation

mastication

elaboration of conscious thought

Frontal Lobe

swallowing

hearing

bilateral vision

visual and auditory recollection

perceptual judgment

Temporal Lobe

Occipital Lobe

posture. It receives information from the inner ear indicating the position of the body and sends impulses to muscles that maintain or restore balance.

> The thalamus is concerned with sensory reception, and the cerebellum is concerned with muscle control.

The Conscious Brain

The cerebrum, the foremost part of the brain, is the largest part of the brain in humans. It consists of 2 large masses called cerebral hemispheres, which are connected by a bridge of nerve fibers called the corpus callosum. The outer portion of the cerebral hemispheres, the cerebral cortex, is highly convoluted and gray in color because it contains cell bodies and short fibers.

A comparative study of vertebrates indicates a progressive increase in the size of the cerebrum from fishes to humans, and the cerebral cortex is more convoluted in humans than other vertebrates (fig. 39.11). The function of the cerebrum has also changed. In fishes and amphibians, the cerebrum largely has an olfactory function, but in reptiles, birds, and mammals, the cerebrum receives information from other parts of the brain and coordinates sensory data and motor functions. Only the cerebrum is responsible for consciousness, and it is the portion of the brain that governs intelligence and reason. These qualities are particularly well developed in humans.

The cerebral cortex of each hemisphere contains 4 surface lobes: frontal, parietal, temporal, and occipital (fig. 39.12). Different functions are associated with each lobe. For example, the *frontal lobe* controls motor functions and permits us to control our muscles consciously. The *parietal lobe* receives information from receptors located in the skin, such as those for touch, pressure, and pain. The *occipital lobe* interprets visual input. The *temporal lobe* has sensory areas for hearing and smelling.

Certain areas of the cerebral cortex have been "mapped" in great detail. For example, we know which portions of the frontal lobe control various parts of the body and which portions of the parietal lobe receive sensory information from these same parts. Each of the 4 lobes of the cerebral cortex contains an association area, which receives information from the other lobes and integrates it into higher, more complex levels of consciousness. These areas are concerned with intellect, artistic and creative ability, learning, and memory.

> The cerebrum is the most highly developed region of the brain and it alone carries on conscious thought processes. The cerebrum interprets incoming sensory data and initiates voluntary muscle movements.

The Limbic System

The **limbic system** involves portions of both the unconscious and conscious brain. It lies just beneath the cortex and contains neural pathways that connect portions of the frontal lobes, temporal lobes, thalamus, and hypothalamus (fig. 39.12). Several masses of gray matter that lie deep within each hemisphere of the cerebrum, termed the *basal nuclei,* are also a part of the limbic system.

Figure 39.13
Individual nerve cells in a snail, *Hermissenda,* are being stimulated by microelectrodes, which produce the signals scientists previously recorded when a snail learns to avoid light. When this snail is freed, it automatically avoids the light and does not need to be taught like other snails. Normally, to teach snails to avoid light, they are placed on a table that rotates every time they venture toward light.

Stimulation of different areas of the limbic system causes the subject to experience rage, pain, pleasure, or sorrow. By causing pleasant or unpleasant feelings about experiences, the limbic system apparently guides the individual into behavior that is likely to increase the chance of survival.

Learning and Memory The limbic system is also involved in the processes of learning and memory. Learning requires memory, but just what permits memory development is not definitely known. Investigators have been working with invertebrates such as slugs and snails because their nervous system is very simple and yet they can be conditioned to perform a particular behavior. To study this simple type of learning, it has been possible to insert electrodes into individual cells and to alter or record the electrochemical responses of these cells (fig. 39.13). This type of research has shown that learning is accompanied by an increase in the number of synapses, while forgetting involves a decrease in the number of synapses. In other words, the nerve-circuit patterns are constantly changing as learning, remembering, and forgetting occur. Within the individual neuron, learning involves a change in gene regulation and nerve protein synthesis and an increased ability to secrete transmitter substances.

Animal Structure and Function

Figure 39.14

Some drug actions at synapses. **a.** Drug stimulates the release of the neurotransmitter. **b.** Drug blocks the release of the neurotransmitter. **c.** Drug combines with the neurotransmitter, preventing its breakdown. **d.** Drug mimics the action of the neurotransmitter. **e.** Drug blocks the receptor so that the neurotransmitter cannot be received. Generally, only one of these actions occurs at a given synapse.

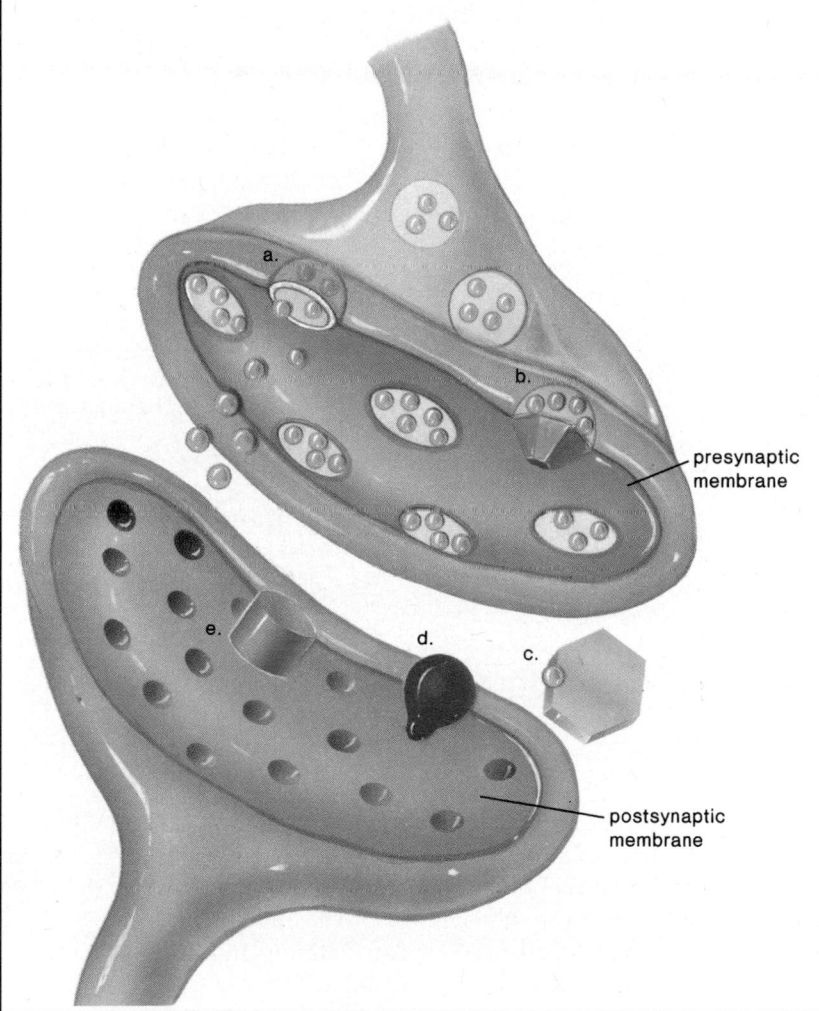

presynaptic membrane

postsynaptic membrane

Table 39.2
Drug Action

Drug Action	Neurotransmitter	Result
Blocks	Excitatory	Depression
Enhances	Excitatory	Stimulation
Blocks	Inhibitory	Stimulation
Enhances	Inhibitory	Depression

At the other end of the spectrum, some investigators have been studying learning and memory in monkeys. This work has led to the conclusion that the limbic system is absolutely essential to both short-term and long-term memory. An example of short-term memory in humans is the ability to recall a telephone number long enough to dial the number; an example of long-term memory is the ability to recall the events of the day. After nerve impulses circulate within the limbic system, the basal nuclei stimulate the sensory areas where memories are stored. The involvement of the limbic system certainly explains why emotionally charged events result in our most vivid memories. The fact that the limbic system communicates with the sensory areas for touch, smell, vision, and so forth accounts for the ability of any particular sensory stimulus to awaken a complex memory.

> The limbic system is particularly involved in the emotions and in memory and learning.

Neurotransmitters in the Brain

As discussed previously, neurotransmitters released at the ends of axons affect the membrane potential of postsynaptic membranes. Some excitatory transmitters are amines, such as acetylcholine (ACh), norepinephrine (NE), serotonin, and dopamine. We have already had occasion to mention that ACh and NE are transmitters for the autonomic nervous system. *Serotonin* and *dopamine* are associated with behavior states, such as mood, sleep, attention, learning, and memory. Increasingly, it appears that feelings of pleasure are accompanied by the release of dopamine. The inhibitory transmitters include the amino acids gamma-aminobutyric acid (GABA) and glycine.

> Both excitatory and inhibitory neurotransmitters are active in the brain.

In addition to the neurotransmitters already mentioned, a number of different types of peptides have been discovered in the CNS. The *endorphins,* molecules that are the body's natural opioids, are of particular interest. They are called natural opioids because the psychoactive drugs morphine and heroin, which are both derived from opium, attach to the receptors for endorphins in the CNS. When endorphins are released, they, like opium, produce a feeling of elation and reduce the sensation of pain. For example, if endorphins are present, neurons do not release substance P, a neurotransmitter that brings about the sensation of pain. Exercise has been associated with their presence, and this may account for the so-called runner's high.

Neurotransmitter Disorders It has been discovered that several neurological illnesses, such as *Parkinson disease* and *Huntington disease,* are due to an imbalance of neurotransmitters.

Five Drugs of Abuse

Alcohol

I t is possible to drink alcoholic beverages in moderation, but they are often abused. Alcohol use becomes "abuse," or an illness, when alcohol ingestion impairs an individual's social relationships, health, job efficiency, or judgment. While it is general knowledge that alcoholics are prone to drink until they become intoxicated, there is much debate as to what causes alcoholism. Some believe that alcoholism is due to an underlying psychological disorder, while others maintain that the condition is due to an inherited physiological disorder.

Alcohol effects on the brain are biphasic. After consuming several drinks, blood alcohol concentration rises rapidly and the drinker reports feeling "high" and happy (euphoric); however, after about 90 minutes and lasting until about 5–6 hours after consumption, the drinker feels depressed and unhappy (dysphoric). On the other hand, if the drinker continues to drink to maintain a high blood level of alcohol, he or she experiences ever-increasing loss of control. Coma and death are even possible if a substantial amount of alcohol is consumed within a limited period.

Marijuana

The dried flowering tops, leaves, and stems of the Indian hemp plant *Cannabis sativa* contain and are covered by a resin that is rich in THC (tetrahydrocannabinol). The names *cannabis* and *marijuana* can apply either to the plant or to THC.

The effects of marijuana differ depending upon the strength and the amount consumed, the expertise of the user, and the setting in which it is taken. Usually, the user reports experiencing a mild euphoria, along with alterations in vision and judgment that result in distortions of space and time. The inability to concentrate or to speak coherently and motor incoordination may also be involved.

Intermittent use of low-potency marijuana is not generally associated with obvious symptoms of toxicity, but heavy use may produce chronic intoxication. Intoxication is recognized by the presence of hallucinations, anxiety, depression, rapid flow of ideas, body image distortions, paranoid reactions, and similar psychotic symptoms. The terms *cannabis psychosis* and *cannabis delirium* refer to such reactions.

The use of marijuana has ill effects, but it does not seem to produce physical dependence. There may be a psychological dependence on the euphoric and sedative effects. Craving or difficulty in stopping may also occur as a part of regular heavy use.

Marijuana has been called a *gateway* drug because adolescents who have used marijuana tend to also try other drugs. For example, in a study of 100 cocaine abusers, 60% had also smoked marijuana for more than 10 years.

Cocaine

Cocaine, currently the second most popular illegal drug after marijuana, is derived from the shrub *Erythroxylon coca.* Cocaine is sold in powder form and in the form of crack, a more potent extract. Users often use the word *rush* to describe the feeling of euphoria that follows intake of the drug. Snorting (inhaling through the nose) produces this effect in a few minutes; injection, within 30 seconds; and smoking in less than 10 seconds. Persons dependent upon the drug are, therefore, most likely to use the last method of intake. The rush only lasts a few seconds and is then replaced by a state of arousal that lasts from 5 minutes–30 minutes. Then the user begins to feel restless, irritable, and depressed. To overcome these symptoms the user is apt to take more of the drug, repeating the cycle until there is no more drug left. A binge of

---

Parkinson disease is a condition characterized by a wide-eyed, unblinking expression, an involuntary tremor of the fingers and thumbs, muscle rigidity, and a shuffling gait. All these symptoms are due to dopamine deficiencies. Huntington disease is characterized by a progressive deterioration of the individual's nervous system; this eventually leads to constant thrashing and writhing movements and finally to insanity and death. The problem is believed to be malfunction of the inhibitory neurotransmitter GABA. *Alzheimer disease*, a severe form of senility with marked memory loss in 5%–10% of all people over age 65, is recognized by the presence of innumerable diffuse betaprotein plaques in the brain. Many different types of neurons and neurotransmitters are adversely affected.

Some neurological illnesses are associated with the deficiency of a particular neurotransmitter in the brain.

Action of Neurological Drugs Drugs that alter the mood of a person either enhance or block the action of a particular neurotransmitter found particularly in the limbic system. There are a number of different ways drugs can influence the action of neurotransmitters, some of which are shown in figure 39.14. It is clear, as outlined in table 39.2 that stimulants can either enhance the action of an excitatory transmitter or block the action of an inhibitory transmitter. On the other hand, depressants can either enhance the action of an inhibitory transmitter or block the action of an excitatory transmitter.

One group of scientists believes that the neurotransmitter dopamine is involved in all forms of pleasure. Substantiating this position is the knowledge that cocaine interferes with the uptake of dopamine from synaptic clefts. Also, neurons producing dopamine make up a significant part of the limbic system. Many new medications developed to counter drug addiction and mental illness affect the release, reception, or breakdown of dopamine. There is some evidence to suggest that those with mental illness and those prone to drug addiction are afflicted with the same genetic defects; perhaps ones affecting dopamine transmission across the synapse.

A wide variety of drugs can be used to alter mood and/or emotional state. This chapter's reading discusses 5 of the most commonly abused drugs: alcohol, marijuana, cocaine, methamphetamine, and heroin. Drug abuse occurs when a person takes

Animal Structure and Function

this sort can go on for days, after which the individual suffers a *crash*. During the binge period, the user is hyperactive and has little desire for food or sleep but has an increased sex drive. During the crash period, the user is fatigued, depressed, and irritable and has memory and concentration problems and displays no interest in sex. Indeed, men are often impotent. Other drugs, such as marijuana, alcohol, or heroin, are often taken to ease the symptoms of the crash.

With continued cocaine use, the body begins to make less dopamine to compensate for a seemingly excess supply. The user, therefore, now experiences *tolerance* (always needing more of the drug for the same effect), *withdrawal* (symptoms such as the crash symptoms described previously when a drug is not taken), and an intense *craving* for cocaine. These are indications that the person is highly dependent upon the drug or, in other words, that cocaine is extremely addictive.

Methamphetamine (Ice)

Methamphetamine is related to amphetamine, a well-known stimulant. Both methamphetamine and amphetamine have been drugs of abuse for some time, but a new form of methamphetamine known as "ice"

is now being used as an alternative to cocaine. Ice is a pure, crystalline hydrochloride salt that has the appearance of sheet-like crystals. Unlike cocaine, ice can be produced in this country in illegal laboratories and does not need to be imported.

Ice, like crack, will vaporize in a pipe and can be smoked, avoiding the complications of intravenous injections. After rapid absorption into the bloodstream, the drug moves quickly to the brain. It has the same stimulatory effect as cocaine, and subjects report they cannot distinguish between the 2 after intravenous administration. Methamphetamine effects, however, persist for hours instead of a few seconds. Therefore, it is the preferred drug of abuse by many.

Heroin

Heroin is derived from morphine, a derivative of *opium*. Heroin is usually injected. After intravenous injection, the onset of action is noticeable within one minute and reaches its peak in 3 minutes–6 minutes. There is a feeling of euphoria along with relief of pain. Side effects can include nausea, vomiting, dysphoria, and respiratory and circulatory depression leading to death.

Heroin binds to receptors meant for the body's own opioids, the endorphins. As mentioned previously, the opioids are be-

lieved to alleviate pain by preventing the release of the neurotransmitter termed substance P from certain sensory neurons in the region of the spinal cord. When substance P is released, pain is felt, and when substance P is not released, pain is not felt. Evidence also indicates that there are opioid receptors in neurons that travel from the spinal cord to the limbic system and that stimulation of these can cause a feeling of pleasure. This explains why opium and heroin not only kill pain but also produce a feeling of tranquility.

Individuals who inject heroin become physically dependent on the drug. With time, the body's production of endorphins decreases; now *tolerance* develops so that the user needs to take more of the drug just to prevent *withdrawal* symptoms. The euphoria originally experienced upon injection is no longer felt.

The withdrawal symptoms include perspiration, dilation of pupils, tremors, restlessness, abdominal cramps, goose flesh, defecation, vomiting, and increase in systolic pressure and respiratory rate. Those who are excessively dependent may experience convulsions, respiratory failure, and death. Infants born to women who are physically dependent also experience these withdrawal symptoms.

excessive amounts of a drug under circumstances that increase the likelihood of a harmful effect. Individuals who are drug abusers are also apt to display a *physical dependence* on the drug (formerly called an addiction to the drug). Dependence is present when the person (1) spends much time thinking about the drug or arranging to get it; (2) often takes more of the drug than was intended; (3) is tolerant to the drug—that is, must increase the

amount of the drug to get the same effect; (4) has withdrawal symptoms when he or she stops taking the drug; and (5) has a repeated desire to cut down on use.

Neurological drugs interfere with normal neurotransmitter function in the brain, particularly in the limbic system. Drug abuse often results in physical dependence on the drug.

Summary

1. In humans, the nervous system, along with the endocrine system, regulates the other systems of the body and coordinates body functions.

2. A comparative study of the invertebrates shows a gradual increase in the complexity of nervous system. The human nervous system, like that of the earthworm, is divided into the central and peripheral nervous systems.

3. Neurons, cells that conduct nerve impulses, have 3 parts: dendrite(s), cell body, and axon. The dendrites and the cell body receive information from the environment, and if stimulation is sufficient, nerve impulses travel down the axon to where neurotransmitters are stored in vesicles.

4. The nerve impulse, which is recognized when the resting potential becomes an action potential, is an electrochemical phenomenon involving

the movement first of sodium and then of potassium ions across the axomembrane.

Resting potential: Sodium-potassium pump at work, inside of the neuron is negative (−65 mV) compared to the outside of the neuron.

Action potential: (1) Sodium ions move to the inside, making it positive compared to outside (+40 mV). (2) Potassium ions move to the outside, making the inside negative again compared to the outside (–65 mV).

5. The nerve impulse is self-propagating because the gated channels that allow sodium and potassium ions to flow down their concentration gradients are sensitive to a nearby decrease in the membrane potential.

6. Saltatory conduction occurs in myelinated fibers—the action potential jumps from node of Ranvier to node of Ranvier. This accounts for the great speed of impulses in these fibers.

7. Transmission across a synapse usually requires neurotransmitters because there is a small space, the synaptic cleft, that separates neuron from neuron. The neurotransmitters released at the ends of axons may be either excitatory or inhibitory.

8. Acetylcholine and norepinephrine are well-known excitatory transmitters. After its release, acetylcholine is destroyed by acetylcholinesterase in the synaptic cleft.

9. Within the human peripheral nervous system, the somatic system includes cranial and spinal nerves. Spinal nerves extend from the spinal cord and contain both sensory and motor fibers. It is possible to use them to trace the path of a reflex arc from receptor to effector, as described here.

Sensory neuron: receptor generates nerve impulses that travel in the dendrite to the cell body and then to the axon, which enters the cord.

Interneuron: transmits impulses from the dorsal to the ventral root of the cord.

Motor neuron: impulses begin in the dendrites and cell body and then pass out of the cord by way of the axon, which innervates an effector.

10. The autonomic nervous system controls internal organs and includes the sympathetic and parasympathetic nervous systems, which are contrasted in table 39.1.

11. The human central nervous system includes the spinal cord and the brain. In the brain, the medulla oblongata and hypothalamus regulate internal organs. The thalamus receives sensory input and passes it to the cerebrum. The cerebellum functions in muscle coordination.

12. A survey of vertebrates shows a continual evolutionary increase in the size of the cerebrum, with its greatest development in humans. The highly convoluted cerebral cortex is divided into lobes, each of which has specific functions. In general, the cerebrum is responsible for consciousness, sensory perception, motor control, and all the higher forms of thought.

13. Research in invertebrates indicates that learning is accompanied by an increase in the number of synapses; research in monkeys indicates that the limbic system is involved. There are short-term and long-term memories; the involvement of the limbic system explains why emotionally charged events result in vivid long-term memories.

14. Various illnesses such as Parkinson, Huntington, and Alzheimer diseases are due to an imbalance in a specific neurotransmitter in the brain. Drugs that people take to alter the mood and/or emotional state often interfere with normal neurotransmitter function in the brain, particularly the limbic system.

Writing Across the Curriculum

In order to practice writing skills, students should write out the answers to any or all of the study questions and the critical thinking questions. The study questions are sequenced in the same order as the text. Suggested answers to the critical thinking questions are in appendix D.

Study Questions

1. What is the overall function of the nervous and endocrine systems? How do they differ in regard to this function?

2. Trace the evolution of the nervous system by contrasting the organization of the nervous system in hydras, planarians, earthworms, and humans.

3. Describe the structure of a neuron and give a function for each part mentioned. Name 3 types of neurons, and give a function for each.

4. What are the major events of an action potential and the ion changes that are associated with each event?

5. Describe the mode of action of a neurotransmitter at a synapse, including how it is stored and how it is destroyed.

6. Contrast the structure and function of the peripheral and the central nervous systems.

7. Trace the path of a spinal reflex.

8. Contrast the sympathetic and parasympathetic divisions of the autonomic nervous system.

9. Name the major parts of the human brain and give a principal function for each part.

10. Define the limbic system and discuss its possible involvement in learning and memory.

11. Name several specific neurotransmitters and, in general, describe how various types of drugs can affect the action of neurotransmitters.

Objective Questions

1. Which is the most complete list of animals that have both a central and a peripheral nervous system?
   a. hydra, planaria, earthworm, rabbit, human
   b. planaria, earthworm, rabbit, human
   c. earthworm, rabbit, human
   d. rabbit, human

2. Which of these are the first element and the last element in a spinal reflex?
   a. axon and dendrite
   b. sense organ and muscle effector
   c. ventral horn and dorsal horn
   d. motor neuron and sensory neuron

Animal Structure and Function

3. Which phrase does not belong with the others?
   a. cerebrum
   b. cerebral cortex
   c. cerebral hemispheres
   d. cerebellum

4. A spinal nerve takes nerve impulses
   a. to the CNS.
   b. away from the CNS.
   c. both to and away from the CNS.
   d. only inside the CNS.

5. Which of these correctly describes the distribution of ions on either side of an axon when it is not conducting a nerve impulse?
   a. Na$^+$ outside and K$^+$ inside
   b. K$^+$ outside and Na$^+$ inside
   c. charged protein outside; Na$^+$ and K$^+$ inside
   d. Na$^+$ and K$^+$ outside and water only inside

6. When the action potential begins, sodium gates open, allowing sodium ions to cross the axomembrane. Now the polarity changes to
   a. negative outside and positive inside.
   b. positive outside and negative inside.
   c. no difference in charge between outside and inside.
   d. Any one of these.

7. Transmission of the nerve impulse across a synapse is accomplished by the
   a. movement of sodium and potassium ions.
   b. release of neurotransmitters.
   c. Both of these.
   d. Neither of these.

8. The autonomic nervous system has 2 divisions called the
   a. CNS and peripheral systems.
   b. somatic and skeletal systems.
   c. efferent and afferent systems.
   d. sympathetic and parasympathetic systems.

9. Synaptic vesicles are
   a. at the ends of dendrites and axons.
   b. at the ends of axons only.
   c. along the length of all long fibers.
   d. All of these.

10. Which of these is mismatched?
    a. cerebrum—consciousness
    b. thalamus—motor and sensory centers
    c. hypothalamus—internal environment regulator
    d. cerebellum—motor coordination

11. Label this diagram of a reflex arc. State a function for each structure labeled:

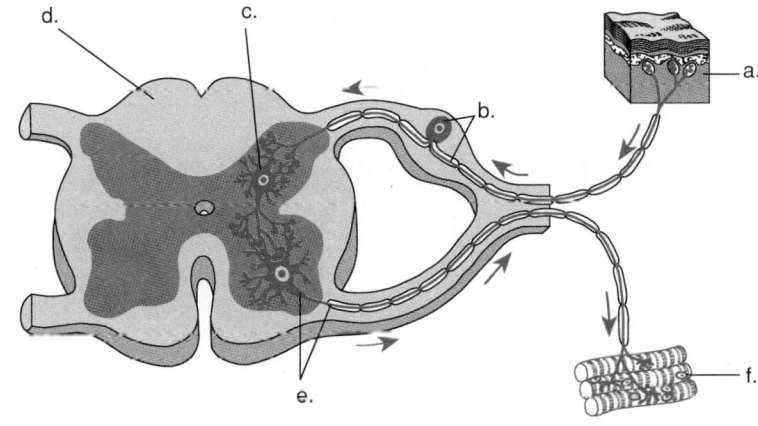

Concepts and Critical Thinking

1. *All systems of an animal's body contribute to homeostasis.*

   Tell several ways in which the nervous system contributes to homeostasis.

2. *Segmentation leads to specialization of parts.*

   How is segmentation along with specialization of parts reflected in the structure of the nervous system?

3. *Structure suits function.*

   How does the structure of a neuron suit its function?

Selected Key Terms

central nervous system (sen'tral ner'vus sis'tem) 634
peripheral nervous system (pĕ-rif'er-al ner'vus sis'tem) 634
neuron (nu'ron) 635
dendrite (den'drīt) 635
axon (ak'son) 635
myelin sheath (mi'ĕ-lin shēth) 636
resting potential (rest'ing po-ten'shal) 637
action potential (ak'shun po-ten'shal) 639

saltatory conduction (sal'tah-to"re kon-duk'shun) 639
synapse (sin'aps) 639
neurotransmitter (nu"ro-trans-mit'er) 639
acetylcholine (as"ĕ-til-ko'lēn) 639
norepinephrine (nor"ep-i-nef'ron) 639
nerve (nerv) 640
ganglion (gang'gle-on) 640
reflex (re'fleks) 641
autonomic nervous system (aw"to-nom'ik ner'vus sis'tem) 641

sympathetic nervous system (sim"pah-thet'ik ner'vus sis'tem) 641
parasympathetic nervous system (par"ah-sim"pah-thet'ik ner'vus sis'tem) 641
medulla oblongata (mĕ-dul'ah ob"long-ga'tah) 644
hypothalamus (hi"po-thal'ah-mus) 644
thalamus (thal'ah-mus) 645
cerebellum (ser"ĕ-bel'um) 645
limbic system (lim'bic sis'tem) 646

40

Sense Organs

The eyes of an octopus and other cephalopods are highly developed and similar to those of humans. Light is brought to a focus on a single array of photo-receptor cells. Predators need good eyesight, and studies indicate that an octopus can see an object as small as 0.5 cm from a distance of 1 m.

Your study of this chapter will be complete when you can

1. state the function of sense organs;
2. describe the receptors for smell and taste in humans;
3. tell why the receptors for smell and taste are categorized as chemoreceptors;
4. in general, contrast the eye of arthropods with the eye of humans;
5. list the structures of the human eye, and give a function for each;
6. contrast the action of the sight receptors, the rods and cones;
7. give examples of various types of mechanoreceptors in animals;
8. list the structures of the human ear, and give a function for each one;
9. explain how the ear serves as an organ for static and dynamic equilibrium;
10. describe the organ of Corti and how it functions to permit hearing.

Sense organs receive external and internal stimuli; therefore, they are called **receptors.** Each type of receptor is designed to respond to a particular stimulus: for example, eyes respond only to light, and ears respond only to sound waves. Receptors do not interpret stimuli—they act merely as transducers that transform the energy of stimuli into nerve impulses. Interpretation is the function of the brain, which has a specific region for receiving nerve impulses from each of the sense organs. Impulses arriving at a particular sensory area of the brain can be interpreted in only one way; for example, those arriving at the visual area result in sight sensation, and those arriving at the olfactory area result in smell sensation. Usually the brain is able to discriminate the type of stimulus, the intensity of the stimulus, and the origin of the stimulus. On occasion, the brain can be fooled, as when a blow to the eyes causes you to "see stars."

Interoceptors located within the body monitor such conditions as blood pressure, expansion of lungs and bladder, and movement of limbs. *Exteroceptors* are located near the surface of the animal and respond to outer stimuli. Both types of receptors send information to the brain, and both are needed for homeostasis and to promote appropriate behavior. Intero[...] cussed later in connection with the internal o[...] limit our coverage to exteroceptors.

Chemoreceptors

The receptors responsible for taste and smell a[...] **ceptors** because they are sensitive to certain chemical substances in food, liquids, and air. Chemoreception is found universally in animals and is therefore believed to be the most primitive sense. Chemoreceptors are present all over the body of planarians, but they are concentrated on the auricles at the sides of the head. In arthropods, chemoreceptors are found on the antennae and mouthparts, but in humans, unlike these animals, sense and smell have distinguishable receptors, as we will now discuss.

Smell

In humans, the receptors for the sense of smell are called the **olfactory cells** (*olfactus* means smell in Latin), located high in the roof of the nasal cavity (fig. 40.1). The olfactory cells are actually modified neurons that synapse with the nerve fibers making up the

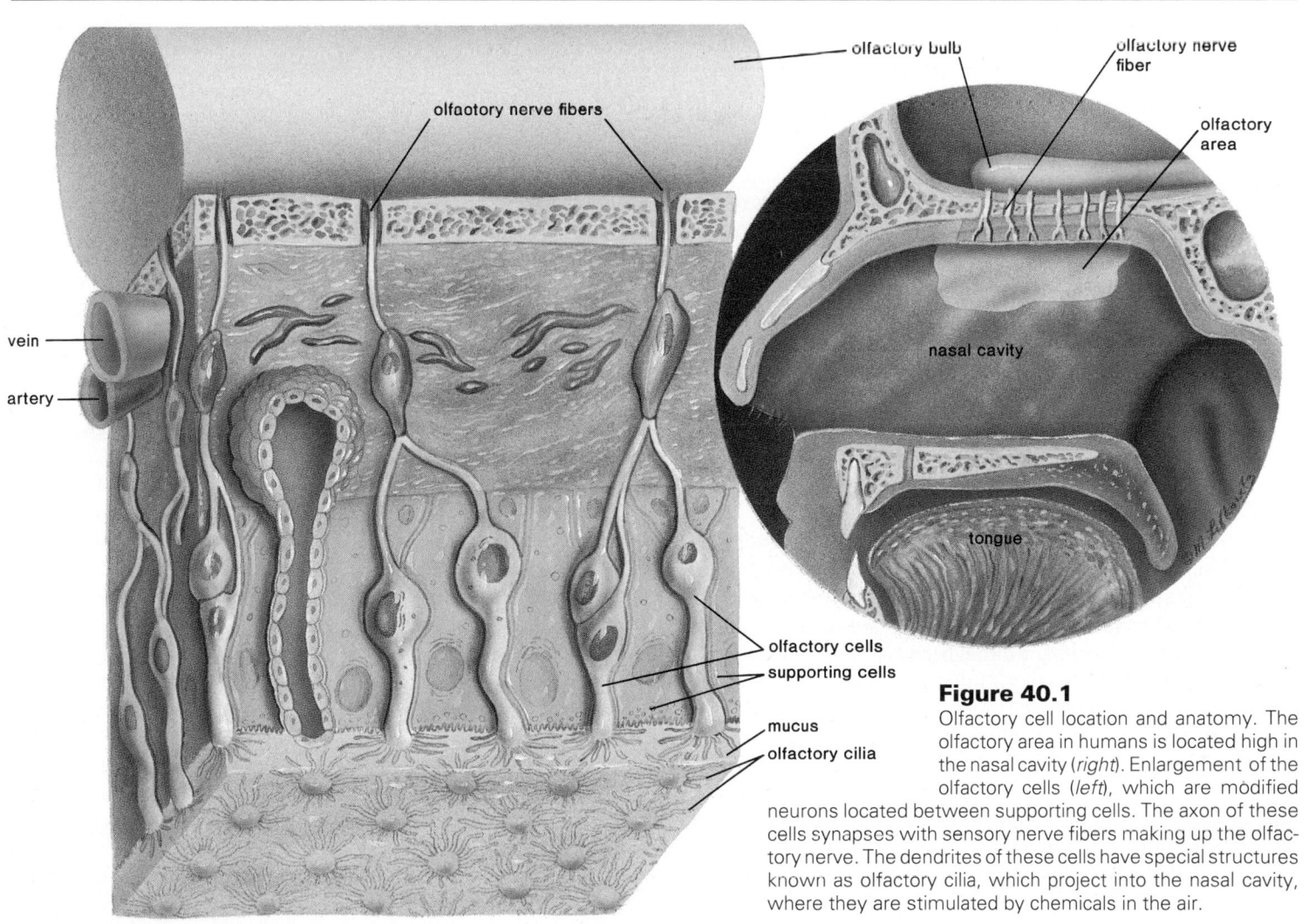

Figure 40.1
Olfactory cell location and anatomy. The olfactory area in humans is located high in the nasal cavity (*right*). Enlargement of the olfactory cells (*left*), which are modified neurons located between supporting cells. The axon of these cells synapses with sensory nerve fibers making up the olfactory nerve. The dendrites of these cells have special structures known as olfactory cilia, which project into the nasal cavity, where they are stimulated by chemicals in the air.

Figure 40.2

Taste buds. **a.** Elevations on the tongue indicate the presence of taste buds. The location of those containing taste buds responsive to sweet, sour, salt, and bitter is indicated. **b.** Enlargement of elevations, called papillae. **c.** The taste buds occur along the walls of the papillae. **d.** Drawing shows the various cells that make up a taste bud. Sensory cells in a bud end in microvilli that have receptors for the chemicals that exhibit the tastes noted in (**a**). When the chemicals combine with the receptors, nerve impulses are generated.

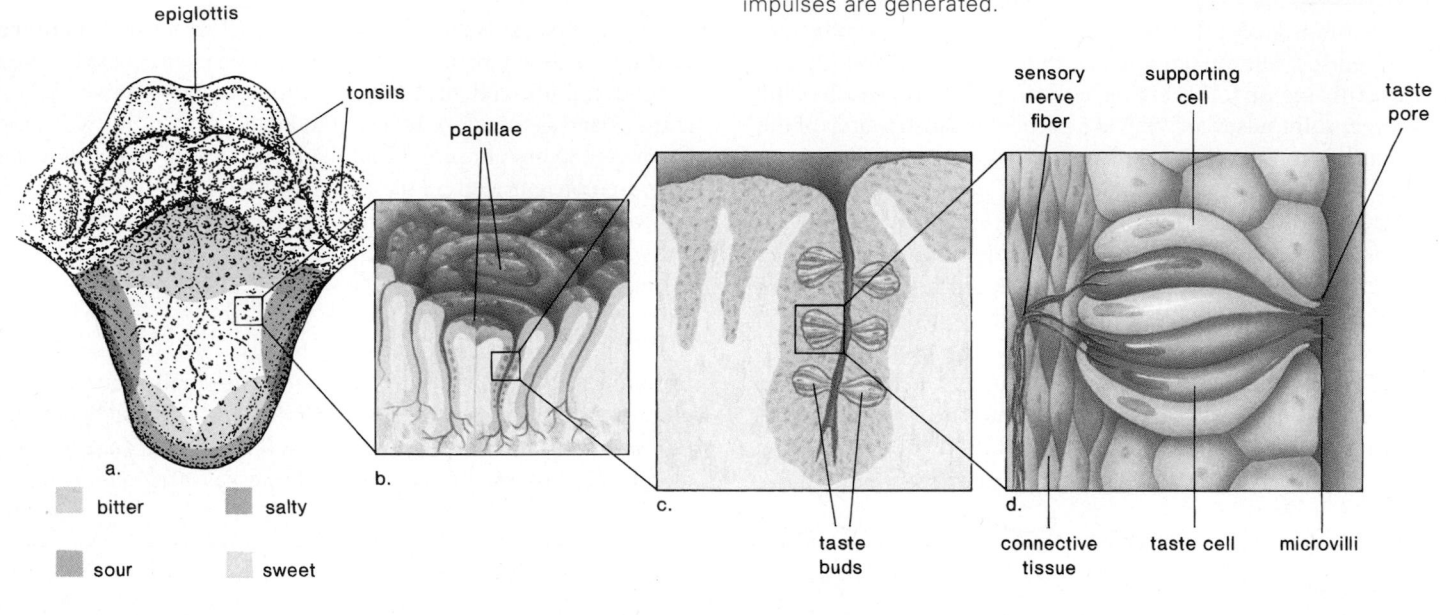

olfactory nerve. Each olfactory cell has cilia, which are stimulated by many chemicals in the air. Research resulting in the stereochemical theory of smell has shown that different smells may be related to the various shapes of the molecules rather than to the atoms that make up the molecules. These shapes fit specific olfactory sites on the olfactory cells' cilia.

The sense of smell is generally much more acute than the sense of taste. The human nose, for example, can detect one 25-millionth of 1 mg of mercaptan, the odoriferous chemical given off by a skunk. This averages out to approximately one molecule per sensory ending. Yet, humans have a weak sense of smell compared to other vertebrates, such as dogs.

Taste

The receptors for the sense of taste in mammals are the **taste buds.** Most of these are located on the tongue, but they are also present on the surface of the soft palate, pharynx, and epiglottis (fig. 40.2). Each taste bud contains a number of elongated cells. These cells have *microvilli*, which project through the taste pore, an opening in the taste bud. It is the microvilli that are stimulated by various chemicals in the environment. Humans are believed to have 4 types of taste buds, each type stimulated by chemicals that result in a bitter, a sour, a salty, or a sweet sensation.

The sense of taste and the sense of smell supplement each other, creating a combined effect when interpreted by the cerebral cortex. For example, when you have a cold, food seems to lose its taste, but actually the ability to sense its smell is temporarily absent.

This may work in reverse also. When we smell something, some of the molecules move from the nose down into the mouth and stimulate certain taste buds. Thus, part of what we refer to as smell is actually taste.

Photoreceptors

Many animals have light-sensitive receptors called **photoreceptors.** In its simplest form, a photoreceptor indicates only the presence of light and its intensity. The "eyespots" of planarians also allow the animal to determine the direction of light. More complex eyes have lenses that focus light on certain receptors. Arthropods have **compound eyes** composed of many independent visual units, each of which has its own lens and views a separate portion of the object (fig. 40.3). How well the brain combines this information to see the entire object is not known, but compound eyes seem especially well suited to detecting motion, as anyone who has tried to catch an insect knows.

Insects have color vision, but they make use of a slightly shorter range of the electromagnetic spectrum compared to humans. They can see the longest of the ultraviolet rays, and this enables them to be especially sensitive to the reproductive parts of flowers, which reflect particular ultraviolet patterns (fig. 40.4). Fishes, reptiles, and most birds are believed to have color vision, but among mammals, only humans and other primates have color vision. It would seem, then, that this trait was adaptive for life in trees, which accounts for its retention in these few mammals.

Figure 40.3

Each visual unit of a compound eye has a cornea and lens that focus light onto photoreceptor cells. The cells generate nerve impulses that are transmitted to the brain, where interpretation produces a mosaic image.

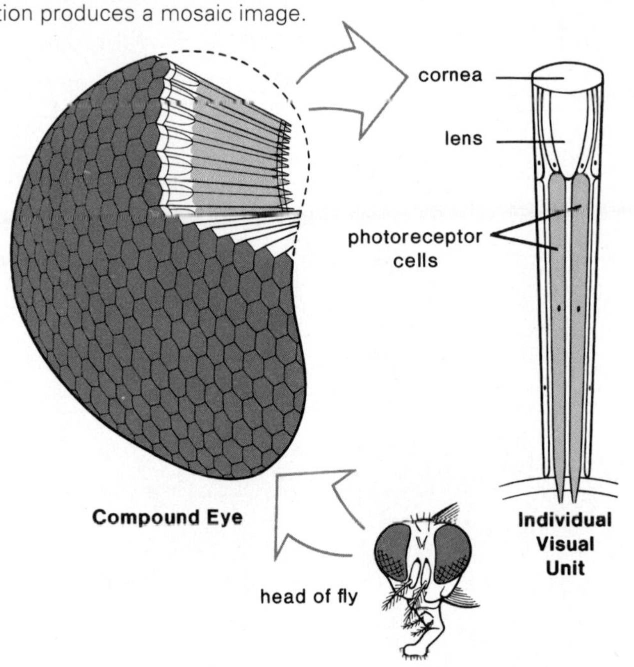

cornea

lens

photoreceptor cells

Compound Eye

Individual Visual Unit

head of fly

Figure 40.4

Marsh marigold as seen by humans (*left*) and insects (*right*). Humans see no markings, but insects see distinct blotches because their eyes respond to ultraviolet rays. These types of markings often highlight the reproductive parts of flowers where insects feed on nectar and pick up pollen at the same time.

Vertebrates and certain mollusks, like the squid, have a *camera type of eye*. A *single* lens focuses an image of the visual field on the photoreceptors, which are closely packed together. All of the photoreceptors taken together can be compared to a piece of film in a camera. The human eye is more complex than a camera, however, as we shall see.

Human Eye

The most important parts of the human eye and their functions are listed in table 40.1. The human eye, which is an elongated sphere about 2.5 cm in diameter, has 3 layers, or coats (fig. 40.5). The outer layer, the **sclera,** is an opaque, white, fibrous layer that surrounds the transparent cornea, the window of the eye. The middle, thin, dark-brown layer, the *choroid,* contains many blood vessels and pigment that absorbs stray light rays. Toward the front of the eye, the choroid thickens and forms the ring-shaped ciliary body and finally becomes a thin, circular, muscular diaphragm, the *iris,* which regulates the size of an opening called the *pupil.* The *lens,* which is attached to the ciliary body by ligaments, divides the cavity of the eye into 2 portions. A basic, watery solution called *aqueous humor* fills the anterior cavity between the cornea and the lens. A viscous, gelatinous material, the *vitreous humor,* fills the large posterior cavity behind the lens.

The inner layer of the eye, the **retina,** contains the receptors for sight: the **rods** and **cones** (fig. 40.6). Nerve impulses initiated by the rods and cones are passed to the bipolar cells, which in turn pass them to the ganglionic cells. The fibers of these cells pass in front of the retina forming the **optic nerve,** which carries the nerve impulses to the brain. Notice in figure 40.6*b* that there are many more rods and cones than nerve fibers leaving ganglionic cells. This means that there is considerable mixing of messages and a certain amount of integration before nerve impulses are sent to the brain. There are no rods or cones at the point where the optic nerve passes through the retina; therefore, this point is called the blind spot.

The center of the retina contains a special region called the fovea centralis, an oval, yellowish area with a depression where there are only cone cells (fig. 40.5). In the fovea centralis or fovea, color vision is most acute in daylight; at night, it is barely sensitive. At this time, the rods in the rest of the retina are active.

The human eye has 3 layers: the outer sclera, the middle choroid, and the inner retina. Only the retina contains receptors for sight.

Vision Process

Light rays entering the eye are bent (refracted) as they pass through the cornea, lens, and humors and brought to a focus on the retina. The lens is relatively flat when viewing distant objects but rounds up for close objects because light rays must be bent to a greater degree when viewing a close object. These changes of the lens shape are called accommodation (fig. 40.7). With

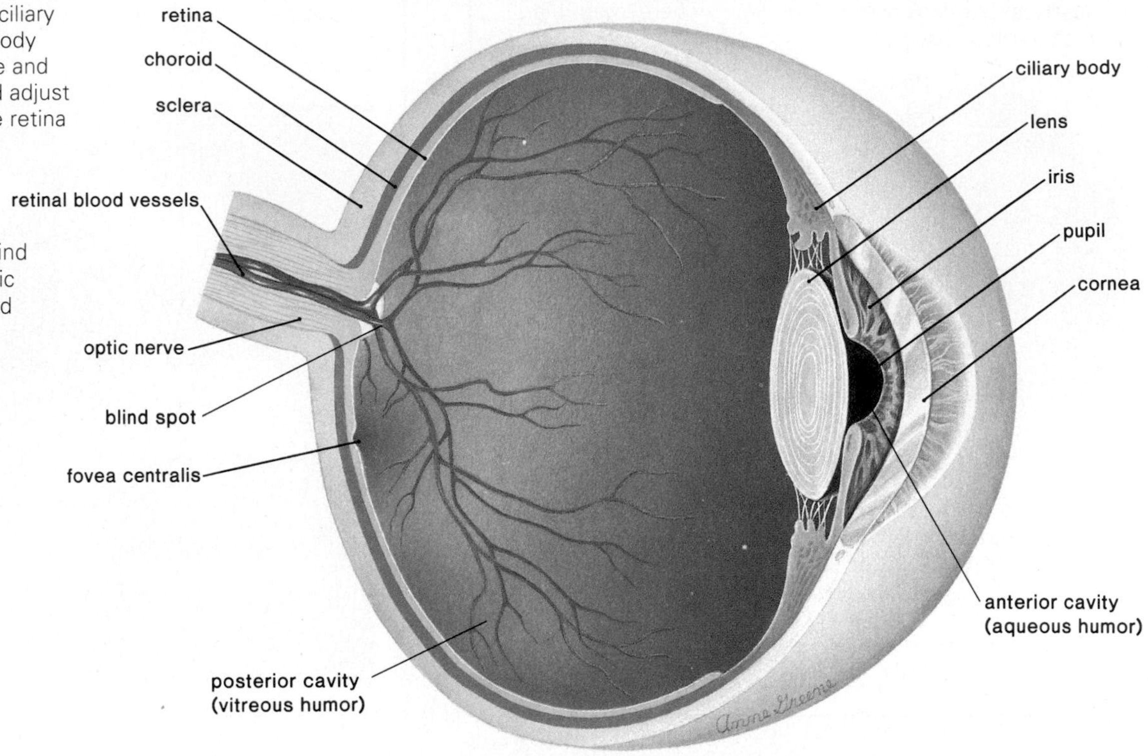

Figure 40.5

Anatomy of the human eye. Notice that the sclera becomes the cornea; the choroid becomes the ciliary body and iris. The ciliary body contains the ciliary muscle and ligaments, which hold and adjust the shape of the lens. The retina contains the receptors for sight, and vision is most acute in the fovea centralis, where there are only cones. A blind spot exists where the optic nerve leaves the retina and where there are no receptors for sight.

retina

choroid

sclera

retinal blood vessels

optic nerve

blind spot

fovea centralis

posterior cavity
(vitreous humor)

ciliary body

lens

iris

pupil

cornea

anterior cavity
(aqueous humor)

Figure 40.6

a. The retina is the inner layer of the eye. *b.* Rods and cones, which are located toward the back of the retina, are preceded by the bipolar cells and the ganglionic cells, whose fibers become the optic nerve. Notice that rods share bipolar cells but cones do not. Cones, therefore, distinguish more detail. *c.* The photosensitive

pigment is located in the disks of the outer segment of rods and cones. A disk is apparently a modified cilium. *d.* Scanning electron micrograph of rods and cones. The cones are responsible for color vision, and the rods are responsible for night vision.

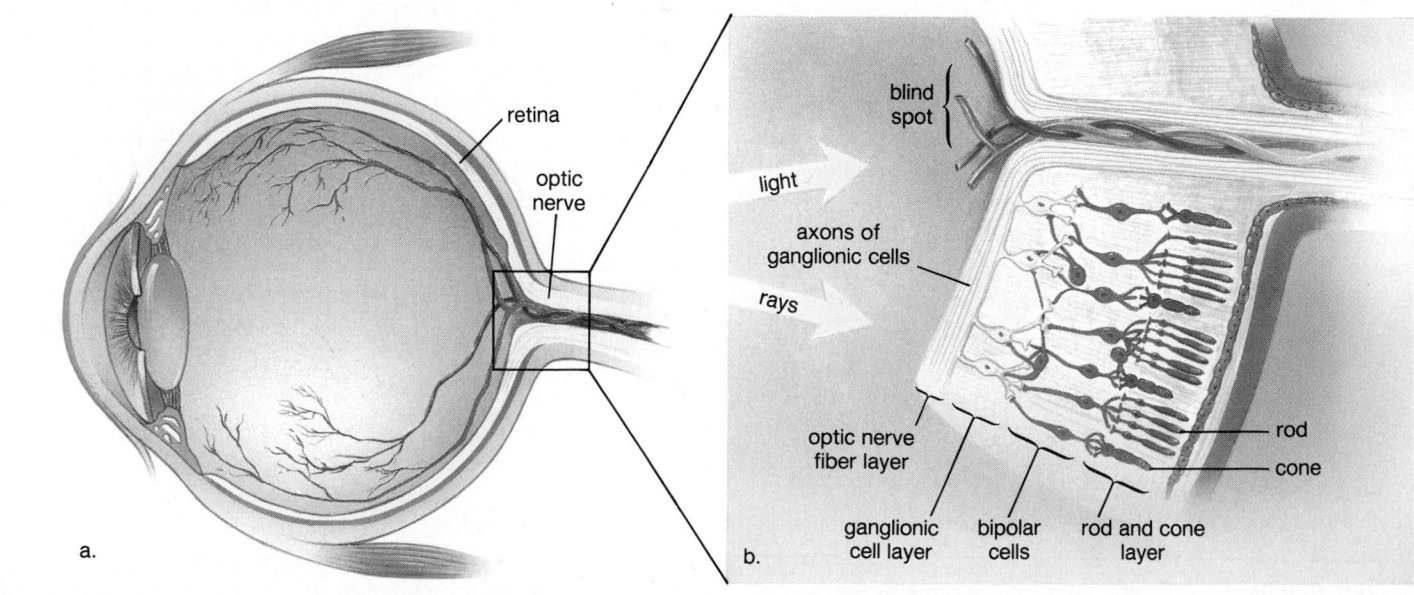

retina

optic
nerve

blind
spot

light

axons of
ganglionic cells

rays

optic nerve
fiber layer

ganglionic
cell layer

bipolar
cells

rod and cone
layer

rod

cone

a.

b.

normal aging, the lens loses its ability to accommodate for close objects; therefore, persons frequently need reading glasses once they reach middle age.

Because of refraction, the image on the retina is rotated 180° from the actual, but it is believed that this image is righted in the brain. In one experiment, scientists wore glasses that inverted and reversed the field. At first, they had difficulty adjusting to the placement of objects, but they soon became accustomed to their inverted world. Experiments such as this suggest that if the retina sees the world "upside down," the brain has learned to see it right side up.

Table 40.1
Function of the Parts of the Eye

Part	Function
Lens	Refracts and focuses light
Iris	Regulates light entrance
Pupil	Admits light
Choroid	Absorbs stray light
Sclera	Protects
Cornea	Refracts light
Humors	Refracts light
Ciliary body	Holds lens in place
Retina	Contains receptors
Rods	Allow black-and-white vision
Cones	Allow color vision
Optic nerve	Transmits impulse
Fovea centralis	Region of cones in retina

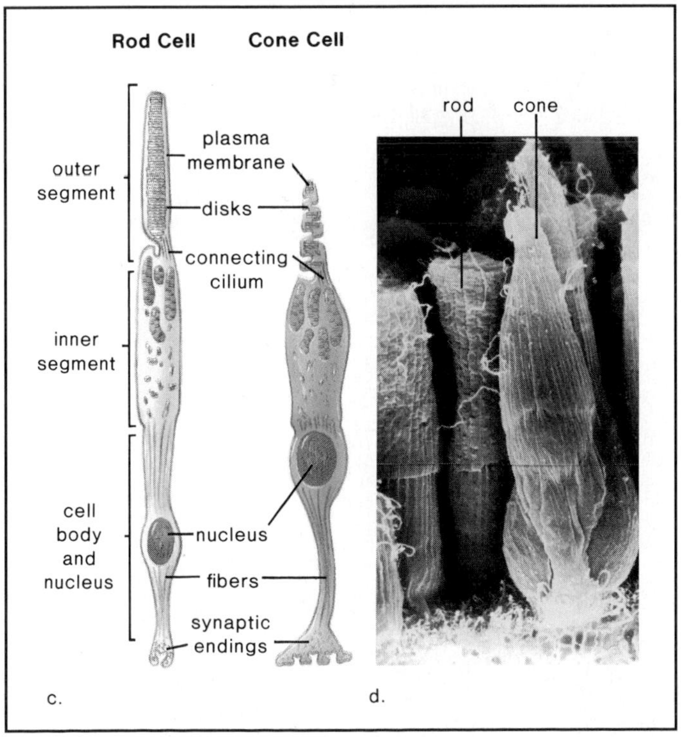

Figure 40.7
Accommodation. *a.* When the eye focuses on a far object, the lens is flat because the eye muscles holding the lens are relaxed and the ligament is taut. *b.* When the eye focuses on a near object, the lens rounds up because the eye muscles contract, causing the ligament to relax. *c.* In nearsighted individuals, the eyeball is too long. They cannot see far because when viewing a distant object, the image is brought to focus in front of the retina. *d.* In farsighted individuals, the eyeball is too short. They cannot see near objects well because when viewing a near object, the image is brought to focus behind the retina.

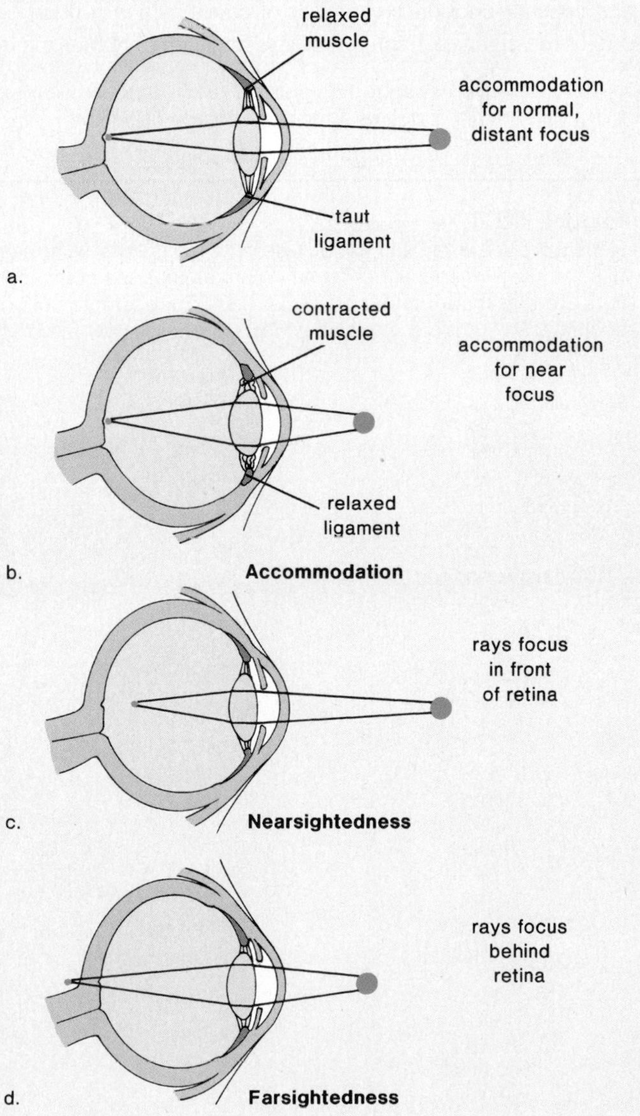

Rods Only dim light is required to stimulate rods; therefore, they are responsible for *night vision*. The numerous rods are also better at detecting motion than cones, but they cannot provide distinct and/or color vision. This causes objects to appear blurred and look gray in dim light. Many molecules of **rhodopsin** are located within the membrane of the disks (lamellae) found in the outer segment of the rods (fig. 40.6c). Rhodopsin is a complex molecule that contains protein (opsin) and a pigment molecule called *retinal,* which is a derivative of vitamin A. When light strikes retinal it changes shape, and opsin is activated. The reactions that follow eventually end after many molecules of GMP (guanosine monophosphate) have been converted to cyclic GMP.[1] Cyclic GMP, in turn, initiates nerve impulses in the rod, which pass through the retina to the optic nerve. Each stimulus generated lasts about one-tenth of a second. This is why we continue to see an image if we close our eyes immediately after looking at an object. It also allows us to see motion if still frames are presented at a rapid rate, as in "movies."

Cones The cones, located primarily in the fovea and activated by bright light, detect the fine detail and the color of an object. *Color vision* depends on 3 different kinds of cones, which contain a blue, green, or red pigment. Each pigment is made up of retinal and opsin,

[1]GMP is a nucleotide that contains the base guanine, the sugar ribose, and one phosphate group. In cyclic GMP, the single phosphate is attached to ribose in 2 places.

but there is a slight difference in the opsin structure of each, which accounts for their individual absorption patterns. Various combinations of cones are believed to be stimulated by in-between shades of color, and the combined nerve impulses are interpreted in the brain as a particular color.

In the human eye, the receptors for sight are the rods and cones. The rods are responsible for vision in dim light, and the cones are responsible for vision in bright light and color vision. When a receptor is stimulated, nerve impulses are transmitted by the optic nerve to the brain.

Mechanoreceptors

Mechanoreceptors are sensitive to mechanical stimuli, such as pressure, sound waves, and gravity. Human skin contains various types of mechanoreceptors, such as *touch receptors* and pressure receptors (fig. 40.8). A pressure receptor, called the Pacinian corpuscle, is shaped like an onion and consists of a series of concentric layers of connective tissue wrapped around the end (dendrite) of a sensory neuron. In contrast, *pain receptors* are only the unmyelinated ("naked") ends (dendrites) of the fibers of sensory neurons. Some pain receptors are especially sensitive to mechanical stimuli; others are most sensitive to temperature or chemicals.

Figure 40.8

Receptors in human skin. The classical view is that each receptor has the function indicated; however, investigations in this century indicate that matters are not so clear-cut. For example, microscopic examination of the skin of the ear shows only free nerve endings (pain receptors), and yet the skin of the ear is sensitive to all sensations. Therefore, it appears that the receptors of the skin are somewhat but not completely specialized. Touch and pressure receptors are considered to be membrane receptors.

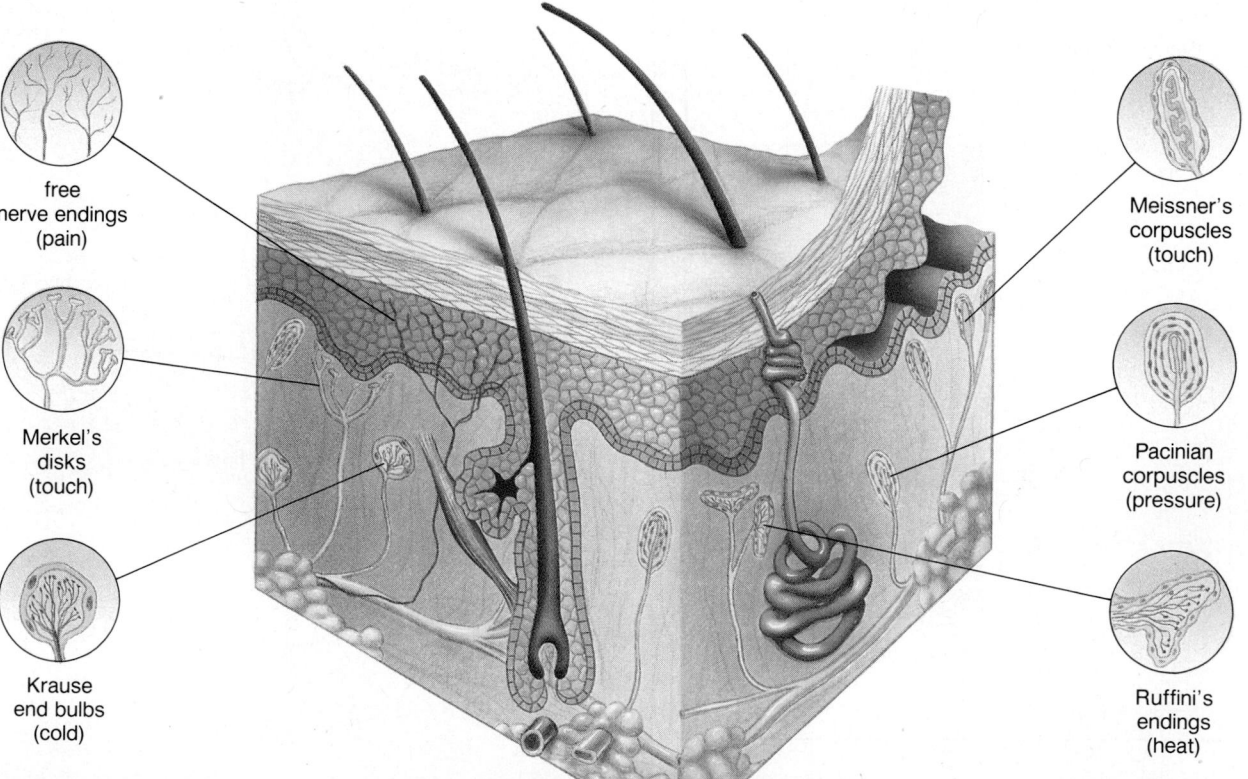

free nerve endings (pain)

Merkel's disks (touch)

Krause end bulbs (cold)

Meissner's corpuscles (touch)

Pacinian corpuscles (pressure)

Ruffini's endings (heat)

Animal Structure and Function

Hair cells, which are named for the cilia they bear, are often mechanoreceptors in various specialized vertebrate sense organs. Fishes and amphibians have a series of such receptors called the lateral line system, which detect water currents and pressure waves from nearby objects. In primitive fishes and aquatic amphibians, the receptors are located on the body surface, but in advanced fishes, they are located within a canal that has openings to the outside (fig. 40.9). A lateral line receptor is a collection of hair cells with cilia embedded in a mass of gelatinous material known as a cupula. When the cupula bends due to pressure waves, the hair cells initiate nerve impulses.

The ears of fishes, which function mainly as equilibrium organs, are derived from a portion of the lateral line system. The evolution of the inner ear of humans can also be traced back to the lateral line system of fishes. The inner ear of humans contains both equilibrium and second receptors.

Human Ear

Table 40.2 lists the most important parts of the human ear and their function.

As figure 40.10 shows, the human ear has an outer, a middle, and an inner portion. The *outer ear* consists of the pinna (external flap) and *auditory canal.* The auditory canal is lined with

Figure 40.9

Lateral line system of fishes. Location of the system (*upper*); longitudinal section of the system (*lower*). A main canal has openings to the exterior. Lining the canal are hair cells (embedded in cupulae) that act as sense receptors for pressure.

Table 40.2
Function of the Parts of the Ear

Part	Function
Outer Ear	
Pinna	Collects sound waves
Middle Ear	
Tympanic membrane and ossicles	Amplify sound waves
Inner Ear	
Oval window	Initiates pressure waves
Cochlea	Transmits pressure waves
Organ of Corti	Receptor for hearing
Utricle and saccule	Static equilibrium
Semicircular canals	Dynamic equilibrium

Figure 40.10

Anatomy of the human ear. In the middle ear, the hammer, the anvil, and the stirrup amplify sound waves. Otosclerosis is a condition in which the stirrup becomes attached to the inner ear and is unable to carry out its normal function. It can be replaced by a plastic piston, and thereafter the individual hears normally because sound waves are transmitted as usual to the cochlea, which contains the receptors for hearing.

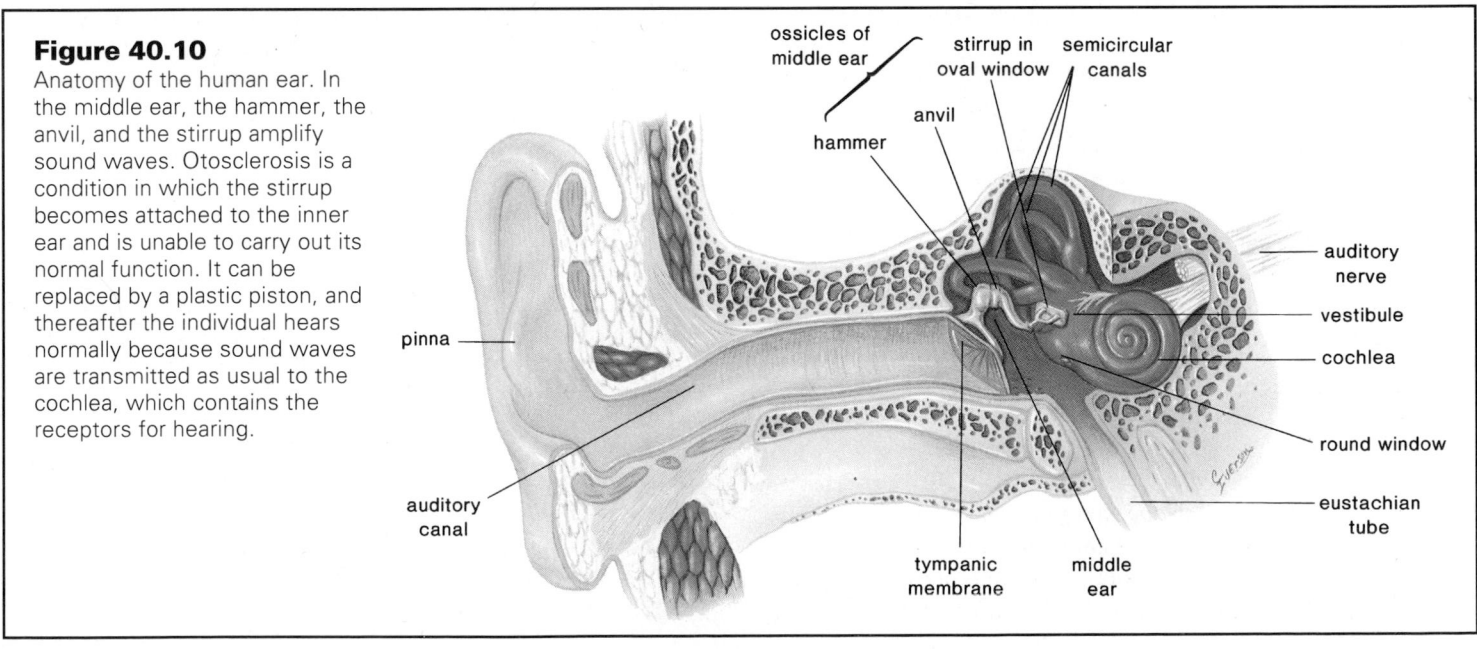

Figure 40.11

Inner ear. **a.** The inner ear contains the semicircular canals, a vestibule, and the cochlea. The cochlea has been cut to show the location of the organ of Corti. **b.** There is an ampulla at the base of each semicircular canal that contains the receptors (hair cells) for dynamic equilibrium. **c.** In a vestibule are the utricle and the saccule, small sacs that contain the receptors (hair cells) for static equilibrium. **d.** The sense organ for hearing, the organ of Corti, is in the cochlea. The organ of Corti consists of hair cells with cilia that touch the tectorial membrane. Pressure waves cause the basilar membrane beneath the hair cells to vibrate. When the cilia touch the tectorial membrane, nerve impulses are initiated and taken up by sensory nerve fibers within the auditory nerve. The intensity of the sound is determined by how many cells are stimulated; the quality of the sound is determined by which particular cells along the entire organ of Corti are stimulated.

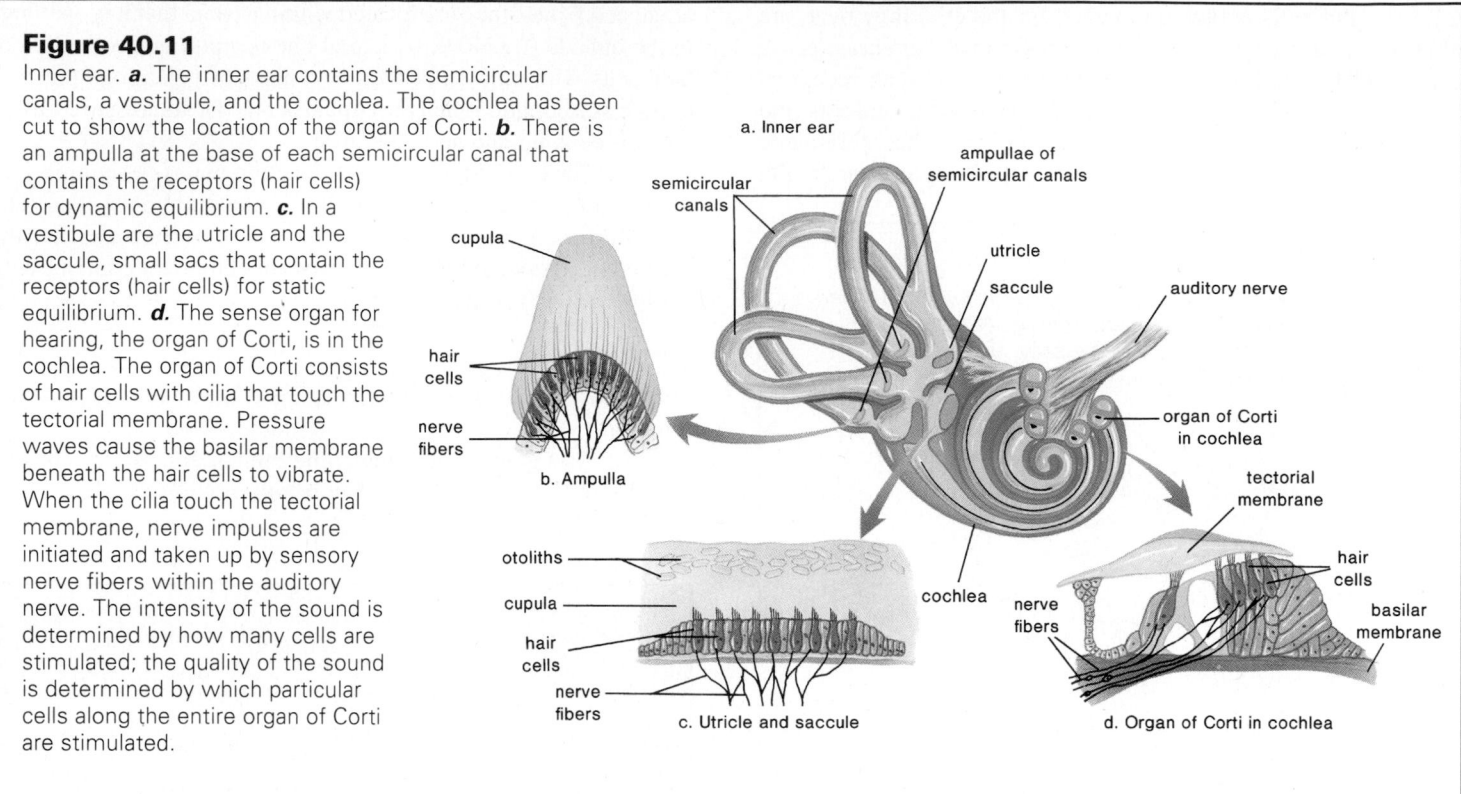

fine hairs, which filter the air. Modified sweat glands are located in the upper wall of the canal; these secrete earwax, which helps guard the ear against entrance of foreign materials.

The *middle ear* begins at the **tympanic membrane** (eardrum) and ends at a bony wall that has small openings covered by membranes, the *oval window* and the *round window*. Three small bones are located between the tympanic membrane and the oval window. Collectively called *ossicles,* individually they are the hammer (malleus), anvil (incus), and stirrup (stapes), named for their structural resemblance to these objects. The Eustachian tube extends from the middle ear to the pharynx and permits the equalization of air pressure between the inside and the outside of the ear. Chewing gum, yawning, and swallowing help move air through the Eustachian tubes during ascent and descent in airplanes or elevators.

The inner ear has 3 anatomic areas: the first 2 consist of a vestibule (chamber) and the semicircular canals and are concerned with balance; the third, the cochlea, is concerned with hearing. Whereas the outer ear and middle ear contain air, the inner ear is filled with fluid.

Balance

The **semicircular canals** are *dynamic equilibrium organs* because they initiate a sensation of movement (fig. 40.11*a*). At their bases are the ampullae, each one having touch-sensitive hair cells (fig. 40.11*b*). As the body moves about, there is first a slight lag, and then the fluid within the canals moves; this causes the cilia of the hair cells to bend. The 3 semicircular canals are each orien-

tated in a different plane to detect movement in any direction. The brain integrates the impulses it receives from each of the 3 canals and therefore can determine the direction and rate of movement. Very rapid or prolonged movements of the head may cause uncomfortable side effects, such as the dizziness and nausea of seasickness.

The vestibule between the semicircular canals and the cochlea contain 2 small sacs, the utricle and saccule (fig. 40.11*c*). Within each of these are groups of little hair cells; calcium carbonate granules called *otoliths* rest on their cilia. When the head tilts, the otoliths press on the cilia in a certain direction, and this initiates nerve impulses that inform the brain of the position of the head. Similar types of organs, called statocysts, are also found in cnidarians, mollusks, and crustacea. These organs only give information about the position of the head and do not result in a sensation of movement (fig. 40.12). They are therefore called *static equilibrium organs.*

The outer ear, middle ear, and cochlea are necessary for hearing in humans. The vestibule, containing the utricle and the saccule, and the semicircular canals are concerned with the sense of balance.

Hearing

The **cochlea** within the inner ear resembles the shell of a snail because it is spiral shaped (fig. 40.11*a*). A cross section of the cochlea shows that it contains 3 canals: the vestibular, cochlear, and tympanic canals. Along the length of the *basilar membrane,*

Animal Structure and Function

W e have an idea of what noise does to the ear," David Lipscomb [of the University of Tennessee Noise Laboratory] says. "There's a pretty clear cause-effect relationship." And these photomicrographs (see fig. 40.A) of the cochlea's tiny structures graphically document noise trauma to the inner ear.

Hair cells transmit the mechanical energy of sound waves into those neural impulses that the brain interprets as sound. Loud noise can damage or destroy hair cells as these scanning electron micrographs illustrate.

Hair cells come in 2 varieties: a single row of inner cells and a triple row of outer ones. "Outer cells degenerate before inner cells," notes Clifton Springs, N.Y.-otolaryngologist Stephen Falk. The most subtle change wrought by noise is a development of vesicles, or blisterlike protrusions along the walls of the hair cells' stereocilia. Continued assault by noise will lead to a rupturing of the vesicles and damage. In addition, the "cuticular plate"—base tissue supporting the stereocilia—may soften, followed by a swelling and ultimate degeneration of hair cells.

But sensory hair cells are not the only structures at risk. Adjacent inner-ear cells … may undergo vacuolation—development

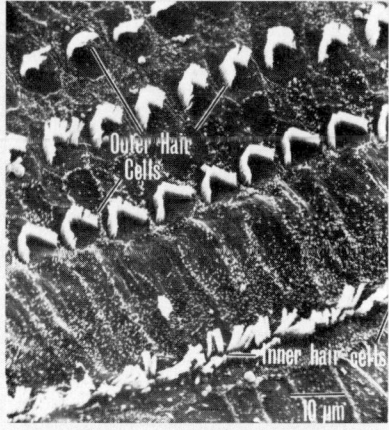
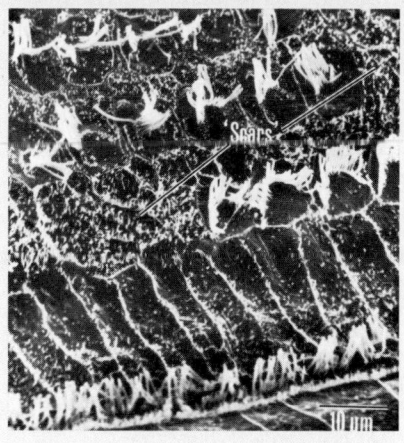

a.   b.

Figure 40.A

Electron micrographs showing damage to the organ of Corti due to loud noise. *a.* Normal organ of Corti. The tectorial membrane has been removed to reveal the hair cells. *b.* Organ of Corti after 24-hour exposure to noise level typical of rock music. Note scars where cilia on hair cells have worn away. Magnification, X1,400.

of degenerative empty spaces in cells. Even nerve fibers synapsing at the hair cells' roots may die. In the final phase of noise-induced cochlear damage, the organ of Corti—of which hair cells and supporting cells are a part—is completely denuded and covered by a layer of scar tissue.

From *Science News,* the weekly newsmagazine of science, copyright 1982 by Science Service, Inc., Washington, DC.

Figure 40.12

Generalized statocysts as found in mollusks and crustaceans. A small particle, the statolith, moves in response to a change in the animal's position. When the statolith stops moving, it stimulates the closest cilia of hair cells. These cilia transmit impulses, indicating the position of the body.

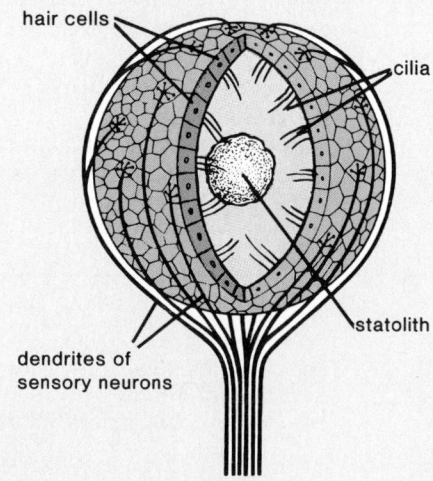

which forms the floor of the cochlear canal, there are at least 24,000 ciliated hair cells. Just above them is another membrane called the *tectorial membrane* (fig. 40.11*d*). The hair cells plus the tectorial membrane form the **organ of Corti,** the sense organ for hearing.

Sound waves reach the organ of Corti by way of the outer and middle ear. Ordinarily, sound waves do not carry much energy, but when a large number of waves strike the eardrum, it moves back and forth (vibrates) very slightly. The hammer transfers the pressure from the inner surface of the eardrum to the anvil, and then to the stirrup (fig. 40.10). The pressure is multiplied about 20-fold as it moves from the eardrum to the stirrup. The stirrup vibrates the oval window, which transmits pressure waves to the fluid in the inner ear. These waves cause the basilar membrane to move up and down and the cilia of the hair cells to rub against the tectorial membrane. Bending of the cilia initiates nerve impulses, which pass by way of the auditory nerve to the brain, where the impulses are interpreted as a sound.

The organ of Corti is narrow at its base, but it widens as it approaches the tip of the cochlear canal. Various cells along it are sensitive to different wave frequencies, or pitches. Near its apex, the organ of Corti responds to low pitches, such as a bass drum, and near the base, it responds to high pitches, such as a bell or whistle. The nerve impulses from each region along the organ lead to slightly different areas in the brain. The pitch sensation we experience depends on which brain area is stimulated. Volume is a function of the *amplitude* of sound waves. Loud noises (measured in decibels, db) cause the fluid of the cochlea to oscillate to a greater degree, and this, in turn, causes the basilar membrane to move up and down to a greater extent. The resulting increased stimulation is interpreted by the brain as loudness. It is believed that tone is interpreted by the brain according to the distribution of hair cells that are stimulated.

The sense receptors for sound are ciliated hair cells on the basilar membrane (the organ of Corti). When the basilar membrane vibrates, the delicate cilia touch the tectorial membrane, producing nerve impulses, which are transmitted by the auditory nerve to the brain.

Summary

1. Sense organs are transducers; they transform the energy of a stimulus to the energy of nerve impulses. It is the brain, not the sense organ, that interprets the stimulus.
2. Human olfactory cells and taste buds are chemoreceptors. They are sensitive to chemicals in water and air.
3. The human eye is a photoreceptor. The compound eye of arthropods is made up of many individual units, whereas the human eye is a camera-type eye with a single lens. Table 40.1 lists the parts of the eye and the function of each part.
4. The receptors for sight in humans are the rods and cones, which are in the retina. The rods work in minimum light and detect motion, but they do not detect color. The cones require bright light and do detect color.
5. When light strikes rhodopsin, a molecule composed of opsin and retinal, retinal changes shape and opsin is activated. Chemical reactions that produce nerve impulses follow. These impulses are eventually picked up by the optic nerve.
6. There are 3 kinds of cones, containing blue, green, or red pigment. Each pigment is also made up of retinal and opsin, but opsin structure varies among the 3.
7. Mechanoreceptors include the touch receptors and pressure receptors in the human skin.
8. Many mechanoreceptors are hair cells with cilia, such as those found in the lateral line of fishes as well as the inner ear of humans. Table 40.2 lists the parts of the ear and the function of each part.
9. The inner ear contains the sense organs for balance. Just like the statocysts of invertebrates, portions of the human inner ear contain calcium carbonate granules resting on hair cells. The movement of these granules gives us a sense of static equilibrium. Movement of fluid past hair cells in the semicircular canals gives us a sense of dynamic equilibrium.
10. Hair cells on the basilar membrane (the organ of Corti) are responsible for hearing. Pressure waves, which begin at the oval window, cause the basilar membrane to vibrate so that the cilia of the hair cells touch the tectorial membrane. This causes the hair cells to initiate nerve impulses, which are carried by the auditory nerve to the brain.

Writing Across the Curriculum

In order to practice writing skills, students should write out the answers to any or all of the study questions and the critical thinking questions. The study questions are sequenced in the same order as the text. Suggested answers to the critical thinking questions are in appendix D.

Study Questions

1. In what ways are all receptors similar; in what ways are they different?
2. Discuss the structure and function of human chemoreceptors.
3. In general, how does the arthropod eye differ from that in humans? What types of animals have eyes that are constructed similarly to the human eye?
4. Name the parts of the eye, and give a function for each part.
5. Contrast the location and function of rods to that of cones.
6. What are the types of mechanoreceptors in human skin?
7. Describe how the lateral line system of fishes works and why it is considered to contain mechanoreceptors.
8. Describe the anatomy of the ear and how we hear.
9. Describe the role of the utricle, saccule, and semicircular canals in balance.

Objective Questions

1. A receptor
   a. is the first portion of a reflex arc.
   b. initiates nerve impulses.
   c. responds to only one type of stimulus.
   d. All of these.
2. Which of these gives the correct path for light rays entering the human eye?
   a. sclera, retina, choroid, lens, cornea
   b. fovea centralis, pupil, aqueous humor, lens
   c. cornea, pupil, lens, vitreous humor, retina
   d. optic nerve, sclera, choroid, retina, humors

Animal Structure and Function

3. Which gives an incorrect function for the structure?
   a. lens—focusing
   b. iris—regulation of amount of light
   c. choroid—location of cones
   d. sclera—protection
4. Which of these contain mechanoreceptors?
   a. human skin
   b. lateral line of fishes
   c. statocysts of arthropods
   d. All of these.
5. Which association is incorrect?
   a. lateral line—fishes
   b. compound eye—arthropods
   c. camera-type eye—squid
   d. statocysts—sea stars
6. Which one of these wouldn't you mention if you were tracing the path of sound vibrations?
   a. auditory canal
   b. tympanic membrane
   c. semicircular canals
   d. cochlea
7. Which one of these correctly describes the location of the organ of Corti?
   a. between the tympanic membrane and the oval window in the inner ear
   b. in the utricle and saccule within the vestibule
   c. between the tectorial membrane and the basilar membrane in the cochlear canal
   d. between the outer and inner ear within the semicircular canals

8. Which of these is mismatched?
   a. semicircular canals—inner ear
   b. utricle and saccule—outer ear
   c. auditory canal—outer ear
   d. ossicles—middle ear
9. Retinal is
   a. sensitive to light energy.
   b. a part of rhodopsin.
   c. found in both rods and cones.
   d. All of these.

10. Both olfactory receptors and sound receptors have cilia, and they both
    a. are chemoreceptors.
    b. are mechanoreceptors.
    c. initiate nerve impulses.
    d. All of these.
11. Label this diagram of the human eye. State a function for each structure labeled:

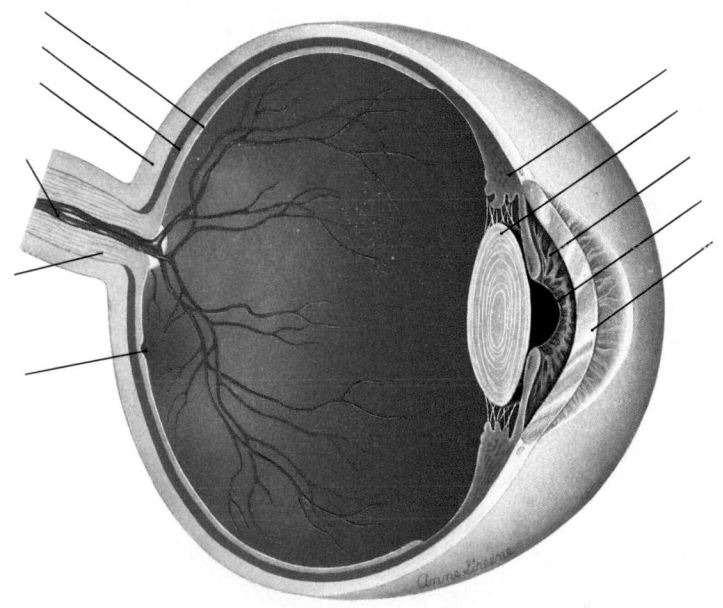

Concepts and Critical Thinking

1. *All organ systems contribute to homeostasis.*

   Tell several ways the sense organs contribute to homeostasis.

2. *In the whole animal, all systems work together and influence one another.*

   How do the nervous system and the sense organs work together?

3. *Animals are sensitive to only certain types of stimuli.*

   Why might animals be sensitive to only certain types of stimuli?

Selected Key Terms

receptor (re-sep'tor) 653
chemoreceptor (ke"mo-re-sep'tor) 653
olfactory cell (ol-fak'to-re sel) 653
taste bud (tāst bud) 654
photoreceptor (fo"to-re-sep'tor) 654
compound eye (kom'pownd i) 654

sclera (skle'rah) 655
retina (ret'ĭ-nah) 655
rod (rod) 655
cone (kōn) 655
rhodopsin (ro-dop'sin) 658
mechanoreceptor (mek"ah-no-re-sep'tor) 658

tympanic membrane (tim-pan'ik mem'brān) 660
semicircular canal (sem"e-ser'ku-lar kah-nal') 660
cochlea (kok'le-ah) 660
organ of Corti (or'gan uv kor'ti) 661

41

Musculoskeletal System

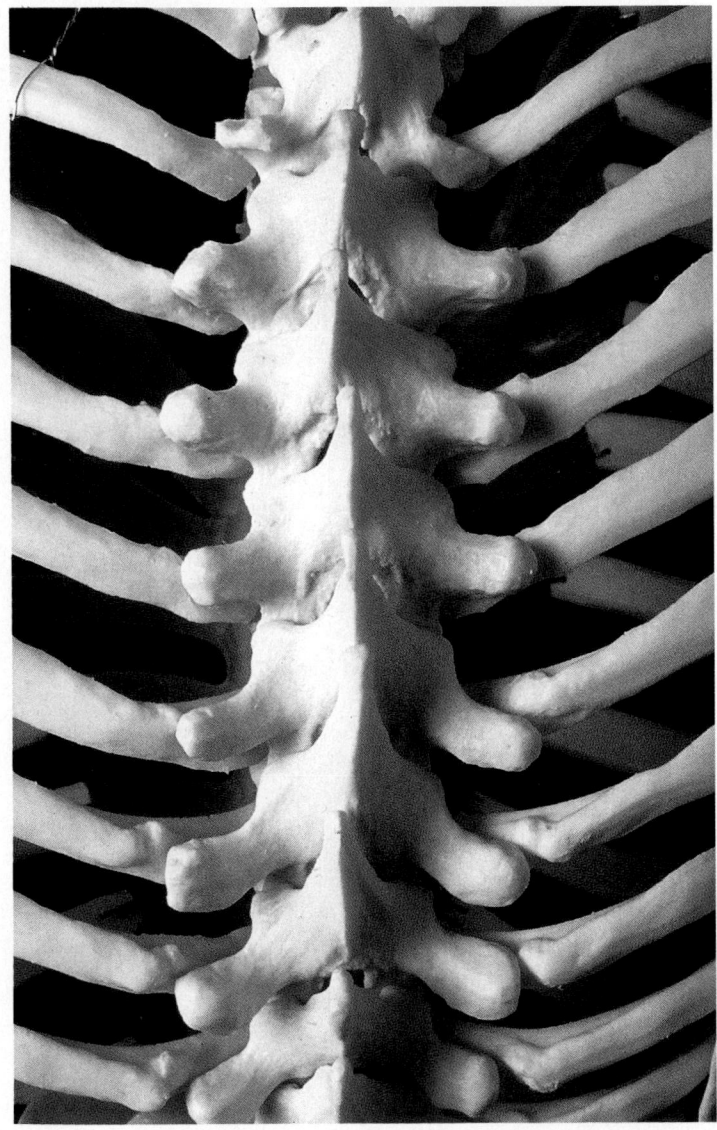

The human vertebral column illustrates that vertebrates have a strong but flexible skeleton. In humans, the vertebral column supports the head and trunk, yet it allows the body to bend forward, backward, and to the side. Only arthropods and vertebrates have a rigid skeleton that is jointed and segmented.

Your study of this chapter will be complete when you can

1. distinguish among the 3 types of skeletons in the animal kingdom, and give examples of animals that have each type;
2. give the functions of a skeletal system in animals;
3. name the 2 parts of the human skeleton, and list the location of the bones for both;
4. describe the anatomy of a long bone and the tissues found in a long bone;
5. explain how bone is continually being broken down and rebuilt;
6. name the different types of joints, and give an example of each;
7. describe the anatomy and location of the 3 types of muscles found in the vertebrate body;
8. describe how whole muscles work in antagonistic pairs;
9. describe the anatomy of a muscle fiber and sarcomere;
10. explain the sliding filament theory of muscle contraction;
11. list the sources of ATP energy for muscle contraction, and explain the occurrence of oxygen debt;
12. list the series of events that occur after a muscle fiber is innervated by nervous stimulation.

The ability to move by means of muscle fibers is one of the distinctive characteristics of animals. Being mobile helps animals obtain food, escape enemies, locate mates, and disperse the species. The title of this chapter—"Musculoskeletal System"—recognizes that the muscular system and the skeletal system work together to provide movement. Nevertheless, we will first consider the skeleton and then show how muscles cause the skeleton to move.

Skeletal System

The skeleton is the framework of the body. It helps protect the internal organs and assists in movement. To produce body movements, the force of muscle contractions must be specifically directed against other parts of the body.

Table 41.1
Classification of Skeletons

Type	Phylum	Example
Hydrostatic skeleton		
gastrovascular cavity	Cnidaria	Hydras, anemones
	Platyhelminthes	Flatworms
coelom*	Annelida	Earthworms
Exoskeleton	Arthropoda	Lobsters, grasshoppers
Endoskeleton	Chordata	Vertebrates

*See figure 26.9.

Figure 41.1
Arthropods have an external skeleton. In order to grow, they must periodically shed this exoskeleton, a process called molting. In this photo, a cicada is just emerging from its discarded skeleton. It will be more vulnerable than usual until its new exoskeleton has hardened.

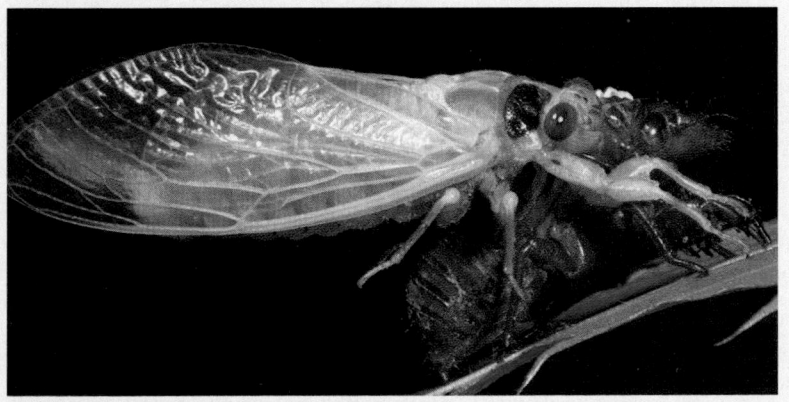

Hydras and planarians use their fluid-filled gastrovascular cavity and annelids use their fluid-filled coelom as a hydrostatic skeleton. In these animals, muscular contraction is applied against the fluid-filled cavity (table 41.1). For example, in the earthworm, the contractions of first the circular muscle and then the longitudinal muscle of the body wall enable the worm alternately to extend and shorten. In this way, the earthworm can move forward.

Other members of the animal kingdom have an exoskeleton. The rigid calcium carbonate exoskeleton of a clam (the bivalve shell) is largely for protection. It grows as the animal grows. The chitinous exoskeleton of arthropods is jointed and movable. Chitin is a strong flexible nitrogenous polysaccharide. Arthropods molt to rid themselves of an exoskeleton that has become too small (fig. 41.1).

Vertebrates have an endoskeleton, composed of bone and cartilage, which grows with the animal. It is also jointed, as is the arthropod skeleton. A rigid but flexible skeleton helped the arthropods and vertebrates successfully colonize the terrestrial environment.

Cnidarians, flatworms, and annelids have a hydrostatic skeleton. Arthropods and vertebrates have a rigid but flexible skeleton that allows complex body movements.

Human Skeleton

Bone and cartilage, the 2 types of tissue found in the human skeleton, were discussed in chapter 33. The human skeleton has many functions. It provides support and protection; it allows movement; it serves as a storage area for calcium and phosphorus salts; and it is the site of blood cell production in adults.

Growth and Development

Most of the bones of the skeleton are cartilaginous during prenatal development. Later, bone-forming cells known as *osteoblasts* replace cartilage with bone. At first, there is only a primary ossification center at the middle of a long bone (fig. 33.4*a*), but later, secondary centers form at the ends of the bones. A *cartilaginous disk* remains between the primary ossification center and each secondary center. The length of a bone is dependent on how long the cartilage cells within the disk continue to divide. Eventually, though, the disks disappear, and the bone stops increasing in size when the individual attains adult height.

In the adult, bone is *continually* being broken down and then built up again. Bone-absorbing cells called *osteoclasts* are derived from cells carried in the bloodstream. As they break down bone, they remove worn cells and deposit calcium in the blood. Apparently, after about 3 weeks, osteoclasts disappear. The destruction caused by the work of osteoclasts is repaired by osteoblasts. As they form new bone, they take calcium from the blood. Eventually some of these cells get caught in the matrix they secrete and are converted to **osteocytes**, the cells found within Haversian systems.

Bones of the Skeleton

The human skeleton can be divided into 2 parts: the **axial skeleton** and the **appendicular skeleton** (fig. 41.2).

Axial Skeleton The axial skeleton includes the skull, the vertebrae, the ribs, and the sternum. The *skull,* or cranium, is composed of many bones fitted tightly together in adults. In newborns, certain bones are not completely formed and instead are joined by membranous regions called *fontanels,* all of which usually close by the age of 16 months. The bones of the skull contain the *sinuses,* air spaces lined by mucous membrane. Two of these, called the mastoid sinuses, drain into the middle ear. Mastoiditis, a condition that can lead to deafness, is an inflammation of these sinuses. Whereas the skull protects the brain, the several bones of the face join together to support and protect the special sense organs and to form the jawbones.

The *vertebral column* extends from the skull to the pelvis and forms a dorsal backbone, which protects the spinal cord (fig. 41.3). Normally, the vertebral column has 4 curvatures, which provide more resiliency and strength than a straight column could. It is composed of many parts called *vertebrae,* which are held together by bony facets, muscles, and strong ligaments. The vertebrae are named according to their location in the body.

Disks between the vertebrae act as padding (fig. 41.3); they prevent the vertebrae from grinding against one another and also absorb shock caused by movements such as running, jumping, and even walking. The disks allow motion between the vertebrae so that we can bend forward, backward, and from side-to-side. Unfortunately, these disks weaken with age and may slip or even rupture. Pain results when the damaged disk presses against the spinal cord and/or spinal nerves. The body may heal itself or the disk can be removed surgically. If surgery occurs, the 2 adjacent vertebrae can be fused together, but this limits the flexibility of the body.

The vertebral column, directly or indirectly, serves as an anchor for all the other bones of the skeleton. All the *ribs* connect directly to the thoracic vertebrae in the back, and all but 2 pairs connect either directly or indirectly via shafts of cartilage to the *sternum* (breastbone) in the front. The lower 2 pairs of ribs are called "floating ribs" because they do not attach to the sternum.

Appendicular Skeleton The appendicular skeleton consists of the bones within the pectoral and pelvic girdles and the attached appendages. The pectoral (shoulder) girdle and appendages (arms and hands) are specialized for flexibility, while the pelvic girdle (innominate bones) and appendages (legs and feet) are specialized for strength.

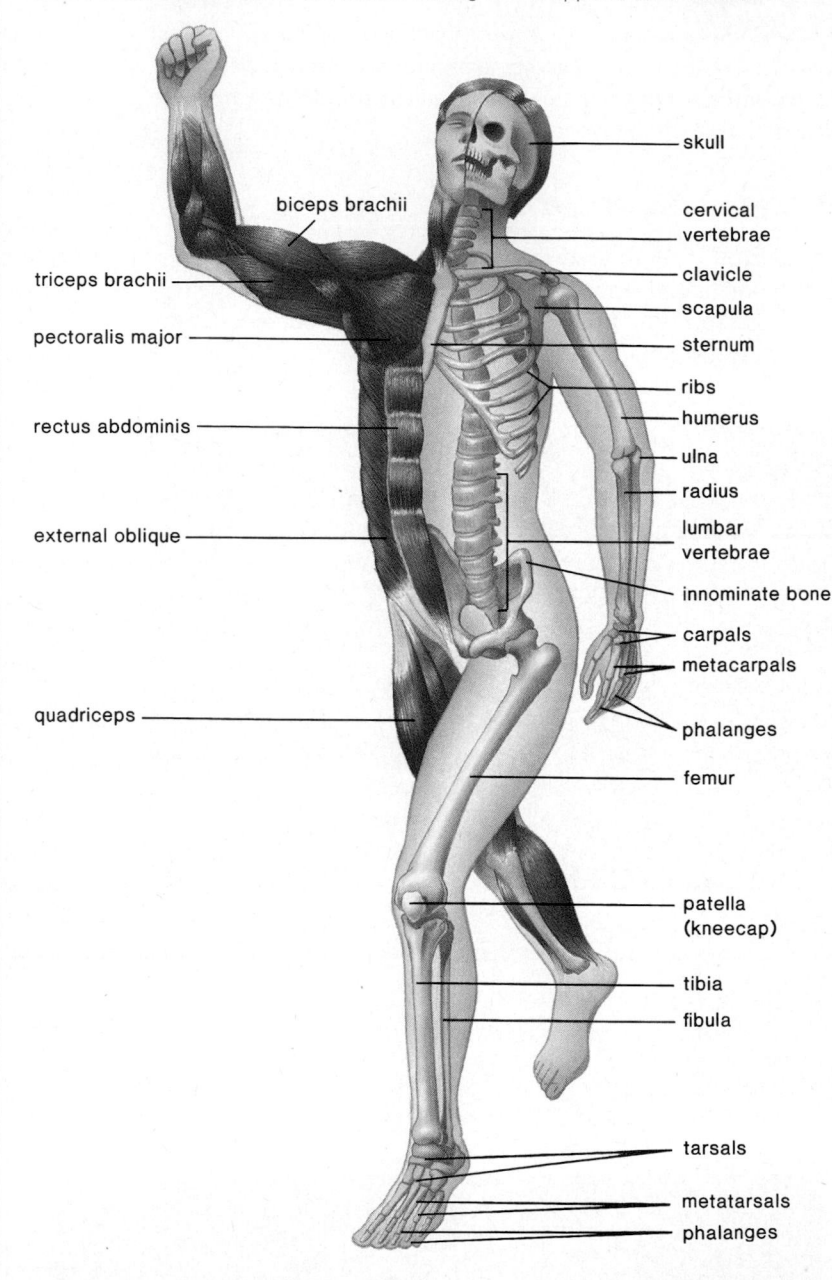

Figure 41.2

Major bones (*right*) and muscles (*left*) of the human body. The axial skeleton, composed of the skull, the vertebral column, the sternum, and the ribs, lies in the midline; the rest of the bones belong to the appendicular skeleton.

biceps brachii

triceps brachii

pectoralis major

rectus abdominis

external oblique

quadriceps

skull

cervical vertebrae

clavicle

scapula

sternum

ribs

humerus

ulna

radius

lumbar vertebrae

innominate bone

carpals

metacarpals

phalanges

femur

patella (kneecap)

tibia

fibula

tarsals

metatarsals

phalanges

The components of the *pectoral girdle* are loosely linked by ligaments rather than firm joints (fig. 41.4). Each *clavicle* (collarbone) connects with the sternum in front and the *scapula* (shoulder blade) behind, but the scapula is freely movable and is held in place by muscles. This allows it to follow the movements of the arm freely. The single long bone in the upper arm, the *humerus,* has a smooth, round head that fits into a socket on the scapula (fig. 41.4). The socket, however, is very shallow and much smaller than the head of the humerus. Although this means that the

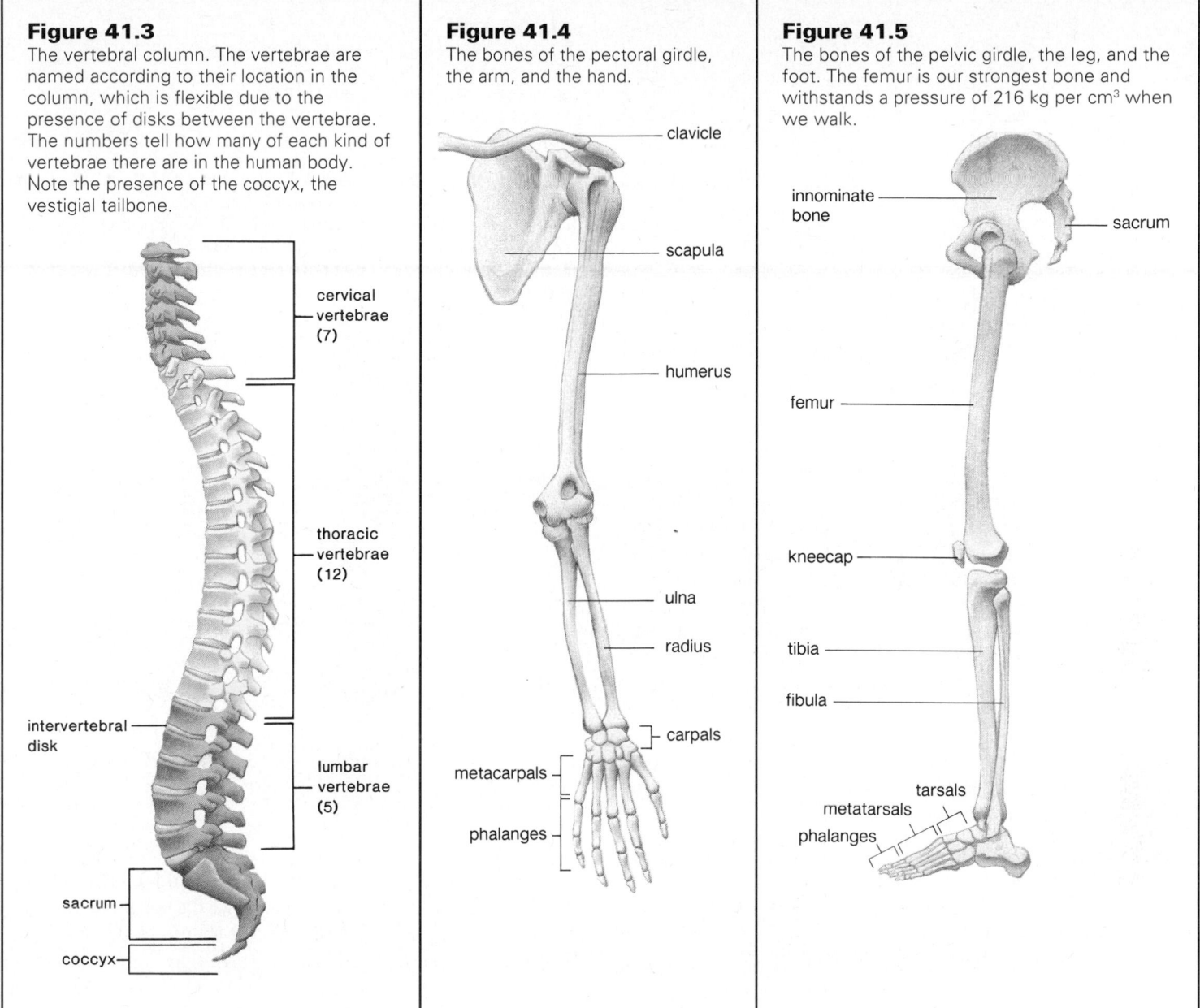

Figure 41.3
The vertebral column. The vertebrae are named according to their location in the column, which is flexible due to the presence of disks between the vertebrae. The numbers tell how many of each kind of vertebrae there are in the human body. Note the presence of the coccyx, the vestigial tailbone.

cervical vertebrae (7)

thoracic vertebrae (12)

intervertebral disk

lumbar vertebrae (5)

sacrum

coccyx

Figure 41.4
The bones of the pectoral girdle, the arm, and the hand.

clavicle

scapula

humerus

ulna

radius

carpals

metacarpals

phalanges

Figure 41.5
The bones of the pelvic girdle, the leg, and the foot. The femur is our strongest bone and withstands a pressure of 216 kg per cm^3 when we walk.

innominate bone

sacrum

femur

kneecap

tibia

fibula

tarsals

metatarsals

phalanges

arm can move in almost any direction, there is little stability. Therefore this is the joint that is most apt to dislocate. The opposite end of the humerus meets the 2 bones of the lower arm, the *ulna* and the *radius,* at the elbow. (The prominent bone in the elbow is the topmost part of the ulna.) When the arm is held so that the palm is turned frontward, the radius and the ulna are about parallel to one another. When the arm is turned so that the palm is next to the body, the radius crosses in front of the ulna, a feature that contributes to the easy twisting motion of the lower arm.

The many bones of the hand increase its flexibility. The wrist has 8 *carpal* bones, which look like small pebbles. From these, 5 *metacarpal* bones fan out to form a framework for the palm. The metacarpal bone that leads to the thumb is placed in such a way that the thumb can reach out and touch the other digits. (*Digits* is a term that refers to either fingers or toes.) Beyond the metacarpals are the *phalanges,* the bones of the fingers and the thumb. The phalanges of the hand are long, slender, and lightweight.

The *pelvic girdle* consists of 2 heavy, large *innominate* bones, or hipbones (fig. 41.5). The innominate bones are anchored to the *sacrum,* and together these bones form a hollow cavity, the pelvis. The weight of the body is transmitted through the pelvis to the legs and then onto the ground. The largest bone in the body is the *femur,* or thigh bone. Although the femur is a strong bone, it is doubtful that the femurs of a fairy-tale giant could support the increase in weight. If a giant were 10 times taller than an ordinary human being, he would also be about 10 times wider and thicker, making him weigh about 1,000 times as much. This amount of weight would break even giant-size femurs.

Table 41.2
The Major Bones of the Skeleton

Part	Bones
Axial skeleton	Skull, vertebral column, sternum, ribs
Appendicular skeleton	
Pectoral girdle	Clavicle, scapula
Arm	Humerus, ulna, radius
Hand	Carpals, metacarpals, phalanges
Pelvic	girdle Innominate bone
Leg	Femur, tibia, fibula
Foot	Tarsals, metatarsals, phalanges

In the lower leg the larger of the 2 bones, the *tibia*, has a ridge we call the shin (fig. 41.5). Both of the bones of the lower leg have a prominence, which contributes to the ankle—the tibia on the inside of the ankle and the *fibula* on the outside of the ankle. Although there are 7 *tarsal* bones in the ankle, only one receives the weight and passes it on to the heel and the ball of the foot. If you wear high-heeled shoes, your weight is thrown even further toward the front of the foot. The *metatarsal* bones form the arches of the foot—a longitudinal arch from the heel to the toes and a transverse arch across the foot. These provide a stable, springy base for the body. If the tissues that bind the metatarsals together become weakened, flatfeet are apt to result. The bones of the toes are called *phalanges,* just like those of the fingers, but in the foot the phalanges are stout and extremely sturdy.

> The axial and appendicular skeletons contain the bones that are listed in table 41.2.

Joints of the Skeleton
Bones are linked at the joints, which are often classified according to the amount of movement they allow. Some bones, such as those that make up the cranium, are sutured together; they are *immovable*. Other joints are *slightly movable,* such as the joints between the vertebrae. The vertebrae are separated by disks, described earlier, which increase their flexibility. Similarly, the 2 hipbones are slightly movable where they are ventrally joined by cartilage. Owing to hormonal changes, this joint becomes more flexible during late pregnancy, which allows the female pelvis to expand during childbirth.

Most joints are *freely movable* or *synovial joints,* in which the 2 bones are separated by a cavity. **Ligaments,** which are composed of fibrous connective tissue, form a capsule, which binds the 2 bones to each other, holding them in place. In a "double-jointed" individual, the ligaments are unusually loose. The joint capsule is lined by synovial membrane, which produces *synovial fluid,* a lubricant for the joint.

Figure 41.6
The knee joint, an example of a freely moveable or synovial joint. Notice that the cavity between the bones is encased by ligaments and lined by synovial membrane. The kneecap protects the joint. Synovial joints are subject to arthritis. In rheumatoid arthritis, the synovial membrane becomes inflamed and thickens. Degenerative changes, which make the joint almost immovable and very painful to use, take place. There is evidence that these effects are brought on by an autoimmune reaction. In old-age arthritis, or osteoarthritis, the cartilage at the ends of the bones disintegrates and the ends of the bones become rough and irregular. This type of arthritis is most likely to affect the joints that have received the greatest use over the years.

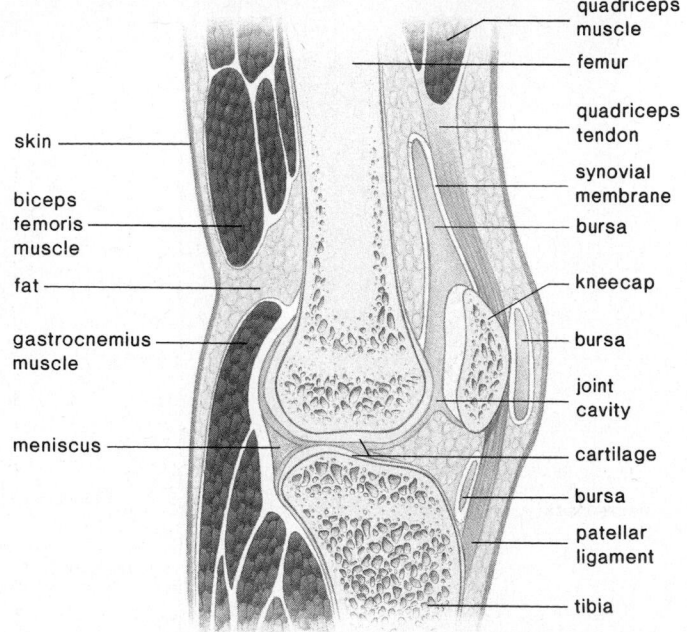

The knee is an example of a synovial joint (fig. 41.6). In the knee, as in other freely movable joints, the bones are capped by cartilage, but the knee also has crescent-shaped pieces of cartilage between the bones, called *menisci* (meniscus, sing.). These add stability, helping to support the weight placed on the knee joint. Unfortunately, athletes often injure the menisci, an injury known as torn cartilage. The knee joint also contains 13 fluid-filled sacs called *bursae* (bursa, sing.), which ease friction between tendons and ligaments and between tendons and bones. Inflammation of bursae is called bursitis. Tennis elbow is a form of bursitis.

There are other types of synovial joints. The knee and elbow joints are *hinge joints* because, like a hinged door, they largely permit movement in one direction only. More movable are the ball-and-socket joints; for example, the ball of the femur fits into a socket on the innominate bone. *Ball-and-socket joints* allow movement in all planes and even a rotational movement.

> Joints are classified according to degree of movement. Some joints are immovable, some are slightly movable, and some are freely movable.

Animal Structure and Function

Figure 41.7

Antagonistic muscle pairs. Muscles can exert force only by shortening. Movable joints are supplied with double sets of muscles, which work in opposite directions. As indicated by the arrows, flexor muscles move the lower limb toward the body and extensor muscles move the lower limb away from the body. **a.** Muscles in an insect's leg as an example of antagonistic muscles attached to the inside of an arthropod exoskeleton. **b.** Muscles in the human leg as an example of antagonistic muscles attached to a vertebrate endoskeleton.

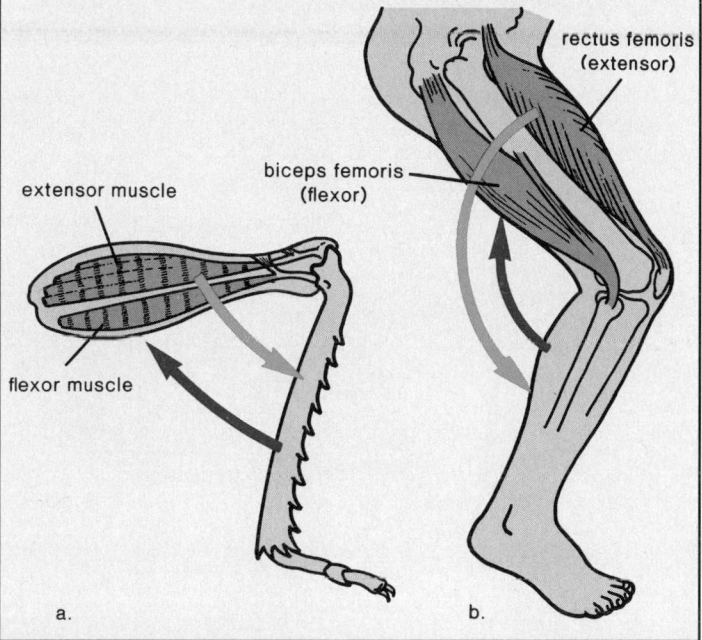

Figure 41.8

Laboratory study of muscle fiber versus whole muscle contraction. **a.** When a muscle fiber is electrically stimulated in the laboratory, it may contract. At first the stimulus may be so weak that no contraction occurs, but as soon as the strength of the stimulus reaches the threshold, the muscle fiber contracts and then relaxes. This action, which is called a muscle fiber twitch, is divided into 3 stages as shown. Increasing the strength of the stimulus does not change the strength of the muscle fiber's contraction; therefore, it is said that a fiber obeys the all-or-none law. If a muscle fiber is given 2 stimuli in quick succession, it does not respond to the second stimulus because it is still recovering from the first stimulus. **b.** A whole muscle behaves quite differently. If the strength of the stimulus is increased, the strength of the response increases. This is because more and more fibers are being stimulated to contract. Also, quickly repeated stimuli to a whole muscle can result in a summation of twitches. This occurs because more fibers can begin to contract while others are relaxing. The muscle as a whole exhibits a greater and greater degree of contraction until tetanus, a sustained contraction, is reached. Eventually the muscle suffers fatigue and begins to relax. Fatigue in isolated muscles is caused by depletion of ATP.

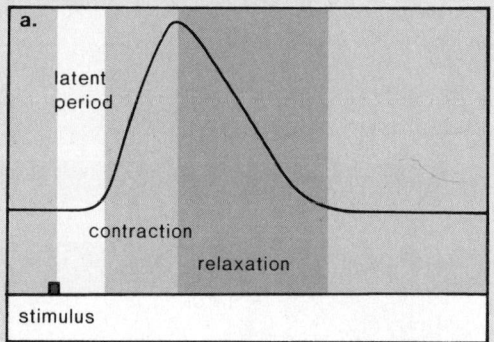

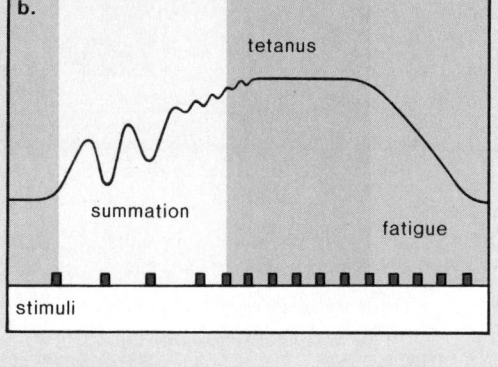

Muscular System

In chapter 33, we discussed the 3 types of muscle in the body of animals. It is skeletal muscle contraction that allows an animal to move, smooth muscle contraction that pushes food along in the digestive tract, and cardiac muscle that allows the heart to pump the blood (fig. 33.6).

Extensive experimental work has been done to understand vertebrate skeletal muscle contraction. It is assumed that the contraction of the other types of muscles is similar to this type of muscle.

Whole Muscle Anatomy and Physiology

Skeletal muscles, which make up over 40% of the body's weight, are attached to the skeleton by **tendons,** made of fibrous connective tissue. When muscles contract, they shorten. Therefore, muscles can only pull; they cannot push. Because of this, skeletal muscles must work in *antagonistic pairs*. For example, one muscle of an antagonistic pair bends the joint and brings the limb toward the body. The other muscle then straightens the joint and extends the limb. Figure 41.7 illustrates this principle and compares the actions of muscles in the limb of an arthropod, in which the muscles are attached to an exoskeleton, with those of a human, in whom the muscles are attached to an endoskeleton.

It is possible to study the contraction of individual whole muscles in the laboratory. Customarily, the calf muscle is removed from a frog and mounted so one end is fixed and the other is movable. The mechanical force of contraction is transduced into an electrical current recorded by an apparatus called a *physiograph* (fig. 41.8). The resulting pattern is called a *myogram*.

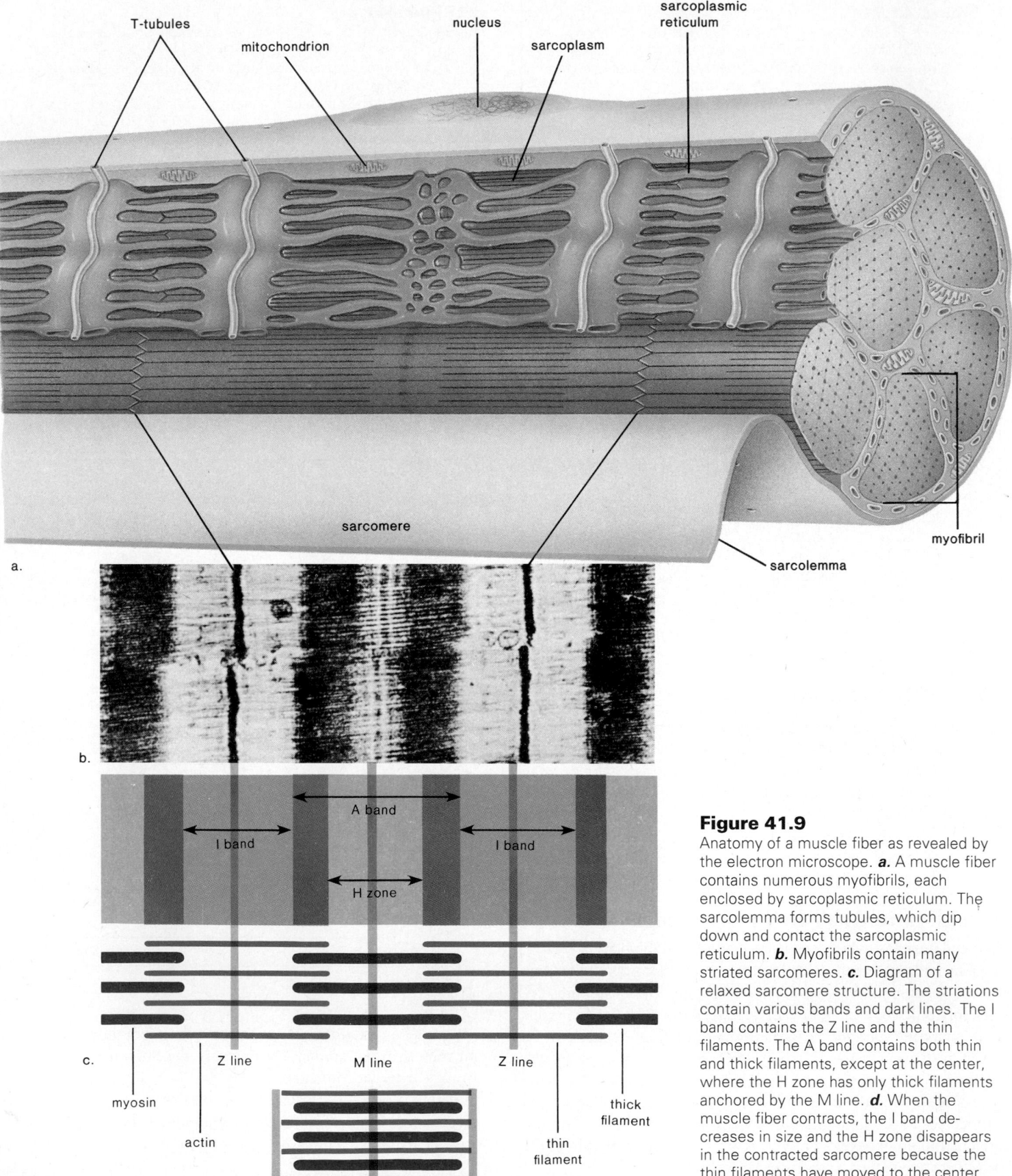

T-tubules **mitochondrion** **nucleus** **sarcoplasm** **sarcoplasmic reticulum**

myofibril

sarcolemma

sarcomere

a.

b.

A band

I band   I band

H zone

Z line   M line   Z line

myosin

actin

thin filament

thick filament

c.

d.

Figure 41.9

Anatomy of a muscle fiber as revealed by the electron microscope. **a.** A muscle fiber contains numerous myofibrils, each enclosed by sarcoplasmic reticulum. The sarcolemma forms tubules, which dip down and contact the sarcoplasmic reticulum. **b.** Myofibrils contain many striated sarcomeres. **c.** Diagram of a relaxed sarcomere structure. The striations contain various bands and dark lines. The I band contains the Z line and the thin filaments. The A band contains both thin and thick filaments, except at the center, where the H zone has only thick filaments anchored by the M line. **d.** When the muscle fiber contracts, the I band decreases in size and the H zone disappears in the contracted sarcomere because the thin filaments have moved to the center and have joined together.

Animal Structure and Function

Muscle Fiber Anatomy and Physiology

Our understanding of the microscopic anatomy and the physiology of muscle contraction is in part due to the research of 2 English scientists. In the late 1950s, A. F. Huxley studied muscle contraction mainly using the light microscope. H. E. Huxley studied the fine structure of skeletal muscle with the electron microscope.

A whole skeletal muscle is composed of a number of bundles of **muscle fibers.** A muscle fiber (cell) has some unique anatomical characteristics (fig. 41.9). For one thing, it has a T (for transverse) system. The **sarcolemma,** or plasma membrane, forms *T-tubules* that penetrate, or dip down into, the cell so that they come into contact, but do not fuse, with expanded portions of modified endoplasmic reticulum, termed the **sarcoplasmic reticulum.** The expanded portions of the sarcoplasmic reticulum store calcium (Ca^{++}) ions, which are essential for muscle contraction. The sarcoplasmic reticulum encases hundreds and sometimes even thousands of myofibrils, the contractile elements of muscle fibers.

Myofibrils

Myofibrils are cylindrical in shape and run the length of the muscle fiber. The light microscope shows that a myofibril has light and dark bands called striations (fig. 33.4*b*). The electron microscope shows that these striations are dependent on the placement of protein filaments within units called **sarcomeres** (fig. 41.9*b*). Notice that a sarcomere extends between 2 dark lines called Z lines. The placement of thick and thin filaments in a sarcomere creates the bands and zones of a sarcomere (fig. 41.9*c*). The I band is light because it contains only thin filaments; the A band is dark because it contains both thin and thick filaments except at the center in the lighter H zone, where only thick filaments are found.

The thick filaments of a sarcomere are made up of a protein called **myosin,** and the thin filaments are made up of a protein called **actin.**

Sarcomere Contraction

A muscle fiber contracts when the sarcomeres within the myofibrils shorten. When sarcomeres shorten, the actin (thin) filaments slide past the myosin (thick) filaments and approach one another (fig. 41.9*d*). This causes the I band to get smaller and the H zone to almost or completely disappear. The movement of actin filaments in relation to myosin filaments is called the *sliding filament theory* of muscle contraction. Notice that after the sarcomere contracts, the filaments are still the same length.

The overall formula for muscle contraction can be represented as follows:

$$\text{actin} + \text{myosin} \xrightarrow[\text{Ca}^{++}]{\text{ATP} \longrightarrow \text{ADP} + \textcircled{P}} \text{actomyosin}$$

The participants in this reaction have the functions listed in table 41.3. Even though it is the actin filaments that slide past myosin filaments, it is the myosin filaments that do the work. In the presence of calcium (Ca^{++}) ions and ATP, portions of a myosin

Table 41.3
Muscle Contraction

Name	Function
Actin filaments	Slide past myosin
Myosin filaments	a. Enzyme that splits ATP b. Pulls actin by means of cross-bridges
Ca^{++}	Needed for actin to bind to myosin
ATP	Supplies energy for bonding between actin and myosin = actomyosin

Figure 41.10

Sliding filament theory. *a.* Relaxed sarcomere. *b.* Contracted sarcomere. Note that during contraction, the I band and H zone decrease in size. This indicates that the thin filaments slide past the thick filaments. Even so, the thick filaments do the work by pulling the thin filaments by means of cross-bridges.

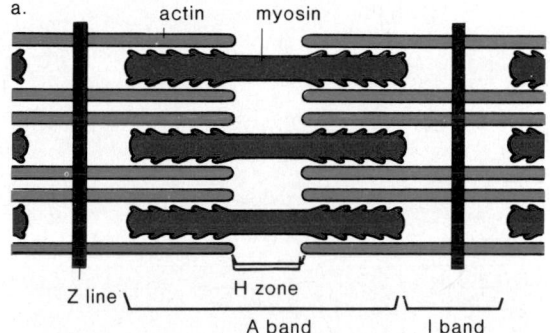

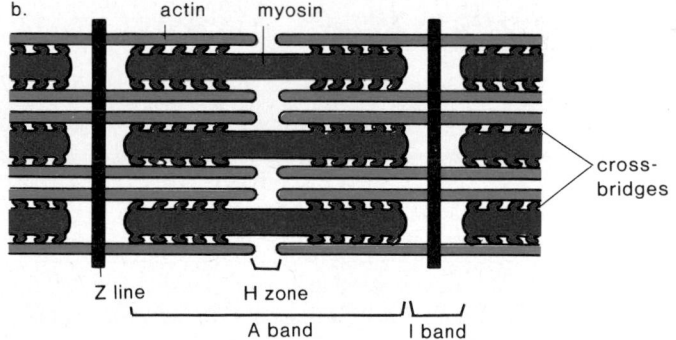

filament called *cross-bridges* bend backward and attach to an actin filament (fig. 41.10). (Each cross-bridge binds to an actin filament at a cross-bridge binding site; actomyosin represents this in the preceding formula.) After attaching, the cross-bridges bend forward and the actin filament is pulled along. Now ATP is broken down by myosin, and detachment occurs. Notice that myosin is not only a structural protein, it is also an ATPase enzyme. The cross-bridges attach and detach some 50–100 times as the thin filaments are pulled to the center of a sarcomere.

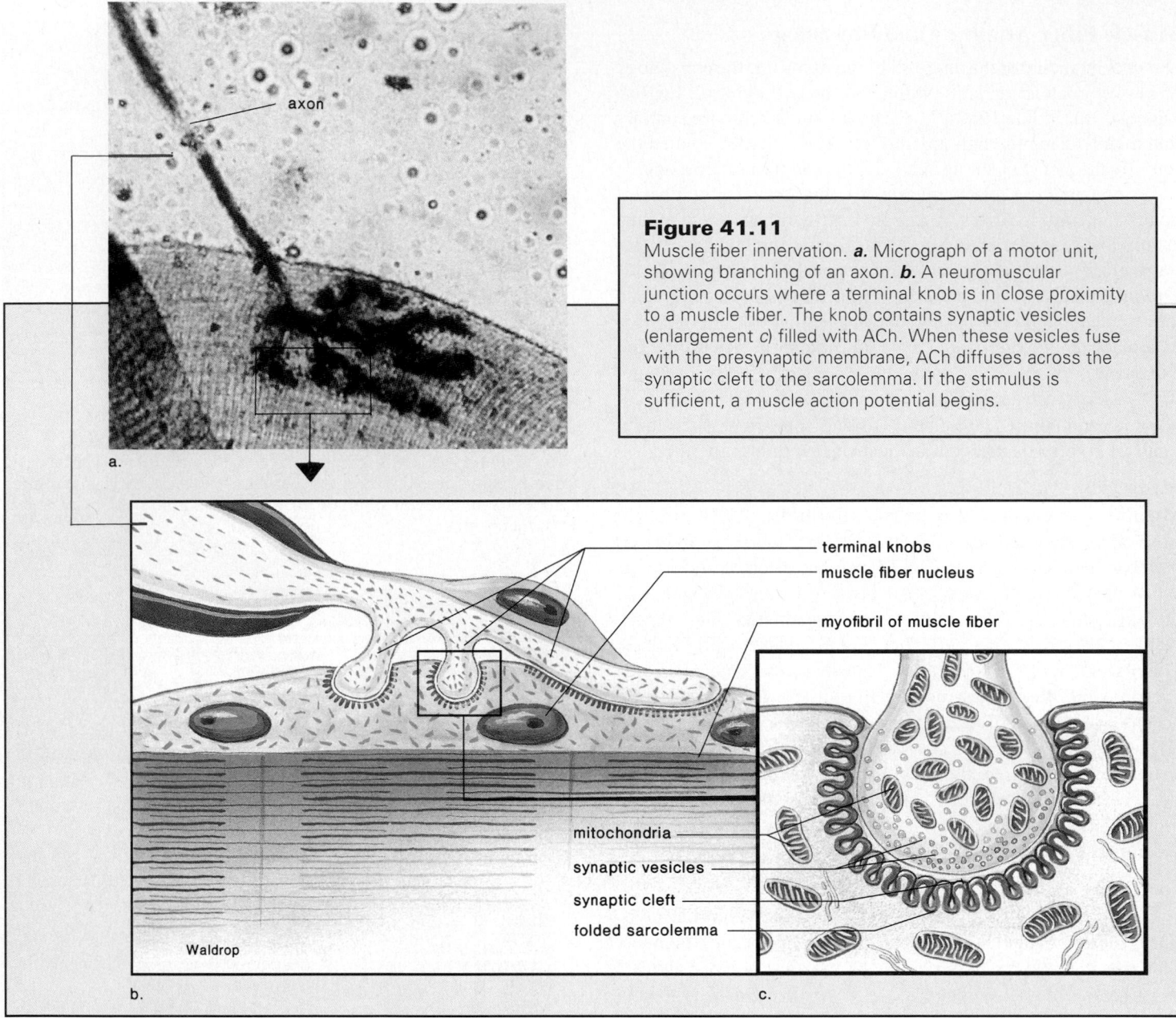

Figure 41.11
Muscle fiber innervation. *a.* Micrograph of a motor unit, showing branching of an axon. *b.* A neuromuscular junction occurs where a terminal knob is in close proximity to a muscle fiber. The knob contains synaptic vesicles (enlargement *c*) filled with ACh. When these vesicles fuse with the presynaptic membrane, ACh diffuses across the synaptic cleft to the sarcolemma. If the stimulus is sufficient, a muscle action potential begins.

axon

a.

terminal knobs
muscle fiber nucleus
myofibril of muscle fiber

mitochondria
synaptic vesicles
synaptic cleft
folded sarcolemma

Waldrop

b.

c.

The sliding filament theory states that actin filaments slide past myosin filaments because myosin has cross-bridges, which pull the actin filaments inward.

It is obvious from our discussion that ATP provides the energy for muscle contraction. To ensure a ready supply of ATP, muscle fibers contain *creatine phosphate* (phosphocreatine), a storage form of high-energy phosphate. Creatine phosphate does not directly participate in muscle contraction. Instead, it is used to regenerate ATP by the following reaction:

$$\text{creatine} \sim \text{P} + \text{ADP} \longrightarrow \text{ATP} + \text{creatine}$$

Oxygen Debt

When all of the creatine phosphate is depleted and no oxygen (O_2) is available for aerobic respiration, a muscle fiber can generate ATP by using fermentation, an anaerobic process. Fermentation, which is apt to occur during strenuous exercise, can supply ATP for only a short time because of lactate buildup. Lactate is a metabolic poison that produces muscle aching and fatigue for a minute or so before it is broken down.

We have all had the experience of having to continue deep breathing for a few minutes following strenuous exercise. We do this because we are in *oxygen debt*—oxygen is needed to complete the metabolism of lactate that has accumulated during exercise. Lactate is transported to the liver, where one-fifth of it is completely broken down to carbon dioxide and water. The

Animal Structure and Function

resulting buildup of ATP by the electron transport system provides the energy to convert the remaining four-fifths of lactate back to glucose.

> Muscle contraction requires a ready supply of ATP. Creatine phosphate is used to generate ATP rapidly. If oxygen is in limited supply, fermentation produces ATP and results in oxygen debt.

Innervation

Muscles are innervated; that is, nerve impulses cause muscles to contract. Each motor axon of a nerve branches to several muscle fibers, and collectively, these muscle fibers are called a motor unit. Each branch has several terminal knobs, where there are synaptic vesicles filled with the neurotransmitter acetylcholine (ACh). The region where a terminal knob lies in close proximity to the sarcolemma of a muscle fiber is called a **neuromuscular junction**

(fig. 41.11). A neuromuscular junction has the same components as a synapse; a presynaptic membrane, a synaptic cleft, and a postsynaptic membrane. In this junction, however, the postsynaptic membrane is a portion of the sarcolemma of a muscle fiber.

Nerve impulses cause synaptic vesicles to merge with the presynaptic membrane and release ACh into the synaptic cleft. When ACh reaches the sarcolemma, it is depolarized. The result is a *muscle action potential* that spreads over the sarcolemma and down the T tubules (fig. 41.9*a*) to where calcium (Ca^{++}) ions are stored in the sarcoplasmic reticulum. Now calcium ions are released, and they diffuse into the sarcoplasm, where they participate in muscle contraction.

It is now necessary to consider the structure of a thin filament in more detail. Figure 41.12 shows the placement of 2 other proteins associated with a thin filament (the double row of twisted globular actin molecules). Threads of tropomyosin wind about a thin filament, and troponin occurs at intervals along the threads. After calcium ions are released, they combine with troponin. After binding occurs, the tropomyosin threads shift their position, and the cross-bridge binding sites are exposed.

The thick filament is a bundle of myosin molecules, each having a globular head. Each head is a cross-bridge that has an ATP-binding site. After ATP attaches to ATP-binding sites, the cross-bridges bend backward and attach to the cross-bridge binding sites on the actin filaments. After attachment occurs, the cross-bridges bend forward, pulling the actin filaments a short distance. Then the myosin heads break down ATP, and detachment of the cross-bridges occurs. The actin filaments move nearer the center of the sarcomere each time the cycle is repeated.

The movement of the actin filaments causes muscle contraction. Contraction ceases when nerve impulses no longer stimulate the muscle fiber. With the cessation of a muscle action potential, calcium ions are pumped back into the sarcoplasmic reticulum by active transport. Relaxation now occurs.

> A neuromuscular junction functions like a synapse except that a muscle action potential causes calcium ions to be released from sarcoplasmic reticulum, and thereafter muscle contraction occurs.

Figure 41.12

Detailed structure and function of sarcomere contraction. After calcium (Ca^{++}) ions are released from the sarcoplasmic reticulum, they combine with troponin, a protein that occurs periodically along tropomyosin threads. This causes the tropomyosin threads to shift their position so that cross-bridge binding sites are revealed along the actin filaments. The myosin filament extends its globular heads, forming cross-bridges, which bind to these sites. The breakdown of ATP by myosin causes the cross-bridges to detach and to reattach farther along the actin. In this way, the actin filaments are pulled along past the myosin filaments.

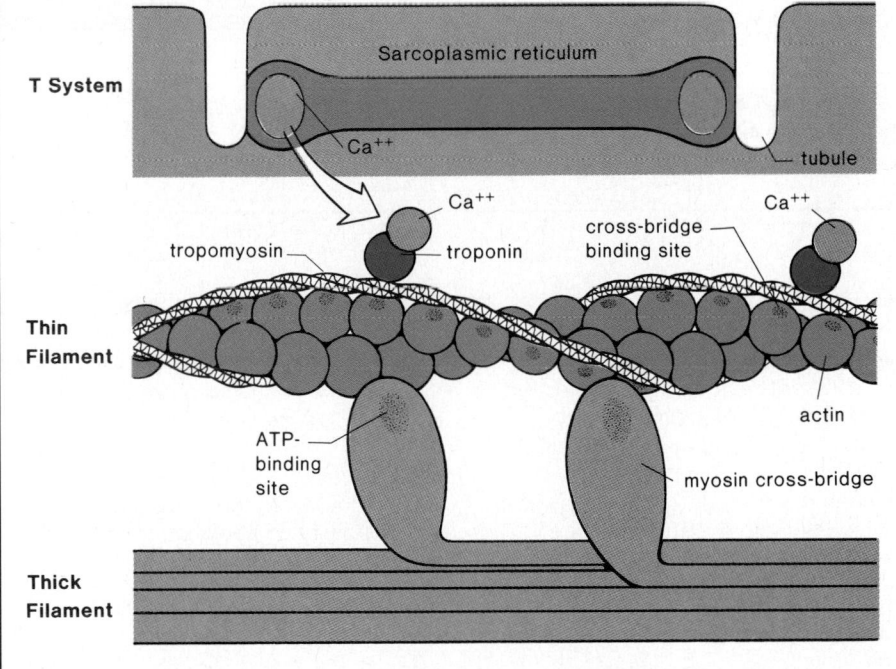

Summary

1. The 3 types of skeletons that are found in the animal kingdom are the hydrostatic skeleton (flatworms and segmented worms), the exoskeleton (arthropods), and the endoskeleton (vertebrates).

2. A rigid skeleton gives support to the body, helps protect internal organs, and assists movement. In humans, the skeleton is also a storage area for calcium and phosphorus salts and the site of blood cell production.

3. Bone is constantly being renewed; osteoclasts break down bone and osteoblasts build new bone. Osteocytes are in the lacunae of Haversian systems.

4. The human skeleton is divided into 2 parts: (1) the axial skeleton, which is made up of the skull, ribs, and vertebrae, and (2) the appendicular skeleton, which is composed of the appendages and their girdles.

5. Joints are classified as immovable, like those of the cranium; slightly movable, like those between the vertebrae; and freely movable or synovial joints, like those in the knee and hip. In synovial joints, ligaments bind the 2 bones together, forming a capsule in which there is synovial fluid.

6. Skeletal muscles can only pull bones; they cannot push. Therefore, they work in antagonistic pairs.

7. Muscle fibers contain a T system: the sarcolemma forms tubules, which contact the sarcoplasmic reticulum. Here calcium ions are stored.

8. The sarcoplasmic reticulum encases myofibrils, long contractile units having longitudinal subunits called sarcomeres. The anatomy of the sarcomere shows the placement of the protein filaments actin and myosin. It is the placement of these that results in the striations of skeletal muscle. Actin (thin) filaments are attached to the Z line, and both actin and myosin (thick) filaments are in the A band. The H zone contains only myosin filaments.

9. The sliding filament theory of muscle contraction says that the myosin filaments have cross-bridges, which attach to and detach from the actin filaments causing them to slide and the sarcomere to shorten. The H zone disappears as the actin filaments approach one another.

10. ATP is the only source of energy for muscle contraction. The globular heads of myosin break down ATP when they detach from the actin filament. Creatine phosphate is a storage form of high-energy phosphate in muscle tissue. When there is no more freely available ATP, fermentation occurs and the organism goes into oxygen debt.

11. Innervation of a muscle fiber begins at a neuromuscular junction. Here synaptic vesicles release ACh into the synaptic cleft. When the sarcolemma receives the ACh, a muscle action potential moves down the T system to the calcium-storage sacs.

12. After calcium ions are released, contraction occurs because calcium combines with troponin, a protein that occurs on tropomyosin threads on the actin filament. This causes the threads to shift their position so that myosin binding sites on actin become available. When calcium ions are actively transported back into the storage sacs, muscle relaxation occurs.

Writing Across the Curriculum

In order to practice writing skills, students should write out the answers to any or all of the study questions and the critical thinking questions. The study questions are sequenced in the same order as the text. Suggested answers to the critical thinking questions are in appendix D.

Study Questions

1. What are the 3 types of skeletons found in the animal kingdom, how do they differ, and what are some animals having each type?
2. Give several functions of the skeletal system in humans.
3. Distinguish between the axial and appendicular skeletons.
4. List the bones that form the pectoral and pelvic girdles.
5. Describe the anatomy of a long bone, and tell how bone is continually being rejuvenated.
6. Name the different types of joints, and give an example of each type.
7. Why do muscles act in antagonistic pairs?
8. Describe the microscopic anatomy of a muscle fiber and the structure of a sarcomere. What is the sliding filament theory?
9. Discuss the availability and the specific role of ATP during muscle contraction. What is oxygen debt and how is it repaid?
10. What causes a muscle action potential? How does the muscle action potential bring about sarcomere and muscle fiber contraction?

Animal Structure and Function

Objective Questions

For questions 1–4, match the following items with the correct locations given in the key.

Key:

- **a.** upper arm
- **b.** lower arm
- **c.** pectoral girdle
- **d.** pelvic girdle
- **e.** upper leg
- **f.** lower leg

1. ulna
2. tibia
3. clavicle
4. femur
5. Spongy bone
   - **a.** contains Haversian systems.
   - **b.** contains red marrow where blood cells are formed.
   - **c.** lends no strength to bones.
   - **d.** All of these.
6. Which of these is mismatched?
   - **a.** slightly movable joint—vertebrae
   - **b.** hinge joint—hip joint
   - **c.** synovial joint—elbow
   - **d.** immovable joint—sutures in cranium
7. In a muscle fiber
   - **a.** the sarcolemma is connective tissue holding the myofibrils together.
   - **b.** the T system contains calcium storage sacs.
   - **c.** both filaments have cross-bridges.
   - **d.** All of these.

8. When muscles contract
   - **a.** sarcomeres increase in size.
   - **b.** myosin slides past actin.
   - **c.** the H zone disappears.
   - **d.** calcium is taken up by calcium storage sacs.
9. Which of these is a source of energy for muscle contraction?
   - **a.** ATP
   - **b.** creatine phosphate
   - **c.** lactic acid
   - **d.** Both a and b.

10. Nervous stimulation of muscles
    - **a.** occurs at a neuromuscular junction.
    - **b.** results in an action potential that travels down the T system.
    - **c.** causes calcium to be released from storage sacs.
    - **d.** All of these.
11. Give the function of each participant in the following reaction.

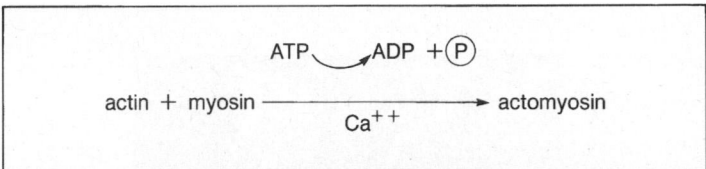

12. Label this diagram of a sarcomere.

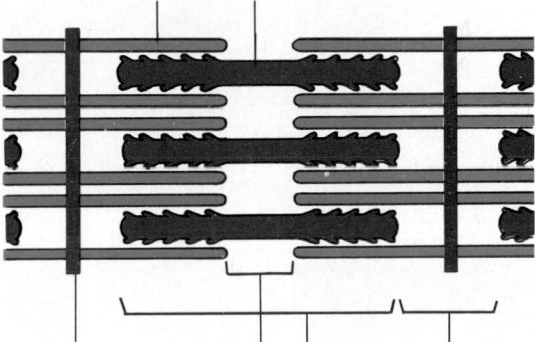

Concepts and Critical Thinking

1. *A bony skeleton determines the shape of an animal.*

   How does the skeleton aid adaptation to the environment?

2. *Movement is fundamental to the nature of animals.*

   Why is movement in animals more essential than in plants?

3. *Locomotion requires energy.*

   How do animals acquire the energy for locomotion, and specifically how is energy used to bring about locomotion?

Selected Key Terms

axial skeleton (ak'se-al skel'ē-ton) 666
appendicular skeleton (ap"en-dik'u-lar skel'ē-ton) 666
muscle fiber (mus'el fi'ber) 671

sarcolemma (sar"ko-lem'ah) 671
sarcoplasmic reticulum (sar"ko-plaz'mik rē-tik'u-lum) 671
myofibril (mi"o-fi'bril) 671
sarcomere (sar"ko-mēr) 671

myosin (mi'o-sin) 671
actin (ak'tin) 671
neuromuscular junction (nu"ro-mus'ku-lar junk'shun) 673

42

Endocrine System

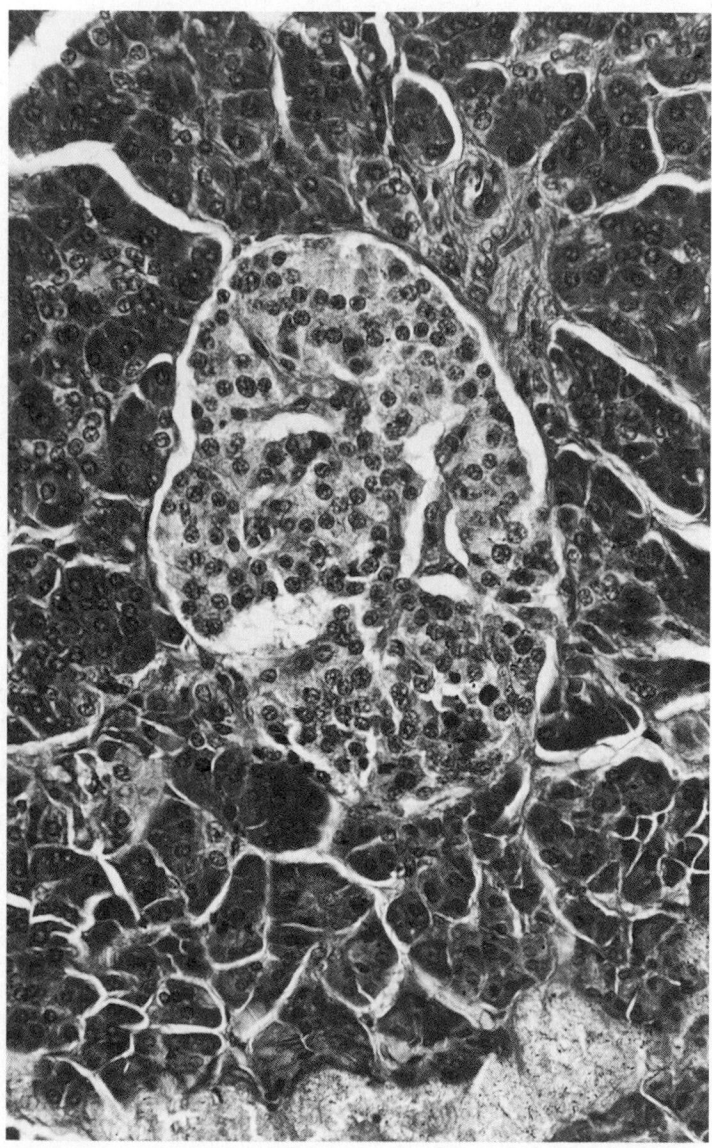

The pancreatic islets are specialized regions of the pancreas, an endocrine gland. The islets secrete the hormone insulin when the blood glucose level is high. Insulin increases the passage of glucose into cells, restoring the normal level. The endocrine glands regulate and coordinate body functions. Magnification, X250.

Your study of this chapter will be complete when you can

1. discuss invertebrate endocrine systems;
2. describe the location of each of the major endocrine glands of the human body;
3. chemically distinguish between the 2 major types of vertebrate hormones, and tell how each type brings about its effect on the cell;
4. contrast the ways in which the hypothalamus controls the posterior and anterior pituitary;
5. list the most important hormones produced by each endocrine gland, and describe their most important effects;
6. give examples of medical conditions associated with the overproduction or underproduction of specific hormones;
7. explain the concept of control by negative feedback, and give an example that involves the hypothalamus, anterior pituitary, and a gland controlled by the anterior pituitary; also give examples that do not involve this 3-tiered system;
8. discuss the relationship between the nervous system and the endocrine system; give examples to show that there are regions of overlap between their functions;
9. discuss the concept of environmental signals in general, and group environmental signals into 3 categories; discuss why the term *hormone* could be broadened to include all these categories.

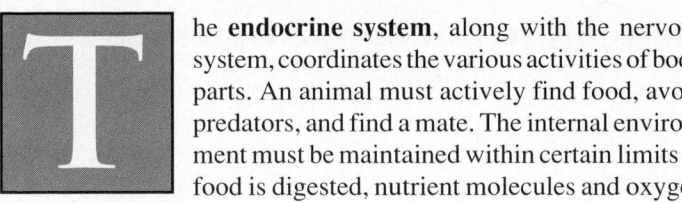

The **endocrine system**, along with the nervous system, coordinates the various activities of body parts. An animal must actively find food, avoid predators, and find a mate. The internal environment must be maintained within certain limits as food is digested, nutrient molecules and oxygen are distributed to the cells, and wastes eliminated. Both the endocrine and nervous systems utilize chemical messengers to coordinate these activities. In chapter 39, we discussed the neurotransmitters that are released by one neuron and influence the excitability of other neurons. In contrast, the endocrine system utilizes hormones, chemical messengers that are typically released directly into the bloodstream.

The nervous system reacts quickly to external and internal stimuli; you rapidly pull your hand away from a hot stove, for example. The endocrine system is somewhat slower because it takes time for a hormone to travel through the circulatory system to its target organ. You might think from the use of this terminology that the hormone is seeking out a particular organ, but quite the contrary, the organ is awaiting the arrival of the hormone. Cells that can react to a hormone have specific receptors, which combine with the hormone in a lock-and-key manner. Therefore, certain cells respond to one hormone and not to another, depending on their receptors.

Sometimes a hormone is defined as a chemical produced by one set of cells that affects a different set. The problem with this definition of it, however, is that it would allow us to categorize all sorts of chemical messengers as hormones, even neurotransmitters! We will have more to say about this later, but a certain amount of overlap between the nervous and endocrine systems is to be expected. The systems evolved together, and at the same time, no doubt making occasional use of the same chemical messengers and communicating not only with other systems but with each other as well. Later, we will give several examples of such associations between the 2 systems.

Both the nervous and endocrine systems function to coordinate body parts and activities. Perhaps since they evolved at the same time, a certain amount of overlap between the chemical messengers utilized and the types of activities performed by the 2 systems is to be expected.

Invertebrate Endocrine Systems

In the first half of this century, biologists showed that many invertebrates have an endocrine system. In invertebrates, hormones are produced only by neurosecretory cells, specialized neurons capable of synthesizing and secreting hormones. Neurosecretions are discharged directly into the blood. This lends support to the hypothesis that hormones first evolved as neurosecretions and only later did nonnervous endocrine glands appear, especially among the vertebrates.

Neurosecretions have been extensively studied in insects. Insects, being arthropods, have an exoskeleton. Typically, they undergo a series of larval stages marked by molting before

Figure 42.1

Three hormones are involved in insect development. Brain hormone controls the secretion of ecdysone, the molt and maturation hormone. Ecdysone causes the insect to pupate if the level of juvenile hormone is low, and to metamorphose if juvenile hormone is absent.

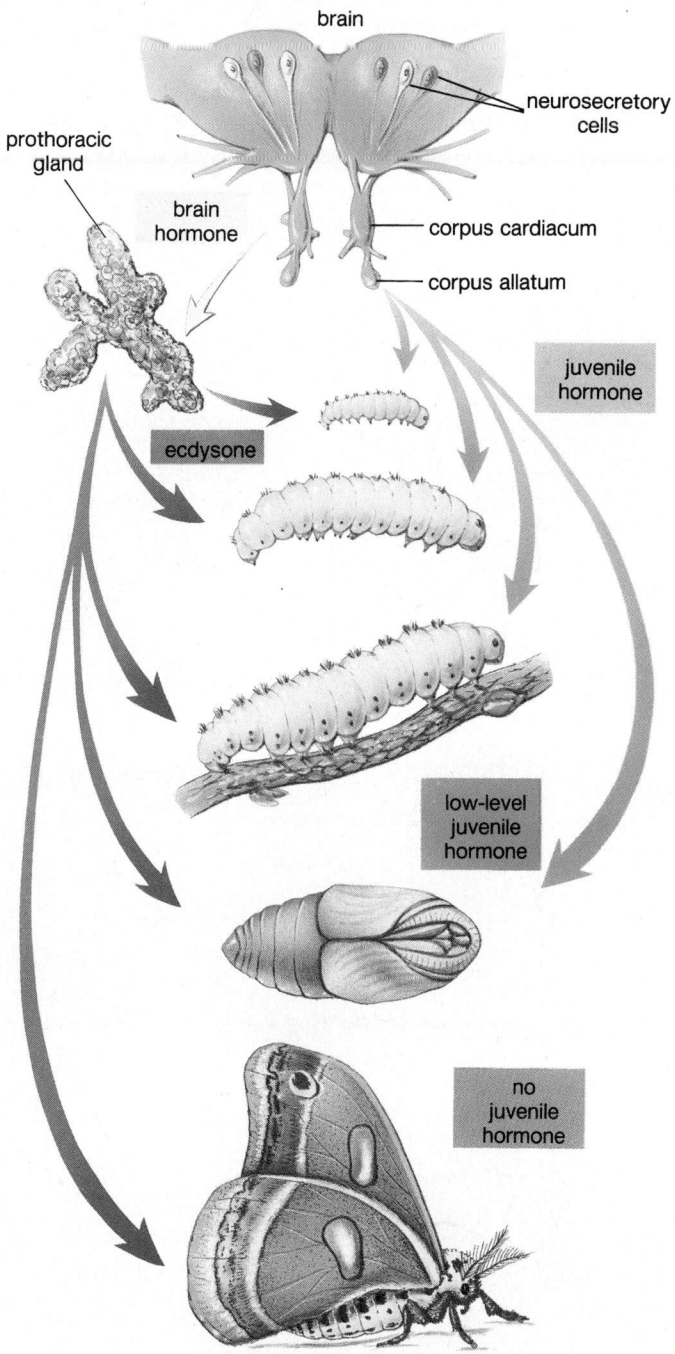

they pupate. V. B. Wigglesworth, who performed experiments on a bloodsucking bug (*Rhodnius*) in the 1930s, showed that the brain is necessary for maturation to take place because it produces a hormone appropriately called *brain hormone* (fig. 42.1). Brain hormone stimulates the prothoracic gland, which lies in the region just behind the head, to secrete *ecdysone,* a steroid hormone. This hormone is also called the molt and maturation hormone because it promotes both molting and maturation. During larval stages, however, response to ecdysone is modified by the action of *juvenile hormone.* If juvenile hormone is present, the insect molts into another larval form; if it is minimally present, the insect pupates; and if it is absent, the insect undergoes metamorphosis into an adult.

Some plants produce compounds that are similar or identical to either ecdysone or juvenile hormone. These compounds apparently protect the plants by disrupting the development of the insects that feed upon them. Work is underway to extract these compounds and to use them as insecticides.

Vertebrate Endocrine Systems

Vertebrate endocrine glands can be contrasted with exocrine glands, which have ducts and secrete their products into these ducts for transport into body cavities; for example, the salivary glands send saliva into the mouth by way of the salivary ducts. Endocrine glands are ductless; they secrete their hormones directly into the bloodstream for distribution throughout the body.

The human endocrine system can be used to exemplify the vertebrate endocrine system (fig. 42.2). Even so, there are marked differences in the vertebrate hormones. For example, prolactin in humans stimulates female breasts to secrete milk, but

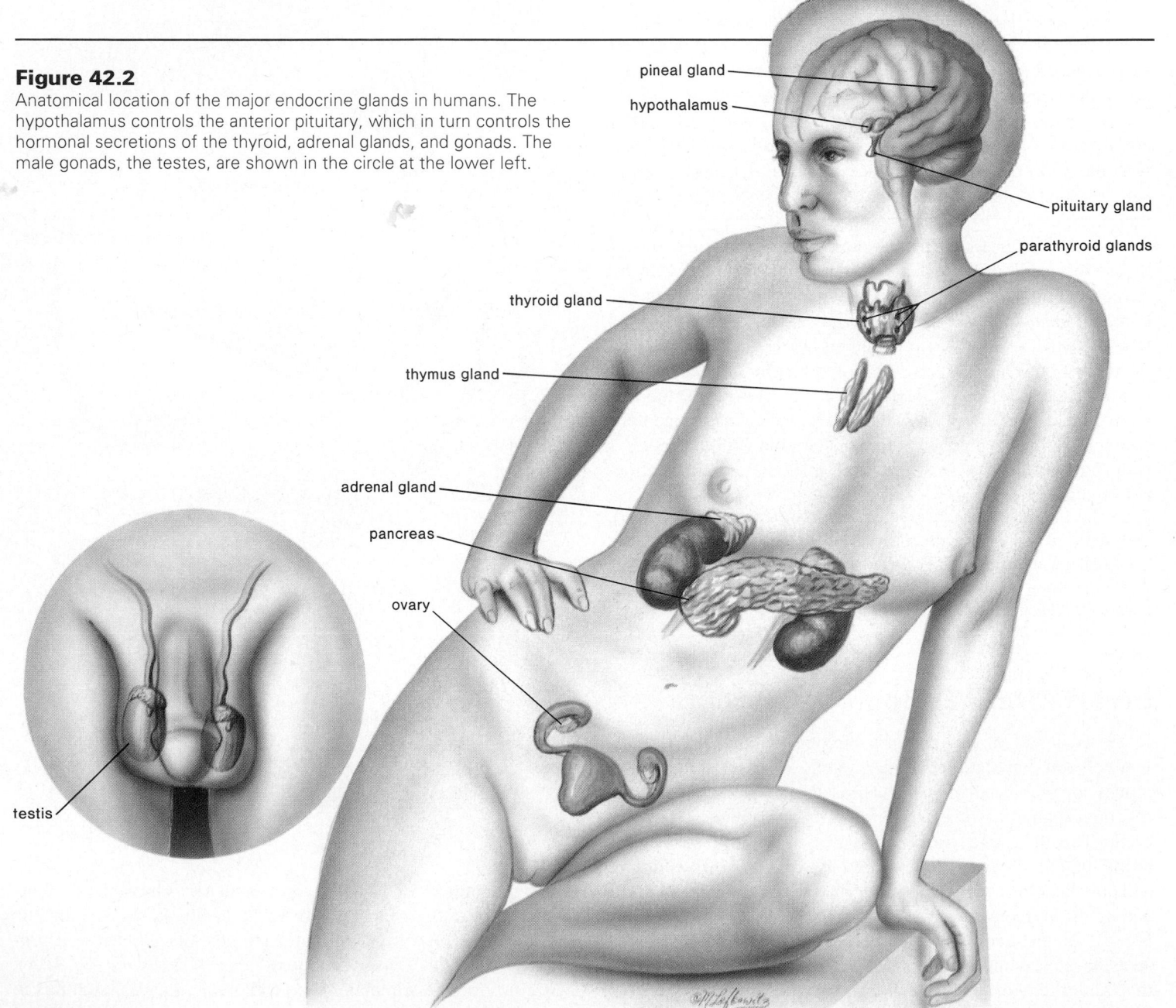

Figure 42.2
Anatomical location of the major endocrine glands in humans. The hypothalamus controls the anterior pituitary, which in turn controls the hormonal secretions of the thyroid, adrenal glands, and gonads. The male gonads, the testes, are shown in the circle at the lower left.

pineal gland

hypothalamus

pituitary gland

parathyroid glands

thyroid gland

thymus gland

adrenal gland

pancreas

ovary

testis

Animal Structure and Function

in pigeons it stimulates the secretion of crop milk, a product of the gut. Some hormones have an entirely different function; for example, thyroxin in humans stimulates metabolism, but in frogs it induces metamorphosis or the transformation of tadpole to adult. Other hormones are species specific, that is, they function only in one species.

Mechanism of Hormonal Action

Table 42.1 indicates that many of the human hormones are proteins or peptides. These molecules are coded for by genes and are synthesized at the ribosomes (see fig. 16.13). Eventually, they are packaged into vesicles at the Golgi apparatus and are secreted at the plasma membrane. There are some other hormones, called catecholamines, that are derived from the amino acid tyrosine; their production requires only a series of metabolic reactions within the cytoplasm. From the viewpoint of hormonal action, we can group the protein, peptide, and catecholamines together and refer to them as *peptide hormones*.

The hormones produced by the adrenal cortex, the ovaries, and the testes are steroids. *Steroid hormones* are derived from cholesterol (see fig. 4.13) by a series of metabolic reactions. These hormones are stored in fat droplets in the cell cytoplasm until their release at the plasma membrane.

Peptide Hormones

The mode of action of peptide hormones was discovered in the 1950s by Earl W. Sutherland, who studied the effects of epinephrine on liver cells. Sutherland received a Nobel Prize for his hypothesis that when this hormone binds to a cell-surface receptor, the resulting complex leads to the activation of an enzyme that produces **cyclic AMP** (fig. 42.3a). Cyclic AMP (cAMP) is a compound made from ATP, but it contains only one phosphate group, which is attached to adenosine at 2 locations.

Since peptide hormones never enter the cell, they are sometimes called the first messengers, while cAMP, which sets the metabolic machinery in motion, is called the second messenger (fig. 42.3). Other messengers have been discovered in cells. Inositol triphosphate (IP_3) is a second messenger that causes the release of calcium (Ca^{++}) ions in muscle cells, for example. Since calcium goes on to activate other proteins, it is called the third messenger.

cAMP operates in a second messenger system that sets an *enzyme cascade* into motion. cAMP activates only one particular enzyme in the cell, but this enzyme in turn activates another, and so forth. The activated enzymes, of course, can be used repeatedly. Therefore, at every step in the enzyme cascade, more and more reactions take place—the binding of a single hormone molecule eventually results in a 1,000-fold response.

Steroid Hormones

Steroid hormones do not bind to cell-surface receptors; they can enter the cell freely (fig. 42.3b). For example, scans of the reproductive organs clearly show that the hormones estrogen and progesterone enter the cells of the reproductive organs. Once inside, steroid hormones bind to receptors in the cytoplasm. The hormone-receptor complex then enters the nucleus, where it binds

Figure 42.3

Cellular activity of hormones. **a.** Peptide hormones combine with receptors located in the plasma membrane. This promotes the production of cAMP, which in turn leads to activation of a particular enzyme. **b.** Steroid hormones pass through the plasma membrane to combine with receptors; the complex activates certain genes, leading to protein synthesis.

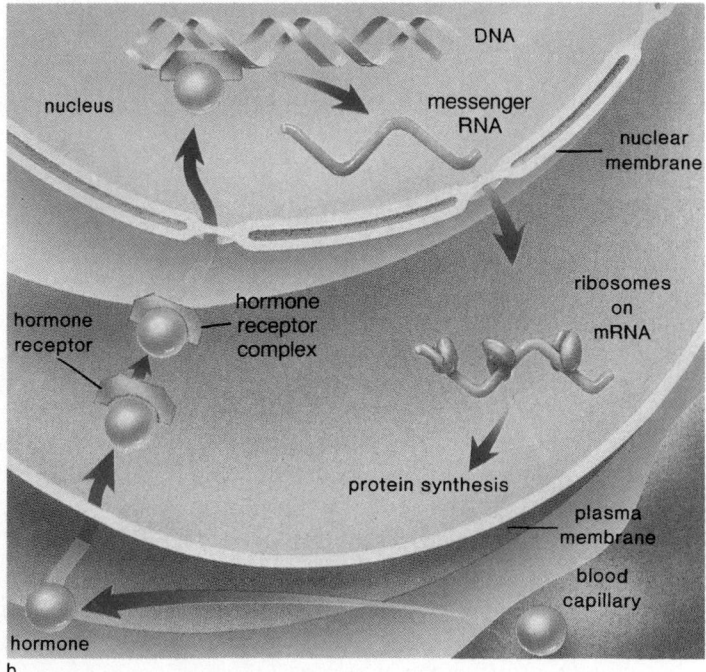

a.

b.

Table 42.1
Human Endocrine Glands and Their Hormones

Endocrine Gland	Hormone Released	Chemical Structure of Hormone	Chief Function of Hormone
Hypothalamus	Hypothalamic-releasing hormones Hypothalamic-release-inhibiting hormones	Peptides	Regulate anterior pituitary hormones
Anterior pituitary	Thyroid-stimulating (TSH, thyrotropic)	Glycoprotein	Stimulates thyroid
	Adrenocorticotropic (ACTH)	Polypeptide	Stimulates adrenal cortex
	Gonadotropic follicle-stimulating (FSH)	Glycoprotein	Controls egg and sperm production
	luteinizing (LH)		Controls sex hormone production
	Prolactin (Pr)	Protein	Stimulates milk production and secretion
	Growth (GH, somatotropic)	Protein	Stimulates cell division, protein synthesis, and bone growth
	Melanocyte-stimulating (MSH)	Polypeptide	Unknown function in humans; regulates skin color in lower vertebrates
Posterior pituitary (storage of hypothalamic hormones)	Antidiuretic (ADH, vasopressin)	Peptide	Stimulates water reabsorption by kidneys
	Oxytocin	Peptide	Stimulates uterine muscle contraction and release of milk by mammary glands
Thyroid	Thyroxin and triiodothyronine	Iodinated amino acid	Increases metabolic rate; helps regulate growth and development
	Calcitonin	Peptide	Lowers blood calcium level
Parathyroids	Parathyroid (PTH)	Polypeptide	Raises blood calcium level
Adrenal cortex	Glucocorticoids (cortisol)	Steroids	Raise blood glucose level; stimulate breakdown of protein
	Mineralocorticoids (aldosterone)	Steroids	Stimulate kidneys to reabsorb sodium and to excrete potassium
	Sex hormones	Steroids	Stimulate development of secondary sex characteristics (particularly in male)
Adrenal medulla	Epinephrine and norepinephrine	Catecholamines	Stimulate fight or flight reactions; raise blood glucose level
Pancreas	Insulin	Polypeptide	Lowers blood glucose level; promotes formation of glycogen, proteins, and fats
	Glucagon	Polypeptide	Raises blood glucose level; promotes breakdown of glycogen, proteins, and fats
Gonads			
Testes	Androgens (testosterone)	Steroids	Stimulate spermatogenesis; develop and maintain secondary male sex characteristics
Ovaries	Estrogen and progesterone	Steroids	Stimulate growth of uterine lining; develop and maintain secondary female sex characteristics
Thymus	Thymosins	Peptides	Stimulate production and maturation of T lymphocytes
Pineal gland	Melatonin	Catecholamine	Involved in circadian and circannual rhythms; possibly involved in maturation of sex organs
Gut*	Gastrin	Peptide	Stimulates secretion of gastric juices
	Secretin	Peptide	Stimulates secretion of pancreatic juices
	Cholecystokinin-pancreozymin (CCK-PZ)	Peptide	Stimulates secretion of pancreatic juices and flow of bile from gallbladder

*The hormones produced by the gut are discussed in chapter 36.

Animal Structure and Function

with chromatin at a location that promotes activation of particular genes. Transcription of DNA and translation of mRNA follow. In this manner, steroid hormones lead to protein synthesis.

Steroids act more slowly than peptides because it takes more time to synthesize new proteins than to activate enzymes that are already present in the cell. But steroids have a more *sustained* effect on the metabolism of the cell than do peptide hormones.

Hormones are chemical messengers that influence the metabolism of the recipient cell. Peptide hormones activate existing enzymes in the cell, and steroid hormones bring about the synthesis of new proteins.

Hypothalamus and the Pituitary Gland

The hypothalamus is the portion of the brain (fig. 39.10) that regulates the internal environment; for example, it helps to control the heart rate, body temperature, and water balance, as well as the glandular secretions of the **pituitary gland.** The pituitary, a small gland about 1 cm in diameter, lies just below the hypothalamus and is divided into 2 portions called the posterior pituitary and the anterior pituitary.

Posterior Pituitary

The posterior pituitary is connected to the hypothalamus by means of a stalklike structure. There are neurons in the hypothalamus that are called neurosecretory cells because they both respond to neurotransmitter substances and produce the hormones that are stored in and released from the posterior pituitary. The hormones pass from the hypothalamus through axons that terminate in the posterior pituitary (fig. 42.4).

The axon endings in the posterior pituitary store **antidiuretic hormone (ADH),** sometimes called vasopressin, and oxytocin. ADH, as discussed in chapter 38, promotes the reabsorption of water from the collecting duct, a portion of the nephrons within the kidneys. The hypothalamus contains other nerve cells that act as a sensor because they are sensitive to the osmolarity of the blood. When these cells determine that the blood is too concentrated, ADH is released into the bloodstream from the axon endings in the posterior pituitary. As the blood becomes dilute, the hormone no longer is released. This is an example of control by negative feedback because the effect of the hormone (dilute blood) acts to shut down the release of the hormone. Negative feedback mechanisms regulate the activities of most endocrine glands.

Oxytocin is the other hormone that is made in the hypothalamus and stored in the posterior pituitary. Oxytocin causes the uterus to contract and is used to artificially induce labor. It also stimulates the release of milk from the mammary glands when a baby is nursing.

The posterior pituitary stores 2 hormones, ADH and oxytocin, both of which are produced by and released from neurosecretory cells in the hypothalamus.

Figure 42.4

The hypothalamus produces ADH and oxytocin, 2 hormones that are stored in and secreted by the posterior pituitary.

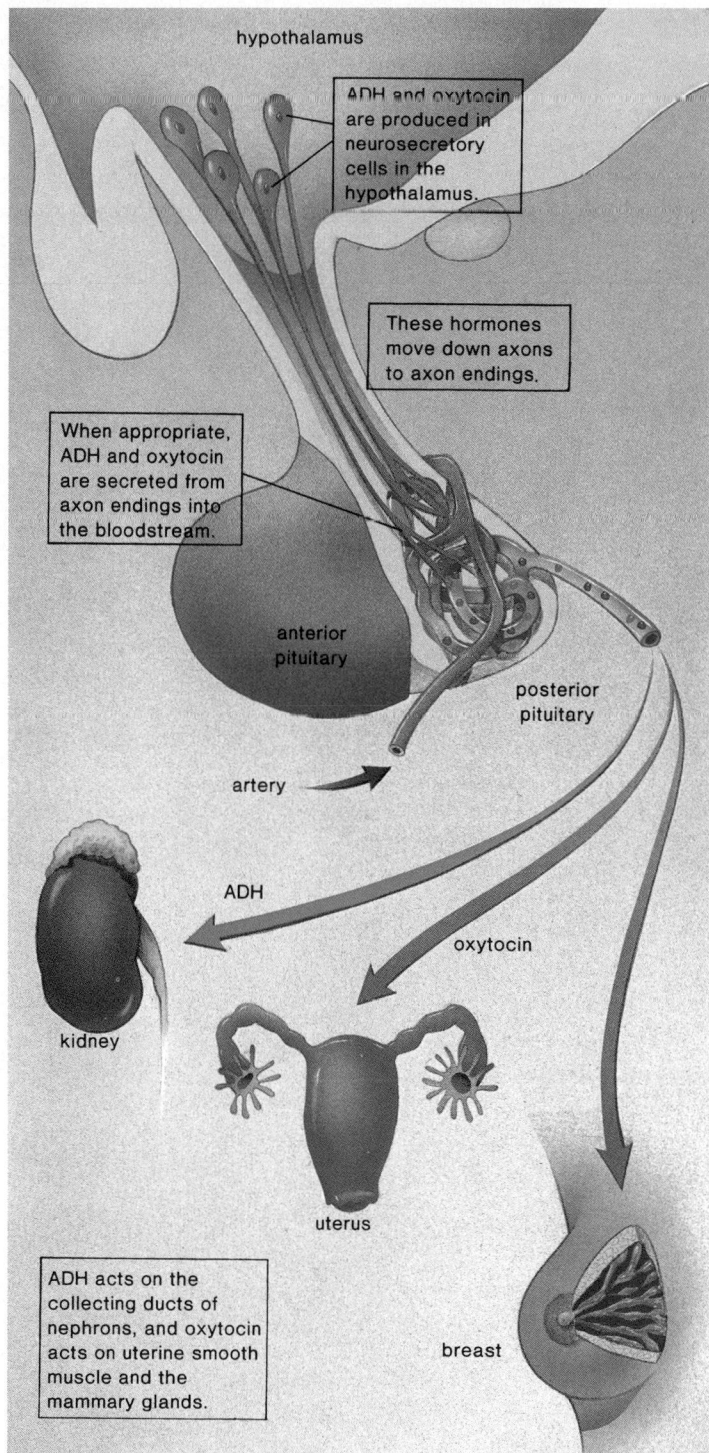

hypothalamus

ADH and oxytocin are produced in neurosecretory cells in the hypothalamus.

These hormones move down axons to axon endings.

When appropriate, ADH and oxytocin are secreted from axon endings into the bloodstream.

anterior pituitary

posterior pituitary

artery

ADH

oxytocin

kidney

uterus

breast

ADH acts on the collecting ducts of nephrons, and oxytocin acts on uterine smooth muscle and the mammary glands.

Figure 42.5

The hypothalamus also controls the secretions of the anterior pituitary.

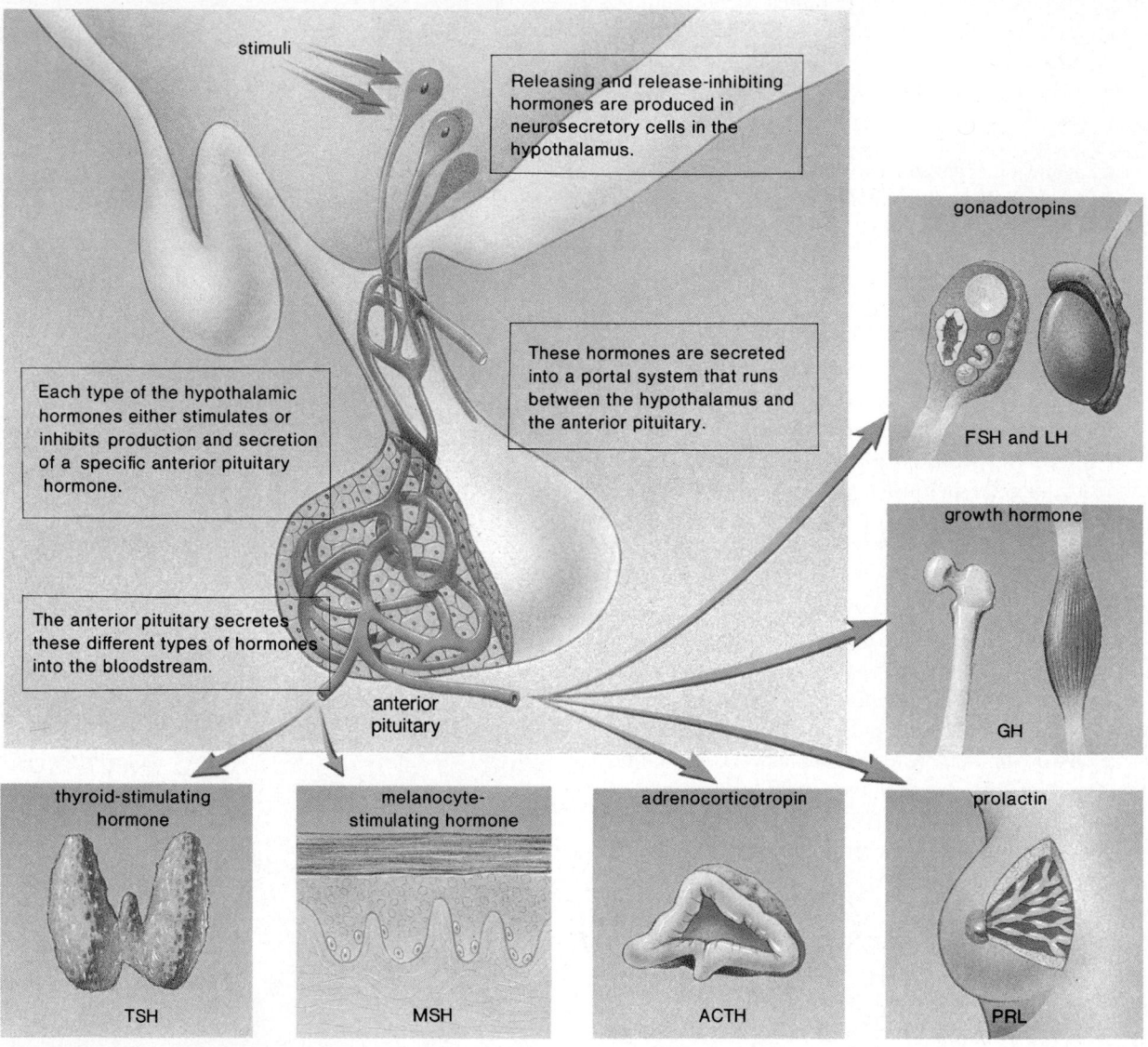

stimuli

Releasing and release-inhibiting hormones are produced in neurosecretory cells in the hypothalamus.

Each type of the hypothalamic hormones either stimulates or inhibits production and secretion of a specific anterior pituitary hormone.

These hormones are secreted into a portal system that runs between the hypothalamus and the anterior pituitary.

The anterior pituitary secretes these different types of hormones into the bloodstream.

anterior pituitary

gonadotropins

FSH and LH

growth hormone

GH

thyroid-stimulating hormone

TSH

melanocyte-stimulating hormone

MSH

adrenocorticotropin

ACTH

prolactin

PRL

Anterior Pituitary

By the 1930s, it had been determined that the hypothalamus also controls the release of the anterior pituitary hormones. For example, electrical stimulation of the hypothalamus causes the release of anterior pituitary hormones, but *direct* stimulation of the anterior pituitary itself has no effect. Detailed anatomical studies showed that there is a *portal* system consisting of blood vessels connecting a capillary bed in the hypothalamus with one in the anterior pituitary (fig. 42.5). The hypothesis was put forward that a different set of neurosecretory cells in the hypothalamus synthesize a number of different releasing hormones that are sent to the anterior pituitary by way of the vascular portal system. There began a long and competitive struggle between 2 groups of investigators to identify one of these *hypothalamic-releasing hormones,* which

cause the anterior pituitary to release hormones. The group headed by R. Guillemin processed nearly 2 million sheep hypothalami, and the group headed by A. V. Schally processed more than 1 million pig hypothalami until each group announced, in November of 1969, that it had isolated and determined the structure of one of these hormones, a peptide containing only 3 amino acids.

Later, it was found that some of the hypothalamic hormones inhibit the release of anterior pituitary hormones. Therefore, today, it is customary to speak of *hypothalamic-releasing hormones* and *hypothalamic-release-inhibiting hormones.* For example, there is a **gonadotropic-releasing hormone (GNRH)** and a gonadotropic-release-inhibiting hormone (GnRIH). The first hormone stimulates the anterior pituitary to release gonadotropic hormones, and the second inhibits the anterior pituitary from releasing the same hormones.

Figure 42.6

The difference in size between a giant and a dwarf sometimes can be explained by a difference in production of growth hormone by the anterior pituitary. Sandy Allen is one of the world's tallest women due to a higher than usual amount of growth hormone produced by the anterior pituitary.

The anterior pituitary is controlled by the hypothalamic-releasing hormones. These hormones are produced in neurosecretory cells in the hypothalamus and pass to the anterior pituitary by way of a vascular portal system.

The anterior pituitary produces at least 6 different types of hormones, each by a distinct cell type (fig. 42.5). Three of these hormones have a direct effect on the body. **Growth hormone (GH),** or somatotropic hormone, dramatically affects physical appearance since it determines the height of the individual (fig. 42.6). If too little GH is produced during childhood, the individual becomes a pituitary dwarf, and if too much is produced, the individual is a pituitary giant. In both instances, the individual has normal body proportions. On occasion, however, there is overproduction of growth hormone in the adult and a condition called *acromegaly* results. Since only the feet, hands, and face (particularly chin, nose, and eyebrow ridges) can respond, these portions of the body become overly large.

Growth hormone promotes cell division, protein synthesis, and bone growth. It stimulates the transport of amino acids into cells and increases the activity of ribosomes, both of which are essential to protein synthesis. In bones, it promotes growth of the cartilaginous plates and causes osteoblasts to form bone. Evidence suggests that the effects on cartilage and bone may actually be due to hormones called somatomedins, which are released by the liver. Growth hormone causes the liver to release somatomedins.

Prolactin (Pr), is produced in quantity only after childbirth. It causes the mammary glands in the breasts to develop and produce milk. It also plays a role in carbohydrate and fat metabolism.

Melanocyte-stimulating hormone (MSH) causes skin color changes in many fishes, amphibians, and reptiles who have melanophores, special skin cells that produce color variations. The hormone is present in humans, but no one knows what its function might be. It is derived, however, from a molecule that is also the precursor for both ACTH and the anterior pituitary endorphins. These endorphins are structurally and functionally similar to the endorphins produced in brain nerve cells.

Regulation of the Anterior Pituitary

The anterior pituitary is sometimes called the *master gland* because it controls the secretion of some other endocrine glands (fig. 42.5). As indicated in table 42.1, the anterior pituitary secretes the following hormones, which have an effect on other glands:

1. **TSH,** thyroid-stimulating hormone
2. **ACTH,** adrenocorticotropic hormone, which stimulates the adrenal cortex
3. **Gonadotropic hormones** (FSH and LH), which stimulate the gonads—the testes in males and the ovaries in females

TSH causes the thyroid to produce thyroxin; ACTH causes the adrenal cortex to produce cortisol; and gonadotropic hormones cause the gonads to secrete sex hormones. A 3-tiered relationship exists among the hypothalamus, anterior pituitary, and other endocrine glands; the hypothalamus produces releasing hormones, which control the anterior pituitary, and the anterior pituitary produces hormones that control the thyroid, adrenal cortex, and gonads. Figure 42.7 illustrates the negative feedback mechanism that controls the activity of these glands.

The hypothalamus, the anterior pituitary, and the other endocrine glands controlled by the anterior pituitary are all involved in a self-regulating negative feedback loop.

Thyroid Gland

The **thyroid gland** is a large gland located in the neck, where it is attached to the trachea just below the larynx (fig. 42.8a). The 2 hormones produced by the thyroid both contain iodine: **thyroxin,** or T_4, contains 4 atoms of iodine; it is secreted in greater amounts but is less potent than triiodothyronine, or T_3, which has only 3 atoms of iodine. Iodine is actively transported into the thyroid gland, and it may reach a concentration as much as 25 times greater than that of the blood (see fig. 3.4).

Even before the structure of thyroxin was known, it was surmised that the hormone contained iodine because when iodine is lacking in the diet, the thyroid gland enlarges, producing a **goiter** (fig. 42.9). The cause of the enlargement becomes clear if we refer to figure 42.7. When there is a low level of thyroxin in the blood,

Figure 42.7

TRH (thyroid-releasing hormone) stimulates the anterior pituitary, and TSH (thyroid-stimulating hormone) stimulates the thyroid to secrete thyroxin. The level of thyroxin in the body is controlled by negative feedback: *a.* the level of TSH exerts negative feedback control over the hypothalamus; *b.* the level of thyroxin exerts feedback control over the anterior pituitary; and *c.* the level of thyroxin exerts feedback control over the hypothalamus. In this way, thyroxin controls its own secretion. Cortisol and sex hormone levels are controlled in similar ways.

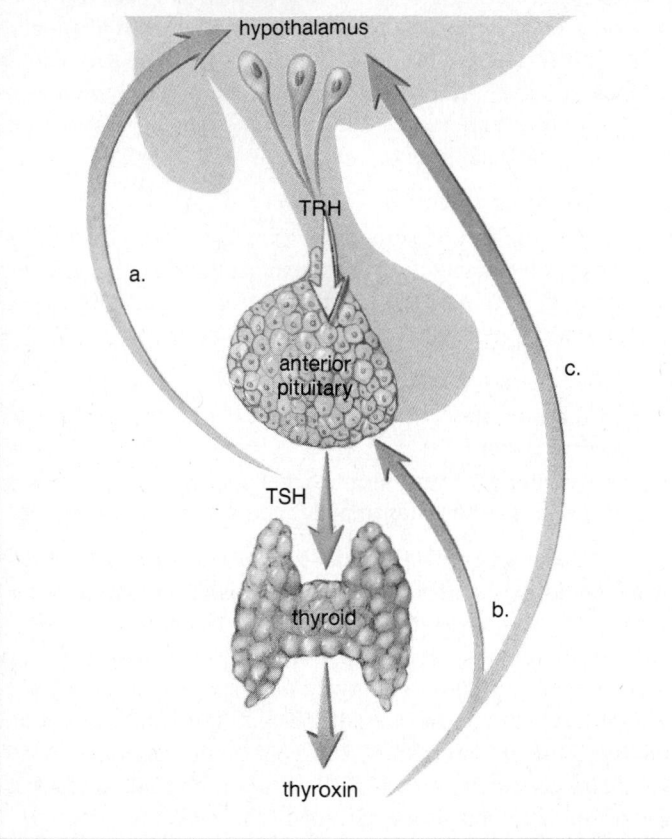

Figure 42.8

a. The thyroid gland is located in the neck in front of the trachea. *b.* The 4 parathyroid glands are embedded in the posterior surface of the thyroid gland, yet the parathyroids and thyroid glands have no anatomical or physiological connection with one another. *c.* Regulation of parathyroid hormone secretion. A low blood level of calcium causes the parathyroids to secrete PTH, which causes the kidneys and gut to retain calcium and osteoclasts to break down bone. The end result is an increased level of calcium in the blood. A high blood level of calcium inhibits secretion of PTH.

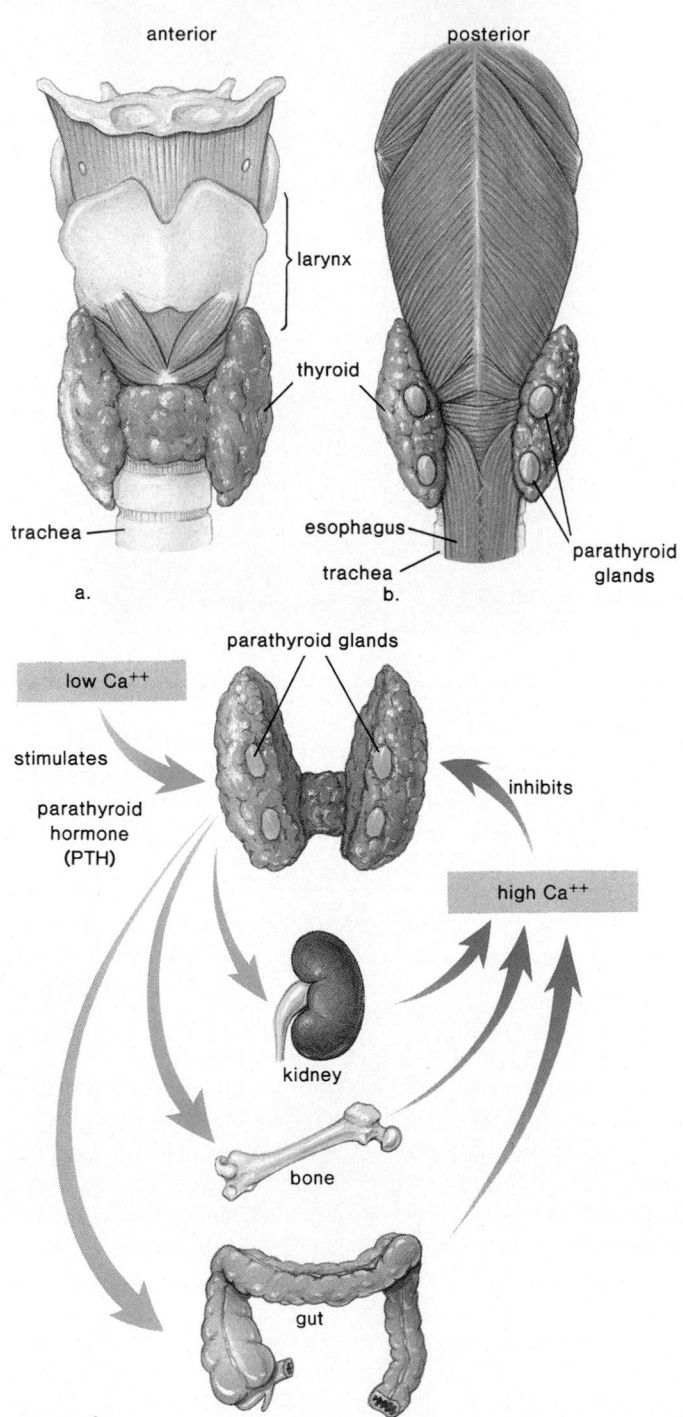

the anterior pituitary is stimulated to produce TSH (thyroid-stimulating hormone). The thyroid responds by increasing in size, but this increase in size is ineffective because active thyroxin cannot be produced without iodine. Later, it was found that goiter can be prevented by consumption of iodized salt.

Thyroxin is necessary in vertebrates for proper growth and development. For example, without thyroxin, frogs do not metamorphose properly and humans do not mature properly. Cretinism occurs in individuals who have suffered from *hypothyroidism* (low thyroid function) since birth. They show reduced skeletal growth, sexual immaturity, and abnormal protein metabolism, which leads to mental retardation.

In general, thyroxin increases the metabolic rate in all cells; the number of respiratory enzymes increases, as does oxygen uptake. Hypothyroidism (too little thyroxin) in adults produces the condition known as myxedema, which is characterized by leth-

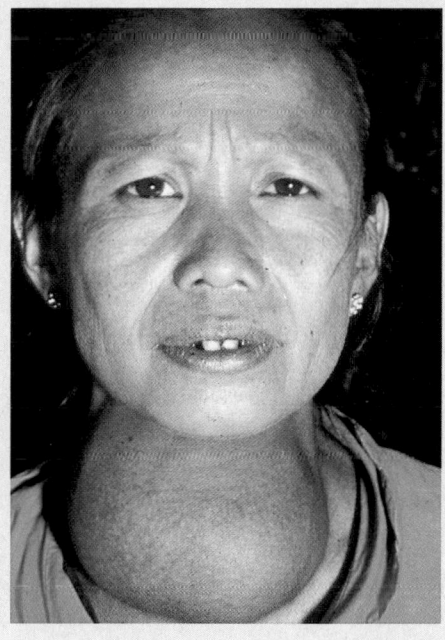

Figure 42.9
A woman suffering from goiter, an enlargement of the thyroid gland that often accompanies hypothyroidism. Continued TSH stimulation of a thyroid that lacks iodine causes massive overgrowth of thyroid tissue.

argy, weight gain, loss of hair, slow pulse, and decreased body temperature. In hyperthyroidism (too much thyroxin), the thyroid gland is enlarged and overactive, causing a goiter to form and the eyes to protrude because of edema in the tissues of the eye sockets and swelling of muscles that move the eyes. This type of goiter is called *exophthalmic goiter*. The patient usually becomes hyperactive, nervous, irritable, and suffers from insomnia. Removal or destruction of a portion of the thyroid by means of radioactive iodine sometimes is effective in curing the condition.

In addition to thyroxin, the thyroid gland also produces the hormone *calcitonin*. This hormone lowers the level of calcium in the blood and opposes the action of the parathyroid hormone, which we will discuss in the next section.

The anterior pituitary produces TSH, a hormone that promotes the production of thyroxin (and triiodothyronine) by the thyroid. Thyroxin helps regulate growth and development in immature animals and speeds up metabolism in all animals.

Parathyroid Glands

The 4 **parathyroid glands** are embedded in the posterior surface of the thyroid gland, as shown in figure 42.8*b*. They produce a hormone called **parathyroid hormone (PTH).** Under the influence of PTH, the calcium (Ca^{++}) level in the blood increases and the phosphate (PO_4^{---}) level decreases. PTH stimulates the *absorption* of calcium from the gut by activating vitamin D, the *retention* of

calcium (the excretion of phosphate) by the kidneys, and the *demineralization* of the bone by promoting the activity of the osteoclasts, the bone-reabsorbing cells. When the blood calcium level reaches the appropriate level, the parathyroid glands no longer produce PTH (fig. 42.8*c*).

If PTH is not produced in response to low blood calcium, *tetany* results because calcium plays an important role in both nervous conduction and muscle contraction. In tetany, the body shakes from continuous muscle contraction. The effect is brought about by increased excitability of the nerves, which fire spontaneously and without rest.

Parathyroid hormone (PTH) maintains a high blood calcium level. Its actions are opposed by calcitonin, which is produced by the thyroid.

Adrenal Glands

Each of the 2 **adrenal glands,** as the name implies (*ad*—near; *renal*—kidney), lies atop a kidney (fig. 42.2). Each consists of an inner portion called the medulla and an outer portion called the cortex. These portions, like the anterior pituitary and the posterior pituitary, have no connection with one another.

Adrenal Medulla

The adrenal medulla secretes *epinephrine* and *norepinephrine* under conditions of stress. They bring about all those responses we associate with the "fight or flight" reaction: the blood glucose level and the metabolic rate increase, as do breathing and the heart rate. The blood vessels in the intestine constrict, and those in the muscles dilate. This increased circulation to the muscles causes them to have more stamina than usual. In times of emergency, the sympathetic nervous system *initiates* these responses but they are *maintained* by secretions from the adrenal medulla.

The adrenal medulla is innervated by one set of sympathetic nerve fibers. Recall that usually there are preganglionic and postganglionic nerve fibers for each organ stimulated. In this instance, what happened to the postganglionic neurons? It appears that the adrenal medulla may have evolved from a modification of the postganglionic neurons. Like the neurosecretory neurons in the hypothalamus, these neurons also secrete hormones into the bloodstream.

The adrenal medulla releases epinephrine and norepinephrine into the bloodstream. These hormones help us and other animals cope with situations that threaten survival.

Adrenal Cortex

Although the adrenal medulla may be removed with no ill effects, the adrenal cortex is absolutely necessary to life. The 2 major classes of hormones made by the adrenal cortex are the *glucocorticoids* and the *mineralocorticoids*. The cortex also secretes a small amount of male and an even smaller amount of female sex hormones. All of these hormones are steroids.

Of the various glucocorticoids, the hormone responsible for the greatest amount of activity is *cortisol*. Cortisol promotes the hydrolysis of muscle protein to amino acids that enter the blood. This leads to an increased level of glucose when the liver converts these amino acids to glucose. Cortisol also favors metabolism of fatty acids rather than carbohydrates. In opposition to insulin, therefore, cortisol raises the blood glucose level. Cortisol also counteracts the inflammatory response, which leads to the pain and swelling of joints in arthritis and bursitis. The administration of cortisol eases the symptoms of these conditions because it reduces inflammation.

The secretion of cortisol by the adrenal cortex is under the control of the anterior pituitary hormone ACTH. Using the same negative feedback system shown in figure 42.7, the hypothalamus produces a releasing hormone (CRH) that stimulates the anterior pituitary to release ACTH. ACTH in turn stimulates the adrenal cortex to secrete cortisol, which regulates its own synthesis by negative feedback of both CRH and ACTH synthesis.

The secretion of mineralocorticoids, the most significant of which is **aldosterone,** is not under the control of the anterior pituitary. Aldosterone regulates the level of sodium (Na^+) ions and potassium (K^+) ions in the blood, its primary target organ being the kidney, where it promotes renal absorption of sodium and renal excretion of potassium. The level of sodium is particularly important to the maintenance of blood pressure, and the concentration of this ion indirectly regulates the secretion of aldosterone. When the blood sodium level is low, the kidneys secrete renin (fig. 42.10). Renin is an enzyme that converts the plasma protein angiotensinogen to angiotensin I, which becomes angiotensin II in the lungs. Angiotensin II stimulates the adrenal cortex to release aldosterone. This is called the renin-angiotensin-aldosterone system. The effect of this system raises the blood pressure in 2 ways. First, angiotensin constricts the arteries directly, and second, aldosterone causes the kidneys to reabsorb sodium. When blood sodium level is high, water is reabsorbed, and blood volume and pressure are maintained.

Two medical conditions are associated with malfunctioning of the adrenal cortex. In Cushing syndrome, the adrenal cortex is overactive. The muscles of the arms and legs waste away, and the blood glucose level and the blood pressure rise. Hypertension develops. In Addison disease, the adrenal cortex is underactive. There is a reduction in blood pressure and blood glucose levels, and an unexplained, peculiar bronzing of the skin.

There is another hormone involved in regulating blood sodium levels. This hormone, called atrial natriuretic hormone, is produced by the atria of the heart and causes natriuresis, the excretion of sodium. Once sodium is excreted, so is water; therefore, blood volume and blood pressure decrease.

Cortisol, which raises the blood glucose level, and aldosterone, which raises the blood sodium level, are 2 hormones secreted by the adrenal cortex.

Pancreas

The **pancreas** is a long organ that lies transversely in the abdomen between the kidneys and near the duodenum of the small intestine (fig. 42.11). It is composed of 2 types of tissues—one of these produces and secretes *digestive* juices that go by way of the

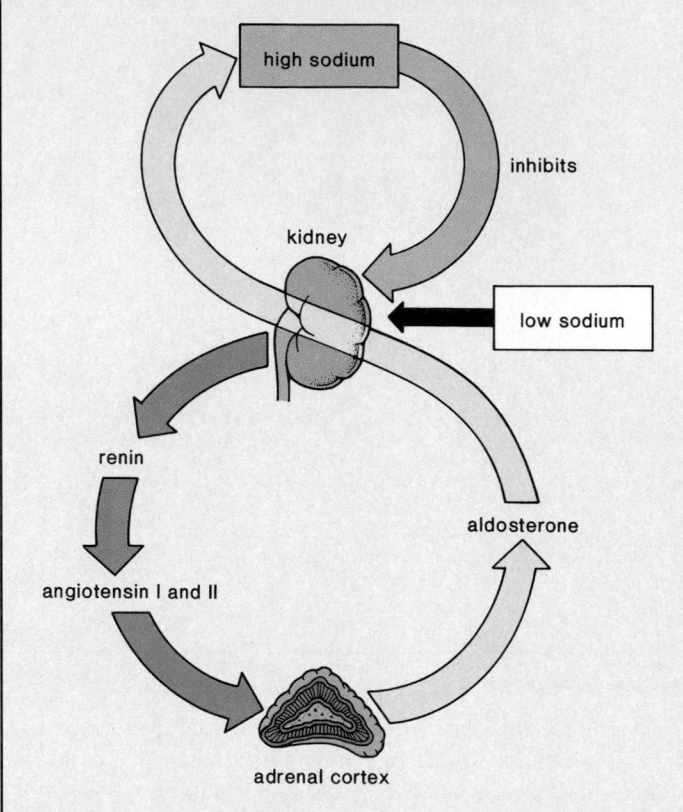

Figure 42.10
Renin-angiotensin-aldosterone system. If the blood level of sodium is low, the kidneys secrete renin. The increased renin acts via the increased production of angiotensin I and II to stimulate aldosterone secretion. Aldosterone promotes reabsorption of sodium by the kidneys; when the sodium level in the blood rises, the kidneys stop secreting renin.

pancreatic duct to the duodenum; the other type, called the *islets of Langerhans* (pancreatic islets), produces and secretes the hormones **insulin** and **glucagon** directly into the blood.

Insulin is secreted when there is a high level of glucose in the blood, which usually occurs just after eating. Insulin has 3 different actions: (1) it stimulates all cells, and in particular fat, liver, and muscle cells to take up and metabolize glucose; (2) it stimulates the liver and muscles to store glucose as glycogen; and (3) it promotes the buildup of fats and proteins and inhibits their use as an energy source. Therefore, these nutrients will be on hand during leaner times. As a result of its activities, insulin lowers the blood glucose level. Glucagon is secreted from the pancreas in between eating, and its effects are opposite to those of insulin; glucagon stimulates the breakdown of stored nutrients and causes the blood glucose level to rise (fig. 42.12).

Diabetes Mellitus

Diabetes mellitus is a fairly common disease caused by a defect in insulin production or utilization. The most common laboratory test for diabetes mellitus (usually called diabetes) is to look for sugar, or glucose, in the urine. Glucose in the urine means

Figure 42.11

The pancreas is both an exocrine and an endocrine gland. It sends digestive juices to the duodenum by way of the pancreatic duct and the hormones insulin and glucagon into the bloodstream. In 1920, Frederick Banting, a physician who occasionally gave physiology lectures, decided to try to isolate insulin. Investigators before him had been unable to do this because the enzymes in the digestive juices destroyed insulin (a protein) during the isolation procedure. He hit upon the idea of tying off the pancreatic duct, which he knew from previous research would lead to the degeneration only of the cells that produce digestive juices and not of the islets of Langerhans, where insulin is made. J. J. Macleod made a laboratory available to him at the University of Toronto and also assigned a graduate student named Charles Best to assist him. Banting and Best had limited funds and spent that summer working, sleeping, and eating in the lab. By the end of the summer, they had obtained pancreatic extracts that did lower the blood glucose level in diabetic dogs. Macleod then brought in biochemists, who purified the extract. Insulin therapy for the first human patient was begun in January 1922 and large-scale production of purified insulin from pigs and cattle followed. Banting and Macleod received a Nobel Prize for their work in 1923. The amino acid sequence of insulin was determined in 1953. Insulin is presently synthesized using recombinant DNA technology (chapter 18). Banting and Best followed the required steps given in the chart to identify a chemical messenger.

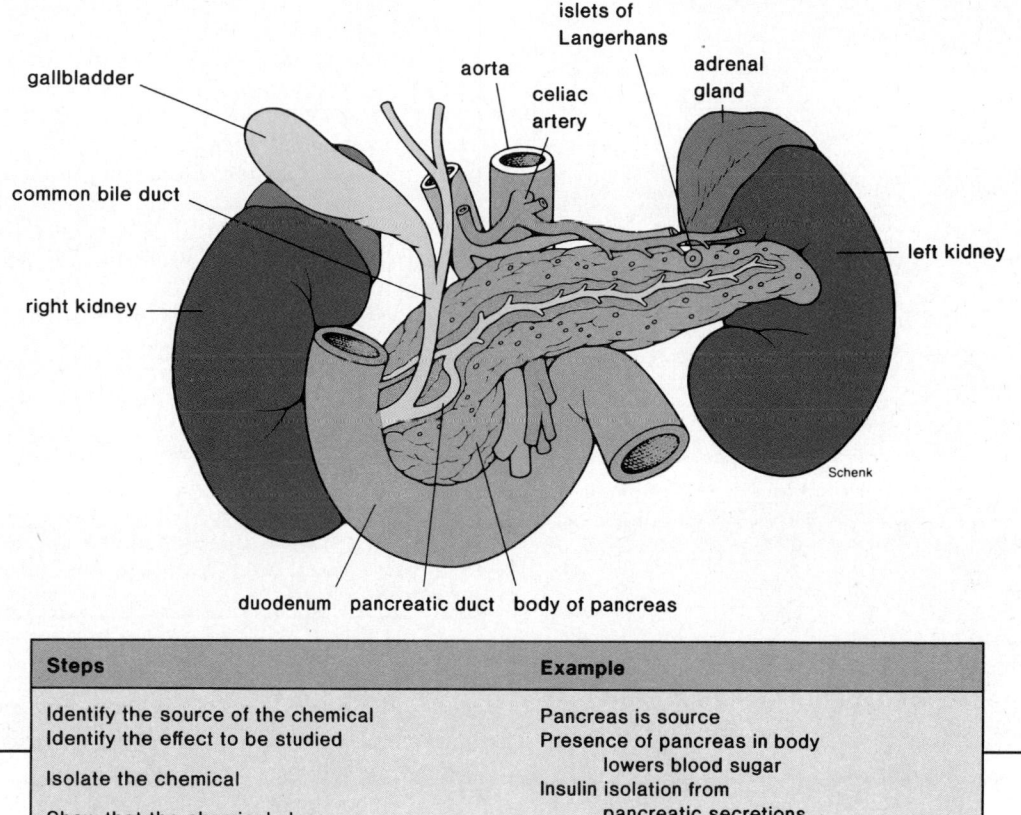

Steps	Example
Identify the source of the chemical	Pancreas is source
Identify the effect to be studied	Presence of pancreas in body lowers blood sugar
Isolate the chemical	Insulin isolation from pancreatic secretions
Show that the chemical alone has the effect	Insulin alone lowers blood sugar

that the blood glucose level is high enough to cause the kidneys to excrete glucose. This indicates that the liver is not storing glucose as glycogen and that cells are using proteins and fats as an energy source due to insufficient production or use of insulin. An individual with untreated diabetes is also extremely thirsty because the solute content of the blood is higher than normal. The breakdown of proteins and fats also leads to rapid weight loss, to the buildup of acids in the blood (acidosis), and to respiratory distress. It is acidosis that can eventually cause coma and death if the diabetic does not receive treatment.

There are 2 forms of diabetes. In Type I diabetes, the pancreas does not produce insulin. This type of diabetes frequently follows a viral infection. Apparently, the infection sets off an autoimmune reaction that destroys the insulin-secreting cells in the islets of Langerhans. Patients with Type I diabetes must either have daily insulin injections or undergo a pancreas transplant. In Type II diabetes, the most common type, the pancreas is producing insulin, but the cells of the body do not respond to it. At first, the cells lack the receptors necessary to detect the presence of insulin, and then later they become incapable of taking up glucose. The best treatment for this type of diabetes is a low-fat and low-sugar diet and regular exercise. If this treatment fails, oral drugs that make the cells more sensitive to the effects of insulin or stimulate the pancreas to make more of it can be prescribed.

Figure 42.12

Insulin and glucagon have contrary effects. When blood glucose level is high, the pancreas secretes insulin. Insulin promotes the storage of glucose as glycogen, the use of glucose as an energy source, and the synthesis of proteins and fats as opposed to their use as energy sources. Therefore, insulin lowers the blood glucose level. When the blood glucose level is low, the pancreas secretes glucagon. Glucagon acts in opposition to insulin in all respects; therefore, glucagon raises the blood glucose level.

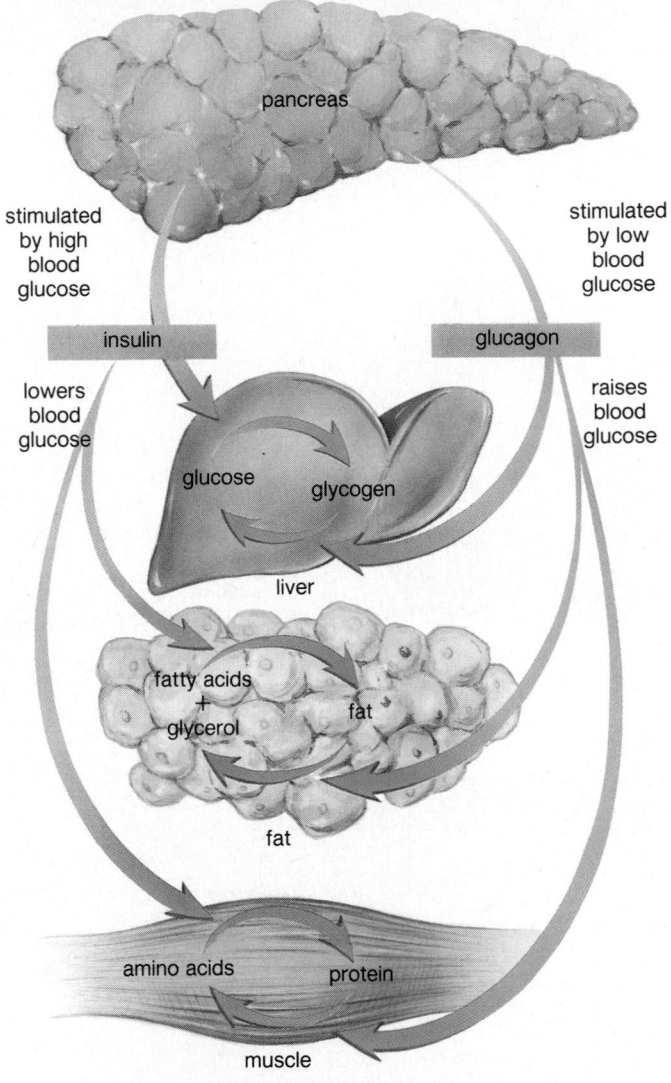

The pancreas secretes insulin, which acts to promote storage of nutrients, thereby lowering the blood glucose level. In diabetes mellitus, either insulin is not produced or it cannot be utilized; the excess glucose in the blood is excreted by the kidneys.

Other Endocrine Glands

Gonads

The gonads are the endocrine glands that produce the hormones that determine sexual characteristics. As we will discuss in detail in the following chapter, the *testes* produce the male sex hormones, the androgens—the most important of which is testosterone—and

Anabolic Steroid Use

Anabolic steroids are essentially the male sex hormone testosterone and its synthetic derivatives. The latter were developed in the 1930s to prevent the atrophy of muscle tissue in patients with debilitating illnesses. In some cases, they are also given to burn victims and surgical patients to speed recovery. Steroids were first used for nonmedical reasons in World War II, when German doctors gave them to soldiers, perhaps to make them more aggressive. Following the war, the Soviet Union began dispensing steroids to athletes, and when the results were observed by a U.S. physician, he recommended that the U.S. weight-lifting team use them. In the 1960s, when their dangerous side effects became apparent, steroid therapy fell into disfavor, and in 1973 the Olympic Committee added anabolic steroids to its list of banned substances (fig. 42.A). Although the FDA (Food and Drug Administration) has also banned most steroids except for limited medical use, they are not *controlled* substances like cocaine and heroin. Presently, they are smuggled into the United States from Mexico and to a lesser degree, Eastern Europe; this illicit trafficking has become a $100-million-a-year business. Congress has passed an antidrug bill that makes the selling of steroids a felony offense with jail terms of 3 years for dispensing to adults and 6 years for dispensing to children.

There is great concern because these drugs, originally taken primarily by professional athletes, are now being more widely used, even by children, to increase muscle size. The latest estimate is that 1–3 million Americans use steroids, a figure that has increased steadily since the early 1970s. No one has really conclusively determined the long-term harmful effects of steroid use because researchers are reluctant to do such studies on humans. The accompanying figure indicates the major side effects that have been observed in users. Often these side effects can be understood in view of the fact that testosterone

the *ovaries* produce estrogen and progesterone, the female sex hormones. The secretion of these hormones is under the control of the anterior pituitary.

The sex hormones bring about the secondary sex characteristics of males and females. Among other traits, males have greater muscle strength than do females. As discussed in this chapter's reading, athletes and others sometimes take so-called *anabolic steroids*, which are synthetic steroids that mimic the action of testosterone, to improve their strength and physique. Unfortunately, this practice is accompanied by harmful side effects.

Thymus Gland

The *thymus* gland extends from below the thyroid gland in the neck into the thoracic cavity. This organ grows during childhood but gradually decreases in size after puberty. Lymphocytes that have passed through the thymus are transformed into T cells. Colonies of T cells in the lymph nodes and other organs are apparently able to produce new T cells when stimulated by various thymus hormones called *thymosins*. There is hope that these hormones can be used in conjunction with lymphokine therapy to restore or stimulate T cell function in patients suffering from AIDS or cancer.

Animal Structure and Function

is the male sex hormone that maintains the secondary sex characteristics of males, such as the development of the testes and penis, a deep voice, and facial hair. Anabolic steroid use can also apparently cause death. Steroids have been implicated in the deaths of young athletes from liver cancer and a type of kidney tumor. Furthermore, because steroids cause the body to retain fluid, users often take diuretics; excessive amounts of these can rob the body of its proper electrolyte balance, resulting in heart attack.

It has come to light that steroids may be addictive and might cause psychological problems. Psychotic side effects, sometimes called "roid mania," have been observed in abusers who experience wild aggression and delusions. For example, it has been reported that one user had a friend videotape him while he deliberately drove a car into a tree at 35 mph. Psychiatrists feel that the magnitude of the mental effects of steroids is only now being realized.

Figure 42.A
The effects of anabolic steroid use.

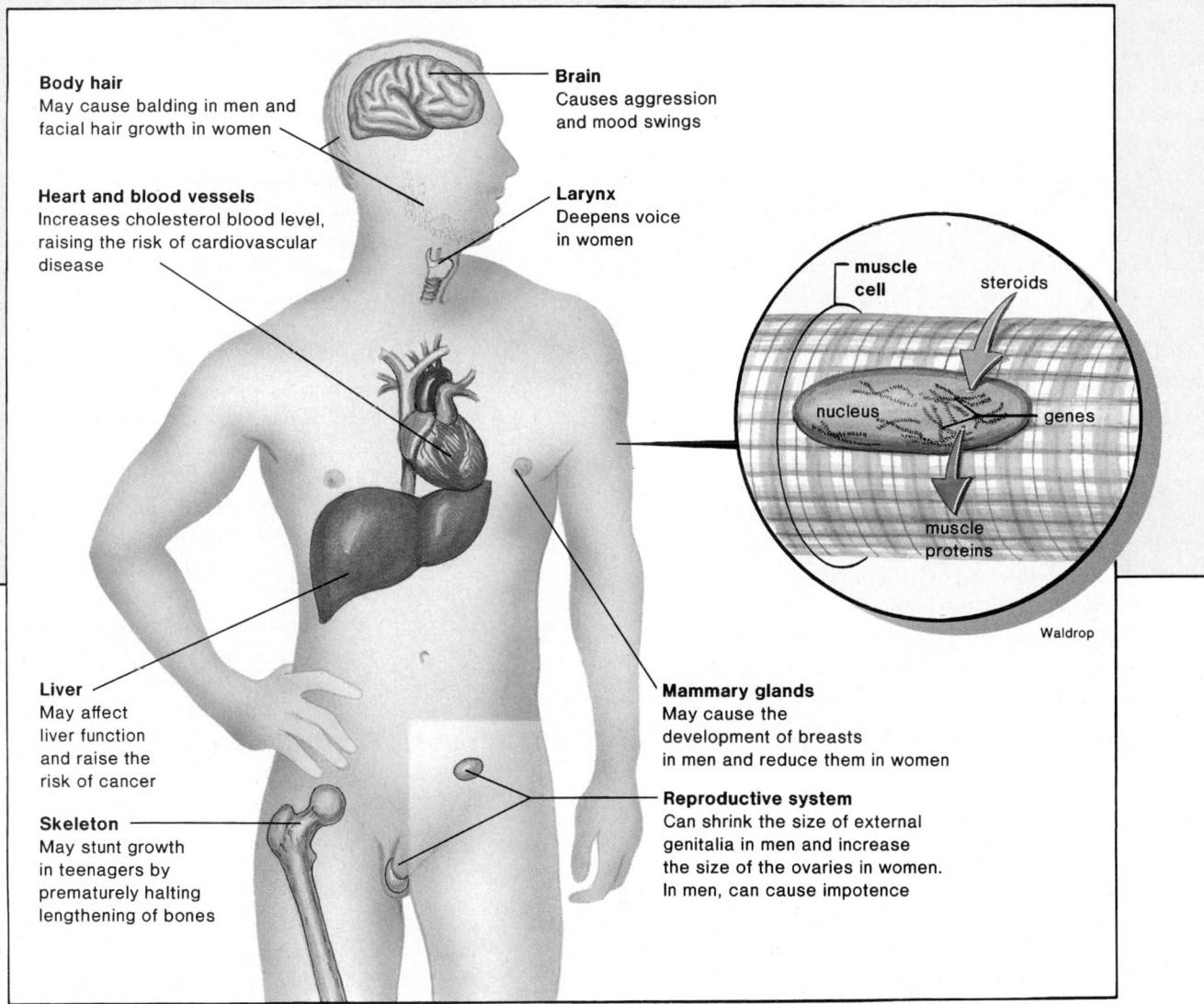

Body hair
May cause balding in men and facial hair growth in women

Brain
Causes aggression and mood swings

Heart and blood vessels
Increases cholesterol blood level, raising the risk of cardiovascular disease

Larynx
Deepens voice in women

muscle cell

steroids

nucleus

genes

muscle proteins

Waldrop

Liver
May affect liver function and raise the risk of cancer

Skeleton
May stunt growth in teenagers by prematurely halting lengthening of bones

Mammary glands
May cause the development of breasts in men and reduce them in women

Reproductive system
Can shrink the size of external genitalia in men and increase the size of the ovaries in women. In men, can cause impotence

Pineal Gland

The *pineal gland* in amphibians and fishes is located near the skin surface, where it functions as a "third eye," transforming light signals into nerve impulses sent to the brain. In birds and mammals, the pineal gland is located in the brain; in birds it still receives light signals directly, but in mammals the eyes communicate with the pineal gland. In any case, at night when light is absent, the pineal gland produces melatonin, a hormone that is involved in both circadian and circannual rhythms. For example, in amphibians it causes a blanching of the skin and in birds it causes roosting, which are nighttime events. Daily cycles such as these are called circadian rhythms.

The reproductive cycles of most temperate zone animals are related to the seasons. Melatonin causes the reproductive organs to shrink in size. Apparently the reduced production of melatonin during the shorter nights of summer promotes enlargement of reproductive organs. Therefore, mating occurs in the fall and young are born in the spring when food is available to nourish them. Yearly rhythms such as these are called circannual rhythms.

The function of melatonin in humans is not known; however, children with brain tumors that destroy the pineal gland experience early puberty. Therefore, it's possible that melatonin is also involved in human sexual development.

Still Other Glands

Organs traditionally not considered to be endocrine glands have been found to secrete hormones into the bloodstream. For example, as mentioned on page 630, the heart produces atrial natriuretic hormone, a peptide that helps to regulate the water and salt balance in kidneys. Also, in chapter 36, we discussed the hormones produced by the stomach and the small intestine.

A number of different types of organs and cells have been discovered to produce peptide *growth factors* that stimulate cell division and mitosis. They are like hormones in that they act on cell types that have specific receptors to receive them. Some, including lymphokines (p. 579) and blood cell growth factors (p. 563), are released into the blood; others diffuse to nearby neighboring cells, and still others are self-stimulating. The latter are of interest because they play a role in the formation of cancer, as discussed on page 271. Other growth factors are as follows:

Platelet-derived growth factor is released from platelets and also many other cell types. It helps in wound healing and causes an increase in the number of fibroblasts, smooth muscle cells, and certain cells of the nervous system.

Epidermal growth factor and *nerve growth factor* stimulate the cells indicated by their names and many others also. These growth factors are also important in wound healing.

Tumor angiogenesis factor stimulates the formation of capillary networks and is released by tumor cells. One treatment for cancer is to prevent the activity of this growth factor.

Growth factors that simply diffuse to nearby cells and act locally do not fit the standard definition of a hormone. **Prostaglandins (PG)** are another class of chemical messengers that also are produced and act locally. They are derived from fatty acids stored in plasma membranes as phospholipids. When a cell is stimulated by reception of a hormone or even by trauma, a series of synthetic reactions takes place in the plasma membrane, and PG first is released into the cytoplasm and then secreted from the cell. There are many different types of prostaglandins produced by many different tissues. In the uterus, certain prostaglandins cause muscles to contract; therefore, they are implicated in the pain and discomfort of menstruation in some women. (Antiprostaglandin therapy is useful in these cases.) On the other hand, certain prostaglandins are used to treat ulcers because they reduce gastric secretion, to treat hypertension because they lower blood pressure, and to prevent thrombosis because they inhibit platelet aggregation. Because the different prostaglandins can have contrary effects, however, it has been very difficult to standardize their use, and in most instances, prostaglandin therapy still is considered experimental.

Environmental Signals

In this chapter, we concentrated on describing the functions of the human endocrine glands and their hormonal secretions. We already know that hormones are only one type of chemical messenger or environmental signal between cells. In fact, the concept of the environmental signal now has been broadened to include at least the following 3 different categories of messengers (fig. 42.13):

1. *Environmental signals that act at a distance between individuals.* Many organisms release chemical messengers, called *pheromones,* into the air or in externally deposited body fluids. These are intended to be messages for other members of the species. For example, ants lay down a pheromone trail to direct other ants to food, and the female silkworm moth releases bombykol, a sex attractant that is received by male moth antennae even several miles away. This chemical is so potent that it has been estimated that only 40 out of 40,000 receptors on the male antennae need to be activated in order for the male to respond. Mammals, too, release pheromones; the urine of dogs serves as a territorial marker, for example. Studies are being conducted to determine if humans also have pheromones.

2. *Environmental signals that act at a distance between body parts.* This category includes the endocrine secretions, which traditionally have been called hormones. It also includes the secretions of the neurosecretory cells in the hypothalamus—the production and action of ADH and oxytocin illustrate the close relationship between the nervous system and the endocrine system. Neurosecretory cells produce these hormones, which are released when these cells receive nerve impulses. As another example of the overlap between the nervous and endocrine systems, consider that endorphins on occasion travel in the bloodstream, but they act on nerve cells to alter their membrane potential. Also, norepinephrine is both a neurotransmitter and hormone secreted by the adrenal medulla.

3. *Environmental signals that act locally between adjacent cells.* Neurotransmitter substances belong in this category, as do substances like prostaglandins and growth factors, which are sometimes called local hormones. Also, when the skin is cut, histamine is released by mast cells and promotes the inflammatory response.

Redefinition of a Hormone

Traditionally, a hormone was considered to be a secretion of an endocrine gland that was carried in the bloodstream to a target organ. In recent years, some scientists have broadened the definition of a hormone to include *all* types of chemical messengers. This change seemed necessary because those chemicals traditionally considered to be hormones now have been found in all sorts of tissues in the human body. For example, it is impossible for insulin produced by the pancreas to enter the brain because of the blood-brain barrier—a tight fusion of endothelial cells of the capillary walls that prevents passage of larger molecules like peptides. Yet, insulin has been found in the brain. It now appears that the brain cells themselves can produce insulin, which is used locally to influence the metabolism of adjacent cells. Also, some chemicals identical to the hormones of the endocrine system have been found in lower organisms, even in bacteria! A moment's thought about the evolutionary process helps to explain this; these regulatory chemicals may have been present in the earliest cells and only became specialized as hormones as evolution proceeded.

Figure 42.13

The 3 categories of environmental signals. Pheromones are chemical messengers that act at a distance between individuals. Endocrine hormones and neurosecretions are typically carried in the bloodstream and act at a distance within the body of a single organism. Some chemical messengers have local effects only; they pass between cells that are adjacent to one another. This, of course, includes neurotransmitters.

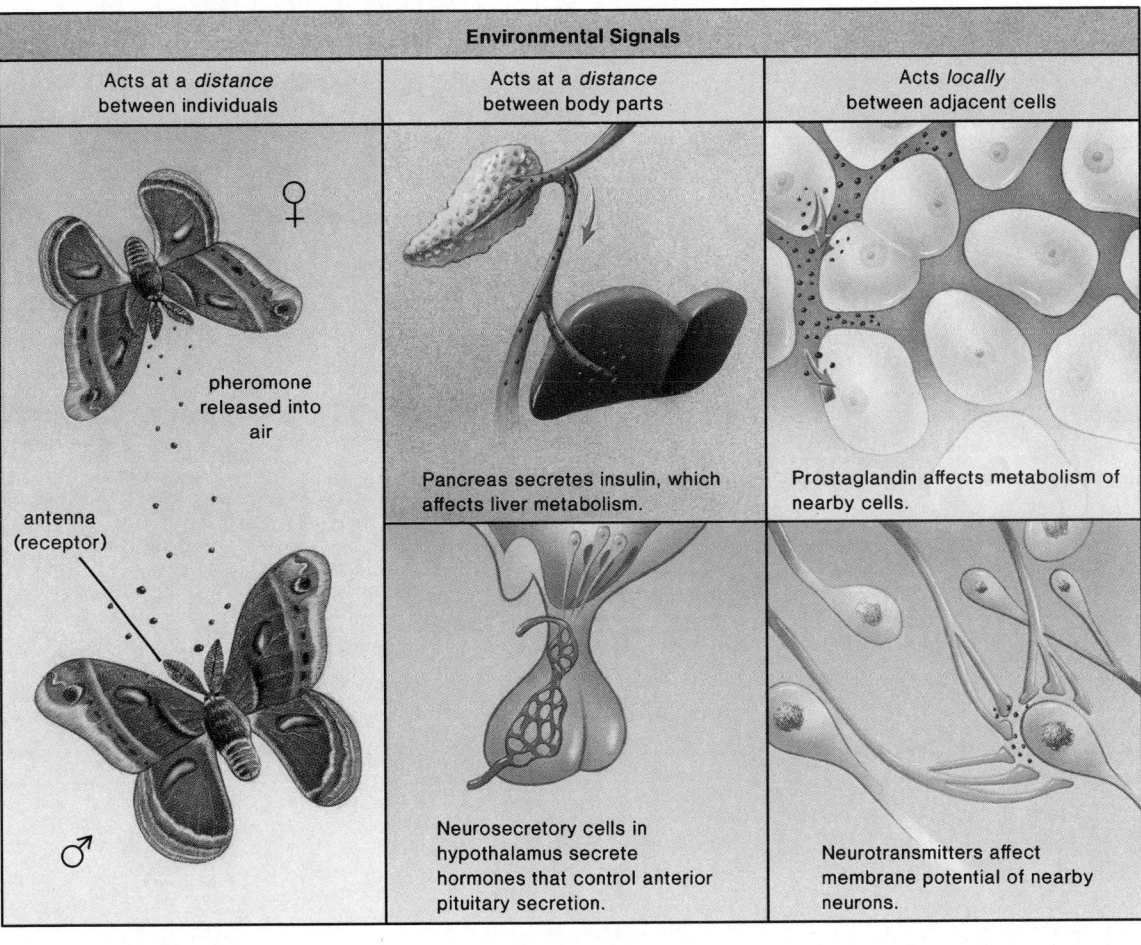

Environmental Signals

Acts at a *distance* between individuals	Acts at a *distance* between body parts	Acts *locally* between adjacent cells

♀

pheromone released into air

antenna (receptor)

♂

Pancreas secretes insulin, which affects liver metabolism.

Neurosecretory cells in hypothalamus secrete hormones that control anterior pituitary secretion.

Prostaglandin affects metabolism of nearby cells.

Neurotransmitters affect membrane potential of nearby neurons.

Summary

1. The endocrine glands in humans, which are shown in figure 42.2, produce hormones that are secreted into the bloodstream and circulate about the body until they are received by their target organs.
2. There are 2 major types of compounds used as hormones; most hormones are classified as peptides; a few hormones are steroids.
3. The peptide protein hormones are usually received by a receptor located in the plasma membrane. Most often their reception activates an enzyme that changes ATP to cyclic AMP. Cyclic AMP then activates another enzyme that activates another and so forth. There is an amplification of effect at each step of this enzyme cascade.
4. The steroid hormones are lipid soluble and can pass through plasma membranes. Once inside the cytoplasm, they combine with a receptor

molecule, and the complex moves into the nucleus, where it combines with chromatin. Transcription and translation lead to protein synthesis; in this way, steroid hormones alter the m abolism of the cell.

5. Neurosecretory cells in the hypothalamus produce ADH and oxytocin, which are stored in axon endings in the posterior pituitary until they are released. ADH secretion is controlled by a negative feedback mechanism in which concentrated blood stimulates ADH release and dilute blood inhibits its release. ADH causes the collecting ducts of the kidney to reabsorb water. Oxytocin causes the uterus to contract and milk to be released from the mammary glands.

6. The hypothalamus produces hypothalamic-releasing and hypothalamic-release-inhibiting hormones, which pass to the anterior pituitary by way of a portal system and either stimulate or inhibit the release of a particular anterior pituitary hormone.

7. The anterior pituitary produces at least 6 types of hormones (table 42.1). GH promotes body growth; Pr promotes breast development and production of milk following pregnancy; MSH causes skin-color changes in lower vertebrates; TSH stimulates the thyroid; ACTH stimulates the adrenal cortex to produce corticoids; gonadotropic hormones (FSH and LH) stimulate the gonads. Because the anterior pituitary also stimulates certain other hormonal glands, it is sometimes called the master gland.

8. The thyroid gland produces thyroxin and triiodothyronine, hormones that contain iodine. These hormones play a role in growth and development of immature forms; in mature individuals they increase the metabolic rate. A goiter may develop in individuals who have inadequate iodine in the diet. The thyroid gland also produces calcitonin, which helps lower the blood calcium level.

9. The parathyroid glands raise the level of calcium in the blood by (1) activating vitamin D, which is involved in absorption of calcium from the gut, (2) stimulating retention of calcium by the kidneys, and (3) stimulating calcium removal from the bones. PTH also decreases the blood phosphate level.

10. The adrenal medulla secretes epinephrine and norepinephrine, which bring about responses we associate with the fight or flight reaction.

11. The adrenal cortex produces the glucocorticoids (cortisol) and the mineralocorticoids (aldosterone). Cortisol stimulates hydrolysis of proteins to amino acids that are converted to glucose; in this way, it raises the blood glucose level. It also counteracts the inflammatory reaction. Aldosterone causes the kidneys to reabsorb sodium: when the blood sodium level is low, the kidneys secrete renin, which converts angiotensinogen to angiotensin, a molecule that stimulates the release of aldosterone. The main effect of this renin-angiotensin-aldosterone system is to raise the blood pressure.

12. The pancreas secretes insulin, which lowers the blood glucose level, and glucagon, which has opposite effects to insulin. In Type I diabetes mellitus, insulin-secreting cells in the islets of Langerhans have been destroyed; in Type II diabetes, the cells lack receptors for insulin or do not respond to it. Only Type I diabetes requires daily injections of insulin, while Type II requires a special diet.

13. The gonads produce the sex hormones, which are discussed in chapter 43; the thymus secretes thymosins, which stimulate T lymphocyte production and maturation; the pineal gland produces melatonin, whose function in mammals is uncertain—it may affect development of the reproductive organs.

14. There are 3 categories of chemical messengers: those that act at a distance between individuals (pheromones); those that act at a distance within the individual (traditional endocrine hormones and secretions of neurosecretory cells); and local messengers (such as prostaglandins, growth factors, and neurotransmitters). Since there is great overlap between these categories, perhaps the definition of a hormone should now be expanded to include all of them.

Writing Across the Curriculum

In order to practice writing skills, students should write out the answers to any or all of the study questions and the critical thinking questions. The study questions are sequenced in the same order as the text. Suggested answers to the critical thinking questions are in appendix D.

Study Questions

1. Give a definition of endocrine hormones that includes their most likely source, how they are transported in the body, and how they are received. What does "target" organ mean?
2. Categorize endocrine hormones according to their chemical makeup.
3. Tell how the 2 major types of hormones influence the metabolism of the cell.
4. Give the location in the human body of all the major endocrine glands. Name the hormones secreted by each gland, and describe their chief functions.
5. Explain the relationship of the hypothalamus to the posterior pituitary and to the anterior pituitary.
6. Explain the concept of negative feedback and give an example involving ADH.
7. Give an example of the 3-tiered relationship among the hypothalamus, the anterior pituitary, and other endocrine glands. Explain why the anterior pituitary can be called the master gland.
8. Draw a diagram to explain the contrary actions of insulin and glucagon. Use your diagram to explain the symptoms of Type I diabetes mellitus.
9. Categorize chemical messengers into 3 groups, and give examples of each group.
10. Give examples to show that there is an overlap between the mode of operation of the nervous system and that of the endocrine system. Explain why the traditional definition of a hormone may need to be expanded.

Animal Structure and Function

Objective Questions

Match the hormone in numbers 1–5 to the correct gland in the key.

Key:

a. pancreas
b. anterior pituitary
c. posterior pituitary
d. thyroid
e. adrenal medulla
f. adrenal cortex
1. cortisol
2. GH
3. oxytocin storage
4. insulin
5. epinephrine
6. The anterior pituitary controls the secretion(s) of
   a. both the adrenal medulla and the adrenal cortex.
   b. both cortisol and aldosterone.
   c. thyroxin.
   d. All of these.
7. Peptide hormones
   a. are received by a receptor located in the plasma membrane.
   b. are received by a receptor located in the cytoplasm.
   c. bring about the transcription of DNA.
   d. Both b and c.
8. Aldosterone causes the
   a. kidneys to release renin.
   b. kidneys to reabsorb sodium.
   c. blood volume to increase.
   d. All of these.

9. Diabetes mellitus is associated with
   a. too much insulin in the blood.
   b. too high a level of glucose in the blood.
   c. blood that is too dilute.
   d. All of these.
10. The level of cortisol in the blood controls the secretion of
    a. a releasing hormone from the hypothalamus.
    b. ACTH from the anterior pituitary.
    c. cortisol from the adrenal cortex.
    d. All of these.
11. It is now clear that
    a. there is a discrete distinction between the activities of the nervous and endocrine systems.

b. both the nervous and endocrine systems coordinate the activities of body parts.
c. the nervous system controls the secretions of all endocrine glands.
d. All of these.
12. One of the chief differences between pheromones and local hormones is
    a. one is a chemical messenger and the other is not.
    b. the distance over which they act.
    c. one is made by invertebrates and the other is made by vertebrates.
    d. All of these.
13. Label this diagram and explain how a negative feedback system keeps the hormone level constant in the body:

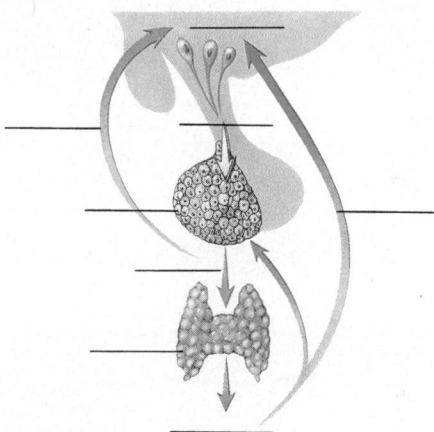

Concepts and Critical Thinking

1. *Hormone levels are maintained by feedback control.*

   Contrast control of neurotransmitter level in the nervous system with control of hormone level in the endocrine system.

2. *The nervous system is fast acting, and the endocrine system is fairly slow moving.*

   Contrast the manner in which the nervous system delivers its message with the way the endocrine system delivers a message.

3. *Hormone levels greatly affect the phenotype.*

   Use the effect of sex hormones to substantiate this concept.

Selected Key Terms

pituitary gland (pi-tu′ĭ-tār″e gland) 681
thyroid gland (thi′roid gland) 683
goiter (goi′ter) 683

parathyroid gland (par″ah-thi′roid gland) 685
adrenal gland (ah-dre′nal gland) 685

pancreas (pan′kre-as) 686
diabetes mellitus (di″ah-be′tēz me-lī′tus) 686

43

Reproduction in Animals

Many animals sexually reproduce and invest much energy in the rearing of offspring. Newly hatched birds usually have to be fed before they are able to fly away and seek food for themselves. Complex hormonal and neural regulation are involved in reproductive behavior of parental birds.

Your study of this chapter will be complete when you can

1. contrast asexual reproduction with sexual reproduction, and relate their advantages to environmental conditions;
2. explain why aquatic animals are apt to practice external fertilization, whereas land animals are apt to practice internal fertilization;
3. name and give a function for the organs of the human male reproductive system; trace the path of sperm from the testes to the penis;
4. describe the microscopic anatomy of the testes, including the stages of spermatogenesis; relate the action of the hormones FSH (ICSH) and LH to the functions of the testes;
5. describe hormonal control in the male, mentioning the hypothalamus, anterior pituitary, and testosterone; give several actions of testosterone;
6. name and give a function for the organs of the human female reproductive system;
7. describe the microscopic anatomy of the ovary, including a description of the ovarian cycle;
8. describe the uterine cycle; explain the relationship among the hormones GnRH, FSH, LH, and the female sex hormones; give several functions of estrogen and progesterone;
9. explain the cessation of the ovarian and uterine cycles in the pregnant female;
10. list several means of birth control, and compare their effectiveness in preventing pregnancy;
11. describe the cause and symptoms of the following sexually transmitted diseases: AIDS, genital herpes, genital warts, gonorrhea, chlamydia, and syphilis.

 nimals expend a considerable amount of energy on reproduction. After all, reproduction ensures that the animals' genes are passed on to the next generation. The life cycle of any particular animal comes to an end, but its genes can be perpetuated as long as reproduction has taken place. In evolutionary terms, the most fit organisms are the ones that have produced the most offspring. They are the best adapted to the environment, and it is beneficial to the species that they reproduce more than other members of their cohort.

Patterns of Reproduction

There are 2 fundamental patterns of reproduction: asexual and sexual. We will examine the methodology and particular advantage of each pattern.

Asexual Reproduction

In asexual reproduction, there is only one parent, and the offspring have the same phenotype and genotype as that parent (fig. 43.1). This lack of variation is not a disadvantage as long as the environment stays the same. Organisms that reproduce asexually often have the tremendous advantage of being able to produce a large number of offspring within a limited amount of time.

Only certain methods of asexual reproduction are found among animals. Many flatworms can constrict into 2 halves, each of which can become a new individual. This form of **regeneration** is also seen among sponges and echinoderms. Chopping up a sea star does not kill it; instead, each fragment grows into another animal, and the total number of sea stars increases in the process. Some cnidarians, such as hydras, reproduce by **budding** (see fig. 26.5), during which the new individual arises as an outgrowth (bud) of the parent. Some insects and several other types of arthropods have the ability to reproduce parthenogenetically (fig. 43.1b). **Parthenogenesis** is a modification of sexual reproduction in which an unfertilized egg develops into a complete individual.

Sexual Reproduction

Sexual reproduction involves gametes (sex cells) (fig. 43.1). The gametes may be specialized into eggs or sperm, which are produced by the same or separate individuals. Earthworms (see fig. 27.8) practice cross-fertilization, even though they are *hermaphroditic*—they have both male and female sex organs. Among vertebrates, the sexes tend to be separate, and it is often easy to tell whether an animal is an egg-producing female or a sperm-producing male (fig. 43.2a). When animals reproduce sexually, an offspring inherits half its genes from one parent and the other half from the other parent. Therefore, an offspring often has a different combination of genes than either parent. In this way, variation may be introduced and maintained. Such variation is an advantage to the species if the environment is changing, because an offspring might be better adapted to the new environment than is either parent.

Figure 43.1

Asexual reproduction versus sexual reproduction. **a.** In asexual reproduction (*left*), there is only one parent and the offspring tend to be identical to each other and to the parent. Asexual reproduction occurs in various ways: by regeneration, by budding, or by parthenogenesis. In sexual reproduction (*right*), there are most often 2 parents. Each parent contributes one-half of the genes to the offspring either by way of the sperm or the egg. Therefore, the offspring tend to be genetically and phenotypically different from either parent. **b.** Stem covered with aphids. In the summer, many generations of aphid females produce up to 100 young, without prior fertilization. In the autumn, however, aphids reproduce sexually and the zygotes overwinter.

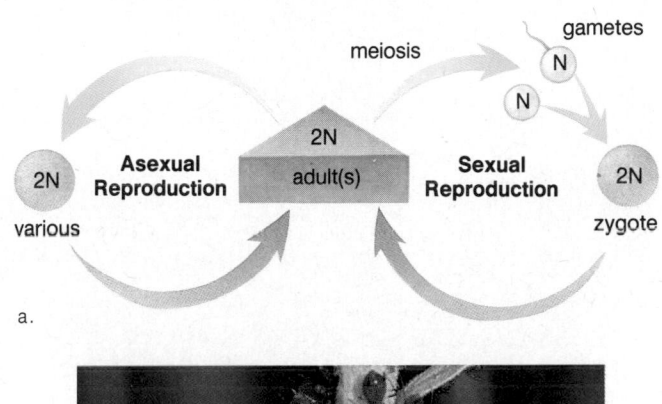

a.

b.

Reproductive Timing

The majority of many sexually reproducing animals has a single breeding period each year. Mating is timed so that the young are born when conditions are favorable for their growth. Many temperate zone mammals mate in the fall and deliver young in the spring, for example. Such reproductive cycles are tied to day-length changes. The length of the day reliably indicates the proper time for reproductive behavior, including migration to distant places. Re-

Figure 43.2

Sexual reproduction among vertebrates. **a.** During mating in terrestrial vertebrates, the male usually passes sperm to the female by use of a copulatory organ such as a penis. Note also that it is often possible to tell one sex from the other; in this case because the male lion has a mane. **b.** During mating in aquatic vertebrates, both male and female sexes often shed their gametes directly into the water. The watery environment protects the gametes and zygotes from drying out.

a.

b.

searchers have found that melatonin is produced by the pineal gland during the night, and that this hormone shrinks reproductive organs in seasonally breeding animals. The nights lengthen in the winter, and the increased production of melatonin causes reproductive behavior to cease.

External versus Internal Fertilization

Many aquatic animals practice external fertilization. They shed a large number of gametes into the water at the same time and place (fig. 43.2b). External fertilization is possible because the water protects the zygote and embryo from drying out. Often the embryo is a swimming larva capable of acquiring its own food.

Internal fertilization is a mechanism practiced by animals that lay a shelled egg or in which the embryo develops for a time within the female parent. Even some aquatic animals practice internal fertilization. In certain sharks, skates, and rays, the pelvic fins are specialized to pass sperm to the female, and in most of these animals, the young develop internally and are born alive. On land, internal fertilization is a necessity because sperm and eggs dry quickly when exposed to air. Often the male has a copulatory organ to transfer sperm to the female. Reptiles and birds lay shelled eggs (see fig. 28.10). In placental mammals, the fetus develops within the uterus, where it receives nourishment by way of the placenta. Humans can be used as an example of reproduction in placental mammals.

Some animals practice asexual reproduction, especially when environmental conditions are favorable. Most practice sexual reproduction, timed so that the young are born when food is plentiful. Aquatic animals are apt to practice external fertilization, and internal fertilization is common in land animals.

Human Reproduction

The manner in which humans reproduce is similar to that of other mammals and represents an adaptation to the land environment. When the male penis passes sperm to the female, desiccation of not only the male and female gametes but also the subsequent embryo is prevented.

Male Reproductive System

The male reproductive system includes the organs pictured in figure 43.3 and listed in table 43.1. The male gonads are paired testes, which are suspended within the *scrotal sacs* of the scrotum. The testes begin their development inside the abdominal cavity, but they descend into the scrotal sacs as development proceeds. If the testes do not descend, and the male is not treated by administration of hormones or operated on to place the testes in the scrotum, sterility (the inability to produce offspring) results. Sterility in this case results because normal sperm production cannot occur at body temperature; a cooler temperature is required.

Figure 43.3

Side view of the male reproductive system. Notice that the urethra in males carries either urine from the bladder or semen from the testes.

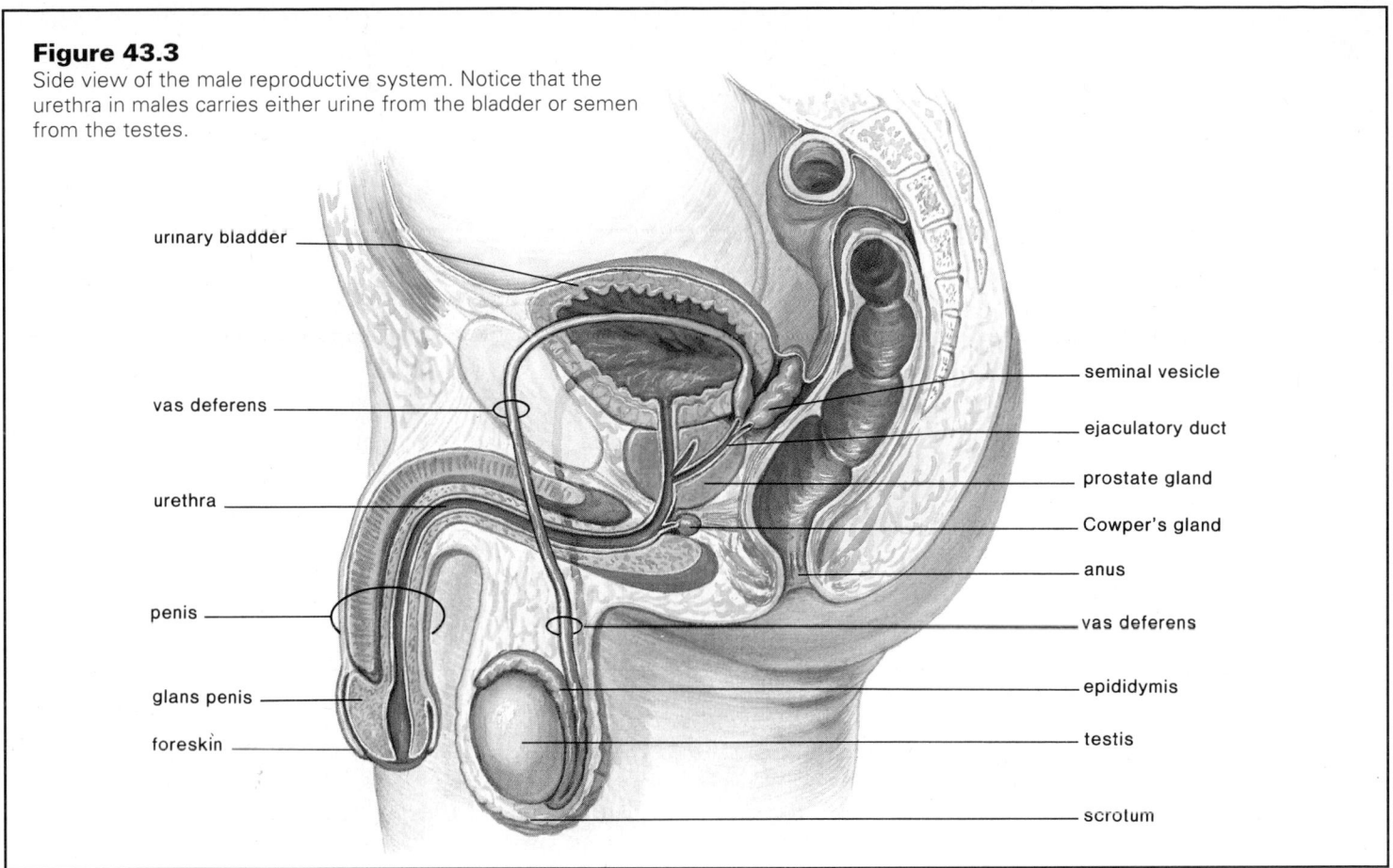

Labels (left side, top to bottom): urinary bladder, vas deferens, urethra, penis, glans penis, foreskin

Labels (right side, top to bottom): seminal vesicle, ejaculatory duct, prostate gland, Cowper's gland, anus, vas deferens, epididymis, testis, scrotum

Table 43.1
Male Reproductive System

Organ	Function
Testes	Produce sperm and sex hormones
Epididymides	Maturation and some storage of sperm
Vas deferentia	Conduct and store sperm
Seminal vesicles	Contribute to seminal fluid
Prostate gland	Contributes to seminal fluid
Urethra	Conducts sperm
Bulbourethral glands	Contribute to seminal fluid
Penis	Organ of copulation

Path of Sperm

Sperm produced by the testes mature within the *epididymides* (epididymis, sing.), which are tightly coiled tubules lying just outside the testes. Maturation seems to be required for the sperm to swim to the egg. Once the sperm have matured, they are propelled into the **vas deferentia** (vas deferens, sing.) by muscular contrac-

tions. Sperm are stored in both the epididymides and the vas deferentia. When a male becomes sexually aroused, sperm enter the urethra, part of which is located within the **penis.**

The penis is a cylindrical organ that usually hangs in front of the scrotum. Spongy, erectile tissue containing distensible blood spaces extends through the shaft of the penis (fig. 43.4). During sexual arousal, nervous reflexes cause an increase in arterial blood flow to the penis. This increased blood flow fills the blood space in the erectile tissue, and the penis, which is normally limp (flaccid), stiffens and increases in size. These changes are called *erection*. If the penis fails to become erect, the condition is called *impotency*.

Orgasm in both sexes is characterized by a release of neuromuscular tension, particularly in the genital area. In males, orgasm is obvious because rhythmic muscle contractions compress the urethra and expel semen. This is termed *ejaculation,* after which a male typically experiences a *refractory period* when stimulation does not bring about an erection.

Semen **Seminal fluid** (semen) is a thick, whitish fluid that contains sperm and fluids. As table 43.1 indicates, 3 types of glands contribute fluids to semen. The *seminal vesicles* lie between the bladder and the rectum. Each joins a vas deferens to form an ejaculatory duct that enters the urethra. As sperm pass from the vas deferentia, these vesicles secrete a thick, viscous fluid containing

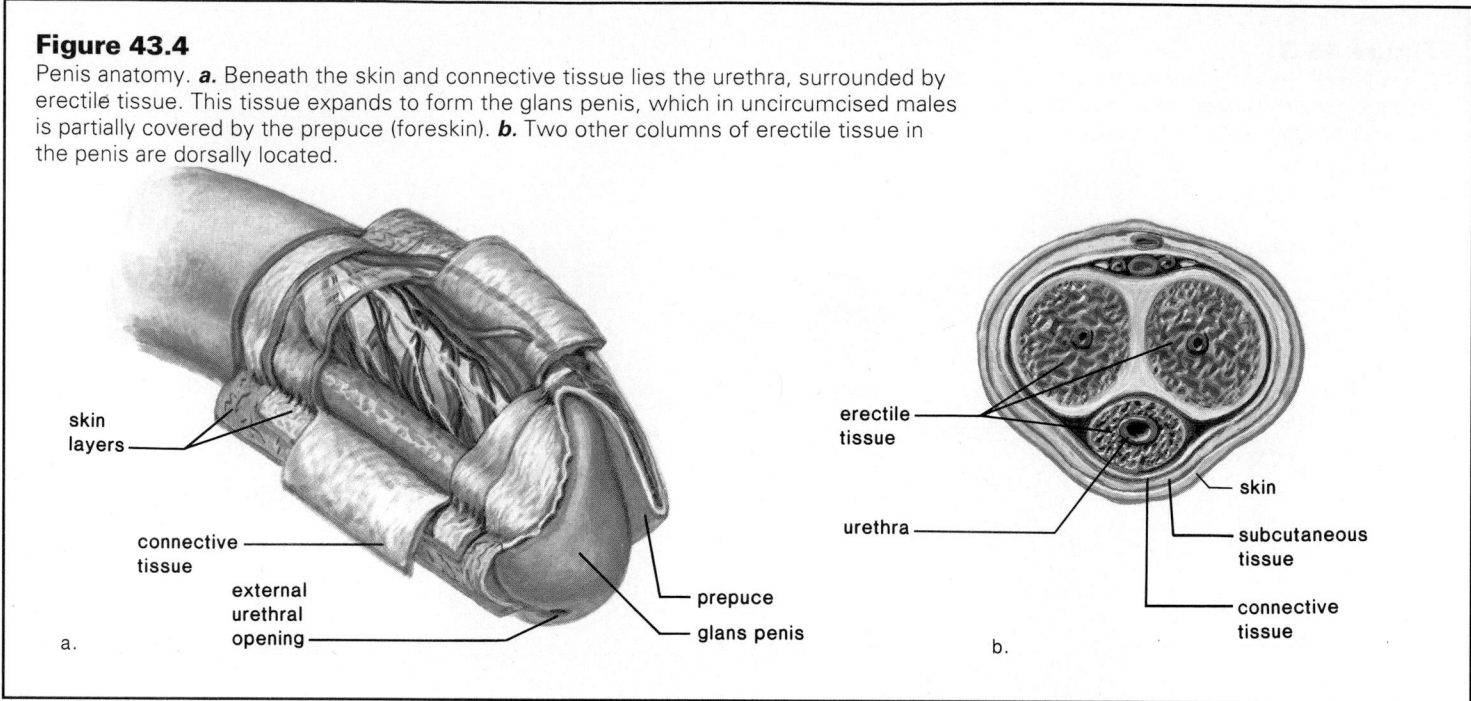

Figure 43.4

Penis anatomy. **a.** Beneath the skin and connective tissue lies the urethra, surrounded by erectile tissue. This tissue expands to form the glans penis, which in uncircumcised males is partially covered by the prepuce (foreskin). **b.** Two other columns of erectile tissue in the penis are dorsally located.

nutrients for possible use by the sperm. Just below the bladder is the *prostate gland,* which secretes a milky alkaline fluid believed to activate or increase the motility of the sperm. In older men, the prostate gland frequently becomes enlarged, thereby constricting the urethra and making urination difficult. Slightly below the prostate gland, on either side of the urethra, is a pair of small glands called *bulbourethral glands,* which have mucous secretions with a lubricating effect. Notice from figure 43.3 that at different times the urethra carries urine from the bladder and semen from the vas deferentia.

> Sperm produced by the testes mature in the epididymides and pass from the vas deferentia to the urethra, where certain glands add seminal fluid prior to ejaculation.

Sperm Production and Hormonal Regulation

A longitudinal section of a testis shows that each testis is composed of compartments called lobules, each of which contains one to 3 tightly coiled *seminiferous tubules* (fig. 43.5*a*). Altogether, these tubules have a combined length of approximately 250 m. A microscopic cross section of a seminiferous tubule shows that it is packed with cells undergoing *spermatogenesis* (fig. 43.5*b*), which involves the process of meiosis (see fig. 11.6). Also present are the Sertoli (or nurse) cells, which support, nourish, and regulate the spermatogenic cells (fig. 43.5*c*).

Mature **sperm,** or spermatozoans, have 3 distinct parts: a head, a middle piece, and a tail (fig. 43.5*d*). The middle piece and tail contain microtubules, in the characteristic 9 + 2 pattern of cilia and flagella. In the middle piece, mitochondria are wrapped around the microtubules and provide the energy for movement. The head

contains a nucleus covered by a cap called the *acrosome,* which stores enzymes needed for fertilization. The human egg is surrounded by several layers of cells and a thick membrane—these acrosome enzymes play a role in allowing a sperm to reach the surface of the egg. The normal human male usually produces several hundred million sperm per day, assuring an adequate number for fertilization to take place. Fewer than 100 ever reach the vicinity of the egg and only one sperm enters an egg.

Hormonal Regulation in Males The hypothalamus has ultimate control of the testes' sexual function because it secretes a releasing hormone (GnRH—gonadotropic-releasing hormone) that stimulates the anterior pituitary to produce the gonadotropic hormones. There are 2 gonadotropic hormones, FSH and LH, in both males and females. In males, FSH promotes spermatogenesis in the seminiferous tubules, which also release the hormone inhibin.

LH in males is sometimes given the name *interstitial cell-stimulating hormone (ICSH)* because it controls the production of testosterone by the interstitial cells, scattered in the spaces between the seminiferous tubules (fig. 43.5*b*). All these hormones are involved in a feedback relationship that maintains the fairly constant production of sperm and testosterone (fig. 43.6).

Male Sex Hormones Testosterone is the main sex hormone in males. It is essential for the normal development and functioning of the organs listed in table 43.1. Testosterone is also necessary for the maturation of sperm.

Testosterone also brings about and maintains the secondary sex characteristics in males that develop at the time of puberty. Testosterone causes growth of a beard, axillary (underarm) hair, and pubic hair. It prompts the larynx and vocal cords to enlarge,

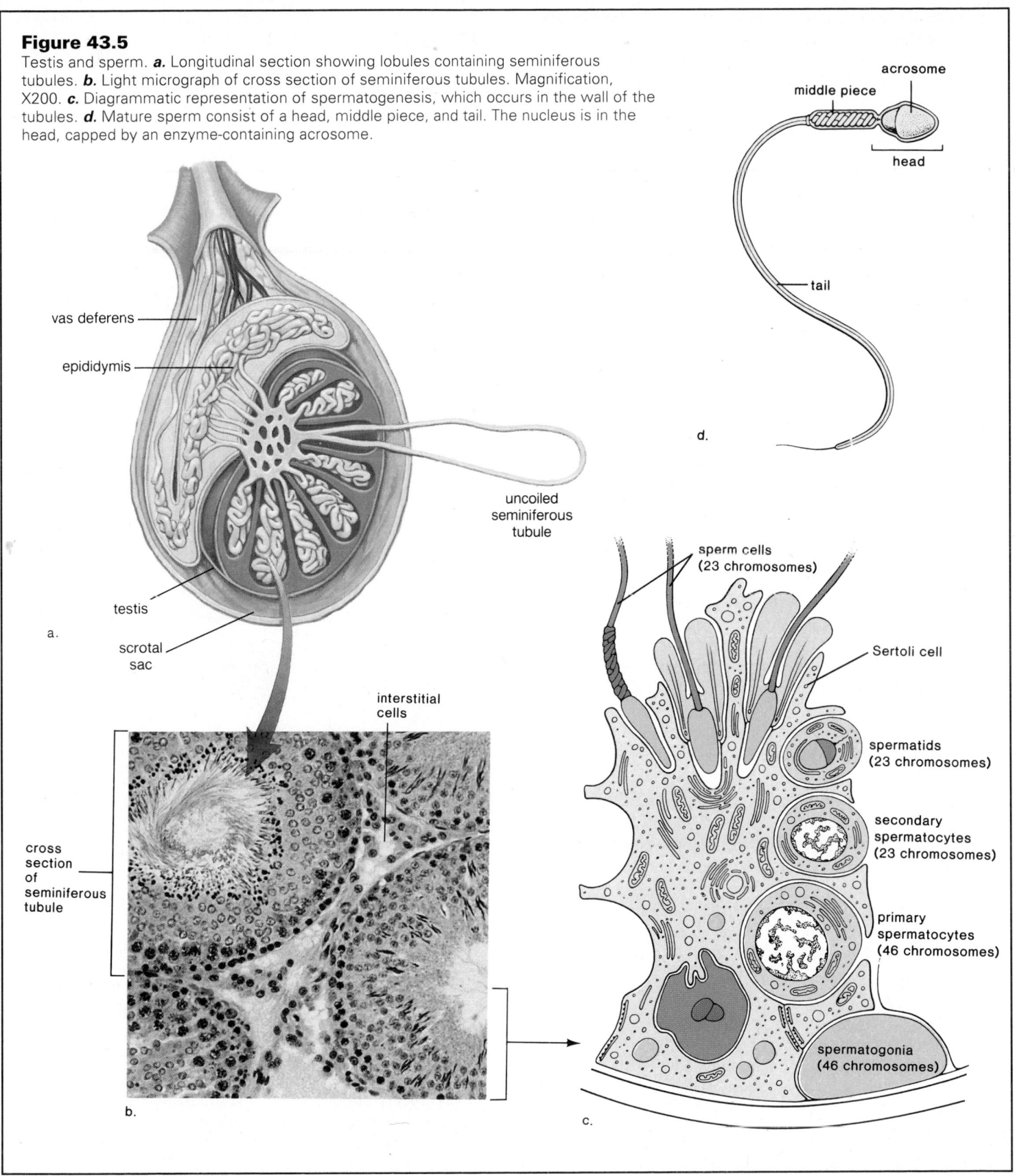

Figure 43.5

Testis and sperm. ***a.*** Longitudinal section showing lobules containing seminiferous tubules. ***b.*** Light micrograph of cross section of seminiferous tubules. Magnification, X200. ***c.*** Diagrammatic representation of spermatogenesis, which occurs in the wall of the tubules. ***d.*** Mature sperm consist of a head, middle piece, and tail. The nucleus is in the head, capped by an enzyme-containing acrosome.

acrosome

middle piece

head

tail

d.

vas deferens

epididymis

uncoiled
seminiferous
tubule

testis

scrotal
sac

a.

interstitial
cells

cross
section
of
seminiferous
tubule

b.

sperm cells
(23 chromosomes)

Sertoli cell

spermatids
(23 chromosomes)

secondary
spermatocytes
(23 chromosomes)

primary
spermatocytes
(46 chromosomes)

spermatogonia
(46 chromosomes)

c.

Reproduction in Animals

causing the voice to change. It is responsible for the greater muscle strength of males, and this is the reason some athletes take supplemental amounts of anabolic steroids, which are either testosterone or related chemicals. The contraindications of taking anabolic steroids are discussed in this chapter's reading. Testosterone is believed to be largely responsible for the sex drive and may even contribute to the supposed aggressiveness of males.

Testosterone also causes oil and sweat glands in the skin to secrete; therefore, it is largely responsible for acne and body odor. Another side effect of testosterone activity is baldness, as discussed on page 231.

> The hypothalamus, anterior pituitary, and testes are engaged in a feedback system that promotes spermatogenesis and maintains the level of testosterone in the male. Testosterone stimulates development of the male sex organs and secondary sex characteristics.

Female Reproductive System

The female reproductive system includes the ovaries, oviducts, uterus, and vagina (fig. 43.7 and table 43.2). The **ovaries,** which produce an egg each month, lie in shallow depressions, one on each side of the upper pelvic cavity. The **oviducts,** also called uterine or fallopian tubes, extend from the ovaries to the uterus; however, the oviducts are not attached to the ovaries. Instead, they have fingerlike projections called *fimbriae* that sweep over the ovaries. When an egg bursts from an ovary during ovulation, it usually is swept into an oviduct by the combined action of the fimbriae and the beating of cilia that line the oviducts. Fertilization, if it occurs, takes place in an oviduct, and the developing embryo is propelled slowly by cilia movement and tubular muscle contraction to the uterus. The **uterus** is a thick-walled muscular organ about the size and shape of an inverted pear. When an embryo embeds itself in the uterine lining, called the **endometrium,** the female is pregnant. A small opening at the cervix leads to the vaginal canal. The **vagina** is a tube that makes a 45-degree angle with the small of the back. The mucosal lining of the vagina lies in folds and can extend. This is especially important when the vagina serves as the birth canal, and it also can facilitate intercourse, when the vagina receives the penis during copulation.

The external genital organs of the female are known collectively as the *vulva* (fig. 43.7b). The *mons pubis* and 2 folds of skin called *labia minora* and *labia majora* are on either side of the urethral and vaginal openings. At the juncture of the labia minora is the *clitoris,* which is analogous to the penis in males. The clitoris has a shaft of erectile tissue and is capped by a pea-shaped glans. The many sense receptors of the clitoris allow it to function as a sexually sensitive organ. Orgasm in the female is a release of neuromuscular tension in the muscles of the genital area, vagina, and uterus.

Egg Production and Hormonal Regulation

The ovaries are responsible for egg production, and they also produce the female sex hormones, **estrogen** and **progesterone,** during the ovarian cycle.

Figure 43.6

The hypothalamus-pituitary-testis control relationship. Testosterone acts on various body tissues and also regulates the amount of hypothalamic GnRH being sent to the pituitary. GnRH affects gonadotropic hormone production by the pituitary. LH regulates the amount of testosterone produced, and FSH controls spermatogenesis. The seminiferous tubules release a substance called inhibin, which is also involved in feedback inhibition.

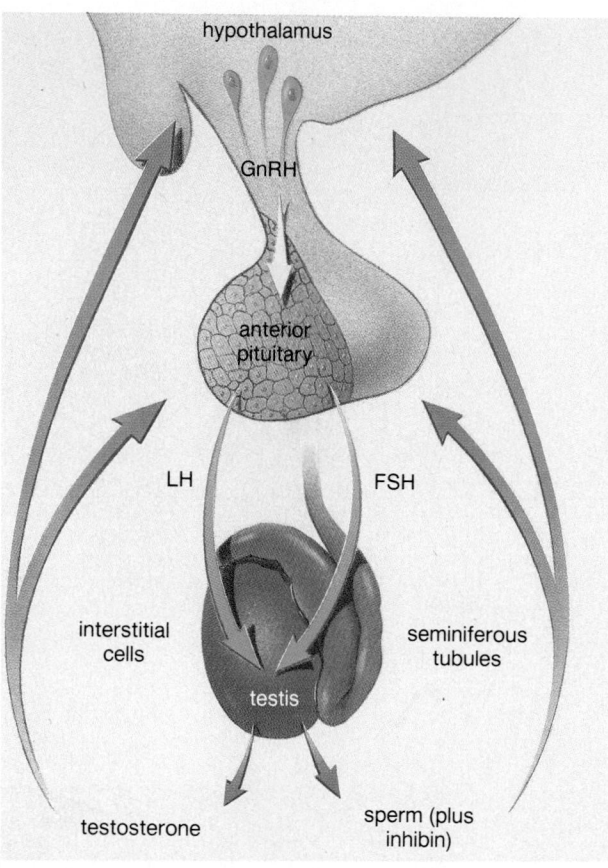

The Ovarian Cycle   A longitudinal section through an ovary shows many **follicles,** each containing an oocyte (fig. 43.8). A female is born with as many as 2 million follicles, but the number is reduced to 300,000–400,000 by the time of puberty. Only a small number of follicles (about 400) ever mature, because a female usually produces only one egg per month during her reproductive years.

As the follicle undergoes maturation, it develops from a primary follicle to a secondary follicle to a Graafian follicle. *Oogenesis* is occurring (see fig. 11.6), and a secondary follicle contains a secondary oocyte pushed to one side of a fluid-filled cavity. In a *Graafian follicle,* the fluid-filled cavity increases to the point that the follicle wall balloons out on the surface of the ovary and bursts, releasing the secondary oocyte surrounded by a clear membrane and follicular cells. This is referred to as **ovulation,** and for the sake of convenience, the released oocyte is often called an ovum, or egg. Actually, the second meiotic

Animal Structure and Function

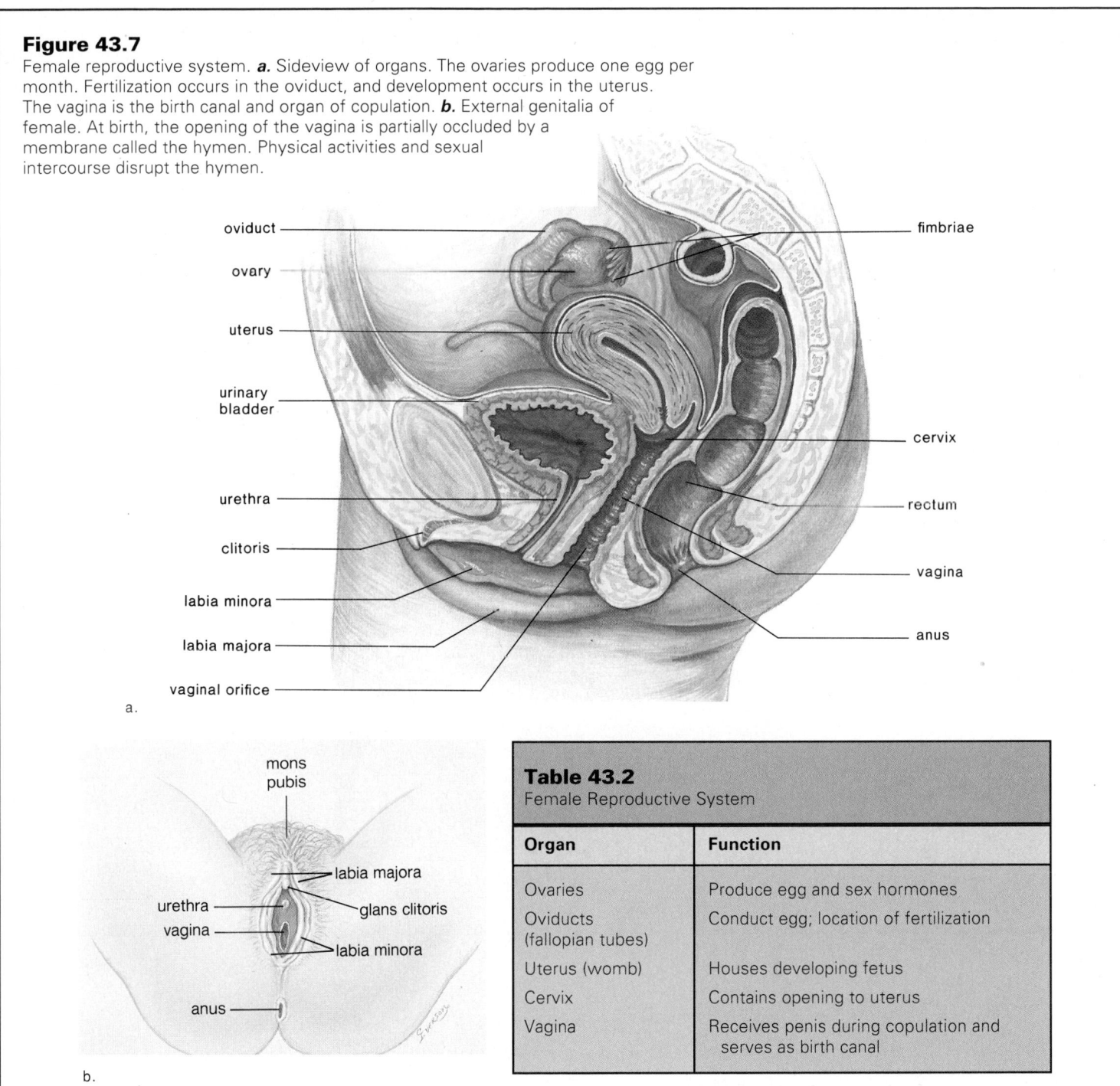

Figure 43.7
Female reproductive system. **a.** Sideview of organs. The ovaries produce one egg per month. Fertilization occurs in the oviduct, and development occurs in the uterus. The vagina is the birth canal and organ of copulation. **b.** External genitalia of female. At birth, the opening of the vagina is partially occluded by a membrane called the hymen. Physical activities and sexual intercourse disrupt the hymen.

oviduct — — fimbriae

ovary

uterus

urinary bladder

— cervix

urethra

— rectum

clitoris

labia minora — vagina

labia majora

— anus

vaginal orifice

a.

mons pubis

labia majora
urethra
glans clitoris
vagina
labia minora

anus

b.

Table 43.2
Female Reproductive System

Organ	Function
Ovaries	Produce egg and sex hormones
Oviducts (fallopian tubes)	Conduct egg; location of fertilization
Uterus (womb)	Houses developing fetus
Cervix	Contains opening to uterus
Vagina	Receives penis during copulation and serves as birth canal

division does not take place unless fertilization occurs. In the meantime, the follicle is developing into the **corpus luteum.** If pregnancy does not occur, the corpus luteum begins to degenerate after about 10 days.

The ovarian cycle is under the control of the gonadotropic hormones, **follicle-stimulating hormone (FSH)** and **luteinizing hormone (LH)** (fig. 43.9 and table 43.3). The gonadotropic

hormones are not present in constant amounts and instead are secreted at different rates during the cycle. For simplicity's sake, it is convenient to emphasize that during the first half, or *follicular phase,* of the cycle, FSH promotes the development of a follicle, which secretes estrogen. As the estrogen level in the blood rises, it exerts feedback control over the anterior pituitary secretion of FSH so that the follicular phase comes to an end.

Figure 43.8

Anatomy of ovary and follicle. **a.** As a follicle matures, the oocyte enlarges and is surrounded by a mantle of follicular cells and fluid. Eventually, ovulation occurs, the mature follicle ruptures, and the secondary oocyte is released. A single follicle actually goes through all stages in one place within the ovary. **b.** Scanning electron micrograph of a secondary follicle. Magnification, X80.

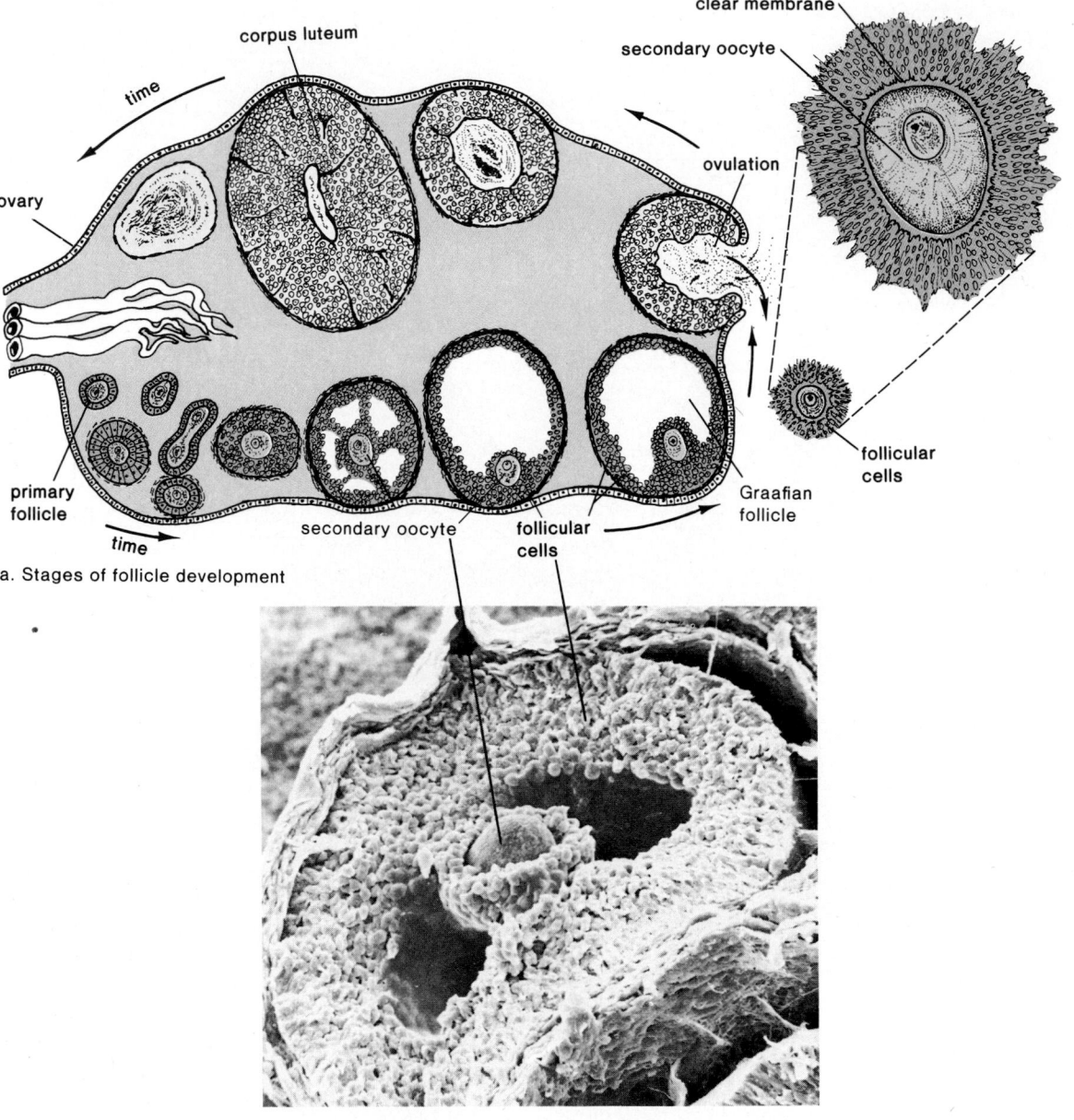

a. Stages of follicle development

b. Secondary follicle

Presumably, the high level of estrogen in the blood also causes the hypothalamus suddenly to secrete a large amount of GnRH. This leads to a surge of LH production by the anterior pituitary and ovulation at about the fourteenth day of a 28-day cycle (fig. 43.10).

During the second half, or *luteal phase*, of the ovarian cycle, it is convenient to emphasize that LH promotes the development of the corpus luteum, which secretes progesterone. As the blood level of progesterone rises, it exerts feedback control over anterior pituitary secretion of LH so that the corpus luteum begins to degenerate. As the luteal phase comes to an end, menstruation occurs.

A negative feedback system involving the hypothalamus and anterior pituitary causes one ovarian follicle per month to produce a secondary oocyte. Following ovulation, the follicle, which secretes estrogen before ovulation, develops into the corpus luteum, which produces progesterone after ovulation.

Animal Structure and Function

Ovarian Cycle	Events	Uterine Cycle	Events
Follicular phase—Days 1-13	FSH Follicle maturation Estrogen	Menstruation—Days 1-5 Proliferative phase—Days 6-13	Endometrium breaks down Endometrium rebuilds
Ovulation—Day 14*			
Luteal phase—Days 15-28	LH Corpus luteum Progesterone	Secretory phase—Days 15-28	Endometrium thickens and glands are secretory

*Assuming a 28-day cycle

Figure 43.9
Hypothalamic-pituitary-gonad system (simplified) as it functions in the female. GnRH is a hypothalamic-releasing hormone that stimulates the anterior pituitary to secrete FSH and LH. These gonadotropic hormones act on the ovaries. FSH promotes the development of the follicle that later, under the influence of LH, becomes the corpus luteum. Negative feedback controls the level of all hormones involved.

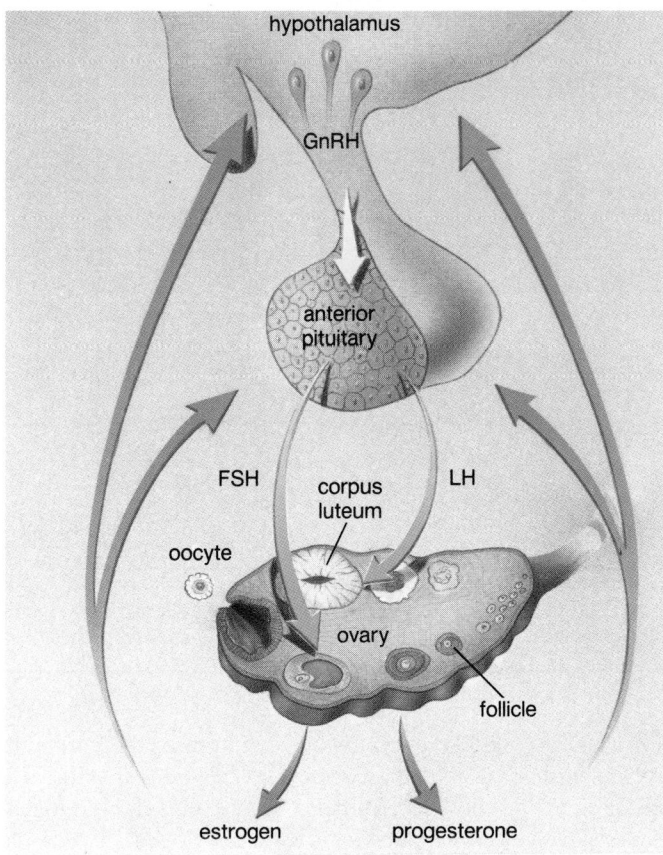

The Uterine Cycle The female sex hormones, estrogen and progesterone, have numerous functions. The effect these hormones have on the endometrium of the uterus causes the uterus to undergo a cyclical series of events known as the **uterine cycle** (table 43.3). Cycles that last 28 days are divided as follows.

During *days 1–5*, there is a low level of female sex hormones in the body, causing the uterine lining to disintegrate and its blood vessels to rupture. A flow of blood, known as the *menses*, passes out of the vagina during a period of **menstruation**, also known as the menstrual period.

During *days 6–13*, increased production of estrogen by an ovarian follicle causes the endometrium to thicken and to become vascular and glandular. This is called the proliferative phase of the uterine cycle.

Ovulation usually occurs on the fourteenth day of the 28-day cycle.

During *days 15–28*, increased production of progesterone by the corpus luteum causes the endometrium to double in thickness and the uterine glands to mature, producing a thick mucoid secretion. This is called the secretory phase of the uterine cycle. The endometrium now is prepared to receive the developing embryo, but if pregnancy does not occur, the corpus luteum degenerates and the low level of sex hormones in the female body causes the uterine lining to break down. This is evident, due to the menstrual discharge that begins at this time. Even while menstruation is occurring, the anterior pituitary begins to increase its production of FSH and a new follicle begins to mature. Table 43.3 indicates how the ovarian cycle controls the uterine cycle.

At about age 45–50, the ovaries gradually cease to respond to the anterior pituitary hormones. Eventually, no more follicles are produced. Following this occurrence, called menopause, menstruation ceases entirely.

The female sex hormones, estrogen and progesterone, regulate the uterine cycle, in which the endometrium first builds up, becomes secretory, and then is shed (menstruation).

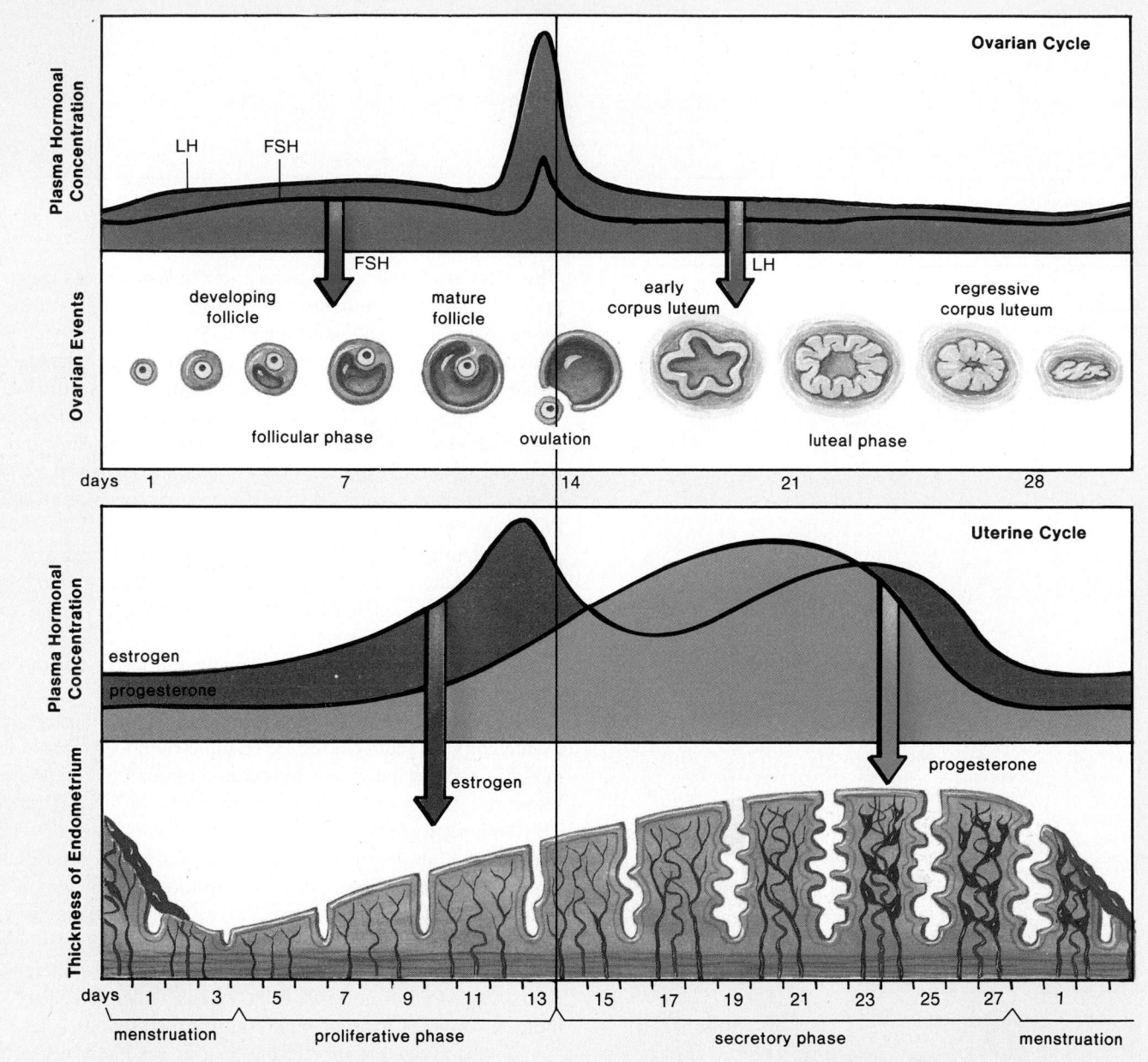

Figure 43.10

Plasma hormonal levels associated with the ovarian and uterine cycles. During the follicular phase, FSH produced by the anterior pituitary promotes the maturation of a follicle in the ovary. The structure produces increasing levels of estrogen, which causes the endometrial lining of the uterus to thicken. After ovulation and during the luteal phase, LH promotes the development of the corpus luteum. This structure produces increasing levels of progesterone, which causes the endometrial lining to become secretory. Menstruation begins when progesterone production declines to a low level.

Pregnancy

If fertilization does occur, an embryo begins development even as it travels down the oviduct to the uterus. The endometrium is now prepared to receive the developing embryo, which becomes embedded in the lining several days following fertilization. This process, called *implantation,* causes the female to become pregnant. During implantation, an outer layer of cells surrounding the embryo produces a gonadotropic hormone (*HCG,* or human chori-

onic gonadotropin) that prevents degeneration of the corpus luteum and instead causes it to secrete even larger quantities of progesterone. The corpus luteum may be maintained for as long as 6 months, even after the placenta is fully developed.

The *placenta* originates from both maternal and fetal tissues. It is the region of exchange of molecules between fetal and maternal blood, although there is rarely any mixing of the 2 types of blood. After its formation, the placenta continues production of

HCG and begins production of progesterone and estrogen. Progesterone and estrogen have 2 effects. They shut down the anterior pituitary so no new follicles mature, and they maintain the lining of the uterus so the corpus luteum is not needed. There is no menstruation during pregnancy.

HCG is produced in such large quantities that pregnant women excrete considerable amounts of it in their urine. The chemical tests for pregnancy are based on the detection of this hormone in the urine. HCG is so readily available that it is routinely used in many teaching and research laboratories. For example, it is even used to induce ovulation in female frogs to obtain eggs for embryological studies.

Illness of the mother or a failure of the placenta to continue adequate HCG production may occasionally cause corpus luteum activity to decrease. Due to a resultant drop in progesterone levels, menstruation-like breakdown of the endometrium and loss of the implanted embryo is initiated. Although the blood flow is somewhat heavier and longer than normal as miscarriage occurs, the terminated pregnancy may be mistaken for a somewhat delayed menstrual period.

Female Sex Hormones

The female sex hormones, estrogen and progesterone, have many other affects in the body, in addition to their affect on the uterus. Estrogen is largely responsible for the secondary sex characteristics in females, including body hair and fat distribution. In general, females have a more rounded appearance than males because of a greater accumulation of fat beneath the skin. Also, the pelvic girdle enlarges in females, and the pelvic cavity has a larger relative size compared to males. This means that females have wider hips. Both estrogen and progesterone are also required for breast development.

Breasts A female breast contains 15–25 lobules, each with its own mammary duct (fig. 43.11). This duct begins at the nipple and divides into numerous other ducts, which end in blind sacs called *alveoli*. In a nonlactating breast, the ducts far outnumber the alveoli because alveoli are made up of cells that can produce milk.

Milk is not produced during pregnancy. Prolactin is needed for lactation (milk production) to begin, and production of this hormone is suppressed by the feedback inhibition estrogen and progesterone have on the anterior pituitary during pregnancy. It takes a couple of days after delivery of a baby for milk production to begin. In the meantime, the breasts produce a watery, yellowish white fluid called *colostrum*, which has a similar composition to milk but contains more protein and less fat.

Reproductive Concerns

Two primary reproductive concerns are control of pregnancy and avoidance of sexually transmitted diseases.

Control of Pregnancy

Sterility causes a person to have no children, and *infertility* causes a person to have fewer children than he or she desires despite frequent intercourse. The 2 major causes of these conditions in females are blocked oviducts, possibly due to a sexually transmit-

Figure 43.11
Anatomy of breast. The female breast contains lobules consisting of ducts and alveoli. The alveoli are lined by milk-producing cells in the lactating (milk-producing) breast.

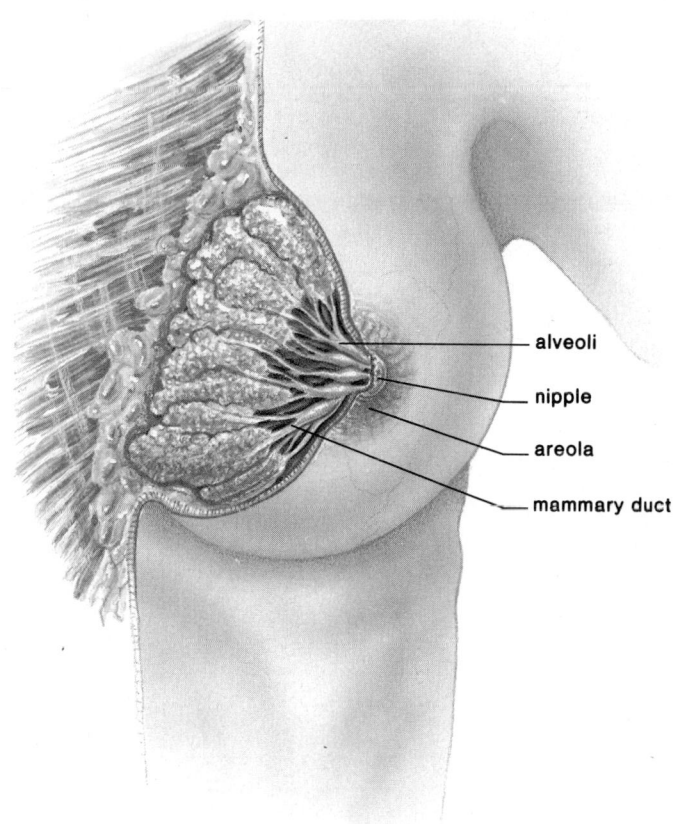

- alveoli
- nipple
- areola
- mammary duct

ted disease, and failure to ovulate due to low body weight. Endometriosis, the spread of uterine tissue beyond the uterus, is also a cause. If no obstruction is apparent and body weight is normal, it is possible to give females HCG extracted from the urine of postmenopausal women. This treatment causes multiple ovulations and sometimes multiple pregnancies.

The most frequent causes of sterility and infertility in males are low sperm count and/or a large proportion of abnormal sperm. Disease, radiation, chemical mutagens, too much heat near the testes, and the use of psychoactive drugs can contribute to this condition.

When reproduction does not occur in the usual manner, couples often seek alternative reproductive methods, which include artificial insemination (sperm are placed in the vagina by a physician), in vitro fertilization (fertilization takes place in laboratory glassware and the zygote is inserted into the uterus of the woman), and surrogate motherhood (a woman has another woman's child).

Sometimes couples wish to prevent a possible pregnancy. Several means of *birth control* are readily available in the United States, and the most common are listed in table 43.4. Effectiveness of the method refers to the number of women per year who will not get pregnant even though they are regularly engaging in sexual

Table 43.4
Common Birth-Control Methods

Name	Procedure	Methodology	Effectiveness*	Action Needed	Risk
Vasectomy	Vas deferentia are cut and tied	No sperm in semen	Almost 100%	Sexual freedom	Irreversible sterility
Tubal ligation	Oviducts are cut and tied	No eggs in oviduct	Almost 100%	Sexual freedom	Irreversible sterility
Pill	Medication is taken daily	Shuts down anterior pituitary	Almost 100%	Sexual freedom	Thromboembolism
Norplant	Progesterone implant is placed under skin by physician	Shuts down anterior pituitary	99.7%	Sexual freedom	_____
IUD	Inserted into uterus by physician	Prevents implantation	More than 90%	Sexual freedom	Infection
Sponge and diaphragm	Inserted into vagina at cervix	Blocks entrance of sperm into uterus	With jelly about 90%	Must be inserted each time before intercourse	_____
Condom	Sheath is placed over erect penis	Traps sperm	About 85%	Must be placed on penis at time of intercourse	_____
Coitus interruptus (withdrawal)	Male withdraws penis before ejaculation	Prevents sperm from entering vagina	About 80%	Intercourse must be interrupted before ejaculation	_____
Jellies, creams, foams	Inserted into vagina	Spermicidal chemicals kill a large number of sperm	About 75%	Must be inserted before intercourse	_____
Rhythm method	Determine day of ovulation by record keeping; testing by various methods	Avoid day of ovulation	About 70%	Limits sexual activity	_____

*Effectiveness is the average percentage of women who did not become pregnant in a population of 100 sexually active women using the technique for one year.

intercourse. For example, with the least effective method given in table 43.4, 70 out of 100, or 70% of sexually active women will not get pregnant, while 30 women will get pregnant, within a year. If no birth-control method is used, it is expected that 80 out of 100 women will be pregnant within a year.

Sexually Transmitted Diseases

Sexually transmitted diseases (STDs) are caused by organisms ranging from viruses to arthropods; however, we will discuss only certain STDs caused by viruses and bacteria. Unfortunately, for unknown reasons, humans cannot develop good immunity to any of the STDs. Therefore, prompt and proper medical treatment should be received when exposed to an STD. A condom serves as protection against spreading STDs. The concomitant use of a spermicide containing nonoxynol-9 gives added protection.

STDs of Viral Origin
AIDS (acquired immunodeficiency syndrome), genital herpes, and genital warts are all viral infections. AIDS is discussed later in this chapter.

Genital herpes is caused by herpes simplex virus, of which type 1 usually causes cold sores and fever blisters, while type 2 more often causes genital herpes. Many times an infected person has no symptoms, but if symptoms are present, there are painful ulcers on the genitals that heal and then reappear. The ulcers may be accompanied by fever, pain upon urination, and swollen lymph nodes. At this time, the individual has an increased risk of acquiring an AIDS infection. Exposure in the birth canal can cause an infection in the newborn, which leads to neurological disorders and even death. Birth by cesarean section prevents this possibility.

Genital warts are caused by the human papilloma viruses (HPVs). Many times, carriers do not have any sign of warts or only flat lesions may be present. If visible warts are removed, they may recur. HPVs are now associated with cancer of the cervix, as well as tumors of the vulva, the vagina, the anus, and the penis. Some researchers believe that the viruses are involved in 90%–95% of all cases of cancer of the cervix.

STDs of Bacterial Origin
Gonorrhea, chlamydia, and syphilis are STDs caused by bacteria. Therefore, they are curable by antibiotic therapy.

Animal Structure and Function

Gonorrhea is caused by the bacterium *Neisseria gonorrheae*. Diagnosis in the male is not difficult, as long as he displays the typical symptoms of pain upon urination and a thick, greenish yellow urethral discharge. In males and females, a latent infection leads to pelvic inflammatory disease (PID), in which the vas deferentia or oviducts are affected. As the inflamed tubes heal, they may become partially or completely blocked by scar tissue, resulting in sterility or infertility. If a baby is exposed during the process of birth, an eye infection leading to blindness can result. All newborns are given eye drops to prevent this possibility.

Chlamydia is name for the tiny bacterium that causes it (*Chlamydia trachomatis*). Chlamydia is the most common cause of nongonococcal urethritis (NGU), which is often difficult to distinguish from gonococcal urethritis. Since an infection can also cause PID, physicians routinely prescribe medicines for both gonorrhea and chlamydia at the same time. Chlamydia also causes cervical ulcerations, which increase the risk of acquiring AIDS. If a baby comes in contact with chlamydia during birth, inflammation of the eyes or pneumonia can result.

Syphilis, which is caused by the bacterium *Treponema pallidum,* has 3 stages, which are typically separated by latent periods. In the primary stage, a hard chancre (ulcerated sore with hard edges) appears. In the secondary stage, a rash appears all over the body—even on the palms of the hands and on the soles of the feet. During the tertiary stage, syphilis may affect the cardiovascular and/or nervous system. An infected person may become mentally retarded, blind, walk with a shuffle, or show signs of insanity. *Gummas,* which are large destructive ulcers, may develop on the skin or within the internal organs. Syphilitic bacteria can cross the placenta, causing a baby to be stillborn or have many and various anatomical malformations. Unlike the other STDs discussed, there is a blood test to diagnose syphilis.

Sexually transmitted diseases include AIDS; herpes, which can recur; genital warts, which lead to cancer of the cervix; gonorrhea and chlamydia, which cause PID; and syphilis, which has cardiovascular and neurological complications if untreated.

AIDS

AIDS is caused by a group of related retroviruses known as HIV (human immunodeficiency viruses). The World Health Organization now estimates that in 9 years, over 40 million persons worldwide will be infected with HIV. Presently in the United States, 90% of those infected are males and 10% are female. In Africa, where heterosexual transmission is more prevalent, an equal number of men and women are infected.

Transmission of AIDS

After HIV enter the blood, they infect helper T lymphocytes (p. 579) of a type known as T4 cells. The T4 cells die and their number declines as the disease progresses. About 28% of AIDS cases are intravenous drug abusers who have contracted the disease after using a contaminated needle. About 3% of AIDS cases are persons who have received infected blood during a blood transfusion.

Figure 43.12
Balance of power between HIV and the immune system during the course of an HIV infection. During the time that a person is an asymptomatic carrier, the concentration of HIV in the body is high, but then the immune system becomes active. While an infected person has ARC, the immune system begins to lose the battle and eventually the battle is lost when a person develops full-blown AIDS. Source. Data from R. R. Redfield and D. S. Burke, "HIV Infection: The Clinical Picture" in *Scientific American*, October 1988.

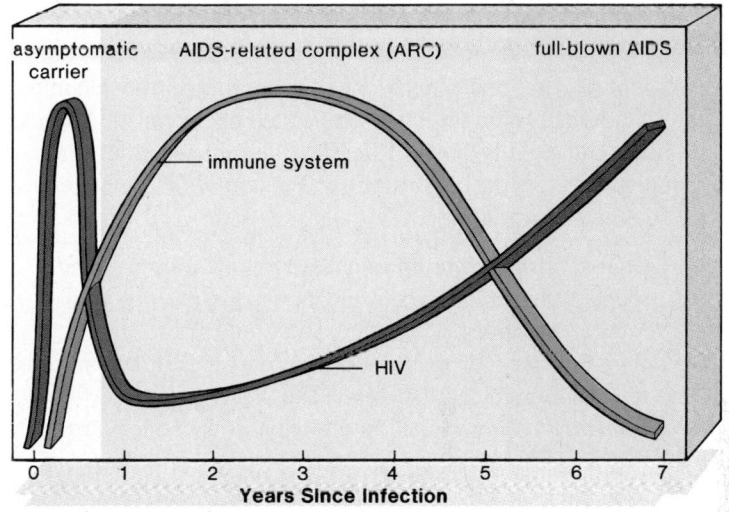

AIDS, however, is primarily a sexually transmitted disease. Semen can contain the virus, but more likely an infected lymphocyte does. About 64% of the total cases in the United States are homosexual men who practice anal intercourse. (About 6% of these are also IV drug abusers.) Unlike the vagina, the epithelial lining of the rectum is a thin, single-celled layer that is easily torn during intercourse. Nevertheless, heterosexual transmission does occur and may become more prevalent as more females become infected. One unhappy side effect to female infection is the fact that viruses and infected lymphocytes can pass to a fetus via the placenta or to an infant via the mother's milk. Presently, infected infants account for about 1% of all AIDS cases.

Symptoms of AIDS

An HIV infection can be divided into 3 stages (fig. 43.12).

Asymptomatic Carrier

Only 1%–2% of those newly infected have mononucleosis-like symptoms that may include fever, chills, aches, swollen lymph glands, and an itchy rash. These symptoms disappear, and there are no other symptoms for 9 months or longer. Although the individual exhibits no symptoms during this stage, he or she is highly infectious. The standard HIV blood test for the presence of antibody becomes positive during this stage.

AIDS Related Complex (ARC)

The most common symptom of ARC is swollen lymph glands in the neck, armpits, or groin that persist for 3 months or more. There is severe fatigue unrelated to exercise or drug use; unexplained persistent or recurrent fevers, often with night sweats; persistent cough not associated with smoking, a cold, or the flu; and persistent diarrhea. Also possible are signs of nervous system impairment, including loss of memory, inability to think clearly, loss of judgment, and/or depression.

When the individual develops non-life-threatening and recurrent infections such as thrush or herpes simplex, it is a signal that full-blown AIDS will occur shortly.

Full-Blown AIDS

In this final stage, there is severe weight loss and weakness due to persistent diarrhea and usually one of several opportunistic infections is present. These infections are called opportunistic because the body can usually prevent them—only an impaired immune system gives them the opportunity to get started. These infections include the following:

Pneumocystis carinii **pneumonia.** There is not a single documented case of this type of pneumonia in persons with normal immunity.

Toxoplasmic encephalitis. In AIDS patients, this infection leads to loss of brain cells, seizures, and weakness.

Myobacterium avium. This is an infection of the bone marrow that leads to a decrease in red blood cells, white blood cells, and platelets.

Kaposi's sarcoma. A cancer of the blood vessels that causes reddish purple, coin-size spots and lesions on the skin.

Treatment

The drug zidovudine (also called azidothymidine, or AZT) and dideoxyinosine (DDI) prevent HIV reproduction in cells. Proteases are enzymes HIV needs to bud from the host cell; researchers are hopeful that a protease inhibitor drug will soon be available.

A number of different types of vaccines are in, or are expected to be in, human trials. Several of these are subunit vaccines that utilize genetically engineered proteins that resemble those found in HIV. For example, HIV-1, the cause of most AIDS cases in the United States, has an outer envelope molecule called GP120 (fig. 43.13). When GP120 combines with a CD4 molecule that projects from a helper T lymphocyte, the virus enters the cell. There are subunit vaccines that make use of GP120. An entirely different approach is being taken by Jonas Salk, who developed the polio vaccine. His vaccine utilizes whole HIV-1 killed by treatment with chemicals and radiation. So far, this vaccine has been found to be effective against experimental HIV-1 infection in chimpanzees, and clinical trials will occur soon.

Figure 43.13

An HIV-1 has an envelope molecule called GP120 that allows it to attach to CD4 molecules that project from a T4 cell. Infection of the T4 cell follows, and HIV eventually buds from the infected T4 cell. If the immune system can be trained by the use of a vaccine to attack and destroy all cells that bear GP120, a person would not be able to be infected with HIV-1.

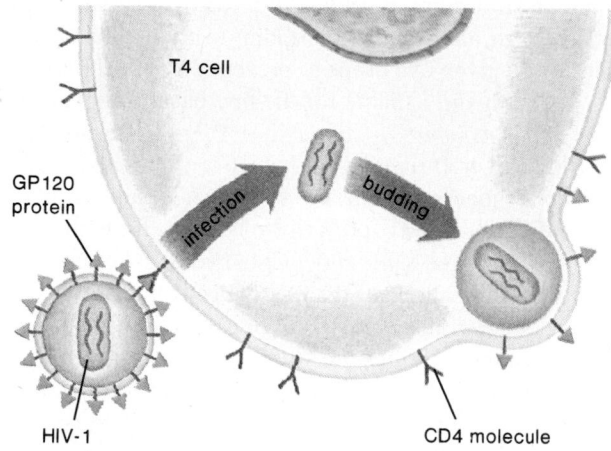

AIDS Prevention

Shaking hands, hugging, social kissing, coughing or sneezing, and swimming in the same pool do not transmit the AIDS virus. You cannot get AIDS from inanimate objects such as toilets, doorknobs, telephones, office machines, or household furniture.

The following behaviors will help prevent the spread of AIDS:

1. Do not use alcohol or drugs in a way that prevents you from being in control of your behavior. Especially, do not inject drugs into veins, but if you are an intravenous drug user and cannot stop your behavior, always use a sterile needle for injection or one cleansed by bleach.

2. Refrain from multiple sex partners, especially with homosexual or bisexual men or intravenous drug users of either sex. Either abstain from sexual intercourse or develop a long-term monogamous (always the same partner) sexual relationship with a partner who is free of HIV and is not an intravenous drug user.

3. If you are uncertain about your partner, always use a latex condom. Follow the directions, and also use a spermicide containing nonoxynol-9, which kills viruses and virus-infected lymphocytes. The risk of contracting AIDS is greater in persons who already have a sexually transmitted disease.

Summary

1. Asexual reproduction may quickly produce a large number of offspring exactly like the single parent and is advantageous when environmental conditions are relatively unchanging.

2. Sexual reproduction involves the use of gametes and produces offspring that are slightly different from the parents. This may be advantageous if the environment is changing.

3. Drying out of gametes and zygotes is not a threat for aquatic animals, and they tend to practice external fertilization. Drying out is a problem for land

Animal Structure and Function

animals, and they tend to practice internal fertilization.

4. In human males, sperm are produced in the testes, mature in the epididymides, and may be stored in the vas deferens before entering the urethra, along with seminal fluid (produced by the seminal vesicles, prostate gland, and Cowper's gland), prior to ejaculation during male orgasm, when the penis becomes erect.

5. Spermatogenesis occurs in the seminiferous tubules of the testes, which also produce testosterone in interstitial cells. Testosterone maintains the secondary sex characteristics of males, such as low voice, facial hair, and increased muscle strength.

6. FSH (also called ICSH) from the anterior pituitary stimulates spermatogenesis, and LH stimulates testosterone production. A hypothalamic releasing hormone, GnRH, controls anterior pituitary production and FSH and LH release. The level of testosterone in the blood controls the secretion of GnRH and the anterior pituitary hormones by a negative feedback system.

7. In females, an egg produced by an ovary enters an oviduct, which leads to the uterus. The uterus opens into the vagina. The genital area of women includes the vaginal opening, clitoris, and labia minora and labia majora.

8. In either ovary, one follicle a month matures, produces a secondary oocyte, and becomes a corpus luteum. This is called the ovarian cycle. The follicle and corpus luteum produce estrogen and progesterone, the female sex hormones.

9. A uterine cycle occurs concurrently with the ovarian cycle. In the first half of these cycles (days 1–13 before ovulation), the anterior pituitary produces FSH and the follicle produces estrogen. Estrogen causes the uterine lining to increase in thickness. In the second half of these cycles (days 15–28 after ovulation), the anterior pituitary produces LH and the follicle produces progesterone. Progesterone causes the uterine lining to become secretory. Feedback control of the hypothalamus and anterior pituitary causes the level of estrogen and progesterone to fluctuate. When they are at a low level, menstruation begins.

10. If fertilization occurs, the corpus luteum is maintained because of HCG production. Progesterone production does not cease, and the zygote implants itself in the thick uterine lining.

11. Estrogen and progesterone maintain the secondary sex characteristics of females, including less body hair than males, a wider pelvic girdle, more rounded appearance, and development of breasts.

12. Infertile couples are increasingly resorting to alternative methods of reproduction. Numerous birth-control methods and devices are available for those who wish to prevent pregnancy.

13. Sexually transmitted diseases include AIDS; herpes, which can recur; genital warts, which lead to cancer of the cervix; gonorrhea and chlamydia, which cause PID; and syphilis, which has cardiovascular and neurological complications if untreated.

14. AIDS is caused by HIV, which infect helper T4 cells. As the number of T4 cells declines, the symptoms of AIDS related complex and then full-blown AIDS appear. Drugs are available that prevent HIV reproduction in cells, and it is hoped that a vaccine will one day be available. In the meantime, all persons should take the proper steps to prevent the spread of AIDS.

Writing Across the Curriculum

In order to practice writing skills, students should write out the answers to any or all of the study questions and the critical thinking questions. The study questions are sequenced in the same order as the text. Suggested answers to the critical thinking questions are in appendix D.

Study Questions

1. Give examples of asexual and sexual reproduction among animals. Relate these practices to environmental conditions.
2. Discuss the human reproductive system as an adaptation to life on land.
3. Discuss the anatomy and physiology of the testes. Describe the structure of sperm.
4. Give the path of sperm. What glands contribute fluids to semen?

5. Name the endocrine glands involved in maintaining the sex characteristics of males and the hormones produced by each.
6. Discuss the anatomy and physiology of the ovaries. Describe ovulation.
7. Give the path of the egg. Where do fertilization and implantation occur? Name 2 functions of the vagina.
8. Discuss hormonal regulation in the female by giving the events of the

uterine cycle and relating these to the ovarian cycle. In what way is menstruation prevented if pregnancy occurs?
9. Describe at least 3 common sexually transmitted diseases.
10. What means of birth control help prevent the spread of AIDS? What other measures can be taken to protect oneself from AIDS?

Objective Questions

1. Which of these is a requirement for sexual reproduction?
   a. male and female parents
   b. production of gametes
   c. optimal environmental conditions
   d. aquatic habitat

2. Internal fertilization
   a. prevents the drying out of gametes and zygotes.
   b. must take place on land.
   c. is practiced by humans.
   d. Both a and c.

3. Which of these is mismatched?
   a. interstitial cells—testosterone
   b. seminiferous tubules—sperm production
   c. vas deferens—seminal fluid production
   d. penis—erection

4. FSH
   a. occurs in females but not males.
   b. stimulates the seminiferous tubules to produce sperm.
   c. secretion is controlled by GnRH.
   d. Both b and c.

5. Which of these combinations is most likely to be present before ovulation occurs?
   a. FSH, corpus luteum, estrogen, secretory uterine lining
   b. LH, follicle, progesterone, thick uterine lining
   c. FSH, follicle, estrogen, uterine lining becoming thick
   d. LH, corpus luteum, progesterone, secretory uterine lining

6. In tracing the path of sperm, you would mention vas deferens before
   a. testes.
   b. epididymis.
   c. urethra.
   d. uterus.

7. An oocyte is fertilized in the
   a. vagina.
   b. uterus.
   c. oviduct.
   d. ovary.

8. During pregnancy,
   a. the ovarian cycle and uterine cycle occur more quickly than before.
   b. GnRH is produced at a higher level than before.
   c. the ovarian cycle and uterine cycle do not occur.
   d. the female secondary sex characteristics are not maintained.

9. Which of the following means of birth control is most effective in preventing AIDS?
   a. condom
   b. pill
   c. diaphragm
   d. spermicidal jelly

10. Which of these sexually transmitted diseases is mismatched with its cause?
   a. AIDS—bacterial infection of red blood cells
   b. gonorrhea—bacterial infection of genital tract
   c. chlamydia—bacterial infection of genital tract
   d. syphilis—systemic bacterial infection

11. Label this diagram of the male reproductive system and trace the path of sperm:

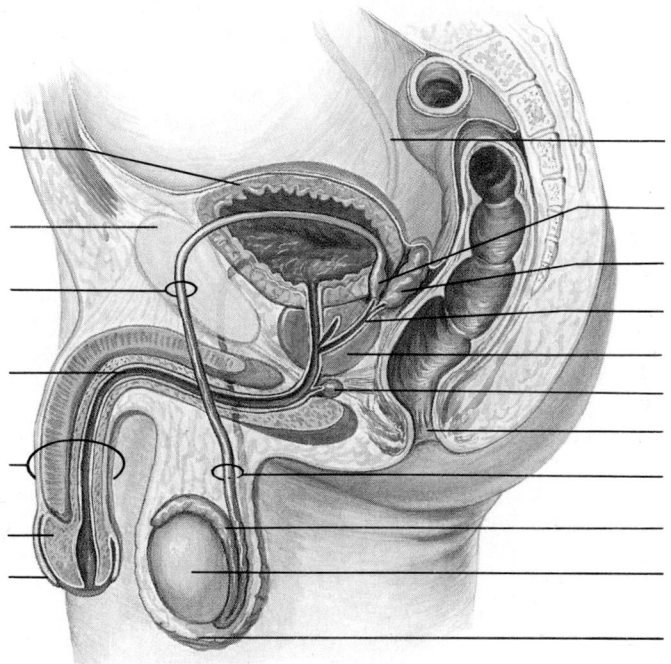

Concepts and Critical Thinking

1. *Successful reproduction on land requires certain adaptations.*

   Contrast the manner in which reptiles are adapted to reproduce on land with the manner in which humans are adapted to reproduce on land.

2. *Reproduction is under hormonal rather than nervous control.*

   Why would you have predicted hormonal rather than nervous control of reproduction?

Selected Key Terms

regeneration (re-jen″er-a′shun) 695
budding (bud′ing) 695
parthenogenesis (par″thē-no-jen′-ē-sis) 695
penis (pe′nis) 697

testosterone (tes-tos′tĕ-rōn) 698
ovary (o′vah-re) 700
oviduct (o′vĭ-dukt) 700
uterus (u′ter-us) 700
vagina (vah-ji′nah) 700

follicle (fol′ĭ-k′l) 700
ovulation (o″vu-la′shun) 700
corpus luteum (kor′pus lu′te-um) 701
uterine cycle (u′ter-īn si′k′l) 703
menstruation (men″stroo-a′shun) 703

Animal Structure and Function

44

Development in Animals

An iguana emerging from its shell, a structure that allows reptiles to develop on land. Within the shelled egg, the extraembryonic membranes perform many necessary functions, and they provide a watery environment. In effect, all vertebrates are surrounded by water as they develop.

Your study of this chapter will be complete when you can

1. name and define 3 processes that occur whenever there is a developmental change;
2. compare the early developmental stages of the lancelet, frog, and chick;
3. explain the germ layer theory of development, and give examples;
4. draw and label a cross section of a typical vertebrate embryo at the neurula stage of development;
5. give evidence that differentiation probably begins soon after formation of the zygote;
6. give evidence that morphogenesis can be accounted for by interactions between tissues;
7. state the stages of embryonic development in humans, and compare these to the development of the chick;
8. list the extraembryonic membranes, and give their function in chicks and in humans;
9. briefly outline the developmental changes in humans from the fetus to adult;
10. define aging, and discuss 3 theories of the aging process.

The study of development concerns the events and processes that occur as a single cell becomes a complex organism. These same processes are also seen as the newly born or hatched organism matures, as lost parts regenerate, as a wound heals, and even during aging. Therefore, it is customary to stress that the study of development encompasses not only embryology (development of the embryo) but these other events as well.

Development requires growth, differentiation, and morphogenesis. When an organism increases in size, we say that it has grown. During *growth,* cells divide, get larger, and divide once again. **Differentiation** occurs when cells become specialized in structure and function. A muscle cell looks and acts quite differently than a nerve cell, for example. **Morphogenesis** goes one step beyond growth and differentiation. It occurs when body parts become shaped and patterned into a certain form. There is a great deal of difference between your arm and leg for example, even though they contain the same types of tissues.

We will discuss these processes as they apply to development of the embryo, but keep in mind that they also occur whenever an organism goes through any developmental change.

Growth, differentiation, and morphogenesis are 3 processes that are seen whenever a developmental change occurs.

Early Developmental Stages

Embryological development begins when the sperm fertilizes the egg. Each gamete has a haploid number of chromosomes, therefore the resulting zygote has the diploid number. Gene expression is required for development to proceed normally, and we will stress this again in later sections.

All chordate embryos go through the same early developmental stages of cleavage, blastulation, gastrulation, and neurulation (fig. 44.1). The presence of *yolk,* which is dense, nutrient material, however, affects the manner in which embryonic cells complete the first 3 stages and hence the appearance of the embryo at the end of each stage. Varying amounts of yolk will result in embryos with different appearances. Table 44.1 indicates the amount of yolk in the 4 embryos discussed in this chapter and relates the amount of yolk to the environment in which the animal develops. The 2 animals (lancelet and frog) that develop in water

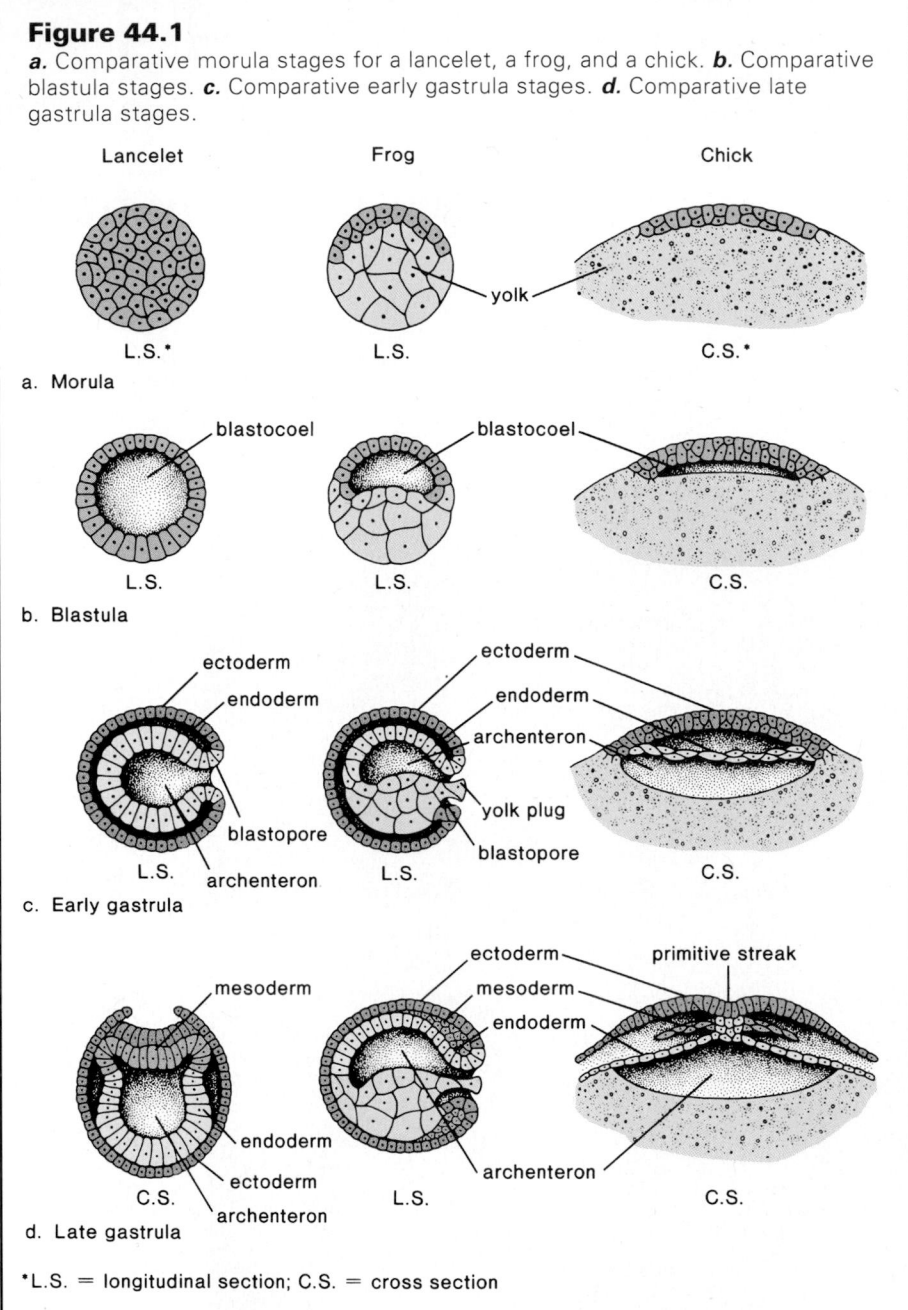

Figure 44.1

a. Comparative morula stages for a lancelet, a frog, and a chick. *b.* Comparative blastula stages. *c.* Comparative early gastrula stages. *d.* Comparative late gastrula stages.

*L.S. = longitudinal section; C.S. = cross section

Table 44.1
Amount of Yolk in Eggs versus Location of Development

Animal	Yolk	Location of Development
Lancelet	Little	External in water
Frog	Some	External in water
Chick	Much	Within hard shell
Human	Little	Inside mother

have less yolk than the chick because development in these 2 animals proceeds quickly to a swimming larval stage that can feed itself. But the chick is representative of animals that have solved the problem of reproduction on land, in part, by providing a great deal of yolk within a hard shell. Development continues in the shell until there is an offspring capable of land existence.

Early stages of human development resemble those of the chick embryo, yet this resemblance cannot be related to the amount of yolk because the human egg contains little yolk. But the evolutionary history of these 2 animals can provide an answer for this similarity. Both birds (e.g., chicks) and mammals (e.g., humans) are related to reptiles, and this explains why all 3 groups develop similarly, despite a difference in the amount of yolk in the eggs.

The amount of yolk affects the manner in which lancelets, frogs, and chicks complete the first 3 stages of development.

Cleavage and Formation of Blastula

Cell division without growth occurs during **cleavage** (fig. 44.2). DNA replication and mitosis occur repeatedly, and the cells get smaller with each division. Chordates, being deuterostomes, have a pattern of cleavage that is radial and indeterminate (see fig. 27.2). The term *radial* means that any plane passing through the major axis will divide the embryo into 2 symmetrical halves. The term *indeterminate* means that the cleavage cells have not differentiated, and therefore their developmental fate is not yet set.

In a lancelet, the cell divisions are equal, and the cells are of uniform size, whereas in a frog, the upper cells are smaller than the lower cells. This difference in size occurs because the upper cells at the animal pole contain little yolk, whereas the lower cells at the vegetal pole contain a large amount of yolk. Cells containing yolk cleave more slowly than those without yolk. The presence of extensive yolk in the chick egg causes cleavage to be incomplete, and only those cells lying on top of the yolk cleave. This means that although cleavage in a lancelet and a frog results in a ball of cells called the morula, no such ball is seen in a chick (fig. 44.1a). Instead, during the morula stage the cells spread out on a portion of the yolk.

The morula is a solid mass of cells, but then a cavity called the **blastocoel** develops (fig. 44.2). This "hollow-ball stage" of development, called the **blastula,** is best exhibited by the lancelet. In the frog, the blastocoel is formed at the animal pole only. The heavily laden yolk cells of the vegetal pole do not participate in this step. In a chick, the blastocoel is created when the cells lift up from the yolk and leave a space between the cells and the yolk (fig. 44.1b).

Cleavage results in a ball of cells, which becomes the blastula when an internal cavity develops. In lancelets and frogs, the blastula is a hollow ball. In the chick, the cavity is found beneath cells that lie flat atop the great mass of yolk.

Gastrulation and Formation of Germ Layers

The **gastrula** stage is evident in a lancelet when certain cells begin to push, or invaginate, into the blastocoel, creating a double layer of cells (fig. 44.2). The outer layer is called the **ectoderm,** and the inner layer is called the **endoderm.** The space created by invagi-

Figure 44.2

Early development in a lancelet. A lancelet has little yolk as an embryo, and it can be used to exemplify the early stages of development in such animals. Cleavage produces a number of cells that form a cavity. Invagination during gastrulation produces the germ layers, ectoderm, and endoderm. Mesoderm arises from pouches that pinch off from the endoderm.

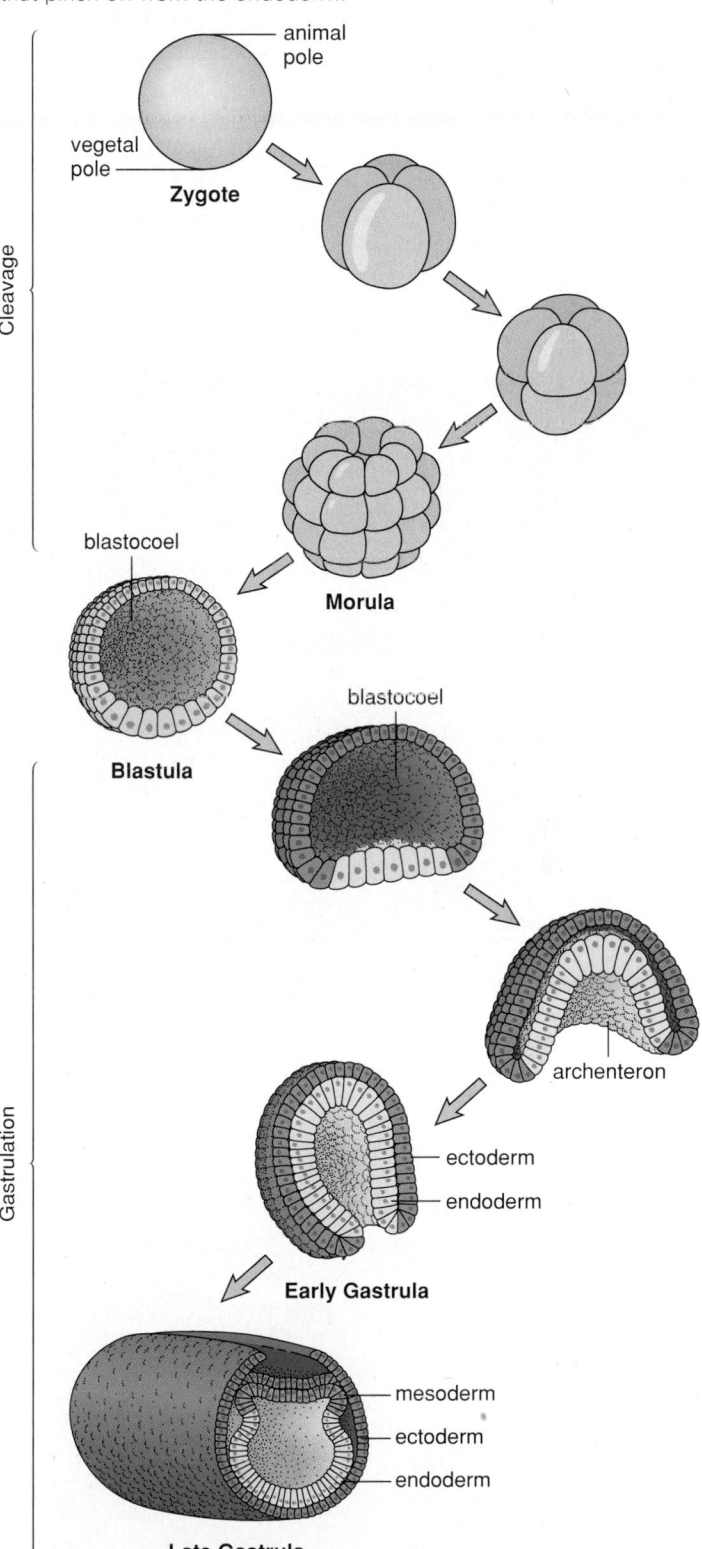

nation becomes the gut and is called either the primitive gut or the archenteron. The pore, or hole, created by invagination is called the blastopore, and in a lancelet, as well as in the other animals discussed here, it eventually becomes the anus (see fig. 27.2).

In the frog, the cells containing yolk do not participate in gastrulation and therefore do not invaginate. Instead, a slitlike blastopore is formed when the animal pole cells begin to invaginate from above. Following this, other animal pole cells move down over the yolk, and the blastopore becomes rounded when these cells also invaginate from below. At this stage, there are some yolk cells temporarily left in the region of the pore; these are called the yolk plug. In the chick, there is so much yolk that endoderm formation does not occur by invagination. Instead, an upper layer of cells differentiates into ectoderm, and a lower layer differentiates into endoderm (fig. 44.1c).

Gastrulation is not complete until 3 layers of cells have formed (fig. 44.2). The third, or middle, layer of cells is called the

mesoderm. In a lancelet, this layer begins as outpocketings from the primitive gut (see fig. 27.2). These outpocketings grow in size until they meet and fuse. In effect, then, 2 layers of mesoderm are formed, and the space between them is the coelom.

In the frog, cells from the dorsal lip of the blastopore migrate between the ectoderm and endoderm, forming the mesoderm. Later, a splitting of the mesoderm creates the coelom. In the chick, the mesoderm layer arises by an invagination of cells along the edges of a longitudinal furrow in the midline of the embryo. Because of its appearance this furrow is called the *primitive streak* (fig. 44.1d). Later, the newly formed mesoderm will split to give a coelomic cavity.

Ectoderm, mesoderm, and endoderm are called the primary **germ layers** of the embryo, and no matter how gastrulation takes place, the end result is the same: 3 germ layers are formed. It is possible to relate the development of future organs to these germ layers, as is done in table 44.2. Karl E. Von Baer, the nineteenth-century embryologist, first related later development to the early formation of germ layers. This is called the *germ layer theory*.

The 3 embryonic germ layers arise during gastrulation, when cells invaginate into the blastocoel. The development of organs can be related to the 3 germ layers: ectoderm, mesoderm, and endoderm.

Neurula

In chordate animals, newly formed mesoderm cells that lie along the main longitudinal axis of the animal coalesce to form a dorsal supporting rod called the **notochord.** The notochord persists in lancelets (see fig. 27.15), but in frogs, chicks, and humans, it is later replaced by the vertebral column.

The nervous system develops from ectoderm located just above the notochord. At first, a thickening of cells called the *neural plate* is seen along the dorsal surface of the embryo. Then, *neural folds* develop on either side of a neural groove, which becomes the *neural tube* when these folds fuse. Figure 44.3 shows cross sections of frog development to illustrate the formation of the neural tube. At this point, the embryo is called a *neurula*. Later, the anterior end of the neural tube develops into the brain.

Table 44.2
Organs Developed from the 3 Primary Germ Layers

Ectoderm	Mesoderm	Endoderm
Skin epidermis, including hair, nails, and sweat glands	All muscles	Lining of digestive tract, trachea, bronchi, lungs, gallbladder, and urethra
Nervous system, including brain, spinal cord, ganglia, nerves, and sense receptors	Dermis of skin	Liver
Lens and cornea of eye	All connective tissue, including bone, cartilage, and blood	Pancreas
Lining of nose, mouth, and anus	Blood vessels	Thyroid, parathyroid, and thymus glands
Tooth enamel	Kidneys	Urinary bladder
	Reproductive organs	

Figure 44.3
Development of neural tube and coelom in a frog embryo.
a. Ectoderm cells that lie above the future notochord (called presumptive notochord) thicken to form a neural plate. *b.* The neural groove and folds are noticeable as the neural tube begins to form. *c.* A splitting of the mesoderm produces a coelom, which is completely lined by mesoderm. *d.* A neural tube and coelom have now developed.

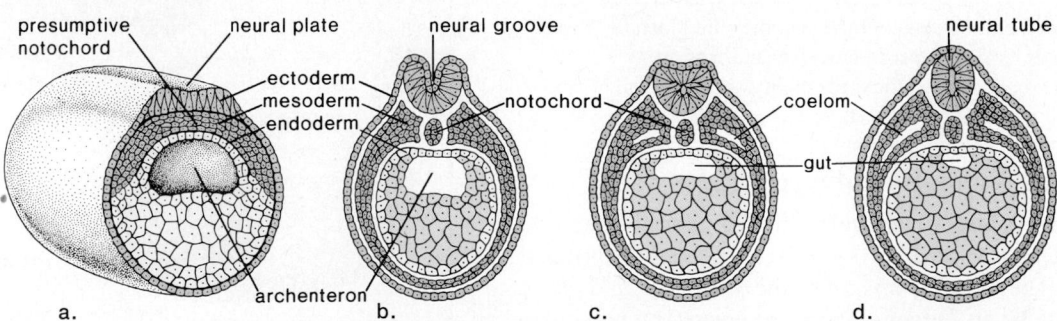

Midline mesoderm cells that did not contribute to the formation of the notochord now become 2 longitudinal masses of tissue. These 2 masses become blocked off into the somites, which give rise to segmental muscles in all chordates. In vertebrates, the somites also produce the vertebral bones.

With the formation of the nervous system, it is possible to show a generalized diagram (fig. 44.4) of a chordate embryo to illustrate the location of parts. Consideration of this figure with table 44.2 will help you relate the formation of chordate structures and organs to the 3 embryonic layers of cells: the ectoderm, the mesoderm, and the endoderm.

During neurulation, the neural tube develops just above the notochord. At the neurula stage of development, a cross section of all chordate embryos is similar in appearance.

Differentiation and Morphogenesis

Differentiation accounts for specialization of tissues that have specific functions, and morphogenesis accounts for the formation of organs that have an overall pattern of shape and form. The process of differentiation must start long before different types of cells are recognizable, most likely even with the very first cleavage of the egg.

The reading in chapter 17 (p. 267) describes an experiment in which a tadpole cell nucleus is placed in an enucleated egg. Development proceeds normally, showing that tadpole nuclei are totipotent—they contain all the genetic information required to bring about complete development of the organism. Therefore, differentiation cannot be due to a parceling out of genes into the various embryonic cells. Instead, it must be due to the expression of particular genes, controlled at first by *ooplasmic segregation,* which is the distribution of maternal cytoplasmic contents to the cells of the morula:

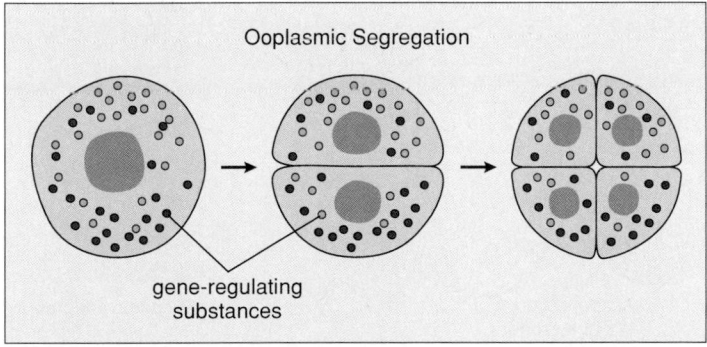

Ooplasmic Segregation

gene-regulating substances

The diagram proposes that the contents of the egg cytoplasm are not distributed uniformly. Following cleavage, therefore, embryonic cells will differ according to the cytoplasmic contents they receive.

The fact that the cytoplasm of an egg is not uniform can be substantiated by considering the egg of a frog. As we mentioned earlier, a frog's egg is polar: there is an animal pole and a vegetal pole (fig. 44.5a). The vegetal pole, which is distinguishable by the presence of yolk, becomes the germ layer called

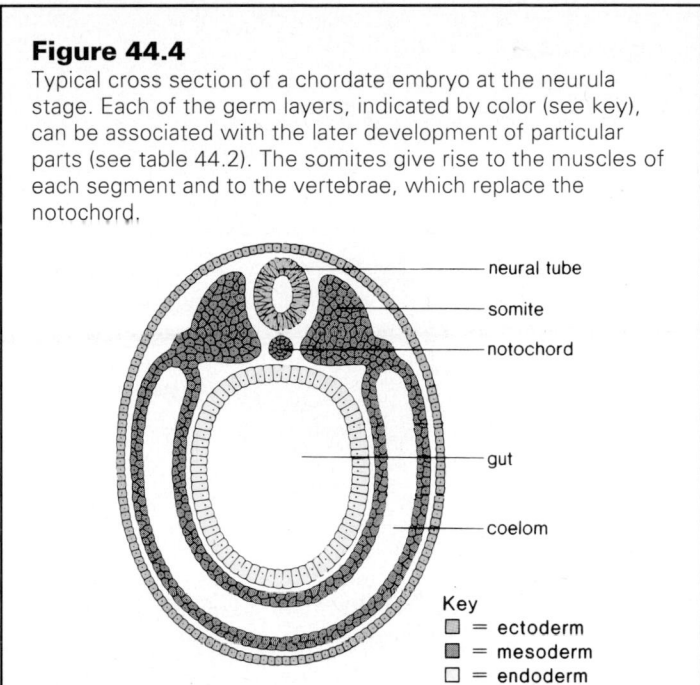

Figure 44.4

Typical cross section of a chordate embryo at the neurula stage. Each of the germ layers, indicated by color (see key), can be associated with the later development of particular parts (see table 44.2). The somites give rise to the muscles of each segment and to the vertebrae, which replace the notochord.

neural tube

somite

notochord

gut

coelom

Key
☐ = ectoderm
◼ = mesoderm
☐ = endoderm

endoderm. Researchers working with the frog *Xenopus* have been able to identify a particular mRNA (called Vg 1) that is localized at the vegetal pole during the process of oogenesis. It is translated into a particular peptide growth factor that perhaps induces the formation of endoderm. **Induction** is the ability of a chemical or a tissue to influence the development of another tissue. The inducing chemical is called a *signal.*

After a frog's egg is fertilized, contents of the egg shift position, and a *gray crescent* appears on the egg opposite the point where the sperm entered (fig. 44.5a). The gray crescent marks the dorsal side of the embryo. It most likely represents a combination of animal and vegetal pole contents, and it is speculated that the gray crescent might contain growth factors from each pole. The dorsal side of the embryo develops a notochord and nervous system—perhaps due to the action of these growth factors.

This line of reasoning is substantiated by an experiment performed by Hans Spemann, who received a Nobel Prize in 1935 for his extensive work in embryology. Spemann showed that if the gray crescent is divided equally by the first cleavage, each experimentally separated daughter cell develops into a complete embryo. If he caused the egg to divide so that only one daughter cell receives the gray crescent, however, only that cell becomes a complete embryo (fig. 44.5b). The other cell gives rise to a tissue mass that lacks any sign of a notochord and nervous system.

Spemann later showed that the gray crescent becomes the dorsal lip of the blastopore, where gastrulation begins. Since this region is necessary for complete development, he called it the primary organizer. The cells closest to Spemann's primary organizer become endoderm, those farther away become mesoderm,

Figure 44.5

Importance of the gray crescent in frog development **a.** The position of the gray crescent can be correlated with the anterior/posterior and dorsal/ventral axes of the body. **b.** The first cleavage normally divides the gray crescent in half and each daughter cell is capable of developing into a complete tadpole. But if only one daughter cell receives the gray crescent, then only that cell can become a complete embryo.

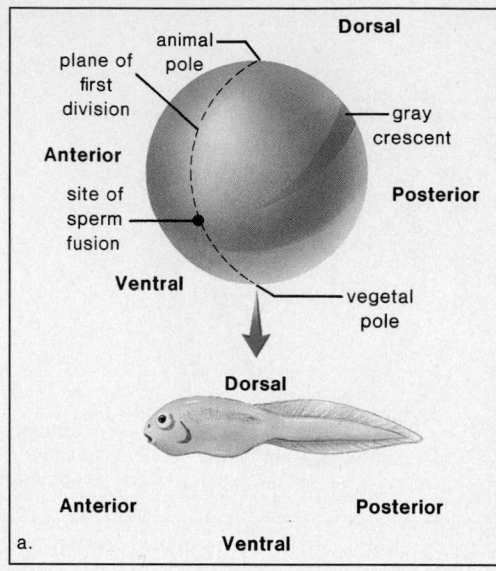

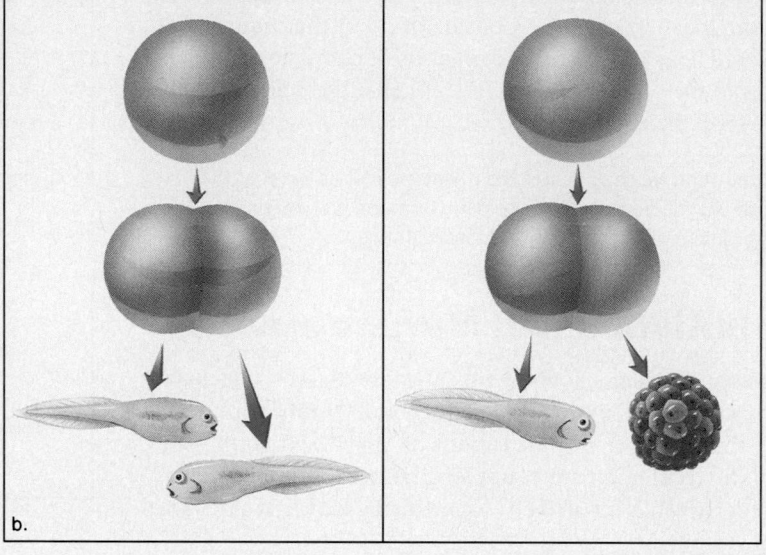

and those farthest away become ectoderm. This suggests that there may be a molecular concentration gradient that acts as a signal to induce germ layer differentiation. A recent experiment performed by Jim Smith of the National Institute of Medical Research in London indicates that the peptide growth factor called activin may play a role in the signaling process. At low concentrations of activin, animal pole cells become epidermis, an ectoderm derived tissue, and at higher concentrations they become muscle and notochord, a mesoderm derived tissue.

Later Spemann and his colleague Hilde Mangold discovered other instances of induction. For example, presumptive (potential) notochord tissue induces the formation of the nervous system (fig. 44.6). If presumptive nervous system tissue, located just above the presumptive notochord, is cut out and transplanted to the belly region of the embryo, it does not form a neural tube. On the other hand, if presumptive notochord tissue is cut out and transplanted beneath what would be belly ectoderm, this ectoderm differentiates into neural tissue. Still other examples of induction are now known. In 1905, Warren Lewis studied the formation of the eye in frog embryos (fig. 44.7). He found that an optic vesicle, which is a lateral outgrowth of developing brain tissue, induces overlying ectoderm to thicken and become a lens. The developing lens in turn induces an optic vesicle to form an optic cup, where the retina develops.

Figure 44.6

Experiment performed by Hans Spemann in 1924. **a.** The presumptive nervous system (blue) does not develop into the neural plate if moved from the normal location. **b.** The presumptive notochord (red) can cause the belly ectoderm to develop into the neural plate (blue). This shows that the notochord induces ectoderm cells to become a neural plate most likely by sending out chemical signals.

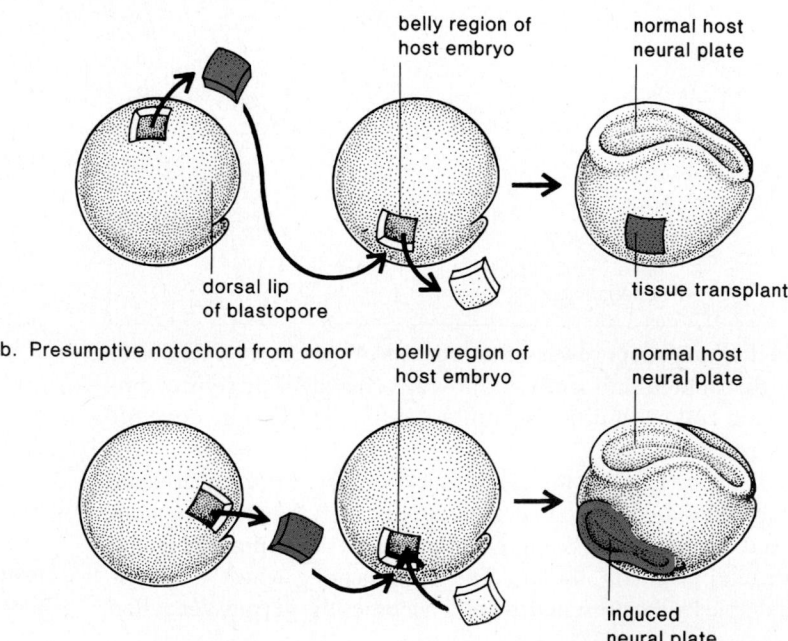

Animal Structure and Function

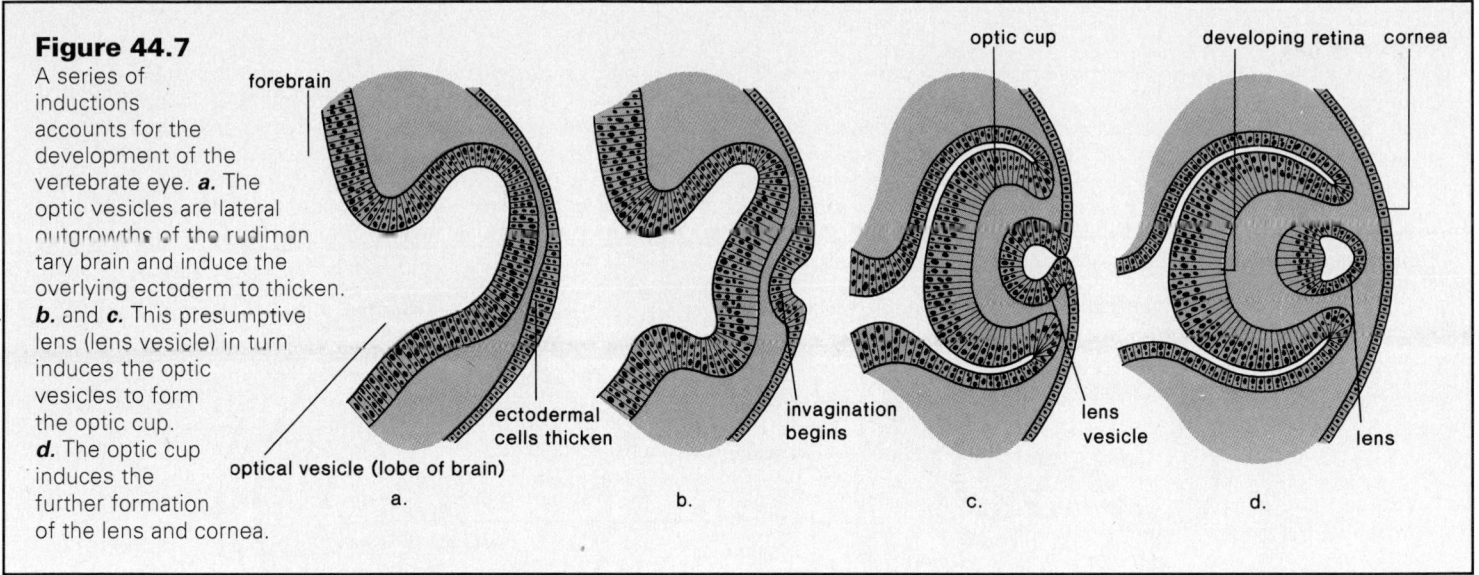

Figure 44.7
A series of inductions accounts for the development of the vertebrate eye. *a.* The optic vesicles are lateral outgrowths of the rudimentary brain and induce the overlying ectoderm to thicken. *b.* and *c.* This presumptive lens (lens vesicle) in turn induces the optic vesicles to form the optic cup. *d.* The optic cup induces the further formation of the lens and cornea.

forebrain

optic cup

developing retina   cornea

ectodermal cells thicken

invagination begins

lens vesicle

lens

optical vesicle (lobe of brain)

a.   b.   c.   d.

All these various experiments allow us to envision that as development proceeds, a series of inductive events occurs. A signal activates certain genes, which in turn encode other signals, which activate new genes, which encode new signals, and so forth:

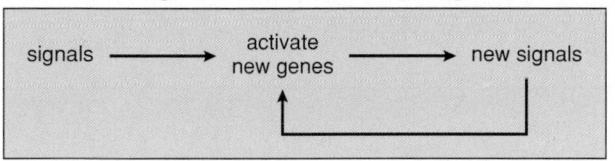

signals ⟶ activate new genes ⟶ new signals

The end result is the orderly process of development from the fertilized egg to the adult.

Homeobox Genes and Pattern Formation

A *homeobox gene,* also called a homeotic gene, is a master gene that controls the activity of many other genes necessary for embryonic development. Each organism has several of these genes. One portion of a homeobox gene (called the homeobox) contains a constant sequence of 180 nucleotides. The sequence of nucleotides in the variable region of a homeobox gene varies from gene to gene. A homeobox gene codes for a homeodomain protein. Each of these has a homeodomain, a sequence of 60 amino acids that is found in all other homeodomain proteins.

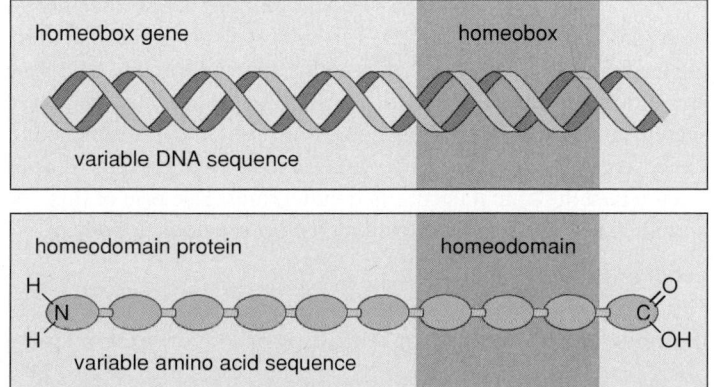

homeobox gene

homeobox

variable DNA sequence

homeodomain protein

homeodomain

variable amino acid sequence

Since homeobox genes and their products contain variable regions, each is different from the other. Their similarity is limited to just the homeobox and homeodomain, respectively.

Homeobox DNA can be radioactively labeled and used as a DNA probe (p. 283) to locate other homeobox genes. This methodology has allowed researchers to determine that almost all eukaryotic organisms contain homeobox genes. The conclusion is that all homeoboxes are derived from an original DNA sequence that has been largely conserved (maintained from generation to generation) because of its importance in the regulation of animal development.

Homeoboxes were discovered by researchers studying homeotic mutations in *Drosophila.* A homeotic mutation causes body parts to be misplaced—a homeotic mutant fly can have 2 pairs of wings or extra legs where antennae should be. Homeobox genes are clearly involved in pattern formation, that is, the shaping of an embryo so that the adult has a normal appearance. Homeobox genes are arranged in a definite order on a chromosome, and the gene closest to the DNA 5' end is mainly involved in determining the anterior portion of the embryo, while the gene closest to the 3' end is mainly involved in determining the most posterior portion of the embryo. In *Xenopus,* if the expression levels of the homeobox genes are altered, headless and tailless embryos are produced.

In keeping with the diagram explaining sequential development events above, researchers envision that a homeodomain protein produced by one homeobox gene binds to and turns on the next homeobox gene, and this orderly process determines the overall pattern of the embryo. It appears that homebox genes also establish homeodomain protein gradients that affect the pattern development of specific parts, such as the limbs. Many laboratories are engaged in discovering much more about homeobox genes and homeodomain proteins. We need to know, for example, how homeobox genes are turned on, if cells have receptors for homeodomain proteins, and how their gradient is maintained

Figure 44.8

Evolution of the extraembryonic membranes in the reptile made development on land possible. If an embryo develops in water, the water supplies oxygen, takes away waste, prevents desiccation, and provides a protective cushion. On land some of these functions are performed by the extraembryonic membranes, which must be protected from drying out by an outer shell or by internal development. ***a.*** Chick within its hard shell. The chorion lies just beneath the shell and carries on gas exchange. The allantois collects nitrogenous wastes. The yolk sac provides nourishment, and the amnion provides a watery environment. ***b.*** In humans, only the chorion and amnion have comparable functions. The chorion forms the fetal half of the placenta, where exchange with the mother's blood occurs, and the amnion provides a watery environment. The allantoic blood vessels become the umbilical blood vessels, and the yolk sac is the first site of blood cell formation.

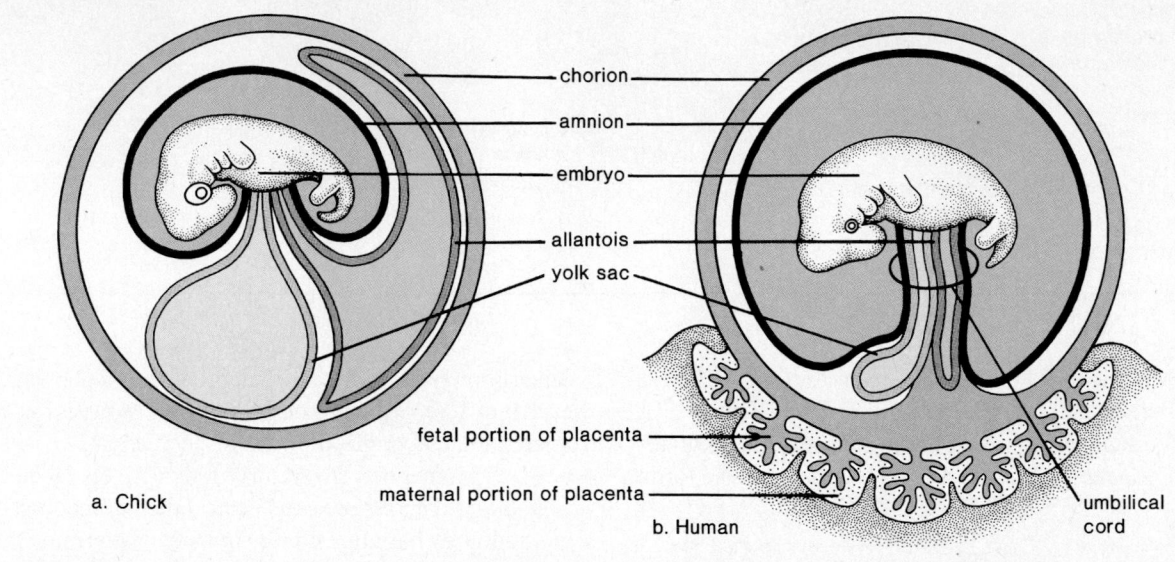

chorion
amnion
embryo
allantois
yolk sac
fetal portion of placenta
maternal portion of placenta
umbilical cord

a. Chick

b. Human

in tissues. The roles of the cytoskeleton and extracellular matrix are also being explored.

> Differentiation and morphogenesis are dependent upon signals (in many cases peptide growth hormones) from neighboring cells. The signals activate homeobox genes, which produce homeodomain proteins, which are themselves signals.

Human Development

Human development is often divided into embryonic development (first 2 mon) and fetal development (3–9 mon). The embryonic period consists of early formation of the major organs, and fetal development is a refinement of these structures.

Of the 3 animals we have previously discussed, human development most resembles that of the chick, no doubt because reptiles are a common ancestor of both birds and mammals. In the chick, the germ layers extend out over the yolk and develop into the *extraembryonic membranes,* so called because they lie outside the embryo. Their placement and function are described in figure 44.8. In humans, the extraembryonic membranes develop earlier than in the chick. In contrast to the chick, the human egg has little yolk, and the nutritional needs of the embryo are supplied by the chorion, which becomes part of the placenta.

Embryonic Development

Cleavage

Following fertilization in the oviduct, the human zygote begins to cleave as it travels down the oviduct to the uterus (fig. 44.9). The first cleavage is complete about a day after fertilization, and subsequent divisions occur at intervals of 8–10 hours. Passage of the embryo through the oviduct takes from 3–3½ days. While the embryo lies free in the uterine cavity, the morula becomes the *blastocyst* (the human equivalent of the blastula), consisting of the *inner cell mass* and a single layer of surrounding cells known as the *trophoblast.* Later, the trophoblast, reinforced by a layer of mesoderm, gives rise to the **chorion,** an extraembryonic membrane. The inner cell mass eventually becomes the fetus.

At the end of the first week, the embryo begins the process of implanting itself in the wall of the uterus. During implantation, the trophoblast secretes enzymes to digest away some of the tissue and blood vessels of the uterine wall. The trophoblast begins to secrete HCG (human chorionic gonadotropin), the hormone that is the basis for the pregnancy test and serves to maintain the corpus luteum past the time it normally disintegrates. Because of this, the endometrium is maintained and menstruation does not occur.

Gastrulation

Gastrulation occurs during the second week. The inner cell mass becomes the *embryonic disk,* consisting of a layer of endodermal

Figure 44.9

Human development before implantation. Structures and events proceed counterclockwise. At ovulation, the secondary oocyte leaves the ovary. Fertilization occurs in the oviduct. As the zygote moves along the oviduct, it undergoes cleavage to produce a morula. The blastocyst forms and implants itself in the uterine lining.

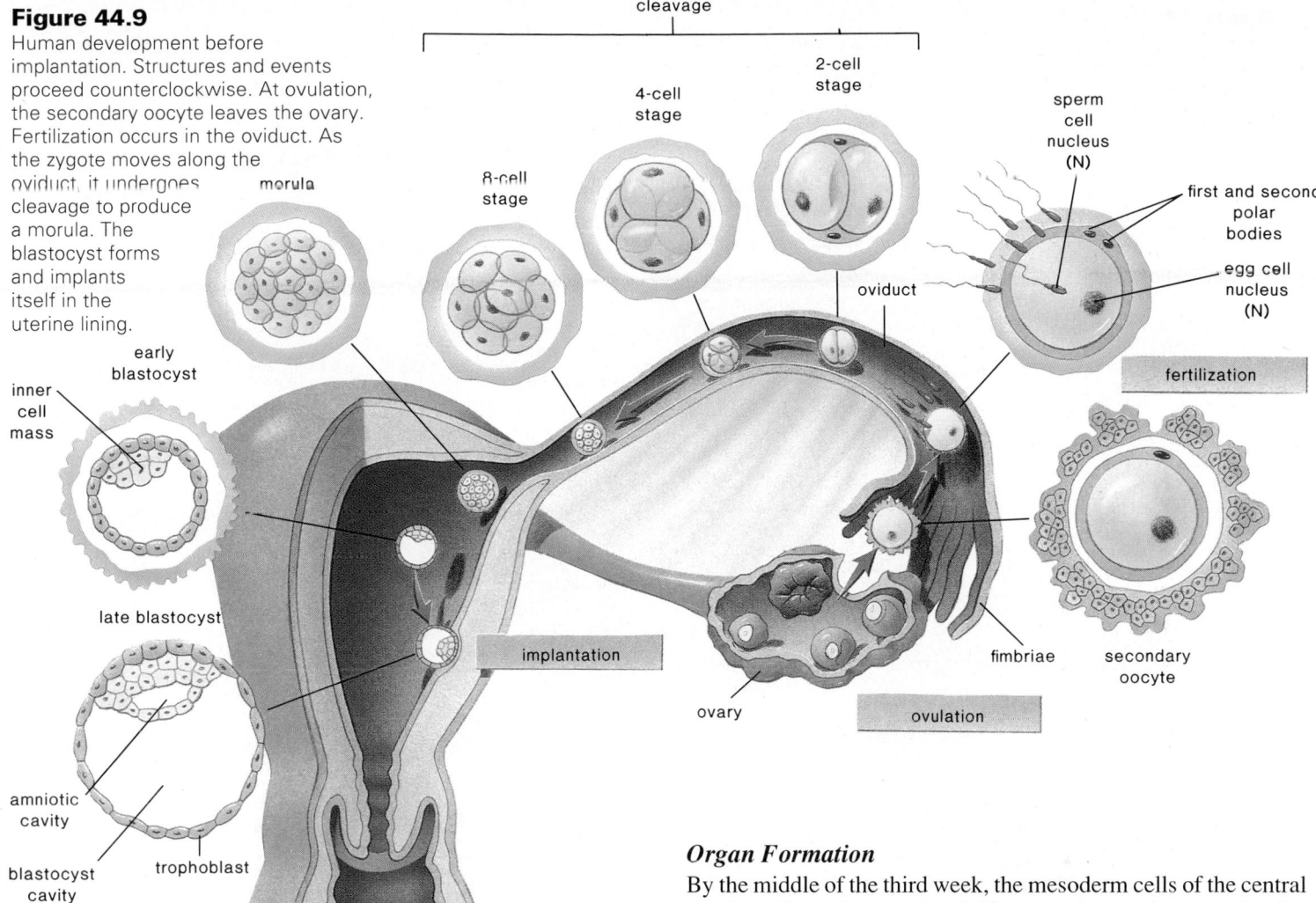

cleavage

2-cell stage

4-cell stage

8-cell stage

morula

early blastocyst

inner cell mass

sperm cell nucleus (N)

first and second polar bodies

egg cell nucleus (N)

oviduct

fertilization

late blastocyst

implantation

amniotic cavity

blastocyst cavity

trophoblast

fimbriae

secondary oocyte

ovary

ovulation

cells beneath a layer of ectodermal cells (fig. 44.10). The **amnion,** an extraembryonic membrane that contains a cavity filled with amniotic fluid, is now visible above the embryo. The amniotic fluid cushions the embryo and keeps it from drying out. At this time we also see the **yolk sac,** which has no nutritive function but is the first site of blood cell formation.

As gastrulation continues, mesoderm forms along the length of a primitive streak by invagination of cells between the ectoderm and endoderm. When the edges of the embryonic disk fold under, the embryo is converted into a tubular, 3-layered body. The ectoderm is on the outside, the endoderm is on the inside, and the mesoderm is between these 2. It is possible to relate the development of future organs of these germ layers (table 44.2).

Organ Formation

By the middle of the third week, the mesoderm cells of the central axis have formed a notochord. The central nervous system begins to develop above the notochord. Neural folds arise from the ectoderm and fuse to form the neural tube, which swells anteriorly to form the brain (fig. 44.11). The somites, which become the dermis of the skin, the muscle tissue, and the vertebrae and other bones, appear on both sides of the neural tube.

The heart and blood vessels also arise from the mesoderm. At first, the heart is 2 tiny tubes. These fuse to form a single chamber as the third week ends. The primitive heart bulges and is easily observed as it begins to pump blood through simple arteries and veins. These vessels take blood to the developing esophagus, stomach, liver, pancreas, and intestine.

At 4 weeks, the embryo is barely larger than the height of this print. There is a bridge of mesoderm called the body stalk, which connects the caudal end of the embryo with the chorion (fig. 44.10c). The fourth extraembryonic membrane, the **allantois,** is contained within this stalk, and its blood vessels become the umbilical blood vessels. Then the head and the tail lift up and body stalk moves anteriorly by constriction (fig. 44.10d). Once this process is complete, the umbilical cord that connects the developing embryo to the placenta is fully formed (fig. 44.10e).

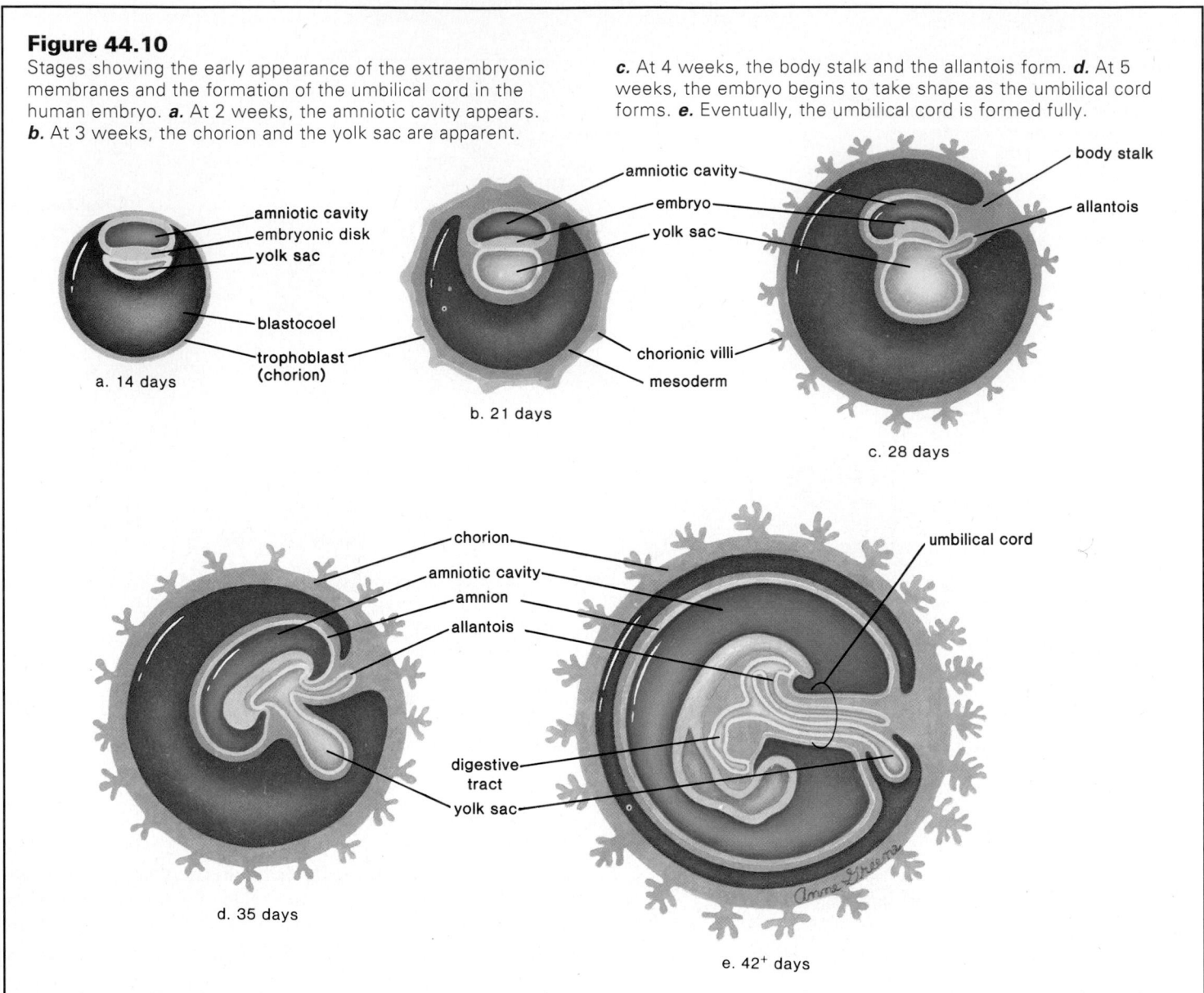

Figure 44.10

Stages showing the early appearance of the extraembryonic membranes and the formation of the umbilical cord in the human embryo. *a.* At 2 weeks, the amniotic cavity appears. *b.* At 3 weeks, the chorion and the yolk sac are apparent. *c.* At 4 weeks, the body stalk and the allantois form. *d.* At 5 weeks, the embryo begins to take shape as the umbilical cord forms. *e.* Eventually, the umbilical cord is formed fully.

a. 14 days

amniotic cavity
embryonic disk
yolk sac
blastocoel
trophoblast (chorion)

b. 21 days

amniotic cavity
embryo
yolk sac
chorionic villi
mesoderm

c. 28 days

body stalk
allantois

d. 35 days

chorion
amniotic cavity
amnion
allantois
digestive tract
yolk sac

e. 42⁺ days

umbilical cord

Little flippers called *limb buds* appear (fig. 44.12); later, the arms and the legs develop from the limb buds, and even the hands and the feet become apparent. At the same time, during the fifth week, the head becomes larger, and the sense organs become more prominent. It is possible to make out the developing eyes, ears, and even nose.

At first the skeleton is composed of cartilage, but the cartilage begins to be replaced by bone during the sixth week. There is a remarkable change in external appearance during the sixth through eighth weeks of development from a form that is difficult to recognize as human to one that is easily recognizable as human. Concurrent with brain development, the head achieves its normal relationship with the body as a neck region develops. The nervous system is developed well enough to permit reflex actions, such as a startle response to being touched. At the end of this period, the embryo is about 38 mm long and weighs no more than an aspirin tablet, even though all organ and systems are established.

Placenta Formation

The placenta begins formation once the embryo is implanted fully. Treelike extensions of the chorion called chorionic villi project into the maternal tissues (fig. 44.10*c–e*). Later, these disappear in all areas except where the placenta develops. By the tenth week, the placenta is fully formed and begins to produce progesterone and estrogen (fig. 44.13). These maintain the corpus luteum even though production of HCG by the placenta drops off. There is no follicle production or menstruation during pregnancy.

The umbilical cord contains the blood vessels that take fetal blood to and from the **placenta.** Here the chorionic villi are surrounded by maternal blood sinuses, yet the blood of the mother

Figure 44.11

Human embryo at 21 days. The neural folds still need to close at the anterior and posterior ends of the embryo. The pericardial area contains the primitive heart, and the somites are the precursors of the muscles.

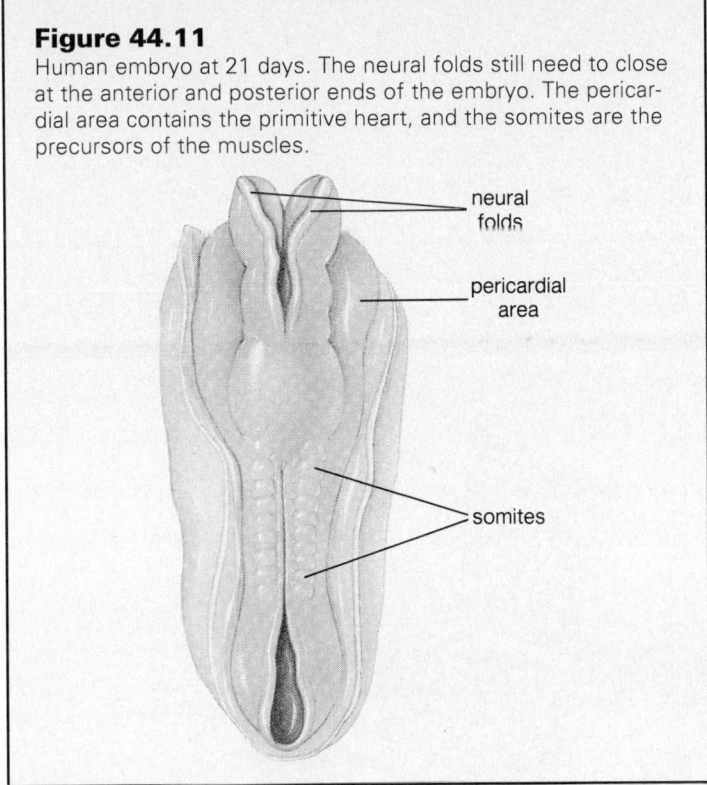

neural folds

pericardial area

somites

Figure 44.12

Human embryo at beginning of fifth week. **a.** Scanning electron micrograph. **b.** The embryo is curled so that the head touches the heart, 2 organs whose development is further along than the rest of the body. The organs of the gastrointestinal tract are forming, and the arms and legs develop from the bulges that are called limb buds. The presence of the tail is an evolutionary remnant; its bones regress and become those of the coccyx. The pharyngeal pouches only become functioning gills in fishes and amphibians; in humans the first pair of pharyngeal pouches becomes the auditory cavity of the middle ear and the eustachian tube. The second pair becomes the tonsils, while the third and fourth become the thymus and the parathyroids.

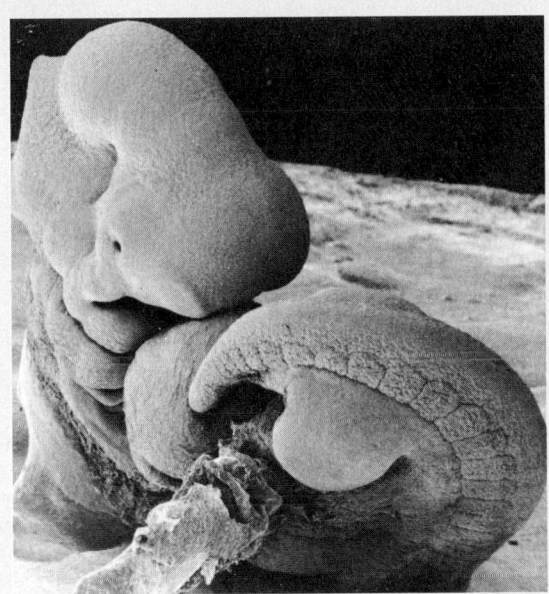

a.

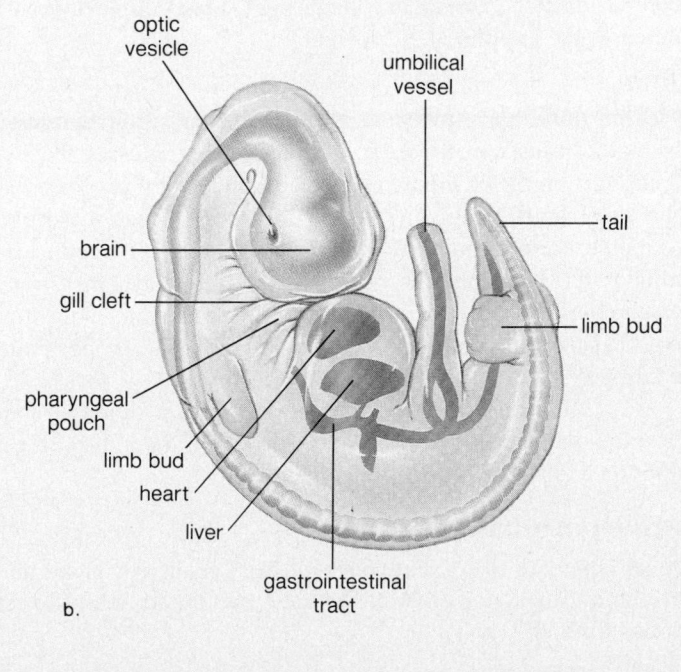

optic vesicle

umbilical vessel

brain

tail

gill cleft

limb bud

pharyngeal pouch

limb bud

heart

liver

gastrointestinal tract

b.

docs not mix with the blood of the fetus. Instead, exchange takes place across plasma membranes. Nitrogenous wastes and carbon dioxide move from the fetal side to the maternal side of the placenta, and nutrients and oxygen move from the maternal side to the fetal side of the placenta.

Harmful chemicals can also cross the placenta, and this is of particular concern during the embryonic period, when various structures are first forming. Each organ or part seems to have a sensitive period during which a substance can alter its normal function.

Fetal Development

Fetal development is marked by an extreme increase in size. Weight multiplies 600 times, going from less than 28 g to 3 kg. In this time, too, the fetus grows to about 50 cm in length. The genitalia finally make their appearance, and it is possible to tell if the fetus is male or female.

Soon, hair, eyebrows, and eyelashes add finishing touches to the face and head. In the same way, fingernails and toenails complete the hands and feet. A fine, downy hair (lanugo) covers the limbs and trunk, only to later disappear. The fetus looks like an old man because the skin is growing so fast that it wrinkles. A waxy, almost cheeselike substance (vernix caseosa) protects the wrinkly skin from the watery amniotic fluid.

The fetus at first only flexes its limbs and nods its head, but later it can move its limbs vigorously to avoid discomfort. The mother feels these movements from about the fourth month on. The other systems of the body also begin to function. The fetus begins to suck its thumb, swallow amniotic fluid, and urinate.

Development in Animals

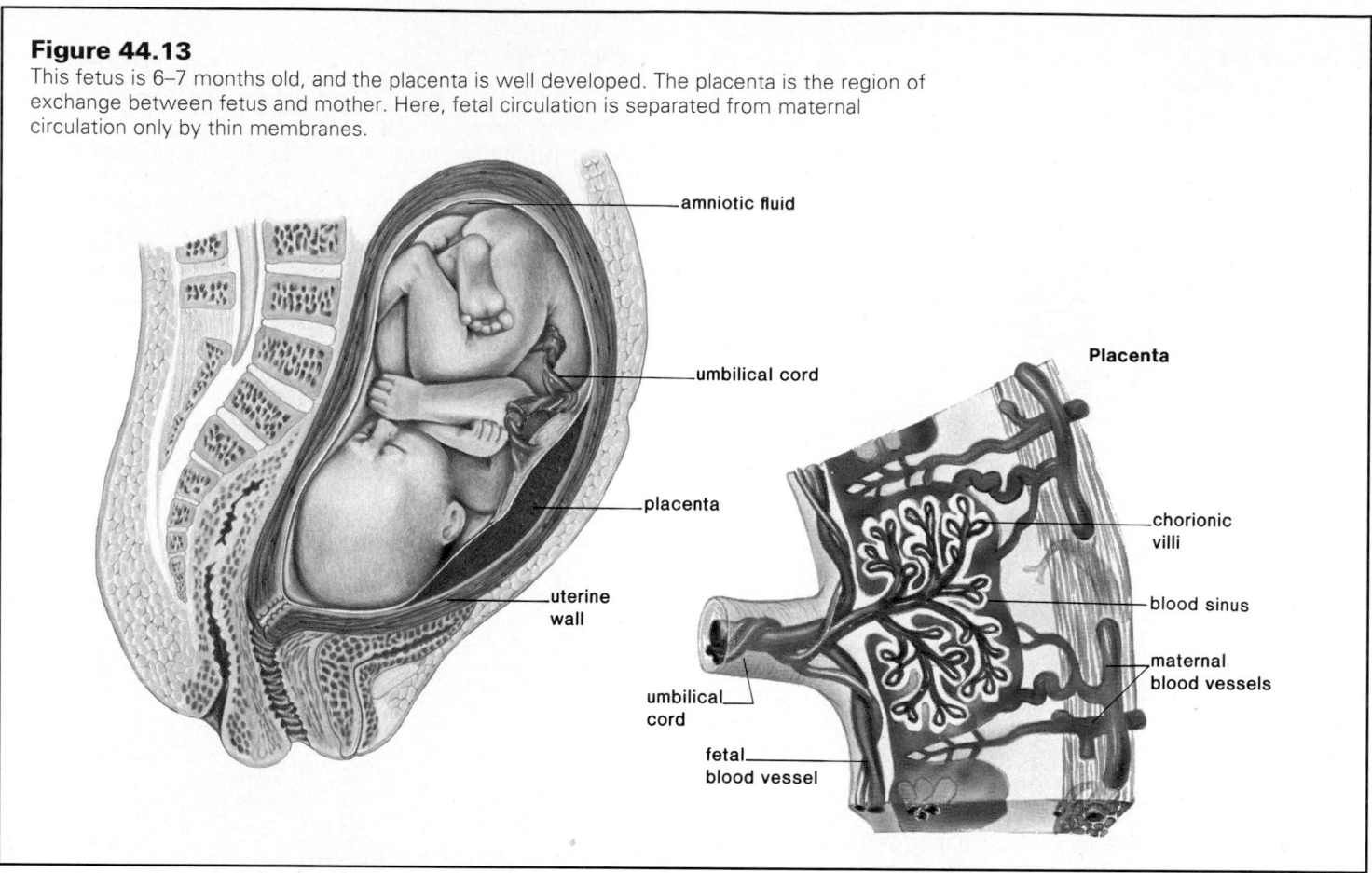

Figure 44.13
This fetus is 6–7 months old, and the placenta is well developed. The placenta is the region of exchange between fetus and mother. Here, fetal circulation is separated from maternal circulation only by thin membranes.

After 16 weeks, the fetal heartbeat is heard through a stethoscope. A fetus born at 24 weeks has a chance of surviving, although the lungs are still immature and often cannot capture oxygen adequately. Weight gain during the last couple of months increases the likelihood of survival.

Birth

The time of birth is usually calculated at 280 days from the start of the mother's last menstruation. As pregnancy progresses, the level of estrogen in the bloodstream exceeds the level of progesterone. This may help bring on birth of the fetus, because estrogen promotes uterine irritability. It is also possible that the fetus itself initiates the birth process by releasing a chemical messenger. Researchers have found in studies with sheep that if the fetal hypothalamus or pituitary is destroyed, birth can be postponed indefinitely.

The process of birth includes 3 stages: dilation of the cervix, expulsion of the fetus, and delivery of the afterbirth (the placenta and the extraembryonic membranes).

Development after Birth

Development does not cease once birth has occurred; it continues throughout the stages of life: infancy, childhood, adolescence, and adulthood.

Infancy lasts until about 2 years of age. It is characterized by tremendous growth and sensorimotor development. During *childhood,* the individual grows, and the body proportions change. *Adolescence* begins with *puberty,* when the secondary sex characteristics appear and the sexual organs become functional. At this time, there is an acceleration of growth leading to changes in height, weight, fat distribution, and body proportions. Males commonly experience a growth spurt later than females; therefore, they grow for a longer period of time. Males are generally taller than females and have broader shoulders and longer legs relative to their trunk length.

Young adults are at their physical peak in muscle strength, reaction time, and sensory perception. The organ systems at this time are best able to respond to altered circumstances in a homeostatic manner. From now on, however, there is an almost imperceptible, gradual loss in certain of the body's abilities.

Aging

Aging encompasses the progressive changes that contribute to an increased risk of infirmity, disease, and death. Figure 44.14 compares the percentage of organ function in a 75- to 80-year-old person to that of a 20-year-old person whose organs are assumed to function at 100% capacity. When making this comparison, we may note that the body has a vast functional reserve so that it can still perform well, even when not at 100% capacity.

Animal Structure and Function

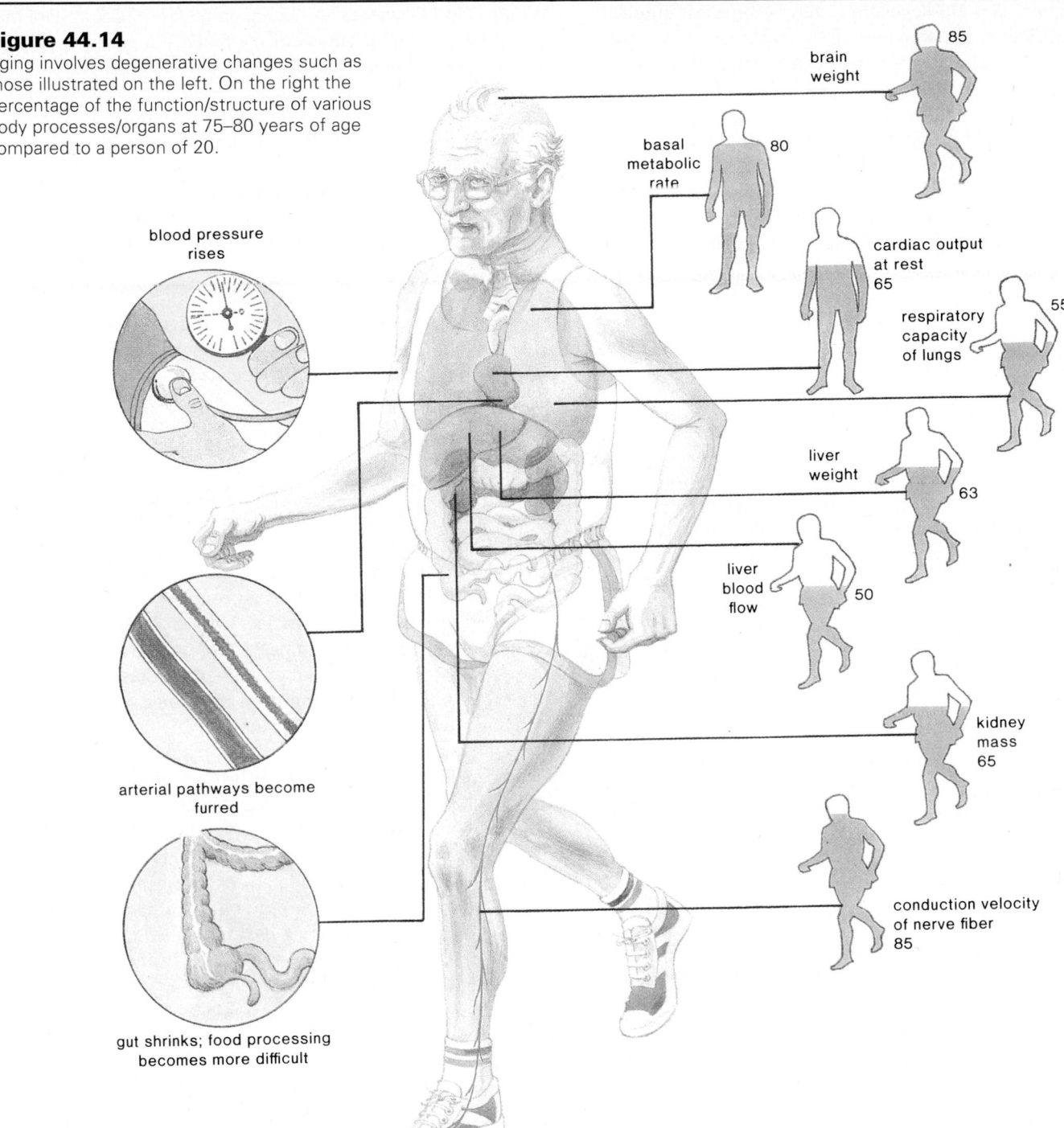

Figure 44.14
Aging involves degenerative changes such as those illustrated on the left. On the right the percentage of the function/structure of various body processes/organs at 75–80 years of age compared to a person of 20.

blood pressure rises

arterial pathways become furred

gut shrinks; food processing becomes more difficult

brain weight 85

basal metabolic rate 80

cardiac output at rest 65

respiratory capacity of lungs 55

liver weight 63

liver blood flow 50

kidney mass 65

conduction velocity of nerve fiber 85

There are many theories about what causes aging. We will consider 4 of them.

Genetic Origin Several lines of evidence indicate that we are genetically programmed to age. The maximum life span of animals is species specific; for humans it is about 110 years. But the children of long-lived parents tend to live longer than those of short-lived parents. Longevity may be related to the number of times the cells divide. A human cell divides a maximum of about

50 times. There could be a genetic off switch that causes cells to lose the ability to divide. David Danner, of the National Institutes of Aging, microinjected a mRNA for the protein prohibin into fibroblasts and they stopped dividing. The gene coding for this protein could be the off switch. It's also possible that other genes that normally block cell division are turned on in aging cells. Other researchers have found that the retinoblastoma (RB) gene is a tumor suppressor that growth factors fail to inactivate in older cells, for example.

Cellular Repair As cells degenerate, they become less efficient at self-repair. Ordinarily, whenever DNA replicates, any errors that occur are enzymatically repaired. Due to environmental insults, the DNA repair system may falter so that errors in replication are not corrected. Mutations may result that lead to the production of nonfunctional proteins. Eventually, the number of inadequately functioning cells may build up and contribute to an aging process.

It's also noted that as cells age, they begin to accumulate metabolic by-products that could damage DNA. For example, senescent cells accumulate free radicals, which are molecules that have an extra electron and can readily bond with cellular macromolecules. Housekeeping enzymes known as antioxidants ordinarily prevent the formation of free radicals. Longer-lived species are known to have more antioxidants than shorter-lived species, and it is hypothesized that older cells also have fewer copies of these beneficial enzymes.

Glucose is a destructive agent that attaches itself to protein. The protein-sugar complex starts a cross-linking process that produces a yellowish brown mass known as advanced glycosylation end products (AGEs). Like free radicals, AGEs may interact with DNA, causing mutations that interfere with the cell's ability to replicate and repair DNA molecules. The presence of AGE-derived cross-links explains why cataracts develop in older persons, and it may also help explain the development of atherosclerosis and the inefficiency of the kidneys in diabetics and older individuals. Researchers are presently experimenting with the drug aminoguanidine, which can prevent the development of AGEs.

Werner syndrome is a rare disorder in which those affected show signs of advanced aging in their 20s. Werner cells have an excess of collagen and fibronectin, as do aging cells in general. Collagen makes up the white fibers in connective tissues, and increasingly collagen cross-linking occurs as people age. Undoubtedly, this cross-linking contributes to the stiffening and loss of elasticity characteristic of aging tendons and ligaments. It may also account for the inability of organs such as the blood vessels, heart, and lungs to function as they once did.

Researchers have not decided what comes first: accumulation of metabolic by-products or a genetic defect that affects the housekeeping enzymes that prevent their formation. Ultimately, the hope is to identify the genetic and cellular events that lead to aging so that cells can live longer while maintaining their normal functions.

Whole Body Processes A decline in the hormonal system may affect many different organs of the body. For example, Type II diabetes is a common disorder in older individuals. The pancreas makes insulin, but the cells lack the receptors that enable them to respond. Menopause in women occurs for a similar reason. There is plenty of FSH in the bloodstream, but the ovaries do not respond. Perhaps aging results from the loss of hormonal activity and a decline in the functions they control.

With age, the immune system, too, no longer performs as it once did, and this may affect the body as a whole. The thymus gland gradually decreases in size, and eventually most of it is replaced by fat and connective tissue. The incidence of cancer increases among the elderly, which may signify that the immune system is no longer functioning as it should. This idea is further substantiated by the increased incidence of autoimmune diseases in older individuals.

Extrinsic Factors The present data on the effects of aging are often based on comparing the characteristics of the elderly with younger age groups. But today's elderly perhaps were not as aware of the importance of diet and exercise for general health. It's possible, then, that much of what we attribute to aging is instead due to years of poor health habits. For example, osteoporosis is associated with a progressive decline in bone density in both males and females. Fractures are more likely to occur after only minimal trauma. Osteoporosis is common in the elderly—by age 65, one-third of women will have had vertebral fractures, and by age 81, one-third of women and one-sixth of men will have suffered a hip fracture. While there is no denying that a decline in bone mass results from aging, certain extrinsic factors are also important. The occurrence of osteoporosis itself is associated with cigarette smoking, heavy alcohol intake, and inadequate calcium intake. Not only is it possible to eliminate these negative factors by personal choice, it is also possible to add a positive factor: a moderate exercise program has been found to slow down the progressive loss of bone mass.

Rather than collecting data on the average changes observed between different age groups, it may be more useful to note the differences within any particular age group so that any extrinsic factors contributing to a decline or promoting the health of an organ can be identified.

Summary

1. Any developmental change requires 3 processes: growth, differentiation, and morphogenesis.

2. The development of 3 types of animals (lancelet, frog, and chick) is compared. The first 3 stages (cleavage, blastulation, and gastrulation) differ according to the amount of yolk in the egg.

3. During cleavage, the zygote divides, but there is no overall growth. The result is a morula, which becomes the blastula when an internal cavity (the blastocoel) appears. During the gastrula stage, invagination of cells into the blastocoel results in formation of the germ layers: ectoderm, mesoderm, and endoderm. Later development of organs can be related to these layers.

4. During neurulation, the nervous system develops from mid-line ectoderm, just above the notochord. At this point, it is possible to draw a typical cross section of a vertebrate embryo (fig. 44.4).

5. Differentiation begins with cleavage, when the egg's cytoplasm is partitioned among the numerous cells. The cytoplasm is not uniform in content, and presumably each of the first few cells differ as to their cytoplasmic contents. Some probably contain substances that can influence gene activity—turning some genes on and others off.

6. After the first cleavage of a frog embryo, only a daughter cell that receives a portion of the gray crescent is able to develop into a complete embryo. This illustrates the importance of cytoplasmic inheritance to early development.
7. Morphogenesis involves the process of induction. The notochord induces the formation of the neural tube in frog embryos. The reciprocal induction that occurs between the lens and the optic vesicle is another good example of this phenomenon.
8. Today we envision induction as always present because cells are believed to constantly give off signals that influence the genetic activity of neighboring cells.
9. Studies of homeobox genes and pattern formation further indicate that morphogenesis involves genetic and environmental influences.
10. Human development can be divided into embryonic development (first 2 mon) and fetal development (3–9 mon). The early stages in human development resemble those of the chick. The similarities are probably due to their evolutionary relationship, not the amount of yolk the eggs contain, because the human egg has little yolk.
11. Fertilization occurs in the oviduct, and cleavage occurs as the embryo moves toward the uterus. The morula becomes the blastocyst before implanting in the uterine lining. Human gastrulation occurs as in the chick, and the three germ layers form.
12. The extraembryonic membranes appear early in human development. The trophoblast of the blastocyst is the first sign of the chorion, which goes on to become the fetal part of the placenta. The placenta is where exchange occurs between fetal and maternal blood. The amnion contains the amniotic fluid, which cushions and protects the embryo. The yolk sac and allantois are largely vestigial.
13. Organ development begins with neural tube and heart formation. There follows a steady progression of organ formation during embryonic development. During fetal development, refinement of features occurs, and the fetus adds weight. Birth occurs about 280 days after the start of the mother's last menstruation.
14. Development after birth consists of infancy, childhood, adolescence, and adulthood. Young adults are at their prime, and then the aging process begins. Aging may be genetic in origin, due to cellular repair changes, whole body processes, or extrinsic factors.

Study Questions

1. State, define, and give examples of the 3 processes that occur whenever a developmental change occurs.
2. Compare the process of cleavage and the formation of the blastula and gastrula in lancelets, frogs, and chicks.
3. State the germ layer theory, and tell which organs are derived from each of the germ layers.
4. Draw a cross section of a typical chordate embryo at the neurula stage, and label your drawing.
5. Give reasons for suggesting that differentiation and morphogenesis are dependent upon signals given off by neighboring cells. What do the signals bring about in the receiving cells?
6. Explain an experiment performed by Spemann that suggests the notochord induces formation of the neural tube. Give another well-known example of induction between tissues.
7. List the human extraembryonic membranes, give a function for each, and compare this function to that in the chick.
8. Tell where the stages of fertilization, cleavage, morula, and blastula occur in humans. What happens to the blastula in the uterus?
9. Describe the structure and function of the placenta in humans.
10. Outline the developmental changes that occur in humans from fetal development to adulthood.
11. Discuss 3 theories of aging.

Objective Questions

1. Which of these stages is the first one out of sequence?
   a. cleavage
   b. blastula
   c. morula
   d. gastrula
2. Which of these stages is mismatched?
   a. cleavage—cell division
   b. blastula—gut formation
   c. gastrula—three germ layers
   d. neurula—nervous system
3. Which of the germ layers is best associated with development of the heart?
   a. ectoderm
   b. mesoderm
   c. endoderm
   d. All of these.
4. Differentiation begins at what stage?
   a. cleavage
   b. blastula
   c. gastrula
   d. neurula
5. Morphogenesis is best associated with
   a. overall growth.
   b. induction of one tissue by another.
   c. genetic mutations.
   d. All of these.
6. In humans, the placenta develops from the chorion. This indicates that human development
   a. resembles that of the chick.
   b. is dependent upon extraembryonic membranes.
   c. cannot be compared to lower animals.
   d. only begins upon implantation.
7. In humans, the fetus
   a. is surrounded by 4 extraembryonic membranes.

b. has developed organs and is recognizably human.

c. is dependent upon the placenta for excretion of wastes and acquisition of nutrients.

d. Both b and c.

8. Developmental changes

    a. require growth, differentiation, and morphogenesis.

    b. stop occurring as soon as one is grown.

    c. are dependent upon a parceling out of genes into daughter cells.

    d. Both a and c.

9. Mesoderm forms by invagination of cells in the

    a. lancelet.

    b. frog.

    c. chick.

    d. All of these.

10. Which of these is mismatched?

    a. brain—ectoderm

    b. gut—endoderm

    c. bone—mesoderm

    d. lens—endoderm

11. Label this diagram illustrating the placement of the extraembryonic membranes, and give a function for each membrane in humans:

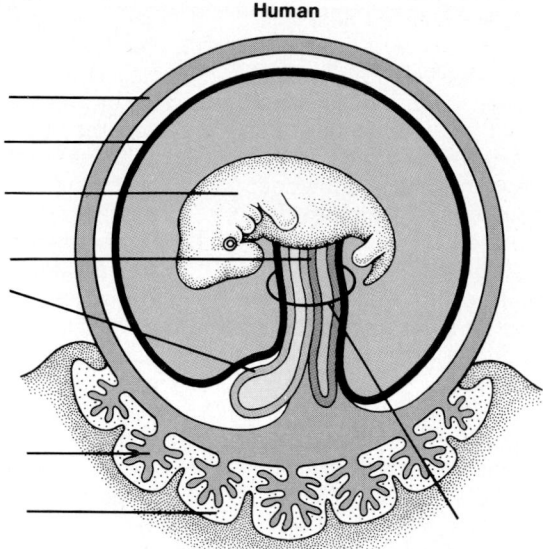

Human

Concepts and Critical Thinking

1. *Development occurs throughout the life of an animal.*

   Why is senescence considered a part of development?

2. *Chemical signals are involved in development.*

   Should these chemical signals be considered hormones? Why or why not?

3. *The genes control development.*

   Use the diagram on page 717 to substantiate a genetic theory of aging.

Selected Key Terms

differentiation (dif"er-en"she-a'shun) 712
morphogenesis (mor"fo-jen'ĕ-sis) 712
cleavage (klēv'ij) 713
blastocoel (blas'to-sēl) 713
blastula (blas'tu-lah) 713

gastrula (gas'troo-lah) 713
ectoderm (ek'to-derm) 713
endoderm (en'do-derm) 713
mesoderm (mes'o-derm) 714
germ layers (jerm la'erz) 714

chorion (kor'e-on) 718
amnion (am'ne-on) 719
yolk sac (yōk sak) 719
allantois (ah-lan'to-is) 719
placenta (plah-sen'tah) 720

PART 5 *Animal Structure and Function*

CRITICAL THINKING CASE STUDY

Skin Breathing

Most animals have specialized respiratory organs like gills or lungs for gas exchange. Many animals, especially amphibians, are also capable of skin breathing, or exchanging gases across the skin. Physiologists decided to test the hypothesis that *amphibians have physiological adaptations that favor skin breathing under appropriate conditions.*

Skin capillaries can be open (carrying blood) or closed (not carrying blood). The amount of open capillaries was determined in normal frogs and in frogs prevented from using their lungs. What do you predict about the amount of open capillaries in the skin of frogs prevented from using their lungs compared to those that use their lungs?

Prediction 1 There will be more open capillaries in the skin of frogs prevented from using their lungs.

Result 1 The predicted results were observed. Apparently amphibians are capable of regulating the amount of open capillaries in the skin to make more blood available for skin breathing.

I In another experiment, frogs that had been submerged and were skin breathing in the water were transferred to air. Give 2 reasons why you would expect the amount of open capillaries in the skin to decrease.

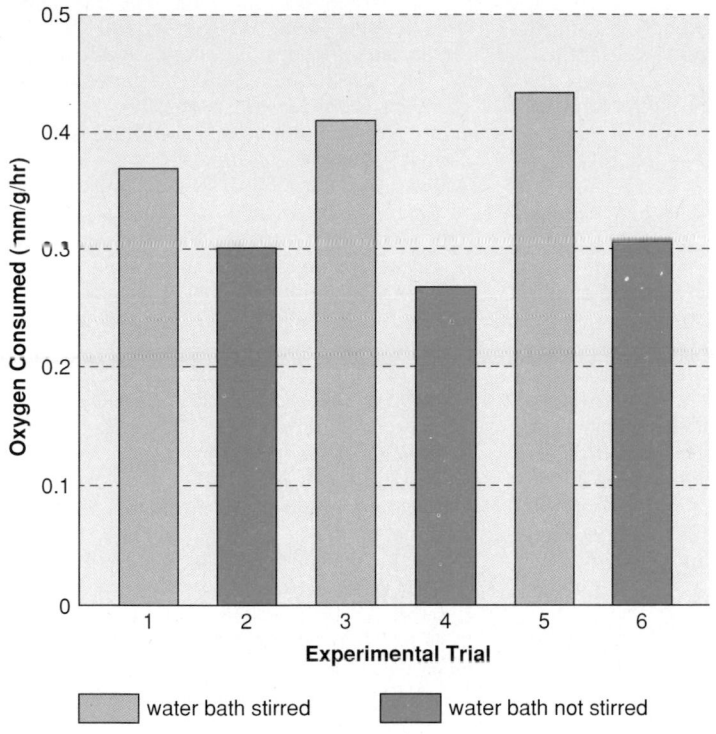

 water bath stirred   water bath not stirred

Figure A
Oxygen consumed by frog via skin-breathing. Adapted from "Skin Breathing in Vertebrates" by Martin E. Feder and Warren W. Burggran. Copyright © 1985 Scientific American, Inc. All rights reserved.

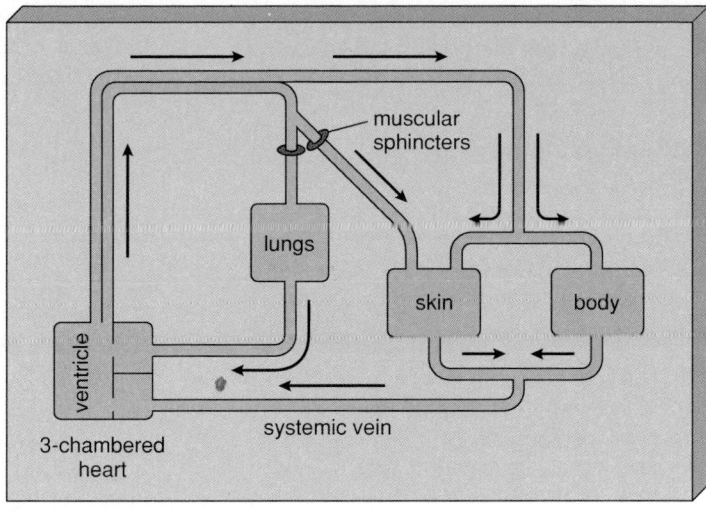

Figure B
Circulatory system in amphibian showing the 3-chambered heart and circulation to lungs, skin, and body. The muscle sphincters can shunt deoxygenated blood to skin or lungs depending on the environmental conditions.

Decreased distance between skin capillaries and the external medium speeds the process of diffusion, and this facilitates skin breathing. One group of tadpoles was raised in oxygen-rich water and another group was raised in oxygen-poor water. What do you predict about the distance of skin capillaries from the external environment in these 2 groups of tadpoles?

Prediction 2 Tadpoles raised in oxygen-poor water will have skin capillaries closer to the external environment than tadpoles raised in oxygen-rich water.

Result 2 Examination revealed that tadpoles raised in the oxygen-poor environment had skin capillaries 20 micrometers closer to the environment than tadpoles raised in an oxygen-rich environment. Moreover, these skin capillaries were finer and grew in a more dense network.

II What are the relative chances of effective skin breathing if tadpoles raised in an oxygen-poor environment are placed in an oxygen-rich environment? if tadpoles raised in an oxygen-rich environment are placed in an oxygen-poor environment?

Flowing water as opposed to stagnant water makes more oxygen available to the skin and therefore favors skin breathing. Amphibians are sometimes observed to have rhythmic body movements when submerged in water. Such movement may facilitate skin breathing. A frog immobilized in a wire cage was immersed in a water bath so that only its mouth and nostrils were exposed to air. Oxygen uptake from the water was determined when the water was stirred and when the water was not stirred. (Oxygen uptake was determined by periodically measuring the oxygen concentration in the water bath. A layer of mineral oil was placed on the surface of the water to prevent atmospheric oxygen from replacing the oxygen consumed.) What do you predict regarding oxygen consumption when the water was stirred and when it was not stirred?

Prediction 3 There will be more oxygen consumption when the water is stirred as opposed to when it is not stirred.

Result 3 The results of this experiment are shown in figure A. These data indicate that oxygen consumption is greater when the water bath is stirred than when it is not stirred. These results suggest that the body movements of amphibians are a behavioral adaptation to promote skin breathing.

III What would be the results given in figure A if mineral oil hadn't been placed on the water?
What would be the results in figure A if the experiment was repeated with no frog? Could this experiment serve as a control? Why?
Would a frog have more skin capillaries open when the water was stirred or not stirred? Why?

Figure B diagrams the circulatory system of an amphibian. Note the muscular sphincters, which apparently can control blood flow to the lungs or the skin. It is possible to measure blood flow in the pulmonary artery, which carries blood to the lungs, and in the cutaneous artery, which takes blood to the skin. Which blood vessel do you predict will carry more blood when a toad is prevented from using its lungs? when a toad is using its lungs?

Prediction 4 If a toad is prevented from using its lungs, the cutaneous artery will carry more blood. If a toad is using its lungs, the pulmonary artery will carry more blood.

Result 4 Subsequent observations were consistent with the prediction.

IV Suppose the data show that there is always more blood in the cutaneous artery even when a frog is using its lungs. What would be the implication regarding skin breathing in amphibians? Would this refute the hypothesis? Why?

A diversity of experiments indicates support for the hypothesis that amphibians have physiological adaptations that favor skin breathing under appropriate conditions.

Other Questions

1. Would you expect to find skin breathing in reptiles and mammals? Why or why not?
2. Effective gas exchange is dependent upon speed of diffusion. Order the following circumstances according to effective gas exchange. Explain your answer.
   a. capillaries close to external medium; oxygen-poor water
   b. capillaries close to external medium; oxygen-rich water
   c. capillaries far from external medium; oxygen-poor water
   d. capillaries far from external medium; oxygen-rich water

References

Feder, M. E., and Burggren, W. W. 1985. Cutaneous gas exchange in vertebrates: Design, patterns, control and implications. *Biological Reviews.* 60:1–45.
———. 1985. Skin breathing in vertebrates. *Scientific American,* 126–43.

Suggested Readings for Part 5

Alkon, D. L. July 1989. Memory storage and neural systems. *Scientific American.*

Aral, S. O., and Holmes, K. K. February 1991. Sexually transmitted diseases in the AIDS era. *Scientific American.*

Atkinson, M. A., and Maclaren, N. K. July 1990. What causes diabetes? *Scientific American.*

Barlow, R. B., Jr. April 1990. What the brain tells the eye. *Scientific American.*

Berns, M. B. June 1991. Laser surgery. *Scientific American.*

Cantin, M., and Genest, J. February 1986. The heart as an endocrine gland. *Scientific American.*

Cerami, A., et al. May 1987. Glucose and aging. *Scientific American.*

Cohen, I. R. April 1988. The self, the world and autoimmunity. *Scientific American.*

Crapo, L. 1985. *Hormones, the messengers of life.* New York: W. H. Freeman and Co.

DeRobertis, E. M., Oliver, G., and Wright, C. V. E. July 1990. Homeobox genes and the vertebrate body plan. *Scientific American.*

Freeman, W. J. February 1991. The physiology of perception. *Scientific American.*

Golde, D. W. December 1991. The stem cell. *Scientific American.*

Grey, H. M., Sette, A., and Buus, S. November 1989. How T cells see antigen. *Scientific American.*

Hole, J. W. 1990. *Human anatomy and physiology.* 5th ed. Dubuque, Iowa: Wm. C. Brown Publishers.

Holloway, M. March 1991. Rx for addiction. *Scientific American.*

Kalil, R. E. December 1989. Synapse formation in the developing brain. *Scientific American.*

Koretz, J. F., and Handelman, G. H. July 1988. How the human eye focuses. *Scientific American.*

Mills, J., and Masur, H. August 1990. AIDS-related infections. *Scientific American.*

Nathans, J. February 1989. The genes for color vision. *Scientific American.*

Orci, L., et al. September 1988. The insulin factor. *Scientific American.*

Powell, C. P. June 1991. Peering inward. *Scientific American.*

Rasmussen, H. October 1989. The cycling of calcium as an intracellular messenger. *Scientific American.*

Rennie, J. December 1990. The body against itself. *Scientific American.*

Scientific American. October 1988. Entire issue is devoted to articles on AIDS.

Scrimshaw, N. S. October 1991. Iron deficiency. *Scientific American.*

Selkoe, D. J. November 1991. Amyloid protein and Alzheimer's disease. *Scientific American.*

Smith, K. A. March 1990. Interleukin-2. *Scientific American.*

Ulmann, A., Teutsch, G., and Philibert, D. June 1990. RU 486. *Scientific American.*

Uvnas-Moberg, K. July 1989. The gastrointestinal tract in growth and reproduction. *Scientific American.*

von Boehmer, H., and Kisielow, P. October 1991. How the immune system learns about self. *Scientific American.*

Wassarman, P. M. December 1988. Fertilization in mammals. *Scientific American.*

Young, J., and Cohn, Z. January 1988. How killer cells kill. *Scientific American.*

Zivin, J. A., and Choi, D. W. July 1991. Stroke therapy. *Scientific American.*

Animal Structure and Function

PART 6

Behavior and Ecology

This part will show that to understand a species completely, you have to know where it lives and how it interacts with other species. White-tailed deer live in woodlands, where they feed by grazing on grass or browsing on shrubs and trees. About once a year, females bear young, which remain with them and are nursed for about 7 months. The males live in bachelor groups, and the 2 sexes only come together in the fall when mating occurs. Ever on guard against predators, the tail is raised and a white underside is revealed as they flee from any possible danger.

45

Animal Behavior

Green-backed herons hide among the branches of an evergreen tree. The behavior of an animal is directed toward surviving and reproducing. Making detection more difficult is an antipredator defense that helps these birds live another day.

Your study of this chapter will be complete when you can

1. discuss the immediate (proximate) control of behavior and the evolutionary (ultimate) control of behavior;
2. discuss behavior as having a genetic basis and under the influence of lifetime experiences, including learning;
3. list 4 components of most behavior patterns;
4. name 5 types of learned behavior and give examples;
5. show that feeding behavior, finding and defending a territory, and reproductive behavior are adaptive for the individual;
6. list and discuss ways in which animals communicate with one another;
7. list and discuss the benefits and disadvantages to living in a society;
8. discuss altruism, kin selection, and inclusive fitness as they apply to sociobiology.

A nimals carry on many activities that help them get food, avoid predators, find shelter, seek mates, and reproduce. *Behavior* encompasses all those activities that allow an animal to meet its needs.

Ethologists (biologists who study behavior) try to determine what are the proximate causes and what are the ultimate causes of behavior. *Proximate causes* are the immediate causes. For example, we know that the nervous system and endocrine system control behavior and that these systems are a part of the phenotype and therefore have a genetic basis. The *ultimate cause* is determined by natural selection—behavior is expected to increase the fitness of the individual. Recall that David P. Barash explained the results of his mountain bluebird experiment by saying that it is adaptive for males to be aggressive toward a rival before the eggs are laid but maladaptive after the eggs are laid (p. 18).

Mechanisms of Behavior

Mechanisms of behavior pertain to the proximate cause of behavior without considering the ultimate (evolutionary) cause of behavior. Genes determine the phenotype, which includes the nervous system and the endocrine system. Behavioral patterns are neuromotor responses to internal and external stimuli.

Physiological Control of Behavior

The roles of the nervous system and hormones in animal behavior are exemplified very well by a study of ringdove reproductive behavior. When male and female ringdoves are put together in a cage, the male begins courting by repeatedly bowing and cooing (fig. 45.1). Because castrated males do not do this, it can be reasoned that the hormone testosterone readies the male for this behavior. The sight of the male courting causes the pituitary gland in the female to release FSH. Then her ovaries produce eggs and release estrogen into the bloodstream. Now both male and female construct a nest, during which time copulation takes place. The hormone progesterone will cause birds of either sex to incubate the eggs; the hormone prolactin causes crop growth so that both parents are capable of feeding their young crop milk.

To determine whether the incubation behavior of the female is actually controlled by this sequence of events, investigators studied 3 experimental groups: (1) females housed alone, (2) females housed with males only, and (3) females housed with a male and nesting material. Each group of females was presented with a nest containing eggs. All females housed with a male and nesting material incubated eggs but not so for the other groups. In another study, male ringdoves were allowed to court and mate with females and to participate in nest construction. After the eggs had been laid, and incubation had begun, each male was placed behind a partition so that he could see his mate sitting on the eggs. Only this visual experience was then needed for the male to undergo normal crop development.

Complex interactions between external cues, the nervous system, and hormones control the reproductive cycle of ringdoves.

Rhythmic or Cyclic Behavior

Certain behavior patterns of animals naturally reoccur at regular intervals. Behavior that occurs on a daily basis is said to have a **circadian rhythm.** For example, some animals, like humans, are usually active during the day and sleep at night. Others, such as bats, sleep during the day and hunt at night. There are also behavior

Figure 45.1
Ringdove mating behavior. *a.* Male and female responding to one another. *b.* Copulation takes place. *c.* Male and female create a nest. *d.* Chicks are fed "crop milk." The reproductive behavior of ringdoves is controlled by the nervous and endocrine systems.

a.

b.

c.

d.

Figure 45.2

Biological clock system. Biological clock systems have 3 components: (**a**) an internal timekeeper, (**b**) a means of detecting light/dark periods, (**c**) a method of communication that eventually brings about a behavior response.

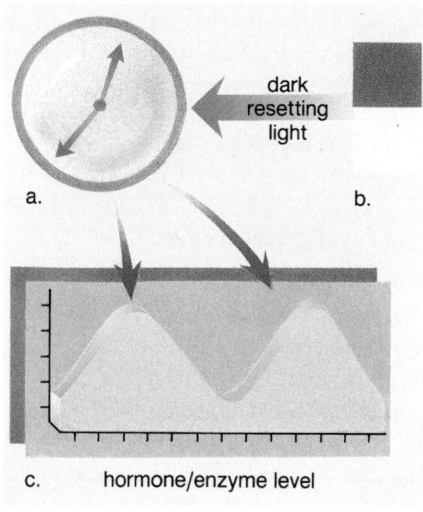

Figure 45.3

Innate types of behavior (taxes and reflexes) are more frequently used by invertebrates, and learned types are more frequently used by vertebrates. V. G. Dethier/Eliot Stellar, *Animal Behavior,* Third Edition, © 1970, p. 91. Adapted by permission of Prentice-Hall, Inc., Englewood Cliffs, NJ.

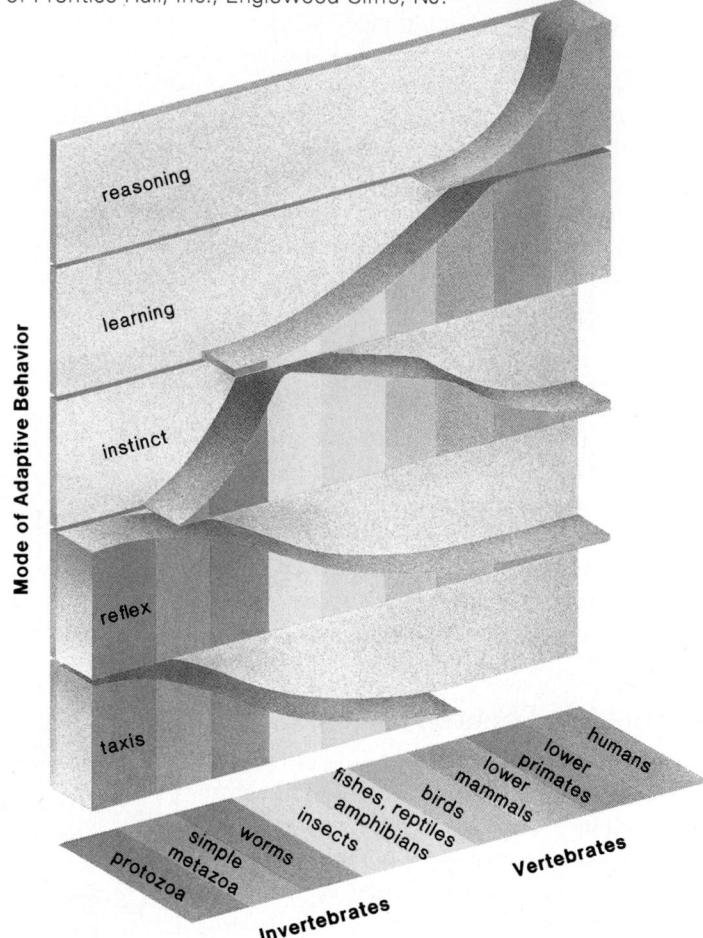

patterns that occur on a yearly basis. For example, birds migrate south in the fall, and the young of many animals are born in the spring. These are called circannual rhythms.

Originally, it was assumed that environmental changes, such as the coming of night or the length of day, directly controlled cyclical behavior in animals. But it is now known that cyclical behavior occurs even when the associated stimulus (daylight or darkness) is lacking. For example, fiddler crabs are dark in color during the day and light in color at night. The cycle continues even when they are kept in a constant environment indefinitely, but the timing of the daily changes tends to drift out of synchronization with the natural cycle. For this reason, it has been suggested that rhythmic behavior is under the control of an innate, internal **biological clock** that runs on its own but is reset by external stimuli. Aside from keeping time, a biological clock must also be able to trigger a change in behavior. In the fiddler crab, for example, the behavior must be able to stimulate the processes that cause the shell to change color. Therefore, a biological clock system needs the following (fig. 45.2):

1. A timekeeping mechanism that *keeps time* independently of external stimuli (i.e., a minute is always a minute);
2. A receptor that is sensitive to light/dark periods and *can reset* the clock for circadian rhythms or can indicate a change in the length of the day/night for circannual rhythms;
3. A communication mechanism by which the clock induces the appropriate behavior.

In animals, we know that melatonin is produced at night by the pineal gland; therefore, the amount produced begins to increase in the fall and to decrease in the spring. Melatonin is believed to inhibit development of the reproductive organs, which would explain why mammals often mate in the fall, when sex organs are most developed. For animals in which the pineal gland is the "third eye," not only the clock but also the receptor is presumed to be in the pineal gland.

For some behavior patterns it appears that only the timekeeping mechanism is required. For example, bees and some birds use the sun and the stars as compasses, even though these celestial bodies change their positions throughout the day and night.

Biological clock systems control behavior that reoccurs at regular intervals. Again the nervous system and/or hormones are involved in controlling these behavior patterns.

Genetic Bases and Learned Behavior

Two early investigators, Konrad Lorenz and Niko Tinbergen, were ethologists who first studied animal behavior in the wild. They were particularly interested in "innate" versus "learned" behavior

Figure 45.4

Robin in breeding condition attacks a tuft of red feathers. The color red is a sign stimulus for a fixed-action pattern.

tuft of red feathers

model without red breast

robin

a.

b.

Most behavior probably contains all of these components:

Preprogramming: the genetic makeup of an organism predisposes it to respond to certain stimuli in a certain way.

Practice: experience affects the development of the behavior.

Potentiation: the physiological state of the animal readies it to perform the behavior (sometimes called motivation or drive).

Performance: the resulting series of acts and internal changes that accompany the behavior.

Learning is a change in behavior as a result of experience. The capacity to learn is inherited and allows an organism to change its behavior to suit the environment. There are different types of learning.

Imprinting

Konrad Lorenz observed that birds become attached to and follow the first moving object they see. He termed this behavior **imprinting** and suggested that it is a means by which organisms learn to recognize their own species. Ordinarily, the object followed is the mother; however, goslings became imprinted on him and chose him over their own mother when given the opportunity (fig. 45.6).

Lorenz found that if goslings are totally isolated for the first 2 days after hatching, they failed to imprint. Apparently, there is a *critical period* of time during which imprinting is possible. Since Lorenz's time, it has been discovered that there are critical periods for other types of learning in birds. A male white-crowned sparrow reared in isolation sings a song, but the song is less complicated than the normal song. If the bird is permitted to hear another adult white-crowned sparrow sing during a critical period of 10 days–50 days following hatching, however, it learns to sing the normal song. White-crowned sparrows never learn to sing the song of another species of birds.

Today, the term *imprinting* is used in a larger context to refer to any type of learning that has a critical period, even when older animals are involved. For example, adult birds will accept any chick, even one of another species, if the chick is put in the nest during a critical period soon after a hatching.

Habituation

When presented with the same stimuli repeatedly, animals will eventually cease to respond. For example, at first baby chicks crouch in fear even when a leaf flutters overhead, but then they learn to disregard these and even long-necked birds. A possible explanation for this is that they have grown accustomed to and habituated to seeing long-necked birds. *Habituation* is useful because it prevents animals from wasting time on unnecessary responses.

and said that the degree of genetic programming and flexibility of responses reflect an evolutionary sequence from invertebrates to vertebrates (fig. 45.3). By innate these investigators meant that the behavior was entirely under genetic control; whereas learned meant that the behavior pattern was solely due to a life experience.

Tinbergen suggested that innate behaviors are like automatic reflexes because the members of a species all perform the exact same sequence under the same environmental conditions. In humans, when a knee is hit by a mallet, the lower leg jerks in a characteristic manner. He used the term **fixed-action pattern (FAP)** for a behavior that occurs automatically, as if it were a composite of reflex actions. And he called a stimulus that initiates a fixed-action pattern a *sign stimulus*. For example, robins attack a red tuft of feathers rather than an exact replica of a male robin without a red breast (fig. 45.4). The color red, not the entire bird, acts as a stimulus that releases the attack behavior.

Today we know that any behavior pattern has a genetic basis and is also influenced by lifetime experiences, including learning. The genes provide a program for the organism, but this program can be modified by environmental influences as it unfolds. Therefore, it can be observed that some behaviors with a strong genetic bases can still be modified by experience. For example, baby gulls peck at the parent's beak to induce the parent to feed them. Research with various models has shown that the chicks are not very discriminatory at first and choose a model that has a large red spot (the sign stimulus), even if the model does not resemble the parent. With time, however, chicks become progressively more selective and choose a model that more nearly resembles the parent (fig. 45.5).

Operant Conditioning

Operant conditioning is trial-and-error learning. An animal faced with several alternatives is rewarded (positively reinforced) for making the proper choice and learns to make this response without hesitation. This type of learning, employed by animal trainers, can be used successfully to make animals learn all sorts of tricks. Using *positive reinforcement*, B. F. Skinner, a well-known Harvard University behaviorist, even taught pigeons to play a form of Ping-Pong. Skinner believes that humans, too, are controlled primarily by positive reinforcement mechanisms, although many disagree with this idea.

Figure 45.5

Modification of behavior. *a.* Chicks peck at the red spot on a parent's bill when they want to be fed. *b.* With time, chicks peck at a more exact replica of a parent's bill (first model). In each example, the top bar represents the pecking frequency of newly hatched chicks, and the bottom bar represents the pecking frequency of chicks 2–5 days old.

a.

b.   **Pecks/30 Sec**

0   5   10   15

Figure 45.6

a. Normally, goslings are imprinted on their mother. *b.* Goslings are following Konrad Lorenz, who, like Niko Tinbergen, performed experiments in the field. Lorenz found that goslings learn to follow the first moving object they see and could be imprinted on an improper object.

a.

b.

Behavior and Ecology

Figure 45.7

Apes are capable of insight learning—reasoning out a solution to a problem without trial and error. ***a.*** A chimpanzee is unable to reach a banana. ***b.*** It stands on boxes to reach the banana. If an animal can reason, does it have a cognitive awareness of itself and its surroundings? Researchers are now wrestling with this question.

a. b.

Table 45.1
Review of Terms Used in the Study of Behavior

Term	Definition
Fitness	The ability of an organism to survive and reproduce in its local (immediate) environment
Natural selection	The environment selects for survival and reproduction certain members of a population; in the end, adaptive traits become more prevalent in the population (p. 324)
Altruistic behavior	Unselfish acts that help others in a group so that they and not the individual increase their chances of surviving and reproducing
Selfish behavior	Acts that increase the likelihood of only the individual surviving and reproducing offspring
Inclusive fitness	The ability of kin (relatives who share common genes) to survive and to reproduce

Insight Learning

Insight learning, or reasoning, is the ability to solve a problem using previous experiences to think through to a solution. To the observer, it seems as if the animal employing insight needs no practice to reach a goal successfully. For example, apes can devise the means to get to bananas placed out of their arms' reach. They will pile up boxes or use a pole in order to reach the food (fig. 45.7).

Natural Selection and Behavior

Behavior patterns enhance fitness—the ability of an organism to survive and to reproduce (table 45.1). This premise applies to such behavior as feeding, finding and defending a territory, and mating.

Feeding Behavior

Some animals are herbivores. Among these are grazers, which feed on grasses, or browsers, which feed on shrubs and trees. Other animals are carnivores: some go out alone and actively seek prey, and others lie in wait. Predators such as lions and wolves sometimes prey on animals larger than themselves and hunt as a group.

Regardless of how an animal feeds, the energy benefit of the food itself should be greater than the energy cost of getting the food. Using this as a principle, we can see why great blue herons have a more varied diet in less productive northern lakes than in the more productive waters of Florida. Still, there is a genetic component to food preference. Garter snakes reared in isolation in the Florida Everglades readily accepted fish as food, but garter snakes from Massachusetts did not.

Characteristically, conditioned learning is maintained only as long as a reward is given for the behavior. If the reward is withdrawn, the behavior also disappears. This is called extinction. Extinction is also a type of learning.

Classical Conditioning

In **classical conditioning,** also called association learning, an animal learns to give a response to an irrelevant stimulus. Pavlov's dogs learned to expect food and began to salivate when a bell rang because they learned to associate the ringing of the bell with food. Associative learning can explain behavior that seems out of place; for example, the ringing of a bell in and of itself does not have anything to do with food. Even so, associative learning can be useful. For example, it has been suggested that mothers can instill love of learning by holding their children on their laps while reading to them. On the other hand, advertisers make use of associative learning when they attempt to associate power and prestige with their particular product.

Animal Behavior

Natural selection not only promotes efficient feeding techniques, it also operates to reduce the likelihood of being eaten. Animals have antipredator defenses, including concealment, fright, warning coloration, and vigilance (see fig. 47.6).

Home Territory

As we shall see in chapter 48, specific habitats have their own particular mix of plants and animals. In general, animals tend to select those habitats that increase their reproductive success. For example, among plant parasites called aphids (p. 502), the first females to hatch in the spring settle on large leaves, leaving the small leaves for latecomers. Females on the large leaves have higher reproductive success than females on small leaves.

Orientation and Navigation

Orientation of the body toward or away from a stimulus is termed **taxis.** *Phototaxis* is movement in relation to light, and *chemotaxis* is movement in relation to a chemical. Salmon are hatched in a tributary of a river but grow to maturity in the open sea. At spawning time, mature salmon travel back up the river to the same spot where they hatched. Experiments have shown that the fish find the spawning ground by following the chemical scent of their first home.

Only a few animals are able to orient themselves by means of sonar, or *echolocation*. Bats send out a series of sound pulses and listen for the echoes that come back. The time it takes for an echo to return indicates the location of both inanimate and animate objects and enables a bat to find its way through dark caves (and also to locate moths for food).

Birds and other animals are able to migrate long distances between their summer breeding range and their winter feeding range (fig. 45.8). Honeybees and homing pigeons are able to use the sun during the day as a compass to determine direction (north, south, east, or west). A biological clock even allows them to compensate for east-to-west movement of the sun during the course of the day. Pigeons can navigate, however, even when it is cloudy. Investigators have discovered iron oxide crystals in the bodies of pigeons and honeybees. Apparently the crystals align themselves with the earth's magnetic field and allow them to determine the direction of home. Other birds have been found to use the stars as compasses during the night.

Territoriality

Territoriality means that an animal defends a certain area, preventing certain other members of the same species from utilizing it. Even the aphid females mentioned earlier defend their leaves against other females, and those with the largest leaves reproduce the most. Among great tits (a type of bird), John Krebs found that 92% of these with woodland nests had offspring, but only 22% of birds with hedgerow nests had offspring. He removed 6 pairs of birds in the woodlands, and within a few days the hedgerow birds moved in. Territoriality in birds improves reproductive success because the better territories provide more food for offspring.

Reproductive Behavior

The reproductive behavior of an organism is expected to increase individual fitness. Songbirds are often monogamous—having only

Figure 45.8
A land bird that breeds in North American clover fields, the bobolink migrates to the grasslands of Argentina by island hopping through the Caribbean. The small bobolink colonies in the Northwest do not use the most direct route to the south through Mexico; instead, they make a detour east to fly with other bobolinks along an ancestral flyway.

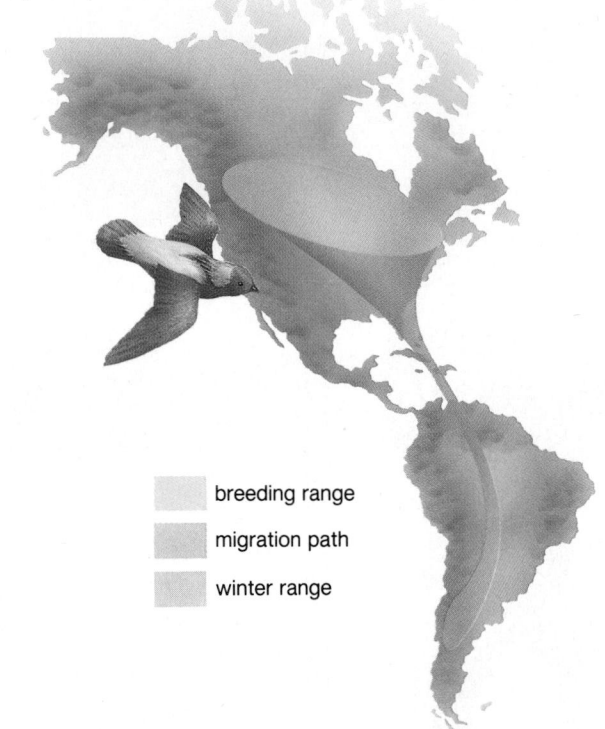

breeding range

migration path

winter range

one mate. In the case of birds, incubating the eggs/or gathering enough food requires the effort of both parents. If they did not cooperate, neither would have any offspring. On the other hand, mammals tend to be polygamous—having more than one mate. A mammalian female produces only a few hundred eggs during her lifetime, but a male produces millions of sperm. A female is certain in a way no male can be certain that the offspring is her own. Under these circumstances, it is more adaptive for a female to invest a great deal of energy in the rearing of offspring, and it is adaptive for males to impregnate as many females as possible.

In polygamous systems, males often compete intensely for the right to inseminate females. In elephant seals, males fight with one another to establish a dominance hierarchy. A **dominance hierarchy** exists when animals form a relationship in which a higher-ranking animal has access to resources before a lower-ranking animal. One study found that the top 5 elephant seal males do at least 50% of the copulating. Since pups are born a year after copulation, males have no way of knowing if an offspring is theirs. Males have been observed to trample pups, possibly their own, when striving to inseminate a female.

Females participate in the selection process. Since the female produces few offspring, which she alone rears, we would expect her to mate preferentially with a male she judges to be genetically superior. Female elephant seals scream loudly during copulation,

Behavior and Ecology

How Do You Track an Animal in the Wild?

Miniature radio transmitters that emit a radio signal allow you to track an animal in the wild. The transmitter is encapsulated along with a battery in a protective epoxy resin covering, and the package is either attached to an animal with a collar or a clip or implanted surgically into the main body cavity of the animal. In either case, it is important to capture the animal carefully. When some form of sedation or anesthesia is used to calm the animal during tagging or surgical implantation, all of the necessary precautions are observed and the animal is permitted to recover fully prior to its release. A radio transmitter device is generally no more than 10% of the animal's body weight, and therefore it should not interfere with normal activities.

To track the animal, a researcher needs a radio receiver equipped with an antenna and earphones. The strongest signal comes from the direction of the animal as the antenna is rotated above the head. The data obtained allow the researcher to obtain a series of fixes (to determine where the animal is) and calculate its rate of movement. Plotting the sequence of fixes on a map of the appropriate scale gives information on the area that is used by the animal in the course of a night or over several nights (fig. 45.A).

The distance over which the signal travels varies with the size of the transmitter and the strength of the battery. For small rodents it may be necessary to be within 5 m–10 m of the animal to hear the signal, whereas for larger collars like some of those used on elk or caribou, the signal can actually be received by a satellite orbiting above the earth. Later, the signals are sent in a radio beam by antenna to the ground. Satellite transmission allows researchers to work with animals that are far away and prevents any possibility of interfering with the animal's activities.

A number of important findings have been made utilizing radiotelemetry and a few examples provide a flavor of the kinds of results obtained with this technology. Elk in Yellowstone National Park in Wyoming have summer home range areas that are 5–10 times larger than in winter. Areas occupied by many mice and voles get larger as the animal ages between the time the animals are weaned and until they are adults, but these increases are greater for males than for females. Winter dens of rattlesnakes (and other snakes) are often some distance from the areas where they spend their summer months. Some crows migrate each spring and fall, whereas others remain as residents in the same locale throughout the year.

Radio transmitters can also be used to obtain data on physiological functions (heart rate or body temperature) in conjunction with observations of behavior while animals continue to engage in normal activities. When physiological measurements are taken, a device is included in the package that is sensitive to temperature or heart rate and can translate that information to the transmitter. Altogether, the use of radio transmitters has added a new dimension to the ability to explore the lives of animals.

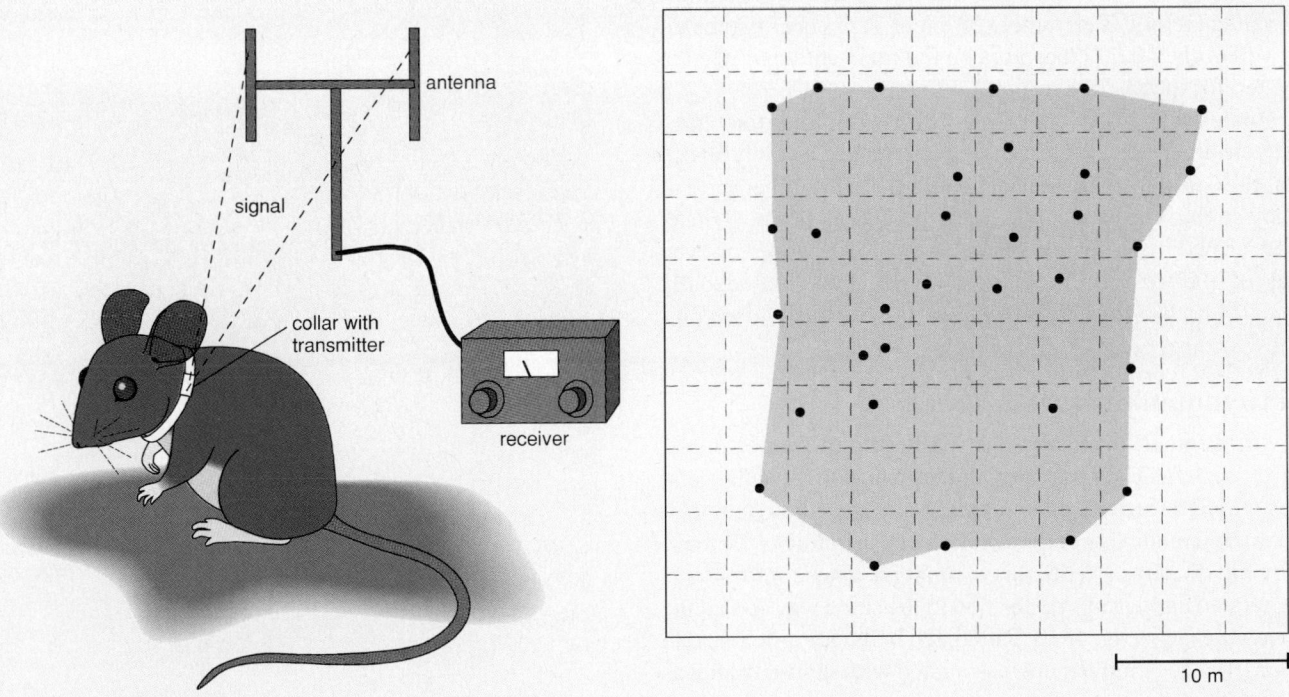

a. Radiotransmitter system

b. Plot of activity range

Figure 45.A
a. By using a miniature radio transmitter in a neck collar, an antenna, and a receiver it is possible to record an animal's location approximately every 15 minutes throughout a 24-hr period. **b.** These data are then used to plot its activity range for that day. (Some data in the plot represent multiple fixes at the same location.)

alerting dominant males, who chase away less dominant males. At times, female selection even seems to be responsible for controlling male attributes, such as the multicolored face of male mandril baboons, the splendid tail of male peacocks, or the long tail of male widowbirds (see fig. 20.11). Malte Andersson allowed females to choose between 3 experimental groups of male widow birds: males with tails of normal length of 50 cm; males with tails clipped to 14 cm; and males with their tails lengthened to 75 cm (feathers from the clipped group were glued onto the tails of this group). At the end of the experiment, more females were nesting in the territories of males with the longest tails. Courtship rituals also probably evolved in part as a way for females to make a choice among males.

> Animal behavior encompasses the way an animal gets food, finds and occupies a home territory, and carries on reproduction. Behavior patterns are expected to increase individual fitness.

Communication

Communication is an action by a sender that influences the behavior of the recipient(s).

Chemical Communication

The term **pheromone** is used to designate chemical signals that are passed between members of the same species. For example, female moths secrete chemicals from special abdominal glands. These chemicals are detected downwind by receptors on male antennae. This signaling method is extremely efficient. It has been estimated that only 40 of the 40,000 receptors on the male antennae need to be activated for the male to respond.

Male mice produce a substance that can alter the reproductive cycle of females. When a male and female are newly placed together, this substance can cause the female to abort her present pregnancy so that she can then be impregnated by her new mate. Similarly, crowding of female mice causes disturbances or even blockage of their estrus (those times when they are sexually receptive). In both instances, removal of the olfactory lobes prevents these occurrences.

Visual Communication

Visual communication includes many sign stimuli. For example, male birds and fishes sometimes undergo a color change that indicates they are ready to mate. Female baboons show that they are in estrus by a reddening of the sex flesh on the buttocks. Defense and courtship displays are ritualized behavior whose movements are exaggerated and always performed in the same way so that the sign stimuli are clear (fig. 45.9). Ritualized behavior is believed to be derived from body movements associated with such activities as locomotion, feeding, and caring for the young.

Tactile Communication

Tactile communication occurs when one animal touches another. For example, baby gulls peck at the parent's beak in order to induce the parent to feed them. In primates, grooming—one animal cleans the coat and skin of another—helps cement social bonds within a group.

Figure 45.9
Courtship ritual of 3-spined stickleback fish is based on sign stimuli (social releasers) and fixed-action patterns.

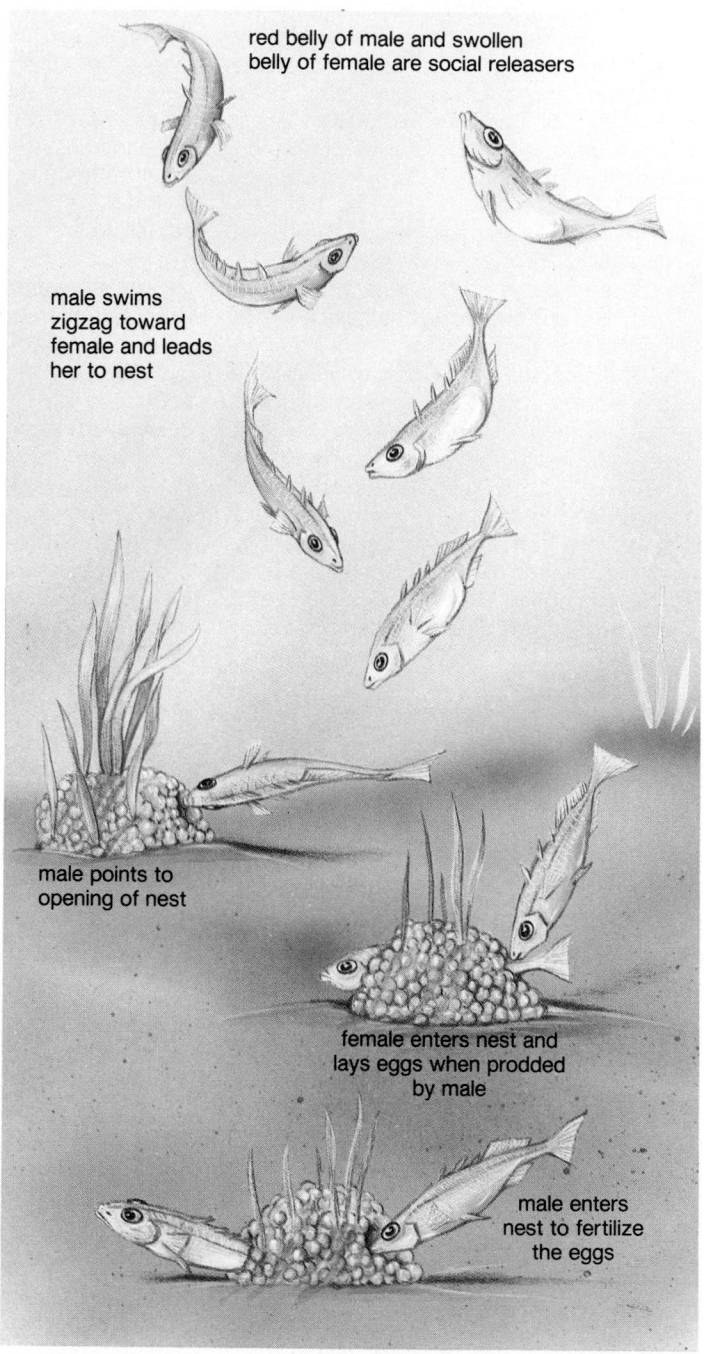

red belly of male and swollen belly of female are social releasers

male swims zigzag toward female and leads her to nest

male points to opening of nest

female enters nest and lays eggs when prodded by male

male enters nest to fertilize the eggs

The communication of honeybees is remarkable because the so-called language of the bees uses a variety of stimuli to impart information about the environment. Karl von Frisch, another famous ethologist, carried out many detailed bee experiments in the 1940s and was able to determine that when a foraging bee returns to the hive, it performs a waggle dance (fig. 45.10). The

Figure 45.10

Tactile communication and auditory communication among bees relate information not about the bee but about the environment. Honeybees do a waggle dance to indicate the direction of food. *a.* If the dance is done outside the hive on a horizontal surface, the straight run of the dance will point to the food source. *b.* If the dance is done inside the hive on a vertical surface, the angle of the straightaway to that of the direction of gravity is the same as the angle of the food source to the sun.

the direction of gravity is the same as the angle of the food source to the sun. In other words, a 40° angle to the left of vertical means that food is 40° to the left of the sun.

As mentioned, honeybees can use the sun as a compass because their biological clock (p. 732) allows them to compensate for the movement of the sun in the sky. In the dark hive, bees use a combination of tactile and auditory communication. Through touch, bees can determine the direction and waggles of the dance. Not only the waggles but also the buzzing noises of the dancer tell the distance to the food source.

Auditory Communication

Auditory (sound) *communication* has some advantages over other kinds of communication. It is faster than chemical communication, and unlike visual communication, it can be sent and received even in the dark. Sound stimuli are shown to be more important than visual stimuli for birds that live in dense woods, where vision is obstructed. In experiments, a male wood thrush attacked models of an unrelated species as long as these models were silent. But, if the unrelated species' song was played via a loudspeaker, the wood thrush paid the models no heed. The intensity with which a wood thrush attacked a model of its own species was directly proportional to the loudness of the song that was played.

Auditory communication can be modified not only by loudness but also by its pattern, its duration, and its repetition. In an experiment with rats, a researcher discovered that an intruder can avoid attack by increasing the frequency with which it makes an appeasement sound. Rats isolated from birth could make the sound but had not learned to increase the frequency.

Male crickets have calls, and male birds have songs for a number of different occasions. For example, birds may have one song for distress, another for courting, and still another for marking territories. Sailors have long heard the songs of humpback whales because they are transmitted through the hull of a ship. But only recently has it been shown that the song has 6 basic themes, each with its own phrases, that can vary in length and be interspersed with sundry cries and chirps. The purpose of the song is probably sexual and serves to advertise the availability of the singer. Individual whales can be recognized by their particular voice characteristics.

Language is the ultimate auditory communication, but only humans have the biological ability to produce a large number of different sounds and to put them together in many different ways. Nonhuman primates have at most only 40 different vocalizations, each having a definite meaning, such as the one meaning "baby on the ground," which is uttered by a baboon when a baby baboon falls out of a tree. Although chimpanzees can be taught to use an artificial language, they never progress beyond the capability level of a 2-year-old child (fig. 45.11). It has also been difficult to prove that chimps understand the concept of grammar or can use their language to reason. It still seems as if humans possess a communication ability unparalleled by other animals.

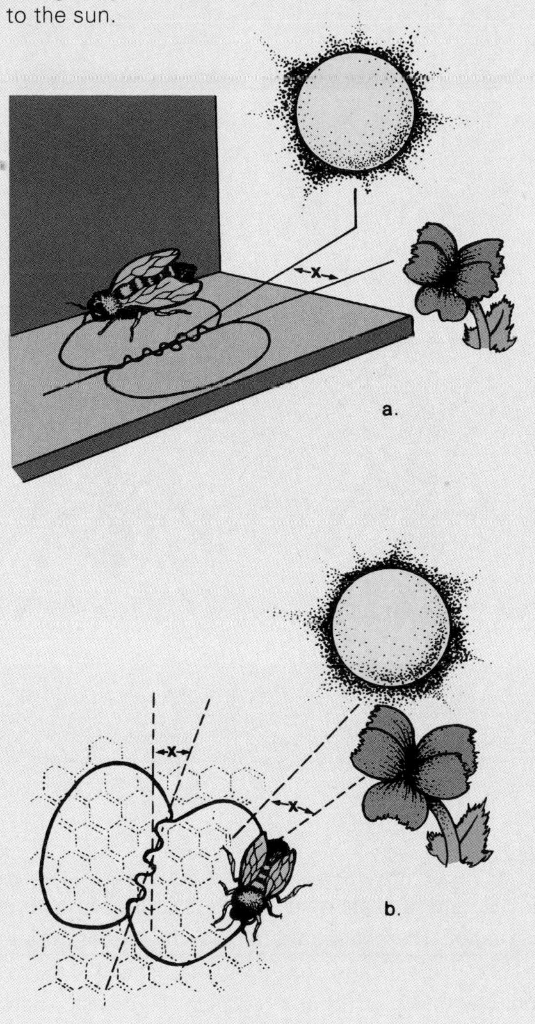

dance, which indicates the distance and the direction of a food source, has a figure-8 pattern. As the bee moves between the 2 loops of the figure 8, it buzzes noisily and shakes its entire body in so-called waggles. Distance to the food source is believed to be indicated by the number of waggles and/or the amount of time taken to complete the straight run. The straight run also indicates the location of the food. Outside the hive, the dance is done on a horizontal surface and the straightaway indicates the exact direction of the food. Inside the hive, the dance is performed on the comb, which is vertical, and the angle of the straightaway to that of

Figure 45.11

A chimpanzee with a researcher. Chimpanzees are unable to speak but can learn to use a visual language consisting of symbols. Some researchers believe that chimps are capable of creating their own sentences, but others believe that they only mimic their teachers and never understand the cognitive use of a language. Here the experimenter shows Nim the sign for "drink." Nim copies.

Societies

A **society** is a group of individuals that belongs to the same species and is organized in a cooperative manner extending beyond sexual and parental behavior. **Sociobiology** applies the principles of evolutionary biology to the study of social behavior in animals.

There are benefits and disadvantages to living in a group. Some of the possible benefits are as follows:

Protection against predators. Members of the society give alarm signals and group together for defense.
Finding and procuring food. Members of a society signal each other as to the location of food and sometimes work together to procure food.
Division of labor among specialists. In insect societies, members have specialized roles to perform.

Some of the possible disadvantages of living in a group are as follows:

Increased competition for resources. Members compete with one another for shelter, food, and mates.
Increased chance of disease and parasites. Members give diseases and spread parasites from one to the other.
Interference with reproduction. Members of a society are sometimes aggressive toward the young of other members.

Animals socialize when the benefits of the behavior outweigh the disadvantages. Benefit is judged by increased fitness of the individual member.

Figure 45.12

Sociobiology and the behavior of ants. It is genetically more advantageous for female ants and bees to assist in caring for siblings rather than produce their own offspring. Offspring possess one-half of the queen's genes, but because the male parent is haploid, siblings share, on the average, three-fourths of their genes.

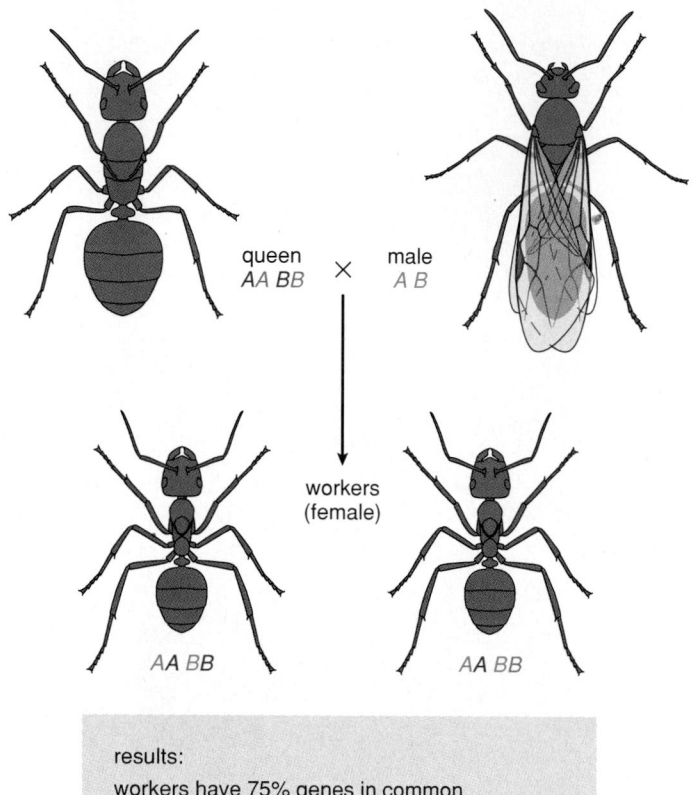

results:
workers have 75% genes in common
worker and mother have 50% genes in common

Altruism

At times, it may seem as if an animal is acting altruistically, that is, for the good of others rather than from self-interest. For example, in ants and bees, only the queen lays eggs, and most of the hive consists of female workers without offspring. How can such a society increase the fitness of the workers? The answer may lie in the fact that the male parent is haploid; therefore, siblings have on the average three-fourths of their genes in common (fig. 45.12). Because offspring have only one-half of their genes in common with their parents, it may actually increase the fitness of the worker ants to raise siblings rather than offspring.

Altruism, exemplified by the willingness of daughters to help the queen rather than have their own offspring, can be explained by noting that *kin selection* increases the animal's inclusive fitness. In other words, survival of close relatives (kin) increases the frequency of an animal's genes in the next generation. Therefore, the animal's overall (inclusive) fitness is also increased.

Under circumstances of ordinary sexual reproduction, the relatedness (r) of 2 individuals is dependent on the fraction of genes they have in common. For parents and offspring, $r = 1/2$; for siblings, $r = 1/2$; for individuals and uncles or aunts, $r = 1/4$; for first cousins, $r = 1/8$.

Communal Groups

The concept of kin selection is exemplified in acorn woodpeckers. These birds received their name because they store acorns in holes drilled in trees. In California and New Mexico, acorn woodpeckers usually live in social groups of about 15 birds. Each group defends its storage tree from birds that are not members of the group.

Researchers have noted that only certain members of a group are *breeders* and they share mates. Mate sharing, which is rarely seen in birds, in this instance increases the inclusive fitness of the individual bird. Groups have more offspring per adult than do single pairs. For a group of three, investigators found there were 1.16 young per adult male on the average; for pairs there were only 0.92 young per male. Secondly, among the sharers, the males are usually brothers and the females are usually sisters.

In a large group, there are several *nonbreeders,* birds who only help the breeders raise their young. Because nonbreeders are offspring of the breeders, it is possible that they are also increasing their inclusive fitness by their behavior. The researchers also note, however, that the adjoining territories are already occupied and birds have a very difficult time establishing themselves as breeders. Under these circumstances, it may simply be more advantageous for the nonbreeders to remain in a group where they can feed on the stored acorns and wait until there is a vacancy among the breeding birds.

Alarm Callers

Some animals that live in groups give alarm calls when a predator approaches, even when such behavior puts the caller at greater risk. The concept of inclusive fitness can explain this behavior when it saves close relatives, which go on to reproduce because of the warning.

Sociobiology is the study of social behavior according to the tenets of evolutionary theory. While it may seem as if members of a group are altruistic, it can be shown that their motives are most likely selfish and their behavior serves to increase their own fitness. Sometimes it is necessary to utilize the concept of inclusive fitness when interpreting an animal's behavior.

Humans

Sociobiologists interpret human behavior according to the same principles used to interpret animal behavior. For example, parental love is clearly selfish in that it promotes the likelihood that an individual's genes will be present in the next generation's gene pool. People tend to be less aggressive toward blood relatives than nonrelatives. We would expect small towns to have less crime than large cities because small towns have a greater percentage of related residents.

Human reproductive behavior can be similarly examined. Human infants are born helpless and have a much better chance of developing properly if both parents contribute to the effort. The human female, unlike other mammals, is continuously receptive to sexual intercourse. (In other mammals, females are receptive only during estrus, a physiological stage that recurs periodically, sometimes only once a year.) The fact that sex is continuously available may help assure that the human male will remain and help the female raise the young.

Human beings are usually monogamous; however, there are exceptions that appear to be adaptations to the environment. Among African tribes, for example, one man may have several wives. This is reproductively advantageous to the male, but is also advantageous to the woman. By this arrangement, she has fewer children and they are thereby assured a more nutritious diet. In Africa, sources of protein are scarce and early weaning poses a threat to the health of the child. In contrast, among the Bihari hill people of India, brothers have the same wife. Here the environment is hostile and it takes 2 men to provide the everyday necessities for one family. Because the men are brothers, they are actually helping each other look after common genes.

Not all scientists and laypeople are in favor of applying sociobiological principles to human behavior. Many are concerned that it may lead to the perception that human reproductive behavior is fixed and cannot be changed. Particularly today, when people are exploring new avenues of behavior, some do not wish to stress unduly any potential obstacles to bringing about a new social order.

Summary

1. All behavior has a genetic basis, but no behavior is genetically determined. Rather, an interaction between inheritance and the environment determines behavior.

2. The nervous system and the endocrine system immediately control behavior. This is illustrated by considering ringdove reproductive behavior.

3. In innate (instinctive) behaviors, the genetic component is believed to be more influential, and in learned behaviors, the environmental component is believed to be more influential.

4. Fixed-action patterns, which appear to be a series of reflexes, occur as a response to a sign stimulus. FAPs are largely inherited, although they sometimes increase in efficiency with practice.

5. Most behavior requires preprogramming (genetic input), practice, potentiation (physiological state), and performance.

6. Learning includes imprinting, habituation, operant conditioning, classical conditioning, and insight learning.

7. The role of natural selection in behavior is exemplified by feeding behavior, selecting a home, and reproductive behavior.

8. The energy cost of getting food must be less than the energy benefit of the food. What an animal eats is a result of interaction between inheritance and environmental influences. Animals have antipredator defenses.

9. Animals have to orientate themselves to find suitable eating and/or reproductive places. For example, some are orientated by chemicals (salmon),

some by echolocation (bats), and some by the sun (honeybees and homing pigeons). A biological clock keeps track of the relationship of the sun to direction.

10. Some animals have territories they defend. The energy investment allows them to have more offspring because the territory provides more food for offspring.

11. The reproductive system of an animal strongly correlates with individual fitness. Mammals are usually polygamous. It is beneficial for a male to

impregnate as many females as possible and for females to have the best possible mate.

12. Animals use various means to communicate with one another including chemical, visual, tactile, and auditory communication.

13. Living in a society has benefits and drawbacks. Animals socialize when the benefits outweigh the drawbacks.

14. Social behavior can be studied from an evolutionary perspective. According to sociobiologists, although it may seem

as if an individual is altruistic, actually the individual is benefitted by the behavior.

Writing Across the Curriculum

In order to practice writing skills, students should write out the answers to any or all of the study questions and the critical thinking questions. The study questions are sequenced in the same order as the text. Suggested answers to the critical thinking questions are in appendix D.

Study Questions

1. Describe the roles of the nervous system and the endocrine system in the reproductive behavior of ringdoves.
2. What is a biological clock system, and what are the functions of its components?
3. Give an example to show that a fixed-action pattern can be modified by learning.

4. What are the 4 components of most behavior patterns? Explain each one.
5. Explain the concept of critical period in relation to imprinting.
6. Name 5 types of learning, and explain the significance of each.
7. What premise underlies the various ways animals get food, establish a territory, and carry on reproductive behavior?

8. List and describe the ways that animals communicate with one another.
9. What are the benefits and disadvantages of living in a group?
10. Explain altruism from the sociobiologist's point of view.

Objective Questions

1. Male robins in breeding plumage attack a tuft of red cotton. This is an example of
   a. habituation.
   b. innate behavior.
   c. fixed-action pattern.
   d. Both b and c.
2. Baby chicks no longer cringe and hide when a long neck bird goes overhead. This is an example of
   a. habituation.
   b. imprinting.
   c. insight learning.
   d. Both b and c.
3. Animals exhibit a wide range feeding behavior patterns. This is because
   a. genes control behavior.
   b. the environment controls behavior.
   c. animal phenotypes and environmental influences differ.
   d. animals need an energy source to live.
4. A student concludes that robins have cones sensitive to the color red. The student is considering
   a. a proximate cause of behavior.
   b. an ultimate cause of behavior.

   c. the role of learning in behavior.
   d. the role of natural selection in behavior.
5. A student concludes that it is adaptive for robins to attack any other object with a red breast. The student is considering
   a. a proximate cause of behavior.
   b. the role of learning in behavior.
   c. the role of natural selection in behavior.
   d. the role of ecology in behavior.
6. Animals are capable of orienting themselves. This helps them find
   a. food.
   b. a mate.
   c. a home.
   d. All of these.
7. Animals in the best territories have
   a. to defend them against intrusion by others.
   b. more food for offspring.
   c. the best chance of getting a mate.
   d. All of these.
8. Males don't have the same assurance as females that the offspring are theirs; therefore, it is beneficial for males to
   a. be aggressive to females.

   b. defend territories.
   c. impregnate as many females as possible.
   d. All of these.
9. Altruism
   a. seems to be self-sacrifice for the benefit of the group.
   b. may increase inclusive fitness.
   c. fosters kin selection.
   d. All of these.
10. Some birds help others raise their young. Most likely the helpers
    a. are related to the parents.
    b. cannot get their own territory.
    c. are poor food gatherers.
    d. Both a and b.
11. For each item, indicate whether it is a form of learning or a form of communication, and then tell what specific form it is. The first is done for you as an example.
    a. pheromone *communication chemical*
    b. critical period _____
    c. language _____
    d. positive reward _____
    e. waggle dance _____
    f. courtship display _____

Concepts and Critical Thinking

1. *All behavior has both a genetic basis and is influenced by lifetime experiences, including learning.*

   Explain why you would expect this statement to apply even to human reasoning.

2. *Behavior increases the individual fitness of the individual.*

   Why would you expect feeding behavior patterns, territoriality, and reproductive behavior patterns to increase individual fitness?

3. *Individuals live in a society when the benefits to the individual outweigh the costs.*

   Show that altruistic acts are actually selfish acts.

Selected Key Terms

fixed-action pattern (FAP) (fikst ak′shun pat′ern) 733
learning (lern′ing) 733
imprinting (im′print-ing) 733
operant conditioning (op′e-rant kon-dish′un-ing) 734

insight learning (in′sīt lern′ing) 735
taxis (tak′sis) 736
territoriality (ter″i-tor″i-al′i-te) 736
dominance hierarchy (dom′i-nans hi′er-ar″ke) 736

communication (ko-mu′ni-ka-shun) 738
pheromone (fer′o-mon) 738
society (so-si′ē-te) 740
sociobiology (so″se-o-bi-ol′ō-je) 740

46

Ecology of Populations

There are many populations of organisms in a community. In this community, we see populations of wild flowers and evergreens, for example. The members of a population share a common gene pool because they reproduce with one another.

Your study of this chapter will be complete when you can

1. calculate the rate of natural increase for a population when provided with the number of individuals in the population, the birthrate, and the death rate;
2. contrast a **J**-shaped growth curve with an **S**-shaped growth curve;
3. indicate which parts of an **S**-shaped growth curve represent biotic potential, environmental resistance, and carrying capacity;
4. give examples of density-independent and density-dependent effects of environmental resistance;
5. contrast 3 likely types of survivorship curves for a population;
6. contrast 3 common age structure diagrams for a population;
7. contrast an r-strategist population with a K-strategist population with regard to growth curves, survivorship curves, and age structure diagrams;
8. describe the growth curve for the world's human population;
9. contrast the demography of more developed and less developed countries; tell why both types of countries contribute heavily to resource consumption;
10. explain why replacement reproduction does not automatically bring about zero population growth.

he German zoologist Ernst Haeckel (1834–1919) coined the word *ecology* from 2 Greek roots: *oikos*, meaning household and *ology*, meaning study of. He recognized that organisms do not exist alone; rather, they are part of a household, a functional unit that includes other types of organisms and their physical surroundings. Today, we define **ecology** as the study of the interactions of organisms with both their *biotic* (living) environment and their *abiotic* (physical) environment.

Ecologists may study these interactions at a particular locale. Various ecological factors can affect the size of a **population, all the organisms of the same species at the locale.** We will begin by taking a look at some of these factors in this and the next chapter.

Ecology is the study of interactions of organisms with each other and with the physical environment.

Population Growth

Populations increase in size whenever the number of births exceeds the number of deaths. Suppose, for example, a population size (N) of 1,000 individuals, a *birthrate* (b) of 30 per year, and a *death rate* (d) of 10 per year. The annual *rate of natural increase* (r) would be

$$r = \frac{b - d}{N} = \frac{30 - 10}{1,000} = 0.02 = 2\% \text{ per year.}$$

The rate of natural increase does not include any change in population size from either immigration or emigration, which for the purpose of our discussion can be assumed to be equivalent.

Populations with a positive rate of natural increase grow larger each year. The expected increase (I) can be calculated by multiplying the rate of natural increase (r) by the current population size (N):

$$I = rN$$

This formula indicates that the population is undergoing **exponential growth:** at the end of each year, N is larger and therefore I is also larger, as demonstrated in table 46.1. This means population size increases by an ever-larger amount each year. This is apparent when population size is plotted against time. When an arithmetic scale is used for population size, the resulting growth curve resembles that in figure 46.1. It starts off slowly and then increases dramatically, giving a *J-shaped curve*.

Biotic Potential

The **biotic potential** of a population is the maximum rate of natural increase (r_{max}) that can possibly occur under ideal circumstances. These circumstances include plenty of room for each member of the population, unlimited resources, and no hindrances, such as predators or parasites. Because these conditions rarely occur, it is very difficult to determine the r_{max} for most species. A few approximations, however, are given in table 46.2.

Figure 46.1
The exponential growth curve is J shaped, indicating that the population size is experiencing a steady rise.

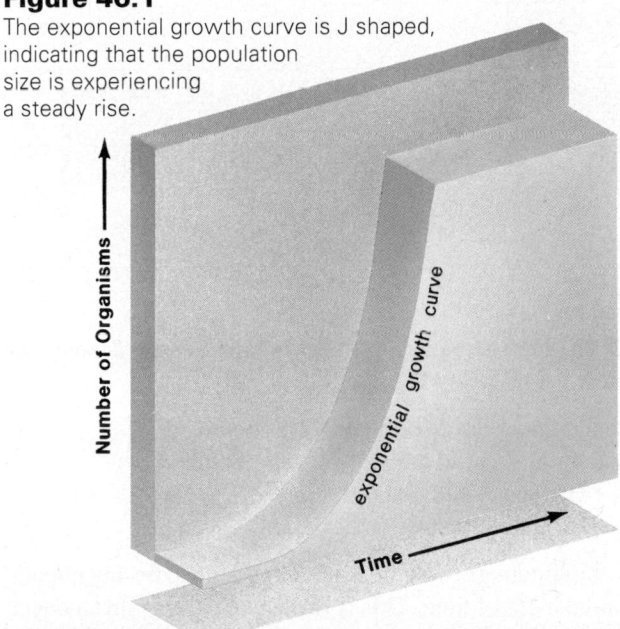

Table 46.1
Exponential Growth (r = 2.0%)

Unit Time	N	I	Unit Time	N	I
0	10,000.00	200.00	7	11,486.86	229.74
1	10,200.00	204.00	8	11,716.60	234.33
2	10,404.00	208.08	9	11,950.93	239.02
3	10,612.08	212.24	10	12,189.95	243.80
4	10,824.32	216.49	11	12,433.75	248.68
5	11,040.81	220.82	12	12,682.43	253.65
6	11,261.63	225.23			

Table 46.2
Biotic Potential

Kind of Organism	Approximate Biotic Potential (r_{max} per year)
Large mammals	0.02-0.5
Birds	0.05-1.5
Small mammals	0.3-8
Larger invertebrates	10-30
Insects	4-50
Small invertebrates (including large protozoans)	30-800
Protozoans and unicellular algae	600-2,000
Bacteria	3,000-20,000

Table from *Principles of Ecology* by Richard Brewer, copyright © 1979 by Saunders College Publishing, reprinted by permission of the publisher.

Figure 46.2

The number of fruit flies in a colony was counted every other day. When these numbers were plotted, the result was an S-shaped growth curve. Data from *Laboratory Studies in Biology: Observations and Their Implications,* Chester A. Lawson, Ralph W. Lewis, Mary Alice Burmester, and Garrett Hardin. Copyright © 1955 W. H. Freeman and Company, New York. Reprinted by permission.

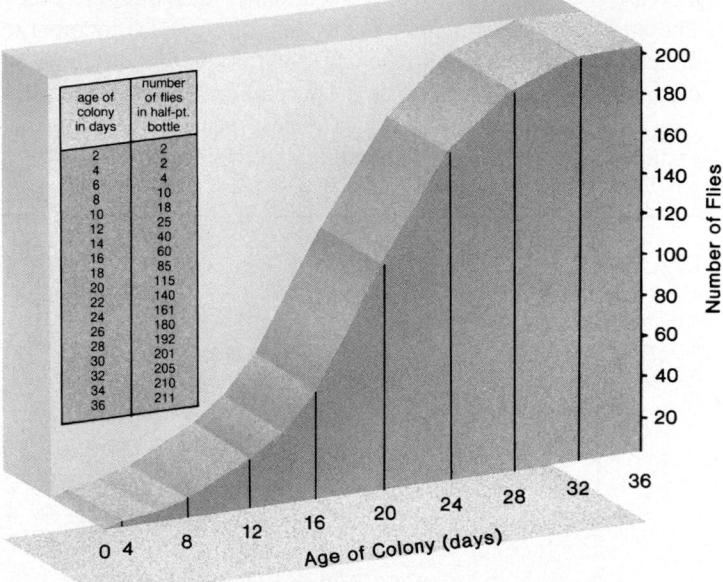

age of colony in days	number of flies in half-pt. bottle
2	2
4	2
6	4
8	10
10	18
12	25
14	40
16	60
18	85
20	115
22	140
24	161
26	180
28	192
30	201
32	205
34	210
36	211

Whether the biotic potential is high or low depends on such factors as the following:

1. usual number of offspring per reproduction;
2. chances of survival until age of reproduction;
3. how often each individual reproduces;
4. age at which reproduction begins.

Maximum growth rate is not experienced by any population for any length of time. This is readily apparent when considering that if a single female pig had her first litter at 9 months, and produced 2 litters a year, each of which contained an average of 4 females (which, in turn, reproduced at the same rate), there would be 2,220 pigs by the end of 3 years. Because of exponential growth, even species with a low biotic potential would soon cover the face of the earth with their own kind. For example, as Charles Darwin calculated, a single pair of elephants would have over 19 million live descendants after 750 years.

Environmental Resistance and Carrying Capacity

Populations do not undergo exponential growth for very long. Even in the laboratory setting, where crowding may be the only limiting factor, growth of populations eventually levels off. Figure 46.2 gives the actual growth curve for a fruit fly population reared in a culture bottle. This type of curve is often called a sigmoid or *S-shaped growth curve.* During an acceleration phase, when births exceed deaths, *r* is positive; during a deceleration phase, when births and deaths are more nearly equal, *r* is declining. Finally, when births exactly equal deaths, *r* is zero, and the population maintains a steady state (fig. 46.3).

Figure 46.3

S-shaped growth curve. The population initially grows exponentially but begins to slow down as resources become limited. Population size remains stable when the carrying capacity of the environment is reached. *a.* The rate of natural increase of a population can be determined by this formula. *b.* When a population is undergoing accelerated growth, r is positive. When a population is undergoing decelerated growth, r is declining. When a population is in a steady stage, r is zero.

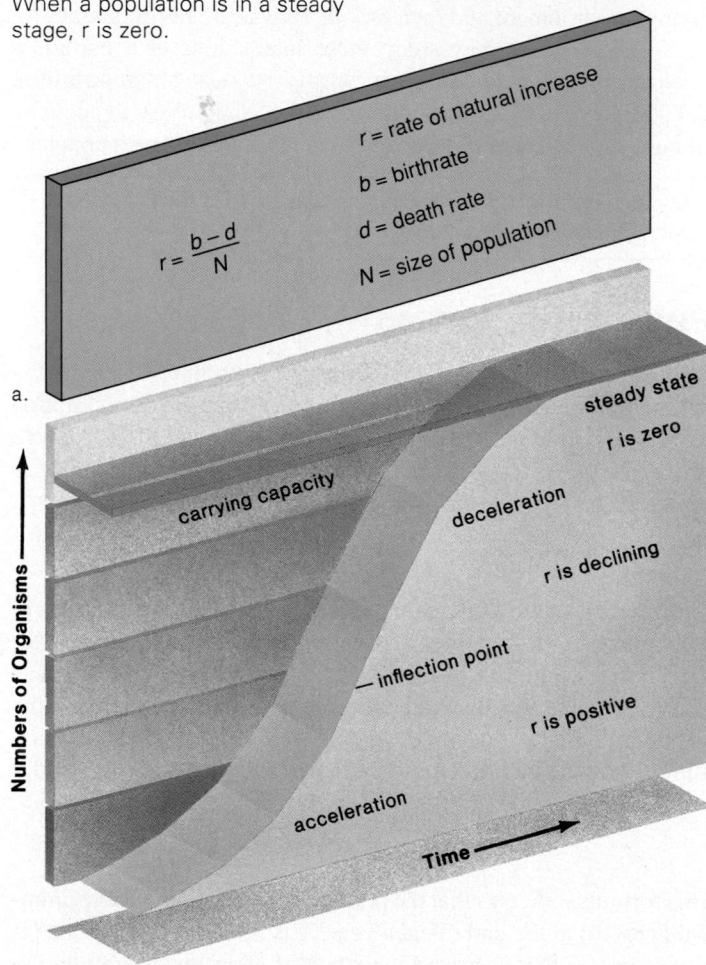

$$r = \frac{b - d}{N}$$

r = rate of natural increase
b = birthrate
d = death rate
N = size of population

a.

b.

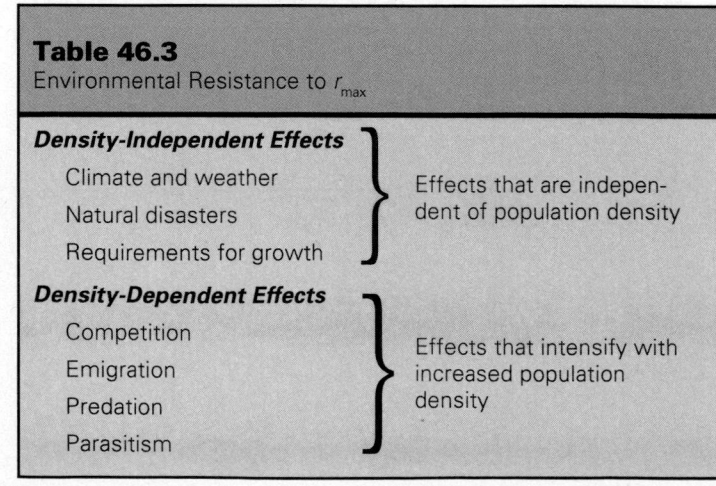

Table 46.3
Environmental Resistance to r_{max}

Density-Independent Effects	
Climate and weather	
Natural disasters	Effects that are independent of population density
Requirements for growth	
Density-Dependent Effects	
Competition	
Emigration	Effects that intensify with increased population density
Predation	
Parasitism	

Figure 46.4

a. St. Matthew Island, Alaska, reindeer. *b.* The herd grew exponentially for several decades and then underwent a sharp decline as a result of overgrazing the available range.

a.

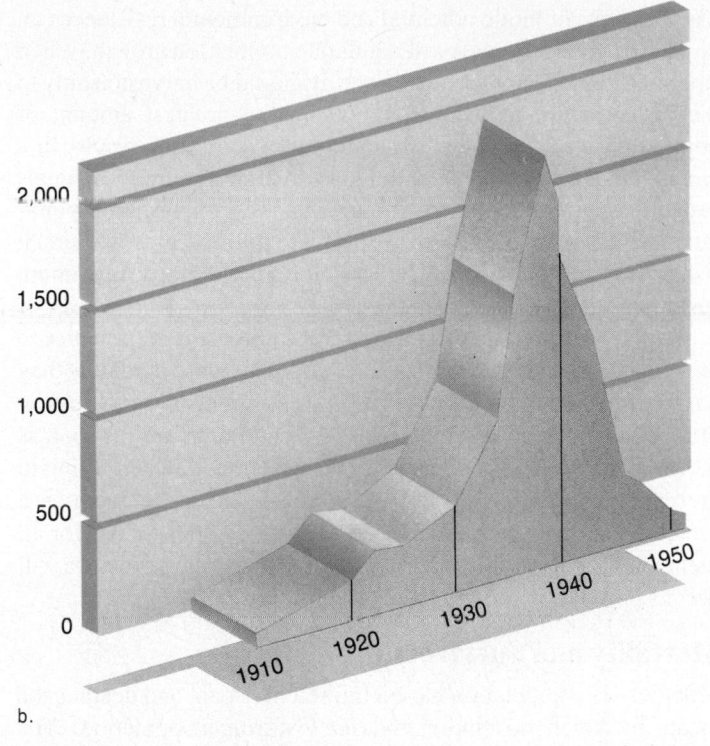

b.

Population size is believed to level off at the **carrying capacity** (*K*) of the environment. In other words, the environment is capable of sustaining a population of a limited size. As the population increases in size, there will be more and more competition for available space and food. Eventually, this will affect population size. It is said that the environment increasingly resists the biotic potential of the population. **Environmental resistance** contains the effects given in table 46.3. The expression

$$\frac{K - N}{K}$$

pertains to density-dependent resistance. It also tells us what proportion of the carrying capacity still remains. If *N* is small, the expression approximates 1, and most of the carrying capacity still remains. If *N* is large, the expression approaches 0, which indicates that the carrying capacity has almost been reached. Insert real numbers in the equation to show that this is the case.

If our argument to this point is sound, then insertion of the expression

$$\frac{K - N}{K}$$

into the equation for exponential growth should result in an **S**-shaped curve.

$$I = r \left(\frac{K - N}{N} \right) N$$

At low population densities, when the expression approximates 1, growth would be nearly equal to *rN;* therefore, growth is exponential. But as the density increases, the expression approaches 0, and

growth levels off. Therefore, it seems that the concept of carrying capacity explains the logistic growth curve. Further, we can predict that if a population happens to overshoot the carrying capacity, it will have to decline once again because the expression

$$\frac{K - N}{K}$$

then will have a negative value.

Fluctuations are often seen in population sizes. For example, fluctuations can be due to an abrupt change in carrying capacity from season to season. During the summer, when resources are plentiful, a population increases in size. In winter, when the environment is less favorable, a population declines. As a result, population size can fluctuate from year to year. The mean of the increases and decreases over time represents the average carrying capacity of the environment.

Among some species, notably insects and certain plants, populations grow rapidly for a limited period of time and then quickly die out. The growth curve for these populations is at first **J**-shaped, but the exponential growth is followed by a sharp decline. For example, a reindeer herd introduced on St. Matthew Island, Alaska, underwent a rapid expansion and then declined just as quickly when food resources were depleted by the overgrazing of the herd, which had grown too large (fig. 46.4).

Every population has a biotic potential for increase in size. Biotic potential is normally held in check by environmental resistance so that the growth curve is **S**-shaped and levels off at the carrying capacity of the environment. **J**-shaped curves are also known.

Practical Applications

The concepts of biotic potential and environmental resistance can be used to devise the means of regulating population growth. When a population is a human food source, it should be harvested only to the inflection point (fig. 46.3), where the greatest amount of population growth occurs. This amount of cropping results in a *maximum sustained yield,* as long as enough mature individuals capable of reproduction remain. Unfortunately, many marine populations, such as whales, have been overexploited by commercial fisheries to the point where it will be many years before a maximum sustained yield will again be possible.

For pests, such as rats and mosquitoes, it is far better to reduce breeding sites and food supply or to increase predators than it is to try to kill off the animals. Killing the animals does not restrict their growth potential, which may come into full force as soon as the population size is reduced. On the other hand, when trying to preserve an endangered species, it is best to place them in a protective environment, such as a reserve, which provides for all their needs. In this case, reduction in environmental resistance will permit population growth to occur.

Mortality and Survivorship

Members of a population have a limited life span, and death is the means by which most individuals are lost from a population. (The other means is emigration—departure.) The number of members of a population that are dying per unit of population per unit time is the death rate. The death rate is usually determined by dividing the number of individuals that have died during a given time period by the number alive at the beginning of that period of time.

Rates of natural increase in population size are a function of additions by birth and losses by death. The rate of increase sometimes becomes greater, not because of a growing birthrate but because of a decline in the death rate. This decline in deaths is one reason for the explosive growth of the human population in certain countries of the world.

Because living members of the population are more important to the population than dead ones, another way of looking at mortality is **survivorship,** which is the number of individuals in a given population alive at the beginning of each age interval. A survivorship curve can be drawn by plotting the number of survivors against time.

Three types of survivorship curves are shown in figure 46.5. The type-I curve shows that most individuals within a designated group live out their allotted life span and then die of old age. Under very favorable environmental conditions, certain mammals, such as humans, and annual plants, which live one season, tend to approach this curve. The type-II curve results when organisms die at a constant rate through time. Such curves are typical of birds, rodents, and some perennial plants that live more than one season. The type-III curve indicates that many organisms tend to die early in life. This curve is typical of fishes, many invertebrates, and many perennial plants.

Age Structure

Populations have 3 age groups: *dependency, reproductive,* and *postreproductive.* One way of characterizing populations is by

Figure 46.5
The 3 basic types of survivorship curves. Type-I curve occurs when the members of a species usually live the full physiological life span. Type-II curve occurs when the rate of mortality is fairly constant at all age levels. Type-III curve occurs when there is a high mortality early in life.

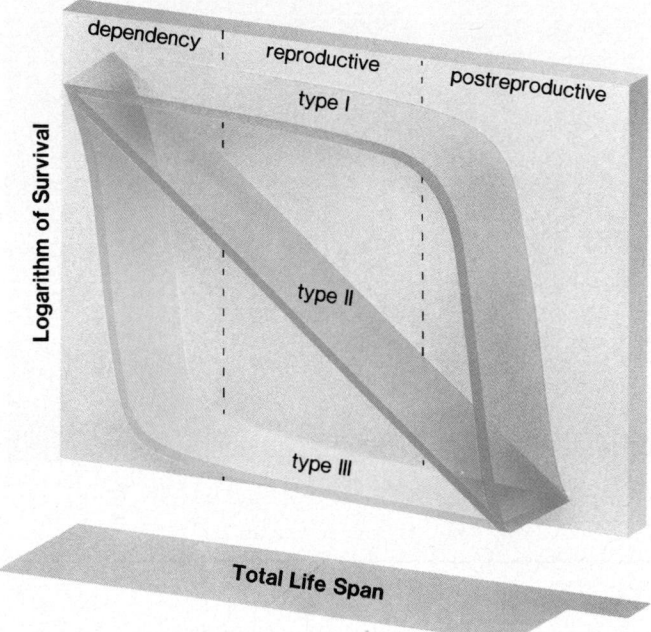

these age groups. This is best visualized when the proportion of individuals in each group is plotted on a bar graph, thereby producing an **age structure diagram.** If the birthrate is high, and the population is undergoing exponential growth, a *pyramid-shaped* figure results, because each successive generation is larger than the previous one (fig. 46.6). When the birthrate equals the death rate, a *bell-shaped* age diagram is expected, because the dependency and reproductive age groups become more nearly equal. The postreproductive group is still the smallest, however, because of mortality. If the birthrate falls below the death rate, the dependency group will become smaller than the reproductive group. The age structure diagram will then be *urn shaped* because the postreproductive group is now the largest.

In a study of a honeybee hive, these changes in the age structure diagram were observed in a single season. At the beginning of the season, when the population was expanding, the age structure diagram was pyramid shaped; in midseason, when the population size was stationary, the age structure diagram was bell shaped; and finally, as the season came to a close and the population was dying off, the age structure diagram was urn shaped.

Strategies for Existence

Populations do not increase in size year after year because environmental resistance, including both *density-independent* and *density-dependent* effects (table 46.3), regulates number. Some populations are regulated primarily by density-independent effects. These tend to have **J**-shaped growth curves until

Figure 46.6

Age structure diagrams for 3 different populations. **a.** Age structure diagram is a pyramid shape, indicating that the population is undergoing exponential growth. **b.** Age structure diagram is tending to be bell shaped, indicating that population size is fairly stationary. **c.** Age structure diagram is tending toward an urn shape, indicating that the population is on the decline. From "The Human Population" by Ronald Freedman and Bernard Berelson. Copyright © 1974 by Scientific American, Inc. All rights reserved.

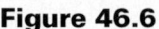

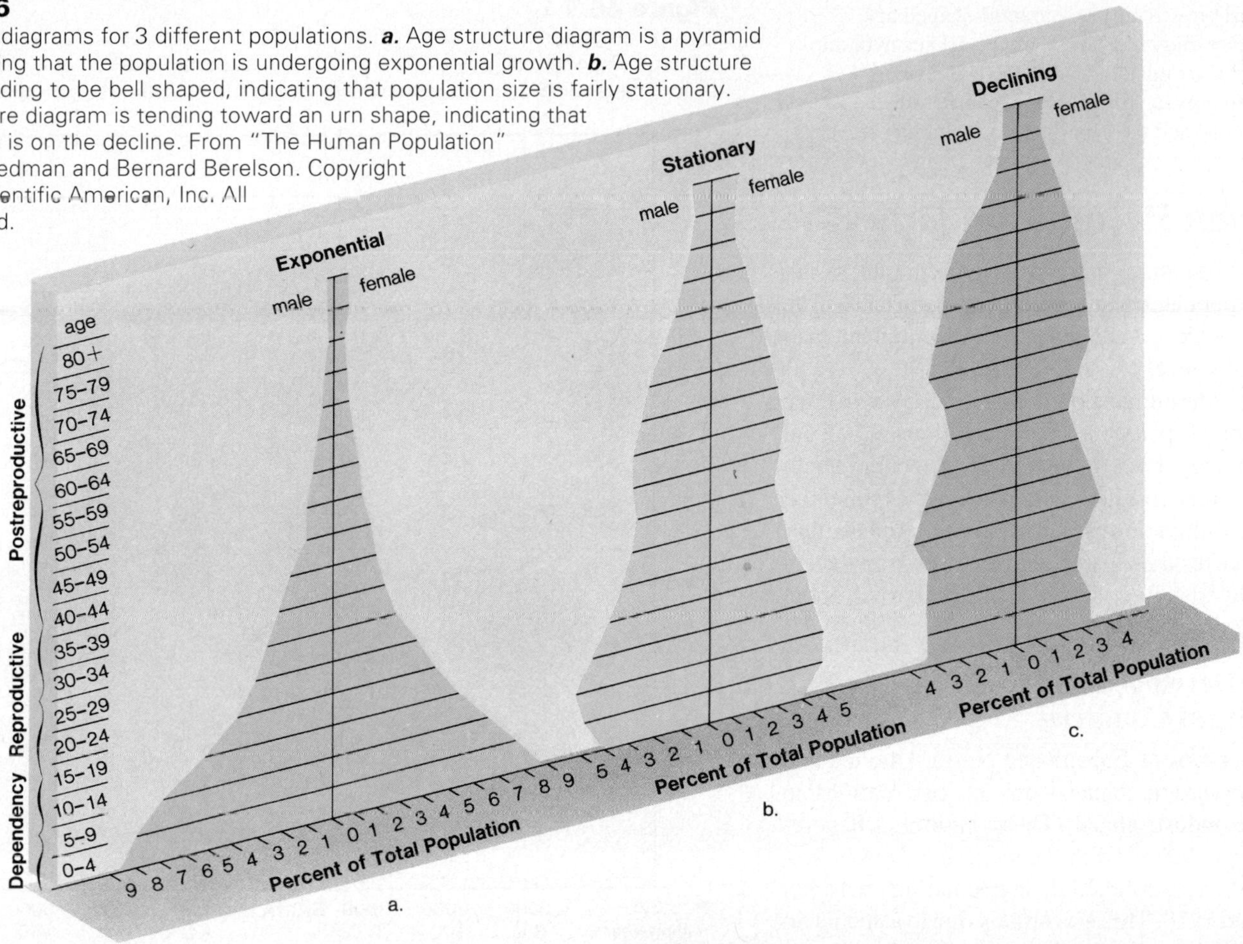

some environmental change causes them to decline, usually within a short time. Most likely these populations are *opportunistic species,* which move in and occupy new environments. Dandelions in lawns and along roadsides and tent caterpillars in cherry and apple trees are opportunistic species. During a short period of time, they produce many offspring, which require little care. Therefore, these populations usually have a survivorship curve similar to type-III in figure 46.5. But chances are, some of these offspring will quickly disperse to colonize new habitats because opportunistic species usually have efficient dispersal mechanisms. From an evolutionary point of view, such species have undergone selection to maximize their rate of natural increase and for this reason are said to be *r*-selected or to be *r*-strategists (table 46.4).

Other populations are regulated primarily by density-dependent effects. These tend to be **S**-shaped growth curves. Such populations are termed *equilibrium species.* For example, whales are relatively large, long-lived, and slow to mature. They produce a small number of offspring and expend considerable energy in producing each individual because resources are limited and competitive superiority is a necessity for survival. These organisms tend to have a survivorship curve like type-I in figure 46.5. Equilibrium species are strong competitors and,

Table 46.4
r-strategists versus *K*-strategists

r-strategists	*K*-strategists
Small individuals	Large individuals
Short life span	Long life span
Fast to mature	Slow to mature
Many offspring	Few offspring
Little or no care of offspring	Much care of offspring

once established, can dominate or exclude opportunistic species. They are specialists rather than colonizers, and they tend to become extinct when their normal way of life is destroyed. When undisturbed, the population size remains stable at or near the carrying capacity of the environment. Equilibrium species are said to be *K*-selected or to be **K-strategists.**

Most populations cannot be characterized as being either *r*- or *K*-strategists because they have intermediate characteristics. Such populations tend to have the type-II survivorship curve.

Ecology of Populations

749

Populations of *r*-strategists tend to have **J**-shaped growth curves, pyramid-shaped age structure diagrams, and the type-III survivorship curve. *K*-strategists tend to have **S**-shaped growth curves, urn-shaped age structure diagrams, and the type-I survivorship curve.

Human Population Growth

The rate of natural increase or growth rate, of the human population of the world is now 1.8%. The growth curve is **J**-shaped and exponential but may eventually level off (fig. 46.7).

The human population is believed to have undergone 3 phases of exponential growth. Tool making may have been the first technological advance that resulted in a phase of exponential growth. Cultivation of plants and animal husbandry may have resulted in a second phase of growth; and the Industrial Revolution, which occurred about 1850, promoted the third phase.

More Developed and Less Developed Countries

The countries of Europe and North America and also Russia and Japan were the first nations to become industrialized. These nations, often referred to collectively as the *more developed countries (MDCs)*, doubled their populations between 1850 and 1950. This was largely due to a decline in the death rate caused by the rise of modern medicine and improved socioeconomic conditions. The death rate decline was followed shortly thereafter by a decline in the birthrate, so the populations in the MDCs showed only modest growth between 1950 and 1975. This sequence of events is termed a **demographic transition.**

The growth rate for the MDCs as a whole has now stabilized at about 0.6%. A few of the MDCs—Italy, Denmark, Germany, Hungary, and Sweden—are not growing or are actually losing population. The United States has a higher growth rate (0.7%) than average, because many people immigrate to the United States each year. In addition, a baby boom between 1947 and 1964 resulted in an unusually large number of women who are still of reproductive age.

Most countries in Africa, Asia, and Latin America are collectively known as the *less developed countries (LDCs)* because they are not yet fully industrialized. Although mortality began to decline steeply in these countries following World War II with the importation of modern medicine from the MDCs, the birthrate did not decline to the same extent. Therefore, the populations of the LDCs began to increase dramatically. The LDCs were unable to cope adequately with such rapid population expansion, and many people in these countries are underfed, ill-housed, unschooled, and living in abject poverty.

Figure 46.7

Growth curve for world human population. The human population is now undergoing rapid exponential growth. The population size may level off at 8, 10.5, or 14.2 billion, depending upon the speed with which the growth rate may decline.

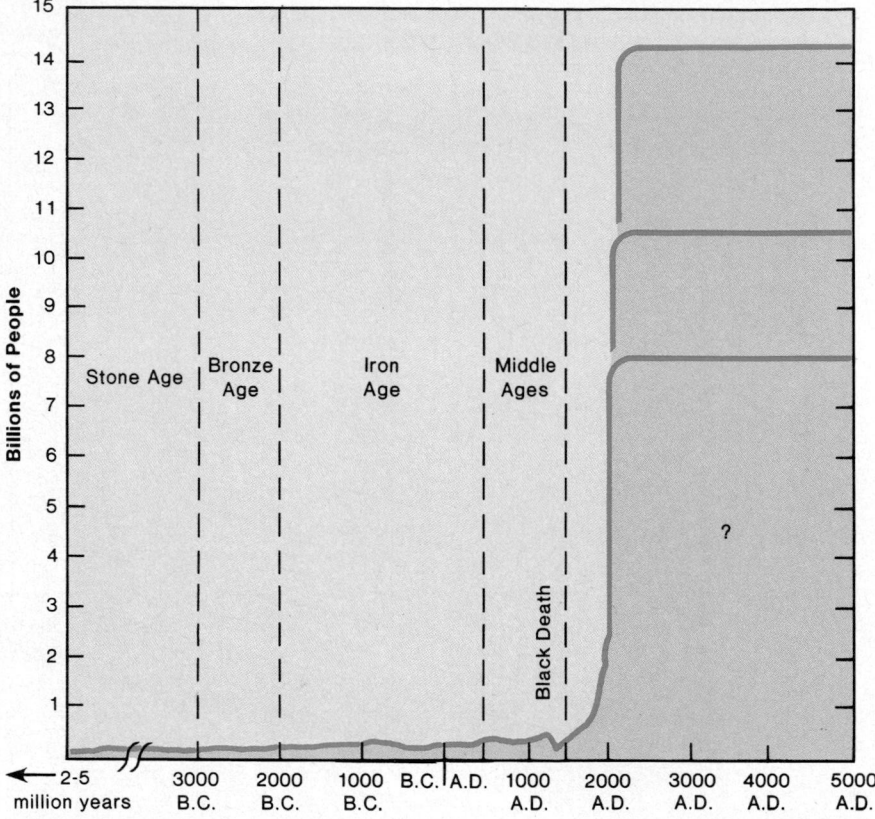

The growth rate of the LDCs peaked at 2.4% between 1960 and 1965. Since that time, the mortality decline has slowed and the birthrate has fallen. The growth rate is expected to be 1.8% by the end of the century. At that time, however, more than two-thirds of the world population will be in the LDCs.

The reason for a decline in the growth rate in the LDCs is not clear. Previously, it was argued that decline would only be seen when socioeconomic conditions had improved. It was believed that as soon as the populace enjoyed the benefits of an industrialized society, the birthrate would decline. It is now apparent, however, that the LDCs with the greatest decline are those with good family-planning programs supported by community leaders. Such programs may bring about stable population sizes in the LDCs even before socioeconomic conditions have improved. Nevertheless, certain socioeconomic factors have also probably contributed to the decline in the LDC's growth rate. A relatively high gross national product (GNP), urbanization, low infant mortality, increased life expectancy, literacy, and education, especially of women, have all had a dampening effect on the *fertility rate*—the average number of children born to each woman.

The history of the world population shows that the more developed countries underwent a demographic transition between 1950 and 1975; the less developed countries are just now undergoing demographic transition.

Figure 46.8

Age structure diagram for less developed and more developed countries, 1989. The diagrams illustrate that the more developed countries are approaching stabilization, whereas the less developed countries will expand rapidly due to the shape of their age structure diagram. Source: Data from *World Population Profile: 1989,* WP–89.

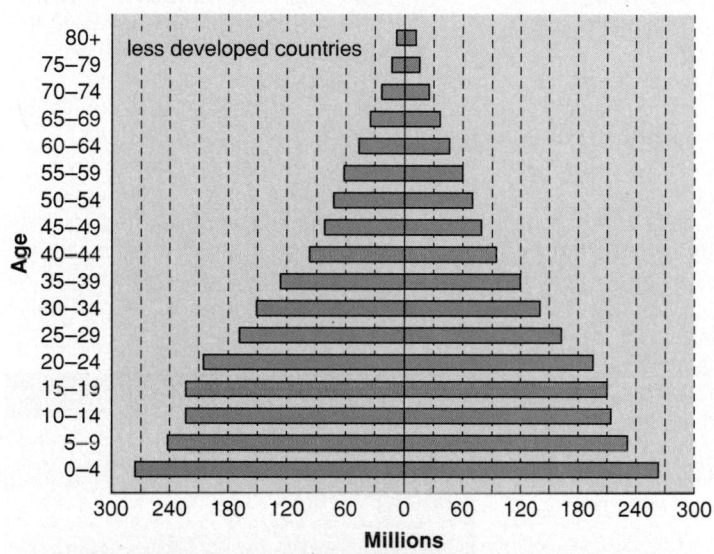

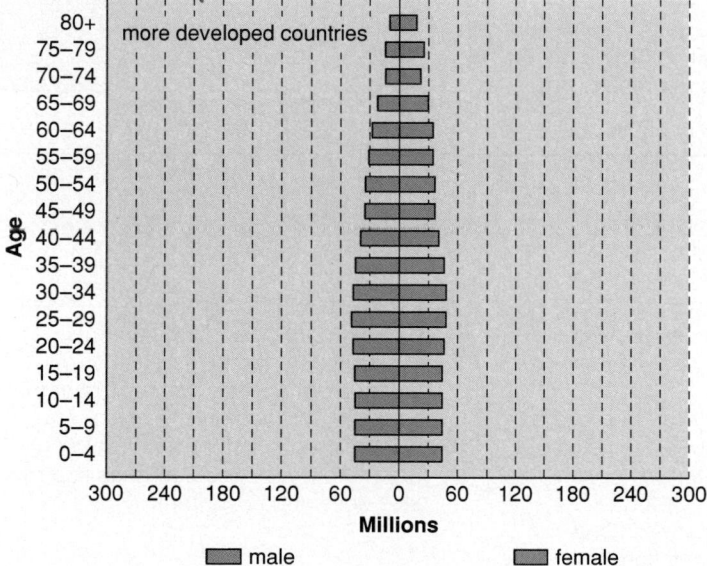

Age Structure Comparison

Laypersons are sometimes under the impression that if each couple has 2 children, **zero population growth** will immediately take place. *Replacement reproduction,* as it is called, however, still cause most countries to continue growth because of the age structure of the population. If there are more young women entering the reproductive years than there are older women leaving them, replacement reproduction will still result in a positive growth rate.

Reproduction is at or below replacement level in some 20 MDCs, including the United States. Even so, some of these countries will continue to grow modestly, in part because of the baby boom after World War II. These women are still in their reproductive years, and even if each one has fewer than 2 children, the population will have grown. Also keep in mind that even the smallest growth rate can add a considerable number of individuals to a large country.

Whereas many MDCs are tending toward a stabilized age structure, most LDCs have a youthful profile—a large proportion of the population is younger than age 15 (fig. 46.8). Because there are so many young women entering their reproductive years, the population will still greatly expand, even after replacement reproduction is attained. The more quickly replacement reproduction is achieved, however, the sooner zero population growth will result.

Consequences of Population Growth

Increasing world population is putting extreme pressure on the earth's resources, environment, and social organization. Although population increases in LDCs might seem to be of the gravest concern, this is not necessarily the case. Each person in a MDC consumes more resources, and is therefore responsible for a greater amount of pollution than a person in a LDC. Environmental impact (EI) is measured not only in terms of population size but also in terms of the resources used and the pollution caused by each person in the population. Therefore, there are 2 possible types of overpopulation. The first type is due to increased population, and the second is due to increased resource consumption. The first type of overpopulation is more obvious in LDC, and the second type is more obvious in MDC.

Resource consumption is dependent on population size and on consumption per individual. The less developed countries are responsible for the first type of environmental impact, and the more developed countries are responsible for the second type.

Summary

1. Ecology is the study of the interactions of organisms with the physical environment and with each other. Organisms of the same species living in one locale form a population.

2. Populations with a positive rate of natural increase (r) have an increase in size expressed as $I = rN$, where (N) is the number of individuals in the population. Because (N) is larger each year, so is (I). This means the population is increasing in size exponentially, and the growth curve is a **J**-shaped, or exponential one.

3. The maximum rate of natural increase (biotic potential) for a population (r_{max}) occurs only when there is no environmental resistance, which includes density-independent and density-dependent effects.

4. An r_{max} may be observed for a short time during exponential growth, but eventually environmental resistance causes the growth curve to become sigmoidal as the population size levels off at the carrying capacity of the environment.

5. Sometimes populations exhibit a **J**-shaped curve followed by a sharp decline. Opportunistic species often colonize an area and then are cut back by the first frost, for example. Or unrestricted growth is often followed by a crash, as the population greatly exceeds the environment's carrying capacity.

6. Populations tend to have one of 3 types of survivorship curves depending on whether most individuals live out the normal life span, die at a constant rate regardless of age, or die early.

7. Each population also has a particular type of age structure diagram, which indicates the number of individuals in the dependency, reproductive, and postreproductive age groups. The shape of the diagram tells whether the population is a young, expanding one, a stable one, or a dying one.

8. Populations seem to have a particular overall strategy for continued existence. Those that are opportunists produce many young within a short period of time and rely on rapid dispersal to new, unoccupied environments. These populations tend to have a **J**-shaped growth curve and to be regulated by density-independent effects. Such populations belong to species that are r-selected.

9. Other populations produce a limited number of young, which they nurture for a long time. These populations tend to be regulated by density-dependent effects and belong to species that are K-selected.

10. The human population is now in an exponential part of its growth curve. More developed countries underwent a demographic transition between 1950 and 1975, with the result that their growth rate is now only about 0.6%. Less developed countries are just now undergoing demographic transition. Because this transition was delayed, their growth rate at one time was as high as 2.4%, but it is declining slowly now.

11. Replacement reproduction will not bring about zero population growth even in more developed countries if the age structure is biased toward those in their early reproductive years. In less developed countries, where the average age is under 15, it will be many years before replacement reproduction will mean zero population growth.

12. Resource consumption depends on population size and on consumption per individual. Less developed countries are responsible for the first type of environmental impact, and more developed countries are responsible for the second type.

Writing Across the Curriculum

In order to practice writing skills, students should write out the answers to any or all of the study questions and the critical thinking questions. The study questions are sequenced in the same order as the text. Suggested answers to the critical thinking questions are in appendix D.

Study Questions

1. Draw a sigmoid growth curve and relate the concepts of biotic potential, environmental resistance, and carrying capacity to the curve.
2. Draw a J-shaped curve and discuss why populations characterized by such curves may suddenly decline in size.
3. Show that a population of 10,000 persons will add more individuals with each generation, even if the growth rate remains constant. Is it possible that even with a declining growth rate the same number of individuals could be added each generation?
4. Name 4 types of density-dependent effects, and explain what is meant by this term.
5. Draw 3 different survivorship curves, and tell what each curve indicates about the expected time of death.
6. Draw 3 different age structure diagrams and tell what each one indicates about a population.
7. What is an "r-strategist"? a "K-strategist"?
8. Define demographic transition. When did the more developed countries undergo demographic transition? When did the less developed countries undergo demographic transition?
9. Give at least 3 differences between the more developed countries and the less developed countries.

Objective Questions

1. A population contains all the
   a. organisms that live at one locale.
   b. organisms that are members of the same species at one locale.
   c. similar types of organisms at one locale.
   d. organisms that are interacting with each other at one locale.
2. In a population $N = 1,500$, if the birthrate is 42 per year and the death rate is 27 per year, the rate of natural increase will be
   a. 42%.
   b. 27%.
   c. 1%.
   d. 0.1%.
3. Whether the biotic potential for a species is high or low depends on such factors as
   a. usual number of offspring per reproduction.
   b. chances of survival until age of reproduction.
   c. age at which reproduction begins.
   d. All of these.
4. Environmental resistance
   a. prevents exponential growth from occurring.
   b. includes density-independent effects.
   c. helps determine the carrying capacity.
   d. Both b and c.
5. Populations termed *r*-strategists
   a. have **J**-shaped growth curves.
   b. have type-III survivorship curves.
   c. are usually pioneer species.
   d. All of these.
6. Replacement reproduction cannot bring about zero population growth unless
   a. all diseases such as cholera, typhus, and diphtheria are brought under control.
   b. there are many fatalities due to a world war.
   c. there are as many young women entering their reproductive years as older women leaving them.
   d. Any one of these.
7. Once the demographic transition has occurred
   a. both the death rate and birthrate are high.
   b. both the death rate and birthrate are low.
   c. the death rate is high but the birthrate is low.
   d. the death rate is low but the birthrate is high.
8. Which shape best describes the age structure for a country whose population has stabilized?
   a. pyramid shaped
   b. urn shaped
   c. bell shaped
   d. exponential
9. Label this diagram of an **S**-shaped growth curve. What does *r* stand for?

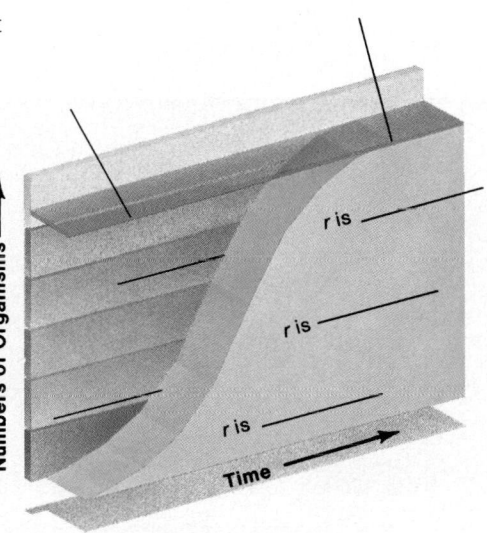

Concepts and Critical Thinking

1. *Population growth is always finite.*

   Under what circumstances could population growth possibly be infinite?

2. *Each population has a life strategy.*

   How is the human life strategy like that of an *r*-strategist? like that of a *K*-strategist?

Selected Key Terms

ecology (e-kol'o-je) 745
population (pop"u-la'shun) 745
exponential growth (ex"po-nen'shal groth) 745
biotic potential (bi-ot'ik po-ten'shal) 745
carrying capacity (kar'e-ing kah-pas'i-te) 747

environmental resistance (en-vi"ron-men'tal re-zis'tans) 747
survivorship (sur-vi'vor-ship) 748
age structure diagram (aj struk'tūr di'ah-gram) 748
K-**strategist** (ka strat'e-jist) 749

demographic transition (dē"mo-graf'ik tran-zish'un) 750
zero population growth (ze'ro pop"u-la'shun grōth) 751

47

Interaction within Communities

A bat swoops down out of the sky to capture a mouse. The predator-prey interaction is an important one among animal populations. Over time, the prey evolves better protective mechanisms and the predator evolves better capture mechanisms. This coevolutionary process continues, and the end result is that a predator continues to capture its prey.

Your study of this chapter will be complete when you can

1. discuss the effect that interspecific competition can have on population size;
2. state the competitive exclusion principle, and relate this principle to the diversity of organisms;
3. state several characteristics of a niche, and distinguish between niche and the habitat of an organism;
4. discuss the effect that predation can have on the size of the prey population and diversity of the community;
5. give examples to show that human intervention can upset the natural balance of a community;
6. give examples of plant and animal antipredator defenses;
7. explain the principle of mimicry, and give examples of the 2 kinds of mimicry;
8. give examples of the 3 types of symbiotic relationships, and tell what effect they can have on population size.

Whenever we walk into a forest or stop by the side of a pond, we are in a **community** of populations interacting with one another. The types of interactions are various, but we will primarily consider interactions involving competition, predation, and symbiosis. These interactions are important in controlling the diversity and size of populations. Only when populations remain at an appropriate size—one to the other—can a community of populations continue to sustain itself.

Competition

In chapter 46, we saw that *intraspecific competition*—competition for resources among members of a population—is a part of environmental resistance helping to limit population growth. Some animals live in groups with *dominance hierarchies* that determine the social position and importance of each animal. Dominance is determined by a contest—either actual fighting or simply a ritual contest of who can frighten whom. The dominant animals eat first and may have a chance to mate before lower ranking animals. The other animals either assist the dominant individuals or leave the group. In either case, they may not produce offspring.

Another social interaction that may affect population size is *territoriality,* in which successful competitors establish territories that they defend against other animals. Many animals mate and produce offspring within the territory. The best territory defenders often have the best territories—those with the best food resources and, among birds, the best nesting sites. The less successful defenders must make do with less desirable territories, and some have no territories at all. Therefore, these animals produce fewer offspring than those in better territories, or they produce no offspring at all.

Similar types of species with the same needs are likely to compete with one another for resources, such as water, food, sunlight, and space. The outcome of this *interspecific competition* is often the predominance of one species and the virtual elimination of the other. This was demonstrated in the laboratory by G. F. Gause, who grew 2 species of paramecia in one test tube containing a fixed amount of bacterial food. Although each population survived when grown separately, only one survived when they were grown together (fig. 47.1). The successful paramecium population had a higher rate of natural increase than the unsuccessful population. This is an example of local extinction resulting from interspecific competition.

Instead of extinction in natural ecosystems, there may be restriction in the range of one of the species, as observed by Joseph Connell when he studied the distribution of barnacles on the Scottish coast. *Chthamalus stellatus* lives on the high part of the intertidal zone, and *Balanus balanoides* lives on the lower part (fig. 47.2). Free-swimming larvae of both species attach themselves to rocks at any point in the intertidal zone, where they develop into the

Figure 47.1

Competition between 2 species of paramecium. When grown alone in pure culture, *Paramecium caudatum* and *Paramecium aurelia* exhibit sigmoidal growth. When the 2 species are grown together in mixed culture, *P. aurelia* is the better competitor and its population increases. *P. caudatum* cannot grow in the presence of a growing population of *P. aurelia,* so it dies out. Source: Data from G. F. Gause, *The Struggle for Existence.* Copyright © 1934 Williams & Wilkins Company, Baltimore, MD.

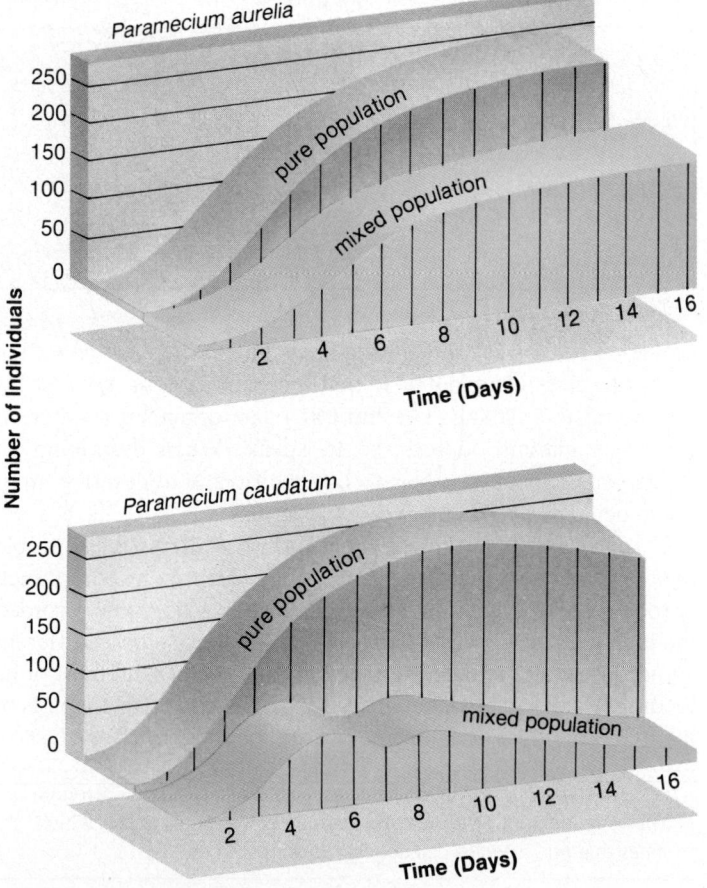

Figure 47.2

Effects of competition. Competition prevents 2 species of barnacles from occupying as much of the intertidal zone as possible. Both exist in the area of competition between *Chthamalus* and *Balanus.* Above this area only *Chthamalus* survives and below it only *Balanus* survives.

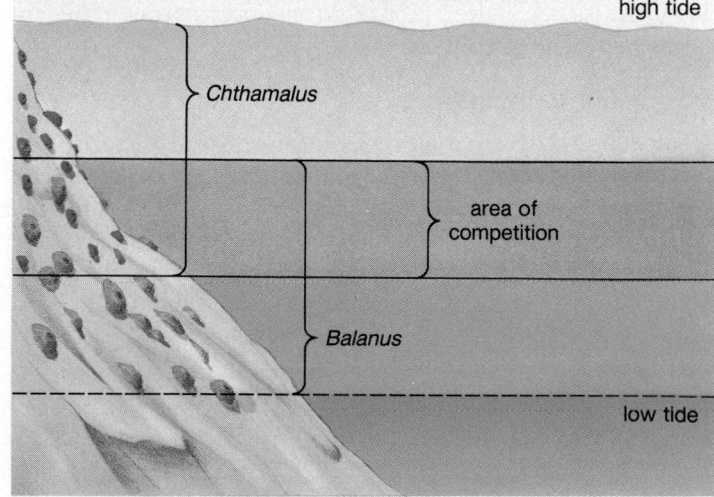

Figure 47.3

Resource partitioning among 5 species of coexisting warblers. The diagrams that appear beside the birds represent spruce trees. The time each species spent in various portions of the trees was determined, and it was discovered that each species spends more than half its time in the blue regions.

| Cape May warbler | Black-throated green warbler | Yellow-rumped warbler | Bay-breasted warbler | Blackburnian warbler |

sessile adult form. In the lower zone, the faster-growing *Balanus* individuals either force *Chthamalus* individuals off the rocks or grow over them. *Balanus,* however, is not as resistant to drying out as is *Chthamalus* and does not survive well in the upper intertidal zone, thus allowing *Chthamalus* to grow there.

Human Intervention

The fact that successful competition can cause one species to increase in size at the expense of another inadvertently has been demonstrated by human intervention. The carp, a fish imported to the United States from the Orient, is able to tolerate polluted water. Therefore, this fish now is often more prevalent than our own native fishes. Melaleuca, an ornamental tree that thrives in wet habitats, became a pest in the Everglades National Park after being introduced into Florida. The burro, which originated in Ethiopia and Somalia and is adapted to a dry environment, now is threatening the existence of deer, pronghorn antelope, and desert bighorn sheep in the Grand Canyon.

Exclusion Principle

Competition between species does not always lead to expansion of one population and restriction of another. Gause showed in another laboratory experiment that when 2 different species of paramecia occupied the same tube, both survived because one species fed on bacteria on the bottom and the other fed on bacteria suspended in solution. Such experiments have led some biologists to formulate and support a **competitive exclusion principle,** which states that no 2 species can occupy the same niche at the same time.

Niche is the term used to refer to the role a species plays in a community of organisms. To describe the niche of a species, it is necessary to state all the requirements and the activities of the

Table 47.1
Aspects of Niche

Plants	Animals
Season of year for growth and reproduction	Time of day for feeding and season of year for reproduction
Sunlight, water, and soil requirements	Habitat and food requirements
Relationships with other organisms	Relationships with other organisms
Effect on abiotic environment	Effect on abiotic environment

species. Table 47.1 lists factors to be included when describing the niche of a particular plant or animal species. One aspect of niche is the organism's habitat. The **habitat** of an organism is where it lives—its mailing address, so to speak. When describing an organism's habitat, it is best to be specific and to describe, in as much detail as possible, the exact location.

The concept of niche suggests that similar species are able to coexist in one ecosystem because they partition resources such as food and living space. For example, Robert MacArthur recorded the length of time each of 5 species of warblers spent in different spruce tree zones to determine where each species did most of its feeding. He discovered that different species use different parts of the tree canopy, although some overlap does occur (fig. 47.3).

Several species living in the same area partition available resources. This is in keeping with the competitive exclusion principle, which states that no 2 species can occupy the same niche.

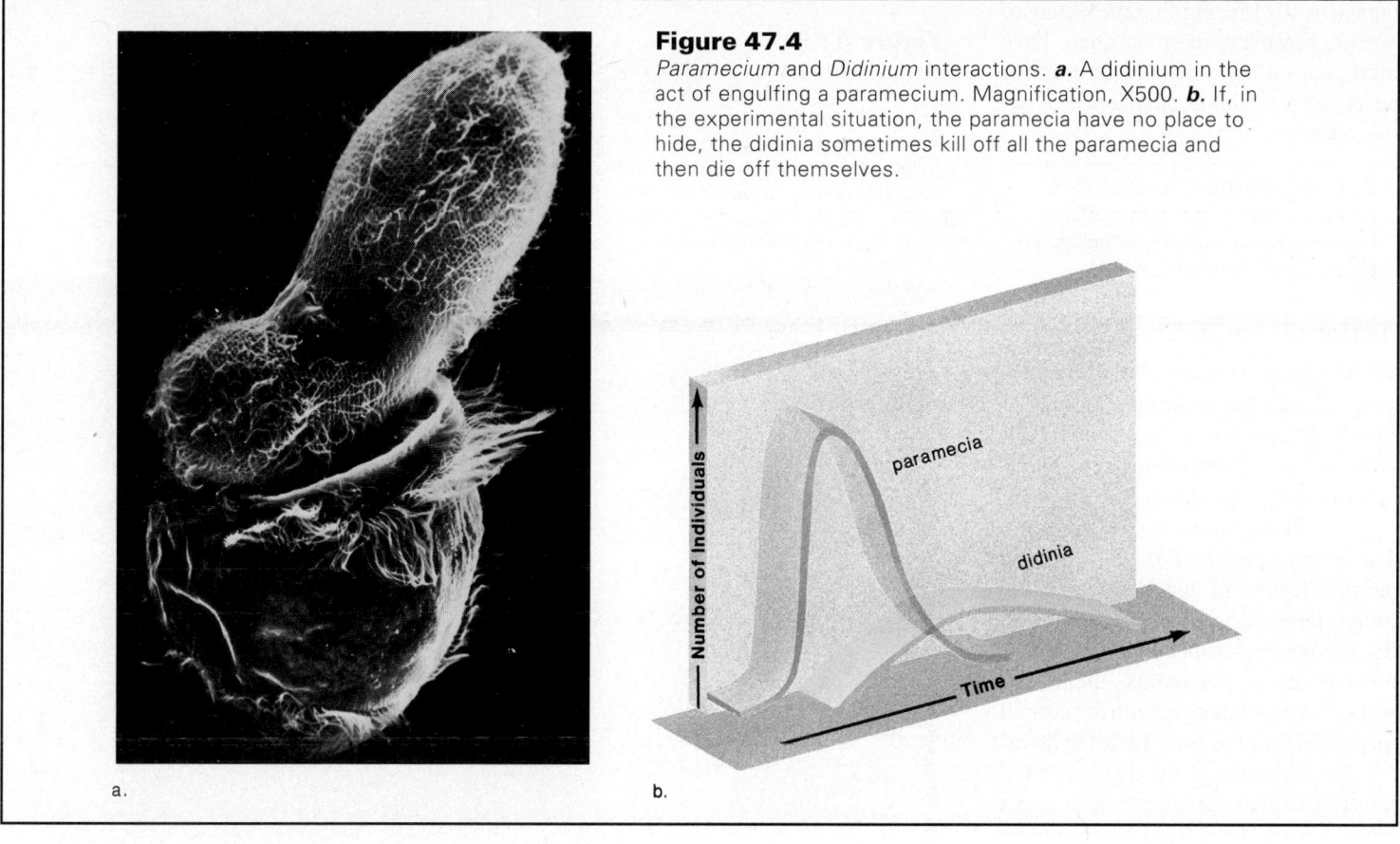

Figure 47.4
Paramecium and *Didinium* interactions. *a.* A didinium in the act of engulfing a paramecium. Magnification, X500. *b.* If, in the experimental situation, the paramecia have no place to hide, the didinia sometimes kill off all the paramecia and then die off themselves.

a.

b.

Predation

Predation, simply defined, is one organism feeding on another, called the *prey*. Through the evolutionary process, predators become adapted to capturing their prey. The anatomy of browsers enables them to reach high into trees—giraffes have long necks and elephants have long trunks. Birds of prey have talons and sharp beaks, which enable them to capture, hold, and kill their prey. Carnivores also need keen eyesight. Hawk eyes have a resolving power 8 times that of the human eye due to the concentration of cones in 2 foveas instead of one.

Effects of Predation

In one laboratory experiment, Gause placed didinia and paramecia in a test tube. In this experimental environment, the didinia actually captured all the paramecia and then died out from starvation (fig. 47.4). If Gause provided refuge for the prey, however, in this case sediment on the bottom of the laboratory glassware, part of the prey population survived. This suggests that predator-prey relationships can stabilize when predator pressure eases as the prey population decreases in size. In nature, this can be the result of prey being able to hide and/or of predators switching to a more abundant, and hence more economical, prey species.

In a famous Canadian field study, the size of the lynx population was observed to fluctuate with the size of the snowshoe hare population (fig. 47.5). Ecologists are not convinced, however,

that these oscillations are due to the predator-prey relationship. These changes may be due to a periodic decline in the availability of food for the hares. As their population size declines, the lynx population also declines. This explanation seems reasonable because the size of hare populations has fluctuated even in the absence of lynxes.

Robert Paine demonstrated in the 1960s that if the large starfish *Piaster* is removed from an intertidal community of the Washington Coast, the mollusk *Mytilus* outcompetes all the other species on the rocks. He said that *Piaster* was a *keystone predator* because it preyed on the competitively dominant species of the area. A keystone predator helps increase the diversity of a community by keeping the size of its prey population in check. Expanding on Paine's concept, other researchers use the term *keystone species* to mean one whose removal causes the extinction of other species.

Human Intervention

Sometimes humans have neglected to take into consideration that predators help to keep prey populations in check. For example, in the past, coyotes have been killed indiscriminately in the West without regard for the fact that they help to keep the prairie dog population under control. Similarly, when the dingo, a wild dog in Australia, was killed off because it attacked sheep, the rabbit and wallaby populations greatly increased. In contrast, humans formerly kept the burro population in the Grand Canyon within reasonable limits by killing them for meat. Since a federal law was passed in 1971 forbidding the killing of burros, however, their numbers have increased until they now are destructive pests and threaten the existence of native animals.

> Predators help to control the size of prey populations. Whether oscillations in predator and prey populations occur still is being investigated.

Antipredator Defenses

While predators have evolved strategies to secure the maximum amount of food with minimal expenditure of energy, prey organisms have evolved strategies to escape predation.

Plants try to avoid predation. The sharp spines of the cactus, the pointed leaves of the holly, and the tough, leathery leaves of oak trees all discourage predation by insects. Plants even produce poisonous chemicals, some of which interfere with the normal metabolism of the adult insect and others that act as hormone analogues and interfere with the development of insect larvae. Some insects still can inhabit trees that produce toxins because the insects have evolved either detoxifying enzymes or a method of keeping the toxin within their body in a manner that is not harmful to them.

Animals have varied *antipredator defenses*. Some of the more effective defenses are concealment, fright, warning coloration, and vigilance (fig. 47.6).

Mimicry

Mimicry occurs when one species resembles another for its own benefit. Mimicry can help a predator capture food or it can help a prey avoid capture. For example, snapping turtles have tongues and angler fishes have lures that resemble worms for the purpose of bringing small fish within reach.

There are many examples of mimicry among prey animals. Aside from those given in figure 47.6, there are inch worms that resemble twigs and caterpillars that can transform themselves into shapes that resemble snakes, for example. *Batesian mimicry* (named for Henry Bates, who discovered it) occurs when a prey mimics another species that has a successful antipredator defense. Many examples of Batesian mimicry involve warning coloration. Among flies of the family Syrphidae, which feed on the nectar and pollen of flowers, one species resembles the wasp *Vespula arenaria* so closely it is difficult to tell them apart (fig. 47.7). Once a predator

Figure 47.5

a. Montana Lynx, a solitary predator. A long, strong forelimb grabs its main prey, the snowshoe hare, with sharp claws. A lynx lives in northern forests. Its brownish gray coat blends in well against the trunk of trees, and its long legs help walking through deep snow. ***b.*** Example of predator-prey cycling. As the hare population increases in size so does the lynx population, but then the hare population, followed by the lynx population, suffers a dramatic decrease in size. While it may seem as if this cycling is due to "overkill" by the lynx population, other factors could be involved.
(These data are based on the number of lynx and snowshoe hare pelts received by the Hudson Bay Company in the years indicated.)

a.

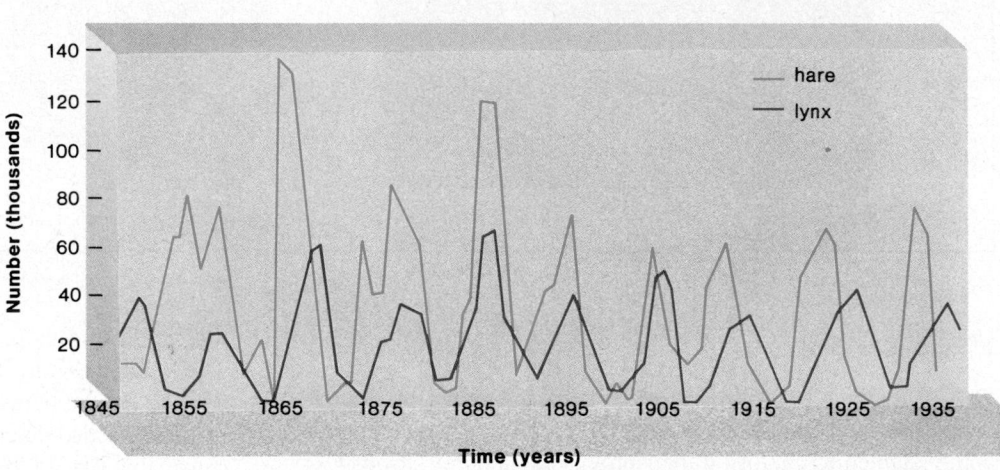

b.

experiences the defense of a wasp it remembers the coloration and avoids all animals that look similarly.[1]

There are also examples of species that have the same defense and resemble each other. For example, many coral snake species have brilliant red, black, and yellow body rings. And the stinging insects, bees, wasps, and hornets all have the familiar black and yellow color bands. Mimics that share the same protective defense are called *Müllerian mimics* after Fritz Müller, who discovered this form of mimicry.

> Antipredator defenses are varied. Sometimes one species mimics another that has a successful defense (Batesian mimic), and sometimes several different species with the same defense mimic each other (Müllerian mimic).

[1]It has long been said that the viceroy butterfly is a mimic of the monarch butterfly, which is poisonous to birds. This is being questioned because (1) birds don't often eat monarch butterflies in the wild; (2) when flycatchers are fed monarch butterflies in the laboratory, they don't get sick; and (3) the viceroy butterfly is toxic when its larvae have fed on willow plants.

Behavior and Ecology

Figure 47.6

Antipredator defenses. ***a.*** Concealment. Flouder can take on the same coloration as their background. ***b.*** Fright. The South American lantern fly has a large false head that resembles an alligator. This may frighten a predator into thinking it is facing a dangerous animal. ***c.*** Warning coloration. The skin secretions of poison-arrow frogs are so poisonous that they were used by natives to make their arrows instant lethal weapons. The coloration of these frogs warns others to beware.

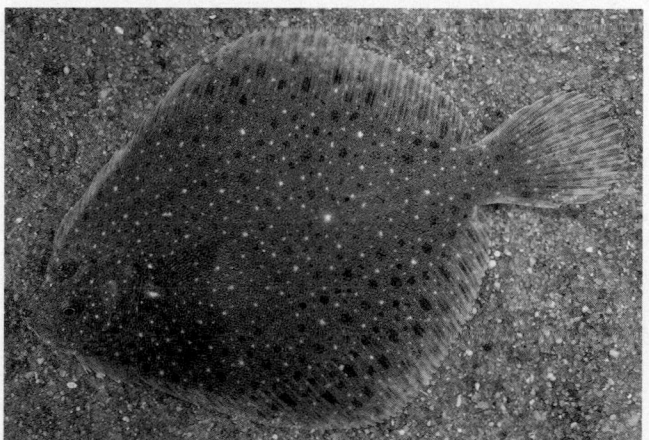

a.

b.

c.

Figure 47.7

Flies of the family Syrphidae are called flower flies because they are likely to be found on flowers where they drink nectar and eat pollen. Some species, like this one, mimic a wasp, which is protected from predation by its sting.

Table 47.2
Symbiosis

	Species 1	Species 2
Parasitism	+	−
Commensalism	+	0
Mutualism	+	+

Key: + = benefited
− = harmed
0 = no effect

Symbiosis

Symbiosis is an intimate relationship between 2 different species (table 47.2). As with predation, coevolution occurs, and the species become closely adapted to one another. In parasitism, the parasite benefits, but the host is harmed; in commensalism, one species benefits, but the other is unaffected; and in mutualism, both species benefit.

Parasitism

Parasitism is similar to predation in that the *parasite* derives nourishment from the host (just as the predator derives nourishment from its prey). Usually, however, the host is larger than the parasite, and an efficient parasite does not kill the host, at least until its own life cycle is complete. Viruses are always parasites, as are a number of bacteria, protists, plants, and animals. The smaller parasites tend to be endoparasites, which live in the bodies of the hosts. Larger parasites tend to be ectoparasites and remain attached to the exterior of the host by means of specialized organs and appendages.

Just as predators can reduce dramatically the size of a prey population that lacks a suitable defense, so parasites can reduce the size of a host population that has no defense. Thousands of elm trees have died in the United States from the inadvertent introduction of Dutch elm disease, which is caused by a parasitic fungus. A new method of treating Dutch elm disease utilizes bacteria that produce a fungicide when injected into the tree. Another means of curbing this fungal infection of trees is to control the bark beetle that spreads the disease.

Many other parasites use a secondary host for dispersal or completion of the stages of development. For example, the deer tick has 3 different hosts in its life cycle. The adult tick feeds on white-tailed deer, the larva feeds primarily on white-footed mice, and the nymph occasionally feeds on humans. As discussed in this chapter's reading, the deer tick is a carrier for a bacterium that causes Lyme disease.

In parasitism, the host species is harmed. Parasites, which sometimes help to control the size of their host population, may require a secondary host for dispersal.

Commensalism

Commensalism is a relationship between 2 species in which one species is benefited and the other is neither benefited or harmed. Often, the host species provides a home and/or transportation for the other species. Barnacles that attach themselves to the backs of whales and the shells of horseshoe crabs are provided with both a home and transportation. Remoras are fishes that attach themselves to the bellies of sharks by means of a modified dorsal fin acting as a suction cup. The remoras obtain a free ride and also feed on the remains of the shark's meals. Epiphytes grow in the branches of trees, where they receive light, but they take no nourishment from the trees. Instead, their roots obtain nutrients and water

Lyme Disease, a Parasitic Infection

W ith the exception of AIDS, Lyme disease is the most newly recognized serious infectious disease in the United States in terms of number of cases and potential for severe illness. Lyme disease has been reported in 43 states, but most cases occur in the Northeast (New York, New Jersey, Connecticut, and Massachusetts), the upper Midwest (Minnesota and Wisconsin),

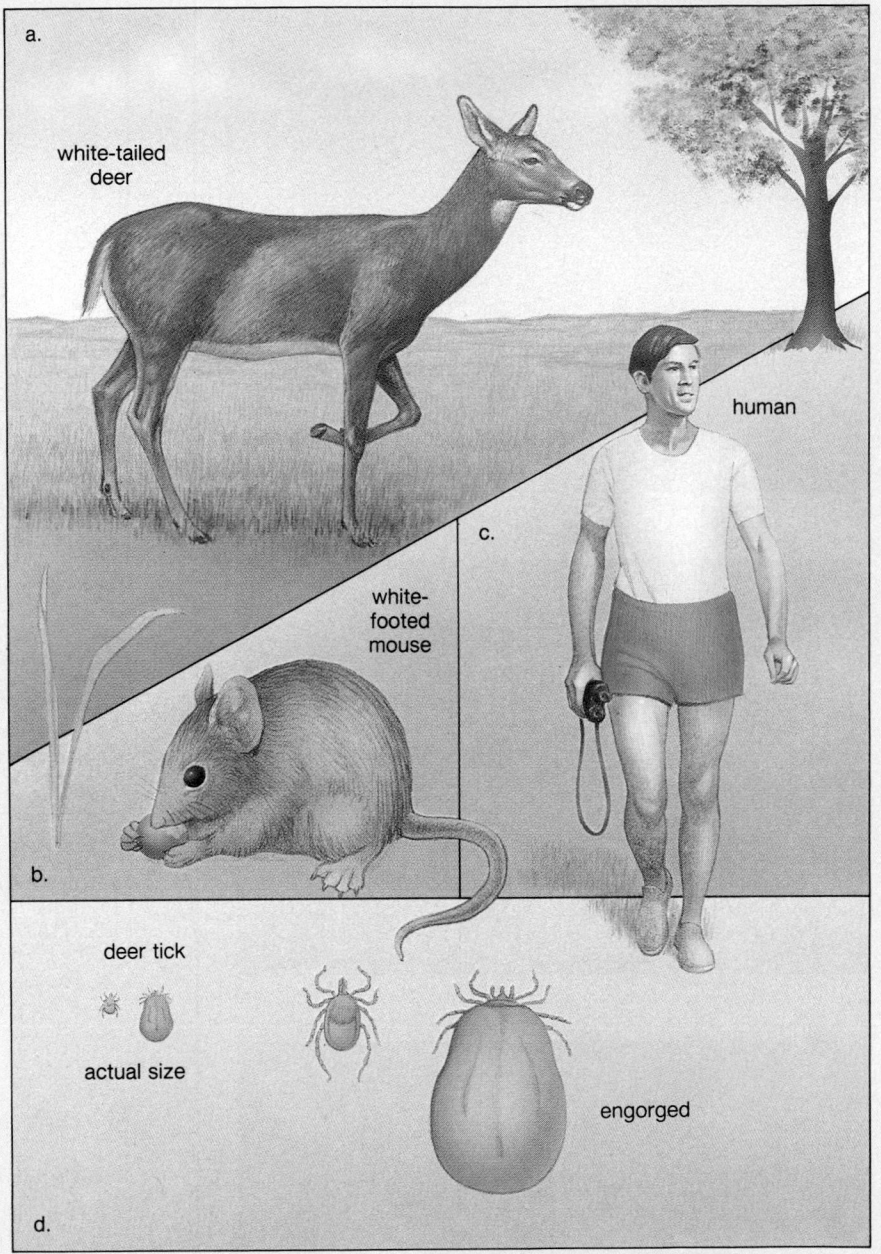

Figure 47.A

The life cycle of a deer tick. **a.** Adult ticks feed and mate on white-tailed deer, which accounts for the common name of the ticks. Female adults lay their eggs in soil and then die. **b.** In the spring and summer, larvae feed mainly on white-footed mice, they then overwinter. **c.** During the next summer, nymphs feed on white-footed mice or other animals, including humans. If a nymph is infected with the bacterium, humans get Lyme disease. **d.** Deer tick before feeding and after feeding. Actual size is shown along with enlarged size.

Behavior and Ecology

and the Pacific Northwest (California and Oregon). Lyme disease takes its name from Lyme, Connecticut, where the disease first was recognized in 1975.

The cause of Lyme disease is a spirochete *(Borrelia burgdorferi)*, a coiled bacterium that is spread by the bite of a deer tick *(Ixodes dammini* in the East and *Ixodes ricinus* in the West). Adult deer ticks are only about two-thirds the size of dog ticks. They are so named because adult ticks feed and mate on white-tailed deer. Ticks feed on the blood of their host and triple in size following a meal (see fig. 47.A). Ticks are arthropods that go through a number of states (egg, larva, nymph, adult) as they develop; molting occurs between the stages.

After a female tick has fed and mated in the fall, she drops down onto the ground and lays her eggs. When these eggs hatch in the spring, they become larvae that feed primarily on white-footed mice. These mice are an important reservoir for the Lyme disease bacterium; when mice are infected, the tick becomes infected. The fed larvae overwinter and molt the next spring to become nymphs that are not much larger than the head of a pin. It is during the nymph stage, between May and September, that ticks are most apt to feed on humans and to pass on to them the bacterium that causes Lyme disease. The nymph is dark brown and hard bodied and is mistaken easily for a scab or a piece of dirt on the skin. Other likely hosts for nymphs are white-footed mice, raccoons, and skunks. After feeding, the nymph develops into an adult, and the life cycle repeats itself.

Altogether, ticks remain on their hosts only about 2 1/2 weeks. The unfed adults, larvae, and nymphs spend most of their time attached to grasses or shrubs waiting for an appropriate host to come by. Infestation is greatest in wooded areas, but people also have been bitten in their own backyards, especially if the yards are adjacent to wooded areas.

The effects of Lyme disease vary from mild to severe. First, some victims, but not all, get a rash that is called a bull's-eye rash because it has a red center encircled by a light area and then a red area again. The rash can get as large as a dinner plate. At this time, infected persons possibly feel like they have a mild case of flu. In about a month or so, nausea and neck and joint pain similar to that caused by arthritis may develop. Finally, a few may experience heart blockage, sometimes even requiring the installation of a pacemaker, and neurological disorders that mimic the symptoms of meningitis and encephalitis. Fierce headaches, facial paralysis, depression, and a temporary loss of memory also can occur. Loss of muscular coordination sometimes makes the disease seem like multiple sclerosis. Lyme disease is called "the great imitator" because it easily is misdiagnosed as a number of other illnesses.

It is never too late to be treated for Lyme disease. A blood test can detect antibodies to the bacterium in the system. If the blood test is positive, the administration of an antibiotic, like penicillin, tetracycline, or erythromycin, cures the condition. Antitick sprays, which discourage ticks from choosing a human host in the first place, are available. Some home owners are trying to control the tick population in their vicinity by killing off the mice and ticks with chemicals. A more natural solution can be attempted by local governments. It is possible to release into the environment tiny parasitic wasps that lay their eggs only in ticks. The developing wasp larvae kill the ticks.

Figure 47.8

Clownfishes often live among a sea anemone's tentacles but are not seized and eaten as prey. This is a form of commensalism in which the clownfish species gets a place to live and the anemone species is not harmed.

from the air. Clownfishes live within the waving mass of tentacles of sea anemones (fig. 47.8). Because most fishes avoid the poisonous tentacles of the anemones, the clownfishes are protected from predators. Perhaps this relationship borders on mutualism because the clownfishes actually may attract other fishes on which the anemone can feed. The sea anemone's tentacles quickly paralyze and seize other fishes as prey.

In commensalism, one species is benefited and the other is unaffected. Often the host simply provides a home and/or transportation.

Mutualism

Mutualism is a symbiotic relationship in which both members of the association benefit. Mutualistic relationships often allow organisms to obtain food or to avoid predation.

Bacteria that reside in the human intestinal tract are provided with food, but they also provide humans with vitamins, which are molecules we are unable to synthesize for ourselves. Termites would not even be able to digest wood if not for the protozoans that inhabit their intestinal tract. The bacteria in the

Figure 47.9

The bullhorn acacia is adapted to provide a home for ants of the species *Pseudomyrmex ferruginea*. ***a.*** The thorns are hollow and the ants live inside. ***b.*** The base of leaves have nectaries (openings) where ants can feed. ***c.*** Leaves have bodies at the tips that ants harvest for larval food.

b.

a.

c.

protozoans digest cellulose, which termites cannot digest. Mycorrhizae, also called fungal roots, are symbiotic associations between the roots of plants and fungal hyphae. Mycorrhizal hyphae improve the uptake of nutrients for the plant, protect the plant's roots against pathogens, and produce plant growth hormones. In return, the fungus obtains carbohydrates from the plant. As we discussed on page 518, flowers and their pollinators have coevolved and are dependent upon one another. The flower is benefited when the pollinator carries pollen to another flower, ensuring cross-fertilization, and the flower provides food for the pollinator.

In tropical America, the bullhorn acacia is adapted to provide a home for ants of the species *Pseudomyrmex ferruginea* (fig. 47.9). Unlike other acacias, this species has swollen thorns with a hollow interior, where ant larvae can grow and develop. In addition to housing the ants, the acacias provide them with food. The ants feed from nectaries at the base of the leaves and eat fat- and protein-containing nodules called Beltian bodies, which are found at the tips of some of the leaves. Bullhorn acacias have leaves throughout the year, while related acacia species lose their leaves during the dry season. The ants constantly protect the plant from herbivores and other plants that might shade it because unlike other ants, they are active 24 hours a day.

In mutualism, both species benefit and the 2 species often are closely adapted to one another.

Summary

1. A community is a group of populations that are interacting with one another.
2. Two very important concepts in ecology are habitat and niche. Habitat is where a population lives, and niche is the role that it plays in an ecosystem. We find that when there are several similar species living in the same environment, they occupy different niches.
3. Competition may be intraspecific or interspecific. Interspecific competition occurs only for a short period of time, because either one population replaces the other or the 2 species evolve to occupy different niches. Although it may seem that similar species are occupying the same niche, actually there are slight differences.
4. Evidence suggests that predation tends to control the size of prey population, but the relationship may involve other factors as well. For example, the availability of food affects population sizes.

5. Just as predators are adapted to capture prey, prey are adapted to escape predators. Plants have both physical and chemical defenses. Animals have defenses that include concealment, fright of the predator, warning coloration, and vigilance.
6. Some animals mimic other animals that have a defense, especially those others that are poisonous to eat.
7. Symbiotic relationships include parasitism, commensalism, and mutualism. Parasitism may control the size of the host populations.
8. In commensalism, one species is benefited, and the other is unaffected. Often, the host simply provides a home and/or transportation.
9. In mutualism, both species benefit. The 2 species are often closely adapted to one another, as are flowers and their pollinators. There are also examples of mutualistic species that live together in the same locale, such as ants and acacia trees.

Study Questions

1. Why doesn't interspecific competition between similar populations always lead to the extinction of one of them?
2. What is the competitive exclusion principle, and how is it related to diversity of organisms?
3. Explain what is meant by the niche of an organism and why it is possible for similar types of organisms to be found in the same general area.
4. What evidence is there that predators might overkill their prey? that they do not do so?
5. Why would you recommend that humans not kill animals, such as coyotes, that might sometimes prey on farm animals?
6. Describe some prey defenses of plants and animals.
7. What is mimicry? What is the difference between Batesian and Müllerian mimicry?
8. Give examples of parasitism, and state the effects on the host organism.
9. Give examples of commensalism, and state the effects on the host organism.
10. Give examples of mutualism, and state the benefits to each organism involved in the relationship.

Objective Questions

1. Six species of monkeys are found in a tropical forest. Most likely, they
   a. occupy the same niche.
   b. eat different foods and occupy different ranges.
   c. spend much time fighting each other.
   d. All of these.
2. Leaf cutter ants keep fungal gardens. The ants provide food for the fungus on which they feed. This is an example of
   a. competition.
   b. predation.
   c. commensalism.
   d. mutualism.
3. Clownfishes live among sea anemone tentacles, where they are protected. If the clownfish provides no service to the anemone, this is an example of
   a. competition.
   b. predation.
   c. commensalism.
   d. mutualism.
4. Two species of barnacles vie for space in the intertidal zone. The one that remains is
   a. the better competitor.
   b. better adapted to the area.
   c. the better predator on the other.
   d. Both a and b.
5. Territoriality occurs as a result of
   a. predation.
   b. parasitism.
   c. competition.
   d. cooperation.
6. Bullhorn acacia provides a home and nutrients for ants. Which statement is correct?
   a. The plant is under the control of pheromones produced by the ants.
   b. The ants protect the plant.
   c. They have coevolved to occupy different niches.
   d. All of these.
7. The niche of an organism
   a. is the same as its habitat.
   b. includes how it competes and acquires food.
   c. is specific to the organism.
   d. Both b and c.
8. The frilled lizard of Australia suddenly opened its mouth wide and unfurled folds of skin around its neck. Most likely this was a way to
   a. conceal itself.
   b. warn that it was noxious to eat.
   c. scare a predator.
   d. All of these.
9. When one species mimics another species, the mimic sometimes
   a. lacks the defense of the model.
   b. possesses the defense of the model.
   c. is brightly colored.
   d. All of these.
10. Under each example, write either parasitism, commensalism, or mutualism.
    a. A heterotroph provides a home for an autotroph, who supplies it with organic molecules.

    _____

    b. A fish attaches to and receives nutrients directly from another fish.

    _____

    c. An organism lives in the tube along with a tubeworm.

    _____

Concepts and Critical Thinking

1. *Communities vary in their diversity.*

   Why do more diverse communities have more niches?

2. *Population sizes in a community are in part controlled by predation.*

   If predation acts as a check on prey population size, what are the counterchecks?

3. *Symbiotic relationships are important to the fabric of a community.*

   Why could parasitism affect both population sizes and diversity within a community?

Selected Key Terms

community (kŏ-mu'nĭ-te) 755
competitive exclusion principle (kom-pet'ĭ-tiv eks-kloo'zhun prin'sĭ-p'l) 756

niche (nich) 756
habitat (hab'ĭ-tat) 756
mimicry (mim'ik-re) 758
symbiosis (sim"bi-o'sis) 759

parasitism (par'ah-si"tizm) 759
commensalism (kŏ-men'sal-izm") 760
mutualism (mu'tu-al-izm") 761

48

The Biosphere

The tundra is a beautiful, vast, treeless plain located about the arctic circle. The average annual temperature is 5° C, and the area is slow to repair itself after a disruption. Yet, there is always intense pressure to develop the area for mineral exploitation and the building of pipelines to carry oil to the contiguous states.

Your study of this chapter will be complete when you can

1. distinguish between freshwater communities and saltwater communities;
2. list and describe the life zones of a lake and the types of organisms that reside in these zones;
3. distinguish between coastal communities and the oceans;
4. explain why coastal communities are more productive than the open ocean;
5. describe the life zones of the ocean and the types of organisms that reside in these zones;
6. describe the process of primary and secondary succession on land; contrast the productivity of the early stages of succession to the climax stage of succession;
7. describe the characteristics of the treeless biomes: deserts, tundra, grasslands, and chaparral;
8. describe the characteristics of forests: taiga, temperate deciduous forests, and tropical rain forests;
9. relate the location of the various terrestrial biomes to latitude and altitude;
10. contrast the productivity of the various terrestrial biomes, and explain why the tropical rain forest is the most productive of these.

The earth is enveloped by the **biosphere**—a thin realm composed of water, land, and air—where organisms are found. The biosphere contains communities of populations whose interactions were discussed in the previous chapter. The largest communities on land are called **biomes.** There is no equivalent term for large aquatic communities, which is unfortunate because both the aquatic environment and the land environment contain large communities that have unique, defined characteristics.

Aquatic Communities

Aquatic communities can be divided into 2 types: freshwater (inland) communities and saltwater (oceanic or marine) communities.

Freshwater Communities

The freshwater communities are lakes, ponds, rivers, and streams. Here, we consider only the composition of a *lake,* a body of water that has some depths that are always dark. This study is sufficient because the other freshwater communities have a similar mix of organisms.

Lakes

Lakes have 3 life zones: the littoral zone, which is closest to the shore, the limnetic zone, which is the sunlit main body of the lake; and the profundal zone, which is the deep part where light does not penetrate.

Aquatic plants are rooted in the shallow littoral zone of a lake (fig. 48.1). Here, there are microscopic organisms that cling to plants and rocks. In the limnetic zone, there are a few types of organisms, like the water strider, that are found on the surface of the lake. In the water, there are microscopic, floating organisms known as **plankton.** Phytoplankton (*phyto*—plant; *plankton*—drifting) are photosynthesizing algae, and zooplankton (*zoo*—animal; *plankton*—drifting) are protozoans and tiny crustaceans. Small and large fishes also are found in the limnetic zone. In the profundal zone, there are mollusks, crustaceans, and worms that feed on debris that falls from above.

Saltwater Communities

Saltwater or marine communities begin along the shoreline. These coastal communities are discussed before the oceans themselves are taken up.

Coastal Communities

An **estuary** forms where a large river flows into the ocean (fig. 48.2). The river brings fresh water into the estuary. Because of the tides, the sea brings salt water into this same area. Therefore, an estuary is characterized by a mixture of fresh water and salt water; this is called brackish water. Most estuaries are rather shallow with good sunlight penetration. The temperature of estuarine waters is generally higher than that of the surrounding marine or freshwater environments. Further, an estuary acts as a nutrient trap; the tides bring nutrients from the sea, and at the same time, they prevent the seaward escape of nutrients brought by the river.

Figure 48.1

The life zones of a lake. Rooted plants and clinging organisms occur in the littoral zone. Phytoplankton, zooplankton, and fishes are in the sunlit limnetic zone. Bottom-dwelling organisms, like crayfishes and mollusks, are in the profundal zone.

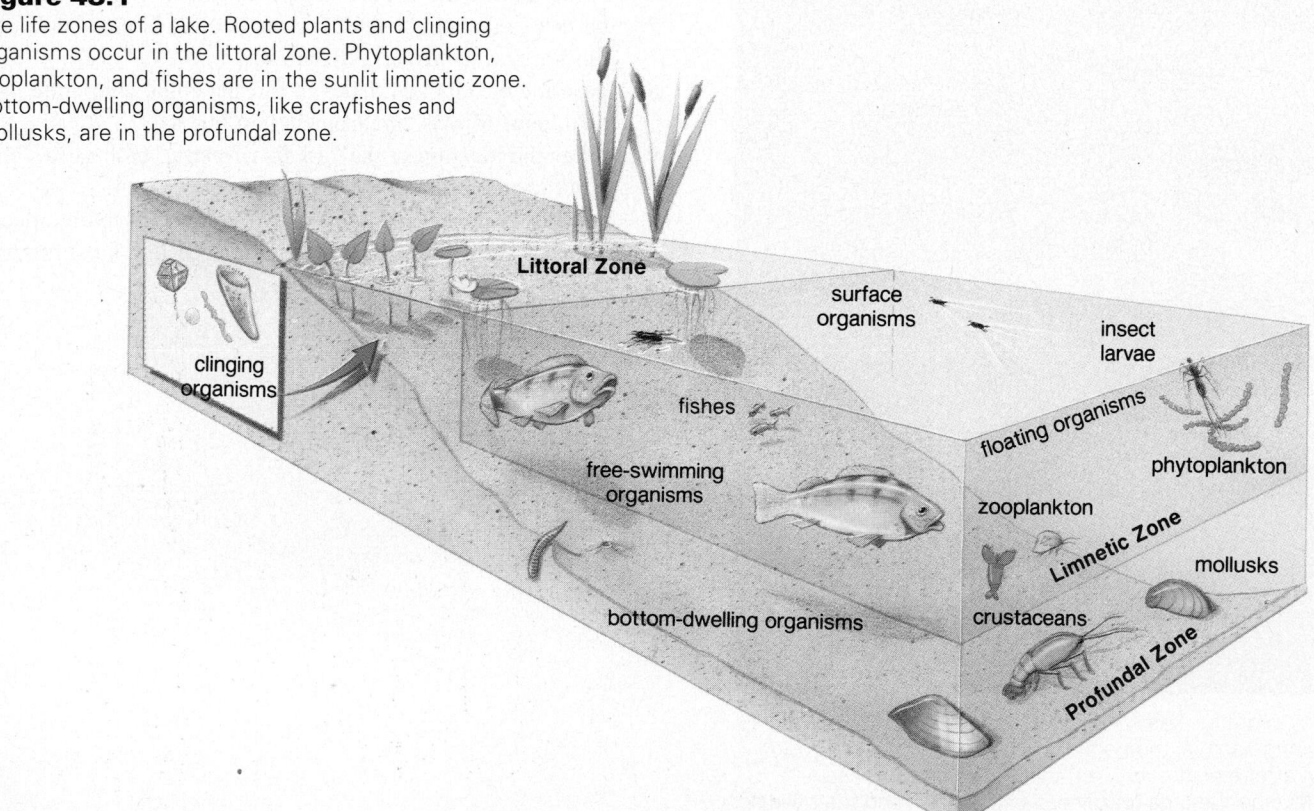

Behavior and Ecology

Figure 48.2

Estuary structure and function. Since an estuary is located where a river flows into the ocean, it receives nutrients from land and also from the sea by way of the tides. Decaying grasses also provide nutrients. Estuaries serve as a nursery for the spawning and rearing of the young for many species of fishes, mollusks, shrimp, and other crustaceans.

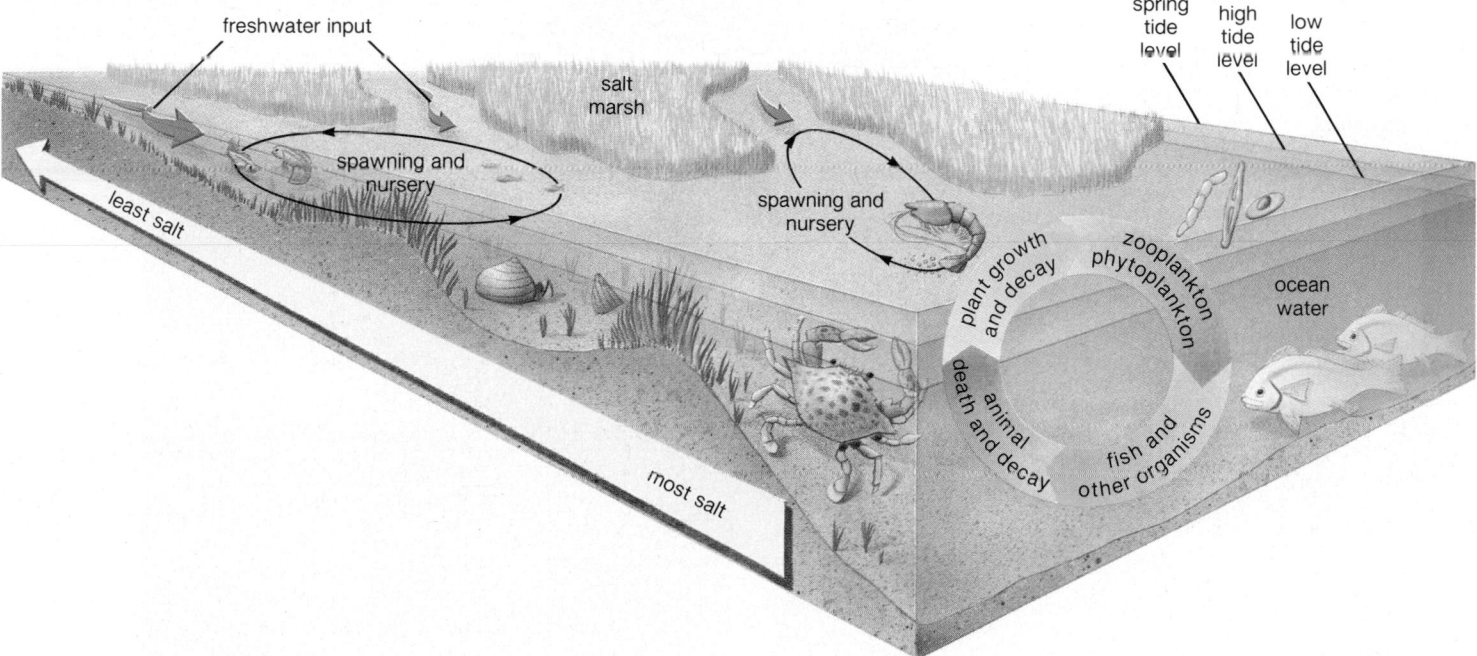

Only a few types of small fishes live permanently in an estuary, but many types develop there; therefore, both larval and immature fishes are present in large numbers. It has been estimated that well over half of all marine fishes develop in the protective environment of an estuary; this is why estuaries are called the nurseries of the sea. Shrimps and mollusks, too, use the estuary as a nursery.

Salt marshes dominated by salt-marsh cord grass are often adjacent to estuaries. Salt marshes contribute nutrients and detritus to estuaries, in addition to being an important habitat for wildlife.

The *rocky shore* offers a firm substratum to which organisms can attach. Large macroscopic brown algae cling to the rocks and offer support for other algae, protists, and small crustacea. Barnacles, mussels, and snails can cling firmly to rocks or seaweeds. These close up tightly during the periods the tide is out.

The shifting, unstable sands of a *sandy beach* do not provide a suitable substratum for the attachment of organisms. Many organisms, like ghost crabs and sand hoppers, burrow above high tide and feed at night whenever the tide is out. Sandworms and sand (ghost) shrimps remain within their burrows and feed on detritus. Shorebirds that feed on the burrowers are links with land communities.

Coral reefs are found in warm tropical seas, where the water is still shallow (fig. 48.3). Their chief constituents are stony corals (phylum Cnidaria), which have a calcium carbonate exoskeleton, and carbonaceous red and green algae. Most of the solid part of a coral reef is composed of the skeletons of dead coral; the outer layer contains living organisms.

Figure 48.3

A close-up of a coral reef. A coral reef is a unique community of marine organisms. The abundance of life is the result of the optimal environmental conditions of tropical seas.

Figure 48.4

The ocean and the oceanic floor. The waters of the ocean and the oceanic floor are divided as shown.

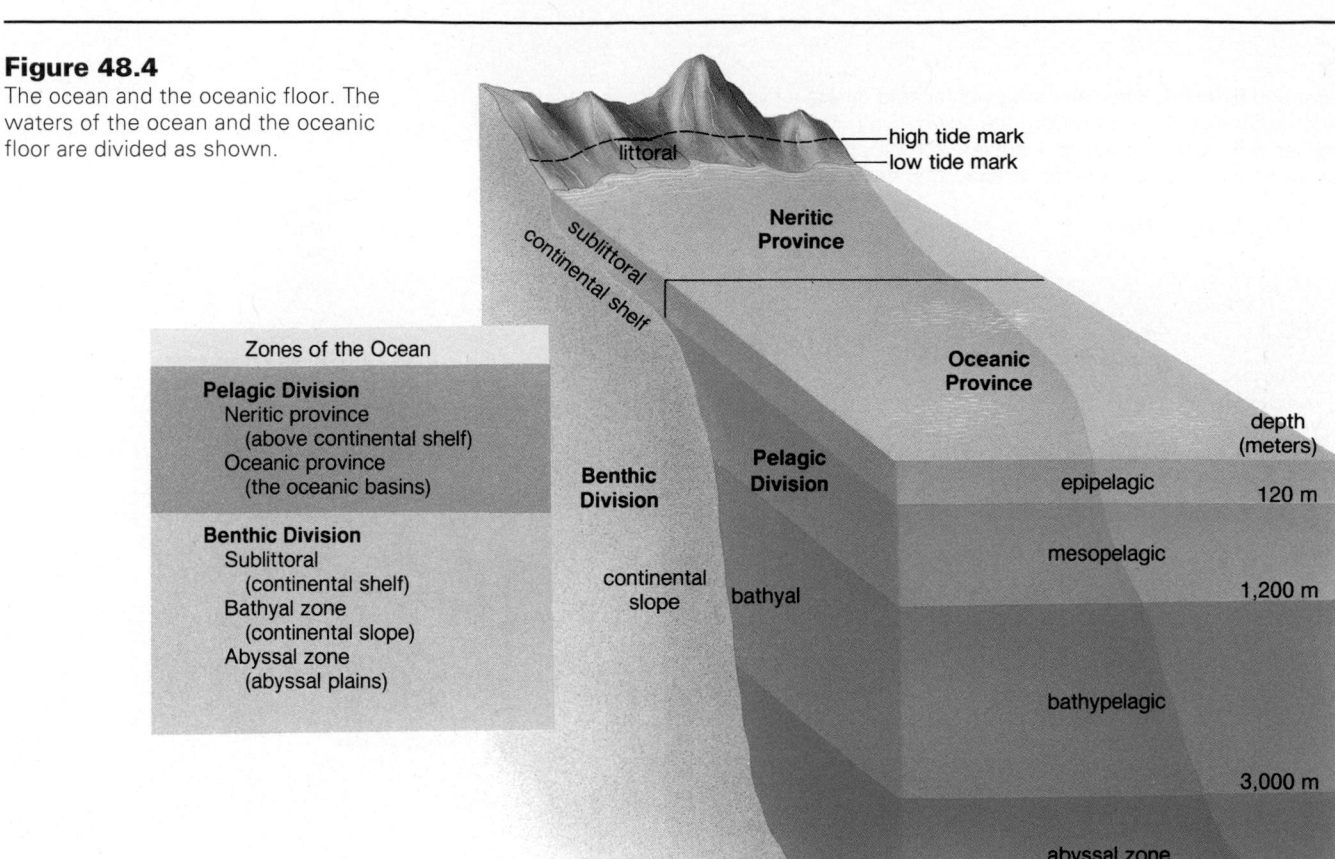

A reef is densely populated with diverse animal life. There are many types of small fishes, all of which are beautifully colored. In addition, the numerous crevices and caves provide shelter for filter feeders (sponges, sea squirts, fan worms) and for scavengers (crabs and sea urchins). The barracuda and the moray eel prey on these animals.

Oceans

The oceans cover approximately three-quarters of our planet. The geographic areas and zones of an ocean are shown in figure 48.4. It is customary to place the organisms of the oceans into either the **pelagic division** (open waters) or the **benthic division** (ocean floor).

The Pelagic Division The pelagic division includes the neritic province and the oceanic province. There is a greater concentration of organisms in the neritic province than in the oceanic province because sunlight penetrates the waters of the neritic province and this is where you find the most nutrients. Phytoplankton is food not only for zooplankton but also for small fishes. These small fishes in turn are food for commercial fishes—herring, cod, and flounder.

In the oceanic province, the epipelagic zone does not have a high concentration of phytoplankton because of the lack of nutrients (fig. 48.5). These photosynthesizers, however, still sup-port a large assembly of zooplankton because the oceans are so large. The zooplankton are food for herrings and bluefishes, which in turn are eaten by larger mackerels, tunas, and sharks. Flying fishes, which glide above the surface, are preyed upon by dolphins, not to be confused with mammalian porpoises, which also are present. Whales are other mammals found in the epipelagic zone. Baleen whales strain krill (small crustacea) from the water, and the toothed sperm whales feed primarily on the common squid.

Animals in the mesopelagic zone are carnivores, are adapted to the absence of light, and tend to be translucent, red colored, or even luminescent. There are luminescent shrimps, squids, and fishes, such as lantern and hatchet fishes.

The bathypelagic zone is in complete darkness except for an occasional flash of bioluminescent light. Carnivores and scavengers are found in this zone. Strange-looking fishes with distensible mouths and abdomens and small, tubular eyes feed on infrequent prey.

The Benthic Division The benthic division includes organisms that live on the continental shelf, the continental slope, and the abyssal plain (fig. 48.4). These are the organisms of the sublittoral, bathyal, and abyssal zones.

Figure 48.5

The zones of the oceanic province (pelagic division). The epipelagic zone contains the most organisms because this is where you find the phytoplankton and the zooplankton. Only carnivores are found in the mesopelagic zone. Both carnivores and scavengers are found in the bathypelagic zone. The abyssal zone is a zone of the benthic division, but it is shown here because it lies beneath the bathypelagic zone. The abyssal zone is characterized especially by certain echinoderms (sea cucumbers and sea lilies) and tubeworms. A few kinds of fishes, such as the tripod fish, also are found here.

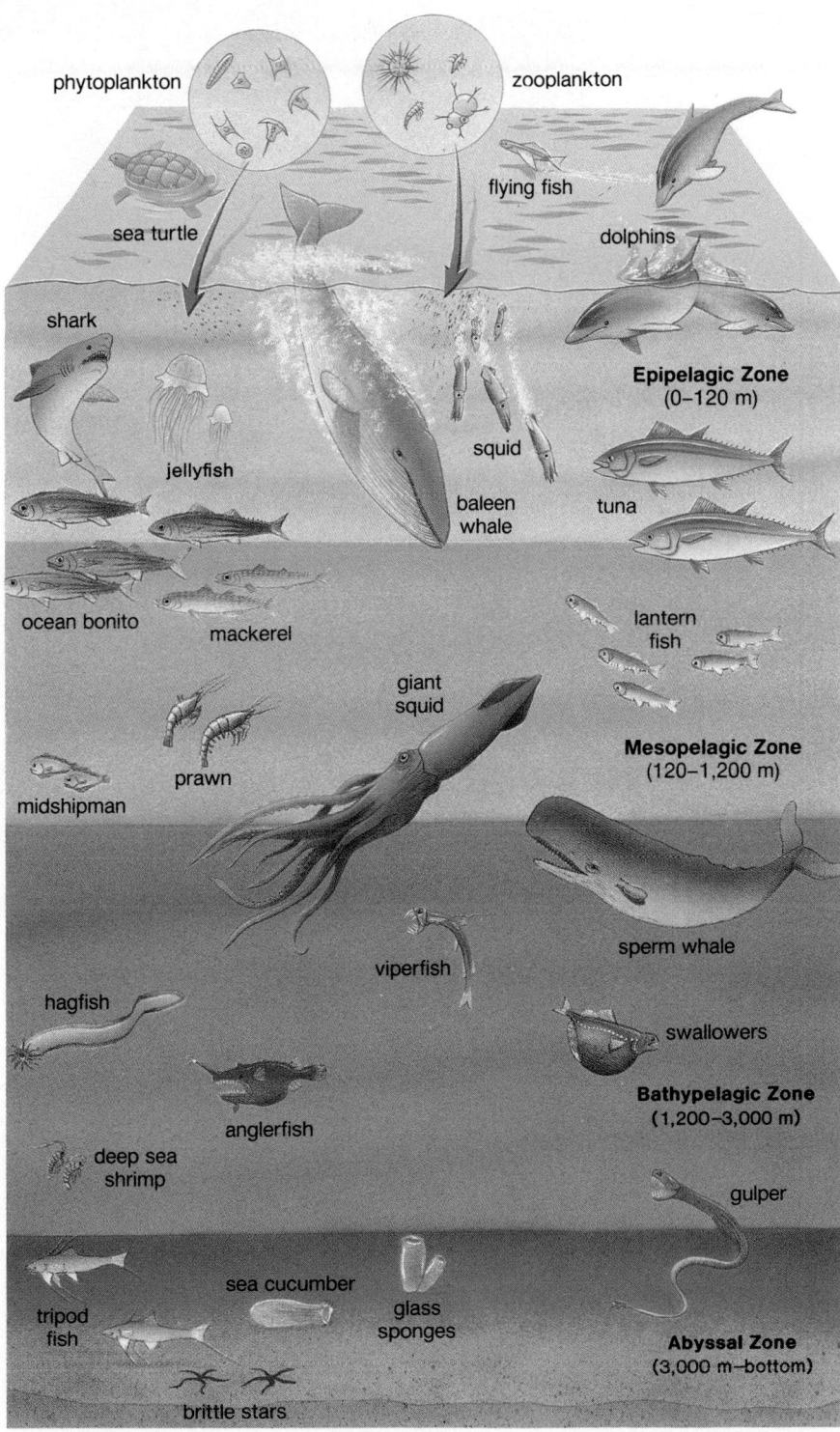

Seaweed grows in the first part of the sublittoral zone, and it can be found in batches on outcroppings as the water gets deeper. Nearly all benthic organisms, however, are dependent on the slow rain of plankton and detritus from the sunlit waters above. There is more diversity of life in the sublittoral and bathyal zones than in the abyssal zone. In these first 2 zones, clams, worms, and sea urchins are preyed upon by starfishes, lobsters, crabs, and brittle stars.

The abyssal life zone is inhabited by animals that live just above and in the dark abyssal plain. It once was thought that few animals exist in this zone because of the intense pressure and the extreme cold. Yet, many invertebrates live here by feeding on debris floating down from the mesopelagic zone. Sea lilies rise above the seafloor; sea cucumbers and sea urchins crawl around on the sea bottom; and tubeworms burrow in the mud.

The flat abyssal plain is interrupted by enormous underwater mountain chains called oceanic ridges. Along the axes of the ridges, crustal plates spread apart and molten magma rises to fill the gap. At specific sites of these tectonic spreading zones, seawater percolates through cracks and is heated to about 350° C, causing sulfate to react with water and form hydrogen sulfide (H_2S). The water temperature-hydrogen sulfide combination has been found to support communities that contain huge tubeworms and clams. Chemosynthetic bacteria that obtain energy from oxidizing hydrogen sulfide live within these organisms.

In the pelagic division of the ocean, only the neritic province and the epipelagic zone of the oceanic province receive sunlight and contain the organisms with which we are most familiar. The organisms of the benthic division are dependent upon debris that floats down from above.

Terrestrial Biomes

The locations of terrestrial biomes studied in this text are indicated in figure 48.6. Before we begin our discussion, however, we will consider how these biomes arose.

Succession

Succession is an orderly sequence of communities; each one prepares the way for a more complex community that follows. The complete sequence of communities is called a *sere,* and each community is a seral stage. The biomes are final seral stages (sometimes called climax communities), which can be associated with definite regions of the world.

Figure 48.6

a. Temperature and rainfall determine the nature of the biome to a large extent. For example, rain forests are found in tropical regions where temperatures are warm and rainfall plentiful year-round. Deserts, on the other hand, are found in tropical and temperate regions where rainfall is minimal. ***b.*** A map showing the major biomes of the world as they would be distributed if undisturbed by humans.

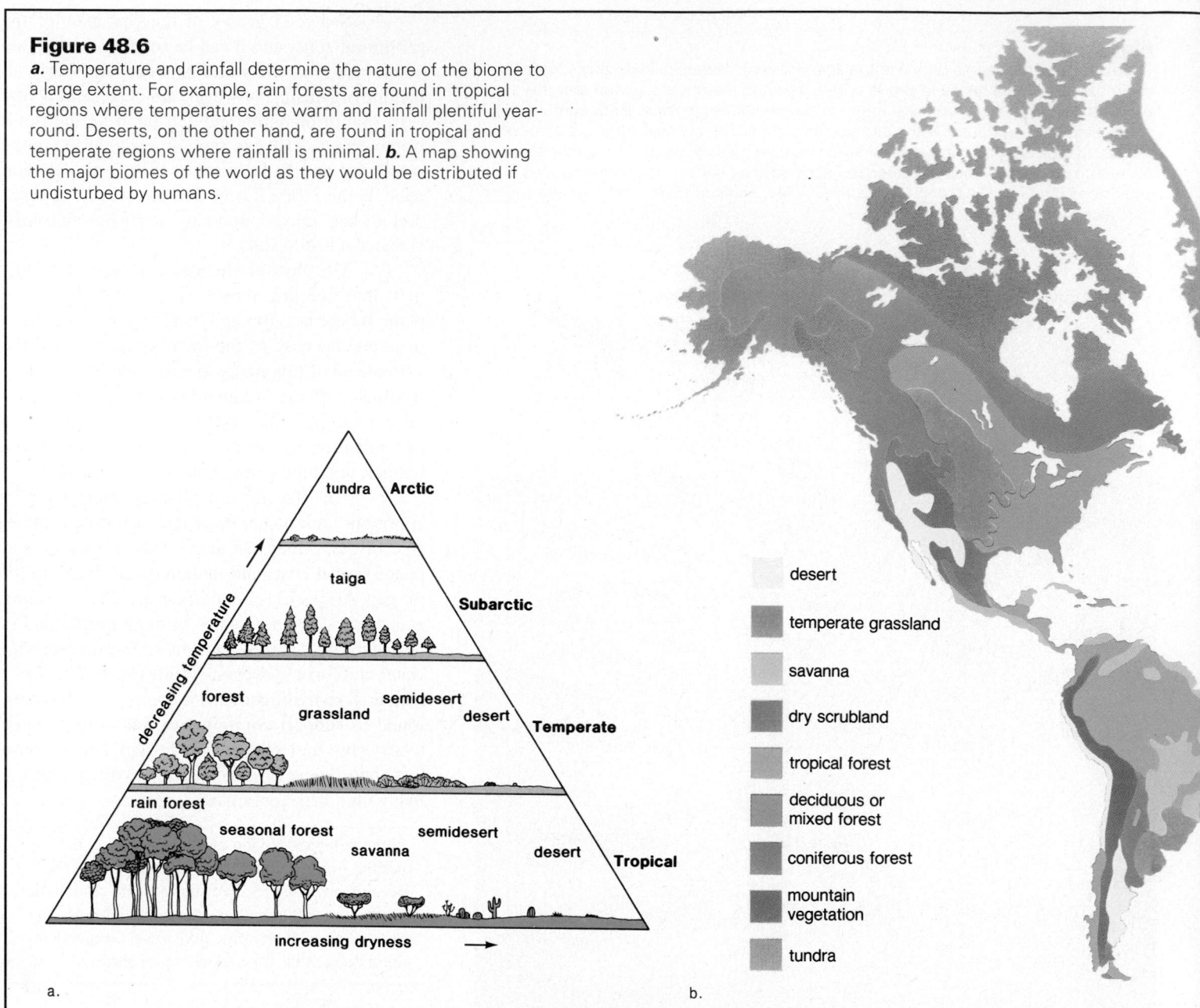

Primary succession begins with bare rock (fig. 48.7). At first, the rock is subjected to weathering as wind and rain act on it. Then lichens begin growing on the rock. Their acid secretions further break down the rock, and mosses and ferns take hold after a while. As soil begins to build up, longer-growing native plants begin to grow in the area, and eventually these may be followed by shrubs and trees.

Secondary succession occurs in regions that have been disturbed. The first seral stage, called the pioneer community, contains plants that can colonize disturbed areas because of tolerance to full sun and limited nutrient and moisture requirements.

This colonization prepares the way for native plants of the area that most likely have a longer growing period. Again the final seral stage may be shrubs and trees.

Succession is an undisputed observed phenomenon. Unlike former views, however, most modern ecologists do not believe that each seral stage in a particular region has to have the same mix of plants and animals. After all, chance plays a role as to which organisms will invade and begin the process of succession, for example. Also, the final seral stage can be dependent on other factors aside from climate. It is often pointed out that the Midwest of the United States could most likely support trees; however, frequent fires play a role in preventing succession to this point.

Behavior and Ecology

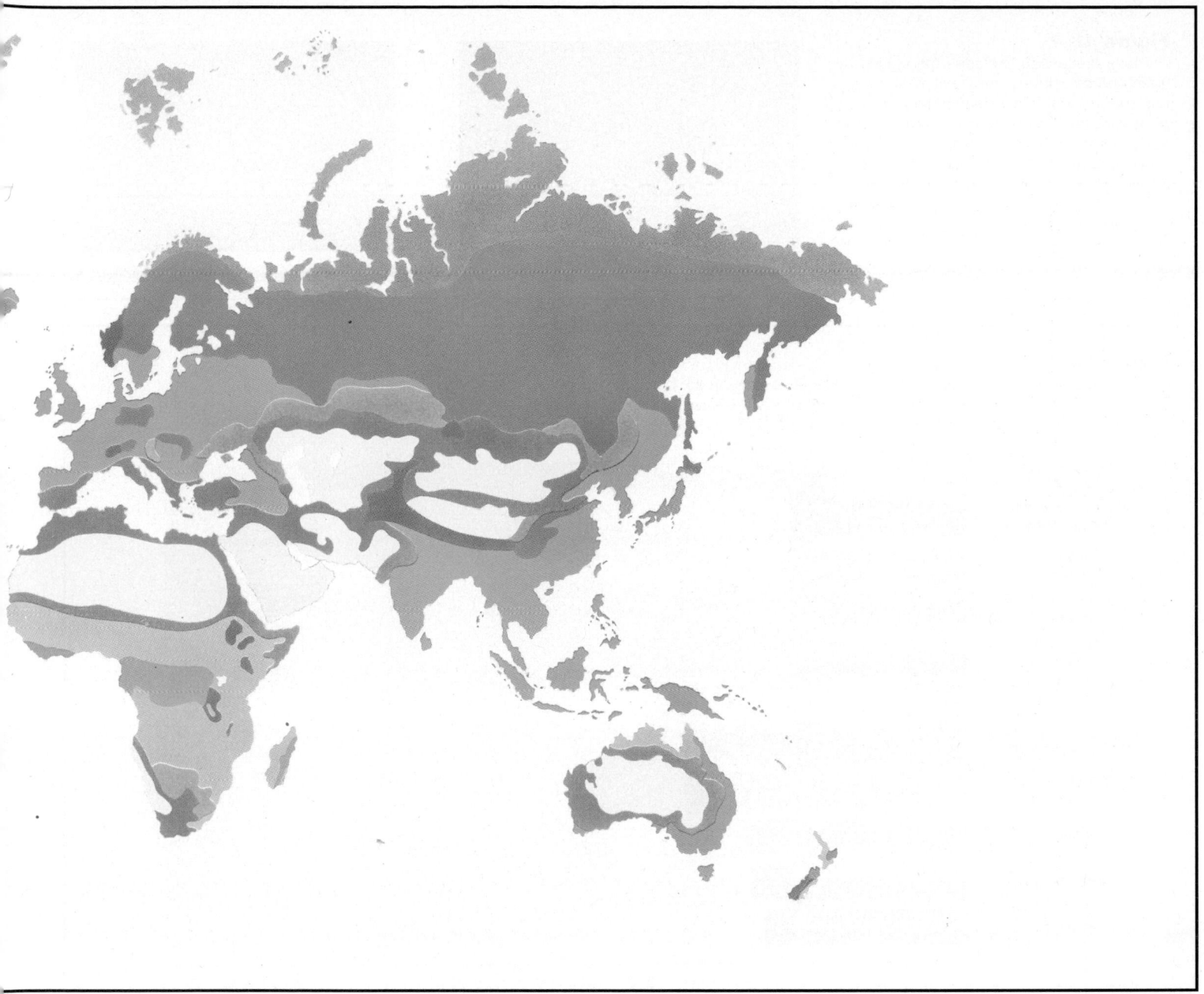

Productivity of Seral Stages

The early stages of succession show the most growth and therefore are the most productive; however, the inputs into the system in terms of energy and nutrients are high. The outputs (heat and loss of nutrients), however, are also high, because the community lacks the diversity to make full use of all the inputs.

As succession proceeds, diversity usually increases, and nutrients are recycled so that there is more efficient utilization of energy within the community. Figure 48.8 compares the productivity of the producers found within the climax communities studied in this chapter. These are the communities that are believed to use chemical and energy inputs efficiently. They are capable of maintaining themselves and therefore are considered stable.

Humans often replace climax communities, which tend to have the characteristics listed in table 48.1, with simpler communities, which tend to have the same characteristics as a pioneer community. Agricultural fields and managed forest show an increase in biomass and have high yields because they are provided with large inputs, such as fertilizer applications. The price paid for this growth potential, however, is a certain amount of waste and possible loss of stability.

Succession is a process in which a series of communities (a sere) replace one another until there is a final seral stage. The early, less diverse communities do not use inputs (chemical and energy) as efficiently as the later, more diverse communities.

Figure 48.7

Primary succession includes the stages by which bare rock becomes a climax community within a particular biome. These photographs show a possible sequence of early events: (**a**) lichens growing on bare rock; (**b**) individual plants taking hold; (**c**) perennial herbs spread out over the area. (**d**) trees are possible in the final stage of succession.

a.

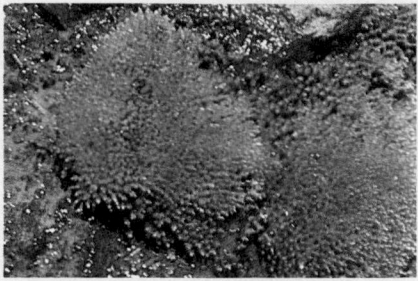

b.

Figure 48.8

The average net primary productivity for several communities. It is influenced by such factors as temperature and rainfall. Also important is the availability of sunlight and nutrients and the length of the growing season.

estuaries, swamps, and marshes

tropical rain forest

temperate deciduous forest

agricultural land

temperate grassland (prairie)

lakes and streams

rocky beach and sandy beach

tundra

open ocean

desert

800 1,600 2,400 3,200 4,000 4,800 5,600 6,400 7,200 8,000 8,800 9,600

Average Net Primary Productivity (kcal/m²/yr)

Gross primary productivity (GPP): Total rate of photosynthesis in a specified area usually expressed as kilocalories of energy produced per square meter per year (kcal/m²/yr).

Autotrophs (mostly photosynthetic plants and algae) absorb the energy of the sun and produce food for all organisms in a community.

Net primary productivity (NPP): The rate at which plants produce usable food or chemical energy (also expressed as kilocalories per square meters per year).

$$NPP = GPP - RS$$

where RS is the energy used by the autotrophs for respiration.

Since plants respire they use some of the food they produce (GPP) for respiration, and what is left (NPP) is available for growth and storage and can be used as food for heterotrophs. Usually, between 50% and 90% of GPP remains as NPP.

Behavior and Ecology

c.   d.

Table 48.1	
Properties of Communities	
Pioneer Community	**Climax Community**
Harsh environment	Most favorable environment
Biomass* increasing	Biomass stable
Energy consumption inefficient	Energy consumption efficient
Some nutrient loss	Nutrient cycling
Low species diversity	High species diversity
Little stability	High degree of stability

*Biomass is the weight of living material.

Treeless Biomes

Deserts

Deserts occur in regions where annual rainfall is less than 25 cm (fig. 48.9). The rain that does fall is subject to rapid runoff and evaporation. The days are hot because the lack of cloud cover allows the sun's rays to penetrate easily, but the nights are cold because the heat escapes easily into the atmosphere.

Some deserts, like the Sahara in Africa, have little or no vegetation, but most have a variety of plants. The best-known desert perennials in this country are the succulent, leafless cacti, which have stems that store water and also carry on photosynthesis. All cacti have extensive root systems that can absorb great quantities of water during brief periods of rainfall. Their spines provide a means of defense against desert herbivores. Desert vegetation also includes small-leaved, nonsucculent shrubs, such as sagebrush, a densely branched evergreen, and creosote, with leaves that turn brown and drop off. The deciduous mesquite tree, which has deep roots, also is common. Desert annuals exist most of the year as seeds; they burst into flower during the limited period of time when moisture and temperature are favorable.

Some animals are adapted to the desert environment. A desert has numerous insects, some of which, like the annual plants, have a compressed life cycle. They pass through the stages of development from pupa to pupa within a very short time because the pupa can remain inactive until it rains again. Reptiles, especially

Figure 48.9
Desert vegetation. These plants are adapted to retain water so that they can live in areas with infrequent rain.

lizards and snakes, are perhaps the most characteristic group of vertebrates found in deserts, but running birds (e.g., the roadrunner) and rodents (e.g., the kangaroo rat) are also well known. Larger mammals, like the coyote, prey on the rodents, as do the hawks.

Most deserts are located in high-temperature areas that receive less than 25 cm of rainfall a year. Plants and animals living in deserts have adaptations that protect them from the sun's rays and the shortage of water.

Tundra

The arctic **tundra** biome encircles the earth just south of ice-covered polar seas in the Northern Hemisphere (fig. 48.10). (A similar community, called the alpine tundra, occurs above the

Figure 48.10

Vegetation of the tundra. Notice the lack of trees, which are unable to root successfully because the ground is always frozen beneath the surface.

Figure 48.11

Soil profiles. *a.* Soil has horizontal layers called horizons. The top layer, the A horizon, contains most of the organic matter consisting of litter and humus. It also has a zone of leaching through which dissolved materials move downward. The middle layer, the B horizon, accumulates minerals and particles as water moves from the A to B horizon. The last layer, the C horizon, consists of weathered rock. This layer determines the pH of the soil and its ability to absorb and retain moisture. *b.* Grassland profile. Soils formed in grasslands have a deep A horizon built up from decaying grasses over many years. The shallow B horizon does not have sufficient nutrients to support growth because the low amount of rainfall in the grassland areas limits the amount of leaching from the topsoil. *c.* Forest profile. In forest soils, both the A and B horizons possess enough nutrients to allow for root growth. In tropical rain forests, the A horizon is more shallow than this generalized profile and the B horizon is deeper, signifying that leaching is more extensive. Since the topsoil of rain forests lacks nutrients, it only has a limited ability to support crops.

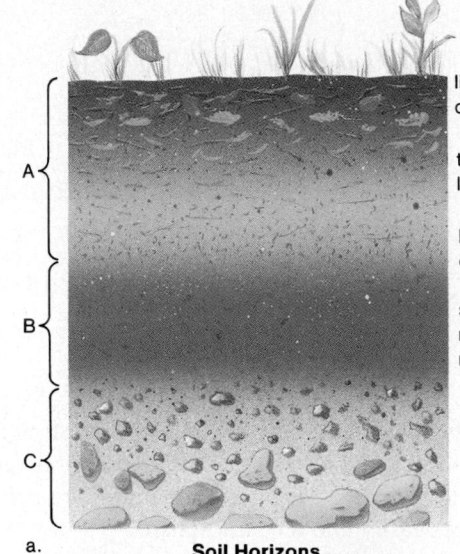

litter: leaves and other debris

topsoil: humus plus living organisms

leaching: removal of nutrients

subsoil: accumulation of minerals and organic materials

parent material: weathered rock

a. **Soil Horizons**

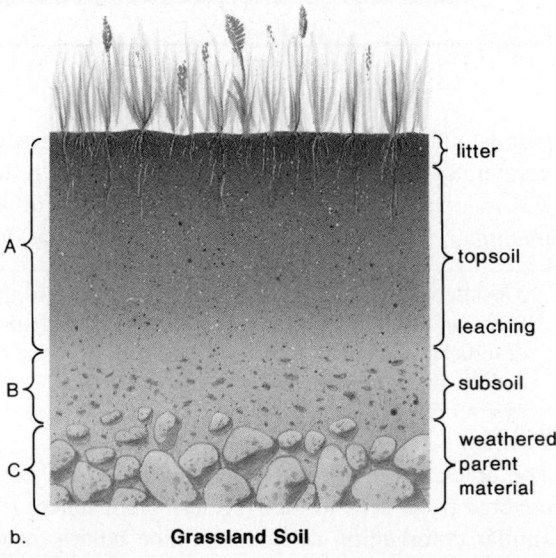

litter

topsoil

leaching

subsoil

weathered parent material

b. **Grassland Soil**

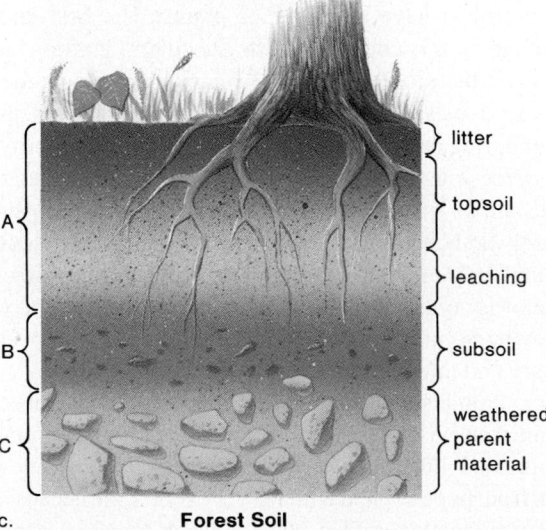

litter

topsoil

leaching

subsoil

weathered parent material

c. **Forest Soil**

Behavior and Ecology

timberline on mountain ranges.) The arctic tundra is cold and dark much of the year. Because rainfall amounts to only about 20 cm a year, the tundra possibly could be considered a desert, but melting snow makes water plentiful in the summer, especially because so little evaporates. Only the topmost layer of earth thaws; the *permafrost* beneath this layer is always frozen.

Trees are not found in the tundra because the growing season is too short, their roots cannot penetrate the permafrost, and they cannot become anchored in the boggy soil of summer. In the summer, the ground is just about covered with sedges and short-grasses, but there also are numerous patches of lichens and mosses. Dwarf woody shrubs flower and seed quickly while there is plentiful sun for photosynthesis.

A few animals live in the tundra year-round. For example, the ratlike lemming stays beneath the snow; the ptarmigan, a grouse, burrows in the snow during storms; and the musk-ox conserves heat because of its thick coat and short, squat body. In the summer, the tundra is alive with numerous insects and birds, particularly shorebirds and waterfowl that migrate inland. Caribou and reindeer also come, along with the wolves that prey upon them. Polar bears are common near the coast.

Grasslands

Grasslands occur where rainfall is greater than 25 cm but is generally insufficient to support trees. The extensive root system of grasses allows them to recover quickly from drought and fire, which occur frequently in this biome. The matted roots also efficiently absorb surface water and prevent invasion by most trees. Over the years, organic matter builds up in the rich soil, which often is exploited for agriculture (fig. 48.11).

In the tropics, particularly in much of Africa, there is a tropical grassland called the **savanna** (fig. 48.12). The savanna has few trees because of a severe dry season. One tree that is found here is the flat-topped acacia, which sheds its leaves during a drought.

The African savanna supports the greatest variety and number of large herbivores of all the biomes. Elephants and giraffes are browsers that feed on tree vegetation. Antelopes, zebras, wildebeests, water buffalo, and rhinoceroses are grazers that feed on grasses. Any plant litter that is not consumed by grazers is attacked by a variety of small organisms, among them termites. Termites build towering nests in which they tend fungal gardens, their source of food. The herbivores support a large population of carnivores. Lions and hyenas hunt in packs, cheetahs hunt singly by day, and leopards hunt singly by night.

A **prairie** is a temperate grassland. When traveling east to west across the U.S. Midwest, the tall-grass prairie gradually gives way to the short-grass prairie. Although grasses dominate both types of prairie, they are interspersed by forbs, herbs other than grasses. These often catch the eye because they have colorful flowers, whereas grasses do not.

Because there are so few trees, only grazers are found in the prairie. Small mammals, such as mice, prairie dogs, and rabbits, typically burrow in the ground, although they usually feed above ground. Hawks, snakes, badgers, coyotes, and foxes feed on these

Figure 48.12
Vegetation and animal life of the African savanna. There are more types of grazers in the savanna than in any other biome.

mammals. Large herds of buffalo—estimated at hundreds of thousands—once roamed the prairies and plains, as did herds of pronghorn antelope.

Grasslands occur where rainfall is greater than 20 cm but is insufficient to support many trees. The savanna supports the greatest number of different types of herbivores, which serve as food for carnivores. Much of the U.S. Midwest was once a prairie.

Chaparral

In parts of South Africa, western Australia, and central Chile, around the Mediterranean Sea, and in California, most of the rain falls in winter and the summers are very dry. There is a dense scrubland known as *chaparral* in this country. The shrubs have small but thick evergreen leaves that often are coated with a waxy material that prevents loss of moisture from the leaves. Their thick underground stems survive the dry summers and frequent fires and can sprout new growth. Rodents and reptiles abound in this biome.

Forests

Taiga

The **taiga** is a coniferous forest extending in a broad belt across northern Eurasia and North America. The climate is characterized by cold winters and cool summers, with a growing season of about 130 days. Rainfall ranges between 40 cm and 100 cm per year, with much

Figure 48.13

The taiga biome stretches round the globe in the northern temperature zone and contains narrow-leaved evergreen trees that are adapted to harsh conditions.

of it in the form of heavy snows. The great stands of evergreen, narrow-leaved trees, such as spruce, fir, and pine, are interrupted by many lakes and swamps (fig. 48.13). (*Taiga* is Russian for swampland.) Beneath the dense tree canopy (upper layer of leaves, which is the first to receive sunlight) there is little light, but a ground cover of lichens, mosses, and ferns usually is found.

Compared to other forests, the taiga has relatively few consumer species. In the summer, insects attack the trees and are eaten by warblers and flycatchers. Other birds, such as crossbills and grosbeaks, extract seeds from the cones of the trees. Birds of prey and other carnivores—weasel, lynx, and wolf—feed on small animals, such as the plentiful rodents. The large herbivores—moose and bear—are apt to be found in clearings or near the water's edge, where small trees and shrubs are found.

Temperate Deciduous Forests

A temperate deciduous forest is found south of the taiga in eastern North America, eastern Asia, and much of Europe. The climate is moderate, with relatively high rainfall (75 cm–150 cm per year). The seasons are well defined, and the growing season ranges between 140 and 300 days. The trees, such as oak, beech, and maple, are *deciduous;* they lose their leaves in the fall and

The Urgency of Tropical Conservation

Sharing the planet earth with us are at least 4–5 million species of animals, plants, and microorganisms, most of which are poorly known. About two-thirds of them have not even been given scientific names.

Of all the world's species, roughly 10% occur in the United States and Canada together. We can state with confidence that most of these plants, vertebrates, and larger invertebrates and fungi have been studied and classified. Furthermore, we now know that some 5%–10% of North America's species are threatened or endangered, and we are making advances to protect them.

To think that by "taking care of our own" we are saving a large proportion of the world's organisms is to delude ourselves. It is to the tropics that we must turn. For instance, about one-third of the total number of earth's plants, animals, and microorganisms occur in Latin America. Of these, only about one-sixth have been catalogued. Of the remainder, amounting to nearly 30% of all the planet's organisms, we know absolutely nothing. Some scientists, including Terry Erwin of the U.S. National Museum of Natural History, have estimated that the total number of species in Latin America might be much higher. His calculations indicate a regional total of perhaps 15 million species. Whatever the final figure, the number of unknown species is staggering.

For example, South America's fresh waters are inhabited by an estimated 5,000 fish species, only about 3,000 of which have been named. (This amounts to about one-eighth of all the world's fish species.) On the eastern slopes of the Andes, 80 or more species of frogs and toads often exist within a single square mile—almost as many as in all of temperate North America. Approximately the size of the state of Colorado, Ecuador harbors more than 1,300 species of birds—roughly twice as many as those inhabiting the United States and Canada.

Moreover, hundreds of new species of plants—including many dozen tree species—are being discovered in Latin America every year, as are dozens of new species of terrestrial vertebrates and fishes. Over one-third of the nearly 800 species of reptiles and amphibians known to occur in Ecuador have been discovered since 1970, and many more still are being found there. In fact, the northern Andean countries of Colombia, Ecuador, and Peru combined are home to about one-sixth of all the earth's biota—some 750,000 species that are also the most poorly known organisms on our planet. . . .

The actual and potential importance of species for human welfare cannot be overstated. About 85% of our food is derived directly or indirectly from 20 species of plants, and some 60% comes from only 3: corn, wheat, and rice. It stands to reason that among the world's remaining estimated 235,000 known flowering plant species, there must be many more that

grow them in the spring. Enough sunlight penetrates the canopy for the development of a well-developed understory: a layer of shrubs is followed by herbaceous plants and then often a ground cover of mosses and ferns. Animal life is plentiful. Birds and rodents provide food for bobcats, wolves, and foxes. The white-tailed deer has increased in number of late, while the black bear, an omnivore, has decreased in number but still is found commonly in some parts of the country.

Behavior and Ecology

could provide important sources of food—not to mention medicines, oils, chemicals, and renewable fuels. Our continued survival depends on our ability to use plants, animals, and microorganisms extensively and wisely. Yet, we know absolutely nothing about most tropical organisms, and we certainly have not examined them for their potential value to humans. By saving the tropical forests and their myriad plant and animal species, we also will be investing wisely in the future survival of our own species.

Despite its riches, tropical vegetation worldwide is being altered or eradicated at an alarming rate. In 1981, the Tropical Forest Resources Assessment Project of the United Nation's Food and Agriculture Organization (FAO) estimated that 44% of the tropical rain forests already had been degraded or destroyed by 1980. Recent estimates based on satellite imagery and field surveys show that most deforestation is now occurring in Brazil and India. India is losing 3.7 million acres per year and Brazil is losing about 10 times that number of acres per year. Brazil's loss alone is more than 50 acres per minute. If present trends continue, the rich tropical moist forests of many developing countries will disappear in the next 2–3 decades....

The greatest consequence of all, however, is that of biological extinction—the extermination of a major fraction of earth's plants, animals, and microorganisms during the lifetimes of most people living today (fig. 48.A). Many tropical organisms have a very narrow geographical range and highly specific ecological requirements. At least 40% of the world's species occur in the very tropical forests that may not survive the next few decades. The loss of only half these organisms would amount to the permanent disappearance from our planet of at least 750,000 species—a far greater number than the total number of species found in the United States and Canada. If we wish to preserve the world's biota, can it be doubted that our major concern in the closing years of the twentieth century and the early years of the twenty-first century must be with the tropics? . . .

To find an extinction event of this magnitude in the geological record, we need to go back some 65 million years to the end of the Cretaceous period, when—possibly spurred by a cloud thrown up by the collision of a gigantic meteorite with the earth—a large proportion of the earth's organisms, including the dinosaurs, became extinct. The extinction that is taking place now will occur during our life and that of our children. But we can ameliorate its effects by learning about tropical organisms and using our knowledge to save them. Our contribution to preserving natural diversity will be minute if we do not look beyond our own borders to the tropics. This task is one of extreme urgency. If we succeed, we shall play an important role in shaping the quality of life for all who come after us.

Reprinted from *Nature Conservancy News*, January–March 1986. Copyright 1986 The Nature Conservancy, Arlington, VA. Reprinted by permission.

a.   b.   c.

Figure 48.A
Diversity of life in tropical rain forests. Destroying the tropical rain forests will greatly reduce the variety of life on earth. Already vulnerable are (**a**) the Green iguana, (**b**) South American ocelot, and (**c**) the blue-and-yellow macaw.

Tropical Rain Forests

Tropical rain forests[1] in South America, Africa, and the Indo-Malayan region occur at the equator, where it is always warm (between 20° C and 25° C) and rainfall is plentiful (in excess of 250 cm per year). This is the richest biome, both in the different kinds of species found and the total amount of living matter. The need to preserve tropical rain forests is discussed in this chapter's reading.

The rain forest has a complex structure, with many layers of canopy. Some of the broad-leaved evergreen trees grow to 50 m or taller, some to 35 m, and some to 15 m. These tall trees often have trunks buttressed at ground level to prevent their toppling over. Lianas, or woody vines, which encircle the tree as it grows, also

[1]In some tropical regions, there are tropical deciduous forests. Here, broad-leaved trees lose their leaves because of a dry season.

help to strengthen the trunk. There are many understory layers in a tropical rain forest; however, the lowest understory layer usually allows one to walk fairly freely except in open areas or clearings, where the presence of light may produce a thick jungle of vines and secondary growth.

Although there is animal life on the ground (the paca, the agouti, the peccary, the armadillo, and the coati), most animals live in the trees. Insect life is so abundant that the majority of species have not been identified yet. The various birds, such as the hummingbird, the parakeet, the parrot, and the toucan, are often beautifully colored. Amphibians and reptiles also are well represented by many types of snakes, lizards, and frogs. Monkeys are well-known primates that feed on the fruits of the trees. The largest carnivores are the big cats—the jaguars in South America and the leopards in the Old World.

Not only do many animals spend their entire lives in the canopy, some plants do also. **Epiphytes** are plants that grow on tall trees but do not parasitize them. Instead, some have roots that take moisture and minerals leached from the canopy and others catch daily rain and debris in special hollow leaves. The most common epiphytes belong to the pineapple, orchid, and fern families.

Whereas the soil of temperate deciduous forests is rich enough for agricultural purposes, the soil of a tropical rain forest is not. Numerous organisms quickly break down any litter, and nutrients are recycled immediately to the plants. Of the minerals, only aluminum and iron sometimes remain near the surface, producing a red-colored soil known as laterite. When the trees are cleared, laterite bakes in the hot sun to a bricklike consistency and

will not support crops. Slash-and-burn agriculture is the only type that thus far has been successful in the tropics. Trees are felled and burned, and the ashes provide enough nutrients for several harvests. Thereafter, the forest is allowed to regrow, and a new section is utilized for agriculture. These matters are discussed further on page 795.

Forests require adequate rainfall. The taiga has the least amount of rainfall. The temperate deciduous forest has trees that grow and shed their leaves with the seasons. The tropical rain forest is the least studied and the most complex of all the biomes.

Latitude versus Altitude

Latitude

If you travel from the Southern Hemisphere to the Northern Hemisphere, it is possible to observe first a tropical rain forest, followed by a temperate deciduous forest, and then the taiga and the tundra, in that order. This shows that the location of the biomes is influenced by temperature.

Altitude

It is also possible to observe a similar sequence of biomes by traveling from the bottom to the top of a mountain (fig. 48.6a). These transitions are largely due to decreasing temperature as the altitude increases, but soil conditions and rainfall are also important.

In general, the same series of biomes can be observed according to both latitude and altitude.

Summary

1. Aquatic communities are divided into freshwater communities and saltwater communities.
2. Lakes are freshwater communities with 3 life zones and a different mix of organisms in each zone.
3. Saltwater communities are divided into coastal communities and the oceans. The coastal communities, especially estuaries, are more productive than the oceans.
4. The rocky shore, with organisms that cling to rocks, have more obvious signs of life than the sandy seashore, where animals burrow in the sand. Coral reefs are extremely productive communities found in shallow tropical waters.
5. An ocean is divided into the pelagic division and the benthic division.
6. The pelagic division (open waters) has 3 zones. The epipelagic zone receives adequate sunlight and supports the most life. The mesopelagic zone and bathypelagic zone contain organisms adapted to minimum and no light, respectively.
7. The benthic division (ocean floor) includes the seaweeds of the sublittoral zone and the bottom-dwelling organisms of the bathyal and abyssal zones.
8. Terrestrial communities (biomes) arise by a process called succession. The first stage of succession, called the pioneer community, produces new growth but tends to be unstable. The final stage, called the climax community, produces little new growth but tends to be stable.
9. The treeless biomes include the deserts, tundra, grasslands, and chaparral.
10. The organisms of a desert are adapted to heat and minimum rainfall. Those in the tundra are adapted to cold; there is water only because melting snow does not rapidly evaporate in the summer.
11. Among grasslands, the savanna, a tropical grassland, supports the greatest number of different types of large herbivores. The prairie, found in the United States, has a limited variety of vegetation and animal life.
12. Forests require adequate rainfall. The taiga, a coniferous forest, has the least amount of rainfall. The temperate deciduous forest has trees that gain and lose their leaves because of the alternating seasons of summer and winter. Tropical rain forests are the most complex and productive of all biomes.

Writing Across the Curriculum

In order to practice writing skills, students should write out the answers to any or all of the study questions and the critical thinking questions. The study questions are sequenced in the same order as the text. Suggested answers to the critical thinking questions are in appendix D.

Study Questions

1. Describe the life zones of a lake and the organisms you would expect to find in each zone.
2. Describe the coastal communities, including coral reefs, and discuss the importance of estuaries to the productivity of the ocean.
3. Describe the zones of the open ocean and the organisms you would expect to find in each zone.
4. Describe the stages of succession by which an abandoned field becomes a forest.
5. Describe the climate and the populations of a desert in North America.
6. Describe the location, the climate, and the populations of the arctic tundra.
7. Describe the climate and the populations of the African savanna.
8. Describe the location and the climate of the North American prairie. What are its populations of organisms?
9. Describe the location, the climate, and the populations of the taiga.
10. Describe the location, the climate, and the populations of temperate deciduous forests in North America.
11. Describe the location, the climate, and the populations of tropical rain forests.
12. Name the terrestrial biomes you would expect to find when going from the base to the top of a mountain.

Objective Questions

1. Phytoplankton are more likely to be found in which zone of the ocean?
   a. epipelagic zone
   b. mesopelagic zone
   c. bathypelagic zone
   d. continental slope
2. An estuary acts as a nutrient trap because of the
   a. action of rivers and tides.
   b. depth at which photosynthesis can occur.
   c. amount of rainfall received.
   d. height of the water table.
3. Which area of an ocean has the greatest concentration of nutrients?
   a. epipelagic zone and benthic zone
   b. epipelagic zone only
   c. benthic zone only
   d. neritic province
4. The forest with a multilevel understory is the
   a. tropical rain forest.
   b. coniferous forest.

   c. tundra.
   d. temperate deciduous forest.
5. All of these phrases describe a tropical rain forest, except
   a. nutrient-rich soil.
   b. many arboreal plants and animals.
   c. canopy composed of many layers.
   d. broad-leaved evergreen trees.
6. Which type of animal would you be least likely to find in a grassland biome?
   a. hoofed herbivore
   b. active carnivore
   c. arboreal primate
   d. flying insect
7. All of these phrases are true of the tundra, except
   a. low-lying vegetation.
   b. northernmost biome.
   c. short growing season.
   d. many different types of species.
8. Which of these is mismatched?
   a. tundra—permafrost
   b. savanna—acacia trees

   c. prairie—epiphytes
   d. coniferous forest—evergreen trees
9. Which of these lists the least productive to the most productive?
   a. ocean, rocky beach, temperate forest, estuary
   b. agricultural land, tundra, tropical rain forest, desert
   c. lakes and streams, tropical rain forest, open ocean
   d. estuary, agricultural land, lakes and streams, tundra
10. Use plus signs to rank order these biomes for temperature and rainfall.

	Temperature	Rainfall
a. tropical rain forest		
b. savanna		
c. taiga		
d. tundra		

Concepts and Critical Thinking

1. *The productivity of communities vary.*

   Use figure 48.4 to explain why a neritic province is more productive but more polluted than the oceanic province.

2. *Climax communities arise by the process of succession.*

   Does the concept of succession require a set sequence of communities? Why or why not?

3. *The gross and net primary productivity of communities vary.*

   Explain why you might see this sequence of communities from the base to the top of a mountain: temperate forest, coniferous forest, tundra.

Selected Key Terms

biosphere (bi′o-sfēr) 766
biome (bi′ōm) 766
plankton (plank′ton) 766
estuary (es′tu-ar″e) 766

pelagic division (pe-laj′ik de-vizh′en) 768
benthic division (ben′thik de-vizh′en) 768
succession (suk-sesh′un) 769
tundra (tun′drah) 773
savanna (sah-van′ah) 775

prairie (prare) 775
taiga (ti′gah) 775
tropical rain forest (trop′i-kal rān for′est) 777

49

Ecosystems: The Flow of Energy and the Cycling of Materials

Large birds, like this roadrunner eating an American collared lizard, are top predators whose population size is usually limited by the quantity of available food. When top predators die, the chemicals making up their bodies are used by photosynthesizers, who produce food for themselves and indirectly for all other populations within the community.

Your study of this chapter will be complete when you can

1. give examples of biotic components within an ecosystem;
2. name 2 types of food chains, and give an example of each type;
3. give an example of a food web, and define trophic level;
4. contrast the human food chain with a natural food web;
5. construct a generalized pyramid of energy using it as a basis to explain why energy flows through an ecosystem;
6. use a pyramid of biomass to explain biological magnification;
7. name and give a function for each part of a generalized chemical cycle;
8. describe the carbon and nitrogen cycles;
9. describe the human influence on each of these cycles.

In previous chapters, we have studied the ecology of individual populations and communities. Now we will consider the interactions that take place between the community and the environment. Table 49.1 lists various ecological terms that have already been introduced.

An **ecosystem** comprises all of the populations in a given area, together with the physical environment. Therefore, an ecosystem possesses both living (biotic) and nonliving (abiotic) components. The nonliving components include soil, water, light, inorganic nutrients, and weather variables. The living components of the ecosystem can be categorized as either producers or consumers. **Producers** are autotrophic organisms with the capability of carrying on photosynthesis and making food for themselves (and indirectly for the other populations as well). In terrestrial ecosystems, the producers are predominantly green plants, while in freshwater and marine ecosystems, the dominant producers are various species of algae.

Consumers are heterotrophic organisms that use preformed food. It is possible to distinguish 4 types of consumers, depending on their food source. **Herbivores** feed directly on green plants; they are termed primary consumers. **Carnivores** feed only on other animals and are thus secondary, or tertiary, consumers. **Omnivores** feed on both plants and animals. Therefore, a caterpillar feeding on a leaf is a herbivore; a green heron feeding on a fish is a carnivore; a human being eating both leafy green vegetables and beef is an omnivore.

Detritus is the remains of plants and animals following their death. The fourth type of consumer is decomposers, which feed on detritus. The bacteria and fungi of decay are important detritus feeders, but so are other soil organisms, such as earthworms and various small arthropods. The importance of the latter can be demonstrated by placing leaf litter in bags with mesh too fine to allow soil animals to enter; the leaf litter does not decompose well, even though bacteria and fungi are present. Small soil organisms precondition the detritus so that bacteria and fungi can break it down to inorganic matter that producers can use again.

When we diagram all the biotic components of an ecosystem, as in figure 49.1, it is possible to illustrate that every ecosystem is characterized by 2 fundamental phenomena: energy flow and chemical cycling. Energy flow begins when producers absorb solar energy, and chemical cycling begins when producers take in inorganic nutrients from the physical environment. Energy flow occurs because all the energy content of organic food is eventually lost to the environment as heat. The original inorganic nutrients, however—the individual chemical elements that make up organisms—are cycled through the abiotic environment and eventually cycled back to the producers.

Energy Flow

Energy flow in an ecosystem is a consequence of the 2 laws of thermodynamics we discussed in chapter 7. The first law states that energy cannot be created or destroyed, and the second law states

Table 49.1
Ecological Terms

Term	Definition
Ecology	Study of the interactions of organisms with the physical environment and with each other.
Population	All the members of the same species that inhabit a particular area.
Community	All the populations that are found in a particular area.
Ecosystem	A community, along with its physical environment; has a living (biotic) component and a nonliving (abiotic) component.
Biosphere	That portion of the surface of the earth (air, water, and land) where living things exist.

Figure 49.1
Energy flow and chemical cycling through an ecosystem. Energy does not cycle because all the energy that is derived from the sun eventually dissipates as heat.

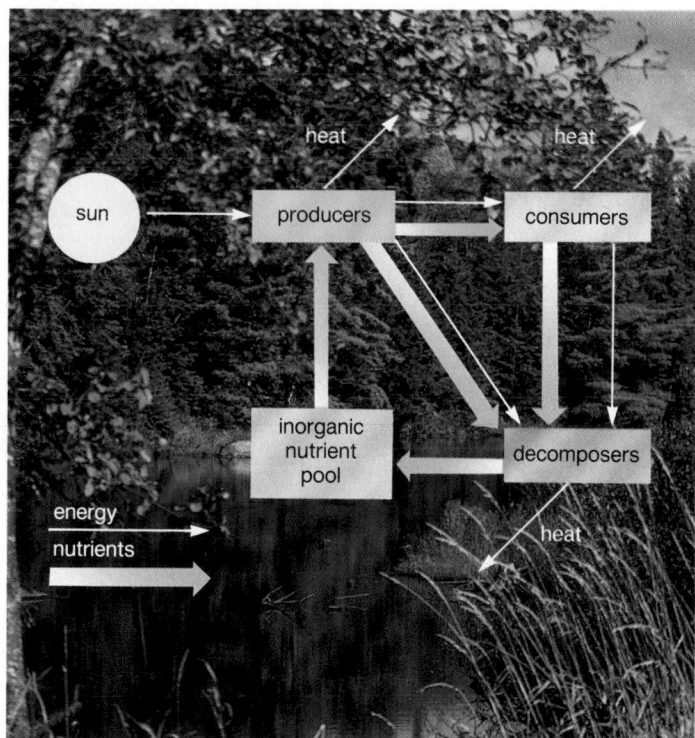

that when energy is transformed from one form to another, there is always a loss of some usable energy as heat. This means that ecosystems are unable to function unless there is a constant energy input from an external source, such as the sun, the ultimate source of energy for our planet.

Food Chains and Food Webs

Energy flows through an ecosystem as the individuals of one population feed on those of another. A **food chain** indicates who eats whom in an ecosystem. Figure 49.2 depicts a possible terrestrial food chain and a possible aquatic food chain. It is important to realize that each represents just one path of energy flow through an ecosystem. Natural ecosystems have numerous food chains, each linked to others to form a complex food web. For example, figure 49.3 shows a deciduous forest **food web** in which plants are eaten by a variety of insects, and these, in turn, are eaten by several different birds, while any one of the latter may be eaten by a larger bird, such as a hawk. Therefore, energy flow is better described in terms of **trophic** (feeding) **levels,** each one further removed from the producer population, the first (photosynthetic) level. All animals acting as primary consumers are part of a second trophic level; all animals acting as secondary consumers are part of the third trophic level; and so on.

Figure 49.2
Examples of food chains. **a.** Terrestrial. **b.** Aquatic.

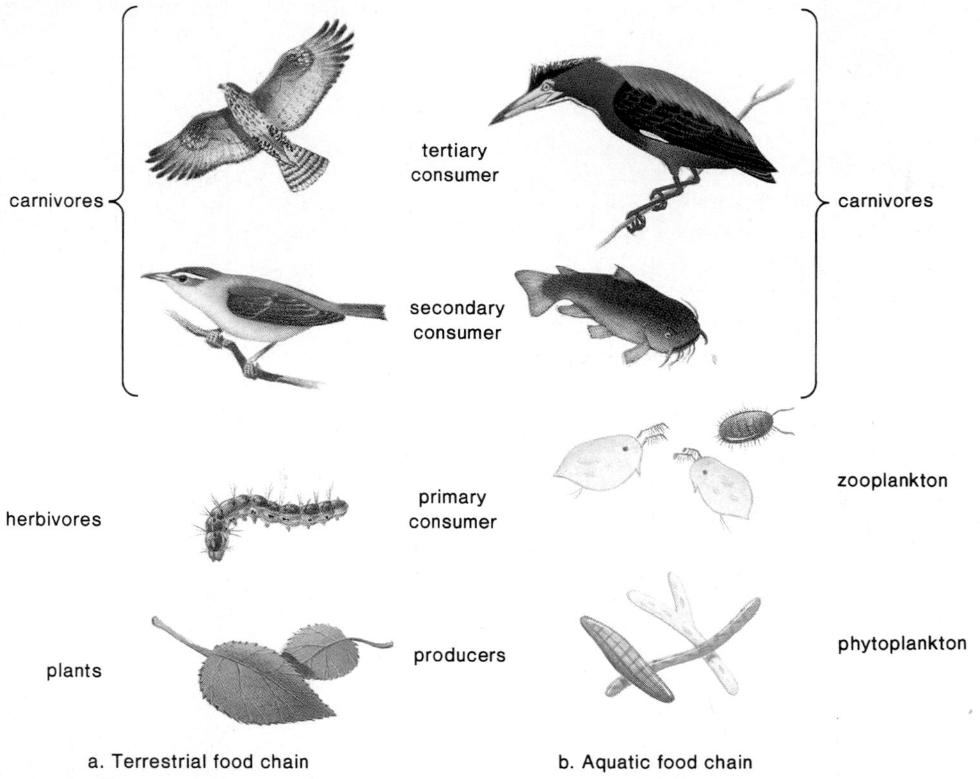

a. Terrestrial food chain

b. Aquatic food chain

The populations in an ecosystem form food chains in which the producers produce food for the other populations, which are consumers. While it is convenient to study food chains, the populations in an ecosystem actually form a food web in which food chains join and overlap with one another.

One of the food chains depicted in figure 49.2 is part of the forest food web shown in figure 49.3, and the other is part of the aquatic food web shown in figure 49.4. Both these food chains are called *grazing food chains* because the primary consumer feeds on a photosynthesizer. In some ecosystems (forests, rivers, and marshes) the primary consumer feeds mostly on detritus (dead organisms). This *detritus food chain* accounts for more energy flow than the grazing food chain because most organisms die without having been eaten. In the forest, an example of a detritus food chain is

detritus soil → bacteria → earthworms

A detritus food chain is often connected to a grazing food chain, as when earthworms are eaten by a tufted titmouse. Eventually, however, as dead organisms decompose, all the solar energy that was taken up by the producer populations dissipates as heat. Therefore, energy does not cycle.

Pyramids of Energy and Biomass

The trophic structure of an ecosystem can be summarized in the form of an **ecological pyramid.** The base of the pyramid represents the producer trophic level; the apex is the tertiary or some higher level consumer, and the other consumer trophic levels are in between. There are 3 possible kinds of pyramids. One is a *pyramid of numbers,* based on the number of organisms at each trophic level. A second is the *pyramid of biomass.* Biomass is the weight of living material at some particular time. To calculate the biomass for each trophic level, first an average weight for the organisms at each level is determined, and then the number of organisms at each level is estimated. Now, multiplying the average weight by the estimated number gives the approximate biomass for each trophic level. A third pyramid, the *pyramid of energy,* illustrates the fact that each succeeding trophic level is smaller than the previous level (fig. 49.5). Less energy is found in each succeeding trophic level for the following reasons:

1. Of the food available, only a certain amount is captured and eaten by the next trophic level. After all, prey are adapted in many ways to escaping their predators.
2. Some of the food that is eaten cannot be digested and exits the digestive tract as waste.

Behavior and Ecology

Figure 49.3

A deciduous forest ecosystem. The arrows indicate the flow of energy in a food web.

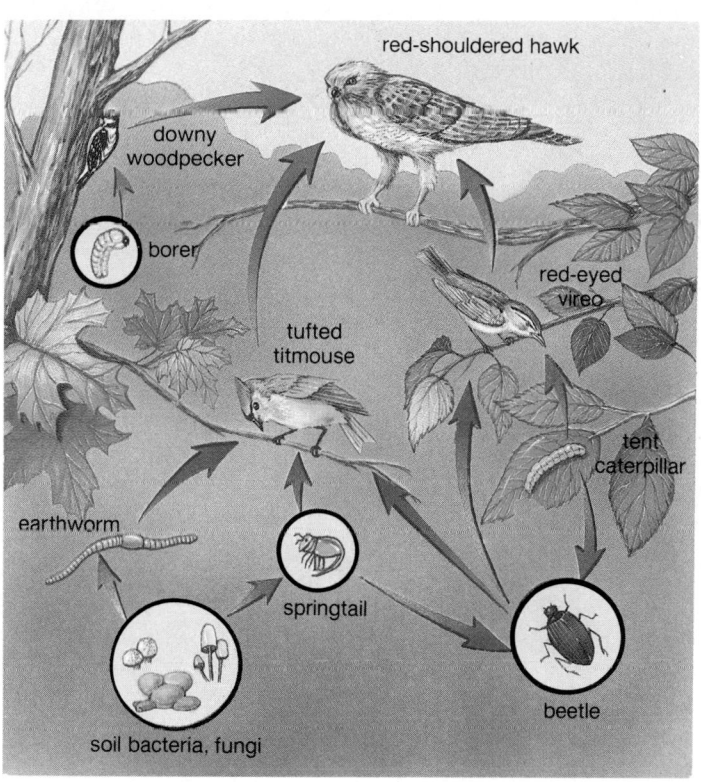

Figure 49.4

A freshwater pond ecosystem. The arrows indicate the flow of energy in a food web.

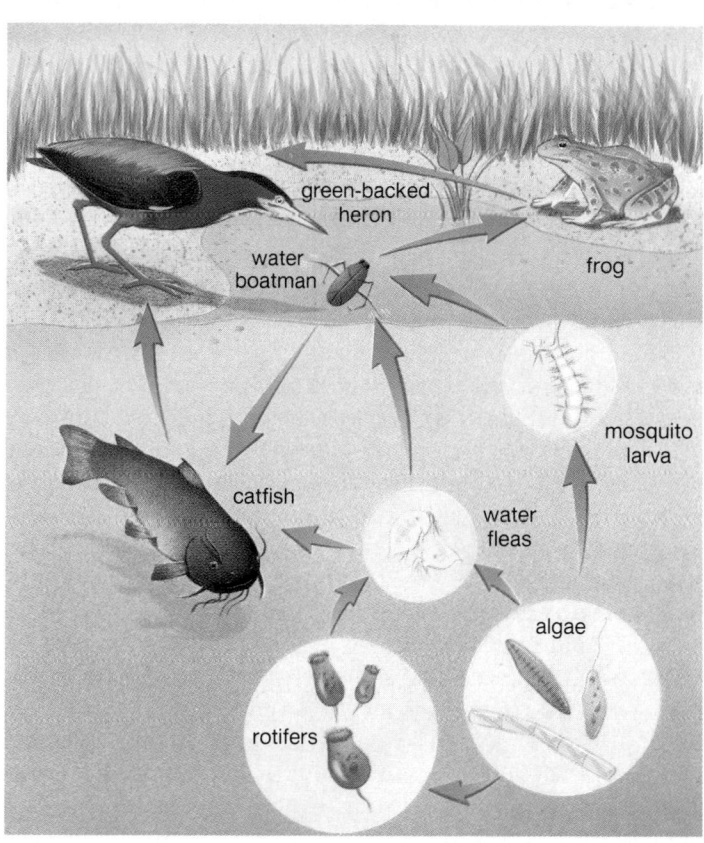

Figure 49.5

Pyramid of energy. At each succeeding trophic level in the pyramid, an appreciable portion of energy originally trapped by the producer is dissipated as heat. Accordingly, organisms in each trophic level pass on less energy than they received.

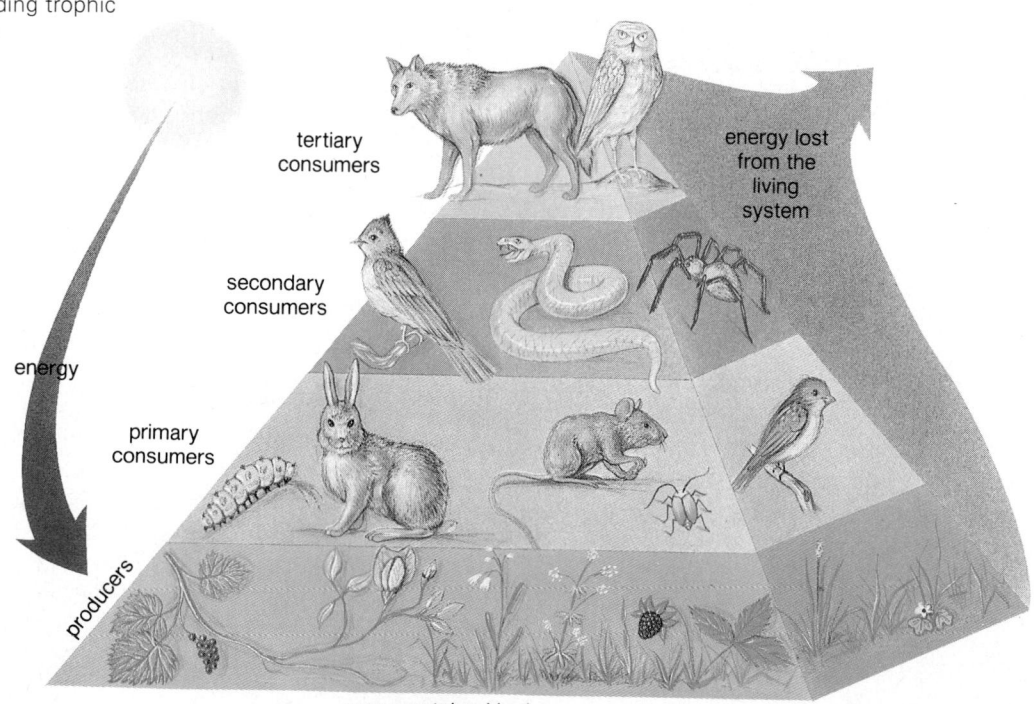

Ecosystems: The Flow of Energy and the Cycling of Materials

3. Only a portion of the food that is digested becomes part of the organisms's body. The rest is used as a source of energy.

In regard to the last point, consider that even when food molecules are used for growth and repair, energy is needed for these molecules to react. Synthetic reactions do not occur spontaneously, and when the body builds proteins, carbohydrates, and fats, energy in the form of ATP must be supplied. ATP is also needed by the body for such activities as muscle contraction and nerve conduction. Altogether, a significant portion of food molecules supplies the energy needed to form ATP by means of cellular respiration within mitochondria. Also remember, as discussed on page 106, that one form of energy can never be transformed completely into another form. There is always some loss of usable energy between trophic levels for this same reason.

The shaded areas in figure 49.5 represent the amount of energy that is unavailable to the next trophic level. The unavoidable loss of usable energy between feeding levels explains why food chains are relatively short—at most, 4 or 5 links—and why mice (herbivores) are more common than weasels, foxes, or hawks (carnivores). The population sizes are appropriate to the amount of energy available at each trophic level, until finally there is an insufficient amount of energy to support another level. The largest populations in an ecological pyramid are the producer populations at the base of the pyramid, and the smallest populations are the top predators. As with all organisms, the population size of top predators is controlled by the amount of food energy available to them.

The energy considerations associated with ecological pyramids have implications for the human population. It is generally stated that only about 10% of the energy available at a particular trophic level is incorporated into the tissues of the next level. This being the case, it can be estimated that 100 kg of grain would, if consumed directly, result in 10 human kg, but if fed to cattle, it would result in only 1 human kg. Therefore, a larger human population can be sustained by eating grain than by eating grain-consuming animals. Humans generally need some meat in their diet, however, because meat provides all the essential amino acids, as discussed in chapter 36.

In a food web, each successive trophic level has less total energy content because when energy is transferred from one level to the next, some energy is lost to the environment as heat.

Biological Magnification

Humans add all sorts of pollutants to the environment; some of these are hazardous substances that cannot be degraded by decomposers. These include heavy metals such as lead and mercury, synthetic organic chemicals, such as PCB and DDT,[1] and radioactive materials. Since decomposers are unable to break down these substances, they remain in the body. Once they enter a food chain, they become more concentrated at each trophic level. Notice in figure 49.6 that the number of dots representing DDT becomes more concentrated as the chemical is passed along from producer

[1]PCB =polychlorinated biphenyl, a chemical used in manufacturing plastics. DDT= dichlorodiphenyltrichloroethane, a pesticide no longer in use in the United States.

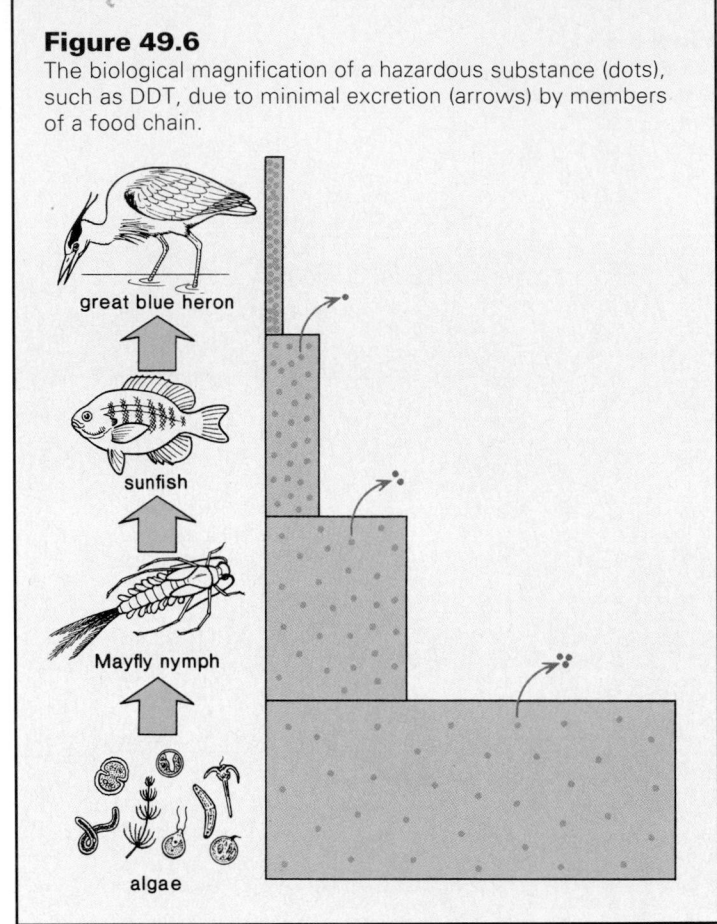

Figure 49.6
The biological magnification of a hazardous substance (dots), such as DDT, due to minimal excretion (arrows) by members of a food chain.

great blue heron

sunfish

Mayfly nymph

algae

to tertiary consumer. These dots could also represent the toxic by-products from the burning of petroleum. This so-called petroleum poisoning will be an effect from the oil well fires that were set during the Gulf War. **Biological magnification** is most apt to occur in aquatic food chains; there are more trophic levels in aquatic food chains than there are in terrestrial food chains. Humans are the final consumers in both types of food chains, and in some areas, human milk contains detectable amounts of DDT and PCB.

The use of DDT as a pesticide is now banned for most uses in this country. The first indication that the substance was harmful to living things was the reduction in the number of birds of prey. DDT interfered with the deposition of calcium in their eggshells, which were then too fragile to allow proper development of the enclosed chick.

Human Food Chain

Complex food webs tend to be stable because species' diversity provides alternate pathways by which individuals in the community can obtain energy and nutrients. To put this another way, if a large number of species are present, then when one species declines or disappears, its place and function can be assumed in part by another. If the red-eyed vireo population shown in figure 49.3 declined in size, the tufted titmouse population would most likely increase in size, and in this way the hawk population would be

Figure 49.7

The natural food chain versus human food chain. **a.** The cyclic nature of natural food chains and therefore of natural ecosystems. Notice, too, that the inputs and outputs remain constant, as does the size of the populations. **b.** The noncyclic nature of the human food chain and therefore the human ecosystem. Notice that the use of supplemental fossil fuel energy and of substances that do not cycle results in pollution. The ever-increasing size of the human population and the concomitant use of more artificial supplements cause an ever-greater amount of pollution.

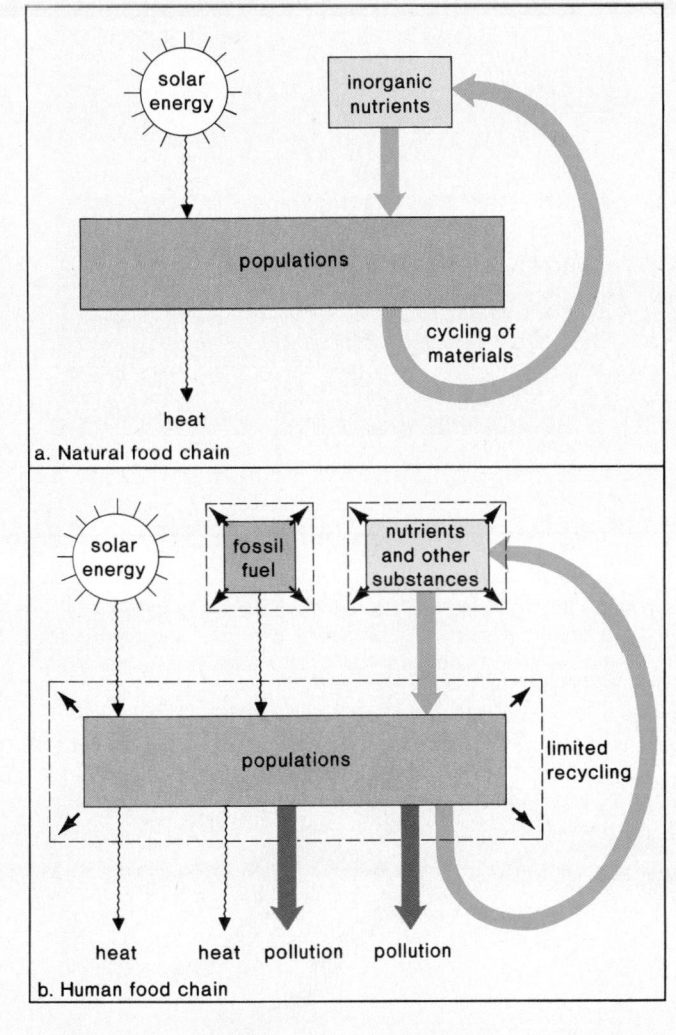

a. Natural food chain

b. Human food chain

maintained. If the hawk population disappeared, competition between its prey populations would keep their numbers in check.

Generally, population sizes remain constant in mature natural communities primarily due to biotic factors, such as competition and predation. Under normal circumstances, most of the solar energy utilized goes into maintenance of the community rather than into the growth of certain populations. The same amounts of energy and materials are required by the system, and it is said that the inputs are constant. Also, because of nutrient cycling, there is little output except heat (fig. 49.7).

When humans were hunters and gatherers, human populations, like a few native tribes today, were members of food webs, such as the one depicted in figure 49.3. With the advent of the Agricultural Revolution (about 8000 B.C.) and the Industrial Revolution (about A.D. 1650), human populations were able to become largely independent of natural food webs. Instead, humans created their own captive populations of producers and herbivores, which they have kept separate from natural food chains even to the present time. To maintain this independence from nature, the domesticated plants or animals have to be kept in highly unnatural conditions, with a minimum of species diversity and a high degree of energy input in the form of human care and fossil fuel.

Whereas natural food webs are remarkable for their inner stability, the human food chain is highly unstable. First, note that the human food chain is very short and usually has at most 3 links; for example,

$$\text{grain} \rightarrow \text{cattle} \rightarrow \text{humans}$$

Adding to the lack of diversity is the tendency of farmers to plant the same hybrid variety of wheat or corn year after year because it gives the highest yield. This so-called monoculture farming contributes to instability, since any one environmental factor, either biological or physical, can cause a crop failure in all regions.

One frequently cited example is the failure of the potato crop in Ireland in the 1800s due to a water mold infection. A more recent example occurred in 1970, when 15% of the corn crop in the United States was damaged by a corn blight. The damage could have been far worse because 80% of the crop planted that year was of the same susceptible hybrid variety. The fungus causing this particular corn blight is endemic to the South, but it happened to undergo a mutation that made it more lethal than before, and it spread to the Corn Belt, where warm, moist weather helped it flourish. The next year, resistant hybrid seed was available for planting, but this occurrence still shows that planting the same type of crop over a wide area produces a potentially unstable ecological situation.

The size of the populations in the human food chain are not constant. The human population has tripled since 1900, and the overall size of the population continues to increase dramatically (see fig. 46.7). More food must be produced to sustain the ever-growing human population, although this population encroaches on agricultural land. A half-million hectares (one hectare = 2.471 acres) of farmland are taken over each year for other human uses, such as highways and housing. Nevertheless, food production (but not necessarily food distribution) has kept up with population increases. The newly developed hybrid plants are designed to give a high yield as long as they receive adequate supplements of fertilizers, pesticides, and water. The production of fertilizers and pesticides utilizes energy, and the application of these substances as well as water requires the use of mechanized farm machinery run on fossil fuel energy.

Fertilizers, pesticides, water, and fossil fuels are visible supplements to the U.S. agricultural system, but there is another supplement as well. For example, many farmers are not using methods to minimize soil erosion, such as contour farming (fig. 49.8). The U.S. Department of Agriculture estimates that erosion is causing a steady drop in the productivity of land that is equivalent to the loss of .5 million hectares of tillable land per year. Ever more fertilizers, pesticides, and energy supplements will be required to maintain yield.

Supplemental fossil fuel energy also contributes to animal husbandry yields because at least 50% of all cattle are kept in feedlots, where they are fed grain. Production of the grain is enhanced by great expenditures of fossil fuel energy, and more fossil fuel energy must then be used to make the grain available to the cattle. Chickens are raised in a completely artificial environment, where the climate is controlled and each chicken has its own cage to which food is delivered on a conveyor belt.

From this discussion, it is evident that the human food chain is largely noncyclic (fig. 49.7b). There is a large input of fossil fuel energy (in addition to solar energy) and other substances that do not cycle back to the producers. Unfortunately, this means there is also a large output of wastes. Excess fertilizers and pesticides from farmlands and home lawns are often transported by water and wind to aquatic ecosystems. Animal and human sewage are also added to natural waters, where the bacteria of decay are often unable immediately to break down the quantity that is disposed. Therefore, it is customary for each community to pretreat sewage in sewage treatment plants before the effluent is added to nearby bodies of water.

It is clear that the human food chain cannot be isolated completely from the natural communities. In fact, it is dependent on them not only because they by themselves lend stability to the biosphere, but also because they help stabilize the human food chain by absorbing by-products, such as excess fertilizers, pesticides, and fuel breakdown products. Unfortunately, because natural communities have decreased in size as human populations have increased, the natural communities are often overwhelmed. It is then that humans recognize that something is amiss and realize that pollutants[2] have been added to the environment.

To control pollution, it is necessary to control the amount of inputs and/or to control the amount and the treatment of the outputs. For example, a few farmers have stopped using artificial additives, such as fertilizers and pesticides, to increase crop yields.

[2]Pollutant: Material added to the environment by human or natural activities that produces undesirable effects.

Figure 49.8
In contour farming, crops are planted according to the lay of the land in order to reduce soil erosion. This farmer has planted alfalfa in between the strips of corn in order to replenish the nitrogen content of the soil. Alfalfa, a legume, has root nodules that contain nitrogen-fixing bacteria.

Figure 49.9
Components of a chemical cycle. The reservoir stores the chemical and the exchange pool makes it available to producers. The chemical then cycles through food chains. Decomposition returns the chemical to the exchange pool once again if it has not already returned by another process.

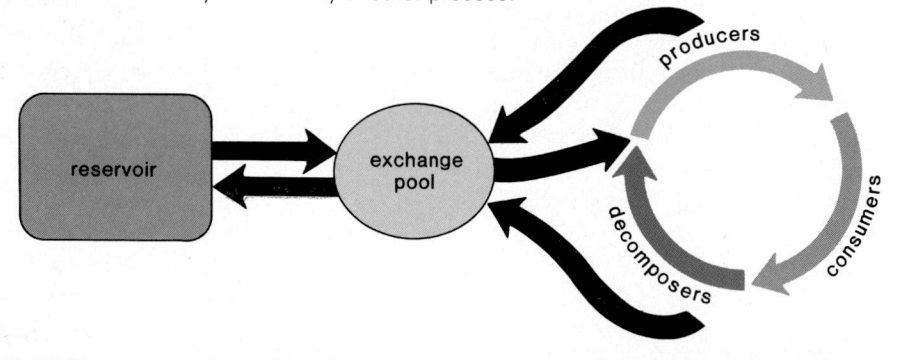

A study indicates that this method of farming, labeled organic farming by the investigators, does have reduced yields but can be equally profitable to the farmer because of the reduced costs. Other citizens are attempting to reduce pollution by making the human food chain more cyclic. For example, at a sewage treatment system functioning outside Atlanta, Georgia, wastewater is sprinkled on a forest area, providing nutrients for tree growth and purifying the water at the same time. Not only is the water then fit for drinking, but the trees can be harvested for human use on a 4-year cycle.

Behavior and Ecology

Figure 49.10

Carbon cycle. Photosynthesizers take up carbon dioxide (CO_2) from the air or bicarbonate (HCO_3^-) from the water. They and all other organisms return carbon dioxide to the environment. The carbon dioxide level is also increased when volcanoes erupt and fossil fuels are burned. Presently, the oceans are a primary reservoir for carbon in the form of limestone and carbonaceous shells.

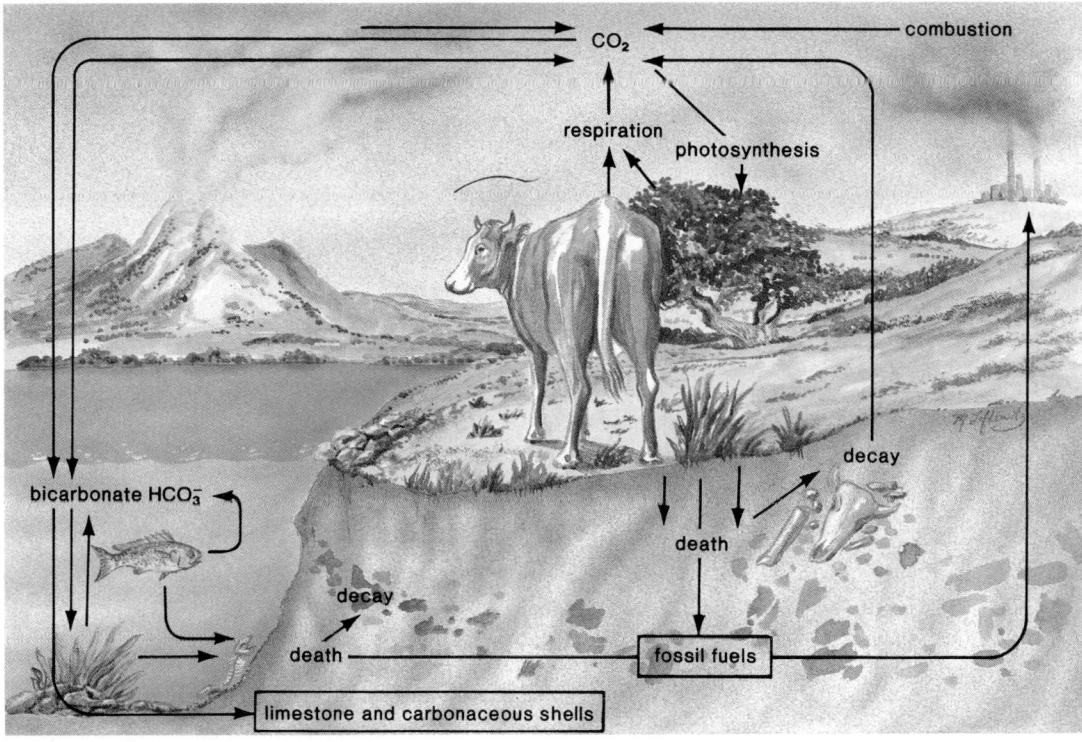

These examples demonstrate that it is possible to find more natural ways to fulfill our needs. This is often called "working with nature rather than against nature." On the other hand, if we choose to continue practices that create environmental problems, we must ask ourselves if we are willing to bear the cost. As ecologists are fond of saying, "There is no free lunch." By this they mean that for every artificial benefit, there is a price. For example, if we kill off coyotes because they prey on chickens, we then have to realize that the mouse population (mice are a common food for coyotes) will probably increase. In an ecosystem, as we have seen, "everything is connected to everything else."

The human food chain is shorter and less stable than those found in natural food webs. As the human population increases in size, a greater quantity of materials and fossil fuel energy is required to sustain the population, and a greater quantity of outputs (e.g., pollutants) results.

Chemical Cycling

In contrast to energy, inorganic nutrients do cycle through large natural ecosystems. Because there is minimal input from the outside, the various elements essential to life are used over and over. Since the pathways by which chemicals circulate through ecosystems involve both the living (biosphere) and nonliving (geological) areas, they are known as **biogeochemical cycles.** For each element, the cycling process involves (1) a reservoir—that portion of the earth that acts as a storehouse for the element; (2) an exchange pool—that portion of the environment from which the producers take their nutrients; and (3) the biotic community—through which chemicals move along food chains to and from the exchange pool (fig. 49.9).

The Carbon Cycle

In the carbon cycle, organisms in both terrestrial and aquatic ecosystems exchange carbon dioxide with the atmosphere (fig. 49.10). On land, plants take up carbon dioxide from the air, and through photosynthesis, they incorporate carbon into food that is used by autotrophs and heterotrophs alike. When organisms respire, a portion of this carbon is returned to the atmosphere as carbon dioxide.

In aquatic ecosystems, the exchange of carbon dioxide with the atmosphere is indirect. Carbon dioxide from the air combines with water to give bicarbonate (HCO_3^-), a source of carbon for algae that produce food for themselves and for heterotrophs. Similarly, when aquatic organisms respire, the carbon dioxide they give off becomes bicarbonate. The amount of bicarbonate in the water is in equilibrium with the amount of carbon dioxide in the air.

Carbon Reservoirs

Living and dead organisms contain organic carbon and serve as one of the reservoirs for the carbon cycle. The world's biota, particularly trees, contain 800 billion tons of organic carbon, and an additional 1,000–3,000 billion metric tons are estimated to be held in the remains of plants and animals in the soil. Before decomposition can occur, some of these remains are subjected to physical processes that transform them into coal, oil, and natural gas. We call these materials the fossil fuels. Most of the fossil fuels were formed during the Carboniferous period, 280–350 million years ago, when an exceptionally large amount of organic matter was buried before decomposing. Another reservoir is the inorganic carbonate that accumulates in limestone and in carbonaceous shells. The oceans abound in organisms, some microscopic, that are composed of carbonaceous shells and accumulate in ocean bottom sediments. Limestone is formed from these sediments by geological transformation.

Human Influence on the Carbon Cycle

The activities of human beings have increased the amount of carbon dioxide (CO_2) and other gases in the atmosphere. Data from monitoring stations record an increase of 20 ppm (parts per million) in carbon dioxide in only 22 years. (This is equivalent to 42 billion metric tons of carbon.) This buildup is primarily attributed to the burning of fossil fuels and wood. The oil well fires that were set during the Gulf War added carbon dioxide to the atmosphere but the on-going burning of tropical rain forests is of even more concern (fig. 49.11). When we do away with forests, we reduce a reservoir that takes up excess carbon dioxide. At this time, the oceans are believed to be taking up most of the excess carbon dioxide; the burning of fossil fuels in the last 22 years has probably released 78 billion metric tons of carbon, yet the atmosphere registers an increase of "only" 42 billion metric tons.

As discussed on page 800, there is much concern that an increased amount of carbon dioxide (and other gases) in the atmosphere is causing a global warming. These gases allow the sun's rays to pass through, but they absorb and reradiate heat back to the earth, a phenomenon called the *greenhouse effect*.

> In the carbon cycle, carbon dioxide is removed from the atmosphere by photosynthesis but is returned by respiration. Living things and dead matter are carbon reservoirs. The oceans, because they abound with carbonaceous shells and accumulate limestone, are also major carbon reservoirs.

The Nitrogen Cycle

Nitrogen is an abundant element in the atmosphere. Nitrogen gas (N_2) makes up about 78% of the atmosphere by volume, yet nitrogen deficiency commonly limits plant growth. Plants cannot incorporate nitrogen gas into organic compounds and therefore depend on various types of bacteria to make nitrogen available to them (fig. 49.12).

Nitrogen Fixation

The reduction of nitrogen gas (N_2) prior to the incorporation of nitrogen into organic compounds is called **nitrogen fixation.** For example, some cyanobacteria in aquatic ecosystems and some

Figure 49.11
Burning of trees in tropical rain forest. When trees are burned, carbon dioxide (CO_2) is released from one of the reservoirs in the carbon cycle.

free-living bacteria in the soil reduce nitrogen gas to ammonium (NH_4^+), which then is subject to processes leading to organic compounds. Other *nitrogen-fixing bacteria* infect and live in nodules on the roots of legumes (fig. 49.12). They fix atmospheric nitrogen gas and incorporate nitrogen into organic compounds that not only they themselves use but the host plant as well.

Nitrogen fixation also occurs after plants take up nitrates (NO_3^-) from the soil. Plants use nitrates as their source of nitrogen when they make amino acids and nucleic acids, for example.

Nitrification and Denitrification

Nitrogen gas (N_2) is converted to nitrate (NO_3^-) in the atmosphere when cosmic radiation, meteor trails, and lightning provide the high energy needed for nitrogen to react with oxygen. Also, humans make a most significant contribution to the nitrogen cycle when they convert nitrogen gas to nitrate for use in fertilizers.

Ammonium (NH_4^+) in the soil is also converted to nitrate by bacteria in a process called nitrification. Ammonium gets into the soil in 2 ways. As mentioned, some free-living bacteria in the soil reduce nitrogen gas to ammonium. Decomposers also produce ammonium when they decompose waste and dead organic matter. Other types of bacteria carry out nitrification. First, nitrite-producing bacteria convert ammonium to nitrite (NO_2^-), and then nitrate-producing bacteria convert nitrite to nitrate. These 2 groups of bacteria are called the *nitrifying bacteria*. Notice that there is a subcycle in the nitrogen cycle that involves only nitrates, ammonium, and nitrites. This subcycle does not depend on the presence of nitrogen gas at all (fig. 49.12).

Figure 49.12

Nitrogen cycle. Several types of bacteria are at work: nitrogen-fixing bacteria reduce nitrogen gas (N_2); nitrifying bacteria, which include both nitrite-producing and nitrate-producing bacteria, convert ammonium (NH^+_4) to nitrate; and the denitrifying bacteria convert nitrate back to nitrogen gas. Humans contribute to the cycle by using nitrogen gas to produce nitrate for fertilizers.

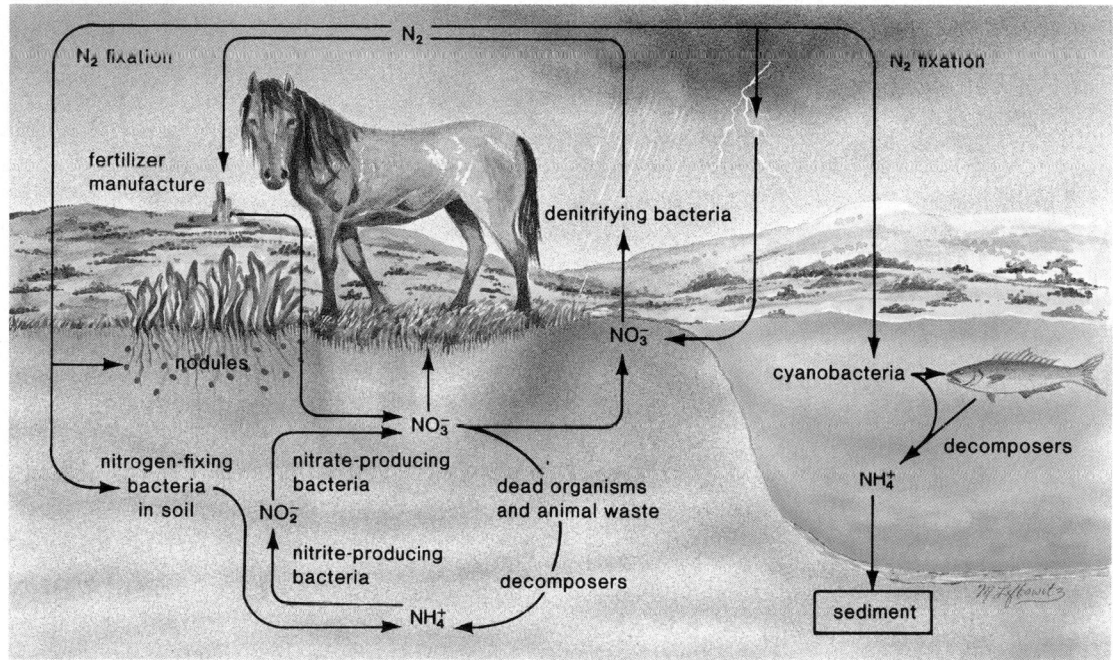

Denitrification is the conversion of nitrate to nitrogen gas. There are *denitrifying bacteria* in both aquatic and terrestrial ecosystems. Denitrification counterbalances nitrogen fixation but not completely. There is more nitrogen fixation, especially due to fertilizer production.

In the nitrogen cycle, nitrogen-fixing bacteria (in nodules and in the soil) reduce nitrogen gas, and thereafter nitrogen can be incorporated into organic compounds; nitrifying bacteria convert ammonium to nitrate; denitrifying bacteria convert nitrate back to nitrogen gas.

Summary

1. Ecosystems contain both abiotic (physical) and biotic (living) components. Biotic components are either producers or consumers. Consumers may be herbivores, carnivores, omnivores, or decomposers.
2. Energy flow and chemical cycling are 2 phenomena observed in ecosystems.
3. Food chains are paths of energy flow through an ecosystem. Grazing food chains always begin with a producer population, which is capable of producing organic food, followed by a series of consumer populations. Detritus food chains begin with dead organic matter, which is consumed by decomposers, including bacteria and fungi as well as various soil animals. Eventually, all members of food chains die and decompose. The very same chemicals are then made available to the producer population again, but the energy has been dissipated as heat. Therefore, energy does not cycle through an ecosystem.
4. If we consider the various food chains together, we realize that they form an intricate food web in which there are various trophic (feeding) levels. All producers are on the first level, all primary consumers are on the second level, and so forth. To illustrate that energy does not cycle, it is customary to arrange the various trophic levels to form an energy pyramid. Then we see that each level contains less energy than the previous level. It is estimated that only 10% of the available energy is actually incorporated into the body tissues of the next level.
5. In mature natural food webs, usually the same, constant amount of energy is required and additional material inputs are minimal because matter cycles. In contrast, the human food chain tends to be unstable. The human food chain is quite short, usually containing at most only 3 links. Adding to this lack of diversity is the tendency of farmers to use only certain high-yield varieties of agricultural plants, which require such supplements as fertilizers, pesticides, and water. The human food chain is noncyclic, and worse yet, more fossil fuel energy and material supplements are needed each year to produce more food for the human population, which

continues to increase in size. Pollution is a side effect; it would be beneficial for us to find ways to fulfill our needs in a more natural way.

6. Chemical cycling through an ecosystem involves not only the biotic components of an ecosystem but also the physical environment. Each cycle involves a reservoir where the element is stored, an exchange pool from which the populations take and return nutrients, and the populations themselves.

7. In the carbon cycle, the reservoir is organic matter, carbonaceous shells, and limestone. The exchange pool is the atmosphere: photosynthesis removes carbon dioxide, and respiration and combustion add carbon dioxide.

8. In the nitrogen cycle, the reservoir is the atmosphere, but nitrogen (N_2) gas must be converted to nitrate (NO_3^-) for use by producers. Nitrogen-fixing bacteria, particularly in root nodules, make organic nitrogen available to plants. Other bacteria active in the nitrogen cycle are the nitrifying bacteria, which convert ammonia to nitrates, and the denitrifying bacteria, which reconvert nitrate to N_2.

Writing Across the Curriculum

In order to practice writing skills, students should write out the answers to any or all of the study questions and the critical thinking questions. The study questions are sequenced in the same order as the text. Suggested answers to the critical thinking questions are in appendix D.

Study Questions

1. Name 4 different types of consumers found in natural ecosystems.
2. Give an example of a grazing and a detritus food chain for a terrestrial and for an aquatic food chain.
3. Draw an energy pyramid and explain why such a pyramid can be used to verify that energy does not cycle.
4. Give an example of biological magnification and tell why it occurs.
5. Compare and contrast the basic characteristics of a natural food web to the characteristics of the human food chain.
6. Discuss in detail the carbon and nitrogen biogeochemical cycles. Indicate the manner in which humans disturb the equilibrium of these cycles.

Objective Questions

1. Compare this food chain: algae → water fleas → catfish → green herons to this food chain: trees → tent caterpillars → red-eyed vireos → hawks. Both water fleas and tent caterpillars are
   a. carnivores.
   b. primary consumers.
   c. detritus feeders.
   d. Both a and b.
2. Consider the components of a food chain: producers → herbivores → carnivores → top carnivores. Which level contains the most energy?
   a. primary producers
   b. herbivores
   c. carnivores
   d. top carnivores
3. Consider the components of a food chain: producers → herbivores → carnivores → top carnivores. Eventually what happens to all the energy passed from one element to the next?
   a. It recycles back to the producers.
   b. It results in a much larger decomposer population.
   c. It is dissipated into the environment.
   d. It is recaptured by another food chain.

4. Which of the following contribute to the carbon cycle?
   a. respiration
   b. photosynthesis
   c. fossil fuel combustion
   d. All of these.
5. A natural food web
   a. contains only grazing food chains.
   b. contains several trophic levels.
   c. is usually unstable.
   d. All of these.
6. How do plants contribute to the carbon cycle?
   a. When they respire, they release CO_2 into the atmosphere.
   b. When they photosynthesize, they consume carbon dioxide from the atmosphere.
   c. They do not contribute to the carbon cycle.
   d. Both a and b.
7. How do nitrogen-fixing bacteria contribute to the nitrogen cycle?
   a. They return N_2 to the atmosphere.
   b. They change ammonia to nitrate.
   c. They change N_2 to ammonia.
   d. They withdraw nitrate from the soil.

8. In what way are decomposers like producers?
   a. Either one may be the first member of a grazing food chain.
   b. Both produce oxygen for other forms of life.
   c. Both require a source of nutrient molecules and energy.
   d. Both supply organic food for the biosphere.
9. How is a first-order consumer like a second-order consumer?
   a. Both pass on less energy to the next trophic level than they received.
   b. Both pass on the same amount of energy to the next trophic level.
   c. Both tend to be herbivores that produce nutrients for plants.
   d. Both are able to convert organic compounds to ATP without the loss of energy.
10. Which statement is true concerning this food chain: grass → rabbits → snakes → hawks?
    a. Each predator population has a greater biomass than its prey population.

b. Each prey population has a greater biomass than its predator population.

c. Each population is omnivorous.

d. Both a and c.

11. Add these trophic levels to this diagram representing a pyramid of energy: algae, large fishes, humans, small fishes, zooplankton:

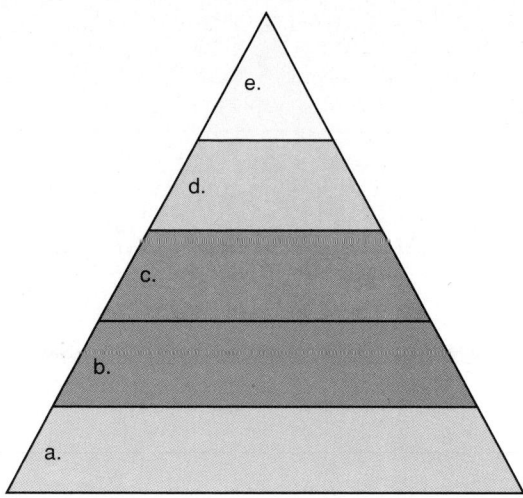

Concepts and Critical Thinking

1. *Chemicals cycle through an ecosystem.*

How does chemical cycling involve both the abiotic and biotic components of an ecosystem?

2. *Energy flows through an ecosystem.*

Relate energy flow through an ecosystem to the laws of thermodynamics (see chap. 7).

Selected Key Terms

producer (pro-dūs'er) 781
consumer (kon-sūm'er) 781
herbivore (her'bĭ-vōr) 781
carnivore (kar'nĭ-vōr) 781
omnivore (om'nĭ-vōr) 781

detritus (de-tri'tus) 781
food chain (fōōd chān) 782
food web (fōōd web) 782
trophic level (trof'ik lev'el) 782
ecological pyramid (e"ko-log'ĭ-kal pir'ah-mid) 782

biological magnification (bi"o-loj'ĭ-kal mag"nĭ-fi-ka'shun) 784
nitrogen fixation (ni'tro-jen fik-sa'shun) 788
denitrification (de-ni"trĭ-fi-ka'shun) 789

50

Environmental Concerns

Proper management allows land to be productive for many years. Mismanagement can lead to desert conditions and the loss of land fertility. In this photograph, crops are being covered by sand, and therefore, land productivity will be reduced.

Your study of this chapter will be complete when you can

1. give 3 general reasons why the quality of the land is being degraded today;
2. give 3 reasons why tropical rain forests are being destroyed, and explain why this destruction should be halted immediately;
3. give reasons why underground water supplies are being polluted, and tell how the ocean receives pollutants;
4. list the substances that contribute to air pollution; tell their sources and associate each one with either development of photochemical smog, acid deposition, global warming, or destruction of the ozone shield.
5. explain how smog develops, how acids develop in the atmosphere, how chlorofluorocarbons break down the ozone shield, and how the greenhouse gases bring about global warming;
6. list the environmental effects of global warming, and tell what can be done to prevent global warming;
7. review the energy sources available to humans, and indicate which of these would best prevent the occurrence of the greenhouse effect;
8. describe the characteristics of a human society that is in a steady state.

As the human population increases in size, the space allotted to natural ecosystems is reduced in size. Natural ecosystems are then no longer able to process and rid the biosphere of wastes, which accumulate and are called pollutants. **Pollutants** are substances added to the environment, particularly by human activities, that lead to undesirable effects for all living things. Human beings add pollutants to all parts of the biosphere—land, water, and air.

Land

Whereas 20% of the world's population lived in cities in 1950, it is predicted that 50% will live in cities by the year 2000. Near cities, the development of new housing areas has led to urban sprawl (fig. 50.1). Such areas tend to take over agricultural land or simply further degrade the land in the area.

Soil Erosion and Desertification

Soil erosion causes the productivity of agricultural lands to decline. It occurs when wind blows and rain washes away the topsoil, leaving the land exposed. The U.S. Department of Agriculture estimates that erosion is causing a steady drop in the productivity of farmland equivalent to the loss of .5 million hectares per year. To maintain the productivity of eroding land, more fertilizers, more pesticides, and more energy must be used.

One answer to the problem of erosion is to adopt soil conservation measures. Unfortunately, at the present time many farmers are failing to use any form of soil conservation, and it is estimated that soil erosion is removing topsoil at 3 times the rate at which it can be reformed by natural means. There are, however, both new and old measures that farmers can adopt. Some farmers have turned to no-tillage farming as a means to prevent soil erosion. In no-tillage farming, the remains of the previous crop are left in place rather than plowed under. At planting time, a disc is used to cut furrows for planting new seeds. There is one drawback to this method: it requires the use of potent herbicides to prevent weed growth, and it may also require a more liberal application of pesticides, since pests thrive in the undisturbed soil and litter. For this reason, some experts prefer the older methods of preventing soil erosion, such as strip and contour farming (see fig. 49.8).

The seriousness of land degradation is made clear by some recent trends. People are abandoning marginal cropland and rangeland worldwide because of severe soil erosion. Another process, **desertification,** leads to the transformation of marginal lands to desert conditions. Desertification has been particularly evident along the southern edge of the Sahara Desert in Africa, where it is estimated that 350,000 square miles of once-productive grazing land has become desert in the last 50 years. Desertification, however, also occurs in this country. The U.S. Bureau of Land Management, which opens up federal lands for grazing, reports that much of the rangeland it manages is in poor or bad condition, with much of its topsoil gone and with greatly reduced ability to support forage plants.

Figure 50.1
Urban sprawl now occurs worldwide as the cities expand due to newly arrived residents.

Humans often use land for their cities and for agriculture. Agricultural land quality is threatened by soil erosion, which can lead to desertification.

Land Spoilage Due to Wars

Many plants and animals are killed because of wars. For example, during a civil war in Angola rhinos and elephants were shot, and their tusks and horns were sold to buy uniforms and weapons. During a war between Uganda and Tanzania, hippos were used as target practice, and other animals were killed for food and for their ivory tusks.

However, habitat destruction caused by wars probably leads to more loss of life than direct killing. Vietnamese scientists estimate that 5.45 million acres of tropical forests were destroyed due to bombing, bulldozing, and the spraying of defoliates such as Agent Orange. Also, about half of south Vietnam's wetlands are now devoid of their mangrove trees due to the effects of war. The Gulf War was similarly destructive. Although the sands of Kuwait may seem lifeless, they are actually home to a variety of spiders, snakes, and scorpions as well as sheep, gazelles, and camels. Water and air pollution due to the Gulf War is discussed later.

Exploitation of Tropical Rain Forest

In developed countries, much of the hardwood forests were long ago converted to rapidly growing softwood forest plantations to provide humans with a source of wood. There are still virgin rain

a.

Figure 50.2
Tropical rain forest sustainability. **a.** In its natural state, a tropical rain forest is immensely rich in vegetation and breathtakingly beautiful. **b.** If the trees are not cut down, rubber tappers can earn a sustainable living by tapping the same trees year after year. **c.** Slash-and-burn agriculture is the first step toward destruction of a portion of the rain forest. With this method, agriculture cannot be sustained for more than a few years because the soil is infertile and does not hold moisture.

b.

c.

forests, however, in Southeast Asia and Oceania, Central and South America, and Africa (fig. 50.2). These forests are severely threatened by human exploitation. The people living in developed countries want all sorts of things made from beautiful and costly tropical woods and provide a market for such lumber. Most of the loss due to logging is in Southeast Asia, and in Malaysia and the Philippines, for example, the best commercial forests are already gone. Indonesia is cutting heavily to feed its new wood-exporting business, and Brazil is racing to catch up. Much of this wood goes to Japan, the United States, and Europe.

Tropical rain forests are undergoing destruction also because of the needs of the people who live there. For example, in Brazil there are large numbers of persons who have no means to support their families. To ease social unrest, the government allows citizens to own any land they clear in the Amazon forest (along the Amazon River). In tropical rain forests, it is the custom to practice what is called **slash-and-burn agriculture,** in which trees are cut down and burned to provide space to raise crops. Unfortunately, however, the land is fertile for only a few years. In tropical forests the inorganic nutrients immediately cycle back to producers (see fig. 48.11) and do not accumulate in the soil. Therefore, despite the massive amount of growth above ground, the soil itself is nutrient-poor. Once the cleared land is unable to sustain crops, the farmer moves on to another part of the forest to slash and burn again.

Cattle ranchers are the greatest beneficiaries of deforestation, and ranching expansion is therefore another reason for tropical rain forest destruction. The ranchers in Brazil are so aggressive that they actually force colonists to sell them newly cleared land at gunpoint. The cattle that are raised provide beef for export, and much of this beef is bought by United States fast-food chains. Because the imported beef is cheaper, hamburgers can be sold for 5 cents less than those with beef purchased in the United States. The destruction of tropical forests on this account has been termed the "hamburger connection."

A pig-iron industry has also begun in the Amazon and indirectly results in further exploitation of the rain forest. Deposits of iron ore found in the Carajas Mountains are processed before being exported. The pig-iron smelters are fired by charcoal bought from small, private producers. The largest pig-iron company acknowledges having paid for construction of 1,500 small, make-shift ovens used by peasants who burn trees from the forests to produce the charcoal.

There are currently 3 primary reasons for tropical rain forest destruction: (1) logging to provide hardwoods for export, (2) slash-and-burn agriculture, and (3) cattle ranching. Industrialization in countries having extensive tropical rain forests will no doubt become another major reason in the future.

Loss of Biological Diversity

Tropical rain forests are much more biologically diverse than temperate forests. For example, temperate forests across the entire United States contain about 400 tree species. In the rain forest, a typical 10-hectare area holds as many as 750 types of trees. The fresh waters of South America are inhabited by an estimated 5,000 fish species; on the eastern slopes of the Andes, there are 80 or more species of frogs and toads, and in Ecuador, there are more than 1,200 species of birds—roughly twice as many as those inhabiting all of the United States and Canada. Therefore, a very serious side effect of deforestation in tropical countries is the loss of biological diversity.

Altogether, about half of the world's species are believed to live in tropical forests. A National Academy of Sciences study estimated that a million species of plants and animals are in danger of disappearing within 20 years as a result of deforestation in tropical countries. Many of these life forms have never been studied and yet many could possibly be useful. At present, the entire domesticated crop production around the world relies on fewer than 30 species of plants and animals that have been domesticated during the last 10,000 years! It's quite possible that many additional species of wild plants and animals now living in tropical forests could be domesticated. As matters now stand, the clearing of tropical forests will very likely prevent humans from ever having the possibility of utilizing more than a tiny fraction of the earth's biological diversity.

While it may seem like an either-or situation—either biological diversity or human survival—there is a growing recognition that this is not the case. Natives who harvest rubber from the trees can have a sustainable income from the same trees year after year (fig. 50.2b). A recent study calculated the market value of rubber and exotic produce, like the aguaje palm fruit, that can be harvested continually from the Amazon forest. It concluded that selling these products would yield more than twice the income of either lumbering or cattle ranching.

There is much worldwide concern about the loss of biological diversity due to the destruction of tropical rain forests. The myriad of plants and animals that live there could possibly benefit human beings.

Waste Disposal

Every year, the United States population discards billions of metric tons of solid wastes, much of it on land (fig. 50.3). Solid wastes include not only household trash but also sewage sludge, agricultural residues, mining refuse, and industrial wastes. Some of these solid wastes contain substances that cause human illness and sometimes even death; they are called *hazardous wastes*.

Hazardous wastes, such as heavy metals, chlorinated hydrocarbons, or organochlorides, and nuclear wastes, are subject to biological magnification (see fig. 49.6). Biological magnification is most apt to occur in aquatic food chains; there are more links in aquatic food chains than there are in terrestrial food chains. Humans are the final consumers in both types of food chains, and in some areas, human milk contains detectable amounts of DDT and PCBs, which are organochlorides.

The dumping of hazardous wastes directly endangers public health. Chemical wastes buried over a quarter of a century ago in Love Canal, near Niagara Falls, have seriously damaged the health of some residents there. Similarly, the town of Times Beach,

Missouri, is abandoned because workers spread an organochloride (dioxin)-laced oil on the city streets, leading to a myriad of illnesses among its citizens. In other places, such as in Holbrook, Massachusetts, manufacturers have left thousands of drums in abandoned or uncontrolled sites, where toxic chemicals are oozing out into the ground and are contaminating the water supply. Illnesses, especially forms of cancer, are quite common not only in Holbrook but also in adjoining towns.

Water Pollution

Pollution of surface water, groundwater, and the oceans is of major concern today.

Surface Water Pollution

All sorts of pollutants from various sources enter surface waters, as depicted in figure 50.4 and listed in table 50.1. Sewage treatment plants can help degrade organic wastes, which otherwise can cause oxygen depletion in lakes and rivers. As the oxygen level decreases, the diversity of life is greatly reduced. Also, human feces can contain pathogenic microorganisms that cause cholera, typhoid fever, and dysentery. In less developed countries, where the population is growing and where waste treatment is practically nonexistent, many children die each year from these diseases.

   Typically, sewage treatment plants use bacteria to break down organic matter to inorganic nutrients, like nitrates and phosphates, which then enter surface waters. These types of nutrients, which also can enter waters by fertilizer runoff and soil erosion, lead to *cultural eutrophication,* an acceleration of the natural process by which bodies of water fill in and disappear. First, the nutrients cause overgrowth of algae. Then, when the algae die, oxygen is used up by the decomposers, and the water's capacity to support life is reduced. Massive fish kills are sometimes the result of cultural eutrophication.

   Industrial wastes can include heavy metals and organochlorides, such as pesticides. These materials are not degraded readily under natural conditions nor in conventional sewage treatment plants. Sometimes, they accumulate in the bottom mud of deltas and estuaries of highly polluted rivers and cause environmental problems if they are disturbed. Industrial pollution is being addressed in many industrialized countries but usually has low priority in less developed countries.

   Some pollutants enter bodies of water from the atmosphere. Acid deposition has caused many lakes to become sterile in the industrialized world because acid leaches aluminum and iron out of the soil. A high concentration of these ions kills fishes and other forms of aquatic life. Lime is sometimes helpful against acidification of a lake.

Groundwater Pollution

Figure 50.5 shows the ways—both intentionally and unintentionally—that pollutants can reach underground rivers called aquifers. In areas of intensive animal farming or where there are many septic tanks, ammonium (NH_4^+) released from animal and human waste

Figure 50.3
Dump sites not only cause land and air pollution, they also allow chemicals to enter groundwater, which may later become part of human drinking water.

is converted by soil bacteria to soluble nitrate, which moves down through the soil (percolates) into underground water supplies. Between 5% and 10% of all wells examined in the United States have nitrate levels higher than the recommended maximum.

   Industry also pollutes aquifers. Previously, industry ran wastewater into a pit, from which pollutants could seep into the ground. Wastewater and chemical wastes were also injected into deep wells, from which pollutants constantly discharge. Both of these customs have been or are in the process of being phased out. It is very difficult for industry to find other ways to dispose of wastes, especially since citizens do not wish to live near waste treatment plants. The emphasis today, therefore, is on prevention of wastes in the first place. Industry is trying to use processes that do not create wastes and/or to recycle the wastes they do generate.

Ocean Water Pollution

Coastal regions are not only the immediate receptors for local pollutants, they are also the final receptors for pollutants carried by rivers that empty at a coast. Waste dumping also occurs at sea, and ocean currents sometimes transport both trash and pollutants back to shore. Examples are the nonbiodegradable plastic bottles, pellets, and containers that now commonly litter beaches and the oceans' surfaces. Some of these, such as the plastic that holds a 6-pack of cans, cause the death of birds, fishes, and marine mammals that mistake them for food and get entangled in them.

   Offshore mining and shipping add pollutants to the oceans. Some 5 million metric tons of oil a year, or more than one gram per 100 square meters of the oceans' surfaces, end up in the oceans. Large oil spills kill plankton, fish larvae, and shellfishes, as well as

Behavior and Ecology

Figure 50.4

Sources of surface water pollution. So many different pollutants enter rivers (and also lakes) that they are in danger of dying.

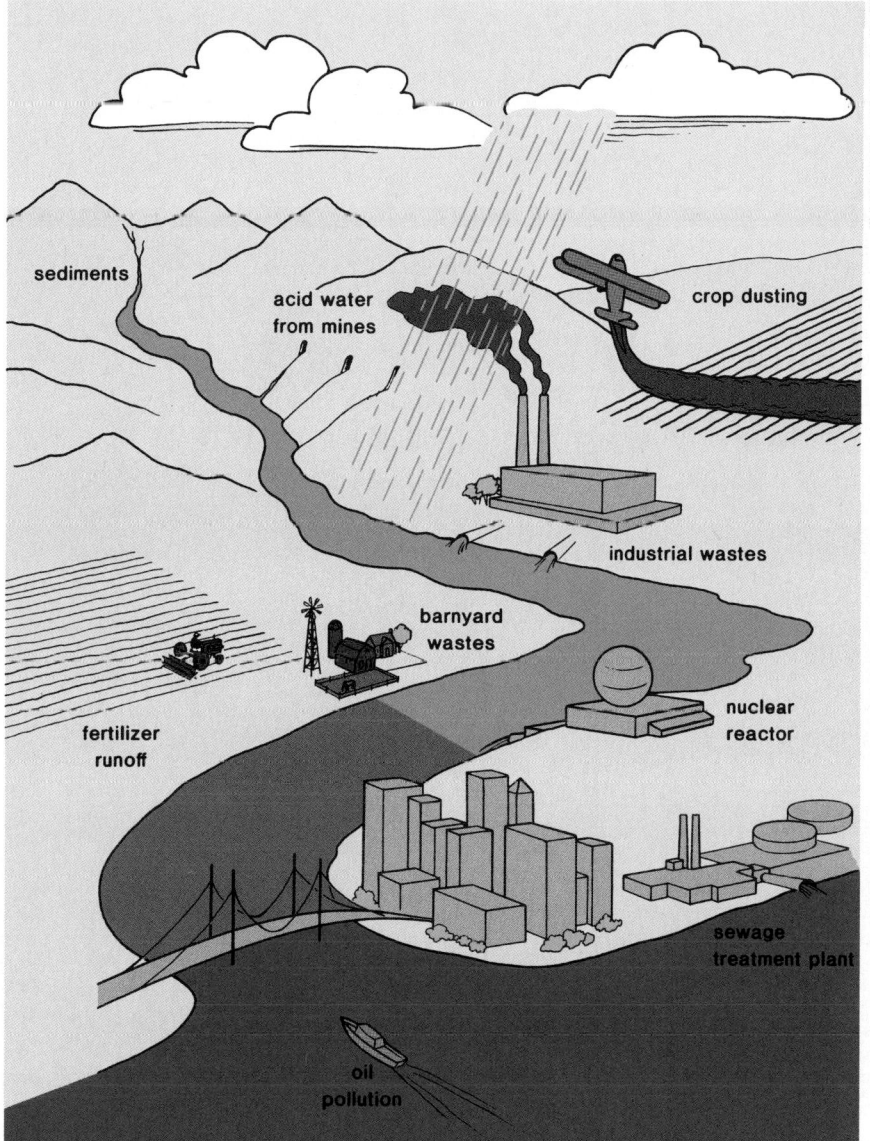

Table 50.1
Sources of Water Pollution

Leading to Cultural Eutrophication	
Oxygen-demanding wastes	Biodegradable organic compounds (e.g., sewage, wastes from food processing plants, paper mills, and tanneries)
Plant nutrients	Nitrates and phosphates from detergents, fertilizers, and sewage treatment plants
Sediments	Enriched soil in water due to soil erosion
Thermal discharges	Heated water from power plants
Health Hazards	
Disease-causing agents	Bacteria and viruses from sewage (e.g., food poisoning and hepatitis)
Synthetic organic compounds	Pesticides, industrial chemicals (e.g., PCBs)
Inorganic chemicals and minerals	Acids from mines and air pollution; dissolved salts; heavy metals (e.g., mercury) from industry
Radioactive substances	From nuclear power plants, medical and research facilities, and nuclear weapons testing

birds and marine mammals. The largest spill may have occurred on March 24, 1989, when the tanker *Exxon Valdez* struck a reef in Alaska's Prince William Sound and leaked 44 million liters of crude oil. During the war with Iraq, 120 million liters were released from damaged onshore storage tanks into the Persian Gulf—an event that was called environmental terrorism. Although petroleum is biodegradable, the process takes a long time because the low-nutrient content of seawater does not support a large bacterial population. Once the oil washes up onto beaches, it takes many hours of work and millions of dollars to clean it up.

Adequate sewage treatment and waste disposal are necessary to prevent the pollution of rivers and oceans. New methods also are needed to prevent pollution of underground water supplies.

Air Pollution

The atmosphere has 2 layers, the stratosphere and the troposphere. The stratosphere is a layer that lies 15 km–50 km above the surface of the earth. Here, the energy of the sun splits some molecular oxygen (O_2) molecules. These individual oxygen (O) atoms then combine with intact molecular oxygen to give ozone (O_3). This ozone layer is called a shield because it absorbs the ultraviolet rays of the sun, preventing them from striking the earth. If these rays did penetrate the atmosphere, life on earth would not be possible because living things cannot tolerate heavy doses of ultraviolet radiation. The troposphere is the atmospheric layer closest to the earth's surface; it ordinarily contains the gases nitrogen (N_2)—78%—oxygen (O_2)—21%—and carbon dioxide (CO_2)—0.3%.

Figure 50.5

Sources of groundwater pollution. Discontinuance of these means of industrial waste disposal has been difficult to achieve because citizens do not want to have waste disposal plants located near them. Source: U.S. Environmental Protection Agency, Office of Water Supply and Solid Waste Management Programs, "Waste Disposal Practices and Their Effects on Ground Water" in *Executive Summary*. U.S. Government Printing Office, Washington, DC, 1977.

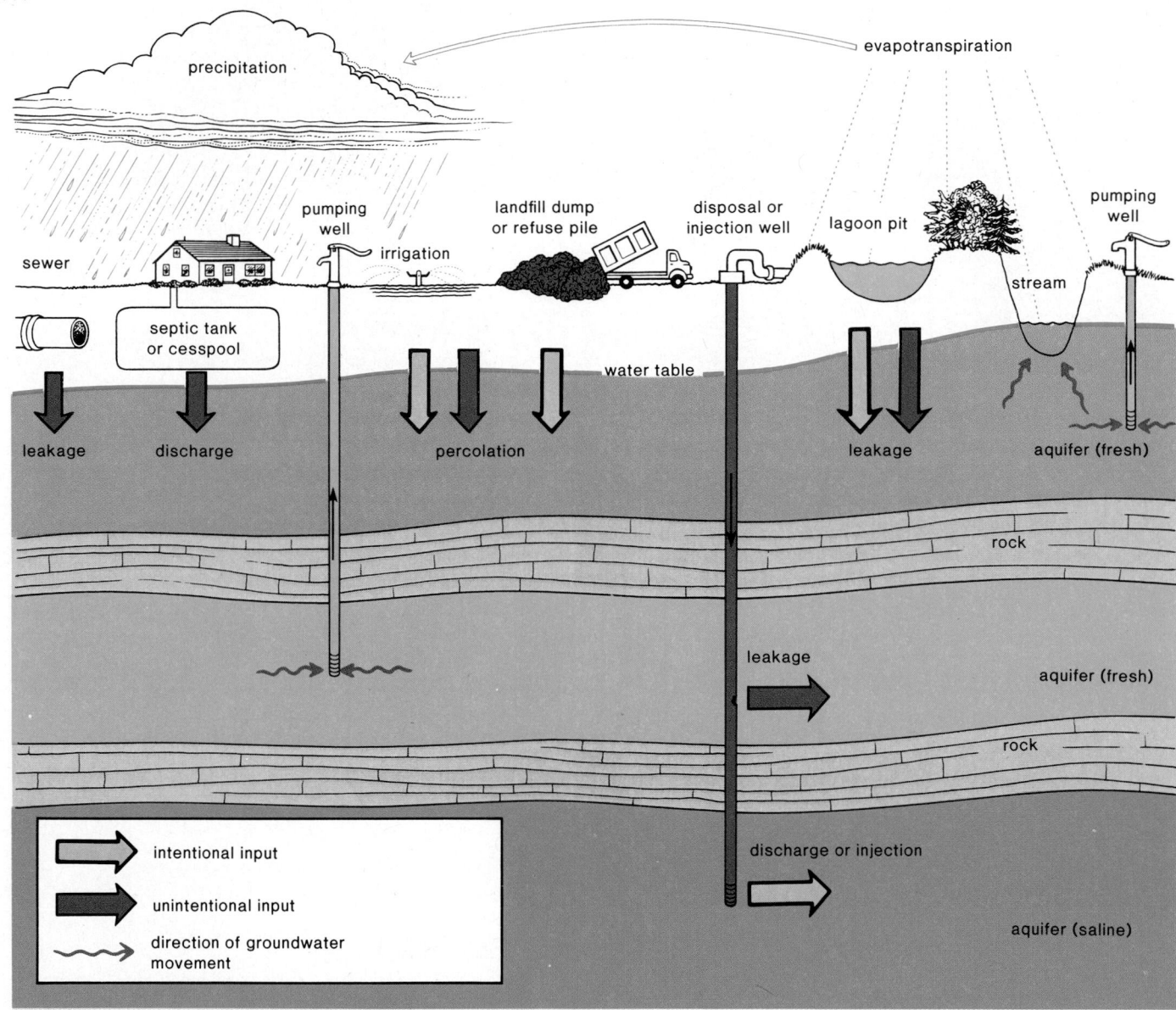

Outdoor Air Pollution

Four major concerns (photochemical smog, acid deposition, global warming, and destruction of the ozone shield) are associated with the air pollutants listed in figure 50.6. You can see that fossil fuel burning and vehicle exhaust are primary causes of air pollution gases. These 2 are related because gasoline is derived from petroleum (oil), a fossil fuel. The fossil fuels (oil, coal, and natural gas) are burned in the home to provide heat and in power

plants to generate electricity. In other words, civilization's need for energy creates the major ecological problems we will be discussing.

Photochemical Smog

Photochemical smog contains 2 air pollutants—nitrogen oxides (NO_x) and hydrocarbons (HC)—that react with one another in the presence of sunlight to produce ozone (O_3) and PAN (peroxylacetyl nitrate). Both nitrogen oxides and hydrocarbons

Figure 50.6

Air pollutants. These are the gases, along with their sources, that contribute to 4 environmental effects of major concern: photochemical smog, acid deposition, the greenhouse effect, and ozone shield destruction. An examination of the sources of these gases shows that vehicle exhaust and fossil fuel burning are the chief contributors.

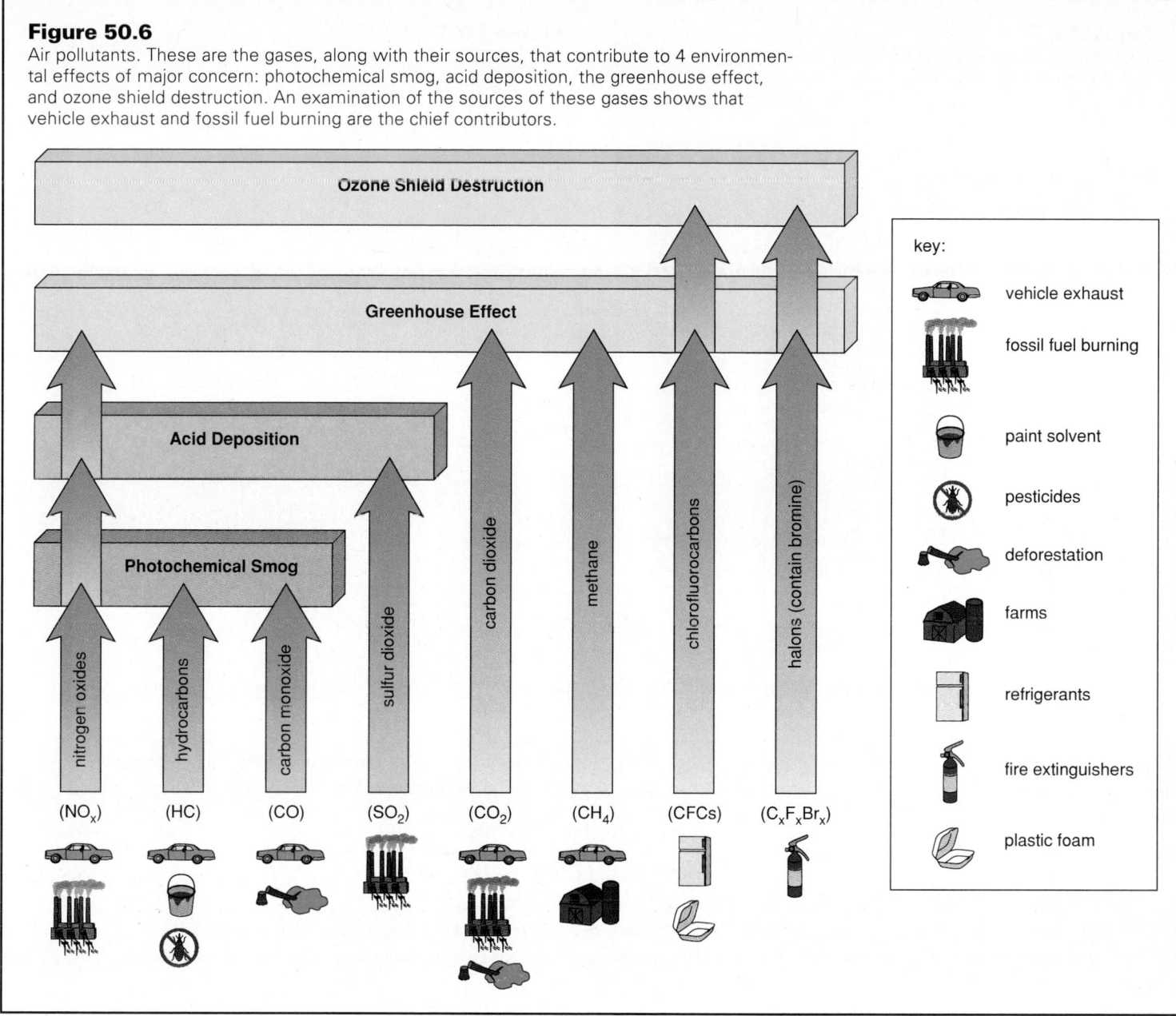

come from fossil fuel combustion, but additional hydrocarbons come from various other sources as well, including paint solvents and pesticides.

Ozone and PAN are commonly referred to as oxidants. Breathing ozone affects the respiratory and nervous systems, resulting in respiratory distress, headache, and exhaustion. These symptoms are particularly apt to appear in young people; therefore, in Los Angeles, where ozone levels are often high, schoolchildren must remain inside the school building whenever the ozone level reaches 0.35 ppm (parts per million by weight). Ozone is especially damaging to plants, resulting in leaf mottling and reduced growth (fig. 50.7).

Carbon monoxide (CO) is another gas that comes from the burning of fossil fuels in the industrial Northern Hemisphere. High levels of carbon monoxide increase the formation of ozone. Carbon monoxide also combines preferentially with hemoglobin and thereby prevents hemoglobin from carrying oxygen. Breathing large quantities of automobile exhaust can even result in death because of this effect. Of late, it has been discovered that the amount of carbon monoxide over the Southern Hemisphere is equal to the amount over the Northern Hemisphere. Here it comes from the burning of tropical forests.

Normally, warm air near the ground is able to escape into the atmosphere. Sometimes, however, air pollutants, including smog and soot, are trapped near the earth due to a long-lasting

Figure 50.7
The milkweed in (**a**) was exposed to ozone and appears unhealthy—leaf mottling is apparent. The milkweed in (**b**) was grown in an enclosure with filtered air and appears healthy.

a.   b.

Figure 50.8
Acid deposition in the United States. The numbers in this map are average pH recordings of deposition in parts of the United States. The northeast is the hardest hit area because the winds carry the acid pollutants from other parts of the country in this direction.

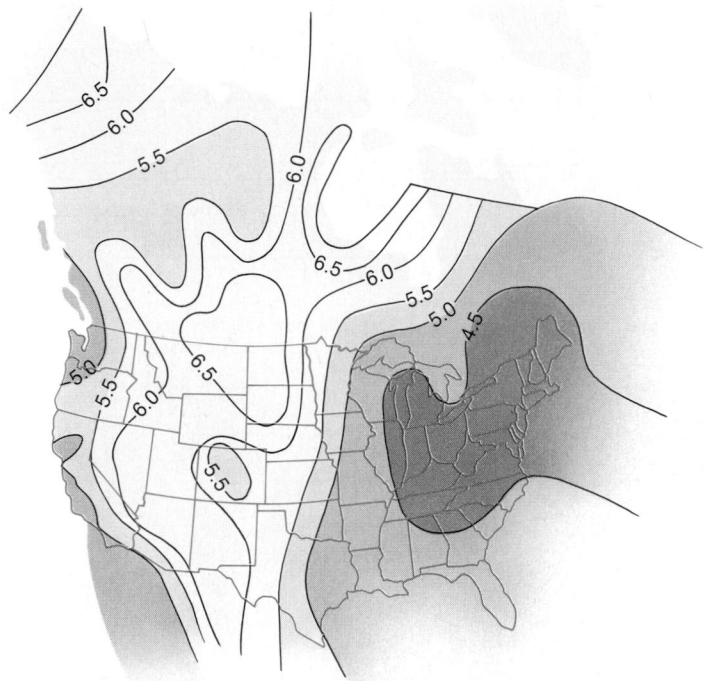

thermal inversion. During a **thermal inversion,** there is the cold air at ground level beneath a layer of warm stagnant air above. This often occurs at sunset, but turbulence usually mixes these layers during the day. Some areas surrounded by hills are particularly susceptible to the effects of a temperature inversion because the air tends to stagnate and there is little turbulent mixing.

Acid Deposition

The coal and oil routinely burned by power plants release sulfur dioxide into the air. Kuwait oil has a high sulfur content, and therefore, the oil well fires started during the Gulf War released much sulfur dioxide into the atmosphere. Automobile exhaust routinely puts nitrogen oxides in the air. Both sulfur dioxide and nitrogen oxides are converted to acids when they combine with water vapor in the atmosphere, a reaction that is promoted by ozone in smog. These acids return to earth as either wet deposition (acid rain or snow) or dry deposition (sulfate and nitrate salts).

As discussed in the reading on page 37, **acid deposition** is now associated with dead or dying lakes and forests, particularly in North America and Europe (fig. 50.8). Acid deposition also corrodes marble, metal, and stonework, an effect that is noticeable in cities. It can also degrade our water supply by leaching heavy metals from the soil into drinking-water supplies. Similarly, acid water dissolves copper from pipes and from lead solder, which is used to join pipes.

Global Warming

Certain air pollutants allow the sun's rays to pass through but then absorb and reradiate the heat back toward the earth (fig. 50.9*a*). This is called the **greenhouse effect** because the glass of a greenhouse allows sunlight to pass through and then traps the resulting heat inside the structure. The air pollutants responsible for the greenhouse effect are known as the greenhouse gases. They are as follows:

Carbon dioxide (CO_2) from fossil fuel and wood burning
Nitrous oxide (NO_2), primarily from fertilizer use and animal wastes
Methane (CH_4) from biogas, bacterial decomposition, particularly in the guts of animals, sediments, and flooded rice paddies
Chlorofluorocarbons (CFCs) from Freon, a refrigerant
Halons (halocarbons, $C_xF_xBr_x$) from fire extinguishers

If nothing is done to control the level of greenhouse gases in the atmosphere, a rise in global temperature is expected. Figure 50.9 predicts a rise of over .05°C by 2020, but some authorities predict the rise in temperature could be as high as 5°C by 2050. It's possible that global warming has already begun; the 4 hottest years on record have occurred during the 1980s: there has been greater warming in the winters than in the summers; there has been greater warming at high latitudes than near the equator; the stratosphere is cooler and the lower atmosphere is warmer than before. All of these effects had been predicted by computer models of the greenhouse effect.

The ecological effects of a 5°C rise in global temperature would be severe. The sea level would rise—melting of the polar ice caps would add more water to the sea, and in any case water expands when it heats up. There would be coastal flooding and the

Figure 50.9

Greenhouse effect. ***a.*** The greenhouse effect is caused by the accumulation of certain gases, such as CO_2, in the atmosphere. These gases allow the rays of the sun to pass through but absorb and reradiate heat back to the earth. ***b.*** The greenhouse gases. This graph shows the fraction of warming caused by the gases carbon dioxide (CO_2), chlorofluorocarbons (CFCs), methane (CH_4), and nitrous oxide (NO_2) for the decades from 1950 to 2020. There was no accumulation of CFCs in the 1950s because they were not being manufactured to any degree then. By 2020, the CO_2 and the other gases combined will each contribute about 50% to the projected global warming.

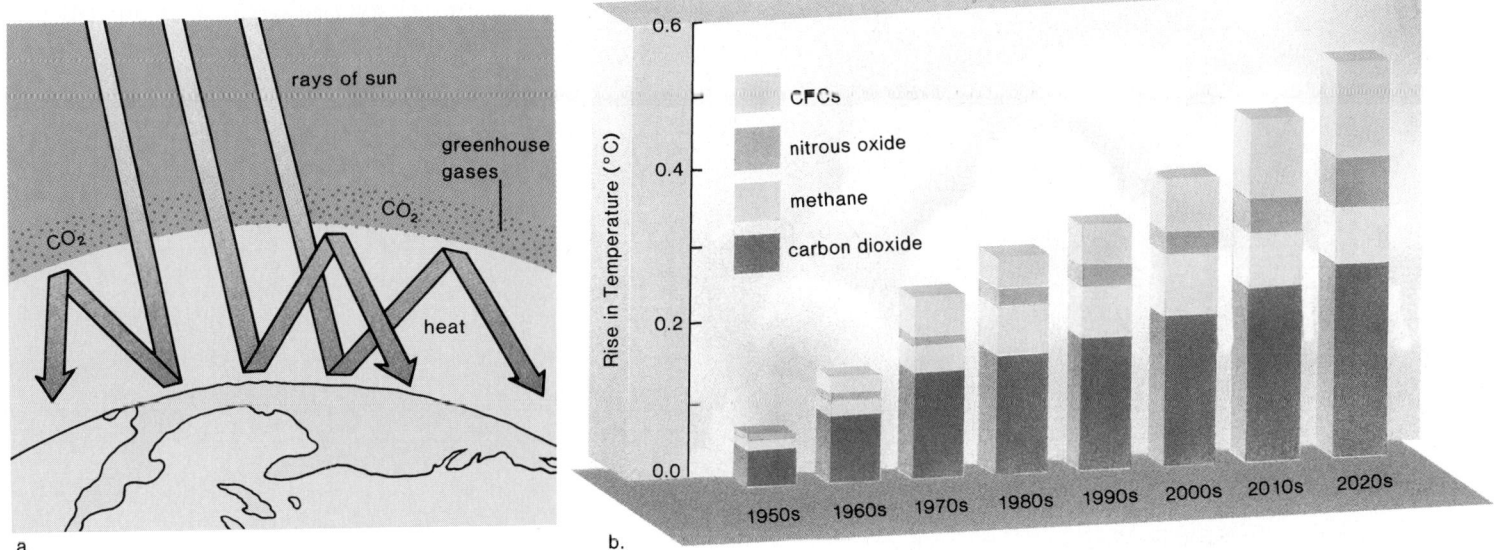

possible loss of many cities, like New York, Boston, Miami, and Galveston in the United States. Coastal ecosystems, such as marshes, swamps, and bayous, would normally move inland to higher ground as the sea level rises, but many of these ecosystems are blocked by artificial structures and may be unable to move inland. If so, the loss of fertility would be immense.

There may also be food loss because of regional changes in climate. Because of greater heat and also drought in the midwestern United States, the suitable climate for growing wheat and corn may shift as far as Canada, where the soil is not as suitable.

It is clear from figure 50.9*b* that carbon dioxide accounts for at least 50% of the predicted rise in global temperature. Therefore, a sharp decrease in the consumption of fossil fuels is recommended, and we must find more efficient ways to acquire energy from cleaner fuels, such as natural gas. We must also use alternative energy sources, such as solar and geothermal energy and even perhaps nuclear power, more aggressively. In addition to fossil fuel consumption, deforestation is a major contributor to the rise in carbon dioxide. Burning one acre of primary forest releases 200,000 kg of carbon dioxide into the air; moreover, the trees are no longer available to act as a sink to take up carbon dioxide during photosynthesis. Therefore, tropical rain forest deforestation should be halted and extensive reforestation all over the globe should take place.

The other greenhouse gases combined account for the other 50% predicted rise in global temperature. A complete phase-out of chlorofluorocarbon use would be most beneficial. Fortunately, in an effort to arrest ozone shield destruction, the United States and the European countries have agreed to reduce CFC production by 85% as soon as possible and to stop their production altogether by the end of the century.

Ozone Shield Destruction

Ozone shield destruction is primarily caused by chlorofluorocarbons (CFCs) and halons (halocarbons, $C_xF_xBr_x$). CFCs are heat-transfer agents used in refrigerators and air conditioners. They are also used as foaming agents in such products as styrofoam cups and egg cartons. Formerly, they were used as propellants in spray cans, but this application is now banned in the United States. Halons are antifire agents used in fire extinguishers.

Scientists knew that CFCs would drift up into the stratosphere, but it was believed that they would be nonreactive there during their 150-year life span. It is now apparent, however, that when the temperature drops, these compounds react chemically on frozen particle surfaces within stratospheric clouds and release active chlorine, which can react with the ozone in the **ozone shield.** (Similarly, halons release bromine, which reacts with ozone.) Once freed, a single atom of chlorine destroys about 100,000 molecules of ozone before settling to the earth's surface as chloride years later. At the South Pole of the Antarctic, up to 60% of the ozone is destroyed each spring (fig. 50.10). There is similar reduction called an *ozone hole,* but not as severe, over the North Pole of the Arctic as well. Also, it now appears that northern midlatitudes (roughly between Seattle and New Orleans) are losing ozone at the rate of 4%–5% per decade. This is twice as fast as had been originally assumed.

Figure 50.10

Ozone depletion has been confirmed by observations of a drastic decrease in ozone over Antarctica each spring. The so-called ozone hole appears as a large lavender area at the center of this picture, taken by the total ozone mapping spectrometer aboard NASA's *Nimbus 7* satellite on 5 October 1987, when the ozone loss reached nearly 60%.

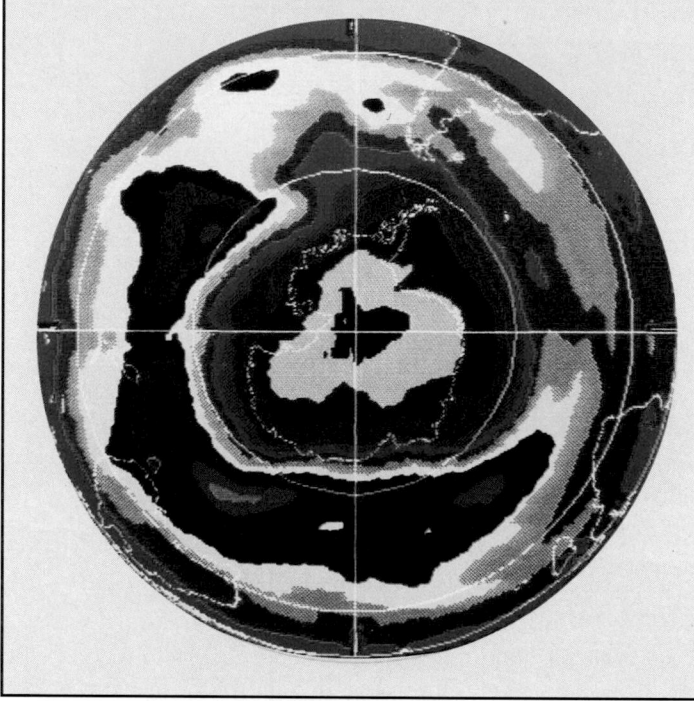

Table 50.2
Renewable and Nonrenewable Energy Sources

	Advantages	Disadvantages
Nonrenewable	*Technology Well Established*	*Finite Fuel Supply*
Fossil fuels		
Coal	Plentiful supply	Surface mining
		Air and water pollution
Petroleum	Cleaner burning	Limited supply
Natural gas	Cleanest burning	Limited supply
Nuclear		
Fission	Fuel availability	Radiation pollution
Renewable	*Fuel Supply*	*Technology under Development*
Nuclear		
Breeder	Fuel availability	Radiation pollution
		Nuclear weapons proliferation
Fusion	Fuel availability	Radiation pollution
Geothermal	Less pollution	Availability limited
Solar and wind	Nonpolluting Large and small scale possible	Expensive
Ocean	Nonpolluting	Applicable only in certain areas
Biomass	Utilizes wastes	Air and water pollution

Depletion of the ozone layer will allow more ultraviolet rays to enter the troposphere. The incidence of human cancer, especially skin cancer, is expected to increase, and plants and animals living in the top microlayer of the ocean will begin to die. Increased ultraviolet radiation also will hasten the rate at which smog is formed. Many believe that the current plans to reduce CFCs and halon emissions are not sufficient, and more stringent measures are called for.

Outdoor air pollutants are involved in causing 4 major environmental effects: photochemical smog, acid deposition, detrimental global warming, and ozone shield destruction. Each pollutant may be involved in more than one of these.

While each of these environmental effects is bad enough when considered separately, they actually feed on one another, making the total effect much worse than is predicted for each separately.

Indoor Air Pollution

While much attention has been given to the quality of outdoor air, indoor air is often more polluted. Nitrous oxide (NO_2) traced to gas combustion in stoves has been found indoors at twice the outdoor level; carbon monoxide (CO) in offices, repair garages, and hockey rinks is routinely in excess of the governmental standard, and hydrocarbons from myriad sources appear in high concentrations. Radioactive radon gas, emitted naturally from radioactive decay in uranium-bearing materials in the earth, as well as those incorporated into such building materials as concrete blocks, has been detected indoors at levels that exceed those outdoors by factors of 2-20.

Because of these findings, some have questioned the advisability of tightening up residential buildings to save energy. Reduced ventilation unquestionably increases concentrations of pollutants. The EPA is now recommending that ventilation rates not be reduced below one complete air change per hour.

Energy Choices

The threat of global warming due to the buildup of gases that bring about the greenhouse effect has renewed the desire to use sources of energy that will not add carbon dioxide to the atmosphere. The energy sources currently available or being investigated are listed in table 50.2. **Nonrenewable resources** are those with a supply that can be used up or exhausted. **Renewable resources** are those that can be replenished by either physical or biological means.

Figure 50.11

Present energy sources in the United States. Oil became the favored fuel during the 1900s because it is cleaner burning than coal and because it can be converted to gasoline. In recent years, use of coal has increased because it is more plentiful than oil and natural gas in the United States, but it may fall into disfavor once again because SO_2 from coal contributes most heavily to air pollution acid deposition. There is also much concern about the possibility of global warming because of the CO_2 produced by burning coal.

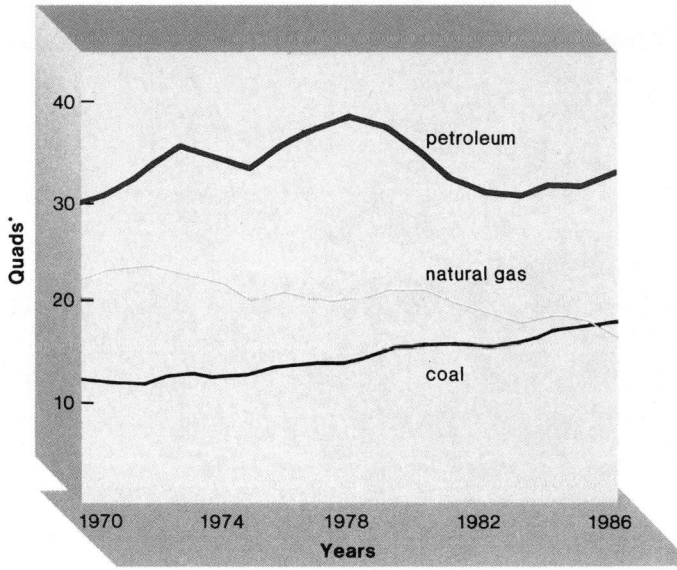

*Quadrillion BTUs, the same amount of energy as in 8 billion gallons of gasoline.

Nonrenewable Energy Sources

Although it is difficult to realize, supplies of the cleaner-burning petroleum and natural gas are not expected to last more than another 30 years. Petroleum became the favored fuel in the United States during the 1900s not only because it is cleaner burning than coal, but also because it is transported easily and because it is the raw material for gasoline and many organic compounds (fig. 50.11).

Although coal has been projected to be the primary energy source in the United States in the future, this is now doubtful because it contributes more than other fossil fuels to sulfur dioxide pollution. Also, to supply coal, thousands of acres of land are strip-mined, and this results in huge piles of residue with the potential to leach acids and hazardous chemicals into underground aquifers.

While all fossil fuels release carbon dioxide in addition to other pollutants, **nuclear power** does not contribute to air pollution. Thus far, only fission power plants, which split [235]uranium, have been utilized in the United States. In recent years, the nuclear power industry has declined in favor because coal-fired power plants are cheaper to build and because the public is concerned about nuclear power dangers, such as the meltdown that occurred in 1986 at the Chernobyl nuclear power plant in Russia. At a proposed plant to be built in Idaho, designers plan to use a series of 4 small-scale modular reactors, which would use fuel in such small

quantities their cores could not achieve meltdown temperatures. The fuel would be packed inside tiny heat-resistant ceramic spheres and cooled by inert helium gas.

Even if the possibility of a meltdown can be eliminated, radioactive wastes that have accumulated since the start of the nuclear age must still be stored. Another source of radioactive waste, not often mentioned, occurs when uranium is mined. Uranium mill tailings are the sand that remains after uranium has been removed from mined material. The tailings contain [226]radium, which has a half-life[1] of 1,620 years. As [226]radium decays, it gives off radon gas, which has been associated with the development of lung cancer.

Nuclear fusion requires the fusion of 2 atoms, usually deuterium and tritium. The energy needed for this is so great that there is no conventional container for the reaction. Scientists have been perfecting laser-beam ignition and magnetic containment for the reaction, which thus far uses up more energy than it produces. Since the fusion reaction gives off neutrons that can change uranium to plutonium, the best use of fusion plants may be to provide fuel for breeder reactors. One idea is to develop hybrid fusion plants, which would combine both fusion and breeder fission plants. But, in any case, the fusion process is still experimental and is not expected to be ready for production any time soon.

Renewable Energy Sources

Solar energy is diffuse energy, which must be collected and concentrated before utilization. Solar heating has been used primarily in homes and offices. Passive solar systems consist of specially constructed glass for the south wall of a building and building materials that can collect the sun's energy during the day and release it at night. Solar collectors placed on rooftops absorb radiant energy but do not release the resulting heat. A fluid within the solar collector heats up and is pumped to other parts of the building for space heating, cooling, or electricity generation.

Another possibility is the use of photovoltaic (solar) cells, which produce electricity directly from sunlight. However, there is great environmental impact because of mining for the rare metals used in their construction.

Solar energy does not produce air or water pollution, nor does it add heat to the atmosphere, since radiant energy is eventually converted to heat anyway. The problem of storage can be overcome in a number of ways, including the use of solar energy to produce hydrogen by means of hydrolysis of water. Hydrogen as either a gas or a liquid can be piped into existing pipelines and used as fuel for automobiles and airplanes. When it is burned, it forms fog, not smog:

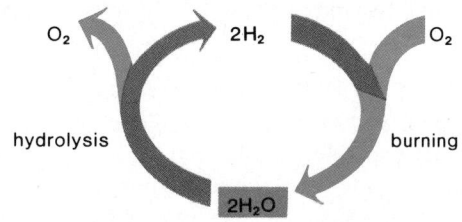

[1]Length of time required for one-half of the radiation to dissipate.

Figure 50.12

In the steady state, environmental preservation will be an important consideration.

Certain types of renewable energy sources have been utilized for quite some time. Falling water is used to produce electricity in hydroelectric plants. Geothermal energy is derived from the heat of the magma in the earth's core. Water converted to steam by this heat can be pumped up and used to heat buildings or to generate electricity. Wind power provides enough force to turn vanes, blades, or propellers attached to shafts, which, in turn, spin generator motors that produce electricity. The government has allocated a small amount of money for wind research, particularly the promotion of large windmills, such as those now in operation in central and southern California.

> The threat of global warming makes it imperative to use a source of energy that does not pollute the atmosphere. Nuclear power and several of the renewable sources such as solar and wind power are possibilities.

The Steady State

A stable population size would be a new experience for humans. Such a population has many benefits, as discussed here:

1. Over 40% of the people would be fairly youthful, with only about 15% being in the senior citizen category. There would be proportionately fewer children and teenagers than in a rapidly expanding population.
2. The quality of life for children might increase substantially since fewer unwanted babies might be born and the opportunity would exist for children to receive the loving attention each needs.
3. There might be increased employment opportunities for women and a generally less competitive workplace because newly qualified workers would enter the job market at a more moderate rate.
4. Creativity need not be impaired. A study of Nobel Prize winners showed that the average age at which the prize-winning work was done was over 30.

Environmental preservation would be the most important consideration in a steady-state world. Renewable energy sources, such as solar energy, would play a greater role in meeting energy needs. Pollution would be minimized. Ecological diversity would be maintained, and overexploitation would cease (fig. 50.12). Ecological principles would serve as guidelines for specialists in all fields, creating a unified approach to the environment. In a steady-state world, all people would strive to be aware of the environmental consequences of their actions, consciously working toward achieving balance in the ecosystems of their planet.

In a **steady-state society,** there would be no yearly increase in population or resource consumption. It is forecast that under these circumstances, the quality of life would improve.

What would our culture be like if we had steady-state manufacturing and a steady-state population? Perhaps it would be greatly improved. Certainly, there are no limits to growth in knowledge, education, art, music, scientific research, human rights, justice, and cooperative human interactions. In a steady-state world, the general sense of fearful competition among peoples might diminish, allowing human compassion and creativity to prosper as never before.

Behavior and Ecology

Summary

1. An increasing human population is causing land, water, and air pollution.
2. Most people of the world now live in cities that are increasing in size and taking over land formerly used for agriculture and other purposes.
3. Soil erosion reduces the quality of land and leads to desertification.
4. The tropical rain forests in Southeast Asia and Oceania, Central and South America, and Africa are being reduced in size for several reasons: (1) to provide wood for export, (2) slash-and-burn agriculture, and (3) cattle ranching. Industrialization may soon be a fourth reason.
5. The loss of biological diversity due to the destruction of tropical rain forests will be immense. Many of these threatened organisms could possibly be of benefit to humans if we had time to study and domesticate them.
6. Solid wastes, including hazardous wastes, are deposited on land. The latter, including metals, organochlorides, and nuclear wastes, may contaminate water supplies. The oceans are the final recipients of wastes deposited in rivers and along the coasts.
7. Two layers of the atmosphere are of interest: (1) the troposphere, where ozone is a pollutant; and (2) the stratosphere, where ozone forms a shield against ultraviolet radiation.

8. Various substances are associated with air pollution, such as sulfur dioxide, hydrocarbons, nitrogen oxides, methane, CFCs, carbon dioxide, and carbon monoxide. These have various sources but most come from vehicle exhaust and fossil fuel burning.
9. Hydrocarbons and nitrogen oxides (NO_x) react to form smog, which contains ozone and PAN. These oxidants are harmful to animal and plant life.
10. Sulfur dioxide (SO_2) and nitrogen oxide react with water vapor to form acids that contribute to acid deposition. Acid deposition is killing lakes and forests and also corrodes marble, metal, and stonework.
11. Carbon dioxide, nitrous oxide, methane, and CFCs trap solar heat just like the panes of a greenhouse, leading to the so-called greenhouse effect. It is predicted that a buildup in these "greenhouse gases" will lead to a global warming. The effects of global warming could be a rise in sea level and a change in climate patterns. Food shortages could follow.
12. To control global warming, it is recommended that we reduce our consumption of fossil fuel energy and halt tropical rain forest destruction. Also, we should increase the amount of land devoted to forests. A ban on the manufacture of CFCs has already begun.

13. Ozone shield destruction is particularly associated with CFCs. CFCs rise into the stratosphere and react on frozen particles within clouds to release chlorine. Chlorine causes ozone to break down. A 60% reduction in ozone was found above the Antarctic and dubbed an "ozone hole." Since ozone prevents harmful ultraviolet radiation from reaching the surface of the earth, reduction in the amount of ozone will lead to skin cancer in humans and decreased productivity of the oceans.
14. A review of the energy sources available indicates that cleaner burning petroleum and natural gas are not expected to be available many more years. It has been suggested that nuclear power and renewable sources such as solar and wind power are the best possibilities to prevent global warming.

Writing Across the Curriculum

In order to practice writing skills, students should write out the answers to any or all of the study questions and the critical thinking questions. The study questions are sequenced in the same order as the text. Suggested answers to the critical thinking questions are in appendix D.

Study Questions

1. Give 3 reasons why the quality of the land is being degraded today. What is desertification?
2. Give 3 reasons why tropical rain forests are being destroyed. What is another possible reason in the future?
3. What are the primary ecological concerns associated with the destruction of rain forests?
4. What are the 3 types of hazardous wastes that contribute to pollution on land?

5. What are several ways in which underground water supplies can be polluted?
6. What substances contribute to air pollution? What are their sources? Which ones are associated with photochemical smog, acid deposition, global warming, and destruction of the ozone shield?
7. How does smog develop? How do acids develop in the atmosphere? How do CFCs break down the ozone shield?

How do the greenhouse gases bring about global warming?
8. What are the environmental effects of global warming, and what can be done to prevent global warming?
9. What are the various energy sources available to humans? What are the drawbacks of each one? Which of these would not contribute to the greenhouse effect?
10. Describe the characteristics of a human society that is in a steady state.

Objective Questions

For questions 1–4, match the terms with those in the key:

Key:

  a. sulfur dioxide
  b. ozone
  c. carbon dioxide
  d. CFCs

1. acid deposition
2. ozone shield destruction
3. greenhouse effect
4. photochemical smog
5. Which of these is a true statement?
   a. More developed countries do not contribute to the destruction of tropical rain forests.
   b. Carbon dioxide from fossil fuel combustion is the primary greenhouse gas.
   c. Water from underground sources is never subject to contamination.
   d. Sulfur dioxide is given off to the same degree by the burning of all fossil fuels.

6. Which of these is a true statement?
   a. Global warming is of no immediate concern.
   b. Global warming is so imminent that nothing can be done.
   c. Reduction in fossil fuel burning will lessen the greenhouse effect.
   d. Since gases not derived from fossil fuel combustion are involved, reduction in fossil fuel burning will not help the greenhouse effect.

7. Tropical rain forest destruction is extremely serious because
   a. it will lead to a severe reduction in biological diversity.
   b. tropical soils cannot support agriculture for long.
   c. large tracts of forest absorb carbon dioxide, reducing the threat of global warming.
   d. All of these.

8. Which of these is mismatched?
   a. fossil fuel burning—carbon dioxide given off
   b. nuclear power—radioactive wastes

   c. solar energy—greenhouse effect
   d. biomass burning—carbon dioxide given off

9. Acid deposition causes
   a. lakes and forests to die.
   b. acid indigestion in humans.
   c. the greenhouse effect to lessen.
   d. All of these.

10. Water is a renewable resource, and
   a. there will always be a plentiful supply.
   b. the oceans can never become polluted.
   c. it is still subject to pollution.
   d. primary sewage treatment plants assure clean drinking water.

Concepts and Critical Thinking

1. *Tropical rain forests play a major role in the biosphere.*

   Tropical rain forest destruction might increase the net primary productivity of which biomes listed in figure 48.8?

2. *Air pollutants have far-reaching effects.*

   Use figure 50.6 to substantiate that air pollutants have combined effects.

3. *All ecosystems require an outside energy source.*

   Show that the size of the human population has been dependent upon the use of nonrenewable energy sources.

Selected Key Terms

pollutant (po-lut'ant) 793
desertification (dez"ert-i-fi-ka'shun) 793
slash-and-burn agriculture (slash and burn ag'ri-kul"tur) 795
photochemical smog (fo"to-kem'i-kal smog) 798

thermal inversion (ther'mal in-ver'zhun) 800
acid deposition (as'id dep"o-zish'un) 800
greenhouse effect (gren'hows e-fekt) 800
deforestation (de"for-es-ta'shun) 801
ozone shield (o'zon sheld) 801
nonrenewable resource (non"re-nu'ah-b'l re-sors') 802

renewable resource (re-nu'ah-b'l re-sors') 802
nuclear power (nu'kle-ar pow'er) 803
solar energy (so'lar en'er-je) 803
steady-state society (sted'e stat so-si'e-te) 804

CRITICAL THINKING CASE STUDY

Deep-Sea Ecosystems

Researchers have found a great diversity of organisms packed around occasional thermal vents 2,600 m below the surface of the ocean. There are giant tubeworms, clams, mussels, shrimps, and crabs. Most ecosystems have an outside energy source, producers, consumers, and decomposers (see fig. 49.1). Therefore, the researchers put forward the hypothesis that *these deep-sea assemblages of organisms are ecosystems that will have these same elements.*

Initial investigations revealed that thermal vents release large amounts of hydrogen sulfide (H_2S). It's well known that some chemosynthetic bacteria have the capability of oxidizing sulfur, with the concomitant buildup of ATP molecules. These ATP molecules are then used to drive the Calvin cycle, so that carbohydrates are produced. Do you predict that chemosynthetic bacteria capable of oxidizing hydrogen sulfide will be found in the vicinity of the thermal vents?

Prediction 1 Bacteria will be found that can use hydrogen sulfide as an energy source.

Result 1 Bacteria that can utilize hydrogen sulfide as an energy source have been found near the vents.

I What does the term *producer* mean? Is it possible that these sulfur-oxidizing bacteria are the producers in deep-sea ecosystems? Is hydrogen sulfide equivalent to the sun in deep-sea ecosystems? Why or why not?

Do you predict that there will be consumers that feed on the sulfur-oxidizing bacteria if this is a deep-sea ecosystem?

Prediction 2 There will be consumers that feed on the sulfur-oxidizing bacteria.

Result 2 Contrary to the prediction, no organisms appear to utilize the sulfur-oxidizing bacteria as a major food source. Even the clams that normally feed on bacteria did not appear to feed on these bacteria.

Also, it was found that the large, 2-m-long tubeworms don't have a mouth or any other sign of a digestive system! Microscopic examination of the tubeworm body, however, indicated that the internal cells contain enormous numbers of a type of sulfur-oxidizing bacteria. A similar examination of the clams revealed that another type of

sulfur-oxidizing bacteria had colonized the gills of these organisms. This suggested that there is a mutualistic relationship between the bacteria and their hosts.

II In a mutualistic relationship, each organism contributes to the livelihood of the other. What could the bacteria be contributing to their hosts?
Design an experiment that would test the hypothesis that tubeworms and clams are dependent on sulfur-oxidizing bacteria to supply their nutritional needs.
What could the hosts be contributing to the bacteria?
Since the bacteria are deep inside the tissues of the worms, what system in the body of the tubeworms would be transporting H_2S to the bacteria?

What do you predict about the circulatory system of the tubeworms?

Prediction 3 The circulatory system of a tubeworm is able to transport H_2S to the bacteria.

Result 3 It was discovered that the hemoglobin of tubeworms, which is in plasma and not in red blood cells, is a very special type—it contains a binding site for H_2S. In the clam, where hemoglobin is in red blood cells, there is a special transport protein that transports H_2S to the bacteria.

III What does the term *consumer* mean? Should the tubeworm and clams, which harbor sulfur-oxidizing bacteria, be considered consumers? Why or why not?
Can we assume that decomposers are present at thermal vents?
Does it seem, then, that our title—deep-sea ecosystem—is suitable?

The hemoglobin of tubeworms transports both hydrogen sulfide and oxygen but in such a way that hydrogen sulfide is not oxidized before it reaches the bacteria. Hydrogen sulfide is a powerful inhibitor of cytochrome oxidase, equivalent to that of cyanide. Do you predict the hemoglobin of tubeworms keeps hydrogen sulfide from diffusing into mitochondria where it would poison cytochrome oxidase?

Prediction 4 The hemoglobin is able to prevent hydrogen sulfide from diffusing into mitochondria.

Result 4 If tubeworm hemoglobin is added to a test tube containing cytochrome oxidase and hydrogen sulfide, cytochrome oxidase is not inhibited. This shows that tubeworm hemoglobin has a higher affinity for hydrogen sulfide than does cytochrome oxidase.

IV How would the hemoglobin's "higher affinity for hydrogen sulfide" protect cytochrome oxidase from inhibition by hydrogen sulfide?
What do you predict about the bacteria's affinity for H_2S?
Rank order from highest to lowest the expected affinity of hemoglobin, cytochrome oxidase, and bacteria for hydrogen sulfide in the body of the worm.

These observations and experiments show how hydrogen sulfide provides usable energy for deep-sea ecosystems. In spite of a markedly different environment, the ecosystem is characterized by energetic relationships and a biotic structure similar to those found in other ecosystems.

Childress, J. J., Felbeck, H., and Somero, G. N. May 1987. Symbiosis in the deep sea. *Scientific American.*

Jannasch, H. W., and Mottl, M. J. August 1985. Geomicrobiology of deep-sea hydrothermal vents. *Science.*

Other Questions

1. Some animals in the abyssal zone (see fig. 48.5) feed on detritus, which rains down from above. These organisms belong to an ecosystem that has what outside energy source? Why?
2. A parasite of giant tubeworms would be which of these: a producer, a consumer, or a decomposer? Why?

Suggested Readings for Part 6

Alcock, J. 1988. *Animal behavior: An evolutionary approach.* 4th ed. Sunderland, Mass.: Sinauer, Associates.

Bazzaz, F. A., and Fajer, E. D. January 1992. Plant life in a CO_2 rich world. *Scientific American.*

Begon, J., Harper, J., and Townsend, C. 1986. *Ecology: individuals, populations, and communities.* Sunderland, Mass.: Sinauer, Associates.

Berner, R. A., and Lasaga, A. C. March 1989. Modeling the geochemical carbon cycle. *Scientific American.*

Brown, L. R., et al. 1992. *State of the world: 1992*. New York: W. W. Norton and Co.

Colinvaux, P. A. May 1989. The past and future Amazon. *Scientific American.*

Davies, N. B., and Brooke, M. January 1991. Coevolution of the cukoo and its hosts. *Scientific American.*

Ghiglieri, M. P. June 1985. The social ecology of chimpanzees. *Scientific American.*

Holloway, M., and Horgan, J. October 1991. Soiled shores. *Scientific American.*

Horn, M. H., and Gibson, R. N. January 1988. Intertidal fishes. *Scientific American.*

Houghton, R. A., and Woodwell, G. M. April 1989. Global climatic change. *Scientific American.*

Jones, P. D., and Wigley, T. M. L. August 1990. Global warming trends. *Scientific American.*

Lohmann, K. J. January 1992. How sea turtles navigate. *Scientific American.*

Miller, J. T. 1991. *Living in the environment.* 6th ed. Belmont, Calif.: Wadsworth.

Mohnen, V. A. August 1988. The challenge of acid rain. *Scientific American.*

Nebel, B. J. 1987. *Environmental science: The way the world works.* 2d ed. Englewood Cliffs, N.J.: Prentice Hall.

Newell R. E., Reichle, H. G., Jr., and Seiler, W. October 1989. Carbon monoxide and the burning earth. *Scientific American.*

Odum, H. T. 1989. *Ecology: Our Endangered Life-Support Systems.* Sunderland, Mass.: Sinauer, Associates.

O'Leary, P. R., Walsh, P. W., and Ham, R. H. December 1988. Managing solid waste. *Scientific American.*

Reganold, J. P., Papendik, R. I., and Parr, J. F. June 1990. Sustainable agriculture. *Scientific American.*

Rennie, J. January 1992. Living together. *Scientific American.*

Repetto, R. April 1990. Deforestation in the tropics. *Scientific American.*

Ricklefs, R. E. 1989. *Ecology.* 4th ed. New York: Chiron Press.

Scientific American. September 1990. Energy for planet earth. Special Issue.

Scientific American. September 1989. Managing planet earth. Special Issue.

Smith, R. L. 1985. *Ecology and field biology.* 3d ed. New York: Harper & Row.

Strobel, G. July 1991. Biological control of weeds. *Scientific American.*

Sumich, J. 1992. *Biology of marine life.* 5th ed. Dubuque, Iowa: Wm. C. Brown Publishers.

White, R. M. July 1990. The great climate debate. *Scientific American.*

Wursig, B. April 1988. The behavior of baleen whales. *Scientific American.*

Classification of Organisms

The classification system given here is a simplified one, containing all the major kingdoms, as well as the major divisions (called phyla in the kingdom Protista and the kingdom Animalia).

Kingdom Monera

Prokaryotic, unicellular organisms. Nutrition principally by absorption, but some are photosynthetic or chemosynthetic.

Division Archaebacteria: methanogens, halophiles, and thermoacidophiles

Division Eubacteria: all other bacteria, including cyanobacteria (formerly called blue-green algae)

Kingdom Protista

Eukaryotic, unicellular organisms (and the most closely related multicellular forms). Nutrition by photosynthesis, absorption, or ingestion.

Phylum Sarcodina: amoeboid protozoans
Phylum Ciliophora: ciliated protozoans
Phylum Zoomastigina: flagellated protozoans
Phylum Sporozoa: parasitic protozoans
Phylum Chlorophyta: green algae
Phylum Pyrrophyta: dinoflagellates
Phylum Euglenophyta: *Euglena* and relatives
Phylum Chrysophyta: diatoms
Phylum Rhodophyta: red algae
Phylum Phaeophyta: brown algae
Phylum Myxomycota: slime molds
Phylum Oomycota: water molds

Kingdom Fungi

Eukaryotic organisms, usually having haploid or multinucleated hyphal filaments. Spore formation during both asexual and sexual reproduction. Nutrition principally by absorption.

Division Zygomycota: black bread molds
Division Ascomycota: sac fungi
Division Basidiomycota: club fungi
Division Deuteromycota: imperfect fungi (means of sexual reproduction not known).

Kingdom Plantae

Eukaryotic, terrestrial, multicellular organisms with rigid cellulose cell walls and chlorophylls *a* and *b*. Nutrition principally by photosynthesis. Starch is the food reserve.

Division Bryophyta: mosses and liverworts
Division Psilophyta: whisk ferns
Division Lycophyta: club mosses
Division Sphenophyta: horsetails
Division Pterophyta: ferns
Division Cycadophyta: cycads
Division Ginkgophyta: ginkgo
Division Gnetophyta: gnetae
Division Coniferophyta: conifers
Division Anthophyta: flowering plants
Class Dicotyledonae: dicots
Class Monocotyledonae: monocots

Kingdom Animalia

Eukaryotic, usually motile, multicellular organisms without cell walls or chlorophyll. Nutrition principally ingestive, with digestion in an internal cavity.

Phylum Porifera: sponges
Phylum Cnidaria: radially symmetrical aquatic animals
Class Hydrozoa: hydras, Portuguese man-of-war
Class Scyphozoa: jellyfishes
Class Anthozoa: sea anemones and corals
Phylum Platyhelminthes: flatworms
Class Turbellaria: free-living flatworms
Class Trematoda: parasitic flukes
Class Cestoda: parasitic tapeworms
Phylum Nematoda: roundworms
Phylum Rotifera: rotifers
Phylum Mollusca: soft-bodied, unsegmented animals
Class Polyplacophora: chitons
Class Monoplacophora: *Neopilina*
Class Gastropoda: snails and slugs
Class Bivalvia: clams and mussels
Class Cephalopoda: squids and octopuses
Phylum Annelida: segmented worms
Class Polychaeta: sandworms
Class Oligochaeta: earthworms
Class Hirudinea: leeches

Phylum Arthropoda: chiton exoskeleton, jointed appendages
  Class Crustacea: lobsters, crabs, barnacles
  Class Arachnida: spiders, scorpions, ticks
  Class Chilopoda: centipedes
  Class Diplopoda: millipedes
  Class Insecta: grasshoppers, termites, beetles
Phylum Onychophora: small, sluglike animals with legs; internal features of both annelids and arthropods
Phylum Echinodermata: marine; spiny, radially symmetrical animals
  Class Crinoidea: sea lilies and feather stars
  Class Asteroidea: sea stars
  Class Ophiuroidea: brittle stars
  Class Echinoidea: sea urchins and sand dollars
  Class Holothuroidea: sea cucumbers
Phylum Chordata: dorsal supporting rod (notochord) at some stage; dorsal hollow nerve cord; pharyngeal pouches or gill slits
  Subphylum Urochordata: tunicates
  Subphylum Cephalochordata: lancelets
  Subphylum Vertebrata: vertebrates
    Class Agnatha: jawless fishes (lampreys, hagfishes)
    Class Chondrichthyes: cartilaginous fishes (sharks, rays)
    Class Osteichthyes: bony fishes
      Subclass Dipnoi: lungfishes
      Subclass Crossopterygii: lobe-finned fishes
      Subclass Actinopterygii: ray-finned fishes
    Class Amphibia: frogs, toads, salamanders
    Class Reptilia: snakes, lizards, turtles
    Class Aves: birds
    Class Mammalia: mammals

Subclass Prototheria: egg-laying mammals
  Order Monotremata: duckbilled platypus, spiny anteater
Subclass Metatheria: marsupial mammals
  Order Marsupialia: opossums, kangaroos
Subclass Eutheria: placental mammals
  Order Insectivora: shrews, moles
  Order Chiroptera: bats
  Order Edentata: anteaters, armadillos
  Order Rodentia: rats, mice, squirrels
  Order Lagomorpha: rabbits, hares
  Order Cetacea: whales, dolphins, porpoises
  Order Carnivora: dogs, bears, weasels, cats, skunks
  Order Proboscidea: elephants
  Order Sirenia: manatees
  Order Perissodactyla: horses, hippopotamuses, zebras
  Order Artiodactyla: pigs, deer, cattle
  Order Primates: lemurs, monkeys, apes, humans
    Suborder Prosimii: lemurs, tree shrews, tarsiers, lorises, pottos
    Suborder Anthropoidea: monkeys, apes, humans
      Superfamily Ceboidea: New World monkeys
      Superfamily Cercopithecoidea: Old World monkeys
      Superfamily Hominoidea: apes and humans
        Family Hylobatidae: gibbons
        Family Pongidae: chimpanzees, gorillas, orangutans
        Family Hominidae: *Australopithecus,* * *Homo erectus,* * *Homo sapiens sapiens*

*extinct

APPENDIX B

Table of Chemical Elements

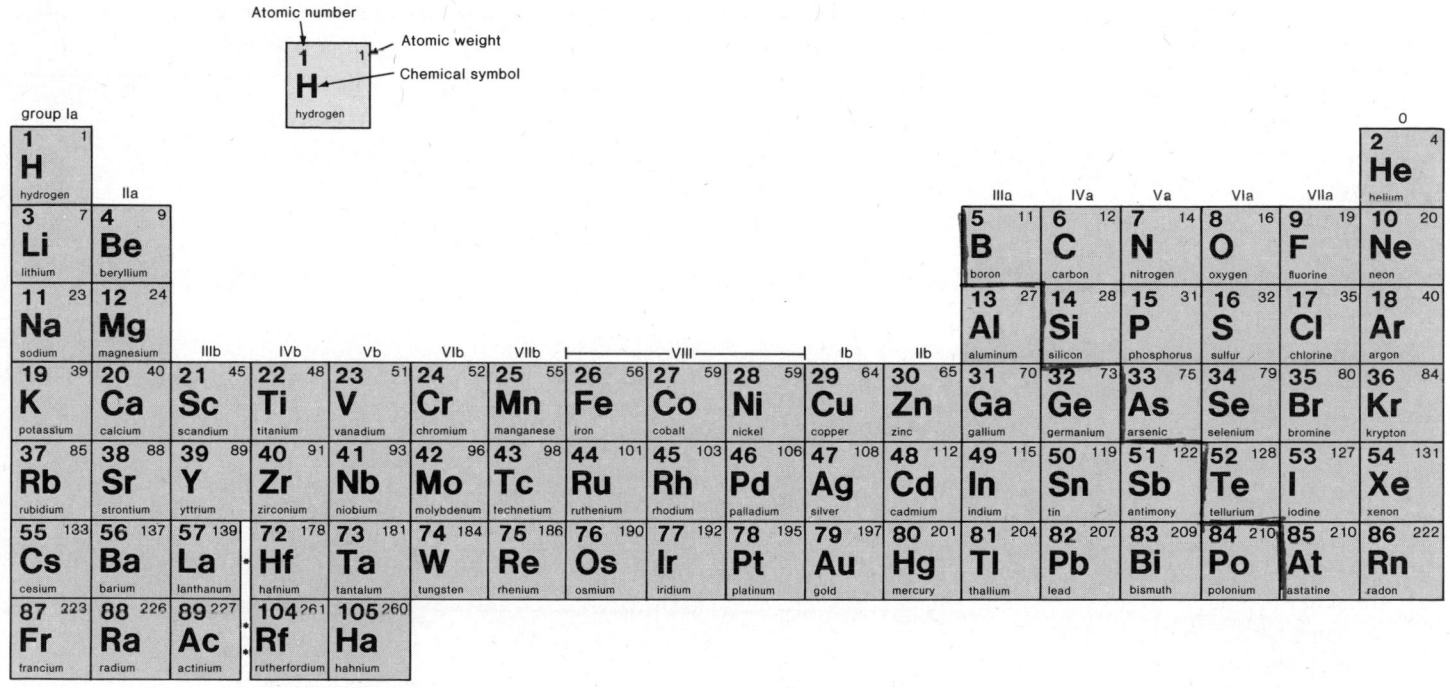

group Ia																	0
1 1 **H** hydrogen	IIa											IIIa	IVa	Va	VIa	VIIa	**2** 4 **He** helium
3 7 **Li** lithium	**4** 9 **Be** beryllium											**5** 11 **B** boron	**6** 12 **C** carbon	**7** 14 **N** nitrogen	**8** 16 **O** oxygen	**9** 19 **F** fluorine	**10** 20 **Ne** neon
11 23 **Na** sodium	**12** 24 **Mg** magnesium	IIIb	IVb	Vb	VIb	VIIb		VIII		Ib	IIb	**13** 27 **Al** aluminum	**14** 28 **Si** silicon	**15** 31 **P** phosphorus	**16** 32 **S** sulfur	**17** 35 **Cl** chlorine	**18** 40 **Ar** argon
19 39 **K** potassium	**20** 40 **Ca** calcium	**21** 45 **Sc** scandium	**22** 48 **Ti** titanium	**23** 51 **V** vanadium	**24** 52 **Cr** chromium	**25** 55 **Mn** manganese	**26** 56 **Fe** iron	**27** 59 **Co** cobalt	**28** 59 **Ni** nickel	**29** 64 **Cu** copper	**30** 65 **Zn** zinc	**31** 70 **Ga** gallium	**32** 73 **Ge** germanium	**33** 75 **As** arsenic	**34** 79 **Se** selenium	**35** 80 **Br** bromine	**36** 84 **Kr** krypton
37 85 **Rb** rubidium	**38** 88 **Sr** strontium	**39** 89 **Y** yttrium	**40** 91 **Zr** zirconium	**41** 93 **Nb** niobium	**42** 96 **Mo** molybdenum	**43** 98 **Tc** technetium	**44** 101 **Ru** ruthenium	**45** 103 **Rh** rhodium	**46** 106 **Pd** palladium	**47** 108 **Ag** silver	**48** 112 **Cd** cadmium	**49** 115 **In** indium	**50** 119 **Sn** tin	**51** 122 **Sb** antimony	**52** 128 **Te** tellurium	**53** 127 **I** iodine	**54** 131 **Xe** xenon
55 133 **Cs** cesium	**56** 137 **Ba** barium	**57** 139 **La** lanthanum	**72** 178 **Hf** hafnium	**73** 181 **Ta** tantalum	**74** 184 **W** tungsten	**75** 186 **Re** rhenium	**76** 190 **Os** osmium	**77** 192 **Ir** iridium	**78** 195 **Pt** platinum	**79** 197 **Au** gold	**80** 201 **Hg** mercury	**81** 204 **Tl** thallium	**82** 207 **Pb** lead	**83** 209 **Bi** bismuth	**84** 210 **Po** polonium	**85** 210 **At** astatine	**86** 222 **Rn** radon
87 223 **Fr** francium	**88** 226 **Ra** radium	**89** 227 **Ac** actinium	**104** 261 **Rf** rutherfordium	**105** 260 **Ha** hahnium													

58 140 **Ce** cerium	**59** 141 **Pr** praseodymium	**60** 144 **Nd** neodymium	**61** 147 **Pm** promethium	**62** 150 **Sm** samarium	**63** 152 **Eu** europium	**64** 157 **Gd** gadolinium	**65** 159 **Tb** terbium	**66** 163 **Dy** dysprosium	**67** 165 **Ho** holmium	**68** 167 **Er** erbium	**69** 169 **Tm** thulium	**70** 173 **Yb** ytterbium	**71** 175 **Lu** lutetium
90 232 **Th** thorium	**91** 231 **Pa** protactinium	**92** 238 **U** uranium	**93** 237 **Np** neptunium	**94** 242 **Pu** plutonium	**95** 243 **Am** americium	**96** 247 **Cm** curium	**97** 247 **Bk** berkelium	**98** 249 **Cf** californium	**99** 254 **Es** einsteinium	**100** 253 **Fm** fermium	**101** 256 **Md** mendelevium	**102** 254 **No** nobelium	**103** 257 **Lr** lawrencium

APPENDIX C

Metric System

Metric System		
Standard Metric Units		**Abbreviations**
Standard unit of mass	gram	g
Standard unit of length	meter	m
Standard unit of volume	liter	l
Common Prefixes		**Examples**
kilo (k)	1,000	a kilogram is 1,000 grams
centi (c)	0.01	a centimeter is 0.01 of a meter
milli (m)	0.001	a milliliter is 0.001 of a liter
micro (µ)	one-millionth	a micrometer is 0.000001 (one-millionth) of a meter
nano (n)	one-billionth	a nanogram is 10^{-9} (one billionth) of a gram
pico (p)	one-trillionth	a picogram is 10^{-12} (one trillionth) of a gram

Think Metric
Length

1. The speed of a car is 60 miles/hr or 100 km/hr.
2. A man who is 6 feet tall is 180 centimeters tall.
3. A 6-inch ruler is 15 centimeters long.
4. One yard is almost a meter (0.9 m).

Units of Length		
Unit	**Abbreviation**	**Equivalent**
meter	m	approximately 39 in
centimeter	cm	10^{-2} m
millimeter	mm	10^{-3} m
micrometer	µm	10^{-6} m
nanometer	nm	10^{-9} m
angstrom	Å	10^{-10} m

Length Conversions

1 in = 2.5 cm
1 ft = 30 cm
1 yd = 0.9 m
1 mi = 1.6 km

1 mm = 0.039 in
1 cm = 0.39 in
1 m = 39 in

1 m = 1.094 yd
1 km = 0.6 mi

To Convert	Multiply By	To Obtain
inches	2.54	centimeters
feet	30	centimeters
centimeters	0.39	inches
millimeters	0.039	inches

Think Metric

Volume

1. One can of beer (12 oz) contains 360 milliliters.
2. The average human body contains between 10-12 pints of blood or between 4.7-5.6 liters.
3. One cubic foot of water (7.48 gal) is 28.426 liters.
4. If a gallon of unleaded gasoline costs $1.00, a liter costs 26¢.

Units of Volume

Unit	Abbreviation	Equivalent
liter	l	approximately 1.06 qt
milliliter	ml	10-3 l (1 ml = 1 cm^3 = 1 cc)
microliter	μl	10-6 l

Volume Conversions

1 tsp = 5 ml	1 pt – 0.47 l	1 ml – 0.03 fl oz
1 tbsp = 15 ml	1 qt = 0.95 l	1 l = 2.1 pt
1 fl oz = 30 ml	1 gal = 3.8 l	1 l = 1.06 qt
1 cup = 0.24 l		1 l = 0.26 gal

To Convert	Multiply by	To Obtain
fluid ounces	30	milliliters
quarts	0.95	liters
milliliters	0.03	fluid ounces
liters	1.06	quarts

Think Metric

Weight

1. One pound of hamburger is 448 grams.
2. The average human male brain weighs 1.4 kilograms (3 lb 1.7 oz).
3. A person who weighs 154 pounds weighs 70 kilograms.
4. Lucia Zarate weighed 5.85 kilograms (13 lbs) at age 20.

Units of Weight

Unit	Abbreviation	Equivalent
kilogram	kg	10^3 g (approximately 2.2 lb)
gram	g	approximately 0.035 oz
milligram	mg	10-3 g
microgram	μg	10-6 g
nanogram	ng	10-9 g
picogram	pg	10-12 g

Weight Conversions

1 oz = 28.3 g	1 g = 0.035 oz
1 lb = 453.6 g	1 kg = 2.2 lb
1 lb = 0.45 kg	

To Convert	Multiply By	To Obtain
ounces	28.3	grams
pounds	453.6	grams
pounds	0.45	kilograms
grams	0.035	ounces
kilograms	2.2	pounds

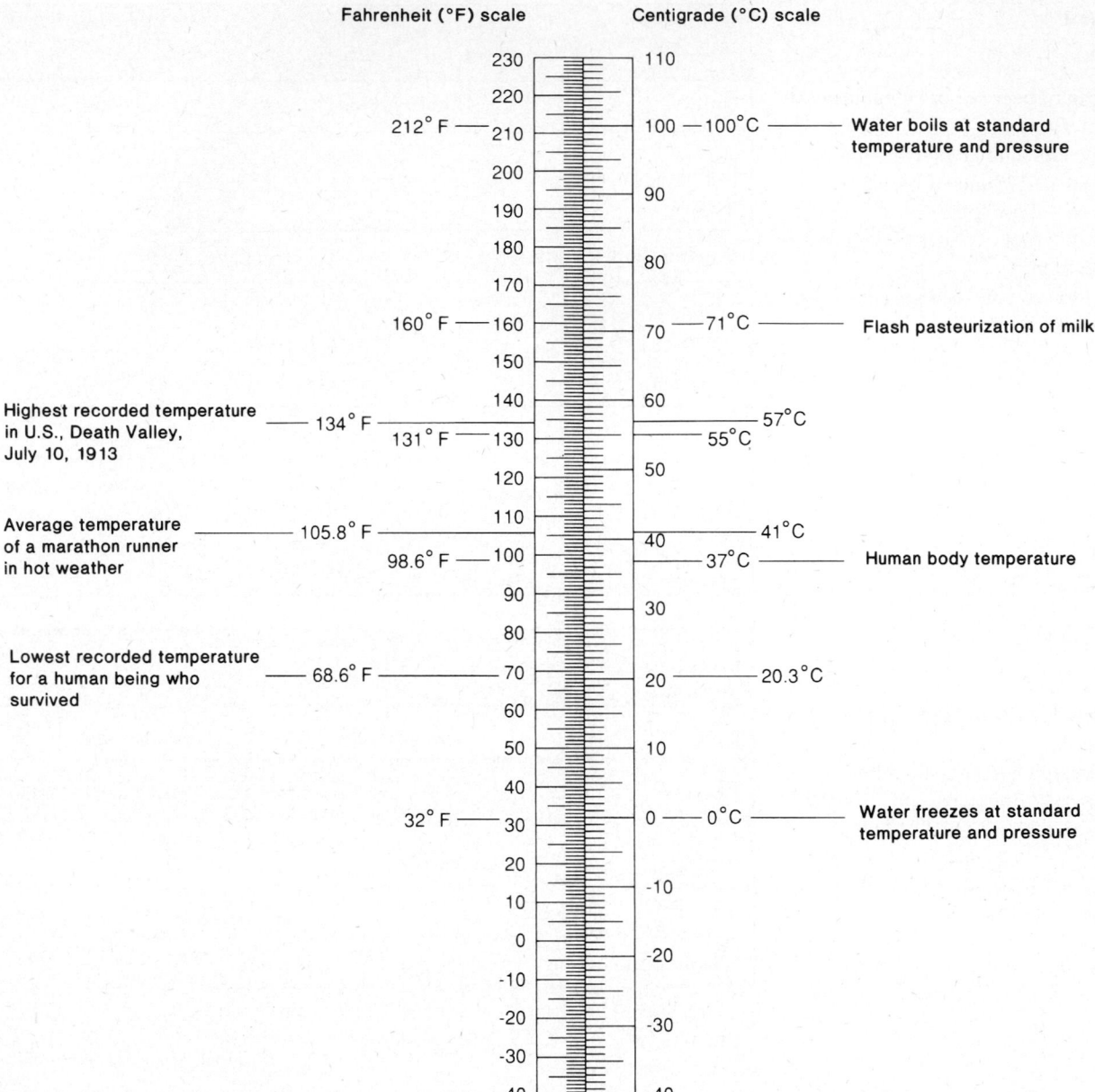

230 — 110
220
212° F — 210   100 — 100°C ——— Water boils at standard
200   temperature and pressure
190   90
180   80
170
160° F — 160   70 — 71°C ——— Flash pasteurization of milk
150
Highest recorded temperature   140   60   57°C
in U.S., Death Valley, — 134° F
July 10, 1913   131° F — 130   55°C
120   50
Average temperature — 105.8° F   110   41°C
of a marathon runner   98.6° F — 100   40 — 37°C ——— Human body temperature
in hot weather   90
80   30
Lowest recorded temperature   70   20 — 20.3°C
for a human being who — 68.6° F
survived   60
50   10
40
32° F — 30   0 — 0°C ——— Water freezes at standard
20   temperature and pressure
10
0 — -10
-10
-20 — -20
-30
-40 — -40

APPENDIX D

Answers

Chapter 1
Objective Questions

1. d
2. c
3. a
4. b
5. e
6. c
7. b
8. c
9. a. plants; b. animals; c. nutrients for plants; d. death and decay.

Concepts and Critical Thinking

1. A common ancestor has passed on to all living things common characteristics, such as using DNA genes.
2. Ways of life are diverse; therefore, organisms are diverse.
3. Nonliving things may have levels of organization, but they do not demonstrate emergent properties, where the whole is greater than the sum of the parts. For example, a cell is alive, but the structures, which make up a cell, are not alive.

Chapter 2
Objective Questions

1. d
2. a
3. d
4. a
5. b
6. c
7. b
8. d
9. a. Bacteria exposed to sunlight do not die if medium contains dye. b. Bacteria can be protected from UV light by dye. c. Experimental plate contains bacteria and dye; control contains only bacteria; exposure to UV light causes all bacteria to die. d. Hypothesis refuted.

Concepts and Critical Thinking

1. Even though the results of experimentation or observation support a hypothesis, it could be that the further studies will prove it false.
2. In the everyday sense, a theory means a supposition; in the scientific sense, a theory means a hypothesis that has been supported by many experiments and observations.
3. A scientist does studies to understand the natural world; citizens make decisions how these results should be used.

Chapter 3
Objective Questions

1. c
2. b
3. c
4. d
5. d
6. d
7. b
8. c
9. a
10. 7 p and 7 n in nucleus; 2 electrons in inner shell, and 5 electrons in outer shell. This means that nitrogen needs 3 more electrons to complete its outer shell; therefore the formula for ammonia is NH_3.

Concepts and Critical Thinking

1. For example, people take supplemental calcium to keep their bones strong.
2. Cells have a complex structure that is built upon the atoms and molecules in cells.
3. For example, if water was not slow to heat up and slow to cool down, living things might be subject to the killing effects of rapid heating and cooling.

Chapter 4
Objective Questions

1. c
2. a
3. d
4. b
5. c
6. b
7. c
8. c
9. d
10. c
11. a
12. a
13. a. monomer; b. condensation; c. polymer; d. hydrolysis. The diagram shows the manner in which macromolecules are synthesized and degraded in cells.
14. a. primary level; b. secondary level; c. tertiary level; d. quaternary level.
15. AGTTCGGCATGC

Concepts and Critical Thinking

1. They are unified in the use of carbohydrate as a structural molecule, but they are diversified as to the particular carbohydrate.
2. Butter is solid, and an oil is a liquid at room temperature because a fat containing saturated hydrocarbon chains melts at a higher temperature than one containing unsaturated chains.

3. Phospholipids have a polar head and nonpolar tails; they make up the plasma membrane, which is selectively permeable. Keratin, a fibrous protein that contains only helix polypeptides, is found in tough structures such as hair and nails.

Chapter 5
Objective Questions

1.	c	2.	c
3.	d	4.	a
5.	c	6.	c
7.	a	8.	d
9.	d	10.	d

11. a. Golgi apparatus further modifies; b. smooth ER modifies; c. chromatin—DNA directs; d. nucleolus—RNA helps; e. rough ER produces.

12. a. See text. b. Mitochondria and chloroplasts are a pair because they are both membranous structures involved in energy metabolism. c. Centrioles and flagella are a pair because they both contain microtubules; centrioles give rise to the basal bodies of flagella. d. Endoplasmic reticulum and ribosomes are a pair because together they are rough ER, which produces proteins.

Concepts and Critical Thinking

1. Show them microscopic slides of all sorts of tissues from different organisms. You would have to find an organism whose tissues did not contain cells.

2. The various organelles listed in table 5.1 show that the cell is compartmentalized; each organelle has a separate function.

3. All the structures labeled in question 11 plus mitochondria supply ATP energy. All the other parts of a cell assist to some degree also.

Chapter 6
Objective Questions

1.	b	2.	b
3.	a	4.	c
5.	c	6.	d
7.	d	8.	b
9.	b		

10. See figure 6.3. Phospholipid tails have no polar groups to interact with water's polar groups.

11. a. hypertonic—cell shrinks due to loss of water; b. hypotonic—cell has swelled due to gain of water.

Concepts and Critical Thinking

1. Structurally, the plasma membrane is the outer boundary of the cell. Functionally, the plasma membrane regulates what enters and leaves a cell.

2. A cell can die in either a severely hypertonic or severely hypotonic solution.

3. The plasma membrane secretes chemical messengers that communicate with distant cells; the plasma membrane participates in the structure of junctions between adjacently located cells.

Chapter 7
Objective Questions

1.	a	2.	d
3.	c	4.	a
5.	d	6.	c
7.	d	8.	c

9. a. macromolecules; b. degradative reactions; c. ADP + P → ATP; d. NADP → NADPH; e. small molecules; f. synthetic reactions; g. macromolecules.

10. See figure 7.13.

Concepts and Critical Thinking

1. Both the rough ER and Golgi apparatus produce molecules that are used for structural purposes. It takes energy to produce molecules.

2. When glucose energy is converted to ATP, there is a loss, and the energy released when ATP is broken down eventually becomes heat.

3. The temperature of the body is not high, and enzymes are needed to bring reactants together.

Chapter 8
Objective Questions

1.	d	2.	d
3.	a	4.	d
5.	c	6.	d
7.	c	8.	d

9. See figure 8.3.

10. a. water; b. oxygen; c. carbon dioxide; d. carbohydrate; e. ADP + P → ATP; f. NADP → NADPH.

Concepts and Critical Thinking

1. Show that solar energy is needed for photosynthesis, and photosynthetic organisms are the food for the biosphere, including humans. The bodies of plants became the fossil fuels, which we use to produce electricity and heat houses and convert to gasoline for cars.

2. Chlorophyll captures solar energy, which is converted to ATP energy and used to produce NADPH. ATP and NADPH are used to reduce CO_2 to a carbohydrate.

3. The thylakoid space is separated from the stroma by the thylakoid membrane; the build-up of hydrogen ions in the thylakoid space leads to ATP production.

Chapter 9
Objective Questions

1.	b	2.	c
3.	a	4.	c
5.	c	6.	c
7.	a	8.	b
9.	c	10.	d
11.	c	12.	a
13.	b	14.	d
15.	b	16.	a
17.	d	18.	a
19.	c	20.	b

21. See figure 9.11a.

22. a. 32; b. electron transport system; c. NADH; d. NADH; e. NADH; f. pyruvate; g. acetyl CoA; h. Krebs cycle; i. 2; j. lactate; k. CO_2; l. 2; m. CO_2.

Concepts and Critical Thinking

1. Flow of energy; photosynthesis converts the energy of the sun into carbohydrates that are converted to ATP energy by cellular respiration. Recycling of matter: Carbohydrates and oxygen from photosynthesis participate in cellular respiration, whose end product carbon dioxide reenters plants again.
2. Nutrient molecules are broken down to molecules that participate in cellular respiration, with the concomitant build-up of ATP. It is the ATP that is used by cells.
3. Glycolysis is a metabolic pathway that is almost universally found in organisms; it must have evolved in an ancestor common to all other organisms.

Chapter 10
Objective Questions

1.	c	2.	b
3.	d	4.	d
5.	a	6.	c
7.	b	8.	c
9.	b	10.	b
11.	See figure 10.5		

Concepts and Critical Thinking

1. As cells grow, they become ready for reproduction by producing more plasma membrane, cytoplasm, and organelles.
2. The daughter cells inherit DNA along with its inherent regulation from the parent cell.
3. Mitosis involves the use of a spindle, a structure that ensures that each daughter cell receives a copy of each chromosome.

Chapter 11
Objective Questions

1.	b	2.	c
3.	c	4.	b
5.	a	6.	c
7.	d	8.	d
9.	c		
10.	a—It shows bivalents at equators.		

Concepts and Critical Thinking

1. Asexual reproduction and sexual reproduction begin with cells, usually single cells. This shows how fundamental the cell is to the life of an organism.
2. Variation during asexual reproduction is limited to the occurrence of mutations. Variation during sexual reproduction is introduced due to mutations, crossing-over, and recombination (from independent assortment and fertilization).
3. Advantage: favorable mutations may be of immediate advantage. Disadvantage: unfavorable mutations may be of immediate disadvantage.

Chapter 12
Practice Problems 1

1. a. 100% W; b. 50% W, 50% w; c. 50% T, 50% t; d. 100% T
2. a. gamete; b. genotype; c. gamete

Practice Problems 2

1. bb
2. $Tt \times tt$, tt
3. 3/4 or 75%
4. Yy and yy

Practice Problems 3

1. a. 100% tG; b. 50% TG, 50% tG; c. 25% TG, 25% Tg; 25% tG; 25% tg; d. 50% TG, 50% Tg
2. a. genotype; b. gamete; c. genotype; d. gamete

Practice Problems 4

1. $BbTt$ only
2. a. $LlGg \times llgg$
   b. $LlGg \times LlGg$
3. 9/16

Objective Questions

1.	b	2.	a
3.	c	4.	d
5.	c	6.	d
7.	a	8.	b
9.	b	10.	b
11.	c	12.	d

Additional Genetics Problems

1. 100% chance for widow's peak and 0% chance for continuous hairline
2. Ee

3. 50%
4. 210 gray body and 70 black body; 140 = heterozygous; cross fly with recessive
5. F_1 = all black with short hair; F_2 = 9:3:3:1; offspring would be 1 brown long: 1 brown short: 1 black long: 1 black short
6. $Bbtt \times bbTt$ and $bbtt$
7. $GGLl$
8. 25%

Concepts and Critical Thinking

1. You inherit chromosomes containing the same types of genes from parents, but each parent contributes one-half of the particular genes inherited.
2. Plants and animals both have chromosomes containing the genetic material, DNA.
3. The individual's genes were not modified by the accident.

Chapter 13
Practice Problems 1

1. Cc^h (wild); $c^{ch}c^h$ (light gray); Cc^h (wild); c^hc^h (Himalaya)
2. 1 pink:1 white
3. 7
4. pleiotropy, epistasis

Practice Problems 2

1. females: X^RX^R, X^RX^r, X^rX^r, males: X^RY (gametes X^R, Y); X^rY (gametes Xr, Y)
2. b; 1:1
3. 100%, none, 100%
4. mother: X^BX^bRr; father: X^BYRr; son: X^bYrr

Practice Problems 3

1. 9:3:3:1, linkage
2. 54.5 − 13.0 = 41.5%
3. 12
4. bar eye, scalloped wings, garnet eye

Objective Questions

1.	b	2.	c
3.	e	4.	a
5.	b	6.	b
7.	d	8.	b
9.	c	10.	d

Additional Genetics Problems

1. pleiotropy
2. codominance, F^BF^B, F^WF^W, F^BF^W
3. 50%
4. 6, yes
5. linkage, 10
6. Both males and females are 1:1.
7. $X^BX^B \times X^bY$ = both males and females bar-eyed;
   $X^bX^b \times X^BY$ = females bar-eyed; males all normal
8. $XbXbWw$
9. females—1 short-haired tortoise shell: 1 short-haired yellow; males—1 short-haired black: 1 short-haired yellow

Concepts and Critical Thinking

1. Most human traits, such as height and color of skin, are examples of polygenic inheritance.
2. The genes behave similarly to the chromosomes (i.e., they come in pairs, they segregate during meiosis, and there is only one in the gametes).
3. Almost all males have an Y chromosome; XO individuals are females. This shows that most likely sex determining alleles on the Y chromosome have some degree of dominance to sex-determining alleles on the X chromosome.

Chapter 14

Practice Problem 1

1. 25%
2. 50%
3. heterozygous
4. heterozygous—unless a new mutation, which is a fairly common cause of NF.

Practice Problem 2

1. Hb^AHb^S; either Hb^SHb^S or Hb^AHb^S
2. light
3. white
4. Doe—baby #1
   Jones—baby #2

Practice Problem #3

1. mother; mother X^HX^b, father X^HY, son X^bY

2. 0%, 0%, 100%
3. rrX^bX^b and RrX^BX^b, rrX^BX^b, RrX^bX^b
4. colorblind

Objective Questions

1. c
2. a
3. b
4. c
5. All are consistent.
6. c
7. b
8. d
9. c
10. a
11. autosomal dominant condition

Additional Genetics Problems

1. 0%, 0%, 100%
2. 25%
3. 50%, 50%
4. AB, Yes, A, B, AB, or O
5. light, white
6. X^cYww, X^CX^cWw, X^CYWw; normal vision with widow's peak
7. a. recessive, Aa
   b. dominant, Aa
   c. sex-linked recessive, X^AX^a

Concepts and Critical Thinking

1. Yes, the concept still holds, because any differences don't negate the basic similarities.
2. Divide the cells of an early embryo to produce as large a number of identical children as possible. (This solves the nature question.) Raise each child under entirely different circumstances and then see how similar the children are.
3. This is a question that each person must decide on their own. Genetic counseling and genetic testing can sometimes tell parents their chances of having a child with a genetic disease.

Chapter 15

Objective Questions

1. b
2. d
3. a
4. c
5. d
6. c
7. a
8. a
9. b
10. d

11. The parental helix is heavy-heavy, and each daughter helix is heavy-light. (The cell was supplied with ^{14}N nucleotides as replication began.) This shows that each daughter helix is composed of one template strand and one new strand. This is consistent with semiconservative replication.

Concepts and Critical Thinking

1. ^{35}S would have been found with the bacterial cells, showing that it was needed for viral replication.
2. Replication is a part of duplication of the chromosomes. Replication before mitosis ensures that each daughter cell will have the diploid number of chromosomes, and replication before meiosis ensures that the sperm and egg will have the haploid number of chromosomes.
3. Mutation is the ultimate source of genetic variation. Without variation, new organisms would never evolve.

Chapter 16

Objective Questions

1. b
2. d
3. a
4. c
5. a
6. a
7. c
8. b
9. c
10. d
11. a. ACU CCU GAA UGC AAA;
    b. UGA GGA CUU ACG UUU;
    c. threonine-proline-glutamate-cysteine-lysine.

Concepts and Critical Thinking

1. The genetic information stored by DNA is the sequence of amino acids in a protein; this is stored in the sequence of DNA bases.
2. A universal code means that the ancestry of all types of organisms can be traced to a common origin.
3. The sequence of bases of DNA ultimately determines the sequence of bases in a protein. Figure 16.5 shows how the molecules are colinear.

Chapter 17

Objective Questions

1.	c	**2.**	a
3.	d	**4.**	a
5.	b	**6.**	b
7.	b	**8.**	d
9.	d	**10.**	d

11. a. See figure 17.2*a*. b. For the trp operon to be in the off position, a corepressor has to be attached to the repressor.

Concepts and Critical Thinking

1. Only certain mRNA molecules are present in cells at any particular time.
2. Gene expression necessitates a functioning protein product. Posttranslational control involves regulation of the activity of the protein product.
3. Checks and balances are often seen in biological systems. This is a safety feature that allows greater control.

Chapter 18

Objective Questions

1.	a	**2.**	c
3.	d	**4.**	a
5.	d	**6.**	d
7.	b	**8.**	c
9.	d	**10.**	a

11. a. AATT; b. TTAA.
12. 2. reverse transcription occurs; 4. most of the viral genes are removed, etc.; 5. transcription of DNA occurs; 6. recombinant RNA is repackaged; 8. reverse transcription occurs.

Concepts and Critical Thinking

1. It is possible to make recombinant DNA, to modify natural DNA in the laboratory, to perform PCR analysis, to do DNA fingerprinting, and to determine the sequence of the bases of DNA. This is disturbing to people who resist thinking of humans as physical and chemical machines.
2. Human genes can be placed in a bacterium, where they function normally.

Chapter 19

Objective Questions

1.	d	**2.**	b
3.	b	**4.**	d
5.	d	**6.**	d
7.	d	**8.**	b

9. a. All the continents of today were one continent during the Triassic period, and the reptiles spread throughout the land. b. All vertebrates share a common ancestor, who had pharyngeal pouches during development. c. Two different continents can have similar environments and therefore unrelated organisms that are similarly adapted. d. This demonstrates that diversification has occurred in the nightshade family.

Concepts and Critical Thinking

1. Both bacteria and humans use DNA as the hereditary material and have similar metabolic pathways.
2. If adaptation were purposeful, organisms wouldn't have structures that don't serve a current function.
3. It is possible to formulate a hypothesis about evolution and then make observations to see if the hypothesis is correct.

Chapter 20

Objective Questions

1.	c	**2.**	b
3.	c	**4.**	c
5.	c	**6.**	c
7.	d	**8.**	c
9.	b		

10. See figure 20.7*b*.

Population Genetics Problems

1. 99%
2. recessive allele = 0.2; dominant allele = 0.8; homozygous recessive = 0.04; homozygous dominant = 0.64; heterozygous = 0.32

Concepts and Critical Thinking

1. Traits are passed from one generation to the next via the gametes, and no mechanism has been found by which phenotype changes can affect the genes in the gametes.

2. The more fit organisms have more offspring, and this is the way by which adaptive traits accumulate in a population.
3. Increase variation between populations: genetic drift, natural selection, mutation, nonrandom mating. Decrease variation between populations: gene flow. These mechanisms alter gene frequencies in populations.

Chapter 21

Objective Questions

1.	d	**2.**	d
3.	c	**4.**	b
5.	b	**6.**	d
7.	a	**8.**	d
9.	b		

10. Do away with 5 middle arrows so that 2 sets of circles remain. The circles of each set are connected by arrows.

Concepts and Critical Thinking

1. The higher taxa show how species are related. Species in different kingdoms are distantly related, and those in the same genus are closely related, for example.
2. All new species have evolved from previous species; therefore all organisms have a phylogenetic history.
3. One group gives rise to many groups, each adapted to a particular environment.

Chapter 22

Objective Questions

1.	d	**2.**	c
3.	b	**4.**	b
5.	a	**6.**	c
7.	c	**8.**	b
9.	b	**10.**	d

11. The diagram suggests that the eukaryotic cell acquired mitochondria and, if present, chloroplasts by engulfing particular prokaryotic cells. This hypothesis is called the endosymbiotic theory.

Concepts and Critical Thinking

1. All 3 hypotheses suggest that a chemical evolution produced the first cell(s).
2. The increase in variability among the offspring of sexually reproducing organisms (as compared to asexual ones) provides more opportunity for change.
3. Under the conditions of the primitive earth, a chemical evolution produced the first cell(s). Under the conditions of today's earth, life comes only from life.

Chapter 23
Objective Questions

1.	b	2.	b
3.	c	4.	b
5.	c	6.	c
7.	d	8.	c
9.	b	10.	d
11.	d		

12. See figure 23.3.

Concepts and Critical Thinking

1. Viruses are unable to carry on reproduction outside a living cell. They reproduce by taking over the machinery of the cell; therefore, they are obligate parasites.
2. Monerans are metabolically diverse, as witnessed by the fact that there is probably no single type of organic compound that some Moneran cannot digest.
3. This statement refers to structure, not biochemistry.

Chapter 24
Objective Questions

1.	d	2.	a
3.	b	4.	c
5.	d	6.	a
7.	a	8.	d

9. a. Protista, Chlorophyta (green algae); b. Fungi, Ascomycota (sac fungi); c. Protista, Sarcodina (amoeboid protozoans).
10. a. locomotion; b. pigmentation.
11. See figure 24.9b. Sexual reproduction involves the union of gametes.

Concepts and Critical Thinking

1. The protozoans are heterotrophic; the algae are photosynthetic; and the slime molds are saprophytic.
2. In the haplontic cycle, the zygote undergoes meiosis, and therefore the adult is haploid. In alternation of generations, the sporophyte produces haploid spores that mature into the gametophyte. In the diplontic cycle, meiosis produces haploid gametes, the only part of the cycle that is haploid.
3. Fungi live on the organic material they digest. Plants produce their own organic food, and animals go out and find it.

Chapter 25
Objective Questions

1.	d	2.	c
3.	c	4.	c
5.	d	6.	b
7.	b	8.	b
9.	d		

10. See figure 25.3b.
11. a. leafy shoot; b. heart-shaped structure; c. stalk with sporangium; d. rhizome with fronds, roots. The fern has the sporophyte dominant. There are sporangia on the underside of the fronds, and this is the generation that persists longer.
12. a. underside of scales of male cone; b. in the anther of stamens; c. upper surface of scales of female cone; d. in ovary of pistil. The flowering plant produces fruit, which develops from the ovary.

Concepts and Critical Thinking

1. The 5-kingdom system of classification is based in part on mode of nutrition. From this standpoint, it would seem that all photosynthesizers should be in the same kingdom. On the other hand, it would seem that we need some other criteria to go by; for example, all organisms in the plant kingdom are multicellular with some degree of complexity. In this text, all organisms in the plant kingdom protect the zygote.

2. Green algae at the base could lead to nonvascular plants (bryophytes) and vascular plants. The primitive vascular plants as a group could lead to gymnosperms (4 divisions) and angiosperms (2 classes). The tree goes from the first evolved to the latest evolved according to the fossil record.
3. Roots enable a plant to take water out of the soil; stems hold the leaves aloft to catch solar energy, and the leaves also take in carbon dioxide at stomata.

Chapter 26
Objective Questions

1.	d	2.	c
3.	d	4.	d
5.	b	6.	b
7.	a		

8. See figure 26.4.
9. Sponges (no items)
   Cnidarians
       radial symmetry
       tissue level of organization
   Flatworms
       3 germ layers
   Roundworms
       pseudocoelom
       3 germ layers
       tube-within-a-tube
10. Sponges
        collar cells
    Planarians
        eyespots
        gastrovascular cavity
    Tapeworms
        proglottids
    Roundworms
        uterus

Concepts and Critical Thinking

1. For a contrast of the plant cell and the animal cell, see table 5.2. Note that animal cells cannot make their own food (lack chloroplasts) and are motile (have flagella), 2 distinct characteristics of animals.
2. A common ancestor would have to have those features shared by both types of organisms: bilateral symmetry, 3 germ layers, and organs.

3. Internal parasites need to have a means of protecting themselves from host attack, absorbing nutrients from the host, and reproducing that allows dispersal to new hosts.

Chapter 27
Objective Questions

1.	a	**2.**	a
3.	c	**4.**	c
5.	a	**6.**	d
7.	d	**8.**	b
9.	d		

10. Mollusks
    organ system level of
        organization
    coelom
    cephalization in some
        representatives
    complete gut
    Annelids
    organ system level of
        organization
    segmentation
    coelom
    cephalization in some
        representatives
    complete gut
    Arthropods
    organ system level of
        organization
    segmentation
    coelom
    cephalization in some
        representatives
    complete gut
    Chordates
    organ system level of
        organization
    segmentation
    coelom
    cephalization in some
        representatives
    complete gut
11. Clam
    Mollusk
    3 ganglia
    gills
    open circulatory system
    hatchet foot

Earthworm
    annelid
    ventral nerve cord
    closed circulatory system
    hydrostatic skeleton
    setae

12. Label wings (for flying in the air), spiracle (for breathing air), tympanum (for picking up sound waves), ovipositor (for laying eggs in earth), digestive system (for eating grass), Malpigian tubules (for excretion without loss of water), tracheae (for breathing air), vagina and seminal receptacle (for reception and storage of sperm so they do not dry out). See figure 27.12 for placement of labels.

Concepts and Critical Thinking

1. Body rings indicate that the earthworm's body is segmented and there are setae on each segment. In the grasshopper, segmentation is not so obvious because there is a head, thorax with legs, and abdomen—that is, specialization of parts.
2. See table 27.2, which lists all the features that arthropods and chordates have in common. The fact that both lineages have these features must mean that these features increase the fitness of organisms having them.
3. See answer to objective question 12.

Chapter 28
Objective Questions

1.	a	**2.**	d
3.	a	**4.**	b
5.	b	**6.**	d
7.	d	**8.**	b
9.	b	**10.**	c

11. a. mammals; b. birds; c. reptiles; d. amphibians; e. modern bony fishes; f. cartilagenous fishes; g. modern jawless fishes.
12. Primate (all), Prosimian (only lemurs), Anthropoid (all but lemurs), Hominoid (all but lemurs and monkeys), Hominid (only Australopithecines and humans).

13. Kingdom Animalia, phylum Chordata, subphylum Vertebrata, class Mammalia, order Primate, superfamily Hominoidea, family Hominidae, genus and species *Homo sapiens*.

Concepts and Critical Thinking

1. The number of species in varied habitats is a good criterion. Vertebrates are chordates; therefore table 27.2 lists the features they have in common. In addition, they both have a rigid, but jointed, skeleton.
2. Gametes are protected from drying out: reptiles pass sperm directly from male to female; angiosperms rely on the pollen grain. Embryo is protected from drying out: reptiles develop in a shelled egg; angiosperms produce seeds that protect the embryo until germination under favorable conditions occurs.
3. No living organism evolved from another living organism. Instead, 2 living organisms may have shared a common ancestor.

Chapter 29
Objective Questions

1.	c	**2.**	b
3.	c	**4.**	c
5.	b	**6.**	b
7.	b	**8.**	c
9.	c	**10.**	b

11. a. epidermis; b. cortex; c. endodermis; d. phloem; e. xylem.
12. a. cork; b. phloem; c. vascular cambium; d. xylem; e. pith; f. bark.
13. a. upper epidermis; b. palisade mesophyll; c. leaf vein; d. spongy mesophyll; e. lower epidermis.

Concepts and Critical Thinking

1. A leaf that is broad and flat is well shaped to catch the sun's rays; the cells contain chloroplasts; the leaf veins bring water from the roots; the stomata admit carbon dioxide into the air spaces of the spongy layer.
2. Figure 29.1 and table 29.1 show how a plant is organized.

3. Wood helps a plant resist the pull of gravity. It serves as an internal skeleton and allows the plant to lift the leaves, exposing them to the rays of the sun.

Chapter 30
Objective Questions

1. d
2. a
3. c
4. b
5. c
6. c
7. c
8. d
9. The diagram shows that gravity pushing down on mercury in the pan can only raise a column of mercury to 76 cm. When water above the column is transpired, it pulls on the mercury and raises it higher. This suggests that transpiration would be able to raise water to the top of trees.
10. See figure 30.6*b* and *c*. After K^+ enters guard cells, water follows by osmosis and the stoma opens.
11. There is more solute in (1) than (2); therefore water enters (1); this creates a pressure that causes water, along with solute, to flow toward (2).

Concepts and Critical Thinking

1. A physical process accounts for the transport of water in xylem; causes a stoma to open; and accounts for the transport of organic substances in phloem.
2. Because vessel cells are hollow and nonliving, they form a continuous pipeline from the roots to the leaves. This allows plants to transport water and minerals from the roots to the leaves.

Chapter 31
Objective Questions

1. c
2. a
3. b
4. a
5. a
6. b
7. a
8. d
9. c
10. c
11. a. sporophyte; b. microsporangium; c. megasporangium (in ovule); d. meiosis; e. microspore;

f. megaspore; g. male gametophyte (pollen grain); h. female gametophyte; i. egg and sperm; j. fertilization; k. zygote in seed.

Concepts and Critical Thinking

1. Flowering plants follow the alternation of generations life cycle, but the gametophyte is reduced to a microscopic size and is protected by the sporophyte. The gametes are also protected; the egg stays within the ovule within the ovary and awaits the sperm. A pollen grain is transported to the pistil by wind or animals and then germinates, allowing the sperm to move down the style to the egg. The zygote (new sporophyte) develops into an embryo within the ovule, which becomes the seed enclosed by fruit. Fruit is dispersed in various ways.
2. As stated, the gametophyte is protected by the sporophyte and the embryonic sporophyte is protected by the seed.
3. Coevolution results in one type of pollinator gathering food from a particular plant. This reduces competition and assures the pollinator of food from this particular source.

Chapter 32
Objective Questions

1. d
2. b
3. e
4. a
5. c
6. d
7. b
8. d
9. d
10. a
11. The arrows show the movement of auxin from the shady side to the sunny side of the stem. Auxin causes the cells on the shady side to elongate; therefore, the oat seedling bends toward the light.
12. The arrows (K^+ ions) leave guard cells under the influence of abscisic acid (ABA) when a plant is water stressed. Water osmotically follows the movement of K^+ and the stoma closes.

Concepts and Critical Thinking

1. The best example is that the ratio of auxin to cytokinin determines whether a plant tissue will form an undifferentiated mass or form roots, vegetative shoots and leaves, or flowers.
2. It is adaptive, for example, for leaves to unfold and stomata to open during the day, when photosynthesis can occur. A biological clock system needs a receptor that is sensitive to light and dark; a timekeeper, that is, a biological clock; and a means of communication within the body of the organism. In regard to flowering, phytochrome is the receptor and hormones are probably the means of communication. The possible timekeeper is not known.

Chapter 33
Objective Questions

1. b
2. b
3. a
4. d
5. d
6. d
7. b
8. d
9. d
10. c
11. a. smooth muscular tissue; b. blood cells—connective tissue; c. nervous tissue; d. ciliated columnar epithelium.

Concepts and Critical Thinking

1. The cell is alive, but the organelles making up a cell are not alive; a tissue performs functions that the individual cell in the tissue cannot perform and so forth from level to level.
2. Acquiring food is located to the right and left of the diagram. Cellular respiration is a degradative pathway requiring an exchange of gases. Excretion of waste accompanies degradation.
3. Acquiring food provides materials and energy; the musculoskeletal system helps us get food; exchange of gases gets rid of carbon dioxide and brings in oxygen; transporting materials brings nutrients to and takes wastes away from cells;

protecting the body from disease maintains its integrity; coordinating body activities makes sure that all the other functions are performed.

Chapter 34
Objective Questions

1. b   2. b
3. a   4. d
5. d   6. c
7. d   8. b
9. b   10. b
11. d
12. a (See figure 34.6.)

Concepts and Critical Thinking

1. Tissue fluid remains relatively constant because materials are constantly being added and removed at the capillaries.
2. The cells are far removed from the exterior, and therefore they have to have nutrients brought to them and wastes removed from them.
3. ATP energy keeps the heart pumping (arterial flow) and the skeletal muscles contracting (venous flow). ATP energy is produced by mitochondria, where chemical energy is transformed into usable energy.

Chapter 35
Objective Questions

1. d   2. d
3. a   4. b
5. c   6. b
7. a   8. a
9. b   10. d

Concepts and Critical Thinking

1. Death occurs when immunity fails to prevent microorganisms from taking over the body.
2. Immunity consists of immediate defense mechanisms and specific defense mechanisms, which work more slowly.
3. Bones are of course part of the skeletal system. Red bone marrow is the site of red and white blood cell formation; therefore, it is part of the circulatory and lymphatic systems.

Chapter 36
Objective Questions

1. d   2. b
3. d   4. b
5. c   6. a
7. c   8. c
9. d   10. c
11. d   12. c
13. See figure 36.6.
14. Test tube 1: No digestion
    No enzyme and no HCl
    Test tube 2: Some digestion
    No HCl
    Test tube 3: No digestion
    No enzyme
    Test tube 4: Digestion
    Both enzyme and HCl are present

Concepts and Critical Thinking

1. In humans, the digestive system takes in nutrients, excretes certain metabolites like heavy metals, and prevents bacterial invasion by the low pH of the stomach.
2. See figure 36.3.
3. Only autotrophs are able to produce their own food. Heterotrophs feed on autotrophs and acquire material and energy. Autotrophs use wastes from heterotrophs (fertilizer, CO_2).

Chapter 37
Objective Questions

1. a   2. b
3. b   4. d
5. c   6. b
7. c   8. b
9. d   10. b
11. See figure 37.7.

Concepts and Critical Thinking

1. The respiratory system carries out exchange of gases and helps maintain the pH of the blood.
2. Gills are external extensions and lungs are internal cavities. Both are minutely divided and highly vascularized for the exchange of gases.
3. In humans, the lungs expand when the chest moves up and out and deflate when the chest moves down and in. The lungs contain many alveoli (air sacs) and are highly

vascularized. Oxygen enters the blood following inhalation, and carbon dioxide leaves the body upon exhalation.

Chapter 38
Objective Questions

1. d   2. a
3. c   4. b
5. a   6. c
7. b   8. d
9. a   10. b
11. See figure 38.10.

Concepts and Critical Thinking

1. When molecules undergo degradation, wastes result. For example, following glucose metabolism, carbon dioxide must be excreted. Following amino acid breakdown, urea must be excreted.
2. Water contains a lower amount of oxygen than air. The countercurrent mechanism in the gills of fishes helps them extract oxygen out of water. Mammals evolved on land; a countercurrent mechanism in the kidneys helps them conserve water.
3. Reabsorption, in part by active transport, takes place in the proximal convoluted tubule. An increased surface area (microvilli) helps reabsorption, and mitochondria supply the ATP energy for active transport.

Chapter 39
Objective Questions

1. c   2. b
3. d   4. c
5. a   6. a
7. b   8. d
9. b   10. b
11. a. receptor (initiates nerve impulse); b. sensory neuron (takes impulses to cord); c. interneuron (passes impulses to motor neuron and other interneurons in the cord); d. white matter (contains tracts that take impulses up and down the cord); e. motor neuron (takes impulses to effector); f. effector (brings about adaptive response).

Concepts and Critical Thinking

1. Controls internal organs like the beating of the heart and skeletal muscles, which allow an animal to seek environments compatible with a constant internal environment.
2. There is a spinal nerve in each segment, but the brain is an obvious specialization of part of the spinal cord.
3. A neuron has long processes that conduct nerve impulses. The postsynaptic membrane is part of a dendrite, and the presynaptic membrane is part of an axon. Impulses always flow in this direction because only an axon contains synaptic vesicles.

Chapter 40
Objective Questions

1.	d	2.	c
3.	c	4.	d
5.	d	6.	c
7.	c	8.	b
9.	d	10.	c

11. See figure 40.5 and table 40.1.

Concepts and Critical Thinking

1. The sense organs help animals find favorable environments and avoid unfavorable ones; they help animals find food and avoid being eaten; they help animals communicate with others for the purpose of cooperation.
2. The sense organs are receptors for external and internal stimuli; they generate nerve impulses, which supply information to the central nervous system.
3. Relative fitness determines what structures characterize a species. Therefore, animals have those sense organs that are most adaptive.

Chapter 41
Objective Questions

1.	b	2.	f
3.	c	4.	e
5.	b	6.	b
7.	b	8.	c
9.	a	10.	d

11. See table 41.3.
12. See figure 41.10a.

Concepts and Critical Thinking

1. Consult, for example, figure 21.19 and note how the skeleton of a horse was adapted at first to a forest and then later to a grassland habitat.
2. Animals have to go out and get their food; therefore, locomotion is essential to them.
3. Nutrient molecules from the digestive system travel in the blood to muscle cells. Here, glucose energy is converted to ATP energy, largely within mitochondria. ATP energy is used by myosin filaments to pull actin filaments, and muscle contraction occurs.

Chapter 42
Objective Questions

1.	f	2.	b
3.	c	4.	a
5.	e	6.	c
7.	a	8.	d
9.	b	10.	d
11.	b	12.	b

13. See figure 42.7. In a negative feedback system, the output (i.e., thyroxin) cancels the input (e.g., thyroid growth hormone from the pituitary); therefore, the production of thyroxin is shut off when it gets too high.

Concepts and Critical Thinking

1. Neurotransmitters are released when needed and then reabsorbed by the presynaptic membrane or broken down within the synaptic cleft by an enzyme, for example. The level of a hormone is controlled typically by using negative feedback to regulate its production.
2. Specialized structures (i.e., neurons) deliver messages in the nervous system, whereas the endocrine system uses the blood system.
3. The human body undergoes a dramatic change following puberty.

Chapter 43
Objective Questions

1.	b	2.	d
3.	c	4.	d
5.	c	6.	c
7.	c	8.	c
9.	a	10.	a

11. See figure 43.3 and table 43.1. Testis, epididymis, vas deferens, urethra (in penis).

Concepts and Critical Thinking

1. They are similarly adapted, except the embryos of reptiles develop in shelled eggs while the embryos of placental mammals develop within the uterus of the female. Even so, the same extraembryonic membranes surround the embryo.
2. Reproduction is a process that takes some time; it is not like movement, which takes place quickly.

Chapter 44
Objective Questions

1.	b	2.	b
3.	b	4.	a
5.	b	6.	b
7.	d	8.	a
9.	d	10.	d

11. See figure 44.8. Chorion exchanges wastes for nutrients with the mother; amnion protects and prevents desiccation; blood vessels of allantois become umbilical blood vessels; yolk sac is the first site of blood cell formation.

Concepts and Critical Thinking

1. Senescence is a change that is part of the normal life cycle from fertilization to death.
2. They could be considered local hormones because they are messengers that are produced by one cell and act on another cell.
3. Perhaps as individuals age, genes are turned off (instead of on), and this leads to degenerative changes.

Chapter 45

Objective Questions

1.	d	2.	a
3.	d	4.	a
5.	c	6.	d
7.	d	8.	c
9.	d	10.	a

11. a. communication, chemical; b. learning, imprinting; c. communication, sound; d. learning, operant conditioning; e. communication, touch; f. communication, visual.

Concepts and Critical Thinking

1. The genes control the development of the brain and therefore even insight learning has a genetic basis.
2. Feeding behavior patterns affect fitness because an animal needs a body in good working order for reproduction to be successful. Territoriality affects fitness because an animal needs enough food to nourish its offspring. Reproductive patterns affect fitness because an animal needs to have a mate to reproduce.
3. Altruistic acts can be shown to increase inclusive fitness.

Chapter 46

Objective Questions

1.	b	2.	c
3.	d	4.	d
5.	d	6.	c
7.	b	8.	c

9. See figure 46.3b. The r stands for rate of natural increase.

Concepts and Critical Thinking

1. When there are unlimited resources, such as space, food, and shelter, and where there are no deaths due to natural disasters, population growth could be infinite.
2. Human life strategy is only like that of a K-strategist. See table 46.4.

Chapter 47

Objective Questions

1.	b	2.	d
3.	c	4.	d
5.	c	6.	b
7.	d	8.	c
9.	d		

10. a. mutualism; b. parasitism; c. commensalism.

Concepts and Critical Thinking

1. The organisms themselves provide niches for other organisms. For example, several types of warblers find niches in spruce trees (fig. 47.3).
2. As prey populations decrease in size, it is less likely that a predator will find that type of prey, and prey populations have defenses against predation.
3. Parasites keep down population sizes when they weaken or directly kill off their hosts. Parasites increases diversity because, for example, even parasites have parasites. As mentioned in number 1, organisms themselves provide niches for other organisms.

Chapter 48

Objective Questions

1.	a	2.	a
3.	d	4.	a
5.	a	6.	c
7.	d	8.	c
9.	a		

10. a. 4+ for both; b. 3+ for both; c. 2+ for both; d. 1+ for both.

Concepts and Critical Thinking

1. The neritic province is more productive because it receives both nutrients and sunlight. In contrast, only the epipelagic zone of the oceanic province receives adequate sunlight. This zone is not as nutrient rich as the neritic province.
2. Succession does not require a definite sequence, but there is a sequence of communities from the simple to the complex.
3. It gets colder as you go from the base of a mountain to the top; therefore, the communities will change as noted.

Chapter 49

Objective Questions

1.	b	2.	a
3.	c	4.	d
5.	b	6.	d
7.	c	8.	c
9.	a	10.	b

11. a. algae; b. zooplankton; c. small fishes; d. large fishes; e. humans.

Concepts and Critical Thinking

1. For example, the atmosphere is the abiotic component of the nitrogen cycle, and the passage of nitrogen compounds from producer to consumers is the biotic component of the nitrogen cycle.
2. It is impossible to create energy to keep an ecosystem going (first law), and as energy is passed from one trophic level to the next, there is always some loss of energy (second law). Therefore, there is a flow of energy from the sun through an ecosystem.

Chapter 50

Objective Questions

1.	a	2.	d
3.	c	4.	b
5.	b	6.	c
7.	d	8.	c
9.	a	10.	c

11. a. acid deposition; b. ozone shield destruction; c. greenhouse effect; d. photochemical smog.

Concepts and Critical Thinking

1. The primary productivity of the other terrestrial biomes might increase due to carbon dioxide buildup following tropical rain forest destruction.
2. The figure shows, for example, that nitrous dioxide (NO_2), nitrous oxide, carbon dioxide, methane, CFCs, and halons all contribute to the greenhouse effect, and of these CFCs and halons also caused ozone shield destruction.
3. Fossil fuel energy contributes to agricultural yield and therefore helps support the increased human population size of today.

GLOSSARY

A

abscisic acid (ABA) (ab-sis'ik) a plant hormone causing stomata to close and initiating and maintaining dormancy *528*

abscission (ab-sizh'un) the dropping of leaves, fruits, or flowers from a plant body *528*

acetylcholine (as"ĕ-til-ko'lēn) a neurotrans-mitter used within both the peripheral and central nervous systems *639*

acetyl CoA (as"ĕ-til-ko-A) molecule made up of a 2-carbon acetyl group attached to coenzyme A, which enters the Krebs cycle for further oxidation *143*

acid a compound tending to dissociate and yield hydrogen ions in a solution and to lower its pH numerically *36*

acid deposition the return to earth as rain or snow of the sulfate or nitrate salts of acids produced by commercial and industrial activities on earth *800*

acoelomate (a-sēl'o-māt) condition where an organism does not develop a coelom or true body cavity *424*

ACTH adrenocorticotrophic hormone, which is secreted by the anterior pituitary and stimulates activity in the adrenal cortex *683*

actin a muscle protein making up the thin filaments in a sarcomere; its movement shortens the sarcomere, yielding muscle contraction *671*

action potential change in the polarity of a neuron caused by opening and closing of ion channels in the axon plasma mem-brane *639*

active site that part of an enzyme mol-ecule where the substrate fits and the chemical reaction occurs *109*

active transport use of a plasma mem-brane carrier molecule to move particles from a region of lower to higher concentra-tion; it opposes an equilibrium and requires energy *94*

adaptation an organism's modification in structure, function, or behavior to increase the likelihood of continued existence *8, 307*

adaptive radiation formation of a large number of species from a common ancestor *336*

adenine (A) (ad'ē-nīn) one of 4 organic bases in the nucleotides composing the structure of DNA and RNA *54, 238*

adenosine diphosphate (ADP) (ah-den'o-sēn di-fos'fāt) one of the products of the hydrolysis of ATP, a process that liberates energy *116*

adenosine triphosphate (ATP) (ah-den'o-sēn tri-fos'fāt) a compound containing adenine, ribose, and 3 phosphates, 2 of which are high-energy phosphates. The breakdown of ATP to ADP makes energy available for energy-requiring processes in cells *115*

adipose tissue type of loose connective tissue in which fibroblasts enlarge and store fat *541*

ADP *see* adenosine diphosphate

adrenal gland lies atop a kidney and secretes the stress hormones epinephrine, norepinephrine, and the corticoid hor-mones *685*

aerobic process in which oxygen is required *145*

age structure diagram a representation of the number of individuals in each age group in a population *748*

agglutination (ah-glūt-en-ā'shun) clumping of red blood cells, as when incompatible red cell antigens and antibodies mix *568*

AIDS acquired immunodeficiency syn-drome, caused by the HIV virus in humans; the virus attacks T4 lymphocytes, making them ineffective and leading to the development of other infections *707*

aldosterone (al"do-ster'on) hormone secreted by the adrenal gland that main-tains the sodium and potassium balance of the blood *630, 686*

alga (pl., algae) aquatic organisms carrying out photosynthesis and belonging to the kingdom Protista *383*

allantois (ah-lan'to-is) an extraembryonic membrane that accumulates nitrogenous wastes in the embryo of birds and reptiles and contributes to the formation of umbilical blood vessels in mammals *719*

allele (ah-lēl') alternative forms of a gene, for example, an allele for long wings and an allele for short wings in fruit flies *186*

allopatric speciation origin of new species in populations that are separated geo-graphically *334*

alternation of generations cycle life cycle in which a haploid generation alternates with a diploid generation *384*

alveolus (al-ve'o-lus) terminal, micro-scopic, grapelike air sacs found in verte-brate lungs *612*

altruism form of social behavior found in insect societies, exemplified by the willing-ness of the daughters to help the queen rather than reproduce their own offspring *740*

amino acid organic subunits each having an amino group and an acid group, that covalently bond to produce protein molecules *50*

amnion an extraembryonic membrane of the bird, reptile, and mammal embryo that forms an enclosing fluid-filled sac *719*

amphibian class of terrestrial vertebrates that includes frogs, toads, and sala-manders; still tied to a watery environment for reproduction *452*

anaerobic a process that does not require oxygen *142*

angiosperm flowering plant having double fertilization that results in development of seed-bearing fruit *404*

annelid (ah'nel-id) a phylum of segmented worms characterized by a tube-within-a-tube body plan, bilateral symmetry, and segmentation of some organ systems *434*

anther part of stamen where pollen grains develop *407, 510*

antheridium (an"ther-id'i-um) male structures in nonseed plants where swimming sperm are produced *398*

antibody a protein molecule usually formed naturally in an organism to combat antigens or foreign substances *563, 577*

anticodon 3 nucleotides on a tRNA molecule attracted to a complementary codon on mRNA *257*

antidiuretic hormone (ADH) (an"tī-di'u-ret'ik) hormone secreted by the posterior pituitary that increases the permeability of the collecting duct in a nephron *628, 681*

antigen substances, usually proteins, that are not normally found in the body *563, 577*

aorta largest systemic artery, which transports blood from the heart to all other systemic arteries *560*

appendicular skeleton part of the skeleton forming the upper appendages, shoulder girdle, lower appendages, and hip girdle *666*

archaebacterium (ar"ke-bak-te're-um) probably the earliest prokaryote; the cell wall and plasma membrane differ from other bacteria, and many live in extreme environments *374*

archegonium (ar"kē-go'ne-um) female structure in nonseed plants where an egg is produced *398*

artery a blood vessel that transports blood away from the heart toward arterioles *555*

arthropod a diverse phylum of animals that have jointed appendages; includes lobsters, insects, and spiders *437*

asexual reproduction reproduction involving only mitosis which produces offspring genetically identical to the parent *166*

aster short, radiating fibers produced by the centrioles; important during mitosis and meiosis *160*

atom the smallest particle of an element that displays its properties *26*

ATP *see* adenosine triphosphate

autonomic nervous system a branch of the peripheral nervous system administering motor control over internal organs *641*

autosome any chromosome other than the sex-determining pair *204*

autotroph self-nourishing organism; referring to producers starting food chains that make organic molecules from inorganic nutrients *356*

auxin plant hormone regulating growth, particularly cell elongation; also called indoleacetic acid *523*

axial skeleton part of the skeleton forming the vertical support or axis, including the skull, rib cage, and vertebral column *666*

axon the part of a neuron that conducts the impulse from cell body to synapse *635*

B

B lymphocyte white blood cell that produces and secretes antibodies that can combine with antigens *577*

bacteriophage a virus that parasitizes a bacterial cell as its host, destroying it by lytic action *237*

bacterium (pl., bacteria) one-celled organism that lacks a nucleus and cytoplasmic organelles other than ribosomes; reproduces by binary fission and occurs in one of 3 shapes (rod, sphere, spiral) *66, 370*

Barr body X chromosome in females that is inactive *269*

basal body structure in the cell cytoplasm, located at the base of a cilium or flagellum *78*

base a compound tending to lower the hydrogen ion concentration in a solution and raise its pH numerically *36*

behavior all responses made by an organism to changes in the environment *5*

benthic division the ocean floor with a unique set of organisms in contrast to the pelagic division or open waters *768*

bicarbonate ion form in which most of the carbon dioxide is transported in the bloodstream *614*

bilaterally symmetrical a body having 2 corresponding or complementary halves *419*

bile a substance released from the gallbladder to small intestine to emulsify fats prior to chemical digestion *594*

binary fission splitting of a parent cell into 2 daughter cells; serves as an asexual form of reproduction in single-celled organisms *156, 371*

biogeography the study of the geographical distribution of organisms *303*

biological clock internal mechanism that maintains a biological rhythm in the absence of environmental stimuli *530, 732*

biological magnification process by which substances become more concentrated in organisms in the higher trophic levels of the food chain *784*

biome a major biotic community having well-recognized life forms and a typical climax species *766*

biosphere a thin shell of air, land, and water around the earth that supports life *8, 766*

biotic potential maximum rate of natural increase of a population that can occur under ideal circumstances *745*

bivalent (tetrad) (bi-va'lent) homologous chromosomes each having sister chromatids that are joined by a nucleoprotein lattice during meiosis *172*

blastocoel (blas'to-sēl) the fluid-filled cavity of a blastula in the early embryo *713*

blastula a hollow, fluid-filled ball of cells prior to gastrula formation of the early embryo *713*

blood type of connective tissue in which cells are separated by a liquid called plasma *541*

blood pressure force of flowing blood pushing against the inside wall of an artery *561*

bone type of connective tissue in which a matrix of calcium salts is deposited around protein fibers *541*

bronchus (pl., bronchi) one of 2 main branches of the trachea in vertebrates that have lungs *612*

bronchiole small tube that conducts air from a bronchus to the alveoli *612*

bryophyte a plant of a group that includes mosses and liverworts *398*

budding in animals an asexual form of reproduction whereby a new organism develops as an outgrowth of the body of the parent *695*

buffer a substance or group of substances that tend to resist pH changes in a solution, thus stabilizing its relative acidity *36*

bundle sheath cells that surround the xylem and phloem of leaf veins *131, 491*

C

calorie 1/1,000 of a kilocalorie; a unit used to express the quantity of energy of carbohydrates, fats, and proteins *34, 109*

C_3 plant plant using the enzyme rubisco to fix carbon dioxide to RuBP; the first detected molecule after fixation is PGAL, a 3-carbon molecule *131*

C_4 plant plant that does not directly use the Calvin cycle and produces an immediate 4-carbon molecule during photosynthesis *131*

Calvin cycle series of photosynthetic reactions in which carbon dioxide is reduced in the chloroplast *129*

cambium meristematic, or growth, tissue of a plant; for example, vascular cambium between the xylem and phloem in a stem or the cork cambium beneath the epidermis of the stem *484*

CAM plant plant that uses PEP carboxylase to fix carbon dioxide at night; CAM stands for crassulacean acid metabolism *133*

cancer condition by which cells grow and divide uncontrollably, detaching from tumors and spreading throughout the body *271*

capillary microscopic blood vessel for gas and nutrient exchange with body cells *555*

carbohydrate a family of organic compounds consisting of carbon, hydrogen, and oxygen atoms; includes subfamilies monosaccharides, disaccharides, and polysaccharides *44*

carbon cycle circulation of carbon between the biotic and abiotic component of both terrestrial and aquatic ecosystems *787*

carbon dioxide (CO₂) fixation photosynthetic reaction in which carbon dioxide is attached to an organic compound *130*

carbonic anhydrase an enzyme in red blood cells that speeds up the formation of carbonic acid from water and carbon dioxide *614*

carcinogen agent that contributes to the development of cancer *273*

cardiovascular system animal organ system consisting of the blood, heart, and a series of blood vessels that distribute blood under the pumping action of the heart *555*

carnivore a secondary consumer in a food chain that eats other animals *781*

carpel simple plant reproductive unit of a simple pistil; consisting of 3 parts, the stigma, style, and ovary *509*

carrier an individual who is capable of transmitting an infectious or genetic disease; a protein that combines with and transports a molecule across the plasma membrane *94, 219*

carrying capacity the maximum size of a population that can be supported by the environment in a particular locale *747*

cartilage a type of connective tissue having a matrix of protein and proteinaceous fibers that is found in animal skeletal systems *541*

Casparian strip a waxy layer bordering 4 sides of root endodermal cells; prevents water and solute transport between adjacent cells *481*

catastrophism (kah-tas'tro-fizm) belief espoused by Cuvier that periods of catastrophic extinctions occurred, after which repopulation of surviving species took place and gave the appearance of change through time *301*

cell the smallest unit that displays properties of life; composed of cytoplasmic regions, possibly organelles, and surrounded by a plasma membrane *3, 61*

cell plate (sel plāt) structure across a dividing plant cell that signals the location of new plasma membranes and cell walls *164*

cell theory statement that all organisms are made of cells and that cells are produced only from preexisting cells *61*

cellular respiration metabolic reactions that provide energy to cells by the step-by-step oxidation of carbohydrates *114, 138*

cellulose polysaccharide consisting of covalently bonded glucose chains, hydrogen-bonded within fibrils; important in plant cell walls *45*

cell wall a relatively rigid structure composed mostly of polysaccharides that surrounds the plasma membrane of plants, fungi, and bacteria *64*

central nervous system the brain and nerve (spinal) cord in animals *634*

centriole cell organelle, existing in pairs, that possibly organizes a mitotic spindle for chromosome movement during cell division *78*

centromere a constriction where duplicates (chromatids) of chromosomes are held together *157*

cephalization (sef'al-i-za'shun) having a well-recognized anterior head with concentrated nerve masses and receptors *421*

cerebellum portion of the brain that is dorsal to the pons and medulla oblongata; functioning in muscle coordination to produce smooth graceful motions *645*

cerebrum foremost part of the brain consisting of 2 large masses—cerebral hemispheres; the largest part of the brain in humans *644*

chemical evolution an increase in the complexity of chemicals that could have led to first cells *355*

chemiosmotic phosphorylation (kem"ĭ-os-mot'ik fos"for-ĭ-la'shun) production of ATP by utilizing the energy released when H⁺ ions flow through an ATP synthase complex in mitochondria and chloroplasts *117*

chemoreceptor a receptor that is sensitive to chemical stimulation, for example, receptors for taste and smell *653*

chemoautotroph bacterium capable of oxidizing inorganic compounds, such as ammonia, nitrites, and sulfides, to gain energy to synthesize carbohydrates *373*

chiasma during synapsis in meiosis, the region of attachment between nonsister chromatids of a bivalent due to crossing-over *173*

chitin (ki'tin) a strong but flexible, nitrogenous polysaccharide that is found in the exoskeleton of arthropods *46, 437*

chlorophyll green pigment that absorbs sunlight energy and is important in photosynthesis *125*

chloroplast a membrane-bounded organelle with membranous grana that contain chlorophyll and where photosynthesis takes place *74, 125*

cholesterol one of the major lipids found in animal plasma membranes; makes the membrane impermeable to many molecules *50, 88*

chordate phylum of animals that includes lancelets, tunicates, fishes, amphibians, reptiles, birds, and mammals; characterized by a notochord, dorsal nerve cord, and gill pouches *443*

chorion (kor'e-on) an extraembryonic membrane functioning for respiratory exchange in the eggs of birds and reptiles; contributes to placenta formation in mammals *718*

chromatid one of 2 identical parts of a chromosome, produced through DNA replication; within the same chromosome, one chromatid is the sister chromatid of the other *157*

chromatin the mass of DNA and associated proteins observed within a nucleus that is not dividing *68, 157, 269*

chromosome association of DNA and proteins arranged linearly in genes and visible only during cell division *68, 157*

chromosome theory of inheritance theory stating that the genes are on chromosomes accounting for their similar behavior *202*

cilium (pl., cilia) short, hairlike projection from the cell membrane, found as one of a group (cilia) *78*

circadian rhythm (ser"kah-de'an) a biological rhythm with a 24-hour cycle *530, 731*

citric acid cycle a cycle of reactions in mitochondria, that begins and ends with citric acid; produces CO₂, ATP, NADH, and FADH₂; also called the Krebs cycle *144*

cladistics (clah-dis'tiks) means of classification that places a common ancestor, and all organisms evolved from it, into one taxon; based on studying homologous structures and other similarities among organisms *342*

classical conditioning one type of learning in which an animal learns to give a response to an irrelevant stimulus; also called associative learning *735*

cleavage earliest division of the zygote developmentally, without cytoplasmic addition or enlargement *163, 713*

cleavage furrowing indenting of the plasma membrane that leads to division of the cytoplasm during cell division *163*

climax community mature stage of a community that can sustain itself indefinitely as long as it is not stressed *771*

cloned (klōnd) the production of identical copies; in genetic engineering, the production of many, identical copies of a gene 279

cnidarian (ni-dah′re-an) phylum of animals with a gastrovascular cavity; includes sea anemones, corals, hydras, and jellyfishes 419

cochlea (kok′le-ah) spiral-shaped structure of the inner ear containing the receptor hair cells for hearing 660

codominance a condition in heredity in which both alleles in the gene pair of an organism are expressed 197

codon 3 nucleotides of DNA or mRNA, codes for a particular amino acid 253

coelomate (sēl′o-māt) characterizes animals with a true coelom which is completely lined with mesoderm 424

coenzyme a nonprotein organic part of an enzyme structure, often with a vitamin as a subpart 112

cohesion-tension model explanation for the transport of water to great heights in a plant by water molecules clinging together as transpiration occurs 497

collecting duct the final portion of a nephron where reabsorption of water can be controlled 625

commensalism a symbiotic relationship in which one species is benefited and the other is neither harmed nor benefited 760

communication an action by a sender that influences the behavior of the recipient 738

community many different populations of organisms that interact with each other 755

competitive exclusion principle theory that no 2 species can occupy the same niche 756

complementary base pairing bonding between particular purines and pyrimidines in DNA 55, 240

complement system series of proteins, produced by the liver and present in the plasma, that produces a cascade of reactions as part of the immune system 576

compound (kom′pownd) a chemical substance having 2 or more different elements in fixed ratio 29

compound eye a type of eye found in arthropods composed of many independent visual units 437, 654

condensation chemical change producing the covalent bonding of 2 monomers with the accompanying loss of a water molecule 44

cone photoreceptors in vertebrate eyes that respond to bright light and allow color vision 655

conifer conifers are one of the 4 groups of gymnosperm plants; cone-bearing trees that include pine, cedar, and spruce 405

conjugation the transfer of genetic material from one cell to another through a cytoplasmic bridge 385

connective tissue animal tissue type that binds structures together, provides support and protection, fills spaces, stores fat, and forms blood cells 540

consumer an organism that feeds on another organism in a food chain; primary consumers eat plants, and secondary consumers eat animals 781

control group a sample that goes through all the steps of an experiment except the one being tested; a standard against which results of an experiment are checked 15

corepressor molecule that binds to a repressor, allowing the repressor to bind to an operator in a repressible operon 268

cork outer covering of bark of trees; made up of dead cells that may be sloughed off 477

corpus luteum (kor′pus lu′te-um) a follicle that has released an egg and increases its secretion of progesterone 701

cortex layer of the young plant root or stem beneath the epidermis; consisting of large, thin-walled parenchyma cells 481

cotyledon (kot-el-ēd′en) seed leaf for embryonic plant, providing nutrient molecules for the developing plant before its mature leaves begin photosynthesis 478

covalent bond (ko-va′lent) a chemical bond in which the atoms share one pair of electrons 31

cristae shelflike folds of the inner membrane of the mitochondrion, projecting into the inner space of the organelle 75

Cro-Magnon hominid who lived 40,000 years ago; accomplished hunters, made compound stone tools, and possibly had language 467

crossing-over an exchange of segments between nonsister chromatids of a bivalent during meiosis 172

cultural eutrophication human activities result in an acceleration of the process by which nutrient poor lakes and ponds become nutrient rich; speeds up aquatic succession, in which a pond or lake eventually fills in and disappears 796

cyanobacterium (si″ah-no-bak-te′re-um) formerly called blue-green alga; photosynthetic prokaryote that contains chlorophyll and releases oxygen 66, 375

cyclic AMP an ATP-related compound that promotes chemical reactions in the body; the "second messenger" in peptide hormone activity, it initiates activity of the metabolic machinery 679

cyclic photophosphorylation (fo″to-fos″for-ĭ-la′shun) photosynthetic pathway in which electrons from photosystem P700 pass through the electron transport system, produce ATP, and eventually return to P700 128

cystic fibrosis a lethal genetic disease affecting the mucus in the lungs and digestive tract 221

cytochrome system a series of electron carrier molecules, part of the electron transport systems of mitochondria and chloroplasts 128, 144

cytokinesis (si″to-ki-ne′sis) the division of the cytoplasm following mitosis and meiosis 157

cytokinin (si″to-ki′nin) plant hormone that promotes cell division; often works in combination with auxin during organ development in plant embryos 526

cytoplasm contents of a cell between the nucleus (nucleoid) and the plasma membrane 64

cytosine (C) (si′to-sin) one of 4 organic bases in the nucleotides composing structure of DNA and RNA 238

cytoskeleton internal framework of the cell, consisting of microtubules, actin filaments, and intermediate filaments 76

cytosol fluid medium that bathes the contents of the cytoplasm 64

D

datum (pl., data) fact or information collected through observation and/or experimentation 14

decomposer organism, usually a bacterial or fungal species, that breaks down large organic molecules into elements that can be recycled in the environment 373

deductive reasoning a process of logic and reasoning, using "if . . . then" statements 14

demographic transition a decline in death rate, followed shortly by a decline in birthrate, resulting in slower population growth 750

denatured condition of an enzyme when its shape is changed so that its active site cannot bind substrate molecules 111

dendrite the part of a neuron that sends impulses toward the cell body 635

denitrification (de-ni″trĭ-fi-ka′shun) the process of converting nitrogen compounds to atmospheric nitrogen; is a part of the nitrogen cycle 781

deoxyribonucleic acid (DNA) an organic molecule produced from covalent bonding of nucleotide subunits; the chemical composition of genes on chromosomes 54, 235

dependent variable result or change that occurs when the experimental variable is manipulated 15

dermis deeper, thicker layer of the skin that consists of fibrous connective tissue and contains various structures such as sense organs 544

desert treeless biome where the annual rainfall is less than 25 cm; the rain that does fall is subject to rapid runoff and evaporation 773

desertification (dez"ert-ĭ-fi-ka'shun) transformation of marginal lands to desert 793

detritus (de-tri'tus) falling, settled remains of plants and animals on the land-floor or water bed 781

deuterostome (du'ter-o-stōm") a group of coelomate animals in which the second embryonic opening becomes the mouth; the first embryonic opening, the blasto-pore, becomes the anus 430

diabetes mellitus (di"ah-be'tēz mě-li'tus) a disease caused by a lack of insulin production by the pancreas 686

diaphragm a dome-shaped muscle separating the thoracic cavity from the abdominal cavity in the animal body and involved in respiration 609

diastole (di-as'to-le) relaxation period of a heart during the cardiac cycle 558

dicotyledon a flowering plant group; members show 2 embryonic leaves, net-veined leaves, cylindrical arrangement of vascular bundles, and other characteristics 407

dictyosome organelle in a plant cell that is called a Golgi apparatus in animal cells 72

differentiation specialization of early embryonic cells with regard to structure and function 712

diffusion the movement of molecules from a region of high to low concentration, requiring no energy and tending toward an equal distribution 89

dihybrid a genetic cross involving 2 traits 190

diploid (2N) number (dip'loid) the condition in which cells have 2 of each type of chromosome 157

diplontic cycle (dip-lon'tik) life cycle in which the adult is diploid 384

directional selection outcome of natural selection in which an extreme phenotype is favored, usually in a changing environment 326

disruptive selection outcome of natural selection in which extreme phenotypes are favored over the average phenotype and can lead to polymorphic forms 325

distal convoluted tubule the portion of a nephron between the loop of the nephron and the collecting duct, where tubular secretion occurs 625

DNA see deoxyribonucleic acid

DNA ligase (li'gas) an enzyme that links DNA fragments; used in genetic engineering to join foreign DNA to the vector DNA 280

DNA marker genetic variation found within highly repetitive DNA and inherited in a Mendelian fashion; the distinctive pattern of repetition in a person serves as a DNA fingerprint or means of identification 290

DNA polymerase (pol-im'er-ās) enzyme that joins complementary nucleotides to form double-stranded DNA 241

DNA probe known sequences of DNA that are used to find complementary DNA strands; can be used diagnostically to determine the presence of particular genes 283

dominance hierarchy organization of animals in a group that determines the order in which the animals have access to resources 736

dominant allele (ah-lēl') allele of a given gene that hides the effect of the recessive allele in a heterozygous condition 186

Down syndrome a genetic disease caused by trisomy 21 that is characterized by mental retardation and a specific set of physical features 216

dryopithecine (dri"o-pith'e-sin) a possible hominoid ancestor; a forest-dwelling primate with characteristics somewhat like those of living apes 461

duodenum (dū-ah-dē'-num) first part of the small intestine where chyme enters from the stomach 593

Duchenne muscular dystrophy (du-shen') an X-linked, recessive disorder that causes muscular weakness and eventually death at a young age 230

E

echinoderm (e-kin'o-derm) a phylum of marine animals that includes sea stars, sea urchins, and sand dollars; characterized by radial symmetry and a water vascular system 440

ecological pyramid pictorial graph representing the biomass, organism number, or energy content of each trophic level in a food web from the producer to the final consumer populations 782

ecology the study of the interactions of organisms with their living and physical environment 745

ecosystem a biological community together with the associated abiotic environment 8, 781

ectoderm outermost of an animal's primary germ layers 713

effector organ that makes a response when signaled by a motor neuron 641

electromagnetic spectrum solar radiation divided on the basis of wavelength, with gamma rays having the shortest wave-length and radio waves having the longest wavelength 122

electron negative subatomic particle, moving about in energy levels around the nucleus of the atom 26

electron transport system a mechanism whereby electrons are passed along a series of carrier molecules to produce energy for the synthesis of ATP 114, 144

element the simplest of substances consisting of only one type of atom; for example, carbon, hydrogen, oxygen 26

embryo early developmental stage of a plant or animal, produced from a zygote 510

emigrate (n., emigration) to move from a geographical area 748

endocrine system one of the major systems involved in the coordination of body activities; uses messengers called hormones which are secreted into the bloodstream 677

endocytosis (en"do-si-to'sis) moving particles or debris into the cell from the environment by phagocytosis (cellular eating) or pinocytosis (cellular drinking) 97

endoderm innermost of an animal's primary germ layers 713

endodermis internal plant root tissue forming a boundary between the cortex and vascular cylinder 481

endometrium (en-do-me'tre-um) a mucous membrane lining the inside free surface of the uterus 700

endoplasmic reticulum (en"do-plas'mik rě-tik'u-lum) a system of membranous saccules and channels in the cytoplasm 71

endosperm nutritive tissue in a seed; often triploid from a fusion of sperm cell with 2 polar nuclei 510

endospore a bacterium that has shrunk its cell, rounded up within the former plasma membrane, and secreted a new and thicker cell wall in the face of unfavorable environmental conditions 371

endosymbiotic theory statement that chloroplasts and mitochondria (and perhaps cilia) were originally prokaryotes that evolved a mutualistic relationship with the eukaryotic cell 75

energy capacity to do work and bring about change; occurs in a variety of forms *28, 105*

environment surroundings with which a cell or organism interacts *8*

environmental resistance the opposing force of the environment on the biotic potential of a population *747*

enzyme a protein that acts as an organic catalyst to speed up reaction rates in living systems *50, 108*

epicotyl (ep″ĭ-kot′il) plant embryo portion above the cotyledon that contributes to shoot development *515*

epidermis covering tissue of roots and leaves of plants, plus stems of nonwoody organisms; outer, protective layer of the skin *476, 544*

epiglottis a flaplike covering, hinged to the back of the larynx and capable of covering the glottis or air-tract opening *593, 612*

epiphyte (ep′ĭ-fīt) a plant that takes its nourishment from the air because its attachment to other plants gives it an aerial position *503, 778*

epistatic gene (epistasis) (ĕp″ĭ-stat′ik jēn) a gene that interferes with the expression of alleles that are at a different locus *198*

epithelial tissue (ep-ah-thē′lē-al) animal tissue type forming a continuous layer over most body surfaces (i.e., skin) and inner cavities *539*

erythrocyte (e-rith′ro-sit) red blood cell; important for oxygen transport in blood *563*

esophagus a muscular tube for moving swallowed food from pharynx to stomach *593*

essential amino acid an amino acid that is required by the body but must be obtained in the diet because the body is unable to produce it *149, 598*

estrogen female ovarian sex hormone that has numerous effects on the endometrium of the uterus throughout the ovarian cycle *700*

estuary the end of a river where fresh water and salt water mix as they meet *766*

ethylene plant hormone that causes ripening of fruit and is also involved in abscission *527*

eubacterium most common type of bacteria, including the photosynthetic bacteria *374*

euchromatin diffuse chromatin, which is being transcribed *269*

eukaryotic cell typical of organisms, except bacteria and cyanobacteria, having organelles and a well-defined nucleus *64*

evolution genetic and phenotypic changes that occur in populations of organisms with the passage of time, often resulting in increased adaptation of organisms to the prevailing environment *300*

exhalation stage during respiration when air is pushed out of the lungs *609*

exocytosis a process by which cells expel particles or debris in vesicles passing through a plasma membrane to the extracellular environment *97*

exon in a gene, the portion of the DNA code that is expressed as the result of polypeptide formation *254*

exoskeleton a protective, external skeleton, as in arthropods *437*

experimental variable condition that is tested in an experiment by varying it and observing the results *15*

exponential referring to a geometrically multiplying, rapid population growth rate *745*

F

facilitated transport process whereby molecules pass through a plasma membrane, utilizing a carrier molecule but without the expenditure of energy *94*

FAD (flavin adenine dinucleotide) (fla′vin ad′ĕ-nīn di-nu′kle-o-tīd) coenzyme that functions as an electron acceptor in cellular oxidation-reduction reactions *114*

fat organic compound consisting of 3 fatty acids covalently bonded to one glycerol molecule *47*

fatty acid subunit of a fat molecule, characterized by a carboxyl acid group at one end of a long hydrocarbon chain *47*

feedback inhibition process by which a substance, often an end product of a reaction or a metabolic pathway, controls its own continued production by binding with the enzyme which produced it *112*

fermentation anaerobic breakdown of carbohydrates that results in products such as alcohol and lactate *139*

fibroblast cell type of loose and fibrous connective tissue with cells at some distance from one another and separated by a jellylike substance *540*

filament the elongated stalk of a stamen bearing the anther at the tip *510*

fitness ability of an organism to survive and reproduce in its local environment *305*

fixed-action pattern (FAP) a stereotyped behavior pattern that occurs automatically *733*

flower reproductive organ of a flowering plant, consisting of several kinds of modified leaves arranged in concentric rings and attached to a modified stem called the receptacle *407*

fluid-mosaic model model for the plasma membrane based on the changing location and pattern of protein molecules in a fluid phospholipid bilayer *86*

follicle structures in the ovary of animals that contain oocytes; site of egg production *700*

follicle-stimulating-hormone (FSH) gonadotrophic hormone secreted by anterior pituitary that controls the events of the ovarian cycle by being secreted at different rates throughout the cycle *701*

food chain a succession of organisms in an ecosystem that are linked by an energy flow and the order of who eats whom *782*

food web a complex pattern of interlocking and crisscrossing food chains *782*

founder effect the tendency for a new, small population to experience genetic drift after it has separated from an original, larger population *323*

fruit flowering plant structure consisting of one or more ripened ovaries that usually contain seeds *407, 511*

fruiting body a spore-bearing structure found in certain types of fungi, such as mushrooms *391*

FSH *see* follicle-stimulating hormone

fungus saprophytic decomposer; body is made up of filaments called hyphae that form a mass called a mycelium *391*

G

gallbladder attached to the liver; serves as a storage organ for bile *594*

gamete (gam′et) haploid sex cell *157*

gametophyte haploid generation of the alternation of generations life cycle of a plant; it produces gametes that unite to form a diploid zygote *398*

ganglion (sing., ganglia) (gang′gle-on) a knot or bundle of neuron cell bodies outside the central nervous system *640*

gastrovascular cavity a blind, branched digestive cavity that also serves a circulatory (transport) function in animals that lack a circulatory system *419*

gastrula (gas′troo-lah) cup-shaped early embryo with 2 primary germ layers, ectoderm and endoderm, enclosing a primitive digestive tract *713*

gel electrophoresis (ĭ-lek-tra-fa-rē′sus) technique that allows DNA fragments or proteins to be separated according to differences in the charge and size *290*

gene the unit of heredity occupying a particular locus on the chromosome and passed on to offspring *5*

gene flow sharing of genes between 2 populations through interbreeding *320*

gene locus the specific location of a particular gene on a chromosome *186*

gene mutation (jēn mu-ta'shun) a change in the base sequence of DNA such that the sequence of amino acids is changed in a protein *316*

gene pool the total of all the genes of all the individuals in a population *318*

gene therapy use of transplanted genes to overcome an inborn error in metabolism *288*

genetic drift change in the genetic make-up of a population due to chance (random) events; important in small populations or when only a few individuals mate *322*

genetic engineering alteration of the genome of an organism by technological processes *279*

genome (jē'nōm) all of the genes of an organism *244*

genotype (jen'o-tīp) the genes of an organism for a particular trait or traits; for example, *BB* or *Aa 187*

germination resumption of growth by a seed or any other reproductive structure of a plant or protist *515*

germ layer developmental layer of body; that is, ectoderm, mesoderm, and endoderm *714*

gibberellin (gib-ber-el'in) plant hormone producing increased stem growth by cell division and enlargement; also involved in flowering and seed germination *526*

gill respiratory organ in most aquatic animals; most common in fish as an outward extension of the pharynx *607*

girdling removing a strip of bark from around a tree *503*

glomerular capsule a cuplike structure that is the initial portion of a nephron; where pressure filtration occurs *625*

glomerulus (glo-mer'u-lus) a capillary network within a glomerular capsule of a nephron *625*

glottis opening for airflow in the larynx *612*

glucagon (glū'kah-gon) hormone secreted by the pancreas that stimulates the breakdown of stored nutrients and increases the level of glucose in the blood *686*

glucose monosaccharide with the molecular formula of $C_6H_{12}O_6$ found in the blood of animals, serving as an energy source in most organisms and transported in the blood of animals *45*

glycogen polysaccharide consisting of covalently bonded glucose molecules, characterized by many side branches, energy storage product in animals *45*

glycolysis (gli-kol'i-sis) pathway of metabolism converting a sugar, usually glucose, to pyruvate; resulting in a net gain of 2 ATP and 2 NADH molecules *139*

goiter an enlargement of the thyroid gland caused by a lack of iodine in the diet *683*

Golgi apparatus an organelle consisting of a central region of saccules and vesicles that modifies and/or activates proteins and produces lysosomes *72*

gonadotrophic hormone (GNRH) (gō-nad-ah-trō'fik) either FSH or LH, a substance secreted by the anterior pituitary that stimulates the gonads *482*

granum (pl., grana) (gra'num) in a chloroplast, a stack of flattened membrane sacs, thylakoids, that contain chlorophyll *74, 123*

gravitropism (grav"ĭ-tro'pizm) directional growth in response to the earth's gravity; roots demonstrate positive gravitropism, stems demonstrate negative gravitropism *529*

greenhouse effect the reradiation of heat toward the earth caused by gases in the atmosphere *800*

growth hormone (GH) substance stored and secreted by the anterior pituitary; promotes cell division, protein synthesis, and bone growth *683*

guanine (G) (gwan'in) one of 4 organic bases in the structure of nucleotides composing the structure of DNA and RNA *54, 238*

guard cell plant cell type, found in pairs, with one on each side of a leaf stoma; changes in the turgor pressure of these cells regulate the size and passage of gases through the stoma *498*

guttation (gut-ta'shun) liberation of water droplets from the edges and tips of leaves *497*

gymnosperm a vascular plant producing naked seeds, as in conifers *405*

H

habitat region of an ecosystem occupied by an organism *756*

haploid (N) number a cell condition in which only one of each type of chromosome is present *157*

haplontic cycle life cycle in which the adult is haploid *384*

Hardy-Weinberg law a law stating that the frequency of an allele in a population remains stable under certain assumptions, such as random mating; therefore, no change or evolution occurs *318*

helper T cell cell that secretes lymphokines, which stimulate all kinds of immune cells *579*

hemoglobin red respiratory pigment of erythrocytes for transport of oxygen *554, 563, 614*

hemophilia a disease in which the blood fails to clot; caused by insufficient clotting factor VIII *229*

herbaceous plant a plant that lacks persistent woody tissue *407*

herbivore a primary consumer in a food chain; a plant eater *781*

hermaphroditic animal (her'maf"ro-dit'ik) characterizes an animal having both male and female sex organs *422*

heterochromatin highly compacted chromatin during interphase *269*

heterogamete (het"er-o-gam'ēt) a nonidentical gamete, for example, large, nonmotile egg and small, flagellated sperm *385*

heterotroph (het'er-o-trōf") an organism that cannot synthesize organic compounds from inorganic substances and therefore must take in preformed food *356*

heterozygous possessing 2 different alleles for a particular trait *187*

homeostasis the maintenance of internal conditions in cell or organisms, for example, relatively constant temperature, pH, blood sugar *546*

hominid referring to humans; ancestors of humans who had diverged from the ape lineage *461*

Homo erectus hominid who lived during the Pleistocene; with a posture and locomotion similar to modern humans *466*

Homo habilis (ho'mo hah'bĭ-lis) fossil hominid of 2 million years ago; possible first direct ancestor of modern humans *465*

homologous chromosome a pair of chromosomes that carry genes for the same traits and synapse during prophase of the first meiotic division *171*

homologous structure in evolution, structure derived from a common ancestor *310*

homologue (hom'o-log) member of homologous pair of chromosomes *171*

homozygous possessing 2 identical alleles for a particular trait *186*

hormone a chemical secreted into the bloodstream in one part of the body that controls the activity of other parts *523*

Huntington disease (HD) a lethal, genetic disease affecting one in 10,000 people in the United States; affects neurological functioning and generally does not manifest itself until middle age *223*

hydra a freshwater cnidarian with only a polyp stage that reproduces both sexually and asexually *420*

hydrogen bond a weak bond that arises between a partially positive hydrogen and a partially negative oxygen, often on different molecules or separated by some distance *33*

hydrolysis splitting of a compound into parts in a reaction that involves addition of water, with the H^+ ion being incorporated in one fragment and the OH^- ion in the other *44*

hydrophilic a type of molecule that interacts with water, such as polar molecules, which dissolve in water or form hydrogen bonds with water molecules *34, 43*

hydrophobic a type of molecule that does not interact with water, such as nonpolar molecules, which do not dissolve in water or form hydrogen bonds with water molecules *43*

hydroponics water culture method of growing plants that allows an experimenter to vary the nutrients and minerals provided so as to determine specific growth requirements *500*

hydrostatic skeleton characteristic of animals in which muscular contractions are applied to fluid-filled body compartments *434*

hypertonic solution lesser water concentration, greater solute concentration than the cytoplasm of a particular cell; situation in which cell tends to lose water by osmosis *92*

hypha a filament of the vegetative body of a fungus *391*

hypocotyl (hi″po-kot′il) plant embryo portion below the cotyledon that contributes to development of stem *515*

hypothalamus portion of the brain that regulates the internal environment of the body; involved in control of heart rate, body temperature, water balance, and glandular secretions of the stomach and pituitary gland *644*

hypothesis a supposition that is established by reasoning after consideration of available evidence and that can be tested by obtaining more data, often by experimentation *14*

hypotonic solution greater water concentration, lesser solute concentration than the cytoplasm of a particular cell; situation in which a cell tends to gain water by osmosis *92*

I

immune system includes many leukocytes and various organs and provides specific defense against foreign antigens *577*

immunity the ability of the body to protect itself from foreign substances and cells, including infectious microbes *575*

imprinting a behavior pattern with both learned and innate components that is acquired during a specific and limited period, usually when very young *733*

incomplete dominance a pattern of inheritance in which the offspring shows characteristics intermediate between 2 extreme parental characteristics, for example, a red and a white flower producing pink offspring *197*

inducer a molecule that brings about activity of an operon by joining with a repressor and preventing it from binding to the operator *268*

inducible operon an operon that is normally inactive but is turned on by an inducer *268*

induction the ability of a chemical or tissue to influence the development of another tissue *715*

inductive reasoning a process of logic and reasoning, using specific observations to arrive at a hypothesis *14*

inhalation stage during respiration when air is drawn into the lungs *609*

inheritance of acquired characteristics Lamarckian belief that organisms become adapted to their environment during their lifetime and pass on these adaptations to their offspring *301*

inorganic referring to molecules that are not compounds of carbon and hydrogen *42*

insight learning ability to respond correctly to a new, different situation the first time it is encountered by using reason *735*

insulin hormone secreted by the pancreas that stimulates fat, liver, and muscle cells to take up and metabolize glucose, stimulates the conversion of glucose into glycogen in muscle and liver cells, and promotes the buildup of fats and proteins *686*

interferon an antiviral agent produced by an infected cell that blocks the infection of another cell *577*

interneuron neuron, within the central nervous system, conveying messages within parts of the central nervous system *636*

interphase stage of cell cycle during which DNA synthesis occurs and the nucleus is not actively dividing *159*

intron noncoding segments of DNA that are transcribed but removed before mRNA leaves nucleus *254*

invertebrate referring to an animal without a serial arrangement of vertebrae or a backbone *416*

ion charged derivative of an atom, positive if the atom loses electrons and negative if the atom gains electrons *30*

ionic bond an attraction between charged particles (ions) through their opposite charge that involves a transfer of electrons from one atom to another *31*

isomer molecules with the same molecular formula but different structure and, therefore, shape *43*

isotonic solution a solution that is equal in solute and water concentration to that of the cytoplasm of a particular cell; situation in which cell neither loses nor gains water by osmosis *91*

isotope forms of an element having the same atomic number but a different atomic weight due to a different number of neutrons *27*

J

jointed appendage freely moving appendages of arthropods *437*

K

karyotype (kar′e-o-tīp) chromosomes are cut from a photograph of the nucleus just prior to cell division and arranged in groups for display *216*

kidney bean-shaped, reddish brown organ in humans that regulates the chemical composition of the blood and produces a waste product called urine *624*

killer T cell (kil′er te sel) cell that attacks and destroys cells that bear a foreign antigen *579*

kingdom a taxonomic category grouping related phyla (animals) or divisions (plants) *3*

kin selection an explanation for the evolution of altruistic behavior that benefits one's relatives *740*

Klinefelter syndrome a condition caused by the inheritance of a chromosome abnormality in number; an XXY (trisomy) condition where normally a chromosome pair exists *218*

Krebs cycle *see* citric acid cycle

K-strategist a species that has evolved characteristics that keep its population size near carrying capacity, for example, few offspring, longer generation time *749*

L

lacteal (lak'te-al) a lymphatic vessel in an intestinal villus; aids in the absorption of fats *596*

larynx voicebox, cartilage-made box for airflow between the glottis and trachea *612*

leaf usually broad, flat structure of a plant shoot system, containing cells that carry out photosynthesis *482*

learning change in behavior as a result of experience; there are 3 kinds of learning—imprinting, operant conditioning, and insight learning *733*

leukocyte (lu'ko-sit) white blood cell; important for immune response *563*

LH *see* luteinizing hormone

lichen (li'ken) a symbiotic relationship between certain fungi and algae which has long been thought to be mutualistic, the fungi providing inorganic food and the algae providing organic food *375, 392*

life cycle entire sequence of developmental phases from zygote formation to gamete formation *384*

ligament structure consisting of fibrous connective tissue that connects bones at joints *541, 668*

light-dependent reaction a reaction of photosynthesis that requires light energy to proceed; involved in the production of ATP and NADH *125*

light-independent reaction photosynthetic reaction that does not directly require light energy; involved primarily in the conversion of carbon dioxide to carbohydrate utilizing the products of the light-dependent reaction *125*

limbic system part of the brain beneath the cortex that contains pathways that connect various portions of the brain; allows the experience of many emotions *646*

linkage group genes that are located on the same chromosome and tend to be inherited together *206*

lipid (lip'id) a family of organic compounds that tend to be soluble in nonpolar solvents such as alcohol; includes fats and oils *46*

liposome (lip'o-som) droplet of phospholipid molecules formed in a liquid environment *356*

loop of the nephron the portion of a nephron between the proximal and distal convoluted tubules where some water reabsorption occurs *625*

lung respiratory organ mainly found in terrestrial vertebrates, evolving as a vascularized outgrowth of the lower pharyngeal region *608*

luteinizing hormone (LH) (lūt'ē-ah-nīz-ing) gonadotrophic hormone that controls events of the ovarian cycle by being secreted at different rates throughout the cycle *701*

lymph tissue fluid collected by the lymphatic system and returned to the general systemic circuit *574*

lymphatic system mammalian organ system consisting of lymphatic vessels and lymphoid organs *753*

lymphocyte specialized white blood cell, produced by lymphoid tissue, that fights infection; occurs in 2 forms—T lymphocyte and B lymphocyte *563*

lymphokine (lim'fo-kin) chemicals secreted by T cells that stimulate immune cells *579*

lysogenic cycle one of the bacteriophage life cycles in which the virus incorporates its DNA into that of the bacterium; only later does it begin a lytic cycle which ends with the destruction of the bacterium *367*

lysosome membrane-bounded vesicles that contain hydrolytic enzymes to digest macromolecules *73*

lytic cycle (lit'ik) one of the bacteriophage life cycles in which the virus takes over the operation of the bacterium immediately upon entering it and subsequently destroys the bacterium *367*

M

macroevolution large-scale evolutionary change, for example, the formation of new species *340*

macrophage (mak'ro-faj) large, phagocytotic white blood cell *563*

Malpighian tubule (mal-pig'i-an) blind, threadlike excretory tubule near anterior end of insect hindgut *438, 623*

mammal a class of vertebrates characterized especially by the presence of hair and mammary glands *455*

marsupial a mammal bearing immature young nursed in a marsupium, or pouch; for example, kangaroo, opossum *457*

mass extinction recorded in the fossil record as a very large number of extinctions over a relatively short geologic time *343*

matrix inner space of the mitochondrion, filled with a gellike fluid *75*

mechanoreceptor a cell that is sensitive to mechanical stimulation, such as that from pressure, sound waves, and gravity *658*

medulla oblongata (mĕ-dul'ah ob"long-ga'tah) a brain region at the base of the brainstem controlling heartbeat, blood pressure, breathing, and other vital functions *644*

megaspore reproductive structure produced by the megasporangium of seed plants; of the 4 produced, one develops into a female gametophyte *405*

meiosis (mi-o'sis) type of nuclear division that occurs as part of sexual reproduction in which the daughter cells receive the haploid number of chromosomes *157, 171*

meiosis I (mi-o'sis) first division of meiosis, which ends when duplicated homologous chromosomes separate and move into different cells *173*

meiosis II (mi-o'sis) second division of meiosis in which sister chromatids separate and move into different cells, resulting in the formation of haploid cells *175*

melanocyte-stimulating hormone (mah-lan'ah-sīt) substance stored and secreted by the anterior pituitary; causes color changes in many fish, amphibians, and reptiles *683*

memory B cell a B lymphocyte that remains in the bloodstream after an immune response ceases and is capable of producing antibodies to an antigen *578*

memory T cell a T lymphocyte that remains in the bloodstream after an immune response ceases *579*

menstruation periodic shedding of tissue and blood from the inner lining of the uterus *703*

meristem tissue undifferentiated, embryonic tissue in the active growth regions of plants *475*

mesoderm middle layer of an animal's primary germ layers *714*

mesoglea (mes"-o-gle'ah) a jellylike layer between the epidermis and endodermis of cnidarians *419*

mesophyll inner, thickest layer of a leaf consisting of a palisade layer of elongated cells and a spongy layer of irregularly spaced cells *491*

messenger RNA (mRNA) a nucleic acid (ribonucleic acid) complementary to genetic DNA and bearing a message to direct cell protein synthesis at the ribosome *251*

metabolic pool metabolites that are the products of and/or the substrates for key reactions in cells allowing one type of molecule to be changed into another type, such as the conversion of carbohydrates to fats 148

metabolism all of the chemical reactions that occur in a cell during growth and repair involve and produce metabolites 4, 107

metafemale a female with 3 X chromosomes; most show no obvious physical abnormalities 218

metamorphosis change in shape and form that some animals, such as insects, undergo during development 438

MHC protein (major histocompatibility complex) protein on the plasma membrane of a macrophage that aids in the stimulation of T cells 580

microbody membrane-bounded vesicle containing specific enzymes involved in lipid and alcohol metabolism, photosynthesis, and germination 73

microsphere formed from proteinoids exposed to water; has properties similar to today's cells 356

microspore reproductive structure produced by a microsporangium of a seed plant; it develops into a pollen grain 510

microtubule small cylindrical organelle that is believed to be involved in maintaining the shape of the cell and directing various cytoplasmic structures 76

mimicry superficial resemblance of an organism to one of another species; often used to avoid predation 758

mitochondrion membrane-bounded organelle of the cell known as the power-house because it transforms the energy of carbohydrates and fats to ATP energy 74

mitosis a process in which a parent nucleus reproduces 2 daughter nuclei, each identical to the parent nucleus; this division is necessary for growth and development 157

model a suggested explanation for experimental results that can help direct future research 17

molecule a unit of a chemical substance formed by the union of 2 or more atoms by covalent or ionic bonding; smallest part of a compound that retains the properties of the compound 29

mollusk a vertebrate phylum that includes squids, clams, snails, and chitons and is characterized by a visceral mass, a mantle, and a foot 431

molt periodic shedding of the exoskeleton in arthropods 437

Monera kingdom that includes bacteria, which are microscopic single-celled prokaryotes 370

monocotyledon a flowering plant group; members show one embryonic leaf, parallel leaf veins, scattered vascular bundles, and other characteristics 407

monohybrid a genetic cross involving one trait 185

monotreme an egg-laying mammal; for example, duckbill platypus or spiny anteater 457

morphogenesis (mor"fo-jen'ĕ-sis) movement of early embryonic cells to establish body outline and form 712

multiple allele more than 2 alleles for a particular trait 200

muscle fiber a cell with myofibrils containing actin and myosin filaments arranged within sarcomeres; a group of muscle fibers is a muscle 671

muscular (contractile) tissue animal tissue type, composed of fibers with actin and myosin filaments, that can shorten its length and produce movements 542

mutagen an agent, such as radiation or a chemical, that brings about a mutation in DNA 260

mutation a change in the composition of DNA, due to either a chromosomal or a genetic alteration 7, 208, 235

mutualism a symbiotic relationship in which both species benefit 761

mycelium (mi-se'le-um) a tangled mass of hyphal filaments composing the vegetative body of a fungus 391

mycorrhiza (mi"ko-ri'zah) a "fungus root" composed of a fungus growing around a plant root, which assists the growth and development of the plant 502

myelin sheath (mi'ĕ-lin) covering on neurons that is formed from Schwann cell membrane and aids in transmission of nervous impulses 636

myofibril a specific muscle cell organelle containing a linear arrangement of sarcomeres which shorten to produce muscle contraction 671

myosin (mi'o-sin) a muscle protein making up the thick filaments in a sarcomere; it pulls actin to shorten the sarcomere, yielding muscle contraction 671

N

NAD⁺ (nicotinamide-adenine dinucleotide) (nik"o-tin'ah-mīd ad'ĕ-nīn di-nu'kle-o-tīd) a coenzyme that functions as an electron and hydrogen ion carrier in cellular oxidation-reduction reactions of glycolysis and cellular respiration 113

NADP⁺ (nicotinamide-adenine dinucleotide phosphate) (nik"o-tin'ah-mīd ad'ĕ-nīn di-nu'kle-o-tīd fos'fāt) a coenzyme that functions as an electron and hydrogen ion carrier in cellular oxidation-reduction reactions of photosynthesis 114

natural selection the guiding force of evolution caused by environmental selection of organisms most fit to reproduce, resulting in adaptation 305

Neanderthal hominid who lived during the last ice age in Europe and the Middle East; made stone tools, hunted large game, and lived together in a kind of society 466

negative feedback loop pattern of homeostatic response in which the output is counter to and cancels the input 548

nematocyst (nem'ah-to-sist) in the cnidaria, a threadlike fiber enclosed within the capsule of a stinging cell; when released, aids in the capture of prey 419

nephridium (nĕ-frid'e-um) for invertebrates, a tubular structure for excretion; its contents are released outside through a nephridiopore 435

nephron microscopic kidney unit that regulates blood composition by filtration and reabsorption; over one million per human kidney 625

nerve bundle of axons and/or dendrites 640

neurofibromatosis (nu"ro-fi-bro'mah-to-sus) disease caused by an autosomal dominant allele, characterized by the proliferation of nerve cells into benign tumors under the skin 223

neuromuscular junction region where a nerve end comes into contact with a muscle fiber; contains a presynaptic membrane, synaptic cleft, and postsynaptic membrane 673

neuron a nerve cell; composed of dendrite(s), a cell body, and axon 543, 635

neurotransmitter a chemical made at the ends of axons that is responsible for transmission across a synapse or neuromuscular junction 639

neutron neutral subatomic particle, located in the nucleus of the atom 26

neutrophil (nu'tro-fil) white blood cell with granules 563

niche total description of an organism's functional role in an ecosystem, from activities to reproduction 756

nitrogen fixation process whereby free atmospheric nitrogen is converted into compounds, such as ammonia and nitrates, usually by soil bacteria 373, 788

nitrogen-fixing bacterium a bacterium that can convert atmospheric nitrogen into compounds such as ammonia and nitrates; free living in the soil or symbiotic with plants such as legumes 373

nodule (nod′ul) structure on plant roots that contains nitrogen-fixing bacteria 502

noncyclic photophosphorylation (fo″to-fos″for-ĭ-la′shun) photosynthetic pathway in which electrons from photosystem P680 pass through the electron transport system and move on to another photosystem (P700) 128

nonrenewable resource resource found in limited supplies that can be used up 803

norepinephrine (nor″ep-ah-nef′ren) a hormone released by the adrenal medulla as a reaction to stress; also a neurotransmitter released by neurons of the central and peripheral nervous systems 639

notochord (no′to-kord) a dorsal, elongated, supporting structure in chordates beneath the neural tube or spinal cord; present in the embryo of all vertebrates 714

nuclear envelope double membrane that separates the nucleus from the cytoplasm 69

nuclear power energy source generation by atomic reactions 803

nucleic acid polymer of nucleotides; both DNA and RNA are nucleic acids 54, 235

nucleoid area in prokaryotic cell where DNA is found 64, 156

nucleolus (pl., nucleoli) dark-staining, spherical body in the cell nucleus that contains ribosomal RNA 68

nucleotide the building block subunit of DNA and RNA consisting of a 5-carbon sugar bonded to a nitrogenous base and phosphorus group 54

nucleus (nu′kle-us) region of a eukaryotic cell, containing chromosomes, that controls the metabolic function of the cell 66

O

oil lipid containing triglycerides having unsaturated hydrocarbon chains; liquid at room temperature 47

olfactory cell modified neuron that is a receptor for the sense of smell 653

omnivore a food chain organism feeding on both plants and animals 781

oncogene (ong′ko-jen) a cancer-causing gene 273

oogenesis (o″o-jen′ĕ-sis) production of eggs in females by the process of meiosis and maturation 175

operant conditioning trial-and-error learning in which behavior is reinforced with a reward or a punishment 734

operator the sequence of DNA in an operon to which the repressor protein binds 266

operon a group of structural and regulating genes that functions as a single unit 267

optic nerve nerve that carries impulses from the retina of the eye to the brain 655

orbital volume of space around a nucleus where electrons can be found most of the time 28

organ combination of 2 or more different tissues performing a common function 544

organ of Corti specialized region of the cochlea containing the hair cells for sound detection and discrimination 661

organelle small, membranous structure in the cytoplasm having a specific function 64

organic containing carbon and hydrogen; such molecules usually also have oxygen attached to the carbon(s) 42

organic soup accumulation of simple organic compounds in the early oceans 355

osmosis the diffusion of water through a selectively permeable membrane 89

osmotic pressure measure of the tendency of water to move across a selectively permeable membrane into a solution; visible as an increase in liquid on the side of the membrane with higher solute concentration 91

osteocyte bone cell found within the Haversian system of bone tissue 665

ovary in flowering plants, enlarged, base portion of the pistil that eventually develops into the fruit; in animals, an egg-producing organ 407, 509, 700

oviduct tubular portion of the female reproductive tract from ovary to the uterus 700

ovulation bursting of a follicle at the release of an egg from the ovary 700

ovule (o′vul) in seed plants, a structure that contains the megasporangium, where meiosis occurs and the female gametophyte is produced; develops into the seed 407, 510

oxidation a chemical reaction that results in removal of one or more electrons from an atom, ion, or compound; oxidation of one substance occurs simultaneously with reduction of another 33

oxidative phosphorylation (fos″for-ĭ-la′shun) the process of building ATP by an electron transport system that uses oxygen as the final receptor; occurs in mitochondria 145

oxidizing atmosphere an atmosphere with free oxygen; for example, the atmosphere of today 355, 358

oxygen debt use of oxygen to convert lactate, which builds up due to anaerobic conditions, to pyruvate 142

oxytocin hormone made by the hypothalamus and stored in the posterior pituitary; causes the uterus to contract and stimulates the release of milk from mammary glands 681

ozone shield formed from oxygen in the upper atmosphere; protects the earth from ultraviolet radiation 358, 801

P

paleontology study of fossils that results in knowledge about the history of life 301

palisade layer layer in a plant leaf containing elongated cells with many chloroplasts; along with the spongy layer, it is the site of most of photosynthesis 491

pancreas an abdominal organ that produces digestive enzymes and the hormones insulin and glucagon 686

parasitism (par′ah-si″tizm) a symbiotic relationship in which one species (parasite) benefits for growth and reproduction to the harm of the other species (host) 759

parasympathetic nervous system a division of the autonomic nervous system that is involved in normal activities; uses acetylcholine as a neurotransmitter 641

parathyroid gland embedded in the posterior surface of the thyroid gland; produces parathyroid hormone, which is involved in the regulation of calcium and phosphorus balance 685

parathyroid hormone (PTH) substance secreted from the 4 parathyroid glands that increases the calcium level and decreases the phosphate level in the blood 685

parenchyma (pah-ren′kah-mah) least specialized of all plant cell or tissue types; found in all organs of a plant with many of the cells containing chloroplasts 477

parthenogenesis (par″thĕ-no-jen′-ĕ-sis) development of an egg cell into a whole organism without fertilization 695

pelagic division (pe-laj′ik) the open portion of the sea 768

penis male copulatory organ 697

peptide 2 or more amino acids joined together by covalent bonding 50

peptide bond a covalent bond between two amino acids resulting from a condensation reaction between the carboxyl group of one and the amino group of another 50

pericycle external layer of cells in the vascular cylinder of a plant root, producing branch and secondary roots 481

peripheral nervous system (PNS) made up of the nerves that branch off of the central nervous system 634

peristalsis the rhythmic, wavelike contraction that moves food through the digestive tract 593

petal plant structure belonging to an inner whorl of the flower; colored and internal to the outer whorl of green sepals 509

phagocytosis (fag-ah-si-to'sis) transport process by which amoeboid-type cells engulf large material, forming an intracellular vacuole 97, 381

pharynx a common passageway (throat) for both food intake and air movement 612

phenotype the visible expression of a genotype; for example, brown eyes, height 187

phenylketonuria (PKU) (fen"il-ke"to-nu're-ah) a genetic disorder characterized by the absence of an enzyme that metabolizes phenylalanine; results in severe mental retardation 222

pheromone (fer'o-mōn) substance produced and discharged into the environment by an organism that influences the behavior of another organism 738

phloem (flo'em) vascular tissue conducting organic solutes in plants; contains sieve-tube cells and companion cells 399, 477

phospholipid a molecule having the same structure as a neutral fat except one of the bonded fatty acids is replaced by a phosphorous-containing group; an important component of plasma membranes 48, 87

photochemical smog air pollution that contains nitrogen oxides and hydrocarbons which react to produce ozone and peroxylacetyl nitrate 798

photon discrete packet of light energy; the amount of energy in a photon is inversely related to the wavelength of light in the photon 122

photoperiodism relative lengths of daylight and darkness that affect the physiology and behavior of an organism 530

photoreceptor light sensitive receptor cell 654

photosynthesis a process whereby chlorophyll-containing organisms trap sunlight energy to build a sugar from carbon dioxide and water 114

photosystem a photosynthetic unit where light is absorbed; contains an antenna (photosynthetic pigments) and an electron acceptor 125

phototropism (fo-tot'ro-pizm) movement in response to light; in plants, growth toward or away from light 529

pH scale measurement scale for the relative concentrations of hydrogen (H^+) and hydroxyl (OH^-) ions in a solution 36

phylogenetic tree (fī-lō-jah-net'ik) diagram with branches illustrating common evolutionary ancestors and the descendants produced from them through evolutionary divergence 340

phytochrome a plant photoreversible pigment; this reversion rate of 2 molecular forms is believed to measure the photoperiod 531

pinocytosis (pi"no-si-to'sis) process by which vesicles form around and bring macromolecules into the cell 97

pistil a female flower structure consisting of an ovary, style, and stigma 407, 509

pituitary gland a small gland that lies just below the hypothalamus and is important for its hormone storage and production activities 681

placenta a structure formed from an extraembryonic membrane, the chorion, and uterine tissue through which nutrient and waste exchange occur for the embryo and fetus 720

placental mammal a mammalian subclass that is characterized by a placenta, which is an organ of exchange between maternal and fetal blood that supplies nutrients to the growing offspring 458

plankton fresh- and saltwater organisms that float on or near the surface of the water 766

plasma liquid portion of blood; contains nutrients, wastes, salts, and proteins 562

plasma cell an activated B cell that is currently secreting antibodies 577

plasma membrane membrane that separates the contents of a cell from the environment and regulates the passage of molecules into and out of the cell 64

plasmid a self-duplicating ring of accessory DNA in the cytoplasm of bacteria 279

plasmodesmata (plaz-mah-dez'mah-tah) in plants, cytoplasmic channels bounded by the plasma membrane and connecting the cytoplasm of adjacent cells 101

platelet cell-like disks formed from fragmentation of megakaryocytes that are involved in blood clotting in vertebrate blood 563

plumule (ploō'mūl) embryonic plant shoot that bears young leaves 515

polar body a nonfunctional product of oogenesis; 3 of the 4 meiotic products are of this type 176

polar covalent bond (ko-va'lent) a bond in which the sharing of electrons between atoms is unequal 33

pollen grain a male gametophyte in seed plants 404

pollination transfer of pollen from an anther to a stigma 404, 510

pollutant substance that is added to the environment and leads to undesirable effects for living organisms 793

polygenic inheritance a pattern of inheritance in which a trait is controlled by several gene pairs that segregate independently 200

polymer macromolecule consisting of covalently bonded monomers; for example, a protein is a polymer of monomers called amino acids 43

polypeptide chain of many amino acids, covalently bonded together by peptide bonds 50

polyploid (polyploidy) a condition in which an organism has more than 2 sets of chromosomes 208

polysaccharide carbohydrate macromolecule consisting of many monosaccharides covalently bonded together 45

polysome a string of ribosomes, simultaneously translating different regions of the same mRNA strand during protein synthesis 257

population a group of organisms of the same species occupying a certain area and sharing a common gene pool 8, 745

portal system circulatory pathway that begins and ends in capillaries, such as the one found between the small intestine and liver 560

positive feedback loop pattern of homeostatic response in which the output intensifies and increases the likelihood of response instead of countering it and canceling it 548

prairie terrestrial biome that is a temperate grassland, changing from tall-grass prairies to short-grass prairies when traveling east to west across the Midwest of the United States 775

pressure-flow model model explaining transport through sieve-tube cells of phloem, with leaves serving as a source and roots as a sink in the summer, and vice versa in the spring 503

primate an animal order of mammals having hands and feet with 5 distinct digits and some having an opposable thumb 458

primitive atmosphere earliest atmosphere of the earth after the planet's formation 355

producer organism at the start of a food chain that makes its own food (e.g., green plants on land and algae in water) 781

progesterone female ovarian sex hormone that causes the endometrium of the uterus to become secretory during the ovarian cycle; along with estrogen maintains secondary sex characteristics in females 700

prokaryote first type of cell to evolve; found in the kingdom Monera and lacking a well-defined nucleus and organelles 64, 370

prokaryotic cell the first, primitive cells on earth, exemplified today by bacteria, which lack a defined nucleus and most organelles 64

promoter a sequence of DNA in an operon where RNA polymerase begins transcription 266

prostaglandins (pros-tah-glan'dins) class of chemical messengers, acting locally in the body, with widespread effects; derived from fatty acids and stored in the plasma membrane 690

protein macromolecule having, as its primary structure, a sequence of amino acids united through covalent bonding; many kinds, important in structure and metabolism 50

proteinoid (pro'te-in-oid") abiotically polymerized amino acids that are joined in a preferred manner; possible early step in cell evolution 356

protist organism belonging to the kingdom Protista; for example, protozoans, algae, or slime molds 380

protocell a cell forerunner developed from cell-like microspheres 356

proton positive subatomic particle, located in the nucleus of the atom 26

protostome a group of coelomate animals in which the blastopore (the first embryonic opening) becomes the mouth 430

protozoan animal-like, heterotrophic, unicellular organisms 381

proximal convoluted tubule the portion of a nephron following the Bowman's capsule where selective reabsorption of filtrate occurs 625

pseudocoelom (su"do-se'lom) a coelom that is not completely lined by mesoderm 423

pseudopodia (sud-ah-pod-e-ah) cytoplasmic extensions of amoeboid protists; used for locomotion by these organisms 381

pulmonary circuit pathway of bloodflow between the lungs and heart 560

Punnett square a gridlike graph that enables one to calculate the results of simple genetic crosses by lining gametic genotypes of 2 parents on the outside margin and their recombination in boxes inside the grid 187

purine (pu'rin) a type of nitrogenous base, such as adenine and guanine, having a double-ring structure 54

pyrimidine (pi-rim'i-din) a type of nitrogenous base, such as cytosine, thymine, and uracil, having a single-ring structure 54

pyruvate the end product of glycolysis; its further fate involving fermentation or Krebs cycle incorporation depending on oxygen availability 139

R

radially symmetrical body plan in which similar parts are arranged around a central axis like spokes of a wheel 419

radicle part of the plant embryo that contains the root apical meristem and becomes the first root of the seedling 515

receptor a cell, often in groups, that detects change or stimulus and initiates a nerve impulse 641, 653

recessive allele (ah-lēl') form of a gene whose effect is hidden by the dominant allele in a heterozygote 186

recombinant DNA (rDNA) DNA that contains genes from more than one source 279

reducing atmosphere an atmosphere with little if any free oxygen, for example, the primitive atmosphere 355

reduction a chemical reaction that results in addition of one or more electrons to an atom, ion, or compound. Reduction of one substance occurs simultaneously with oxidation of another 33

reflex an automatic, involuntary response of an organism to a stimulus 641

regeneration reforming a lost body part from the remaining body mass by active cell division and growth 695

regulator gene in an operon, a gene that codes for a repressor 266

renewable resource resource that can be replenished by physical or biological means 803

repetitive DNA DNA with the same set of base pairs repeated many times, separating genes that direct the synthesis of cytoplasmic proteins 244

replication fork V-shaped region of a eukaryotic chromosome wherever DNA is being replicated 243

repressible operon an operon that is normally active because the repressor must combine with a corepressor before the complex can bind to the operator 268

repressor protein molecule that binds to an operator site, preventing RNA polymerase from binding to the promotor site of an operon 266

reproduce make a copy similar to oneself; for example, bacteria dividing to produce more bacteria, or egg and sperm joining to produce offspring in more advanced organisms 5

reptile class of terrestrial vertebrates with internal fertilization, scaly skin, and an egg with a leathery shell; includes snakes, lizards, turtles, and crocodiles 453

resting potential state of an axon when the membrane potential is about -65 mV and no impulse is being conducted 637

restriction enzyme enzyme that stops viral reproduction by cutting viral DNA; used in genetic engineering to cut DNA at specific points 279

retina innermost layer of the eyeball containing the light receptors—rods and cones 655

retrovirus RNA virus containing the enzyme reverse transcriptase that carries out RNA/DNA transcription; for example, viruses that carry cancer-causing oncogenes and the AIDS virus 368

rhizoid rootlike hairs that anchor a plant and absorb minerals and water from the soil 398

rhizome a rootlike, underground stem 485

rhodopsin (ro-dop'sin) a light-absorbing molecule in rods and cones that contains a pigment and its attached protein 658

ribonucleic acid (RNA) a nucleic acid with a sequence of subunits dictated by DNA; involved in protein synthesis 54, 235

ribosomal RNA (rRNA) type of RNA found in ribosomes; sometimes called structural RNA 251

ribosome cytoplasmic organelle; site of protein synthesis 70

RNA see ribonucleic acid

RNA polymerase (pol-im'er-ās) enzyme that links ribonucleotides together during transcription 254

rod photoreceptor in vertebrate eyes that responds to dim light 655

root structure of a plant root system; serving to anchor the plant, absorb water and minerals from the soil, and store products of photosynthesis 479

root pressure a force generated by an osmotic gradient that serves to elevate sap through xylem a short distance 497

rough ER complex organelle that is continuous with the nuclear envelope, consisting of membranous channels and studded with ribosomes 71

r-strategist a species that has evolved characteristics that maximize its rate of natural increase, for example, high birthrate 749

RuBP (ribulose biphosphate) 5-carbon compound that combines with carbon dioxide during the Calvin cycle and is later regenerated by the same cycle 130

S

SA node (es a nod) mass of specialized tissue in right atrium wall of heart initiates a heartbeat; the "pacemaker" 559

sac body plan a body with a digestive cavity that has only one opening, as in cnidarians and flatworms 421

saltatory conduction (sal'tah-to"re) movement of a nervous impulse from one node of Ranvier to another along a myelinated axon 639

saprophyte organism that carries out external digestion and absorbs the resulting nutrients across the plasma membrane of its cells; indicative of fungi and slime molds 373, 380

sarcolemma (sar"ko-lem'ah) plasma membrane of a muscle fiber that forms the tubules of the T system involved in muscular contraction 671

sarcomere (sar"ko-mer) a unit of a myofibril, many of which are arranged linearly along its length, that shortens to give muscle contraction 671

sarcoplasmic reticulum (sar"ko-plaz'mik re-tik'u-lum) the modified endoplasmic reticulum of a muscle fiber that stores calcium ions, whose release initiates muscle contraction 671

savanna terrestrial biome that is a tropical grassland in Africa, characterized by a few trees and a severe dry season 775

scientific method a step-by-step process for discovery and generation of knowledge, ranging from observation and hypothesis to theory and principle 14

sclera outer, white, fibrous layer of the eye that surrounds the transparent cornea 655

secondary oocyte (sek'on-dary o'o-sīt) the largest functional product of meiosis; becomes the egg 176

seed a mature ovule that contains an embryo with stored food enclosed in a protective coat 404, 511

seed coat covering around the embryonic sporophyte and stored foods in a mature seed; derived from the integument of the ovule 513

segmented having repeating body units as is seen in the earthworm 434

selectively permeable ability of plasma membranes to regulate the passage of substances into and out of the cell, allowing some to pass through the membrane and preventing the passage of others 89

semicircular canal one of 3 half-circle-shaped canals of the inner ear that are fluid filled for registering changes in motion 660

semiconservative replication duplication of DNA resulting in a double helix having one parental and one new strand 241

seminal fluid (sem'in-al) thick, whitish fluid consisting of sperm and secretions from several glands of the male reproductive tract 697

sepal protective leaflike structure enclosing the flower when in bud 509

sessile filter feeder an organism that stays in one place and filters its food from the water 417

sex chromosome the chromosome that determines the biological sex of an organism 204

sex-influenced trait a genetic trait that is expressed differently in the 2 sexes but is not controlled by alleles on the sex chromosomes, for example, pattern baldness 231

sex-linked gene unit of heredity, a gene, located on a sex chromosome 204

sexual reproduction reproduction involving meiosis and gamete formation; produces offspring with genes inherited from each parent 166, 170

sexual selection type of natural selection by which one organism specifically prefers another organism for mating 327

sickle-cell disease a genetic disorder in which sickling of red blood cells clogs blood vessels; individuals who are heterozygous for the allele have a greater resistance to malaria; more common among Blacks, but even then quite rare 226

sister chromatid one of 2 genetically identical chromosomal units that are the result of DNA replication and are attached to each other at the centromere 157

slash-and-burn agriculture practice of cutting and burning forest to use the land for agriculture 795

smooth ER complex organelle that is continuous with the nuclear envelope; consists of membranous channels and saccules but not studded with ribosomes 71

society a group of individuals of the same species that shows cooperative behavior 740

sociobiology the biological study of social behavior 740

sodium-potassium pump a transport protein in the plasma membrane that moves sodium out of and potassium into animal cells; important in nerve and muscle cells 94

solar energy energy from the sun; light energy 803

speciation (spe"se-a'shun) the process whereby a new species is produced or originates 334

species a taxonomic category that is the subdivision of a genus; its members can breed successfully with each other but not with organisms of another species 8

sperm male sex cell with 3 distinct parts at maturity: head, middle piece, and tail 698

spermatogenesis (sper"mah-to-jen'ĕ-sis) production of sperm in males by the process of meiosis and maturation 175

sphincter circular muscle in the wall of a tubular structure, such as an artery or the digestive tract, that can open and close the vessel, therefore regulating the amount of material moving through it 555

spicule (spik'ūl) skeletal structures of sponges composed of calcium carbonate or silicate 418

spinal cord part of the central nervous system, continuous with the base of the brain and housed within the vertebral column 640

spindle microtubule structure that brings about chromosome movement during cell division 160

spongy layer layer of loosely packed cells in a plant leaf that increases the amount of surface area for gas exchange; along with the palisade layer, it is the site of most of photosynthesis 491

spore usually a haploid reproductive structure that develops into a haploid generation; characteristic product of meiosis in plants 384

sporophyte diploid generation of the alternation of generations life cycle of a plant; it produces haploid spores, by meiosis, that develop into the haploid generation 398

stabilizing selection outcome of natural selection in which extreme phenotypes are eliminated and the average phenotype is conserved 325

stamen a pollen-producing flower structure consisting of an anther on a filament tip 407, 510

starch polysaccharide consisting of covalently bonded glucose molecules, characterized by few side branches; typical storage product in plants 45

steady-state society a society with no yearly increase in population or resource consumption 804

stem upright, vertical portion of a plant shoot system, transporting substances to and from the leaves 482

steroid biologically active lipid molecule having 4 interlocking rings; examples are cholesterol, progesterone, testosterone 48

stigma enlarged, sticky knob at one end of the pistil where pollen grains are received during pollination *407, 509*

stolon modified, horizontal stem of a plant that is aboveground and produces new plants where its nodes touch the ground *485*

stoma (stō-ma) small opening with 2 guard cells on the underside of leaf epidermis; their opening controls the rate of gas exchange *476*

stroma (stro'mah) large, central space in a chloroplast that is fluid filled and contains enzymes used in photosynthesis *74, 123*

structural gene gene that codes for proteins in metabolic pathways *267*

style the tubular part of the pistil of a flower where a pollen tube develops from a transferred pollen grain *407, 509*

substrate the reactant in an enzymatic reaction; each enzyme has a specific substrate *108*

substrate-level phosphorylation (fos"fōr-ĭ-la'shun) process in which ATP is formed by transferring a phosphate from a metabolic substrate to ADP *117*

succession an orderly sequence of community replacement, one following the other, to an eventual climax community *769*

suppressor T cell T lymphocyte that increases in number more slowly than other T cells and eventually brings about an end to the immune response *579*

survivorship usually shown graphically to depict death rates or percentage of remaining survivors of a population over time *748*

symbiosis (sim"bi-o'sis) a relationship that occurs when 2 different species live together in a unique way; may be beneficial, neutral, or detrimental to one and/or the other species *759*

symbiotic (sim"bi-ot'ik) one species having a close relationship with another species; includes parasitism, mutualism, and commensalism *373*

sympathetic nervous system a division of the autonomic nervous system that is involved in fight or flight responses; uses norepinephrine as a neurotransmitter *641*

sympatric speciation (spe"se-a'shun) origin of new species in populations that overlap geographically *334*

synapse a junction between neurons consisting of presynaptic (axon) membrane, the synaptic cleft, and the postsynaptic (usually dendrite) membrane *639*

synapsis pairing of homologous chromosomes during meiosis I *173*

systematics means to classify organisms based on their phylogeny or evolutionary history *340*

systemic circuit pathway of bloodflow ranging from the heart to the tissues throughout the body and back to the heart *560*

systolc (sis'to-le) contraction period of a heart during the cardiac cycle *558*

T

taiga (tī'gah) terrestrial biome that is a coniferous forest extending in a broad belt across northern Eurasia and North America *775*

taste bud an elongated cell that functions as a taste receptor *654*

taxis movement of an organism toward or away from a stimulus *736*

taxonomy a system to classify organisms meaningfully into groups based on similarities and differences and using morphology and evolution *340*

Tay-Sachs disease (ta saks') a lethal, genetic lysosomal storage disease, best known among the U.S. Jewish population, which results from the lack of a particular enzyme; nervous system damage leads to early death *222*

tendon structure consisting of fibrous connective tissue that connects skeletal muscles to bones *541, 669*

territoriality behavior used to guarantee exclusive use of a given space for reproduction, feeding, etc. *736*

testcross a genetic mating in which a possible heterozygote is crossed with an organism homozygous recessive for the characteristic(s) in question in order to determine its genotype *189*

testosterone male sex hormone produced from interstitial cells in the testis *698*

tetrad the set of 4 chromatids of a synapsed homologous chromosome pair, visible during prophase of meiosis I; also called bivalent *172*

thalamus (thal'ah-mus) a lower forebrain region in vertebrates involved in crude sensory perception and screening messages intended for the higher forebrain or cerebrum *645*

theory a conceptual scheme arrived at by the scientific method and supported by innumerable observations and experimentations *17*

thermal inversion temperature inversion that traps cold air and its pollutants near the earth with the warm air above it *800*

thigmotropism (thig-mot'ro-pizm) movement in response to touch; in plants, the coiling of tendrils *529*

thoracic cavity (tho-ras'ik) internal body space of some animals that contains the lungs, protecting them from desiccation; the chest *610*

thrombocyte cell fragment in the blood that initiates the process of blood clotting *563*

thylakoid (thi'lah-koid) flattened disk of a granum; its membrane contains the photosynthetic pigments *74, 125*

thymine (T) (thi'min) one of 4 organic bases in the nucleotides composing the structure of DNA *54, 238*

thyroid gland large gland in the neck that produces several important hormones, including thyroxin and calcitonin *683*

thyroxin a substance, T4, secreted from the thyroid gland, that promotes growth and development in vertebrates; in general, it increases the metabolic rate in cells *683*

tissue group of similar cells combined to perform a common function *539*

tissue fluid derivative of the blood plasma from capillaries that bathes the cells throughout the body *566*

T lymphocyte white blood cell that directly attacks antigen-bearing cells *577*

tonicity (tō-nis'ĭ-tē) referring to the solute concentration of a solution *91*

trachea 1. an air tube (windpipe) of the respiratory tract in vertebrates; 2. an air tube in insects *438, 593, 612*

transcription the process whereby the DNA code determines (is transcribed into) the sequence of codons in mRNA *252*

transfer RNA (tRNA) active in protein synthesis, and transfers a particular amino acid to a ribosome; at one end it binds to the amino acid and at the other end it has an anticodon that binds to an mRNA codon *251*

transgenic organism a multicellular organism that contains a transplanted gene *284*

transition reaction a reaction involving the removal of CO_2 from pyruvate; connects glycolysis to the Krebs cycle *143*

translation the process whereby the sequence of codons in mRNA determines (is translated into) the sequence of amino acids in a polypeptide *252*

transpiration the plant's loss of water to the atmosphere, mainly through evaporation at leaf stomata *497*

transposon (trans-po'zun) movable segment of DNA *245*

triplet code genetic code (mRNA, tRNA) in which sets of 3 bases call for specific amino acids in the formation of polypeptides *253*

trophic level (trof'ik) feeding level of one or more populations in a food web *782*

tropical rain forest forest found in warm climates with plentiful rainfall *777*

tropism (tro'pizm) in plants, a growth response toward or away from an external stimulus *529*

TSH thyroid stimulating hormone, secreted by the anterior pituitary; stimulates activity in the thyroid gland *683*

tube-within-a-tube body plan a body with a digestive tract that has both a mouth and an anus *421*

tumor suppressor gene unit of heredity (gene) that codes for proteins that ordinarily suppress cell division, thereby promoting organized cell growth *273*

tundra treeless terrestrial biome of cold climates; occurs on high mountains and in polar regions *773*

turgor pressure the pressure of the plasma membrane against the cell wall, determined by the water content of the plant cell vacuole; gives internal support to the cell *92*

Turner syndrome a condition that results from the inheritance of a single X chromosome; a second sex chromosome is absent: XO *218*

tympanic membrane eardrum; membranous region that receives air vibrations in an auditory organ *660*

U

uracil one of 4 nucleotides composing the structure of RNA *54*

urea main nitrogenous waste of terrestrial amphibians and mammals *621*

ureter (u-re'ter) a tubular structure conducting urine from kidney to urinary bladder *624*

urethra (u-re'thrah) tubular structure that receives urine from the bladder and carries it to the outside of the body *624*

uric acid main nitrogenous waste of insects, reptiles, birds, and some dogs *621*

urinary bladder organ where urine is stored *624*

urine liquid waste product made by the nephrons of the kidney through the processes of pressure filtration and selective reabsorption *624*

uterine cycle a cycle that runs concurrently with the ovarian cycle; prepares the uterus to receive a developing zygote *703*

uterus pear-shaped portion of female reproductive tract that lies between the oviducts and the vagina; site of embryo development *700*

V

vaccine a substance that wakes the immune response without causing illness *580*

vacuole organelle that is a membranous sac, particulary prominent in plant cells *74*

vagina muscular tube leading from the uterus; the female copulatory organ and the birth canal *700*

vas deferentia part of the male reproductive tract that transports sperm cells from the epididymis to the penis *697*

vascular plant any organism of the plant kingdom that contains the vascular tissues xylem and phloem as part of its structure *399*

vector in genetic engineering, a means to transfer foreign genetic material into a cell, for example, a plasmid *279*

vein a blood vessel that arises from venules and transports blood toward the heart *555*

vena cava (pl., venae cavae) (ve'nah ka'vah) one of 2 largest veins in the body; returns blood to the right atrium in a 4-chambered heart *560*

vertebra (pl., vertebrae) one of many bones of the vertebral column; held to other vertebrae by bony facets, muscles, and strong ligaments *666*

vertebrate referring to a chordate animal with a serial arrangement of vertebrae, or backbone *416*

villus (pl., villi) (vil'us) small, fingerlike projection of the inner small intestine wall *594*

viroid unusual infectious particle consisting of a short chain of naked RNA *368*

visible light a portion of the electromagnetic spectrum of light that is visible to the human eye *122*

vitamin an organic molecule that is required in small quantities for various biological processes and must be in an organism's diet because it cannot be synthesized by the organism; becomes part of enzyme structure *112, 600*

W

white matter regions of the brain and spinal cord, consisting of myelinated nerve fibers *642*

woody angiosperm characterizes angiosperm trees with hardwood *407*

X

X-linked gene gene located on the X chromosome that does not control a sexual feature of the organism *204*

xylem a vascular tissue that transports water and mineral solutes upward through the plant body *399, 477*

XYY male a genetic condition in which affected males have 2 Y chromosomes, are taller than average, and may exhibit learning difficulties *218*

Y

yolk sac one of 4 extraembryonic membranes in vertebrates; important in the development of fishes, amphibians, reptiles, and birds, it is largely vestigial in mammals *719*

Z

zero population growth no growth in population size *751*

zygote the diploid (2N) cell formed by the union of 2 gametes, the product of fertilization *510*

CREDITS

Photographs

History of Biology
Leeuwenhoek: Bettmann Newsphotos; **Darwin**: © 78 John Moss/ Black Star; **Pasteur**: The Bettmann Archive; **Koch**: Bettmann Newsphotos; **Pavlov**: UPI/Bettmann; **Lorenz**: UPI/Bettmann; **McClintock**: AP/Wide World Photos, Inc.; **Jarvik**: UPI/Bettmann Newsphotos; **Gallo**: Rueters/Bettmann Newsphotos

Part Openers
Part 1: © Carolina Biological Supply Co.; **Part 2:** © Gunter Ziesler/Peter Arnold, Inc.; **Part 3:** © Douglas Faulkner/Photo Researchers, Inc.; **Part 4:** © Astrid and Hahns-Frieder Michler/Science Photo Library/Photo Researchers, Inc.; **Part 5:** © Runk/Schoenberger/Grant Heilman; **Part 6:** © Carl R. Sams II/Peter Arnold, Inc.

Chapter 1
Opener: © Gerard Lacz/Peter Arnold, Inc.; **1.1a:** © Rod Allin/Tom Stack & Associates; **1.1b,c:** © Kjell B. Sanved; **1.Aa:** © Eric Grave/Photo Researchers, Inc., **1.Ab:** © John Cunningham/ Visuals Unlimited; **1.Ac:** © Rod Planck/Tom Stack & Associates; **1.Ad:** © Farrell Grehan/ Photo Researchers, Inc.; **1.Ae:** © Leonard Lee Rue Enterprises; **1.2:** © John Gerlach/Animals Animals; **1.3a:** © Stephen Kraseman/Photo Researchers, Inc.; **1.3b:** © Mitch Reardon/Photo Researchers, Inc.; **1.3c:** © David Fritts/Animals Animals; **1.4a:** © Biophoto Associates/Photo Researchers, Inc.; **1.4b:** © Philip Zito; **1.4c:** © Breck Kent/Animals Animals; **1.4d:** © Simion/ IKAN/Peter Arnold, Inc.; **1.6a:** © Ed Robinson/ Tom Stack & Associates; **1.6b:** © F. Stuart Westmorland/Tom Stack & Associates; **1.6c:** © Dave Fleetham/Tom Stack & Associates; **1.6d:** © C. Ressler/Animals Animals; **1.6e:** © Tom Stack/Tom Stack & Associates; **1.7a:** © Richard Thom/Tom Stack & Associates; **1.7b:** © Zig Lesczynski/Animals Animals; **1.7c:** © C.W. Schwartz/Animals Animals; **1.7d:** © Oxford Scientific Films/Animals Animals; **1.7e:** © Wendy Neefus/Animals Animals; **1.7f:** © Tom McHugh/Photo Researchers, Inc.

Chapter 2
Opener: © Jim Olive/Peter Arnold, Inc.; **2.5:** © Kenneth E. Glander, Ph.D.

Chapter 3
Opener: © Werner Muller/Peter Arnold, Inc.; **3.4:** © SIU/Journalism Services; **3.8c:** © Charles Kinger/Micro/Macro Stock Photography; **3.9:** © Douglas Faulkner/Sally Faulkner Collection; **3.12a:** © Clyde Smith/Peter Arnold, Inc.; **3.12b:** © Holt Confer/Grant Heilman; **3.12c:** © JoAnn Kalish/Peter Arnold, Inc.

Chapter 4
Opener: © CNRI/Science Photo Library/Photo Researchers, Inc.; **4.1a:** © John Gerlach/Tom Stack & Associates; **4.1b:** © Leonard Lee Rue III/Photo Researchers, Inc.; **4.1c:** © H. Pol/CNRI/ Science Photo Library/Photo Researchers, Inc.; **4.7b:** © Stanley Flegler/Visuals Unlimited; **4.7c:** © Don Fawcett/Photo Researchers, Inc.; **4.8b:** © Biophoto Associates/Photo Researchers, Inc.; **4.11:** © S. Maslowski/Visuals Unlimited; **4.Ba:** © James Karales/Peter Arnold, Inc.; **4.Cb:** Courtesy Dr. Wolfgang Beerman.

Chapter 5
Opener: © Dr. Jeremy Burgess/Science Photo Library/Photo Researchers, Inc.; **5.1a:** © Bob Coyle; **5.1b:** © Carolina Biological Supply; **5.1c:** © B. Miller/Biological Photo Service; **5.1d:** © Ed Reschke; **5.Aa:** © M. Abbey/Visuals Unlimited; **5.Ab:** © M. Schliwa/Visuals Unlimited; **5.Ac:** © Drs. Kessel/Shih/Peter Arnold, Inc.; **5.Ba-c:** © David Phillips/Visuals Unlimited; **5.3a:** © Henry Aldrich/Visuals Unlimited; **5.4b:** © Richard Rodewald/Biological Photo Service; **5.5b:** © W. P. Wergin, University of Wisconsin, courtesy E. A. Newcomb/Biological Photo Service; **5.6b:** © L. L. Sims/Visuals Unlimited; **5.7b:** © D. W. Fawcett/Photo Researchers, Inc.; **5.8a:** © W. Rosenberg/Biological Photo Service; **5.9a:** © David Phillips/Visuals Unlimited; **5.11:** © K. G. Murti/Visuals Unlimited; **5.12:** © S. E. Frederick, courtesy of E. H. Newcomb/Biological Photo Service; **5.13a:** © Dr. Herbert Israel/Cornell University; **5.14a:** © Dr. Keith Porter; **5.16b,c:** © Don Fawcett/Photo Researchers, Inc.; **5.16d:** © Don Fawcett/John Heuser/Photo Researchers, Inc.; **5.17b:** © Photo Researchers, Inc.; **5.18a:** © Biophoto Associates/Science Source/Photo Researchers, Inc.

Chapter 6
Opener: © CNRI/Science Photo Library/Photo Researchers, Inc.; **6.2a:** © W. Rosenberg/ Biological Photo Service; **6.2d:** © Don Fawcett/ Photo Researchers, Inc.; **6.10a:** © Eric Grave/ Photo Researchers, Inc.; **6.10b,c:** © Thomas Eisner; **6.13:** © Francis Leroy; **6.14b:** © Mark S. Bretscher; **6.17a:** © Ray F. Evert.

Chapter 7
Opener: © William Weber/Visuals Unlimited; **7.1:** © Joe DiStefano/Photo Researchers, Inc.; **7.5:** Courtesy Dr. Alfonso Tramontano; **7.8:** © Doolittle, R. F., University of California, from Scientific American, "Proteins," Oct. 1985, p. 95.

Chapter 8
Opener: © John Kaprielian/Photo Researchers, Inc.; **8.1:** © Jeff Lepore/Photo Researchers, Inc.; **8.3b,c:** © Gordon Leedale/Biophoto Associates; **8.8a:** © Dr. Melvin Calvin; **8.11a:** © Ray F. Evert; **8.13:** © Pat Armstrong/Visuals Unlimited.

Chapter 9
Opener: © Gunter Ziesler/Peter Arnold, Inc.; **9.1a:** © Dieter Blum/Peter Arnold, Inc.

Chapter 10
Opener: © Carolina Biological Supply; **10.1a:** © Brian Parker/Tom Stack & Associates; **10.1b:** © John Cunningham/Visuals Unlimited; **10.2b:** © John J. Cardamone Jr./University of Pittsburg, Medical Microbiology/Biological Photo Service; **10.B:** © Dr. Stephen Wolfe; **10.3a:** BioPhoto Associates/Science Source/Photo Researchers, Inc.; **10.5b, 10.6b, 10.7b:** © Ed Reschke; **10.8a:** © Francis Leroy; **10.9b:** © Ed Reschke; **10.10:** R. G. Kessel and C. Y. Shih: *Scanning Electron Microscopy* © 1976, Springer Verlag, Berlin; **10.11a-d:** © Dr. Andrew Bajer; **10.12a:** © B. A. Palevity and E. H. Newcomb, University of Wisconsin/Biological Photo Service

25.10b: © John Cunningham/Visuals Unlimited; 25.10c: © Field Museum of Natural History, Chicago; 25.11a: © Karlene Schwartz; 25.11b: © Milton Rand/Tom Stack & Associates; 25.12b: © Cabisco/Visuals Unlimited; 25.12c: © John Cunningham/Visuals Unlimited; 25.A: Courtesy of the Center for Plant Conservation/Dr. Charles McDonald

Chapter 26
Opener: © E. Robinson/Tom Stack & Associates; 26.1a: © BioMedia Associates; 26.1b: © Stephen Krasemann/Peter Arnold, Inc.; 26.6a: © Bill Carlsinger/Photo Researchers, Inc.; 26.6b: © Ron Taylor/Bruce Coleman, Inc.; 26.6c: © Carolina Biological Supply Company; 26.8a: © Dr. Fred Whittaker; 26.10b: © Fred Marsik/Visuals Unlimited; 26.11: © Jim Solliday/Biological Photo Service; 26.12: © Markell, E.K. and Voge, M.: Medical Paristology, 4th ed., W.B. Saunders Co. 1981

Chapter 27
Opener: © J. Alcock/Visuals Unlimited; 27.3b: © William Jorgenson/Visuals Unlimited; 27.4a: © William Ferguson; 27.4c: © William Jorgenson/Visuals Unlimited; 27.6a: © Michael DiSpezio/Images; 27.6c: © Geral Corsi/Tom Stack & Associates; 27.7a: © Michael DiSpezio/Images; 27.7c: © Gary Milburn/Tom Stack & Associates; 27.9: © Michael DiSpezio/Images; 27.10a: © Robert Evans/Peter Arnold, Inc.; 27.10b: © Tom McHugh/Photo Researchers, Inc.; 27.10c: © Dwight Kuhn; 27.10d: © John McGregor/Peter Arnold, Inc.; 27.10e, f: © John Fowler/Valan Photos; 27.11a: © James H. Carmichael/Peter Arnold, Inc.; 27.11c: © Fred Bavendam/Peter Arnold, Inc.; 27.Ab: © Edward S. Ross; 27.14a: © Michael DiSpezio/Images; 27.15b: © Oxford Scientific Films, Ltd.

Chapter 28
Opener: © John Cancalosi/Peter Arnold, Inc.; 28.3: © Field Museum of Natural History, Chicago; 28.4a: © Tom Stack/Tom Stack & Associates; 28.4b: © Douglas Faulkner/Sally Faulkner Collection; 28.5a: © Paul L. Janosi/Valan Photos; 28.5b: © Thomas Kitchin/Valan Photos; 28.6a: © Tom McHugh/Steinhart Aquarium; 28.7a-d: © Jane Burton/Bruce Coleman, Inc.; 28.8a: © American Museum of Natural History; 28.8b: © John Cunningham/Visuals Unlimited; 28.9a: © Andrew Odum/Peter Arnold, Inc.; 28.9b: © E.R. Degginger/Bruce Coleman, Inc.; 28.10a: © Wolfgang Bayer Productions, Inc.; 28.11: © American Museum of Natural History; 28.12: © R. Austing/Photo Researchers, Inc.; 28.13: © Jim David/Photo Researchers, Inc.; 28.14: © Renee Stockdale/Animals Animals; 28.15: © Stouffer Enterprises, Inc./Animals Animals; 28.18a: © Martha Reeves/Photo Researchers, Inc.; 28.18b: © Bios/Peter Arnold, Inc.; 28.18c: © George Holton/Photo Researchers, Inc.; 28.18d: © Tom McHugh/Photo Researchers, Inc.; 28.23: © American Museum of Natural History

Chapter 29
Opener: © M.I. Walker/Science Source/Photo Researchers, Inc.; 29.2a: © Ed Reschke/Peter Arnold, Inc.; 29.2b: © John Cunningham/Visuals Unlimited; 29.2c: © Biophoto Associates/Photo Researchers, Inc.; 29.3a: © Biological Photo Service; 29.3b, c: © Biophoto Associates/Photo Researchers, Inc.; 29.4b: © Biological Photo Service; 29.5b: © George Wilder/Visuals Unlimited; 29.8b: © Carolina Biological Supply Company; 29.9: © Dwight Kuhn; 29.10: © John Cunningham/Visuals Unlimited; 29.11a: © G.R. Roberts; 29.11b: © E.R. Degginger/Color-Pic; 29.11c: © David Newman/Visuals Unlimited; 29.12: © J.R. Waaland, University of Washington/Biological Photo Service; 29.13a: © Carolina Biological Supply Company; 29.13b, c: © Runk/Schoenberger/Grant Heilman; 29.14: © Carolina Biological Supply Company; 29.16b: © Carolina Biological Supply Company; 29.B: © Earl Roberge/Photo Researchers, Inc.; 29.19b: © J.H. Troughton and F.B. Sampson; 29.21a: © Martha Cooper/Peter Arnold, Inc.; 29.21b: © Larry Mellichamp/Visuals Unlimited; 29.21c: © Carolina Biological Supply Company

Chapter 30
Opener: © Cabisco/Visuals Unlimited; 30.6a: © David Phillips/Visuals Unlimited; 30.8a-c: © Dr. Mary E. Doohan; 30.10a: © Gordon Leedale/BioPhoto Associates; 30.10b: © Donald Marx/USDA Forest Service

Chapter 31
Opener: © D. Newman/Visuals Unlimited; 31.3: © J. Robert Waaland/Biological Photo Service; 31.4a: © BioPhoto Associates/Photo Researchers, Inc.; 31.4b: © Ed Reschke/Peter Arnold, Inc.; 31.4c: © David Scharf/Peter Arnold, Inc.; 31.Aa: © Nicholas Smythe/Photo Researchers, Inc.; 31.Ab: © H. Eisenbeiss/Photo Researchers, Inc.; 31.Ac: © Anthony Mercieca/Photo Researchers, Inc.; 31.Ad: © Donna Howell; 31.6a (1): © Ralph A. Reinhold/Animals Animals; 31.6a (2): © W. Ormerod/Visuals Unlimited; 31.6b (1): © C.S. Lobban/Biological Photo Service; 31.6b (2): © Biological Photo Service; 31.6c (both): © Dwight Kuhn; 31.7: © Ted Levin/Earth Scenes

Chapter 32
Opener: © D. Newman/Visuals Unlimited; 32.8: Courtesy R.J. Weaver; 32.4: © Tom McHugh/Photo Researchers, Inc.; 32.5: © Robert E. Lyons/Visuals Unlimited; 32.6a-d: © Kiem Tran Thanh Van and her colleagues; 32.7: © Runk/Schoenberger; 32.10: © John Cunningham/Visuals Unlimited; 32.11a, b: © Tom McHugh/Photo Researchers, Inc.; 32.14: © Frank B. Salisbury

Chapter 33
Opener: © Michael Gabridge/Visuals Unlimited; 33.2a-e: © Ed Reschke; 33.2f: © Ed Reschke/Peter Arnold, Inc.; 33.3a,b, 33.4b,c, 33.6a-c, 33.7: © Ed Reschke

Chapter 34
Opener: © Manfred Kage/Peter Arnold, Inc.; 34.1a: © Eric Grave/Photo Researchers, Inc.; 34.1b: © Carolina Biological Supply Company; 34.1c: © Michael DiSpezio/Images; 34.A: The Bettmann Archive; 34.Ba: © Lewis Lainey; 34.12b: © Science Photo Library/Photo Researchers, Inc.; 34.15a: © Stuart I. Fox

Chapter 35
Opener: © Manfred Kage/Peter Arnold, Inc.; 35.2: © Keetin, Biological Science, Norton Publishing; 35.4: © Boehringer Ingleheim International GmbH, Courtesy of Lennart Nilsson; 35.7: © R. Feldman/Dan McCoy/Rainbow; 35.8: © Boehringer Ingleheim International GmbH, courtesy Lennart Nilsson

Chapter 36
Opener: © Matt Meadows/Peter Arnold, Inc.; 36.1a: © David Dennis/Tom Stack & Associates; 36.1b: © Harry Rogers/Photo Researchers, Inc.; 36.1c: © Grant Heilman Photography; 36.14a,b: Courtesy of the World Health Organizations; 36.14c,d: Courtesy of the Atlanta Centers for Disease Control

Chapter 37
Opener: © CNRI/Science Photo Library/Photo Researchers, Inc.; 37.10a: © Lennart Nilsson: Behold Man, Little Brown and Company, Boston; 37.A, 37.B: © Martin Rotker

Chapter 38
Opener: © CNRI/Science Photo Library/Photo Researchers, Inc.; 38.2b: © Robert Myers/Visuals Unlimited

Chapter 39
Opener: © John Allison/Peter Arnold, Inc.; 39.13: © Dan McCoy/Rainbow

Chapter 40
Opener: © Runk/Schoenberger/Grant Heilman; 40.4: © T. Norman Tait/BioPhoto Associates; 40.6d: © Dr. Frank Werblin, University of California at Berkeley; 40.A (both): Courtesy Professor J.E. Hawkins, Robert S. Preston, Kresage

Chapter 41
Opener: © SIU/Visuals Unlimited; 41.1: © Joe McDonald/Bruce Coleman, Inc.; 41.9b: © H.E. Huxley; 41.11a: © Victor Eichler/Bio Art

Chapter 42
Opener: © Manfred Kage/Peter Arnold, Inc.; 42.6: © Bettina Cirone/Photo Researchers, Inc.; 42.9: © Lester Bergman and Associates

Chapter 43
Opener: © John Mitchell/Photo Researchers, Inc.; 43.1b: © John Shaw/Bruce Coleman, Inc.; 43.2a: © Manfred Kage/Peter Arnold, Inc.; 43.2b: © Matt Meadows/Peter Arnold, Inc.; 43.5: © BioPhoto Associates/Photo Researchers, Inc.; 43.8b: © Baganvandoss/Photo Researchers, Inc.

Chapter 44
Opener: © Karl Switak/Photo Researchers, Inc.; 44.12a: © Edelmann/First Days of Life/Black Star

Chapter 37
37.4, 37.7, 37.8: Copyright © Mark Lefkowitz.
37.9: From John W. Hole, Jr., *Human Anatomy and Physiology,* 5th ed. Copyright © 1990 Wm. C. Brown Communications, Inc., Dubuque, Iowa. All Rights Reserved. Reprinted by permission.
37.11: From A. J. Vander, et al., *Human Physiology.* Copyright © McGraw-Hill, Inc., New York, NY. Reprinted by permission.
page 605: Copyright © Mark Lefkowitz.

Chapter 38
38.7, 38.8, 38.9: Copyright © Mark Lefkowitz.

Chapter 39
39.6, 39.7, 39.8, 39.10, 39.12: Copyright © Mark Lefkowitz.

Chapter 40
40.1: Copyright © Mark Lefkowitz.

Chapter 41
41.11b,c: From John W. Hole, *Human Anatomy and Physiology,* 5th ed. Copyright © 1990 Wm. C. Brown Communications, Inc., Dubuque, Iowa. All Rights Reserved. Reprinted by permission.

Chapter 42
42.2: Copyright © Mark Lefkowitz.

Chapter 43
43.3: From John W. Hole, Jr., *Human Anatomy and Physiology,* 5th ed. Copyright © 1990 Wm. C. Brown Communications, Inc., Dubuque, Iowa. All Rights Reserved. Reprinted by permission.
43.4, 43.5c: From Kent M. Van De Graaff and Stuart Ira Fox, *Concepts of Human Anatomy and Physiology,* 2d ed. Copyright © 1989 Wm. C. Brown Communications, Inc., Dubuque, Iowa. All Rights Reserved. Reprinted by permission.
43.7, 43.8a, 43.10: From John W. Hole, *Human Anatomy and Physiology,* 5th ed. Copyright © 1990 Wm. C. Brown Communications, Inc., Dubuque, Iowa. All Rights Reserved. Reprinted by permission.
43.11: From Kent M. Van De Graaff and Stuart Ira Fox, *Concepts of Human Anatomy and Physiology,* 2d ed. Copyright © 1989 Wm. C. Brown Communications, Inc., Dubuque, Iowa. All Rights Reserved. Reprinted by permission.
page 713: From John W. Hole, Jr., *Human Anatomy and Physiology,* 5th ed. Copyright © 1990 Wm. C. Brown Communications, Inc., Dubuque, Iowa. All Rights Reserved. Reprinted by permission.

Chapter 44
44.13: From John W. Hole, Jr., *Human Anatomy and Physiology,* 5th ed. Copyright © 1990 Wm. C. Brown Communications, Inc., Dubuque, Iowa. All Rights Reserved. Reprinted by permission.
44.14: From *The Human Body: Growth and Development.* Copyright © Marshall Editions Limited, London, England. Reprinted by permission.

Chapter 45
45.5b: From J. P. Hailman, *Behavior Supplement #15.* Copyright © E. J. Brill, Publisher, Leiden, Holland. Reprinted by permission.

Chapter 46
46.2: From *Laboratory Studies in Biology: Observations and Their Implications* by Chester A. Lawson, et al. Copyright © 1955 W. H. Freeman and Company, New York, NY. Reprinted by permission.
46.4b: From Victor B. Scheffer, "The Rise and Fall of a Reindeer Herd" in *Scientific Monthly,* Vol. 73:356-362. Copyright © 1951 by the AAAS.

Chapter 47
47.5b: Figure from *Fundamentals of Ecology,* Third Edition, by Eugene P. Odum, copyright © 1971 by Saunders College Publishing, reprinted by permission of the publisher.

Illustrators

Part 1
Kathleen Hagelston:
5.16a.

Illustrious, Inc.:
text art, pages 20, 39, 47; 4.9, 4.B, 4.C.a, text art, page 50; 4.16a-d, text art, pages 58, 72; 5.18b-e, 5.19a-b, 6.11, 6.12, 7.9, 7.13, 7.14, text art, page 119; 8.9, text art, pages 130, 133, 136, 140, 142; 9.7, text art, page 144; 9.10, text art, page 151.

Carlyn Iverson:
2.7a, text art, page 82; 8.2, 8.3a, 8.7, 9.1b, 9.2.

Laurie O'Keefe:
5.10.

Mark Lefkowitz:
2.3, 3.1.

Rolin Graphics:
text art, page 15; 2.2, 2.4, 2.6, text art, pages 33, 34, 36; 3.2, 3.5, 3.6, 3.7, 3.8a-b, 3.A, 3.10, 3.11, 3.13, 3.14, 3.15, text art, page 43; 4.2, 4.3, 4.4, text art, page 48; 4.5, 4.6, text art, pages 54, 58; 4.10, 4.13, 4.14, 4.15, 4.17, 4.18, 4.20, text art, page 65; 5.2a-b, 5.3b, text art, page 74; 5.4b, 5.5a, 5.7a, 5.7c, 5.9b, 5.15, 5.17a, text art, page 85; 6.1a-b, text art, page 86; 6.2b-c, 6.3, text art, page 87; 6.4, text art, page 88; 6.5, 6.6, 6.8, 6.9, text art, page 97; 6.A.a-f, 6.17b, text art, pages 102, 103, 106; 7.3a-b, text art, page 109; 7.6a-b, 7.7, text art, pages 110, 113; 7.10, text art, page 114; 7.11, text art, pages 117, 122; 8.4, text art, pages 127, 128; 8.8b, text art, page 131; 8.10, 8.12, text art, pages 138, 140; 9.3, 9.4, text art, page 142; 9.5, 9.6, 9.8, text art, page 145; 9.9, text art, pages 146, 150; 9.11, 9.12.

Rolin/Iverson:
4.19.

Part 2
Molly Babich:
text art, page 184

Kathleen Hagelston:
18.3

Hagelston/Margorie Leggitt:
14.2B

Illustrious, Inc.:
11.1, text art, page 171; 13.3, 13.5, 13.8, 13.11, text art, page 209; 14.12A, 15.5B, 15.6, 15.7, 15.8, 16.2, 16.3B&C, 16.4, text art, page 253; 16.7, 16.11A, 16.15, text art pages 269, 274; 18.2, text art pages 280, 289, 18.10, 18.11, text art, page 294.

Illustrious, Inc./Leonard Morgan:
10.5A, 10.6A, 10.7A, 10.9A, 11.3, 11.7, text art, page 167.

Carlyn Iverson:
14.1, 16.10A, 18.B.

Laurie O'Keefe:
14.8A-C, 17.7, 17.8

Mark Lefkowitz:
10.8B, 10.A, 10.12B, 12.2, 13.6, 15.2A, 17.4B&C

Precision Graphics:
14.9B

Rolin Graphics:
text art, pages 155, 157; 10.2A, text art, page 160; 10.4, 11.2, text art, page 174; 11.4B-D, 11.6, 12.3, text art, page 187; 12.4, 12.5, text art, page 189; 12.6, 12.7, 12.8, 12.9, text art, page 192; 13.1, text art, page 201; 13.9, 13.12, 13.16, 14.3, 14.6, 14.7, 14.B.A-C, 14.14, text art, page 236; 15.3, text art, page 237; 15.4A, text art, page 240; 15.A, text art, pages 243, 249; 16.6, 16.8B, 16.9A-E, text art, page 266; 16.14, 17.1, text art, pages 258, 266; 17.A, 17.B, 17.2, 17.5, 18A, 18.8, 18.12B.

Rolin/Iverson:
16.1

Part 3
Molly Babich:
22.6, 24.14, text art, page 401; 27.A, 28.17

Illustrious, Inc.:
text art, page 309; 20.2A&B, text art, page 318; 20.5, 20.8, 20.12, 21.3, 21.5, text art, page 340; Tbl. 21.3, text art, pages 364, 446, 470; fig. A.

Carlyn Iverson:
21.6A&B, 21.7, 21.16B

Margorie Leggitt:
28.16

Mark Lefkowitz:
19.6A&B, 19.13, 21.17, 21.19, 23.1C1, 24.2, 24.4, 26.2, 26.3, 26.5, 26.7, 27.1, 27.5, 27.12, 28.19, 28.20, 28.21, 28.22.

Laurie O'Keefe:
21.8, 21.9, 21.10, 21.12, text art, page 450

Rolin Graphics:
19.1A, 20.3A&B, 20.9A, text art, page 344; 21.11, 21.14, 21.15, 21.16A, 22.1, 22.5, 23.3, 23.5A, text art, page 377; 24.8, 24.9, 24.12, 24.16, 24.18A, 24.19A, text art, page 395; 25.4A, 25.5, 25.7, text art, page 405; 25.12A, text art, page 414; 27.2, 27.4B, 27.6B.

Rolin/Iverson:
25.3

Part 4

Kathleen Hagelston:
29.16A

Illustrious, Inc.:
29.A, text arts, page 493; Fig. A

Mark Lefkowitz:
31.2, 31.8A-F

Rolin Graphics:
29.1, 29.3A2, 29.3B2, 29.3C2, 29.6, 29.14B, 29.15, 29.17A-B, 29.19A&C, 29.20, 30.1, 30.2, 30.3A-B, 30.5B, text art, page 498; 30.6B&C, text art, page 500; 30.9, 30.11, 30.12, text art, pages 504-505; 31.1, text art, pages 509, 521; 32.1, 32.2A-C, 32.3, 32.8, text art, page 529; 32.12A&B, 32.13.

Rolin/Iverson:
31.5

Part 5

Molly Babich:
34.2, 38.6B

Chris Creek:
35.3, 38.1A

Anne Greene:
405, text art, page 663; 44.10A-E

Kathleen Hagelston:
43.12, 43.13

Hagelston/Leggitt:
36.13

Hagelston/O'Keefe:
35.5

Illustrious, Inc.:
text art, page 539; 33.1, 33.2A, text art, page 545, 558; 34.11B, 35.A, text art, page 596; 44.2, text arts, page 717; Figs. A&B

Carlyn Iverson:
33.8, 33.12, 34.2, 34.3, 34.11A, 35.6, 36.7, 38.11, 39.14A-E, 40.10, 40.11A-D, 41.9A, 43.7B, 44.11, 44.12B

Laurie O'Keefe:
33.5, 35.1

Mark Lefkowitz:
34.5, 34.6, 36.2, 36.6, text art, page 604; 37.4, 37.5, 37.7, 37.8, text art, page 618; 38.5, 38.6A, 38.7, 38.8, 38.9, 39.1, 39.6, 39.7, 39.8, 39.10, 39.12, 40.1, 42.2

Ron McClean:
43.8A

Steve Moon:
33.9, 43.11

Diane Nelson:
34.B.B, 44.13

Rolin Graphics:
33.6A.2, 33.6B.2, 33.6C.2, 33.11, 34.7, 34.8,

34.12A, 34.13, 34.14, 35.9A-B, 35.10, text art, pages 591, 593, 594, 595; 36.12, text art, page 605; 37.10B&C, text art, pages 614, 616; 38.1B, 38.2A, 38.3, 38.10, 38.12, text art, page 631; 39.3, 39.4, 39.5, 40.2A-D, 40.6C, 40.7A-D, 40.8, 41.2, 41.3, 41.4, 41.5, 41.6, text art, pages 671, 672, 675; 42.3A&B; 42.4, 42.5, 42.7A-C, 42.8A-C, 42.12, 42.13, text art, page 693; 43.1A, 43.6, 43.9, 44.5, 44.9

Mike Schenk:
42.11, 43.5C

Tom Waldrop:
text art, page 550; 39.2A, 39.9, 41.11B&C, 42.A, 43.3, 43.4A&B, 43.7A, 43.10, text art, page 710

Part 6

Illustrious, Inc.:
45.A, 45.12, 46.3A, 46.8, 50.6

Carlyn Iverson:
49.5

Mark Lefkowitz:
49.10, 49.12

Rolin Graphics:
45.2A-C, 45.3, 45.4, 45.5B, 45.7A&B, 45.8, 45.9, 46.1, 46.2, 46.3B, 46.4B, 46.5, 46.6A-C, text art, page 753; 47.1, 47.2, 47.3, 47.4B, 47.5B, 47.A.A-D, 48.1, 48.2, 48.4, 48.5, 48.6B, 48.8, 48.11A-C, 49.1A, 49.2A&B, 49.3, 49.4, 50.8, 50.9B, 50.11

How Do You Do That? Boxes

p. 292, Holly Ahern; p. 345, Steve Miller; p. 487, Dr. Kingsley Stern; p. 581, Dr. Andy Anderson; p. 737, Lee Drickamer

INDEX

shapes of, 371
and STD, 706–7
transformation of, 235–37, 282, 283
and transgenic bacteria, 284–86
Bacteriophages, 236, 237, 367–68
Balance, 659, 660
Balanus balanoides, 755–56
Baldness, 700
Ball-and-socket joints, 668
Baltimore oriole, 334, 335
Bananas, 209
Banting, Frederick, 687
Barash, David P., 18–19, 731
Barnacles, 755–56
Barn owl, 10
Barr, M. L., 220
Barracuda, 9
Barr body, 220, 269
Basal body, 67
Basal nuclei, 646
Base pairing, 55, 56, 288, 292
Base-pair map, 288, 292
Bases, 36, 37–38, 54, 238, 240
Basidia, 392, 393
Basidiomycetes, 392–93
Basidiospores, 393
Basilar membrane, 660
Basophils, 565
Bass, 645
Bates, Henry Walter, 308, 758
Batesian mimicry, 758
Bateson, William, 198
Bats, and pollination, 519
B cells, 563, 577–79, 583
B-DNA, 243–44
Beach, 767
Beadle, George, 249, 250
Beagle, 306
Beagle, 299, 302, 307
Bean, 517
Beaumont, William, 593, 594
Bees, and pollination, 518
Beets, 481
Behavior, 5
    animal behavior, 730–43
    and communication, 738–40
    and ecology, 729–808, 807–8
    genetic bases and learned behavior, 732–35
    innate types of, 732
    mechanisms of, 731–32
    modification of, 734
    and natural selection, 735–38
    physiological control of, 731
    and reporduction, 736–38
Behavioral inheritance, 224–25
Behavioral isolation, 333
Bell-shaped curve, 748
Beltian bodies, 762
Beneden, Pierre-Joseph van, 170
Bengal tiger, 1
Benthic division, 768–69
Bernard, Claude, 18, 546
Bernstein, Julius, 637
Best, Charles, 687
Bicarbonate ion, 614
Bilaterally symmetrical, 419, 421, 426, 435, 443, 448
Bilateral symmetry, 422
Bile, 594

Binary fission, 154–68, 371
Binnig, Gerd, 63
Binomial names, 4
Bioavailability, 603
Biochemistry, and comparative biochemistry, 311–12
Biogenesis theory, 18
Biogeochemical cycle, 787
Biogeography, 303–5, 309–10
Biological clock, 530, 732
Biological diversity, loss of, 795
Biological evolution, 357–60
Biological magnification, 784
Biology
    and cell biology, 62–64
    and measurements, 65–66
    and molecular biology, 250–52
    sociobiology, 740–41
    unifying theories of, 18
Biomass, pyramid of, 782–84
Biomes, 773–75, 769–78
Bioremediation, 286
Biosphere, 8, 765–79, 781
Biosynthesis, and metabolic pool, 148–49
Biotechnology, and rDNA, 278–95
Biotic components, 781
Biotic environment, 745
Biotic potential, 745–46
Biotin, 602
Bipolar cells, 656
Birds, 2, 5, 299, 309, 325, 327, 332, 334, 335, 343, 443, 449, 454–55, 456, 519, 694, 733, 734, 736, 756, 759
Birds of prey, 454
Birth, 722
Birth control, 705, 706
Birth defects, detecting, 226
Birthrate, 745
Birth weight, 325
Bisphosphoglycerate, 141, 143
Biston betularia, 326
Bivalent, 171, 172, 174, 186, 208
Bivalves, 431, 433
Bivalve shells, 431
Black bread molds, 391
Blackman, F. F., 125, 126
Blade, 489
Blastocoel, 713
Blastocyst, 718, 719
Blastopore, 430, 431
Blastula, 712, 713
Blending theory of inheritance, 183
Blind spot, 655
Blood, 31, 36, 541–42, 543, 555
    of humans, 562–69
    pH, 629
    and pregnancy, 704–5
    and Rh blood factor, 226
    and systemic blood flow, 561–62
    and trypanosome infection, 383
    vessels, 560–61
Blood cells, 541, 563–65
Blood clotting, 563–65
Blood groups, 569
Bloodhound, 306
Blood pressure, 284, 561, 562, 566
Blood Rh factor, 468
Blood Rh system, 567, 568–69
Blood types, 225–26, 567–69

Bluebirds, 18
B lymphocytes, 574, 577
Bobcat, 10, 416
Bobolink, 736
Body, temperature control, 548–49
Body cavities, 426, 546, 548
Body cells, 220
Body plans, 421, 422, 423, 426, 434, 435, 443, 448
Bog moss, 399
Boiling, 33, 35
Bonds
    chemical, 29
    covalent, 31–33, 53
    hydrogen, 33, 35, 47, 53
    ionic, 30–31, 53
    peptide, 50
    polar covalent, 32–33
Bone, and bones, 541, 542, 666, 668
Bone marrow, 564, 574
Bonner, J., 531
Bony fishes, 448–52
Book lungs, 438
Booster, 582
Boston terrier, 306
Bottleneck effect, 324
Botulism, 373
Boveri, Theodor, 202
Bovine growth hormone, (bGH), 282–83, 287
Bowman's capsule, 625, 626
Brachydactyly, 223
Bracken fern, 397, 410–11
Bracket fungi, 392, 393
Brain
    conscious brain, 646
    and drugs, 647
    human, 644–49
    and neurotransmitters, 647–49
    sizes among vertebrates, 645
    unconscious brain, 644–46
Brain hormone, 677–78
Branchiostoma, 444
Brassica oleracea, 307
Bread molds, 391
Breasts, 705
Breathing, 610, 612, 726–28
Breeders, 741
Bright-field light microscope, 62
Bristlecone pines, 407
Brittle stars, 442
Broad-bean plant, 499
Broad billed birds, 454
Bromine, 501
Bronchi, 611
Bronchioles, 611
Brown algae, 388, 389
Bryophytes, 398–99, 410–11
Bubble baby, 288
Budding, 368, 695
Buds, 482–84, 591, 654, 720
Buffers, 36
Buffon, Comte de, 300
Bulbourethral glands, 697
Bulbs, 485, 491
Bulk, 45, 46
Bullhorn acacia, 762
Bullock's oriole, 335
Bundle sheath, 491
Bundle sheath cells, 131

Chick, 311, 712
Chiggers, 438
Chihuahua, 306
Childhood, 722
Chilopoda, 438, 439
Chimpanzee, 312, 342, 461, 463, 645, 735, 740
Chinese cabbage, 307
Chipmunk, 10
Chironomus, 270, 295–96
Chironomus pallidivittatus, 295–96
Chironomus tentans, 295–96
Chiroptera, 459
Chitin, 45–46, 47, 437, 665
Chlamydia, 706, 707
Chlamydomonas, 384, 385, 386
Chlorella, 130
Chlorides, 627
Chlorine, 26, 30
Chlorofluorocarbons (CFCs), 801
Chlorophyll, 124, 125, 126, 371, 384
Chlorophyta, 384
Chloroplasts, 68, 69, 74–75, 81, 123–25, 476, 489
CHNOPS, 26, 42
Cholecystokinin-pancreozymin (CCK-PZ), 596, 680
Cholesterol, 50, 57, 87, 223, 566–67, 599–600
Cholesterol blood level, 566–67
Chondrichthyes, 448
Chordae tendineae, 557
Chordates, 4, 417, 429, 431, 443–44, 448, 665
Chorion, 455, 718
Chorionic villi sampling, 226, 227
Choroid, 655, 657
Chromatids, 157, 159, 160, 208
Chromatin, 67, 68, 69, 70, 157, 269–70
Chromosome mapping, 206–8
Chromosome mutations, 208–11, 316–17
Chromosome number
    changes in, 208–9
    and organism types, 157
Chromosome puffs, 269, 295
Chromosomes, 66, 68, 69, 157
    and autosomal chromosome abnormalities, 216–18, 222–24
    and chromosome mapping, 206–8
    and chromosome number, 208–9
    and chromosome puffs, 269, 295
    comparison of meiosis and mitosis, 178, 179
    contents of, 158, 159, 160
    duplicated, 159, 160
    eukaryotic, 157–59
    and genes, 196–214
    and genetic mapping, 288–92
    and human chromosomes, 216–19
    independent assortment of, 176
    and life cycles of algae, 385
    and meiosis, 171–76
    and mutations, 208–11
    and pangenesis, 177
    and polytene chromosomes, 269
    prokaryotic, 156–57
    and sex chromosomes, 204, 218–19, 220, 222, 228–31
    structure changes, 209–11
Chromosome sex determination, 204
Chromosome theory of inheritance, 202
Chrysanthemum, 471
Chrysophyta, 388–89
Chthamalus stellatus, 755–56

Chyme, 593
Cicada, 665
Cigarettes, 615
Cilia, 69, 78, 79, 81
Ciliary body, 657
Ciliates, 381–82
Ciliophora, 381
Cilium, 67
Circadian rhythms, 530, 731
Circulatory system, 546, 552–71
    closed circulatory system, 448, 552, 553–54
    invertebrates, 553–54
    and open circulatory system, 431, 553–54
    vertebrates, 555–57
Cirrhosis, 597
Citric acid cycle, 144
Cladistics, 342–43
Cladogenesis, 340, 347–48
Cladogram, 342, 343
Clams, 417, 431, 433, 590
Class, 4, 340
Classical conditioning, 735
Classifications
    of bacteria, 374–76
    and five kingdoms system of classification, 360–62
    hierarchy of, 340
    of leaves, 490
    of organisms, 4–5, A-1–A-2
    of primates, 459
    of skeletons, 665
Clathrin, 98
Clavicle, 666
Cleavage, 713, 716, 718
Climax community, 769
Clitellum, 435
Clitoris, 700, 701
Cloaca, 621
Clonal selection theory, 577, 578
Cloning, 279–81, 286–87, 288, 577, 578
Closed circulatory system, 448, 552, 553–54
Clostridium botulinum, 373
Clotting, 284, 563–65
Clotting factor, 284
Clover, 409
Clownfishes, 761
Club fungi, 392
Club mosses, 405, 410–11
Cnidarians, 417, 419–21, 426, 553, 665
Coastal communities, 766–68
Cobamide coenzymes, 112
Cocaine, and abuse, 648–49
Coccolithophore, 359
Coccus, 371, 372
Coccyx, 667
Cochlea, 659, 660
Code, and genes, 252–54
Codominance, 197
Codon, 257, 259
Coelacanth, 451, 452
Coelom, 424, 434, 443, 665, 714
Coelomates, 431
Coenzyme A, 143, 145
Coenzymes, 57, 112–14, 143, 145
Coevolution, of plants and pollinators, 518–19
Cofactor, 112
Cohesion, 34, 35, 497–98
Cohesion-tension model of xylem transport, 497–98

Coitus interruptus, 706
Cole, K. S., 637
Coleoptile, 517, 523, 525
Collagen, 57, 540, 724
Collagen fibers, 540
Collar cells, 417
Collecting duct, 625, 626
Collenchyma cells, 476, 477
Colon, 597
Colonial green algae, 385
Color, of skin, 224, 225
Color blindness, 228–29
Color vision, 658
Colostrum, 705
Columbia University, 204
Columnar epithelium, 539
Commelinales, 4
Commensalism, 373, 374, 759, 760–61
Commensalistic bacteria, 373, 374
Common descent, 308–10
Communication, and behavior, 738–40
Community, 3, 744, 754–64, 766–69, 773, 781
Companion cells, 478
Comparative anatomy, 310–12
Comparative biochemistry, 311–12
Comparative embryology, 311
Compartmentalization, 66
Competition, 755–56
Competitive exclusion principle, 756
Competitive inhibition, 111–12
Complementary base pairing, 55, 56, 239, 240
Complementary DNA (cDNA), 280, 289, 368, 369
Complement system, 576–77
Complete gut, 590–91
Complete linkage, 207
Complete proteins, 598
Complete ventilation, 610
Complex carbohydrates, 598
Compound eye, 437, 440, 654, 655
Compound fruits, 514
Compounds, and molecules, 29–33
Concentration, and substrates, 111
Concentration gradient, 89
Concepts and critical thinking, answers to, A-7–A-18
Conclusions, 14
Condensation, and hydrolysis, 44
Conditioning, classical and operant, 734–35
Condom, 706
Condor, 412
Conduction, 639
Cones, 397, 406, 655, 656, 657, 658
Configurations, 29
Conidia, 391
Coniferophyta, 410–11
Conifers, 397, 405–7, 410–11
Conjugation, 282, 283, 371, 382, 385–86
Connective tissue, 539, 540–42
Connell, Joseph, 755
Conscious brain, 646
Conservation, 412, 776–77
Constant regions, 578
Consumers, 8, 781, 783
Contact inhibition, 271
Continental drift, 344
Continuous feeders, 589
Contour farming, 786
Contour feathers, 454

Contractile tissue, 542–43
Contractile vacuoles, 381
Contraction, 669, 671
Contrast microscopy, 63
Contributions to the Theory of Natural Selection,
    308
Control
    of body temperature, 548–49
    and gene action, 265
    physiological control of behavior, 731
    of pregnancy, 705–6
    of water excretion, 626–30
Control group, 15
Controlled experiment, 15, 16
Convergent evolution, 348
Convoluted tubules, 625, 626
Cooling, and origin of first cell, 355
Copepods, 437
Copper, 26
Coprinus comatus, 5
Coral, 9, 415, 420, 421, 451, 767
Coral reefs, 9, 767
Cordgrass, 335
Corepressor, 268
Corey, Robert, 51
Cork, 476, 479, 484
Cork cambium, 479, 484
Corms, 485, 486
Corn, 60, 61, 238, 245, 481, 483, 517
Cornea, 656, 657
Cornell University, 117
Coronary arteries, 560
Corpus luteum, 703, 704
Correns, Karl, 202
Cortex
    adrenal, 624, 680, 685–86
    of root, 481
Cortisol, 680, 685
Cotyledons, 516
Countercurrent flow, 607
Countercurrent mechanism, 626–30
Country. *See* Less developed country (LDC) *and*
       More developed country (MDC)
Courtship, 738
Covalent bonding, 31–33, 53
Covered seeds, 407
C plants, 131–32, 133
Crabs, 42, 417, 439, 440
Cranial nerves, 640
Cranium, 666
Crassulaceae, 133
Craving, 649
Crayfishes, 440
Creatine phosphate, 672, 673
Creatinine, 627
Creation, 300
Cretaceous, 401, 450
Cretinism, 684
CRH, 686
Crick, Francis H. C., 18, 238, 239, 240, 241, 242,
    243
Cri du chat syndrome, 210, 218
Crill, 437
Crinoidea, 442
Cristae, 75
Critical period, 733
Crocodiles, 453, 455
Cro-Magnon, 467
Crop, 590

Cross-bridges, 671
Crossing-over, 172, 174, 176, 178, 207, 208, 317
Crossopterygii, 451
Crotalus atrox, 305
Crustaceans, 437, 439, 440, 661
Cuboidal epithelium, 539
Cucumbers, 490
Cultural eutrophication, 796
Culture, of plant, 516
Cup fungi, 392–93
Curtis, J. J., 637
Cushing syndrome, 686
Cuticle, 57
Cuticular plate, 661
Cuttings, 516
Cuvier, George, 301
Cyanobacteria, 5, 66, 358, 371, 372, 375–76, 788
Cycadophyta, 410–11
Cycads, 405, 406, 410–11
Cycles
    biogeochemical cycle, 787
    Calvin cycle, 807
    carbon cycle, 787–88
    chemical cycle, 787–89
    nitrogen cycle, 788–89
    of materials and ecosystems, 780–91
Cyclic AMP, 679
Cyclic behavior, 731–32
Cyclic electron pathway, 126–29
Cyclic GMP, 658
Cyclic photophosphorylation, 128
Cypress, 323
Cysteine, 51, 253
Cysticercus, 422
Cystic fibrosis, 221–22, 226, 284
Cystic fibrosis transmembrane conductance
      regulator (CFTR), 221–22
Cytochrome, 125, 128, 144
Cytochrome system, 128, 144
Cytokinesis, 157, 160
    in animal cells, 163, 164
    and mitosis, 160–64
    in plant cells, 164, 165
Cytokinins, 523, 524, 526–27
Cytoplasm, 64, 66, 543, 544, 553, 591
Cytoplasmic cleavage, 163
Cytosine, 238, 239, 240, 251, 253
Cytoskeleton, 69, 76, 77, 81
Cytosol, 64, 67, 139
Cytotoxic T cells, 579, 580

D

Dacrydium, 338
Dalmation, 306
Dalton, John, 26
Dandelion, 515
Danielli, J., 85
Danner, David, 723
Daphne Major, 340
Dark-field illumination, 63
Darter, 412
Darwin, Charles, 18, 176, 177, 183, 186, 298–314,
      316, 324, 523
Darwin, Erasmus, 300
Dasyurus, 310
Data, 14, 17
Dating of fossils, 344–45
Daughter cells, 160, 179, 716

Daughter molecules, 240
Daughter nuclei, 160, 179
Daughter strands, 242
Davson, H., 85
Day-neutral plants, 531
DDI. *See* Dideoxyinosine (DDI)
DDT, 326, 784, 795
Deamination, 597
Decapods, 440
Decay series, 345
Decibels (db), 662
Deciduous forests, 776, 783
Decomposers, 8, 373
Deductive reasoning, 14
Deep-sea ecosystems, 807–8
Deer, 341, 729
Deer tick, life cycle of, 760–61
Defenses, 758, 759, 575–80
Deforestation, 801
Degradative reaction, 107, 114, 115
Degrees of dominance, 197–98, 222, 226
Dehydration synthesis, 44, 45
Dehydrogenase, 108
Deletion, 210, 211
Delirium, 648
Demographic transition, 750
Denaturation, 53, 111
Dendrites, 543, 544, 635–40
Denitrification, 788–89
Denitrifying bacteria, 789
Density, 35
Density-dependent effects, 746
Density-independent effects, 746
Deoxyribonucleic acid (DNA), 5, 7, 19, 45, 54–56,
      57, 64, 66, 67–68, 75, 81, 234–47, 248–
      63, 264–77, 278–95
    and aging, 724
    and apes, 461
    and autoimmune diseases, 585
    and bacteria classification, 374
    B-DNA and Z-DNA, 243–44
    and binary fission, 156–57
    and biological evolution, 357–60
    and blastula, 713
    cDNA, 280, 289, 368, 369
    and chromosome contents, 158, 163, 196
    and chromosome mutations, 316–17
    and comparative biochemistry, 311–12
    compared to RNA, 252
    and cytokinins, 526
    DNA animal viruses, 368
    DNA fingerprinting, 291
    DNA ligase, 280
    DNA markers, 290–92
    DNA polymerase, 241
    DNA probe, 283, 284, 291
    DNA sequencing, 292
    and fruit fly chromosomes, 211
    gDNA, 281
    gel electrophoresis, 290, 292
    and gene expression, 261
    and gene mutations, 316
    and genetic code, 252–54
    and HIV, 581
    homeobox DNA, 717
    and hormones, 681, 687
    and meiosis and mitosis, 172, 173, 179
    mitochondrial DNA, 467–68
    and modern humans, 467–68

and molecular biology, 250–52
and molecular systematics, 340–41
and pangenesis, 177
and plants, 412
plasmid DNA or viral DNA, 279–80, 282–83, 289
and prokaryotic reproduction, 371
rDNA and biotechnology, 278–95
recombinant DNA, 412
and regulation of gene activity, 264–77
and repetitive DNA, 244–45
replication, 240, 241–43
RNA. See Ribonucleic acid (RNA)
and RNA-first hypothesis, 356
and search for genetic material, 235–37
structure of, 237–40, 249
target DNA, 285
viral DNA, 368, 369
and viruses, 366, 367, 368, 369
Deoxyribose, 45, 54
Dependence, 649
Dependency age group, 748, 749
Dependent variables, 15
Depolarization, 637, 639
Depolymerize, 76
Dermal tissue, 475, 476–77
Dermatitis, 601
Dermis, 544, 545
Descent, 300, 308–10
Descent with modification, 300
Descriptive research, 15–17
Desertification, 793
Deserts, 773
Desiccation, 398
Desmosomes, 99, 101, 540
Detritus, 781, 782
Detritus food chains, 782
Deuteromycota, 393
Deuterostomes, 430, 431
Development
    after birth, 722–24
    in animals, 711–26
    of embryo, 513–15
    of fetus, 721–22
    of gametophytes, 510
    of humans, 665, 718–24
    of human skeleton, 665
    in plants, 522–34
    of pollen grain, 511
Devonian, 401, 448, 450, 452
de Vries, Hugo, 202
Diabetes mellitus, 284, 625, 686–88
Dialysis, 629
Diaphragm, 609, 610, 655, 706
Diarrhea, 597
Diastole, 558
Diastolic pressure, 561
Diatoms, 388–89, 390
Dicot, 407, 410–11, 478, 479, 479–81, 516
Dictyosome, 68, 69, 72
Dideoxyinosine (DDI), 370, 708
Didinium, 757
Diet
    and cancer, 274–75, 600
    recommended daily, 599
    and sodium, 603
Differential reproduction, 307, 324
Differentiation, 712, 715–18
Diffusion, 89–92, 93, 99

Digestive enzymes, 594, 595
Digestive juice, 686
Digestive system, 546, 587
    animal feeding modes, 588–89
    and digestive tracts, 589–97
    and hormones, 596
    and nutrition, 587–605
Digestive tracts, 589–97
Digits, 667
Dihybrid crosses, 190–93
Dihybrid genetics problems, 190–92
Dihybrid inheritance, 190–93
Dihybrid testcross, 192–93
Dihydroxyacetone, 43
Dimorphism, 327
Dinoflagellates, 388
Dinosaurs, 449, 450, 454
Diploid, 403, 409, 471
Diploid chromosomes, 157
Diploid number, 157
Diploidy, 327–28
Diplontic cycle, 384, 385
Diplopoda, 438, 439
Dipnoi, 451
Directional selection, 324, 326
Disaccharides, 45, 57
Discontinuous feeders, 589
Disease, STD. See Sexually transmitted disease
    (STD)
Disks, 666
Disruptive selection, 324, 325–26
Distal convoluted tubule, 625, 626
Distribution of phenotypes, 224
Divergent evolution, 340, 347–48
Divering birds, 454
Diversity
    and diatoms, 390
    and evolution, 297–472, 471–72
    of fruits, 514–15
    of life, 11
    and loss of biological diversity, 795
    of plants, 396–414
    in rain forests, 777
    and reptiles, 454
    and roots, 481–82
    and vertebrates, 447–71
Division, 4, 340
Dixon, Henry, 497
DNA. See Deoxyribonucleic acid (DNA)
Dogs, 306, 457
Dogwood, 7
Dolichotis patagonium, 303
Domesticated horse, 341
Dominance, 184, 185, 197, 222, 223–24, 226
    allele, 186
    degrees of, 197–98, 222, 226
    hierarchies, 736, 755
    modification, 198
    simple, 193
Donor, and electrons, 30
Dopamine, 288, 647
Dormancy, 528
Double fertilization, 407, 510, 513
Double helix, 239, 240, 244, 252, 254
Double jointed, 668
Douglass, A. E., 487
Doves, 731
Down, John L., 216
Down feathers, 454

Down syndrome, 216–18, 231
Dragonfly, 3
Drosophila melanogaster, 190–92, 204–7, 210, 211, 235, 238, 244, 245, 269, 270, 282, 290, 318, 322, 717, 746
Drosophila pseudoobscura, 319–20
Drugs
    and abuse, 648–49
    antibiotics and antivirals, 370
    and brain, 647
Dryopithecines, 461, 462
Duchenne muscular dystrophy, 220, 222, 230–31, 284
Ducks, 454
Dump sites, 796
Duodenum, 593
Duplicated chromosomes, 159, 160
Duplication, 210, 211
Dwarfism, 223, 323, 683
Dynamic equilibrium, 660
Dynein, 79
Dystrophin, 230

E

Ear, of human, 659–62
Eardrum, 661
Earlobes, 187–88
Earthworms, 417, 435–36, 554, 588, 591, 609, 623, 782
Ear wax, 57
Ecdysone, 677, 678
Echinoderms, 417, 431, 440–43
Echinoidea, 442
Echolocation, 736
Ecological niche, 299
Ecological pyramid, 782–84
Ecology, 781
    and behavior, 729–808, 807–8
    and bioremediation, 286
    and ecological pyramid, 782–84
    of populations, 744–53
EcoRI, 280
Ecosystems, 3, 8–11, 780–91, 807–8
Ectoderm, 421, 713–14
Edward syndrome, 217, 218
Edward VII, King, 229
Eel, 9
Effectors, 547, 641
Efficiency, and energy transformation, 148
Egg, 7, 169, 177, 204, 386, 695
    hard-shelled, 455
    and hormones, 700–704
    and reptiles, 455
    yolk in, 712–13
EI. See Environmental impact (EI)
Ejaculation, 697
Elaphe obsoleta, 320, 321, 334
Eldredge, Niles, 339
Electrocardiogram (ECG), 559
Electrochemical gradient, 94
Electromagnetic spectrum, 122, 123
Electron model, of water, 32
Electrons, 26, 27, 28
    and acceptor and donor, 30
    configurations, 29
    and oxidation-reduction reaction, 33
    and photosynthesis, 126–29
    and sharing, 31

Lobules, 705
Local osmosis, 95
Locus, 186, 220
Loligo, 434, 637
Loligo pedlei, 434
Long bone, anatomy of, 542
Long-day plants, 531
Loons, 454
Loop of Henle, 625, 626
Loop of the nephron, 625, 626
Loose connective tissue, 540–41, 541
Lorenz, Konrad, 732–35, 734
Love Canal, 795
Low-density lipoproteins (LDL), 98, 599–600
Lucy, 465
Lungfishes, 451
Lungs, 606, 612
    amphibians and reptilians, 609, 610
    and book lungs, 438
    and cancer from smoking, 615
    disorders of, 616
    and gas exchange and transport, 612–16
    saclike lungs, 448
Luteal phase, 702, 703
Luteinizing hormones (LH), 680, 683, 698, 700, 701, 702, 703, 704
Lycophyta, 410–11
Lycopodium, 402
Lycopod trees, 402
Lyell, Charles, 302, 308
Lyme disease, 284, 438, 760–61
Lymph, 574
Lymphatic duct, 574
Lymphatic system, 547, 573–75
Lymphatic vessels, 568, 573–74
Lymph nodes, 574
Lymphocytes, 563, 565, 577
Lymphoid organs, 574–75
Lymphokines, 579
Lynx, 5, 758
Lyon, Mary, 220
Lysine, 51
Lysogenic cells, 368
Lysogenic cycle, 367, 368
Lysosomes, 67, 69, 73, 81, 222, 223
Lytic cycle, 367, 368

M

MacArthur, Robert, 756
Macaw, 777
McClintock, Barbara, 244, 245
Macleod, J. J., 687
Macrapus, 310
Macrocystis, 389
Macroevolution, 340–47
Macromolecules, 43, 55, 99, 356
Macrophages, 97, 99, 563, 565, 575, 576
Madagascar periwinkle, 412
Magnesium, 26, 29, 30, 112, 494, 500, 627
Magnification, biological, 784
Major histocompatibility complex (MHC) protein, 580, 582, 583, 584
Malaria, 226, 284, 326, 328, 383, 384
Malay Archipelago, 308
Male gametophyte, 405, 508
Male reproductive system, 696–700
Malignant tumors, 272
Malleus, 660

Malpighi, Marcello, 503, 624
Malpighian tubules, 438, 441, 623–24
Maltase, 108
Malthus, Thomas, 305, 308
Maltose, 45, 108
Mammals, 4, 5, 7, 297, 310, 343, 349, 417, 443, 449, 450, 455–58, 459
    body cavities of, 548
    and respiration, 609, 610
    spermatogenesis and oogenesis in, 176
Mammary glands, 455, 456
Manatees, 297, 307
Mangold, Hilde, 716
Manic depressive psychosis, 228
Man-of-war, 420
Mantle, 431, 434
Maples, 333
Mapping
    and chromosome mapping, 206–8
    and genetic mapping, 288–90
Maranta leuconeura, 530
Marchantia, 398
Marfan syndrome, 223
Marigolds, 10, 655
Marijuana, 409, 648
Marine annelids, 435
Marine Iguana, 299
Markers, and DNA markers, 290–92
Marrow, 564, 574
Marshes, 767
Marsh marigold, 655
Marsupials, 310, 457
Maryland darter, 412
Masked messengers, 271
Mass extinctions, 343–44
Mast cells, 584
Master gland, 683
Mastodon, 301
Materials, and ecosystems, 780–91
Mating, 321–22, 332–34
Matrix, 75, 99
Matter, 26
Matthei, J. Heinrich, 253
Maturation promoting factor (MPF), 160
Mature mRNA, 254
Maturing face, 72
Maximum sustained yield, 748
Mayr, Ernst, 334
M. D. Anderson Hospital and Tumor Institute, 274
MDC. *See* More developed country (MDC)
Measurements, and biology, 65–66
Mechanical isolation, 333
Mechanoreceptors, 658–62
Medicago, 484
Medulla, adrenal, 624, 680, 685
Medulla oblongata, 644
Medusae, 419, 421
Megaspores, 404, 405, 407, 509, 510
Meiosis, 157, 169–81, 203, 207, 208, 209, 210, 219, 317, 335, 385, 386, 393, 400, 405, 513, 695
Meissner's corpuscles, 658
Melanocyte-stimulating hormone (MSH), 680, 683
Melatonin, 680, 689–90
Membranes, structure and function, 84–103
Memory, 645, 646–47
Memory B cells, 578
Memory T cells, 579
Mendel, Gregor, 18, 182–95, 197, 198, 199, 202, 282, 316

Mendelian patterns of inheritance, 182–95
Mendel's law of independent assortment, 190, 192, 203
Mendel's law of segregation, 186, 188, 203
Meninges, 642
Menisci, 668
Menopause, 724
Menses, 703
Menstruation, 702, 703
Mental retardation, 217, 218, 221, 228, 231
Menthol, 409
Meristem tissue, 475
Meristem tissues, 479
Merkel's disks, 658
Mermaid's wineglass, 69
Merrick, Joseph, 223
Meselson, Matthew, 241–43
Mesoderm, 713–14
Mesoglea, 419, 421, 553
Mesophyll, 131, 489, 491
Mesozoic, 344, 401, 404, 450, 454, 457
Messenger RNA (mRNA), 251, 252, 253, 254, 255, 261, 270, 271, 368, 681, 715, 723
Metabolic pathways, 108, 139
Metabolic pool, and biosynthesis, 148–49
Metabolism, 4, 6, 8
    and cellular metabolism, 107–15
    and glucose metabolism, 139
    inborn error of metabolism, 249
    and prokaryotic metabolism, 372–74
Metabolites, 249
Metacarpals, 667
Metafemale syndrome, 217, 218, 219, 220
Metamorphosis, 438, 452, 453, 677
Metaphase, 161, 163, 164, 172, 173, 174, 175
Metastasis, 272
Metatarsals, 668
Metatheria, 457
Methamphetamine (ice), and abuse, 649
Methyl group, 112
Metochondrion, 139
Metric system, A-4–A-6
MHC protein. *See* Major histocompatibility complex (MHC) protein
Mice, 10, 16
Microbodies, 67, 69, 73–74, 81
Microevolution diagram, 320
Microfilaments, 78
Micrometer, 65
Micropyle, 515
Microscopes, 62–64
Microspheres, 356, 357
Microsporangium, 404
Microspores, 404, 405, 509, 510
Microtubule organizing center (MTOC), 76–78, 160
Microtubules, 67, 68, 76–78, 79, 160, 161, 163, 174
Microvilli, 67, 654
Midbrain, 645
Middle ear, 659
Middle lamella, 68
Midges, 270, 295–96
Miescher, Friedrich, 235
Milk, 45, 705
Miller, Stanley, 355
Millimeter, 65
Millipedes, 10, 438–40
Mimicry, 758
Mimosa pudica, 530

Mineralocorticoids, 680, 685
Minerals, 494, 495
    requirements and uptake in plants, 499–503
    and symbiosis, 502
    and vitamins, 600–603
Miocene, 461
Mitchell, Peter, 117
Mites, 438
Mitochondria, 67, 68, 69, 74, 75, 81
Mitochondrial DNA, 467–68
Mitosis, 154–68, 177–79
Mitotic spindle, 154, 160, 161, 162, 166
Mixed nerves, 640
Models
    of origin of life, 355
    scientific, 17
    of water molecule, 32
Moderately repetitive DNA, 245
Modern humans, 467–68
Modification of behavior, 734
Molars, 588
Molds, 5, 380, 389–91, 391
Molecular biology, 250–52
Molecular clock, 341
Molecular systematics, 340–41
Molecules, 25
    and compounds, 29–33, 34
    evolution of organic molecules, 355
    inorganic. See Inorganic molecules
    and macromolecules, 43, 356
    movement into and out of cells, 89–99
    organic. See Organic molecules
Mollusks, 417, 429, 431–34, 437, 443, 661
Molt, 437, 677
Monera, 3, 5, 370–74
    and five kingdoms system, 360–62
    and viruses, 365–78
Moniliasis, 393
Monkeys, 16–17, 312, 342, 459–61
Monoclonal antibodies, 582–84
Monocot roots, 481
Monocots, 4, 407, 410–11, 478, 479, 483
Monocytes, 565, 575
Monod, Jacques, 267, 268
Monohybrid cross, 185
Monohybrid genetics problems, 187–88
Monohybrid inheritance, 185–90
Monohybrid testcross, 188–90
Monomers, 43, 44
Monosaccharides, 43, 45, 57
Monosomy, 209, 210
Monotremata, 457
Monotremes, 457
Moray eel, 9
More developed country (MDC), 750–51
Morgan, Thomas Hunt, 204, 235, 282
Morning glory, 529
Morphine, 409, 649
Morphogenesis, 712, 715–18
Mortality, 748
Morula, 712, 715, 719
Mosquito, 326, 383, 384
Mosses, 5, 397, 398–99, 400, 405, 409, 410–11
Moth, 326
Mother Earth, 359
Moths, 14, 15, 429, 518–19
Motor neurons, 636
Mount St. Helens, 344
Mouse, 312

Mouth, 590, 591–93
Movements
    and cells, 76, 77
    of molecules into and out of cells, 89–99
    of plants, 528–30
MPF. See Maturation promoting factor (MPF)
M phase, 159
mRNA. See Messenger RNA (mRNA)
MS. See Multiple sclerosis (MS)
MSH. See Melanocyte-stimulating hormone
    (MSH)
MTOC. See Microtubule organizing center
    (MTOC)
Mucous membrane layer, 592
Mucus, 435
Muller, Fritz, 758
Mullerian mimics, 758
Multicellular green algae, 386
Multicellular heterotrophic organisms, 416
Multicellularity, 360
Multiple alleles, 200, 225–26
Multiple sclerosis (MS), 584–85
Muscle action potential, 673
Muscle contraction, 669
Muscle fiber, 542, 669, 670, 671–72, 673
Muscles, 539, 542–43, 546, 669–73
    anatomy and physiology, 669–72
    of human skeleton, 666
Muscular dystrophy, 220, 222, 230–31, 284
Muscular system, 546, 669–73
Muscular tissue, 539, 542–43
Musculoskeletal system, 664–75
Mushrooms, 5, 379, 392, 393
Mussels, 433
Mutagens, 260
Mutations, 7, 371
    and chromosomes, 208–11, 316–17
    and DNA, 235
    and evolution, 319–20
    and frameshift mutations, 259
    and gene mutations, 259–61, 316
    and point mutations, 259–61
Mutualism, 75, 373, 374, 759, 761–62
Mutualistic bacteria, 373, 374
Mycelium, 391
Mycena, 392
Mycorrhizae, 393, 502
Myelin sheath, 636
Myobacterium avium, 708
Myocardium, 557
Myofibrils, 671
Myoglobin, 616
Myogram, 669
Myosin, 57, 671
Myxomycota, 389

N

NAD. See Nicotinamide adenine dinucleotide
    (NAD)
NADP. See Nicotinamide adenine dinucleotide
    phosphate (NADP)
Nanometer, 65
Nasal cavity, 611, 612, 653
National Heart, Lung, and Blood Institute, 566
National Institute of Medical Research, 716
National Institutes of Aging, 723
Native cat, 310
Natural food chain, 784–87

Natural selection, 305–8, 324–27, 735–38
Nautilus, 95, 434
Navigation, 736
NE. See Norepinephrine (NE)
Neanderthals, 466–67
Neanthes, 435
Negative feedback loop, 548–49
Negative pressure, and breathing, 610
Neisseria gonorrheae, 707
Nematocysts, 419, 420
Nematoda, 423
Neopilina gelatheae, 437
Nephridia, 435, 623
Nephrons, 625, 626
Nereocystis, 389
Nerve branches, 441
Nerve cords, 431, 436, 437
Nerve fibers, 543, 544
Nerve growth factor, 690
Nerve impluse, 637–39
Nerve net, 419
Nerve ring, 431, 441
Nerves, 640, 655
Nervous system, 547, 633–51
    evolution of, 634–35
    of humans, 635
    ladder-type nervous system, 421–22
Nervous tissue, 539, 543–44
Nesting cycle, 18
Neural tube, 714
Neurofibromatosis (NF), 222, 226
Neuromuscular junction, 673
Neurons, 543, 544, 633, 635–40
Neurosecretions, 677
Neurospora, 238, 249, 250, 392
Neurospora crassa, 238, 249, 250
Neurotransmitters, 639, 647–49, 690, 691
Neurula, 714–15
Neutrons, 26, 27, 28
Neutrophils, 563, 565
Newborn, 284, 569
Newsprint, 488
New World monkeys, 312, 459–61
NF. See Neurofibromatosis (NF)
NGU. See Nongonoccal urethritis (NGU)
Niacin, 112, 601, 602
Niche, 299, 756
Nicolson, G., 86
Nicotinamide adenine dinucleotide (NAD), 112,
        113–14, 140, 141, 142, 143, 144, 145,
        146, 147–48, 600
Nicotinamide adenine dinucleotide phosphate
        (NADP), 114, 124, 125, 127, 128, 129
Night vision, 658
Nilsson-Ehle, H., 201
Nirenberg, Marshall, 253
Nitrates, 373
Nitrification, 788–89
Nitrifying bacteria, 788
Nitrites, 373
Nitrogen, 26, 28, 345, 494, 500
Nitrogen cycle, 788–89
Nitrogen fixation, 373, 374, 502, 788
Nitrogenous wastes, 620–21
Nitrogen oxides, 798
Nocturnal birds, 454
Nodes, 482, 502, 574, 639
Nodes of Ranvier, 639
Nodules, 502

Rat, 312
Rat snake, 320, 321, 334
Rattlesnakes, 305
Ray-finned fishes, 450, 451
Rays, 448, 451
RB gene. *See* Retinoblastoma (RB) gene
RDA. *See* Required dietary (daily) allowance (RDA)
Reabsorption, 625
Reactants, and products, 33
Reactions
    chemical, 33
    light-dependent and light-independent, 125–36
    and transition reactions, 143, 144, 145, 147
Reactivation, of proteins, 53
Reasoning, 14, 735
Receptacle, 509
Receptor-mediated endocytic cycle, 98–99
Receptor protein, 90
Receptors, 88, 547, 612, 641, 653–62
Receptor sites, 639
Recessive, 184, 185, 197, 222
Recessive allele, 186
Recessive inheritance, autosomal, 221–23
Reciprocal crosses, 185
Recombinant DNA (rDNA), 278–95, 412
Recombinant gametes, 207, 208
Recombinant phenotypes, 207, 208
Recombination, 175, 278–95, 317–18, 412
Recycling, of NAD and FAD, 146
Red algae, 388
Red blood cells, 541, 563–65, 614
Red bone marrow, 564
Red chow, 306
Red deer, 341
Red grouper, 9
Red maple, 333
Red tide, 388
Reducing atmosphere, 355
Reduction, 33, 130
Redwoods, 407, 410–11
Reefs, 9, 451, 767
Reflex arc, 641
Reflexes, 641
Refraction, 657
Refractory period, 697
Regeneration, 131, 695
Regulation
    of enzyme activity, 111
    of gene action, 264–77
    osmotic regulation, 621–23
Regulator gene, 266, 267
Reindeer, 341, 747
Reinforcement, 734
Rejection, of tissue, 584
Relative dating method, 345
Relaxation, 669
Remediation, 286
Renal pelvis, 624
Renewable energy sources, 802–4
Renin, 628, 629, 686
Renin-angiotensin-aldosterone system, 686
Repair enzymes, 261
Repetitive DNA, 244–45
Replacement reproduction, 751
Replica-plating technique, 317
Replication, 203, 240
    and first replication system, 357
    and replication fork, 243

Repolarization, 639
Reporting of experiments, 14
Repressible operon, 268
Repressor, 266, 267
Reproduction, 5, 7, 8
    in animals, 694–710
    and asexual reproduction, 166, 384, 386, 387, 515–19, 695
    and behavior, 736–38
    cell. *See* Cell reproduction
    and comparisons among plants, 410–11
    and differential reproduction, 307, 324
    female reproductive system, 700–704
    of humans, 696–705
    patterns of, 695–96
    in plants, 507–21
    and prokaryotic reproduction, 371–72, 373
    replacement reproduction, 751
    and sexual reproduction, 166, 169, 170, 384, 386, 387, 695–96
Reproductive age group, 748, 749
Reproductive behavior, 736–38
Reproductive isolating mechanism, 332–34
Reproductive system, 547
    female, 700–704
    male, 696–700
Reptiles, 5, 309, 343, 349, 417, 443, 449, 453–54, 455, 609, 610, 718
Required dietary (daily) allowance (RDA), 598, 600, 601
Research, descriptive, 15–17
Resistance, environmental, 746–48
Resolution, 62
Respiration
    and cellular respiration, 114, 137–52
    and glycolysis, 137–52
    and health, 616
    and origin of first cell, 355
    and photorespiration, 132–33
    and tracheae, 441
Respiratory distress of newborn, 284
Respiratory gas-exchange surfaces, 607–10
Respiratory system, 546, 606–18
Responses, 4–5, 8
Responsibility, and science, 19
Resting potential, 637–39
Restricted fragment length polymorphism (RFLP), 290–92
Restriction enzymes, 279–80
Retardation, 217, 218, 221, 228, 231
Reticulitermes, 333
Retina, 222, 655, 656, 657
Retinal, 658
Retinoblastoma (RB) gene, 274, 723
Retroviruses
    RNA, 368, 369
    as vectors, 289
Reverse transcriptase, 368
Revision, 15
RFLP. *See* Restricted fragment length polymorphism (RFLP)
R groups, 51, 52, 53
Rhea, 299
Rhesus monkey, 342
Rh factor, and system, 226, 468, 567, 568–69
Rhizobium, 374
Rhizobium, 502
Rhizoids, 398, 403
Rhizomes, 401, 485, 486, 515

Rhizopus, 391
Rhodnius, 677
Rhodopsin, 658
Rhynia, 400, 401, 409
Rhyniophytes, 400
Rhythmic behavior, 731–32
Rhythm method, 706
Rhythms, of plants, 530
Riboflavin, 112, 601, 602
Ribonuclease, 53, 108
Ribonucleic acid (RNA), 45, 54–56, 57, 68, 69, 108, 235, 245, 248, 251, 252, 253, 254, 255, 264–77
    and aging, 723
    antisense RNA, 268–69
    and bacteria classification, 374
    and biological evolution, 357–60
    compared to DNA, 252
    and cytokinins, 526
    and genetic code, 252–54
    and leaf senescence, 535
    and molecular biology, 251–52
    mRNA, 251, 252, 253, 254, 255, 261, 270, 271, 368, 681, 715, 723
    and regulation of gene control, 264–77
    RNA animal viruses, 368
    RNA-first hypothesis, 356
    RNA polymerase, 254
    RNA processing or RNA splicing, 254–56
    RNA replicase and RNA transcriptase, 368
    RNA retroviruses, 368, 369
    rRNA, 245, 251, 256–57, 261, 264
    tRNA, 251, 254, 257–58, 261
    and viruses, 366, 367, 368, 369
Ribose, 45, 54
Ribosomal RNA (rRNA), 68, 245, 251, 256–57, 261, 264
Ribosomes, 66, 67, 68, 69, 70, 81
Ribozymes, 108
Ribs, 666
Ribulose bisphosphate (RuBP), 130, 131, 132
Rich, Alexander, 243
Rickets, 601
Ringdove, 731
RNA. *See* Ribonucleic acid (RNA)
Robertson, J. D., 86
Robins, 18
Rockhopper penguins, 2
Rocks, dating methods, 345
Rocky Mountain spotted fever, 438
Rocky shore, 767
Rodentia, 459
Rods, 655, 656, 657, 658
Rohrer, Heinrich, 63
Roid mania, 689
Romania, 316
Root epidermis, 476
Root hairs, 494
Root pressure, 497
Roots, 494
    and diversity, 481–82
    and mineral uptake, 502–3
    types of, 481–82
    vegetative organs and major tissues, 492
    woody roots, 486
Root system, 475, 479–82
Root tip, 60, 480
Rosa, 5
Rostrum, 440

Skin
    breathing of, 726–28
    color, 224, 225
    of humans, 544–45, 658
Skin gills, 441
Skinner, B. F., 734
Skull, 666
Slash-and-burn agriculture, 795
SLE. *See* Systemic lupus erythematosus (SLE)
Sliding filament theory, 671
Slime molds, 380, 389
Slugs, 431, 432
Small intestine, 592, 593, 594–96
Small nectar feedering birds, 454
Smell, 653–54
Smith, Hamilton, 279
Smith, Jim, 716
Smog, 798–800
Smoking, 615
Smooth endoplasmic reticulum (ER), 67, 68, 69, 71–74
Smooth muscle fibers, 542
Smooth muscle layer, 592
Smuts, 393
Snails, 302, 325, 417, 431, 537, 646
Snakes, 320, 321, 334, 453, 587
Snowshoe hare, 758
Social responsibility, 19
Society, 740–41, 804
Sociobiology, 740–41
Sodium, 26, 30, 603, 627
Sodium chloride, 30, 33
Sodium gates, 639
Sodium ions, 639
Sodium-potassium pump, 94, 95, 96
Soft palate, 654
Soil erosion, 793
Soil pores, 496
Soil profiles, 774
Soil water, 496
Solar energy, 114, 803
Solar radiation, 122, 123
Solutions, 91–92, 93
Solvent, 33
Somatic nervous system, 641
Somatotropic hormone, 680
Sonar, 736
Songbirds, 454
Sori, 402, 403, 404
Sound, 661–62, 739–40
Southern Illinois University, 356
Soybean plant, 374
Space-filling model, of water, 32
Spartina, 335
Spawning, 736
Special creation, 300
Speciation, 332–40
Species, 4, 8, 332
    category in classification, 340
    keystone species, 757
    and strategies for existence, 749
Specific defense, 577–80
Specimen, 63
Spectrum, 122, 123
Spemann, Hans, 715–16
Sperm, 79, 169, 177, 204, 386, 403, 695, 698–700
Spermatogenesis, 175, 176, 219, 698
Sphagnum, 399
S phase, 159

Sphenophyta, 410–11
Sphincters, 555, 593
Spicules, 418
Spiders, 417, 438, 588
Spinal cord, 640, 641
Spinal nerves, 640
Spinal reflex, 641
Spindle, 154, 160, 161, 162, 166
Spines, 490
Spiracles, 441
Spirillum, 371, 372
Spirogyra, 385–86, 387
Spleen, 574–75
Sponges, 5, 391, 399, 403, 404, 415, 417–19, 426
Spongy layer, 489, 491
Spontaneous, 106
Sporazoans, 383, 384
Spores, 384, 385, 386, 391, 393, 396, 404, 509
Sporophyte, 398, 399, 400, 403, 409, 410–11, 509, 513
Sporozoans, 381, 383
Spot desmosomes, 99, 101, 540
Spruces, 410–11
Squamous epithelium, 539
Squids, 417, 431, 433–34, 590, 637
Squirrel, 10, 122
Sry gene, 220
S-shaped growth curve, 746
Stabilizing selection, 324, 325, 326
Staghorn fern, 410–11
Stahl, Franklin, 241–43
Staining, 216, 220, 374
Stalk, of plant, 399
Stamens, 407, 409, 510
Stanley, W. M., 366
Stapes, 660
Staphylococcus aureus, 370, 374
Starch, 45, 46, 57, 68, 126, 596, 597
Starch grain, 68
Starfishes, 441, 757
Stasis, 348
Static equilibrium organs, 660
Statocysts, 661
Statolith, 661
STD. *See* Sexually transmitted disease (STD)
Steady state, 804
Steady-state society, 804
Steam, 34, 35
Stearate, 48
Stem cell growth factor (SCF), 563
Stem elongation, 526
Stem reptiles, 449
Stems, 482
    herbaceous stems, 483–84
    modifications of, 486
    monocot stems, 483
    primary growth of, 482–84
    secondary growth of, 484–85
    types and uses, 485–89
    vegetative organs and major tissues, 492
    woody stems, 485, 486
Stereocilia, 661
Sterility, 334, 705
Sterilization, 373
Sternum, 666
Steroids, 47, 48–50, 57, 678, 679–81, 688–89
Stickleback, 451, 738
Sticky ends, 280
Stigma, 407, 409

Stimuli, 4–5, 8, 733
Sting ray, 451
Stirrup, 660, 661
STM. *See* Scanning tunneling microscope (STM)
Stohr, Oskar, 225
Stolons, 485, 486
Stomach, 592, 593–94
Stomata, 474, 476, 489, 498–99, 528
Stop codon, 259
Strategy for existence, 748–50
Stratified epithelium, 539
Streptococcus pneumoniae, 235
Streptomyces, 370
Streptomycin, 370
Stress, 566
Striated muscle fibers, 542
Strobili, 396
Stroma, 74, 123, 124, 125
Stromatolites, 357, 358
Strontium, 345, 500
Structural gene, 266, 267
Structures
    and comparative anatomy, 310
    homologous and analogous, 340
Struggle for existence, 305
STS. *See* Sequence-tagged site (STS)
Style, 407, 409, 509
Stylet, 502
Subatomic particles, 27
Subcutaneous, 544, 545
Substrate concentration, 111
Substrate-level phosphorylation, 117, 140, 141, 142, 147
Substrates, and enzymes, 108–10
Succession, 769–72
Suckers, 516
Sucrose, 45, 57
Sudoriferous glands, 544
Sugar diabetes, 625
Sugar maple, 333
Sugars, 43, 57
    and diet, 599
    simple sugars and polysaccharides, 44–46
Sulfa, 370
Sulfates, 627
Sulfhydryl, 43
Sulfur, 26, 500
Sulfur dioxide, 37–38
Summation, and integration, 639
Sun, 4, 8
Sunflowers, 316, 501
Sunlight, 122–23
Suppressor T cells, 579
Surface water pollution, 796, 797
Surroundings, 106
Survival of the fittest, 308
Survivorship, 748
Sustained yield, 748
Sutherland, Earl W., 679
Sutton, Walter S., 202, 206
Swallowing, 593
Swallowtail butterfly, 439
Swamps, 401, 402, 405
Sweat glands, 544
Sweetener, 15, 16
Swim bladder, 450
Swimmerets, 440
Swordfish, 451
Symbiosis, 360, 373, 502, 759–62

Wars, and land spoilage, 793
Warts, 706
Wasp, 7, 758
Wastes, 620–21, 795–96, 803
Water, 24, 31, 32, 33–36, 122, 140, 494, 495–96
    and acid deposition, 37–38
    and condensation and hydrolysis, 44, 45
    control of excretion, 626–30
    ionization of, 36
    pollution, 796–97, 798
    properties of, 33–35
    and salt balance of fishes, 622
    and soil water, 496
    uptake and transport, 495–98
Water buttercup, 202
Water culture, 500
Watermelons, 209
Water molds, 389–91
Water-soluble vitamins, 602
Water vascular system, 441
Watson, James, 18, 238, 239, 240, 241, 242, 243
Waves, of ECG, 559
Waxes, 48, 57
Web footed birds, 454
Wegener, Alfred, 344
Weight
    and birth weight, 325
    units of, A-5
Weinberg, W., 318
Welwitschia, 405, 406, 410–11
Welwitschia mirabilis, 406
Went, Frits W., 523, 526
Werner syndrome, 724
Whale, 31
Wheat, 46, 201
Whisk ferns, 400, 402, 410–11
White blood cells, 541, 563–65, 572, 576

White matter, 642
White-tailed deer, 729
Whooping cough, 284
Whorled pogonia, 412
Widowbird, 327
Wigglesworth, V. B., 677
Wilkins, M. H. F., 238
Willow, 481
Wilson, H. V., 418
Windows, 659, 660
Windpipe, 611, 612
Winge, O., 471
Withdrawal, 649, 706
Wolf. *See* Wolves
Wollman, Elie, 282
Wolves, 10, 306, 307, 310
Womb, 701
Wombat, 310
Wood, 488
Wood Hole, 637
Woodpecker, 299
Woody, 407
Woody roots, 486
Woody stems, 485, 486
World War II, 751
Worms, 5, 9, 417, 434, 436

X

X chromosomes, inactive, 220
Xenopus, 270, 715, 717
Xeroderma pigmentosum, 221
X-linked genes, 204–6, 222, 228–31
X-ray diffraction, 238–40
Xylem, 399, 477–78, 488, 497
XYY male syndrome, 217, 218, 219

Y

Yale University, 356
Y chromosomes, active, 220
Yeasts, 391, 393
Yews, 410–11
Yield, 748
Yolk, 65, 712–13, 718, 719, 720
Yufe, Jack, 225

Z

Z-DNA, 243–44
Zea mays, 4, 238, 245, 481, 483, 517
Zeatin, 523, 526
Zebra, 6, 285, 341, 457, 775
Zero population growth, 751
Zidovudine (AZT), 370
Zinc, 26
Zinder, Norton, 283
Z lines, 671
Zone of cell division, 479
Zone of elongation, 479
Zone of maturation, 479
Zones of the ocean, 768–69
Zooflagellates, 381, 382–83
Zoomastigina, 382
Zooplankton, 421, 766
Zygospores, 385, 391
Zygotes, 334, 510, 516, 695
Zytomycota, 391
Zytospore, 386